CHILTON®

FORD
DIAGNOSTIC SERVICE
2006 EDITION

THOMSON

DELMAR LEARNING

Australia • Canada • Mexico • Singapore • Spain • United Kingdom • United States

CHILTON®

FORD
DIAGNOSTIC SERVICE
2006 Edition

Vice President,
Technology Professional Business Unit:
Gregory L. Clayton

Publisher,
Professional Business Unit:
David Koontz

Production Director:
Mary Ellen Black

Marketing Director:
Beth A. Lutz

Marketing Specialist:
Brian McGrath

Marketing Coordinator:
Marissa Mariella

Marketing Assistant:
Jennifer Stall

Sr. Production Editor:
Elizabeth Hough

Editorial Assistant:
Christine Wade

Editor:
Dennis Bailey

Publishing Coordinator:
Paula Baillie

Cover Design:
Melinda Possinger

© 2006 Thomson Delmar Learning, a part of The Thomson Corporation. Thomson, the Star logo, and Delmar Learning are trademarks used herein under license.

Printed in the United States of America
1 2 3 4 5 XX 07 06 05

For more information contact:

Thomson Delmar Learning
Executive Woods
5 Maxwell Drive, PO Box 8007,
Clifton Park, NY 12065-8007

Or find us on the World Wide Web at
www.delmarlearning.com

ISBN: 1-4180-2119-9

NOTICE TO THE READER

Publisher does not warrant or guarantee any of the products described herein or perform any independent analysis in connection with any of the product information contained herein. Publisher does not assume, and expressly disclaims, any obligation to obtain and include information other than that provided to it by the manufacturer.

The reader is expressly warned to consider and adopt all safety precautions that might be indicated by the activities herein and to avoid all potential hazards. By following the instructions contained herein, the reader willingly assumes all risks in connection with such instructions.

The publisher makes no representation or warranties of any kind, including but not limited to, the warranties of fitness for particular purpose or merchantability, nor are any such representations implied with respect to the material set forth herein, and the publisher takes no responsibility with respect to such material. The publisher shall not be liable for any special, consequential, or exemplary damages resulting, in whole or part, from the readers' use of, or reliance upon, this material.

TABLE OF CONTENTS

7 - PIN CHARTS

USING THIS INFORMATION

Organization

To find where a particular model section or procedure is located, look in the Table of Contents. Main topics are listed with the page number on which they may be found. Following the main topics is a listing of all of the subjects within the section and their page numbers.

Manufacturer and Model Coverage

This product covers 1990-2005 Ford Motor Company models that are produced in sufficient quantities to warrant coverage, and which have technical content available from the vehicle manufacturers before our publication date. Although this information is as complete as possible at the time of publication, some manufacturers may make changes which cannot be included here. While striving for total accuracy, the publisher cannot assume responsibility for any errors, changes, or omissions that may occur in the compilation of this data.

Part Numbers & Special Tools

Part numbers and special tools are recommended by the publisher and vehicle manufacturer to perform specific jobs. Before substituting any part or tool for the one recommended, you must be completely satisfied that neither your personal safety, nor the performance of the vehicle will be endangered.

ACKNOWLEDGEMENT

This product contains material that is reproduced and distributed under a license from Ford Motor Company. No further reproduction or distribution of the Ford Motor Company material is allowed without the express written permission of Ford Motor Company. The publisher would like to express appreciation to Ford Motor Company for its assistance in producing this product.

PRECAUTIONS

Before servicing any vehicle, please be sure to read all of the following precautions, which deal with personal safety, prevention of component damage, and important points to take into consideration when servicing a motor vehicle:

• Always wear safety glasses or goggles when drilling, cutting, grinding or prying.

• Steel-toed work shoes should be worn when working with heavy parts. Pockets should not be used for carrying tools. A slip or fall can drive a screwdriver into your body.

• Work surfaces, including tools and the floor should be kept clean of grease, oil or other slippery material.

• When working around moving parts, don't wear loose clothing. Long hair should be tied back under a hat or cap, or in a hair net.

• Always use tools only for the purpose for which they were designed. Never pry with a screwdriver.

• Keep a fire extinguisher and first aid kit handy.

• Always properly support the vehicle with approved stands or lift.

• Always have adequate ventilation when working with chemicals or hazardous material.

• Carbon monoxide is colorless, odorless and dangerous. If it is necessary to operate the engine with vehicle in a closed area such as a garage, always use an exhaust collector to vent the exhaust gases outside the closed area.

• When draining coolant, keep in mind that small children and some pets are attracted by ethylene glycol antifreeze, and are quite likely to drink any left in an open container, or in puddles on the ground. This will prove fatal in sufficient quantity. Always drain the coolant into a sealable container.

• To avoid personal injury, do not remove the coolant pressure relief cap while the engine is operating or hot. The cooling system is under pressure; steam and hot liquid can come out forcefully when the cap is loosened slightly. Failure to follow these instructions may result in personal injury. The coolant must be recovered in a suitable, clean container for reuse. If the coolant is contaminated it must be recycled or disposed of correctly.

• When carrying out maintenance on the starting system be aware that heavy gauge leads are connected directly to the battery. Make sure the protective caps are in place when maintenance is completed. Failure to follow these instructions may result in personal injury.

• Do not remove any part of the engine emission control system. Operating the engine without the engine emission control system will reduce fuel economy and engine ventilation. This will weaken engine performance and shorten engine life. It is also a violation of Federal law.

• Due to environmental concerns, when the air conditioning system is drained, the refrigerant must be collected using refrigerant recovery/recycling equipment. Federal law requires that refrigerant be recovered into appropriate recovery equipment and the process be conducted by qualified technicians who have been certified by an approved organization, such as MACS, ASI, etc. Use of a recovery machine dedicated to the appropriate refrigerant is necessary to reduce the possibility of oil and refrigerant incompatibility concerns. Refer to the instructions provided by the equipment manufacturer when removing refrigerant from or charging the air conditioning system.

• Always disconnect the battery ground when working on or around the electrical system.

• Batteries contain sulfuric acid. Avoid contact with skin, eyes, or clothing. Also, shield your eyes when working near batteries to protect against possible splashing of the acid solution. In case of acid contact with skin or eyes, flush immediately with water for a minimum of 15 minutes and get prompt medical attention. If acid is swallowed, call a physician immediately. Failure to follow these instructions may result in personal injury.

• Batteries normally produce explosive gases. Therefore, do not allow flames, sparks or lighted substances to come near the battery. When charging or working near a battery, always shield your face and protect your eyes. Always provide ventilation. Failure to follow these instructions may result in personal injury.

• When lifting a battery, excessive pressure on the end walls could cause acid to spew through the vent caps, resulting in personal injury, damage to the vehicle or battery. Lift with a battery carrier or with your hands on opposite corners. Failure to follow these instructions may result in personal injury.

• Observe all applicable safety precautions when working around fuel. Whenever servicing the fuel system, always work in a well-ventilated area. Do not allow fuel spray or vapors to come in contact with a spark, open flame, or excessive heat (a hot drop light, for example). Keep a dry chemical fire extinguisher near the work area. Always keep fuel in a container specifically designed for fuel storage; also, always properly seal fuel containers to avoid the possibility of fire or explosion. Do not smoke or carry lighted tobacco or open flame of any type when working on or near any fuel-related components.

• Fuel injection systems often remain pressurized, even after the engine has been turned OFF. The fuel system pressure must be relieved before disconnecting any fuel lines. Failure to do so may result in fire and/or personal injury.

• The evaporative emissions system contains fuel vapor and condensed fuel vapor. Although not present in large quantities, it still presents the danger of explosion or fire. Disconnect the battery ground cable from the battery to minimize the possibility of an electrical spark occurring, possibly causing a fire or explosion if fuel vapor or liquid fuel is present in the area. Failure to follow these instructions can result in personal injury.

• The EPA warns that prolonged contact with used engine oil may cause a number of skin disorders, including cancer! You should make every effort to minimize your exposure to used engine oil. Protective gloves should be worn when changing oil. Wash your hands and any other exposed skin areas as soon as possible after exposure to used engine oil. Soap and water, or waterless hand cleaner should be used.

• Some vehicles are equipped with an air bag system, often referred to as a Supplemental Restraint System (SRS) or Supplemental Inflatable Restraint (SIR) system. The system must be disabled before performing service on or around system components, steering column, instrument panel components, wiring and sensors. Failure to follow safety and disabling procedures could result in accidental air bag deployment, possible personal injury and unnecessary system repairs.

• Always wear safety goggles when working with, or around, the air bag system. When carrying a non-deployed air bag, be sure the bag and trim cover are pointed away from your body. When placing a non-deployed air bag on a work surface, always face the bag and trim cover upward, away from the surface. This will reduce the motion of the module if it is accidentally deployed.

• Electronic modules are sensitive to electrical charges. The ABS module can be damaged if exposed to these charges.

• Brake pads and shoes may contain asbestos, which has been determined to be a cancer-causing agent. Never clean brake surfaces with compressed air. Avoid inhaling brake dust. Clean all brake surfaces with a commercially available brake cleaning fluid.

• When replacing brake pads, shoes, discs or drums, replace them as complete axle sets.

• When servicing drum brakes, disassemble and assemble one side at a time, leaving the remaining side intact for reference.

• Brake fluid often contains polyglycol ethers and polyglycols. Avoid contact with the eyes and wash your hands thoroughly after handling brake fluid. If you do get brake fluid in your eyes, flush your eyes with clean, running water for 15 minutes. If eye irritation persists, or if you have taken brake fluid internally, immediately seek medical assistance.

• Clean, high quality brake fluid from a sealed container is essential to the safe and proper operation of the brake system. You should always buy the correct type of brake fluid for your vehicle. If the brake fluid becomes contaminated, completely flush the system with new fluid. Never reuse any brake fluid. Any brake fluid that is removed from the system should be discarded. Also, do not allow any brake fluid to come in contact with a painted or plastic surface; it will damage the paint.

• Never operate the engine without the proper amount and type of engine oil; doing so will result in severe engine damage.

• Timing belt maintenance is extremely important! Many models utilize an interference-type, non-freewheeling engine. If the timing belt breaks, the valves in the cylinder head may strike the pistons, causing potentially serious (also time-consuming and expensive) engine damage.

• Disconnecting the negative battery cable on some vehicles may interfere with the functions of the on-board computer system(s) and may require the computer to undergo a relearning process once the negative battery cable is reconnected.

• Steering and suspension fasteners are critical parts because they affect performance of vital components and systems and their failure can result in major service expense. They must be replaced with the same grade or part number or an equivalent part if replacement is necessary. Do not use a replacement part of lesser quality or substitute design. Torque values must be used as specified during reassembly to ensure proper retention of these parts.

Contents

OBD II Vehicle Applications

FORD

Aerostar
1996-1997
3.0L V6 MPI .... VIN U
4.0L V6 MPI .... VIN X

Aspire
1996-1997
1.3L I4 MPI .... VIN H

Crown Victoria
1996-2005
4.6L V8 MPI .... VIN 6, 9, W

Bronco
1996
5.0L V8 MPI .... VIN N
5.8L V8 MPI .... VIN H

E-Series Van
1996-2005
4.2L V6 MPI .... VIN 2
4.9L I6 MPI .... VIN Y
4.6L V8 MPI .... VIN 6, W
5.0L V8 MPI .... VIN N
5.4L V8 MPI .... VIN L, M, Z
5.8L V8 MPI .... VIN H
6.0L V8 Diesel .... VIN P
6.8L V10 MPI .... VIN 2
7.3L V8 Diesel .... VIN F, G

Escape
2001-2005
2.0L I4 MPI .... VIN B
2.0L V6 MPI .... VIN 1

Escort, ZX2
1996-2003
1.8L I4 MPI .... VIN 8
1.9L I4 MPI .... VIN J
2.0L I4 MPI .... VIN P, 3

Excursion
2000-2005
5.4L V8 MPI .... VIN L
6.0L V8 Diesel .... VIN P
6.8L V10 MPI .... VINS
7.3L V8 Diesel .... VIN F

Expedition
1997-2005
4.6L V8 MPI .... VIN 6, W
5.4L V8 MPI .... VIN L

Explorer
1996-2005
4.0L V6 MPI .... VIN E, K, X
4.6L V8 MPI .... VIN W
5.0L V8 MPI .... VIN P

Focus
2000-2005
    2.0L I4 MPI ....VIN 3, 5, P
    2.3L I4 MPI ....VIN Z

Freestar
2004-2005
    3.9L V6 MPI ....VIN 6
    4.2L V6 MPI ....VIN 2

Freestyle
2005
    3.0L V6 MPI ....VIN 1

Five Hundred
2005
    3.0L V6 MPI ....VIN 1

Mustang
1996-2005
    3.8L V6 MPI ....VIN 4
    3.9L V6 MPI ....VIN 6
    4.6L V8 MPI .... VIN R, V, W, X, Y

F-Series Pickup
1996-2005
    4.2L V6 MPI ....VIN 2
    4.6L V8 MPI ....VIN 6, W
    4.9L I6 MPI ....VIN Y
    5.0L V8 MPI ....VIN N
    5.4L V8 MPI ....3, 5, L, M, Z
    6.0L V8 Diesel ....VIN P
    6.8L V10 MPI ....VIN S
    7.3L V8 Diesel ....VIN F
    7.5L V8 MPI ....VIN G

Probe
1996-1997
    2.0L I4 MPI ....VIN A
    2.5L V6 MPI ....VIN B

Ranger
1996-2005
    2.3L I4 MPI ....VIN A, D
    2.5L I4 MPI ....VIN C
    3.0L V6 MPI ....VIN U, V
    4.0L V6 MPI ....VIN E, X

Taurus
1996-2004
    3.0L V6 MPI ....VIN 1, 2, U, S
    3.4L V8 MPI ....VIN N

Thunderbird
1996-1997, 2002-2005
 3.8L V6 MPI...VIN 4
 3.9L V8 MPI...VIN A
 4.6L V8 MPI...VIN W

Windstar
1996-2003
 3.0L V6 MPI...VIN U
 3.8L V6 MPI...VIN 4

LINCOLN

Aviator
2003-2005
 4.6L V8 MPI...VIN H

Blackwood
2002-2003
 5.4L V8 MPI...VIN A

Continental
1996-2002
 4.6L V8 MPI...VIN V

LS
2000-2005
 3.0L V6 MPI...VIN S
 3.9L V8 MPI...VIN A

Mark VIII
1996-1998
 4.6L V8 MPI...VIN V

Navigator
1998-2005
 5.4L V8 MPI...VIN A, L, R

Town Car
1996-2005
 4.6L V8 MPI...VIN W

MERCURY

Cougar XR7
1996-1997
 3.8L V6 MPI...VIN 4
 4.6L V8 MPI...VIN W

Cougar
1999-2002
 2.0L I4 MPI..VIN 3
 2.5L V6 MPI...VIN G, L

Grand Marquis
1996-2005
 4.6L V8 MPI..VIN 6, W

Marauder
2003-2004
 4.6L V8 MPI...VIN V

Montego
2005
 3.0L V6 MPI...VIN 1

Monterey
2004-2005
    4.2L V6 MPI ....VIN 2

Mountaineer
1997-2005
    4.0L V6 MPI .... VIN E, K
    4.6L V8 MPI ....VIN W
    5.0L V8 MPI ....VIN P

Mystique
1996-2000
    2.0L I4 MPI ....VIN 3
    2.5L V6 MPI ....VIN L

Sable
1996-2004
    3.0L V6 MPI .... VIN U, S

Tracer
1996-1999
    1.8L I4 MPI ....VIN 8
    1.9L I4 MPI ....VIN J
    2.0L I4 MPI ....VIN P

Villager
1996-2002
    3.0L V6 MPI ....VIN 1, W
    3.3L V6 MPI .... VIN T

NOTES & CAUTIONS

Before servicing any vehicle, please be sure to read all of the following precautions, which deal with personal safety, prevention of component damage, and important points to take into consideration when servicing a motor vehicle:

- Observe all applicable safety precautions when working around fuel. Whenever servicing the fuel system, always work in a well-ventilated area. Do NOT allow fuel spray or vapors to come in contact with a spark, open flame, or excessive heat (a hot drop light, for example). Keep a dry chemical fire extinguisher near the work area. Always keep fuel in a container specifically designed for fuel storage; also, always properly seal fuel containers to avoid the possibility of fire or explosion. Refer to the additional fuel system precautions later in this section.
- Fuel injection systems often remain pressurized, even after the engine has been turned **OFF**. The fuel system pressure must be relieved before disconnecting any fuel lines. Failure to do so may result in fire and/or personal injury.
- Brake fluid often contains Polyglycol Ethers and Polyglycols. Avoid contact with the eyes and wash your hands thoroughly after handling brake fluid. If you do get brake fluid in your eyes, flush your eyes with clean, running water for 15 minutes. If eye irritation persists, or if you have taken brake fluid internally, IMMEDIATELY seek medical assistance.
- The EPA warns that prolonged contact with used engine oil may cause a number of skin disorders, including cancer. You should make every effort to minimize your exposure to used engine oil. Protective gloves should be worn when changing oil. Wash your hands and any other exposed skin areas as soon as possible after exposure to used engine oil. Soap and water, or waterless hand cleaner should be used.
- The air bag system must be disabled (negative battery cable disconnected and/or air bag system main fuse removed) for at least 30 seconds before performing service on or around system components, steering column, instrument panel components, wiring and sensors. Failure to follow safety and disabling procedures could result in accidental air bag deployment, possible personal injury and unnecessary system repairs.
- Always wear safety goggles when working with, or around, the air bag system. When carrying a non-deployed air bag, be sure the bag and trim cover are pointed away from your body. When placing a non-deployed air bag on a work surface, always face the bag and trim cover upward, away from the surface. This will reduce the motion of the module if it is accidentally deployed. Refer to the additional air bag system precautions later in this section.
- Disconnecting the negative battery cable on some vehicles may interfere with the functions of the on-board computer system(s) and may require the computer to undergo a relearning process once the negative battery cable is reconnected.
- It is critically important to observe all instructions regarding ground disconnects, ignition switch positions, etc., in each diagnostic routine provided. Ignoring these instructions can result in false readings, damage to electronic components or circuits, or personal injury.

Preliminary Diagnostics

HISTORY OF OBD SYSTEMS

Starting in 1978, several vehicle manufacturers introduced a new type of control for several vehicle systems and computer control of engine management systems. These computer-controlled systems included programs to test for problems in the engine mechanical area, electrical fault identification and tests to help diagnose the computer control system. Early attempts at diagnosis involved expensive and specialized diagnostic testers that hooked up externally to the computer in series with the wiring connector and monitored the input/output operations of the computer.

By early 1980, vehicle manufacturers had designed systems in which the onboard computer incorporated programs to monitor selected components, and to store a trouble code in its memory that could be retrieved at a later time. These trouble codes identified failure conditions that could be used to refer a technician to diagnostic repair charts or test procedures to help pinpoint the problem area.

OBD I SYSTEM DIAGNOSTICS

One of the most important things to understand about the automotive repair industry is the fact that you have to continually learn new systems and new diagnostic routines (the test procedures designed to isolate a problem on a vehicle system). For OBD I and II systems, a diagnostic routine can be defined as a procedure (a series of steps) that you follow to find the cause of a problem, make a repair and then verify the problem is fixed.

CHANGES IN DIAGNOSTIC ROUTINES

In some cases, a new Engine Control system may be similar to an earlier system, but it can have more indepth control of vehicle emissions, input and output devices and it may include a diagnostic "monitor" embedded in the engine controller designed to run a thorough set of emission control system tests.

OBD I Diagnostic Flowchart

The OBD I Diagnostic Flowchart on this page can be used to find the cause of problems related to Engine Control system trouble codes or driveability symptoms detected on OBD I systems. It includes a step-by-step procedure to use to repair these systems. To compare this flowchart with the one used on OBD II systems, refer to the next page.

The steps in this flow chart should be followed as described below (from top to bottom).

- Do the Pre-Computer Checks.
- Check for any trouble codes stored in memory.
- Read the trouble codes - If trouble codes are set, record them and then clear the codes.
- Start the vehicle and see if the trouble code(s) reset. If they do, then use the correct trouble code repair chart to make the repair.
- If the codes do not reset, than the problem may be intermittent in nature. In this case, refer to the test steps used to find the cause of an intermittent fault (wiggle test).
- In no trouble codes are found at the initial check, then determine if a driveability symptom is present. If so, then refer to the approriate driveability symptom repair chart to make the repair. If the first symptom chart does not isolate the cause of the condition, then go on to another driveability symptom and follow that procedure to conclusion.
- If the problem is intermittent in nature, then refer to the special intermittent tests. Follow all available intermittent tests to determine the cause of this type of fault (usually an electrical connection problem).

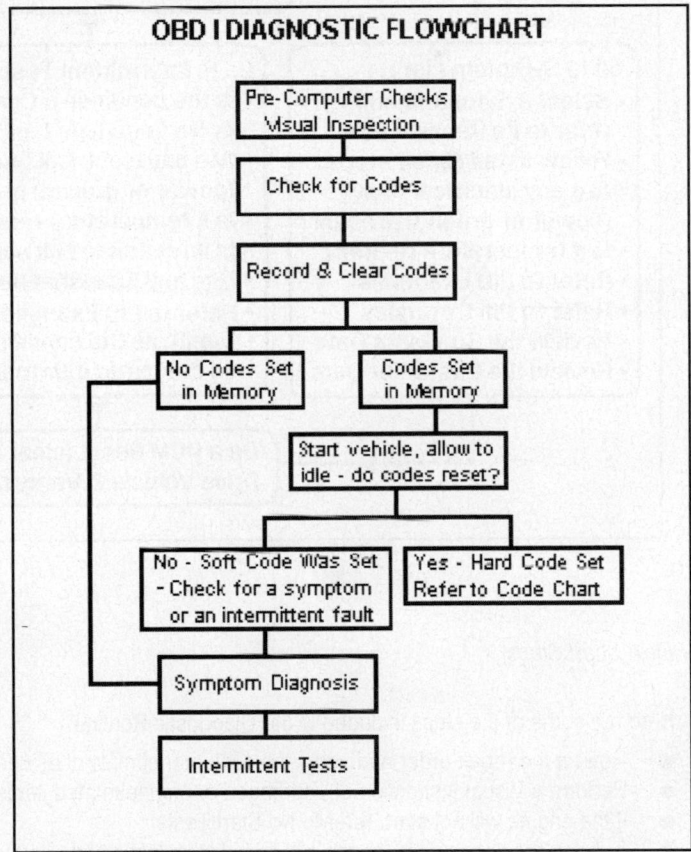

OBD I DIAGNOSTIC FLOWCHART

Pre-Computer Checks
Visual Inspection

Check for Codes

Record & Clear Codes

No Codes Set in Memory — Codes Set in Memory

Start vehicle, allow to idle - do codes reset?

No - Soft Code Was Set - Check for a symptom or an intermittent fault

Yes - Hard Code Set Refer to Code Chart

Symptom Diagnosis

Intermittent Tests

OBD II System Diagnostics

The diagnostic approach used in OBD II systems is more complex than that of the one for OBD I systems. This complexity will effect how you approach diagnosing the vehicle. On an OBD II system, the onboard diagnostics will identify sensor faults (i.e., open, shorted or grounded circuits) as well as those that lose calibration. Another new test that arrived with OBD II is the rationality test (a test that checks whether the value for one input makes rational sense when compared against other sensor input values). The changes plus the use of OBD II Monitors have dramatically changed OBD II diagnostics.

The use of a repeatable test routine can help you quickly get to the root cause of a customer complaint, save diagnostic time and result in a higher percentage of properly repaired vehicles. You can use this Diagnostic Flow Chart to keep on track as you diagnose an Engine Control problem or a base engine fault on vehicles with OBD II.

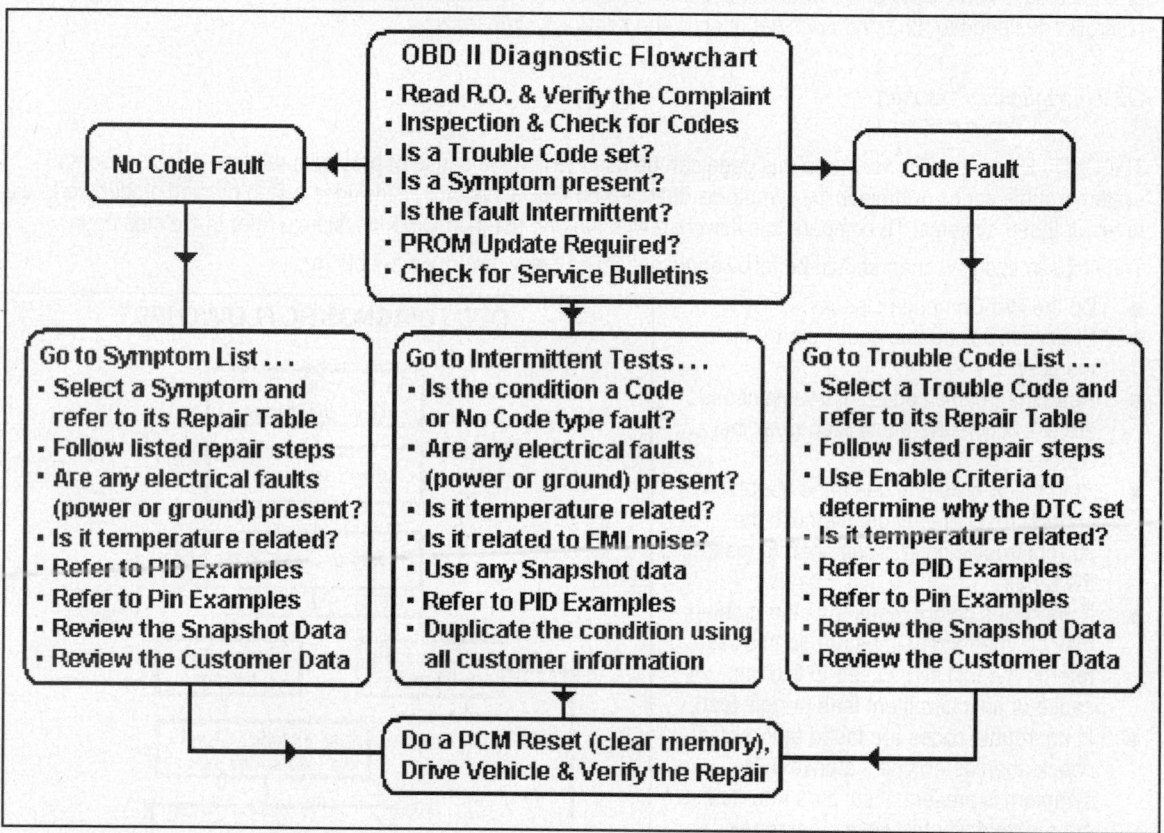

Flow Chart Steps

Here are some of the steps included in the Diagnostic Routine:

● Review the repair order and verify the customer complaint as described
● Perform a Visual Inspection of underhood or engine related items
● If the engine will not start, refer to No Start Tests
● If codes are set, refer to the trouble code list, select a code and use the repair chart
● If no codes are set, and a symptom is present, refer to the Symptom List
● Check for any related technical service bulletins (for both Code and No Code Faults)
● If the problem is intermittent in nature, refer to the special Intermittent Tests

OBD II SYSTEM OVERVIEW

The OBD II system was developed as a step toward compliance with California and Federal regulations that set standards for vehicle emission control monitoring for all automotive manufacturers. The primary goal of this system is to detect when the degradation or failure of a component or system will cause emissions to rise by 50%. Every manufacturer must meet OBD II standards by the 1996 model year. Some manufacturers began programs that were OBD II mandated as early as 1992, but most manufacturers began an OBD II phase-in period starting in 1994.

The changes to On-Board Diagnostics influenced by this new program include:

- Common Diagnostic Connector
- Expanded Malfunction Indicator Light Operation
- Common Trouble Code and Diagnostic Language
- Common Diagnostic Procedures
- New Emissions-Related Procedures, Logic and Sensors
- Expanded Emissions-Related Monitoring

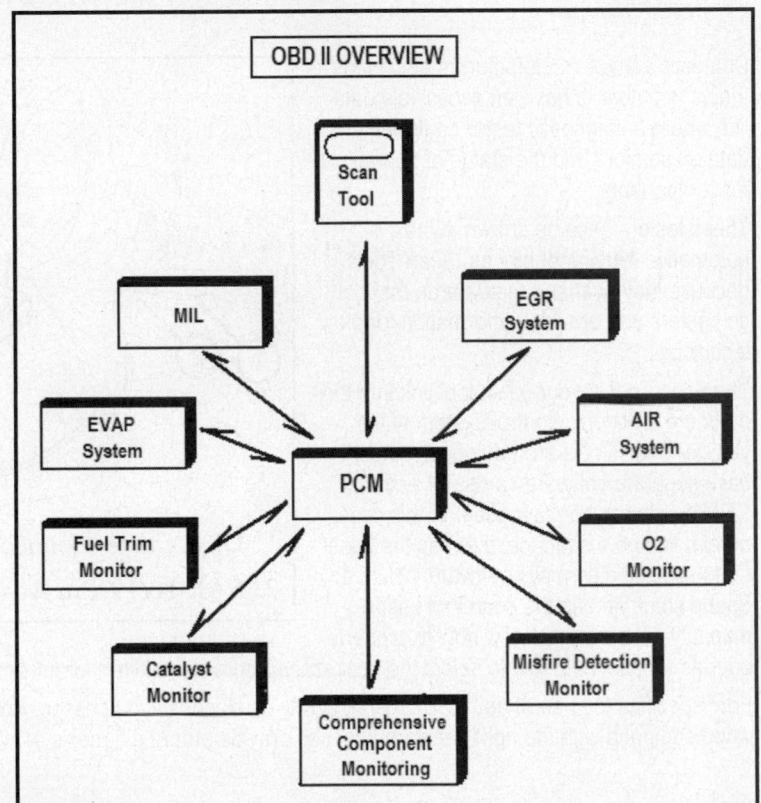

COMMON TERMINOLOGY

OBD II introduces common terms, connectors, diagnostic language and new emissions-related monitoring procedures. The most important benefit of OBD II is that all vehicles will have a common data output system with a common connector. This allows equipment Scan Tool manufacturers to read data from every vehicle and pull codes with common names and similar descriptions of fault conditions. In the future, emissions testing will require the use of an OBD II certifiable Scan Tool.

Diagnostic Tools & Circuit Testing

HAND TOOLS & METER OPERATION

To effectively use this or any diagnostic information, you should have a solid understanding of how to operate required tools and test equipment.

SCAN TOOLS

Domestic vehicle manufacturers designed their computers to have an accessible data line where a diagnostic tester could retrieve data on sensors and the status of operation for components.

These testers became known in the automotive repair industry as "Scan Tools" because they scanned the data on the computers and provided information for the technician.

The Scan Tool is your basic tool link into the on-board electronic control system of the vehicle. Scan Tools are equipped with, or have separate software cards, for each OEM needed to be diagnosed. In this case, always secure a scan tool that has the latest OEM-specific diagnostic software included. Spend some time in the scan tool user's manual to ensure you know how to properly operate the tool and how to select the necessary programs required for full and proper diagnostics.

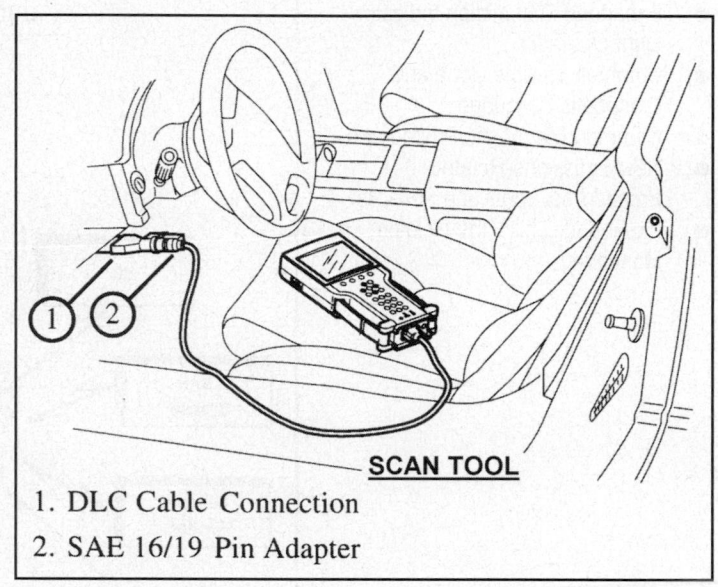

SCAN TOOL

1. DLC Cable Connection
2. SAE 16/19 Pin Adapter

Ford specifies the use of an "NGS" scan tool with its diagnostic processes. However, there are aftermarket scan tools, when equipped with the right software, that can provide proper diagnosis as well.

MALFUNCTION INDICATOR LAMP

Emission regulations require that a Malfunction Indicator Lamp (MIL) be illuminated when an emissions related fault is detected and that a Diagnostic Trouble Code be stored in the vehicle controller (PCM) memory.

When the MIL is illuminated, it is an indication of a problem within one of the electronic components or circuits. When the scan tool is attached to the Data Link Connector (DLC) in the vehicle, it can access the DTCs. In some situations, without the use of a scan tool, the MIL can be activated to flash a series of long and short flashes, which correspond to the numbering of the DTC.

OBD II guidelines define *when* an emissions-related fault will cause the MIL to activate and set a Diagnostic Trouble Code (DTC). There are some DTCs that will not cause the MIL to illuminate. OBD II guidelines determine how quickly the onboard diagnostics must be able to identify a fault, set the trouble code in memory and activate the MIL (lamp).

ELECTRONIC CONTROLS

You should have a basic knowledge of electronic controls when performing test procedures to keep from making an incorrect diagnosis or damaging components. Do NOT attempt to diagnose an electronic control problem without this basic knowledge!

ELECTRICITY & ELECTRICAL CIRCUITS

You should understand basic electricity and know the meaning of voltage (volts), current (amps), and resistance (ohms). You should understand what happens in an electrical circuit when it is open or shorted, and you should be able to identify an open circuit or shorted circuit using a DVOM. You should also be able to read and understand automotive electrical wiring diagrams and schematics.

CIRCUIT TESTING TOOLS

You should know when to use and when NOT to use a 12-volt test light during diagnosis of electronic controls (Do NOT use this tester unless specifically instructed to do so by a test procedure). Instead of using a 12-volt test light, you should use a DVOM or Lab Scope with a breakout box whenever a diagnostic procedure calls for a measurement at a PCM connector or component wiring harness.

Effective Diagnostics

GETTING STARTED

If you are reasonably certain that the problem is related to a particular electronic control system, the first step is to check for any stored trouble codes in that controller.

On vehicles with more than one vehicle controller (i.e., PCM, BCM, MIC, TCM, etc.), if you are unsure whether the problem is Powertrain related, start by checking for codes in the other controllers to determine if the problem is related to another vehicle system.

If there are no codes set, and you are certain which Powertrain subsystem has a problem, you can start by checking one of the subsystems. The subsystems include the Charging, Cooling, Fuel, Ignition and Speed Control systems.

If a wiring problem is found during testing, you will need to refer to wiring diagrams in the appropriate information resource. Using a wiring schematic can help you determine:

● Wiring circuits, circuit numbers, and wire colors
● Electrical component connector and component relationships within a circuit
● Power, ground, and splice locations within a circuit
● Related circuits connected into the circuit you are reviewing

Once you decide how to repair the vehicle, in addition to performing the repair, it is a good idea to clear any trouble codes that were set and to verify they do not reset.

To verify a repair, you should confirm that the Check Engine Light is operational and goes out after the 4-second key-on bulb check. Then, you need to duplicate the conditions present when the customer complaint occurred or when a trouble code set; these are the actual code conditions that caused a code to set. The individual code conditions and possible causes are included in **Section 3**. You can use this information to find out how to drive a vehicle for problem verification.

Contents

INTRODUCTION

SYSTEM CONTROL MODULES

Before attempting diagnosis of the Electronic Engine Control system, familiarize yourself with the basics of how the system is designed to operate. It consists of a central processing unit: Powertrain Control Module (PCM), Engine Control Module (ECM), Transmission Control Module (TCM) and/or the Body Control Module (BCM). These units are the "heart" of the electronic control systems on the vehicle. In some cases, these units are integral with one another, and on some applications, they are separate. As you get deeper into actual diagnostic testing, you will find out which units are used on the vehicle you are testing.

The PCM is a digital computer that contains a microprocessor. The PCM receives input signals from various sensors and switches that are referred to as PCM inputs. Based on these inputs, the PCM adjusts various engine and vehicle operations through devices that are referred to as PCM outputs. Examples of the input and output devices are shown in the graphic below.

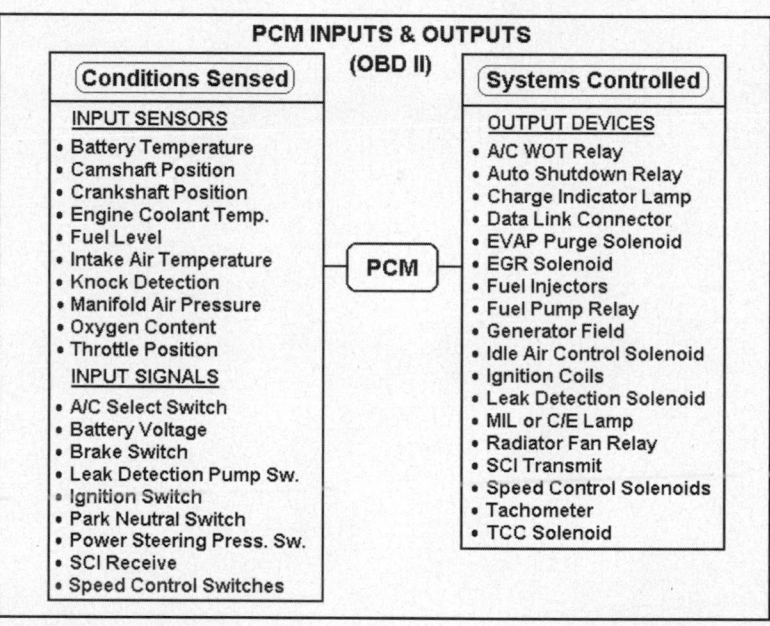

PCM INPUTS & OUTPUTS (OBD II)

Conditions Sensed

INPUT SENSORS
- Battery Temperature
- Camshaft Position
- Crankshaft Position
- Engine Coolant Temp.
- Fuel Level
- Intake Air Temperature
- Knock Detection
- Manifold Air Pressure
- Oxygen Content
- Throttle Position

INPUT SIGNALS
- A/C Select Switch
- Battery Voltage
- Brake Switch
- Leak Detection Pump Sw.
- Ignition Switch
- Park Neutral Switch
- Power Steering Press. Sw.
- SCI Receive
- Speed Control Switches

Systems Controlled

OUTPUT DEVICES
- A/C WOT Relay
- Auto Shutdown Relay
- Charge Indicator Lamp
- Data Link Connector
- EVAP Purge Solenoid
- EGR Solenoid
- Fuel Injectors
- Fuel Pump Relay
- Generator Field
- Idle Air Control Solenoid
- Ignition Coils
- Leak Detection Solenoid
- MIL or C/E Lamp
- Radiator Fan Relay
- SCI Transmit
- Speed Control Solenoids
- Tachometer
- TCC Solenoid

Input & Output Device Graphic (Example)

POWERTRAIN SUBSYSTEMS

A key to the diagnosis of the PCM and its subsystems is to determine which subsystems are on a vehicle. Examples of typical subsystems appear below:

- Cranking & Charging System
- Emission Control Systems
- Engine Cooling System
- Engine Air/Fuel Controls
- Exhaust System
- Ignition System
- Speed Control System
- Transaxle Controls

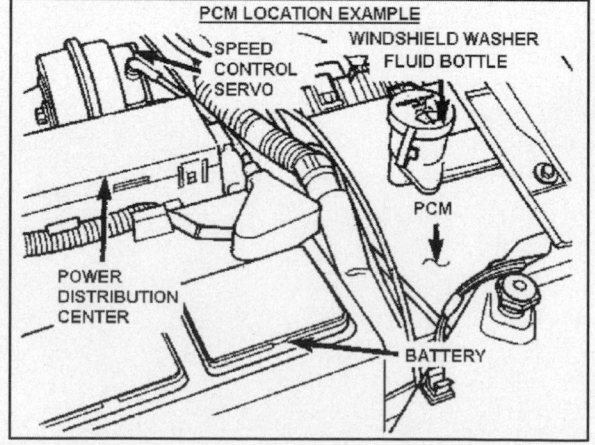

PCM LOCATION EXAMPLE
SPEED CONTROL SERVO
WINDSHIELD WASHER FLUID BOTTLE
PCM
POWER DISTRIBUTION CENTER
BATTERY

<u>WHERE TO BEGIN</u>

Diagnosis of engine performance or drivability problems on a vehicle with an onboard computer requires that you have a logical plan on how to approach the problem. The "Six Step Test Procedure" is designed to provide a uniform approach to repair any problems that occur in one or more of the vehicle subsystems.

The diagnostic flow built into this test procedure has been field-tested for several years at dealerships - *it is the starting point when a repair is required!*

It should be noted that a commonly overlooked part of the "Problem Resolution" step is to check for any related Technical Service Bulletins.

Six-Step Test Procedure

The steps outlined on this page were defined to help you determine how to perform a proper diagnosis. Refer to the flow chart that outlines the Six Step Test Procedure on the previous page as needed. The recommended steps include:

Verify the Complaint & Check for TSBs
To verify the customer complaint, the technician should understand the normal operation of the system. Conduct a thorough visual and operational inspection, review the service history, detect unusual sounds or odors, and gather diagnostic trouble code (DTC) information resources to achieve an effective repair.

This check should include videos, newsletters, and any other information in the form of TSBs or Dealer Service Bulletins. Analyze the complaint and then use the recommended Six Step Test Procedure. Utilize the wiring diagrams and theory of operation articles. Combine your own knowledge with efficient use of the available service information.

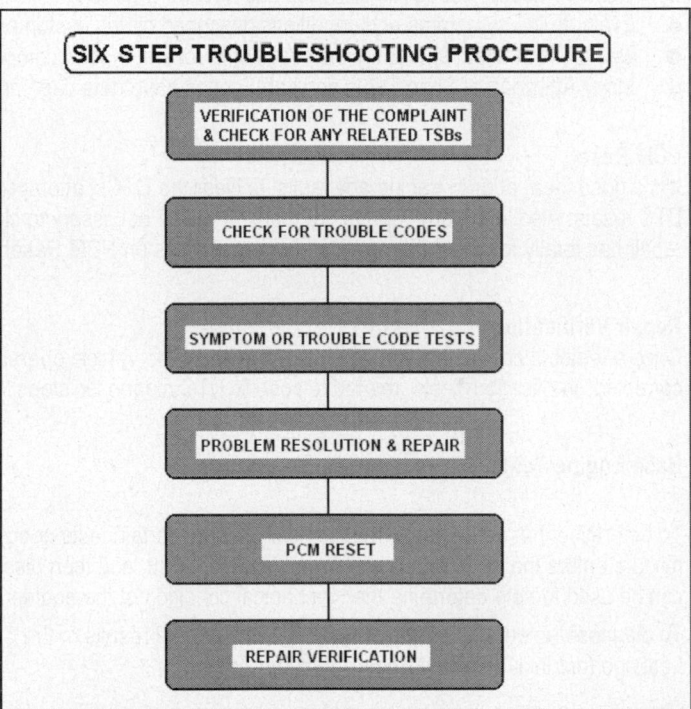

Verify the cause of any related symptoms that may or may not be supported by one or more trouble codes. There are various checks that can be performed to Engine Controls that will help verify the cause of a related symptom. This step helps to lead you in an organized diagnostic approach.

Check for Trouble Codes or Symptoms
Determine if the problem is a Code or a No Code Fault. Then refer to the appropriate published service diagnostic information to make the repair.

Problem Resolution & Repair

Once the problem component or circuit has been properly identified and verified using published diagnostic procedures, make any needed repairs or replacement to restore the vehicle to proper working order. If the condition has set a DTC, follow the designated repair chart to make an effective repair. If there is not a DTC set, but you can determine specific symptoms that are evident during the failure, select the symptom from the symptom tables and follow the diagnostic paths or suggestions to complete the repair or refer to the applicable component or system in service information.

If the vehicle does not set a DTC and has only intermittent operating failures or concerns, to resolve an intermittent fault, perform the following steps:

- Observe trouble codes, DTC modes and freeze frame data.
- Evaluate the symptoms and conditions described by the customer.
- Use a check sheet to identify the circuit or electrical system component.
- Many Aftermarket Scan Tools and Lab Scopes have data capturing features.

PCM Reset

It is a good idea, prior to tracing any faults, to clear the DTCs, attempt to replicate the condition and see if the same DTC resets. Also, once any repairs are made, it will be necessary to clear the DTC(s) – PCM Reset – to ensure the repair has totally resolved the problem. For procedures on PCM Reset, see DIAGNOSTIC TROUBLE CODES section.

Repair Verification

Once a repair is completed, the next step is to verify the vehicle operates properly and that the original symptom was corrected. Verification Tests, related to specific DTC diagnostic steps, can be used to verify a repair.

Base Engine Tests

To determine that an engine is mechanically sound, certain tests need to be performed to verify that the correct A/F mixture enters the engine, is compressed, ignited, burnt, and then discharged out of the exhaust system. These tests can be used to help determine the mechanical condition of the engine.

To diagnose an engine-related complaint, compare the results of the Compression, Cylinder Balance, Engine Cylinder Leakage (not included) and Engine Vacuum Tests.

Engine Compression Test

The Engine Compression Test is used to determine if each cylinder is contributing its equal share of power. The compression readings of all the cylinders are recorded and then compared to each other and to the manufacturer's specification (if available).

Cylinders that have low compression readings have lost their ability to seal. It this type of problem exists, the location of the compression leak must be identified. The leak can be in any of these areas: piston, head gasket, spark plugs, and exhaust or intake valves.

The results of this test can be used to determine the overall condition of the engine and to identify any problem cylinders as well as the most likely cause of the problem.

CAUTION: *Prior to starting this procedure, set the parking brake, place the gear selector in P/N and block the drive wheels for safety. The battery must be fully charged.*

Compression Test Procedure

1. Allow the engine to run until it is fully warmed up.

2. Remove the spark plugs and disable the Ignition system and the Fuel system for safety. Disconnecting the CKP sensor harness connector will disable both fuel and ignition (except on NGC vehicles).

3. Carefully block the throttle to the wide-open position.

4. Insert the compression gauge into the cylinder and tighten it firmly by hand.

5. Use a remote starter switch or ignition key and crank the engine for 3-5 complete engine cycles. If the test is interrupted for any reason, release the gauge pressure and retest. Repeat this test procedure on all cylinders and record the readings.

The lowest cylinder compression reading should not be less than 70% of the highest cylinder compression reading and no cylinder should read less than 100 psi.

Evaluating the Test Results

To determine why an individual cylinder has a low compression reading, insert a small amount of engine oil (3 squirts) into the suspect cylinder. Reinstall the compression gauge and retest the cylinder and record the reading. Review the explanations below.

Reading is higher - If the reading is higher at this point, oil inserted into the cylinder helped to seal the piston rings against the cylinder walls. Look for worn piston rings.

Reading did not change - If the reading didn't change, the most likely cause of the low cylinder compression reading is the head gasket or valves.

Low readings on companion cylinders - If low compression readings were recorded from cylinders located next to each other, the most likely cause is a blown head gasket.

Readings are higher than normal - If the compression readings are higher than normal, excessive carbon may have collected on the pistons and in the exhaust areas. One way to remove the carbon is with an approved brand of "Top Engine Cleaner."

Note: *Always clean spark plug threads and seat with a spark plug thread chaser and seat cleaning tool prior to reinstallation. Use anti-seize compound on aluminum heads.*

Engine Vacuum Tests

An engine vacuum test can be used to determine if each cylinder is contributing an equal share of power. Engine vacuum, defined as any pressure lower than atmospheric pressure, is produced in each cylinder during the intake stroke. If each cylinder produces an equal amount of vacuum, the measured vacuum in the intake manifold will be even during engine cranking, at idle speed, and at off-idle speeds.

Engine vacuum is measured with a vacuum gauge calibrated to show the difference between engine vacuum (the lack of pressure in the intake manifold) and atmospheric pressure. Vacuum gauge measurements are usually shown in inches of Mercury (in. Hg).

Note: *In the tests described in this article, connect the vacuum gauge to an intake manifold vacuum source at a point below the throttle plate on the throttle body.*

Engine Cranking Vacuum Test Procedure

The Engine Cranking Vacuum Test can be used to verify that low engine vacuum is not the cause of a No Start, Hard Start, Starts and Dies or Rough Idle condition (symptom).

The vacuum gauge needle fluctuations that occur during engine cranking are indications of individual cylinder problems. If a cylinder produces less than normal engine vacuum, the needle will respond by fluctuating between a steady high reading (from normal cylinders) and a lower reading (from the faulty cylinder). If more than one cylinder has a low vacuum reading, the needle will fluctuate very rapidly.

1. Prior to starting this test, set the parking brake, place the gearshift in P/N and block the drive wheels for safety. Then block the PCV valve and disable the idle air control device.

2. Disable the fuel and/or ignition system to prevent the vehicle from starting during the test (while it is cranking).

3. Close the throttle plate and connect a vacuum gauge to an intake manifold vacuum source. Crank the engine for three seconds (do this step at least twice).

The test results will vary due to engine design characteristics, the type of PCV valve and the position of the AIS or IAC motor and throttle plate. However, the engine vacuum should be steady between 1.0-4.0 in. Hg during normal cranking.

Engine Running Vacuum Test Procedure

1. Allow the engine to run until fully warmed up. Connect a vacuum gauge to a clean intake manifold source. Connect a tachometer or Scan Tool to read engine speed.

2. Start the engine and let the idle speed stabilize. Raise the engine speed rapidly to just over 2000 rpm. Repeat the test (3) times. Compare the idle and cruise readings.

Evaluating the Test Results

If the engine wear is even, the gauge should read over 16 in. Hg and be steady. Test results can vary due to engine design and the altitude above or below sea level.

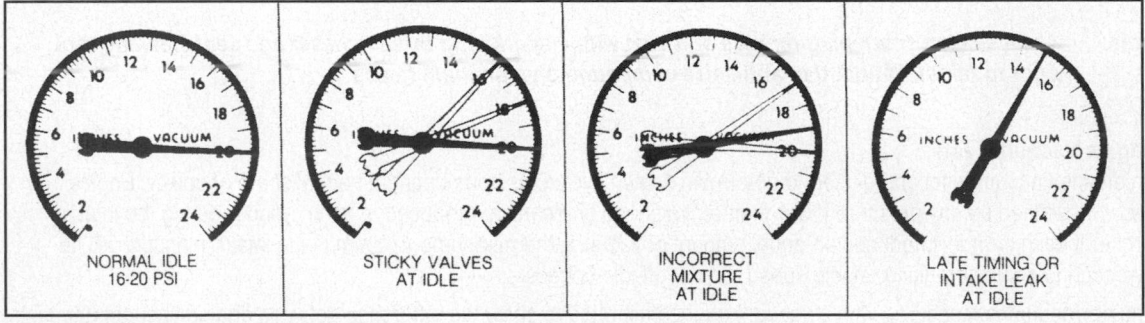

NORMAL IDLE 16-20 PSI   STICKY VALVES AT IDLE   INCORRECT MIXTURE AT IDLE   LATE TIMING OR INTAKE LEAK AT IDLE

Engine Running Vacuum Test Graphic

Ignition System Tests - Distributor

This next section provides an overview of ignition tests with examples of Engine Analyzer patterns for a Distributor Ignition System.

Preliminary Inspection

1. Perform these checks prior to connecting the Engine Analyzer:

2. Check the battery condition (verify that it can sustain a cranking voltage of 9.6v).

3. Inspect the ignition coil for signs of damage or carbon tracking at the coil tower.

4. Remove the coil wire and check for signs of corrosion on the wire or tower.

5. Test the coil wire resistance with a DVOM (it should be less than 7 k/ohm per foot).

6. Connect a *low* output spark tester to the coil wire and engine ground. Verify that the ignition coil can sustain adequate spark output while cranking for 3-6 seconds.

7. Connect the Engine Analyzer to the Ignition System, and choose Parade display. Run the engine at 2000 RPM, and note the display patterns, looking for any abnormalities.

Ignition System Tests - Distributorless

Perform the following checks prior to connecting the Engine Analyzer:

1. Check the battery condition (verify that it can sustain a cranking voltage of 9.6v).

2. Inspect the ignition coils for signs of damage or carbon tracking at the coil towers.

3. Remove the secondary ignition wires and check for signs of corrosion.

4. Test the plug wire resistance with a DVOM (specification varies from 15-30 k/ohm).

5. Connect a *low* output spark tester to a plug wire and to engine ground. Verify that the ignition coil can sustain adequate spark output for 3-6 seconds.

Secondary Ignition System Scope Patterns (V6 Engine)

1. Connect the Engine Analyzer to the ignition system.

2. Turn the scope selector to view the "Parade Display" of the ignition secondary.

3. Start the engine in Park or Neutral and slowly increase the engine speed from idle to 2000 rpm.

4. Compare actual display to the examples below.

Secondary Ignition System Graphic (V6 Engine)

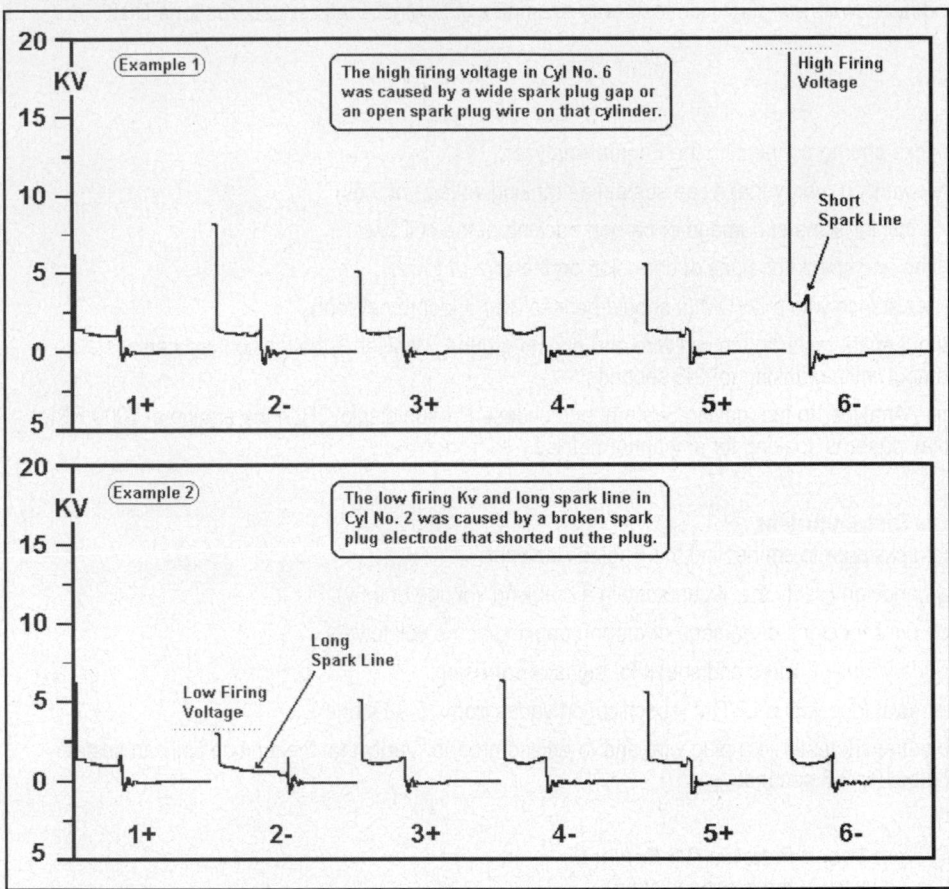

Symptom Diagnosis

To determine whether vehicle problems are identified by a set Diagnostic Trouble Code, you will first have to connect a proper scan tool to the Data Link Connector and retrieve any set codes. See DIAGNOSTIC TROUBLE CODES section for information on retrieving and reading codes.

If no codes are set, the problem must be diagnosed using only vehicle operating symptoms. A complete set of "No Code" symptoms is found in the SYMPTOM DIAGNOSIS (NO CODES) section.

Do NOT attempt to diagnose driveability symptoms without having a logical plan to use to determine which engine control system is the cause of the symptom - this plan should include a way to determine which systems do NOT have a problem! Remember, there are 2 kinds of NO CODE conditions:

● Symptom diagnosis, in which a continuous problem exists, but no DTC is set as a result. Therefore, only the operating symptoms of the vehicle can be used to pinpoint the root cause of the problem.
● Intermittent problem diagnosis, in which the problem does not occur all the time and does not set any DTCs.
Both of these NO CODE conditions are covered in Section 6: SYMPTOM DIAGNOSIS.

Accessing Components & Circuits

Every vehicle and every diagnostic situation is different. It is a good idea to first determine the best diagnostic path to follow using flow charts, wiring diagrams, TSBs, etc. Part of choosing steps is to determine how time-consuming and effective each step will be. It may be easy to access a component or circuit in one vehicle, but difficult in another. Many circuits are integrated into a large harness and are difficult to test. Many components are inaccessible without disassembly of unrelated systems.

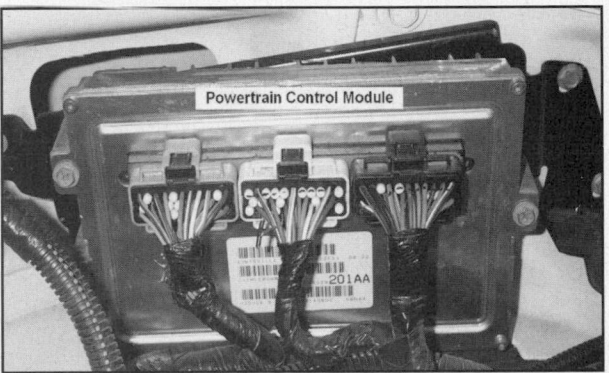

In the graphic, you will note that the protective covers have been removed from the PCM connectors, and any circuit can be easily identified and back probed. In other cases, PCM access is difficult, and it may be easier to access circuits at the component side of the harness.

Another important point to remember is that any circuit or component controlled by a relay or fused circuit can be monitored from the appropriate fuse box.

There is generally more than one of each type of relay or fuse. Therefore, swapping a suspect relay from another system may be more efficient than testing the relay itself. Relays and fuses may also be removed and replaced with fused jumper wires for testing circuits. Jumper wires can also provide a loop for inductive amperage tests.

Choosing the easiest way has its limitations, however. Remember that an appropriate signal on a PCM controlled circuit at an actuator means that the signal at the PCM is also good. However, a sensor signal at the sensor does not necessarily mean that the PCM is receiving the same signal. Think about the direction flow through a circuit, and not just what signal is appropriate, to save time without making costly assumptions.

Ford OBD II Diagnostics Contents

OBD II SYSTEMS

The California Air Resources Board (CARB) began regulating On-Board Diagnostic (OBD) systems for vehicles sold in California beginning with the 1988 model year. The initial requirements, known as OBD I, required the identification of the likely area of a fault with regard to the fuel metering system, EGR system, emission-related components and the PCM. Implementation of this new vehicle emission control monitoring regulation was done in several phases.

OBD I Systems

A malfunction indicator lamp (MIL) labeled *Check Engine Lamp* or *Service Engine Soon* was required to illuminate and alert the driver of a fault, and the need to service the emission controls. A diagnostic trouble code (DTC) was required to assist in identifying the system or component associated with the fault. If the fault that caused the MIL goes away, the MIL will go out and the code associated with the fault will disappear after a predetermined number of ignition cycles.

Following extensive research, CARB determined that by the time an Emission System component failed and caused the MIL to illuminate, that the vehicle could have emitted excess emissions over a long period of time. CARB also concluded that semi-annual or annual tailpipe tests were not catching enough of the vehicles with Emission Control systems operating at less than normal efficiency.

To take advantage of improvements in vehicle manufacturer adaptive and failsafe strategies, CARB developed new requirements designed to monitor the performance of Emission Control components, as well as to detect circuit and component hard faults. The new diagnostics were designed to operate under normal driving conditions, and the results of its tests would be viewable without any special equipment.

Enhanced OBD Systems

Beginning in the 1994 model year, both CARB and the EPA mandated Enhanced OBD systems, commonly known as OBD II. The objectives of OBD II were to improve air quality by reducing high in-use emissions caused by emission-related faults, reduce the time between the occurrence of a fault and its detection and repair, and assist in the diagnosis and repair of an emissions-related fault.

Differences Between OBD I & OBD II

As with OBD I, if an emission related problem is detected on a vehicle with OBD II, the MIL is activated and a code is set. However, that is the only real similarity between these systems. OBD II procedures that define emissions component and system tests, code clearing and drive cycles are more comprehensive than tests in the OBD I system.

Powertrain Control Module

The PCM in the OBD II system monitors almost all Emission Control systems that affect tailpipe or evaporative emissions. In most cases, the fault must be detected before tailpipe emissions exceed 1.5 times applicable 50K or 100K-mile FTP standards. If a component exceeds emission levels or fails to operate within the design specifications, the MIL is illuminated and a code is stored within two OBD II drive cycles.

The OBD II test runs continuously or once per trip (it depends on the driving mode requirement). Tests are run once per drive cycle during specific drive patterns called trips. Codes are stored in the PCM memory when a fault is first detected. In most cases, the MIL is turned on after two trips with a fault present. If the MIL is "on", it will go off after three consecutive trips if the same fault does not reappear. If the same fault is not detected after 40 engine warmup periods, the code will be erased (Fuel and Misfire faults require 80 warmup cycles).

OBD II Standardization

OBD II diagnostics require the use of a standardized diagnostic link connector (DLC), standard communication protocol and messages, and standardized trouble codes and terminology. Examples of this standardization are Freeze Frame Data and I/M Readiness Monitors.

Changes in MIL Operation

An important change for OBD II involves when to activate the MIL. The MIL must be activated by at least the second trip if vehicle emissions could exceed 1.5 times the FTP standard. If any single component or system failure would allow the emissions to exceed this level, the MIL is activated and a related code is stored in the PCM.

1994 OBD II Phase-In Systems

Starting in 1994, Ford began to "phase-in" the OBD II system on certain Mustang models (equipped with 3.8L V6 engines) and certain Thunderbird and Cougar models (equipped with 4.6L V8 engines). The OBD II "phase-in" system on these vehicles included the use of a Misfire Monitor that operated with a "lower threshold" Misfire Detection system designed to monitor misfires without setting any codes. In addition, the EVAP Monitor was not operational on these vehicles.

1996 & Later OBD II Systems

By the 1996 model year, all California passenger cars and trucks up to 14,000 lb. GVWR, and all Federal passenger cars and trucks up to 8,600 lb. GWVR were required to comply with the CARB-OBD II or EPA OBD requirements. The requirements applied to diesel and gasoline vehicles, and were phased in on alternative-fuel vehicles.

Flash EEPROM

The PCM on an EEC-V system includes a flash electrically erasable and programmable read-only memory (EEPROM) module. This software, in the form of an integrated circuit, contains the program used by the PCM to control the vehicle Powertrain. The EEPROM can be updated (reprogrammed) at a Ford dealership through the DLC and SBDS without removing the PCM. It can also be updated if it is removed and then taken to the parts counter. Changes to vehicle calibration are performed as directed by Recall Letters and Technical Service Bulletins. Refer to TSB 98-26-3 for an explanation of the complete procedure.

EEC-V Powertrain Controller

The purpose of the EEC-V system is to provide optimum control of the engine and transmission while meeting the objectives of the OBD II regulations. The PCM connects to various input and output devices through a wiring harness via a 104-pin connector (88 pins on the Villager and 150 pins on LS models). The PCM receives inputs from various sensors and switches, performs calculations based on data stored in an integrated circuit called Keep Alive Memory (KAM), and controls various output devices (i.e., actuators, relays, and solenoids).

EEC-V Hardware & Software

The EEC-V system hardware components include:

- All related actuators, relays, solenoids, sensors and switches
- The CCRM, PCM and VLCM (modules) and connecting wiring

The EEC-V system software components include:

- Programs that make up the strategies used by the PCM to control operation of the engine, electronic transmission, Failure Mode Effects Management, idle speed and fuel delivery systems.
- The EEC-V system includes backup or fail-safe circuitry should the Central Processing Unit (CPU) or EEPROM in the PCM fail.

Diagnostic Test Modes

The "test mode" messages available on a Scan Tool are listed below:

- Mode $01: Used to display Powertrain Data (PID data)
- Mode $02: Used to display any stored Freeze Frame data
- Mode $03: Used to request any trouble codes stored in memory
- Mode $04: Used to request that any trouble codes be cleared
- Mode $05: Used to monitor the Oxygen sensor test results
- Mode $06: Used to monitor Non-Continuous Monitor test results
- Mode $07: Used to monitor the Continuous Monitor test results
- Mode $08: Used to request control of a special test (EVAP Leak)
- Mode $09: Used to request vehicle information (INFO MENU)

Onboard Diagnostics

OBD II Systems incorporate the dedicated Ford test procedures built into the system. In effect, the key on, engine off (KOEO) and key on, engine running (KOER) Self-Tests are still an important functional part of the diagnostics as with earlier Ford diagnostics for OBD I systems.

Trouble codes associated with OBD II System are linked to the Ford code repair charts (Pinpoint Tests) using the customary CONT or MEM, KOEO, and KOER designators. In addition, the OBD II Main Monitors frequently run as part of the dedicated Ford Self-Tests.

Diagnostic Procedure

The Diagnostic Repair Chart on this page should be used as follows:

- Trouble Code Diagnosis - Refer to the Code List (in this section) or electronic media for a repair chart for a particular trouble code.
- Driveability Symptoms - Refer to the Driveability Symptom List in other manuals or in electronic media.
- Intermittent Faults - Refer to the Intermittent Test Procedures.
- OBD II Drive Cycles - Refer to the Comprehensive Component Monitor or a Main Monitor drive cycle articles in this section.

OBD II Repair Chart Graphic

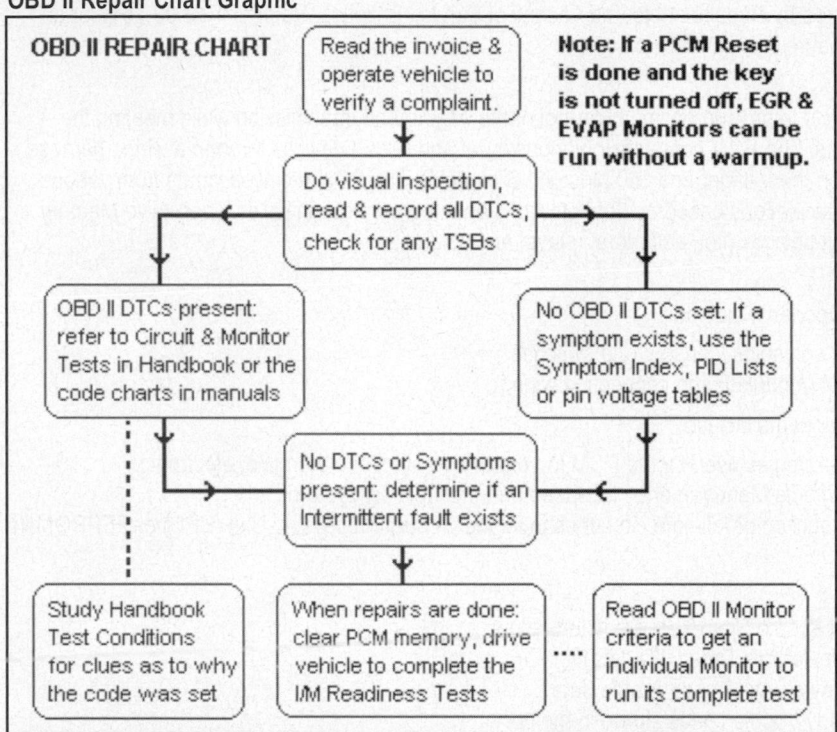

OBD SYSTEM TERMINOLOGY

It is very important that service technicians understand terminology related to OBD II test procedures. Several of the essential OBD II terms and definitions are discussed on the next few pages.

Two-Trip Detection

Frequently, an emission system or component must fail a Monitor test more than once before the MIL is activated. In these cases, the first time an OBD II Monitor detects a fault during any drive cycle it sets a pending code in the PCM memory.

A pending code, which is read by selecting DDL from the Scan Tool menu, appears when Memory or Continuous codes are read. In order for a pending code to cause the MIL to activate, the original fault must be repeated under *similar conditions*.

This is a critical issue to understand as a pending code could remain in the PCM for a long time before the conditions that caused the code to set reappear. This type of OBD II trouble code logic is frequently referred to as the "Two-Trip Detection Logic".

NOTE: *Codes related to a Misfire fault and Fuel Trim can cause the PCM to activate the MIL after <u>one</u> trip because   these codes are related to critical emission systems that could cause emissions to exceed the federally mandated limits.*

Similar Conditions

If a pending code is set because of a Misfire or Fuel System Monitor fault, the vehicle must meet *similar conditions* for a second trip before the code matures the PCM activates the MIL and stores the code in memory. Refer to Note above for exceptions to this rule. The meaning of *similar conditions* is important when attempting to repair a fault detected by a Misfire or Fuel System Monitor.

To achieve *similar conditions*, the vehicle must reach the following engine running conditions simultaneously:

●   Engine speed must be within 375 rpm of the speed when the trouble code set.
●   Engine load must be within 10% of the engine load when the trouble code set.
●   Engine warmup state must match a previous cold or warm state.

Summary - Similar conditions are defined as conditions that match the conditions recorded in Freeze Frame when the fault was first detected and the trouble code was set in the PCM memory.

OBD II Warmup Cycle

The meaning of the expression *warmup cycle* is important. Once the fault that caused an OBD II trouble code to set is gone and the MIL is turned off, the PCM will not erase that code until after 40 warmup cycles. *This is the purpose of the warmup cycle - to help clear stored codes.*

OBD II Warmup Cycle Graphic

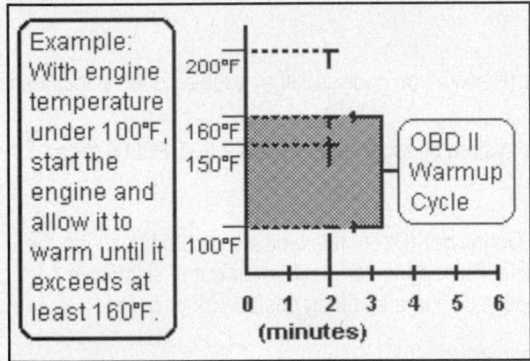

However, trouble codes related to a Fuel system or Misfire fault require that 80 warmup cycles occur without the fault reappearing before codes related to these monitors will be erased from the PCM memory.

NOTE: *A warmup cycle is defined as vehicle operation (after an engine off and cool-down period) when the engine temperature rises to at least 40°F and reaches at least 160°F.*

Malfunction Indicator Lamp

If the PCM detects an emission related component or system fault for two consecutive drive cycles on OBD II systems, the MIL is turned on and a trouble code is stored. The MIL is turned off if three consecutive drive cycles occur without the same fault being detected.

Most trouble codes related to a MIL are erased from KAM after 40 warmup periods if the same fault is not repeated. The MIL can be turned off after a repair by using the Scan Tool PCM Reset function.

Freeze Frame Data

The term *Freeze Frame* is used to describe the engine conditions that are recorded in PCM memory at the time a Monitor detects an emissions related fault. These conditions include fuel control state, spark timing, engine speed and load.

Freeze Frame data is recorded when a system fails the first time for two-trip type faults. The Freeze Frame Data will only be overwritten by a different fault with a "higher emission priority."

Scan Tool Freeze Frame Graphic

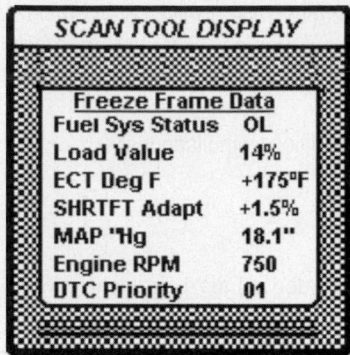

SCAN TOOL DISPLAY	
Freeze Frame Data	
Fuel Sys Status	OL
Load Value	14%
ECT Deg F	+175°F
SHRTFT Adapt	+1.5%
MAP "Hg	18.1"
Engine RPM	750
DTC Priority	01

Diagnostic Trouble Codes

The OBD II system uses a *Diagnostic Trouble Code (DTC)* identification system established by the Society of Automotive Engineers (SAE) and the EPA. The first letter of a DTC is used to identify the type of computer system that has failed as shown below:

- The letter 'P' indicates a Powertrain related device
- The letter 'C' indicates a Chassis related device
- The letter 'B' indicates a Body related device
- The letter 'U' indicates a Data Link or Network device code.

The first DTC number indicates a generic (P0xxx) or manufacturer (P1xxx) type code. A list of trouble codes is included in this section.

The number in the hundreds position indicates the specific vehicle system or subgroup that failed (i.e., P0300 for a Misfire code, P0400 for an emission system code, etc.).

Data Link Connector

Ford vehicles equipped with OBD II use a standardized Data Link Connector (DLC). It is typically located between the left end of the instrument panel and 12 inches past vehicle centerline. The connector is mounted out of sight from vehicle passengers, but should be easy to see from outside by a technician in a kneeling position (door open). *However, not all of the connectors are located in this exact area.*

The DLC is rectangular in design and capable of accommodating up to 16 terminals. It has keying features to allow easy connection to the Scan Tool. Both the DLC and Scan Tool have latching features used to ensure that the Scan Tool will remain connected to the vehicle during testing.

Once the Scan Tool is connected to the DLC, it can be used to:

- Display the results of the most current I/M Readiness Tests
- Read and clear any diagnostic trouble codes
- Read the Parameter ID (PID) data from the PCM
- Perform Enhanced Diagnostic Tests (manufacturer specific)

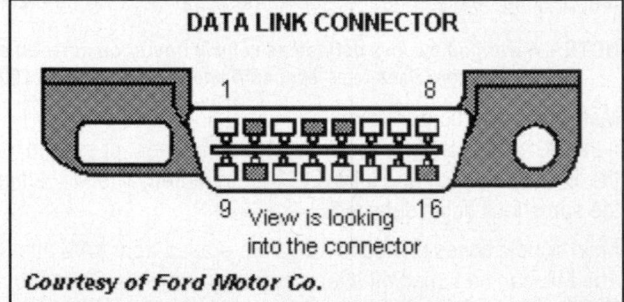

DATA LINK CONNECTOR

1   8

9 View is looking 16
into the connector

Courtesy of Ford Motor Co.

Standard Corporate Protocol

On vehicles equipped with OBD II, a Standard Corporate Protocol (SCP) communication language is used to exchange bi-directional messages between stand-alone modules and devices. With this type of system, two or more messages can be sent over one circuit.

OBD II Monitor Software

The Diagnostic Executive contains software designed to allow the PCM to organize and prioritize the Main Monitor tests and procedures, and to record and display test results and diagnostic trouble codes.

The functions controlled by this software include:

- To control the diagnostic system so the vehicle continues to operate in a normal manner during testing.
- To ensure the OBD II Monitors run during the first two sample periods of the Federal Test Procedure.
- To ensure that all OBD II Monitors and their related tests are sequenced so that required inputs (enable criteria) for a particular Monitor are present prior to running that particular Monitor.
- To sequence the running of the Monitors to eliminate the possibility of different Monitor tests interfering with each other or upsetting normal vehicle operation.
- To provide a Scan Tool interface by coordinating the operation of special tests or data requests.

Cylinder Bank Identification

Engine sensors are identified on each engine cylinder bank as explained next.

Bank - A specific group of engine cylinders that share a common control sensor (e.g., Bank 1 identifies the location of Cyl 1 while Bank 2 identifies the cylinders on the opposite bank).

An example of the cylinder bank configuration for a Ford Taurus with FWD and a 3.0L V6 (VIN U) engine is shown in the Graphic on this page.

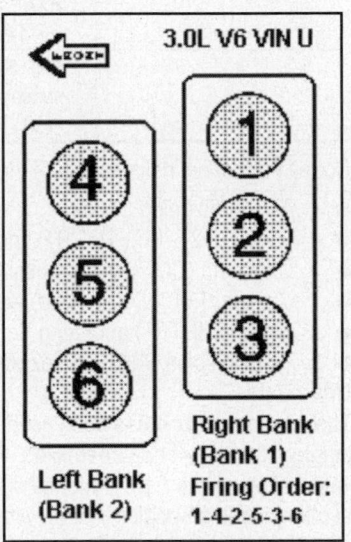

Oxygen Sensor Identification

Oxygen sensors are identified in each cylinder bank as the front O2S (pre-catalyst) or rear O2S (post-catalyst). The acronym HO2S-11 identifies the front oxygen sensor located (Bank 1) while the HO2S-21 identifies the front oxygen sensor in Bank 2 of the engine, and so on.

OBD II Monitor Test Results

Generally, when an OBD II Monitor runs and fails a particular test during a trip, a pending code is set. If the same Monitor detects a fault for two consecutive trips, the MIL is activated and a code is set in PCM memory. The results of a particular Monitor test indicate that an emission system or component failed - not the circuit that failed!

To determine where the fault is located, follow the correct code repair chart, symptom diagnosis or intermittent test. The code and symptom repair charts are the most efficient way to repair an OBD II system.

NOTE:: *Two important pieces of information that can help speed up a diagnosis are code conditions (including all enable criteria), and the parameter information (PID) stored in the Freeze Frame at the time a trouble code is set and stored in memory.*

Adaptive Fuel Control Strategy

The PCM incorporates an Adaptive Fuel Control Strategy that includes an adaptive fuel control table stored in KAM to compensate for normal changes in fuel system devices due to age or engine wear.

During closed loop operation, the Fuel System Monitor has two methods of attempting to maintain an ideal A/F ratio of 14:7 to 1 (they are referred to as short term fuel trim and long term fuel trim).

NOTE: *If a fuel injector, fuel pressure regulator or oxygen sensor is replaced the KAM in the PCM should be cleared by a PCM Reset step so that the PCM will not use a previously learned strategy.*

Short Term Fuel Trim

Short term fuel trim (SHRTFT) is an engine operating parameter that indicates the amount of short term fuel adjustment made by the PCM to compensate for operating conditions that vary from the ideal A/F ratio condition. A SHRTFT number that is negative (-15%) means that the HO2S is indicating a richer than normal condition to the PCM, and that the PCM is attempting to lean the A/F mixture. If the A/F ratio conditions are near ideal, the SHRTFT number will be close to 0%.

Long Term Fuel Trim

Long term fuel trim (LONGFT) is an engine parameter that indicates the amount of long term fuel adjustment made by the PCM to correct for operating conditions that vary from ideal A/F ratios. A LONGFT number that is positive (+15%) means that the HO2S is indicating a leaner than normal condition, and that it is attempting to add more fuel to the A/F mixture. If A/F ratio conditions are near ideal, the LONGFT number will be close to 0%. The PCM adjusts the LONGFT in a range from -35 to +35%. The values are in percentage on a Scan Tool.

Enable Criteria

The term *enable criteria* describe the conditions necessary for any of the OBD II Monitors to run their diagnostic tests. Each Monitor has specific conditions that must be met before it will run its test.

Examples of trouble code conditions for DTC P0460 and P1168 are shown below:

DTC	Trouble Code Title & Conditions
	EVAP System Small Leak Conditions: Cold startup, engine running at off-idle conditions, then the PCM detected a small leak (a leak of more than 0.040") in the EVAP system.
	FRP Sensor in Range but Low Conditions: Engine running, then the PCM detected that the FRP sensor signal was out-of-range low. *Scan Tool Tip: Monitor the FRP PID for a value below 80 psi (551 kPa).*

Code information includes any of the following examples:

- Air Conditioning Status
- BARO, ECT, IAT, TFT, TP and Vehicle Speed sensors
- Camshaft (CMP) and Crankshaft (CKP) sensors
- Canister Purge (duty cycle) and Ignition Control Module Signals
- Short (SHRTFT) and Long Term (LONGFT) Fuel Trim Values
- Transmission Shift Solenoid On/Off Status

Drive Cycle

The term *drive cycle* has been used to describe a drive pattern used to verify that a trouble code, driveability symptom or intermittent fault had been fixed. With OBD II systems, this term is used to describe a vehicle drive pattern that would allow all the OBD II Monitors to initiate and run their diagnostic tests. For OBD II purposes, a minimum *drive cycle* includes an engine startup with continued vehicle operation that exceeds the amount of time required to enter closed loop fuel control.

OBD II Trip

The term *OBD II Trip* describes a method of driving the vehicle so that one or more of the following OBD II Monitors complete their tests:

- Comprehensive Component Monitor (completes anytime in a trip)
- Fuel System Monitor (completes anytime during a trip)
- EGR System Monitor (completes after accomplishing a specific idle and acceleration period)
- Oxygen Sensor Monitor (completes after accomplishing a specific steady state cruise speed for a certain amount of time)

OBD II Drive Cycle

The ambient or inlet air temperature must be from 40-100°F to initiate the OBD II drive cycle. Allow the engine to warm to 130°F prior to starting the test (except for Escort and Tracer models that require the engine be less than 100°F to run the EVAP Monitor).

Connect the Scan Tool prior to beginning the drive cycle. Some tools are designed to emit a three-pulse beep when all of the OBD II Monitors complete their tests and DTC P1000 has been erased.

NOTE: *The IAT PID must be from 50-100°F to start the drive cycle. If it is less than 50°F at any time during the highway part of the drive cycle, the EVAP Monitor may not complete. The engine should reach 130°F before starting the trip except on Escort &Tracer models where a cold engine startup of less than 100°F is used before attempting to verify an EVAP system fault. Disengage the PTO before proceeding (PTO PID will show OFF) if applicable. For the EVAP Running Loss system, verify FLI PID is at 15-85%. Some Monitors require very specific idle and acceleration steps.*

Drive Cycle Procedure

The primary intention of the Ford OBD II drive cycle is to clear a DTC P1000. The drive cycle can also be used to assist in identifying any OBD II concerns present through total Monitor testing. Perform all of the Vehicle Preparation steps. Then refer to the Drive Cycle Table and Graphic below for details on how to run a Ford OBD II Drive Cycle.

Connect a Scan Tool and have an assistant watch the Scan Tool I/M Readiness Status to determine when the Catalyst, EGR, EVAP, Fuel System, O2 Sensor, Secondary AIR and Misfire Monitors complete.

OBD II Drive Cycle Graphic

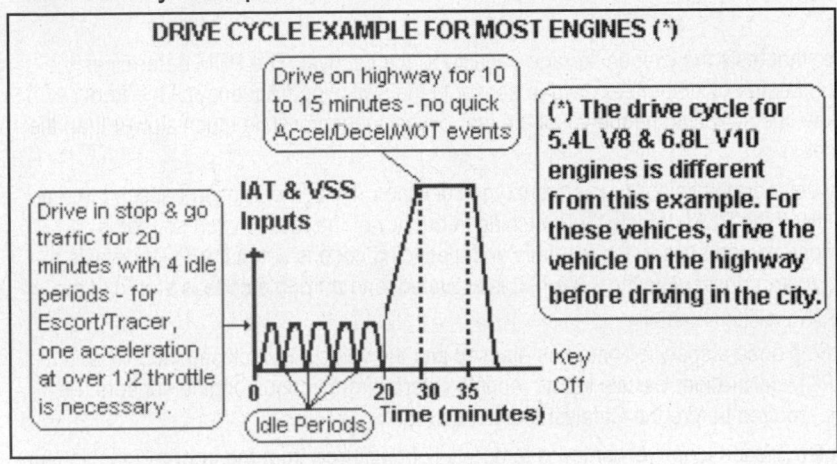

OBD II SYSTEM MONITORS

Comprehensive Component Monitor

OBD II regulations require that all emission related circuits and components controlled by the PCM that could affect emissions are monitored for circuit continuity and out-of-range faults. The Comprehensive Component Monitor (CCM) consists of four different monitoring strategies: two for inputs and two for output signals. *The CCM is a two trip Monitor for emission faults on Ford vehicles.*

Input Strategies

One input strategy is used to check devices with analog inputs for opens, shorts, or out-of-range values. The CCM accomplishes this task by monitoring A/D converter input voltages. The analog inputs monitored include the ECT, IAT, MAF, TP and Transmission Range Sensors signals.

A second input strategy is used to check devices with digital and frequency inputs by performing rationality checks. The PCM uses other sensor readings and calculations to determine if a sensor or switch reading is correct under existing conditions. Some tests run continuously, some only after actuation. The inputs monitored by the PCM include the CMP, IDM, PIP, OSS, TSS and VSS signals.

Output Strategies

An Output State Monitor in the PCM checks outputs for opens or shorts by observing the control voltage level of the related device. The control voltage is low with it on, and high with the device off. Monitored outputs include the EPC, SS1, SS2, SS3, TCC, HFC, VMV, WOT A/C Cutout and the HO2S Heater.

Component Monitor Graphic

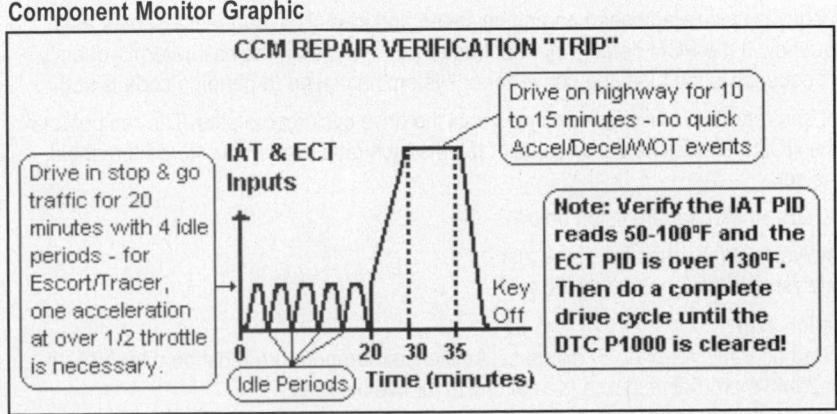

IAC Motor Test

The PCM monitors the IAC system in order to "learn" the closed loop correlation it needs to reposition the IAC solenoid (a rationality check).

Catalyst Efficiency Monitor

The Catalyst Monitor is a PCM diagnostic run once per drive cycle that uses the downstream heated Oxygen Sensor (HO2S-12) to determine if a catalyst falls below a minimum level of effectiveness in its ability to control exhaust emissions. The PCM uses a program to determine the catalyst efficiency based on the oxygen storage capacity of the catalytic converter.

Catalyst Monitor Operation

The Catalyst Monitor is a diagnostic that tests the oxygen storage capacity of the catalyst. The PCM determines the capacity by comparing the switching frequency of the rear oxygen sensor to the switching frequency of the front oxygen sensor. If the catalyst is okay, the switching frequency of the rear oxygen sensor will be much slower than the frequency of the front oxygen sensor.

However, as the catalyst efficiency deteriorates its ability to store oxygen declines. This deterioration causes the rear oxygen sensor to switch more rapidly. If the PCM detects the switching frequency of the rear oxygen sensor is approaching the frequency of the front oxygen sensor, the test fails and a pending code is set. If the PCM detects a fault on consecutive trips *(from two to six consecutive trips)* the MIL is activated, and a trouble code is stored in the PCM memory.

The Catalyst Monitor runs after startup once a specified time has elapsed and the vehicle is in closed loop. The amount of time is subject to each PCM calibration. Certain inputs (enable criteria) from various engine sensors (i.e., CKP, ECT, IAT, TPS and VSS) are required before the Catalyst Monitor can run.

Once the Catalyst Monitor is activated, closed loop fuel control is temporarily transferred from the front oxygen sensor to the rear oxygen sensor. During the test, the Monitor analyzes the switching frequency of both sensors to determine if a catalyst has degraded.

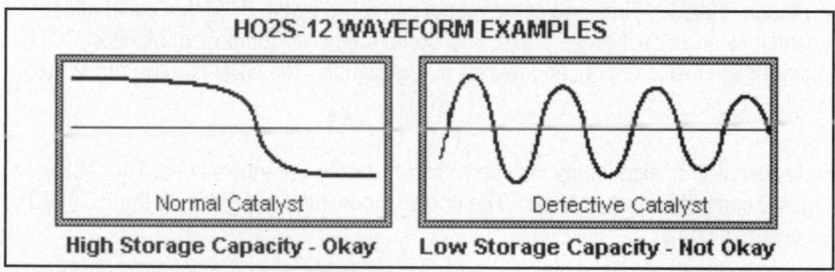

HO2S-12 WAVEFORM EXAMPLES

Normal Catalyst — High Storage Capacity - Okay

Defective Catalyst — Low Storage Capacity - Not Okay

Rear Oxygen Sensor Waveform Graphic

Catalyst Efficiency Monitor

Catalyst Test - Steady State Catalyst Efficiency Test

The PCM transfers the input for closed loop fuel control from the front HO2S-11 to the rear HO2S-21 during this test. The PCM measures the output frequency of the rear HO2S. This "test frequency" indicates the current oxygen storage capacity of the converter. The slower the frequency of the test result, the higher the efficiency of the converter.

Catalyst Test - Calibrated Frequency Test

In Part 2 of the test a second frequency is calculated based on engine speed and load. This frequency serves as a high limit threshold for the test frequency. If the PCM detects the test frequency is less than the calibrated frequency the catalyst passes the test. If the frequency is too high, the converter or system has failed (a pending code is set).

The sequence of counting the front and rear O2S switches continues until the drive cycle completes. The ratio of total HO2S-21 switches to the total of the HO2S-11 switches is calculated. If the switch ratio is over the stored threshold, the catalyst has failed and a code is set.

Trouble Codes associated with this OBD II Monitor are listed below:

- DTC P0420, P0421 - The catalyst in Bank 1 has failed the test
- DTC P0430, P0431 - The catalyst in Bank 2 has failed the test

Catalytic Monitor Repair Verification Trip

Start the engine, and drive in stop and go traffic for over 20 minutes. (Ambient air temperature must be over 50°F to run this test). Drive at speeds from 25-40 mph (6 times) and then at cruise for five minutes.

Catalyst Monitor Graphic

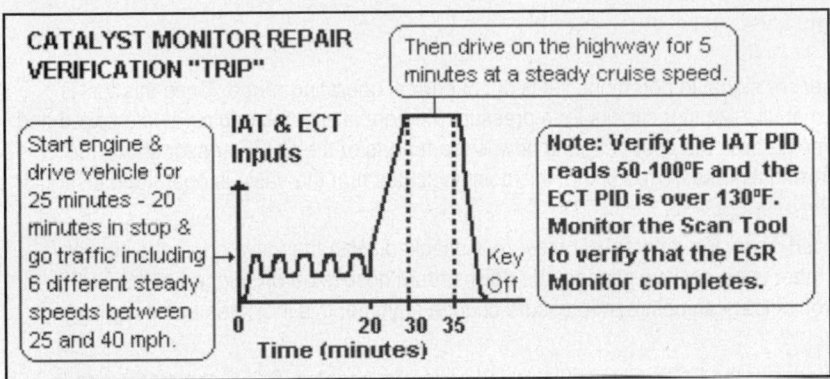

Possible Causes of a Catalyst Efficiency Fault
- Base Engine faults (engine mechanical)
- Exhaust leaks or contaminated fuel

EGR System Monitor

The EGR System Monitor is a PCM diagnostic run once per trip that monitors EGR system component functionality and components for faults that could cause vehicle tailpipe levels to exceed 1.5 times the FTP Standard. A series of sequenced tests is used to test the system.

Differential Pressure Feedback EGR System

This system includes a DPFE sensor, vacuum regulator solenoid, EGR valve, orifice tube assembly, the PCM and related wiring/hoses.

EGR DPFE System Graphic

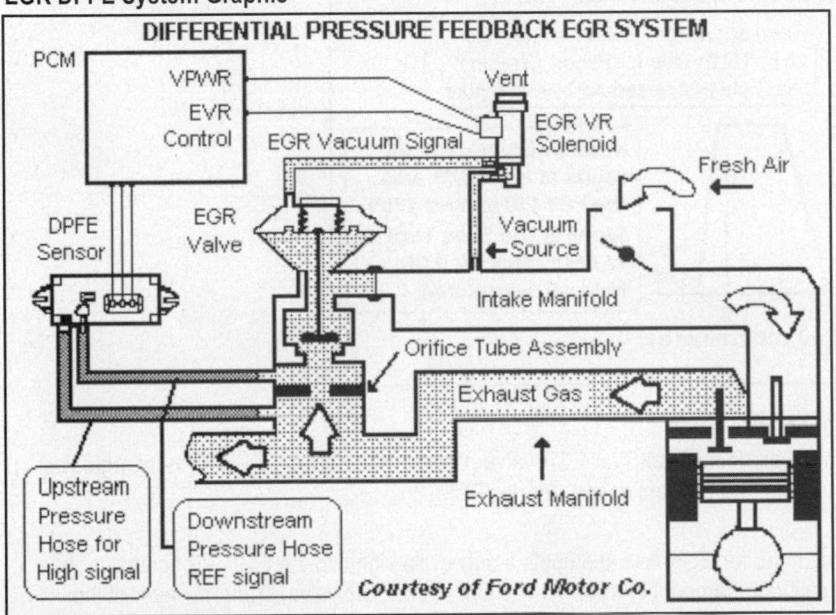

When PCM strategy dictates that EGR flow should be enabled, it sends a duty cycle command to the EGR Vacuum Regulator (VR) solenoid. The VR solenoid responds by delivering a portion of its manifold vacuum signal to the EGR valve. Once the vacuum applied to the EGR valve is sufficient to overcome the EGR valve spring force, the valve opens allowing exhaust gases to enter the intake system through a metering orifice located in the orifice tube assembly. A pressure drop is created across this orifice that is proportional to the rate of EGR flow entering the intake manifold.

The DPFE sensor senses the differential pressure signal across the orifice and sends an analog signal to the PCM that indicates the rate of EGR flow. This feedback signal is used to adjust the EVR duty cycle to achieve the desired amount of EGR flow.

EGR System Monitor (Continued)

DPFE EGR Valve - Test One

First, the Monitor tests the DPFE sensor signal to determine if it is out of normal operating range. Once this test is passed, and all enable criteria are met, the Monitor checks for a pressure differential with the engine at idle speed and with the EGR valve closed. At this point, both the upstream and downstream ports of the DPFE sensor should be reading exhaust pressure. Any differential pressure reading at this point indicates that the valve is open (and it should not be open).

Next, with the EGR valve commanded open, the differential pressure is checked. With the valve open, the sensor should signal a positive change. If there is no positive change, the downstream hose is off or plugged at the DPFE sensor. If the DPFE sensor does not indicate an upstream pressure change anytime, this indicates that it is plugged.

DPFE EGR Valve - Test Two

Next, the EGR DPFE sensor is tested with the EGR valve commanded open. If a negative DPFE sensor reading is detected, the indication is that the downstream and upstream hoses are reversed on the EGR valve. Then, the EGR valve flow is checked.

Once the engine reaches cruise speed with the ECT, MAP and TP sensor signals constant, the PCM compares the actual and expected DPFE sensor values. If the actual reading is lower than the expected value due to a restriction, the PCM fails the test and sets a code.

Possible Causes of an EGR System Failure

- Leaks or disconnects in upstream or downstream vacuum hoses
- Damaged DPFE or EGR EVP sensor
- Plugged or restricted DPFE or EGR VP sensor or orifice assembly

EGR Monitor Graphic

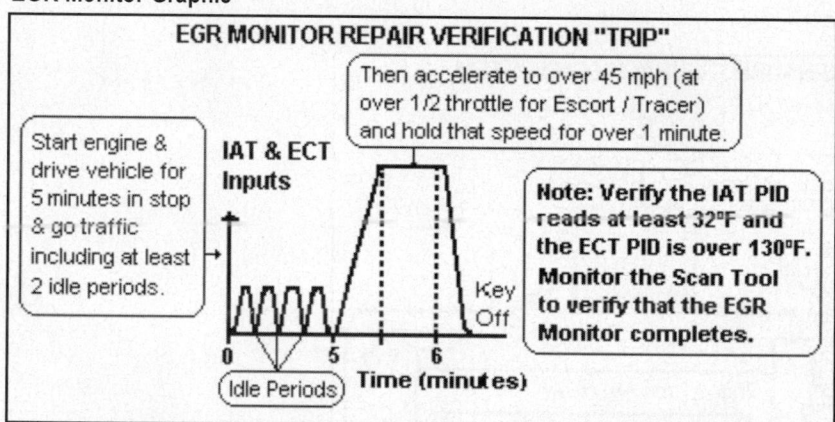

Sonic EGR System (Explorer 5.0L V8)

The Sonic EGR system used on these models includes an EGR valve position (VP) sensor, EGR vacuum regulator (VR) solenoid, EGR valve, system wiring and vacuum hoses, and the PCM.

System Operation

When the PCM strategy dictates the need for EGR flow, it outputs a duty cycle signal to the EGR VR solenoid. The solenoid responds by delivering a calibrated amount of manifold vacuum to the EGR valve. The remainder of the source vacuum is vented into the atmosphere.

At some point, engine vacuum applied to the EGR valve is sufficient to overcome the spring force of the valve, and the valve begins to open. This action allows exhaust gases to enter the intake manifold. As the EGR valve pintle lifts upward, it causes the EGR VP sensor to also lift upward in direct proportion to the amount of EGR opening.

In this Sonic EGR system, the PCM uses the VP sensor signal as an indication of the amount of EGR flow into the engine. It adjusts the VR solenoid duty cycle signal to achieve the desired amount of EGR flow.

If a fault is detected during the EGR Monitor test that could cause the tailpipe emissions to exceed 1.5 times the FTP Standard, the test fails and a pending code is set. If the EGR test fails on two consecutive trips, the MIL is activated and a hard code is set.

Purge Flow System Graphic

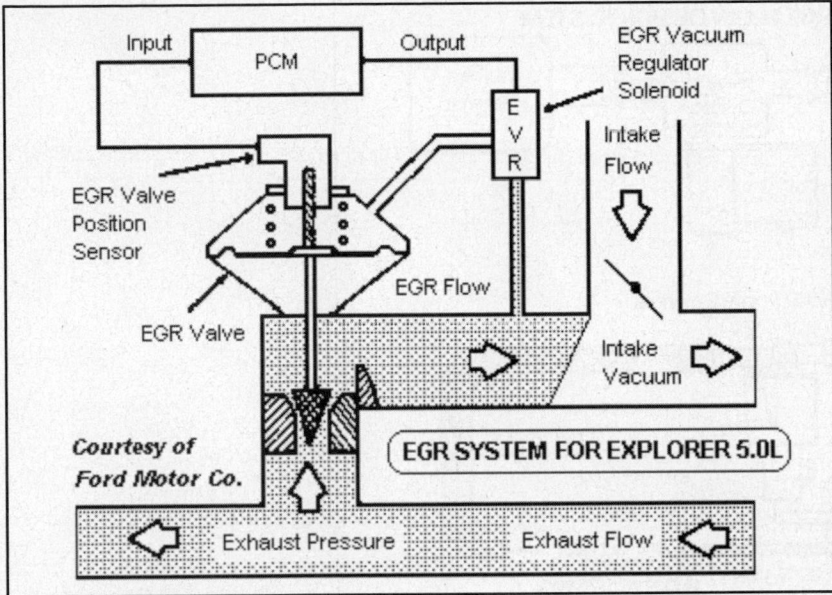

EGR SYSTEM FOR EXPLORER 5.0L

Courtesy of Ford Motor Co.

EVAP System Monitor

The EVAP System Monitor is a PCM diagnostic run once per trip that monitors the EVAP system in order to detect a loss of system integrity or leaks in the system (anywhere from 0.020" to 0.040" in diameter).

The Ford vehicles included in this article are equipped with three different EVAP systems: Purge Flow, Vapor Management Valve and On Board Refueling Vapor Recovery.

Purge Flow System (Explorer 5.0L V8 Engine)

The Purge Flow system consists of a purge flow (PF) sensor, purge solenoid, fuel vapor valve, a gas tank, charcoal canister (with internal atmospheric vent) and related wiring and fuel vapor hoses.

In this EVAP system, the PCM commands a normally closed (N.C.) purge solenoid "on" and "off" to purge fuel vapors from the charcoal canister into the intake manifold during various engine conditions. The PCM monitors changes in the fuel pressure (FP) sensor analog signal to detect fuel vapor flow through the purge solenoid during the EVAP diagnostic test. Changes are based on "flow" or "no flow" conditions.

Purge Flow System Graphic (Explorer)

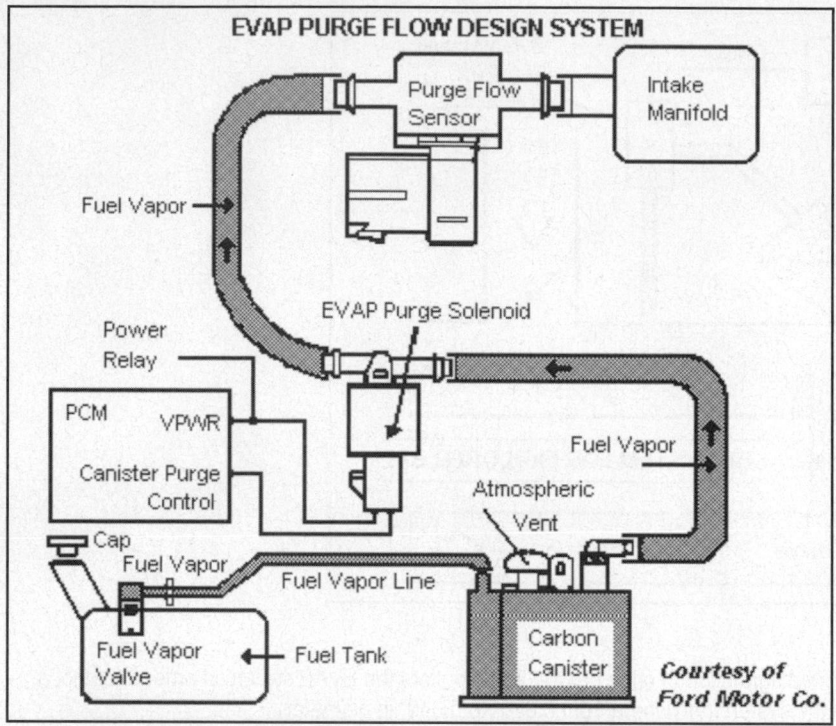

Purge Flow System (Probe 2.0L I4)

The Purge Flow system used on these models includes a canister purge solenoid, fuel line check valve, charcoal canister, fuel tank, related wiring and vacuum hoses, and the PCM.

System Operation

On this system, the PCM commands a normally closed (N.C.) canister purge solenoid "on" and "off" to control purging of fuel vapors from the charcoal canister into the intake manifold under various engine conditions.

Purge Flow System Graphic (Probe)

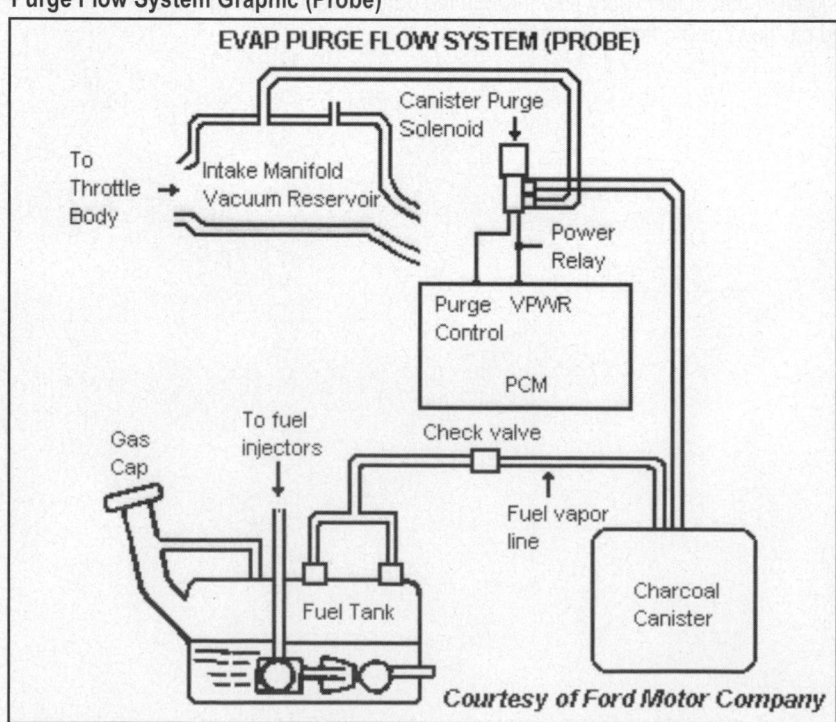

Possible Causes of an EVAP System Failure
● Cracks, leaks or disconnected hoses in the fuel vapor lines, components or plastic connectors or lines
● Backed-out or loose connectors to the Canister Purge solenoid
● Fuel filler cap (gas cap) loose or missing
● PCM has failed

EVAP VMV Design System

This system consists of a Vapor Management Valve (VMV), fuel vapor valve, gas tank, the charcoal canister with internal atmospheric vent, related wiring and fuel vapor hoses, and the PCM. The PCM commands a normally closed (N.C.) valve "on" and "off" to control when to purge fuel vapors from the canister to the intake manifold.

Fuel vapors trapped in the sealed fuel tank are vented through a vapor valve assembly on top of fuel tank. The vapors leave the valve assembly through a single vapor line and continue on to the carbon canister for storage until they are purged to the engine for recycling.

The VMV and vent solenoid control signals are cycled "on" and "off" at a frequency of 10 hertz with a variable duty cycle. The duty cycle is ramped-up to slowly draw the canister vapors into the intake manifold.

VMV System Graphic

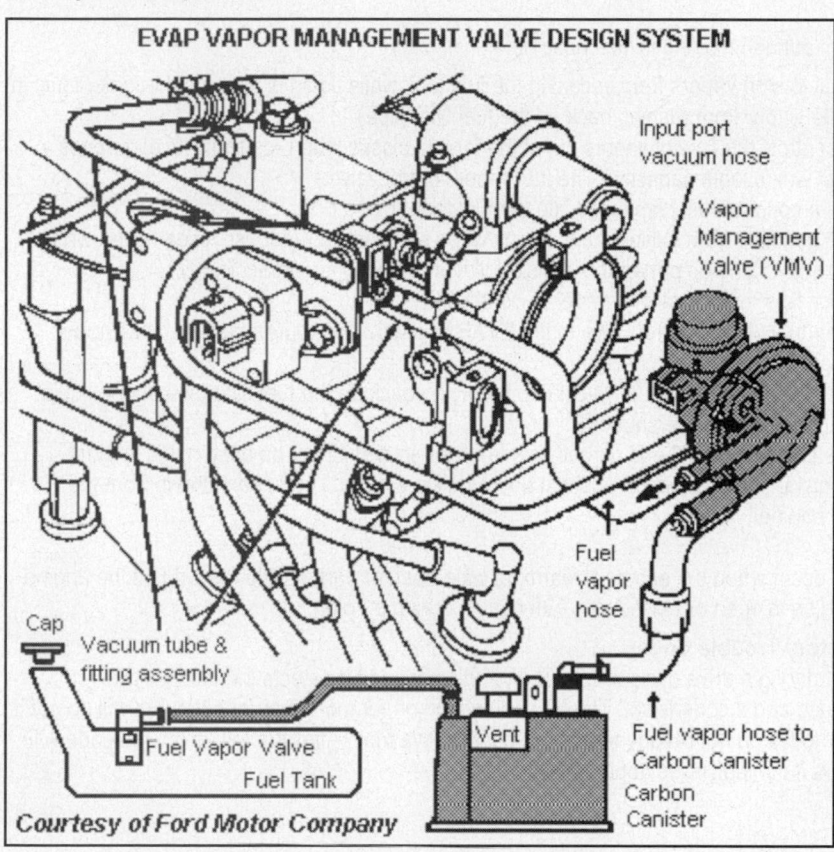

EVAP VAPOR MANAGEMENT VALVE DESIGN SYSTEM

Input port vacuum hose

Vapor Management Valve (VMV)

Fuel vapor hose

Cap

Vacuum tube & fitting assembly

Fuel Vapor Valve

Fuel Tank

Vent

Fuel vapor hose to Carbon Canister

Carbon Canister

Courtesy of Ford Motor Company

On-Board Refueling Vapor Recovery System

An On-Board Refueling Vapor Recovery (ORVR) system is used on late model vehicles to recover fuel vapors during vehicle refueling.

ORVR System Graphic

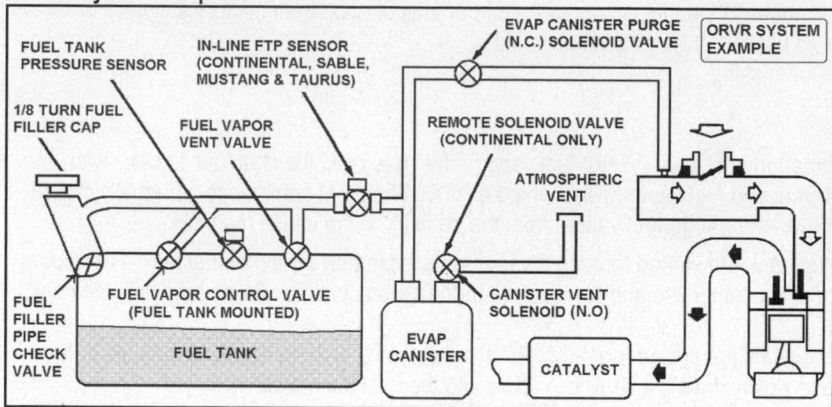

System Operation

The operation of the ORVR system during refueling is described next:

- The fuel filler pipe forms a seal to stop vapors from escaping the fuel tank while liquid is entering the tank (liquid in the 1" diameter tube blocks fuel vapor from rushing back up the fuel filler pipe).
- The fuel vapor control valve controls the flow of vapors out of the tank (it closes when the liquid level reaches a height associated with the fuel tank usable capacity). The fuel vapor control valve:
  - Limits the total amount of fuel dispensed into the fuel tank.
  - Prevents liquid gasoline from exiting the fuel tank when submerged (and also when tipped well beyond a horizontal plane as part of the vehicle rollover protection in an accident).
  - Minimizes vapor flow resistance in a refueling condition.
- Fuel vapor tubing connects the fuel vapor control valve to the EVAP canister. This routes the fuel tank vapors (that are displaced by the incoming fuel) to the canister.
- A check valve in the bottom of the pipe prevents any liquid from rushing back up the fuel filler pipe during liquid flow variations associated with the filler nozzle shut-off.
- Between refueling events, the charcoal canister is purged with fresh air so that it may be used again to store vapors accumulated during engine soak periods or subsequent refueling events. The vapors drawn from the canister are consumed in the engine.

EVAP Monitor Test Conditions

The PCM allows canister purge to occur when the engine is warm, at wide open or part throttle (as long as the engine is not overheated). The engine can be in open or closed loop fuel control during purging.

MIL Operation, How to Clear History Trouble Codes

If the EVAP Monitor detects a fault during a drive cycle, it will set a pending code. If it detects the fault for two consecutive trips, the MIL is activated and a code is set. The MIL will remain on for more than one trip, but will go out if conditions that caused the Monitor to fail do not reappear on three consecutive trips. After the MIL is off, the code will be erased after 40 consecutive trips if the fault does not reappear.

EVAP Running Loss Monitor Graphic

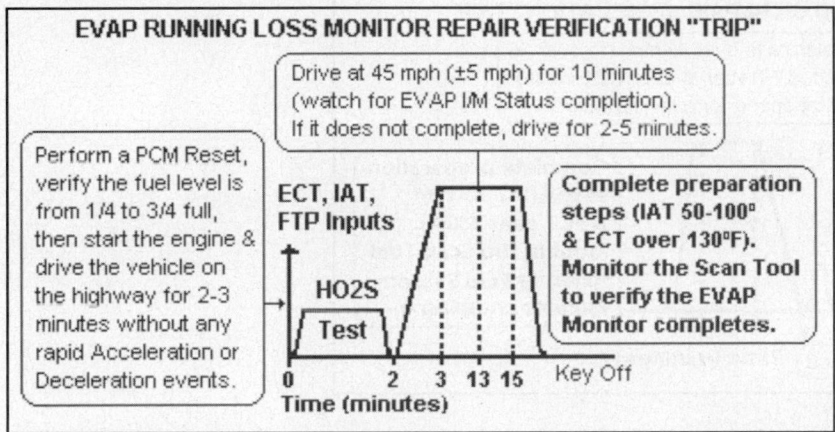

EVAP Leak Check Monitor Graphic

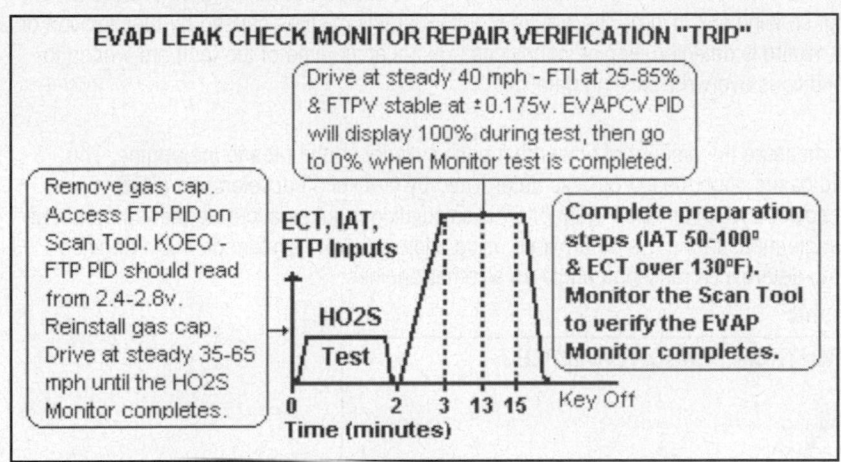

Fuel System Monitor

The Fuel System Monitor is a PCM diagnostic that monitors the Adaptive Fuel Control system. The PCM uses adaptive fuel tables that are updated constantly and stored in long term memory (KAM) to compensate for wear and aging in the Fuel system components.

Fuel System Monitor Operation

Once the PCM determines all the enable criteria has been are met (ECT, IAT and MAF PIDs in range and closed loop enabled), the PCM uses its adaptive strategy to "learn" changes needed to correct a Fuel system that is biased either rich or lean. The PCM accomplishes this task by monitoring Short Term and Long Term fuel trim in closed loop mode.

Long and Short Term Fuel Trim

Short Term fuel trim is a PCM parameter identification (PID) used to indicate Short Term fuel adjustments. This parameter is expressed as a percentage and its range of authority is from -10% to +10%. Once the engine enters closed loop, if the PCM receives a HO2S signal that indicates the A/F mixture is richer than desired, it moves the SHRTFT command to a more negative range to correct for the rich condition.

If the PCM detects the SHRTFT is adjusting for a rich condition for too long a time, the PCM will "learn" this fact, and move LONGFT into a negative range to compensate so that SHRTFT can return to a value close to 0%. Once a change occurs to LONGFT or SHRTFT, the PCM adds a correction factor to the injector pulsewidth calculation to adjust for variations. If the change is too large, the PCM will detect a fault.

NOTE: *If a fuel injector, fuel pressure regulator, etc. is replaced, clear the KAM and then drive the vehicle through the Fuel System Monitor drive pattern to reset the fuel control table in the PCM.*

Fuel System Monitor Graphic

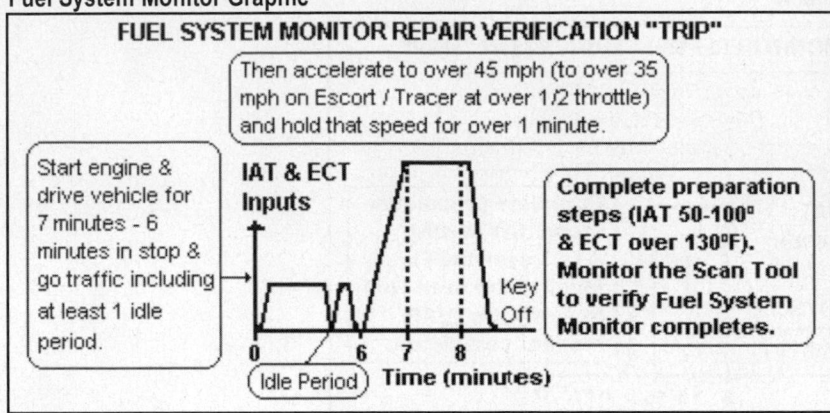

FUEL SYSTEM MONITOR REPAIR VERIFICATION "TRIP"

Then accelerate to over 45 mph (to over 35 mph on Escort / Tracer at over 1/2 throttle) and hold that speed for over 1 minute.

Start engine & drive vehicle for 7 minutes - 6 minutes in stop & go traffic including at least 1 idle period.

IAT & ECT Inputs

Key Off

Complete preparation steps (IAT 50-100° & ECT over 130°F). Monitor the Scan Tool to verify Fuel System Monitor completes.

Idle Period   Time (minutes)

Misfire Detection Monitor

The Misfire Monitor is a PCM diagnostic that continuously monitors for engine misfires under all engine positive load and speed conditions (accelerating, cruising and idling). The Misfire Monitor detects misfires caused by fuel, ignition or mechanical misfire conditions. If a misfire is detected, engine conditions present at the time of the fault are written to the Freeze Frame Data. These conditions overwrite existing data.

Misfire Monitor Operation

The Misfire Monitor is designed to measure the amount of power that each cylinder contributes to the engine. The amount of contribution is calculated based upon measurements determined by crankshaft acceleration (TDC of compression stroke to BDC of the power stroke) for each cylinder. This calculation requires accurate measurement of the crankshaft angle. Crankshaft angle measurement is determined using a low data rate system on 4-Cyl engines. The high data rate system is used to determine crankshaft angle on all other engines.

Crankshaft Position Sensor Graphic

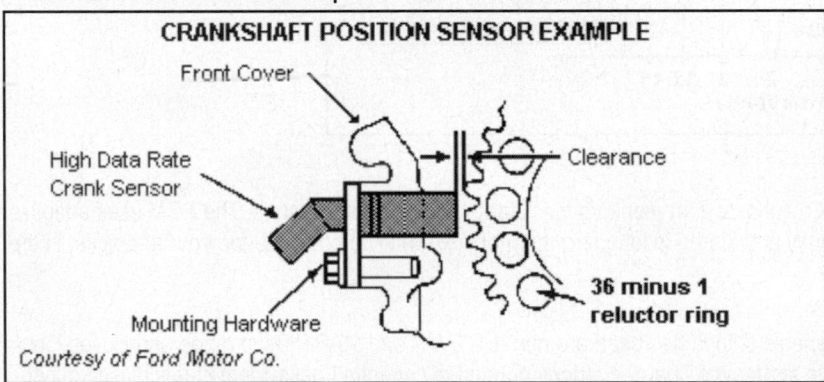

CRANKSHAFT POSITION SENSOR EXAMPLE

Front Cover

High Data Rate Crank Sensor

Clearance

Mounting Hardware

36 minus 1 reluctor ring

Courtesy of Ford Motor Co.

Catalyst Damaging Misfire (One-Trip Detection)

If the PCM detects a Catalyst Damaging Misfire, the MIL will flash once per second within 200 engine revolutions from the point where misfire is detected. The MIL will stop flashing and remain on if the engine stops misfiring in a manner that could damage the catalyst.

High Emissions Misfire (Two-Trip Detection)

A High Emissions Misfire is set if a misfire condition is present that could cause the tailpipe emissions to exceed the FTP emissions standard by 1.5 times. If this fault is detected for two consecutive trips under similar engine speed, load and temperature conditions, the MIL is activated. It is also activated if a misfire is detected under *similar conditions* for two non-consecutive trips that are not 80 trips apart.

State Emissions Failure Misfire (Two-Trip Detection)

A State Emissions Failure Misfire is set if the misfire is sufficient to cause the vehicle to fail a State Inspection or Maintenance (I/M) Test. This fault is determined by identifying misfire percentages that would cause a "durability demonstration vehicle" to fail an Inspection Maintenance (I/M) Test. If the Misfire Monitor detects the fault for two consecutive trips with the engine at similar engine speed, load and temperature conditions, the MIL is activated and a code is set. The MIL is also activated if this type of misfire is detected under similar conditions for two non-consecutive trips of not more than 80 trips apart.

NOTE: *Some vehicles set Misfire codes because of an early version of OBD II hardware and software. If a misfire code is set and the cause of the fault is not found, clear the code and retest. Search the TSB list for possible answers or contact the dealer.*

Misfire Detection

The Misfire Monitor uses the CKP sensor signals to detect an engine misfire. The amount of contribution is calculated based upon measurements determined by crankshaft acceleration from each cylinder's power stroke.

The PCM performs various calculations to detect individual cylinder acceleration rates. If acceleration for a cylinder deviates beyond the average variation of acceleration for all cylinders, a misfire is detected.

Misfire Detection Monitor Graphic

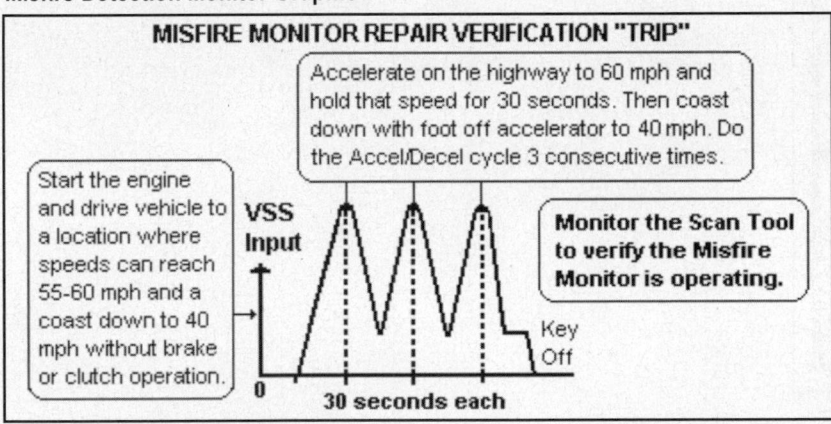

Faults detected by the Misfire Monitor

- Engine mechanical faults, restricted intake or exhaust system
- Dirty or faulty fuel injectors, loose or damaged injector connectors
- The vehicle has been run low on fuel or run until it ran out of fuel

Oxygen Sensor Monitor

The Oxygen Sensor Monitor is a PCM diagnostic designed to monitor the front and rear oxygen sensor for faults or deterioration that could cause tailpipe emissions to exceed 1.5 times the FTP standard. The front oxygen sensor voltage and response time are also monitored.

HO2S Monitor Operation

Fuel System and Misfire Monitors must be run and complete before the PCM will start the HO2S Monitor. Additionally, parts of the HO2S Sensor Monitor are enabled during the KOER Self-Test. The HO2S Monitor is run during each drive cycle after the CKP, ECT, IAT and MAF sensor signals are within a predetermined range.

Fixed Frequency Closed Loop Test

The HO2S Monitor constantly monitors the sensor voltage and frequency. The PCM detects a high voltage condition by comparing the HO2S signal to a preset level.

A Fixed Frequency Closed Loop Test is used to check the HO2S voltage and frequency. A sample of the HO2S signal is checked to determine if the sensor is capable of switching properly or has a slow response time (referred to as a lazy sensor).

Fixed Frequency Test Graphic

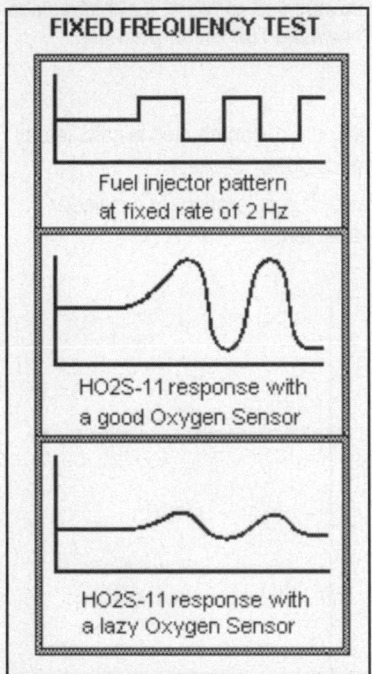

Oxygen Sensor Monitor Graphic

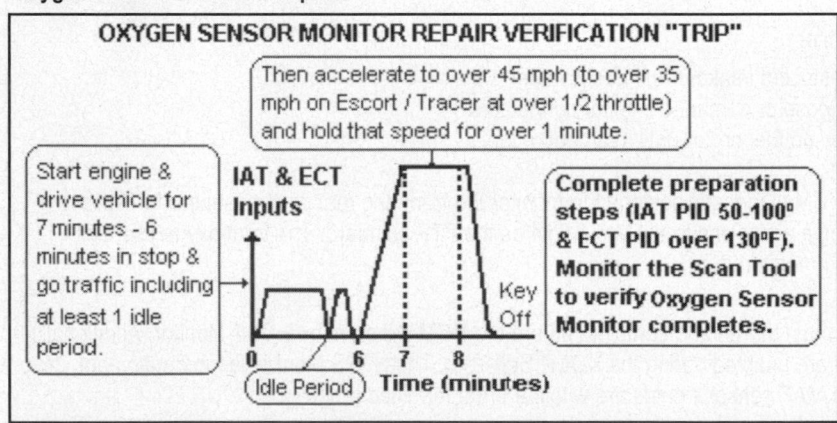

Oxygen Sensor Heater Monitor

The Oxygen Sensor Heater Monitor is a PCM diagnostic designed to monitor the Oxygen Sensor Heater and its related circuits for faults.

Oxygen Sensor Heater Monitor Operation

The Oxygen Sensor Heater Monitor performs its task by detecting whether the proper amount of O2 sensor voltage change occurred as the HO2S Heater is turned from "on" to "off" with the engine in closed loop. The time it takes for the HO2S-11 and HO2S-12 signal to switch (the response time) is constantly monitored by the Oxygen Sensor Monitor. Once the Oxygen Sensor Heater Monitor is enabled, if the switch time for the HO2S-11 or HO2S-12 signal is too long, the PCM fails the test, the MIL is activated and a trouble code is set.

Note: *Response time is defined as the amount of time it takes for a HO2S signal to switch from Rich to Lean, and then Lean to Rich.*

Front and Rear Oxygen Sensor Heater Operation

Both upstream and downstream Oxygen sensors are used on the OBD II system. These sensors are designed with additional protection around the ceramic core to protect them from condensation that could crack them if the heater is turned on with condensation present.

The HO2S heaters are not turned on until the ECT sensor signal indicates that the engine is warm. The delay period can last for as long as 5 minutes from startup. The delay allows any condensation in the Exhaust system to evaporate.

Oxygen Sensor Monitor Graphic

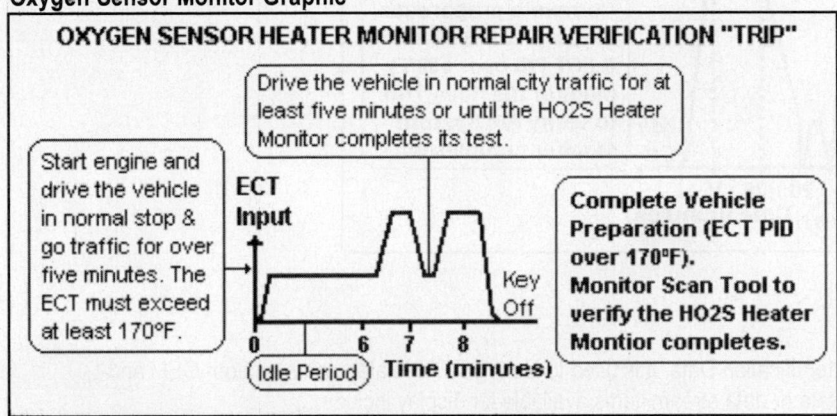

Faults detected by the HO2S or HO2S Heater Monitor
- A fault in the HO2S, the HO2S heater or its related circuits
- A fault in the HO2S connectors (look for moisture tracking)
- A defective Power Control Module

Air Injection System Monitor
The Air Injection System Monitor is an OBD diagnostic controlled by the PCM that monitors the Air Injection (AIR) system. The Oxygen Sensor Monitor must run and complete before the PCM will run this test. The PCM enables this test during AIR system operation after certain engine conditions are met and these enable criteria are met:

- Crankshaft Position sensor signal must be present
- ECT and IAT sensor input signals must be within limits

AIR Monitor - Electric Pump Design
The AIR Monitor consists of these Solid State Monitor tests:

- A check of the Solid State relay for electrical faults.
- A check of the secondary side of the relay for electrical faults.
- A test to determine if the AIR system can inject additional air.

AIR Monitor - Mechanical Pump Design
The AIR Monitor for the mechanical (belt-driven air pump) design uses two Output State Monitor configurations to perform two different circuit tests. One test is used to check for faults in the Secondary Air Bypass (AIRB) solenoid circuit. The normal function of the AIRB solenoid and valve assembly is to dump air into the atmosphere.

A second test is used to check for electrical faults in the Secondary Air Divert (AIRD) solenoid. The normal function of the AIRD solenoid and valve assembly is to direct the air either upstream or downstream.

Functional Check
An AIR system functional check is done at startup with the AIR pump on or during a hot idle period if the startup part of the test was not performed. A flow test is included that uses the HO2S signal to indicate the presence of extra air injected into the exhaust stream.

Secondary AIR Monitor Graphic

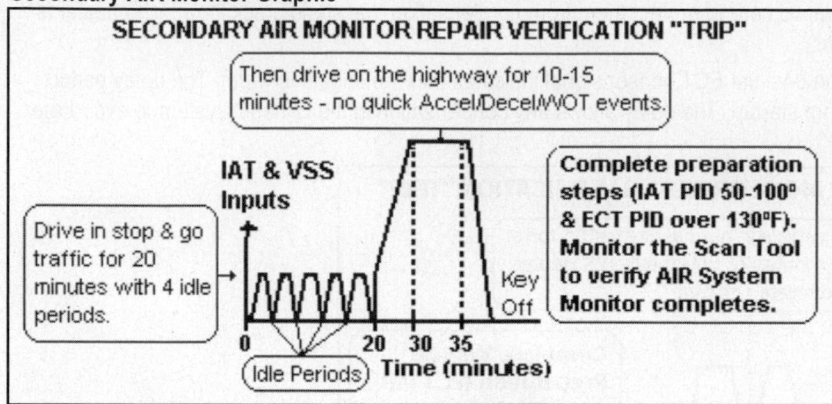

SECONDARY AIR MONITOR REPAIR VERIFICATION "TRIP"

Then drive on the highway for 10-15 minutes - no quick Accel/Decel/WOT events.

IAT & VSS Inputs

Drive in stop & go traffic for 20 minutes with 4 idle periods.

Complete preparation steps (IAT PID 50-100° & ECT PID over 130°F). Monitor the Scan Tool to verify AIR System Monitor completes.

Key Off

0   20   30 35

Idle Periods   Time (minutes)

REFERENCE INFORMATION

PID Data Explanation

PID is an acronym for Parameter Identification Data. It is used to identify PCM related items on both OEM and aftermarket Scan Tools. The PID Data or data stream items available for display include:

- PCM analog & digital input signals (ACT, ECT, IAT, MAP, TP)
- PCM analog & digital output signals (EVR, FUELPW1, MIL)
- PCM calculated values (VACUUM, LOAD, MISF, SPARK)
- PCM system status information (OBD II I/M Readiness Status)

OBD I & OBD II Data Stream Information

PID Data is separated by OBD System type as listed below:

- OBD I System PID Data for Ford Cars from 1990-95
- OBD II System PID Data for Ford Cars from 1994-2002

How to View PID Data

To view PID Data on a Scan Tool, connect the Scan Tool to the vehicle connector; then identify the vehicle by following the setup instructions, and select Data Stream or Parameter from the menu.

NOTE: *If a Scan Tool does not power up or read PID Data, verify that the power, ground and tool cable connections are all okay. To test if the tool is working properly, try it on another vehicle.*

Example of Scan Tool Connection

In this example, the Scan Tool is connected to the DLC in order to view "live" HO2S PID Data once it has been converted in the PCM.

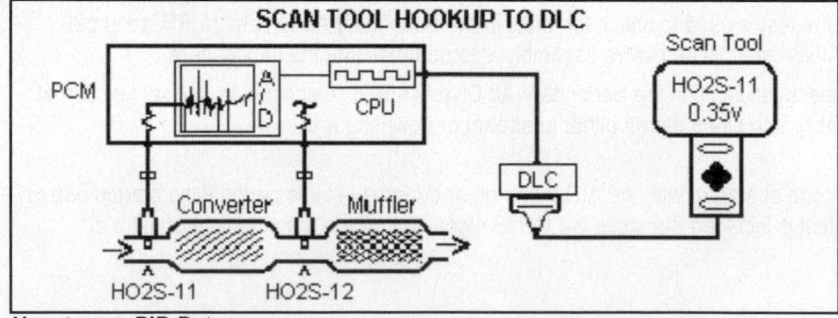

SCAN TOOL HOOKUP TO DLC

PCM

CPU

Converter   Muffler

HO2S-11   HO2S-12

DLC

Scan Tool

HO2S-11
0.35v

How to use PID Data

Information contained within the PID Data Charts can be used to:

- Validate a repair procedure
- Check the operation of a component before or after a repair
- Check the operation of a component or system by viewing "live" data from the vehicle computer data stream

OBD II PID Mode

Vehicles equipped with OBD II diagnostics have a unique PID Mode that can be selected to allow access to vehicle Parameter Data information. Part of the PID Data is generic in nature and can be accessed with an OBD II compatible Scan Tool in Generic Mode.

However, some of the PID Data can only be accessed using an OEM New Generation Star (NGS) Scan Tool or an Aftermarket Scan Tool with the capability to read Ford Enhanced Data. To learn how to view Generic or Enhanced Data, refer to the Scan Tool operating manual.

PID Data Display

An example of Enhanced Data captured with a Scan Tool is shown in the table below. This data is from a 1996 Ford 3.0L V6 Taurus VIN U.

PID #	Acronym	Description	Measurement Units
000F	IAT	Intake Air Temperature	110°F
0001	DTC CNTM	Continuous DTC Counter	Numbers
0003	FUEL SYS1	Fuel System Feedback (BK1)	CL (Closed Loop)
0005	ECT	Engine Coolant Temperature	188°F
0014	O2S-11	Bank 1 Front Oxygen Sensor	0.635v (varies at cruise)
0015	O2S-12	Bank 1 Rear Oxygen Sensor	750 mv (varies at cruise)

Scan Tool Graphing Mode

Most Scan Tools include a function called Scan Tool Graphing Mode that can be used to capture signals from various devices (TP sensor, IAC Motor, etc.). Note how this feature was used to capture changes in the TP sensor signal and IAC Motor control signals in the Graphic below. This is a good test to use to detect an intermittent fault.

Scan Tool Graphing Mode Examples

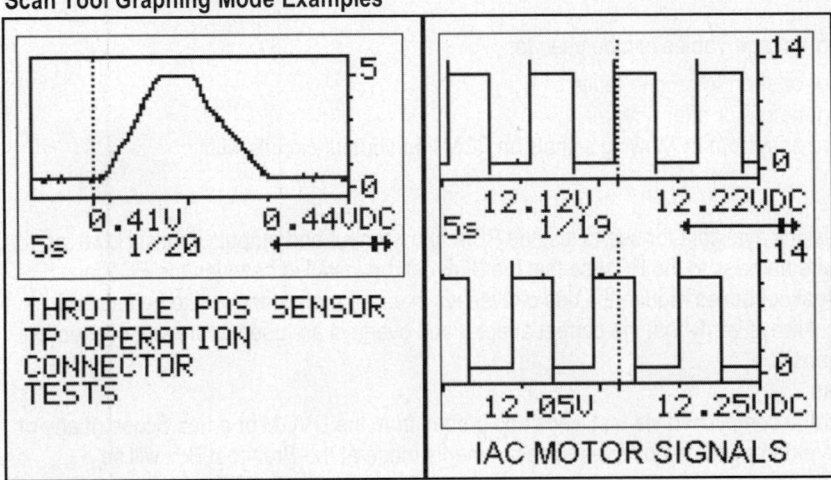

Pin Voltage Table Explanation

A Pin Voltage Table is a term used to describe a table that identifies PCM and Breakout Box (BOB) pins, wire colors of the PCM circuits and "known good" values for devices that connect to the PCM. These tables include the following information:

- Signals from various sensors (ECT, IAT, MAP, TOT, TP, VAF)
- Signals from various switches (BOO, CES, IDL, NDS, PSP)
- Signals from oxygen sensors (O2S-11, HO2S-11, HO2S-12)
- Signals from output devices (CANP, EVR, IAC, INJ, TCC)
- Power & ground signals (KAPWR, PWR GND, VPWR, VREF)

NOTE *Acronyms shown above are listed in the Glossary in this section.*

OBD I & OBD II Pin Voltage Tables

Pin Voltages are separated by OBD System type as listed below:

- OBD I System pin voltage values for Ford Cars from 1990-95
- OBD II System pin voltage values for Ford Cars from 1994-2002

How to Connect a Breakout Box

To use pin voltage information with a DVOM, a Breakout Box (BOB) should be installed. To connect a BOB, first turn the ignition off and remove the wire harness at the engine controller (PCM). Next, connect the correct wiring adapter to the PCM and BOB connectors. This places the BOB between the PCM and wiring so that circuit measurements can be made at the pin connections on the BOB.

NOTE: *Be sure to read and record all trouble code and freeze frame data in the PCM before connecting the BOB as all codes and data are lost when the PCM connector is removed.*

Example of a BOB Connected to the PCM

In this example, the BOB is connected to the PCM connector and wire harness so that measurements can be made with a DVOM (or Lab Scope) of the Oxygen Sensor circuits with the engine running.

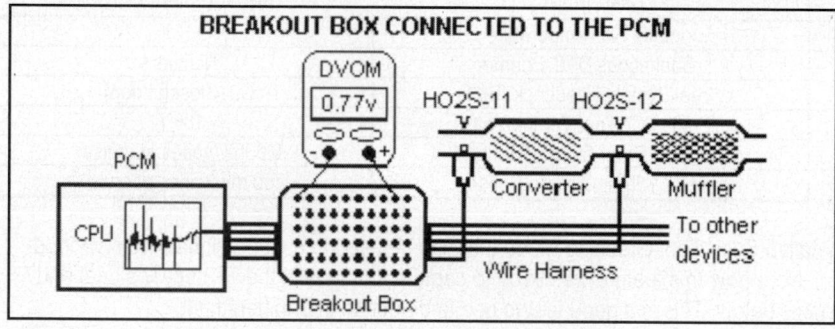

Pin Voltage Tables

Information contained within the Pin Voltage Tables can be used to:

Test circuits for open, short to power or short to ground faults
Check the operation of a component before or after a repair
Check the operation of a component or system by viewing signals on PCM input/output circuits with
 a DVOM or Lab Scope

Using a Breakout Box

There are several Breakout Box designs available for use to test the PCM and its input and output circuits. However, all of them require removal of the wire harness to the PCM so that the BOB can be installed between the PCM and wire harness connector. Several breakout boxes require the use of overlays in order to allow the tool to be used on more than one year or engine type. Always verify that the correct adapter and overlays are used to prevent connection to the wrong circuits and a misdiagnosis.

Power and Ground Circuit Checks

Measurements made at the BOB are accomplished via test leads and probes from the DVOM or a Lab Scope. If any of the terminals on the PCM or BOB are damaged or loose, test measurements made at the Breakout Box will be inaccurate. To verify the PCM battery power (KAPWR and VPWR) and ground circuits (PWR GND) are normal (correct) at the BOB, test the condition of the circuit between the battery negative (-) post and these circuits prior to starting a test sequence.

NOTE: *With the key on, the voltage drop from the battery positive (+) terminal to KAPWR or VPWR at the BOB should be less than 0.1v. The voltage drop from the battery negative (-) post to PWR GND at the BOB should be less   than 0.1v (key on).*

Diagnosis with Pin Voltage Tables

Once an actual PCM pin voltage reading is recorded, it can be compared to an example from a vehicle with "known good" values. In the example below (from a 1990 Crown Victoria 5.0L V8), the Value at Hot Idle for the EVP sensor signal (0.4v) is the "known good" value.

PCM Pin #	W/Color	Circuit Description (60-Pin)	Value at Hot Idle
27	BN/LG	EVP Sensor Signal	0.4v

WIRE COLOR CHANGES

Every effort has been made to obtain and list the correct circuit wire colors for all vehicles. However, running changes from the vehicle manufacturer can cause the wrong colors to be listed.

PCM Actuator & Sensor Tables
Application: Association Vehicles

- Aspire 1.3L I4
- Escort/Tracer 1.8L I4
- Probe 2.5L V6

ACT, ECT, IAT & TFT Sensor Conversion Chart

Temperature		Voltage Resistance	
Degrees F	Degrees C	Volts	Kohms
194°F	90°C	0.90v	0.2 Kohms
122°F	50°C	2.20v	0.8 Kohms
68°F	20°C	3.50v	2.5 Kohms
14°F	-10°C	4.40v	9.2 Kohms

Application: Vehicles with EEC-IV & EEC-V Systems

ACT, ECT, IAT, TFT & TOT Sensor Conversion Chart

Temperature		Voltage Resistance	
Degrees F	Degrees C	Volts	Kohms
302	160	0.12	0.54
267	131	0.20	0.80
248	120	0.28	1.18
230	110	0.36	1.55
212	100	0.47	2.07
194	90	0.61	2.8
176	80	0.8	3.84
158	70	1.04	5.37
140	60	1.35	7.6
104	40	2.16	16.15
86	30	2.62	24.27
68	20	3.06	37.3
50	10	3.52	58.75
32	0	3.97	65.85
14	-10	4.42	78.19
-4	-20	4.87	90.54
-22	-30	4.89	102.88
-40	-40	4.91	115.23

Differential Pressure Feedback EGR Sensor

Pressure			Voltage
PSI	Inches Hg	KPa	Volts
4.34	8.83	29.81	4.56
3.25	6.62	22.36	3.54
2.17	4.41	14.90	2.51
1.08	2.21	7.46	1.48
0	0	0	0.45

Application: All vehicles with a BARO/MAP Sensor

Barometric Pressure Sensor Chart

Approximate Altitude in Feet	BARO Hertz Rate
0 feet (sea level)	159 Hz
1000 feet	156 Hz
2000 feet	153 Hz
3000 feet	150 Hz
4000 feet	147 Hz
5000 feet	144 Hz
6000 feet	141 Hz
7000 feet	139 Hz
8000 feet	136 Hz

Manifold Absolute Pressure Sensor Chart

Manifold Vacuum (" Hg)	MAP Hertz Rate
KOEO at Sea Level	159 Hz
5" Hg	146 Hz
8" Hg	134 Hz
10" Hg	129 Hz
11" Hg	126 Hz
12" Hg	124 Hz
13" Hg	121 Hz
14" Hg	119 Hz
15" Hg	116 Hz
16" Hg	114 Hz
17" Hg	111 Hz
18" Hg	109 Hz
19" Hg	107 Hz
20" Hg	104 Hz
21" Hg	102 Hz
22" Hg	99 Hz
23" Hg	97 Hz
24" Hg	95 Hz

Application: Domestic Vehicles
- Aspire 1.3L I4, Probe 2.5L V6 Engine
- Vehicles with EEC-V and 5.0L V8 (VIN P) Engine

EGR Valve Position (EVP) Sensor

Percent of EGR	Voltage	Code Set Values
0% (valve closed)	0.40v	Hot Idle: A code sets if the EVP
10% (valve open)	0.75v	Sensor voltage is less than 0.24v or
30% (valve open)	1.45v	If the EVP signal is more than 0.67v
50% (valve open)	2.15v	
70% (valve open)	2.85v	
90% (valve open)	3.55v	

Application: Domestic Vehicles

● All applicable EEC-IV & EEC-V Vehicles

Pressure Feedback EGR (PFE) Sensor

PSI	Inches HG	Volts	Code Set Values
1.82	3.70	4.75	Hot idle: A code sets if the PFE
1.36	2.79	4.38	Sensor signal is under 2.60v or if
0.91	1.85	4.00	the PFE signal is more than 4.20v
0.46	0.94	3.63	
0	0	3.25	
-2.47	-5.03	1.22	
-3.63	-7.40	0.25	

EGR Vacuum Regulator (EVR) Solenoid Vacuum Examples

Duty	Vacuum				Output	
Cycle (%)	Min		Nom		Max	
	In-Hg	KPa	In-Hg	KPa	In-Hg	KPa
0	0	0	0.38	1.28	0.75	2.53
33	0.55	1.86	1.3	4.39	2.05	6.90
90	5.69	19.2	6.32	21.3	6.95	23.47

NOTE: *EVR resistance values - for 4.9L and 5.8L engines: 20-45 ohms, for 7.5L engines: 100 ohms. For all other engines: 30-70 ohms.*

Output Device Resistance Chart for OBD I and OBD II Vehicles

Component	Ohm Value	Component	Ohm Value
AM1 Valve	40-80 ohms	AM2 Valve	40-80 ohms
Canister Purge Valve	40-80 ohms	Idle Air Control Motor	7-13 ohms
CCRM (module)	1K minimum	IRCM (module) - FP	65-110 ohms
EPC Solenoid	36-43 ohms	High Pressure Injector	12-16 ohms
EGR VR Solenoid	26-40 ohms	Shift Solenoid 1 (SS1)	15-25 ohms
Electro Drive Fan (EDF)	65-110 ohms	Shift Solenoid 2 (SS2)	15-25 ohms
Fan or Low Fan Control	65-110 ohms	Shift Solenoid 3 (SS3)	15-25 ohms
Fuel Pump (FP) Relay	40-90 ohms	TCC Solenoid	0.9-1.9 ohms
High Electro Drive Fan	50-100 ohms	AC WOT Cutout Relay	1K minimum

Output Device Resistance Chart for Association Vehicles

Component	Ohm Value	Component	Ohm Value
Canister Purge Valve	30-34 ohms	A/C WOT Cutout Relay	1K minimum
EPC Solenoid	36-43 ohms	Shift Solenoid 1, 2, 3	11-27 ohms
EGR Control, Vent	30-45 ohms	VRIS Solenoid 1, 2	35-40 ohms
EGR Boost Solenoid	90-110 ohms	Shift Solenoid 1, 2, 3	11-27 ohms
Idle Air Control Valve	10.7-12 ohms	TCC Solenoid	11-27 ohms

NOTE *An Amp Draw Test should not be done on output devices with a resistance of less than 25 ohms (or duty cycle controlled).*

Voltage to Mass Airflow Conversation Table (1996-98 Models)

Vehicle & Engine Application	MAF Volts 0.34v =	MAF Volts 0.39v =	MAF Volts 0.60v =	MAF Volts 1.00v =
Aerostar with 3.0L Engines	0.94 gm/sec	1.07 gm/sec	1.68 gm/sec	4.77 gm/sec
Aerostar with 4.0L Engines	1.46 gm/sec	1.68 gm/sec	2.52 gm/sec	7.64 gm/sec
Continental, Mark VII & Mustang with 4.6L Engine	1.39 gm/sec	1.60 gm/sec	2.69 gm/sec	8.36 gm/sec
Contour, Mystique with 2.0L	0.39 gm/sec	0.45 gm/sec	0.72 gm/sec	2.10 gm/sec
Contour & Mystique with 2.5L	0.33 gm/sec	0.95 gm/sec	1.69 gm/sec	5.53 gm/sec
Cougar, Thunderbird with 3.8L & 4.6L Engines	1.44 gm/sec	1.66 gm/sec	2.73 gm/sec	8.06 gm/sec
Crown Victoria 4.6L, Crown Victoria CNG 4.6L	1.28 gm/sec	1.48 gm/sec	2.37 gm/sec	6.97 gm/sec
Escort and Tracer with 1.9L Engines	0.94 gm/sec	1.07 gm/sec	1.68 gm/sec	4.77 gm/sec
Explorer and Ranger with 4.0L Engines	1.46 gm/sec	1.68 gm/sec	2.52 gm/sec	7.64 gm/sec
E-Series Van with 4.9L, 5.0L & 5.8L	1.24 gm/sec	1.42 gm/sec	2.98 gm/sec	8.38 gm/sec
F-Series Truck 4.6L & 7.5L Engines	1.35 gm/sec	1.56 gm/sec	2.66 gm/sec	8.31 gm/sec
F-Series Truck 4.9L and Bronco with 5.0L & 5.8L	1.68 gm/sec	1.93 gm/sec	3.32 gm/sec	8.43 gm/sec
Grand Marquis, Town Car with 4.6L Engines	1.28 gm/sec	1.48 gm/sec	2.37 gm/sec	6.97 gm/sec
Mustang 3.8L and Aerostar, Explorer and Ranger with 4.0L	1.46 gm/sec	1.68 gm/sec	2.52 gm/sec	7.64 gm/sec
Probe 2.0L and Windstar with 3.0L Engines	0.82 gm/sec	0.88 gm/sec	1.25 gm/sec	3.44 gm/sec
Ranger Pickup with 2.3L & 3.0L Engines	0.93 gm/sec	1.07 gm/sec	1.71 gm/sec	4.58 gm/sec
Sable, Taurus 3.0L 2v and Taurus FF 3.0L Engines	1.31 gm/sec	1.11 gm/sec	1.72 gm/sec	4.82 gm/sec
Sable/Taurus 3.0L 4v, 3.8L Windstar and F-Series Pickup with 4.2L	1.43 gm/sec	1.64 gm/sec	2.58 gm/sec	7.53 gm/sec
Taurus SHO 3.4L Engines	1.39 gm/sec	1.60 gm/sec	2.69 gm/sec	8.36 gm/sec

NOTE: *To use this table, connect a Scan Tool and select the MAF PID. Read the value and then compare it against the known good grams per second (gm/sec) reading for the vehicle and engine in this table. A faulty MAF sensor will read out of normal range.*

Voltage to Mass Airflow Conversation Table (1999-2005 Models)

Vehicle & Engine Application	MAF Volts 0.23v =	MAF Volts 0.27v =	MAF Volts 0.46v =	MAF Volts 2.44v =
Continental & Mustang with 4.6L Engine	1.07 gm/sec	1.22 gm/sec	2.08 gm/sec	59.52 gm/sec
Contour, Mystique & Cougar with 2.0L Engine	0.70 gm/sec	0.80 gm/sec	1.36 gm/sec	30.86 gm/sec
Contour, Mystique & Cougar 2.5L and Taurus 3.0L 4v Engines	1.00 gm/sec	1.14 gm/sec	1.95 gm/sec	52.50 gm/sec
Crown Victoria, Grand Marquis, & Town Car with 4.6L Engines	1.01 gm/sec	1.16 gm/sec	1.98 gm/sec	50.26 gm/sec
Escort & Tracer 2.0L 2v Engines	0.67 gm/sec	0.77 gm/sec	1.32 gm/sec	30.55 gm/sec
Escort & Tracer 2.0L 42v Engines	0.62 gm/sec	0.71 gm/sec	1.22 gm/sec	30.66 gm/sec
Expedition & Navigator 4.6L & 5.4L Engines	1.68 gm/sec	1.93 gm/sec	3.32 gm/sec	58.33 gm/sec
Explorer & Mountaineer 4.0L & 5.0L Engines	0.97 gm/sec	1.11 gm/sec	1.90 gm/sec	49.97 gm/sec
E-Series Van with 4.2L Engines	0.97 gm/sec	1.11 gm/sec	1.90 gm/sec	49.97 gm/sec
E-Series Van with 4.6L & 5.4L Engines	1.15 gm/sec	1.31 gm/sec	2.24 gm/sec	63.37 gm/sec
F-150 Truck with 4.2L, 4.6L & 5.4L Engines	1.68 gm/sec	1.93 gm/sec	3.32 gm/sec	58.43 gm/sec
F-250, F-350 Truck 4.2L, 5.4L & 6.8L Engines	1.06 gm/sec	1.21 gm/sec	2.07 gm/sec	58.33 gm/sec
Lighting 5.4L Engine	1.17 gm/sec	1.34 gm/sec	1.28 gm/sec	62.52 gm/sec
LS6, LS8 3.0L & 3.9L Engines	0.90 gm/sec	1.03 gm/sec	1.76 gm/sec	52.27 gm/sec
Mustang 3.8L, Ranger 4.0L Engines	1.00 gm/sec	1.14 gm/sec	1.94 gm/sec	52.44 gm/sec
Ranger 2.5L and 3.0L Engines	0.67 gm/sec	0.77 gm/sec	1.32 gm/sec	30.55 gm/sec
Sable & Taurus 3.0L 2v, Taurus FFV 3.0L Engines	0.67 gm/sec	0.77 gm/sec	1.32 gm/sec	30.55 gm/sec
Taurus SHO 3.4L Engine	1.06 gm/sec	1.21 gm/sec	2.07 gm/sec	58.33 gm/sec
Windstar 3.0L Engine	0.62 gm/sec	0.71 gm/sec	1.90 gm/sec	30.66 gm/sec
Windstar 3.8L Engine	0.62 gm/sec	0.71 gm/sec	1.90 gm/sec	30.66 gm/sec

NOTE: *Connect a Scan Tool and select the MAF PID. Read the value and compare it against the known good grams per second reading of that vehicle in the table. A faulty MAF sensor will be out of range.*

PID Data & Pin Chart Glossary

Glossary of Terms and Acronyms	
(<): Indicates less than the value	(>): Indicates more than the value
ACCS: A/C Cycling Clutch Switch	ACP: Air Conditioning Pressure Sensor
AODE: Automatic Overdrive Electronic	A4R70W: Automatic Overdrive Electronic
BOO: Brake On/Off Switch	CANP: EVAP Canister Purge Solenoid
CES: Clutch Engage Switch	CCS: Coast Clutch Solenoid
CKP: Crankshaft Position	CMP: Camshaft Position
COP: Coil On Plug Electronic Ignition	CTM: Central Timer Module
CTO: Clean Tachometer Output	DI: Distributor Ignition
DIS: Direct Ignition (Waste Spark)	DLC: Data Link Connector
DPFE: Differential Pressure Feedback	DRL: Daytime Running Lights
EBCM: Electronic Brake Control Module	EBTCM: Electronic Brake T/C Module
EFI: Electronic Fuel Injection	EGR: Exhaust Gas Recirculation
EGRC: EGR Control Solenoid	EGR Monitor: OBD II EGR Test
EGRV: EGR Vent Solenoid	EI: Electronic Ignition System
EPC: Electronic Pressure Control	EPT: EGR Pressure Transducer
EVAP: Evaporative Emission System	EVAP CP: Evaporative Canister Purge
EVAP CV: Evaporative Canister Vent	EVP: EGR Valve Position Sensor
EVR: EGR Vacuum Regulator	FAN: Cooling Fan (Low or High Speed)
FF: Flexible Fuel Vehicle	FIC: Fast Idle Control
FEM: Front Electronic Module	FMEM: Failure Mode Effect Management
FDPM: Fuel Pump Driver Module	FTO: Filtered Tachometer Output
FPM: Fuel Pump Monitor (in the PCM)	FTP: Fuel Tank Pressure
GEM: Generic Electronic Module	GOOSE: Brief throttle open/close
GND: Electrical ground connection	GVW: Gross Vehicle Weight
HDR-CKP: High Data Rate Crankshaft Position Sensor	HFC: High Fan Control, controls the high speed cooling fan
HFP: High Fuel Pump (Relay) Control	HLOS: Hardware Limited Operating Strategy
HO: High Output	HO2S: Heated Oxygen Sensor
HO2S-11 (Bank 1 Sensor 1) Signal	HO2S-12 (Bank 1 Sensor 2) Signal
HO2S-21 (Bank 2 Sensor 1) Signal	HO2S-22 (Bank 2 Sensor 2) Signal
HSC: High Swirl Combustion	IAC: Idle Air Control (solenoid)
IAT: Intake Air Temperature	IC: Integrated Circuit
ICM: Ignition Control Module	IDL: Idle Position Switch
IDM: Ignition Diagnostic Monitor	IFS: Inertia Fuel Switch
IGN GND: Ignition Ground	IMRC: Intake Manifold Runner Control
IMRC: Intake Manifold Runner Control	INJ 1 to INJ 10: Fuel Injectors 1 to 10
EI: Integrated-Electronic Ignition	ISO: International Standards Org.
ITS: Idle Tracking Switch	KAM: Keep Alive Memory
KAPWR: Direct Battery Power	Kg/cm^2: Kilograms/Cubic Centimeters
KOEC: Key On, Engine Cranking	KOEO: Key On, Engine Off
KOER: Key On, Engine Running	KS: Knock Sensor (Bank 1 or Bank 2)
LAMBSE: Short Term Fuel Trim	LCD: Liquid Crystal Display
LFC: Low Fan Control	LFP: Low Speed Fuel Pump (control)
LONGFT: Long Term Fuel Trim	LOS: Limited Operating Strategy
LOOP: Engine Operating Loop Status	LPG: Liquid Petroleum Gas

PID Data & Pin Chart Glossary

Glossary of Terms and Acronyms	
M/T: Manual Transmission	MAF: Mass Airflow
MAF RTN: Mass Airflow Sensor Return	MAP: Manifold Absolute Pressure
MCU: Microprocessor Control Unit	MFI: Multiport Fuel Injection
MIL: Malfunction Indicator Lamp	MLP: Manual Lever Position (Sensor)
MPH: Miles Per Hour	N.C.: Normally Closed
NDS: Neutral Drive (Switch)	N/V: Input shaft speed to vehicle speed
NGV: Natural Gas Vehicles	N.O.: Normally Open
OASIS: Online Automotive Service Information System	OBD I: On Board Diagnostics Version I OBD II: On Board Diagnostics Version II
OCIL: Overdrive Cancel Indicator Lamp	OCS: Overdrive Cancel Switch
OCT ADJ: Octane Adjust Fuel (Switch)	OHC: Overhead Cam Engine
OSM: Output State Monitor	OSS: Output Speed Shaft
O2S-11 (Bank 1 Sensor 1) Signal	O2S-21 (Bank 2 Sensor 1) Signal
PAS: Passive Anti-Theft System	PCM: Powertrain Control Module
PCV: Positive Crankcase Ventilation	PFE: Pressure Feedback EGR System
PFI: Port Fuel Injection	PID: Parameter Identification Location
PIP: Profile Ignition Pickup Signal	PMI: Programmable Module Installation
PNP: Park Neutral Position	PSP: Power Steering Pressure (switch)
PSOM: Programmable Speedometer Odometer Module	RAM: Random Access Memory
RCM: Restraint Control Module	REM: Rear Electronic Module
ROM: Read Only Memory	RPM: Revolutions Per Minute
RTN: Dedicated sensor ground circuit	RWD: Rear Wheel Drive
SC: Supercharged Engine	SCP: Standard Corporate Protocol
SFI: Sequential Fuel Injection	SHRTFT1: Short Term Fuel Trim Bank 1
SHO: Super High Output Engine	SIG RTN: Signal Return
SIL: Shift Indicator Lamp	SPOUT: Spark Output signal
SS: Shift Solenoid	STAR: Self Test Automatic Readout
STI: Self Test Input	STO: Self Test Output
TAB: Thermactor Air Bypass	TAD: Thermactor Air Diverter
TC: Turbo Charged	TCC: Torque Converter Clutch
TCC: Torque Converter Clutch Control	TCS: Traction Control Switch
TCIL: Transmission Control Indicator Lamp	TFT: Transmission Fluid Temperature
TKS: Throttle Kicker Solenoid	TOT: Transmission Oil Temperature
TP: Throttle Position (Sensor)	TP Mode: Throttle Position Mode
TR: Transmission Range sensor	TSB: Technical Service Bulletin
TSS: Transmission Shaft Speed Sensor	TWC: Three Way Catalyst
TWC + OC: 3-Way Catalyst	UBP: UART Based Protocol Network
VAF: Vane Airflow Meter or sensor	VAT: Vane Air Temperature sensor
VID: Vehicle Information Block (memory)	VIM: Vehicle Interface Module
VBAT: Vehicle Battery Voltage	VLCM: Variable Load Control Module
VPWR: Ignition switched Power	VREF: Reference Voltage (from PCM)
VRS: Variable Reluctance Sensor	VMV: Vapor Management Valve (EVAP)
VCT: Variable Camshaft Timing	WAC: WOT A/C Cutout Relay
WACA- A/C WOT Cutout Relay Monitor	WOT: Wide Open Throttle

GAS ENGINE TROUBLE CODE LIST

Introduction

To use this information, first read and record all codes in memory along with any Freeze Frame data. *If the PCM reset function is done prior to recording any data, all codes and freeze frame data will be lost!* Look up the desired code by DTC number, Code Title and Conditions (enable criteria) that indicate why a code set, and how to drive the vehicle. **1T and 2T** indicate a 1-trip or 2-trip fault and the Monitor type.

Gas Engine OBD II Trouble Code List (P0xxx Codes)

DTC	Trouble Code Title, Conditions & Possible Causes
P0010 **2T CCM, MIL: Yes** Cougar, Focus, LS, Five Hundred, Freestyle, Montego, Thunderbird & ZX2 Transmissions: All	**Variable Cam Timing Solenoid 'A' Circuit Malfunction Conditions:** Key on or engine running; and the PCM detected an unexpected high voltage or low voltage condition on the Variable Cam Timing (VCT) Solenoid 'A' control circuit during testing. **Possible Causes** • VCT 'A' solenoid connector is damaged, loose or shorted • VCT 'A' solenoid control circuit is open, shorted to ground or shorted to power • VCT 'A' solenoid is damaged or the PCM has failed
P0011 **2T CCM, MIL: Yes** Cougar, Focus, LS, Five Hundred, Freestyle, Montego,, Thunderbird & ZX2 Transmissions: All	**Variable Cam Timing Over Advanced (Bank 1) Conditions:** Engine started; and the PCM detected the camshaft timing exceeded the maximum calibrated advance value, or the camshaft remained in an advanced position during the CCM test. **Possible Causes** • Camshaft timing improperly set, or continuous oil flow to the VCT piston chamber • Camshaft advance mechanism (the VCT unit) is sticking or binding mechanically • VCT solenoid valve is stuck in open position
P0012 **2T CCM, MIL: Yes** Cougar, Focus, LS, Five Hundred, Freestyle, Montego,, Thunderbird & ZX2 Transmissions: All	**Variable Cam Timing Over Retarded (Bank 1) Conditions:** Engine started; and the PCM detected the camshaft timing exceeded the maximum calibrated retard value, or the camshaft remained in a retarded position during the CCM test. **Possible Causes** • Camshaft timing improperly set, or continuous oil flow to the VCT piston chamber • Camshaft advance mechanism (the VCT unit) is sticking or binding mechanically • VCT solenoid valve is stuck in open position
P0020 **2T CCM, MIL: Yes** Cougar, Focus, LS, Five Hundred, Freestyle, Montego,, Thunderbird & ZX2 Transmissions: All	**Variable Cam Timing Solenoid 'B' Circuit Malfunction Conditions:** Key on or engine running; and the PCM detected an unexpected high or low voltage condition on the Variable Cam Timing (VCT) Solenoid 'B' control circuit during testing. **Possible Causes** • VCT 'B' solenoid connector is damaged, loose or shorted • VCT 'B' solenoid control circuit is open, shorted to ground or shorted to power • VCT 'B' solenoid is damaged or the PCM has failed
P0020 **2T CCM, MIL: Yes** Cougar, Focus, LS, Five Hundred, Freestyle, Montego,, Thunderbird & ZX2 Transmissions: All	**Variable Cam Timing Over Advanced (Bank 2) Conditions:** Engine started; and the PCM detected the camshaft timing exceeded the maximum calibrated advance value, or the camshaft remained in an advanced position during the CCM test. **Possible Causes** • Camshaft timing improperly set, or continuous oil flow to the VCT piston chamber • Camshaft advance mechanism (the VCT unit) is sticking or binding mechanically • VCT solenoid valve is stuck in open position
P0022 **2T CCM, MIL: Yes** Cougar, Focus, LS, Five Hundred, Freestyle, Montego,, Thunderbird & ZX2 Transmissions: All	**Variable Cam Timing Over Retarded (Bank 2) Conditions:** Engine started; and the PCM detected the camshaft timing exceeded the maximum calibrated retard value, or the camshaft remained in a retarded position during the CCM test. **Possible Causes** • Camshaft timing improperly set, or continuous oil flow to the VCT piston chamber • Camshaft advance mechanism (the VCT unit) is sticking or binding mechanically • VCT solenoid valve is stuck in open position
P0040 **2T CCM, MIL: Yes** Crown Victoria, Marauder & Grand Marquis Transmissions: All	**Upstream Oxygen Sensors Swapped From Bank To Bank Conditions:** Engine started; and after the PCM performed a brief fuel shift in engine Bank 1, it did not detect the correct response in the HO2S-11 and/or HO2S-21 (sensors) during testing. **Possible Causes** • HO2S-11 and HO2S-21 harness connectors are swapped • HO2S-11 and HO2S-21 wiring is crossed inside the harness • HO2S-11 and HO2S-21 wires are crossed at 104-pin connector

Gas Engine OBD II Trouble Code List (P0xxx Codes)

DTC	Trouble Code Title, Conditions & Possible Causes
P0041 **2T CCM, MIL: Yes** Crown Victoria, Marauder & Grand Marquis Transmissions: All	**Downstream Oxygen Sensors Swapped Bank To Bank Conditions:** Engine started; and after the PCM performed a brief fuel shift in engine Bank 1, it did not detect the correct response in the HO2S-12 and/or HO2S-22 (sensors) during testing. **Possible Causes** • HO2S-12 and HO2S-22 harness connectors are swapped • HO2S-12 and HO2S-22 wiring is crossed inside the harness • HO2S-12 and HO2S-22 wires are crossed at 104-pin connector
P0041 **2T CCM, MIL: Yes** Crown Victoria, Marauder & Grand Marquis Transmissions: All	**TP Sensor, ETC TP Sensor Inconsistent With MAF Sensor Conditions:** Engine started, KOER Self Test enabled, and the PCM detected the MAF and ETC TP sensor signals were not consistent with the calibrated values expected for these sensors in the test. **TP Sensor:** Drive the vehicle and then monitor the TPNS PID in all gears. A TP PID under 0.24v (4.82%) with a LOAD PID over 55%, or a TP V PID over 2.44v (49.05%) with a LOAD PID under 30% will set this code. **ETC TP:** Drive the vehicle and then monitor the TPNS PID in all gears. A TPNS PID under 0.24v (4.82%) with a LOAD PID over 55% or TPNS PID over 2.44v (49.05%) with LOAD PID under 30% will set a code. **Possible Causes** • Air leak exists between MAF sensor and the throttle body • MAF sensor is out-of-calibration or it has failed • TP sensor is not seated properly or the TP sensor is damaged
P0102 **2T CCM, MIL: Yes** All Models Transmissions: All	**MAF Sensor Circuit Low Input Conditions:** DTC P0505 not set, engine started, and the PCM detected the MAF sensor signal was less than 0.23v during the CCM test period. **Possible Causes** • Check for leaks at air outlet tube • Sensor power circuit open, sensor ground circuit open • Sensor signal circuit open (may be disconnected) • Check for loose tube clamps near the MAF sensor
P0103 **2T CCM, MIL: Yes** All Models Transmissions: All	**MAF Sensor Circuit High Input Conditions:** DTC P0505 not set, engine started, and the PCM detected the MAF sensor signal was more than 4.60v during the CCM test period. **Possible Causes** • Check for a restricted inlet screen on the MAF sensor. • MAF sensor signal circuit is shorted to system power (B+) • MAF sensor is damaged or the PCM has failed
P0106 **2T CCM, MIL: Yes** All Models Transmissions: All	**Barometric Pressure Sensor Circuit Performance Conditions:** Engine started, and the PCM detected the BARO sensor was out of range during the CCM test. The BARO sensor signal should be in a range of 4.0-6.0v. The Scan Tool displays the sensor reading as a frequency. Note: The sensor VREF should be less than 6.0v at all times. **Possible Causes** • Sensor has deteriorated (response time too slow) or has failed • PCM has failed
P0107 **2T CCM, MIL: Yes** All Models Transmissions: All	**Barometric Pressure Sensor Circuit Low Input Conditions:** Engine started, and the PCM detected the BARO sensor indicated less than the minimum calibrated parameter. The BARO sensor is a variable capacitance unit used to detect altitude. **Possible Causes** • BARO sensor signal circuit is shorted to ground • BARO sensor VREF circuit (5v) is open • BARO sensor is damaged or it has failed • PCM has failed
P0107 **2T CCM, MIL: Yes** 2.0L I4 VIN Z CNG engine Contour Transmissions: All	**MAP/BARO Sensor (Compuvalve) Circuit Low Input Conditions:** Engine started, vehicle driven under heavy load conditions, and the PCM detected the Compuvalve MAP/BARO sensor signal indicated too low a value. Note: The MAP/BARO sensor is mounted (internal) in the Compuvalve assembly. The PCM monitors this signal. **Possible Causes** • Vacuum hoses from the intake manifold and the Compuvalve may be open or restricted • Compuvalve MAP/BARO sensor has failed

Gas Engine OBD II Trouble Code List (P0xxx Codes)

DTC	Trouble Code Title, Conditions & Possible Causes
P0108 **2T CCM, MIL: Yes** All Models Transmissions: All	**Barometric Sensor Circuit High Input Conditions:** Engine started, and the PCM detected the BARO sensor signal was more than the maximum calibrated parameter of 5.0v. Note: The sensor VREF should be 4.0v to 6.0v at all times. **Possible Causes** • BARO sensor connector is damaged, open or shorted • BARO sensor signal circuit is open or shorted to VREF (5v) • BARO sensor is damaged or it has failed • PCM has failed
P0108 **2T CCM, MIL: Yes** 2.0L I4 VIN Z CNG engine Contour Transmissions: All	**MAP/BARO Sensor (Compuvalve) Circuit High Input Conditions:** Engine started, vehicle driven under heavy load conditions, and the PCM detected the Compuvalve MAP/BARO sensor signal indicated too high a value during the CCM test. The MAP/BARO sensor is mounted (internal) in the Compuvalve assembly on this application. **Possible Causes** • Vacuum hoses from the intake manifold and Compuvalve may be open and/or restricted • Compuvalve MAP/BARO sensor has failed
P0109 **2T CCM, MIL: Yes** All Models Transmissions: All	**Barometric Sensor Circuit Intermittent Input Conditions:** Engine started, and the PCM detected the BARO sensor signal had an intermittent failure during normal engine operation. **Possible Causes** • BARO sensor signal circuit is open or shorted to ground (intermittent) • BARO sensor VREF circuit (5v) is open • BARO sensor signal circuit is shorted to ground (Intermittent) • BARO sensor is damaged or it has failed
P0112 **2T CCM, MIL: Yes** All Models Transmissions: All	**Intake Air Temperature Sensor Circuit Low Input Conditions:** Key on or engine running; and the PCM detected the IAT sensor signal was less than the self-test minimum of 0.20v (Scan Tool reads over 250ºF). This is a thermistor-type sensor with a variable resistance that changes when exposed to different temperatures. **Possible Causes** • IAT sensor signal circuit is grounded (check wiring & connector) • IAT sensor is damaged or it has failed • PCM has failed
P0113 **2T CCM, MIL: Yes** All Models Transmissions: All	**Intake Air Temperature Sensor Circuit High Input Conditions:** Key on or engine running; and the PCM detected the IAT sensor signal was more than the self-test maximum 4.60v (Scan Tool reads under -46ºF). This is a thermistor-type sensor with a variable resistance that changes when exposed to different temperatures. **Possible Causes** • IAT sensor signal circuit is open (inspect wiring & connector) • IAT sensor signal circuit is shorted to VREF (5v) • IAT sensor is damaged or it has failed • PCM has failed
P0116 **2T ECT, MIL: Yes** All Models Transmissions: All	**ECT Sensor / CHT Sensor Signal Range/Performance Conditions:** Engine off for a calibrated period of time after an engine cold soak period over 6 hours, engine started, and the PCM detected the ECT sensor exceeded the IAT sensor by more than a calibrated value (e.g., 30ºF), or the ECT sensor was more than a calibrated value of 225ºF; the Catalyst, Fuel System, HO2S and Misfire Monitor did not complete, or the timer expired. **Possible Causes** • Check for low coolant level or incorrect coolant mixture • CHT sensor is out-of-calibration or it has failed • ECT sensor is out-of-calibration or it has failed
P0117 **2T CCM, MIL: Yes** All Models Transmissions: All	**ECT Sensor Circuit Low Input Conditions:** Key on or engine running; and the PCM detected the ECT sensor signal was less than the self-test minimum of 0.20v (Scan Tool reads over 250ºF). This is a thermistor-type sensor with a variable resistance that changes when exposed to different temperatures **Possible Causes** • ECT sensor signal circuit is grounded in the wiring harness • ECT sensor is damaged or the PCM has failed

Gas Engine OBD II Trouble Code List (P0xxx Codes)

DTC	Trouble Code Title, Conditions & Possible Causes
P0118 **2T CCM, MIL: Yes** All Models Transmissions: All	**ECT Sensor Circuit High Input Conditions:** Key on or engine running; and the PCM detected the ECT sensor signal was more than the self-test maximum of 4.60v (Scan Tool reads under -46ºF). This is a thermistor-type sensor with a variable resistance that changes when exposed to different temperatures **Possible Causes** • ECT sensor signal circuit is open (inspect wiring & connector) • ECT sensor signal circuit is shorted to VREF (5v) • ECT sensor is damaged or it has failed • PCM has failed
P0121 **2T CCM, MIL: Yes** All Models Transmissions: All	**TP Sensor Signal Range/Performance Conditions:** Engine started; then immediately following a condition where the engine was running under at off-idle, the PCM detected the TP sensor signal indicated the throttle did not return to its previous closed position during the CCM Rationality test. **Possible Causes** • Throttle plate is binding, dirty or sticking • TP sensor signal circuit open (inspect wiring & connector) • TP sensor ground circuit open (inspect wiring & connector) • TP Sensor is damaged or has failed
P0121 **2T CCM, MIL: Yes** Crown Victoria, Marauder & Grand Marquis Transmissions: All	**ETC Throttle Position Sensor Signal Range/Performance Conditions:** Key on or engine running; then immediately after the throttle is open or following a closed throttle (deceleration) event, the PCM detected the TP2 PID (TP Sensor 2) signal indicated more than 93% (4.65v) from its previous closed position during the CCM Rationality test. **Possible Causes** • ECT TP sensor connector is damaged, loose or shorted • ETC TP sensor signal circuit is open • ETC TP sensor ground circuit is open • ETC TP sensor circuit VREF (5v) is open • ECT TP sensor is not seated properly, or it is sticking • ETC TP Sensor is damaged or it has failed
P0122 **2T CCM, MIL: Yes** All Models Transmissions: All	**TP Sensor Circuit Low Input Conditions:** Key on or engine running; and the PCM detected the TP sensor was less than 0.17v (Scan Tool TP PID reads under 3.42%) in the test. **Possible Causes** • TP sensor signal circuit open (inspect wiring & connector) • TP sensor signal shorted to ground (inspect wiring & connector) • TP sensor VREF circuit is open (between the sensor and PCM) • TP sensor is damaged or has failed • PCM has failed
P0122 **2T CCM, MIL: Yes** Crown Victoria, Marauder & Grand Marquis Transmissions: All	**ETC Throttle Position Sensor Circuit Low Input Conditions:** Key on or engine running; and the PCM detected the ETC TP sensor was below the self-test minimum of 0.17v (Scan Tool below 3.42%). **Possible Causes** • TP sensor signal circuit open (inspect wiring & connector) • TP sensor signal shorted to ground (inspect wiring & connector) • TP sensor VREF circuit is open (between the sensor and PCM) • TP sensor is damaged or has failed • PCM has failed
P0123 **2T CCM, MIL: Yes** All Models Transmissions: All	**TP Sensor Circuit High Input Conditions:** Engine started, and the PCM detected the TP sensor signal was more than the self-test maximum of 4.65v (equivalent to a Scan Tool TP PID of more than 93%) during testing. **Possible Causes** • TP sensor not seated correctly in housing (may be damaged) • TP sensor signal is circuit shorted to VREF or system voltage • TP sensor ground circuit is open (check the wiring harness) • Perform a "sensor sweep test" and monitor for any glitches • PCM has failed

Gas Engine OBD II Trouble Code List (P0xxx Codes)

DTC	Trouble Code Title, Conditions & Possible Causes
P0122 **2T CCM, MIL: Yes** Crown Victoria, Marauder & Grand Marquis Transmissions: All	**ETC Throttle Position Sensor Circuit High Input Conditions:** Key on or engine running; and the PCM detected the ETC TP sensor was less than the self-test maximum of 4.65v (Scan Tool TP PID reads more than 93%) during the CCM test. **Possible Causes** • TP sensor connector is damaged, loose or shorted • TP sensor signal circuit is open or shorted to ground • TP sensor VREF circuit is open (between the sensor and PCM) • TP sensor is damaged or has failed • PCM has failed
P0125 **2T ECT, MIL: Yes** All Models Transmissions: All	**Insufficient Coolant Temperature For Closed Loop Conditions:** Engine runtime at road load more than 6 minutes, and the PCM detected that the ECT sensor (or CHT sensor) signal did not indicate the required engine temperature value to enter closed loop within a specified amount of time. The amount of time is calculated from the point at which the engine is started, and depends upon the ECT or CHT sensor signal value at startup. **Possible Causes** • Check the coolant mixture for an incorrect mixture • Check the operation of the thermostat (it may be stuck open) • ECT sensor (or the CHT sensor) has failed • Inspect for low coolant level
P0127 **2T CCM, MIL: Yes** F150 Series Truck, Mustang with a 4.6L VIN Y & 5.4L VIN 3 engines Transmissions: All	**IAT Sensor 2 Circuit High Input Conditions:** Engine started, engine running for a calibrated period of time, and the PCM detected the IAT Sensor 2 (PID IAT2) signal was too high. This code indicates a potential fault present in the Intercooler system (the Supercharger Boost is bypassed when this code is set). **Possible Causes** • Blockage present in the Heat Exchangers • Low fluid level, or a fluid leak is present • Intercooler pump or relay has failed • Intercooler coolant lines may be crossed
P0128 **2T CCM, MIL: Yes** All Models Transmissions: All	**Intake Air Temperature Sensor 2 Circuit High Input Conditions:** Engine started, vehicle driven for over 10 minutes, and the PCM detected the engine did not reach an engine operating temperature of 160°F after an additional runtime of 2 minutes. **Possible Causes** • Check the operation of the thermostat (it may be stuck open) • ECT sensor or CHT sensor is out-of-calibration, or has failed • Inspect for low coolant level or an incorrect coolant mixture
P0131 **2T CCM, MIL: Yes** All Models Transmissions: All	**HO2S-11 (Bank 1 Sensor 1) Circuit Low Input Conditions:** Engine running for more than 5 minutes, and the PCM detected the HO2S signal was in a negative voltage range referred to as "character shift downward". This code sets when the HO2S signal remains in a low state (usually less than 156 mv). In effect, it does not switch properly between 0.1v and 1.1v in closed loop operation. **Possible Causes** • HO2S is contaminated (due to presence of silicone in fuel) • HO2S signal and ground circuit wires crossed in wiring harness • HO2S signal circuit is shorted to sensor or chassis ground • HO2S element has failed (internal short condition) • PCM has failed
P0132 **2T CCM, MIL: Yes** All Models Transmissions: All	**HO2S-11 (Bank 1 Sensor 1) Circuit High Input Conditions:** Engine running for more than 5 minutes, and the PCM detected the HO2S signal remained in a high state (i.e., more than 1.5v). Note: The HO2S signal circuit may be shorted to the heater power circuit due to "tracking inside of the HO2S connector. Remove the connector and visually inspect the connector for signs of oil or water. **Possible Causes** • HO2S signal shorted to heater power circuit inside connector • HO2S signal circuit shorted to VREF or to system voltage • PCM has failed

Gas Engine OBD II Trouble Code List (P0xxx Codes)

DTC	Trouble Code Title, Conditions & Possible Causes
P0133 **2T O2S, MIL: Yes** All Models Transmissions: All	**HO2S-11 (Bank 1 Sensor 1) Circuit Slow Response Conditions:** Engine started, engine running in closed loop for over 5 minutes, and the PCM detected the HO2S amplitude and frequency were out of the normal range (e.g., the HO2S rich to lean switch time was more than 100 ms) during the HO2S Monitor test. **Possible Causes** • HO2S is contaminated (due to presence of silicone in fuel) • HO2S signal circuit open • Leaks present in the exhaust manifold or exhaust pipes • HO2S is damaged or has failed • PCM has failed
P0135 **2T O2HTR, MIL: Yes** All Models Transmissions: All	**HO2S-11 (Bank 1 Sensor 1) Heater Circuit Malfunction Conditions:** Engine started, engine running for 5 minutes, and the PCM detected an unexpected voltage condition, or it detected excessive current draw in the heater circuit during the CCM test. **Possible Causes** • HO2S heater power circuit is open or heater ground circuit open • HO2S signal tracking (due to oil or moisture in the connector) • HO2S is damaged or has failed • PCM has failed
P0136 **2T O2S, MIL: Yes** All Models Transmissions: All	**HO2S-12 (Bank 1 Sensor 1) Circuit No Activity Conditions:** Engine started, engine running in closed loop for over 5 minutes, and the PCM detected the HO2S signal failed to meet the maximum or minimum voltage levels (i.e., it failed the voltage range check). **Possible Causes** • Leaks present in the exhaust manifold or exhaust pipes • HO2S signal wire and ground wire crossed in connector • HO2S element is fuel contaminated or has failed • PCM has failed
P0138 **2T CCM, MIL: Yes** All Models Transmissions: All	**HO2S-12 (Bank 1 Sensor 2) Circuit High Input Conditions:** Engine running for more than 5 minutes, and the PCM detected the HO2S signal remained in a high state (i.e., more than 1.5v). *Note: The HO2S signal circuit may be shorted to the heater power circuit due to "tracking inside of the HO2S connector. Remove the connector and visually inspect the connector for signs of oil or water.* **Possible Causes** • HO2S signal shorted to heater power circuit in the connector • HO2S signal circuit shorted to VREF or to system voltage • PCM has failed
P0141 **2T O2HTR, MIL: Yes** All Models Transmissions: All	**HO2S-12 (Bank 1 Sensor 2) Heater Circuit Malfunction Conditions:** Engine running for 5 minutes, and the PCM detected an open or shorted condition, or excessive current draw in the heater circuit. **Possible Causes** • HO2S heater power circuit is open or heater ground circuit open • HO2S signal tracking (due to oil or moisture in the connector) • HO2S is damaged or has failed • PCM has failed
P0148 **2T CCM, MIL: Yes** All Models Transmissions: All	**Fuel Delivery Error Conditions:** Engine started, engine running for a specified period of time in closed loop, then after at least one WOT event was recorded, the PCM detected a lean air/fuel condition in at least one engine bank during the wide-open throttle event. **Possible Causes** • Severely restricted fuel filter • Severely pinched or restricted fuel delivery line

Gas Engine OBD II Trouble Code List (P0xxx Codes)

DTC	Trouble Code Title, Conditions & Possible Causes
P0151 **2T CCM, MIL: Yes** Models with a 4-Cyl, V6 or V8 engine Transmissions: All	**HO2S-21 (Bank 2 Sensor 1) Circuit Low Input Conditions:** Engine started, engine running for over 5 minutes, and the PCM detected the HO2S signal was in a negative voltage range referred to as "character shift downward". This code sets when the HO2S signal remains in a low state (less than 156 mv). In effect, the sensor did not switch properly between 0.1v and 1.1v in closed loop. **Possible Causes** • HO2S connector is damaged or shorted • HO2S signal and ground circuit wires crossed in wiring harness • HO2S signal circuit is shorted to sensor or chassis ground • HO2S element has failed (internal short condition) • PCM has failed
P0152 **2T CCM, MIL: Yes** Models with a 4-Cyl, V6 or V8 engine Transmissions: All	**HO2S-21 (Bank 2 Sensor 1) Circuit High Input Conditions:** Engine started, engine runtime over 5 minutes, and the PCM detected the HO2S signal remained in a high state (more than 1.5v). Note: The HO2S signal circuit may be shorted to the heater power circuit due to "tracking inside of the HO2S connector. Remove the connector and visually inspect the connector for signs of oil or water. **Possible Causes** • HO2S is contaminated (due to presence of silicone in fuel) • HO2S signal tracking (due to oil or moisture in the connector) • HO2S signal circuit is open or shorted to VREF • PCM has failed
P0153 **2T O2S, MIL: Yes** Models with a 4-Cyl, V6 or V8 engine Transmissions: All	**HO2S-21 (Bank 2 Sensor 1) Circuit Slow Response Conditions:** Engine started, engine running in closed loop for over 5 minutes, and the PCM detected the HO2S amplitude and frequency were out of the normal range (e.g., the HO2S rich to lean switch time was more than 100 ms) during the HO2S Monitor test. **Possible Causes** • HO2S is contaminated (due to presence of silicone in fuel) • Leaks present in the exhaust manifold or exhaust pipes • HO2S is damaged or has failed • PCM has failed
P0155 **2T O2HTR, MIL: Yes** Models with a 4-Cyl, V6 or V8 engine Transmissions: All	**HO2S-21 (Bank 2 Sensor 1) Heater Circuit Malfunction Conditions:** Engine running for 5 minutes, and the PCM detected an open or shorted condition, or excessive current draw in the heater circuit. **Possible Causes** • HO2S heater power circuit is open • HO2S heater ground circuit is open • HO2S signal tracking (due to oil or moisture in the connector) • HO2S is damaged or has failed • PCM has failed
P0156 **2T O2S, MIL: Yes** Models with a 4-Cyl, V6 or V8 engine Transmissions: All	**HO2S-22 (Bank 2 Sensor 2) Circuit No Activity Conditions:** Engine running in closed loop for more than 5 minutes, and the PCM detected the HO2S signal failed to meet the maximum or minimum voltage (i.e., it failed the voltage check). **Possible Causes** • Leaks present in the exhaust manifold or exhaust pipes • HO2S signal wire and ground wire crossed in connector • HO2S element is fuel contaminated or has failed • PCM has failed
P0158 **2T CCM, MIL: Yes** Models with a 4-Cyl, V6 or V8 engine Transmissions: All	**HO2S-22 (Bank 2 Sensor 2) Circuit High Input Conditions:** Engine running for more than 5 minutes, and the PCM detected the HO2S signal remained in a high state (i.e., more than 1.5v). Note: The HO2S signal circuit may be shorted to the heater power circuit due to "tracking inside of the HO2S connector. Remove the connector and visually inspect the connector for signs of oil or water. **Possible Causes** • HO2S signal shorted to the heater power circuit (due to oil or moisture in the connector) • HO2S signal circuit shorted to VREF or to system voltage • PCM has failed

Gas Engine OBD II Trouble Code List (P0xxx Codes)

DTC	Trouble Code Title, Conditions & Possible Causes
P0161 **2T O2HTR, MIL: Yes** Models with a 4-Cyl, V6 or V8 engine Transmissions: All	**HO2S-22 (Bank 2 Sensor 2) Heater Circuit Malfunction Conditions:** Engine running for 5 minutes, and the PCM detected an open or shorted condition, or excessive current draw in the heater circuit. **Possible Causes** • HO2S heater power circuit or the heater ground circuit is open • HO2S signal tracking (due to oil or moisture in the connector) • HO2S has failed, or the PCM has failed
P0171 **2T FUEL, MIL: Yes** Some Models Transmissions: All	**Fuel System Too Lean (Cylinder Bank 1) Conditions:** Engine started, engine running at cruise speed for 3 to 4 minutes, and the PCM detected the Bank 1 Adaptive Fuel Control System reached its rich correction limit (a lean A/F condition). **Possible Causes** • Air leaks after the MAF sensor, or leaks in the PCV system • Exhaust leaks before or near where the HO2S is mounted • Fuel injector(s) restricted or not supplying enough fuel • Fuel pump not supplying enough fuel during high fuel demand conditions • Leaking EGR gasket, or leaking EGR valve diaphragm • MAF sensor dirty (causes PCM to underestimate airflow) • Vehicle running out of fuel or engine oil dip stick not seated • TSB 01-20-5 contains a repair procedure for this trouble code
P0171 **2T FUEL, MIL: Yes** Windstar Models Transmissions: All	**Fuel System Too Lean (Cylinder Bank 1) Conditions:** Engine started, engine running at cruise speed for 3 to 4 minutes, and the PCM detected the Bank 1 Adaptive Fuel Control System reached its rich correction limit (a lean A/F condition). **Possible Causes** • Air leaks after the MAF sensor, or leaks in the PCV system • Exhaust leaks before or near where the HO2S is mounted • Fuel injector(s) restricted or not supplying enough fuel • Fuel system not supplying enough fuel during high fuel demand conditions (e.g., the fuel pump may not supply enough fuel) • Leaking EGR gasket, or leaking EGR valve diaphragm • MAF sensor dirty (causes PCM to underestimate airflow) • Vehicle running out of fuel or engine oil dip stick not seated • TSB 03-16-1 contains a repair procedure for this trouble code
P0172 **2T FUEL, MIL: Yes** All Models Transmissions: All	**Fuel System Too Rich (Cylinder Bank 1) Conditions:** Engine started, engine running at cruise speed for 3 to 4 minutes, and the PCM detected the Bank 1 Adaptive Fuel Control System reached its rich correction limit (a rich A/F condition). **Possible Causes** • Camshaft timing is incorrect, or the engine has an oil overfill condition • EVAP vapor recovery system failure (may be pulling vacuum) • Fuel pressure regulator is damaged or leaking • HO2S element is contaminated with alcohol or water • MAF or MAP sensor values are incorrect or out-of-range • One of more fuel injectors is leaking
P0174 **2T FUEL, MIL: Yes** Some Models (V6, V8) Transmissions: All	**Fuel System Too Lean (Cylinder Bank 2) Conditions:** Engine started, engine running at cruise speed for 3 to 4 minutes, and the PCM detected the Bank 2 Adaptive Fuel Control System reached its rich correction limit (a lean A/F condition). **Possible Causes** • Air leaks after the MAF sensor, or leaks in the PCV system • Exhaust leaks before or near where the HO2S is mounted • Fuel injector(s) restricted or not supplying enough fuel • Fuel system not supplying enough fuel during high fuel demand conditions (e.g., the fuel pump may not supply enough fuel) • Leaking EGR gasket, or leaking EGR valve diaphragm • MAF sensor dirty (causes PCM to underestimate airflow) • Vehicle running out of fuel or engine oil dip stick not seated • TSB 1-20-5 contains a repair procedure for this trouble code

Gas Engine OBD II Trouble Code List (P0xxx Codes)

DTC	Trouble Code Title, Conditions & Possible Causes
P0174 **2T FUEL, MIL: Yes** Windstar Models Transmissions: All	**Fuel System Too Lean (Cylinder Bank 2) Conditions:** Engine started, engine running at cruise speed for 3 to 4 minutes, and the PCM detected the Bank 2 Adaptive Fuel Control System reached its rich correction limit (a lean A/F condition). **Possible Causes** • Air leaks after the MAF sensor, or leaks in the PCV system • Exhaust leaks before or near where the HO2S is mounted • Fuel injector(s) restricted or not supplying enough fuel • Fuel system not supplying enough fuel during high fuel demand conditions (e.g., the fuel pump may not supply enough fuel) • Leaking EGR gasket, or leaking EGR valve diaphragm • MAF sensor dirty (causes PCM to underestimate airflow) • Vehicle running out of fuel or engine oil dip stick not seated • TSB 03-16-1 contains a repair procedure for this trouble code
P0175 **2T FUEL, MIL: Yes** Some Models (V6, V8) Transmissions: All	**Fuel System Too Rich (Cylinder Bank 2) Conditions:** Engine started, engine running at cruise speed for 3 to 4 minutes, and the PCM detected the Bank 2 Adaptive Fuel Control System reached its rich correction limit (a rich A/F condition). **Possible Causes** • Camshaft timing is incorrect • Engine oil overfill condition • EVAP vapor recovery system failure (may be pulling vacuum) • Fuel pressure regulator is damaged or leaking • HO2S element is contaminated with alcohol or water • MAF or MAP sensor values are incorrect or out-of-range • One of more fuel injectors is leaking
P0176 **2T CCM, MIL: Yes** Taurus with a 3.0L VIN 1 or VIN 2 engine Transmissions: All	**Flexible Fuel Sensor Circuit Malfunction Conditions:** Engine started, and the PCM detected the Flexible Fuel sensor (FFV) signal was not within its maximum or minimum calibrated range (the normal calibrated range is from 40-160 Hz). Note: Prefix 610C indicates an ethanol vehicle and prefix 610G indicates a methanol vehicle. **Possible Causes** • FFV sensor VPWR circuit is open or shorted to ground • FFV sensor ground circuit open or circuit shorted to ground • FFV sensor has a fuel contamination condition • FFV sensor has a fuel separation condition • PCM has failed
P0180 **2T CCM, MIL: Yes** Crown Victoria, Grand Marquis, E & F Series Vans & Trucks with a 4.6L VIN 9, 5.4L VIN M, 5.4L VIN Z engine Transmissions: All	**Engine Fuel Temperature Sensor 'A' Circuit Malfunction Conditions:** Engine runtime over 2 minutes, and the PCM detected the Engine Fuel Temperature (EFT) sensor 'A' signal was out-of-range (i.e., it was more than 4.54v [-46°F] or less than 0.21v [275°F]. Note: Monitor the EFT PID value to identify an open or short circuit. **Possible Causes** • Engine operating under "low" ambient temperature conditions • EFT sensor signal circuit open or shorted in the wiring harness • EFT sensor is damaged or it has failed • PCM has failed
P0181 **2T CCM, MIL: Yes** Crown Victoria, Grand Marquis, E & F Series Vans & Trucks with a 4.6L VIN 9, 5.4L VIN M, 5.4L VIN Z engine Transmissions: All	**Fuel Temperature Sensor 'A' Signal Performance Conditions:** Engine runtime over 2 minutes, and the PCM detected the Engine Fuel Temperature (EFT) sensor 'A' signal was more than 4.50v [-46°F] or less than 0.21v [275°F] the calibrated limit in the self-test. **Possible Causes** • Engine operating under "low" ambient temperature conditions • EFT sensor signal circuit open or shorted in the wiring harness • EFT sensor is damaged or it has failed • PCM has failed

Gas Engine OBD II Trouble Code List (P0xxx Codes)

DTC	Trouble Code Title, Conditions & Possible Causes
P0182 **2T CCM, MIL: Yes** Crown Victoria, Grand Marquis, E & F Series Vans & Trucks with a 4.6L VIN 9, 5.4L VIN M, 5.4L VIN Z engine Transmissions: All	**Engine Fuel Temperature Sensor 'A' Circuit Low Input Conditions:** Key on or engine running; and the PCM detected the Engine Fuel Temperature (EFT) Sensor 'A' signal was over 260ºF in the self-test. **Possible Causes** • EFT sensor connector is damaged or shorted • EFT sensor VREF circuit is open or shorted to ground • EFT sensor circuit is shorted to chassis or to sensor ground • EFT sensor is damaged or it has failed • PCM has failed
P0182 **2T CCM, MIL: Yes** 2.0L I4 VIN Z CNG engine Contour Transmissions: All	**Engine Fuel Temperature Sensor 'A' Circuit Low Input Conditions:** Key on or engine running; and the PCM detected the EFT sensor 'A' signal from the Compuvalve was less than -38ºF during the self-test. **Possible Causes** • EFT sensor connector is damaged or shorted • EFT sensor VREF circuit is open or shorted to ground • EFT sensor circuit is shorted to chassis or to sensor ground • EFT sensor is damaged or it has failed • PCM has failed
P0183 **2T CCM, MIL: Yes** Crown Victoria, Grand Marquis, E & F Series Vans & Trucks with a 4.6L VIN 9, 5.4L VIN M, 5.4L VIN Z engine Transmissions: All	**Engine Fuel Temperature Sensor 'A' Circuit High Input Conditions:** Key on or engine running; and the PCM detected the EFT sensor 'A' signal was more than the limit of 4.50v [-46ºF] during the self-test. **Possible Causes** • EFT sensor connector is damaged, loose or open • EFT sensor signal circuit is shorted to VREF (5v) • EFT sensor signal circuit is open • EFT sensor is damaged or it has failed • PCM has failed
P0183 **2T CCM, MIL: Yes** 2.0L I4 VIN Z CNG engine Contour Transmissions: All	**Engine Fuel Temperature Sensor 'A' Circuit High Input Conditions:** Key on or engine running; and the PCM detected the EFT sensor 'A' signal from the Compuvalve was more than 260ºF in the self-test. **Possible Causes** • EFT sensor connector is damaged, loose or open • EFT sensor signal circuit is open or it is shorted to VREF (5v) • EFT sensor is damaged or it has failed • PCM has failed
P0186 **2T CCM, MIL: Yes** Crown Victoria, Grand Marquis, E & F Series Vans & Trucks with a 4.6L VIN 9, 5.4L VIN M, 5.4L VIN Z engine Transmissions: All	**Engine Fuel Temperature Sensor 'B' Signal Performance Conditions:** Engine started, engine runtime over 2 minutes, and the PCM detected the EFT sensor 'B' signal was more than 4.50v (Scan Tool reads less than -46ºF) or less than 0.21v (Scan Tool reads more than 275ºF) the calibrated limit during the self-test. **Possible Causes** • Engine operating under "low" ambient temperature conditions • EFT sensor signal circuit open or shorted in the wiring harness • EFT sensor is damaged or it has failed • PCM has failed
P0187 **2T CCM, MIL: Yes** Crown Victoria, Grand Marquis, E & F Series Vans & Trucks with a 4.6L VIN 9, 5.4L VIN M, 5.4L VIN Z engine Transmissions: All	**Engine Fuel Temperature Sensor 'B' Circuit Low Input Conditions:** Engine started, engine runtime over 2 minutes, and the PCM detected the EFT Sensor 'B' signal was less than the calibrated limit of 0.21v (Scan Tool reads less than 275ºF) during the self-test. **Possible Causes** • Engine operating under "low" ambient temperature conditions • EFT sensor signal circuit shorted to ground in the harness • EFT sensor is damaged or it has failed • PCM has failed

Gas Engine OBD II Trouble Code List (P0xxx Codes)

DTC	Trouble Code Title, Conditions & Possible Causes
P0187 **2T CCM, MIL: Yes** 2.0L I4 VIN Z CNG engine Contour Transmissions: All	**Fuel Tank Temperature Sensor Circuit Low Input Conditions:** Engine started, and the PCM detected the Fuel Tank Temperature (FTT) sensor voltage was too low (Scan Tool reads less than 140ºF) during the self-test. Note that when this fault is detected, the PCM sets the FFT sensor default value to a reading of 32ºF. **Possible Causes** • FTT sensor signal circuit is shorted to ground in the harness • FFT sensor is damaged or it has failed • PCM has failed
P0188 **2T CCM, MIL: Yes** Crown Victoria, Grand Marquis, E & F Series Vans & Trucks with a 4.6L VIN 9, 5.4L VIN M, 5.4L VIN Z engine Transmissions: All	**Engine Fuel Temperature Sensor 'B' Circuit High Conditions:** Engine running for more than 2 minutes, and the PCM detected the EFT sensor 'A' signal was more than the calibrated limit of 4.50v [-46ºF] during the self-test. **Possible Causes** • EFT sensor signal circuit open in the wiring harness • EFT sensor signal circuit shorted to VREF in the wiring harness • EFT sensor is damaged or it has failed • PCM has failed
P0188 **2T CCM, MIL: Yes** 2.0L I4 VIN Z CNG engine Contour Transmissions: All	**Fuel Tank Temperature Sensor Circuit High Input Conditions:** Engine started, and the PCM detected the Fuel Tank Temperature (FTT) sensor voltage was too high (Scan Tool read less than -38ºF) during the self-test. Note that when this fault is detected, the PCM sets the FFT sensor default value to a reading of 32ºF. **Possible Causes** • FTT sensor signal circuit is open in the wiring harness • FTT sensor signal circuit is shorted to VREF or to power • FFT sensor is damaged or it has failed • PCM has failed
P0190 **2T CCM, MIL: Yes** Crown Victoria, Grand Marquis, E & F Series Vans & Trucks with a 4.6L VIN 9, 5.4L VIN M, 5.4L VIN Z engine Transmissions: All	**Fuel Rail Pressure Sensor Circuit Malfunction Conditions:** Key on, and the PCM detected that the FRP sensor VREF was less than its acceptable range (minimum value of 4.0v) during the CCM test). The FRP sensor signal should be in a range of 4.0-6.0v during testing. *Note: The sensor VREF should be between 4.0 to 6.0v at all times.* **Possible Causes** • Sensor VREF circuit open in the wiring harness • Sensor VREF circuit open at the sensor • Sensor VREF circuit open between the sensor and the PCM • PCM has failed
P0191 **2T CCM, MIL: Yes** Crown Victoria, Grand Marquis, E & F Series Vans & Trucks with a 4.6L VIN 9, 5.4L VIN M, 5.4L VIN Z engine Transmissions: All	**Fuel Rail Pressure Sensor Range/Performance Conditions:** Engine started, and the PCM detected the FRP sensor signal was less than the minimum acceptable range or was more than the maximum acceptable range. With the engine running, the FRP PID should read between 20 psi [138 kPa] and 60 psi [413 kPa] for gasoline powered vehicles, or between 85 psi [586 kPa] and 105 psi [725 kPa] for natural gas (NG) powered vehicles. **Possible Causes** • Fuel pressure too high or too low during engine operation • FRP sensor signal circuit has high resistance • FRP sensor is damaged or has failed
P0192 **2T CCM, MIL: Yes** Crown Victoria, Grand Marquis, E & F Series Vans & Trucks with a 4.6L VIN 9, 5.4L VIN M, 5.4L VIN Z engine Transmissions: All	**Fuel Rail Pressure Sensor Circuit Low Conditions:** Engine started, and the PCM detected the FRP sensor signal was less than 0.3v for gasoline vehicles or less than 0.5v for NG vehicles during the self-test. **Possible Causes** • FRP sensor signal shorted to chassis ground or sensor ground • FRP sensor signal circuit open (NG usage only) • FRP sensor is damaged or has failed • PCM has failed

Gas Engine OBD II Trouble Code List (P0xxx Codes

DTC	Trouble Code Title, Conditions & Possible Causes
P0192 **2T CCM, MIL: Yes** 2.0L I4 VIN Z CNG engine Contour Transmissions: All	**Fuel Absolute Pressure Sensor Circuit Low Input Conditions:** Engine started, and the PCM detected the Fuel Absolute Pressure (FAP) sensor inside the Compuvalve signal was too low (less than 4.9 psi). The FRPPREAB PID displays the actual PSI value. Bleed the system (to 0 psi), the FRPPREAB and MAP PID should agree. **Possible Causes** • FAP sensor signal circuit is shorted to ground in the harness • FAP sensor is damaged or it has failed • PCM has failed
P0193 **2T CCM, MIL: Yes** Crown Victoria, Grand Marquis, E & F Series Vans & Trucks with a 4.6L VIN 9, 5.4L VIN M, 5.4L VIN Z engine Transmissions: All	**Fuel Absolute Pressure Sensor Circuit High Input Conditions:** Engine started, and the PCM detected the FRP sensor was more than 4.5v for gasoline vehicles or more than 4.8v for NG vehicles. **Possible Causes** • FRP sensor signal shorted to VREF or system voltage • FRP sensor signal circuit open (gasoline usage only) • Low fuel pressure (NG vehicle only) • FRP sensor is damaged or has failed • PCM has failed
P0193 **2T CCM, MIL: Yes** 2.0L I4 VIN Z CNG engine Contour Transmissions: All	**Fuel Rail Pressure Sensor Circuit High Input Conditions:** Engine started, and the PCM detected the Fuel Absolute Pressure (FAP) sensor in the Compuvalve unit was too high (over 167 psi). Note: The FRPPREAB PID displays the actual PSI value. Bleed the system to 0 psi, and the FRPPREAB and MAP PID should agree). **Possible Causes** • FAP sensor signal circuit is open in the wiring harness • FAP sensor is damaged or it has failed • PCM has failed
P0196 **2T CCM, MIL: Yes** Crown Victoria, Marauder & Grand Marquis Transmissions: All	**Engine Oil Temperature Sensor Signal Performance Conditions:** Engine started, KOER Self Test enabled, and the PCM detected the Engine Oil Temperature (EOT) sensor signal was not within a calibrated amount of the ECT sensor signal during the test. The EOT sensor value should be close to the engine oil temperature. **Possible Causes** • Cooling system malfunction, or the thermostat is stuck • Engine not operating at normal operating temperature • EOT sensor is damaged or it has failed • PCM has failed
P0197 **2T CCM, MIL: Yes** Crown Victoria, Marauder & Grand Marquis Transmissions: All	**Engine Oil Temperature Sensor Circuit Low Input Conditions:** Key on or engine running; and the PCM detected that the Engine Oil Temperature (EOT) sensor was less than 0.20v during the test. The EOT sensor value should be close to the engine oil temperature. **Possible Causes** • EOT sensor connector is damaged or shorted • EOT sensor signal circuit shorted to chassis or sensor ground • EOT sensor is damaged or it has failed • PCM has failed
P0198 **2T CCM, MIL: Yes** Crown Victoria, Marauder & Grand Marquis Transmissions: All	**Engine Oil Temperature Sensor Circuit High Input Conditions:** Key on or engine running; and the PCM detected that the Engine Oil Temperature (EOT) sensor was less than 4.50v during the test. The EOT sensor value should be close to the engine oil temperature. **Possible Causes** • EOT sensor connector is damaged or open • EOT sensor signal circuit is shorted to VREF (5v) • EOT sensor is damaged or it has failed • PCM has failed

Gas Engine OBD II Trouble Code List (P0xxx Codes)

DTC	Trouble Code Title, Conditions & Possible Causes
P0201 **2T CCM, MIL: Yes** All Models Transmissions: All	**Cylinder 1 Injector Circuit Malfunction Conditions:** Engine started, and the PCM detected the fuel injector "1" control circuit was in a high state when it should have been low, or in a low state when it should have been high (wiring harness & injector okay). Note: Monitor the INJIF PID Fault "flags" with the Scan Tool. The appropriate INJF PID "flag" will read Yes when this code is set. **Possible Causes** • Injector 1 connector is damaged, open or shorted • Injector 1 control circuit is open, shorted to ground or to power • PCM has failed (the injector driver circuit may be damaged)
P0202 **2T CCM, MIL: Yes** All Models Transmissions: All	**Cylinder 2 Injector Circuit Malfunction Conditions:** Engine started, and the PCM detected the fuel injector "2" control circuit was in a high state when it should have been low, or in a low state when it should have been high (wiring harness & injector okay). Note: Monitor the INJIF PID Fault "flags" with the Scan Tool. The appropriate INJF PID "flag" will read Yes when this code is set. **Possible Causes** • Injector 2 connector is damaged, open or shorted • Injector 2 control circuit is open, shorted to ground or to power • PCM has failed (the injector driver circuit may be damaged)
P0203 **2T CCM, MIL: Yes** All Models Transmissions: All	**Cylinder 3 Injector Circuit Malfunction Conditions:** Engine started, and the PCM detected the fuel injector "3" control circuit was in a high state when it should have been low, or in a low state when it should have been high (wiring harness & injector okay). Note: Monitor the INJIF PID Fault "flags" with the Scan Tool. The appropriate INJF PID "flag" will read Yes when this code is set. **Possible Causes** • Injector 3 connector is damaged, open or shorted • Injector 3 control circuit is open, shorted to ground or to power • PCM has failed (the injector driver circuit may be damaged)
P0204 **2T CCM, MIL: Yes** All Models Transmissions: All	**Cylinder 4 Injector Circuit Malfunction Conditions:** Engine started, and the PCM detected the fuel injector "4" control circuit was in a high state when it should have been low, or in a low state when it should have been high (wiring harness & injector okay). Note: Monitor the INJIF PID Fault "flags" with the Scan Tool. The appropriate INJF PID "flag" will read Yes when this code is set. **Possible Causes** • Injector 4 connector is damaged, open or shorted • Injector 4 control circuit is open, shorted to ground or to power • PCM has failed (the injector driver circuit may be damaged)
P0205 **2T CCM, MIL: Yes** Models equipped with V6, V8 or V10 engine Transmissions: All	**Cylinder 5 Injector Circuit Malfunction Conditions:** Engine started, and the PCM detected the fuel injector "5" control circuit was in a high state when it should have been low, or in a low state when it should have been high (wiring harness & injector okay). Note: Monitor the INJIF PID Fault "flags" with the Scan Tool. The appropriate INJF PID "flag" will read Yes when this code is set. **Possible Causes** • Injector 5 connector is damaged, open or shorted • Injector 5 control circuit is open, shorted to ground or to power • PCM has failed (the injector driver circuit may be damaged)
P0206 **2T CCM, MIL: Yes** Models equipped with V6, V8 or V10 engine Transmissions: All	**Cylinder 6 Injector Circuit Malfunction Conditions:** Engine started, and the PCM detected the fuel injector control circuit was in a high state when it should have been low, or in a low state when it should have been high (wiring harness & injector okay). Note: Monitor the INJIF PID Fault "flags" with the Scan Tool. The appropriate INJF PID "flag" will read Yes when this code is set. **Possible Causes** • Injector 6 connector is damaged, open or shorted • Injector 6 control circuit is open, shorted to ground or to power • PCM has failed (the injector driver circuit may be damaged)

Gas Engine OBD II Trouble Code List (P0xxx Codes)

DTC	Trouble Code Title, Conditions & Possible Causes
P0207 **2T CCM, MIL: Yes** Models equipped with a V8 or V10 engine Transmissions: All	**Cylinder 7 Injector Circuit Malfunction Conditions:** Engine started, and the PCM detected the fuel injector "7" control circuit was in a high state when it should have been low, or in a low state when it should have been high (wiring harness & injector okay). Note: Monitor the INJIF PID Fault "flags" with the Scan Tool. The appropriate INJF PID "flag" will read Yes when this code is set. **Possible Causes** • Injector 7 connector is damaged, open or shorted • Injector 7 control circuit is open, shorted to ground or to power • PCM has failed (the injector driver circuit may be damaged)
P0208 **2T CCM, MIL: Yes** a V8 or V10 engine Transmissions: All	**Cylinder 8 Injector Circuit Malfunction Conditions:** Engine started, and the PCM detected the fuel injector "8" control circuit was in a high state when it should have been low, or in a low state when it should have been high (wiring harness & injector okay). Note: Monitor the INJIF PID Fault "flags" with the Scan Tool. The appropriate INJF PID "flag" will read Yes when this code is set. **Possible Causes** • Injector 8 connector is damaged, open or shorted • Injector 8 control circuit is open, shorted to ground or to power • PCM has failed (the injector driver circuit may be damaged)
P0209 **2T CCM, MIL: Yes** E-Van, F-Series Truck & Excursion with 6.8L VIN S or VIN Z engine Transmissions: All	**Cylinder 9 Injector Circuit Malfunction Conditions:** Engine started, and the PCM detected the fuel injector "9" control circuit was in a high state when it should have been low, or in a low state when it should have been high (wiring harness & injector okay). Note: Monitor the INJIF PID Fault "flags" with the Scan Tool. The appropriate INJF PID "flag" will read Yes when this code is set. **Possible Causes** • Injector 9 connector is damaged, open or shorted • Injector 9 control circuit is open, shorted to ground or to power • PCM has failed (the injector driver circuit may be damaged)
P0210 **2T CCM, MIL: Yes** E-Van, F-Series Truck & Excursion with 6.8L VIN S or VIN Z engine Transmissions: All	**Cylinder 10 Injector Circuit Malfunction Conditions:** Engine started, and the PCM detected the fuel injector "10" control circuit was in a high state when it should have been low, or in a low state when it should have been high (wiring harness & injector okay). Note: Monitor the INJIF PID Fault "flags" with the Scan Tool. The appropriate INJF PID "flag" will read Yes when this code is set. **Possible Causes** • Injector 10 connector is damaged, open or shorted • Injector 10 control circuit is open, shorted to ground or to power • PCM has failed (the injector driver circuit may be damaged)
P0217 **2T CCM, MIL: Yes** F-Series Truck with a 5.4L VIN 3 SC engine Transmissions: All	**Engine Coolant Over-Temperature Condition Conditions:** Engine started, vehicle driven, and the PCM detected a signal from the ECT or CHT sensor (depends on how the vehicle is equipped) that indicated an engine overheat condition. The PCM will cause the boost from the Supercharger to be bypassed when this code is set. **Possible Causes** • Base engine problems that could cause the engine to overheat • Cooling system faults that could cause the engine to overheat • Low engine coolant or incorrect mixture of engine coolant
P0219 **2T CCM, MIL: Yes** All Models Transmissions: All	**Engine Over-Speed Condition Conditions:** Engine started, and the PCM determined the vehicle had been driven in a manner that caused the engine to over-speed, and to exceed the engine speed calibration limit stored in memory. **Possible Causes** • Engine operated in the wrong transmission gear position • Excessive engine speed with gear selector in Neutral position • Wheel slippage due to wet, muddy or snowing conditions

Gas Engine OBD II Trouble Code List (P0xxx Codes)

DTC	Trouble Code Title, Conditions & Possible Causes
P0221 **2T CCM, MIL: Yes** Crown Victoria, Grand Marquis, LS & Thunderbird equipped with a 3.0L VIN S, 3.9L VIN A, 4.6L VIN 9, 4.6L VIN W engine Transmissions: All	**Throttle Position Sensor 'B' Signal Performance Conditions:** Engine started; and the PCM detected the TP Sensor 'B' circuit was out of its normal operating range during a condition with the throttle wide open, or with it completely closed. **Possible Causes** • Throttle body is damaged • Throttle linkage is binding or sticking • ETC TP Sensor 'B' signal circuit to the PCM is open • ETC TP Sensor 'B' ground circuit is open • ETC TP Sensor 'B' is damaged or it has failed
P0222 **2T CCM, MIL: Yes** Crown Victoria, Grand Marquis, LS & Thunderbird equipped with a 3.0L VIN S, 3.9L VIN A, 4.6L VIN 9, 4.6L VIN W engine Transmissions: All	**Throttle Position Sensor 'B' Circuit Low Input Conditions:** Key on or engine running; and the PCM detected the TP Sensor 'B' indicated less than 0.17v (Scan Tool reads less than 3.42%). **Possible Causes** • ETC TP Sensor 'B' connector is damaged or shorted • ETC TP Sensor 'B' signal circuit is shorted to ground • ETC TP Sensor 'B' is damaged or it has failed • PCM has failed
P0223 **2T CCM, MIL: Yes** Crown Victoria, Grand Marquis, LS & Thunderbird equipped with a 3.0L VIN S, 3.9L VIN A, 4.6L VIN 9, 4.6L VIN W engine Transmissions: All	**Throttle Position Sensor 'B' Circuit High Input Conditions:** Key on or engine running; and the PCM detected the TP Sensor 'B' indicated more than 4.65v (Scan Tool reads more than 93%) during the CCM test period. **Possible Causes** • ETC TP Sensor 'B' connector is damaged or open • ETC TP Sensor 'B' signal circuit is open • ETC TP Sensor 'B' signal circuit is shorted to VREF (5v) • ETC TP Sensor 'B' is damaged or it has failed
P0230 **2T CCM, MIL: Yes** All Models Transmissions: All	**Fuel Pump Primary Circuit Malfunction Conditions:** Key on, and the PCM detected high current in fuel pump or fuel shutoff valve (FSV) circuit (NG only), or it detected voltage with the valve off, or it did not detect voltage on the circuit. The circuit is used to energize the fuel pump relay for 20 seconds at key on or while running. **Possible Causes** • FP or FSV circuit is open or shorted • Fuel pump relay VPWR circuit open • Fuel pump relay is damaged or has failed • PCM has failed
P0231 **2T CCM, MIL: Yes** All Models Transmissions: All	**Fuel Pump Primary Circuit Low Input Conditions:** Key on, and the PCM detected a lack of voltage on the FP Monitor circuit with the fuel pump commanded on. The fuel pump control circuit is used by the PCM to energize the fuel pump relay. At key on, the relay is energized for 20 seconds, and all the time the engine is running. **Possible Causes** • FP or FSV circuit is open or shorted to ground • Fuel pump relay VPWR circuit open or fuel pump relay failed • PCM has failed
P0232 **2T CCM, MIL: Yes** All Models Transmissions: All	**Fuel Pump Secondary Circuit High Input Conditions:** Key on, and the PCM detected voltage on the FP Monitor circuit with fuel pump "off". The PCM uses the fuel pump control circuit to energize the fuel pump relay. At key on, the relay is "on" for 20 seconds or while running. This circuit is used to check voltage to the pump. **Possible Causes** • Fuel pump relay contacts always closed • Fuel pump ground circuit has high resistance • Fuel pump secondary circuit is shorted to power • Low speed fuel pump relay damaged or related circuit problem

Gas Engine OBD II Trouble Code List (P0xxx Codes)

DTC	Trouble Code Title, Conditions & Possible Causes
P0234 **2T CCM, MIL: Yes** F-Series Truck with a 5.4L VIN 3 SC engine Transmissions: All	**Supercharger Overboost Condition Conditions:** Engine started, vehicle driven, and the PCM detected an operating condition that could harm the engine or automatic transmission. **Possible Causes** • Brake "on" with throttle under wide open throttle condition • Ignition misfire condition exceeds the calibrated threshold • Knock sensor circuit has failed, or excessive knock detected • Low speed fuel pump relay not switching properly • Transmission oil temperature beyond the calibrated threshold
P0243 **2T CCM, MIL: Yes** F-Series Truck with a 5.4L VIN 3 SC engine Transmissions: All	**Supercharger Boost Bypass Solenoid Circuit Malfunction Conditions:** Key on or engine running; and the PCM detected an unexpected low or high voltage condition on the S/C Bypass Solenoid control circuit **Possible Causes** • SCB power supply circuit (VPWR) is open • SCB control circuit is open, shorted to ground or system power • SCB assembly is damaged or has failed • PCM has failed
P0297 **2T CCM, MIL: Yes** All Models Transmissions: All	**Vehicle Over-Speed Condition Conditions:** Engine started, vehicle driven at a very high engine speed, and the PCM detected the vehicle speed exceeded the calibration limit, and then enabled the High Vehicle Speed Strategy to control the speed. **Possible Causes** • The code indicates the vehicle was driven at very high engine speed (rpm) for too long. The PCM temporarily prohibits high engine speed by disabling the fuel injectors with this code set.
P0298 **2T CCM, MIL: Yes** LS Models all engines Transmissions: All	**Engine Oil Over-Temperature Condition Conditions:** Engine started, engine running for several minutes, and the PCM detected an engine overheating condition, and then enabled the Engine Oil Temperature Protection Strategy (injectors off). **Possible Causes** • Check for signs of base engine concern or engine overheating • EOP sensor or related circuit fault • Very high engine speed (rpm) for an extended period of time
P0300 **2T MISFIRE** **MIL: Yes** All Models Transmissions: All	**Random Misfire Detected Conditions:** DTC P0136, P0156, P0171, P0172, P0175, P1130 and P1150 not set, engine running under positive torque conditions, and the PCM detected a misfire in 1000 revolution (High Emissions) or the 200 revolution (Catalyst Damaging 1T) range in two or more cylinders. Note: If the misfire is severe, the MIL will flash on/off on the 1st trip! **Possible Causes** • Base engine mechanical fault that affects two or more cylinders • Fuel metering fault that affects two or more cylinders • Fuel pressure too low or too high, fuel supply contaminated • EVAP system problem or the EVAP canister is fuel saturated • EGR valve is stuck open or the PCV system has a vacuum leak • Ignition system fault (coil, plug) affecting two or more cylinders • MAF sensor contamination (it can cause a very lean condition) • Vehicle driven while very low on fuel (less than 1/8 of a tank) • TSB 03-14-4 contains repair help for this code for COP ignition

Gas Engine OBD II Trouble Code List (P0xxx Codes)

DTC	Trouble Code Title, Conditions & Possible Causes
P0301 **2T MISFIRE** **MIL: Yes** All Models Transmissions: All	**Cylinder Number 1 Misfire Detected Conditions:** DTC P0136, P0156, P0171, P0172, P0175, P1130 and P1150 not set, engine started, engine running under positive torque conditions, and the PCM detected a misfire a misfire in Cylinder 1 during the 200 revolution (Catalyst) or 1000 revolution (High Emissions) period. Note: If the misfire is severe, the MIL will flash on/off on the 1st trip! **Possible Causes** • Air leak in the intake manifold, or in the EGR or PCM system • Base engine mechanical problem that affects only Cylinder 1 • Fuel delivery component problem that affects only Cylinder 1 (i.e., a contaminated, dirty or sticking fuel injector) • Ignition system problem (coil, plug) that affects only Cylinder 1 • TSB 02-16-2 contains repair help for this code (LS & T-Bird) • TSB 03-14-4 contains repair help for this code for COP ignition
P0302 **2T MISFIRE** **MIL: Yes** All Models Transmissions: All	**Cylinder Number 2 Misfire Detected Conditions:** DTC P0136, P0156, P0171, P0172, P0175, P1130 and P1150 not set, engine started, engine running under positive torque conditions, and the PCM detected a misfire a misfire in Cylinder 2 during the 200 revolution (Catalyst) or 1000 revolution (High Emissions) period. Note: If the misfire is severe, the MIL will flash on/off on the 1st trip! **Possible Causes** • Air leak in the intake manifold, or in the EGR or PCM system • Base engine mechanical problem that affects only Cylinder 2 • Fuel delivery component problem that affects only Cylinder 2 (i.e., a contaminated, dirty or sticking fuel injector) • Ignition system problem (coil, plug) that affects only Cylinder 2 • TSB 02-16-2 contains repair help for this code (LS & T-Bird) • TSB 03-14-4 contains repair help for this code for COP ignition
P0303 **2T MISFIRE** **MIL: Yes** All Models Transmissions: All	**Cylinder Number 3 Misfire Detected Conditions:** DTC P0136, P0156, P0171, P0172, P0175, P1130 and P1150 not set, engine started, engine running under positive torque conditions, and the PCM detected a misfire a misfire in Cylinder 3 during the 200 revolution (Catalyst) or 1000 revolution (High Emissions) period. Note: If the misfire is severe, the MIL will flash on/off on the 1st trip! **Possible Causes** • Air leak in the intake manifold, or in the EGR or PCM system • Base engine mechanical problem that affects only Cylinder 3 • Fuel delivery component problem that affects only Cylinder 3 (i.e., a contaminated, dirty or sticking fuel injector) • Ignition system problem (coil, plug) that affects only Cylinder 3 • TSB 02-16-2 contains repair help for this code (LS & T-Bird) • TSB 03-14-4 contains repair help for this code for COP ignition
P0304 **2T MISFIRE** **MIL: Yes** All Models Transmissions: All	**Cylinder Number 4 Misfire Detected Conditions:** DTC P0136, P0156, P0171, P0172, P0175, P1130 and P1150 not set, engine started, engine running under positive torque conditions, and the PCM detected a misfire a misfire in Cylinder 4 during the 200 revolution (Catalyst) or 1000 revolution (High Emissions) period. Note: If the misfire is severe, the MIL will flash on/off on the 1st trip! **Possible Causes** • Air leak in the intake manifold, or in the EGR or PCM system • Base engine mechanical problem that affects only Cylinder 4 • Fuel delivery component problem that affects only Cylinder 4 (i.e., a contaminated, dirty or sticking fuel injector) • Ignition system problem (coil, plug) that affects only Cylinder 4 • TSB 02-16-2 contains repair help for this code (LS & T-Bird) • TSB 03-14-4 contains repair help for this code for COP ignition

Gas Engine OBD II Trouble Code List (P0xxx Codes)

DTC	Trouble Code Title, Conditions & Possible Causes
P0305 **2T MISFIRE** **MIL: Yes** Models equipped with V6, V8 or V10 engine Transmissions: All	**Cylinder Number 5 Misfire Detected Conditions:** DTC P0136, P0156, P0171, P0172, P0175, P1130 and P1150 not set, engine started, engine running under positive torque conditions, and the PCM detected a misfire a misfire in Cylinder 5 during the 200 revolution (Catalyst) or 1000 revolution (High Emissions) period. Note: If the misfire is severe, the MIL will flash on/off on the 1st trip! **Possible Causes** • Air leak in the intake manifold, or in the EGR or PCM system • Base engine mechanical problem that affects only Cylinder 5 • Fuel delivery component problem that affects only Cylinder 5 (i.e., a contaminated, dirty or sticking fuel injector) • Ignition system problem (coil, plug) that affects only Cylinder 5 • TSB 02-16-2 contains repair help for this code (LS & T-Bird) • TSB 03-14-4 contains repair help for this code for COP ignition
P0306 **2T MISFIRE** **MIL: Yes** Models equipped with V6, V8 or V10 engine Transmissions: All	**Cylinder Number 6 Misfire Detected Conditions:** DTC P0136, P0156, P0171, P0172, P0175, P1130 and P1150 not set, engine started, engine running under positive torque conditions, and the PCM detected a misfire a misfire in Cylinder 6 during the 200 revolution (Catalyst) or 1000 revolution (High Emissions) period. Note: If the misfire is severe, the MIL will flash on/off on the 1st trip! **Possible Causes** • Air leak in the intake manifold, or in the EGR or PCM system • Base engine mechanical problem that affects only Cylinder 6 • Fuel delivery component problem that affects only Cylinder 6 (i.e., a contaminated, dirty or sticking fuel injector) • Ignition system problem (coil, plug) that affects only Cylinder 6 • TSB 02-16-2 contains repair help for this code (LS & T-Bird) • TSB 03-14-4 contains repair help for this code for COP ignition
P0307 **2T MISFIRE** **MIL: Yes** Models equipped with V8 or V10 engine Transmissions: All	**Cylinder Number 7 Misfire Detected Conditions:** DTC P0136, P0156, P0171, P0172, P0175, P1130 and P1150 not set, engine started, engine running under positive torque conditions, and the PCM detected a misfire a misfire in Cylinder 7 during the 200 revolution (Catalyst) or 1000 revolution (High Emissions) period. Note: If the misfire is severe, the MIL will flash on/off on the 1st trip! **Possible Causes** • Air leak in the intake manifold, or in the EGR or PCM system • Base engine mechanical problem that affects only Cylinder 7 • Fuel delivery component problem that affects only Cylinder 7 (i.e., a contaminated, dirty or sticking fuel injector) • Ignition system problem (coil, plug) that affects only Cylinder 7 • TSB 02-16-2 contains repair help for this code (LS & T-Bird) • TSB 03-14-4 contains repair help for this code for COP ignition
P0308 **2T MISFIRE** **MIL: Yes** Models equipped with V8 or V10 engine Transmissions: All	**Cylinder Number 8 Misfire Detected Conditions:** DTC P0136, P0156, P0171, P0172, P0175, P1130 and P1150 not set, engine started, engine running under positive torque conditions, and the PCM detected a misfire a misfire in Cylinder 8 during the 200 revolution (Catalyst) or 1000 revolution (High Emissions) period. Note: If the misfire is severe, the MIL will flash on/off on the 1st trip! **Possible Causes** • Air leak in the intake manifold, or in the EGR or PCM system • Base engine mechanical problem that affects only Cylinder 8 • Fuel delivery component problem that affects only Cylinder 8 (i.e., a contaminated, dirty or sticking fuel injector) • Ignition system problem (coil, plug) that affects only Cylinder 8 • TSB 02-16-2 contains repair help for this code (LS & T-Bird) • TSB 03-14-4 contains repair help for this code for COP ignition

Gas Engine OBD II Trouble Code List (P0xxx Codes)

DTC	Trouble Code Title, Conditions & Possible Causes
P0309 **2T MISFIRE** **MIL: Yes** Models equipped with a V10 engine Transmissions: All	**Cylinder Number 9 Misfire Detected Conditions:** DTC P0136, P0156, P0171, P0172, P0175, P1130 and P1150 not set, engine started, engine running under positive torque conditions, and the PCM detected a misfire a misfire in Cylinder 9 during the 200 revolution (Catalyst) or 1000 revolution (High Emissions) period. Note: If the misfire is severe, the MIL will flash on/off on the 1st trip! **Possible Causes** • Air leak in the intake manifold, or in the EGR or PCM system • Base engine mechanical problem that affects only Cylinder 9 • Fuel delivery component problem that affects only Cylinder 9 (i.e., a contaminated, dirty or sticking fuel injector) • Ignition system problem (coil, plug) that affects only Cylinder 9 • TSB 03-14-4 contains repair help for this code for COP ignition
P0310 **2T MISFIRE** **MIL: Yes** Models equipped with a V10 engine Transmissions: All	**Cylinder Number 10 Misfire Detected Conditions:** DTC P0136, P0156, P0171, P0172, P0175, P1130 and P1150 not set, engine started, engine running under positive torque conditions, and the PCM detected a misfire a misfire in Cylinder 10 during the 200 revolution (Catalyst) or 1000 revolution (High Emissions) period. Note: If the misfire is severe, the MIL will flash on/off on the 1st trip! **Possible Causes** • Air leak in the intake manifold, or in the EGR or PCM system • Base engine mechanical problem that affects only Cylinder 10 • Fuel delivery component problem that affects only Cylinder 10 i.e., a contaminated, dirty or sticking fuel injector) • Ignition system problem (coil, plug) that affects only Cylinder 10 • TSB 03-14-4 contains repair help for this code for COP ignition
P0315 **2T CCM, MIL: Yes** All Models Transmissions: All	**Unable to Learn Crankshaft Variation Conditions:** Engine started, and the PCM determined that it was unable to correct for mechanical inaccuracies in the CKP wheel tooth spacing. Note: The Misfire Monitor will be disabled. **Possible Causes** • Inspect the CKP sensor for damage • Inspect the CKP sensor for debris on the rotor • Inspect the crankshaft pulse wheel for damaged teeth • Inspect the crankshaft pulse wheel for wobble (loose condition)
P0316 **2T MISFIRE** **MIL: Yes** All Models Transmissions: All	**Misfire in the First 1000 Revolutions Conditions:** Engine started, and the PCM detected a severe misfire within the first 1000 engine revolutions. **Possible Causes** • Check for CMC DTC P0136, P0156, P0171, P0172, P0175, P1130 and P1150. Repair these adaptive fuel and HO2S codes • Check for any other CMC in memory. Repair these codes first! • Ignore P1000 codes that set during KOEO and KOER Self-Test
P0320 **2T CCM, MIL: Yes** E Van, F Series Truck equipped with a 4.9L VIN Y, 5.0L VIN N, 5.8L VIN H, 7.5L VIN G engine Transmissions: All	**Ignition Engine Speed Input Circuit Malfunction Conditions:** Engine started, and the PCM detected two or more successive erratic PIP signals during the self-test. **Possible Causes** • Verify that the vehicle Antitheft system is operational • Inspect for any Aftermarket 2-way radio problems • Inspect for signs of "arcing" at one or more of the ignition coils • Verify that the Inertia Fuel Switch (IFS) is set properly (reset)
P0320 **2T CCM, MIL: Yes** F Series Truck equipped with a 7.5L VIN G engine Transmissions: All	**Ignition Engine Speed Input Circuit Malfunction Conditions:** Engine started, and the PCM detected 2 or more successive erratic PIP signals while testing. **Possible Causes** • Verify that the vehicle Antitheft system is operational • Inspect for any Aftermarket 2-way radio problems • Inspect for signs of "arcing" at one or more of the ignition coils • Verify that the Inertia Fuel Switch (IFS) is set properly (reset)

Gas Engine OBD II Trouble Code List (P0xxx Codes)

DTC	Trouble Code Title, Conditions & Possible Causes
P0320 **2T CCM, MIL: Yes** All Models Transmissions: All	**Ignition Engine Speed Input Circuit Malfunction Conditions:** Engine started, and the PCM detected 2 or more successive erratic PIP signals during testing. **Possible Causes** • Inspect for problems with an Aftermarket 2-way radio • Inspect for signs of "arcing" at one or more of the ignition coils • Inspect the Profile Ignition Pickup (PIP) unit inside distributor (check for damage or corrosion at the PIP sensor connector) • PIP sensor is damaged or it has failed (distributor models) • Ignition control module (ICM) has failed (Distributorless models)
P0325 **2T CCM, MIL: Yes** All Models Transmissions: All	**Knock Sensor 1 Circuit Malfunction Conditions:** Key on or engine running; and the PCM detected the knock Sensor 1 (KS1) signal was more than 0.5v at key on, engine off, or the KS1 signal was out of normal range (engine running). **Possible Causes** • Knock sensor circuit is open • Knock sensor circuit is shorted to ground, or shorted to power • Knock sensor is damaged or it has failed • PCM has failed
P0326 **2T CCM, MIL: Yes** All Models Transmissions: All	**Knock Sensor 1 Signal Range/Performance Conditions:** Engine started, vehicle driven, and the PCM detected the Knock Sensor 1 (KS1) signal was more than the calibrated value. This code can set at key on, engine off, if the KS 1 signal is more than 0.5v. **Possible Causes** • Knock sensor circuit is open • Knock sensor circuit is shorted to ground, or shorted to power • Knock sensor is damaged or it has failed • PCM has failed
P0330 **2T CCM, MIL: Yes** All Models Transmissions: All	**Knock Sensor 2 Circuit Malfunction Conditions:** Key on or engine running; and the PCM detected the knock Sensor 2 (KS2) signal was more than 0.5v at key on, engine off, or that the KS2 signal was out of the normal range with the engine running. **Possible Causes** • Knock sensor circuit is open • Knock sensor circuit is shorted to ground, or shorted to power • Knock sensor is damaged or it has failed • PCM has failed
P0331 **2T CCM, MIL: Yes** All Models Transmissions: All	**Knock Sensor 2 Signal Range/Performance Conditions:** Engine started, vehicle driven, and the PCM detected that the Knock Sensor 2 (KS2) signal was more than the calibrated value. This code can set at key on, engine off, if the KS2 signal is more than 0.5v. **Possible Causes** • Knock sensor circuit is open • Knock sensor circuit is shorted to ground, or shorted to power • Knock sensor is damaged or it has failed • PCM has failed
P0340 **2T CCM, MIL: Yes** All Models Transmissions: All	**Camshaft Position Sensor Circuit Malfunction Conditions:** Engine started, and the PCM detected the CMP sensor signal was missing or it was erratic. **Possible Causes** • CMP sensor circuit is open or shorted to ground • CMP sensor circuit is shorted to power • CMP sensor ground (return) circuit is open • CMP sensor installation incorrect (Hall-effect type) • CMP sensor is damaged or CMP sensor shielding damaged • PCM has failed • TSB 02-22-1 contains repair information for this trouble code

Gas Engine OBD II Trouble Code List (P0xxx Codes)

DTC	Trouble Code Title, Conditions & Possible Causes
P0350 **2T CCM, MIL: Yes** Some Models Transmissions: All	**Ignition Coil (Undetermined) Primary/Secondary Circuit Malfunction Conditions:** Engine started, and the PCM did not receive valid IDM pulses from the ignition module. The PCM did not identify the coil with a problem. **Possible Causes** • Ignition START/RUN circuit is open or shorted to ground • Ignition coil driver circuit is open or shorted to ground • Ignition coil circuit is shorted to power • Ignition coil damaged or it has failed • PCM has failed
P0350 **2T CCM, MIL: Yes** Crown Victoria, Grand Marquis, Excursion, LS, Five Hundred, Freestyle, Montego,, E & F Series Vans & Trucks & Town Car equipped with a 3.0L VIN S, 3.9L VIN A, 4.6L VIN 9, 4.6L VIN W, 6.8L VIN S engine Transmissions: All	**Ignition Coil Primary/Secondary Circuit Malfunction Conditions:** Engine started, and the PCM did not receive valid IDM pulses from the ignition module. The PCM did not identify the coil with a problem. **Possible Causes** • Ignition START/RUN circuit is open or shorted to ground • Ignition coil driver circuit is open • Ignition coil driver circuit is shorted to ground • Ignition coil driver circuit is shorted to system power (B+) • Ignition coil damaged or it has failed • PCM has failed • TSB 01-1-6 contains a repair procedure for this trouble code
P0351-P0354 **2T CCM, MIL: Yes** Escort, ZX2 & Focus equipped with 2.0L VIN 3, 2.0L VIN P engine (w/Coilpack) Transmissions: All	**Ignition Coilpack 1-4 Primary/Secondary Circuit Malfunction Conditions:** Engine started, and the PCM did not receive any valid IDM pulses from the ignition module for the Ignition Coilpack 1-4 primary circuit. **Possible Causes** • Ignition START/RUN circuit is open or shorted to ground • Ignition coilpack 1-4 control circuit is open or shorted to ground • Ignition coilpack 1-4 control circuit is shorted to power • Ignition coilpack 1-4 is damaged or it has failed • PCM has failed
P0351-P0356 **2T CCM, MIL: Yes** Focus equipped with a 3.0L VIN Z engine (Coil on Plug design) Transmissions: All	**Ignition COP 1-6 Primary/Secondary Circuit Malfunction Conditions:** Engine started, and the PCM did not detect a valid IDM pulse from the ignition module for the Ignition Coil on Plug 1-6 primary circuit. **Possible Causes** • Ignition START/RUN circuit is open or shorted to ground • Ignition COP driver 1-6 control circuit open or shorted to ground • Ignition COP driver 1-6 control circuit is shorted to power • Ignition COP 1-6 is damaged or it has failed • PCM has failed
P0351-P0310 **2T CCM, MIL: Yes** Aviator, Blackwood, Crown Victoria, Grand Marquis, Explorer, E & F Series Van & Truck, Explorer, Town Car, Mark VII and Ranger equipped with 2.5L VIN C, 4.6L VIN 9, 4.6L VIN B, 4.6L VIN H, 4.6L VIN W, 5.4L VIN A, 5.4L VIN L, 5.4L VIN R, 6.8L VIN S engine (COP) Transmissions: All	**Ignition Coil 1-10 Primary/Secondary Circuit Malfunction Conditions:** Engine started, and the PCM did not receive any valid IDM pulses from the ignition module for the Ignition Coil 1-10 primary circuit. **Possible Causes** • Ignition START/RUN circuit is open or shorted to ground • Ignition coil driver 1-10 circuit is open or shorted to ground • Ignition coil 1-10 circuit is shorted to power • Ignition coil -101 damaged or it has failed • PCM has failed

Gas Engine OBD II Trouble Code List (P0xxx Codes)

DTC	Trouble Code Title, Conditions & Possible Causes
P0351-P0358 **2T CCM, MIL: Yes** Escape, LS, Five Hundred, Freestyle, Montego,, Ranger & Thunderbird equipped with a 2.0L VIN B, 3.0L (VIN 1, VIN 2, VIN S, VIN U), 3.9L VIN A, 4.0L VIN E, 4.0L VIN K, 4.6L VIN W, 5.4L VIN S engine (Coil on Plug) Transmissions: All	**Ignition Coil 1-8 Primary/Secondary Circuit Malfunction Conditions:** Engine started, and the PCM did not receive any valid IDM pulses from the ignition module for the Ignition Coil 1-8 primary circuit. **Possible Causes** • Ignition START/RUN circuit is open or shorted to ground • Ignition coil driver 1-8 circuit is open or shorted to ground • Ignition coil 1-8 circuit is shorted to power • Ignition coil 1-8 damaged or it has failed • PCM has failed
P0351-P0310 **2T CCM, MIL: Yes** Aviator, Explorer, Mountaineer, E & F Series Vans & Trucks equipped with a 4.0L VIN E, 4.0L VIN K, 4.2L VIN 2, 4.2L VIN E, 4.6L VIN 6, 4.6L VIN W, 5.4L VIN 3, 5.4L VIN E, 5.4L VIN L, 5.4L VIN M, 5.4L VIN Z, 6.8L VIN S, 6.8L VIN Z engine Transmissions: All	**Ignition Coil 1-10 Primary/Secondary Circuit Malfunction Conditions:** Engine started, and the PCM did not receive any valid IDM pulses from the ignition module for the Ignition Coil 1-10 primary circuit. **Possible Causes** • Ignition START/RUN circuit is open or shorted to ground • Ignition coilpack or COP 1-10 circuit is open or shorted to ground • Ignition coilpack or COP 1-10 circuit is shorted to power • Ignition coilpack or COP 1-10 damaged or it has failed • PCM has failed
P0351-P0358 **2T CCM, MIL: Yes** Expedition, Navigator equipped with 4.6L VIN 6, 4.6L VIN W engine (w/Coilpack) Transmissions: All	**Ignition Coilpack 1-8 Primary/Secondary Circuit Malfunction Conditions:** Engine started, and the PCM did not receive any valid IDM pulses from the ignition module for the Ignition Coilpack 1-8 primary circuit. **Possible Causes** • Ignition START/RUN circuit is open or shorted to ground • Ignition coilpack 1-8 control circuit is open or shorted to ground • Ignition coilpack 1-8 control circuit is shorted to power • Ignition coilpack 1-8 is damaged or it has failed • PCM has failed
P0351-P0358 **2T CCM, MIL: Yes** Expedition, Navigator equipped with 4.6L VIN 6, 4.6L VIN W engine (Coil on Plug) Transmissions: All	**Ignition COP 1-8 Primary/Secondary Circuit Malfunction Conditions:** Engine started, and the PCM did not detect a valid IDM pulse from the Ignition Module on the Ignition Coil on Plug 1-8 primary circuit. **Possible Causes** • Ignition START/RUN circuit is open or shorted to ground • Ignition COP driver 1-8 circuit is open or shorted to ground • Ignition COP driver 1-8 circuit is shorted to power • Ignition COP 1-8 is damaged or it has failed • PCM has failed
P0351-P0358 **2T CCM, MIL: Yes** Expedition, Navigator with a 5.4L VIN A, 5.4L VIN L & VIN R engine (Coil on Plug) Transmissions: All	**Ignition COP 1-8 Primary/Secondary Circuit Malfunction Conditions:** Engine started, and the PCM did not detect a valid IDM pulse from the Ignition Module on the Ignition Coil on Plug 1 primary circuit. **Possible Causes** • Ignition START/RUN circuit is open or shorted to ground • Ignition COP driver 1-8 circuit is open or shorted to ground • Ignition COP driver 1-8 circuit is shorted to power • Ignition COP 1-8 is damaged or it has failed • PCM has failed

Gas Engine OBD II Trouble Code List (P0xxx Codes)

DTC	Trouble Code Title, Conditions & Possible Causes
P0351-P0358 **2T CCM, MIL: Yes** Crown Victoria equipped with a 4.6L VIN 9, 4.6L VIN W engine (Coilpack) Transmissions: All	**Ignition Coil 1-8 Primary/Secondary Circuit Malfunction Conditions:** Engine started, and the PCM did not receive any valid IDM pulses from the ignition module for the Ignition Coil 1-8 primary circuit. **Possible Causes** • Ignition START/RUN circuit is open or shorted to ground • Ignition coil driver 1-8 circuit open, shorted to ground or to power • Ignition coil 1-8 is damaged or it has failed • PCM has failed • TSB 01-1-6 contains a repair procedure for this trouble code
P0351-P0356 **2T CCM, MIL: Yes** E Series Vans with a 3.4L VIN N engine Transmissions: All	**Ignition Coil 1-6 Primary/Secondary Circuit Malfunction Conditions:** Engine started, and the PCM did not receive any valid IDM pulses from the ignition module for the Ignition Coil 1-6 primary circuit. **Possible Causes** • Ignition START/RUN circuit is open or shorted to ground • Ignition coil driver 1-6 circuit is open or shorted to ground • Ignition coil 1-6 circuit is shorted to power • Ignition coil 1-6 damaged or it has failed • PCM has failed
P0351-P0354 **2T CCM, MIL: Yes** Contour & Mystique equipped with a 2.0L engine (Coil on Plug) Engines: All Transmissions: All	**Ignition Coil 1-4 Primary/Secondary Circuit Malfunction Conditions:** Engine started, and the PCM did not receive any valid IDM pulses from the ignition module for the Ignition Coil 1-4 primary circuit. **Possible Causes** • Ignition START/RUN circuit is open or shorted to ground • Ignition coil driver 1-4 circuit open, shorted to ground or to power • Ignition coil 1-4 is damaged or it has failed • PCM has failed • TSB 99-21-7 contains a repair procedure for this trouble code
P0400 **2T EGR, MIL: Yes** Probe equipped with a 2.0L I4 VIN A or 2.5L V6 VIN B engine Transmissions: All	**EGR Flow Malfunction Conditions:** DTC P0106, P0107, P0108, P1400, P1401 and P1408 not set, engine started, vehicle driven at a steady speed in closed loop for over 1 minute, and the PCM detected the EGR flow rate was less than or more than the calibrated flow rate limits during the test. **Possible Causes** • EGR atmospheric solenoid connector is damaged or loose • EGR atmospheric solenoid is damaged or it has failed • EGR check solenoid connector is damaged, loose or shorted • EGR check solenoid is damaged or it has failed • EGR vacuum regulator connector is damaged, loose or shorted • EGR vacuum regulator solenoid is damaged or it has failed • EGR valve position sensor connector is damaged or loose • EGR valve position sensor is damaged or it has failed • Exhaust system is restricted • PCM has failed
P0400 **2T EGR, MIL: Yes** Aviator, Escort, Focus, Explorer, Mountaineer, E Van, F Series Truck, Mustang, Sable, Taurus Navigator, Escape, Explorer, Expedition, Navigator, ZX2 Transmissions: All	**Exhaust EGR System Malfunction Conditions:** DTC P0102, P0103, P0107, P0108, P1100 and P1101 not set, vehicle driven at over 48 mph at a steady speed in closed loop for 1 minute, and the PCM detected the EGR flow rate was less than or more than the calibrated flow rate limits during the EGR test period. **Possible Causes** • DPFE EGR sensor connector is damaged, loose or shorted • EGR valve is sticking, damaged or it has failed • MAP sensor is damaged or out of calibration • PCM has failed • TSB 03-31-3 contains repair information for this trouble code

Gas Engine OBD II Trouble Code List (P0xxx Codes)

DTC	Trouble Code Title, Conditions & Possible Causes
P0400 **2T EGR, MIL: Yes** Focus & Ranger equipped with a 2.3L VIN D or 2.3L VIN Z engine Transmissions: All	**Electronic EGR System Malfunction Conditions:** DTC P0102, P0103, P0107, P0108, P1100 and P1101 not set, engine started, vehicle driven at over 48 mph at a steady speed in closed loop for 1 minute, and the PCM detected the EGR flow rate was less than or more than the calibrated limits during the EGR test. **Possible Causes** • EEGR valve connector is damaged, loose or shorted • EEGR valve is sticking, damaged or it has failed • MAP or TMAP sensor is damaged or not properly seated • PCM has failed
P0400 **2T EGR, MIL: Yes** Crown Victoria, Grand Marquis, Aviator, LS, Five Hundred, Freestyle, Montego,, Marauder, Town Car & Thunderbird Models Transmissions: All	**Exhaust EGR System Malfunction (ESM System) Conditions:** DTC P0102, P0103, P0107, P0108, P1100 and P1101 not set, vehicle driven at over 48 mph at a steady speed in closed loop for 1 minute, and the PCM detected the EGR flow rate was less than or more than the calibrated flow rate limits during the EGR test period. **Possible Causes** • DPFE EGR sensor connector is damaged, loose or shorted • EGR valve is sticking, damaged or it has failed • MAP sensor is damaged or out of calibration • PCM has failed
P0401 **2T EGR, MIL: Yes** Explorer, Mountaineer equipped with a 5.0L VIN P engine Transmissions: All	**Insufficient EGR Flow Detected Conditions:** Engine running at a steady cruise speed in closed loop, and the PCM detected insufficient EGR gas flow. The EGR valve actuator and EGR valve position sensor are in one unit. **Possible Causes** • EVP sensor is damaged or has failed • EGR valve source vacuum supply problem (hose off or leaking) • EGR valve is stuck partially open or closed (check for carbon) • EGR valve may be leaking vacuum (the diaphragm is broken) • VR solenoid is damaged or the VR control circuit is open • PCM has failed
P0401 **2T EGR, MIL: Yes** All Models Transmissions: All	**Insufficient EGR Flow Detected Conditions:** Engine started, engine running in closed loop under steady cruise conditions, and the PCM detected the DPFE sensor input indicated insufficient EGR gas flow. Run the KOER Self-Test, and if DTC P1408 is present, the fault is currently present. **Possible Causes** • DPFE sensor signal circuit is shorted to ground • DPFE sensor VREF circuit is open between sensor and PCM • DPFE sensor downstream hose off or plugged • DPFE sensor hoses both off, loose or damaged • DPFE sensor hoses connected wrong (reversed) • EGR orifice tube is damaged or restricted • TSB 03-31-3 contains repair information for this trouble code
P0401 **2T EGR, MIL: Yes** Escort, Escape, Focus, E Van, F Series Truck, Excursion, Expedition, Explorer, Mustang, Ranger, Navigator, Sable, Mountaineer Transmissions: All	**Exhaust Gas Recirculation Malfunction Conditions:** Engine started, engine running under at cruise speed in closed loop, and the PCM detected a problem in the EGR system. Run the KOER self-test. If DTC P1406 is set, test the EGR valve operation. **Possible Causes** • DPFE EGR valve hoses are damaged, leaking or restricted • DPFE EGR valve hoses may be reversed at the sensor • EGR valve connector is damaged, loose or shorted • EGR valve is damaged or it has failed • PCM has failed • TSB 4-3-1 contains repair information for this trouble code

Gas Engine OBD II Trouble Code List (P0xxx Codes)

DTC	Trouble Code Title, Conditions & Possible Causes
P0401 **2T EGR, MIL: Yes** Crown Victoria, Grand Marquis, Aviator, LS, Five Hundred, Freestyle, Montego,, Marauder, Town Car & Thunderbird Models Transmissions: All	**Exhaust Gas Recirculation Malfunction (ESM System) Conditions:** Engine running under at cruise speed in closed loop, and the PCM detected a problem in the EGR ESM system. Run the KOER self-test. If DTC P1408 is present, inspect the EGR valve. **Possible Causes** • DPFE EGR valve hoses are damaged, leaking or restricted • EGR valve connector is damaged, loose or shorted • EGR valve is damaged or it has failed • PCM has failed
P0402 **2T EGR, MIL: Yes** All Models Transmissions: All	**Excessive EGR Flow Detected Conditions:** Engine started, engine running in hot idle speed, and the PCM detected the Actual DPFE sensor value indicated more than the KOEO DPFE sensor value stored in the PCM memory. **Possible Causes** • DPFE EGR valve source hoses loose or connected wrong • DPFE sensor slow to respond or sluggish (it may have failed) • DPFE sensor signal circuit is open or shorted to ground • DPFE EGR sensor is damaged or the PCM has failed
P0402 **2T EGR, MIL: Yes** Explorer, Mountaineer equipped with a 5.0L VIN P engine Transmissions: All	**Excessive EGR Flow Detected Conditions:** Engine running under at a steady cruise speed in closed loop, and the PCM detected the EGR valve position did not match the Desired EGR position based on engine speed and load. **Possible Causes** • EGR valve actuator is damaged or sticking (perform a KOEO or KOER Self Test to test the actuator for an On Demand code) • EGR position sensor signal circuit shorted to bias voltage • EP sensor signal circuit shorted to bias voltage • PCM has failed
P0402 **2T EGR, MIL: Yes** Crown Victoria, Grand Marquis, Aviator, LS, Five Hundred, Freestyle, Montego,, Marauder, Town Car & Thunderbird Models Transmissions: All	**EGR Flow At Idle Speed Detected (ESM System) Conditions:** Engine started, engine running in hot idle speed, and the PCM detected the Actual DPFE sensor value indicated more than the KOEO DPFE sensor value stored in the memory. If DTC P1405 is set, repair the cause of that trouble code prior to repairing P0402. **Possible Causes** • DPFE EGR sensor is damaged • DPFE EGR valve source hoses loose or connected wrong • DPFE sensor slow to respond or sluggish (it may have failed) • DPFE sensor signal circuit is open or shorted to ground • PCM has failed • TSB 03-31-3 contains repair information for this trouble code
P0402 **2T EGR, MIL: Yes** Escort, Escape, Focus, E Van, F Series Truck, Excursion, Expedition, Explorer, Mustang, Ranger, Navigator, Sable, Mountaineer Transmissions: All	**EGR Flow At Idle Speed Detected (ESM System) Conditions:** Engine started, engine running in hot idle speed, and the PCM detected the Actual DPFE sensor value indicated more than the KOEO DPFE sensor value stored in the memory. If DTC P1405 is set, repair the cause of that trouble code prior to repairing P0402. **Possible Causes** • DPFE EGR sensor is damaged • DPFE EGR valve source hoses loose or connected wrong • DPFE sensor slow to respond or sluggish (it may have failed) • DPFE sensor signal circuit is open or shorted to ground • PCM has failed • TSB 03-31-3 contains repair information for this trouble code
P0403 **2T CCM, MIL: Yes** Ranger equipped with a 2.3L VIN D engine Transmissions: All	**EGR Solenoid Circuit Malfunction Conditions:** Engine started, and the PCM detected an unexpected "high" or "low" voltage condition on the EEGR solenoid control circuit at idle speed. **Possible Causes** • EEGR solenoid control circuit is open, or shorted to ground • EEGR solenoid power circuit is open (check power to relay) • EEGR motor winding is open or shorted to power • EEGR solenoid connector not seated correctly or the EGR solenoid has failed • PCM has failed

Gas Engine OBD II Trouble Code List (P0xxx Codes)

DTC	Trouble Code Title, Conditions & Possible Causes
P0403 **2T CCM, MIL: Yes** Crown Victoria, Grand Marquis, Aviator, LS, Five Hundred, Freestyle, Montego,, Marauder, Town Car & Thunderbird Models Transmissions: All	**EGR Solenoid Circuit Malfunction (ESM System) Conditions:** Engine started, and the PCM detected an unexpected high or low voltage condition on the ESM EGR solenoid control circuit at idle. **Possible Causes** • EGR solenoid connector is damaged, loose or shorted • EGR solenoid control circuit is open, or shorted to ground • EGR solenoid power circuit is open (check power to relay) • EGR solenoid is damaged or the PCM has failed
P0403 **2T CCM, MIL: Yes** Aviator, Escort, Focus, Explorer, Mountaineer, E Van, F Series Truck, Mustang, Sable, Taurus Navigator, Escape, Explorer, Expedition, Navigator, ZX2 Transmissions: All	**EGR Solenoid Circuit Malfunction Conditions:** Engine started, and the PCM detected an unexpected high or low voltage condition on the EGR solenoid control circuit at idle speed. **Possible Causes** • EGR solenoid connector is damaged, loose or shorted • EGR solenoid control circuit is open, or shorted to ground • EGR solenoid power circuit is open (check power to relay) • EGR motor winding is open or shorted to power • EGR solenoid is damaged or has failed • PCM has failed
P0405 **2T CCM, MIL: Yes** Crown Victoria, Grand Marquis, Aviator, LS, Five Hundred, Freestyle, Montego,, Marauder, Town Car & Thunderbird Models Transmissions: All	**DPFE Sensor Circuit Low Input (ESM System) Conditions:** Engine started, and the PCM detected an unexpected low voltage condition (less than 0.20v) on the ESM DPFE sensor circuit. **Possible Causes** • DPFE sensor connector is damaged or shorted • DPFE sensor power supply circuit is open or shorted to ground • DPFE sensor signal circuit is shorted to ground • DPFE sensor is damaged or the PCM has failed
P0405 **2T CCM, MIL: Yes** Aviator, Escort, Focus, Explorer, Mountaineer, E Van, F Series Truck, Mustang, Sable, Taurus Navigator, Escape, Explorer, Expedition, Navigator, ZX2 Transmissions: All	**DPFE Sensor Circuit Low Input Conditions:** Engine started, and the PCM detected an unexpected low voltage condition (less than 0.20v) on the ESM DPFE sensor circuit. **Possible Causes** • DPFE sensor connector is damaged or shorted • DPFE sensor power supply circuit is open or shorted to ground • DPFE sensor signal circuit is shorted to ground • DPFE sensor is damaged or has failed • PCM has failed
P0406 **2T CCM, MIL: Yes** Crown Victoria, Grand Marquis, Aviator, LS, Five Hundred, Freestyle, Montego,, Marauder, Town Car & Thunderbird Models Transmissions: All	**DPFE Sensor Circuit High Input (ESM System) Conditions:** Key on or engine running at idle speed, and the PCM detected an unexpected high voltage condition (more than 4.00v) on the ESM DPFE sensor circuit during the CCM test period. **Possible Causes** • DPFE sensor connector is damaged or open • DPFE sensor signal circuit is open or it is shorted to VREF (5v) • DPFE sensor is damaged or has failed • PCM has failed
P0406 **2T CCM, MIL: Yes** Aviator, Escort, Focus, Explorer, Mountaineer, E Van, F Series Truck, Mustang, Sable, Taurus Navigator, Escape, Explorer, Expedition, Navigator, ZX2 Transmissions: All	**DPFE Sensor Circuit High Input Conditions:** Key on or engine running at idle speed, and the PCM detected an unexpected high voltage condition (more than 4.00v) on the ESM DPFE sensor circuit during the CCM test period. **Possible Causes** • DPFE sensor connector is damaged or open • DPFE sensor signal circuit is shorted to VREF (5v) • DPFE sensor signal circuit is open • DPFE sensor is damaged or has failed • PCM has failed

Gas Engine OBD II Trouble Code List (P0xxx Codes)

DTC	Trouble Code Title, Conditions & Possible Causes
P0411 **2T AIR, MIL: Yes** Some Models Transmissions: All	**Secondary AIR System Incorrect Upstream Flow Detected Conditions:** Engine started, engine running at idle speed in closed loop, and the PCM detected the secondary AIR pump airflow was not diverted correctly when requested during the self-test. **Possible Causes** • Air pump output is blocked • AIR bypass solenoid leaking or blocked • AIR bypass solenoid is stuck open or stuck closed • Air injection pump hose(s) leaking • PCM has failed
P0411 **2T AIR, MIL: Yes** LS, Five Hundred, Freestyle, Montego,, Mustang, Taurus equipped with a 3.0L VIN 2, 3.0L VIN S or 3.8L VIN 4 engine Transmissions: All	**Secondary Air Injection System Upstream Flow Detected Conditions:** Engine started, engine runtime from 20-120 seconds at any speed, and the PCM detected the Secondary AIR pump airflow was not diverted correctly when requested during the self-test. **Possible Causes** • Air pump output is blocked or restricted • AIR bypass solenoid is leaking or it is restricted • AIR bypass solenoid is stuck open or stuck closed • Check valve (one or more) is damaged or leaking • Electric air injection pump hose(s) leaking • PCM has failed
P0412 **2T CCM, MIL: Yes** Some Models Transmissions: All	**Secondary Air Injection Solenoid Circuit Malfunction Conditions:** Engine started, and the PCM detected an unexpected low or high voltage condition on the AIR solenoid control circuit during testing. **Possible Causes** • AIR solenoid power circuit (B+) is open (check dedicated fuse) • AIR bypass solenoid control circuit is open or shorted to ground • AIR diverter solenoid control circuit open or shorted to ground • AIR pump control circuit is open or shorted to ground • Check valve (one or more) is damaged or leaking • Solid State relay is damaged or it has failed • PCM has failed
P0412 **2T CCM, MIL: Yes** LS, Five Hundred, Freestyle, Montego,, Mustang, Taurus equipped with a 3.0L VIN 2, 3.0L VIN S or 3.8L VIN 4 engine Transmissions: All	**Secondary AIR Solenoid Control Circuit Malfunction Conditions:** Engine started, and the PCM detected an unexpected low or high voltage condition on the AIR solenoid control circuit during testing. **Possible Causes** • AIR solenoid power circuit (B+) is open (check dedicated fuse) • AIR bypass solenoid control circuit is open or shorted to ground • AIR diverter solenoid control circuit open or shorted to ground • Electric AIR pump control circuit is open or shorted to ground • PCM has failed
P0413 **2T CCM, MIL: Yes** Some Models Transmissions: All	**Secondary AIR System Switching Valve 'A' Circuit Malfunction Conditions:** Engine started; Air Injection solenoid commanded "on", and the PCM detected an unexpected "high" voltage condition on the Secondary AIR control circuit during the CCM test. **Possible Causes** • AIR solenoid control circuit is shorted to system power • AIR pump solenoid is damaged or has failed • PCM has failed
P0414 **2T CCM, MIL: Yes** Some Models Transmissions: All	**Secondary AIR System Switching Valve 'A' Circuit Malfunction Conditions:** Engine started; AIR solenoid disabled, and the PCM detected an unexpected low voltage condition on the AIR Solenoid control circuit **Possible Causes** • AIR solenoid control circuit is shorted to ground • AIR solenoid power circuit is open (no power to the solenoid) • AIR pump solenoid is damaged or has failed • Solid State relay is damaged or the PCM has failed

Gas Engine OBD II Trouble Code List (P0xxx Codes)

DTC	Trouble Code Title, Conditions & Possible Causes
P0416 **2T CCM, MIL: Yes** Some Models Transmissions: All	**Secondary AIR System Switching Valve 'B' Circuit Fault Conditions:** Engine started, AIR solenoid enabled, and the PCM detected an unexpected high voltage condition on AIR solenoid control circuit. **Possible Causes** • AIR solenoid control circuit is open or it is shorted to system power • AIR pump is damaged or the Solid State relay is damaged or has failed • PCM has failed
P0417 **2T CCM, MIL: Yes** Some Models Transmissions: All	**Secondary AIR System Switching Valve 'B' Circuit Fault Conditions:** Engine started; Air Injection solenoid commanded "on", and the PCM detected an unexpected "low" voltage condition on the AIR Solenoid control circuit during the CCM test. **Possible Causes** • AIR solenoid control circuit is shorted to ground or there is no power to the circuit • AIR pump is damaged or the Solid State relay is damaged or has failed • PCM has failed
P0420 **2T CAT, MIL: Yes** All Models Transmissions: All	**Catalyst System Efficiency Bank 1 Below Threshold Conditions:** Vehicle driven at steady cruise speed for 5 minutes, and the PCM detected the switch rate of the rear HO2S-12 was close to the switch rate of front HO2S (it should be much slower). **Possible Causes** • Air leaks at the exhaust manifold or in the exhaust pipes • Catalytic converter is damaged, contaminated or it has failed • ECT/CHT sensor has lost its calibration (the signal is incorrect) • Engine cylinders misfiring, or the ignition timing is over retarded • Engine oil is contaminated • Front HO2S or rear HO2S is contaminated with fuel or moisture • Front HO2S and/or the rear HO2S is loose in the mounting hole • Front HO2S much older than the rear HO2S (HO2S-11 is lazy) • Fuel system pressure is too high (check the pressure regulator) • Rear HO2S wires improperly connected or the HO2S has failed
P0421 **2T CAT, MIL: Yes** Probe equipped with a 2.5L V6 VIN B engine Transmissions: All	**Catalyst System Efficiency Bank 1 Below Threshold Conditions:** Vehicle driven at steady cruise speed for 5 minutes, and the PCM detected the switch rate of the rear HO2S-12 was close to the switch rate of front HO2S (it should be much slower). **Possible Causes** • Air leaks at the exhaust manifold or in the exhaust pipes • Catalytic converter is damaged, contaminated or it has failed • ECT/CHT sensor has lost its calibration (the signal is incorrect) • Engine cylinders misfiring, or the ignition timing is over retarded • Front HO2S or rear HO2S is contaminated with fuel or moisture • Front HO2S and/or the rear HO2S is loose in the mounting hole • Front HO2S much older than the rear HO2S (HO2S-11 is lazy) • Fuel system pressure is too high (check the pressure regulator) • Rear HO2S wires improperly connected or the HO2S has failed
P0430 **2T CAT, MIL: Yes** All Models Transmissions: All	**Catalyst System Efficiency Bank 2 Below Threshold Conditions:** Vehicle driven at steady cruise speed for 5 minutes, and the PCM detected the switch rate of the rear HO2S-12 was close to the switch rate of front HO2S (it should be much slower). **Possible Causes** • Air leaks at the exhaust manifold or in the exhaust pipes • Catalytic converter is damaged, contaminated or it has failed • ECT/CHT sensor has lost its calibration (the signal is incorrect) • Engine cylinders misfiring, or the ignition timing is over retarded • Engine oil is contaminated • Front HO2S or rear HO2S is contaminated with fuel or moisture • Front HO2S and/or the rear HO2S is loose in the mounting hole • Front HO2S much older than the rear HO2S (HO2S-11 is lazy) • Fuel system pressure is too high (check the pressure regulator) • Rear HO2S wires improperly connected or the HO2S has failed

Gas Engine OBD II Trouble Code List (P0xxx Codes)

DTC	Trouble Code Title, Conditions & Possible Causes
P0431 **2T CAT, MIL: Yes** Probe equipped with a 2.5L V6 VIN B engine Transmissions: All	**Catalyst System Efficiency Bank 2 Below Threshold Conditions:** Vehicle driven at steady cruise speed for 5 minutes, and the PCM detected the switch rate of the rear HO2S-12 was close to the switch rate of front HO2S (it should be much slower). **Possible Causes** • Air leaks at the exhaust manifold or in the exhaust pipes • Catalytic converter is damaged, contaminated or it has failed • ECT/CHT sensor has lost its calibration (the signal is incorrect) • Engine cylinders misfiring, or the ignition timing is over retarded • Front HO2S or rear HO2S is contaminated with fuel or moisture • Front HO2S and/or the rear HO2S is loose in the mounting hole • Front HO2S much older than the rear HO2S (HO2S-11 is lazy) • Fuel system pressure is too high (check the pressure regulator) • Rear HO2S wires improperly connected or the HO2S has failed
P0440 **2T EVAP, MIL: Yes** Probe equipped with a 2.0L VIN A or 2.5L V6 VIN B engine Transmissions: All	**EVAP System Malfunction Conditions:** ECT sensor less than 90°F at startup (cold engine), engine running in closed loop at a steady cruise speed, and the PCM detected a problem in the EVAP system during the EVAP System Monitor test. Note: If DTC P0443 is set, repair that diagnostic trouble code first. **Possible Causes** • Canister Purge valve is damaged (DTC P0443 not set) • Vapor line between Purge solenoid and intake manifold vacuum reservoir is damaged, or vapor line between EVAP Canister Purge solenoid and charcoal canister is damaged • Vapor line between charcoal canister and check valve, or vapor line between check valve and fuel vapor valves is damaged • PCM has failed
P0442 **2T EVAP, MIL: Yes** All Models Transmissions: All	**EVAP Control System Small Leak Detected Conditions:** ECT sensor less than 90°F at startup (cold engine), engine running in closed loop at a steady cruise speed, and the PCM detected a leak in the EVAP system as small as 0.040" in the test. **Possible Causes** • Aftermarket EVAP parts that do not conform to specifications • CV solenoid stays partially open when commanded to close • EVAP component seals leaking (i.e., leaks in the Purge valve, fuel vapor control valve tube assembly or fuel vapor vent valve) • Fuel filler cap damaged, cross-threaded or loosely installed • Loose fuel vapor hose/tube connections to EVAP components • Small holes or cuts in fuel vapor hoses or EVAP canister tubes
P0442 **2T EVAP, MIL: Yes** All Models Transmissions: All	**EVAP Control System Small Leak (0.040") Detected Conditions:** ECT sensor less than 90°F and within 10°F of the IAT sensor at startup (cold engine), engine started, the with the engine running in closed loop at a steady cruise speed, the PCM detected a leak in the EVAP system as small as 0.040" during the EVAP Monitor Test. **Possible Causes** • Aftermarket EVAP parts that do not conform to specifications • CV solenoid remains partially open when commanded to close • EVAP component seals leaking (i.e., leaks in the Purge valve, fuel tank pressure sensor, canister vent solenoid, fuel vapor control valve tube assembly or fuel vapor vent valve). • Fuel filler cap damaged, cross-threaded or loosely installed • Loose fuel vapor hose/tube connections to EVAP components • Small holes or cuts in fuel vapor hoses or EVAP canister tubes • TSB 99-23-4, TSB 3-9-8 & TSB 3-20-3 contain a repair procedure for this trouble code
P0443 **2T CCM, MIL: Yes** All Models Transmissions: All	**EVAP Canister Purge Solenoid Circuit Malfunction Conditions:** Engine started, and the PCM detected an unexpected high or low voltage condition on the Purge solenoid control circuit when the device was cycled "on" and "off" during testing. **Possible Causes** • EVAP purge solenoid supply circuit is open • EVAP purge solenoid control circuit open, shorted to ground • EVAP purge solenoid control circuit is shorted to power (B+) • EVAP canister purge solenoid valve is damaged or the PCM has failed

Gas Engine OBD II Trouble Code List (P0xxx Codes)

DTC	Trouble Code Title, Conditions & Possible Causes
P0443 **2T CCM, MIL: Yes** Contour & Mystique Models Transmissions: All	**EVAP Vapor Management Valve Circuit Malfunction Conditions:** Engine started, and the PCM detected an unexpected high or low voltage condition on the Vapor Management Valve (VMV) circuit when the device was cycled On/Off during testing. **Possible Causes** • EVAP VMV power supply circuit is open • EVAP VMV solenoid control circuit is open or shorted to ground • EVAP VMV solenoid control circuit is shorted to power (B+) • EVAP VMV solenoid valve is damaged or it has failed • PCM has failed
P0443 **2T CCM, MIL: Yes** E Van, F Series Truck Models equipped with a 5.8L VIN H or 7.5L VIN G engine Transmissions: All	**EVAP Vapor Management Valve Circuit Malfunction Conditions:** Engine started, and the PCM detected an unexpected high or low voltage condition on the Vapor Management Valve (VMV) circuit after the device was cycled On/Off during testing. **Possible Causes** • EVAP VMV power supply circuit is open • EVAP VMV solenoid control circuit is open or shorted to ground • EVAP VMV solenoid control circuit is shorted to power (B+) • EVAP VMV solenoid valve is damaged or it has failed • PCM has failed
P0443 **2T CCM, MIL: Yes** All Models Transmissions: All	**EVAP Canister Purge Solenoid Circuit Malfunction Conditions:** Engine started, and the PCM detected an unexpected high or low voltage condition on the EVAP Purge solenoid control circuit when the device was cycled On/Off during testing. **Possible Causes** • EVAP purge solenoid supply circuit is open • EVAP purge solenoid control circuit open, shorted to ground • EVAP purge solenoid control circuit is shorted to power (B+) • EVAP canister purge solenoid valve is damaged or it has failed • PCM has failed
P0446 **2T EVAP, MIL: Yes** All Models Transmissions: All	**EVAP Canister Vent System Performance Conditions:** ECT sensor less than 90ºF, engine started, engine running at a steady cruise speed, and the PCM detected excessive vacuum was present in the EVAP system during the test period. **Possible Causes** • Canister vent (CV) solenoid is stuck closed (partially or fully) • EVAP canister purge outlet tube blocked or kinked between the canister purge valve and the EVAP canister, or EVAP canister tube blocked between the fuel tank and canister • EVAP canister restricted, or plugged CV solenoid filter unit • Plugged or contaminated CV solenoid filter • EVAP canister purge valve stuck open • Fuel filler cap stuck closed (no vacuum relief) • FTP sensor VREF circuit open, or FTP sensor is damaged • Fuel vapor elbow at the EVAP canister contaminated
P0446 **2T EVAP, MIL: Yes** All Models Transmissions: All	**EVAP Canister Vent Solenoid Circuit Malfunction Conditions:** Engine started, and the PCM detected an unexpected high or low voltage condition on the EVAP Canister Vent solenoid control circuit after the device was cycled On/Off in the test. **Possible Causes** • Canister vent solenoid supply circuit is open • Canister vent solenoid control circuit open, shorted to ground • Canister vent solenoid control circuit is shorted to power (B+) • Canister vent solenoid valve is damaged or the PCM has failed
P0451 **2T CCM, MIL: Yes** All Models Transmissions: All	**Fuel Tank Pressure Sensor Intermittent Signal Conditions:** Engine started, and the PCM detected the FTP sensor signal changed from over +15" H2O to under -15" H2O within 100 ms. **Possible Causes** • FTP sensor signal circuit has an intermittent open condition • FTP sensor signal circuit has an intermittent shorted condition • FTP sensor is damaged or it has failed

Gas Engine OBD II Trouble Code List (P0xxx Codes)

DTC	Trouble Code Title, Conditions & Possible Causes
P045 **2T CCM, MIL: Yes** All Models Transmissions: All	**FTP Sensor Circuit Low Input Conditions:** Key on or engine running; and the PCM detected the FTP sensor indicated less than the minimum calibrated limit of 0.22v in the test. **Possible Causes** • FTP sensor connector has internal damage or contamination • FTP sensor signal circuit is shorted to chassis or signal ground • FTP sensor is damaged • PCM has failed
P0453 **2T CCM, MIL: Yes** All Models Transmissions: All	**FTP Sensor Circuit High Input Conditions:** Key on or engine running; and the PCM detected the FTP sensor indicated more than the maximum calibrated limit (4.50v) in the test. **Possible Causes** • FTP sensor signal circuit is open or the ground circuit is open • FTP sensor signal circuit is shorted to VREF (5v) • FTP sensor is damaged or the PCM has failed
P0455 **2T EVAP, MIL: Yes** All Models Transmissions: All	**EVAP Control System Large Leak Detected Conditions:** ECT sensor less than 90ºF at startup, engine running, and the PCM detected several small fuel vapor leaks or a large leak in the system. **Possible Causes** • Aftermarket EVAP hardware non-conforming to specifications • Canister vent (CV) solenoid stuck open • EVAP canister purge valve stuck closed, or canister damaged • EVAP canister tube, EVAP canister purge outlet tube or EVAP return tube disconnected or cracked, or canister is damaged • Fuel filler cap missing, loose (not tightened) or the wrong part • Loose fuel vapor hose/tube connections to EVAP components • Purge sensor or FTP sensor is out of calibration or has failed
P0455 **2T EVAP, MIL: Yes** All Models Transmissions: All	**EVAP Control System Large Leak (0.080") Detected Conditions:** ECT sensor less than 90ºF, engine running at a steady cruise speed, and the PCM detected multiple small fuel vapor leaks; or it detected a large leak in the system during the leak test. **Possible Causes** • Aftermarket EVAP hardware non-conforming to specifications • EVAP canister tube, EVAP canister purge outlet tube or EVAP return tube disconnected or cracked, or canister is damaged • EVAP canister purge valve stuck closed, or canister damaged • Fuel filler cap missing, loose (not tightened) or the wrong part • Loose fuel vapor hose/tube connections to EVAP components • Canister vent (CV) solenoid stuck open • Fuel tank pressure (FTP) sensor has failed mechanically • TSB 99-23-4 contains a repair procedure for this trouble code • TSB 03-9-8 contains a repair procedure for this trouble code • TSB 3-20-3 contains a repair procedure for this trouble code
P0456 **2T EVAP, MIL: Yes** All Models Transmissions: All	**EVAP Control System Very Small Leak (0.020") Detected Conditions:** ECT sensor less than 90ºF (cold engine), engine started, engine running at a steady cruise speed, and the PCM detected a very small fuel vapor leak (0.020") during the leak test. **Possible Causes** • Canister tube, EVAP canister purge outlet tube or return tube disconnected or cracked • EVAP canister purge valve stuck closed, or canister damaged • Fuel vapor hoses/tubes that have very small holes and/or cuts • Fuel vapor hose/tube connections are loose or damaged • EVAP component seals are leaking (i.e., Purge valve, fuel tank pressure sensor, canister vent solenoid, fuel vapor control valve tube assembly or fuel vapor vent valve assembly) • TSB 03-9-8 contains a repair procedure for this trouble code • TSB 3-20-3 contains a repair procedure for this trouble code

Gas Engine OBD II Trouble Code List (P0xxx Codes)

DTC	Trouble Code Title, Conditions & Possible Causes
P0457 **2T EVAP, MIL: Yes** All Models Transmissions: All	**EVAP Control System Leak Detected (Fuel Cap Missing) Conditions:** ECT sensor less than 90ºF at startup, engine running at a steady state cruise speed, and the PCM detected the fuel tank pressure changed more than minus (-) 7" H2O in 30 seconds, or excessive purge flow (over 0.06 pounds per minute) occurred in the EVAP Running Loss Monitor Test ("Check Fuel Cap" Lamp may be "on"). **Possible Causes** • Fuel filler cap not installed after refueling (CMC P0457 is set) • Fuel filler cap missing, loose or cross-threaded • TSB 03-9-8 contains a repair procedure for this trouble code • TSB 3-20-3 contains a repair procedure for this trouble code
P0460 **2T CCM, MIL: Yes** All Models Transmissions: All	**Fuel Level Sensor Signal Range/Performance Conditions:** Engine started, and the PCM detected the FLI sensor did not match the fuel level (e.g., FLI V PID below 0.90v with FLI PID at 25%, or FLIV PID more than 2.45v with FLI PID at 75%). **Possible Causes** • Fuel tank is empty • FP module is stuck open • Fuel gauge is incorrectly installed • Instrument cluster damaged • PCM Case ground circuit open • Fuel level indicator (FLI) circuit is shorted to power, or is open • Fuel tank has been overfilled, or fuel gauge is damaged • Fuel pump (FP) module is stuck closed, or is stuck open • Fuel level indicator circuit shorted to Case or to power ground • PCM Case ground shorted to VPWR (shorted to system power) • TSB 03-1-7 contains repair help for this code (LS & T-Bird)
P0462 **2T CCM, MIL: Yes** All Models Transmissions: All	**Fuel Level Sensor Circuit Low Input Conditions:** Key on or engine running; and the PCM detected the FLI sensor indicated less than 0.20v at any time during the CCM test period. **Possible Causes** • Fuel tank is empty • FLI signal circuit is open • FLI signal circuit is shorted to case or chassis ground • PCM has failed
P0463 **2T CCM, MIL: Yes** All Models Transmissions: All	**Fuel Level Sensor Circuit High Input Conditions:** Key on or engine running; and the PCM detected that the FLI sensor indicated more than 4.50v at any time during the CCM test period. **Possible Causes** • Fuel level sensor connector is damaged or shorted • Fuel tank has been over-filled • FLI signal circuit is shorted to VREF (5v or 12v) • PCM has failed
P0480 **2T CCM, MIL: Yes** Crown Victoria, Grand Marquis, Town Car, LS Thunderbird, Mustang, Expedition, Navigator, ZX2 Models Transmissions: All	**Visctronic Drive Fan Primary Circuit Malfunction Conditions:** Key on or engine running; and the PCM detected an unexpected high or low voltage condition on the Visctronic Drive Fan (VDF) primary circuit during the CCM test period. **Possible Causes** • VDF variable control circuit is open • VDF variable control circuit is shorted to chassis ground • VDF variable control circuit shorted to Fan Speed Sensor circuit • VDF clutch power supply (VPWR) circuit is open • VDF clutch is damaged or it has failed • PCM has failed

Gas Engine OBD II Trouble Code List (P0xxx Codes)

DTC	Trouble Code Title, Conditions & Possible Causes
P0480 **2T CCM, MIL: Yes** Focus, E Van, Escape, Taurus, Sable, Ranger Models equipped with a 2.0L VIN B, 2.0L VIN 3, 2.0L VIN 5, 2.0L VIN P, 2.3L VIN D, 2.3L VIN Z, 3.0L VIN 1, 3.0L VIN 2, 3.0L VIN S, 3.0L VIN U, 3.0L VIN V, 4.0L VIN E engine Transmissions: All	**Fan Control Relay Circuit Malfunction Conditions:** Key on or engine running; and the PCM detected an unexpected high or low voltage condition on the Fan Control relay control circuit. **Possible Causes** • High/Low/Medium FC relay control circuit is open • High/Low/Medium FC relay control circuit is shorted to ground • High/Low/Medium FC relay VPWR circuit is open • High/Low/Medium FC relay direct battery (B+) circuit is open • High/Low/Medium FC relay is damaged or it has failed • PCM has failed
P0481 **2T CCM, MIL: Yes** Escort, ZX2, Mustang equipped with a 2.0L VIN 3, 3.8L VIN 4, 4.6L VIN X, 4.6L VIN X, 4.6L VIN Y Transmissions: All	**Constant Control Relay Module Circuit Malfunction Conditions:** Key on or engine running; and the PCM detected an unexpected high or low voltage condition on the High Fan Control (HFC) relay control circuit (located inside the CCRM). **Possible Causes** • HFC relay control circuit is open • HFC relay control circuit is shorted to chassis ground • HFC relay power supply (VPWR) circuit is open • HFC relay direct battery (B+) circuit is open • HFC relay (inside the CCRM) is damaged or it has failed • PCM has failed
P0481 **2T CCM, MIL: Yes** Escape, Focus, Taurus, Sable, Ranger equipped with a 2.0L (VIN B, VIN 3, VIN 5, VIN P), 2.3L VIN D, 2.3L VIN Z, 3.0L VIN 1, 3.0L VIN 2, 3.0L VIN S, 3.0L VIN U, 3.0L VIN V, 4.0L VIN E engine Transmissions: All	**Fan Control Relay Circuit Malfunction Conditions:** Key on or engine running; and the PCM detected an unexpected high or low voltage condition on the Fan Control relay control circuit. **Possible Causes** • FC relay control circuit is open • FC relay control circuit is shorted to chassis ground • FC relay power supply (VPWR) circuit is open • FC relay direct battery (B+) circuit is open • FC relay is damaged or it has failed • PCM has failed
P0482 **2T CCM, MIL: Yes** Aviator, Escape, Focus, Taurus, Sable, Ranger, Explorer, Mountaineer, Expedition, Navigator equipped with a 2.0L VIN B, 2.0L VIN 3, 2.0L VIN 5, 2.0L VIN P, 2.3L VIN D, 2.3L VIN Z, 3.0L VIN 1, 3.0L VIN 2, 3.0L VIN S, 3.0L VIN U, 3.0L VIN V, 3.8L VIN 4, 4.0L VIN E, 4.6L VIN H engine Transmissions: All	**Constant Control Relay Module Circuit Malfunction Conditions:** Key on or engine running; and the PCM detected an unexpected high or low voltage condition on the High Fan Control (HFC) relay control circuit (located inside the CCRM) during the CCM test period. **Possible Causes** • FC relay control circuit is open • FC relay control circuit is shorted to chassis ground • FC relay power supply (VPWR) circuit is open • FC relay direct battery (B+) circuit is open • PCM has failed

Gas Engine OBD II Trouble Code List (P0xxx Codes)

DTC	Trouble Code Title, Conditions & Possible Causes
P0500 **2T CCM, MIL: Yes** All Models Transmissions: All	**Vehicle Speed Sensor Malfunction Conditions:** Engine running, then with the engine speed more than the TCC stall speed, the PCM detected a lack of vehicle speed data occurred. Note: The PCM receives vehicle speed data from the VSS, TCSS, ABS module, CTM or GEM controller, depending up the application. **Possible Causes** • Modules connected to VSC/VSS harness circuits are damaged • Mechanical drive mechanism for the VSS or TCSS is damaged • VSS+ or VSS- harness circuit is open • TCSS signal or TCSS signal return harness circuit is open • VSS harness circuit, TCSS harness circuit is shorted to ground • VSS harness circuit, CSS harness circuit is shorted to power • VSS circuit open between the PCM and related control module • VSS or TCSS, or wheel speed sensors circuits are damaged • TSB 01-21-13 contains a repair procedure for this trouble code
P0500 **2T CCM, MIL: Yes** All Models Transmissions: All	**Vehicle Speed Sensor Circuit Malfunction Conditions:** Engine running, then with the engine speed more than the TCC stall speed, the PCM detected a lack of vehicle speed data occurred. Note: The PCM receives vehicle speed data from the Vehicle Speed Sensor on these vehicle applications. **Possible Causes** • VSS signal circuit is open or shorted to ground • VSS ground circuit is open or VSS power circuit is open • VSS is damaged or it has failed • PSOM is damaged or it has failed (some models) • PCM has failed
P0500 **2T CCM, MIL: Yes** Continental, Town Car, LS, Five Hundred, Freestyle, Montego,, Windstar Models Transmissions: All	**Vehicle Speed Sensor Circuit Malfunction Conditions:** Engine running, then with the engine speed more than the TCC stall speed, the PCM detected a lack of vehicle speed data occurred. Note: The PCM receives vehicle speed data from the ABS module. **Possible Causes** • The vehicle speed information on this vehicle application is provided to the PCM by the Antilock Brake System module. • Refer to the ABS diagnostics and trouble codes to diagnose this particular trouble code.
P0500 **2T CCM, MIL: Yes** Contour, Cougar & Mystique Models Transmissions: All	**Vehicle Speed Sensor Circuit Malfunction Conditions:** Engine running, then with the engine speed more than the TCC stall speed, the PCM detected a lack of vehicle speed data occurred. Note: The PCM receives vehicle speed data from the VSS sensor. **Possible Causes** • VSS+ signal circuit is open or shorted to ground • VSS- signal circuit is open • VSS power supply (VPWR) circuit is open • VSS is damaged or it has failed • PCM has failed
P0500 **2T CCM, MIL: Yes** F Series Trucks equipped a 4.6L VIN W, 5.4L VIN 3, 5.4L VIN L, 5.4L VIN M, 5.4L VIN Z engine Transmissions: All	**Vehicle Speed Sensor Circuit Malfunction Conditions:** Engine running, then with the engine speed more than the TCC stall speed, the PCM detected a lack of vehicle speed data for a period of time. Note: The PCM receives vehicle speed data from the Transfer Case Speed Sensor on these vehicle applications. **Possible Causes** • TCSS signal circuit is open or shorted to ground • TCSS ground circuit is open • TCSS is damaged or it has failed • PCM has failed

Gas Engine OBD II Trouble Code List (P0xxx Codes)

DTC	Trouble Code Title, Conditions & Possible Causes
P0500 **2T CCM, MIL: Yes** Aviator, Explorer, Mountaineer, Mustang, F Series Truck, Ranger equipped with a 2.5L VIN C, 3.0L (VIN U, VIN V), 4.0L (VIN E, VIN K, VIN X), 5.0L VIN P, 5.4L (VIN 3, VIN L, VIN M, VIN Z) 6.8L VIN S engines Transmissions: All	**Vehicle Speed Sensor Circuit Malfunction Conditions:** Engine started; then with the engine speed more than the TCC stall speed, the PCM detected a lack of vehicle speed data occurred. Note: The PCM receives vehicle speed data from the Rear Wheel ABS (RABS) or 4-Wheel ABS (4WABS) on these applications. **Possible Causes** • VSC positive signal circuit is open or shorted to ground • VSC negative signal circuit is open • RABS or 4WABS control unit is damaged or has failed • One of the other modules (CTM or GEM) may be the cause of this trouble code. Diagnose other codes from these modules.
P0501 **1T CCM, MIL: Yes** All Models Transmissions: All	**Vehicle Speed Sensor or PSOM Range/Performance Conditions:** Engine started; engine speed above the TCC stall speed, and the PCM detected a loss of the VSS signal over a period of time. Note: The PCM receives vehicle speed data from the VSS, TCSS, ABS module, CTM or GEM controller, depending up the application. **Possible Causes** • VSS+ or VSS- signal circuit is open or shorted to ground • TCSS signal or TCSS signal return harness circuit is open • VSS harness circuit, TCSS harness circuit is shorted to ground • VSS harness circuit, CSS harness circuit is shorted to power • VSS circuit open between the PCM and related control module • VSS or TCSS, or wheel speed sensors circuits are damaged • Modules connected to VSC/VSS harness circuits are damaged • Mechanical drive mechanism for the VSS or TCSS is damaged
P0501 **1T CCM, MIL: Yes** All Models Transmissions: All	**Vehicle Speed Sensor Range/Performance Conditions:** Engine started; then with the engine speed more than the TCC stall speed, the PCM detected a problem with the vehicle speed data. Note: The PCM receives vehicle speed data from the Vehicle Speed Sensor on these vehicle applications. **Possible Causes** • VSS signal circuit is open or shorted to ground • VSS ground circuit is open • VSS power circuit (VPWR) is open • VSS is damaged or it has failed • PCM has failed
P0500 **2T CCM, MIL: Yes** Continental, Town Car, LS, Five Hundred, Freestyle, Montego,, Windstar Models Transmissions: All	**Vehicle Speed Sensor Signal Range/Performance Conditions:** Engine started; then with the engine speed more than the TCC stall speed, the PCM detected a problem with the vehicle speed data. Note: The PCM receives vehicle speed data from the ABS module. **Possible Causes** • The vehicle speed information on this vehicle application is provided to the PCM by the Antilock Brake System module. • Refer to the ABS diagnostics and trouble codes to diagnose this particular trouble code.
P0501 **2T CCM, MIL: Yes** Contour, Cougar & Mystique Models Transmissions: All	**Vehicle Speed Sensor Signal Range/Performance Conditions:** Engine started; then with the engine speed more than the TCC stall speed, the PCM detected a problem with the vehicle speed data. Note: The PCM receives vehicle speed data from the VSS sensor. **Possible Causes** • VSS+ signal circuit is open or shorted to ground • VSS- signal circuit is open • VSS power supply (VPWR) circuit is open • VSS is damaged or it has failed • PCM has failed

Gas Engine OBD II Trouble Code List (P0xxx Codes)

DTC	Trouble Code Title, Conditions & Possible Causes
P0501 **2T CCM, MIL: Yes** F Series Trucks equipped a 4.6L VIN W, 5.4L VIN 3, 5.4L VIN L, 5.4L VIN M, 5.4L VIN Z engine Transmissions: All	**Vehicle Speed Sensor Signal Range/Performance Conditions:** Engine running, then with the engine speed more than the TCC stall speed, the PCM detected a problem with the vehicle speed data. Note: The PCM receives vehicle speed data from the Transfer Case Speed Sensor on these vehicle applications. **Possible Causes** • TCSS signal circuit is open or shorted to ground • TCSS ground circuit is open • TCSS is damaged or it has failed • PCM has failed
P0501 **2T CCM, MIL: Yes** Aviator, Explorer, Mountaineer, Mustang, F Series Truck, Ranger equipped with a 2.5L VIN C, 3.0L (VIN U, VIN V), 4.0L (VIN E, VIN K, VIN X), 5.0L VIN P, 5.4L (VIN 3, VIN L, VIN M, VIN Z) 6.8L VIN S engines Transmissions: All	**Vehicle Speed Sensor Signal Range/Performance Conditions:** Engine running, then with the engine speed more than the TCC stall speed, the PCM detected a problem with the vehicle speed data. Note: The PCM receives vehicle speed data from the Rear Wheel ABS (RABS) or 4-Wheel ABS (4WABS) on these applications. **Possible Causes** • VSC positive signal circuit is open or shorted to ground • VSC negative signal circuit is open • RABS or 4WABS control unit is damaged or has failed • One of the other modules (CTM or GEM) may be the cause of this trouble code. Diagnose other codes from these modules.
P0503 **2T CCM, MIL: Yes** All Models Transmissions: A/T	**Vehicle Speed Sensor Signal Intermittent Conditions:** Engine started, engine speed above the TCC stall speed, and the PCM detected the vehicle speed data was "noisy" or intermittent. Note: The PCM receives vehicle speed data from the VSS, TCSS, ABS module, CTM or GEM controller, depending up the application. **Possible Causes** • Module or circuits connected to VSS/TCSS circuit are damaged • VSS/TCSS wiring harness or connector is damaged or loose • VSS/TCSS signal is "noisy" due to RFI or EMI interference from sources such as ignition components or charging system • VSS/TCSS gears are damaged or there is debris on the sensor
P0503 **2T CCM, MIL: Yes** Aviator, Contour, Mystique, Explorer, Mountaineer, Mustang, F Series Truck, Ranger equipped with a 2.0L (VIN 3, VIN Z, VIN B) 2.5L VIN L, 2.5L VIN G, 4.2L VIN 2, 4.6L VIN 6, 4.6L VIN W, 5.4L VIN L engine Transmissions: M/T	**Vehicle Speed Sensor Signal Intermittent Conditions:** Engine started, engine speed above the TCC stall speed, and the PCM detected the vehicle speed data was "noisy" or intermittent. Note: The PCM receives vehicle speed data from the VSS or TCSS. **Possible Causes** • TCSS or VSS signal circuit is open or shorted to ground • TCSS or VSS ground circuit is open (an intermittent problem) • TCSS or VSS power supply (VREF) circuit is open (intermittent) • TCSS or VSS is damaged or it has failed (intermittent problem) • PCM has failed

Gas Engine OBD II Trouble Code List (P0xxx Codes)

DTC	Trouble Code Title, Conditions & Possible Causes
P0505 **2T CCM, MIL: Yes** All Models Transmissions: All	**Idle Air Control System Malfunction Conditions:** Engine started, engine running at hot idle speed, and the PCM detected the Actual Idle Speed was too low or too high when compared to the Target Idle Speed during the KOER self-test. Specification: The IAC valve resistance is 6-13 ohms at 68ºF. **Possible Causes** • Air inlet dirty, restricted or the air cleaner is severely restricted • IAC solenoid control circuit is open, shorted to ground or to B+ • IAC solenoid power circuit (VPWR) is open from the relay • IAC valve is damaged or has failed • PCM has failed • TSB 03-3-5 contains repair information for this trouble code
P0506 **2T CCM, MIL: Yes** All Models Transmissions: All	**Idle Air Control System RPM Lower Than Expected Conditions:** DTC P0402 not set, engine started, engine running in closed loop, and the PCM detected it could not control the idle speed correctly. **Possible Causes** • Air inlet is plugged or the air filter element is severely clogged • IAC circuit is open or shorted to the VPWR circuit • IAC circuit VPWR circuit is open • IAC solenoid is damaged or has failed • PCM has failed • TSB 03-3-5 contains repair information for this trouble code
P0507 **2T CCM, MIL: Yes** All Models Transmissions: All	**Idle Air Control System RPM Higher Than Expected Conditions:** DTC P0402 not set, engine started, engine running in closed loop, and the PCM detected it could not control the idle speed correctly. **Possible Causes** • Air intake leak located somewhere after the throttle body • IAC control circuit is shorted to chassis ground • IAC solenoid is damaged or has failed • PCM has failed • TSB 03-3-5 contains repair information for this trouble code
P0511 **2T CCM, MIL: Yes** All Models Transmissions: All	**Idle Air Control Valve Circuit Malfunction Conditions:** DTC P0402 not set, engine started, engine running in closed loop, and the PCM detected it could not control the idle speed correctly. **Possible Causes** • IAC control circuit is open • IAC control circuit is shorted to power (B+) • IAC power supply circuit (VPWR) is open • IAC solenoid is damaged or the PCM has failed
P0528 **2T CCM, MIL: Yes** Crown Victoria, Grand Marquis, Town Car, Mustang, Thunderbird, Escort, Expedition, LS, Five Hundred, Freestyle, Montego,, Navigator & ZX2 Models Transmissions: All	**Visctronic Drive Fan Speed Sensor Circuit Malfunction Conditions:** Engine started, Visctronic Drive Fan (VDF) commanded to a 100% duty cycle position, and the PCM detected the VDF Speed Sensor signal was less than a calibrated value in the test. **Possible Causes** • VDF fan motor has a mechanical interference fault or is binding • VDF speed sensor circuit is open or shorted to ground • Vehicle Buffered Power (VBPWR) circuit is open or shorted • VDF speed sensor power ground circuit is open • VDF speed sensor is damaged or the PCM has failed
P0534 **2T CCM, MIL: Yes** All Models Transmissions: All	**Low Air Conditioning Cycle Period Conditions:** Engine started; A/C enabled, and the PCM detected frequent A/C compressor clutch cycling during the CCM test period. Note that this trouble code and test was designed to protect the transmission. In some cases, the PCM will unlock TCC operation. **Possible Causes** • A/C cycling pressure switch signal to PCM open (intermittent) • A/C cycling pressure switch IGN (B+) circuit open (intermittent) • A/C mechanical problem (low A/C refrigerant charge or a damaged A/C cycling switch)

Gas Engine OBD II Trouble Code List (P0xxx Codes)

DTC	Trouble Code Title, Conditions & Possible Causes
P0537 **2T CCM, MIL: Yes** All Models Transmissions: All	**A/C Evaporator Temperature Circuit Sensor Low Input Conditions:** Engine started; A/C enabled, and the PCM detected an unexpected low voltage condition on the A/C Evaporator Temperature (ACET) sensor circuit during the CCM test period. **Possible Causes** • ACET sensor signal circuit shorted to sensor or chassis ground • ACET sensor is damaged or it has failed • PCM has failed
P0538 **2T CCM, MIL: Yes** All Models Transmissions: All	**A/C Evaporator Temperature Sensor Circuit High Input Conditions:** Engine started; A/C enabled, and the PCM detected an unexpected high voltage condition on the A/C Evaporator Temperature (ACET) sensor circuit during the CCM test period. **Possible Causes** • ACET sensor signal circuit is open • ACET sensor signal circuit is shorted to VREF (5v) • ACET sensor ground circuit is open • ACET sensor is damaged or it has failed • PCM has failed
P0552 **2T CCM, MIL: Yes** All Models Transmissions: All	**Power Steering Pressure Sensor Circuit Low Input Conditions:** Engine started, and the PCM detected an unexpected low voltage condition on the Power Steering Pressure (PSP) sensor circuit. **Possible Causes** • PSP sensor signal circuit is shorted to sensor ground • PSP sensor signal circuit is shorted to chassis ground • PSP sensor VREF (5v) circuit is open • PSP sensor is damaged or it has failed • PCM has failed
P0553 **2T CCM, MIL: Yes** All Models Transmissions: All	**Power Steering Pressure Sensor Circuit High Input Conditions:** Engine started, and the PCM detected an unexpected high voltage condition on the Power Steering Pressure (PSP) sensor circuit. **Possible Causes** • PSP sensor ground circuit is open • PSP sensor ground circuit is shorted to VREF (5v) • PSP sensor signal circuit is shorted to VREF (5v) • PSP sensor is damaged or the PCM has failed
P0597 **2T CCM, MIL: Yes** Ranger equipped with a 2.3L I4 VIN D engine Transmissions: All	**Thermostat Heater Control Circuit Malfunction Conditions:** Engine started, and the PCM detected an unexpected low or high voltage condition on the Thermostat Heater Control (THTRC) circuit. **Possible Causes** • THTRC circuit is open or shorted to ground • THTRC power (VPWR) circuit is open • Thermostat assembly is damaged or the PCM has failed
P0602 **1T PCM, MIL: Yes** All Models Transmissions: All	**Control Module Programming Error Conditions:** Key on, and the PCM detected a programming error in the VID block. This fault requires that the VID Block be reprogrammed, or that the EEPROM be re-flashed. **Possible Causes** • During the VID reprogramming function, the Vehicle ID (VID) data block failed during reprogramming wit the Scan Tool.
P0603 **1T PCM, MIL: Yes** All Models Transmissions: All	**PCM Keep Alive Memory Test Error Conditions:** Key on, and the PCM detected an internal memory fault. This code will set if KAPWR to the PCM is interrupted (at the initial key on). **Possible Causes** • Battery terminal corrosion, or loose battery connection • KAPWR to PCM interrupted, or the circuit has been opened • Reprogramming error has occurred • PCM has failed and needs replacement. Remember to check for Aftermarket Performance Products before replacing a PCM.

Gas Engine OBD II Trouble Code List (P0xxx Codes)

DTC	Trouble Code Title, Conditions & Possible Causes
P0605 **1T PCM, MIL: Yes** All Models Transmissions: All	**PCM Read Only Memory Test Error Conditions:** Key on, and the PCM detected a ROM test error (ROM inside PCM is corrupted). The PCM is normally replaced if this code has set. **Possible Causes** • An attempt was made to change the module calibration, or a Module programming error may have occurred • Clear the trouble codes and then check for this trouble code. If it resets, the PCM has failed and needs replacement. • Remember to check for signs of Aftermarket Performance Products installation before replacing the PCM.
P0606 **1T PCM, MIL: Yes** All Models Transmissions: All	**PCM Internal Communication Error Conditions:** Key on, and the PCM detected an internal communications register read back error during the initial key on check period. **Possible Causes** • Clear the trouble codes and then check for this trouble code. If it resets, the PCM has failed and needs replacement. • Remember to check for signs of Aftermarket Performance Products installation before replacing the PCM.
P0622 **1T CCM, MIL: Yes** All Models Transmissions: All	**Generator Regulator System Malfunction Conditions:** Engine started; and the PCM detected an unexpected voltage condition on the Generator control circuit. **Possible Causes** • Generator belt is loose or worn out • Generator or regulator is damaged or has failed • PCM has failed
P0645 **1T CCM, MIL: Yes** All Models Transmissions: All	**Wide Open Throttle A/C Output Primary Circuit Malfunction Conditions:** Key on or engine running; and the PCM detected an unexpected low or high voltage condition WAC output primary circuit during the test. **Possible Causes** • WAC relay control circuit is open or shorted to ground • WAC relay power circuit (VPWR) is open • WAC relay is damaged or it has failed • PCM has failed
P0645 **1T CCM, MIL: Yes** Escort, Mustang, ZX2 equipped with a 2.0L VIN 3, 3.8L VIN 4, 4.6L VIN X, 4.6L VIN X, 4.6L VIN Y engine Transmissions: All	**Wide Open Throttle A/C Output Primary Circuit Malfunction Conditions:** Key on or engine running; and the PCM detected an unexpected low or high voltage condition on the WAC output circuit (in the CCRM). **Possible Causes** • WAC relay control circuit is open or shorted to ground • WAC relay power circuit (VPWR) is open • WAC relay is damaged or it has failed • PCM has failed
P0660 **1T CCM, MIL: Yes** All Models Transmissions: All Escort, Focus, E Van & F Series Truck, ZX2 & Windstar Models equipped with a 2.0L VIN P, 3.8L VIN 4, 4.2L VIN 2 engine Transmissions: All	**Intake Manifold Runner Control Valve Circuit Malfunction Conditions:** Key on or engine running; and the PCM detected an unexpected low or high voltage condition on the Intake Manifold Runner Control (IMRC) signal circuit during the CCM test. **Possible Causes** • IMRC signal circuit is open • IMRC signal circuit is shorted to chassis ground • IMRC actuator assembly is damaged or failed • PCM has failed

Gas Engine OBD II Trouble Code List (P0xxx Codes)

DTC	Trouble Code Title, Conditions & Possible Causes
P0660 **1T CCM, MIL: Yes** Blackwood, Expedition Navigator, LS, Five Hundred, Freestyle, Montego,, Taurus, Sable, Mustang Models equipped with a 3.0L VIN S, 4.6L VIN W, 5.4L VIN A, 5.4L VIN L, 5.4L VIN R engine Transmissions: All	**Intake Manifold Tuning Valve Circuit Malfunction Conditions:** Key on or engine running; and the PCM detected an unexpected low or high voltage condition on the Intake Manifold Tuning Valve (ITMV) signal circuit during the CCM test. **Possible Causes** • ITMV signal circuit is open • ITMV signal circuit is shorted to chassis ground • ITMV electric actuator assembly is damaged or failed • PCM has failed
P0660 **1T CCM, MIL: Yes** Ranger Models equipped with a 2.3L VIN D engine Transmissions: All	**Intake Manifold Swirl Control Actuator Circuit Malfunction Conditions:** Key on or engine running; and the PCM detected an unexpected low or high voltage condition on the Intake Manifold Swirl Control (IMSC) signal circuit during the CCM test. **Possible Causes** • IMSC signal circuit is open • IMSC signal circuit is shorted to chassis ground • IMSC actuator assembly is damaged or failed • PCM has failed
P0660 **1T CCM, MIL: Yes** Aviator equipped with a 4.6L VIN H engine Transmissions: All	**Intake Manifold Communication Control Circuit Malfunction Conditions:** Key on or engine running; and the PCM detected an unexpected low or high voltage condition on the Intake Manifold Communication Control (IMCC) signal circuit in the test. **Possible Causes** • IMCC signal circuit is open • IMCC signal circuit is shorted to chassis ground • Long / Short actuator assembly is damaged or failed • PCM has failed
P0703 **2T CCM, MIL: Yes** Continental, Town Car & Windstar Models Transmissions: A/T	**Brake Switch Circuit Malfunction Conditions:** Engine started, and the PCM did not detect any change in the Brake Pedal Position switch status, or with the vehicle running at Cruise speed, followed by a short deceleration periods, the PCM did not detect any change in the Brake Pedal Position switch status. **Possible Causes** • BPP switch circuit is open • BPP switch is damaged or it is out of adjustment • BPP switch power circuit is open (test switch inline fuse) • Module(s) connected the BPP switch circuit have a problem (e.g., Rear Electronic Module on Windstar or LS6/LS8, or the Lighting Control Module (LCM) Continental and Town Car)
P0703 **2T CCM, MIL: Yes** LS Models Transmissions: A/T	**Brake Switch Circuit Malfunction Conditions:** Engine started, and the PCM did not detect any change in the Brake Pedal Position switch status, or with the vehicle running at Cruise speed, followed by a short deceleration periods, the PCM did not detect any change in the Brake Pedal Position switch status. **Possible Causes** • BPA/ BPP switch circuit is open • BPA/ BPP switch is damaged or it is out of adjustment • BPA/ BPP switch power circuit is open (test switch inline fuse) • One or more of the Module(s) that connect to the BPA or the BPP switch circuits have a problem (e.g., Rear Electronic Module or the Vehicle Speed Control)
P0703 **2T CCM, MIL: Yes** Some Models Transmissions: A/T	**Brake Switch Circuit Malfunction Conditions:** Engine started, and the PCM did not detect any change in the Brake Pedal Position (BPP) switch status, or with the vehicle at Cruise speed, followed by one or more short deceleration periods, the PCM did not detect any change in the Brake Pedal Position switch status. **Possible Causes** • BPP switch circuit is open • BPP switch is damaged or it is out of adjustment • BPP switch power circuit is open (check the switch inline fuse)

Gas Engine OBD II Trouble Code List (P0xxx Codes)

DTC	Trouble Code Title, Conditions & Possible Causes
P0703 **2T CCM, MIL: Yes** Thunderbird Models Transmissions: A/T	**Brake Switch Circuit Malfunction Conditions:** Engine started, and the PCM did not detect any change in the Brake Pedal Position switch status, or with the vehicle running at Cruise speed, followed by a short deceleration periods, the PCM did not detect any change in the Brake Pedal Position switch status. **Possible Causes** • BPP switch circuit open, stop lamp switch circuit open/shorted • BPP/SLW switch is damaged or it is out of adjustment • BPP/SLW switch power circuit is open (test switch inline fuse) • One or more of the Module(s) that connect to the BPA or Stop Lamp switch circuits have a problem (e.g., Generic Electronic Module, ABS or Shift Lock Actuator or Vehicle Speed Control)
P0704 **1T CCM, MIL: No** Contour, Mystique, LS, Five Hundred, Freestyle, Montego,, Escort, Focus, E Van & F Series Truck, ZX2 & Windstar Models Transmissions: M/T	**Clutch Pedal Position Switch Circuit Malfunction Conditions:** Engine running in gear, followed by several gearshift changes, and the PCM did not detect any change in the clutch switch status. *Note: The CCP PID should change (5v to 0v) with clutch depressed.* **Possible Causes** • CPP switch signal circuit shorted to power • CPP switch ground (return) circuit is open • CPP switch is damaged or out of adjustment • PCM has failed
P0705 **2T CCM, MIL: Yes** All Models Transmissions: A/T	**DTR Sensor / TR Sensor Circuit Malfunction Conditions:** Key on or engine running; and the PCM detected that one or more of the Digital Transmission Range (DTR) or Transmission Range sensor (TR) signals (TR4, TR3, TR2 and TR1) were invalid (e.g., two TR or DR sensor signals received at the same time). **Possible Causes** • DTR or TR sensor connector is damaged or shorted • DTR or TR sensor signal circuit is open or shorted to ground • DTR or TR sensor signal circuit is shorted to VREF (5v) • DTR or TR sensor damaged • PCM has failed
P0707 **2T CCM, MIL: Yes** All Models Transmissions: A/T	**DTR Sensor / TR Sensor Circuit Low Input Conditions:** Key on or engine running; and the PCM detected the Digital Transmission Range (DTR) or Transmission Range sensor (TR) signal was less than the self-test minimum value in the test. **Possible Causes** • DTR or TR sensor connector is damaged or it is shorted • DTR or TR sensor signal circuit is shorted to sensor ground • DTR or TR sensor damaged • PCM has failed
P0708 **2T CCM, MIL: Yes** All Models Transmissions: A/T	**DTR Sensor or TR Sensor Circuit High Input Conditions:** Key on or engine running; and the PCM detected the Digital Transmission Range (DTR) or Transmission Range sensor (TR) input was more than the self-test maximum range in the test. **Possible Causes** • DTR or TR sensor connector is damaged or open • DTR or TR sensor signal circuit is open • DTR or TR sensor is shorted to VREF (5v) • DTR or TR sensor is damaged or the PCM has failed
P0711 **2T CCM, MIL: No** All Models Transmissions: A/T	**TFT Sensor Signal Range/Performance Conditions:** Engine started, KOER Self-Test enabled, engine running for over 10 minutes, and the PCM detected the Transmission Fluid Temperature (TFT) sensor value was not close its normal operating temperature. **Possible Causes** • ATF is low, contaminated, dirty or burnt • TFT sensor signal circuit has a high resistance condition • TFT sensor is out-of-calibration ("skewed") or it has failed • PCM has failed

Gas Engine OBD II Trouble Code List (P0xxx Codes)

DTC	Trouble Code Title, Conditions & Possible Causes
P0712 **2T CCM, MIL: No** All Models Transmissions: A/T	**TFT Sensor Circuit Low Input Conditions:** Key on or engine running; and the PCM detected the Transmission Fluid Temperature (TFT) sensor was less than its minimum self-test range (Scan Tool reads below -40ºF) in the test. **Possible Causes** • TFT sensor signal circuit is shorted to chassis ground • TFT sensor signal circuit is shorted to sensor ground • TFT sensor is damaged, or out-of-calibration, or has failed • PCM has failed
P0713 **2T CCM, MIL: No** All Models Transmissions: A/T	**TFT Sensor Circuit High Input Conditions:** Key on or engine running; and the PCM detected the Transmission Fluid Temperature (TFT) sensor was more than its maximum self-test range (Scan Tool reads over 315ºF) in the test. **Possible Causes** • TFT sensor signal circuit is open between the sensor and PCM • TFT sensor ground circuit is open between sensor and PCM • TFT sensor is damaged or has failed • PCM has failed
P0715 **2T CCM, MIL: No** All Models Transmissions: A/T	**Transmission Speed Shaft Sensor Circuit Malfunction Conditions:** Engine started, vehicle driven with the vehicle speed sensor indicating more than 1 mph, and the PCM detected the TSS signals were erratic, or that they were missing for a period of time. **Possible Causes** • TSS signal circuit is open • TSS signal is shorted to chassis ground • TSS signal is shorted to sensor ground • TSS assembly is damaged or it has failed • PCM has failed
P0717 **2T CCM, MIL: No** All Models Transmissions: A/T	**Transmission Speed Shaft Sensor Signal Intermittent Conditions:** Engine started, vehicle speed sensor indicating over 1 mph, and the PCM detected an intermittent loss of TSS signals (i.e., the TSS signals were erratic, irregular or missing). **Possible Causes** • TSS connector is damaged, loose or shorted • TSS signal circuit has an intermittent open condition • TSS signal circuit has an intermittent short to ground condition • TSS assembly is damaged or is has failed • PCM has failed
P0718 **2T CCM, MIL: No** All Models Transmissions: A/T	**Transmission Speed Shaft Sensor Signal Noisy Conditions:** Engine started, vehicle speed sensor signal over 1 mph, and the PCM detected the "noise" interference on the TSS signal circuit. **Possible Causes** • TSS signal is "noisy" due to RFI or EMI interference from sources such as ignition components or charging system • TSS signal wiring is damaged or contacting other signal wiring • PCM has failed
P0718 **2T CCM, MIL: Yes** All Models Transmissions: A/T	**A/T Output Shaft Speed Sensor Insufficient Input Conditions:** Engine started, VSS signal more than 1 mph, and the PCM detected the Output Shaft Speed signal did not correlate to the incoming signals received from the VSS or TCSS devices or related modules. **Possible Causes** • OSS sensor signal circuit is shorted to ground or • OSS sensor signal circuit is open • OSS sensor circuit is shorted to power • OSS sensor is damaged or it has failed • PCM has failed

Gas Engine OBD II Trouble Code List (P0xxx Codes)

DTC	Trouble Code Title, Conditions & Possible Causes
P0721 2T CCM, MIL: No All Models Transmissions: A/T	**A/T Output Shaft Speed Sensor Noise Interference Conditions:** Engine started, VSS signal more than 1 mph, and the PCM detected "noise" interference on the Output Shaft Speed (OSS) sensor circuit. **Possible Causes** • After market add-on devices interfering with the OSS signal • OSS connector is damaged, loose or shorted, or the wiring is misrouted or it is damaged • OSS assembly is damaged or it has failed • PCM has failed
P072 2T CCM, MIL: No All Models Transmissions: A/T	**A/T Output Speed Sensor No Signal Conditions:** Engine started, and the PCM did not detect any Output Shaft Speed (OSS) sensor signals upon initial vehicle movement. **Possible Causes** • After market add-on devices interfering with the OSS signal • OSS sensor wiring is misrouted or damaged, or the OSS sensor is damaged • PCM has failed
P0723 2T CCM, MIL: No All Models Transmissions: A/T	**A/T Output Speed Sensor Signal Intermittent Conditions:** Engine started, and the PCM detected the Output Shaft Speed (OSS) sensor signal was interrupted or irregular during testing. **Possible Causes** • OSS harness connector is damaged, loose or shorted, or the connector is not seated • OSS signal is open or it is shorted to ground (intermittent fault) • OSS assembly is damaged or it has failed
P0731 2T CCM, MIL: No Some Models Transmissions: A/T	**Incorrect First Gear Ratio Conditions:** Engine started, vehicle operating with 1st Gear commanded "on", and the PCM detected an incorrect 1st gear ratio during the test. **Possible Causes** • 1st Gear solenoid harness connector not properly seated • 1st Gear solenoid signal shorted to ground, or open • 1st Gear solenoid wiring harness connector is damaged • 1st Gear solenoid is damaged or not properly installed
P0731 2T CCM, MIL: No Escape, Contour, Probe Cougar, Mystique Transmissions: A/T	**Incorrect First Gear Ratio Conditions:** Engine started, vehicle operating with 1st Gear commanded "on", and the PCM detected an incorrect 1st gear ratio during the test. **Possible Causes** • 1st Gear solenoid harness connector not properly seated • 1st Gear solenoid signal shorted to ground, or open • 1st Gear solenoid wiring harness connector is damaged • 1st Gear solenoid is damaged or not properly installed • TSB 02-2-4 contains a repair procedure for this trouble code
P0732 2T CCM, MIL: No Some Models Transmissions: A/T	**Incorrect Second Gear Ratio Conditions:** Engine started, vehicle operating with 2nd Gear commanded "on", and the PCM detected an incorrect 2nd gear ratio during the test. **Possible Causes** • 2nd Gear solenoid harness connector not properly seated • 2nd Gear solenoid signal shorted to ground, or open • 2nd Gear solenoid wring harness connector is damaged • 2nd Gear solenoid is damaged or not properly installed
P0732 2T CCM, MIL: No Escape, Contour, Probe Cougar, Mystique Transmissions: A/T	**Incorrect Second Gear Ratio Conditions:** Engine started, vehicle operating with 2nd Gear commanded "on", and the PCM detected an incorrect 2nd gear ratio during the test. **Possible Causes** • 2nd Gear solenoid harness connector not properly seated • 2nd Gear solenoid signal shorted to ground, or open • 2nd Gear solenoid wring harness connector is damaged • 2nd Gear solenoid is damaged or not properly installed • TSB 02-2-4 contains a repair procedure for this trouble code

Gas Engine OBD II Trouble Code List (P0xxx Codes)

DTC	Trouble Code Title, Conditions & Possible Causes
P0733 **2T CCM, MIL: No** All Models Transmissions: A/T	**Incorrect Third Gear Ratio Conditions:** Engine started, vehicle operating with 3rd Gear commanded "on", and the PCM detected an incorrect 3rd gear ratio during the test. **Possible Causes** • 3rd Gear solenoid harness connector not properly seated • 3rd Gear solenoid signal shorted to ground, or open • 3rd Gear solenoid wiring harness connector is damaged • 3rd Gear solenoid is damaged or not properly installed
P0734 **2T CCM, MIL: No** Some Models Transmissions: A/T	**Incorrect Fourth Gear Ratio Conditions:** Engine started, vehicle operating with 4th Gear commanded "on", and the PCM detected an incorrect 4th gear ratio during the test. **Possible Causes** • 4th Gear solenoid harness connector not properly seated • 4th Gear solenoid signal shorted to ground, or open • 4th Gear solenoid wiring harness connector is damaged • 4th Gear solenoid is damaged or not properly installed
P0734 **2T CCM, MIL: No** Escape, Contour, Probe Cougar, Mystique Transmissions: A/T	**Incorrect Fourth Gear Ratio Conditions:** Engine started, vehicle operating with 4th Gear commanded "on", and the PCM detected an incorrect 4th gear ratio during the test. **Possible Causes** • 4th Gear solenoid harness connector not properly seated • 4th Gear solenoid signal shorted to ground, or open • 4th Gear solenoid wiring harness connector is damaged • 4th Gear solenoid is damaged or not properly installed • TSB 02-2-4 contains a repair procedure for this trouble code
P0735 **2T CCM, MIL: No** All Models Transmissions: A/T	**Incorrect Fifth Gear Ratio Conditions:** Engine started, vehicle operating with 5th Gear commanded "on", and the PCM detected an incorrect 5th gear ratio during the test. **Possible Causes** • 5th Gear solenoid harness connector not properly seated • 5th Gear solenoid signal shorted to ground, or open • 5th Gear solenoid wiring harness connector is damaged • 5th Gear solenoid is damaged or not properly installed
P0736 **2T CCM, MIL: No** All Models Transmissions: A/T	**Incorrect Reverse Gear Ratio Conditions:** Engine started, vehicle operating with Reverse Gear commanded "on", and the PCM detected an incorrect reverse gear ratio occurred. **Possible Causes** • Reverse Gear solenoid harness connector not properly seated • Reverse Gear solenoid signal shorted to ground, or open • Reverse Gear solenoid wiring harness connector is damaged • Reverse Gear solenoid is damaged or not properly installed
P0740 **2T CCM, MIL: No** All Models Transmissions: A/T	**TCC Solenoid Circuit Malfunction Conditions:** Engine started, KOER Self-Test enabled, vehicle driven at cruise speed, and the PCM did not detect any voltage drop across the TCC solenoid circuit during the test period. **Possible Causes** • TCC solenoid control circuit is open or shorted to ground • TCC solenoid wiring harness connector is damaged • TCC solenoid is damaged or has failed • PCM has failed
P0741 **2T CCM, MIL: No** All Models Transmissions: A/T	**TCC Mechanical System Range/Performance Conditions:** Engine started, vehicle driven in gear with VSS signals received, and the PCM detected excessive slippage while in normal operation. **Possible Causes** • TCC solenoid has a mechanical failure • TCC solenoid has a hydraulic failure • PCM has failed

Gas Engine OBD II Trouble Code List (P0xxx Codes)

DTC	Trouble Code Title, Conditions & Possible Causes
P0741 **2T CCM, MIL: No** Continental, Sable, Taurus, Windstar Transmissions: A/T	**TCC Mechanical System Range/Performance Conditions:** Engine started, vehicle driven in gear with VSS signals received, and the PCM detected excessive slippage while in normal operation. **Possible Causes** • TCC solenoid has a mechanical failure • TCC solenoid has a hydraulic failure • PCM has failed • TSB 3-12-3 contains a repair procedure for this trouble code
P0741 **2T CCM, MIL: No** Escape, Contour, Probe Cougar, Mystique Transmissions: A/T	**TCC Mechanical System Range/Performance Conditions:** Engine started, vehicle driven in gear with VSS signals received, and the PCM detected excessive slippage while in normal operation. **Possible Causes** • TCC solenoid has a mechanical failure • TCC solenoid has a hydraulic failure • PCM has failed • TSB 02-2-4 contains a repair procedure for this trouble code
P0743 **2T CCM, MIL: Yes** All Models Transmissions: A/T	**TCC Solenoid Circuit Malfunction Conditions:** Key on, KOEO Self-Test enabled and the PCM did not detect any voltage drop across the TCC solenoid circuit during the test period. **Possible Causes** • TCC solenoid control circuit is open • TCC solenoid control circuit is shorted to ground • TCC solenoid wiring harness connector is damaged • TCC solenoid is damaged or it has failed • PCM has failed
P0746 **2T CCM, MIL: No** All Models Transmissions: A/T	**A/T EPC Solenoid Circuit Malfunction Conditions:** Key on, KOEO Self-Test enabled and the PCM did not detect any voltage drop across the EPC solenoid circuit during the test period. **Possible Causes** • EPC solenoid control circuit is open • EPC solenoid control circuit is shorted to ground • EPC solenoid wiring harness connector is damaged • EPC solenoid is damaged or it has failed • PCM has failed
P0750 **2T CCM, MIL: Yes** All Models Transmissions: A/T	**A/T Shift Solenoid 1/A Circuit Malfunction Conditions:** Engine started, vehicle driven with the solenoid applied, and the PCM detected an unexpected voltage condition on the SS1/A solenoid circuit was incorrect during the test. **Possible Causes** • SS1/A solenoid control circuit is open • SS1/A solenoid control circuit is shorted to ground • SS1/A solenoid wiring harness connector is damaged • SS1/A solenoid is damaged or has failed • PCM has failed
P0751 **2T CCM, MIL: No** All Models Transmissions: A/T	**A/T Shift Solenoid 1/A Function Range/Performance Conditions:** Engine started, vehicle driven with the solenoid applied, and the PCM detected a mechanical failure while operating the Shift Solenoid 1/A during the CCM test period. **Possible Causes** • SS1/A solenoid is stuck in the "off" position • SS1/A solenoid has a mechanical failure • SS1/A solenoid has a hydraulic failure • PCM has failed

Gas Engine OBD II Trouble Code List (P0xxx Codes)

DTC	Trouble Code Title, Conditions & Possible Causes
P0752 **1T CCM, MIL: No** All Models Transmissions: A/T	**A/T Shift Solenoid 1/A Function Range/Performance Conditions:** Engine started, vehicle driven with the solenoid applied, and the PCM detected a mechanical failure while operating the Shift Solenoid 1/A during the CCM test period. **Possible Causes** • SS1/A solenoid is stuck in the "on" position • SS1/A solenoid has a mechanical failure • SS1/A solenoid has a hydraulic failure • PCM has failed
P0753 **1T CCM, MIL: Yes** All Models Transmissions: A/T	**A/T Shift Solenoid 1/A Circuit Malfunction Conditions:** Engine started, vehicle driven with the solenoid applied, and the PCM detected an unexpected voltage condition on the SS1/A solenoid circuit was incorrect during the test. **Possible Causes** • SS1/A solenoid control circuit is open • SS1/A solenoid control circuit is shorted to ground • SS1/A solenoid wiring harness connector is damaged • SS1/A solenoid is damaged or has failed • PCM has failed
P0755 **1T CCM, MIL: Yes** All Models Transmissions: A/T	**A/T Shift Solenoid 2/B Circuit Malfunction Conditions:** Engine started, vehicle driven with the solenoid applied, and the PCM detected an unexpected voltage condition on the SS2/B solenoid circuit was incorrect during the test. **Possible Causes** • SS2/B solenoid control circuit is open • SS2/B solenoid control circuit is shorted to ground • SS2/B solenoid wiring harness connector is damaged • SS2/B solenoid is damaged or has failed • PCM has failed
P0756 **1T CCM, MIL: Yes** All Models Transmissions: A/T	**A/T Shift Solenoid 2/B Function Range/Performance Conditions:** Engine started, vehicle driven with the solenoid applied, and the PCM detected a mechanical failure while operating the Shift Solenoid 2/B during the CCM test period. **Possible Causes** • SS2/B solenoid is stuck in the "on" position • SS2/B solenoid has a mechanical failure • SS2/B solenoid has a hydraulic failure • PCM has failed
P0757 **1T CCM, MIL: Yes** All Models Transmissions: A/T	**A/T Shift Solenoid 2/B Function Range/Performance Conditions:** Engine started, vehicle driven with the solenoid applied, and the PCM detected a mechanical failure while operating the Shift Solenoid 2/B during the CCM test period. **Possible Causes** • SS2/B solenoid is stuck in the "on" position • SS2/B solenoid has a mechanical failure • SS2/B solenoid has a hydraulic failure • PCM has failed
P0758 **1T CCM, MIL: Yes** All Models Transmissions: A/T	**A/T Shift Solenoid 2/B Circuit Malfunction Conditions:** Key on, KOEO Self-Test enabled, Shift Solenoid 2/B applied, and the PCM detected an unexpected voltage condition on the Shift Solenoid 2/B circuit during the CCM test period. **Possible Causes** • Shift Solenoid 2/B connector is damaged, open or shorted • Shift Solenoid 2/B control circuit is open • Shift Solenoid 2/B control circuit is shorted to ground • Shift Solenoid 2/B is damaged or it has failed • PCM has failed

Gas Engine OBD II Trouble Code List (P0xxx Codes)

DTC	Trouble Code Title, Conditions & Possible Causes
P0760 **1T CCM, MIL: Yes** All Models Transmissions: A/T	**A/T Shift Solenoid 3/C Circuit Malfunction Conditions:** Engine started, vehicle driven with Shift Solenoid 3/C applied, and the PCM detected an unexpected voltage condition on the Shift Solenoid 3/C circuit during the CCM test period. **Possible Causes** • Shift Solenoid 3/C connector is damaged, open or shorted • Shift Solenoid 3/C control circuit is open • Shift Solenoid 3/C control circuit is shorted to ground • Shift Solenoid 3/C is damaged or it has failed • PCM has failed
P0761 **1T CCM, MIL: No** All Models Transmissions: A/T	**A/T Shift Solenoid 3/C Function Range/Performance Conditions:** Engine started, vehicle driven with Shift Solenoid 3/C applied, and the PCM detected a mechanical failure occurred (stuck "off") while operating Shift Solenoid 3/C during the test. **Possible Causes** • SS3/C solenoid may be stuck "off" • SS3/C solenoid has a mechanical failure • SS3/C solenoid has a hydraulic failure • PCM has failed
P0762 **1T CCM, MIL: No** All Models Transmissions: A/T	**A/T Shift Solenoid 3/C Function Range/Performance Conditions:** Engine started, vehicle driven with Shift Solenoid 3/C applied, and the PCM detected a mechanical failure occurred (stuck "on") while operating Shift Solenoid 3/C during the test. **Possible Causes** • SS3/C solenoid may be stuck "on" • SS3/C solenoid has a mechanical failure • SS3/C solenoid has a hydraulic failure • PCM has failed
P0765 **1T CCM, MIL: Yes** All Models Transmissions: A/T	**A/T Shift Solenoid 4/D Circuit Malfunction Conditions:** Engine started, vehicle driven with Shift Solenoid 4/D applied, and the PCM detected an unexpected voltage condition on Shift Solenoid 4/D circuit during the CCM continuous test. **Possible Causes** • Shift Solenoid 4/D wiring harness or connector is damaged • Shift Solenoid 4/D control circuit is open or shorted to ground • Shift Solenoid 4/D is damaged or it has failed • PCM has failed
P0781 **1T CCM, MIL: No** All Models Transmissions: A/T	**A/T 1 to 2 Shift Error Conditions:** Engine started, vehicle driven in gear with VSS signals received, and the PCM detected the engine speed (rpm) did not decrease properly (i.e., an incorrect 1-2 gear ratio was detected during a shift event). **Possible Causes** • SS1/A solenoid may be stuck • SS1/A solenoid has a hydraulic problem • SS2/B solenoid may be stuck • SS2/B has a hydraulic problem • Transmission may have damaged friction material • Transmission has internal damage and needs replacement
P0782 **1T CCM, MIL: No** All Models Transmissions: A/T	**A/T 2 to 3 Shift Error Conditions:** Engine started, vehicle driven in gear with VSS signals received, and the PCM detected the engine speed (rpm) did not decrease properly (i.e., an incorrect 2-3 gear ratio was detected during a shift event). **Possible Causes** • SS1/A solenoid may be stuck • SS1/A solenoid has a hydraulic problem • SS2/B solenoid may be stuck • SS2/B has a hydraulic problem • Transmission may have damaged friction material • Transmission has internal damage and needs replacement

Gas Engine OBD II Trouble Code List (P0xxx Codes)

DTC	Trouble Code Title, Conditions & Possible Causes
P0783 **1T CCM, MIL: No** All Models Transmissions: A/T	**A/T 3 to 4 Shift Error Conditions:** Engine started, vehicle driven in gear with VSS signals received, and the PCM detected the engine speed (rpm) did not change properly (i.e., an incorrect 3-4 gear ratio was detected during the shift event). **Possible Causes** • SS1/A solenoid may be stuck, or a hydraulic failure exists • SS2/B solenoid may be stuck, or a hydraulic failure exists • Transmission may have damaged friction material
P0784 **1T CCM, MIL: No** All Models Transmissions: A/T	**A/T 4 to 5 Shift Error Conditions:** Engine started, vehicle driven in gear with VSS signals received, and the PCM detected the engine speed (rpm) did not change properly (i.e., an incorrect 4-5 gear ratio was detected during a shift event). **Possible Causes** • SS2/B solenoid may be stuck, or a hydraulic failure exists • SS3/C solenoid may be stuck, or a hydraulic failure exists • Transmission may have damaged friction material
P0812 **1T CCM, MIL: No** All Models Transmissions: A/T	**A/T Reverse Switch Circuit Malfunction Conditions:** Key on, engine off, KOEO Self Test enabled, and the PCM detected the reverse switch signal did not change as the selector was shifted in or out of reverse gear. Note: The RS PID should change from ON to OFF while shifting. **Possible Causes** • Transmission shift not indicating neutral during the self-test • RS switch circuit shorted to VREF or VPWR • RS switch circuit is open or shorted to ground (signal return) • Reverse switch is damaged • PCM has failed
P0813 **1T CCM, MIL: Yes** All Models Transmissions: A/T	**Transmission Control System Malfunction Conditions:** Engine started, vehicle speed more than 1 in gear, and the PCM detected a problem in the Transmission Control System operation. **Possible Causes** • Refer to the information in the Transmission Section of the appropriate Workshop Repair manual (i.e., the information for the particular vehicle that set this trouble code).
P0815 **1T CCM, MIL: Yes** All Models Transmissions: A/T	**Transmission Control System Malfunction Conditions:** Key on, engine off, KOEO Self Test enabled, and the PCM detected the reverse switch input did not change as the selector was shifted in or out of reverse (i.e., it was high when it should have been low). Note: The RS PID should change from ON to OFF while shifting. **Possible Causes** • Refer to the information in the Transmission Section of the appropriate Workshop Repair manual (i.e., the information for the particular vehicle that set this trouble code).

Gas Engine OBD II Trouble Code List (P1xxx Codes)

DTC	Trouble Code Title, Conditions & Possible Causes
P1000 **1T PCM, MIL: No** All Models Transmissions: All	**OBD II Monitor Testing Not Complete Conditions:** Key on or engine running; and the PCM detected one the conditions shown under Possible Causes (i.e., this code cannot be cleared manually - it must clear itself after all of the OBD II Monitors complete). Note: This code must be cleared to pass an Inspection/Maintenance Test required to register a vehicle in certain states. **Possible Causes** • Battery keep alive power (KAPWR) was removed to the PCM • One or more OBD II Monitors did not complete during an official OBD II Drive Cycle • PCM Reset step was performed with an OBD II Scan Tool
P1001 **1T CCM, MIL: No** All Models Transmissions: All	**KOER Self-Test Not Completed, KOER Test Aborted Conditions:** Key on, engine running self-test not completed during the normal allowable time period. **Possible Causes** • Engine speed (rpm) out of specification during the KOER test • Incorrect Self-Test Procedure • Scan Tool has a communication problem • Unexpected response from Self-Test monitors
P1100 **2T CCM, MIL: Yes** All Models Transmissions: All	**MAF Sensor Signal Intermittent Conditions:** Engine started, engine running at idle or cruise speed, and the PCM detected the MAF sensor signal above or below the calibrated limit. **Possible Causes** • MAF sensor continuity problems at the connector • MAF sensor continuity through the wiring harness • MAF sensor circuit intermittent open inside the sensor • PCM has failed
P1101 **2T CCM, MIL: Yes** All Models Transmissions: All	**MAF Sensor Out Of Self-Test Range Conditions:** Key on and engine off, and the PCM detected the MAF sensor was more than 0.27v, or with the engine running, the MAF sensor voltage was not within a normal range of 0.46v to 2.44v. **Possible Causes** • Low battery charge • MAF sensor partially connected, or the sensor is contaminated • MAF sensor power ground circuit or sensor signal (return) open • MAF sensor is damaged or it has failed • PCM has failed
P1112 **2T CCM, MIL: Yes** All Models Transmissions: All	**IAT Sensor Circuit Intermittent Conditions:** Engine started, and the PCM detected an intermittent condition in the IAT sensor signal during the self-test. Note: Select the IAT PID and monitor the signal for sudden changes. **Possible Causes** • IAT sensor wiring harness is damaged (wire may be open) • IAT sensor harness connector is damaged • IAT sensor is damaged or the PCM has failed
P1114 **2T CCM, MIL: Yes** All Models Transmissions: All	**IAT Sensor Circuit Low Input Conditions:** Engine started, and the PCM detected the IAT sensor signal was less than the self-test minimum of 0.20v (equivalent to 250°F). Monitor the IAT PID for very low signal. **Possible Causes** • IAT sensor wiring harness is damaged (wire may be grounded) • IAT sensor harness connector is damaged (may be grounded) • IAT sensor is damaged or the PCM has failed
P1115 **2T CCM, MIL: Yes** All Models Transmissions: All	**IAT Sensor 2 Circuit High Input Conditions:** Engine started, and the PCM detected the IAT Sensor 2 signal was more than the self-test maximum of 4.60v (equivalent to 250°F). Monitor the IAT PID for very high signal. **Possible Causes** • IAT sensor wiring harness or harness connector is damaged (wire may be open) • IAT sensor signal circuit is open, or the ground circuit is open • IAT sensor is damaged or has failed • PCM has failed

Gas Engine OBD II Trouble Code List (P1xxx Codes)

DTC	Trouble Code Title, Conditions & Possible Causes
P1116 **1T CCM, MIL: Yes** All Models Transmissions: All	**CHT or ECT Sensor Out Of Self-Test Range Conditions:** Key on, KOEO Self-Test enabled, and the PCM detected the ECT sensor was more than the expected range (50ºF), or engine running, KOER Self-Test enabled, and the PCM detected the ECT senor signal was less than 180ºF during the self test period. The ECT PID must be above 50ºF in the KOEO test or above 180ºF in the KOER self-test to pass these parameters. **Possible Causes** • ECT sensor harness connector is damaged, loose or shorted • ECT sensor is damaged • KOER or KOER Self-Test performed with the engine "too cold"
P1117 **2T CCM, MIL: Yes** All Models Transmissions: All	**CHT or ECT Sensor Signal Intermittent Conditions:** Engine started, and the PCM detected an intermittent loss of the CHT or ECT sensor signal (it may have an open circuit condition). Note: Select the CHT or IAT PID and monitor the signal for sudden changes while wiggling the CHT or IAT sensor connector. On the 5.4L V8, if the temperature exceeds 258ºF, the PCM disables four fuel injectors at a time. It alternates which four fuel injectors are disabled every 32-engine cycles. The cylinders that are disabled do not inject fuel, so they act as air pumps to aid in cooling the engine. If the temperature exceeds 310ºF, the PCM disables all of the fuel injectors until the engine temperature drops below 310ºF. **Possible Causes** • ECT sensor harness connector is damaged, loose or shorted • ECT sensor is damaged or it has failed • Engine overheating condition present • Thermostat is faulty, or engine coolant level is low
P1120 **2T CCM, MIL: Yes** All Models Transmissions: All	**TP Sensor Signal Out-of-Range Low Conditions:** Key on or engine running; and the PCM detected the TP sensor signal was between 0.17-0.49v (3.42-9.85%) with the signal within the calibrated self-test range. **Possible Causes** • ECT sensor harness connector is damaged • ECT sensor is damaged • Engine coolant level is low • PCM has failed
P1121 **2T CCM, MIL: Yes** All Models Transmissions: All	**TP Sensor Inconsistent With MAF Sensor Conditions:** Engine started; and the PCM detected the MAF and TP sensor signals were not consistent the calibrated values expected for these two sensors during the self-test. Note: Drive the vehicle and monitor the TP PID in all gears. A TP PID of less than 0.24v (4.82%) with a LOAD PID over 55%, or a TP PID over 2.44v (49.05%) with a LOAD PID under 30% will set this code. **Possible Causes** • Air leak exists between MAF sensor and the throttle body • MAF sensor is damaged or it has failed • TP sensor is not seated properly • TP sensor is damaged
P1124 **1T CCM, MIL: Yes** All Models Transmissions: All	**TP Sensor Out of Self-Test Range Conditions:** Key on, KOEO Self-Test enabled, and the PCM detected the TP sensor signal was less than 0.66v (13.27%), or with the engine running, KOER Self-Test enabled, the PCM detected the TP sensor signal was approximately 1.17v (23.52%). Note: A TP V PID less than 4.82 % (0.24 volt) with a LOAD PID more than 55%; or the TP V PID more than 49.05% (2.44 volts) with a LOAD PID less than 30% indicates a hard fault is present. **Possible Causes** • Throttle linkage is binding, or TP sensor is not seated properly • Throttle plate below closed throttle position • Throttle plate screw is misadjusted • TP sensor is damaged or it has failed • PCM has failed

Gas Engine OBD II Trouble Code List (P1xxx Codes)

DTC	Trouble Code Title, Conditions & Possible Causes
P1125 **2T CCM, MIL: Yes** All Models Transmissions: All	**TP Sensor Circuit Malfunction (Intermittent) Conditions:** Engine started, and the PCM detected the TP sensor rotational angle changed beyond the minimum or maximum calibrated limit. Note: Monitor the TP V PID, and tap lightly on the TP sensor housing and wiggle the wiring harness. Watch for the value to suddenly go below 0.49v or over 4.65v. **Possible Causes** • TP sensor wiring harness or connector has an intermittent open • TP sensor has an intermittent open or shorted condition
P1127 **2T CCM, MIL: Yes** All Models Transmissions: All	**Exhaust Not Warm, Downstream Sensor Not Tested Conditions:** Engine started, KOER Self-Test enabled, and the PCM detected the inferred exhaust temperature was less than a minimum value. Note: Monitor the HO2S Heater PID to determine their ON/OFF status (the heaters must work properly in order to pass this test). **Possible Causes** • Engine not operating long enough prior to the KOER Self-Test • Exhaust system temperature too cold to run the self-test
P1128 **2T CCM, MIL: Yes** All Models Transmissions: All	**Upstream Oxygen Sensors Swapped From Bank-to-Bank Conditions:** Engine started, KOER Self-Test enabled, and the PCM detected the HO2S signal response to a related fuel shift did not correspond to the correct engine cylinder bank (e.g., the HO2S-11 and the HO2S-21 wires were crossed) during the test period. **Possible Causes** • Upstream HO2S-11, HO2S-21 wiring crossed at the connector • Upstream HO2S-11, HO2S-21 crossed in the wiring harness • Upstream HO2S-11, HO2S-21 crossed at PCM pin connector
P1129 **2T CCM, MIL: Yes** All Models Transmissions: All	**Downstream Oxygen Sensors Swapped From Bank-to-Bank Conditions:** Engine started, KOER Self-Test enabled, and the PCM detected the HO2S signal response to a related fuel shift did not correspond to the correct engine cylinder bank (e.g., the HO2S-12 and HO2S-22 wires were crossed) during the test period. **Possible Causes** • Upstream HO2S-12, HO2S-21 wiring crossed at the connector • Upstream HO2S-12, HO2S-21 crossed in the wiring harness • Upstream HO2S-12, HO2S-21 crossed at PCM pin connector
P1130 **2T O2S, MIL: Yes** All Models Transmissions: All	**Lack of HO2S-11 Switching, Fuel Trim at Rich/Lean Limit Conditions:** DTC P0300-P0310 not set, engine running in closed loop, and the PCM detected the HO2S circuit was too lean or too rich, or that it could no longer change Fuel Trim because it was at its rich limit or its lean limit. **Possible Causes** • Air intake system leaking, vacuum hoses leaking or damaged • Air leaks located after the MAF sensor mounting location • EGR valve sticking, EGR diaphragm leaking, or gasket leaking • EVAP vapor recovery system has failed • Excessive fuel pressure, leaking or contaminated fuel injectors • Exhaust leaks before or near the HO2S(s) mounting location • Fuel pressure regulator is leaking or damaged • HO2S circuits wet or oily, corroded, or poor terminal contact • HO2S is damaged or it has failed • HO2S signal circuit open, shorted to ground, shorted to power • Low fuel pressure or vehicle driven until it was out of fuel • Oil dipstick not seated or engine oil level too high (overfilled)

Gas Engine OBD II Trouble Code List (P1xxx Codes)

DTC	Trouble Code Title, Conditions & Possible Causes
P1131 **2T O2S, MIL: Yes** All Models Transmissions: All	**Lack of HO2S-11 Switching, HO2S Signal Low Input Conditions:** DTC P0300-P0310 not set, engine started, engine running in closed loop, and the PCM detected the HO2S-11 was not switching (i.e., the HO2S-11 indicated a lean A/F mixture). **Possible Causes** • Air intake system leaking, vacuum hoses leaking or damaged • Air leaks located after the MAF sensor mounting location • Base engine mechanical fault (i.e., compression, valve timing) • HO2S circuits wet or oily, corroded, or poor terminal contact • HO2S signal circuit open, shorted to ground, shorted to power, or the sensor has failed • Low fuel pressure or vehicle driven until it was out of fuel • Possible air leaks at the PCV valve or at the related hoses
P1132 **2T O2S, MIL: Yes** All Models Transmissions: All	**Lack of HO2S-11 Switching, HO2S Signal High Input Conditions:** DTC P0300-P0310 not set, engine started, engine running in closed loop, and the PCM detected the HO2S-11 was not switching (i.e., the HO2S-11 indicated a rich A/F mixture). **Possible Causes** • Check air cleaner element and air cleaner housing for blockage • EVAP vapor recovery system has failed (canister full of fuel) • Fuel pressure too high, contaminated or leaking fuel injectors • HO2S is fuel contaminated, or coated with silicone or moisture
P1137 **2T O2S, MIL: Yes** All Models Transmissions: All	**Lack of HO2S-12 Switching, HO2S Signal Low Input Conditions:** DTC P0300-P0310 not set, engine started, engine running in closed loop, and the PCM detected the HO2S-12 was not switching (i.e., the HO2S-12 indicated a lean A/F mixture). **Possible Causes** • Air intake system leaking, vacuum hoses leaking or damaged • Air leaks located after the MAF sensor mounting location • Base engine mechanical fault (i.e., compression, valve timing) • HO2S circuits wet or oily, corroded, or poor terminal contact • HO2S is damaged or it has failed • HO2S signal circuit open, shorted to ground, shorted to power • Low fuel pressure or vehicle driven until it was out of fuel • Possible air leaks at the PCV valve or at the related hoses
P1138 **2T O2S, MIL: Yes** All Models Transmissions: All	**Lack of HO2S-12 Switching, HO2S Signal High Input Conditions:** DTC P0300-P0310 not set, engine started, engine running in closed loop, and the PCM detected the HO2S-12 was not switching (i.e., the HO2S-12 indicated a rich A/F mixture). **Possible Causes** • Check air cleaner element and air cleaner housing for blockage • EVAP vapor recovery system has failed (canister full of fuel) • Fuel pressure too high, contaminated or leaking fuel injectors • HO2S is fuel contaminated, or coated with silicone or moisture
P1150 **2T O2S, MIL: Yes** All Models Transmissions: All	**Lack of HO2S-21 Switching, Fuel Trim At Rich/Lean Limit Conditions:** DTC P0300-P0310 not set, engine running in closed loop, and the PCM detected the HO2S circuit was too lean or too rich, or that it could no longer correct Fuel Trim (i.e., the Fuel Trim was at its calibrated rich limit or its calibrated lean limit). **Possible Causes** • Air intake system leaking, vacuum hoses leaking or damaged • Air leaks located after the MAF sensor mounting location • EGR valve sticking, EGR diaphragm leaking, or gasket leaking • EVAP vapor recovery system has failed • Excessive fuel pressure, leaking or contaminated fuel injectors • Exhaust leaks before or near the HO2S(s) mounting location • Fuel pressure regulator is leaking or damaged • HO2S circuits wet or oily, corroded, or poor terminal contact • HO2S signal circuit open, shorted to ground, shorted to power, or the sensor has failed • Low fuel pressure or vehicle driven until it was out of fuel • Oil dipstick not seated or engine oil level too high (overfilled)

Gas Engine OBD II Trouble Code List (P1xxx Codes)

DTC	Trouble Code Title, Conditions & Possible Causes
P1151 **2T O2S, MIL: Yes** All Models Transmissions: All	**Lack of HO2S-21 Switching, HO2S Signal Low Input Conditions:** DTC P0300-P0310 not set, engine started, engine running in closed loop, and the PCM detected the HO2S-21 was not switching (i.e., the HO2S-21 indicated a lean A/F mixture). **Possible Causes** • Air intake system leaking, vacuum hoses leaking or damaged • Air leaks located after the MAF sensor mounting location or in the PCV system • Base engine mechanical fault (i.e., compression, valve timing) • HO2S circuits wet or oily, corroded, or poor terminal contact • HO2S signal circuit open, shorted to ground, shorted to power, or the sensor has failed • Low fuel pressure or vehicle driven until it was out of fuel
P1152 **2T O2S, MIL: Yes** All Models Transmissions: All	**Lack of HO2S-21 Switching, HO2S Signal High Input Conditions:** DTC P0300-P0310 not set, engine started, engine running in closed loop, and the PCM detected the HO2S-21 was not switching (i.e., the HO2S-21 indicated a rich A/F mixture). **Possible Causes** • Check air cleaner element and air cleaner housing for blockage • EVAP vapor recovery system has failed (canister full of fuel) • Fuel pressure too high, contaminated or leaking fuel injectors • HO2S is fuel contaminated, or coated with silicone or moisture
P1157 **2T O2S, MIL: Yes** All Models Transmissions: All	**Lack of HO2S-22 Switching, HO2S Signal Low Input Conditions:** DTC P0300-P0310 not set, engine started, engine running in closed loop, and the PCM detected the HO2S-22 was not switching (i.e., the HO2S-22 indicated a lean A/F mixture). **Possible Causes** • Air intake system leaking, vacuum hoses leaking or damaged • Air leaks located after the MAF sensor mounting location • Base engine mechanical fault (i.e., compression, valve timing) • HO2S circuits wet or oily, corroded, or poor terminal contact • HO2S is damaged or it has failed • HO2S signal circuit open, shorted to ground, shorted to power • Low fuel pressure or vehicle driven until it was out of fuel • Possible air leaks at the PCV valve or at the related hoses
P1158 **2T O2S, MIL: Yes** All Models Transmissions: All	**Lack of HO2S-22 Switching, HO2S Signal High Input Conditions:** DTC P0300-P0310 not set, engine started, engine running in closed loop, and the PCM detected the HO2S-22 was not switching (i.e., the HO2S-22 indicated a rich A/F mixture). **Possible Causes** • Check air cleaner element and air cleaner housing for blockage • EVAP vapor recovery system has failed (canister full of fuel) • Fuel pressure too high, contaminated or leaking fuel injectors • HO2S is fuel contaminated, or coated with silicone or moisture
P1168 **2T CCM, MIL: Yes** Crown Victoria, E Van & F Series Truck with a 4.6L VIN 9 or 5.4L VIN M CNG engine Transmissions: All	**Fuel Rail Pressure Sensor In Range But Low Conditions:** DTC P0230, P0231 and P0232 not set, Engine started; and the PCM detected the FRP sensor signal was less than the normal self-test range for these operating conditions. Note: A FRP PID value below 551 kPa (80 psi) indicates a failure. **Possible Causes** • Low fuel level or the vehicle is out of fuel • Low fuel pressure • FRP sensor is damaged • FRP sensor signal circuit has a high resistance condition
P1169 **2T CCM, MIL: Yes** Crown Victoria, E Van & F Series Truck with a 4.6L VIN 9 or 5.4L VIN M CNG engine Transmissions: All	**Fuel Rail Pressure Sensor In Range But High Conditions:** DTC P0230, P0231 and P0232 not set, Engine started; and the PCM detected the FRP sensor signal was more than the normal self-test range for these operating conditions. Note: A FRP PID value above 896 kPa (130 psi) indicates a failure. **Possible Causes** • Low fuel level, no fuel or low fuel pressure • FRP sensor is damaged • FRP sensor signal circuit has a high resistance condition

Gas Engine OBD II Trouble Code List (P1xxx Codes)

DTC	Trouble Code Title, Conditions & Possible Causes
P1180 **2T CCM, MIL: Yes** Crown Victoria, E Van & F Series Truck with a 4.6L VIN 9 or 5.4L VIN M CNG engine Transmissions: All	**Fuel Delivery System Low Conditions:** Engine started, and the PCM detected the FTP sensor signal from the NG module data indicated an inferred fuel deliver pressure that was less than a minimum calibrated value. **Possible Causes** • Fuel line is restricted • Fuel filter is plugged or restricted
P1181 **2T CCM, MIL: Yes** Crown Victoria, E Van & F Series Truck with a 4.6L VIN 9 or 5.4L VIN M CNG engine Transmissions: All	**Fuel Delivery System High Conditions:** Engine started, and the PCM detected the FTP sensor signal from the NG module data indicated an inferred fuel deliver pressure that was higher than a maximum calibrated value. **Possible Causes** • Fuel pressure regulator is damaged or has failed
P1183 **2T CCM, MIL: Yes** All Models Transmissions: All	**Engine Oil Temperature Sensor Circuit Malfunction Conditions:** Engine started, and the PCM detected the engine oil temperature (EOT) sensor circuit was open or shorted to ground (i.e., this fault is usually caused by an interruption of the signal - intermittent fault). **Possible Causes** • EOT sensor circuit is open or shorted to ground • EOT sensor has failed • PCM has failed
P1184 **2T CCM, MIL: Yes** All Models Transmissions: All	**Engine Oil Temperature Sensor Out Of Self-Test Range Conditions:** Engine started, and the PCM detected the engine oil temperature (EOT) sensor circuit was open or shorted to ground (i.e., this fault can be caused by an intermittent loss of this signal). **Possible Causes** • EOT sensor circuit is open or shorted to ground (intermittent) • EOT sensor is corroded, damaged or it has failed • PCM has failed
P1220 **1T CCM, MIL: Yes** Continental with 4.6L DOHC VIN V engine Transmissions: All	**Series Throttle Control System Malfunction Conditions:** Engine started and the PCM detected a malfunction in the Series Throttle Control system. **Possible Causes** • Series Throttle Stepper Motor circuit open, ground or shorted • STC module VPWR circuit open (no power) • STC module ground or power circuit open • TP-B sensor ground (return) circuit is open • TAPW circuit open, shorted to ground, or shorted to VPWR • Series Throttle (ST), Stepper Motor or TP-B sensor damaged • STC Module has failed
P1224 **2T CCM, MIL: Yes** Continental with 4.6L DOHC VIN V engine Transmissions: All	**TP Sensor 'B' Out Of Self-Test Range Conditions:** Key on or engine running; no other Traction Control codes set, and the PCM detected the TP-B signal was out-of-range during the Continuous self test. **Possible Causes** • TP-B sensor binding/sticking, throttle stop screws misadjusted • TP-B sensor VREF circuit out of range • TP-B sensor is damaged • Series Throttle is damaged

Gas Engine OBD II Trouble Code List (P1xxx Codes)

DTC	Trouble Code Title, Conditions & Possible Causes
P1229 **2T CCM, MIL: Yes** F Series Truck with a 5.4L VIN 3 SC engine Transmissions: All	**Supercharger Intercooler Pump Not Operating Conditions:** Engine running at Cruise speed, then after the PCM commanded the Intercooler pump (ICP) to operate it did not detect any current flow in the ICP circuit (with the intercooler pump energized). Note: The Scan Tool ICP PID should change from ON to OFF. **Possible Causes** • Pump motor windings are open, or the pump relay coil is open • The circuit between the relay and the pump is open, or the circuit between the relay and the PCM is open • Pump motor is shorted, or the motor ground connection is open • PCM has failed
P1230 **1 CCM, MIL: Yes** Mark VIII Transmissions: All	**Low Speed Fuel Pump Malfunction Conditions:** Key on, Low Speed fuel pump energized; and the PCM detected an open condition in the Power-to-Pump circuit between the VLCM and the FPM splice to this fuel pump circuit. **Possible Causes** • An open condition exists in the Power-To-Pump circuit between the VLCM and the FPM splice to the circuit.
P1231 **1T CCM, MIL: No** All Models Transmissions: All	**Fuel Pump Secondary Low, High Speed Pump On Conditions:** Key on, KOEO Self-Test enabled; High Speed Fuel Pump (HFP) relay energized, fuel pump driver in VLCM off (to VLCM Pin 7) off, the PCM detected voltage on the FPM circuit. **Possible Causes** • HFP relay circuit to battery power (B+) is open • HFP relay is damaged or it has failed • Power-To-Pump circuit between HFP relay and splice is open
P1232 **2T CCM, MIL: Yes** All Models Transmissions: All	**Low Speed Fuel Pump Primary Circuit Malfunction Conditions:** Engine started, Low Speed Fuel Pump (LFP) relay energized, the PCM detected excessive current on the LFP circuit; or with LFP commanded off it detected power on the LFP circuit. **Possible Causes** • Low fuel pump (LFP) circuit open or shorted • Low speed fuel pump relay VPWR circuit open • Low speed fuel pump relay is damaged • PCM has failed
P1233 **2T CCM, MIL: Yes** All Models Transmissions: All	**Fuel System Disabled Or Offline Conditions:** Key on or engine running; and the PCM did not receive any diagnostic information (via duty cycle signals) from the FPDM. **Possible Causes** • Inertia fuel shutoff (IFS) switch needs to be reset • FPDM ground circuit is open, or FPM circuit is open or shorted • Mark VIII: FPDM PWR circuit is open, the FPDM power supply relay VPWR circuit is opened or grounded, or power relay failed • Escort/Tracer: FPDM PWR circuit is open, or the CCRM pin 11 is open to power (B+), or the CCRM (relay) is damaged • Continental: FPDM circuit to VPWR is open, or the FPDM or the IFS is damaged • Refer to the GEM or REM controllers for related trouble codes
P1234 **1T CCM, MIL: Yes** All Models Transmissions: All	**Fuel System Disabled Or Offline Conditions:** Key on, and the PCM did not receive any diagnostic information from the FPDM. **Possible Causes** • Inertia fuel shutoff (IFS) switch needs to be reset or has failed • FPDM ground circuit is open, or FPM circuit is open or shorted • Mark VIII: FPDM PWR circuit is open, the FPDM power supply relay VPWR circuit is opened or grounded, or power relay failed • Escort/Tracer: FPDM PWR circuit is open, or the CCRM pin 11 is open to power (B+), or the CCRM (relay) is damaged • Continental: FPDM circuit to VPWR is open, or the FPDM or the IFS is damaged • LS6, LS8 Models: This code indicates the PCM is not receiving data about the fuel level on the SCP data line from the Rear Electronics Module (REM). Test the REM first! • PCM has failed

Gas Engine OBD II Trouble Code List (P1xxx Codes)

DTC	Trouble Code Title, Conditions & Possible Causes
P1235 **2T CCM, MIL: Yes** Some Models Transmissions: All	**Fuel Pump Control Out Of Range Conditions:** Key on or engine running; and the PCM received a signal from the FPM over the SCP bus that the FPDM had received an invalid or missing fuel pump command from the PCM. **Possible Causes** • FP circuit is open or shorted • FPDM is damaged • PCM has failed
P1235 **2T CCM, MIL: Yes** LS Models Transmissions: All	**Fuel Pump Control Out Of Range Conditions:** Key on or engine running; and the PCM received a signal from the FPM over the SCP bus that the FPDM had received an invalid or missing fuel pump command from the PCM. Note that the FPDM commands to the Rear Electronics Module (REM) are sent over the SCP bus communication circuits. **Possible Causes** • FP circuit is open or it is shorted • FPDM is damaged • REM is damaged or it has failed • PCM has failed
P1236 **2T CCM, MIL: No** Some Models Transmissions: All	**Fuel Pump Control Out Of Range Conditions:** Key on or engine running; and the PCM received a signal (from the FPM over the SCP bus) that the FPDM had received an invalid or missing fuel pump command from the PCM. **Possible Causes** • FP circuit is open or it is shorted • FPDM is damaged or the PCM has failed
P1236 **1T CCM, MIL: No** LS Models Transmissions: All	**Fuel Pump Control Out Of Range Conditions:** Key on or engine running; and the PCM received a signal (from the FPM over the SCP bus) that the FPDM had received an invalid or missing fuel pump command from the PCM. Note that the FPDM commands to the Rear Electronics Module (REM) are sent over the SCP bus. **Possible Causes** • FP circuit is open or it is shorted • FPDM is damaged or the PCM has failed
P1237 **2T CCM, MIL: Yes** Some Models Transmissions: All	**Fuel Pump Secondary Circuit Malfunction Conditions:** Key on or engine running; and the PCM received a signal from the FPDM that it had detected a fault in the fuel pump secondary circuit. **Possible Causes** • FP PWR circuit is open or shorted • FPDM fuel pump return circuit is open • Fuel pump windings are open or shorted, or the rotor is locked • FPDM is damaged
P1237 **2T CCM, MIL: Yes** LS Models Transmissions: All	**Fuel Pump Secondary Circuit Malfunction Conditions:** Key on or engine running; and the PCM received a signal from the FPDM that it had detected a fault in the fuel pump secondary circuit. Note that the FPDM commands to the Rear Electronics Module (REM) are sent over the SCP bus communication circuits. **Possible Causes** • FP PWR circuit is open or shorted • FPDM fuel pump return circuit is open • Fuel pump windings are open or shorted, or the rotor is locked • FPDM is damaged
P1238 **2T CCM, MIL: Yes** Some Models Transmissions: All	**Fuel Pump Secondary Circuit Malfunction Conditions:** Key on or engine running; and the PCM received a signal from the FPDM that it had detected a fault in the fuel pump secondary circuit. **Possible Causes** • FP PWR circuit is open or shorted • FPDM fuel pump return circuit is open • Fuel pump windings are open or shorted • Fuel pump rotor is locked • FPDM is damaged

Gas Engine OBD II Trouble Code List (P1xxx Codes)

DTC	Trouble Code Title, Conditions & Possible Causes
P1238 **1T CCM, MIL: Yes** LS Models Transmissions: All	**Fuel Pump Secondary Circuit Malfunction Conditions:** Key on or engine running; and the PCM received a signal from the FPDM that it had detected a fault in the fuel pump secondary circuit. Note that the FPDM commands to the Rear Electronics Module (REM) are sent over the SCP bus communication circuits. **Possible Causes** • FP PWR circuit is open or shorted • FPDM fuel pump return circuit is open • Fuel pump windings are open or shorted • Fuel pump rotor is locked • FPDM is damaged
P1244 **2T CCM, MIL: Yes** Some Models Transmissions: All	**Generator Load Circuit Low Input Conditions:** Engine started, and the PCM detected the GLI signal was less than the calibrated limit for a calibrated amount of time. **Possible Causes** • GLI circuit is open or shorted • Voltage regulator/generator is damaged • PCM has failed
P1245 **2T CCM, MIL: Yes** Some Models Transmissions: All	**Generator Load circuit High Input Conditions:** Engine started, and the PCM detected the GLI signal was more than the calibrated limit for a calibrated amount of time. **Possible Causes** • GLI circuit is open or shorted • Voltage regulator/generator is damaged • PCM has failed
P1246 **2T CCM, MIL: Yes** Some Models Transmissions: All	**Generator Load Circuit Malfunction Conditions:** Engine started, and the PCM detected the GLI was more than or less than a calibrated amount for too long a period of time. **Possible Causes** • Generator circuit is open, shorted to ground or shorted to power • Generator drive mechanism has failed • Generator/regulator assembly is damaged or the PCM has failed
P1246 **2T CCM, MIL: Yes** Crown Victoria, Grand Marquis & Town Car Transmissions: All	**Generator Load Circuit Malfunction Conditions:** Engine started, and the PCM detected the GLI was more than or less than a calibrated amount for too long a period of time. **Possible Causes** • Generator circuit is open, shorted to ground or shorted to power • Generator drive mechanism has failed • Generator/regulator assembly is damaged or the PCM has failed • TSB 03-14-2 contains repair information for this trouble code
P1250 **2T CCM, MIL: Yes** Probe Models with a 2.0L I4 VIN A, 2.5L V6 VIN B engine Transmissions: All	**Fuel Pressure Regulator Control Circuit Malfunction Conditions:** KOEO or KOER Self-Test enabled, and the PCM detected a lack of power (VPWR) to the Fuel Pressure Regulator Control (FPRC) solenoid circuit. **Possible Causes** • FPRC solenoid valve harness circuits are open or shorted • FPRC input port or output port vacuum lines are damaged • FRPC solenoid is damaged • PCM has failed
P1260 **2T PCM, MIL: Yes** All Models Transmissions: All	**Theft Detected, Vehicle Immobilized Conditions:** Key on, and the PCM received a signal from the Anti-Theft System that a theft condition had occurred. The theft indicator on the dash will flash rapidly or remain on "solid" with the ignition switch in the "on" position. The engine may "start and stall", or may not crank if the vehicle is equipped with the PATS starter disable feature. **Possible Causes** • A Previous theft condition has occurred • Anti-Theft System is damaged or has failed • TSB 01-6-2 (superseded from 76-65-4) contains an updated repair procedure

Gas Engine OBD II Trouble Code List (P1xxx Codes)

DTC	Trouble Code Title, Conditions & Possible Causes
P1270 1T CCM, MIL: Yes All Models Transmissions: All	**Engine Speed/Vehicle Speed Limiter Fault Conditions:** Engine started, and after the PCM monitored the engine speed and VSS signals), it detected the vehicle was operated in a manner where the engine or vehicle speed to exceeded its limit. **Possible Causes** • Excessive wheel slippage due to water, ice, mud and snow • Excessive engine speed (rpm) with the gearshift in Neutral • Vehicle driven at a high rate of speed
P1285 2T CCM, MIL: Yes All Models with CHT Transmissions: All	**Cylinder Head Over-Temperature Sensed Conditions:** Key on or engine running; and the PCM detected an engine overheat condition through inputs from the cylinder head temperature sensor. Engine started, and the PCM detected the CHT or ECT sensor signal was intermittent (it may have an intermittent open condition). Note: Select the CHT or IAT PID and monitor the signal for sudden changes while wiggling the CHT or IAT sensor connector. On the 5.4L V8, if the temperature exceeds 258ºF, the PCM disables four fuel injectors at a time. It alternates which four fuel injectors are disabled every 32-engine cycles. The cylinders that are disabled do not inject fuel, so they act as air pumps to aid in cooling the engine. If the temperature exceeds 310ºF, the PCM disables all of the fuel injectors until the engine temperature drops below 310ºF. **Possible Causes** • Base engine problems or related concerns • CHT sensor has deteriorated or it has failed • Engine coolant level is too low • Engine cooling system has a problem • TSB 10-29-1 contains repair help for this code (LS & T-Bird)
P1288 **2T CCM, MIL: Yes** All Models with CHT Transmissions: All	**Cylinder Head Temperature Sensor Out of Self-Test Range Conditions:** Key on and KOEO Self-Test enabled, or engine running with the KOER Self-Test enabled, and the PCM detected the CHT sensor was out of its self-test range (i.e., the engine was too hot or it did not warm to its normal operating temperature) during the test period. **Possible Causes** • CHT sensor harness connector is damaged • CHT sensor is damaged • Engine coolant level is too low • Engine is cold, or the engine is overheated
P1289 **2T CCM, MIL: Yes** All Models with CHT Transmissions: All	**Cylinder Head Temperature Sensor Circuit High Input Conditions:** Key on or engine running; and the PCM detected a Cylinder Head Temperature (CHT) sensor signal that was more than 4.60v. This code may be due to an intermittent fault. Wiggle the CHT sensor wiring and connector while monitoring the CHT PID for a sudden change in voltage. DTC P0118 may also be reported when this code is set, and either code will cause the PCM to activate the MIL. **Possible Causes** • CHT sensor circuit is open in the wiring harness, or an open circuit exists in the CHT sensor circuit at the harness connector • CHT sensor is damaged or has failed • Engine coolant level is too low or the thermostat has failed • PCM has failed
P1290 **2T CCM, MIL: Yes** All Models with CHT Transmissions: All	**Cylinder Head Temperature Sensor Circuit Low Input Conditions:** Key on or engine running; and the PCM detected a Cylinder Head Temperature (CHT) sensor signal that was less than 0.2v. Note that this trouble code may be due to an intermittent type of fault. Wiggle the CHT sensor wiring and connector while monitoring the CHT V PID for signs of a sudden change in the voltage. DTC P0118 may also set along with this code (both codes will cause a MIL to be on). **Possible Causes** • CHT sensor connector is damaged or a short circuit exists • CHT sensor signal circuit is shorted to sensor ground • CHT sensor is damaged or the PCM has failed

Gas Engine OBD II Trouble Code List (P1xxx Codes)

DTC	Trouble Code Title, Conditions & Possible Causes
P1299 **2T CCM, MIL: Yes** All Models Transmissions: All	**Cylinder Head Over-Temperature Protection Active Conditions:** Engine started, and after a period of time with the engine running, the PCM detected the engine was in an overheated condition. Note: The PCM enables the Fail-Safe Cooling whenever this code is set to cool the engine (a Failure Mode Effects Strategy or FMEM). **Possible Causes** • Cooling system has a problem • Engine coolant level is too low • A Base Engine problem may be present • TSB 10-29-1 contains repair help for this code (LS & T-Bird)
P1309 **1T MISFIRE** **MIL: Yes** All Models Transmissions: All	**Misfire Monitor Disabled Conditions:** DTC P0136, P0156, P0171, P0172, P0174, P0175, P1130 and P1150 not set, engine started, and the PCM disabled the Misfire Monitor in order to verify that the CMP sensor is synchronized. Note that this code can be caused by an incorrect input from the CMP sensor (i.e., it senses the passage of teeth from the CMP wheel). **Possible Causes** • Camshaft position sensor is damaged or it has failed • CKP, ECT or MAF sensors may be out-of-calibration or failed • PCM has failed • TSB 02-22-1 contains repair information for this trouble code
P1336 **2T CCM, MIL: Yes** All Models Transmissions: All	**CKP or CMP Signal Malfunction Conditions:** Engine started, and the PCM detected an erratic signal from CKP sensor or the CMP sensor. It is possible for EMI/RFI interference to cause this code when they occur on these circuits. **Possible Causes** • Base Engine problem or concern exists • CKP sensor or CMP signal circuit is open or shorted to ground • CKP sensor or CMP sensor is damaged or failed (check for EMI/RFI on this circuit). • PCM has failed • TSB 02-22-1 contains repair information for this trouble code
P1351 **2T CCM, MIL: Yes** All Models Transmissions: All	**Ignition Diagnostic Monitor Circuit Malfunction Conditions:** Engine started, and the PCM detected a loss of the IDM circuit from the ignition module in the distributor (the fault may be intermittent). Note: If DTC P0350, P0351, P0352, P0353 or P0354 is set, repair these trouble codes and then recheck to see if DTC P1351 resets. **Possible Causes** • Camshaft position sensor may have failed • IDM signal circuit may be open or grounded • CKP, ECT or MAF sensors may be damaged or have failed • PCM has failed
P1356 **1T CCM, MIL: No** All Models Transmissions: All	**PIP Signals Present With Engine Off Conditions:** Key on, and the PCM detected the presence of PIP signals, yet the Ignition Diagnostic Monitor (IDM) signals indicated the engine was not turning. **Possible Causes** • Ignition Module has failed • PCM has failed
P1358 **2T CCM, MIL: Yes** All Models Transmissions: All	**IDM Signals Out Of Self-Test Range Conditions:** Engine started; and the PCM detected PIP signals that indicated that the Ignition Diagnostic Monitor signals were out of the self-test range under these operating conditions. **Possible Causes** • Ignition Module has failed • PCM has failed
P1359 **2T CCM, MIL: Yes** All Models Transmissions: All	**Spark Output Circuit Malfunction Conditions:** Engine started, and the PCM did not detect any change in the Spark Output (SPOUT) signals during the test period. **Possible Causes** • SPOUT signal circuit may be open (check the connector) • SPOUT signal circuit may be grounded • Ignition Module is damaged or has failed

Gas Engine OBD II Trouble Code List (P1xxx Codes)

DTC	Trouble Code Title, Conditions & Possible Causes
P1380 **2T CCM, MIL: Yes** Cougar, Focus, Escort, LS, Five Hundred, Freestyle, Montego,, Thunderbird, ZX2 Equipped with a 2.0L (VIN 3, VIN 5, VIN P) 2.3L VIN Z, 3.0L VIN S, 3.9L VIN A engine Transmissions: All	**Variable Cam Timing Solenoid 'A' Circuit Malfunction (Bank 1) Conditions:** Key on or engine running; and the PCM detected an unexpected voltage condition on the Variable Cam Timing signal circuit. Note that this code is due to an electrical fault (not a mechanical fault). **Possible Causes** • VCT solenoid circuit is open or shorted • VCT solenoid is damaged or has failed • PCM has failed
P1381 **2T CCM, MIL: Yes** Cougar, Focus, Escort, LS, Five Hundred, Freestyle, Montego,, Thunderbird, ZX2 Equipped with a 2.0L (VIN 3, VIN 5, VIN P) 2.3L VIN Z, 3.0L VIN S, 3.9L VIN A engine Transmissions: All	**Variable Cam Timing Over-Advanced (Cylinder Bank 1) Conditions:** Engine started, and the PCM detected the Variable Cam Timing position indicated the camshaft timing was over-advanced when compared to a maximum calibrated limit in an advanced position. Note: This code is a mechanical problem - not an electrical problem. The engine may be hard to start or idle rough when this code is set. **Possible Causes** • Cam timing improperly set • No oil flow to VCT piston chamber, or low engine oil pressure • VCT solenoid valve stuck in closed position • Camshaft advance mechanism binding (inside the VCT unit)
P1383 **2T CCM, MIL: Yes** Cougar, Focus, Escort, LS, Five Hundred, Freestyle, Montego,, Thunderbird, ZX2 Equipped with a 2.0L (VIN 3, VIN 5, VIN P) 2.3L VIN Z, 3.0L VIN S, 3.9L VIN A engine Transmissions: All	**Variable Cam Timing Over-Retarded (Cylinder Bank 1) Conditions:** Engine started, and the PCM detected the Variable Cam Timing position indicated the camshaft timing was over-retarded when compared to a maximum calibrated limit in a retarded position. Note that this code is a mechanical problem - not an electrical problem. The engine may be hard to start or idle rough when this code is set. **Possible Causes** • Cam timing improperly set • Low oil pressure, VCT piston chamber not receiving any oil flow • VCT solenoid valve stuck in closed position • Camshaft advance mechanism binding (inside the VCT unit)
P1385 **2T CCM, MIL: Yes** LS & Thunderbird Models with 3.0L VIN S, 3.9L VIN A engine Transmissions: All	**Variable Cam Timing Solenoid 'A' Circuit Malfunction (Cylinder Bank 2) Conditions:** Key on or engine running; and the PCM detected an unexpected voltage condition on the Variable Cam Timing signal circuit. Note that this code is due to an electrical fault (not a mechanical fault). **Possible Causes** • VCT solenoid circuit is open or shorted • VCT solenoid is damaged or has failed • PCM has failed
P1386 **2T CCM, MIL: Yes** LS & Thunderbird Models with 3.0L VIN S, 3.9L VIN A engine Transmissions: All	**Variable Cam Timing Over-Advanced (Cylinder Bank 2) Conditions:** Engine started, and the PCM detected the Variable Cam Timing position indicated the camshaft timing was over-advanced when compared to a maximum calibrated limit in an advanced position. Note: This code indicates the presence of a mechanical problem - not an electrical problem. The engine may be hard to start or idle rough when this code is set. **Possible Causes** • Cam timing improperly set • No oil flow to VCT piston chamber, or low engine oil pressure • VCT solenoid valve stuck in closed position • Camshaft advance mechanism binding (inside the VCT unit)

Gas Engine OBD II Trouble Code List (P1xxx Codes)

DTC	Trouble Code Title, Conditions & Possible Causes
P1388 **2T CCM, MIL: Yes** LS & Thunderbird Models with 3.0L VIN S, 3.9L VIN A engine Transmissions: All	**Variable Cam Timing Over-Retarded (Cylinder Bank 2) Conditions:** Engine started, and the PCM detected the Variable Cam Timing position indicated the camshaft timing was over-retarded when compared to a maximum calibrated limit in a retarded position. This code indicates a mechanical problem - not an electrical problem. The engine may be hard to start or idle rough when this code is set. **Possible Causes** • Cam timing improperly set • Low oil pressure, VCT piston chamber not receiving any oil flow • VCT solenoid valve stuck in closed position • Camshaft advance mechanism binding (inside the VCT unit)
P1390 **1T CCM, MIL: No** All Models Transmissions: All	**Octane Adjust Circuit Malfunction Conditions:** Key on, KOEO Self-Test enabled, and with the octane adjust software activated, the PCM detected a malfunction in the OCT circuit. **Possible Causes** • OCT shorting bar removed • OCT circuit open • PCM has failed
P1400 **2T CCM, MIL: Yes** Some Models Transmissions: All	**DPFE Sensor Circuit Low Input Conditions:** Key on, and the PCM detected the DPF EGR sensor signal was less than the minimum calibrated value of 0.2v. Note: The DPF EGR PID will read less than 0.2v with this code set. **Possible Causes** • DPF EGR signal circuit shorted to ground • DPF EGR signal VREF circuit open • DPF EGR sensor is damaged or has failed • PCM has failed • TSB 4-3-1 contains repair information for this trouble code
P1400 **2T CCM, MIL: Yes** Crown Victoria, Grand Marquis, Aviator, LS, Five Hundred, Freestyle, Montego,, Marauder, Town Car & Thunderbird Models Transmissions: All	**DPFE Sensor Circuit Low Input Conditions:** Engine started, and the PCM detected the DPF EGR sensor signal was less than the minimum calibrated value of 0.2v. The DPFE, EGR valve and EVR solenoid are integrated into the ESM assembly. **Possible Causes** • DPFE sensor signal circuit is shorted to ground • DPFE sensor VREF circuit (5v) is open • DPFE sensor is damaged or it has failed • PCM has failed
P1400 **2T CCM, MIL: Yes** Explorer, Mountaineer equipped with a 5.0L VIN P engine Transmissions: All	**EVP EGR Sensor Circuit Low Input Conditions:** Key on or engine running; and the PCM detected the EVP sensor signal was below the self-test minimum value of 0.2v. The EVP EGR PID will read less than 0.2v with this code set. **Possible Causes** • EVP sensor circuit is open or shorted to ground • EVP sensor VREF circuit is open • EGR valve position sensor is damaged or has failed • PCM has failed
P1401 **2T CCM, MIL: Yes** Some Models Transmissions: All	**DPFE Sensor Circuit High Input Conditions:** Key on; and the PCM detected the DPF EGR sensor signal was more than the maximum calibrated value of 4.5v. The DPF EGR PID will read more than 4.5v with this code set. **Possible Causes** • DPF EGR signal circuit open, or sensor ground circuit open • DPF EGR signal shorted to VREF or to power • DPF EGR sensor is damaged or has failed • PCM has failed • TSB 4-3-1 contains repair information for this trouble code

Gas Engine OBD II Trouble Code List (P1xxx Codes)

DTC	Trouble Code Title, Conditions & Possible Causes
P1401 **2T CCM, MIL: Yes** Crown Victoria, Grand Marquis, Aviator, LS, Five Hundred, Freestyle, Montego,, Marauder, Town Car & Thunderbird Models Transmissions: All	**DPFE Sensor Circuit High Input Conditions:** Key on or engine running; and the PCM detected the DPF EGR sensor signal was more than the maximum calibrated value of 4.5v. On this vehicle application, the DPFE, EGR valve and EVR solenoid are integrated into the ESM assembly. **Possible Causes** • DPFE sensor signal circuit is open • DPFE sensor ground circuit is open • DPFE sensor signal is shorted to VREF (5v) • DPFE sensor is damaged or it has failed • PCM has failed
P1405 **2T CCM, MIL: Yes** Some Models Transmissions: All	**DPFE Sensor Upstream Hose Off Or Plugged Conditions:** Engine started; and the PCM detected the DPF EGR sensor indicated EGR flow in a negative direction (a closed EGR valve). Check for signs of icing in the hose, or wrong hose routing. **Possible Causes** • DPF EGR sensor upstream hose is disconnected • DPF EGR sensor upstream hose is plugged (ice) • EGR tube is plugged or damaged
P1406 **2T CCM, MIL: Yes** Some Models Transmissions: All	**DPFE Sensor Downstream Hose Off Or Plugged Conditions:** Engine started; and the PCM detected the DPF EGR sensor signal indicated EGR flow existed with the EGR valve commanded closed. **Possible Causes** • Check for signs of icing in the hose, or for a restricted tube • DPF EGR sensor downstream hose is disconnected • DPF EGR sensor downstream hose is plugged (ice) • EGR tube is plugged or damaged
P1406 **2T CCM, MIL: Yes** Crown Victoria, Grand Marquis, Aviator, LS, Five Hundred, Freestyle, Montego,, Marauder, Town Car & Thunderbird Models Transmissions: All	**DPFE Sensor Downstream Hose Off Or Plugged Conditions:** Engine started, and the PCM detected the DPFE sensor indicated that EGR flow was present with the EGR valve commanded closed. On this vehicle application, the DPFE, EGR valve and EVR solenoid are integrated into the ESM assembly. **Possible Causes** • Check for signs of icing in the hose, or for a restricted tube • DPF EGR sensor downstream hose is disconnected • DPF EGR sensor downstream hose is plugged (ice) • EGR tube is plugged or damaged
P1407 **2T EGR, MIL: Yes** Probe Models with a 2.0L I4 VIN A, 2.5L V6 VIN B engine Transmissions: All	**EGR Flow Out Of Self-Test Range Conditions:** DTC P0107, P0108, P1400, P1401 and P1408 not set, engine running, and the PCM detected the EGR valve did not move. Note: The EGR valve may be stuck in the closed position. **Possible Causes** • EGR VR solenoid circuit open or VR solenoid is damaged • EGR Atmospheric solenoid circuit open or solenoid is damaged • EGR Valve Position sensor circuit open or sensor is damaged • EGR Check solenoid circuit open or solenoid is damaged • Exhaust system is restricted • PCM has failed
P1408 **1T EGR, MIL: Yes** Some Models Transmissions: All	**EGR Flow Out Of Self-Test Range Conditions:** KOER Self-Test enabled, and the PCM detected the EGR flow was out of the self-test range during the self-test with the engine running. **Possible Causes** • EGR vacuum regulator solenoid vacuum supply problem • EVR valve stuck closed or iced up, or the flow path is restricted • EGR valve diaphragm leaking, hose is off, plugged or leaking • EGR VR solenoid open or the VPWR circuit is open • DPF EGR sensor pressure hoses connected wrong (reversed) • DPF EGR downstream hose connection leaking or plugged • EGR Orifice tube assembly is damaged • DPF EGR sensor or the EGR VR solenoid is damaged, or the PCM has failed

Gas Engine OBD II Trouble Code List (P1xxx Codes)

DTC	Trouble Code Title, Conditions & Possible Causes
P1408 **1T EGR, MIL: No** Probe Models with a 2.0L I4 VIN A, 2.5L V6 VIN B engine Transmissions: All	**EGR Flow Out Of Self-Test Range Conditions:** DTC P1400, P1401 and P1407 not set, engine started, KOER Self-Test enabled, and the PCM detected a lack of EGR valve movement while monitoring the EGR EVP sensor signal. **Possible Causes** • EGR valve vacuum lines are loose or damaged • EGR valve is damaged or has failed • EGR VR solenoid circuit open or the solenoid is damaged • EGR Atmospheric solenoid circuit open or solenoid is damaged • PCM has failed
P1408 **1T EGR, MIL: No** Focus, Ranger Models with a 2.3L VIN D, 2.3L VIN Z engine Transmissions: All	**EGR Flow Out Of Self-Test Range Conditions:** Engine started, KOER Self-Test enabled, and the PCM detected the EGR flow was out of the self-test range. There is no DPFE sensor or orifice tube assembly on this vehicle application. **Possible Causes** • E-EGR valve connector is damaged, loose or shorted • E-EGR valve is sticking, damaged or it has failed • MAP or TMAP sensor is damaged or out of calibration • PCM has failed
P1408 **1T EGR, MIL: No** Crown Victoria, Grand Marquis, Aviator, LS, Five Hundred, Freestyle, Montego,, Marauder, Town Car & Thunderbird Models Transmissions: All	**EGR Flow Out Of Self-Test Range Conditions:** Engine started, KOER Self-Test enabled, and the PCM detected the EGR flow was out of the self-test range during the self-test. On this vehicle application, the DPFE, EGR valve and EVR solenoid are integrated into the ESM assembly. **Possible Causes** • EGR vacuum regulator solenoid vacuum supply problem • EVR valve stuck closed or iced up, or the flow path is restricted • EGR valve diaphragm leaking, hose is off, plugged or leaking • EGR VR solenoid open or the VPWR circuit is open • DPF EGR sensor pressure hoses connected wrong (reversed) • DPF EGR sensor VREF circuit is open • DPF EGR downstream hose connection leaking or plugged • EGR Orifice tube assembly is damaged • DPF EGR sensor or the EGR VR solenoid is damaged • PCM has failed
P1409 **2T CCM, MIL: Yes** Some Models Transmissions: All	**EGR Vacuum Regulator Solenoid Circuit Malfunction Conditions:** Engine started, and the PCM detected a fault in the EGR VR solenoid circuit (i.e., the VR circuit was too high or low when compared to the expected range with the solenoid enabled). **Possible Causes** • EGR VR solenoid circuit is open, or shorted to ground • EGR VR circuit is shorted to power or the VPWR circuit is open • EGR vacuum regulator solenoid is damaged or the PCM has failed
P1409 **1T CCM, MIL: Yes** Crown Victoria, Grand Marquis, Aviator, LS, Five Hundred, Freestyle, Montego,, Marauder, Town Car & Thunderbird Models Transmissions: All	**EGR Vacuum Regulator Solenoid Circuit Malfunction Conditions:** Engine started, and the PCM detected a fault in the EGR VR solenoid circuit (i.e., the VR circuit was too high or low when compared to its expected range with the solenoid enabled). **Possible Causes** • EGR VR solenoid circuit is open, or shorted to ground • EGR VR circuit is shorted to power or the VPWR circuit is open • EGR vacuum regulator solenoid is damaged or the PCM has failed
P1410 **2T CCM, MIL: Yes** Probe Models with a 2.0L I4 VIN A, 2.5L V6 VIN B engine Transmissions: All	**EGR Check Or Fuel Pressure Solenoid Circuit Malfunction Conditions:** Key on or engine running; and the PCM detected a fault in the EGR Check solenoid or the EGR Fuel Pressure Control Solenoid circuits. **Possible Causes** • EGR check solenoid control circuit is open or shorted to ground • FPC solenoid control circuit is open or shorted to ground • Inspect the EGR Check Solenoid and Fuel Pressure Control solenoid harness connectors to determine if they are swapped • PCM has failed

Gas Engine OBD II Trouble Code List (P1xxx Codes)

DTC	Trouble Code Title, Conditions & Possible Causes
P1411 **2T AIR, MIL: Yes** All Models Transmissions: All	**Secondary Air Injection System Downstream Flow Conditions:** Engine started, engine running with AIR system "on", and the PCM detected the HO2S signal did not go lean with the AIR system "on". **Possible Causes** • Secondary AIR System Electric Pump is damaged • Secondary AIR System Mechanical Pump is damaged • Secondary AIR pump hose is leaking • Secondary AIR pump hose is blocked • Secondary AIR Bypass solenoid passage leaking or blocked • Secondary AIR Bypass solenoid stuck open or stuck closed
P1413 **2T CCM, MIL: Yes** All Models Transmissions: All	**Secondary AIR System Monitor Circuit Low Input Conditions:** Engine started, engine running with AIR system "off", and the PCM detected an unexpected "low" voltage condition on the Secondary AIR monitor during the CCM test. **Possible Causes** • AIR solenoid control circuit is open or it is shorted to ground • AIR pump is damaged or it has failed • Solid State relay is damaged or it has failed • Solid State relay battery power circuit (B+) is open • PCM has failed
P1414 **2T CCM, MIL: Yes** All Models Transmissions: All	**Secondary AIR System Monitor Circuit High Input Conditions:** Engine started, AIR system not active, and the PCM detected a high voltage signal present on the Secondary AIR monitor signal circuit. **Possible Causes** • AIR Monitor circuit from the pump to the PCM is open • AIR solenoid control circuit is shorted to power • Solid State relay is damaged or has failed • AIR pump ground circuit is open • AIR pump is damaged or has failed • PCM has failed
P1432 **1T CCM, MIL: Yes** All Models Transmissions: All	**Thermostat Heater Control Circuit Malfunction Conditions:** Engine started; and the PCM detected the Thermostat Heater Control circuit was less than or more than a calibrated limit for too long a period of time during the CCM self-test. **Possible Causes** • Thermostat Heater Control (THTRC) circuit open or shorted • Thermostat Heater Control (THTRC) VPWR circuit open • Thermostat Heater assembly is damaged or has failed • PCM has failed
P1436 **2T CCM, MIL: Yes** All Models Transmissions: All	**A/C Evaporator Temperature (ACET) Circuit Low Input Conditions:** Key on or engine running; and the PCM detected the ACET signal was less than the self-test minimum amount of 0.13v in the self-test. **Possible Causes** • ACET signal circuit shorted to sensor ground (return) • ACET signal circuit shorted to chassis ground • ACET sensor is damaged or has failed • PCM has failed
P1437 **2T CCM, MIL: Yes** All Models Transmissions: All	**A/C Evaporator Temperature (ACET) Circuit High Input Conditions:** Key on or engine running; and the PCM detected the ACET signal was more than the self-test maximum amount of 4.5v in the self-test. **Possible Causes** • ACET signal circuit is open, or the ground circuit is open • ACET signal is shorted to VREF • ACET sensor is damaged or has failed • PCM has failed

Gas Engine OBD II Trouble Code List (P1xxx Codes)

DTC	Trouble Code Title, Conditions & Possible Causes
P1442 **2T EVAP, MIL: Yes** All Models Transmissions: All	**EVAP System Small Leak (0.040") Detected Conditions:** Cold startup requirement met, engine running at Cruise speed in closed loop for 2-3 minutes, and the PCM detected a leak (as small as 0.040") in the EVAP system. Note: Inspect the CV solenoid for contamination (as contamination can hold the CV open set DTC P0442 and also plugs the port to atmosphere enough to keep system from being vented quickly). **Possible Causes** • Fuel filler cap damaged, cross-threaded or loosely installed • Aftermarket EVAP parts that do not conform to specifications • Small holes or cuts in fuel vapor hoses or EVAP canister tubes • CV solenoid stays partially open when commanded closed • Loose fuel vapor hose/tube connections to EVAP components • EVAP component seals leaking (i.e., leaks in the Purge valve, fuel tank pressure sensor, canister vent solenoid, fuel vapor control valve tube assembly or fuel vapor vent valve) • TSB 03-9-8 contains a repair procedure for this trouble code • TSB 3-20-3 contains a repair procedure for this trouble code
P1443 **2T EVAP, MIL: Yes** All Models Transmissions: All	**EVAP Canister Purge System Malfunction Conditions:** Engine started, engine warmup completed engine running at a steady cruise speed, and the PCM detected a leak or blockage was present somewhere in the vapor line between the intake manifold, EVAP purge valve and the charcoal canister during the Continuous self test. **Possible Causes** • EVAP canister purge valve is damaged or has failed • PF sensor is out-of-calibration or it is "skewed" • PCM has failed
P1443 **2T EVAP, MIL: Yes** Contour & Mystique Models Transmissions: All	**EVAP VMV System Malfunction Conditions:** Engine started, engine warmup completed engine running at a steady cruise speed, and the PCM detected a leak or blockage was present somewhere in the vapor line between the intake manifold, VMV valve and the charcoal canister during the CCM self-test. **Possible Causes** • VMV (valve) is sticking or has failed mechanically • PF sensor is out-of-calibration or it is "skewed" • PCM has failed
P1443 **2T EVAP, MIL: Yes** All Models Transmissions: All	**Low Purge Flow Or No Purge Flow Condition Detected Conditions:** ECT sensor less than 90°Fat startup (cold engine), engine running at a steady cruise speed, and the PCM detected a fuel tank pressure change occurred of more than -7" H2O within 30 seconds with the purge flow less than 0.02 pounds per minute during testing. **Possible Causes** • EVAP canister purge valve stuck closed (mechanically) • Fuel vapor hose blocked between EVAP purge valve and FTP sensor, or blocked between purge valve and intake manifold, or vacuum hose blocked between purge valve and intake manifold
P1444 **2T CCM, MIL: Yes** All Models Transmissions: All	**Purge Flow Sensor Circuit Low Input Conditions:** Key on or engine running; and the PCM detected the Purge Flow (PF) sensor signal was less than the minimum calibrated limit of 0.40v during the Continuous self test. **Possible Causes** • PF sensor signal circuit is shorted to sensor or chassis ground • PF sensor is damaged or has failed • PCM has failed
P1445 **2T CCM, MIL: Yes** All Models Transmissions: All	**Purge Flow Sensor Circuit High Input Conditions:** Key on or engine running; and the PCM detected the Purge Flow (PF) sensor was more than the maximum calibrated limit of 4.80v. **Possible Causes** • PF sensor signal circuit shorted to VREF or power (VPWR) • PF sensor signal circuit open or sensor ground circuit open • PF sensor is damaged or has failed • PCM has failed

Gas Engine OBD II Trouble Code List (P1xxx Codes)

DTC	Trouble Code Title, Conditions & Possible Causes
P1450 **2T EVAP, MIL: Yes** All Models Transmissions: All	**Unable to Bleed Up Fuel Tank Vacuum Conditions:** ECT sensor less than 90°F at startup (cold engine), engine running at a steady cruise speed, and the PCM detected a high fuel tank vacuum condition was present during the EVAP test. **Possible Causes** • CV solenoid is stuck partially or fully open or filter is plugged • EVAP canister tube or EVAP canister purge outlet tube blocked or kinked between fuel tank, purge valve and EVAP canister • Fuel filler cap stuck closed (vacuum relief cannot occur) • Contaminated fuel vapor elbow at the EVAP canister, or the EVAP canister is restricted or canister purge valve stuck open
P1450 **2T EVAP, MIL: Yes** All Models Transmissions: All	**Unable to Bleed Up Fuel Tank Vacuum Conditions:** ECT sensor less than 90°F at startup (cold engine), engine running at a steady cruise speed, and the PCM detected a high fuel tank vacuum condition was present during the EVAP test. **Possible Causes** • CV solenoid is stuck partially or fully open or filter is plugged • EVAP canister tube or EVAP canister purge outlet tube blocked or kinked between fuel tank, purge valve and EVAP canister • Fuel filler cap stuck closed (vacuum relief cannot occur) • Contaminated fuel vapor elbow at the EVAP canister, or the EVAP canister is restricted or canister purge valve stuck open • FTP sensor is damaged
P1451 **2T CCM, MIL: Yes** All Models Transmissions: All	**EVAP System Canister Vent Solenoid Circuit Malfunction Conditions:** Engine started, engine running at a steady cruise speed, canister vent solenoid enabled, and the PCM detected an unexpected voltage condition on the Canister Vent solenoid circuit. **Possible Causes** • CV solenoid circuit is open, shorted to ground or system power • CV solenoid is damaged or has failed • PCM has failed
P1452 **2T CCM, MIL: Yes** All Models Transmissions: All	**Fuel Tank Pressure Sensor Circuit Malfunction Conditions:** Key on or engine running; and the PCM detected that the Fuel Tank Pressure (FTP) sensor signal was less than or more than the calibrated amount during the self-test. Note that the FTP V PID should read from 2.40-2.80 with the cap off. **Possible Causes** • FTP sensor signal circuit is open or shorted to ground • FTP sensor ground return circuit is open • FTP sensor is damaged or has failed • PCM has failed
P1455 **2T EVAP, MIL: Yes** All Models Transmissions: All	**EVAP System Gross Leak Detected Conditions:** ECT sensor less than 90°F at startup (cold engine), engine running at a steady cruise speed for 2-3 minutes, and the PCM detected a gross leak in the EVAP system during the test. **Possible Causes** • Fuel filler cap missing, loose (not tightened) or the wrong part • FTP sensor signal circuit open or sensor ground circuit open • FTP sensor ground circuit open • FTP sensor is damaged or has failed • PCM has failed
P1460 **1T CCM, MIL: Yes** All Models Transmissions: All	**Wide Open Throttle A/C Cutout Relay Circuit Malfunction Conditions:** Key on, and the PCM detected a malfunction in the A/C wide-open throttle (WOT) circuit during the test. Note: If this code sets on vehicles without an A/C system, ignore this code. **Possible Causes** • WOT A/C Relay control circuit is open or shorted to ground • WOT A/C Relay VREF circuit is open • WOT A/C Relay is damaged or has failed • PCM has failed

Gas Engine OBD II Trouble Code List (P1xxx Codes)

DTC	Trouble Code Title, Conditions & Possible Causes
P1461 **2T CCM, MIL: Yes** All Models Transmissions: All	**A/C Pressure Sensor Circuit High Input Conditions:** Engine started, and the PCM detected the A/C Pressure sensor signal was over the test limit. **Possible Causes** • ACP sensor circuit shorted to VREF or to power (VPWR) • ACP sensor circuit is open, or the ground circuit is open • ACP sensor is damaged or has failed • PCM has failed
P1462 **2T CCM, MIL: Yes** All Models Transmissions: All	**A/C Pressure Sensor Circuit Low Input Conditions:** Engine started, and the PCM detected the A/C Pressure sensor signal was under the test limit. **Possible Causes** • ACP sensor circuit shorted to VREF or to power (VPWR) • ACP sensor circuit is open, or the ground circuit is open • ACP sensor is damaged or has failed • PCM has failed
P1463 **2T CCM, MIL: Yes** All Models Transmissions: All	**A/C Pressure Sensor Insufficient Pressure Change Conditions:** Engine started, and with the A/C compressor operating, the PCM detected the A/C refrigerant pressure did not change as the compressor cycled during the self-test period. **Possible Causes** • A/C system mechanical failure, or A/C clutch always engaged • ACP sensor signal open, or sensor ground circuit open • A/C sensor is damaged or the PCM has failed
P1464 **1T CCM, MIL: No** All Models Transmissions: All	**A/C Demand Out of Self-Test Range Conditions:** Key on, KOEO Self-Test enabled, or with the engine running, KOER Self-Test enabled, and the PCM detected the A/C demand switch signal was high during the self-test period. **Possible Causes** • A/C switch was left "on" during the KOER self-test • A/C PWR circuit is shorted to power (N/C WAC relay contacts) • ACCS circuit is shorted to power • A/C Demand Switch, WAC relay or CCRM is damaged
P1469 **2T CCM, MIL: Yes** All Models Transmissions: All	**Low A/C Cycling Period Conditions:** Engine started, and with the A/C selected, PCM detected frequent cycling of the A/C compressor clutch. This test was designed to protect the transmission. In some strategies, the PCM will unlock the torque converter during A/C clutch engagement. If a concern is present that results in frequent A/C clutch cycling, damage could occur if the torque converter was cycled at these intervals. This test will detect this condition, set the code and prevent the torque converter from excessive cycling. **Possible Causes** • Cycling pressure switch circuit open between pin 41 (ACCS) and the PCM, or the IGN RUN circuit is open to the cycling pressure switch circuit (if applicable) • Mechanical A/C system concern (i.e., low refrigerant charge, damaged A/C switch)
P1473 **2T CCM, MIL: Yes** All Models Transmissions: All	**Fan Secondary High with Fan(s) Off Conditions:** Key on, KOEO Self-Test enabled, and the PCM detected an unexpected voltage condition on the Power-To-Cooling fan circuit **Possible Causes** • Power-to-Cooling fan circuit open in the wiring harness • Power-to-Cooling fan circuit shorted to power in wiring harness • Cooling fan motor windings open, or fan ground circuit is open • VLCM is damaged or has failed
P1474 **2T CCM, MIL: Yes** All Models Transmissions: All	**Hydraulic Cooling Fan Primary Circuit Malfunction Conditions:** Key on or engine running; low cooling fan enabled, and the PCM detected the voltage to the Hydraulic Cooling Fan (HFC) motor was higher or lower than the expected range for the fan primary circuit. **Possible Causes** • FC circuit is open or shorted to power • LFC circuit is open or shorted to power (CCRM models) • FC or LFC relay VPWR circuit is open (Start/Run on Probe)

Gas Engine OBD II Trouble Code List (P1xxx Codes)

DTC	Trouble Code Title, Conditions & Possible Causes
P1474 **2T CCM, MIL: Yes** LS Models with a 3.0L VIN S, 3.9L VIN A Transmissions: All	**Hydraulic Cooling Fan Primary Circuit Malfunction Conditions:** Key on or engine running; hydraulic cooling fan "on", and the PCM detected an unexpected voltage (too high or too low) condition on the Hydraulic Cooling Fan (HFC) motor circuit. **Possible Causes** • HCF control circuit is open or shorted to ground • HCF control circuit is shorted to power (VPWR) • HCF motor is damaged or has failed • PCM has failed
P1477 **2T CCM, MIL: Yes** All Models Transmissions: All	**Medium Fan Control Primary Circuit Malfunction Conditions:** Key on, medium cooling fan (MFC) enabled; and the PCM detected excessive current draw in the circuit; or with the MFC disabled (off), it detected voltage present on the MFC circuit. **Possible Causes** • MFC circuit is open • MFC relay circuit to IGN START/RUN is open • MFC relay is damaged or has failed • PCM has failed
P1479 **2T CCM, MIL: Yes** All Models Transmissions: All	**High Fan Control Primary Circuit Malfunction Conditions:** Key on, high cooling fan (HFC) enabled, and the PCM detected excessive current draw in the circuit; or with the HFC commanded off, it detected voltage present on the HFC circuit. **Possible Causes** • HFC circuit is open • HFC circuit is shorted to ground • HFC relay power circuit (VPWR) Is open • High speed FC relay is damaged or it has failed • PCM has failed
P1481 **2T CCM, MIL: Yes** All Models Transmissions: All	**Fan Secondary Low With High Fan On Conditions:** Key on or engine running; high speed cooling fan enabled, and the PCM detected the fan secondary circuit was low with the High Speed cooling fan commanded "on" during testing. **Possible Causes** • High speed cooling fan circuit is open • High speed cooling fan circuit is shorted to ground • High speed cooling fan relay power circuit (VPWR) is open • High speed FC relay is damaged or has failed • PCM has failed
P1483 **2T CCM, MIL: Yes** All Models Transmissions: All	**Power To Fan Circuit Over-Current Detected Conditions:** Key on or engine running; cooling fan enabled, and the PCM detected the current in the Fan PWR circuit exceeded the limit. **Possible Causes** • Power-to-Cooling Fan circuit shorted to ground • Cooling fan motor is damaged or has failed • VLCM is damaged or has failed • PCM has failed
P1487 **2T CCM, MIL: Yes** Mark VIII Models Transmissions: All	**VLCM Power Ground Circuit Malfunction Conditions:** Key on or engine running; and the PCM detected an unexpected voltage condition on the Variable Load Control Module (VLCM) power ground circuit during the CCM test period. **Possible Causes** • VLCM power ground circuit is open • VLCM is damaged or it has failed
P1487 **2T CCM, MIL: Yes** Probe Models with a 2.0L I4 VIN A, 2.5L V6 VIN B engines Transmissions: All	**EGR Check Solenoid Circuit Malfunction Conditions:** Key on or engine running; and the PCM detected an unexpected voltage condition on the EGR Check solenoid circuit during testing. **Possible Causes** • EGR Check solenoid circuit is open or shorted in the harness • EGR Check solenoid circuit is open or shorted in the connector • EGR Check solenoid power circuit (VREF) is open • EGR Check solenoid is damaged or the PCM has failed

Gas Engine OBD II Trouble Code List (P1xxx Codes)

DTC	Trouble Code Title, Conditions & Possible Causes
P1500 **2T CCM, MIL: Yes** All Models Transmissions: All	**VSS Signal Or POSM Signal Intermittent Conditions:** Engine running in gear with a VSS signal present, and the PCM detected that the VSS signal was intermittent Note: The VSS signal is received from the VSS, transfer case speed sensor, ABS Control module, the GEM or the Central Timer module (CTM), depending upon the vehicle application. **Possible Causes** • VSS pins damaged, loose or pushed in at the connector • VSS circuit open or shorted in the wiring harness (insulation) • VSS wiring harness routing incorrect or VSS mounting incorrect • TSB 01-21-13 contains a repair procedure for this trouble code
P1501 **1T CCM, MIL: No** All Models Transmissions: All	**VSS Signal Out Of Self-Test Range Conditions:** Engine started, KOER Self-Test enabled, and the PCM detected a VSS signal during the self-test (i.e., with the vehicle not moving). **Possible Causes** • VSS signal is noisy due to Radio Frequency Interference/ Electro-Magnetic Interference (RFI/EMI) from outside devices (ignition wires, charging circuit or aftermarket devices)
P1501 **1T CCM, MIL: Yes** All Models Transmissions: All	**VSS Signal Intermittent Conditions:** Engine started, and the PCM detected the VSS signal dropped out. The TCIL will flash on the first trip this code sets. The speed signal is received from the VSS, transfer case speed sensor, ABS Control module, GEM or the Central Timer module (depends upon the vehicle). **Possible Causes** • VSS signal is noisy due to Radio Frequency Interference/ Electro-Magnetic Interference (RFI/EMI) from outside devices (ignition wires, charging circuit or aftermarket devices)
P1502 **1T CCM, MIL: Yes** All Models Transmissions: All	**VSS Signal Intermittent Conditions:** Engine started, and the PCM detected an intermittent VSS signal. The TCIL will flash on the first trip that this code is set. The VSS signal is received from the VSS, transfer case speed sensor, ABS Control module, GEM or the Central Timer module (depends upon the vehicle). **Possible Causes** • VSS+ or VSS- harness circuit is open • TCSS signal or TCSS signal return harness circuit is open • VSS harness circuit, TCSS harness circuit is shorted to ground • VSS harness circuit, CSS harness circuit is shorted to power • VSS circuit open between the PCM and related control module • VSS or TCSS, or wheel speed sensors circuits are damaged • Modules connected to VSC/VSS harness circuits are damaged • Mechanical drive mechanism for the VSS or TCSS is damaged
P1504 **2T CCM, MIL: Yes** All Models Transmissions: All	**Idle Air Control Circuit Malfunction Conditions:** Engine started, engine running for 1 minute, and the PCM detected an electrical load failure on the IAC motor circuit during the self-test. **Possible Causes** • IAC circuit is open, shorted to ground or to the VPWR circuit • IAC solenoid VPWR circuit is open • IAC valve is damaged or has failed • PCM has failed
P1505 **2T CCM, MIL: Yes** Models Transmissions: All	**Idle Air Control System At Adaptive Clip Conditions:** Engine running for over one minute, and the PCM detected the idle speed control had reached its "idle air trim limit" during the Continuous self test. **Possible Causes** • Base engine air leaks are present • Air cleaner element is dirty, plugged or restricted • Throttle body/linkage is binding • IAC valve body is damaged or contaminated • Throttle body is damaged

Gas Engine OBD II Trouble Code List (P1xxx Codes)

DTC	Trouble Code Title, Conditions & Possible Causes
P1506 **2T CCM, MIL: Yes** Crown Victoria, Grand Marquis Models Transmissions: All	**Idle Air Control Overspeed Error Conditions:** Engine started, engine running for 1 minute, and the PCM detected the idle speed was more than the desired engine Target Idle Speed. **Possible Causes** • Base engine vacuum leaks present • EVAP system has a problem • IAC circuit shorted to ground • IAC valve is stuck open, or it is damaged • Throttle body or throttle plate is contaminated or very dirty • TSB 98-25-19 contains a repair procedure for this trouble code
P1506 **2T CCM, MIL: Yes** Some Models Transmissions: All	**Idle Air Control Overspeed Error Conditions:** Engine started, engine running for 1 minute, and the PCM detected the idle speed was more than the desired engine Target Idle Speed. **Possible Causes** • Base engine vacuum leaks present • EVAP system has a problem • IAC circuit shorted to ground • IAC valve is stuck open, or it is damaged • Throttle body or throttle plate is contaminated or very dirty
P1507 **2T CCM, MIL: Yes** Some Models Transmissions: All	**Idle Air Control Underspeed Error Conditions:** Engine started, engine running for 1 minute, and the PCM detected the idle speed was less than the desired engine Target Idle Speed. **Possible Causes** • Air inlet is plugged or the air filter element is severely clogged • IAC circuit is open, or shorted to the VPWR circuit • IAC circuit VPWR circuit is open • IAC solenoid is damaged or has failed • Throttle body or throttle plate is contaminated or very dirty
P1512 **2T CCM, MIL: Yes** Some Models Transmissions: All	**Intake Manifold Runner Control System Malfunction Conditions:** Engine started, and the PCM detected the IMRC Monitor indicated that the IMRC was stuck closed during the Continuous self test. **Possible Causes** • Leaky vacuum reservoir, vacuum lines loose or damaged • Vacuum solenoid or vacuum actuator is damaged • IMRC actuator cable/gears are seized, or the cables are improperly routed or seized • IMRC housing return springs are damaged or disconnected • Lever/shaft return stop may be obstructed or bent, or the lever/shaft wide open stop may be obstructed or bent, or the IMRC lever/shaft may be sticking, binding or disconnected • IMRC control circuit open, shorted or the VPWR circuit is open • PCM has failed
P1513 **2T CCM, MIL: Yes** Some Models Transmissions: All	**Intake Manifold Runner Control Malfunction (Bank 1) Conditions:** Engine started, and the PCM detected the IMRC Monitor indicated the IMRC was not functioning correctly during the self-test period. **Possible Causes** • IMRC actuator cable/gears are seized, or the cables are improperly routed or seized • IMRC control circuit open, shorted or the VPWR circuit is open • IMRC housing return springs are damaged or disconnected • Leaky vacuum reservoir, vacuum lines loose or damaged • Lever/shaft return stop may be obstructed or bent, or the lever/shaft wide open stop may be obstructed or bent, or the IMRC lever/shaft may be sticking, binding or disconnected • Vacuum solenoid or vacuum actuator is damaged • PCM has failed

Gas Engine OBD II Trouble Code List (P1xxx Codes)

DTC	Trouble Code Title, Conditions & Possible Causes
P1516 **2T CCM, MIL: Yes** Some Models Transmissions: All	**Intake Manifold Runner Control Input Error (Bank 1) Conditions:** Key on or engine running; and the PCM detected the IMRC Monitor signal for Bank 1 was outside of its expected calibrated range during the Continuous self test. **Possible Causes** • IMRC mechanical fault - the linkage may be bound or seized • Inspect for binding or improper routing. The cable core wire at the IMRC/IMSC housing attachment must have slack and lever must contact close plate stop screw
P1517 **2T CCM, MIL: Yes** Some Models Transmissions: All	**Intake Manifold Runner Control Input Error (Bank 2) Conditions:** Key on or engine running; and the PCM detected the IMRC Monitor signal for Bank 2 was outside of its expected calibrated range during the Continuous self test. **Possible Causes** • IMRC mechanical fault - the linkage may be bound or seized • Visually inspect for binding or improper routing. The cable core wire at the IMRC or IMSC housing attachment must have slack and lever must contact close plate stop screw
P1518 **2T CCM, MIL: Yes** Some Models Transmissions: All	**Intake Manifold Runner Control Malfunction (Stuck Open) Conditions:** Engine started, and the PCM detected the IMRC Monitor signal was less than its expected calibrated range at closed throttle. An IMRCM PID of 1v at closed throttle indicates a fault. **Possible Causes** • IMRC monitor signal circuit shorted to power ground • IMRC Monitor signal circuit shorted to signal ground (return) • IMRC actuator is damaged or has failed • PCM has failed
P1519 **2T CCM, MIL: Yes** Some Models Transmissions: All	**Intake Manifold Runner Control Stuck Closed Conditions:** Key on, and the PCM detected the IMRC Monitor was more than the expected calibrated range at closed throttle. Note: An IMRCM PID of VREF at 3000 rpm may indicate a fault. **Possible Causes** • IMRC monitor signal circuit shorted to power ground • IMRC Monitor signal circuit shorted to signal ground (return) • IMRC actuator is damaged or has failed (e.g., there may be a small leak in the vacuum diaphragm of the actuator) • PCM has failed
P1520 **2T CCM, MIL: Yes** Some Models Transmissions: All	**Intake Manifold Runner Control Input Error Conditions:** Key on or engine running; and the PCM detected the IMRC Monitor signal for was outside of its expected calibrated range. Use the Active Command or Output State Control on a Generic Scan Tool to help determine if an electrical fault is present. **Possible Causes** • IMRC control circuit is open or shorted to ground • IMRC Monitor VREF circuit is open • IMRC is damaged or the PCM has failed
P1530 **2T CCM, MIL: Yes** All Models Transmissions: All	**Air Conditioning Clutch Circuit Malfunction Conditions:** Key on or engine running; and the PCM detected a circuit fault in the A/C Clutch power (Power To Clutch) circuit. **Possible Causes** • A/C Clutch power circuit open or shorted to VPWR in harness • A/C clutch ground circuit is open • A/C clutch is open • VLCM is damaged or has failed
P1537 **2T CCM, MIL: Yes** Some Models Transmissions: All	**Intake Manifold Runner Control Malfunction (Bank 1 Stuck Open) Conditions:** Key on or engine running; and the PCM detected the Bank 1 IMRC Monitor signal was less than its expected calibrated range at closed throttle (it may be stuck in open position). An IMRCM PID of 1v at closed throttle may indicate a fault is present. **Possible Causes** • IMRC monitor signal circuit shorted to power ground • IMRC Monitor signal circuit shorted to signal ground (return) • IMRC actuator is damaged or the PCM has failed

Gas Engine OBD II Trouble Code List (P1xxx Codes)

DTC	Trouble Code Title, Conditions & Possible Causes
P1538 **2T CCM, MIL: Yes** Some Models Transmissions: All	**Intake Manifold Runner Control Stuck Open (Bank 2) Conditions:** Key on or engine running; and the PCM detected the Bank 2 IMRC Monitor signal was more than its expected calibrated range at closed throttle (it may be stuck in open position). An IMRCM PID of VREF at 3000 rpm may indicate a fault is present. **Possible Causes** • IMRC monitor signal circuit shorted to power ground • IMRC Monitor signal circuit shorted to signal ground (return) • IMRC actuator is damaged or has failed • PCM has failed
P1539 **2T CCM, MIL: No** All Models Transmissions: All	**Power To A/C Clutch Circuit Over-Current Conditions:** Key on or engine running; and with the A/C switch "on", the PCM detected the current in the A/C Clutch power (PWR) circuit exceeded the normal current level during the self-test. **Possible Causes** • A/C Clutch power circuit open or shorted to VPWR in harness • A/C clutch ground circuit is open • A/C clutch is open • VLCM is damaged or has failed
P1549 **1T CCM, MIL: No** Escort, Focus, E Van & F Series Truck, Ranger, LS, Five Hundred, Freestyle, Montego,, Windstar, ZX2 Models equipped with a 2.0L (VIN 3, VIN 5, VIN P) 2.3L VIN D, 4.2L VIN 2 engine Transmissions: All	**Intake Manifold Tuning Valve Circuit Malfunction Conditions:** Key on or engine running and the PCM detected an unexpected voltage on the Intake Manifold Tuning Valve (IMTV) during testing. An IMT valve (IMTVF) PID of YES status indicates a fault is present. **Possible Causes** • IMTV circuit is open or shorted to ground • IMTV power circuit (VPWR) is open • IMTV assembly is damaged or it has failed • PCM has failed
P1549 **1T CCM, MIL: No** Blackwood, LS, Five Hundred, Freestyle, Montego,, Sable, Taurus, LS, Five Hundred, Freestyle, Montego,, Expedition & Navigator Models with a 3.0L VIN S, 4.6L VIN W, 5.4L VIN A, 5.4L VIN L, 5.4L VIN R engine Transmissions: All	**Intake Manifold Runner Control Circuit Malfunction Conditions:** Key on or engine running and the PCM detected an unexpected voltage on the Intake Manifold Runner Control (IMRC) during testing. An IMRC valve PID of YES status indicates a fault is present. **Possible Causes** • IMRC circuit is open or shorted to ground • IMRC power circuit (VPWR) is open • IMRC assembly is damaged or it has failed • PCM has failed
P1549 **1T CCM, MIL: No** Aviator, Explorer & Mountaineer with a 4.6L VIN H engine Transmissions: All	**Long / Short Runner Control Circuit Malfunction Conditions:** Key on or engine running and the PCM detected an unexpected voltage on the Long / Short Runner Control (LSRC) circuit. An LSRC valve PID of YES status indicates a fault exists. **Possible Causes** • LSRC circuit is open or shorted to ground • LSRC power circuit (VPWR) is open • LSRC assembly is damaged or it has failed • PCM has failed
P1549 **1T CCM, MIL: No** Ranger Models with a 2.3L VIN D engine Transmissions: All	**Intake Manifold Swirl Control Circuit Malfunction Conditions:** Key on or engine running and the PCM detected an unexpected voltage on the Intake Manifold Swirl Control (IMSC) circuit. An IMSC valve PID of YES status indicates a fault. **Possible Causes** • IMSC circuit is open or shorted to ground • IMSC power circuit (VPWR) is open • IMSC assembly is damaged or the PCM has failed

Gas Engine OBD II Trouble Code List (P1xxx Codes)

DTC	Trouble Code Title, Conditions & Possible Causes
P1550 1T CCM, MIL: No All Models Transmissions: All	**Power Steering Pressure Switch Circuit Malfunction Conditions:** KOER Self-Test enabled, and the PCM detected the PSP switch signal did not change during the self-test. This code indicates the PSP input is out of its self-test range. **Possible Causes** • PSP switch circuit open or shorted, or the ground circuit open • Steering wheel was not rotated during the KOER Self-Test • PCM has failed
P1572 2T CCM, MIL: Yes All Models Transmissions: All	**Brake Pedal Switch Circuit Malfunction Conditions:** Engine started, and the PCM detected the Brake Pedal switch and Brake Pressure switch inputs failed the Rationality test (i.e., one or both of these inputs did not change as expected). DTC P1572 is set when the PCM does not see the proper sequence of the brake pedal input signal from both the BPP and BPA when the brake pedal is pressed and released. **Possible Causes** • BPP or BPA switches are out of adjustment (one or both) • Blown fuse to switch power circuit • BPP switch or BPA switch is damaged (one or both) • BPP or BPA switch circuit is open or shorted • PCM has failed
P1605 1T PCM, MIL: Yes All Models Transmissions: All	**PCM Keep Alive Memory Test Error Conditions:** Key on, and the PCM detected an internal memory fault. This code can be set if KAPWR to the PCM is interrupted. This trouble code will set at first key on if a battery circuit is opened. **Possible Causes** • Battery terminals loose or corroded (high resistance in circuit) • Keep Alive Memory circuit to PCM interrupted or open • Reprogramming function not performed • PCM has failed
P1625 2T CCM, MIL: Yes Mark VIII Transmissions: All	**Battery Supply to VLCM Fan Circuit Malfunction Conditions:** Key on or engine running; and the PCM detected no battery power at the VLCM fan circuit. **Possible Causes** • B+ circuit to VLCM (pins 4 and 5) has been open (check for an intermittent condition)
P1626 2T CCM, MIL: Yes Mark VIII Transmissions: All	**Battery Supply To VLCM A/C Circuit Malfunction Conditions:** Key on or engine running; and the PCM detected no battery power at the VLCM fan circuit. **Possible Causes** • B+ circuit to VLCM (pins 4 and 5) is open (check for an intermittent condition)
P1633 1T PCM, MIL: Yes All Models Transmissions: All	**PCM Keep Alive Memory Voltage Too Low Conditions:** Key on, and the PCM detected that the KAM power circuit to the battery was interrupted. **Possible Causes** • KAPWR circuit has been interrupted (this problem may be an intermittent condition) • PCM has failed
P1635 1T PCM, MIL: Yes All Models Transmissions: All	**Tire Axle/Ratio Out Of Acceptable Range Conditions:** Key on, and the PCM detected the tire and axle information in the VID Block does not match the vehicle hardware. Note: This code indicates that the PCM needs to be reprogrammed. **Possible Causes** • Incorrect tire size or Incorrect axle ratio • Incorrect VID configuration parameters • PCM need to be reprogrammed • TSB 02-23-4 contains repair information for this trouble code
P1636 1T PCM, MIL: Yes All Models Transmissions: All	**Inductive Signature Chip Communication Error Conditions:** Key on, and the PCM determined it had lost communication with the Inductive Signature Chip. The PCM has internal damage when this trouble code is present. **Possible Causes** • PCM has failed and needs to be replaced

Gas Engine OBD II Trouble Code List (P1xxx Codes)

DTC	Trouble Code Title, Conditions & Possible Causes
P1639 **1T PCM, MIL: Yes** All Models Transmissions: All	**Vehicle ID Block Not Programmed Or Is Corrupt Conditions:** Key on, and the PCM determined the Vehicle ID Block information was incorrect. **Possible Causes** • PCM may not be the correct application • PCM may need to be reprogrammed • VID configuration may not be correct • TSB 02-23-4 contains repair information for this trouble code
P1640 **1T PCM, MIL: Yes** All Models Transmissions: All	**PCM Trouble Codes Available In Another Module Conditions:** Engine started, and the PCM received a request from another module to turn on the MIL due to a fault that could affect emissions. Note: Vehicles using a secondary Engine Control Module can request that the PCM turn on the Check Engine Light when a failure occurs that could affect emissions. Request PID 0946 to determine which module made the request. Then select that module to read the related trouble code(s). **Possible Causes** • Trouble codes are stored in a secondary module, which in turn, requested that the PCM turn on the MIL when this code is set.
P1650 **2T CCM, MIL: Yes** All Models Transmissions: All	**Power Steering Pressure Switch Circuit Malfunction Conditions:** Engine started, and the PCM detected the PSP switch signal did not change after a certain number of vehicle speed transitions. The PCM counts the number of times that the vehicle speed transitions from 0 mph to a calibrated speed. The PCM expects the PSP switch input to change after a certain number of transitions. **Possible Causes** • Steering wheel must be turned during the KOER Self-Test • PSP switch/shorting bar is damaged • PSP signal circuit is open or shorted to ground • PSP switch ground (return) circuit is open • PCM has failed
P1651 **2T CCM, MIL: Yes** All Models Transmissions: All	**Power Steering Pressure Switch Circuit Malfunction Conditions:** Engine started, and the PCM detected the PSP switch signal did not change after a certain number of vehicle speed transitions. Note: The PCM counts the number of times that the vehicle speed transitions from 0 mph to a calibrated speed. The PCM expects the PSP switch input to change after a certain number of transitions. **Possible Causes** • Steering wheel must be turned during the KOER Self-Test • PSP switch/shorting bar is damaged • PSP signal circuit is open or shorted to ground • PSP switch ground (return) circuit is open • PCM has failed
P1700 **1T CCM, MIL: No** All Models Transmissions: A/T	**Transaxle Mechanical Malfunction Conditions:** Engine started, vehicle driven in gear, and the PCM detected a transmission mechanical fault. **Possible Causes** • This code can set due to low transmission fluid level • Refer to the appropriate Transmission Repair Manual or information in electronic media to perform a complete diagnosis of the automatic transmission when this code is set
P1701 **1T CCM, MIL: No** All Models Transmissions: A/T	**Reverse Engagement Error Conditions:** Engine started, and the PCM detected a Transmission Range (TR) sensor signal that indicated a reverse engagement error. **Possible Causes** • Refer to the appropriate Transmission Repair Manual or information in electronic media to perform a complete diagnosis of the automatic transmission when this code is set
P1702 **1T CCM, MIL: No** All Models Transmissions: A/T	**TR Sensor Signal Intermittent Conditions:** Key on or engine running; and the PCM detected the failure Trouble Code Conditions for DTC P0705 or P0708 were met intermittently. **Possible Causes** • Refer to the appropriate Transmission Repair Manual or information in electronic media to perform a complete diagnosis of the automatic transmission when this code is set

Gas Engine OBD II Trouble Code List (P1xxx Codes)

DTC	Trouble Code Title, Conditions & Possible Causes
P1703 **1T CCM, MIL: No** All Models Transmissions: A/T	**Brake Switch Circuit Out of Self-Test Range Conditions:** Key on, KOEO Self-Test enabled; and the PCM detected the brake switch signal was high, or with the KOER Self-Test enabled, the PCM detected the switch signal did not cycle On / Off. **Possible Causes** • BPP switch circuit open or shorted • Brake Switch is misadjusted, damaged or has failed • Stop lamp circuits open or shorted • Malfunction in the module(s) connected to BPP circuit (i.e., the Rear Electronic Module on Windstar and LS, or the Lighting Control Module on the Continental and Town Car) • PCM has failed
P1704 **1T CCM, MIL: No** All Models Transmissions: A/T	**Transmission Range Sensor Circuit Out Of Self-Test Range Conditions:** Key on, KOEO Self Test enabled, and the PCM detected a Transmission Range (TR) sensor signal occurred in between gear positions. **Possible Causes** • Digital TR sensor or shift cable misadjusted • Digital TR sensor circuit is open or shorted to ground • Digital TR sensor has failed
P1705 **1T CCM, MIL: No** All Models Transmissions: A/T	**Transmission Range Sensor Out of Self-Test Range Conditions:** Key on, KOEO Self Test enabled, and the PCM detected it did not receive a Transmission Range (TR) sensor signal in Park or Neutral position. **Possible Causes** • Gear selector not in Park or Neutral during the self-test • Digital TR sensor circuit is open or shorted to ground • Digital TR sensor has failed • PCM has failed
P1705 **1T CCM, MIL: Yes** Villager Models Transmissions: A/T	**Throttle Position Sensor Circuit Malfunction Conditions:** Key on, KOEO Self-Test enabled, or engine running, KOER Self-Test enabled, and the PCM detected the TP sensor was too high or low. **Possible Causes** • TP sensor signal circuit is open • TP sensor signal circuit is shorted to ground • TP sensor circuit is shorted to VREF or power • TP switch is open, shorted to ground or power • TP sensor or the TP switch has failed, or the PCM has failed
P1708 **1T CCM, MIL: Yes** All Models Transmissions: A/T	**Digital Transmission Range Sensor Circuit Malfunction Conditions:** Engine started, and the PCM detected it did not receive a change in the Digital Transmission Range (TR) sensor signal after the vehicle was driven in gear. **Possible Causes** • Digital TR sensor circuit open • Digital TR sensor ground circuit open • Digital TR sensor is damaged or has failed • PCM has failed
P1709 **1T CCM, MIL: No** All Models Transmissions: A/T	**PNP Switch Out Of Self-Test Range Conditions:** Key on, KOEO Self-Test enabled, and the PCM detected the PNP switch was high when is should have been low (wrong gearshift position). **Possible Causes** • PNP switch ground circuit is open • PNP switch circuit short to power (VPWR) • PNP switch is damaged or has failed • PCM has failed
P1710 **2T CCM, MIL: Yes** All Models Transmissions: A/T	**TFT Sensor In-Range Circuit Malfunction Conditions:** Engine started, vehicle driven to a speed over 1 mph, TFT sensor signal in-range, and the PCM did not detect any change in the TFT signal in the self-test. **Possible Causes** • Refer to the appropriate Transmission Repair Manual or information in electronic media to perform a complete diagnosis of the automatic transmission when this code is set

Gas Engine OBD II Trouble Code List (P1xxx Codes)

DTC	Trouble Code Title, Conditions & Possible Causes
P1711 1T CCM, MIL: No All Models Transmissions: A/T	**TFT Sensor Out of Self-Test Range Conditions:** Key on, KOER Self Test enabled; or engine running with the KOER Self Test enabled, and the PCM detected the Transmission Fluid Temperature (TFT) sensor was more than or less than the calibrated range (25ºF to 240ºF) during the self-test. **Possible Causes** • Refer to the appropriate Transmission Repair Manual or information in electronic media to perform a complete diagnosis of the automatic transmission when this code is set
P1712 1T CCM, MIL: No All Models Transmissions: A/T	**TFT Sensor Circuit Low Input Conditions:** **Engine started, and the PCM detected the TFT sensor signal was less than 0.2v (equivalent to a temperature of more than 357ºF).** **Possible Causes** • Refer to the appropriate Transmission Repair Manual or information in electronic media to perform a complete diagnosis of the automatic transmission when this code is set
P1713 1T CCM, MIL: No All Models Transmissions: A/T	**TFT Sensor No Activity or TFT Sensor Circuit Low Input Conditions:** **Engine started, VSS over 1 mph, and the PCM did not detect any change in the TFT low range circuit during the self-test.** **Possible Causes** • Refer to the appropriate Transmission Repair Manual or information in electronic media to perform a complete diagnosis of the automatic transmission when this code is set
P1714 1T CCM, MIL: Yes All Models Transmissions: A/T	**Transmission Control System Malfunction Conditions:** **Engine started, VSS over 1 mph, and the PCM did not detect any change in the TFT low range circuit during the self-test.** **Possible Causes** • Refer to the appropriate Transmission Repair Manual or information in electronic media to perform a complete diagnosis of the automatic transmission when this code is set
P1715 1T CCM, MIL: Yes All Models Transmissions: A/T	**Transmission Control System Malfunction Conditions:** **Engine started, VSS over 1 mph, and the PCM detected a mechanical problem in the Shift Solenoid 'B' (SSB) during the test.** **Possible Causes** • Refer to the appropriate Transmission Repair Manual or information in electronic media to perform a complete diagnosis of the automatic transmission when this code is set
P1716 2T CCM, MIL: No All Models Transmissions: A/T	**Transmission Control System Malfunction Conditions:** **Engine started, VSS over 1 mph, and the PCM detected a problem in the Transmission Control system during the self-test.** **Possible Causes** • Refer to the appropriate Transmission Repair Manual or information in electronic media to perform a complete diagnosis of the automatic transmission when this code is set
P1717 1T CCM, MIL: No All Models Transmissions: A/T	**Transmission Control System Malfunction Conditions:** **Engine started, VSS over 1 mph, and the PCM detected a problem in the Transmission Control system during the self-test.** **Possible Causes** • Refer to the appropriate Transmission Repair Manual or information in electronic media to perform a complete diagnosis of the automatic transmission when this code is set
P1718 1T CCM, MIL: No All Models Transmissions: A/T	**TFT Sensor No Activity Or TFT Sensor Circuit High Input Conditions:** **Engine started, VSS over 1 mph, and the PCM did not detect any change in the TFT high range circuit during the self-test.** **Possible Causes** • Refer to the appropriate Transmission Repair Manual or information in electronic media to perform a complete diagnosis of the automatic transmission when this code is set
P1719 1T CCM, MIL: No All Models Transmissions: A/T	**Transmission Control System Malfunction Conditions:** **Engine started, VSS over 1 mph, and the PCM detected a problem in the Transmission Control system during the self-test.** **Possible Causes** • Refer to the appropriate Transmission Repair Manual or information in electronic media to perform a complete diagnosis of the automatic transmission when this code is set

Gas Engine OBD II Trouble Code List (P1xxx Codes)

DTC	Trouble Code Title, Conditions & Possible Causes
P1727 **1T CCM, MIL: No** All Models Transmissions: A/T	**Transmission Coast Clutch Solenoid Slip Malfunction Conditions:** Engine started, VSS over 1 mph in gear, and the PCM detected a signal that indicated the coast clutch solenoid had a slippage fault. **Possible Causes** • Refer to the appropriate Transmission Repair Manual or information in electronic media to perform a complete diagnosis of the automatic transmission when this code is set
P1728 **1T CCM, MIL: No** All Models Transmissions: A/T	**Transmission Slip Malfunction Conditions:** Engine started, VSS over 1 mph in gear, and the PCM detected a signal that indicated the transmission was slipping while in gear. **Possible Causes** • Refer to the appropriate Transmission Repair Manual or information in electronic media to perform a complete diagnosis of the automatic transmission when this code is set
P1729 **1T CCM, MIL: No** All Models Transmissions: A/T	**4 x 4 Low Switch Circuit Malfunction Conditions:** Engine started and the PCM detected the 4x4 switch did not go low after the switch was on. **Possible Causes** • Speedometer out of calibration • 4x4L wiring harness is open or shorted, 4x4L switch is damaged or has failed • Electronic Shift Control Module is damaged or has failed, or the PCM has failed
P1740 **1T CCM, MIL: Yes** All Models Transmissions: A/T	**TCC Solenoid Mechanical Malfunction Conditions:** Engine started, vehicle speed more than 20 mph, and the PCM detected that TCC lockup did not occur (the lockup event is inferred from other inputs). **Possible Causes** • Refer to the appropriate Transmission Repair Manual or information in electronic media to perform a complete diagnosis of the automatic transmission when this code is set
P1741 **1T CCM, MIL: No** Some Models Transmissions: A/T	**TCC Engagement Error Conditions:** Engine started, vehicle in gear at Cruise speed, and the PCM detected an error due to excessive TCC engagement. Note: This problem can cause speed changes or vehicle surges. **Possible Causes** • Refer to the appropriate Transmission Repair Manual or information in electronic media to perform a complete diagnosis of the automatic transmission when this code is set
P1741 **1T CCM, MIL: No** Escape, MF, Contour, Mystique, Probe Transmissions: A/T	**TCC Engagement Error Conditions:** Engine started, vehicle in gear at Cruise speed, and the PCM detected an error due to excessive TCC engagement. Note: This problem can cause speed changes or vehicle surges. **Possible Causes** • Refer to the appropriate Transmission Repair Manual for more information on this code • TSB 02-2-4 contains a repair procedure for this trouble code
P1742 **1T CCM, MIL: Yes** Some Models Transmissions: A/T	**TCC Solenoid Failed On (Electrical Or Mechanical Fault) Conditions:** Engine started, vehicle in gear at Cruise speed, and the PCM detected that the Torque Converter Clutch system had failed "on". **Possible Causes** • Refer to the appropriate Transmission Repair Manual or information in electronic media to perform a complete diagnosis of the automatic transmission when this code is set.
P1744 **1T CCM, MIL: Yes** Some Models Transmissions: A/T	**TCC System Mechanically Stuck In Off Position Conditions:** Engine started, vehicle in gear at Cruise speed, and the PCM detected the Torque Converter Clutch system had failed with the TCC in the mechanically "off" position. **Possible Causes** • Refer to the appropriate Transmission Repair Manual or information in electronic media to perform a complete diagnosis of the automatic transmission when this code is set.
P1744 **2T CCM, MIL: No** Continental, Taurus, Sable, Windstar models Transmissions: A/T	**EPC Solenoid Circuit Malfunction Conditions:** Engine started, vehicle in gear, and the PCM detected the Electronic Pressure Control (EPC) solenoid control circuit was "open". This fault can cause maximum can cause harsh shifts. **Possible Causes** • Refer to the appropriate Transmission Repair Manual for more information on this code • PCM has failed • TSB 3-12-3 contains a repair procedure for this trouble code

Gas Engine OBD II Trouble Code List (P1xxx Codes)

DTC	Trouble Code Title, Conditions & Possible Causes
P1746 **1T CCM, MIL: No** Some Models Transmissions: A/T	**EPC Solenoid Circuit Malfunction Conditions:** Engine started, vehicle in gear, and the PCM detected the Electronic Pressure Control (EPC) solenoid circuit indicated "open". This fault can cause harsh engagements and shifts. **Possible Causes** • Refer to the appropriate Transmission Repair Manual or information in electronic media to perform a complete diagnosis of the automatic transmission when this code is set
P1747 **1T CCM, MIL: No** Some Models Transmissions: A/T	**A/T EPC Solenoid Circuit Malfunction Conditions:** Engine started, vehicle in gear at Cruise speed, and the PCM detected a shorted output driver or the TCC solenoid was shorted. **Possible Causes** • Refer to the appropriate Transmission Repair Manual or information in electronic media to perform a complete diagnosis of the automatic transmission when this code is set.
P1749 **1T CCM, MIL: No** Some Models Transmissions: A/T	**A/T EPC Solenoid Failed Low Conditions:** Engine started, vehicle in gear at Cruise speed, and the PCM detected the Torque Converter Clutch solenoid had failed "low". **Possible Causes** • Refer to the appropriate Transmission Repair Manual or information in electronic media to perform a complete diagnosis of the automatic transmission when this code is set.
P1751 **1T CCM, MIL: No** Some Models Transmissions: A/T	**A/T Shift Solenoid 1 Performance Conditions:** Engine started, vehicle in gear at Cruise speed, and the PCM detected a mechanical fault in the Shift Solenoid 1 (SS1) operation. **Possible Causes** • Refer to the appropriate Transmission Repair Manual or information in electronic media to perform a complete diagnosis of the automatic transmission when this code is set.
P1754 **1T CCM, MIL: No** Some Models Transmissions: A/T	**A/T Coast Clutch Solenoid Circuit Malfunction Conditions:** Engine started, vehicle in gear at Cruise speed, and the PCM detected an unexpected voltage condition on the Coast Clutch Solenoid (CCS) circuit during the CCM test period. **Possible Causes** • Refer to the appropriate Transmission Repair Manual or information in electronic media to perform a complete diagnosis of the automatic transmission when this code is set.
P1756 **1T CCM, MIL: No** Some Models Transmissions: A/T	**A/T Shift Solenoid 2 Performance Conditions:** Engine started, vehicle in gear at Cruise speed, and the PCM detected a mechanical fault in the Shift Solenoid 2 (SS2) operation. **Possible Causes** • Refer to the appropriate Transmission Repair Manual or information in electronic media to perform a complete diagnosis of the automatic transmission when this code is set
P1760 **1T CCM, MIL: No** Some Models Transmissions: A/T	**A/T EPC Solenoid Circuit Malfunction Conditions:** Engine started, vehicle in gear at Cruise speed, and the PCM detected a shorted output driver or the TCC solenoid was shorted. **Possible Causes** • Refer to the appropriate Transmission Repair Manual or information in electronic media to perform a complete diagnosis of the automatic transmission when this code is set.
P1761 **1T CCM, MIL: No** Some Models Transmissions: A/T	**A/T Shift Solenoid 3 Performance Conditions:** Engine started, vehicle in gear at Cruise speed, and the PCM detected a malfunction in the Shift Solenoid 3 (SS3) operation. **Possible Causes** • Refer to the appropriate Transmission Repair Manual or information in electronic media to perform a complete diagnosis of the automatic transmission when this code is set.
P1762 **1T CCM, MIL: No** Some Models Transmissions: A/T	**Transmission System Malfunction Conditions:** Engine started, vehicle in gear at Cruise speed, and the PCM detected a malfunction in the Transmission System operation. **Possible Causes** • Refer to the appropriate Transmission Repair Manual or information in electronic media to perform a complete diagnosis of the automatic transmission when this code is set.

Gas Engine OBD II Trouble Code List (P1xxx Codes)

DTC	Trouble Code Title, Conditions & Possible Causes
P1767 1T CCM, MIL: No Some Models Transmissions: A/T	**A/T Shift Solenoid Performance Conditions:** Engine started, vehicle in gear at Cruise speed, and the PCM detected a malfunction in the Shift Solenoid operation. **Possible Causes** • Refer to the appropriate Transmission Repair Manual or information in electronic media to perform a complete diagnosis of the automatic transmission when this code is set.
P1780 1T CCM, MIL: No Some Models Transmissions: A/T	**Transmission Control Switch Out of Self-Test Range Conditions:** Engine started, KOER Self-Test enabled, and the PCM detected the Transmission Control Switch (TCS) was out of range during the test. **Possible Causes** • TCS circuit open or shorted in the wiring harness • TCS not cycled during the self-test • TCS is damaged, or the PCM has failed
P1781 1T CCM, MIL: No Some Models Transmissions: All	**4 x 4 Low Switch Out Of Self-Test Range Conditions:** Key on, KOEO Self-Test enabled, and the PCM detected the 4x4 switch input was not low with the switch engaged or "on". **Possible Causes** • 4x4L switch circuit is open or shorted in the wiring harness • Electronic Shift Module is damaged or has failed • PCM has failed
P1783 1T CCM, MIL: No Some Models Transmissions: A/T	**Transmission Over-Temperature Malfunction Conditions:** Engine started, engine runtime more than 5 minutes, vehicle in gear at Cruise speed, and the PCM detected the TFT sensor signal was more than 300ºF during the CCM test period. **Possible Causes** • Refer to the appropriate Transmission Repair Manual or information in electronic media to perform a complete diagnosis of the automatic transmission when this code is set.
P1784 1T CCM, MIL: No Some Models Transmissions: A/T	**Transmission System First Or Reverse Gear Malfunction Conditions:** Engine started, vehicle speed over 1 mph in gear, shift command received for First or Reverse gear, and the PCM detected a problem in the Transmission Control system. **Possible Causes** • Refer to the appropriate Transmission Repair Manual or information in electronic media to perform a complete diagnosis of the automatic transmission when this code is set.
P1785 1T CCM, MIL: No Some Models Transmissions: A/T	**Transmission System First Or Second Gear Malfunction Conditions:** Engine started, vehicle speed over 1 mph in gear, shift command received for First or Second gear, and the PCM detected a problem in the Transmission Control system during the test. **Possible Causes** • Refer to the appropriate Transmission Repair Manual or information in electronic media to perform a complete diagnosis of the automatic transmission when this code is set
P1786 1T CCM, MIL: No Some Models Transmissions: A/T	**Transmission System Second Or Third Gear Malfunction Conditions:** Engine started, vehicle speed over 1 mph in gear, shift command received for Second or Third gear, and the PCM detected a problem in the Transmission Control system. **Possible Causes** • Refer to the appropriate Transmission Repair Manual or information in electronic media to perform a complete diagnosis of the automatic transmission when this code is set.
P1787 1T CCM, MIL: No Some Models Transmissions: A/T	**Transmission System Third Or Fourth Gear Malfunction Conditions:** Engine started, vehicle speed over 1 mph in gear, shift command received for Third or Fourth gear, and the PCM detected a problem in the Transmission Control system during the test. **Possible Causes** • Refer to the appropriate Transmission Repair Manual or information in electronic media to perform a complete diagnosis of the automatic transmission when this code is set.
P1788 1T CCM, MIL: No Some Models Transmissions: A/T	**3-2 Timing/Coast Clutch Solenoid Signal High Input Conditions:** Engine started, vehicle in gear at Cruise speed, and the PCM detected the malfunction 3-2 Timing or Coast Clutch solenoid circuit. **Possible Causes** • 3-2 Timing or Coast Clutch solenoid circuit open or grounded, or the solenoid has failed • Coast Clutch solenoid is damaged or has failed

Gas Engine OBD II Trouble Code List (P1xxx Codes) - (P2xxx Codes)

DTC	Trouble Code Title, Conditions & Possible Causes
P1789 **1T CCM, MIL: No** Some Models Transmissions: A/T	**3-2 Timing/Coast Clutch Solenoid Signal Low Input Conditions:** Engine started, vehicle in gear at Cruise speed, and the PCM detected the malfunction 3-2 Timing or Coast Clutch solenoid circuit. **Possible Causes** • 3-2 Timing or Coast Clutch solenoid circuit is shorted • 3-2 Timing solenoid is damaged or has failed • Coast Clutch solenoid is damaged or has failed
P1900 **1T CCM, MIL: No** Some Models Transmissions: A/T	**Transmission System Malfunction Conditions:** Engine started, vehicle in gear at Cruise speed, and the PCM detected a malfunction in the Transmission System operation. **Possible Causes** • Refer to the appropriate Transmission Repair Manual or information in electronic media to perform a complete diagnosis of the automatic transmission when this code is set
P1901 **1T CCM, MIL: No** Some Models Transmissions: A/T	**Transmission System Malfunction Conditions:** Engine started, vehicle in gear at Cruise speed, and the PCM detected a malfunction in the Transmission System operation. **Possible Causes** • Refer to the appropriate Transmission Repair Manual or information in electronic media to perform a complete diagnosis of the automatic transmission when this code is set.
P2004 **1T CCM, MIL: No** Escort, Focus, Aviator, Mountaineer, Explorer, LS, Five Hundred, Freestyle, Montego,, Ranger, Blackwood E Van, F Series Truck, Expedition, Navigator, Windstar, ZX2 with a 2.0L (VIN 3, VIN 5, 2.0L VIN P), 2.3L VIN D, 3.0L VIN S, 4.2L VIN 2, 4.6L VIN H, 4.6L VIN W, 5.4L VIN A, 5.4L VIN L, 5.4L VIN R engine Transmissions: All	**Intake Air System Malfunction Conditions:** Engine started, engine running at hot idle speed for one minute, and the PCM detected a problem in the Intake Air System operation. It should be noted that the throttle bore cannot be cleaned as any attempt to clean it will damage the throttle bore and plate. **Possible Causes** • Test for a sticking Accelerator or speed control cable condition: Turn the key off and disconnect accelerator and speed control cable from the throttle body. Rotate the throttle body linkage to determine if it rotates freely (the throttle body may have failed). • Check the air cleaner and air inlet assembly for restrictions • Check the IAC motor response (it may be damaged or sticking) • Check the PCV system (valve and hoses) for leaks or plugging • Check for signs of vacuum leaks in the engine or components • Test TP sensor signal (due a sweep test at key on, engine off)
P2005 **1T CCM, MIL: No** Escort, Focus, Aviator, Mountaineer, Explorer, LS, Five Hundred, Freestyle, Montego,, Ranger, Blackwood E Van, F Series Truck, Expedition, Navigator, Windstar, ZX2 with a 2.0L (VIN 3, VIN 5, 2.0L VIN P), 2.3L VIN D, 3.0L VIN S, 4.2L VIN 2, 4.6L VIN H, 4.6L VIN W, 5.4L VIN A, 5.4L VIN L, 5.4L VIN R engine Transmissions: All	**Intake Air System Malfunction Conditions:** Engine started, engine running at hot idle speed for one minute, and the PCM detected a problem in the Intake Air System operation. It should be noted that the throttle bore cannot be cleaned as any attempt to clean it will damage the throttle bore and plate. **Possible Causes** • Test for a sticking Accelerator or speed control cable condition: Turn the key off and disconnect accelerator and speed control cable from the throttle body. Rotate the throttle body linkage to determine if it rotates freely (the throttle body may have failed). • Check the air cleaner and air inlet assembly for restrictions • Check the IAC motor response (it may be damaged or sticking) • Check the PCV system (valve and hoses) for leaks or plugging • Check for signs of vacuum leaks in the engine or components • Test TP sensor signal (due a sweep test at key on, engine off)

Gas Engine OBD II Trouble Code List (P2xxx Codes)

DTC	Trouble Code Title, Conditions & Possible Causes
P2006 **1T CCM, MIL: No** Escort, Focus, Aviator, Mountaineer, Explorer, LS, Five Hundred, Freestyle, Montego,, Ranger, Blackwood E Van, F Series Truck, Expedition, Navigator, Windstar, ZX2 with a 2.0L (VIN 3, VIN 5, 2.0L VIN P), 2.3L VIN D, 3.0L VIN S, 4.2L VIN 2, 4.6L VIN H, 4.6L VIN W, 5.4L VIN A, 5.4L VIN L, 5.4L VIN R engine Transmissions: All	**Intake Air System Malfunction Conditions:** Engine started, engine running at hot idle speed for one minute, and the PCM detected a problem in the Intake Air System operation. It should be noted that the throttle bore cannot be cleaned as any attempt to clean it will damage the throttle bore and plate. **Possible Causes** • Test for a sticking Accelerator or speed control cable condition: Turn the key off and disconnect accelerator and speed control cable from the throttle body. Rotate the throttle body linkage to determine if it rotates freely (the throttle body may have failed). • Check the air cleaner and air inlet assembly for restrictions • Check the IAC motor response (it may be damaged or sticking) • Check the PCV system (valve and hoses) for leaks or plugging • Check for signs of vacuum leaks in the engine or components • Test TP sensor signal (due a sweep test at key on, engine off)
P2008 **1T CCM, MIL: No** Escort, Focus, Aviator, Mountaineer, Explorer, LS, Five Hundred, Freestyle, Montego,, Ranger, Blackwood E Van, F Series Truck, Expedition, Navigator, Windstar, ZX2 with a 2.0L (VIN 3, VIN 5, 2.0L VIN P), 2.3L VIN D, 3.0L VIN S, 4.2L VIN 2, 4.6L VIN H, 4.6L VIN W, 5.4L VIN A, 5.4L VIN L, 5.4L VIN R engine Transmissions: All	**Intake Air System Malfunction Conditions:** Engine started, engine running at hot idle speed for one minute, and the PCM detected a problem in the Intake Air System operation. It should be noted that the throttle bore cannot be cleaned as any attempt to clean it will damage the throttle bore and plate. **Possible Causes** • Accelerator or speed control cable sticking or binding. To test for this condition, turn the key off. Then disconnect the accelerator and speed control cable from the throttle body. Then rotate the throttle body linkage to determine if it rotates freely. If it is sticking, the throttle body may need replacement. • Check the air cleaner and air inlet assembly for restrictions • Check the IAC motor response (it may be damaged or sticking) • Check the PCV system (valve and hoses) for leaks or plugging • Check for signs of vacuum leaks in the engine or components • Test TP sensor signal (due a sweep test at key on, engine off)
P2014 **1T CCM, MIL: No** Escort, Focus, Aviator, Mountaineer, Explorer, LS, Five Hundred, Freestyle, Montego,, Ranger, Blackwood E Van, F Series Truck, Expedition, Navigator, Windstar, ZX2 with a 2.0L (VIN 3, VIN 5, 2.0L VIN P), 2.3L VIN D, 3.0L VIN S, 4.2L VIN 2, 4.6L VIN H, 4.6L VIN W, 5.4L VIN A, 5.4L VIN L, 5.4L VIN R engine Transmissions: All	**Intake Air System Malfunction Conditions:** Engine started, engine running at hot idle speed for one minute, and the PCM detected a problem in the Intake Air System operation. It should be noted that the throttle bore cannot be cleaned as any attempt to clean it will damage the throttle bore and plate. **Possible Causes** • Accelerator or speed control cable sticking or binding. To test for this condition, turn the key off. Then disconnect the accelerator and speed control cable from the throttle body. Then rotate the throttle body linkage to determine if it rotates freely. If it is sticking, the throttle body may need replacement. • Check the air cleaner and air inlet assembly for restrictions • Check the IAC motor response (it may be damaged or sticking) • Check the PCV system (valve and hoses) for leaks or plugging • Check for signs of vacuum leaks in the engine or components • Test TP sensor signal (due a sweep test at key on, engine off)

Gas Engine OBD II Trouble Code List (P2xxx Codes)

DTC	Trouble Code Title, Conditions & Possible Causes
P2019 **1T CCM, MIL: No** Escort, Focus, Aviator, Mountaineer, Explorer, LS, Five Hundred, Freestyle, Montego,, Ranger, Blackwood E Van, F Series Truck, Expedition, Navigator, Windstar, ZX2 with a 2.0L (VIN 3, VIN 5, 2.0L VIN P), 2.3L VIN D, 3.0L VIN S, 4.2L VIN 2, 4.6L VIN H, 4.6L VIN W, 5.4L VIN A, 5.4L VIN L, 5.4L VIN R engine Transmissions: All	**Intake Air System Malfunction Conditions:** Engine started, engine running at hot idle speed for one minute, and the PCM detected a problem in the Intake Air System operation. It should be noted that the throttle bore cannot be cleaned as any attempt to clean it will damage the throttle bore and plate. **Possible Causes** • Accelerator or speed control cable sticking or binding. To test these devices, turn the key off and disconnect the accelerator and speed control cable from the throttle body. Then rotate the throttle body linkage to determine if it rotates freely. • Check the air cleaner and air inlet assembly for restrictions • Check the IAC motor response (it may be damaged or sticking) • Check the PCV system (valve and hoses) for leaks or plugging • Check for signs of vacuum leaks in the engine or components • Test TP sensor signal (due a sweep test at key on, engine off)
P2070 **1T CCM, MIL: No** Escort, Focus, E Van, F Series Truck, Windstar, ZX2 engines with a 2.0L (VIN 3, VIN 5, VIN P), 2.3L VIN D, 4.2L VIN 2 engine Transmissions: All	**Intake Manifold Tuning Valve Malfunction (Stuck Open) Conditions:** Key on or engine running; and the PCM detected an unexpected low voltage condition on the Intake Manifold Tuning Valve circuit during the CCM test period (i.e., the valve may be stuck open). **Possible Causes** • IMTV signal circuit shorted to chassis ground • IMTV signal circuit shorted to sensor ground • IMTV actuator is damaged or has failed • PCM has failed
P2070 **2T CCM, MIL: No** E Van, F Series Truck, Blackwood, Expedition Navigator & LS with a 3.0L VIN S, 4.6L VIN W, 5.4L VIN A, 5.4L VIN L, 5.4L VIN R Transmissions: All	**Intake Manifold Runner Control Malfunction (Stuck Open) Conditions:** Key on or engine running; and the PCM detected an unexpected low voltage condition on the Intake Manifold Runner Control circuit during the CCM test period (i.e., the valve may be stuck open). **Possible Causes** • IMRC signal circuit shorted to chassis ground • IMRC signal circuit shorted to sensor ground • IMRC actuator is damaged or has failed • PCM has failed
P2070 **2T CCM, MIL: No** Aviator, Mountaineer, Explorer equipped with a 4.6L VIN H engine Transmissions: All	**Long / Short Runner Control Circuit Malfunction (Open) Conditions:** Key on or engine running; and the PCM detected an unexpected low voltage condition on the Long / Short Runner Control (LSRC) signal circuit during the test (i.e., the valve may be stuck open). The PCM uses the Intake Manifold Communication Control (IMCC) signal to monitor the status of the Long / Short Runner Control position. **Possible Causes** • LSRC signal circuit is shorted to chassis ground • LSRC signal circuit is shorted to sensor ground • LSRC actuator assembly is damaged or failed • PCM has failed
P2070 **2T CCM, MIL: No** Ranger equipped with a 2.3L VIN D engine Transmissions: All	**Intake Manifold Swirl Control Actuator Circuit Malfunction Conditions:** Key on or engine running; and the PCM detected an unexpected low voltage condition on the Intake Manifold Swirl Control (IMSC) circuit (i.e., the valve may be stuck open). **Possible Causes** • IMSC signal circuit is shorted to chassis ground • IMSC signal circuit is shorted to sensor ground • IMSC actuator assembly is damaged or failed • PCM has failed

Gas Engine OBD II Trouble Code List (P2xxx Codes)

DTC	Trouble Code Title, Conditions & Possible Causes
P2071 **2T CCM, MIL: No** Escort, Focus, E Van, F Series Truck, Windstar, ZX2 engines with a 2.0L (VIN 3, VIN 5, VIN P), 2.3L VIN D, 4.2L VIN 2 engine Transmissions: All	**Intake Manifold Tuning Valve Circuit Malfunction (Stuck Closed) Conditions:** Key on or engine running; and the PCM detected an unexpected high voltage condition on the Intake Manifold Tuning Valve circuit during the CCM test period (i.e., the valve may be stuck closed). **Possible Causes** • IMTV signal circuit is open • IMTV power circuit (VPWR) is open • IMTV actuator is damaged or has failed • PCM has failed
P2071 **2T CCM, MIL: No** E Van, F Series Truck, Blackwood, Expedition Navigator & LS with a 3.0L VIN S, 4.6L VIN W, 5.4L VIN A, 5.4L VIN L, 5.4L VIN R Transmissions: All	**Intake Manifold Runner Control Circuit Malfunction (Stuck Closed) Conditions:** Key on or engine running; and the PCM detected an unexpected high voltage condition on the Intake Manifold Runner Control (IMRC) circuit during the CCM test (i.e., the valve may be stuck closed). **Possible Causes** • IMRC monitor signal circuit is open • IMRC power circuit (VPWR) is open • IMRC actuator is damaged or has failed • PCM has failed
P2071 **2T CCM, MIL: No** Aviator, Mountaineer, Explorer equipped with a 4.6L VIN H engine Transmissions: All	**Long / Short Runner Control Circuit Malfunction (Stuck Closed) Conditions:** Key on or engine running; and the PCM detected an unexpected high voltage condition on the Long / Short Runner Control (LSRC) circuit (i.e., the valve may be stuck closed). The PCM uses the IMCC signal to monitor the status of the Long / Short Runner Control valve. **Possible Causes** • LSRC control (signal) circuit is open • LSRC power circuit (VPWR) is open • LSRC actuator assembly is damaged or failed • PCM has failed
P2071 **2T CCM, MIL: No** Ranger equipped with a 2.3L VIN D engine Transmissions: All	**Intake Manifold Swirl Control Circuit Malfunction (Stuck Closed) Conditions:** Key on or engine running; and the PCM detected an unexpected high voltage condition on the Intake Manifold Swirl Control (IMSC) Monitor circuit (i.e., the valve is stuck closed). **Possible Causes** • IMSC control (signal) circuit is open • IMSC power circuit (VPWR) is open • IMSC actuator assembly is damaged or failed • PCM has failed
P2075 **12T CCM, MIL: No** Escort, Focus, E Van, F Series Truck, Windstar, ZX2 engines with a 2.0L (VIN 3, VIN 5, VIN P), 2.3L VIN D, 4.2L VIN 2 engine Transmissions: All	**Intake Manifold Tuning Valve Monitor Circuit Malfunction Conditions:** Key on or engine running; and the PCM detected an unexpected low or high voltage condition on the Intake Manifold Tuning Valve Monitor circuit during the CCM test period. **Possible Causes** • IMTV monitor signal circuit is open • IMTV monitor signal circuit shorted to chassis ground • IMTV actuator is damaged or has failed • PCM has failed
P2075 **2T CCM, MIL: No** E Van, F Series Truck, Blackwood, Expedition Navigator & LS with a 3.0L VIN S, 4.6L VIN W, 5.4L VIN A, 5.4L VIN L, 5.4L VIN R Transmissions: All	**Intake Manifold Runner Control Monitor Circuit Malfunction Conditions:** Key on or engine running; and the PCM detected an unexpected high voltage condition on the Intake Manifold Runner Control (IMRC) Monitor circuit during the CCM test period. **Possible Causes** • IMRC monitor signal circuit is open • IMRC monitor signal circuit shorted to chassis ground • IMRC actuator is damaged or has failed • PCM has failed

Gas Engine OBD II Trouble Code List (P2xxx Codes)

DTC	Trouble Code Title, Conditions & Possible Causes
P2075 **2T CCM, MIL: No** Aviator, Mountaineer, Explorer equipped with a 4.6L VIN H engine Transmissions: All	**Long / Short Runner Control Monitor Circuit Malfunction Conditions:** Key on or engine running; and the PCM detected an unexpected low or high voltage condition on the Intake Manifold Communication Control (IMCC) Monitor circuit during the CCM test period. **Possible Causes** • LSRC control (signal) circuit is open • LSRC control (signal) circuit is shorted to chassis ground • LSRC actuator assembly is damaged or failed • PCM has failed
P2075 **2T CCM, MIL: No** Ranger equipped with a 2.3L VIN D engine Transmissions: All	**Intake Manifold Swirl Control Monitor Circuit Malfunction Conditions:** Key on or engine running; and the PCM detected an unexpected high voltage condition on the Intake Manifold Swirl Control (IMSC) Actuator monitor circuit during the CCM test period. **Possible Causes** • IMSC monitor circuit is open • IMSC monitor circuit is shorted to chassis ground • IMSC actuator assembly is damaged or failed • PCM has failed
P2075 **2T PCM, MIL: No** LS & Thunderbird Models Transmissions: All	**Throttle Actuator Control Motor Circuit Malfunction (Open) Conditions:** Key on or engine running; and the PCM detected an unexpected voltage condition on the Throttle Actuator Control Motor (TACM) circuit during the CCM test period. **Possible Causes** • TACM wiring harness connector is damaged or open • TACM (motor) circuit is open • TACM assembly is damaged or it has failed (an open circuit) • PCM has failed
P2101 **2T PCM, MIL: No** LS & Thunderbird Models Transmissions: All	**Throttle Actuator Control Motor Range/Performance Conditions:** Key on or engine running; and the PCM detected an unexpected low or high voltage condition on the Throttle Actuator Control Motor (TACM) circuit during the CCM test. **Possible Causes** • TACM wiring harness connector is damaged or open • TACM wiring may be crossed in the wire harness assembly • TACM (motor) circuit is open, or TACM assembly is damaged (possible open circuit) • PCM has failed
P2104 **2T PCM, MIL: No** LS & Thunderbird Models Transmissions: All	**Throttle Actuator Control System - Forced Idle Mode Conditions:** Key on, and the PCM detected the Throttle Actuator Control Motor (TACM) system was in Forced Idle mode while operating in Failure Mode Effect Management (FMEM). **Possible Causes** • PCM is damaged. Clear the codes (do a PCM reset), and if the same trouble code resets, the PCM will have to be replaced.
P2105 **2T PCM, MIL: No** LS & Thunderbird Models Transmissions: All	**Throttle Actuator Control System - Forced Engine Shutdown Conditions:** Key on, and the PCM detected the Throttle Actuator Control Motor (TACM) system was in Forced Engine Shutdown mode while in Failure Mode Effect Management (FMEM). **Possible Causes** • A total system failure has occurred • PCM is damaged. Clear the codes (do a PCM reset), and if the same trouble code resets, the PCM will have to be replaced.
P2106 **2T PCM, MIL: No** LS & Thunderbird Models Transmissions: All	**Title: Throttle Actuator Control System - Forced Limited Power** **Trouble Code Conditions** Key on, and the PCM detected the Throttle Actuator Control Motor (TACM) system was in Forced Limited Power mode while operating in Failure Mode Effect Management (FMEM). **Possible Causes** • TACM (motor) wiring harness is disconnected or loose • TACM (motor) circuits are shorted to power • PCM may have failed. Clear the codes (do a PCM reset), and if the same code resets, the PCM will have to be replaced.

Gas Engine OBD II Trouble Code List (P2xxx Codes)

DTC	Trouble Code Title, Conditions & Possible Causes
P2107 **2T PCM, MIL: No** LS & Thunderbird Models Transmissions: All	**Throttle Actuator Control Motor Processor Malfunction Conditions:** Key on or engine running; and the PCM detected the Throttle Actuator Control Motor processor received an invalid command or the TACM processor did not execute a command. **Possible Causes** • TACM (motor) wiring harness is shorted, or TACM signal wires are shorted together • TACM (motor) circuits are shorted to power • PCM is damaged. Clear the codes. If the same code resets, the PCM needs replacement.
P2110 **2T PCM, MIL: No** LS & Thunderbird Models Transmissions: All	**Throttle Actuator Control System - Forced Limited RPM Conditions:** Key on or engine running; and the PCM detected the Throttle Actuator Control System was operating in Forced Limited RPM mode while in Failure Mode Effect Management (FMEM). **Possible Causes** • PCM is damaged. Clear the codes. If the same code resets, the PCM needs replacement.
P2111 **2T PCM, MIL: No** LS & Thunderbird Models Transmissions: All	**Throttle Actuator Control System - Stuck Open Conditions:** Key on or engine running; and the PCM detected the throttle plate angle (opening) was more the commanded amount during testing. **Possible Causes** • PCM is damaged. Clear the codes. If the same code resets, the PCM needs replacement.
P2112 **2T PCM, MIL: No** LS & Thunderbird Models Transmissions: All	**Throttle Actuator Control System - Stuck Closed Conditions:** Key on or engine running; and the PCM detected the throttle plate angle (opening) was less the commanded amount during testing. **Possible Causes** • TACM wiring may be crossed in the wire harness assembly • Throttle body is binding - the throttle is stuck closed when these conditions are present • PCM has failed
P2119 **2T PCM, MIL: No** LS & Thunderbird Models Transmissions: All	**Throttle Actuator Control Throttle Body Range/Performance Conditions:** Key on or engine running; and the PCM detected a signal that indicated the throttle return spring was damaged or it had failed. **Possible Causes** • Throttle body is binding or sticking • Throttle return spring is damaged or it is broken • PCM has failed
P2121 **2T PCM, MIL: No** LS & Thunderbird Models Transmissions: All	**Accelerator Pedal Position Sensor 'D' Signal Range/Performance Conditions:** Key on or engine running; and the PCM detected the Accelerator Pedal Position Sensor 'D' signal circuit was out of the normal operating range during the CCM test. **Possible Causes** • APP sensor signal circuits are shorted together • APP sensor is damaged or the PCM has failed
P2122 **2T PCM, MIL: No** LS & Thunderbird Models Transmissions: All	**Accelerator Pedal Position Sensor 'D' Circuit Low Input Conditions:** Key on or engine running; and the PCM detected the Accelerator Pedal Position Sensor 'D' signal circuit was less than the normal range during the test period. **Possible Causes** • APP sensor signal circuit is open • APP sensor signal circuit is shorted to ground • APP sensor is damaged or it has failed • PCM has failed
P2123 **2T PCM, MIL: No** LS & Thunderbird Models Transmissions: All	**Accelerator Pedal Position Sensor 'D' Circuit High Input Conditions:** Key on or engine running; and the PCM detected the Accelerator Pedal Position Sensor 'D' signal circuit was more than the normal range during the test period. **Possible Causes** • APP sensor connector is damaged or shorted • APP sensor signal circuit is shorted to VREF (5v) • APP sensor is damaged or it has failed • PCM has failed

Gas Engine OBD II Trouble Code List (P2xxx Codes)

DTC	Trouble Code Title, Conditions & Possible Causes
P2126 **2T PCM, MIL: No** LS & Thunderbird Models Transmissions: All	**Accelerator Pedal Position Sensor 'E' Signal Range/Performance Conditions:** Key on or engine running; and the PCM detected the Accelerator Pedal Position Sensor 'E' signal circuit was more than the normal range during the test period. **Possible Causes** • APP sensor connector is damaged or shorted • APP sensor signal circuit is shorted to VREF (5v) • APP sensor is damaged or the PCM has failed
P212 **2T PCM, MIL: No** LS & Thunderbird Models Transmissions: All	**Accelerator Pedal Position Sensor 'E' Circuit Low Input Conditions:** Key on or engine running; and the PCM detected the Accelerator Pedal Position Sensor 'D' signal circuit was less than the normal range during the test period. **Possible Causes** • APP sensor signal circuit is open • APP sensor signal circuit is shorted to ground • APP sensor is damaged or the PCM has failed
P2128 **2T PCM, MIL: No** LS & Thunderbird Models Transmissions: All	**Accelerator Pedal Position Sensor 'E' Circuit High Input Conditions:** Key on or engine running; and the PCM detected the Accelerator Pedal Position Sensor 'E' signal circuit was more than the normal range during the test period. **Possible Causes** • APP sensor connector is damaged or shorted • APP sensor signal circuit is shorted to VREF (5v) • APP sensor is damaged or it has failed • PCM has failed
P2131 **2T PCM, MIL: No** LS & Thunderbird Models Transmissions: All	**Accelerator Pedal Position Sensor 'F' Signal Range/Performance Conditions:** Key on or engine running; and the PCM detected the Accelerator Pedal Position Sensor 'F' signal circuit was more than the normal range during the test period. **Possible Causes** • APP sensor connector is damaged or shorted • APP sensor signal circuit is shorted to VREF (5v) • APP sensor is damaged or it has failed • PCM has failed
P2132 **2T PCM, MIL: No** LS & Thunderbird Models Transmissions: All	**Accelerator Pedal Position Sensor 'F' Circuit Low Input Conditions:** Key on or engine running; and the PCM detected the Accelerator Pedal Position Sensor 'F' signal circuit was less than the normal range during the test period. **Possible Causes** • APP sensor signal circuit is open • APP sensor signal circuit is shorted to ground • APP sensor is damaged or it has failed • PCM has failed
P2133 **2T PCM, MIL: No** LS & Thunderbird Models Transmissions: All	**Accelerator Pedal Position Sensor 'F' Circuit High Input Conditions:** Key on or engine running; and the PCM detected the Accelerator Pedal Position Sensor 'F' signal circuit was more than the normal range during the test period. **Possible Causes** • APP sensor connector is damaged or shorted • APP sensor signal circuit is shorted to VREF (5v) • APP sensor is damaged or it has failed • PCM has failed
P2135 **2T PCM, MIL: No** LS & Thunderbird Models Transmissions: All	**ETC Throttle Position Sensor A/B Voltage Correlation Conditions:** Key on or engine running; and the PCM detected the Throttle Position 'A' (TPA) and Throttle Position 'B' (TPB) sensors disagreed, or that the TPA sensor should not be in its detected position, or that the TPB sensor should not be in its detected position during testing. **Possible Causes** • ETC TP sensor connector is damaged or shorted • ETC TP sensor circuits shorted together in the wire harness • ETC TP sensor signal circuit is shorted to VREF (5v) • ETC TP sensor is damaged or the PCM has failed

Gas Engine OBD II Trouble Code List (P2xxx Codes)

DTC	Trouble Code Title, Conditions & Possible Causes
P2195 **2T CCM, MIL: No** All Models Transmissions: All	**Lack of HO2S-11 Switching, Sensor Indicates Lean Conditions:** DTC P0300-P0310 not set, engine running in closed loop, and the PCM detected the HO2S indicated a lean signal, or it could no longer control Fuel Trim because it was at lean limit. **Possible Causes** • Base engine problems: engine oil level high, camshaft timing error, cylinder compression low, exhaust leaks in front of HO2S • EGR System problem: EGR valve is stuck open, the gasket is leaking, or the EVR diaphragm is leaking • Fuel System problem: damaged fuel pressure regulator or extremely low fuel pressure • HO2S problems: HO2S circuit is open or shorted in the wiring harness or the HO2S is damaged or it has failed • Induction System problems: air leaks after the MAF sensor, PCV system leaks, engine vacuum leaks or dip stick not seated
P2196 **2T CCM, MIL: No** All Models Transmissions: All	**Lack of HO2S-21 Switching, Sensor Indicates Rich Conditions:** DTC P0300-P0310 not set, engine running in closed loop, and the PCM detected the HO2S indicated a rich signal, or it could no longer control Fuel Trim because it was at its rich limit. **Possible Causes** • Base engine problems: engine oil level high, camshaft timing error, cylinder compression low, exhaust leaks in front of HO2S • Fuel System problem: excessive fuel pressure, leaking fuel injectors, fuel pressure regulator leaking • HO2S problems: HO2S circuit is open or shorted in the wiring harness, the HO2S signal circuit is contacting moisture in harness connector, or the HO2S is damaged or it has failed
P2197 **2T CCM, MIL: No** All Models Transmissions: All	**Lack of HO2S-21 Switching, Sensor Indicates Lean Conditions:** DTC P0300-P0310 not set, engine running in closed loop, and the PCM detected the HO2S indicated a lean signal, or it could no longer control Fuel Trim because it was at lean limit. **Possible Causes** • Base engine problems: engine oil level high, camshaft timing error, cylinder compression low, exhaust leaks in front of HO2S • EGR System problem: EGR valve is stuck open, the gasket is leaking, or the EVR diaphragm is leaking • Fuel System problem: damaged fuel pressure regulator or extremely low fuel pressure • HO2S problems: HO2S circuit is open or shorted in the wiring harness or the HO2S is damaged or it has failed • Induction System problems: air leaks after the MAF sensor, PCV system leaks, engine vacuum leaks or dip stick not seated
P2198 **2T CCM, MIL: No** All Models Transmissions: All	**Lack of HO2S-21 Switching, Sensor Indicates Rich Conditions:** DTC P0300-P0310 not set, engine running in closed loop, and the PCM detected the HO2S indicated a rich signal, or it could no longer control Fuel Trim because it was at its rich limit. **Possible Causes** • Base engine problems: engine oil level high, camshaft timing error, cylinder compression low, exhaust leaks in front of HO2S • Fuel System problem: excessive fuel pressure, leaking fuel injectors, fuel pressure regulator leaking • HO2S problems: HO2S circuit is open or shorted in the wiring harness, the HO2S signal circuit is contacting moisture in harness connector, or the HO2S is damaged or it failed

Gas Engine OBD II Trouble Code List (P2xxx Codes) – (U1xxx)

DTC	Trouble Code Title, Conditions & Possible Causes
P2270 **2T CCM, MIL: No** All Models Transmissions: All	**Lack of HO2S-12 Switching, Sensor Indicates Lean Conditions:** DTC P0300-P0310 not set, engine running in closed loop, and the PCM detected the HO2S indicated a lean signal, or it could no longer control Fuel Trim because it was at lean limit. **Possible Causes** • Base engine problems: engine oil level high, camshaft timing error, cylinder compression low, exhaust leaks in front of HO2S • EGR System problem: EGR valve is stuck open, the gasket is leaking, or the EVR diaphragm is leaking • Fuel System problem: damaged fuel pressure regulator or extremely low fuel pressure • HO2S problems: HO2S circuit is open or shorted in the wiring harness or the HO2S is damaged or it has failed • Induction System problems: air leaks after the MAF sensor, PCV system leaks, engine vacuum leaks or dip stick not seated
P2271 **2T CCM, MIL: No** All Models Transmissions: All	**Lack of HO2S-12 Switching, Sensor Indicates Rich Conditions:** DTC P0300-P0310 not set, engine running in closed loop, and the PCM detected the HO2S indicated a rich signal, or it could no longer control Fuel Trim because it was at its rich limit. **Possible Causes** • Base engine problems: engine oil level high, camshaft timing error, cylinder compression low, exhaust leaks in front of HO2S • Fuel System problem: excessive fuel pressure, leaking fuel injectors, fuel pressure regulator leaking • HO2S problems: HO2S circuit is open or shorted in the wiring harness, the HO2S signal circuit is contacting moisture in harness connector, or the HO2S is damaged or it failed
P2272 **2T CCM, MIL: No** All Models Transmissions: All	**Lack of HO2S-22 Switching, Sensor Indicates Lean Conditions:** DTC P0300-P0310 not set, engine running in closed loop, and the PCM detected the HO2S indicated a lean signal, or it could no longer control Fuel Trim because it was at lean limit. **Possible Causes** • Base engine problems: engine oil level high, camshaft timing error, cylinder compression low, exhaust leaks in front of HO2S • EGR System problem: EGR valve is stuck open, the gasket is leaking, or the EVR diaphragm is leaking • Fuel System problem: damaged fuel pressure regulator or extremely low fuel pressure • HO2S problems: HO2S circuit is open or shorted in the wiring harness or the HO2S is damaged or it has failed • Induction System problems: air leaks after the MAF sensor, PCV system leaks, engine vacuum leaks or dip stick not seated
P2273 **2T CCM, MIL: No** All Models Transmissions: All	**Title: Lack of HO2S-22 Switching, Sensor Indicates Rich** **Trouble Code Conditions** DTC P0300-P0310 not set, engine running in closed loop, and the PCM detected the HO2S indicated a rich signal, or it could no longer control Fuel Trim because it was at its rich limit. **Possible Causes** • Base engine problems: engine oil level high, camshaft timing error, cylinder compression low, exhaust leaks in front of HO2S • Fuel System problem: excessive fuel pressure, leaking fuel injectors, fuel pressure regulator leaking • HO2S problems: HO2S circuit is open or shorted in the wiring harness, the HO2S signal circuit is contacting moisture in harness connector, or the HO2S is damaged or it failed
U1011 **1T PCM, MIL: No** All Models Transmissions: All	**Data Circuit Message Conditions:** Key on, and the PCM detected that invalid or Missing Data from the Engine Air Intake system was received on the SCP data bus. Note: Network codes occur during module-to-module communication failures. Invalid and Missing data network faults are outlined below. **Possible Causes** • Invalid Data: Data transferred in normal inter-module messages with known invalid data. Transmitting module will set the code. • Missing Network Data: Missing message fault logged by a module upon failure to receive a message from another module within a defined retry period.

Gas Engine OBD II Trouble Code List (U1xxx Codes)

DTC	Trouble Code Title, Conditions & Possible Causes
U1020 **1T PCM, MIL: No** All Models Transmissions: All	**Data Circuit Message Conditions:** Key on, and the PCM detected that invalid or Missing Data from the Air Conditioning system was received on the SCP data bus. Note: Network codes occur during module-to-module communication failures. Invalid and Missing data network faults are outlined below. **Possible Causes** • Invalid Data: Data transferred in normal inter-module messages with known invalid data. Transmitting module will set the code. • Missing Network Data: Missing message fault logged by a module upon failure to receive a message from another module within a defined retry period.
U1021 **1T PCM, MIL: No** All Models Transmissions: All	**Data Circuit Message Conditions:** Key on, and the PCM detected that invalid or Missing Data from the Air Conditioning Clutch status was received on the SCP data bus. Note: Network codes occur during module-to-module communication failures. Invalid and Missing data network faults are outlined below. **Possible Causes** • Invalid Data: Data transferred in normal inter-module messages with known invalid data. Transmitting module will set the code. • Missing Network Data: Missing message fault logged by a module upon failure to receive a message from another module within a defined retry period.
U1037 **1T PCM, MIL: No** All Models Transmissions: All	**Data Circuit Message Conditions:** Key on, and the PCM detected that invalid or Missing Data from the Telltale Lamp Module was received on the SCP data bus. Note: Network codes occur during module-to-module communication failures. Invalid and Missing data network faults are outlined below. **Possible Causes** • Invalid Data: Data transferred in normal inter-module messages with known invalid data. Transmitting module will set the code. • Missing Network Data: Missing message fault logged by a module upon failure to receive a message from another module within a defined retry period.
U1039 **1T PCM, MIL: No** All Models Transmissions: All	**Data Circuit Message Conditions:** Key on, and the PCM detected that invalid or Missing Data from the Vehicle Speed Sensor was received on the SCP data bus. Note: Network codes occur during module-to-module communication failures. Invalid and Missing data network faults are outlined below. **Possible Causes** • Invalid Data: Data transferred in normal inter-module messages with known invalid data. Transmitting module will set the code. • Missing Network Data: Missing message fault logged by a module upon failure to receive a message from another module within a defined retry period. • TSB 01-21-13 contains a repair procedure for this trouble code
U1041 **1T PCM, MIL: No** All Models Transmissions: All	**Data Circuit Message Conditions:** Key on, and the PCM detected that invalid or Missing Data from the Vehicle Speed Sensor was received on the SCP data bus. Note: Network codes occur during module-to-module communication failures. Invalid and Missing data network faults are outlined below. **Possible Causes** • Invalid Data: Data transferred in normal inter-module messages with known invalid data. Transmitting module will set the code. • Missing Network Data: Missing message fault logged by a module upon failure to receive a message from another module within a defined retry period.
U1051 **1T PCM, MIL: No** All Models Transmissions: All	**Data Circuit Message Conditions:** Key on, and the PCM detected that invalid or Missing Data from the Antilock Brake System was received on the SCP data bus. Note: Network codes occur during module-to-module communication failures. Invalid and Missing data network faults are outlined below. **Possible Causes** • Invalid Data: Data transferred in normal inter-module messages with known invalid data. Transmitting module will set the code. • Missing Network Data: Missing message fault logged by a module upon failure to receive a message from another module within a defined retry period.

Gas Engine OBD II Trouble Code List (U1xxx Codes)

DTC	Trouble Code Title, Conditions & Possible Causes
U1071 **1T PCM, MIL: No** All Models Transmissions: All	**Data Circuit Message Conditions:** Key on, and the PCM detected that invalid or Missing Data from the Engine Sensor was received on the SCP data bus. Note: Network codes occur during module-to-module communication failures. Invalid and Missing data network faults are outlined below. **Possible Causes** • Invalid Data: Data transferred in normal inter-module messages with known invalid data. Transmitting module will set the code. • Missing Network Data: Missing message fault logged by a module upon failure to receive a message from another module within a defined retry period.
U1073 **1T PCM, MIL: No** All Models Transmissions: All	**Data Circuit Message Conditions:** Key on, and the PCM detected that invalid or Missing Data from the Engine Coolant Fan Status was received on the SCP data bus. Note: Network codes occur during module-to-module communication failures. Invalid and Missing data network faults are outlined below. **Possible Causes** • Invalid Data: Data transferred in normal inter-module messages with known invalid data. Transmitting module will set the code. • Missing Network Data: Missing message fault logged by a module upon failure to receive a message from another module within a defined retry period.
U1089 **1T PCM, MIL: No** All Models Transmissions: All	**Data Circuit Message Conditions:** Key on, and the PCM detected that invalid or Missing Data from the Suspension System Module was received on the SCP data bus. Note: Network codes occur during module-to-module communication failures. Invalid and Missing data network faults are outlined below. **Possible Causes** • Invalid Data: Data transferred in normal inter-module messages with known invalid data. Transmitting module will set the code. • Missing Network Data: Missing message fault logged by a module upon failure to receive a message from another module within a defined retry period.
U1098 **1T PCM, MIL: No** All Models Transmissions: All	**Data Circuit Message Conditions:** Key on, and the PCM detected that invalid or Missing Data from the Vehicle Speed Control Module was received on the SCP data bus. Note: Network codes occur during module-to-module communication failures. Invalid and Missing data network faults are outlined below. **Possible Causes** • Invalid Data: Data transferred in normal inter-module messages with known invalid data. Transmitting module will set the code. • Missing Network Data: Missing message fault logged by a module upon failure to receive a message from another module within a defined retry period.
U1130 **1T PCM, MIL: No** All Models Transmissions: All	**Data Circuit Message Conditions:** Key on, and the PCM detected that invalid or Missing Data from the Fuel System was received on the SCP data bus. Note: Network codes occur during module-to-module communication failures. Invalid and Missing data network faults are outlined below. **Possible Causes** • Invalid Data: Data transferred in normal inter-module messages with known invalid data. Transmitting module will set the code. • Missing Network Data: Missing message fault logged by a module upon failure to receive a message from another module within a defined retry period.
U1131 **1T PCM, MIL: No** All Models Transmissions: All	**Data Circuit Message Conditions:** Key on, and the PCM detected that invalid or Missing Data from the Fuel System was received on the SCP data bus. Note: Network codes occur during module-to-module communication failures. Invalid and Missing data network faults are outlined below. **Possible Causes** • Invalid Data: Data transferred in normal inter-module messages with known invalid data. Transmitting module will set the code. • Missing Network Data: Missing message fault logged by a module upon failure to receive a message from another module within a defined retry period.

Gas Engine OBD II Trouble Code List (U1xxx Codes)

DTC	Trouble Code Title, Conditions & Possible Causes
U1135 **1T PCM, MIL: No** All Models Transmissions: All	**Data Circuit Message Conditions:** Key on, and the PCM detected that invalid or Missing Data from the Ignition Switch Signal was received on the SCP data bus. Note: Network codes occur during module-to-module communication failures. Invalid and Missing data network faults are outlined below. **Possible Causes** • Invalid Data: Data transferred in normal inter-module messages with known invalid data. Transmitting module will set the code. • Missing Network Data: Missing message fault logged by a module upon failure to receive a message from another module within a defined retry period.
U1147 **1T PCM, MIL: No** All Models Transmissions: All	**Data Circuit Message Conditions:** Key on, and the PCM detected that invalid or Missing Data from the Vehicle Security System was received on the SCP data bus. Note: Network codes occur during module-to-module communication failures. Invalid and Missing data network faults are outlined below. **Possible Causes** • Invalid Data: Data transferred in normal inter-module messages with known invalid data. Transmitting module will set the code. • Missing Network Data: Missing message fault logged by a module upon failure to receive a message from another module within a defined retry period.
U1243 **1T PCM, MIL: No** All Models Transmissions: All	**Data Circuit Message Conditions:** Key on, and the PCM detected that invalid or Missing Data from the Exterior Environment System was received on the SCP data bus. Note: Network codes occur during module-to-module communication failures. Invalid and Missing data network faults are outlined below. **Possible Causes** • Invalid Data: Data transferred in normal inter-module messages with known invalid data. Transmitting module will set the code. • Missing Network Data: Missing message fault logged by a module upon failure to receive a message from another module within a defined retry period.
U1256 **1T PCM, MIL: No** All Models Transmissions: All	**Data Circuit Message Conditions:** Key on, and the PCM detected a signal indicating a communication error had occurred with another module over the SCP data bus. Note: Network codes occur during module-to-module communication failures. Invalid and Missing data network faults are outlined below. **Possible Causes** • Invalid Data: Data transferred in normal inter-module messages with known invalid data. Transmitting module will set the code. • Missing Network Data: Missing message fault logged by a module upon failure to receive a message from another module within a defined retry period.
U1260 **1T PCM, MIL: No** All Models Transmissions: All	**Data Circuit Message Conditions:** Key on, and the PCM detected a signal that indicated an open or shorted condition was present in the SCP (+) bus circuit. Note: Network codes occur during module-to-module communication failures. Invalid and Missing data network faults are outlined below. **Possible Causes** • Invalid Data: Data transferred in normal inter-module messages with known invalid data. Transmitting module will set the code. • Missing Network Data: Missing message fault logged by a module upon failure to receive a message from another module within a defined retry period.
U1261 **1T PCM, MIL: No** All Models Transmissions: All	**Data Circuit Message Conditions:** Key on, and the PCM detected a signal that indicated an open or shorted condition was present in the SCP (-) bus circuit. Note: Network codes occur during module-to-module communication failures. Invalid and Missing data network faults are outlined below. **Possible Causes** • Invalid Data: Data transferred in normal inter-module messages with known invalid data. Transmitting module will set the code. • Missing Network Data: Missing message fault logged by a module upon failure to receive a message from another module within a defined retry period.

Gas Engine OBD II Trouble Code List (U1xxx Codes) – (U2xxx)

DTC	Trouble Code Title, Conditions & Possible Causes
U1262 **1T PCM, MIL: No** All Models Transmissions: All	**Data Circuit Message Conditions:** Key on, and the PCM detected a signal that indicated a fault was present in the SCP bus (perform the network communication tests). Note: Network codes occur during module-to-module communication failures. Invalid and Missing data network faults are outlined below. **Possible Causes** • Invalid Data: Data transferred in normal inter-module messages with known invalid data. Transmitting module will set the code. • Missing Network Data: Missing message fault logged by a module upon failure to receive a message from another module within a defined retry period.
U1341 **1T PCM, MIL: No** All Models Transmissions: All	**Data Circuit Message Conditions:** Key on, and the PCM detected that invalid or Missing Data from the Function Read Vehicle Speed was received on the SCP data bus. Note: Network codes occur during module-to-module communication failures. Invalid and Missing data network faults are outlined below. **Possible Causes** • Invalid Data: Data transferred in normal inter-module messages with known invalid data. Transmitting module will set the code. • Missing Network Data: Missing message fault logged by a module upon failure to receive a message from another module within a defined retry period.
U1451 **1T PCM, MIL: No** All Models Transmissions: All	**Data Circuit Message Conditions:** Key on, and the PCM detected that invalid or Missing Data from the Vehicle Antitheft Module was received on the SCP data bus. Note: Network codes occur during module-to-module communication failures. Invalid and Missing data network faults are outlined below. **Possible Causes** • Invalid Data: Data transferred in normal inter-module messages with known invalid data. Transmitting module will set the code. • Missing Network Data: Missing message fault logged by a module upon failure to receive a message from another module within a defined retry period.
U2015 **1T PCM, MIL: No** All Models Transmissions: All	**Data Circuit Message Conditions:** Key on, and the PCM detected that invalid or Missing Data from the Function Read Vehicle Speed was received on the SCP data bus. Note: Network codes occur during module-to-module communication failures. Invalid and Missing data network faults are outlined below. **Possible Causes** • Invalid Data: Data transferred in normal inter-module messages with known invalid data. Transmitting module will set the code. • Missing Network Data: Missing message fault logged by a module upon failure to receive a message from another module within a defined retry period.
U2195 **1T PCM, MIL: No** All Models Transmissions: All	**Data Circuit Message Conditions:** Key on, and the PCM detected an open or shorted condition in the Signal Link circuit (not on the SCP data bus circuits). Note: Network codes occur during module-to-module communication failures. Invalid and Missing data network faults are outlined below. **Possible Causes** • Invalid Data: Data transferred in normal inter-module messages with known invalid data. Transmitting module will set the code. • Missing Network Data: Missing message fault logged by a module upon failure to receive a message from another module within a defined retry period.
U2243 **1T PCM, MIL: No** All Models Transmissions: All	**Data Circuit Message Conditions:** Key on, and the PCM detected that invalid or Missing Data from the SCLM Status was received on the SCP data bus. Note: Network codes occur during module-to-module communication failures. Invalid and Missing data network faults are outlined below. **Possible Causes** • Invalid Data: Data transferred in normal inter-module messages with known invalid data. Transmitting module will set the code. • Missing Network Data: Missing message fault logged by a module upon failure to receive a message from another module within a defined retry period.

<u>DIESEL ENGINE TROUBLE CODE LIST</u>

Introduction

To use this information, first read and record all codes in memory along with any Freeze Frame data. *If the PCM reset function is done prior to recording any data, all codes and freeze frame data will be lost!* Look up the desired code by DTC number, Code Title and Conditions (enable criteria) that indicate why a code set, and how to drive the vehicle. **1T and 2T** indicate a 1-trip or 2-trip fault and the Monitor type.

Diesel Engine OBD II Trouble Code List (B1xxx Codes)

DTC	Trouble Code Title, Conditions & Possible Causes
B1213 **1T PCM, MIL: No** F-Series Truck & Excursion with a 6.0L VIN P Diesel engine Transmissions: All	**Less Than Two Keys Programmed To The System Conditions:** Key on, and the PCM received a signal that indicated there were less than two (2) keys programmed to the system. **Possible Causes** • Refer to test procedures Section 419 of the Workshop Manual. Then reprogram the correct amount of keys.
B1342 **1T PCM, MIL: No** F-Series Truck & Excursion with a 6.0L VIN P Diesel engine Transmissions: All	**ECU Damaged (EEPROM Inside PCM Not Working) Conditions:** Key on, and the PCM received a signal that indicated there were less than two (2) keys programmed to the system. **Possible Causes** • Refer to test procedures Section 419 of the Workshop Manual. • Replace the PCM. Follow the correct procedures to "flash" new PCM. Then recheck the system for trouble codes.
B1600 **1T PCM, MIL: No** F-Series Truck & Excursion with a 6.0L VIN P Diesel engine Transmissions: All	**PATS Ignition Key Transponder Signal Not Received Conditions:** Key on, and the PATS did not receive Ignition Key Transponder Signal was not received. **Possible Causes** • Refer to test procedures Section 419 of the Workshop Manual to determine why the Passive Antitheft System (PATS) did not receive the Ignition Key Transponder signal correctly.
B1601 **1T PCM, MIL: No** F-Series Truck & Excursion with a 6.0L VIN P Diesel engine Transmissions: All	**PATS Received Incorrect Key Code From Ignition Key Transponder Conditions:** Key on, and the PATS received an incorrect Key Code from the Ignition Key Transponder. **Possible Causes** • Refer to test procedures Section 419 of the Workshop Manual to determine why the Passive Antitheft System (PATS) received an incorrect Key Code from the Ignition Key Transponder.
B1602 **1T PCM, MIL: No** F-Series Truck & Excursion with a 6.0L VIN P Diesel engine Transmissions: All	**PATS Received Invalid Format of Key Code From Ignition Key Transponder Conditions:** Key on, and the PATS received an invalid format of Key Code from the Ignition Key Transponder. **Possible Causes** • Refer to test procedures Section 419 of the Workshop Manual to determine why the Passive Antitheft System (PATS) received an invalid format of Key Code from the Ignition Key Transponder.
B1681 **1T PCM, MIL: No** F-Series Truck & Excursion with a 6.0L VIN P Diesel engine Transmissions: All	**PATS Receiver Module Signal Not Received Conditions:** Key on, and the PATS Receiver Module Signal was not received. **Possible Causes** • Refer to test procedures Section 419 of the Workshop Manual to determine why the Passive Antitheft System (PATS) Receiver Module Signal was not received.
B2103 **1T PCM, MIL: No** F-Series Truck & Excursion with a 6.0L VIN P Diesel engine Transmissions: All	**Antenna Not Connected Conditions:** Key on, and the PATS detected that the Antenna was not connected. **Possible Causes** • Refer to test procedures Section 419 of the Workshop Manual to determine why the Passive Antitheft System (PATS) detected the Antenna was not connected.

Diesel Engine OBD II Trouble Code List (B2xxx Codes)

DTC	Trouble Code Title, Conditions & Possible Causes
B2103 **1T PCM, MIL: No** F-Series Truck & Excursion with a 6.0L VIN P Diesel engine Transmissions: All	**Transponder Programming Failed Conditions:** Key on, and the PATS detected that the Transponder Programming failed. **Possible Causes** • Refer to test procedures Section 419 of the Workshop Manual to determine why the Passive Antitheft System (PATS) detected the Transponder Programming failed.
P0046 **1T CCM, MIL: No** F-Series Truck & Excursion with a 6.0L VIN P Diesel engine Transmissions: All	**Turbo/Supercharger Boost Solenoid Signal Performance Conditions:** Key on or engine running; and the PCM detected a high or low voltage on the Variable Geometry Turbo (VGT) control circuit. The VGT actuator is a variable position valve that controls the vane position in the turbine housing. The VGT is located on top of the turbocharger. The valve position signal is controlled inside the PCM. **Possible Causes** • VGT solenoid connector is damaged, open or shorted • VGT solenoid control circuit is open or shorted • VGT solenoid is damaged, or it has failed • PCM has failed
P0069 **1T CCM, MIL: Yes** F-Series Truck & Excursion with a 6.0L VIN P Diesel engine Transmissions: All	**MAP/BARO Sensor Signal Correlation Conditions:** Engine started, engine running at idle speed, and the PCM detected the difference between the BARO sensor and MAP sensor signal was more than a specified value during the CCM continuous test. **Possible Causes** • MAP/BARO sensor connector is damaged, open or shorted • MAP/BARO sensor circuit is open or shorted • MAP/BARO sensor is damaged, or it has failed • PCM has failed
P0096 **1T CCM, MIL: Yes** F-Series Truck & Excursion with a 6.0L VIN P Diesel engine Transmissions: All	**Intake Air Temperature Sensor 2 Signal Performance Conditions:** Engine started, and the PCM detected an unexpected voltage condition on the IAT Sensor 2 signal during the CCM test period. **Possible Causes** • IAT2 assembly connector is damaged, open or shorted • IAT2 assembly signal circuit is open or shorted • IAT2 assembly is damaged, or it has failed • PCM has failed
P0097 **1T CCM, MIL: Yes** F-Series Truck & Excursion with a 6.0L VIN P Diesel engine Transmissions: All	**Intake Air Temperature Sensor 2 Circuit Low Input Conditions:** Key on or engine running; and the PCM detected an unexpected low voltage condition on the IAT Sensor 2 signal during the CCM test. **Possible Causes** • IAT2 assembly connector is damaged or shorted • IAT2 assembly signal circuit is shorted to ground • IAT2 assembly is damaged, or it has failed • PCM has failed
P0098 **1T CCM, MIL: Yes** F-Series Truck & Excursion with a 6.0L VIN P Diesel engine Transmissions: All	**Intake Air Temperature Sensor 2 Circuit High Input Conditions:** Key on or engine running; and the PCM detected an unexpected high voltage condition on the IAT Sensor 2 signal during the test. **Possible Causes** • IAT2 sensor connector is damaged, open or shorted • IAT sensor signal circuit is open or shorted to 5v VREF • IAT2 sensor ground circuit is open • IAT2 assembly is damaged, or it has failed • PCM has failed

Diesel Engine OBD II Trouble Code List (P0xxx Codes)

DTC	Trouble Code Title, Conditions & Possible Causes
P0101 **1T CCM, MIL: No** F-Series Truck & Excursion with a 6.0L VIN P Diesel engine Transmissions: All	**MAF Sensor Or VAF Sensor Signal Range/Performance Conditions:** DTC P0102 and DTC P0103 not set, engine started, and the PCM detected the Actual MAF sensor value was not within a preset range of the Calculated MAF sensor value while testing. **Possible Causes** • Base engine vacuum leak, PCV valve leaking or stuck open • Check for leaks at air outlet tube, or loose tube clamps • Engine oil dipstick missing or not fully seated • MAF sensor element (wire) is contaminated or dirty • MAF sensor signal or ground circuit fault or sensor has failed
P0102 **2T CCM, MIL: Yes** F-Series Truck & Excursion with a 6.0L VIN P Diesel engine Transmissions: All	**MAF Sensor Circuit Low Input Conditions:** DTC P0505 not set, engine started, and the PCM detected the MAF sensor signal was less than 0.23v during the CCM test period. **Possible Causes** • Check for leaks at air outlet tube • Sensor power circuit open, sensor ground circuit open • Sensor signal circuit open (may be disconnected) • Check for loose tube clamps near the MAF sensor
P0103 **2T CCM, MIL: Yes** F-Series Truck & Excursion with a 6.0L VIN P Diesel engine Transmissions: All	**MAF Sensor Circuit High Input Conditions:** DTC P0505 not set, engine started, and the PCM detected the MAF sensor signal was more than 4.60v during the CCM test period. **Possible Causes** • Check for a restricted inlet screen on the MAF sensor. • MAF sensor signal circuit is shorted to system power (B+) • MAF sensor is damaged or has failed • PCM has failed
P0107 **2T CCM, MIL: Yes** E-Van, F-Series Truck & Excursion with 6.0L or 7.3L Diesel engine Transmissions: All	**Barometric Pressure Sensor Circuit Low Input Conditions:** Key on, KOEO Self-Test enabled, or engine running, and the PCM detected the BARO sensor signal was less than the minimum calibrated parameter. The BARO sensor is a variable capacitance sensor used to determine altitude. Prior to the 1999 1/2 model year, this sensor was mounted under the Instrument Cluster. This sensor is internal to the PCM on later models, and it is serviced separately. **Possible Causes** • BARO sensor signal circuit is shorted to ground • BARO sensor VREF circuit is open • BARO sensor is damaged or it has failed • PCM has failed
P0108 **2T CCM, MIL: Yes** E-Van, F-Series Truck & Excursion with 6.0L or 7.3L Diesel engine Transmissions: All	**Barometric Pressure Sensor Circuit High Input Conditions:** Key on, KOEO Self-Test enabled, or engine running, and the PCM detected the BARO sensor signal was more than the minimum calibrated parameter. The BARO sensor is a variable capacitance sensor used to determine altitude. Prior to the 1999 1/2 model year, this sensor was mounted under the Instrument Cluster. This sensor is internal to the PCM on later models, and it is serviced separately. **Possible Causes** • BARO ground signal circuit is open • BARO sensor signal circuit is shorted to VREF (5v) • BARO sensor is damaged or it has failed • PCM has failed
P0112 **2T CCM, MIL: No** E-Van, F-Series Truck & Excursion with 6.0L or 7.3L Diesel engine Transmissions: All	**IAT Sensor Circuit Low Input Conditions:** Key on or engine running; and the PCM detected the IAT sensor signal was less than the self-test minimum of 0.20v (Scan Tool reads over 250°F). This is a thermistor-type sensor with a variable resistance that changes when exposed to different temperatures. **Possible Causes** • IAT sensor signal circuit is grounded (check wiring & connector) • IAT sensor is damaged or it has failed • PCM has failed

Diesel Engine OBD II Trouble Code List (P0xxx Codes)

DTC	Trouble Code Title, Conditions & Possible Causes
P0113 **2T CCM, MIL: No** E-Van, F-Series Truck & Excursion with 6.0L or 7.3L Diesel engine Transmissions: All	**IAT Sensor Circuit High Input Conditions:** Key on or engine running; and the PCM detected the IAT sensor was more than the self-test maximum of 4.60v (Scan Tool reads -46ºF) during the self-test. This is a thermistor-type sensor with a variable resistance that changes as the temperature changes. **Possible Causes** • Sensor signal circuit open (check wiring harness/connector) • Sensor signal circuit shorted to VREF or system voltage • Sensor ground circuit is open • Sensor is damaged or it has failed • PCM has failed
P0117 **2T CCM, MIL: No** E-Van, F-Series Truck & Excursion with 6.0L or 7.3L Diesel engine Transmissions: All	**ECT Sensor Circuit Low Input Conditions:** Key on or engine running; and the PCM detected the ECT sensor signal was less than the self-test minimum of 0.20v (Scan Tool reads over 250ºF). This is a thermistor-type sensor with a variable resistance that changes when exposed to different temperatures **Possible Causes** • ECT sensor signal circuit is grounded in the wiring harness • ECT sensor is damaged or it has failed • PCM has failed
P0118 **2T CCM, MIL: No** E-Van, F-Series Truck & Excursion with 6.0L or 7.3L Diesel engine Transmissions: All	**ECT Sensor Circuit High Input Conditions:** Key on or engine running; and the PCM detected the ECT sensor signal was more than the self-test maximum of 4.60v (Scan Tool reads under -46ºF). This is a thermistor-type sensor with a variable resistance that changes when exposed to different temperatures **Possible Causes** • ECT sensor signal circuit is open (inspect wiring & connector) • ECT sensor signal circuit is shorted to VREF (5v) • ECT sensor is damaged or it has failed • PCM has failed
P0122 **2T CCM, MIL: Yes** E-Van, F-Series Truck & Excursion with 6.0L or 7.3L Diesel engine Transmissions: All	**Accelerator Position Sensor Circuit Low Input Conditions:** Key on or engine running; and the PCM detected the APP sensor was less than the minimum calibrated limit during the CCM test. **Possible Causes** • APP sensor signal circuit is open (inspect wiring & connector) • APP sensor signal is shorted to ground (check the connector) • APP sensor VREF circuit (5v) is open (from sensor to PCM) • APP sensor is damaged or it has failed • PCM has failed
P0123 **2T CCM, MIL: Yes** E-Van, F-Series Truck & Excursion with 6.0L or 7.3L Diesel engine Transmissions: All	**Accelerator Position Sensor Circuit High Input Conditions:** Key on or engine running; and the PCM detected the APP sensor was more than the minimum calibrated limit during the CCM test. **Possible Causes** • APP sensor ground circuit is open (inspect wiring & connector) • APP sensor signal is shorted to the VREF circuit (5v) • APP sensor is damaged or it has failed • PCM has failed
P0148 **2T CCM, MIL: Yes** F-Series Truck & Excursion with a 6.0L VIN P Diesel engine Transmissions: All	**FICM Fuel Delivery Circuit Error Conditions:** Engine started, and the PCM detected a Fuel Delivery Error related to the FICM circuit. The FICM monitor input line informs the Diesel Engine Power Monitor (DEPM) when the injectors are turned "on" and "off". If the FICM line is open or shorted, the monitor strategy assumes that the fuel injectors are always turned "on", and it sets DTC P2552. In some cases, the DEPM may also set DTC P0148. **Possible Causes** • Severely restricted fuel filter • Severely pinched or restricted fuel delivery line

Diesel Engine OBD II Trouble Code List (P0xxx Codes)

DTC	Trouble Code Title, Conditions & Possible Causes
P00196 **2T CCM, MIL: Yes** F-Series Truck & Excursion with a 6.0L VIN P Diesel engine Transmissions: All	**Engine Oil Temperature Sensor Signal Performance Conditions:** Engine started, KOER Self Test enabled, and the PCM detected that the Engine Oil Temperature (EOT) sensor signal was not within a calibrated amount of the ECT sensor signal. The EOT sensor signal is used to determine the timing and quality of the fuel required to optimize engine startup over all temperature conditions. **Possible Causes** • Cooling system malfunction, or the thermostat is stuck • Engine not operating at normal operating temperature • EOT sensor is damaged or it has failed • PCM has failed
P0197 **2T CCM, MIL: Yes** E-Van, F-Series Truck & Excursion with 6.0L or 7.3L Diesel engine Transmissions: All	**Engine Oil Temperature Sensor Circuit Low Input Conditions:** Key on or engine running; and the PCM detected that the Engine Oil Temperature (EOT) sensor was less than 0.20v. The EOT sensor signal is used to determine the timing and quality of the fuel required to optimize engine startup over all temperature conditions. **Possible Causes** • EOT sensor connector is damaged or shorted to ground • EOT sensor signal circuit shorted to sensor ground • EOT sensor is damaged or it has failed • PCM has failed
P0198 **2T CCM, MIL: Yes** E-Van, F-Series Truck & Excursion with 6.0L or 7.3L Diesel engine Transmissions: All	**Engine Oil Temperature Sensor Circuit High Input Conditions:** Key on or engine running; and the PCM detected that the Engine Oil Temperature (EOT) sensor was less than 4.50v. The EOT sensor signal is used to determine the timing and quality of the fuel required to optimize engine startup over all temperature conditions. **Possible Causes** • EOT sensor connector is damaged or open • EOT sensor signal circuit is shorted to VREF (5v) • EOT sensor is damaged or it has failed • PCM has failed
P0219 **1T CCM, MIL: Yes** F-Series Truck & Excursion with a 6.0L VIN P Diesel engine Transmissions: All	**Engine Overspeed Condition Conditions:** Engine started, and the PCM detected an Engine Overspeed condition had occurred. **Possible Causes** • Repeat the KOEO and KOER Self Test • Refer to the Symptom Charts for further information
P0220 **2T CCM, MIL: Yes** E-Van, F-Series Truck & Excursion with 6.0L or 7.3L Diesel engine Transmissions: All	**Throttle Position Sensor 'B' Circuit Malfunction Conditions:** Key on or engine running; and the PCM detected an unexpected voltage condition on the TP Sensor 'B' circuit during the CCM test. The Idle Validation Switch (IVS) provides the PCM with a signal to verify when the accelerator is in the idle position. **Possible Causes** • ETC TP Sensor 'B' connector is damaged or shorted • ETC TP Sensor 'B' signal circuit is shorted to ground • ETC TP Sensor 'B' is damaged or it has failed
P0221 **2T CCM, MIL: Yes** E-Vans & F-Series Trucks & Excursion with a 7.3L Diesel engine Transmissions: All	**Idle Validation/Throttle Switch 'B' Performance Conditions:** Engine started, KOER Self Test enabled, and the PC'M detected the TP-B sensor was less than or more than its calibrated minimum. Note: Wait 5 seconds before starting the KOER switch test after first pressing the trigger to start running the driver-operated controls. **Possible Causes** • Blown fuse, or open in power circuit to IVS switch • IVS signal circuit open • IVS has failed, or the IVS transition is out of range

Diesel Engine OBD II Trouble Code List (P0xxx Codes)

DTC	Trouble Code Title, Conditions & Possible Causes
P0222 **2T CCM, MIL: Yes** E-Van, F-Series Truck & Excursion with 6.0L or 7.3L Diesel engine Transmissions: All	**Throttle Position Sensor 'B' Circuit Low Input Conditions:** Key on or engine running; and the PCM detected the TP Sensor 'B' circuit was out of its normal operating range during the CCM test. The Idle Validation Switch (IVS) provides the PCM with a signal to verify when the accelerator is in the idle position. **Possible Causes** • Throttle body is damaged • Throttle linkage is binding or sticking • ETC TP Sensor 'B' signal circuit to the PCM is open • ETC TP Sensor 'B' ground circuit is open • ETC TP Sensor 'B' is damaged or it has failed
P0223 **2T CCM, MIL: Yes** E Van, F-Series Truck & Excursion with 7.3L VIN F Diesel engine Transmissions: All	**Throttle Position Sensor 'B' Circuit High Input Conditions:** Key on or engine running; and the PCM detected the TP Sensor 'B' signal indicated more than 4.65v (Scan Tool reads more than 93%). **Possible Causes** • ETC TP Sensor 'B' connector is damaged or open • ETC TP Sensor 'B' signal circuit is open • ETC TP Sensor 'B' signal circuit is shorted to VREF (5v) • ETC TP Sensor 'B' is damaged or it has failed
P0230 **2T CCM, MIL: Yes** E-Van, F-Series Truck & Excursion with 6.0L or 7.3L Diesel engine Transmissions: All	**Fuel Pump Relay Driver Circuit Malfunction Conditions:** Key on or engine running; and the PCM detected an unexpected voltage condition on the Fuel Pump Relay Driver control circuit. The fuel pump control circuit is used by the PCM to energize the fuel pump relay. At key on, the relay is energized for 20 seconds, and all the time the engine is running. The Fuel Pump Monitor (FPM) circuit is after the inertia switch. It is used to monitor voltage to the pump. **Possible Causes** • Fuel pump relay driver control circuit is open • Fuel pump relay driver control circuit is shorted to ground • Fuel pump relay is damaged or it has failed • PCM has failed
P0231 **2T CCM, MIL: Yes** E-Van, F-Series Truck & Excursion with 6.0L or 7.3L Diesel engine Transmissions: All	**Fuel Pump Relay Driver Failed On Conditions:** Key on or engine running; and the PCM did not detect any voltage on the FP Monitor circuit with the fuel pump commanded on. The fuel pump control circuit is used by the PCM to energize the fuel pump relay. At key on, the relay is energized for 20 seconds, and all the time the engine is running. The Fuel Pump Monitor (FPM) circuit is after the inertia switch. **Possible Causes** • FPM circuit is open or it is shorted to ground • Fuel pump relay output circuit (VPWR) is open • Fuel pump relay is damaged or it has failed (no output voltage) • PCM has failed
P0232 **2T CCM, MIL: Yes** E-Van, F-Series Truck & Excursion with 6.0L or 7.3L Diesel engine Transmissions: All	**Fuel Pump Relay Driver Failed Off Conditions:** Key on or engine running; and the PCM detected a voltage on the FP Monitor circuit with fuel pump commanded off. The fuel pump control circuit is used by the PCM to energize the fuel pump relay. At key on, the relay is energized for 20 seconds, and all the time the engine is running. The Fuel Pump Monitor (FPM) circuit is after the inertia switch. **Possible Causes** • Fuel pump relay contacts always closed • Fuel pump ground circuit has high resistance • Fuel pump secondary circuit shorted to power • Low speed fuel pump relay damaged or related circuit problem

Diesel Engine OBD II Trouble Code List (P0xxx Codes)

DTC	Trouble Code Title, Conditions & Possible Causes
P0236 **2T CCM, MIL: Yes** E-Van, F-Series Truck & Excursion with 6.0L or 7.3L Diesel engine Transmissions: All	**Turbo Boost Sensor 'A' Signal Range/Performance Conditions:** Key on or engine running; and the PCM detected an unexpected low or high voltage condition for the specific operating conditions on the MAP sensor circuit. The MAP sensor is a variable capacitance sensor that, when supplied with a 5v reference signal, produces an analog signal that indicates pressure. The MAP sensor is used to control (diesel) smoke by limiting fuel quality during acceleration until a specified boost pressure is obtained. **Possible Causes** • MAP sensor wiring harness or connector is damaged or open • MAP sensor signal circuit is open or shorted to ground • MAP sensor VREF circuit (5v) is open or shorted to ground • MAP sensor is damaged or it has failed • PCM has failed
P0237 **2T CCM, MIL: Yes** E-Van, F-Series Truck & Excursion with 6.0L or 7.3L Diesel engine Transmissions: All	**Turbo Boost Sensor 'A' Circuit Low Input Conditions:** Key on or engine running; and the PCM detected an unexpected low voltage on the MAP sensor circuit. The MAP sensor is a variable capacitance sensor supplied with a 5v reference signal. It produces an analog signal that indicates pressure. The MAP sensor is used to control (diesel) smoke by limiting fuel quality during acceleration until a specified boost pressure is obtained. **Possible Causes** • MAP sensor wiring harness or connector is shorted • MAP sensor signal circuit is shorted to ground • MAP sensor VREF circuit (5v) is open • MAP sensor is damaged or it has failed • PCM has failed
P0238 **2T CCM, MIL: Yes** E-Van, F-Series Truck & Excursion with 6.0L or 7.3L Diesel engine Transmissions: All	**Turbo Boost Sensor 'A' Circuit High Input Conditions:** Key on or engine running; and the PCM detected an unexpected high voltage on the MAP sensor circuit. The MAP sensor is a variable capacitance sensor produces an analog signal that indicates pressure. The signal from this sensor is used to control (diesel) smoke by limiting fuel quality during acceleration until a specified boost pressure is obtained. **Possible Causes** • MAP sensor wiring harness or connector is open • MAP sensor signal circuit is open • MAP sensor signal circuit is shorted to VREF (5v) • MAP sensor is damaged or it has failed • PCM has failed
P0261 **2T CCM, MIL: Yes** E-Van, F-Series Truck & Excursion with 6.0L or 7.3L Diesel engine Transmissions: All	**Fuel Injector Circuit Low Input - Cylinder 1 Conditions:** Key on or engine running; and the PCM detected an unexpected low voltage condition on the Injector 1 control circuit. The High Side driver output function is to distribute energy to the correct bank based on cylinder identification and to provide regulated current to the unit injectors, based on fuel delivery command signal from the injector driver module internal 115v supply. Injector timing and duration is commanded by the PCM in the FDCS module. **Possible Causes** • Injector 1 control circuit is open • Injector 1 power circuit (B+) is open • Injector 1 control circuit is shorted to chassis ground • Injector 1 is damaged or has failed • PCM has failed
P0262 **2T CCM, MIL: No** E-Van, F-Series Truck & Excursion with 6.0L or 7.3L Diesel engine Transmissions: All	**Fuel Injector Circuit High Input - Cylinder 1 Conditions:** Key on or engine running; and the PCM detected an unexpected high voltage condition on the Injector 1 control circuit. The High Side driver output function is to distribute energy to the correct bank based on cylinder identification and to provide regulated current to the unit injectors, based on fuel delivery command signal from the injector driver module internal 115v supply. Injector timing and duration are commanded by the PCM in the FDCS module. **Possible Causes** • Injector 1 control circuit is shorted to system power (B+) • PCM has failed

Diesel Engine OBD II Trouble Code List (P0xxx Codes)

DTC	Trouble Code Title, Conditions & Possible Causes
P0263 **2T CCM, MIL: No** E-Van, F-Series Truck & Excursion with 6.0L or 7.3L Diesel engine Transmissions: All	**Fuel Injector Cylinder 1 Contribution/Balance Malfunction Conditions:** Engine started, and the PCM detected an engine condition with low cylinder contribution or "no" cylinder contribution condition on Engine Cylinder 1 during the CCM test period. **Possible Causes** • Base engine problem affecting only Cylinder 1 • Fuel injection delivery problem affecting only Cylinder 1 • PCM has failed
P0264 **2T CCM, MIL: Yes** E-Van, F-Series Truck & Excursion with 6.0L or 7.3L Diesel engine Transmissions: All	**Fuel Injector Circuit Low Input - Cylinder 2 Conditions:** Key on or engine running; and the PCM detected an unexpected low voltage condition on the Injector 2 control circuit. The High Side driver output function is to distribute energy to the correct bank based on cylinder identification and to provide regulated current to the unit injectors, based on fuel delivery command signal from the injector driver module internal 115v supply. Injector timing and duration is commanded by the PCM in the FDCS module. **Possible Causes** • Injector 2 power circuit (B+) is open • Injector 2 control circuit is open or shorted to chassis ground • Injector 2 is damaged or has failed • PCM has failed
P0265 **2T CCM, MIL: No** E-Van, F-Series Truck & Excursion with 6.0L or 7.3L Diesel engine Transmissions: All	**Fuel Injector Circuit High Input - Cylinder 2 Conditions:** Key on or engine running; and the PCM detected an unexpected high voltage condition on the Injector 2 control circuit. The High Side driver output function is to distribute energy to the correct bank based on cylinder identification and to provide regulated current to the unit injectors, based on fuel delivery command signal from the injector driver module internal 115v supply. Injector timing and duration are commanded by the PCM in the FDCS module. **Possible Causes** • Injector 2 control circuit is shorted to system power (B+) • PCM has failed
P0266 **, T CCM** **MIL: No** E-Van, F-Series Truck & Excursion with 6.0L or 7.3L Diesel engine Transmissions: All	**Fuel Injector 2 Cylinder Contribution/Balance Malfunction Conditions:** Engine started, and the PCM detected an engine condition with low cylinder contribution or "no" cylinder contribution condition on Engine Cylinder 2 during the CCM test period. **Possible Causes** • Base engine problem affecting only Cylinder 2 • Fuel injection delivery problem affecting only Cylinder 2 • PCM has failed
P0267 **2T CCM, MIL: Yes** E-Van, F-Series Truck & Excursion with 6.0L or 7.3L Diesel engine Transmissions: All	**Fuel Injector Circuit Low Input - Cylinder 3 Conditions:** Key on or engine running; and the PCM detected an unexpected low voltage condition on the Injector 3 control circuit. The High Side driver output function is to distribute energy to the correct bank based on cylinder identification and to provide regulated current to the unit injectors, based on fuel delivery command signal from the injector driver module internal 115v supply. Injector timing and duration is commanded by the PCM in the FDCS module. **Possible Causes** • Injector 3 control circuit is open • Injector 3 power circuit (B+) is open • Injector 3 control circuit is shorted to chassis ground
P0268 **2T CCM, MIL: No** E-Van, F-Series Truck & Excursion with 6.0L or 7.3L Diesel engine Transmissions: All	**Fuel Injector Circuit High Input - Cylinder 3 Conditions:** Key on or engine running; and the PCM detected an unexpected high voltage condition on the Injector 3 control circuit. The High Side driver output function is to distribute energy to the correct bank based on cylinder identification and to provide regulated current to the unit injectors, based on fuel delivery command signal from the injector driver module internal 115v supply. Injector timing and duration is commanded by the PCM in the FDCS module. **Possible Causes** • Injector 3 control circuit is shorted to system power (B+) • PCM has failed

Diesel Engine OBD II Trouble Code List (P0xxx Codes)

DTC	Trouble Code Title, Conditions & Possible Causes
P0269 **2T CCM, MIL: No** E-Van, F-Series Truck & Excursion with 6.0L or 7.3L Diesel engine Transmissions: All	**Fuel Injector Cylinder 3 Contribution/Balance Malfunction Conditions:** Engine started, and the PCM detected an engine condition with low cylinder contribution or "no" cylinder contribution condition on Engine Cylinder 3 during the CCM test period. **Possible Causes** • Base engine problem affecting only Cylinder 3 • Fuel injection delivery problem affecting only Cylinder 3 • PCM has failed
P0270 **2T CCM, MIL: Yes** E-Van, F-Series Truck & Excursion with 6.0L or 7.3L Diesel engine Transmissions: All	**Fuel Injector Circuit Low Input - Cylinder 4 Conditions:** Key on or engine running; and the PCM detected an unexpected low voltage condition on the Injector 4 control circuit. The High Side driver output function is to distribute energy to the correct bank based on cylinder identification and to provide regulated current to the unit injectors, based on fuel delivery command signal from the injector driver module internal 115v supply. Injector timing and duration is commanded by the PCM in the FDCS module. **Possible Causes** • Injector 4 control circuit is open • Injector 4 power circuit (B+) is open • Injector 4 control circuit is shorted to chassis ground • PCM has failed
P0271 **2T CCM, MIL: No** E-Van, F-Series Truck & Excursion with 6.0L or 7.3L Diesel engine Transmissions: All	**Fuel Injector Circuit High Input - Cylinder 4 Conditions:** Key on or engine running; and the PCM detected an unexpected high voltage condition on the Injector 4 control circuit. The High Side driver output function is to distribute energy to the correct bank based on cylinder identification and to provide regulated current to the unit injectors, based on fuel delivery command signal from the injector driver module internal 115v supply. Injector timing and duration are commanded by the PCM in the FDCS module. **Possible Causes** • Injector 4 control circuit is shorted to system power (B+) • PCM has failed
P0272 **2T CCM, MIL: No** E-Van, F-Series Truck & Excursion with 6.0L or 7.3L Diesel engine Transmissions: All	**Fuel Injector Cylinder 4 Contribution/Balance Malfunction Conditions:** Engine started, and the PCM detected an engine condition with low cylinder contribution or "no" cylinder contribution condition on Engine Cylinder 4 during the CCM test period. **Possible Causes** • Base engine problem affecting only Cylinder 4 • Fuel injection delivery problem affecting only Cylinder 4.
P0273 **2T CCM, MIL: Yes** E-Van, F-Series Truck & Excursion with 6.0L or 7.3L Diesel engine Transmissions: All	**Fuel Injector Circuit Low Input - Cylinder 5 Conditions:** Key on or engine running; and the PCM detected an unexpected low voltage condition on the Injector 5 control circuit. The High Side driver output function is to distribute energy to the correct bank based on cylinder identification and to provide regulated current to the unit injectors, based on fuel delivery command signal from the injector driver module internal 115v supply. Injector timing and duration are commanded by the PCM in the FDCS module. **Possible Causes** • Injector 5 control circuit is open • Injector 5 power circuit (B+) is open • Injector 5 control circuit is shorted to chassis ground
P0274 **2T CCM, MIL: No** E-Van, F-Series Truck & Excursion with 6.0L or 7.3L Diesel engine Transmissions: All	**Fuel Injector Circuit High Input - Cylinder 5 Conditions:** Key on or engine running; and the PCM detected an unexpected high voltage condition on the Injector 5 control circuit. The High Side driver output function is to distribute energy to the correct bank based on cylinder identification and to provide regulated current to the unit injectors, based on fuel delivery command signal from the injector driver module internal 115v supply. Injector timing and duration are commanded by the PCM in the FDCS module. **Possible Causes** • Injector 5 control circuit is shorted to system power (B+) • PCM has failed

Diesel Engine OBD II Trouble Code List (P0xxx Codes)

DTC	Trouble Code Title, Conditions & Possible Causes
P0275 **2T CCM, MIL: No** E-Van, F-Series Truck & Excursion with 6.0L or 7.3L Diesel engine Transmissions: All	**Fuel Injector Cylinder 5 Contribution/Balance Malfunction Conditions:** Engine started, and the PCM detected an engine condition with low cylinder contribution or "no" cylinder contribution condition on Engine Cylinder 5 during the CCM test period. **Possible Causes** • Base engine problem affecting only Cylinder 5 • Fuel injection delivery problem affecting only Cylinder 5 • PCM has failed
P0276 **2T CCM, MIL: Yes** E-Van, F-Series Truck & Excursion with 6.0L or 7.3L Diesel engine Transmissions: All	**Fuel Injector Circuit Low Input - Cylinder 6 Conditions:** Key on or engine running; and the PCM detected an unexpected low voltage condition on the Injector 6 control circuit. The High Side driver output function is to distribute energy to the correct bank based on cylinder identification and to provide regulated current to the unit injectors, based on fuel delivery command signal from the injector driver module internal 115v supply. Injector timing and duration are commanded by the PCM in the FDCS module. **Possible Causes** • Injector 6 control circuit is open • Injector 6 power circuit (B+) is open • Injector 6 control circuit is shorted to chassis ground • PCM has failed
P0277 **2T CCM, MIL: No** E-Van, F-Series Truck & Excursion with 6.0L or 7.3L Diesel engine Transmissions: All	**Fuel Injector Circuit High Input - Cylinder 6 Conditions:** Key on or engine running; and the PCM detected an unexpected high voltage condition on the Injector 6 control circuit. The High Side driver output function is to distribute energy to the correct bank based on cylinder identification and to provide regulated current to the unit injectors, based on fuel delivery command signal from the injector driver module internal 115v supply. Injector timing and duration are commanded by the PCM in the FDCS module. **Possible Causes** • Injector 6 control circuit is shorted to system power (B+) • PCM has failed
P0278 **2T CCM, MIL: No** E-Van, F-Series Truck & Excursion with 6.0L or 7.3L Diesel engine Transmissions: All	**Fuel Injector Cylinder 6 Contribution/Balance Malfunction Conditions:** Engine started, and the PCM detected an engine condition with low cylinder contribution or "no" cylinder contribution condition on Engine Cylinder 6 during the CCM test period. **Possible Causes** • Base engine problem affecting only Cylinder 6 • Fuel injection delivery problem affecting only Cylinder 6
P0279 **2T CCM, MIL: Yes** E-Van, F-Series Truck & Excursion with 6.0L or 7.3L Diesel engine Transmissions: All	**Fuel Injector Circuit Low Input - Cylinder 7 Conditions:** Key on or engine running; and the PCM detected an unexpected low voltage condition on the Injector 7 control circuit. The High Side driver output function is to distribute energy to the correct bank based on cylinder identification and to provide regulated current to the unit injectors, based on fuel delivery command signal from the injector driver module internal 115v supply. Injector timing and duration are commanded by the PCM in the FDCS module. **Possible Causes** • Injector 7 control circuit is open • Injector 7 power circuit (B+) is open • Injector 7 control circuit is shorted to chassis ground
P0280 **2T CCM, MIL: No** E-Van, F-Series Truck & Excursion with 6.0L or 7.3L Diesel engine Transmissions: All	**Fuel Injector Circuit High Input - Cylinder 7 Conditions:** Key on or engine running; and the PCM detected an unexpected high voltage condition on the Injector 7 control circuit. The High Side driver output function is to distribute energy to the correct bank based on cylinder identification and to provide regulated current to the unit injectors, based on fuel delivery command signal from the injector driver module internal 115v supply. Injector timing and duration are commanded by the PCM in the FDCS module. **Possible Causes** • Injector 7 control circuit is shorted to system power (B+) • PCM has failed

Diesel Engine OBD II Trouble Code List (P0xxx Codes)

DTC	Trouble Code Title, Conditions & Possible Causes
P0281 **2T CCM, MIL: Yes** E-Van, F-Series Truck & Excursion with 6.0L or 7.3L Diesel engine Transmissions: All	**Fuel Injector Cylinder 7 Contribution/Balance Malfunction Conditions:** Engine started, and the PCM detected an engine condition with low cylinder contribution or "no" cylinder contribution condition on Engine Cylinder 7 during the CCM test period. **Possible Causes** • Base engine problem affecting only Cylinder 7 • Fuel injection delivery problem affecting only Cylinder 7
P0282 **2T CCM, MIL: Yes** E-Van, F-Series Truck & Excursion with 6.0L or 7.3L Diesel engine Transmissions: All	**Fuel Injector Circuit Low Input - Cylinder 8 Conditions:** Key on or engine running; and the PCM detected an unexpected low voltage condition on the Injector 8 control circuit. The High Side driver output function is to distribute energy to the correct bank based on cylinder identification and to provide regulated current to the unit injectors, based on fuel delivery command signal from the injector driver module internal 115v supply. Injector timing and duration is commanded by the PCM in the FDCS module. **Possible Causes** • Injector 8 control circuit is open • Injector 8 power circuit (B+) is open • Injector 8 control circuit is shorted to chassis ground
P0283 **2T CCM, MIL: Yes** E-Van, F-Series Truck & Excursion with 6.0L or 7.3L Diesel engine Transmissions: All	**Fuel Injector Circuit High Input - Cylinder 8 Conditions:** Key on or engine running; and the PCM detected an unexpected high voltage condition on the Injector 8 control circuit. The High Side driver output function is to distribute energy to the correct bank based on cylinder identification and to provide regulated current to the unit injectors, based on fuel delivery command signal from the injector driver module internal 115v supply. Injector timing and duration is commanded by the PCM in the FDCS module. **Possible Causes** • Injector 8 control circuit is shorted to system power (B+) • PCM has failed
P0284 **2T CCM, MIL: Yes** E-Van, F-Series Truck & Excursion with 6.0L or 7.3L Diesel engine Transmissions: All	**Fuel Injector Cylinder 8 Contribution/Balance Malfunction Conditions:** Engine started, and the PCM detected an engine condition with low cylinder contribution or "no" cylinder contribution condition on Engine Cylinder 8 during the CCM test period. **Possible Causes** • Base engine problem affecting only Cylinder 8 • Fuel injection delivery problem affecting only Cylinder 8 • PCM has failed
P0301 **2T Misfire, MIL: Yes** E-Van, F-Series Truck & Excursion with 6.0L or 7.3L Diesel engine Transmissions: All	**Fault Cylinder 'A' - Misfire Detected Conditions:** DTP P0263 not set, engine started, engine running under positive torque conditions, and the PCM detected a low cylinder contribution (misfire) condition related to Engine Cylinder 1 or Engine Cylinder 'A'. **Possible Causes** • Base engine problems: broken compression rings, leaking or bent valves, bent push rod, broken rocker arm bolts or bent connecting rod • Fuel metering problem on Cylinder 'A' (a fault with the injector)
P0302 **2T Misfire, MIL: Yes** E-Van, F-Series Truck & Excursion with 6.0L or 7.3L Diesel engine Transmissions: All	**Fault Cylinder 'B' - Misfire Detected Conditions:** DTP P0266 not set, engine started, engine running under positive torque conditions, and the PCM detected a low cylinder contribution (misfire) condition related to Engine Cylinder 2 or Engine Cylinder 'B'. **Possible Causes** • Base engine problems: broken compression rings, leaking or bent valves, bent push rod, broken rocker arm bolts or bent connecting rod • Fuel metering problem on Cylinder 'B' (a fault with the injector)

Diesel Engine OBD II Trouble Code List (P0xxx Codes)

DTC	Trouble Code Title, Conditions & Possible Causes
P0303 **2T Misfire, MIL: Yes** E-Van, F-Series Truck & Excursion with 6.0L or 7.3L Diesel engine Transmissions: All	**Fault Cylinder 'C' - Misfire Detected Conditions:** DTP P0269 not set, engine started, engine running under positive torque conditions, and the PCM detected a low cylinder contribution (misfire) condition related to Engine Cylinder 3 or Engine Cylinder 'C'. **Possible Causes** • Base engine problems: broken compression rings, leaking or bent valves, bent push rod, broken rocker arm bolts or bent connecting rod • Fuel metering problem on Cylinder 'C' (a fault with the injector)
P0304 **2T Misfire, MIL: Yes** E-Van, F-Series Truck & Excursion with 6.0L or 7.3L Diesel engine Transmissions: All	**Fault Cylinder 'D' - Misfire Detected Conditions:** DTP P0272 not set, engine started, engine running under positive torque conditions, and the PCM detected a low cylinder contribution (misfire) condition related to Engine Cylinder 4 or 'D'. **Possible Causes** • Base engine problems: broken compression rings, leaking or bent valves, bent push rod, broken rocker arm bolts or bent connecting rod • Fuel metering problem on Cylinder 'D' (a fault with the injector)
P0305 **2T Misfire, MIL: Yes** E-Van, F-Series Truck & Excursion with 6.0L or 7.3L Diesel engine Transmissions: All	**Fault Cylinder 'E' - Misfire Detected Conditions:** DTP P0275 not set, engine started, engine running under positive torque conditions, and the PCM detected a low cylinder contribution (misfire) condition related to Engine Cylinder 5 or 'E'. **Possible Causes** • Base engine problems: broken compression rings, leaking or bent valves, bent push rod, broken rocker arm bolts or bent connecting rod • Fuel metering problem on Cylinder 'E' (a fault with the injector)
P0306 **2T Misfire, MIL: Yes** E-Van, F-Series Truck & Excursion with 6.0L or 7.3L Diesel engine Transmissions: All	**Fault Cylinder 'F' - Misfire Detected Conditions:** DTP P0278 not set, engine started, engine running under positive torque conditions, and the PCM detected a low cylinder contribution (misfire) condition related to Engine Cylinder 6 or 'F'. **Possible Causes** • Base engine problems: broken compression rings, leaking or bent valves, bent push rod, broken rocker arm bolts or bent connecting rod • Fuel metering problem on Cylinder 'F' (a fault with the injector)
P0307 **2T Misfire, MIL: Yes** E-Van, F-Series Truck & Excursion with 6.0L or 7.3L Diesel engine Transmissions: All	**Fault Cylinder 'G' - Misfire Detected Conditions:** DTP P0281 not set, engine started, engine running under positive torque conditions, and the PCM detected a low cylinder contribution (misfire) condition related to Engine Cylinder 7 or 'G'. **Possible Causes** • Base engine problems: broken compression rings, leaking or bent valves, bent push rod, broken rocker arm bolts or bent connecting rod • Fuel metering problem on Cylinder 'G' (a fault with the injector)
P0308 **2T Misfire, MIL: Yes** E-Van, F-Series Truck & Excursion with 6.0L or 7.3L Diesel engine Transmissions: All	**Fault Cylinder 'H' - Misfire Detected Conditions:** DTP P0284 not set, engine started, engine running under positive torque conditions, and the PCM detected a low cylinder contribution (misfire) condition related to Engine Cylinder 8 or 'H'. **Possible Causes** • Base engine problems: broken compression rings, leaking or bent valves, bent push rod, broken rocker arm bolts or bent connecting rod • Fuel metering problem on Cylinder 'H' (a fault with the injector)

Diesel Engine OBD II Trouble Code List (P0xxx Codes)

DTC	Trouble Code Title, Conditions & Possible Causes
P0335 **1T CCM, MIL: Yes** F-Series Truck & Excursion with a 6.0L VIN P Diesel engine Transmissions: All	**Crankshaft Position Sensor 'A' Circuit Malfunction Conditions:** Engine cranking or running; and the PCM did not detect any Crankshaft Position Sensor 'A' circuit signals during the test. **Possible Causes** • CKP sensor signal circuit is open or shorted to ground • CKP sensor ground (return) circuit is open • CKP sensor is damaged or CKP sensor shielding is damaged • PCM has failed
P0336 **1T CCM, MIL: Yes** F-Series Truck & Excursion with a 6.0L VIN P Diesel engine Transmissions: All	**Crankshaft Position Sensor 'A' Circuit Performance Conditions:** Engine started, and the PCM detected an erratic Crankshaft Position Sensor 'A' signal or an intermittent loss of the CKP Sensor 'A' signal. **Possible Causes** • CKP sensor signal circuit is open or shorted to ground • CKP sensor ground (return) circuit is open • CKP sensor is damaged or CKP sensor shielding is damaged • PCM has failed
P0340 **2T CCM, MIL: No** E-Van, F-Series Truck & Excursion with 6.0L or 7.3L Diesel engine Transmissions: All	**Camshaft Position Sensor Circuit Malfunction Conditions:** Engine started, and the PCM did not detect any CMP sensor signals. This device is a Hall Effect design sensor that generates a digital frequency, as windows in a target wheel pass through its magnetic field. The frequency of the windows passing by the sensor as well as the width of the selected windows allows the PCM to detect the engine speed and position. The engine speed is determined by counting the 12 the windows on the cam gear in each camshaft revolution. The position of cylinders 1 and 4 is determined by distinguishing a narrow or wide window on the camshaft gear. **Possible Causes** • CMP (Hall effect) sensor circuit is open or shorted to ground • CMP (Hall effect) sensor power (B+) circuit is open • CMP (Hall effect) sensor is incorrectly installed or damaged • PCM is damaged
P0341 **2T CCM, MIL: Yes** E-Van, F-Series Truck & Excursion with 6.0L or 7.3L Diesel engine Transmissions: All	**Camshaft Position Sensor Signal Range/Performance Conditions:** Engine started, and the PCM detected an irregular or out-of-phase CMP sensor signal during the CCM test. The target wheel spokes are each 15 degrees apart except for a narrow spoke that identifies Cylinder 1 and a wide spoke that identifies Cylinder 4. As the camshaft rotates, this sensor generates a digital frequency. **Possible Causes** • CMP sensor connector is damaged, open or shorted • CMP sensor signal circuit is open or shorted to ground • CMP sensor signal circuit is erratic (check for EMI or RFI) • CMP sensor is damaged or it has failed • PCM has failed
P0340 **2T CCM, MIL: Yes** E-Van, F-Series Truck & Excursion with 6.0L or 7.3L Diesel engine Transmissions: All	**Camshaft Position Sensor Circuit Malfunction (Intermittent) Conditions:** Engine started, and the PCM detected that the CMP sensor signal was interrupted during testing. The target wheel spokes are each 15 degrees apart except for a narrow spoke that identifies Cylinder 1 and a wide spoke that identifies Cylinder 4. As the camshaft rotates, this sensor generates a digital frequency. **Possible Causes** • CMP (Hall effect) sensor circuit is open (an intermittent fault) • CMP (Hall effect) sensor circuit shorted to ground (intermittent) • CMP (Hall effect) sensor is damaged or it has failed • PCM has failed

Diesel Engine OBD II Trouble Code List (P0xxx Codes)

DTC	Trouble Code Title, Conditions & Possible Causes
P0380 **2T CCM, MIL: Yes** E-Van, F-Series Truck & Excursion with 6.0L or 7.3L Diesel engine Transmissions: All	**Glow Plug Relay Circuit Malfunction Conditions:** Key on, and the PCM detected an unexpected voltage condition on the Glow Plug Relay control circuit during the CCM test. The Glow Plug Lamp remains "on" for 1-12 seconds (depending on the Glow Plug relay on-time which can vary from 1 and 120 seconds). **Possible Causes** • Glow plug relay control circuit is open or shorted to ground • Glow plug relay power circuit is open (test the 12GA fuse link) • Glow plug relay is damaged or it has failed
P0381 **2T CCM, MIL: Yes** E-Van, F-Series Truck & Excursion with 6.0L or 7.3L Diesel engine Transmissions: All	**Glow Plug Indicator Circuit Malfunction Conditions:** Key on, and the PCM detected an unexpected voltage condition on the Glow Plug Lamp circuit during the CCM test. The Glow Plug Lamp remains "on" for 1-12 seconds (depending on the Glow Plug relay on-time which can vary from 1 and 120 seconds). **Possible Causes** • Glow plug lamp circuit is open or shorted to ground • Glow plug relay control circuit is open or shorted to ground • Glow plug relay power circuit is open (test the 12GA fuse link) • Glow plug relay is damaged or it has failed
P0401 **2T EGR, MIL: Yes** F-Series Truck & Excursion with a 6.0L VIN P Diesel engine Transmissions: All	**Insufficient EGR Flow Detected Conditions:** Engine started, engine running at steady cruise speed, and the PCM detected the EGR valve position did not match the Desired EGR position based on engine speed and load. **Possible Causes** • EGR valve actuator is damaged or sticking (perform a KOEO or KOER Self Test to test the actuator for an On Demand code) • EGR position sensor signal circuit shorted to bias voltage • EP sensor signal circuit shorted to bias voltage • PCM has failed
P0402 **2T EGR, MIL: Yes** F-Series Truck & Excursion with a 6.0L VIN P Diesel engine Transmissions: All	**Excessive EGR Flow Detected Conditions:** Engine started, engine running at steady cruise speed, and the PCM detected excessive EGR gas flow. The EGR valve actuator and EGR valve position sensor are in one unit. **Possible Causes** • EVP sensor is damaged or has failed • EGR valve source vacuum supply problem (hose off or leaking) • EGR valve is stuck partially open or closed (check for carbon) • EGR valve may be leaking vacuum (the diaphragm is broken) • VR solenoid is damaged or the VR control circuit is open • PCM has failed
P0403 **2T CCM, MIL: Yes** F-Series Truck & Excursion with a 6.0L VIN P Diesel engine Transmissions: All	**EGR Solenoid Control Circuit Malfunction Conditions:** Engine started, and the PCM detected an unexpected high or low voltage condition on the EGR solenoid control circuit at idle speed. **Possible Causes** • EGR solenoid connector is damaged, loose or shorted • EGR solenoid control circuit is open, shorted to ground or B+ • EGR solenoid is damaged or has failed • PCM has failed
P0404 **2T CCM, MIL: Yes** F-Series Truck & Excursion with a 6.0L VIN P Diesel engine Transmissions: All	**EGR Solenoid Control Signal Range/Performance Conditions:** Engine started, and the PCM detected and EGRP error ($\pm$ 0.10v) from the Actual to the Commanded EGR valve position at cruise. **Possible Causes** • EGR position sensor circuit is open or shorted (intermittent) • EGR position sensor is damaged or it has failed • EGR valve is damaged or it has failed • PCM has failed

Diesel Engine OBD II Trouble Code List (P0xxx Codes)

DTC	Trouble Code Title, Conditions & Possible Causes
P0405 **2T CCM, MIL: Yes** F-Series Truck & Excursion with a 6.0L VIN P Diesel engine Transmissions: All	**EGR Position Sensor 'A' Circuit Low Input Conditions:** Engine started, and the PCM detected an unexpected low voltage condition (less than 0.30v) on the EGR valve position sensor circuit. **Possible Causes** • EGR position sensor connector is damaged, open or shorted • EGR position sensor circuit is open or shorted to ground • EGR position sensor is damaged or it has failed • PCM has failed
P0406 **2T CCM, MIL: Yes** F-Series Truck & Excursion with a 6.0L VIN P Diesel engine Transmissions: All	**EGR Position Sensor 'A' Circuit High Input Conditions:** Engine started, and the PCM detected an unexpected high voltage condition (more than 4.9v) on the EGR valve position sensor circuit. **Possible Causes** • EGR position sensor connector is damaged or shorted • EGR position sensor circuit is shorted to VREF or power (B+) • EGR position sensor is damaged or it has failed • PCM has failed
P0460 **2T CCM, MIL: Yes** E-Van, F-Series Truck & Excursion with 6.0L or 7.3L Diesel engine Transmissions: All	**Fuel Level Sensor Circuit Malfunction Conditions:** Engine started, and the PCM detected the FLI signal did not match the fuel level during the CCM test. For example, a FLI V PID below 0.90v (Scan Tool reads FLI PID = 25%), or a FLI over 2.45v (Scan Tool reads FLI PID = 75). **Possible Causes** • Fuel tank is empty, or fuel pump (FP) module is stuck open • Fuel gauge incorrectly installed or Instrument cluster damaged • PCM Case ground circuit is open • Fuel level indicator (FLI) circuit shorted to VPWR, or is open • Fuel tank has been overfilled, or fuel gauge is damaged • Fuel pump (FP) module is stuck closed, or is stuck open • Fuel level indicator circuit shorted to Case or to power ground • PCM Case ground shorted to VPWR (shorted to system power)
P0470 **2T CCM, MIL: Yes** E-Van, F-Series Truck & Excursion with 6.0L or 7.3L Diesel engine Transmissions: All	**Exhaust Backpressure Sensor Circuit Malfunction Conditions:** Key on or engine running; and the PCM detected an unexpected low or high voltage condition on the Exhaust Backpressure Sensor (EBP) signal circuit. The EBP sensor is a variable capacitance sensor that, when supplied with a 5v reference signal, produces a linear analog signal that indicates pressure. The primary function of this sensor is to measure exhaust backpressure so that the PCM can control the Exhaust Back Pressure Regulator (EBP) operation. **Possible Causes** • EBP sensor signal circuit is open or shorted to ground • EBP sensor signal circuit is shorted to VREF (5v) • EBP sensor power circuit (VREF) is open from the PCM • EBP sensor is damaged or it has failed • PCM is damaged
P0471 **2T CCM, MIL: Yes** E-Van, F-Series Truck & Excursion with 6.0L or 7.3L Diesel engine Transmissions: All	**Exhaust Backpressure Sensor Signal Range/Performance Conditions:** Key on or engine running; and the PCM detected an unexpected voltage condition on the Exhaust Backpressure Sensor signal circuit. The EBP sensor is a variable capacitance sensor that, when supplied with a 5v reference signal, produces a linear analog signal that indicates pressure. The primary function of this sensor is to measure exhaust backpressure so that the PCM can control the Exhaust Back Pressure Regulator (EBP) operation. **Possible Causes** • EBP sensor is damaged • EBP sensor pressure supply tube is damaged or clogged • EPR linkage is damaged or sticking, or the butterfly is damaged • PCM is damaged

Diesel Engine OBD II Trouble Code List (P0xxx Codes)

DTC	Trouble Code Title, Conditions & Possible Causes
P0472 **2T CCM, MIL: Yes** E-Van, F-Series Truck & Excursion with 6.0L or 7.3L Diesel engine Transmissions: All	**Exhaust Backpressure Sensor Circuit Low Input Conditions:** Key on or engine running; and the PCM detected an unexpected low voltage condition on the Exhaust Backpressure Sensor signal circuit. This sensor is used to measure exhaust backpressure so the PCM can control the Exhaust Back Pressure Regulator (EBP) operation. **Possible Causes** • EBP sensor signal circuit is shorted to ground • EBP sensor power circuit (VREF) is open • EBP sensor is damaged or it has failed • PCM is damaged
P0473 **2T CCM, MIL: Yes** E-Van, F-Series Truck & Excursion with 6.0L or 7.3L Diesel engine Transmissions: All	**Exhaust Backpressure Sensor Circuit High Input Conditions:** Key on or engine running; and the PCM detected an unexpected high voltage condition on the Exhaust Backpressure Sensor signal circuit. This sensor is used to measure exhaust backpressure so the PCM can control the Exhaust Back Pressure Regulator operation. **Possible Causes** • EBP sensor is shorted to VREF (5v) • EBP sensor ground circuit is open • EBP sensor is damaged or it has failed • PCM is damaged
P0475 **2T CCM, MIL: Yes** E-Van, F-Series Truck & Excursion with 6.0L or 7.3L Diesel engine Transmissions: All	**Exhaust Backpressure Control Valve Malfunction Conditions:** Key on, KOEO Self-Test enabled and the PCM detected an unexpected voltage condition on the Exhaust Backpressure Control Valve circuit. The EPR is a variable position valve that is used (along with signals from the IAT sensor and engine load) to control exhaust backpressure during cold ambient temperature to increase cab heat and decrease the amount of time needed to defrost the windshield. **Possible Causes** • EPR (regulator) valve circuit is open or shorted to ground • EPR (regulator) valve circuit is shorted to system power (B+) • EPR (regulator) valve is damaged or it has failed • PCM is damaged
P0476 **2T CCM, MIL: Yes** E-Van, F-Series Truck & Excursion with 6.0L or 7.3L Diesel engine Transmissions: All	**Exhaust Backpressure Regulator Malfunction Conditions:** Engine started, KOER Self-Test enabled, and the PCM detected a problem in the operation of the Exhaust Backpressure Regulator. The EPR is a variable position valve that is used (along with signals from the IAT sensor and engine load) to control exhaust backpressure during cold ambient temperature to increase cab heat and decrease the amount of time needed to defrost the windshield. **Possible Causes** • EPR (regulator) control valve is binding or sticking • EPR (regulator) control valve is damaged or it has failed • EPR (regulator) resistance is 2.5-12.0 ohms at 68°F • PCM is damaged
P0478 **2T CCM, MIL: Yes** E-Van, F-Series Truck & Excursion with 6.0L or 7.3L Diesel engine Transmissions: All	**Exhaust Backpressure Control Valve Excessive Conditions:** Engine started, vehicle driven under normal conditions, and the PCM detected an excessive backpressure condition. The EPR is a variable position valve that is used (along with signals from the IAT sensor and engine load) to control exhaust backpressure during cold ambient temperature to increase cab heat and decrease the amount of time needed to defrost the windshield. **Possible Causes** • EBP butterfly valve is stuck • EBP sensor line (or exhaust system) is clogged or restricted • EPR linkage is adjusted incorrectly or it is binding • Wastegate turbo may be in Overboost mode • PCM is damaged

Diesel Engine OBD II Trouble Code List (P0xxx Codes)

DTC	Trouble Code Title, Conditions & Possible Causes
P0480 **1T CCM, MIL: Yes** F-Series Truck & Excursion with a 6.0L VIN P Diesel engine Transmissions: All	**VDF Fan Control Circuit Malfunction Conditions:** Engine started, and the PCM detected an unexpected voltage condition on the Visctronic Drive Fan (VDF) control circuit. The VDF is a viscous coupling. The viscous drag should be smooth during fan rotation. The amount of resistance is dependant upon the final VDF operational state before engine shutdown. Check the PCM to determine if it has the latest calibration. **Possible Causes** • VDF assembly is damaged or it has failed (the VDF resistance is 6-10 ohms at 68°F) • PCM has failed
P0500 **2T CCM, MIL: Yes** E-Van, F-Series Truck & Excursion with 6.0L or 7.3L Diesel engine Transmissions: All	**Vehicle Speed Sensor Malfunction Conditions:** Engine running, then with the engine speed more than the TCC stall speed, the PCM detected a lack of vehicle speed data for a period of time. The PCM receives vehicle speed data from the ABS, VSS, TCSS, GEM, or CTM controller, depending up the application. **Possible Causes** • Mechanical drive mechanism for the VSS is damaged • VSS (+) or (-) signal circuit is open, shorted to ground or power • VSS is damaged or it has failed • PCM has failed • TSB 01-21-13 contains a repair procedure for this code
P0501 **2T CCM, MIL: Yes** E-Vans & F-Series Trucks with a 7.3L Diesel engine Transmissions: All	**Vehicle Speed Sensor Signal Range/Performance Conditions:** Engine running, then with the engine speed more than the TCC stall speed, the PCM detected a problem with the VSS signal data during testing. The PCM receives vehicle speed data from the VSS (sensor) on this vehicle application. **Possible Causes** • Mechanical drive mechanism for the VSS is damaged • VSS (+) or (-) signal circuit is open, shorted to ground or power • VSS is damaged or it has failed • PCM has failed
P0502 **2T CCM, MIL: Yes** E-Vans & F-Series Trucks with a 7.3L Diesel engine Transmissions: All	**Vehicle Speed Sensor Intermittent Signal Conditions:** Engine running, then with the engine speed more than the TCC stall speed, the PCM detected an intermittent loss of the VSS signal. The PCM receives vehicle speed data from a vehicle speed sensor on this vehicle application. **Possible Causes** • Mechanical drive mechanism for the VSS is damaged • VSS (+) or (-) signal circuit is open, shorted to ground or power • VSS is damaged or it has failed • PCM has failed
P0503 **2T CCM, MIL: Yes** E-Van, F-Series Truck & Excursion with 6.0L or 7.3L Diesel engine Transmissions: All	**Vehicle Speed Sensor Intermittent Signal Conditions:** Engine running, then with the engine speed more than the TCC stall speed, the PCM detected an intermittent or "noisy" signal on the VSS signal circuit. The PCM receives vehicle speed data from the ABS, VSS, TCSS, GEM, or CTM controller, depending up the vehicle. **Possible Causes** • An Aftermarket add-on device causing "noise" on the circuit • Module or circuits connected to VSS/TCSS circuit are damaged • VSS/TCSS signal "noise" due to RFI or EMI interference from sources such as ignition components or charging system • VSS/TCSS gears are damaged • VSS/TCSS wiring harness or connectors are damaged
P0528 **1T CCM, MIL: Yes** F-Series Truck & Excursion with a 6.0L VIN P Diesel engine Transmissions: All	**VDF Fan Control Circuit Malfunction Conditions:** Engine started, and the PCM detected an unexpected voltage condition on the Visctronic Drive Fan (VDF) control circuit. The VDF is a viscous coupling. The viscous drag should be smooth during fan rotation. The amount of resistance is dependant upon the final VDF operational state before engine shutdown. Check PCM calibration. **Possible Causes** • VDF assembly is damaged or it has failed (the VDF resistance is 6-10 ohms at 68°F) • PCM has failed

Diesel Engine OBD II Trouble Code List (P0xxx Codes)

DTC	Trouble Code Title, Conditions & Possible Causes
P0541 **2T CCM, MIL: Yes** E-Van, F-Series Truck & Excursion with a 7.3L Diesel engine Transmissions: All	**Manifold Intake Air Heater Circuit Low Input Conditions:** Engine started, IAT sensor under 32ºF, EOT sensor less than 131ºF, system voltage at 11.8-15.0v, and the PCM detected an unexpected low voltage condition on the Manifold Intake Air Heater circuit. **Possible Causes** • MIAH circuit is open or shorted to ground from relay to the PCM • MIAH heater relay power (B+) circuit is open to the Alternator • MIAH is damaged or it has failed • PCM has failed
P0542 **2T CCM, MIL: No** E-Van, F-Series Truck & Excursion with a 7.3L Diesel engine Transmissions: All	**Manifold Intake Air Heater Circuit High Input Conditions:** Engine started, IAT sensor less than 32ºF, system voltage from 11.8-15.0v, EOT sensor less than 131ºF, and the PCM detected an unexpected high voltage condition on the Manifold Intake Air Heater circuit during the CCM test period. **Possible Causes** • MIAH circuit shorted to power between MIAH relay and heater • MIAH is damaged or it has failed • PCM has failed
P0560 **2T CCM, MIL: No** E-Van, F-Series Truck & Excursion with 6.0L or 7.3L Diesel engine Transmissions: All	**System Voltage Malfunction Conditions:** Engine started, engine running at idle or cruise speed, and the PCM detected the system voltage was too high or too low during the test. **Possible Causes** • Battery connections corroded (high resistance) or loose • Generator is damaged or it has failed (output is too high or low) • Ignition system voltage circuit is open at the PCM terminals • PCM has failed
P0562 **2T CCM, MIL: Yes** E-Van, F-Series Truck & Excursion with 6.0L or 7.3L Diesel engine Transmissions: All	**System Voltage Low Input Conditions:** Engine started, engine running at idle or cruise speed, and the PCM detected the system voltage was too low during the CCM test. This code can set due to a low battery voltage condition during cranking. **Possible Causes** • Battery connections corroded (high resistance) or loose • Generator is damaged or it has failed (output is too low) • Ignition system voltage circuit is open at the PCM terminals • PCM has failed
P0563 **2T CCM, MIL: No** E-Van, F-Series Truck & Excursion with 6.0L or 7.3L Diesel engine Transmissions: All	**System Voltage High Input Conditions:** Engine started, engine running at idle or cruise speed, and the PCM detected the system voltage was too high during the CCM test. This code can set due to a 24v battery jump-start event. **Possible Causes** • Battery connections corroded (high resistance) or loose • Generator is damaged or it has failed (output is too high) • Ignition system voltage circuit is open at the PCM terminals • PCM has failed
P0565 **2T CCM, MIL: No** E-Van, F-Series Truck & Excursion with 6.0L or 7.3L Diesel engine Transmissions: All	**Speed Control ON, Not Pressed - KOER Switch Test Conditions:** Engine started, KOER Self-Test enabled, and the PCM detected the Cruise Control "On" switch was not pressed during the KOER test. The Speed Control function is integrated into the PCM. The speed command switches are "momentary" devices located on the face of the steering wheel. These are On/Off switches and one 3-position SET/ACCEL-COAST-RESUME switch. When these switches are pressed, they select one of several resistance values to the PCM. **Possible Causes** • Cruise Control "On" switch not pressed during the Self-Test • Cruise Control "On" switch is damaged, or the circuit has failed • PCM has failed

Diesel Engine OBD II Trouble Code List (P0xxx Codes)

DTC	Trouble Code Title, Conditions & Possible Causes
P0566 **2T CCM, MIL: No** E-Van, F-Series Truck & Excursion with 6.0L or 7.3L Diesel engine Transmissions: All	**Speed Control OFF, Not Pressed - KOER Switch Test Conditions:** Engine started, KOER Self-Test enabled, and the PCM detected the Cruise Control "Off" switch was not pressed. The speed command switches are "momentary" devices located on the face of the steering wheel. These are On/Off switches and one 3-position SET/ACCEL-COAST-RESUME switch. When these switches are pressed, they select one of several resistance values to the PCM. **Possible Causes** • Cruise Control "Off" switch not pressed during the Self-Test • Cruise Control "Off" switch is damaged, or the circuit has failed • PCM has failed
P0567 **2T CCM, MIL: No** E-Van, F-Series Truck & Excursion with 6.0L or 7.3L Diesel engine Transmissions: All	**Speed Control RESUME, Not Pressed - KOER Switch Test Conditions:** Engine started, KOER Self-Test enabled, and the PCM detected the Cruise Control "RESUME" switch was not pressed. The speed command switches are "momentary" devices located on the face of the steering wheel. These are On/Off switches and one 3-position SET/ACCEL-COAST-RESUME switch. When these switches are pressed, they select one of several resistance values to the PCM. **Possible Causes** • Cruise Control "Resume" switch not pressed in the Self-Test • Cruise Control "Resume" switch is damaged has a circuit fault • PCM has failed
P0568 **2T CCM, MIL: No** E-Van, F-Series Truck & Excursion with 6.0L or 7.3L Diesel engine Transmissions: All	**Speed Control "Set", Not Pressed - KOER Switch Test Conditions:** Engine started, KOER Self-Test enabled, and the PCM detected the Cruise Control "Set" switch was not pressed. The speed command switches are "momentary" devices located on the face of the steering wheel. These are On/Off switches and one 3-position SET/ACCEL-COAST-RESUME switch. As a switch is pressed, a resistance value is selected in the PCM. **Possible Causes** • Cruise Control "Set" switch not pressed in the Self-Test • Cruise Control "Set" switch is damaged has a circuit fault • PCM has failed
P0569 **2T CCM, MIL: No** E-Van, F-Series Truck & Excursion with 6.0L or 7.3L Diesel engine Transmissions: All	**Cruise Control "Coast", Not Pressed - KOER Switch Test Conditions:** Engine started, KOER Self-Test enabled, and the PCM detected the Cruise Control "Coast" switch was not pressed. The speed command switches are "momentary" devices located on the face of the steering wheel. These are On/Off switches and one 3-position SET/ACCEL-COAST-RESUME switch. As a switch is pressed, a resistance value is selected in the PCM. **Possible Causes** • Cruise Control "Coast" switch not pressed in the Self-Test • Cruise Control "Coast" switch is damaged has a circuit fault • PCM has failed
P0571 **2T CCM, MIL: No** E-Van, F-Series Truck & Excursion with 6.0L or 7.3L Diesel engine Transmissions: All	**Brake Pressure Applied Switch Circuit Malfunction Conditions:** Engine started, KOER Self-Test enabled, and the PCM detected the Brake Pressure Applied (BPA) switch did not change status after the brake was pressed. This is a pressure switch that senses brake pressure, and is redundant to the Brake On/Off (BOO) switch to provide a backup signal to deactivate the Speed Control system. **Possible Causes** • Brake Pedal switch is open between switch and the PCM • Brake Pedal switch shorted to ground between switch and PCM • Brake Pedal switch power circuit is open from J/Box to switch • Brake Pedal switch is damaged or it has failed
P0602 **1T PCM, MIL: Yes** E-Van, F-Series Truck & Excursion with 6.0L or 7.3L Diesel engine Transmissions: All	**Control Module Programming Error Conditions:** Key on, and the PCM detected a programming error in the VID block. This fault requires that the VID Block be reprogrammed, or that the EEPROM be re-flashed. **Possible Causes** • During the VID reprogramming function, the Vehicle ID (VID) data block failed during reprogramming wit the Scan Tool.

Diesel Engine OBD II Trouble Code List (P0xxx Codes)

DTC	Trouble Code Title, Conditions & Possible Causes
P0603 **1T PCM, MIL: Yes** E-Van, F-Series Truck & Excursion with 6.0L or 7.3L Diesel engine Transmissions: All	**PCM Keep Alive Memory Test Error Conditions:** Key on or engine running; and the PCM detected an internal memory fault. This code can be set if KAPWR to the PCM is interrupted. The code will set the first time during the initial key "on". **Possible Causes** • Battery terminal corrosion, or loose battery connection • KAPWR to PCM interrupted, or circuit opened • Reprogramming error occurred • PCM has failed
P0603 **1T PCM, MIL: Yes** E-Van, F-Series Truck & Excursion with 6.0L or 7.3L Diesel engine Transmissions: All	**PCM Read Only Memory Test Error Conditions:** Key on, and the PCM detected a ROM test error (the ROM is corrupted). The PCM is normally replaced if this code has set. **Possible Causes** • An attempt was made to change the control module calibration • Module programming error has occurred • PCM has failed
P0606 **1T PCM, MIL: Yes** E-Van, F-Series Truck & Excursion with 6.0L or 7.3L Diesel engine Transmissions: All	**PCM Internal Communication Error Conditions:** Key on, and the PCM detected an internal communications register read back error at initial key on. **Possible Causes** • PCM has failed
P0611 **1T CCM, MIL: Yes** F-Series Truck & Excursion with a 6.0L VIN P Diesel engine Transmissions: All	**Fuel Injector Control Module Performance Conditions:** Key on or engine cranking; and the PCM detected an invalid signal from the Fuel Injector Control Module (FICM) during the test period. **Possible Causes** • FICM is damaged or it has failed. Replace the FICM and then retest the system for trouble codes. • PCM has failed
P0615 **2T CCM, MIL: Yes** F-Series Truck & Excursion with a 6.0L VIN P Diesel engine Transmissions: All	**Starter Relay Signal Circuit Malfunction Conditions:** Key on or engine cranking; and the PCM detected an invalid signal from the Starter Relay signal under these conditions. **Possible Causes** • Starter relay connector is damaged, loose or shorted • Starter relay signal circuit is open, shorted to ground or power • Starter relay is damaged or it has failed • PCM has failed
P0620 **2T CCM, MIL: Yes** F-Series Truck & Excursion with a 6.0L VIN P Diesel engine Transmissions: All	**Generator 1 Control Circuit Malfunction Conditions:** Engine started; and the PCM detected an unexpected voltage condition on the Generator 1 control circuit during the CCM test. **Possible Causes** • Generator 1 connector is damaged, loose or shorted • Generator 1 control circuit is open or shorted to ground • Generator or regulator function is damaged or has failed • PCM has failed

Diesel Engine OBD II Trouble Code List (P0xxx Codes)

DTC	Trouble Code Title, Conditions & Possible Causes
P0623 **2T CCM, MIL: Yes** F-Series Truck & Excursion with a 6.0L VIN P Diesel engine Transmissions: All	**Generator Lamp Control Circuit Malfunction Conditions:** Engine started; and the PCM detected an unexpected voltage condition on the Generator lamp control circuit during the CCM test. **Possible Causes** • Generator lamp control circuit is open or shorted to ground • Generator or regulator function is damaged or has failed • PCM has failed
P0640 **2T PCM, MIL: Yes** E-Van, F-Series Truck & Excursion with a 7.3L Diesel engine Transmissions: All	**Manifold Intake Air Heater Relay Circuit Malfunction Conditions:** Engine started, IAT sensor less than 32ºF, Engine Oil Temperature (EOT) sensor less than 131ºF, system voltage from 11.8-15.0v, and the PCM detected an unexpected voltage condition on the Manifold Intake Air Heater (MIAH) Relay control circuit. The PCM activates the MIAH relay under the conditions listed to reduce the amount of white smoke during long idle periods at low ambient temperatures. **Possible Causes** • MIAH relay circuit is open, shorted to ground or to power (B+) • MIAH heater relay power (B+) circuit is open to the Alternator • MIAH heater relay is damaged or it has failed • PCM has failed
P0645 **1T CCM, MIL: Yes** F-Series Truck & Excursion with a 6.0L VIN P Diesel engine Transmissions: All	**Air Conditioning Clutch Relay Control Circuit Malfunction Conditions:** Key on or engine running; and the PCM detected an unexpected low or high voltage condition A/C clutch relay control circuit. **Possible Causes** • A/C clutch relay control circuit is open or shorted to ground • A/C clutch relay power circuit (VPWR) is open • A/C clutch relay is damaged or it has failed • PCM has failed
P0649 **2T CCM, MIL: Yes** F-Series Truck & Excursion with a 6.0L VIN P Diesel engine Transmissions: All	**Cruise Control Lamp Control Circuit Malfunction Conditions:** Engine started, vehicle driven at cruise with the Cruise switch set, and the PCM detected a fault in the C/C lamp control circuit. The Cruise Control function is integrated into the PCM. The Speed Control (S/C) command switches are momentary switches located in the face of the steering wheel. They include a 3-position SET/ACCEL/COAST/RESUME switch and On/Off toggle switch. When a switch is depressed, a resistance values is selected in the PCM. **Possible Causes** • Cruise Control lamp control circuit is open or shorted • Cruise Control lamp is damaged or it has failed • PCM has failed
P0670 **2T CCM, MIL: Yes** E-Van, F-Series Truck & Excursion with 6.0L or 7.3L Diesel engine Transmissions: All	**Glow Plug Control Circuit Malfunction Conditions:** Engine started, IAT sensor less than 32ºF, Engine Oil Temperature (EOT) sensor less than 131ºF, system voltage from 11.8-15.0v, and the PCM detected an unexpected voltage condition on the Glow Plug control circuit. Glow Plug "on" time, controlled by the PCM, is a function of the inputs listed here. The Glow Plug "on" time varies from 1-120 seconds. **Possible Causes** • Glow Plug control circuit is open or shorted to ground • PCM has failed • TSB 99-25-10 contains a repair procedure for this code
P0671 **2T CCM, MIL: Yes** E-Van, F-Series Truck & Excursion with 6.0L or 7.3L Diesel engine Transmissions: All	**Glow Plug No. 1 Circuit Malfunction Conditions:** Engine started, system voltage from 11-15v, and the PCM detected an unexpected voltage condition on the Glow Plug 1 circuit. The Glow Plug "on" time varies from 1-120 seconds. **Possible Causes** • Glow Plug No. 1 power circuit is open or shorted to ground • Glow Plug No. 1 is damaged or it has failed • Glow Plug Green and Black connectors could be mismatched • PCM has failed • TSB 99-25-10 contains a repair procedure for this code

Diesel Engine OBD II Trouble Code List (P0xxx Codes)

DTC	Trouble Code Title, Conditions & Possible Causes
P0672 **2T CCM, MIL: Yes** E-Van, F-Series Truck & Excursion with 6.0L or 7.3L Diesel engine Transmissions: All	**Glow Plug No. 2 Circuit Malfunction Conditions:** Engine started, system voltage from 11-15v, and the PCM detected an unexpected voltage condition on the Glow Plug 2 circuit. The Glow Plug "on" time varies from 1-120 seconds. **Possible Causes** • Glow Plug No. 2 power circuit is open or shorted to ground • Glow Plug No. 2 is damaged or it has failed • Glow Plug Green and Black connectors could be mismatched • PCM has failed • TSB 99-25-10 contains a repair procedure for this code
P0673 **T CCM MIL: Yes** E-Van, F-Series Truck & Excursion with 6.0L or 7.3L Diesel engine Transmissions: All	**Glow Plug No. 3 Circuit Malfunction Conditions:** Engine started, system voltage from 11-15v, and the PCM detected an unexpected voltage condition on the Glow Plug 3 circuit. The Glow Plug "on" time varies from 1-120 seconds. **Possible Causes** • Glow Plug No. 3 power circuit is open or shorted to ground • Glow Plug No. 3 is damaged or it has failed • Glow Plug Green and Black connectors could be mismatched • PCM has failed • TSB 99-25-10 contains a repair procedure for this code
P0674 **2T CCM, MIL: Yes** E-Van, F-Series Truck & Excursion with 6.0L or 7.3L Diesel engine Transmissions: All	**Glow Plug No. 4 Circuit Malfunction Conditions:** Engine started, system voltage from 11-15v, and the PCM detected an unexpected voltage condition on the Glow Plug 4 circuit. The Glow Plug "on" time varies from 1-120 seconds. **Possible Causes** • Glow Plug No. 4 power circuit is open or shorted to ground • Glow Plug No. 4 is damaged or it has failed • Glow Plug Green and Black connectors could be mismatched • PCM has failed • TSB 99-25-10 contains a repair procedure for this code
P0675 **2T CCM, MIL: Yes** E-Van, F-Series Truck & Excursion with 6.0L or 7.3L Diesel engine Transmissions: All	**Glow Plug No. 5 Circuit Malfunction Conditions:** Engine started, system voltage from 11-15v, and the PCM detected an unexpected voltage condition on the Glow Plug 5 circuit. The Glow Plug "on" time varies from 1-120 seconds. **Possible Causes** • Glow Plug No. 5 power circuit is open or shorted to ground • Glow Plug No. 5 is damaged or it has failed • Glow Plug Green and Black connectors could be mismatched • PCM has failed • TSB 99-25-10 contains a repair procedure for this code
P0676 **2T CCM, MIL: Yes** E-Van, F-Series Truck & Excursion with 6.0L or 7.3L Diesel engine Transmissions: All	**Glow Plug No. 6 Circuit Malfunction Conditions:** Engine started, system voltage from 11-15v, and the PCM detected an unexpected voltage condition on the Glow Plug 6 circuit. The Glow Plug "on" time varies from 1-120 seconds. **Possible Causes** • Glow Plug No. 6 power circuit is open or shorted to ground • Glow Plug No. 6 is damaged or it has failed • Glow Plug Green and Black connectors could be mismatched • PCM has failed • TSB 99-25-10 contains a repair procedure for this code
P0677 **2T CCM, MIL: Yes** E-Van, F-Series Truck & Excursion with 6.0L or 7.3L Diesel engine Transmissions: All	**Glow Plug No. 7 Circuit Malfunction Conditions:** Engine started, system voltage from 11-15v, and the PCM detected an unexpected voltage condition on the Glow Plug 7 circuit. The Glow Plug "on" time varies from 1-120 seconds. **Possible Causes** • Glow Plug No. 7 power circuit is open or shorted to ground • Glow Plug No. 7 is damaged or it has failed • Glow Plug Green and Black connectors could be mismatched • PCM has failed • TSB 99-25-10 contains a repair procedure for this code

Diesel Engine OBD II Trouble Code List (P0xxx Codes)

DTC	Trouble Code Title, Conditions & Possible Causes
P0678 **2T CCM, MIL: Yes** E-Van, F-Series Truck & Excursion with 6.0L or 7.3L Diesel engine Transmissions: All	**Glow Plug No. 8 Circuit Malfunction Conditions:** Engine started, system voltage from 11-15v, and the PCM detected an unexpected voltage condition on the Glow Plug 8 circuit. The Glow Plug "on" time varies from 1-120 seconds). **Possible Causes** • Glow Plug No. 8 power circuit is open or shorted to ground • Glow Plug No. 8 is damaged or it has failed • Glow Plug Green and Black connectors could be mismatched • PCM has failed • TSB 99-25-10 contains a repair procedure for this code
P0683 **2T CCM, MIL: Yes** E-Van, F-Series Truck & Excursion with 6.0L or 7.3L Diesel engine Transmissions: All	**Glow Plug Diagnostic Communication Circuit Malfunction Conditions:** Engine started, system voltage at 11-15.0v, and the PCM detected an unexpected voltage condition on the Glow Plug data line circuit. **Possible Causes** • Glow Plug diagnostic circuit is open or shorted to ground • Glow Plug diagnostic circuit is shorted to system power (B+) • Glow Plug control module is damaged or it has failed • PCM has failed • TSB 99-25-10 contains a repair procedure for this code
P0684 **2T CCM, MIL: Yes** E-Van, F-Series Truck & Excursion with 6.0L or 7.3L Diesel engine Transmissions: All	**Glow Plug Module Communication Signal Performance Conditions:** Engine started, system voltage at 11-15.0v, and the PCM detected an unexpected voltage condition on the Glow Plug Communication Module (GPCM) to PCM signal circuit. **Possible Causes** • Glow Plug Module communication circuit to the PCM is open or shorted to ground • GPCM circuit is picking up EMI or RFI "noise"
P0645 **1T CCM, MIL: Yes** F-Series Truck & Excursion with a 6.0L VIN P Diesel engine Transmissions: A/T	**Transmission Control System MIL Request Detected Conditions:** Engine started; and the PCM received a message from the TCM to turn "on" the MIL after it detected a fault in a monitored function. **Possible Causes** • Read and record any trouble codes stored in the TCM as this is only an indicator code. There will be other TCM related codes stored along with DTC P0700.
P0703 **1T CCM, MIL: No** E-Van, F-Series Truck & Excursion with 6.0L or 7.3L Diesel engine Transmissions: A/T	**Brake Switch Circuit Malfunction Conditions:** Engine started, KOER Test enabled, and the PCM did not detect a change in the Brake Switch signal status after the brake pedal was pressed. This BPP switch is wired to the stoplamp switch. The PCM uses the BBP switch signal to detect when the brake is applied to determine when to disengage Speed Control and Auxiliary devices. **Possible Causes** • Brake pedal not depressed correctly during the KOER Self-Test • Brake switch circuit is open or shorted to ground • Brake switch is damaged or it has failed • PCM has failed
P0704 **1T CCM, MIL: No** E-Van, F-Series Truck & Excursion with 6.0L or 7.3L Diesel engine Transmissions: A/T	**Clutch Pedal Position Switch Circuit Malfunction Conditions:** Engine started, KOER Self-Test enabled, and the PCM did not detect any change in the CPP switch status during the Self-Test. Note: The CCP PID should change (5v to 0v) with clutch depressed. **Possible Causes** • Clutch pedal was not depressed during the KOER Self-Test • Clutch pedal switch circuit is open or shorted to ground • Clutch pedal switch is damaged or it has failed • PCM has failed

Diesel Engine OBD II Trouble Code List (P0xxx Codes)

DTC	Trouble Code Title, Conditions & Possible Causes
P0705 **2T CCM, MIL: Yes** E-Van, F-Series Truck & Excursion with 6.0L or 7.3L Diesel engine Transmissions: A/T	**DTR Sensor / TR Sensor Circuit Malfunction Conditions:** Engine started, and the PCM detected the Transmission Range (TR) sensor signal was out of its normal operating range during the test. **Possible Causes** • TR sensor signal circuit open or shorted to ground or to VREF • TR sensor is damaged or it has failed • PCM has failed
P0707 **2T CCM, MIL: Yes** E-Van, F-Series Truck & Excursion with 6.0L or 7.3L Diesel engine Transmissions: A/T	**DTR Sensor / TR Sensor Circuit Low Input Conditions:** Engine started, and the PCM detected the Transmission Range (TR) sensor signal was less than its self-test minimum voltage. **Possible Causes** • TR sensor signal circuit is shorted to ground • TR sensor is damaged or it has failed • PCM has failed
P0708 **2T CCM, MIL: Yes** E-Van, F-Series Truck & Excursion with 6.0L or 7.3L Diesel engine Transmissions: A/T	**DTR Sensor / TR Sensor Circuit High Input Conditions:** Engine running, and the PCM detected the Digital Transmission Range (DTR) or Transmission Range sensor (TR) signal was more than its self-test maximum range. **Possible Causes** • DTR or TR sensor ground circuit is open • DTR or TR sensor signal circuit is open or shorted to power • DTR or TR sensor damaged • PCM has failed
P0712 **2T CCM, MIL: No** E-Van, F-Series Truck & Excursion with 6.0L or 7.3L Diesel engine Transmissions: A/T	**TFT Sensor Circuit Low Input Conditions:** Key on or engine running; and the PCM detected the Transmission Fluid Temperature (TFT) sensor signal voltage was less than its self-test minimum range. Note: The TFT PID will read -40°F when this problem is present. **Possible Causes** • TFT sensor signal circuit is shorted to sensor or chassis ground • TFT sensor is damaged, or out-of-calibration, or it has failed • PCM has failed
P0713 **2T CCM, MIL: No** E-Van, F-Series Truck & Excursion with 6.0L or 7.3L Diesel engine Transmissions: A/T	**TFT Sensor Circuit High Input Conditions:** Key on or engine running; and the PCM detected the Transmission Fluid Temperature (TFT) sensor signal voltage was more than the maximum self-test value. Note: The TFT PID will read 315°F when this fault is present. **Possible Causes** • TFT sensor signal circuit is open between the sensor and PCM • TFT sensor ground circuit is open between sensor and PCM • TFT sensor is damaged or it has failed • PCM has failed
P0715 **2T CCM, MIL: No** E-Van, F-Series Truck & Excursion with 6.0L or 7.3L Diesel engine Transmissions: A/T	**Transmission Speed Shaft Sensor Circuit Malfunction Conditions:** Engine started, vehicle driven to a speed of over 1 mph, and the PCM detected the TSS signals were missing or erratic. **Possible Causes** • TSS signal circuit open or shorted to ground • TSS is damaged • PCM has failed

Diesel Engine OBD II Trouble Code List (P0xxx Codes)

DTC	Trouble Code Title, Conditions & Possible Causes
P0717 **2T CCM, MIL: No** E-Van, F-Series Truck & Excursion with 6.0L or 7.3L Diesel engine Transmissions: A/T	**Transmission Speed Shaft Sensor Circuit Intermittent Conditions:** Engine running, vehicle speed sensor signal more than 1 mph, and the PCM detected the TSS signals were erratic, irregular or missing. **Possible Causes** • TSS signal circuit has an intermittent open or short to ground • TSS is damaged • PCM has failed
P0718 **2T CCM, MIL: No** E-Van, F-Series Truck & Excursion with 6.0L or 7.3L Diesel engine Transmissions: A/T	**Transmission Speed Shaft Sensor Circuit Noisy Conditions:** Engine started, vehicle driven to a speed of over 1 mph, and the PCM detected "noise" interference on the TSS signal circuit. **Possible Causes** • TSS signal is "noisy" due to RFI or EMI interference from sources such as ignition components or charging system • TSS signal wiring is damaged or contacting other signal wiring • PCM has failed
P0720 **2T CCM, MIL: Yes** E-Van, F-Series Truck & Excursion with 6.0L or 7.3L Diesel engine Transmissions: A/T	**Output Shaft Speed Sensor Circuit Low Input Conditions:** Engine started, VSS signal over 1 mph, and the PCM detected the Output Shaft Speed signal did not correlate to the incoming signals received from the VSS or TCSS devices/modules. **Possible Causes** • OSS sensor circuit shorted to ground or open • OSS sensor circuit shorted to VPWR • OSS sensor is damaged • PCM has failed
P0721 **2T CCM, MIL: No** E-Van, F-Series Truck & Excursion with 6.0L or 7.3L Diesel engine Transmissions: A/T	**Output Shaft Speed Sensor Noise Interference Conditions:** Engine started, VSS input more than 1 mph, and the PCM detected "noise" interference on the Output Shaft Speed (OSS) sensor circuit. **Possible Causes** • OSS sensor wiring misrouted or damaged • After market add-on devices interfering with the OSS signal • OSS sensor damaged • PCM has failed
P0721 **2T CCM, MIL: No** E-Van, F-Series Truck & Excursion with 6.0L or 7.3L Diesel engine Transmissions: A/T	**Output Speed Sensor No Signal Conditions:** Engine started, and the PCM did not detect any OSS sensor signals upon initial vehicle movement during the test. **Possible Causes** • OSS sensor wiring misrouted or damaged • After market add-on devices interfering with the OSS signal • OSS sensor damaged • PCM has failed
P0732 **2T CCM, MIL: No** E-Van, F-Series Truck & Excursion with 6.0L or 7.3L Diesel engine Transmissions: A/T	**Incorrect Second Gear Ratio Conditions:** Engine started, vehicle operating with 2nd Gear commanded "on", and the PCM detected an incorrect 2nd gear ratio. **Possible Causes** • 2nd Gear solenoid harness connector not properly seated • 2nd Gear solenoid wring harness connector damaged • 2nd Gear solenoid signal shorted to ground, or open • 2nd Gear solenoid is damaged or not properly installed
P0733 **2T CCM, MIL: No** E-Van, F-Series Truck & Excursion with 6.0L or 7.3L Diesel engine Transmissions: A/T	**Incorrect Third Gear Ratio Conditions:** Engine started, vehicle operating with 3rd Gear commanded "on", and the PCM detected an incorrect 3rd gear ratio. **Possible Causes** • 3rd Gear solenoid harness connector not properly seated • 3rd Gear solenoid wiring harness connector damaged • 3rd Gear solenoid signal shorted to ground, or open • 3rd Gear solenoid is damaged or not properly installed

Diesel Engine OBD II Trouble Code List (P0xxx Codes)

DTC	Trouble Code Title, Conditions & Possible Causes
P0741 **2T CCM, MIL: No** E-Van, F-Series Truck & Excursion with 6.0L or 7.3L Diesel engine Transmissions: A/T	**A/T TCC Mechanical System Range/Performance Conditions:** Engine started, vehicle driven in gear with VSS signals received, and the PCM detected excessive TCC slippage while in normal vehicle operation. **Possible Causes** • TCC solenoid has a mechanical failure • TCC solenoid has a hydraulic failure • PCM has failed
P0743 **1T CCM, MIL: Yes** E-Van, F-Series Truck & Excursion with 6.0L or 7.3L Diesel engine Transmissions: A/T	**A/T TCC Solenoid Circuit Malfunction Conditions:** KOEO Self-Test enabled, and the PCM did not detect any voltage drop from the TCC solenoid circuit during testing. **Possible Causes** • TCC solenoid control circuit open or shorted to ground • TCC solenoid wiring harness connector damaged • TCC solenoid is damaged or it has failed • PCM has failed
P0750 **1T CCM, MIL: Yes** E-Van, F-Series Truck & Excursion with 6.0L or 7.3L Diesel engine Transmissions: A/T	**A/T Shift Solenoid 1/A Circuit Malfunction Conditions:** Engine started, vehicle driven in 1st gear, and the PCM detected an unexpected voltage condition on the SS1/A solenoid circuit with the SS1/A solenoid applied during testing. **Possible Causes** • SS1/A solenoid control circuit is open or shorted to ground • SS1/A solenoid wiring harness connector damaged • SS1/A solenoid is damaged or it has failed • PCM has failed
P0755 **1T CCM, MIL: Yes** E-Van, F-Series Truck & Excursion with 6.0L or 7.3L Diesel engine Transmissions: A/T	**Shift Solenoid 2/B Circuit Malfunction Conditions:** Engine started, vehicle driven in 1st gear, and the PCM detected an unexpected voltage condition on the SS1/A solenoid circuit with the SS2/B solenoid applied during testing. **Possible Causes** • SS2/B solenoid control circuit open or shorted to ground • SS2/B solenoid wiring harness connector damaged • SS2/B solenoid is damaged or it has failed • PCM has failed
P0781 **1T CCM, MIL: No** E-Van, F-Series Truck & Excursion with 6.0L or 7.3L Diesel engine Transmissions: A/T	**A/T 1 to 2 Shift Error Conditions:** Engine started, vehicle driven with VSS signals received, and the PCM detected the engine speed (rpm) did not change properly (i.e., an incorrect 1-2 gear ratio was detected during shifting). **Possible Causes** • SS1/A solenoid may be stuck, or a hydraulic failure exists • SS2/B solenoid may be stuck, or a hydraulic failure exists • Transmission may have damaged friction material
P0782 **1T CCM, MIL: No** E-Van, F-Series Truck & Excursion with 6.0L or 7.3L Diesel engine Transmissions: A/T	**A/T 2 to 3 Shift Error Conditions:** Engine started, vehicle driven with VSS signals received, and the PCM detected the engine speed (rpm) did not change properly (an incorrect 2-3 gear ratio was detected while shifting). **Possible Causes** • SS1/A solenoid may be stuck, or a hydraulic failure exists • SS2/B solenoid may be stuck, or a hydraulic failure exists • Transmission may have damaged friction material
P0783 **1T CCM, MIL: No** E-Van, F-Series Truck & Excursion with 6.0L or 7.3L Diesel engine Transmissions: A/T	**A/T 3 to 4 Shift Error Conditions:** Engine started, vehicle driven with VSS signals received, and the PCM detected the engine speed (rpm) did not change properly (an incorrect 3-4 gear ratio was detected while shifting). **Possible Causes** • SS1/A solenoid may be stuck, or a hydraulic failure exists • SS2/B solenoid may be stuck, or a hydraulic failure exists • Transmission may have damaged friction material

Diesel Engine OBD II Trouble Code List (P1xxx Codes)

DTC	Trouble Code Title, Conditions & Possible Causes
P1000 **1T PCM, MIL: No** E-Van, F-Series Truck & Excursion with 6.0L or 7.3L Diesel engine Transmissions: All	**OBD II Monitor Testing Not Complete Conditions:** Key on or engine running; and the PCM detected one the conditions shown under Possible Causes (i.e., this code cannot be cleared manually - it must clear itself after all the OBD II Monitors complete). Note: This code must be cleared to pass an I/M Test for registration. **Possible Causes** • Battery keep alive power (KAPWR) was removed to the PCM • One or more OBD II Monitors did not complete during an official OBD II Drive Cycle • PCM Reset step was performed with an OBD II Scan Tool
P1001 **1T CCM, MIL: No** E-Van, F-Series Truck & Excursion with 6.0L or 7.3L Diesel engine Transmissions: All	**KOER Self-Test Not Completed, KOER Test Aborted Conditions:** Key on, engine running self-test not completed during the normal allowable time period. **Possible Causes** • Engine speed (rpm) out of specification during the KOER test • Incorrect Self-Test Procedure • Scan Tool has a communication problem • Unexpected response from Self-Test monitors
P1105 **1T CCM, MIL: No** E-Van, F-Series Truck & Excursion with 6.0L or 7.3L Diesel engine Transmissions: All	**Dual Alternator Upper Fault (Monitor) Conditions:** Key on or engine running; and the PCM detected an unexpected voltage condition on the Dual Alternator Upper circuit during the test. **Possible Causes** • Dual Alternator Upper circuit connection problem (corrosion) • Dual Alternator Upper monitor circuit open or shorted to ground • Dual Alternator is damaged or it has failed • PCM has failed
P1106 **1T CCM, MIL: No** E-Van, F-Series Truck & Excursion with 6.0L or 7.3L Diesel engine Transmissions: All	**Dual Alternator Lower Fault (Control) Conditions:** Key on or engine running; and the PCM detected an unexpected voltage condition on the Dual Alternator Lower circuit during the test. **Possible Causes** • D Dual Alternator Lower circuit connection problem (corrosion) • Dual Alternator Lower control circuit open or shorted to ground • Dual Alternator is damaged or it has failed • PCM has failed
P1107 **1T CCM, MIL: Yes** E-Van, F-Series Truck & Excursion with 6.0L or 7.3L Diesel engine Transmissions: All	**Dual Alternator Lower Circuit Malfunction Conditions:** Key on or engine running; and the PCM detected an unexpected voltage condition on the Dual Alternator Lower circuit during the test. **Possible Causes** • Dual Alternator Lower circuit (IGN) is open or shorted to ground • Dual Alternator is damaged or it has failed • PCM has failed
P1108 **1T CCM, MIL: No** E-Van, F-Series Truck & Excursion with 6.0L or 7.3L Diesel engine Transmissions: All	**Dual Alternator BATT Lamp Circuit Malfunction Conditions:** Key on or engine running; and the PCM detected an unexpected voltage condition on the Dual Alternator Lower circuit during the test. **Possible Causes** • Dual Alternator BATT Lamp circuit is open or shorted to ground • Instrument Panel fuse is open (check for cause of the short) Instrument Panel is damaged or it has failed • PCM has failed
P1118 **1T CCM, MIL: Yes** E-Van, F-Series Truck & Excursion with 6.0L or 7.3L Diesel engine Transmissions: All	**Manifold Air Temperature Sensor Low Input Conditions:** Key on or engine running; and the PCM detected an unexpected "low" voltage condition on the MAT sensor circuit during the test. This is a thermistor-type sensor with a variable resistance that changes when exposed to changes in the air temperature. **Possible Causes** • MAT sensor signal circuit is shorted to sensor ground • MAT sensor signal circuit is shorted to chassis ground • MAT sensor is damaged or it has failed • PCM has failed

Diesel Engine OBD II Trouble Code List (P1xxx Codes)

DTC	Trouble Code Title, Conditions & Possible Causes
P1119 **1T CCM, MIL: Yes** E-Van, F-Series Truck & Excursion with 6.0L or 7.3L Diesel engine Transmissions: All	**Manifold Air Temperature Sensor High Input Conditions:** Key on or engine running; and the PCM detected an unexpected "high" voltage condition on the MAT sensor circuit during the test. This is a thermistor-type sensor with a variable resistance that changes when exposed to changes in the air temperature. **Possible Causes** • MAT sensor signal circuit is open between sensor and the PCM • MAT sensor ground circuit is open between sensor and PCM • MAT sensor signal circuit is shorted to VREF or system power • MAT sensor is damaged or it has failed • PCM has failed
P1139 **1T CCM, MIL: No** E-Van, F-Series Truck & Excursion with 6.0L or 7.3L Diesel engine Transmissions: All	**Water-In-Fuel Indicator Circuit Malfunction Conditions:** Key on or engine running; and the PCM detected an unexpected voltage condition on the MAT sensor circuit during the test. **Possible Causes** • Water-In-Fuel circuit is open between Indicator and the PCM • Water-In-Fuel circuit is shorted to ground • Water-In-Fuel power circuit is open (check fuse in I/P panel) • Water is present in the fuel filter housing (drain it and retest) • PCM has failed
P1140 **1T CCM, MIL: No** E-Van, F-Series Truck & Excursion with 6.0L or 7.3L Diesel engine Transmissions: All	**Water-In-Fuel Condition Conditions:** Key on or engine running; and the PCM detected an unexpected voltage condition on the MAT sensor circuit during the test. **Possible Causes** • Drain the water from the fuel filter into a 1-quart clear container • Inspect the fuel that was previously drained. If there is no water or contaminants in the container, check for an intermittent fault • PCM has failed
P1148 **2T CCM, MIL: No** F-Series Truck & Excursion with a 6.0L VIN P Diesel engine Transmissions: All	**Generator 2 Control Circuit Malfunction Conditions:** Engine started, and the PCM detected an unexpected voltage condition on the Generator 2 control circuit during the CCM test. **Possible Causes** • Generator 2 control circuit is open or shorted, or the connector is damaged • Generator has failed • PCM has failed
P1149 **2T CCM, MIL: No** F-Series Truck & Excursion with a 6.0L VIN P Diesel engine Transmissions: All	**Generator 2 Control Circuit High Input Conditions:** Engine started, and the PCM detected an unexpected high voltage condition on the Generator 2 control circuit during the CCM test. **Possible Causes** • Generator 2 control circuit is shorted to system power, or the connector is damaged • Generator has failed • PCM has failed
P1184 **2T CCM, MIL: No** F-Series Truck & Excursion with a 6.0L VIN P Diesel engine Transmissions: All	**Engine Oil Temperature Sensor Out Of Self Test Range Conditions:** Key on, KOEO Self Test enabled, and the PCM detected that the engine temperature was below the self test minimum - test aborted. **Possible Causes** • Engine temperature to cold to run the KOEO Self Test • Thermostat is damaged or leaking (engine will not warm up) • EOT sensor signal circuit is open or shorted, or the PCM has failed
P1209 **2T CCM, MIL: Yes** E-Van, F-Series Truck & Excursion with 6.0L or 7.3L Diesel engine Transmissions: All	**Injection Control System Pressure Peak Malfunction Conditions:** Engine cranking and the PCM detected an unexpected voltage condition on the Injection Control Pressure circuit during the test. Note: The engine may not start if this code is set. **Possible Causes** • ICP sensor signal circuit is open, shorted to ground or shorted to system power • ICP sensor is damaged or it has failed • PCM has failed

Diesel Engine OBD II Trouble Code List (P1xxx Codes)

DTC	Trouble Code Title, Conditions & Possible Causes
P1210 **2T CCM, MIL: Yes** E-Van, F-Series Truck & Excursion with 6.0L or 7.3L Diesel engine Transmissions: All	**Injection Control System Pressure Above Expected Level Conditions:** Key on, and the PCM detected an unexpected "high" pressure condition on the Injection Control Pressure circuit during the test. **Possible Causes** • ICP sensor signal circuit is open or shorted to VREF or system power • ICP sensor ground circuit is open • ICP sensor is damaged or it has failed • PCM has failed
P1211 **1T CCM, MIL: Yes** E-Van, F-Series Truck & Excursion with 6.0L or 7.3L Diesel engine Transmissions: All	**Injection Control System Pressure Not Controllable - Pressure Above/Below Desired Code Conditions:** Engine started, and the PCM detected the engine was operating in open loop with the Injection Control Pressure too "high" or too "low". **Possible Causes** • ICP sensor circuit is open or shorted to ground (intermittent) • ICP sensor is damaged or it has failed • ICP system has failed, or the PCM has failed
P1212 **1T CCM, MIL: No** E-Van, F-Series Truck & Excursion with 6.0L or 7.3L Diesel engine Transmissions: All	**Injection Control System Pressure Not At Expected Level Conditions:** Engine started, and the PCM detected the engine was operating in open loop with the Injection Control Pressure too "high" or too "low". **Possible Causes** • ICP sensor circuit is open or shorted to ground (intermittent) • ICP sensor is damaged or it has failed • ICP system is damaged or it has failed • PCM has failed
P1218 **2T CCM, MIL: Yes** E-Van, F-Series Truck & Excursion with 6.0L or 7.3L Diesel engine Transmissions: All	**Cylinder ID Signal Stuck High Conditions:** Engine started, and the PCM detected the engine was operating in open loop with the Injection Control Pressure too "high" or too "low". **Possible Causes** • CID signal circuit to the IDM is open or shorted to system power • CID signal return circuit is open between the IDM and the PCM • IDM is damaged or it has failed • PCM has failed
P1219 **2T CCM, MIL: Yes** E-Van, F-Series Truck & Excursion with 6.0L or 7.3L Diesel engine Transmissions: All	**Cylinder ID Signal Stuck Low Conditions:** Engine started, and the PCM detected the engine was operating in open loop with the Injection Control Pressure too "high" or too "low". **Possible Causes** • CID signal circuit to the IDM is shorted to ground • IDM is damaged or it has failed • PCM has failed
P1247 **2T CCM, MIL: Yes** E-Van, F-Series Truck & Excursion with 6.0L or 7.3L Diesel engine Transmissions: All	**Turbo Boost Pressure Circuit Low Input Conditions:** Engine started, and the PCM detected a signal from the Injection Pressure sensor indicating the Boost Pressure was too "low". **Possible Causes** • ICP sensor signal circuit is shorted to ground (intermittent fault) • Injection Pressure system is damaged or it has failed • PCM has failed
P1248 **2T CCM, MIL: Yes** E-Van, F-Series Truck & Excursion with 6.0L or 7.3L Diesel engine Transmissions: All	**Turbo Boost Pressure Not Detected Conditions:** Engine started, and the PCM detected a signal from the Injection Pressure sensor indicating a lack of Boost Pressure was present. **Possible Causes** • Wastegate Control hose going to the actuator is leaking, loose or damaged • Wastegate solenoid is damaged, sticking or it has failed • PCM has failed

Diesel Engine OBD II Trouble Code List (P1xxx Codes)

DTC	Trouble Code Title, Conditions & Possible Causes
P1249 **2T CCM, MIL: Yes** E-Van, F-Series Truck & Excursion with 6.0L or 7.3L Diesel engine Transmissions: All	**Wastegate Fail Steady State Test Conditions:** Engine started, and the PCM detected an Injection Pressure sensor signal indicating the Wastegate failed the Steady state test. **Possible Causes** • WGC circuit is shorted to ground • WGC hose going to the actuator is leaking, loose or damaged • WGC actuator or valve is damaged or it has failed • Wastegate solenoid is damaged or the PCM has failed
P1250 **1T PCM, MIL: No** E-Van, F-Series Truck & Excursion with 6.0L or 7.3L Diesel engine Transmissions: All	**Theft Detected, Vehicle Immobilized Conditions:** Key on, and the PCM received a signal from the Anti-Theft System that a theft condition had occurred. The theft indicator on the dash will flash rapidly or remain on "solid" with the ignition switch in the "on" position. The engine may "start and stall", or may not crank if the vehicle is equipped with the PATS starter disable feature. **Possible Causes** • Anti-Theft System is damaged or it has failed • Previous theft condition may have occurred and set this code • TSB 01-6-02 contains a repair procedure for this trouble code
P1260 **1T PCM, MIL: No** F-Series Truck & Excursion with a 6.0L VIN P Diesel engine Transmissions: All	**Theft Detected, Vehicle Immobilized Conditions:** Key on or engine cranking; and the PCM received a message that indicated an invalid key had been inserted during a start sequence. **Possible Causes** • Ignition key is invalid or not coded properly • Theft detection module or circuit is damaged or it has failed • PCM has failed
P1261 **1T CCM, MIL: No** E-Van, F-Series Truck & Excursion with a 7.3L Diesel engine Transmissions: All	**Cylinder 1 - High To Low Side Short Detected Conditions:** Engine started, and the PCM detected a shorted condition on the High to Low side of the Injector 1 control circuit. The High side driver output function provides power and regulated current to the correct injector bank based on the CID signal and Fuel Delivery Command Signal from the Injector Driver Module that controls the 115v supply. **Possible Causes** • Injector 1 "high" side to "low" side shorted circuit detected • IDM or the PCM is damaged or it has failed
P1262 **1T CCM, MIL: No** E-Van, F-Series Truck & Excursion with a 7.3L Diesel engine Transmissions: All	**Cylinder 2 - High To Low Side Short Detected Conditions:** Engine started, and the PCM detected a shorted condition on the High to Low side of the Injector 2 control circuit. The High side driver output function provides power and regulated current to the correct injector bank based on the CID signal and Fuel Delivery Command Signal from the Injector Driver Module that controls the 115v supply. **Possible Causes** • Injector 2 "high" side to "low" side shorted circuit detected • IDM or the PCM is damaged or it has failed
P1263 **1T CCM, MIL: No** E-Van, F-Series Truck & Excursion with a 7.3L Diesel engine Transmissions: All	**Cylinder 3 - High To Low Side Short Detected Conditions:** Engine started, and the PCM detected a shorted condition on the High to Low side of the Injector 3 control circuit. The High side driver output function provides power and regulated current to the correct injector bank based on the CID signal and Fuel Delivery Command Signal from the Injector Driver Module that controls the 115v supply. **Possible Causes** • Injector 3 "high" side to "low" side shorted circuit detected • IDM or the PCM is damaged or it has failed
P1264 **1T CCM, MIL: No** E-Van, F-Series Truck & Excursion with a 7.3L Diesel engine Transmissions: All	**Cylinder 4 - High To Low Side Short Detected Conditions:** Engine started, and the PCM detected a shorted condition on the High to Low side of the Injector 4 control circuit. The High side driver output function provides power and regulated current to the correct injector bank based on the CID signal and Fuel Delivery Command Signal from the Injector Driver Module that controls the 115v supply. **Possible Causes** • Injector 4 "high" side to "low" side shorted circuit detected • IDM or the PCM is damaged or it has failed

Diesel Engine OBD II Trouble Code List (P1xxx Codes)

DTC	Trouble Code Title, Conditions & Possible Causes
P1265 **1T CCM, MIL: No** E-Van, F-Series Truck & Excursion with a 7.3L Diesel engine Transmissions: All	**Cylinder 5 - High To Low Side Short Detected Conditions:** Engine started, and the PCM detected a shorted condition on the High to Low side of the Injector 5 control circuit. The High side driver output function provides power and regulated current to the correct injector bank based on the CID signal and Fuel Delivery Command Signal from the Injector Driver Module that controls the 115v supply. **Possible Causes** • Injector 5 "high" side to "low" side shorted circuit detected • IDM or the PCM is damaged or it has failed
P1266 **1T CCM, MIL: No** E-Van, F-Series Truck & Excursion with a 7.3L Diesel engine Transmissions: All	**Cylinder 6 - High To Low Side Short Detected Conditions:** Engine started, and the PCM detected a shorted condition on the High to Low side of the Injector 6 control circuit. The High side driver output function provides power and regulated current to the correct injector bank based on the CID signal and Fuel Delivery Command Signal from the Injector Driver Module that controls the 115v supply. **Possible Causes** • Injector 6 "high" side to "low" side shorted circuit detected • IDM or the PCM is damaged or it has failed
P1267 **1T CCM, MIL: No** E-Van, F-Series Truck & Excursion with a 7.3L Diesel engine Transmissions: All	**Cylinder 7 - High To Low Side Short Detected Conditions:** Engine started, and the PCM detected a shorted condition on the High to Low side of the Injector 7 control circuit. The High side driver output function provides power and regulated current to the correct injector bank based on the CID signal and Fuel Delivery Command Signal from the Injector Driver Module that controls the 115v supply. **Possible Causes** • Injector 7 "high" side to "low" side shorted circuit detected • IDM or the PCM is damaged or it has failed
P1268 **1T CCM, MIL: No** E-Van, F-Series Truck & Excursion with a 7.3L Diesel engine Transmissions: All	**Cylinder 8 - High To Low Side Short Detected Conditions:** Engine started, and the PCM detected a shorted condition on the High to Low side of the Injector 8 control circuit. The High side driver output function provides power and regulated current to the correct injector bank based on the CID signal and Fuel Delivery Command Signal from the Injector Driver Module that controls the 115v supply. **Possible Causes** • Injector8 "high" side to "low" side shorted circuit detected • IDM or the PCM is damaged or it has failed
P1271 **1T CCM, MIL: No** E-Van, F-Series Truck & Excursion with a 7.3L Diesel engine Transmissions: All	**Cylinder 1 - High To Low Side Open Detected Conditions:** Engine started, and the PCM detected an open condition on the High to Low side of the Injector 1 control circuit. The High side driver output function provides power and regulated current to the correct injector bank based on the CID signal and Fuel Delivery Command Signal from the Injector Driver Module that controls the 115v supply. **Possible Causes** • Injector 1 high side circuit is open between the feed and injector • IDM or the PCM is damaged or it has failed
P1272 **1T CCM, MIL: No** E-Van, F-Series Truck & Excursion with a 7.3L Diesel engine Transmissions: All	**Cylinder 2 - High To Low Side Open Detected Conditions:** Engine started, and the PCM detected an open condition on the High to Low side of the Injector 2 control circuit. The High side driver output function provides power and regulated current to the correct injector bank based on the CID signal and Fuel Delivery Command Signal from the Injector Driver Module that controls the 115v supply. **Possible Causes** • Injector 2 high side circuit is open between the feed and injector • IDM or the PCM is damaged or it has failed

Diesel Engine OBD II Trouble Code List (P1xxx Codes)

DTC	Trouble Code Title, Conditions & Possible Causes
P1273 **1T CCM, MIL: No** E-Van, F-Series Truck & Excursion with a 7.3L Diesel engine Transmissions: All	**Cylinder 3 - High To Low Side Open Detected Conditions:** Engine started, and the PCM detected an open condition on the High to Low side of the Injector 3 control circuit. The High side driver output function provides power and regulated current to the correct injector bank based on the CID signal and Fuel Delivery Command Signal from the Injector Driver Module that controls the 115v supply. **Possible Causes** • Injector 3 high side circuit is open between the feed and injector • IDM or the PCM is damaged or it has failed
P1274 **1T CCM, MIL: No** E-Van, F-Series Truck & Excursion with a 7.3L Diesel engine Transmissions: All	**Cylinder 4 - High To Low Side Open Detected Conditions:** Engine started, and the PCM detected an open condition on the High to Low side of the Injector 4 control circuit. The High side driver output function provides power and regulated current to the correct injector bank based on the CID signal and Fuel Delivery Command Signal from the Injector Driver Module that controls the 115v supply. **Possible Causes** • Injector 4 high side circuit is open between the feed and injector • IDM or the PCM is damaged or it has failed
P1275 **1T CCM, MIL: No** E-Van, F-Series Truck & Excursion with a 7.3L Diesel engine Transmissions: All	**Cylinder 5 - High To Low Side Open Detected Conditions:** Engine started, and the PCM detected an open condition on the High to Low side of the Injector 5 control circuit. The High side driver output function provides power and regulated current to the correct injector bank based on the CID signal and Fuel Delivery Command Signal from the Injector Driver Module that controls the 115v supply. **Possible Causes** • Injector 5 high side circuit is open between the feed and injector • IDM or the PCM is damaged or it has failed
P1276 **1T CCM, MIL: No** E-Van, F-Series Truck & Excursion with a 7.3L Diesel engine Transmissions: All	**Cylinder 6 - High To Low Side Open Detected Conditions:** Engine started, and the PCM detected an open condition on the High to Low side of the Injector 6 control circuit. The High side driver output function provides power and regulated current to the correct injector bank based on the CID signal and Fuel Delivery Command Signal from the Injector Driver Module that controls the 115v supply. **Possible Causes** • Injector 6 high side circuit is open between the feed and injector • IDM or the PCM is damaged or it has failed
P1277 **1T CCM, MIL: No** E-Van, F-Series Truck & Excursion with a 7.3L Diesel engine Transmissions: All	**Cylinder 7 - High To Low Side Open Detected Conditions:** Engine started, and the PCM detected an open condition on the High to Low side of the Injector 7 control circuit. The High side driver output function provides power and regulated current to the correct injector bank based on the CID signal and Fuel Delivery Command Signal from the Injector Driver Module that controls the 115v supply. **Possible Causes** • Injector 7 high side circuit is open between the feed and injector • IDM or the PCM is damaged or it has failed
P1278 **1T CCM, MIL: No** E-Van, F-Series Truck & Excursion with a 7.3L Diesel engine Transmissions: All	**Cylinder 8 - High To Low Side Open Detected Conditions:** Engine started, and the PCM detected an open condition on the High to Low side of the Injector 8 control circuit. The High side driver output function provides power and regulated current to the correct injector bank based on the CID signal and Fuel Delivery Command Signal from the Injector Driver Module that controls the 115v supply. **Possible Causes** • Injector 8 high side circuit is open between the feed and injector • IDM or the PCM is damaged or it has failed

Diesel Engine OBD II Trouble Code List (P1xxx Codes)

DTC	Trouble Code Title, Conditions & Possible Causes
P1280 **1T CCM, MIL: Yes** E-Van, F-Series Truck & Excursion with a 7.3L Diesel engine Transmissions: All	**Injection Control System Pressure Out-Of-Range Low Conditions:** Engine started, and the PCM detected the Injector Pressure Sensor (ICP) indicated an out-of-range "low" condition. The ICP sensor is a variable capacitance sensor that, when supplied with a 5v reference signal, produces an analog signal that indicates pressure. It is designed to provide a feedback signal that indicates the fuel rail pressure so the PCM can command injector timing and pressure. **Possible Causes** • ICP sensor circuit is shorted to ground • ICP sensor is "skewed" to the "low" side of its range, or it has failed • PCM has failed
P1281 **1T CCM, MIL: Yes** E-Van, F-Series Truck & Excursion with a 7.3L Diesel engine Transmissions: All	**Injection Control Pressure Circuit Out-Of-Range High Conditions:** Engine started, and the PCM detected the Injector Pressure Sensor (ICP) indicated an out-of-range "high" condition. The ICP sensor is a variable capacitance sensor that, when supplied with a 5v reference signal, produces an analog signal that indicates pressure. It is designed to provide a feedback signal that indicates the fuel rail pressure so the PCM can command injector timing and pressure. **Possible Causes** • ICP sensor circuit is open or shorted to VREF (5v) • ICP sensor is "skewed" to the "high" side of its range, or it has failed • PCM has failed
P1282 **1T CCM, MIL: Yes** E-Van, F-Series Truck & Excursion with a 7.3L Diesel engine Transmissions: All	**Injection Pressure Regulator Excessive Pressure Conditions:** Engine started, and the PCM detected the Injector Pressure Sensor signal was more than 3,675 psi for 1.5 seconds during the CCM test. **Possible Causes** • IPR control circuit is shorted to ground • ICP sensor is damaged or it has failed • PCM has failed
P1283 **1T CCM, MIL: No** E-Van, F-Series Truck & Excursion with a 7.3L Diesel engine Transmissions: All	**Injection Pressure Regulator Circuit Malfunction Conditions:** Key on, KOEO Self-Test enabled and the PCM detected an unexpected voltage condition on the Injector Pressure Regulator circuit during the test period. **Possible Causes** • IPR control circuit is open or shorted to ground • IPR control circuit is shorted to system power (B+) • IPR assembly is damaged or it has failed • PCM has failed
P1284 **1T CCM, MIL: No** E-Van, F-Series Truck & Excursion with a 7.3L Diesel engine Transmissions: All	**Injection Pressure Regulator Failure - KOER Test Aborted Conditions:** Engine started, KOER Self-Test enabled, and the PCM detected an unexpected voltage condition on the Injector Pressure Regulator and "aborted" the rest of the KOER test. **Possible Causes** • IPR control circuit is open or shorted to ground • IPR control circuit is shorted to system power • IPR is damaged or it has failed • PCM has failed
P1291 **1T CCM, MIL: No** E-Van, F-Series Truck & Excursion with a 7.3L Diesel engine Transmissions: All	**High Side No. 1 (Right) Short To B+ or Ground Detected Conditions:** Engine started, and the PCM detected the Injector Bank 1 "high" side circuit was shorted to ground or shorted to system power (B+). Note: The "high" side driver output function is to provide power to the proper bank based on a Fuel Delivery Command Signal (FDCS) and Injector Driver Module (IDM) which controls the 115v supply. **Possible Causes** • Injector 1 "high" side circuit is shorted to ground or to power • IDM or the PCM is damaged or it has failed

Diesel Engine OBD II Trouble Code List (P1xxx Codes)

DTC	Trouble Code Title, Conditions & Possible Causes
P1292 **1T CCM, MIL: No** E-Van, F-Series Truck & Excursion with a 7.3L Diesel engine Transmissions: All	**High Side No. 2 (Left) Short To Ground Or B+ Detected Conditions:** Engine started, and the PCM detected the Injector Bank 2 "high" side circuit was shorted to ground or shorted to system power (B+). Note: The "high" side driver output function is to provide power to the proper bank based on a Fuel Delivery Command Signal (FDCS) and Injector Driver Module (IDM) which controls the 115v supply. **Possible Causes** • Injector 2 "high" side circuit is shorted to ground or to power • IDM or the PCM is damaged or it has failed
P1293 **1T CCM, MIL: No** E-Van, F-Series Truck & Excursion with a 7.3L Diesel engine Transmissions: All	**High Side Bank No. 1 (Right) Open Conditions:** Engine started, and the PCM detected an open circuit condition on Injector Bank 1 (right bank). Note: The "high" side driver output function is to provide power to the proper bank based on a Fuel Delivery Command Signal (FDCS) and Injector Driver Module (IDM) which controls the 115v supply. **Possible Causes** • Injector Bank 1 "high" side circuit is open (the right side bank) • IDM or the PCM is damaged or it has failed
P1294 **1T CCM, MIL: No** E-Van, F-Series Truck & Excursion with a 7.3L Diesel engine Transmissions: All	**High Side Bank No. 2 (Left) Open Conditions:** Engine started, and the PCM detected an open circuit condition on Injector Bank 2 (left bank). Note: The "high" side driver output function is to provide power to the proper bank based on a Fuel Delivery Command Signal (FDCS) and Injector Driver Module (IDM) which controls the 115v supply. **Possible Causes** • Injector "high" side circuit is open (the left side bank) • IDM or the PCM is damaged or it has failed
P1295 **1T CCM, MIL: Yes** E-Van, F-Series Truck & Excursion with a 7.3L Diesel engine Transmissions: All	**Multiple Faults On Bank No. 1 (Right) Conditions:** Engine started, and the PCM detected multiple circuit faults on Cylinder Bank 1 (right side bank). Note: The "high" side driver output function is to provide power to the proper bank based on a Fuel Delivery Command Signal (FDCS) and Injector Driver Module (IDM) which controls the 115v supply. **Possible Causes** • Multiple circuit faults exist on Cylinder Bank 1 (the right side) • IDM or the PCM is damaged or it has failed
P1296 **1T CCM, MIL: Yes** E-Van, F-Series Truck & Excursion with a 7.3L Diesel engine Transmissions: All	**Multiple Faults On Bank No. 2 (Left) Conditions:** Engine started, and the PCM detected multiple circuit faults on Cylinder Bank 2 (left side bank). Note: The "high" side driver output function is to provide power to the proper bank based on a Fuel Delivery Command Signal (FDCS) and Injector Driver Module (IDM) which controls the 115v supply. **Possible Causes** • Multiple circuit faults exist on Cylinder Bank 2 (the left side) • IDM or the PCM is damaged or it has failed
P1297 **1T CCM, MIL: No** E-Van, F-Series Truck & Excursion with a 7.3L Diesel engine Transmissions: All	**High Sides Shorted Together Conditions:** Engine started, and the PCM detected the "high" side circuit were shorted together. Note: The "high" side driver output function is to provide power to the proper bank based on a Fuel Delivery Command Signal (FDCS) and Injector Driver Module (IDM) which controls the 115v supply. **Possible Causes** • Injector "high" side circuits are shorted together • IDM or the PCM is damaged or it has failed

Diesel Engine OBD II Trouble Code List (P1xxx Codes)

DTC	Trouble Code Title, Conditions & Possible Causes
P1298 **1T CCM, MIL: No** E-Van, F-Series Truck & Excursion with a 7.3L Diesel engine Transmissions: All	**Injector Driver Module Malfunction Conditions:** Engine started, and the PCM detected a signal from the Injector Driver Module (IDM) indicating that it had failed during the CCM test. The "high" side driver output function is to provide power to the proper bank based on a Fuel Delivery Command Signal (FDCS) and Injector Driver Module (IDM) which controls the 115v supply. **Possible Causes** • Injector Driver Module is damaged or it has failed
P1316 **1T CCM, MIL: Yes** E-Van, F-Series Truck & Excursion with a 7.3L Diesel engine Transmissions: All	**Injector Circuit/IDM Codes Detected Conditions:** Engine started, and the PCM detected a signal from the Injector Driver Module (IDM) indicating that IDM codes were in its memory. The "high" side driver output function is to provide power to the proper bank based on a Fuel Delivery Command Signal (FDCS) and Injector Driver Module (IDM) which controls the 115v supply. **Possible Causes** • IDM contains one or more IDM related trouble codes
P1378 **1T CCM, MIL: Yes** F-Series Truck & Excursion with a 6.0L VIN P Diesel engine Transmissions: All	**FICM Supply Voltage Low Input Conditions:** Key on or engine started, and the PCM detected an unexpected low voltage condition on the FICM supply voltage circuit during the test. **Possible Causes** • FICM supply voltage connector is damaged or open • FICM supply voltage circuit is open or shorted to ground • FICM is damaged or it has failed • PCM has failed
P1379 **1T CCM, MIL: Yes** F-Series Truck & Excursion with a 6.0L VIN P Diesel engine Transmissions: All	**FICM Supply Voltage High Input Conditions:** Key on or engine started, and the PCM detected an unexpected high voltage condition on the FICM supply voltage circuit during the test. **Possible Causes** • FICM supply voltage connector is damaged or shorted • FICM supply voltage circuit is shorted to system power • FICM is damaged or it has failed • PCM has failed
P1391 **1T CCM, MIL: Yes** E-Van, F-Series Truck & Excursion with a 7.3L Diesel engine Transmissions: All	**Glow Plug Circuit Low Input On Bank 1 (Right Side) Conditions:** Engine started, engine running and the PCM detected all four Glow Plug circuits were open on Cylinder Bank 1 (the right side bank). The "high" side driver output function is to provide power to the proper bank based on a Fuel Delivery Command Signal (FDCS) and Injector Driver Module (IDM) which controls the 115v supply. **Possible Causes** • Glow Plugs circuits or fusible links open on Bank 1 (right side) • Glow plugs are open on Cylinder Bank 1 (right side) • Glow plug relay circuit is open • IDM or PCM is damaged or it has failed
P1393 **1T CCM, MIL: Yes** E-Van, F-Series Truck & Excursion with a 7.3L Diesel engine Transmissions: All	**Glow Plug Circuit Low Input On Bank 2 (Left Side) Conditions:** Engine started, and the PCM detected that all four Glow Plug circuits were open on Cylinder Bank 2 (the left side bank). The "high" side driver output function is to provide power to the proper bank based on a Fuel Delivery Command Signal (FDCS) and Injector Driver Module (IDM) which controls the 115v supply. **Possible Causes** • Glow Plugs circuits or fusible links open on Bank 2 (left side) • Glow plugs are open on Cylinder Bank 2 (left side) • Glow plug relay circuit is open • PCM is damaged or it has failed

Diesel Engine OBD II Trouble Code List (P1xxx Codes)

DTC	Trouble Code Title, Conditions & Possible Causes
P1395 **1T CCM, MIL: Yes** E-Van, F-Series Truck & Excursion with a 7.3L Diesel engine Transmissions: All	**Glow Plug Monitor Fault On Bank 1 (Right Side) Conditions:** Key on, glow plugs enabled, system voltage over 11.4v, and the PCM detected the Bank 1 Glow Plug current flow was less than 39 amps. The "high" side driver output function is to provide power to the proper bank based on a Fuel Delivery Command Signal (FDCS) and Injector Driver Module (IDM) which controls the 115v supply. **Possible Causes** • Glow plug circuit or one or two Glow Plugs with high resistance • One or more Glow Plugs shorted (low resistance) on Bank 2
P1396 **1T CCM, MIL: Yes** E-Van, F-Series Truck & Excursion with a 7.3L Diesel engine Transmissions: All	**Glow Plug Monitor Fault On Bank 2 (Left Side) Conditions:** Key on, glow plugs enabled, system voltage over 11.4v, and the PCM detected the Bank 1 Glow Plug current flow was less than 39 amps. The "high" side driver output function is to provide power to the proper bank based on a Fuel Delivery Command Signal (FDCS) and Injector Driver Module (IDM) which controls the 115v supply. **Possible Causes** • Glow plug circuit or one or two Glow Plugs with high resistance • One or more Glow Plugs shorted (low resistance) on Bank 1
P1397 **1T CCM, MIL: Yes** E-Van, F-Series Truck & Excursion with 6.0L or 7.3L Diesel engine Transmissions: All	**Glow Plug Voltage Out Of Self-Test Range Conditions:** Engine started, and the PCM detected the system voltage (to the Glow Plugs) was out of its normal operating range. The "high" side driver output function is to provide power to the proper bank based on a Fuel Delivery Command Signal (FDCS) and Injector Driver Module (IDM) which controls the 115v supply. **Possible Causes** • Glow Plug system voltage is out of its self-test range
P1408 **1T EGR, MIL: No** F-Series Truck & Excursion with a 6.0L VIN P Diesel engine Transmissions: All	**EGR Flow Out Of Self-Test Range Conditions:** Engine started, KOER Self-Test enabled, and the PCM detected the EGR flow was out of the self-test range during the self-test. The EGR valve and EGR sensor are included in one unit. **Possible Causes** • EVR valve stuck closed or iced up, or the flow path is restricted • EGR valve diaphragm leaking, hose is off, plugged or leaking • EGR sensor or the EGR solenoid is damaged or has failed • PCM has failed
P1464 **1T CCM, MIL: No** E-Van, F-Series Truck & Excursion with 6.0L or 7.3L Diesel engine Transmissions: All	**A/C Demand Out of Self-Test Range Conditions:** Engine started, KOER Self-Test enabled, and the PCM detected the ACCS input was high during the self-test. This code can set if the A/C is turned "on" during the KOER Self-Test. **Possible Causes** • A/C switch was "on" during self-test • A/C PWR circuit shorted to VPWR (N/C WAC relay contacts) • ACCS circuit shorted to power • A/C Demand Switch, WAC relay or CCRM is damaged
P1501 **1T CCM, MIL: No** E-Van, F-Series Truck & Excursion with 6.0L or 7.3L Diesel engine Transmissions: All	**VSS Signal Out of Self-Test Range Conditions:** Engine started, KOER Self-Test enabled, and the PCM detected a VSS signal during the KOER test (i.e., with the vehicle not moving). **Possible Causes** • VSS signal is noisy due to Radio Frequency Interference/ Electro-Magnetic Interference (RFI/EMI) • Check for RFI or EMI "noise" from external sources such as ignition wires, charging circuit or aftermarket devices
P1502 **1T CCM, MIL: No** E-Van, F-Series Truck & Excursion with 6.0L or 7.3L Diesel engine Transmissions: All	**Invalid Self-Test - APCM Functioning Conditions:** Engine started, KOER Self-Test enabled, and the PCM detected the APCM was "on" during the test period. **Possible Causes** • APCM is "on" during the KOER Self-Test • Repeat the KOER Self-Test with the APCM turned "off"

Diesel Engine OBD II Trouble Code List (P1xxx Codes)

DTC	Trouble Code Title, Conditions & Possible Causes
P1531 **1T CCM, MIL: No** E-Van, F-Series Truck & Excursion with 6.0L or 7.3L Diesel engine Transmissions: All	**Invalid Self-Test - Accelerator Pedal Movement Conditions:** Engine started, KOER Self-Test enabled, and the PCM detected the accelerator pedal was moved during the test period. **Possible Causes** • Accelerator pedal movement during the KOER Self-Test • Repeat KOER Self-Test without moving the accelerator pedal
P1536 **1T CCM, MIL: No** E-Van, F-Series Truck & Excursion with 6.0L or 7.3L Diesel engine Transmissions: All	**Parking Brake Applied Failure Conditions:** Engine started, KOER Self-Test enabled, and the PCM detected the Parking Brake application failed during the PTO / Raised Idle Mode portion of the KOER test. **Possible Causes** • Parking brake switch signal is open or shorted to ground • Parking brake switch is damaged or it has failed • PCM has failed
P1610 **1T PCM, MIL: No** F-Series Truck & Excursion with a 6.0L VIN P Diesel engine Transmissions: All	**Interactive Reprogramming Code - Replace The PCM Conditions:** Key on, and the PCM detected an error that indicated the PCM had failed during the initial test phase. **Possible Causes** • Perform the correct procedures to replace the PCM. Once a new PCM is installed, perform the correct procedures to "flash" the new PCM. Then retest the system for any trouble codes.
P1611 **1T PCM, MIL: No** F-Series Truck & Excursion with a 6.0L VIN P Diesel engine Transmissions: All	**Interactive Reprogramming Code - Diagnose Further Conditions:** Key on, and the PCM detected an error that indicated further diagnosis of the PCM is required to complete the test phase. **Possible Causes** • Perform the correct procedures to "flash" the control module. Then retest the system for any trouble codes.
P1615 **1T PCM, MIL: No** F-Series Truck & Excursion with a 6.0L VIN P Diesel engine Transmissions: All	**Interactive Reprogramming Code - Flash Erase Error Conditions:** Key on, and the PCM detected a Flash Error during its initial startup. **Possible Causes** • Perform the correct procedures to "flash" the control module. Then retest the system for any trouble codes.
P1616 **1T PCM, MIL: No** F-Series Truck & Excursion with a 6.0L VIN P Diesel engine Transmissions: All	**Interactive Reprogramming Code - Flash Erase Error, Low Voltage Conditions:** Key on, and the PCM detected a Flash Erase Error due to a low voltage condition during a previous "flash" event. **Possible Causes** • Perform the correct procedures to "flash" the control module. Then retest the system for any trouble codes.
P1617 **1T PCM, MIL: No** F-Series Truck & Excursion with a 6.0L VIN P Diesel engine Transmissions: All	**Interactive Reprogramming Code - Block Programming Error Conditions:** Key on, and the PCM detected a Block Programming Error existed. **Possible Causes** • Perform the correct procedures to "flash" the control module. Then retest the system for any trouble codes.

Diesel Engine OBD II Trouble Code List (P1xxx Codes)

DTC	Trouble Code Title, Conditions & Possible Causes
P1618 **1T PCM, MIL: No** F-Series Truck & Excursion with a 6.0L VIN P Diesel engine Transmissions: All	**Interactive Reprogramming Code - Block Programming Error, Low Voltage Conditions:** Key on, and the PCM detected a Block Programming Error existed due to a low voltage condition during a previous reflash event. **Possible Causes** • Perform the correct procedures to "flash" the control module. Then retest the system for any trouble codes.
P1660 **1T CCM, MIL: No** E-Van, F-Series Truck & Excursion with 7.3L VIN F Diesel engine Transmissions: All	**OCC Signal High Conditions:** Key on, KOEO Self-Test enabled, and the PCM detected high system voltage during the test due to an internal PCM failure. **Possible Causes** • PCM is damaged or it has failed. Clear the codes, and then rerun the KOEO Self-Test. If the same code resets, the PCM has failed and must be replaced.
P1660 **1T CCM, MIL: No** E-Van, F-Series Truck & Excursion with 7.3L VIN F Diesel engine Transmissions: All	**OCC Signal Low Conditions:** Key on, KOEO Self-Test enabled, and the PCM detected low system voltage during the test due to an internal PCM failure. **Possible Causes** • PCM is damaged or it has failed. Clear the codes, and then rerun the KOEO Self-Test. If the same code resets, the PCM has failed and must be replaced.
P1662 **1T CCM, MIL: No** E-Van, F-Series Truck & Excursion with 6.0L or 7.3L Diesel engine Transmissions: All	**EF Feedback Signal Not Detected Conditions:** Engine started, and the PCM did not detect an Electronic Feedback signal from the IDM during the CCM test. **Possible Causes** • EF signal circuit is open or shorted to ground • EF ground circuit is open • IDM is damaged or it has failed • PCM has failed
P1663 **1T CCM, MIL: No** E-Van, F-Series Truck & Excursion with 6.0L or 7.3L Diesel engine Transmissions: All	**Fuel Delivery Command Signal Circuit Malfunction Conditions:** Key on, KOEO Self-Test enabled and the PCM detected an unexpected low or high voltage condition on the Fuel Delivery Command Signal (FDCS) circuit during the initial test period. **Possible Causes** • FDCS signal circuit is open • FDCS signal circuit is shorted to ground • FDCS ground circuit is open • PCM has failed
P1667 **1T CCM, MIL: No** E-Van, F-Series Truck & Excursion with 6.0L or 7.3L Diesel engine Transmissions: All	**CID Circuit Malfunction Conditions:** Engine started, KOEO Self-Test enabled, and the PCM detected an unexpected low or condition on the Cylinder Identification (CID) circuit during the test period. **Possible Causes** • CID signal circuit is open • CID signal circuit is shorted to ground • CID ground circuit is open • PCM has failed
P1668 **1T CCM, MIL: No** E-Van, F-Series Truck & Excursion with 6.0L or 7.3L Diesel engine Transmissions: All	**PCM TO IDM Diagnostic Communication Error Conditions:** Engine started, and the PCM detected an unexpected low or high voltage condition on the Electronic Feedback signal circuit during the test period. **Possible Causes** • EF signal circuit is open or shorted to ground • EF ground circuit is open • PCM has failed

Diesel Engine OBD II Trouble Code List (P1xxx Codes)

DTC	Trouble Code Title, Conditions & Possible Causes
P1670 **2T CCM, MIL: Yes** E-Van, F-Series Truck & Excursion with 6.0L or 7.3L Diesel engine Transmissions: All	**EF Feedback Signal Not Detected Conditions:** Engine started, and the PCM did not detect any signals on the EF feedback signal circuit during the test. **Possible Causes** • EF signal circuit is open or shorted to ground • EF ground circuit is open • PCM has failed
P1690 **2T CCM, MIL: Yes** E-Van, F-Series Truck & Excursion with 6.0L or 7.3L Diesel engine Transmissions: All	**Wastegate Control Valve Malfunction Conditions:** Engine started, and the PCM did not detect any signals on the EF feedback signal circuit during the test. A Wastegate type of turbo is designed to reach maximum boost sooner than a conventional turbo. However, over-boosting will cause damage to the turbo assembly. The PCM controls the boost pressure with a duty cycle signal to the solenoid to maximize boost performance (no more than 16.5 psi). When pressure is supplied on the Red hose to the actuator (with the solenoid not energized), the valve will open, dumping boost. When low or no pressure is supplied to the Red hose to the actuator (solenoid is being energized), the actuator valve will stay closed. **Possible Causes** • EF signal circuit is open or shorted to ground • EF ground circuit is open • PCM has failed
P1702 **1T CCM, MIL: No** E-Van, F-Series Truck & Excursion with a 7.3L Diesel engines Transmissions: A/T	**Transmission Range Sensor Intermittent Signal Conditions:** Key on or engine running; and the PCM detected the failure conditions were met intermittently for DTC P0705 or P0708 due to a malfunction of the Transmission Range (TR) Sensor or its circuit. **Possible Causes** • Refer to the appropriate Transmission Repair Manual or information in electronic media to perform a complete diagnosis of the automatic transmission when this code is set
P1703 **1T CCM, MIL: No** F-Series Truck & Excursion with a 6.0L VIN P Diesel engine Transmissions: A/T	**Brake Switch Out Of Self Test Range Conditions:** Key on with KOEO Self Test enabled, or engine started with KOER Self Test enabled, and the PCM detected an invalid Brake Switch signal during the CCM test period. Rerun the appropriate self test. **Possible Causes** • Brake switch not cycled properly during the self test • Brake switch connector is damaged, loose or shorted • Brake switch is damaged or it has failed
P1704 **1T CCM, MIL: No** E-Van, F-Series Truck & Excursion with a 7.3L Diesel engines Transmissions: A/T	**Transmission Range Sensor Out Of Self-Test Range Conditions:** Key on or engine running; and the PCM detected a Transmission Range (TR) sensor signal occurred in between gear positions. **Possible Causes** • Digital TR sensor or shift cable misadjusted • Digital TR sensor circuit open or shorted to ground • Digital TR sensor is damaged or it has failed
P1705 **1T CCM, MIL: No** E-Van, F-Series Truck & Excursion with 6.0L or 7.3L Diesel engine Transmissions: A/T	**Transmission Range Sensor Out Of Self-Test Range Conditions:** Key on, KOEO Self-Test enabled, or engine running, KOER Self-Test enabled, and the PCM detected it did not receive a Transmission Range (TR) sensor signal while in Park or Neutral. **Possible Causes** • Gear selector not in Park or Neutral during the self-test • Digital TR sensor circuit open or shorted to ground • Digital TR sensor is damaged or it has failed

Diesel Engine OBD II Trouble Code List (P1xxx Codes)

DTC	Trouble Code Title, Conditions & Possible Causes
P1711 **1T CCM, MIL: No** E-Van, F-Series Truck & Excursion with a 7.3L Diesel engines Transmissions: A/T	**TFT Sensor Out of Self-Test Range Conditions:** **KOEO or KOER Self-Test Enabled** Key on, KOEO Self-Test enabled, or engine running, KOER Self-Test enabled, and the PCM detected the Transmission Fluid Temperature (TFT) sensor was more than or less than the calibrated range (25°F to 240°F) during the self-test period. **Possible Causes** • Refer to the appropriate Transmission Repair Manual or information in electronic media to perform a complete diagnosis of the automatic transmission when this code is set.
P1713 **2T CCM, MIL: No** E-Van, F-Series Truck & Excursion with a 7.3L Diesel engines Transmissions: A/T	**TFT Sensor Stuck In Low Range Conditions:** Engine started, vehicle driven to over 1 mph for 2-3 minutes, and the PCM detected the TFT sensor signal was stuck in low range (less than 50°F) during the CCM test. **Possible Causes** • Refer to the appropriate Transmission Repair Manual or information in electronic media to perform a complete diagnosis of the automatic transmission when this code is set.
P1714 **2T CCM, MIL: Yes** E-Van, F-Series Truck & Excursion with 6.0L or 7.3L Diesel engine Transmissions: A/T	**A/T Shift Solenoid 'A' Inductive Signature Malfunction Conditions:** Engine started, VSS over 1 mph, and the PCM detected a problem with the Shift Solenoid 'A' Inductive signature during the CCM test. **Possible Causes** • Refer to the appropriate Transmission Repair Manual or information in electronic media to perform a complete diagnosis of the automatic transmission when this code is set.
P1715 **2T CCM, MIL: Yes** E-Van, F-Series Truck & Excursion with 6.0L or 7.3L Diesel engine Transmissions: A/T	**A/T Shift Solenoid 'B' Inductive Signature Malfunction Conditions:** Engine running, VSS over 1 mph, and the PCM detected a problem with the Shift Solenoid 'B' Inductive signature during the CCM test. **Possible Causes** • Refer to the appropriate Transmission Repair Manual or information in electronic media to perform a complete diagnosis of the automatic transmission when this code is set
P1718 **1T CCM, MIL: No** E-Van, F-Series Truck & Excursion with a 7.3L Diesel engines Transmissions: A/T	**TFT Sensor Stuck In High Range Conditions:** Engine started, vehicle driven to over 1 mph for 2-3 minutes, and the PCM detected the TFT sensor signal was stuck in high range (more than 250°F) during the CCM test. **Possible Causes** • Refer to the appropriate Transmission Repair Manual or information in electronic media to perform a complete diagnosis of the automatic transmission when this code is set.
P1725 **1T CCM, MIL: No** F-Series Truck & Excursion with a 6.0L VIN P Diesel engine Transmissions: A/T	**Insufficient Engine Speed Increase During The Self Test Conditions:** Engine started, KOER Self Test enabled, and the PCM detected an insufficient amount of engine speed increase during the self test. **Possible Causes** • Check for other trouble codes in the PCM • Rerun the Self Test as directed
P1726 **1T CCM, MIL: No** F-Series Truck & Excursion with a 6.0L VIN P Diesel engine Transmissions: A/T	**Insufficient Engine Speed Decrease During The Self Test Conditions:** Engine started, KOER Self Test enabled, and the PCM detected an insufficient amount of engine speed decrease during the self test. **Possible Causes** • Check for other trouble codes in the PCM • Rerun the self test as directed

Diesel Engine OBD II Trouble Code List (P1xxx Codes)

DTC	Trouble Code Title, Conditions & Possible Causes
P1727 **1T CCM, MIL: No** E-Van, F-Series Truck & Excursion with a 7.3L Diesel engines Transmissions: A/T	**Coast Clutch Solenoid Inductive Signature Malfunction Conditions:** Engine started, VSS over 1 mph in gear, and the PCM detected a signal that indicated a problem had been detected in the Coast Clutch Solenoid Inductive Signature value during the test period. **Possible Causes** • Refer to the appropriate Transmission Repair Manual or information in electronic media to perform a complete diagnosis of the automatic transmission when this code is set.
P1728 **1T CCM, MIL: No** E-Van, F-Series Truck & Excursion with 6.0L or 7.3L Diesel engine Transmissions: A/T	**Transmission Slip Malfunction Conditions:** Engine running, VSS over 1 mph in gear, and the PCM detected a signal that indicated the transmission was slipping while in gear. **Possible Causes** • Refer to the appropriate Transmission Repair Manual or information in electronic media to perform a complete diagnosis of the automatic transmission when this code is set.
P1729 **1T CCM, MIL: No** E-Van, F-Series Truck & Excursion with 6.0L or 7.3L Diesel engine Transmissions: A/T	**4 X 4 Low Switch Circuit Malfunction Conditions:** Key on or engine running; and the PCM detected the 4x4 switch input did not go low with the switch cycled "on" during the self-test. **Possible Causes** • Speedometer out of calibration • 4x4L wiring harness is open or shorted • 4x4L switch is damaged or it has failed • Electronic Shift Control Module is damaged or it has failed • PCM has failed
P1744 **1T CCM, MIL: Yes** E-Van, F-Series Truck & Excursion with 6.0L or 7.3L Diesel engine Transmissions: A/T	**TCC System Mechanical Malfunction (Stuck Off) Conditions:** Engine started, vehicle in gear at Cruise speed, and the PCM detected the Torque Converter Clutch system had failed with the TCC in the mechanically "off" position. **Possible Causes** • Refer to the appropriate Transmission Repair Manual or information in electronic media to perform a complete diagnosis of the automatic transmission when this code is set.
P1746 **1T CCM, MIL: Yes** E-Van, F-Series Truck & Excursion with 6.0L or 7.3L Diesel engine Transmissions: A/T	**A/T EPC Solenoid Circuit Malfunction (Open) Conditions:** Engine started, vehicle in gear, and the PCM detected an unexpected high voltage condition on the Electronic Pressure Control (EPC) solenoid circuit. This fault causes maximum EPC pressure and results in harsh engagements and shifts. **Possible Causes** • EPC solenoid control circuit open between solenoid and PCM • Refer to the appropriate Transmission Repair Manual or information in electronic media to perform a complete diagnosis of the automatic transmission when this code is set
P1747 **1T CCM, MIL: Yes** E-Van, F-Series Truck & Excursion with 6.0L or 7.3L Diesel engine Transmissions: A/T	**A/T EPC Solenoid Circuit Malfunction (Shorted) Conditions:** Engine started, vehicle driven at a steady cruise speed, and the PCM detected an unexpected low voltage condition on the TCC solenoid driver circuit during the CCM test. **Possible Causes** • EPC solenoid control circuit is shorted to ground or power (B+) • EPC solenoid power circuit is open • Refer to the appropriate Transmission Repair Manual or information in electronic media to perform a complete diagnosis of the automatic transmission when this code is set
P1748 **1T CCM, MIL: Yes** E-Van, F-Series Truck & Excursion with 6.0L or 7.3L Diesel engine Transmissions: A/T	**A/T EPC Solenoid Circuit Malfunction Conditions:** Engine started, vehicle driven at a steady cruise speed, and the PCM detected an unexpected voltage condition on the TCC solenoid driver circuit during the CCM test. **Possible Causes** • EPC solenoid control circuit is open • EPC solenoid power circuit is open • Refer to the appropriate Transmission Repair Manual or information in electronic media to perform a complete diagnosis of the automatic transmission when this code is set

Diesel Engine OBD II Trouble Code List (P1xxx Codes)

DTC	Trouble Code Title, Conditions & Possible Causes
P1754 1T CCM, MIL: No E-Van, F-Series Truck & Excursion with 6.0L or 7.3L Diesel engine Transmissions: A/T	**A/T Coast Clutch Solenoid Circuit Malfunction Conditions:** Key on, KOEO Self-Test enabled and the PCM detected an unexpected low or high voltage condition on the Coast Clutch Solenoid (CCS) circuit during the test period. **Possible Causes** • CCS control circuit is open or shorted to ground • Refer to the appropriate Transmission Repair Manual or information in electronic media to perform a complete diagnosis of the automatic transmission when this code is set
P1760 1T CCM, MIL: No E-Van, F-Series Truck & Excursion with 6.0L or 7.3L Diesel engine Transmissions: A/T	**A/T EPC Solenoid Circuit Malfunction (Intermittent) Conditions:** Engine started, vehicle driven at a steady cruise speed, and the PCM detected an interruption of the TCC solenoid driver signal (the TCC control circuit) during the CCM test. **Possible Causes** • EPC solenoid control circuit is open or shorted (intermittent) • Refer to the appropriate Transmission Repair Manual or information in electronic media to perform a complete diagnosis of the automatic transmission when this code is set
P1779 1T CCM, MIL: No E-Van, F-Series Truck & Excursion with 7.3L Diesel engine Transmissions: A/T	**Transmission Control Switch Out Of Self-Test Range Conditions:** Key on, KOEO Self-Test enabled and the PCM detected an unexpected voltage condition on the Transmission Control Switch (TCS) circuit during the test period. **Possible Causes** • TCS not cycled during the self-test • TCS connector is damaged, open or shorted • TCS circuit is open • TCS circuit is shorted to ground • TCS is damaged • PCM has failed
P1780 1T CCM, MIL: No E-Van, F-Series Truck & Excursion with 6.0L or 7.3L Diesel engine Transmissions: A/T	**Transmission Control Switch Out Of Self-Test Range Conditions:** Engine started, KOER Self-Test enabled, and the PCM detected the Transmission Control Switch (TCS) was out of range in the self-test. **Possible Causes** • TCS not cycled during the self-test • TCS connector is damaged, open or shorted • TCS circuit is open • TCS circuit is shorted to ground • TCS is damaged • PCM has failed
P1781 1T CCM, MIL: No E-Van, F-Series Truck & Excursion with 6.0L or 7.3L Diesel engine Transmissions: All	**4 X 4 Low Switch Out Of Self-Test Range Conditions:** Key on, KOEO Self-Test enabled, and the PCM detected the 4x4 switch input was not low with the switch engaged ("on"). **Possible Causes** • 4x4L switch circuit is open or shorted in the wiring harness • Electronic Shift Module is damaged or it has failed • PCM has failed
P1783 1T CCM, MIL: No E-Van, F-Series Truck & Excursion with 6.0L or 7.3L Diesel engine Transmissions: A/T	**Transmission Over-Temperature Conditions:** Engine runtime over 5 minutes, vehicle in gear at Cruise speed, and the PCM detected the TFT sensor signal was more than 300°F. **Possible Causes** • Refer to the appropriate Transmission Repair Manual or information in electronic media to perform a complete diagnosis of the automatic transmission when this code is set

Diesel Engine OBD II Trouble Code List (P2xxx Codes)

DTC	Trouble Code Title, Conditions & Possible Causes
P2121 **1T CCM, MIL: No** F-Series Truck & Excursion with a 6.0L VIN P Diesel engine Transmissions: All	**Accelerator Pedal Position Sensor 'D' Signal Range/Performance Conditions:** Key on or engine running; and the PCM detected the Accelerator Pedal Position Sensor 'D' signal circuit was out of the normal operating range during the CCM test. **Possible Causes** • APP sensor signal circuits are shorted together • APP sensor is damaged or it has failed • PCM has failed
P2122 **1T CCM, MIL: No** F-Series Truck & Excursion with a 6.0L VIN P Diesel engine Transmissions: All	**Accelerator Pedal Position Sensor 'D' Circuit Low Input Conditions:** Key on or engine running; and the PCM detected the Accelerator Pedal Position Sensor 'D' signal circuit was less than the normal range during the test period. **Possible Causes** • APP sensor signal circuit is open • APP sensor signal circuit is shorted to ground • APP sensor is damaged or it has failed • PCM has failed
P2123 **1T CCM, MIL: No** F-Series Truck & Excursion with a 6.0L VIN P Diesel engine Transmissions: All	**Accelerator Pedal Position Sensor 'D' Circuit High Input Conditions:** Key on or engine running; and the PCM detected the Accelerator Pedal Position Sensor 'D' signal circuit was more than the normal range during the test period. **Possible Causes** • APP sensor connector is damaged or shorted • APP sensor signal circuit is shorted to VREF (5v) • APP sensor is damaged or it has failed • PCM has failed
P2124 **1T CCM, MIL: No** F-Series Truck & Excursion with a 6.0L VIN P Diesel engine Transmissions: All	**Accelerator Pedal Position Sensor 'D' Signal Intermittent Conditions:** Key on or engine running; and the PCM detected an intermittent signal from the Accelerator Pedal Position Sensor 'D' during the test. **Possible Causes** • APP sensor connector is damaged or shorted (intermittent) • APP sensor signal circuit is open or shorted (intermittent fault) • APP sensor is damaged or it has failed (intermittent fault) • PCM has failed
P2126 **1T CCM, MIL: No** F-Series Truck & Excursion with a 6.0L VIN P Diesel engine Transmissions: All	**Accelerator Pedal Position Sensor 'E' Signal Range/Performance Conditions:** Key on or engine running; and the PCM detected the Accelerator Pedal Position Sensor 'E' signal circuit was more than the normal range during the test period. **Possible Causes** • APP sensor connector is damaged or shorted • APP sensor signal circuit is shorted to VREF (5v) • APP sensor is damaged or it has failed • PCM has failed
P2127 **1T CCM, MIL: No** F-Series Truck & Excursion with a 6.0L VIN P Diesel engine Transmissions: All	**Accelerator Pedal Position Sensor 'E' Circuit Low Input Conditions:** Key on or engine running; and the PCM detected the Accelerator Pedal Position Sensor 'D' signal circuit was less than the normal range during the test period. **Possible Causes** • APP sensor signal circuit is open • APP sensor signal circuit is shorted to ground • APP sensor is damaged or it has failed • PCM has failed
P2128 **1T CCM, MIL: No** F-Series Truck & Excursion with a 6.0L VIN P Diesel engine Transmissions: All	**Accelerator Pedal Position Sensor 'E' Circuit High Input Conditions:** Key on or engine running; and the PCM detected the Accelerator Pedal Position Sensor 'E' signal circuit was more than the normal range during the test period. **Possible Causes** • APP sensor connector is damaged or shorted • APP sensor signal circuit is shorted to VREF (5v) • APP sensor is damaged or it has failed • PCM has failed

Diesel Engine OBD II Trouble Code List (P2xxx Codes)

DTC	Trouble Code Title, Conditions & Possible Causes
P2129 **1T CCM, MIL: No** F-Series Truck & Excursion with a 6.0L VIN P Diesel engine Transmissions: All	**Accelerator Pedal Position Sensor 'E' Signal Intermittent Conditions:** Key on or engine running; and the PCM detected an intermittent signal from the Accelerator Pedal Position Sensor 'E' during the test. **Possible Causes** • APP sensor connector is damaged or shorted (intermittent) • APP sensor signal circuit is open or shorted (intermittent fault) • APP sensor is damaged or it has failed (intermittent fault) • PCM has failed
P2131 **1T CCM, MIL: No** F-Series Truck & Excursion with a 6.0L VIN P Diesel engine Transmissions: All	**Accelerator Pedal Position Sensor 'F' Signal Range/Performance Conditions:** Key on or engine running; and the PCM detected the Accelerator Pedal Position Sensor 'F' signal circuit was more than the normal range during the test period. **Possible Causes** • APP sensor connector is damaged or shorted • APP sensor signal circuit is shorted to VREF (5v) • APP sensor is damaged or it has failed • PCM has failed
P2132 **1T CCM, MIL: No** F-Series Truck & Excursion with a 6.0L VIN P Diesel engine Transmissions: All	**Accelerator Pedal Position Sensor 'F' Circuit Low Input Conditions:** Key on or engine running; and the PCM detected the Accelerator Pedal Position Sensor 'F' signal circuit was less than the normal range during the test period. **Possible Causes** • APP sensor signal circuit is open • APP sensor signal circuit is shorted to ground • APP sensor is damaged or it has failed • PCM has failed
P2133 **1T CCM, MIL: No** F-Series Truck & Excursion with a 6.0L VIN P Diesel engine Transmissions: All	**Accelerator Pedal Position Sensor 'F' Circuit High Input Conditions:** Key on or engine running; and the PCM detected the Accelerator Pedal Position Sensor 'F' signal circuit was more than the normal range during the test period. **Possible Causes** • APP sensor connector is damaged or shorted • APP sensor signal circuit is shorted to VREF (5v) • APP sensor is damaged or it has failed • PCM has failed
P2134 **1T CCM, MIL: No** F-Series Truck & Excursion with a 6.0L VIN P Diesel engine Transmissions: All	**Accelerator Pedal Position Sensor 'F' Signal Intermittent Conditions:** Key on or engine running; and the PCM detected an intermittent signal from the Accelerator Pedal Position Sensor 'F' during the test. **Possible Causes** • APP sensor connector is damaged or shorted (intermittent) • APP sensor signal circuit is open or shorted (intermittent fault) • APP sensor is damaged or it has failed (intermittent fault) • PCM has failed
P2138 **1T CCM, MIL: No** F-Series Truck & Excursion with a 6.0L VIN P Diesel engine Transmissions: All	**ETC Throttle Position Sensor D/E Voltage Correlation Conditions:** Key on or engine running; and the PCM detected the Throttle Position 'D' (TPD) and Throttle Position 'E' (TPE) sensors disagreed, or that the TPD sensor should not be in its detected position, or that the TPE sensor should not be in its detected position during testing. **Possible Causes** • ETC TP sensor connector is damaged or shorted • ETC TP sensor circuits shorted together in the wire harness • ETC TP sensor signal circuit is shorted to VREF (5v) • ETC TP sensor is damaged or it has failed • PCM has failed

Diesel Engine OBD II Trouble Code List (P2xxx Codes)

DTC	Trouble Code Title, Conditions & Possible Causes
P2139 **1T CCM, MIL: No** F-Series Truck & Excursion with a 6.0L VIN P Diesel engine Transmissions: All	**ETC Throttle Position Sensor D/F Voltage Correlation Conditions:** Key on or engine running; and the PCM detected the Throttle Position 'D' (TPD) and Throttle Position 'F' (TPF) sensors disagreed, or that the TPD sensor should not be in its detected position, or that the TPF sensor should not be in its detected position during testing. **Possible Causes** • ETC TP sensor connector is damaged or shorted • ETC TP sensor circuits shorted together in the wire harness • ETC TP sensor signal circuit is shorted to VREF (5v) • ETC TP sensor is damaged or it has failed • PCM has failed
P2140 **1T CCM, MIL: No** F-Series Truck & Excursion with a 6.0L VIN P Diesel engine Transmissions: All	**ETC Throttle Position Sensor E/F Voltage Correlation Conditions:** Key on or engine running; and the PCM detected the Throttle Position 'E' (TPE) and Throttle Position 'F' (TPF) sensors disagreed, or that the TPE sensor should not be in its detected position, or that the TPF sensor should not be in its detected position during testing. **Possible Causes** • ETC TP sensor connector is damaged or shorted • ETC TP sensor circuits shorted together in the wire harness • ETC TP sensor signal circuit is shorted to VREF (5v) • ETC TP sensor is damaged or it has failed • PCM has failed
P2199 **1T CCM, MIL: No** F-Series Truck & Excursion with a 6.0L VIN P Diesel engine Transmissions: All	**IAT Sensor 1-2 Voltage Correlation Conditions:** Key on or engine running; and the PCM detected the Intake Air Temperature Sensor 1-2 signals did not agree during the CCM test. **Possible Causes** • IAT Sensor 1 connector is damaged or has high resistance • IAT Sensor 1 signal circuit has a high resistance condition • IAT Sensor 1 is damaged or it has failed • IAT Sensor 2 connector is damaged or has high resistance • IAT Sensor 2 signal circuit has a high resistance condition • IAT Sensor 2 is damaged or it has failed
P2262 **1T CCM, MIL: No** F-Series Truck & Excursion with a 6.0L VIN P Diesel engine Transmissions: All	**Turbo/Super Charger Boost Pressure Not Detected Conditions:** Engine started, vehicle driven under boost requirement conditions, and the PCM did not detect sufficient boost pressure during the test. **Possible Causes** • CAC system is leaking (cold air cooler assembly) • EP sensor is damaged or it has failed • Exhaust system is leaking or severely restricted • Intake system has air leaks • MAP sensor vacuum hose is disconnected, restricted or the MAP sensor has failed
P2263 **1T CCM, MIL: No** F-Series Truck & Excursion with a 6.0L VIN P Diesel engine Transmissions: All	**Turbo/Super Charger System Performance Conditions:** Engine started, vehicle driven under boost requirement conditions, and the PCM did not detect sufficient boost pressure during the test. **Possible Causes** • CAC system is leaking (cold air cooler assembly) • EP sensor is damaged or it has failed • Exhaust system is leaking or severely restricted • Intake system has air leaks • MAP sensor vacuum hose is disconnected, restricted or the MAP sensor has failed
P2269 **1T CCM, MIL: No** F-Series Truck & Excursion with a 6.0L VIN P Diesel engine Transmissions: All	**Water In Fuel Condition Conditions:** Key on or engine running; and the PCM detected a signal from the Water In Fuel (WIF) sensor that there was water in the Fuel system. **Possible Causes** • Drain the water from the fuel separator and retest the system • WIF sensor circuit is damaged, open or shorted • WIF sensor is damaged or it has failed

Diesel Engine OBD II Trouble Code List (P2xxx Codes)

DTC	Trouble Code Title, Conditions & Possible Causes
P2284 **1T CCM, MIL: No** F-Series Truck & Excursion with a 6.0L VIN P Diesel engine Transmissions: All	**Injector Control Pressure Sensor Signal Performance Conditions:** Engine started; and the PCM detected an unexpected voltage condition on the Injector Control Pressure sensor circuit. **Possible Causes** • ICP sensor connector is damaged, loose or shorted • ICP sensor signal circuit is open or shorted • ICP sensor is damaged or it has failed • PCM has failed
P2285 **1T CCM, MIL: No** F-Series Truck & Excursion with a 6.0L VIN P Diesel engine Transmissions: All	**Injector Control Pressure Sensor Circuit Low Input Conditions:** Engine started; and the PCM detected an unexpected low voltage condition (less than 0.04v) on the Injector Control Pressure sensor circuit during the CCM test. **Possible Causes** • ICP sensor connector is damaged or shorted • ICP sensor signal circuit is shorted to ground • ICP sensor is damaged or it has failed • PCM has failed
P2286 **1T CCM, MIL: No** F-Series Truck & Excursion with a 6.0L VIN P Diesel engine Transmissions: All	**Injector Control Pressure Sensor Circuit High Input Conditions:** Engine started; and the PCM detected an unexpected high voltage condition (more than 4.91v) on the Injector Control Pressure sensor circuit during the CCM test. **Possible Causes** • ICP sensor signal circuit is open or shorted to VREF (5v) • ICP sensor ground circuit is open • ICP sensor is damaged or it has failed • PCM has failed
P2288 **1T CCM, MIL: No** F-Series Truck & Excursion with a 6.0L VIN P Diesel engine Transmissions: All	**Injector Control Pressure Too High Conditions:** Engine started; and the PCM detected the Injector Control Pressure was too high (more than 3675 psi) during the CCM test period. **Possible Causes** • Refer to ICP system diagnosis information • ICP sensor is damaged or it has failed • PCM has failed
P2289 **1T CCM, MIL: No** F-Series Truck & Excursion with a 6.0L VIN P Diesel engine Transmissions: All	**Injector Control Pressure Too High - Engine Off Conditions:** Key on, KOEO Self Test enabled; and the PCM detected the Injector Control Pressure was too high (more than 3675 psi) during the test. **Possible Causes** • ICP sensor ground circuit is open • ICP sensor signal circuit is open • ICP sensor is damaged or it has failed • PCM has failed
P2290 **1T CCM, MIL: No** F-Series Truck & Excursion with a 6.0L VIN P Diesel engine Transmissions: All	**Injector Control Pressure Too Low Conditions:** Engine started; and the PCM detected the Injector Control Pressure was less than the Desired pressure during the CCM test period. **Possible Causes** • Refer to ICP system diagnosis information • ICP sensor is damaged or it has failed • PCM has failed
P2291 **1T CCM, MIL: No** F-Series Truck & Excursion with a 6.0L VIN P Diesel engine Transmissions: All	**Injector Control Pressure Too Low - Engine Cranking Conditions:** Engine cranking; and the PCM detected the Injector Control Pressure was too low (less than 725 psi) during the CCM test period. **Possible Causes** • Refer to ICP system diagnosis information • ICP sensor is damaged or it has failed • PCM has failed

Diesel Engine OBD II Trouble Code List (P2xxx Codes)

DTC	Trouble Code Title, Conditions & Possible Causes
P2552 **1T CCM, MIL: No** F-Series Truck & Excursion with a 6.0L VIN P Diesel engine Transmissions: All	**FICMM Circuit - Throttle/Fuel Inhibit Circuit Malfunction Conditions:** Key on or engine running; and the PCM did not detect a signal from the FICM during the CCM test. **Possible Causes** • FICM connector is damaged, open or shorted • FICMM circuit is damaged, open or shorted • FICM is damaged or it has failed • PCM has failed
P2614 **1T CCM, MIL: No** F-Series Truck & Excursion with a 6.0L VIN P Diesel engine Transmissions: All	**Camshaft Position Output Circuit Intermittent Conditions:** Key on or engine running; and the PCM detected an intermittent loss of the Camshaft Position (CMP) sensor circuit during the CCM test. **Possible Causes** • CMP sensor connector is damaged, open or shorted • CMP sensor circuit is damaged, open or shorted • CMP is damaged or it has failed • PCM has failed
P2617 **1T CCM, MIL: No** F-Series Truck & Excursion with a 6.0L VIN P Diesel engine Transmissions: All	**Crankshaft Position Output Circuit Intermittent Conditions:** Key on or engine running; and the PCM detected an intermittent loss of the Crankshaft Position (CKP) sensor circuit during the CCM test. **Possible Causes** • CKP sensor connector is damaged, open or shorted • CKP sensor circuit is damaged, open or shorted • CKP is damaged or it has failed • PCM has failed
P2623 **1T CCM, MIL: No** F-Series Truck & Excursion with a 6.0L VIN P Diesel engine Transmissions: All	**Injector Control Pressure Regulator Circuit Malfunction Conditions:** Key on or engine running; and the PCM detected an unexpected voltage condition on the Injector Pressure regulator circuit. **Possible Causes** • IPR (regulator) connector is damaged, open or shorted • IPR (regulator) circuit is damaged, open or shorted • IPR (regulator) is damaged, sticking or it has failed • PCM has failed

Diesel Engine OBD II Trouble Code List (U0xxx Codes)

DTC	Trouble Code Title, Conditions & Possible Causes
U0101 **1T PCM, MIL: No** F-Series Truck & Excursion with a 6.0L VIN P Diesel engine Transmissions: All	**Lost Communication With TCM Conditions:** Key on, and the PCM detected that it has lost communication with the Transmission Control Module (TCM) during its initial startup. **Possible Causes** • Replace the PCM and retest the system for codes.
U0105 **1T PCM, MIL: No** F-Series Truck & Excursion with a 6.0L VIN P Diesel engine Transmissions: All	**Lost Communication With FICM Conditions:** Key on, and the PCM detected that it has lost communication with the Fuel Injection Control Module (FICM) during its initial startup. **Possible Causes** • Replace the PCM and retest the system for codes.
U0155 **1T PCM, MIL: No** F-Series Truck & Excursion with a 6.0L VIN P Diesel engine Transmissions: All	**Lost Communication With Instrument Cluster Conditions:** Key on, and the PCM detected that it has lost communication with the Instrument Cluster Panel (I/P) during its initial startup. **Possible Causes** • Refer to the test procedures found in the Workshop Manual in Section 418.
U0306 **1T PCM, MIL: No** F-Series Truck & Excursion with a 6.0L VIN P Diesel engine Transmissions: All	**Software Incompatibility With Fuel Injector Control Module Conditions:** Key on, and the PCM detected a software incompatibility condition with the Fuel Injector Control Module (FICM) during its initial startup. **Possible Causes** • Refer to the test procedures found in the Workshop Manual in Section 418.

Contents

Aerostar

OPERATION

Electronic Engine Control IV (EEC-IV) System

The Electronic Engine Control IV system is used on the 2.3L EFI, 3.0L and 4.0L fuel injected engines. Minimal variations may occur between years and models. This system is designed to improve emission control, fuel economy, driveability, and engine performance. This is achieved by the means of an on-board control assembly which reads the inputs from various sensors and makes computations based on these inputs and then sends controlling outputs to various engine components in order to provide the optimum air/fuel ratio.

The electronic control assembly is calibrated to optimize emissions, fuel economy and driveability. The system controls the fuel injectors for air fuel mixture, spark timing, deceleration fuel cut-off, EGR function, curb and fast idle speed, evaporative emission purge, air condition cut-off during wide open throttle, cold engine start and enrichment, electric fuel pump and self test engine diagnostics.

OXYGEN SENSOR

Operation

The exhaust gas oxygen sensor supplies the electronic control assembly with a signal which indicates either a rich or lean mixture condition, during the engine operation. This sensor is screwed into the exhaust manifold before the catalytic converter.

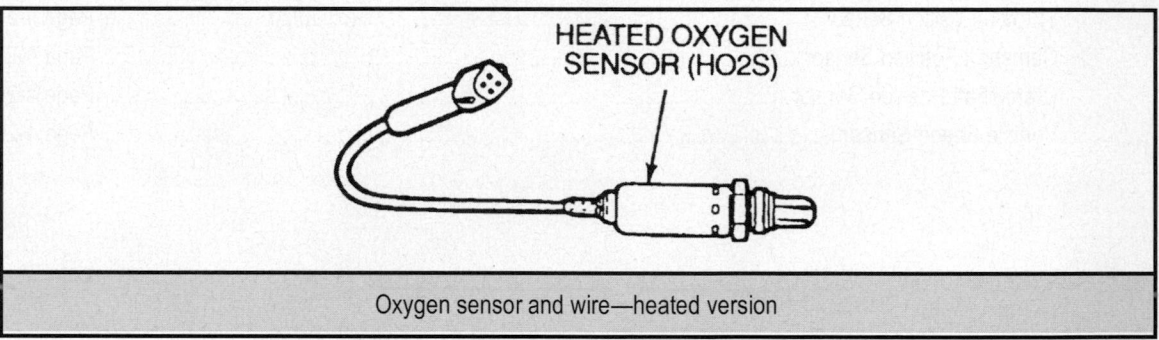

HEATED OXYGEN
SENSOR (HO2S)

Oxygen sensor and wire—heated version

Testing

1. Perform a visual inspection of the heated oxygen sensor as follows:

2. Disconnect the electrical lead, then remove the sensor from the exhaust system.

3. If the sensor tip has a black/sooty deposit, this may indicate a rich fuel mixture.

4. If the sensor tip has a white, gritty deposit, this may indicate an internal coolant leak.

5. If the sensor tip has a brown deposit, this could indicate oil consumption.

6. Reinstall the heated oxygen sensor, but do not connect its electrical lead.

7. Measure resistance between the PWR and GND (heater) terminals of the sensor. If the reading is about 6 ohms at 68°F (20°C) the sensor's heater element is okay. Connect the sensor's electrical lead.

With the heated oxygen sensor connected and the engine running, measure voltage with a DVOM by backprobing the SIG RTN wire of the HO2S connector. The voltage readings at idle should stay below 1.0 volts. When the speed is increased or decreased, the voltage should fluctuate between 0 and 1.0 volts; if so, the sensor is okay.

AIR BYPASS SOLENOID VALVE

Operation

The air bypass solenoid or idle air control valve is used to control the engine idle speed. This component is operated by the electronic control module. The valve allows air to pass around the throttle plates in order to control cold engine idle and promote easier starting and dashpot operation.

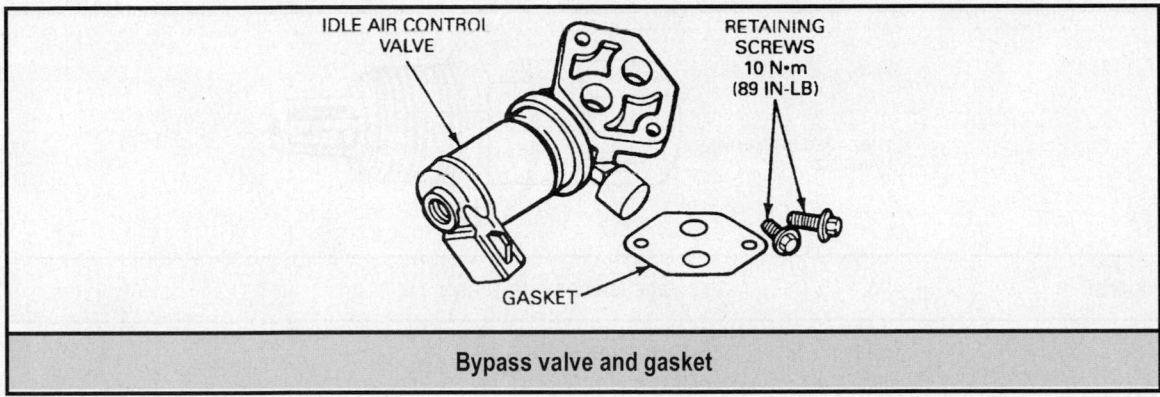

IDLE AIR CONTROL VALVE

RETAINING SCREWS 10 N•m (89 IN-LB)

GASKET

Bypass valve and gasket

Testing

1. With the ignition in the ON position, use a multimeter and check for current at the valve. The meter should register approximately one volt.

2. Connect a tachometer using the manufacturer's instructions.

3. With the engine running at normal operating temperature, apply 12 volts to the valve and watch the tachometer.

4. If the bypass valve is functioning correctly, the idle speed will increase 700–1000 rpm. If the idle speed does not change, or increases less than 500 rpm, replace the bypass valve.

<u>AIR CHARGE TEMPERATURE SENSOR</u>
Operation

The air charge temperature sensor provides the electronic fuel injection system with air/fuel mixture information. This sensor is used as both a density corrector to calculate air flow and to proportion cold enrichment fuel flow. This sensor is similar in construction to the engine coolant temperature sensor.

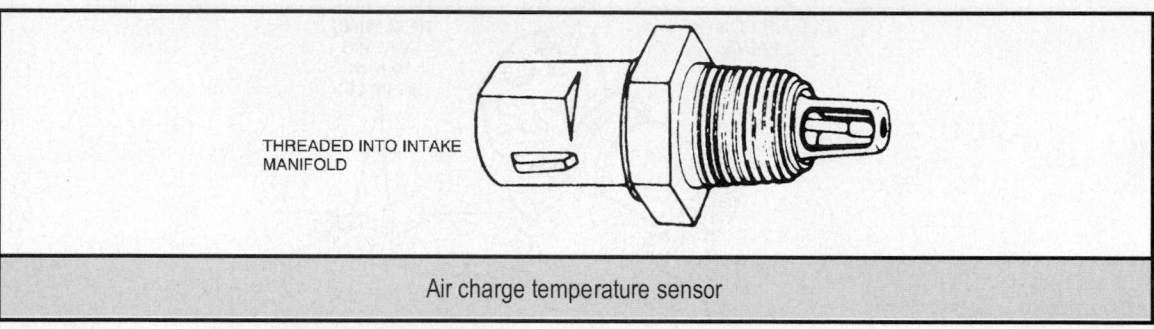

THREADED INTO INTAKE MANIFOLD

Air charge temperature sensor

Testing

1. To check the Idle Air Control (IAC) valve resistance, unplug the IAC valve connector.

2. Connect a high impedance Digital Volt Ohmmeter (DVOM) to the terminals.

3. Measure the resistance.

4. The resistance should be 7.7-9.3 ohms.

5. If the resistance does not meet specification, replace the IAC assembly.

Unplug the IAC valve electrical connector

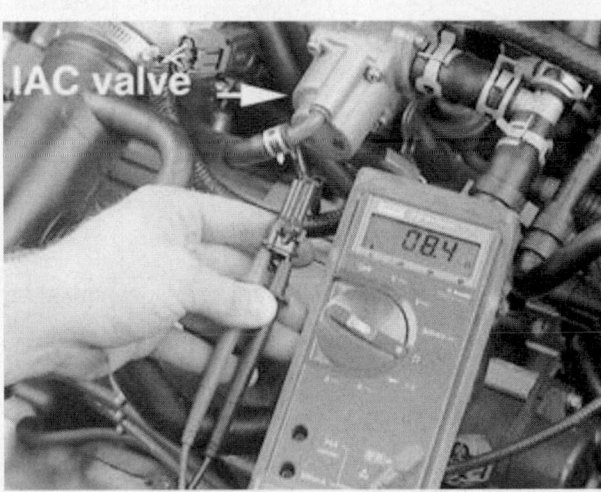

Connect an ohmmeter to the IAC valve; the resistance should be 7.7-9.3 ohms

6. To test the solenoid, connect a scan tool to the Data Link Connector (DLC).

7. Turn the key ON and enter the simulation test.

8. Access the IACV Parameter Identification (PID) and turn the dial to cycle at 75 degrees.

9. Listen for the IAC solenoid to click when using the simulation test.

COOLANT TEMPERATURE SENSOR
Operation

This component detects the temperature of the engine coolant and relays the information to the electronic control assembly. The sensor is located by the heater outlet fitting or in a cooling passage on the engine, depending upon the particular type vehicle. The function of the sensor is to modify ignition timing, control EGR flow and regulate the air/fuel mixture. On vehicles equipped with the electronic instrument cluster, the sensor is also used to control the coolant temperature indicator.

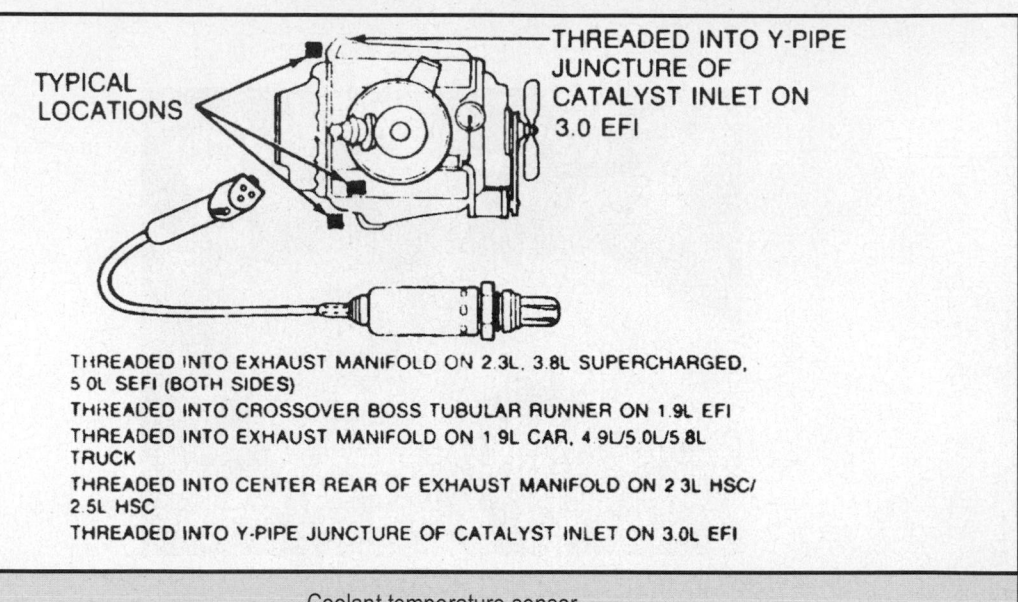

TYPICAL LOCATIONS

THREADED INTO Y-PIPE JUNCTURE OF CATALYST INLET ON 3.0 EFI

THREADED INTO EXHAUST MANIFOLD ON 2.3L, 3.8L SUPERCHARGED, 5 0L SEFI (BOTH SIDES)
THREADED INTO CROSSOVER BOSS TUBULAR RUNNER ON 1.9L EFI
THREADED INTO EXHAUST MANIFOLD ON 1 9L CAR, 4 9L/5 0L/5 8L TRUCK
THREADED INTO CENTER REAR OF EXHAUST MANIFOLD ON 2 3L HSC/ 2.5L HSC
THREADED INTO Y-PIPE JUNCTURE OF CATALYST INLET ON 3.0L EFI

Coolant temperature sensor

Testing

Under correct operating conditions, the coolant switch is closed in the normal operating temperature range above 180°F. The switch remains open below the 180°F range. If a switch is defective, the electronic control module is able to determine this and alert the driver.

This switch can also be tested as follows;

1. Remove the switch from the engine while the engine is cool.

2. Using a multimeter, test for continuity at the switch. If the switch is functioning correctly, the meter will register open.

3. Place the probe end of the switch into boiling water for at least two minutes. Use the meter and check for continuity. If the switch is function correctly, there should be continuity.

INTAKE AIR TEMPERATURE SENSOR
Operation

The intake air temperature sensor, mounted in the intake air tube to the airflow meter and throttle body assembly, changes resistance in response to changes in intake air temperature. The resistance will decrease as the air temperature increased, and will increase as the air temperature decrease. The sensor provides a signal to the electronic control module of incoming air temperature. This information is used to adjust the injector pulse width and in turn the air/fuel ratio.

Testing

1. Remove the intake air temperature sensor from the intake tube on the vehicle.

2. Using a hair dryer with adjustable temperature controls, start on the lowest setting and measure the sensor resistance with a multimeter. Record each reading. Do not place the hair dryer directly in front of the sensor. Make sure there is at least a 12 in. (30cm) gap between the sensor and hair dryer.

3. After the reading has been recorded, adjust the temperature controls of the hair dryer to increase the heat, and point at the sensor. Continue this through the entire heat range of the hair dryer, recording each reading.

4. If the sensor is working correctly, as the heat from the hair dryer increased, the resistance recorded from the sensor should have decreased. If the multimeter reading did not decrease, or decreased, then stopped decreasing, replace the sensor.

AIRFLOW METER SENSOR

Operation

The air flow sensor, mounted on the outside of the air cleaner assembly, uses a hot wire sensing element to measure the amount of air entering the engine. Air passing over the hot wire causes the wire to cool, which in turn creates an analog signal which is relayed to the electronic control unit where a intake mass air calculation is made. The electronic control module uses this information to determine the required fuel injector pulse width in order to provided the optimum air/fuel ratio.

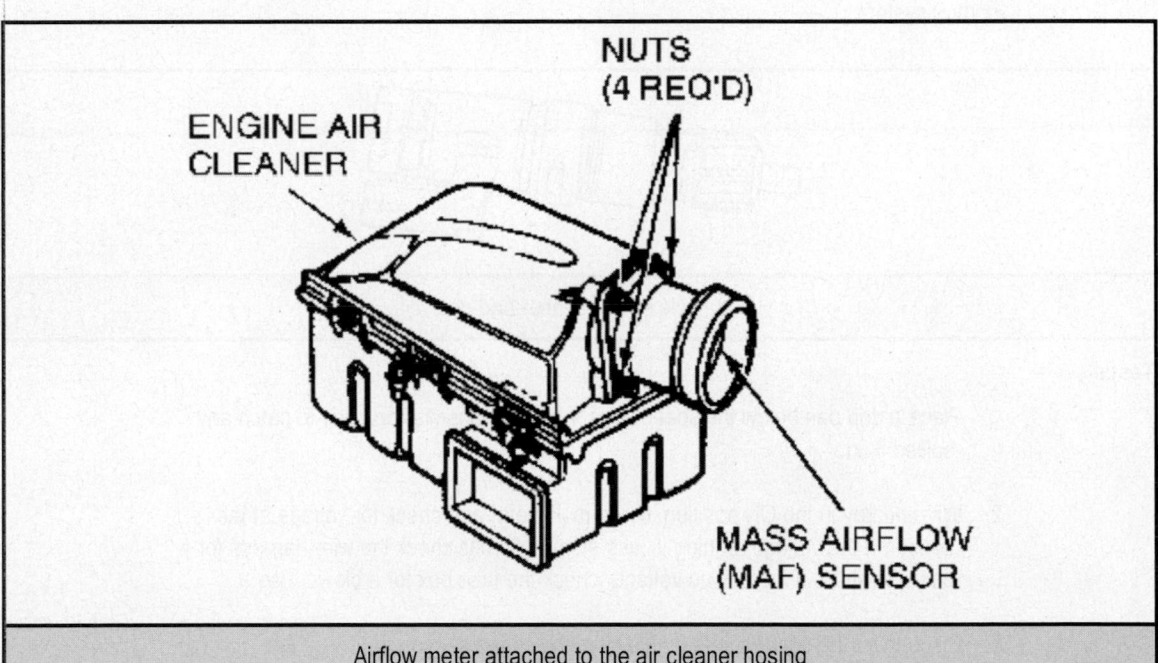

Airflow meter attached to the air cleaner hosing

Testing

1. With the key OFF, check and make sure all engine control components are installed.

2. Unfasten the airflow sensor harness and connect a breakout box to the sensor. Connect the wire harness to the breakout box.

3. With the wheels blocked, start the vehicle and place in PARK or NEUTRAL. Allow the vehicle to reach normal operating temperature.

4. Using a multimeter, measure the voltage between test pin 14 at the breakout box, and the positive terminal of the battery or starter solenoid.

5. If the sensor is working correctly, at idle, the voltage reading should be 0.6–0.7 volts.

6. If the reading is below 0.6, perform the same test while applying throttle. If the voltage is below 0.8 volts while at throttle, replace the airflow sensor.

VEHICLE SPEED SENSOR
Operation

The vehicle speed sensor, mounted to the speedometer cable at the transmission assembly, is a magnetic pickup that sends a signal to the electronic control module of the vehicle's speed. The electronic module uses this information to control transmission shift patterns.

The vehicle speed sensor is also used to send a speed signal to the speed control servo of the cruise control system.

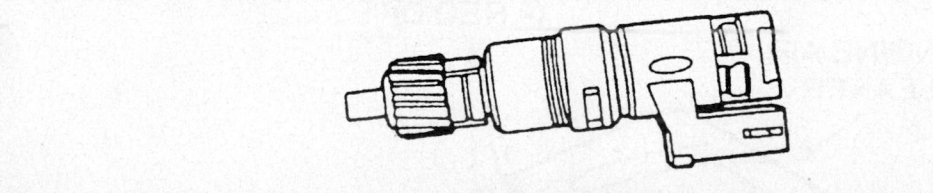

Vehicle speed sensor and gear

Testing

1. Place a drip pan below the speed sensor on the transmission body to catch any spilled fluid.

2. With the key in the ON position, use a multimeter and check for voltage at the sensor. If the voltage reading is less than 10.5 volts check the wire harness for a possible short. If there is no voltage, check the fuse box for a blown fuse.

3. Disconnect the speedometer cable from the speed sensor.

4. Disconnect the speed sensor from the transmission body.

5. Connect a multimeter to the speed sensor terminals and rotate the gear. If the resistance is between 190–250 ohms, the sensor is functioning correctly. If the reading is above or below the specified level, replace the switch.

THROTTLE POSITION SENSOR (TPS)
Operation

The throttle position sensor, mounted to the throttle body assembly, is a potentiometer type sender which provides a signal to the electronic control module that is related to the relative throttle plate position. As the throttle plate moves in relation to driving conditions, a signal is sent to the control unit which adjusts the injector pulse width and air/fuel ratio. As the throttle plate is opened further, more air is taken into the combustion chambers, and as a result the relative fuel demand of the engine changes. The throttle position sensor relays this information to the control unit which in alters the fuel amount.

Testing

1. Visually inspect the throttle linkage and throttle for binding and sticking.

2. With the key in the ON position, use a multimeter and test for voltage at the two outside wires of the harness. Voltage should be between 4–6 volts. If there is no voltage present, check the wire harness for a break, or the fuse box for a blown fuse.

3. Connect a breakout box between the powertrain control module and the powertrain control module wire harness.

4. Start the engine and allow to reach normal operating temperature. Record the voltage between the signal and return terminals at the breakout box while slowly opening the throttle plate. Depending on the throttle plate position, the voltage will be between 0.17–0.49 volts. If the voltage level is below or above the specified amount, replace the switch.

CAMSHAFT POSITION SENSOR
Operation

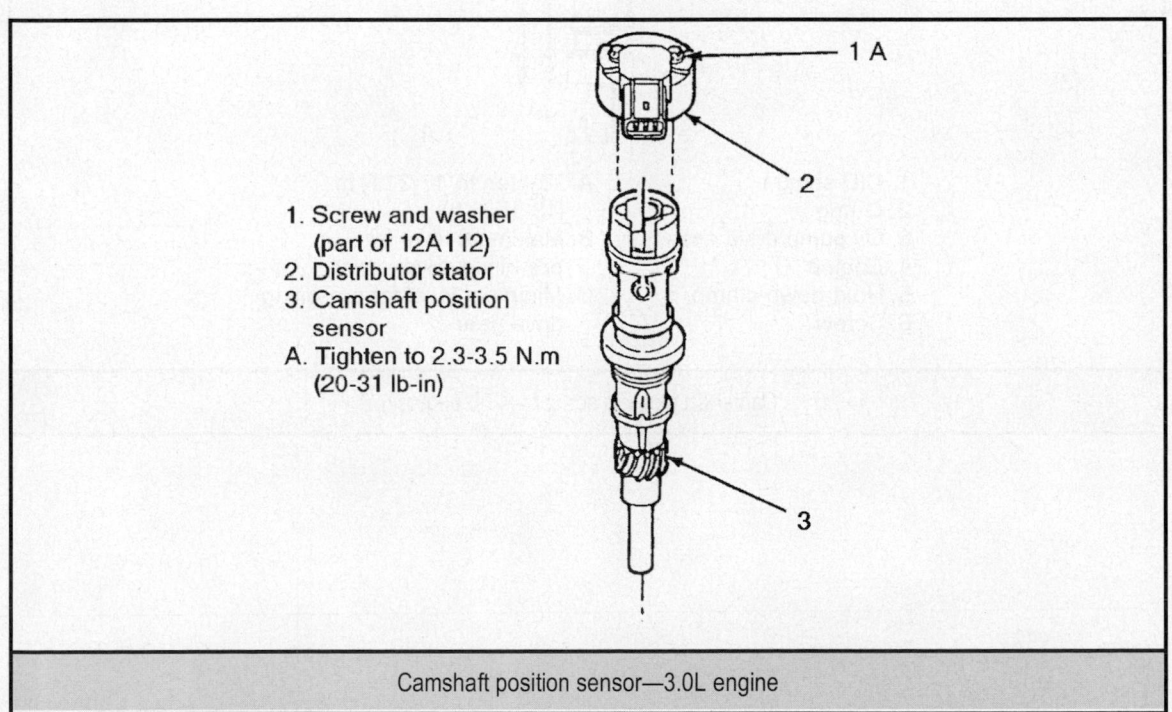

1. Screw and washer
   (part of 12A112)
2. Distributor stator
3. Camshaft position
   sensor
A. Tighten to 2.3-3.5 N.m
   (20-31 lb-in)

Camshaft position sensor—3.0L engine

The camshaft position sensor relays the relative camshaft position to the powertrain control module for determining the position and stroke of the No. 1 cylinder.

This information is required for proper fuel injection functioning.

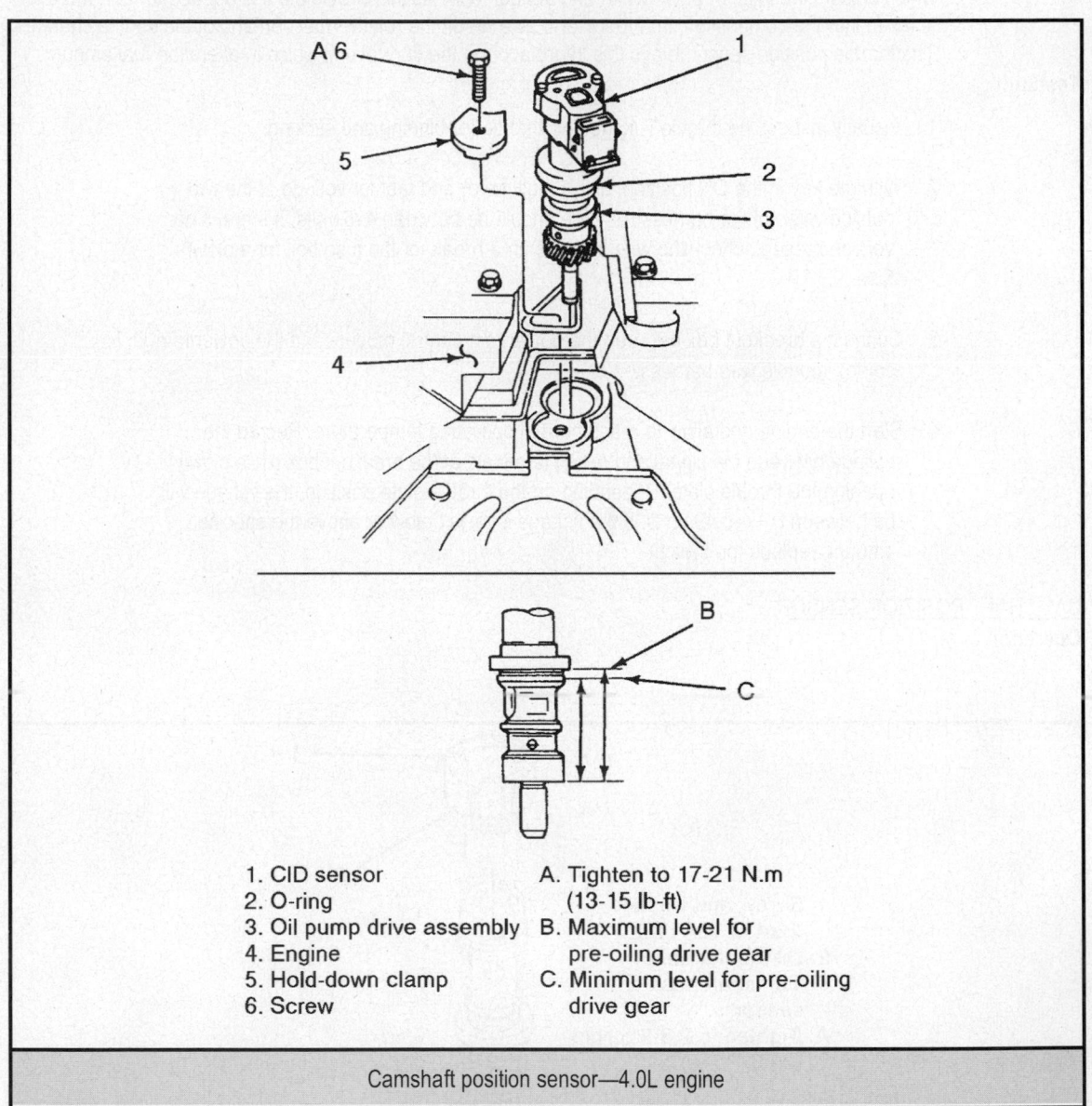

1. CID sensor
2. O-ring
3. Oil pump drive assembly
4. Engine
5. Hold-down clamp
6. Screw

A. Tighten to 17-21 N.m (13-15 lb-ft)
B. Maximum level for pre-oiling drive gear
C. Minimum level for pre-oiling drive gear

Camshaft position sensor—4.0L engine

Testing

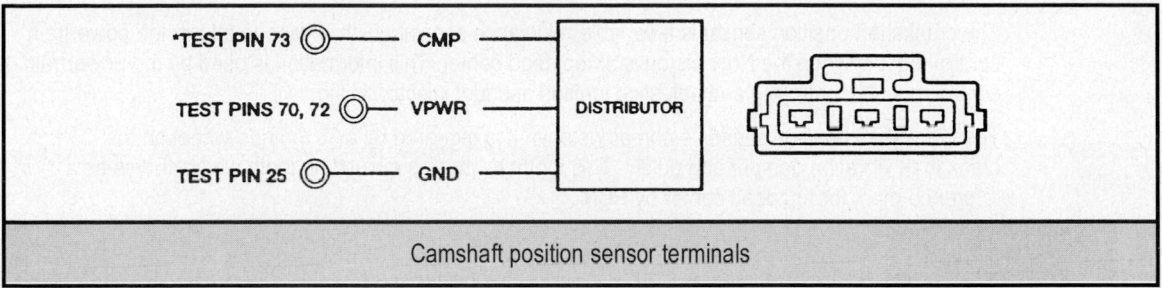

*TEST PIN 73	— CMP —	
TEST PINS 70, 72	— VPWR —	DISTRIBUTOR
TEST PIN 25	— GND —	

Camshaft position sensor terminals

1. Make sure the camshaft position sensor harness is connected to the sensor. Inspect the wire for any breaks.

2. Turn the key to the ON position.

3. Disconnect the wire harness from the sensor. And test for current between the center terminal and the negative terminal of the battery. Voltage should register greater than 10.5 volts. If there is no current check the fuse box for a blown fuse, or a broken wire in the harness.

4. Connect a breakout box between the powertrain control module and control module wire harness.

5. Start the engine and allow it to reach normal operating temperature.

6. Record the voltage between the sensor and vehicle ground at different idle speeds. If the voltage varies more than 0.1 volts the sensor is functioning correctly.

CRANKSHAFT POSITION SENSOR
Operation

The crankshaft position sensor is a variable reluctance sensor which is used to inform the powertrain control module when the No.1 piston is at top dead center. This information is used by the powertrain control module controlling and adjusting ignition and fuel injector timing.

The crankshaft position sensor is non-adjustable. It is triggered by a 36 toothed wheel on the crankshaft vibration damper and pulley. This toothed wheel is minus one tooth which serves as a reference mark for top dead center by PCM.

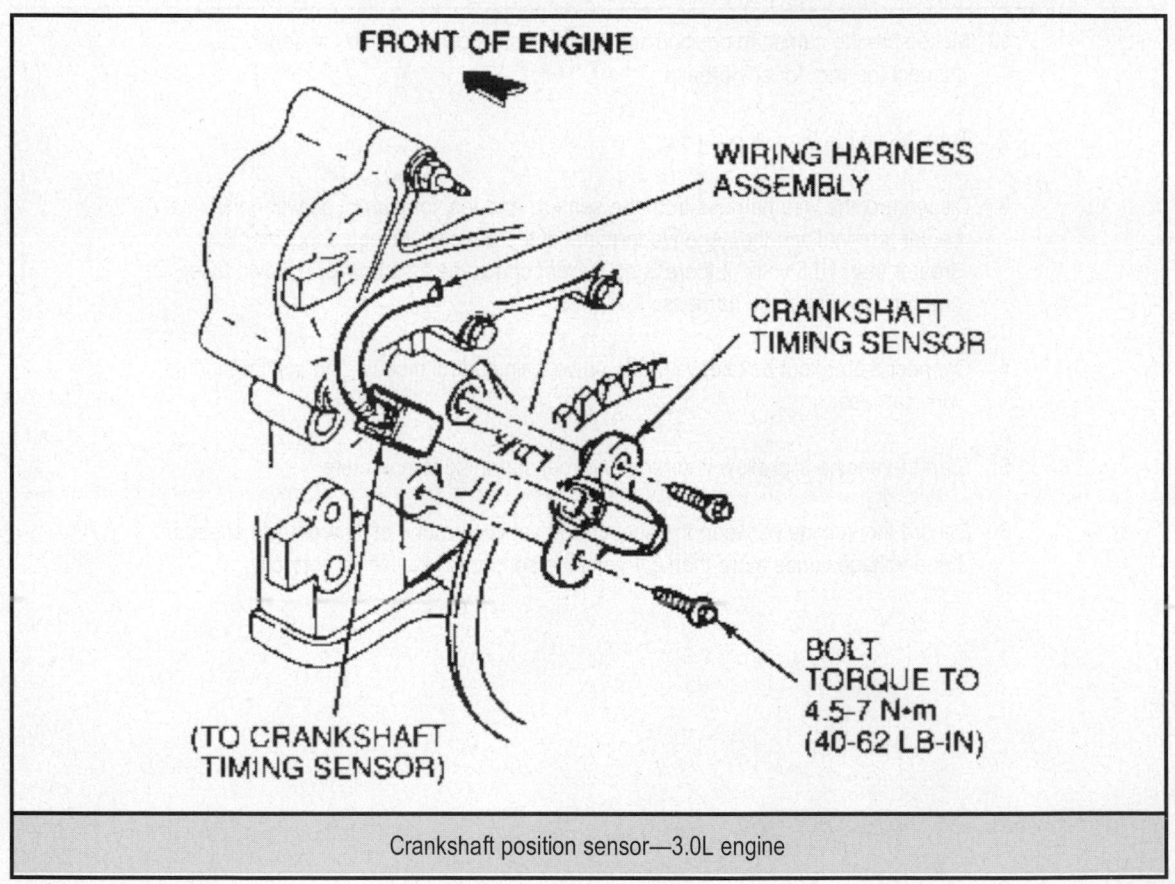

Crankshaft position sensor—3.0L engine

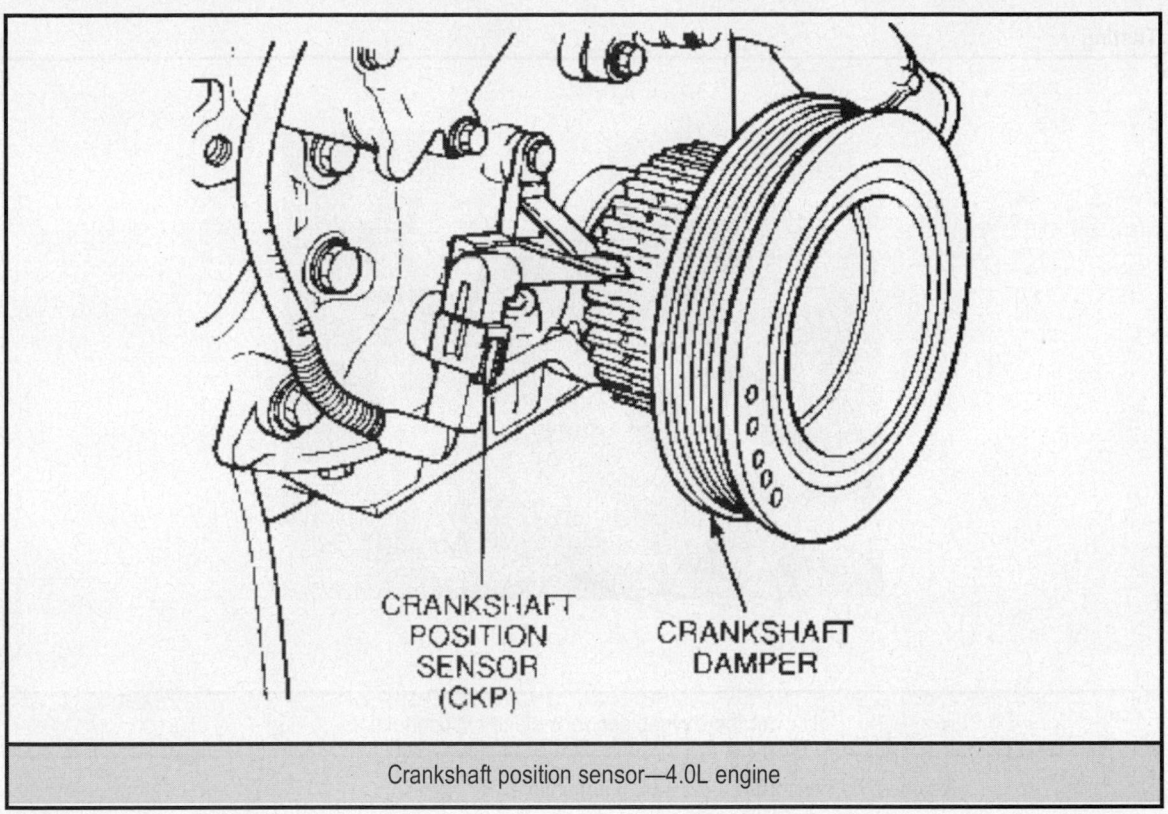

CRANKSHAFT
POSITION
SENSOR
(CKP)

CRANKSHAFT
DAMPER

Crankshaft position sensor—4.0L engine

Aspire

POWERTRAIN CONTROL MODULE
Operation

The models covered by this manual employ the fourth and fifth generation Electronic Engine Control systems, commonly designated EEC-IV and EEC-V, to manage fuel, ignition and emissions on vehicle engines.

Typically, the EEC-IV systems were used on 1994-95 models and the EEC-V systems were used on 1996-97 models.

The Powertrain Control Module (PCM) detects engine operating and driving conditions, along with the exhaust gas oxygen content. Various switches, sensors and components provide the PCM with information that allows it to control the air/fuel ratio (mixture). The PCM can also control some evaporative emission, ignition and deceleration systems.

The PCM is located underneath the left-hand side of the instrument panel.

OXYGEN SENSOR
Operation

An Oxygen Sensor (O2S) is used on all engines, and is mounted in the exhaust manifold. The sensor protrudes into the exhaust stream and monitors the oxygen content of the exhaust gases. The difference between the oxygen content of the exhaust gases and that of the outside air generates a voltage signal to the PCM. The PCM monitors this voltage and, depending upon the value of the signal received, issues a command to adjust for a rich or lean condition.

Testing

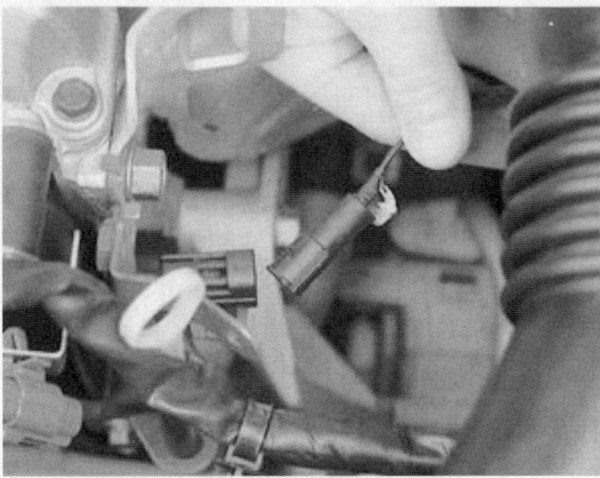

Unplug the oxygen sensor's electrical connection

1. Remove the sensor from the exhaust manifold.

2. If the sensor tip has a black/sooty deposit, this may indicate a rich fuel mixture.

3. If the sensor tip has a white, gritty deposit, this may indicate an internal coolant leak.

4. If the sensor tip has a brown deposit, this could indicate oil consumption. All these contaminants can destroy the sensor; if the problem is not repaired, the new sensor will also be damaged.

5. Reinstall the sensor, but do not engage its electrical connection.

6. Connect jumper wires from the sensor connector to the wiring harness. This permits the engine to operate normally while you check the sensor.

7. Start the engine and allow it to reach normal operating temperature. This will take around ten minutes.

8. Connect the positive lead of a high impedance Digital Volt Ohmmeter (DVOM) to the sensor signal wire and the negative lead to a good known engine ground, such as the battery negative terminal.

9. The voltage reading should fluctuate as the sensor detects varying levels of oxygen in the exhaust stream.

10. If the sensor voltage does not fluctuate, the sensor may be defective, or the fuel mixture could be extremely out of range.

11. If the sensor reads below 550 millivolts constantly, the fuel mixture may be too lean, or you could have an exhaust leak near the sensor.

12. Under normal conditions, the sensor should fluctuate high and low. Prior to condemning the sensor, try forcing the system to have a rich fuel mixture by restricting the air intake, or lean by removing a vacuum line. If this causes the sensor to respond, look for problems in other areas of the system.

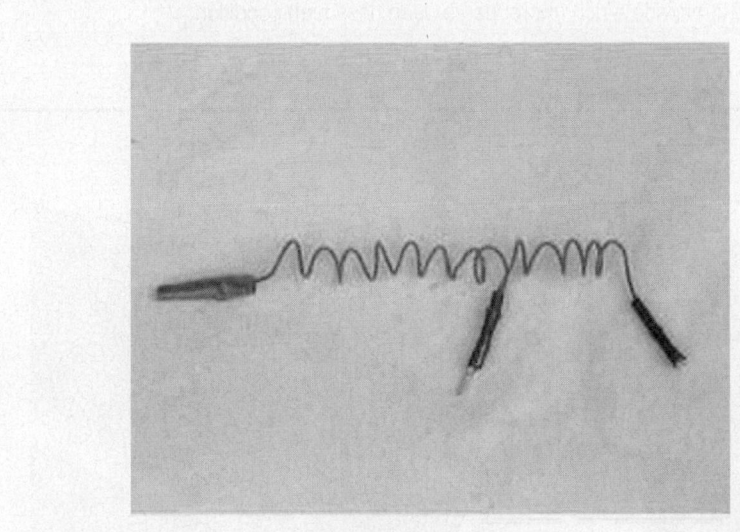

Jumper wires that will make sensor testing easier and safer
(for the sensor) are available from your local auto parts store

<u>HEATED OXYGEN SENSOR</u>
Operation

A Heated Oxygen Sensor (HO2S) is mounted on the muffler inlet pipe, just below the three-way catalytic converter. The sensor monitors the oxygen content in the three-way catalytic converter and then transmits this information to the Powertrain Control Module (PCM). The PCM then adjusts the air/fuel ratio to provide a rich (more fuel) or lean (less fuel) condition.

Testing

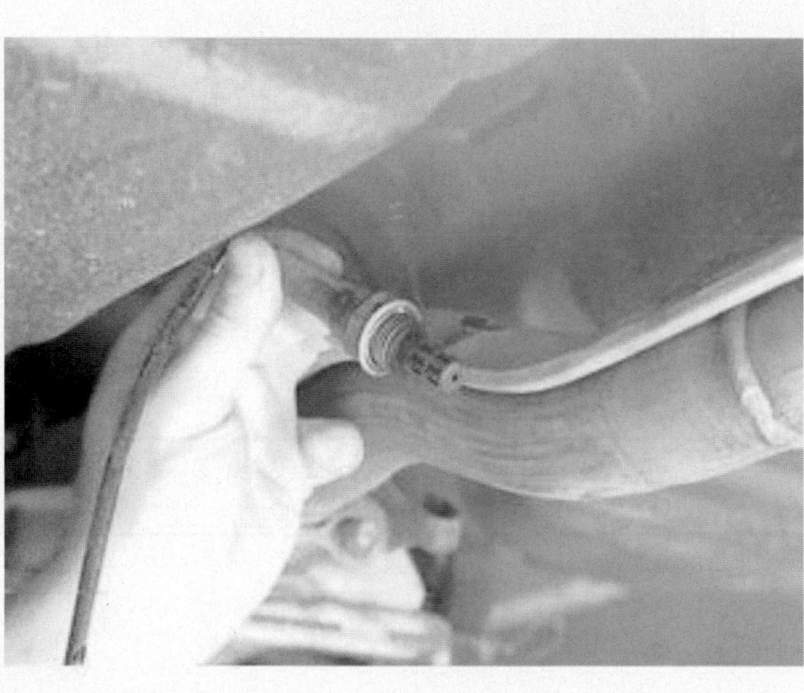

Check the tip of the heated oxygen sensor for contamination

1. Disconnect the electrical lead, then remove the sensor from the exhaust system.

2. If the sensor tip has a black/sooty deposit, this may indicate a rich fuel mixture.

3. If the sensor tip has a white, gritty deposit, this may indicate an internal coolant leak.

4. If the sensor tip has a brown deposit, this could indicate oil consumption.

5. Reinstall the heated oxygen sensor, but do not connect its electrical lead.

6.   Measure resistance between the PWR and GND (heater) terminals of the sensor. If the reading is about 6 ohms at 68°F (20°C) the sensor's heater element is okay. Connect the sensor's electrical lead.

7.   With the heated oxygen sensor connected and the engine running, measure voltage with a DVOM by backprobing the SIG RTN wire of the HO2S connector. The voltage readings at idle should stay below 1.0 volts. When the speed is increased or decreased, the voltage should fluctuate between 0 and 1.0 volts; if so, the sensor is okay.

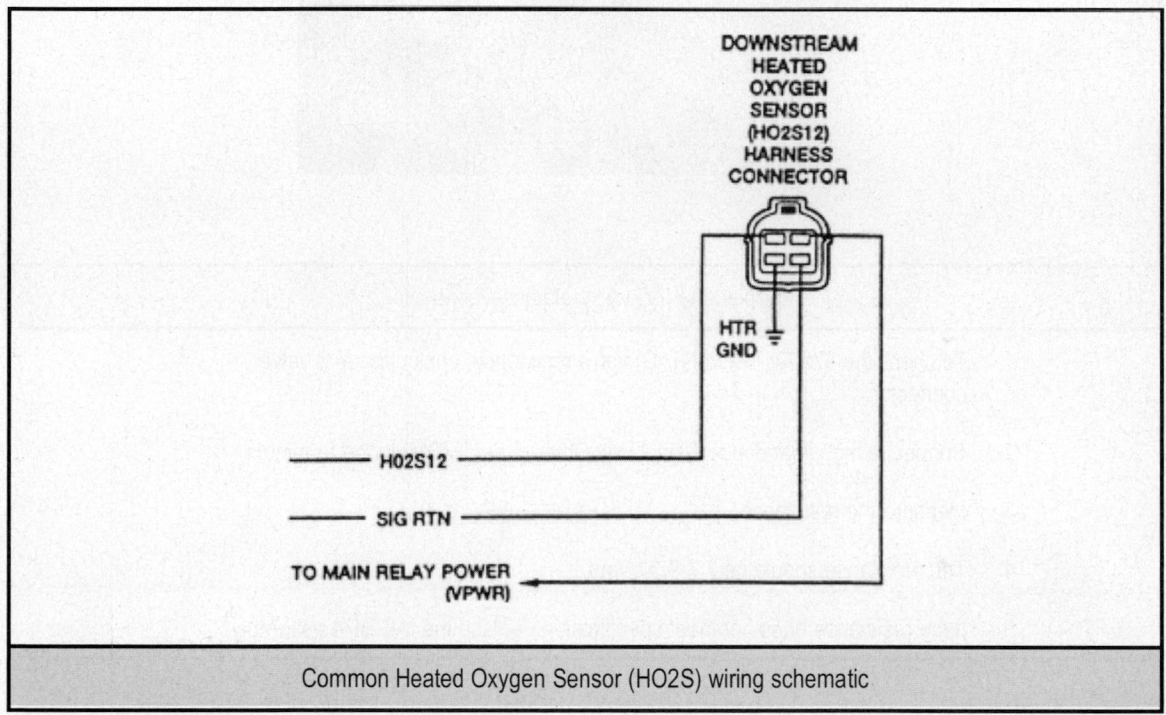

Common Heated Oxygen Sensor (HO2S) wiring schematic

IDLE AIR CONTROL BYPASS AIR VALVE
Operation

The Idle Air Control Bypass Air (IAC BPA) valve consists of an idle air control valve and a bypass air valve.

The bypass air valve functions during cold engine conditions to increase engine idle speed. It consists of a thermowax bead and a valve.

Engine coolant is directed around the thermowax, which opens and closes the valve. During cold engine operation below 140°F (60°C), the thermowax is contracted enough to allow the valve to open. As the coolant heats, the thermowax begins to expand. When the coolant reaches temperatures above 140°F (60°C), the thermowax expands and closes the valve.

The valve controls the amount of throttle valve bypass, which ensures a smooth idle under all engine operating conditions.

When the engine is cold, air flows through the valve during all modes of engine operation, to maintain the factory set idle speed.

Testing

Unplug the IAC valve electrical connector

1   To check the Idle Air Control (IAC) valve resistance, unplug the IAC valve connector.

2   Connect a high impedance Digital Volt Ohmmeter (DVOM) to the terminals.

3   Measure the resistance.

4   The resistance should be 7.7-9.3 ohms.

5   If the resistance does not meet specification, replace the IAC BPA assembly.

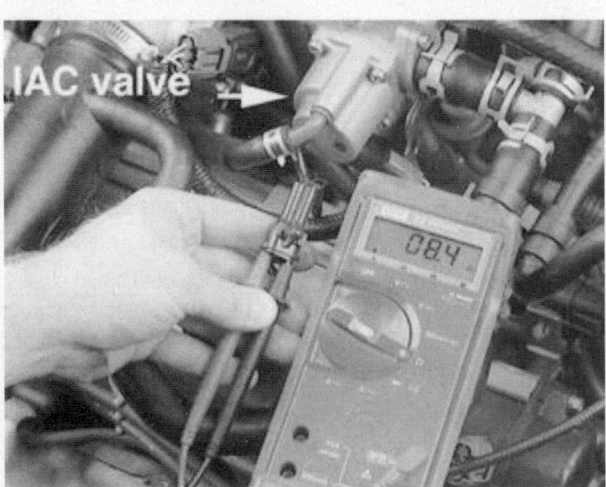

Connect an ohmmeter to the IAC valve; the resistance should be 7.7-9.3 ohms

6   To check the BPA valve function, remove the BPA valve from the engine.

7   Wait until the valve reaches room temperature. When the valve is cold, blow through the air inlet and verify that air flows freely through the valve.

8   Heat the valve with a hair dryer; the air valve should move out and block the passage.

9   If the valve does not function as specified, replace the IAC BPA assembly.

10   To test the solenoid, connect a scan tool to the Data Link Connector (DLC).

11   Turn the key ON and enter the simulation test.

12   Access the IACV Parameter Identification (PID) and turn the dial to cycle at 75 degrees.

13   Listen for the IAC solenoid to click when using the simulation test.

14   If the solenoid clicks it is working properly.

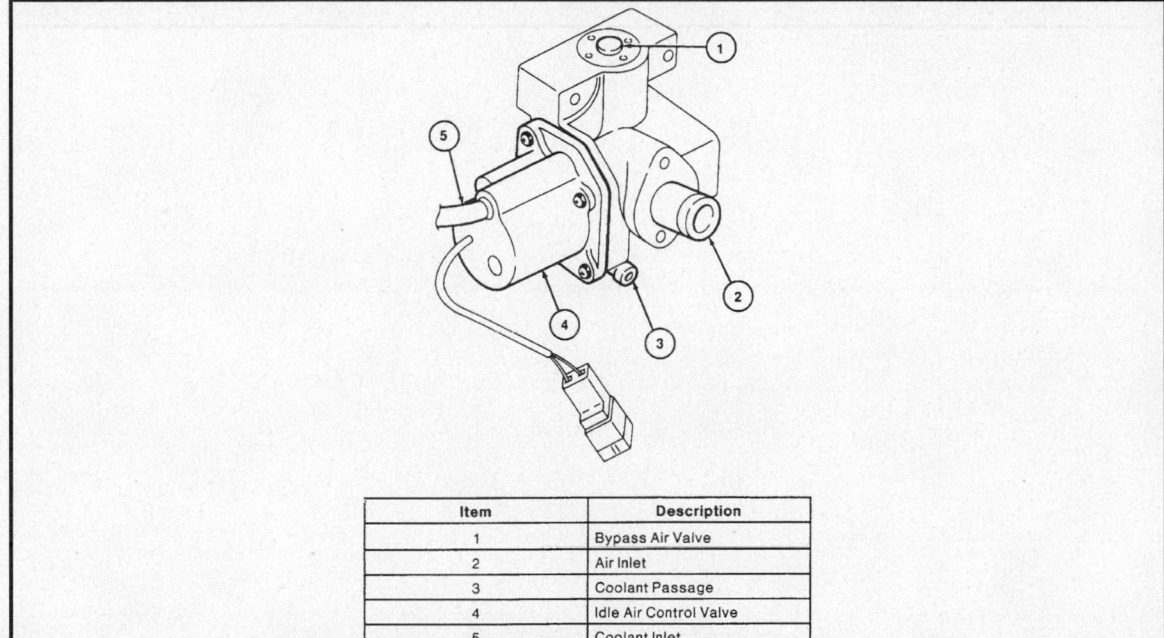

Item	Description
1	Bypass Air Valve
2	Air Inlet
3	Coolant Passage
4	Idle Air Control Valve
5	Coolant Inlet

Idle Air Control Bypass Air (IAC BPA) valve assembly components

COOLANT TEMPERATURE SENSOR
Operation

The Engine Coolant Temperature (ECT) sensor is mounted in the intake manifold coolant passage. The ECT sensor is a thermistor (a device which changes resistance as temperature changes). This sensor detects the temperature of engine coolant and provides a corresponding signal to the Powertrain Control Module (PCM).

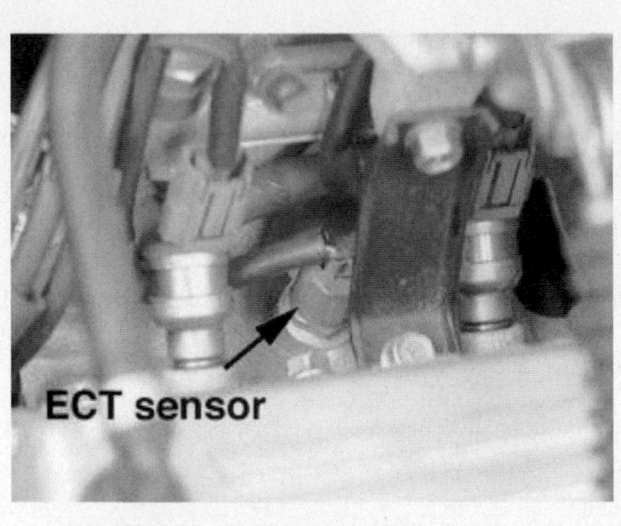

Location of the ECT sensor

Testing
1994-95 MODELS

1. Disengage the engine coolant temperature sensor's electrical connection.

2. Attach jumper wires to the sensor's terminals.

Do not attach to the wiring harness connector, since the sensor's resistance, rather than voltage, will be measured.

3. Connect a Digital Volt Ohmmeter (DVOM), set to the kilohms scale, to the jumper wires.

4. Measure the resistance with the engine OFF and cool, and also with the engine warmed up. Compare the temperature and resistance values obtained with those in the chart.

5. Remove the jumper wires.

6. Replace the sensor if any readings are incorrect.

1996-97 MODELS

1. Disengage the engine coolant temperature sensor's electrical connection.

2. Attach jumper wires from the sensor terminals to the wiring harness connector. This permits the engine to operate normally while you check the sensor.

3. Connect a Digital Volt Ohmmeter (DVOM) between the jumper wires.

4. Measure the voltage with the engine OFF and cool, and also with the engine running and warmed up. Compare the temperature and voltage values obtained with those in the chart.

5. Remove the jumper wires.

6. Replace the sensor if any readings are incorrect.

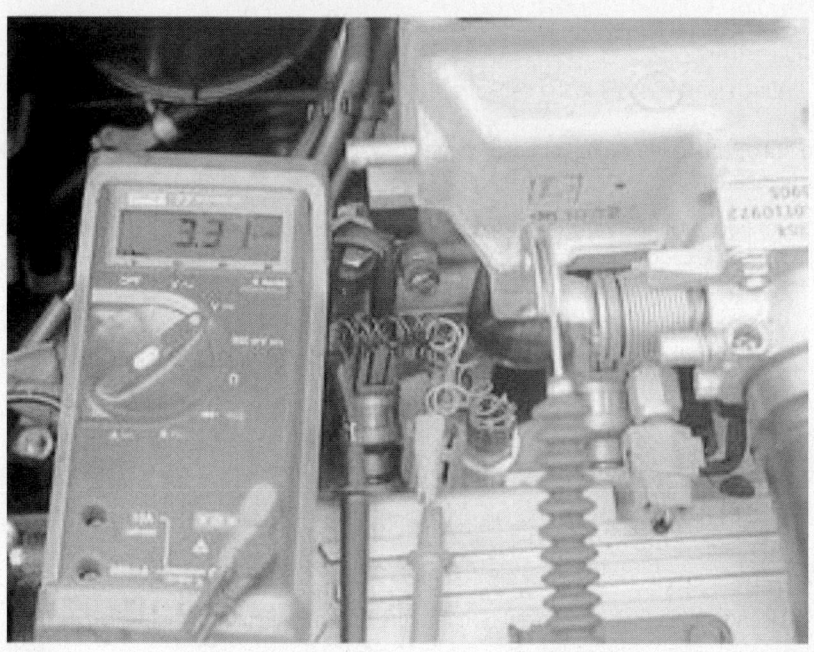

Use jumper wires to backprobe the ECT sensor terminals and
measure the voltage with a DVOM-OBD-II models

| Temperature | | Voltage* |
°F	°C	Volts
-4	-20	4.7
32	0	3.4
68	20	2.5
104	40	2.0
140	60	1.2
176	80	0.7
194	90	0.45
203	95	0.33
212	100	0.2

ECT and IAT sensors' temperature versus voltage chart

INTAKE AIR TEMPERATURE SENSOR
Operation

The Intake Air Temperature (IAT) sensor is a thermistor that changes its resistance in response to the intake air temperature. This sensor is mounted on the upper side of the upper engine air cleaner, where it senses the intake air temperature, then sends this information to the Powertrain Control Module (PCM). The PCM uses this information to calculate the correct amount of fuel injection.

Testing
1994-95 MODELS

1. Disengage the intake air temperature sensor's electrical connection.

2. Attach jumper wires to the sensor's terminals.

Do not attach to the wiring harness connector, since the sensor's resistance, rather than voltage, will be measured.

3. Connect a Digital Volt Ohmmeter (DVOM), set to the kilohms scale, to the jumper wires.

4. Measure the resistance with the engine OFF and cool, and also with the engine warmed up. Compare the temperature and resistance values obtained with those in the chart.

5. Remove the jumper wires.

6. Replace the sensor if any readings are incorrect.

Temperature °C (°F)	Resistance (kohms)
0 (32)	72.1 - 79.4
13 (55)	54.3 - 58.6
25 (77)	29.7 - 36.3
43 (110)	17.9 - 19.3
85 (185)	3.3 - 3.7

IAT sensor temperature versus resistance chart-OBD-I systems

1996-97 MODELS

1. Disengage the Intake Air Temperature (IAT) sensor electrical connection.

2. Connect jumper wires from the sensor terminal to the wiring harness connector. This permits the engine to operate normally while you check the sensor.

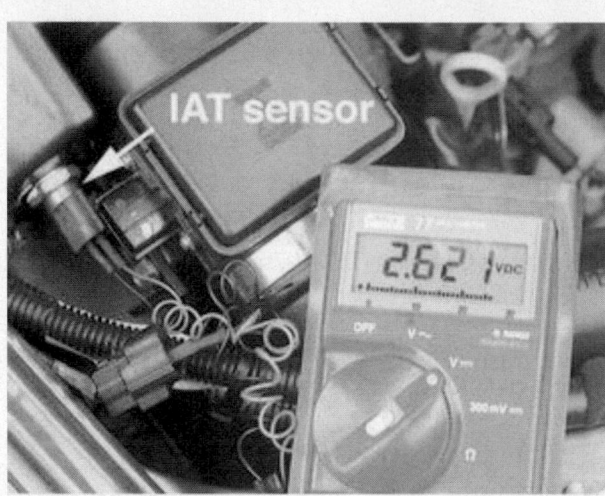

Use jumper wires to backprobe the IAT sensor terminals and measure the voltage with a DVOM-OBD-II models

3. Connect a Digital Volt Ohmmeter (DVOM) between the jumper wires.

4. Measure the voltage with the engine OFF and cool, and also with the engine running and warmed up. Compare the temperature versus voltage values obtained with those in the chart.

5. Remove the jumper wires.

6. Replace the sensor if any readings are incorrect.

MASS AIR FLOW SENSOR

Operation

The Mass Air Flow (MAF) sensor detects the intake air quantity and converts the measurement to a voltage reading by way of a heated resistor. The voltage signal is sent to the Powertrain Control Module (PCM) which, in turn, determines such things as fuel injection quantities and engine speed.

Testing

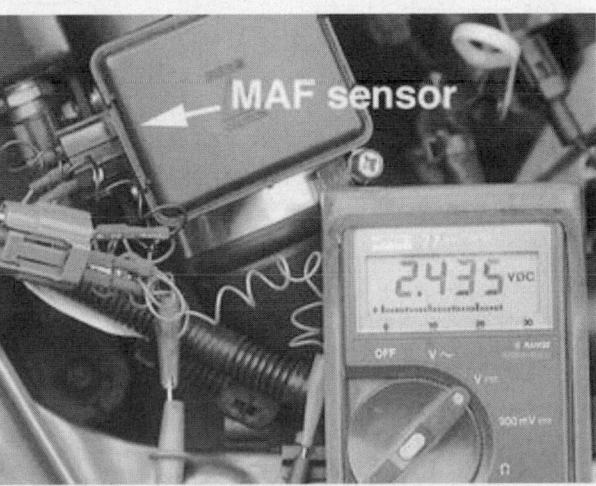

Measuring the voltage of the MAF sensor using jumper wires and a DVOM

1. Make sure the ignition is OFF.

2. Connect jumper wires from the sensor terminal to the wiring harness connector. This permits the engine to operate normally while you check the sensor.

3. Connect a Digital Volt Ohmmeter (DVOM) between the jumper wires.

4. Check for voltage at the MAF sensor connections. On OBD-I systems, test terminals MAF and SIG RTN. On OBD-II systems, check MAF and GND. Refer to the wiring illustrations.

5. Turn the ignition switch to the ON position, but do not start the engine. The voltage should be 1.0-1.5 volts.

6. Start the engine, then check the voltage, which should be 1.5-5.0 volts.

7. If the voltage readings do not meet specifications, replace the sensor.

8. Remove the jumper wires.

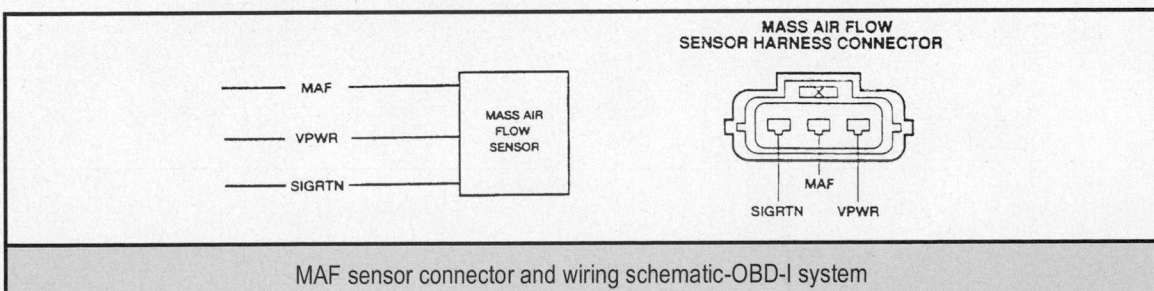

MAF sensor connector and wiring schematic-OBD-I system

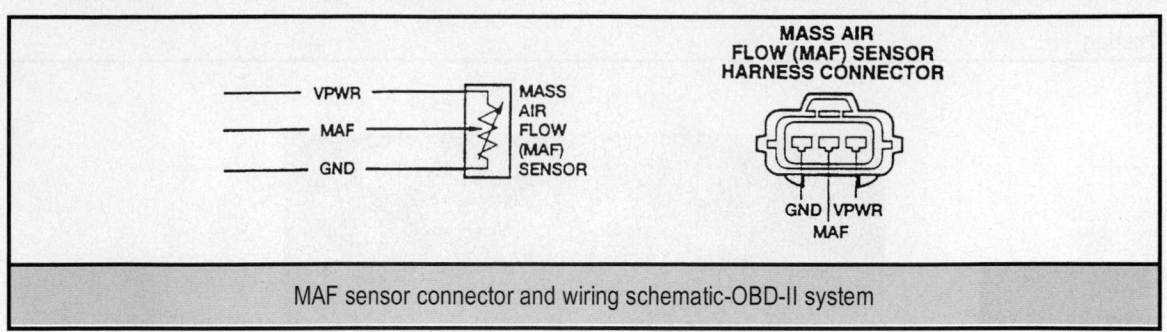

MAF sensor connector and wiring schematic-OBD-II system

BAROMETRIC PRESSURE SENSOR
Operation

The Barometric Pressure (BARO) sensor is located inside the Powertrain Control Module (PCM). The BARO sensor provides the PCM with atmospheric pressure information.

Testing
OBD-I SYSTEM

1. Connect a scan tool to the Data Link Connector (DLC).

2. Operate the scan tool and check for trouble codes.

3. If code 14 is present, clear the codes and repeat the scan tool test.

4. If code 14 is still present, the Barometric Pressure (BARO) sensor is defective and the Powertrain Control Module (PCM) must be replaced.

THROTTLE POSITION SENSOR

Operation

The Throttle Position (TP) sensor is a variable resistor type sensor. The sensor is mounted on the left-hand side of the throttle body, and detects the angle which the throttle valve has been opened. The sensor then relays this information to the Powertrain Control Module (PCM) which, after analyzing the data, regulates the air/fuel mixture.

Testing

1. Backprobe the TP terminal at the TP sensor connector, with the positive lead of a Digital Volt Ohmmeter (DVOM).

2. Backprobe the SIG RTN terminal with the negative lead.

3. Have an assistant turn the ignition key ON, then depress the accelerator pedal (or manually rotate the throttle lever) so that the throttle lever attains $1/4$ open, $1/2$ open, $3/4$ open and wide-open positions.

4. Observe the voltmeter at each of these positions and note the voltage displayed. Compare your readings with the figures in the accompanying charts.

Backprobe the TP sensor terminals with a DVOM, move the throttle lever and observe the voltage reading

Move the throttle lever through different opening positions and note the voltage readings

5. If the TP sensor readings match the values in the chart, the sensor is working properly. If, however, the TP sensor readings are not as specified, continue with the test procedure.

6. Turn the ignition switch OFF.

7. Unplug the TP sensor electrical connector, then attach the positive lead of the DVOM to the VREF terminal of the wiring harness connector.

8. Turn the ignition switch ON and observe the voltage reading on the DVOM; it should be between 4.5 and 5.5 volts DC.

9. If the VREF voltage is within specifications, and the TP voltage readings do not agree with the values displayed in the chart, replace the TP sensor.

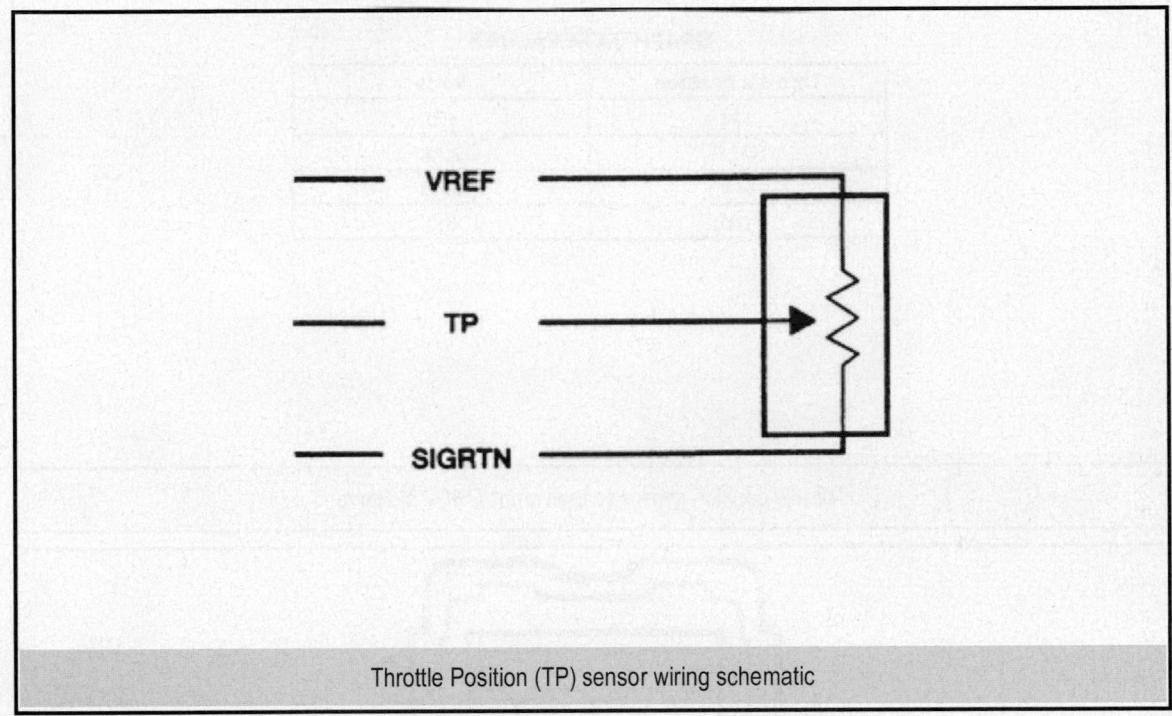

Throttle Position (TP) sensor wiring schematic

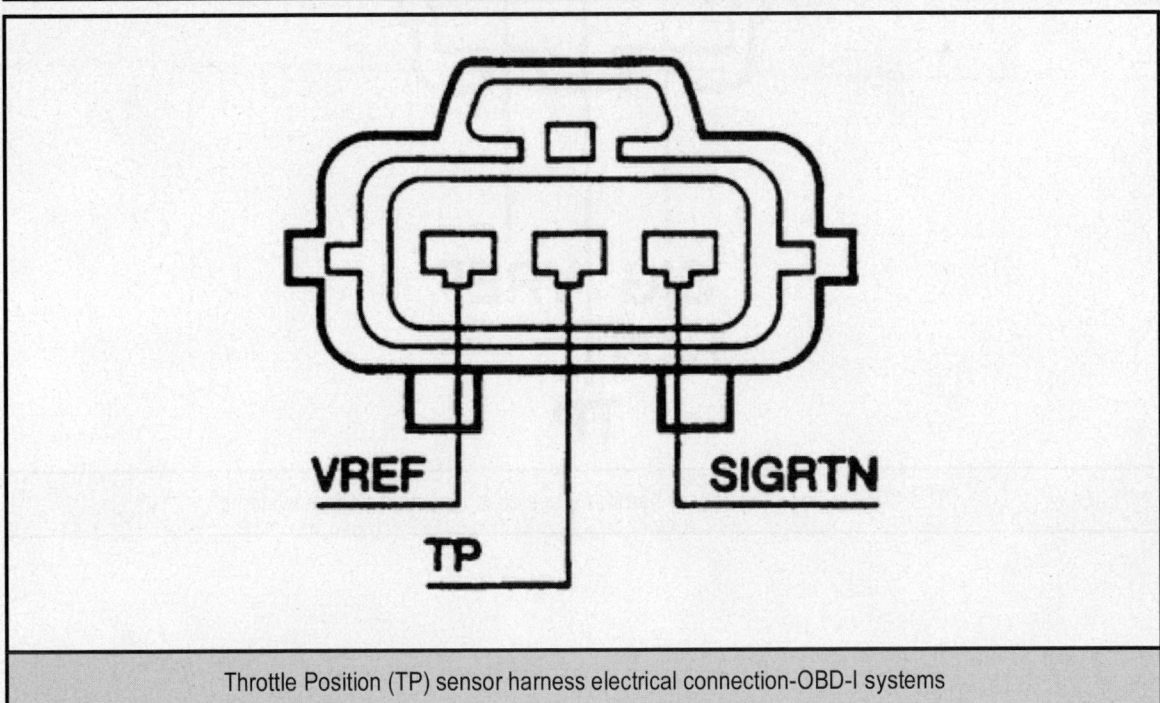

Throttle Position (TP) sensor harness electrical connection-OBD-I systems

GRAPH DATA VALUES

Throttle Position	Volts
1/4	0.5
HALF	2.75
3/4	3.88
FULL	5.0

Throttle position sensor voltage chart-OBD-I systems

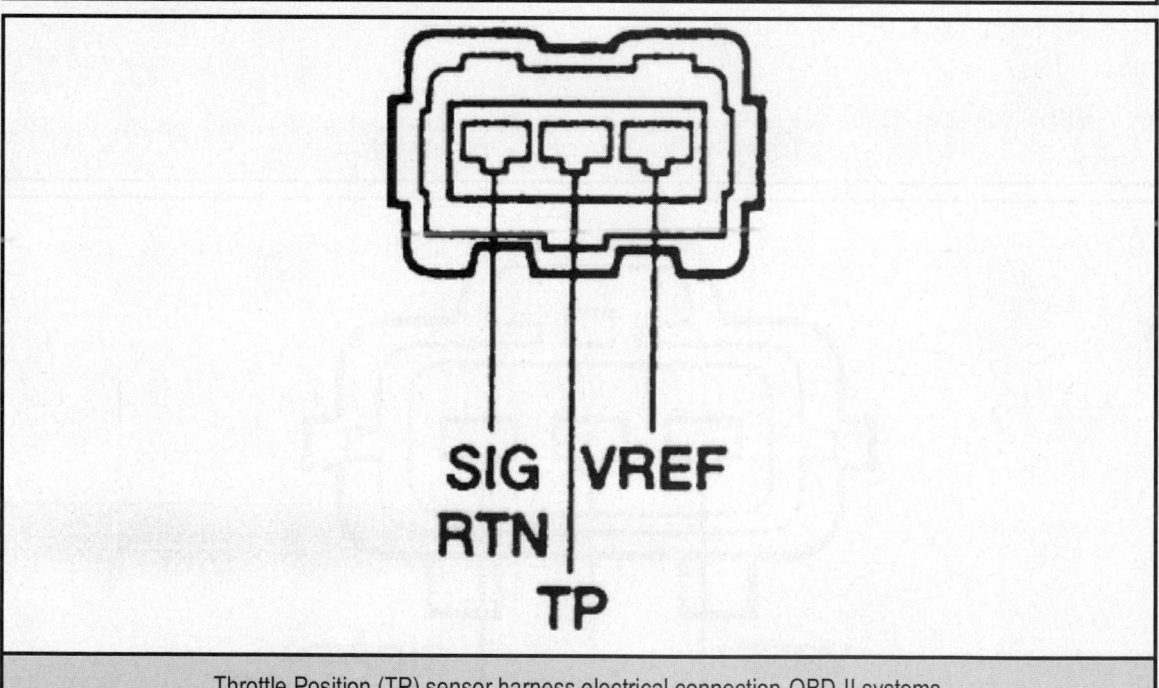

Throttle Position (TP) sensor harness electrical connection-OBD-II systems

Throttle Position	Voltage (Volts)
0	0.5
1/4	1.3
Half	2.2
3/4	2.9
Full	3.7

Throttle position sensor voltage chart-OBD-II systems

CAMSHAFT POSITION SENSOR

Operation

On OBD-I systems, the Camshaft Position (CMP) sensor detects the No.1 cylinder when it reaches Top Dead Center (TDC) and signals the Powertrain Control Module (PCM) to control fuel injection.

On OBD-II systems, there are two Camshaft Position Sensors. One CMP detects the No.1 cylinder when it reaches Top Dead Center (TDC) and signals the Powertrain Control Module (PCM) to control fuel injection. The other CMP signals the PCM when the camshaft turns a quarter of a revolution.

Testing
OBD-I SYSTEM

1. Turn the ignition switch OFF.

2. Set a DVOM to a low DC voltage range (under 20 volts).

3. Backprobe the ground terminal of the distributor connector with the negative lead.

4. Backprobe the VPWR terminal of the distributor connector with the positive lead.

5. Turn the ignition switch to the ON position. The DVOM should read battery voltage. If not, check the wiring to the ECM.

6. Turn the ignition switch to the OFF position.

7. Backprobe the CMP terminal of the distributor connector with the positive lead.

8. Turn the ignition switch to the ON position.

9. Bump the starter to rotate the engine, but do not start the engine.

10. As the vanes of the distributor rotor pass through the Hall effect sensor, the voltage should switch back and forth between 5 volts and less than 1 volt.

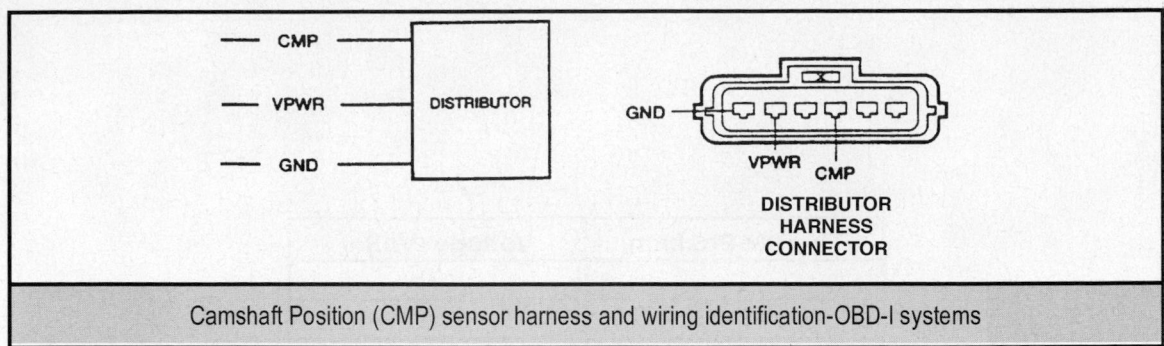

Camshaft Position (CMP) sensor harness and wiring identification-OBD-I systems

OBD-II SYSTEM

1. Turn the ignition switch OFF.

2. Set a DVOM to a low DC voltage range (under 20 volts).

3. Backprobe the ground terminal of the distributor connector with the negative lead.

4. Backprobe the VPWR terminal of the distributor connector with the positive lead.

5. Turn the ignition switch to the ON position. The DVOM should read battery voltage. If not, check the wiring to the ECM.

6. Turn the ignition switch to the OFF position.

Backprobe the GND and the CMP 1 and CMP 2 terminals with a DVOM, and observe the readings

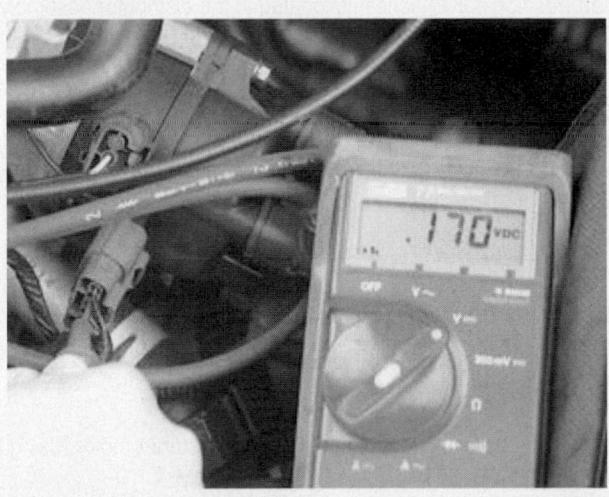

The voltage readings should change as the vanes in the distributor pass through the Hall effect sensor

7. Backprobe the CMP1 terminal of the distributor connector with the positive lead.

8. Turn the ignition switch to the ON position.

9. Bump the starter to rotate the engine, but do not start the engine.

10. As the vanes in the distributor pass through the Hall effect sensor, the voltage should switch back and forth between 5 volts and less than 1 volt.

11. Perform the same test on the CMP 2 terminal.

12. If the voltage doesn't vary, check that the distributor is rotating. If the distributor is rotating, the CMP sensor may be defective.

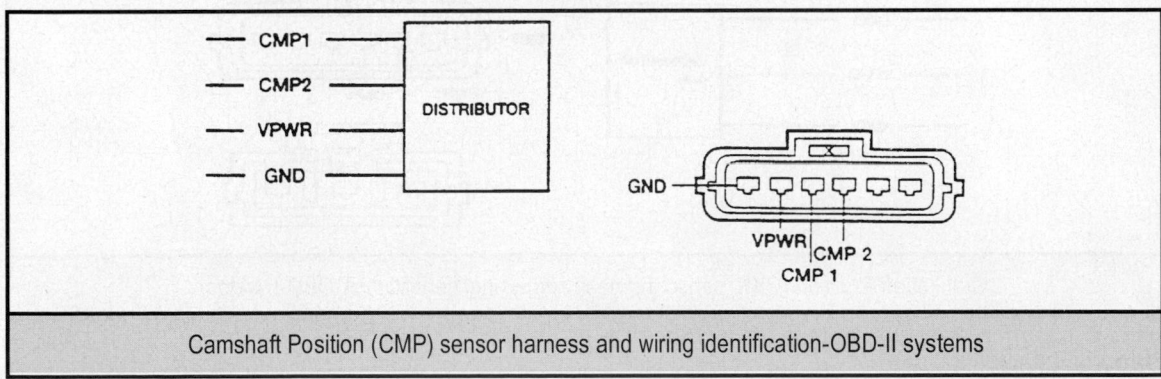

Camshaft Position (CMP) sensor harness and wiring identification-OBD-II systems

CRANKSHAFT POSITION SENSOR

The Crankshaft Position (CKP) sensor is located in the distributor assembly on 1994-95 models, and on the lower front of the oil pump on 1996-97 models.

Operation

The Crankshaft Position (CKP) sensor relays engine speed and crankshaft position to the Powertrain Control Module (PCM).

Testing
OBD-I SYSTEM

1. Turn the ignition switch OFF.

2. Using a Digital Volt Ohmmeter (DVOM), backprobe the CKP1 terminal of the CKP sensor connector with the DVOM's positive lead. Attach the DVOM's negative lead to a good engine ground.

3. While an assistant bumps over the engine with the starter motor (using the ignition switch), observe the voltage displayed on the DVOM.

4. The CKP sensor voltage should fluctuate between approximately 0 and 5 volts. If the CKP sensor performs as specified, it is functioning properly. If, however, the voltage is not as specified, continue with the test procedure.

5. Turn the ignition switch OFF.

6. Disengage the wiring harness connector from the distributor, then attach the positive lead of the DVOM to the VPWR terminal on the wiring harness connector. Attach the negative lead to a good ground.

7. Turn the ignition key ON and observe the voltage displayed on the DVOM. The voltage should be greater than 10 volts. If not, the CKP sensor is not receiving the proper voltage from the PCM; the problem lies elsewhere in the electrical system.

8. If the VPWR voltage is satisfactory and the CKP output is not as specified, replace the CKP sensor.

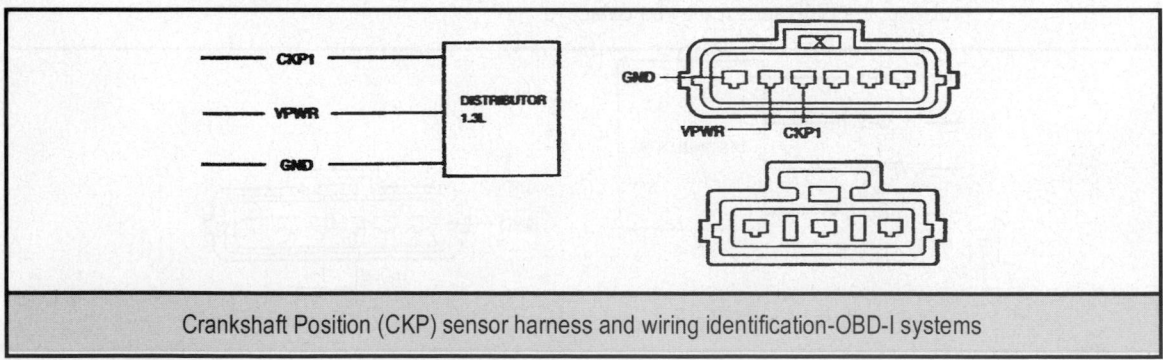

Crankshaft Position (CKP) sensor harness and wiring identification-OBD-I systems

OBD-II SYSTEM

To perform this test accurately, the ambient air temperature should be 68°F (20°C).

1. Turn the ignition switch OFF.

2. Unplug the CKP electrical connection.

3. Using a Digital Volt Ohmmeter (DVOM) set to read resistance, probe the CKP (+) terminal of the CKP sensor connector with the DVOM's positive lead. Attach the DVOM's negative lead to a good engine ground. Note the reading.

Backprobe the CKP sensor connector with the DVOM and note the readings

4. Backprobe the CKP + terminal of the CKP (-) sensor connector with the DVOM's positive lead in the same manner and note the resistance reading.

5. The resistance should be 520-580 ohms at 68°F (20°C). If readings are not as specified, replace the CKP sensor.

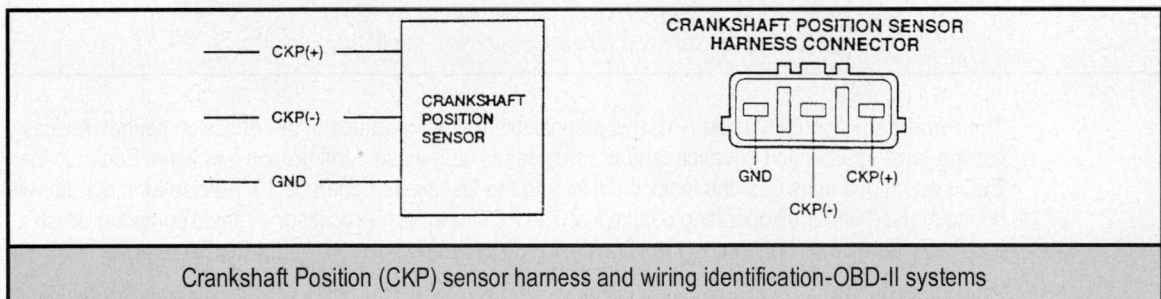

Crankshaft Position (CKP) sensor harness and wiring identification-OBD-II systems

Bronco

POWERTRAIN CONTROL MODULE
Operation

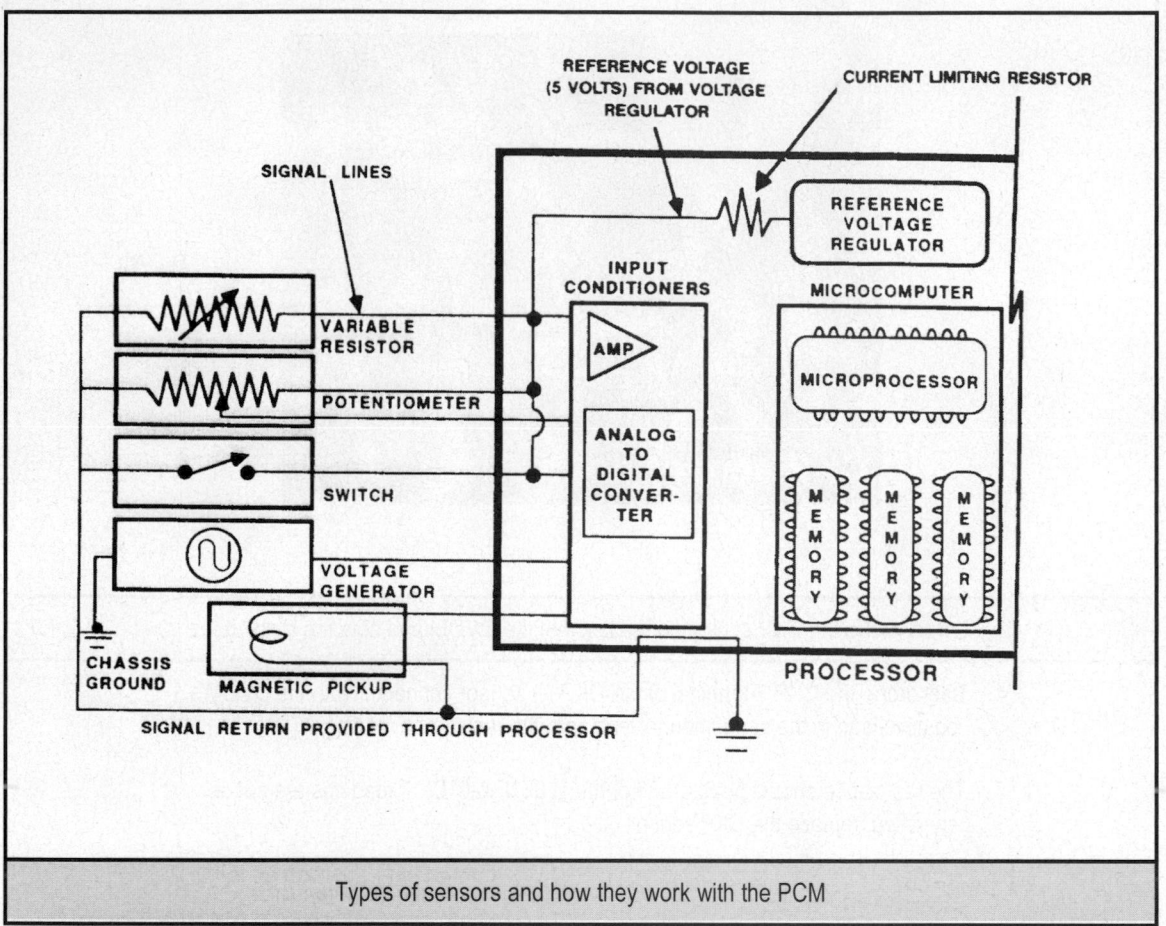

Types of sensors and how they work with the PCM

The Powertrain Control Module (PCM) is responsible for the operation of the emission control devices, cooling fans, ignition and advance, and in some cases, automatic transmission functions. Because the EEC system oversees both the ignition timing and the fuel injector operation, a precise air/fuel ratio will be maintained under all operating conditions. The PCM is a microprocessor or small computer which receives electrical inputs from several sensors, switches and relays on and around the engine.

PCM's for EEC-IV systems use a 60-pin connector. For the EEC-V PCM, a 104-pin connector is used.

Based on combinations of these inputs, the PCM controls outputs to various devices concerned with engine operation and emissions. The engine control assembly relies on the signals to form a correct picture of current vehicle operation. If any of the input signals is incorrect, the PCM reacts to what ever picture is painted for it. For example, if the coolant temperature sensor is inaccurate and reads too low, the PCM may see a picture of the engine never warming up. Consequently, the engine settings will be maintained as if the engine were cold. Because so many inputs can affect one output, correct diagnostic procedures are essential on these systems.

The EEC system employs adaptive fuel logic. This process is used to compensate for normal wear and variability within the fuel system. Once the engine enters steady-state operation, the engine control assembly watches the oxygen sensor signal for a bias or tendency to run slightly rich or lean. If such a bias is detected, the adaptive logic corrects the fuel delivery to bring the air/fuel mixture towards a centered or 14.7:1 ratio. This compensating shift is stored in a non-volatile memory which is retained by battery power even with the ignition switched off. The correction factor is then available the next time the vehicle is operated.

The Powertrain Control Module (PCM) is usually located under the instrument panel or passenger's seat and is usually covered by a kick panel. A multi-pin connector links the PCM with all system components. The processor provides a continuous reference voltage to the B/MAP, EVP and TP sensors. EEC systems use a 5 volt reference signal. Different calibration information is used in different vehicle applications, such as California or Federal models. For this reason, careful identification of the engine, year, model and type of electronic control system is essential to ensure correct component replacement.

If the battery cable(s) is disconnected for longer than 5 minutes, the adaptive fuel factor will be lost. After repair it will be necessary to drive the truck at least 10 miles to allow the processor to relearn the correct factors. The driving period should include steady-throttle open road driving if possible. During the drive, the vehicle may exhibit driveability symptoms not noticed before. These symptoms should clear as the PCM computes the correction factor. The PCM will also store Code 19 indicating loss of power to the controller.

Electronic Engine Control

The electronic engine control subsystem consists of the PCM and various sensors and actuators. The PCM reads inputs from engine sensors, then outputs a voltage signal to various components (actuators) to control engine functions. The period of time that the injectors are energized ("ON" time or "pulse width") determines the amount of fuel delivered to each cylinder. The longer the pulse width, the richer the fuel mixture.

The operating reference voltage (Vref) between the PCM and its sensors and actuators is 5 volts. This allows these components to work during the crank operation even though the battery voltage drops.

In order for the PCM to properly control engine operation, it must first receive current status reports on various operating conditions. The control unit constantly monitors crankshaft position, throttle plate position, engine coolant temperature, exhaust gas oxygen level, air intake volume and temperature, air conditioning (On/Off), spark knock and barometric pressure.

OXYGEN SENSOR

Operation

An Oxygen Sensor (O_2S) or heated Oxygen Sensor (HO_2S) is used on all engines. The sensor is mounted in the right side exhaust manifold on some V8 engines, while other V8 engines use a sensor in both right and left manifolds. The sensor protrudes into the exhaust stream and monitors the oxygen content of the exhaust hoses. The difference between the oxygen content of the exhaust gases and that of the outside air generates a voltage signal to the PCM. The PCM monitors this voltage and, depending upon the value of the signal received, issues a command to adjust for a rich or a lean condition.

Testing

8. Perform a visual inspection of the heated oxygen sensor as follows:

9. Disconnect the electrical lead, then remove the sensor from the exhaust system.

10. If the sensor tip has a black/sooty deposit, this may indicate a rich fuel mixture.

11. If the sensor tip has a white, gritty deposit, this may indicate an internal coolant leak.

12. If the sensor tip has a brown deposit, this could indicate oil consumption.

13. Reinstall the heated oxygen sensor, but do not connect its electrical lead.

14. Measure resistance between the PWR and GND (heater) terminals of the sensor. If the reading is about 6 ohms at 68°F (20°C) the sensor's heater element is okay. Connect the sensor's electrical lead.

With the heated oxygen sensor connected and the engine running, measure voltage with a DVOM by backprobing the SIG RTN wire of the HO2S connector. The voltage readings at idle should stay below 1.0 volts. When the speed is increased or decreased, the voltage should fluctuate between 0 and 1.0 volts; if so, the sensor is okay.

HEATED OXYGEN SENSOR

Operation

Heated oxygen sensors are located in the exhaust pipes below the exhaust manifolds. The sensors react with the oxygen in the exhaust gasses and generates a voltage based on this reaction. A low voltage indicates too much oxygen or a lean condition. Where as a high voltage indicates not enough oxygen or a rich condition.

Testing

1. Disconnect the Oxygen Sensor (O_2S). Measure resistance between PWR and GND (heater) terminals of the sensor. If the reading is about 6 ohms at 68°F (20°C). the sensor's heater element is okay.

2. With the O_2S connected and engine running, measure voltage with DVOM between terminals HO2S and SIG RTN (GND) of the oxygen sensor connector. If the voltage readings are about equal to those in the table, the sensor is okay.

IDLE AIR CONTROL VALVE (IAC)

Operation

The idle air control valve (IAC) control the engine idle speed and dashpot functions. The valve is located on the throttle body on. This valve allows air to bypass the throttle plate. The amount of air is determined by the Powertrain Control Module (PCM) and controlled by a duty cycle signal.

Testing

1. Make sure the ignition key is OFF.

2. Disconnect the air control valve.

3. Use an ohmmeter to measure the resistance between the terminals of the valve solenoid.

    Due to the diode in the solenoid, place the ohmmeter positive lead on the VPWR pin and the negative lead on the ISC pin.

4. If the resistance is not 7-13 ohms replace the air control valve.

ENGINE COOLANT TEMPERATURE (ECT) SENSOR

Operation

The ECT sensor is located either in the heater supply tube at the rear of the engine, or in the lower intake manifold. The ECT sensor is a thermistor (changes resistance as temperature changes). The sensor detects the temperature of engine coolant and provides a corresponding signal to the PCM. From this signal, the PCM will modify the air/fuel ratio (mixture), idle speed, spark advance, EGR and Canister purge control. When the engine coolant is cold, the ECT sensor signal causes the PCM to provide enrichment to the air/fuel ratio for good cold drive away as engine coolant warms up, the voltage will drop.

Testing

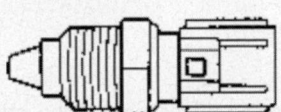 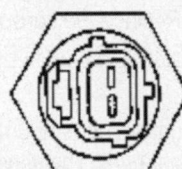

There are two prongs behind the sensor for testing

Submerge the end of the temperature sensor in cold or hot water and check resistance

1   Disconnect the temperature sensor.

2   Connect an ohmmeter between the sensor terminals and set the ohmmeter scale on 200,000 ohms.

3   Measure the resistance with the engine off and cool and with the engine running and warmed up. Compare the resistance values obtained with the chart.

4   Replace the sensor if the readings are incorrect.

INTAKE AIR TEMPERATURE (IAT) SENSOR

Operation

The intake air temperature sensor (IAT) changes the resistance in response to the intake air temperature. The sensor resistance decreases as the surrounding air temperature increases. This provides a signal to the PCM indicating the temperature of the incoming air intake.

Testing

With ignition OFF, disconnect the IAT sensor. Measure the resistance across the sensor connector terminals. If the reading for a given temperature is about that shown in the table, the IAT sensor is okay.

MASS AIR FLOW (MAF) SENSOR
Operation

The Mass Air Flow (MAF) sensor directly measures the mass of the air flowing into the engine. The sensor output is an analog signal ranging from about 0.5-5.0 volts. The signal is used by the PCM to calculate the injector pulse width. The sensing element is a thin platinum wire wound on a ceramic bobbin and coated with glass. This "hot wire" is maintained at 200°C above the ambient temperature as measured by a constant "cold wire". The MAF sensor is located in the outlet side of the air cleaner lid assembly.

Testing

1. Make sure the ignition key is OFF.

2. Connect Breakout Box T83L-50-EEC-IV or equivalent, to the PCM harness and connect the PCM.

3. Start the engine and let it idle.

4. Use a voltmeter to measure the voltage between test pin 50 of the Breakout Box and the battery negative post.

5. Replace the MAF sensor if the voltage is not 0.36-1.50 volts.

BAROMETRIC/MANIFOLD ABSOLUTE PRESSURE SENSORS (B/MAP)
Operation

The B/MAP sensor used on some engines is separate from the barometric sensor and is located on the left fender panel in the engine compartment. The barometric sensor signals the PCM of changes in atmospheric pressure and density to regulate calculated air flow into the engine. The MAP sensor monitors and signals the PCM of changes in intake manifold pressure which result from engine load, speed and atmospheric pressure changes.

The Manifold Absolute Pressure (MAP) sensor measures the pressure in the intake manifold and sends a variable frequency signal to the PCM. When the ignition is ON and the engine OFF, the MAP sensor will indicate the barometric pressure in the intake manifold.

Testing

1. Connect MAP/BARO tester to sensor connector and sensor harness connector. With ignition ON and engine OFF, use a DVOM to measure voltage across tester terminals. If the tester's 4-6V indicator is ON, the reference voltage input to the sensor is okay.

2. If the DVOM voltage reading is as indicated in the table, the sensor is okay.

THROTTLE POSITION (TP) SENSOR

Operation

The Throttle Position (TP) sensor is a potentiometer that provides a signal to the PCM that is directly proportional to the throttle plate position. The TP sensor is mounted on the side of the throttle body and is connected to the throttle plate shaft.

The TP senses the throttle movement and position and transmits an appropriate electrical signal to the PCM. These signals are used by the PCM to adjust the air/fuel mixture, spark timing and EGR operation according to engine load at idle, part throttle, or full throttle. The TP sensor has 2 versions, an adjustable and a non-adjustable; the difference being elongated mounting holes that allow the rotary sensor to be turned slightly to adjust the output voltage. The rotary TP sensor with round mounting holes are not adjustable.

Testing

With ignition OFF, disconnect the TP sensor connector. Measure resistance between sensor connector terminals SIG RTN and TP. If resistance readings are about equal to the values in the tables, the sensor is okay.

CAMSHAFT POSITION (CMP) SENSOR

Operation

The Camshaft Position Sensor, is a Hall-effect sensor that generates a digital frequency while windows in a target wheel pass through its magnetic field. The frequency of the windows passing by the sensor, as well as the width of selected windows, allows the PCM to detect engine speed and position.

Testing
TWO-WIRE SENSORS

1. With the ignition OFF, install a Breakout Box.

2. Connect CMP sensor and ECM.

3. Using DVOM on AC scale and set to monitor less than 5V, measure voltage between Breakout Box terminals 24 and 46 with the engine running at varying RPM. If the voltage reading varies more than 0.1V AC, the sensor is okay.

THREE-WIRE SENSORS

1. With the ignition OFF, disconnect the CMP sensor. With the ignition ON and the engine OFF, measure the voltage between sensor harness connector VPWR and PWR GND terminals (refer to the figure). If the reading is greater than 10.5V, the power circuit to the sensor is okay.

2. With the ignition OFF, install a Breakout Box. Connect CMP sensor and PCM. Using DVOM on AC and scale set to monitor less than 5V, measure voltage between Breakout Box terminals 24 and 40 with the engine running at varying RPM. If the voltage reading varies more than 0.1V AC, the sensor is okay.

CRANKSHAFT POSITION (CP) SENSOR

Operation

The CP sensor is mounted on the right front of some V8 engines, is a variable reluctance sensor triggered by a trigger pulse wheel (36 minus 1 tooth). Its purpose is to provide the PCM with an accurate ignition timing reference (when the piston reaches 10 degrees BTDC) and injector operation information (twice each crankshaft revolution). The crankshaft vibration damper is fitted with a 4 lobe pulse ring. As the crankshaft rotates, the pulse ring lobes interrupt the magnetic field at the tip of the CP sensor.

Testing

Using DVOM on the AC scale and set to monitor less than 5V, measure voltage between the sensor Cylinder Identification (CID) terminal and ground. The sensor is okay if the voltage reading varies more than 0.1V AC with the engine running at varying RPM.

KNOCK SENSOR (KS)

Operation

This sensor is used on some 4.9L engines. The KS detects engine vibrations caused by preignition or detonation and provides information to the PCM, which then retards the timing to eliminate detonation.

Testing

1. With ignition ON and engine OFF, measure voltage between KS connector terminals. If voltage reading is 2.4-2.6V, the circuit between the ECM and KS is okay.

2. With engine running at idle and 3000 rpm, measure voltage using a DVOM on the AC setting between the KS terminals. If the AC voltage reading increases as the rpm increases, the sensor is okay.

EGR VALVE POSITION (EVP) SENSOR EEC-IV MANAGEMENT SYSTEM

Operation

The Exhaust Gas Recirculation (EGR) Valve Position (EVP) system uses an electronic EGR valve to control the flow of exhaust gases. The Engine Control Module (ECM) monitors the flow by means of an EVP sensor and regulates the electronic EGR valve accordingly. The valve is operated by a vacuum signal from the EGR Vacuum Regulator (EVR) solenoid which actuates the valve diaphragm.

As the supply vacuum overcomes the spring load, the diaphragm is actuated. This lifts the pintle off its seat and allows exhaust gases to flow. The amount of flow is proportional to the pintle position. The EVP sensor, mounted on the valve, sends an electronic signal representing pintle position to the ECM.

Testing

1. Disconnect the EVP sensor connector. With the ignition ON and the engine OFF, measure the voltage between VREF and SIG RTN terminals of EVP sensor harness connector. If the voltage is 4.0-6.0V, the power circuits to the sensor are okay.

2. Reconnect the EVP sensor. With the ignition ON and the engine OFF, measure the voltage between EVP sensor terminals EVP and SIG RTN. If the voltage reading is 0.67V or less, the sensor is okay.

ENGINE COMPONENT, AND CONNECTOR LOCATIONS

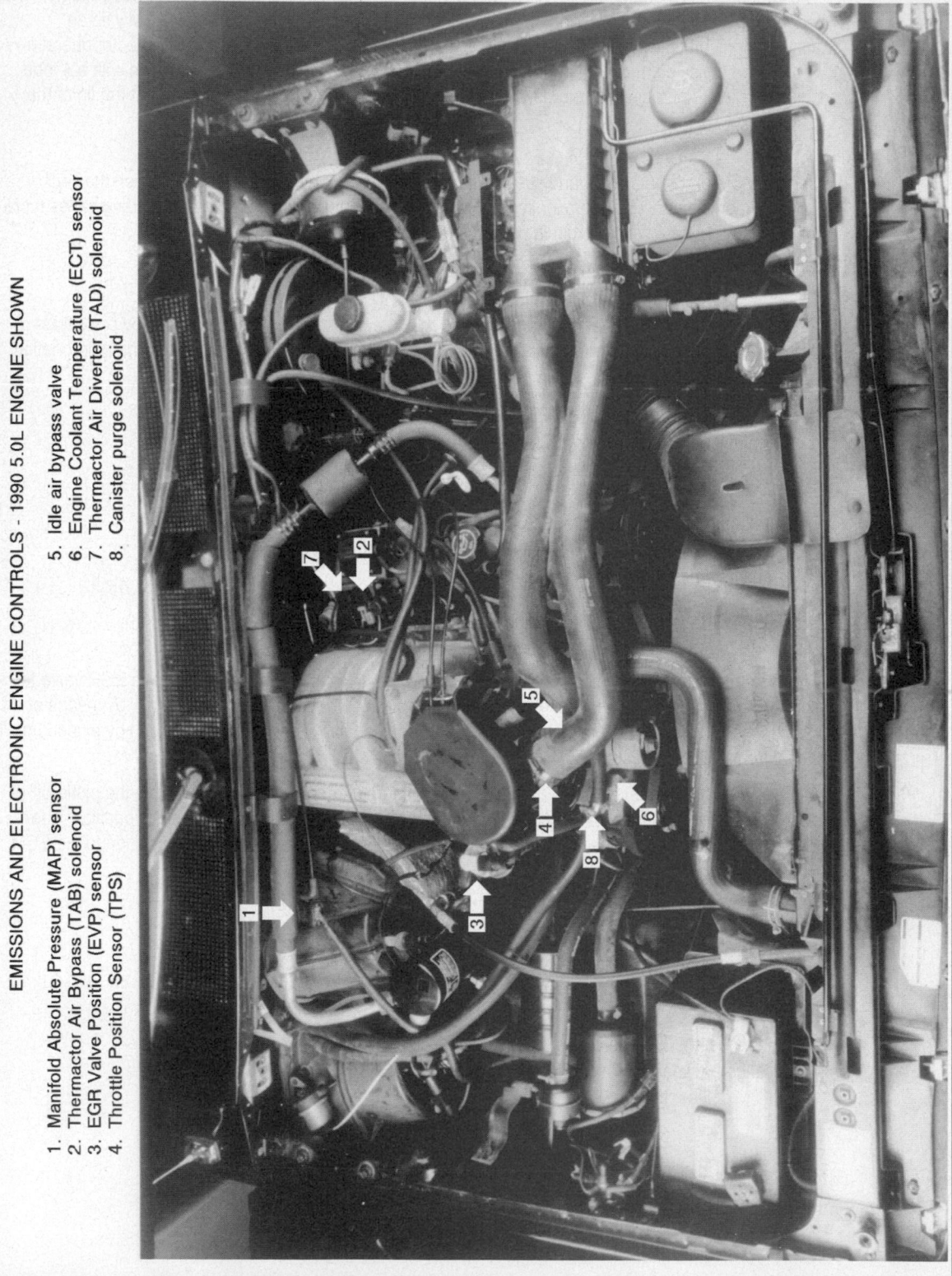

EMISSIONS AND ELECTRONIC ENGINE CONTROLS - 1990 5.0L ENGINE SHOWN

1. Manifold Absolute Pressure (MAP) sensor
2. Thermactor Air Bypass (TAB) solenoid
3. EGR Valve Position (EVP) sensor
4. Throttle Position Sensor (TPS)
5. Idle air bypass valve
6. Engine Coolant Temperature (ECT) sensor
7. Thermactor Air Diverter (TAD) solenoid
8. Canister purge solenoid

EMISSION AND ELECTRONIC ENGINE CONTROL LOCATIONS

Contour, Mystique & Cougar
POWERTRAIN CONTROL MODULE
Operation

The Powertrain Control Module (PCM) performs many functions on your vehicle. The module accepts

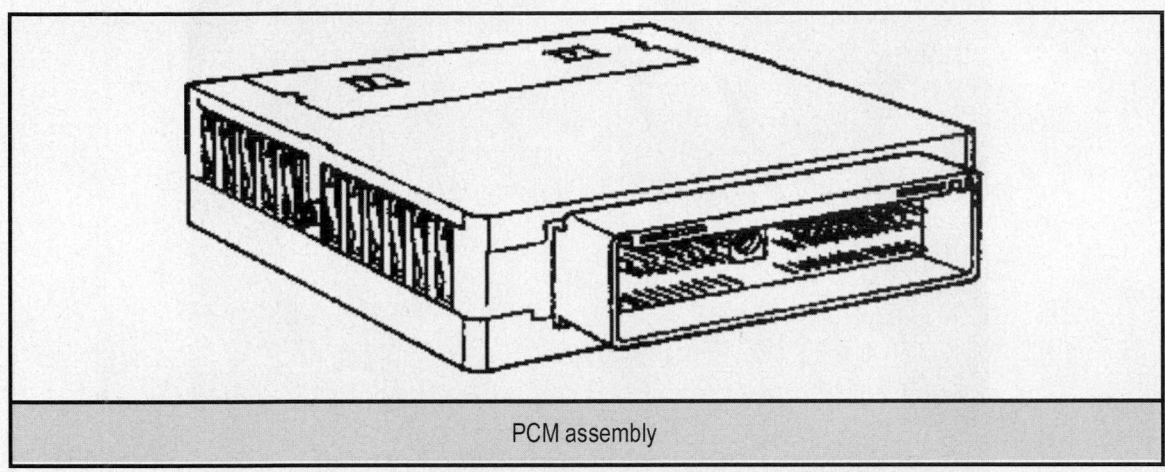

PCM assembly

information from various engine sensors and computes the required fuel flow rate necessary to maintain the correct amount of air/fuel ratio throughout the entire engine operational range.

Based on the information that is received and programmed into the PCM's memory, the PCM generates output signals to control relays, actuators and solenoids. The PCM also sends out a command to the fuel injectors that meters the appropriate quantity of fuel. The module automatically senses and compensates for any changes in altitude when driving your vehicle.

HEATED OXYGEN SENSOR (HO2S)
Operation

The oxygen (O2) sensor is a device which produces an electrical voltage when exposed to the oxygen present in the exhaust gases. The sensor is mounted in the exhaust system, usually in the manifold or a boss located on the down pipe before the catalyst.. The oxygen sensors used on the Ford Contour/Mercury Mystique/Mercury Cougar are electrically heated internally for faster switching when the engine is started cold. The oxygen sensor produces a voltage within 0 and 1 volt. When there is a large amount of oxygen present (lean mixture), the sensor produces a low voltage (less than 0.4v). When there is a lesser amount present (rich mixture) it produces a higher voltage (0.6-1.0v). The stoichiometric or correct fuel to air ratio will read between 0.4 and 0.6v. By monitoring the oxygen content and converting it to electrical voltage, the sensor acts as a rich-lean switch. The voltage is transmitted to the PCM.

Some models have two sensors, one before the catalyst and one after. This is done for a catalyst efficiency monitor that is a part of the OBD-II engine controls. The one before the catalyst measures the exhaust emissions right out of the engine, and sends the signal to the PCM about the state of the mixture as previously talked about. The second sensor reports the difference in the emissions after the exhaust gases have gone through the catalyst. This sensor reports to the PCM the amount of emissions reduction the catalyst is performing.

The oxygen sensor will not work until a predetermined temperature is reached, until this time the PCM is running in what as known as OPEN LOOP operation. OPEN LOOP means that the PCM has not yet begun to correct the air-to-fuel ratio by reading the oxygen sensor. After the engine comes to operating temperature, the PCM will monitor the oxygen sensor and correct the air/fuel ratio from the sensor's readings. This is what is known as CLOSED LOOP operation.

A heated oxygen sensor (HO2S) has a heating element that keeps the sensor at proper operating temperature during all operating modes. Maintaining correct sensor temperature at all times allows the system to enter into CLOSED LOOP operation sooner.

In CLOSED LOOP operation the PCM monitors the sensor input (along with other inputs) and adjusts injector pulse width accordingly. During OPEN LOOP operation the PCM ignores the sensor input and adjusts the injector pulse to a preprogrammed value based on other inputs.

Testing

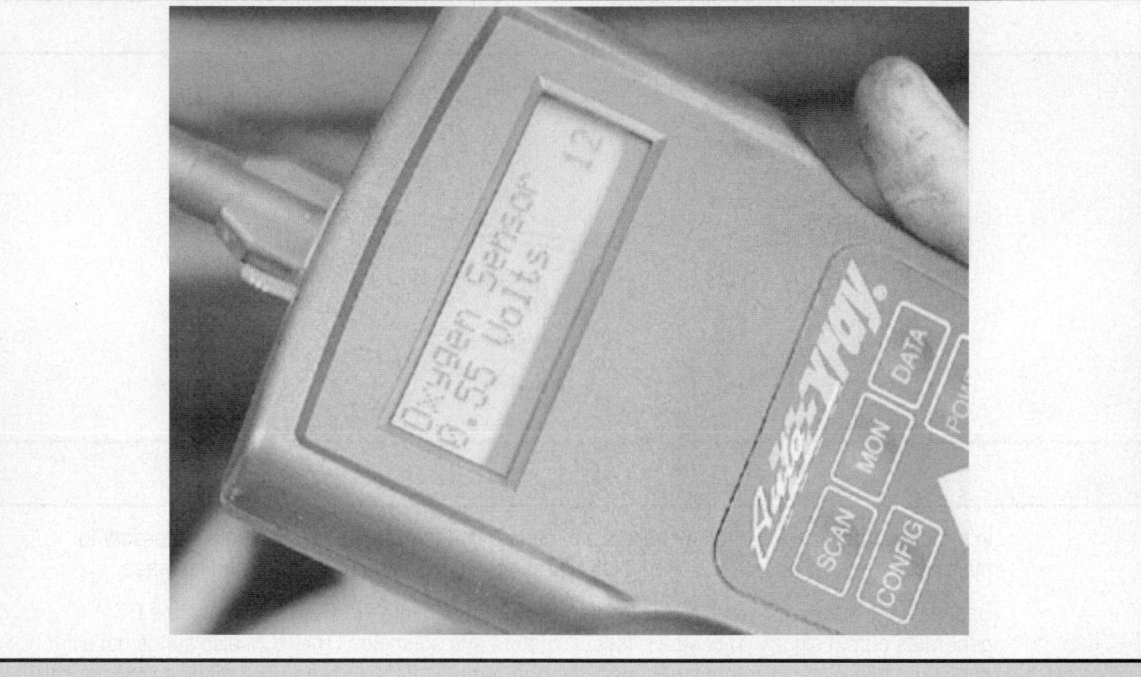

The HO2S can be monitored with an appropriate and Data-stream capable scan tool

1. Disconnect the HO2S.

2. Measure the resistance between PWR and GND terminals of the sensor. Resistance should be approximately 6 ohms at 68°F (20°C). If resistance is not within specification, the sensor's heater element is faulty.

3. With the HO2S connected and engine running, measure the voltage with a Digital Volt-Ohmmeter (DVOM) between terminals HO2S and SIG RTN (GND) of the oxygen sensor connector. Voltage should fluctuate between 0.01-1.1 volts. If voltage fluctuation is slow or voltage is not within specification, the sensor may be faulty.

<u>IDLE AIR CONTROL VALVE</u>
Operation

The Idle Air Control (IAC) valve adjusts the engine idle speed. The valve is located on the side of the throttle body. The valve is controlled by a duty cycle signal from the PCM and allows air to bypass the throttle plate in order to maintain the proper idle speed.

The IAC is located at the top of the upper intake manifold adjacent to the throttle body.

Do NOT attempt to clean the IAC valve. Carburetor tune-up cleaners or any type of solvent cleaners will damage the internal components of the valve.

Testing

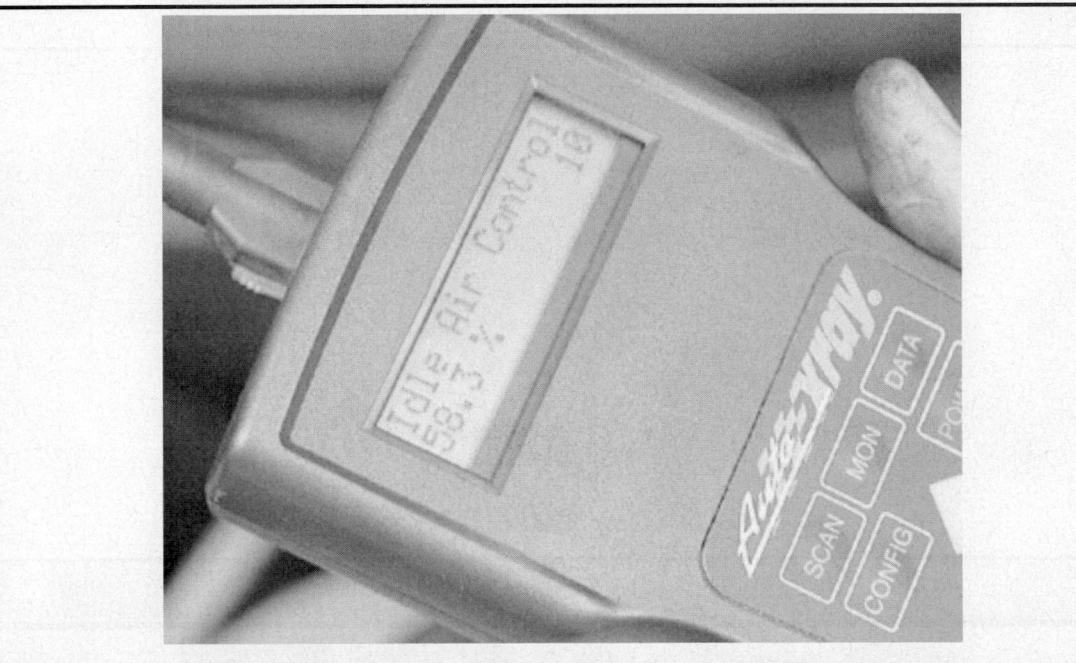

The IAC can be monitored with an appropriate and Data-stream capable scan tool

1. Turn the ignition switch to the OFF position.

2. Disconnect the wiring harness from the IAC valve.

3. Measure the resistance between the terminals of the valve.

Due to the diode in the solenoid, place the ohmmeter positive lead on the VPWR terminal and the negative lead on the ISC terminal.

4. Resistance should be 6-13 ohms.

5. If resistance is not within specification, the valve may be faulty.

<u>COOLANT TEMPERATURE SENSOR</u>
Operation

The Engine Coolant Temperature (ECT) sensor resistance changes in response to engine coolant temperature. The sensor resistance decreases as the coolant temperature increases, and increases as the coolant temperature decreases. This provides a reference signal to the PCM, which indicates engine coolant temperature. The signal sent to the PCM by the ECT sensor helps the PCM to determine spark advance, EGR flow rate, air/fuel ratio, and engine temperature. The ECT also is used for temperature gauge operation by sending its signal to the instrument cluster.

The ECT is a two wire sensor, a 5-volt reference signal is sent to the sensor and the signal return is based upon the change in the measured resistance due to temperature.

Testing

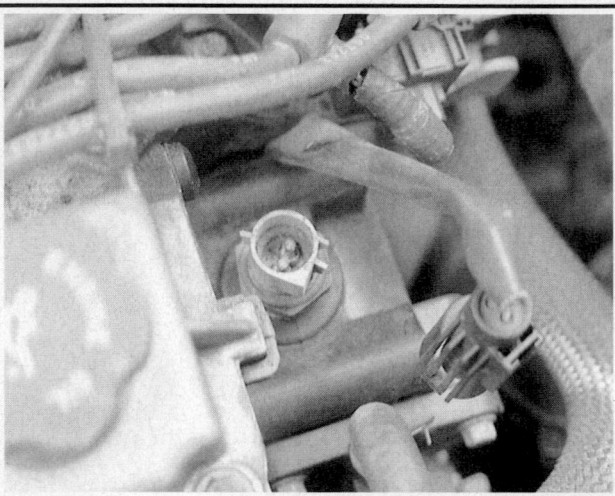

Unplug the ECT sensor to access the sensor

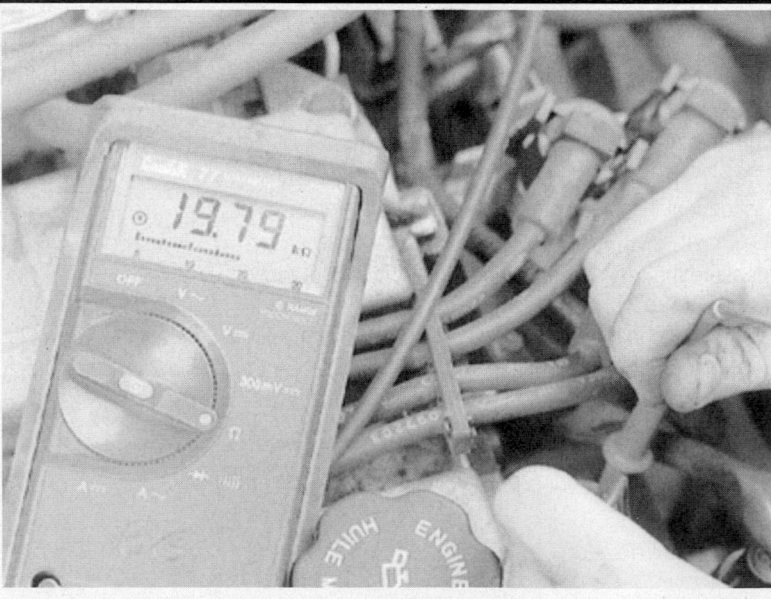

Test the resistance of the ECT sensor across the two sensor pins

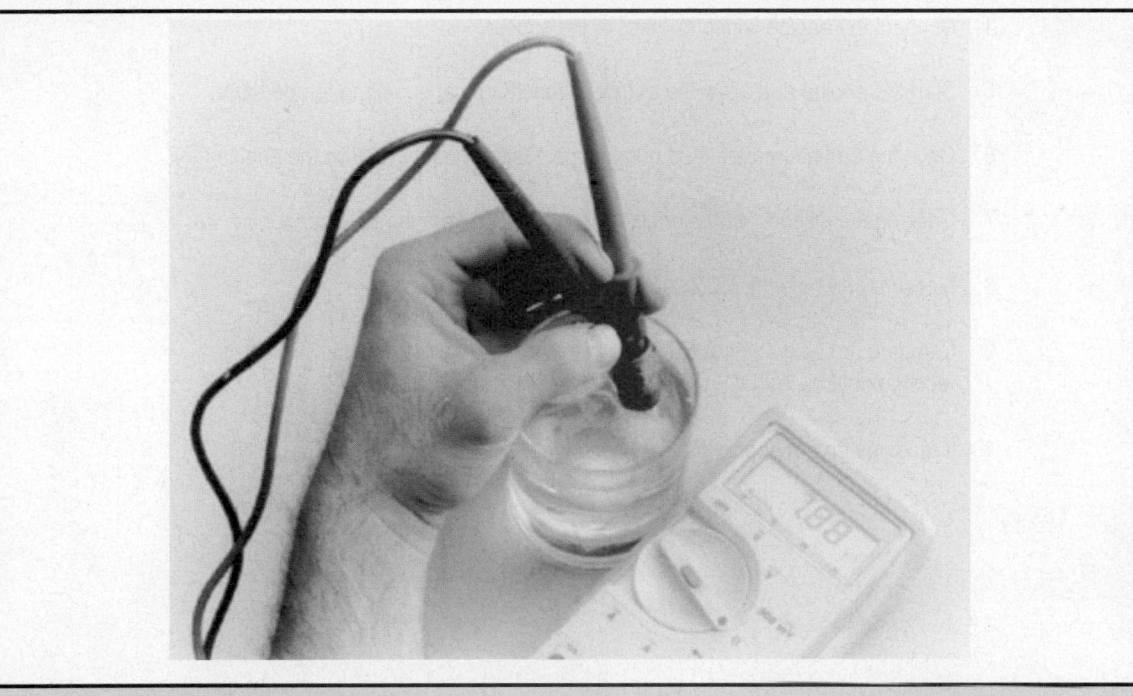

Another method of testing the ECT is to submerge it in cold or hot water and check resistance

The ECT can be monitored with an appropriate and Data-stream capable scan tool

1. Disconnect the engine wiring harness from the ECT sensor.

2. Connect an ohmmeter between the ECT sensor terminals.

3. With the engine cold and the ignition switch in the OFF position, measure and note the ECT sensor resistance.

4. Connect the engine wiring harness to the sensor.

5. Start the engine and allow the engine to reach normal operating temperature.

6. Once the engine has reached normal operating temperature, turn the engine OFF.

7. Once again, disconnect the engine wiring harness from the ECT sensor.

8. Measure and note the ECT sensor resistance with the engine hot.

9. Compare the cold and hot ECT sensor resistance measurements with the accompanying chart.

10. If readings do not approximate those in the chart, the sensor may be faulty.

INTAKE AIR TEMPERATURE SENSOR
Operation

The tip of the IAT sensor has an exposed thermistor that changes the
resistance of the sensor based upon the force of the air rushing past it

The Intake Air Temperature (IAT) sensor determines the air temperature inside the intake manifold. Resistance changes in response to the ambient air temperature. The sensor has a negative temperature coefficient. As the temperature of the sensor rises, the resistance across the sensor decreases. This provides a signal to the PCM indicating the temperature of the incoming air charge. This sensor helps the PCM to determine spark timing and air/fuel ratio. Information from this sensor is added to the pressure sensor information to calculate the air mass being sent to the cylinders. The IAT is a two wire sensor, a 5-volt reference signal is sent to the sensor and the signal return is based upon the change in the measured resistance due to temperature.

Testing

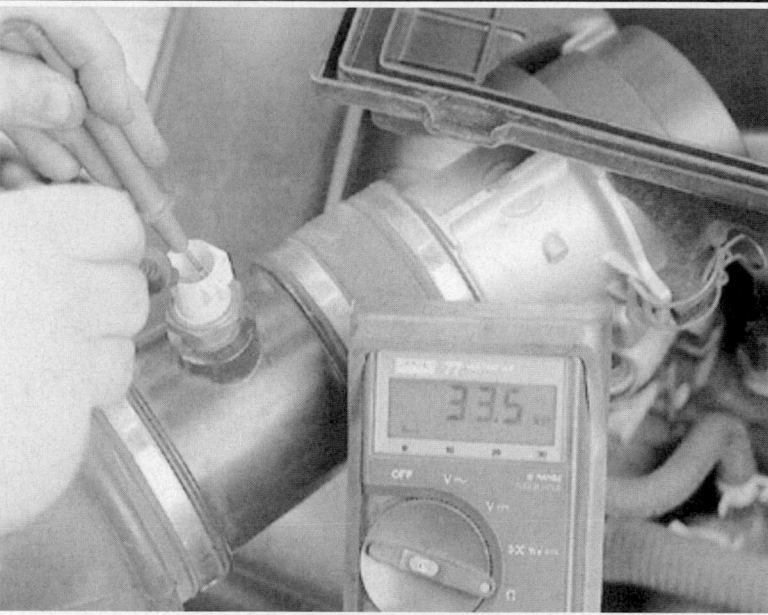

Test the resistance of the IAT sensor across the two sensor pins

The IAT sensor can be monitored with an appropriate and Data-stream capable scan tool

1. Turn the ignition switch OFF.

2. Disconnect the wiring harness from the IAT sensor.

3. Measure the resistance between the sensor terminals.

4. Compare the resistance reading with the accompanying chart.

5. If the resistance is not within specification, the IAT may be faulty.

6. Connect the wiring harness to the sensor.

MASS AIR FLOW SENSOR
Operation

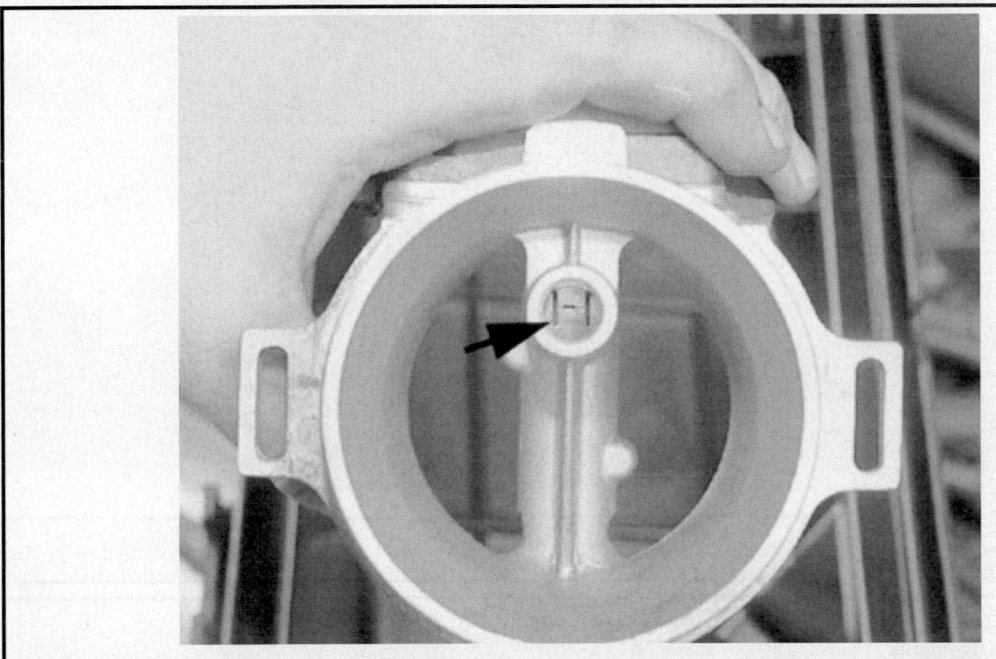

The exposed "hot wire" of the MAF sensor

The Mass Air Flow (MAF) sensor directly measures the mass of air being drawn into the engine. The sensor output is used to calculate injector pulse width. The MAF sensor is what is referred to as a "hot-wire sensor". The sensor uses a thin platinum wire filament, wound on a ceramic bobbin and coated with glass, that is heated to 200°C (417°F) above the ambient air temperature and subjected to the intake airflow stream. A "cold-wire" is used inside the MAF sensor to determine the ambient air temperature.

Battery voltage from the EEC power relay, and a reference signal and a ground signal from the PCM are supplied to the MAF sensor. The sensor returns a signal proportionate to the current flow required to keep the "hot-wire" at the required temperature. The increased airflow across the "hot-wire" acts as a cooling fan, lowering the resistance and requiring more current to maintain the temperature of the wire. The increased current is measured by the voltage in the circuit, as current increases, voltage increases. As the airflow increases the signal return voltage of a normally operating MAF sensor will increase.

Testing

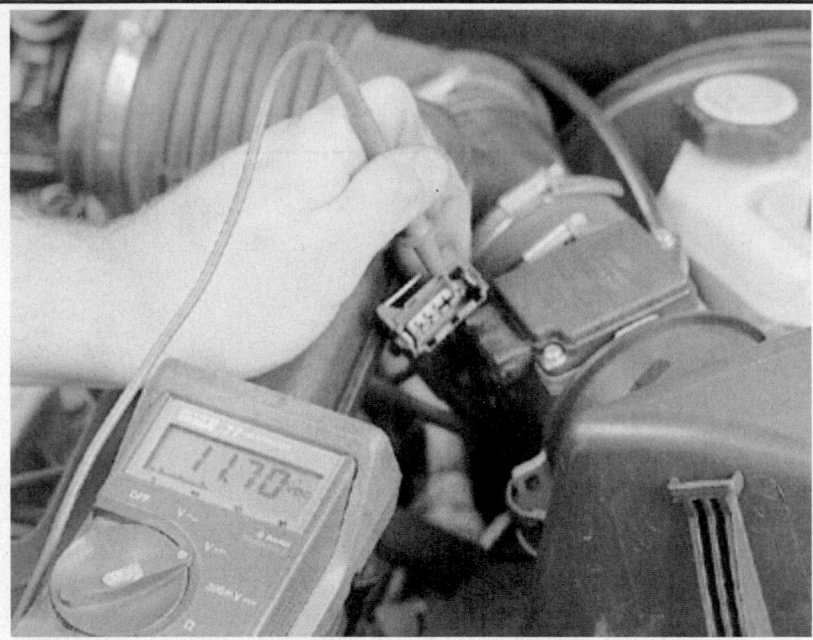

Testing the VPWR circuit to the MAF sensor

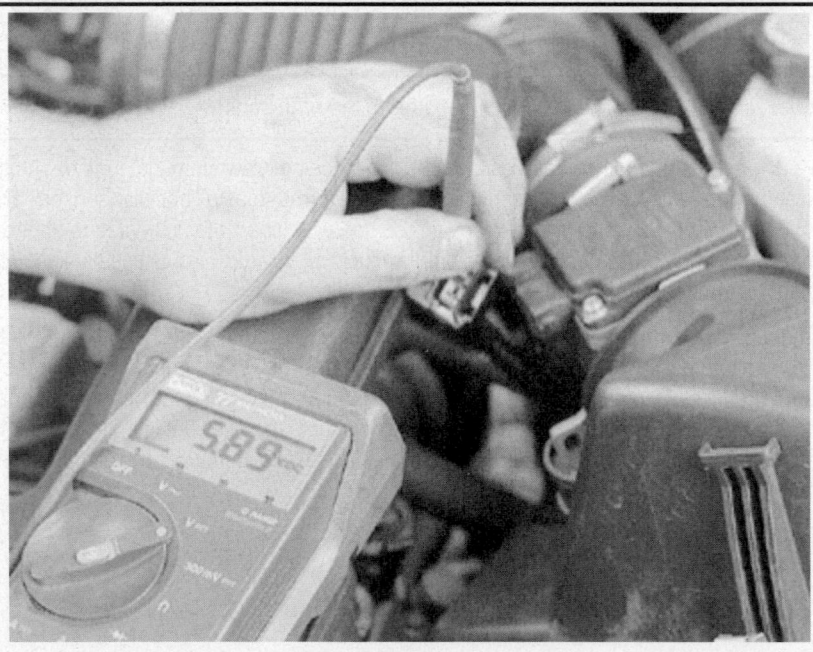

Testing the SIG circuit to the MAF sensor

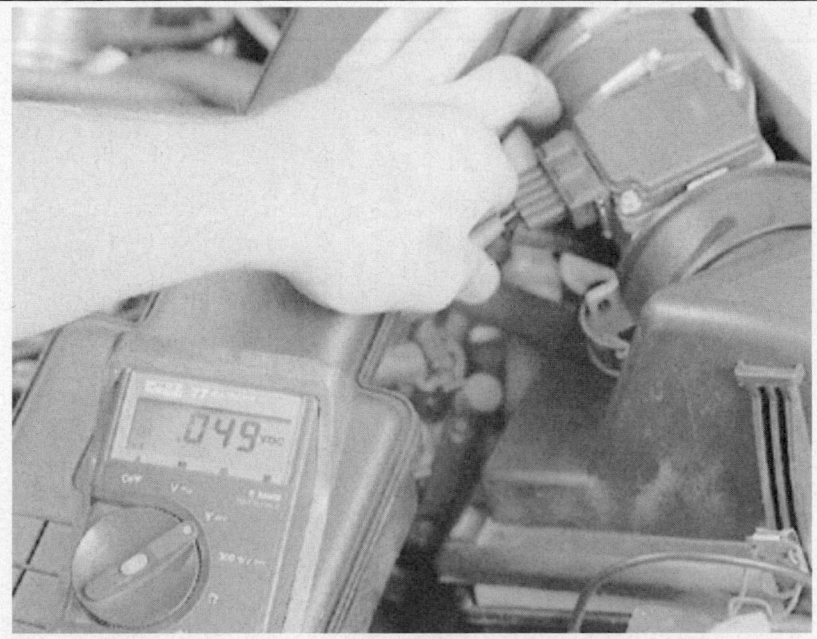

Testing the SIG RTN circuit to the MAF sensor

1. Using a multimeter, check for voltage by backprobing the MAF sensor connector.

2. With the key ON, and the engine OFF, verify that there is at least 10.5 volts between the VPWR and GND terminals of the MAF sensor connector. If voltage is not within specification, check power and ground circuits and repair as necessary.

3. With the key ON, and the engine ON, verify that there is at least 4.5 volts between the SIG and GND terminals of the MAF sensor connector. If voltage is not within specification, check power and ground circuits and repair as necessary.

4. With the key ON, and the engine ON, check voltage between GND and SIG RTN terminals. Voltage should be approximately 0.34-1.96 volts. If voltage is not within specification, the sensor may be faulty.

THROTTLE POSITION SENSOR
Operation

The Throttle Position (TP) sensor is a potentiometer that provides a signal to the PCM that is directly proportional to the throttle plate position. The TP sensor is mounted on the side of the throttle body and is connected to the throttle plate shaft. The TP sensor monitors throttle plate movement and position, and transmits an appropriate electrical signal to the PCM. These signals are used by the PCM to adjust the air/fuel mixture, spark timing and EGR operation according to engine load at idle, part throttle, or full throttle. The TP sensor is not adjustable.

The TP sensor receives a 5 volt reference signal and a ground circuit from the PCM. A return signal circuit is connected to wiper that runs on a resistor internally on the sensor. The further the throttle is opened, the wiper moves along the resistor, at wide open throttle, the wiper essentially creates a loop between the reference signal and the signal return returning the full or nearly full 5 volt signal back to the PCM. At idle the signal return should be approximately 0.9 volts.

Testing

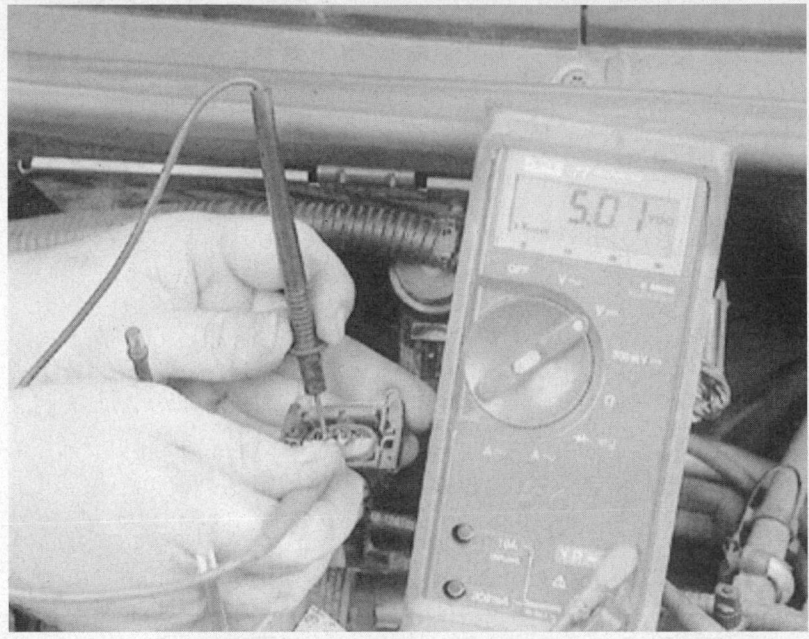

Testing the SIG circuit to the TP sensor

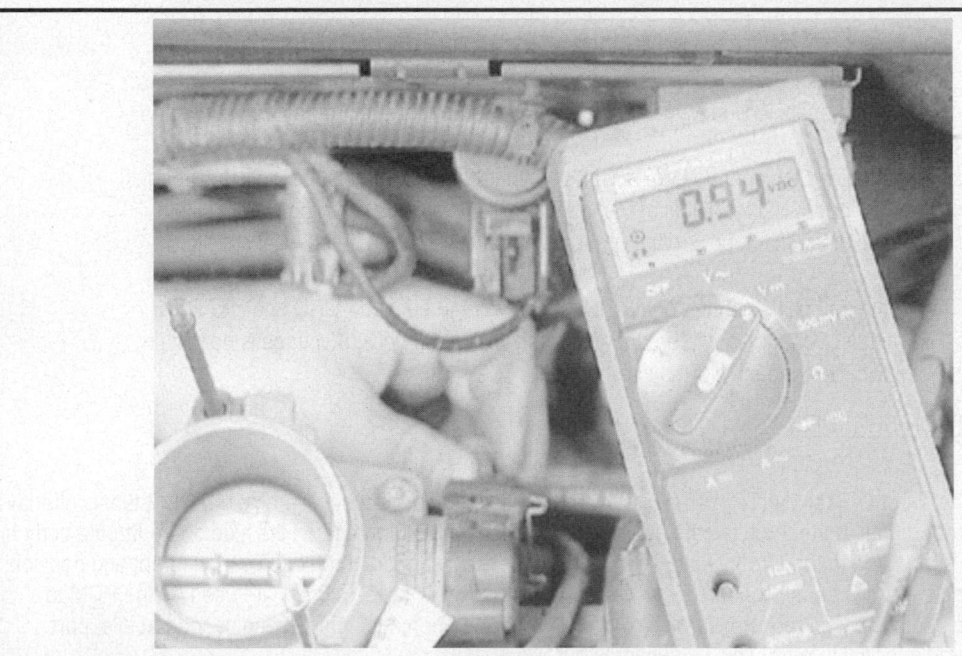

Testing the SIG RTN circuit of the TP sensor

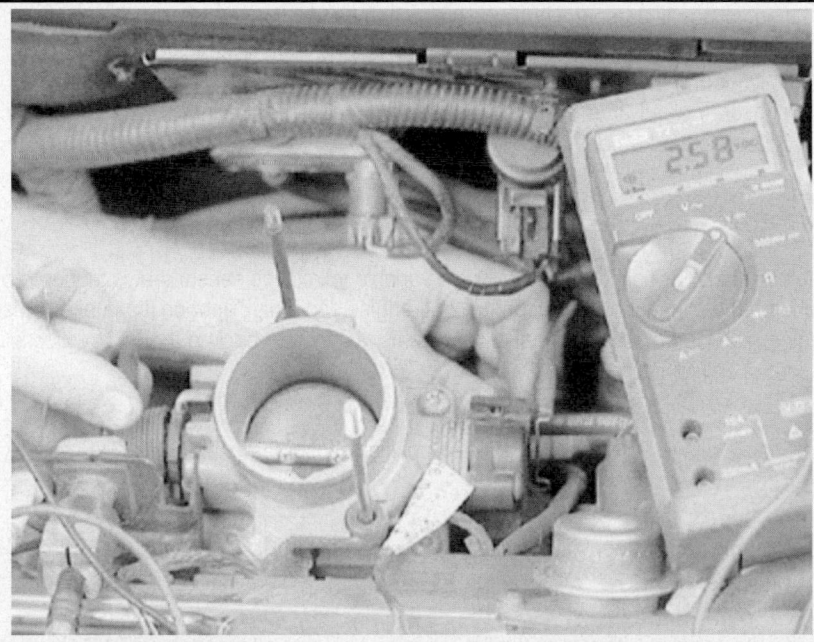

Testing the operation of the potentiometer inside the TP sensor while slowly opening the throttle

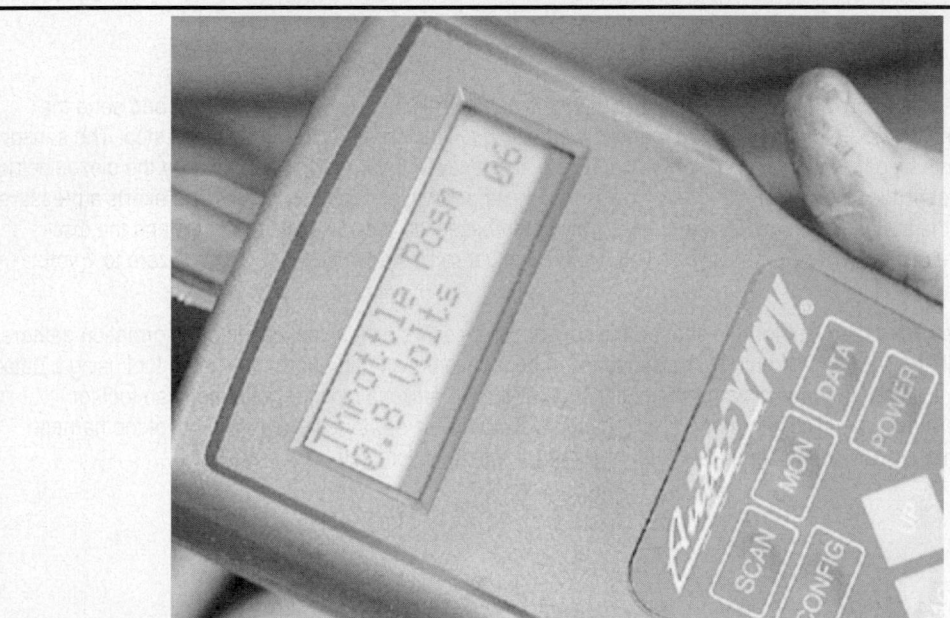

The TP sensor can be monitored with an appropriate and Data-stream capable scan tool

1. With the engine OFF and the ignition ON, check the voltage at the signal return
   circuit of the TP sensor by carefully backprobing the connector using a DVOM.

2. Voltage should be between 0.2 and 1.4 volts at idle.

3. Slowly move the throttle pulley to the wide open throttle (WOT) position and watch
   the voltage on the DVOM. The voltage should slowly rise to slightly less than 4.8v
   at Wide Open Throttle (WOT).

4. If no voltage is present, check the wiring harness for supply voltage (5.0v) and ground (0.3v or less), by referring to your corresponding wiring guide. If supply voltage and ground are present, but no output voltage from TP, replace the TP sensor. If supply voltage and ground do not meet specifications, make necessary repairs to the harness or PCM.

CAMSHAFT POSITION SENSOR
Operation

The camshaft position sensor (CMP) is a variable reluctance sensor that is triggered by a high point on the left-hand exhaust camshaft on the 2.5L engine and a high spot on the intake camshaft on the 2.0L engine. The CMP sends a signal relating camshaft position back to the PCM which is used by the PCM to control engine timing.

Testing

1. Check voltage between the camshaft position sensor terminals PWR GND and CID.

2. With engine running, voltage should be greater than 0.1 volt AC and vary with engine speed.

3. If voltage is not within specification, check for proper voltage at the VPWR terminal.

4. If VPWR voltage is greater than 10.5 volts, sensor may be faulty.

KNOCK SENSOR
Operation

The operation of the Knock Sensor (KS) is to monitor preignition or "engine knocks" and send the signal to the PCM. The PCM responds by adjusting ignition timing until the "knocks" stop. The sensor works by generating a signal produced by the frequency of the knock as recorded by the piezoelectric ceramic disc inside the KS. The disc absorbs the shock waves from the knocks and exerts a pressure on the metal diaphragm inside the KS. This compresses the crystals inside the disc and the disc generates a voltage signal proportional to the frequency of the knocks ranging from zero to 1 volt.

Testing

There is real no test for this sensor, the sensor produces it's own signal based on information gathered while the engine is running. The sensors can be monitored with an appropriate scan tool using a data display or other data stream information. Follow the instructions included with the scan tool for information on accessing the data. The only tests available are to test the continuity of the harness from the PCM to the sensor, and the resistance of the sensor itself.

VEHICLE SPEED SENSOR
Operation

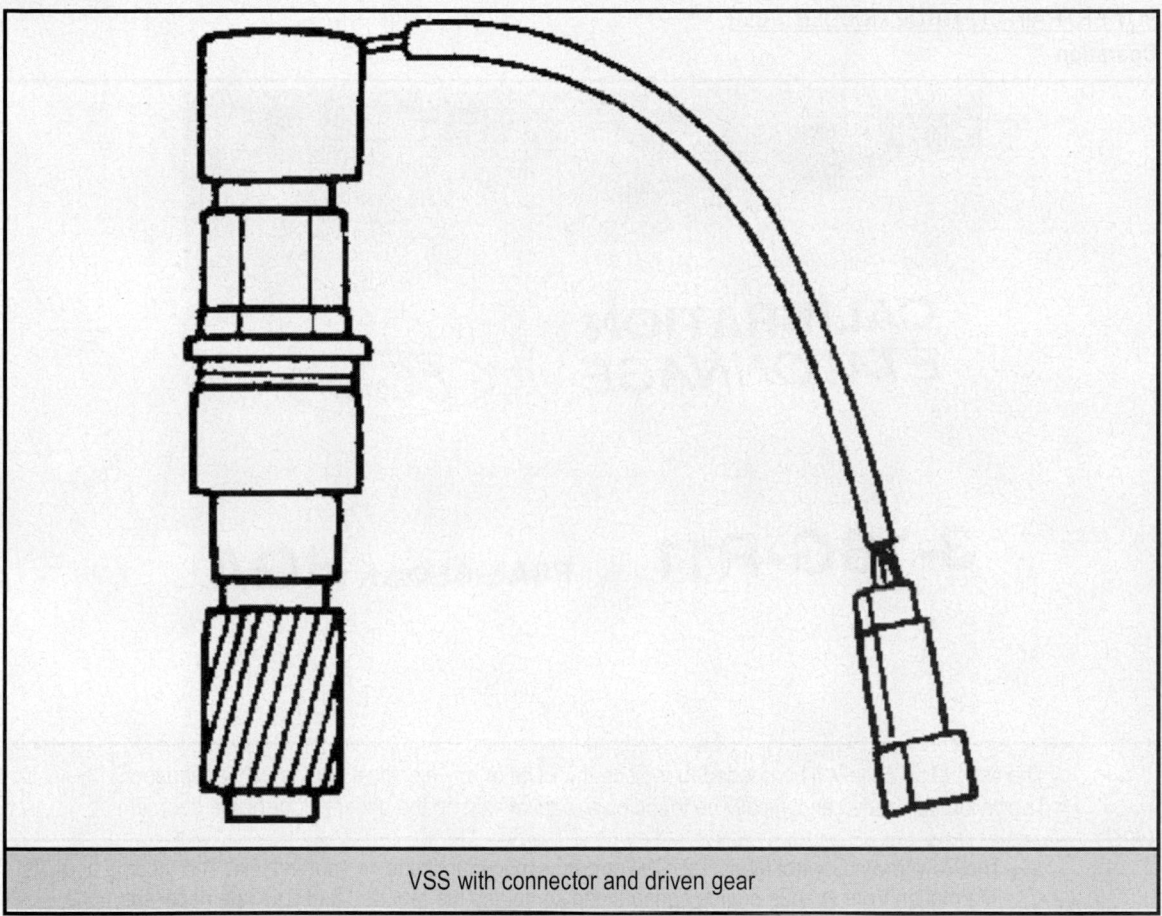

VSS with connector and driven gear

The Vehicle Speed Sensor (VSS) is a magnetic pick-up sensor that sends a signal to the Powertrain Control Module (PCM) and the speedometer. The sensor measures the rotation of the output shaft on the transaxle and sends an AC voltage signal to the PCM which determines the corresponding vehicle speed.

Testing

1. Disconnect the negative battery cable.

2. Disengage the wiring harness connector from the VSS.

3. Using a Digital Volt-Ohmmeter (DVOM), measure the resistance (ohmmeter function) between the sensor terminals. If the resistance is 190-250 ohms, the sensor is okay.

Crown Victoria, Grand Marquis & Town Car

POWERTRAIN CONTROL MODULE (PCM)
Operation

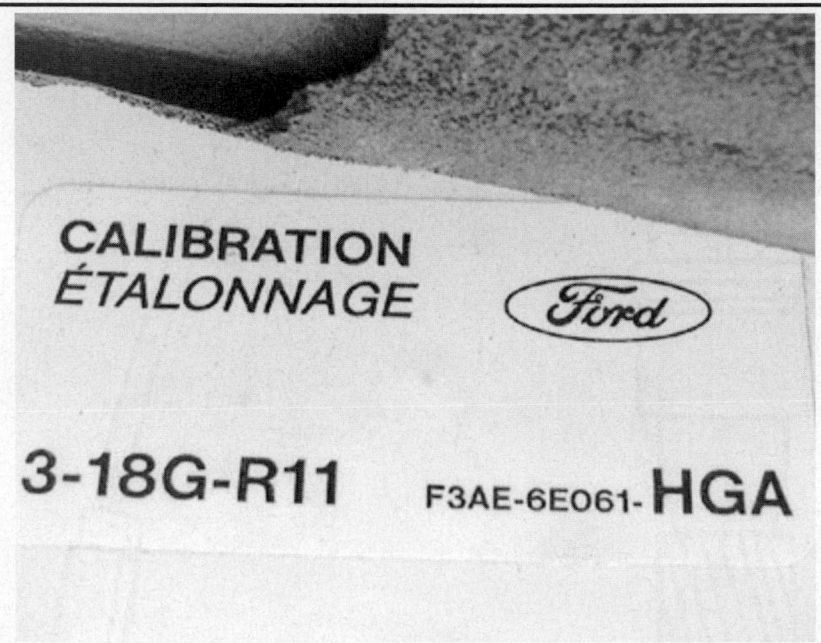

CALIBRATION
ÉTALONNAGE   *Ford*

3-18G-R11   F3AE-6E061-**HGA**

The calibration code will be needed to replace the PCM or any electrical engine control sensors.
The calibration code can typically be found on a label affixed on the driver's door or the door jam

The Powertrain Control Module (PCM) performs many functions on your vehicle. The module accepts information from various engine sensors and computes the required fuel flow rate necessary to maintain the correct amount of air/fuel ratio throughout the entire engine operational range.

Based on the information that is received and programmed into the PCM's memory, the PCM generates output signals to control relays, actuators and solenoids. The PCM also sends out a command to the fuel injectors that meters the appropriate quantity of fuel. The module automatically senses and compensates for any changes in altitude when driving your vehicle.

OXYGEN SENSOR
Operation

The oxygen (O2) sensor is a device which produces an electrical voltage when exposed to the oxygen present in the exhaust gases. The sensor is mounted in the exhaust system, usually in the manifold or a boss located on the down pipe before the catalyst.. Some of the oxygen sensors used on the Ford Crown Victoria/Mercury Grand Marquis are electrically heated internally for faster switching when the engine is started cold. The oxygen sensor produces a voltage within 0 and 1 volt. When there is a large amount of oxygen present (lean mixture), the sensor produces a low voltage (less than 0.4v). When there is a lesser amount present (rich mixture) it produces a higher voltage (0.6 -1.0v).The stoichiometric or correct fuel to air ratio will read between 0.4 and 0.6v. By monitoring the oxygen content and converting it to electrical voltage, the sensor acts as a rich-lean switch. The voltage is transmitted to the PCM.

Some models have two or more sensors, before the catalyst and after. This is done for a catalyst efficiency monitor that is a part of the OBD-II engine controls that are on all models from the 1995 model year on. The sensor before the catalyst measures the exhaust emissions right out of the engine, and sends the signal to the PCM about the state of the mixture as previously talked about. The second sensor reports the difference in the emissions after the exhaust gases have gone through the catalyst. This sensor reports to the PCM the amount of emissions reduction the catalyst is performing.

The oxygen sensor will not work until a predetermined temperature is reached, until this time the PCM is running in what as known as OPEN LOOP operation. OPEN LOOP means that the PCM has not yet begun to correct the air-to-fuel ratio by reading the oxygen sensor. After the engine comes to operating temperature, the PCM will monitor the oxygen sensor and correct the air/fuel ratio from the sensor's readings. This is what is known as CLOSED LOOP operation.

A heated oxygen sensor (HO2S) has a heating element that keeps the sensor at proper operating temperature during all operating modes. Maintaining correct sensor temperature at all times allows the system to enter into CLOSED LOOP operation sooner.

In CLOSED LOOP operation the PCM monitors the sensor input (along with other inputs) and adjusts the injector pulse width accordingly. During OPEN LOOP operation the PCM ignores the sensor input and adjusts the injector pulse to a preprogrammed value based on other inputs.

Testing

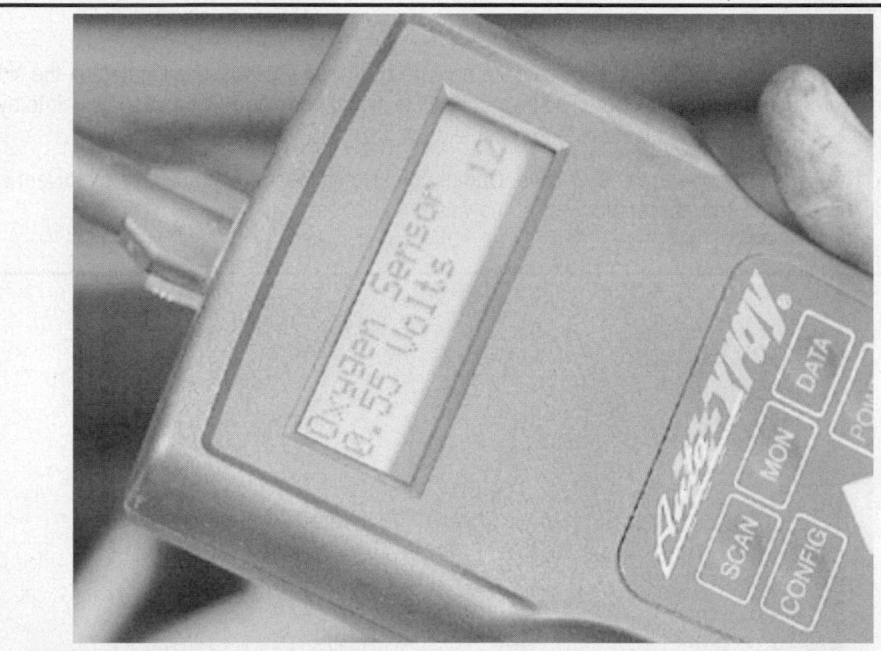

The HO2S can be monitored with an appropriate and Data-stream capable scan tool

1. Disconnect the HO2S.

2. Measure the resistance between PWR and GND terminals of the sensor. Resistance should be approximately 6 ohms at 68 °F (20 °C). If resistance is not within specification, the sensor's heater element is faulty.

3. With the HO2S connected and engine running, measure the voltage with a Digital Volt-Ohmmeter (DVOM) between terminals HO2S and SIG RTN (GND) of the oxygen sensor connector. Voltage should fluctuate between 0.01 -1.0 volts. If voltage fluctuation is slow or voltage is not within specification, the sensor may be faulty.

IDLE AIR CONTROL VALVE

Operation

The Idle Air Control (IAC) valve adjusts the engine idle speed. The valve is located on the side of the throttle body. The valve is controlled by a duty cycle signal from the PCM and allows air to bypass the throttle plate in order to maintain the proper idle speed.

Do NOT attempt to clean the IAC valve. Carburetor tune-up cleaners or any type of solvent cleaners will damage the internal components of the valve.

Testing

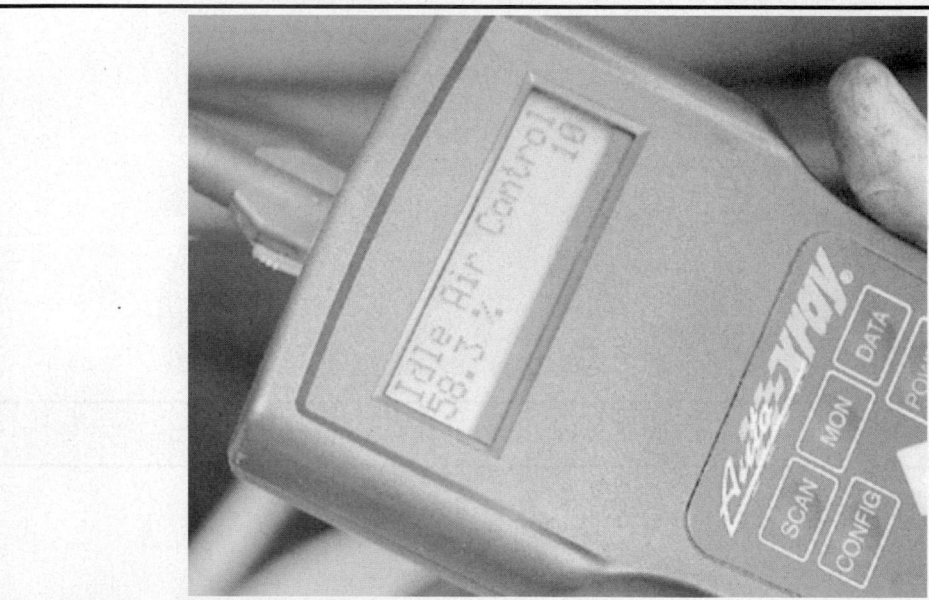

The IAC can be monitored with an appropriate and Data-stream capable scan tool

1. Turn the ignition switch to the OFF position.

2. Disconnect the wiring harness from the IAC valve.

3. Measure the resistance between the terminals of the valve.

Due to the diode in the solenoid, place the ohmmeter positive lead on the VPWR terminal and the negative lead on the ISC terminal.

4. Resistance should be 6-13 ohms.

5. If resistance is not within specification, the valve may be faulty.

ENGINE COOLANT TEMPERATURE (ECT) SENSOR
Operation

The Engine Coolant Temperature (ECT) sensor resistance changes in response to engine coolant temperature. The sensor resistance decreases as the coolant temperature increases, and increases as the coolant temperature decreases. This provides a reference signal to the PCM, which indicates engine coolant temperature. The signal sent to the PCM by the ECT sensor helps the PCM to determine spark advance, EGR flow rate, air/fuel ratio, and engine temperature. The ECT is a two wire sensor, a 5-volt reference signal is sent to the sensor and the signal return is based upon the change in the measured resistance due to temperature.

Testing

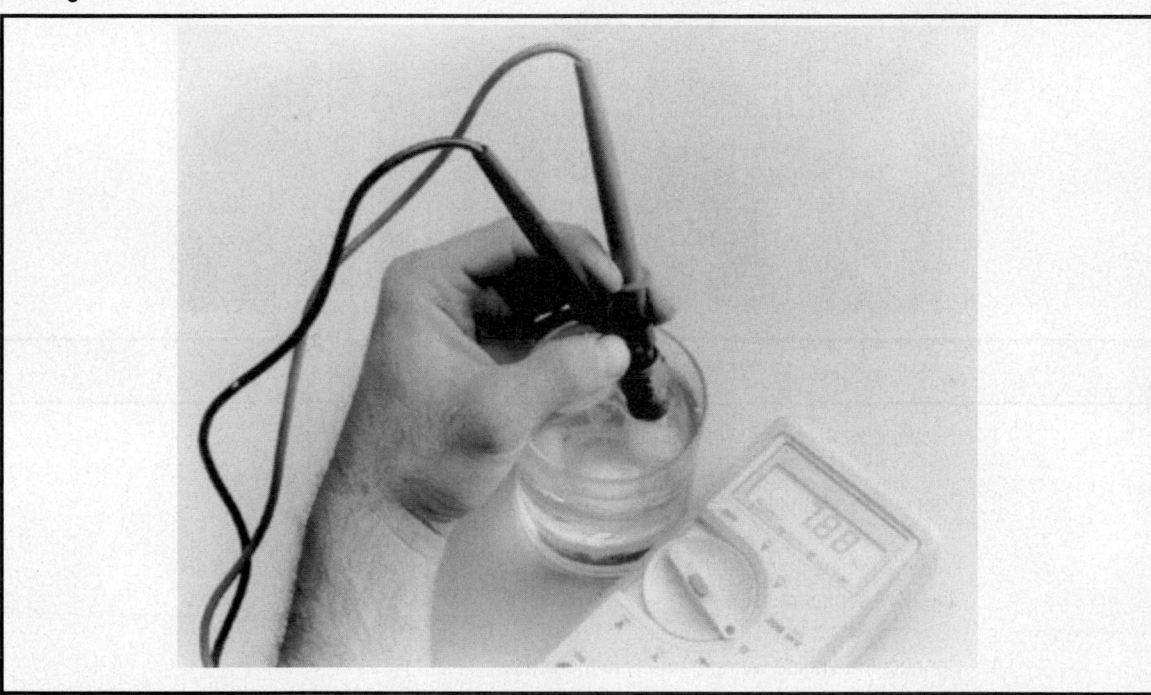

Another method of testing the ECT is to submerge it in cold or hot water and check resistance

Test the ECT resistance across the two sensor terminals

1. Disconnect the engine wiring harness from the ECT sensor.

2. Connect an ohmmeter between the ECT sensor terminals.

3. With the engine cold and the ignition switch in the OFF position, measure and note the ECT sensor resistance.

4. Connect the engine wiring harness to the sensor.

5. Start the engine and allow the engine to reach normal operating temperature.

6. Once the engine has reached normal operating temperature, turn the engine OFF.

7. Once again, disconnect the engine wiring harness from the ECT sensor.

8. Measure and note the ECT sensor resistance with the engine hot.

9. Compare the cold and hot ECT sensor resistance measurements with the accompanying chart.

10. If readings do not approximate those in the chart, the sensor may be faulty.

INTAKE AIR TEMPERATURE SENSOR
Operation

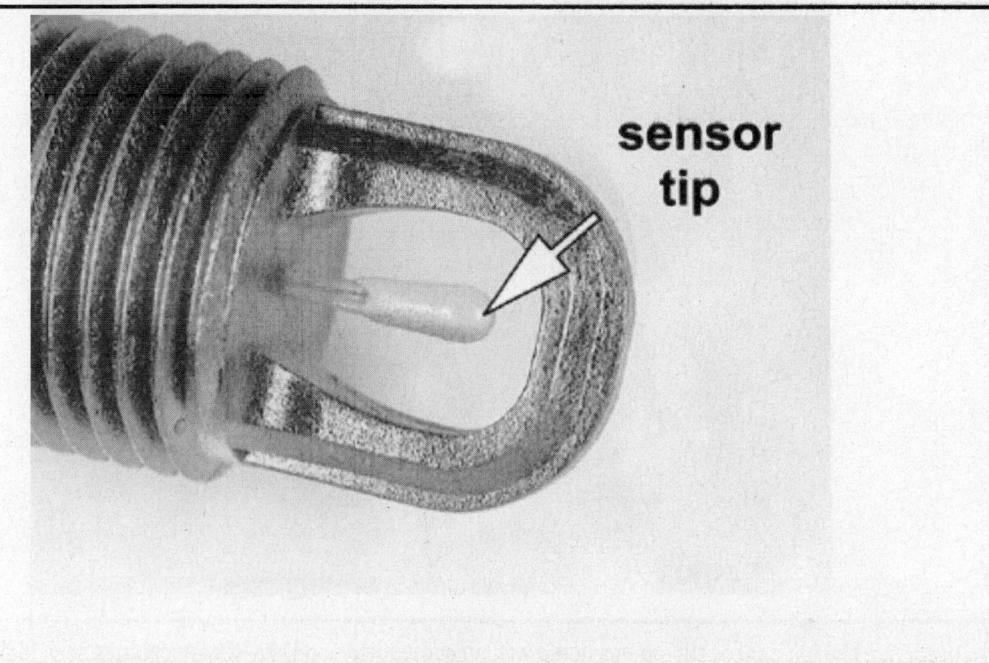

The tip of the IAT sensor has an exposed thermistor that changes the resistance of the sensor based upon the force of the air rushing past it

The Intake Air Temperature (IAT) sensor determines the air temperature inside the intake manifold. Resistance changes in response to the ambient air temperature. The sensor has a negative temperature coefficient. As the temperature of the sensor rises the resistance across the sensor decreases. This provides a signal to the PCM indicating the temperature of the incoming air charge. This sensor helps the PCM to determine spark timing and air/fuel ratio. Information from this sensor is added to the pressure sensor information to calculate the air mass being sent to the cylinders. The IAT is a two wire sensor, a 5-volt reference signal is sent to the sensor and the signal return is based upon the change in the measured resistance due to temperature.

Testing

The IAT sensor can be monitored with an appropriate and Data-stream capable scan tool

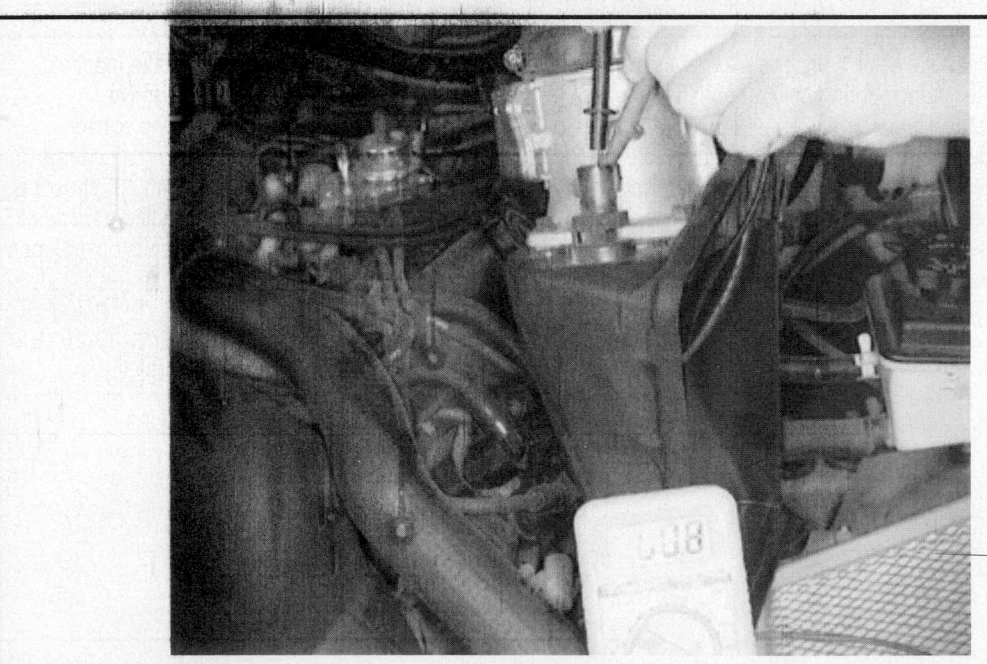

Measure the resistance of the IAT sensor across the two sensor pins

1. Turn the ignition switch OFF.

2. Disconnect the wiring harness from the IAT sensor.

3. Measure the resistance between the sensor terminals.

4. Compare the resistance reading with the accompanying chart.

5. If the resistance is not within specification, the IAT may be faulty.

6. Connect the wiring harness to the sensor.

MASS AIRFLOW SENSOR
Operation

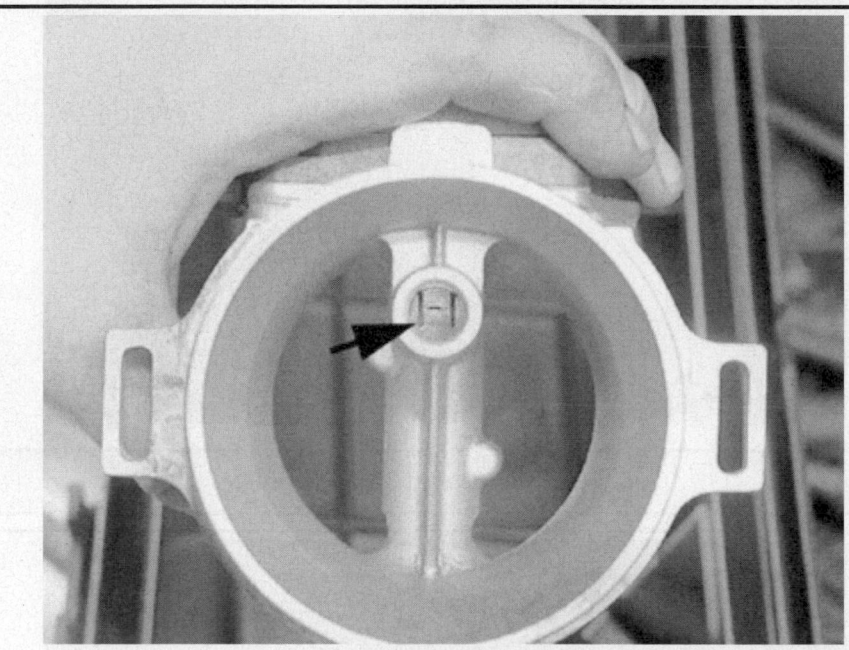

The exposed "hot wire" of the MAF sensor

The Mass Air Flow (MAF) sensor directly measures the mass of air being drawn into the engine. The sensor output is used to calculate injector pulse width. The MAF sensor is what is referred to as a "hot-wire sensor". The sensor uses a thin platinum wire filament, wound on a ceramic bobbin and coated with glass, that is heated to 200°C (392°F) above the ambient air temperature and subjected to the intake airflow stream. A "cold-wire" is used inside the MAF sensor to determine the ambient air temperature.

Battery voltage from the EEC power relay, and a reference signal and a ground signal from the PCM are supplied to the MAF sensor. The sensor returns a signal proportionate to the current flow required to keep the "hot-wire" at the required temperature. The increased airflow across the "hot-wire" acts as a cooling fan, lowering the resistance and requiring more current to maintain the temperature of the wire. The increased current is measured by the voltage in the circuit, as current increases, voltage increases. As the airflow increases the signal return voltage of a normally operating MAF sensor will increase.

Testing

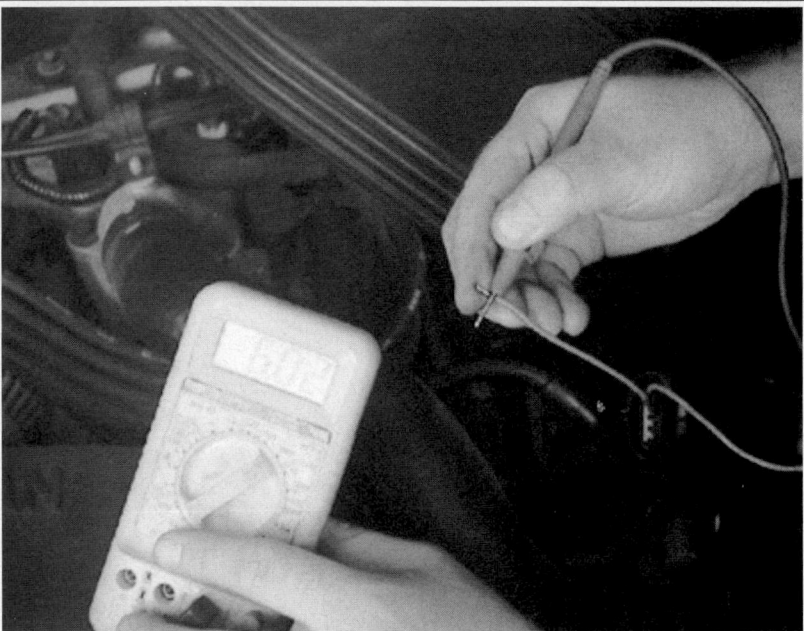

Testing the SIG circuit of the MAF sensor

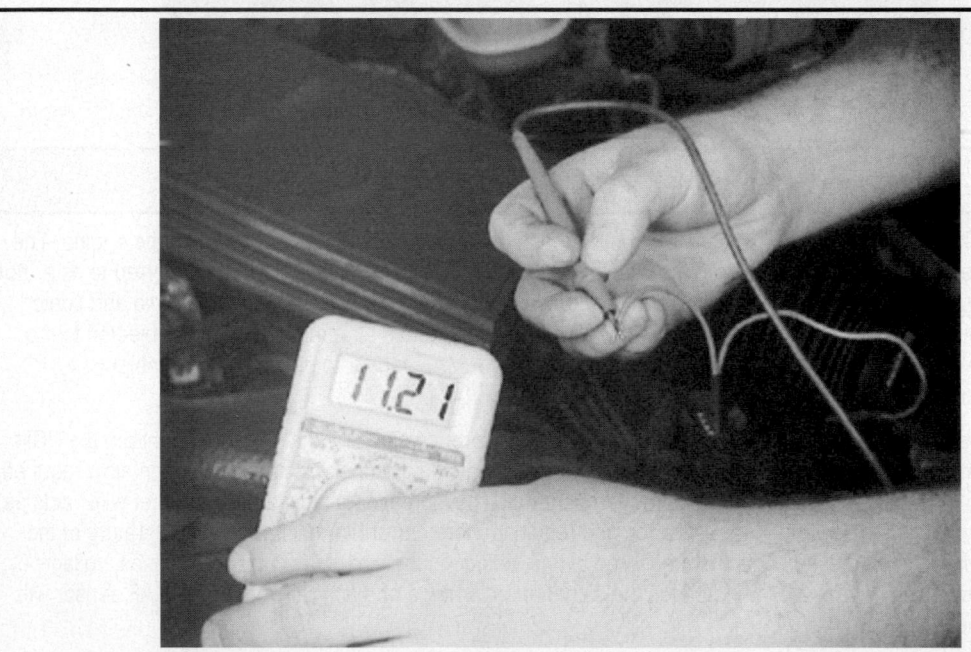

Testing the VPWR circuit of the MAF sensor

1. Using a multimeter, check for voltage by backprobing the MAF sensor connector.

2. With the key ON, and the engine OFF, verify that there is at least 10.5 volts between the VPWR and GND terminals of the MAF sensor connector. If voltage is not within specification, check power and ground circuits and repair as necessary.

3. With the key ON, and the engine ON, verify that there is at least 4.5 volts between the SIG and GND terminals of the MAF sensor connector. If voltage is not within specification, check power and ground circuits and repair as necessary.

4. With the key ON, and the engine ON, check voltage between GND and SIG RTN terminals. Voltage should be approximately 0.34-1.96 volts. If voltage is not within specification, the sensor may be faulty.

MANIFOLD ABSOLUTE PRESSURE (MAP) SENSOR

Operation

The Manifold Absolute Pressure (MAP) sensor is used on the 5.0L engine, except on 1990-91 sedans sold in California.

The most important information for measuring engine fuel requirements comes from the pressure sensor. Using the pressure and temperature data, the PCM calculates the intake air mass. It is connected to the engine intake manifold through a hose and takes readings of the absolute pressure. A piezoelectric crystal changes a voltage input to a frequency output which reflects the pressure in the intake manifold.

Atmospheric pressure is measured both when the engine is started and when driving fully loaded, then the pressure sensor information is adjusted accordingly.

Testing

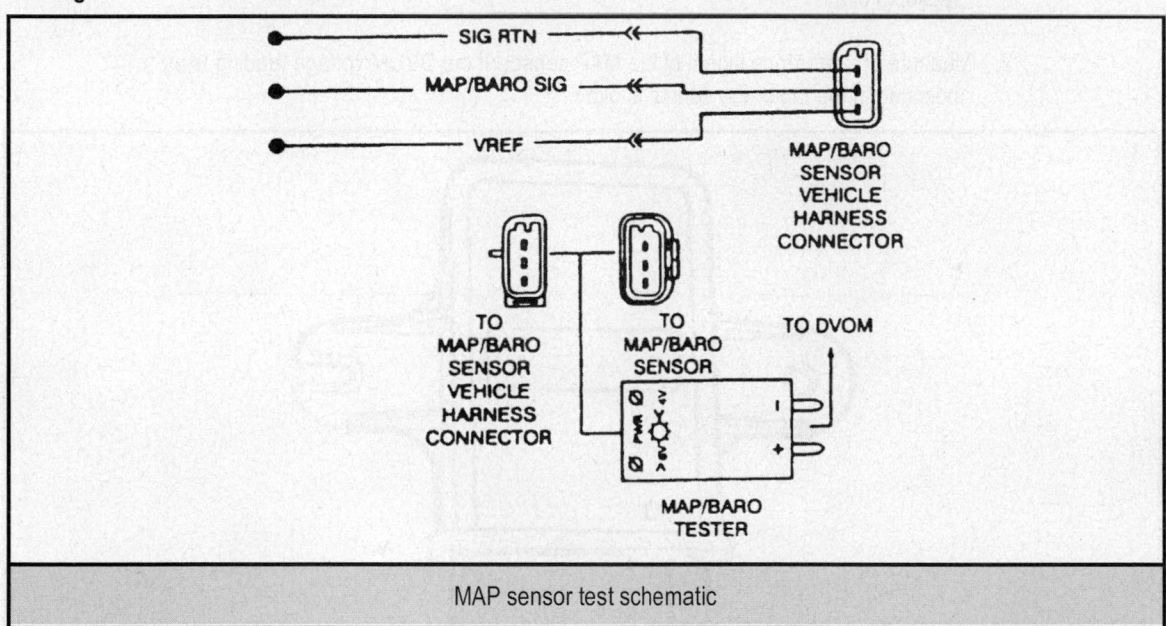

MAP sensor test schematic

Approximate Altitude (Feet)	Signal Voltage (±0.04V)
0	1.59
1000	1.56
2000	1.53
3000	1.50
4000	1.47
5000	1.44
6000	1.41
7000	1.39

MAP sensor altitude/voltage output relationship

1. Connect MAP/BARO tester to the sensor connector and sensor harness connector. With ignition ON and engine OFF, use DVOM to measure voltage across tester terminals. If the tester's 4-6V indicator is ON, the reference voltage input to the sensor is okay.

2. Measure the reference signal of the MAP sensor. If the DVOM voltage reading is as indicated in the table, the sensor is okay.

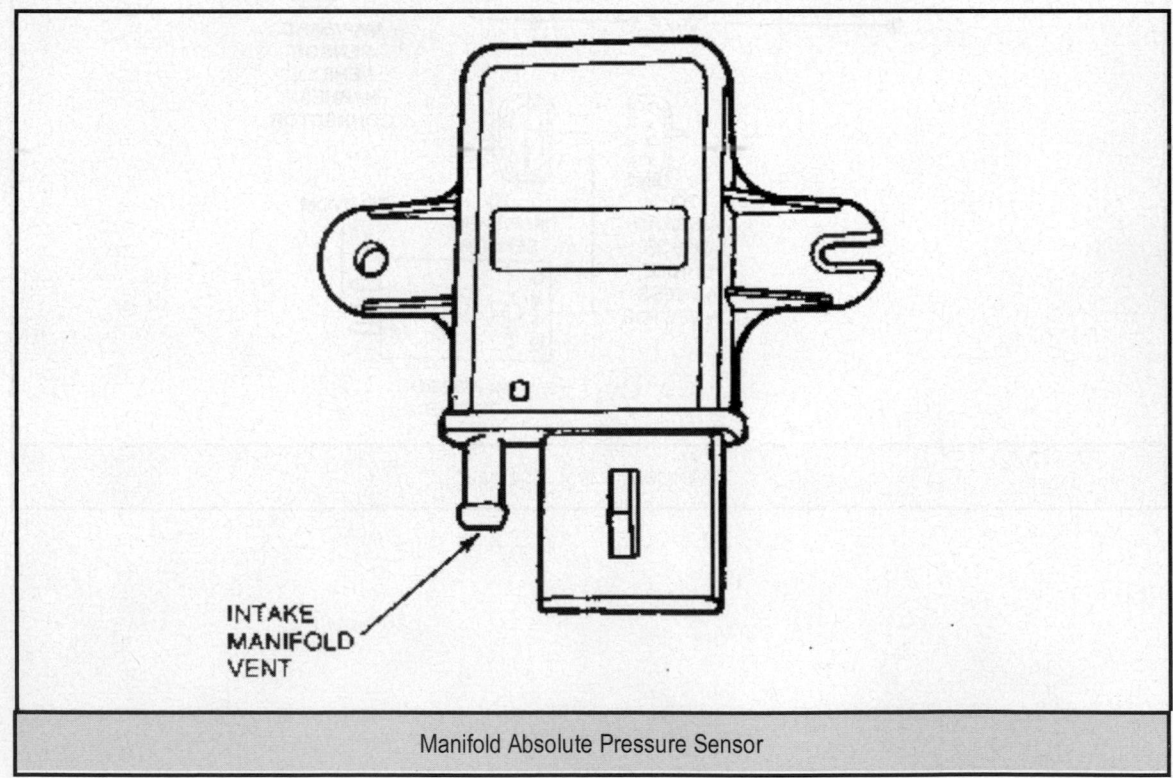

INTAKE MANIFOLD VENT

Manifold Absolute Pressure Sensor

Approximate Altitude (Ft.)	Voltage Output (±.04 Volts)
0	1.59
1000	1.56
2000	1.53
3000	1.50
4000	1.47
5000	1.44
6000	1.41
7000	1.39

MAP sensor altitude/voltage output relationship

THROTTLE POSITION SENSOR

Operation

The Throttle Position (TP) sensor is a potentiometer that provides a signal to the PCM that is directly proportional to the throttle plate position. The TP sensor is mounted on the side of the throttle body and is connected to the throttle plate shaft. The TP sensor monitors throttle plate movement and position, and transmits an appropriate electrical signal to the PCM. These signals are used by the PCM to adjust the air/fuel mixture, spark timing and EGR operation according to engine load at idle, part throttle, or full throttle. The TP sensor is not adjustable.

The TP sensor receives a 5 volt reference signal and a ground circuit from the PCM. A return signal circuit is connected to a wiper that runs on a resistor internally on the sensor. The further the throttle is opened, the wiper moves along the resistor, at wide open throttle, the wiper essentially creates a loop between the reference signal and the signal return returning the full or nearly full 5 volt signal back to the PCM. At idle the signal return should be approximately 0.9 volts.

Testing

Testing the TP sensor signal return voltage at idle

Test the operation of the TP sensor by gently opening the throttle while observing the
signal return voltage. The voltage should move smoothly according to the amount the throttle is opened

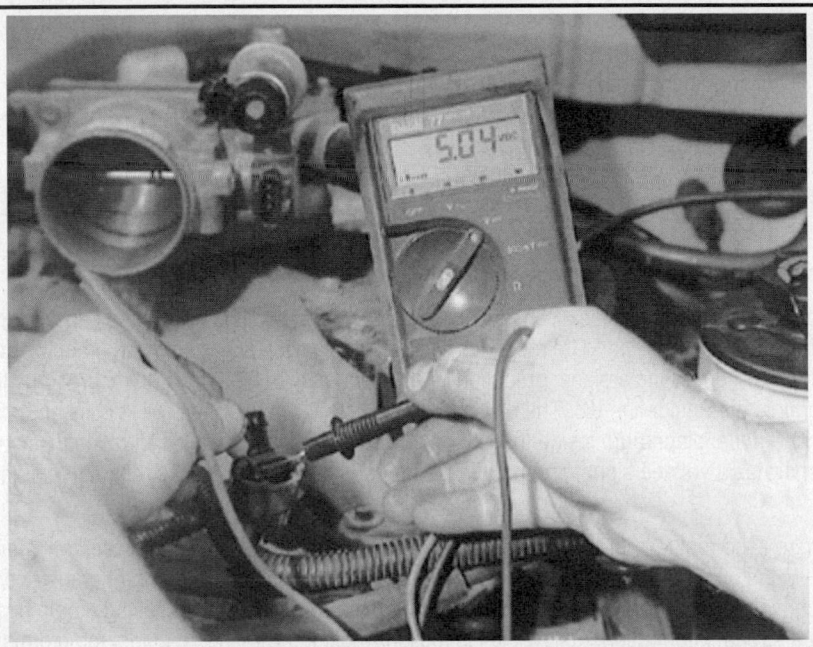

Testing the supply voltage at the TP sensor connector

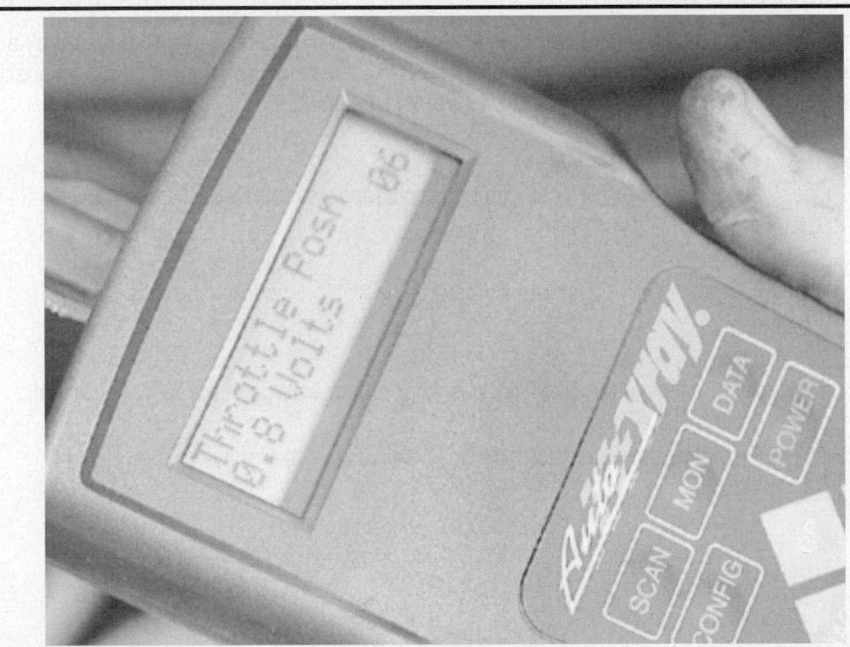

The TP sensor can be monitored with an appropriate and Data-stream capable scan tool

1. With the engine OFF and the ignition ON, check the voltage at the signal return circuit of the TP sensor by carefully backprobing the connector using a DVOM.

2. Voltage should be between 0.2 and 1.4 volts at idle.

3. Slowly move the throttle pulley to the wide open throttle (WOT) position and watch the voltage on the DVOM. The voltage should slowly rise to slightly less than 4.8v at Wide Open Throttle (WOT).

4. If no voltage is present, check the wiring harness for supply voltage (5.0v) and ground (0.3v or less), by referring to your corresponding wiring guide. If supply voltage and ground are present, but no output voltage from TP, replace the TP sensor. If supply voltage and ground do not meet specifications, make necessary repairs to the harness or PCM.

KNOCK SENSOR

Operation

The Knock Sensor (KS) is only outfitted on the 4.6L engine.

The operation of the Knock Sensor (KS) is to monitor preignition or "engine knocks" and send the signal to the PCM. The PCM responds by adjusting ignition timing until the "knocks" stop. The sensor works by generating a signal produced by the frequency of the knock as recorded by the piezoelectric ceramic disc inside the KS. The disc absorbs the shock waves from the knocks and exerts a pressure on the metal diaphragm inside the KS. This compresses the crystals inside the disc and the disc generates a voltage signal proportional to the frequency of the knocks ranging from zero to 1 volt.

Testing

There is no real test for this sensor, the sensor produces it's own signal based on information gathered while the engine is running. The sensors also are usually inaccessible without major component removal. The sensors can be monitored with an appropriate scan tool using a data display or other data stream information. Follow the instructions included with the scan tool for information on accessing the data. The only test available is to test the continuity of the harness from the PCM to the sensor.

CAMSHAFT POSITION SENSOR

The camshaft position sensor (CMP) is a variable reluctance sensor that is triggered by a high point on the left-hand exhaust camshaft sprocket. The CMP sends a signal relating camshaft position back to the PCM which is used by the PCM to control engine timing.

Testing

1. Check voltage between the camshaft position sensor terminals PWR GND and CID.

2. With engine running, voltage should be greater than 0.1 volt AC and vary with engine speed.

3. If voltage is not within specification, check for proper voltage at the VPWR terminal.

4. If VPWR voltage is greater than 10.5 volts, sensor may be faulty.

CRANKSHAFT POSITION SENSOR
Operation

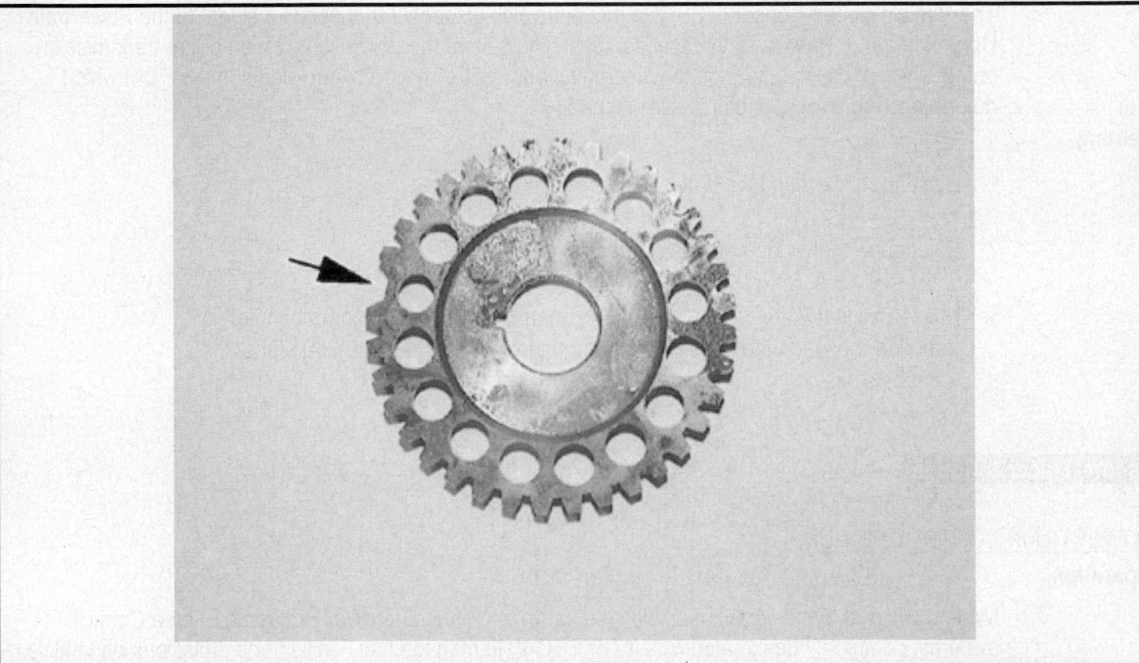

The CKP sensor trigger wheel rides on the front of the crankshaft.
The missing tooth creates a fluctuation of voltage in the sensor

The Crankshaft Position (CKP) sensor is a variable reluctance sensor that uses a trigger wheel to induce voltage. The CKP sensor is a fixed magnetic sensor mounted to the engine block and monitors the trigger or "pulse" wheel which is attached to the crank pulley/damper. As the pulse wheel rotates by the CKP sensor, teeth on the pulse wheel induce voltage inside the sensor through magnetism. The pulse wheel has a missing tooth that changes the reading of the sensor. This is used for the Cylinder Identification (CID) function to properly monitor and adjust engine timing by locating the number 1 cylinder. The voltage created by the CKP sensor is alternating current (A/C). This voltage reading is sent to the PCM and is used to determine engine RPM, engine timing, and is used to fire the ignition coils.

Testing

1. Measure the voltage between the sensor CKP sensor terminals by backprobing the sensor connector.

   If the connector cannot be backprobed, fabricate or purchase a test harness.

2. Sensor voltage should be more than 0.1 volt AC with the engine running and should vary with engine RPM.

3. If voltage is not within specification, the sensor may be faulty.

VEHICLE SPEED SENSOR

Operation

The Vehicle Speed Sensor (VSS) is a magnetic pick-up sensor that sends a signal to the Powertrain Control Module (PCM) and the speedometer. The sensor measures the rotation of the transmission output shaft or the ring gear on the differential and sends an AC voltage signal to the PCM which determines the corresponding vehicle speed.

Testing

1. Disconnect the negative battery cable.

2. Disengage the wiring harness connector from the VSS.

3. Using a Digital Volt-Ohmmeter (DVOM), measure the resistance (ohmmeter function) between the sensor terminals. If the resistance is 190-250 ohms, the sensor is okay.

Escort, Tracer & Cougar

POWERTRAIN CONTROL MODULE

Operation

Models covered by this manual employ the fourth and fifth generation Electronic Engine Control systems, commonly designated EEC-IV or EEC-V, to manage fuel, ignition and emissions on vehicle engines.

Typically the EEC-IV system was used on 1991-95 models and the EEC-V system was used on 1996-03 models.

The Powertrain Control Module (PCM) detects the engine operating and driving conditions, along with the exhaust gas oxygen content. Various switches, sensors and components provide the PCM with information which allows it to control the air/fuel ratio (mixture), emissions and drivability.

OXYGEN SENSOR

Operation

An Oxygen Sensor (O2S) is used on 1991-94 1.8L engines. The sensor is mounted in the exhaust manifold. The sensor protrudes into the exhaust stream and monitors the oxygen content of the exhaust gases. The difference between the oxygen content of the exhaust gases and that of the outside air generates a voltage signal to the PCM. The PCM monitors this voltage and, depending upon the value of the signal received, issues a command to adjust for a rich or a lean condition.

Testing

1. Perform a visual inspection on the sensor as follows:

    1. Remove the sensor from the exhaust.

    2. If the sensor tip has a black/sooty deposit, this may indicate a rich fuel mixture.

    3. If the sensor tip has a white gritty deposit, this may indicate an internal anti-freeze leak.

    4. If the sensor tip has a brown deposit, this could indicate oil consumption.

All these contaminants can destroy the sensor; if the problem is not repaired, the new sensor will also be damaged.

2. Reinstall the sensor and disengage its electrical connection.

3. Connect jumper wires from the sensor connector to the wiring harness. This permits the engine to operate normally while you check the engine.

4. Start the engine and allow it to reach normal operating temperature. This will take around ten minutes.

5. Engage the positive lead of a high impedance Digital Volt Ohmmeter (DVOM) to the sensor signal wire and the negative lead to a good known engine ground, such as the negative battery terminal.

6. The voltage reading should fluctuate as the sensor detects varying levels of oxygen in the exhaust stream.

7. If the sensor voltage does not fluctuate, the sensor may be defective or the fuel mixture could be extremely out of range.

8. If the sensor reads above 550 millivolts constantly, the fuel mixture may be too lean or you could have an exhaust leak near the sensor.

9. Under normal conditions, the sensor should fluctuate high and low. Prior to condemning the sensor, try forcing the system to have a rich fuel mixture by restricting the air intake, or a lean mixture by removing a vacuum line. If this causes the sensor to respond, look for problems in other areas of the system.

HEATED OXYGEN SENSOR
Operation

A Heated Oxygen Sensor (HO2S) monitors the oxygen content in the three-way catalytic converter and transmits this information to the Powertrain Control Module (PCM). The PCM then adjusts the air/fuel ratio to provide a rich (more fuel) or lean (less fuel) condition.

Testing

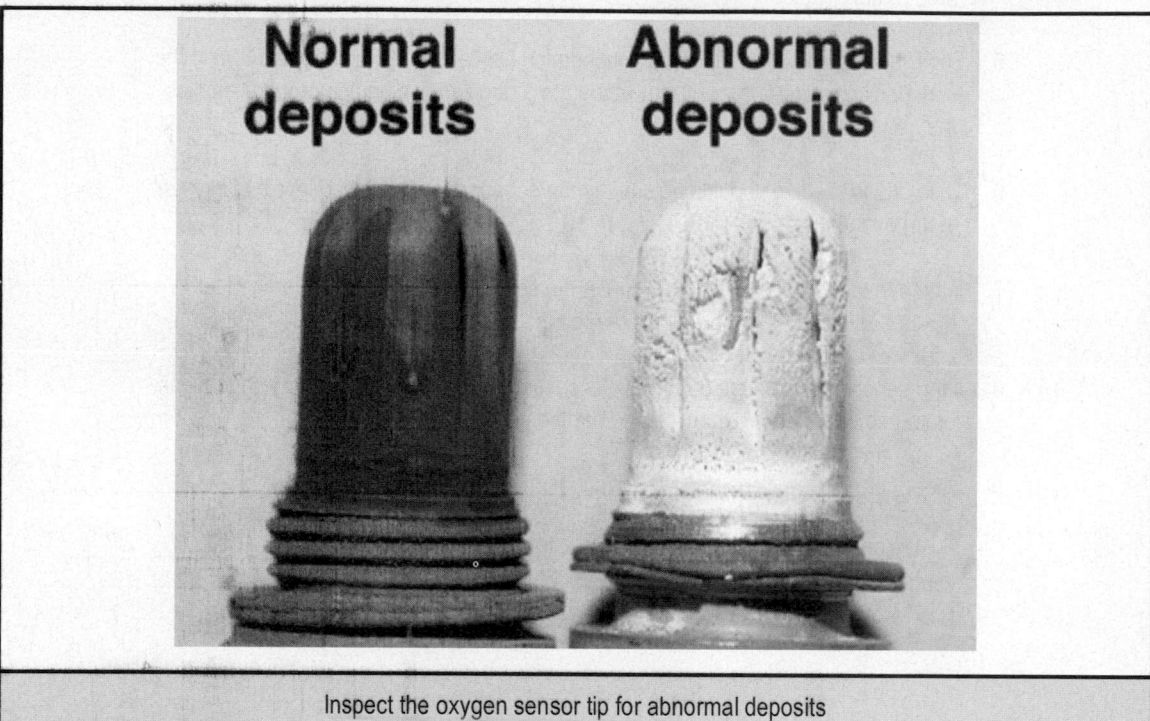

Inspect the oxygen sensor tip for abnormal deposits

1. Perform a visual inspection on the sensor as follows:

    1. Remove the sensor from the exhaust.

    2. If the sensor tip has a black/sooty deposit, this may indicate a rich fuel mixture.

    3. If the sensor tip has a white gritty deposit, this may indicate an internal anti-freeze leak.

    4. If the sensor tip has a brown deposit, this could indicate oil consumption.

All these contaminates can destroy the sensor, if the problem is not repaired the new sensor will also be damaged.

2. Reinstall the sensor.

3. Disconnect the Heated Oxygen Sensor (HO2S). Measure resistance between PWR and GND (heater) terminals of the sensor.

4. With the engine hot-to-warm, the resistance should be 5-30 ohms.

5. At room temperature, the resistance should be 2-5 ohms. If the readings are within specification, the sensor's heater element is okay.

6   With the O₂S connected and engine running, measure voltage with a DVOM by backprobing the SIG RTN wire of the oxygen sensor connector. The voltage readings at idle should stay below 1 volt. When the speed is increased or decreased, the voltage should fluctuate between 0 and 1 volt; if so, the sensor is okay.

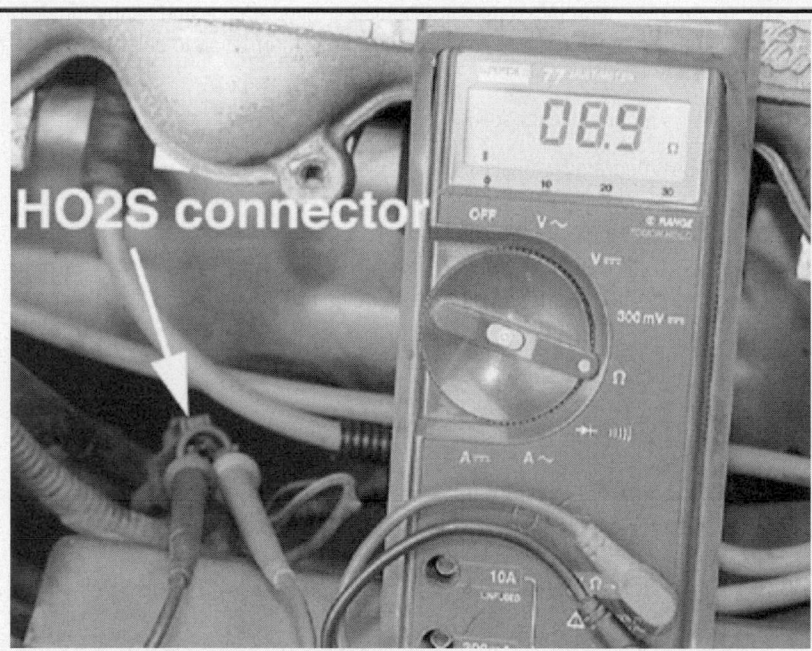

Probe the terminals of the Heated Oxygen Sensor (HO2S) harness to test the heating element resistance

COOLANT TEMPERATURE SENSOR

Operation

The Engine Coolant Temperature (ECT) sensor is a thermistor (a device which changes resistance or voltage as temperature changes). The sensor detects the temperature of engine coolant and provides a corresponding signal to the PCM.

The ECT sensor is mounted in the following locations:

● 1.8L engines: in the cylinder head

● 1.9L engines: threaded into the heater hose inlet pipe

● 2.0L engines: at the left rear of the engine

Testing

1. Disengage the temperature sensor electrical connection.

2. Connect jumper wires from the sensor connector to the wiring harness. This permits the engine to operate normally while you check the sensor.

3. Connect a Digital Volt Ohmmeter (DVOM) between the sensor terminals.

4. Measure the voltage or resistance with the engine off and cool and with the engine running and warmed up. Compare the temperature versus voltage or resistance values obtained with the chart(s).

5. Replace the sensor if the readings are incorrect.

INTAKE AIR TEMPERATURE SENSOR

Operation

The Intake Air Temperature (IAT) sensor is a thermistor that changes the resistance in response to the intake air temperature. The IAT senses the intake air temperature and sends that information to the Powertrain Control Module (PCM). The PCM uses this information to calculate the correct amount of fuel injection.

The IAT sensor is mounted in the following locations:

- 1.8L engines: integrated with the Vane Air Flow (VAF) meter

- 1.9L engines: threaded into the air cleaner lid

- 2.0L engines: at the top left front of the engine compartment, on the front of the air cleaner

Testing

1. Disengage the Intake Air Temperature (IAT) sensor electrical connection.

2. Connect jumper wires from the sensor connector to the wiring harness. This permits the engine to operate normally while you check the sensor.

3. Connect a Digital Volt Ohmmeter (DVOM) between the sensor terminals.

4. Measure the voltage or ohms with the engine off and cool, as well as with the engine running and warmed up. Compare the temperature versus voltage or resistance values obtained with those in the chart.

5. Replace the sensor if the readings are incorrect.

6. If the reading for a given temperature is about that shown in the table, the IAT sensor is okay.

Temperature °C (°F)	Resistance (kohms)
-20 (-4)	10.0 - 20.0
0 (32)	4.0 - 7.0
20 (68)	2.0 - 3.0
40 (104)	0.9 - 1.3
60 (140)	0.4 - 0.7

IAT sensor temperature versus resistance chart-1.8L engines

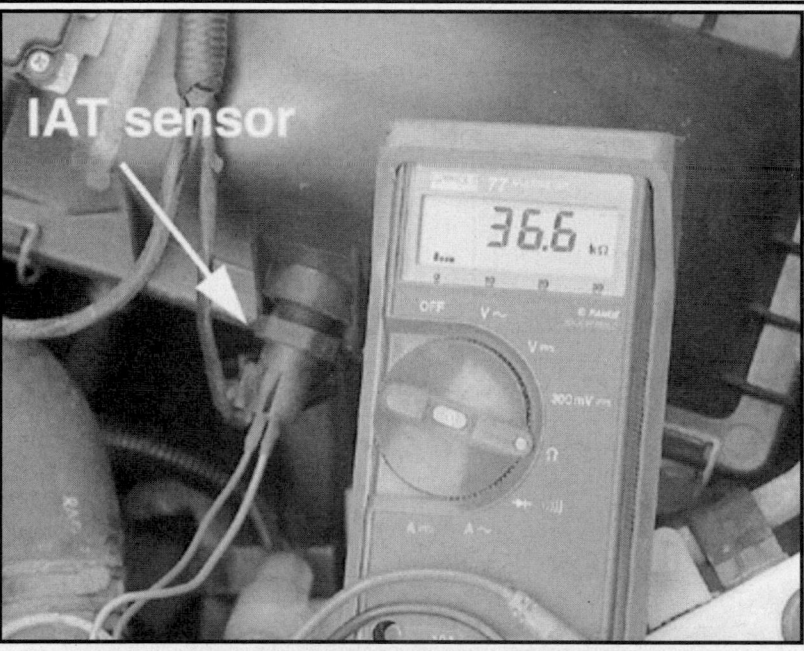

Probe the terminals of the IAT sensor and measure the voltage or resistance,
then compare the readings to the appropriate chart

MASS AIR FLOW SENSOR

Operation

The Mass Air Flow (MAF) sensor detects the intake air quantity and converts the measurement to a voltage reading by way of a heated resistor. The voltage signal is sent to the Powertrain Control Module (PCM) which, in turn, determines such things as injection quantities and engine speed.

The MAF sensor is mounted in the following locations:

● 1.9L engines: on the outside of the air cleaner assembly lid

● 2.0L engines: at the top left-hand front of the engine compartment, on the inside of the air cleaner assembly

Testing
OBD-I SYSTEMS

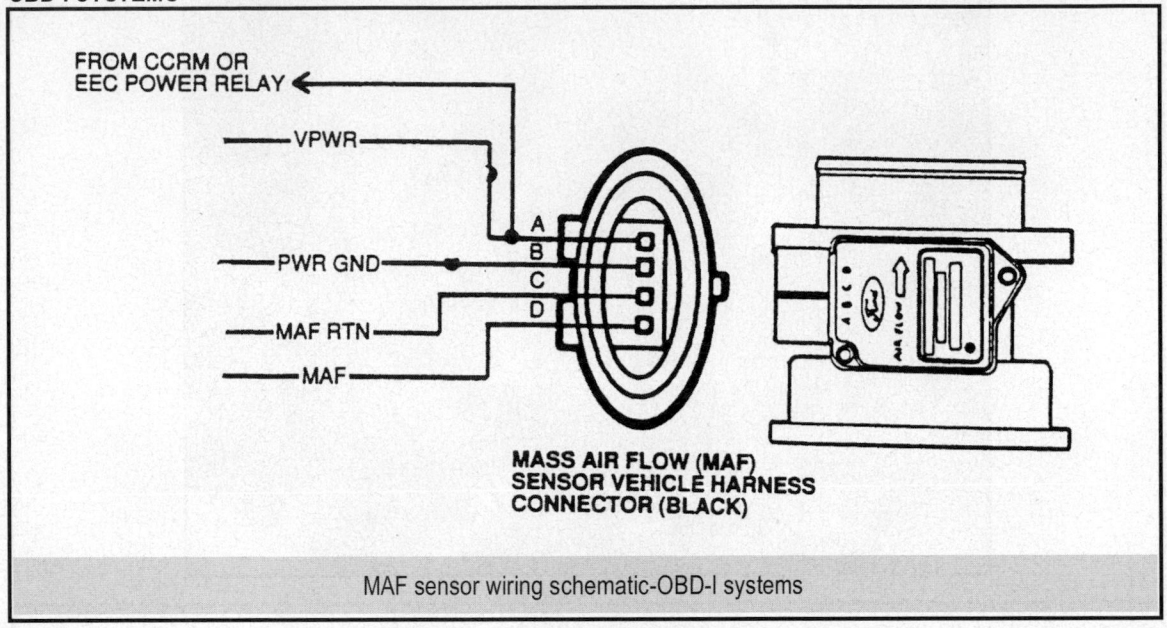

MAF sensor wiring schematic-OBD-I systems

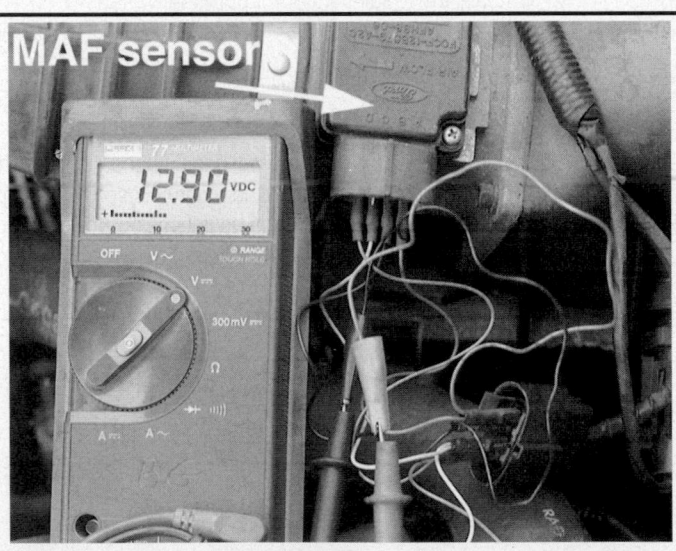

Connect a DVOM between MAF sensor harness terminal VPWR and the
negative battery terminal; the voltage should exceed 10.5 volts

1. Make sure the ignition key is OFF.

2. Connect jumper wires from the sensor connector to the wiring harness. This
   permits the engine to operate normally while you check the sensor.

3. Turn the ignition key ON, but do not start the engine.

4. Connect a Digital Volt Ohmmeter (DVOM), set to read voltage, between harness
   terminal VPWR and the negative battery terminal.

5. The voltage reading should be more than 10.5 volts.

6. If the reading is incorrect, check for an open in the VPWR circuit.

7. Connect the DVOM between harness terminals VPWR and PWR GND.

8. The voltage reading should be more than 10.5 volts.

9. If the reading is incorrect, check for an open in the PWR GND circuit.

10. Start the engine.

11. Connect the DVOM between the MAF terminal and the negative battery terminal.

12. The voltage reading should be between 0.36 and 1.50 volts.

13. Connect the DVOM between the MAF and MAF RTN terminals.

14. The voltage reading should be between 0.36 and 1.50 volts.

15. If the voltage readings are not within specifications, replace the sensor.

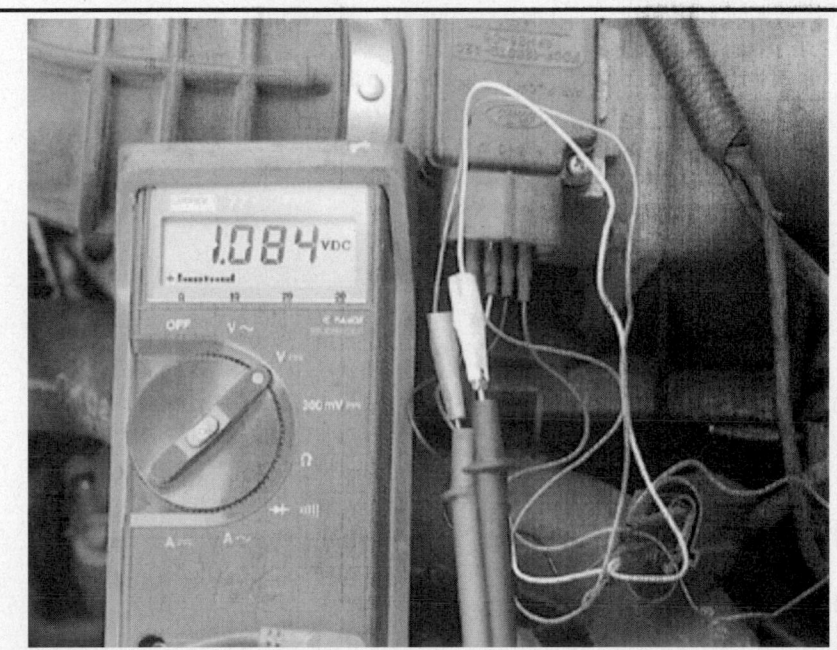

With the DVOM connected between the MAF terminal and negative battery
terminal, the voltage reading should be 0.36-1.50 volts

MAF sensor wiring schematic-OBD-II systems

1. Make sure the ignition key is OFF.

2. Connect jumper wires from the sensor connector to the wiring harness. This permits the engine to operate normally while you check the sensor.

3. Connect a Digital Volt Ohmmeter (DVOM), set to read voltage, between harness terminal VPWR and the negative battery terminal.

4. The voltage reading should be approximately 10.5 volts.

5. If the reading is incorrect, check for an open in the VPWR circuit.

6. Connect the DVOM between harness terminals VPWR and PWR GND.

7. The voltage reading should be 10.5 volts.

8. If the reading is incorrect, check for an open in the PWR GND circuit.

9. Start the engine.

10. Connect the DVOM between the MAF terminal and the negative battery terminal.

11. The voltage reading should be between 0.34 and 1.96 volts.

12. Connect the DVOM between the MAF and MAF RTN terminals.

13. The voltage reading should be between 0.34 and 1.96 volts.

14. If the voltage readings are not within specifications, replace the sensor.

VANE AIR FLOW (VAF) METER
Operation

The Vane Air Flow (VAF) meter, also known as a Volume Air Flow Meter, measures the air flowing into the engine. The meter contains a movable vane which is connected to a potentiometer. As air flows through the meter, the movable vane and potentiometer change position and provide an input to the PCM. The PCM can then translate the vane position into the volume of air flowing into the engine.

The VAF meter is mounted in the following locations:

- 1.8L engines: between the air cleaner and throttle body

Testing

VAF meter wiring schematic and connector terminal identification-1.8L engines

1. Make sure the ignition key is OFF.

2. Connect jumper wires from the sensor connector to the wiring harness. This permits the engine to operate normally while you check the sensor.

3. Connect a Digital Volt Ohmmeter (DVOM), set to read voltage, between sensor terminals VAF and SIG RTN.

4. Turn the key ON and access the VAF meter measuring vane.

5. With the vane fully closed, the voltage reading should be 4.5-5 volts.

6. Slowly move the vane to the fully open position. As the vane moves through its travel, the voltage should drop slowly and smoothly. When the vane reaches the end of its travel, the voltage reading should be 0.5-1.5 volts.

7. If the voltage readings are not within specifications, replace the VAF meter.

THROTTLE POSITION SENSOR
Operation

The Throttle Position (TP) sensor detects the throttle plate opening angle and supplies the PCM with an input signal indicating throttle position.

On 1991-95 1.8L engines with a manual transaxle, the TP sensor consists of a two-position switch sensing only closed or open throttle positions. These two positions are monitored by the Idle (IDL) switch and the Wide Open Throttle (WOT) switch.

On 1991-951.8L engines with automatic transaxles, the TP sensor consists of a combination switch and idle switch. The TP sensor sends a range of signals to the PCM indicating throttle plate position. This range of signals could range from idle to off idle.

On 1996 1.8L engines, the TP sensor consists of a combination switch and idle switch. The TP sensor sends a range of signals to the PCM indicating throttle plate position. This range of signals could range from idle to off idle.

On 1.9L and 2.0L engines, the TP sensor is a rotary potentiometer which consists of either a rigid or flexible thick film resistive substrate, and a moving wiper that is mounted on a rotor. As the TP sensor is rotated by the throttle shaft blade, the PCM receives information on the throttle plate position.

Testing
1.8L ENGINES
1991-95 MODELS

1. Connect jumper wires from the sensor connector to the wiring harness. This permits the engine to operate normally while you check the sensor.

2. Using the jumper wires, probe the TP terminal at the TP sensor connector, with the positive lead of a Digital Volt Ohmmeter (DVOM).

3. Using the jumper wire, probe the SIG RTN terminal with the negative lead.

4. Have an assistant turn the ignition key ON and depress the accelerator pedal in $1/8$ increments to the wide open position.

5. Observe the voltmeter at each of these positions and note the voltage displayed. Compare your readings with the accompanying charts.

6. If the TP sensor readings match the values in the chart, the sensor is working properly. If, however, the TP sensor readings are not as specified, continue with the test procedure.

7. Turn the ignition key OFF.

8. Unplug the TP sensor electrical connector, then attach the positive lead of the DVOM to the VREF terminal of the wiring harness connector.

9. Turn the ignition key ON and observe the voltage reading on the DVOM; it should be between 4.5 and 5.5 volts.

10. If the VREF voltage is within specifications and the TP voltage readings do not agree with the values displayed in the chart, replace the TP sensor.

THROTTLE POSITION	VOLTS
1/8	.998
2/8	1.60
3/8	2.37
4/8	2.74
5/8	3.15
6/8	3.43
7/8	3.60
8/8	4.02

NOTE: Voltage and Resistance values may vary ± 15%.

Throttle position sensor voltage chart-1991-95 1.8L engines

1996 MODELS

1. Connect jumper wires from the sensor connector to the wiring harness. This permits the engine to operate normally while you check the sensor.

2. Using the jumper wires, probe the TP terminal at the TP sensor connector, with the positive lead of a Digital Volt Ohmmeter (DVOM).

3. Probe the SIG RTN terminal with the negative lead.

4. Have an assistant turn the ignition key ON and depress the accelerator pedal to the $1/4$ open, $1/2$ open, $3/4$ open and wide open positions.

5. Observe the voltmeter at each of these positions and note the voltage displayed. Compare your readings with the accompanying charts.

6. If the TP sensor readings match the values in the chart, the sensor is working properly. If, however, the TP sensor readings are not as specified, continue with the test procedure.

7. Turn the ignition key OFF.

8. Unplug the TP sensor electrical connector, then attach the positive lead of the DVOM to the VREF terminal of the wiring harness connector.

9. Turn the ignition key ON and observe the voltage reading on the DVOM; it should be between 4.5 and 5.5 volts.

10. If the VREF voltage is within specifications and the TP voltage readings do not agree with the values displayed in the chart, replace the TP sensor.

Throttle Position	Volts
1/4	0.5
HALF	2.75
3/4	3.88
FULL	5.0

NOTE: Voltage and Resistance values may vary ± 15%.

Throttle position sensor voltage chart-1996 1.8L engines

1.9L AND 2.0L ENGINES

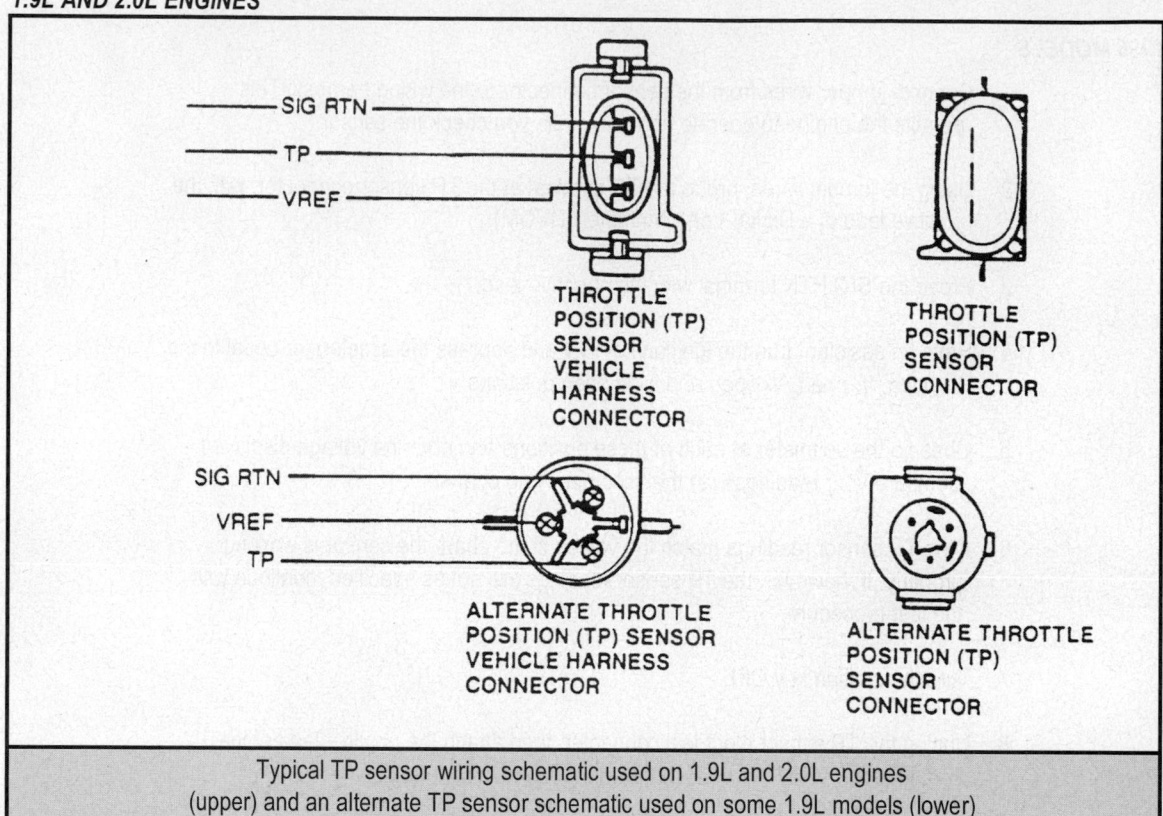

Typical TP sensor wiring schematic used on 1.9L and 2.0L engines
(upper) and an alternate TP sensor schematic used on some 1.9L models (lower)

1. Connect jumper wires from the sensor connector to the wiring harness. This permits the engine to operate normally while you check the sensor.

2. Using the jumper wires, probe the TP terminal at the TP sensor connector, with the positive lead of a Digital Volt Ohmmeter (DVOM).

3. Probe the SIG RTN terminal with the negative lead.

4. Have an assistant turn the ignition key ON and depress the accelerator pedal in small, smooth increments from the fully closed to the wide open position.

5. Observe the voltmeter at each of these positions and note the voltage displayed.

6. The voltage at the closed position should be approximately 0 volts. As the throttle is moved slowly through its travel, the voltage should rise evenly until you reach the fully open position, where the voltage should be approximately 4.5 volts.

7. If the voltage does not meet specifications, or jumps around erratically when moving the throttle through its travel, the sensor may be defective and should be replaced.

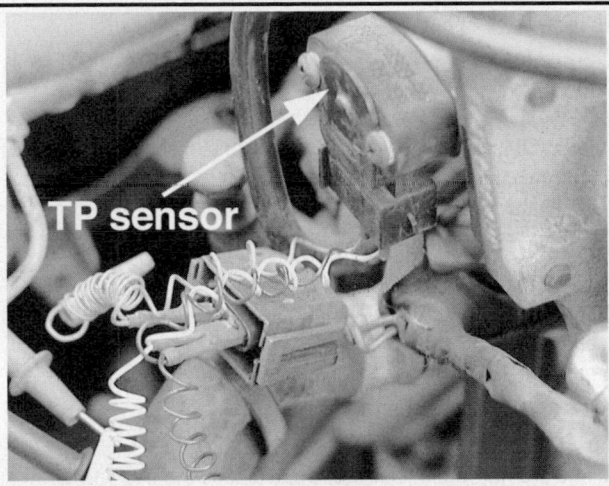

Attach jumper wires from the TP sensor connector to the wiring harness.
This permits the engine to operate normally while you check the sensor

IDLE AIR CONTROL VALVE

Operation

The Idle Air Control (IAC) valve is used to control idle speed. The amount of air allowed to bypass the throttle plate is determined by the PCM.

When the engine is cold, air flows through the valve during all modes of engine operation to maintain the factory set idle speed.

Testing
1.8L ENGINES

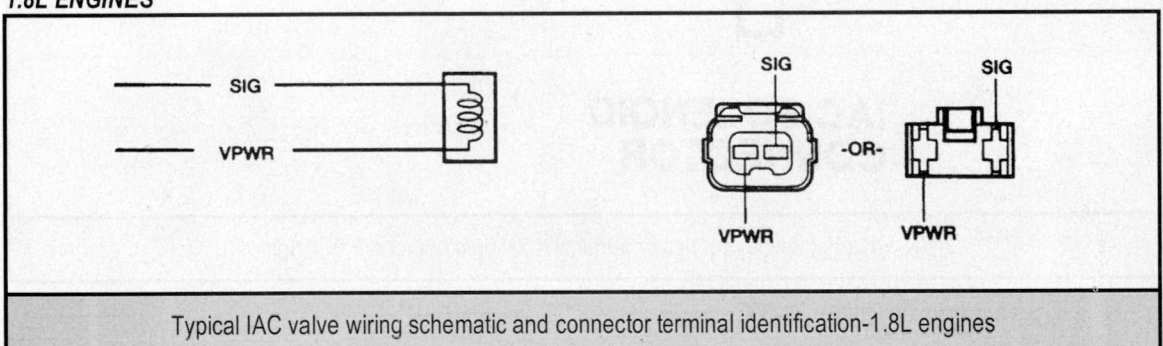

Typical IAC valve wiring schematic and connector terminal identification-1.8L engines

1. Turn the ignition key OFF.

2. Connect jumper wires from the sensor connector to the wiring harness. This permits the engine to operate normally while you check the sensor.

3. Connect a Digital Volt Ohmmeter (DVOM), set to read resistance, between sensor terminal SIG and the negative battery cable.

4. The resistance should not be greater than 10,000 ohms.

5. Connect the DVOM, set to read resistance, between sensor terminal VPWR and the negative battery cable.

6. The resistance should not be greater than 10,000 ohms.

7. Turn the ignition key ON.

8. Connect the DVOM, set to read voltage, between sensor terminal VPWR and the negative battery cable.

9. The voltage should be more than 1 volt.

10. Connect the DVOM, set to read voltage, between sensor terminal SIG and the negative battery cable.

11. The voltage should be more than 1 volt.

12. If the readings are not within specifications, replace the IAC valve.

1.9L AND 2.0L ENGINES

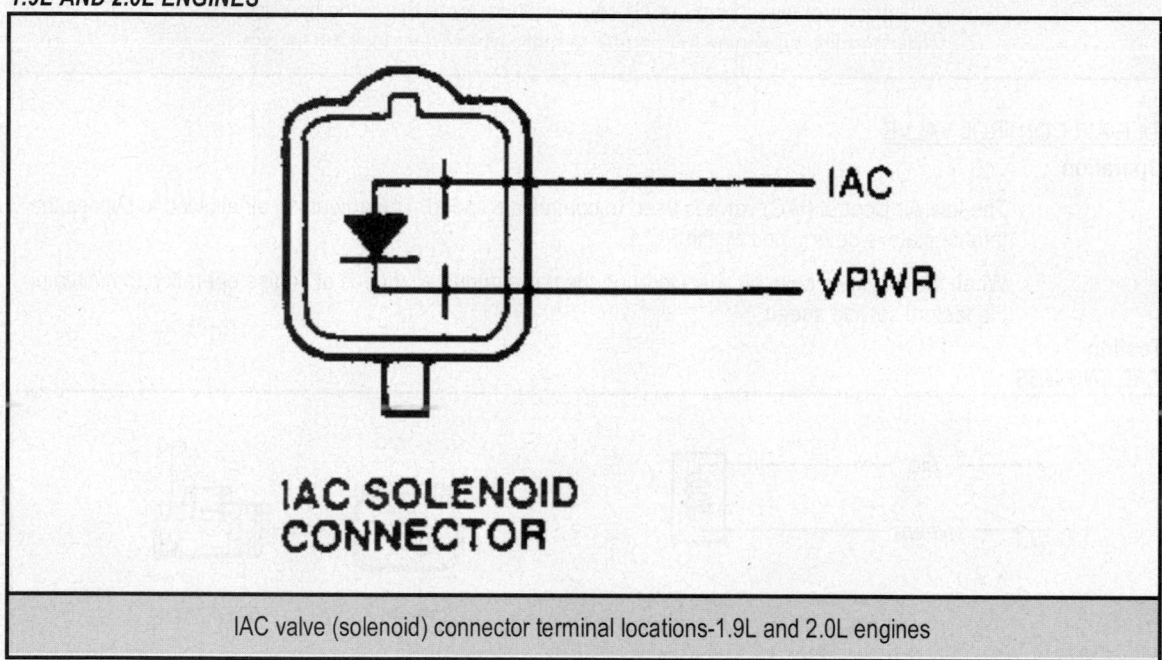

IAC valve (solenoid) connector terminal locations-1.9L and 2.0L engines

1. Turn the ignition key OFF.

2. Unplug the sensor electrical connection.

Due to the diode in the IAC valve, be sure to observe the correct polarity when connecting the DVOM.

3. Connect a Digital Volt Ohmmeter (DVOM), set to read resistance, between the sensor terminals as follows. Connect the positive lead to the VPWR terminal and the negative lead to the IAC terminal.

4. The resistance should be between 6 and 13 ohms.

5. Measure the resistance between either of the valve terminals and the valve body. The resistance should be greater than 10,000 ohms.

6. If the resistances are not within specification, replace the sensor.

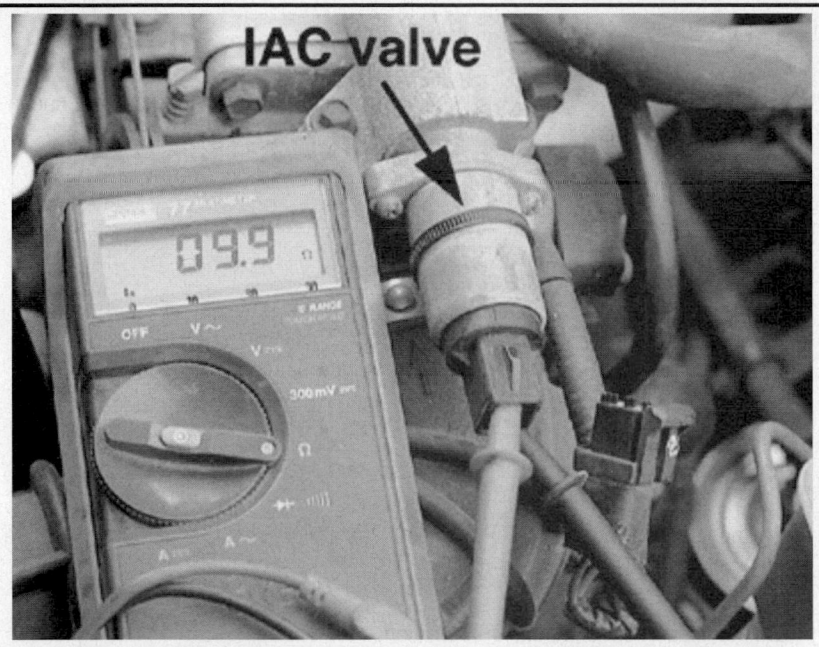

Connect the DVOM positive lead to the VPWR terminal and the negative
lead to the IAC terminal. The resistance should be between 6 and 13 ohms

Measure resistance between either of the valve terminals and the valve body.
The resistance should be greater than 10,000 ohms

CAMSHAFT POSITION SENSOR
Operation

The Camshaft Position (CMP) sensor provides camshaft position and rpm information to the PCM when the No. 1 piston is on the compression stroke.

Testing
1.8L ENGINES
OBD-I SYSTEM

1. Turn the ignition switch OFF.

2. Set a DVOM to a low DC voltage range (under 20 volts).

3. Backprobe the ground terminal of the distributor connector with the negative lead.

4. Backprobe the VPWR terminal of the distributor connector with the positive lead.

5. Turn the ignition switch to the ON position. The DVOM should read battery voltage. If not, check the wiring to the PCM.

6. Turn the ignition switch to the OFF position.

7. Back probe the CMP terminal of the distributor connector with the positive lead.

8. Turn the ignition switch to the ON position.

9. Bump the starter to rotate the engine, but do not start the engine.

10. As the vanes of the pass through the Hall effect sensor the voltage should switch back and forth between 5 volts and less than 1 volt.

OBD-II SYSTEM

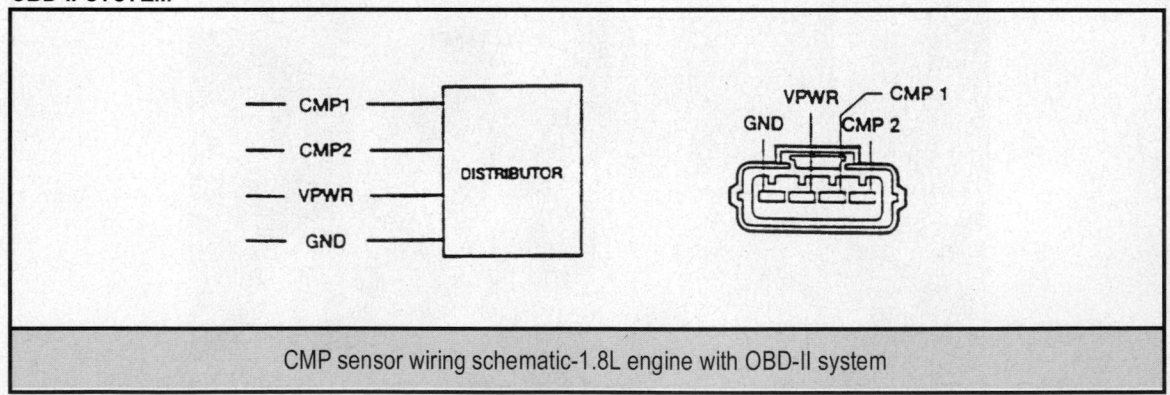

CMP sensor wiring schematic-1.8L engine with OBD-II system

1. Turn the ignition switch OFF.

2. Set a DVOM to a low DC voltage range (under 20 volts).

3. Backprobe the ground terminal of the distributor connector with the negative lead.

4. Backprobe the VPWR terminal of the distributor connector with the positive lead.

5. Turn the ignition switch to the ON position. The DVOM should read battery voltage. If not, check the wiring to the PCM.

6. Turn the ignition switch to the OFF position.

7. Back probe the CMP1 terminal of the distributor connector with the positive lead.

8. Turn the ignition switch to the ON position.

9. Bump the starter to rotate the engine, but do not start the engine.

10. As the vanes in the distributor pass through the Hall effect sensor the voltage should switch back and forth between 5 volts and less than 1 volt.

11. Perform the same test on the CMP 2 terminal.

12. If the voltage doesn't vary, check that the distributor is rotating. If the distributor is rotating, the CMP sensor may be defective.

1.9L AND 2.0L ENGINES

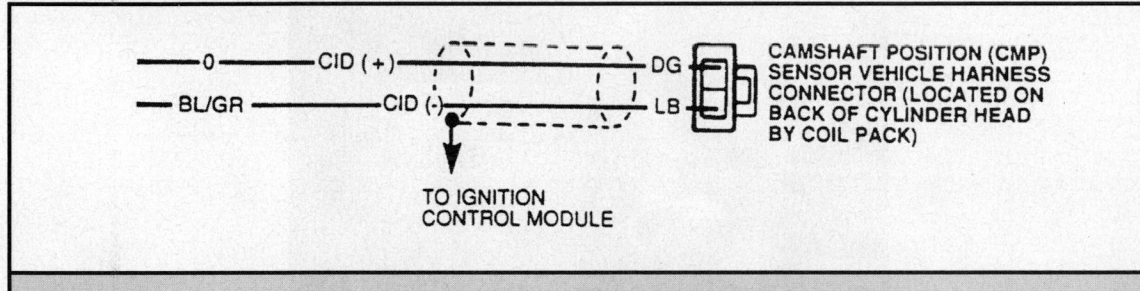

Common CMP sensor wiring schematic-1.9L and 2.0L engines

1. Turn the ignition switch OFF.

2. Set a DVOM on the AC scale to read less than 5 volts.

3. Connect jumper wires from the CMP sensor connector to the wiring harness. This permits the engine to operate normally while you check the sensor.

4. Backprobe the CID (-) or CMP (-) terminal of the sensor with the negative lead.

5. Backprobe the CID (+) or CMP (+) terminal of the sensor with the positive lead.

6. Vary the engine rpm by turning the throttle assembly.

7. The voltage reading should vary by more than 0.1 volt AC.

8. If the voltage is not as specified, replace the CMP sensor.

Measure the CMP sensor voltage with the throttle assembly closed ...

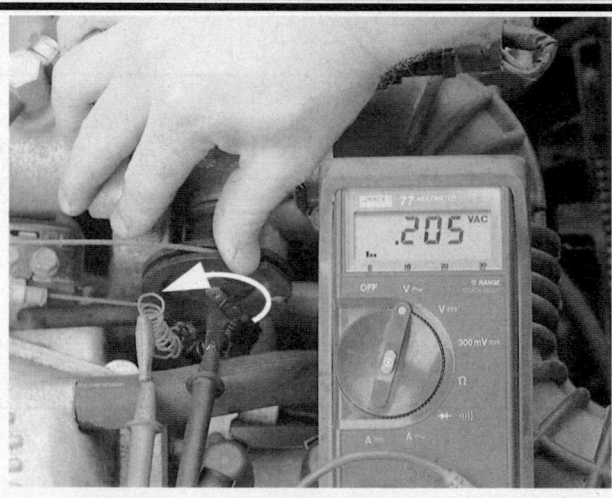

... then move the throttle assembly to vary the engine rpm and observe the CMP sensor voltage reading

CRANKSHAFT POSITION SENSOR

The Crankshaft Position (CKP) sensor relays engine speed and crankshaft position to the Powertrain Control Module (PCM).

Testing
1.8L ENGINES
OBD-I SYSTEM

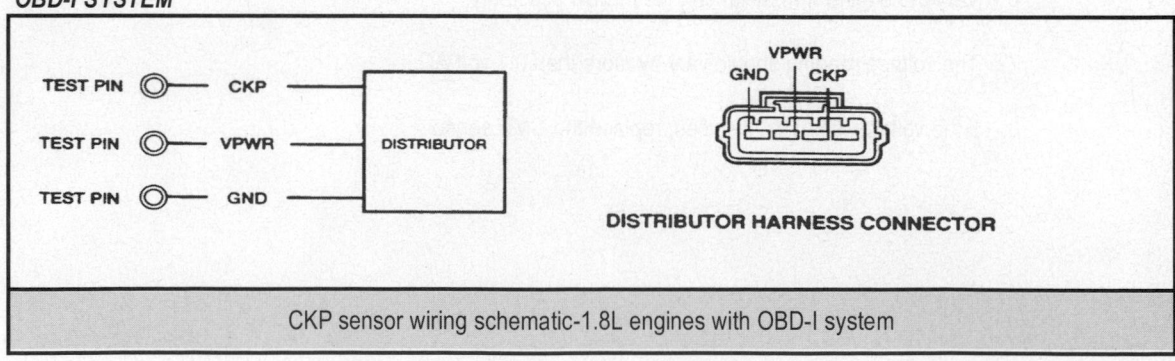

CKP sensor wiring schematic-1.8L engines with OBD-I system

1. Turn the ignition key OFF.

2. Using a Digital Volt Ohmmeter (DVOM), backprobe the CKP terminal of the CKP sensor connector with the DVOM's positive lead. Attach the DVOM's negative lead to a good engine ground.

3. While an assistant cranks the engine with the starter motor (using the ignition switch), observe the voltage displayed on the DVOM.

4. The CKP sensor voltage should fluctuate between approximately 0 and 5 volts. If the CKP sensor performs as specified, it is functioning properly. If, however, the voltage is not as specified, continue with the test procedure.

5. Turn the ignition key OFF.

6. Disengage the wiring harness connector from the distributor, then attach the positive lead of the DVOM to the VPWR terminal on the wiring harness connector. Attach the negative lead to a good ground.

7. Turn the ignition key ON and observe the voltage displayed on the DVOM. The voltage should be greater than 10 volts. If not, the CKP sensor is not receiving the proper voltage from the PCM; the problem lies elsewhere in the electrical system.

8. If the VPWR voltage is satisfactory and the CKP output is not as specified, replace the CKP sensor.

OBD-II SYSTEM

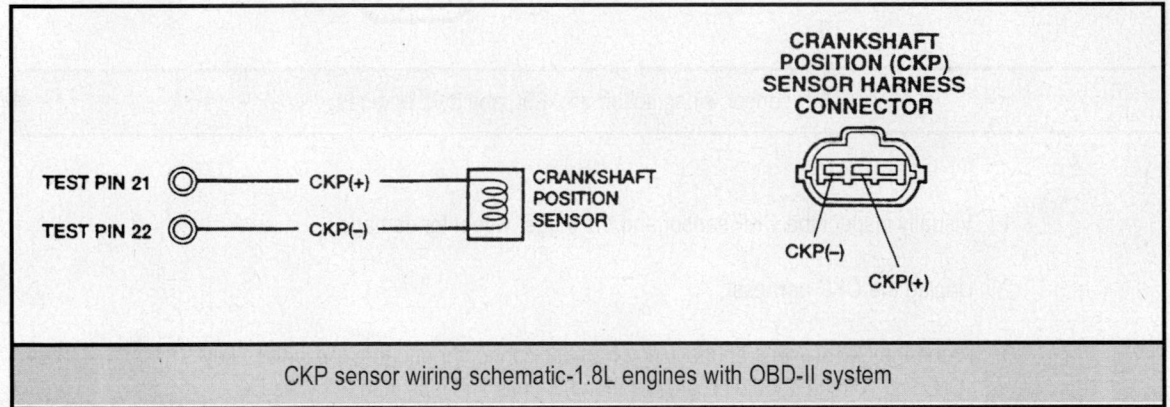

CKP sensor wiring schematic-1.8L engines with OBD-II system

1. Turn the ignition key OFF.

2. Using a Digital Volt Ohmmeter (DVOM), backprobe the CKP (-) terminal of the harness and ground.

3. The resistance should be less than 5 volts. If not check for a short in the GND circuit.

4. Measure the resistance between CKP (+) and the negative battery terminal.

5. The resistance should be approximately 10, 000 ohms. If the resistance is greater than specified resistance, replace the sensor. If the resistance is less than 10,000 ohms, check for a wiring harness for short circuits.

6. Measure the resistance between CKP (-) and the negative battery terminal.

7. The resistance should be approximately 10, 000 ohms. If the resistance is greater than specified resistance, replace the sensor. If the resistance is less than 10,000 ohms, check for a wiring harness for short circuits.

To perform this test accurately, the ambient air temperature should be 68°F (20°C).

8. Turn the ignition key OFF.

9. Unplug the CKP electrical connection.

10. Using a Digital Volt Ohmmeter (DVOM), set to read resistance, probe the CKP (+) terminal of the CKP sensor connector with the DVOM's positive lead. Attach the DVOM's negative lead to a good engine ground. Note the reading.

11. Backprobe the CKP (+) terminal of the CKP sensor connector with the DVOM's positive lead in the same manner and note the resistance reading.

12. The resistance should be 520-580 ohms at 68°F (20°C). If readings are not as specified, replace the CKP sensor.

1.9L AND 2.0L ENGINES

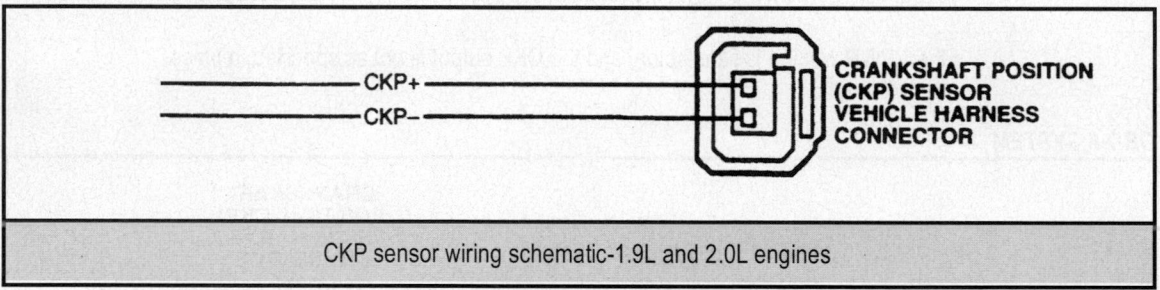

CKP sensor wiring schematic-1.9L and 2.0L engines

1. Visually inspect the CKP sensor and the trigger wheel for damage.

2. Unplug the CKP harness.

3. Probe the CKP (-) wire on the wiring harness connector and turn the ignition key ON. The voltage should be greater than 0.8 volts and less than 2.2 volts DC.

4. If no voltage is present, check the continuity of the CKP (-) wire between the wiring connector and the EDIS module. If the wire has continuity, check the EDIS module is getting power, If power is present, replace the EDIS module and retest the system.

5. Attach jumper wires from the CKP sensor connector to the wiring harness. This permits the engine to operate normally while you check the sensor.

6. Set a Digital Volt Ohmmeter (DVOM) on the AC voltage scale.

7. Start the engine and observe the voltage reading while varying the rpm, the voltage should also vary.

8. If there is no signal, unplug the wiring harness from the sensor and check the harness for continuity and shorts to ground. Repair as necessary and attach the harness to the sensor.

Attach jumper wires from the CKP sensor connector to the wiring harness, connect a voltmeter, start the engine and observe the voltage reading

VARIABLE CAMSHAFT TIMING OIL CONTROL SOLENOID
Operation

The oil control solenoid is part of the Variable Camshaft Timing (VCT) unit.

The VCT unit used on the 2.0L DOHC engine is connected to the exhaust camshaft, and the rotation of the exhaust camshaft takes place in the VCT unit. The unit is oil filled to move the pistons inside the unit. Oil and piston movement inside the unit can retard the exhaust camshaft without affecting the intake camshaft.

Adjustment of the VCT unit permits every piston position between maximum advance and maximum retardation.

Testing

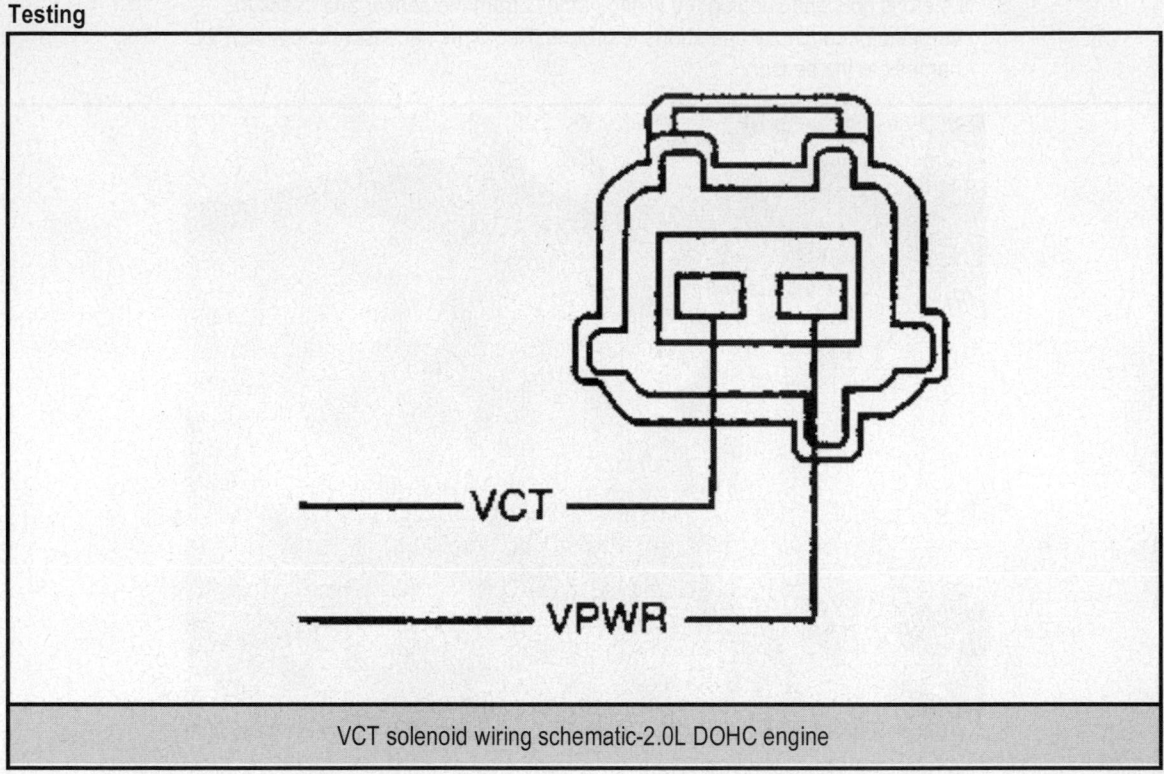

VCT solenoid wiring schematic-2.0L DOHC engine

1. Turn the ignition switch OFF.

2. Unplug the wiring harness from the sensor and check the harness for continuity and shorts to ground. Repair as necessary and attach the harness to the sensor.

3. Using a Digital Volt Ohmmeter (DVOM), set to read resistance, probe the terminals of the solenoid.

4. The resistance should be between 3.0 and 6.0 volts.

5. If the resistance is not within specification, replace the sensor.

Expedition & Navigator

POWERTRAIN CONTROL MODULE (PCM)

Operation

The Powertrain Control Module (PCM) performs many functions on your vehicle. The module accepts information from various engine sensors and computes the required fuel flow rate necessary to maintain the correct amount of air/fuel ratio throughout the entire engine operational range.

Based on the information that is received and programmed into the PCM's memory, the PCM generates output signals to control relays, actuators and solenoids. The PCM also sends out a command to the fuel injectors that meters the appropriate quantity of fuel. The module automatically senses and compensates for any changes in altitude when driving your vehicle.

HEATED OXYGEN SENSOR

Operation

The oxygen (O2) sensor is a device which produces an electrical voltage when exposed to the oxygen present in the exhaust gases. The sensor is mounted in the exhaust system, usually in the manifold or a boss located on the down pipe before the catalyst.. The oxygen sensors used on the Ford F-series, Expedition and the Lincoln Navigator are electrically heated internally for faster switching when the engine is started cold. The oxygen sensor produces a voltage within 0 and 1 volt. When there is a large amount of oxygen present (lean mixture), the sensor produces a low voltage (less than 0.4v). When there is a lesser amount present (rich mixture) it produces a higher voltage (0.6-1.0v).The stoichiometric or correct fuel to air ratio will read between 0.4 and 0.6v. By monitoring the oxygen content and converting it to electrical voltage, the sensor acts as a rich-lean switch. The voltage is transmitted to the PCM.

Some models have two sensors, one before the catalyst and one after. This is done for a catalyst efficiency monitor that is a part of the OBD-II engine controls that are on all models covered by this manual except for those with the 5.8L or 7.5L engines. The sensor before the catalyst measures the exhaust emissions right out of the engine, and sends the signal to the PCM about the state of the mixture as previously talked about. The second sensor reports the difference in the emissions after the exhaust gases have gone through the catalyst. This sensor reports to the PCM the amount of emissions reduction the catalyst is performing.

The oxygen sensor will not work until a predetermined temperature is reached, until this time the PCM is running in what as known as OPEN LOOP operation. OPEN LOOP means that the PCM has not yet begun to correct the air-to-fuel ratio by reading the oxygen sensor. After the engine comes to operating temperature, the PCM will monitor the oxygen sensor and correct the air/fuel ratio from the sensor's readings. This is what is known as CLOSED LOOP operation.

A heated oxygen sensor (HO2S) has a heating element that keeps the sensor at proper operating temperature during all operating modes. Maintaining correct sensor temperature at all times allows the system to enter into CLOSED LOOP operation sooner.

In CLOSED LOOP operation the PCM monitors the sensor input (along with other inputs) and adjusts the injector pulse width accordingly. During OPEN LOOP operation the PCM ignores the sensor input and adjusts the injector pulse to a preprogrammed value based on other inputs.

Testing

The HO2S can be monitored with an appropriate and Data-stream capable scan tool

1. Disconnect the HO2S.

2. Measure the resistance between PWR and GND terminals of the sensor. Resistance should be approximately 6 ohms at 68°F (20°C). If resistance is not within specification, the sensor's heater element is faulty.

3. With the HO2S connected and engine running, measure the voltage with a Digital Volt-Ohmmeter (DVOM) between terminals HO2S and SIG RTN (GND) of the oxygen sensor connector. Voltage should fluctuate between 0.01-1.0 volts. If voltage fluctuation is slow or voltage is not within specification, the sensor may be faulty.

IDLE AIR CONTROL VALVE

Operation

The Idle Air Control (IAC) valve adjusts the engine idle speed. The valve is located on the side of the throttle body. The valve is controlled by a duty cycle signal from the PCM and allows air to bypass the throttle plate in order to maintain the proper idle speed.

Do not attempt to clean the IAC valve. Carburetor tune-up cleaners or any type of solvent cleaners will damage the internal components of the valve.

Testing

The IAC can be monitored with an appropriate and Data-stream capable scan tool

1. Turn the ignition switch to the OFF position.

2. Disconnect the wiring harness from the IAC valve.

3. Measure the resistance between the terminals of the valve.

Due to the diode in the solenoid, place the ohmmeter positive lead on the VPWR terminal and the negative lead on the ISC terminal.

4. Resistance should be 6-13 ohms.

5. If resistance is not within specification, the valve may be faulty.

ENGINE COOLANT TEMPERATURE SENSOR
Operation

The Engine Coolant Temperature (ECT) sensor resistance changes in response to engine coolant temperature. The sensor resistance decreases as the coolant temperature increases, and increases as the coolant temperature decreases. This provides a reference signal to the PCM, which indicates engine coolant temperature. The signal sent to the PCM by the ECT sensor helps the PCM to determine spark advance, EGR flow rate, air/fuel ratio, and engine temperature. The ECT is a two wire sensor, a 5-volt reference signal is sent to the sensor and the signal return is based upon the change in the measured resistance due to temperature.

Testing

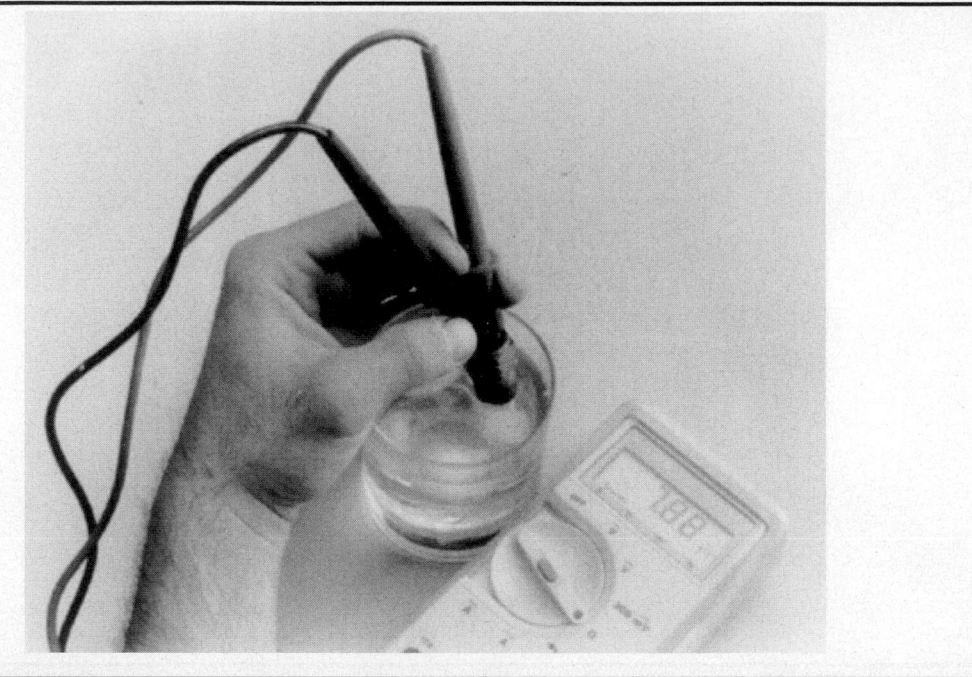

Another method of testing the ECT is to submerge it in cold or hot water and check resistance

Test the ECT resistance across the two sensor terminals

1. Disconnect the engine wiring harness from the ECT sensor.

2. Connect an ohmmeter between the ECT sensor terminals.

3. With the engine cold and the ignition switch in the OFF position, measure and note the ECT sensor resistance.

4. Connect the engine wiring harness to the sensor.

5. Start the engine and allow the engine to reach normal operating temperature.

6. Once the engine has reached normal operating temperature, turn the engine OFF.

7. Once again, disconnect the engine wiring harness from the ECT sensor.

8. Measure and note the ECT sensor resistance with the engine hot.

9. Compare the cold and hot ECT sensor resistance measurements with the accompanying chart.

10. If readings do not approximate those in the chart, the sensor may be faulty.

INTAKE AIR TEMPERATURE SENSOR
Operation

The tip of the IAT sensor has an exposed thermistor that changes the resistance of the sensor based upon the force of the air rushing past it

The Intake Air Temperature (IAT) sensor determines the air temperature inside the intake manifold. Resistance changes in response to the ambient air temperature. The sensor has a negative temperature coefficient. As the temperature of the sensor rises the resistance across the sensor decreases. This provides a signal to the PCM indicating the temperature of the incoming air charge. This sensor helps the PCM to determine spark timing and air/fuel ratio. Information from this sensor is added to the pressure sensor information to calculate the air mass being sent to the cylinders. The IAT is a two wire sensor, a 5-volt reference signal is sent to the sensor and the signal return is based upon the change in the measured resistance due to temperature.

Testing

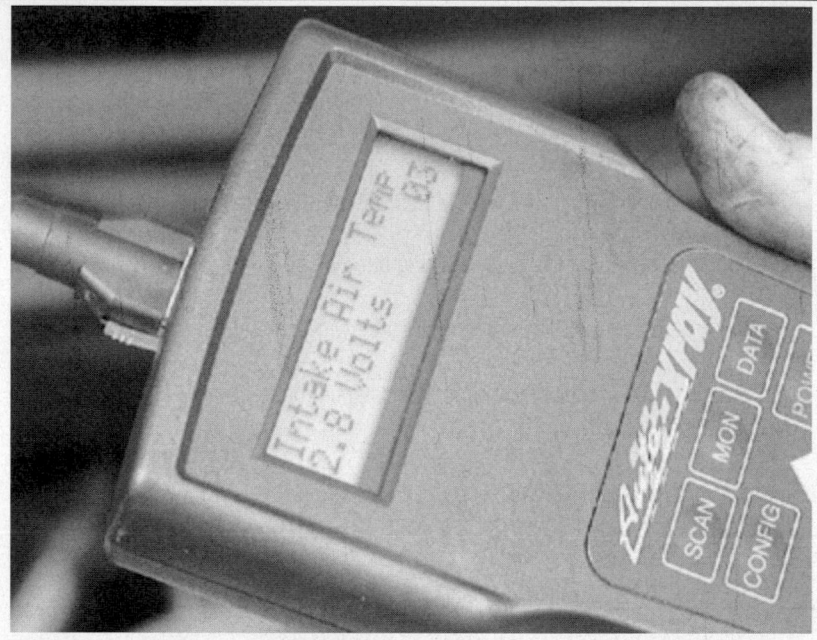

The IAT sensor can be monitored with an appropriate and Data-stream capable scan tool

Measure the resistance of the IAT sensor across the two sensor pins

1. Turn the ignition switch OFF.

2. Disconnect the wiring harness from the IAT sensor.

3. Measure the resistance between the sensor terminals.

4. Compare the resistance reading with the accompanying chart.

5. If the resistance is not within specification, the IAT may be faulty.

6. Connect the wiring harness to the sensor.

MASS AIRFLOW SENSOR
Operation

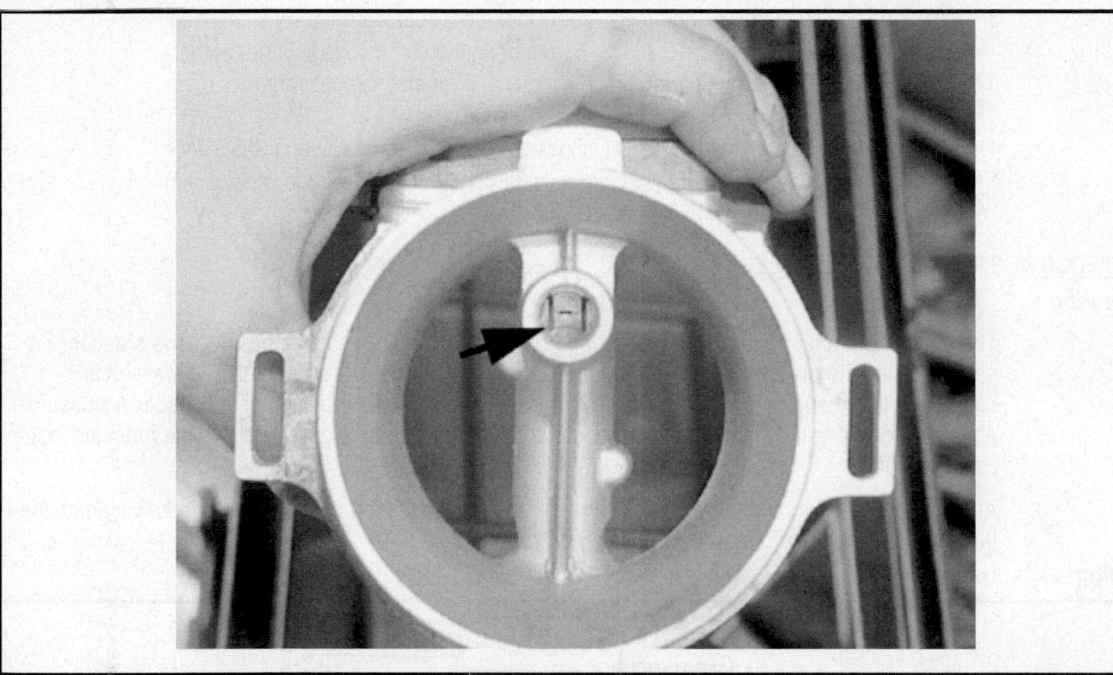

The exposed "hot wire" of the MAF sensor

The Mass Air Flow (MAF) sensor directly measures the mass of air being drawn into the engine. The sensor output is used to calculate injector pulse width. The MAF sensor is what is referred to as a "hot-wire sensor". The sensor uses a thin platinum wire filament, wound on a ceramic bobbin and coated with glass, that is heated to 200°C (417°F) above the ambient air temperature and subjected to the intake airflow stream. A "cold-wire" is used inside the MAF sensor to determine the ambient air temperature.

Battery voltage from the EEC power relay, and a reference signal and a ground signal from the PCM are supplied to the MAF sensor. The sensor returns a signal proportionate to the current flow required to keep the "hot-wire" at the required temperature. The increased airflow across the "hot-wire" acts as a cooling fan, lowering the resistance and requiring more current to maintain the temperature of the wire. The increased current is measured by the voltage in the circuit, as current increases, voltage increases. As the airflow increases the signal return voltage of a normally operating MAF sensor will increase.

Testing

1. Using a multimeter, check for voltage by backprobing the MAF sensor connector.

2. With the key ON, and the engine OFF, verify that there is at least 10.5 volts between the VPWR and GND terminals of the MAF sensor connector. If voltage is not within specification, check power and ground circuits and repair as necessary.

3. With the key ON, and the engine ON, verify that there is at least 4.5 volts between the SIG and GND terminals of the MAF sensor connector. If voltage is not within specification, check power and ground circuits and repair as necessary.

4. With the key ON, and the engine ON, check voltage between GND and SIG RTN terminals. Voltage should be approximately 0.34-1.96 volts. If voltage is not within specification, the sensor may be faulty.

MANIFOLD ABSOLUTE PRESSURE SENSOR

Operation

The most important information for measuring engine fuel requirements comes from the pressure sensor. Using the pressure and temperature data, the PCM calculates the intake air mass. It is connected to the engine intake manifold through a hose and takes readings of the absolute pressure. A piezoelectric crystal changes a voltage input to a frequency output which reflects the pressure in the intake manifold.

Atmospheric pressure is measured both when the engine is started and when driving fully loaded, then the pressure sensor information is adjusted accordingly.

Testing

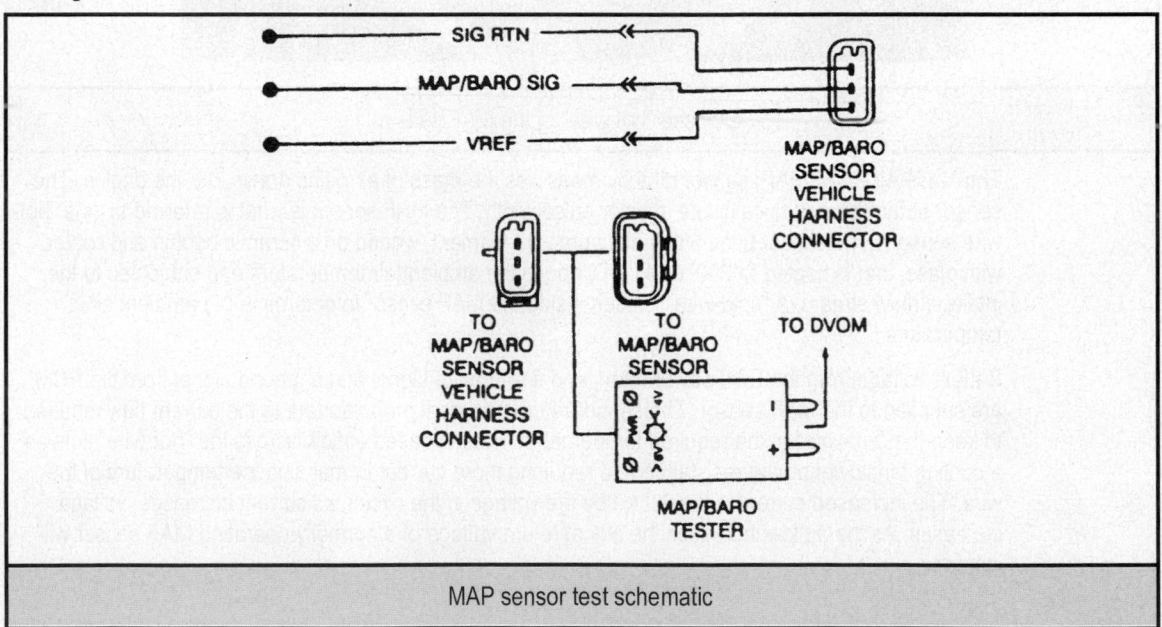

MAP sensor test schematic

Approximate Altitude (Feet)	Signal Voltage (±0.04V)
0	1.59
1000	1.56
2000	1.53
3000	1.50
4000	1.47
5000	1.44
6000	1.41
7000	1.39

MAP sensor altitude/voltage output relationship

1. Connect MAP/BARO tester to the sensor connector and sensor harness connector. With ignition ON and engine OFF, use DVOM to measure voltage across tester terminals. If the tester's 4-6V indicator is ON, the reference voltage input to the sensor is okay.

2. Measure the reference signal of the MAP sensor. If the DVOM voltage reading is as indicated in the table, the sensor is okay.

BAROMETRIC PRESSURE SENSOR

Operation

The barometric pressure sensor (BARO sensor) is a variable capacitance sensor that when supplied with a %-volt reference signal from the PCM, produces an analog voltage that indicates barometric pressure. The sensor signal is used to determine altitude to adjust the ignition timing and quantity of fuel to optimize engine operation. The output of the barometric sensor is one of the variables used to calculate glow plug on time on diesel engines.

Testing

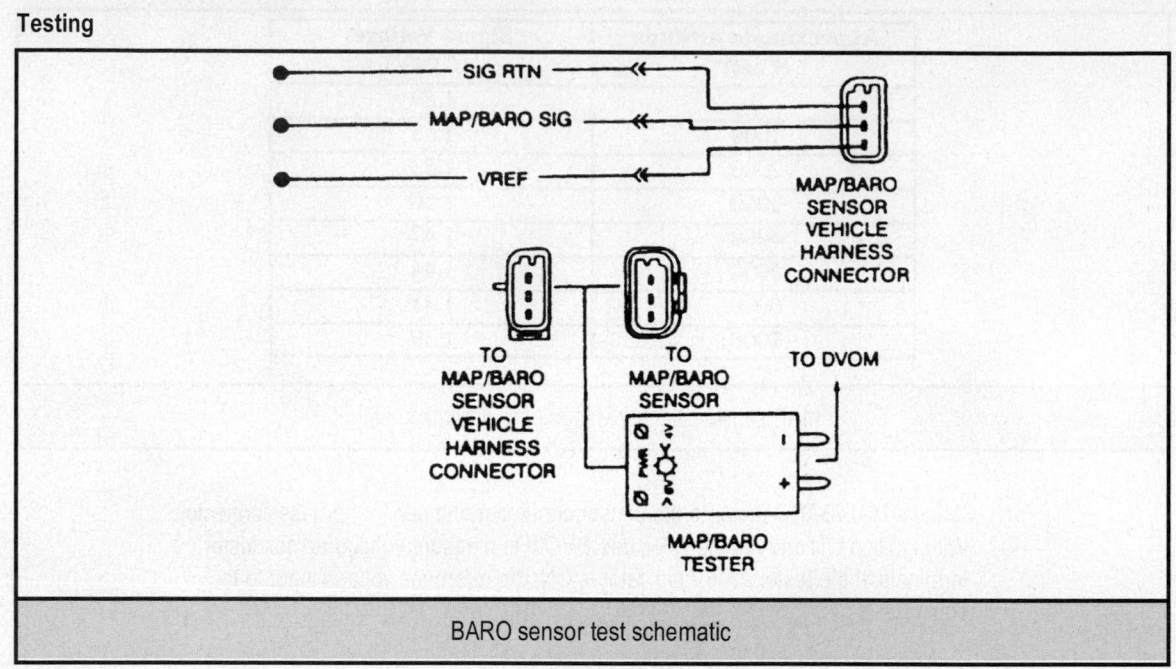

BARO sensor test schematic

Approximate Altitude (Feet)	Signal Voltage (±0.04V)
0	1.59
1000	1.56
2000	1.53
3000	1.50
4000	1.47
5000	1.44
6000	1.41
7000	1.39

BARO sensor altitude/voltage output relationship

1. Connect MAP/BARO tester to the sensor connector and sensor harness connector. With ignition ON and engine OFF, use DVOM to measure voltage across tester terminals. If the tester's 4-6V indicator is ON, the reference voltage input to the sensor is okay.

2. Measure the reference signal of the BARO sensor. If the DVOM voltage reading is as indicated in the table, the sensor is okay.

THROTTLE POSITION SENSOR
Operation

The Throttle Position (TP) sensor is a potentiometer that provides a signal to the PCM that is directly proportional to the throttle plate position. The TP sensor is mounted on the side of the throttle body and is connected to the throttle plate shaft. The TP sensor monitors throttle plate movement and position, and transmits an appropriate electrical signal to the PCM. These signals are used by the PCM to adjust the air/fuel mixture, spark timing and EGR operation according to engine load at idle, part throttle, or full throttle. The TP sensor is not adjustable.

The TP sensor receives a 5 volt reference signal and a ground circuit from the PCM. A return signal circuit is connected to a wiper that runs on a resistor internally on the sensor. The further the throttle is opened, the wiper moves along the resistor, at wide open throttle, the wiper essentially creates a loop between the reference signal and the signal return returning the full or nearly full 5 volt signal back to the PCM. At idle the signal return should be approximately 0.9 volts.

Testing

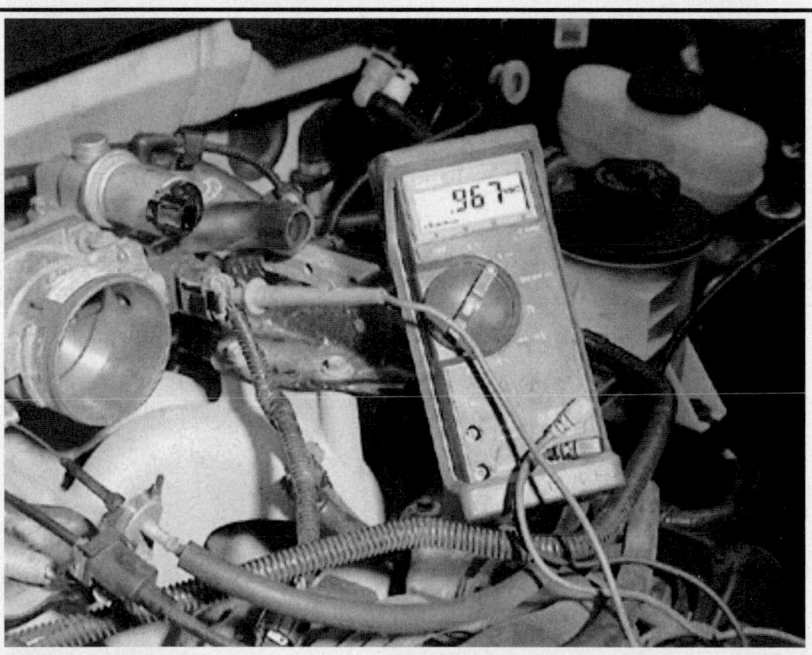

Testing the TP sensor signal return voltage at idle

Test the operation of the TP sensor by gently opening the throttle while observing the signal return voltage. The voltage should move smoothly according to the amount the throttle is opened

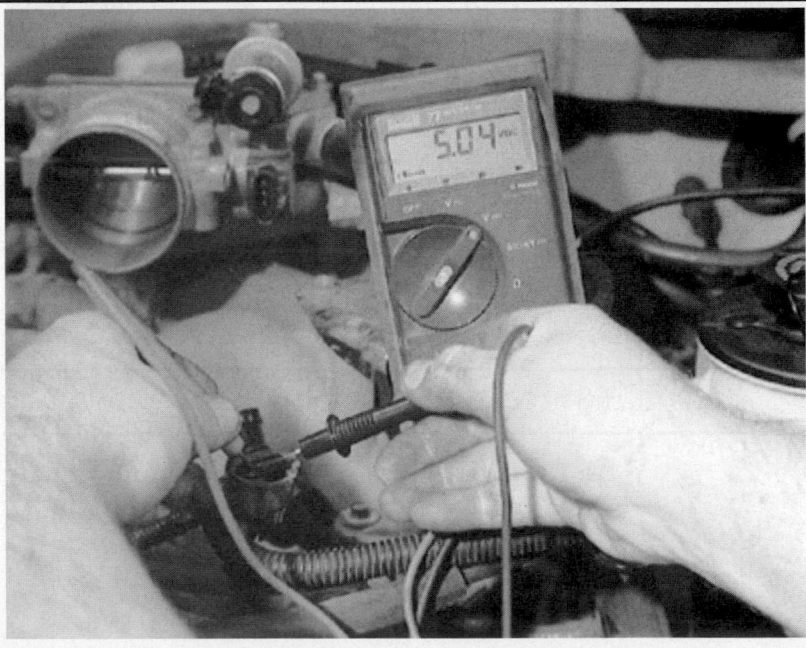

Testing the supply voltage at the TP sensor connector

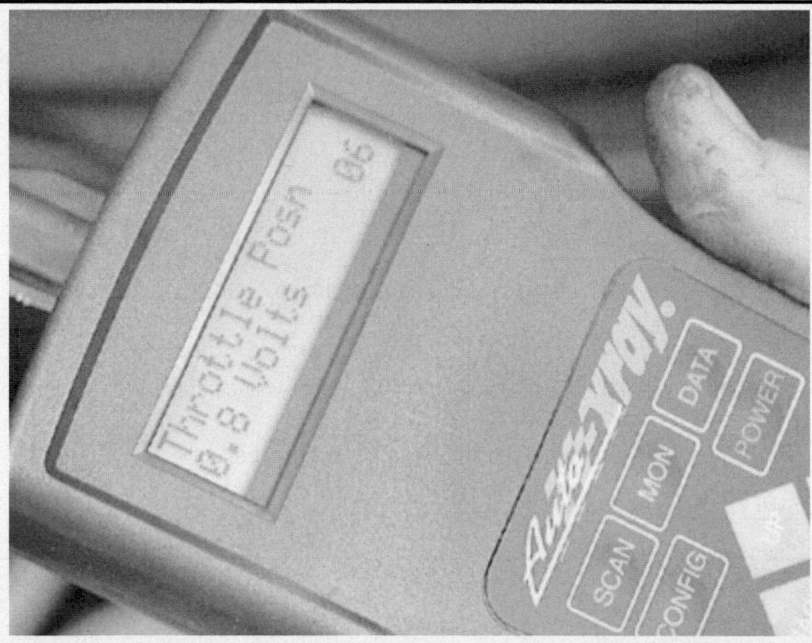

The TP sensor can be monitored with an appropriate and Data-stream capable scan tool

1. With the engine OFF and the ignition ON, check the voltage at the signal return circuit of the TP sensor by carefully backprobing the connector using a DVOM.

2. Voltage should be between 0.2 and 1.4 volts at idle.

3. Slowly move the throttle pulley to the wide open throttle (WOT) position and watch the voltage on the DVOM. The voltage should slowly rise to slightly less than 4.8v at Wide Open Throttle (WOT).

4. If no voltage is present, check the wiring harness for supply voltage (5.0v) and ground (0.3v or less), by referring to your corresponding wiring guide. If supply voltage and ground are present, but no output voltage from TP, replace the TP sensor. If supply voltage and ground do not meet specifications, make necessary repairs to the harness or PCM.

CAMSHAFT POSITION SENSOR

Operation
4.2L ENGINE

The camshaft position sensor (CMP) is a single hall-effect magnetic switch that is triggered by a single vane which is driven by the camshaft. The CMP sends a signal relating camshaft position back to the PCM which is used by the PCM to control engine timing.

4.6L, 5.4L AND 6.8L ENGINES

The camshaft position sensor (CMP) is a variable reluctance sensor that is triggered by a high point on the left-hand exhaust camshaft sprocket. The CMP sends a signal relating camshaft position back to the PCM which is used by the PCM to control engine timing.

DIESEL ENGINE

The Camshaft Position Sensor, is a Hall-effect sensor that generates a digital frequency while windows in a target wheel pass through its magnetic field. The frequency of the windows passing by the sensor, as well as the width of selected windows, allows the PCM to detect engine speed and position.

Testing

1. Check voltage between the camshaft position sensor terminals PWR GND and CID.

2. With engine running, voltage should be greater than 0.1 volt AC and vary with engine speed.

3. If voltage is not within specification, check for proper voltage at the VPWR terminal.

4. If VPWR voltage is greater than 10.5 volts, sensor may be faulty.

CRANKSHAFT POSITION SENSOR
Operation

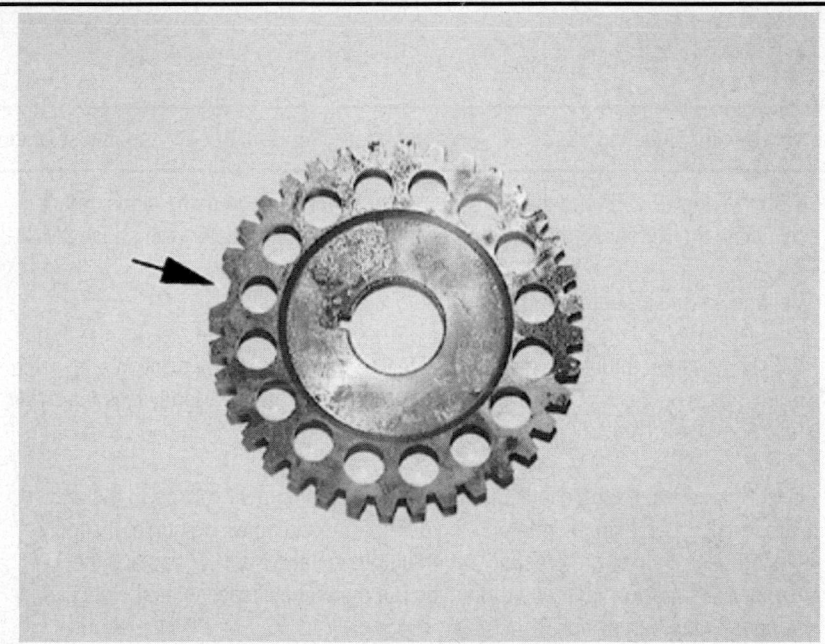

The CKP sensor trigger wheel rides on the front of the crankshaft. The missing tooth creates a fluctuation of voltage in the sensor

The Crankshaft Position (CKP) sensor is a variable reluctance sensor that uses a trigger wheel to induce voltage. The CKP sensor is a fixed magnetic sensor mounted to the engine block and monitors the trigger or "pulse" wheel which is attached to the crank pulley/damper. As the pulse wheel rotates by the CKP sensor, teeth on the pulse wheel induce voltage inside the sensor through magnetism. The pulse wheel has a missing tooth that changes the reading of the sensor. This is used for the Cylinder Identification (CID) function to properly monitor and adjust engine timing by locating the number 1 cylinder. The voltage created by the CKP sensor is alternating current (A/C). This voltage reading is sent to the PCM and is used to determine engine RPM, engine timing, and is used to fire the ignition coils.

Testing

1. Measure the voltage between the sensor CKP sensor terminals by backprobing the sensor connector.

If the connector cannot be backprobed, fabricate or purchase a test harness.

2. Sensor voltage should be more than 0.1 volt AC with the engine running and should vary with engine RPM.

3. If voltage is not within specification, the sensor may be faulty.

KNOCK SENSOR
Operation

The operation of the Knock Sensor (KS) is to monitor preignition or "engine knocks" and send the signal to the PCM. The PCM responds by adjusting ignition timing until the "knocks" stop. The sensor works by generating a signal produced by the frequency of the knock as recorded by the piezoelectric ceramic disc inside the KS. The disc absorbs the shock waves from the knocks and exerts a pressure on the metal diaphragm inside the KS. This compresses the crystals inside the disc and the disc generates a voltage signal proportional to the frequency of the knocks ranging from zero to 1 volt.

Testing

There is real no test for this sensor, the sensor produces it's own signal based on information gathered while the engine is running. The sensors also are usually inaccessible without major component removal. The sensors can be monitored with an appropriate scan tool using a data display or other data stream information. Follow the instructions included with the scan tool for information on accessing the data. The only test available is to test the continuity of the harness from the PCM to the sensor.

VEHICLE SPEED SENSOR
Operation

The Vehicle Speed Sensor (VSS) is a magnetic pick-up sensor that sends a signal to the Powertrain Control Module (PCM) and the speedometer. The sensor measures the rotation of the transmission output shaft or the ring gear on the differential and sends an AC voltage signal to the PCM which determines the corresponding vehicle speed.

Testing

1. Disconnect the negative battery cable.

2. Disengage the wiring harness connector from the VSS.

3. Using a Digital Volt-Ohmmeter (DVOM), measure the resistance (ohmmeter function) between the sensor terminals. If the resistance is 190-250 ohms, the sensor is okay.

INJECTOR DRIVER MODULE
Operation

The injector driver module (IDM) applies 115volts DC to all of the fuel injectors on the 7.3L DIT diesel engine. The IDM receives signals from the PCM for cylinder identification and fuel delivery command. The PCM controls when the timing of the injectors should start and how long the injector is open. The IDM controls the injector firing sequence through output drivers. The IDM has an output driver for each injector: one low side driver for each injector and one high side injector for each bank of injectors. The injector is fired when the output driver closes the circuit to ground.

ACCELERATOR PEDAL POSITION SENSOR
Operation

The accelerator pedal position (APP) sensor is a potentiometer that provides a signal to the PCM proportional to the accelerator pedal position. This sensor is only on models equipped with the 7.3L DIT diesel engine. The 7.3L DIT diesel engine incorporates a "drive by wire" system in which there is no cable to open the throttle plate. The APP sensor sends a signal to the PCM and the PCM control the air/fuel mixture accordingly. The accelerator pedal position sensor acts similar to the TP sensor on a gasoline engine. It provides the input to the PCM directly proportional to the engine load.

Testing

The accelerator pedal position sensor is tested just like the throttle position sensor. See throttle position sensor for testing.

INJECTION CONTROL PRESSURE SENSOR
Operation

The injection control pressure sensor (ICP) is a variable capacitance sensor that when supplied with a 5-volt reference signal from the PCM, produces a linear analog voltage signal that indicates oil pressure in the high pressure oil pump on the 7.3L DIT diesel engine. The ICP sensor provides a feedback signal to indicate high oil pressure so that the PCM can command the correct injector timing, pulse width and injection control pressure for proper fuel delivery at all speed and load conditions.

IDLE VALIDATION SWITCH
Operation

The idle validation switch (IDS) provides the PCM with a redundant signal to verify when the accelerator pedal is in the idle position. Any detected malfunction of the idle validation switch will illuminate the check engine light and cause the engine to operate at idle only.

ENGINE OIL TEMPERATURE SENSOR
Operation

The engine oil temperature (EOT) sensor changes resistance in response to the changing temperature of the engine oil. This sensor is located on the high pressure oil pump used to fire the fuel injectors on the 7.3L DIT diesel engine. The EOT sensor is a standard thermistor type sensor like an ECT or IAT. The EOT resistance decreases as the oil temperature increases providing the PCM with a signal relevant to the temperature of the engine oil.

Testing

1. Turn the ignition switch OFF.

2. Disconnect the wiring harness from the EOT sensor.

3. Measure the resistance between the sensor terminals.

4. Compare the resistance reading with the accompanying chart.

5. If the resistance is not within specification, the EOT may be faulty.

6. Connect the wiring harness to the sensor.

EXHAUST BACK PRESSURE SENSOR
Operation

The exhaust back pressure (EBP) sensor is a variable capacitance sensor that when supplied with a 5-volt reference signal from the PCM, produces a linear analog voltage signal that indicates exhaust back pressure. The EBP sensor is the input to the PCM used to control the exhaust back pressure regulator.

EXHAUST BACK PRESSURE REGULATOR
Operation

The exhaust back pressure regulator (EPR) is a variable position valve that control the exhaust back pressure during cold ambient temperatures to decrease the amount of time required to bring the engine to normal operating temperature. The PCM measures the exhaust back pressure, ambient air temperature, and the engine oil temperature to determine the desired exhaust back pressure.

INJECTION PRESSURE REGULATOR
Operation

The injection pressure regulator (IPR) is a variable position valve that controls injection control pressure on the 7.3L DIT diesel engine. Battery voltage is supplied to the IPR when the ignition switch is in the ON position. The valve position is controlled by switching the output signal circuit to ground inside the PCM. The injection control pressure is controlled by the PCM based on inputs received from the IAT, EOT, MAP/BARO, and accelerator pedal position sensors.

Explorer & Mountaineer
ELECTRONIC ENGINE CONTROL (EEC)

All Sequential Fuel Injection (SFI) systems use the EEC system. The heart of the EEC system is a micro-processor called the Powertrain Control Module (PCM). The PCM receives data from a number of sensors and other electronic components (switches, relay, etc.). Based on information received and information programmed in the PCM's memory, it generates output signals to control various relay, solenoids and other actuators. The PCM in the EEC system has calibration modules located inside the assembly that contain calibration specifications for optimizing emissions, fuel economy and drive ability. The calibration module is called a PROM.

The following are the electronic engine controls used by Ranger/Explorer and Mountaineers:

● Powertrain Control Module (PCM)

● Throttle Position (TP) sensor

● Mass Air Flow (MAF) sensor

● Intake Air Temperature (IAT) sensor

● Idle Air Control (IAC) valve

- Engine Coolant Temperature (ECT) sensor

- Heated Oxygen Sensor (HO2S)

- Camshaft Position (CMP) sensor

- Knock Sensor (KS)

- Vehicle Speed Sensor (VSS)

- Crankshaft Position (CKP) sensor

The MAF sensor (a potentiometer) senses the position of the airflow in the engine's air induction system and generates a voltage signal that varies with the amount of air drawn into the engine. The IAT sensor (a sensor in the area of the MAF sensor) measures the temperature of the incoming air and transmits a corresponding electrical signal. Another temperature sensor (the ECT sensor) inserted in the engine coolant tells if the engine is cold or warmed up. The TP sensor, a switch that senses throttle plate position, produces electrical signals that tell the PCM when the throttle is closed or wide open. A special probe (the HO2S) in the exhaust manifold measures the amount of oxygen in the exhaust gas, which is in indication of combustion efficiency, and sends a signal to the PCM. The sixth signal, camshaft position information, is transmitted by the CMP sensor, installed in place of the distributor (engines with distributorless ignition), or integral with the distributor.

The EEC microcomputer circuit processes the input signals and produces output control signals to the fuel injectors to regulate fuel discharged to the injectors. It also adjusts ignition spark timing to provide the best balance between driveability and economy, and controls the IAC valve to maintain the proper idle speed.

Because of the complicated nature of the Ford system, special tools and procedures are necessary for testing and troubleshooting.

POWERTRAIN CONTROL MODULE (PCM)
Operation

The Powertrain Control Module (PCM) performs many functions on your car. The module accepts information from various engine sensors and computes the required fuel flow rate necessary to maintain the correct amount of air/fuel ratio throughout the entire engine operational range.

Based on the information that is received and programmed into the PCM's memory, the PCM generates output signals to control relays, actuators and solenoids. The PCM also sends out a command to the fuel injectors that meters the appropriate quantity of fuel. The module automatically senses and compensates for any changes in altitude when driving your vehicle.

HEATED OXYGEN SENSORS (HO2S)
Operation

The oxygen sensor supplies the computer with a signal which indicates a rich or lean condition during engine operation. The input information assists the computer in determining the proper air/fuel ratio. A low voltage signal from the sensor indicates too much oxygen in the exhaust (lean condition) and, conversely, a high voltage signal indicates too little oxygen in the exhaust (rich condition).

The oxygen sensors are threaded into the exhaust manifold and/or exhaust pipes on all vehicles. Heated oxygen sensors are used on all models to allow the engine to reach the closed loop faster.

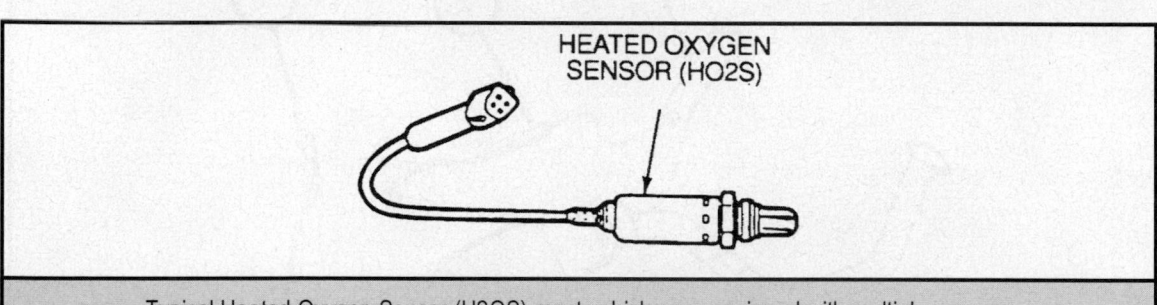

Typical Heated Oxygen Sensor (H2OS)-most vehicles are equipped with multiple sensors

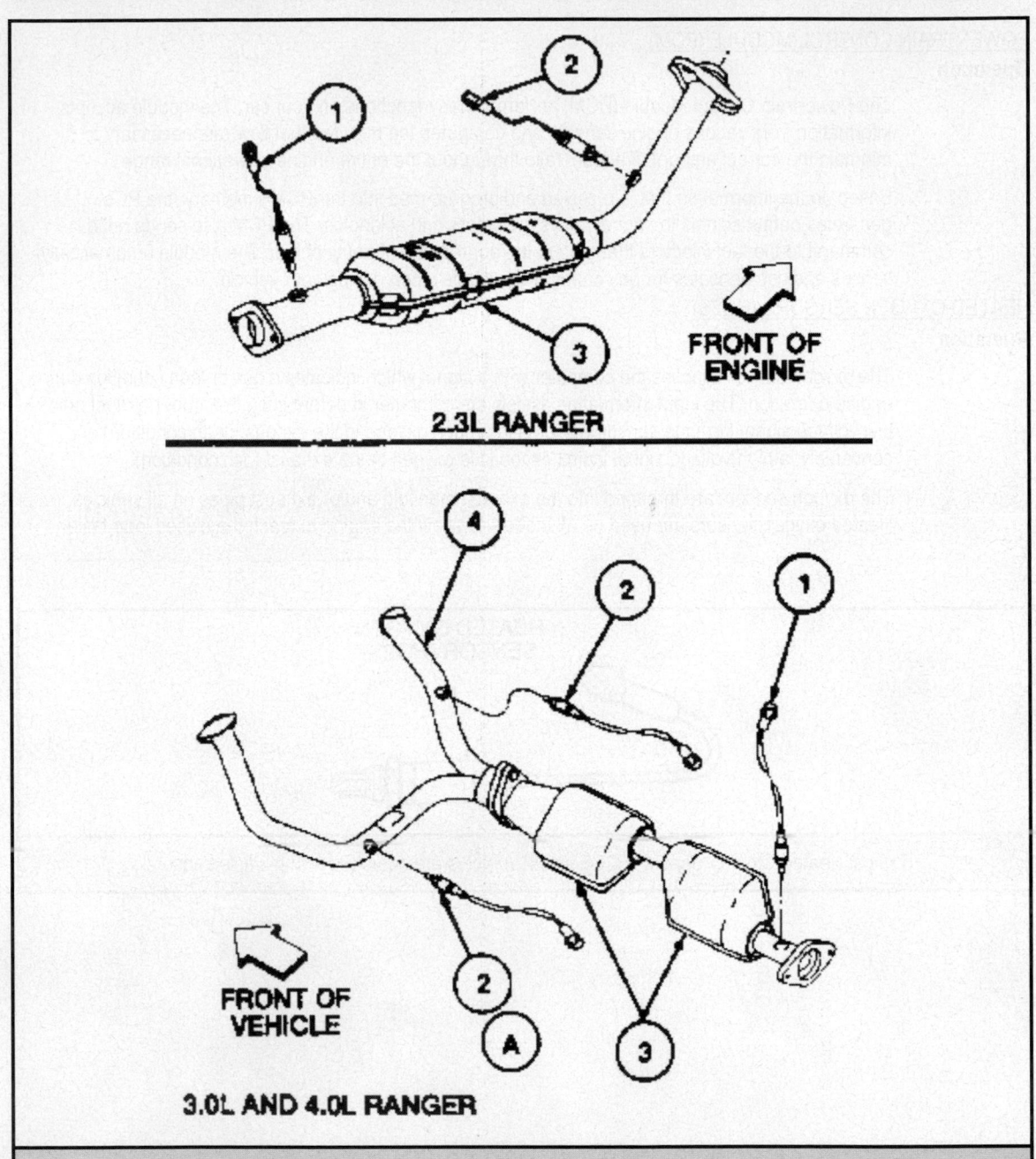

2.3L RANGER

FRONT OF ENGINE

3.0L AND 4.0L RANGER

FRONT OF VEHICLE

Typical Heated Oxygen (HO2S) sensor mounting locations-Ranger shown, other models are similar

Testing

1. Disconnect the HO2S.

2. Measure the resistance between PWR and GND terminals of the sensor. If the reading is approximately 6 ohms at 68°F (1°C). the sensor's heater element is in good condition.

3. With the HO2S connected and engine running, measure the voltage with a Digital Volt-Ohmmeter (DVOM) between terminals HO2Sand SIG RTN(GND) of the oxygen sensor connector. If the voltage readings are swinging rapidly between 0.01-1.1 volts, the sensor is probably okay.

IDLE AIR CONTROL (IAC) VALVE
Operation

The Idle Air Control (IAC) valve controls the engine idle speed and dashpot functions. The valve is located on the side of the throttle body. This valve allows air, determined by the Powertrain Control Module (PCM) and controlled by a duty cycle signal, to bypass the throttle plate in order to maintain the proper idle speed.

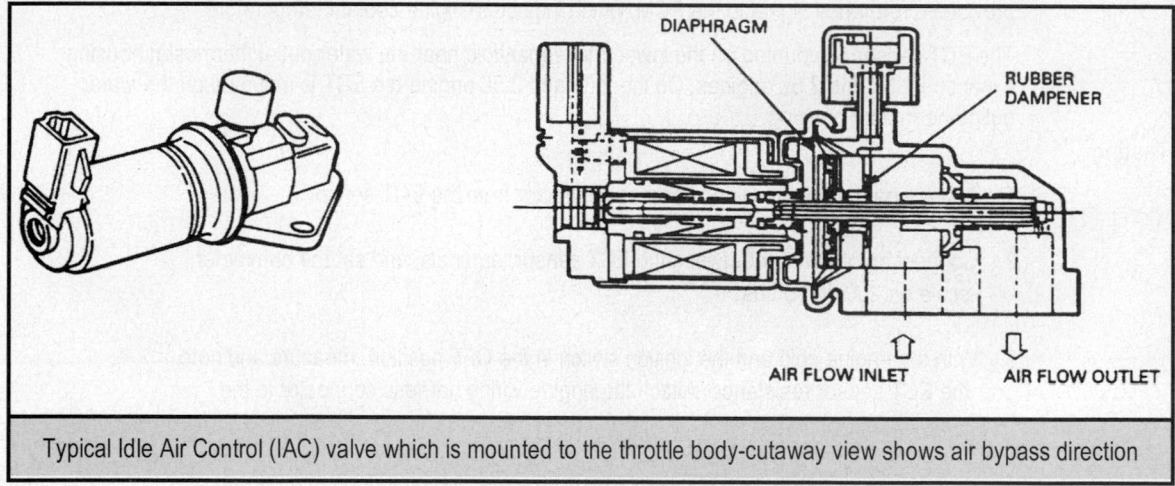

Typical Idle Air Control (IAC) valve which is mounted to the throttle body-cutaway view shows air bypass direction

Testing

1. Turn the ignition switch to the OFF position.

2. Disengage the wiring harness connector from the IAC valve .

3. Using an ohmmeter, measure the resistance between the terminals of the valve.

Due to the diode in the solenoid, place the ohmmeter positive lead on the VPWR terminal and the negative lead on the ISC terminal.

4. If the resistance is not 7-13 ohms, replace the IAC valve.

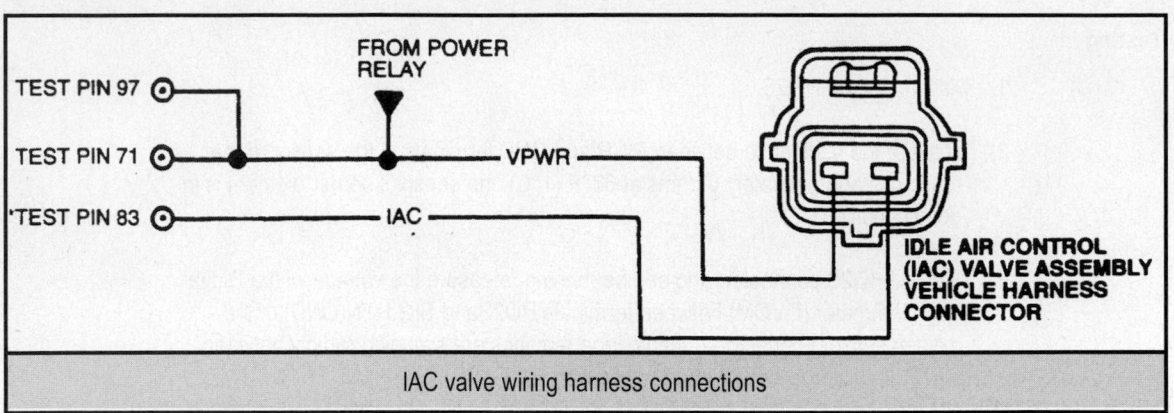

TEST PIN 97

TEST PIN 71

TEST PIN 83

FROM POWER RELAY

VPWR

IAC

IDLE AIR CONTROL (IAC) VALVE ASSEMBLY VEHICLE HARNESS CONNECTOR

IAC valve wiring harness connections

ENGINE COOLANT TEMPERATURE (ECT) SENSOR

Operation

The engine coolant temperature sensor resistance changes in response to engine coolant temperature. The sensor resistance decreases as the surrounding temperature increases. This provides a reference signal to the PCM, which indicates engine coolant temperature.

The ECT sensor is mounted on the lower intake manifold near the water outlet/thermostat housing, except on 2.3L and 2.5L engines. On the 2.3L and 2.5L engine the ECT is mounted on the water outlet/thermostat housing.

Testing

1. Disengage the engine wiring harness connector from the ECT sensor.

2. Connect an ohmmeter between the ECT sensor terminals, and set the ohmmeter scale on 200,000 ohms.

3. With the engine cold and the ignition switch in the OFF position, measure and note the ECT sensor resistance. Attach the engine wiring harness connector to the sensor.

4. Start the engine and allow the engine to warm up to normal operating temperature.

5. Once the engine has reached normal operating temperature, turn the engine OFF

6. Once again, detach the engine wiring harness connector from the ECT sensor.

7. Measure and note the ECT sensor resistance, then compare the cold and hot ECT sensor resistance measurements with the accompanying chart.

8. Replace the ECT sensor if the readings do not approximate those in the chart, otherwise reattach the engine wiring harness connector to the sensor.

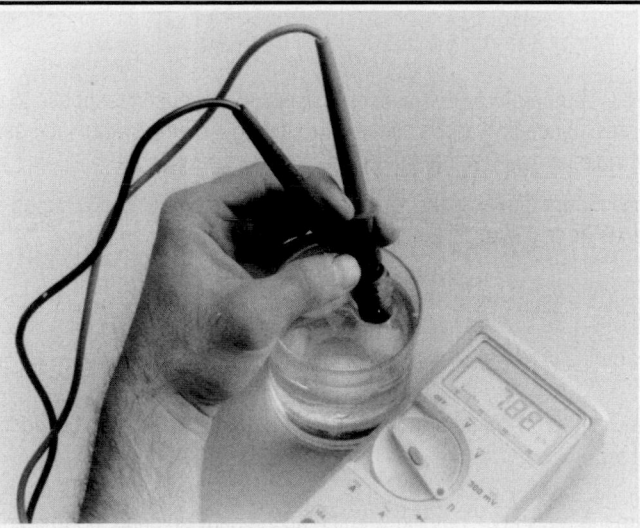

Submerge the end of the temperature sensor in cold or hot water and check resistance

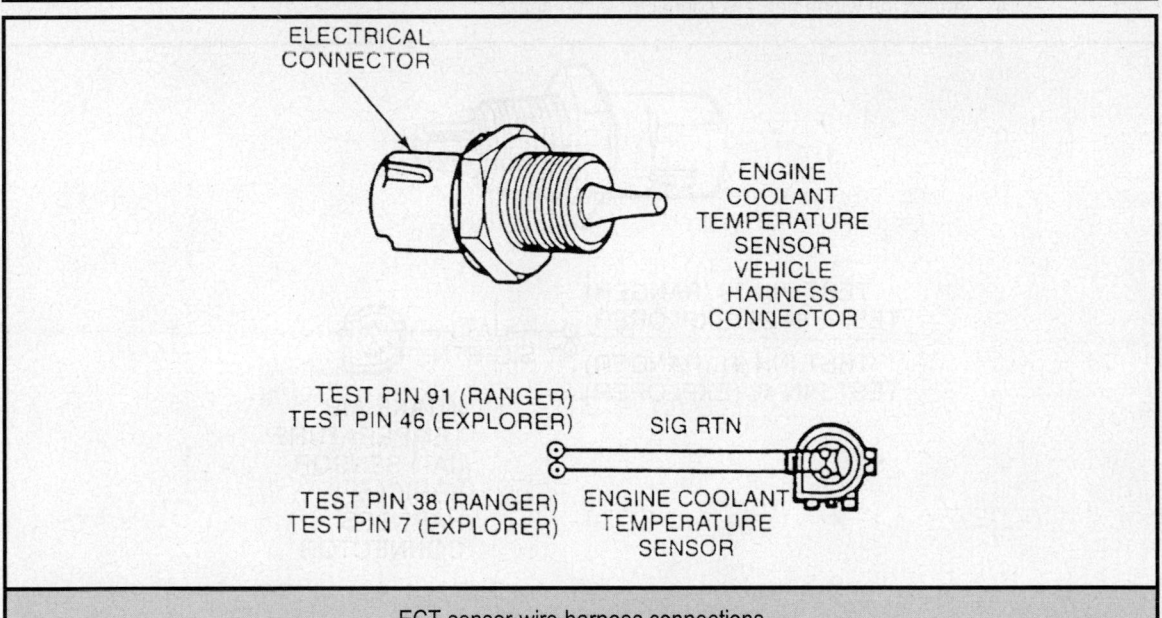

ELECTRICAL
CONNECTOR

ENGINE
COOLANT
TEMPERATURE
SENSOR
VEHICLE
HARNESS
CONNECTOR

TEST PIN 91 (RANGER)
TEST PIN 46 (EXPLORER)   SIG RTN

TEST PIN 38 (RANGER)   ENGINE COOLANT
TEST PIN 7 (EXPLORER)   TEMPERATURE
SENSOR

ECT sensor wire harness connections

INTAKE AIR TEMPERATURE (IAT) SENSOR

Operation

The Intake Air Temperature (IAT) sensor resistance changes in response to the intake air temperature. The sensor resistance decreases as the surrounding air temperature increases. This provides a signal to the PCM indicating the temperature of the incoming air charge.

Most engines mount the IAT sensor in the air cleaner-to-throttle body supply tube. However, some earlier engines have it mounted to the upper intake manifold.

Testing

Turn the ignition switch OFF.

1. Disengage the wiring harness connector from the IAT sensor.

2. Using a Digital Volt-Ohmmeter (DVOM), measure the resistance between the two sensor terminals.

3. Compare the resistance reading with the accompanying chart. If the reading for a given temperature is approximately that shown in the table, the IAT sensor is okay.

4. Attach the wiring harness connector to the sensor.

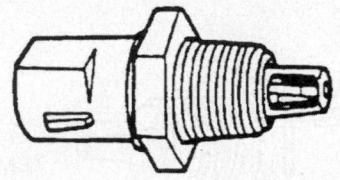

TEST PIN 39 (RANGER)
TEST PIN 25 (EXPLORER)   IAT
                               SIG RTN
TEST PIN 91 (RANGER)
TEST PIN 46 (EXPLORER)

INTAKE AIR
TEMPERATURE
(IAT) SENSOR
VEHICLE
HARNESS
CONNECTOR

IAT sensor wiring harness connections

MASS AIR FLOW (MAF) SENSOR

Operation

The Mass Air Flow (MAF) sensor directly measures the amount of the air flowing into the engine. The sensor is mounted between the air cleaner assembly and the air cleaner outlet tube.

The sensor utilizes a hot wire sensing element to measure the amount of air entering the engine. The sensor does this by sending a signal, generated by the sensor when the incoming air cools the hot wire down, to the PCM. The signal is used by the PCM to calculate the injector pulse width, which controls the air/fuel ratio in the engine. The sensor and plastic housing are integral and must be replaced if found to be defective.

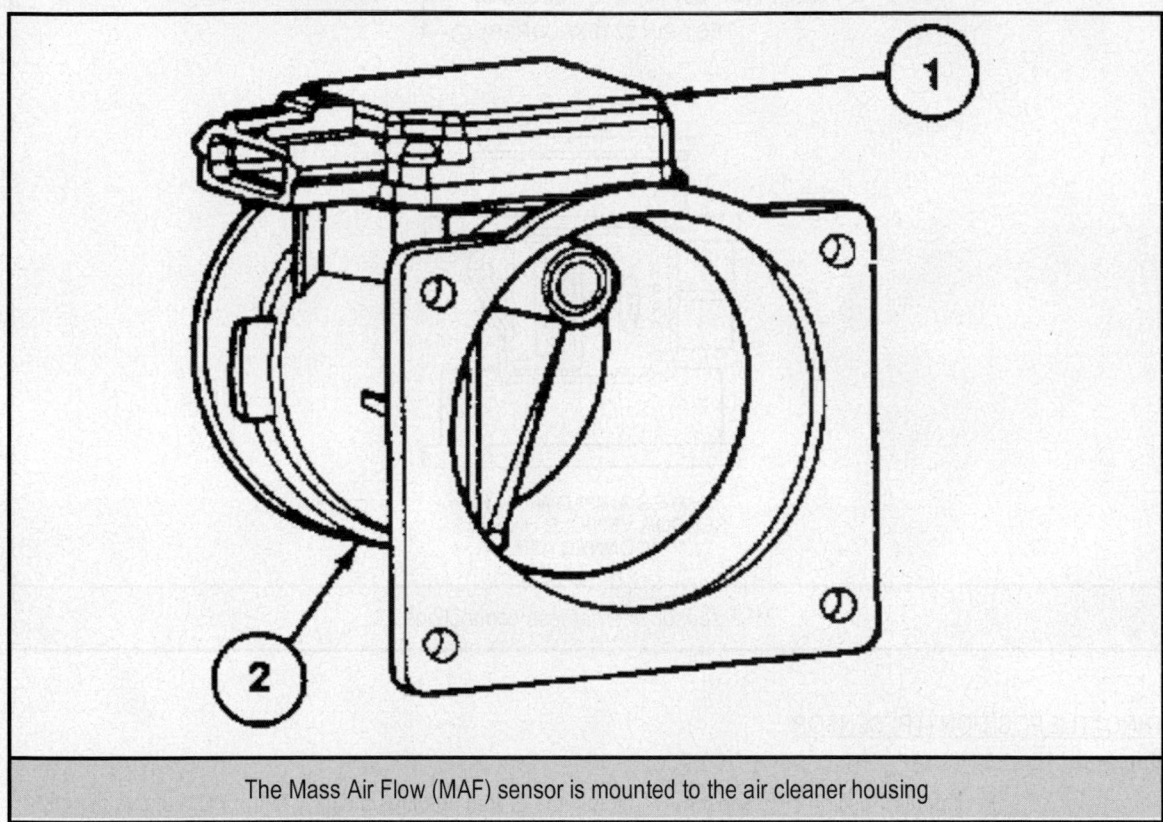

The Mass Air Flow (MAF) sensor is mounted to the air cleaner housing

The sensing element (hot wire) is a1 thin platinum wire wound on a ceramic bobbin and coated with glass. This hot wire is maintained at 392°F (200°C) above the ambient temperature as measured by a constant "cold wire".

Testing

1. With the engine running at idle, use a DVOM to verify there is at least 10.5 volts between terminals A and B of the MAF sensor connector. This indicates the power input to the sensor is correct. Then, measure the voltage between MAF sensor connector terminals C and D. If the reading is approximately 0.34-1.96 volts, the sensor is functioning properly.

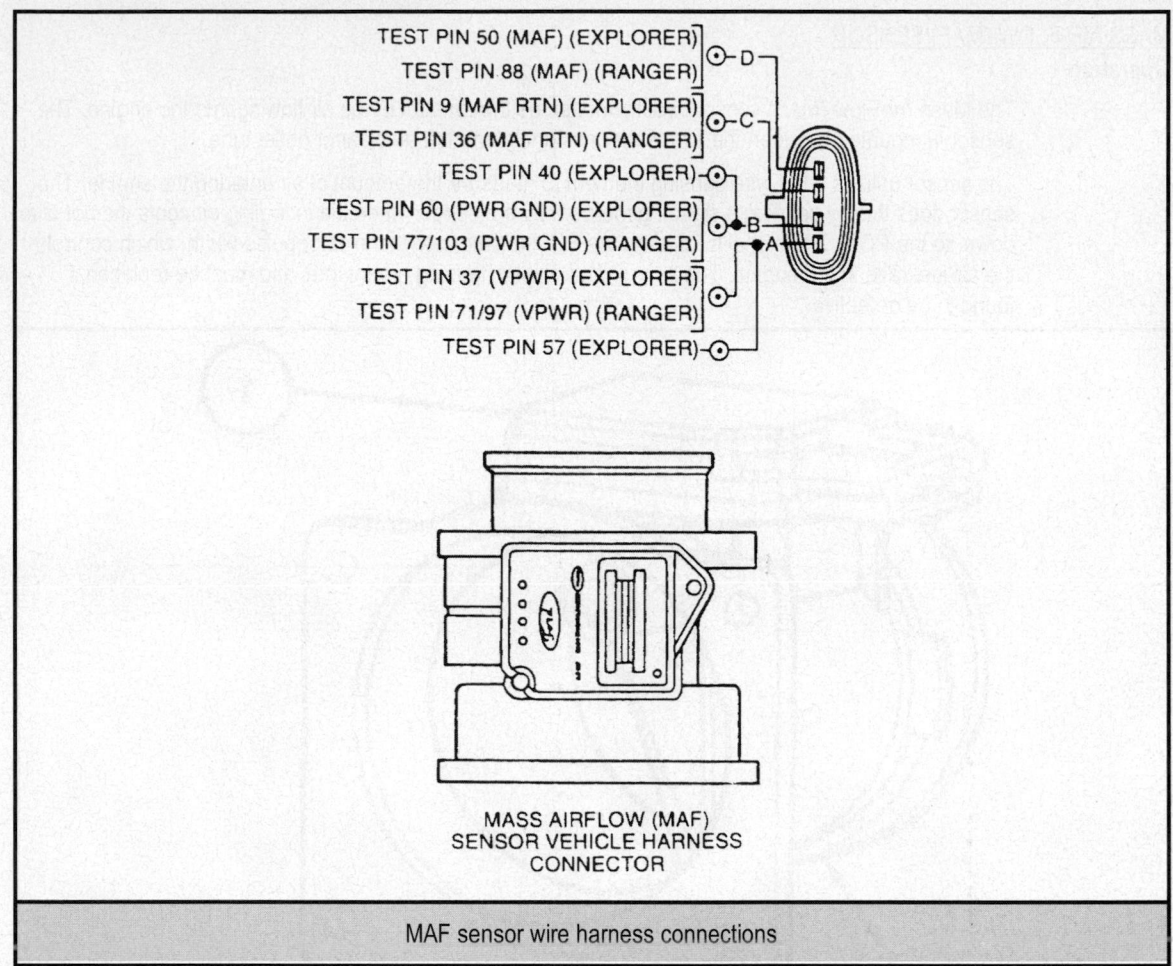

TEST PIN 50 (MAF) (EXPLORER)
TEST PIN 88 (MAF) (RANGER)
TEST PIN 9 (MAF RTN) (EXPLORER)
TEST PIN 36 (MAF RTN) (RANGER)
TEST PIN 40 (EXPLORER)
TEST PIN 60 (PWR GND) (EXPLORER)
TEST PIN 77/103 (PWR GND) (RANGER)
TEST PIN 37 (VPWR) (EXPLORER)
TEST PIN 71/97 (VPWR) (RANGER)
TEST PIN 57 (EXPLORER)

MASS AIRFLOW (MAF)
SENSOR VEHICLE HARNESS
CONNECTOR

MAF sensor wire harness connections

THROTTLE POSITION (TP) SENSOR

Operation

The Throttle Position (TP) sensor is a potentiometer that provides a signal to the PCM that is directly proportional to the throttle plate position. The TP sensor is mounted on the side of the throttle body and is connected to the throttle plate shaft. The TP sensor monitors throttle plate movement and position, and transmits an appropriate electrical signal to the PCM. These signals are used by the PCM to adjust the air/fuel mixture, spark timing and EGR operation according to engine load at idle, part throttle, or full throttle. The TPS is not adjustable.

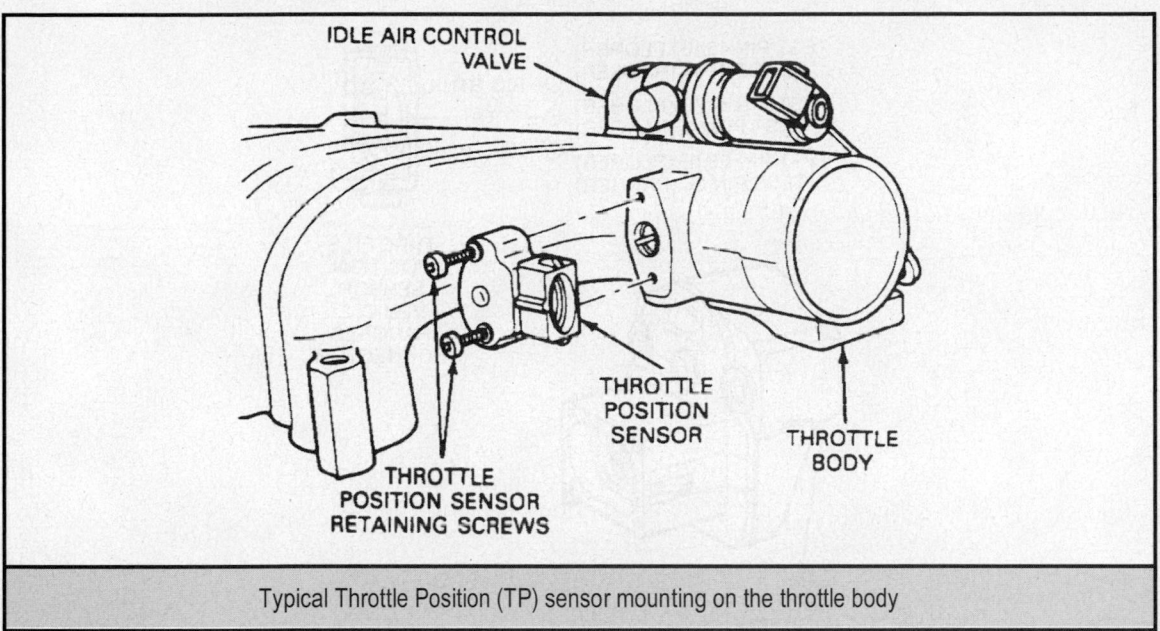

IDLE AIR CONTROL VALVE

THROTTLE POSITION SENSOR

THROTTLE BODY

THROTTLE POSITION SENSOR RETAINING SCREWS

Typical Throttle Position (TP) sensor mounting on the throttle body

Testing

1. Disconnect the negative battery cable.

2. Disengage the wiring harness connector from the TP sensor.

3. Using a Digital Volt-Ohmmeter (DVOM) set on ohmmeter function, probe the terminals, which correspond to the Brown/White and the Gray/White connector wires, on the TP sensor. Do not measure the wiring harness connector terminals, rather the terminals on the sensor itself.

4. Slowly rotate the throttle shaft and monitor the ohmmeter for a continuous, steady change in resistance. Any sudden jumps, or irregularities (such as jumping back and forth) in resistance indicates a malfunctioning sensor.

5. Reconnect the negative battery cable.

6. Turn the DVOM to the voltmeter setting.

7. Detach the wiring harness connector from the PCM (located behind the lower right-hand kick panel in the passengers' compartment), then install a break-out box between the wiring harness connector and the PCM connector.

8. Turn the ignition switch ON and using the DVOM on voltmeter function, measure the voltage between terminals 89 and 90 of the breakout box. The specification is 0.9 volts.

9. If the voltage is outside the standard value or if it does not change smoothly, inspect the circuit wiring and/or replace the TP sensor.

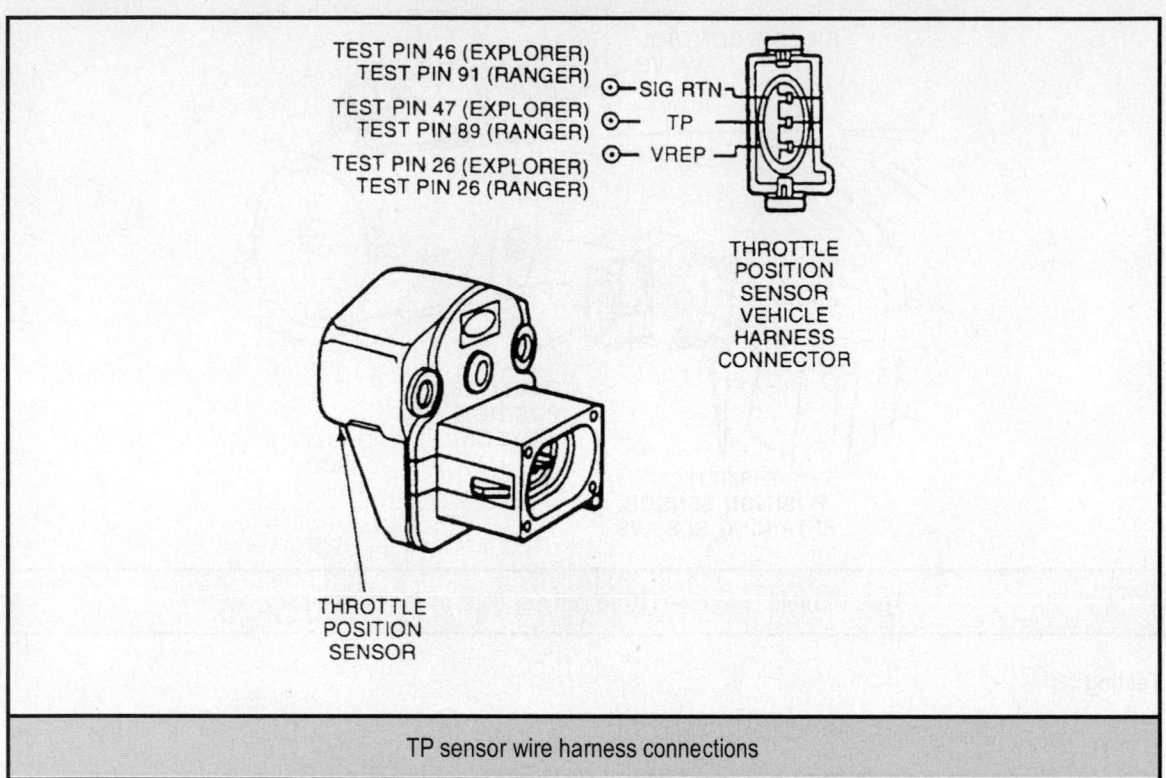

TEST PIN 46 (EXPLORER)
TEST PIN 91 (RANGER)

TEST PIN 47 (EXPLORER)
TEST PIN 89 (RANGER)

TEST PIN 26 (EXPLORER)
TEST PIN 26 (RANGER)

SIG RTN
TP
VREP

THROTTLE
POSITION
SENSOR
VEHICLE
HARNESS
CONNECTOR

THROTTLE
POSITION
SENSOR

TP sensor wire harness connections

CAMSHAFT POSITION (CMP) SENSOR
Operation

The CMP sensor provides the camshaft position information, called the CMP signal, which is used by the Powertrain Control Module (PCM) for fuel synchronization.

1991-93 2.3L and 1991-95 4.0L (VIN X) engines did not use CMP sensors.

1994 2.3L California only and 1995-99 2.3L and 2.5L engines utilize CMP sensors. The 1994 CMP is located on the oil pump drive assembly, on the left-hand lower side of the engine block. 1995-99 models CMP sensor is located and triggered by the auxiliary shaft drive sprocket.

On the 1991-94 2.9L and 3.0L engine, the distributor stator is the Camshaft Position (CMP) sensor, and it is a Hall effect magnetic switch. On the 1995-99 3.0L engine, the CMP is mounted on the oil pump drive assembly, located towards the rear of the block. it is also a single hall effect magnetic switch and it is activated by a single vane, and is driven by the camshaft.

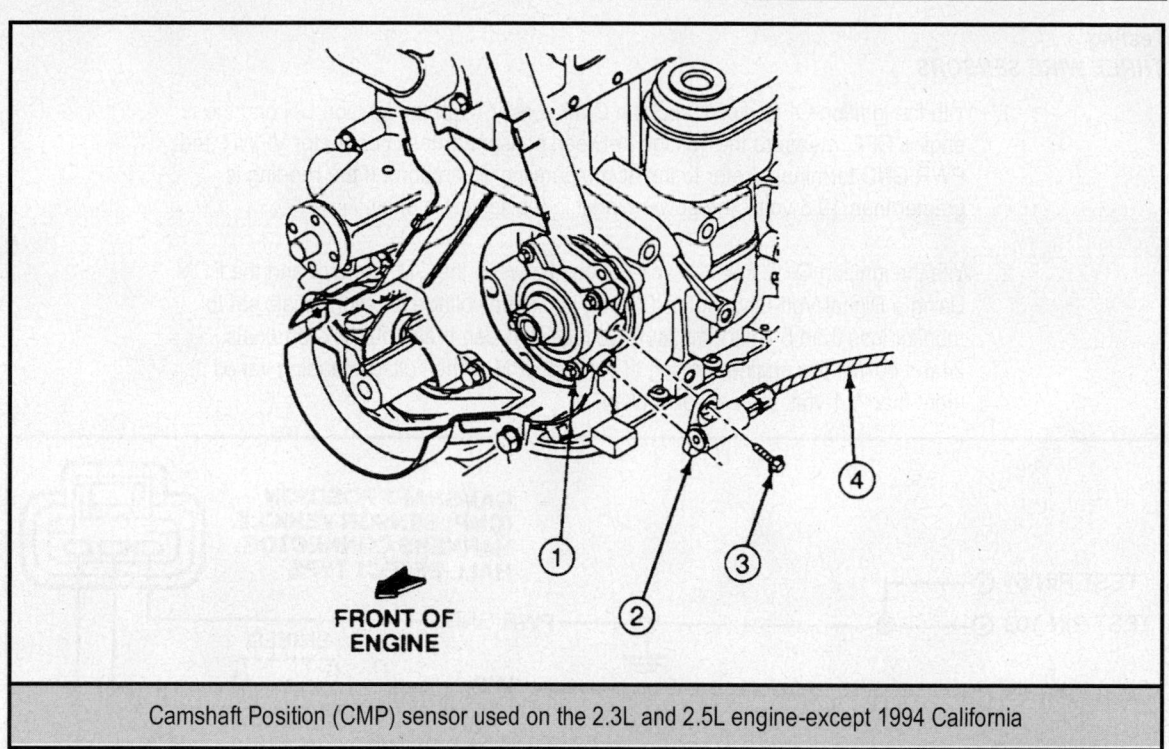

FRONT OF
ENGINE

Camshaft Position (CMP) sensor used on the 2.3L and 2.5L engine-except 1994 California

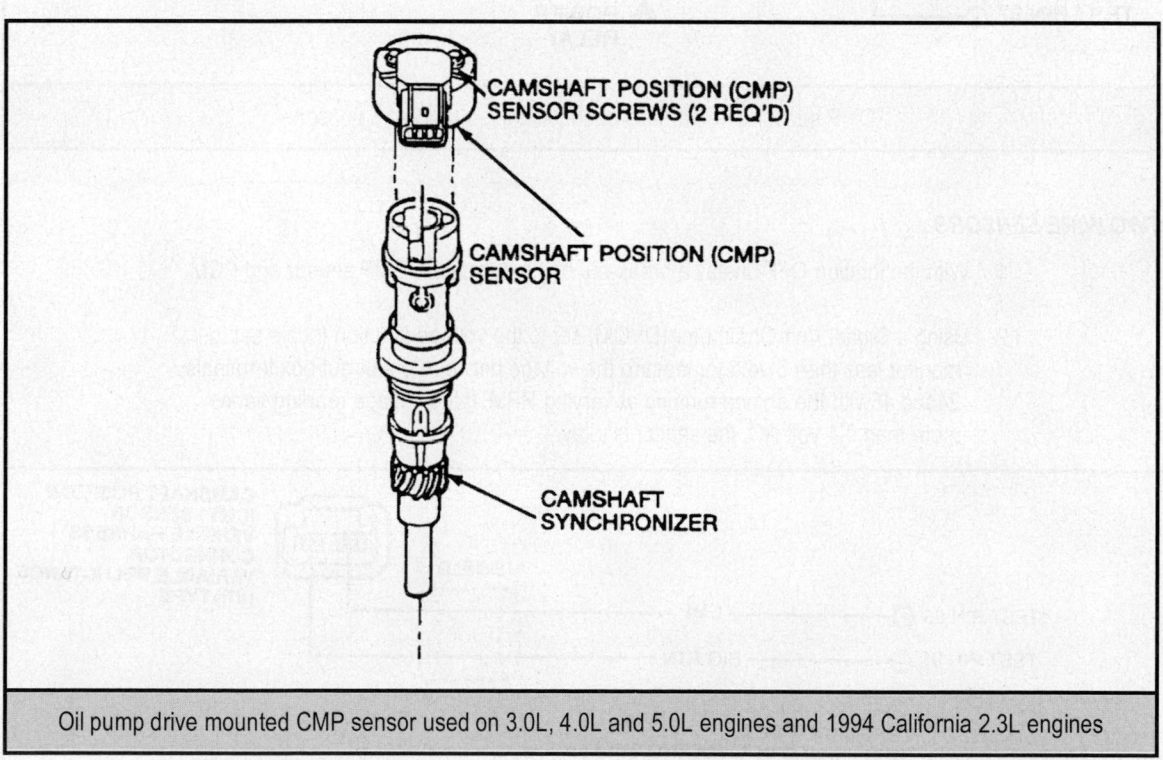

CAMSHAFT POSITION (CMP)
SENSOR SCREWS (2 REQ'D)

CAMSHAFT POSITION (CMP)
SENSOR

CAMSHAFT
SYNCHRONIZER

Oil pump drive mounted CMP sensor used on 3.0L, 4.0L and 5.0L engines and 1994 California 2.3L engines

On the 4.0L SOHC engine (VIN E), the Camshaft Position (CMP) sensor is a variable reluctance sensor, which is triggered by the high-point mark on the left-hand camshaft. It is mounted to the valve cover.

The 4.0L and 5.0L engines use a separate CMP sensor mounted to the oil pump drive. The drive assembly is located toward the rear of the engine on the 4.0L, and towards the front on the 5.0L engine.

Testing
THREE WIRE SENSORS

1. With the ignition OFF, disconnect the CMP sensor. With the ignition ON and the engine OFF, measure the voltage between sensor harness connector VPWR and PWR GND terminals (refer to the accompanying illustration). If the reading is greater than 10.5 volts, the power circuit to the sensor is okay.

2. With the ignition OFF, install break-out box between the CMP sensor and the PCM. Using a Digital Volt-Ohmmeter (DVOM) set to the voltage function (scale set to monitor less than 5 volts), measure voltage between break-out box terminals 24and 40with the engine running at varying RPM. If the voltage reading varies more than 0.1 volt, the sensor is okay.

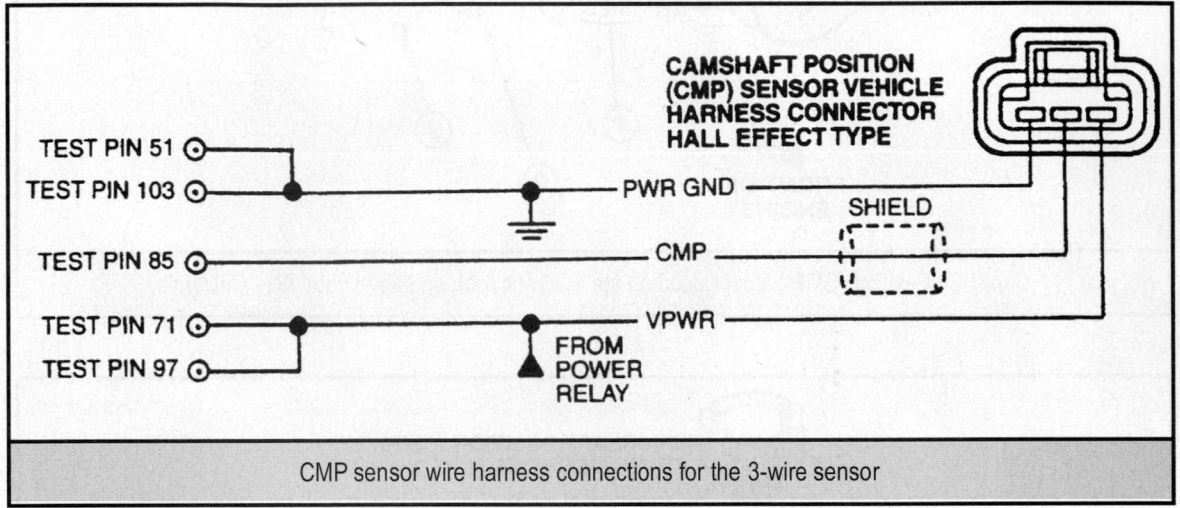

CMP sensor wire harness connections for the 3-wire sensor

TWO WIRE SENSORS

1. With the ignition OFF, install a break-out box between the CMP sensor and PCM.

2. Using a Digital Volt-Ohmmeter (DVOM) set to the voltage function (scale set to monitor less than 5 volts), measure the voltage between break-out box terminals 24and 46with the engine running at varying RPM. If the voltage reading varies more than 0.1 volt AC, the sensor is okay.

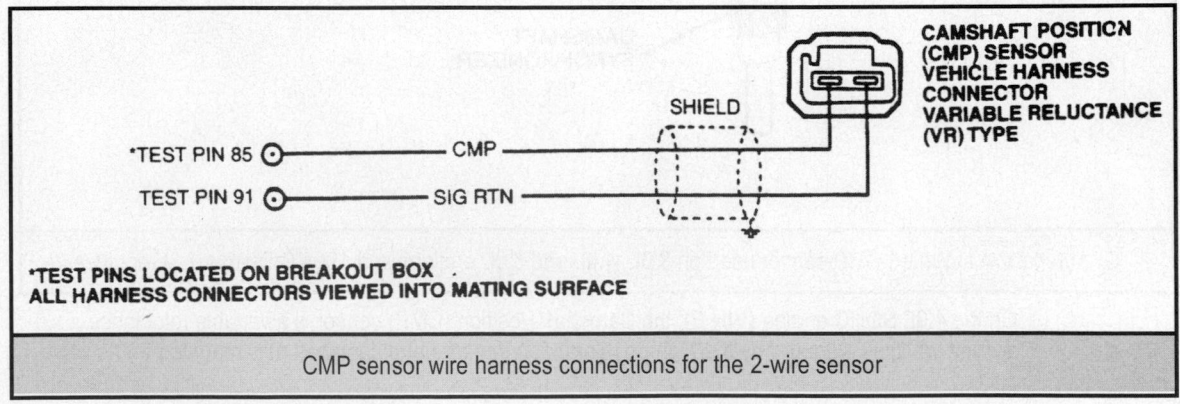

CMP sensor wire harness connections for the 2-wire sensor

CRANKSHAFT POSITION (CKP) SENSOR

Operation

The Crankshaft Position (CKP) sensor, located on the front cover (near the crankshaft pulley) is used to determine crankshaft position and crankshaft rpm. The CKP sensor is a reluctance sensor which senses the passing of teeth on a sensor ring because the teeth disrupt the magnetic field of the sensor. This disruption creates a voltage fluctuation, which is monitored by the PCM.

The CKP sensor (A) is triggered by the teeth which are machined into the crankshaft damper (B)

Testing

Using a DVOM set to the DC scale to monitor less than 5 volts, measure the voltage between the sensor Cylinder Identification (CID) terminal and ground by backprobing the sensor connector. If the connector cannot be backprobed, fabricate or purchase a test harness. The sensor is okay if the voltage reading varies more than 0.1 volt with the engine running at varying RPM.

VEHICLE SPEED SENSOR (VSS)

Operation

The Vehicle Speed Sensor (VSS) is a magnetic pick-up that sends a signal to the Powertrain Control Module (PCM). The sensor measures the rotation of the transmission and the PCM determines the corresponding vehicle speed.

Testing

1. Turn the ignition switch to the OFF position.

2. Disengage the wiring harness connector from the VSS.

3. Using a Digital Volt-Ohmmeter (DVOM), measure the resistance (DVOM ohmmeter function) between the sensor terminals. If the resistance is 190-250 ohms, the sensor is okay.

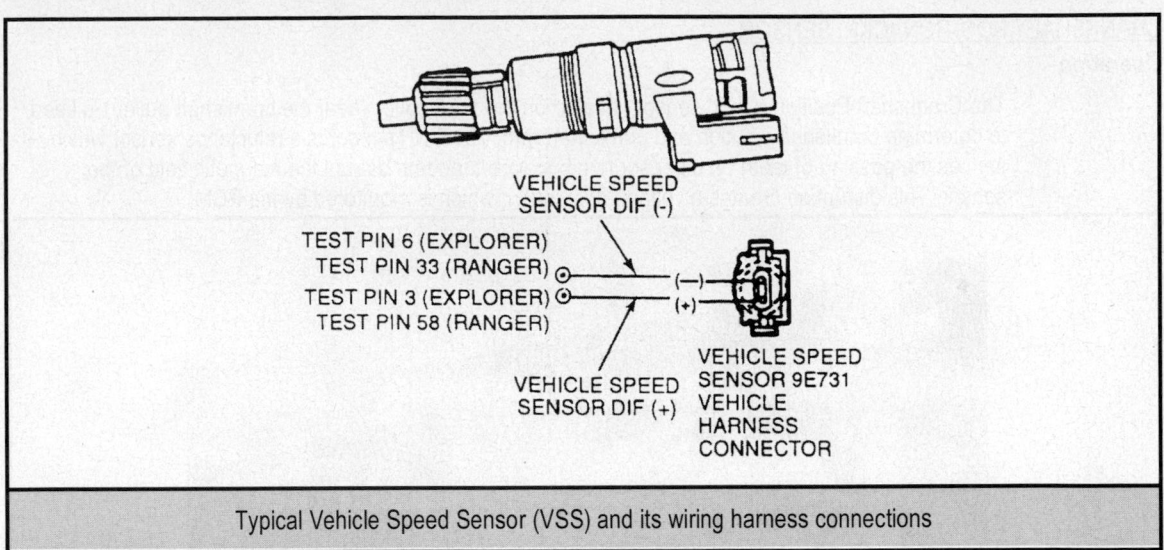

VEHICLE SPEED
SENSOR DIF (-)

TEST PIN 6 (EXPLORER)
TEST PIN 33 (RANGER)
TEST PIN 3 (EXPLORER)
TEST PIN 58 (RANGER)

(—)
(+)

VEHICLE SPEED
SENSOR 9E731
VEHICLE
HARNESS
CONNECTOR

VEHICLE SPEED
SENSOR DIF (+)

Typical Vehicle Speed Sensor (VSS) and its wiring harness connections

F-Series Trucks

ELECTRONIC ENGINE CONTROLS

Operation

The Powertrain Control Module (PCM) performs many functions on your vehicle. The module accepts information from various engine sensors and computes the required fuel flow rate necessary to maintain the correct amount of air/fuel ratio throughout the entire engine operational range.

Based on the information that is received and programmed into the PCM's memory, the PCM generates output signals to control relays, actuators and solenoids. The PCM also sends out a command to the fuel injectors that meters the appropriate quantity of fuel. The module automatically senses and compensates for any changes in altitude when driving your vehicle.

HEATED OXYGEN SENSOR

Operation

The oxygen (O2) sensor is a device which produces an electrical voltage when exposed to the oxygen present in the exhaust gases. The sensor is mounted in the exhaust system, usually in the manifold or a boss located on the down pipe before the catalyst.. The oxygen sensors used on the Ford F-series, Expedition and the Lincoln Navigator are electrically heated internally for faster switching when the engine is started cold. The oxygen sensor produces a voltage within 0 and 1 volt. When there is a large amount of oxygen present (lean mixture), the sensor produces a low voltage (less than 0.4v). When there is a lesser amount present (rich mixture) it produces a higher voltage (0.6-1.0v).The stoichiometric or correct fuel to air ratio will read between 0.4 and 0.6v. By monitoring the oxygen content and converting it to electrical voltage, the sensor acts as a rich-lean switch. The voltage is transmitted to the PCM.

Some models have two sensors, one before the catalyst and one after. This is done for a catalyst efficiency monitor that is a part of the OBD-II engine controls that are on all models covered by this manual except for those with the 5.8L or 7.5L engines. The sensor before the catalyst measures the exhaust emissions right out of the engine, and sends the signal to the PCM about the state of the mixture as previously talked about. The second sensor reports the difference in the emissions after the exhaust gases have gone through the catalyst. This sensor reports to the PCM the amount of emissions reduction the catalyst is performing.

The oxygen sensor will not work until a predetermined temperature is reached, until this time the PCM is running in what as known as OPEN LOOP operation. OPEN LOOP means that the PCM has not yet begun to correct the air-to-fuel ratio by reading the oxygen sensor. After the engine comes to operating temperature, the PCM will monitor the oxygen sensor and correct the air/fuel ratio from the sensor's readings. This is what is known as CLOSED LOOP operation.

A heated oxygen sensor (HO2S) has a heating element that keeps the sensor at proper operating temperature during all operating modes. Maintaining correct sensor temperature at all times allows the system to enter into CLOSED LOOP operation sooner.

In CLOSED LOOP operation the PCM monitors the sensor input (along with other inputs) and adjusts the injector pulse width accordingly. During OPEN LOOP operation the PCM ignores the sensor input and adjusts the injector pulse to a preprogrammed value based on other inputs.

Testing

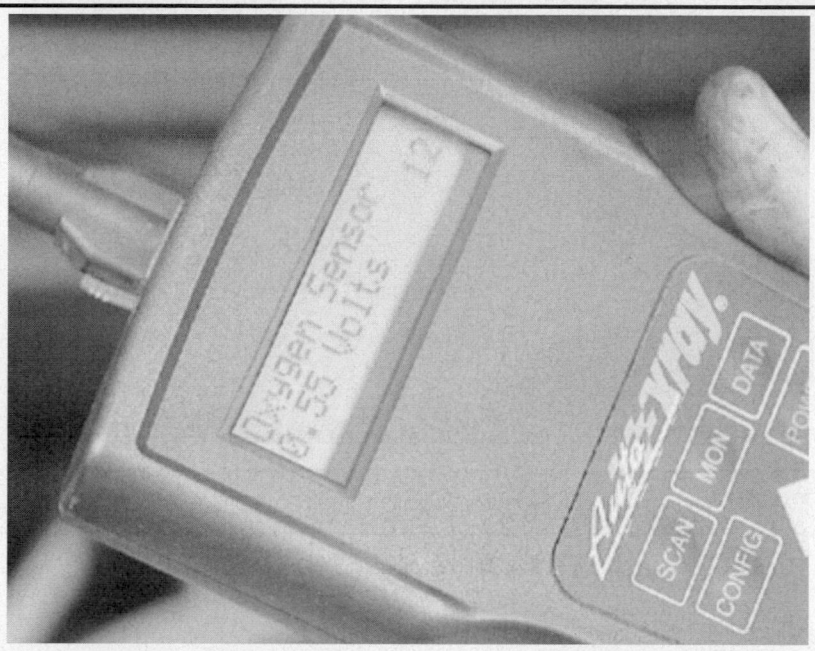

The HO2S can be monitored with an appropriate and Data-stream capable scan tool

1. Disconnect the HO2S.

2. Measure the resistance between PWR and GND terminals of the sensor. Resistance should be approximately 6 ohms at 68°F (20°C). If resistance is not within specification, the sensor's heater element is faulty.

3. With the HO2S connected and engine running, measure the voltage with a Digital Volt-Ohmmeter (DVOM) between terminals HO2S and SIG RTN (GND) of the oxygen sensor connector. Voltage should fluctuate between 0.01-1.0 volts. If voltage fluctuation is slow or voltage is not within specification, the sensor may be faulty.

IDLE AIR CONTROL VALVE

Operation

The Idle Air Control (IAC) valve adjusts the engine idle speed. The valve is located on the side of the throttle body. The valve is controlled by a duty cycle signal from the PCM and allows air to bypass the throttle plate in order to maintain the proper idle speed.

Do not attempt to clean the IAC valve. Carburetor tune-up cleaners or any type of solvent cleaners will damage the internal components of the valve.

Testing

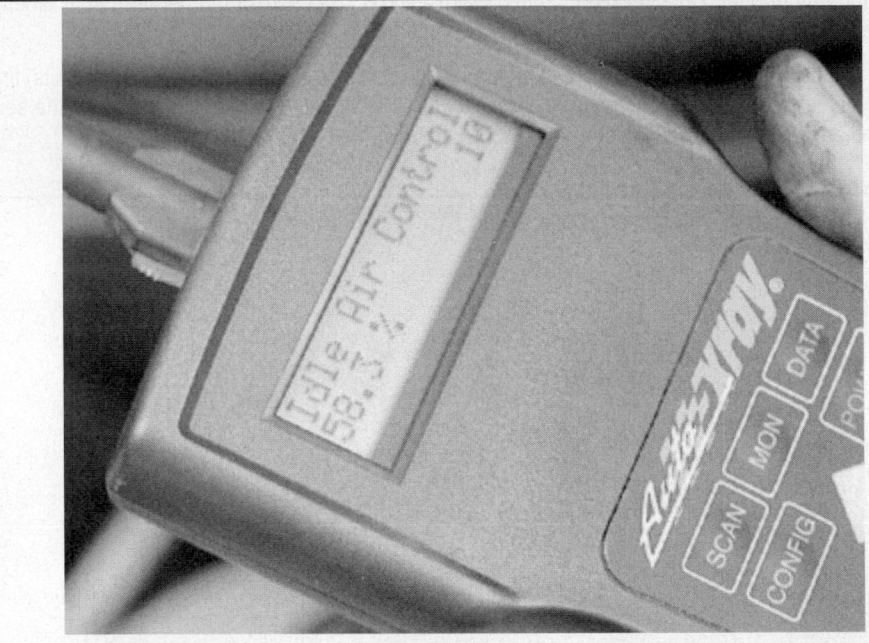

The IAC can be monitored with an appropriate and Data-stream capable scan tool

1. Turn the ignition switch to the OFF position.

2. Disconnect the wiring harness from the IAC valve.

3. Measure the resistance between the terminals of the valve.

Due to the diode in the solenoid, place the ohmmeter positive lead on the VPWR terminal and the negative lead on the ISC terminal.

4. Resistance should be 6-13 ohms.

5. If resistance is not within specification, the valve may be faulty.

ENGINE COOLANT TEMPERATURE SENSOR

Operation

The Engine Coolant Temperature (ECT) sensor resistance changes in response to engine coolant temperature. The sensor resistance decreases as the coolant temperature increases, and increases as the coolant temperature decreases. This provides a reference signal to the PCM, which indicates engine coolant temperature. The signal sent to the PCM by the ECT sensor helps the PCM to determine spark advance, EGR flow rate, air/fuel ratio, and engine temperature. The ECT is a two wire sensor, a 5-volt reference signal is sent to the sensor and the signal return is based upon the change in the measured resistance due to temperature.

Testing

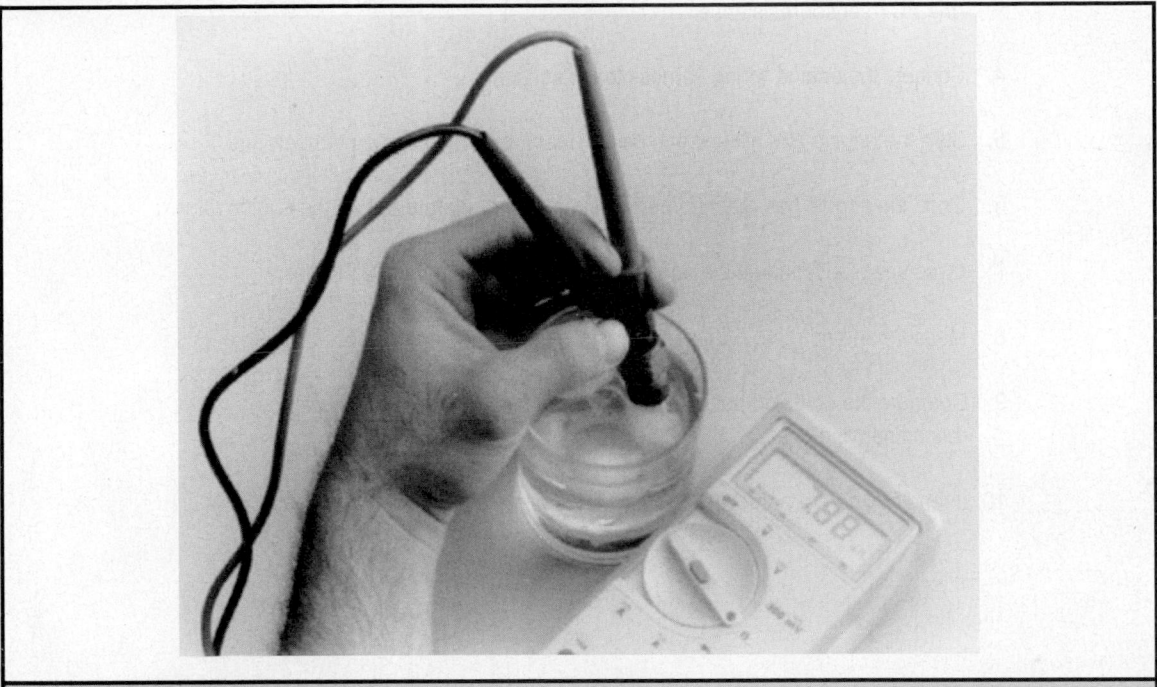

Another method of testing the ECT is to submerge it in cold or hot water and check resistance

Test the ECT resistance across the two sensor terminals

1. Disconnect the engine wiring harness from the ECT sensor.

2. Connect an ohmmeter between the ECT sensor terminals.

3. With the engine cold and the ignition switch in the OFF position, measure and note the ECT sensor resistance.

4. Connect the engine wiring harness to the sensor.

5. Start the engine and allow the engine to reach normal operating temperature.

6. Once the engine has reached normal operating temperature, turn the engine OFF.

7. Once again, disconnect the engine wiring harness from the ECT sensor.

8. Measure and note the ECT sensor resistance with the engine hot.

9. Compare the cold and hot ECT sensor resistance measurements with the accompanying chart.

10. If readings do not approximate those in the chart, the sensor may be faulty.

INTAKE AIR TEMPERATURE SENSOR
Operation

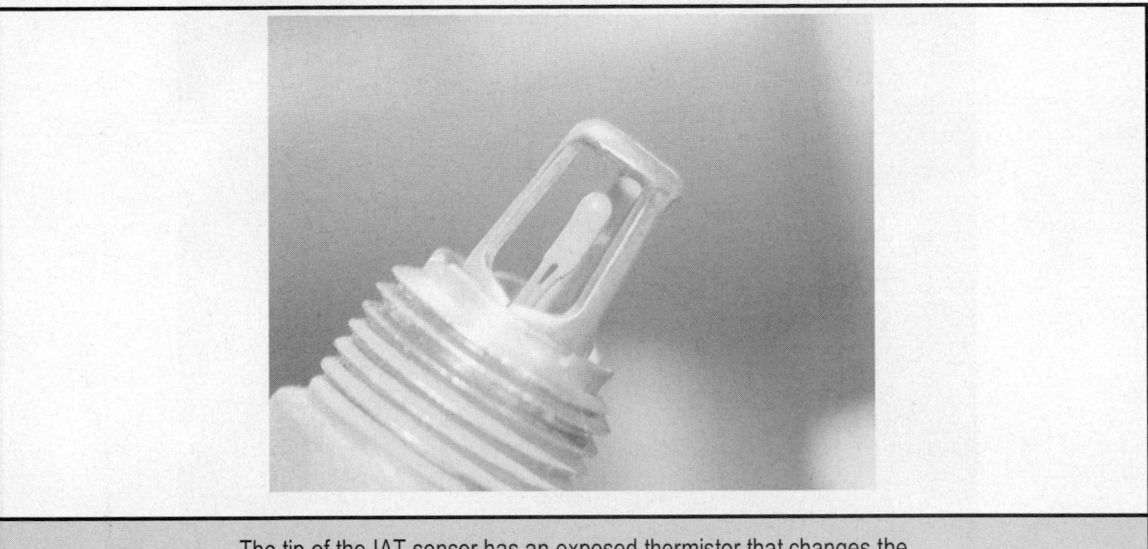

The tip of the IAT sensor has an exposed thermistor that changes the
resistance of the sensor based upon the force of the air rushing past it

The Intake Air Temperature (IAT) sensor determines the air temperature inside the intake manifold.
Resistance changes in response to the ambient air temperature. The sensor has a negative
temperature coefficient. As the temperature of the sensor rises the resistance across the sensor
decreases. This provides a signal to the PCM indicating the temperature of the incoming air charge.
This sensor helps the PCM to determine spark timing and air/fuel ratio. Information from this sensor is
added to the pressure sensor information to calculate the air mass being sent to the cylinders. The IAT
is a two wire sensor, a 5-volt reference signal is sent to the sensor and the signal return is based upon
the change in the measured resistance due to temperature.

Testing

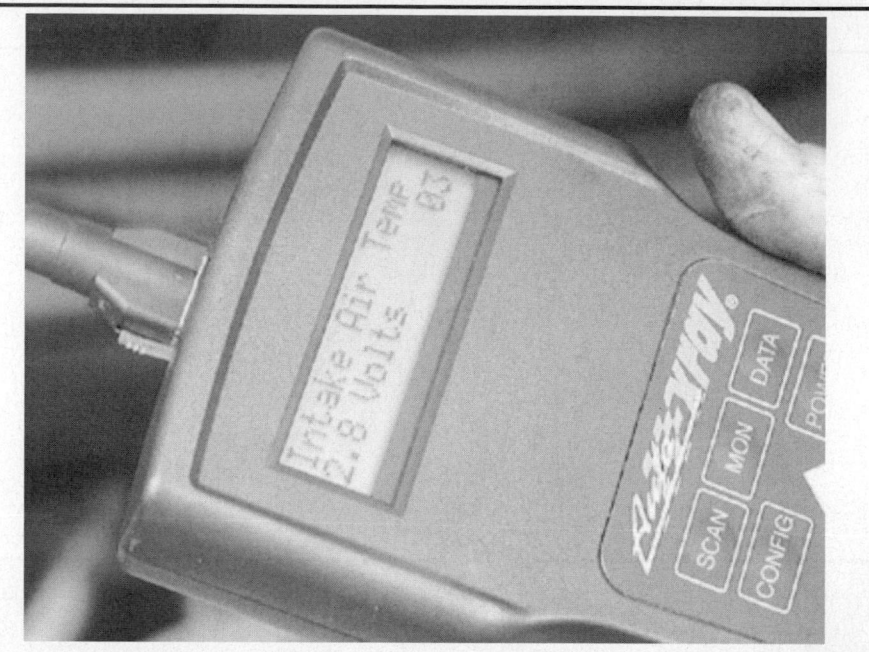

The IAT sensor can be monitored with an appropriate and Data-stream capable scan tool

Measure the resistance of the IAT sensor across the two sensor pins

1. Turn the ignition switch OFF.

2. Disconnect the wiring harness from the IAT sensor.

3. Measure the resistance between the sensor terminals.

4. Compare the resistance reading with the accompanying chart.

5. If the resistance is not within specification, the IAT may be faulty.

6. Connect the wiring harness to the sensor.

MASS AIRFLOW SENSOR
Operation

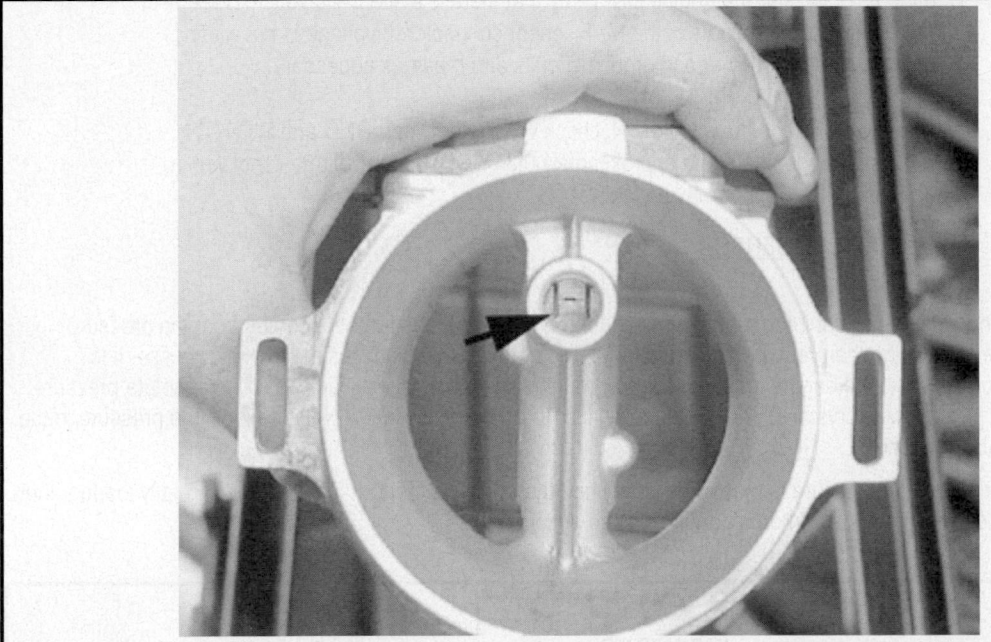

The exposed "hot wire" of the MAF sensor

The Mass Air Flow (MAF) sensor directly measures the mass of air being drawn into the engine. The sensor output is used to calculate injector pulse width. The MAF sensor is what is referred to as a "hot-wire sensor". The sensor uses a thin platinum wire filament, wound on a ceramic bobbin and coated with glass, that is heated to 200°C (417°F) above the ambient air temperature and subjected to the intake airflow stream. A "cold-wire" is used inside the MAF sensor to determine the ambient air temperature.

Battery voltage from the EEC power relay, and a reference signal and a ground signal from the PCM are supplied to the MAF sensor. The sensor returns a signal proportionate to the current flow required to keep the "hot-wire" at the required temperature. The increased airflow across the "hot-wire" acts as a cooling fan, lowering the resistance and requiring more current to maintain the temperature of the wire. The increased current is measured by the voltage in the circuit, as current increases, voltage increases. As the airflow increases the signal return voltage of a normally operating MAF sensor will increase.

Testing

1. Using a multimeter, check for voltage by backprobing the MAF sensor connector.

2. With the key ON, and the engine OFF, verify that there is at least 10.5 volts between the VPWR and GND terminals of the MAF sensor connector. If voltage is not within specification, check power and ground circuits and repair as necessary.

3. With the key ON, and the engine ON, verify that there is at least 4.5 volts between the SIG and GND terminals of the MAF sensor connector. If voltage is not within specification, check power and ground circuits and repair as necessary.

4. With the key ON, and the engine ON, check voltage between GND and SIG RTN terminals. Voltage should be approximately 0.34-1.96 volts. If voltage is not within specification, the sensor may be faulty.

MANIFOLD ABSOLUTE PRESSURE SENSOR
Operation

The most important information for measuring engine fuel requirements comes from the pressure sensor. Using the pressure and temperature data, the PCM calculates the intake air mass. It is connected to the engine intake manifold through a hose and takes readings of the absolute pressure. A piezoelectric crystal changes a voltage input to a frequency output which reflects the pressure in the intake manifold.

Atmospheric pressure is measured both when the engine is started and when driving fully loaded, then the pressure sensor information is adjusted accordingly.

Testing

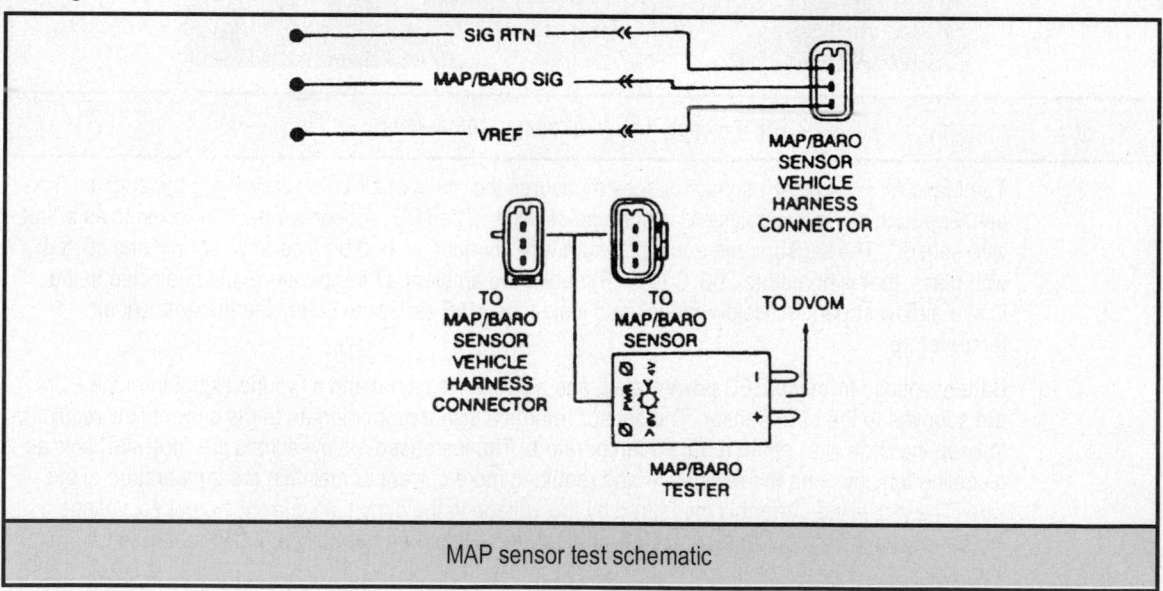

MAP sensor test schematic

Approximate Altitude (Feet)	Signal Voltage (±0.04V)
0	1.59
1000	1.56
2000	1.53
3000	1.50
4000	1.47
5000	1.44
6000	1.41
7000	1.39

MAP sensor altitude/voltage output relationship

1. Connect MAP/BARO tester to the sensor connector and sensor harness connector. With ignition ON and engine OFF, use DVOM to measure voltage across tester terminals. If the tester's 4-6V indicator is ON, the reference voltage input to the sensor is okay.

2. Measure the reference signal of the MAP sensor. If the DVOM voltage reading is as indicated in the table, the sensor is okay.

BAROMETRIC PRESSURE SENSOR

Operation

The barometric pressure sensor (BARO sensor) is a variable capacitance sensor that when supplied with a %-volt reference signal from the PCM, produces an analog voltage that indicates barometric pressure. The sensor signal is used to determine altitude to adjust the ignition timing and quantity of fuel to optimize engine operation. The output of the barometric sensor is one of the variables used to calculate glow plug on time on diesel engines.

Testing

BARO sensor test schematic

Approximate Altitude (Feet)	Signal Voltage (±0.04V)
0	1.59
1000	1.56
2000	1.53
3000	1.50
4000	1.47
5000	1.44
6000	1.41
7000	1.39

BARO sensor altitude/voltage output relationship

1. Connect MAP/BARO tester to the sensor connector and sensor harness connector. With ignition ON and engine OFF, use DVOM to measure voltage across tester terminals. If the tester's 4-6V indicator is ON, the reference voltage input to the sensor is okay.

2. Measure the reference signal of the BARO sensor. If the DVOM voltage reading is as indicated in the table, the sensor is okay.

THROTTLE POSITION SENSOR

Operation

The Throttle Position (TP) sensor is a potentiometer that provides a signal to the PCM that is directly proportional to the throttle plate position. The TP sensor is mounted on the side of the throttle body and is connected to the throttle plate shaft. The TP sensor monitors throttle plate movement and position, and transmits an appropriate electrical signal to the PCM. These signals are used by the PCM to adjust the air/fuel mixture, spark timing and EGR operation according to engine load at idle, part throttle, or full throttle. The TP sensor is not adjustable.

The TP sensor receives a 5 volt reference signal and a ground circuit from the PCM. A return signal circuit is connected to a wiper that runs on a resistor internally on the sensor. The further the throttle is opened, the wiper moves along the resistor, at wide open throttle, the wiper essentially creates a loop between the reference signal and the signal return returning the full or nearly full 5 volt signal back to the PCM. At idle the signal return should be approximately 0.9 volts.

Testing

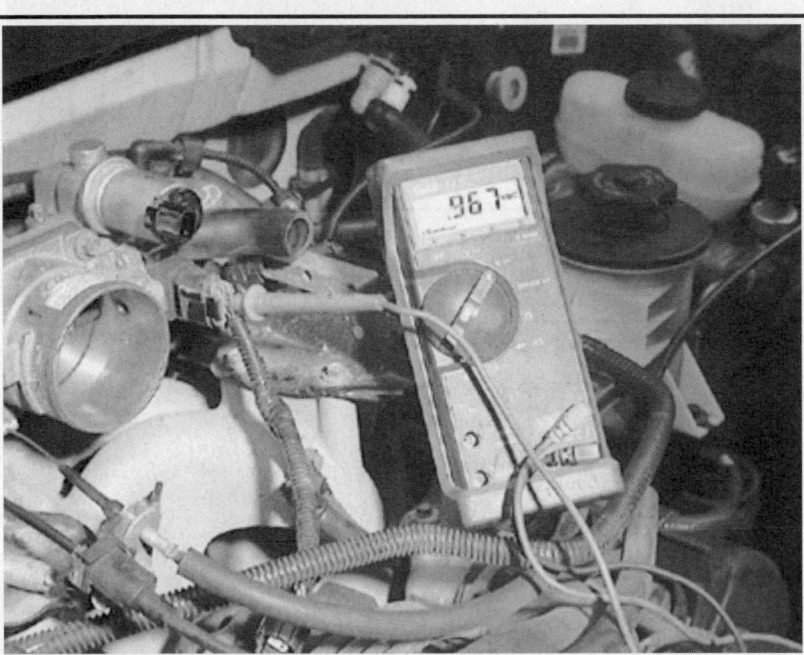

Testing the TP sensor signal return voltage at idle

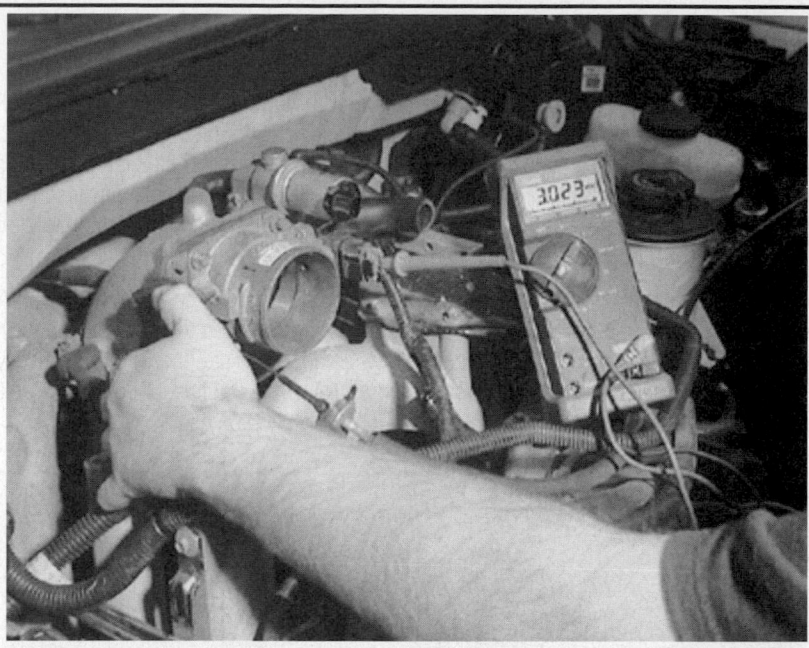

Test the operation of the TP sensor by gently opening the throttle while observing the signal return voltage. The voltage should move smoothly according to the amount the throttle is opened

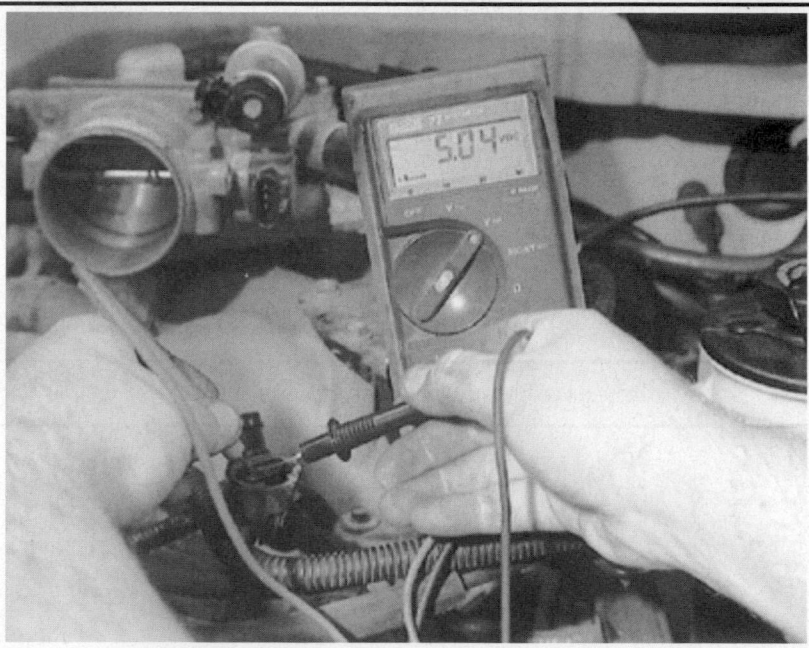

Testing the supply voltage at the TP sensor connector

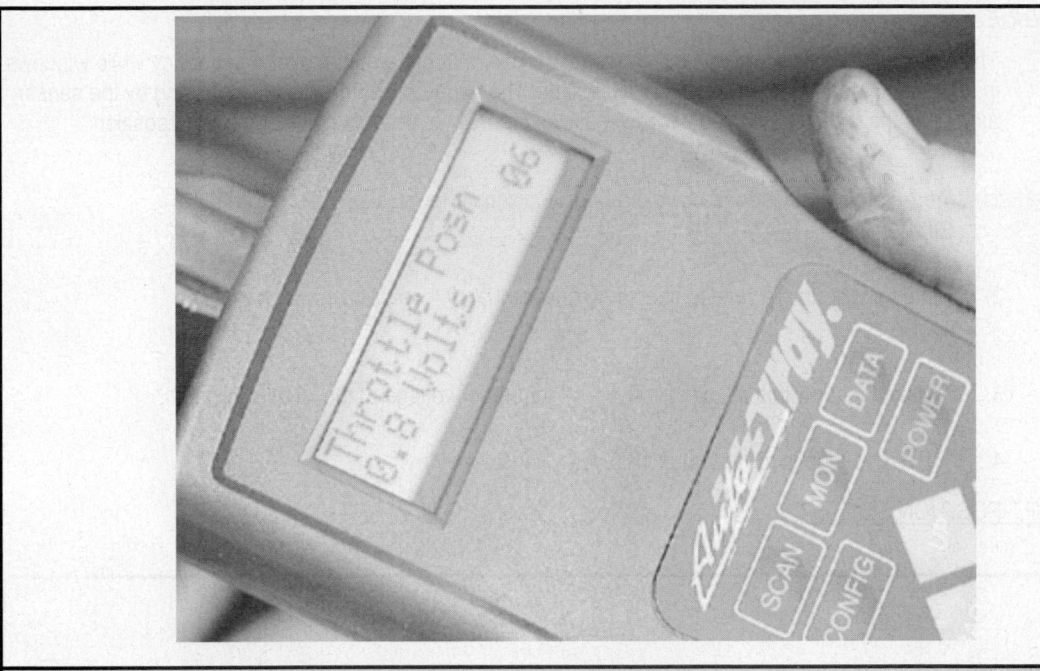

The TP sensor can be monitored with an appropriate and Data-stream capable scan tool

1. With the engine OFF and the ignition ON, check the voltage at the signal return circuit of the TP sensor by carefully backprobing the connector using a DVOM.

2. Voltage should be between 0.2 and 1.4 volts at idle.

3. Slowly move the throttle pulley to the wide open throttle (WOT) position and watch the voltage on the DVOM. The voltage should slowly rise to slightly less than 4.8v at Wide Open Throttle (WOT).

4. If no voltage is present, check the wiring harness for supply voltage (5.0v) and ground (0.3v or less), by referring to your corresponding wiring guide. If supply voltage and ground are present, but no output voltage from TP, replace the TP sensor. If supply voltage and ground do not meet specifications, make necessary repairs to the harness or PCM.

CAMSHAFT POSITION SENSOR
Operation
4.2L ENGINE

The camshaft position sensor (CMP) is a single hall-effect magnetic switch that is triggered by a single vane which is driven by the camshaft. The CMP sends a signal relating camshaft position back to the PCM which is used by the PCM to control engine timing.

4.6L, 5.4L AND 6.8L ENGINES

The camshaft position sensor (CMP) is a variable reluctance sensor that is triggered by a high point on the left-hand exhaust camshaft sprocket. The CMP sends a signal relating camshaft position back to the PCM which is used by the PCM to control engine timing.

DIESEL ENGINE

The Camshaft Position Sensor, is a Hall-effect sensor that generates a digital frequency while windows in a target wheel pass through its magnetic field. The frequency of the windows passing by the sensor, as well as the width of selected windows, allows the PCM to detect engine speed and position.

Testing

1. Check voltage between the camshaft position sensor terminals PWR GND and CID.

2. With engine running, voltage should be greater than 0.1 volt AC and vary with engine speed.

3. If voltage is not within specification, check for proper voltage at the VPWR terminal.

4. If VPWR voltage is greater than 10.5 volts, sensor may be faulty.

CRANKSHAFT POSITION SENSOR
Operation

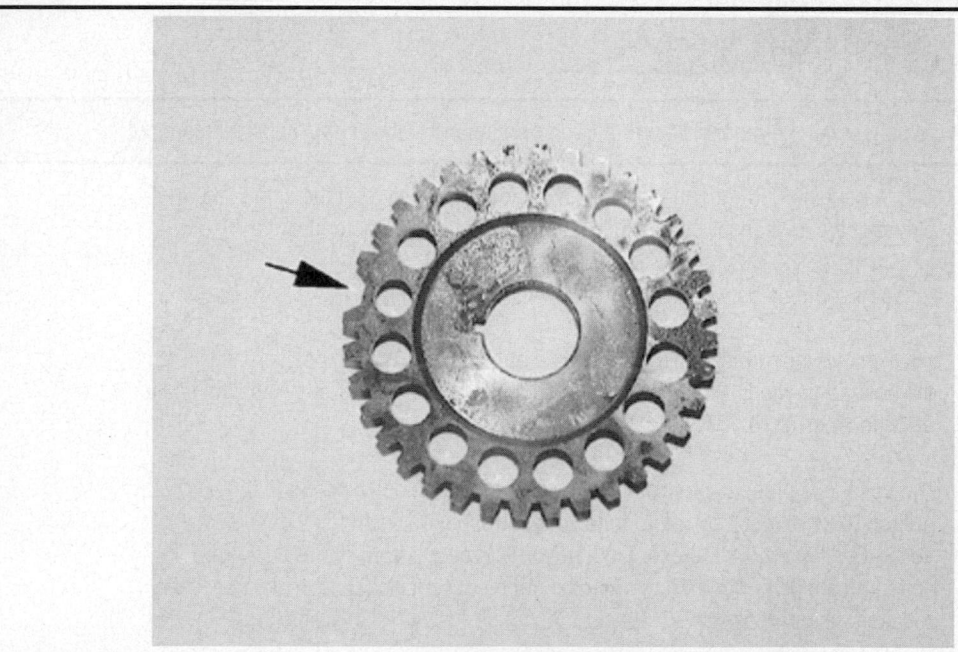

The CKP sensor trigger wheel rides on the front of the crankshaft. The missing tooth creates a fluctuation of voltage in the sensor

The Crankshaft Position (CKP) sensor is a variable reluctance sensor that uses a trigger wheel to induce voltage. The CKP sensor is a fixed magnetic sensor mounted to the engine block and monitors the trigger or "pulse" wheel which is attached to the crank pulley/damper. As the pulse wheel rotates by the CKP sensor, teeth on the pulse wheel induce voltage inside the sensor through magnetism. The pulse wheel has a missing tooth that changes the reading of the sensor. This is used for the Cylinder Identification (CID) function to properly monitor and adjust engine timing by locating the number 1 cylinder. The voltage created by the CKP sensor is alternating current (A/C). This voltage reading is sent to the PCM and is used to determine engine RPM, engine timing, and is used to fire the ignition coils.

Testing

1. Measure the voltage between the sensor CKP sensor terminals by backprobing the sensor connector.

If the connector cannot be backprobed, fabricate or purchase a test harness.

2. Sensor voltage should be more than 0.1 volt AC with the engine running and should vary with engine RPM.

3. If voltage is not within specification, the sensor may be faulty.

KNOCK SENSOR
Operation

The operation of the Knock Sensor (KS) is to monitor preignition or "engine knocks" and send the signal to the PCM. The PCM responds by adjusting ignition timing until the "knocks" stop. The sensor works by generating a signal produced by the frequency of the knock as recorded by the piezoelectric ceramic disc inside the KS. The disc absorbs the shock waves from the knocks and exerts a pressure on the metal diaphragm inside the KS. This compresses the crystals inside the disc and the disc generates a voltage signal proportional to the frequency of the knocks ranging from zero to 1 volt.

Testing

There is real no test for this sensor, the sensor produces it's own signal based on information gathered while the engine is running. The sensors also are usually inaccessible without major component removal. The sensors can be monitored with an appropriate scan tool using a data display or other data stream information. Follow the instructions included with the scan tool for information on accessing the data. The only test available is to test the continuity of the harness from the PCM to the sensor.

VEHICLE SPEED SENSOR
Operation

The Vehicle Speed Sensor (VSS) is a magnetic pick-up sensor that sends a signal to the Powertrain Control Module (PCM) and the speedometer. The sensor measures the rotation of the transmission output shaft or the ring gear on the differential and sends an AC voltage signal to the PCM which determines the corresponding vehicle speed.

Testing

1. Disconnect the negative battery cable.

2. Disengage the wiring harness connector from the VSS.

3. Using a Digital Volt-Ohmmeter (DVOM), measure the resistance (ohmmeter function) between the sensor terminals. If the resistance is 190-250 ohms, the sensor is okay.

INJECTOR DRIVER MODULE
Operation

The injector driver module (IDM) applies 115volts DC to all of the fuel injectors on the 7.3L DIT diesel engine. The IDM receives signals from the PCM for cylinder identification and fuel delivery command. The PCM controls when the timing of the injectors should start and how long the injector is open. The IDM controls the injector firing sequence through output drivers. The IDM has an output driver for each injector: one low side driver for each injector and one high side injector for each bank of injectors. The injector is fired when the output driver closes the circuit to ground.

ACCELERATOR PEDAL POSITION SENSOR
Operation

The accelerator pedal position (APP) sensor is a potentiometer that provides a signal to the PCM proportional to the accelerator pedal position. This sensor is only on models equipped with the 7.3L DIT diesel engine. The 7.3L DIT diesel engine incorporates a "drive by wire" system in which there is no cable to open the throttle plate. The APP sensor sends a signal to the PCM and the PCM control the air/fuel mixture accordingly. The accelerator pedal position sensor acts similar to the TP sensor on a gasoline engine. It provides the input to the PCM directly proportional to the engine load.

Testing

The accelerator pedal position sensor is tested just like the throttle position sensor. See throttle position sensor for testing.

INJECTION CONTROL PRESSURE SENSOR
Operation

The injection control pressure sensor (ICP) is a variable capacitance sensor that when supplied with a 5-volt reference signal from the PCM, produces a linear analog voltage signal that indicates oil pressure in the high pressure oil pump on the 7.3L DIT diesel engine. The ICP sensor provides a feedback signal to indicate high oil pressure so that the PCM can command the correct injector timing, pulse width and injection control pressure for proper fuel delivery at all speed and load conditions.

IDLE VALIDATION SWITCH
Operation

The idle validation switch (IDS) provides the PCM with a redundant signal to verify when the accelerator pedal is in the idle position. Any detected malfunction of the idle validation switch will illuminate the check engine light and cause the engine to operate at idle only.

ENGINE OIL TEMPERATURE SENSOR
Operation

The engine oil temperature (EOT) sensor changes resistance in response to the changing temperature of the engine oil. This sensor is located on the high pressure oil pump used to fire the fuel injectors on the 7.3L DIT diesel engine. The EOT sensor is a standard thermistor type sensor like an ECT or IAT. The EOT resistance decreases as the oil temperature increases providing the PCM with a signal relevant to the temperature of the engine oil.

Testing

1. Turn the ignition switch OFF.

2. Disconnect the wiring harness from the EOT sensor.

3. Measure the resistance between the sensor terminals.

4. Compare the resistance reading with the accompanying chart.

5. If the resistance is not within specification, the EOT may be faulty.

6. Connect the wiring harness to the sensor.

EXHAUST BACK PRESSURE SENSOR
Operation

The exhaust back pressure (EBP) sensor is a variable capacitance sensor that when supplied with a 5-volt reference signal from the PCM, produces a linear analog voltage signal that indicates exhaust back pressure. The EBP sensor is the input to the PCM used to control the exhaust back pressure regulator.

EXHAUST BACK PRESSURE REGULATOR
Operation

The exhaust back pressure regulator (EPR) is a variable position valve that control the exhaust back pressure during cold ambient temperatures to decrease the amount of time required to bring the engine to normal operating temperature. The PCM measures the exhaust back pressure, ambient air temperature, and the engine oil temperature to determine the desired exhaust back pressure.

INJECTION PRESSURE REGULATOR
Operation

The injection pressure regulator (IPR) is a variable position valve that controls injection control pressure on the 7.3L DIT diesel engine. Battery voltage is supplied to the IPR when the ignition switch is in the ON position. The valve position is controlled by switching the output signal circuit to ground inside the PCM. The injection control pressure is controlled by the PCM based on inputs received from the IAT, EOT, MAP/BARO, and accelerator pedal position sensors.

Mustang
ELECTRONIC ENGINE CONTROL (EEC) SYSTEM

All Sequential Fuel Injection (SFI) systems use the EEC system. The heart of the EEC system is a microprocessor called the Powertrain Control Module (PCM). The PCM receives data from a number of sensors and other electronic components (switches, relays, etc.). Based on information received and information programmed in the PCM's memory, it generates output signals to control various relays, solenoids and other actuators. The PCM in the EEC system has calibration modules located inside the assembly that contain calibration specifications for optimizing emissions, fuel economy and drivability. The calibration module is called a PROM.

- Powertrain Control Module (PCM)

- Throttle Position (TP) sensor

- Mass Air Flow (MAF) sensor

- Intake Air Temperature (IAT) sensor

- Idle Air Control (IAC) valve

- Engine Coolant Temperature (ECT) sensor

- Heated Oxygen Sensor (HO2S)

- Camshaft Position (CMP) sensor

- Knock Sensor (KS)

- Vehicle Speed Sensor (VSS)

- Crankshaft Position (CKP) sensor

The MAF sensor (a potentiometer) senses the quantity of airflow in the engine's air induction system and generates a voltage signal that varies with the amount of air drawn into the engine. The IAT sensor (a sensor in the area of the MAF sensor) measures the temperature of the incoming air and transmits a corresponding electrical signal. Another temperature sensor (the ECT sensor) inserted in the engine coolant tells if the engine is cold or warmed up. The TP sensor, a switch that senses throttle plate position, produces electrical signals that tell the PCM when the throttle is closed or wide open. A special probe (the HO2S) in the exhaust manifold measures the amount of oxygen in the exhaust gas, which is in indication of combustion efficiency, and sends a signal to the PCM. The sixth signal, camshaft position information, is transmitted by the CMP sensor, installed in place of the distributor (except 5.0L engines), or integral with the distributor (5.0L engines).

The EEC microcomputer circuit processes the input signals and produces output control signals to the fuel injectors to regulate fuel discharged to the injectors. It also adjusts ignition spark timing to provide the best balance between drivability and economy, and controls the IAC valve to maintain the proper idle speed.

Because of the complicated nature of the Ford system, special tools and procedures are necessary for testing and troubleshooting.

POWERTRAIN CONTROL MODULE (PCM)
Operation

The Powertrain Control Module (PCM) performs many functions on your car. The module accepts information from various engine sensors and computes the required fuel flow rate necessary to maintain the correct amount of air/fuel ratio throughout the entire engine operational range.

Based on the information that is received and programmed into the PCM's memory, the PCM generates output signals to control relays, actuators and solenoids. The PCM also sends out a command to the fuel injectors that meters the appropriate quantity of fuel. The module automatically senses and compensates for any changes in altitude when driving your vehicle.

HEATED OXYGEN SENSORS (HO2S)
Operation

The oxygen sensors supply the computer with a signal that indicates a rich or lean condition during engine operation. This input information assists the computer in determining the proper air/fuel ratio. A low voltage signal from one or more sensors indicates too much oxygen in the exhaust (lean condition) and, conversely, a high voltage signal indicates too little oxygen in the exhaust (rich condition).

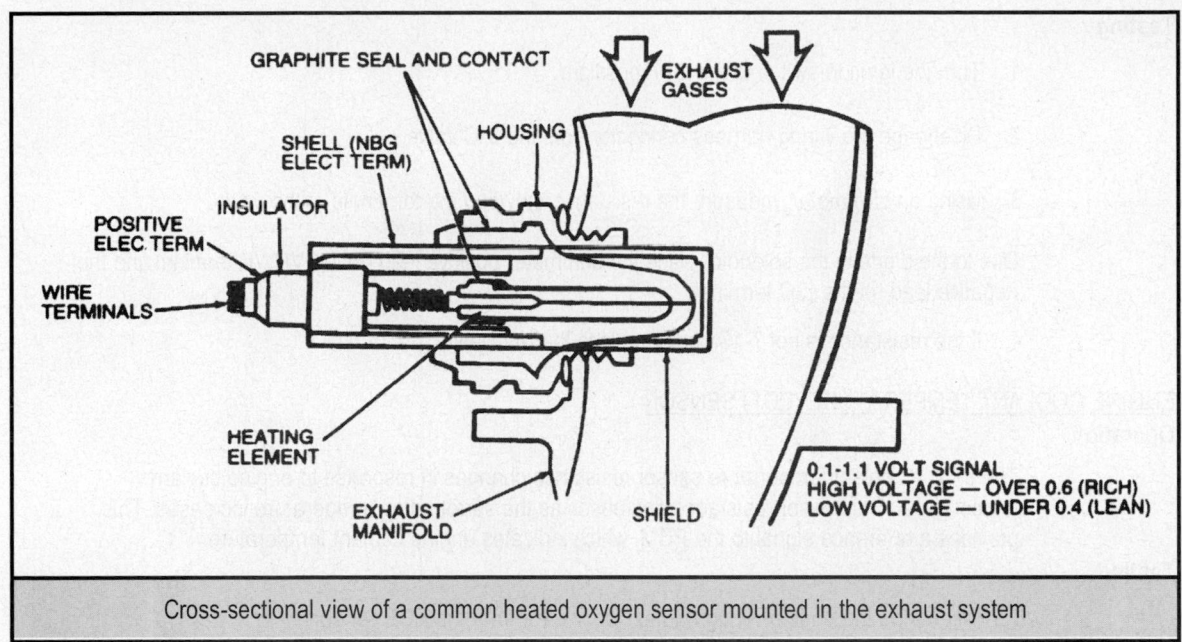

Cross-sectional view of a common heated oxygen sensor mounted in the exhaust system

The oxygen sensors are threaded into the exhaust manifold and/or exhaust pipes on all vehicles. Heated oxygen sensors are used on all models to allow the engine to reach the closed loop faster.

Testing

Condition	Voltage—Between Terminals HO2S and SIG RTN
Ignition ON, engine OFF	0V
Idle (cold)	0V
Idle (warm)	0–1.0V
Acceleration	0.5–1.0V
Deceleration	0–0.5V

When testing the oxygen sensor voltage signal, it should correspond to the values shown in this chart

1. Disconnect the HO2S.

2. Measure the resistance between PWR and GND terminals of the sensor. If the reading is approximately 6 ohms at 68°F (20°C), the sensor's heater element is in good condition.

3. With the HO2S connected and engine running, measure the voltage with a Digital Volt-Ohmmeter (DVOM) between terminals HO2S and SIG RTN (GND) of the oxygen sensor connector. If the voltage readings are approximately equal to those in the table, the sensor is okay.

IDLE AIR CONTROL (IAC) VALVE
Operation

The Idle Air Control (IAC) valve controls the engine idle speed and dashpot functions. The valve is located on the side of the throttle body. This valve allows the necessary amount of air, as determined by the Powertrain Control Module (PCM) and controlled by a duty cycle signal, to bypass the throttle plate in order to maintain the proper idle speed.

Testing

1. Turn the ignition switch to the OFF position.

2. Disengage the wiring harness connector from the IAC valve .

3. Using an ohmmeter, measure the resistance between the terminals of the valve.

   Due to the diode in the solenoid, place the ohmmeter positive lead on the VPWR terminal and the negative lead on the ISC terminal.

4. If the resistance is not 7-13 ohms, replace the IAC valve.

ENGINE COOLANT TEMPERATURE (ECT) SENSOR
Operation

The engine coolant temperature sensor resistance changes in response to engine coolant temperature. The sensor resistance decreases as the surrounding temperature increases. This provides a reference signal to the PCM, which indicates engine coolant temperature.

Testing

1. Disengage the engine wiring harness connector from the ECT sensor.

2. Connect an ohmmeter between the ECT sensor terminals, and set the ohmmeter scale on 200,000 ohms.

3. With the engine cold and the ignition switch in the OFF position, measure and note the ECT sensor resistance. Attach the engine wiring harness connector to the sensor.

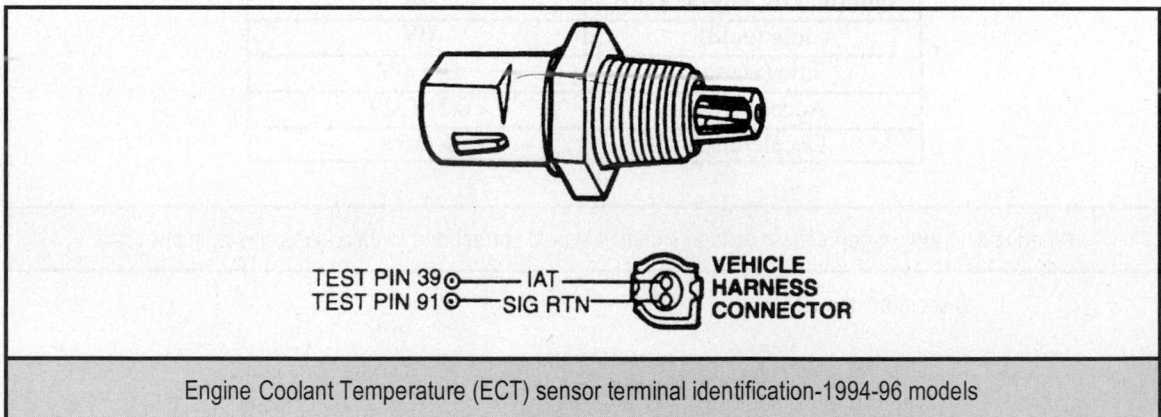

Engine Coolant Temperature (ECT) sensor terminal identification-1994-96 models

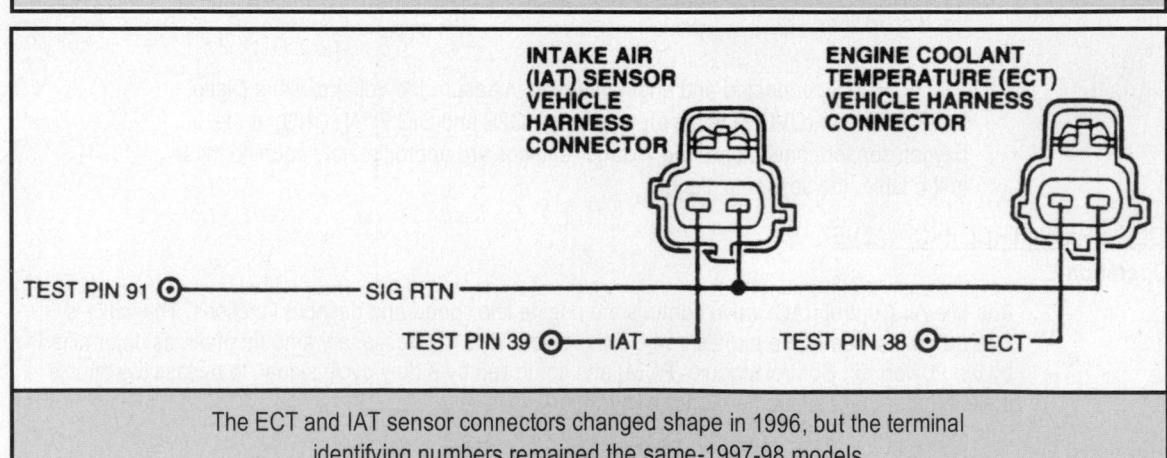

The ECT and IAT sensor connectors changed shape in 1996, but the terminal identifying numbers remained the same-1997-98 models

Temperature		Engine Coolant/Air Charge Temperature Sensor Values	
°F	°C	Voltage (volts)	Resistance (K ohms)
248	120	.27	1.18
230	110	.35	1.55
212	100	.46	2.07
194	90	.60	2.80
176	80	.78	3.84
158	70	1.02	5.37
140	60	1.33	7.70
122	50	1.70	10.97
104	40	2.13	16.15
86	30	2.60	24.27
68	20	3.07	37.30
50	10	3.51	58.75

A properly functioning Engine Coolant Temperature (ECT) sensor
should exhibit the resistance values indicated in this chart

4. Start the engine and allow it to warm up to normal operating temperature.

5. Once the engine has reached normal operating temperature, turn it OFF.

6. Once again, detach the engine wiring harness connector from the ECT sensor.

7. Measure and note the ECT sensor resistance, then compare the cold and hot ECT sensor resistance measurements with the accompanying chart.

8. Replace the ECT sensor if the readings do not approximate those in the chart; otherwise, reattach the engine wiring harness connector to the sensor.

INTAKE AIR TEMPERATURE (IAT) SENSOR

Operation

The Intake Air Temperature (IAT) sensor resistance changes in response to the intake air temperature. The sensor resistance decreases as the surrounding air temperature increases. This provides a signal to the PCM, indicating the temperature of the incoming air charge.

Testing

Turn the ignition switch OFF before testing the sensor.

1. Disengage the wiring harness connector from the IAT sensor.

2. Using a Digital Volt-Ohmmeter (DVOM), measure the resistance between the two sensor terminals.

TEST PIN 39 — IAT —
TEST PIN 91 — SIG RTN —

VEHICLE
HARNESS
CONNECTOR

Intake Air Temperature (IAT) sensor connector terminal identification-1994-96 models

Temperature		Engine Coolant/Air Charge Temperature Sensor Values	
°F	°C	Voltage (volts)	Resistance (K ohms)
248	120	.27	1.18
230	110	.35	1.55
212	100	.46	2.07
194	90	.60	2.80
176	80	.78	3.84
158	70	1.02	5.37
140	60	1.33	7.70
122	50	1.70	10.97
104	40	2.13	16.15
86	30	2.60	24.27
68	20	3.07	37.30
50	10	3.51	58.75

A properly functioning Intake Air Temperature (IAT) sensor
should exhibit the resistance values indicated in this chart

3. Compare the resistance reading with the accompanying chart. If the reading for a
    given temperature is approximately that shown in the table, the IAT sensor is okay.

4. Attach the wiring harness connector to the sensor.

MASS AIR FLOW (MAF) SENSOR

Operation

The Mass Air Flow (MAF) sensor directly measures the amount of air flowing into the engine. The sensor is mounted between the air cleaner assembly and the air cleaner outlet tube.

The sensor utilizes a hot wire sensing element to measure the amount of air entering the engine. The sensor does this by sending a signal, which is generated by the sensor when the incoming air cools the hot wire, to the PCM. The signal is used by the PCM to calculate the injector pulse width, which controls the air/fuel ratio in the engine. The sensor and plastic housing are integral and must be replaced if found to be defective.

The sensing element (hot wire) is a thin platinum wire wound on a ceramic bobbin and coated with glass. This hot wire is maintained at 392°F (200°C) above the ambient temperature as measured by a constant "cold wire".

Testing

1. With the engine running at idle, use a DVOM to verify that there are at least 10.5 volts between terminals A and B of the MAF sensor connector. Such a reading indicates that the power input to the sensor is correct. Then, measure the voltage between MAF sensor connector terminals C and D. If the reading is approximately 0.34-1.96 volts, the sensor is functioning properly.

Engine Condition	Signal Voltage—Sensor Terminals C and D
Idle	0.60V
20 mph	1.10V
40 mph	1.70V
60 mph	2.10V

The Mass Air Flow (MAF) sensor should exhibit the same voltage specifications as indicated in this chart; otherwise, replace the MAF sensor

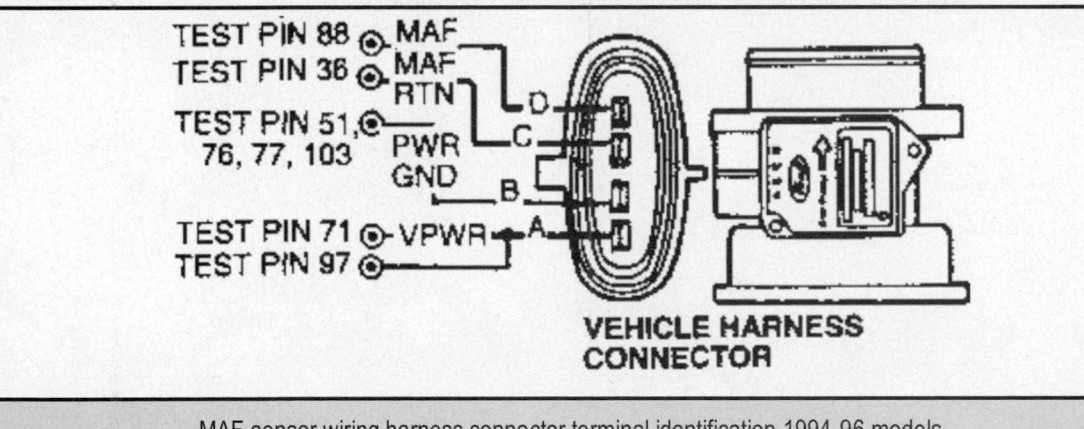

MAF sensor wiring harness connector terminal identification-1994-96 models

```
*TEST PIN 24/51/76/77/103 ⊙————————— PWR GND ————————
         TEST PIN 36 ⊙————————————— MAF RTN ———————
         TEST PIN 88 ⊙————————————— MAF ————————
      TEST PIN 71/97 ⊙————————————— VPWR ————————
```

MASS AIR FLOW (MAF) SENSOR VEHICLE HARNESS CONNECTOR AT MAF EXTENSION

MAF sensor wiring harness connector terminal identification-1997-98 models

THROTTLE POSITION (TP) SENSOR

Operation

The Throttle Position (TP) sensor is a potentiometer which provides a signal to the PCM that is directly proportional to the throttle plate position. The TP sensor is mounted on the side of the throttle body and is connected to the throttle plate shaft. The TP sensor monitors throttle plate movement and position, and transmits an appropriate electrical signal to the PCM. These signals are used by the PCM to adjust the air/fuel mixture, spark timing and EGR operation according to engine load at idle, part throttle, or full throttle. The TPS is not adjustable.

Testing

1. Disconnect the negative battery cable.

2. Disengage the wiring harness connector from the TP sensor.

3. Using a Digital Volt-Ohmmeter (DVOM) set on ohmmeter function, probe the terminals which correspond to the brown/white and gray/white connector wires on the TP sensor. Do not measure the wiring harness connector terminals, but rather the terminals on the sensor itself.

4. Slowly rotate the throttle shaft and monitor the ohmmeter for a continuous, steady change in resistance. Any sudden jumps or irregularities in resistance (such as jumping back and forth) indicates a malfunctioning sensor.

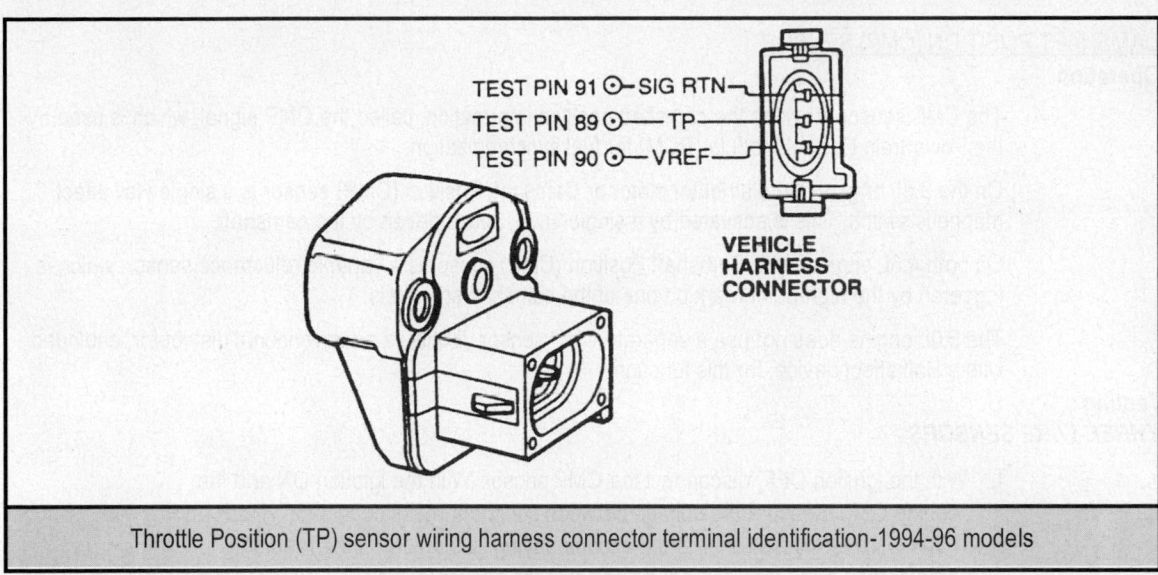

TEST PIN 91 ⊙—SIG RTN
TEST PIN 89 ⊙—— TP
TEST PIN 90 ⊙— VREF

VEHICLE HARNESS CONNECTOR

Throttle Position (TP) sensor wiring harness connector terminal identification-1994-96 models

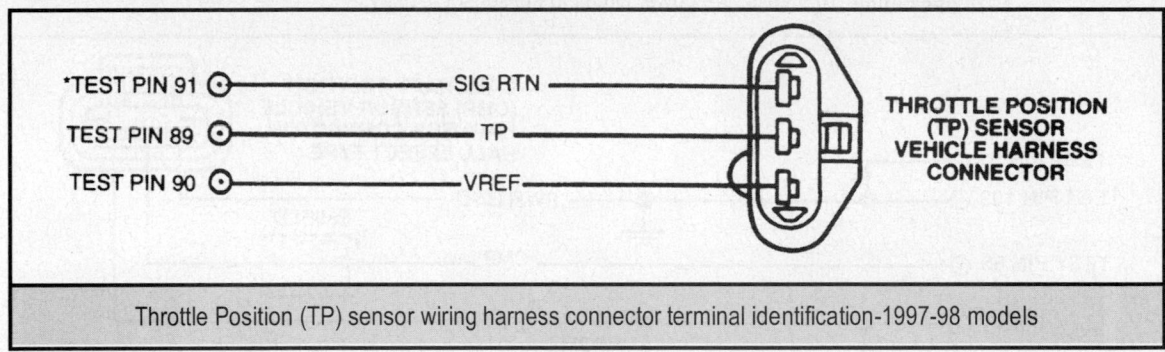

*TEST PIN 91 ⊙——————— SIG RTN —————
TEST PIN 89 ⊙——————— TP ——————
TEST PIN 90 ⊙——————— VREF —————

THROTTLE POSITION (TP) SENSOR VEHICLE HARNESS CONNECTOR

Throttle Position (TP) sensor wiring harness connector terminal identification-1997-98 models

5. Reconnect the negative battery cable.

6. Turn the DVOM to the voltmeter setting.

7. Detach the wiring harness connector from the PCM (located behind the lower right-hand kick panel in the passenger compartment), then install a break-out box between the wiring harness connector and the PCM connector.

8. Turn the ignition switch ON and using the DVOM set to its voltmeter function, measure the voltage between terminals 89 and 90 of the breakout box. The specification is 0.9 volts.

9. If the voltage is outside the standard value or if it does not change smoothly, inspect the circuit wiring and/or replace the TP sensor.

CAMSHAFT POSITION (CMP) SENSOR

Operation

The CMP sensor provides the camshaft position information, called the CMP signal, which is used by the Powertrain Control Module (PCM) for fuel synchronization.

On the 3.8L engine, the distributor stator or Camshaft Position (CMP) sensor is a single Hall effect magnetic switch. This is activated by a single vane, and is driven by the camshaft.

On both 4.6L engines, the Camshaft Position (CMP) sensor is a variable reluctance sensor, which is triggered by the high-point mark on one of the camshaft sprockets.

The 5.0L engine does not use a separate CMP sensor. It utilizes a conventional distributor, equipped with a Hall effect device, for this function.

Testing

THREE-WIRE SENSORS

1. With the ignition OFF, disconnect the CMP sensor. With the ignition ON and the engine OFF, measure the voltage between sensor harness connector VPWR and PWR GND terminals (refer to the accompanying illustration). If the reading is greater than 10.5 volts, the power circuit to the sensor is okay.

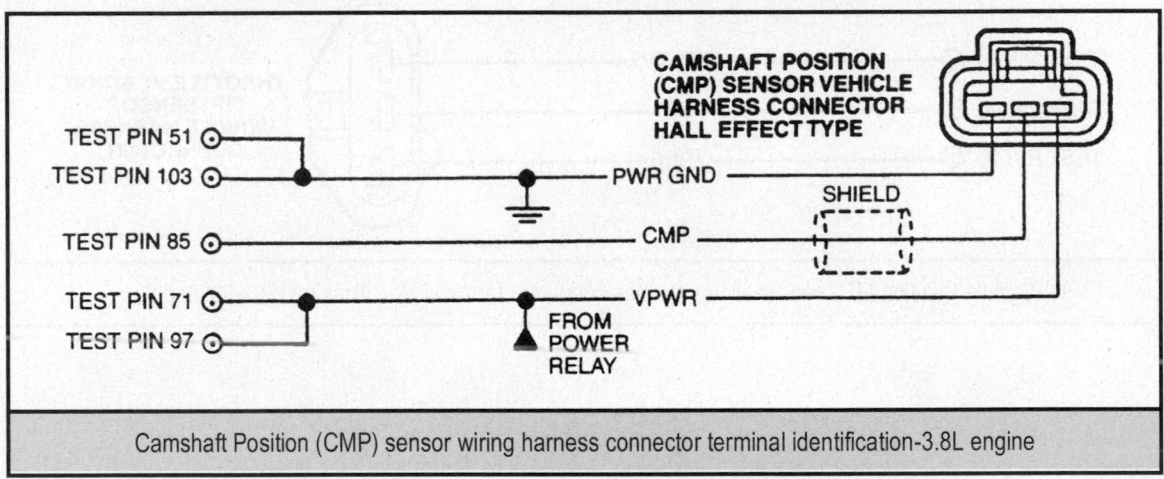

Camshaft Position (CMP) sensor wiring harness connector terminal identification-3.8L engine

2. With the ignition OFF, install a break-out box between the CMP sensor and the PCM. Using a Digital Volt-Ohmmeter (DVOM) set to the voltage function (scale set to monitor less than 5 volts), measure the voltage between break-out box terminals 24 and 4 with the engine running at varying RPM. If the voltage reading varies more than 0.1 volt, the sensor is okay.

TWO-WIRE SENSORS

1. With the ignition OFF, install a break-out box between the CMP sensor and PCM.

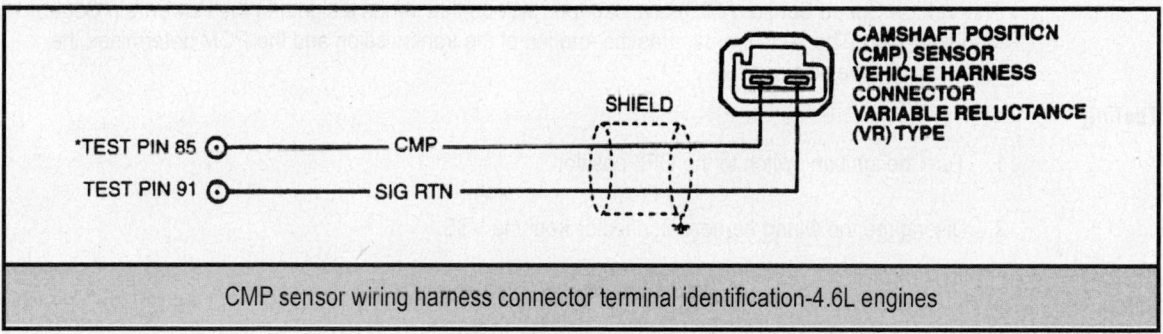

CMP sensor wiring harness connector terminal identification-4.6L engines

2. Using a Digital Volt-Ohmmeter (DVOM) set to the voltage function (scale set to monitor less than 5 volts), measure the voltage between break-out box terminals 24 and 46 with the engine running at varying RPM. If the voltage reading varies more than 0.1 volt AC, the sensor is okay.

CRANKSHAFT POSITION (CKP) SENSOR

Operation

The Crankshaft Position (CKP) sensor, located on the front cover (near the crankshaft pulley) is used to determine crankshaft position and crankshaft rpm. The CKP sensor is a reluctance sensor which senses the passing of teeth on the sensor ring because the teeth disrupt the magnetic field of the sensor. This disruption creates a voltage fluctuation, which is monitored by the PCM.

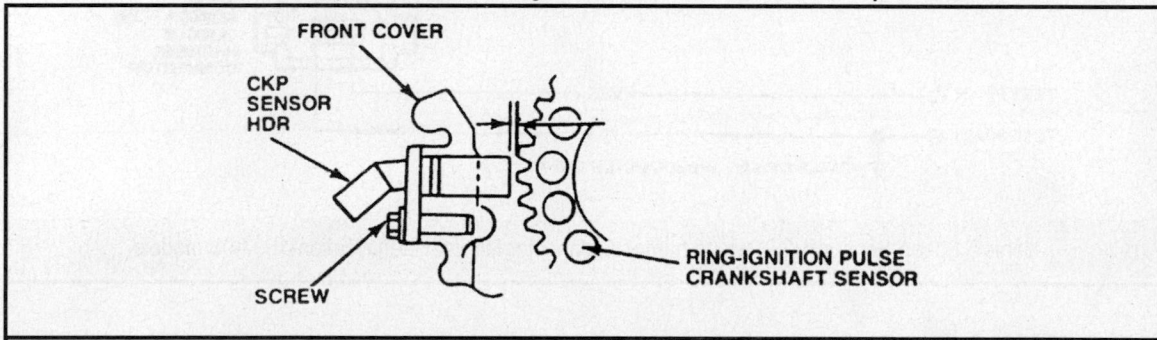

The Crankshaft Position (CKP) sensor is mounted on the engine front cover, and senses the teeth on the sensor ring as they pass through its magnetic field

Testing

Using a DVOM set to the AC scale to monitor less than 5 volts, measure the voltage between the sensor Cylinder Identification (CID) terminal and ground by back probing the sensor connector. If the connector cannot be backprobe, fabricate or purchase a test harness. The sensor is okay if the voltage reading varies more than 0.1 volt with the engine running at varying RPM.

VEHICLE SPEED SENSOR (VSS)

Operation

The Vehicle Speed Sensor (VSS) is a magnetic pick-up that sends a signal to the Powertrain Control Module (PCM). The sensor measures the rotation of the transmission and the PCM determines the corresponding vehicle speed.

Testing

1. Turn the ignition switch to the OFF position.

2. Disengage the wiring harness connector from the VSS.

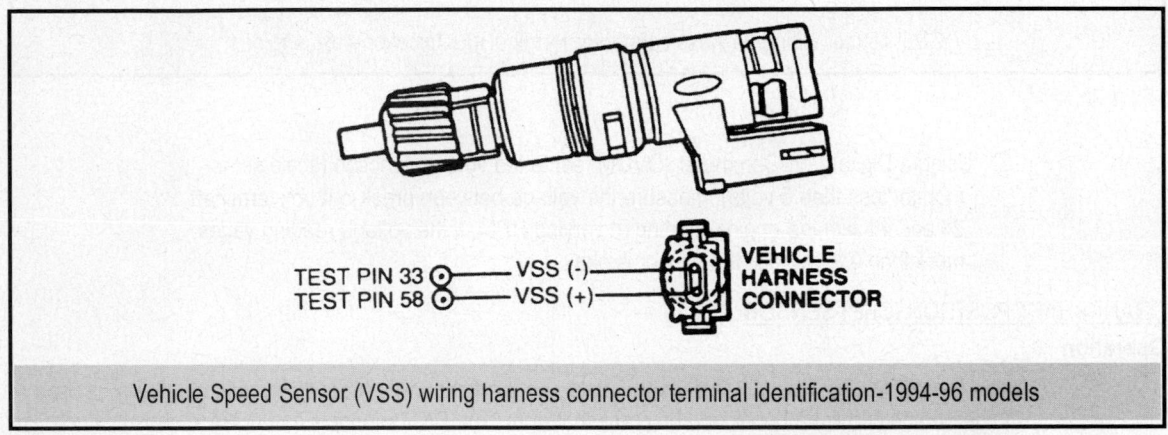

Vehicle Speed Sensor (VSS) wiring harness connector terminal identification-1994-96 models

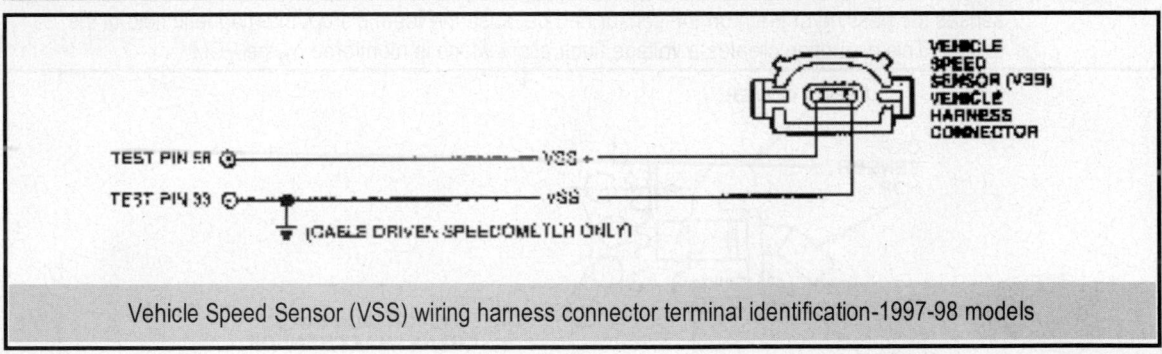

Vehicle Speed Sensor (VSS) wiring harness connector terminal identification-1997-98 models

3. Using a Digital Volt-Ohmmeter (DVOM), measure the resistance (ohmmeter function) between the sensor terminals. If the resistance is 190-250 ohms, the sensor is okay.

Ranger
ELECTRONIC ENGINE CONTROLS
Electronic Engine Control (EEC)

All Sequential Fuel Injection (SFI) systems use the EEC system. The heart of the EEC system is a micro-processor called the Powertrain Control Module (PCM). The PCM receives data from a number of sensors and other electronic components (switches, relay, etc.). Based on information received and information programmed in the PCM's memory, it generates output signals to control various relay, solenoids and other actuators. The PCM in the EEC system has calibration modules located inside the assembly that contain calibration specifications for optimizing emissions, fuel economy and drive ability. The calibration module is called a PROM.

The following are the electronic engine controls used by Ranger/Explorer and Mountaineers:

- Powertrain Control Module (PCM)

- Throttle Position (TP) sensor

- Mass Air Flow (MAF) sensor

- Intake Air Temperature (IAT) sensor

- Idle Air Control (IAC) valve

- Engine Coolant Temperature (ECT) sensor

- Heated Oxygen Sensor (HO2S)

- Camshaft Position (CMP) sensor

- Knock Sensor (KS)

- Vehicle Speed Sensor (VSS)

- Crankshaft Position (CKP) sensor

The MAF sensor (a potentiometer) senses the position of the airflow in the engine's air induction system and generates a voltage signal that varies with the amount of air drawn into the engine. The IAT sensor (a sensor in the area of the MAF sensor) measures the temperature of the incoming air and transmits a corresponding electrical signal. Another temperature sensor (the ECT sensor) inserted in the engine coolant tells if the engine is cold or warmed up. The TP sensor, a switch that senses throttle plate position, produces electrical signals that tell the PCM when the throttle is closed or wide open. A special probe (the HO2S) in the exhaust manifold measures the amount of oxygen in the exhaust gas, which is in indication of combustion efficiency, and sends a signal to the PCM. The sixth signal, camshaft position information, is transmitted by the CMP sensor, installed in place of the distributor (engines with distributorless ignition), or integral with the distributor.

The EEC microcomputer circuit processes the input signals and produces output control signals to the fuel injectors to regulate fuel discharged to the injectors. It also adjusts ignition spark timing to provide the best balance between driveability and economy, and controls the IAC valve to maintain the proper idle speed.

Because of the complicated nature of the Ford system, special tools and procedures are necessary for testing and troubleshooting.

POWERTRAIN CONTROL MODULE (PCM)
Operation

The Powertrain Control Module (PCM) performs many functions on your car. The module accepts information from various engine sensors and computes the required fuel flow rate necessary to maintain the correct amount of air/fuel ratio throughout the entire engine operational range.

Based on the information that is received and programmed into the PCM's memory, the PCM generates output signals to control relays, actuators and solenoids. The PCM also sends out a command to the fuel injectors that meters the appropriate quantity of fuel. The module automatically senses and compensates for any changes in altitude when driving your vehicle.

HEATED OXYGEN SENSORS (HO2S)
Operation

The oxygen sensor supplies the computer with a signal which indicates a rich or lean condition during engine operation. The input information assists the computer in determining the proper air/fuel ratio. A low voltage signal from the sensor indicates too much oxygen in the exhaust (lean condition) and, conversely, a high voltage signal indicates too little oxygen in the exhaust (rich condition).

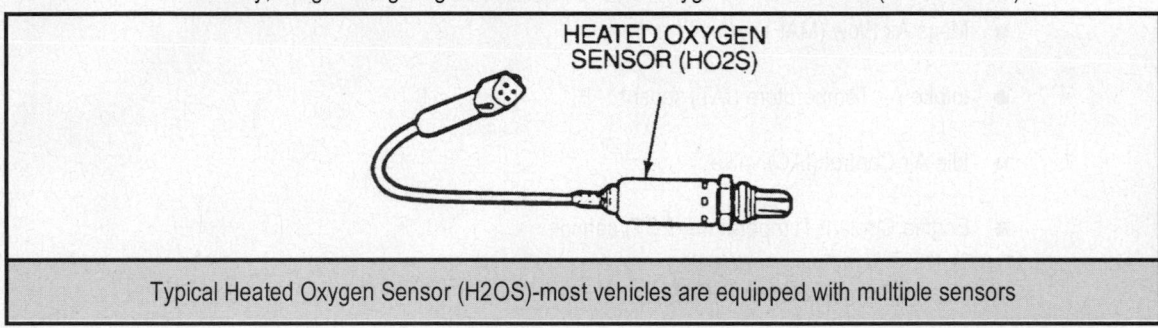

Typical Heated Oxygen Sensor (H2OS)-most vehicles are equipped with multiple sensors

The oxygen sensors are threaded into the exhaust manifold and/or exhaust pipes on all vehicles. Heated oxygen sensors are used on all models to allow the engine to reach the closed loop faster.

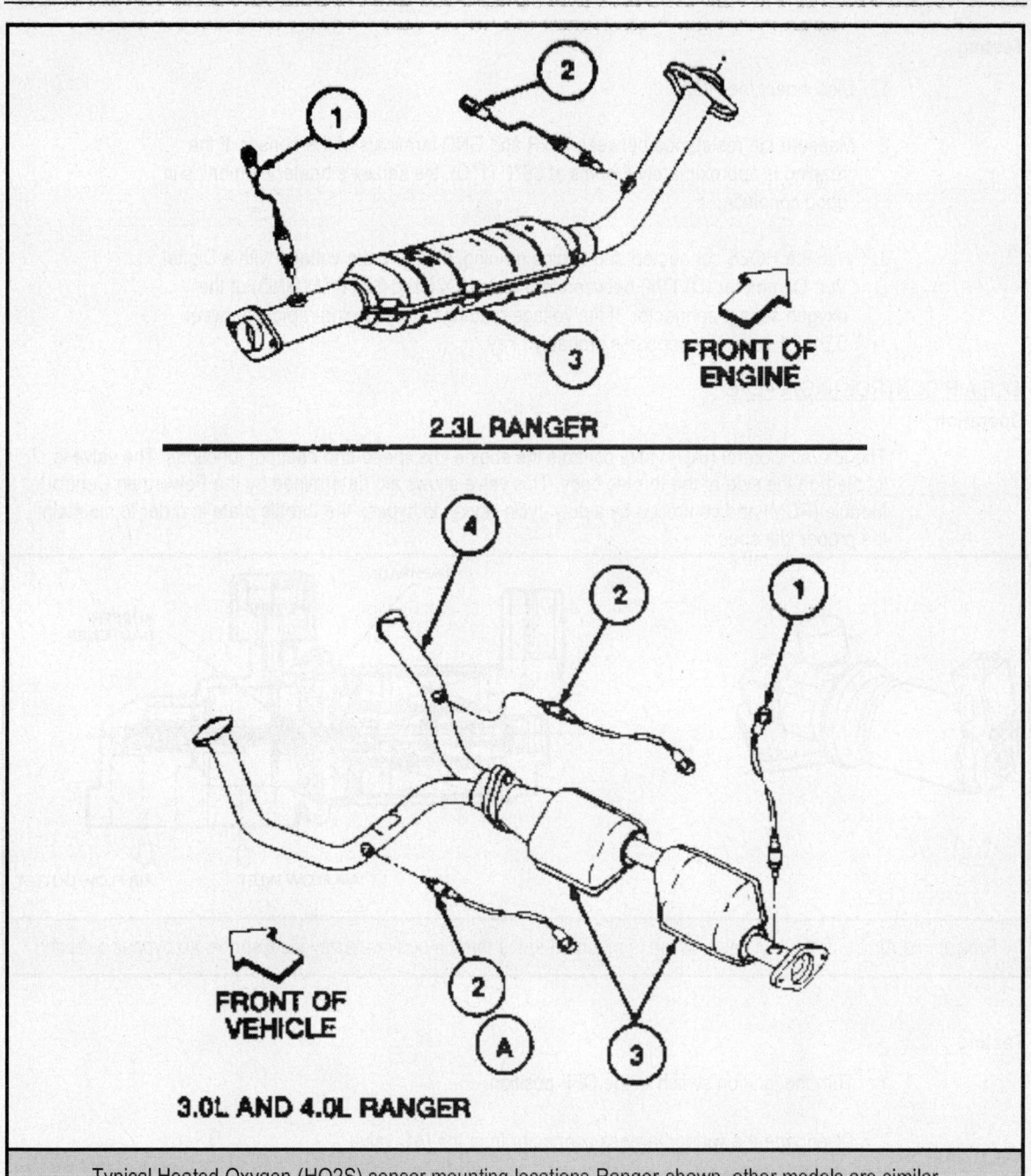

2.3L RANGER

FRONT OF ENGINE

FRONT OF VEHICLE

3.0L AND 4.0L RANGER

Typical Heated Oxygen (HO2S) sensor mounting locations-Ranger shown, other models are similar

Testing

1. Disconnect the HO2S.

2. Measure the resistance between PWR and GND terminals of the sensor. If the reading is approximately 6 ohms at 68°F (1°C). the sensor's heater element is in good condition.

3. With the HO2S connected and engine running, measure the voltage with a Digital Volt-Ohmmeter (DVOM) between terminals HO2Sand SIG RTN(GND) of the oxygen sensor connector. If the voltage readings are swinging rapidly between 0.01-1.1 volts, the sensor is probably okay.

IDLE AIR CONTROL (IAC) VALVE
Operation

The Idle Air Control (IAC) valve controls the engine idle speed and dashpot functions. The valve is located on the side of the throttle body. This valve allows air, determined by the Powertrain Control Module (PCM) and controlled by a duty cycle signal, to bypass the throttle plate in order to maintain the proper idle speed.

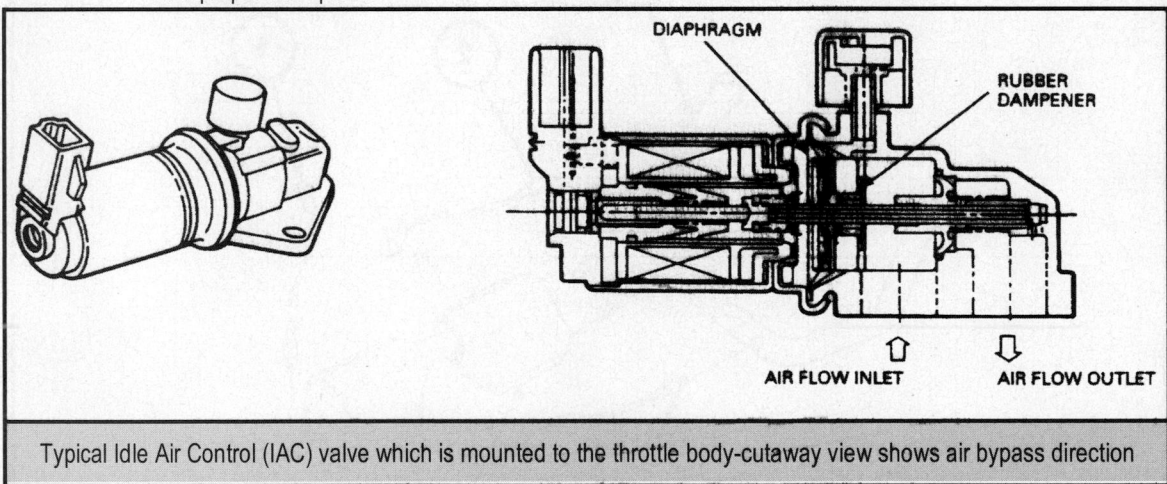

Typical Idle Air Control (IAC) valve which is mounted to the throttle body-cutaway view shows air bypass direction

Testing

1. Turn the ignition switch to the OFF position.

2. Disengage the wiring harness connector from the IAC valve .

3. Using an ohmmeter, measure the resistance between the terminals of the valve.

Due to the diode in the solenoid, place the ohmmeter positive lead on the VPWR terminal and the negative lead on the ISC terminal.

4. If the resistance is not 7-13 ohms, replace the IAC valve.

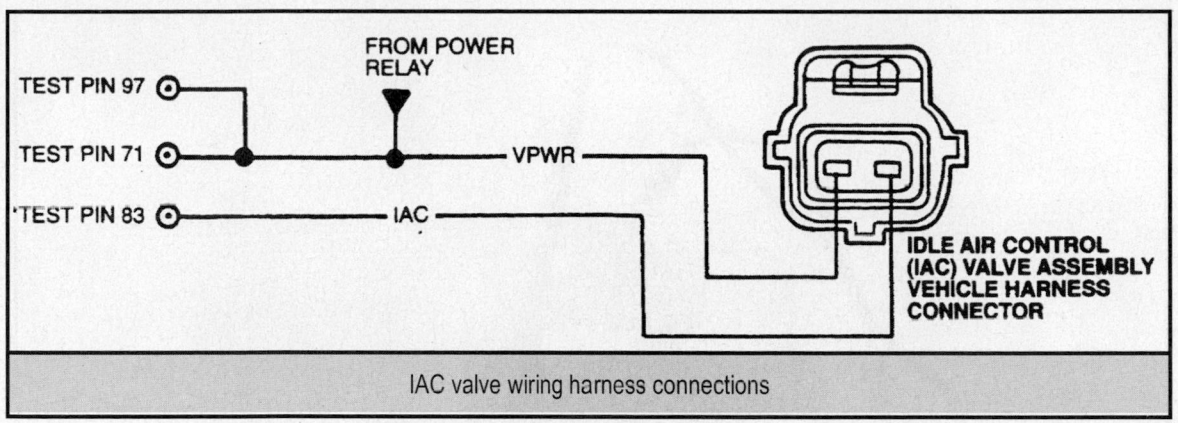

TEST PIN 97

FROM POWER RELAY

TEST PIN 71 — VPWR

TEST PIN 83 — IAC

IDLE AIR CONTROL (IAC) VALVE ASSEMBLY VEHICLE HARNESS CONNECTOR

IAC valve wiring harness connections

ENGINE COOLANT TEMPERATURE (ECT) SENSOR

Operation

The engine coolant temperature sensor resistance changes in response to engine coolant temperature. The sensor resistance decreases as the surrounding temperature increases. This provides a reference signal to the PCM, which indicates engine coolant temperature.

The ECT sensor is mounted on the lower intake manifold near the water outlet/thermostat housing, except on 2.3L and 2.5L engines. On the 2.3L and 2.5L engine the ECT is mounted on the water outlet/thermostat housing.

Testing

1. Disengage the engine wiring harness connector from the ECT sensor.

2. Connect an ohmmeter between the ECT sensor terminals, and set the ohmmeter scale on 200,000 ohms.

3. With the engine cold and the ignition switch in the OFF position, measure and note the ECT sensor resistance. Attach the engine wiring harness connector to the sensor.

4. Start the engine and allow the engine to warm up to normal operating temperature.

5. Once the engine has reached normal operating temperature, turn the engine OFF

6. Once again, detach the engine wiring harness connector from the ECT sensor.

7. Measure and note the ECT sensor resistance, then compare the cold and hot ECT sensor resistance measurements with the accompanying chart.

8. Replace the ECT sensor if the readings do not approximate those in the chart, otherwise reattach the engine wiring harness connector to the sensor.

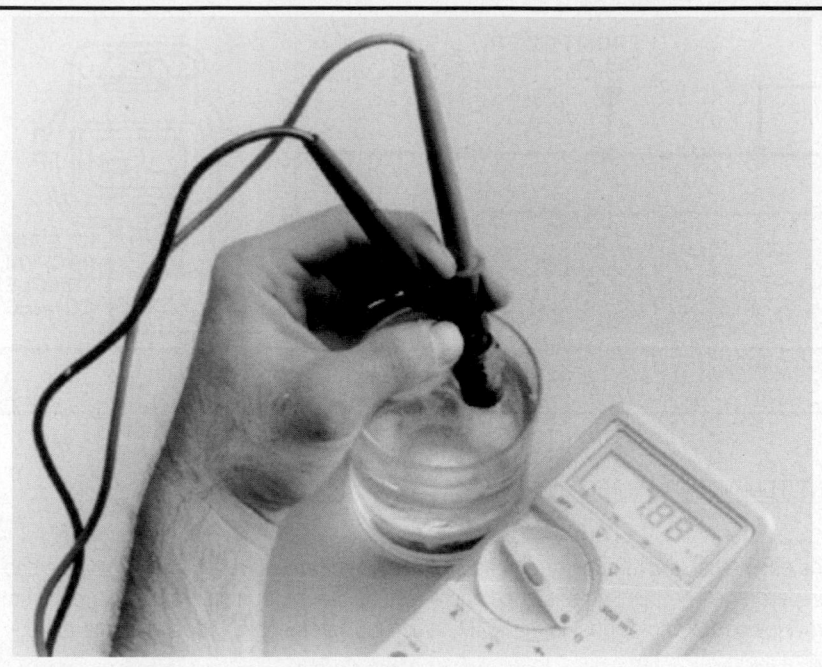

Submerge the end of the temperature sensor in cold or hot water and check resistance

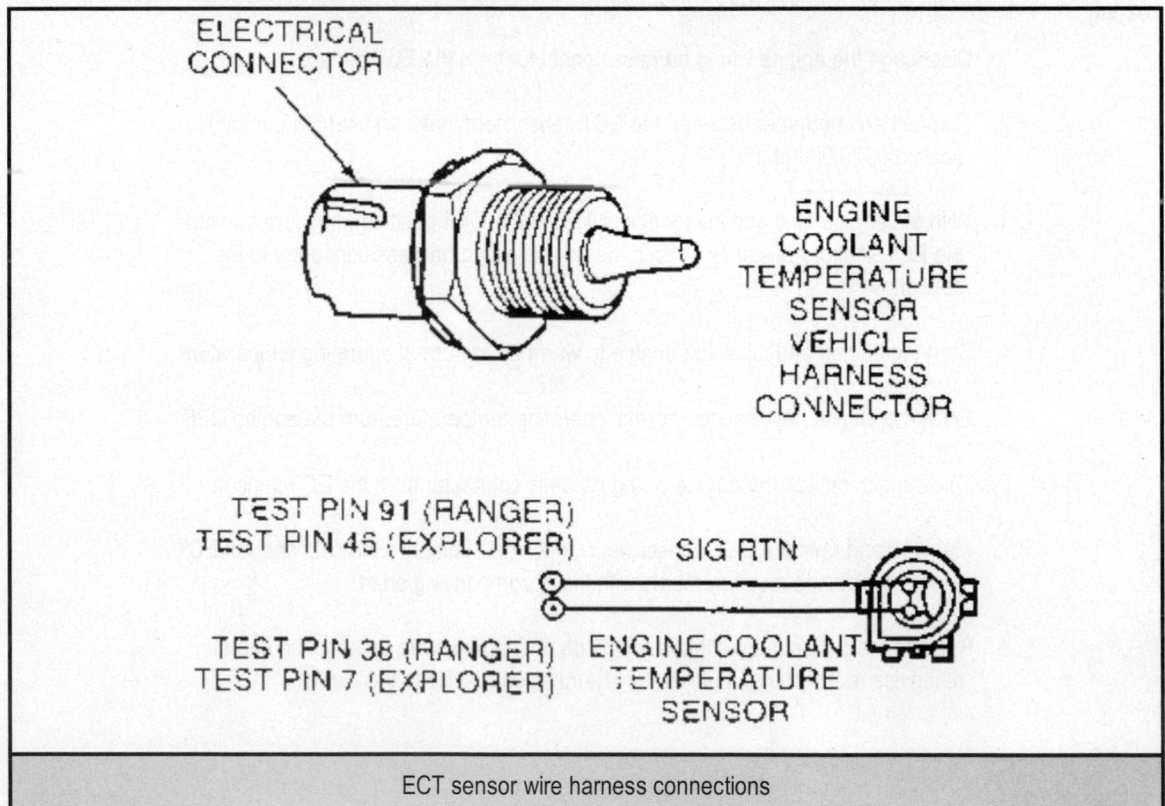

ELECTRICAL
CONNECTOR

ENGINE
COOLANT
TEMPERATURE
SENSOR
VEHICLE
HARNESS
CONNECTOR

TEST PIN 91 (RANGER)
TEST PIN 46 (EXPLORER)

SIG RTN

TEST PIN 36 (RANGER)
TEST PIN 7 (EXPLORER)

ENGINE COOLANT
TEMPERATURE
SENSOR

ECT sensor wire harness connections

INTAKE AIR TEMPERATURE (IAT) SENSOR

Operation

The Intake Air Temperature (IAT) sensor resistance changes in response to the intake air temperature. The sensor resistance decreases as the surrounding air temperature increases. This provides a signal to the PCM indicating the temperature of the incoming air charge.

Most engines mount the IAT sensor in the air cleaner-to-throttle body supply tube. However, some earlier engines have it mounted to the upper intake manifold.

Testing

Turn the ignition switch OFF.

1. Disengage the wiring harness connector from the IAT sensor.

2. Using a Digital Volt-Ohmmeter (DVOM), measure the resistance between the two sensor terminals.

3. Compare the resistance reading with the accompanying chart. If the reading for a given temperature is approximately that shown in the table, the IAT sensor is okay.

4. Attach the wiring harness connector to the sensor.

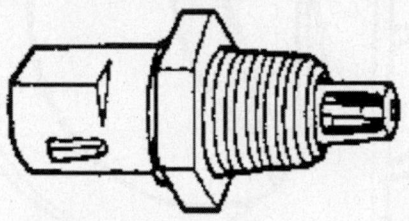

TEST PIN 39 (RANGER)
TEST PIN 25 (EXPLORER)
TEST PIN 91 (RANGER)
TEST PIN 46 (EXPLORER)

IAT
SIG RTN

INTAKE AIR
TEMPERATURE
(IAT) SENSOR
VEHICLE
HARNESS
CONNECTOR

IAT sensor wiring harness connections

MASS AIR FLOW (MAF) SENSOR
Operation

The Mass Air Flow (MAF) sensor directly measures the amount of the air flowing into the engine. The sensor is mounted between the air cleaner assembly and the air cleaner outlet tube.

The sensor utilizes a hot wire sensing element to measure the amount of air entering the engine. The sensor does this by sending a signal, generated by the sensor when the incoming air cools the hot wire down, to the PCM. The signal is used by the PCM to calculate the injector pulse width, which controls the air/fuel ratio in the engine. The sensor and plastic housing are integral and must be replaced if found to be defective.

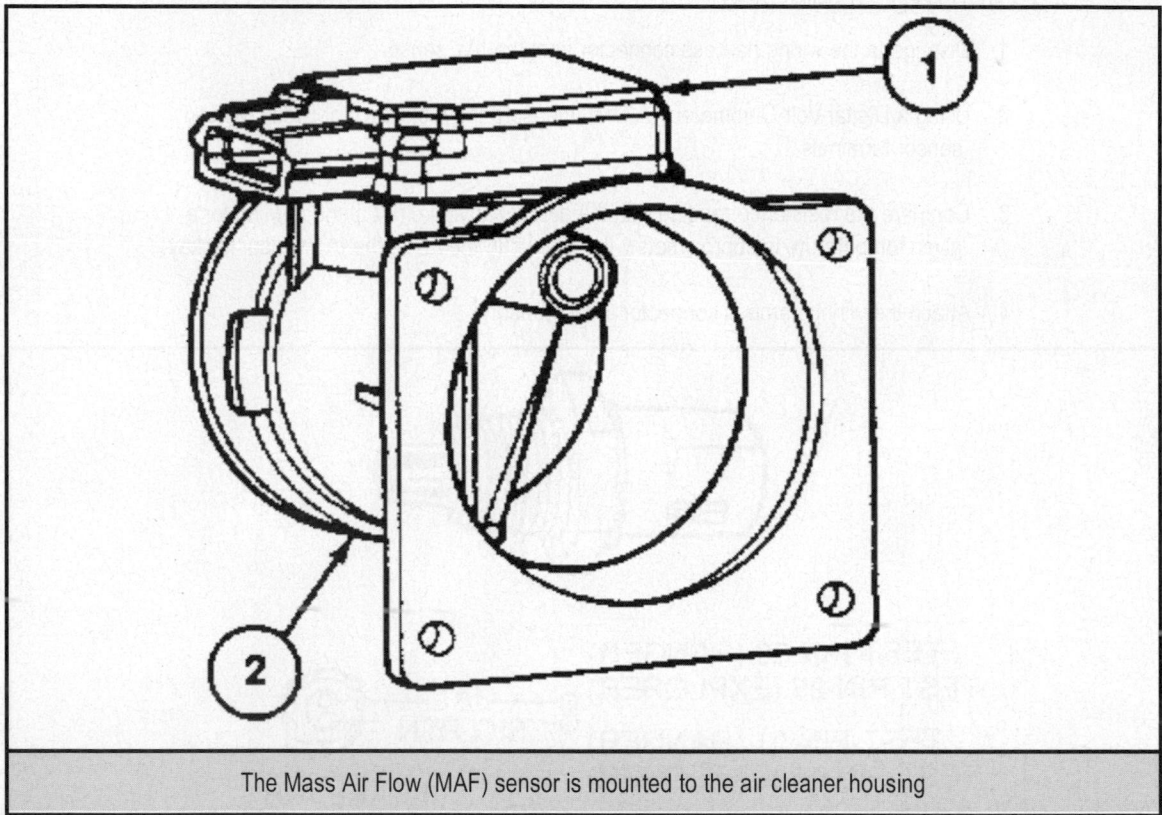

The Mass Air Flow (MAF) sensor is mounted to the air cleaner housing

The sensing element (hot wire) is a1 thin platinum wire wound on a ceramic bobbin and coated with glass. This hot wire is maintained at 392°F (200°C) above the ambient temperature as measured by a constant "cold wire".

Testing

1. With the engine running at idle, use a DVOM to verify there is at least 10.5 volts between terminals A and B of the MAF sensor connector. This indicates the power input to the sensor is correct. Then, measure the voltage between MAF sensor connector terminals C and D. If the reading is approximately 0.34-1.96 volts, the sensor is functioning properly.

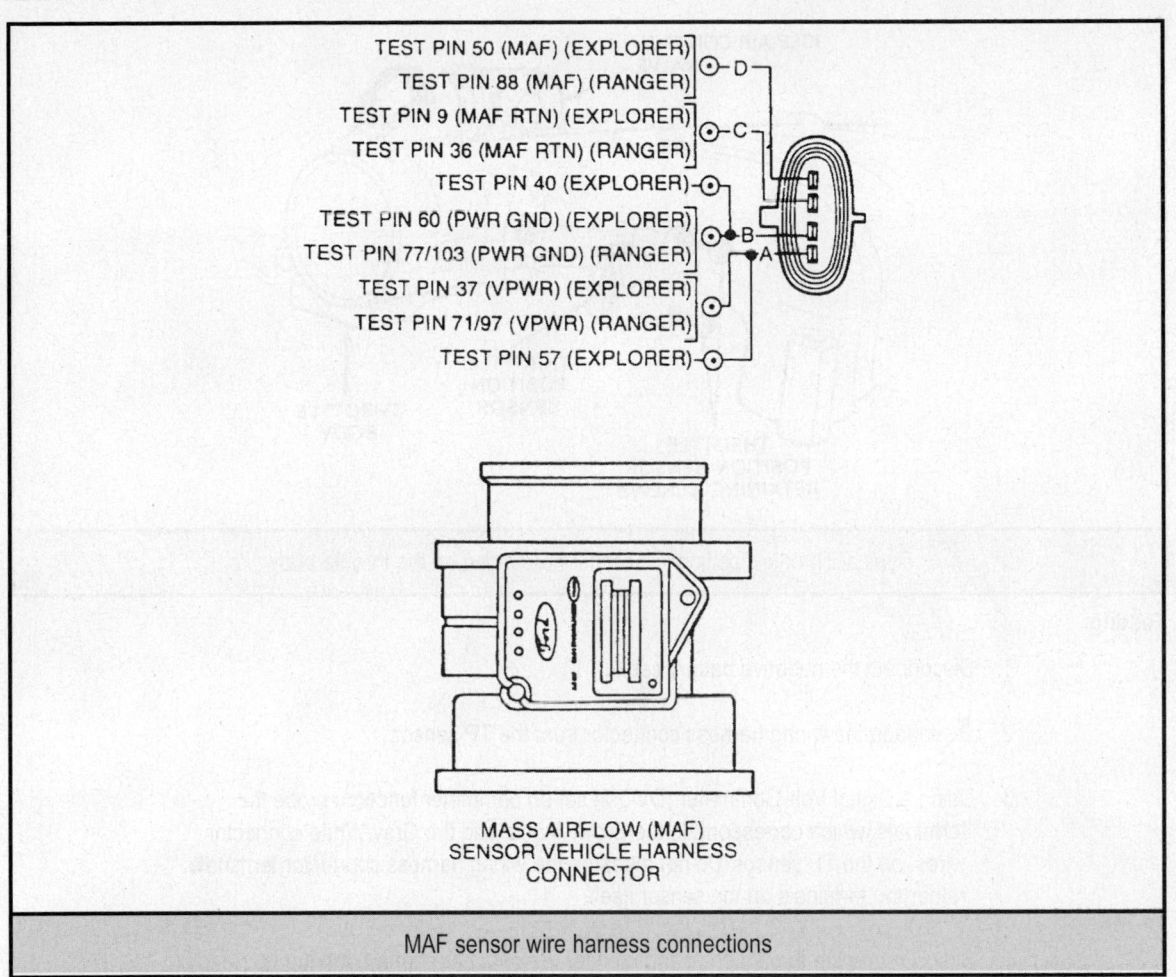

TEST PIN 50 (MAF) (EXPLORER)
TEST PIN 88 (MAF) (RANGER)
TEST PIN 9 (MAF RTN) (EXPLORER)
TEST PIN 36 (MAF RTN) (RANGER)
TEST PIN 40 (EXPLORER)
TEST PIN 60 (PWR GND) (EXPLORER)
TEST PIN 77/103 (PWR GND) (RANGER)
TEST PIN 37 (VPWR) (EXPLORER)
TEST PIN 71/97 (VPWR) (RANGER)
TEST PIN 57 (EXPLORER)

MASS AIRFLOW (MAF)
SENSOR VEHICLE HARNESS
CONNECTOR

MAF sensor wire harness connections

THROTTLE POSITION (TP) SENSOR

Operation

The Throttle Position (TP) sensor is a potentiometer that provides a signal to the PCM that is directly proportional to the throttle plate position. The TP sensor is mounted on the side of the throttle body and is connected to the throttle plate shaft. The TP sensor monitors throttle plate movement and position, and transmits an appropriate electrical signal to the PCM. These signals are used by the PCM to adjust the air/fuel mixture, spark timing and EGR operation according to engine load at idle, part throttle, or full throttle. The TPS is not adjustable.

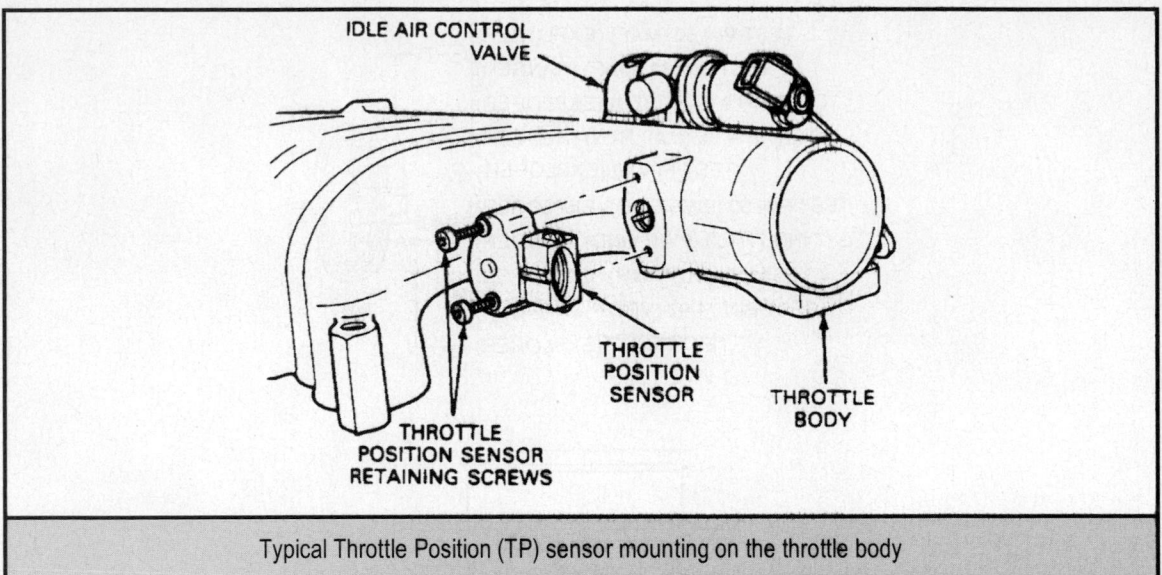

Typical Throttle Position (TP) sensor mounting on the throttle body

Testing

1. Disconnect the negative battery cable.

2. Disengage the wiring harness connector from the TP sensor.

3. Using a Digital Volt-Ohmmeter (DVOM) set on ohmmeter function, probe the terminals, which correspond to the Brown/White and the Gray/White connector wires, on the TP sensor. Do not measure the wiring harness connector terminals, rather the terminals on the sensor itself.

4. Slowly rotate the throttle shaft and monitor the ohmmeter for a continuous, steady change in resistance. Any sudden jumps, or irregularities (such as jumping back and forth) in resistance indicates a malfunctioning sensor.

5. Reconnect the negative battery cable.

6. Turn the DVOM to the voltmeter setting.

7. Detach the wiring harness connector from the PCM (located behind the lower right-hand kick panel in the passengers' compartment), then install a break-out box between the wiring harness connector and the PCM connector.

8. Turn the ignition switch ON and using the DVOM on voltmeter function, measure the voltage between terminals 89 and 90 of the breakout box. The specification is 0.9 volts.

9. If the voltage is outside the standard value or if it does not change smoothly, inspect the circuit wiring and/or replace the TP sensor.

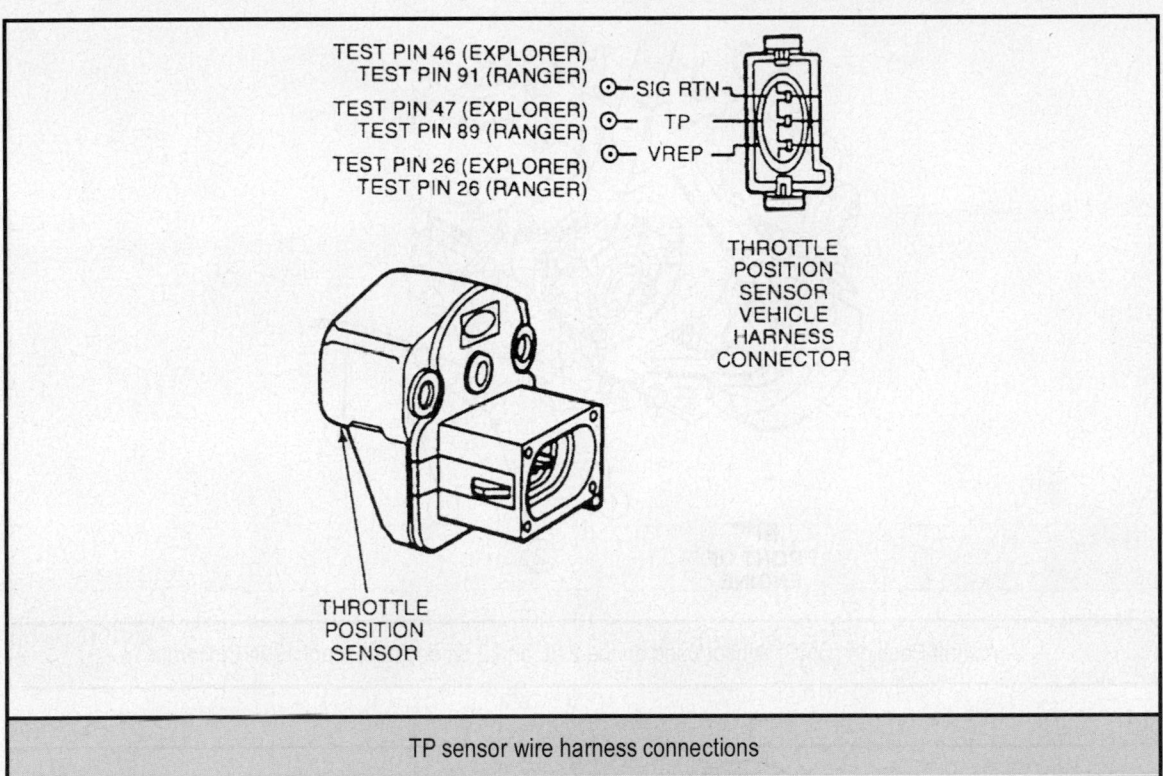

TEST PIN 46 (EXPLORER)
TEST PIN 91 (RANGER)

TEST PIN 47 (EXPLORER)
TEST PIN 89 (RANGER)

TEST PIN 26 (EXPLORER)
TEST PIN 26 (RANGER)

SIG RTN

TP

VREP

THROTTLE
POSITION
SENSOR
VEHICLE
HARNESS
CONNECTOR

THROTTLE
POSITION
SENSOR

TP sensor wire harness connections

CAMSHAFT POSITION (CMP) SENSOR

Operation

The CMP sensor provides the camshaft position information, called the CMP signal, which is used by the Powertrain Control Module (PCM) for fuel synchronization.

1991-93 2.3L and 1991-95 4.0L (VIN X) engines did not use CMP sensors.

1994 2.3L California only and 1995-99 2.3L and 2.5L engines utilize CMP sensors. The 1994 CMP is located on the oil pump drive assembly, on the left-hand lower side of the engine block. 1995-99 models CMP sensor is located and triggered by the auxiliary shaft drive sprocket.

On the 1991-94 2.9L and 3.0L engine, the distributor stator is the Camshaft Position (CMP) sensor, and it is a Hall effect magnetic switch. On the 1995-99 3.0L engine, the CMP is mounted on the oil pump drive assembly, located towards the rear of the block. it is also a single hall effect magnetic switch and it is activated by a single vane, and is driven by the camshaft.

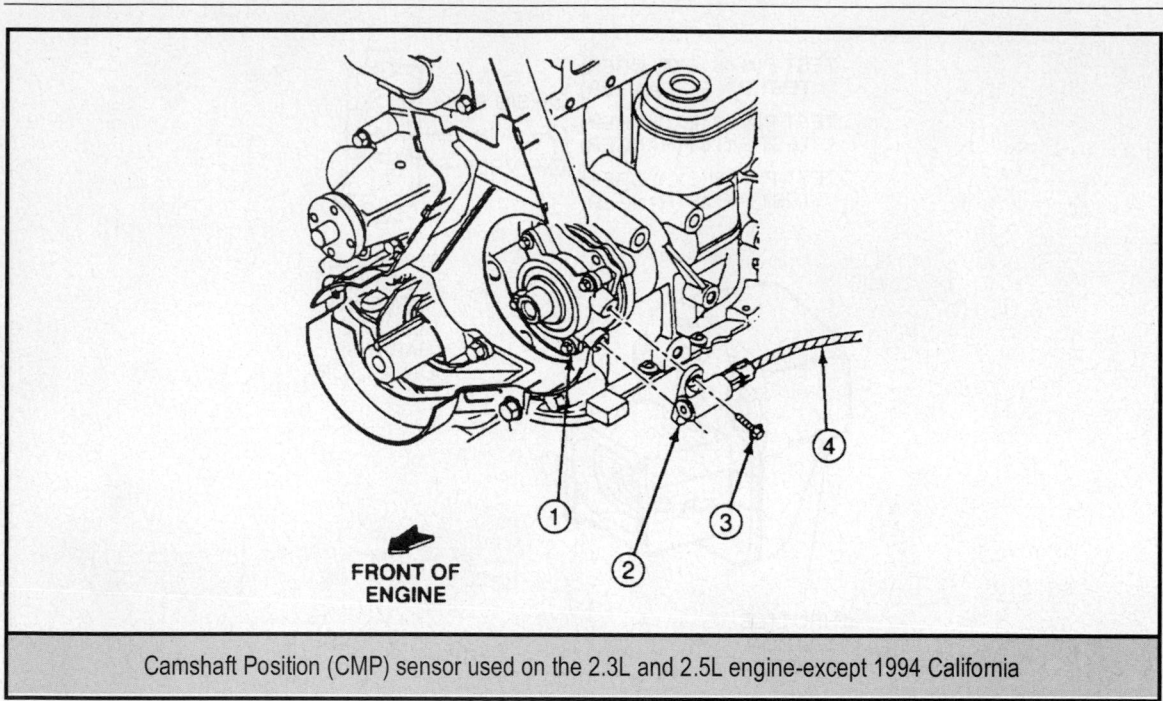

FRONT OF
ENGINE

Camshaft Position (CMP) sensor used on the 2.3L and 2.5L engine-except 1994 California

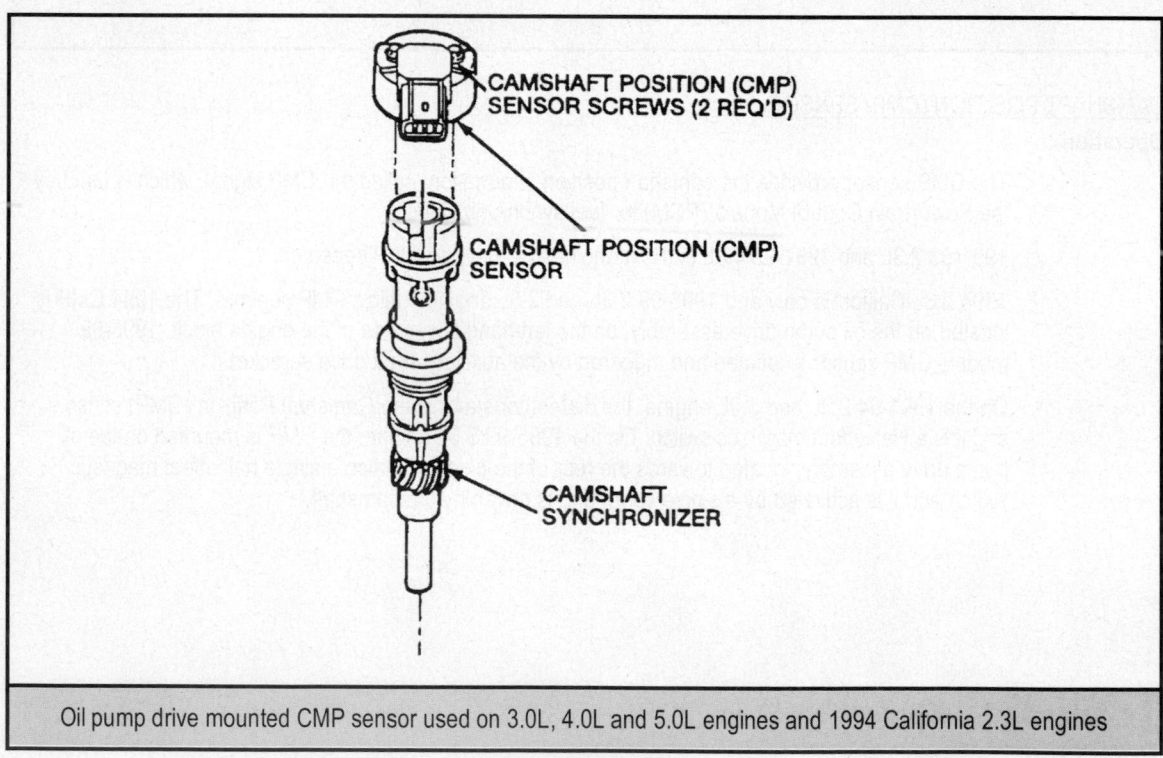

CAMSHAFT POSITION (CMP)
SENSOR SCREWS (2 REQ'D)

CAMSHAFT POSITION (CMP)
SENSOR

CAMSHAFT
SYNCHRONIZER

Oil pump drive mounted CMP sensor used on 3.0L, 4.0L and 5.0L engines and 1994 California 2.3L engines

On the 4.0L SOHC engine (VIN E), the Camshaft Position (CMP) sensor is a variable reluctance sensor, which is triggered by the high-point mark on the left-hand camshaft. It is mounted to the valve cover.

The 4.0L and 5.0L engines use a separate CMP sensor mounted to the oil pump drive. The drive assembly is located toward the rear of the engine on the 4.0L, and towards the front on the 5.0L engine.

Testing
THREE WIRE SENSORS

1. With the ignition OFF, disconnect the CMP sensor. With the ignition ON and the engine OFF, measure the voltage between sensor harness connector VPWR and PWR GND terminals (refer to the accompanying illustration). If the reading is greater than 10.5 volts, the power circuit to the sensor is okay.

2. With the ignition OFF, install break-out box between the CMP sensor and the PCM. Using a Digital Volt-Ohmmeter (DVOM) set to the voltage function (scale set to monitor less than 5 volts), measure voltage between break-out box terminals 24 and 40 with the engine running at varying RPM. If the voltage reading varies more than 0.1 volt, the sensor is okay.

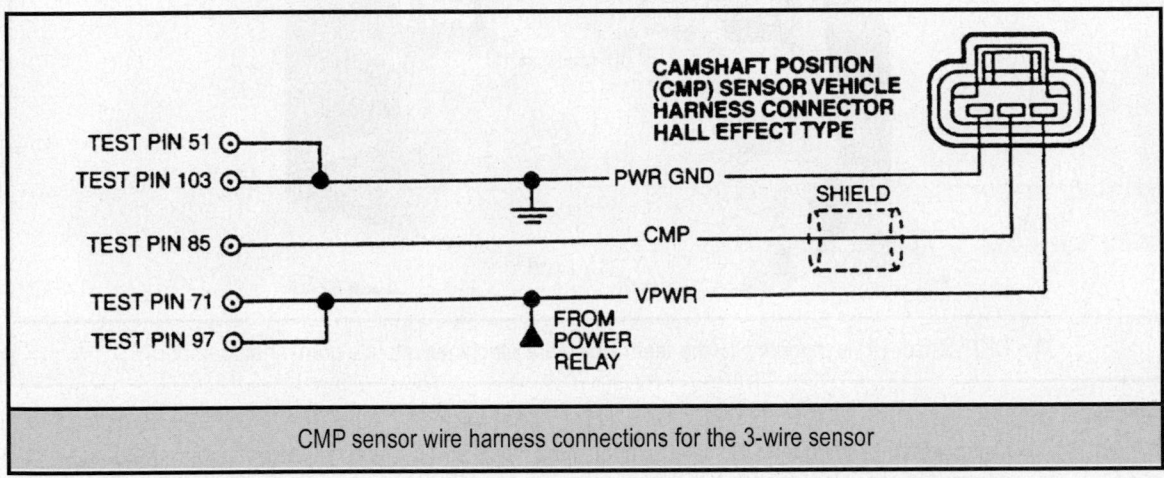

CMP sensor wire harness connections for the 3-wire sensor

TWO WIRE SENSORS

1. With the ignition OFF, install a break-out box between the CMP sensor and PCM.

2. Using a Digital Volt-Ohmmeter (DVOM) set to the voltage function (scale set to monitor less than 5 volts), measure the voltage between break-out box terminals 24 and 46 with the engine running at varying RPM. If the voltage reading varies more than 0.1 volt AC, the sensor is okay.

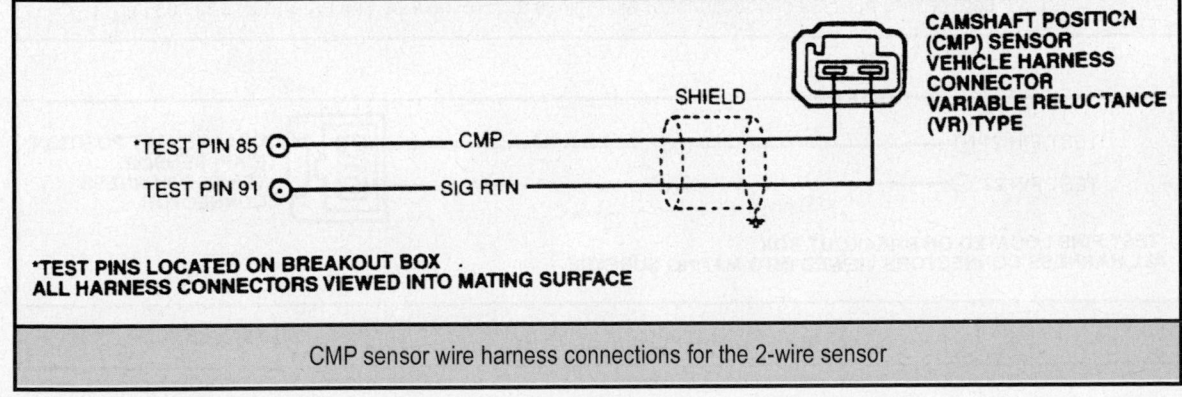

CMP sensor wire harness connections for the 2-wire sensor

CRANKSHAFT POSITION (CKP) SENSOR

Operation

The Crankshaft Position (CKP) sensor, located on the front cover (near the crankshaft pulley) is used to determine crankshaft position and crankshaft rpm. The CKP sensor is a reluctance sensor which senses the passing of teeth on a sensor ring because the teeth disrupt the magnetic field of the sensor. This disruption creates a voltage fluctuation, which is monitored by the PCM.

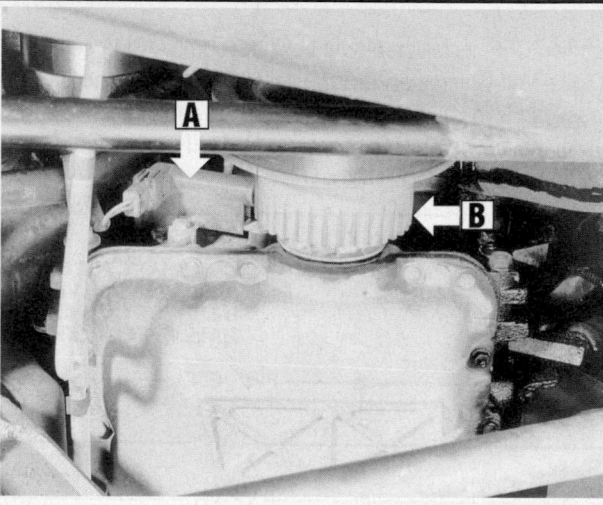

The CKP sensor (A) is triggered by the teeth which are machined into the crankshaft damper (B)

Testing

Using a DVOM set to the DC scale to monitor less than 5 volts, measure the voltage between the sensor Cylinder Identification (CID) terminal and ground by backprobing the sensor connector. If the connector cannot be backprobed, fabricate or purchase a test harness. The sensor is okay if the voltage reading varies more than 0.1 volt with the engine running at varying RPM.

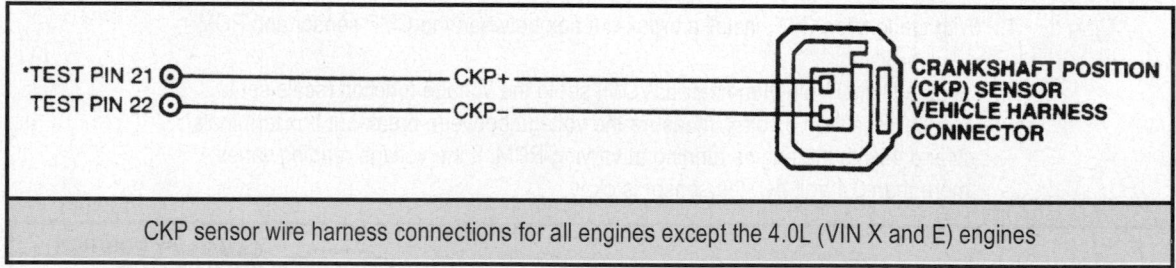

CKP sensor wire harness connections for all engines except the 4.0L (VIN X and E) engines

*TEST PIN 21 ○ ———— CKP+ ————
TEST PIN 22 ○ ———— CKP– ————

CRANKSHAFT POSITION (CKP) SENSOR VEHICLE HARNESS CONNECTOR

*TEST PINS LOCATED ON BREAKOUT BOX
ALL HARNESS CONNECTORS VIEWED INTO MATING SURFACE

CKP sensor wire harness connections for the 4.0L (VIN X and E) engines

VEHICLE SPEED SENSOR (VSS)

Operation

The Vehicle Speed Sensor (VSS) is a magnetic pick-up that sends a signal to the Powertrain Control Module (PCM). The sensor measures the rotation of the transmission and the PCM determines the corresponding vehicle speed.

Testing

1. Turn the ignition switch to the OFF position.

2. Disengage the wiring harness connector from the VSS.

3. Using a Digital Volt-Ohmmeter (DVOM), measure the resistance (DVOM ohmmeter function) between the sensor terminals. If the resistance is 190-250 ohms, the sensor is okay.

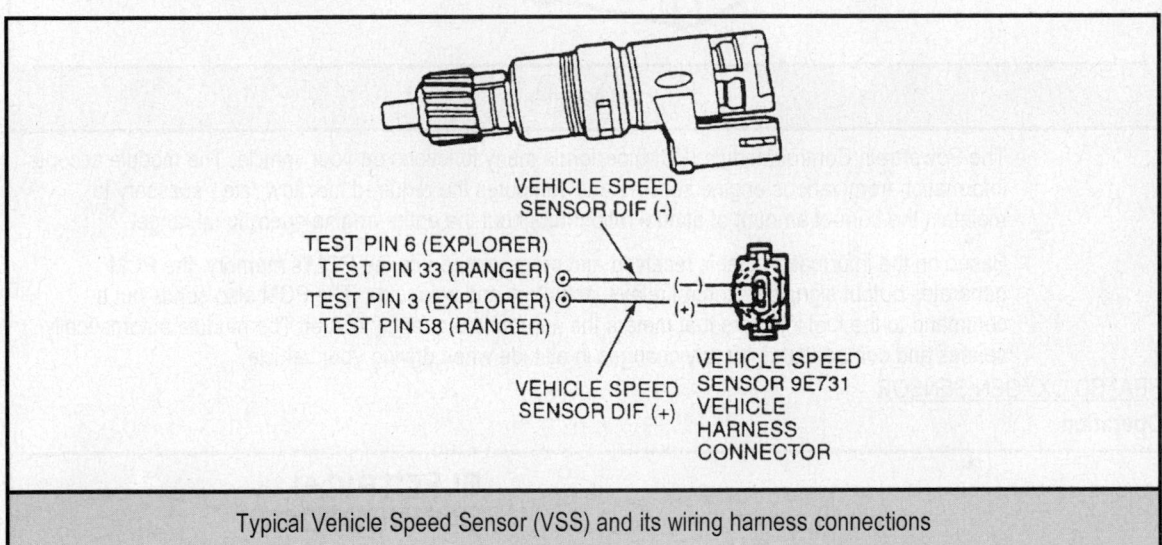

VEHICLE SPEED
SENSOR DIF (-)

TEST PIN 6 (EXPLORER)
TEST PIN 33 (RANGER)
TEST PIN 3 (EXPLORER)
TEST PIN 58 (RANGER)

(—)
(+)

VEHICLE SPEED
SENSOR DIF (+)

VEHICLE SPEED
SENSOR 9E731
VEHICLE
HARNESS
CONNECTOR

Typical Vehicle Speed Sensor (VSS) and its wiring harness connections

POWERTRAIN CONTROL MODULE (PCM)
Operation

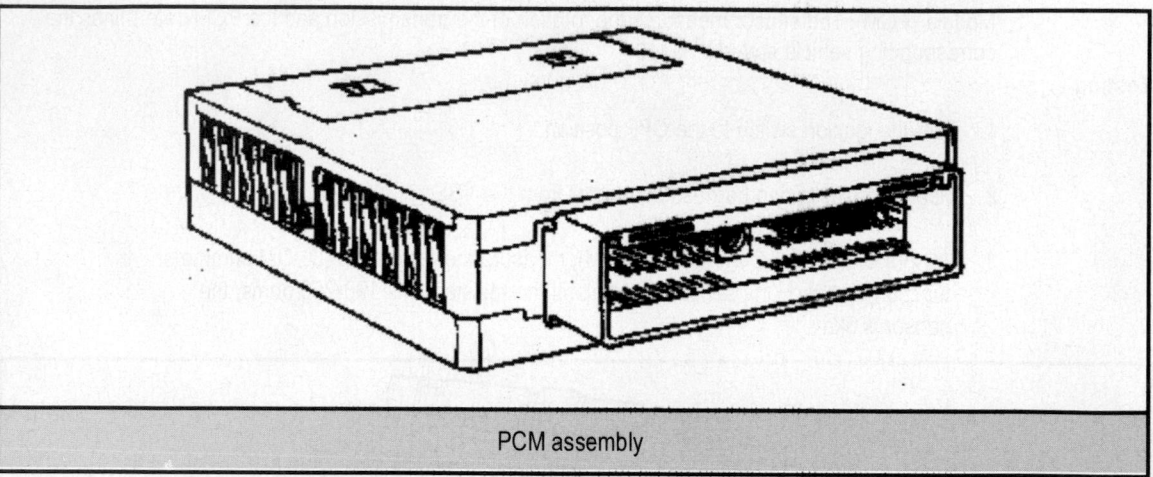

PCM assembly

The Powertrain Control Module (PCM) performs many functions on your vehicle. The module accepts information from various engine sensors and computes the required fuel flow rate necessary to maintain the correct amount of air/fuel ratio throughout the entire engine operational range.

Based on the information that is received and programmed into the PCM's memory, the PCM generates output signals to control relays, actuators and solenoids. The PCM also sends out a command to the fuel injectors that meters the appropriate quantity of fuel. The module automatically senses and compensates for any changes in altitude when driving your vehicle.

HEATED OXYGEN SENSOR
Operation

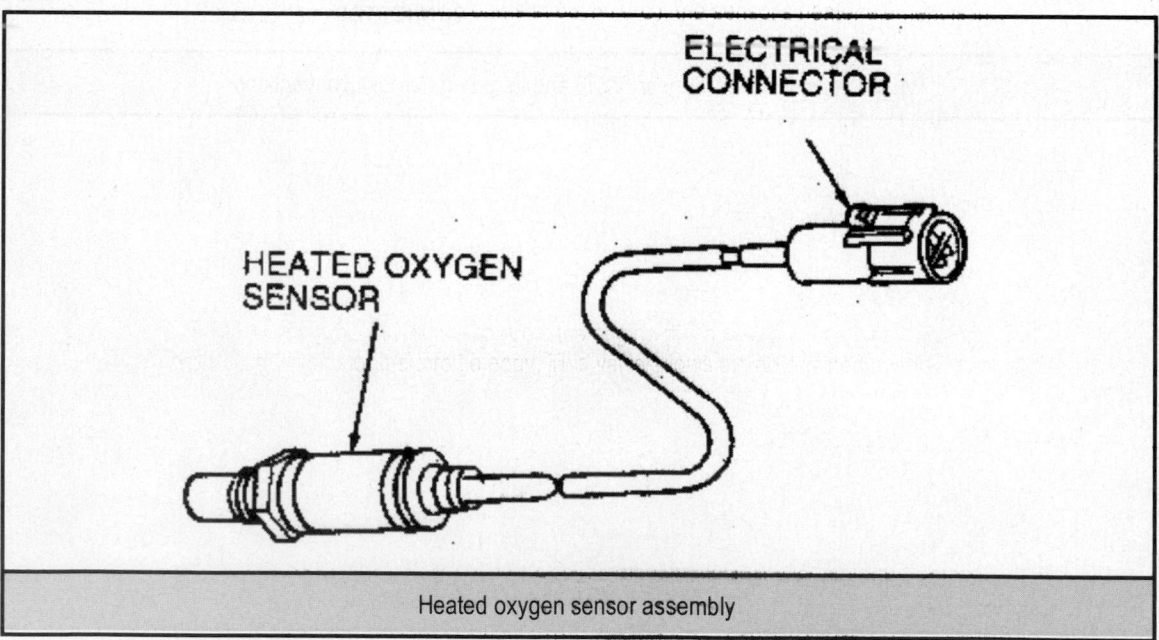

Heated oxygen sensor assembly

The oxygen (O2) sensor is a device which produces an electrical voltage when exposed to the oxygen present in the exhaust gases. The sensor is mounted in the exhaust system, usually in the manifold or a boss located on the down pipe before the catalyst.. The oxygen sensors used on the Ford Contour/Mercury Mystique/Mercury Cougar are electrically heated internally for faster switching when the engine is started cold. The oxygen sensor produces a voltage within 0 and 1 volt. When there is a large amount of oxygen present (lean mixture), the sensor produces a low voltage (less than 0.4v). When there is a lesser amount present (rich mixture) it produces a higher voltage (0.6-1.0v). The stoichiometric or correct fuel to air ratio will read between 0.4 and 0.6v. By monitoring the oxygen content and converting it to electrical voltage, the sensor acts as a rich-lean switch. The voltage is transmitted to the PCM.

Some models have two sensors, one before the catalyst and one after. This is done for a catalyst efficiency monitor that is a part of the OBD-II engine controls. The one before the catalyst measures the exhaust emissions right out of the engine, and sends the signal to the PCM about the state of the mixture as previously talked about. The second sensor reports the difference in the emissions after the exhaust gases have gone through the catalyst. This sensor reports to the PCM the amount of emissions reduction the catalyst is performing.

The oxygen sensor will not work until a predetermined temperature is reached, until this time the PCM is running in what as known as OPEN LOOP operation. OPEN LOOP means that the PCM has not yet begun to correct the air-to-fuel ratio by reading the oxygen sensor. After the engine comes to operating temperature, the PCM will monitor the oxygen sensor and correct the air/fuel ratio from the sensor's readings. This is what is known as CLOSED LOOP operation.

A heated oxygen sensor (HO2S) has a heating element that keeps the sensor at proper operating temperature during all operating modes. Maintaining correct sensor temperature at all times allows the system to enter into CLOSED LOOP operation sooner.

In CLOSED LOOP operation the PCM monitors the sensor input (along with other inputs) and adjusts the injector pulse width accordingly. During OPEN LOOP operation the PCM ignores the sensor input and adjusts the injector pulse to a preprogrammed value based on other inputs.

Testing

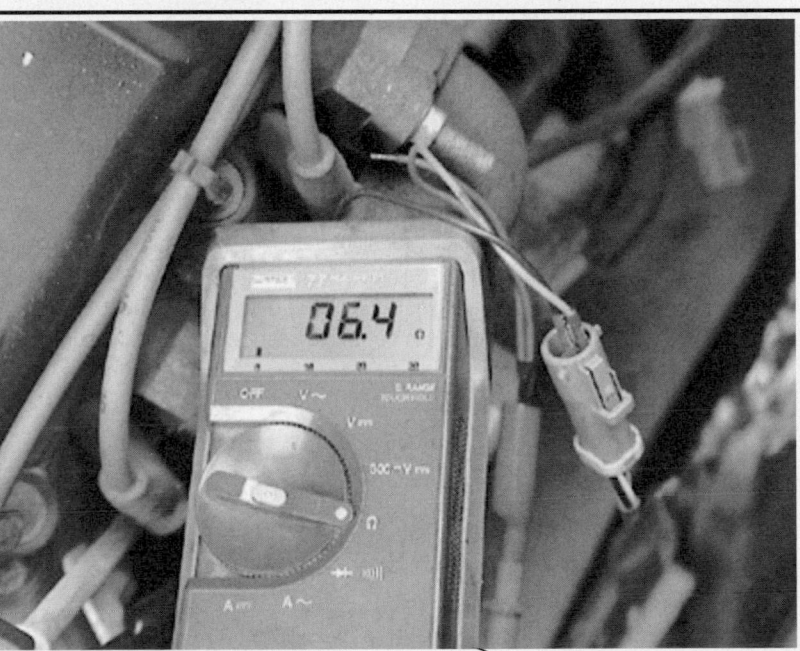

The voltage of the HO2S should be between 0.1 and 1.0 volts DC with the engine running

The HO2S can be monitored with an appropriate and Data-stream capable scan tool

1. Disconnect the HO2S.

2. Measure the resistance between PWR and GND terminals of the sensor. Resistance should be approximately 6 ohms at 68°F (20°C). If resistance is not within specification, the sensor's heater element is faulty.

3. With the HO2S connected and engine running, measure the voltage with a Digital Volt-Ohmmeter (DVOM) between terminals HO2S and SIG RTN (GND) of the oxygen sensor connector. Voltage should fluctuate between 0.01-1.0 volts. If voltage fluctuation is slow or voltage is not within specification, the sensor may be faulty.

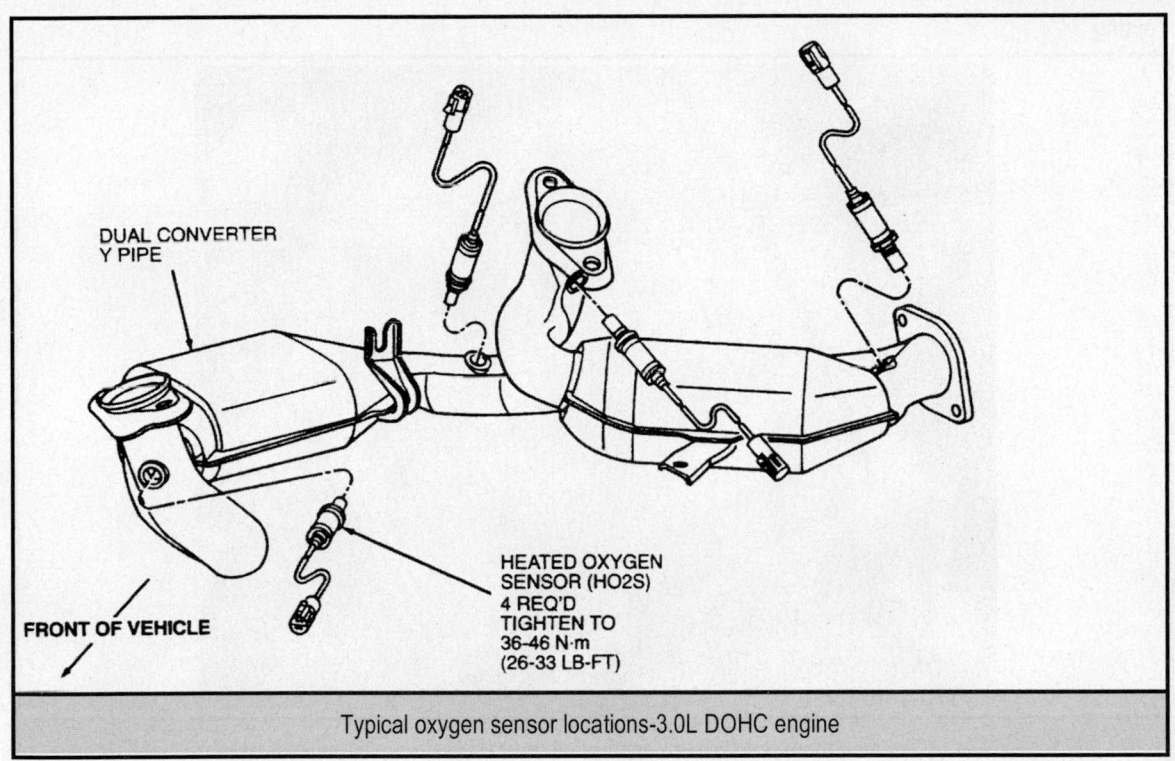

DUAL CONVERTER
Y PIPE

HEATED OXYGEN
SENSOR (HO2S)
4 REQ'D
TIGHTEN TO
36-46 N·m
(26-33 LB-FT)

FRONT OF VEHICLE

Typical oxygen sensor locations-3.0L DOHC engine

IDLE AIR CONTROL VALVE
Operation

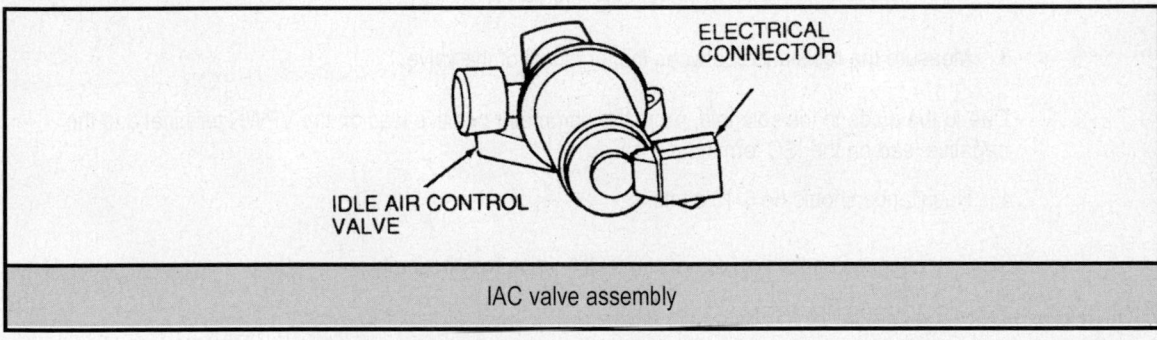

ELECTRICAL
CONNECTOR

IDLE AIR CONTROL
VALVE

IAC valve assembly

The Idle Air Control (IAC) valve adjusts the engine idle speed. The valve is located on the side of the throttle body. The valve is controlled by a duty cycle signal from the PCM and allows air to bypass the throttle plate in order to maintain the proper idle speed.

Do not attempt to clean the IAC valve. Carburetor tune-up cleaners or any type of solvent cleaners will damage the internal components of the valve.

Testing

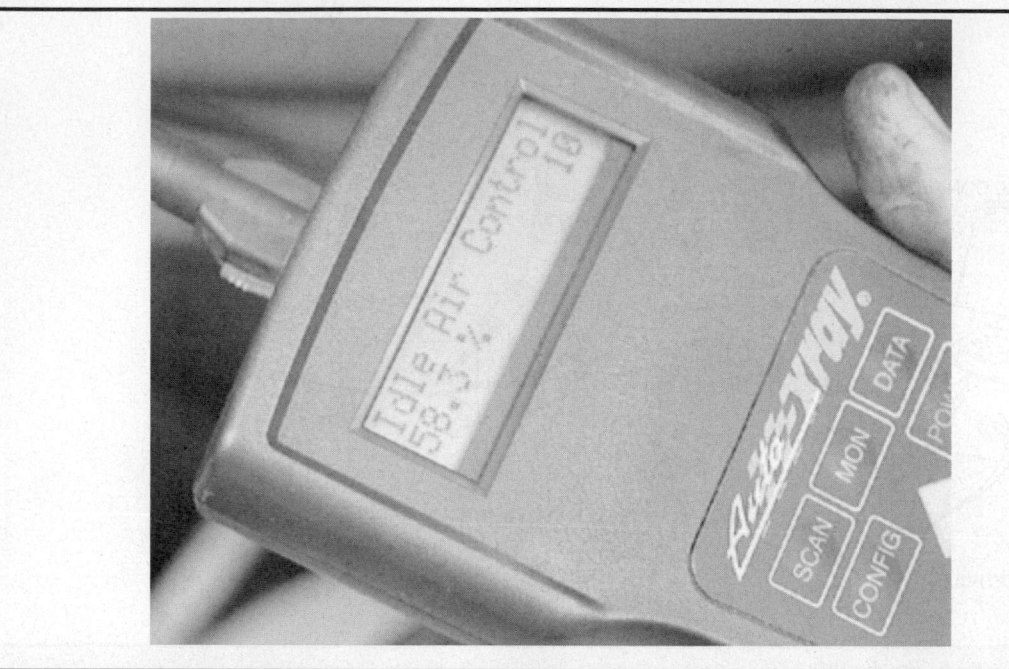

The IAC can be monitored with an appropriate and Data-stream capable scan tool

1. Turn the ignition switch to the OFF position.

2. Disconnect the wiring harness from the IAC valve.

3. Measure the resistance between the terminals of the valve.

   Due to the diode in the solenoid, place the ohmmeter positive lead on the VPWR terminal and the negative lead on the ISC terminal.

4. Resistance should be 6-13 ohms.

5. If resistance is not within specification, the valve may be faulty.

ENGINE COOLANT TEMPERATURE SENSOR

Operation

The Engine Coolant Temperature (ECT) sensor resistance changes in response to engine coolant temperature. The sensor resistance decreases as the coolant temperature increases, and increases as the coolant temperature decreases. This provides a reference signal to the PCM, which indicates engine coolant temperature. The signal sent to the PCM by the ECT sensor helps the PCM to determine spark advance, EGR flow rate, air/fuel ratio, and engine temperature. The ECT is a two wire sensor, a 5-volt reference signal is sent to the sensor and the signal return is based upon the change in the measured resistance due to temperature.

Testing

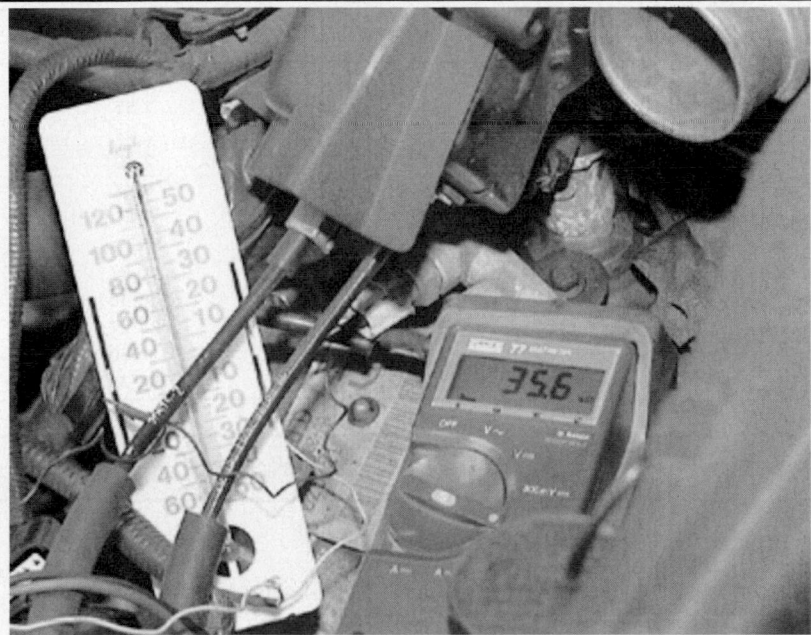

Measure the resistance of the ECT sensor across the two sensor terminals using a suitable DVOM. Compare the readings to the temp/resistance chart

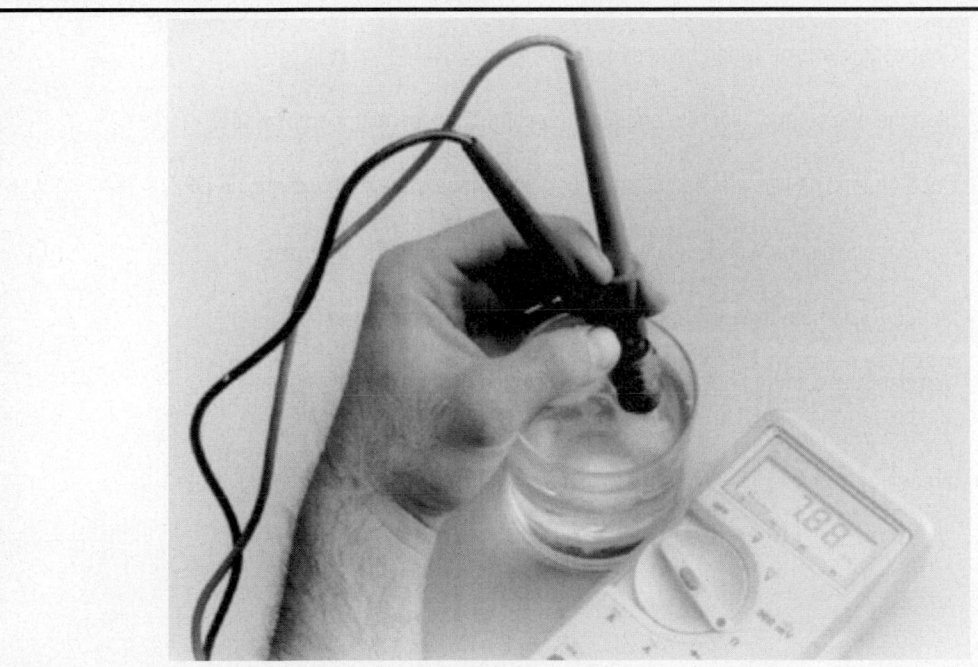

Another method of testing the ECT is to submerge it in cold or hot water and check resistance

Temperature		Engine Coolant/Intake Air Temperature Sensor Values
°F	°C	Resistance (K ohms)
248	120	1.18
230	110	1.55
212	100	2.07
194	90	2.80
176	80	3.84
158	70	5.37
140	60	7.70
122	50	10.97
104	40	16.15
86	30	24.27
68	20	37.30
50	10	58.75

ECT resistance-to-temperature specifications

1. Disconnect the engine wiring harness from the ECT sensor.

2. Connect an ohmmeter between the ECT sensor terminals.

3. With the engine cold and the ignition switch in the OFF position, measure and note the ECT sensor resistance.

4. Connect the engine wiring harness to the sensor.

5. Start the engine and allow the engine to reach normal operating temperature.

6. Once the engine has reached normal operating temperature, turn the engine OFF.

7. Once again, disconnect the engine wiring harness from the ECT sensor.

8. Measure and note the ECT sensor resistance with the engine hot.

9. Compare the cold and hot ECT sensor resistance measurements with the accompanying chart.

10. If readings do not approximate those in the chart, the sensor may be faulty.

INTAKE AIR TEMPERATURE SENSOR
Operation

The tip of the IAT sensor has an exposed thermistor that changes the
resistance of the sensor based upon the force of the air rushing past it

The Intake Air Temperature (IAT) sensor determines the air temperature inside the intake manifold. Resistance changes in response to the ambient air temperature. The sensor has a negative temperature coefficient. As the temperature of the sensor rises the resistance across the sensor decreases. This provides a signal to the PCM indicating the temperature of the incoming air charge. This sensor helps the PCM to determine spark timing and air/fuel ratio. Information from this sensor is added to the pressure sensor information to calculate the air mass being sent to the cylinders. The IAT is a two wire sensor. A 5-volt reference signal is sent to the sensor and the signal return is based upon the change in the measured resistance due to temperature.

Testing

Measure the resistance of the IAT sensor across the two sensor terminals using a suitable DVOM.
Compare the readings to the temp/resistance chart

The IAT sensor can be monitored with an appropriate and Data-stream capable scan tool

Temperature		Engine Coolant/Intake Air Temperature Sensor Values
°F	°C	Resistance (K ohms)
248	120	1.18
230	110	1.55
212	100	2.07
194	90	2.80
176	80	3.84
158	70	5.37
140	60	7.70
122	50	10.97
104	40	16.15
86	30	24.27
68	20	37.30
50	10	58.75

ECT/IAT resistance-to-temperature specifications

1. Turn the ignition switch OFF.

2. Disconnect the wiring harness from the IAT sensor.

3. Measure the resistance between the sensor terminals.

4. Compare the resistance reading with the accompanying chart.

5. If the resistance is not within specification, the IAT may be faulty.

6. Connect the wiring harness to the sensor.

MASS AIRFLOW SENSOR
Operation

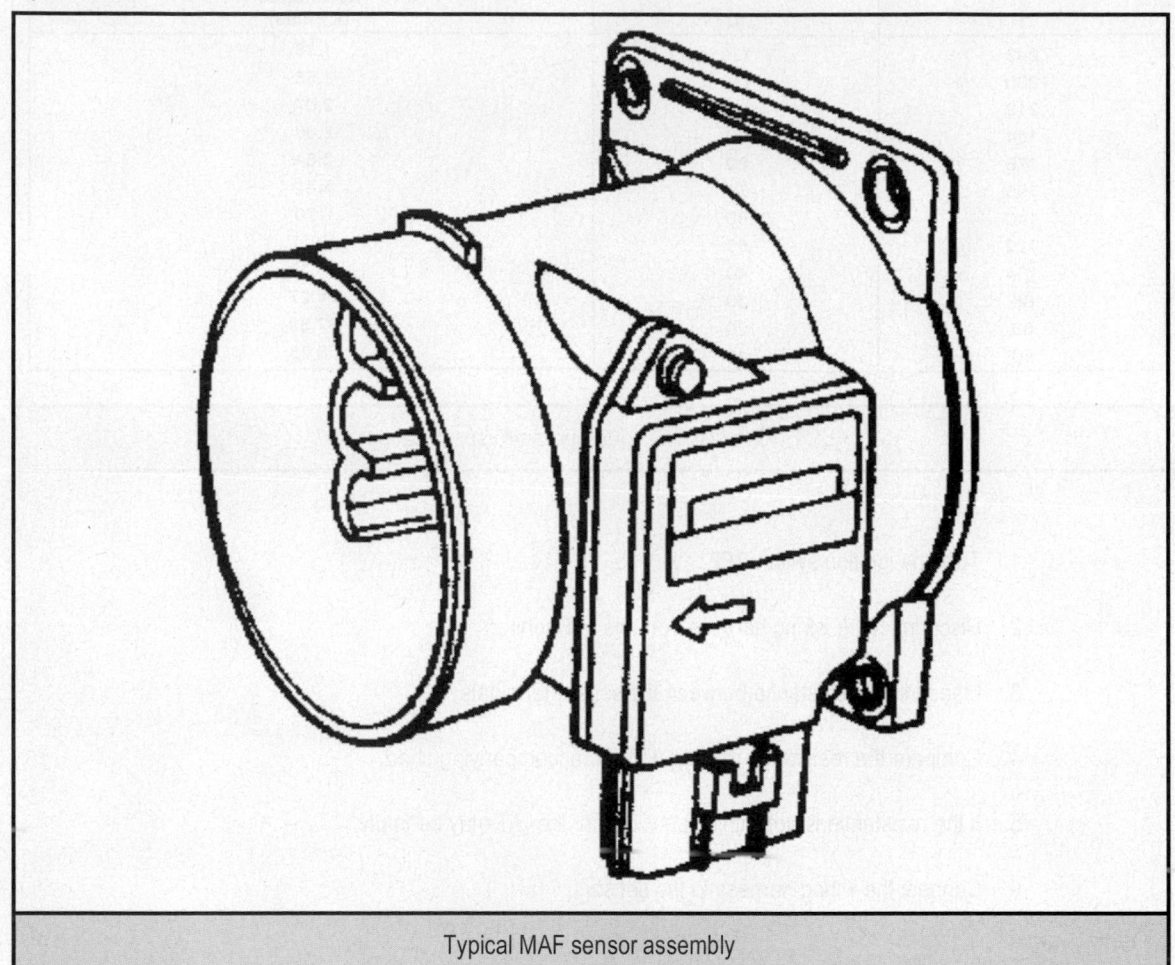

Typical MAF sensor assembly

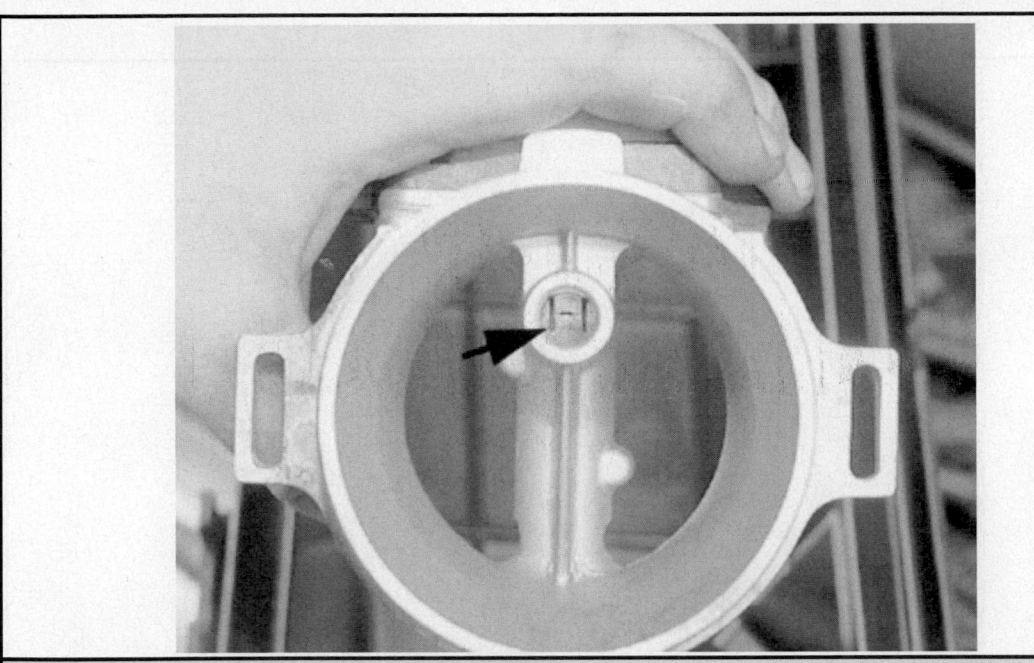

The exposed "hot wire" of the MAF sensor

The Mass Air Flow (MAF) sensor directly measures the mass of air being drawn into the engine. The sensor output is used to calculate injector pulse width. The MAF sensor is what is referred to as a "hot-wire sensor". The sensor uses a thin platinum wire filament, wound on a ceramic bobbin and coated with glass, that is heated to 200°C (417°F) above the ambient air temperature and subjected to the intake airflow stream. A "cold-wire" is used inside the MAF sensor to determine the ambient air temperature.

Battery voltage from the EEC power relay, and a reference signal and a ground signal from the PCM are supplied to the MAF sensor. The sensor returns a signal proportionate to the current flow required to keep the "hot-wire" at the required temperature. The increased airflow across the "hot-wire" acts as a cooling fan, lowering the resistance and requiring more current to maintain the temperature of the wire. The increased current is measured by the voltage in the circuit, as current increases, voltage increases. As the airflow increases the signal return voltage of a normally operating MAF sensor will increase.

Testing

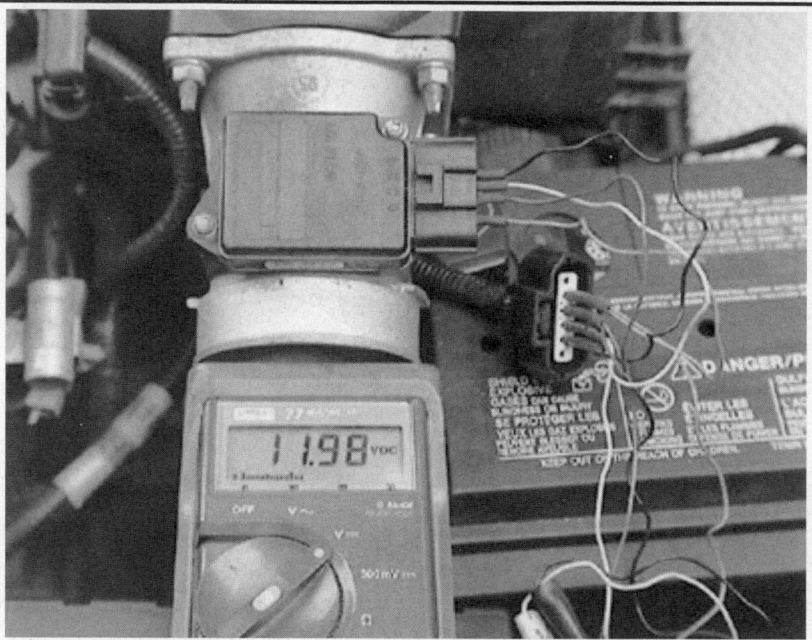

Testing the VPWR circuit of the MAF sensor

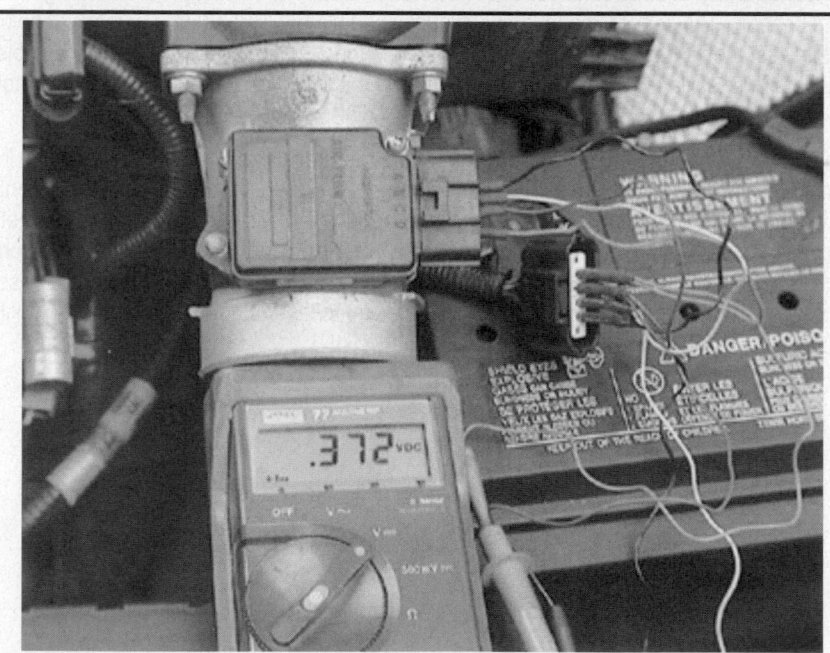

Testing the SIG RTN circuit of the MAF sensor

1. Using a multimeter, check for voltage by backprobing the MAF sensor connector.

2. With the key ON, and the engine OFF, verify that there is at least 10.5 volts between the VPWR and GND terminals of the MAF sensor connector. If voltage is not within specification, check power and ground circuits and repair as necessary.

3. With the key ON, and the engine ON, verify that there is at least 4.5 volts between the SIG and GND terminals of the MAF sensor connector. If voltage is not within specification, check power and ground circuits and repair as necessary.

4. With the key ON, and the engine ON, check voltage between GND and SIG RTN terminals. Voltage should be approximately 0.34-1.96 volts. If voltage is not within specification, the sensor may be faulty.

THROTTLE POSITION SENSOR
Operation

The Throttle Position (TP) sensor is a potentiometer that provides a signal to the PCM that is directly proportional to the throttle plate position. The TP sensor is mounted on the side of the throttle body and is connected to the throttle plate shaft. The TP sensor monitors throttle plate movement and position, and transmits an appropriate electrical signal to the PCM. These signals are used by the PCM to adjust the air/fuel mixture, spark timing and EGR operation according to engine load at idle, part throttle, or full throttle. The TP sensor is not adjustable.

The TP sensor receives a 5 volt reference signal and a ground circuit from the PCM. A return signal circuit is connected to wiper that runs on a resistor internally on the sensor. The further the throttle is opened, the wiper moves along the resistor, at wide open throttle, the wiper essentially creates a loop between the reference signal and the signal return returning the full or nearly full 5 volt signal back to the PCM. At idle the signal return should be approximately 0.9 volts.

Testing

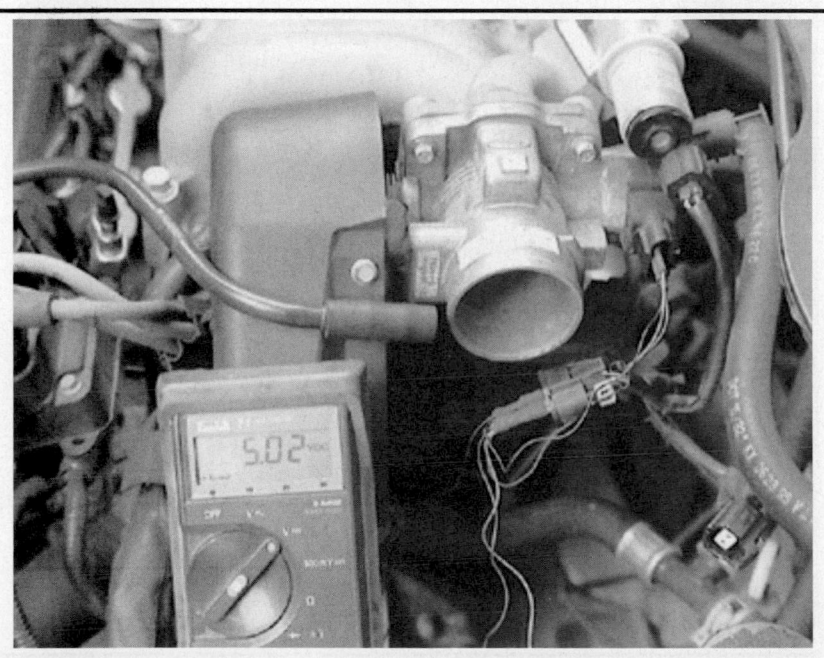

Testing the SIG circuit to the TP sensor

Testing the SIG RTN circuit of the TP sensor

Testing the operation of the potentiometer inside the TP sensor while slowly opening the throttle

The TP sensor can be monitored with an appropriate and Data-stream capable scan tool

1. With the engine OFF and the ignition ON, check the voltage at the signal return circuit of the TP sensor by carefully backprobing the connector using a DVOM.

2. Voltage should be between 0.2 and 1.4 volts at idle.

3. Slowly move the throttle pulley to the wide open throttle (WOT) position and watch the voltage on the DVOM. The voltage should slowly rise to slightly less than 4.8v at Wide Open Throttle (WOT).

4. If no voltage is present, check the wiring harness for supply voltage (5.0v) and ground (0.3v or less), by referring to your corresponding wiring guide. If supply voltage and ground are present, but no output voltage from TP, replace the TP sensor. If supply voltage and ground do not meet specifications, make necessary repairs to the harness or PCM.

CAMSHAFT POSITION SENSOR
Operation

The camshaft position sensor (CMP) on the 3.0L DOHC and 3.4L DOHC engines is a variable reluctance sensor that is triggered by a high point on the left-hand exhaust camshaft sprocket. The CMP sends a signal relating camshaft position back to the PCM which is used by the PCM to control engine timing.

The camshaft position sensor (CMP) on the 3.0L OHV engine is a single hall-effect magnetic switch that is triggered by a single vane which is driven by the camshaft. The CMP sends a signal relating camshaft position back to the PCM which is used by the PCM to control engine timing.

Testing

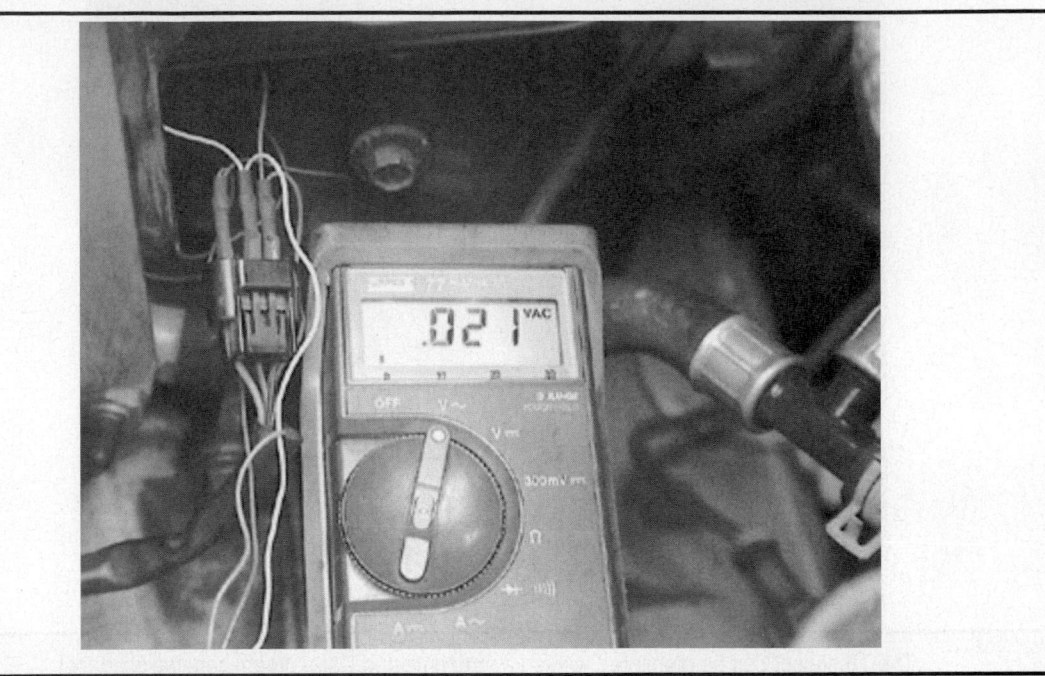

Testing the CMP sensor for voltage as the engine is running

Testing the VPWR circuit of the CMP sensor

1   Check voltage between the camshaft position sensor terminals PWR GND and CID.

2   With engine running, voltage should be greater than 0.1 volt AC and vary with engine speed.

3   If voltage is not within specification, check for proper voltage at the VPWR terminal.

4   If VPWR voltage is greater than 10.5 volts, sensor may be faulty.

CRANKSHAFT POSITION SENSOR

Operation

The Crankshaft Position (CKP) sensor is a variable reluctance sensor that uses a trigger wheel to induce voltage. The CKP sensor is a fixed magnetic sensor mounted to the engine block and monitors the trigger or "pulse" wheel which is attached to the crank pulley/damper. As the pulse wheel rotates by the CKP sensor, teeth on the pulse wheel induce voltage inside the sensor through magnetism. The pulse wheel has a missing tooth that changes the reading of the sensor. This is used for the Cylinder Identification (CID) function to properly monitor and adjust engine timing by locating the number 1 cylinder. The voltage created by the CKP sensor is alternating current (A/C). This voltage reading is sent to the PCM and is used to determine engine RPM, engine timing, and is used to fire the ignition coils.

Testing

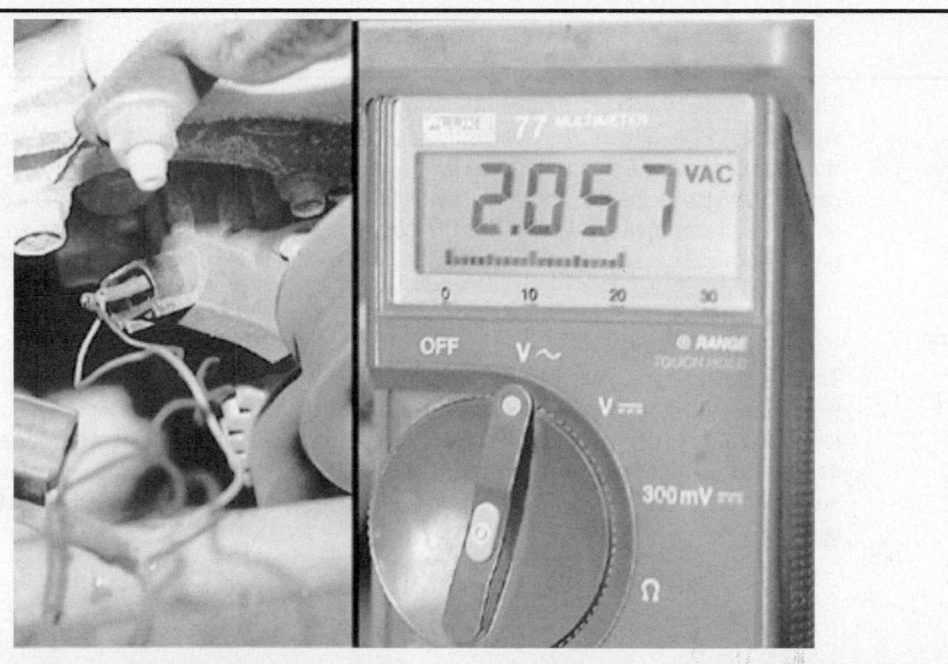

Test the CKP sensor for output voltage while the engine is cranked o r running. Voltage should be more than 0.1 volts A/C

1. Measure the voltage between the sensor CKP sensor terminals by backprobing the sensor connector.

    If the connector cannot be backprobed, fabricate or purchase a test harness.

2. Sensor voltage should be more than 0.1 volt AC with the engine running and should vary with engine RPM.

3. If voltage is not within specification, the sensor may be faulty.

<u>KNOCK SENSOR</u>
Operation

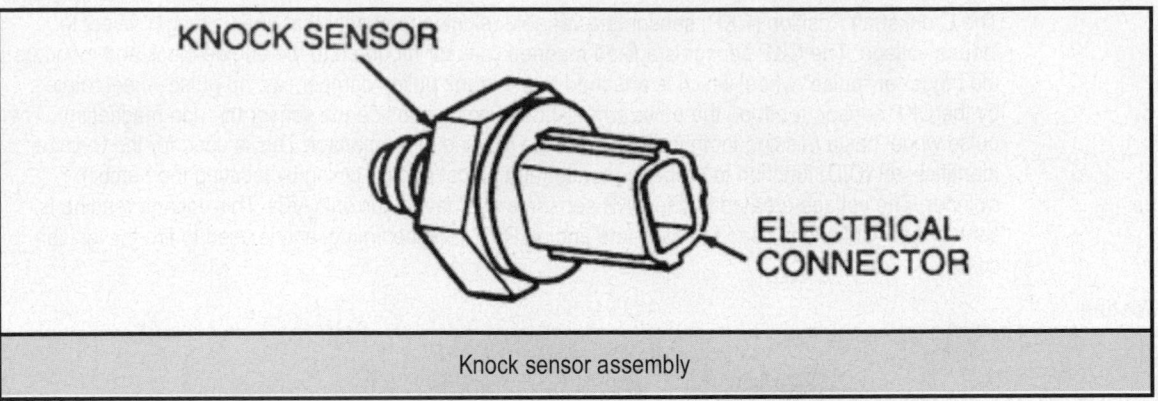

Knock sensor assembly

The operation of the Knock Sensor (KS) is to monitor preignition or "engine knocks" and send the signal to the PCM. The PCM responds by adjusting ignition timing until the "knocks" stop. The sensor works by generating a signal produced by the frequency of the knock as recorded by the piezoelectric ceramic disc inside the KS. The disc absorbs the shock waves from the knocks and exerts a pressure on the metal diaphragm inside the KS. This compresses the crystals inside the disc and the disc generates a voltage signal proportional to the frequency of the knocks ranging from zero to 1 volt.

Testing

There is real no test for this sensor, the sensor produces it's own signal based on information gathered while the engine is running. The sensors also are usually inaccessible without major component removal. The sensors can be monitored with an appropriate scan tool using a data display or other data stream information. Follow the instructions included with the scan tool for information on accessing the data. The only test available is to test the continuity of the harness from the PCM to the sensor.

VEHICLE SPEED SENSOR
Operation

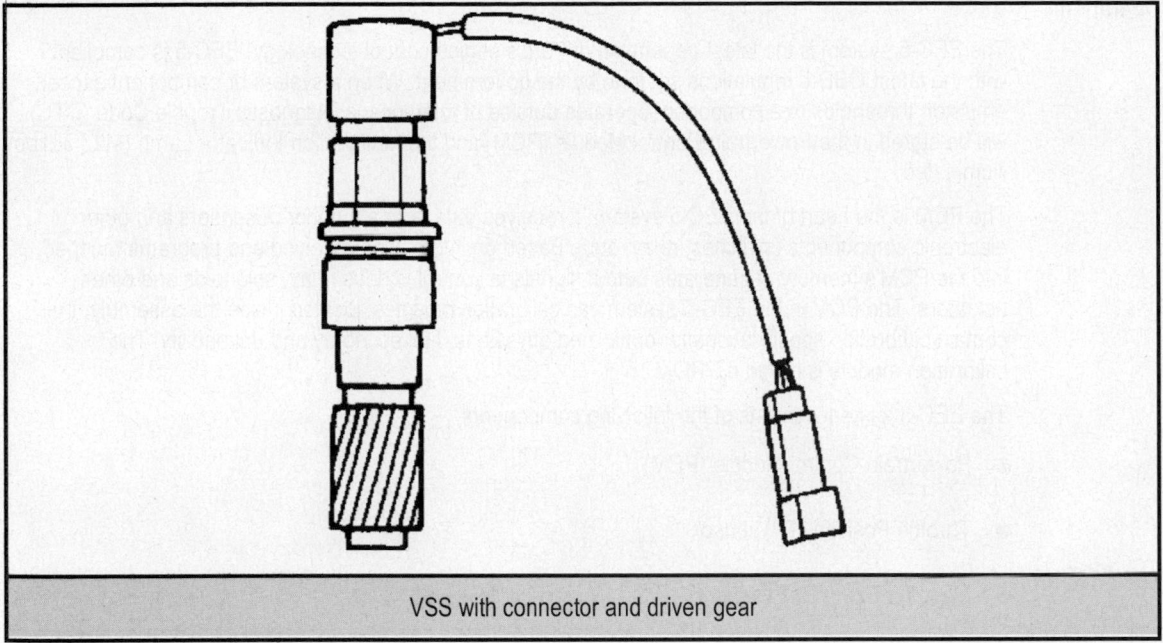

VSS with connector and driven gear

The Vehicle Speed Sensor (VSS) is a magnetic pick-up sensor that sends a signal to the Powertrain Control Module (PCM) and the speedometer. The sensor measures the rotation of the output shaft on the transaxle and sends an AC voltage signal to the PCM which determines the corresponding vehicle speed.

Testing

1. Disconnect the negative battery cable.

2. Disengage the wiring harness connector from the VSS.

3. Using a Digital Volt-Ohmmeter (DVOM), measure the resistance (ohmmeter function) between the sensor terminals. If the resistance is 190-250 ohms, the sensor is okay.

ELECTRONIC ENGINE CONTROL 5 (EEC-V) SYSTEM

Operation

The EEC-5 system is the latest generation of Ford's engine control technology. EEC-5 is compliant with the latest OBD-II regulations set forth by the government. When a system or component exceeds emission thresholds or a component operates outside of tolerance, a Diagnostic Trouble Code (DTC) will be stored in the Powertrain Control Module (PCM) and the Malfunction Indicator Lamp (MIL) will be illuminated.

The PCM is the heart of the EEC-5 system. It receives data from a number of sensors and other electronic components (switches, relay, etc.). Based on information received and programs mapped into the PCM's memory, it generates output signals to control various relay, solenoids and other actuators. The PCM in the EEC-5 system has calibration modules, located inside the assembly, that contain calibration specifications for optimizing emissions, fuel economy and driveability. The calibration module is called a PROM.

The EEC-5 system consists of the following components:

- Powertrain Control Module (PCM)

- Throttle Position (TP) sensor

- Mass Air Flow (MAF) sensor

- Intake Air Temperature (IAT) sensor

- Idle Air Control (IAC) valve

- Engine Coolant Temperature (ECT) sensor

- Heated Oxygen Sensor (HO2S)

- Camshaft Position (CMP) sensor

- Knock Sensor (KS)

- Vehicle Speed Sensor (VSS)

- Crankshaft Position (CKP) sensor

The MAF sensor (a potentiometer) senses the position of the airflow in the engine's air induction system and generates a voltage signal that varies with the amount of air drawn into the engine. The IAT sensor (a sensor in the area of the MAF sensor) measures the temperature of the incoming air and transmits a corresponding electrical signal. Another temperature sensor (the ECT sensor) inserted in the engine coolant tells if the engine is cold or warmed up. The TP sensor, a switch that senses throttle plate position, produces electrical signals that tell the PCM when the throttle is closed or wide open. A special probe (the HO2S) in the exhaust manifold measures the amount of oxygen in the exhaust gas, which is in indication of combustion efficiency, and sends a signal to the PCM. The sixth signal, camshaft position information, is transmitted by the CMP sensor, installed in place of the distributor.

The EEC-5 microcomputer circuit processes the input signals and produces output control signals to the fuel injectors to regulate fuel discharged to the injectors. It also adjusts ignition spark timing to provide the best balance between driveability and economy, and controls the IAC valve to maintain the proper idle speed.

Because of the complicated nature of the EEC-5 OBD II system, special tools such as a Break Out Box (BOB) and a Generic Scan Tool (GST) are necessary for testing and troubleshooting.

POWERTRAIN CONTROL MODULE (PCM)

Operation

The Powertrain Control Module (PCM) performs many functions on your vehicle. The module accepts information from various engine sensors and computes the required fuel flow rate necessary to maintain the correct amount of air/fuel ratio throughout the entire engine operational range.

Based on the information that is received and programmed into the PCM's memory, the PCM generates output signals to control relays, actuators and solenoids. The PCM also sends out a command to the fuel injectors that meters the appropriate quantity of fuel. The module automatically senses and compensates for any changes in altitude when driving your vehicle.

The PCM is located behind the passenger's side of the firewall.

HEATED OXYGEN SENSORS (HO2S)

Operation

The heated oxygen sensor supplies the PCM with a signal which indicates a rich or lean condition during engine operation. The input information assists the computer in determining the proper air/fuel ratio. A low voltage signal from the sensor indicates too much oxygen in the exhaust (lean condition) and a high voltage signal indicates too little oxygen in the exhaust (rich condition).

The sensors are threaded into the dual converter Y-pipe. Heated oxygen sensors are used on all models to allow the engine to reach the closed loop state faster.

Heated oxygen sensors are located in the dual converter Y-pipe both before and after the catalyst.

Testing

1. Disconnect the HO2S.

2. Measure the resistance between PWR and GND terminals of the sensor.
   Resistance should be approximately 6 ohms at 68°F (20°C). If resistance is not within specification, the sensor's heater element is faulty.

3. With the HO2S connected and engine running, measure the voltage with a Digital Volt-Ohmmeter (DVOM) between terminals HO2S and SIG RTN (GND) of the oxygen sensor connector. Voltage should fluctuate between 0.01-1.1 volts. If voltage fluctuation is slow or voltage is not within specification, the sensor may be faulty.

IDLE AIR CONTROL (IAC) VALVE

Operation

The IAC valve adjusts the engine idle speed. The valve is located on the side of the throttle body. The valve is controlled by a duty cycle signal from the PCM and allows air to bypass the throttle plate in order to maintain the proper idle speed.

The IAC is located at the top of the upper intake manifold adjacent to the throttle body.

Do not attempt to clean the IAC valve. Carburetor tune-up cleaners or any type of solvent cleaners will damage the internal components of the valve.

Testing

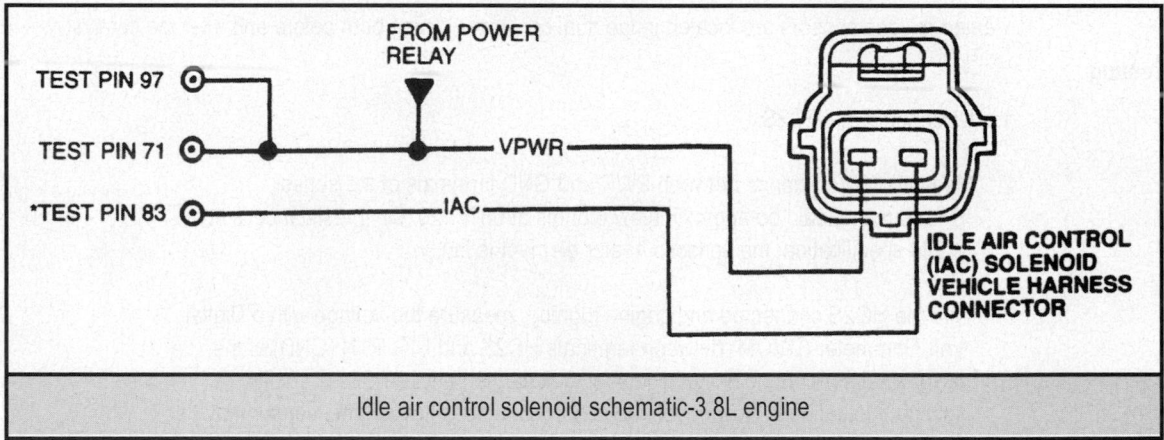

Idle air control solenoid schematic-3.0L engine

Idle air control solenoid schematic-3.8L engine

1. Turn the ignition switch to the OFF position.

2. Disconnect the wiring harness from the IAC valve .

3. Measure the resistance between the terminals of the valve.

Due to the diode in the solenoid, place the ohmmeter positive lead on the VPWR terminal and the negative lead on the ISC terminal.

4. Resistance should be 6-13 ohms.

5. If resistance is not within specification, the valve may be faulty.

ENGINE COOLANT TEMPERATURE (ECT) SENSOR
Operation

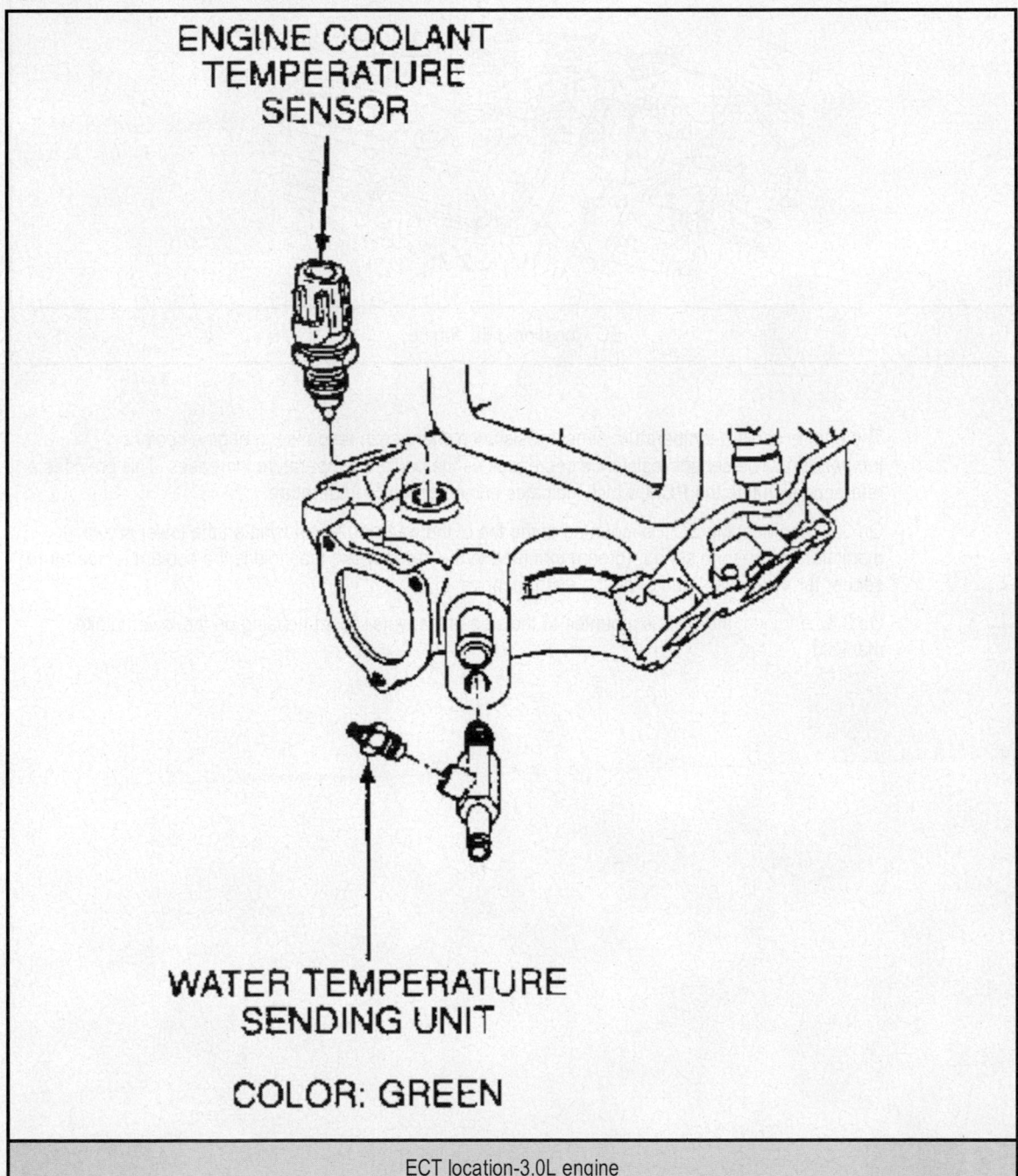

ENGINE COOLANT
TEMPERATURE
SENSOR

WATER TEMPERATURE
SENDING UNIT

COLOR: GREEN

ECT location-3.0L engine

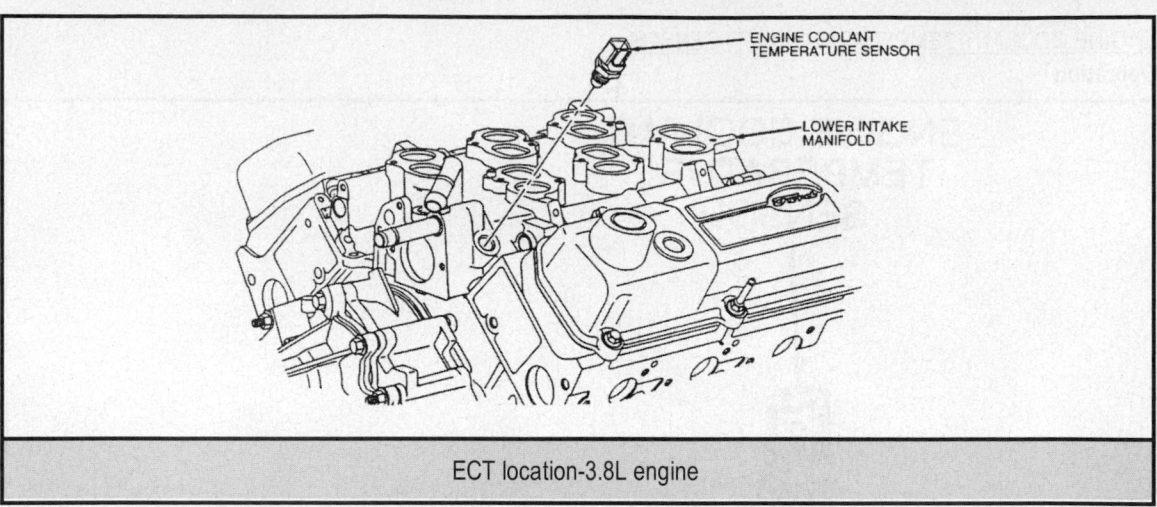

ECT location-3.8L engine

The engine coolant temperature sensor resistance changes in response to engine coolant temperature. The sensor resistance decreases as the coolant temperature increases. This provides a reference signal to the PCM, which indicates engine coolant temperature.

On 3.0L engines, the ECT is mounted at the top of the water outlet housing on the lower intake manifold. The second sensor (green) mounted to the lower intake manifold is the coolant temperature sender for the coolant gauge on the instrument panel.

On 3.8L engines, the ECT is mounted to the side of the water outlet housing on the lower intake manifold.

Testing

Temperature		Engine Coolant/Intake Air Temperature Sensor Values
°F	°C	Resistance (K ohms)
248	120	1.18
230	110	1.55
212	100	2.07
194	90	2.80
176	80	3.84
158	70	5.37
140	60	7.70
122	50	10.97
104	40	16.15
86	30	24.27
68	20	37.30
50	10	58.75

ECT and IAT resistance-to-temperature specifications

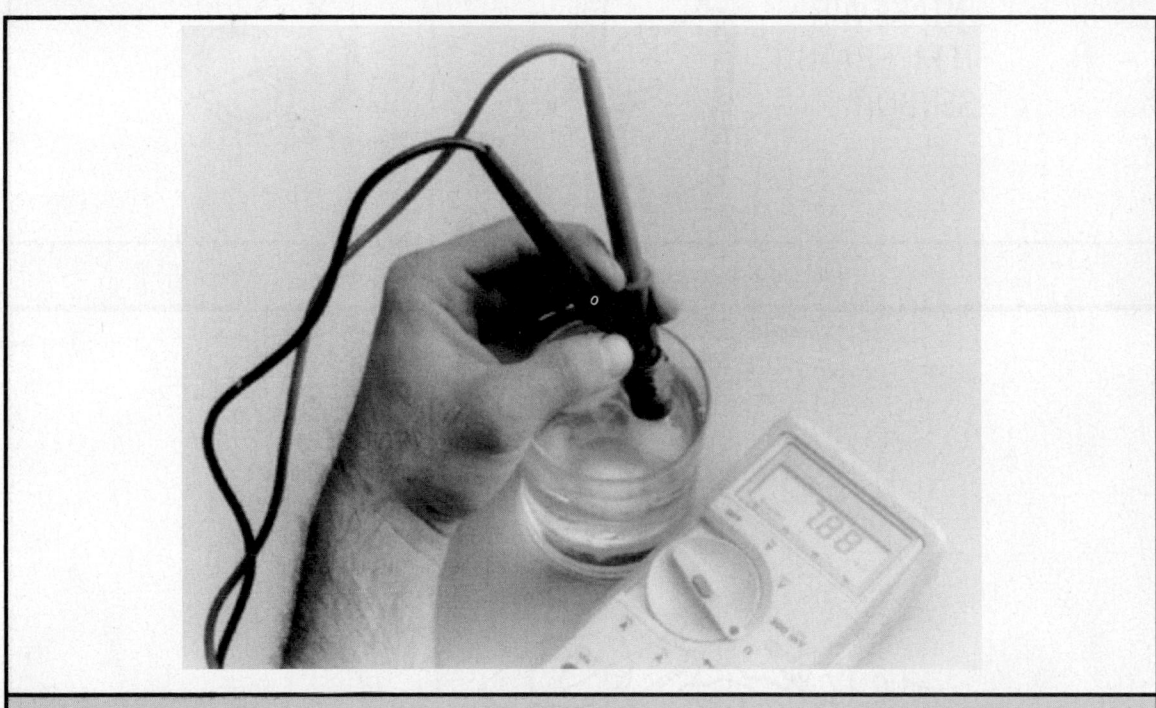

Another method of testing the ECT is to submerge it in cold or hot water and check resistance

1. Disconnect the engine wiring harness from the ECT sensor.

2. Connect an ohmmeter between the ECT sensor terminals.

3. With the engine cold and the ignition switch in the OFF position, measure and note the ECT sensor resistance.

4. Connect the engine wiring harness to the sensor.

5. Start the engine and allow the engine to reach normal operating temperature.

6. Once the engine has reached normal operating temperature, turn the engine OFF.

7. Once again, disconnect the engine wiring harness from the ECT sensor.

8. Measure and note the ECT sensor resistance with the engine hot.

9. Compare the cold and hot ECT sensor resistance measurements with the accompanying chart.

10. If readings do not approximate those in the chart, the sensor may be faulty.

INTAKE AIR TEMPERATURE (IAT) SENSOR

Operation

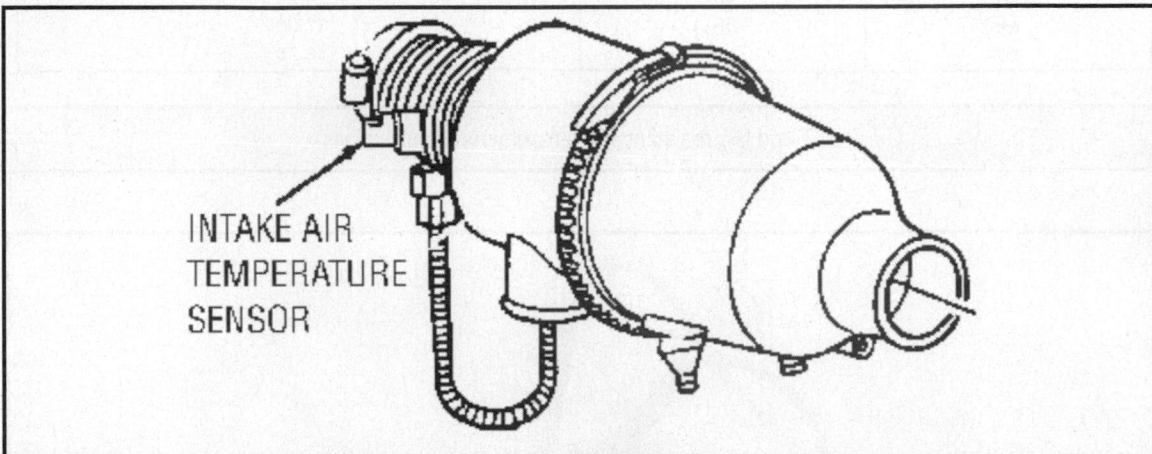

Intake air temperature sensor location-3.8L engine

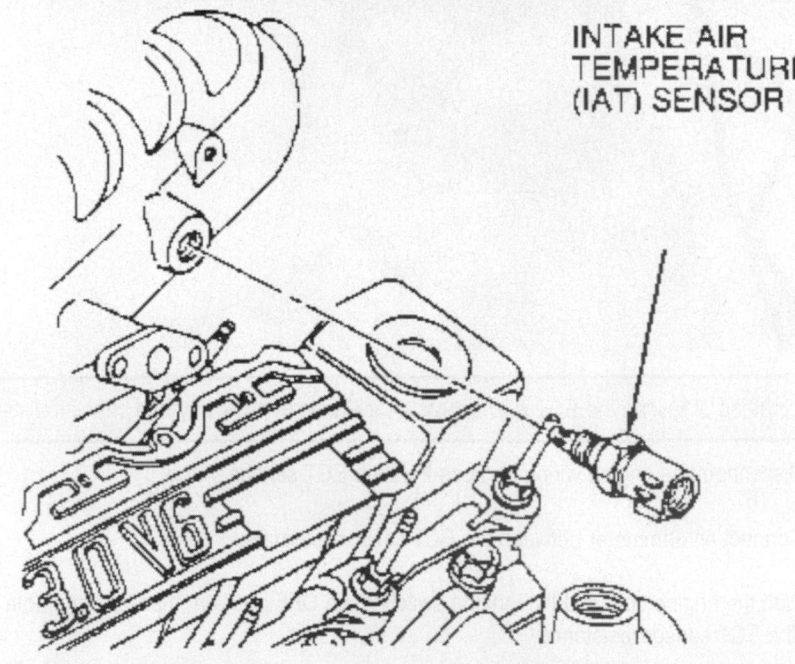

Intake air temperature sensor location-3.0L engine

The Intake Air Temperature (IAT) sensor resistance changes in response to the ambient air temperature. The sensor resistance decreases as the air temperature increases. This provides a signal to the PCM indicating the temperature of the incoming air charge.

The 3.0L engine sensor is mounted in the air cleaner-to-throttle body supply tube, while the 3.8L engine sensor is mounted in the air cleaner box.

Testing

1. Turn the ignition switch OFF.

2. Disconnect the wiring harness from the IAT sensor.

3. Measure the resistance between the sensor terminals.

4. Compare the resistance reading with the accompanying chart.

5. If the resistance is not within specification, the IAT may be faulty.

6. Connect the wiring harness to the sensor.

MASS AIR FLOW (MAF) SENSOR
Operation

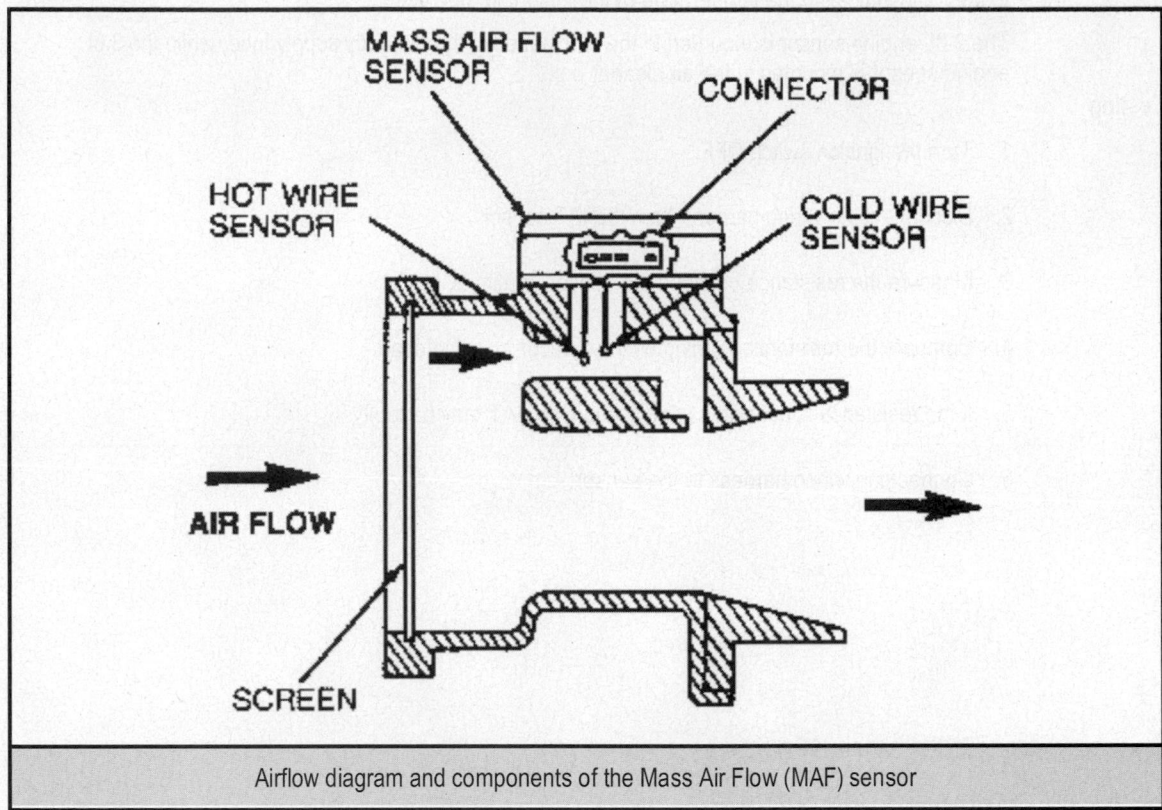

Airflow diagram and components of the Mass Air Flow (MAF) sensor

The MAF sensor directly measures the amount of the air flowing into the engine. The sensor is mounted between the air cleaner assembly and the air cleaner outlet tube.

The sensor utilizes a hot wire sensing element to measure the amount of air entering the engine. The sensor does this by sending a signal, generated by the sensor when the incoming air cools the hot wire down, to the PCM. The signal is used by the PCM to calculate the injector pulse width, which controls the air/fuel ratio in the engine. The sensor and plastic housing are integral and must be replaced if found to be defective.

The sensing element (hot wire) is one thin platinum wire wound on a ceramic bobbin and coated with glass. This hot wire is maintained at 392°F (200°C) above the ambient temperature as measured by a constant "cold wire".

Testing

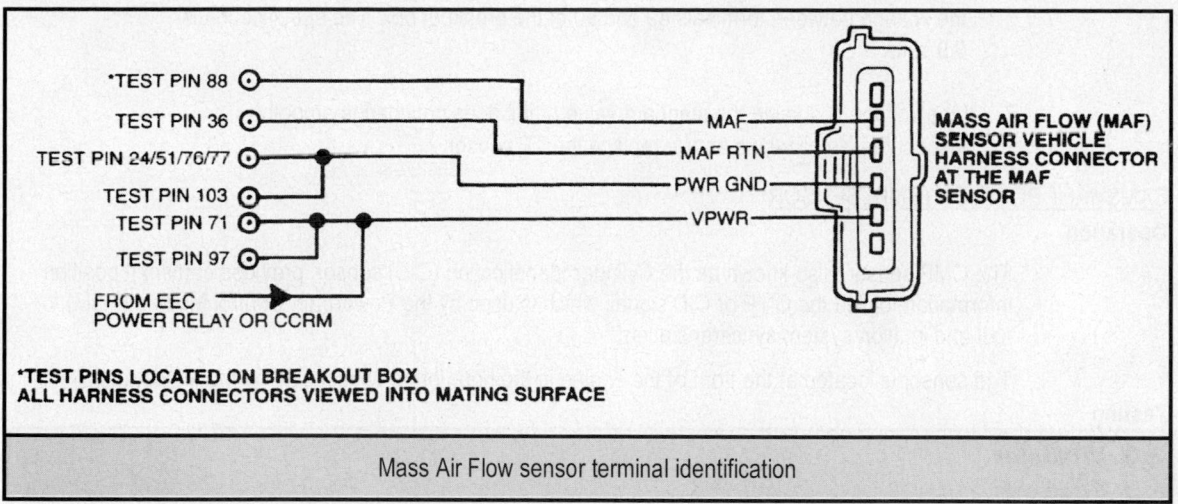

*TEST PIN 88
TEST PIN 36
TEST PIN 24/51/76/77
TEST PIN 103
TEST PIN 71
TEST PIN 97
FROM EEC
POWER RELAY OR CCRM

MAF
MAF RTN
PWR GND
VPWR

MASS AIR FLOW (MAF)
SENSOR VEHICLE
HARNESS CONNECTOR
AT THE MAF
SENSOR

*TEST PINS LOCATED ON BREAKOUT BOX
ALL HARNESS CONNECTORS VIEWED INTO MATING SURFACE

Mass Air Flow sensor terminal identification

1. Using a multimeter, check for voltage by backprobing the MAF sensor connector.

2. With the engine running at idle, verify that there is at least 10.5 volts between the VPWR and PWR GND terminals of the MAF sensor connector. If voltage is not within specification, check power and ground circuits and repair as necessary.

3. Check voltage between the MAF and MAF RTN terminals. Voltage should be approximately 0.34-1.96 volts. If voltage is not within specification, the sensor may be faulty.

THROTTLE POSITION (TP) SENSOR
Operation

The TP sensor is a potentiometer that provides a signal to the PCM that is directly proportional to the throttle plate position. The TP sensor is mounted on the side of the throttle body and is connected to the throttle plate shaft. The TP sensor monitors throttle plate movement and position, and transmits an appropriate electrical signal to the PCM. These signals are used by the PCM to adjust the air/fuel mixture, spark timing and EGR operation according to engine load at idle, part throttle, or full throttle. The TP sensor is not adjustable.

Testing

1. Disconnect the negative battery cable.

2. Disconnect the wiring harness from the sensor.

3. Check resistance between terminals TP (br/w wire) and the VREF (gy/w wire), on the TP sensor.

Do not measure the wiring harness connector terminals, rather the terminals on the sensor itself.

4. Slowly rotate the throttle shaft and monitor the ohmmeter for a continuous, steady change in resistance. Any sudden jumps, or irregularities (such as jumping back and forth) in resistance indicates a malfunctioning sensor.

5. Reconnect the negative battery cable.

6. Turn the ignition switch ON and using the DVOM on voltmeter function, measure the voltage between terminals 89 and 90 of the breakout box. The specification is 0.9 volts.

7. If the voltage is outside the standard value or if it does not change smoothly, inspect the circuit wiring and/or replace the TP sensor.

CAMSHAFT POSITION (CMP) SENSOR

Operation

The CMP sensor, also known as the Cylinder Identification (CID) sensor, provides camshaft position information, called the CMP or CID signal, which is used by the Powertrain Control Module (PCM) for fuel and ignition system synchronization.

The sensor is located at the front of the engine in the bore formerly occupied by the distributor.

Testing

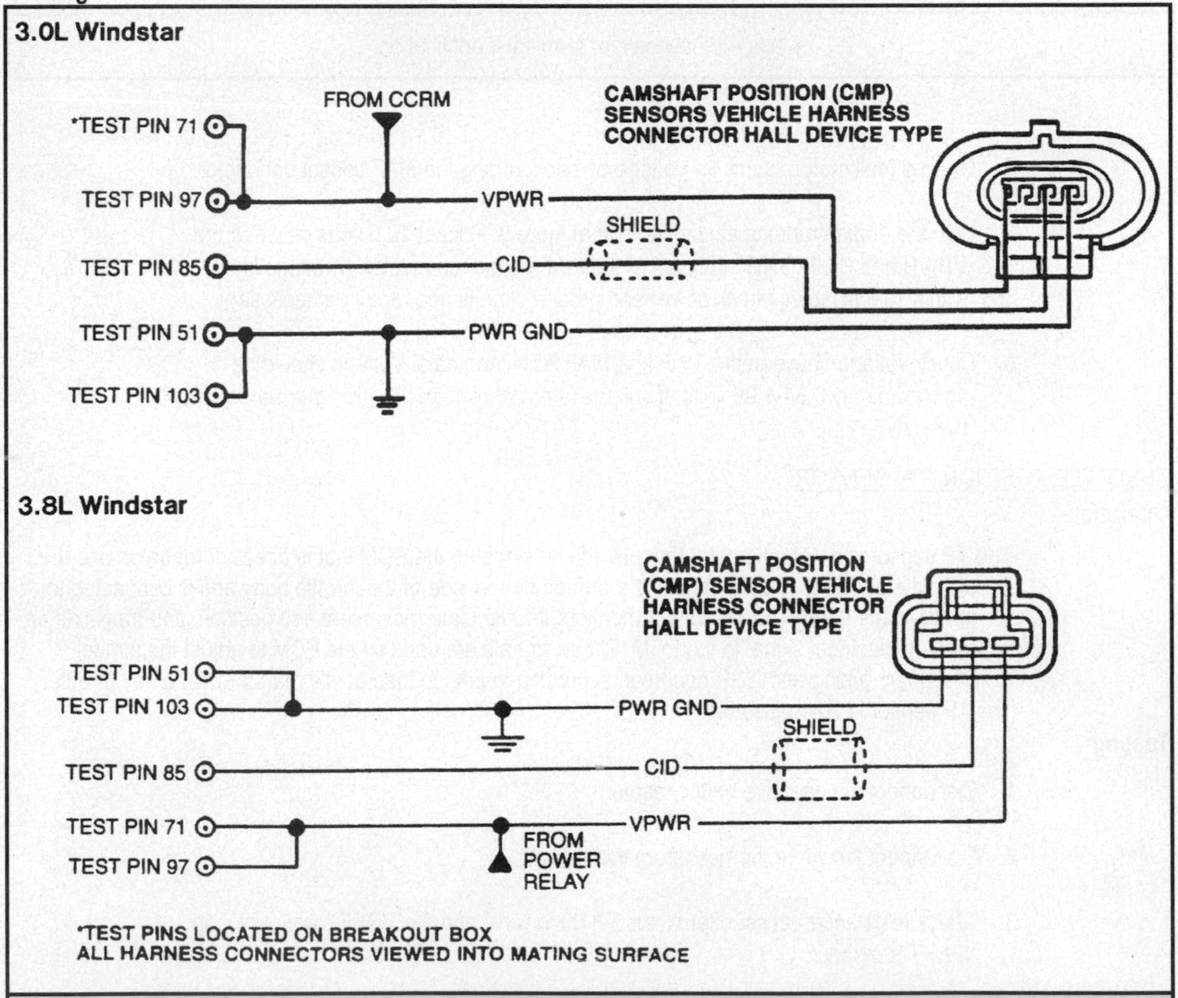

3.0L Windstar

FROM CCRM

CAMSHAFT POSITION (CMP) SENSORS VEHICLE HARNESS CONNECTOR HALL DEVICE TYPE

*TEST PIN 71

TEST PIN 97 — VPWR

SHIELD

TEST PIN 85 — CID

TEST PIN 51 — PWR GND

TEST PIN 103

3.8L Windstar

CAMSHAFT POSITION (CMP) SENSOR VEHICLE HARNESS CONNECTOR HALL DEVICE TYPE

TEST PIN 51
TEST PIN 103 — PWR GND

SHIELD

TEST PIN 85 — CID

TEST PIN 71 — VPWR
TEST PIN 97

FROM POWER RELAY

*TEST PINS LOCATED ON BREAKOUT BOX
ALL HARNESS CONNECTORS VIEWED INTO MATING SURFACE

Camshaft position sensor electrical schematic

1. Check voltage between the camshaft position sensor terminals PWR GND and CID.

2. With engine running, voltage should be greater than 0.1 volt AC and vary with engine speed.

3. If voltage is not within specification, check for proper voltage at the VPWR terminal.

4. If VPWR voltage is greater than 10.5 volts, sensor may be faulty.

CRANKSHAFT POSITION (CKP) SENSOR
Operation

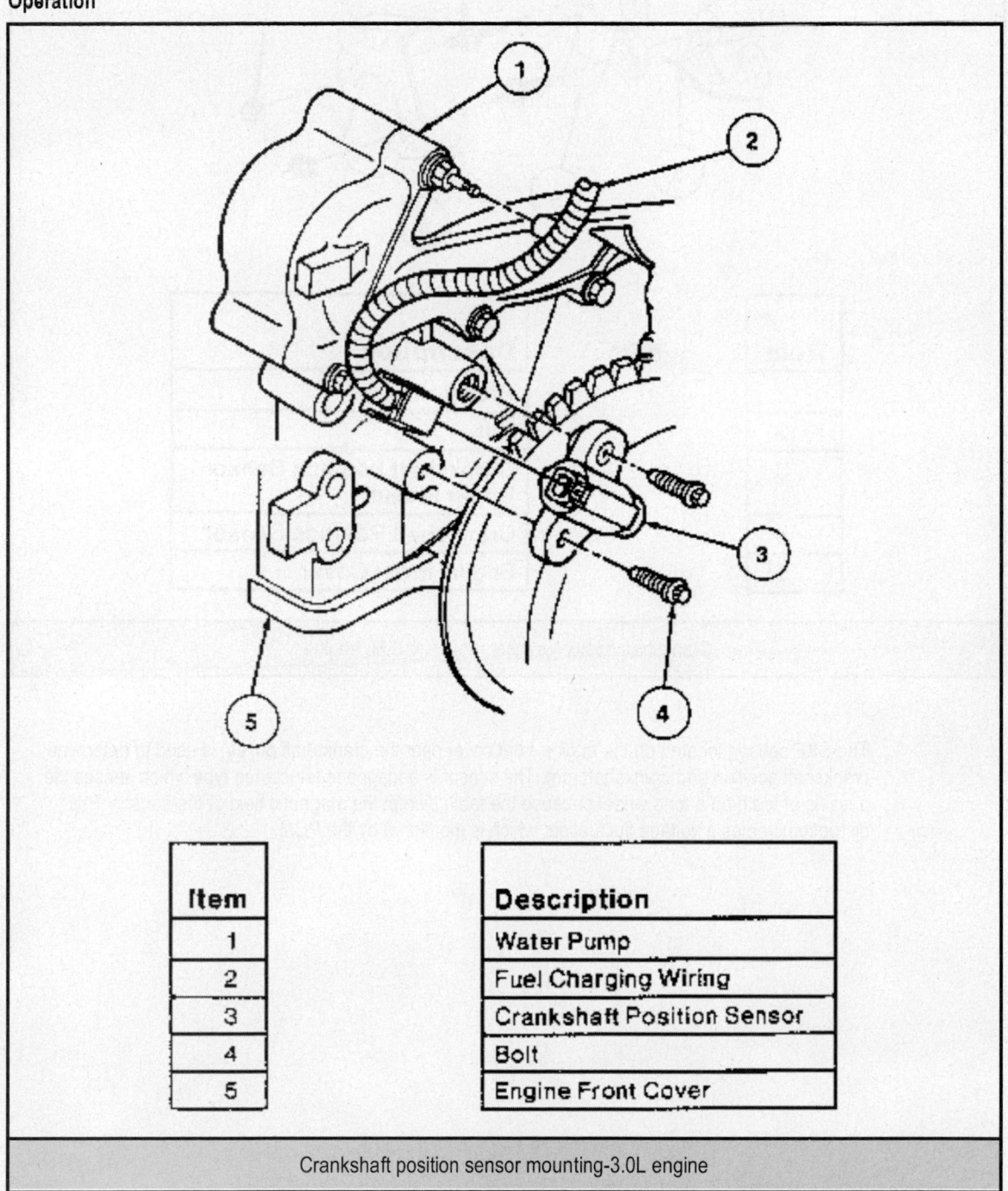

Item	Description
1	Water Pump
2	Fuel Charging Wiring
3	Crankshaft Position Sensor
4	Bolt
5	Engine Front Cover

Crankshaft position sensor mounting-3.0L engine

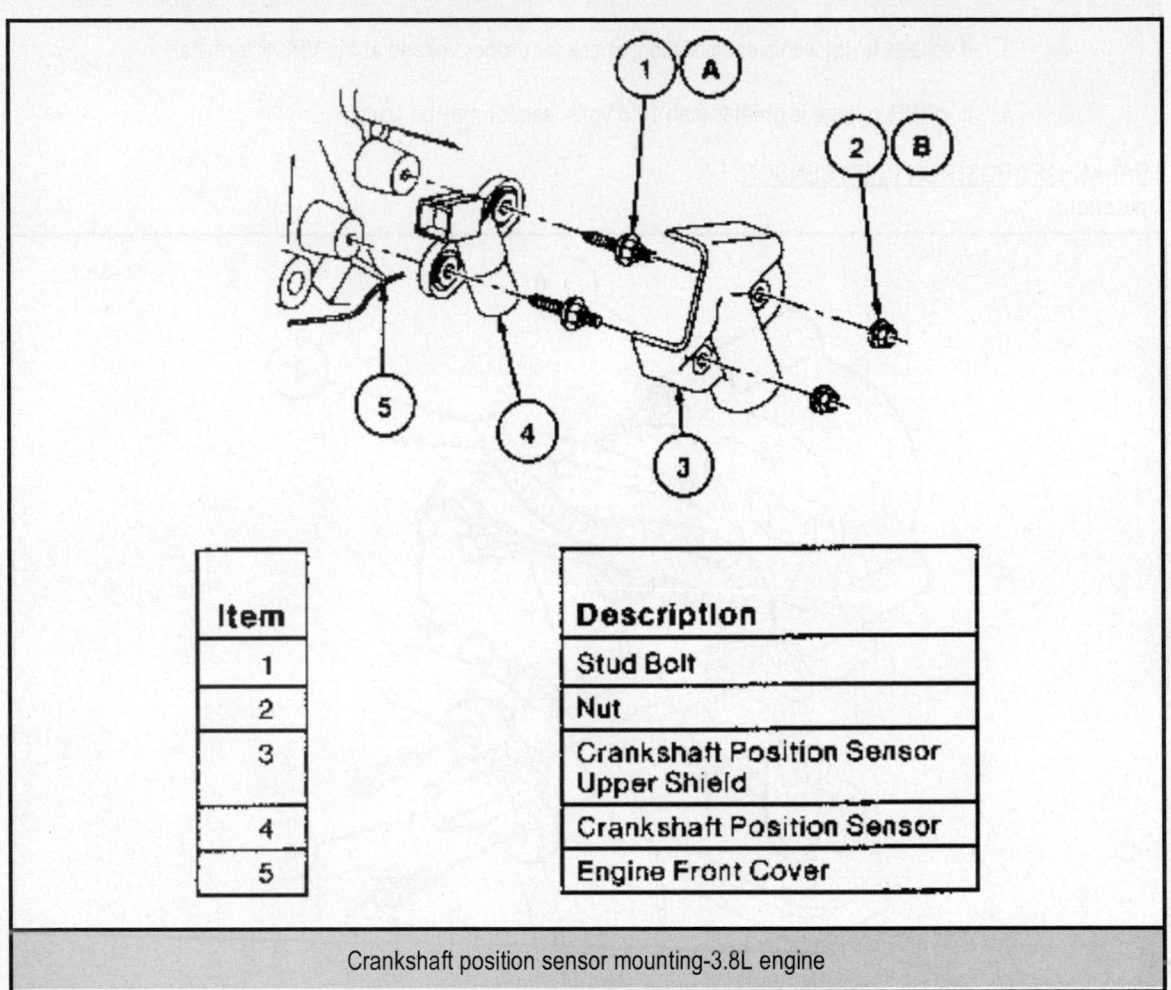

Item	Description
1	Stud Bolt
2	Nut
3	Crankshaft Position Sensor Upper Shield
4	Crankshaft Position Sensor
5	Engine Front Cover

Crankshaft position sensor mounting-3.8L engine

The CKP sensor, located on the engine front cover near the crankshaft pulley, is used to determine crankshaft position and crankshaft rpm. The sensor is a magnetic reluctance type which senses the passing of teeth on a tone wheel because the teeth disrupt the magnetic field of the sensor. This disruption creates a voltage fluctuation, which is monitored by the PCM.

Testing

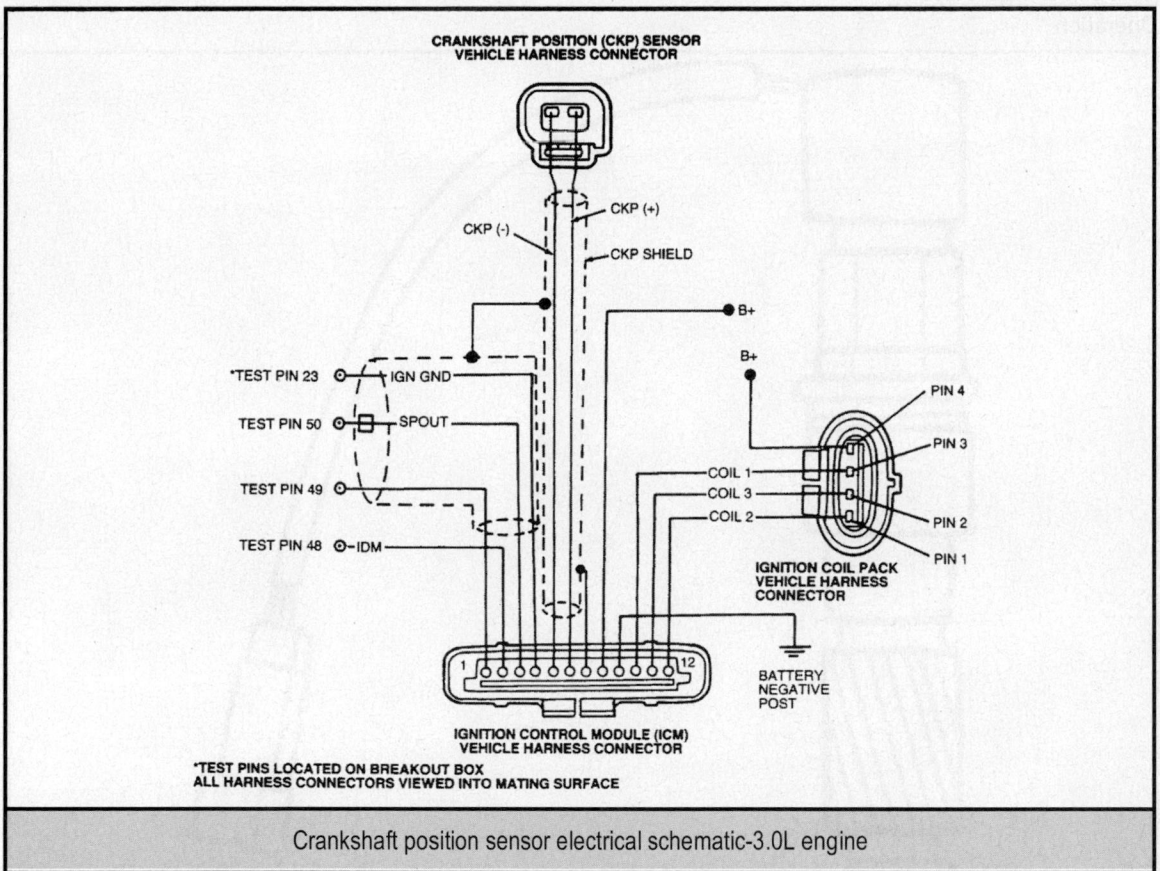

Crankshaft position sensor electrical schematic-3.0L engine

1. Measure the voltage between the sensor CKP sensor terminals by backprobing the sensor connector.

   If the connector cannot be backprobed, fabricate or purchase a test harness.

2. Sensor voltage should be more than 0.1 volt DC with the engine running and should vary with engine RPM.

3. If voltage is not within specification, the sensor may be faulty.

VEHICLE SPEED SENSOR
Operation

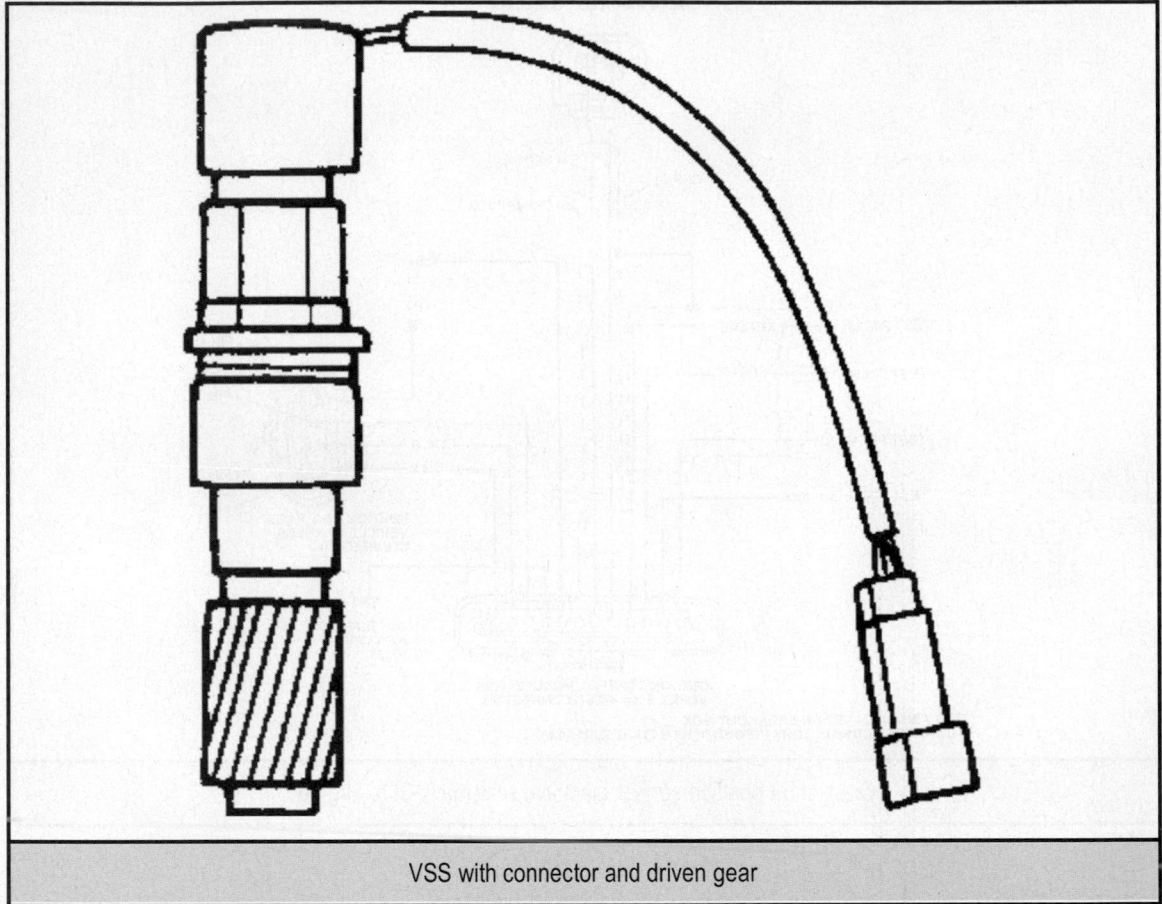

VSS with connector and driven gear

The Vehicle Speed Sensor (VSS) is a magnetic pick-up sensor that sends a signal to the Powertrain Control Module (PCM) and the speedometer. The sensor measures the rotation of the output shaft on the transaxle and sends an AC voltage signal to the PCM which determines the corresponding vehicle speed.

Testing

4. Disconnect the negative battery cable.

5. Disengage the wiring harness connector from the VSS.

Using a Digital Volt-Ohmmeter (DVOM), measure the resistance (ohmmeter function) between the sensor terminals. If the resistance is 190-250 ohms, the sensor is okay.

Contents

<u>WHAT TO DO WHEN THERE ARE NO DTCS</u>

Do not attempt to diagnose a Drivability Symptoms without having a logical plan to use to determine which Engine Control system is the cause of the symptom - this plan should include a way to determine which systems do not have a problem! *Drivability symptom diagnosis is a part of an organized approach to problem solving and repair.*

Drivability Symptom Index Table

To use this list, locate the symptom that matches a particular problem and refer to the areas to test. The items listed under each symptom may not apply to all models, engines or vehicle systems. The repair steps indicate what vehicle component or system to test.

Note: *The Drivability Symptoms in this list are intended to be generic. While they apply to most vehicles, some vehicles may not have all of the components listed. Refer to other Chilton repair manuals and electronic media for specific tests.*

Symptom Test Table

Symptom Description	Suggested Areas to Test
Test 1 - No Start, Hard Start Condition ● No Crank ● Hard Start, Long Crank, Erratic Crank ● Stall After Start ● No Start, Normal Crank ● No Start, MIL is off (if the VREF shorts to ground)	- Check battery, battery circuits to starter - Check for a damaged flywheel, engine compression, base timing and minimum air rate - Check for a failed fuel pump relay - Check for distributor rotor "punch-through" - Check for a faulty ignition control module (ICM) - Check for a VREF circuit shorted to ground - Check SKIM (security system) with a Scan Tool
Test 2 - Rough Idle or Stalls Condition ● Low or slow idle speed ● Fast idle speed ● Hunting or rolling idle speed ● Slow return to idle speed ● Stalls or almost stalls	- Check for engine vacuum leaks - Check the condition of the PCV valve and lines - Check for excessive carbon buildup - Check for a restricted exhaust (in Section 2) - Check base idle speed, check for low fuel pressure - Check the throttle linkage for sticking or binding
Test 3 - Runs Rough Condition ● At idle speed ● During acceleration ● At cruise speed ● During deceleration	- Check for engine vacuum leaks at intake manifold - Check condition of ignition secondary components - Check base timing and idle speed settings - Check for low or high fuel pressure - Check for dirty, leaking or shorted fuel injectors - Check for excessive carbon buildup on valves
Test 4 - Cuts-out, Misses Condition ● At idle speed ● During acceleration ● At cruise speed ● During deceleration	- Check for engine vacuum leaks at intake manifold - Check condition of ignition secondary components - Check that spark timing advance is available - Check for low or high fuel pressure - Check for dirty, leaking or shorted fuel injectors - Check for excessive carbon buildup on valves
Test 5 - Bucks, Jerks Condition ● During acceleration ● At cruise speed ● During deceleration	- Check for engine vacuum leaks at intake manifold - Check condition of ignition secondary components - Check that spark timing advance is available - Check for low or high fuel pressure - Check for dirty, leaking or shorted fuel injectors - Check operation of the TCC solenoid, brake switch

Symptom Diagnosis - Test 1

No Start, Hard Start Condition

Note: *If there is no spark output or fuel pressure available, check for a failed fuel pump relay, no power to the PCM, or loss of the ignition reference signal to the PCM.*

PRELIMINARY CHECKS

Prior to starting this symptom test routine, inspect these underhood items:

1. Check battery charge and condition, starter current draw.
2. Verify the starter relay operation and that the engine cranks (turns over).
3. Verify the check engine light (MIL) operation - if it does not activate, check the PCM power and ground circuits, and check for 5v supply at the MAP or TP sensor.
4. Check Air Intake system for restrictions (inspect air inlet tubes, air filter for dirt, etc.).
5. Check the status of the Smart Key Immobilizer System (SKIM) with the Scan Tool.

Test 1 Chart

Step	Action	Yes	No
1	**Step Description: No Start Condition Only** » Check battery cables, state of charge. » If the engine does not rotate, inspect for a locked engine (hydrostatic lockup condition). » Does the engine crank normally?	Go to Step 2.	Repair the fault in the battery, starter, or Base Engine. Retest for the symptom when all repairs are done.
2	**Step Description: Check the Fuel System** » Verify that the pump operates at key on. » Check the fuel pump relay operation. If the relay does not operate, check for blown fuse. » Inspect pump for a leak-down condition » Test fuel pressure, volume and quality. » Test the operation of the fuel regulator. » Are there any faults in the Fuel system?	Make needed repairs. Fuel Pressure Gauge Fuel Rail Test Port	Go to Step 3.
3	**Step Description: Check the Ignition System** » Inspect ignition secondary components for damage (look for rotor "punch-through"). » Inspect the coils for signs of spark leakage at coil towers or primary connections. » Check the spark output with a spark tester. » Test Ignition system with an engine analyzer. » Are there any faults in the Ignition system?	Make repairs to the Ignition system. Then retest the symptom. CABLE SPARK TESTER	Go to Step 4.
4	**Step Description: Check the Exhaust System** » Check Exhaust system for leaks or damage. » Check the Exhaust system for a restriction using the Vacuum or Pressure Gauge Test (e.g., exhaust backpressure reading should not exceed 1.5 psi at cruise speeds). » Are there any faults in the Exhaust system?	Make repairs to the Exhaust system. Then retest the symptom. Inspect for Damage	Go to Step 5.
5	**Step Description: Check the MAP Sensor** » Disconnect the MAP sensor and attempt to start the engine. » Does the engine start and run normally?	Replace the MAP sensor. Retest for the symptom when repairs are completed.	Go to Step 6.

No Start, Hard Start Condition (Continued)

Test 1 Chart (Continued)

Step	Action (Hard Start Only)	Yes	No
6	Step Description: Check for a Hot Engine » Check for signs of an engine overheating condition related to a Hard Start Symptom. » Does the engine appear to be overheated?	Make the repairs to correct the hot engine and then retest for the symptom when done.	Go to Step 7.
7	Step Description: Check ECT Sensor PID » Connect a Scan Tool and turn the key to on. » Read the ECT sensor (compare to chart). » Has the ECT sensor shifted out of range?	Replace the ECT sensor. Then retest for the symptom when all repairs are completed.	Go to Step 8.
8	Step Description: Check the PCV System » Inspect the PCV system components for broken parts or loose connections. » Test the operation of the PCV valve. » Are there any faults in the PCV system?	Repair the PCV system. Refer to the PCV system tests in this manual. Retest the symptom when all repairs are done.	Go to Step 9.
9	Step Description: Check the EVAP System » Inspect for damaged or disconnected EVAP system components. » Inspect for a fuel saturated charcoal canister. » Are there any faults in the EVAP system?	Refer to the EVAP system tests in this manual. Retest for the symptom when all repairs are completed.	Go to Step 10.
10	Step Description: Test the Base Engine » Check the engine compression. » Test valve timing and timing chain condition. » Check for a worn camshaft or valve train. » Check for any large intake manifold leaks. » Are there any faults in the Base Engine?	Repair the Base Engine. Refer to the Base Engine Tests in this manual. Retest symptom when done.	Return to Step 2 to repeat the test steps in this series to locate and repair the "No Start, Hard Start" condition.

Symptom Diagnosis - Test 2

Rough, Low or High Idle Speed Condition

Note: *If the vehicle has a rough idle and the base timing, idle speed and the IAC (or AIS) motor operates properly, check the engine for excessive carbon buildup.*

Preliminary Checks

Prior to starting this symptom test routine, inspect these underhood items:

1. All related vacuum lines for proper routing and integrity.
2. All related electrical connectors and wiring harnesses for faults (Wiggle Test).
3. Check the throttle linkage for a sticking or binding condition.
4. Air Intake system for restrictions (air inlet tubes, dirty air filter, etc.).
5. Search for any technical service bulletins related to this symptom.
6. Turn the key to off. Unplug the MAP sensor connection and restart the engine to recheck for the idle concern. If the condition is gone, replace the MAP sensor.

Test 2 Chart

Step	Action	Yes	No
1	Step Description: Verify the rough idle or stall » Does the engine have a warm engine rough idle, low idle or high Idle condition in P or N?	Go to Step 2.	Fault is intermittent. Return to the Symptom List and select another fault.
2	Step Description: Verify idle speed & timing » Verify the base timing is within specifications » Verify that the base idle speed is set properly » Are the timing and idle speed set properly?	Go to Step 3.	Set the base idle speed and timing to the specifications and then retest for the symptom.
3	Step Description: Check AIS / IAC Operation » Check the AIS or IAC motor operation » Inspect the AIS/IAC housing in throttle body for restricted passages. Clean as needed. » Set the parking brake, block the drive wheels and turn the A/C off. Install the Scan Tool. » IAC Motor Tester - Turn the key off and then connect the IAC tester to the IAC valve. » Start the engine and use the IAC tester to extend and retract the IAC valve. » ATM Test - Start the engine. Use the tool to change the speed from min-idle to 1500 rpm. » Did the idle speed change as commanded?	Install an Aftermarket Noid light and check the operation of the PCM and AIS or IAC motor circuits. Check the motor for signs of open or shorted circuits. Replace the IAC motor or PCM as needed or make repairs to the IAC motor wiring. If all are okay, go to Step 4.	If the AIS/IAC motor passages are clean and engine speed did not change as described when the AIS/IAC motor was extended and retracted, replace the AIS/IAC motor. Then retest for the condition.
4	Step Description: Check/compare PID values » Connect Scan Tool & turn off all accessories. » Start the engine and allow it to fully warmup. » Monitor all related PIDs on the Scan Tool. » Verify the P/N switch input in gear and Park. » Check the O2S operation with a Lab Scope. » Are all PIDs within normal range?	Go to Step 5. Note: An IAC motor count of over 80 indicates the pintle is extended and an IAC count of (0) indicates the pintle is retracted.	One or more of the PIDs are out of range when compared to "known good" values. Make repairs to the system that is out of range, then retest for the symptom.
5	Step Description: Check the Ignition System » Inspect the coils for signs of spark leakage at coil towers or primary connections. » Check the spark output with a spark tester. » Test Ignition system with an engine analyzer. » Were any faults found in the Ignition system?	Make repairs as needed 	Go to Step 6.
6	Step Description: Check the Fuel System » Inspect the Fuel delivery system for leaks. » Test the fuel pressure, quality and volume. » Test the operation of the pressure regulator. » Were any faults found in the Fuel system?	Make repairs as needed esz	Go to Step 7.
7	Step Description: Check the Exhaust System » Check Exhaust system for leaks or damage. » Check the Exhaust system for a restriction using the Vacuum or Pressure Gauge Test (e.g., exhaust backpressure reading should not exceed 1.5 psi at cruise speeds). » Were any faults found in Exhaust System?	Make repairs to the Exhaust system. Then retest the symptom. Inspect for Damage 	Go to Step 8.

Test 2 Chart (Continued)

Step	Action	Yes	No
8	**Step Description: Check the PCV System** » Inspect the PCV system components for broken parts or loose connections. » Test the operation of the PCV valve. » Were any faults found in the PCV system?	Make repairs to the PCV system. Refer to the PCV system tests in this manual. Then retest for the condition.	Go to Step 9.
9	**Step Description: Check the EVAP System** » Inspect for damaged or disconnected EVAP system components or a saturated canister. » Were any faults found in the EVAP system?	Make repairs to EVAP system (use the EVAP tests in this manual). Retest for the condition.	Go to Step 10.
10	**Step Description: Check the Base Engine** » Test the engine compression. » Test valve timing and timing chain condition. » Check for a worn camshaft or valve train. » Check for any large intake manifold leaks. » Were any faults found in the Base Engine?	Make repairs as needed to the Base Engine. Refer to the Base Engine tests in this manual. Then retest for the condition when repairs are completed.	Go to Step 2 and repeat the tests from the beginning to locate and repair the cause of the "Rough, Low or High Idle Speed" condition.

Symptom Diagnosis - Test 3

Runs Rough Condition

<u>PRELIMINARY CHECKS</u>

Prior to starting this symptom test routine, inspect these underhood items:

1. All related vacuum lines for proper routing and integrity
2. Air Intake system for restrictions (air inlet tubes, dirty air filter, etc.)
3. Search for any technical service bulletins related to this symptom.

Test 3 Chart

Step	Action	Yes	No
1	**Step Description: Verify engine runs rough** » Start the engine and allow it to idle in P or N. » Does the engine run rough when warm in Park or Neutral position?	Check for any stored codes. If codes are set, repair codes and retest. If no codes are set, go to Step 3.	Go to Step 2.
2	**Step Description: Condition does not exist!** » Inspect various underhood items that could cause an intermittent Runs Rough condition (i.e., dirt in the throttle body, vacuum leaks, IAC motor connections, etc.). » Were any problems located in this step?	Correct the problems. Do a PCM reset and engine "idle relearn" procedure. Then verify the "runs rough" condition is repaired.	The problem is not present at this time. It may be an intermittent problem.
3	**Step Description: Check/compare PID values** » Connect a Scan Tool to the test connector. » Turn off all accessories. » Start the engine and allow it to fully warmup. » Monitor all related PIDs on the Scan Tool. » Were all PIDs within their normal range?	Go to Step 4. Note: The IAC motor should read from 5-50 counts. Check the LONGFT reading for a large shift into the negative range (due to a rich condition).	One or more of the PIDs are out of range when compared to "known good" values. Make repairs to the system that is out of range, then retest for the symptom.

Test 3 Chart (Continued)

Step	Action	Yes	No
4	**Step Description: Check the Ignition System** » Inspect the coils for signs of spark leakage at coil towers or primary connections. » Check the spark output with a spark tester. » Test Ignition system with an engine analyzer. » Were any faults found in the Ignition system?	Make repairs as needed	Go to Step 5.
5	**Step Description: Check the Fuel System** » Inspect the Fuel delivery system for leaks. » Test the fuel pressure, quality and volume. » Test the operation of the pressure regulator. » Were any faults found in the Fuel system?	Make repairs as needed	Go to Step 6.
6	**Step Description: Check the Exhaust System** » Check Exhaust system for leaks or damage. » Check the Exhaust system for a restriction using the Vacuum or Pressure Gauge Test (e.g., exhaust backpressure reading should not exceed 1.5 psi at cruise speeds). » Were any faults found in Exhaust System?	Make repairs to the Exhaust system. Then retest the symptom.	Go to Step 7.
7	**Step Description: Check the PCV System** » Inspect the PCV system components for broken parts or loose connections. » Test the operation of the PCV valve. » Were any faults found in the PCV system?	Make repairs to the PCV system. Refer to the PCV system tests in this manual. Then retest for the condition.	Go to Step 9.
8	**Step Description: Check the EVAP System** » Inspect for damaged or disconnected EVAP system components or a saturated canister. » Were any faults found in the EVAP system?	Make repairs to EVAP system (use the EVAP tests in this manual). Retest for the condition.	Go to Step 10.
9	**Step Description: Check Engine Condition** » Test the engine compression. » Test valve timing and timing chain condition. » Check for a worn camshaft or valve train. » Check for any large intake manifold leaks. » Were any faults found in the Base Engine?	Make repairs as needed to the Base Engine. Refer to the Base Engine tests in this manual. Then retest for the condition when repairs are completed.	Return to Step 2 and repeat the tests from the beginning to locate and repair the cause of the "Runs Rough" condition.

Example EVAP System Graphic

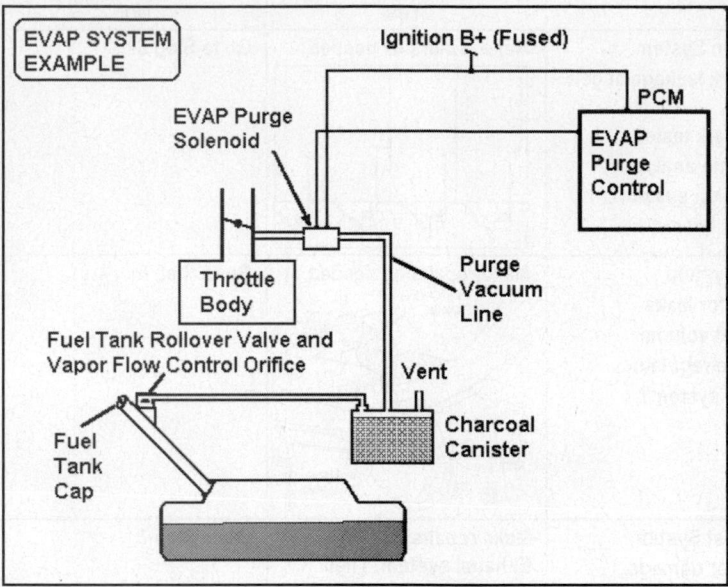

Symptom Diagnosis - Test 4
Cuts-out or Misses Condition

<u>PRELIMINARY CHECKS</u>

Prior to starting this symptom test routine, inspect these underhood items:

1. All related vacuum lines for proper routing and integrity
2. Search for any technical service bulletins related to this symptom.

Test 4 Chart

Step	Action	Yes	No
1	**Step Description: Verify Cuts-out condition** » Start the engine and attempt to verify the Cuts-out or misses condition. » Does the engine have a cuts-out condition?	Check for any stored codes. If codes are set, repair codes and retest. If no codes are set, go to Step 3.	Go to Step 2.
2	**Step Description: Condition does not exist!** » Inspect various underhood items that could cause an intermittent Cuts-out condition (i.e., EVAP, Fuel or Ignition system components). » Were any problems located in this step?	Correct the problems. Do a PCM reset and "Fuel Trim Relearn" procedure. Then verify condition is repaired.	The problem is not present at this time. It may be an intermittent problem.
3	**Step Description: Check/compare PID values** » Connect a Scan Tool to the test connector. » Turn off all accessories. » Start the engine and allow it to fully warmup. » Monitor all related PIDs on the Scan Tool (i.e., ECT IAC Counts and LONGFT at idle). » Were all PIDs within their normal range?	Go to Step 4. Note: The IAC motor should be from 5-50 counts. Watch fuel trim (%) for a large shift into the negative (-) range (due to a rich condition).	One or more of the PIDs are out of range when compared to "known good" values. Make repairs to the system that is out of range, then retest for the symptom.

Test 4 Chart (Continued)

Step	Action	Yes	No
4	**Step Description: Check the Ignition System** » Inspect the coils for signs of spark leakage at coil towers or primary connections. » Check the spark output with a spark tester. » Test Ignition system with an engine analyzer. » Were any faults found in the Ignition system?	Make repairs as needed	Go to Step 5.
5	**Step Description: Check the Fuel System** » Inspect the Fuel delivery system for leaks. » Test the fuel pressure, quality and volume. » Test the operation of the pressure regulator. » Were any faults found in the Fuel system?	Make repairs as needed	Go to Step 6.
6	**Step Description: Check the Exhaust System** » Check Exhaust system for leaks or damage. » Check the Exhaust system for a restriction using the Vacuum or Pressure Gauge Test (e.g., exhaust backpressure reading should not exceed 1.5 psi at cruise speeds). » Were any faults found in Exhaust System?	Make repairs to the Exhaust system. Then retest the symptom.	Go to Step 7.
7	**Step Description: Check the PCV System** » Inspect the PCV system components for broken parts or loose connections. » Test the operation of the PCV valve. » Were any faults found in the PCV system?	Make repairs to the PCV system. Refer to the PCV system tests in this manual. Then retest for the condition.	Go to Step 8.
8	**Step Description: Check the EVAP System** » Inspect for damaged or disconnected EVAP system components » Check for a saturated EVAP canister. » Were any faults found in the EVAP system?	Make repairs to EVAP system (use the EVAP tests in this manual). Retest for the condition.	Go to Step 9.
9	**Step Description: Check the AIR system** » Inspect AIR system for broken parts, leaking valves or disconnected hoses (see graphic). » Test the operation of Secondary AIR system. » Were any faults found in the AIR system?	Make repairs as needed. Refer to the Secondary AIR system tests in this manual. Retest for the condition.	Go to Step 10.
10	**Step Description: Check Engine Condition** » Test the engine compression. » Test valve timing and timing chain condition. » Check for a worn camshaft or valve train. » Check for any large intake manifold leaks. » Were any faults found in the Base Engine?	Make repairs as needed to the Base Engine. Refer to the Base Engine tests in this manual. Then retest for the condition when repairs are completed.	Go to Step 2 and repeat the tests from the beginning to locate and repair the cause of the "Cuts Out or Misses" condition.

Typical Secondary Air System Graphic

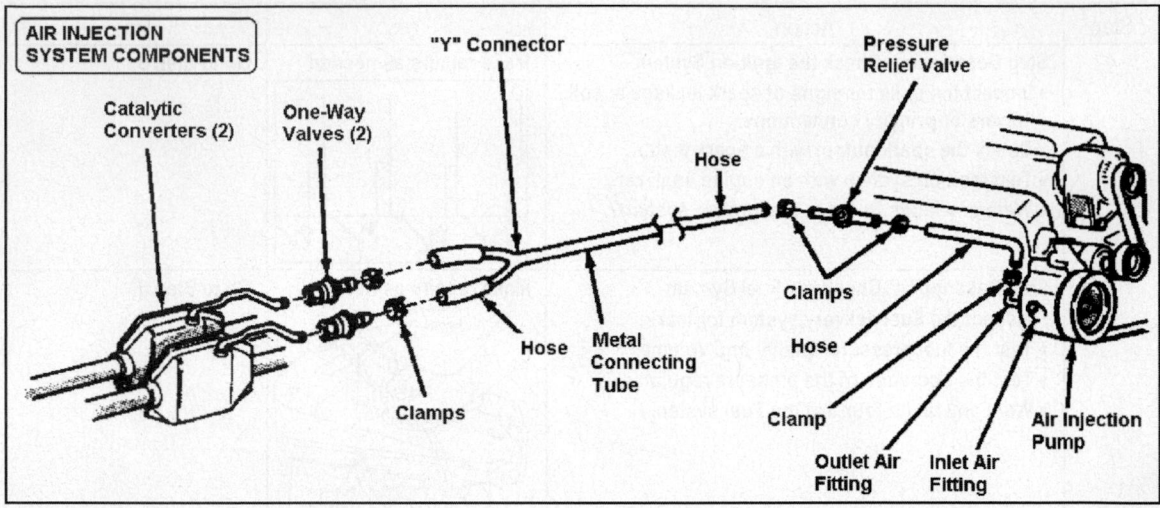

Symptom Diagnosis - Test 5

Surge Condition

PRELIMINARY CHECKS

1. Discuss how the operation of the torque converter clutch (TCC) or air conditioning compressor can affect the "feel" of the vehicle during normal operation. Refer to the information in the Owner's Manual to explain how these devices normally operate.
2. Search for any technical service bulletins related to this symptom.

Test 5 Chart

Step	Action	Yes	No
1	**Step Description: Verify the surge condition** » Drive the vehicle and attempt to verify that the vehicle surges at cruise speeds. » Does the engine have a surge condition?	Check for any stored codes. If codes are set, repair codes and retest. If no codes are set, go to Step 3.	Go to Step 2.
2	**Step Description: Condition does not exist!** » Inspect various underhood items that could cause an intermittent surge condition (check for leaks in the MAP sensor vacuum lines). » Were any problems located in this step?	Correct the problems. Do a PCM reset and "Fuel Trim Relearn" procedure. Then verify condition is repaired.	The problem is not present at this time. It may be an intermittent problem.
3	**Step Description: Check/compare PID values** » Connect a Scan Tool to the test connector. » Start the engine and allow it to fully warmup. » Monitor all related PIDs on Scan Tool (HO2S switching, LONGFT, and the TCC operation) » Compare VSS PID reading to speedometer. » Were all PIDs within their normal range?	Go to Step 4. Note: Verify that the front HO2S responds quickly to throttle changes. Check for silicon contamination on the front HO2S (this can cause a rich A/F signal).	One or more of the PIDs are out of range when compared to "known good" values. Make repairs to the system that is out of range, then retest for the symptom.
4	Step Description: Check the Ignition System » Inspect the coils for signs of spark leakage at coil towers or primary connections. » Check the spark output with a spark tester. » Test Ignition system with an engine analyzer. » Were any faults found in the Ignition system?	Make repairs as needed	Go to Step 5.

Test 5 Chart (Continued)

Step	Action	Yes	No
5	**Step Description: Check the Fuel System** » Inspect the Fuel delivery system for leaks. » Test the fuel pressure, quality and volume. » Test the operation of the pressure regulator. » Were any faults found in the Fuel system?	Make repairs as needed Fuel Pressure Gauge Fuel Rail Test Port	Go to Step 6.
6	**Step Description: Check the Exhaust System** » Check Exhaust system for leaks or damage. » Check the Exhaust system for a restriction using the Vacuum or Pressure Gauge Test (e.g., exhaust backpressure reading should not exceed 1.5 psi at cruise speeds). » Were any faults found in Exhaust System?	Make repairs to the Exhaust system. Then retest the symptom. Inspect for Damage	Return to Step 2 and repeat the tests from the beginning to locate and repair the cause of the "Surge" condition.

INTERMITTENT TESTS

Many trouble code repair charts end with a result that reads "Fault Not Present at this Time." What this expression means is that the conditions that were present when a code set or drivability symptom occurred are no longer there or were not met. In effect, the problem was present at least once, but is not present at this time. However, it is likely to return in the future, so it should be diagnosed and repaired if at all possible.

One way to find an intermittent problem is to gather the information that was present when the problem occurred. In the case of a Code Fault, this can be done in two ways: by capturing the data in Snapshot or Movie mode or by driver observations.

The PCM has to detect the fault for a specific period of time before a trouble code will set. While intermittent problems may appear to be occasional in nature, they usually occur under specific conditions. Therefore, you should identify and duplicate these conditions. Since intermittent faults are difficult to duplicate, a logical routine (checklist) must be followed when attempting to find the faulty component, system or circuit. The tests on the next page can be used to help find the cause of an intermittent fault.

Some intermittent faults occur due to a loose connection, wiring problem or warped circuit board. An intermittent fault can also be caused by poor test techniques that cause damage to the male or female ends of a connector.

Test for Loose Connectors
To test for a loose or damaged connection, take the male end of a connector from another wiring harness and carefully push it into the "suspect" female terminal to verify that the opening is tight. There should be some resistance felt as the male connector is inserted in the terminal connection.

The Wiggle Test
A wiggle test can be used to locate the cause of some intermittent faults. The sensor, switch or the PCM wiring can be back-probed, as shown, while the test is done.

During testing, move or wiggle the suspect device, connector or wiring while watching for a change.
If the DVOM has a Min/Max record mode, use this mode during the test.

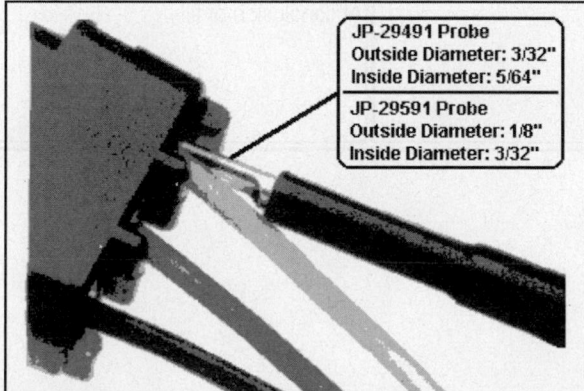

JP-29491 Probe
Outside Diameter: 3/32"
Inside Diameter: 5/64"

JP-29591 Probe
Outside Diameter: 1/8"
Inside Diameter: 3/32"

Back-Probing Connector

Wiggle Test Example

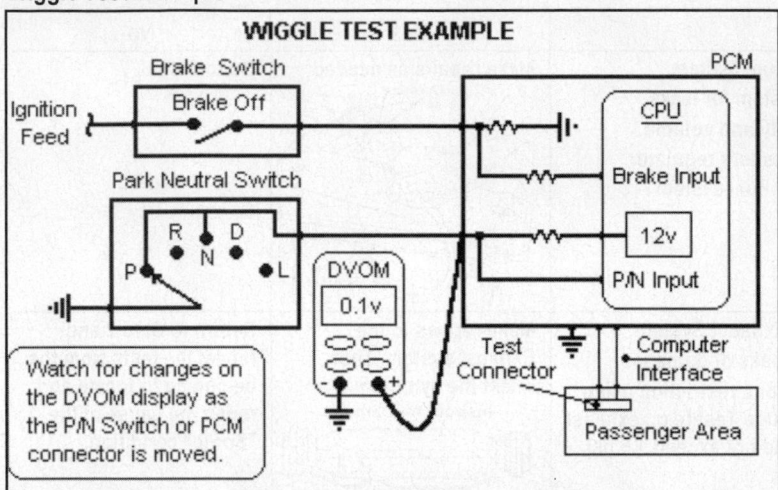

WIGGLE TEST EXAMPLE

Diagnosis And Testing - Vehicle Does Not Fill

CONDITION	POSSIBLE CAUSES	CORRECTION
Pre-Mature Nozzle Shut-Off	Defective fuel tank assembly components.	Fill tube improperly installed (sump)
		Fill tube hose pinched.
		Check valve stuck shut.
		Control valve stuck shut.
	Defective vapor/vent components.	Vent line from control valve to canister pinched.
		Vent line from canister to vent filter pinched.
		Canister vent valve failure (requires double failure, plugged to NVLD and atmosphere).
		Leak detection pump failed closed.
		Leak detection pump filter plugged.
	On-Board diagnostics evaporative system leak test just conducted.	Canister vent valve vent plugged to atmosphere.
		Engine still running when attempting to fill (System designed not to fill).
	Defective fill nozzle.	Try another nozzle.
Fuel Spits Out Of Filler Tube.	During fill.	See Pre-Mature Shut-Off.
	At conclusion of fill.	Defective fuel handling component. (Check valve stuck open).
		Defective vapor/vent handling component.
		Defective fill nozzle.

FORD PARAMETER ID (PID) CONTENTS

About This Section

Introduction

This section of the Domestic Car Handbook contains Parameter ID (PID) tables for Ford, Lincoln and Mercury vehicles from 1990-2005. It can be used to assist in the repair of both Code & No Code problems.

VEHICLE COVERAGE

- Aspire Applications (1996-97)
- Continental Applications (1990-2002)
- Contour & Mystique Applications (1996-2000)
- Cougar & Mustang Applications (1991-2005)
- Crown Victoria & Grand Marquis Applications (1992-2005)
- Escort & Tracer Applications (1991-2002)
- Escort ZX2 Applications (2003)
- LS6 & LS8 Applications (1999-2005)
- Mark VIII Applications (1993-98)
- Probe Applications (1990-97)
- Sable & Taurus Applications (1990-2004)
- Tempo Applications (1992-94)
- Thunderbird Applications (1990-97)
- Town Car Applications (1990-2005)

How to Use This Section

This Section of the Handbook can be used to look up PID Data so that you can compare the "known good" values to the actual values you see on the Scan Tool display. To locate the PID Data, find the model, correct engine size (with VIN Code) and finally the year of the vehicle.

For example, to look up the PID Data for a 1999 Taurus 3.0L VIN U, go to Contents Page 2 and find the text string shown below:

Then turn to Page 6-59 to find the following PCM PID information.

1996-2002 Taurus/Sable 3.0L V6 MFI VIN U (A/T) - Inputs

PCM PID Acronym	Parameter Identification	PID Range	PID Value at Hot Idle	PID Value at 30	PID Value at 55
FTP	FTP Sensor (1998-02)	0-5.1v or 0-10" H20	2.6v at 0" of H20	2.6v at 0" of H20	2.6v at 0" of H20

In this example, the Fuel Tank Pressure Sensor should read near 2.6v with the cap removed. The actual PID value from the vehicle can be used to check the sensor calibration by comparing it against the known good PID value in pin voltage tab

ASPIRE PID DATA - OBD II

1996-97 Aspire 1.3L I4 MFI VIN H (All) - Inputs / Outputs

PCM PID Acronym	Parameter Identification	PID Value Range	PID Value at Hot Idle	PID Value at 30	PID Value at 55
ACR	AC Relay	ON / OFF	ON (if on)	OFF	OFF
ACCS	AC Relay	ON / OFF	ON (if on)	OFF	OFF
ATP	Atmospheric Press.	0-5.1v	3.9	3.9	3.9
B+	Battery Power	0-25.5v	12-14	12-14	12-14
BARO	BARO Pressure	0-30 in. Hg	29.4	29.4	29.4
BLT SW	Blower Motor SW	ON / OFF	ON (if on)	OFF	OFF
DEF SW	Defrost Switch	ON / OFF	ON (if on)	OFF	OFF
DRL SW	Daytime Running	ON / OFF	ON (if on)	OFF	OFF
ECT (°F)	ECT Sensor	-40-304°F	177	177	177
ECT (V)	ECT Sensor	0-5.1v	0.8	0.8	0.8
EGRS (V)	EGR Sensor Volts	0-5.1v	0.7	0.7-1.6	2.5-3.2
EGR-A	EGR 'A' Solenoid	0-100 ms	6400	0	2.7
EGR-V	EGR Vent Solenoid	0-25.5v	14.3	0.1	0.1
HDP SW	Headlamp Switch	ON / OFF	ON (if on)	OFF	OFF
HOS2-11	HO2S-11 (Bank 1)	0.1-1.1v	0.1-1.1	0.1-1.1	0.1-1.1
HOS2-12	HO2S-12 (Bank 1)	0.1-1.1v	0.1-1.1	0.1-1.1	0.1-1.1
IAT (°F)	IAT Sensor	-40-304°F	77	86	68
IAT (V)	IAT Sensor	0-5.1v	2.4	2.4	2.7
IDL SW	Throttle Idle Switch	ON / OFF	ON	OFF	OFF
IGN SW	Ignition Switch	ON / OFF	ON-KOEC	OFF	OFF
INJ1	Fuel Injector 1	0-99.9 ms	4.7	4.8	6.4
INJ2	Fuel Injector 2	0-99.9 ms	4.7	4.8	6.4
ISC	Idle Speed Control	0-100 ms	2.2	2.4	2.1
LOAD	Engine Load	0-100%	26	27	56
LONGFT	Long Term F/T B1	-20 to 20%	-5 to +5	-5 to +5	-5 to +5
MAF	MAF Sensor LB/M	0-10 LBM	0.0	0.85	1.43
MAF (V)	MAF Sensor	0-5.1v	2.0v	2.4	3.2
MIL	MIL Control	ON / OFF	ON (if on)	OFF	OFF
PSP	PSP Switch	ON / OFF	ON turning	OFF	OFF
PUMP	Fuel Pump	ON / OFF	ON	ON	ON
Purge	EVAP Purge	0-100%	0	40-80	84-90
RPM	Engine Speed	0-10K rpm	750	1690	2713
DES RPM	Desired Engine Speed	0-10K rpm	750	750	750
SHRTFT	Short Term F/T B1	-5 to +5	-5 to +5	-5 to +5	-5 to +5
SPARK	Spark Advance (°)	-90° to 90°	10	12	20
STP SW	Stop Lamp Switch	ON / OFF	ON (if on)	OFF	OFF
TPV	TP Sensor	0-5.1v	0.5	0.6	0.4v
VSS	Vehicle Speed	0-159 mph	0	30	55

CONTOUR & MYSTIQUE PID DATA - OBD II

1996-2000 Contour/Mystique 2.0L I4 MFI VIN 3 (All) Inputs

PCM PID Acronym	Parameter Identification	PID Range	PID Value at Hot Idle	PID Value at 30	PID Value at 55
ACCS	A/C Clutch Switch	ON / OFF	ON (if on)	OFF	OFF
ACP	A/C Press. Switch	CL/OPEN	Open	OPEN	OPEN
BPP	Brake Pedal Switch	ON / OFF	ON (if on)	OFF	OFF
CID	CID Sensor	0-1 KHz	5-7	11-15	17-21
CKP	CKP Sensor	0-10 KHz	470-530	990-1200	1350-1460
CPP	Clutch Pedal Switch	ON / OFF	ON (if in)	OFF	OFF
DPFE	DPFE Sensor	0-5.1v	0.4-0.6	0.7-0.9	0.7-1.2v
ECT (°F)	ECT Sensor	-40-304°F	160-200	160-200	160-200
FEPS	Flash EEPROM	0-5.1v	0.1	0.1	0.1
FLI	Fuel Level Indicator (1998-2002)	0-100%	50 (1.7v)	50 (1.7v)	50 (1.7v)
FPM	Fuel Pump Monitor	ON / OFF	ON	ON	ON
FRP	Fuel Rail Pressure (1999-2002)	0-100 psi	39	39	39
FTP	Fuel Tank Pressure (1998-2002)	0-5.1v or 0-10" H2O	2.6v / 0" H2O (with cap off)	2.6v / 0" H2O (with cap off)	2.6v / 0" H2O (with cap off)
HO2S-11	HO2S-11 (Bank 1)	0.1-1.1v	0.1-1.1	0.1-1.1	0.1-1.1
HO2S-12	HO2S-12 (Bank 1)	0.1-1.1v	0.1-1.1	0.1-1.1	0.1-1.1
IAT (°F)	IAT Sensor	-40-304°F	50-120	50-120	50-120
KS1	Knock Sensor 1	2.5v	0	0	0
GEAR	Gear Position	1-2-3-4	1	3	4
LOAD	Engine Load	0-100%	10-20	20-28	34-42
MAFV	MAF Sensor	0-5.1v	0.6-0.9	0.8-1.5	1.2-2.5
MISF	Misfire Detection	ON / OFF	OFF	OFF	OFF
OCTADJ	Octane Adjustment	CL/OPEN	CLOSED	CLOSED	CLOSED
PNP	PNP Switch	ON / OFF	ON (in 'P')	OFF	OFF
PSP	PSP Switch	HI / LOW	HI: turning	LOW	LOW
RPM	Engine Speed	0-10K rpm	790-900	1785-1835	2390-2430
TCS	TCS Switch	ON / OFF	OFF	ON	ON
TFT (°F)	TFT Sensor	-40-304°F	110-210	110-210	110-210
TPV	TP Sensor	0-5.1v	0.5-0.9	1.0-1.3	1.1-1.4
TR/V	TR Sensor	0-5.1v	4.4	2.1	2.1
TSS	TSS Sensor	0-10K rpm	680-710	1100-1220	2000-2800
Vacuum	Engine Vacuum	0-30" Hg	19-20	18-20	8-15
VPWR	Vehicle Power	0-25.5v	12.8-14.0	12.8-14.0	12.8-14.0
VSS	Vehicle Speed (MPH & Hertz Rate)	0-159 mph	0 = 0 Hz	30 = 65 Hz	55 = 125 Hz

1996-2000 Contour/Mystique 2.0L I4 MFI VIN 3 (All) Outputs

PCM PID Acronym	Parameter Identification	PID Range	PID Value at Hot Idle	PID Value at 30	PID Value at 55
CD1	Coil 1 Driver	0-90° dwell	7	10	12
CD2	Coil 2 Driver	0-90° dwell	7	10	12
CTO	Clean Tach Output	0-10 KHz	23-31	54-68	73-90
EGRVR	Vacuum Regulator	0-100%	0	0-40	0-40
EPC	EPC Solenoid	0-500 psi	15-18	23-28	30-35
EVAP DC	EVAP Purge Valve	0-100%	0-100	80	90
EVAPCV	EVAP CV Valve	0-100%	0	0	0
FP	Fuel Pump Relay	0-100%	100	100	100
FUELPW	INJ 1 Pulsewidth	0-99.9 ms	2.3-2.9	2.9-4.7	4.6-6.5
HFC	High Fan Control	ON / OFF	ON (if on)	OFF	OFF
HTR11	HO2S-11 Heater	ON / OFF	ON	ON	ON
HTR12	HO2S-12 Heater	ON / OFF	ON	ON	ON
IAC	Idle Air Control	0-100%	33-35	40-48	40-55
INJ1	INJ 1 Pulsewidth	0-999 ms	2.3-2.9	2.9-4.7	4.6-6.5
INJ2	INJ 2 Pulsewidth	0-999 ms	2.3-2.9	2.9-4.7	4.6-6.5
INJ3	INJ 3 Pulsewidth	0-999 ms	2.3-2.9	2.9-4.7	4.6-6.5
INJ4	INJ 4 Pulsewidth	0-999 ms	2.3-2.9	2.9-4.7	4.6-6.5
LFC	Low Fan Control	ON / OFF	ON (if on)	OFF	OFF
LONGFT	Long Term F/T B1	-20 to 20%	-5 to +5	-5 to +5	-5 to +5
MIL	MIL Control	ON / OFF	ON (if on)	OFF	OFF
SHRTFT	Short Term F/T B1	-20 to 20%	-5 to +5	-5 to +5	-5 to +5
SPARK	Spark Advance	-90° to 90°	15-25	19-30	25-36
SS1	Shift Solenoid 1	ON / OFF	OFF	ON	ON
SS2	Shift Solenoid 2	ON / OFF	ON	OFF	OFF
SS3	Shift Solenoid 3	0-25.5v	7-9.5v	8.3-9.5	8.3-9.5
TCC	TCC Solenoid	0-100%	0	50-100	95-100
TCIL	TCIL (lamp)	ON / OFF	ON (if on)	OFF	OFF
VCT	VCT Solenoid (1998-99)	0-25.5v	12-14	10.5-14	10.5-14
VREF	Voltage Reference	0-5.1v	4.9-5.1	4.9-5.1	4.9-5.1
WAC	A/C WOT Cutout Relay	ON / OFF	ON (if on)	OFF	OFF

1996-2000 Contour/Mystique 2.5L V6 MFI VIN L (All) Inputs

PCM PID Acronym	Parameter Identification	PID Range	PID Value at Hot Idle	PID Value at 30	PID Value at 55
ACCS	A/C Clutch Switch	ON / OFF	ON (if on)	OFF	OFF
ACPSW	A/C Press. Switch	CL/OPEN	OPEN	OPEN	OPEN
BPP	Brake Pedal Switch	ON / OFF	ON (if on)	OFF	OFF
CID	CID Sensor	0-1 KHz	5-7	7-9	11-13
CKP	CKP Sensor	0-10 KHz	410-440	770-830	1190-1120
CPP	Clutch Pedal Switch	ON / OFF	ON (if in)	OFF	OFF
DPFE	DPFE Sensor	0-5.1v	0.4-0.6	0.7-1.1v	1.1-1.7
ECT (°F)	ECT Sensor	-40-304°F	160-200	160-200	160-200
FEPS	Flash EEPROM	0-5.1v	0.1	0.1	0.1
FLI	Fuel Level Indicator (1998-2002)	0-100%	50 (1.7v)	50 (1.7v)	50 (1.7v)
FPM	Fuel Pump Monitor	0-100%	50	50	50
FRP	Fuel Rail Pressure (1999-2000)	0-100 psi	39	39	39
FTP	Fuel Tank Pressure (1998-2002)	0-5.1v or 0-10" H2O	2.6v / 0" H2O (with cap off)	2.6v / 0" H2O (with cap off)	2.6v / 0" H2O (with cap off)
Gear	Transmission Gear	1-2-3-4	1	3	4
HO2S-11	HO2S-11 (Bank 1)	0.1-1.1v	0.1-1.1	0.1-1.1	0.1-1.1
HO2S-12	HO2S-12 (Bank 1)	0.1-1.1v	0.1-1.1	0.1-1.1	0.1-1.1
HO2S-21	HO2S-21 (Bank 2)	0.1-1.1v	0.1-1.1	0.1-1.1	0.1-1.1
HO2S-22	HO2S-22 (Bank 2)	0.1-1.1v	0.1-1.1	0.1-1.1	0.1-1.1
IAT (°F)	IAT Sensor	-40-304°F	50-120	50-120	50-120
IMRCM	IMRC Monitor	ON / OFF	OFF	OFF	OFF
KS1	Knock Sensor 1	0-5.1v	2.5v	2.5v	2.5v
Load	Engine Load	0-100%	10-20	20-28	34-46
MAF (V)	MAF Sensor	0-5.1v	0.6-0.9	0.9-1.1	1.2-2.6v
MISF	Misfire Detection	ON / OFF	OFF	OFF	OFF
OCTADJ	Octane Adjustment	CL/OPEN	CLOSED	CLOSED	CLOSED
PNP	PNP Switch	ON / OFF	ON (in 'P')	OFF	OFF
PSP	PSP Switch	HI / LOW	HI: turning	LOW	LOW
RPM	Engine Speed	0-10K rpm	700-760	1340-1380	1950-2100
TCS	TCS (switch)	ON / OFF	ON (if on)	OFF	OFF
TFT (°F)	TFT Sensor	-40-304°F	110-210	110-210	110-210
TPV	TP Sensor	0-5.1v	0.5-0.9	1.0-1.3	1.1-1.4
TRS	TR Sensor	0-5.1v	4.4	2.1	2.1
TSS	TSS Sensor	0-10 KHz	680-710	1150-1240	2030-3020
Vacuum	Engine Vacuum	0-30" Hg	19-20	18-20	8-15
VPWR	Vehicle Power	0-25.5v	12-14	12-14	12-14
VSS	Vehicle Speed (MPH & Hertz Rate)	0-159 mph	0 = 0 Hz	30 = 65 Hz	55 = 125 Hz

1996-99 Contour/Mystique 2.5L V6 MFI VIN L (All) Outputs

PCM PID Acronym	Parameter Identification	PID Range	PID Value at Hot Idle	PID Value at 30	PID Value at 55
CD1-CD4	Coil Driver 1-4	0-60° dwell	7	10	11
CTO	Clean Tach Output	0-10 KHz	25-42	62-73	88-115
EGRVR	VR Solenoid	0-100%	0	0-40	35-45
EPC	EPC Solenoid	0-500 psi	11-18	23-28	30-35
EVAPCV	EVAP CV Valve	0-100%	0	0	0
EVAPDC	EVAP Purge Valve	0-100%	0-100	0-100	0-100
FP	Fuel Pump Relay	0-100%	100	100	100
FuelPW1	INJ Pulsewidth B1	0-99.9 ms	2.1-2.5	3.5-5.5	3.3-5.9
FuelPW2	INJ Pulsewidth B2	0-99.9 ms	2.1-2.5	3.5-5.5	3.3-5.9
HFC	High Fan Control	ON / OFF	ON (if on)	OFF	OFF
HTR11	HO2S-11 Heater	ON / OFF	ON	ON	ON
HTR12	HO2S-12 Heater	ON / OFF	ON	ON	ON
HTR21	HO2S-21 Heater	ON / OFF	ON	ON	ON
HTR22	HO2S-22 Heater	ON / OFF	ON	ON	ON
IAC	Idle Air Control	0-100%	32-40	30-55	40-60
IMRC	IMRC Solenoid	ON / OFF	OFF	OFF	OFF
INJ1	INJ 1 Pulsewidth	0-99.9 ms	2.1-2.5	3.5-5.5	3.3-5.9
INJ2	INJ 2 Pulsewidth	0-99.9 ms	2.1-2.5	3.5-5.5	3.3-5.9
INJ3	INJ 3 Pulsewidth	0-99.9 ms	2.1-2.5	3.5-5.5	3.3-5.9
INJ4	INJ 4 Pulsewidth	0-99.9 ms	2.1-2.5	3.5-5.5	3.3-5.9
INJ5	INJ 5 Pulsewidth	0-99.9 ms	2.1-2.5	3.5-5.5	3.3-5.9
INJ6	INJ 6 Pulsewidth	0-99.9 ms	2.1-2.5	3.5-5.5	3.3-5.9
LFC	Low Fan Control	ON / OFF	ON (if on)	OFF	OFF
LongFT1	Long Term F/T B1	-20 to 20%	-5 to +5	-5 to +5	-5 to +5
LongFT2	Long Term F/T B2	-20 to 20%	-5 to +5	-5 to +5	-5 to +5
MIL	MIL Control	ON / OFF	ON (if on)	OFF	OFF
SHTFT1	Short Term F/T B1	-20 to 20%	-5 to +5	-5 to +5	-5 to +5
SHTFT2	Short Term F/T B2	-20 to 20%	-5 to +5	-5 to +5	-5 to +5
SPARK	Spark Advance (°)	-90° to 90°	4-7	19-30	25-36
SS1	Shift Solenoid 1	ON / OFF	OFF	ON	ON
SS2	Shift Solenoid 2	ON / OFF	ON	OFF	OFF
SS3	Shift Solenoid 3	7.0-8.1v	7.0-9.5	8.3-9.5	8.3-9.5
TCC	TCC Solenoid	0-100%	0	50-100	95-100
TCIL	TCIL (lamp)	ON (if on)	OFF	OFF	OFF
VREF	Voltage Reference	0-5.1v	4.9-5.1	4.9-5.1	4.9-5.1
WAC	A/C WOT Cutout Relay	ON / OFF	ON (if on)	OFF	OFF

COUGAR PID DATA - OBD II

1999-2002 Cougar 2.0L I4 4v MFI VIN 3 (All) Inputs

PCM PID Acronym	Parameter Identification	PID Range	PID Value at Hot Idle	PID Value at 30	PID Value at 55
ACCS	Cycling Clutch SW	ON / OFF	ON (if on)	OFF	OFF
ACP	A/C Press. Switch	CL/OPEN	OPEN	OPEN	OPEN
BOO	BPP Switch	ON / OFF	ON (if on)	OFF	OFF
CID	Cylinder ID Signal	0-1 KHz	5-7	11-15	17-21
CKP	CKP Sensor	0-10 KHz	470-650	995-1210	1345-1455
CPP	Clutch Switch	ON / OFF	ON (if in)	OFF	OFF
ECT (°F)	ECT Sensor	-40-304°F	160-200	160-200	160-200
FEPS	Flash EEPROM	0-5.1v	0.1	0.1	0.1
FLI	Fuel Level Indicator	0-100%	50 (1.7v)	25-75	25-75
FPM	Fuel Pump Monitor	0-100%	87-100	87-100	87-100
FRP	Fuel Rail Pressure	0-100 psi	39	39	39
FTP	Fuel Tank Pressure	0-5.1v or 0-10" H2O	2.6v / 0" H2O (with cap off)	2.6v / 0" H2O (with cap off)	2.6v / 0" H2O (with cap off)
IAT (°F)	IAT Sensor	-40-304°F	50-120	50-120	50-120
GEAR	Gear Position	1-2-3-4	1	3	4
GLI	Generator Load	0-25.5v	8.2 (high)	7.6	7.9
HO2S-11	HO2S-11 (Bank 1)	0.1-1.1v	0.1-1.1	0.1-1.1	0.1-1.1
HO2S-12	HO2S-12 (Bank 1)	0.1-1.1v	0.1-1.1	0.1-1.1	0.1-1.1
KS1	Knock Sensor 1	0-5.1v	0	0	0
LOAD	Engine Load	0-100%	10-20	19-30	40-47
MAFV	MAF Sensor	0-5.1v	0.6-0.9	0.8-1.5	1.2-2.5
MISF	Misfire Detection	ON / OFF	OFF	OFF	OFF
OCTADJ	Octane Adjustment	CL/OPEN	CLOSED	CLOSED	CLOSED
PATSIN	PAT System	See PATS	See PATS	See PATS	See PATS
PNP	PNP Switch	ON / OFF	ON (in 'P')	OFF	OFF
PSP	PSP Switch	HI / LOW	HI: turning	LOW	LOW
RPM	Engine Speed	0-10K rpm	790-900	1680-1840	2250-2385
TFT (°F)	TFT Sensor	-40-304°F	110-210	110-210	110-210
TPV	TP Sensor	0-5.1v	0.53-1.27	1.0-1.3	1.1-1.4
Vacuum	Engine Vacuum	0-30" Hg	19-20	18-20	8-15
VPWR	Vehicle Power	0-25.5v	12.8-14.0	12.8-14.0	12.8-14.0
VSS	Vehicle Speed (MPH & Hertz Rate)	0-159 mph	0 = 0 Hz	30 = 65 Hz	55 = 125 Hz

1999-2002 Cougar 2.0L I4 4v MFI VIN 3 (All) Outputs

PCM PID Acronym	Parameter Identification	PID Range	PID Value at Hot Idle	PID Value at 30	PID Value at 55
CDA	Coil Driver 'A'	0-90° dwell	7	10	11
CD2	Coil Driver 'B'	0-90° dwell	7	10	11
CTO	Clean Tach Output	0-10K KHz	23-31	54-68	73-90
EPC	EPC Solenoid	0-500 psi	0	23-28	35-48
EVAPCV	EVAP CV Valve	0-100%	0	0	0
EVAPDC	EVAP Purge Valve	0-100%	0-100	80	90
FP	Fuel Pump Relay	0-100%	100	100	100
FuelPW1	INJ Pulsewidth B1	0-999 ms	2.9-2.4.0	2.9-5.6	6.5-9.3
HFC	High Fan Control	ON / OFF	ON (if on)	OFF	OFF
HTR11	HO2S-11 Heater	ON / OFF	ON	ON	ON
HTR12	HO2S-12 Heater	ON / OFF	ON	ON	ON
IAC	Idle Air Control	0-100%	32	35-48	40-55
INJ1	INJ 1 Pulsewidth	0-999 ms	2.9-2.4.0	2.9-5.6	6.5-9.3
INJ2	INJ 2 Pulsewidth	0-999 ms	2.9-2.4.0	2.9-5.6	6.5-9.3
INJ3	INJ 3 Pulsewidth	0-999 ms	2.9-2.4.0	2.9-5.6	6.5-9.3
INJ4	INJ 4 Pulsewidth	0-999 ms	2.9-2.4.0	2.9-5.6	6.5-9.3
LFC	Low Fan Control	ON / OFF	ON (if on)	OFF	OFF
LONGFT	Long Term F/T B1	-20 to 20%	-5 to +5	-5 to +5	-5 to +5
MIL	MIL Control	ON / OFF	ON (if on)	OFF	OFF
PATSIL	PAT System	See PATS	See PATS	See PATS	See PATS
PATSOUT	PAT System	See PATS	See PATS	See PATS	See PATS
PATSTRT	PAT System	See PATS	See PATS	See PATS	See PATS
SHRTFT	Short Term F/T B1	-20 to 20%	-5 to +5	-5 to +5	-5 to +5
SPARK	Spark Advance	-90° to 90°	15-25	19-33	25-36
SS1	Shift Solenoid 1	ON / OFF	OFF	ON	ON
SS2	Shift Solenoid 2	ON / OFF	OFF	ON	ON
SS3	Shift Solenoid 3	0-25.5v	7.0-9.5	8.3-9.5	8.3-9.5
TCC	TCC Solenoid	0-100%	0	0	95-100
TCIL	TCIL Lamp	ON / OFF	OFF	OFF	OFF
VCT	VCT Solenoid	0-25.5v	14	10.5-14	10.5-14
VSO	VSO Signal	0-1K Hz	0	65	125
VREF	Voltage Reference	0-5.1v	4.9-5.1	4.9-5.1	4.9-5.1
WAC	A/C WOT Cutout Relay	ON / OFF	ON (if on)	OFF	OFF

1999-2002 Cougar 2.5L V6 MFI VIN L (All) Inputs

PCM PID Acronym	Parameter Identification	PID Range	PID Value at Hot Idle	PID Value at 30	PID Value at 55
ACCS	A/C Clutch Switch	ON / OFF	ON (if on)	OFF	OFF
ACP	A/C Press. Switch	CL/OPEN	OPEN	OPEN	OPEN
BPP	Brake Pedal Switch	ON / OFF	ON (if on)	OFF	OFF
CID	CID Sensor	0-1 KHz	5-7	10-14	15-21
CKP	CKP Sensor	0-10 KHz	410-480	830-1000	1220-1400
CPP	Clutch Pedal Switch	ON / OFF	ON (if in)	OFF	OFF
DPFE	DPFE Input (99-'00)	0-5.1v	0.2-1.3	0.2-4.5	0.2-4.5
DPFE	DPFE Input (01-'02)	0-5.1v	0.95-1.05	0.95-4.65	0.95-4.65
ECT (°F)	ECT Sensor	-40-304°F	160-200	160-200	160-200
FEPS	Flash EEPROM	0-5.1v	0.1	0.1	0.1
FLI	Fuel Level Indicator	0-5.1v	50 (1.7v)	50 (1.7v)	50 (1.7v)
FPM	Fuel Pump Monitor	0-100%	100	100	100
FRP	Fuel Rail Pressure	0-100 psi	39	39	39
FTP	Fuel Tank Pressure	0-5.1v or 0-10" H2O	2.6v / 0" H2O (with cap off)	2.6v / 0" H2O (with cap off)	2.6v / 0" H2O (with cap off)
Gear	Transmission Gear	1-2-3-4	1	3	4
HO2S-11	HO2S-11 (Bank 1)	0.1-1.1v	0.1-1.1	0.1-1.1	0.1-1.1
HO2S-12	HO2S-12 (Bank 1)	0.1-1.1v	0.1-1.1	0.1-1.1	0.1-1.1
HO2S-21	HO2S-21 (Bank 2)	0.1-1.1v	0.1-1.1	0.1-1.1	0.1-1.1
HO2S-22	HO2S-22 (Bank 2)	0.1-1.1v	0.1-1.1	0.1-1.1	0.1-1.1
IMRCM	IMRC Monitor	ON / OFF	OFF	OFF	OFF
IAT (°F)	IAT Sensor	-40-304°F	50-120	50-120	50-120
KS1	Knock Sensor 1	0-5.1v	0	0	0
Load	Engine Load	0-100%	10-20	20-28	30-42
MAF (V)	MAF Sensor	0-5.1v	0.6-0.9	1.1-1.3	1.2-1.6
MISF	Misfire Detection	ON / OFF	OFF	OFF	OFF
OCTADJ	Octane Adjustment	CL/OPEN	CLOSED	CLOSED	CLOSED
OSS	OSS Sensor Hertz	0-10 KHz	0	590	1050
PATSIN	PAT System	See PATS	See PATS	See PATS	See PATS
PNP	PNP Switch	ON / OFF	ON (in 'P')	OFF	OFF
PSP	PSP Switch	HI / LOW	HI: turning	LOW	LOW
RPM	Engine Speed	0-10K rpm	700-760	1700-1800	2100-2390
TCS	TCS (switch)	ON / OFF	ON (if on)	OFF	OFF
TFT (°F)	TFT Sensor	-40-304°F	110-210	110-210	110-210
TPV	TP Sensor	0-5.1v	0.53-1.27	1.0-1.3	1.1-1.4
TRV	TR Sensor	0-5.1v	4.4 (Park)	2.1 (O/D)	2.1 (O/D)
TSS	TSS Sensor	0-10 KHz	46-50	70-95	130-145
VPWR	Vehicle Power	0-25.5v	12-14	12-14	12-14
VSS (+)	Vehicle Speed (MPH & Hertz Rate)	0-159 mph	0 = 0 Hz	30 = 65 Hz	55 = 125 Hz

1999-2002 Cougar 2.5L V6 MFI VIN L (All) Outputs

PCM PID Acronym	Parameter Identification	PID Range	PID Value at Hot Idle	PID Value at 30	PID Value at 55
CDA	Coil Driver 'A'	0-60° dwell	7	10	11
CDB	Coil Driver 'B'	0-60° dwell	7	10	11
CDC	Coil Driver 'C'	0-60° dwell	7	10	11
CTO	Clean Tach Output	0-10 KHz	35-42	73-86	88-120
EGRVR	VR Solenoid	0-100%	0	0-40	0-40
EPC	EPC Solenoid	0-500 psi	0	23-28	30-35
EVAPCV	EVAP CV Valve	0-100%	0	0	0
EVAPDC	EVAP Purge Valve	0-100%	0-100	0-100	0-100
FP	Fuel Pump Relay	0-100%	30	30	30
FuelPW1	Injector on time B1	0-99.9 ms	3.5-4.0	3.0-6.0	3.3-7.3
FuelPW2	Injector on time B2	0-99.9 ms	3.5-4.0	3.0-6.0	3.3-7.3
HFC	High Fan Control	ON / OFF	ON (if on)	OFF	OFF
HTR11	HO2S-11 Heater	ON / OFF	ON	ON	ON
HTR12	HO2S-12 Heater	ON / OFF	ON	ON	ON
HTR21	HO2S-21 Heater	ON / OFF	ON	ON	ON
HTR22	HO2S-22 Heater	ON / OFF	ON	ON	ON
IAC	Idle Air Control	0-100%	33	40-48	40-55
IMRC	IMRC Solenoid	ON / OFF	OFF	OFF	OFF
INJ1	INJ 1 Pulsewidth	0-99.9 ms	3.5-4.0	3.0-6.0	3.3-7.3
INJ2	INJ 2 Pulsewidth	0-99.9 ms	3.5-4.0	3.0-6.0	3.3-7.3
INJ3	INJ 3 Pulsewidth	0-99.9 ms	3.5-4.0	3.0-6.0	3.3-7.3
INJ4	INJ 4 Pulsewidth	0-99.9 ms	3.5-4.0	3.0-6.0	3.3-7.3
INJ5	INJ 5 Pulsewidth	0-99.9 ms	3.5-4.0	3.0-6.0	3.3-7.3
INJ6	INJ 6 Pulsewidth	0-99.9 ms	3.5-4.0	3.0-6.0	3.3-7.3
LFC	Low Fan Control	ON / OFF	ON (if on)	OFF	OFF
LongFT1	Long Term F/T B1	-20 to 20%	-5 to +5	-5 to +5	-5 to +5
LongFT2	Long Term F/T B2	-20 to 20%	-5 to +5	-5 to +5	-5 to +5
MIL	MIL Control	ON / OFF	ON (if on)	OFF	OFF
PATSIL	PAT System	See PATS	See PATS	See PATS	See PATS
PATSOUT	PAT System	See PATS	See PATS	See PATS	See PATS
PATSTRT	PAT System	See PATS	See PATS	See PATS	See PATS
SHTFT1	Short Term F/T B1	-20 to 20%	-5 to +5	-5 to +5	-5 to +5
SHTFT2	Short Term F/T B2	-20 to 20%	-5 to +5	-5 to +5	-5 to +5
SPARK	Spark Advance (°)	-90° to 90°	4-7	19-30	25-36
SS1	Shift Solenoid 1	ON / OFF	OFF	ON	ON
SS2	Shift Solenoid 2	ON / OFF	ON	OFF	OFF
SS3	Shift Solenoid 3	0-25.5v	7.0-9.5	8.3-9.5	8.3-9.5
TCC	TCC Solenoid	0-100%	0	0	95-100
TCIL	TCIL (lamp)	ON (if on)	OFF	OFF	OFF
VREF	Voltage Reference	0-5.1v	4.9-5.1	4.9-5.1	4.9-5.1
VSO	VSO Signal	0-1K Hz	0	65	125
WAC	A/C WOT Cutout Relay	ON / OFF	ON (if on)	OFF	OFF

CROWN VICTORIA & GRAND MARQUIS PID DATA - OBD I

1992 Crown Victoria/Grand Marquis 4.6L V8 VIN W (A/T - AOD)

PCM PID Acronym	Parameter Identification	PID Range	PID Value at Hot Idle	PID Value at 30	PID Value at 55
CANP	EVAP Purge Valve	ON / OFF	OFF	ON	ON
ECT (°F)	ECT Sensor	-40-304°F	160-200	160-200	160-200
ECT (V)	ECT Sensor	0-5.1v	0.6	0.6	0.6
EVR	VR Solenoid	0-100%	0	0-40	0-60
FuelPW1	INJ Pulsewidth B1	0-99.9 ms	6.4-6.8	8.2-8.8	8.4-9.0
FuelPW2	INJ Pulsewidth B2	0-99.9 ms	6.4-6.8	8.2-8.8	8.4-9.0
HO2S-11	HO2S-11 (Bank 1)	0.1-1.1v	0.1-1.1	0.1-1.1	0.1-1.1
HO2S-21	HO2S-21 (Bank 2)	0.1-1.1v	0.1-1.1	0.1-1.1	0.1-1.1
IAC	Idle Air Control	0-100%	20-40	34-40	45-55
IAT (°F)	IAT Sensor	-40-304°F	50-120	50-120	50-120
IAT (V)	IAT Sensor	-40-304°F	1.7-3.5	1.7-3.5	1.7-3.5
LOOP	Loop Status	CL or OL	CL	CL	CL
LongFT1	Long Term F/T B1	-20 to 20%	-5 to +5	-5 to +5	-5 to +5
LongFT2	Long Term F/T B2	-20 to 20%	-5 to +5	-5 to +5	-5 to +5
MAF (V)	MAF Sensor	0-5.1v	0.6	0.9-1.1	1.1-1.3
MLP	MLP Sensor	Park/Drive	PARK	DRIVE	DRIVE
MLP (V)	MLP Sensor	0-5.1v	0v	5v	5v
PFE	PFE Sensor	0-5.1v	3.2	3.3	3.4
PNP	PNP Switch	Neutral/DR	NEUTRAL	DRIVE	DRIVE
RPM	Engine Speed	0-10K rpm	750-820	1450-1630	1750-2100
SHTFT1	Short Term F/T B1	-20 to 20%	-5 to +5	-5 to +5	-5 to +5
SHTFT2	Short Term F/T B2	-20 to 20%	-5 to +5	-5 to +5	-5 to +5
SPARK	Spark Advance (°)	-90° to 90°	15-22	28-35	25-35
TP	TP Sensor	0-5.1v	0.7	0.8-1.0	1.1-1.3
TP Mode	TP Mode	C/T or P/T	C/T	P/T	P/T
VPWR	Vehicle Power	0-25.5v	12-14	12-14	12-14
VREF	Voltage Reference	0-5.1v	4.9-5.1	4.9-5.1	4.9-5.1
VSS	Vehicle Speed	0-159 mph	0	30	55
WAC	A/C WOT Cutout Relay	ON / OFF	ON (if on)	OFF	OFF

1992-94 Crown Victoria/Grand Marquis 4.6L VIN W (A/T AODE)

PCM PID Acronym	Parameter Identification	PID Range	PID Value at Hot Idle	PID Value at 30	PID Value at 55
ACCS	A/C Clutch Switch	ON / OFF	ON (if on)	OFF	ON
BOO	Brake Switch	ON / OFF	ON (if on)	OFF	OFF
DPFE	DPFE Sensor	0-5.1v	0.4-0.6	0.6-1.8	1.9-4.8
ECT (°F)	ECT Sensor	-40-304°F	160-200	160-200	160-200
ECT (V)	ECT Sensor	0-5.1v	0.6	0.6	0.6
EPC	EPC Solenoid	0-500 psi	20	12	30
EVR	VR Solenoid	0-100%	0	0-40	0-60
FP	Fuel Pump Relay	ON / OFF	ON	ON	ON
GEAR	Gear Position	P-R-N-D	P-R-N-D	DRIVE 3	DRIVE 4
FuelPW1	INJ Pulsewidth B1	0-99.9 ms	5.5-5.8	6.4-8.4	9.4-12
FuelPW2	INJ Pulsewidth B2	0-99.9 ms	5.5-5.8	6.4-8.4	9.4-12
HO2S-11	HO2S-11 (Bank 1)	0.1-1.1v	0.1-1.1	0.1-1.1	0.1-1.1
HO2S-21	HO2S-21 (Bank 1)	0.1-1.1v	0.1-1.1	0.1-1.1	0.1-1.1
IAC	Idle Air Control	0-100%	20-40	34-40	45-55
IAT (°F)	IAT Sensor	-40-304°F	50-120	50-120	50-120
IAT (V)	IAT Sensor	0-5.1v	1.7-3.5	1.7-3.5	1.7-3.5
LOOP	Loop Status	CL or OL	CL	CL	CL
LongFT1	Long Term F/T B1	-20 to 20%	-5 to +5	-5 to +5	-5 to +5
LongFT2	Long Term F/T B2	-20 to 20%	-5 to +5	-5 to +5	-5 to +5
MAF	MAF Sensor	0-5.1v	1.0v	1.3-1.5	1.6-2.0
MLP	MLP Switch	DRIVE or PARK	PARK	DRIVE	DRIVE
MLP (V)	MLP Switch	0-5.1v	0	5	5
PNP	PNP Switch	Neutral/DR	NEUTRAL	DRIVE	DRIVE
RPM	Engine Speed	0-10K rpm	750-820	1450-1630	1750-2100
SHTFT1	Short Term F/T B1	-20 to 20%	-5 to +5	-5 to +5	-5 to +5
SHTFT2	Short Term F/T B2	-20 to 20%	-5 to +5	-5 to +5	-5 to +5
SPARK	Spark Advance	-90° to 90°	15-22	28-35	25-35
TCC	TCC Solenoid	0-100%	0	42-44	92-100
TOT (°F)	TOT Sensor	-40-304°F	110-210	110-210	110-210
TOT (V)	TOT Sensor	0-5.1v	1.7-3.5	1.7-3.5	1.7-3.5
TPV	TP Sensor	0-5.1v	0.7	0.8-1.0	1.1-1.3
TP MIN	TP Minimum	0-5.1v	0.7	0.8-1.0	1.1-1.3
TP Mode	TP Mode	C/T or P/T	C/T	P/T	P/T
VPWR	Vehicle Power	0-25.5v	12-14	12-14	12-14
VREF	Voltage Reference	0-5.1v	4.9-5.1	4.9-5.1	4.9-5.1
VSS	Vehicle Speed MPH	0-159 mph	0	30	55
WAC	A/C WOT Cutout Relay	OFF	ON (if on)	OFF	OFF

1995-2005 Crown Victoria/Grand Marquis 4.6L V8 VIN W (A/T)

PCM PID Acronym	Parameter Identification	PID Range	PID Value at Hot Idle	PID Value at 30	PID Value at 55
ACCS	A/C Clutch Switch	ON / OFF	ON (if on)	OFF	OFF
ACP	A/C Pressure	OPEN or CLOSED	CLOSED (if low)	OPEN	OPEN
BOO	BPP Switch	ON / OFF	ON (if on)	OFF	OFF
CHT (°F)	CHT Sensor (°F)	-40-304°F	190-194	190-194	190-194
CID	CID Sensor Signal	0-1 KHz	6-7	10-11	13-14
CKP	CKP Sensor Signal	0-10 KHz	440-490	550-780	900-1100
DPFE	DPFE Sensor	0-5.1v	0.25-0.6	0.25-4.65	0.25-4.65
ECT (°F)	ECT Sensor	-40-304°F	160-200	160-200	160-200
FEPS	Flash EEPROM	0-5.1v	0.1	0.1	0.1
FLI	Fuel Level Indicator	0-100%	50 (1.7v)	25-75	25-75
FPM	Fuel Pump Monitor	ON / OFF	ON	ON	ON
FTP	Fuel Tank Pressure	0-5.1v or 0-10" H2O	2.6v / 0" H2O (with cap off)	2.6v / 0" H2O (with cap off)	2.6v / 0" H2O (with cap off)
GEAR	Transmission Gear	1-2-3-4	1	3	4
HO2S-11	HO2S-11 (Bank 1)	0.1-1.1v	0.1-1.1	0.1-1.1	0.1-1.1
HO2S-12	HO2S-12 (Bank 1)	0.1-1.1v	0.1-1.1	0.1-1.1	0.1-1.1
HO2S-21	HO2S-21 (Bank 2)	0.1-1.1v	0.1-1.1	0.1-1.1	0.1-1.1
HO2S-22	HO2S-22 (Bank 2)	0.1	0.1-1.1	0.1-1.1	0.1-1.1
IAT (°F)	IAT Sensor	-40-304°F	50-120	50-120	50-120
LOAD	Engine Load	0-100%	15-19	20-26	28-38
MAFV	MAF Sensor	0-5.1v	0.6-0.9	0.9-1.5	1.4-2.1
MISF	Misfire Detection	ON / OFF	OFF	OFF	OFF
OCTADJ	Octane Adjustment	CL/OPEN	CLOSED	CLOSED	CLOSED
OSS	OSS Sensor	0-10 KHz	0 rpm	1260-1330	2265-2400
PNP	PNP Switch	ON / OFF	ON (in 'P')	OFF	OFF
RPM	Engine Speed	0-10K rpm	790-815	1250-1400	1540-1620
TCS	TCS (Momentary ON Switch)	ON when depressed	OFF	OFF	OFF
TFT (°F)	TFT Sensor	110-210	110-210	110-210	110-210
TPV	TP Sensor	0.5-0.9	0.5-0.9	1.0-1.2	1.0-1.3
TR1	TR1 Sensor (2003)	0-11.5v	0	11.5	11.5
TR2 (V)	TR2 Sensor	0-25.5v	0	11.5	11.5
TR4 (V)	TR4 Sensor	0-25.5v	0	11.5	11.5
TR/V	TR Sensor	0-5v	4.4	2.1	2.1
Vacuum	Engine Vacuum	0-30" Hg	19-20	18-20	8-15
VPWR	Vehicle Power	12-14v	12-14	12-14	12-14
VSS/VSO	Vehicle Speed (MPH & Hertz Rate)	0-159 mph	0 = 0 Hz	30 = 65 Hz	55 = 125 Hz
WACA	A/C WOT Cutout Relay Monitor	ON / OFF	ON (if on)	OFF	OFF

1995-2005 Crown Victoria/Grand Marquis 4.6L V8 VIN W (A/T)

PCM PID Acronym	Parameter Identification	PID Range	PID Value at Hot Idle	PID Value at 30	PID Value at 55
CD1-CD8	Coil Driver 1-8	0-45° dwell	8	10	12
CHTIL	Cylinder Head Temperature Lamp	ON / OFF	OFF	OFF	OFF
CTO	Clean Tach Output	0-10 KHz	50-57	82-90	100-110
EGRVR	VR Solenoid	0-100%	0	0-40	40-50
EPC	EPC Solenoid	0-500 psi	20	22	22
EVAPCV	EVAP CV Valve	0-100%	0	0	0
EVAPDC	EVAP Purge Valve	0-100%	0-100	80	90
FP	Fuel Pump Monitor	0-100%	100	100	100
FuelPW1	INJ Pulsewidth B1	0-99.9 ms	3.4-3.7	3.7-6.0	5.5-9.0
FuelPW2	INJ Pulsewidth B2	0-99.9 ms	3.4-3.7	3.7-6.0	5.5-9.0
HTR-11	HO2S-11 Heater	ON / OFF	Switch	Switch	Switch
HTR-12	HO2S-12 Heater	ON / OFF	ON	ON	ON
HTR-21	HO2S-21 Heater	ON / OFF	Switch	Switch	Switch
HTR-22	HO2S-22 Heater	ON / OFF	ON	ON	ON
IAC	Idle Air Control	0-100%	32-40	30-55	40-60
INJ1	INJ 1 Pulsewidth	0-99.9 ms	3.4-3.7	3.7-6.0	5.5-9.0
INJ2	INJ 2 Pulsewidth	0-99.9 ms	3.4-3.7	3.7-6.0	5.5-9.0
INJ3	INJ 3 Pulsewidth	0-99.9 ms	3.4-3.7	3.7-6.0	5.5-9.0
INJ4	INJ 4 Pulsewidth	0-99.9 ms	3.4-3.7	3.7-6.0	5.5-9.0
INJ5	INJ 5 Pulsewidth	0-99.9 ms	3.4-3.7	3.7-6.0	5.5-9.0
INJ6	INJ 6 Pulsewidth	0-99.9 ms	3.4-3.7	3.7-6.0	5.5-9.0
INJ7	INJ 7 Pulsewidth	0-99.9 ms	3.4-3.7	3.7-6.0	5.5-9.0
INJ8	INJ 8 Pulsewidth	0-99.9 ms	3.4-3.7	3.7-6.0	5.5-9.0
LFC	Low Fan Control	ON / OFF	ON (if on)	OFF	OFF
LongFT1	Long Term F/T B1	-20 to 20%	-5 to +5	-5 to +5	-5 to +5
LongFT2	Long Term F/T B2	-20 to 20%	-5 to +5	-5 to +5	-5 to +5
MIL	MIL Control	ON / OFF	OFF	OFF	OFF
SHTFT1	Short Term F/T B1	-20 to 20%	-5 to +5	-5 to +5	-5 to +5
SHTFT2	Short Term F/T B2	-20 to 20%	-5 to +5	-5 to +5	-5 to +5
SPARK	Spark Advance	-90° to 90°	15-18	33-36	32-38
SS1	Shift Solenoid 1	ON / OFF	ON	OFF	ON
SS2	Shift Solenoid 2	ON / OFF	OFF	ON	ON
TCC	TCC Solenoid	0-100%	0	42-44	90-100
TCIL	TCIL (lamp)	ON (if on)	OFF	OFF	OFF
VREF	Voltage Reference	0-5.1v	4.9-5.1	4.9-5.1	4.9-5.1
WAC	A/C WOT Cutout Relay	ON / OFF	ON (if on)	OFF	OFF

1996-2005 Crown Victoria/Grand Marquis 4.6L V8 NGV VIN 9

PCM PID Acronym	Parameter Identification	PID Range	PID Value at Hot Idle	PID Value at 30	PID Value at 55
ACCS	A/C Clutch Switch	ON / OFF	ON (if on)	OFF	OFF
BPP	Brake Pedal Switch	ON / OFF	ON (if on)	OFF	OFF
CID	CID Sensor	0-1 KHz	6-7	8-9	16-17
CKP	CKP Sensor	0-10 KHz	440-490	580-770	850-1100
DPFE	DPFE Sensor	0-5.1v	0.25-1.30	0.25-4.65	0.25-4.65
ECT (°F)	ECT Sensor	-40-304°F	160-200	160-200	160-200
EFTA (°F)	EFT Sensor B1	-40-304°F	50-120	50-120	50-120
EFTB (°F)	EFT Sensor B2	-40-304°F	50-120	50-120	50-120
FEPS	Flash EEPROM	0-5.1v	0.5-0.6	0.5-0.6	0.5-0.6
FSV M	Fuel Shutoff Valve Monitor	ON / OFF	ON	ON	ON
FRP (psi)	Fuel Pressure	0-500 psi	105-130	105-130	105-130
FRP	Fuel Rail Pressure	0-5v	2.7-3.7	2.7-3.7	2.7-3.7
GEAR	Transmission Gear	1-2-3-4	1	3	4
HO2S-11	HO2S-11 (Bank 1)	0.1-1.1v	0.1-1.1	0.1-1.1	0.1-1.1
HO2S-12	HO2S-12 (Bank 1)	0.1-1.1v	0.1-1.1	0.1-1.1	0.1-1.1
HO2S-21	HO2S-21 (Bank 2)	0.1-1.1v	0.1-1.1	0.1-1.1	0.1-1.1
HO2S-22	HO2S-22 (Bank 2)	0.1-1.1v	0.1-1.1	0.1-1.1	0.1-1.1
IAT (°F)	IAT Sensor	-40-304°F	50-120	50-120	50-120
IP ('96)	Injector Pressure	0-5v	2.7-3.7	2.7-3.7	2.7-3.7
LOAD	Engine Load	0-100%	15-19	22-30	31-46
MAF (V)	MAF Sensor	0-5.1v	0.6-0.9	0.9-1.5	1.2-2.1
MISF	Misfire Detection	ON / OFF	OFF	OFF	OFF
OCTADJ	Octane Adjustment	CL/OPEN	CLOSED	CLOSED	CLOSED
OSS	Output Speed Shaft	0-10K rpm	0	950-1100	1750-1900
PNP	PNP Switch	ON / OFF	ON (in 'P')	OFF	OFF
RPM	Engine Speed	0-10K rpm	790-815	925-1125	1320-1350
TCS	TCS (Switch)	ON / OFF	ON (if OD & TCS on)	OFF	OFF
TFT (°F)	TFT Sensor	-40-304°F	110-210	110-210	110-210
TPV	TP Sensor	0-5.1v	0.5-0.9	0.8-1.1	0.9-1.3
TRV/TR	TR Sensor	0-5.1v	0-PARK	1.7-OD	1.7-OD
VPWR	Vehicle Power	12-14v	12-14	12-14	12-14
VSS	Vehicle Speed (MPH & Hertz Rate)	0-159 mph	0 / 0	30 = 65 Hz	55 = 125 Hz
WACA	A/C WOT Cutout Relay Monitor	ON / OFF	ON (if AC is on)	OFF	OFF

1996-2005 Crown Victoria/Grand Marquis 4.6L V8 NGV VIN 9

PCM PID Acronym	Parameter Identification	PID Range	PID Value at Hot Idle	PID Value at 30	PID Value at 55
CD1-CD8	Coil Driver 1-8	0-45° dwell	8	10	12
CHTIL	Cylinder Head Temperature Lamp	ON / OFF	ON (if on)	OFF	OFF
CTO	Clean Tach Output	0-10 KHz	50-57	60-65	85-95
EGRVR	VR Solenoid	0-100%	0	0-40	0-50
EPC	EPC Solenoid	0-500 psi	20	12	30
FSV	Fuel Shutoff Valve	ON / OFF	ON	ON	ON
FuelPW1	INJ Pulsewidth B1	0-99.9 ms	3.9-4.6	4.7-12	4.7-12.2
FuelPW2	INJ Pulsewidth B2	0-99.9 ms	3.9-4.6	4.7-12	4.7-12.2
HTR-11	HO2S-11 Heater	ON / OFF	ON	ON	ON
HTR-12	HO2S-12 Heater	ON / OFF	ON	ON	ON
HTR-21	HO2S-21 Heater	ON / OFF	ON	ON	ON
HTR-22	HO2S-22 Heater	ON / OFF	ON	ON	ON
IAC	Idle Air Control	0-100%	32-36	30-55	40-60
INJ1	INJ 1 Pulsewidth	0-99.9 ms	3.9-4.6	4.7-12.0	4.7-12.2
INJ2	INJ 2 Pulsewidth	0-99.9 ms	3.9-4.6	4.7-12.0	4.7-12.2
INJ3	INJ 3 Pulsewidth	0-99.9 ms	3.9-4.6	4.7-12.0	4.7-12.2
INJ4	INJ 4 Pulsewidth	0-99.9 ms	3.9-4.6	4.7-12.0	4.7-12.2
INJ5	INJ 5 Pulsewidth	0-99.9 ms	3.9-4.6	4.7-12.0	4.7-12.2
INJ6	INJ 6 Pulsewidth	0-99.9 ms	3.9-4.6	4.7-12.0	4.7-12.2
INJ7	INJ 7 Pulsewidth	0-99.9 ms	3.9-4.6	4.7-12.0	4.7-12.2
INJ8	INJ 8 Pulsewidth	0-99.9 ms	3.9-4.6	4.7-12.0	4.7-12.2
LFC	Low Fan Control	ON / OFF	ON (if on)	OFF	OFF
LongFT1	Long Term F/T B1	-20 to 20%	-5 to 5%	-5 to 5%	-5 to 5%
LongFT2	Long Term F/T B2	-20 to 20%	-5 to 5%	-5 to 5%	-5 to 5%
MIL	MIL Control	ON / OFF	ON (if on)	OFF	OFF
SHTFT1	Short Term F/T B1	-20 to 20%	-5 to 5%	-5 to 5%	-5 to 5%
SHTFT2	Short Term F/T B2	-20 to 20%	-5 to 5%	-5 to 5%	-5 to 5%
SPARK	Spark Advance	-90° to 90°	5-10	20-31	20-28
SS1	Shift Solenoid 1	ON / OFF	ON	OFF	ON
SS2	Shift Solenoid 2	ON / OFF	OFF	ON	ON
TCC	TCC Solenoid	0-100%	0	0-50	80-100
TCIL	TCIL (lamp)	ON / OFF	ON (if on)	OFF	OFF
VREF	Voltage Reference	0-5.1v	4.9-5.1	4.9-5.1	4.9-5.1
WAC	A/C WOT Cutout Relay	ON / OFF	ON (if on)	OFF	OFF

ESCORT & TRACER PID DATA - OBD II

1996 Escort/Tracer 1.8L I4 MFI VIN 8 (All) Inputs / Outputs

PCM PID Acronym	Parameter Identification	PID Range	PID Value at Hot Idle	PID Value at 30	PID Value at 55
ACR	A/C Relay (On with A/C blower at 3-4)	ON / OFF	ON (if on)	OFF	OFF
ACS	A/C Relay (On with A/C blower at 3-4)	ON / OFF	ON (if on)	OFF	OFF
ATP	Atmospheric Press.	0-5.1v	3.9	3.9	3.9
BARO	BARO Sensor " Hg	0-30" Hg	29.4	29.4	29.4
BLMTSW	Blower Motor SW	ON / OFF	ON (if on)	OFF	OFF
DEF SW	Defrost Switch	ON / OFF	ON (if on)	OFF	OFF
DRL SW	Daytime Run Lamp	ON / OFF	ON (if on)	OFF	OFF
ECT (°F)	ECT Sensor	-40-304°F	185	180	192
ECT (V)	ECT Sensor	0-5.1v	0.7	0.7	0.6
HO2S-11	HO2S-11 (Bank 1)	0.1-1.1v	0.1-1.1	0.1-1.1	0.1-1.1
HO2S-12	HO2S-12 (Bank 1)	0.1-1.1v	0.1-1.1	0.1-1.1	0.1-1.1
HTR1	HO2S-11 Heater	ON / OFF	ON	ON	ON
IAT (°F)	IAT Sensor	-40-304°F	69	69	71
IAT (V)	IAT Sensor	0-5.1v	2.2	2.2	2.2
IDL SW	Throttle Idle Switch	ON / OFF	ON	OFF	OFF
IGN SW	Ignition Switch	ON-KOEC	OFF	OFF	OFF
INJ1 & 2	Fuel Injector B1, B2	0-99.9 ms	2.3	3.4	4.2
ISC	Idle Speed Control	0-99.9 ms	2.8	3.2	3.3
LOAD	Engine Load	0-100%	18	21	38
LongFT1	Long Term F/T B1	-20 to 20%	-5 to 5%	-5 to 5%	-5 to 5%
LUP	Lockup Solenoid	ON / OFF	OFF	OFF	OFF
MAF LB	MAF Sensor LB/M	0-25 LB/M	0.8	2.35	3.0
MAF (V)	MAF Sensor	0-5.1v	1.3	1.9	1.9
MIL	MIL Control	ON / OFF	ON (if on)	OFF	OFF
PRC	Pressure Regulator	ON / OFF	ON (if on)	OFF	OFF
PSP	PSP Switch	ON / OFF	ON turning	OFF	OFF
PUMP	Fuel Pump	ON / OFF	ON	ON	ON
Purge	EVAP Purge	0-100%	0	0-50	34-67
RPM	Engine Speed	0-10K rpm	765	1690	1850
RPMDES	Desired RPM	0-10K rpm	750	750	750
SOL1-2-3	Shift Solenoid 1-2-3	ON / OFF	OFF	ON	ON
SHTFT1	Short Term F/T B1	-20 to 20%	-5 to +5	-5 to +5	-5 to +5
SPARK	Spark Advance	-90° to 90°	10	11	20
STP SW	Stop Lamp Switch	ON / OFF	ON (if on)	OFF	OFF
TFT	TFT Sensor Volts	0-5.1v	4.0	3.9	3.1
TPV	TP Sensor	0-5.1v	0.5	0.7	0.8
NS SW	Neutral Start Switch	ON / OFF	ON (in 'P')	OFF	OFF
TSS	TSS Sensor	0-10 KHz	712	1575	1712
VICS	VICS Solenoid	ON / OFF	ON	ON	ON
VSS	Vehicle Speed	0-159 mph	0	30	55

1991-95 Escort/Tracer 1.9L I4 MFI VIN J (All) Inputs / Outputs

PCM PID Acronym	Parameter Identification	PID Range	PID Value at Hot Idle	PID Value at 30	PID Value at 55
BOO	Brake Switch (1993-95)	ON / OFF	ON (if on)	OFF	OFF
CANP	EVAP Purge Valve (1992-95)	0-100%	0	0-40	40-50
ECT (°F)	ECT Sensor	-40-304°F	160-200	160-200	160-200
ECT (V)	ECT Sensor	0-5.1v	0.6	0.6	0.6
EVR	VR Solenoid	0-100%	0%	0-40	39-50%
HFC	High Fan Control	ON / OFF	ON (if on)	ON	ON
INJPW1	INJ Pulsewidth B1	0-99.9 ms	4.8-5.0	7.8-8.4	10-11
IAC	Idle Air Control	0-100%	20-40	34-40	45-55
IAT (°F)	IAT Sensor	-40-304°F	50-120	50-120	50-120
IAT (V)	IAT Sensor	0-5.1v	1.7-3.5	1.7-3.5	1.7-3.5
LFC	Low Fan Control	ON / OFF	ON (if on)	OFF	OFF
LOOP	Loop Status	CL or OL	CL	CL	CL
MAF (V)	MAF Sensor	0-5.1v	0.7	1.3-1.5	1.8-2.2
HO2S-11	HO2S-11 (Bank 1)	0.1-1.1v	0.1-1.1	0.1-1.1	0.1-1.1
LongFT1	Long Term F/T B1	-20 to 20%	-5 to +5	-5 to +5	-5 to +5
PFE (V)	PFE Sensor	0-5.1v	3.2	3.3	3.4
PNP	PNP Switch	DRIVE or NEUTRAL	NEUTRAL	DRIVE	DRIVE
RPM	Engine Speed	0-10K rpm	750-820	1450-1630	1750-2100
SHTFT1	Short Term F/T B1	-20 to +20	-5 to +5	-5 to +5	-5 to +5
SPARK	Spark Advance	-90° to 90°	15-22	28-35	25-35
TP	TP Sensor	0-5.1v	0.7	0.8-1.0	1.1-1.3
TP Mode	TP Mode	C/T or P/T	C/T	P/T	P/T
VPWR	Vehicle Power	0-25.5v	12-14	12-14	12-14
VREF	Voltage Reference	0-5.1v	4.9-5.1	4.9-5.1	4.9-5.1
VSS	Vehicle Speed	0-159 mph	0	30	55
WAC	A/C WOT Cutout Relay	ON / OFF	ON (if on)	OFF	OFF

1996 Escort/Tracer 1.9L I4 MFI VIN J (A/T) - Inputs

PCM PID Acronym	Parameter Identification	PID Range	PID Value at Hot Idle	PID Value at 30	PID Value at 55
1ST	First Position	OFF / 1ST	1ST	1ST	1ST
2ND	Second Position	OFF / 2ND	2ND	2ND	2ND
ACCS	A/C Clutch Switch	ON / OFF	ON (if on)	OFF	OFF
ACPSW	AC Pressure Switch	0-5.1v	1.5-1.9v	0.6-0.8	0.6-0.8
BPP	Brake Pedal Switch	ON / OFF	ON (if on)	OFF	OFF
CID	CID Sensor	0-1 KHz	5-7	11-15	17-21
CKP	CKP Sensor	0-10 KHz	435-475	770-900	1200-1400
DRV	Drive Position	OFF / DR	OFF	DR	DR
DPFE	DPFE Sensor	0-5.1v	0.4-0.6	0.6-1.0v	1.1-3.8v
ECT (°F)	ECT Sensor	-40-304°F	160-200	160-200	160-200
FEPS	Flash EEPROM	0.1	0.1	0.1	0.1
FRP PSI	Fuel Pressure	29-32 psi	30-32	30-32	30-35
FPM	Fuel Pump Monitor	ON / OFF	ON	ON	ON
GEAR	Transmission Gear	P-R-N-D	P-R-N-D	DRIVE 3	DRIVE 4
HO2S-11	HO2S-11 (Bank 1)	0.1-1.1v	0.1-1.1	0.1-1.1	0.1-1.1
HO2S-12	HO2S-12 (Bank 1)	0.1-1.1v	0.1-1.1	0.1-1.1	0.1-1.1
IAT (°F)	IAT Sensor	-40-304°F	50-120	50-120	50-120
LOAD	Engine Load	0-100%	10-20	20-31%	36-52%
MAF (V)	MAF Sensor	0-5.1v	0.6-0.9	1.0-1.6v	1.3-1.9v
MISF	Misfire Detection	ON / OFF	OFF	OFF	OFF
OCTADJ	Octane Adjustment	CL/OPEN	CLOSED	CLOSED	CLOSED
PF (V)	Purge Flow Sensor	0-5.1v	1.55	2.10	2.30
PNP	PNP Switch	ON / OFF	ON (in 'P')	OFF	OFF
RPM	Engine Speed	0-10K rpm	760-820	1450-1630	1750-2100
REV	Reverse Position	OFF / REV	REV	OFF	OFF
TFT (°F)	TFT Sensor	-14-304°F	110-210	110-210	110-210
TPV	TP Sensor	0-5.1v	0.5-0.9	1.0-1.2	1.0-1.3
TSS	TSS Sensor	0-10 KHz	340-380	620-680	1090-1150
Vacuum	Engine Vacuum	0-30" Hg	19-20	18-20	8-15
VPWR	Vehicle Power	0-25.5v	12-14	12-14	12-14
VSS	Vehicle Speed (MPH & Hertz Rate)	0-159 mph	0 = 0 Hz	30 = 65 Hz	55 = 125 Hz

1996 Escort/Tracer 1.9L I4 MFI VIN J (A/T) - Outputs

PCM PID Acronym	Parameter Identification	PID Range	PID Value at Hot Idle	PID Value at 30	PID Value at 55
CD1	Coil 1 Driver	0-90° dwell	7	10	12
CD2	Coil 2 Driver	0-90° dwell	7	10	12
CTO	Clean Tach Output	0-10 KHz	25-38	40-48	55-85 Hz
EGRVR	VR Solenoid	0-100%	0	0-40	0-60
EVAPDC	EVAP Purge Valve	0-100%	0-100	0-100	0-100
FP	Fuel Pump Relay	ON / OFF	ON	ON	ON
FuelPW1	INJ Pulsewidth B1	0-99.9 ms	4.0-4.5	4.1-8.0	4.4-10
HFC	High Fan Control	ON / OFF	ON (if on)	OFF	OFF
HTR-11	HO2S-11 Heater	ON / OFF	ON	ON	ON
HTR-12	HO2S-12 Heater	ON / OFF	ON	ON	ON
IAC	Idle Air Control	0-100%	33	40-52	66
INJ1	INJ 1 Pulsewidth	0-99.9 ms	4.0-4.5	4.1-8.0	4.4-10
INJ2	INJ 2 Pulsewidth	0-99.9 ms	4.0-4.5	4.1-8.0	4.4-10
INJ3	INJ 3 Pulsewidth	0-99.9 ms	4.0-4.5	4.1-8.0	4.4-10
INJ4	INJ 4 Pulsewidth	0-99.9 ms	4.0-4.5	4.1-8.0	4.4-10
LFC	Low Fan Control	ON / OFF	ON (if on)	OFF	OFF
LongFT1	Long Term F/T B1	-20 to 20%	-5 to +5	-5 to +5	-5 to +5
MIL	MIL Control	ON / OFF	ON (if on)	OFF	OFF
SHTFT1	Short Term F/T B1	-20 to 20%	-5 to +5	-5 to +5	-5 to +5
SPARK	Spark Advance	-90° to 90°	15-22	28-35	25-35
SS1	Shift Solenoid 1	ON / OFF	OFF	ON	ON
SS2	Shift Solenoid 2	ON / OFF	ON	OFF	OFF
SS3	Shift Solenoid 3	ON / OFF	ON	ON	ON
TCC	TCC Solenoid	0-100%	0	55-100	95-100
VREF	Voltage Reference	0-5.1v	4.9-5.1	4.9-5.1	4.9-5.1
WAC	A/C WOT Cutout Relay	ON / OFF	ON (if on)	OFF	OFF

1996 Escort/Tracer 1.9L I4 MFI VIN J (M/T) - Inputs

PCM PID Acronym	Parameter Identification	PID Range	PID Value at Hot Idle	PID Value at 30	PID Value at 55
ACCS	A/C Clutch Switch	ON / OFF	ON (if on)	OFF	OFF
ACPSW	AC Pressure Switch	0-5.1v	1.5-1.9v	0.6-0.8	0.6-0.8
BPP	Brake Switch	ON / OFF	ON (if on)	OFF	OFF
CID	CID Sensor	0-1 KHz	5-7	11-15	17-21
CKP	CKP Sensor	0-10 KHz	435-475	770-900	1200-1400
CPP	CPP Switch	ON / OFF	ON (if in)	OFF	OFF
DPFE	DPFE Sensor	0-5.1v	0.4-0.6	0.6-1.0v	1.1-3.8v
ECT (°F)	ECT Sensor	-40-304°F	160-200	160-200	160-200
FEPS	Flash EEPROM	0-5.1v	0.1	0.1	0.1
FPM	Fuel Pump Monitor	ON / OFF	ON	ON	ON
FRP	Fuel Rail Pressure	0-100 psi	30-32	30-32	30-35
HO2S-11	HO2S-11 (Bank 1)	0.1-1.1v	0.1-1.1	0.1-1.1	0.1-1.1
HO2S-12	HO2S-12 (Bank 1)	0.1-1.1v	0.1-1.1	0.1-1.1	0.1-1.1
IAT (°F)	IAT Sensor	-40-304°F	50-120	50-120	50-120
LOAD	Engine Load	0-100%	10-20	20-31%	36-52%
MAF (V)	MAF Sensor	0-5.1v	0.6-0.9	1.0-1.6v	1.3-1.9v
MISF	Misfire Detection	OFF	OFF	OFF	OFF
OCTADJ	Octane Adjustment	CL or OP	CLOSED	CLOSED	CLOSED
PF (V)	Purge Flow Sensor	0-5.1v	1.55	2.10	2.30
RPM	Engine Speed	0-10K rpm	760-820	1450-1630	1750-2100
TPV	TP Sensor	0-5.1v	0.5-0.9	1.0-1.2	1.0-1.3
Vacuum	Engine Vacuum	0-30" Hg	19-20	18-20	8-15
VPWR	Vehicle Power	0-25.5v	12-14	12-14	12-14

1996 Escort/Tracer 1.9L I4 MFI VIN J (M/T) - Outputs

PCM PID Acronym	Parameter Identification	PID Range	PID Value at Hot Idle	PID Value at 30	PID Value at 55
CD1-CD2	Coil Driver 1-2	0-90° dwell	7	10	12
CTO	Clean Tach Output	0-10 KHz	25-38	40-48	55-85 Hz
EGRVR	VR Solenoid	0-100%	0%	0-40	0-60
EVAPDC	EVAP Purge Valve	0-100%	0-100%	0-100%	0-100%
FP	Fuel Pump Relay	ON / OFF	ON	ON	ON
FUELPW	INJ 1 Pulsewidth	0-99.9 ms	4.0-4.5	4.1-8.0	4.4-10
HFC	High Fan Control	ON / OFF	ON (if on)	OFF	OFF
HTR-11	HO2S-11 Heater	ON / OFF	ON	ON	ON
HTR-12	HO2S-12 Heater	ON / OFF	ON	ON	ON
IAC	Idle Air Control	0-100%	30%	40-52%	60-66%
INJ1-4	INJ 1-4 On-time	0-99.9 ms	4.0-4.5	4.1-8.0	4.4-10
LFC	Low Fan Control	ON / OFF	ON (if on)	OFF	OFF
LongFT1	LONGFT (Bank 1)	-20 to 20%	-5 to +5	-5 to +5	-5 to +5
MIL	MIL Control	ON / OFF	ON (if on)	OFF	OFF
SIL	Shift Indicator Lamp	ON / OFF	OFF	ON (if on)	OFF
SHTFT1	SHRTFT (Bank 1)	-20 to 20%	-5 to +5	-5 to +5	-5 to +5
SPARK	Spark Advance	-90° to 90°	15-22	28-35	25-35
WAC	A/C WOT Relay	ON / OFF	ON (if on)	OFF	OFF

1997-2002 Escort/Tracer 2.0L 2v I4 MFI VIN P (A/T) - Inputs

PCM PID Acronym	Parameter Identification	PID Range	PID Value at Hot Idle	PID Value at 30	PID Value at 55
ACCS	A/C Clutch Switch	ON / OFF	ON (if on)	OFF	OFF
ACP	AC Press. Sensor	0-5.1v	1.5-2.6	0.0-0.8	0.0-0.8
BPP	Brake Pedal Switch	ON / OFF	ON (if on)	OFF	OFF
CID	CID Sensor	0-1 KHz	5-7	11-15	17-21
CKP	CKP Sensor	0-10 KHz	400-425	770-900	1200-1400
DPFE	DPFE Sensor	0-5.1v	0.4-0.6	0.6-1.5	0.5-3.8
ECT (°F)	ECT Sensor	-14-304°F	160-200	160-200	160-200
EFTA (°F)	EFT 'A' Sensor	-14-304°F	50-120	50-120	50-120
FEPS	Flash EEPROM	0-5.1v	0.1	0.1	0.1
FLI	Fuel Level Indicator (1998-99)	0-100%	50 (1.7v)	25-75	25-75
FPM	Fuel Pump Monitor	ON / OFF	ON	ON	ON
FRP	Fuel Rail Pressure (1998-99)	0-100 psi	39	39	39
FTPV	Fuel Tank Pressure (1998-99)	0-5.1v or 0-10" H2O	2.6v / 0" H2O (with cap off)	2.6v / 0" H2O (with cap off)	2.6v / 0" H2O (with cap off)
GEAR	Gear Position	1-2-3-4	1	3	4
HO2S-11	HO2S-11 (Bank 1)	0.1-1.1v	0.1-1.1	0.1-1.1	0.1-1.1
HO2S-12	HO2S-12 (Bank 1)	0.1-1.1v	0.1-1.1	0.1-1.1	0.1-1.1
IAT (°F)	IAT Sensor	-14-304°F	50-120	50-120	50-120
IMRCM	IMRC Monitor	0-5.1v	5	5	5
KS1	Knock Sensor	0-5.1v	0	0	0
LOAD	Engine Load	0-100%	10-20	20-31	25-52
MAF (V)	MAF Sensor	0-5.1v	0.6-0.9	1.0-1.6	1.3-2.3
MISF	Misfire Detection	OFF	OFF	OFF	OFF
OCTADJ	Octane Adjustment	CL/OPEN	CLOSED	CLOSED	CLOSED
PF (V)	Purge Flow Sensor	0-5.1v	1.55	2.10	2.30
PNP	PNP Switch	ON / OFF	ON (in 'P')	OFF	OFF
PSP (V)	PSP Sensor	0-5.1v	1.5-1.8	0.5-0.8	0.5-0.8
RPM	Engine Speed	0-10K rpm	730-790	1450-1630	1750-2100
TFT (°F)	TFT Sensor	-14-304°F	110-210	110-210	110-210
TPV	TP Sensor	0-5.1v	0.5-0.9	1.0-1.3	1.1-1.9
TRD/TR	TR Sensor	N / OD	N	OD	OD
TRL/TR	TR Sensor Low	MAN1 /OD	MAN1	OD	OD
TROD/TR	TR Overdrive	N / OD	OD	OD	OD
TRR/TR	TR Reverse	R / OD	R	OD	OD
TSS	TSS Sensor	0-10 KHz	340-380	620-680	1090-1250
Vacuum	Engine Vacuum	0-30" Hg	19-20	18-20	8-15
VPWR	Vehicle Power	0-25.5v	12-14	12-14	12-14
VSS	Vehicle Speed (MPH & Hertz Rate)	0-159 mph	0 = 0 Hz	30 = 65 Hz	55 = 125 Hz

1997-2002 Escort/Tracer 2.0L 2v I4 MFI VIN P (A/T) - Outputs

PCM PID Acronym	Parameter Identification	PID Range	PID Value at Hot Idle	PID Value at 30	PID Value at 55
CD1-CD2	Coil Driver 1-2	0-90° dwell	7	10	11
CTO	Clean Tach Output	0-10 KHz	25-38	40-48	72-85
EGRVR	VR Solenoid	0-100%	0	0-40	0-60
EVAPCV	EVAP CV Valve	0-100%	0-100	0-100	0-100
EVAPDC	EVAP Purge Valve	0-100%	0-100	0-100	0-100
FP	Fuel Pump Relay	0-100%	100	100	100
FuelPW1	INJ Pulsewidth B1	0-99.9 ms	3.3-3.7	4.1-8.0	4.4-10
HFC	High Fan Relay	ON / OFF	ON (if on)	OFF	OFF
HTR-11	HO2S-11 Heater	ON / OFF	ON	ON	ON
HTR-12	HO2S-12 Heater	ON / OFF	ON	ON	ON
IAC	Idle Air Control	0-100%	20-40	34-40	45-55
IMRC	IMRC Solenoid	ON / OFF	OFF	OFF	OFF
INJ1	INJ 1 Pulsewidth	0-99.9 ms	3.3-3.7	4.1-8.0	4.4-10
INJ2	INJ 2 Pulsewidth	0-99.9 ms	3.3-3.7	4.1-8.0	4.4-10
INJ3	INJ 3 Pulsewidth	0-99.9 ms	3.3-3.7	4.1-8.0	4.4-10
INJ4	INJ 4 Pulsewidth	0-99.9 ms	3.3-3.7	4.1-8.0	4.4-10
LFC	Low Fan Relay	ON / OFF	ON (if on)	OFF	OFF
LongFT1	Long Term F/T B1	-20 to 20%	-5 to +5%	-5 to +5%	-5 to +5%
MIL	MIL Control	ON / OFF	ON (if on)	OFF	OFF
SS1	Shift Solenoid 1	ON / OFF	ON	OFF	OFF
SS2	Shift Solenoid 2	ON / OFF	OFF	ON	ON
SS3	Shift Solenoid 3	0-25.5v	7-9.5v	8.3-9.5	8.3-9.5
SHTFT1	Short Term F/T B1	-20 to 20%	-5 to +5	-5 to +5	-5 to +5
SPARK	Spark Advance	-90° to 90°	15-22	28-35	25-35
TCC	TCC Solenoid	0-100%	0	55-100	95-100
VREF	Voltage Reference	0-5.1v	4.9-5.1	4.9-5.1	4.9-5.1
WAC	A/C WOT Cutout Relay	ON / OFF	ON (if on)	OFF	OFF

1997-2002 Escort/Tracer 2.0L 2v I4 MFI VIN P (M/T) - Inputs

PCM PID Acronym	Parameter Identification	PID Range	PID Value at Hot Idle	PID Value at 30	PID Value at 55
ACCS	A/C Clutch Switch	ON / OFF	ON (if on)	OFF	OFF
ACP	A/C Press. Switch	0-5.1v	1.5-2.6	0.0-0.8	0.0-0.8
CID	CID Sensor	0-1 KHz	5-7	11-15	17-21
CKP	CKP Sensor	0-10 KHz	400-425	770-900	1200-1400
CPP	Clutch Pedal Switch	ON / OFF	ON (if in)	OFF	OFF
DPFE	DPFE Sensor	0-5.1v	0.4-0.6	0.6-1.5	0.5-3.8
ECT (°F)	ECT Sensor	-40-304°F	160-200	160-200	160-200
FEPS	Flash EEPROM	0-5.1v	0.1	0.1	0.1
FLI	Fuel Level Indicator (1998-99)	0-100%	50 (1.7v)	25-75	25-75
FPM	Fuel Pump Monitor	ON / OFF	ON	ON	ON
FRP (V)	Fuel Rail Pressure (1998-99)	0-5.1v or 0-100 psi	2.8v / 39	2.8v / 39	2.8v / 39
FTPV	Fuel Tank Pressure (1998-99)	0-5.1v or 0-10" H2O	2.6v / 0" H2O (with cap off)	2.6v / 0" H2O (with cap off)	2.6v / 0" H2O (with cap off)
HO2S-11	HO2S-11 (Bank 1)	0.1-1.1v	0.1-1.1	0.1-1.1	0.1-1.1
HO2S-12	HO2S-12 (Bank 1)	0.1-1.1v	0.1-1.1	0.1-1.1	0.1-1.1
IAT (°F)	IAT Sensor	-40-304°F	50-120	50-120	50-120
LOAD	Engine Load	0-100%	10-20	20-31	25-52
IMRCM	IMRC Monitor	0-5.1v	5	5	5
MAF (V)	MAF Sensor	0-5.1v	0.6-0.9	1.0-1.6	1.3-2.3
MISF	Misfire Detection	OFF	OFF	OFF	OFF
OCT ADJ	Octane Adjustment	CL/OPEN	CLOSED	CLOSED	CLOSED
PF (V)	Purge Flow Sensor (1997 only)	0-5.1v	1.55	2.10	2.30
PSPV	PSP Switch	0-5.1v	1.5-1.8	0.5-0.8	0.5-0.8
RPM	Engine Speed	0-10K rpm	750-820	1450-1630	1750-2100
TPV	TP Sensor	0-5.1v	0.5-0.9	1.0-1.3	1.1-1.9
Vacuum	Engine Vacuum	0-30" Hg	19-20	18-20	8-15
VPWR	Vehicle Power	0-25.5v	12-14	12-14	12-14
VSS	Vehicle Speed (MPH & Hertz Rate)	0-159 mph	0 = 0 Hz	30 = 65 Hz	55 = 125 Hz

1997-2002 Escort/Tracer 2.0L 2v I4 MFI VIN P (M/T) - Outputs

PCM PID Acronym	Parameter Identification	PID Range	PID Value at Hot Idle	PID Value at 30	PID Value at 55
CD1	Coil 1 Driver	0-90° dwell	7	10	11
CD2	Coil 2 Driver	0-90° dwell	7	10	11
CTO	Clean Tach Output	0-10 KHz	25-38	40-48	72-85
EGRVR	VR Solenoid	0-100%	0	0-40	0-60
EVAPCV	EVAP CV Valve	0-100%	0-100	0-100	0-100
EVAPDC	EVAP Purge Valve	0-100%	0-100	0-100	0-100
FP	Fuel Pump Relay	0-100%	100	100	100
FuelPW1	INJ Pulsewidth B1	0-99.9 ms	3.3-3.7	4.1-8.0	4.4-10
HFC	High Fan Relay	ON / OFF	ON (if on)	OFF	OFF
HTR-11	HO2S-11 Heater	ON / OFF	ON	ON	ON
HTR-12	HO2S-12 Heater	ON / OFF	ON	ON	ON
IAC	Idle Air Control	0-100%	20-40	34-40	45-55
INJ1	INJ 1 Pulsewidth	0-99.9 ms	3.3-3.7	4.1-8.0	4.4-10
INJ2	INJ 2 Pulsewidth	0-99.9 ms	3.3-3.7	4.1-8.0	4.4-10
INJ3	INJ 3 Pulsewidth	0-99.9 ms	3.3-3.7	4.1-8.0	4.4-10
INJ4	INJ 4 Pulsewidth	0-99.9 ms	3.3-3.7	4.1-8.0	4.4-10
IMRC	IMRC Solenoid	ON / OFF	OFF	OFF	OFF
LFC	Low Fan Relay	ON / OFF	ON (if on)	OFF	OFF
LongFT1	Long Term F/T B1	-20 to 20%	-5 to +5%	-5 to +5%	-5 to +5%
MIL	MIL Control	ON / OFF	ON (if on)	OFF	OFF
SHTFT1	Short Term F/T B1	-20 to 20%	-5 to +5	-5 to +5	-5 to +5
SPARK	Spark Advance	-90° to 90°	15-22	28-35	25-35
VREF	Voltage Reference	0-5.1v	4.9-5.1	4.9-5.1	4.9-5.1
WAC	A/C WOT Cutout Relay	ON / OFF	ON (if on)	OFF	OFF

1998-2003 Escort/Tracer/ZX2 2.0L 4v I4 MFI VIN 3 (A/T) - Inputs

PCM PID Acronym	Parameter Identification	PID Range	PID Value at Hot Idle	PID Value at 30	PID Value at 55
ACCS	A/C Clutch Switch	ON / OFF	ON (if on)	OFF	OFF
ACP	A/C Press. Sensor	CLOSED or OPEN	OPEN	OPEN	OPEN
BPP	Brake Pedal Switch	ON / OFF	ON (if on)	OFF	OFF
CID	CID Sensor	0-1 KHz	5-7	11-15	17-21
CKP	CKP Sensor	0-10 KHz	400-425	770-900	1200-1400
DPFE	DPFE Sensor	0-5.1v	0.4-0.6	0.6-1.5	0.5-3.8
ECT (°F)	ECT Sensor	-40-304°F	160-200	160-200	160-200
EFTA (°F)	EFT 'A' Sensor	-40-304°F	50-120	50-120	50-120
FEPS	Flash EEPROM	0-5.1v	0.5-0.6	0.5-0.6	0.5-0.6
FLI	Fuel Level Indicator (1998-99)	0-100%	50 (1.7v)	50 (1.7v)	50 (1.7v)
FPM	Fuel Pump Monitor	ON / OFF	ON	ON	ON
FRP (V)	Fuel Rail Pressure (1998-99)	0-100 psi	39	39	39
FTPV	Fuel Tank Pressure (1998-99)	0-5.1v or 0-10" H2O	2.6v / 0" H2O (with cap off)	2.6v / 0" H2O (with cap off)	2.6v / 0" H2O (with cap off)
GEAR	Gear Position	1-2-3-4	1	3	4
HO2S-11	HO2S-11 (Bank 1)	0.1-1.1v	0.1-1.1	0.1-1.1	0.1-1.1
HO2S-12	HO2S-12 (Bank 1)	0.1-1.1v	0.1-1.1	0.1-1.1	0.1-1.1
IAT (°F)	IAT Sensor	-40-304°F	50-120	50-120	50-120
KS1	Knock Sensor	0-5.1v	0v	0v	0v
LOAD	Engine Load	0-100%	10-20	20-31	25-52
MAF (V)	MAF Sensor	0-5.1v	0.6-0.9	1.0-1.6	1.3-2.3
MISF	Misfire Detection	ON / OFF	OFF	OFF	OFF
OCTADJ	Octane Adjustment	CL/OPEN	CLOSED	CLOSED	CLOSED
PF (V)	Purge Flow Sensor (1997 only)	0-5.1v	1.55	2.10	2.30
PNP	PNP Switch	ON / OFF	ON	OFF	OFF
PSP (V)	PSP Sensor	0-5.1v	1.5-1.8	0.5-0.8	0.5-0.8
RPM	Engine Speed	0-10K rpm	750-820	1450-1630	1750-2100
TFT (°F)	TFT Sensor	-40-304°F	110-210	110-210	110-210
TPV	TP Sensor	0-5.1v	0.5-0.9	1.0-1.3	1.1-1.9
TRD/TR	TR Sensor	N / OD	N	OD	OD
TRL/TR	TR Sensor Low	MAN1 /OD	MAN1	OD	OD
TROD/TR	TR Overdrive	N / OD	OD	OD	OD
TRR/TR	TR Reverse	R / OD	R	OD	OD
TSS	TSS Sensor	0-10 KHz	340-380	620-680	1090-1250
VPWR	Vehicle Power	0-25.5V	12-14	12-14	12-14
VSS	Vehicle Speed (MPH & Hertz Rate)	0-159 mph	0 = 0 Hz	30 = 65 Hz	55 = 125 Hz

1998-2003 Escort/Tracer/ZX2 2.0L 4v I4 MFI VIN 3 (A/T) - Outputs

PCM PID Acronym	Parameter Identification	PID Range	PID Value at Hot Idle	PID Value at 30	PID Value at 55
CD1-CD2	Coil Driver 1-2	0-90° dwell	7	10	11
CTO	Clean Tach Output	0-10 KHz	25-38	40-48	72-85
EGRVR	VR Solenoid	0-100%	0	0-40	0-60
EVAPCV	EVAP CV Valve	0-100%	0-100	0-100	0-100
EVAPDC	EVAP Purge Valve	0-100%	0-100	0-100	0-100
FP	Fuel Pump Relay	0-100%	100	100	100
FuelPW1	INJ Pulsewidth B1	0-99.9 ms	3.3-3.7	4.1-8.0	4.4-10
HFC	High Fan Relay	ON / OFF	ON (if on)	OFF	OFF
HTR-11	HO2S-11 Heater	ON / OFF	ON	ON	ON
HTR-12	HO2S-12 Heater	ON / OFF	ON	ON	ON
IAC	Idle Air Control	0-100%	20-40	34-40	45-55
INJ1	INJ 1 Pulsewidth	0-99.9 ms	3.3-3.7	4.1-8.0	4.4-10
INJ2	INJ 2 Pulsewidth	0-99.9 ms	3.3-3.7	4.1-8.0	4.4-10
INJ3	INJ 3 Pulsewidth	0-99.9 ms	3.3-3.7	4.1-8.0	4.4-10
INJ4	INJ 4 Pulsewidth	0-99.9 ms	3.3-3.7	4.1-8.0	4.4-10
LFC	Low Fan Relay	ON / OFF	ON (if on)	OFF	OFF
LongFT1	Long Term F/T B1	-20 to 20%	-5 to +5%	-5 to +5%	-5 to +5%
MIL	MIL Control	ON / OFF	OFF	OFF	OFF
SHTFT1	Short Term F/T B1	-20 to 20%	-5 to +5	-5 to +5	-5 to +5
SPARK	Spark Advance	-90° to 90°	15-22	28-35	25-35
SS1	Shift Solenoid 1	ON / OFF	ON	OFF	OFF
SS2	Shift Solenoid 2	ON / OFF	OFF	ON	ON
SS3	Shift Solenoid 3	0-25.5v	7.0-9.5	8.3-9.5	8.3-9.5
TCC	TCC Solenoid	0-100%	0	55-100	95-100
VCT	Variable Cam Timing Solenoid	0-25.5v	12-14	10.5-14	10.5-14
VREF	Voltage Reference	0-5.1v	4.9-5.1	4.9-5.1	4.9-5.1
WAC	A/C WOT Cutout Relay	ON / OFF	ON (if on)	OFF	OFF

1998-2003 Escort/Tracer/ZX2 2.0L 4v I4 MFI VIN 3 (M/T) - Inputs

PCM PID Acronym	Parameter Identification	PID Range	PID Value at Hot Idle	PID Value at 30	PID Value at 55
ACCS	A/C Clutch Switch	ON / OFF	ON (if on)	OFF	OFF
ACP	A/C Press. Switch	CL/OPEN	OPEN	OPEN	OPEN
CID	CID Sensor	0-1 KHz	5-7	11-15	17-21
CKP	CKP Sensor	0-10 KHz	400-425	770-900	1200-1400
CPP	Clutch Pedal Switch	ON / OFF	ON (if in)	OFF	OFF
DPFE	DPFE Sensor	0-5.1v	0.4-0.6	0.6-1.5	0.5-3.8
ECT (°F)	ECT Sensor	-40-304°F	160-200	160-200	160-200
FEPS	Flash EEPROM	0-5.1v	0.5-0.6	0.5-0.6	0.5-0.6
FLI	Fuel Level Indicator (1998-99)	0-100%	50 (1.7v)	50 (1.7v)	50 (1.7v)
FPM	Fuel Pump Monitor	ON / OFF	ON	ON	ON
FRP (V)	Fuel Rail Pressure (1998-99)	0-100 psi	39	39	39
FTPV	Fuel Tank Pressure (1998-99)	0-5.1v or 0-10" H2O	2.6v / 0" H2O (with cap off)	2.6v / 0" H2O (with cap off)	2.6v / 0" H2O (with cap off)
HO2S-11	HO2S-11 (Bank 1)	0.1-1.1v	0.1-1.1	0.1-1.1	0.1-1.1
HO2S-12	HO2S-12 (Bank 1)	0.1-1.1v	0.1-1.1	0.1-1.1	0.1-1.1
IAT (°F)	IAT Sensor	-40-304°F	50-120	50-120	50-120
LOAD	Engine Load	0-100%	10-20	20-31	25-52
MAF (V)	MAF Sensor	0-5.1v	0.6-0.9	1.0-1.6	1.3-2.3
MISF	Misfire Detection	ON / OFF	OFF	OFF	OFF
OCT ADJ	Octane Adjustment	CL/OPEN	CLOSED	CLOSED	CLOSED
PF (V)	Purge Flow Sensor (1997 only)	0-5.1v	1.55	2.10	2.30
PSP	PSP Switch	0-5.1v	1.5-1.8	0.5-0.8	0.5-0.8
RPM	Engine Speed	0-10K rpm	750-820	1450-1630	1750-2100
TPV	TP Sensor	0-5.1v	0.5-0.9	1.0-1.3	1.1-1.9
VPWR	Vehicle Power	0-25.5V	12-14	12-14	12-14
VSS	Vehicle Speed (MPH & Hertz Rate)	0-159 mph	0 = 0 Hz	30 = 65 Hz	55 = 125 Hz

1998-2003 Escort/Tracer/ZX2 2.0L 4v I4 MFI VIN 3 (M/T) - Outputs

PCM PID Acronym	Parameter Identification	PID Range	PID Value at Hot Idle	PID Value at 30	PID Value at 55
CD1	Coil 1 Driver	0-90° dwell	7	10	11
CD2	Coil 2 Driver	0-90° dwell	7	10	11
CTO	Clean Tach Output	0-10 KHz	25-38	40-48	72-85
EGRVR	VR Solenoid	0-100%	0	0-40	0-60
EVAPCV	EVAP CV Valve	0-100%	0	0	0
EVAPDC	EVAP Purge Valve	0-100%	0-100	0-100	0-100
FP	Fuel Pump Relay	0-100%	100	100	100
FuelPW1	INJ Pulsewidth B1	0-99.9 ms	3.3-3.7	4.1-8.0	4.4-10
HFC	High Fan Relay	ON / OFF	ON (if on)	OFF	OFF
HTR-11	HO2S-11 Heater	ON / OFF	ON	ON	ON
HTR-12	HO2S-12 Heater	ON / OFF	ON	ON	ON
IAC	Idle Air Control	0-100%	20-40	34-40	45-55
INJ1	INJ 1 Pulsewidth	0-99.9 ms	3.3-3.7	4.1-8.0	4.4-10
INJ2	INJ 2 Pulsewidth	0-99.9 ms	3.3-3.7	4.1-8.0	4.4-10
INJ3	INJ 3 Pulsewidth	0-99.9 ms	3.3-3.7	4.1-8.0	4.4-10
INJ4	INJ 4 Pulsewidth	0-99.9 ms	3.3-3.7	4.1-8.0	4.4-10
LFC	Low Fan Relay	ON / OFF	ON (if on)	OFF	OFF
LongFT1	Long Term F/T B1	-20 to 20%	-5 to +5	-5 to +5	-5 to +5
MIL	MIL Control	ON / OFF	ON (if on)	OFF	OFF
SHTFT1	Short Term F/T B1	-20 to 20%	-5 to +5	-5 to +5	-5 to +5
SPARK	Spark Advance	-90° to 90°	15-22	28-35	25-35
VCT	Variable Cam Timing Solenoid	0-25.5v	12-14	10.5-14	10.5-14
VREF	Voltage Reference	0-5.1v	4.9-5.1	4.9-5.1	4.9-5.1
WAC	A/C WOT Cutout Relay	ON / OFF	ON (if on)	OFF	OFF

FIVE HUNDRED & FREESTYLE PID DATA

2005 Five Hundred, Freestyle 3.0L 4v V6 MFI VIN 1 - Inputs

PCM PID Acronym	Parameter Identification	PID Range	PID Value at Hot Idle	PID Value at 30	PID Value at 55
AIR-M	AIR Monitor	ON / OFF	OFF	OFF	OFF
ACP	A/C Press. Switch	CL/OPEN	OPEN	OPEN	OPEN
AFS	AFS Signal Hertz	0-1 KHz	130	130	130
BPP	BPP Switch	ON / OFF	ON (if on)	OFF	OFF
BPS	Brake Pedal Switch	0-25.5v	0.1 (if on)	12-14	12-14
CHT (°F)	CHT Sensor (°F)	-40-304°F	160-200	160-200	160-200
CID	CID Sensor	0-1 KHz	6.6	10	17
CKP	CKP Sensor	0-10 KHz	435	780-710	1170-1180
CSTT	Clutch Switch	0-25.5v	12v (if in)	0.1	0.1
DPFE	DPFE input (99-'00)	0-5.1v	0.2-1.3	0.2-4.5	0.2-4.5
DPFE	DPFE input (01-'02)	0-5.1v	0.95-1.05	0.95-4.65	0.95-4.65
ECT (°F)	ECT Sensor	-40-304°F	160-200	160-200	160-200
EFTA (°F)	EFT Sensor (°F)	-40-304°F	50-120	50-120	50-120
FEPS	Flash EEPROM	0-5.1v	0.1	0.1	0.1
FLI	Fuel Level Indicator	0-100%	50 (1.7v)	25-75	25-75
FP-M	Fuel Pump Monitor	0-100%	50	50	50
FRP	Fuel Rail Pressure	0-100 psi	39	39	39
FTP	FTP Sensor	0-5.1v or 0-10" H20	2.6v at 0" of H20	2.6v at 0" of H20	2.6v at 0" of H20
GEAR	Transmission Gear	1-2-3-4-5	1	4	5
HO2S-11	HO2S-11 (Bank 1)	0.1-1.1v	0.1-1.1	0.1-1.1	0.1-1.1
HO2S-12	HO2S-12 (Bank 1)	0.1-1.1v	0.1-1.1	0.1-1.1	0.1-1.1
HO2S-21	HO2S-21 (Bank 2)	0.1-1.1v	0.1-1.1	0.1-1.1	0.1-1.1
HO2S-22	HO2S-22 (Bank 2)	0.1-1.1v	0.1-1.1	0.1-1.1	0.1-1.1
IAT (°F)	IAT Sensor	-40-304°F	50-120	50-120	50-120
KS1	Knock Sensor B1	0-5.1v	0	0	0
KS2	Knock Sensor B2	0-5.1v	0	0	0
LOAD	Engine Load	0-100%	18.6	35	40
MAF (V)	MAF Sensor	0-5.1v	0.7	1.6-1.8	2.1-2.3
OCTADJ	Octane Adjustment	CLOSED	CLOSED	CLOSED	CLOSED
OSS	OSS Sensor	0-10 KHz	0 rpm	1500	2660
PSP	PSP Switch	HI / LOW	HI: turning	LOW	LOW
PS1	PS1 Switch	0-25.5v	11.7	11.7	11.7
REV	Reverse Signal	0-25.5v	0.1 (in 'R')	12-14	12-14
SCCS	S/C Switch	0-5.1v	0.1 (if on)	4.6	4.6
TCS	TCS (switch)	ON / OFF	ON (if on)	OFF	OFF
TSS	TSS Sensor	0-10 KHz	680	1080	2060
TFT (°F)	TFT Sensor	-40-304°F	110-210	110-210	110-210
TPV	TP Sensor	0-5.1v	0.5-0.9	1.1v	1.3v
TR1	TR Sensor 1 Volts	0-12v	0	10.7	10.7
TR2	TR2 Sensor	0-12v	0	10.7	10.7
TR4	TR4 Sensor	0-12v	0	10.7	10.7
TRV/TR	TR Sensor	0-5.1v	0v / PN	1.7 / OD	1.7 / OD
WACA	A/C WOT Relay	ON / OFF	ON (if on)	OFF	OFF

2005 Five Hundred, Freestyle 3.0L 4v V6 MFI VIN 1 - Outputs

PCM PID Acronym	Parameter Identification	PID Range	PID Value at Hot Idle	PID Value at 30	PID Value at 55
AIR	AIR Solenoid	ON / OFF	OFF	OFF	OFF
ALDFDC	Alternator Control	0-1 KHz	0-130	0 -10	0-10
CD1-CD6	Coil Driver 1-6	0-60° dwell	7	10	11
CHTIL	CHTIL (lamp)	ON / OFF	ON (if on)	OFF	OFF
CTO	Clean Tach Output	0-10 KHz	39	89	100
EGRVR	VR Solenoid	0-100%	0	0-40	55
EPC	EPC Solenoid	0-25.5v	9.3	10.4	10.5
EPC2	EPC2 Solenoid PSI	0-25.5v	11.1	10.4	10.5
EPC3	EPC3 Solenoid PSI	0-25.5v	7.5	12-14	12-14
EVAPCV	EVAP CV Valve	0-100%	0-100	0-100	0-100
EVAPDC	EVAP Purge Valve	0-100%	0-100	0-100	0-100
FP	Fuel Pump Control	0-100%	25	25	25
HTR-11	HO2S-11 Heater	ON / OFF	ON	ON	ON
HTR-12	HO2S-12 Heater	ON / OFF	ON	ON	ON
HTR-21	HO2S-21 Heater	ON / OFF	ON	ON	ON
HTR-22	HO2S-22 Heater	ON / OFF	ON	ON	ON
IAC	Idle Air Control	0-100%	34	53	67
INJ1	INJ 1 Pulsewidth	0-99.9 ms	3.2	4.9	6.7-7.1
INJ2	INJ 2 Pulsewidth	0-99.9 ms	3.2	4.9	6.7-7.1
INJ3	INJ 3 Pulsewidth	0-99.9 ms	3.2	4.9	6.7-7.1
INJ4	INJ 4 Pulsewidth	0-99.9 ms	3.2	4.9	6.7-7.1
INJ5	INJ 5 Pulsewidth	0-99.9 ms	3.2	4.9	6.7-7.1
INJ6	INJ 6 Pulsewidth	0-99.9 ms	3.2	4.9	6.7-7.1
LongFT1	Long Term F/T B1	-20 to 20%	-5 to +5	-5 to +5	-5 to +5
LongFT2	Long Term F/T B2	-20 to 20%	-5 to +5	-5 to +5	-5 to +5
MIL	MIL Control	ON / OFF	ON (if on)	OFF	OFF
SCC	S/C C Signal	0-25.5v	12-14	12-14	12-14
SCMA	S/C MA Signal	0-25.5v	12-14	12-14	12-14
SCMB	S/C MB Signal	0-25.5v	12-14	12-14	12-14
SCMC	S/C MC Signal	0-25.5v	12-14	12-14	12-14
SHTFT1	Short Term F/T B1	-20 to 20%	-5 to +5	-5 to +5	-5 to +5
SHTFT2	Short Term F/T B2	-20 to 20%	-5 to +5	-5 to +5	-5 to +5
SPARK	Spark Advance	-90° to 90°	12-17	34	40
S1	Shift Solenoid 1	ON / OFF	ON	OFF	ON
S2	Shift Solenoid 2	ON / OFF	OFF	OFF	OFF
S3	Shift Solenoid 3	ON / OFF	OFF	ON	ON
SS4	Shift Solenoid 4	ON / OFF	ON	ON	ON
TCC	TCC Solenoid	0-100%	0	0	100
VREF	Voltage Reference	0-5.1v	4.9-5.1	4.9-5.1	4.9-5.1
WAC	WOT Cutout Relay	ON / OFF	ON (if on)	OFF	OFF

MONTEGO PID DATA
2005 Montego 3.0L 4v V6 MFI VIN 1 - Inputs

PCM PID Acronym	Parameter Identification	PID Range	PID Value at Hot Idle	PID Value at 30	PID Value at 55
AIR-M	AIR Monitor	ON / OFF	OFF	OFF	OFF
ACP	A/C Press. Switch	CL/OPEN	OPEN	OPEN	OPEN
AFS	AFS Signal Hertz	0-1 KHz	130	130	130
BPP	BPP Switch	ON / OFF	ON (if on)	OFF	OFF
BPS	Brake Pedal Switch	0-25.5v	0.1 (if on)	12-14	12-14
CHT (°F)	CHT Sensor (°F)	-40-304°F	160-200	160-200	160-200
CID	CID Sensor	0-1 KHz	6.6	10	17
CKP	CKP Sensor	0-10 KHz	435	780-710	1170-1180
CSTT	Clutch Switch	0-25.5v	12v (if in)	0.1	0.1
DPFE	DPFE input (99-'00)	0-5.1v	0.2-1.3	0.2-4.5	0.2-4.5
DPFE	DPFE input (01-'02)	0-5.1v	0.95-1.05	0.95-4.65	0.95-4.65
ECT (°F)	ECT Sensor	-40-304°F	160-200	160-200	160-200
EFTA (°F)	EFT Sensor (°F)	-40-304°F	50-120	50-120	50-120
FEPS	Flash EEPROM	0-5.1v	0.1	0.1	0.1
FLI	Fuel Level Indicator	0-100%	50 (1.7v)	25-75	25-75
FP-M	Fuel Pump Monitor	0-100%	50	50	50
FRP	Fuel Rail Pressure	0-100 psi	39	39	39
FTP	FTP Sensor	0-5.1v or 0-10" H20	2.6v at 0" of H20	2.6v at 0" of H20	2.6v at 0" of H20
GEAR	Transmission Gear	1-2-3-4-5	1	4	5
HO2S-11	HO2S-11 (Bank 1)	0.1-1.1v	0.1-1.1	0.1-1.1	0.1-1.1
HO2S-12	HO2S-12 (Bank 1)	0.1-1.1v	0.1-1.1	0.1-1.1	0.1-1.1
HO2S-21	HO2S-21 (Bank 2)	0.1-1.1v	0.1-1.1	0.1-1.1	0.1-1.1
HO2S-22	HO2S-22 (Bank 2)	0.1-1.1v	0.1-1.1	0.1-1.1	0.1-1.1
IAT (°F)	IAT Sensor	-40-304°F	50-120	50-120	50-120
KS1	Knock Sensor B1	0-5.1v	0	0	0
KS2	Knock Sensor B2	0-5.1v	0	0	0
LOAD	Engine Load	0-100%	18.6	35	40
MAF (V)	MAF Sensor	0-5.1v	0.7	1.6-1.8	2.1-2.3
OCTADJ	Octane Adjustment	CLOSED	CLOSED	CLOSED	CLOSED
OSS	OSS Sensor	0-10 KHz	0 rpm	1500	2660
PSP	PSP Switch	HI / LOW	HI: turning	LOW	LOW
PS1	PS1 Switch	0-25.5v	11.7	11.7	11.7
REV	Reverse Signal	0-25.5v	0.1 (in 'R')	12-14	12-14
SCCS	S/C Switch	0-5.1v	0.1 (if on)	4.6	4.6
TCS	TCS (switch)	ON / OFF	ON (if on)	OFF	OFF
TSS	TSS Sensor	0-10 KHz	680	1080	2060
TFT (°F)	TFT Sensor	-40-304°F	110-210	110-210	110-210
TPV	TP Sensor	0-5.1v	0.5-0.9	1.1v	1.3v
TR1	TR Sensor 1 Volts	0-12v	0	10.7	10.7
TR2	TR2 Sensor	0-12v	0	10.7	10.7
TR4	TR4 Sensor	0-12v	0	10.7	10.7
TRV/TR	TR Sensor	0-5.1v	0v / PN	1.7 / OD	1.7 / OD
WACA	A/C WOT Relay	ON / OFF	ON (if on)	OFF	OFF

2005 Montego 3.0L 4v V6 MFI VIN 1 - Outputs

PCM PID Acronym	Parameter Identification	PID Range	PID Value at Hot Idle	PID Value at 30	PID Value at 55
AIR	AIR Solenoid	ON / OFF	OFF	OFF	OFF
ALDFDC	Alternator Control	0-1 KHz	0-130	0 -10	0-10
CD1-CD6	Coil Driver 1-6	0-60° dwell	7	10	11
CHTIL	CHTIL (lamp)	ON / OFF	ON (if on)	OFF	OFF
CTO	Clean Tach Output	0-10 KHz	39	89	100
EGRVR	VR Solenoid	0-100%	0	0-40	55
EPC	EPC Solenoid	0-25.5v	9.3	10.4	10.5
EPC2	EPC2 Solenoid PSI	0-25.5v	11.1	10.4	10.5
EPC3	EPC3 Solenoid PSI	0-25.5v	7.5	12-14	12-14
EVAPCV	EVAP CV Valve	0-100%	0-100	0-100	0-100
EVAPDC	EVAP Purge Valve	0-100%	0-100	0-100	0-100
FP	Fuel Pump Control	0-100%	25	25	25
HTR-11	HO2S-11 Heater	ON / OFF	ON	ON	ON
HTR-12	HO2S-12 Heater	ON / OFF	ON	ON	ON
HTR-21	HO2S-21 Heater	ON / OFF	ON	ON	ON
HTR-22	HO2S-22 Heater	ON / OFF	ON	ON	ON
IAC	Idle Air Control	0-100%	34	53	67
INJ1	INJ 1 Pulsewidth	0-99.9 ms	3.2	4.9	6.7-7.1
INJ2	INJ 2 Pulsewidth	0-99.9 ms	3.2	4.9	6.7-7.1
INJ3	INJ 3 Pulsewidth	0-99.9 ms	3.2	4.9	6.7-7.1
INJ4	INJ 4 Pulsewidth	0-99.9 ms	3.2	4.9	6.7-7.1
INJ5	INJ 5 Pulsewidth	0-99.9 ms	3.2	4.9	6.7-7.1
INJ6	INJ 6 Pulsewidth	0-99.9 ms	3.2	4.9	6.7-7.1
LongFT1	Long Term F/T B1	-20 to 20%	-5 to +5	-5 to +5	-5 to +5
LongFT2	Long Term F/T B2	-20 to 20%	-5 to +5	-5 to +5	-5 to +5
MIL	MIL Control	ON / OFF	ON (if on)	OFF	OFF
SCC	S/C C Signal	0-25.5v	12-14	12-14	12-14
SCMA	S/C MA Signal	0-25.5v	12-14	12-14	12-14
SCMB	S/C MB Signal	0-25.5v	12-14	12-14	12-14
SCMC	S/C MC Signal	0-25.5v	12-14	12-14	12-14
SHTFT1	Short Term F/T B1	-20 to 20%	-5 to +5	-5 to +5	-5 to +5
SHTFT2	Short Term F/T B2	-20 to 20%	-5 to +5	-5 to +5	-5 to +5
SPARK	Spark Advance	-90° to 90°	12-17	34	40
S1	Shift Solenoid 1	ON / OFF	ON	OFF	ON
S2	Shift Solenoid 2	ON / OFF	OFF	OFF	OFF
S3	Shift Solenoid 3	ON / OFF	OFF	ON	ON
SS4	Shift Solenoid 4	ON / OFF	ON	ON	ON
TCC	TCC Solenoid	0-100%	0	0	100
VREF	Voltage Reference	0-5.1v	4.9-5.1	4.9-5.1	4.9-5.1
WAC	WOT Cutout Relay	ON / OFF	ON (if on)	OFF	OFF

MUSTANG PID DATA

1991-93 Mustang 2.3L I4 MFI VIN M (All) Inputs / Outputs

PCM PID Acronym	Parameter Identification	PID Range	PID Value at Hot Idle	PID Value at 30	PID Value at 55
BOO	Brake Switch	ON / OFF	ON (if on)	OFF	OFF
BARO	Barometric Pressure Sensor	0-1 KHz	159 Hz at Sea Level	159 Hz at Sea Level	159 Hz at Sea Level
CANP	EVAP Purge Valve	0-100%	0%	0-40	40-50
ECT (°F)	ECT Sensor	-40-304°F	160-200	160-200	160-200
ECT (V)	ECT Sensor	0-5.1v	0.6	0.6	0.6
EVP (V)	EGR EVP Sensor	0-5.1v	0.3	0.4-1.0	0.9-2.0
EVR	VR Solenoid	0-100%	0%	0-40	30-50
HO2S-11	HO2S-11 (Bank 1)	0.1-1.1v	0.1-1.1	0.1-1.1	0.1-1.1
IAC	Idle Air Control	0-100%	20-40	34-40	45-55
IAT (°F)	IAT Sensor	-40-304°F	50-120	50-120	50-120
IAT (V)	IAT Sensor	0-5.1v	1.7-3.5	1.7-3.5	1.7-3.5
INJPW1	INJ Pulsewidth B1	0-99.9 ms	3.0-3.8 ms	4.5-5.5 ms	6.0-6.8 mv
LFAN	Low Speed Fan	ON / OFF	ON (if on)	OFF	OFF
LOOP	Loop Status	CL or OL	CL	CL	CL
MAF (V)	Mass Airflow	0-5.1v	0.5	1.3-1.4v	1.5-2.0v
PNP	PNP Switch	Neutral/DR	NEUTRAL	DRIVE	DRIVE
RPM	Engine Speed	0-10K rpm	750-820	1450-1630	1750-2100
SHTFT1	Short Term F/T B1	-20 to 20%	-5 to +5	-5 to +5	-5 to +5
SPARK	Spark Advance	-90° to 90°	15-22	28-35	25-35
TPV	TP Sensor	0-5.1v	0.7	0.8-1.0	1.1-1.3
TP Mode	TP Mode	C/T or P/T	C/T	P/T	P/T
VPWR	Vehicle Power	0-25.5v	12-14	12-14	12-14
VREF	Voltage Reference	0-5.1v	4.9-5.1	4.9-5.1	4.9-5.1
VSS	Vehicle Speed	0-159 mph	0	30	55
WAC	A/C WOT Cutout Relay	ON / OFF	ON (if on)	OFF	OFF

1994-2005 Mustang 3.8L 2v V6 MFI VIN 4 (A/T) - Inputs

PCM PID Acronym	Parameter Identification	PID Range	PID Value at Hot Idle	PID Value at 30	PID Value at 55
ACCS	A/C Clutch Switch	ON / OFF	ON (if on)	OFF	OFF
AIR-M	AIR System Monitor	ON / OFF	OFF	OFF	OFF
BPP	Brake Pedal Switch	ON / OFF	ON (if on)	OFF	OFF
CID	CID Sensor	0-1 KHz	5-7	9-11	10-15
CKP	CKP Sensor	0-10 KHz	390-450	650-700	875-1000
DPFE	DPFE Sensor	0-5.1v	0.25-1.30	0.25-4.65	0.25-4.65
ECT (°F)	ECT Sensor	-40-304°F	160-200	160-200	160-200
EFT (°F)	EFT Sensor (°F)	-40-304°F	50-120	50-120	50-120
FEPS	Flash EEPROM	0-5.1v	0.5-0.6	0.5-0.6	0.5-0.6
FLI	Fuel Level Indicator	0-100%	50 (1.7v)	25-75	25-75
FPM	Fuel Pump Monitor	0-100%	100	100	100
FRP	Fuel Rail Pressure (1998-2001)	0-100 psi	39	39	39
FTPV	Fuel Tank Pressure (1998-2001)	0-5.1v or 0-10" H2O	2.6v / 0" H2O (with cap off)	2.6v / 0" H2O (with cap off)	2.6v / 0" H2O (with cap off)
GEAR	Transmission Gear	1-2-3-4	1	3	4
HO2S-11	HO2S-11 (Bank 1)	0.1-1.1v	0.1-1.1	0.1-1.1	0.1-1.1
HO2S-12	HO2S-12 (Bank 1)	0.1-1.1v	0.1-1.1	0.1-1.1	0.1-1.1
HO2S-21	HO2S-21 (Bank 2)	0.1-1.1v	0.1-1.1	0.1-1.1	0.1-1.1
HO2S-22	HO2S-22 (Bank 2)	0.1-1.1v	0.1-1.1	0.1-1.1	0.1-1.1
IAT (°F)	IAT Sensor	-40-304°F	50-120	50-120	50-120
LOAD	Engine Load	0-100%	10-20	16-36	25-35
MAF (V)	MAF Sensor	0-5.1v	0.6-0.9	0.8-1.6	1.1-2.3
MISF	Misfire Detection	ON / OFF	OFF	OFF	OFF
OCTADJ	Octane Adjustment	CL/OPEN	CLOSED	CLOSED	CLOSED
OSS	Output Speed Shaft Sensor RPM	0-10 KHz rpm	0	1150-1300	2400
PF (V)	Purge Flow Sensor Volts (1996-97)	0-5.1v	1.55	2.10	2.30
PNP	PNP Switch	ON/OFF	ON in Park	OFF	OFF
RPM	Engine Speed	0-10K rpm	700-730	1000-1200	1500-1700
TCS	TCS (Switch)	ON / OFF	ON (if on)	OFF	OFF
TFT (°F)	TFT Sensor	-40-304°F	110-210	110-210	110-210
TPV	TP Sensor	0-5.1v	0.5-1.27	0.8-1.1	0.8-1.2
TR1	TR1 Sensor	0-25.5v	0	11.5	11.5
TRV/TR	TR Sensor	0-5.1v	0v (PN)	1.7v (OD)	1.7v (OD)
VPWR	Vehicle Power	0-25.5V	12-14	12-14	12-14
VSS	Vehicle Speed (MPH & Hertz Rate)	0-159 mph	0 = 0 Hz	30 = 65 Hz	55 = 125 Hz

1994-2005 Mustang 3.8L 2v V6 MFI VIN 4 (A/T) - Outputs

PCM PID Acronym	Parameter Identification	PID Range	PID Value at Hot Idle	PID Value at 30	PID Value at 55
AIR-GS	AIR Solenoid	ON / OFF	OFF	OFF	OFF
CD1-CD3	Coil Driver 1-3	0-60° dwell	7	11	12
CTO	Clean Tach Output	0-10 KHz	33-37	57-63	68-74
EGRVR	VR Solenoid	0-100%	0	0-40	35-45
EPC	EPC Solenoid	0-500 psi	8	12-18	18-22
EVAPCV	EVAP CV Valve	0-100%	0-100	0-100	0-100
EVAPDC	EVAP Purge Valve	0-100%	0-100	0-100	0-100
FLI	Fuel Level Indicator (1998-99)	0-100%	50 (1.7v)	25-75	25-75
FP	Fuel Pump Control	0-100%	50	50	50
FuelPW1	INJ Pulsewidth B1	0-99.9 ms	4.9-5.1	5.3-10	6.5-12
FuelPW2	INJ Pulsewidth B2	0-99.9 ms	4.9-5.1	5.3-10	6.5-12
HTR-11	HO2S-11 Heater	ON / OFF	ON	ON	ON
HTR-12	HO2S-12 Heater	ON / OFF	ON	ON	ON
HTR-21	HO2S-21 Heater	ON / OFF	ON	ON	ON
HTR-22	HO2S-22 Heater	ON / OFF	ON	ON	ON
IAC	Idle Air Control	0-100%	34-39	44-73	50-75
INJ1	INJ 1 Pulsewidth	0-99.9 ms	4.9-5.1	5.3-10	6.5-12
INJ2	INJ 2 Pulsewidth	0-99.9 ms	4.9-5.1	5.3-10	6.5-12
INJ3	INJ 3 Pulsewidth	0-99.9 ms	4.9-5.1	5.3-10	6.5-12
INJ4	INJ 4 Pulsewidth	0-99.9 ms	4.9-5.1	5.3-10	6.5-12
INJ5	INJ 5 Pulsewidth	0-99.9 ms	4.9-5.1	5.3-10	6.5-12
INJ6	INJ 6 Pulsewidth	0-99.9 ms	4.9-5.1	5.3-10	6.5-12
LFC	Low Fan Relay	ON / OFF	ON (if on)	OFF	OFF
LongFT1	Long Term F/T B1	-20 to 20%	-5 to +5	-5 to +5	-5 to +5
LongFT2	Long Term F/T B2	-20 to 20%	-5 to +5	-5 to +5	-5 to +5
MIL	MIL Control	ON / OFF	ON (if on)	OFF	OFF
SHTFT1	Short Term F/T B1	-20 to 20%	-5 to +5	-5 to +5	-5 to +5
SHTFT2	Short Term F/T B2	-20 to 20%	-5 to +5	-5 to +5	-5 to +5
SPARK	Spark Advance	-90° to 90°	15-20	25-35	31-40
SS1	Shift Solenoid 1	ON / OFF	ON	OFF	ON
SS2	Shift Solenoid 2	ON / OFF	OFF	ON	ON
TCC	TCC Solenoid	0-100%	0	0-45	95-100
TCIL	TCIL (lamp)	ON / OFF	ON (if on)	OFF	OFF
VREF	Voltage Reference	0-5.1v	4.9-5.1	4.9-5.1	4.9-5.1
VSO	VSS Output Signal	0-1 KHz	0	55	125
WAC	A/C WOT Cutout Relay	ON / OFF	ON (if on)	OFF	OFF

1994-2005 Mustang 3.8L 2v V6 MFI VIN 4 (M/T) - Inputs

PCM PID Acronym	Parameter Identification	PID Range	PID Value at Hot Idle	PID Value at 30	PID Value at 55
ACCS	A/C Clutch Switch	ON / OFF	ON (if on)	OFF	OFF
CID	CID Sensor	0-1 KHz	5-7	9-11	10-15
CKP	CKP Sensor	0-10 KHz	390-450	700-740	825-875
CPP/NDS	Clutch Pedal Position Switch	ON/OFF	ON when clutch depressed	OFF	OFF
DPFE	DPFE Sensor	0-5.1v	0.25-1.30	0.25-4.65	0.25-4.65
ECT (°F)	ECT Sensor	-40-304°F	160-200	160-200	160-200
FEPS	Flash EEPROM	0-5.1v	0.5-0.6	0.5-0.6	0.5-0.6
FLI	Fuel Level Indicator	0-100%	50 (1.7v)	25-75	25-75
FPM	Fuel Pump Monitor	0-100%	100	100	100
FRP	Fuel Rail Pressure (1998-2001)	0-100 psi	39	39	39
FTPV	Fuel Tank Pressure (1998-2001)	0-5.1v or 0-10" H2O	2.6v / 0" H2O (with cap off)	2.6v / 0" H2O (with cap off)	2.6v / 0" H2O (with cap off)
HO2S-11	HO2S-11 (Bank 1)	0.1-1.1v	0.1-1.1	0.1-1.1	0.1-1.1
HO2S-12	HO2S-12 (Bank 1)	0.1-1.1v	0.1-1.1	0.1-1.1	0.1-1.1
HO2S-21	HO2S-21 (Bank 2)	0.1-1.1v	0.1-1.1	0.1-1.1	0.1-1.1
HO2S-22	HO2S-22 (Bank 2)	0.1-1.1v	0.1-1.1	0.1-1.1	0.1-1.1
IAT (°F)	IAT Sensor	-40-304°F	50-120	50-120	50-120
MAF (V)	MAF Sensor	0-5.1v	0.6-0.9	0.8-1.6	1.1-2.3v
MISF	Misfire Detection	ON / OFF	OFF	OFF	OFF
OCTADJ	Octane Adjustment	CL/OPEN	CLOSED	CLOSED	CLOSED
LOAD	Engine Load	0-100%	10-20	16-36	25-35
PF (V)	Purge Flow Sensor Volts (1996-97)	0-5.1v	1.55	2.10	2.30
RPM	Engine Speed	0-10K rpm	700-780	1000-1200	1500-1700
TPV	TP Sensor	0-5.1v	0.53-1.27	0.8-1.1	0.8-1.2
VPWR	Vehicle Power	0-25.5v	12-14	12-14	12-14
VSS	Vehicle Speed (MPH & Hertz Rate)	0-159 mph	0 = 0 Hz	30 = 65 Hz	55 = 125 Hz

1994-2005 Mustang 3.8L 2v V6 MFI VIN 4 (M/T) - Outputs

PCM PID Acronym	Parameter Identification	PID Range	PID Value at Hot Idle	PID Value at 30	PID Value at 55
AIR-GS	AIR Solenoid	ON / OFF	OFF	OFF	OFF
CD1-CD3	Coil Driver 1-3	0-60° dwell	7	11	12
CTO	Clean Tach Output	0-10 KHz	33-37	57-63	68-74
EGRVR	VR Solenoid	0-100%	0	0-40	35-45
EVAPCV	EVAP CV Valve	0-100%	0-100	0-100	0-100
EVAPDC	EVAP Purge Valve	0-100%	0-100	0-100	0-100
FP	Fuel Pump Control	0-100%	26	26	26
FuelPW1	INJ Pulsewidth B1	0-99.9 ms	4.9-5.1	5.3-10	6.5-12
FuelPW2	INJ Pulsewidth B2	0-99.9 ms	4.9-5.1	5.3-10	6.5-12
HTR-11	HO2S-11 Heater	ON / OFF	ON	ON	ON
HTR-12	HO2S-12 Heater	ON / OFF	ON	ON	ON
HTR-21	HO2S-21 Heater	ON / OFF	ON	ON	ON
HTR-22	HO2S-22 Heater	ON / OFF	ON	ON	ON
IAC	Idle Air Control	0-100%	34-39	44-73	50-75
INJ1	INJ 1 Pulsewidth	0-99.9 ms	4.9-5.1	5.3-10	6.5-12
INJ2	INJ 2 Pulsewidth	0-99.9 ms	4.9-5.1	5.3-10	6.5-12
INJ3	INJ 3 Pulsewidth	0-99.9 ms	4.9-5.1	5.3-10	6.5-12
INJ4	INJ 4 Pulsewidth	0-99.9 ms	4.9-5.1	5.3-10	6.5-12
INJ5	INJ 5 Pulsewidth	0-99.9 ms	4.9-5.1	5.3-10	6.5-12
INJ6	INJ 6 Pulsewidth	0-99.9 ms	4.9-5.1	5.3-10	6.5-12
LFC	Low Fan Control	ON / OFF	ON (if on)	OFF	OFF
LongFT1	Long Term F/T B1	-20 to 20%	-5 to +5	-5 to +5	-5 to +5
LongFT2	Long Term F/T B2	-20 to 20%	-5 to +5	-5 to +5	-5 to +5
MIL	MIL Control	ON / OFF	ON (if on)	OFF	OFF
SHTFT1	Short Term F/T B1	-20 to 20%	-5 to +5	-5 to +5	-5 to +5
SHTFT2	Short Term F/T B2	-20 to 20%	-5 to +5	-5 to +5	-5 to +5
SPARK	Spark Advance	-90° to 90°	15-20	25-35	31-40
VREF	Voltage Reference	0-5.1v	4.9-5.1	4.9-5.1	4.9-5.1
VSO	VSS Output Signal	0-1 KHz	0	65	125
WAC	A/C WOT Cutout Relay	ON / OFF	ON (if on)	OFF	OFF

1996-2005 Mustang 4.6L 4v V8 MFI VIN V, Y (All) - Inputs

PCM PID Acronym	Parameter Identification	PID Range	PID Value at Hot Idle	PID Value at 30	PID Value at 55
ACCS	A/C Clutch Switch	ON / OFF	ON (if on)	OFF	OFF
ACP	A/C Press. Sensor	CL/OPEN	CLOSED (if low)	OPEN	OPEN
AIR-M	AIR System Monitor	ON / OFF	OFF	OFF	OFF
BPP	Brake Pedal Switch	ON / OFF	ON (if on)	OFF	OFF
CID	CID Sensor	0-1 KHz	5-7	9-11	11-14
CKP	CKP Sensor	0-10 KHz	360-420	680-800	860-910
DPFE	DPFE Sensor	0-5.1v	0.25-1.30	0.25-4.65	0.25-4.65
ECT (°F)	ECT Sensor	-40-304°F	160-200	160-200	160-200
EFTA (°F)	EFTA Sensor (°F)	-40-304°F	50-120	50-120	50-120
FEPS	Flash EEPROM	0-5.1v	0.5-0.6	0.5-0.6	0.5-0.6
FLI	Fuel Level Indicator (1998-99)	0-100%	50 (1.7v)	25-75	25-75
FPM	Fuel Pump Monitor	0-100%	87-100	87-100	87-100
FTPV	Fuel Tank Pressure (1998-99)	0-5.1v or 0-10" H2O	2.6v / 0" H2O (with cap off)	2.6v / 0" H2O (with cap off)	2.6v / 0" H2O (with cap off)
HO2S-11	HO2S-11 (Bank 1)	0.1-1.1v	0.1-1.1	0.1-1.1	0.1-1.1
HO2S-12	HO2S-12 (Bank 1)	0.1-1.1v	0.1-1.1	0.1-1.1	0.1-1.1
HO2S-21	HO2S-21 (Bank 2)	0.1-1.1v	0.1-1.1	0.1-1.1	0.1-1.1
HO2S-22	HO2S-22 (Bank 2)	0.1-1.1v	0.1-1.1	0.1-1.1	0.1-1.1
IAT (°F)	IAT Sensor	-40-304°F	50-120	50-120	50-120
IMRCM	IMRC Monitor	0-5.1v	5	5	5
KS1	Knock Sensor B1	0-5.1v	0	0	0
KS2	Knock Sensor B2	0-5.1v	0	0	0
LOAD	Engine Load	0-100%	10-20	16-30	20-30
MAF (V)	MAF Sensor	0-5.1v	0.5-0.8	0.8-1.3	1.3-1.7
MISF	Misfire Detection	ON / OFF	OFF	OFF	OFF
OCTADJ	Octane Adjustment	CL/OPEN	CLOSED	CLOSED	CLOSED
OSS	Output Speed Shaft Sensor RPM	0-10K rpm	0 rpm	1365 rpm	2440 rpm
PF (V)	Purge Flow Sensor Volts (1996-97)	0-5.1v	1.55	2.10	2.30
PNP	PNP Switch	Neutral/DR	NEUTRAL	DRIVE	DRIVE
RPM	Engine Speed	0-10K rpm	630-750	1180-1360	1530-1750
TPV	TP Sensor	0-5.1v	0.5-0.9	0.9-1.0	1.0-1.1
VPWR	Vehicle Power	0-25.5V	12-14	12-14	12-14
VSS	Vehicle Speed (MPH & Hertz Rate)	0-159 mph	0 = 0 Hz	30 = 65 Hz	55 = 125 Hz

1996-2005 Mustang 4.6L 4v V8 MFI VIN V, Y (All) - Outputs

PCM PID Acronym	Parameter Identification	PID Range	PID Value at Hot Idle	PID Value at 30	PID Value at 55
AIR	AIR Solenoid	ON / OFF	OFF	OFF	OFF
CD1-CD8	Coil Driver 1-8	0-90° dwell	7	11	12
CTO	Clean Tach Output	0-10 KHz	37-44	86-92	87-115
EGRVR	VR Solenoid	0-100%	0	0-40	0-40
EVAP CP	EVAP CV Valve	0-100%	0-100	0-100	0-100
EVAPPC	EVAP Purge Valve	0-100%	0-100	0-100	0-100
FP	Fuel Pump Control	0-100%	33	28	100
FPL	Fuel Pump Low	0-25.5v	0.1	0.1	0.1
FuelPW1	INJ Pulsewidth B1	0-99.9 ms	2.4-2.8	1.6-5.0	3.3-5.0
FuelPW2	INJ Pulsewidth B2	0-99.9 ms	2.4-2.8	1.6-5.0	3.3-5.0
HFC	High Fan Control	ON / OFF	ON (if on)	OFF	OFF
HTR-11	HO2S-11 Heater	ON / OFF	ON	ON	ON
HTR-12	HO2S-12 Heater	ON / OFF	ON	ON	ON
HTR-21	HO2S-21 Heater	ON / OFF	ON	ON	ON
HTR-22	HO2S-22 Heater	ON / OFF	ON	ON	ON
IAC	Idle Air Control	0-100%	32	34-46	34-46
IMRC	IMRC Solenoid	ON / OFF	OFF	OFF	OFF
INJ1	INJ 1 Pulsewidth	0-99.9 ms	2.4-2.8	1.6-5.0	3.3-5.0
INJ2	INJ 2 Pulsewidth	0-99.9 ms	2.4-2.8	1.6-5.0	3.3-5.0
INJ3	INJ 3 Pulsewidth	0-99.9 ms	2.4-2.8	1.6-5.0	3.3-5.0
INJ4	INJ 4 Pulsewidth	0-99.9 ms	2.4-2.8	1.6-5.0	3.3-5.0
INJ5	INJ 5 Pulsewidth	0-99.9 ms	2.4-2.8	1.6-5.0	3.3-5.0
INJ6	INJ 6 Pulsewidth	0-99.9 ms	2.4-2.8	1.6-5.0	3.3-5.0
INJ7	INJ 7 Pulsewidth	0-99.9 ms	2.4-2.8	1.6-5.0	3.3-5.0
INJ8	INJ 8 Pulsewidth	0-99.9 ms	2.4-2.8	1.6-5.0	3.3-5.0
LFC	Low Fan Control	ON / OFF	ON (if on)	OFF	OFF
LongFT1	Long Term F/T B1	-20 to 20%	-5 to +5	-5 to +5	-5 to +5
LongFT2	Long Term F/T B2	-20 to 20%	-5 to +5	-5 to +5	-5 to +5
MIL	MIL Control	ON / OFF	ON (if on)	OFF	OFF
SHTFT1	Short Term F/T B1	-20 to 20%	-5 to +5	-5 to +5	-5 to +5
SHTFT2	Short Term F/T B2	-20 to 20%	-5 to +5	-5 to +5	-5 to +5
SPARK	Spark Advance	-90° to 90°	11-15	17-20	19-24
SS1	Shift Solenoid 1	ON / OFF	ON	OFF	ON
SS2	Shift Solenoid 2	ON / OFF	OFF	ON	ON
VREF	Voltage Reference	0-5.1v	4.9-5.1	4.9-5.1	4.9-5.1
VSO	VSS Output	0-1 KHz	0	65	125
WAC	A/C WOT Cutout Relay	ON / OFF	ON (if on)	OFF	OFF

1996-2005 Mustang 4.6L 2v V8 MFI VIN W, X (All) - Inputs

PCM PID Acronym	Parameter Identification	PID Range	PID Value at Hot Idle	PID Value at 30	PID Value at 55
ACCS	A/C Clutch Switch	ON / OFF	ON (if on)	OFF	OFF
ACP	A/C Press. Sensor	CL/OPEN	OPEN	OPEN	OPEN
AIR-M	AIR System Monitor	ON / OFF	OFF	OFF	OFF
BPP	Brake Switch	ON / OFF	ON (if on)	OFF	OFF
CID	CID Sensor	0-1 KHz	5-7	10-12	12-16
CKP	CKP Sensor	0-10 KHz	390-450	650-760	980-1020
CPP/NDS	Clutch Pedal Switch	ON/OFF	ON with clutch depressed	OFF	OFF
DPFE	DPFE Sensor	0-5.1v	0.25-1.30	0.25-4.65	0.25-4.65
ECT (°F)	ECT Sensor	-40-304°F	160-200	160-200	160-200
EFTA (°F)	EFTA Sensor (°F)	-40-304°F	50-120	50-120	50-120
FEPS	Flash EEPROM	0-5.1v	0.5-0.6	0.5-0.6	0.5-0.6
FLI	Fuel Level Indicator (1998-99)	0-100%	50 (1.7v)	50 (1.7v)	50 (1.7v)
FPM	Fuel Pump Monitor	0-100%	87-100	87-100	87-100
FTP	Fuel Tank Pressure (1998-99)	0-5.1v or 0-10" H2O	2.6v / 0" H2O (with cap off)	2.6v / 0" H2O (with cap off)	2.6v / 0" H2O (with cap off)
FRP	Fuel Rail Pressure	0-100 psi	39	39	39
GEAR	Transmission Gear	1-2-3-4	1	3	4
HO2S-11	HO2S-11 (Bank 1)	0.1-1.1v	0.1-1.1	0.1-1.1	0.1-1.1
HO2S-12	HO2S-12 (Bank 1)	0.1-1.1v	0.1-1.1	0.1-1.1	0.1-1.1
HO2S-21	HO2S-21 (Bank 2)	0.1-1.1v	0.1-1.1	0.1-1.1	0.1-1.1
HO2S-22	HO2S-22 (Bank 2)	0.1-1.1v	0.1-1.1	0.1-1.1	0.1-1.1
IAT (°F)	IAT Sensor	-40-304°F	50-120	50-120	50-120
LOAD	Engine Load	0-100%	10-20	16-30	20-30
MAF (V)	MAF Sensor	0-5.1v	0.6-0.9	0.8-1.2	1.4-1.9
MISF	Misfire Detection	ON / OFF	OFF	OFF	OFF
OCTADJ	Octane Adjustment	CL/OPEN	CLOSED	CLOSED	CLOSED
OSS	Output Speed Shaft	0-10 KHz	0	1385-1420	2475-2500
PF (V)	Purge Flow Sensor (1996-97)	0-5.1v	1.55	2.10	2.30
PNP	PNP Switch	Neutral/DR	NEUTRAL	DRIVE	DRIVE
RPM	Engine Speed	0-10K rpm	660-700	1380-1420	1700-1740
TCS	TCS (switch)	ON / OFF	ON (if on)	OFF	OFF
TFT (°F)	TFT Sensor	-40-304°F	110-210	110-210	110-210
TPV	TP Sensor	0-5.1v	0.5-0.9	1.0-1.2	1.2-1.5
TRV/TR	TR Sensor	0-5.1v	4.4 / PN	2.1 / OD	2.1 / OD
VPWR	Vehicle Power	0-25.5V	12-14	12-14	12-14
VSS	Vehicle Speed (MPH & Hertz Rate)	0-159 mph	0 = 0 Hz	30 = 65 Hz	55 = 125 Hz

1996-2005 Mustang 4.6L 2v V8 MFI VIN W, X (All) - Outputs

PCM PID Acronym	Parameter Identification	PID Range	PID Value at Hot Idle	PID Value at 30	PID Value at 55
AIR-GS	AIR Solenoid	ON / OFF	OFF	OFF	OFF
CD1-CD8	Coil Driver 1-8	0-90° dwell	7	11	12
CTO	Clean Tach Output	0-10 KHz	37-44	86-92	87-115
EGRVR	VR Solenoid	0-100%	0	0-40	0-40
EPC	EPC Solenoid	0-100 psi	15-18	12-16	40
EVAP CP	EVAP CV Valve	0-100%	0-100%	0-100%	0-100%
EVAPDC	EVAP Purge Valve	0-100%	0-100%	0-100%	0-100%
FP	Fuel Pump Control	0-100%	33	28	100
FPL	Fuel Pump Low	0-25.5v	0.1	0.1	0.1
FuelPW1	INJ Pulsewidth B1	0-99.9 ms	3.5-3.7	3.8-5.5	4.5-9.6
FuelPW2	INJ Pulsewidth B2	0-99.9 ms	3.5-3.7	3.8-5.5	4.5-9.6
HFC	High Fan Control	ON / OFF	ON (if on)	OFF	OFF
HTR-11	HO2S-11 Heater	ON / OFF	ON	ON	ON
HTR-12	HO2S-12 Heater	ON / OFF	ON	ON	ON
HTR-21	HO2S-21 Heater	ON / OFF	ON	ON	ON
HTR-22	HO2S-22 Heater	ON / OFF	ON	ON	ON
IAC	Idle Air Control	0-100%	37	50-57	60-65
INJ1	INJ 1 Pulsewidth	0-99.9 ms	3.5-3.7	3.8-5.5	4.5-9.6
INJ2	INJ 2 Pulsewidth	0-99.9 ms	3.5-3.7	3.8-5.5	4.5-9.6
INJ3	INJ 3 Pulsewidth	0-99.9 ms	3.5-3.7	3.8-5.5	4.5-9.6
INJ4	INJ 4 Pulsewidth	0-99.9 ms	3.5-3.7	3.8-5.5	4.5-9.6
INJ5	INJ 5 Pulsewidth	0-99.9 ms	3.5-3.7	3.8-5.5	4.5-9.6
INJ6	INJ 6 Pulsewidth	0-99.9 ms	3.5-3.7	3.8-5.5	4.5-9.6
INJ7	INJ 7 Pulsewidth	0-99.9 ms	3.5-3.7	3.8-5.5	4.5-9.6
INJ8	INJ 8 Pulsewidth	0-99.9 ms	3.5-3.7	3.8-5.5	4.5-9.6
LFC	Low Fan Control	ON / OFF	ON (if on)	OFF	OFF
LongFT1	Long Term F/T B1	-20 to 20%	-5 to +5	-5 to +5	-5 to +5
LongFT2	Long Term F/T B2	-20 to 20%	-5 to +5	-5 to +5	-5 to +5
MIL	MIL Control	ON / OFF	ON (if on)	OFF	OFF
SHTFT1	Short Term F/T B1	-20 to 20%	-5 to +5	-5 to +5	-5 to +5
SHTFT2	Short Term F/T B2	-20 to 20%	-5 to +5	-5 to +5	-5 to +5
SPARK	Spark Advance	-90° to 90°	15-20	29-38	25-34
SS1	Shift Solenoid 1	ON / OFF	ON	OFF	ON
SS2	Shift Solenoid 2	ON / OFF	OFF	ON	ON
TCC	TCC Solenoid	0-100%	0	37-43	95-100
TCIL	TCIL (lamp)	ON / OFF	ON (if on)	OFF	OFF
VREF	Voltage Reference	0-5.1v	4.9-5.1	4.9-5.1	4.9-5.1
VSO	VSS Output	0-1 KHz	0	65	125 Hz
WAC	A/C WOT Cutout Relay	ON / OFF	ON (if on)	OFF	OFF

1994-95 Mustang 5.0L V8 MFI VIN T (All) Inputs / Outputs

PCM PID Acronym	Parameter Identification	PID Range	PID Value at Hot Idle	PID Value at 30	PID Value at 55
ACCS	A/C Clutch Switch	ON / OFF	ON (if on)	OFF	OFF
BOO	Brake Switch	ON / OFF	ON (if on)	OFF	OFF
CANP	EVAP Purge	0-100%	0	0-40	40-50
ECT (°F)	ECT Sensor	-40-304°F	160-200	160-200	160-200
ECT (V)	ECT Sensor	0-5.1v	0.6	0.6	0.6
EPC	EPC Solenoid	0-100 psi	20	12	30
EVP (V)	EGR EVP Sensor	0-5.1v	0.3	0.9-1.0	2.0-3.0v
EVR	VR Solenoid	0-100%	0%	0-40	0-60
GEAR	Gear Position	P-R-N-D	P-R-N-D	DRIVE 3	DRIVE 4
HO2S-11	HO2S-11 (Bank 1)	0.1-1.1v	0.1-1.1	0.1-1.1	0.1-1.1
HO2S-12	HO2S-12 (Bank 1)	0.1-1.1v	0.1-1.1	0.1-1.1	0.1-1.1
IAC	Idle Air Control	0-100%	20-40	34-40	45-55
IAT (°F)	IAT Sensor	-40-314F	50-120	50-120	50-120
IAT (V)	IAT Sensor	0-5.1v	1.7-3.5	1.7-3.5	1.7-3.5
INJPW1	INJ Pulsewidth B1	0-99.9 ms	5.5-5.8	6.4-8.4	9.4-12
INJPW2	INJ Pulsewidth B2	0-99.9 ms	5.5-5.8	6.4-8.4	9.4-12
LOOP	Loop Status	CL or OL	CL	CL	CL
LongFT1	Long Term F/T B1	-20 to 20%	-5 to +5	-5 to +5	-5 to +5
LongFT2	Long Term F/T B2	-20 to 20%	-5 to +5	-5 to +5	-5 to +5
MAF	MAF Sensor	0-5.1v	1.0v	1.3-1.5	1.6-2.0
MLP	MLP Sensor	Park / OD	PARK	OD	OD
MLP (V)	MLP Sensor	0-5.1v	0v	5v	5v
PNP	PNP Switch	Neutral/DR	NEUTRAL	DRIVE	DRIVE
RPM	Engine Speed	0-10K rpm	750-820	1450-1630	1750-2100
SHTFT1	Short Term F/T B1	-20 to 20%	-5 to +5	-5 to +5	-5 to +5
SHTFT2	Short Term F/T B2	-20 to 20%	-5 to +5	-5 to +5	-5 to +5
SPARK	Spark Advance	-90° to 90°	15-22	28-35	25-35
TCC	TCC Solenoid	0-100%	0%	42-44	92-100
TOT (°F)	TOT Sensor	-40-304°F	110-210	110-210	110-210
TOT (V)	TOT Sensor	0-5.1v	1.7-3.5	1.7-3.5	1.7-3.5
TP	TP Sensor	0-5.1v	0.7	0.8-1.0	1.1-1.3
TP C/T	TP Closed Throttle	0-5.1v	0.7	0.8-1.0	1.1-1.3
TP Mode	TP Mode	C/T or P/T	C/T	P/T	P/T
VPWR	Vehicle Power	0-25.5v	12-14	12-14	12-14
VREF	Voltage Reference	0-5.1v	4.9-5.1	4.9-5.1	4.9-5.1
VSS	Vehicle Speed	0-159 mph	0	30	55
WAC	A/C WOT Cutout Relay	ON / OFF	ON (if on)	OFF	OFF

PROBE PID DATA

1994-95 Probe 2.0L I4 MFI VIN A (CD4E) Inputs / Outputs

PCM PID Acronym	Parameter Identification	PID Range	PID Value at Hot Idle	PID Value at 30	PID Value at 55
ACCS	A/C Clutch Switch	ON / OFF	ON (if on)	OFF	OFF
BOO	Brake Switch	ON / OFF	ON (if on)	OFF	ON
CANP	EVAP Purge Valve	0-100%	0	0-40	40-50
ECT (ºF)	ECT Sensor	-40-304ºF	160-200	160-200	160-200
ECT (V)	ECT Sensor	0-5.1v	0.6	0.6	0.6
EPC	EPC Solenoid	0-100 psi	8-9 psi	24 psi	20-30
EVP (V)	EGR EVP Sensor	0-5.1v	0.4	0.5-1.1	0.9-2.1
EVR	VR Solenoid	0-100%	0	0	0-40
GEAR	Gear Position	P-R-N-DR	P-R-N-D	DRIVE 3	DRIVE 4
HFAN	High Speed Fan	ON / OFF	ON (if on)	OFF	OFF
FuelPW1	INJ Pulsewidth B1	0-99.9 ms	1.7-2.3	2.4-4.6	3.5-8.4
HO2S-11	HO2S-11 (Bank 1)	0.1-1.1v	0.1-1.1	0.1-1.1	0.1-1.1
IAC	Idle Air Control	0-100%	41	20-35	50-75
IAT (ºF)	IAT Sensor	-40-304ºF	50-120	50-120	50-120
IAT (V)	IAT Sensor	0-5.1v	1.7-3.5	1.7-3.5	1.7-3.5
LFAN	Low Speed Fan	ON / OFF	ON (if on)	OFF	OFF
LongFT1	Long Term F/T B1	-20 to 20%	-5 to +5	-5 to +5	-5 to +5
LOOP	Loop Status	CL or OL	CL	CL	CL
MAF (V)	MAF Sensor	0-5.1v	0.4-0.7	0.7-1.6	1.2-2.2
MLP (V)	MLP Sensor	0-5.1v	0	5	5
PNP	PNP Switch	Neutral/DR	NEUTRAL	DRIVE	DRIVE
RPM	Engine Speed	0-10K rpm	680-720	1575-1635	2200-2450
SHTFT1	Short Term F/T B1	-20 to 20%	-5 to +5	-5 to +5	-5 to +5
SPARK	Spark Advance	-90º to 90º	15-20	25-35	28-33
TCC	TCC Solenoid	0-100%	0	0-45	95-100
TOT (ºF)	TOT Sensor	-40-304ºF	110-210	110-210	110-210
TOT (V)	TOT Sensor	0-5.1v	0.5-2.0	0.5-2.0	0.5-2.0
TSS	TSS Sensor	0-10 KHz	680-720	1575-1635	2200-2450
TP	TP Sensor	0-5.1v	0.7	0.8-1.0	1.1-1.3
TP MIN	TP Minimum	0-5.1v	0.5	0.8-1.0	1.1-1.3
TP Mode	TP Mode	C/T or P/T	C/T	P/T	P/T
VPWR	Vehicle Power	0-25.5v	12-14	12-14	12-14
VREF	Voltage Reference	0-5.1v	4.9-5.1	4.9-5.1	4.9-5.1
VSS	Vehicle Speed	0-159 mph	0	30	55
WAC	A/C WOT Cutout Relay	ON / OFF	ON (if on)	OFF	OFF

1996-97 Probe 2.0L I4 MFI VIN A (All) - Inputs

PCM PID Acronym	Parameter Identification	PID Range	PID Value at Hot Idle	PID Value at 30	PID Value at 55
ACCS	A/C Clutch Switch	ON / OFF	ON (if on)	OFF	OFF
ACP	A/C Press. Sensor	0-5.1v	0.8-1.9v	0.8-1.9v	0.8-1.9v
BPP	Brake Pedal Switch	ON / OFF	ON (if on)	OFF	OFF
CID	CID Sensor	0-1 KHz	5-7	11-15	17-21
CKP	CKP Sensor	0-10 KHz	390-450	925-950	1320-1410
CPP	M/T: CPP Switch	ON / OFF	ON (if in)	OFF	OFF
DRL	Daytime Running Lamps	0-5.1v	0.1	0.1	0.1
ECT (°F)	ECT Sensor	-40-304°F	160-200	160-200	160-200
EGRB	EGR BARO Sensor	0-5.1v	0.4-4.8	0.4-4.8	0.4-4.8
EGR VP	EGR EVP Sensor	0-5.1v	0.5-0.8	0.7-1.6	2.5-3.2
FEPS	Flash EEPROM	0-5.1v	0.1	0.1	0.1
FPM	Fuel Pump Monitor	ON / OFF	ON	ON	ON
FRP	Fuel Rail Pressure	0-100 psi	30-32	30-32	32-35
IAT	IAT Sensor	-40-304°F	50-120	50-120	50-120
GEAR	Gear Position	P-N-R-D or DRIVE	P-R-N-D	DRIVE 3	DRIVE 4
HDLMP	Headlamp Switch	ON / OFF	ON (if on)	OFF	OFF
HO2S-11	HO2S-11	0.1-1.1v	0.1-1.1	0.1-1.1	0.1-1.1
HO2S-12	HO2S-12	0.1-1.1v	0.1-1.1	0.1-1.1	0.1-1.1
LOAD	Engine Load	0-100%	10-20	16-26	24-36
MAF (V)	MAF Sensor	0-5.1v	0.4-0.7	0.7-1.6	1.2-2.2
MISF	Misfire Detection	OFF	OFF	OFF	OFF
OCTADJ	Octane Adjustment	CL/OPEN	CLOSED	CLOSED	CLOSED
PNP	A/T: PNP Switch	Neutral/DR	NEUTRAL	DRIVE	DRIVE
PSP	PSP Switch	0-25.5v	0v turning	12-14	12-14
RDEF	Rear Defrost Switch	ON / OFF	ON (if on)	OFF	OFF
RPM	Engine Speed	0-10K rpm	680-720	1575-1635	2200-2450
TCS	TCS (Switch)	ON / OFF	ON (if on)	OFF	OFF
TFT (°F)	TFT Sensor	-40-304°F	110-210	110-210	110-210
TPV	TP Sensor	0-5.1v	0.5-0.9	0.6-0.8	0.7-1.0
TSS	TSS Sensor	0-10 KHz	680-720	1575-1635	2200-2450
VACUUM	Engine Vacuum	0-30" Hg	19-20	18-20	8-15
VPWR	Vehicle Power	0-25.5v	12-14	12-14	12-14
VSS	Vehicle Speed in MPH & Hz	0-159 mph	0 = 0 Hz	30 = 65 Hz	55 = 125 Hz

1996-97 Probe 2.0L I4 MFI VIN A (All) - Outputs

PCM PID Acronym	Parameter Identification	PID Range	PID Value at Hot Idle	PID Value at 30	PID Value at 55
BLWMTR	Blower Motor	ON / OFF	ON (if on)	OFF	OFF
CD1	Coil Driver 1	0-90° dwell	7	10	12
EGRC	EGR Control Solenoid	ON / OFF	OFF	ON	ON
EGRCS	EGR Check Solenoid	ON / OFF	OFF	OFF	OFF
EGRV	EGR Vent Solenoid	ON / OFF	ON	ON	ON
EVAP CP	EVAP Purge Valve	ON / OFF	OFF	ON	ON
FP	Fuel Pump Control	ON / OFF	ON	ON	ON
FPRC	Fuel Pump Regulator Control	ON / OFF	ON (if on)	OFF	OFF
FuelPW1	INJ Pulsewidth B1	0-99.9 ms	1.0-2.3	2.4-4.6	3.5-8.4
HFC	High Fan Control	ON / OFF	ON (if on)	OFF	OFF
HTR-11	HO2S-11 Heater	ON / OFF	ON	ON	ON
HTR-12	HO2S-12 Heater	ON / OFF	ON	ON	ON
IAC	Idle Air Control	0-100%	41	20-35	50-75
INJ1	INJ 1 Pulsewidth	0-99.9 ms	1.0-2.3	2.4-4.6	3.5-8.4
INJ2	INJ 2 Pulsewidth	0-99.9 ms	1.0-2.3	2.4-4.6	3.5-8.4
INJ3	INJ 3 Pulsewidth	0-99.9 ms	1.0-2.3	2.4-4.6	3.5-8.4
INJ4	INJ 4 Pulsewidth	0-99.9 ms	1.0-2.3	2.4-4.6	3.5-8.4
LFC	Low Fan Control	ON / OFF	ON (if on)	OFF	OFF
LongFT1	Long Term F/T B1	-20 to 20%	-5 to +5	-5 to +5	-5 to +5
MIL	MIL Control	ON / OFF	ON (if on)	OFF	OFF
SS1	Shift Solenoid 1	ON / OFF	OFF	ON	ON
SS2	Shift Solenoid 2	ON / OFF	ON	OFF	OFF
SS3	Shift Solenoid 3	0-25.5v	6.7-7.7	6.7-7.7	6.7-7.7
SHTFT1	Short Term F/T B1	-20 to 20%	-5 to +5	-5 to +5	-5 to +5
SPARK	Spark Advance	-90° to 90°	15-20	25-35	28-33
TCC	TCC Solenoid	0-100%	0	0-45	95-100
TCIL	TCIL (lamp)	ON / OFF	ON (if on)	OFF	OFF
VREF	Voltage Reference	0-5.1v	4.9-5.1	4.9-5.1	4.9-5.1
WAC	A/C WOT Cutout Relay	ON / OFF	ON (if on)	OFF	OFF

1996-97 Probe 2.5L V6 MFI VIN B (All) - Inputs / Outputs

PCM PID Acronym	Parameter Identification	PID Range	PID Value at Hot Idle	PID Value at 30	PID Value at 55
ACR	A/C Relay (on with A/C & BLMT at 3-4)	ON / OFF	ON (if on)	OFF	OFF
ACCS	A/C Relay (on with A/C & BLMT at 3-4)	ON / OFF	ON (if on)	OFF	OFF
ATP	Atmospheric Pressure Sensor	0-5.1v	3.9	3.9	3.9
B+	Vehicle Power	0-25.5v	12-14	12-14	12-14
BARO	BARO Sensor " Hg (varies with altitude)	0-30" Hg	29.4	29.4	29.4
BLMT SW	Blower Motor Switch	ON / OFF	ON (if on)	OFF	OFF
DEF SW	Defroster Switch	ON / OFF	ON (if on)	OFF	OFF
DRL SW	Daytime Running Lamp	ON / OFF	ON (if on)	ON	ON
ECT (°F)	ECT Sensor	-40-304°F	177	185°F	186
ECT (V)	ECT (°F) Sensor	0-5.1v	0.6	0.4	0.4v
EGRS	EGR Sensor	0-5.1v	0.8	0.9	1.4v
EGR-A	EGR 'A' Solenoid	0-99.9 ms	64-99.9	64-99.9	19-99.9
HDLPSW	Headlamp Switch	ON / OFF	ON (if on)	OFF	OFF
HOS2-11	HO2S-11 (Bank 1)	0.1-1.1v	0.1-1.1	0.1-1.1	0.1-1.1
HOS2-12	HO2S-12 (Bank 1)	0.1-1.1v	0.1-1.1	0.1-1.1	0.1-1.1
IAT (°F)	IAT Sensor	-40-304°F	77	86	69°F
IAT (V)	IAT Sensor	0-5.1v	2.4	2.6v	2.7
IDL SW	Throttle Idle Switch	ON / OFF	ON	OFF	OFF
IGN SW	Ignition Switch	ON / OFF	ON-KOEC	OFF	OFF
INJ1	Fuel Injector 1	0-99.9 ms	4.0-5.0	5.0-5.8	5.0-6.0
INJ2	Fuel Injector 2	0-99.9 ms	4.0-5.0	5.0-5.8	5.0-6.0
KS	Knock Sensor	0-5.1v	2.5	2.5	2.5
MC-VAF	Measuring Core Vane Airflow sensor	0-5.1v	3.06	1.50	1.62
MIL	MIL Control	ON / OFF	ON (if on)	OFF	OFF
PNP SW	PNP Switch	0-25.5v	0v (in 'P')	14	14
PSP	PSP Switch	ON / OFF	ON turning	OFF	OFF
PUMP	Fuel Pump	ON / OFF	ON	ON	ON
PURGE	EVAP Purge Valve	0-100%	0	40-50	84-90
SPARK	Spark Advance	-90° to 90°	11	20	20
STP SW	Stop Lamp Switch	ON / OFF	ON (if on)	OFF	OFF
TPV	TP Sensor	0-5.1v	0.5	0.6	0.7
VSS	Vehicle Speed	0-159 mph	0	30	55

1990-92 Probe 3.0L V6 MFI VIN U (All) Inputs / Outputs

PCM PID Acronym	Parameter Identification	PID Range	PID Value at Hot Idle	PID Value at 30	PID Value at 55
ARC	Auto Ride Control	ON / OFF	ON	ON	ON
CANP	EVAP Purge Valve	0-100%	0	0-40	85-95
ECT (°F)	ECT Sensor	-40-304°F	160-200	160-200	160-200
ECT (V)	ECT Sensor	0-5.1v	0.5	0.5	0.5
EVR	VR Solenoid	0-100%	0%	0-40	30-50
HFAN	High Fan Control	ON / OFF	ON (if on)	OFF	OFF
HO2S-11	HO2S-11 (Bank 1)	0.1-1.1v	0.1-1.1	0.1-1.1	0.1-1.1
INJPW1	INJ Pulsewidth B1	0-99.9 ms	5.8-6.0 ms	7.4-7.8 ms	9.0-9.9 ms
IAC	Idle Air Control	0-100%	20-40	34-40	45-55
IAT (°F)	IAT Sensor	-40-304°F	50-120	50-120	50-120
IAT (V)	IAT Sensor	0-5.1v	1.7-3.5	1.7-3.5	1.7-3.5
LFAN	Low Fan Control	ON / OFF	ON (if on)	OFF	OFF
LOOP	Loop Status	CL or OL	CL	CL	CL
MAP	MAP Sensor Hertz	0-1 KHz	107-108	108-110	120-128
PFE (V)	PFE Sensor	0-5.1v	3.2	3.3-3.4	3.3-3.5
PNP	PNP Switch	Neutral/DR	NEUTRAL	DRIVE	DRIVE
RPM	Engine Speed	0-10K rpm	750-820	1450-1630	1750-2100
SHTFT1	Short Term F/T B1	-20 to +20	-5 to +5	-5 to +5	-5 to +5
SPARK	Spark Advance	-90° to 90°	15-22	28-35	25-35
TPV	TP Sensor	0-5.1v	0.7	0.8-1.0	1.1-1.3
TP Mode	TP Mode	C/T or P/T	C/T	P/T	P/T
VPWR	Vehicle Power	0-25.5v	12-14	12-14	12-14
VREF	Voltage Reference	0-5.1v	4.9-5.1	4.9-5.1	4.9-5.1
VSS	Vehicle Speed	0-159 mph	0	30	55
WAC	A/C WOT Cutout Relay	ON / OFF	ON (if on)	OFF	OFF

TAURUS PID DATA

1993-95 Taurus 3.0L V6 FFV VIN 1 (A/T) Inputs / Outputs

PCM PID Acronym	Parameter Identification	PID Range	PID Value at Hot Idle	PID Value at 30	PID Value at 55
ACCS	AC Cycling Clutch	ON / OFF	ON (If on)	OFF	OFF
BOO	Brake Switch	ON / OFF	ON (if on)	OFF	OFF
CANP	EVAP Solenoid	0-100%	0	0-40	85-95
ECT (°F)	ECT Sensor	-40-304°F	160-200	160-200	160-200
ECT (V)	ECT Sensor	0-5.1v	0.6	0.6	0.6
EPC	EPC Solenoid	0-100 psi	40	15	42
EVR	VR Solenoid	0-100%	0%	0-40	40-50
FFS	FF Sensor (value at 100% gas mixture)	0-1 KHz	40-60	40-60	40-60
FPM	Fuel Pump Monitor	ON / OFF	ON	ON	ON
FUELP1	INJ On-time B1	0-99.9 ms	3.8-4.5	3.8-8.1	4.1-12
FUELP2	INJ On-time B2	0-99.9 ms	3.8-4.5	3.8-8.1	4.1-12
GEAR	Gear Position	P-R-N-D	P-R-N-D	DRIVE 3	DRIVE 4
HFC	High Fan Control	ON / OFF	ON (if on)	OFF	OFF
HO2S-11	HO2S-11 (Bank 1)	0.1-1.1v	0-1v	0-1v	0-1v
HO2S-21	HO2S-21 (Bank 2)	0.1-1.1v	0-1v	0-1v	0-1v
IAC	Idle Air Control	0-100%	20-40	34-40	45-55
IAT (°F)	IAT Sensor	-40-304°F	50-120	50-120	50-120
IAT (V)	IAT Sensor	0-5.1v	1.7-3.5	1.7-3.5	1.7-3.5
LFC	Low Fan Control	ON / OFF	ON (if on)	OFF	ON
LOOP	Loop Status	CL or OL	CL	CL	CL
LGFT1	L/T Fuel Trim (B1)	-20 to 20%	-5 to +5	-5 to +5	-5 to +5
LGFT2	L/T Fuel Trim (B2)	-20 to 20%	-5 to +5	-5 to +5	-5 to +5
MAF (V)	MAF Sensor	0-5.1v	0.8	0.9-1.4v	1.4-1.9
MLP (V)	MLP Sensor	0-5.1v	0v	5v	5v
PFEV	PFE Sensor volts	0-5.1v	3.2	3.3	3.4
PNP	PNP Switch	NEUTRAL	NEUTRAL	DRIVE	DRIVE
PSP	PSP Switch	ON / OFF	ON turning	OFF	OFF
RPM	Engine Speed	0-10K rpm	700-750	1410-1510	1760-1860
SHFT1	S/T Fuel Trim (B1)	-20 to 20%	-5 to +5	-5 to +5	-5 to +5
SHFT2	S/T Fuel Trim (B2)	-20 to 20%	-5 to +5	-5 to +5	-5 to +5
SPARK	Spark Advance	-90° to 90°	26-30	44-48	46-52
TCC	TCC Solenoid	0-100%	0	90-100	90-100
TOT °F	TOT Sensor	-40-304°F	110-210	110-210	110-210
TOT (V)	TOT Sensor	0-5.1v	0.5-2.0	0.5-2.0	0.5-2.0
TP	TP Sensor	0-5.1v	0.8	0.9-1.1	1.1-1.3
TP MIN	TP Minimum	0-5.1v	0.5	0.9-1.1	1.1-1.3
TP Mode	TP Mode	C/T or P/T	C/T	P/T	P/T
TSS	TSS Sensor	0-10 KHz	790-820	1400-1450	1740-1800
VPWR	Vehicle Power	0-25.5v	12-14	12-14	12-14
VREF	Voltage Reference	0-5.1v	4.9-5.1	4.9-5.1	4.9-5.1
VSS	Vehicle Speed	0-159 mph	0	30	55
WAC	A/C WOT Cutout Relay	ON / OFF	ON (if on)	OFF	OFF

1996-2004 Taurus 3.0L V6 FFV VIN 1, 2 (A/T) - Inputs

PCM PID Acronym	Parameter Identification	PID Range	PID Value at Hot Idle	PID Value at 30	PID Value at 55
ACCS	A/C Clutch Switch	ON / OFF	ON (if on)	OFF	OFF
ACP	A/C Press. Switch	OPEN or CLOSED	OPEN	OPEN	OPEN
AIR-M	AIR System Monitor	ON / OFF	ON	ON	ON
BPP	Brake Pedal Switch	ON / OFF	ON (if on)	OFF	OFF
CID	CID Sensor	0-1 KHz	6-8	11-13	14-17
CKP	CKP Sensor	0-10 KHz	410-510	810-950	1050-1820
DPFE	DPFE Sensor	0-5.1v	0.2-1.3v	0.2-4.5v	0.2-4.5v
ECT (°F)	ECT Sensor	-40-304°F	160-200	160-200	160-200
FEPS	Flash EEPROM	0-5.1v	0.5-0.6	0.5-0.6	0.5-0.6
FFS	Flexible Fuel sensor (100% gas mixture)	0-1 KHz	40-60	40-60	40-60
FP-M	Fuel Pump Monitor	ON / OFF	ON	ON	ON
FRP	Fuel Rail Pressure	0-100 psi	54	42	40
FTP	FTP Sensor	0-5.1v or 0-10" H20	2.6v at 0" of H20	2.6v at 0" of H20	2.6v at 0" of H20
GEAR	Transmission Gear	P-R-N-D	P-R-N-D	DRIVE 3	DRIVE 4
HO2S-11	HO2S-11 (Bank 1)	0.1-1.1v	0.1-1.1	0.1-1.1	0.1-1.1
HO2S-12	HO2S-12 (Bank 1)	0.1-1.1v	0.1-1.1	0.1-1.1	0.1-1.1
HO2S-21	HO2S-21 (Bank 2)	0.1-1.1v	0.1-1.1	0.1-1.1	0.1-1.1
HO2S-22	HO2S-22 (Bank 2)	0.1-1.1v	0.1-1.1	0.1-1.1	0.1-1.1
IAT (°F)	IAT Sensor	-40-304°F	50-120	50-120	50-120
LOAD	Engine Load	0-100%	10-20	16-30	13-50
MAF (V)	MAF Sensor	0-5.1v	0.6-0.9	0.9-1.9	1.5-2.5
MISF	Misfire Detection	OFF	OFF	OFF	OFF
OCTADJ	Octane Adjustment	CL/OPEN	CLOSED	CLOSED	CLOSED
TSS	TSS Sensor	0-10 KHz	790-820	1190-1450	1950-2200
PNP	PNP Switch	ON / OFF	ON	OFF	OFF
PSP	PSP Switch	HI / LOW	HI: turning	LOW	LOW
RPM	Engine Speed	0-10K rpm	660-800	1440-1625	1830-1970
TFT (°F)	TFT Sensor	-40-304°F	110-210	110-210	110-210
TPV	TP Sensor	0-5.1v	0.53-1.27	0.8-1.2	0.8-1.2
TRV/TR	TR Sensor	0-5.1v	4.4 / PN	2.1 / OD	2.1 / OD
Vacuum	Engine Vacuum	0-30" Hg	19-20	18-20	8-15
VPWR	Vehicle Power	0-25.5v	12-14	12-14	12-14
VSS	Vehicle Speed	0-159 mph 0-1 KHz	0 = 0 Hz	30 = 65 Hz	55 = 125 Hz

1996-2004 Taurus 3.0L V6 FFV VIN 1, 2 (A/T) - Outputs

PCM PID Acronym	Parameter Identification	PID Range	PID Value at Hot Idle	PID Value at 30	PID Value at 55
AIR	AIR Solenoid	ON / OFF	OFF	OFF	OFF
CD1	Coil Driver 1	0-60° dwell	7	11	12
CD2	Coil Driver 2	0-60° dwell	7	11	12
CD3	Coil Driver 3	0-60° dwell	7	11	12
CTO	Clean Tach Output	0-10 KHz	42-50	65-78	91-105
EGRVR	VR Solenoid	0-100%	0	0-40	40-50
EPC	EPC Solenoid	0-100 psi	38-42	26-30	28-30
EVAPCV	EVAP CV Valve	0-100%	0-100	0-100	0-100
EVAPDC	EVAP Purge Valve	0-100%	0-100	0-100	0-100
FP	Fuel Pump Control	ON / OFF	ON	ON	ON
FuelPW1	INJ Pulsewidth B1	0-99.9 ms	2.3-2.8	2.5-6.0	3.0-7.0
FuelPW2	INJ Pulsewidth B2	0-99.9 ms	2.3-2.8	2.5-6.0	3.0-7.0
HFC	High Fan Control	ON / OFF	ON (if on)	OFF	OFF
HTR-11	HO2S-11 Heater	ON / OFF	ON	ON	ON
HTR-12	HO2S-12 Heater	ON / OFF	ON	ON	ON
HTR-21	HO2S-21 Heater	ON / OFF	ON	ON	ON
HTR-22	HO2S-22 Heater	ON / OFF	ON	ON	ON
IAC	Idle Air Control	0-100%	40-45	40-60	40-60
INJ1	INJ 1 Pulsewidth	0-99.9 ms	2.3-2.8	2.5-6.0	3.0-7.0
INJ2	INJ 2 Pulsewidth	0-99.9 ms	2.3-2.8	2.5-6.0	3.0-7.0
INJ3	INJ 3 Pulsewidth	0-99.9 ms	2.3-2.8	2.5-6.0	3.0-7.0
INJ4	INJ 4 Pulsewidth	0-99.9 ms	2.3-2.8	2.5-6.0	3.0-7.0
INJ5	INJ 5 Pulsewidth	0-99.9 ms	2.3-2.8	2.5-6.0	3.0-7.0
INJ6	INJ 6 Pulsewidth	0-99.9 ms	2.3-2.8	2.5-6.0	3.0-7.0
LFC	Low Fan Control	ON / OFF	ON (if on)	OFF	OFF
LongFT1	Long Term F/T B1	-20 to 20%	-5 to +5	-5 to +5	-5 to +5
LongFT2	Long Term F/T B2	-20 to 20%	-5 to +5	-5 to +5	-5 to +5
MIL	MIL Control	ON / OFF	ON (if on)	OFF	OFF
SHTFT1	Short Term F/T B1	-20 to 20%	-5 to +5	-5 to +5	-5 to +5
SHTFT2	Short Term F/T B2	-20 to 20%	-5 to +5	-5 to +5	-5 to +5
SPARK	Spark Advance	-90° to 90°	24-30	30-45	33-43
SS1	Shift Solenoid 1	ON / OFF	OFF	OFF	ON
SS2	Shift Solenoid 2	ON / OFF	ON	OFF	OFF
SS3	Shift Solenoid 3	ON / OFF	OFF	ON	ON
TCC	TCC Solenoid	0-100%	0	90-100	95-100
VMV	EVAP VMV Valve (1996 only)	0-100%	0-100	0-100	0-100
VREF	Voltage Reference	0-5.1v	4.9-5.1	4.9-5.1	4.9-5.1
WAC	A/C WOT Cutout Relay	ON / OFF	ON (if on)	OFF	OFF

1993-95 Taurus 3.2L V6 MFI SHO VIN P (A/T) Inputs / Outputs

PCM PID Acronym	Parameter Identification	PID Range	PID Value at Hot Idle	PID Value at 30	PID Value at 55
ACCS	A/C Clutch Switch	ON / OFF	ON (if on)	OFF	OFF
BOO	Brake Switch	ON / OFF	ON (if on)	OFF	OFF
CANP	EVAP Purge Valve	0-100%	0	0-40	85-95
ECT (°F)	ECT Sensor	-40-304°F	160-200	160-200	160-200
ECT (V)	ECT Sensor	0-5.1v	0.6	0.6	0.6
EPC	EPC Solenoid	0-100 psi	40	15	42
EVR	VR Solenoid	0-100%	0	0-40	40-50
FPM	Fuel Pump Monitor	ON / OFF	ON	ON	ON
GEAR	Gear Position	P-R-N-D	P-R-N-D	DRIVE 3	DRIVE 4
HFC	High Fan Control	ON / OFF	ON (if on)	OFF	OFF
FLPW1	INJ On-time (B1)	0-99.9 ms	3.6-3.8	4.5-4.7	5.0-10
FLPW2	INJ On-time (B2)	0-99.9 ms	3.6-3.8	4.5-4.7	5.0-10
IAC	Idle Air Control	0-100%	20-40	34-40	45-55
IAT (°F)	IAT Sensor	-40-304°F	50-120	50-120	50-120
IAT (V)	IAT Sensor	0-5.1v	1.7-3.5	1.7-3.5	1.7-3.5
LFC	Low Fan Control	ON / OFF	ON (if on)	OFF	ON
LOOP	Loop Status	CL or OL	CL	CL	CL
LongFT1	Long Term F/T B1	-20 to 20%	-5 to +5	-5 to +5	-5 to +5
LongFT2	Long Term F/T B2	-20 to 20%	-5 to +5	-5 to +5	-5 to +5
MAF (V)	MAF Sensor	0-5.1v	0.6	1.0-1.8v	1.4-1.8v
MLP (V)	MLP Sensor	0-5.1v	0	5	5
HO2S-11	HO2S-11 (Bank 1)	0.1-1.1v	0.1-1.1	0.1-1.1	0.1-1.1
HO2S-21	HO2S-21 (Bank 2)	0.1-1.1v	0.1-1.1	0.1-1.1	0.1-1.1
PFE	PFE Sensor	0-5.1v	3.2	3.3	3.4
PNP	PNP Switch	NEUTRAL	NEUTRAL	DRIVE	DRIVE
PSP	PSP Switch	ON / OFF	ON turning	OFF	OFF
RPM	Engine Speed	0-10K rpm	700-750	1410-1510	1760-1860
SHTFT1	Short Term F/T B1	-20 to 20%	-5 to +5	-5 to +5	-5 to +5
SHTFT2	Short Term F/T B2	-20 to 20%	-5 to +5	-5 to +5	-5 to +5
SPARK	Spark Advance	-90° to 90°	16-21	32-36	30-36°
TCC	TCC Solenoid	0-100%	0	90-100	90-100
TOT °F	TOT Sensor	-40-304°F	110-210	110-210	110-210
TOT (V)	TOT Sensor	0-5.1v	0.5-2.0	0.5-2.0	0.5-2.0
TP	TP Sensor	0-5.1v	0.8	0.9-1.0	1.1-1.2v
TP MIN	TP Minimum	0-5.1v	0.5	0.9-1.0	1.1-1.2v
TP Mode	TP Mode	C/T or P/T	C/T	P/T	P/T
TSS	TSS Sensor	0-10K rpm	550-600	1350-1440	1640-1795
VPWR	Vehicle Power	0-25.5v	12-14	12-14	12-14
VREF	Voltage Reference	0-5.1v	4.9-5.1	4.9-5.1	4.9-5.1
VSS	Vehicle Speed	0-159 mph	0	30	55
WAC	A/C WOT Cutout Relay	ON / OFF	ON (if on)	OFF	OFF

1996-99 Taurus 3.4L V6 MFI SHO VIN N (A/T) - Inputs

PCM PID Acronym	Parameter Identification	PID Range	PID Value at Hot Idle	PID Value at 30	PID Value at 55
ACCS	A/C Clutch Switch	ON / OFF	ON (if on)	OFF	OFF
ACP	A/C Press. Switch	0-5.1v	0.8-1.9v	0.8-1.9v	0.8-1.9v
AIR-M	AIR System Monitor	ON / OFF	ON	ON	ON
BPP	Brake Pedal Switch	ON / OFF	ON (if on)	OFF	OFF
CID	CID Sensor	0-1 KHz	5-7	11-13	12-15
CKP	CKP Sensor	0-10 KHz	380-420	850-900	950-1130
DPFE	DPFE Sensor	0-5.1v	0.4-0.6	0.4-0.7	0.8-2.7
ECT (°F)	ECT Sensor	-40-304°F	160-200	160-200	160-200
FEPS	Flash EEPROM	0-5.1v	0.1	0.1	0.1
FLI	Fuel Level Indicator (1998-99)	0-100%	50 (1.7v)	25-75	25-75
FTP	FTP Sensor (1998-99)	0-5.1v or 0-10" H20	2.6v at 0" of H20	2.6v at 0" of H20	2.6v at 0" of H20
FP-M	Fuel Pump Monitor	ON / OFF	ON	ON	ON
FRP	Fuel Rail Pressure	0-100 psi	30-32	30-32	32-35
FTP	FTP Sensor	0-5.1 or 0-10" H20	2.6v at 0" of H20	2.6v at 0" of H20	2.6v at 0" of H20
GEAR	Transmission Gear	1-2-3-4	1	3	4
HO2S-11	HO2S-11 (Bank 1)	0.1-1.1v	0.1-1.1	0.1-1.1	0.1-1.1
HO2S-12	HO2S-12 (Bank 1)	0.1-1.1v	0.1-1.1	0.1-1.1	0.1-1.1
HO2S-21	HO2S-21 (Bank 2)	0.1-1.1v	0.1-1.1	0.1-1.1	0.1-1.1
HO2S-22	HO2S-22 (Bank 2)	0.1-1.1v	0.1-1.1	0.1-1.1	0.1-1.1
IAT (°F)	IAT Sensor	-40-304°F	50-120	50-120	50-120
IMRCM	IMRC Monitor	0-5.1v	5v	5v	5v
KS1	Knock Sensor B1	0-5.1v	0	0	0
KS2	Knock Sensor B2	0-5.1v	0	0	0
LOAD	Engine Load	0-100%	10-20	16-30	13-32
MAF (V)	MAF Sensor	0-5.1v	0.6-0.9	0.8-1.4	1.1-1.9
MISF	Misfire Detection	OFF	OFF	OFF	OFF
OCTADJ	Octane Adjustment	CL/OPEN	CLOSED	CLOSED	CLOSED
PNP	PNP Switch	NEUTRAL	NEUTRAL	DRIVE	DRIVE
PSP	PSP Switch	ON / OFF	ON turning	OFF	OFF
RPM	Engine Speed	0-10K rpm	650-730	1360-1495	1780-1900
TCS	TCS (switch)	ON / OFF	ON (if on)	OFF	OFF
TFT (°F)	TFT Sensor	-40-304°F	110-210	110-210	110-210
TPV	TP Sensor	0-5.1v	0.5-0.9	0.8-1.2	0.0-1.3v
TRV/TR	TR Sensor	0-5.1v	4.4 / PN	2.1 / OD	2.1 / OD
TSS	TSS Sensor	0-10 KHz	550-600	1350-1440	1640-1795
Vacuum	Engine Vacuum	0-30" Hg	19-20	18-20	8-15
VPWR	Vehicle Power	0-25.5v	12-14	12-14	12-14
VSS	Vehicle Speed	0-159 mph 0-1 KHz	0 = 0 Hz	30 = 65 Hz	55 = 125 Hz

1996-99 Taurus 3.4L V6 SHO MFI VIN N (A/T) - Outputs

PCM PID Acronym	Parameter Identification	PID Range	PID Value at Hot Idle	PID Value at 30	PID Value at 55
AIR	AIR Solenoid	OFF	OFF	OFF	OFF
CD1-CD6	COP Driver 1-6	0-60° dwell	7	10	11
CTO	Clean Tach Output	0-10 KHz	40-50	85-110	100-140
EGRVR	VR Solenoid	0-100%	0%	0-50	0-50
EPC	EPC Solenoid	0-100 psi	40	15	42
EVAPCV	EVAP CV Valve	0-100%	0%	0%	0%
EVAPDC	EVAP Purge Valve	0-100%	0-100%	0-100%	0-100%
FP	Fuel Pump Control	0-100%	100%	100%	100%
FuelPW1	INJ Pulsewidth B1	0-99.9 ms	2.3-2.8	2.8-5.5	2.9-6.0
FuelPW2	INJ Pulsewidth B2	0-99.9 ms	2.3-2.8	2.8-5.5	2.9-6.0
HFC	High Fan Control	ON / OFF	ON	OFF	OFF
HTR-11	HO2S-11 Heater	ON / OFF	ON	ON	ON
HTR-12	HO2S-12 Heater	ON / OFF	ON	ON	ON
HTR-21	HO2S-21 Heater	ON / OFF	ON	ON	ON
HTR-22	HO2S-22 Heater	ON / OFF	ON	ON	ON
IAC	Idle Air Control	0-100%	34	40-60	40-50
INJ1	INJ 1 Pulsewidth	0-99.9 ms	2.3-2.8	2.8-5.5	2.9-6.0
INJ2	INJ 2 Pulsewidth	0-99.9 ms	2.3-2.8	2.8-5.5	2.9-6.0
INJ3	INJ 3 Pulsewidth	0-99.9 ms	2.3-2.8	2.8-5.5	2.9-6.0
INJ4	INJ 4 Pulsewidth	0-99.9 ms	2.3-2.8	2.8-5.5	2.9-6.0
INJ5	INJ 5 Pulsewidth	0-99.9 ms	2.3-2.8	2.8-5.5	2.9-6.0
INJ6	INJ 6 Pulsewidth	0-99.9 ms	2.3-2.8	2.8-5.5	2.9-6.0
IMRC	IMRC Solenoid	ON / OFF	OFF	OFF	OFF
LFC	Low Fan Control	ON / OFF	ON (if on)	OFF	OFF
LongFT1	Long Term F/T B1	-20 to 20%	-5 to +5	-5 to +5	-5 to +5
LongFT2	Long Term F/T B2	-20 to 20%	-5 to +5	-5 to +5	-5 to +5
MIL	MIL Control	ON / OFF	ON (if on)	OFF	OFF
SHTFT1	Short Term F/T B1	-20 to 20%	-5 to +5	-5 to +5	-5 to +5
SHTFT2	Short Term F/T B2	-20 to 20%	-5 to +5	-5 to +5	-5 to +5
SPARK	Spark Advance	-90° to 90°	5-10	31-39	29-41
SS1	Shift Solenoid 1	ON / OFF	OFF	ON	ON
SS2	Shift Solenoid 2	ON / OFF	ON	OFF	OFF
SS3	Shift Solenoid 3	ON / OFF	OFF	ON	ON
TCC	TCC Solenoid	0-100%	0	50-95	90-95
TCIL	TCIL (lamp)	ON / OFF	On (if on)	OFF	OFF
VMV	EVAP VMV Valve (1996-97)	0-100%	0-100	0-100	0-100
VREF	Voltage Reference	0-5.1v	4.9-5.1	4.9-5.1	4.9-5.1
WAC	A/C WOT Cutout Relay	ON / OFF	ON (if on)	OFF	OFF

TAURUS & SABLE PID DATA - OBD II
1996-2002 Taurus/Sable 3.0L V6 MFI VIN S (A/T) - Inputs

PCM PID Acronym	Parameter Identification	PID Range	PID Value at Hot Idle	PID Value at 30	PID Value at 55
ACCS	A/C Clutch Switch	ON / OFF	ON (if on)	OFF	OFF
ACP	A/C Press. Sensor	CL/OPEN	OPEN	OPEN	OPEN
AIR-M	AIR System Monitor	ON / OFF	OFF	OFF	OFF
BPP	Brake Pedal Switch	ON / OFF	ON (if on)	OFF	OFF
CID	CID Sensor	0-1 KHz	5-7	10-13	14-17
CKP	CKP Sensor	0-10 KHz	390-520	850-1120	1140-1220
DPFE	DPFE Sensor	0-5.1v	0.4-0.6	0.4-1.5v	2.0-2.3v
ECT (°F)	ECT Sensor	-40-304°F	160-200	160-200	160-200
FRP	Fuel Rail Pressure	0-100 psi	39	41	40
FEPS	Flash EEPROM	0-5.1v	0.5-0.6	0.5-0.6	0.5-0.6
FLI	Fuel Level Indicator (1998-99)	0-100%	50 (1.7v)	50 (1.7v)	50 (1.7v)
FTP	FTP Sensor (1998-99)	0-5.1v or 0-10" H20	2.6v at 0" of H20	2.6v at 0" of H20	2.6v at 0" of H20
FP-M	Fuel Pump Monitor	ON / OFF	ON	ON	ON
GEAR	Transmission Gear	1-2-3-4	1	3	4
HO2S-11	HO2S-11 (Bank 1)	0.1-1.1v	0.1-1.1	0.1-1.1	0.1-1.1
HO2S-12	HO2S-12 (Bank 1)	0.1-1.1v	0.1-1.1	0.1-1.1	0.1-1.1
HO2S-21	HO2S-21 (Bank 2)	0.1-1.1v	0.1-1.1	0.1-1.1	0.1-1.1
HO2S-22	HO2S-22 (Bank 2)	0.1-1.1v	0.1-1.1	0.1-1.1	0.1-1.1
IAT (°F)	IAT Sensor	-40-304°F	50-120	50-120	50-120
IMRCM	IMRC Monitor	0-5.1v	5v	5v	5v
KS1	Knock Sensor B1	0-5.1v	0v	0v	0v
LOAD	Engine Load	0-100%	15-20%	20-35	15-35
MAF (V)	MAF Sensor	0-5.1v	0.6-0.9	0.7-1.5v	1.3-2.0
MISF	Misfire Detection	OFF	OFF	OFF	OFF
OCTADJ	Octane Adjustment	CL/OPEN	CLOSED	CLOSED	CLOSED
PNP	PNP	NEUTRAL	NEUTRAL	DRIVE	DRIVE
PSPV	PSP Switch	0-25.5v	14 (wheel turning)	0.1	0.1
RPM	Engine Speed	0-10K rpm	700-730	1350-1650	1800-2060
TCS	TCS (switch)	ON / OFF	ON (if on)	OFF	OFF
TFT (°F)	TFT Sensor	-40-304°F	110-210	110-210	110-210
TPV	TP Sensor	0-5.1v	0.5-0.9	0.8-1.1	1.0-1.5v
TRV/TR	TR Sensor	0-5.1v	4.4 / PN	2.1 / OD	2.1 / OD
TSS	TSS Sensor	0-10 KHz	650	1480-1570	1690-1855
Vacuum	Engine Vacuum	0-30" Hg	19-20	18-20	8-15
VPWR	Vehicle Power	0-25.5v	12-14	12-14	12-14
VSS	Vehicle Speed	0-159 mph 0-1 KHz	0 = 0 Hz	30 = 65 Hz	55 = 125 Hz

1996-2002 Taurus/Sable 3.0L V6 MFI VIN S (A/T) - Outputs

PCM PID Acronym	Parameter Identification	PID Range	PID Value at Hot Idle	PID Value at 30	PID Value at 55
AIR	AIR Solenoid	ON / OFF	OFF	OFF	OFF
CD1-CD3	Coil Driver 1-3	0-90° dwell	7	11	12
CTO	Clean Tach Output	0-10 KHz	33-37	75-85	92-120
EGRVR	VR Solenoid	0-100%	0	0-50	0-60
EPC	EPC Solenoid	0-100 psi	9-15	16-25	27-29
EVAPCV	EVAP CV Valve	0-100%	0-100	0-100	0-100
EVAPDC	EVAP Purge Valve	0-100%	0-100	0-100	0-100
FP	Fuel Pump Control	0-100%	100	100	100
FuelPW1	INJ Pulsewidth B1	0-99.9 ms	2.2-2.7	2.3-5.5	2.0-7.0
FuelPW2	INJ Pulsewidth B2	0-99.9 ms	2.2-2.7	2.3-5.5	2.0-7.0
HFC	High Fan Control	OFF	ON (if on)	OFF	OFF
HTR-11	HO2S-11 Heater	ON / OFF	ON	ON	ON
HTR-12	HO2S-12 Heater	ON / OFF	ON	ON	ON
HTR-21	HO2S-21 Heater	ON / OFF	ON	ON	ON
HTR-22	HO2S-22 Heater	ON / OFF	ON	ON	ON
IAC	Idle Air Control	0-100%	25-35	30-35	50-59
INJ1	INJ 1 Pulsewidth	0-99.9 ms	2.2-2.7	2.3-5.5	2.0-7.0
INJ2	INJ 2 Pulsewidth	0-99.9 ms	2.2-2.7	2.3-5.5	2.0-7.0
INJ3	INJ 3 Pulsewidth	0-99.9 ms	2.2-2.7	2.3-5.5	2.0-7.0
INJ4	INJ 4 Pulsewidth	0-99.9 ms	2.2-2.7	2.3-5.5	2.0-7.0
INJ5	INJ 5 Pulsewidth	0-99.9 ms	2.2-2.7	2.3-5.5	2.0-7.0
INJ6	INJ 6 Pulsewidth	0-99.9 ms	2.2-2.7	2.3-5.5	2.0-7.0
LFC	Low Fan Control	ON / OFF	ON (if on)	OFF	OFF
LongFT1	Long Term F/T B1	-20 to 20%	-5 to +5	-5 to +5	-5 to +5
LongFT2	Long Term F/T B2	-20 to 20%	-5 to +5	-5 to +5	-5 to +5
MIL	MIL Control	ON / OFF	ON (if on)	OFF	OFF
SHTFT1	Short Term F/T B1	-20 to 20%	-5 to +5	-5 to +5	-5 to +5
SHTFT2	Short Term F/T B2	-20 to 20%	-5 to +5	-5 to +5	-5 to +5
SPARK	Spark Advance	-90° to 90°	12-27	25-42	20-40
SS1	Shift Solenoid 1	ON / OFF	OFF	OFF	ON
SS2	Shift Solenoid 2	ON / OFF	ON	OFF	OFF
SS3	Shift Solenoid 3	ON / OFF	OFF	ON	ON
TCC	TCC Solenoid	0-100%	0	0-70	90-100
VMV	EVAP VMV Valve (1996-97)	0-100%	0	0-100	0-100
VREF	Voltage Reference	0-5.1v	4.9-5.1	4.9-5.1	4.9-5.1
WAC	A/C WOT Cutout Relay	ON / OFF	ON (if on)	OFF	OFF

1990 Taurus/Sable 3.0L V6 MFI VIN U (A/T) - California

PCM PID Acronym	Parameter Identification	PID Range	PID Value at Hot Idle	PID Value at 30	PID Value at 55
ARC	Auto Ride Control	ON / OFF	ON	ON	ON
BOO	Brake Switch	ON / OFF	ON (if on)	OFF	OFF
CANP	EVAP Purge Valve	ON / OFF	OFF	ON	ON
ECT (°F)	ECT Sensor	-40-304°F	160-200	160-200	160-200
ECT (V)	ECT Sensor	0-5.1v	0.6	0.6	0.6
EVR	VR Solenoid	0-100%	0	0-40	40-50
GEAR	Gear Position	P-R-N-D	P-R-N-D	DRIVE 3	DRIVE 4
HFC	High Fan Control	ON / OFF	ON (if on)	OFF	OFF
FuelPW1	INJ Pulsewidth B1	0-99.9 ms	5.4-5.6	4.8-5.4	12-13
FuelPW2	INJ Pulsewidth B2	0-99.9 ms	5.4-5.6	4.8-5.4	12-13
HO2S-11	HO2S-11 (Bank 1)	0.1-1.1v	0-1v	0-1v	0-1v
HO2S-21	HO2S-21 (Bank 2)	0.1-1.1v	0-1v	0-1v	0-1v
IAC	Idle Air Control	0-100%	20-40	34-40	45-55
IAT (°F)	IAT Sensor	-40-304°F	50-120	50-120	50-120
IAT (V)	IAT Sensor	0-5.1v	1.7-3.5	1.7-3.5	1.7-3.5
LFC	Low Fan Control	ON / OFF	ON (if on)	OFF	OFF
LOOP	Loop Status	CL or OL	CL	CL	CL
MAP	MAP Sensor Hertz	0-1 KHz	107 Hz	112	128 Hz
PFE (V)	PFE Sensor	0-5.1v	3.2	3.3-3.4	3.1-3.5v
PNP	PNP Switch	NEUTRAL	NEUTRAL	DRIVE	DRIVE
PSP	PSP Switch	ON / OFF	ON turning	OFF	OFF
RPM	Engine Speed	0-10K rpm	700-750	1410-1510	1760-1860
SCCS	S/C Command Switch	0-25.5v	6.7	6.7	6.7
SCVAC	S/C Vacuum Solenoid	ON / OFF	OFF	OFF	OFF
SCVENT	S/C Vent Solenoid	ON / OFF	OFF	ON	ON
SHTFT1	Short Term F/T B1	-20 to +20	-5 to +5	-5 to +5	-5 to +5
SHTFT2	Short Term F/T B2	-20 to +20	-5 to +5	-5 to +5	-5 to +5
SPARK	Spark Advance	-90° to 90°	15-22	28-35	25-35
TCC	TCC Solenoid	ON / OFF	OFF	ON	ON
THS 3-2	Trans. 3-2 Switch	0-25.5v	11v	0.1	0.1
THS 4-3	Trans. 4-3 Switch	0-25.5v	11v	0.1	0.1
TPV	TP Sensor	0-5.1v	0.7	0.9	1.2-1.4
TP Mode	TP Mode	C/T or P/T	C/T	P/T	P/T
VPWR	Vehicle Power	0-25.5v	12-14	12-14	12-14
VREF	Voltage Reference	0-5.1v	4.9-5.1	4.9-5.1	4.9-5.1
VSS	Vehicle Speed	0-159 mph	0	30	55
WAC	A/C WOT Cutout Relay	ON / OFF	ON (if on)	OFF	OFF

1991-95 Taurus/Sable 3.0L V6 MFI VIN U (A/T) Inputs / Outputs

PCM PID Acronym	Parameter Identification	PID Range	PID Value at Hot Idle	PID Value at 30	PID Value at 55
ACCS	A/C Clutch Switch	ON / OFF	ON (if on)	OFF	OFF
BOO	Brake Switch	ON / OFF	ON (if on)	OFF	OFF
CANP	EVAP Purge Valve	0-100%	0	0-40	85-95
ECT (°F)	ECT Sensor	-40-304°F	160-200	160-200	160-200
ECT (V)	ECT Sensor	0-5.1v	0.6	0.6	0.6
EPC	EPC Solenoid	0-100 psi	40	15	42
EVR	VR Solenoid	0-100%	0	0-40	40-50
FPM	Fuel Pump Monitor	ON / OFF	ON	ON	ON
GEAR	Gear Position	P-R-N-D	P-R-N-D	DRIVE 3	DRIVE 4
HFC	High Fan Control	ON / OFF	ON (if on)	OFF	OFF
FUELB1	INJ On-time B1	0-99.9 ms	5.4-5.6	4.8-5.4	12-13
FUELB2	INJ On-time B2	0-99.9 ms	5.4-5.6	4.8-5.4	12-13
HO2S-11	HO2S-11 (Bank 1)	0.1-1.1v	0.1-1.1	0.1-1.1	0.1-1.1
HO2S-21	HO2S-21 (Bank 2)	0.1-1.1v	0.1-1.1	0.1-1.1	0.1-1.1
IAC	Idle Air Control	0-100%	20-40	34-40	45-55
IAT (°F)	IAT Sensor	-40-304°F	50-120	50-120	50-120
IAT (V)	IAT Sensor	0-5.1v	1.7-3.5	1.7-3.5	1.7-3.5
LFC	Low Fan Control	ON / OFF	ON (if on)	OFF	ON
LOOP	Loop Status	CL or OL	CL	CL	CL
LongFT1	Long Term F/T B1	-20 to 20%	-5 to +5	-5 to +5	-5 to +5
LongFT2	Long Term F/T B2	-20 to 20%	-5 to +5	-5 to +5	-5 to +5
MAF (V)	MAF Sensor	0-5.1v	0.7	1.3-1.6	1.9-2.2
MLP (V)	MLP Sensor	0-5.1v	0v	5v	5v
PFE (V)	PFE Sensor volts	0-5.1v	3.2	3.3	3.4
PNP	PNP Switch	NEUTRAL	NEUTRAL	DRIVE	DRIVE
PSP	PSP Switch	ON / OFF	ON turning	OFF	OFF
RPM	Engine Speed	0-10K rpm	700-750	1410-1510	1760-1860
SHTFT1	Short Term F/T B1	-20 to 20%	-5 to +5	-5 to +5	-5 to +5
SHTFT2	Short Term F/T B2	-20 to 20%	-5 to +5	-5 to +5	-5 to +5
SPARK	Spark Advance	-90° to 90°	26-30	44-48	46-52
TCC	TCC Solenoid	0-100%	0	90-100	90-100
TOT °F	TOT Sensor	-40-304°F	110-210	110-210	110-210
TOT (V)	TOT Sensor	0-5.1v	0.5-2.0	0.5-2.0	0.5-2.0
TPV	TP Sensor	0-5.1v	0.7	0.9	1.2-1.4
TP MIN	TP Minimum	0-5.1v	0.5	0.9	1.2-1.4
TP Mode	TP Mode	C/T or P/T	C/T	P/T	P/T
TSS	TSS Sensor	0-10 KHz	790-820	1400-1450	1740-1800
VPWR	Vehicle Power	0-25.5v	12-14	12-14	12-14
VREF	Voltage Reference	0-5.1v	4.9-5.1	4.9-5.1	4.9-5.1
VSS	Vehicle Speed	0-159 mph	0	30	55
WAC	A/C WOT Cutout	ON / OFF	ON (if on)	OFF	OFF

1995 Taurus/Sable 3.0L V6 MFI VIN U (A/T) - California

PCM PID Acronym	Parameter Identification	PID Range	PID Value at Hot Idle	PID Value at 30	PID Value at 55
ACCS	A/C Clutch Switch	ON / OFF	ON (if on)	OFF	OFF
BOO	Brake Switch	ON / OFF	ON (if on)	OFF	OFF
CANP	EVAP Purge Valve	0-100%	0	0-40	85-95
DPFE	DPFE Sensor	0-5.1v	0.4-0.6	0.4-0.9	0.4-2.3
ECT (°F)	ECT Sensor	-40-304°F	160-200	160-200	160-200
ECT (V)	ECT Sensor	0-5.1v	0.6	0.6	0.6
EPC	EPC Solenoid	0-100 psi	40	15	42
EVR	VR Solenoid	0-100%	0%	0-40	40-50
FPM	Fuel Pump Monitor	ON / OFF	ON	ON	ON
GEAR	Gear Position	P-R-N-D	P-R-N-D	DRIVE 3	DRIVE 4
HFC	High Fan Control	ON / OFF	ON (if on)	OFF	OFF
FUELB1	INJ On-time B1	0-99.9 ms	5.4-5.6	4.8-5.4	12-13
FUELB2	INJ On-time B2	0-99.9 ms	5.4-5.6	4.8-5.4	12-13
HO2S-11	HO2S-11 (Bank 1)	0.1-1.1v	0-1v	0-1v	0-1v
HO2S-21	HO2S-21 (Bank 2)	0.1-1.1v	0-1v	0-1v	0-1v
IAC	Idle Air Control	0-100%	20-40	34-40	45-55
IAT (°F)	IAT Sensor	-40-304°F	50-120	50-120	50-120
IAT (V)	IAT Sensor	0-5.1v	1.7-3.5	1.7-3.5	1.7-3.5
LFC	Low Fan Control	ON / OFF	ON (if on)	OFF	ON
LOOP	Loop Status	CL or OL	CL	CL	CL
LongFT1	Long Term F/T B1	-20 to 20%	-5 to +5	-5 to +5	-5 to +5
LongFT2	Long Term F/T B2	-20 to 20%	-5 to +5	-5 to +5	-5 to +5
MAF (V)	MAF Sensor	0-5.1v	0.7	1.3-1.6	1.9-2.2
MLP (V)	MLP Sensor	0-5.1v	0v	5v	5v
PNP	PNP Switch	NEUTRAL	NEUTRAL	DRIVE	DRIVE
PSP	PSP Switch	ON / OFF	ON turning	OFF	OFF
RPM	Engine Speed	0-10K rpm	700-750	1410-1510	1760-1860
SHTFT1	Short Term F/T B1	-20 to 20%	-5 to +5	-5 to +5	-5 to +5
SHTFT2	Short Term F/T B2	-20 to 20%	-5 to +5	-5 to +5	-5 to +5
SPARK	Spark Advance	-90° to 90°	26-30	44-48	46-52
TCC	TCC Solenoid	0-100%	0%	90-100	90-100
TOT °F	TOT Sensor	-40-304°F	110-210	110-210	110-210
TOT (V)	TOT Sensor	0-5v	0.5-2.0	0.5-2.0	0.5-2.0
TPV	TP Sensor	0-5.1v	0.7	0.9	1.2-1.4
TP MIN	TP Minimum	0-5.1v	0.5	0.9	1.2-1.4
TP Mode	TP Mode	C/T or P/T	C/T	P/T	P/T
TSS	TSS Sensor	0-10 KHz	790-820	1400-1450	1740-1800
VPWR	Vehicle Power	0-25.5v	12-14	12-14	12-14
VREF	Voltage Reference	0-5.1v	4.9-5.1	4.9-5.1	4.9-5.1
VSS	Vehicle Speed	0-159 mph	0	30	55
WAC	A/C WOT Cutout	ON / OFF	ON (if on)	OFF	OFF

1996-2004 Taurus/Sable 3.0L V6 MFI VIN U (A/T) - Inputs

PCM PID Acronym	Parameter Identification	PID Range	PID Value at Hot Idle	PID Value at 30	PID Value at 55
ACCS	A/C Clutch Switch	ON / OFF	ON (if on)	OFF	OFF
ACET	A/C Temp. Sensor	0-12v	4.95-5.15	4.95-5.15	4.95-5.15
ACP	A/C Press. Sensor	CL/OPEN	OPEN	OPEN	OPEN
AIR-M	EAM Monitor (CAL)	ON / OFF	ON	ON	ON
BPP	Brake Pedal Switch	ON / OFF	ON (if on)	OFF	OFF
CID	CID Sensor	0-1 KHz	6-8	12-14	13-16
CKP	CKP Sensor	0-10 KHz	410-510	810-950	1050-1820
DPFE	DPFE Input (96-'00)	0-5.1v	0.2-1.3	0.2-4.5	0.2-4.5
DPFE	DPFE Sensor	0-5.1v	0.95-1.05	0.9-4.5	0.9-4.5
ECT (ºF)	ECT Sensor	-40-304ºF	160-200	160-200	160-200
FRP	Fuel Rail Pressure	0-100 psi	54	42	40
FEPS	Flash EEPROM	0-5.1v	0.5-0.6	0.5-0.6	0.5-0.6
FLI	Fuel Level Indicator (1998-02)	0-100%	50 (1.7v)	25-75	25-75
FPM	FP Monitor ('96-'99)	ON / OFF	ON	ON	ON
FPM	FP Monitor ('00-'02)	0-100%	90	90	90
FTP	FTP Sensor (1998-02)	0-5.1v or 0-10" H20	2.6v at 0" of H20	2.6v at 0" of H20	2.6v at 0" of H20
GEAR	Transmission Gear	1-2-3-4	1	3	4
GFS	Generator Field	0-1 KHz	130 (30%)	130 (27%)	130 (33%)
HO2S-11	HO2S-11 (Bank 1)	0.1-1.1v	0.1-1.1	0.1-1.1	0.1-1.1
HO2S-12	HO2S-12 (Bank 1)	0.1-1.1v	0.1-1.1	0.1-1.1	0.1-1.1
HO2S-21	HO2S-21 (Bank 2)	0.1-1.1v	0.1-1.1	0.1-1.1	0.1-1.1
HO2S-22	HO2S-22 (Bank 2)	0.1-1.1v	0.1-1.1	0.1-1.1	0.1-1.1
IAT (ºF)	IAT Sensor	-40-304ºF	50-120	50-120	50-120
KS1	Knock Sensor 1	0-25v	0	0	0
LOAD	Engine Load	0-100%	10-20	16-30	13-50
MAF (V)	MAF Sensor	0-5.1v	0.6-0.9	1.0-1.5	1.1-2.0
MISF	Misfire Detection	OFF	OFF	OFF	OFF
OCTADJ	Octane Adjust	Retard/No	No Retard	No Retard	No Retard
OSS	Output Shaft Speed	0-10 KHz	0	300	500
PATSIN	PAT System	See PATS	See PATS	See PATS	See PATS
PNP	PNP Switch	ON / OFF	ON (in 'P')	OFF	OFF
PSP	PSP Switch	HI / LOW	HI: turning	LOW	LOW
RPM	Engine Speed	0-10K rpm	660-800	1440-1625	1830-1970
TFT (ºF)	TFT Sensor	-40-304ºF	110-210	110-210	110-210
TPV	TP Sensor	0-5.1v	0.53-1.27	0.8-1.2	0.8-1.2
TR1	TR Sensor 1	0-12v	0	10.5	10.5
TR2	TR Sensor 2	0-12v	0	10.5	10.5
TR4	TR Sensor 4	0-12v	0	10.5	10.5
TRV/TR	TR Sensor	0-5.1v	4.4 / PN	1.7 / OD	1.7 / OD
TSS	Turbine shaft speed	0-10 KHz	50-65	82-99	88-120
Vacuum	Engine Vacuum	0-30" Hg	19-20	18-20	8-15
VPWR	Vehicle Power	0-25.5v	12-14	12-14	12-14
VSS	Vehicle Speed	0-159 mph 0-1 KHz	0 = 0 Hz	30 = 65 Hz	55 = 125 Hz

1996-2004 Taurus/Sable 3.0L V6 MFI VIN U (A/T) - Outputs

PCM PID Acronym	Parameter Identification	PID Range	PID Value at Hot Idle	PID Value at 30	PID Value at 55
AIR	EAM Signal (CAL)	ON / OFF	OFF	OFF	OFF
CD1-CD3	Coil Driver 1-3	0-60° dwell	7	10	12
CTO	Clean Tach Output	0-10 KHz	35-50	65-78	91-105
EGRVR	VR Solenoid	0-100%	0	0-40	36-50
EPC AX4N	EPC Solenoid	0-100 psi	40	15	42
EPC AX4S	EPC Solenoid	0-100 psi	15	17	40
EVAPCV	EVAP CV Valve	0-100%	0-100	0-100	0-100
EVAPDC	EVAP Purge Valve	0-100%	0-100	0-100	0-100
FP	Fuel Pump (96-00)	ON / OFF	ON	ON	ON
FP	Fuel Pump (01-02)	0-12v	1v (100%)	1v (100%)	1v (100%)
FUELB1	INJ Pulsewidth B1	0-99.9 ms	3.8-4.7	3.9-8.0	3.0-9.0
FUELB2	INJ Pulsewidth B2	0-99.9 ms	3.8-4.7	3.9-8.0	3.0-9.0
GENFDC	Generator Field	0-1 KHz	0	0	0
HFC	High Fan Control	ON / OFF	ON (if on)	OFF	OFF
HTR-11	HO2S-11 Heater	ON / OFF	ON	ON	ON
HTR-12	HO2S-12 Heater	ON / OFF	ON	ON	ON
HTR-21	HO2S-21 Heater	ON / OFF	ON	ON	ON
HTR-22	HO2S-22 Heater	ON / OFF	ON	ON	ON
IAC	Idle Air Control	0-100%	40	40-60	40-47
INJ1	INJ 1 Pulsewidth	0-99.9 ms	3.8-4.7	3.9-8.0	3.0-9.0
INJ2	INJ 2 Pulsewidth	0-99.9 ms	3.8-4.7	3.9-8.0	3.0-9.0
INJ3	INJ 3 Pulsewidth	0-99.9 ms	3.8-4.7	3.9-8.0	3.0-9.0
INJ4	INJ 4 Pulsewidth	0-99.9 ms	3.8-4.7	3.9-8.0	3.0-9.0
INJ5	INJ 5 Pulsewidth	0-99.9 ms	3.8-4.7	3.9-8.0	3.0-9.0
INJ6	INJ 6 Pulsewidth	0-99.9 ms	3.8-4.7	3.9-8.0	3.0-9.0
LFC	Low Fan Control	ON / OFF	ON (if on)	OFF	OFF
LongFT1	Long Term F/T B1	-20 to 20%	-5 to +5	-5 to +5	-5 to +5
LongFT2	Long Term F/T B2	-20 to 20%	-5 to +5	-5 to +5	-5 to +5
MIL	MIL Control	ON / OFF	ON (if on)	OFF	OFF
PATSOUT	PAT System	See PATS	See PATS	See PATS	See PATS
PATSIL	PAT System	See PATS	See PATS	See PATS	See PATS
PATSTRT	PAT System	See PATS	See PATS	See PATS	See PATS
SHTFT1	Short Term F/T B1	-20 to 20%	-5 to +5	-5 to +5	-5 to +5
SHTFT2	Short Term F/T B2	-20 to 20%	-5 to +5	-5 to +5	-5 to +5
SPARK	Spark Advance	-90° to 90°	24-30	34-42	33-46
SS1	Shift Solenoid 1	ON / OFF	OFF	ON	ON
SS2	Shift Solenoid 2	ON / OFF	ON	OFF	ON
SS3	Shift Solenoid 3	ON / OFF	OFF	ON	ON
TCC	TCC Solenoid	0-100%	0	42	95-100
VMV	EVAP VMV '96-'97	0-100	0-100	0-100	0-100
VREF	Voltage Reference	0-5.1v	4.9-5.1	4.9-5.1	4.9-5.1
VSO	VSS Output Signal	0-1 KHz	0	65	125
WAC	A/C WOT Cutout Relay	ON / OFF	ON (if on)	OFF	OFF

1990 Taurus/Sable 3.8L V6 MFI VIN 4 (A/T) Inputs / Outputs

PCM PID Acronym	Parameter Identification	PID Range	PID Value at Hot Idle	PID Value at 30	PID Value at 55
ARC	Auto Ride Control	ON / OFF	OFF	OFF	OFF
BOO	Brake Switch	ON / OFF	ON (if on)	OFF	OFF
CANP	EVAP Purge Valve	ON / OFF	OFF	ON	ON
ECT (°F)	ECT Sensor	-40-304°F	160-200	160-200	160-200
ECT (V)	ECT Sensor	0-5.1v	0.6	0.6	0.6
EVR	VR Solenoid	0-100%	0	0-40	0-60
GEAR	Gear Position	P-R-N-D	P-R-N-D	DRIVE 3	DRIVE 4
FLPW1	INJ On-time (B1)	0-99.9 ms	5.9-6.1	7.4-8.4	8.0-9.0
FLPW2	INJ On-time (B2)	0-99.9 ms	5.9-6.1	7.4-8.4	8.0-9.0
HFC	High Fan Control	ON / OFF	ON (if on)	OFF	OFF
HO2S-11	HO2S-11 (Bank 1)	0.1-1.1v	0.1-1.1	0.1-1.1	0.1-1.1
HO2S-21	HO2S-21 (Bank 2)	0.1-1.1v	0.1-1.1	0.1-1.1	0.1-1.1
IAC	Idle Air Control	0-100%	20-40	34-40	45-55
IAT (°F)	IAT Sensor	-40-304°F	50-120	50-120	50-120
IAT (V)	IAT Sensor	0-5.1v	1.7-3.5	1.7-3.5	1.7-3.5
LFC	Low Fan Control	ON / OFF	ON (if on)	OFF	OFF
LongFT1	Long Term F/T B1	-20 to 20%	-5 to +5	-5 to +5	-5 to +5
LongFT2	Long Term F/T B2	-20 to 20%	-5 to +5	-5 to +5	-5 to +5
LOOP	Loop Status	CL or OL	CL	CL	CL
MAP	MAP Sensor Hertz	0-1 KHz	107	112	128
PFE (V)	PFE Sensor	0-5.1v	3.2	3.3	3.4
PNP	PNP Switch	NEUTRAL	NEUTRAL	DRIVE	DRIVE
RPM	Engine Speed	0-10K rpm	650-750	1340-1440	2410-2510
SCCS	S/C Command Switch	0-25.5v	6.7	6.7	6.7
SCVAC	S/C Vacuum Solenoid	0-100%	0	0	0
SCVENT	S/C Vent Solenoid	0-100%	0	98	98
SHTFT1	Short Term F/T B1	-20 to 20%	-5 to +5	-5 to +5	-5 to +5
SHTFT2	Short Term F/T B2	-20 to 20%	-5 to +5	-5 to +5	-5 to +5
SPARK	Spark Advance	-90° to 90°	18-22	34-38	41-45
TCC	TCC Solenoid	ON / OFF	OFF	ON	ON
THS 3-2	Trans. 3-2 Switch	0-25.5v	11	0.1	0.1
THS 4-3	Trans. 4-3 Switch	0-25.5v	11	0.1	0.1
TPV	TP Sensor	0-5.1v	0.8	0.9-1.0	1.0-1.1
TP Mode	TP Mode	C/T or P/T	C/T	P/T	P/T
VPWR	Vehicle Power	0-25.5v	12-14	12-14	12-14
VREF	Voltage Reference	0-5.1v	4.9-5.1	4.9-5.1	4.9-5.1
VSS	Vehicle Speed	0-159 mph	0	30	55
WAC	A/C WOT Cutout	ON / OFF	ON (if on)	Off	Off

1991 Taurus/Sable 3.8L V6 MFI VIN 4 (A/T) Inputs / Outputs

PCM PID Acronym	Parameter Identification	PID Range	PID Value at Hot Idle	PID Value at 30	PID Value at 55
BOO	Brake Switch	ON / OFF	ON (if on)	OFF	OFF
CANP	EVAP Purge Valve	ON / OFF	OFF	ON	ON
ECT (ºF)	ECT Sensor	-40-304ºF	160-200	160-200	160-200
ECT (V)	ECT Sensor	0-5.1v	0.6	0.6	0.6
EVR	VR Solenoid	0-100%	0	0-40	0-60
GEAR	Gear Position	P-R-N-D	P-R-N-D	DRIVE 3	DRIVE 4
FuelPW1	INJ Pulsewidth B1	0-99.9 ms	5.9-6.1	7.4-8.4	8.0-9.0
FuelPW2	INJ Pulsewidth B2	0-99.9 ms	5.9-6.1	7.4-8.4	8.0-9.0
HFC	High Fan Control	ON / OFF	ON (if on)	OFF	ON
HO2S-11	HO2S-11 (Bank 1)	0.1-1.1v	0.1-1.1	0.1-1.1	0.1-1.1
HO2S-21	HO2S-21 (Bank 2)	0.1-1.1v	0.1-1.1	0.1-1.1	0.1-1.1
IAC	Idle Air Control	0-100%	20-40	34-40	45-55
IAT (ºF)	IAT Sensor	-40-304ºF	50-120	50-120	50-120
IAT (V)	IAT Sensor	0-5.1v	1.7-3.5	1.7-3.5	1.7-3.5
LFC	Low Fan Control	ON / OFF	ON (if on)	OFF	OFF
LOOP	Loop Status	CL or OL	CL	CL	CL
LongFT1	Long Term F/T B1	-20 to 20%	-5 to +5	-5 to +5	-5 to +5
LongFT2	Long Term F/T B2	-20 to 20%	-5 to +5	-5 to +5	-5 to +5
MAF (V)	MAF Sensor	0-5.1v	0.1	0.6-0.8	1.9-2.2
PFE (V)	PFE Sensor	0-5.1v	3.2	3.3	3.4
PNP	PNP Switch	NEUTRAL	NEUTRAL	DRIVE	DRIVE
RPM	Engine Speed	0-10K rpm	650-750	1340-1440	2410-2510
SHTFT1	Short Term F/T B1	-20 to 20%	-5 to +5	-5 to +5	-5 to +5
SHTFT1	Short Term F/T B2	-20 to 20%	-5 to +5	-5 to +5	-5 to +5
SPARK	Spark Advance	-90º to 90º	18-22	34-38	41-45
TCC	TCC Solenoid	ON / OFF	OFF	ON	ON
TPV	TP Sensor	0-5.1v	0.8	0.9-1.0	1.0-1.1
TP Mode	TP Mode	C/T or P/T	C/T	P/T	P/T
VPWR	Vehicle Power	0-25.5v	12-14	12-14	12-14
VREF	Voltage Reference	0-5.1v	4.9-5.1	4.9-5.1	4.9-5.1
VSS	Vehicle Speed	0-159 mph	0	30	55
WAC	A/C WOT Cutout Relay	ON / OFF	ON (if on)	OFF	OFF

1992-95 Taurus/Sable 3.8L V6 MFI VIN 4 (A/T) Inputs / Outputs

PCM PID Acronym	Parameter Identification	PID Range	PID Value at Hot Idle	PID Value at 30	PID Value at 55
ACCS	A/C Clutch Switch	ON / OFF	ON (if on)	OFF	OFF
BOO	Brake Switch	ON / OFF	ON (if on)	OFF	OFF
CANP	EVAP Purge Valve	0-100%	0	0-40	85-95
ECT (ºF)	ECT Sensor	-40-304ºF	160-200	160-200	160-200
ECT (V)	ECT Sensor	0-5.1v	0.6	0.6	0.6
EPC	EPC Solenoid	0-100 psi	16-18	12-16	18-22
EVR	VR Solenoid	0-100%	0%	0-40	35-45
FPM	Fuel Pump Monitor	ON / OFF	ON	ON	ON
GEAR	Gear Position	P-R-N-D	P-R-N-D	DRIVE 3	DRIVE 4
HFC	High Fan Control	ON / OFF	ON (if on)	OFF	OFF
FUEL B1	INJ On-time B1	0-99.9 ms	5.9-6.1	7.4-8.4	8.0-9.0
FUEL B2	INJ On-time B2	0-99.9 ms	5.9-6.1	7.4-8.4	8.0-9.0
HO2S-11	HO2S-11 (Bank 1)	0.1-1.1v	0-1v	0-1v	0-1v
HO2S-21	HO2S-21 (Bank 2)	0.1-1.1v	0-1v	0-1v	0-1v
IAC	Idle Air Control	0-100%	34	44-74	50-75
IAT (ºF)	IAT Sensor	-40-304ºF	50-120	50-120	50-120
IAT (V)	IAT Sensor	0-5.1v	1.7-3.5	1.7-3.5	1.7-3.5
LFC	Low Fan Control	ON / OFF	ON (if on)	OFF	ON
LOOP	Loop Status	CL or OL	CL	CL	CL
LongFT1	Long Term F/T B1	-20 to 20%	-5 to +5	-5 to +5	-5 to +5
LongFT2	Long Term F/T B2	-20 to 20%	-5 to +5	-5 to +5	-5 to +5
MAF (V)	MAF Sensor	0-5.1v	0.7	1.3-1.6	1.9-2.2
MLP (V)	MLP Sensor	0-5.1v	0v	5	5
PFE (V)	PFE Sensor Volts	0-5.1v	3.2	3.3	3.4
PNP	PNP Switch	NEUTRAL	NEUTRAL	DRIVE	DRIVE
PSP	PSP Switch	ON / OFF	ON turning	OFF	OFF
RPM	Engine Speed	0-10K rpm	650-750	1340-1440	2410-2510
SHTFT1	Short Term F/T B1	-20 to 20%	-5 to +5	-5 to +5	-5 to +5
SHTFT2	Short Term F/T B2	-20 to 20%	-5 to +5	-5 to +5	-5 to +5
SPARK	Spark Advance	-90º to 90º	26-30	44-48	46-52
TCC	TCC Solenoid	0-100%	0	90-100	90-100
TOT (ºF)	TOT Sensor	-40-304ºF	110-210	110-210	110-210
TOT (V)	TOT Sensor	0-5.1v	.5-2.0	0.5-2.0	0.5-2.0
TPV	TP Sensor	0-5.1v	0.7	0.9	1.2-1.4
TP MIN	TP Minimum	0-5.1v	0.5	0.9	1.2-1.4
TP Mode	TP Mode	C/T or P/T	C/T	P/T	P/T
TSS	TSS Sensor	0-10 KHz	790-820	1400-1450	1740-1800
VPWR	Vehicle Power	0-25.5v	12-14	12-14	12-14
VREF	Voltage Reference	0-5.1v	4.9-5.1	4.9-5.1	4.9-5.1
VSS	Vehicle Speed	0-159 mph	0	30	55
WAC	A/C WOT Cutout	ON / OFF	ON (if on)	OFF	OFF

1994-95 Taurus/Sable 3.8L V6 MFI VIN 4 (A/T) - California

PCM PID Acronym	Parameter Identification	PID Range	PID Value at Hot Idle	PID Value at 30	PID Value at 55
ACCS	A/C Clutch Switch	ON / OFF	ON (if on)	OFF	OFF
BOO	Brake Switch	ON / OFF	ON (if on)	OFF	OFF
CANP	EVAP Purge Valve	0-100%	0	0-40	85-95
DPFE	DPFE Sensor	0-5.1v	0.4-0.6	0.4-0.7	0.4-1.6v
ECT (°F)	ECT Sensor	-40-304°F	160-200	160-200	160-200
ECT (V)	ECT Sensor	0-5.1v	0.6	0.6	0.6
EPC	EPC Solenoid	0-100 psi	16-18	12-16	18-22
EVR	VR Solenoid	0-100%	0	0-40	35-45
FPM	Fuel Pump Monitor	ON / OFF	ON	ON	ON
GEAR	Gear Position	P-R-N-D	P-R-N-D	DRIVE 3	DRIVE 4
HFC	High Fan Control	ON / OFF	ON (if on)	OFF	OFF
FUEL B1	INJ On-time B1	0-99.9 ms	5.9-6.1	7.4-8.4	8.0-9.0
FUEL B2	INJ On-time B2	0-99.9 ms	5.9-6.1	7.4-8.4	8.0-9.0
HO2S-11	HO2S-11 (Bank 1)	0.1-1.1v	0.1-1.1	0.1-1.1	0.1-1.1
HO2S-21	HO2S-21 (Bank 2)	0.1-1.1v	0.1-1.1	0.1-1.1	0.1-1.1
IAC	Idle Air Control	0-100%	34	44-74	50-75
IAT (°F)	IAT Sensor	-40-304°F	50-120	50-120	50-120
IAT (V)	IAT Sensor	0-5.1v	1.7-3.5	1.7-3.5	1.7-3.5
LFC	Low Fan Control	ON / OFF	ON (if on)	OFF	ON
LOOP	Loop Status	CL or OL	CL	CL	CL
LongFT1	Long Term F/T B1	-20 to 20%	-5 to +5	-5 to +5	-5 to +5
LongFT2	Long Term F/T B2	-20 to 20%	-5 to +5	-5 to +5	-5 to +5
MAF (V)	MAF Sensor	0-5.1v	0.7	1.3-1.6	1.9-2.2
MLP (V)	MLP Sensor	0-5.1v	0	5	5
PNP	PNP Switch	NEUTRAL	NEUTRAL	DRIVE	DRIVE
PSP	PSP Switch	ON / OFF	ON turning	OFF	OFF
RPM	Engine Speed	0-10K rpm	650-750	1340-1440	2410-2510
SHTFT1	Short Term F/T B1	-20 to 20%	-5 to +5	-5 to +5	-5 to +5
SHTFT2	Short Term F/T B2	-20 to 20%	-5 to +5	-5 to +5	-5 to +5
SPARK	Spark Advance	-90° to 90°	26-30	44-48	46-52
TCC	TCC Solenoid	0-100%	0	90-100	90-100
TOT (°F)	TOT Sensor	-40-304°F	110-210	110-210	110-210
TOT (V)	TOT Sensor	0-5.1v	0.5-2.0	0.5-2.0	0.5-2.0
TPV	TP Sensor	0-5.1v	0.7	0.9	1.2-1.4
TP MIN	TP Minimum	0-5.1v	0.5	0.9	1.2-1.4
TP Mode	TP Mode	C/T or P/T	C/T	P/T	P/T
TSS	TSS Sensor	0-10 KHz	790-820	1400-1450	1740-1800
VPWR	Vehicle Power	0-25.5v	12-14	12-14	12-14
VREF	Voltage Reference	0-5.1v	4.9-5.1	4.9-5.1	4.9-5.1
VSS	Vehicle Speed	0-159 mph	0	30	55
WAC	A/C WOT Cutout	ON / OFF	ON (if on)	OFF	OFF

TEMPO & TOPAZ PID DATA

1992-94 Tempo/Topaz 2.3L VIN X, 3.0L VIN U (All) Inputs/Outputs

PCM PID Acronym	Parameter Identification	PID Range	PID Value at Hot Idle	PID Value at 30	PID Value at 55
ACCS	A/C Clutch Switch	ON / OFF	ON (if on)	OFF	OFF
CANP	EVAP Purge Valve	ON / OFF	OFF	ON	ON
ECT (°F)	ECT Sensor	-40-304°F	160-200	160-200	160-200
ECT (V)	ECT Sensor	0-5.1v	0.6	0.6	0.6
EVR	VR Solenoid	0-100%	0%	0%	30-40
FPM	Fuel Pump Monitor	ON / OFF	ON	ON	ON
FuelPW1	INJ Pulsewidth B1	0-99.9 ms	3.6-6.0	4.0-7.6	4.0-10
HFAN	High Fan Control	ON / OFF	ON (if on)	OFF	OFF
HO2S-11	HO2S-11 (Bank 1)	0.1-1.1v	0.1-1.1	0.1-1.1	0.1-1.1
HO2S-21	HO2S-21 (Bank 2)	0.1-1.1v	0.1-1.1	0.1-1.1	0.1-1.1
IAC	Idle Air Control	0-100%	20-40	34-40	45-55
IAT (°F)	IAT Sensor	-40-304°F	50-120	50-120	50-120
IAT (V)	IAT Sensor	0-5.1v	1.7-3.5	1.7-3.5	1.7-3.5
LFAN	Low Fan Control	ON / OFF	ON (if on)	OFF	OFF
LongFT1	Long Term F/T B1	-20 to 20%	-5 to +5	-5 to +5	-5 to +5
LongFT2	Long Term F/T B2	-20 to 20%	-5 to +5	-5 to +5	-5 to +5
LOOP	Loop Status	CL or OL	CL	CL	CL
MAF (V)	Mass Airflow	0-5.1v	0.5	1.3-1.4	1.5-2.0
PFE (V)	PFE Sensor	0-5.1v	3.3	3.4	2.5-3.5
PNP	PNP Switch	NEUTRAL	NEUTRAL	DRIVE	DRIVE
PSP	PSP Switch	HI / LOW	HI: turning	LOW	LOW
RPM	Engine Speed	0-10K rpm	820-860	1500-1600	2550-2650
SHTFT1	Short Term F/T B1	-20 to 20%	-5 to +5	-5 to +5	-5 to +5
SHTFT2	Short Term F/T B2	-20 to 20%	-5 to +5	-5 to +5	-5 to +5
SPARK	Spark Advance	-90° to 90°	16-30	30-34	34-40
TPV	TP Sensor	0-5.1v	0.8	1.0	1.3-1.4
TP MIN	TP Minimum	0-5.1v	0.8	1.0	1.3-1.4
TP Mode	TP Mode	C/T or P/T	C/T	P/T	P/T
VPWR	Vehicle Power	0-25.5v	12-14	12-14	12-14
VREF	Voltage Reference	0-5.1v	4.9-5.1	4.9-5.1	4.9-5.1
VSS	Vehicle Speed	0-159 mph	0	30	55
WAC	A/C WOT Cutout Relay	ON / OFF	ON (if on)	OFF	OFF

THUNDERBIRD PID DATA

1990-95 Thunderbird 3.8L V6 MFI SC VIN R (A/T) Inputs / Outputs

PCM PID Acronym	Parameter Identification	PID Range	PID Value at Hot Idle	PID Value at 30	PID Value at 55
ACCS	A/C Clutch Switch	ON / OFF	ON (if on)	OFF	OFF
BARO	Barometric Pressure Sensor	0-1 KHz	159 Hz (sea level)	159 Hz (sea level)	159 Hz (sea level)
BOO	Brake Switch	ON / OFF	ON (if on)	OFF	OFF
CANP	EVAP Purge Valve	0-100%	0	0-40	85-95
DPFE	DPFE Sensor	0.5.1v	0.4-0.6	0.4-0.9	0.4-2.3
ECT (°F)	ECT Sensor	-40-304°F	160-200	160-200	160-200
ECT (V)	ECT Sensor	0-5.1v	0.6	0.6	0.6
EPC	EPC Solenoid	0-100 psi	40	15	42
EVR	VR Solenoid	0-100%	0%	0-40	40-50
FPM	Fuel Pump Monitor	ON / OFF	ON	ON	ON
GEAR	Gear Position	P-R-N-D	P-R-N-D	DRIVE 3	DRIVE 4
HFC	High Fan Control	ON / OFF	ON (if on)	OFF	OFF
HO2S-11	HO2S-11 (Bank 1)	0.1-1.1v	0-1v	0-1v	0-1v
FUEL B1	INJ On-time B1	0-99.9 ms	3.6-3.8	4.5-6.2	5.8-7.0
FUEL B2	INJ On-time B2	0-99.9 ms	3.6-3.8	4.5-6.2	5.8-7.0
IAC	Idle Air Control	0-100%	20-40	34-40	45-55
IAT (°F)	IAT Sensor	-40-304°F	50-120	50-120	50-120
IAT (V)	IAT Sensor	0-5.1v	1.7-3.5	1.7-3.5	1.7-3.5
LFC	Low Fan Control	ON / OFF	ON (if on)	OFF	ON
LOOP	Loop Status	OL or CL	CL	CL	CL
LNGFT1	LONGFT (Bank 1)	-20 to 20%	-5 to +5	-5 to +5	-5 to +5
LNGFT2	LONGFT (Bank 2)	-20 to 20%	-5 to +5	-5 to +5	-5 to +5
MAF (V)	MAF Sensor	0-5.1v	0.7	1.0-1.4	1.5-2.0v
MLP (V)	MLP Sensor	0-5.1v	0v	5v	5v
PNP	PNP Switch	NEUTRAL	NEUTRAL	DRIVE	DRIVE
PSP	PSP Switch	ON / OFF	ON turning	OFF	OFF
RPM	Engine Speed	0-10K rpm	770-830	1150-1250	1460-1560
SHTFT1	Short Term F/T B1	-20 to 20%	-5 to +5	-5 to +5	-5 to +5
SHTFT2	Short Term F/T B2	-20 to 20%	-5 to +5	-5 to +5	-5 to +5
SPARK	Spark Advance	-90° to 90°	17-22	28-32	28-32
TCC	TCC Solenoid	0-100%	0%	90-100	90-100
TOT °F	TOT Sensor	-40-304°F	110-210	110-210	110-210
TOT (V)	TOT Sensor	0-5.1v	0.5-2.0	0.5-2.0	0.5-2.0
TPV	TP Sensor	0-5.1v	0.7	0.9	1.2-1.4
TP MIN	TP Minimum	0-5.1v	0.5	0.9	1.2-1.4
TP Mode	TP Mode	C/T or P/T	C/T	P/T	P/T
TSS	TSS Sensor	0-10 KHz	790-820	1400-1450	1740-1800
VPWR	Vehicle Power	0-25.5v	12-14	12-14	12-14
VREF	Voltage Reference	0-5.1v	4.9-5.1	4.9-5.1	4.9-5.1
VSS	Vehicle Speed	0-159 mph	0	30	55
WAC	A/C WOT Cutout	ON / OFF	ON (if on)	OFF	OFF

2002-05 Thunderbird 3.9L 4v V8 MFI VIN A (A/T) - Inputs

PCM PID Acronym	Parameter Identification	PID Range	PID Value at Hot Idle	PID Value at 30	PID Value at 55
ACP	A/C Press. Switch	CL/OPEN	OPEN	OPEN	OPEN
BPP	BPP Switch	ON / OFF	ON (if on)	OFF	OFF
BPS	Brake Pedal Switch	0v or 12v	0v (12 on)	0v	0v
CHT (°F)	CHT Sensor (°F)	-40-304°F	160-200	160-200	160-200
CID	CID Sensor	0-1 KHz	6.6	10	17
CKP	CKP Sensor	0-10 KHz	420	665-800	1160-1200
DPFE	DPFE Sensor	0-5.1v	0.25-1.30	0.25-4.65	0.25-4.65
ECT (°F)	ECT Sensor	-40-304°F	160-200	160-200	160-200
EFTA (°F)	EFT Sensor (°F)	-40-304°F	50-120	50-120	50-120
FEPS	Flash EEPROM	0-5.1v	0.1	0.1	0.1
FP-M	Fuel Pump Monitor	0-100%	50	50	50
FRP	Fuel Rail Pressure	0-100 psi	44	44	44
FTP	FTP Sensor	0-5.1v or 0-10" H20	2.6v at 0" of H20	2.6v at 0" of H20	2.6v at 0" of H20
FuelPW1	Fuel On-time B1	0-99.9 ms	2.9-3.6	5.1	6.5-7.5
FuelPW2	Fuel On-time B2	0-99.9 ms	2.9-3.6	5.1	6.5-7.5
GEAR	Transmission Gear	1-2-3-4-5	1	4	5
GFS	Generator Field	0-1K Hz	130 (45%)	130 (25%)	130 (20%)
HO2S-11	HO2S-11 (Bank 1)	0.1-1.1v	0.1-1.1	0.1-1.1	0.1-1.1
HO2S-12	HO2S-12 (Bank 1)	0.1-1.1v	0.1-1.1	0.1-1.1	0.1-1.1
HO2S-21	HO2S-21 (Bank 2)	0.1-1.1v	0.1-1.1	0.1-1.1	0.1-1.1
HO2S-22	HO2S-22 (Bank 2)	0.1-1.1v	0.1-1.1	0.1-1.1	0.1-1.1
IAT (°F)	IAT Sensor	-40-304°F	50-120	50-120	50-120
ISS	Input Shaft Speed	0-10 KHz	235-380	725	1370
KS1	Knock Sensor B1	0-5.1v	0	0	0
KS2	Knock Sensor B2	0-5.1v	0	0	0
LOAD	Engine Load	0-100%	17-18.6	26-35.7	30-50
MAF (V)	MAF Sensor	0-5.1v	0.7	1.6-1.8	2.1-2.3
MISF	Misfire Detection	ON / OFF	OFF	OFF	OFF
OCTADJ	Octane Adj. Switch	Retard/NO	No Retard	No Retard	No Retard
OSS	OSS Sensor Hertz	0-10 KHz	0	536-595	956-1100
PNP	PNP Switch	ON / OFF	ON	OFF	OFF
PS1	PS1 Switch	0-25.5v	12	11.7	11.7
RPM	Engine Speed	0-10K rpm	660-700	1425	1950
TCS	TCS (switch)	ON / OFF	ON (if on)	OFF	OFF
TSS	TSS Sensor	0-10K rpm	340	539	1025
TFT (°F)	TFT Sensor	-40-304°F	110-210	110-210	110-210
TPV	TP Sensor	0-5.1v	0.53-1.05	1.1-1.3	1.3-1.5
TR 1	TR Sensor 1 Volts	0-12v	0.1	11.5	11.5
TR 2	TR2 Sensor	0-12v	0.1	11.5	11.5
TR 4	TR4 Sensor	0-12v	0.1	11.5	11.5
TRV/TR	TR Sensor	0-5.1v	0.1 / PN	1.7 / OD	1.7 / OD
VPWR	Vehicle Power	0-25.5v	12-14	12-14	12-14
VSS	Vehicle Speed	0-159 mph 0-1 KHz	0 = 0 Hz	30 = 65 Hz	55 = 125 Hz

2002-05 T-Bird 3.9L 4v V8 MFI VIN A (A/T) - Outputs

PCM PID Acronym	Parameter Identification	PID Range	PID Value at Hot Idle	PID Value at 30	PID Value at 55
ALDFDC	Alternator Control	0-1 KHz	0-130	0	0
CD1-8	Coil Driver 1-8	0-45° dwell	8	11	12
CHTIL	CHTIL (lamp)	ON / OFF	ON (if on)	OFF	OFF
EGRVR	VR Solenoid	0-100%	0	40	40
EPC	EPC Solenoid (V)	0-25.5v	8.8	9.3	9.9
EPC2	EPC2 Solenoid (V)	0-25.5v	10.9	9.3	9.9
EPC3	EPC3 Solenoid (V)	0-25.5v	8.5	12-14	12-14
EVAPCV	EVAP CV Valve	0-100%	0-100%	0-100%	0-100%
EVAPDC	EVAP Purge Valve	0-100%	0-100%	0-100%	0-100%
FP	Fuel Pump Control	0-100%	26	29	27
HCFD	Hydraulic Cooling Fan Drive	ON / OFF	ON (if on)	OFF	OFF
HTR-11	HO2S-11 Heater	ON / OFF	ON	ON	ON
HTR-12	HO2S-12 Heater	ON / OFF	ON	ON	ON
HTR-21	HO2S-21 Heater	ON / OFF	ON	ON	ON
HTR-22	HO2S-22 Heater	ON / OFF	ON	ON	ON
IAC	Idle Air Control	0-100%	25	54	72
IMRC	IMRC Solenoid	ON / OFF	OFF	OFF	OFF
INJ1	INJ 1 Pulsewidth	0-99.9 ms	2.9-3.6	5.1	6.5-7.5
INJ2	INJ 2 Pulsewidth	0-99.9 ms	2.9-3.6	5.1	6.5-7.5
INJ3	INJ 3 Pulsewidth	0-99.9 ms	2.9-3.6	5.1	6.5-7.5
INJ4	INJ 4 Pulsewidth	0-99.9 ms	2.9-3.6	5.1	6.5-7.5
INJ5	INJ 5 Pulsewidth	0-99.9 ms	2.9-3.6	5.1	6.5-7.5
INJ6	INJ 6 Pulsewidth	0-99.9 ms	2.9-3.6	5.1	6.5-7.5
INJ7	INJ 7 Pulsewidth	0-99.9 ms	2.9-3.6	5.1	6.5-7.5
INJ8	INJ 8 Pulsewidth	0-99.9 ms	2.9-3.6	5.1	6.5-7.5
LNGFT1	L/T Fuel Trim (B1)	-20 to 20%	-5 to +5	-5 to +5	-5 to +5
LNGFT2	L/T Fuel Trim (B2)	-20 to 20%	-5 to +5	-5 to +5	-5 to +5
MIL	MIL Control	ON / OFF	ON (if on)	OFF	OFF
SCC	S/C Signal	0-25.5v	12-14	12-14	12-14
SCCS	S/C Switch Signal	0-5.1v	0.1	4.6	4.6
SCMA	S/C MA Signal	0-25.5v	12-14	9-12	9-12
SCMB	S/C MB Signal	0-25.5v	12-14	9-12	9-12
SCMC	S/C MC Signal	0-25.5v	12-14	9-12	9-12
SHTFT1	Short Term F/T B1	-20 to 20%	-5 to +5	-5 to +5	-5 to +5
SHTFT2	Short Term F/T B2	-20 to 20%	-5 to +5	-5 to +5	-5 to +5
SPARK	Spark Advance	-90° to 90°	10-17	37	36
SS1	Shift Solenoid 1	ON / OFF	ON	OFF	OFF
SS2	Shift Solenoid 2	ON / OFF	OFF	OFF	OFF
SS3	Shift Solenoid 3	ON / OFF	OFF	ON	ON
SS4	Shift Solenoid 4	ON / OFF	ON	ON	ON
TCC	TCC Solenoid	0-100%	0	0	100
VREF	Voltage Reference	0-5.1v	4.9-5.1	4.9-5.1	4.9-5.1

THUNDERBIRD & COUGAR PID DATA

1990-95 Thunderbird/Cougar 3.8L V6 MFI VIN 4 (A/T)

PCM PID Acronym	Parameter Identification	PID Range	PID Value at Hot Idle	PID Value at 30	PID Value at 55
ACCS	A/C Clutch Switch	ON / OFF	ON (if on)	OFF	OFF
ARC	Auto Ride Control (1990)	ON / OFF	ON	ON	ON
CANP	EVAP Purge Valve	0-100%	0	0-40	85-95
ECT (°F)	ECT Sensor	-40-304°F	160-200	160-200	160-200
ECT (V)	ECT Sensor	0-5.1v	0.6	0.6	0.6
EVR	VR Solenoid	0-100%	0%	0-40	35-45
FPM	Fuel Pump Monitor	ON / OFF	ON	ON	ON
FLPW1	INJ On-time (B1)	0-99.9 ms	5.9-6.1	7.4-8.4	8.0-9.0
FLPW2	INJ On-time (B2)	0-99.9 ms	5.9-6.1	7.4-8.4	8.0-9.0
HFC	High Fan Control	ON / OFF	ON (if on)	OFF	OFF
HO2S-11	HO2S-11 (Bank 1)	0.1-1.1v	0.1-1.1	0.1-1.1	0.1-1.1
HO2S-21	HO2S-21 (Bank 2)	0.1-1.1v	0.1-1.1	0.1-1.1	0.1-1.1
IAC	Idle Air Control	0-100%	34	44-74	50-75
IAT (°F)	IAT Sensor	-40-304°F	50-120	50-120	50-120
IAT (V)	IAT Sensor	0-5.1v	1.7-3.5	1.7-3.5	1.7-3.5
LFC	Low Fan Control	ON / OFF	ON (if on)	OFF	OFF
LOOP	Loop Status	OL or CL	CL	CL	CL
LongFT1	Long Term F/T B1	-20 to 20%	-5 to +5	-5 to +5	-5 to +5
LongFT2	Long Term F/T B2	-20 to 20%	-5 to +5	-5 to +5	-5 to +5
MAF (V)	MAF Sensor	0-5.1v	0.7	1.3-1.6	1.9-2.2
PFE (V)	PFE Sensor volts	0-5.1v	3.2	3.3	3.4
PNP	PNP Switch	NEUTRAL	NEUTRAL	DRIVE	DRIVE
PSP	PSP Switch	ON / OFF	ON turning	OFF	OFF
RPM	Engine Speed	0-10K rpm	650-750	1340-1440	2410-2510
SCCS	S/C Command Switch	0-25.5v	6.7	6.7	6.7
SCVAC	S/C Vacuum Solenoid	0-100%	0	0	0
SCVENT	S/C Vent Solenoid	0-100%	0	98	98
SHTFT1	Short Term F/T B1	-20 to 20%	-5 to +5	-5 to +5	-5 to +5
SHTFT2	Short Term F/T B2	-20 to 20%	-5 to +5	-5 to +5	-5 to +5
SPARK	Spark Advance	-90° to 90°	26-30	44-48	46-52
TPV	TP Sensor	0-5.1v	0.7	0.9	1.2-1.4
TP Mode	TP Mode	C/T or P/T	C/T	P/T	P/T
TSS	TSS Sensor	0-10 KHz	790-820	1400-1450	1740-1800
VPWR	Vehicle Power	0-25.5v	12-14	12-14	12-14
VREF	Voltage Reference	0-5.1v	4.9-5.1	4.9-5.1	4.9-5.1
VSS	Vehicle Speed	0-159 mph	0	30	55
WAC	A/C WOT Cutout Relay	ON / OFF	ON (if on)	OFF	OFF

1996-97 Thunderbird/Cougar 3.8L V6 VIN 4 (A/T) Inputs

PCM PID Acronym	Parameter Identification	PID Range	PID Value at Hot Idle	PID Value at 30	PID Value at 55
ACCS	A/C Clutch Switch	ON / OFF	ON (if on)	OFF	OFF
ACP	A/C Press. Sensor	CL/OPEN	OPEN	OPEN	OPEN
BPP	Brake Pedal Switch	ON / OFF	ON (if on)	OFF	OFF
CID	CID Sensor	0-1 KHz	5-7	9-11	11-15
CKP	CKP Sensor	0-10 KHz	390-450	710-760	875-940
DPFE	DPFE Sensor	0-5.1v	0.4-0.6	0.4-0.9	1.0-1.3
ECT (°F)	ECT Sensor	-40-304°F	160-200	160-200	160-200
FRP	Fuel Rail Pressure	0-100 psi	30-32	30-32	32-35
FEPS	Flash EEPROM	0-5.1v	0.1	0.1	0.1
FPM	Fuel Pump Monitor	ON / OFF	ON	ON	ON
GEAR	Transmission Gear	1-2-3-4	1	3	4
HO2S-11	HO2S-11 (Bank 1)	0.1-1.1v	0.1-1.1	0.1-1.1	0.1-1.1
HO2S-12	HO2S-12 (Bank 1)	0.1-1.1v	0.1-1.1	0.1-1.1	0.1-1.1
HO2S-21	HO2S-21 (Bank 2)	0.1-1.1v	0.1-1.1	0.1-1.1	0.1-1.1
HO2S-22	HO2S-22 (Bank 2)	0.1-1.1v	0.1-1.1	0.1-1.1	0.1-1.1
IAT	IAT Sensor	-40-304°F	50-120	50-120	50-120
LOAD	Engine Load	0-100%	10-20	16-30	25-38
MAF (V)	MAF Sensor	0-5.1v	0.6-0.9	1.0-1.2	1.3-2.0
MISF	Misfire Detection	ON / OFF	OFF	OFF	OFF
OCTADJ	Octane Adjustment	CL/OPEN	CLOSED	CLOSED	CLOSED
OSS	OSS Sensor	0-10K rpm	0	1150-1300	2260-2295
PF (V)	Purge Flow Sensor	0-5.1v	1.55	2.10	2.30
PNP	PNP Switch	ON / OFF	ON (in 'P')	OFF	OFF
RPM	Engine Speed	0-10K rpm	740-760	1200-1325	1580-1640
TCS	TCS (switch)	ON / OFF	ON (if on)	OFF	OFF
TFT (°F)	TFT Sensor	-40-304°F	110-210	110-210	110-210
TPV	TP Sensor	0-5.1v	0.5-0.9	0.8-0.9	0.9-1.2
TRV/TR	TR Sensor	0-5.1v	4.4 / PN	2.1 / OD	2.1 / OD
Vacuum	Engine Vacuum	0-30" Hg	19-20	18-20	8-15
VPWR	Vehicle Power	0-25.5v	12-14	12-14	12-14
VSS	Vehicle Speed	0-159 mph 0-1 KHz	0 = 0 Hz	30 = 65 Hz	55 = 125 Hz

1996-97 Thunderbird/Cougar 3.8L V6 VIN 4 (A/T) Outputs

PCM PID Acronym	Parameter Identification	PID Range	PID Value at Hot Idle	PID Value at 30	PID Value at 55
CD1-CD3	Coil Driver 1-3	0-60° dwell	7	10	12
EGRVR	VR Solenoid	0-100%	0	0-40	40-50
EPC	EPC Solenoid	0-100 psi	20	24-30	25-30
EVAPDC	EVAP Purge Valve	0-100%	0-100	0-100	0-100
FP	Fuel Pump Control	0-100%	100	100	100
FuelPW1	INJ Pulsewidth B1	0-99.9 ms	4.5-4.8	5.0-8.0	6.0-11
FuelPW2	INJ Pulsewidth B2	0-99.9 ms	4.5-4.8	5.0-8.0	6.0-11
HFC	High Fan Control	ON / OFF	ON (if on)	OFF	OFF
HTR-11	HO2S-11 Heater	ON / OFF	ON	ON	ON
HTR-12	HO2S-12 Heater	ON / OFF	ON	ON	ON
HTR-21	HO2S-21 Heater	ON / OFF	ON	ON	ON
HTR-22	HO2S-22 Heater	ON / OFF	ON	ON	ON
IAC	Idle Air Control	0-100%	40	45-55	65-75
INJ1	INJ 1 Pulsewidth	0-99.9 ms	4.5-4.8	5.0-8.0	6.0-11
INJ2	INJ 2 Pulsewidth	0-99.9 ms	4.5-4.8	5.0-8.0	6.0-11
INJ3	INJ 3 Pulsewidth	0-99.9 ms	4.5-4.8	5.0-8.0	6.0-11
INJ4	INJ 4 Pulsewidth	0-99.9 ms	4.5-4.8	5.0-8.0	6.0-11
INJ5	INJ 5 Pulsewidth	0-99.9 ms	4.5-4.8	5.0-8.0	6.0-11
INJ6	INJ 6 Pulsewidth	0-99.9 ms	4.5-4.8	5.0-8.0	6.0-11
LFC	Low Fan Control	ON / OFF	ON (if on)	OFF	OFF
LongFT1	Long Term F/T B1	-20 to 20%	-5 to +5	-5 to +5	-5 to +5
LongFT2	Long Term F/T B2	-20 to 20%	-5 to +5	-5 to +5	-5 to +5
MIL	MIL Control	ON / OFF	ON (if on)	OFF	OFF
SHTFT1	Short Term F/T B1	-20 to 20%	-5 to +5	-5 to +5	-5 to +5
SHTFT2	Short Term F/T B2	-20 to 20%	-5 to +5	-5 to +5	-5 to +5
SPARK	Spark Advance	-90° to 90°	15-20	25-35	31-40
SS1	Shift Solenoid 1	ON / OFF	ON	OFF	ON
SS2	Shift Solenoid 2	OFF	OFF	ON	ON
TCC	TCC Solenoid	0-100%	0	35-45	95-100
TCIL	TCIL (lamp)	ON / OFF	ON (if on)	OFF	OFF
VREF	Voltage Reference	0-5.1v	4.9-5.1	4.9-5.1	4.9-5.1
WAC	A/C WOT Cutout Relay	ON / OFF	ON (if on)	OFF	OFF

1994-97 Thunderbird/Cougar 4.6L V8 VIN W (A/T) Inputs

PCM PID Acronym	Parameter Identification	PID Range	PID Value at Hot Idle	PID Value at 30	PID Value at 55
ACCS	A/C Clutch Switch	ON / OFF	ON (if on)	OFF	OFF
ACP	A/C Press. Sensor	CL/OPEN	OPEN	OPEN	OPEN
BPP	Brake Pedal Switch	ON / OFF	ON (if on)	OFF	OFF
CID	CID Sensor	0-1 KHz	5-9	10-12	11-15
CKP	CKP Sensor	0-10 KHz	450-480	720-750	830-870
DPFE	DPFE Sensor	0-5.1v	0.4-0.6	0.6-1.5	1.1-1.7
ECT (°F)	ECT Sensor	-40-304°F	160-200	160-200	160-200
FEPS	Flash EEPROM	0-5.1v	0.1	0.1	0.1
FRP	Fuel Rail Pressure	0-100 psi	30-32	30-32	32-35
FP-M	Fuel Pump Monitor	ON / OFF	ON	ON	ON
GEAR	Transmission Gear	1-2-3-4	1	3	4
HO2S-11	HO2S-11 (Bank 1)	0.1-1.1v	0.1-1.1	0.1-1.1	0.1-1.1
HO2S-12	HO2S-12 (Bank 1)	0.1-1.1v	0.1-1.1	0.1-1.1	0.1-1.1
HO2S-21	HO2S-21 (Bank 2)	0.1-1.1v	0.1-1.1	0.1-1.1	0.1-1.1
HO2S-22	HO2S-22 (Bank 2)	0.1-1.1v	0.1-1.1	0.1-1.1	0.1-1.1
IAT (°F)	IAT Sensor	-40-304°F	50-120	50-120	50-120
LOAD	Engine Load	0-100%	10-20	15-25	14-31
MAF (V)	MAF Sensor	0-5.1v	0.6-0.9	0.8-1.6	0.6-3.1
MISF	Misfire Detection	ON / OFF	OFF	OFF	OFF
OCTADJ	Octane Adjustment	CL/OPEN	CLOSED	CLOSED	CLOSED
OSS	OSS Sensor	0-10 KHz	0 rpm	1210-1235	2050-2200
PF (V)	Purge Flow Sensor	0-5.1v	1.55	2.10	2.30
PNP	PNP Switch	ON / OFF	ON (in 'P')	OFF	OFF
RPM	Engine Speed	0-10K rpm	660-700	1170-1200	1460-1500
TCS	TCS (switch)	ON / OFF	ON (if on)	OFF	OFF
TFT (°F)	TFT Sensor	-40-304°F	110-210	110-210	110-210
TPV	TP Sensor	0-5.1v	0.5-0.9	0.9-1.1	0.9-1.2
TRV/TR	TR Sensor	0-5.1v	4.4 / PN	2.1 / OD	2.1 / OD
Vacuum	Engine Vacuum	0-30" Hg	19-20	18-20	8-15
VPWR	Vehicle Power	0-25.5v	12-14	12-14	12-14
VSS	Vehicle Speed	0-159 mph 0-1 KHz	0 = 0 Hz	30 = 65 Hz	55 = 125 Hz

1994-97 Thunderbird/Cougar 4.6L V8 VIN W (A/T) Outputs

PCM PID Acronym	Parameter Identification	PID Range	PID Value at Hot Idle	PID Value at 30	PID Value at 55
CD1-CD4	Coil Driver 1-4	0-45° dwell	8	11	12
EGRVR	VR Solenoid	0-100%	0	0-40	40-50
EVAPDC	EVAP Purge Valve	0-100%	0-100	0-100	0-100
FP	Fuel Pump Control	ON / OFF	ON	ON	ON
FuelPW1	INJ Pulsewidth B1	0-99.9 ms	3.1-3.5	3.1-6.6	4.7-7.2
FuelPW2	INJ Pulsewidth B2	0-99.9 ms	3.1-3.5	3.1-6.6	4.7-7.2
HFC	High Fan Control	ON / OFF	ON (if on)	OFF	OFF
HTR-11	HO2S-11 Heater	ON / OFF	ON	ON	ON
HTR-12	HO2S-12 Heater	ON / OFF	ON	ON	ON
HTR-21	HO2S-21 Heater	ON / OFF	ON	ON	ON
HTR-22	HO2S-22 Heater	ON / OFF	ON	ON	ON
IAC	Idle Air Control	0-100%	37	50-57	60-65
INJ1	INJ 1 Pulsewidth	0-99.9 ms	3.1-3.5	3.1-6.6	4.7-7.2
INJ2	INJ 2 Pulsewidth	0-99.9 ms	3.1-3.5	3.1-6.6	4.7-7.2
INJ3	INJ 3 Pulsewidth	0-99.9 ms	3.1-3.5	3.1-6.6	4.7-7.2
INJ4	INJ 4 Pulsewidth	0-99.9 ms	3.1-3.5	3.1-6.6	4.7-7.2
INJ5	INJ 5 Pulsewidth	0-99.9 ms	3.1-3.5	3.1-6.6	4.7-7.2
INJ6	INJ 6 Pulsewidth	0-99.9 ms	3.1-3.5	3.1-6.6	4.7-7.2
INJ7	INJ 7 Pulsewidth	0-99.9 ms	3.1-3.5	3.1-6.6	4.7-7.2
INJ8	INJ 8 Pulsewidth	0-99.9 ms	3.1-3.5	3.1-6.6	4.7-7.2
LFC	Low Fan Control	ON / OFF	ON (if on)	OFF	OFF
LongFT1	Long Term F/T B1	-20 to 20%	-5 to +5	-5 to +5	-5 to +5
LongFT2	Long Term F/T B2	-20 to 20%	-5 to +5	-5 to +5	-5 to +5
MIL	MIL Control	ON / OFF	ON (if on)	OFF	OFF
SHTFT1	Short Term F/T B1	-20 to 20%	-5 to +5	-5 to +5	-5 to +5
SHTFT2	Short Term F/T B2	-20 to 20%	-5 to +5	-5 to +5	-5 to +5
SPARK	Spark Advance	-90° to 90°	15-20	30-41	29-40
SS1	Shift Solenoid 1	ON / OFF	ON	OFF	ON
SS2	Shift Solenoid 2	ON / OFF	OFF	ON	ON
TCC	TCC Solenoid	0-100%	0	37-43	95-100
TCIL	TCIL (lamp)	ON / OFF	ON (if on)	OFF	OFF
VREF	Voltage Reference	0-5.1v	4.9-5.1	4.9-5.1	4.9-5.1
WAC	A/C WOT Cutout Relay	ON / OFF	ON (if on)	OFF	OFF

1991-93 Thunderbird/Cougar 5.0L V8 MFI VIN T (A/T)

PCM PID Acronym	Parameter Identification	PID Range	PID Value at Hot Idle	PID Value at 30	PID Value at 55
ACCS	A/C Clutch Switch	ON / OFF	ON (if on)	OFF	OFF
ARC	Auto Ride Control	ON / OFF	ON	ON	ON
BARO	Barometric Pressure Sensor	0-1 KHz	159 Hz (sea level)	159 Hz (sea level)	159 Hz (sea level)
CANP	EVAP Purge Valve	0-100%	0	0-40	85-95
ECT (°F)	ECT Sensor	-40-304°F	160-200	160-200	160-200
ECT (V)	ECT Sensor	0-5.1v	0.6	0.6	0.6
EVP	EGR EVP Sensor	0-5.1v	0.4v	0.4v	1.3-1.6
EVR	VR Solenoid	0-100%	0%	0-40	35-45
FPM	Fuel Pump Monitor	ON / OFF	ON	ON	ON
FuelPW1	INJ.Pulsewidth B1	0-99.9 ms	4.9-5.2	4.8-6.0	7.4-8.6
FuelPW2	INJ Pulsewidth B2	0-99.9 ms	4.9-5.2	4.8-6.0	7.4-8.6
HFC	High Fan Control	ON / OFF	ON (if on)	OFF	OFF
HO2S-11	HO2S-11 (Bank 1)	0.1-1.1v	0.1-1.1	0.1-1.1	0.1-1.1
HO2S-21	HO2S-21 (Bank 2)	0.1-1.1v	0.1-1.1	0.1-1.1	0.1-1.1
IAC	Idle Air Control	0-100%	34	44-74	50-75
IAT (°F)	IAT Sensor	-40-304°F	50-120	50-120	50-120
IAT (V)	IAT Sensor	0-5.1v	1.7-3.5	1.7-3.5	1.7-3.5
LFC	Low Fan Control	ON / OFF	ON (if on)	OFF	OFF
LOOP	Loop Status	OL or CL	CL	CL	CL
LongFT1	Long Term F/T B1	-20 to 20%	-5 to +5	-5 to +5	-5 to +5
LongFT2	Long Term F/T B2	-20 to 20%	-5 to +5	-5 to +5	-5 to +5
MAF (V)	MAF Sensor	0-5.1v	0.7	1.3-1.6	1.9-2.2
PNP	PNP Switch	ON / OFF	ON (in 'P')	OFF	OFF
PSP	PSP Switch	ON / OFF	ON turning	OFF	OFF
RPM	Engine Speed	0-10K rpm	715-755	1140-1180	1390-1420
SHTFT1	Short Term F/T B1	-20 to 20%	-5 to +5	-5 to +5	-5 to +5
SHTFT2	Short Term F/T B2	-20 to 20%	-5 to +5	-5 to +5	-5 to +5
SPARK	Spark Advance	-90° to 90°	18-22	32-36	40-44
TP	TP Sensor	0-5.1v	1.0	1.1	1.2-1.3
TP Mode	TP Mode	C/T or P/T	C/T	P/T	P/T
VPWR	Vehicle Power	0-25.5v	12-14	12-14	12-14
VREF	Voltage Reference	0-5.1v	4.9-5.1	4.9-5.1	4.9-5.1
VSS	Vehicle Speed	0-159 mph 0-1 KHz	0 = 0 Hz	30 = 65 Hz	55 = 125 Hz
WAC	A/C WOT Cutout Relay	ON / OFF	ON (if on)	OFF	OFF

CONTINENTAL PID DATA

1990-94 Continental 3.8L V6 MFI VIN 4 (A/T) Inputs / Outputs

PCM PID Acronym	Parameter Identification	PID Range	PID Value at Hot Idle	PID Value at 30	PID Value at 55
ACCS	A/C Clutch Switch	ON / OFF	ON (if on)	OFF	OFF
ARC	Auto Ride Control	ON / OFF	ON	ON	ON
BOO	Brake Switch	ON / OFF	ON (if on)	OFF	OFF
CANP	EVAP Purge Valve	0-100%	0	0-40	85-95
ECT (°F)	ECT Sensor	-40-304°F	160-200	160-200	160-200
ECT (V)	ECT Sensor	0-5.1v	0.6	0.6	0.6
EPC	EPC Solenoid	0-100 psi	40	15	42
EVR	VR Solenoid	0-100%	0%	0-40	40-50
FPM	Fuel Pump Monitor	ON / OFF	ON	ON	ON
GEAR	Gear Position	P-R-N-D	P-R-N-D	DRIVE 3	DRIVE 4
HFC	High Fan Control	ON / OFF	ON (if on)	OFF	OFF
FUEL B1	INJ On-time B1	0-99.9 ms	3.6-3.8	4.5-6.2	5.8-7.0
FUEL B2	INJ On-time B2	0-99.9 ms	3.6-3.8	4.5-6.2	5.8-7.0
HO2S-11	HO2S-11 (Bank 1)	0.1-1.1v	0.1-1.1	0.1-1.1	0.1-1.1
HO2S-21	HO2S-21 (Bank 2)	0.1-1.1v	0.1-1.1	0.1-1.1	0.1-1.1
IAC	Idle Air Control	0-100%	20-40	34-40	45-55
IAT (°F)	IAT Sensor	-40-304°F	50-120	50-120	50-120
IAT (V)	IAT Sensor	0-5.1v	1.7-3.5	1.7-3.5	1.7-3.5
LFC	Low Fan Control	ON / OFF	ON (if on)	OFF	ON
LOOP	Loop Status	OL or CL	CL	CL	CL
LGFT1	L/T Fuel Trim (B1)	-20 to 20%	-5 to +5	-5 to +5	-5 to +5
LGFT2	L/T Fuel Trim (B2)	-20 to 20%	-5 to +5	-5 to +5	-5 to +5
MAF (V)	MAF Sensor	0-5.1v	0.7	1.0-1.4	1.5-2.0v
MLP (V)	MLP Sensor	0-5.1v	0v	5v	5v
PFE (V)	PFE Sensor	0-5.1v	3.3	3.2	2.8-3.2
PNP	PNP Switch	NEUTRAL	NEUTRAL	DRIVE	DRIVE
PSP	PSP Switch	ON / OFF	ON turning	OFF	OFF
RPM	Engine Speed	0-10K rpm	700-750	1300-1400	1650-1750
SHTFT1	Short Term F/T B1	-20 to 20%	-5 to +5	-5 to +5	-5 to +5
SHTFT2	Short Term F/T B2	-20 to 20%	-5 to +5	-5 to +5	-5 to +5
SPARK	Spark Advance	-90° to 90°	20-22	42-48	42-48
TCC	TCC Solenoid	0-100%	0	90-100	90-100
TOT °F	TOT Sensor	-40-304°F	110-210	110-210	110-210
TOT (V)	TOT Sensor	0-5.1v	0.5-2.0	0.5-2.0	0.5-2.0
TPV	TP Sensor	0-5.1v	0.7	0.9	1.2-1.4
TP MIN	TP Minimum	0-5.1v	0.5	0.9	1.2-1.4
TP Mode	TP Mode	C/T or P/T	C/T	P/T	P/T
TSS	TSS Sensor	0-10 KHz	790-820	1400-1450	1740-1800
VPWR	Vehicle Power	0-25.5v	12-14	12-14	12-14
VREF	Voltage Reference	0-5.1v	4.9-5.1	4.9-5.1	4.9-5.1
VSS	Vehicle Speed	0-159 mph 0-1 KHz	0 = 0 Hz	30 = 65 Hz	55 = 125 Hz
WAC	A/C WOT Cutout	OFF	ON (if on)	OFF	OFF

1995-2000 Continental 4.6L V8 MFI VIN V (A/T) - Inputs

PCM PID Acronym	Parameter Identification	PID Range	PID Value at Hot Idle	PID Value at 30	PID Value at 55
ACCS	A/C Clutch Switch	ON / OFF	ON (if on)	OFF	OFF
ACP	A/C Press. Switch	OP / CL	OPEN	OPEN	OPEN
AIR-M	AIR System Monitor	ON / OFF	OFF	OFF	OFF
BPP	Brake Switch	ON / OFF	ON (if on)	OFF	OFF
CID	CID Sensor	0-1 KHz	4-6	10-13	12-16
CKP	CKP Sensor	0-10 KHz	330-420	800-850	990-1100
DPFE	DPFE Sensor	0-5.1v	0.2-1.3	0.2-4.5	0.2-4.5
ECT (°F)	ECT Sensor	-40-304°F	160-200	160-200	160-200
FEPS	Flash EEPROM	0-5.1v	0.1	0.1	0.1
FLI	Fuel Level Indicator	0-100%	50 (1.7v)	25-75	25-75
FP M	Fuel Pump Monitor	0-100%	50-100	50-100	50-100
FRP	Fuel Rail Pressure	0-100 psi	39 (2.8v)	39 (2.8v)	39 (2.8v)
FTP	FTP Sensor (1998-2002)	0-5.1v or 0-10" H20	2.6v at 0" of H20	2.6v at 0" of H20	2.6v at 0" of H20
Gear	Gear Position	1 - 5	1	3	4
HO2S-11	HO2S-11 (Bank 1)	0.1-1.1v	0.1-1.1	0.1-1.1	0.1-1.1
HO2S-12	HO2S-12 (Bank 1)	0.1-1.1v	0.1-1.1	0.1-1.1	0.1-1.1
HO2S-21	HO2S-21 (Bank 2)	0.1-1.1v	0.1-1.1	0.1-1.1	0.1-1.1
HO2S-22	HO2S-22 (Bank 2)	0.1-1.1v	0.1-1.1	0.1-1.1	0.1-1.1
IAT (°F)	IAT Sensor	-40-304°F	50-120	50-120	50-120
IMRCM	IMRC Monitor	0-5.1v	5	5	5
KS1	Knock Sensor B1	0-5.1v	0	0	0
KS2	Knock Sensor B2	0-5.1v	0	0	0
LOAD	Engine Load	0-100%	10-20	16-36	23-33
MAF (V)	MAF Sensor	0-5.1v	0.5-0.8	0.8-1.6	1.1-1.9
MISF	Misfire Detection	ON / OFF	OFF	OFF	OFF
OCTADJ	Octane Adjust	Retard/NO	No Retard	No Retard	No Retard
PNP	PNP Switch	ON / OFF	ON	OFF	OFF
PSP	PSP Switch	HI / LOW	HI: turning	LOW	LOW
RPM	Engine Speed	0-10K rpm	695-760	1350-1440	1700-1820
TCS	TCS (switch)	ON / OFF	ON (if on)	OFF	OFF
TFT (°F)	TFT Sensor	-40-304°F	110-210	110-210	110-210
TP-B	SEC TP Sensor	0-5.1v	0.5-0.7	0.5-0.7	0.5-0.7
TPV	TP Sensor	0-5.1v	0.53-1.27	0.8-1.1	0.8-1.2
TRAC	Traction Control	ON / OFF	ON	ON	ON
TR1	TR Sensor 1 Volts	0-12v	0	10.7	10.7
TR2	TR2 Sensor	0-12v	0	10.7	10.7
TR4	TR4 Sensor	0-12v	0	10.7	10.7
TRV/TR	TR Sensor	0-5.1v	4.4 / PN	2.1 / OD	2.1 / OD
TSS	Turbine shaft speed	0-1 KHz	40-45	85-105	110-118
VPWR	Vehicle Power	0-25.5v	12-14	12-14	12-14
VSS	Vehicle Speed	0-159 mph 0-1 KHz	0 = 0 Hz	30 = 65 Hz	55 = 125 Hz
WACA	A/C WOT Relay	ON / OFF	ON (if on)	OFF	OFF

1995-2000 Continental 4.6L V8 MFI VIN V (A/T) - Outputs

PCM PID Acronym	Parameter Identification	PID Range	PID Value at Hot Idle	PID Value at 30	PID Value at 55
AIR	AIR Solenoid	ON / OFF	OFF	OFF	OFF
CD1-8	Coil Driver 1-8	0-45° dwell	8	11	12
CTO	Clean Tach Output	0-10 KHz	33-37	57-63	68-74
EGRVR	VR Solenoid	0-100%	0	0-40	35-55
EPC	EPC Solenoid	0-100 psi	16-18	12-16	18-27
EVAPCV	EVAP CV Valve	0-100%	0-100	0-100	0-100
EVAPDC	EVAP Purge Valve	0-100%	0-100	0-100	0-100
FP	Fuel Pump Control	0-100%	33	33	33
FLPW1	INJ On-time (B1)	0-99.9 ms	2.8-2.9	2.9-5.5	4.5-8.0
FLPW2	INJ On-time (B2)	0-99.9 ms	2.8-2.9	2.9-5.5	4.5-8.0
HFC	High Fan Control	ON / OFF	ON (if on)	OFF	OFF
HTR-11	HO2S-11 Heater	ON / OFF	ON	ON	ON
HTR-12	HO2S-12 Heater	ON / OFF	ON	ON	ON
HTR-21	HO2S-21 Heater	ON / OFF	ON	ON	ON
HTR-22	HO2S-22 Heater	ON / OFF	ON	ON	ON
IAC	Idle Air Control	0-100%	35	44-55%	51-59%
IMRC	IMRC Solenoid	ON / OFF	OFF	OFF	OFF
INJ1	INJ 1 Pulsewidth	0-99.9 ms	2.8-2.9	2.9-5.5	4.5-8.0
INJ2	INJ 2 Pulsewidth	0-99.9 ms	2.8-2.9	2.9-5.5	4.5-8.0
INJ3	INJ 3 Pulsewidth	0-99.9 ms	2.8-2.9	2.9-5.5	4.5-8.0
INJ4	INJ 4 Pulsewidth	0-99.9 ms	2.8-2.9	2.9-5.5	4.5-8.0
INJ5	INJ 5 Pulsewidth	0-99.9 ms	2.8-2.9	2.9-5.5	4.5-8.0
INJ6	INJ 6 Pulsewidth	0-99.9 ms	2.8-2.9	2.9-5.5	4.5-8.0
INJ7	INJ 7 Pulsewidth	0-99.9 ms	2.8-2.9	2.9-5.5	4.5-8.0
INJ8	INJ 8 Pulsewidth	0-99.9 ms	2.8-2.9	2.9-5.5	4.5-8.0
LFC	Low Fan Control	ON / OFF	ON (if on)	OFF	OFF
LongFT1	Long Term F/T B1	-20 to 20%	-5 to +5	-5 to +5	-5 to +5
LongFT2	Long Term F/T B2	-20 to 20%	-5 to +5	-5 to +5	-5 to +5
MIL	MIL Control	ON / OFF	ON (if on)	OFF	OFF
SHTFT1	Short Term F/T B1	-20 to 20%	-5 to +5	-5 to +5	-5 to +5
SHTFT2	Short Term F/T B2	-20 to 20%	-5 to +5	-5 to +5	-5 to +5
SPARK	Spark Advance	-90° to 90°	8-10	35-40	30-40
SS1	Shift Solenoid 1	ON / OFF	ON	OFF	ON
SS2	Shift Solenoid 2	ON / OFF	OFF	ON	ON
SS3	Shift Solenoid 3	ON / OFF	OFF	ON	ON
TCC	TCC Solenoid	0-100%	0	90	90-95
TC SEC	T/C Secondary Throttle	0-100%	0	0	0
TCIL ('96)	TCIL (lamp)	ON / OFF	ON (if on)	OFF	OFF
VREF	Voltage Reference	0-5.1v	4.9-5.1	4.9-5.1	4.9-5.1
WAC	A/C WOT Cutout	ON / OFF	ON (if on)	OFF	OFF

2001-02 Continental 4.6L V8 MFI VIN V (A/T) - Inputs

PCM PID Acronym	Parameter Identification	PID Range	PID Value at Hot Idle	PID Value at 30	PID Value at 55
ACCS	A/C Clutch Switch	ON / OFF	ON (if on)	OFF	OFF
ACP	A/C Press. Switch	OP / CL	OPEN	OPEN	OPEN
BPP	Brake Switch	ON / OFF	ON (if on)	OFF	OFF
CID	CID Sensor	0-1 KHz	4-6	10-13	12-16
CKP	CKP Sensor	0-10 KHz	330-420	800-850	990-1100
DPFE	DPFE Sensor	0-5.1v	0.95-1.05	0.95-4.65	0.95-4.65
ECT (°F)	ECT Sensor	-40-304°F	160-200	160-200	160-200
FEPS	Flash EEPROM	0-5.1v	0.1	0.1	0.1
FLI	Fuel Level Indicator	0-100%	50 (1.7v)	25-75	25-75
FP M	Fuel Pump Monitor	0-100%	50-100	50-100	50-100
FRP	Fuel Rail Pressure	0-100 psi	39 (2.8v)	39 (2.8v)	39 (2.8v)
FTP	FTP Sensor (1998-2002)	0-5.1v or 0-10" H20	2.6v at 0" of H20	2.6v at 0" of H20	2.6v at 0" of H20
Gear	Gear Position	1 - 5	1	3	4
HO2S-11	HO2S-11 (Bank 1)	0.1-1.1v	0.1-1.1	0.1-1.1	0.1-1.1
HO2S-12	HO2S-12 (Bank 1)	0.1-1.1v	0.1-1.1	0.1-1.1	0.1-1.1
HO2S-21	HO2S-21 (Bank 2)	0.1-1.1v	0.1-1.1	0.1-1.1	0.1-1.1
HO2S-22	HO2S-22 (Bank 2)	0.1-1.1v	0.1-1.1	0.1-1.1	0.1-1.1
IAT (°F)	IAT Sensor	-40-304°F	50-120	50-120	50-120
IMRCM	IMRC Monitor	0-5.1v	5	5	5
KS1	Knock Sensor B1	0-5.1v	0	0	0
KS2	Knock Sensor B2	0-5.1v	0	0	0
LOAD	Engine Load	0-100%	10-20	16-36	23-33
MAF (V)	MAF Sensor	0-5.1v	0.5-0.8	0.8-1.6	1.1-1.9
MISF	Misfire Detection	ON / OFF	OFF	OFF	OFF
OCTADJ	Octane Adjust	Retard/NO	No Retard	No Retard	No Retard
PNP	PNP Switch	ON / OFF	ON	OFF	OFF
PSP	PSP Switch	HI / LOW	HI: turning	LOW	LOW
RPM	Engine Speed	0-10K rpm	695-760	1350-1440	1700-1820
TCS	TCS (switch)	ON / OFF	ON (if on)	OFF	OFF
TFT (°F)	TFT Sensor	-40-304°F	110-210	110-210	110-210
TP-B	SEC TP Sensor	0-5.1v	0.5-0.7	0.5-0.7	0.5-0.7
TPV	TP Sensor	0-5.1v	0.53-1.27	0.8-1.1	0.8-1.2
TRAC	Traction Control	ON / OFF	ON	ON	ON
TR1	TR Sensor 1 Volts	0-12v	0	10.7	10.7
TR2	TR2 Sensor	0-12v	0	10.7	10.7
TR4	TR4 Sensor	0-12v	0	10.7	10.7
TRV/TR	TR Sensor	0-5.1v	4.4 / PN	2.1 / OD	2.1 / OD
TSS	Turbine shaft speed	0-1 KHz	40-45	85-105	110-118
VPWR	Vehicle Power	0-25.5v	12-14	12-14	12-14
VSS	Vehicle Speed	0-159 mph 0-1 KHz	0 = 0 Hz	30 = 65 Hz	55 = 125 Hz
WACA	A/C WOT Relay	ON / OFF	ON (if on)	OFF	OFF

2001-02 Continental 4.6L V8 MFI VIN V (A/T) - Outputs

PCM PID Acronym	Parameter Identification	PID Range	PID Value at Hot Idle	PID Value at 30	PID Value at 55
CD1-8	Coil Driver 1-8	0-45° dwell	8	11	12
CTO	Clean Tach Output	0-10 KHz	33-37	57-63	68-74
EGRVR	VR Solenoid	0-100%	0	0-40	35-55
EPC	EPC Solenoid	0-100 psi	16-18	12-16	18-27
EVAPCV	EVAP CV Valve	0-100%	0-100	0-100	0-100
EVAPDC	EVAP Purge Valve	0-100%	0-100	0-100	0-100
FP	Fuel Pump Control	0-100%	33	31	33
FLPW1	INJ On-time (B1)	0-99.9 ms	2.8-2.9	2.9-5.5	4.5-8.0
FLPW2	INJ On-time (B2)	0-99.9 ms	2.8-2.9	2.9-5.5	4.5-8.0
HFC	High Fan Control	ON / OFF	ON (if on)	OFF	OFF
HTR-11	HO2S-11 Heater	ON / OFF	ON	ON	ON
HTR-12	HO2S-12 Heater	ON / OFF	ON	ON	ON
HTR-21	HO2S-21 Heater	ON / OFF	ON	ON	ON
HTR-22	HO2S-22 Heater	ON / OFF	ON	ON	ON
IAC	Idle Air Control	0-100%	28-30	44-55	51-75
IMRC	IMRC Solenoid	ON / OFF	OFF	OFF	OFF
INJ1	INJ 1 Pulsewidth	0-99.9 ms	2.8-2.9	2.9-5.5	4.5-8.0
INJ2	INJ 2 Pulsewidth	0-99.9 ms	2.8-2.9	2.9-5.5	4.5-8.0
INJ3	INJ 3 Pulsewidth	0-99.9 ms	2.8-2.9	2.9-5.5	4.5-8.0
INJ4	INJ 4 Pulsewidth	0-99.9 ms	2.8-2.9	2.9-5.5	4.5-8.0
INJ5	INJ 5 Pulsewidth	0-99.9 ms	2.8-2.9	2.9-5.5	4.5-8.0
INJ6	INJ 6 Pulsewidth	0-99.9 ms	2.8-2.9	2.9-5.5	4.5-8.0
INJ7	INJ 7 Pulsewidth	0-99.9 ms	2.8-2.9	2.9-5.5	4.5-8.0
INJ8	INJ 8 Pulsewidth	0-99.9 ms	2.8-2.9	2.9-5.5	4.5-8.0
LFC	Low Fan Control	ON / OFF	ON (if on)	OFF	OFF
LongFT1	Long Term F/T B1	-20 to 20%	-5 to +5	-5 to +5	-5 to +5
LongFT2	Long Term F/T B2	-20 to 20%	-5 to +5	-5 to +5	-5 to +5
MIL	MIL Control	ON / OFF	ON (if on)	OFF	OFF
SHTFT1	Short Term F/T B1	-20 to 20%	-5 to +5	-5 to +5	-5 to +5
SHTFT2	Short Term F/T B2	-20 to 20%	-5 to +5	-5 to +5	-5 to +5
SPARK	Spark Advance	-90° to 90°	8-10	35-45	30-45
SS1	Shift Solenoid 1	ON / OFF	OFF	ON	ON
SS2	Shift Solenoid 2	ON / OFF	ON	OFF	ON
SS3	Shift Solenoid 3	ON / OFF	OFF	ON	ON
TCC	TCC Solenoid	0-100%	0	0	100
VREF	Voltage Reference	0-5.1v	4.9-5.1	4.9-5.1	4.9-5.1
WAC	A/C WOT Cutout	ON / OFF	ON (if on)	OFF	OFF

LS6 PID DATA

2000-2005 LS6 3.0L 4v V6 MFI VIN S (All) - Inputs

PCM PID Acronym	Parameter Identification	PID Range	PID Value at Hot Idle	PID Value at 30	PID Value at 55
AIR-M	AIR Monitor	ON / OFF	OFF	OFF	OFF
ACP	A/C Press. Switch	CL/OPEN	OPEN	OPEN	OPEN
AFS	AFS Signal Hertz	0-1 KHz	130	130	130
BPP	BPP Switch	ON / OFF	ON (if on)	OFF	OFF
BPS	Brake Pedal Switch	0-25.5v	0.1 (if on)	12-14	12-14
CHT (°F)	CHT Sensor (°F)	-40-304°F	160-200	160-200	160-200
CID	CID Sensor	0-1 KHz	6.6	10	17
CKP	CKP Sensor	0-10 KHz	435	780-710	1170-1180
CSTT	Clutch Switch	0-25.5v	12v (if in)	0.1	0.1
DPFE	DPFE input (99-'00)	0-5.1v	0.2-1.3	0.2-4.5	0.2-4.5
DPFE	DPFE input (01-'02)	0-5.1v	0.95-1.05	0.95-4.65	0.95-4.65
ECT (°F)	ECT Sensor	-40-304°F	160-200	160-200	160-200
EFTA (°F)	EFT Sensor (°F)	-40-304°F	50-120	50-120	50-120
FEPS	Flash EEPROM	0-5.1v	0.1	0.1	0.1
FLI	Fuel Level Indicator	0-100%	50 (1.7v)	25-75	25-75
FP-M	Fuel Pump Monitor	0-100%	50	50	50
FRP	Fuel Rail Pressure	0-100 psi	39	39	39
FTP	FTP Sensor	0-5.1v or 0-10" H20	2.6v at 0" of H20	2.6v at 0" of H20	2.6v at 0" of H20
GEAR	Transmission Gear	1-2-3-4-5	1	4	5
HO2S-11	HO2S-11 (Bank 1)	0.1-1.1v	0.1-1.1	0.1-1.1	0.1-1.1
HO2S-12	HO2S-12 (Bank 1)	0.1-1.1v	0.1-1.1	0.1-1.1	0.1-1.1
HO2S-21	HO2S-21 (Bank 2)	0.1-1.1v	0.1-1.1	0.1-1.1	0.1-1.1
HO2S-22	HO2S-22 (Bank 2)	0.1-1.1v	0.1-1.1	0.1-1.1	0.1-1.1
IAT (°F)	IAT Sensor	-40-304°F	50-120	50-120	50-120
KS1	Knock Sensor B1	0-5.1v	0	0	0
KS2	Knock Sensor B2	0-5.1v	0	0	0
LOAD	Engine Load	0-100%	18.6	35	40
MAF (V)	MAF Sensor	0-5.1v	0.7	1.6-1.8	2.1-2.3
OCTADJ	Octane Adjustment	CLOSED	CLOSED	CLOSED	CLOSED
OSS	OSS Sensor	0-10 KHz	0 rpm	1500	2660
PSP	PSP Switch	HI / LOW	HI: turning	LOW	LOW
PS1	PS1 Switch	0-25.5v	11.7	11.7	11.7
REV	Reverse Signal	0-25.5v	0.1 (in 'R')	12-14	12-14
SCCS	S/C Switch	0-5.1v	0.1 (if on)	4.6	4.6
TCS	TCS (switch)	ON / OFF	ON (if on)	OFF	OFF
TSS	TSS Sensor	0-10 KHz	680	1080	2060
TFT (°F)	TFT Sensor	-40-304°F	110-210	110-210	110-210
TPV	TP Sensor	0-5.1v	0.5-0.9	1.1v	1.3v
TR1	TR Sensor 1 Volts	0-12v	0	10.7	10.7
TR2	TR2 Sensor	0-12v	0	10.7	10.7
TR4	TR4 Sensor	0-12v	0	10.7	10.7
TRV/TR	TR Sensor	0-5.1v	0v / PN	1.7 / OD	1.7 / OD
WACA	A/C WOT Relay	ON / OFF	ON (if on)	OFF	OFF

2000-2005 LS6 3.0L 4v V6 MFI VIN S (All) - Outputs

PCM PID Acronym	Parameter Identification	PID Range	PID Value at Hot Idle	PID Value at 30	PID Value at 55
AIR	AIR Solenoid	ON / OFF	OFF	OFF	OFF
ALDFDC	Alternator Control	0-1 KHz	0-130	0 -10	0-10
CD1-CD6	Coil Driver 1-6	0-60° dwell	7	10	11
CHTIL	CHTIL (lamp)	ON / OFF	ON (if on)	OFF	OFF
CTO	Clean Tach Output	0-10 KHz	39	89	100
EGRVR	VR Solenoid	0-100%	0	0-40	55
EPC	EPC Solenoid	0-25.5v	9.3	10.4	10.5
EPC2	EPC2 Solenoid PSI	0-25.5v	11.1	10.4	10.5
EPC3	EPC3 Solenoid PSI	0-25.5v	7.5	12-14	12-14
EVAPCV	EVAP CV Valve	0-100%	0-100	0-100	0-100
EVAPDC	EVAP Purge Valve	0-100%	0-100	0-100	0-100
FP	Fuel Pump Control	0-100%	25	25	25
HTR-11	HO2S-11 Heater	ON / OFF	ON	ON	ON
HTR-12	HO2S-12 Heater	ON / OFF	ON	ON	ON
HTR-21	HO2S-21 Heater	ON / OFF	ON	ON	ON
HTR-22	HO2S-22 Heater	ON / OFF	ON	ON	ON
IAC	Idle Air Control	0-100%	34	53	67
INJ1	INJ 1 Pulsewidth	0-99.9 ms	3.2	4.9	6.7-7.1
INJ2	INJ 2 Pulsewidth	0-99.9 ms	3.2	4.9	6.7-7.1
INJ3	INJ 3 Pulsewidth	0-99.9 ms	3.2	4.9	6.7-7.1
INJ4	INJ 4 Pulsewidth	0-99.9 ms	3.2	4.9	6.7-7.1
INJ5	INJ 5 Pulsewidth	0-99.9 ms	3.2	4.9	6.7-7.1
INJ6	INJ 6 Pulsewidth	0-99.9 ms	3.2	4.9	6.7-7.1
LongFT1	Long Term F/T B1	-20 to 20%	-5 to +5	-5 to +5	-5 to +5
LongFT2	Long Term F/T B2	-20 to 20%	-5 to +5	-5 to +5	-5 to +5
MIL	MIL Control	ON / OFF	ON (if on)	OFF	OFF
SCC	S/C C Signal	0-25.5v	12-14	12-14	12-14
SCMA	S/C MA Signal	0-25.5v	12-14	12-14	12-14
SCMB	S/C MB Signal	0-25.5v	12-14	12-14	12-14
SCMC	S/C MC Signal	0-25.5v	12-14	12-14	12-14
SHTFT1	Short Term F/T B1	-20 to 20%	-5 to +5	-5 to +5	-5 to +5
SHTFT2	Short Term F/T B2	-20 to 20%	-5 to +5	-5 to +5	-5 to +5
SPARK	Spark Advance	-90° to 90°	12-17	34	40
S1	Shift Solenoid 1	ON / OFF	ON	OFF	ON
S2	Shift Solenoid 2	ON / OFF	OFF	OFF	OFF
S3	Shift Solenoid 3	ON / OFF	OFF	ON	ON
SS4	Shift Solenoid 4	ON / OFF	ON	ON	ON
TCC	TCC Solenoid	0-100%	0	0	100
VREF	Voltage Reference	0-5.1v	4.9-5.1	4.9-5.1	4.9-5.1
WAC	WOT Cutout Relay	ON / OFF	ON (if on)	OFF	OFF

LS8 PID DATA

2000-2005 LS8 3.9L 4v V8 MFI VIN A (A/T) - Inputs

PCM PID Acronym	Parameter Identification	PID Range	PID Value at Hot Idle	PID Value at 30	PID Value at 55
AIR-M	AIR Monitor	ON / OFF	OFF	OFF	OFF
ACP	A/C Press. Switch	CL/OPEN	OPEN	OPEN	OPEN
AFS	AFS Signal Hertz	0-1 KHz	130	130	130
BPP	BPP Switch	ON / OFF	ON (if on)	OFF	OFF
BPS	Brake Pedal Switch	0-25.5v	12v (if on)	0v	0v
CHT (°F)	CHT Sensor (°F)	-40-304°F	160-200	160-200	160-200
CID	CID Sensor	0-1 KHz	6	11	18
CKP	CKP Sensor	0-10 KHz	420	665-800	1160-1200
CTO	Clean Tach Output	0-10 KHz	46	106	123
DPFE	DPFE input	0-5.1v	0.20-1.3	0.2-4.5	0.2-4.5
ECT (°F)	ECT Sensor	-40-304°F	160-200	160-200	160-200
EFTA (°F)	EFT Sensor (°F)	-40-304°F	50-120	50-120	50-120
EOT	Engine Oil Temp.	0-5.1v	1.0	0.9	1.2
FEPS	Flash EEPROM	0-5.1v	0.1	0.1	0.1
FLI	Fuel Level Indicator	0-100%	50 (1.7v)	25-75	25-75
FP-M	Fuel Pump Monitor	0-100%	50	50	50
FRP	Fuel Rail Pressure	0-100 psi	44	44	44
FTP	FTP Sensor	0-5.1v	2.6 (0")	2.6 (0")	2.6 (0")
GEAR	Transmission Gear	1-2-3-4-5	1	4	5
GFS	Generator Field	0-1 KHz	130 (35%)	130 (25%)	130 (20%)
HO2S-11	HO2S-11 (Bank 1)	0.1-1.1v	0.1-1.1	0.1-1.1	0.1-1.1
HO2S-12	HO2S-12 (Bank 1)	0.1-1.1v	0.1-1.1	0.1-1.1	0.1-1.1
HO2S-21	HO2S-21 (Bank 2)	0.1-1.1v	0.1-1.1	0.1-1.1	0.1-1.1
HO2S-22	HO2S-22 (Bank 2)	0.1-1.1v	0.1-1.1	0.1-1.1	0.1-1.1
IAT (°F)	IAT Sensor	-40-304°F	50-120	50-120	50-120
ISS	Input Shaft Speed	0-10 KHz	235-380	725	1370
KS1	Knock Sensor B1	0-5.1v	0	0	0
KS2	Knock Sensor B2	0-5.1v	0	0	0
LOAD	Engine Load	0-100%	17-18.6	26-35.7	30-50
MAF (V)	MAF Sensor	0-5.1v	0.78	1.2-1.4	1.5-1.9
MISF	Misfire Detection	ON / OFF	OFF	OFF	OFF
OSS	Output Shaft Speed	0-10 KHz	0	595	1070
PS1	PS1 Switch	0-25.5v	12	11.7	11.7
PNP	PNP Switch	ON / OFF	ON	OFF	OFF
PSP	PSP Switch	HI / LOW	HI: turning	LOW	LOW
RPM	Engine Speed	0-10K rpm	660-700	1425	1950
TCS	TCS (switch)	ON / OFF	ON (if on)	OFF	OFF
TSS	Turbine shaft speed	0-10 KHz	340	590	1070
TFT (°F)	TFT Sensor	-40-304°F	110-210	110-210	110-210
TPV	TP Sensor	0-5.1v	0.53-1.37	1.1-1.3	1.3-1.5
TR1	TR Sensor 1 Volts	0-12v	0.1	12	12
TR2	TR2 Sensor	0-12v	0.1	12	12
TR4	TR4 Sensor	0-12v	0.1	12	12
TRV/TR	TR Sensor	0-5.1v	0 / PN	1.7 / OD	1.7 / OD

2000-2002 LS8 3.9L 4v V8 MFI VIN A (A/T) - Outputs

PCM PID Acronym	Parameter Identification	PID Range	PID Value at Hot Idle	PID Value at 30	PID Value at 55
AIR	AIR Solenoid	ON / OFF	OFF	OFF	OFF
CD1-8	Coil Driver 1-8	0-45° dwell	8	11	12
CHTIL	CHTIL (lamp)	ON / OFF	ON (if on)	OFF	OFF
EGRVR	VR Solenoid	0-100%	0	40	40
EPC	EPC Solenoid	0-25.5v	7.4	9.3	9.9
EPC2	EPC2 Solenoid PSI	0-25.5v	12-14	9.3	9.9
EPC3	EPC3 Solenoid PSI	0-25.5v	7.5	12-14	12-14
EVAPCV	EVAP CV Valve	0-100%	0-100	0-100	0-100
EVAPDC	EVAP Purge Valve	0-100%	0-100	0-100	0-100
FP	Fuel Pump Control	0-100%	26 (3.5v)	29 (4v)	27 (3.8v)
GENFDC	Alternator Control	0-1 KHz	0-130	0	0
HFC	High Fan Control	ON / OFF	ON (if on)	OFF	OFF
HTR-11	HO2S-11 Heater	ON / OFF	ON	ON	ON
HTR-12	HO2S-12 Heater	ON / OFF	ON	ON	ON
HTR-21	HO2S-21 Heater	ON / OFF	ON	ON	ON
HTR-22	HO2S-22 Heater	ON / OFF	ON	ON	ON
IAC	Idle Air Control	0-100%	32	54	72
IMRC	IMRC Solenoid	ON / OFF	OFF	OFF	OFF
INJ1	INJ 1 Pulsewidth	0-99.9 ms	2.9-3.6	5.0-5.1	6.5-7.5
INJ2	INJ 2 Pulsewidth	0-99.9 ms	2.9-3.6	5.0-5.1	6.5-7.5
INJ3	INJ 3 Pulsewidth	0-99.9 ms	2.9-3.6	5.0-5.1	6.5-7.5
INJ4	INJ 4 Pulsewidth	0-99.9 ms	2.9-3.6	5.0-5.1	6.5-7.5
INJ5	INJ 5 Pulsewidth	0-99.9 ms	2.9-3.6	5.0-5.1	6.5-7.5
INJ6	INJ 6 Pulsewidth	0-99.9 ms	2.9-3.6	5.0-5.1	6.5-7.5
INJ7	INJ 7 Pulsewidth	0-99.9 ms	2.9-3.6	5.0-5.1	6.5-7.5
INJ8	INJ 8 Pulsewidth	0-99.9 ms	2.9-3.6	5.0-5.1	6.5-7.5
LGFT1	L/T Fuel Trim (B1)	-20 to 20%	-5 to +5	-5 to +5	-5 to +5
LGFT2	L/T Fuel Trim (B2)	-20 to 20%	-5 to +5	-5 to +5	-5 to +5
MIL	MIL Control	ON / OFF	ON (if on)	OFF	OFF
SCCS	S/C Switch Signal	0-5.1v	0.1	4.6	4.6
SCMA	S/C MA Signal	0-25.5v	12-14	12-14	12-14
SCMB	S/C MB Signal	0-25.5v	12-14	12-14	12-14
SCMC	S/C MC Signal	0-25.5v	12-14	12-14	12-14
SHTFT1	Short Term F/T B1	-20 to 20%	-5 to +5	-5 to +5	-5 to +5
SHTFT2	Short Term F/T B2	-20 to 20%	-5 to +5	-5 to +5	-5 to +5
SPARK	Spark Advance	-90° to 90°	10-20	36	33
S1	Shift Solenoid 1	ON / OFF	ON	OFF	OFF
S2	Shift Solenoid 2	ON / OFF	OFF	OFF	OFF
S3	Shift Solenoid 3	ON / OFF	OFF	ON	ON
SS4	Shift Solenoid 4	ON / OFF	ON	ON	ON
TCC	TCC Solenoid	0-100%	0	0	100
TCIL	Trans. Control lamp	ON / OFF	ON (if on)	OFF	OFF
WAC	WOT Cutout Relay	ON / OFF	ON (if on)	OFF	OFF

MARK VIII PID DATA

1993-95 Mark VIII 4.6L V8 MFI VIN V (A/T) Inputs / Outputs

PCM PID Acronym	Parameter Identification	PID Range	PID Value at Hot Idle	PID Value at 30	PID Value at 55
ACCS	A/C Clutch Switch	ON / OFF	ON (if on)	OFF	OFF
BOO	Brake Switch	On / OFF	ON (if on)	OFF	OFF
DPFE	DPFE Sensor	0-5.1v	0.4-0.6	0.6-1.8	1.9-4.8
ECT (ºF)	ECT Sensor	-40-304ºF	160-200	160-200	160-200
ECT (V)	ECT Sensor	0-5.1v	0.6	0.6	0.6
EPC	EPC Solenoid	0-100 psi	20	12	30
EVR	VR Solenoid	0-100%	0	0-40	0-60
FP	Fuel Pump Control	ON / OFF	ON	ON	ON
GEAR	Gear Position	P-R-N-D	P-R-N-D	DRIVE 3	DRIVE 4
FuelPW1	INJ Pulsewidth B1	0-99.9 ms	5.5-5.8	6.4-8.4	9.4-12
FuelPW2	INJ Pulsewidth B2	0-99.9 ms	5.5-5.8	6.4-8.4	9.4-12
HO2S-11	HO2S-11 (Bank 1)	0.1-1.1v	0.1-1.1	0.1-1.1	0.1-1.1
HO2S-21	HO2S-21 (Bank 2)	0.1-1.1v	0.1-1.1	0.1-1.1	0.1-1.1
IAC	Idle Air Control	0-100%	20-40	34-40	45-55
IAT (ºF)	IAT Sensor	-40-304ºF	50-120	50-120	50-120
IAT (V)	IAT Sensor	1.7-3.5	1.7-3.5	1.7-3.5	1.7-3.5
LOOP	Loop Status	OL or CL	CL	CL	CL
LongFT1	Long Term F/T B1	-20 to 20%	-5 to +5	-5 to +5	-5 to +5
LongFT2	Long Term F/T B2	-20 to 20%	-5 to +5	-5 to +5	-5 to +5
MAF	MAF Sensor	0-5.1v	1.0v	1.3-1.5	1.6-2.0
MLP	MLP Sensor	PARK	PARK	DRIVE	O/D
MLP (V)	MLP Sensor	0-5.1v	0v	5v	5v
PNP	PNP Switch	NEUTRAL	NEUTRAL	DRIVE	DRIVE
RPM	Engine Speed	0-10K rpm	750-820	1450-1630	1750-2100
SHTFT1	Short Term F/T B1	-20 to 20%	-5 to +5	-5 to +5	-5 to +5
SHTFT2	Short Term F/T B2	-20 to 20%	-5 to +5	-5 to +5	-5 to +5
SPARK	Spark Advance	-90º to 90º	15-22	28-35	25-35
TCC	TCC Solenoid	0-100%	0	42-44	92-100
TOT (V)	TFT Sensor Volts	0-5.1v	1.7-3.5	1.7-3.5	1.7-3.5
TOT (ºF)	TFT Sensor	-40-304ºF	110-210	110-210	110-210
TPV	TP Sensor	0-5.1v	0.7	0.8-1.0	1.1-1.3
TP MIN	TP Minimum	0-5.1v	0.5	0.8-1.0	1.1-1.3
TP Mode	TP Mode	C/T or P/T	C/T	P/T	P/T
VPWR	Vehicle Power	0-25.5v	12-14	12-14	12-14
VREF	Voltage Reference	0-5.1v	4.9-5.1	4.9-5.1	4.9-5.1
VSS	Vehicle Speed	0-159 mph	0	30	55
WAC	A/C WOT Cutout Relay	ON / OFF	ON (if on)	OFF	OFF

1996-98 Mark VIII 4.6L V8 MFI VIN V (A/T) - Inputs

PCM PID Acronym	Parameter Identification	PID Range	PID Value at Hot Idle	PID Value at 30	PID Value at 55
ACCS	A/C Clutch Switch	ON / OFF	ON (if on)	OFF	OFF
ACP	ACP Sensor Volts	0-5.1v	0.8-1.4	0.8-1.4	0.8-1.4
BPP	Brake Pedal Switch	ON / OFF	ON (if on)	OFF	OFF
AIR-M	AIR System Monitor	ON / OFF	ON	ON	ON
CID	CID Sensor	0-1 KHz	5-7	7-9	11-14
CKP	CKP Sensor	0-10 KHz	365-395	540-550	850-895
DPFE	DPFE Sensor	0-5.1v	0.2-1.3v	0.2-4.5v	0.2-4.5v
ECT (°F)	ECT Sensor	-40-304°F	160-200	160-200	160-200
FEPS	Flash EEPROM	0-5.1v	0.1	0.1	0.1
FLI ('98)	Fuel Level Indicator	0-100%	50 (1.7v)	50 (1.7v)	50 (1.7v)
FP-M	Fuel Pump Monitor	0-100%	50%	50%	50%
FTP ('98)	FTP Sensor	0-5.1v or 0-10" H20	2.6v at 0" of H20	2.6v at 0" of H20	2.6v at 0" of H20
GEAR	Gear Position	1-2-3-4	1	3	4
HO2S-11	HO2S-11 (Bank 1)	0.1-1.1v	0.1-1.1	0.1-1.1	0.1-1.1
HO2S-12	HO2S-12 (Bank 1)	0.1-1.1v	0.1-1.1	0.1-1.1	0.1-1.1
HO2S-21	HO2S-21 (Bank 2)	0.1-1.1v	0.1-1.1	0.1-1.1	0.1-1.1
HO2S-22	HO2S-22 (Bank 2)	0.1-1.1v	0.1-1.1	0.1-1.1	0.1-1.1
IAT (°F)	IAT Sensor	-40-304°F	50-120	50-120	50-120
IMRCM	IMRC Monitor	0-5.1v	5	5	5
KS1	Knock Sensor B1	0-5.1v	0	0	0
KS2	Knock Sensor B2	0-5.1v	0	0	0
LOAD	Engine Load	0-100%	10-20	23-35	25-35
MAF (V)	MAF Sensor	0-5.1v	0.6-0.9	0.9-1.1	1.2-1.6v
MISF	Misfire Detection	ON / OFF	OFF	OFF	OFF
OCTADJ	Octane Adjustment	CL/OPEN	CLOSED	CLOSED	CLOSED
OSS	OSS Sensor	0-10K rpm	0	1345-1355	2170-2210
PNP	PNP Switch	ON / OFF	ON	OFF	OFF
RPM	Engine Speed	0-10K rpm	660-700	1380-1420	1700-1740
TCS	TCS (switch)	ON / OFF	ON (if on)	OFF	OFF
TFT (°F)	TFT Sensor	-40-304°F	110-210	110-210	110-210
TP-B	SEC TP Sensor	0-5.1v	0.5-0.7	0.5-0.7	0.5-0.7
TPV	TP Sensor	0-5.1v	0.5-0.9	0.8-1.1	0.8-1.2
TR1	TR Sensor 1 Volts	0-12v	0	11.5	11.5
TR2	TR2 Sensor	0-12v	0	11.5	11.5
TRV/TR	TR Sensor	0-5.1v	0	1.7 / OD	1.7 / OD
TR4	TR4 Sensor	0-12v	0	11.5	11.5
Vacuum	Engine Vacuum	0-30" Hg	19-20	18-20	8-15
VPWR	Vehicle Power	0-25.5v	12-14	12-14	12-14
VSS	Vehicle Speed	0-159 mph 0-1 KHz	0 = 0 Hz	30 = 65 Hz	55 = 125 Hz

1996-98 Mark VIII 4.6L V8 MFI VIN V (A/T) - Outputs

PCM PID Acronym	Parameter Identification	PID Range	PID Value at Hot Idle	PID Value at 30	PID Value at 55
AIR	AIR Solenoid	ON / OFF	OFF	OFF	OFF
CD1-CD8	COP Driver 1-8	0-45° dwell	8	11	12
EGRVR	VR Solenoid	0-100%	0	0-45	45-90
EPC	EPC Solenoid	0-100 psi	15	40	40
EVAPCV	EVAP CV Valve	0-100%	0	0	0
EVAPDC	EVAP Purge Valve	0-100%	0-100	0-100	0-100
FP	Fuel Pump Control	0-100%	35	35	35
FuelPW1	INJ Pulsewidth B1	0-99.9 ms	2.5-3.0	3.5-8.0	3.8-9.0
FuelPW2	INJ Pulsewidth B2	0-99.9 ms	2.5-3.0	3.5-8.0	3.8-9.0
HFC	High Fan Control	ON / OFF	ON (if on)	OFF	OFF
HTR-11	HO2S-11 Heater	ON / OFF	ON	ON	ON
HTR-12	HO2S-12 Heater	ON / OFF	ON	ON	ON
HTR-21	HO2S-21 Heater	ON / OFF	ON	ON	ON
HTR-22	HO2S-22 Heater	ON / OFF	ON	ON	ON
IAC	Idle Air Control	0-100%	32	46-53	60-75
IMRC	IMRC Solenoid	ON / OFF	OFF	OFF	OFF
INJ1	INJ 1 Pulsewidth	0-99.9 ms	2.5-3.0	3.5-8.0	3.8-9.0
INJ2	INJ 2 Pulsewidth	0-99.9 ms	2.5-3.0	3.5-8.0	3.8-9.0
INJ3	INJ 3 Pulsewidth	0-99.9 ms	2.5-3.0	3.5-8.0	3.8-9.0
INJ4	INJ 4 Pulsewidth	0-99.9 ms	2.5-3.0	3.5-8.0	3.8-9.0
INJ5	INJ 5 Pulsewidth	0-99.9 ms	2.5-3.0	3.5-8.0	3.8-9.0
IN6	INJ 6 Pulsewidth	0-99.9 ms	2.5-3.0	3.5-8.0	3.8-9.0
INJ7	INJ 7 Pulsewidth	0-99.9 ms	2.5-3.0	3.5-8.0	3.8-9.0
INJ8	INJ 8 Pulsewidth	0-99.9 ms	2.5-3.0	3.5-8.0	3.8-9.0
LFC	Low Fan Control	ON / OFF	ON (if on)	OFF	OFF
LongFT1	Long Term F/T B1	-20 to 20%	-5 to +5	-5 to +5	-5 to +5
LongFT2	Long Term F/T B2	-20 to 20%	-5 to +5	-5 to +5	-5 to +5
MIL	MIL Control	ON / OFF	ON (if on)	OFF	OFF
SHTFT1	Short Term F/T B1	-20 to 20%	-5 to +5	-5 to +5	-5 to +5
SHTFT2	Short Term F/T B2	-20 to 20%	-5 to +5	-5 to +5	-5 to +5
SPARK	Spark Advance	-90° to 90°	13-19	19-32	25-36
SS1	Shift Solenoid 1	ON / OFF	ON	OFF	ON
SS2	Shift Solenoid 2	ON / OFF	OFF	ON	ON
SS3	Shift Solenoid 3	ON / OFF	OFF	ON	ON
TCC	TCC Solenoid	0-100%	0	90-95	90-95
VREF	Voltage Reference	0-5.1v	4.9-5.1	4.9-5.1	4.9-5.1

TOWN CAR PID DATA

1991-94 Town Car 4.6L V8 MFI VIN W (A/T) Inputs / Outputs

PCM PID Acronym	Parameter Identification	PID Range	PID Value at Hot Idle	PID Value at 30	PID Value at 55
ACCS	A/C Clutch Switch	ON / OFF	ON (if on)	OFF	OFF
BOO	Brake Switch	ON / OFF	ON (if on)	OFF	OFF
CANP	EVAP Purge Valve	ON / OFF	OFF	ON	ON
DPFE	DPFE Sensor	0-5.1v	0.4-0.6	0.6-1.8	1.9-4.8
ECT (°F)	ECT Sensor	-40-304°F	160-200	160-200	160-200
ECT (V)	ECT Sensor	0-5.1v	0.6	0.6	0.6
EPC	EPC Solenoid	0-100 psi	20	12	30
EVR	VR Solenoid	0-100%	0%	0-40	38-57
FP	Fuel Pump Control	ON / OFF	ON	ON	ON
GEAR	Gear Position	NEUTRAL	NEUTRAL	DRIVE	DRIVE
FuelPW1	INJ Pulsewidth B1	0-99.9 ms	4.2-4.4	4.6-5.6	6.4-7.0
FuelPW2	INJ Pulsewidth B2	0-99.9 ms	4.2-4.4	4.6-5.6	6.4-7.0
HO2S-11	HO2S-11 (Bank 1)	0.1-1.1v	0.1-1.1	0.1-1.1	0.1-1.1
HO2S-21	HO2S-21 (Bank 2)	0.1-1.1v	0.1-1.1	0.1-1.1	0.1-1.1
IAC	Idle Air Control	0-100%	32-36	42-47	58-65
IAT (°F)	IAT Sensor	-40-304°F	50-120	50-120	50-120
IAT (V)	IAT Sensor	0-5.1v	1.7-3.5	1.7-3.5	1.7-3.5
LongFT1	Long Term F/T B1	-20 to 20%	-5 to +5	-5 to +5	-5 to +5
LongFT2	Long Term F/T B2	-20 to 20%	-5 to +5	-5 to +5	-5 to +5
LOOP	Loop Status	OL or CL	CL	CL	CL
MAF (V)	MAF Sensor	0-5.1v	0.7	1.1-1.3	1.4-1.6
MLP (V)	MLP Sensor	0-5.1v	0	5	5
PNP	PNP Switch	PARK	PARK	DRIVE	DRIVE
RPM	Engine Speed	0-10K rpm	750-820	1450-1630	1750-2100
SHTFT1	Short Term F/T B1	-20 to 20%	-5 to +5	-5 to +5	-5 to +5
SHTFT2	Short Term F/T B2	-20 to 20%	-5 to +5	-5 to +5	-5 to +5
SPARK	Spark Advance	-90° to 90°	15-22	28-35	25-35
TCC	TCC Solenoid	ON / OFF	OFF	ON	ON
TOT (V)	TOT Sensor	0-5.1v	0.5-2.0v	0.5-2.0	0.5-2.0
TPV	TP Sensor	0-5.1v	0.9	0.9-1.0	1.1-1.3
TP C/T	TP Mode	C/T or P/T	C/T	P/T	P/T
VPWR	Vehicle Power	0-25.5v	12-14	12-14	12-14
VREF	Voltage Reference	0-5.1v	4.9-5.1	4.9-5.1	4.9-5.1
VSS	Vehicle Speed	0-159 mph	0	30	55
WAC	A/C WOT Cutout Relay	ON / OFF	ON (if on)	OFF	OFF

1995-2005 Town Car 4.6L V8 MFI VIN W (A/T) - Inputs

PCM PID Acronym	Parameter Identification	PID Range	PID Value at Hot Idle	PID Value at 30	PID Value at 55
ACCS	A/C Clutch Switch	ON / OFF	ON (if on)	OFF	Off
BPP	Brake Pedal Switch	ON / OFF	ON (if on)	OFF	OFF
CID	CID Sensor	0-1 KHz	6-7	9-10	12-14
CKP	CKP Sensor	0-10 KHz	440-490	680-700	870-885
DPFE	DPFE Sensor	0-5.1v	0.4-0.6	0.5-0.9	1.6-4.4
ECT (°F)	ECT Sensor	-40-304°F	160-200	160-200	160-200
FEPS	Flash EEPROM	0-5.1v	0.1	0.1	0.1
FLI	Fuel Level Indicator (1998-99)	0-100%	50 (1.7v)	50 (1.7v)	50 (1.7v)
FP-M	Fuel Pump Monitor	0-100%	100	100	100
FTPV	FTP Sensor (1998-99)	0-5.1v or 0-10" H20	2.6v at 0" of H20	2.6v at 0" of H20	2.6v at 0" of H20
GEAR	Gear Position	1-2-3-4	1	3	4
IAT (°F)	IAT Sensor	-40-304°F	50-120	50-120	50-120
LOAD	Engine Load	0-100%	12-18	17-23	24-28
MAF (V)	MAF Sensor	0-5.1v	0.6-0.9	0.9-1.3	1.3-2.0
MISF	Misfire Detection	ON / OFF	OFF	OFF	OFF
OCTADJ	Octane Adjustment	CL/OPEN	CLOSED	CLOSED	CLOSED
OSS	OSS Sensor	0-10 KHz	0	1200	2130
HO2S-11	HO2S-11 (Bank 1)	0.1-1.1v	0.1-1.1	0.1-1.1	0.1-1.1
HO2S-12	HO2S-12 (Bank 1)	0.1-1.1v	0.1-1.1	0.1-1.1	0.1-1.1
HO2S-21	HO2S-21 (Bank 2)	0.1-1.1v	0.1-1.1	0.1-1.1	0.1-1.1
HO2S-22	HO2S-22 (Bank 2)	0.1-1.1v	0.1-1.1	0.1-1.1	0.1-1.1
PNP	PNP Switch	ON / OFF	ON (in 'P')	OFF	OFF
RPM	Engine Speed	0-10K rpm	790-815	1150-1180	1480-1530
TCS	TCS (switch)	ON / OFF	ON (if on)	OFF	OFF
TFT (°F)	TFT Sensor	-40-304°F	110-210	110-210	110-210
TPV	TP Sensor	0-5.1v	0.5-0.9	1.0-1.2	1.0-1.4
TRV/TR	TR Sensor	0-5.1v	4.4 / PN	2.1 / OD	2.1 / OD
Vacuum	Engine Vacuum	0-30" Hg	19-20	18-20	8-15
VPWR	Vehicle Power	0-25.5v	12-14	12-14	12-14
VSS	Vehicle Speed	0-159 mph 0-1 KHz	0 = 0 Hz	30 = 65 Hz	55 = 125 Hz
WACA	A/C WOT Cutout Relay Monitor	ON / OFF	ON (if on)	OFF	OFF

1995-2005 Town Car 4.6L V8 MFI VIN W (A/T) - Outputs

PCM PID Acronym	Parameter Identification	PID Range	PID Value at Hot Idle	PID Value at 30	PID Value at 55
CD1-CD8	Coil Driver 1-8	0-45° dwell	8	11	12
EGRVR	VR Solenoid	0-100%	0	0-40	38-57
EPC	EPC Solenoid	0-100 psi	20	12	30
EVAPCV	EVAP CV Valve	0-100%	0-100	0-100	0-100
EVAPDC	EVAP Purge Valve	0-100%	0-100	0-100	0-100
FP	Fuel Pump Control	0-100%	100	100	100
FuelPW1	INJ Pulsewidth B1	0-99.9 ms	3.3-3.5	3.7-4.4	5.2-5.6
FuelPW2	INJ Pulsewidth B2	0-99.9 ms	3.3-3.5	3.7-4.4	5.2-5.6
HTR-11	HO2S-11 Heater	ON / OFF	ON	ON	ON
HTR-12	HO2S-12 Heater	ON / OFF	ON	ON	ON
HTR-21	HO2S-21 Heater	ON / OFF	ON	ON	ON
HTR-22	HO2S-22 Heater	ON / OFF	ON	ON	ON
IAC	Idle Air Control	0-100%	32-36	42-47	58-61
INJ1	INJ 1 Pulsewidth	0-99.9 ms	3.3-3.5	3.7-4.4	5.2-5.6
INJ2	INJ 2 Pulsewidth	0-99.9 ms	3.3-3.5	3.7-4.4	5.2-5.6
INJ3	INJ 3 Pulsewidth	0-99.9 ms	3.3-3.5	3.7-4.4	5.2-5.6
INJ4	INJ 4 Pulsewidth	0-99.9 ms	3.3-3.5	3.7-4.4	5.2-5.6
INJ5	INJ 5 Pulsewidth	0-99.9 ms	3.3-3.5	3.7-4.4	5.2-5.6
INJ6	INJ 6 Pulsewidth	0-99.9 ms	3.3-3.5	3.7-4.4	5.2-5.6
INJ7	INJ 7 Pulsewidth	0-99.9 ms	3.3-3.5	3.7-4.4	5.2-5.6
INJ8	INJ 8 Pulsewidth	0-99.9 ms	3.3-3.5	3.7-4.4	5.2-5.6
LFC	Low Fan Control	ON / OFF	ON (if on)	OFF	OFF
LongFT1	Long Term F/T B1	-20 to 20%	-5 to +5	-5 to +5	-5 to +5
LongFT2	Long Term F/T B2	-20 to 20%	-5 to +5	-5 to +5	-5 to +5
MIL	MIL Control	ON / OFF	ON (if on)	OFF	OFF
SHTFT1	Short Term F/T B1	-20 to 20%	-5 to +5	-5 to +5	-5 to +5
SHTFT2	Short Term F/T B2	-20 to 20%	-5 to +5	-5 to +5	-5 to +5
SPARK	Spark Advance	-90° to 90°	18	33-36	32-36
SS1	Shift Solenoid 1	ON / OFF	ON	OFF	OFF
SS2	Shift Solenoid 2	ON / OFF	OFF	ON	ON
TCC	TCC Solenoid	0-100%	0	40-47	85-93
TCIL	TCIL (lamp)	ON / OFF	ON (if on)	OFF	OFF
VREF	Voltage Reference	0-5.1v	4.9-5.1	4.9-5.1	4.9-5.1
WAC	A/C WOT Cutout Relay	ON / OFF	ON (if on)	OFF	OFF

1990 Town Car 5.0L V8 MFI VIN F (A/T) Inputs / Outputs

PCM PID Acronym	Parameter Identification	PID Range	PID Value at Hot Idle	PID Value at 30	PID Value at 55
ACCS	A/C Clutch Switch	ON / OFF	ON (if on)	OFF	OFF
BARO	Barometric Pressure Sensor	0-1 KHz	159 Hz (sea level)	159 Hz (sea level)	159 Hz (sea level)
BOO	Brake Switch	ON / OFF	ON (if on)	OFF	OFF
CANP	EVAP Purge Valve	ON / OFF	OFF	ON	ON
ECT (°F)	ECT Sensor	-40-304°F	160-200	160-200	160-200
ECT (V)	ECT Sensor	0-5.1v	0.6	0.6	0.6
EVP	EGR EVP Sensor	0-5.1v	0.3	0.3	0.8-1.7
EVR	VR Solenoid	0-100%	0	0-40	0-42
FP	Fuel Pump Control	ON / OFF	ON	ON	ON
FuelPW1	INJ Pulsewidth B1	0-99.9 ms	4.7-5.2	5.5-6.5	7.8-9.5
FuelPW2	INJ Pulsewidth B2	0-99.9 ms	4.7-5.2	5.5-6.5	7.8-9.5
HO2S-11	HO2S-11 (Bank 1)	0.1-1.1v	0.1-1.1	0.1-1.1	0.1-1.1
HO2S-21	HO2S-21 (Bank 2)	0.1-1.1v	0.1-1.1	0.1-1.1	0.1-1.1
IAC	Idle Air Control	0-100%	20-40	34-40	45-55
IAT (°F)	IAT Sensor	-40-304°F	50-120	50-120	50-120
IAT (V)	IAT Sensor	0-5.1v	1.7-3.5	1.7-3.5	1.7-3.5
LOOP	Loop Status	OL or CL	CL	CL	CL
MAP	MAP Sensor Hertz	0-1 KHz	110	112	126
MLP (V)	MLP Sensor	0-5.1v	0	5	5
PNP	PNP Switch	NEUTRAL	NEUTRAL	DRIVE	DRIVE
RPM	Engine Speed	0-10K rpm	650-750	1120-1270	1410-1510
SHTFT1	Short Term F/T B1	-20 to +20	-5 to +5	-5 to +5	-5 to +5
SHTFT2	Short Term F/T B2	-20 to +20	-5 to +5	-5 to +5	-5 to +5
SCCS	S/C Command Switch	0-25.5v	6.7	6.7	6.7
SCVAC	S/C Vacuum Solenoid	0-100%	0	0	0
SCVENT	S/C Vent Solenoid	0-100%	0	98	98
SPARK	Spark Advance	-90° to 90°	15-22	28-35	25-35
TCC	TCC Solenoid	ON / OFF	OFF	ON	ON
TPV	TP Sensor	0-5.1v	0.7	0.8-1.0	1.1-1.3
TP Mode	TP Mode	C/T or P/T	C/T	P/T	P/T
VPWR	Vehicle Power	0-25.5v	12-14	12-14	12-14
VREF	Voltage Reference	0-5.1v	4.9-5.1	4.9-5.1	4.9-5.1
VSS	Vehicle Speed	0-159 mph	0	30	55
WAC	A/C WOT Cutout Relay	ON / OFF	ON (if on)	OFF	OFF

F-SERIES TRUCK PID DATA

1997-2005 F-Series 4.2L V6 VIN 2 (All) Inputs

PCM PID Acronym	Parameter Identification	PID Range	PID Value at Hot Idle	PID Value at 30	PID Value at 55
4x4L	4x4 Switch Signal	ON / OFF	ON (if on)	OFF	OFF
ACCS	A/C Switch Signal	ON / OFF	ON (if on)	OFF	OFF
BPP	Brake Position	ON / OFF	ON (if on)	OFF	OFF
CID	CMP Sensor	0-999 Hz	5-7	10-12	13-17
CKP	CKP Sensor	0-9999 Hz	430-475	900-1000	1140-1300
CPP	CPP Switch Signal	ON / OFF	ON (if in)	OFF	OFF
DPFE	DPFE Sensor	0-5.1v	0.4-0.6	0.4-1.0	0.6-1.1
ECT (°F)	ECT Sensor	-40-304°F	160-200	160-200	160-200
FEPS	Flash EEPROM	0-5.1v	0.1	0.1	0.1
FLI	Fuel Level Indicator	0-100%	50% (1.7)	50% (1.7)	50% (1.7)
FPM	Fuel Pump Monitor	ON / OFF	ON	ON	ON
FTP	Fuel Tank Pressure	0-5.1v or 0-10" H2O	2.6v / 0" H2O (with cap off)	2.6v / 0" H2O (with cap off)	2.6v / 0" H2O (with cap off)
IAT (°F)	IAT Sensor	-40-304°F	50-120	50-120	50-120
IMRC-M	IMRC 1 Monitor	5v / 0	5v	5v	5v
KS1	Knock Sensor 1	0-5.1v	0	0	0
LOAD	Engine Load	0-100%	10-20	20-27	30-45
MAFV	MAF Sensor Signal	0-5.1v	0.6-0.9	1.3-1.7	1.2-2.3
MISF	Misfire Detection	ON / OFF	OFF	OFF	OFF
OCTADJ	Octane Adjustment	CL/OPEN	CLOSED	CLOSED	CLOSED
HO2S-11	HO2S-11 (Bank 1)	0.1-1.1v	0.1-1.1	0.1-1.1	0.1-1.1
HO2S-12	HO2S-12 (Bank 1)	0.1-1.1v	0.1-1.1	0.1-1.1	0.1-1.1
HO2S-21	HO2S-21 (Bank 2)	0.1-1.1v	0.1-1.1	0.1-1.1	0.1-1.1
HO2S-22	HO2S-22 (Bank 2)	0.1-1.1v	0.1-1.1	0.1-1.1	0.1-1.1
PNP	PNP Switch Signal	ON / OFF	ON	OFF	OFF
RPM	Engine Speed	0-10K rpm	680-830	1200-1300	1600-1800
TFT	TFT Sensor Signal	-40-304°F	110-210	110-210	110-210
TPV	TP Sensor	0-5.1v	0.53-1.27	1.0-1.3	1.1-1.6
TRV/TR	TR Sensor	P/N or O/D	P/N	O/D	O/D
OSS	OSS Sensor	0-10K rpm	0	1250-1310	2450-2550
Vacuum	Engine Vacuum	0-30" Hg	19-20	18-20	8-15
VPWR	Vehicle Power	0-25.5v	12-14	12-14	12-14
VSS	Vehicle Speed (MPH & Hertz Rate)	0-159 mph	0 = 0 Hz	30 = 65 Hz	55 = 125 Hz

1997-2005 F-Series 4.2L V6 VIN 2 (All) Outputs

PCM PID Acronym	Parameter Identification	PID Range	PID Value at Hot Idle	PID Value at 30	PID Value at 55
CD1	Coil 1 Driver	0-60°	6	8	12
CD2	Coil 2 Driver	0-60°	6	8	12
CD3	Coil 3 Driver	0-60°	6	8	12
CTO	Clean Tachometer signal	0-9999 Hz	35-49	65-90	90-120
EGR VR	EGR VR Solenoid	0-100%	0-100	0-40	0-50
EPC	EPC Solenoid	0-300 psi	4	20	20
EVAPCV	EVAP CV Valve	0-100%	0-100	0-100	0-100
EVAPPC	EVAP Purge Valve	0-100%	0-100	0-100	0-100
FP	Fuel Pump Control	ON / OFF	ON	ON	ON
FuelPW1	INJ Pulsewidth - Bank 1	0-999 ms	2.7-4.1	4.5-8.0	5.5-11.0
FuelPW2	INJ Pulsewidth - Bank 2	0-999 ms	2.7-4.1	4.5-8.0	5.5-11.0
HTR-11	HO2S-11 Heater	ON / OFF	ON	ON	ON
HTR-12	HO2S-12 Heater	ON / OFF	ON	ON	ON
HTR-21	HO2S-21 Heater	ON / OFF	ON	ON	ON
HTR-22	HO2S-22 Heater	ON / OFF	ON	ON	ON
IAC	Idle Air Control	0-100%	25-32	30-55	60-70
IMRC	IMRC Solenoid	ON / OFF	OFF	OFF	OFF
INJ1	INJ 1 Pulsewidth	0-999 ms	2.7-4.1	4.5-8.0	5.5-11.0
INJ2	INJ 2 Pulsewidth	0-999 ms	2.7-4.1	4.5-8.0	5.5-11.0
INJ3	INJ 3 Pulsewidth	0-999 ms	2.7-4.1	4.5-8.0	5.5-11.0
INJ4	INJ 4 Pulsewidth	0-999 ms	2.7-4.1	4.5-8.0	5.5-11.0
INJ5	INJ 5 Pulsewidth	0-999 ms	2.7-4.1	4.5-8.0	5.5-11.0
INJ6	INJ 6 Pulsewidth	0-999 ms	2.7-4.1	4.5-8.0	5.5-11.0
LongFT1	Long Term FT - Bank 1	-20 to 20%	-5 to +5	-5 to +5	-5 to +5
LongFT2	Long Term FT - Bank 2	-20 to 20%	-5 to +5	-5 to +5	-5 to +5
MIL	MIL (lamp) Control	ON / OFF	OFF	OFF	OFF
SHTFT1	Short Term F/T - Bank 1	-10 to 10%	-5 to +5	-5 to +5	-5 to +5
SHTFT2	Short Term FT - Bank 2	-10 to 10%	-5 to +5	-5 to +5	-5 to +5
SPARK	Spark Advance	-90° to 90°	11-20	15-35	20-39
SS1	Shift Solenoid 1	ON / OFF	ON	OFF	ON
SS2	Shift Solenoid 2	ON / OFF	OFF	ON	ON
TCC	TCC Solenoid	0-100%	0	0-45	90-95
TCIL	TCIL (lamp) Control	ON / OFF	ON (if on)	OFF	OFF
VREF	Vehicle Reference	0-5.1v	4.9-5.1	4.9-5.1	4.9-5.1

1997-2005 F-Series 4.6L V8 VIN W, VIN 6 (All) Inputs

PCM PID Acronym	Parameter Identification	PID Range	PID Value at Hot Idle	PID Value at 30	PID Value at 55
4x4L	4x4 Switch Signal	ON / OFF	ON (if on)	OFF	OFF
ACCS	A/C Switch Signal	ON / OFF	ON (if on)	OFF	OFF
BPP	Brake Position	ON / OFF	ON (if on)	OFF	OFF
CHT	CHT Sensor	-40-304°F	194	194	194
CID	CMP Sensor	0-999 Hz	5-7	10-12	13-17
CKP	CKP Sensor	0-9999 Hz	430-475	900-1100	1140-1220
CPP	CPP Switch Signal	ON / OFF	ON (if in)	OFF	OFF
DPFE	DPFE Sensor	0-5.1v	0.4-0.6	0.4-1.0	0.6-1.1
ECT	ECT Sensor	-40-304°F	160-200	160-200	160-200
FEPS	Flash EEPROM	0-5.1v	0.1	0.1	0.1
FLI	Fuel Level Indicator	0-100%	50% (1.7)	0-100%	0-100%
FPM	Fuel Pump Monitor	ON / OFF	ON	ON	ON
FTP	Fuel Tank Pressure	0-5.1v or 0-10" H2O	2.6v / 0" H2O (with cap off)	2.6v / 0" H2O (with cap off)	2.6v / 0" H2O (with cap off)
IAT	IAT Sensor	-40-304°F	50-120	50-120	50-120
IMRC1-M	IMRC 1 Monitor	0-5.1v	5v	5v	5v
KS1	Knock Sensor 1	0-5.1v	0	0	0
LOAD	Engine Load	0-100%	10-20	20-27	30-45
MAFV	MAF Sensor	0-5.1v	0.6-0.9	0.7-1.0	1.2-2.3
MISF	Misfire Detection	ON / OFF	OFF	OFF	OFF
OCTADJ	Octane Adjustment	CL/OPEN	CLOSED	CLOSED	CLOSED
HO2S-11	HO2S-11 (Bank 1)	0.1-1.1v	0.1-1.1	0.1-1.1	0.1-1.1
HO2S-12	HO2S-12 (Bank 1)	0.1-1.1v	0.1-1.1	0.1-1.1	0.1-1.1
HO2S-21	HO2S-21 (Bank 2)	0.1-1.1v	0.1-1.1	0.1-1.1	0.1-1.1
HO2S-22	HO2S-22 (Bank 2)	0.1-1.1v	0.1-1.1	0.1-1.1	0.1-1.1
PNP	PNP Switch Signal	ON / OFF	ON	OFF	OFF
RPM	Engine Speed	0-10K rpm	680-830	1200-1500	1600-1800
TFT	TFT Sensor Signal	-40-304°F	110-210	110-210	110-210
TPO	Power Take-Off	0-25.5v	0.9	1.4	1.8
TPV	TP Sensor	0-5.1v	0.53-1.27	1.0-1.3	1.1-1.6
TRV/TR	TR Sensor	P/N or O/D	P/N or O/D	O/D	O/D
OSS	OSS Sensor	0-10K rpm	0	1250-1310	2450-2550
Vacuum	Engine Vacuum	0-30" Hg	19-20	18-20	8-15
VPWR	Vehicle Power	0-25.5v	12-14	12-14	12-14
VSS	Vehicle Speed (Hertz or MPH)	0-159 mph 0-999 Hz	0 = 0 Hz	30 = 65 Hz	55 = 125 Hz

1997-2005 F-Series 4.6L V8 VIN W, VIN 6 (All) Outputs

PCM PID Acronym	Parameter Identification	PID Range	PID Value at Hot Idle	PID Value at 30	PID Value at 55
CD1	Coil 1 Driver	0-45	5	6	8
CD2	Coil 2 Driver	0-45	5	6	8
CD3	Coil 3 Driver	0-45	5	6	8
CD4	Coil 4 Driver	0-45	5	6	8
CHTIL	CHT (lamp) Control	ON / OFF	OFF	OFF	OFF
CTO	Clean Tachometer signal	0-9999 Hz	35-49	65-90	90-120
EGR VR	EGR VR Solenoid	0-100%	0-100	0-100	0-100
EVAP CP	EVAP Purge Valve	0-100%	0-100	0-100	0-100
EPC	EPC Solenoid	0-300 psi	4	20	20
FP	Fuel Pump Control	ON / OFF	ON	ON	ON
FuelPW1	INJ Pulsewidth - Bank 1	0-999 ms	2.7-3.5	4.5-8.0	5.5-9.0
FuelPW2	INJ Pulsewidth - Bank 2	0-999 ms	2.7-4.1	4.5-8.0	5.5-9.0
HTR-11	HO2S-11 Heater	ON / OFF	ON	ON	ON
HTR-12	HO2S-12 Heater	ON / OFF	ON	ON	ON
HTR-21	HO2S-21 Heater	ON / OFF	ON	ON	ON
HTR-22	HO2S-22 Heater	ON / OFF	ON	ON	ON
IAC	Idle Air Control	0-100%	25-32	30-55	60-70
IMRC	IMRC Signal	0-5.1v	0.1	0.1	0.1
INJ1	INJ 1 Pulsewidth	0-999 ms	2.7-4.1	4.5-8.0	5.5-9.0
INJ2	INJ 2 Pulsewidth	0-999 ms	2.7-4.1	4.5-8.0	5.5-9.0
INJ3	INJ 3 Pulsewidth	0-999 ms	2.7-4.1	4.5-8.0	5.5-9.0
INJ4	INJ 4 Pulsewidth	0-999 ms	2.7-4.1	4.5-8.0	5.5-9.0
INJ5	INJ 5 Pulsewidth	0-999 ms	2.7-4.1	4.5-8.0	5.5-9.0
INJ6	INJ 6 Pulsewidth	0-999 ms	2.7-4.1	4.5-8.0	5.5-9.0
INJ7	INJ 7 Pulsewidth	0-999 ms	2.7-4.1	4.5-8.0	5.5-9.0
INJ8	INJ 8 Pulsewidth	0-999 ms	2.7-4.1	4.5-8.0	5.5-9.0
LongFT1	Long Term FT - Bank 1	-20 to 20%	-5 to +5	-5 to +5	-5 to +5
LongFT2	Long Term FT - Bank 2	-20 to 20%	-5 to +5	-5 to +5	-5 to +5
MIL	MIL (lamp) Control	ON / OFF	ON (if on)	OFF	OFF
SS1	Shift Solenoid 1	ON / OFF	ON	ON	ON
SS2	Shift Solenoid 2	ON / OFF	OFF	ON	ON
SHTFT1	Short Term F/T - Bank 1	-10 to 10%	-5 to +5	-5 to +5	-5 to +5
SHTFT2	Short Term F/T - Bank 2	-10 to 10%	-5 to +5	-5 to +5	-5 to +5
SPARK	Spark Advance	-90-90°	11-20	15-35	20-39
TCC	TCC Solenoid	0-100%	0	0-45	90-95
TCIL	TCIL (lamp) Control	ON / OFF	ON (if on)	OFF	OFF
VREF	Vehicle Reference	0-5.1v	4.9-5.1	4.9-5.1	4.9-5.1

F-SERIES & BRONCO PID DATA

1990-95 F-Series & Bronco 4.9L I6 VIN Y (A/T, M/T, E4OD)

PCM PID Acronym	Parameter Identification	PID Range	PID Value at Hot Idle	PID Value at 30	PID Value at 55
ACCS	A/C Switch Signal	ON / OFF	ON (if on)	OFF	OFF
CANP	EVAP Purge Valve	ON / OFF	OFF	ON	ON
CPP	Clutch Pedal Switch	ON / OFF	ON (if in)	OFF	OFF
ECT (ºF)	ECT Sensor	-40-304ºF	160-200	160-200	160-200
ECTV	ECT Sensor	0-5.1v	0.6	0.6	0.6
EPC	EPC Solenoid	0-300 psi	5	5	15
EVP	EGR Valve Position	0-5.1v	0.3	1.2-2.0	2.5-3.5
EVR	EGR VR Solenoid	0-100%	0	0-40	0-40
FuelPW1	INJ Pulsewidth - Bank 1	0-999 ms	6.8-7.0	9.5-10	12-13
GEAR	Gear Position	P-R-N-D	P-R-N-D	DRIVE 3	DRIVE 4
HO2S-11	HO2S-11 (Bank 1)	0.1-1.1v	0.1-1.1	0.1-1.1	0.1-1.1
IAC	Idle Air Control	0-100%	35	44-50	59-65
IAT (ºF)	IAT Sensor	-40-304ºF	50-120	50-120	50-120
IATV	IAT Sensor	0-5.1v	1.5-3.5	1.5-3.5	1.5-3.5
LOOP	Loop Status	CL or OL	CL	CL	CL
MLPV	MLP Switch (E4OD)	0-5.1v	0v	5	5
MAP	MAP Sensor	0-999 Hz	107	114-120	120-130
PNP	PNP Switch Signal	Neutral/DR	NEUTRAL	DRIVE	DRIVE
RPM	Engine Speed	0-10K rpm	600-700	1050-1150	1840-1940
SHTFT1	Short Term F/T - Bank 1	-10 to 10%	-5 to +5	-5 to +5	-5 to +5
SPARK	Spark Advance	-90º to 90º	17-20	24-28	24-30
TP	TP Sensor	0-5.1v	1.0	1.2-1.3	1.5-1.6
TCC	TCC Sol. (E4OD)	ON / OFF	OFF	ON	ON
TOT	TOT sensor (E4OD)	0-5.1v	2.10-2.40	2.10-2.40	2.10-2.40
TP	TP Sensor	0-5.1v	1.0	1.2-1.3	1.5-1.6
TP Mode	TP Sensor Mode	C/T or P/T	C/T	P/T	P/T
VPWR	Vehicle Power	0-25.5v	12-14	12-14	12-14
VREF	Vehicle Reference	0-5.1v	4.9-5.1	4.9-5.1	4.9-5.1
VSS	Vehicle Speed	0-159 mph	0	30	55

1996 F-Series & Bronco 4.9L I6 MFI VIN Y Inputs

PCM PID Acronym	Parameter Identification	PID Range	PID Value at Hot Idle	PID Value at 30	PID Value at 55
ACCS	A/C Cycling Clutch	ON / OFF	ON	OFF	OFF
CPP	CPP Switch Signal	ON / OFF	ON (if in)	OFF	OFF
DPFE	DPFE Sensor	0-5.1v	0.4-0.6	0.4-0.9	0.6-1.0
ECT	ECT Sensor	-40-304°F	160-200	160-200	160-200
FEPS	Flash EEPROM	0-5.1v	0.1	0.1	0.1
FPM	Fuel Pump Monitor	ON / OFF	ON	ON	ON
IAT	IAT Sensor	-40-304°F	50-120	50-120	50-120
IDM	IDM Signal	0-9999 Hz	32-38	59-65	88-95
KS1	Knock Sensor 1	0-5.1v	0	0	0
LOAD	Engine Load	0-100%	12-14	16-25	30-40
MAFV	MAF Sensor	0-5.1v	0.5-0.7	1.1-1.5	1.7-2.2
MISF	Misfire Detection	ON / OFF	OFF	OFF	OFF
OCTADJ	Octane Adjustment	OPEN/CL	CLOSED	CLOSED	CLOSED
HO2S-11	HO2S-11 (Bank 1)	0.1-1.1v	0.1-1.1	0.1-1.1	0.1-1.1
HO2S-12	HO2S-12 (Bank 1)	0.1-1.1v	0.1-1.1	0.1-1.1	0.1-1.1
HO2S-21	HO2S-21 (Bank 2)	0.1-1.1v	0.1-1.1	0.1-1.1	0.1-1.1
PIP	PIP Sensor	0-9999 Hz	32-38	59-65	88-95
PNP	PNP Switch Signal	NEUT/DR	NEUTRAL	DRIVE	DRIVE
PTO	Power Takeoff Sig.	ON / OFF	ON (if on)	OFF	OFF
RPM	Engine Speed	0-10K rpm	680-730	1200-1300	1810-2000
TPV	TP Sensor	0-5.1v	0.53-1.27	0.8-1.1	0.9-1.2
Vacuum	Engine Vacuum	0-30" Hg	19-20	18-20	8-15
VPWR	Vehicle Power	0-25.5v	12-14	12-14	12-14
WACA	A/C WOT Monitor	ON / OFF	OFF (if on)	ON	ON

1996 F-Series & Bronco 4.9L I6 MFI VIN Y Outputs

PCM PID Acronym	Parameter Identification	PID Range	PID Value at Hot Idle	PID Value at 30	PID Value at 55
AIRB	AIR System Bypass	0v / VBAT	0v	0v	0v
AIRD	AIR System Divert	0v / VBAT	VBAT	VBAT	VBAT
EGR VR	EGR VR Solenoid	0-100%	0	0-40	0-40
FP	Fuel Pump Control	ON / OFF	ON	ON	ON
PW - Bank 1, 2	INJ Pulsewidth	0-999 ms	4.9-5.1	5.3-10.0	9.4-13.0
HTR-11	HO2S-11 Heater	ON / OFF	ON	ON	ON
HTR-12	HO2S-12 Heater	ON / OFF	ON	ON	ON
HTR-21	HO2S-21 Heater	ON / OFF	ON	ON	ON
IAC	Idle Air Control	0-100%	35	44-50	59-65
INJ1-6	INJ 1 Pulsewidth	0-999 ms	4.9-5.1	5.3-10.0	9.4-13.0
LTB1, - Bank 2	Long Term F/T - Bank 1	-20 to 20%	-5 to +5	-5 to +5	-5 to +5
MIL	MIL (lamp) Control	ON / OFF	OFF	OFF	OFF
STB1, - Bank 2	Short Term F/T - Bank 1	-10 to 10%	-5 to +5	-5 to +5	-5 to +5
SPARK	Spark Advance	-90° to 90°	13-16	17-20	17-22
SPOUT	Spark Output Signal	0-999 Hz	32-38	59-65	88-95
VMV	EVAP Solenoid	0-999 Hz	0-10	0-10	0-10
VREF	Vehicle Reference	0-5.1v	4.9-5.1	4.9-5.1	4.9-5.1
WAC	A/C WOT Relay	ON / OFF	ON (if on)	OFF	OFF

1996 F-Series & Bronco 4.9L I6 MFI VIN Y (E4OD) Inputs

PCM PID Acronym	Parameter Identification	PID Range	PID Value at Hot Idle	PID Value at 30	PID Value at 55
4x4L	4x4 Switch Signal	ON / OFF	ON (if on)	OFF	OFF
ACCS	A/C Switch Signal	ON / OFF	ON (if on)	OFF	OFF
BPP	Brake Position	ON / OFF	ON (if on)	OFF	OFF
DPFE	DPFE Sensor	0-5.1v	0.4-0.6	0.4-0.9	0.4-0.9
ECT	ECT Sensor	-40-304°F	160-200	160-200	160-200
FEPS	Flash EEPROM	0-5.1v	0.1	0.1	0.1
FPM	Fuel Pump Monitor	ON / OFF	ON	ON	ON
IAT	IAT Sensor	-40-304°F	50-120	50-120	50-120
GEAR	Gear Position	1-2-3-4	1	3	4
IDM	IDM Signal	0-9999 Hz	35-42	58-69	71-82
KS1	Knock Sensor 1	0-5.1v	0	0	0
LOAD	Engine Load	0-100%	14-16	21-32	33-42
MAFV	MAF Sensor	0-5.1v	0.7-0.9	1.2-1.8	1.7-2.0
MISF	Misfire Detection	ON / OFF	OFF	OFF	OFF
OCTADJ	Octane Adjustment	CL/OPEN	CLOSED	CLOSED	CLOSED
HO2S-11	HO2S-11 (Bank 1)	0.1-1.1v	0.1-1.1	0.1-1.1	0.1-1.1
HO2S-12	HO2S-12 (Bank 1)	0.1-1.1v	0.1-1.1	0.1-1.1	0.1-1.1
HO2S-21	HO2S-21 (Bank 2)	0.1-1.1v	0.1-1.1	0.1-1.1	0.1-1.1
PIP	PIP Sensor	0-9999 Hz	35-42	58-69	71-82
PNP	PNP Switch Signal	Neutral/DR	NEUTRAL	DRIVE	DRIVE
PTO	PTO Signal	ON / OFF	OFF	OFF	OFF
RPM	Engine Speed	0-10K rpm	760-830	1200-1270	1510-1570
TCS	TCS Switch	ON / OFF	OFF	OFF	OFF
TFT	TFT Sensor Signal	-40-304°F	110-210	110-210	110-210
TPV	TP Sensor	0-5.1v	0.53-1.27	1.0-1.3	1.2-1.7
TRV/TR	TR Sensor	0v / 1.7v	0	1.7	1.7
Vacuum	Engine Vacuum	0-30" Hg	19-20	18-20	8-15
VPWR	Vehicle Power	0-25.5v	12-14	12-14	12-14
VSS	Vehicle Speed (Hertz or MPH)	0-999 Hz (0-159 mph)	0 Hz = 0 mph	65 Hz = 30 mph	125 Hz = 55 mph
WACA	A/C WOT Relay Monitor	ON / OFF	OFF	ON	ON

1996 F-Series & Bronco 4.9L I6 VIN Y (E4OD) Outputs

PCM PID Acronym	Parameter Identification	PID Range	PID Value at Hot Idle	PID Value at 30	PID Value at 55
AIRB	AIR System Bypass	0v / VBAT	0v	0v	0v
AIRD	AIR System Divert	0v / VBAT	VBAT	VBAT	VBAT
CCS	Coast Clutch Solenoid Control	ON / OFF	OFF	OFF	OFF
EGR VR	EGR VR Solenoid	0-100%	0-100	0-40	0-40
EPC	EPC Solenoid	0-300 psi	5	6	7
FP	Fuel Pump Control	ON / OFF	ON	ON	ON
FuelPW1	INJ Pulsewidth - Bank 1	0-999 ms	4.9-5.3	5.5-10.0	9.5-13.0
FuelPW2	INJ Pulsewidth - Bank 2	0-999 ms	4.9-5.3	5.5-10.0	9.5-13.0
HTR-11	HO2S-11 Heater	ON / OFF	ON	ON	ON
HTR-12	HO2S-12 Heater	ON / OFF	ON	ON	ON
HTR-21	HO2S-21 Heater	ON / OFF	ON	ON	ON
IAC	Idle Air Control	0-100%	30-34	30-55	58-61
INJ1	INJ 1 Pulsewidth	0-999 ms	4.9-5.3	5.5-10.0	9.5-13.0
INJ2	INJ 2 Pulsewidth	0-999 ms	4.9-5.3	5.5-10.0	9.5-13.0
INJ3	INJ 3 Pulsewidth	0-999 ms	4.9-5.3	5.5-10.0	9.5-13.0
INJ4	INJ 4 Pulsewidth	0-999 ms	4.9-5.3	5.5-10.0	9.5-13.0
INJ5	INJ 5 Pulsewidth	0-999 ms	4.9-5.3	5.5-10.0	9.5-13.0
INJ6	INJ 6 Pulsewidth	0-999 ms	4.9-5.3	5.5-10.0	9.5-13.0
LongFT1	Long Term FT - Bank 1	-20 to 20%	-5 to +5	-5 to +5	-5 to +5
LongFT2	Long Term FT - Bank 2	-20 to 20%	-5 to +5	-5 to +5	-5 to +5
MIL	MIL (lamp) Control	ON / OFF	OFF	OFF	OFF
SS1	Shift Solenoid 1	ON / OFF	ON	OFF	OFF
SS2	Shift Solenoid 2	ON / OFF	OFF	ON	OFF
SHTFT1	Short Term F/T - Bank 1	-10 to 10%	-5 to +5	-5 to +5	-5 to +5
SHTFT2	Short Term FT - Bank 2	-10 to 10%	-5 to +5	-5 to +5	-5 to +5
SPARK	Spark Advance	-90° to 90°	13-16	17-20	17-22
SPOUT	Spark Output Signal	0-9999 Hz	47-52	82-88	105-110
TCC	TCC Solenoid	ON / OFF	OFF	ON	ON
TCIL	TCIL (lamp) Control	ON / OFF	OFF	OFF	OFF
VMV	EVAP Solenoid	0-10	0-10	0-10	0-10
VREF	Vehicle Reference	0-5.1v	4.9-5.1	4.9-5.1	4.9-5.1
WAC	A/C WOT Relay Control	ON / OFF	ON (AC on at WOT)	OFF	OFF

1990-95 F-Series & Bronco 5.0L V8 VIN N (A/T, M/T, E4OD)

PCM PID Acronym	Parameter Identification	PID Range	PID Value at Hot Idle	PID Value at 30	PID Value at 55
ACCS	A/C Switch Signal	ON / OFF	ON (if on)	OFF	OFF
CANP	EVAP Purge Valve	ON / OFF	OFF	ON	ON
CPP	Clutch Pedal Switch	ON / OFF	ON (if in)	OFF	OFF
DPFE	DPFE Sensor ('95)	0-5.1v	0.4-0.6	0.4-0.9	0.4-0.9
ECT (°F)	ECT Sensor	-40-304°F	160-200	160-200	160-200
ECTV	ECT Sensor	0-5.1v	0.6	0.6	0.6
EPC	EPC Solenoid	0-300 psi	4	4	14
EVP	EGR Valve Position	0-5.1v	0.4	0.4-0	3.5-4.5
EVR	EGR VR Solenoid	0-100%	0-100	0-40	0-40
FuelPW1	INJ Pulsewidth - Bank 1	0-999 ms	3.8-4.8	4.4-7.8	7.5-12.0
GEAR (E4OD)	Gear Selector Position Signal	P-R-N-D	P-R-N-D	DRIVE 3	DRIVE 4
IAC	Idle Air Control	0-100%	30-34	43-48	58-61
IAT (°F)	IAT Sensor	-40-304°F	50-120	50-120	50-120
IATV	IAT Sensor	0-5.1v	1.5-3.5	1.5-3.5	1.5-3.5
LOOP	Loop Status	CL or OL	CL	CL	CL
MLPV	MLP Switch (E4OD)	0-5.1v	0	5	5
MAP	MAP Sensor	0-999 Hz	103-105	112-120	122-140
HO2S-11	HO2S-11 (Bank 1)	0.1-1.1v	0.1-1.1	0.1-1.1	0.1-1.1
PNP	PNP Switch Signal	Neutral/DR	NEUTRAL	DRIVE	DRIVE
RPM	Engine Speed	0-10K rpm	680-780	1240-1340	1650-1750
SHTFT1	Short Term F/T - Bank 1	-10 to 10%	-5 to +5	-5 to +5	-5 to +5
SPARK	Spark Advance	-90° to 90°	14-20	28-36	30-40
TCC	TCC Sol. (E4OD)	ON / OFF	OFF	ON	ON
TOTV	TOT sensor (E4OD)	0-5.1v	2.10-2.40	2.10-2.40	2.10-2.40
TP	TP Sensor	0-5.1v	1.0	1.2-1.3	1.5-1.6
TP Mode	TP Sensor Mode	C/T or P/T	C/T	P/T	P/T
VPWR	Vehicle Power	0-25.5v	12-14	12-14	12-14
VREF	Vehicle Reference	0-5.1v	4.9-5.1	4.9-5.1	4.9-5.1
VSS	Vehicle Speed	0-159 mph	0	30	55

F-SERIES & BRONCO PID DATA - OBD II

1996 F-Series & Bronco 5.0L V8 VIN N (All) Inputs

PCM PID Acronym	Parameter Identification	PID Range	PID Value at Hot Idle	PID Value at 30	PID Value at 55
4x4L	4x4 Switch Signal	ON / OFF	ON (if on)	OFF	OFF
ACCS	A/C Switch Signal	ON / OFF	ON (if on)	OFF	OFF
BPP	Brake Switch Signal	ON / OFF	ON (if on)	OFF	OFF
CPP	CPP Switch Signal	ON / OFF	ON (if in)	OFF	OFF
DPFE	DPFE Sensor	0-5.1v	0.4-0.6	0.4-0.9	0.4-0.9
ECT	ECT Sensor	-40-304°F	160-200	160-200	160-200
FEPS	Flash EEPROM	0-5.1v	0.1	0.1	0.1
FPM	Fuel Pump Monitor	ON / OFF	ON	ON	ON
IAT	IAT Sensor	-40-304°F	50-120	50-120	50-120
GEAR	Gear Position	1-2-3-4	1	3	4
IDM	IDM Signal	0-9999 Hz	47-52	82-88	105-110
KS1	Knock Sensor 1	0	0	0	0
LOAD	Engine Load	0-100%	14-16	19-25	26-35
MAFV	MAF Sensor	0-5.1v	0.7-0.9	1.1-1.6	1.7-2.4
MISF	Misfire Detection	ON / OFF	OFF	OFF	OFF
OCTADJ	Octane Adjustment	CL/OPEN	CLOSED	CLOSED	CLOSED
OSS	OSS Sensor	0-9999 Hz	0	122-133	235-250
HO2S-11	HO2S-11 (Bank 1)	0.1-1.1v	0.1-1.1	0.1-1.1	0.1-1.1
HO2S-12	HO2S-12 (Bank 1)	0.1-1.1v	0.1-1.1	0.1-1.1	0.1-1.1
HO2S-21	HO2S-21 (Bank 2)	0.1-1.1v	0.1-1.1	0.1-1.1	0.1-1.1
PIP	PIP Sensor	0-9999 Hz	47-52	82-88	105-110
PNP	PNP Switch Signal	Neutral/DR	NEUTRAL	DRIVE	DRIVE
PTO	Power Takeoff Switch	ON / OFF	ON (if on)	OFF	OFF
RPM	Engine Speed	0-10K rpm	760-830	1200-1270	1590-1675
TCS	TCS Switch	ON / OFF	ON	ON	ON
TFT	TFT Sensor Signal	-40-304°F	110-210	110-210	110-210
TPV	TP Sensor	0-5.1v	0.53-1.27	0.8-1.0	1.0-1.3
TRV/TR	TR Sensor	P/N or O/D	P/N	O/D	O/D
Vacuum	Engine Vacuum	0-30" Hg	19-20	18-20	8-15
VPWR	Vehicle Power	0-25.5v	12-14	12-14	12-14
VSS	Vehicle Speed (Hertz or MPH)	0-999 or 0-159 mph	0 Hz = 0 mph	65 Hz = 30 mph	125 Hz = 55 mph

1996 F-Series & Bronco 5.0L V8 VIN N (All) Outputs

PCM PID Acronym	Parameter Identification	PID Range	PID Value at Hot Idle	PID Value at 30	PID Value at 55
AIRB	AIR System Bypass	0-25.5v	0.1	0.1	0.1
AIRD	AIR System Divert	0-25.5v	VBAT	VBAT	VBAT
EGR VR	EGR VR Solenoid	0-100%	0	0-40	35-40
EPC	EPC Solenoid	0-300 psi	4	5	5
FP	Fuel Pump Control	ON / OFF	ON	ON	ON
FuelPW1	INJ Pulsewidth - Bank 1	0-999 ms	3.2-3.8	4.1-6.9	6.5-12
FuelPW2	INJ Pulsewidth - Bank 2	0-999 ms	3.2-3.8	4.1-6.9	6.5-12
HTR-11	HO2S-11 Heater	ON / OFF	ON	ON	ON
HTR-12	HO2S-12 Heater	ON / OFF	ON	ON	ON
HTR-21	HO2S-21 Heater	ON / OFF	ON	ON	ON
IAC	Idle Air Control	0-100%	30-34	43-48	58-61
INJ1	INJ 1 Pulsewidth	0-999 ms	3.2-3.8	4.1-6.9	6.5-12
INJ2	INJ 2 Pulsewidth	0-999 ms	3.2-3.8	4.1-6.9	6.5-12
INJ3	INJ 3 Pulsewidth	0-999 ms	3.2-3.8	4.1-6.9	6.5-12
INJ4	INJ 4 Pulsewidth	0-999 ms	3.2-3.8	4.1-6.9	6.5-12
INJ5	INJ 5 Pulsewidth	0-999 ms	3.2-3.8	4.1-6.9	6.5-12
INJ6	INJ 6 Pulsewidth	0-999 ms	3.2-3.8	4.1-6.9	6.5-12
INJ7	INJ 7 Pulsewidth	0-999 ms	3.2-3.8	4.1-6.9	6.5-12
INJ8	INJ 8 Pulsewidth	0-999 ms	3.2-3.8	4.1-6.9	6.5-12
LongFT1	Long Term FT - Bank 1	-20 to 20%	-5 to +5	-5 to +5	-5 to +5
LongFT2	Long Term FT - Bank 2	-20 to 20%	-5 to +5	-5 to +5	-5 to +5
MIL	MIL (lamp) Control	ON / OFF	OFF	OFF	OFF
SS1	Shift Solenoid 1	ON / OFF	ON	OFF	OFF
SS2	Shift Solenoid 2	ON / OFF	OFF	ON	OFF
SHTFT1	Short Term F/T - Bank 1	-10 to 10%	-5 to +5	-5 to +5	-5 to +5
SHTFT2	Short Term FT - Bank 2	-10 to 10%	-5 to +5	-5 to +5	-5 to +5
SPARK	Spark Advance	-90° to 90°	12-17	35-40	28-37
SPOUT	Spark Output Signal	0-9999 Hz	47-52	82-88	105-110
TCC	TCC Solenoid	0-100%	0	0-40	90-100
TCIL	TCIL (lamp) Control	ON / OFF	OFF	OFF	OFF
VMV	EVAP Solenoid	0-999 Hz	0-10	0-10	0-10
VREF	Voltage Reference	0-5.1v	4.9-5.1	4.9-5.1	4.9-5.1

1996 F-Series & Bronco 5.0L V8 MFI VIN N (E4OD) Inputs

PCM PID Acronym	Parameter Identification	PID Range	PID Value at Hot Idle	PID Value at 30	PID Value at 55
4x4L	4x4 Switch Signal	ON / OFF	ON (if on)	OFF	OFF
ACCS	A/C Switch Signal	ON / OFF	ON (if on)	OFF	OFF
BPP	Brake Switch Signal	ON / OFF	ON (if on)	OFF	OFF
DPFE	DPFE Sensor	0-5.1v	0.4-0.6	0.4-0.9	0.4-0.9
ECT	ECT Sensor	-40-304°F	160-200	160-200	160-200
FEPS	Flash EEPROM	0-5.1v	0.1	0.1	0.1
FPM	Fuel Pump Monitor	ON / OFF	ON	ON	ON
IAT	IAT Sensor	-40-304°F	50-120	50-120	50-120
GEAR	Gear Position	1-2-3-4	1	3	4
IDM	IDM Signal	0-9999 Hz	47-52	82-88	105-110
KS1	Knock Sensor 1	0-5.1v	0	0	0
LOAD	Engine Load	0-100%	14-16	19-25	26-35
MAFV	MAF Sensor	0-5.1v	0.7-0.9	1.1-1.6	1.7-2.4
MISF	Misfire Detection	ON / OFF	OFF	OFF	OFF
OCTADJ	Octane Adjustment	CL/OPEN	CLOSED	CLOSED	CLOSED
HO2S-11	HO2S-11 (Bank 1)	0.1-1.1v	0.1-1.1	0.1-1.1	0.1-1.1
HO2S-12	HO2S-12 (Bank 1)	0.1-1.1v	0.1-1.1	0.1-1.1	0.1-1.1
HO2S-21	HO2S-21 (Bank 2)	0.1-1.1v	0.1-1.1	0.1-1.1	0.1-1.1
HO2S-22	HO2S-22 (Bank 2)	0.1-1.1v	0.1-1.1	0.1-1.1	0.1-1.1
PIP	PIP Sensor	0-9999 Hz	47-52	82-88	105-110
PNP	PNP Switch Signal	Neutral/DR	NEUTRAL	DRIVE	DRIVE
PTO	Power Takeoff Switch	ON / OFF	ON (if on)	OFF	OFF
RPM	Engine Speed	0-10K rpm	760-830	1200-1270	1600-1650
TCS	TCS Switch	ON / OFF	OFF	OFF	OFF
TFT	TFT Sensor Signal	-40-304°F	110-210	110-210	110-210
TPV	TP Sensor	0-5.1v	0.53-1.27	0.8-1.0	1.0-1.3
TRV/TR	TR Sensor	P/N or O/D	P/N	O/D	O/D
Vacuum	Engine Vacuum	0-30" Hg	19-20	18-20	8-15
VPWR	Vehicle Power	0-25.5v	12-14	12-14	12-14
VSS	Vehicle Speed (Hertz or MPH)	0-999 Hz (0-159 mph)	0 Hz = 0 mph	65 Hz = 30 mph	125 Hz = 55 mph

1996 F-Series & Bronco 5.0L V8 VIN N (E4OD) Outputs

PCM PID Acronym	Parameter Identification	PID Range	PID Value at Hot Idle	PID Value at 30	PID Value at 55
AIRB	AIR System Bypass	0-25.5v	0.1	0.1	0.1
AIRD	AIR System Divert	0-25.5v	VBAT	VBAT	VBAT
CCS	Coast Clutch Solenoid Control	0-25.5v	VBAT	VBAT	VBAT
EGR VR	EGR VR Solenoid	0-100%	0	0-40	35-40
EPC	EPC Solenoid	0-300 psi	4	5	5
FP	Fuel Pump Control	ON / OFF	ON	ON	ON
FuelPW1	INJ Pulsewidth - Bank 1	0-999 ms	3.2-3.8	4.1-6.9	6.5-12
FuelPW2	INJ Pulsewidth - Bank 2	0-999 ms	3.2-3.8	4.1-6.9	6.5-12
HTR-11	HO2S-11 Heater	ON / OFF	ON	ON	ON
HTR-12	HO2S-12 Heater	ON / OFF	ON	ON	ON
HTR-21	HO2S-21 Heater	ON / OFF	ON	ON	ON
IAC	Idle Air Control	0-100%	30-34	43-48	58-61
INJ1	INJ 1 Pulsewidth	0-999 ms	3.2-3.8	4.1-6.9	6.5-12
INJ2	INJ 2 Pulsewidth	0-999 ms	3.2-3.8	4.1-6.9	6.5-12
INJ3	INJ 3 Pulsewidth	0-999 ms	3.2-3.8	4.1-6.9	6.5-12
INJ4	INJ 4 Pulsewidth	0-999 ms	3.2-3.8	4.1-6.9	6.5-12
INJ5	INJ 5 Pulsewidth	0-999 ms	3.2-3.8	4.1-6.9	6.5-12
INJ6	INJ 6 Pulsewidth	0-999 ms	3.2-3.8	4.1-6.9	6.5-12
INJ7	INJ 7 Pulsewidth	0-999 ms	3.2-3.8	4.1-6.9	6.5-12
INJ8	INJ 8 Pulsewidth	0-999 ms	3.2-3.8	4.1-6.9	6.5-12
LongFT1	Long Term FT - Bank 1	-20 to 20%	-5 to +5	-5 to +5	-5 to +5
LongFT2	Long Term FT - Bank 2	-20 to 20%	-5 to +5	-5 to +5	-5 to +5
MIL	MIL (lamp) Control	ON	OFF	OFF	OFF
SS1	Shift Solenoid 1	ON	ON	OFF	OFF
SS2	Shift Solenoid 2	OFF	OFF	ON	OFF
SHTFT1	Short Term F/T - Bank 1	-10 to 10%	-5 to +5	-5 to +5	-5 to +5
SHTFT2	Short Term FT - Bank 2	-10 to 10%	-5 to +5	-5 to +5	-5 to +5
SPARK	Spark Advance	-90° to 90°	12-17	35-40	28-37
SPOUT	Spark Output Signal	0-9999 Hz	47-52	82-88	105-110
TCC	TCC Solenoid	ON / OFF	OFF	ON	ON
TCIL	TCIL (lamp) Control	ON / OFF	OFF	OFF	OFF
VMV	EVAP Solenoid	0-999 Hz	0-10	0-10	0-10
VREF	Vehicle Reference	0-5.1v	4.9-5.1	4.9-5.1	4.9-5.1

1997-2005 F-Series 5.4L V8 MFI VIN L (E4OD) Inputs

PCM PID Acronym	Parameter Identification	PID Range	PID Value at Hot Idle	PID Value at 30	PID Value at 55
4x4L	4x4 Switch Signal	ON / OFF	ON (if on)	OFF	OFF
ACCS	A/C Switch Signal	ON / OFF	ON (if on)	OFF	OFF
BPP	Brake Switch Signal	ON / OFF	ON (if on)	OFF	OFF
CHT	CHT Sensor	-40-304°F	194	194	194
CID	CMP Sensor	0-999 Hz	6	11	11
CKP	CKP Sensor	0-9999 Hz	41	800-840	900-1125
DPFE	DPFE Sensor	0-5.1v	0.20-1.30	0.9-1.36	0.9-1.4
ECT	ECT Sensor	-40-304°F	160-200	160-200	160-200
FEPS	Flash EEPROM	0-5.1v	0.1	0.1	0.1
FLI	Fuel Level Input	1.7 / 50%	1.7 / 50%	1.7 / 50%	1.7 / 50%
FPM	Fuel Pump Monitor	ON / OFF	ON	ON	ON
FTP	Fuel Tank Pressure	2.6v / 0 psi	2.6v / 0 psi	2.6v / 0 psi	2.6v / 0 psi
IAT	IAT Sensor	-40-304°F	50-120	50-120	50-120
GEAR	Gear Position	1-2-3-4	1	3	4
KS1	Knock Sensor 1	0-5.1v	0	0	0
LOAD	Engine Load	0-100%	14-16	19-25	26-35
MAFV	MAF Sensor	0-5.1v	0.7-0.9	1.1-1.6	1.7-2.4
MISF	Misfire Detection	ON / OFF	OFF	OFF	OFF
OCTADJ	Octane Adjustment	CL/OPEN	CLOSED	CLOSED	CLOSED
HO2S-11	HO2S-11 (Bank 1)	0.1-1.1v	0.1-1.1	0.1-1.1	0.1-1.1
HO2S-12	HO2S-12 (Bank 1)	0.1-1.1v	0.1-1.1	0.1-1.1	0.1-1.1
HO2S-21	HO2S-21 (Bank 2)	0.1-1.1v	0.1-1.1	0.1-1.1	0.1-1.1
HO2S-22	HO2S-22 (Bank 2)	0.1-1.1v	0.1-1.1	0.1-1.1	0.1-1.1
PNP	PNP Switch Signal	Neutral/DR	NEUTRAL	DRIVE	DRIVE
PTO	Power Takeoff Switch	OFF	ON (if on)	OFF	OFF
RPM	Engine Speed	0-10K rpm	760-830	1200-1270	1590-1675
TCS	TCS Switch	ON / OFF	OFF	OFF	OFF
TFT	TFT Sensor Signal	-40-304°F	110-210	110-210	110-210
TPV	TP Sensor	0-5.1v	0.53-1.27	0.8-1.0	1.0-1.3
TR1	TR1 Sensor	0v / 11.5v	0v	11.5	11.5
TR2	TR2 Sensor	0v / 11.5v	0v	11.5	11.5
TR3/TR	TR3A Sensor	0v / 1.7v	0v / PN	O/D	O/D
TR4	TR4 Sensor	0v / 11.5v	0	11.5	11.5
Vacuum	Engine Vacuum	0-30" Hg	19-20	18-20	8-15
VPWR	Vehicle Power	0-25.5v	12-14	12-14	12-14
VSS	Vehicle Speed (Hertz or MPH)	0-999 Hz (0-159 mph)	0 Hz = 0 mph	65 Hz = 30 mph	125 Hz = 55 mph

1997-2005 F-Series 5.4L V8 MFI VIN L (E4OD) Outputs

PCM PID Acronym	Parameter Identification	PID Range	PID Value at Hot Idle	PID Value at 30	PID Value at 55
CD1-CD8	COP Driver 1-8	0-45° dwell	5	6	8
CHTIL	CHT (lamp) Control	ON / OFF	OFF	OFF	OFF
CTO	Clean Tachometer signal	0-9999 Hz	46	90	115
EGR VR	EGR VR Solenoid	0-100%	0	0-40	35-45
EVAP CP	EVAP Purge Valve	0-100%	0-100	0-100	0-100
EVAP CV	EVAP CV Valve	0-100%	0-100	0-100	0-100
EPC	EPC Solenoid	0-300 psi	4	5	5
FP	Fuel Pump Control	ON / OFF	ON	ON	ON
FuelPW1	INJ Pulsewidth - Bank 1	0-999 ms	3.2-3.8	4.2-6.9	6.5-12
FuelPW2	INJ Pulsewidth - Bank 2	0-999 ms	3.2-3.8	4.2-6.9	6.5-12
HTR-11	HO2S-11 Heater	ON / OFF	ON	ON	ON
HTR-12	HO2S-12 Heater	ON / OFF	ON	ON	ON
HTR-21	HO2S-21 Heater	ON / OFF	ON	ON	ON
HTR-22	HO2S-22 Heater	ON / OFF	ON	ON	ON
IAC	Idle Air Control	0-100%	30-34	43-48	58-61
INJ1	INJ 1 Pulsewidth	0-999 ms	3.2-3.8	4.2-6.9	6.5-12
INJ2	INJ 2 Pulsewidth	0-999 ms	3.2-3.8	4.2-6.9	6.5-12
INJ3	INJ 3 Pulsewidth	0-999 ms	3.2-3.8	4.2-6.9	6.5-12
INJ4	INJ 4 Pulsewidth	0-999 ms	3.2-3.8	4.2-6.9	6.5-12
INJ5	INJ 5 Pulsewidth	0-999 ms	3.2-3.8	4.2-6.9	6.5-12
INJ6	INJ 6 Pulsewidth	0-999 ms	3.2-3.8	4.2-6.9	6.5-12
INJ7	INJ 7 Pulsewidth	0-999 ms	3.2-3.8	4.2-6.9	6.5-12
INJ8	INJ 8 Pulsewidth	0-999 ms	3.2-3.8	4.2-6.9	6.5-12
LongFT1	Long Term FT - Bank 1	-20 to 20%	-5 to +5	-5 to +5	-5 to +5
LongFT2	Long Term FT - Bank 2	-20 to 20%	-5 to +5	-5 to +5	-5 to +5
MIL	MIL (lamp) Control	ON / OFF	OFF	OFF	OFF
SS1	Shift Solenoid 1	ON / OFF	ON	OFF	OFF
SS2	Shift Solenoid 2	ON / OFF	OFF	ON	OFF
SHTFT1	Short Term F/T - Bank 1	-10 to 10%	-5 to +5	-5 to +5	-5 to +5
SHTFT2	Short Term FT - Bank 2	-10 to 10%	-5 to +5	-5 to +5	-5 to +5
SPARK	Spark Advance	-90° to 90°	12-17	35-40	28-37
TCC	TCC Solenoid	0-100%	0	0-40	90-95
TCIL	TCIL (lamp) Control	ON / OFF	OFF	OFF	OFF
VREF	Vehicle Reference	0-5.1v	4.9-5.1	4.9-5.1	4.9-5.1

<u>FORD LIGHTNING</u>

1999-05 Lightning 5.4L V8 SC MFI VIN 3 (E4OD) Inputs

PCM PID Acronym	Parameter Identification	PID Range	PID Value at Hot Idle	PID Value at 30	PID Value at 55
ACCS	A/C Switch Signal	ON / OFF	ON (if on)	OFF	OFF
BARO	BARO Sensor	0-999 Hz	159	159	159
BPP	Brake Switch Signal	ON / OFF	ON (if on)	OFF	OFF
CHT	CHT Sensor	-40-304°F	194	194	194
CID	CMP Sensor	0-999 Hz	6-8	10-12	14-17
CKP	CKP Sensor	0-9999 Hz	410	650	1060
DPFE	DPFE Sensor	0-5.1v	0.75-1.25	0.9-1.36	0.9-1.4
FEPS	Flash EEPROM	0-5.1v	0.1	0.1	0.1
FLI	Fuel Level Input	1.7v / 50%	1.7 / 50%	1.7 / 50%	1.7 / 50%
FPM	Fuel Pump Monitor	ON / OFF	ON	ON	ON
FTP	Fuel Tank Pressure	2.6v / 0 psi	2.6v / 0 psi	2.6v / 0 psi	2.6v / 0 psi
IAT1	IAT Sensor 1 (°F)	-40-304°F	50-120	50-120	50-120
IAT2	IAT Sensor 2 (°F)	-40-304°F	50-120	50-120	50-120
GEAR	Gear Position	1-2-3-4	1	3	4
KS1	Knock Sensor 1	0-5.1v	0	0	0
LOAD	Engine Load	0-100%	18-21	27-32	35-45
MAFV	MAF Sensor	0-5.1v	0.7-0.9	1.1-1.6	1.7-2.4
MISF	Misfire Detection	ON / OFF	OFF	OFF	OFF
HO2S-11	HO2S-11 (Bank 1)	0.1-1.1v	0.1-1.1	0.1-1.1	0.1-1.1
HO2S-12	HO2S-12 (Bank 1)	0.1-1.1v	0.1-1.1	0.1-1.1	0.1-1.1
HO2S-21	HO2S-21 (Bank 2)	0.1-1.1v	0.1-1.1	0.1-1.1	0.1-1.1
HO2S-22	HO2S-22 (Bank 2)	0.1-1.1v	0.1-1.1	0.1-1.1	0.1-1.1
OCTADJ	Octane Adjustment	CL/OPEN	CLOSED	CLOSED	CLOSED
PNP	PNP Switch Signal	ON / OFF	ON	OFF	OFF
RPM	Engine Speed	0-10K rpm	700	1270-1330	1575-1670
TCS	TCS Switch	ON / OFF	OFF	OFF	ON
TFT	TFT Sensor Signal	-40-304°F	110-210	110-210	110-210
TPV	TP Sensor	0-5.1v	0.53-1.27	0.8-1.0	1.0-1.3
TR1	TR1 Sensor	0v / 11.5v	0v	11.5	11.5
TR2	TR2 Sensor	0v / 11.5v	0v	11.5	11.5
TR4	TR4 Sensor	0v / 11.5v	0v	11.5	11.5
TRV/TR	TR Sensor	0v / 1.7v	0v / PN	O/D	O/D
TSS	TSS Sensor	0-9999 Hz	110	710	1700
Vacuum	Engine Vacuum	0-30" Hg	19-20	18-20	8-15
VPWR	Vehicle Power	0-25.5v	12-14	12-14	12-14
VSS	Vehicle Speed (Hertz or MPH)	0-999 Hz (0-159 mph)	0 Hz = 0 mph	65 Hz = 30 mph	125 Hz = 55 mph

1999-05 Lightning 5.4L V8 SC MFI VIN 3 (E4OD) Outputs

PCM PID Acronym	Parameter Identification	PID Range	PID Value at Hot Idle	PID Value at 30	PID Value at 55
CD1-CD8	COP Driver 1-8	0-45° dwell	5	6	8
CHTIL	CHT (lamp) Control	ON / OFF	OFF	OFF	OFF
CTO	Clean Tachometer signal	0-9999 Hz	46	90	115
EGR VR	EGR VR Solenoid	0-100%	0	0-40	35-45
EVAP CP	EVAP Purge Valve	0-100%	0-100	0-100	0-100
EVAP CV	EVAP CV Valve	0-100%	0-100	0-100	0-100
EPC	EPC Solenoid	0-300 psi	4	5	5
FP	Fuel Pump Control	ON / OFF	ON	ON	ON
FuelPW1	INJ Pulsewidth - Bank 1	0-999 ms	3.2-3.8	4.2-6.9	6.5-12
FuelPW2	INJ Pulsewidth - Bank 2	0-999 ms	3.2-3.8	4.2-6.9	6.5-12
HTR-11	HO2S-11 Heater	ON / OFF	ON	ON	ON
HTR-12	HO2S-12 Heater	ON / OFF	ON	ON	ON
HTR-21	HO2S-21 Heater	ON / OFF	ON	ON	ON
HTR-22	HO2S-22 Heater	ON / OFF	ON	ON	ON
IAC	Idle Air Control	0-100%	30-34	43-48	58-61
ICP	Injector Control Pressure Solenoid	ON / OFF	ON	ON	ON
INJ1	INJ 1 Pulsewidth	0-999 ms	3.2-3.8	4.2-6.9	6.5-12
INJ2	INJ 2 Pulsewidth	0-999 ms	3.2-3.8	4.2-6.9	6.5-12
INJ3	INJ 3 Pulsewidth	0-999 ms	3.2-3.8	4.2-6.9	6.5-12
INJ4	INJ 4 Pulsewidth	0-999 ms	3.2-3.8	4.2-6.9	6.5-12
INJ5	INJ 5 Pulsewidth	0-999 ms	3.2-3.8	4.2-6.9	6.5-12
INJ6	INJ 6 Pulsewidth	0-999 ms	3.2-3.8	4.2-6.9	6.5-12
INJ7	INJ 7 Pulsewidth	0-999 ms	3.2-3.8	4.2-6.9	6.5-12
INJ8	INJ 8 Pulsewidth	0-999 ms	3.2-3.8	4.2-6.9	6.5-12
LongFT1	Long Term FT - Bank 1	-20 to 20%	-5 to +5	-5 to +5	-5 to +5
LongFT2	Long Term FT - Bank 2	-20 to 20%	-5 to +5	-5 to +5	-5 to +5
MIL	MIL (lamp) Control	ON / OFF	OFF	OFF	OFF
SCB	S/C Bypass Control	ON / OFF	OFF	OFF	ON
SHTFT1	Short Term F/T - Bank 1	-10 to 10%	-5 to +5	-5 to +5	-5 to +5
SHTFT2	Short Term FT - Bank 2	-10 to 10%	-5 to +5	-5 to +5	-5 to +5
SPARK	Spark Advance	-90° to 90°	12-17	35-40	28-37
SS1	Shift Solenoid 1	ON / OFF	ON	OFF	OFF
SS2	Shift Solenoid 2	ON / OFF	OFF	OFF	OFF
TCC	TCC Solenoid	0-100%	0	0-40	90-95
TCIL	TCIL (lamp) Control	ON / OFF	OFF	OFF	OFF
VREF	Vehicle Reference	0-5.1v	4.9-5.1	4.9-5.1	4.9-5.1
VSO	VSO Control	ON / OFF	ON	ON	ON

FORD F-SERIES TRUCK

1997-2005 F-Series 5.4L V8 CNG VIN M (E4OD) Inputs

PCM PID Acronym	Parameter Identification	PID Range	PID Value at Hot Idle	PID Value at 30	PID Value at 55
ACCS	A/C Switch Signal	ON / OFF	ON (if on)	OFF	OFF
BPP	Brake Switch Signal	ON / OFF	ON (if on)	OFF	OFF
CHT	CHT Sensor	-40-304°F	194	194	194
CID	CMP Sensor	0-999 Hz	6-8	9-12	15-17
CKP	CKP Sensor	0-9999 Hz	400-490	790-870	1000-1089
EFTA	EFT Sensor	-40-275F	50-120	50-120	50-120
FEPS	Flash EEPROM	0-5.1v	0.1	0.1	0.1
FPM	Fuel Pump Monitor	ON / OFF	ON	ON	ON
FSV-M	Fuel Solenoid Valve	ON / OFF	ON	ON	ON
IAT	IAT Sensor	-40-304°F	50-120	50-120	50-120
FRP	FRP Sensor	0-150 psi	90-100	90-100	90-100
GEAR	Gear Position	1-2-3-4	1	3	4
LOAD	Engine Load	0-100%	13-19	26-33	38-46
MAFV	MAF Sensor	0-5.1v	0.7-0.9	0.9-1.7	1.2-2.4
MISF	Misfire Detection	ON / OFF	OFF	OFF	OFF
OCTADJ	Octane Adjustment	CL/OPEN	CLOSED	CLOSED	CLOSED
HO2S-11	HO2S-11 (Bank 1)	0.1-1.1v	0.1-1.1	0.1-1.1	0.1-1.1
HO2S-21	HO2S-21 (Bank 2)	0.1-1.1v	0.1-1.1	0.1-1.1	0.1-1.1
PNP	PNP Switch Signal	Neutral/DR	NEUTRAL	DRIVE	DRIVE
PTO	Power Takeoff Switch	ON / OFF	ON (if on)	OFF	OFF
RPM	Engine Speed	0-10K rpm	900-930	1470-1490	1840-1860
TCS	TCS Switch	ON / OFF	OFF	OFF	OFF
TPV	TP Sensor	0-5.1v	0.53-1.27	0.8-1.2	0.9-1.6v
TR1	TR1 Sensor	0v / 11.5v	0	11.5	11.5
TR2	TR2 Sensor	0v / 11.5v	0	11.5	11.5
TR3/TR	TR3A Sensor	0v / 1.7v	0	1.7	1.7
TR4	TR4 Sensor	0v / 11.5v	0	11.5	11.5
Vacuum	Engine Vacuum	0-30" Hg	19-20	18-20	8-15
VPWR	Vehicle Power	0-25.5v	12-14	12-14	12-14
VSS	Vehicle Speed (Hertz or MPH)	0-999 Hz (0-159 mph)	0 Hz = 0 mph	65 Hz = 30 mph	125 Hz = 55 mph

1997-2005 F-Series 5.4L V8 CNG VIN M (E4OD) Outputs

PCM PID Acronym	Parameter Identification	PID Range	PID Value at Hot Idle	PID Value at 30	PID Value at 55
CCS	Coast Clutch Solenoid Control	ON / OFF	OFF	OFF	OFF
CD1-CD8	COP Driver 1-8	0-45° dwell	5	6	8
CHTIL	CHT (lamp) Control	ON / OFF	OFF	OFF	OFF
CTO	Clean Tachometer signal	0-9999 Hz	50-58	90-100	115-125
EPC	EPC Solenoid	0-300 psi	5	6	10 psi
FP	Fuel Pump Control	ON / OFF	ON	ON	ON
FSV	Fuel Solenoid Valve	ON / OFF	ON	ON	ON
FuelPW1	INJ Pulsewidth - Bank 1	0-999 ms	3.9-4.6	4.7-12.0	4.7-12.0
FuelPW2	INJ Pulsewidth - Bank 2	0-999 ms	3.9-4.6	4.7-12.0	4.7-12.0
HTR-11	HO2S-11 Heater	ON / OFF	ON	ON	ON
HTR-21	HO2S-21 Heater	ON / OFF	ON	ON	ON
IAC	Idle Air Control	0-100%	30-34	43-48	58-61
IMTV	Intake Manifold Tuning Valve	0-100%	0	0	0
INJ1	INJ 1 Pulsewidth	0-999 ms	3.9-4.6	4.7-12.0	4.7-12.0
INJ2	INJ 2 Pulsewidth	0-999 ms	3.9-4.6	4.7-12.0	4.7-12.0
INJ3	INJ 3 Pulsewidth	0-999 ms	3.9-4.6	4.7-12.0	4.7-12.0
INJ4	INJ 4 Pulsewidth	0-999 ms	3.9-4.6	4.7-12.0	4.7-12.0
INJ5	INJ 5 Pulsewidth	0-999 ms	3.9-4.6	4.7-12.0	4.7-12.0
INJ6	INJ 6 Pulsewidth	0-999 ms	3.9-4.6	4.7-12.0	4.7-12.0
INJ7	INJ 7 Pulsewidth	0-999 ms	3.9-4.6	4.7-12.0	4.7-12.0
INJ8	INJ 8 Pulsewidth	0-999 ms	3.9-4.6	4.7-12.0	4.7-12.0
LongFT1	Long Term FT - Bank 1	-20 to 20%	-5 to 5%	-5 to 5%	-5 to 5%
LongFT2	Long Term FT - Bank 2	-20 to 20%	-5 to 5%	-5 to 5%	-5 to 5%
MIL	MIL (lamp) Control	ON / OFF	OFF	OFF	OFF
SHTFT1	Short Term F/T - Bank 1	-10 to 10%	-5 to 5%	-5 to 5%	-5 to 5%
SHTFT2	Short Term FT - Bank 2	-10 to 10%	-5 to 5%	-5 to 5%	-5 to 5%
SPARK	Spark Advance	-90° to 90°	15-25	20-35	20-30
SS1	Shift Solenoid 1	ON / OFF	ON	OFF	OFF
SS2	Shift Solenoid 2	ON / OFF	OFF	ON	OFF
TCC	TCC Solenoid	0-100%	0%	0-40	80-100
TCIL	TCIL (lamp) Control	ON / OFF	OFF	OFF	OFF
VREF	Vehicle Reference	0-5.1v	4.9-5.1	4.9-5.1	4.9-5.1

1998-99 F-Series 5.4L V8 NGV VIN Z (A/T, M/T, E4OD)

PCM PID Acronym	Parameter Identification	PID Range	PID Value at Hot Idle	PID Value at 30	PID Value at 55
ACCS	A/C Switch Signal	ON / OFF	ON (if on)	OFF	OFF
BPP	Brake Position	ON / OFF	ON (if on)	OFF	OFF
CHT	CHT Sensor	-40-500°F	194	194	194
CID	CMP Sensor	0-999 Hz	6-7	9-12	15-17.5
CKP	CKP Sensor	0-9999 Hz	440-490	810-855	1088-1089
CPP	Clutch Pedal Switch	ON / OFF	ON (if in)	ON	ON
EFTA	EFT Sensor	-40-275F	50-120	50-120	50-120
FEPS	Flash EEPROM	0-5.1v	0.1	0.1	0.1
FPM	Fuel Pump Monitor	ON / OFF	ON	ON	ON
FRP (psi)	FRP Sensor	0-300 psi	90-100	90-100	90-100
FRP (V)	FRP Sensor	0-5.1v	2.7-3.7	2.7-3.7	2.7-3.7
FSV-M	Fuel Solenoid Valve	ON / OFF	ON	ON	ON
IAT	IAT Sensor	-40-304°F	50-120	50-120	50-120
GEAR	Gear Position	1-2-3-4	1	3	4
LOAD	Engine Load	0-100%	13-19	26-33	38-46
MAFV	MAF Sensor	0-5.1v	0.7-0.9	0.9-1.7	1.2-2.4
MISF	Misfire Detection	ON / OFF	OFF	OFF	OFF
O2S-11	O2S-11 (Bank 1)	0.1-1.1v	0.1-1.1	0.1-1.1	0.1-1.1
PNP	PNP Switch Signal	Neutral/DR	NEUTRAL	DRIVE	DRIVE
PTO	Power Takeoff Switch	OFF	ON (if on)	OFF	OFF
RPM	Engine Speed	0-10K rpm	900-930	1470-1490	1840-1860
TCS	TCS Switch	ON / OFF	OFF	OFF	OFF
TFT	TFT Sensor Signal	-40-304°F	110-210	110-210	110-210
TPV	TP Sensor	0-5.1v	0.53-1.27	0.8-1.0	0.9-1.2
TR1	TR1 Sensor	0v / 11.5v	0	11.5	11.5
TR2	TR2 Sensor	0v / 11.5v	0	11.5	11.5
TR4	TR4 Sensor	0v / 11.5v	0	11.5	11.5
TRV/TR	TR Sensor	0v / 1.7v	0	1.7	1.7
Vacuum	Engine Vacuum	0-30" Hg	19-20	18-20	8-15
VPWR	Vehicle Power	0-25.5v	12-14	12-14	12-14
VSS	Vehicle Speed (Hertz or MPH)	0-999 Hz (0-159 mph)	0 Hz = 0 mph	65 Hz = 30 mph	125 Hz = 55 mph

1998-99 F-Series 5.4L V8 NGV VIN Z (A/T, M/T, E4OD)

PCM PID Acronym	Parameter Identification	PID Range	PID Value at Hot Idle	PID Value at 30	PID Value at 55
CCS	Coast Clutch Solenoid Control	ON / OFF	OFF	OFF	OFF
CD1-CD8	COP Driver 1-8	0-45° dwell	5	6	8
CHTIL	CHT (lamp) Control	ON / OFF	OFF	OFF	OFF
CTO	Clean Tachometer signal	0-9999 Hz	50-57	90-100	115-125
EPC	EPC Solenoid (E4OD)	0-300 psi	5	6	11
FP	Fuel Pump Control	ON / OFF	ON	ON	ON
FSV	Fuel Solenoid Valve	ON / OFF	ON	ON	ON
FuelPW1	INJ Pulsewidth - Bank 1	0-999 ms	3.9-4.6	4.7-12.0	4.7-12.0
FuelPW2	INJ Pulsewidth - Bank 2	0-999 ms	3.9-4.6	4.7-12.0	4.7-12.0
HTR-11	HO2S-11 Heater	ON / OFF	ON	ON	ON
HTR-21	HO2S-21 Heater	ON / OFF	ON	ON	ON
IAC	Idle Air Control	0-100%	30-34	43-48	58-61
IMTV	Intake Manifold Tuning Valve	0-100%	0	0	0
INJ1	INJ 1 Pulsewidth	0-999 ms	3.9-4.6	4.7-12.0	4.7-12.0
INJ2	INJ 2 Pulsewidth	0-999 ms	3.9-4.6	4.7-12.0	4.7-12.0
INJ3	INJ 3 Pulsewidth	0-999 ms	3.9-4.6	4.7-12.0	4.7-12.0
INJ4	INJ 4 Pulsewidth	0-999 ms	3.9-4.6	4.7-12.0	4.7-12.0
INJ5	INJ 5 Pulsewidth	0-999 ms	3.9-4.6	4.7-12.0	4.7-12.0
INJ6	INJ 6 Pulsewidth	0-999 ms	3.9-4.6	4.7-12.0	4.7-12.0
INJ7	INJ 7 Pulsewidth	0-999 ms	3.9-4.6	4.7-12.0	4.7-12.0
INJ8	INJ 8 Pulsewidth	0-999 ms	3.9-4.6	4.7-12.0	4.7-12.0
LongFT1	Long Term FT - Bank 1	-20 to 20%	-5 to 5%	-5 to 5%	-5 to 5%
LongFT2	Long Term FT - Bank 2	-20 to 20%	-5 to 5%	-5 to 5%	-5 to 5%
MIL	MIL (lamp) Control	ON / OFF	OFF	OFF	OFF
SHTFT1	Short Term F/T - Bank 1	-10 to 10%	-5 to 5%	-5 to 5%	-5 to 5%
SHTFT2	Short Term FT - Bank 2	-10 to 10%	-5 to 5%	-5 to 5%	-5 to 5%
SPARK	Spark Advance	-90° to 90°	15-25	20-35	20-30
SS1	Shift Solenoid 1	ON / OFF	ON	OFF	ON
SS2	Shift Solenoid 2	ON / OFF	OFF	ON	OFF
TCC	TCC Solenoid	0-100%	0	50-100	90-100
TCIL	TCIL (lamp) Control	ON / OFF	OFF	OFF	OFF
VPWR	Vehicle Power	0-25.5v	12-14	12-14	12-14
VREF	Vehicle Reference	5v	4.9-5.1	4.9-5.1	4.9-5.1

1990-95 F-Series 5.8L V8 MFI VIN H (A/T, M/T E4OD)

PCM PID Acronym	Parameter Identification	PID Range	PID Value at Hot Idle	PID Value at 30	PID Value at 55
ACCS	A/C Switch Signal	ON / OFF	ON (if on)	OFF	OFF
CANP	EVAP Purge Valve	ON / OFF	OFF	ON	ON
CPP	Clutch Pedal Switch	ON / OFF	ON (if in)	OFF	OFF
ECT (°F)	ECT Sensor	-40-304°F	160-200	160-200	160-200
ECTV	ECT Sensor	0-5.1v	0.6	0.6	0.6
EPC	EPC Solenoid (E4OD)	0-300 psi	4	4	14
EVP	EGR Valve Position	0-5.1v	0.3	0.4	0.6-3.0
EVR	EGR VR Solenoid	0-100%	0	0-40	0-40
FuelPW1	INJ Pulsewidth - Bank 1	0-999 ms	5.0-5.8	6.0-6.8	6.4-7.0
GEAR	Gear Position	P-R-N-D	P-R-N-D	DRIVE 3	DRIVE 4
IAC	Idle Air Control	0-100%	30-34	28-32	63-70
IAT (°F)	IAT Sensor	-40-304°F	50-120	50-120	50-120
IATV	IAT Sensor	0-5.1v	1.5-3.5	1.5-3.5	1.5-3.5
LOOP	Loop Status	CL or OL	CL	CL	CL
MLPV	MLP Switch (E4OD)	0-5.1v	0	5	5
MAF	MAF Sensor (Calif.)	0-5.1v	0.7-0.9	1.2-1.7	1.6-2.4
MAP	MAP Sensor	0-999 Hz	108-112	110-120	130-140
HO2S-11	HO2S-11 (Bank 1)	0.1-1.1v	0.1-1.1	0.1-1.1	0.1-1.1
PNP	PNP Switch Signal	Neutral/DR	NEUTRAL	DRIVE	DRIVE
RPM	Engine Speed	0-10K rpm	750-850	1250-1350	1600-1700
SHTFT1	Short Term F/T - Bank 1	-10 to 10%	-5 to +5	-5 to +5	-5 to +5
SPARK	Spark Advance	-90° to 90°	14-18	30-36	38-44
TP	TP Sensor	0-5.1v	0.9	1.0-1.1	1.1-1.2
TCC	TCC Sol. (E4OD)	ON / OFF	OFF	ON	ON
TOTV	TOT sensor (E4OD)	0-5.1v	2.10-2.40	2.10-2.40	2.10-2.40
TP	TP Sensor	0-5.1v	1.0	1.2-1.3	1.5-1.6
TP Mode	TP Sensor Mode	C/T or P/T	C/T	P/T	P/T
VPWR	Vehicle Power	0-25.5v	12-14	12-14	12-14
VREF	Vehicle Reference	0-5.1v	4.9-5.1	4.9-5.1	4.9-5.1
VSS	Vehicle Speed	0-159 mph	0	30	55

1993-95 F-Series 5.8L V8 MFI VIN R (A/T) Inputs & Outputs

PCM PID Acronym	Parameter Identification	PID Range	PID Value at Hot Idle	PID Value at 30	PID Value at 55
CANP	EVAP Purge Valve	ON / OFF	OFF	ON	ON
ECT (°F)	ECT Sensor	-40-304°F	160-200	160-200	160-200
ECTV	ECT Sensor	0-5.1v	0.6	0.6	0.6
EVP	EGR Valve Position	0-5.1v	0.3	0.4	0.6-3.0
EVR	EGR VR Solenoid	0-100%	0	0-40	0-40
FuelPW1	INJ Pulsewidth - Bank 1	0-999 ms	5.0-5.8	6.0-6.8	6.4-7.0
IAC	Idle Air Control	0-100%	30-34	43-48	58-61
IAT (°F)	IAT Sensor	-40-304°F	50-120	50-120	50-120
IATV	IAT Sensor	0-5.1v	1.5-3.5	1.5-3.5	1.5-3.5
LOOP	Loop Status	CL or OL	CL	CL	CL
MAP	MAP Sensor	0-999 Hz	108-112	110-120	130-140
HO2S-11	HO2S-11 (Bank 1)	0.1-1.1v	0.1-1.1	0.1-1.1	0.1-1.1
PNP	PNP Switch Signal	Neutral/DR	NEUTRAL	DRIVE	DRIVE
RPM	Engine Speed	0-10K rpm	750-850	1250-1350	1600-1700
SHTFT1	Short Term F/T - Bank 1	-10 to 10%	-5 to +5	-5 to +5	-5 to +5
SPARK	Spark Advance	-90° to 90°	14-18	30-36	36-44
TP	TP Sensor	0-5.1v	0.9	1.0-1.1	1.1-1.2
TP Mode	TP Sensor Mode	C/T or P/T	C/T	P/T	P/T
VPWR	Vehicle Power	0-25.5v	12-14	12-14	12-14
VREF	Vehicle Reference	0-5.1v	4.9-5.1	4.9-5.1	4.9-5.1
VSS	Vehicle Speed	0-159 mph	0	30	55

F-SERIES & BRONCO PID DATA

1996-97 F-Series & Bronco 5.8L V8 MFI VIN H (E4OD)

PCM PID Acronym	Parameter Identification	PID Range	PID Value at Hot Idle	PID Value at 30	PID Value at 55
4x4L	4x4 Switch Signal	ON / OFF	ON (if on)	OFF	OFF
ACCS	A/C Switch Signal	ON / OFF	ON (if on)	OFF	OFF
BPP	Brake Switch Signal	ON / OFF	ON (if on)	OFF	OFF
DPFE	DPFE Sensor	0-5.1v	0.4-0.6	0.4-0.9	0.8-1.1
ECTV	ECT Sensor	0-5.1v	0.6	0.6	0.6
ECT (°F)	ECT Sensor	-40-304°F	160-200	160-200	160-200
FEPS	Flash EEPROM	0-5.1v	0.1	0.1	0.1
FLI	Fuel Level Input	0-100%	1.7 / 50%	1.7 / 50%	1.7 / 50%
FPM	Fuel Pump Monitor	ON / OFF	ON	ON	ON
FuelPW1	INJ Pulsewidth - Bank 1	0-999 ms	4.2-4.6	4.4-7.0	7.4-12.0
GEAR	Gear Selector	P-R-N-D	P-R-N-D	DRIVE 3	DRIVE 4
IAC	Idle Air Control	0-100%	30-34	28-32	63-70
IAT (°F)	IAT Sensor	-40-304°F	50-120	50-120	50-120
IDM	IDM Signal	0-9999 Hz	42-50	82-88	105-120
LOAD	Engine Load	0-100%	14-16	20-25	24-35
MAFV	MAF Sensor	0-5.1v	0.7-0.9	1.2-1.7	1.6-2.4
MISF	Misfire Detection	ON / OFF	OFF	OFF	OFF
MLPV	MLP Switch (V)	0-5.1v	0	5	5
OCTADJ	Octane Adjustment	OPEN/CL	CLOSED	CLOSED	CLOSED
HO2S-11	HO2S-11 (Bank 1)	0.1-1.1v	0.1-1.1	0.1-1.1	0.1-1.1
HO2S-12	HO2S-12 (Bank 1)	0.1-1.1v	0.1-1.1	0.1-1.1	0.1-1.1
HO2S-21	HO2S-21 (Bank 2)	0.1-1.1v	0.1-1.1	0.1-1.1	0.1-1.1
PIP	PIP Sensor	0-9999 Hz	42-50	82-88	105-120
PNP	PNP Switch Signal	ON / OFF	ON in P/N	OFF	OFF
PTO	Power Takeoff	ON / OFF	ON (if on)	OFF	OFF
RPM	Engine Speed	0-10K rpm	675-715	1250-1390	1590-1700
TCS	TCS Switch	ON / OFF	OFF	OFF	OFF
TFTV	TFT Sensor Signal	-40-304°F	110-210	110-210	110-210
TPV	TP Sensor	0-5.1v	0.53-1.27	0.8-1.0	0.9-1.2
TP Mode	TP Sensor Mode	C/T or P/T	C/T	P/T	P/T
TRV/TR	TR Sensor	P/N or O/D	P/N	O/D	O/D
Vacuum	Engine Vacuum	0-30" Hg	19-20	18-20	8-15
VPWR	Vehicle Power	0-25.5v	12-14	12-14	12-14
VSS	Vehicle Speed (Hertz or MPH)	0-999 Hz (0-159 mph)	0 Hz = 0 mph	65 Hz = 30 mph	125 Hz = 55 mph

1996-97 F-Series & Bronco 5.8L V8 MFI VIN H (E4OD)

PCM PID Acronym	Parameter Identification	PID Range	PID Value at Hot Idle	PID Value at 30	PID Value at 55
CCS	Coast Clutch Solenoid Control	ON / OFF	OFF	OFF	OFF
EGR VR	EGR VR Solenoid	0-100%	0	0-40	35-45
EVAP CP	EVAP Purge Valve	0-100%	0-100	0-100	0-100
EPC	EPC Solenoid	0-300 psi	5	5	5
FP	Fuel Pump Control	ON / OFF	ON	ON	ON
FuelPW1	INJ Pulsewidth - Bank 1	0-999 ms	4.2-4.7	4.4-7.0	7.4-12.0
FuelPW2	INJ Pulsewidth - Bank 2	0-999 ms	4.2-4.7	4.4-7.0	7.4-12.0
HTR-11	HO2S-11 Heater	ON / OFF	ON	ON	ON
HTR-12	HO2S-12 Heater	ON / OFF	ON	ON	ON
HTR-21	HO2S-21 Heater	ON / OFF	ON	ON	ON
HTR-22	HO2S-22 Heater	ON / OFF	ON	ON	ON
IAC	Idle Air Control	0-100%	30-34	28-32	63-70
INJ1	INJ 1 Pulsewidth	0-999 ms	4.2-4.7	4.4-7.0	7.4-12.0
INJ2	INJ 2 Pulsewidth	0-999 ms	4.2-4.7	4.4-7.0	7.4-12.0
INJ3	INJ 3 Pulsewidth	0-999 ms	4.2-4.7	4.4-7.0	7.4-12.0
INJ4	INJ 4 Pulsewidth	0-999 ms	4.2-4.7	4.4-7.0	7.4-12.0
INJ5	INJ 5 Pulsewidth	0-999 ms	4.2-4.7	4.4-7.0	7.4-12.0
INJ6	INJ 6 Pulsewidth	0-999 ms	4.2-4.7	4.4-7.0	7.4-12.0
INJ7	INJ 7 Pulsewidth	0-999 ms	4.2-4.7	4.4-7.0	7.4-12.0
INJ8	INJ 8 Pulsewidth	0-999 ms	4.2-4.7	4.4-7.0	7.4-12.0
LongFT1	Long Term FT - Bank 1	-20 to 20%	-5 to +5	-5 to +5	-5 to +5
LongFT2	Long Term FT - Bank 2	-20 to 20%	-5 to +5	-5 to +5	-5 to +5
MIL	MIL (lamp) Control	ON / OFF	OFF	OFF	OFF
SS1	Shift Solenoid 1	ON / OFF	ON	OFF	OFF
SS2	Shift Solenoid 2	ON / OFF	OFF	ON	OFF
SHTFT1	Short Term F/T - Bank 1	-10 to 10%	-5 to +5	-5 to +5	-5 to +5
SHTFT2	Short Term FT - Bank 2	-10 to 10%	-5 to +5	-5 to +5	-5 to +5
SPARK	Spark Advance	-90° to 90°	15-18	23-29	26-32
SPOUT	Spark Output Signal	0-9999 Hz	42-50	82-88	105-120
TCC	TCC Solenoid	0-100%	0	90-100	90-100
TCIL	TCIL (lamp) Control	ON / OFF	OFF	OFF	OFF
VREF	Vehicle Reference	0-5.1v	4.9-5.1	4.9-5.1	4.9-5.1

1996-2003 F-Series 7.3L V8 Diesel VIN F (E4OD)

PCM PID Acronym	Parameter Identification	PID Range	PID Value at low idle	PID Value at high idle
4x4L	4x4 Low Switch Input	ON / OFF	ON	OFF
ACCS	A/C Switch Signal	ON / OFF	ON	OFF
AP	Accelerator Pedal Sensor	0-5.1v	0.5-1.6	0.5-4.5
ASMM	A/T Shift Modulator (M)	0v / 12	12	12
BAROV	BARO Sensor (sea level)	0-5.1v	4.75	4.75
BPP	Brake Switch Signal	ON / OFF	OFF	OFF
BPA	Brake Pressure Applied	0v / 12	12	0v
CCS	Coast Clutch Solenoid Control	ON / OFF	OFF	ON
C/S (TCS)	Cancel Switch & TCS On/Off Status (depressed is on)	0v/ VBAT	Off: 0v	Off: 0v
CPP	CPP Switch Signal	0v / 5v	0v (In)	5v
CRUISE	Cruise Control Module	---	---	---
DTC CNT	DTC Count	0-256	0	0
EBP V	Exhaust Back Pressure Actual	0-5.1v	0.8-0.95	0.9-3.0
EOT	Engine Oil Temperature	0-5.1v	0.35-4.7	0.35-4.7
EPC	Electronic Pressure Control	---	---	---
EPC V	EPC Solenoid - Actual	0-25.5v	7.5	12
EPR	Exhaust Back Pressure Regulator	ON / OFF	OFF	ON
FDCS	Fuel Delivery Control Signal	0-999 Hz	49	40-240
GEAR	Gear Position	1-2-3-4	1	4
GPC	Glow Plug Control Duty Cycle	---	---	---
GPMH	Glow Plug Monitor (high side)	ON / OFF	OFF	OFF
GPML	Glow Plug Monitor (left side)	ON / OFF	OFF	OFF
GPMR	Glow Plug Monitor (right side)	ON / OFF	OFF	OFF
GPL	Glow Plug Lamp	ON / OFF	OFF	OFF
IAT	IAT Sensor	0-5.1v	1.5-3.5	1.5-3.5
IAT V	IAT Sensor	-40-304°F	50-120	50-120
ICP	ICP Sensor (startup is 0.83v)	0-5.1v	0.25-0.40	0.25-0.40
ICP	Injector Control Pressure Actual	0-5.1v	0.25-0.40	0.25-0.40
IPR	Injector Control Pressure Regulator	0-100%	35	40-100
IVS	Idle Validation Switch	ON / OFF	ON	OFF
MAP	MAP Sensor Expected (4.6-4.8v)	0-5.1v	1-2	1-3
MAP H	MAP Sensor - Actual	0-5.1v	1-2	1-3
MIAH	Manifold Intake Air Heater	0v / 12	12	12
MIAHM	Manifold Intake Air Heater (M)	ON / OFF	OFF	OFF
MGP	Manifold Gauge Pressure	---	---	---
PBA	Parking Brake Applied	Brake On: 0v	Brake Off: 12	Brake Off: 12
RPM	Engine Speed	0-10K rpm	---	---
SCCS	Speed Control Command Switch	0-25.5v	S/C On: 12	S/C On: 12
SCCS-M	Speed Control Command Switch Mode	---	---	---
TCC	Transmission Converter Clutch	ON / OFF	OFF	ON / OFF
SS1	Shift Solenoid 1 Control	ON / OFF	ON	OFF
SS2	Shift Solenoid 2 Control	ON / OFF	OFF	ON
TCIL	Trans. Control Indicator Lamp	ON / OFF	OFF	OFF
TCS	TCS Switch	TCS On: 0.1	TCS On: 0.1	TCS On: 0.1
TFT V	TFT Sensor	0-5.1v	2.10-2.40	2.10-2.40
TORQUE	Engine Torque	---	---	---

1996-2003 F-Series 7.3L V8 Diesel VIN F (E4OD) (continued)

TPREL	Low Idle TP Sensor	---	---	---
TR2	TR2 Sensor	0 / 10.7	0v	10.7
TR3	TR4 Sensor	0 / 5v	4.5	2.2
TR4	TR4 Sensor	0 / 10.7	0v	10.7
TR V	Transmission Range Sensor Actual	0-5.1v	P/N: 4.45v	OD: 2.87v
VFDES	Volume Flow Desired	---	---	---
VPWR	Vehicle Power Supply	0-25.5v	12-14	12-14
VREF	Vehicle Reference	0-5.1v	4.9-5.1	4.9-5.1
VS SET	Vehicle Speed Setting	---	---	---
VSS	Vehicle Speed (MPH)	0-159 mph	0	Actual Speed

1992-97 F-Series 7.5L V8 MFI VIN G (A/T, M/T, E4OD)

PCM PID Acronym	Parameter Identification	PID Range	PID Value at Hot Idle	PID Value at 30	PID Value at 55
ACCS	A/C Cycling Clutch Switch	ON / OFF	ON (A/C on)	OFF	OFF
CPP	Clutch Pedal Switch	ON / OFF	ON (if in)	OFF	OFF
DPFE	DPFE Sensor	0-5.1v	0.4-0.6	0.4-0.7	0.8-1.1
ECT	ECT Sensor	-40-304°F	160-200	160-200	160-200
FEPS	Flash EEPROM	0-5.1v	0.1	0.1	0.1
FPM	Fuel Pump Monitor	ON / OFF	ON	ON	ON
HO2S-11	HO2S-11 (Bank 1)	0.1-1.1v	0.1-1.1	0.1-1.1	0.1-1.1
HO2S-12	HO2S-12 (Bank 1)	0.1-1.1v	0.1-1.1	0.1-1.1	0.1-1.1
HO2S-21	HO2S-21 (Bank 2)	0.1-1.1v	0.1-1.1	0.1-1.1	0.1-1.1
HO2S-22	HO2S-22 (Bank 2)	0.1-1.1v	0.1-1.1	0.1-1.1	0.1-1.1
IAT	IAT Sensor	-40-304°F	50-120	50-120	50-120
IDM	IDM Signal	0-9999 Hz	45-55	88-120	125-160
LOAD	Engine Load	0-100%	14-16	20-25	25-45
MAFV	MAF Sensor	0-5.1v	0.9-1.1	1.1-1.6	2.0-2.6
MISF	Misfire Detection	ON / OFF	OFF	OFF	OFF
OCTADJ	Octane Adjust Switch	CLOSED or OPEN	CLOSED	CLOSED	CLOSED
PIP	PIP Sensor	0-9999 Hz	45-55	88-120	110-220
PNP	PNP Switch Signal	ON / OFF	ON in P/N	OFF	OFF
PTO	Power Takeoff	ON / OFF	ON (if on)	OFF	OFF
RPM	Engine Speed	0-10K rpm	780-810	1280-1360	2150-2400
TPV	TP Sensor	0-5.1v	0.53-1.27	1.1-1.3v	1.3-1.7
Vacuum	Engine Vacuum	0-30" Hg	19-20	18-20	8-15
VPWR	Vehicle Power	0-25.5v	12-14	12-14	12-14
VSS	Vehicle Speed (Hertz or MPH)	0-999 Hz (0-159 mph)	0 Hz = 0 mph	65 Hz = 30 mph	125 Hz = 55 mph

1992-97 F-Series 7.5L V8 MFI VIN G (A/T, M/T, E4OD)

PCM PID Acronym	Parameter Identification	PID Range	PID Value at Hot Idle	PID Value at 30	PID Value at 55
AIRB	AIR System Bypass	0-25.5v	0.1	0.1	0.1
AIRD	AIR System Divert	0-25.5v	VBAT	VBAT	VBAT
EGR VR	EGR VR Solenoid	0-100%	0-100	40-45	50-60
EVAP CP	EVAP Purge Valve	0-100%	0-100	0-100	0-100
EPC	EPC Solenoid	0-300 psi	5	5	10-15
FP	Fuel Pump Control	ON / OFF	ON	ON	ON
FuelPW1	INJ Pulsewidth - Bank 1	0-999 ms	4.3-4.6	5.2-6.5	6.6-9.0
FuelPW2	INJ Pulsewidth - Bank 2	0-999 ms	4.3-4.6	5.2-6.5	6.6-9.0
HTR-11	HO2S-11 Heater	ON / OFF	ON	ON	ON
HTR-12	HO2S-12 Heater	ON / OFF	ON	ON	ON
HTR-21	HO2S-21 Heater	ON / OFF	ON	ON	ON
HTR-22	HO2S-22 Heater	ON / OFF	ON	ON	ON
IAC	Idle Air Control	0-100%	30-34	60-65	63-70
INJ1	INJ 1 Pulsewidth	0-999 ms	4.3-4.6	5.2-6.5	6.6-9.0
INJ2	INJ 2 Pulsewidth	0-999 ms	4.3-4.6	5.2-6.5	6.6-9.0
INJ3	INJ 3 Pulsewidth	0-999 ms	4.3-4.6	5.2-6.5	6.6-9.0
INJ4	INJ 4 Pulsewidth	0-999 ms	4.3-4.6	5.2-6.5	6.6-9.0
INJ5	INJ 5 Pulsewidth	0-999 ms	4.3-4.6	5.2-6.5	6.6-9.0
INJ6	INJ 6 Pulsewidth	0-999 ms	4.3-4.6	5.2-6.5	6.6-9.0
INJ7	INJ 7 Pulsewidth	0-999 ms	4.3-4.6	5.2-6.5	6.6-9.0
INJ8	INJ 8 Pulsewidth	0-999 ms	4.3-4.6	5.2-6.5	6.6-9.0
LongFT1	Long Term FT - Bank 1	-20 to 20%	-5 to +5	-5 to +5	-5 to +5
LongFT2	Long Term FT - Bank 2	-20 to 20%	-5 to +5	-5 to +5	-5 to +5
MIL	MIL (lamp) Control	ON / OFF	OFF	OFF	OFF
SS1	Shift Solenoid 1	ON / OFF	ON	OFF	OFF
SS2	Shift Solenoid 2	ON / OFF	OFF	ON	OFF
SHTFT1	Short Term F/T - Bank 1	-10 to 10%	-5 to +5	-5 to +5	-5 to +5
SHTFT2	Short Term FT - Bank 2	-10 to 10%	-5 to +5	-5 to +5	-5 to +5
SPARK	Spark Advance	-90° to 90°	22-26	25-28	27-32
SPOUT	Spark Output Signal	0-9999 Hz	45-55	88-120	110-220
TCC	TCC Solenoid	0-100%	0	0-100	90-95
TCIL	TCIL (lamp) Control	ON / OFF	OFF	OFF	OFF
VREF	Vehicle Reference	0-5.1v	4.9-5.1	4.9-5.1	4.9-5.1

RANGER PID DATA

1992-94 Ranger 2.3L I4 VIN A (All) Inputs & Outputs

PCM PID Acronym	Parameter Identification	PID Range	PID Value at Hot Idle	PID Value at 30	PID Value at 55
ACCS	A/C Switch Signal	ON / OFF	ON (if on)	OFF	OFF
BOO	Brake Switch Signal	ON / OFF	ON (if on)	OFF	OFF
ECT (°F)	ECT Sensor	-40-304°F	160-200	160-200	160-200
ECTV	ECT Sensor	0-5.1v	0.7	0.7	0.7
EVP	EVP Sensor	0-5.1v	0.4	0.5-0.7	1.2-1.6
EVR	EGR VR Solenoid	0-100%	0%	0-40	30-50
FPM	Fuel Pump Monitor	ON / OFF	ON	ON	ON
FuelPW1	INJ Pulsewidth - Bank 1	0-999 ms	3.3-3.5	4.7-5.4	6.2-7.2
HO2S-11	HO2S-11 (Bank 1)	0.1-1.1v	0.1-1.1	0.1-1.1	0.1-1.1
IAC	Idle Air Control	0-100%	26	42-48	45-52
IAT (°F)	IAT Sensor	-40-304°F	50-120	50-120	50-120
IATV	IAT Sensor	0-5.1v	1.5-3.5	1.5-3.5	1.5-3.5
MAFV	MAF Sensor	0-5.1v	0.1	1.3-1.4	1.8-2.2
PNP	PNP Switch Signal	Neutral/DR	NEUTRAL	DRIVE	DRIVE
PSP	PSP Switch Signal	ON / OFF	ON: turned	OFF	OFF
RPM	Engine Speed	0-10K rpm	760-820	1500-1630	1930-2100
SHTFT1	Short Term F/T - Bank 1	-10 to 10%	-5 to +5	-5 to +5	-5 to +5
SPARK	Spark Advance	-90° to 90°	18-22	28-32	30-34
TP	TP Sensor	0-5.1v	0.9	1.0-1.1	1.3-1.6v
TPCT	TP sensor minimum	0-5.1v	0.1	1.0-1.1	1.3-1.6v
TP Mode	TP Sensor Mode	C/T or P/T	C/T	P/T	P/T
VPWR	Vehicle Power	0-25.5v	12-14	12-14	12-14
VREF	Vehicle Reference	0-5.1v	4.9-5.1	4.9-5.1	4.9-5.1
VSS	Vehicle Speed	0-159 mph	0	30	55
WAC	A/C WOT Relay Control	ON / OFF	ON (AC on at WOT)	OFF	OFF

1995-97 Ranger 2.3L I4 VIN A (All) Inputs

PCM PID Acronym	Parameter Identification	PID Range	PID Value at Hot Idle	PID Value at 30	PID Value at 55
4x4L	4x4 Low Signal	ON / OFF	ON (if on)	OFF	OFF
ACCS	A/C Switch Signal	ON / OFF	ON (if on)	OFF	OFF
BPP	Brake Switch Signal	ON / OFF	ON (if on)	OFF	OFF
CD1A	Coil 1 Driver	0v / VBAT	VBAT	VBAT	VBAT
CD2A	Coil 2 Driver	0v / VBAT	VBAT	VBAT	VBAT
CID	CMP Sensor	0-999 Hz	6-8	13-15	17-19
CKP	CKP Sensor	0-9999 Hz	518-540	840-850	1180-1250
CPP	Clutch Pedal Switch	ON / OFF	ON (if in)	OFF	OFF
DPFE	DPFE Sensor	0-5.1v	0.2-1.3	0.2-4.5	0.2-4.5
ECT	ECT Sensor	-40-304°F	160-200	160-200	160-200
FEPS	Flash EEPROM	0-5.1v	0.1	0.1	0.1
FPM	Fuel Pump Monitor	ON / OFF	ON	ON	ON
HO2S-11	HO2S-11 (Bank 1)	0.1-1.1v	0.1-1.1	0.1-1.1	0.1-1.1
HO2S-12	HO2S-12 (Bank 1)	0.1-1.1v	0.1-1.1	0.1-1.1	0.1-1.1
IAT	IAT Sensor	-40-304°F	50-120	50-120	50-120
GEAR	Gear Position	1-2-3-4	1	3	4
LOAD	Engine Load	0-100%	10-20	16-36	30-45
MAFV	MAF Sensor	0-5.1v	0.6-0.9	0.8-1.1	1.5-2.8
MISF	Misfire Detection	ON / OFF	OFF	OFF	OFF
OCTADJ	Octane Adjustment	CL/OPEN	CLOSED	CLOSED	CLOSED
PF	Purge Flow Sensor	0-5.1v	1.55	2.10	2.30
PNP	PNP Switch Signal	Neutral/DR	NEUTRAL	DRIVE	DRIVE
PSP	PSP Switch Signal	LOW or HI	HI: turned	LOW	LOW
RPM	Engine Speed	0-10K rpm	760-820	1500-1630	1930-2100
TCS	TCS Switch	ON / OFF	OFF	OFF	OFF
TFT	TFT Sensor Signal	-40-304°F	110-210	110-210	110-210
TPV	TP Sensor	0-5.1v	0.53-1.27	1.0-1.3	1.1-1.9
TRV/TR	TR Sensor	P/N or O/D	P/N	O/D	O/D
TR1	TR1 Sensor	0v / 11.5v	0	11.5	11.5
TR2	TR2 Sensor	0v / 11.5v	0	11.5	11.5
TR4	TR4 Sensor	0v / 11.5v	0	11.5	11.5
TSS	TSS Sensor	0-9999 Hz	115-120	192-196	268-275
Vacuum	Engine Vacuum	0-30" Hg	19-20	18-20	8-15
VPWR	Vehicle Power	0-25.5v	12-14	12-14	12-14
VSS	Vehicle Speed (Hertz or MPH)	0-999 Hz (0-159 mph)	0 Hz = 0 mph	65 Hz = 30 mph	125 Hz = 55 mph

1995-97 Ranger 2.3L I4 VIN A (All) Outputs

PCM PID Acronym	Parameter Identification	PID Range	PID Value at Hot Idle	PID Value at 30	PID Value at 55
CD1	Coil 1 Driver	0-90° dwell	5	6	8
CD2	Coil 2 Driver	0-90° dwell	5	6	8
CCS	Coast Clutch Solenoid Control	0-25.5v	VBAT	VBAT	VBAT
CTO	Clean Tachometer signal	0 Hz	25-38	50-60	60-74
EGR VR	EGR VR Solenoid	0-100%	0	0-40	0-40
EVAP CP	EVAP Purge Valve	0-100%	0-100	0-100	0-100
EPC	EPC Solenoid	0-300 psi	24	27	26-30
FP	Fuel Pump Control	ON / OFF	ON	ON	ON
FuelPW1	INJ Pulsewidth - Bank 1	0-999 ms	4.0-4.5	5.3-10.0	10-18.0
HTR-11	HO2S-11 Heater	ON / OFF	ON	ON	ON
HTR-12	HO2S-12 Heater	ON / OFF	ON	ON	ON
IAC	Idle Air Control	0-100%	33	40-48	40-55
INJ1	INJ 1 Pulsewidth	0-999 ms	4.0-4.5	5.3-10.0	10-18.0
INJ2	INJ 2 Pulsewidth	0-999 ms	4.0-4.5	5.3-10.0	10-18.0
INJ3	INJ 3 Pulsewidth	0-999 ms	4.0-4.5	5.3-10.0	10-18.0
INJ4	INJ 4 Pulsewidth	0-999 ms	4.0-4.5	5.3-10.0	10-18.0
LongFT1	Long Term FT - Bank 1	-20 to 20%	-5 to 5%	-5 to 5%	-5 to 5%
MIL	MIL (lamp) Control	ON / OFF	OFF	OFF	OFF
SS1	Shift Solenoid 1	ON / OFF	ON	OFF	OFF
SS2	Shift Solenoid 2	ON / OFF	OFF	ON	ON
SHTFT1	Short Term F/T - Bank 1	-10 to 10%	-5 to 5%	-5 to 5%	-5 to 5%
SPARK	Spark Advance	-90° to 90°	15-22	25-35	15-30
TCC	TCC Solenoid	0-100%	0	0-100	95-100
TCIL	TCIL (lamp) Control	ON / OFF	OFF	OFF	OFF
VREF	Vehicle Reference	0-5.1v	4.9-5.1	4.9-5.1	4.9-5.1
WAC	A/C WOT Relay Control	ON / OFF	ON (AC on at WOT)	OFF	OFF

2001-05 Ranger 2.3L I4 MFI VIN D (All) Inputs

PCM PID Acronym	Parameter Identification	PID Range	PID Value at Hot Idle	PID Value at 30	PID Value at 55
ACCS	A/C Switch Signal	ON / OFF	ON (if on)	OFF	OFF
ACP	A/C Press. Switch	CL/OPEN	OPEN	OPEN	OPEN
BPP	Brake Switch Signal	ON / OFF	ON (if on)	OFF	OFF
CID	CMP Sensor	0-999 Hz	6-8	12-15	12-18
CD1A	Coil 1 Driver	0-25.5v	VBAT	VBAT	VBAT
CD2A	Coil 2 Driver	0-25.5v	VBAT	VBAT	VBAT
CHT	CHT Sensor	-40-304°F	194	194	194
CKP	CKP Sensor	0-9999 Hz	390-450	1000-1220	1220-1500
CPP	Clutch Pedal Switch	ON / OFF	ON (if in)	OFF	OFF
ECT	ECT Sensor	-40-304°F	160-200	160-200	160-200
EGR MS	EGR Motor Speed	0-60 Steps	3	3-31	3-31
FEPS	Flash EEPROM	0-5.1v	0.1	0.1	0.1
FLI	Fuel Level Indicator	0-100%	1.7 (1/2)	0-100	0-100
FPM	Fuel Pump Monitor	ON / OFF	ON	ON	ON
FTP	Fuel Tank Pressure	0-5.1v or 0-10" H2O	2.6v / 0" H2O (with cap off)	2.6v / 0" H2O (with cap off)	2.6v / 0" H2O (with cap off)
IAT	IAT Sensor	-40-304°F	50-120	50-120	50-120
LOAD	Engine Load	0-100%	10-20	16-36	30-45
MAFV	MAF Sensor	0-5.1v	0.6-0.9	0.8-1.1	1.5-2.8
MAP	MAP Sensor	0-5.1v	1.7	3.0	3.5
O2S-11	HO2S-11 (Bank 1)	0.1-1.1v	0.1-1.1	0.1-1.1	0.1-1.1
O2S-12	HO2S-12 (Bank 1)	0.1-1.1v	0.1-1.1	0.1-1.1	0.1-1.1
PF	Purge Flow Sensor	0-5.1v	1.55	2.10	2.30
PNP	PNP Switch Signal	NEUT/DR	NEUTRAL	DRIVE	DRIVE
RPM	Engine Speed	0-10K rpm	800-950	1400-1760	1930-2150
TFT	TFT Sensor Signal	-40-304°F	110-210	110-210	110-210
TPV	TP Sensor	0-5.1v	0.53-1.27	1.0-1.3	1.1-1.9
OSS	OSS Sensor	0-9999 Hz	0	213	385
Vacuum	Engine Vacuum	0-30" Hg	19-20	18-20	8-15
VPWR	Vehicle Power	0-25.5v	12-14	12-14	12-14
VSS	Vehicle Speed (Hertz or MPH)	0-999 Hz (0-159 mph)	0 Hz = 0 mph	65 Hz = 30 mph	125 Hz = 55 mph

2001-05 Ranger 2.3L I4 MFI VIN D (All) Outputs

PCM PID Acronym	Parameter Identification	PID Range	PID Value at Hot Idle	PID Value at 30	PID Value at 55
CD1	Coil 1 Driver	0-90° dwell	5	6	8
CD2	Coil 2 Driver	0-90° dwell	5	6	8
CHTIL	CHT (lamp) Control	ON / OFF	OFF	OFF	OFF
CTO	Clean Tachometer signal	0 Hz	25-38	50-60	60-74
EGRMC1	EGR Motor Cont. 1	0-25.5v	12-14	12-14	12-14
EGRMC2	EGR Motor Cont. 2	0-25.5v	0.2	0.2	0.2
EGRMC3	EGR Motor Cont. 3	0-25.5v	12-14	12-14	12-14
EGRMC4	EGR Motor Cont. 4	0-25.5v	0.2	0.2	0.2
EVAP CP	EVAP Purge Valve	0-100%	0-100	0-100	0-100
EVAP CV	EVAP CV Valve	0-100%	0-100	0-100	0-100
FP	Fuel Pump Control	ON / OFF	ON	ON	ON
FuelPW1	INJ Pulsewidth - Bank 1	0-999 ms	3.0-4.5	5.3-10.0	10-18.0
HTR-11	HO2S-11 Heater	ON / OFF	ON	ON	ON
HTR-12	HO2S-12 Heater	ON / OFF	ON	ON	ON
IAC	Idle Air Control	0-100%	38	42-48	45-52
INJ1	INJ 1 Pulsewidth	0-999 ms	3.0-4.5	5.3-10.0	10-18.0
INJ2	INJ 2 Pulsewidth	0-999 ms	3.0-4.5	5.3-10.0	10-18.0
INJ3	INJ 3 Pulsewidth	0-999 ms	3.0-4.5	5.3-10.0	10-18.0
INJ4	INJ 4 Pulsewidth	0-999 ms	3.0-4.5	5.3-10.0	10-18.0
LongFT1	Long Term FT - Bank 1	-20 to 20%	-5 to +5	-5 to +5	-5 to +5
MIL	MIL (lamp) Control	ON / OFF	OFF	OFF	OFF
MISF	Misfire Monitor	ON / OFF	ON	ON	ON
SCVM	S/C Vacuum Motor	HI/LOW	LOW	LOW	LOW
SHTFT1	Short Term F/T - Bank 1	-10 to 10%	-5 to +5	-5 to +5	-5 to +5
SPARK	Spark Advance	-90° to 90°	15-22	25-35	15-30
TCIL	TCIL (lamp) Control	ON / OFF	OFF	OFF	OFF
VREF	Vehicle Reference	0-5.1v	4.9-5.1	4.9-5.1	4.9-5.1
WAC	A/C WOT Relay Control	ON / OFF	ON (AC on at WOT)	OFF	OFF

1998-2001 Ranger 2.5L I4 MFI VIN C (All) Inputs

PCM PID Acronym	Parameter Identification	PID Range	PID Value at Hot Idle	PID Value at 30	PID Value at 55
ACCS	A/C Switch Signal	ON / OFF	ON (if on)	OFF	OFF
ACP	A/C Press. Switch	CL/OPEN	OPEN	OPEN	OPEN
BPP	Brake Switch Signal	ON / OFF	ON (if on)	OFF	OFF
CID	CMP Sensor	0-999 Hz	5-7	9-12	10-16
CD1A	Coil 1 Driver	0-25.5v	VBAT	VBAT	VBAT
CD2A	Coil 2 Driver	0-25.5v	VBAT	VBAT	VBAT
CKP	CKP Sensor	0-9999 Hz	390-450	850-950	1160-1220
CPP	Clutch Pedal Switch	ON / OFF	ON (If in)	OFF	OFF
DPFE	DPFE Sensor	0-5.1v	0.2-1.3	0.2-4.5	0.2-4.5
ECT	ECT Sensor	-40-304ºF	160-200	160-200	160-200
FEPS	Flash EEPROM	0-5.1v	0.1	0.1	0.1
FLI	Fuel Level Indicator	0-100%	1.7 (1/2)	0-100%	0-100%
FPM	Fuel Pump Monitor	ON / OFF	ON	ON	ON
FTP	Fuel Tank Pressure	0-5.1v or 0-10" H2O	2.6v / 0" H2O (with cap off)	2.6v / 0" H2O (with cap off)	2.6v / 0" H2O (with cap off)
IAT	IAT Sensor	-40-304ºF	50-120	50-120	50-120
GEAR	Gear Position	1-2-3-4	1	3	4
LOAD	Engine Load	0-100%	10-20	16-36	30-45
MAFV	MAF Sensor	0-5.1v	0.6-0.9	0.8-1.1	1.5-2.8
MISF	Misfire Detection	ON / OFF	OFF	OFF	OFF
OCTADJ	Octane Adjustment	CL or OP	CLOSED	CLOSED	CLOSED
O2S-11	HO2S-11 (Bank 1)	0.1-1.1v	0.1-1.1	0.1-1.1	0.1-1.1
O2S-12	HO2S-12 (Bank 1)	0.1-1.1v	0.1-1.1	0.1-1.1	0.1-1.1
PF	Purge Flow Sensor	0-5.1v	1.55	2.10	2.30
PNP	PNP Switch Signal	NEUT/DR	NEUTRAL	DRIVE	DRIVE
PSP	PSP Switch Signal	LOW or HI	HI: turning	LOW	LOW
RPM	Engine Speed	0-10K rpm	760-820	1400-1630	1930-2100
TCS	TCS Switch	ON / OFF	OFF	OFF	OFF
TFT	TFT Sensor Signal	-40-304ºF	110-210	110-210	110-210
TPV	TP Sensor	0-5.1v	0.53-1.27	1.0-1.3	1.1-1.9
TR1	TR1 Sensor	0v / 11.5v	0v	11.5	11.5
TR2	TR2 Sensor	0v / 11.5v	0v	11.5	11.5
TR3/TR	TR Sensor	P/N or O/D	P/N	O/D	O/D
TR4	TR4 Sensor	0v / 11.5v	0	11.5	11.5
TSS	TSS Sensor	0-9999 Hz	95-120	160-180	260-280
Vacuum	Engine Vacuum	0-30" Hg	19-20	18-20	8-15
VPWR	Vehicle Power	0-25.5v	12-14	12-14	12-14
VSS	Vehicle Speed (Hertz or MPH)	0-999 Hz (0-159 mph)	0 Hz = 0 mph	65 Hz = 30 mph	125 Hz = 55 mph

1998-2001 Ranger 2.5L I4 MFI VIN C (All) Outputs

PCM PID Acronym	Parameter Identification	PID Range	PID Value at Hot Idle	PID Value at 30	PID Value at 55
CD1	Coil 1 Driver	0-90° dwell	5	6	8
CD2	Coil 2 Driver	0-90° dwell	5	6	8
CCS	Coast Clutch Solenoid Control	0-25.5v	VBAT	VBAT	VBAT
CTO	Clean Tachometer signal	0 Hz	25-38	50-60	60-74
EGR VR	EGR VR Solenoid	0-100%	0	0-100	0-40
EVAP CP	EVAP Purge Valve	0-100%	0-100	0-100	0-100
EVAP CV	EVAP CV Valve	0-100%	0-100	0-100	0-100
EPC	EPC Solenoid	0-300 psi	25	27	26-30
FP	Fuel Pump Control	ON / OFF	ON	ON	ON
FuelPW1	INJ Pulsewidth - Bank 1	0-999 ms	4.0-4.5	5.3-10.0	10-18.0
HTR-11	HO2S-11 Heater	ON / OFF	ON	ON	ON
HTR-12	HO2S-12 Heater	ON / OFF	ON	ON	ON
IAC	Idle Air Control	0-100%	38	42-48	45-52
INJ1	INJ 1 Pulsewidth	0-999 ms	4.0-4.5	5.3-10.0	10-18.0
INJ2	INJ 2 Pulsewidth	0-999 ms	4.0-4.5	5.3-10.0	10-18.0
INJ3	INJ 3 Pulsewidth	0-999 ms	4.0-4.5	5.3-10.0	10-18.0
INJ4	INJ 4 Pulsewidth	0-999 ms	4.0-4.5	5.3-10.0	10-18.0
LongFT1	Long Term FT - Bank 1	-20 to 20%	-5 to +5	-5 to +5	-5 to +5
MIL	MIL (lamp) Control	ON / OFF	OFF	OFF	OFF
SHTFT1	Short Term F/T - Bank 1	-10 to 10%	-5 to +5	-5 to +5	-5 to +5
SPARK	Spark Advance	-90° to 90°	15-22	25-35	15-30
SS1	Shift Solenoid 1	ON / OFF	ON	OFF	OFF
SS2	Shift Solenoid 2	ON / OFF	OFF	ON	OFF
SS3	Shift Solenoid 3	ON / OFF	OFF	ON	ON
TCC	TCC Solenoid	0-100%	0%	0-100	95-100
TCIL	TCIL (lamp) Control	ON / OFF	OFF	OFF	OFF
VREF	Vehicle Reference	0-5.1v	4.9-5.1	4.9-5.1	4.9-5.1
WAC	A/C WOT Relay Control	ON / OFF	ON (AC on at WOT)	OFF	OFF

1993-94 Ranger 3.0L V6 VIN U (All) Inputs & Outputs

PCM PID Acronym	Parameter Identification	PID Range	PID Value at Hot Idle	PID Value at 30	PID Value at 55
ACCS	A/C Switch Signal	ON / OFF	ON (if on)	OFF	OFF
BOO	Brake Switch Signal	ON / OFF	ON (if on)	OFF	OFF
CANP	EVAP Purge Valve	ON / OFF	OFF	ON	ON
CPP	Clutch Pedal Switch	ON / OFF	ON (if in)	OFF	OFF
DPFE	DPFE Sensor	0-5.1v	0.4-0.6	0.4-0.8	0.4-0.9
ECT (°F)	ECT Sensor	-40-304°F	160-200	160-200	160-200
ECTV	ECT Sensor	0-5.1v	0.7	0.7	0.7
EVR	EGR VR Solenoid	0-100%	0	0-40	0-40
FPM	Fuel Pump Monitor	ON / OFF	ON	ON	ON
FuelPW1	INJ Pulsewidth - Bank 1	0-999 ms	3.7-3.9	4.4-4.6	5.0-6.0
FuelPW2	INJ Pulsewidth - Bank 2	0-999 ms	3.7-3.9	4.4-4.6	5.0-6.0
HO2S-11	HO2S-11 (Bank 1)	0.1-1.1v	0.1-1.1	0.1-1.1	0.1-1.1
HO2S-21	HO2S-21 (Bank 2)	0.1-1.1v	0.1-1.1	0.1-1.1	0.1-1.1
IAC	Idle Air Control	0-100%	33	35-41	57-68
IAT (°F)	IAT Sensor	-40-304°F	50-120	50-120	50-120
IATV	IAT Sensor	0-5.1v	1.5-3.5	1.5-3.5	1.5-3.5
LongFT1	Long Term F/T - Bank 1	-20 to 20%	-5 to +5	-5 to +5	-5 to +5
LongFT2	Long Term F/T - Bank 1	-20 to 20%	-5 to +5	-5 to +5	-5 to +5
MAF	MAF Sensor	0-5.1v	0.9	1.4-1.6	1.9-2.4
PNP	PNP Switch Signal	NEUT/DR	NEUTRAL	DRIVE	DRIVE
RPM	Engine Speed	0-10K rpm	880-920	1430-1470	1550-1750
SHTFT1	Short Term F/T - Bank 1	-10 to 10%	-5 to +5	-5 to +5	-5 to +5
SHTFT2	Short Term FT - Bank 2	-10 to 10%	-5 to +5	-5 to +5	-5 to +5
SPARK	Spark Advance	-90° to 90°	15-22	28-35	25-35
TP	TP Sensor	0-5.1v	0.7	0.9-1.0	1.0-1.2
TPCT	TP sensor Minimum	0-5.1v	0.1	0.9-1.0	1.0-1.2
TP Mode	TP Sensor Mode	C/T or P/T	C/T	P/T	P/T
VPWR	Vehicle Power	0-25.5v	12-14	12-14	12-14
VREF	Vehicle Reference	0-5.1v	4.9-5.1	4.9-5.1	4.9-5.1
VSS	Vehicle Speed	0-159 mph	0	30	55
WAC	A/C WOT Relay Control	ON / OFF	ON (AC on at WOT)	OFF	OFF

1995-2005 Ranger 3.0L V6 MFI VIN U (All) Inputs

PCM PID Acronym	Parameter Identification	PID Range	PID Value at Hot Idle	PID Value at 30	PID Value at 55
4x4L	4x4 Low Signal	ON / OFF	ON (if on)	OFF	OFF
ACCS	A/C Switch Signal	ON / OFF	ON (if on)	OFF	OFF
BOO	Brake Switch Signal	ON / OFF	ON (if on)	OFF	OFF
CID	CMP Sensor	0-999 Hz	6-8	13-15	17-19
CKP	CKP Sensor	0-9999 Hz	518-540	840-860	1180-1210
CPP	Clutch Pedal Switch	ON / OFF	ON (if in)	OFF	OFF
DPFE	DPFE Sensor	0-5.1v	0.4-0.6	0.4-0.8	0.4-0.9
ECT	ECT Sensor	-40-304°F	160-200	160-200	160-200
FEPS	Flash EEPROM	0-5.1v	0.1	0.1	0.1
FLI	Fuel Level Indicator	0-100%	50% (1.7)	0-100%	0-100%
FPM	Fuel Pump Monitor	ON / OFF	ON	ON	ON
FTP	Fuel Tank Pressure	0-5.1v or 0-10" H2O	2.6v / 0" H2O (with cap off)	2.6v / 0" H2O (with cap off)	2.6v / 0" H2O (with cap off)
IAT	IAT Sensor	-40-304°F	50-120	50-120	50-120
GEAR	Gear Position	1-2-3-4	1	3	4
LOAD	Engine Load	0-100%	17-19	21-27	27-34
MAFV	MAF Sensor	0-5.1v	0.6-0.9	1.4-1.7	2.0-0
MISF	Misfire Detection	ON / OFF	OFF	OFF	OFF
OCTADJ	Octane Adjustment	CL/OPEN	CLOSED	CLOSED	CLOSED
HO2S-11	HO2S-11 (Bank 1)	0.1-1.1v	0.1-1.1	0.1-1.1	0.1-1.1
HO2S-12	HO2S-12 (Bank 1)	0.1-1.1v	0.1-1.1	0.1-1.1	0.1-1.1
HO2S-21	HO2S-21 (Bank 2)	0.1-1.1v	0.1-1.1	0.1-1.1	0.1-1.1
PF	Purge Flow Sensor	0-5.1v	1.55	2.10	2.30
PNP	PNP Switch Signal	Neutral/DR	NEUTRAL	DRIVE	DRIVE
PSP	PSP Switch Signal	LOW or HI	HI: turned	LOW	LOW
RPM	Engine Speed	0-10K rpm	880-920	1430-1470	1550-1750
TCS	TCS Switch	ON / OFF	OFF	OFF	OFF
TFT	TFT Sensor Signal	-40-304°F	110-210	110-210	110-210
TPV	TP Sensor	0-5.1v	0.53-1.27	0.8-1.7	1.2-1.7
TRV/TR	TR Sensor	P/N or O/D	P/N	O/D	O/D
TSS	TSS Sensor	0-9999 Hz	115-120	192-196	268-275
Vacuum	Engine Vacuum	0-30" Hg	19-20	18-20	8-15
VPWR	Vehicle Power	0-25.5v	12-14	12-14	12-14
VSS	Vehicle Speed (Hertz or MPH)	0-999 Hz (0-159 mph)	0 Hz = 0 mph	65 Hz = 30 mph	125 Hz = 55 mph

1995-2005 Ranger 3.0L V6 MFI VIN U (All) Outputs

PCM PID Acronym	Parameter Identification	PID Range	PID Value at Hot Idle	PID Value at 30	PID Value at 55
CD1	Coil 1 Driver	0-60°	6	8	12
CD2	Coil 2 Driver	0-60°	6	8	12
CD3	Coil 3 Driver	0-60°	6	8	12
CTO	Clean Tachometer signal	0-9999 Hz	42-48	73-79	99-115
CCS	Coast Clutch Solenoid Control	0-25.5v	VBAT	VBAT	VBAT
EGR VR	EGR VR Solenoid	0-100%	0	0-40	0-45
EVAP CV	EVAP CV Valve	0-100%	0-100	0-100	0-100
EVAP CP	EVAP Purge Valve	0-100%	0-100	0-100	0-100
EPC	EPC Solenoid	0-300 psi	24	23-28	34-38
FP	Fuel Pump Control	ON / OFF	ON	ON	ON
FuelPW1	INJ Pulsewidth - Bank 1	0-999 ms	4.5-4.8	6.3-8.0	7.0-13.0
FuelPW2	INJ Pulsewidth - Bank 2	0-999 ms	4.5-4.8	6.3-8.0	7.0-13.0
HTR-11	HO2S-11 Heater	ON / OFF	ON	ON	ON
HTR-12	HO2S-12 Heater	ON / OFF	ON	ON	ON
HTR-21	HO2S-21 Heater	ON / OFF	ON	ON	ON
IAC	Idle Air Control	0-100%	33	35-41	57-68
INJ1	INJ 1 Pulsewidth	0-999 ms	4.5-4.8	6.3-8.0	7.0-13.0
INJ2	INJ 2 Pulsewidth	0-999 ms	4.5-4.8	6.3-8.0	7.0-13.0
INJ3	INJ 3 Pulsewidth	0-999 ms	4.5-4.8	6.3-8.0	7.0-13.0
INJ4	INJ 4 Pulsewidth	0-999 ms	4.5-4.8	6.3-8.0	7.0-13.0
INJ5	INJ 5 Pulsewidth	0-999 ms	4.5-4.8	6.3-8.0	7.0-13.0
INJ6	INJ 6 Pulsewidth	0-999 ms	4.5-4.8	6.3-8.0	7.0-13.0
LongFT1	Long Term FT - Bank 1	-20 to 20%	-5 to +5	-5 to +5	-5 to +5
LongFT2	Long Term FT - Bank 2	-20 to 20%	-5 to +5	-5 to +5	-5 to +5
MIL	MIL (lamp) Control	ON / OFF	OFF	OFF	OFF
SHTFT1	Short Term F/T - Bank 1	-10 to 10%	-5 to +5	-5 to +5	-5 to +5
SHTFT2	Short Term FT - Bank 2	-10 to 10%	-5 to +5	-5 to +5	-5 to +5
SPARK	Spark Advance	-90° to 90°	11-15	26-31	25-32
SS1	Shift Solenoid 1	ON / OFF	ON	OFF	OFF
SS2	Shift Solenoid 2	ON / OFF	OFF	ON	ON
SS3	Shift Solenoid 3	ON / OFF	OFF	OFF	ON
TCC	TCC Solenoid	0-100%	0	62-67	85-95
TCIL	TCIL (lamp) Control	ON / OFF	OFF	OFF	OFF
VREF	Vehicle Reference	0-5.1v	4.9-5.1	4.9-5.1	4.9-5.1
WAC	A/C WOT Relay Control	ON / OFF	ON (AC on at WOT)	OFF	OFF

1999-2001 Ranger 3.0L V6 FFV VIN V (All) Inputs

PCM PID Acronym	Parameter Identification	PID Range	PID Value at Hot Idle	PID Value at 30	PID Value at 55
4x4L	4x4 Low Signal	ON / OFF	ON (if on)	OFF	OFF
ACCS	A/C Switch Signal	ON / OFF	ON (if on)	OFF	OFF
BOO	Brake Switch Signal	ON / OFF	ON (if on)	OFF	OFF
CID	CMP Sensor	0-999 Hz	6-8	13-15	17-19
CKP	CKP Sensor	0-9999 Hz	518-540	860-1000	1210-1250
CPP	Clutch Pedal Switch	ON / OFF	ON (if in)	OFF	OFF
DPFE	DPFE Sensor	0-5.1v	0.4-0.6	0.4-0.8	0.4-0.9
ECT	ECT Sensor	-40-304°F	160-200	160-200	160-200
FEPS	Flash EEPROM	0-5.1v	0.1	0.1	0.1
FFS	FF Sensor (value at 100% gas mixture)	0-999 Hz	40-60	40-60	40-60
FLI	Fuel Level Indicator	0-100%	50% (1.7)	0-100%	0-100%
FPM	Fuel Pump Monitor	ON / OFF	ON	ON	ON
FTP	Fuel Tank Pressure	0-5.1v or 0-10" H2O	2.6v / 0" H2O (with cap off)	2.6v / 0" H2O (with cap off)	2.6v / 0" H2O (with cap off)
HO2S-11	HO2S-11 (Bank 1)	0.1-1.1v	0.1-1.1	0.1-1.1	0.1-1.1
HO2S-12	HO2S-12 (Bank 1)	0.1-1.1v	0.1-1.1	0.1-1.1	0.1-1.1
HO2S-21	HO2S-21 (Bank 2)	0.1-1.1v	0.1-1.1	0.1-1.1	0.1-1.1
IAT	IAT Sensor	-40-304°F	50-120	50-120	50-120
GEAR	Gear Position	1-2-3-4	1	3	4
LOAD	Engine Load	0-100%	17-19	21-27	27-34
MAFV	MAF Sensor	0-5.1v	0.6-0.9	1.4-1.7	2.0-0
MISF	Misfire Detection	ON / OFF	OFF	OFF	OFF
OCTADJ	Octane Adjustment	CL/OPEN	CLOSED	CLOSED	CLOSED
PF	Purge Flow Sensor	0-5.1v	1.55	2.10	2.30
PNP	PNP Switch Signal	Neutral/DR	NEUTRAL	DRIVE	DRIVE
PSP	PSP Switch Signal	LOW or HI	HI: turned	LOW	LOW
RPM	Engine Speed	0-10K rpm	880-920	1430-1470	1550-1750
TCS	TCS Switch	ON / OFF	OFF	OFF	OFF
TFT	TFT Sensor Signal	-40-304°F	110-210	110-210	110-210
TPV	TP Sensor	0-5.1v	0.53-1.27	0.8-1.7	1.2-1.7
TRV/TR	TR Sensor	P/N or O/D	P/N	O/D	O/D
TSS	TSS Sensor	0-9999 Hz	115-120	192-196	268-275
Vacuum	Engine Vacuum	0-30" Hg	19-20	18-20	8-15
VPWR	Vehicle Power	0-25.5v	12-14	12-14	12-14
VSS	Vehicle Speed (Hertz or MPH)	0-999 Hz (0-159 mph)	0 Hz = 0 mph	65 Hz = 30 mph	125 Hz = 55 mph

1995-2001 Ranger 3.0L V6 FFV VIN V (All) Outputs

PCM PID Acronym	Parameter Identification	PID Range	PID Value at Hot Idle	PID Value at 30	PID Value at 55
CD1	Coil 1 Driver	0-60°	6	8	12
CD2	Coil 2 Driver	0-60°	6	8	12
CD3	Coil 3 Driver	0-60°	6	8	12
CTO	Clean Tachometer signal	0-9999 Hz	42-48	73-79	99-115
CCS	Coast Clutch Solenoid Control	0-25.5v	VBAT	VBAT	VBAT
EGR VR	EGR VR Solenoid	0-100%	0	0-40	0-45
EVAP CP	EVAP Purge Valve	0-100%	0-100	0-100	0-100
EVAP CV	EVAP CV Valve	0-100%	0-100	0-100	0-100
EPC	EPC Solenoid	0-300 psi	39	23-38	34-38
FP	Fuel Pump Control	ON / OFF	ON	ON	ON
FuelPW1	INJ Pulsewidth - Bank 1	0-999 ms	1.9-2.8	3.6-4.8	4.6-6.0
FuelPW2	INJ Pulsewidth - Bank 2	0-999 ms	1.9-2.8	3.6-4.8	4.6-6.0
HTR-11	HO2S-11 Heater	ON / OFF	ON	ON	ON
HTR-12	HO2S-12 Heater	ON / OFF	ON	ON	ON
HTR-21	HO2S-21 Heater	ON / OFF	ON	ON	ON
IAC	Idle Air Control	0-100%	33	35-50	57-68
INJ1	INJ 1 Pulsewidth	0-999 ms	1.9-2.8	3.6-4.8	4.6-6.0
INJ2	INJ 2 Pulsewidth	0-999 ms	1.9-2.8	3.6-4.8	4.6-6.0
INJ3	INJ 3 Pulsewidth	0-999 ms	1.9-2.8	3.6-4.8	4.6-6.0
INJ4	INJ 4 Pulsewidth	0-999 ms	1.9-2.8	3.6-4.8	4.6-6.0
INJ5	INJ 5 Pulsewidth	0-999 ms	1.9-2.8	3.6-4.8	4.6-6.0
INJ6	INJ 6 Pulsewidth	0-999 ms	1.9-2.8	3.6-4.8	4.6-6.0
LongFT1	Long Term FT - Bank 1	-20 to 20%	-5 to +5	-5 to +5	-5 to +5
LongFT2	Long Term FT - Bank 2	-20 to 20%	-5 to +5	-5 to +5	-5 to +5
MIL	MIL (lamp) Control	ON / OFF	OFF	OFF	OFF
SHTFT1	Short Term F/T - Bank 1	-10 to 10%	-5 to +5	-5 to +5	-5 to +5
SHTFT2	Short Term FT - Bank 2	-10 to 10%	-5 to +5	-5 to +5	-5 to +5
SPARK	Spark Advance	-90° to 90°	11-15	26-31	25-32
SS1	Shift Solenoid 1	ON / OFF	ON	OFF	OFF
SS2	Shift Solenoid 2	ON / OFF	OFF	ON	ON
SS3	Shift Solenoid 3	ON / OFF	OFF	OFF	ON
TCC	TCC Solenoid	0-100%	0	62-67	85-95
TCIL	TCIL (lamp) Control	ON / OFF	OFF	OFF	OFF
VREF	Vehicle Reference	0-5.1v	4.9-5.1	4.9-5.1	4.9-5.1
WAC	A/C WOT Relay Control	ON / OFF	ON (AC on at WOT)	OFF	OFF

2002-05 Ranger 4.0L V6 MFI VIN E (A/T) Inputs

PCM PID Acronym	Parameter Identification	PID Range	PID Value at Hot Idle	PID Value at 30	PID Value at 55
4x4L	4x4 Low Signal	ON / OFF	ON (if on)	OFF	OFF
ACCS	A/C Switch Signal	ON / OFF	ON (if on)	OFF	OFF
ACP	A/C Pressure	OPEN/CL	OPEN	OPEN	OPEN
ARC	Auto Ride Control	ON / OFF	OFF	OFF	OFF
BPP	Brake Position	ON / OFF	ON (if on)	OFF	OFF
CID	CMP Sensor	0-999 Hz	6-8	11-15	16-19
CKP	CKP Sensor	0-9999 Hz	400-475	800-1100	1140-1220
DPFE	DPFE Sensor	0-5.1v	0.4-0.6	0.4-1.0	0.6-1.1
ECT	ECT Sensor	-40-304°F	160-200	160-200	160-200
FEPS	Flash EEPROM	0-5.1v	0.1	0.1	0.1
FLI	Fuel Level Input	1.7 / 50%	1.7 / 50%	1.7 / 50%	1.7 / 50%
FPM	Fuel Pump Monitor	ON / OFF	ON	ON	ON
FTP	Fuel Tank Pressure	0-5.1v at 0-10" H2O	2.6v at 0" H2O (gas cap "off")	2.6v at 0" H2O (gas cap "off")	2.6v at 0" H2O (gas cap "off")
IAT	IAT Sensor	-40-304°F	50-120	50-120	50-120
GEAR	Gear Position	1-2-3-4-5	1	4	5
HO2S-11	HO2S-11 (Bank 1)	0.1-1.1v	0.1-1.1	0.1-1.1	0.1-1.1
HO2S-12	HO2S-12 (Bank 1)	0.1-1.1v	0.1-1.1	0.1-1.1	0.1-1.1
HO2S-21	HO2S-21 (Bank 2)	0.1-1.1v	0.1-1.1	0.1-1.1	0.1-1.1
LOAD	Engine Load	0-100%	14-20	21-27	30-45
MAFV	MAF Sensor	0-5.1v	0.6-0.9	1.3-1.7	1.5-2.3
MISF	Misfire Detection	ON / OFF	OFF	OFF	OFF
OCTADJ	Octane Adjustment	CL/OPEN	CLOSED	CLOSED	CLOSED
PNP	PNP Switch Signal	ON / OFF	ON	OFF	OFF
PSP	PSP Switch Signal	LOW	HI: turning	LOW	LOW
RPM	Engine Speed	0-10K rpm	670-750	1400-1600	1800-2100
TCS	TCS Switch	ON / OFF	OFF	OFF	OFF
TFT	TFT Sensor Signal	-40-304°F	110-210	110-210	110-210
TPV	TP Sensor	0-5.1v	0.53-1.27	0.8-1.7	1.2-1.7
TR1	TR1 Sensor	0v / 11.5v	0v	11.5	11.5
TR2	TR2 Sensor	0v / 11.5v	0v	11.5	11.5
TR4	TR4 Sensor	0v / 11.5v	0v	11.5	11.5
TRV/TR	TR Sensor	0v / 1.7v	0v	1.7	1.7
TSS	TSS Sensor	0-9999 Hz	100-125	185-205	260-280
Vacuum	Engine Vacuum	0-30" Hg	19-20	18-20	8-15
VPWR	Vehicle Power	0-25.5v	12-14	12-14	12-14
VSS	Vehicle Speed (Hertz or MPH)	0-999 Hz (0-159 mph)	0 Hz = 0 mph	65 Hz = 30 mph	125 Hz = 55 mph

2002-05 Ranger 4.0L V6 MFI VIN E (A/T) Outputs

PCM PID Acronym	Parameter Identification	PID Range	PID Value at Hot Idle	PID Value at 30	PID Value at 55
CCS	Coast Clutch Solenoid Control	VBAT	VBAT	VBAT	VBAT
CD1	Coil 1 Driver	0-60°	5	6	8
CD2	Coil 2 Driver	0-60°	5	6	8
CD3	Coil 3 Driver	0-60°	5	6	8
CTO	Clean Tachometer signal	0-10K Hz	35-49	65-90	90-120
EGR VR	EGR VR Solenoid	0-100%	0-100	0-40	0-40
EVAP CP	EVAP Purge Valve	0-100%	0-100	0-100	0-100
EVAP CV	EVAP CV Valve	0-100%	0-100	0-100	0-100
EPC	EPC Solenoid	0-300 psi	26	23-38	34-38
FP	Fuel Pump Control	ON / OFF	ON	ON	ON
FuelPW1	INJ Pulsewidth - Bank 1	0-999 ms	3.4-3.8	3.6-7.5	6.0-9.8
FuelPW2	INJ Pulsewidth - Bank 2	0-999 ms	3.4-3.8	3.6-7.5	6.0-9.8
HTR-11	HO2S-11 Heater	ON / OFF	ON	ON	ON
HTR-12	HO2S-12 Heater	ON / OFF	ON	ON	ON
HTR-21	HO2S-21 Heater	ON / OFF	ON	ON	ON
IAC	Idle Air Control	0-100%	25-32	35-49	30-68
INJ1	INJ 1 Pulsewidth	0-999 ms	3.4-3.8	3.6-7.5	6.0-9.8
INJ2	INJ 2 Pulsewidth	0-999 ms	3.4-3.8	3.6-7.5	6.0-9.8
INJ3	INJ 3 Pulsewidth	0-999 ms	3.4-3.8	3.6-7.5	6.0-9.8
INJ4	INJ 4 Pulsewidth	0-999 ms	3.4-3.8	3.6-7.5	6.0-9.8
INJ5	INJ 5 Pulsewidth	0-999 ms	3.4-3.8	3.6-7.5	6.0-9.8
INJ6	INJ 6 Pulsewidth	0-999 ms	3.4-3.8	3.6-7.5	6.0-9.8
LongFT1	Long Term FT - Bank 1	-20 to 20%	-5 to +5	-5 to +5	-5 to +5
LongFT2	Long Term FT - Bank 2	-20 to 20%	-5 to +5	-5 to +5	-5 to +5
MIL	MIL (lamp) Control	ON / OFF	OFF	OFF	OFF
SHTFT1	Short Term F/T - Bank 1	-10 to 10%	-5 to +5	-5 to +5	-5 to +5
SHTFT2	Short Term FT - Bank 2	-10 to 10%	-5 to +5	-5 to +5	-5 to +5
SPARK	Spark Advance	-90° to 90°	11-20	31-36	32-40
SS1	Shift Solenoid 1	ON / OFF	ON	OFF	OFF
SS2	Shift Solenoid 2	ON / OFF	OFF	OFF	OFF
SS3	Shift Solenoid 3	ON / OFF	OFF	OFF	ON
SS4	Shift Solenoid 4	ON / OFF	OFF	OFF	ON
TCC	TCC Solenoid	0-100%	0	0-100	90-95
TCIL	TCIL (lamp) Control	ON / OFF	OFF	OFF	OFF
VREF	Vehicle Reference	0-5.1v	4.9-5.1	4.9-5.1	4.9-5.1
WAC	A/C WOT Relay Control	ON / OFF	ON (AC on at WOT)	ON / OFF	ON / OFF

1990-94 Ranger 4.0L V6 VIN X (All) Inputs & Outputs

PCM PID Acronym	Parameter Identification	PID Range	PID Value at Hot Idle	PID Value at 30	PID Value at 55
ACCS	A/C Switch Signal	ON / OFF	ON (if on)	OFF	OFF
BARO	BARO Sensor	0-999 Hz	159	159	159
BOO	Brake Switch Signal	ON / OFF	ON (if on)	OFF	OFF
CANP	Purge Solenoid	0-100%	0	0-40	85-95
DPFE	DPFE Sensor	0-5.1v	0.4-0.6	0.5-1.0	0.7-1.1
ECT (°F)	ECT Sensor	-40-304°F	160-200	160-200	160-200
ECTV	ECT Sensor	0-5.1v	0.7	0.7	0.7
EVR	EGR VR Solenoid	0-100%	0	0-40	0-40
FPM	Fuel Pump Monitor	ON / OFF	ON	ON	ON
FuelPW1	INJ Pulsewidth - Bank 1	0-999 ms	3.3-3.5	4.0-4.6	5.0-6.0
FuelPW2	INJ Pulsewidth - Bank 2	0-999 ms	3.3-3.5	4.0-4.6	5.0-6.0
HO2S-11	HO2S-11 (Bank 1)	0.1-1.1v	0.1-1.1	0.1-1.1	0.1-1.1
HO2S-21	HO2S-21 (Bank 2)	0.1-1.1v	0.1-1.1	0.1-1.1	0.1-1.1
IAC	Idle Air Control	0-100%	20-40	34-40	45-55
IAT (°F)	IAT Sensor	-40-304°F	50-120	50-120	50-120
IATV	IAT Sensor	0-5.1v	1.5-3.5	1.5-3.5	1.5-3.5
LongFT1	Long Term F/T - Bank 1	-20 to 20%	-5 to +5	-5 to +5	-5 to +5
LongFT2	Long Term F/T - Bank 2	-20 to 20%	-5 to +5	-5 to +5	-5 to +5
LOOP	Loop Status	CL or OL	CLOSED	CLOSED	CLOSED
MAF	MAF Sensor	0-5.1v	0.7	1.3-1.4	1.7-2.0v
PNP	PNP Switch Signal	NEUT/DR	NEUTRAL	DRIVE	DRIVE
RPM	Engine Speed	0-10K rpm	750-830	1500-1650	1800-2100
SHTFT1	Short Term F/T - Bank 1	-10 to 10%	-5 to +5	-5 to +5	-5 to +5
SHTFT2	Short Term FT - Bank 2	-10 to 10%	-5 to +5	-5 to +5	-5 to +5
SPARK	Spark Advance	-90° to 90°	11-20	26-31	20-32
TP	TP Sensor	0-5.1v	0.9	1.2-1.3	1.4-1.6
TPCT	TP sensor Minimum	0-5.1v	0.1	1.2-1.3	1.4-1.6
TP Mode	TP Sensor	C/T or P/T	C/T	P/T	P/T
VPWR	Vehicle Power	0-25.5v	12-14	12-14	12-14
VREF	Vehicle Reference	0-5.1v	4.9-5.1	4.9-5.1	4.9-5.1
VSS	Vehicle Speed	0-159 mph	0	30	55
WAC	A/C WOT Relay Control	OFF	ON (AC on at WOT)	OFF	OFF

1995-2001 Ranger 4.0L V6 MFI VIN X (All) Inputs

PCM PID Acronym	Parameter Identification	PID Range	PID Value at Hot Idle	PID Value at 30	PID Value at 55
4x4L	4x4 Low Signal	ON / OFF	ON (if on)	OFF	OFF
ACCS	A/C Switch Signal	ON / OFF	ON (if on)	OFF	OFF
BPP	Brake Switch Signal	ON / OFF	ON (if on)	OFF	OFF
CID	CMP Sensor	0-999 Hz	6-8	13-15	17-19
CKP	CKP Sensor	0-10K Hz	430-475	810-870	1180-1230
CPP	Clutch Pedal Switch	ON / OFF	ON (if in)	OFF	OFF
DPFE	DPFE Sensor	0-5.1v	0.4-0.6	0.4-1.0	0.4-1.1
ECT	ECT Sensor	-40-304°F	160-200	160-200	160-200
FEPS	Flash EEPROM	0-5.1v	0.1	0.1	0.1
FLI	Fuel Level Indicator	0-100%	50% (1.7)	50% (1.7)	50% (1.7)
FTP	Fuel Tank Pressure	0-5.1v or 0-10" H2O	2.6v / 0" H2O (with cap off)	2.6v / 0" H2O (with cap off)	2.6v / 0" H2O (with cap off)
FPM	Fuel Pump Monitor	ON / OFF	ON	ON	ON
IAT	IAT Sensor	-40-304°F	50-120	50-120	50-120
GEAR	Gear Position	1-2-3-4	1	3	4
HO2S-11	HO2S-11 (Bank 1)	0.1-1.1v	0.1-1.1	0.1-1.1	0.1-1.1
HO2S-12	HO2S-12 (Bank 1)	0.1-1.1v	0.1-1.1	0.1-1.1	0.1-1.1
HO2S-21	HO2S-21 (Bank 2)	0.1-1.1v	0.1-1.1	0.1-1.1	0.1-1.1
LOAD	Engine Load	0-100%	16-20	21-27	27-34
MAFV	MAF Sensor	0-5.1v	0.8-1.0	1.4-1.8	2.0-2.7
MISF	Misfire Detection	ON / OFF	OFF	OFF	OFF
OCTADJ	Octane Adjustment	CL/OPEN	CLOSED	CLOSED	CLOSED
PF	Purge Flow Sensor	0-5.1v	1.55	2.10	2.30
PNP	PNP Switch Signal	Neutral/DR	NEUTRAL	DRIVE	DRIVE
PSP	PSP Switch Signal	LOW or HI	HI: turned	LOW	LOW
RPM	Engine Speed	0-10K rpm	760-830	1400-1475	1525-1760
TCS	TCS Switch	ON / OFF	OFF	OFF	OFF
TFT	TFT Sensor Signal	-40-304°F	110-210	110-210	110-210
TPV	TP Sensor	0-5.1v	0.53-1.27	0.8-1.0	1.2-1.7
TRV/TR	TR Sensor	P/N or O/D	P/N	O/D	O/D
TSS	TSS Sensor	0-9999 Hz	100-125	185-205	260-280
Vacuum	Engine Vacuum	0-30" Hg	19-20	18-20	8-15
VPWR	Vehicle Power	0-25.5v	12-14	12-14	12-14
VSS	Vehicle Speed (Hertz or MPH)	0-999 Hz (0-159 mph)	0 Hz = 0 mph	65 Hz = 30 mph	125 Hz = 55 mph

1995-2001 Ranger 4.0L V6 MFI VIN X (All) Outputs

PCM PID Acronym	Parameter Identification	PID Range	PID Value at Hot Idle	PID Value at 30	PID Value at 55
CD1	Coil 1 Driver	VBAT	VBAT	VBAT	VBAT
CD2	Coil 2 Driver	VBAT	VBAT	VBAT	VBAT
CD3	Coil 3 Driver	VBAT	VBAT	VBAT	VBAT
CTO	Clean Tachometer signal	0-9999 Hz	35-49	70-82 Hz	90-120
CCS	Coast Clutch Solenoid Control	0-14v	VBAT	VBAT	VBAT
EGR VR	EGR VR Solenoid	0-100%	0	0-40	0-40
EVAP CP	EVAP Purge Valve	0-100%	0-100	0-100	0-100
EVAP CV	EVAP CV Valve	0-100%	0-100	0-100	0-100
EPC	EPC Solenoid	0-300 psi	24	23-28	34-38
FP	Fuel Pump Control	ON / OFF	ON	ON	ON
FuelPW1	INJ Pulsewidth - Bank 1	0-999 ms	4.5-4.8	6.3-8.0	7.0-13.0
FuelPW2	INJ Pulsewidth - Bank 2	0-999 ms	4.5-4.8	6.3-8.0	7.0-13.0
HTR-11	HO2S-11 Heater	ON / OFF	ON	ON	ON
HTR-12	HO2S-12 Heater	ON / OFF	ON	ON	ON
HTR-21	HO2S-21 Heater	ON / OFF	ON	ON	ON
IAC	Idle Air Control	0-100%	33	35-41	57-68
INJ1	INJ 1 Pulsewidth	0-999 ms	4.5-4.8	6.3-8.0	7.0-13.0
INJ2	INJ 2 Pulsewidth	0-999 ms	4.5-4.8	6.3-8.0	7.0-13.0
INJ3	INJ 3 Pulsewidth	0-999 ms	4.5-4.8	6.3-8.0	7.0-13.0
INJ4	INJ 4 Pulsewidth	0-999 ms	4.5-4.8	6.3-8.0	7.0-13.0
INJ5	INJ 5 Pulsewidth	0-999 ms	4.5-4.8	6.3-8.0	7.0-13.0
INJ6	INJ 6 Pulsewidth	0-999 ms	4.5-4.8	6.3-8.0	7.0-13.0
LongFT1	Long Term FT - Bank 1	-20 to 20%	-5 to +5	-5 to +5	-5 to +5
LongFT2	Long Term FT - Bank 2	-20 to 20%	-5 to +5	-5 to +5	-5 to +5
MIL	MIL (lamp) Control	ON / OFF	OFF	OFF	OFF
SHTFT1	Short Term F/T - Bank 1	-10 to 10%	-5 to +5	-5 to +5	-5 to +5
SHTFT2	Short Term FT - Bank 2	-10 to 10%	-5 to +5	-5 to +5	-5 to +5
SPARK	Spark Advance	-90° to 90°	11-15	26-31	25-32
SS1	Shift Solenoid 1	ON / OFF	ON	OFF	OFF
SS2	Shift Solenoid 2	ON / OFF	OFF	OFF	OFF
SS3	Shift Solenoid 3	ON / OFF	OFF	OFF	ON
TCC	TCC Solenoid	0-100%	0	55-72	80-95
TCIL	TCIL (lamp) Control	ON / OFF	OFF	OFF	OFF
VREF	Vehicle Reference	0-5.1v	4.9-5.1	4.9-5.1	4.9-5.1
WAC	A/C WOT Relay Control	ON / OFF	ON (AC on at WOT)	OFF	OFF

AEROSTAR PID DATA

1993-95 Aerostar 3.0L V6 MFI VIN U (A/T) Inputs & Outputs

PCM PID Acronym	Parameter Identification	PID Range	PID Value at Hot Idle	PID Value at 30	PID Value at 55
ACCS	A/C Switch Signal	ON / OFF	ON (if on)	OFF	OFF
BOO	Brake Switch Signal	ON / OFF	ON (if on)	OFF	OFF
CANP	EVAP Purge Valve	OFF	OFF	ON	ON
ECT (°F)	ECT Sensor	-40-304°F	160-200	160-200	160-200
ECTV	ECT Sensor	0-5.1v	0.7	0.7	0.7
FPM	Fuel Pump Monitor	ON / OFF	ON	ON	ON
FuelPW1	INJ Pulsewidth - Bank 1	0-999 ms	3.7-3.9	4.4-4.6	5.0-6.0
HO2S-11	HO2S-11 (Bank 1)	0.1-1.1v	0.1-1.1	0.1-1.1	0.1-1.1
IAC	Idle Air Control	0-100%	33	35-41	57-68
IAT (°F)	IAT Sensor	-40-304°F	50-120	50-120	50-120
IATV	IAT Sensor	0-5.1v	1.5-3.5	1.5-3.5	1.5-3.5
LongFT1	Long Term F/T - Bank 1	-20 to 20%	-5 to +5	-5 to +5	-5 to +5
MAF	MAF Sensor	0-5.1v	0.9	1.4-1.6	1.9-2.4
PNP	PNP Switch Signal	Neutral/DR	NEUTRAL	DRIVE	DRIVE
RPM	Engine Speed	0-10K rpm	880-920	1430-1470	1550-1750
SHTFT1	Short Term F/T - Bank 1	-10 to 10%	-5 to +5	-5 to +5	-5 to +5
SPARK	Spark Advance	-90° to 90°	15-22	28-35	25-35
TP	TP Sensor	0-5.1v	0.7	0.9-1.0	1.0-1.2
TPCT	TP sensor Minimum	0-5.1v	0.1	0.9-1.0	1.0-1.2
TP Mode	TP Sensor Mode	C/T or P/T	C/T	P/T	P/T
VPWR	Vehicle Power	0-25.5v	12-14	12-14	12-14
VREF	Vehicle Reference	0-5.1v	4.9-5.1	4.9-5.1	4.9-5.1
VSS	Vehicle Speed	0-159 mph	0	30	55
WAC	A/C WOT Relay Control	OFF / ON	ON (AC on at WOT)	OFF	OFF

1996-97 Aerostar 3.0L V6 MFI VIN U (A/T) Inputs

PCM PID Acronym	Parameter Identification	PID Range	PID Value at Hot Idle	PID Value at 30	PID Value at 55
ACCS	A/C Switch Signal	ON / OFF	ON (if on)	OFF	OFF
BPP	Brake Position	ON / OFF	ON (if on)	OFF	OFF
CID	CMP Sensor	0-999 Hz	6-8	13-15	17-19
CKP	CKP Sensor	0-9999 Hz	518-540	917-1020	1290-1330
DPFE	DPFE Sensor	0-5.1v	0.4-0.6	0.4-0.9	0.4-0.9
ECT	ECT Sensor	-40-304°F	160-200	160-200	160-200
FEPS	Flash EEPROM	0-5.1v	0.1	0.1	0.1
FPM	Fuel Pump Monitor	ON / OFF	ON	ON	ON
GEAR	Gear Position	1-2-3-4	1	3	4
HO2S-11	HO2S-11 (Bank 1)	0.1-1.1v	0.1-1.1	0.1-1.1	0.1-1.1
HO2S-12	HO2S-12 (Bank 1)	0.1-1.1v	0.1-1.1	0.1-1.1	0.1-1.1
IAT	IAT Sensor	-40-304°F	50-120	50-120	50-120
LOAD	Engine Load	0-100%	13-18	25-35	39-46
MAFV	MAF Sensor	0-5.1v	0.8-1.0	1.5-1.9	2.1-2.9
MISF	Misfire Detection	ON / OFF	OFF	OFF	OFF
OCTADJ	Octane Adjustment	CL/OPEN	CLOSED	CLOSED	CLOSED
PF	Purge Flow Sensor	0-5.1v	1.55	2.10	2.30
PNP	PNP Switch Signal	Neutral/DR	NEUTRAL	DRIVE	DRIVE
PSP	PSP Switch Signal	HI or LOW	HI: turning	LOW	LOW
RPM	Engine Speed	0-10K rpm	880-910	1675-1730	2200-2275
TCS	TCS Switch	ON / OFF	OFF	OFF	OFF
TFT	TFT Sensor Signal	-40-304°F	110-210	110-210	110-210
TPV	TP Sensor	0-5.1v	0.53-1.27	0.8-1.7	1.2-1.7
TRV/TR	TR Sensor	P/N or O/D	P/N or O/D	O/D	O/D
TSS	TSS Sensor	0-9999 Hz	118-122	218-230	2200-2275
Vacuum	Engine Vacuum	0-30" Hg	19-20	18-20	8-15
VPWR	Vehicle Power	0-25.5v	12-14	12-14	12-14
VSS	Vehicle Speed (Hertz or MPH)	0-999 Hz (0-159 mph)	0 Hz = 0 mph	65 Hz = 30 mph	125 Hz = 55 mph

1996-97 Aerostar 3.0L V6 MFI VIN U (A/T) Outputs

PCM PID Acronym	Parameter Identification	PID Range	PID Value at Hot Idle	PID Value at 30	PID Value at 55
CD1	Coil 1 Driver	0-60° dwell	6	8	12
CD2	Coil 2 Driver	0-60° dwell	6	8	12
CD3	Coil 3 Driver	0-60° dwell	6	8	12
CTO	Clean Tachometer signal	0-9999 Hz	42-48	79-87	99-115
CCS	Coast Clutch Solenoid Control	0-25.5v	VBAT	VBAT	VBAT
EGR VR	EGR VR Solenoid	0-100%	0	0-40	0-55
EVAP CP	EVAP Purge Valve	0-100%	0-100	0-100	0-100
EPC	EPC Solenoid	0-300 psi	27	27	27
FP	Fuel Pump Control	ON / OFF	ON	ON	ON
FuelPW1	INJ Pulsewidth - Bank 1	0-999 ms	4.5-4.8	5.7-7.0	8.0-14.0
HTR-11	HO2S-11 Heater	ON / OFF	ON	ON	ON
HTR-12	HO2S-12 Heater	ON / OFF	ON	ON	ON
IAC	Idle Air Control	0-100%	33	35-41	57-68
INJ1	INJ 1 Pulsewidth	0-999 ms	4.5-4.8	5.7-7.0	8.0-14.0
INJ2	INJ 2 Pulsewidth	0-999 ms	4.5-4.8	5.7-7.0	8.0-14.0
INJ3	INJ 3 Pulsewidth	0-999 ms	4.5-4.8	5.7-7.0	8.0-14.0
INJ4	INJ 4 Pulsewidth	0-999 ms	4.5-4.8	5.7-7.0	8.0-14.0
INJ5	INJ 5 Pulsewidth	0-999 ms	4.5-4.8	5.7-7.0	8.0-14.0
INJ6	INJ 6 Pulsewidth	0-999 ms	4.5-4.8	5.7-7.0	8.0-14.0
LongFT1	Long Term FT - Bank 1	-20 to 20%	-5 to +5	-5 to +5	-5 to +5
MIL	MIL (lamp) Control	ON / OFF	OFF	OFF	OFF
SHTFT1	Short Term F/T - Bank 1	-10 to 10%	-5 to +5	-5 to +5	-5 to +5
SPARK	Spark Advance	-90° to 90°	15-22	29-39	25-32
SS1	Shift Solenoid 1	ON / OFF	ON	OFF	OFF
SS2	Shift Solenoid 2	ON / OFF	OFF	ON	ON
SS3	Shift Solenoid 3	ON / OFF	OFF	OFF	ON
TCC	TCC Solenoid	0-100%	0	59-66	72-81
TCIL	TCIL (lamp) Control	ON / OFF	OFF	OFF	OFF
VREF	Vehicle Reference	0-5.1v	4.9-5.1	4.9-5.1	4.9-5.1
WAC	A/C WOT Relay Control	ON / OFF	ON (AC on at WOT)	OFF	OFF

1990-95 Aerostar 4.0L V6 MFI VIN X (A/T)

PCM PID Acronym	Parameter Identification	PID Range	PID Value at Hot Idle	PID Value at 30	PID Value at 55
ACCS	A/C Switch Signal	ON / OFF	ON (if on)	OFF	OFF
BARO	BARO Sensor	0-999 Hz	159	159	159
BOO	Brake Switch Signal	ON / OFF	ON (if on)	OFF	OFF
CANP	EVAP Purge Valve	0-100%	0-100	0-100	0-100
ECT (°F)	ECT Sensor	-40-304°F	160-200	160-200	160-200
ECTV	ECT Sensor	0-5.1v	0.7	0.7	0.7
FPM	Fuel Pump Monitor	ON / OFF	ON	ON	ON
FuelPW1	INJ Pulsewidth - Bank 1	0-999 ms	3.3-3.5	4.0-4.6	5.0-6.0
FuelPW2	INJ Pulsewidth - Bank 2	0-999 ms	3.3-3.5	4.0-4.6	5.0-6.0
HO2S-11	HO2S-11 (Bank 1)	0.1-1.1v	0.1-1.1	0.1-1.1	0.1-1.1
HO2S-21	HO2S-21 (Bank 2)	0.1-1.1v	0.1-1.1	0.1-1.1	0.1-1.1
IAC	Idle Air Control	0-100%	20-40	34-40	45-55
IAT (°F)	IAT Sensor	-40-304°F	50-120	50-120	50-120
IATV	IAT Sensor	0-5.1v	1.5-3.5	1.5-3.5	1.5-3.5
LongFT1	Long Term F/T - Bank 1	-20 to 20%	-5 to +5	-5 to +5	-5 to +5
LongFT2	Long Term F/T - Bank 2	-20 to 20%	-5 to +5	-5 to +5	-5 to +5
LOOP	Loop Status	CL or OL	CLOSED	CLOSED	CLOSED
MAF	MAF Sensor	0-5.1v	0.7	1.3-1.4	1.7-2.0v
PNP	PNP Switch Signal	Neutral/DR	NEUTRAL	DRIVE	DRIVE
RPM	Engine Speed	0-10K rpm	790-820	1600-1660	1990-2100
SHTFT1	Short Term F/T - Bank 1	-20 to 20%	-5 to +5	-5 to +5	-5 to +5
SHTFT2	Short Term FT - Bank 2	-20 to 20%	-5 to +5	-5 to +5	-5 to +5
SPARK	Spark Advance	-90° to 90°	15-22	25-29	32-35
TP	TP Sensor	0-5.1v	0.9	1.2-1.3	1.4-1.6
TPCT	TP sensor Minimum	0-5.1v	0.1	1.2-1.3	1.4-1.6
TP Mode	TP Sensor	C/T or P/T	C/T	P/T	P/T
VPWR	Vehicle Power	0-25.5v	12-14	12-14	12-14
VREF	Vehicle Reference	0-5.1v	4.9-5.1	4.9-5.1	4.9-5.1
VSS	Vehicle Speed	0-159 mph	0	30	55
WAC	A/C WOT Relay Control	OFF	ON (AC on at WOT)	OFF	OFF

1996-97 Aerostar 4.0L V6 MFI VIN X (A/T) Inputs

PCM PID Acronym	Parameter Identification	PID Range	PID Value at Hot Idle	PID Value at 30	PID Value at 55
ACCS	A/C Switch Signal	ON / OFF	ON (if on)	OFF	OFF
BPP	Brake Position	ON / OFF	ON (if on)	OFF	OFF
CID	CMP Sensor	0-999 Hz	5-7	13-15	15-17
CKP	CKP Sensor	0-9999 Hz	450-480	1000-1100	1140-1220
DPFE	DPFE Sensor	0-5.1v	0.4-0.6	0.5-1.0	0.7-1.1
ECT	ECT Sensor	-40-304°F	160-200	160-200	160-200
FEPS	Flash EEPROM	0-5.1v	0.1	0.1	0.1
FPM	Fuel Pump Monitor	ON / OFF	ON	ON	ON
GEAR	Gear Position	1-2-3-4	1	3	4
HO2S-11	HO2S-11 (Bank 1)	0.1-1.1v	0.1-1.1	0.1-1.1	0.1-1.1
HO2S-12	HO2S-12 (Bank 1)	0.1-1.1v	0.1-1.1	0.1-1.1	0.1-1.1
IAT	IAT Sensor	-40-304°F	50-120	50-120	50-120
LOAD	Engine Load	0-100%	13-18	18-33	35-50
MAFV	MAF Sensor	0-5.1v	0.7-0.9	1.1-1.6	1.5-2.3
MISF	Misfire Detection	ON / OFF	OFF	OFF	OFF
OCTADJ	Octane Adjustment	CL/OPEN	CLOSED	CLOSED	CLOSED
PF	Purge Flow Sensor	0-5.1v	1.55	2.10	2.30
PNP	PNP Switch Signal	Neutral/DR	NEUTRAL	DRIVE	DRIVE
PSP	PSP Switch Signal	LOW	HI: turning	LOW	LOW
RPM	Engine Speed	0-10K rpm	790-820	1610-1860	1990-2100
TCS	TCS Switch	ON / OFF	OFF	OFF	OFF
TFT	TFT Sensor Signal	-40-304°F	110-210	110-210	110-210
TPV	TP Sensor	0-5.1v	0.53-1.27	0.8-1.7	1.2-1.7
TRV/TR	TR Sensor	P/N or O/D	P/N or O/D	O/D	O/D
TSS	TSS Sensor	0-9999 Hz	95-110	190-200	260-280
Vacuum	Engine Vacuum	0-30" Hg	19-20	18-20	8-15
VPWR	Vehicle Power	0-25.5v	12-14	12-14	12-14
VSS	Vehicle Speed (Hertz or MPH)	0-999 Hz (0-159 mph)	0 Hz = 0 mph	65 Hz = 30 mph	125 Hz = 55 mph

1996-97 Aerostar 4.0L V6 MFI VIN X (A/T) Outputs

PCM PID Acronym	Parameter Identification	PID Range	PID Value at Hot Idle	PID Value at 30	PID Value at 55
CD1	Coil 1 Driver	0-90° dwell	6	8	12
CD2	Coil 2 Driver	0-90° dwell	6	8	12
CD3	Coil 3 Driver	0-90° dwell	6	8	12
CTO	Clean Tachometer signal	0-999	38-42	79-87	96-105
CCS	Coast Clutch Solenoid Control	0-25.5v	VBAT	VBAT	VBAT
EGR VR	EGR VR Solenoid	0-100%	0	0-40	0-45
EVAP CP	EVAP Purge Valve	0-100%	0-100	0-100	0-100
EPC	EPC Solenoid	0-300 psi	26	30	38
FP	Fuel Pump Control	ON / OFF	ON	ON	ON
FuelPW1	INJ Pulsewidth - Bank 1	0-999 ms	3.9-4.1	4.3-6.9	6.0-12.0
HTR-11	HO2S-11 Heater	ON / OFF	ON	ON	ON
HTR-12	HO2S-12 Heater	ON / OFF	ON	ON	ON
IAC	Idle Air Control	0-100%	39	46-52	49-53
INJ1	INJ 1 Pulsewidth	0-999 ms	3.9-4.1	4.3-6.9	6.0-12.0
INJ2	INJ 2 Pulsewidth	0-999 ms	3.9-4.1	4.3-6.9	6.0-12.0
INJ3	INJ 3 Pulsewidth	0-999 ms	3.9-4.1	4.3-6.9	6.0-12.0
INJ4	INJ 4 Pulsewidth	0-999 ms	3.9-4.1	4.3-6.9	6.0-12.0
INJ5	INJ 5 Pulsewidth	0-999 ms	3.9-4.1	4.3-6.9	6.0-12.0
INJ6	INJ 6 Pulsewidth	0-999 ms	3.9-4.1	4.3-6.9	6.0-12.0
LongFT1	Long Term FT - Bank 1	-20 to 20%	-5 to +5	-5 to +5	-5 to +5
MIL	MIL (lamp) Control	ON / OFF	OFF	OFF	OFF
SHTFT1	Short Term F/T - Bank 1	-10 to 10%	-5 to +5	-5 to +5	-5 to +5
SPARK	Spark Advance	-90° to 90°	15-22	25-29	19-25
SS1	Shift Solenoid 1	ON / OFF	ON	OFF	OFF
SS2	Shift Solenoid 2	ON / OFF	OFF	ON	OFF
SS3	Shift Solenoid 3	ON / OFF	OFF	OFF	ON
TCC	TCC Solenoid	0-100%	0	0-100	75-81
TCIL	TCIL (lamp) Control	ON / OFF	OFF	OFF	OFF
VREF	Vehicle Reference	0-5.1v	4.9-5.1	4.9-5.1	4.9-5.1
WAC	A/C WOT Relay Control	ON / OFF	ON (AC on at WOT)	OFF	OFF

E-SERIES VAN PID DATA

1997-2005 E-Series 4.2L V6 MFI VIN 2 (A/T) Inputs

PCM PID Acronym	Parameter Identification	PID Range	PID Value at Hot Idle	PID Value at 30	PID Value at 55
ACCS	A/C Switch Signal	OFF	ON	OFF	OFF
BPP	Brake Position	ON / OFF	ON (if on)	OFF	OFF
CID	CMP Sensor	0-999 Hz	5-7	10-12	13-17
CHT	CHT Sensor	-40-500ºF	194	194	194
CKP	CKP Sensor	0 Hz	430-4500	700-900	1000-1200
DPFE	DPFE Sensor	0-5.1v	0.4-0.6	0.4-1.0	0.6-1.1
FEPS	Flash EEPROM	0-5.1v	0.1	0.1	0.1
FLI	Fuel Level Indicator	0-100%	50% (1.7)	50% (1.7)	50% (1.7)
FPM	Fuel Pump Monitor	ON / OFF	ON	ON	ON
FTP	Fuel Tank Pressure	0-5.1v or 0-10" H2O	2.6v / 0" H2O (with cap off)	2.6v / 0" H2O (with cap off)	2.6v / 0" H2O (with cap off)
GEAR	Gear Position	1-2-3-4	1	3	4
HO2S-11	HO2S-11 (Bank 1)	0.1-1.1v	0.1-1.1	0.1-1.1	0.1-1.1
HO2S-12	HO2S-12 (Bank 1)	0.1-1.1v	0.1-1.1	0.1-1.1	0.1-1.1
HO2S-21	HO2S-21 (Bank 2)	0.1-1.1v	0.1-1.1	0.1-1.1	0.1-1.1
HO2S-22	HO2S-22 (Bank 2)	0.1-1.1v	0.1-1.1	0.1-1.1	0.1-1.1
IAT	IAT Sensor	-40-304ºF	50-120	50-120	50-120
IMRC-M	IMRC Monitor	5v / 0	5	5	5
KS1	Knock Sensor 1	0-5.1v	0	0	0
LOAD	Engine Load	0-100%	10-20	20-27	30-45
MAFV	MAF Sensor	0-5.1v	0.6-0.9	0.7-1.0	1.2-2.3
MISF	Misfire Detection	ON / OFF	OFF	OFF	OFF
OCTADJ	Octane Adjustment	CL/OPEN	CLOSED	CLOSED	CLOSED
OSS	OSS Sensor	0-9999 Hz	0	125-131	245-255
PNP	PNP Switch Signal	Neutral/DR	NEUTRAL	DRIVE	DRIVE
RPM	Engine Speed	0-10K rpm	680-830	1200-1500	1600-1800
TFT	TFT Sensor Signal	-40-304ºF	110-210	110-210	110-210
TPV	TP Sensor	0-5.1v	0.53-1.27	1.0-1.3	1.1-1.6
TR1	TR1 Sensor	0v / 11.5v	0v	11.5	11.5
TR2	TR2 Sensor	0v / 11.5v	0v	11.5	11.5
TR3/TR	TR Sensor	0v / 1.7v	0v	1.7	1.7
TR4	TR4 Sensor	0v / 11.5v	0v	11.5	11.5
Vacuum	Engine Vacuum	0-30" Hg	19-20	18-20	8-15
VPWR	Vehicle Power	0-25.5v	12-14	12-14	12-14
VSS	Vehicle Speed (Hertz or MPH)	0-999 Hz (0-159 mph)	0 Hz = 0 mph	65 Hz = 30 mph	125 Hz = 55 mph
4x4L	4x4 Switch Signal	ON / OFF	ON (if on)	OFF	OFF

1997-2005 E-Series 4.2L V6 MFI VIN 2 (A/T) Outputs

PCM PID Acronym	Parameter Identification	PID Range	PID Value at Hot Idle	PID Value at 30	PID Value at 55
CD1	Coil 1 Driver	0-60° dwell	6	8	12
CD2	Coil 2 Driver	0-60° dwell	6	8	12
CD3	Coil 3 Driver	0-60° dwell	6	8	12
CHIL	CHIL (lamp) Control	ON / OFF	OFF	OFF	OFF
CTO	Clean Tachometer signal	0-9999 Hz	35-49	65-90	90-120
EGR VR	EGR VR Solenoid	0-100%	0	0-40	0-40
EVAP CP	EVAP Purge Valve	0-100%	0-100	0-100	0-100
EVAP CV	EVAP Purge Valve	0-100%	0-100	0-100	0-100
EPC	EPC Solenoid	0-300 psi	15-20	35-40	40
FP	Fuel Pump Control	ON / OFF	ON	ON	ON
FuelPW1	INJ Pulsewidth - Bank 1	0-999 ms	2.7-3.5	4.5-8.0	5.5-9.0
FuelPW2	INJ Pulsewidth - Bank 2	0-999 ms	2.7-3.5	4.5-8.0	5.5-9.0
HTR-11	HO2S-11 Heater	ON / OFF	ON	ON	ON
HTR-12	HO2S-12 Heater	ON / OFF	ON	ON	ON
HTR-21	HO2S-21 Heater	ON / OFF	ON	ON	ON
HTR-22	HO2S-22 Heater	ON / OFF	ON	ON	ON
IAC	Idle Air Control	0-100%	25-32	30-55	60-70
IMRC	IMRC Solenoid	ON / OFF	OFF	OFF	OFF
INJ1	INJ 1 Pulsewidth	0-999 ms	2.7-3.5	4.5-8.0	5.5-9.0
INJ2	INJ 2 Pulsewidth	0-999 ms	2.7-3.5	4.5-8.0	5.5-9.0
INJ3	INJ 3 Pulsewidth	0-999 ms	2.7-3.5	4.5-8.0	5.5-9.0
INJ4	INJ 4 Pulsewidth	0-999 ms	2.7-3.5	4.5-8.0	5.5-9.0
INJ5	INJ 5 Pulsewidth	0-999 ms	2.7-3.5	4.5-8.0	5.5-9.0
INJ6	INJ 6 Pulsewidth	0-999 ms	2.7-3.5	4.5-8.0	5.5-9.0
LongFT1	Long Term FT - Bank 1	-20 to 20%	-5 to +5	-5 to +5	-5 to +5
LongFT2	Long Term FT - Bank 2	-20 to 20%	-5 to +5	-5 to +5	-5 to +5
MIL	MIL (lamp) Control	ON / OFF	OFF	OFF	OFF
SHTFT1	Short Term F/T - Bank 1	-10 to 10%	-5 to +5	-5 to +5	-5 to +5
SHTFT2	Short Term FT - Bank 2	-10 to 10%	-5 to +5	-5 to +5	-5 to +5
SPARK	Spark Advance	-90° to 90°	11-20	15-35	20-39
SS1	Shift Solenoid 1	ON / OFF	ON	OFF	ON
SS2	Shift Solenoid 2	ON / OFF	OFF	ON	ON
TCC	TCC Solenoid	0-100%	0	0-45	90-95
TCIL	TCIL (lamp) Control	ON / OFF	OFF	OFF	OFF
VREF	Vehicle Reference	0-5.1v	4.9-5.1	4.9-5.1	4.9-5.1
WAC	A/C WOT Relay Control	ON / OFF	ON (AC on at WOT)	OFF	OFF

1997-2001 E-Series 4.6L V8 MFI VIN 6 (A/T) Inputs

PCM PID Acronym	Parameter Identification	PID Range	PID Value at Hot Idle	PID Value at 30	PID Value at 55
4x4L	4x4 Switch Signal	ON / OFF	ON (if on)	OFF	OFF
ACCS	A/C Switch Signal	ON / OFF	ON (if on)	OFF	OFF
BPP	Brake Position	ON / OFF	ON (if on)	OFF	OFF
CHT	CHT Sensor	-40-500°F	194	194	194
CID	CMP Sensor	0-999 Hz	5-7	10-12	13-17
CKP	CKP Sensor	0-9999 Hz	430-475	900-1000	1140-1220
DPFE	DPFE Sensor	0-5.1v	0.4-0.6	0.4-1.0	0.6-1.1
ECT	ECT Sensor	-40-304°F	160-200	160-200	160-200
FEPS	Flash EEPROM	0-5.1v	0.1	0.1	0.1
FLI	Fuel Level Input	1.7 / 50%	1.7 / 50%	1.7 / 50%	1.7 / 50%
FPM	Fuel Pump Monitor	ON / OFF	ON	ON	ON
FTP	Fuel Tank Pressure	2.6v / 0 psi	2.6v / 0 psi	2.6v / 0 psi	2.6v / 0 psi
GEAR	Gear Position	1-2-3-4	1	3	4
HO2S-11	HO2S-11 (Bank 1)	0.1-1.1v	0.1-1.1	0.1-1.1	0.1-1.1
HO2S-12	HO2S-12 (Bank 1)	0.1-1.1v	0.1-1.1	0.1-1.1	0.1-1.1
HO2S-21	HO2S-21 (Bank 2)	0.1-1.1v	0.1-1.1	0.1-1.1	0.1-1.1
HO2S-22	HO2S-22 (Bank 2)	0.1-1.1v	0.1-1.1	0.1-1.1	0.1-1.1
IAT	IAT Sensor	-40-304°F	50-120	50-120	50-120
KS1	Knock Sensor 1	0	0	0	0
LOAD	Engine Load	0-100%	10-20	20-27	30-45
MAFV	MAF Sensor	0-5.1v	0.6-0.9	0.7-1.0	1.2-2.3
MISF	Misfire Detection	OFF	OFF	OFF	OFF
OCTADJ	Octane Adjustment	CLOSED	CLOSED	CLOSED	CLOSED
OSS	OSS Sensor	0-9999 Hz	0	125-131	245-255
PNP	PNP Switch Signal	Neutral/DR	NEUTRAL	DRIVE	DRIVE
RPM	Engine Speed	0-10K rpm	680-830	1200-1500	1600-1800
TFT	TFT Sensor Signal	-40-304°F	110-210	110-210	110-210
TPV	TP Sensor	0-5.1v	0.53-1.27	1.0-1.3	1.1-1.6
TR1	TR1 Sensor	0v / 11.5v	0v	11.5	11.5
TR2	TR2 Sensor	0v / 11.5v	0v	11.5	11.5
TR3/TR	TR Sensor	0v / 1.7v	0v	1.7	1.7
TR4	TR4 Sensor	0v / 11.5v	0v	11.5	11.5
Vacuum	Engine Vacuum	0-30" Hg	19-20	18-20	8-15
VPWR	Vehicle Power	0-25.5v	12-14	12-14	12-14
VSS	Vehicle Speed (Hertz or MPH)	0-999 Hz (0-159 mph)	0 Hz = 0 mph	65 Hz = 30 mph	125 Hz = 55 mph

1997-2001 E-Series 4.6L V8 MFI VIN 6 (A/T) Outputs

PCM PID Acronym	Parameter Identification	PID Range	PID Value at Hot Idle	PID Value at 30	PID Value at 55
CD1	Coil 1 Driver	0-45° dwell	5	6	8
CD2	Coil 2 Driver	0-45° dwell	5	6	8
CD3	Coil 3 Driver	0-45° dwell	5	6	8
CD4	Coil 4 Driver	0-45° dwell	5	6	8
CHTIL	CHT (lamp) Control	ON / OFF	OFF	OFF	OFF
CTO	Clean Tachometer signal	0-9999 Hz	35-49	65-90	90-120
EGR VR	EGR VR Solenoid	0-100%	0	0-40	0-40
EVAP CP	EVAP Purge Valve	0-100%	0-100	0-100	0-100
EVAP CV	EVAP CV Valve	0-100%	0-100	0-100	0-100
EPC	EPC Solenoid	0-300 psi	6	40	40
FP	Fuel Pump Control	ON / OFF	ON	ON	ON
FuelPW1	INJ Pulsewidth - Bank 1	0-999 ms	2.7-4.1	4.5-8.0	5.5-9.0
FuelPW2	INJ Pulsewidth - Bank 2	0-999 ms	2.7-4.1	4.5-8.0	5.5-9.0
FRP	Fuel Rail Pressure	29-32 psi	30-32	30-32	32-35
HTR-11	HO2S-11 Heater	ON / OFF	ON	ON	ON
HTR-12	HO2S-12 Heater	ON / OFF	ON	ON	ON
HTR-21	HO2S-21 Heater	ON / OFF	ON	ON	ON
HTR-22	HO2S-22 Heater	ON / OFF	ON	ON	ON
IAC	Idle Air Control	0-100%	25-32	30-55	60-70
IMTV	Intake Manifold Tuning Valve	0v / VBAT	VBAT	VBAT	VBAT
INJ1	INJ 1 Pulsewidth	0-999 ms	2.7-4.1	4.5-8.0	5.5-9.0
INJ2	INJ 2 Pulsewidth	0-999 ms	2.7-4.1	4.5-8.0	5.5-9.0
INJ3	INJ 3 Pulsewidth	0-999 ms	2.7-4.1	4.5-8.0	5.5-9.0
INJ4	INJ 4 Pulsewidth	0-999 ms	2.7-4.1	4.5-8.0	5.5-9.0
INJ5	INJ 5 Pulsewidth	0-999 ms	2.7-4.1	4.5-8.0	5.5-9.0
INJ6	INJ 6 Pulsewidth	0-999 ms	2.7-4.1	4.5-8.0	5.5-9.0
INJ7	INJ 7 Pulsewidth	0-999 ms	2.7-4.1	4.5-8.0	5.5-9.0
INJ8	INJ 8 Pulsewidth	0-999 ms	2.7-4.1	4.5-8.0	5.5-9.0
LongFT1	Long Term FT (B1)	-20 to 20%	-5 to +5	-5 to +5	-5 to +5
LongFT2	Long Term FT (B2)	-20 to 20%	-5 to +5	-5 to +5	-5 to +5
MIL	MIL (lamp) Control	ON / OFF	OFF	OFF	OFF
SHTFT1	Short Term F/T - Bank 1	-10 to 10%	-5 to +5	-5 to +5	-5 to +5
SHTFT2	Short Term FT - Bank 2	-10 to 10%	-5 to +5	-5 to +5	-5 to +5
SPARK	Spark Advance	-90° to 90°	11-20	15-35	20-39
SS1	Shift Solenoid 1	ON / OFF	ON	ON	ON
SS2	Shift Solenoid 2	ON / OFF	OFF	ON	ON
TCC	TCC Solenoid	0-100%	0%	0-45	90-100
TCIL	TCIL (lamp) Control	ON / OFF	OFF	OFF	OFF
VREF	Vehicle Reference	0-5.1v	4.9-5.1	4.9-5.1	4.9-5.1
WAC	A/C WOT Relay	ON / OFF	ON (if on)	OFF	OFF

1999-2005 E-Series 4.6L V8 MFI VIN W (A/T) Inputs

PCM PID Acronym	Parameter Identification	PID Range	PID Value at Hot Idle	PID Value at 30	PID Value at 55
ACCS	A/C Switch Signal	ON / OFF	ON (if on)	OFF	OFF
BPP	Brake Position	ON / OFF	ON (if on)	OFF	OFF
CHT	CHT Sensor	-40-500°F	194	194	194
CID	CMP Sensor	0-999 Hz	5-7	10-12	13-17
CKP	CKP Sensor	0-9999 Hz	430-475	900-1000	1140-1220
DPFE	DPFE Sensor	0-5.1v	0.4-0.6	0.4-1.0	0.6-1.1
ECT	ECT Sensor	-40-304°F	160-200	160-200	160-200
FEPS	Flash EEPROM	0-5.1v	0.1	0.1	0.1
FLI	Fuel Level Input	1.7 / 50%	1.7 / 50%	1.7 / 50%	1.7 / 50%
FPM	Fuel Pump Monitor	ON / OFF	ON	ON	ON
FTP	Fuel Tank Pressure	2.6v / 0 psi	2.6v / 0 psi	2.6v / 0 psi	2.6v / 0 psi
GEAR	Gear Position	1-2-3-4	1	3	4
HO2S-11	HO2S-11 (Bank 1)	0.1-1.1v	0.1-1.1	0.1-1.1	0.1-1.1
HO2S-12	HO2S-12 (Bank 1)	0.1-1.1v	0.1-1.1	0.1-1.1	0.1-1.1
HO2S-21	HO2S-21 (Bank 2)	0.1-1.1v	0.1-1.1	0.1-1.1	0.1-1.1
HO2S-22	HO2S-22 (Bank 2)	0.1-1.1v	0.1-1.1	0.1-1.1	0.1-1.1
IAT	IAT Sensor	-40-304°F	50-120	50-120	50-120
KS1	Knock Sensor 1	0	0	0	0
LOAD	Engine Load	0-100%	10-20	20-27	30-45
MAFV	MAF Sensor	0-5.1v	0.6-0.9	0.7-1.0	1.2-2.3
MISF	Misfire Detection	OFF	OFF	OFF	OFF
OCTADJ	Octane Adjustment	CLOSED	CLOSED	CLOSED	CLOSED
OSS	OSS Sensor	0-9999 Hz	0	125-131	245-255
PNP	PNP Switch Signal	Neutral/DR	NEUTRAL	DRIVE	DRIVE
RPM	Engine Speed	0-10K rpm	680-830	1200-1500	1600-1800
TFT	TFT Sensor Signal	-40-304°F	110-210	110-210	110-210
TPV	TP Sensor	0-5.1v	0.53-1.27	1.0-1.3	1.1-1.6
TR1	TR1 Sensor	0v / 11.5v	0v	11.5	11.5
TR2	TR2 Sensor	0v / 11.5v	0v	11.5	11.5
TR3/TR	TR Sensor	0v / 1.7v	0v	1.7	1.7
TR4	TR4 Sensor	0v / 11.5v	0v	11.5	11.5
Vacuum	Engine Vacuum	0-30" Hg	19-20	18-20	8-15
VPWR	Vehicle Power	0-25.5v	12-14	12-14	12-14
VSS	Vehicle Speed (Hertz or MPH)	0-999 Hz (0-159 mph)	0 Hz = 0 mph	65 Hz = 30 mph	125 Hz = 55 mph
4x4L	4x4 Switch Signal	ON / OFF	ON (if on)	OFF	OFF

1999-2005 E-Series 4.6L V8 MFI VIN W (A/T) Outputs

PCM PID Acronym	Parameter Identification	PID Range	PID Value at Hot Idle	PID Value at 30	PID Value at 55
CD1	Coil 1 Driver	0-45° dwell	5	6	8
CD2	Coil 2 Driver	0-45° dwell	5	6	8
CD3	Coil 3 Driver	0-45° dwell	5	6	8
CD4	Coil 4 Driver	0-45° dwell	5	6	8
CHTIL	CHT (lamp) Control	ON / OFF	OFF	OFF	OFF
CTO	Clean Tachometer signal	0-9999 Hz	35-49	65-90	90-120
EGR VR	EGR VR Solenoid	0-100%	0	0-40	0-40
EVAP CP	EVAP Purge Valve	0-100%	0-100	0-100	0-100
EVAP CV	EVAP CV Valve	0-100%	0-100	0-100	0-100
EPC	EPC Solenoid	0-300 psi	6	40	40
FP	Fuel Pump Control	ON / OFF	ON	ON	ON
FuelPW1	INJ Pulsewidth - Bank 1	0-999 ms	2.7-4.1	4.5-8.0	5.5-9.0
FuelPW2	INJ Pulsewidth - Bank 2	0-999 ms	2.7-4.1	4.5-8.0	5.5-9.0
FRP	Fuel Rail Pressure	29-32 psi	30-32	30-32	32-35
HTR-11	HO2S-11 Heater	ON / OFF	ON	ON	ON
HTR-12	HO2S-12 Heater	ON / OFF	ON	ON	ON
HTR-21	HO2S-21 Heater	ON / OFF	ON	ON	ON
HTR-22	HO2S-22 Heater	ON / OFF	ON	ON	ON
IAC	Idle Air Control	0-100%	25-32	30-55	60-70
IMTV	Intake Manifold Tuning Valve	0v / VBAT	VBAT	VBAT	VBAT
INJ1	INJ 1 Pulsewidth	0-999 ms	2.7-4.1	4.5-8.0	5.5-9.0
INJ2	INJ 2 Pulsewidth	0-999 ms	2.7-4.1	4.5-8.0	5.5-9.0
INJ3	INJ 3 Pulsewidth	0-999 ms	2.7-4.1	4.5-8.0	5.5-9.0
INJ4	INJ 4 Pulsewidth	0-999 ms	2.7-4.1	4.5-8.0	5.5-9.0
INJ5	INJ 5 Pulsewidth	0-999 ms	2.7-4.1	4.5-8.0	5.5-9.0
INJ6	INJ 6 Pulsewidth	0-999 ms	2.7-4.1	4.5-8.0	5.5-9.0
INJ7	INJ 7 Pulsewidth	0-999 ms	2.7-4.1	4.5-8.0	5.5-9.0
INJ8	INJ 8 Pulsewidth	0-999 ms	2.7-4.1	4.5-8.0	5.5-9.0
LongFT1	Long Term FT (B1)	-20 to 20%	-5 to +5	-5 to +5	-5 to +5
LongFT2	Long Term FT (B2)	-20 to 20%	-5 to +5	-5 to +5	-5 to +5
MIL	MIL (lamp) Control	ON / OFF	OFF	OFF	OFF
SHTFT1	Short Term F/T - Bank 1	-10 to 10%	-5 to +5	-5 to +5	-5 to +5
SHTFT2	Short Term FT - Bank 2	-10 to 10%	-5 to +5	-5 to +5	-5 to +5
SPARK	Spark Advance	-90° to 90°	11-20	15-35	20-39
SS1	Shift Solenoid 1	ON / OFF	ON	ON	ON
SS2	Shift Solenoid 2	ON / OFF	OFF	ON	ON
TCC	TCC Solenoid	0-100%	0	0-45	90-100
TCIL	TCIL (lamp) Control	ON / OFF	OFF	OFF	OFF
VREF	Vehicle Reference	0-5.1v	4.9-5.1	4.9-5.1	4.9-5.1
WAC	A/C WOT Relay	ON / OFF	ON (if on)	OFF	OFF

1990-95 E-Series 4.9L I6 MFI VIN Y (A/T, M/T, E4OD)

PCM PID Acronym	Parameter Identification	PID Range	PID Value at Hot Idle	PID Value at 30	PID Value at 55
ACCS	A/C Switch Signal	ON / OFF	ON (if on)	OFF	OFF
CANP	EVAP Purge Valve	ON / OFF	OFF	ON	ON
DPFE	DPFE Sensor (Cal.)	0-5.1v	0.4-0.6	0.4-0.9	0.4-0.9
ECT (°F)	ECT Sensor	-40-304°F	160-200	160-200	160-200
ECTV	ECT Sensor	0-5.1v	0.6	0.6	0.6
EPC	EPC Solenoid (E4OD)	0-300 psi	5	5	15
EVP	EGR Valve Position	0-5.1v	0.3	1.2-2.0	2.5-3.5
EVR	EGR VR Solenoid	0-100%	0	0-40	0-40
FuelPW1	INJ Pulsewidth - Bank 1	0-999 ms	6.8-7.0	9.5-10	12-13
GEAR	Gear Position	P-R-N-D	P-R-N-D	DRIVE 3	DRIVE 4
HO2S-11	HO2S-11 (Bank 1)	0.1-1.1v	0.1-1.1	0.1-1.1	0.1-1.1
IAC	Idle Air Control	0-100%	35	44-50	59-65
IAT (°F)	IAT Sensor	-40-304°F	50-120	50-120	50-120
IATV	IAT Sensor	0-5.1v	1.5-3.5	1.5-3.5	1.5-3.5
LOOP	Loop Status	CL or OL	CLOSED	CLOSED	CLOSED
MAP	MAP Sensor	0-999 Hz	107	114-120	120-130
MLPV	MLP Switch (E4OD)	0-5.1v	0	5	5
PNP	PNP Switch Signal	Neutral/DR	NEUTRAL	DRIVE	DRIVE
RPM	Engine Speed	0-10K rpm	600-700	1050-1150	1840-1940
SHTFT1	Short Term F/T - Bank 1	-10 to 10%	-5 to +5	-5 to +5	-5 to +5
SPARK	Spark Advance	-90° to 90°	17-20	24-28	24-30
TP	TP Sensor	0-5.1v	1.0	1.2-1.3	1.5-1.6
TCC	TCC Sol. (E4OD)	ON / OFF	OFF	ON	ON
TOTV	TOT sensor (E4OD)	0-5.1v	2.10-2.40	2.10-2.40	2.10-2.40
TP	TP Sensor	0-5.1v	1.0	1.2-1.3	1.5-1.6
TP Mode	TP Sensor Mode	C/T or P/T	C/T	P/T	P/T
VPWR	Vehicle Power	0-25.5v	12-14	12-14	12-14
VREF	Vehicle Reference	0-5.1v	4.9-5.1	4.9-5.1	4.9-5.1
VSS	Vehicle Speed	0-159 mph	0	30	55

1996 E-Series 4.9L I6 MFI VIN Y (All) Inputs

PCM PID Acronym	Parameter Identification	PID Range	PID Value at Hot Idle	PID Value at 30	PID Value at 55
ACCS	A/C Switch Signal	ON / OFF	ON (if on)	OFF	OFF
CPP	Clutch Pedal Switch	ON / OFF	ON (if in)	ON	ON
DPFE	DPFE Sensor	0-5.1v	0.4-0.6	0.4-0.9	0.6-1.0
ECT	ECT Sensor	-40-304°F	160-200	160-200	160-200
FEPS	Flash EEPROM	0-5.1v	0.1	0.1	0.1
FPM	Fuel Pump Monitor	ON / OFF	ON	ON	ON
HO2S-11	HO2S-11 (Bank 1)	0.1-1.1v	0.1-1.1	0.1-1.1	0.1-1.1
HO2S-12	HO2S-12 (Bank 1)	0.1-1.1v	0.1-1.1	0.1-1.1	0.1-1.1
HO2S-21	HO2S-21 (Bank 2)	0.1-1.1v	0.1-1.1	0.1-1.1	0.1-1.1
IAT	IAT Sensor	-40-304°F	50-120	50-120	50-120
IDM	IDM Signal	0-9999 Hz	32-38	59-65	88-95
KS1	Knock Sensor 1	0-5.1v	0	0	0
LOAD	Engine Load	0-100%	12-14	16-25	30-40
MAFV	MAF Sensor	0-5.1v	0.5-0.7	1.1-1.5	1.7-2.2
MISF	Misfire Detection	ON / OFF	OFF	OFF	OFF
OCTADJ	Octane Adjustment	CL/OPEN	CLOSED	CLOSED	CLOSED
PIP	PIP Sensor	0-9999 Hz	32-38	59-65	88-95
PNP	PNP Switch Signal	Neutral/DR	NEUTRAL	DRIVE	DRIVE
PTO	Power Takeoff Switch	ON / OFF	ON (if on)	OFF	OFF
RPM	Engine Speed	0-10K rpm	680-730	1200-1300	1810-2000
TPV	TP Sensor	0-5.1v	0.53-1.27	0.8-1.1	0.9-1.2
Vacuum	Engine Vacuum	0-30" Hg	19-20	18-20	8-15
VPWR	Vehicle Power	0-25.5v	12-14	12-14	12-14
VSS	Vehicle Speed (Hertz or MPH)	0-999 Hz (0-159 mph)	0 Hz = 0 mph	65 Hz = 30 mph	125 Hz = 55 mph
WACA	A/C WOT Relay Monitor	ON / OFF	OFF (with A/C on)	OFF	OFF

1996 E-Series 4.9L I6 MFI VIN Y (All) Outputs

PCM PID Acronym	Parameter Identification	PID Range	PID Value at Hot Idle	PID Value at 30	PID Value at 55
AIRB	AIR System Bypass	0v / VBAT	0.1	0.1	0.1
AIRD	AIR System Divert	0v / VBAT	VBAT	VBAT	VBAT
EGR VR	EGR VR Solenoid	0-100%	0	0-40	0-40
FP	Fuel Pump Control	ON / OFF	ON	ON	ON
FuelPW1	INJ Pulsewidth - Bank 1	0-999 ms	4.9-5.1	5.3-10.0	9.4-13.0
FuelPW2	INJ Pulsewidth - Bank 2	0-999 ms	4.9-5.1	5.3-10.0	9.4-13.0
HTR-11	HO2S-11 Heater	ON / OFF	ON	ON	ON
HTR-12	HO2S-12 Heater	ON / OFF	ON	ON	ON
HTR-21	HO2S-21 Heater	ON / OFF	ON	ON	ON
IAC	Idle Air Control	0-100%	35	44-50	59-65
INJ1	INJ 1 Pulsewidth	0-999 ms	4.9-5.1	5.3-10.0	9.4-13.0
INJ2	INJ 2 Pulsewidth	0-999 ms	4.9-5.1	5.3-10.0	9.4-13.0
INJ3	INJ 3 Pulsewidth	0-999 ms	4.9-5.1	5.3-10.0	9.4-13.0
INJ4	INJ 4 Pulsewidth	0-999 ms	4.9-5.1	5.3-10.0	9.4-13.0
INJ5	INJ 5 Pulsewidth	0-999 ms	4.9-5.1	5.3-10.0	9.4-13.0
INJ6	INJ 6 Pulsewidth	0-999 ms	4.9-5.1	5.3-10.0	9.4-13.0
LongFT1	Long Term FT - Bank 1	-20 to 20%	-5 to +5	-5 to +5	-5 to +5
LongFT2	Long Term FT - Bank 2	-20 to 20%	-5 to +5	-5 to +5	-5 to +5
MIL	MIL (lamp) Control	ON / OFF	OFF	OFF	OFF
SHTFT1	Short Term F/T - Bank 1	-10 to 10%	-5 to +5	-5 to +5	-5 to +5
SHTFT2	Short Term FT - Bank 2	-10 to 10%	-5 to +5	-5 to +5	-5 to +5
SPARK	Spark Advance	-90° to 90°	13-16	17-20	17-22
SPOUT	Spark Output Signal	0-9999 Hz	32-38	59-65	88-95
VMV	EVAP Solenoid	0-10	0-10	0-10	0-10
VREF	Vehicle Reference	0-5.1v	4.9-5.1	4.9-5.1	4.9-5.1
WAC	A/C WOT Relay Control	ON / OFF	ON (AC on at WOT)	OFF	OFF

1996 E-Series 4.9L I6 MFI VIN Y (E4OD) Inputs

PCM PID Acronym	Parameter Identification	PID Range	PID Value at Hot Idle	PID Value at 30	PID Value at 55
4x4L	4x4 Switch Signal	ON / OFF	ON (if on)	OFF	OFF
ACCS	A/C Switch Signal	ON / OFF	ON (if on)	OFF	OFF
BPP	Brake Position	ON / OFF	ON (if on)	OFF	OFF
DPFE	DPFE Sensor	0-5.1v	0.4-0.6	0.4-0.9	0.4-0.9
ECT	ECT Sensor	-40-304°F	160-200	160-200	160-200
FEPS	Flash EEPROM	0-5.1v	0.1	0.1	0.1
FPM	Fuel Pump Monitor	ON / OFF	ON	ON	ON
GEAR	Gear Position	1-2-3-4	1	3	4
HO2S-11	HO2S-11 (Bank 1)	0.1-1.1v	0.1-1.1	0.1-1.1	0.1-1.1
HO2S-12	HO2S-12 (Bank 1)	0.1-1.1v	0.1-1.1	0.1-1.1	0.1-1.1
HO2S-21	HO2S-21 (Bank 2)	0.1-1.1v	0.1-1.1	0.1-1.1	0.1-1.1
IAT	IAT Sensor	-40-304°F	50-120	50-120	50-120
IDM	IDM Signal	0-9999 Hz	35-42	58-69	71-82
KS1	Knock Sensor 1	0-5.1v	0	0	0
LOAD	Engine Load	0-100%	14-16	21-32	33-42
MAFV	MAF Sensor	0-5.1v	0.7-0.9	1.2-1.8	1.7-2.0
MISF	Misfire Detection	ON / OFF	OFF	OFF	OFF
OCTADJ	Octane Adjustment	CL/OPEN	CLOSED	CLOSED	CLOSED
PIP	PIP Sensor	0-9999 Hz	35-42	58-69	71-82
PNP	PNP Switch Signal	Neutral/DR	NEUTRAL	DRIVE	DRIVE
PTO	Power Takeoff Switch	ON / OFF	ON (if on)	OFF	OFF
RPM	Engine Speed	0-10K rpm	760-830	1200-1270	1510-1570
TCS	TCS Switch	ON / OFF	OFF	OFF	OFF
TFT	TFT Sensor Signal	-40-304°F	110-210	110-210	110-210
TPV	TP Sensor	0-5.1v	0.53-1.27	1.0-1.3	1.2-1.7
TRV/TR	TR Sensor	0v / 1.7v	0v	1.7	1.7
Vacuum	Engine Vacuum	0-30" Hg	19-20	18-20	8-15
VPWR	Vehicle Power	0-25.5v	12-14	12-14	12-14
VSS	Vehicle Speed (Hertz or MPH)	0-999 Hz (0-159 mph)	0 Hz = 0 mph	65 Hz = 30 mph	125 Hz = 55 mph
WACA	A/C WOT Relay Monitor	ON / OFF	ON (AC on at WOT)	OFF	OFF

1996 E-Series 4.9L I6 MFI VIN Y (E4OD) Outputs

PCM PID Acronym	Parameter Identification	PID Range	PID Value at Hot Idle	PID Value at 30	PID Value at 55
AIRB	AIR System Bypass	0v / VBAT	0.1	0.1	0.1
AIRD	AIR System Divert	0v / VBAT	VBAT	VBAT	VBAT
CCS	Coast Clutch Solenoid Control	ON / OFF	OFF	OFF	OFF
EGR VR	EGR VR Solenoid	0-100%	0	0-40	0-40
EPC	EPC Solenoid	0-300 psi	5	6	7
FP	Fuel Pump Control	ON / OFF	ON	ON	ON
FuelPW1	INJ Pulsewidth - Bank 1	0-999 ms	4.9-5.3	5.5-10.0	9.5-13.0
FuelPW2	INJ Pulsewidth - Bank 2	0-999 ms	4.9-5.3	5.5-10.0	9.5-13.0
HTR-11	HO2S-11 Heater	ON / OFF	ON	ON	ON
HTR-12	HO2S-12 Heater	ON / OFF	ON	ON	ON
HTR-21	HO2S-21 Heater	ON / OFF	ON	ON	ON
IAC	Idle Air Control	0-100%	30-34	30-55	58-61
INJ1	INJ 1 Pulsewidth	0-999 ms	4.9-5.3	5.5-10.0	9.5-13.0
INJ2	INJ 2 Pulsewidth	0-999 ms	4.9-5.3	5.5-10.0	9.5-13.0
INJ3	INJ 3 Pulsewidth	0-999 ms	4.9-5.3	5.5-10.0	9.5-13.0
INJ4	INJ 4 Pulsewidth	0-999 ms	4.9-5.3	5.5-10.0	9.5-13.0
INJ5	INJ 5 Pulsewidth	0-999 ms	4.9-5.3	5.5-10.0	9.5-13.0
INJ6	INJ 6 Pulsewidth	0-999 ms	4.9-5.3	5.5-10.0	9.5-13.0
LongFT1	Long Term FT - Bank 1	-20 to 20%	-5 to +5	-5 to +5	-5 to +5
LongFT2	Long Term FT - Bank 2	-20 to 20%	-5 to +5	-5 to +5	-5 to +5
MIL	MIL (lamp) Control	ON / OFF	OFF	OFF	OFF
SHTFT1	Short Term F/T - Bank 1	-10 to 10%	-5 to +5	-5 to +5	-5 to +5
SHTFT2	Short Term FT - Bank 2	-10 to 10%	-5 to +5	-5 to +5	-5 to +5
SPARK	Spark Advance	-90° to 90°	13-16	17-20	17-22
SPOUT	Spark Output Signal	0-9999 Hz	47-52	82-88	105-110
SS1	Shift Solenoid 1	ON / OFF	ON	OFF	OFF
SS2	Shift Solenoid 2	ON / OFF	OFF	ON	OFF
TCC	TCC Solenoid	ON / OFF	OFF	ON	ON
TCIL	TCIL (lamp) Control	ON / OFF	OFF	OFF	OFF
VMV	EVAP Solenoid	0-10	0-10	0-10	0-10
VREF	Vehicle Reference	0-5.1v	4.9-5.1	4.9-5.1	4.9-5.1
WAC	A/C WOT Relay Control	ON / OFF	ON (AC on at WOT)	OFF	OFF

1990-95 E-Series 5.0L V8 MFI VIN N (A/T, M/T, E4OD)

PCM PID Acronym	Parameter Identification	PID Range	PID Value at Hot Idle	PID Value at 30	PID Value at 55
ACCS	A/C Switch Signal	ON / OFF	ON (if on)	OFF	OFF
CANP	EVAP Purge Valve	ON / OFF	OFF	ON	ON
CPP	Clutch Pedal Switch	ON / OFF	ON (if in)	ON	ON
DPFE	DPFE Sensor ('95)	0-5.1v	0.4-0.6	0.4-0.9	0.4-0.9
ECT (°F)	ECT Sensor	-40-304°F	160-200	160-200	160-200
ECTV	ECT Sensor	0-5.1v	0.6	0.6	0.6
EPC	EPC Solenoid (E4OD)	0-300 psi	4	4	14
EVP	EGR Valve Position	0-5.1v	0.4	0.4-0	3.5-4.5
EVR	EGR VR Solenoid	0-100%	0	0-40	0-40
FuelPW1	INJ Pulsewidth - Bank 1	0-999 ms	4.4-5.0	6.4-7.8	9.8-12.0
GEAR	Gear Position	P-R-N-D	P-R-N-D	DRIVE 3	DRIVE 4
IAC	Idle Air Control	0-100%	30-34	43-48	58-61
IAT (°F)	IAT Sensor	-40-304°F	50-120	50-120	50-120
IATV	IAT Sensor	0-5.1v	1.5-3.5	1.5-3.5	1.5-3.5
LOOP	Loop Status	CL or OL	CLOSED	CLOSED	CLOSED
MLPV	MLP Switch (E4OD)	0-5.1v	0	5	5
MAP	MAP Sensor	0-999 Hz	103-105	112-120	122-140
HO2S-11	HO2S-11 (Bank 1)	0.1-1.1v	0.1-1.1	0.1-1.1	0.1-1.1
PNP	PNP Switch Signal	Neutral/DR	NEUTRAL	DRIVE	DRIVE
RPM	Engine Speed	0-10K rpm	680-780	1240-1340	1650-1750
SHTFT1	Short Term F/T - Bank 1	-10 to 10%	-5 to +5	-5 to +5	-5 to +5
SPARK	Spark Advance	-90° to 90°	14-20	28-36	30-40
TP	TP Sensor	0-5.1v	1.0	1.2-1.3	1.5-1.6
TCC	TCC Solenoid	ON / OFF	OFF	ON	ON
TOTV	TOT sensor (E4OD)	0-5.1v	2.10-2.40	2.10-2.40	2.10-2.40
TP	TP Sensor	0-5.1v	1.0	1.2-1.3	1.5-1.6
TP Mode	TP Sensor Mode	C/T or P/T	C/T	P/T	P/T
VPWR	Vehicle Power	0-25.5v	12-14	12-14	12-14
VREF	Vehicle Reference	0-5.1v	4.9-5.1	4.9-5.1	4.9-5.1
VSS	Vehicle Speed	0-159 mph	0	30	55

1996 E-Series 5.0L V8 MFI VIN N (All) Inputs

PCM PID Acronym	Parameter Identification	PID Range	PID Value at Hot Idle	PID Value at 30	PID Value at 55
4x4L	4x4 Switch Signal	ON / OFF	ON (if on)	OFF	OFF
ACCS	A/C Switch Signal	ON / OFF	ON (if on)	OFF	OFF
BPP	Brake Position	ON / OFF	ON (if on)	OFF	OFF
CPP	Clutch Pedal Switch	ON / OFF	ON (if in)	ON	ON
DPFE	DPFE Sensor	0-5.1v	0.4-0.6	0.4-0.9	0.4-0.9
ECT	ECT Sensor	-40-304°F	160-200	160-200	160-200
FEPS	Flash EEPROM	0-5.1v	0.1	0.1	0.1
FPM	Fuel Pump Monitor	ON / OFF	ON	ON	ON
IAT	IAT Sensor	-40-304°F	50-120	50-120	50-120
GEAR	Gear Position	1-2-3-4	1	3	4
HO2S-11	HO2S-11 (Bank 1)	0.1-1.1v	0.1-1.1	0.1-1.1	0.1-1.1
HO2S-12	HO2S-12 (Bank 1)	0.1-1.1v	0.1-1.1	0.1-1.1	0.1-1.1
HO2S-21	HO2S-21 (Bank 2)	0.1-1.1v	0.1-1.1	0.1-1.1	0.1-1.1
IDM	IDM Signal	0-9999 Hz	47-52	82-88	105-110
KS1	Knock Sensor 1	0-5.1v	0	0	0
LOAD	Engine Load	0-100%	14-16	19-25	26-35
MAFV	MAF Sensor	0-5.1v	0.7-0.9	1.1-1.6	1.7-2.4
MISF	Misfire Detection	ON / OFF	OFF	OFF	OFF
OCTADJ	Octane Adjustment	CL/OPEN	CLOSED	CLOSED	CLOSED
OSS	OSS Sensor	0-9999 Hz	0	122-133	235-250
PIP	PIP Sensor	0-9999 Hz	47-52	82-88	105-110
PNP	PNP Switch Signal	Neutral/DR	NEUTRAL	DRIVE	DRIVE
PTO	Power Takeoff Switch	ON / OFF	ON (if on)	OFF	OFF
RPM	Engine Speed	0-10K rpm	760-830	1200-1270	1590-1675
TCS	TCS Switch	ON / OFF	OFF	OFF	OFF
TFT	TFT Sensor Signal	-40-304°F	110-210	110-210	110-210
TPV	TP Sensor	0-5.1v	0.53-1.27	0.8-1.0	1.0-1.3
TRV/TR	TR Sensor	0v / 1.7v	0	1.7	1.7
Vacuum	Engine Vacuum	0-30" Hg	19-20	18-20	8-15
VPWR	Vehicle Power	0-25.5v	12-14	12-14	12-14
VSS	Vehicle Speed (Hertz or MPH)	0-999 Hz (0-159 mph)	0 Hz = 0 mph	65 Hz = 30 mph	125 Hz = 55 mph

1996 E-Series 5.0L V8 MFI VIN N (All) Outputs

PCM PID Acronym	Parameter Identification	PID Range	PID Value at Hot Idle	PID Value at 30	PID Value at 55
AIRB	AIR System Bypass	0v / VBAT	0.1	0.1	0.1
AIRD	AIR System Divert	0v / VBAT	VBAT	VBAT	VBAT
EGR VR	EGR VR Solenoid	0-100%	0%	0-40	35-40
EPC	EPC Solenoid	0-300 psi	4	5	5
FP	Fuel Pump Control	ON / OFF	ON	ON	ON
FuelPW1	INJ Pulsewidth - Bank 1	0-999 ms	3.2-3.8	4.1-6.9	6.5-12
FuelPW2	INJ Pulsewidth - Bank 2	0-999 ms	3.2-3.8	4.1-6.9	6.5-12
HTR-11	HO2S-11 Heater	ON / OFF	ON	ON	ON
HTR-12	HO2S-12 Heater	ON / OFF	ON	ON	ON
HTR-21	HO2S-21 Heater	ON / OFF	ON	ON	ON
IAC	Idle Air Control	0-100%	30-34	43-48	58-61
INJ1	INJ 1 Pulsewidth	0-999 ms	3.2-3.8	4.1-6.9	6.5-12
INJ2	INJ 2 Pulsewidth	0-999 ms	3.2-3.8	4.1-6.9	6.5-12
INJ3	INJ 3 Pulsewidth	0-999 ms	3.2-3.8	4.1-6.9	6.5-12
INJ4	INJ 4 Pulsewidth	0-999 ms	3.2-3.8	4.1-6.9	6.5-12
INJ5	INJ 5 Pulsewidth	0-999 ms	3.2-3.8	4.1-6.9	6.5-12
INJ6	INJ 6 Pulsewidth	0-999 ms	3.2-3.8	4.1-6.9	6.5-12
INJ7	INJ 7 Pulsewidth	0-999 ms	3.2-3.8	4.1-6.9	6.5-12
INJ8	INJ 8 Pulsewidth	0-999 ms	3.2-3.8	4.1-6.9	6.5-12
LongFT1	Long Term FT - Bank 1	-20 to 20%	-5 to +5	-5 to +5	-5 to +5
LongFT2	Long Term FT - Bank 2	-20 to 20%	-5 to +5	-5 to +5	-5 to +5
MIL	MIL (lamp) Control	ON / OFF	OFF	OFF	OFF
SHTFT1	Short Term F/T - Bank 1	-10 to 10%	-5 to +5	-5 to +5	-5 to +5
SHTFT2	Short Term FT - Bank 2	-10 to 10%	-5 to +5	-5 to +5	-5 to +5
SPARK	Spark Advance	-90° to 90°	12-17	35-40	28-37
SPOUT	Spark Output Signal	0-9999 Hz	47-52	82-88	105-110
SS1	Shift Solenoid 1	ON / OFF	ON	OFF	OFF
SS2	Shift Solenoid 2	ON / OFF	OFF	ON	OFF
TCC	TCC Solenoid	0-100%	0	0-40	90-100
TCIL	TCIL (lamp) Control	ON / OFF	OFF	OFF	OFF
VMV	EVAP Solenoid	0-10	0-10	0-10	0-10
VREF	Voltage Reference	0-5.1v	4.9-5.1	4.9-5.1	4.9-5.1

1996 E-Series 5.0L V8 MFI VIN N (E4OD) Inputs

PCM PID Acronym	Parameter Identification	PID Range	PID Value at Hot Idle	PID Value at 30	PID Value at 55
4x4L	4x4 Switch Signal	ON / OFF	ON (if on)	OFF	OFF
ACCS	A/C Switch Signal	ON / OFF	ON (if on)	OFF	OFF
BPP	Brake Position	ON / OFF	ON (if on)	OFF	OFF
DPFE	DPFE Sensor	0-5.1v	0.4-0.6	0.4-0.9	0.4-0.9
ECT	ECT Sensor	-40-304ºF	160-200	160-200	160-200
FEPS	Flash EEPROM	0-5.1v	0.1	0.1	0.1
FPM	Fuel Pump Monitor	ON / OFF	ON	ON	ON
IAT	IAT Sensor	-40-304ºF	50-120	50-120	50-120
GEAR	Gear Position	1-2-3-4	1	3	4
HO2S-11	HO2S-11 (Bank 1)	0.1-1.1v	0.1-1.1	0.1-1.1	0.1-1.1
HO2S-12	HO2S-12 (Bank 1)	0.1-1.1v	0.1-1.1	0.1-1.1	0.1-1.1
HO2S-21	HO2S-21 (Bank 2)	0.1-1.1v	0.1-1.1	0.1-1.1	0.1-1.1
HO2S-22	HO2S-22 (Bank 2)	0.1-1.1v	0.1-1.1	0.1-1.1	0.1-1.1
IDM	IDM Signal	0-9999 Hz	47-52	82-88	105-110
KS1	Knock Sensor 1	0-5.1v	0	0	0
LOAD	Engine Load	0-100%	14-16	19-25	26-35
MAFV	MAF Sensor	0-5.1v	0.7-0.9	1.1-1.6	1.7-2.4
MISF	Misfire Detection	ON / OFF	OFF	OFF	OFF
OCTADJ	Octane Adjustment	CL/OPEN	CLOSED	CLOSED	CLOSED
PIP	PIP Sensor	0-9999 Hz	47-52	82-88	105-110
PNP	PNP Switch Signal	Neutral/DR	NEUTRAL	DRIVE	DRIVE
PTO	Power Takeoff Switch	ON / OFF	ON (if on)	OFF	OFF
RPM	Engine Speed	0-10K rpm	760-830	1200-1270	1600-1650
TCS	TCS Switch	ON / OFF	OFF	OFF	OFF
TFT	TFT Sensor Signal	-40-304ºF	110-210	110-210	110-210
TPV	TP Sensor	0-5.1v	0.53-1.27	0.8-1.0	1.0-1.3
TRV/TR	TR Sensor	0v / 1.7v	0v	1.7	1.7
Vacuum	Engine Vacuum	0-30" Hg	19-20	18-20	8-15
VPWR	Vehicle Power	0-25.5v	12-14	12-14	12-14
VSS	Vehicle Speed (Hertz or MPH)	0-999 Hz (0-159 mph)	0 Hz = 0 mph	65 Hz = 30 mph	125 Hz = 55 mph

1996 E-Series 5.0L V8 MFI VIN N (E4OD) Outputs

PCM PID Acronym	Parameter Identification	PID Range	PID Value at Hot Idle	PID Value at 30	PID Value at 55
AIRB	AIR System Bypass	0v / VBAT	0.1	0.1	0.1
AIRD	AIR System Divert	0v / VBAT	VBAT	VBAT	VBAT
CCS	Coast Clutch Solenoid Control	0v / VBAT	VBAT	VBAT	VBAT
EGR VR	EGR VR Solenoid	0-100%	0	0-40	35-40
EVAPDC	EVAP Solenoid	0-100%	0-100	0-100	0-100
EPC	EPC Solenoid	0-300 psi	4	5	5
FP	Fuel Pump Control	ON / OFF	ON	ON	ON
FuelPW1	INJ Pulsewidth - Bank 1	0-999 ms	3.2-3.8	4.1-6.9	6.5-12
FuelPW2	INJ Pulsewidth - Bank 2	0-999 ms	3.2-3.8	4.1-6.9	6.5-12
HTR-11	HO2S-11 Heater	ON / OFF	ON	ON	ON
HTR-12	HO2S-12 Heater	ON / OFF	ON	ON	ON
HTR-21	HO2S-21 Heater	ON / OFF	ON	ON	ON
IAC	Idle Air Control	0-100%	30-34	43-48	58-61
INJ1	INJ 1 Pulsewidth	0-999 ms	3.2-3.8	4.1-6.9	6.5-12
INJ2	INJ 2 Pulsewidth	0-999 ms	3.2-3.8	4.1-6.9	6.5-12
INJ3	INJ 3 Pulsewidth	0-999 ms	3.2-3.8	4.1-6.9	6.5-12
INJ4	INJ 4 Pulsewidth	0-999 ms	3.2-3.8	4.1-6.9	6.5-12
INJ5	INJ 5 Pulsewidth	0-999 ms	3.2-3.8	4.1-6.9	6.5-12
INJ6	INJ 6 Pulsewidth	0-999 ms	3.2-3.8	4.1-6.9	6.5-12
INJ7	INJ 7 Pulsewidth	0-999 ms	3.2-3.8	4.1-6.9	6.5-12
INJ8	INJ 8 Pulsewidth	0-999 ms	3.2-3.8	4.1-6.9	6.5-12
LongFT1	Long Term FT - Bank 1	-20 to 20%	-5 to +5	-5 to +5	-5 to +5
LongFT2	Long Term FT - Bank 2	-20 to 20%	-5 to +5	-5 to +5	-5 to +5
MIL	MIL (lamp) Control	ON / OFF	OFF	OFF	OFF
SHTFT1	Short Term F/T - Bank 1	-10 to 10%	-5 to +5	-5 to +5	-5 to +5
SHTFT2	Short Term FT - Bank 2	-10 to 10%	-5 to +5	-5 to +5	-5 to +5
SPARK	Spark Advance	-90° to 90°	12-17	35-40	28-37
SPOUT	Spark Output Signal	0-9999 Hz	47-52	82-88	105-110
SS1	Shift Solenoid 1	ON / OFF	ON	OFF	OFF
SS2	Shift Solenoid 2	ON / OFF	OFF	ON	OFF
TCC	TCC Solenoid	ON / OFF	OFF	ON	ON
TCIL	TCIL (lamp) Control	ON / OFF	OFF	OFF	OFF
VREF	Vehicle Reference	0-5.1v	4.9-5.1	4.9-5.1	4.9-5.1

1997-2005 E-Series 5.4L V8 VIN L (E4OD) Inputs

PCM PID Acronym	Parameter Identification	PID Range	PID Value at Hot Idle	PID Value at 30	PID Value at 55
AIR-M	Air System Monitor	ON / OFF	OFF	OFF	OFF
ACCS	A/C Switch Signal	OFF	ON	OFF	OFF
BPP	Brake Position	ON / OFF	ON (if on)	OFF	OFF
CHT	CHT Sensor	-40-500°F	194	194	194
CID	CMP Sensor	0-999 Hz	6	11	11
CKP	CKP Sensor	0-9999 Hz	411	800-840	1000
DPFE	DPFE Sensor	0-5.1v	0.2-1.3	0.2-4.5	0.2-4.5
ECT	ECT Sensor	-40-304°F	160-200	160-200	160-200
FEPS	Flash EEPROM	0-5.1v	0.1	0.1	0.1
FLI	Fuel Level Input	1.7 / 50%	1.7 / 50%	1.7 / 50%	1.7 / 50%
FPM	Fuel Pump Monitor	ON / OFF	ON	ON	ON
FTP	Fuel Tank Pressure	2.6v / 0 psi	2.6v / 0 psi	2.6v / 0 psi	2.6v / 0 psi
GEAR	Gear Position	1-2-3-4	1	3	4
HO2S-11	HO2S-11 (Bank 1)	0.1-1.1v	0.1-1.1	0.1-1.1	0.1-1.1
HO2S-12	HO2S-12 (Bank 1)	0.1-1.1v	0.1-1.1	0.1-1.1	0.1-1.1
HO2S-21	HO2S-21 (Bank 2)	0.1-1.1v	0.1-1.1	0.1-1.1	0.1-1.1
HO2S-22	HO2S-22 (Bank 2)	0.1-1.1v	0.1-1.1	0.1-1.1	0.1-1.1
IAT	IAT Sensor	-40-304°F	50-120	50-120	50-120
KS1	Knock Sensor 1	0-5.1v	0	0	0
LOAD	Engine Load	0-100%	14-16	19-25	26-35
MAFV	MAF Sensor	0-5.1v	0.7-0.9	1.1-1.6	1.7-2.4
MISF	Misfire Detection	ON / OFF	OFF	OFF	OFF
OCTADJ	Octane Adjustment	CL/OPEN	CLOSED	CLOSED	CLOSED
PNP	PNP Switch Signal	Neutral/DR	NEUTRAL	DRIVE	DRIVE
RPM	Engine Speed	0-10K rpm	760-830	1200-1270	1600-1650
TCS	TCS Switch	ON / OFF	OFF	OFF	OFF
TFT	TFT Sensor Signal	-40-304°F	110-210	110-210	110-210
TPV	TP Sensor	0-5.1v	0.53-1.27	0.8-1.0	1.0-1.3
TR1	TR1 Sensor	0v / 11.5v	0v	11.5	11.5
TR2	TR2 Sensor	0v / 11.5v	0v	11.5	11.5
TR3/TR	TR3A Sensor	0v / 1.7v	0v	1.7	1.7
TR4	TR4 Sensor	0v / 11.5v	0v	11.5	11.5
Vacuum	Engine Vacuum	0-30" Hg	19-20	18-20	8-15
VPWR	Vehicle Power	0-25.5v	12-14	12-14	12-14
VSS	Vehicle Speed (Hertz or MPH)	0-999 Hz (0-159 mph)	0 Hz = 0 mph	65 Hz = 30 mph	125 Hz = 55 mph

1997-2005 E-Series 5.4L V8 VIN L (E4OD) Outputs

PCM PID Acronym	Parameter Identification	PID Range	PID Value at Hot Idle	PID Value at 30	PID Value at 55
AIRD	AIR System Divert	0v / VBAT	VBAT	VBAT	VBAT
AIR	AIR System Monitor	ON / OFF	OFF	OFF	OFF
CCS	Coast Clutch Solenoid Control	0v / VBAT	VBAT	VBAT	VBAT
CD1-CD8	COP Driver 1-8	0-45	5	6	8
CHTIL	CHT (lamp) Control	ON / OFF	OFF	OFF	OFF
CTO	Clean Tachometer signal	0-9999 Hz	46	90	115
EGR VR	EGR VR Solenoid	0-100%	0	0-40	35-40
EVAP CP	EVAP Purge Valve	0-100%	0-100	0-100	0-100
EVAP CV	EVAP CV Valve	0-100%	0-100	0-100	0-100
EPC	EPC Solenoid	0-300 psi	4	4	5
FP	Fuel Pump Control	ON / OFF	ON	ON	ON
FuelPW1	INJ Pulsewidth - Bank 1	0-999 ms	3.2-3.8	4.1-6.9	6.5-12
FuelPW2	INJ Pulsewidth - Bank 2	0-999 ms	3.2-3.8	4.1-6.9	6.5-12
HTR-11	HO2S-11 Heater	ON / OFF	ON	ON	ON
HTR-12	HO2S-12 Heater	ON / OFF	ON	ON	ON
HTR-21	HO2S-21 Heater	ON / OFF	ON	ON	ON
HTR-22	HO2S-22 Heater	ON / OFF	ON	ON	ON
IAC	Idle Air Control	0-100%	30-34	43-48	58-61
INJ1	INJ 1 Pulsewidth	0-999 ms	3.2-3.8	4.1-6.9	6.5-12
INJ2	INJ 2 Pulsewidth	0-999 ms	3.2-3.8	4.1-6.9	6.5-12
INJ3	INJ 3 Pulsewidth	0-999 ms	3.2-3.8	4.1-6.9	6.5-12
INJ4	INJ 4 Pulsewidth	0-999 ms	3.2-3.8	4.1-6.9	6.5-12
INJ5	INJ 5 Pulsewidth	0-999 ms	3.2-3.8	4.1-6.9	6.5-12
INJ6	INJ 6 Pulsewidth	0-999 ms	3.2-3.8	4.1-6.9	6.5-12
INJ7	INJ 7 Pulsewidth	0-999 ms	3.2-3.8	4.1-6.9	6.5-12
INJ8	INJ 8 Pulsewidth	0-999 ms	3.2-3.8	4.1-6.9	6.5-12
LongFT1	Long Term FT (B1)	-20 to 20%	-5 to +5	-5 to +5	-5 to +5
LongFT2	Long Term FT (B2)	-20 to 20%	-5 to +5	-5 to +5	-5 to +5
MIL	MIL (lamp) Control	ON / OFF	OFF	OFF	OFF
SHTFT1	Short Term F/T - Bank 1	-10 to 10%	-5 to +5	-5 to +5	-5 to +5
SHTFT2	Short Term FT - Bank 2	-10 to 10%	-5 to +5	-5 to +5	-5 to +5
SS1	Shift Solenoid 1	ON / OFF	ON	OFF	OFF
SS2	Shift Solenoid 2	ON / OFF	OFF	ON	OFF
SPARK	Spark Advance	-90° to 90°	12-17	35-40	28-37
TCC	TCC Solenoid	ON / OFF	OFF	ON	ON
VREF	Vehicle Reference	0-5.1v	4.9-5.1	4.9-5.1	4.9-5.1

1997-2005 E-Series 5.4L V8 CNG VIN M (E4OD) Inputs

PCM PID Acronym	Parameter Identification	PID Range	PID Value at Hot Idle	PID Value at 30	PID Value at 55
ACCS	A/C Switch Signal	ON / OFF	ON (if on)	OFF	OFF
BPP	Brake Position	ON / OFF	ON (if on)	OFF	OFF
CHT	CHT Sensor	-40-500°F	194	194	194
CID	CMP Sensor	0-999 Hz	6-8	9-12	15-17
CKP	CKP Sensor	0-9999 Hz	440-490	810-870	1088-1089
EFTA	EFT Sensor	-40-275F	50-120	50-120	50-120
FEPS	Flash EEPROM	0-5.1v	0.1	0.1	0.1
FPM	Fuel Pump Monitor	ON / OFF	ON	ON	ON
FSV-M	Fuel Solenoid Valve	ON / OFF	ON	ON	ON
IAT	IAT Sensor	-40-304°F	50-120	50-120	50-120
FRP	FRP Sensor	0-300 psi	90-100	90-100	90-100
GEAR	Gear Position	1-2-3-4	1	3	4
HO2S-11	HO2S-11 (Bank 1)	0.1-1.1v	0.1-1.1	0.1-1.1	0.1-1.1
HO2S-21	HO2S-21 (Bank 2)	0.1-1.1v	0.1-1.1	0.1-1.1	0.1-1.1
LOAD	Engine Load	0-100%	13-19	26-33	38-46
MAFV	MAF Sensor	0-5.1v	0.7-0.9	0.9-1.7	1.2-2.4
MISF	Misfire Detection	ON / OFF	OFF	OFF	OFF
PNP	PNP Switch Signal	Neutral/DR	NEUTRAL	DRIVE	DRIVE
PTO	Power Takeoff Switch	OFF	ON (if on)	OFF	OFF
RPM	Engine Speed	0-10K rpm	900-930	1470-1490	1840-1860
TCS	TCS Switch	ON / OFF	OFF	OFF	OFF
TPV	TP Sensor	0-5.1v	0.53-1.27	0.8-1.2	0.9-1.6v
TR1	TR1 Sensor	0v / 11.5v	0v	11.5	11.5
TR2	TR2 Sensor	0v / 11.5v	0v	11.5	11.5
TR3/TR	TR3A Sensor	0v / 1.7v	0v	1.7	1.7
TR4	TR4 Sensor	0v / 11.5v	0v	11.5	11.5
Vacuum	Engine Vacuum	0-30" Hg	19-20	18-20	8-15
VPWR	Vehicle Power	0-25.5v	12-14	12-14	12-14
VSS	Vehicle Speed (Hertz or MPH)	0-999 Hz (0-159 mph)	0 Hz = 0 mph	65 Hz = 30 mph	125 Hz = 55 mph

1997-2005 E-Series 5.4L V8 CNG VIN M (E4OD) Outputs

PCM PID Acronym	Parameter Identification	PID Range	PID Value at Hot Idle	PID Value at 30	PID Value at 55
CCS	Coast Clutch Solenoid Control	ON / OFF	OFF	OFF	OFF
CD1-CD8	COP Driver 1-8	0-45° dwell	5	6	8
CHTIL	CHT (lamp) Control	ON / OFF	OFF	OFF	OFF
CTO	Clean Tachometer signal	0-9999 Hz	50-57	90-100	115-125
EPC	EPC Solenoid	0-300 psi	5	6	11
FP	Fuel Pump Control	ON / OFF	ON	ON	ON
FSV	Fuel Solenoid Valve	ON / OFF	ON	ON	ON
FuelPW1	INJ Pulsewidth - Bank 1	0-999 ms	3.9-5.5	4.7-12.0	4.7-12.0
FuelPW2	INJ Pulsewidth - Bank 2	0-999 ms	3.9-5.5	4.7-12.0	4.7-12.0
HTR-11	HO2S-11 Heater	ON / OFF	ON	ON	ON
HTR-21	HO2S-21 Heater	ON / OFF	ON	ON	ON
IAC	Idle Air Control	0-100%	30-34	43-48	58-61
IMRC	Intake Manifold Runner Control	0-100%	0	0	0
INJ1	INJ 1 Pulsewidth	0-999 ms	3.9-4.6	4.7-12.0	4.7-12.0
INJ2	INJ 2 Pulsewidth	0-999 ms	3.9-4.6	4.7-12.0	4.7-12.0
INJ3	INJ 3 Pulsewidth	0-999 ms	3.9-4.6	4.7-12.0	4.7-12.0
INJ4	INJ 4 Pulsewidth	0-999 ms	3.9-4.6	4.7-12.0	4.7-12.0
INJ5	INJ 5 Pulsewidth	0-999 ms	3.9-4.6	4.7-12.0	4.7-12.0
INJ6	INJ 6 Pulsewidth	0-999 ms	3.9-4.6	4.7-12.0	4.7-12.0
INJ7	INJ 7 Pulsewidth	0-999 ms	3.9-4.6	4.7-12.0	4.7-12.0
INJ8	INJ 8 Pulsewidth	0-999 ms	3.9-4.6	4.7-12.0	4.7-12.0
LongFT1	Long Term FT - Bank 1	-20 to 20%	-5 to +5	-5 to +5	-5 to +5
LongFT2	Long Term FT - Bank 2	-20 to 20%	-5 to +5	-5 to +5	-5 to +5
MIL	MIL (lamp) Control	ON / OFF	OFF	OFF	OFF
SHTFT1	Short Term F/T - Bank 1	-10 to 10%	-5 to +5	-5 to +5	-5 to +5
SHTFT2	Short Term FT - Bank 2	-10 to 10%	-5 to +5	-5 to +5	-5 to +5
SPARK	Spark Advance	-90° to 90°	14-20	20-35	20-30
SS1	Shift Solenoid 1	ON / OFF	ON	OFF	OFF
SS2	Shift Solenoid 2	ON / OFF	OFF	ON	OFF
TCC	TCC Solenoid	0-100%	0	90-100	90-100
TCIL	TCIL (lamp) Control	ON / OFF	OFF	OFF	OFF
VREF	Vehicle Reference	5v	5v	5v	5v

1998-99 E-Series 5.4L NGV VIN Z (A/T, M/T, E4OD) Inputs

PCM PID Acronym	Parameter Identification	PID Range	PID Value at Hot Idle	PID Value at 30	PID Value at 55
ACCS	A/C Switch Signal	ON / OFF	ON (if on)	OFF	OFF
BPP	Brake Position	ON / OFF	ON (if on)	OFF	OFF
CHT	CHT Sensor	-40-500°F	194	194	194
CID	CMP Sensor	0-999 Hz	6-7	9-12	15-17.5
CKP	CKP Sensor	0-9999 Hz	440-490	810-855	1088-1089
CPP	Clutch Pedal Switch	ON / OFF	ON (if in)	ON	ON
EFTA	EFT Sensor	-40-275F	50-120	50-120	50-120
FEPS	Flash EEPROM	0-5.1v	0.1	0.1	0.1
FPM	Fuel Pump Monitor	ON / OFF	ON	ON	ON
FRP (psi)	FRP Sensor	0-300 psi	90-100	90-100	90-100
FRP (V)	FRP Sensor	0-5.1v	2.7-3.7	2.7-3.7	2.7-3.7
FSV-M	Fuel Solenoid Valve	ON / OFF	ON	ON	ON
GEAR	Gear Position	1-2-3-4	1	3	4
HO2S-11	HO2S-11 (Bank 1)	0.1-1.1v	0.1-1.1v	0.1-1.1v	0.1-1.1v
IAT	IAT Sensor	-40-304°F	50-120	50-120	50-120
LOAD	Engine Load	0-100%	13-19	26-33	38-46
MAFV	MAF Sensor	0-5.1v	0.7-0.9	0.9-1.7	1.2-2.4
MISF	Misfire Detection	ON / OFF	OFF	OFF	OFF
PNP	PNP Switch Signal	Neutral/DR	NEUTRAL	DRIVE	DRIVE
PTO	Power Takeoff Switch	OFF	ON (if on)	OFF	OFF
RPM	Engine Speed	0-10K rpm	900-930	1470-1490	1840-1860
TCS	TCS Switch	ON / OFF	OFF	OFF	OFF
TFT	TFT Sensor Signal	-40-304°F	110-210	110-210	110-210
TPV	TP Sensor	0-5.1v	0.53-1.27	0.8-1.0	0.9-1.2
TR1	TR1 Sensor	0v / 11.5v	0	11.5	11.5
TR2	TR2 Sensor	0v / 11.5v	0	11.5	11.5
TR4	TR4 Sensor	0v / 11.5v	0	11.5	11.5
TRV/TR	TR Sensor	0v / 1.7v	0	1.7	1.7
Vacuum	Engine Vacuum	0-30" Hg	19-20	18-20	8-15
VPWR	Vehicle Power	0-25.5v	12-14	12-14	12-14
VSS	Vehicle Speed (Hertz or MPH)	0-999 Hz (0-159 mph)	0 Hz = 0 mph	65 Hz = 30 mph	125 Hz = 55 mph

1998-99 E-Series 5.4L NGV VIN Z (A/T, M/T, E4OD) Outputs

PCM PID Acronym	Parameter Identification	PID Range	PID Value at Hot Idle	PID Value at 30	PID Value at 55
CCS	Coast Clutch Solenoid Control	ON / OFF	OFF	OFF	OFF
CD1-CD8	COP Driver 1-8	0-45° dwell	5	6	8
CHTIL	CHT (lamp) Control	ON / OFF	OFF	OFF	OFF
CTO	Clean Tachometer signal	0-9999 Hz	50-57	90-100	115-125
EPC	EPC Solenoid (E4OD)	0-300 psi	5	6	11
FP	Fuel Pump Control	ON / OFF	ON	ON	ON
FSV	Fuel Solenoid Valve	ON / OFF	ON	ON	ON
FuelPW1	INJ Pulsewidth - Bank 1	0-999 ms	3.9-5.5	4.7-12.0	4.7-12.0
FuelPW2	INJ Pulsewidth - Bank 2	0-999 ms	3.9-5.5	4.7-12.0	4.7-12.0
HTR-11	HO2S-11 Heater	ON / OFF	ON	ON	ON
HTR-21	HO2S-21 Heater	ON / OFF	ON	ON	ON
IAC	Idle Air Control	0-100%	30-34	43-48	58-61
IMTV	Intake Manifold Tuning Valve	0-100%	0	0	0
INJ1	INJ 1 Pulsewidth	0-999 ms	3.9-4.6	4.7-12.0	4.7-12.0
INJ2	INJ 2 Pulsewidth	0-999 ms	3.9-4.6	4.7-12.0	4.7-12.0
INJ3	INJ 3 Pulsewidth	0-999 ms	3.9-4.6	4.7-12.0	4.7-12.0
INJ4	INJ 4 Pulsewidth	0-999 ms	3.9-4.6	4.7-12.0	4.7-12.0
INJ5	INJ 5 Pulsewidth	0-999 ms	3.9-4.6	4.7-12.0	4.7-12.0
INJ6	INJ 6 Pulsewidth	0-999 ms	3.9-4.6	4.7-12.0	4.7-12.0
INJ7	INJ 7 Pulsewidth	0-999 ms	3.9-4.6	4.7-12.0	4.7-12.0
INJ8	INJ 8 Pulsewidth	0-999 ms	3.9-4.6	4.7-12.0	4.7-12.0
LongFT1	Long Term FT - Bank 1	-20 to 20%	-5 to +5	-5 to +5	-5 to +5
LongFT2	Long Term FT - Bank 2	-20 to 20%	-5 to +5	-5 to +5	-5 to +5
MIL	MIL (lamp) Control	ON / OFF	OFF	OFF	OFF
SHTFT1	Short Term F/T - Bank 1	-10 to 10%	-5 to +5	-5 to +5	-5 to +5
SHTFT2	Short Term FT - Bank 2	-10 to 10%	-5 to +5	-5 to +5	-5 to +5
SPARK	Spark Advance	-90° to 90°	15-25	20-35	20-30
SS1	Shift Solenoid 1	ON / OFF	ON	OFF	ON
SS2	Shift Solenoid 2	ON / OFF	OFF	ON	OFF
TCC	TCC Solenoid	0-100%	0	50-100	90-100
TCIL	TCIL (lamp) Control	ON / OFF	OFF	OFF	OFF
VPWR	Vehicle Power	0-25.5v	12-14	12-14	12-14
VREF	Vehicle Reference	5v	4.9-5.1	4.9-5.1	4.9-5.1

1990-95 E-Series 5.8L V8 MFI VIN H, R (A/T, M/T, E4OD)

PCM PID Acronym	Parameter Identification	PID Range	PID Value at Hot Idle	PID Value at 30	PID Value at 55
ACCS	A/C Switch Signal	ON / OFF	ON (if on)	OFF	OFF
CANP	EVAP Purge Valve	ON / OFF	OFF	ON	ON
CPP	Clutch Pedal Switch	ON / OFF	ON (if in)	OFF	OFF
DPFE	DPFE Sensor (Cal.)	0-5.1v	0.4-0.6	0.4-0.9	0.8-1.1
ECT (°F)	ECT Sensor	-40-304°F	160-200	160-200	160-200
ECTV	ECT Sensor	0-5.1v	0.6	0.6	0.6
EPC	EPC Solenoid (E4OD)	0-300 psi	4	4	14
EVP	EGR Valve Position	0-5.1v	0.3	0.4	0.6-3.0
EVR	EGR VR Solenoid	0-100%	0	0-40	0-40
FuelPW1	INJ Pulsewidth - Bank 1	0-999 ms	5.0-5.8	6.0-6.8	6.4-7.0
GEAR	Gear Position	P-R-N-D	P-R-N-D	DRIVE 3	DRIVE 4
HO2S-11	HO2S-11 (Bank 1)	0.1-1.1v	0.1-1.1	0.1-1.1	0.1-1.1
IAC	Idle Air Control	0-100%	30-34	28-32	63-70
IAT (°F)	IAT Sensor	-40-304°F	50-120	50-120	50-120
IATV	IAT Sensor	0-5.1v	1.5-3.5	1.5-3.5	1.5-3.5
LOOP	Loop Status	CL or OL	CLOSED	CLOSED	CLOSED
MLPV	MLP Switch (E4OD)	0-5.1v	0	5	5
MAP	MAP Sensor	0-999 Hz	108-112	110-120	130-140
PNP	PNP Switch Signal	Neutral/DR	NEUTRAL	DRIVE	DRIVE
RPM	Engine Speed	0-10K rpm	750-850	1250-1350	1600-1700
SHTFT1	Short Term F/T - Bank 1	-10 to 10%	-5 to +5	-5 to +5	-5 to +5
SPARK	Spark Advance	-90° to 90°	14-18	30-36	38-44
TP	TP Sensor	0-5.1v	0.9	1.0-1.1	1.1-1.2
TCC	TCC Solenoid	ON / OFF	OFF	ON	ON
TOTV	TOT sensor (E4OD)	0-5.1v	2.10-2.40	2.10-2.40	2.10-2.40
TP	TP Sensor	0-5.1v	1.0	1.2-1.3	1.5-1.6
TP Mode	TP Sensor Mode	C/T or P/T	C/T	P/T	P/T
VPWR	Vehicle Power	0-25.5v	12-14	12-14	12-14
VREF	Vehicle Reference	0-5.1v	4.9-5.1	4.9-5.1	4.9-5.1
VSS	Vehicle Speed	0-159 mph	0	30	55

1996 E-Series 5.8L V8 MFI VIN H (E4OD) Inputs

PCM PID Acronym	Parameter Identification	PID Range	PID Value at Hot Idle	PID Value at 30	PID Value at 55
4x4L	4x4 Switch Signal	ON / OFF	ON (if on)	OFF	OFF
ACCS	A/C Switch Signal	ON / OFF	ON (if on)	OFF	OFF
BPP	Brake Position	ON / OFF	ON (if on)	OFF	OFF
DPFE	DPFE Sensor	0-5.1v	0.4-0.6	0.4-0.9	0.8-1.1
ECT	ECT Sensor	-40-304°F	160-200	160-200	160-200
FEPS	Flash EEPROM	0-5.1v	0.1	0.1	0.1
FLI	Fuel Level Input	0-100%	1.7 / 50%	1.7 / 50%	1.7 / 50%
FPM	Fuel Pump Monitor	ON / OFF	ON	ON	ON
GEAR	Gear Position	1-2-3-4	1	3	4
HO2S-11	HO2S-11 (Bank 1)	0.1-1.1v	0.1-1.1	0.1-1.1	0.1-1.1
HO2S-12	HO2S-12 (Bank 1)	0.1-1.1v	0.1-1.1	0.1-1.1	0.1-1.1
HO2S-21	HO2S-21 (Bank 1)	0.1-1.1v	0.1-1.1	0.1-1.1	0.1-1.1
IAT	IAT Sensor	-40-304°F	50-120	50-120	50-120
IDM	IDM Signal	0-9999 Hz	42-50	82-88	105-120
LOAD	Engine Load	0-100%	14-16	20-25	24-35
MAFV	MAF Sensor	0-5.1v	0.7-0.9	1.2-1.7	1.6-2.4
MISF	Misfire Detection	ON / OFF	OFF	OFF	OFF
OCTADJ	Octane Adjustment	CL/OPEN	CLOSED	CLOSED	CLOSED
PIP	PIP Sensor	0-9999 Hz	42-50	82-88	105-120
PNP	PNP Switch Signal	ON / OFF	ON	OFF	OFF
PTO	Power Takeoff Switch	OFF	ON (if on)	OFF	OFF
RPM	Engine Speed	0-10K rpm	675-715	1250-1390	1590-1700
TCS	TCS Switch	ON / OFF	OFF	OFF	OFF
TFT	TFT Sensor Signal	-40-304°F	110-210	110-210	110-210
TPV	TP Sensor	0-5.1v	0.53-1.27	0.8-1.0	0.9-1.2
TRV/TR	TR Sensor	0v / 1.7v	0v	1.7	1.7
Vacuum	Engine Vacuum	0-30" Hg	19-20	18-20	8-15
VPWR	Vehicle Power	0-25.5v	12-14	12-14	12-14
VSS	Vehicle Speed (Hertz or MPH)	0-999 Hz (0-159 mph)	0 Hz = 0 mph	65 Hz = 30 mph	125 Hz = 55 mph

1996 E-Series 5.8L V8 MFI VIN H (E4OD) Outputs

PCM PID Acronym	Parameter Identification	PID Range	PID Value at Hot Idle	PID Value at 30	PID Value at 55
CCS	Coast Clutch Solenoid Control	ON / OFF	OFF	OFF	OFF
EGR VR	EGR VR Solenoid	0-100%	0	0-40	35-45
EVAP CP	EVAP Purge Valve	0-100%	0-100	0-100	0-100
EPC	EPC Solenoid	0-300 psi	5	5	15
FP	Fuel Pump Control	ON / OFF	ON	ON	ON
FuelPW1	INJ Pulsewidth - Bank 1	0-999 ms	4.2-4.7	4.4-7.0	7.4-12.0
FuelPW2	INJ Pulsewidth - Bank 2	0-999 ms	4.2-4.7	4.4-7.0	7.4-12.0
HTR-11	HO2S-11 Heater	ON / OFF	ON	ON	ON
HTR-12	HO2S-12 Heater	ON / OFF	ON	ON	ON
HTR-21	HO2S-21 Heater	ON / OFF	ON	ON	ON
HTR-22	HO2S-22 Heater	ON / OFF	ON	ON	ON
IAC	Idle Air Control	0-100%	30-34	28-32	63-70
INJ1	INJ 1 Pulsewidth	0-999 ms	4.2-4.7	4.4-7.0	7.4-12.0
INJ2	INJ 2 Pulsewidth	0-999 ms	4.2-4.7	4.4-7.0	7.4-12.0
INJ3	INJ 3 Pulsewidth	0-999 ms	4.2-4.7	4.4-7.0	7.4-12.0
INJ4	INJ 4 Pulsewidth	0-999 ms	4.2-4.7	4.4-7.0	7.4-12.0
INJ5	INJ 5 Pulsewidth	0-999 ms	4.2-4.7	4.4-7.0	7.4-12.0
INJ6	INJ 6 Pulsewidth	0-999 ms	4.2-4.7	4.4-7.0	7.4-12.0
INJ7	INJ 7 Pulsewidth	0-999 ms	4.2-4.7	4.4-7.0	7.4-12.0
INJ8	INJ 8 Pulsewidth	0-999 ms	4.2-4.7	4.4-7.0	7.4-12.0
LongFT1	Long Term FT - Bank 1	-20 to 20%	-5 to +5	-5 to +5	-5 to +5
LongFT2	Long Term FT - Bank 2	-20 to 20%	-5 to +5	-5 to +5	-5 to +5
MIL	MIL (lamp) Control	ON / OFF	OFF	OFF	OFF
SHTFT1	Short Term F/T - Bank 1	-10 to 10%	-5 to +5	-5 to +5	-5 to +5
SHTFT2	Short Term FT - Bank 2	-10 to 10%	-5 to +5	-5 to +5	-5 to +5
SPARK	Spark Advance	-90° to 90°	15-18	23-29	26-32
SPOUT	Spark Output Signal	0-9999 Hz	42-50	82-88	105-120
SS1	Shift Solenoid 1	ON / OFF	ON	OFF	OFF
SS2	Shift Solenoid 2	ON / OFF	OFF	ON	OFF
TCC	TCC Solenoid	0-100%	0	90-100	90-100
TCIL	TCIL (lamp) Control	ON / OFF	OFF	OFF	OFF
VREF	Vehicle Reference	0-5.1v	4.9-5.1	4.9-5.1	4.9-5.1

1997-2001 E-Series 6.8L V10 VIN S (E4OD) Inputs

PCM PID Acronym	Parameter Identification	PID Range	PID Value at Hot Idle	PID Value at 30	PID Value at 55
4x4L	4x4 Switch Signal	ON / OFF	ON (if on)	OFF	OFF
ACCS	A/C Switch Signal	ON / OFF	ON (if on)	OFF	OFF
BPP	Brake Position	ON / OFF	ON (if on)	OFF	OFF
CHT	CHT Sensor	-40-500°F	194	194	194
CID	CMP Sensor	0-999 Hz	7-10	10-13	15-17
CKP	CKP Sensor	0-9999 Hz	500-525	750-940	1195-1400
DPFE	DPFE Sensor	0-5.1v	0.2-1.3	0.2-4.5	0.2-4.5
FEPS	Flash EEPROM	0-5.1v	0.1	0.1	0.1
FLI	Fuel Level Input	1.7 / 50%	1.7 / 50%	1.7 / 50%	1.7 / 50%
FPM	Fuel Pump Monitor	ON / OFF	ON	ON	ON
FTP	Fuel Tank Pressure	2.6v / 0 psi	2.6v / 0 psi	2.6v / 0 psi	2.6v / 0 psi
GEAR	Gear Position	1-2-3-4	1	4	4
IAT	IAT Sensor	-40-304°F	50-120	50-120	50-120
HO2S-11	HO2S-11 (Bank 1)	0.1-1.1v	0.1-1.1	0.1-1.1	0.1-1.1
HO2S-12	HO2S-12 (Bank 1)	0.1-1.1v	0.1-1.1	0.1-1.1	0.1-1.1
HO2S-21	HO2S-21 (Bank 1)	0.1-1.1v	0.1-1.1	0.1-1.1	0.1-1.1
KS1	Knock Sensor 1	0-5.1v	0	0	0
LOAD	Engine Load	0-100%	14-16	20-25	24-35
MAFV	MAF Sensor	0-5.1v	0.7-0.9	1.2-1.7	1.6-2.4
MISF	Misfire Detection	ON / OFF	OFF	OFF	OFF
OCTADJ	Octane Adjustment	CL/OPEN	CLOSED	CLOSED	CLOSED
PNP	PNP Switch Signal	ON / OFF	ON	OFF	OFF
PTO	Power Takeoff Switch	OFF	ON (if on)	OFF	OFF
RPM	Engine Speed	0-10K rpm	780-810	1380-1450	1790-1840
TCS	TCS Switch	ON / OFF	OFF	OFF	OFF
TFT	TFT Sensor Signal	-40-304°F	110-210	110-210	110-210
TPV	TP Sensor	0-5.1v	0.53-1.27	0.8-1.1	0.9-1.2
TR1	TR1 Sensor	0v / 11.5v	0	11.5	11.5
TR2	TR2 Sensor	0v / 11.5v	0	11.5	11.5
TR3/TR	TR3A Sensor	0v / 1.7v	0	1.7	1.7
TR4	TR4 Sensor	0v / 11.5v	0	11.5	11.5
Vacuum	Engine Vacuum	0-30" Hg	19-20	18-20	8-15
VPWR	Vehicle Power	0-25.5v	12-14	12-14	12-14
VSS	Vehicle Speed (Hertz or MPH)	0-999 Hz (0-159 mph)	0 Hz = 0 mph	65 Hz = 30 mph	125 Hz = 55 mph

1997-2001 E-Series 6.8L V10 VIN S (E4OD) Outputs

PCM PID Acronym	Parameter Identification	PID Range	PID Value at Hot Idle	PID Value at 30	PID Value at 55
CD1-10	Coil 1-10 Driver	0-90° dwell	6	8	10
CCS	Coast Clutch Solenoid Control	ON / OFF	OFF	OFF	OFF
CHTIL	CHT (lamp) Control	ON / OFF	OFF	OFF	OFF
CTO	Clean Tachometer signal	0-9999 Hz	60-70	90-110	150-185
EGR VR	EGR VR Solenoid	0-100%	0	0-40	35-55
EVAP CP	EVAP Purge Valve	0-100%	0-100	0-100	0-100
EVAP CV	EVAP CV Valve	0-100%	0-100	0-100	0-100
EPC	EPC Solenoid	0-300 psi	5	7	27
FP	Fuel Pump Control	ON / OFF	ON	ON	ON
FuelPW1	INJ Pulsewidth - Bank 1	0-999 ms	3.8-4.6	5.2-6.5	6.6-11.0
FuelPW2	INJ Pulsewidth - Bank 2	0-999 ms	3.8-4.6	5.2-6.5	6.6-11.0
HTR-11	HO2S-11 Heater	ON / OFF	ON	ON	ON
HTR-12	HO2S-12 Heater	ON / OFF	ON	ON	ON
HTR-21	HO2S-21 Heater	ON / OFF	ON	ON	ON
HTR-22	HO2S-22 Heater	ON / OFF	ON	ON	ON
IAC	Idle Air Control	0-100%	25-33	30-40	50-65
INJ1	INJ 1 Pulsewidth	0-999 ms	3.8-4.6	5.2-6.5	6.6-11.0
INJ2	INJ 2 Pulsewidth	0-999 ms	3.8-4.6	5.2-6.5	6.6-11.0
INJ3	INJ 3 Pulsewidth	0-999 ms	3.8-4.6	5.2-6.5	6.6-11.0
INJ4	INJ 4 Pulsewidth	0-999 ms	3.8-4.6	5.2-6.5	6.6-11.0
INJ5	INJ 5 Pulsewidth	0-999 ms	3.8-4.6	5.2-6.5	6.6-11.0
INJ6	INJ 6 Pulsewidth	0-999 ms	3.8-4.6	5.2-6.5	6.6-11.0
INJ7	INJ 7 Pulsewidth	0-999 ms	3.8-4.6	5.2-6.5	6.6-11.0
INJ8	INJ 8 Pulsewidth	0-999 ms	3.8-4.6	5.2-6.5	6.6-11.0
INJ9	INJ 9 Pulsewidth	0-999 ms	3.8-4.6	5.2-6.5	6.6-11.0
INJ10	INJ 10 Pulsewidth	0-999 ms	3.8-4.6	5.2-6.5	6.6-11.0
LongFT1	Long Term FT - Bank 1	-20 to 20%	-5 to +5	-5 to +5	-5 to +5
LongFT2	Long Term FT - Bank 2	-20 to 20%	-5 to +5	-5 to +5	-5 to +5
MIL	MIL (lamp) Control	ON / OFF	OFF	OFF	OFF
SHTFT1	Short Term F/T - Bank 1	-10 to 10%	-5 to +5	-5 to +5	-5 to +5
SHTFT2	Short Term F/T - Bank 2	-10 to 10%	-5 to +5	-5 to +5	-5 to +5
SPARK	Spark Advance	-90° to 90°	15-20	23-34	26-34
SS1	Shift Solenoid 1	ON / OFF	ON	OFF	OFF
SS2	Shift Solenoid 2	ON / OFF	OFF	ON	OFF
TCC	TCC Solenoid	0-100%	0	90-100	90-100
TCIL	TCIL (lamp) Control	ON / OFF	OFF	OFF	OFF
VREF	Vehicle Reference	0-5.1v	4.9-5.1	4.9-5.1	4.9-5.1

1996-2003 E-Series 7.3L V8 Diesel VIN F (E4OD)

PCM PID Acronym	Parameter Identification	PID Range	PID Value at low idle	PID Value at high idle
4x4L	4x4 Low Switch Input	ON / OFF	ON	OFF
ACCS	A/C Switch Signal	ON / OFF	ON	OFF
AP	Accelerator Pedal Position Sensor	0-5.1v	0.5-0.9	3.8-4.2v
ARPMDES	Ancillary Engine Speed Desired	---	---	---
BARO	Barometric Pressure	0-450 kPa	---	---
BAROV	Barometric Pressure	0-5.1v	4.75 (Sea level)	4.75 (Sea level)
BPP	Brake Pedal Position	ON / OFF	ON	OFF
BPA	Brake Pressure Applied	ON / OFF	ON	OFF
CCS	Coast Clutch Solenoid Control	ON / OFF	OFF	OFF
CPP	CPP Switch Signal	0v / 5v	0v (clutch "in")	5v
CPP/TCS	TCS & Cancel Switch	0v / VBAT	Switches On: 12	0v
CRUISE	Cruise Control Module	---	---	---
DTC CNT	DTC Count	0-156	0	0
EBP	Exhaust Back Pressure	---	---	---
EBP V	Exhaust Back Pressure Actual	0-5.1v	0.8-0.95	1.25-1.75
EOT	Engine Oil Temperature	0-5.1v	0.35-4.5	0.35-4.5
EPC	Electronic Pressure Control	0-300 psi	7	15
EPC V	Electronic Pressure Control Actual	0-25.5v	7.5	12
EPR	Exhaust Pressure Regulator	0-25.5v	6-8	0-10
FuelPW1	Fuel Pulsewidth	0-999 ms	---	---
GEAR	Gear Position	1-2-3-4	1	4
GPC	Glow Plug Control Duty Cycle	0-100%	---	---
GPC TM	Glow Plug Control Time	---	---	---
GPL TM	Glow Plug Lamp Time	---	---	---
IAT (°F)	IAT Sensor	-40-304°F	50-120	50-120
IAT V	IAT Sensor Actual	0-5.1v	1.5-3.5	1.5-3.5
ICP	Injector Control Pressure Sensor	Min: 0.83v at startup	0.25-0.40	0.25-0.40
ICP	Injector Control Pressure Actual	---	---	---
IPR	Injector Control Pressure Regulator	0-100%	35	40-100
IVS	Idle Validation Switch	0v / VBAT	0v / Closed	12 / CL or OL
MAP	MAP Sensor	0-999 Hz	---	---
MAP H	MAP Sensor Actual	0-999 Hz	110-190	110-190

1996-2003 E-Series 7.3L V8 Diesel VIN F (E4OD)

PCM PID Acronym	Parameter Identification	PID Range	PID value at low idle	PID value at high idle
MFDES	Mass Fuel Desired	---	---	---
MGP	Manifold Gauge Pressure	---	---	---
PBA	Parking Brake Applied	0v / 12v	Parking Brake Off: 12v	0v
RPM	Engine Speed	0-10K rpm	---	---
SCCS	Speed Control Command Switch	0-25.5v	S/C On: 12v	S/C On: 12v
SCCS-M	Speed Control Command Switch Mode	---	---	---
TCC	Transmission Converter Clutch Solenoid Control	ON / OFF	OFF	OFF
SS1	Shift Solenoid 1	ON / OFF	ON	OFF
SS2	Shift Solenoid 2	ON / OFF	OFF	ON
TCIL	TCIL (lamp) Control	ON / OFF	OFF	OFF
TCS	TCS Switch	0v / 12	0	0
TFT V	TFT Sensor	0-5.1v	2.10-2.40	2.10-2.40
TORQUE	Engine Torque	---	---	---
TPREL	Low Idle TP Sensor	---	---	---
TR	Transmission Range Sensor	0-5.1v	4.45 in 'P'	2.87
TR V	Transmission Range Sensor Actual (Volts)	0-5.1v	4.45 in 'P'	2.87 in O/D
VFDES	Volume Flow Desired	---	---	---
VPWR	Vehicle Power Supply	0-25.5v	12-14	12-14
VREF	Vehicle Reference	0-5.1v	4.9-5.1	4.9-5.1
VS SET	Vehicle Speed Setting	---	---	---
VSS	Vehicle Speed (Hertz or MPH)	0-999 Hz (0-159 mph)	0	125 Hz = 55 mph

1992-95 E-Series 7.5L V8 MFI VIN G (A/T, M/T, E4OD)

PCM PID Acronym	Parameter Identification	PID Range	PID Value at Hot Idle	PID Value at 30	PID Value at 55
ACCS	A/C Switch Signal	ON / OFF	ON (if on)	OFF	OFF
CANP	EVAP Purge Valve	ON / OFF	OFF	ON	ON
ECT (°F)	ECT Sensor	-40-304°F	160-200	160-200	160-200
ECTV	ECT Sensor	0-5.1v	0.6	0.6	0.6
EPC	EPC Solenoid (E4OD)	0-300 psi	4	4	14
EVP	EGR Valve Position	0-5.1v	0.3	0.4	0.4-4.0v
EVR (%)	EGR VR Solenoid	0-100%	0	0-40	0-40
FuelPW1	INJ Pulsewidth - Bank 1	0-999 ms	5.6-6.6	6.6-8.6	9.0-11.0
FuelPW2	INJ Pulsewidth - Bank 2	0-999 ms	5.6-6.6	6.6-8.6	9.0-11.0
GEAR	Gear Position	P-R-N-D	P-R-N-D	DRIVE 3	DRIVE 4
HO2S-11	HO2S-11 (B1 S1)	0.1-1.1v	0-1v	0-1v	0-1v
IAC	Idle Air Control	0-100%	30-34	28-32	63-70
IAT (°F)	IAT Sensor	-40-304°F	50-120	50-120	50-120
IATV	IAT Sensor	0-5.1v	1.5-3.5	1.5-3.5	1.5-3.5
LongFT1	Long Term FT - Bank 1	-20 to 20%	-5 to +5	-5 to +5	-5 to +5
LongFT2	Long Term FT - Bank 2	-20 to 20%	-5 to +5	-5 to +5	-5 to +5
LOOP	Loop Status	CL or OL	CLOSED	CLOSED	CLOSED
MLPV	MLP Switch (E4OD)	0-5.1v	0	5	5
MAP	MAP Sensor	0-999 Hz	108-112	110-120	130-140
PNP	PNP Switch Signal	Neutral/DR	NEUTRAL	DRIVE	DRIVE
RPM	Engine Speed	0-10K rpm	650-750	1300-1400	1730-1830
SHTFT1	Short Term F/T - Bank 1	-10 to 10%	-5 to +5	-5 to +5	-5 to +5
SHTFT2	Short Term FT - Bank 2	-10 to 10%	-5 to +5	-5 to +5	-5 to +5
SPARK	Spark Advance	-90° to 90°	18-28	36-42	38-46
TCC	TCC Solenoid	ON / OFF	OFF	ON	ON
TOTV	TOT sensor (E4OD)	0-5.1v	2.10-2.40	2.10-2.40	2.10-2.40
TP	TP Sensor	0-5.1v	1.0	1.2-1.3	1.5-1.6
TP Mode	TP Sensor Mode	C/T or P/T	C/T	P/T	P/T
VPWR	Vehicle Power	0-25.5v	12-14	12-14	12-14
VREF	Vehicle Reference	0-5.1v	4.9-5.1	4.9-5.1	4.9-5.1
VSS	Vehicle Speed	0-159 mph	0	30	55

1996 E-Series 7.5L V8 MFI VIN G (E4OD) Inputs

PCM PID Acronym	Parameter Identification	PID Range	PID Value at Hot Idle	PID Value at 30	PID Value at 55
4x4L	4x4 Switch Signal	ON / OFF	ON (if on)	OFF	OFF
ACCS	A/C Switch Signal	ON / OFF	ON (if on)	OFF	OFF
BPP	Brake Switch Signal	ON / OFF	ON (if on)	OFF	OFF
DPFE	DPFE Sensor	0-5.1v	0.4-0.6	0.4-0.7	0.8-1.1
ECT	ECT Sensor	-40-304°F	160-200	160-200	160-200
FEPS	Flash EEPROM	0-5.1v	0.1	0.1	0.1
FPM	Fuel Pump Monitor	ON / OFF	ON	ON	ON
GEAR	Gear Position	1-2-3-4	1	3	4
HO2S-11	HO2S-11 (Bank 1)	0.1-1.1v	0.1-1.1	0.1-1.1	0.1-1.1
HO2S-12	HO2S-12 (Bank 1)	0.1-1.1v	0.1-1.1	0.1-1.1	0.1-1.1
HO2S-21	HO2S-21 (Bank 1)	0.1-1.1v	0.1-1.1	0.1-1.1	0.1-1.1
HO2S-22	HO2S-22 (Bank 1)	0.1-1.1v	0.1-1.1	0.1-1.1	0.1-1.1
IAT	IAT Sensor	-40-304°F	50-120	50-120	50-120
IDM	IDM Signal	0-9999 Hz	45-55	88-120	110-220
LOAD	Engine Load	0-100%	14-16	20-25	32-40
MAFV	MAF Sensor	0-5.1v	0.9-1.1	1.6-1.8	2.0-2.6
MISF	Misfire Detection	ON / OFF	OFF	OFF	OFF
OCTADJ	Octane Adjust	CL or OP	CLOSED	CLOSED	CLOSED
PIP	PIP Sensor	0-9999 Hz	45-55	88-120	110-220
PNP	PNP Switch Signal	ON / OFF	ON	OFF	OFF
PTO	Power Takeoff	ON / OFF	ON (if on)	OFF	OFF
RPM	Engine Speed	0-10K rpm	780-810	1420-1510	1700-1790
TCS	TCS Switch	ON / OFF	OFF	OFF	OFF
TFT	TFT Sensor Signal	-40-304°F	110-210	110-210	110-210
TPV	TP Sensor	0-5.1v	0.53-1.27	1.1-1.3v	1.3-1.7
TRV/TR	TR Sensor	0v / 1.7v	0	1.7	1.7
Vacuum	Engine Vacuum	0-30" Hg	19-20	18-20	8-15
VPWR	Vehicle Power	0-25.5v	12-14	12-14	12-14
VSS	Vehicle Speed (Hertz or MPH)	0-999 Hz (0-159 mph)	0 Hz = 0 mph	65 Hz = 30 mph	125 Hz = 55 mph

1996 E-Series 7.5L V8 MFI VIN G (E4OD) Outputs

PCM PID Acronym	Parameter Identification	PID Range	PID Value at Hot Idle	PID Value at 30	PID Value at 55
AIRB	AIR System Bypass	0v / VBAT	0.1	0.1	0.1
AIRD	AIR System Divert	0v / VBAT	VBAT	VBAT	VBAT
EGR VR	EGR VR Solenoid	0-100%	0	40-45	50-60
EVAP CP	EVAP Purge Valve	0-100%	0-100	0-100	0-100
EPC	EPC Solenoid	0-300 psi	5	5	10-15
FP	Fuel Pump Control	ON / OFF	ON	ON	ON
FuelPW1	INJ Pulsewidth - Bank 1	0-999 ms	4.3-4.6	5.2-6.5	6.6-9.0
FuelPW2	INJ Pulsewidth - Bank 2	0-999 ms	4.3-4.6	5.2-6.5	6.6-9.0
HTR-11	HO2S-11 Heater	ON / OFF	ON	ON	ON
HTR-12	HO2S-12 Heater	ON / OFF	ON	ON	ON
HTR-21	HO2S-21 Heater	ON / OFF	ON	ON	ON
HTR-22	HO2S-22 Heater	ON / OFF	ON	ON	ON
IAC	Idle Air Control	0-100%	30-34	60-65	63-70
INJ1	INJ 1 Pulsewidth	0-999 ms	4.3-4.6	5.2-6.5	6.6-9.0
INJ2	INJ 2 Pulsewidth	0-999 ms	4.3-4.6	5.2-6.5	6.6-9.0
INJ3	INJ 3 Pulsewidth	0-999 ms	4.3-4.6	5.2-6.5	6.6-9.0
INJ4	INJ 4 Pulsewidth	0-999 ms	4.3-4.6	5.2-6.5	6.6-9.0
INJ5	INJ 5 Pulsewidth	0-999 ms	4.3-4.6	5.2-6.5	6.6-9.0
INJ6	INJ 6 Pulsewidth	0-999 ms	4.3-4.6	5.2-6.5	6.6-9.0
INJ7	INJ 7 Pulsewidth	0-999 ms	4.3-4.6	5.2-6.5	6.6-9.0
INJ8	INJ 8 Pulsewidth	0-999 ms	4.3-4.6	5.2-6.5	6.6-9.0
LongFT1	Long Term FT - Bank 1	-20 to 20%	-5 to +5	-5 to +5	-5 to +5
LongFT2	Long Term FT - Bank 2	-20 to 20%	-5 to +5	-5 to +5	-5 to +5
MIL	MIL (lamp) Control	ON / OFF	OFF	OFF	OFF
SHTFT1	Short Term F/T - Bank 1	-10 to 10%	-5 to +5	-5 to +5	-5 to +5
SHTFT2	Short Term FT - Bank 2	-10 to 10%	-5 to +5	-5 to +5	-5 to +5
SPARK	Spark Advance	-90° to 90°	22-26	25-28	27-32°
SPOUT	Spark Output Signal	0-9999 Hz	45-55	88-120	110-220
SS1	Shift Solenoid 1	ON / OFF	ON	OFF	OFF
SS2	Shift Solenoid 2	ON / OFF	OFF	ON	OFF
TCC	TCC Solenoid	0-100%	0%	0-100%	90-95
TCIL	TCIL (lamp) Control	ON / OFF	OFF	OFF	OFF
VREF	Vehicle Reference	0-5.1v	4.9-5.1	4.9-5.1	4.9-5.1

VILLAGER PID DATA

1996-98 Villager 3.0L V6 SOHC MFI VIN W, 1 (A/T)

PCM PID Acronym	Parameter Identification	PID Range	PID Value at Hot Idle	PID Value at 30	PID Value at 55
ACR	A/C Relay Contro (with A/C & Blower On or Off)	ON / OFF	ON (if on)	OFF	OFF
ACS	AC Cycling Switch (with AC On & blower at 3-4)	ON / OFF	ON	OFF	OFF
A/F COR	Air Fuel Correction	-20 to 20%	-5 to +5	-5 to +5	-5 to +5
ECT	ECT Sensor	-40-304°F	176-220	176-220	176-220
EGRC	EGR Solenoid	ON / OFF	ON	OFF	OFF
EGRT	EGR Temp. Sensor	0-5.1v	0.3	0.2-1.0	0.2-1.0
FUEL	Fuel System Status	LEAN or RICH	LEAN / RICH	LEAN / RICH	LEAN / RICH
FuelPW1	INJ Pulsewidth - Bank 1	0-999 ms	3.38	4.4-5.5	6.6-7.3
HFAN	High Speed Cooling Fan (on with ECT more than 221°F)	ON / OFF	OFF	OFF	OFF
IAC	Idle Speed Control	0-100%	15-28	71	71
IDL SW	Throttle Idle Switch	ON / OFF	ON	OFF	OFF
LFAN	Low Speed Cooling Fan (ON with ECT at 203-210°F - AC Off)	ON / OFF	OFF	OFF	OFF
HO2S-11	HO2S-11 (Bank 1)	0.1-1.1v	0I.1-1.1	0.1-1.1	0.1-1.1
MAFV	MAF Sensor	0-5.1v	0.7	2.4	2.8
PNP SW	PNP Switch Signal	ON / OFF	ON	OFF	OFF
PSP SW	PSP Switch Signal	ON / OFF	ON: turned	OFF	OFF
RPM	Engine Speed	0-10K rpm	750	1200-1650	1850
SHTFT1	Short Term F/T - Bank 1	-10 to 10%	-5 to +5	-5 to +5	-5 to +5
IGN	Ignition Timing	-90° to 90°	15	36	32
TPV	TP Sensor	0-5.1v	0.46	0.78	1.12
VPWR	Vehicle Power	0-25.5v	12-14	12-14	12-14
VSS	Vehicle Speed	0-159 mph	0	30	55
VST	Vehicle Start Signal (on during cranking)	ON / OFF	OFF	OFF	OFF

1999-2003 Villager 3.3L V6 SOHC MFI VIN T (A/T)

PCM PID Acronym	Parameter Identification	PID Range	PID Value at Hot Idle	PID Value at 30	PID Value at 55
ACR	A/C Relay Control	ON / OFF	ON (A/C & BLMT on)	OFF	OFF
ACS	AC Cycling Switch (AC & Blower/3-4)	ON / OFF	ON	OFF	OFF
A/F COR	Air Fuel Correction	-20 to 20%	-5 to +5	-5 to +5	-5 to +5
ECT	ECT Sensor	-40-304°F	176-220	176-220	176-220
EGRC	EGR Solenoid	ON / OFF	ON	OFF	OFF
EGRT	EGR Temp. Sensor	0-5.1v	0.3	0.2-1.0	0.2-1.0
FUEL	Fuel System Status	LEAN or RICH	LEAN / RICH	LEAN / RICH	LEAN / RICH
FuelPW1	INJ Fuel Pulsewidth Bank 1	0-999 ms	4.0-4.3	4.4-5.5	6.6-7.3
HFAN	High Speed Cooling Fan (on with ECT more than 221°F)	ON / OFF	OFF	OFF	OFF
IAC	Idle Speed Control	0-100%	15-28	71	71
IDL SW	Throttle Idle Switch	ON / OFF	ON	OFF	OFF
LFAN	Low Speed Cooling Fan (on with ECT at 203-210°F - AC Off)	ON / OFF	OFF	OFF	OFF
HO2S-11	HO2S-11 (Bank 1)	0.1-1.1v	0.1-1.1	0.1-1.1	0.1-1.1
MAFV	MAF Sensor	0-5.1v	0.7	2.4	2.8
PNP SW	PNP Switch Signal	ON / OFF	ON	OFF	OFF
PSP SW	PSP Switch Signal	ON / OFF	ON: turned	OFF	OFF
RPM	Engine Speed	0-10K rpm	750	1200-1650	1850
SHTFT1	Short Term F/T - Bank 1	-10 to 10%	-5 to +5	-5 to +5	-5 to +5
IGN	Ignition Timing	-90° to 90°	15 BTDC	36 BTDC	32° BTDC
TPV	TP Sensor	0-5.1v	0.46	0.78	1.12
VPWR	Vehicle Power	0-25.5v	12-14	12-14	12-14
VSS	Vehicle Speed	0-159 mph	0	30	55
VST	Vehicle Start Signal (on during cranking)	ON / OFF	OFF	OFF	OFF

WINDSTAR PID DATA

1996-2000 Windstar 3.0L V6 VIN U (A/T) Inputs

PCM PID Acronym	Parameter Identification	PID Range	PID Value at Hot Idle	PID Value at 30	PID Value at 55
ACCS	A/C Switch Signal	OFF	ON	OFF	OFF
ACP	AC Pressure Switch	0-5.1v	1.5-1.9	0.6-0.8	0.6-0.8
AFS	Airflow Sensor	0-999 Hz	130	130	130
BPP	Brake Position	ON / OFF	ON (if on)	OFF	OFF
CID	CMP Sensor	0-999	5-7	12-15	15-20
CKP	CKP Sensor	0-9999 Hz	518-540	975-1020	1200-1330
DPFE	DPFE Sensor	0-5.1v	0.2-1.3	0.2-4.5	0.2-4.5
ECT	ECT Sensor	-40-304°F	160-200	160-200	160-200
FEPS	Flash EEPROM	0-5.1v	0.1	0.1	0.1
FLI	Fuel Level Indicator	0-100%	50% (1.7)	50% (1.7)	50% (1.7)
FPM	Fuel Pump Monitor	ON / OFF	ON	ON	ON
GEAR	Gear Position	1-2-3-4	1	3	4
HO2S-11	HO2S-11 (Bank 1)	0.1-1.1v	0.1-1.1	0.1-1.1	0.1-1.1
HO2S-12	HO2S-12 (Bank 1)	0.1-1.1v	0.1-1.1	0.1-1.1	0.1-1.1
HO2S-21	HO2S-21 (Bank 2)	0.1-1.1v	0.1-1.1	0.1-1.1	0.1-1.1
HO2S-22	HO2S-22 (Bank 2)	0.1-1.1v	0.1-1.1	0.1-1.1	0.1-1.1
IAT	IAT Sensor	-40-304°F	50-120	50-120	50-120
LOAD	Engine Load	0-100%	15-20	20-27	35-45
MAFV	MAF Sensor	0-5.1v	0.6-0.9	0.8-1.9	1.1-2.3
MISF	Misfire Detection	ON / OFF	OFF	OFF	OFF
OCTADJ	Octane Adjustment	CL/OPEN	CLOSED	CLOSED	CLOSED
PNP	PNP Switch Signal	ON / OFF	ON	OFF	OFF
RPM	Engine Speed	0-10K rpm	680-800	1550-1700	1500-2100
TCS	TCS Switch	ON / OFF	OFF	OFF	OFF
TFT	TFT Sensor Signal	-40-304°F	110-210	110-210	110-210
TPV	TP Sensor	0-5.1v	0.53-1.27	0.9-1.2	0.9-1.9
TR1	TR1 Sensor	0v / 11.5v	0	11.5	11.5
TR2	TR2 Sensor	0v / 11.5v	0	11.5	11.5
TR4	TR4 Sensor	0v / 11.5v	0	11.5	11.5
TRV/TR	TR Sensor	0v / 1.7v	0	1.7	1.7
TSS	TSS Sensor	0-9999 Hz	35-40	47-73	100-125
Vacuum	Engine Vacuum	0-30" Hg	19-20	18-20	8-15
VPWR	Vehicle Power	0-25.5v	12-14	12-14	12-14
VSS	Vehicle Speed (Hertz or MPH)	0-999 Hz (0-159 mph)	0 Hz = 0 mph	65 Hz = 30 mph	125 Hz = 55 mph
WACA	AC WOT Monitor	ON	OFF (with A/C on)	ON	ON

1996-2000 Windstar 3.0L V6 VIN U (A/T) Outputs

PCM PID Acronym	Parameter Identification	PID Range	PID Value at Hot Idle	PID Value at 30	PID Value at 55
ALDFDC	Alternator Field (Hz)	0-999 Hz	0-130	0	0
CD1-3	Coil 1 Driver	0-60° dwell	6	8	12
EGR VR	EGR VR Solenoid	0-100%	0	0-40	35-50
EVAP CP	EVAP Purge Valve	0-100%	0-100	0-100	0-100
EVAP CV	EVAP CV Valve	0-100%	0-100	0-100	0-100
EPC	EPC Solenoid	0-300 psi	13-17	15-21	38-47
FP	Fuel Pump Control	ON / OFF	ON	ON	ON
FTP	Fuel Tank Pressure	0-5.1v at 0-10" H2O	2.6v at 0" H2O (gas cap "off")	2.6v at 0" H2O (gas cap "off")	2.6v at 0" H2O (gas cap "off")
FuelPW1	INJ Pulsewidth - Bank 1	0-999 ms	5.0-5.2	5.9-9.0	6.0-11.0
FuelPW2	INJ Pulsewidth - Bank 2	0-999 ms	5.0-5.2	5.9-9.0	6.0-11.0
HFC	High Speed Fan	ON / OFF	ON (if on)	OFF	OFF
HTR-11	HO2S-11 Heater	ON / OFF	ON	ON	ON
HTR-12	HO2S-12 Heater	ON / OFF	ON	ON	ON
HTR-21	HO2S-21 Heater	ON / OFF	ON	ON	ON
HTR-22	HO2S-22 Heater	ON / OFF	ON	ON	ON
IAC	Idle Air Control	0-100%	17-29	38-53	40-55
INJ1	INJ 1 Pulsewidth	0-999 ms	5.0-5.2	5.9-9.0	6.0-11.0
INJ2	INJ 2 Pulsewidth	0-999 ms	5.0-5.2	5.9-9.0	6.0-11.0
INJ3	INJ 3 Pulsewidth	0-999 ms	5.0-5.2	5.9-9.0	6.0-11.0
INJ4	INJ 4 Pulsewidth	0-999 ms	5.0-5.2	5.9-9.0	6.0-11.0
INJ5	INJ 5 Pulsewidth	0-999 ms	5.0-5.2	5.9-9.0	6.0-11.0
INJ6	INJ 6 Pulsewidth	0-999 ms	5.0-5.2	5.9-9.0	6.0-11.0
LFC	Low Speed Fan	ON / OFF	ON (if on)	OFF	OFF
LongFT1	Long Term FT (B1)	-20 to 20%	-5 to +5	-5 to +5	-5 to +5
LongFT2	Long Term FT (B2)	-20 to 20%	-5 to +5	-5 to +5	-5 to +5
MIL	MIL (lamp) Control	ON / OFF	OFF	OFF	OFF
SHTFT1	Short Term F/T - Bank 1	-10 to 10%	-5 to +5	-5 to +5	-5 to +5
SHTFT2	Short Term FT - Bank 2	-10 to 10%	-5 to +5	-5 to +5	-5 to +5
SPARK	Spark Advance	-90° to 90°	15-20	20-35	20-41
SPOUT	Spark Output Signal	0 Hz	33-37	80-87	95-105
SS1	Shift Solenoid 1	ON / OFF	ON	OFF	OFF
SS2	Shift Solenoid 2	ON / OFF	OFF	ON	ON
SS3	Shift Solenoid 3	ON / OFF	OFF	ON	ON
TCC	TCC Solenoid	0-100%	0-100	59-66	72-81
TCIL	TCIL (lamp) Control	ON / OFF	OFF	OFF	OFF
VREF	Vehicle Reference	0-5.1v	4.9-5.1	4.9-5.1	4.9-5.1
WAC	A/C WOT Relay Control	ON / OFF	ON (AC on at WOT)	OFF	OFF

1996-2005 Windstar 3.8L V6 VIN R (A/T) Inputs

PCM PID Acronym	Parameter Identification	PID Range	PID Value at Hot Idle	PID Value at 30	PID Value at 55
ACCS	A/C Switch Signal	ON / OFF	ON (if on)	OFF	OFF
ACP	AC Pressure Switch	0-5.1v	1.5-1.9	0.6-0.8	0.6-0.8
BPP	Brake Position	ON / OFF	ON (if on)	OFF	OFF
CID	CMP Sensor	0-999 Hz	5-7	10-12	13-15
CKP	CKP Sensor	0-9999 Hz	390-450	700-740	950-1050
DPFE	DPFE Sensor	0-5.1v	0.2-1.3	0.2-4.5	0.2-4.5
ECT	ECT Sensor	-40-304°F	160-200	160-200	160-200
FEPS	Flash EEPROM	0-5.1v	0.1	0.1	0.1
FTP	Fuel Tank Pressure	0-5.1v at 0-10" H2O	2.6v at 0" H2O (gas cap "off")	2.6v at 0" H2O (gas cap "off")	2.6v at 0" H2O (gas cap "off")
FPM	Fuel Pump Monitor	ON / OFF	ON	ON	ON
IAT	IAT Sensor	-40-304°F	50-120	50-120	50-120
GEAR	Gear Position	1-2-3-4	1	3	4
HO2S-11	HO2S-11 (Bank 1)	0.1-1.1v	0.1-1.1	0.1-1.1	0.1-1.1
HO2S-12	HO2S-12 (Bank 1)	0.1-1.1v	0.1-1.1	0.1-1.1	0.1-1.1
HO2S-21	HO2S-21 (Bank 2)	0.1-1.1v	0.1-1.1	0.1-1.1	0.1-1.1
HO2S-22	HO2S-22 (Bank 2)	0.1-1.1v	0.1-1.1	0.1-1.1	0.1-1.1
IMRC-M	IMRC Monitor	5v / 0v	5	5	5
LOAD	Engine Load	0-100%	15-20	19-27	31-35
MAFV	MAF Sensor	0-5.1v	0.6-0.9	0.9-1.4	1.3-2.9
MISF	Misfire Detection	ON / OFF	OFF	OFF	OFF
OCTADJ	Octane Adjustment	CL/OPEN	CLOSED	CLOSED	CLOSED
PNP	PNP Switch Signal	Neutral/DR	NEUTRAL	DRIVE	DRIVE
RPM	Engine Speed	0-10K rpm	700-730	1250-1400	1700-1870
TCS	TCS Switch	ON / OFF	OFF	OFF	OFF
TFT	TFT Sensor Signal	-40-304°F	110-210	110-210	110-210
TPV	TP Sensor	0-5.1v	0.53-1.27	0.8-1.1	0.8-1.2
TRV/TR	TR Sensor	0v / 1.7v	0v	1.7	1.7
TR1	TR1 Sensor	0v / 11.5v	0v	11.5	11.5
TR2	TR2 Sensor	0v / 11.5v	0v	11.5	11.5
TR4	TR4 Sensor	0v / 11.5v	0v	11.5	11.5
TSS	TSS Sensor	0-9999 Hz	43	88-93	99-113
Vacuum	Engine Vacuum	0-30" Hg	19-20	18-20	8-15
VPWR	Vehicle Power	0-25.5v	12-14	12-14	12-14
VSS	Vehicle Speed (Hertz or MPH)	0-999 Hz (0-159 mph)	0 Hz = 0 mph	65 Hz = 30 mph	125 Hz = 55 mph
WACA	A/C WOT Relay Monitor	ON / OFF	ON (AC on at WOT)	OFF	OFF

1996-2005 Windstar 3.8L V6 VIN R (A/T) Outputs

PCM PID Acronym	Parameter Identification	PID Range	PID Value at Hot Idle	PID Value at 30	PID Value at 55
ALDFDC	Alternator Field (Hz)	0-999 Hz	0-130	00-130	00-130
CD1-3	Coil 1 Driver-3	0-60° dwell	6	8	12
EGR VR	EGR VR Solenoid	0-100%	0	0-40	35-50
EVAP CP	EVAP Purge Valve	0-100%	0-100	0-100	0-100
EVAP CV	EVAP CV Valve	0-100%	0-100	0-100	0-100
EPC	EPC Solenoid	0-300 psi	15	16-25	Q42
FP	Fuel Pump Control	ON / OFF	ON	ON	ON
FuelPW1	INJ Pulsewidth - Bank 1	0-999 ms	3.5-3.8	4.9-7.9	6.1-11.0
FuelPW2	INJ Pulsewidth - Bank 2	0-999 ms	3.5-3.8	4.9-7.9	6.1-11.0
HFC	High Fan Control	ON / OFF	ON	OFF	OFF
HTR-11	HO2S-11 Heater	ON / OFF	ON	ON	ON
HTR-12	HO2S-12 Heater	ON / OFF	ON	ON	ON
HTR-21	HO2S-21 Heater	ON / OFF	ON	ON	ON
HTR-22	HO2S-22 Heater	ON / OFF	ON	ON	ON
IAC	Idle Air Control	0-100%	25-35	30-55	50-59
IMRC	IMRC Solenoid	ON / OFF	OFF	OFF	OFF
INJ1	INJ 1 Pulsewidth	0-999 ms	3.5-3.8	4.9-7.9	6.1-11.0
INJ2	INJ 2 Pulsewidth	0-999 ms	3.5-3.8	4.9-7.9	6.1-11.0
INJ3	INJ 3 Pulsewidth	0-999 ms	3.5-3.8	4.9-7.9	6.1-11.0
INJ4	INJ 4 Pulsewidth	0-999 ms	3.5-3.8	4.9-7.9	6.1-11.0
INJ5	INJ 5 Pulsewidth	0-999 ms	3.5-3.8	4.9-7.9	6.1-11.0
INJ6	INJ 6 Pulsewidth	0-999 ms	3.5-3.8	4.9-7.9	6.1-11.0
LFC	Low Fan Control	OFF	ON	OFF	OFF
LongFT1	Long Term FT - Bank 1	-20 to 20%	-5 to +5	-5 to +5	-5 to +5
LongFT2	Long Term FT - Bank 2	-20 to 20%	-5 to +5	-5 to +5	-5 to +5
MIL	MIL (lamp) Control	ON / OFF	OFF	OFF	OFF
SHTFT1	Short Term F/T - Bank 1	-10 to 10%	-5 to +5	-5 to +5	-5 to +5
SHTFT2	Short Term FT - Bank 2	-10 to 10%	-5 to +5	-5 to +5	-5 to +5
SPARK	Spark Advance	-90° to 90°	15-20	25-35	27-36
SS1	Shift Solenoid 1	ON / OFF	ON	OFF	OFF
SS2	Shift Solenoid 2	ON / OFF	OFF	ON	ON
SS3	Shift Solenoid 3	ON / OFF	OFF	ON	ON
TCC	TCC Solenoid	0-100%	0%	0-40	90-95
TCIL	TCIL (lamp) Control	ON / OFF	OFF	OFF	OFF
VREF	Vehicle Reference	0-5.1v	4.9-5.1	4.9-5.1	4.9-5.1
WAC	A/C WOT Relay Control	ON / OFF	ON (AC on at WOT)	OFF	OFF

EXPEDITION PID DATA

1997-2005 Expedition 4.6L V8 SOHC 2v MFI VIN W, 6 (A/T)

PCM PID Acronym	Parameter Identification	PID Range	PID Value at Hot Idle	PID Value at 30	PID Value at 55
4x4L	4x4 Switch Signal	ON / OFF	ON (if on)	OFF	OFF
ACCS	A/C Switch Signal	ON / OFF	ON (if on)	OFF	OFF
BPP	Brake Position	ON / OFF	ON (if on)	OFF	OFF
DPFE	DPFE Sensor	0-5.1v	0.2-1.3	0.2-4.5	0.2-4.5
CHT	CHT Sensor	-40-500ºF	194	194	194
FEPS	Flash EEPROM	0-5.1v	0.1	0.1	0.1
FLI	Fuel Level Input	1.7 / 50%	1.7 / 50%	1.7 / 50%	1.7 / 50%
FPM	Fuel Pump Monitor	ON / OFF	ON	ON	ON
FTP	Fuel Tank Pressure	0-5.1v at 0-10" H2O	2.6v at 0" H2O (gas cap "off")	2.6v at 0" H2O (gas cap "off")	2.6v at 0" H2O (gas cap "off")
GEAR	Gear Position	1-2-3-4	1	3	4
HO2S-11	HO2S-11 (Bank 1)	0.1-1.1v	0.1-1.1	0.1-1.1	0.1-1.1
HO2S-12	HO2S-12 (Bank 1)	0.1-1.1v	0.1-1.1	0.1-1.1	0.1-1.1
HO2S-21	HO2S-21 (Bank 2)	0.1-1.1v	0.1-1.1	0.1-1.1	0.1-1.1
HO2S-22	HO2S-22 (Bank 2)	0.1-1.1v	0.1-1.1	0.1-1.1	0.1-1.1
IAT	IAT Sensor	-40-304ºF	50-120	50-120	50-120
LOAD	Engine Load	0-100%	14-16	20-25	24-35
MAFV	MAF Sensor	0-5.1v	0.7-0.9	1.2-1.7	1.6-2.4
MISF	Misfire Detection	ON / OFF	OFF	OFF	OFF
OCTADJ	Octane Adjustment	CL/OPEN	CLOSED	CLOSED	CLOSED
OSS	OSS Sensor	0-9999 Hz	0	125-131	245-255
PNP	PNP Switch Signal	ON / OFF	ON	OFF	OFF
PTO	Power Takeoff Switch	ON / OFF	ON (if on)	OFF	OFF
RPM	Engine Speed	0-10K rpm	680-830	1200-1500	1600-1800
TCS	TCS Switch	ON / OFF	OFF	OFF	OFF
TFT	TFT Sensor Signal	-40-304ºF	110-210	110-210	110-210
TPV	TP Sensor	0-5.1v	0.53-1.27	0.8-1.0	0.9-1.2
TRV/TR	TR Sensor	0v / 1.7v	0v	1.7	1.7
TR1	TR1 Sensor	0v / 11.5v	0v	11.5	11.5
TR2	TR2 Sensor	0v / 11.5v	0v	11.5	11.5
TR4	TR4 Sensor	0v / 11.5v	0v	11.5	11.5
Vacuum	Engine Vacuum	0-30" Hg	19-20	18-20	8-15
VPWR	Vehicle Power	0-25.5v	12-14	12-14	12-14
VSS	Vehicle Speed (Hertz or MPH)	0-999 Hz (0-159 mph)	0 Hz = 0 mph	65 Hz = 30 mph	125 Hz = 55 mph

1997-2005 Expedition 4.6L V8 SOHC MFI 2v VIN W, 6 (A/T)

PCM PID Acronym	Parameter Identification	PID Range	PID Value at Hot Idle	PID Value at 30	PID Value at 55
CCS	Coast Clutch Solenoid Control	ON / OFF	OFF	OFF	OFF
EGR VR	EGR VR Solenoid	0-100%	0	0-40	35-45
EVAP CP	EVAP Purge Valve	0-100%	0-100	0-100	0-100
EVAP CV	EVAP CV Valve	0-100%	0-100	0-100	0-100
EPC	EPC Solenoid	0-300 psi	6	40	40
FP	Fuel Pump Control	ON / OFF	ON	ON	ON
FuelPW1	INJ Pulsewidth - Bank 1	0-999 ms	2.7-4.1	4.5-8.0	5.5-9
FuelPW2	INJ Pulsewidth - Bank 2	0-999 ms	2.7-4.1	4.5-8.0	5.5-9
HTR-11	HO2S-11 Heater	ON / OFF	ON	ON	ON
HTR-12	HO2S-12 Heater	ON / OFF	ON	ON	ON
HTR-21	HO2S-21 Heater	ON / OFF	ON	ON	ON
HTR-22	HO2S-22 Heater	ON / OFF	ON	ON	ON
IAC	Idle Air Control	0-100%	30-34	28-32	63-70
INJ1	INJ 1 Pulsewidth	0-999 ms	2.7-4.1	4.5-8.0	5.5-9.0
INJ2	INJ 2 Pulsewidth	0-999 ms	2.7-4.1	4.5-8.0	5.5-9.0
INJ3	INJ 3 Pulsewidth	0-999 ms	2.7-4.1	4.5-8.0	5.5-9.0
INJ4	INJ 4 Pulsewidth	0-999 ms	2.7-4.1	4.5-8.0	5.5-9.0
INJ5	INJ 5 Pulsewidth	0-999 ms	2.7-4.1	4.5-8.0	5.5-9.0
INJ6	INJ 6 Pulsewidth	0-999 ms	2.7-4.1	4.5-8.0	5.5-9.0
INJ7	INJ 7 Pulsewidth	0-999 ms	2.7-4.1	4.5-8.0	5.5-9.0
INJ8	INJ 8 Pulsewidth	0-999 ms	2.7-4.1	4.5-8.0	5.5-9.0
LongFT1	Long Term FT - Bank 1	-20 to 20%	-5 to +5	-5 to +5	-5 to +5
LongFT2	Long Term FT - Bank 2	-20 to 20%	-5 to +5	-5 to +5	-5 to +5
MIL	MIL (lamp) Control	ON / OFF	OFF	OFF	OFF
SHTFT1	Short Term F/T - Bank 1	-10 to 10%	-5 to +5	-5 to +5	-5 to +5
SHTFT2	Short Term FT - Bank 2	-10 to 10%	-5 to +5	-5 to +5	-5 to +5
SPARK	Spark Advance	-90° to 90°	15-18	23-29	26-32
SS1	Shift Solenoid 1	ON / OFF	ON	OFF	OFF
SS2	Shift Solenoid 2	ON / OFF	OFF	ON	OFF
TCC	TCC Solenoid	0-100%	0	90-100	90-100
TCIL	TCIL (lamp) Control	ON / OFF	OFF	OFF	OFF
VREF	Vehicle Reference	0-5.1v	4.9-5.1	4.9-5.1	4.9-5.1

BLACKWOOD PID DATA

2002-05 Blackwood 5.4L V8 DOHC 4v MFI VIN A (A/T) Inputs

PCM PID Acronym	Parameter Identification	PID Range	PID Value at Hot Idle	PID Value at 30	PID Value at 55
4x4L	4x4 Switch Signal	ON / OFF	ON (if on)	OFF	OFF
ACCS	A/C Switch Signal	ON / OFF	ON (if on)	OFF	OFF
BPP	Brake Position	ON / OFF	ON (if on)	OFF	OFF
CHT	CHT Sensor	-40-500°F	194	194	194
CID	CMP Sensor	0-999 Hz	6-8	10-12	14-17
CKP	CKP Sensor	0-9999 Hz	410	800-850	900-1125
DPFE	DPFE Sensor	0-5.1v	0.2-1.3	0.2-4.5	0.2-4.5
ECT	ECT Sensor	-40-304°F	160-200	160-200	160-200
FEPS	Flash EEPROM	0-5.1v	0.1	0.1	0.1
FLI	Fuel Level Input	1.7 / 50%	1.7 / 50%	1.7 / 50%	1.7 / 50%
FPM	Fuel Pump Monitor	ON / OFF	ON	ON	ON
FTP	Fuel Tank Pressure	0-5.1v at 0-10" H2O	2.6v at 0" H2O (gas cap "off")	2.6v at 0" H2O (gas cap "off")	2.6v at 0" H2O (gas cap "off")
GEAR	Gear Position	1-2-3-4	1	3	4
HO2S-11	HO2S-11 (Bank 1)	0.1-1.1v	0.1-1.1	0.1-1.1	0.1-1.1
HO2S-12	HO2S-12 (Bank 1)	0.1-1.1v	0.1-1.1	0.1-1.1	0.1-1.1
HO2S-21	HO2S-21 (Bank 2)	0.1-1.1v	0.1-1.1	0.1-1.1	0.1-1.1
HO2S-22	HO2S-22 (Bank 2)	0.1-1.1v	0.1-1.1	0.1-1.1	0.1-1.1
IAT	IAT Sensor	-40-304°F	50-120	50-120	50-120
KS1	Knock Sensor 1	0-5.1v	0	0	0
LOAD	Engine Load	0-100%	14-16	19-25	26-35
MAFV	MAF Sensor	0-5.1v	0.7-0.9	1.1-1.6	1.7-2.4
MISF	Misfire Detection	ON / OFF	OFF	OFF	OFF
OCTADJ	Octane Adjustment	CL/OPEN	CLOSED	CLOSED	CLOSED
OSS	OSS Sensor	0-9999 Hz	0	125	350
PNP	PNP Switch Signal	ON / OFF	ON	OFF	OFF
PTO	Power Takeoff Switch	ON / OFF	ON (if on)	OFF	OFF
RPM	Engine Speed	0-10K rpm	790-710	1200-1270	1590-1675
TCS	TCS Switch	ON / OFF	OFF	OFF	OFF
TFT	TFT Sensor Signal	-40-304°F	110-210	110-210	110-210
TPO	TPO TR Sensor	0-5.1v	0.1	0.8	1.0
TPV	TP Sensor	0-5.1v	0.53-1.27	0.8-1.0	1.0-1.3
TR1	TR1 Sensor	0v / 11.5v	0	11.5	11.5
TR2	TR2 Sensor	0v / 11.5v	0	11.5	11.5
TR3/TR	TR3A Sensor	0v / 1.7v	0	1.7	1.7
TR4	TR4 Sensor	0v / 11.5v	0	11.5	11.5
Vacuum	Engine Vacuum	0-30" Hg	19-20	18-20	8-15
VPWR	Vehicle Power	0-25.5v	12-14	12-14	12-14
VSS	Vehicle Speed (Hertz or MPH)	0-999 Hz (0-159 mph)	0 Hz = 0 mph	65 Hz = 30 mph	125 Hz = 55 mph

2002-05 Blackwood 5.4L V8 DOHC 4v MFI VIN A (A/T) Outputs

PCM PID Acronym	Parameter Identification	PID Range	PID Value at Hot Idle	PID Value at 30	PID Value at 55
ARC	Automatic Ride Sig.	ON / OFF	OFF	OFF	OFF
CD1-CD8	COP Driver 1-8	0-45	5	6	8
CHTIL	CHT (lamp) Control	ON / OFF	OFF	OFF	OFF
EGR VR	EGR VR Solenoid	0-100%	0	0-40	35-45
EVAP CP	EVAP Purge Valve	0-100%	0-100	0-100	0-100
EVAP CV	EVAP CV Valve	0-100%	0-100	0-100	0-100
EPC	EPC Sol. Control	7.4v / 4	4	5	5
FP	Fuel Pump Control	OFF	ON	ON	ON
FuelPW1	INJ Pulsewidth - Bank 1	0-999 ms	3.2-3.8	4.0-6.9	6.5-12
FuelPW2	INJ Pulsewidth - Bank 2	0-999 ms	3.2-3.8	4.0-6.9	6.5-12
HTR-11	HO2S-11 Heater	ON / OFF	ON	ON	ON
HTR-12	HO2S-12 Heater	ON / OFF	ON	ON	ON
HTR-21	HO2S-21 Heater	ON / OFF	ON	ON	ON
HTR-22	HO2S-22 Heater	ON / OFF	ON	ON	ON
IAC	Idle Air Control	0-100%	30-34	43-48	58-61
INJ1	INJ 1 Pulsewidth	0-999 ms	3.2-3.8	4.0-6.9	6.5-12
INJ2	INJ 2 Pulsewidth	0-999 ms	3.2-3.8	4.0-6.9	6.5-12
INJ3	INJ 3 Pulsewidth	0-999 ms	3.2-3.8	4.0-6.9	6.5-12
INJ4	INJ 4 Pulsewidth	0-999 ms	3.2-3.8	4.0-6.9	6.5-12
INJ5	INJ 5 Pulsewidth	0-999 ms	3.2-3.8	4.0-6.9	6.5-12
INJ6	INJ 6 Pulsewidth	0-999 ms	3.2-3.8	4.0-6.9	6.5-12
INJ7	INJ 7 Pulsewidth	0-999 ms	3.2-3.8	4.0-6.9	6.5-12
INJ8	INJ 8 Pulsewidth	0-999 ms	3.2-3.8	4.0-6.9	6.5-12
IMTV	Intake Manifold Tune Valve	0-100%	0	0	0
LongFT1	Long Term FT - Bank 1	-20 to 20%	-5 to +5	-5 to +5	-5 to +5
LongFT2	Long Term FT - Bank 2	-20 to 20%	-5 to +5	-5 to +5	-5 to +5
MIL	MIL (lamp) Control	ON / OFF	OFF	OFF	OFF
SHTFT1	Short Term F/T - Bank 1	-10 to 10%	-5 to +5	-5 to +5	-5 to +5
SHTFT2	Short Term FT - Bank 2	-10 to 10%	-5 to +5	-5 to +5	-5 to +5
SPARK	Spark Advance	-90° to 90°	12-17	35-40	28-37
SS1	Shift Solenoid 1	ON / OFF	ON	OFF	OFF
SS2	Shift Solenoid 2	ON / OFF	OFF	ON	OFF
TCC	TCC Solenoid	0-100%	0	0-40	90-95
TCIL	TCIL (lamp) Control	ON / OFF	OFF	OFF	OFF
TPO	TPO TR Sensor	0-5.1v	0.1	0.8	1.0
VREF	Vehicle Reference	0-5.1v	4.9-5.1	4.9-5.1	4.9-5.1

EXPEDITION & NAVIGATOR PID DATA

1997-2001 Expedition/Navigator 5.4L V8 VIN L (E4OD)

PCM PID Acronym	Parameter Identification	PID Range	PID Value at Hot Idle	PID Value at 30	PID Value at 55
4x4L	4x4 Switch Signal	ON / OFF	ON (if on)	OFF	OFF
ACCS	A/C Switch Signal	ON / OFF	ON (if on)	OFF	OFF
BPP	Brake Position	ON / OFF	ON (if on)	OFF	OFF
CHT	CHT Sensor	-40-500°F	194	194	194
CID	CMP Sensor	0-999 Hz	6-8	10-12	14-17
CKP	CKP Sensor	0-9999 Hz	411	800-840	1000
DPFE	DPFE Sensor	0-5.1v	0.2-1.3	0.2-4.5	0.2-4.5
ECT	ECT Sensor	-40-304°F	160-200	160-200	160-200
FEPS	Flash EEPROM	0-5.1v	0.1	0.1	0.1
FLI	Fuel Level Input	1.7 / 50%	1.7 / 50%	1.7 / 50%	1.7 / 50%
FPM	Fuel Pump Monitor	ON / OFF	ON	ON	ON
FTP	Fuel Tank Pressure	0-5.1v at 0-10" H2O	2.6v at 0" H2O (gas cap "off")	2.6v at 0" H2O (gas cap "off")	2.6v at 0" H2O (gas cap "off")
GEAR	Gear Position	1-2-3-4	1	3	4
HO2S-11	HO2S-11 (Bank 1)	0.1-1.1v	0.1-1.1	0.1-1.1	0.1-1.1
HO2S-12	HO2S-12 (Bank 1)	0.1-1.1v	0.1-1.1	0.1-1.1	0.1-1.1
HO2S-21	HO2S-21 (Bank 1)	0.1-1.1v	0.1-1.1	0.1-1.1	0.1-1.1
HO2S-22	HO2S-22 (Bank 1)	0.1-1.1v	0.1-1.1	0.1-1.1	0.1-1.1
IAT	IAT Sensor	-40-304°F	50-120	50-120	50-120
KS1	Knock Sensor 1	0-5.1v	0	0	0
LOAD	Engine Load	0-100%	14-16	19-25	26-35
MAFV	MAF Sensor	0-5.1v	0.7-0.9	1.1-1.6	1.7-2.4
MISF	Misfire Detection	ON / OFF	OFF	OFF	OFF
OCTADJ	Octane Adjustment	CL/OPEN	CLOSED	CLOSED	CLOSED
OSS	OSS Sensor	0-9999 Hz	0	400	700
PNP	PNP Switch Signal	ON / OFF	ON	OFF	OFF
PTO	Power Takeoff Switch	ON / OFF	ON (if on)	OFF	OFF
RPM	Engine Speed	0-10K rpm	760-830	1200-1270	1590-1675
TCS	TCS Switch	ON / OFF	OFF	OFF	OFF
TFT	TFT Sensor Signal	-40-304°F	110-210	110-210	110-210
TPO	TPO TR Sensor	0-5.1v	0.1	0.8	1.0
TPV	TP Sensor	0-5.1v	0.53-1.27	0.8-1.0	1.0-1.3
TR1	TR1 Sensor	0v / 11.5v	0	11.5	11.5
TR2	TR2 Sensor	0v / 11.5v	0	11.5	11.5
TR3/TR	TR3A Sensor	0v / 1.7v	0	1.7	1.7
TR4	TR4 Sensor	0v / 11.5v	0	11.5	11.5
Vacuum	Engine Vacuum	0-30" Hg	19-20	18-20	8-15
VPWR	Vehicle Power	0-25.5v	12-14	12-14	12-14
VSS	Vehicle Speed (Hertz or MPH)	0-999 Hz (0-159 mph)	0 Hz = 0 mph	65 Hz = 30 mph	125 Hz = 55 mph

1997-2001 Expedition/Navigator 5.4L V8 VIN L (E4OD)

PCM PID Acronym	Parameter Identification	PID Range	PID Value at Hot Idle	PID Value at 30	PID Value at 55
ARC	Auto Ride Control	ON / OFF	OFF	OFF	OFF
CD1-CD8	COP Driver 1-8	0-45	5	6	8
CHTIL	CHT (lamp) Control	ON / OFF	OFF	OFF	OFF
EGR VR	EGR VR Solenoid	0-100%	0	0-40	35-45
EVAP CP	EVAP Purge Valve	0-100%	0-100	0-100	0-100
EVAP CV	EVAP CV Valve	0-100%	0-100	0-100	0-100
EPC	EPC Solenoid	7.4v / 4	4	5	5
FP	Fuel Pump Control	OFF	ON	ON	ON
FuelPW1	INJ Pulsewidth - Bank 1	0-999 ms	3.2-4.0	4.1-6.9	6.5-12
FuelPW2	INJ Pulsewidth - Bank 2	0-999 ms	3.2-4.0	4.1-6.9	6.5-12
HTR-11	HO2S-11 Heater	ON / OFF	ON	ON	ON
HTR-12	HO2S-12 Heater	ON / OFF	ON	ON	ON
HTR-21	HO2S-21 Heater	ON / OFF	ON	ON	ON
HTR-22	HO2S-22 Heater	ON / OFF	ON	ON	ON
IAC	Idle Air Control	0-100%	30-34	43-48	58-61
INJ1	INJ 1 Pulsewidth	0-999 ms	3.2-4.0	4.1-6.9	6.5-12
INJ2	INJ 2 Pulsewidth	0-999 ms	3.2-4.0	4.1-6.9	6.5-12
INJ3	INJ 3 Pulsewidth	0-999 ms	3.2-4.0	4.1-6.9	6.5-12
INJ4	INJ 4 Pulsewidth	0-999 ms	3.2-4.0	4.1-6.9	6.5-12
INJ5	INJ 5 Pulsewidth	0-999 ms	3.2-4.0	4.1-6.9	6.5-12
INJ6	INJ 6 Pulsewidth	0-999 ms	3.2-4.0	4.1-6.9	6.5-12
INJ7	INJ 7 Pulsewidth	0-999 ms	3.2-4.0	4.1-6.9	6.5-12
INJ8	INJ 8 Pulsewidth	0-999 ms	3.2-4.0	4.1-6.9	6.5-12
LongFT1	Long Term FT - Bank 1	-20 to 20%	-5 to +5	-5 to +5	-5 to +5
LongFT2	Long Term FT - Bank 2	-20 to 20%	-5 to +5	-5 to +5	-5 to +5
MIL	MIL (lamp) Control	ON / OFF	OFF	OFF	OFF
SHTFT1	Short Term F/T - Bank 1	-10 to 10%	-5 to +5	-5 to +5	-5 to +5
SHTFT2	Short Term FT - Bank 2	-10 to 10%	-5 to +5	-5 to +5	-5 to +5
SPARK	Spark Advance	-90º to 90º	12-17	35-40	28-37
SS1	Shift Solenoid 1	ON / OFF	ON	OFF	OFF
SS2	Shift Solenoid 2	ON / OFF	OFF	ON	OFF
TCC	TCC Solenoid	0-100%	0	0-40	90-95
TCIL	TCIL (lamp) Control	ON / OFF	OFF	OFF	OFF
TPO	TPO TR Sensor	0-5.1v	0.1	0.8	1.0
VREF	Vehicle Reference	0-5.1v	4.9-5.1	4.9-5.1	4.9-5.1

NAVIGATOR PID DATA

1999-2001 Navigator 5.4L V8 DOHC VIN A (A/T) Inputs

PCM PID Acronym	Parameter Identification	PID Range	PID Value at Hot Idle	PID Value at 30	PID Value at 55
4x4L	4x4 Switch Signal	ON / OFF	ON (if on)	OFF	OFF
ACCS	A/C Switch Signal	ON / OFF	ON (if on)	OFF	OFF
BPP	Brake Position	ON / OFF	ON (if on)	OFF	OFF
CHT	CHT Sensor	-40-500°F	194	194	194
CID	CMP Sensor	0-999 Hz	6-8	10-12	14-17
CKP	CKP Sensor	0-9999 Hz	410	800-850	900-1125
DPFE	DPFE Sensor	0-5.1v	0.2-1.3	0.2-4.5	0.2-4.5
ECT	ECT Sensor	-40-304°F	160-200	160-200	160-200
FEPS	Flash EEPROM	0-5.1v	0.1	0.1	0.1
FLI	Fuel Level Input	1.7 / 50%	1.7 / 50%	1.7 / 50%	1.7 / 50%
FPM	Fuel Pump Monitor	ON / OFF	ON	ON	ON
FTP	Fuel Tank Pressure	0-5.1v at 0-10" H2O	2.6v at 0" H2O (gas cap "off")	2.6v at 0" H2O (gas cap "off")	2.6v at 0" H2O (gas cap "off")
GEAR	Gear Position	1-2-3-4	1	3	4
HO2S-11	HO2S-11 (Bank 1)	0.1-1.1v	0.1-1.1	0.1-1.1	0.1-1.1
HO2S-12	HO2S-12 (Bank 1)	0.1-1.1v	0.1-1.1	0.1-1.1	0.1-1.1
HO2S-21	HO2S-21 (Bank 2)	0.1-1.1v	0.1-1.1	0.1-1.1	0.1-1.1
HO2S-22	HO2S-22 (Bank 2)	0.1-1.1v	0.1-1.1	0.1-1.1	0.1-1.1
IAT	IAT Sensor	-40-304°F	50-120	50-120	50-120
KS1	Knock Sensor 1	0-5.1v	0	0	0
LOAD	Engine Load	0-100%	14-16	19-25	26-35
MAFV	MAF Sensor	0-5.1v	0.7-0.9	1.1-1.6	1.7-2.4
MISF	Misfire Detection	ON / OFF	OFF	OFF	OFF
OCTADJ	Octane Adjustment	CL/OPEN	CLOSED	CLOSED	CLOSED
OSS	OSS Sensor	0-9999 Hz	0	125	350
PNP	PNP Switch Signal	ON / OFF	ON	OFF	OFF
PTO	Power Takeoff Switch	ON / OFF	ON (if on)	OFF	OFF
RPM	Engine Speed	0-10K rpm	790-710	1200-1270	1590-1675
TCS	TCS Switch	ON / OFF	OFF	OFF	OFF
TFT	TFT Sensor Signal	-40-304°F	110-210	110-210	110-210
TPO	TPO TR Sensor	0-5.1v	0.1	0.8	1.0
TPV	TP Sensor	0-5.1v	0.53-1.27	0.8-1.0	1.0-1.3
TR1	TR1 Sensor	0v / 11.5v	0	11.5	11.5
TR2	TR2 Sensor	0v / 11.5v	0	11.5	11.5
TR3/TR	TR3A Sensor	0v / 1.7v	0	1.7	1.7
TR4	TR4 Sensor	0v / 11.5v	0	11.5	11.5
Vacuum	Engine Vacuum	0-30" Hg	19-20	18-20	8-15
VPWR	Vehicle Power	0-25.5v	12-14	12-14	12-14
VSS	Vehicle Speed (Hertz or MPH)	0-999 Hz (0-159 mph)	0 Hz = 0 mph	65 Hz = 30 mph	125 Hz = 55 mph

1999-2001 Navigator 5.4L V8 DOHC VIN A (A/T) Outputs

PCM PID Acronym	Parameter Identification	PID Range	PID Value at Hot Idle	PID Value at 30	PID Value at 55
ARC	Automatic Ride Sig.	ON / OFF	OFF	OFF	OFF
CD1-CD8	COP Driver 1-8	0-45	5	6	8
CHTIL	CHT (lamp) Control	ON / OFF	OFF	OFF	OFF
EGR VR	EGR VR Solenoid	0-100%	0	0-40	35-45
EVAP CP	EVAP Purge Valve	0-100%	0-100	0-100	0-100
EVAP CV	EVAP CV Valve	0-100%	0-100	0-100	0-100
EPC	EPC Sol. Control	7.4v / 4	4	5	5
FP	Fuel Pump Control	OFF	ON	ON	ON
FuelPW1	INJ Pulsewidth - Bank 1	0-999 ms	3.2-3.8	4.0-6.9	6.5-12
FuelPW2	INJ Pulsewidth - Bank 2	0-999 ms	3.2-3.8	4.0-6.9	6.5-12
HTR-11	HO2S-11 Heater	ON / OFF	ON	ON	ON
HTR-12	HO2S-12 Heater	ON / OFF	ON	ON	ON
HTR-21	HO2S-21 Heater	ON / OFF	ON	ON	ON
HTR-22	HO2S-22 Heater	ON / OFF	ON	ON	ON
IAC	Idle Air Control	0-100%	30-34	43-48	58-61
INJ1	INJ 1 Pulsewidth	0-999 ms	3.2-3.8	4.0-6.9	6.5-12
INJ2	INJ 2 Pulsewidth	0-999 ms	3.2-3.8	4.0-6.9	6.5-12
INJ3	INJ 3 Pulsewidth	0-999 ms	3.2-3.8	4.0-6.9	6.5-12
INJ4	INJ 4 Pulsewidth	0-999 ms	3.2-3.8	4.0-6.9	6.5-12
INJ5	INJ 5 Pulsewidth	0-999 ms	3.2-3.8	4.0-6.9	6.5-12
INJ6	INJ 6 Pulsewidth	0-999 ms	3.2-3.8	4.0-6.9	6.5-12
INJ7	INJ 7 Pulsewidth	0-999 ms	3.2-3.8	4.0-6.9	6.5-12
INJ8	INJ 8 Pulsewidth	0-999 ms	3.2-3.8	4.0-6.9	6.5-12
IMTV	Intake Manifold Tune Valve	0-100%	0	0	0
LongFT1	Long Term FT - Bank 1	-20 to 20%	-5 to +5	-5 to +5	-5 to +5
LongFT2	Long Term FT - Bank 2	-20 to 20%	-5 to +5	-5 to +5	-5 to +5
MIL	MIL (lamp) Control	ON / OFF	OFF	OFF	OFF
SHTFT1	Short Term F/T - Bank 1	-10 to 10%	-5 to +5	-5 to +5	-5 to +5
SHTFT2	Short Term FT - Bank 2	-10 to 10%	-5 to +5	-5 to +5	-5 to +5
SPARK	Spark Advance	-90° to 90°	12-17	35-40	28-37
SS1	Shift Solenoid 1	ON / OFF	ON	OFF	OFF
SS2	Shift Solenoid 2	ON / OFF	OFF	ON	OFF
TCC	TCC Solenoid	0-100%	0	0-40	90-95
TCIL	TCIL (lamp) Control	ON / OFF	OFF	OFF	OFF
TPO	TPO TR Sensor	0-5.1v	0.1	0.8	1.0
VREF	Vehicle Reference	0-5.1v	4.9-5.1	4.9-5.1	4.9-5.1

2001-05 Navigator 5.4L V8 4v DOHC VIN R (A/T) Inputs

PCM PID Acronym	Parameter Identification	PID Range	PID Value at Hot Idle	PID Value at 30	PID Value at 55
4x4L	4x4 Switch Signal	ON / OFF	ON (if on)	OFF	OFF
ACCS	A/C Switch Signal	ON / OFF	ON (if on)	OFF	OFF
BPP	Brake Position	ON / OFF	ON (if on)	OFF	OFF
CHT	CHT Sensor	-40-500°F	194	194	194
CID	CMP Sensor	0-999 Hz	6-8	10-12	14-17
CKP	CKP Sensor	0-9999 Hz	410	800-850	900-1125
DPFE	DPFE Sensor	0-5.1v	0.2-1.3	0.2-4.5	0.2-4.5
ECT	ECT Sensor	-40-304°F	160-200	160-200	160-200
FEPS	Flash EEPROM	0-5.1v	0.1	0.1	0.1
FLI	Fuel Level Input	1.7 / 50%	1.7 / 50%	1.7 / 50%	1.7 / 50%
FPM	Fuel Pump Monitor	ON / OFF	ON	ON	ON
FTP	Fuel Tank Pressure	0-5.1v at 0-10" H2O	2.6v at 0" H2O (gas cap "off")	2.6v at 0" H2O (gas cap "off")	2.6v at 0" H2O (gas cap "off")
GEAR	Gear Position	1-2-3-4	1	3	4
HO2S-11	HO2S-11 (Bank 1)	0.1-1.1v	0.1-1.1	0.1-1.1	0.1-1.1
HO2S-12	HO2S-12 (Bank 1)	0.1-1.1v	0.1-1.1	0.1-1.1	0.1-1.1
HO2S-21	HO2S-21 (Bank 2)	0.1-1.1v	0.1-1.1	0.1-1.1	0.1-1.1
HO2S-22	HO2S-22 (Bank 2)	0.1-1.1v	0.1-1.1	0.1-1.1	0.1-1.1
IAT	IAT Sensor	-40-304°F	50-120	50-120	50-120
KS1	Knock Sensor 1	0-5.1v	0	0	0
LOAD	Engine Load	0-100%	14-16	19-25	26-35
MAFV	MAF Sensor	0-5.1v	0.7-0.9	1.1-1.6	1.7-2.4
MISF	Misfire Detection	ON / OFF	OFF	OFF	OFF
OCTADJ	Octane Adjustment	CL/OPEN	CLOSED	CLOSED	CLOSED
OSS	OSS Sensor	0-9999 Hz	0	125	350
PNP	PNP Switch Signal	ON / OFF	ON	OFF	OFF
PTO	Power Takeoff Switch	ON / OFF	ON (if on)	OFF	OFF
RPM	Engine Speed	0-10K rpm	790-710	1200-1270	1590-1675
TCS	TCS Switch	ON / OFF	OFF	OFF	OFF
TFT	TFT Sensor Signal	-40-304°F	110-210	110-210	110-210
TPO	TPO TR Sensor	0-5.1v	0.1	0.8	1.0
TPV	TP Sensor	0-5.1v	0.53-1.27	0.8-1.0	1.0-1.3
TR1	TR1 Sensor	0v / 11.5v	0	11.5	11.5
TR2	TR2 Sensor	0v / 11.5v	0	11.5	11.5
TR3/TR	TR3A Sensor	0v / 1.7v	0	1.7	1.7
TR4	TR4 Sensor	0v / 11.5v	0	11.5	11.5
Vacuum	Engine Vacuum	0-30" Hg	19-20	18-20	8-15
VPWR	Vehicle Power	0-25.5v	12-14	12-14	12-14
VSS	Vehicle Speed (Hertz or MPH)	0-999 Hz (0-159 mph)	0 Hz = 0 mph	65 Hz = 30 mph	125 Hz = 55 mph

2001-05 Navigator 5.4L V8 4v DOHC VIN R (A/T) Outputs

PCM PID Acronym	Parameter Identification	PID Range	PID Value at Hot Idle	PID Value at 30	PID Value at 55
ARC	Automatic Ride Sig.	ON / OFF	OFF	OFF	OFF
CD1-CD8	COP Driver 1-8	0-45	5	6	8
CHTIL	CHT (lamp) Control	ON / OFF	OFF	OFF	OFF
EGR VR	EGR VR Solenoid	0-100%	0	0-40	35-45
EVAP CP	EVAP Purge Valve	0-100%	0-100	0-100	0-100
EVAP CV	EVAP CV Valve	0-100%	0-100	0-100	0-100
EPC	EPC Sol. Control	7.4v / 4	4	5	5
FP	Fuel Pump Control	OFF	ON	ON	ON
FuelPW1	INJ Pulsewidth - Bank 1	0-999 ms	3.2-3.8	4.0-6.9	6.5-12
FuelPW2	INJ Pulsewidth - Bank 2	0-999 ms	3.2-3.8	4.0-6.9	6.5-12
HTR-11	HO2S-11 Heater	ON / OFF	ON	ON	ON
HTR-12	HO2S-12 Heater	ON / OFF	ON	ON	ON
HTR-21	HO2S-21 Heater	ON / OFF	ON	ON	ON
HTR-22	HO2S-22 Heater	ON / OFF	ON	ON	ON
IAC	Idle Air Control	0-100%	30-34	43-48	58-61
INJ1	INJ 1 Pulsewidth	0-999 ms	3.2-3.8	4.0-6.9	6.5-12
INJ2	INJ 2 Pulsewidth	0-999 ms	3.2-3.8	4.0-6.9	6.5-12
INJ3	INJ 3 Pulsewidth	0-999 ms	3.2-3.8	4.0-6.9	6.5-12
INJ4	INJ 4 Pulsewidth	0-999 ms	3.2-3.8	4.0-6.9	6.5-12
INJ5	INJ 5 Pulsewidth	0-999 ms	3.2-3.8	4.0-6.9	6.5-12
INJ6	INJ 6 Pulsewidth	0-999 ms	3.2-3.8	4.0-6.9	6.5-12
INJ7	INJ 7 Pulsewidth	0-999 ms	3.2-3.8	4.0-6.9	6.5-12
INJ8	INJ 8 Pulsewidth	0-999 ms	3.2-3.8	4.0-6.9	6.5-12
IMTV	Intake Manifold Tune Valve	0-100%	0	0	0
LongFT1	Long Term FT - Bank 1	-20 to 20%	-5 to +5	-5 to +5	-5 to +5
LongFT2	Long Term FT - Bank 2	-20 to 20%	-5 to +5	-5 to +5	-5 to +5
MIL	MIL (lamp) Control	ON / OFF	OFF	OFF	OFF
SHTFT1	Short Term F/T - Bank 1	-10 to 10%	-5 to +5	-5 to +5	-5 to +5
SHTFT2	Short Term FT - Bank 2	-10 to 10%	-5 to +5	-5 to +5	-5 to +5
SPARK	Spark Advance	-90° to 90°	12-17	35-40	28-37
SS1	Shift Solenoid 1	ON / OFF	ON	OFF	OFF
SS2	Shift Solenoid 2	ON / OFF	OFF	ON	OFF
TCC	TCC Solenoid	0-100%	0	0-40	90-95
TCIL	TCIL (lamp) Control	ON / OFF	OFF	OFF	OFF
TPO	TPO TR Sensor	0-5.1v	0.1	0.8	1.0
VREF	Vehicle Reference	0-5.1v	4.9-5.1	4.9-5.1	4.9-5.1

EXPLORER PID DATA

1991-95 Explorer 4.0L V6 VIN X (All) Inputs & Outputs

PCM PID Acronym	Parameter Identification	PID Range	PID Value at Hot Idle	PID Value at 30	PID Value at 55
ACCS	A/C Switch Signal	ON / OFF	ON (if on)	OFF	OFF
BARO	BARO Sensor	0-999 Hz	159	159	159
BOO	Brake Switch Signal	ON / OFF	ON (if on)	OFF	OFF
CANP	EVAP Purge Valve	0-100%	0-100	0-100	0-100
DPFE	DPFE Sensor	0-5.1v	0.4-0.6	0.5-1.0	0.7-1.1
ECT (°F)	ECT Sensor	-40-304°F	160-200	160-200	160-200
ECTV	ECT Sensor	0-5.1v	0.7	0.7	0.7
EVR	EGR VR Solenoid	0-100%	0%	0-40	0-40
FPM	Fuel Pump Monitor	ON / OFF	ON	ON	ON
FuelPW1	INJ Pulsewidth - Bank 1	0-999 ms	3.3-3.5	4.0-4.6	5.0-6.0
FuelPW2	INJ Pulsewidth - Bank 2	0-999 ms	3.3-3.5	4.0-4.6	5.0-6.0
IAC	Idle Air Control	0-100%	20-40	34-40	45-55
IAT (°F)	IAT Sensor	-40-304°F	50-120	50-120	50-120
IATV	IAT Sensor	0-5.1v	1.5-3.5	1.5-3.5	1.5-3.5
LongFT1	Long Term F/T - Bank 1	-20 to 20%	-5 to +5	-5 to +5	-5 to +5
LongFT2	Long Term F/T - Bank 2	-20 to 20%	-5 to +5	-5 to +5	-5 to +5
LOOP	Loop Status	CL or OL	CLOSED	CLOSED	CLOSED
MAF	MAF Sensor	0-5.1v	0.7	1.3-1.4	1.7-2.0v
HO2S-11	HO2S-11 (B1 S1)	0.1-1.1v	0.1-1.1	0.1-1.1	0.1-1.1
HO2S-21	HO2S-21 (B1 S1)	0.1-1.1v	0.1-1.1	0.1-1.1	0.1-1.1
PNP	PNP Switch Signal	Neutral/DR	NEUTRAL	DRIVE	DRIVE
RPM	Engine Speed	0-10K rpm	750-830	1500-1650	1800-2100
SHTFT1	Short Term F/T - Bank 1	-10 to 10%	-5 to +5	-5 to +5	-5 to +5
SHTFT2	Short Term FT - Bank 2	-10 to 10%	-5 to +5	-5 to +5	-5 to +5
SPARK	Spark Advance	-90° to 90°	11-20	26-31	20-32
TP	TP Sensor	0-5.1v	0.9	1.2-1.3	1.4-1.6
TPCT	TP sensor Minimum	0-5.1v	0.1	1.2-1.3	1.4-1.6
TP Mode	TP Sensor	C/T or P/T	C/T	P/T	P/T
VPWR	Vehicle Power	0-25.5v	12-14	12-14	12-14
VREF	Vehicle Reference	0-5.1v	4.9-5.1	4.9-5.1	4.9-5.1
VSS	Vehicle Speed	0-159 mph	0	30	55
WAC	A/C WOT Relay Control	ON / OFF	ON (AC on at WOT)	OFF	OFF

1996-2001 Explorer/Mountaineer 4.0L V6 MFI VIN X Inputs

PCM PID Acronym	Parameter Identification	PID Range	PID Value at Hot Idle	PID Value at 30	PID Value at 55
4x4L	4x4 Low Signal	ON / OFF	ON (if on)	OFF	OFF
ACCS	A/C Switch Signal	ON / OFF	ON (if on)	OFF	OFF
ARC	Auto Ride Control	ON / OFF	OFF	OFF	OFF
BPP	Brake Position	ON / OFF	ON (if on)	OFF	OFF
CID	CMP Sensor	0-999 Hz	6-8	13-15	16-19
CKP	CKP Sensor	0-9999	400-475	900-1100	1140-1220
CPP	CPP Switch Signal	ON / OFF	ON (if in)	OFF	OFF
DPFE	DPFE Sensor	0-5.1v	0.4-0.6	0.4-1.0	0.6-1.1
ECT	ECT Sensor	-40-304°F	160-200	160-200	160-200
FEPS	Flash EEPROM	0-5.1v	0.1	0.1	0.1
FLI	Fuel Level Input	1.7 / 50%	1.7 / 50%	1.7 / 50%	1.7 / 50%
FPM	Fuel Pump Monitor	ON / OFF	ON	ON	ON
FTP	Fuel Tank Pressure	0-5.1v at 0-10" H2O	2.6v at 0" H2O (gas cap "off")	2.6v at 0" H2O (gas cap "off")	2.6v at 0" H2O (gas cap "off")
GEAR	Gear Position	1-2-3-4	1	3	4
HO2S-11	HO2S-11 (Bank 1)	0.1-1.1v	0.1-1.1	0.1-1.1	0.1-1.1
HO2S-12	HO2S-12 (Bank 1)	0.1-1.1v	0.1-1.1	0.1-1.1	0.1-1.1
HO2S-21	HO2S-21 (Bank 2)	0.1-1.1v	0.1-1.1	0.1-1.1	0.1-1.1
IAT	IAT Sensor	-40-304°F	50-120	50-120	50-120
LOAD	Engine Load	0-100%	14-20	21-27	30-34
MAFV	MAF Sensor	0-5.1v	0.6-0.9	1.3-1.7	1.5-2.3
MISF	Misfire Detection	ON / OFF	OFF	OFF	OFF
OCTADJ	Octane Adjustment	CL/OPEN	CLOSED	CLOSED	CLOSED
PNP	PNP Switch Signal	ON / OFF	ON	OFF	OFF
PSP	PSP Switch Signal	LOW	HI: turning	LOW	LOW
RPM	Engine Speed	0-10K rpm	750-830	1500-1650	1800-2100
. TCS	TCS Switch	ON / OFF	OFF	OFF	OFF
TFT	TFT Sensor Signal	-40-304°F	110-210	110-210	110-210
TPV	TP Sensor	0-5.1v	0.53-1.27	0.8-1.7	1.2-1.7
TRV/TR	TR Sensor	0v / 1.7v	0	1.7	1.7
TR1	TR1 Sensor	0v / 11.5v	0	11.5	11.5
TR2	TR2 Sensor	0v / 11.5v	0	11.5	11.5
TR4	TR4 Sensor	0v / 11.5v	0	11.5	11.5
TSS	TSS Sensor	0-9999 Hz	85-100	185-205	260-280
Vacuum	Engine Vacuum	0-30" Hg	19-20	18-20	8-15
VPWR	Vehicle Power	0-25.5v	12-14	12-14	12-14
VSS	Vehicle Speed (Hertz or MPH)	0-999 Hz (0-159 mph)	0 Hz = 0 mph	65 Hz = 30 mph	125 Hz = 55 mph

1996-2001 Explorer/Mountaineer 4.0L V6 MFI VIN X outputs

PCM PID Acronym	Parameter Identification	PID Range	PID Value at Hot Idle	PID Value at 30	PID Value at 55
CD1	Coil 1 Driver	0-60°	5	6	8
CD2	Coil 2 Driver	0-60°	5	6	8
CD3	Coil 3 Driver	0-60°	5	6	8
CTO	Clean Tachometer signal	0 -9999	35-49	65-90	90-120
CCS	Coast Clutch Solenoid Control	VBAT	VBAT	VBAT	VBAT
EGR VR	EGR VR Solenoid	0-100%	0	0-40	0-40
VMV ('96)	EVAP Solenoid	0-100%	0-100	0-100	0-100
EVAP CP	EVAP Purge Valve	0-100%	0-100	0-100	0-100
EVAP CV	EVAP CV Valve	0-100%	0-100	0-100	0-100
EPC	EPC Solenoid	0-300 psi	27	36	35
FP	Fuel Pump Control	ON / OFF	ON	ON	ON
FuelPW1	INJ Pulsewidth - Bank 1	0-999 ms	3.4-3.8	3.6-7.5	6.0-9.8
FuelPW2	INJ Pulsewidth - Bank 2	0-999 ms	3.4-3.8	3.6-7.5	6.0-9.8
HTR-11	HO2S-11 Heater	ON / OFF	ON	ON	ON
HTR-12	HO2S-12 Heater	ON / OFF	ON	ON	ON
HTR-21	HO2S-21 Heater	ON / OFF	ON	ON	ON
IAC	Idle Air Control	0-100%	33	35-41	57-68
INJ1	INJ 1 Pulsewidth	0-999 ms	3.4-3.8	3.6-7.5	6.0-9.8
INJ2	INJ 2 Pulsewidth	0-999 ms	3.4-3.8	3.6-7.5	6.0-9.8
INJ3	INJ 3 Pulsewidth	0-999 ms	3.4-3.8	3.6-7.5	6.0-9.8
INJ4	INJ 4 Pulsewidth	0-999 ms	3.4-3.8	3.6-7.5	6.0-9.8
INJ5	INJ 5 Pulsewidth	0-999 ms	3.4-3.8	3.6-7.5	6.0-9.8
INJ6	INJ 6 Pulsewidth	0-999 ms	3.4-3.8	3.6-7.5	6.0-9.8
LongFT1	Long Term FT - Bank 1	-20 to 20%	-5 to +5	-5 to +5	-5 to +5
LongFT2	Long Term FT - Bank 2	-20 to 20%	-5 to +5	-5 to +5	-5 to +5
MIL	MIL (lamp) Control	ON / OFF	OFF	OFF	OFF
SHTFT1	Short Term F/T - Bank 1	-10 to 10%	-5 to +5	-5 to +5	-5 to +5
SHTFT2	Short Term FT - Bank 2	-10 to 10%	-5 to +5	-5 to +5	-5 to +5
SPARK	Spark Advance	-90° to 90°	11-20	26-31	20-32
SS1	Shift Solenoid 1	ON / OFF	ON	OFF	OFF
SS2	Shift Solenoid 2	ON / OFF	OFF	OFF	OFF
SS3	Shift Solenoid 3	ON / OFF	OFF	OFF	ON
TCC	TCC Solenoid	0-100%	0	0	90-95
TCIL	TCIL (lamp) Control	ON / OFF	OFF	OFF	OFF
VREF	Vehicle Reference	0-5.1v	4.9-5.1	4.9-5.1	4.9-5.1
WAC	A/C WOT Relay Control	ON / OFF	ON (AC on at WOT)	ON / OFF	ON / OFF

1997-2005 Explorer/Mountaineer 4.0L V6 VIN E, K Inputs

PCM PID Acronym	Parameter Identification	PID Range	PID Value at Hot Idle	PID Value at 30	PID Value at 55
4x4L	4x4 Low Signal	ON / OFF	ON (if on)	OFF	OFF
ACCS	A/C Switch Signal	ON / OFF	ON (if on)	OFF	OFF
ACP	A/C Pressure	OPEN/CL	OPEN	OPEN	OPEN
ARC	Auto Ride Control	ON / OFF	OFF	OFF	OFF
BPP	Brake Position	ON / OFF	ON (if on)	OFF	OFF
CID	CMP Sensor	0-999 Hz	6-8	11-15	16-19
CKP	CKP Sensor	0-9999	400-475	800-1100	1140-1220
DPFE	DPFE Sensor	0-5.1v	0.4-0.6	0.4-1.0	0.6-1.1
ECT	ECT Sensor	-40-304°F	160-200	160-200	160-200
FEPS	Flash EEPROM	0-5.1v	0.1	0.1	0.1
FLI	Fuel Level Input	1.7 / 50%	1.7 / 50%	1.7 / 50%	1.7 / 50%
FPM	Fuel Pump Monitor	ON / OFF	ON	ON	ON
FTP	Fuel Tank Pressure	0-5.1v at 0-10" H2O	2.6v at 0" H2O (gas cap "off")	2.6v at 0" H2O (gas cap "off")	2.6v at 0" H2O (gas cap "off")
IAT	IAT Sensor	-40-304°F	50-120	50-120	50-120
GEAR	Gear Position	1-2-3-4-5	1	4	5
HO2S-11	HO2S-11 (Bank 1)	0.1-1.1v	0.1-1.1	0.1-1.1	0.1-1.1
HO2S-12	HO2S-12 (Bank 1)	0.1-1.1v	0.1-1.1	0.1-1.1	0.1-1.1
HO2S-21	HO2S-21 (Bank 2)	0.1-1.1v	0.1-1.1	0.1-1.1	0.1-1.1
LOAD	Engine Load	0-100%	14-20	21-27	30-45
MAFV	MAF Sensor	0-5.1v	0.6-0.9	1.3-1.7	1.5-2.3
MISF	Misfire Detection	ON / OFF	OFF	OFF	OFF
OCTADJ	Octane Adjustment	CL/OPEN	CLOSED	CLOSED	CLOSED
PNP	PNP Switch Signal	ON / OFF	ON	OFF	OFF
PSP	PSP Switch Signal	LOW	HI: turning	LOW	LOW
RPM	Engine Speed	0-10K rpm	670-750	1400-1600	1800-2100
TCS	TCS Switch	ON / OFF	OFF	OFF	OFF
TFT	TFT Sensor Signal	-40-304°F	110-210	110-210	110-210
TPV	TP Sensor	0-5.1v	0.53-1.27	0.8-1.7	1.2-1.7
TR1	TR1 Sensor	0v / 11.5v	0v	11.5	11.5
TR2	TR2 Sensor	0v / 11.5v	0v	11.5	11.5
TR4	TR4 Sensor	0v / 11.5v	0v	11.5	11.5
TRV/TR	TR Sensor	0v / 1.7v	0v	1.7	1.7
TSS	TSS Sensor	0-9999 Hz	100-125	185-205	260-280
Vacuum	Engine Vacuum	0-30" Hg	19-20	18-20	8-15
VPWR	Vehicle Power	0-25.5v	12-14	12-14	12-14
VSS	Vehicle Speed (Hertz or MPH)	0-999 Hz (0-159 mph)	0 Hz = 0 mph	65 Hz = 30 mph	125 Hz = 55 mph

1997-2005 Explorer/Mountaineer 4.0L V6 VIN E, K Outputs

PCM PID Acronym	Parameter Identification	PID Range	PID Value at Hot Idle	PID Value at 30	PID Value at 55
CCS	Coast Clutch Solenoid Control	VBAT	VBAT	VBAT	VBAT
CD1	Coil 1 Driver	0-60°	5	6	8
CD2	Coil 2 Driver	0-60°	5	6	8
CD3	Coil 3 Driver	0-60°	5	6	8
CTO	Clean Tachometer signal	0 -9999	35-49	65-90	90-120
EGR VR	EGR VR Solenoid	0-100%	0	0-40	0-40
VMV ('96)	EVAP Solenoid	0-100%	0-100	0-100	0-100
EVAP CP	EVAP Purge Valve	0-100%	0-100	0-100	0-100
EVAP CV	EVAP CV Valve	0-100%	0-100	0-100	0-100
EPC	EPC Solenoid	0-300 psi	26	23-38	34-38
FP	Fuel Pump Control	ON / OFF	ON	ON	ON
FuelPW1	INJ Pulsewidth - Bank 1	0-999 ms	3.4-3.8	3.6-7.5	6.0-9.8
FuelPW2	INJ Pulsewidth - Bank 2	0-999 ms	3.4-3.8	3.6-7.5	6.0-9.8
HTR-11	HO2S-11 Heater	ON / OFF	ON	ON	ON
HTR-12	HO2S-12 Heater	ON / OFF	ON	ON	ON
HTR-21	HO2S-21 Heater	ON / OFF	ON	ON	ON
IAC	Idle Air Control	0-100%	25-32	35-49	30-68
INJ1	INJ 1 Pulsewidth	0-999 ms	3.4-3.8	3.6-7.5	6.0-9.8
INJ2	INJ 2 Pulsewidth	0-999 ms	3.4-3.8	3.6-7.5	6.0-9.8
INJ3	INJ 3 Pulsewidth	0-999 ms	3.4-3.8	3.6-7.5	6.0-9.8
INJ4	INJ 4 Pulsewidth	0-999 ms	3.4-3.8	3.6-7.5	6.0-9.8
INJ5	INJ 5 Pulsewidth	0-999 ms	3.4-3.8	3.6-7.5	6.0-9.8
INJ6	INJ 6 Pulsewidth	0-999 ms	3.4-3.8	3.6-7.5	6.0-9.8
LongFT1	Long Term FT - Bank 1	-20 to 20%	-5 to +5	-5 to +5	-5 to +5
LongFT2	Long Term FT - Bank 2	-20 to 20%	-5 to +5	-5 to +5	-5 to +5
MIL	MIL (lamp) Control	ON / OFF	OFF	OFF	OFF
SHTFT1	Short Term F/T - Bank 1	-10 to 10%	-5 to +5	-5 to +5	-5 to +5
SHTFT2	Short Term FT - Bank 2	-10 to 10%	-5 to +5	-5 to +5	-5 to +5
SPARK	Spark Advance	-90° to 90°	11-20	31-36	32-40
SS1	Shift Solenoid 1	ON / OFF	ON	OFF	OFF
SS2	Shift Solenoid 2	ON / OFF	OFF	OFF	OFF
SS3	Shift Solenoid 3	ON / OFF	OFF	OFF	ON
SS4	Shift Solenoid 4	ON / OFF	OFF	OFF	ON
TCC	TCC Solenoid	0-100%	0	0	90-95
TCIL	TCIL (lamp) Control	ON / OFF	OFF	OFF	OFF
VREF	Vehicle Reference	0-5.1v	4.9-5.1	4.9-5.1	4.9-5.1
WAC	A/C WOT Relay Control	ON / OFF	ON (AC on at WOT)	ON / OFF	ON / OFF

2002-05 Explorer/Mountaineer 4.6L V8 MFI VIN W (A/T) Inputs

PCM PID Acronym	Parameter Identification	PID Range	PID Value at Hot Idle	PID Value at 30	PID Value at 55
4x4L	4x4 Switch Signal	ON / OFF	ON (if on)	OFF	OFF
ACCS	A/C Switch Signal	ON / OFF	ON (if on)	OFF	OFF
BPP	Brake Position	ON / OFF	ON (if on)	OFF	OFF
DPFE	DPFE Sensor	0-5.1v	0.2-1.3	0.2-4.5	0.2-4.5
CHT	CHT Sensor	-40-500°F	194	194	194
FEPS	Flash EEPROM	0-5.1v	0.1	0.1	0.1
FLI	Fuel Level Input	1.7 / 50%	1.7 / 50%	1.7 / 50%	1.7 / 50%
FPM	Fuel Pump Monitor	ON / OFF	ON	ON	ON
FTP	Fuel Tank Pressure	0-5.1v at 0-10" H2O	2.6v at 0" H2O (gas cap "off")	2.6v at 0" H2O (gas cap "off")	2.6v at 0" H2O (gas cap "off")
GEAR	Gear Position	1-2-3-4	1	3	4
HO2S-11	HO2S-11 (Bank 1)	0.1-1.1v	0.1-1.1	0.1-1.1	0.1-1.1
HO2S-12	HO2S-12 (Bank 1)	0.1-1.1v	0.1-1.1	0.1-1.1	0.1-1.1
HO2S-21	HO2S-21 (Bank 2)	0.1-1.1v	0.1-1.1	0.1-1.1	0.1-1.1
HO2S-22	HO2S-22 (Bank 2)	0.1-1.1v	0.1-1.1	0.1-1.1	0.1-1.1
IAT	IAT Sensor	-40-304°F	50-120	50-120	50-120
LOAD	Engine Load	0-100%	14-16	20-25	24-35
MAFV	MAF Sensor	0-5.1v	0.7-0.9	1.2-1.7	1.6-2.4
MISF	Misfire Detection	ON / OFF	OFF	OFF	OFF
OCTADJ	Octane Adjustment	CL/OPEN	CLOSED	CLOSED	CLOSED
OSS	OSS Sensor	0-9999 Hz	0	125-131	245-255
PNP	PNP Switch Signal	ON / OFF	ON	OFF	OFF
PTO	Power Takeoff Switch	ON / OFF	ON (if on)	OFF	OFF
RPM	Engine Speed	0-10K rpm	680-830	1200-1500	1600-1800
TCS	TCS Switch	ON / OFF	OFF	OFF	OFF
TFT	TFT Sensor Signal	-40-304°F	110-210	110-210	110-210
TPV	TP Sensor	0-5.1v	0.53-1.27	0.8-1.0	0.9-1.2
TRV/TR	TR Sensor	0v / 1.7v	0	1.7	1.7
TR1	TR1 Sensor	0v / 11.5v	0	11.5	11.5
TR2	TR2 Sensor	0v / 11.5v	0	11.5	11.5
TR4	TR4 Sensor	0v / 11.5v	0	11.5	11.5
Vacuum	Engine Vacuum	0-30" Hg	19-20	18-20	8-15
VPWR	Vehicle Power	0-25.5v	12-14	12-14	12-14
VSS	Vehicle Speed (Hertz or MPH)	0-999 Hz (0-159 mph)	0 Hz = 0 mph	65 Hz = 30 mph	125 Hz = 55 mph

2002-05 Explorer/Mountaineer 4.6L V8 MFI VIN W A/T Outputs

PCM PID Acronym	Parameter Identification	PID Range	PID Value at Hot Idle	PID Value at 30	PID Value at 55
CCS	Coast Clutch Solenoid Control	ON / OFF	OFF	OFF	OFF
EGR VR	EGR VR Solenoid	0-100%	0	0-40	35-45
EVAP CP	EVAP Purge Valve	0-100%	0-100	0-100	0-100
EVAP CV	EVAP CV Valve	0-100%	0-100	0-100	0-100
EPC	EPC Solenoid	0-300 psi	6	40	40
FP	Fuel Pump Control	ON / OFF	ON	ON	ON
FuelPW1	INJ Pulsewidth - Bank 1	0-999 ms	2.7-4.1	4.5-8.0	5.5-9
FuelPW2	INJ Pulsewidth - Bank 2	0-999 ms	2.7-4.1	4.5-8.0	5.5-9
HTR-11	HO2S-11 Heater	ON / OFF	ON	ON	ON
HTR-12	HO2S-12 Heater	ON / OFF	ON	ON	ON
HTR-21	HO2S-21 Heater	ON / OFF	ON	ON	ON
HTR-22	HO2S-22 Heater	ON / OFF	ON	ON	ON
IAC	Idle Air Control	0-100%	30-34	28-32	63-70
INJ1	INJ 1 Pulsewidth	0-999 ms	2.7-4.1	4.5-8.0	5.5-9
INJ2	INJ 2 Pulsewidth	0-999 ms	2.7-4.1	4.5-8.0	5.5-9
INJ3	INJ 3 Pulsewidth	0-999 ms	2.7-4.1	4.5-8.0	5.5-9
INJ4	INJ 4 Pulsewidth	0-999 ms	2.7-4.1	4.5-8.0	5.5-9
INJ5	INJ 5 Pulsewidth	0-999 ms	2.7-4.1	4.5-8.0	5.5-9
INJ6	INJ 6 Pulsewidth	0-999 ms	2.7-4.1	4.5-8.0	5.5-9
INJ7	INJ 7 Pulsewidth	0-999 ms	2.7-4.1	4.5-8.0	5.5-9
INJ8	INJ 8 Pulsewidth	0-999 ms	2.7-4.1	4.5-8.0	5.5-9
LongFT1	Long Term FT - Bank 1	-20 to 20%	-5 to +5	-5 to +5	-5 to +5
LongFT2	Long Term FT - Bank 2	-20 to 20%	-5 to +5	-5 to +5	-5 to +5
MIL	MIL (lamp) Control	ON / OFF	OFF	OFF	OFF
SHTFT1	Short Term F/T - Bank 1	-10 to 10%	-5 to +5	-5 to +5	-5 to +5
SHTFT2	Short Term FT - Bank 2	-10 to 10%	-5 to +5	-5 to +5	-5 to +5
SPARK	Spark Advance	-90° to 90°	15-18	23-29	26-32
SS1	Shift Solenoid 1	ON / OFF	ON	OFF	OFF
SS2	Shift Solenoid 2	ON / OFF	OFF	ON	OFF
TCC	TCC Solenoid	0-100%	0	90-100	90-100
TCIL	TCIL (lamp) Control	ON / OFF	OFF	OFF	OFF
VREF	Vehicle Reference	0-5.1v	4.9-5.1	4.9-5.1	4.9-5.1

1996-2001 Explorer/Mountaineer 5.0L V8 VIN P (A/T) Inputs

PCM PID Acronym	Parameter Identification	PID Range	PID Value at Hot Idle	PID Value at 30	PID Value at 55
ACCS	A/C Switch Signal	ON / OFF	ON (if on)	OFF	OFF
ARC	Auto Ride Control	ON / OFF	OFF	OFF	OFF
BPP	Brake Position	ON / OFF	ON (if on)	OFF	OFF
CID	CMP Sensor	0-999	5-10	10-15	11-17
CKP	CKP Sensor	0-9999 Hz	460-500	650-720	950-1050
ECT	ECT Sensor	-40-304°F	160-200	160-200	160-200
EGR VP (1996-97)	EGR Valve Position	0-5.1v	0.1	0.5-1.5	0.5-1.8
DPFE ('98-'00)	DPFE Sensor Input	0-5.1v	0.20-1.30	0.20-2.75	0.20-2.75
DPFE ('01-'02)	DPFE Sensor Input	0-5.1v	0.95-1.05	0.96-4.65	0.95-4.65
FEPS	Flash EEPROM	0-5.1v	0.1	0.1	0.1
FLI	Fuel Level Input	1.7 / 50%	1.7 / 50%	1.7 / 50%	1.7 / 50%
FPM	Fuel Pump Monitor	ON / OFF	ON	ON	ON
FTP	Fuel Tank Pressure	0-5.1v at 0-10" H2O	2.6v at 0" H2O (gas cap "off")	2.6v at 0" H2O (gas cap "off")	2.6v at 0" H2O (gas cap "off")
IAT	IAT Sensor	-40-304°F	50-120	50-120	50-120
GEAR	Gear Position	1-2-3-4	1	3	4
LOAD	Engine Load	0-100%	10-20	19-33	22-36
MAFV	MAF Sensor	0-5.1v	0.7-0.9	0.9-1.7	1.5-0
MISF	Misfire Detection	ON / OFF	OFF	OFF	OFF
OCTADJ	Octane Adjustment	CL/OPEN	CLOSED	CLOSED	CLOSED
HO2S-11	HO2S-11 (Bank 1)	0.1-1.1v	0.1-1.1	0.1-1.1	0.1-1.1
HO2S-12	HO2S-12 (Bank 1)	0.1-1.1v	0.1-1.1	0.1-1.1	0.1-1.1
HO2S-21	HO2S-21 (Bank 2)	0.1-1.1v	0.1-1.1	0.1-1.1	0.1-1.1
HO2S-22	HO2S-22 (Bank 2)	0.1-1.1v	0.1-1.1	0.1-1.1	0.1-1.1
PNP	PNP Switch Signal	ON / OFF	ON	OFF	OFF
RPM	Engine Speed	0-10K rpm	760-830	1180-1290	1570-1695
TCS	TCS Switch	ON / OFF	OFF	OFF	OFF
TFT	TFT Sensor Signal	-40-304°F	110-210	110-210	110-210
TPV	TP Sensor	0-5.1v	0.53-1.27	0.8-1.1	1.0-1.4
TRV/TR	TR Sensor	0v / 1.7v	0	1.7	1.7
TR1	TR1 Sensor	0v / 11.5v	0	11.5	11.5
TR2	TR2 Sensor	0v / 1.7v	0	11.5	11.5
TR4	TR4 Sensor	0v / 1.7v	0	11.5	11.5
OSS	Output Shaft Speed	0-9999 Hz	0	115-145	230-280
Vacuum	Engine Vacuum	0-30" Hg	19-20	18-20	8-15
VPWR	Vehicle Power	0-25.5v	12-14	12-14	12-14
VSS	Vehicle Speed (Hertz or MPH)	0-999 Hz (0-159 mph)	0 Hz = 0 mph	65 Hz = 30 mph	125 Hz = 55 mph

1996-2001 Explorer/Mountaineer 5.0L V8 VIN P A/T Outputs

PCM PID Acronym	Parameter Identification	PID Range	PID Value at Hot Idle	PID Value at 30	PID Value at 55
CD1	Coil 1 Driver	0-90°	5	6	8
CD2	Coil 2 Driver	0-90°	5	6	8
CD3	Coil 3 Driver	0-90°	5	6	8
CD4	Coil 4 Driver	0-90°	5	6	8
CTO	Clean Tachometer signal	0-9999 Hz	40-55	80-95	100-125
EGR VR	EGR VR Solenoid	0-100%	0	0-40	0-40
EVAP CP	EVAP Purge Valve	0-100%	0-100	0-100	0-100
EVAP VMV	EVAP Purge Valve (1996-97)	0-100%	0-100	0-100	0-100
EVAP CV	EVAP CV Valve	0-100%	0	0	0
EPC	EPC Solenoid	0-300 psi	10	23-38	34-38
FP	Fuel Pump Control	ON / OFF	ON	ON	ON
FuelPW1	INJ Pulsewidth - Bank 1	0-999 ms	3.2-4.5	4.1-8.0	5.5-12.0
FuelPW2	INJ Pulsewidth - Bank 2	0-999 ms	3.2-4.5	4.1-8.0	5.5-12.0
HTR-11	HO2S-11 Heater	ON / OFF	ON	ON	ON
HTR-12	HO2S-12 Heater	ON / OFF	ON	ON	ON
HTR-21	HO2S-21 Heater	ON / OFF	ON	ON	ON
IAC	Idle Air Control	0-100%	33	35-41	57-68
INJ1	INJ 1 Pulsewidth	0-999 ms	3.2-4.5	4.1-8.0	5.5-12.0
INJ2	INJ 2 Pulsewidth	0-999 ms	3.2-4.5	4.1-8.0	5.5-12.0
INJ3	INJ 3 Pulsewidth	0-999 ms	3.2-4.5	4.1-8.0	5.5-12.0
INJ4	INJ 4 Pulsewidth	0-999 ms	3.2-4.5	4.1-8.0	5.5-12.0
INJ5	INJ 5 Pulsewidth	0-999 ms	3.2-4.5	4.1-8.0	5.5-12.0
INJ6	INJ 6 Pulsewidth	0-999 ms	3.2-4.5	4.1-8.0	5.5-12.0
INJ7	INJ 7 Pulsewidth	0-999 ms	3.2-4.5	4.1-8.0	5.5-12.0
INJ8	INJ 8 Pulsewidth	0-999 ms	3.2-4.5	4.1-8.0	5.5-12.0
LongFT1	Long Term FT - Bank 1	-20 to 20%	-5 to +5	-5 to +5	-5 to +5
LongFT2	Long Term FT - Bank 2	-20 to 20%	-5 to +5	-5 to +5	-5 to +5
MIL	MIL (lamp) Control	ON / OFF	OFF	OFF	OFF
SHTFT1	Short Term F/T - Bank 1	-10 to 10%	-5 to +5	-5 to +5	-5 to +5
SHTFT2	Short Term F/T - Bank 2	-10 to 10%	-5 to +5	-5 to +5	-5 to +5
SPARK	Spark Advance	-90° to 90°	12-17	32-40°	25-37°
SS1	Shift Solenoid 1	ON / OFF	ON	OFF	ON
SS2	Shift Solenoid 2	ON / OFF	OFF	ON	ON
TCC	TCC Solenoid	0-100%	0	0-40	90-95
TCIL	TCIL (lamp) Control	ON / OFF	OFF	OFF	OFF
VREF	Vehicle Reference	0-5.1v	4.9-5.1	4.9-5.1	4.9-5.1
WAC	A/C WOT Relay Control	ON / OFF	ON (AC on at WOT)	ON / OFF	ON / OFF

ESCAPE PID DATA

2001-05 Escape 2.0L I4 DOHC 4v MFI VIN B (All) Inputs

PCM PID Acronym	Parameter Identification	PID Range	PID Value at Hot Idle	PID Value at 30	PID Value at 55
ACCS	A/C Clutch Switch	ON / OFF	ON (if on)	OFF	OFF
ACP	A/C Press. Switch	CL/OPEN	Open	OPEN	OPEN
BPP	Brake Pedal Switch	ON / OFF	ON (if on)	OFF	OFF
CID	CID Sensor Hertz	0-1 KHz	5-7	13-16	20-23
CKP	CKP Sensor Hertz	0-10 KHz	400-450	985-1100	1450-1550
CHT (°F)	CHT Sensor	-40-304°F	160-200	160-200	160-200
CPP	CPP Switch	ON / OFF	ON (if in)	OFF	OFF
DPFE	DPFE Sensor Input	0-5.1v	0.95-1.05	0.96-4.65	0.95-4.65
FEPS	Flash EEPROM	0-5.1v	0.1	0.1	0.1
FLI	Fuel Level Indicator	0-100%	50% (1.7)	50% (1.7)	50% (1.7)
FPM	Fuel Pump Monitor	ON / OFF	ON	ON	ON
FTP	Fuel Tank Pressure	0-5.1v or 0-10" H2O	2.6v / 0" H2O (with cap off)	2.6v / 0" H2O (with cap off)	2.6v / 0" H2O (with cap off)
GEAR	Gear Position	1-2-3-4	1	3	4
GFDC	Generator Field DC	0-100%	0	0	0
HO2S-11	HO2S-11 (Bank 1)	0.1-1.1v	0.1-1.1	0.1-1.1	0.1-1.1
HO2S-12	HO2S-12 (Bank 1)	0.1-1.1v	0.1-1.1	0.1-1.1	0.1-1.1
IAT (°F)	IAT Sensor	-40-304°F	50-120	50-120	50-120
KS1	Knock Sensor 1	0-5.1v	0	0	0
LOAD	Engine Load	0-100%	10-20	19-30	35-44
MAFV	MAF Sensor Volts	0-5.1v	0.6-0.9	0.8-1.5	1.2-2.5
MISF	Misfire Detection	ON / OFF	OFF	OFF	OFF
OCTADJ	Octane Adjustment	Retard/NO	NO RET	NO RET	NO RET
OSS	OSS Sensor	0-9999 Hz	0	400	700-740
PATSIN	PATS Input	0-5.1v	0	0	0
PNP	PNP Switch	ON / OFF	ON (in 'P')	OFF	OFF
PSP	PSP Switch	HI / LOW	HI: turning	LOW	LOW
RPM	Engine Speed	0-10K rpm	700-800	1750-1800	2500-2660
TCS	TCS Switch	ON / OFF	OFF	ON	ON
TFT (°F)	TFT Sensor (°F)	-40-304°F	110-210	110-210	110-210
TP (V)	TP Sensor Volts	0-5.1v	0.53-1.27	0.8-1.2	1.0-1.5
Vacuum	Engine Vacuum	0-30" Hg	19-20	18-20	8-15
VPWR	Vehicle Power	0-25.5v	12-14	12-14	12-14
VSS	Vehicle Speed (MPH & Hertz Rate)	0-999 Hz (0-159 mph)	0 = 0 Hz	30 = 65 Hz	55 = 125 Hz

2001-05 Escape 2.0L I4 DOHC 4v MFI VIN B (All) Outputs

PCM PID Acronym	Parameter Identification	PID Range	PID Value at Hot Idle	PID Value at 30	PID Value at 55
CD1	Coil 1 Driver	0-90° dwell	7	10	12
CD2	Coil 2 Driver	0-90° dwell	7	10	12
EGRVR	VR Solenoid	0-100%	0	0-40	0-60
EPC	EPC Solenoid	0-500 psi	15-18	23-28	30-35
EVAP DC	EVAP Purge	0-100%	0-100	0-100	0-100
EVAPCV	EVAP CV	0-100%	0-100	0-100	0-100
FP	Fuel Pump Relay	ON / OFF	ON	ON	ON
FUELPW	INJ 1 Pulsewidth	0-99.9 ms	2.5-3.0	3.3-5.2	4.6-6.5
HFC	High Fan Control	ON / OFF	ON (if on)	OFF	OFF
HTR11	HO2S-11 Heater	ON / OFF	ON	ON	ON
HTR12	HO2S-12 Heater	ON / OFF	ON	ON	ON
IAC	Idle Air Control	0-100%	33-35	40-48	40-55
INJ1	INJ 1 Pulsewidth	0-999 ms	2.5-3.0	3.3-5.2	4.6-6.5
INJ2	INJ 2 Pulsewidth	0-999 ms	2.5-3.0	3.3-5.2	4.6-6.5
INJ3	INJ 3 Pulsewidth	0-999 ms	2.5-3.0	3.3-5.2	4.6-6.5
INJ4	INJ 4 Pulsewidth	0-999 ms	2.5-3.0	3.3-5.2	4.6-6.5
LFC	Low Fan Control	ON / OFF	ON (if on)	OFF	OFF
LONGFT	LONGFT Bank 1	-20 to 20%	-5 to +5	-5 to +5	-5 to +5
MAF	MAF Sensor	0-5.1v	0.6-0.9	1.0-1.7	1.2-2.5
MFC	Medium Fan control	ON / OFF	ON (if on)	OFF	OFF
MIL	MIL Control	ON / OFF	ON (if on)	OFF	OFF
SS1	Shift Solenoid 1	ON / OFF	OFF	ON	ON
SS2	Shift Solenoid 2	ON / OFF	ON	OFF	OFF
SHRTFT	SHRTFT Bank 1	-20 to 20%	-5 to +5	-5 to +5	-5 to +5
SPARK	Spark Advance	-90° to 90°	23-35	30-45	25-36
TCC	TCC Solenoid	0-100%	0	0	95-100
VSO	Vehicle Speed	0-1 KHz	0	65	125
VREF	Voltage Reference	0-5.1v	4.9-5.1	4.9-5.1	4.9-5.1
WAC	A/C WOT Cutout Relay	ON / OFF	ON (if on)	OFF	OFF

2001-05 Escape 3.0L V6 DOHC 4v MFI VIN 1 (All) Inputs

PCM PID Acronym	Parameter Identification	PID Range	PID Value at Hot Idle	PID Value at 30	PID Value at 55
ACCS	A/C Clutch Switch	ON / OFF	ON (if on)	OFF	OFF
ACP	A/C Press. Switch	CL/OPEN	Open	OPEN	OPEN
BPP	Brake Pedal Switch	ON / OFF	ON (if on)	OFF	OFF
CID	CID Sensor Hertz	0-1 KHz	5-7	12-15	14-16
CKP	CKP Sensor Hertz	0-10 KHz	400-450	850-1050	1050-1150
DPFE	DPFE Sensor Input	0-5.1v	0.95-1.05	0.96-4.65	0.95-4.65
ECT (°F)	ECT Sensor	-40-304°F	160-200	160-200	160-200
FEPS	Flash EEPROM	0-5.1v	0.1	0.1	0.1
FLI	Fuel Level Indicator	0-100%	50% (1.7)	50% (1.7)	50% (1.7)
FPM	Fuel Pump Monitor	ON / OFF	ON	ON	ON
FTP	Fuel Tank Pressure (1998-2000)	0-5.1v or 0-10" H2O	2.6v / 0" H2O (with cap off)	2.6v / 0" H2O (with cap off)	2.6v / 0" H2O (with cap off)
GEAR	Gear Position	1-2-3-4	1	3	4
HO2S-11	HO2S-11 (Bank 1)	0.1-1.1v	0.1-1.1	0.1-1.1	0.1-1.1
HO2S-12	HO2S-12 (Bank 1)	0.1-1.1v	0.1-1.1	0.1-1.1	0.1-1.1
HO2S-22	HO2S-22 (Bank 2)	0.1-1.1v	0.1-1.1	0.1-1.1	0.1-1.1
IAT (°F)	IAT Sensor	-40-304°F	50-120	50-120	50-120
KS1	Knock Sensor 1	0	0	0	0
LOAD	Engine Load	0-100%	17-21	21-30	33-40
MAFV	MAF Sensor Volts	0-5.1v	0.8-1.1	1.4-1.7	2.0-2.5
MISF	Misfire Detection	ON / OFF	OFF	OFF	OFF
OCTADJ	Octane Adjustment	CL/OPEN	CLOSED	CLOSED	CLOSED
OSS	OSS Sensor	0-9999 Hz	0	400	700-740
PATSIN	PATS Input	0-5.1v	0	0	0
PNP	PNP Switch	ON / OFF	ON (in 'P')	OFF	OFF
PSP	PSP Switch	HI / LOW	HI: turning	LOW	LOW
RPM	Engine Speed	0-10K rpm	730-750	1550-1700	1800-2000
TCS	TCS Switch	ON / OFF	OFF	ON	ON
TFT (°F)	TFT Sensor (°F)	-40-304°F	110-210	110-210	110-210
TP (V)	TP Sensor Volts	0-5.1v	0.53-1.27	0.8-1.2	1.0-1.5
TR (V)	TR Sensor Volts	0-5.1v	4.4	2.1	2.1
TSS	TSS Sensor	0-9999 Hz	45-50	90-100	110-120
Vacuum	Engine Vacuum	0-30" Hg	19-20	18-20	8-15
VPWR	Vehicle Power	0-25.5v	12-14	12-14	12-14
VSS	Vehicle Speed (MPH & Hertz Rate)	0-159 mph	0 = 0 Hz	30 = 65 Hz	55 = 125 Hz

2001-05 Escape 3.0L V6 DOHC 4v MFI VIN 1 (All) Outputs

PCM PID Acronym	Parameter Identification	PID Range	PID Value at Hot Idle	PID Value at 30	PID Value at 55
COP1	COP 1 Driver	0-90° dwell	7	10	12
COP2	COP 2 Driver	0-90° dwell	7	10	12
COP3	COP 3 Driver	0-90° dwell	7	10	12
COP4	COP 4 Driver	0-90° dwell	7	10	12
COP5	COP 5 Driver	0-90° dwell	7	10	12
COP6	COP 6 Driver	0-90° dwell	7	10	12
EGRVR	VR Solenoid	0-100%	0	0-40	0-60
EPC	EPC Solenoid	0-500 psi	25-37	42-51	50-79
EVAP DC	EVAP Purge	0-100%	0-100	80	90
EVAPCV	EVAP CV	0-100%	0-100	0-100	0-100
FP	Fuel Pump Relay	0-100%	100	100	100
FUELPW	INJ 1 Pulsewidth	0-99.9 ms	2.6-3.2	2.5-5.5	3.5-8.5
HFC	High Fan Control	ON / OFF	ON (if on)	OFF	OFF
HTR11	HO2S-11 Heater	ON / OFF	ON	ON	ON
HTR12	HO2S-12 Heater	ON / OFF	ON	ON	ON
HTR22	HO2S-22 Heater	ON / OFF	ON	ON	ON
IAC	Idle Air Control	0-100%	25-35	30-55	50-79
INJ1	INJ 1 Pulsewidth	0-999 ms	2.6-3.2	2.5-5.5	3.5-8.5
INJ2	INJ 2 Pulsewidth	0-999 ms	2.6-3.2	2.5-5.5	3.5-8.5
INJ3	INJ 3 Pulsewidth	0-999 ms	2.6-3.2	2.5-5.5	3.5-8.5
INJ4	INJ 4 Pulsewidth	0-999 ms	2.6-3.2	2.5-5.5	3.5-8.5
INJ5	INJ 5 Pulsewidth	0-999 ms	2.6-3.2	2.5-5.5	3.5-8.5
INJ6	INJ 6 Pulsewidth	0-999 ms	2.6-3.2	2.5-5.5	3.5-8.5
LFC	Low Fan Control	ON / OFF	ON (if on)	OFF	OFF
LONGFT	LONGFT Bank 1	-20 to 20%	-5 to +5	-5 to +5	-5 to +5
MIL	MIL Control	ON / OFF	ON (if on)	OFF	OFF
PATSOUT	PATS Output	0-5.1v	0.8	0.8	0.8
PATSTRT	PATS Output	0-5.1v	0	0	0
SHRTFT	SHRTFT Bank 1	-20 to 20%	-5 to +5	-5 to +5	-5 to +5
SPARK	Spark Advance	-90° to 90°	16-18	20-36	25-35
SS1	Shift Solenoid 1	ON / OFF	OFF	ON	OFF
SS2	Shift Solenoid 2	ON / OFF	ON	OFF	OFF
SS3	Shift Solenoid 3	ON / OFF	OFF	OFF	ON
TCC	TCC Solenoid	0-100%	0	50-90	95-100
VSO	Vehicle Speed	0-10 KHz	0	65	125
VREF	Voltage reference	0-5.1v	4.9-5.1	4.9-5.1	4.9-5.1
WAC	A/C WOT Cutout Relay	ON / OFF	ON (if on)	OFF	OFF

EXCURSION PID DATA

2000-05 Excursion 5.4L V8 VIN L (E4OD) Inputs

PCM PID Acronym	Parameter Identification	PID Range	PID Value at Hot Idle	PID Value at 30	PID Value at 55
4x4L	4x4 Switch Signal	ON / OFF	ON (if on)	OFF	OFF
ACCS	A/C Switch Signal	ON / OFF	ON (if on)	OFF	OFF
BPP	Brake Position	ON / OFF	ON (if on)	OFF	OFF
CHT	CHT Sensor	-40-500°F	194	194	194
CID	CMP Sensor	0-999 Hz	6-8	10-12	14-17
CKP	CKP Sensor	0-9999 Hz	411	800-840	1000
DPFE	DPFE Sensor	0-5.1v	0.2-1.3	0.2-4.5	0.2-4.5
ECT	ECT Sensor	-40-304°F	160-200	160-200	160-200
FEPS	Flash EEPROM	0-5.1v	0.1	0.1	0.1
FLI	Fuel Level Input	1.7 / 50%	1.7 / 50%	1.7 / 50%	1.7 / 50%
FPM	Fuel Pump Monitor	ON / OFF	ON	ON	ON
FTP	Fuel Tank Pressure	0-5.1v at 0-10" H2O	2.6v at 0" H2O (gas cap "off")	2.6v at 0" H2O (gas cap "off")	2.6v at 0" H2O (gas cap "off")
GEAR	Gear Position	1-2-3-4	1	3	4
HO2S-11	HO2S-11 (Bank 1)	0.1-1.1v	0.1-1.1	0.1-1.1	0.1-1.1
HO2S-12	HO2S-12 (Bank 1)	0.1-1.1v	0.1-1.1	0.1-1.1	0.1-1.1
HO2S-21	HO2S-21 (Bank 1)	0.1-1.1v	0.1-1.1	0.1-1.1	0.1-1.1
HO2S-22	HO2S-22 (Bank 1)	0.1-1.1v	0.1-1.1	0.1-1.1	0.1-1.1
IAT	IAT Sensor	-40-304°F	50-120	50-120	50-120
KS1	Knock Sensor 1	0-5.1v	0	0	0
LOAD	Engine Load	0-100%	14-16	19-25	26-35
MAFV	MAF Sensor	0-5.1v	0.7-0.9	1.1-1.6	1.7-2.4
MISF	Misfire Detection	ON / OFF	OFF	OFF	OFF
OCTADJ	Octane Adjustment	CL/OPEN	CLOSED	CLOSED	CLOSED
OSS	OSS Sensor	0-9999 Hz	0	400	700
PNP	PNP Switch Signal	ON / OFF	ON	OFF	OFF
PTO	Power Takeoff Switch	ON / OFF	ON (if on)	OFF	OFF
RPM	Engine Speed	0-10K rpm	760-830	1200-1270	1590-1675
TCS	TCS Switch	ON / OFF	OFF	OFF	OFF
TFT	TFT Sensor Signal	-40-304°F	110-210	110-210	110-210
TPO	TPO TR Sensor	0-5.1v	0.1	0.8	1.0
TPV	TP Sensor	0-5.1v	0.53-1.27	0.8-1.0	1.0-1.3
TR1	TR1 Sensor	0v / 11.5v	0	11.5	11.5
TR2	TR2 Sensor	0v / 11.5v	0	11.5	11.5
TR3/TR	TR3A Sensor	0v / 1.7v	0	1.7	1.7
TR4	TR4 Sensor	0v / 11.5v	0	11.5	11.5
Vacuum	Engine Vacuum	0-30" Hg	19-20	18-20	8-15
VPWR	Vehicle Power	0-25.5v	12-14	12-14	12-14
VSS	Vehicle Speed (Hertz or MPH)	0-999 Hz (0-159 mph)	0 Hz = 0 mph	65 Hz = 30 mph	125 Hz = 55 mph

2000-05 Excursion 5.4L V8 VIN L (E4OD) Outputs

PCM PID Acronym	Parameter Identification	PID Range	PID Value at Hot Idle	PID Value at 30	PID Value at 55
ARC	Auto Ride Control	ON / OFF	OFF	OFF	OFF
CD1-CD8	COP Driver 1-8	0-45	5	6	8
CHTIL	CHT (lamp) Control	ON / OFF	OFF	OFF	OFF
EGR VR	EGR VR Solenoid	0-100%	0	0-40	0-60
EVAP CP	EVAP Purge Valve	0-100%	0-100	0-100	0-100
EVAP CV	EVAP CV Valve	0-100%	0-100	0-100	0-100
EPC	EPC Solenoid	7.4v / 4	4	5	5
FP	Fuel Pump Control	OFF	ON	ON	ON
FuelPW1	INJ Pulsewidth - Bank 1	0-999 ms	3.2-4.0	4.1-6.9	6.5-12
FuelPW2	INJ Pulsewidth - Bank 2	0-999 ms	3.2-4.0	4.1-6.9	6.5-12
HTR-11	HO2S-11 Heater	ON / OFF	ON	ON	ON
HTR-12	HO2S-12 Heater	ON / OFF	ON	ON	ON
HTR-21	HO2S-21 Heater	ON / OFF	ON	ON	ON
HTR-22	HO2S-22 Heater	ON / OFF	ON	ON	ON
IAC	Idle Air Control	0-100%	30-34	43-48	58-61
INJ1	INJ 1 Pulsewidth	0-999 ms	3.2-4.0	4.1-6.9	6.5-12
INJ2	INJ 2 Pulsewidth	0-999 ms	3.2-4.0	4.1-6.9	6.5-12
INJ3	INJ 3 Pulsewidth	0-999 ms	3.2-4.0	4.1-6.9	6.5-12
INJ4	INJ 4 Pulsewidth	0-999 ms	3.2-4.0	4.1-6.9	6.5-12
INJ5	INJ 5 Pulsewidth	0-999 ms	3.2-4.0	4.1-6.9	6.5-12
INJ6	INJ 6 Pulsewidth	0-999 ms	3.2-4.0	4.1-6.9	6.5-12
INJ7	INJ 7 Pulsewidth	0-999 ms	3.2-4.0	4.1-6.9	6.5-12
INJ8	INJ 8 Pulsewidth	0-999 ms	3.2-4.0	4.1-6.9	6.5-12
LongFT1	Long Term FT - Bank 1	-20 to 20%	-5 to +5	-5 to +5	-5 to +5
LongFT2	Long Term FT - Bank 2	-20 to 20%	-5 to +5	-5 to +5	-5 to +5
MIL	MIL (lamp) Control	ON / OFF	OFF	OFF	OFF
SHTFT1	Short Term F/T - Bank 1	-10 to 10%	-5 to +5	-5 to +5	-5 to +5
SHTFT2	Short Term FT - Bank 2	-10 to 10%	-5 to +5	-5 to +5	-5 to +5
SPARK	Spark Advance	-90° to 90°	12-17	35-40	28-37
SS1	Shift Solenoid 1	ON / OFF	ON	OFF	OFF
SS2	Shift Solenoid 2	ON / OFF	OFF	ON	OFF
TCC	TCC Solenoid	0-100%	0	0-40	90-95
TCIL	TCIL (lamp) Control	ON / OFF	OFF	OFF	OFF
TPO	TPO TR Sensor	0-5.1v	0.1	0.8	1.0
VREF	Vehicle Reference	0-5.1v	4.9-5.1	4.9-5.1	4.9-5.1

2000-05 Excursion 6.8L V10 VIN S (E4OD) Inputs

PCM PID Acronym	Parameter Identification	PID Range	PID Value at Hot Idle	PID Value at 30	PID Value at 55
ACCS	A/C Switch Signal	ON / OFF	ON (if on)	OFF	OFF
BPP	Brake Position	ON / OFF	ON (if on)	OFF	OFF
CHT	CHT Sensor	-40-500ºF	194	194	194
CID	CMP Sensor	0-999 Hz	7-10	10-13	15-17
CKP	CKP Sensor	0-9999 Hz	500-525	750-940	1195-1400
DPFE	DPFE Sensor	0-5.1v	0.2-1.3	0.2-4.5	0.2-4.5
FEPS	Flash EEPROM	0-5.1v	0.1	0.1	0.1
FLI	Fuel Level Input	1.7 / 50%	1.7 / 50%	1.7 / 50%	1.7 / 50%
FPM	Fuel Pump Monitor	ON / OFF	ON	ON	ON
FTP	Fuel Tank Pressure	2.6v / 0 psi	2.6v / 0 psi	2.6v / 0 psi	2.6v / 0 psi
GEAR	Gear Position	1-2-3-4	1	4	4
HO2S-11	HO2S-11 (Bank 1)	0.1-1.1v	0.1-1.1	0.1-1.1	0.1-1.1
HO2S-12	HO2S-12 (Bank 1)	0.1-1.1v	0.1-1.1	0.1-1.1	0.1-1.1
HO2S-21	HO2S-21 (Bank 1)	0.1-1.1v	0.1-1.1	0.1-1.1	0.1-1.1
IAT	IAT Sensor	-40-304ºF	50-120	50-120	50-120
KS1	Knock Sensor 1	0-5.1v	0	0	0
LOAD	Engine Load	0-100%	14-16	20-25	24-35
MAFV	MAF Sensor	0-5.1v	0.7-0.9	1.2-1.7	1.6-2.4
MISF	Misfire Detection	ON / OFF	OFF	OFF	OFF
OCTADJ	Octane Adjustment	CL/OPEN	CLOSED	CLOSED	CLOSED
PNP	PNP Switch Signal	ON / OFF	ON	OFF	OFF
PTO	Power Takeoff Switch	OFF	ON (if on)	OFF	OFF
RPM	Engine Speed	0-10K rpm	780-810	1380-1450	1790-1840
TCS	TCS Switch	ON / OFF	OFF	OFF	OFF
TFT	TFT Sensor Signal	-40-304ºF	110-210	110-210	110-210
TPV	TP Sensor	0-5.1v	0.53-1.27	0.8-1.1	0.9-1.2
TR1	TR1 Sensor	0v / 11.5v	0	11.5	11.5
TR2	TR2 Sensor	0v / 11.5v	0	11.5	11.5
TR3/TR	TR3A Sensor	0v / 1.7v	0	1.7	1.7
TR4	TR4 Sensor	0v / 11.5v	0	11.5	11.5
Vacuum	Engine Vacuum	0-30" Hg	19-20	18-20	8-15
VPWR	Vehicle Power	0-25.5v	12-14	12-14	12-14
VSS	Vehicle Speed (Hertz or MPH)	0-999 Hz (0-159 mph)	0 Hz = 0 mph	65 Hz = 30 mph	125 Hz = 55 mph
4x4L	4x4 Switch Signal	ON / OFF	ON (if on)	OFF	OFF

2000-05 Excursion 6.8L V10 VIN S (E4OD) Outputs

PCM PID Acronym	Parameter Identification	PID Range	PID Value at Hot Idle	PID Value at 30	PID Value at 55
CD1-10	Coil 1-10 Driver	0-36	6	8	10°
CCS	Coast Clutch Solenoid Control	ON / OFF	OFF	OFF	OFF
CHTIL	CHT (lamp) Control	ON / OFF	OFF	OFF	OFF
CTO	Clean Tachometer signal	0-9999 Hz	60-70	90-110	150-185
EGR VR	EGR VR Solenoid	0-100%	0	0-40	0-60
EVAP CP	EVAP Purge Valve	0-100%	0-100	0-100	0-100
EVAP CV	EVAP CV Valve	0-100%	0-100	0-100	0-100
EPC	EPC Solenoid	0-300 psi	5	7	27
FP	Fuel Pump Control	ON / OFF	ON	ON	ON
FuelPW1	INJ Pulsewidth - Bank 1	0-999 ms	3.8-4.6	5.2-6.5	6.6-11.0
FuelPW2	INJ Pulsewidth - Bank 2	0-999 ms	3.8-4.6	5.2-6.5	6.6-11.0
HTR-11	HO2S-11 Heater	ON / OFF	ON	ON	ON
HTR-12	HO2S-12 Heater	ON / OFF	ON	ON	ON
HTR-21	HO2S-21 Heater	ON / OFF	ON	ON	ON
HTR-22	HO2S-22 Heater	ON / OFF	ON	ON	ON
IAC	Idle Air Control	0-100%	25-33	30-40	50-65
INJ1	INJ 1 Pulsewidth	0-999 ms	3.8-4.6	5.2-6.5	6.6-11.0
INJ2	INJ 2 Pulsewidth	0-999 ms	3.8-4.6	5.2-6.5	6.6-11.0
INJ3	INJ 3 Pulsewidth	0-999 ms	3.8-4.6	5.2-6.5	6.6-11.0
INJ4	INJ 4 Pulsewidth	0-999 ms	3.8-4.6	5.2-6.5	6.6-11.0
INJ5	INJ 5 Pulsewidth	0-999 ms	3.8-4.6	5.2-6.5	6.6-11.0
INJ6	INJ 6 Pulsewidth	0-999 ms	3.8-4.6	5.2-6.5	6.6-11.0
INJ7	INJ 7 Pulsewidth	0-999 ms	3.8-4.6	5.2-6.5	6.6-11.0
INJ8	INJ 8 Pulsewidth	0-999 ms	3.8-4.6	5.2-6.5	6.6-11.0
INJ9	INJ 9 Pulsewidth	0-999 ms	3.8-4.6	5.2-6.5	6.6-11.0
INJ10	INJ 10 Pulsewidth	0-999 ms	3.8-4.6	5.2-6.5	6.6-11.0
LongFT1	Long Term FT - Bank 1	-20 to 20%	-5 to +5	-5 to +5	-5 to +5
LongFT2	Long Term FT - Bank 2	-20 to 20%	-5 to +5	-5 to +5	-5 to +5
MIL	MIL (lamp) Control	ON / OFF	OFF	OFF	OFF
SHTFT1	Short Term F/T - Bank 1	-10 to 10%	-5 to +5	-5 to +5	-5 to +5
SHTFT2	Short Term F/T - Bank 2	-10 to 10%	-5 to +5	-5 to +5	-5 to +5
SPARK	Spark Advance	-90° to 90°	15-20	23-34	26-34
SS1	Shift Solenoid 1	ON / OFF	ON	OFF	OFF
SS2	Shift Solenoid 2	ON / OFF	OFF	ON	OFF
TCC	TCC Solenoid	0-100%	0	90-100	90-100
TCIL	TCIL (lamp) Control	ON / OFF	OFF	OFF	OFF
VREF	Vehicle Reference	0-5.1v	4.9-5.1	4.9-5.1	4.9-5.1

2000-03 Excursion 7.3L V8 Turbo Diesel VIN F (E4OD)

PCM PID Acronym	Parameter Identification	PID Range	PID Value at low idle	PID Value at high idle
4x4L	4x4 Low Switch Input	ON / OFF	ON	OFF
ACCS	A/C Switch Signal	ON / OFF	ON	OFF
AP	Accelerator Pedal Position Sensor	0-5.1v	0.5-0.9	3.8-4.2v
ARPMDES	Ancillary Engine Speed Desired	---	---	---
BARO	Barometric Pressure	0-450 kPa	---	---
BAROV	Barometric Pressure	0-5.1v	4.75 (Sea level)	4.75 (Sea level)
BPP	Brake Pedal Position	ON / OFF	ON	OFF
BPA	Brake Pressure Applied	ON / OFF	ON	OFF
CCS	Coast Clutch Solenoid Control	ON / OFF	OFF	OFF
CPP	CPP Switch Signal	0v / 5v	0v (clutch "in")	5v
CPP/TCS	TCS & Cancel Switch	0v / VBAT	Switches On: 12	0v
CRUISE	Cruise Control Module	---	---	---
DTC CNT	DTC Count	0-156	0	0
EBP	Exhaust Back Pressure	---	---	---
EBP V	Exhaust Back Pressure Actual	0-5.1v	0.8-0.95	1.25-1.75
EOT	Engine Oil Temperature	0-5.1v	0.35-4.5	0.35-4.5
EPC	Electronic Pressure Control	0-300 psi	7	15
EPC V	Electronic Pressure Control Actual	0-25.5v	7.5	12
EPR	Exhaust Pressure Regulator	0-25.5v	6-8	0-10
FuelPW1	Fuel Pulsewidth	0-999 ms	---	---
GEAR	Gear Position	1-2-3-4	1	4
GPC	Glow Plug Control Duty Cycle	0-100%	---	---
GPC TM	Glow Plug Control Time	---	---	---
GPL TM	Glow Plug Lamp Time	---	---	---
IAT (ºF)	IAT Sensor	-40-304ºF	50-120	50-120
IAT V	IAT Sensor Actual	0-5.1v	1.5-3.5	1.5-3.5
ICP	Injector Control Pressure Sensor	Min: 0.83v at startup	0.25-0.40	0.25-0.40
ICP	Injector Control Pressure Actual	---	---	---
IPR	Injector Control Pressure Regulator	0-100%	35	40-100
IVS	Idle Validation Switch	0v / VBAT	0v / Closed	12 / CL or OL
MAP	MAP Sensor	0-999 Hz	---	---
MAP H	MAP Sensor Actual	0-999 Hz	110-190	110-190

2000-03 Excursion 7.3L V8 Turbo Diesel VIN F (E4OD)

PCM PID Acronym	Parameter Identification	PID Range	PID value at low idle	PID value at high idle
MFDES	Mass Fuel Desired	---	---	---
MGP	Manifold Gauge Pressure	---	---	---
PBA	Parking Brake Applied	0v / 12	Parking Brake Off: 12	0v
RPM	Engine Speed	0-10K rpm	---	---
SCCS	Speed Control Command Switch	0-25.5v	S/C On: 12	S/C On: 12
SCCS-M	Speed Control Command Switch Mode	---	---	---
TCC	Transmission Converter Clutch Solenoid Control	ON / OFF	OFF	OFF
SS1	Shift Solenoid 1	ON / OFF	ON	OFF
SS2	Shift Solenoid 2	ON / OFF	OFF	ON
TCIL	TCIL (lamp) Control	ON / OFF	OFF	OFF
TCS	TCS Switch	0v / 12	0	0
TFT V	TFT Sensor	0-5.1v	2.10-2.40	2.10-2.40
TORQUE	Engine Torque	---	---	---
TPREL	Low Idle TP Sensor	---	---	---
TR	Transmission Range Sensor	0-5.1v	4.45 in P	2.87
TR V	Transmission Range Sensor Actual (Volts)	0-5.1v	4.45 in P	2.87 in O/D
VFDES	Volume Flow Desired	---	---	---
VPWR	Vehicle Power Supply	0-25.5v	12-14	12-14
VREF	Vehicle Reference	0-5.1v	4.9-5.1	4.9-5.1
VS SET	Vehicle Speed Setting	---	---	---
VSS	Vehicle Speed (Hertz or MPH)	0-999 Hz (0-159 mph)	0	125 Hz = 55 mph

FORD, LINCOLN & MERCURY PIN TABLES

About This Section

About This Section

INTRODUCTION

This section contains Pin Voltage Tables for Ford , Lincoln & Mercury from 1990-2005 that can be used to assist in the repair of Trouble Code and No Code faults related to the PCM.

VEHICLE COVERAGE

*   1995-97 Aspire Applications
*   1996-2000 Contour Applications
*   1990-2005 Crown Victoria Applications
*   1990-2003 Escort & ZX2 Applications
*   2000-05 Focus Applications
*   1990-2005 Mustang Applications
*   1990-97 Probe Applications
*   1990-2004 Taurus Applications
*   1990-94 Tempo Applications
*   1990-97, 2000-05 Thunderbird Applications

HOW TO USE THIS SECTION

This section can be used to look up the location of a particular pin, a Wire Color or a "known good" value of a PCM circuit. To locate the PCM information for a particular vehicle, find the model, correct engine size (with VIN Code) and finally the year of the vehicle.

For example, to look up the PCM terminals for a 1999 Mustang 4.6L VIN W, go to Contents Page 2 and find the text string shown below:

Then turn to Page 7-123 to find the following PCM related information.

1998-99 Mustang 4.6L 2v V8 MFI VIN W (All) 104 Pin Connector

PCM Pin #	Wire Color	Circuit Description (104 Pin)	Value at Hot Idle
1	PK/OR	Shift Solenoid 2 Control	12v, 55 mph: 12v
2	PK/LG	MIL (lamp) Control	MIL On: 1v, Off: 12v
3	YL/BK	Digital TR1 Sensor	0v, 55 mph: 11v
12	YL/WT	Fuel Level Indicator Signal	1.7v (1/2 full)
21	DB	CKP Sensor (+) Signal	390-450 Hz

In this example, the Fuel Level Indicator circuit is connected to Pin 12 of the 104 Pin Connector by a Yellow/White wire. The value at Hot Idle shown here is the normal value for the fuel level with the tank 1/2 full.

The "All" that appears in the Title of the table indicates the information is for both automatic and manual transmission vehicle applications.

ASPIRE PIN TABLES

1994-95 Aspire 1.3L I4 MFI VIN H (All) 'C200' 26 Pin Connector

PCM Pin #	Wire Color	Circuit Description (26 Pin)	Value at Hot Idle
C200-2	BK/OR	Power Ground	<0.1v
C200-2	BK/LG	MAF Ground	<0.050v
C200-3	GN/BK	CID Sensor Signal	5-7 Hz
C200-4	GN/RD	CKP Sensor Signal	400-425 Hz
C200-5	PK	High Pressure Switch Signal	A/C On: 0v, Off: 12v
C200-6	LG/RD	Reference Voltage	4.9-5.1v
C200-7	LG/WT	TP Sensor Signal	Idle: 0.4-1.2v
C200-8	GN/BK	MAF Sensor Signal	2.0v
C200-9	---	Not Used	---
C200-10	RD/YL	EGR Control Solenoid	12v, 55 mph: 1v
C200-11	GN/YL	Injector 1 Control	4.5 ms
C200-12	RD/WT	IAC Motor Control	2.2 ms
C200-13	GN/RD	Injector 3 Control	4.5 ms
C200-14	BK/OR	Power Ground	<0.1v
C200-15	YL/GN	Analog Signal Return	<0.050v
C200-16	BK/RD	Spark Output Signal	0.2v, 55 mph: 0.3v1
C200-17	RD/BL	ECT Sensor Signal	0.5-0.6v
C200-18	YL	EGR EVP Sensor Signal	0.67v, 55 mph: 2.2v
C200-19	GN/RD	IAT Sensor Signal	1.5-2.5v
C200-20	WT	O2S-11 (B1 S1) Signal	0.1-1.1v
C200-21	LG	Condenser Fan Control	On: 0v, Off: 12v
C200-22	BL	EGR Vent Solenoid Control	12v, 55 mph: 1v
C200-23	---	Not Used	---
C200-24	GN/BK	Injector 2 Control	4.5 ms
C200-25	RD/BL	EVAP Purge Solenoid	12v, 55 mph: 1v
C200-26	GN/BL	Injector 4 Control	4.5 ms

Pin Connector Graphic

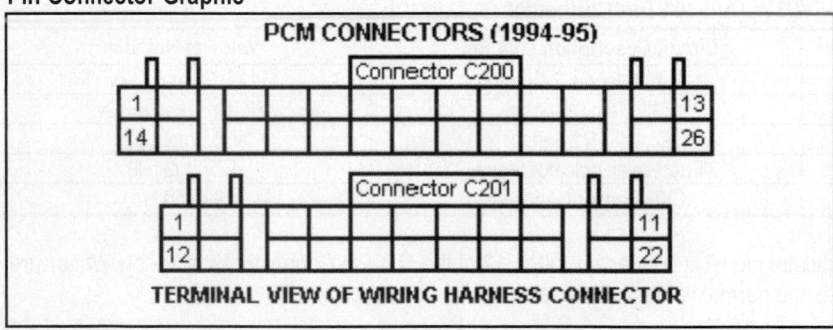

1994-95 Aspire 1.3L I4 MFI VIN H (All) 'C201' 22 Pin Connector

PCM Pin #	Wire Color	Circuit Description (22 Pin)	Value at Hot Idle
C201-1	BL/RD	Keep Alive Power	12-14v
C201-2	BK/WT	Vehicle Start Signal	KOEC: 9-11v
C201-3	BL	MIL (lamp) Control	MIL On: 1v, Off: 12v
C201-4	BK	Power Ground (M/T only)	<0.1v
C201-5	LG	Daytime Running Lamps	DRL On: 0v, Off: 12v
C201-6	BL	Self-Test Input Signal	STI On: 0v, Off: 5v
C201-7	GN/RD	VSS in Instrument Cluster	Vehicle moving: 0-5v
C201-8	GN	Brake Pedal Position Switch	Brake Off: 0v, On: 12v
C201-9	GN/WT	A/C Cycling Clutch Switch	A/C Off: 0v, On: 12v
C201-10	BR	Cooling Fan Control	On: 1v, Off: 12v
C201-11	RD/GN	Headlamp Relay Control	HDLP On: 12v, Off: 0v
C201-12	YL/WT	Vehicle Power (Main Relay)	12-14v
C201-13	BL/BK	Data Link Connector	Digital Signals
C201-14	WT/BK	Self-Test Output Signal	STO On: 0v, Off: 12v
C201-15	WT/YL	Fuel Pump Control	On: 1v, Off: 12v
C201-16	BL/OR	A/C Relay Control	ACR On: 1v, Off: 12v
C201-17	BK/RD	Rear Window Defroster	DEF On: 12v, Off: 0v
C201-18	RD	Idle Throttle Switch Signal	Closed: 0v, Open: 12v
C201-19	BL/YL	PSP Switch (ATX only)	Straight: 12v, Turned: 0v
C201-20	OR/BL	Blower Motor Switch Signal	Motor On: 12v, Off: 0v
C201-21	BL/WT	M/T: Shift Indicator Light	SIL On: 1v, Off: 12v
C201-22	GN/BK	M/T: Clutch Pedal Position	Clutch In: 0v, out: 5v
C201-22	WT	A/T: Park Neutral Position	In 'P': 0v, Others: 5v

Pin Connector Graphic

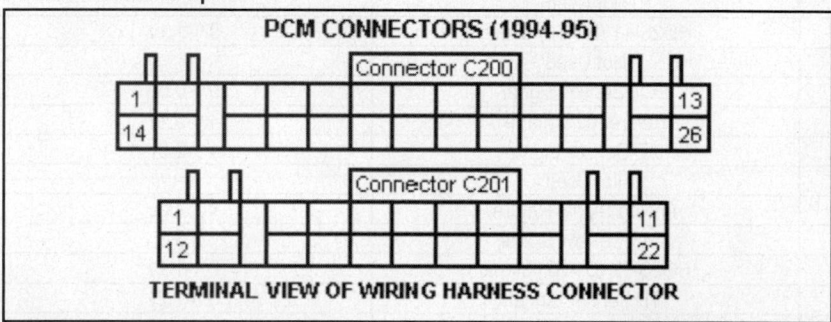

Standard Colors and Abbreviations

Abbreviation	Color	Abbreviation	Color	Abbreviation	Color
BK	Black	GY	Gray	PK	Purple
BL	Blue	GN	Green	RD	Red
BR	Brown	LG	LT Green	TN	Tan
DB	Dark Blue	OR	Orange	WT	White
DG	DK Green	PK	Pink	YL	Yellow

1996-97 Aspire 1.3L I4 MFI Engine VIN H (All) 22 Pin Connector

PCM Pin #	Wire Color	Circuit Description (22 Pin)	Value at Hot Idle
C201-1	LG	Condenser Fan Control	On: 1v, Off: 12v
C201-2	BK/WT	Vehicle Start Signal	KOEC: 9-11v
C201-3	BL	MIL (lamp) Control	MIL On: 1v, Off: 12v
C201-4	BL/OR	Air Conditioning Relay Control	A/C On: 0v, Off: 12v
C201-5	BL	Self-Test Input Signal	STI On: 0v, Off: 12v
C201-6	GN/WT	A/C Cycling Clutch Signal	A/C & BLMT on: 12v
C201-7	GN/RD	Vehicle Speed Sensor	Varies: 0-5v
C201-8	---	Not Used	---
C201-9	GN	Brake Pedal Position Switch	Brake Off: 0v, On: 12v
C201-10	BK	M/T: Power Ground	<0.1v
C201-11	WT/YL	Fuel Pump Control	On: 1v, Off: 12v
C201-12	BK	Fan Control Relay	On: 1v, Off: 12v
C201-13	PK	ISO K-Line for Scan Tool	No Scan Tool: 7v
C201-14	---	Not Used	---
C201-15	RD/GN	Headlamp Switch Signal	Lamps On: 12v, Off: 0v
C201-16	BK/RD	Rear Window Defrost Switch	DEF On: 12v, Off: 0v
C201-17	WT	Clutch or P/N Position Switch	In 'P': 0v, Others: 5v
C201-18	LB	DRL (lamps) Canada only	On: 1v, Off: 12v
C201-19	OR/BL	Blower Motor Control Switch	Motor On: 12v, Off: 0v
C201-20	PK	A/C High Pressure Switch	Open: 12v, closed: 0v
C201-21, 22	---	Not Used	---

1996-97 Aspire 1.3L I4 MFI Engine VIN H (All) 'C257' 16 Pin Connector

PCM Pin #	Wire Color	Circuit Description (16 Pin)	Value at Hot Idle
C257-1	---	Not Used	---
C257-2	WT	HO2S-11 (B1 S1) Signal	0.1-1.1v
C257-3	---	Not Used	---
C257-4	RD/BL	ECT Sensor Signal	0.5-0.6v
C257-5	LG/RD	Reference Voltage	4.9-5.1v
C257-6	GN/RD	IAT Sensor Signal	1.5-2.5v
C257-7	---	Not Used	---
C257-8	YL/GN	Analog Signal Return	<0.050v
C257-9	GN/BK	MAF Sensor Signal	2.0v
C257-10	OR/WT	HO2S-12 (B1 S2) Signal	0.1-1.1v
C257-11	LG/WT	TP Sensor Signal	0.5-1.2v
C257-12	PK/BK	Atmospheric Pressure Sensor	4.3v (28" Hg)
C257-13	YL	EGR EVP Sensor	0.7v
C257-14	RD	Idle Throttle Switch Signal	Closed: 0v, Open: 12v
C257-15	BR	EGR Boost Sensor	3.9v
C257-16	BL/YL	PSP Switch Signal	Straight: 12v, Turned: 0v

Standard Colors and Abbreviations

Abbreviation	Color	Abbreviation	Color	Abbreviation	Color
BK	Black	GY	Gray	PK	Purple
BL	Blue	GN	Green	RD	Red
BR	Brown	LG	LT Green	TN	Tan
DB	Dark Blue	OR	Orange	WT	White
DG	DK Green	PK	Pink	YL	Yellow

1996-97 Aspire 1.3L I4 MFI Engine VIN H (All) C200 26 Pin Connector

PCM Pin #	Wire Color	Circuit Description (26 Pin)	Value at Hot Idle
C200-1	BK/LG	Power Ground	<0.1v
C200-2	BK/OR	Power Ground	<0.1v
C200-3	BL/OR	CKP Sensor (-) Signal	400-425 Hz
C200-4	GN/RD	CMP 1 Sensor Signal	5-7 Hz
C200-5	BL/RD	Keep Alive Power	12-14v
C200-6	---	Not Used	---
C200-7	---	Not Used	---
C200-8	BL	EGR Vent Solenoid Control	12v, 55 mph: 1v
C200-9	RD/WT	IAC Motor Control	2.2 ms
C200-10	---	Not Used	---
C200-11	GN/YL	Injector 1 Control	4.5 ms
C200-12	GN/RD	Injector 3 Control	4.5 ms
C200-13	---	Not Used	---
C200-14	YL/WT	Vehicle Power	12-14v
C200-15	BK/OR	Power Ground	<0.1v
C200-16	GN/BK	CMP 2 Sensor Signal	20-28 Hz
C200-17	LG	CKP Sensor (+) Signal	400-425 Hz
C200-18	---	Not Used	---
C200-19	---	Not Used	---
C200-20	BK/RD	Ignition Control Module Signal	0.2v, 55 mph: 0.3v
C200-21	RD/YL	EGR Control Solenoid	12v, 55 mph: 1v
C200-22	---	Not Used	---
C200-23	BL/RD	EVAP Purge Solenoid	12v, Cruise: 9.1v
C200-24	GN/BK	Injector 2 Control	4.5 ms
C200-25	GN/BL	Injector 4 Control	4.5 ms
C200-26	---	Not Used	---

Pin Connector Graphic

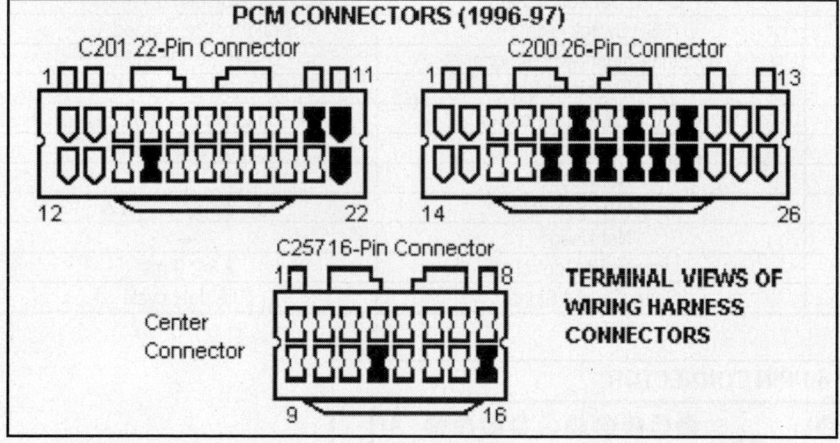

CONTOUR PIN TABLES

1995 Contour 2.0L I4 MFI VIN 3 (All) 60 Pin Connector

PCM Pin #	Wire Color	Circuit Description (60 Pin)	Value at Hot Idle
1	OR/YL	Keep Alive Power	12- 14v
2	OR	Brake On/Off Signal	Brake Off: 0v, On: 12v
3	WT/PK	VSS (+) Signal	0 Hz, 55 mph: 125 Hz
4	WT/GN	Ignition Diagnostic Monitor	20-31 Hz
5	WT/PK	A/T: TSS Sensor (+) Signal	42-50 Hz (680-720 rpm)
6	---	Not Used	---
7	WT/GN	ECT Sensor Signal	0.5-0.6v
8	PK/BK	Fuel Pump Monitor	Pump On: 12v, Off; 0v
9	BR/BL	MAF Sensor Return	<0.050v
10	PK/BL	A/C Cycling Clutch Switch	A/C On: 12v, Off: 0v
11	BK/OR	EVAP Purge Solenoid	12v, 55 mph: 1v
12	---	Not Used	---
13	BK/BL	Low Speed Fan Control	On: 1v, Off: 12v
14	---	Not Used	---
15	---	Not Used	---
16	BK/BL	Ignition Ground	<0.050v
17	BK/OR	STI Output, MIL Control	MIL On: 1v, Off: 12v
18	WT/BK	Data Bus (+) Signal	Digital Signals
19	BR/BK	Data Bus (-) Signal	Digital Signals
20	BK/RD	Case Ground	<0.050v
21	BK/OR	IAC Solenoid Control	Varies: 8-11v
22	BK/BL	Fuel Pump Control	On: 1v, Off: 12v
23	WT/BK	Knock Sensor Signal	0v
24	WT/PK	Camshaft Position Sensor	5-7 Hz
25	WT/PK	IAT Sensor Signal	1-5-2.5v
26	YL	Reference Voltage	4.9-5.1v
27	WT/BL	DPFE Sensor Signal	Hot Idle: 0.1v
28	WT/PK	PSP Switch Signal	Straight: 12v, Turned: 0v
29	WT/BK	Octane Adjust Switch Signal	Closed: 0v, Open: 9.3v
30	WT	A/T: PNP Sensor Signal	In 'P': 0v, 55 mph: 1.7v
30	WT	M/T: Clutch Pedal Position	Clutch In: 0v, Out: 5v
31	BK/WT	High Speed Fan Control	On: 1v, Off: 12v
32	---	Not Used	---
33	GN/WT	EGR VR Solenoid	12v, 30 mph: 10-12v
34	---	Not Used	---
35	BK/OR	Injector 4 Control	2.3-2.9 ms
36	WT/PK	Spark Output Signal	50% duty cycle

Pin Connector Graphic

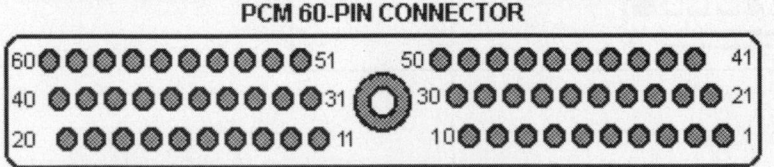

1995 Contour 2.0L I4 MFI VIN 3 (All) 60 Pin Connector

PCM Pin #	Wire Color	Circuit Description (60 Pin)	Value at Hot Idle
37	GN/YL	Vehicle Power	12-14v
38	BK/RD	A/T: EPC Solenoid Control	7-8v (8-9 psi)
39	BK/BL	Injector 3 Control	2.3-2.9 ms
40	BK/YL	Power Ground	<0.1v
41	WT/BK	A/T: TCS (switch) Signal	TCS & O/D On: 12v
42	---	Not Used	---
43	---	Not Used	---
44	WT	HO2S-11 (B1 S1) Signal	0.1-1.1v
45	---	Not Used	---
46	BR	Analog Signal Return	<0.050v
47	WT	TP Sensor Signal	0.5-1.1v
48	WT/BK	Self-Test Input	STI On: 0.1v, Off: 5v
49	WT/RD	A/T: TFT Sensor Signal	2.10-2.40v
50	WT/BL	MAF Sensor Signal	1.9v
51	BK/YL	A/T: Shift Solenoid 1 Control	12v, 55 mph: 1v
52	BK/BL	A/T: Shift Solenoid 2 Control	1v, 55 mph: 12v
53	BK/WT	A/T: TCC Solenoid Control	TCC On: 1v, Off: 12v
54	BK/YL	A/C WOT Relay Control	On: 1v, Off: 12v
55	BK/OR	A/T: Shift Solenoid 3 Control	12v, 55 mph: 1v
56	WT/BK	PIP Sensor Signal	50% duty cycle
57	GN/YL	Vehicle Power	12-14v
58	BK/WT	Injector 1 Control	2.3-2.9 ms
59	BK/YL	Injector 2 Control	2.3-2.9 ms
60	BK/YL	Power Ground	<0.1v

Pin Connector Graphic

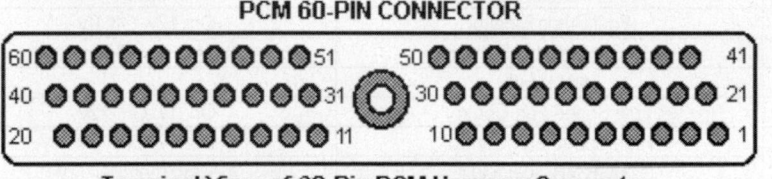

Standard Colors and Abbreviations

Abbreviation	Color	Abbreviation	Color	Abbreviation	Color
BK	Black	GY	Gray	PK	Purple
BL	Blue	GN	Green	RD	Red
BR	Brown	LG	LT Green	TN	Tan
DB	Dark Blue	OR	Orange	WT	White
DG	DK Green	PK	Pink	YL	Yellow

1996-97 Contour 2.0L I4 MFI VIN 3 (All) 104 Pin Connector

PCM Pin #	Wire Color	Circuit Description (104 Pin)	Value at Hot Idle
1	BK/BL	Shift Solenoid 2 Control	12v, 55 mph: 1v
2	BK/OR	MIL (lamp) Control	MIL On: 1v, Off: 12v
3-12	---	Not Used	---
13	WT/BL	Flash EEPROM Power	0.1v
14	---	Not Used	---
15	BR/BL	Data Bus (-) Signal	Digital Signals
16	WT/BL	Data Bus (+) Signal	Digital Signals
17-20	---	Not Used	---
21	WT/RD	CKP Sensor (+) Signal	450-480 Hz
22	BR/WT	CKP Sensor (-) Signal	450-480 Hz
23	---	Not Used	---
24	BK/YL	Power Ground	<0.1v
25	BK/RD	Power Ground	<0.1v
26	BR/BL	Coil Driver 1 Control	5° dwell
27	BK/YL	Shift Solenoid 1 Control	1v, 55 mph: 12v
28	---	Not Used	---
29	PK/BK	TCS (switch) Signal	TCS & O/D On: 12v
30	WT/BK	Octane Adjust Switch	Closed: 0v, Open: 9.3v
31 ('96)	WT/PK	PSP Sensor Signal	Straight: 0v, Turning: 12v
31 ('97)	WT	PSP Sensor Signal	Straight: 0v, Turning: 12v
32-34	---	Not Used	---
35	WT/BL	HO2S-12 (B1 S2) Signal	0.1-1.1v
36	BR/BL	MAF Sensor Return	<0.050v
37	WT/RD	TFT Sensor Signal	2.10-2.40v
38	WT/GN	ECT Sensor Signal	0.5-0.6v
39	WT/PK	IAT Sensor Signal	1.5-2.5v
40	PK/BK	Fuel Pump Monitor	On: 12v, Off: 0v
41	PK/BL	A/C Cycling Clutch Switch	Closed: 12v, Open: 0v
42-44	---	Not Used	---
45	BK/BL	Low Speed Fan Control	On: 1v, Off: 12v
46	BK/WT	High Speed Fan Control	On: 1v, Off: 12v
47	BK/GN	EGR VR Solenoid	12v, 55 mph: 10-12v
48	WT/BK	Clean Tachometer Output	20-31 Hz
49-50	---	Not Used	---
51	BK/YL	Power Ground	<0.1v
52	BR/GN	Coil Driver 2 Control	5° dwell
53	---	Not Used	---
54	BK/WT	TCC Solenoid Control	0%, 55 mph: 50-90%
55	OR/YL	Keep Alive Power	12-14v
56	BK/OR	EVAP VMV Solenoid	0-10 Hz (0-100%)
57	WT/BK	Knock Sensor 1 Signal	0v
58	WT/PK	VSS (+) Signal	0 Hz, 55 mph: 125 Hz
59	---	Not Used	---
60	WT	HO2S-11 (B1 S1) Signal	0.1-1.1v
61-63	---	Not Used	---
64	WT/GN	A/T: TR Sensor	In 'P': 0v, 55 mph: 1.7v
64	WT	M/T: Clutch Pedal Position	Clutch In: 1v, Out: 5v

1996-97 Contour 2.0L I4 MFI VIN 3 (All) 104 Pin Connector

PCM Pin #	Wire Color	Circuit Description (104 Pin)	Value at Hot Idle
65	WT/BL	DPFE Sensor Signal	0.20-1.30v
66-68	---	Not Used	---
69	BK/YL	A/C WOT Relay Control	On: 1v, Off: 12v
70	BK/YL	ZETECT VCT Control	VCT Off: 12v, On: 1v
71	GN/YL	Vehicle Power	12-14v
72-73	---	Not Used	---
74	BK/BL	Injector 3 Control	2.3-2.9 ms
75	BK/WT	Injector 1 Control	2.3-2.9 ms
76	BK/YL	CMP and TSS Ground	<0.050v
77	BK/YL	Power Ground	<0.1v
78	---	Not Used	---
79	WT/BL	TCIL (lamp) Control	On: 1v, Off: 12v
80	BK/BL	Fuel Pump Control	On: 1v, Off: 12v
81	BK/RD	EPC Solenoid Control	17 psi
82	---	Not Used	---
83	BK/YL	IAC Motor Control	9v (33% duty cycle)
84	WT/PK	TSS Sensor Signal	40-56 Hz (680-710 rpm)
85	WT/PK	CID Sensor Signal	5-7 Hz
86	BK/WT	A/C Pressure Switch	A/C On: 12v (Open)
87	---	Not Used	---
88	WT/BL	MAF Sensor Signal	0.6-0.9v
89	WT	TP Sensor Signal	0.53-1.27v
90	YL	Reference Voltage	4.9-5.1v
91	BR	Analog Signal Return	<0.050v
92	OR	Brake Pedal Position Switch	Brake Off: 0v, On: 12v
93	BK/YL	HO2S-11 (B1 S1) Heater	On: 1v, Off: 12v
94	---	Not Used	---
95	BK/OR	HO2S-12 (B1 S2) Heater	On: 1v, Off: 12v
96	---	Not Used	---
97	GN/YL	Vehicle Power	12-14v
98-99	---	Not Used	---
100	BK/OR	Injector 4 Control	2.3-2.9 ms
101	BK/YL	Injector 2 Control	2.3-2.9 ms
102	BK/OR	Shift Solenoid 3 Control	8v, 55 mph: 8-9v
103	BK/YL	Power Ground	<0.1v
104	---	Not Used	---

Pin Connector Graphic

PCM 104-PIN CONNECTOR

Terminal View of 104-Pin PCM Wiring Harness Connector

Standard Colors and Abbreviations

Abbreviation	Color	Abbreviation	Color	Abbreviation	Color
BK	Black	GY	Gray	PK	Purple
BL	Blue	GN	Green	RD	Red
BR	Brown	LG	LT Green	TN	Tan
DB	Dark Blue	OR	Orange	WT	White
DG	DK Green	PK	Pink	YL	Yellow

1998-99 Contour 2.0L I4 MFI VIN 3 (All) 104 Pin Connector

PCM Pin #	Wire Color	Circuit Description (104 Pin)	Value at Hot Idle
1	BK/BL	Shift Solenoid 2 Control	12v, 55 mph: 1v
2	BK/OR	MIL (lamp) Control	MIL On: 1v, Off: 12v
3-11	---	Not Used	---
12	WT	Fuel Level Indicator Signal	1.7v (1/2 full)
13	WT/BL	Flash EEPROM Power	0.1v
15	BL	Data Bus (-) Signal	Digital Signals
16	GN	Data Bus (+) Signal	Digital Signals
17-20	---	Not Used	---
21	WT/RD	CKP Sensor (+) Signal	450-480 Hz
22	BR/WT	CKP Sensor (-) Signal	450-480 Hz
23	---	Not Used	---
24	BK/YL	Power Ground	<0.1v
25	BK/RD	Power Ground	<0.1v
26	BR/BL	Coil Driver 1 Control	5° dwell
27	BK/YL	Shift Solenoid 1 Control	1v, 55 mph: 12v
28	---	Not Used	---
29	PK/BK	TCS (switch) Signal	TCS & O/D On: 12v
30	WT/BK	Octane Adjust Switch	Closed: 0v, Open: 9.3v
31	WT/PK	PSP Sensor Signal	Straight: 0v, Turning: 12v
32-34	---	Not Used	---
35	WT/BL	HO2S-12 (B1 S2) Signal	0.1-1.1v
36	BR/BL	MAF Sensor Return	<0.050v
37	WT/RD	TFT Sensor Signal	2.10-2.40v
38	WT/GN	ECT Sensor Signal	0.5-0.6v
39	WT/PK	IAT Sensor Signal	1.5-2.5v
40	PK/BK	Fuel Pump Monitor	On: 12v, Off: 0v
41	PK/BL	A/C Cycling Clutch Switch	Closed: 12v, Open: 0v
42-43	---	Not Used	---
44	BK/RD	VCT Actuator	VCT Off: 12v, On: 1v
45	BK/BL	Low Speed Fan Control	On: 1v, Off: 12v
46	BK/WT	High Speed Fan Control	On: 1v, Off: 12v
47	BK/GN	EGR VR Solenoid	12v, 55 mph: 10-12v
48	WT/BK	Clean Tachometer Output	20-31 Hz
49-51	---	Not Used	---
51	BK/YL	Power Ground	<0.1v
52	BR/GN	Coil Driver 2 Control	5° dwell
53	---	Not Used	---
54	BK/WT	TCC Solenoid Control	0%, 55 mph: 50-90%
55	OR/YL	Keep Alive Power	12-14v
56	BK/OR	EVAP Purge Solenoid	0-10 Hz (0-100%)
57	WT/BK	Knock Sensor 1 (+) Signal	0v
58	WT/PK	VSS (+) Signal	0 Hz, 55 mph: 125 Hz
60	WT	HO2S-11 (B1 S1) Signal	0.1-1.1v
61	---	Not Used	---
62	WT/PK	FTP Sensor Signal	2.6v (cap off)
63 ('99)	WT/PK	FRP Sensor Signal	2.8v (39 psi)
64	WT/GN	A/T: TR Sensor Signal	In 'P': 0v, 55 mph: 1.7v
64	WT	M/T: Clutch or Neutral Switch	Clutch In: 1v, Out: 5v

1998-99 Contour 2.0L I4 MFI VIN 3 (All) 104 Pin Connector

PCM Pin #	Wire Color	Circuit Description (104 Pin)	Value at Hot Idle
65	WT/BL	DPFE Sensor Signal	0.20-1.30v
67	BK/OR	EVAP CV Solenoid	0-10 Hz (0-100%)
69	BK/YL	A/C WOT Relay Control	On: 1v, Off: 12v
70	BK/YL	VCT Control (ZETECT)	VCT Off: 12v, On: 1v
71	GN/YL	Vehicle Power	12-14v
72-73	---	Not Used	---
74	BK/BL	Injector 3 Control	2.3-2.9 ms
75	BK/WT	Injector 1 Control	2.3-2.9 ms
76	BR	CMP and TSS Ground	<0.050v
77	BK/YL	Power Ground	<0.1v
79	WT/BL	TCIL (lamp) Control	On: 1v, Off: 12v
80	BK/BL	Fuel Pump Control	On: 1v, Off: 12v
81	BK/RD	EPC Solenoid Control	17 psi
83	BK/YL	IAC Motor Control	9v (33% duty cycle)
84	WT/PK	TSS Sensor Signal	40-56 Hz (680-710 rpm)
85	WT/PK	CMP Sensor (+) Signal	5-7 Hz
86	BK/WT	A/C Pressure Switch	A/C On: 12v (Open)
87	---	Not Used	---
88	WT/BL	MAF Sensor Signal	0.6-0.9v
89	WT	TP Sensor Signal	0.53-1.27v
90	YL	Reference Voltage	4.9-5.1v
91	BR	Analog Signal Return	<0.050v
92	OR	Brake Pedal Position Switch	Brake Off: 0v, On: 12v
93	BK/YL	HO2S-11 (B1 S1) Heater	On: 1v, Off: 12v
95	BK/OR	HO2S-12 (B1 S2) Heater	On: 1v, Off: 12v
97	GN/YL	Vehicle Power	12-14v
100	BK/OR	Injector 4 Control	2.3-2.9 ms
101	BK/YL	Injector 2 Control	2.3-2.9 ms
102	BK/OR	Shift Solenoid 3 Control	8v, 55 mph: 8-9v
103	BK/YL	Power Ground	<0.1v

Pin Connector Graphic

PCM 104-PIN CONNECTOR

Terminal View of 104-Pin PCM Wiring Harness Connector

Standard Colors and Abbreviations

Abbreviation	Color	Abbreviation	Color	Abbreviation	Color
BK	Black	GY	Gray	PK	Purple
BL	Blue	GN	Green	RD	Red
BR	Brown	LG	LT Green	TN	Tan
DB	Dark Blue	OR	Orange	WT	White
DG	DK Green	PK	Pink	YL	Yellow

2000 Contour 2.0L I4 MFI VIN 3 (All) 104 Pin Connector

PCM Pin #	Wire Color	Circuit Description (104 Pin)	Value at Hot Idle
1	BK/BL	Shift Solenoid 2 Control	12v, 55 mph: 1v
2	BK/OR	MIL (lamp) Control	MIL On: 1v, Off: 12v
12	WT	Fuel Level Indicator Signal	1.7v (1/2 full)
13	WT/BL	Flash EEPROM Power	0.1v
15	BL	Data Bus (-) Signal	Digital Signals
16	GN	Data Bus (+) Signal	Digital Signals
17-20	---	Not Used	---
20	WT/BK	Gearshift Indicator Signal	N/A
21	WT/RD	CKP Sensor (+) Signal	450-480 Hz
22	BR/RD	CKP Sensor (-) Signal	450-480 Hz
24	BK/YL	Power Ground	<0.1v
25	BK/RD	Power Ground	<0.1v
26	BR/BL	Coil Driver 1 Control	5° dwell
27	BK/YL	Shift Solenoid 1 Control	1v, 55 mph: 12v
28	WT/BL	Speedometer Indicator Signal	Digital Signals
29	GN/BK	TCS (switch) Signal	TCS & O/D On: 12v
30	WT/BK	Octane Adjust Switch	Closed: 0v, Open: 9.3v
31	WT	PSP Sensor Signal	Straight: 0v, Turning: 12v
33	WT/PK	VSS (-) Signal	0 Hz, 55 mph: 125 Hz
34	WT/PK	A/T: TSS Sensor Signal	40-56 Hz (680-710 rpm)
35	WT/BL	HO2S-12 (B1 S2) Signal	0.1-1.1v
36	BR/BL	MAF Sensor Return	<0.050v
37	WT/RD	TFT Sensor Signal	2.10-2.40v
38	WT/GN	ECT Sensor Signal	0.5-0.6v
39	WT/PK	IAT Sensor Signal	1.5-2.5v
40	GN/BK	Fuel Pump Monitor	On: 12v, Off: 0v
41	GN/BL	A/C Cycling Clutch Switch	Closed: 12v, Open: 0v
44	BK/RD	VCT Actuator	VCT Off: 12v, On: 1v
45	BK/BL	Low Speed Fan Control	On: 1v, Off: 12v
46	BK/WT	High Speed Fan Control	On: 1v, Off: 12v
47	BK/GN	EGR VR Solenoid	12v, 55 mph: 10-12v
48	WT/BK	Clean Tachometer Output	20-31 Hz
51	BK/YL	Power Ground	<0.1v
52	BR/GN	Coil Driver 2 Control	5° dwell
54	BK/WT	TCC Solenoid Control	0%, 55 mph: 50-90%
55	OR/YL	Keep Alive Power	12-14v
56	BK/OR	EVAP Purge Solenoid	0-10 Hz (0-100%)
57	WT/BK	Knock Sensor 1 (+) Signal	0v
58	WT/PK	VSS (+) Signal	0 Hz, 55 mph: 125 Hz
59	WT/GN	Generator Load Indicator	1.5-10.v (40-250 Hz)
60	WT	HO2S-11 (B1 S1) Signal	0.1-1.1v
62	WT/PK	FTP Sensor Signal	2.6v (cap off)
63	WT/GN	FRP Sensor Signal	2.8v (39 psi)
64	WT/GN	A/T: TR Sensor Signal	In 'P': 0v, 55 mph: 1.7v
64	WT	M/T: Clutch or Neutral Switch	Clutch In: 1v, Out: 5v

1998-2000 Contour 2.0L I4 MFI VIN 3 (All) 104 Pin Connector

PCM Pin #	Wire Color	Circuit Description (104 Pin)	Value at Hot Idle
65	WT/BL	DPFE Sensor Signal	0.20-1.30v
67	BK/OR	EVAP CV Solenoid	0-10 Hz (0-100%)
69	BK/YL	A/C WOT Relay Control	On: 1v, Off: 12v
70	BK/YL	VCT Control (ZETECT)	VCT Off: 12v, On: 1v
71	GN/YL	Vehicle Power	12-14v
72	---	Not Used	---
73	BK/YL	Shift Solenoid 1 Control	12v, 55 mph: 1v
74	BK/BL	Injector 3 Control	2.3-2.9 ms
75	BK/WT	Injector 1 Control	2.3-2.9 ms
76	BR	CMP Sensor (-) Signal	5-7 Hz
77	BK/YL	Power Ground	<0.1v
79	WT/BL	TCIL (lamp) Control	On: 1v, Off: 12v
80	BK/RD	Fuel Pump Control	On: 1v, Off: 12v
81	BK/RD	EPC Solenoid Control	17 psi
82	---	Not Used	---
83	BK/YL	IAC Motor Control	9v (33% duty cycle)
84	WT/PK	TSS Sensor Signal	40-56 Hz (680-710 rpm)
85	WT/PK	CMP Sensor (+) Signal	5-7 Hz
86	BK/WT	A/C Pressure Switch	A/C On: 12v (Open)
87	BR/YL	Knock Sensor 1 (-) Signal	0v
88	WT/BL	MAF Sensor Signal	0.6-0.9v
89	WT	TP Sensor Signal	0.53-1.27v
90	YL	Reference Voltage	4.9-5.1v
91	BR	Analog Signal Return	<0.050v
92	GN/RD	Brake Pedal Position Switch	Brake Off: 0v, On: 12v
93	BK/YL	HO2S-11 (B1 S1) Heater	On: 1v, Off: 12v
95	BK/OR	HO2S-12 (B1 S2) Heater	On: 1v, Off: 12v
96	---	Not Used	---
97	GN/YL	Vehicle Power	12-14v
98	---	Not Used	---
99	BK/WT	Modulated Lockup Solenoid	N/A
100	BK/OR	Injector 4 Control	2.3-2.9 ms
101	BK/YL	Injector 2 Control	2.3-2.9 ms
102	BK/OR	Shift Solenoid 3 Control	8v, 55 mph: 8-9v
103	BK/YL	Power Ground	<0.1v
104	---	Not Used	---

Pin Connector Graphic

PCM 104-PIN CONNECTOR

Terminal View of 104-Pin PCM Wiring Harness Connector

Standard Colors and Abbreviations

Abbreviation	Color	Abbreviation	Color	Abbreviation	Color
BK	Black	GY	Gray	PK	Purple
BL	Blue	GN	Green	RD	Red
BR	Brown	LG	LT Green	TN	Tan
DB	Dark Blue	OR	Orange	WT	White
DG	DK Green	PK	Pink	YL	Yellow

1998-99 Contour SVT 2.5L 24v V6 VIN G (M/T) 104 Pin Connector

PCM Pin #	Wire Color	Circuit Description (104 Pin)	Value at Hot Idle
1	---	Not Used	---
2	BK/OR	MIL (lamp) Control	MIL On: 1v, Off: 12v
3-7	---	Not Used	---
8	BK/BL	IMRC Solenoid Control	5v
9-11	---	Not Used	---
12	WT	Fuel Level Indicator Signal	1.7v (1/2 full)
13	WT/BL	Flash EEPROM Power	0.1v
14	---	Not Used	---
15	BL	Data Bus (-) Signal	Digital Signals
16	GY	Data Bus (+) Signal	Digital Signals
17	GY/OR	Passive Antitheft System	Digital Signals
18-19	---	Not Used	---
20	BK/BL	Injector 3 Control	2.1-2.5 ms
21	WT/RD	CKP Sensor (+) Signal	410-440 Hz
22	BR/RD	CKP Sensor (-) Signal	410-440 Hz
23	---	Not Used	---
24	BK/YL	Power Ground	<0.1v
25	BK/RD	Power Ground	<0.1v
26	BR/BL	Coil Driver 1 Control	5° dwell
27-28	---	Not Used	---
29	PK/BK	Instrument Interface Module	Digital Signals
30	WT/BK	Octane Adjust Switch	Closed: 0v, Open: 9.3v
31	WT	PSP Sensor Signal	Straight: 0v, Turning: 12v
32-34	---	Not Used	---
34	WT/GN	Passive Antitheft System	Digital Signals
35	WT/BL	HO2S-12 (B1 S2) Signal	0.1-1.1v
36	BR/BL	MAF Sensor Return	<0.050v
37	---	Not Used	---
38	WT/GN	ECT Sensor Signal	0.5-0.6v
39	WT/PK	IAT Sensor Signal	1.5-2.5v
40	PK/BK	Fuel Pump Monitor	On: 12v, Off: 0v
41	PK/BL	A/C Cycling Clutch Switch	A/C On: 12v, Off: 0v
42	BK/RD	IMRC Solenoid Control	12v, 55 mph: 12v
43-44	---	Not Used	---
45	BK/BL	Low Speed Fan Control	On: 1v, Off: 12v
46	BK/WT	High Speed Fan Control	On: 1v, Off: 12v
47	BK/GN	EGR VR Solenoid	0%, 55 mph: 40%
48	WT/BK	Clean Tachometer Output	35-42 Hz
49-50	---	Not Used	---
51	BK/YL	Power Ground	<0.1v
52	BR/GN	Coil Driver 2 Control	5° dwell
53	---	Not Used	---
54	BK/BL	Fuel Pump Control	On: 1v, Off: 12v
55	OR/YL	Keep Alive Power	12-14v
56	BK/OR	EVAP Purge Solenoid	0-10 Hz (0-100%)
57	WT/BK	Knock Sensor 1 Signal	0v
58	WT/PK	VSS (+) Signal	0 Hz, 55 mph: 125 Hz
59	---	Not Used	---
60	WT	HO2S-11 (B1 S1) Signal	0.1-1.1v
61	WT/GN	HO2S-22 (B2 S2) Signal	0.1-1.1v
62	WT/PK	FTP Sensor Signal	2.6v (cap off)
63 ('99)	WT/PK	FRP Sensor Signal	2.8v (39 psi)
64	WT/GN	M/T: Clutch Position Switch	Clutch In: 0.1v, Out: 5v

1998-99 Contour SVT 2.5L 24v V6 VIN G (M/T) 104 Pin Connector

PCM Pin #	Wire Color	Circuit Description (104 Pin)	Value at Hot Idle
65	WT/BL	DPFE Sensor Signal	0.20-1.30v
66	---	Not Used	---
67	BK/OR	EVAP CV Solenoid	0-10 Hz (0-100%)
68	---	Not Used	---
69	BK/YL	A/C WOT Relay Control	On: 1v, Off: 12v
70	BK/WT	Injector 1 Control	2.1-2.5 ms
71	GN/YL	Vehicle Power	12-14v
72	---	Not Used	---
73	BK/YL	HO2S-11 (B1 S1) Heater	On: 1v, Off: 12v
74	BK/BL	Injector 3 Control	2.1-2.5 ms
75	---	Not Used	---
76	BK/YL	CMP Sensor Ground	<0.050v
77	BK/YL	Power Ground	<0.1v
78	BR/YL	Coil Driver 3 Control	5° dwell
79	WT/BL	Instrument Interface Module	Digital Signals
80-82	---	Not Used	---
83	BK/YL	IAC Motor Control	9v (33% duty cycle)
84	---	Not Used	---
85	WT/PK	CMP Sensor Signal	5-7 Hz
86	BK/WT	A/C Pressure Switch	A/C On: 12v (Open)
87	WT/RD	HO2S-21 (B2 S1) Signal	0.1-1.1v
88	WT/BL	MAF Sensor Signal	0.6-0.9v
89	WT	TP Sensor Signal	0.53-1.27v
90	YL	Reference Voltage	4.9-5.1v
91	BR	Analog Signal Return	<0.050v
92	OR	Brake Pedal Position Switch	Brake Off: 0v, On: 12v
93	BK/GN	Injector 5 Control	2.1-2.5 ms
94	BK/RD	Injector 6 Control	2.1-2.5 ms
95	BK/OR	Injector 4 Control	2.1-2.5 ms
96	BK/GN	Injector 2 Control	2.1-2.5 ms
97	GN/YL	Vehicle Power	12-14v
98	---	Not Used	---
99	BK/BL	HO2S-21 (B2 S1) Heater	On: 1v, Off: 12v
100	BK/OR	HO2S-21 (B2 S1) Heater	On: 1v, Off: 12v
101	BK/GN	HO2S-22 (B2 S2) Heater	On: 1v, Off: 12v
102	---	Not Used	---
103	BK/YL	Power Ground	<0.1v
104	---	Not Used	---

Pin Connector Graphic

Terminal View of 104-Pin PCM Wiring Harness Connector

Standard Colors and Abbreviations

Abbreviation	Color	Abbreviation	Color	Abbreviation	Color
BK	Black	GY	Gray	PK	Purple
BL	Blue	GN	Green	RD	Red
BR	Brown	LG	LT Green	TN	Tan
DB	Dark Blue	OR	Orange	WT	White
DG	DK Green	PK	Pink	YL	Yellow

1995 Contour 2.5L V6 MFI VIN L (All) 22 Pin Connector

PCM Pin #	Wire Color	Circuit Description (22 Pin)	Value at Hot Idle
22-A	BL/RD	Keep Alive Power	12- 14v
22-B	RD/BK	Vehicle Power	12-14v
22-C	BK/RD	Vehicle Start Signal	KOEC: 9-11v
22-D	WT/RD	Switch Monitor Lamp Control	Lamp Off: 12v, On: 1v
22-E	BL	STO & MIL (lamp) Control	STO On: 5v, Off: 12v
22-F	---	Not Used	---
22-G	BL/OR	Ignition Control Module	Varies
22-H	WT	Headlamp Switch Signal	Switch On: 12v, Off: 1v
22-I	RD/WT	Self-Test Input	STI On: 1v, STI Off: 12v
22-J	PK	Rear Window Defroster	Switch On: 1v, Off: 12v
22-K	WT/BK	Torque Reduce & ECT Signal	Digital Signals
22-L	GN/BK	A/C Relay Control	On: 1v, Off: 12v
22-M	GN/RD	Vehicle Speed Sensor	0 Hz, 55 mph: 125 Hz
22-N	BL/YL	PSP Switch Signal	Straight: 12v, Turned: 0v
22-O	PK/BK	A/C Cycling Pressure Switch	Switch On: 12v, Off: 1v
22-P	OR/BK	Blower Motor Control Switch	Off or 1st: 12v, 2nd on: 1v
22-Q	WT/GN	Brake On/ Off Switch Signal	Brake Off: 0v, On: 12v
22-R	LG/BK	A/T: PNP Switch Signal	In 'P', 0v, Others: 12v
22-R	LG/BK	M/T: CPP Switch Signal	Clutch In: 0v, Out: 12v
22-S	GN	Torque Reduce Signal #1	Digital Signals
22-T	BR	Idle Switch Signal	Closed: 0v, Open: 12v
22-U	---	Not Used	---
22-V	LG/WT	Torque Reduce Signal #2	Digital Signals

1995 Contour 2.5L V6 MFI VIN L (All) 16 Pin Connector

PCM Pin #	Wire Color	Circuit Description (16 Pin)	Value at Hot Idle
16-A	GN/OR	Barometric Pressure Sensor	3.9v
16-B	RD	Measuring Core VAF Sensor	Idle: 3.06v
16-C	BK/YL	HO2S-11 (B1 S1) Signal	0.1-1.1v
16-D	BL/WT	HO2S-12 (B1 S2) Signal	0.1-1.1v
16-E	RD/GN	ECT Sensor Signal	0.5-0.6v
16-F	YL	Throttle Position Sensor	0.4-1.0v
16-G	-	Not Used	---
16-H	PK/YL	A/C High Pressure Switch	Closed: 1v, open: 12v
16-I	PK	Reference Voltage	4.9-5.1v
16-J	RD/BK	EGR Valve Position Sensor	0.4v
16-K	BK/RD	IAT Sensor Signal	1.5-2.5v
16-L	GN	DRL Signal - Canada	DRL On: 2v, Off: 12v
16-M	WT	Knock Sensor Signal	0v
16-N	-	Not Used	---
16-O	BL/BK	EVAP Purge Solenoid	12v, 55 mph: 1v
16-P	BL/GN	High Speed Fan Control	On: 1v, Off: 12v

Standard Colors and Abbreviations

Abbreviation	Color	Abbreviation	Color	Abbreviation	Color
BK	Black	GY	Gray	PK	Purple
BL	Blue	GN	Green	RD	Red
BR	Brown	LG	LT Green	TN	Tan
DB	Dark Blue	OR	Orange	WT	White
DG	DK Green	PK	Pink	YL	Yellow

1995 Contour 2.5L V6 MFI VIN L (All) 26 Pin Connector

PCM Pin #	Wire Color	Circuit Description (26 Pin)	Value at Hot Idle
26-A	BK	Power Ground	<0.1v
26-B	BK	Power Ground	<0.1v
26-C	BK/RD	Power Ground	<0.1v
26-D	BK/RD	Power Ground	<0.1v
26-E	LG/OR	CKP Sensor 1 Signal	2.5v
26-F	BL	CKP Sensor Ground	0.1v
26-G	BL/PK	CMP Sensor 1 Signal	2.5v
26-H	GN	CKP Sensor 2 Signal	2.5v
26-I	WT/GN	Variable Resonance Induction	VRIS1 On: 1v, Off: 12v
26-J	BL/RD	Variable Resonance Induction	VRIS2 On: 1v, Off: 12v
26-K	---	Not Used	---
26-L	RD/WT	Low Speed Fan Control	On: 1v, Off: 12v
26-M	GN/BK	Fuel Pressure Regulator	Startup: 3v, others: 12v
26-N	BL/OR	Condenser Fan Control	On: 1v, Off: 12v
26-O	WT/BL	EGR Vent Solenoid	12v, off-idle: 1v
26-P	GN/WT	EGR Control Solenoid	12v, off-idle: 1v
26-Q	LG/BK	Idle Air Control Valve	8-10v
26-R	---	Not Used	---
26-S	---	Not Used	---
26-T	LG	Fuel Pump Control	On: 1v, Off: 12v
26-U	RD/LG	Injector 1 Control	3.6 ms
26-V	BL/WT	Injector 2 Control	3.6 ms
26-W	BR	Injector 3 Control	3.6 ms
26-X	RD/YL	Injector 4 Control	3.6 ms
26-Y	WT	Injector 5 Control	3.6 ms
26-Z	WT/BK	Injector 6 Control	3.6 ms

Pin Connector Graphic

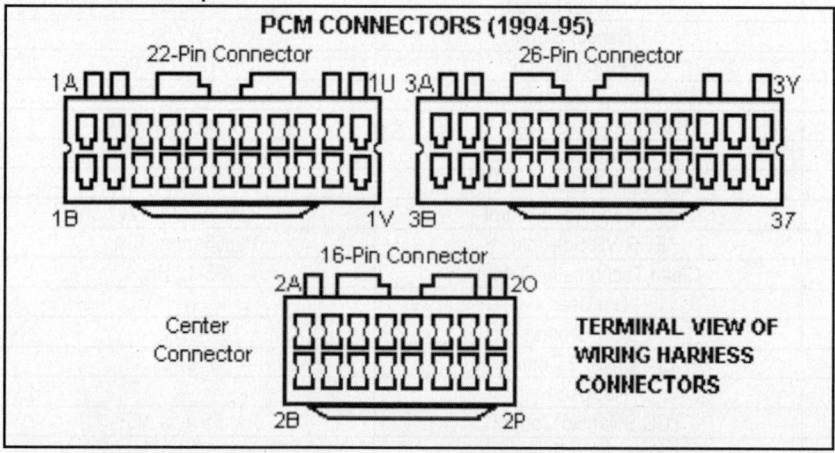

Standard Colors and Abbreviations

Abbreviation	Color	Abbreviation	Color	Abbreviation	Color
BK	Black	GY	Gray	PK	Purple
BL	Blue	GN	Green	RD	Red
BR	Brown	LG	LT Green	TN	Tan
DB	Dark Blue	OR	Orange	WT	White
DG	DK Green	PK	Pink	YL	Yellow

1996-97 Contour 2.5L V6 MFI VIN L (All) 104 Pin Connector

PCM Pin #	Wire Color	Circuit Description (104 Pin)	Value at Hot Idle
1	BK/BL	Shift Solenoid 2 Control	1v, 55 mph: 12v
2	BK/OR	MIL (lamp) Control	MIL On: 1v, Off: 12v
3-7	---	Not Used	---
8	BK/BL	IMRC Solenoid Control	5v (Off)
9-12	---	Not Used	---
13	WT/BL	Flash EEPROM Power	0.1v
14	---	Not Used	---
15	BR/BL	Data Bus (-) Signal	Digital Signals
16	WT/BL	Data Bus (+) Signal	Digital Signals
17	GY/OR	Passive Antitheft System	Digital Signals
18-20	---	Not Used	---
21	WT/RD	CKP Sensor (+) Signal	410-440 Hz
22	BR/RD	CKP Sensor (-) Signal	410-440 Hz
23	---	Not Used	---
24	BK/YL	Power Ground	<0.1v
25	BK/RD	Power Ground	<0.1v
26	BR/BL	Coil Driver 1 Control	5° dwell
27	BK/YL	Shift Solenoid 1 Control	12v, 55 mph: 1v
28	---	Not Used	---
29	PK/BK	TCS (switch) Signal	TCS & O/D On: 12v
30	WT/BK	Octane Adjust Switch	Closed: 0v, Open: 9.3v
31	WT	PSP Sensor Signal	Straight: 0v, Turning: 12v
32-34	---	Not Used	---
34	WT/GN	Passive Antitheft System	Digital Signals
35	WT/BL	HO2S-12 (B1 S2) Signal	0.1-1.1v
36	BR/BL	MAF Sensor Return	<0.050v
37	WT/RD	TFT Sensor Signal	2.10-2.40v
38	WT/GN	ECT Sensor Signal	0.5-0.6v
39	WT/PK	IAT Sensor Signal	1.5-2.5v
40	PK/BK	Fuel Pump Monitor	On: 12v, Off: 0v
41	PK/BL	A/C Cycling Clutch Switch	A/C On: 12v, Off: 0v
42	BK/RD	IMRC Solenoid Control	12v, 55 mph: 12v
43-44	---	Not Used	---
45	BK/BL	Low Speed Fan Control	On: 1v, Off: 12v
46	BK/WT	High Speed Fan Control	On: 1v, Off: 12v
47	BK/GN	EGR VR Solenoid	0%, 55 mph: 40%
48	WT/BK	Clean Tachometer Output	35-42 Hz
49-50	---	Not Used	---
51	BK/YL	Power Ground	<0.1v
52	BR/GN	Coil Driver 2 Control	5° dwell
53	---	Not Used	---
54	BK/WT	TCC Solenoid Control	0%, 55 mph: 95%
55	OR/YL	Keep Alive Power	12-14v
56	BK/OR	EVAP VMV Solenoid	0-10 Hz (0-100%)
57	WT/BK	Knock Sensor Signal	0v
58	WT/PK	VSS (+) Signal	0 Hz, 55 mph: 125 Hz
59	---	Not Used	---
60	WT	HO2S-11 (B1 S1) Signal	0.1-1.1v
61	WT/GN	HO2S-22 (B2 S2) Signal	0.1-1.1v
62-63	---	Not Used	---
64	WT/GN	TR Sensor Signal	In 'P': 0v, 55 mph: 1.7v
64	WT	M/T: Clutch Position Switch	Clutch In: 0.1v, Out: 5v

1996-97 Contour 2.5L V6 MFI VIN L (All) 104 Pin Connector

PCM Pin #	Wire Color	Circuit Description (104 Pin)	Value at Hot Idle
65	WT/BL	DPFE Sensor Signal	0.20-1.30v
66	---	Not Used	---
67	BK/OR	EVAP CV Solenoid	0-10 Hz (0-100%)
68	---	Not Used	---
69	BK/YL	A/C WOT Relay Control	On: 1v, Off: 12v
70	---	Not Used	---
71	GN/YL	Vehicle Power	12-14v
72	---	Not Used	---
73	BK/GN	Injector 5 Control	2.1-2.5 ms
74	BK/BL	Injector 3 Control	2.1-2.5 ms
75	BK/WT	Injector 1 Control	2.1-2.5 ms
76	BK/YL	CMP & TSS Ground	<0.050v
77	BK/YL	Power Ground	<0.1v
78	BR/YL	Coil Driver 3 Control	5° dwell
79	WT/BL	TCIL (lamp) Control	On: 1v, Off: 12v
80	BK/BL	Fuel Pump Control	On: 1v, Off: 12v
81	BK/RD	EPC Solenoid Control	9v (17 psi)
82	---	Not Used	---
83	BK/YL	IAC Motor Control	9v (33% duty cycle)
84	WT/PK	TSS Sensor Signal	46-50 Hz (700-730 rpm)
85	WT/PK	CMP Sensor Signal	5-7 Hz
86	BK/WT	A/C Pressure Switch	A/C On: 12v (Open)
87	WT/RD	HO2S-21 (B2 S1) Signal	0.1-1.1v
88	WT/BL	MAF Sensor Signal	0.6-0.9v
89	WT	TP Sensor Signal	0.53-1.27v
90	YL	Reference Voltage	4.9-5.1v
91	BR	Analog Signal Return	<0.050v
92	OR	Brake Pedal Position Switch	Brake Off: 0v, On: 12v
93	BK/YL	HO2S-11 (B1 S1) Heater	On: 1v, Off: 12v
94	BK/BL	HO2S-21 (B2 S1) Heater	On: 1v, Off: 12v
95	BK/OR	HO2S-12 (B1 S2) Heater	On: 1v, Off: 12v
96	BK/GN	HO2S-22 (B2 S2) Heater	On: 1v, Off: 12v
97	GN/YL	Vehicle Power	12-14v
98	---	Not Used	---
99	BK/RD	Injector 6 Control	2.1-2.5 ms
100	BK/OR	Injector 4 Control	2.1-2.5 ms
101	BK/YL	Injector 2 Control	2.1-2.5 ms
102	BK/OR	Shift Solenoid 3 Control	7-9v, 55 mph: 8-9v
103	BK/YL	Power Ground	<0.1v
104	---	Not Used	---

Pin Connector Graphic

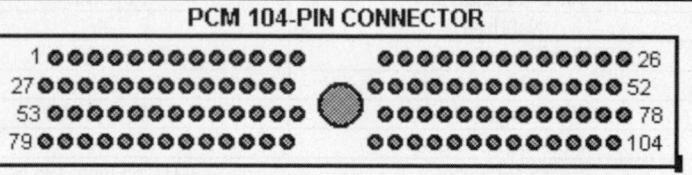

PCM 104-PIN CONNECTOR

Terminal View of 104-Pin PCM Wiring Harness Connector

1998-1999 Contour 2.5L V6 MFI VIN L (All) 104 Pin Connector

PCM Pin #	Wire Color	Circuit Description (104 Pin)	Value at Hot Idle
1	BK/BL	Shift Solenoid 2 Control	1v, 55 mph: 12v
2	BK/OR	MIL (lamp) Control	MIL On: 1v, Off: 12v
3-7	---	Not Used	---
8	BK/BL	IMRC Solenoid Control	5v (Off)
9-11	---	Not Used	---
12	WT	Fuel Level Indicator Signal	1.7v (1/2 full)
13	WT/BL	Flash EEPROM Power	0.1v
14	---	Not Used	---
15	BL	Data Bus (-) Signal	Digital Signals
16	GY	Data Bus (+) Signal	Digital Signals
17	GY/OR	Passive Antitheft System	Digital Signals
18-19	---	Not Used	---
20	BK/BL	Injector 3 Control	2.1-2.5 ms
21	WT/RD	CKP Sensor (+) Signal	410-440 Hz
22	BR/RD	CKP Sensor (-) Signal	410-440 Hz
23	---	Not Used	---
24	BK/YL	Power Ground	<0.1v
25	BK/RD	Power Ground	<0.1v
26	BR/BL	Coil Driver 1 Control	5° dwell
27	BK/YL	Shift Solenoid 1 Control	12v, 55 mph: 1v
28	---	Not Used	---
29	PK/BK	CD4E TCS (switch) Signal	TCS & O/D On: 12v
30	WT/BK	Octane Adjust Switch	Closed: 0v, Open: 9.3v
31	WT	PSP Sensor Signal	Straight: 0v, Turning: 12v
32-34	---	Not Used	---
34	WT/GN	Passive Antitheft System	Digital Signals
35	WT/BL	HO2S-12 (B1 S2) Signal	0.1-1.1v
36	BR/BL	MAF Sensor Return	<0.050v
37	WT/RD	TFT Sensor Signal	2.10-2.40v
38	WT/GN	ECT Sensor Signal	0.5-0.6v
39	WT/PK	IAT Sensor Signal	1.5-2.5v
40	PK/BK	Fuel Pump Monitor	On: 12v, Off: 0v
41	PK/BL	A/C Cycling Clutch Switch	A/C On: 12v, Off: 0v
42	BK/RD	IMRC Solenoid Motor Control	12v, 55 mph: 12v
43-44	---	Not Used	---
45	BK/BL	Low Speed Fan Control	On: 1v, Off: 12v
46	BK/WT	High Speed Fan Control	On: 1v, Off: 12v
47	BK/GN	EGR VR Solenoid	0%, 55 mph: 40%
48	WT/BK	Clean Tachometer Output	35-42 Hz
49-50	---	Not Used	---
51	BK/YL	Power Ground	<0.1v
52	BR/GN	Coil Driver 2 Control	5° dwell
53, 59	---	Not Used	---
54	BK/BL	Fuel Pump Control	On: 1v, Off: 12v
55	OR/YL	Keep Alive Power	12-14v
56	BK/OR	EVAP Purge Solenoid	0-10 Hz (0-100%)
57	WT/BK	Knock Sensor Signal	0v
58	WT/PK	VSS (+) Signal	0 Hz, 55 mph: 125 Hz
60	WT	HO2S-11 (B1 S1) Signal	0.1-1.1v
61	WT/GN	HO2S-22 (B2 S2) Signal	0.1-1.1v
62	WT/PK	FTP Sensor Signal	2.6v (cap off)
63 ('99)	WT/PK	FRP Sensor Signal	2.8v (39 psi)
64	WT/GN	TR Sensor Signal	In 'P': 0v, 55 mph: 1.7v
64	WT	M/T: Clutch Position Switch	Clutch In: 0.1v, Out: 5v

1998-1999 Contour 2.5L V6 MFI VIN L (All) 104 Pin Connector

PCM Pin #	Wire Color	Circuit Description (104 Pin)	Value at Hot Idle
65	WT/BL	DPFE Sensor Signal	0.20-1.30v
66	---	Not Used	---
67	BK/OR	EVAP CV Solenoid	0-10 Hz (0-100%)
68	---	Not Used	---
69	BK/YL	A/C WOT Relay Control	On: 1v, Off: 12v
70	BK/YL	ZETECT VCT Control	VCT Off: 12v, On: 1v
71	GN/YL	Vehicle Power	12-14v
72	---	Not Used	---
73	BK/YL	HO2S-11 (B1 S1) Heater	On: 1v, Off: 12v
74	BK/BL	Injector 3 Control	2.1-2.5 ms
75	BK/WT	Injector 1 Control	2.1-2.5 ms
76	BK/YL	CMP & TSS Ground	<0.050v
77	BK/YL	Power Ground	<0.1v
78	BR/YL	Coil Driver 3 Control	5° dwell
79	WT/BK	TCIL (lamp) Control	On: 1v, Off: 12v
80	BK/WT	TCC Solenoid Control	0%, 55 mph: 95%
81	BK/RD	EPC Solenoid Control	17 psi
82	---	Not Used	---
83	BK/YL	IAC Motor Control	9v (33% duty cycle)
84	WT/PK	TSS Sensor Signal	46-50 Hz (700-730 rpm)
85	WT/PK	CMP Sensor (+) Signal	5-7 Hz
86	BK/WT	A/C Pressure Switch	A/C On: 12v (Open)
87	WT/RD	HO2S-21 (B2 S1) Signal	0.1-1.1v
88	WT/BL	MAF Sensor Signal	0.6-0.9v
89	WT	TP Sensor Signal	0.53-1.27v
90	YL	Reference Voltage	4.9-5.1v
91	BR	Analog Signal Return	<0.050v
92	OR	Brake Pedal Position Switch	Brake Off: 0v, On: 12v
93	BK/YL	Injector 5 Control	2.1-2.5 ms
94	BK/BL	Injector 6 Control	2.1-2.5 ms
95	BK/OR	Injector 4 Control	2.1-2.5 ms
96	BK/GN	Injector 2 Control	2.1-2.5 ms
97	GN/YL	Vehicle Power	12-14v
98	---	Not Used	---
99	BK/BL	HO2S-21 (B2 S1) Heater	On: 1v, Off: 12v
100	BK/OR	HO2S-21 (B2 S1) Heater	On: 1v, Off: 12v
101	BK/GN	HO2S-22 (B2 S2) Heater	On: 1v, Off: 12v
102	BK/OR	Shift Solenoid 3 Control	7-9v, 55 mph: 8-9v
103	BK/YL	Power Ground	<0.1v
104	---	Not Used	---

Pin Connector Graphic

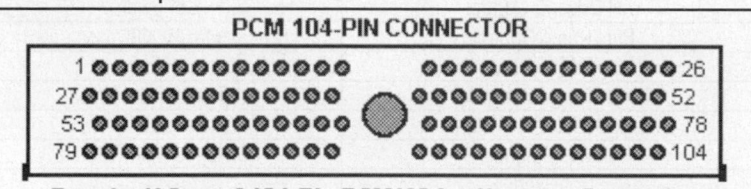

PCM 104-PIN CONNECTOR

Terminal View of 104-Pin PCM Wiring Harness Connector

2000 Contour 2.5L V6 MFI VIN L (All) 104 Pin Connector

PCM Pin #	Wire Color	Circuit Description (104 Pin)	Value at Hot Idle
1	---	Not Used	---
2	BK/OR	MIL (lamp) Control	MIL On: 1v, Off: 12v
3-5	---	Not Used	---
6	BK/YL	Shift Solenoid 1 Control	1v, 55 mph: 12v
7, 10, 14	---	Not Used	---
8	BK/BL	IMRC Solenoid Control	5v (Off)
9	WT	Fuel Level Indicator Signal	1.7v (1/2 full)
11	BK/BL	Shift Solenoid 2 Control	1v, 55 mph: 12v
12	WT/BK	Gearshift Indicator Signal	N/A
13	WT/BL	Flash EEPROM Power	0.1v
15	BL	Data Bus (-) Signal	Digital Signals
16	GY	Data Bus (+) Signal	Digital Signals
17	BK/WT	Hi Speed Fan Relay Control	On: 1v, Off: 12v
18-19	---	Not Used	---
20	BK/BL	Injector 3 Control	2.1-2.5 ms
21	WT/RD	CKP Sensor (+) Signal	410-440 Hz
22	BR/RD	CKP Sensor (-) Signal	410-440 Hz
23	---	Not Used	---
24	BK/YL	Power Ground	<0.1v
25	BK/RD	Power Ground	<0.1v
26	BR/BL	Coil Driver 1 Control	5° dwell
27	---	Not Used	---
28	WT/BL	Speedometer Indicator Signal	N/A
29	GN/BK	TCS (switch) Signal	TCS & O/D On: 12v
30	---	Not Used	---
31	WT	PSP Sensor Signal	Straight: 0v, Turning: 12v
32	BK/YL	Knock Sensor 1 (-) Signal	0v
33	BR/BL	VSS (-) Signal	0 Hz, 55 mph: 125 Hz
34, 43-46	---	Not Used	---
35	WT/BL	HO2S-12 (B1 S2) Signal	0.1-1.1v
36	BR/BL	MAF Sensor Return	<0.050v
37	WT/RD	TFT Sensor Signal	2.10-2.40v
38	WT/GN	ECT Sensor Signal	0.5-0.6v
39	WT/PK	ACT Sensor Signal	1.5-2.5v
40	GN/BK	Fuel Pump Monitor	On: 12v, Off: 0v
41	GN/BL	A/C Cycling Clutch Switch	A/C On: 12v, Off: 0v
42	BK/RD	IMRC Solenoid Motor Control	12v, 55 mph: 12v
47	GN/BK	EGR VR Solenoid	0%, 55 mph: 40%
48	WT/BK	Clean Tachometer Output	35-42 Hz
49-50, 53	---	Not Used	---
51	BK/YL	Power Ground	<0.1v
52	BR/GN	Coil Driver 2 Control	5° dwell
54	BK/RD	Fuel Pump Control	On: 1v, Off: 12v
55	OR/YL	Keep Alive Power	12-14v
56	BK/OR	EVAP Purge Solenoid	0-10 Hz (0-100%)
57	WT/BK	Knock Sensor 1 (+) Signal	0v
58	WT/BL	VSS (+) Signal	0 Hz, 55 mph: 125 Hz
59	---	Not Used	---
60	WT	HO2S-11 (B1 S1) Signal	0.1-1.1v
61	WT/GN	HO2S-22 (B2 S2) Signal	0.1-1.1v
62	WT/PK	FTP Sensor Signal	2.6v (cap off)
63	WT/GN	FRP Sensor Signal	2.8v (39 psi)
64	WT/GN	A/T: TR Sensor Signal	In 'P': 0v, 55 mph: 1.7v
64	WT	M/T: Clutch Position Switch	Clutch In: 0.1v, Out: 5v

2000 Contour 2.5L V6 MFI VIN L (All) 104 Pin Connector

PCM Pin #	Wire Color	Circuit Description (104 Pin)	Value at Hot Idle
65	WT/BL	EPT Sensor Signal	N/A
66	---	Not Used	---
67	BK/OR	EVAP CV Solenoid	0-10 Hz (0-100%)
68	BK/BL	Cooling Fan Relay Control	On: 1v, Off: 12v
69	BK/YL	A/C WOT Relay Control	On: 1v, Off: 12v
70	BK/WT	Injector 1 Control	2.1-2.5 ms
71	GN/YL	Vehicle Power	12-14v
72	---	Not Used	---
73	BK/YL	HO2S-11 (B1 S1) Heater	On: 1v, Off: 12v
74	BK/BL	Injector 3 Control	2.1-2.5 ms
75	---	Not Used	---
76	BK/YL	CMP Sensor (-) Signal	5-7 Hz
77	BK/YL	Power Ground	<0.1v
78	BR/YL	Coil Driver 3 Control	5° dwell
79	---	Not Used	---
80	BK/WT	Modulated Lockup Solenoid	N/A
81	BK/RD	EPC Solenoid Control	17 psi
82	---	Not Used	---
83	BK/YL	IAC Motor Control	9v (33% duty cycle)
84	WT/PK	TSS Sensor Signal	46-50 Hz (700-730 rpm)
85	WT/PK	CMP Sensor (+) Signal	5-7 Hz
86	BK/BL	A/C Pressure Switch	A/C On: 12v (Open)
87	WT/RD	HO2S-21 (B2 S1) Signal	0.1-1.1v
88	WT/BL	MAF Sensor Signal	0.6-0.9v
89	WT	TP Sensor Signal	0.53-1.27v
90	YL	Reference Voltage	4.9-5.1v
91	BR	Analog Signal Return	<0.050v
92	OR	Brake Pedal Position Switch	Brake Off: 0v, On: 12v
93	BK/GN	Injector 5 Control	2.1-2.5 ms
94	BK/RD	Injector 6 Control	2.1-2.5 ms
95	BK/OR	Injector 4 Control	2.1-2.5 ms
96	BK/YL	Injector 2 Control	2.1-2.5 ms
97	GN/YL	Vehicle Power	12-14v
98	---	Not Used	---
99	BK/BL	HO2S-21 (B2 S1) Heater	On: 1v, Off: 12v
100	BK/OR	HO2S-21 (B2 S1) Heater	On: 1v, Off: 12v
101	BK/GN	HO2S-22 (B2 S2) Heater	On: 1v, Off: 12v
102	BK/DG	Shift Solenoid 3 Control	7-9v, 55 mph: 8-9v
103	BK/YL	Power Ground	<0.1v
104	---	Not Used	---

Pin Connector Graphic

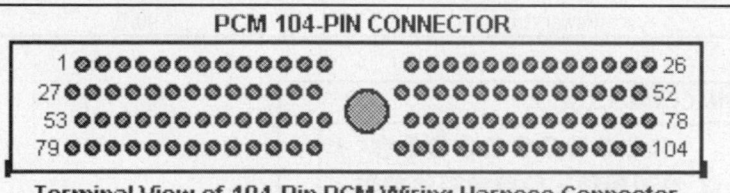

PCM 104-PIN CONNECTOR

Terminal View of 104-Pin PCM Wiring Harness Connector

CROWN VICTORIA PIN TABLES

1992-95 Crown Victoria 4.6L V8 MFI VIN W (A/T) 60 Pin Connector

PCM Pin #	Wire Color	Circuit Description (60 Pin)	Value at Hot Idle
1	YL	Keep Alive Power	12-14v
2 ('93-'95)	LG	Brake Pedal Position Switch	Brake Off: 0v, On: 12v
3	GY/BK	VSS (+) Signal	0 Hz, 55 mph: 125 Hz
4	TN/YL	Ignition Diagnostic Monitor	20-31 Hz
5	PK/BL	TSS Sensor (+) Signal	55 mph: 126-136 Hz
6	PK/OR	VSS (-) Signal	0 Hz, 55 mph: 125 Hz
7	LG/RD	ECT Sensor Signal	0.5-0.6v
8	DG/YL	Fuel Pump Monitor	On: 12v, Off: 0v
9	TN/BL	MAF Sensor Return	<0.050v
10	DG/OR	A/C Cycling Clutch Switch	A/C C/C on: 12v, off: 0v
11	GY/YL	Air Management 2 Solenoid	AM2 On: 1v, Off: 12v
12	LG/OR	Injector 6 Control	4.0-4.4 ms
13	TN/RD	Injector 7 Control	4.0-4.4 ms
14	BL	Injector 8 Control	4.0-4.4 ms
15	TN/BK	Injector 5 Control	4.0-4.4 ms
16	OR/RD	Ignition System Ground	<0.050v
17	TN/RD	STI Output, MIL Control	MIL On: 1v, Off: 12v
18	TN/OR	Data Bus (+) Signal	Digital Signals
19	PK/BL	Data Bus (-) Signal	Digital Signals
20	BK	PCM Case Ground	<0.050v
21	WT/BL	IAC Motor Control	8.3-11.5v
22	BL/OR	Fuel Pump Control	On: 1v, Off: 12v
23	---	Not Used	---
24	DG	CID Sensor Signal	6-7 Hz
25	GY	ACT Sensor Signal	1.5-2.5v
25	LG/PK	ACT Sensor Signal	1.5-2.5v
26	BR/WT	Reference Voltage	4.9-5.1v
27	BR/LG	DPFE EGR Sensor Signal	0.4v
28	---	Not Used	---
29	WT/RD	Octane Adjust Switch	Closed: 0v, Open: 9.1v
30 ('93-'95)	WT/PK	Neutral Drive or MLP Switch	In 'P': 0v, Others: 5v
30 ('93-'95)	BL/YL	Neutral Drive or MLP Switch	In 'P': 0v, Others: 5v
31-32	---	Not Used	---
33	BR/PK	EGR VR Solenoid	0%, 55 mph: 45%
34	DB/LG	Data Output Link	Digital Signals
35	BR/BL	Injector 4 Control	4.0-4.4 ms
36	PK	Spark Angle Word Signal	50% duty cycle
37, 57	RD	Vehicle Power	12-14v
38 ('93-'95)	WT/YL	EPC Solenoid Control	9.5v (20 psi)
39	BR/YL	Injector 4 Control	4.0-4.4 ms
40	BK/WT	Power Ground	<0.1v

Pin Connector Graphic

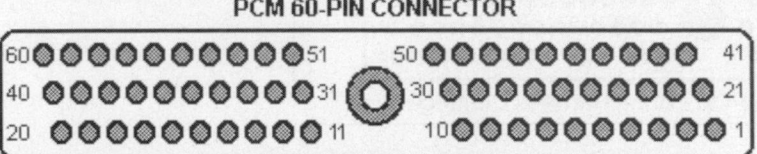

1992-95 Crown Victoria 4.6L V8 MFI VIN W (A/T) 60 Pin Connector

PCM Pin #	Wire Color	Circuit Description (60 Pin)	Value at Hot Idle
41 ('93-'95)	TN/WT	TCS (switch) Signal	TCS & O/D On: 12v
42	---	Not Used	---
43	RD/BK	HO2S-11 (B1 S1) Signal	0.1-1.1v
44	GY/BL	HO2S-21 (B2 S1) Signal	0.1-1.1v
45	---	Not Used	---
46	GY/RD	Analog Signal Return	<0.050v
47	GY/WT	TP Sensor Signal	0.51-1.27v
48	WT/PK	Self-Test Indicator Signal	STI Open: 5v, Closed: 1v
49 ('93-'95)	OR/BK	TOT Sensor Signal	2.10-2.40v
50	BL/RD	MAF Sensor Signal	0.6-0.9v
51 ('93-'94)	OR/YL	Shift Solenoid 1 Control	1v, 55 mph: 12v
52 ('93-'94)	PK/OR	Shift Solenoid 2 Control	12v, 55 mph: 1v
53	BR/OR	TCC Solenoid Signal	12v, 55 mph: 9-10v
53 ('93-'95)	PK/YL	TCC Solenoid Signal	12v, 55 mph: 9-10v
54	OR/BL	A/C WOT Relay Control	On: 1v, Off: 12v
55	DB/WT	TCIL (lamp) Control	On: 1v, Off: 12v
55 ('93-'95)	WT/LG	TCIL (lamp) Control	On: 1v, Off: 12v
56	GY/OR	PIP Sensor Signal	50% dwell
58	TN	Injector 1 Control	4.0-4.4 ms
59	WT	Injector 2 Control	4.0-4.4 ms
60	BK/WT	Power Ground	<0.1v

Standard Colors and Abbreviations

Abbreviation	Color	Abbreviation	Color	Abbreviation	Color
BK	Black	GY	Gray	PK	Purple
BL	Blue	GN	Green	RD	Red
BR	Brown	LG	LT Green	TN	Tan
DB	Dark Blue	OR	Orange	WT	White
DG	DK Green	PK	Pink	YL	Yellow

1996-97 Crown Victoria 4.6L V8 VIN W (A/T) 104 Pin Connector

PCM Pin #	Wire Color	Circuit Description (104 Pin)	Value at Hot Idle
1	PK/OR	Shift Solenoid 2 Control	12v, 55 mph: 1v
2	PK/LG	MIL (lamp) Control	MIL On: 1v, Off: 12v
3-12	---	Not Used	---
13	LG/YL	Flash EEPROM Power	0.1v
14	---	Not Used	---
15	PK/LB	Data Bus (-) Signal	Digital Signals
16	TN/OR	Data Bus (+) Signal	Digital Signals
17-20	---	Not Used	---
21	DB	CKP Sensor (+) Signal	850-1120 Hz
22	GY	CKP Sensor (-) Signal	850-1120 Hz
23	---	Not Used	---
24	BK/WT	Power Ground	<0.1v
25	BK	PCM Case Ground	<0.050v
26	TN/WT	Coil Driver 1 Control	5° dwell
27	OR/YL	Shift Solenoid 1 Control	1v, 55 mph: 12v
28	---	Not Used	---
29	TN/WT	TCS (switch) Signal	TCS & O/D On: 12v
30	WT/RD	Octane Adjust Switch	Closed: 0v, Open: 9.3v
31-32	---	Not Used	---
33	PK/OR	VSS (-) Signal	0 Hz, 55 mph: 125 Hz
34	---	Not Used	---
35	RD/LG	HO2S-11 (B1 S1) Signal	0.1-1.1v
36	TN/LB	MAF Sensor Return	<0.050v
37	OR/BK	TFT Sensor Signal	2.10-2.40v
38	LG/RD	ECT Sensor Signal	0.5-0.6v
39	GY	IAT Sensor Signal	1.5-2.5v
40	DG/YL	Fuel Pump Monitor	On: 12v, Off: 0v
41	DG/OR	A/C Cycling Clutch Switch	A/C On: 12v, Off: 0v
42-44	---	Not Used	---
45	RD/OR	Low Speed Fan Control	On: 1v, Off: 12v
46	---	Not Used	---
47	BR/PK	EGR VR Solenoid	0%, 55 mph: 45%
48-50	---	Not Used	---
51	BK/WT	Power Ground	<0.1v
52	TN/OR	Coil Driver 2 Control	5° dwell
53	---	Not Used	---
54	RD/LB	TCC Solenoid Control	TCC Off Idle: 0%
55	YL/BK	Keep Alive Power	12-14v
56	LG/BK	EVAP VMV Solenoid	0-10 Hz (0-100%)
57	---	Not Used	---
58	GY/BK	VSS (+) Signal	0 Hz, 55 mph: 125 Hz
59	---	Not Used	---
60	GY/LB	HO2S-11 (B1 S1) Signal	0.1-1.1v
61	PK/LG	HO2S-22 (B2 S2) Signal	0.1-1.1v
62	RD/PK	FTP Sensor Signal	2.6v (0" H2O - cap off)
63	---	Not Used	---
64	LB/YL	TR Sensor Signal	In 'P': 0v, 55 mph: 1.7v
65	BR/LG	DPFE Sensor Signal	0.20-1.30v
66	---	Not Used	---
67	PK/WT	EVAP Purge Solenoid	0-10 Hz (0-100%)
68	---	Not Used	---
69	OR/LB	A/C WOT Relay Control	On: 1v, Off: 12v
70	---	Not Used	---
71	RD	Vehicle Power	12-14v

1996-97 Crown Victoria 4.6L V8 VIN W (A/T) 104 Pin Connector

PCM Pin #	Wire Color	Circuit Description (104 Pin)	Value at Hot Idle
72	TN/RD	Injector 7 Control	3.4-3.7 ms
73	TN/BK	Injector 5 Control	3.4-3.7 ms
74	BR/YL	Injector 3 Control	3.4-3.7 ms
75	TN	Injector 1 Control	3.4-3.7 ms
76-77	BK/WT	Power Ground	<0.1v
78	TN/LG	Coil Driver 3 Control	5° dwell
79	WT/LG	TCIL (lamp) Control	On: 1v, Off: 12v
80	LB/OR	Fuel Pump Control	On: 1v, Off: 12v
81	WT/YL	EPC Solenoid Control	9.5v (20 psi)
82	---	Not Used	---
83	WT/LB	IAC Motor Control	34% duty cycle
84	PK/LB	OSS Sensor Signal	0 Hz, 55 mph: 131 Hz
85	DG	CMP Sensor Signal	5-7 Hz
86	---	Not Used	---
87	RD/BK	HO2S-21 (B2 S1) Signal	0.1-1.1v
88	LB/RD	MAF Sensor Signal	0.6v
89	GY/WT	TP Sensor Signal	0.53-1.27v
90	BR/WT	Reference Voltage	4.9-5.1v
91	GY/RD	Analog Signal Return	<0.050v
92	LG	Brake Pedal Position Switch	Brake Off: 0v, On: 12v
93	RD/WT	HO2S-11 (B1 S1) Heater	On: 1v, Off: 12v
94	YL/LB	HO2S-21 (B2 S1) Heater	On: 1v, Off: 12v
95	WT/BK	HO2S-12 (B1 S2) Heater	On: 1v, Off: 12v
96	TN/YL	HO2S-22 (B2 S2) Heater	On: 1v, Off: 12v
97	RD	Vehicle Power	12-14v
98	LB	Injector 8 Control	3.4-3.7 ms
99	LG/OR	Injector 6 Control	3.4-3.7 ms
100	BR/LB	Injector 4 Control	3.4-3.7 ms
101	WT	Injector 2 Control	3.4-3.7 ms
102	---	Not Used	---
103	BK/WT	Power Ground	<0.1v
104	RD/YL	Coil Driver 4 Control	5° dwell

Pin Connector Graphic

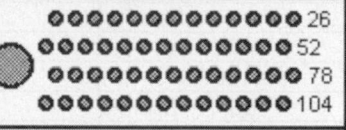

PCM 104-PIN CONNECTOR

Terminal View of 104-Pin PCM Wiring Harness Connector

Standard Colors and Abbreviations

Abbreviation	Color	Abbreviation	Color	Abbreviation	Color
BK	Black	GY	Gray	PK	Purple
BL	Blue	GN	Green	RD	Red
BR	Brown	LG	LT Green	TN	Tan
DB	Dark Blue	OR	Orange	WT	White
DG	DK Green	PK	Pink	YL	Yellow

1998-99 Crown Victoria 4.6L V8 VIN W (A/T) 104 Pin Connector

PCM Pin #	Wire Color	Circuit Description (104 Pin)	Value at Hot Idle
1	OR/YL	COP 6 Driver Control	5° dwell
2	PK/LG	MIL (lamp) Control	MIL On: 1v, Off: 12v
3	BK/WT	Power Ground	<0.1v
4-5	---	Not Used	---
6	OR/YL	Shift Solenoid 'A' Control	1v, 55 mph: 12v
7-8	---	Not Used	---
9	YL/WT	Fuel Level Indicator Signal	1.7v (1/2 full)
10	---	Not Used	---
11	PK/OR	Shift Solenoid 'B' Control	12v, Cruise: 1v
12	WT/LG	TCIL (lamp) Control	Lamp On: 1v, Off: 5v
13	PK	Flash EEPROM Power	0.1v
15	PK/LB	Data Bus (-) Signal	Digital Signals
16	TN/OR	Data Bus (+) Signal	Digital Signals
17-20	---	Not Used	---
21	BK/PK	CKP Sensor (+) Signal	440-490 Hz
22	GY/YL	CKP Sensor (-) Signal	440-490 Hz
23	---	Not Used	---
24	BK/WT	Power Ground	<0.1v
25	BK	PCM Case Ground	<0.050v
26	LG/WT	COP 1 Driver Control	5° dwell
27	LG/YL	COP 5 Driver Control	1v, 55 mph: 12v
28	RD/OR	Low Speed Fan Control	On: 1v, Off: 12v
29	TN/WT	TCS (switch) Signal	TCS & O/D On: 12v
30	WT/RD	Octane Adjust Switch	Closed: 0v, Open: 9.3v
31-32	---	Not Used	---
33	PK/OR	VSS (-) Signal	0 Hz, 55 mph: 125 Hz
34	YL/BK	Digital TR1 Sensor	In 'P': 0v, 55 mph: 11v
35	RD/LG	HO2S-12 (B1 S2) Signal	0.1-1.1v
36	TN/LB	MAF Sensor Return	<0.050v
37	OR/BK	TFT Sensor Signal	2.10-2.40v
38	---	Not Used	---
39	GY	IAT Sensor Signal	1.5-2.5v
40	DG/YL	Fuel Pump Monitor	On: 12v, Off: 0v
41	DG/OR	A/C Cycling Clutch Switch	AC On: 12v, Off: 0v
42	RD/WT	ECT Sensor Signal	0.5-0.6v
43	DB/LG	Fuel Flow Rate Signal	Digital Signals
45	OR/RD	CHTIL (lamp) Control	Lamp On: 1v, Off: 5v
46	LG/PK	High Speed Fan Control	On: 1v, Off: 12v
47	BR/PK	EGR VR Solenoid	0%, 55 mph: 45%
48	---	Not Used	---
49	RD/LB	Digital TR2 Sensor	In 'P': 0v, 55 mph: 11v
50	WT/BK	Digital TR4 Sensor	In 'P': 0v, 55 mph: 11v
51	BK/WT	Power Ground	<0.1v
52	WT/PK	COP 3 Driver Control	5° dwell
53	DG/PK	COP 4 Driver Control	5° dwell
54	PK/YL	TCC Solenoid Control	0%, 55 mph: 95%
55	YL/BK	Keep Alive Power	12-14v
56	LG/BK	EVAP Purge Solenoid	0-10 Hz (0-100%)
57	---	Not Used	---
58	GY/BK	VSS + Sensor Signal	0 Hz, 55 mph: 125 Hz
59	---	Not Used	---
60	GY/LB	HO2S-11 (B1 S1) Signal	0.1-1.1v
61	PK/LG	HO2S-22 (B2 S2) Signal	0.1-1.1v

1998-99 Crown Victoria 4.6L V8 VIN W (A/T) 104 Pin Connector

PCM Pin #	Wire Color	Circuit Description (104 Pin)	Value at Hot Idle
62	RD/PK	FTP Sensor Signal	2.6v (0" H2O - cap off)
63	---	Not Used	---
64	LB/YL	Digital TR3 Sensor	In 'P': 0v, in O/D: 1.7v
65	BR/LG	DPFE Sensor Signal	0.20-1.30v
66	YL/LG	CHT Sensor Signal	0.7v (194°F)
67	PK/WT	EVAP CV Solenoid	0-10 Hz (0-100%)
68	---	Not Used	---
69	PK/YL	A/C WOT Relay Control	On: 1v, Off: 12v
70	---	Not Used	---
71	RD	Vehicle Power	12-14v
72	TN/RD	Injector 7 Control	3.4-3.7 ms
73	TN/BK	Injector 5 Control	3.4-3.7 ms
74	BR/YL	Injector 3 Control	3.4-3.7 ms
75	TN	Injector 1 Control	3.4-3.7 ms
77	BK/WT	Power Ground	<0.1v
78	PK/LB	COP 7 Driver Control	5° dwell
79	WT/RD	COP 8 Driver Control	5° dwell
80	LB/OR	Fuel Pump Control	On: 1v, Off: 12v
81	WT/YL	EPC Solenoid Control	9.5v (20 psi)
82	---	Not Used	---
83	WT/LB	IAC Motor Control	34% duty cycle
84	DB/YL	OSS Sensor Signal	0 Hz
85	DB/OR	CMP Sensor Signal	6-7 Hz
86	WT/BK	A/C High Pressure Switch	Open: 12v, Closed: 0v
87	RD/BK	HO2S-21 (B2 S1) Signal	0.1-1.1v
88	LB/RD	MAF Sensor Signal	0.6v
89	GY/WT	TP Sensor Signal	0.53-1.27v
90	BR/WT	Reference Voltage	4.9-5.1v
91	GY/RD	Analog Signal Return	<0.050v
92	LG	Brake Pedal Position Switch	Brake Off: 0v, On: 12v
93	RD/WT	HO2S-11 (B1 S1) Heater	On: 1v, Off: 12v
94	YL/LB	HO2S-21 (B2 S1) Heater	On: 1v, Off: 12v
95	WT/BK	HO2S-12 (B1 S2) Heater	On: 1v, Off: 12v
96	TN/YL	HO2S-22 (B2 S2) Heater	On: 1v, Off: 12v
97 ('98 only)	RD	Vehicle Power	12-14v
98	LB	Injector 8 Control	3.4-3.7 ms
99	LG/OR	Injector 6 Control	3.4-3.7 ms
100	BR/LB	Injector 4 Control	3.4-3.7 ms
101	WT	Injector 2 Control	3.4-3.7 ms
102	---	Not Used	---
103	BK/WT	Power Ground	<0.1v
104	PK/WT	COP 2 Driver Control	5° dwell

Pin Connector Graphic

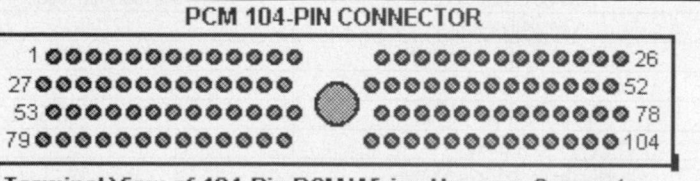

Terminal View of 104-Pin PCM Wiring Harness Connector

2000-02 Crown Victoria 4.6L V8 VIN W (A/T) 104 Pin Connector

PCM Pin #	Wire Color	Circuit Description (104 Pin)	Value at Hot Idle
1	OR/YL	COP 6 Driver Control	5° dwell
2	PK/LG	MIL (lamp) Control	MIL On: 1v, Off: 12v
3	BK/WT	Power Ground	<0.1v
4-5	---	Not Used	---
6	OR/YL	Shift Solenoid 'A' Control	1v, 55 mph: 12v
7-8	---	Not Used	---
9	YL/WT	Fuel Level Indicator Signal	1.7v (1/2 full)
10	LB/RD	Not Used	---
11	PK/OR	Shift Solenoid 'B' Control	12v, Cruise: 1v
12	WT/LG	TCIL (lamp) Control	Lamp On: 1v, Off: 5v
13	PK	Flash EEPROM Power	0.1v
15	PK/LB	Data Bus (-) Signal	Digital Signals
16	TN/OR	Data Bus (+) Signal	Digital Signals
17-20	---	Not Used	---
21	BK/PK	CKP Sensor (+) Signal	440-490 Hz
22	GY/YL	CKP Sensor (-) Signal	440-490 Hz
23-24	---	Not Used	---
25	BK	PCM Case Ground	<0.050v
26	LG/WT	COP 1 Driver Control	5° dwell
27	LG/YL	COP 5 Driver Control	1v, 55 mph: 12v
28	RD/OR	Low Speed Fan Control	On: 1v, Off: 12v
29	TN/WT	TCS (switch) Signal	TCS & O/D On: 12v
30-32	---	Not Used	---
33	PK/OR	VSS (-) Signal	0 Hz, 55 mph: 125 Hz
34	YL/BK	Digital TR1 Sensor	In 'P': 0v, 55 mph: 11v
35	RD/LG	HO2S-12 (B1 S2) Signal	0.1-1.1v
36	TN/LB	MAF Sensor Return	<0.050v
37	OR/BK	TFT Sensor Signal	2.10-2.40v
38	---	Not Used	---
39	GY	IAT Sensor Signal	1.5-2.5v
40	DG/YL	Fuel Pump Monitor	On: 12v, Off: 0v
41	BK/YL	A/C Pressure Cutout switch	AC On: 12v, Off: 0v
42	RD/WT	ECT Sensor Signal	0.5-0.6v
43	DB/LG	Fuel Flow Rate Signal	Digital Signals
45	OR/RD	CHTIL (lamp) Control	Lamp On: 1v, Off: 5v
46	LG/PK	High Speed Fan Control	On: 1v, Off: 12v
47	BR/PK	EGR VR Solenoid	0%, 55 mph: 45%
48	---	Not Used	---
49	LB/BK	Digital TR2 Sensor	In 'P': 0v, 55 mph: 11v
50	WT/BK	Digital TR4 Sensor	In 'P': 0v, 55 mph: 11v
51	BK/WT	Power Ground	<0.1v
52	WT/PK	COP 3 Driver Control	5° dwell
53	DG/PK	COP 4 Driver Control	5° dwell
54	VT/YL	TCC Solenoid Control	0%, 55 mph: 95%
55	YL/BK	Keep Alive Power	12-14v
56	LG/BK	EVAP Purge Solenoid	0-10 Hz (0-100%)
57	---	Not Used	---
58	GY/BK	VSS + Sensor Signal	0 Hz, 55 mph: 125 Hz
59	---	Not Used	---
60	GY/BK	HO2S-11 (B1 S1) Signal	0.1-1.1v
61	PK/LG	HO2S-22 (B2 S2) Signal	0.1-1.1v
62	RD/PK	FTP Sensor Signal	2.6v (0" H2O - cap off)

2000-02 Crown Victoria 4.6L V8 VIN W (A/T) 104 Pin Connector

PCM Pin #	Wire Color	Circuit Description (104 Pin)	Value at Hot Idle
63	---	Not Used	---
64	LB/YL	Digital TR3 Sensor	In 'P': 0v, in O/D: 1.7v
65 ('00)	BR/LG	DPFE Sensor Signal	0.20-1.30v
65 ('01-'02)	BR/LG	DPFE Sensor Signal	0.95-1.05v
66	YL/LG	CHT Sensor Signal	0.7v (194°F)
67	VT/WT	EVAP CV Solenoid	0-10 Hz (0-100%)
68	---	Not Used	---
69	PK/YL	A/C WOT Relay Control	On: 1v, Off: 12v
70	---	Not Used	---
71	RD	Vehicle Power	12-14v
72	TN/RD	Injector 7 Control	3.4-3.7 ms
73	TN/BK	Injector 5 Control	3.4-3.7 ms
74	BR/YL	Injector 3 Control	3.4-3.7 ms
75	TN	Injector 1 Control	3.4-3.7 ms
77	BK/WT	Power Ground	<0.1v
78	PK/LB	COP 7 Driver Control	5° dwell
79	WT/RD	COP 8 Driver Control	5° dwell
80	LB/OR	Fuel Pump Control	On: 1v, Off: 12v
81	WT/YL	EPC Solenoid Control	9.5v (20 psi)
82	---	Not Used	---
83	WT/LB	IAC Motor Control	34% duty cycle
84	DB/YL	OSS Sensor Signal	0 Hz
85	DB/OR	CMP Sensor Signal	6-7 Hz
86	WT/BK	A/C High Pressure Switch	Open: 12v, Closed: 0v
87	RD/BK	HO2S-21 (B2 S1) Signal	0.1-1.1v
88	LB/RD	MAF Sensor Signal	0.6v
89	GY/WT	TP Sensor Signal	0.53-1.27v
90	BR/WT	Reference Voltage	4.9-5.1v
91	GY/RD	Analog Signal Return	<0.050v
92	LG	Brake Pedal Position Switch	Brake Off: 0v, On: 12v
93	RD/WT	HO2S-11 (B1 S1) Heater	On: 1v, Off: 12v
94	YL/LB	HO2S-21 (B2 S1) Heater	On: 1v, Off: 12v
95	WT/BK	HO2S-12 (B1 S2) Heater	On: 1v, Off: 12v
96	TN/YL	HO2S-22 (B2 S2) Heater	On: 1v, Off: 12v
98	LB	Injector 8 Control	3.4-3.7 ms
99	LG/OR	Injector 6 Control	3.4-3.7 ms
100	BR/LB	Injector 4 Control	3.4-3.7 ms
101	WT	Injector 2 Control	3.4-3.7 ms
102	---	Not Used	---
103	BK/WT	Power Ground	<0.1v
104	PK/WT	COP 2 Driver Control	5° dwell

Pin Connector Graphic

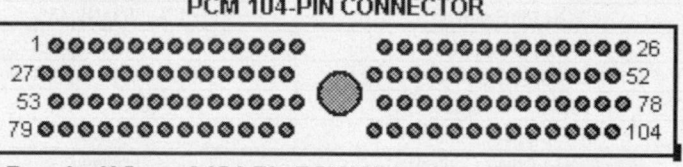

PCM 104-PIN CONNECTOR

Terminal View of 104-Pin PCM Wiring Harness Connector

2003-05 Crown Victoria 4.6L V8 VIN W (A/T) 104 Pin Connector

PCM Pin #	Wire Color	Circuit Description (104 Pin)	Value at Hot Idle
1	OR/YL	COP 6 Driver Control	5° dwell
2	PK/LG	MIL (lamp) Control	MIL On: 1v, Off: 12v
3	BK/WT	Power Ground	<0.1v
4-5, 14, 23	---	Not Used	---
6	OR/YL	Shift Solenoid 'A' Control	1v, 55 mph: 12v
7	YL/LG	Generator Regulator 'S' Terminal	0-130 Hz
8	RD/PK	Fuel Rail Temperature Sensor	1.7-3.5v (50-120°F)
9	YL/WT	Fuel Level Indicator Signal	1.7v (1/2 full)
10	LB/RD	MAP Sensor Signal	107 Hz
11	VT/OR	Shift Solenoid 'B' Control	12v, Cruise: 1v
12	WT/LG	TCIL (lamp) Control	Lamp On: 1v, Off: 5v
13	VT	Flash EEPROM Power	0.1v
15	PK/LB	SCP Bus (-) Signal	Digital Signals
16	TN/OR	SCP Bus (+) Signal	Digital Signals
17	GY/OR	RX Signal	Digital Signals
18	WT/LG	TX Signal	Digital Signals
19	OR/RD	Cylinder Head Temperature Lamp Control	Lamp Off: 12v, On: 1v
20	WT/LG	Fuel Door Release Solenoid Indicator	Solenoid Off: 12v, On: 1v
21	BK/PK	CKP Sensor (+) Signal	440-490 Hz
22	GY/YL	CKP Sensor (-) Signal	440-490 Hz
24, 51	BK/WT	Power Ground	<0.1v
25	BK	PCM Case Ground	<0.050v
26	LG/WT	COP 1 Driver Control	5° dwell
27	LG/YL	COP 5 Driver Control	1v, 55 mph: 12v
28	RD/OR	Low Speed Fan Control	On: 1v, Off: 12v
29	TN/WT	TCS (switch) Signal	TCS & O/D On: 12v
30	DG/RD	Light Emitting Diode Signal Ground	<0.050v
31	YL/LG	Power Steering Pressure Sensor	Straight: 0v, Turned: 12v
32	DG/VT	Knock Sensor Ground	<0.050v
33	---	Not Used	---
34	YL/BK	Digital TR1 Sensor	In 'P': 0v, 55 mph: 11v
35	RD/LG	HO2S-12 (B1 S2) Signal	0.1-1.1v
36	TN/LB	MAF Sensor Return	<0.050v
37	OR/BK	TFT Sensor Signal	2.10-2.40v
38	BK/WT	A/C Pressure Sensor Discharge Temp.	A/C On: 1.5-1.9v
39	GY	IAT Sensor Signal	1.5-2.5v
40	DG/YL	Fuel Pump Monitor	On: 12v, Off: 0v
41	PK/LB	A/C Pressure Cutout Switch	AC On: 12v, Off: 0v
42	RD/WT	ECT Sensor Signal	0.5-0.6v
43	DB/LG	Low Fuel Indicator	Digital Signals
44	GY/RD	Starter Relay Control	Relay Off: 12v, On: 9v
45	LG/RD	Generator Common	Digital Signals
46	---	Not Used	---
47	BR/PK	EGR VR Solenoid	0%, 55 mph: 45%
48	WT	Tachometer Output	DC pulse signals
49	LB/BK	Digital TR2 Sensor	In 'P': 0v, 55 mph: 11v
50	WT/BK	Digital TR4 Sensor	In 'P': 0v, 55 mph: 11v
52	WT/PK	COP 3 Driver Control	5° dwell
53	DG/VT	COP 4 Driver Control	5° dwell
54	VT/YL	TCC Solenoid Control	0%, 55 mph: 95%
55	YL/BK	Keep Alive Power	12-14v
56	LG/BK	EVAP Purge Solenoid Control	0-10 Hz (0-100%)
57	YL/RD	Knock Sensor Signal	0v
58-59	---	Not Used	---

2003-05 Crown Victoria 4.6L V8 VIN W (A/T) 104 Pin Connector

PCM Pin #	Wire Color	Circuit Description (104 Pin)	Value at Hot Idle
60	GY/BK	HO2S-11 (B1 S1) Signal	0.1-1.1v
61	VT/LG	HO2S-22 (B2 S2) Signal	0.1-1.1v
62	RD/PK	FTP Sensor Signal	2.6v (0" H2O - cap off)
63	OR/LG	Injection Pressure Sensor	2.8v (39 psi)
64	LB/YL	Digital TR3 Sensor	In 'P': 0v, in O/D: 1.7v
65	BR/LG	DPFE Sensor Signal	0.95-1.05v
66	YL/LG	Cylinder Head Temperature Sensor	0.7v (194°F)
67	VT/WT	EVAP Canister Vent Solenoid	0-10 Hz (0-100%)
68	GY/BK	Vehicle Speed Sensor (+) Signal	0 Hz, 55 mph: 125 Hz
69	PK/YL	A/C WOT Relay Control	On: 1v, Off: 12v
70	YL	Generator Battery Indicator Control	Lamp Off: 12v, On: 1v
71	RD	Vehicle Power (Start-Run)	12-14v
72	TN/RD	Injector 7 Control	3.4-3.7 ms
73	TN/BK	Injector 5 Control	3.4-3.7 ms
74	BR/YL	Injector 3 Control	3.4-3.7 ms
75	TN	Injector 1 Control	3.4-3.7 ms
76	BK/WT	Power Ground	<0.1v
77	---	Not Used	---
78	PK/LB	COP 7 Driver Control	5° dwell
79	WT/RD	COP 8 Driver Control	5° dwell
80	LB/OR	Fuel Pump Control	On: 1v, Off: 12v
81	WT/YL	EPC Solenoid Control	9.5v (20 psi)
82	---	Not Used	---
83	WT/LB	IAC Motor Control	34% duty cycle
84	DB/YL	OSS Sensor Signal	0 Hz
85	DB/OR	CMP Sensor Signal	6-7 Hz
86	WT/BK	A/C High Pressure Switch	Open: 12v, Closed: 0v
87	RD/BK	HO2S-21 (B2 S1) Signal	0.1-1.1v
88	LB/RD	MAF Sensor Signal	0.6v
89	GY/WT	TP Sensor Signal	0.53-1.27v
90	BR/WT	Reference Voltage	4.9-5.1v
91	GY/RD	Analog Signal Return	<0.050v
92	DG	Brake Pedal Position Switch	Brake Off: 0v, On: 12v
93	RD/WT	HO2S-11 (B1 S1) Heater	On: 1v, Off: 12v
94	YL/LB	HO2S-21 (B2 S1) Heater	On: 1v, Off: 12v
95	WT/BK	HO2S-12 (B1 S2) Heater	On: 1v, Off: 12v
96	TN/YL	HO2S-22 (B2 S2) Heater	On: 1v, Off: 12v
98	LB	Injector 8 Control	3.4-3.7 ms
99	LG/OR	Injector 6 Control	3.4-3.7 ms
100	BR/LB	Injector 4 Control	3.4-3.7 ms
101	WT	Injector 2 Control	3.4-3.7 ms
102	---	Not Used	---
103	BK/WT	Power Ground	<0.1v
104	PK/WT	COP 2 Driver Control	5° dwell

Pin Connector Graphic

PCM 104-PIN CONNECTOR

Terminal View of 104-Pin PCM Wiring Harness Connector

1996-97 Crown Victoria 4.6L V8 CNG VIN 9 104 Pin Connector

PCM Pin #	Wire Color	Circuit Description (104 Pin)	Value at Hot Idle
1	PK/OR	Shift Solenoid 2 Control	12v, 55 mph: 1v
2	PK/LG	MIL (lamp) Control	MIL On: 1v, Off: 12v
3-9	---	Not Used	---
10	LB/W	EFT Sensor 'B' Signal	1.75-3.50v (50-120ºF)
13	LG/YL	Flash EEPROM Power	0.1v
14	---	Not Used	---
15	PK/LB	Data Bus (-) Signal	Digital Signals
16	TN/OR	Data Bus (+) Signal	Digital Signals
17-20	---	Not Used	---
21	DB	CKP Sensor (+) Signal	440-490 Hz
22	GY	CKP Sensor (-) Signal	440-490 Hz
23	---	Not Used	---
24	BK/WT	Power Ground	<0.1v
25	BK	PCM Case Ground	<0.050v
26	TN/WT	Coil Driver 1 Control	5º dwell
27	OR/YL	Shift Solenoid 1 Control	1v, 55 mph: 12v
28	---	Not Used	---
29	TN/WT	TCS (switch) Signal	TCS & O/D On: 12v
30	WT/RD	Octane Adjust Switch	Closed: 0v, Open: 9.3v
31-32	---	Not Used	---
33	PK/OR	VSS (-) Signal	0 Hz, 55 mph: 125 Hz
34	---	Not Used	---
35	RD/LG	HO2S-12 (B1 S2) Signal	0.1-1.1v
36	TN/LB	MAF Sensor Return	<0.050v
37	OR/BK	TFT Sensor Signal	2.10-2.40v
38	LG/RD	ECT Sensor Signal	0.5-0.6v
39	GY	IAT Sensor Signal	1.5-2.5v
40	DG/YL	Fuel Solenoid Valve Control	On: 12v, Off: 0v
41	DG/OR	A/C Cycling Clutch Switch	A/C On: 12v, Off: 0v
42-44	---	Not Used	---
45	RD/OR	Low Speed Fan Control	On: 1v, Off: 12v
46	---	Not Used	---
47	BR/BK	EGR VR Solenoid	0%, 55 mph: 45%
48-50	---	Not Used	---
51	BK/WT	Power Ground	<0.1v
52	TN/OR	Coil Driver 2 Control	5º dwell
53	---	Not Used	---
54	RD/LB	TCC Solenoid Control	0%, 55 mph: 50%
55	YL/BK	Keep Alive Power	12-14v
56-57	---	Not Used	---
58	GY/BK	VSS (+) Signal	0 Hz, 55 mph: 125 Hz
59	---	Not Used	---
60	GY/LB	HO2S-11 (B1 S1) Signal	0.1-1.1v
61	PK/LG	HO2S-22 (B2 S2) Signal	0.1-1.1v
62	BK/YL	EFT Sensor 'A' Signal	1.75-3.50v (50-120ºF)
63	RD/PK	Injector Pressure Sensor	2.7-3.7v (105-130 psi)
64	LB/YL	TR Sensor Signal	In 'P': 0v, 55 mph: 2.1v
65	BR/LG	DPFE Sensor Signal	0.20-1.30v
66-68	---	Not Used	---
69	OR/LB	A/C WOT Relay Control	On: 1v, Off: 12v
70	---	Not Used	---
71	RD	Vehicle Power	12-14v
72	TN/RD	Injector 7 Control	3.9-4.6 ms

1996-97 Crown Victoria 4.6L V8 CNG VIN 9 104 Pin Connector

PCM Pin #	Wire Color	Circuit Description (104 Pin)	Value at Hot Idle
73	TN/BK	Injector 5 Control	3.9-4.6 ms
74	BR/YL	Injector 3 Control	3.9-4.6 ms
75	TN	Injector 1 Control	3.9-4.6 ms
76-77	BK/WT	Power Ground	<0.1v
78	TN/LG	Coil Driver 3 Control	5° dwell
79	WT/LG	TCIL (lamp) Control	On: 1v, Off: 12v
80	LB/OR	Fuel Shutoff Valve Control	Idle: 0.1v (On)
81	WT/YL	EPC Solenoid Control	9.5v (20 psi)
82	---	Not Used	---
83	WT/LB	IAC Motor Control	9.3v (34%)
84	PK/LB	OSS Sensor Signal	55 mph: 1260-1330 rpm
85	DG	CMP Sensor Signal	6-7 Hz
87	RD/BK	HO2S-21 (B2 S1) Signal	0.1-1.1v
88	LB/RD	MAF Sensor Signal	0.6-0.9v
89	GY/WT	TP Sensor Signal	0.53-1.27v
90	BR/WT	Reference Voltage	4.9-5.1v
91	GY/RD	Analog Signal Return	<0.050v
92	LG	Brake Pedal Position Switch	Brake Off: 0v, On: 12v
93	RD/WT	HO2S-11 (B1 S1) Heater	On: 1v, Off: 12v
94	YL/LB	HO2S-21 (B2 S1) Heater	On: 1v, Off: 12v
95	WT/BK	HO2S-12 (B1 S2) Heater	On: 1v, Off: 12v
96	TN/YL	HO2S-22 (B2 S2) Heater	On: 1v, Off: 12v
97	RD	Vehicle Power	12-14v
98	LB	Injector 8 Control	3.9-4.6 ms
99	LG/OR	Injector 6 Control	3.9-4.6 ms
100	BR/LB	Injector 4 Control	3.9-4.6 ms
101	WT	Injector 2 Control	3.9-4.6 ms
102	---	Not Used	---
103	BK/WT	Power Ground	<0.1v
104	TN/LB	Coil Driver 4 Control	5° dwell

Pin Connector Graphic

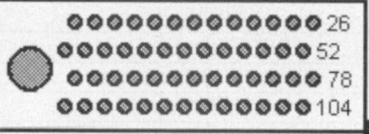

PCM 104-PIN CONNECTOR

Terminal View of 104-Pin PCM Wiring Harness Connector

Standard Colors and Abbreviations

Abbreviation	Color	Abbreviation	Color	Abbreviation	Color
BK	Black	GY	Gray	PK	Purple
BL	Blue	GN	Green	RD	Red
BR	Brown	LG	LT Green	TN	Tan
DB	Dark Blue	OR	Orange	WT	White
DG	DK Green	PK	Pink	YL	Yellow

1998-2002 Crown Victoria 4.6L V8 CNG VIN 9 104 Pin Connector

PCM Pin #	Wire Color	Circuit Description (104 Pin)	Value at Hot Idle
1	OR/YL	COP 6 Driver Control	5° dwell
2	PK/LG	MIL (lamp) Control	MIL On: 1v, Off: 12v
3	BK/WT	Power Ground	<0.1v
4-5	---	Not Used	---
6	OR/YL	Shift Solenoid 'A' Control	1v, 55 mph: 12v
7-10	---	Not Used	---
11	VT/OR	Shift Solenoid 'B' Control	12v, Cruise: 1v
12	WT/LG	TCIL (lamp) Control	Lamp On: 1v, Off: 5v
13	VT	Flash EEPROM Power	0.1v
14	---	Not Used	---
15	PK/LB	SCP Data Bus (-) Signal	Digital Signals
16	TN/OR	SCP Data Bus (+) Signal	Digital Signals
17-20	---	Not Used	---
21	BK/VT	CKP Sensor (+) Signal	440-490 Hz
22	GY/YL	CKP Sensor (-) Signal	440-490 Hz
23	---	Not Used	---
24	BK/WT	Power Ground	<0.1v
25	BK	PCM Case Ground	<0.050v
26	LG/WT	COP 1 Driver Control	5° dwell
27	LG/YL	COP 5 Driver Control	5° dwell
28	RD/OR	Low Speed Fan Control	On: 1v, Off: 12v
29	TN/WT	TCS (switch) Signal	TCS & O/D On: 12v
30-32	---	Not Used	---
33	PK/OR	VSS (-) Signal	0 Hz, 55 mph: 125 Hz
34	YL/BK	Digital TR1 Sensor	In 'P': 0v, 55 mph: 11v
35	RD/LG	HO2S-12 (B1 S2) Signal	0.1-1.1v
36	TN/LB	MAF Sensor Return	<0.050v
37	OR/BK	TFT Sensor Signal	2.10-2.40v
38	---	Not Used	---
39	GY	IAT Sensor Signal	1.5-2.5v
40	DG/YL	Fuel Solenoid Valve Control	On: 12v, Off: 0v
41	BK/YL	A/C Cycling Clutch Switch	AC On: 12v, Off: 0v
42	RD/WT	ECT Sensor Signal	0.5-0.6v
43-44	---	Not Used	---
45	OR/RD	CHTIL (lamp) Control	Lamp On: 1v, Off: 3.5v
46	LG/VT	High Speed Fan Control	On: 1v, Off: 12v
47	BR/VT	EGR VR Solenoid	0%, 55 mph: 45%
48	---	Not Used	---
49	LB/BK	Digital TR2 Sensor	In 'P': 0v, 55 mph: 11v
50	WT/BK	Digital TR4 Sensor	In 'P': 0v, 55 mph: 11v
51	BK/WT	Power Ground	<0.1v
52	WT/PK	COP 3 Driver Control	5° dwell
53	DG/VT	COP 4 Driver Control	5° dwell
54	VT/YL	TCC Solenoid Control	0%, 55 mph: 95%
55	YL/BK	Keep Alive Power	12-14v
56-57	---	Not Used	---
58	GY/BK	VSS (-) Signal	0 Hz, 55 mph: 125 Hz
59	---	Not Used	---
60	GY/LB	HO2S-11 (B1 S1) Signal	0.1-1.1v
61	VT/LG	HO2S-22 (B2 S2) Signal	0.1-1.1v
62	BK/YL	EFT Sensor 'A' Signal	1.7-3.5v (50-120°F)
63	R/VT	Injection Pressure Sensor	2.7-3.7v (105-130 psi)
64	LB/YL	Digital TR Sensor (TR3)	In 'P': 0v, in O/D: 1.7v

1998-2002 Crown Victoria 4.6L V8 CNG VIN 9 104 Pin Connector

PCM Pin #	Wire Color	Circuit Description (104 Pin)	Value at Hot Idle
65 ('98-'00)	BR/LG	DPFE Sensor Signal	0.20-1.30v
65 ('01-'02)	BR/LG	DPFE Sensor Signal	0.95-1.05v
66	YL/LG	CHT Sensor Signal	0.7v (194°F)
67-68	---	Not Used	---
69	PK/YL	A/C WOT Relay Control	On: 1v, Off: 12v
70	---	Not Used	---
71	RD	Vehicle Power	12-14v
72	TN/RD	Injector 7 Control	3.9-5.2 ms
73	TN/BK	Injector 5 Control	3.9-5.2 ms
74	BR/YL	Injector 3 Control	3.9-5.2 ms
75	TN	Injector 1 Control	3.9-5.2 ms
76	---	Not Used	---
77	BK/WT	Power Ground	<0.1v
78	PK/LB	COP 7 Driver Control	5° dwell
79	WT/RD	COP 8 Driver Control	5° dwell
80	LB/OR	Fuel Pump Control	On: 1v, Off: 12v
81	WT/YL	EPC Solenoid Control	9.5v (20 psi)
82	---	Not Used	---
83	WT/LB	IAC Motor Control	34% duty cycle
84	DB/YL	OSS Sensor Signal	0 Hz
85	DB/OR	CMP Sensor Signal	6-7 Hz
86	WT/BK	A/C High Pressure Switch	Open: 12v, Closed: 0v
87	RD/BK	HO2S-21 (B2 S1) Signal	0.1-1.1v
88	LB/RD	MAF Sensor Signal	0.6v
89	GY/WT	TP Sensor Signal	0.53-1.27v
90	BR/WT	Reference Voltage	4.9-5.1v
91	GY/RD	Analog Signal Return	<0.050v
92	LG	Brake Pedal Position Switch	Brake Off: 0v, On: 12v
93	RD/WT	HO2S-11 (B1 S1) Heater	On: 1v, Off: 12v
94	YL/LB	HO2S-21 (B2 S1) Heater	On: 1v, Off: 12v
95	WT/BK	HO2S-12 (B1 S2) Heater	On: 1v, Off: 12v
96	TN/YL	HO2S-22 (B2 S2) Heater	On: 1v, Off: 12v
97	RD	Vehicle Power	12-14v
98	LB	Injector 8 Control	3.9-5.2 ms
99	LG/OR	Injector 6 Control	3.9-5.2 ms
100	BR/LB	Injector 4 Control	3.9-5.2 ms
101	WT	Injector 2 Control	3.9-5.2 ms
102	---	Not Used	---
103	BK/WT	Power Ground	<0.1v
104	PK/WT	COP 2 Driver Control	5° dwell

Pin Connector Graphic

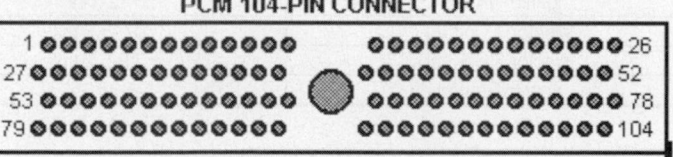

PCM 104-PIN CONNECTOR

Terminal View of 104-Pin PCM Wiring Harness Connector

1998-2002 Crown Victoria 4.6L V8 CNG Module 60 Pin Connector

PCM Pin #	Wire Color	Circuit Description (60 Pin)	Value at Hot Idle
1	YL/BK	Keep Alive Power	12-14v
2	---	Not Used	---
3	TN	Injector 1 Signal from PCM	3.9-5.2 ms
4	WT	Injector 2 Signal from PCM	3.9-5.2 ms
5	BR/YL	Injector 3 Signal from PCM	3.9-5.2 ms
6	---	Not Used	---
7	RD/PK	Fuel Tank Pressure Signal	2.6v (0" HG - cap off)
8-11	---	Not Used	---
12	BK/YL	Instrument Cluster Power	12-14v
13-17	---	Not Used	---
18	PK/LB	SCP Data Bus (-)	Digital Signals
19	TN/OR	SCP Data Bus (+)	Digital Signals
20-22	---	Not Used	---
23	BR/LB	Injector 4 Signal from PCM	3.9-5.2 ms
24	TN/BK	Injector 5 Signal from PCM	3.9-5.2 ms
25	LG/OR	Injector 6 Signal from PCM	3.9-5.2 ms
26	WT/RD	Reference Voltage	4.9-5.1v
27	---	Not Used	---
28	LB/VT	Fuel Tank Temp. Sensor #1	0.1-4.9v
29-30	---	Not Used	---
31	BK/WT	Instrument Cluster Ground	<0.1v
32	---	Not Used	---
33	TN/BK	Injector 5 Control	3.9-5.2 ms
34	---	Not Used	---
35	BR/LB	Injector 4 Control	3.9-5.2 ms
36	---	Not Used	---
37	RD	Ignition Power	12-14v
38	YL/WT	Fuel Display Output	Varies
39	BR/YL	Injector 3 Control	3.9-6.5 ms
40	BK/WT	Power Ground	<0.1v
41	---	Not Used	---
42	LG/OR	Injector 6 Control	3.9-5.2 ms
43	TN/RD	Injector 7 Signal from PCM	3.9-5.2 ms
44	LB	Injector 8 Signal from PCM	3.9-5.2 ms
45	---	Not Used	---
46	BK/LB	Sensor Signal Return	<0.050v
47	WT/YL	Fuel Tank Temp. Sensor #1	0.1-4.9v
48-52	---	Not Used	---
53	TN/RD	Injector 7 Control	3.9-5.2 ms
54	LB	Injector 8 Control	3.9-5.2 ms
55-56	---	Not Used	---
57	RD	Ignition Power	12-14v
58	TN	Injector 1 Control	3.9-5.2 ms
59	WT	Injector 2 Control	3.9-5.2 ms
60	BK/WT	Power Ground	<01v

Pin Connector Graphic

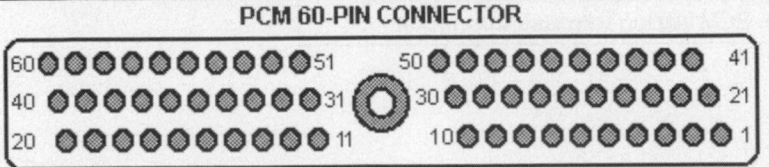

1998-2002 Crown Victoria 4.6L V8 CNG Module Wiring Diagram

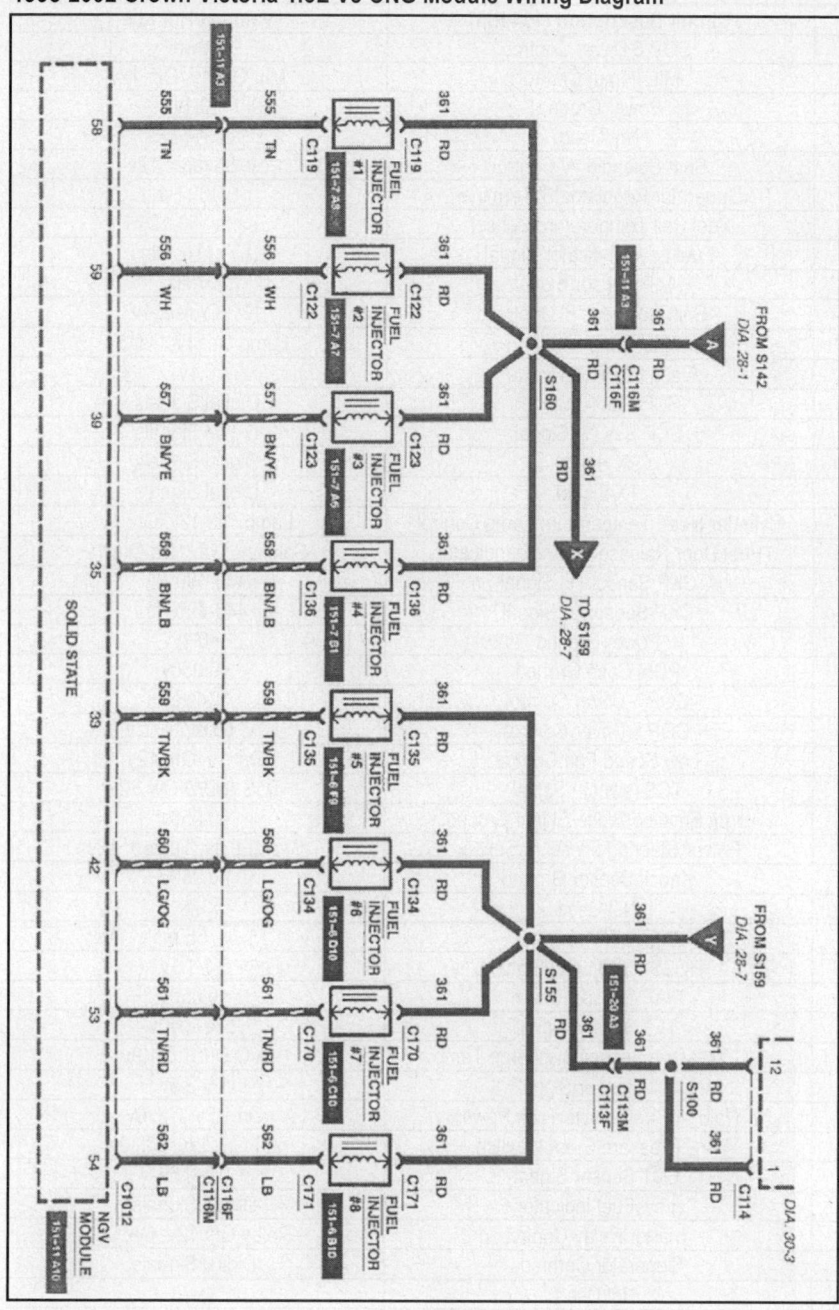

2003-05 Crown Victoria 4.6L V8 CNG VIN 9 (A/T) 104 Pin Connector

PCM Pin #	Wire Color	Circuit Description (104 Pin)	Value at Hot Idle
1	OR/YL	COP 6 Driver Control	5° dwell
2	PK/LG	MIL (lamp) Control	MIL On: 1v, Off: 12v
3	BK/WT	Power Ground	<0.1v
4-5, 14, 23	---	Not Used	---
6	OR/YL	Shift Solenoid 'A' Control	1v, 55 mph: 12v
7	YL/LG	Generator Regulator 'S' Terminal	0-130 Hz
8	RD/PK	Fuel Rail Temperature Sensor	1.7-3.5v (50-120ºF)
9	YL/WT	Fuel Level Indicator Signal	1.7v (1/2 full)
10	LB/RD	MAP Sensor Signal	107 Hz
11	VT/OR	Shift Solenoid 'B' Control	12v, Cruise: 1v
12	WT/LG	TCIL (lamp) Control	Lamp On: 1v, Off: 5v
13	VT	Flash EEPROM Power	0.1v
15	PK/LB	SCP Bus (-) Signal	Digital Signals
16	TN/OR	SCP Bus (+) Signal	Digital Signals
17	GY/OR	RX Signal	Digital Signals
18	WT/LG	TX Signal	Digital Signals
19	OR/RD	Cylinder Head Temperature Lamp Control	Lamp Off: 12v, On: 1v
20	WT/LG	Fuel Door Release Solenoid Indicator	Solenoid Off: 12v, On: 1v
21	BK/PK	CKP Sensor (+) Signal	440-490 Hz
22	GY/YL	CKP Sensor (-) Signal	440-490 Hz
24, 51	BK/WT	Power Ground	<0.1v
25	BK	PCM Case Ground	<0.050v
26	LG/WT	COP 1 Driver Control	5° dwell
27	LG/YL	COP 5 Driver Control	1v, 55 mph: 12v
28	RD/OR	Low Speed Fan Control	On: 1v, Off: 12v
29	TN/WT	TCS (switch) Signal	TCS & O/D On: 12v
30	DG/RD	Light Emitting Diode Signal Ground	<0.050v
31	YL/LG	Power Steering Pressure Sensor	Straight: 0v, Turned: 12v
32	DG/VT	Knock Sensor Ground	<0.050v
33	---	Not Used	---
34	YL/BK	Digital TR1 Sensor	In 'P': 0v, 55 mph: 11v
35	RD/LG	HO2S-12 (B1 S2) Signal	0.1-1.1v
36	TN/LB	MAF Sensor Return	<0.050v
37	OR/BK	TFT Sensor Signal	2.10-2.40v
38	BK/WT	A/C Pressure Sensor Discharge Temp.	A/C On: 1.5-1.9v
39	GY	IAT Sensor Signal	1.5-2.5v
40	DG/YL	Natural Gas Vehicle Tank Power	On: 12v, Off: 0v
41	PK/LB	A/C Pressure Cutout Switch	AC On: 12v, Off: 0v
42	RD/WT	ECT Sensor Signal	0.5-0.6v
43	DB/LG	Low Fuel Indicator	Digital Signals
44	GY/RD	Starter Relay Control	Relay Off: 12v, On: 9v
45	LG/RD	Generator Common	Digital Signals
46	---	Not Used	---
47	BR/PK	EGR VR Solenoid	0%, 55 mph: 45%
48	WT	Tachometer Output	DC pulse signals
49	LB/BK	Digital TR2 Sensor	In 'P': 0v, 55 mph: 11v
50	WT/BK	Digital TR4 Sensor	In 'P': 0v, 55 mph: 11v
52	WT/PK	COP 3 Driver Control	5° dwell
53	DG/VT	COP 4 Driver Control	5° dwell
54	VT/YL	TCC Solenoid Control	0%, 55 mph: 95%
55	YL/BK	Keep Alive Power	12-14v
56	LG/BK	EVAP Purge Solenoid Control	0-10 Hz (0-100%)
57	YL/RD	Knock Sensor Signal	0v
58-59	---	Not Used	---

2003-05 Crown Victoria 4.6L V8 CNG VIN W (A/T) 104 Pin Connector

PCM Pin #	Wire Color	Circuit Description (104 Pin)	Value at Hot Idle
60	GY/BK	HO2S-11 (B1 S1) Signal	0.1-1.1v
61	VT/LG	HO2S-22 (B2 S2) Signal	0.1-1.1v
62	RD/PK	FTP Sensor Signal	2.6v (0" H2O - cap off)
63	OR/LG	Injection Pressure Sensor	2.8v (39 psi)
64	LB/YL	Digital TR3 Sensor	In 'P': 0v, in O/D: 1.7v
65	BR/LG	DPFE Sensor Signal	0.95-1.05v
66	YL/LG	Cylinder Head Temperature Sensor	0.7v (194°F)
67	VT/WT	EVAP Canister Vent Solenoid	0-10 Hz (0-100%)
68	GY/BK	Vehicle Speed Sensor (+) Signal	0 Hz, 55 mph: 125 Hz
69	PK/YL	A/C WOT Relay Control	On: 1v, Off: 12v
70	YL	Generator Battery Indicator Control	Lamp Off: 12v, On: 1v
71	RD	Vehicle Power (Start-Run)	12-14v
72	TN/RD	Injector 7 Control	3.4-3.7 ms
73	TN/BK	Injector 5 Control	3.4-3.7 ms
74	BR/YL	Injector 3 Control	3.4-3.7 ms
75	TN	Injector 1 Control	3.4-3.7 ms
76	BK/WT	Power Ground	<0.1v
77	---	Not Used	---
78	PK/LB	COP 7 Driver Control	5° dwell
79	WT/RD	COP 8 Driver Control	5° dwell
80	LB/OR	Fuel Pump Control	On: 1v, Off: 12v
81	WT/YL	EPC Solenoid Control	9.5v (20 psi)
82	---	Not Used	---
83	WT/LB	IAC Motor Control	34% duty cycle
84	DB/YL	OSS Sensor Signal	0 Hz
85	DB/OR	CMP Sensor Signal	6-7 Hz
86	WT/BK	A/C High Pressure Switch	Open: 12v, Closed: 0v
87	RD/BK	HO2S-21 (B2 S1) Signal	0.1-1.1v
88	LB/RD	MAF Sensor Signal	0.6v
89	GY/WT	TP Sensor Signal	0.53-1.27v
90	BR/WT	Reference Voltage	4.9-5.1v
91	GY/RD	Analog Signal Return	<0.050v
92	DG	Brake Pedal Position Switch	Brake Off: 0v, On: 12v
93	RD/WT	HO2S-11 (B1 S1) Heater	On: 1v, Off: 12v
94	YL/LB	HO2S-21 (B2 S1) Heater	On: 1v, Off: 12v
95	WT/BK	HO2S-12 (B1 S2) Heater	On: 1v, Off: 12v
96	TN/YL	HO2S-22 (B2 S2) Heater	On: 1v, Off: 12v
98	LB	Injector 8 Control	3.4-3.7 ms
99	LG/OR	Injector 6 Control	3.4-3.7 ms
100	BR/LB	Injector 4 Control	3.4-3.7 ms
101	WT	Injector 2 Control	3.4-3.7 ms
102	---	Not Used	---
103	BK/WT	Power Ground	<0.1v
104	PK/WT	COP 2 Driver Control	5° dwell

Pin Connector Graphic

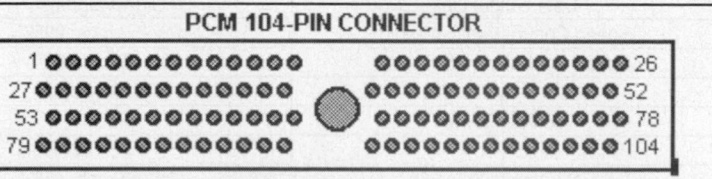

PCM 104-PIN CONNECTOR

```
 1 ●●●●●●●●●●●●●        ●●●●●●●●●●●●● 26
27 ●●●●●●●●●●●●●        ●●●●●●●●●●●●● 52
53 ●●●●●●●●●●●●●   ⬡    ●●●●●●●●●●●●● 78
79 ●●●●●●●●●●●●●        ●●●●●●●●●●●●● 104
```

Terminal View of 104-Pin PCM Wiring Harness Connector

1990-91 Crown Victoria 5.0L V8 MFI VIN F (A/T) 60 Pin Connector

PCM Pin #	Wire Color	Circuit Description (60 Pin)	Value at Hot Idle
1 ('90)	BK/OR	Keep Alive Power	12-14v
1 ('91)	YL	Keep Alive Power	12-14v
2	LG	Brake Pedal Position Switch	Brake On: 12, Off: 0v
3	DG/WT	VSS (+) Signal	0 Hz, 55 mph: 125 Hz
4	DG/YL	Ignition Diagnostic Monitor	20-31 Hz
4	BL/PK	Ignition Diagnostic Monitor	20-31 Hz
4	TN/YL	Ignition Diagnostic Monitor	20-31 Hz
5	---	Not Used	---
6 ('90)	BK/WT	VSS (-) Signal	0 Hz, 55 mph: 125 Hz
6 ('91)	GY/RD	VSS (-) Signal	0 Hz, 55 mph: 125 Hz
7 ('90)	GY/YL	ECT Sensor Signal	0.5-0.6v
7 ('91)	LG/RD	ECT Sensor Signal	0.5-0.6v
8-9	---	Not Used	---
10	LG/PK	A/C Cycling Clutch Switch	A/C On: 12v, Off: 0v
10	PK/BL	A/C Cycling Clutch Switch	A/C On: 12v, Off: 0v
11	LG/BK	Air Management 2 Solenoid	AM2 On: 1v, Off: 12v
12	TN/YL	Injector 3 Control	5.7-6.2 ms
12	BR/YL	Injector 3 Control	5.7-6.2 ms
13	TN/BK	Injector 4 Control	5.7-6.2 ms
13	BR/BL	Injector 4 Control	5.7-6.2 ms
14 ('90)	TN/BL	Injector 5 Control	5.7-6.2 ms
14 ('91)	TN	Injector 5 Control	5.7-6.2 ms
15	TN/LG	Injector 6 Control	5.7-6.2 ms
15	LG	Injector 6 Control	5.7-6.2 ms
15	LG/OR	Injector 6 Control	5.7-6.2 ms
16 ('90)	BK/OR	Ignition System Ground	<0.1v
16 ('91)	OR	Ignition System Ground	<0.1v
17	YL/BK	Self-Test Indicator & MIL	MIL On: 1v, Off: 12v
17	TN/RD	Self-Test Indicator & MIL	MIL On: 1v, Off: 12v
18-19	---	Not Used	---
20	BK	PCM Case Ground	<0.050v
21	WT/BL	IAC Motor Control	9.0-11.5v
22	TN/LG	Fuel Pump Control	On: 1v, Off: 12v
23-24, 32	---	Not Used	---
25	LG/PK	ACT Sensor Signal	1.5-2.5v
26	OR/WT	Reference Voltage	4.9-5.1v
27	BR/LG	EGR EVP Sensor	0.4v
28	---	Not Used	---
29	DG/PK	HO2S-11 (B1 S1) Signal	0.1-1.1v
29	RD/BK	HO2S-11 (B1 S1) Signal	0.1-1.1v
30	WT/PK	Neutral Drive Switch Signal	In 'P': 0v, Others: 5v
30	BL/YL	Neutral Drive Switch Signal	In 'P': 0v, Others: 5v
31	GY/YL	EVAP Purge Solenoid	12v, 55 mph: 1v
33	DG	EGR VR Solenoid	0%, 55 mph: 45%
33	BR/PK	EGR VR Solenoid	0%, 55 mph: 45%
34	BL/PK	Data Output Link	Digital Signals
35 ('90)	WT/PK	Speed Control Vent Solenoid	0%, 55 mph: 98%
35 ('91)	GY/BK	Speed Control Vent Solenoid	0%, 55 mph: 98%
36	YL/LG	Spark Output Signal	50% duty cycle
37 ('90)	RD	Vehicle Power	12-14v
37 ('91)	BK/YL	Vehicle Power	12-14v
38	GY/BK	S/C Vacuum Solenoid	0%, 55 mph: 45%
39	GY/BK	Speed Control Switch Ground	<0.050v
40	BK/LG	Power Ground	<0.1v

1990-91 Crown Victoria 5.0L V8 MFI VIN F (A/T) 60 Pin Connector

PCM Pin #	Wire Color	Circuit Description (60 Pin)	Value at Hot Idle
41	---	Not Used	---
42	TN/OR	Injector 7 Control	5.7-6.2 ms
42	TN/RD	Injector 7 Control	5.7-6.2 ms
43	DB/LG	HO2S-21 (B2 S1) Signal	0.1-1.1v
45	DB/LG	MAP Sensor	107 Hz
45	LG/BK	MAP Sensor	107 Hz
46	BK/WT	Analog Signal Return	<0.050v
46	GY/RD	Analog Signal Return	<0.050v
47	DG/LG	TP Sensor Signal	0.5-1.2v
47	GY/WT	TP Sensor Signal	0.5-1.2v
48	WT/RD	Self-Test Indicator Signal	STI Open: 5v, Closed: 1v
48	WT/BK	Self-Test Indicator Signal	STI Open: 5v, Closed: 1v
48	BR	Self-Test Indicator Signal	STI Open: 5v, Closed: 1v
49	OR	HO2S-11 (Bank 1) Ground	<0.050v
50	OR/RD	S/C Command Switch	6.7v
50	BL/BK	S/C Command Switch	6.7v
51	WT/RD	Air Management 1 Solenoid	AM1 On: 1v, Off: 12v
52 ('90)	BL	Injector 8 Control	5.7-6.2 ms
52 ('91)	TN/RD	Injector 8 Control	5.7-6.2 ms
54	OR/BL	A/C WOT Relay Control	On: 1v, Off: 12v
55	---	Not Used	---
56	DB	PIP Sensor Signal	50% dwell
57 ('90)	RD	Vehicle Power	12-14v
57 ('91)	BK/YL	Vehicle Power	12-14v
58	TN	Injector 1 Control	5.7-6.2 ms
59 ('90)	TN/WT	Injector 2 Control	5.7-6.2 ms
59 ('91)	WT	Injector 2 Control	5.7-6.2 ms
60	BK/LG	Power Ground	<0.1v

Pin Connector Graphic

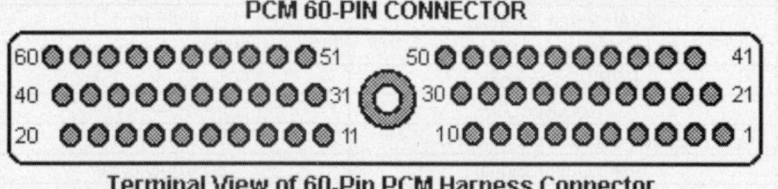

Standard Colors and Abbreviations

Abbreviation	Color	Abbreviation	Color	Abbreviation	Color
BK	Black	GY	Gray	PK	Purple
BL	Blue	GN	Green	RD	Red
BR	Brown	LG	LT Green	TN	Tan
DB	Dark Blue	OR	Orange	WT	White
DG	DK Green	PK	Pink	YL	Yellow

1990-91 Crown Victoria 5.0L V8 VIN F (California) 60 Pin Connector

PCM Pin #	Wire Color	Circuit Description (60 Pin)	Value at Hot Idle
1 ('90)	BK/OR	Keep Alive Power	12-14v
1 ('91)	YL	Keep Alive Power	12-14v
2	---	Not Used	---
3	DG/WT	VSS (+) Signal	0 Hz, 55 mph: 125 Hz
4	DG/YL	Ignition Diagnostic Monitor	20-31 Hz
5	LG	Brake Pedal Position Switch	Brake Off: 0v, On: 12v
6	BK/WT	VSS (-) Signal	0 Hz, 55 mph: 125 Hz
7	LG/YL	ECT Sensor Signal	0.5-0.6v
8	BK/OR	Data Bus (-) Signal	Digital Signals
9	TN/BL	MAF Sensor Return	<0.050v
10	PK/BL	A/C Cycling Clutch Switch	A/C On: 12v, Off: 0v
11	---	Not Used	---
12	BR/YL	Injector 3 Control	5.7-6.2 ms
13	BR/BL	Injector 4 Control	5.7-6.2 ms
14	TN/BL	Injector 5 Control	5.7-6.2 ms
15	LG	Injector 6 Control	5.7-6.2 ms
16	BK/OR	Ignition System Ground	<0.050v
17	TN/RD	STI Output, MIL Control	MIL On: 1v, Off: 12v
18	---	Not Used	---
19	OR	Fuel Pump Monitor	On: 12v, Off: 0v
20	BK	PCM Case Ground	<0.050v
21	WT/BL	IAC Motor Control	9.0-11.5v
22	TN/LG	Fuel Pump Control	On: 1v, Off: 12v
23-24	---	Not Used	---
25	LG/PK	ACT Sensor Signal	1.5-2.5v
26	OR/WT	Reference Voltage	4.9-5.1v
27	BR/LG	EGR EVP Sensor Signal	0.4v
28	BL/BK	S/C Command Switch	6.7v
29	DG/PK	HO2S-11 (B1 S1) Signal	0.1-1.1v
30	WT/PK	Neutral Drive Switch Signal	In 'P': 0v, Others: 5v
31	GY/YL	EVAP Purge Solenoid	12v, at 55 mph: 1v
32	BL	Air Management 2 Solenoid	AM2 On: 1v, Off: 12v
33	DG	EGR VR Solenoid	0%, 55 mph: 45%
34	---	Not Used	---
35	WT/PK	Speed Control Vent Solenoid	0%, 55 mph: 98%
36	YL/LG	Spark Output Signal	50% duty cycle
37	RD	Vehicle Power	12-14v
38	YL	Air Management 1 Solenoid	AM1 On: 1v, Off: 12v
39	GY/BK	Speed Control Switch Ground	<0.050v
40	BK/LG	Power Ground	<0.1v

Pin Connector Graphic

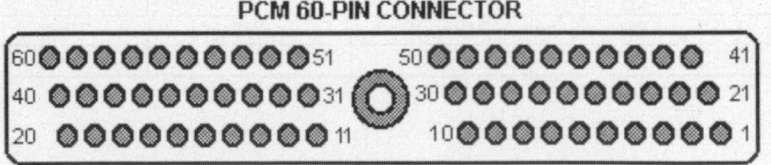

PCM 60-PIN CONNECTOR

Terminal View of 60-Pin PCM Harness Connector

1990-91 Crown Victoria 5.0L V8 VIN F (Cal) 60 Pin Connector

PCM Pin #	Wire Color	Circuit Description (60 Pin)	Value at Hot Idle
41	---	Not Used	---
42	TN/OR	Injector 7 Control	5.7-6.2 ms
43	DB/LG	HO2S-21 (B2 S1) Signal	0.1-1.1v
44	OR/BK	Data Bus (+) Signal	Digital Signals
45	LG/BK	MAP Sensor	107 Hz
46	BK/WT	Analog Signal Return	<0.050v
47	DG/LG	TP Sensor Signal	0.5-1.2v
48	WT/BK	Self-Test Indicator Signal	STI Open: 5v, Closed: 1v
49	OR	HO2S-21 (Bank 2) Ground	<0.050v
50	DB/OR	S/C Command Switch	6.7v
51	---	Not Used	---
52	BL	Injector 8 Control	5.7-6.2 ms
53	---	Not Used	---
54	OR/BL	A/C WOT Relay Control	On: 1v, Off: 12v
55	---	Not Used	---
56	DB	PIP Sensor Signal	50% dwell
57	RD	Vehicle Power	12-14v
58	TN	Injector 1 Control	5.7-6.2 ms
59	WT	Injector 2 Control	5.7-6.2 ms
60	BK/LG	Power Ground	<0.1v

Pin Connector Graphic

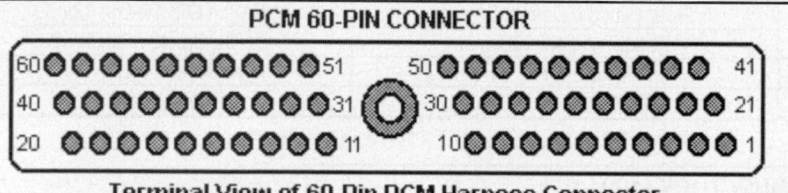

PCM 60-PIN CONNECTOR

Terminal View of 60-Pin PCM Harness Connector

Standard Colors and Abbreviations

Abbreviation	Color	Abbreviation	Color	Abbreviation	Color
BK	Black	GY	Gray	PK	Purple
BL	Blue	GN	Green	RD	Red
BR	Brown	LG	LT Green	TN	Tan
DB	Dark Blue	OR	Orange	WT	White
DG	DK Green	PK	Pink	YL	Yellow

ESCORT PIN TABLES

1991-92 Escort 1.8L I4 MFI VIN 8 (All) 22 Pin Connector

PCM Pin #	Wire Color	Circuit Description (22 Pin)	Value at Hot Idle
22-A	BL/RD	Keep Alive Power	12-14v
22-B	WT/RD	Vehicle Power	12-14v
22-C	PK	Start Signal	KOEC: 9-11v
22-D	WT/YL	Switch Monitor Lamp Signal	SML On: 5v, Off: 12v
22-E	YL/BK	MIL (lamp) Control	MIL On: 2.5v, Off: 12v
22-F	WT/BK	Self-Test Output Signal	STO On: 5v, Off: 12v
22-G	GN/WT	Spark Output Signal	50% duty cycle
22-H	---	Not Used	---
22-I	---	Not Used	---
22-J	BL/BK	A/C WOT Relay Control	A/C On: 12v, Off: 2.5v
22-K	LG/YL	Self-Test Input Signal	STI On: 0v, Off: 5v
22-L	BR/WT	DRL Signal	DRL Off: 12.5v, On: 2.5v
22-M	---	Not Used	---
22-N	RD/WT	Idle Throttle Switch Signal	Closed: 0v, Open: 12v
22-O	GN	M/T: BPP Switch Signal	Brake Off: 0v, On: 12v
22-O	GN/RD	A/T: Down Shift DSS to 4EAT	DSS Off: 1v, On: 12v
22-P	BL/YL	PSP Switch Signal	Straight: 0v, Turned: 12v
22-Q	GN/BK	Air Conditioning Switch Signal	A/C On: 2.5v, Off: 12v
22-R	BK/GN	Low Speed Fan Control	On: 1v, Off: 12v
22-S	OR/BL	Blower Motor Switch Signal	Off or 1st: 12v, 2nd>: 1v
22-T	BK/BL	Rear Window Defrost Switch	Switch Off: 1v, On: 12v
22-U	RD/BK	Headlamp Switch Signal	HDL On: 12v,Off: 1v
22-V	BR/YL	M/T: Neutral Drive Switch	In N: 0v, all others: 11v
22-V	BK/BL	A/T: MLP Sensor Signal	In 'N': 0v, Others: 12v

Pin Connector Graphic

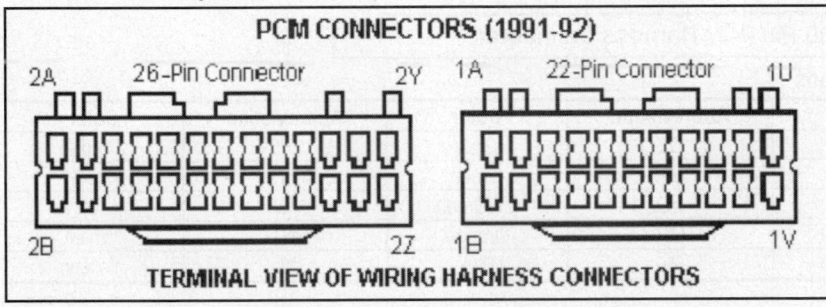

Standard Colors and Abbreviations

Abbreviation	Color	Abbreviation	Color	Abbreviation	Color
BK	Black	GY	Gray	PK	Purple
BL	Blue	GN	Green	RD	Red
BR	Brown	LG	LT Green	TN	Tan
DB	Dark Blue	OR	Orange	WT	White
DG	DK Green	PK	Pink	YL	Yellow

1991-92 Escort 1.8L I4 MFI VIN 8 (All) 26 Pin Connector

PCM Pin #	Wire Color	Circuit Description (26 Pin)	Value at Hot Idle
26-A	BK/OR	Power Ground	<0.1v
26-B	BK/OR	Power Ground	<0.1v
26-C	BK/LG	Power Ground	<0.1v
26-D	BK/WT	Analog Signal Return	<0.050v
26-E	WT	CKP Sensor Signal	400-425 Hz
26-F	---	Not Used	---
26-G	YL/BL	CID Sensor Signal	5-7 Hz
26-H	BK	Power Ground (California)	1v
26-H	BK/YL	Vehicle Power (Canada)	12-14v
26-I	---	Not Used	---
26-J	WT/RD	A/T: Vehicle Power	12-14v
26-K	LG/RD	Reference Voltage	4.9-5.1v
26-L	LG/WT	M/T: WOT Switch	At WOT: 1v
26-M	LG/WT	TP Sensor Signal to 4EAT	0.5-2.1v
26-N	RD/BL	HO2S-11 (B1 S1) Signal	0.1-1.1v
26-O	RD	Vane Airflow Meter Sensor	3.3v
26-P	RD/BK	VAT Sensor Signal	2.5v at 68°F
26-Q	BL/WT	ECT Sensor Signal	2.5v at 68°F
26-R	---	Not Used	---
26-S	BK/RD	High Speed Inlet Air Control	Below 5000 rpm: 1.5v
26-T	GN/OR	FPRC Solenoid Control	Hot engine: startup: <1.5v
26-U	YL	Fuel Injectors 1 & 3 - Bank 1	3.6 ms
26-V	YL/BK	Fuel Injectors 2 & 4 - Bank 2	3.6 ms
26-W	BL/OR	IAC Motor Control	8-10v
26-X	WT/BL	EVAP Purge Solenoid	At off-idle speeds: 1v
26-Y	---	Not Used	---
26-Z	BL/GN	ECT Sensor Signal to 4EAT	ECT at <162°F: <2.5v

Pin Connector Graphic

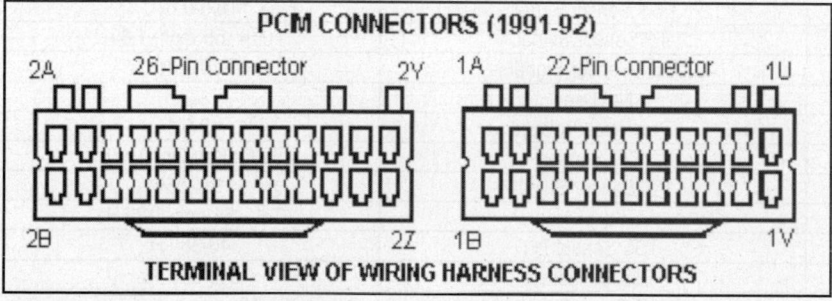

Standard Colors and Abbreviations

Abbreviation	Color	Abbreviation	Color	Abbreviation	Color
BK	Black	GY	Gray	PK	Purple
BL	Blue	GN	Green	RD	Red
BR	Brown	LG	LT Green	TN	Tan
DB	Dark Blue	OR	Orange	WT	White
DG	DK Green	PK	Pink	YL	Yellow

1993 Escort 1.8L I4 Engine VIN 8 (A/T) 22 Pin Connector

PCM Pin #	Wire Color	Circuit Description (22 Pin)	Value at Hot Idle
22-A	DB/RD	Keep Alive Power	12- 14v
22-B	WT/RD	Vehicle Power	12-14v
22-C	PK	Vehicle Start Signal	Cranking: 9-11v
22-D	WT/YL	System Monitor Lamp	Lamp On: 5v, Off: 12v
22-E	YL/BK	MIL (lamp) Control	MIL On: 1v, Off: 12v
22-F	WT/BK	Self-Test Output Signal	STO On: 5v, Off: 12v
22-G	DG/WT	Ignition Control Module Signal	0.1-0.3v
22-H	RD/BK	Headlamp Switch	Switch On: 12v, Off: 1v
22-I	LG/YL	Self-Test Input Signal	STI On: 0v, Off: 5v
22-J	BK/DB	Rear Defroster Control	Switch Off: 1v, On: 12v
22-K	BK	Power Ground	<0.1v
22-L	DB/BK	A/C Heater Relay	On: 1v, Off: 12v
22-M	DG	VSS Signal	0 Hz, 55 mph: 125 Hz
22-N	DB/YL	PSP Switch Signal	Straight: 12v, Turned: 0v
22-O	DG/DB	A/C Switch Signal	Switch On: 2.5v, Off: 12v
22-P	OR/DB	Blower Motor Switch	In 1st: 12v, Others: 1v
22-Q	DG	Brake Pedal Position Switch	Brake Off: 0v, On: 12v
22-R	BK/DB	MLP Sensor Switch	In 'P': 0v, Others: 12v
22-S	---	Not Used	---
22-T	RD/WT	Idle Switch Signal	Closed: 0v, Open: 12v
22-U	---	Not Used	---
22-V	---	Not Used	---

1993 Escort 1.8L I4 Engine VIN 8 (A/T) 16 Pin Connector

PCM Pin #	Wire Color	Circuit Description (16 Pin)	Value at Hot Idle
16-A	WT	CKP Sensor	2.5v (at Key On: 0 or 5v)
16-B	RD	Vane Airflow Meter	0.2v
16-C	RD/DB	O2S-11 (B1 S1) Signal	0.1-1.1v
16-D	BK/DG	Cooling Fan Control	On: 1v, Off: 12v
16-E	DB/WT	ECT Sensor Signal	0.5-0.6v
16-F	LG/WT	TP Sensor Signal	0.6v, 55 mph: 1.8v
16-G	WT/BK	TOT Sensor Signal	2.10-2.40v
16-H	---	Not Used	---
16-I	LG/RD	Reference Voltage	4.9-5.1v
16-J	YL/DB	CID Sensor	2.5v (at Key On: 0 or 5v)
16-K	RD/BK	IAT Sensor Signal	1.5-2.5v
16-L	DB	VSS (-) Signal	0 Hz, 55 mph: 125 Hz
16-M	WT/DB	TSS Sensor (+)	340-380 Hz
16-N	YL/DB	TSS Sensor (-)	340-380 Hz
16-O	WT/DB	EVAP Purge Solenoid	12v, 55 mph: 1v
16-P	---	Not Used	---

Standard Colors and Abbreviations

Abbreviation	Color	Abbreviation	Color	Abbreviation	Color
BK	Black	GY	Gray	PK	Purple
BL	Blue	GN	Green	RD	Red
BR	Brown	LG	LT Green	TN	Tan
DB	Dark Blue	OR	Orange	WT	White
DG	DK Green	PK	Pink	YL	Yellow

1993 Escort 1.8L I4 Engine VIN 8 (A/T) 26 Pin Connector

PCM Pin #	Wire Color	Circuit Description (26 Pin)	Value at Hot Idle
26-A	BK/OR	Power Ground	<0.1v
26-B	BK/OR	Power Ground	<0.1v
26-C	BK/LG	Power Ground	<0.1v
26-D	BK/WT	Analog Signal Return	<0.050v
26-E	YL	MLP Overdrive Switch Signal	OD Switch on: 1v, off: 12v
26-F	BR/WT	Daylight Running Lights	DRL On: 2v, Off: 12v
26-G	YL/WT	MLP Low Switch Signal	Low: 12v, all others: 1v
26-H	YL/RD	MLP Drive Switch Signal	Drive: 12v, all others: 0v
26-I	BK/RD	High Speed Inlet Air Control	Over 4900 rpm: 12v
26-K	---	Not Used	---
26-L	---	Not Used	---
26-M	DG/OR	Fuel Pressure Regulator	Startup: <3v, others: 12v
26-N	---	Not Used	---
26-O	---	Not Used	---
26-P	---	Not Used	---
26-Q	DB/OR	Idle Air Control Valve	8-10v
26-R	---	Not Used	---
26-S	---	Not Used	---
26-T	---	Not Used	---
26-U	YL	Fuel Injectors 1 & 3 (Bank 1)	3.6 ms
26-V	YL/BK	Fuel Injectors 2 & 4 (Bank 2)	3.6 ms
26-W	DB/OR	Shift Solenoid 1 Control	Off: 12v, On: 1v
26-X	DB/YL	Shift Solenoid 2 Control	Off: 12v, On: 1v
26-Y	OR	Shift Solenoid 3 Control	Off: 12v, On: 1v
26-Z	DB	TCC Solenoid Control	Shifting: 12v, Others: 1v

Pin Connector Graphic

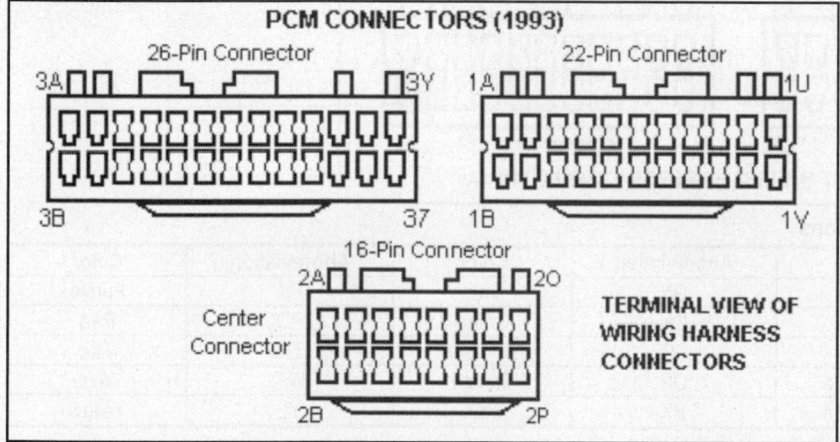

Standard Colors and Abbreviations

Abbreviation	Color	Abbreviation	Color	Abbreviation	Color
BK	Black	GY	Gray	PK	Purple
BL	Blue	GN	Green	RD	Red
BR	Brown	LG	LT Green	TN	Tan
DB	Dark Blue	OR	Orange	WT	White
DG	DK Green	PK	Pink	YL	Yellow

1993 Escort 1.8L I4 MFI VIN 8 (M/T) 22 Pin Connector

PCM Pin #	Wire Color	Circuit Description (22 Pin)	Value at Hot Idle
22-1	DB/RD	Keep Alive Power	12- 14v
22-2	WT/RD	Vehicle Power (Main Relay)	12-14v
22-3	PK	Vehicle Start Signal	Cranking: 9-11v
22-4	WT/YL	System Monitor Lamp	Lamp On: 5v, Off: 12v
22-5	YL/BK	MIL (lamp) Control	MIL On: 1v, Off: 12v
22-6	WT/BK	Self-Test Output Signal	Buzzer On: 5v, Off: 12v
22-7	DG/WT	Spark Output Signal	50% duty cycle
22-8, 22-9	---	Not Used	---
22-10	DB/BK	A/C WOT Relay Control	A/C On: 2.5v, Off: 12v
22-11	LG/YL	Self-Test Input Signal	STI On: 0v, STI Off: 5v
22-12	BR/WT	Daytime Running Lamps	DRL Off: 12.5v, On: 2.5v
22-13	---	Not Used	---
22-14	RD/WT	Idle Switch (IDL) Input	Closed: 0v, Open: 12v
22-15	DG	Brake Pedal Position Switch	Brake Off: 0v, On: 12v
22-16	DB/YL	PSP Switch Signal	Straight: 12v, Turned: 0v
22-17	DB/DB	Air Conditioning Switch Signal	A/C On: 2.5v, Off: 12v
22-18	BK/DG	Cooling Fan Control	On: 1v, Off: 12v
22-19	OR/DB	Blower Motor Switch Signal	1st: 12v, 2nd on up: 1v
22-20	BK/DB	Rear Defroster Signal	Switch On: 12v, Off: 0.1v
22-21	RD/BK	Headlamp Switch Signal	Switch On: 12v, Off: 1v
22-22	BR/YL	A/T: Neutral Drive Switch	In 'N': 0v, Others: 12v
22-22	BR/YL	M/T: Clutch Pedal Switch	Clutch In: 0v, Out: 12v

Pin Connector Graphic

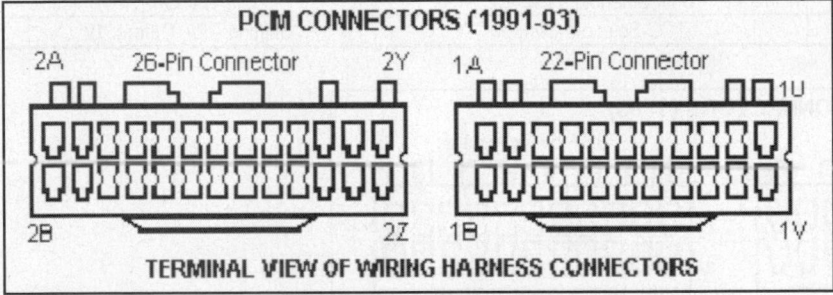

Standard Colors and Abbreviations

Abbreviation	Color	Abbreviation	Color	Abbreviation	Color
BK	Black	GY	Gray	PK	Purple
BL	Blue	GN	Green	RD	Red
BR	Brown	LG	LT Green	TN	Tan
DB	Dark Blue	OR	Orange	WT	White
DG	DK Green	PK	Pink	YL	Yellow

1993 Escort 1.8L I4 MFI VIN 8 (M/T) 26 Pin Connector

PCM Pin #	Wire Color	Circuit Description (26 Pin)	Value at Hot Idle
26-1	BK/OR	Power Ground	<0.1v
26-2	BK/OR	Power Ground	<0.1v
26-3	BK/LG	Power Ground	<0.1v
26-4	BK/WT	Analog Signal Return	<0.050v
26-5	WT	CKP Sensor Signal	2.5v (at Key On: 0 or 5v)
26-6	---	Not Used	---
26-7	YL/DB	CID Sensor Signal	2.5v (at Key On: 0 or 5v)
26-8	BK	Ground (GND)	<0.050v
26-9, 26-10	---	Not Used	---
26-11	LG/RD	Reference Voltage	4.9-5.1v
26-12	LG/WT	A/C WOT or Heater Relay	A/C On: 12v, Off: 2.5v
26-13	---	Not Used	---
26-14	RD/DB	HO2S-11 (B1 S1) Signal	0.1-1.1v
26-15	RD	Vane Airflow Meter Signal	0.1-0.3v
26-16	RD/BK	IAT Sensor Signal	1.5-2.5v
26-17	DB/WT	ECT Sensor Signal	0.5-0.6v
26-18	---	Not Used	---
26-19	BK/RD	High Speed Inlet Air Control	Over 5000 rpm: 12v
26-20	DG/OR	Fuel Pressure Regulator	Startup: <3v, others: 12v
26-21	YL	Fuel Injectors 1 & 3 - Bank 1	3.6 ms
26-22	YL/BK	Fuel Injectors 2 & 4 - Bank 2	3.6 ms
26-23	DB/OR	Idle Air Control Valve	8-10v
26-24	WT/DB	EVAP Purge Solenoid	Idle: 12-14v, off-idle: 1v
26-25	---	Not Used	---
26-26	---	Not Used	---

Pin Connector Graphic

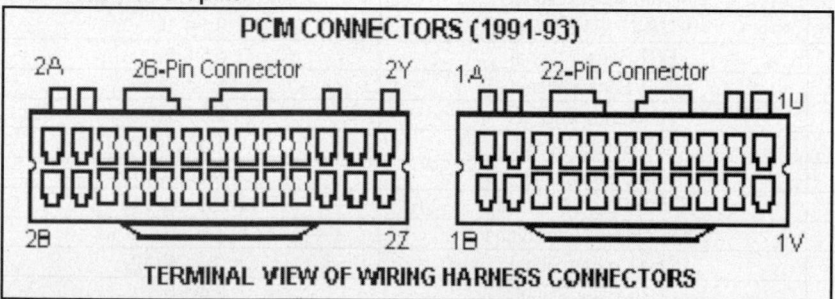

Standard Colors and Abbreviations

Abbreviation	Color	Abbreviation	Color	Abbreviation	Color
BK	Black	GY	Gray	PK	Purple
BL	Blue	GN	Green	RD	Red
BR	Brown	LG	LT Green	TN	Tan
DB	Dark Blue	OR	Orange	WT	White
DG	DK Green	PK	Pink	YL	Yellow

1994-95 Escort 1.8L I4 MFI VIN 8 (A/T) 22 Pin Connector

PCM Pin #	Wire Color	Circuit Description (22 Pin)	Value at Hot Idle
22-1	DB/RD	Keep Alive Power	12- 14v
22-2	PK	Vehicle Start Signal	Cranking: 9-11v
22-3	YL/BK	MIL (lamp) Control	MIL On: 1v, Off: 12v
22-4	DG/WT	CKP Sensor Signal	2.5v (at Key On: 0v or 5v)
22-5	LG/YL	Data Link Connector	Digital Signals
22-6	LG/YL	Data Link Connector	Digital Signals
22-6	BK/YL	Ignition On Power (Canada)	12-14V
22-7	DG	VSS (+) Signal	Vehicle moving: 0-5v
22-8	DGN/BK	A/C Switch Signal	Switch On: 2.5v, Off: 12v
22-9	DG	Brake Pedal Position Switch	Brake Off: 0v, On: 12v
22-10	---	Not Used	---
22-11	---	Not Used	---
22-12	WT/RD	Vehicle Power (PCM Relay)	12-14v
22-13	WT/YL	Data Link Connector	Digital Signals
22-14	WT/BK	Data Link Connector	Digital Signals
22-15	RD/BK	Park Lamp Signal	Switch On: 12v, Off: 1v
22-16	BK/DB	Rear Defroster Signal	Switch On: 12v, Off: 0.1v
22-17	DB/BK	AC/Heater WOT Relay	On: 1v, Off: 12v
22-18	DB/YL	PSP Switch Signal	Wheel Turning: 0v
22-19	OR/DB	Cooling Fan Control	On: 1v, Off: 12v
22-20	BK/DB	Ignition (Hot in Start)	KOEC: 9-11v
22-21	RD/WT	TP Idle Switch Signal	Closed: 0v, Open: 12v
22-22	---	Not Used	---

1994-95 Escort 1.8L I4 MFI VIN 8 (A/T) 16 Pin Connector

PCM Pin #	Wire Color	Circuit Description (16 Pin)	Value at Hot Idle
16-1	WT	CKP Sensor Signal	2.5v (at Key On: 0v or 5v)
16-2	RD/DB	HO2S-11 (B1 S1) Signal	0.1-1.1v
16-3	DB/WT	ECT Sensor Signal	0.5-0.6v
16-4	WT/BK	TOT Sensor Signal	3.5v at 68°F
16-5	LG/RD	Reference Voltage	4.9-5.1v
16-6	RD/BK	IAT Sensor Signal	1.5-2.5v
16-7	WT/DB	TSS Sensor (+)	340-380 rpm
16-8	WT/DB	EVAP Purge Solenoid	12v, 55 mph: 1v
16-9	RD	Vane Airflow Meter Signal	0.2v
16-10	BK/DG	Cooling Fan Signal	On: 1v, Off: 12v
16-11	LG/WT	TP Sensor Signal	0.4-1.0v
16-12	---	Not Used	---
16-13	YL/DB	CID Sensor	2.5v (at Key On: 0v or 5v)
16-14	DB	VSS (-) Signal	Vehicle moving: 0-5v
16-15	YL/DB	TSS Sensor (-)	340-380 rpm
16-16	---	Not Used	---

Standard Colors and Abbreviations

Abbreviation	Color	Abbreviation	Color	Abbreviation	Color
BK	Black	GY	Gray	PK	Purple
BL	Blue	GN	Green	RD	Red
BR	Brown	LG	LT Green	TN	Tan
DB	Dark Blue	OR	Orange	WT	White
DG	DK Green	PK	Pink	YL	Yellow

1994-95 Escort 1.8L I4 MFI VIN 8 (A/T) 26 Pin Connector

PCM Pin #	Wire Color	Circuit Description (26 Pin)	Value at Hot Idle
26-1	BK/OR	Power Ground	<0.1v
26-2	BK/LG	Power Ground	<0.1v
26-3	YL	TR Overdrive Signal	O/D On: 1v, Off: 12v
26-4	YL/WT	TR Sensor Low	In Low: 12v, Others: 1v
26-5	BK/RD	High Speed Inlet Air Control	Over 4900 rpm: 12v
26-6	---	Not Used	---
26-7	DG/OR	Fuel Pressure Regulator	Startup: <3v, others: 12v
26-8	---	Not Used	---
26-9	DB/OR	Idle Air Control Valve	8-10v
26-10	---	Not Used	---
26-11	YL	Fuel Injectors 1 & 3 (Bank 1)	3.6 ms
26-12	DB/OR	Shift Solenoid 1 Control	SS1 On: 1v, Off: 12v
26-13	OR	Shift Solenoid 3 Control	SS3 On: 1v, Off: 12v
26-14	BK/OR	Power Ground	<0.1v
26-15	BK/WT	Analog Signal Return	<0.050v
26-16	BR/WT	Daylight Running Lamps	DRL On: 1v, DRL Off: 12v
26-17	YL/RD	TR Sensor Drive	In 'D': 12v, Others: 1v
26-18	---	Not Used	---
26-19	---	Not Used	---
26-20	---	Not Used	---
26-21	---	Not Used	---
26-22	---	Not Used	---
26-23	---	Not Used	---
26-24	YL/BK	Fuel Injectors 2 & 4 (Bank 2)	3.6 ms
26-25	DB/YL	Shift Solenoid 2 Control	SS2 On: 1v, Off: 12v
26-26	DB	TCC Solenoid	Shifting 12v, Others: 1v

Pin Connector Graphic

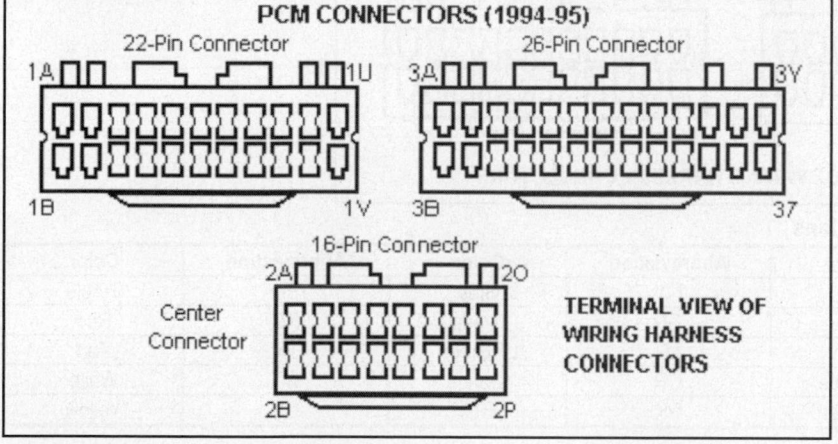

Standard Colors and Abbreviations

Abbreviation	Color	Abbreviation	Color	Abbreviation	Color
BK	Black	GY	Gray	PK	Purple
BL	Blue	GN	Green	RD	Red
BR	Brown	LG	LT Green	TN	Tan
DB	Dark Blue	OR	Orange	WT	White
DG	DK Green	PK	Pink	YL	Yellow

1994-95 Escort 1.8L I4 MFI VIN 8 (M/T) 22 Pin Connector

PCM Pin #	Wire Color	Circuit Description (22 Pin)	Value at Hot Idle
22-A	DB/RD	Keep Alive Power	12- 14v
22-B	PK	Vehicle Start Signal	Cranking: 9-11v
22-C	YL/BK	MIL (lamp) Control	MIL On: 1v, Off: 12v
22-D	DG/WT	Spark Output Signal	50% duty cycle
22-E	LG/YL	Self-Test Input Signal	STI On: 0v, STI Off: 5v
22-F	---	Not Used	---
22-G	---	Not Used	---
22-H	DG	Brake Pedal Position Switch	Brake Off: 0v, On: 12v
22-I	DGN/BK	A/C Switch Signal	Switch On: 12v, Off: 1.5v
22-J	OR/DB	Blower Motor Switch Signal	1st: 12v, others: 1v
22-K	RD/BK	Headlamp Switch Signal	Switch On: 12v, Off: 1v
22-L	WT/RD	Vehicle Power (PCM Relay)	12-14v
22-M	WT/YL	Data Link Connector	Digital Signals
22-N	WT/BK	Data Link Connector	Digital Signals
22-O	---	Not Used	---
22-P	DB/BK	A/C & Heater WOT Relay	On: 1v, Off: 12v
22-Q	BR/WT	Daytime Running Lamp	DRL Off: 12v, On: 2.5v
22-R	RD/WT	Idle Switch Signal	Closed: 0V, Open: 12v
22-S	DB/YL	PSP Switch Signal	Straight: 12v, Turned: 0v
22-T	BK/DG	Cooling Fan Control	On: 1v, Off: 12v
22-U	BK/DB	Rear Defroster Switch Signal	AC On: 12v, Off: 0v
22-V	BR/YL	A/T: Neutral Drive Switch	In 'N': 0v, Others: 12v
22-V	BR/YL	M/T: Clutch Engage Switch	Clutch In: 0v, Out: 12v

Pin Connector Graphic

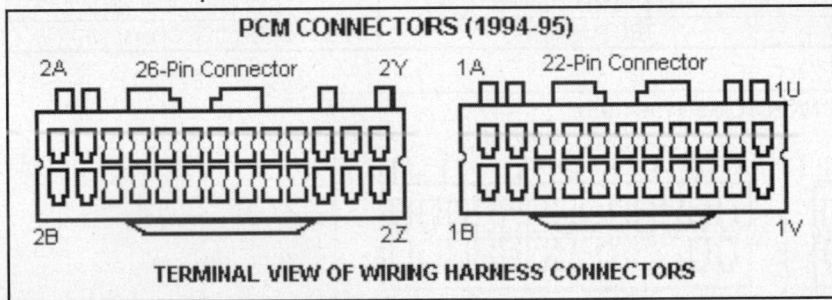

Standard Colors and Abbreviations

Abbreviation	Color	Abbreviation	Color	Abbreviation	Color
BK	Black	GY	Gray	PK	Purple
BL	Blue	GN	Green	RD	Red
BR	Brown	LG	LT Green	TN	Tan
DB	Dark Blue	OR	Orange	WT	White
DG	DK Green	PK	Pink	YL	Yellow

1994-95 Escort 1.8L I4 MFI VIN 8 (M/T) 26 Pin Connector

PCM Pin #	Wire Color	Circuit Description (26 Pin)	Value at Hot Idle
26-A	BK/OR	Power Ground	<0.1v
26-B	BK/LG	Power Ground	<0.1v
26-C	WT	CKP Sensor Signal	2.5v (at Key On: 0-5v)
26-D	YL/DB	CID Sensor Signal	2.5v (at Key On: 0-5v)
26-E	---	Not Used	---
26-F	LG/RD	Reference Voltage	4.9-5.1v
26-G	---	Not Used	---
26-H	RD	Vane Airflow Meter Signal	0.1-0.3v
26-I	DB/WT	ECT Sensor Signal	0.5-0.6v
26-J	BK/RD	High Speed Inlet Air Control	Over 5000 rpm: 12v
26-K	YL	Fuel Injectors 1 & 3 (Bank 1)	3.6 ms
26-L	DB/OR	Idle Air Control Valve	8-10v
26-M	---	Not Used	---
26-N	BK/OR	Power Ground	<0.1v
26-O	BK/WT	Analog Signal Return	<0.050v
26-P	---	Not Used	---
26-Q ('94)	BK	Power Ground	<0.1v
26-Q ('95)	BK/YL	Ignition Power	12-14v
26-R	---	Not Used	---
26-S	LG/WT	TP Sensor Signal	0.4-1.0v
26-T	RD/DB	HO2S-11 (B1 S1) Signal	0.1-1.1v
26-U	RD/BK	IAT Sensor Signal	1.5-2.5v
26-V	---	Not Used	---
26-W	DG/OR	Fuel Pressure Regulator	Startup: <3v, others: 12v
26-X	YL/BK	Fuel Injectors 2 & 4 (Bank 2)	3.6 ms
26-Y	WT/DB	EVAP Purge Solenoid	12v, 55 mph: 1v
26-Z	---	Not Used	---

Pin Connector Graphic

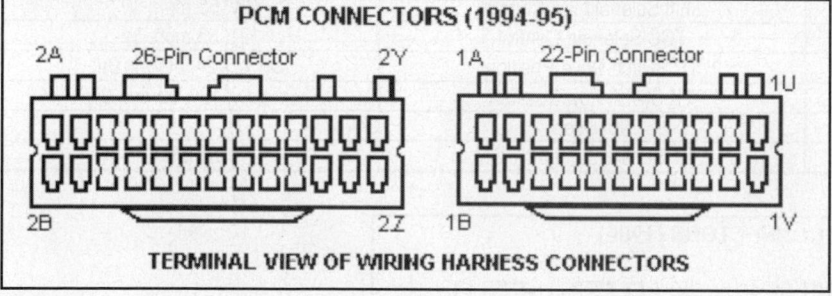

1996 Escort 1.8L I4 MFI VIN 8 (A/T) 22 Pin Connector

PCM Pin #	Wire Color	Circuit Description (22 Pin)	Value at Hot Idle
C1	---	Not Used	---
C1-2	PK	Vehicle Start	KOEC: 9-11v
C1-3	YL/BK	MIL (lamp) Control	MIL On: 0v, Off: 12v
C1-4	BL/BK	A/C WOT Cutout Relay	A/C On: 2.5v, Off: 12v
C1-5	LG/YL	Self-Test Input Signal	STI On: 0v, Off: 5v
C1-6	GN/BK	A/C High Pressure Switch	On: 1v, Off: 12v
C1-7	GN	VSS (+) Signal	Vehicle moving: 0-5v
C1-8	BL	VSS (-) Signal	Vehicle moving: 0-5v
C1-9	GN	Brake Pedal Position Switch	Brake Off: 0v, On: 12v
C10	---	Not Used	---
C1-11	LG	Fuel Pump Control	Relay On: 0v, Off: 12v
C12	---	Not Used	---
C1-13	BL/WT	ISO K-Line (to Scan Tool)	Digital Signals
C1-14	---	Not Used	---
C1-15	RD/BK	Headlamp Switch Input	Switch On: 12v, Off: 1v
C1-16	GY/RD	Rear Defroster Switch Signal	Switch On: 12v, Off: 1v
C1-17	BL/GN	M/T: Cooling Fan Control	Relay On: 0v, Off: 12v
C1-18	BR/WT	DRL Switch	Switch On: 0v, Off: 12v
C1-19	OR/BL	Blower Motor Control Switch	Off or 1st: 12v, 2nd on: 1v
C1-20	BK/GN	Cooling Fan Control	Relay On: 0v, Off: 12v
C21-22	---	Not Used	---
PCM Pin #	**Wire Color**	**Circuit Description (12 Pin)**	**Value at Hot Idle**
C2-1	BL/OR	Shift Solenoid 1-2 Control	12v, 55 mph: 12v
C2-2	OR	Shift Solenoid 2-3 Control	12, 55 mph: 12v
C2-3	---	Not Used	---
C2-4	YL/RD	TR Switch 2nd Position	In 2nd: 12v, others: 0v
C2-5	---	Not Used	---
C2-6	WT/BL	TSS Sensor (+)	340-380 Hz
C2-7	BL/YL	Shift Solenoid 3-4 Control	12v, 55 mph: 1v
C2-8	BL	TCC Solenoid Control	12v, 55 mph: 1v
C2-9	YL	TR Switch Drive Position	Drive: 12v, others: 0v
C2-10	BR/WT	TR Switch Low Position	In Low: 12v, others: 0v
C2-11	---	Not Used	---
C2-12	YL/BL	ECT Sensor Signal to TCM	0.5-0.6v

Pin Connector Graphic

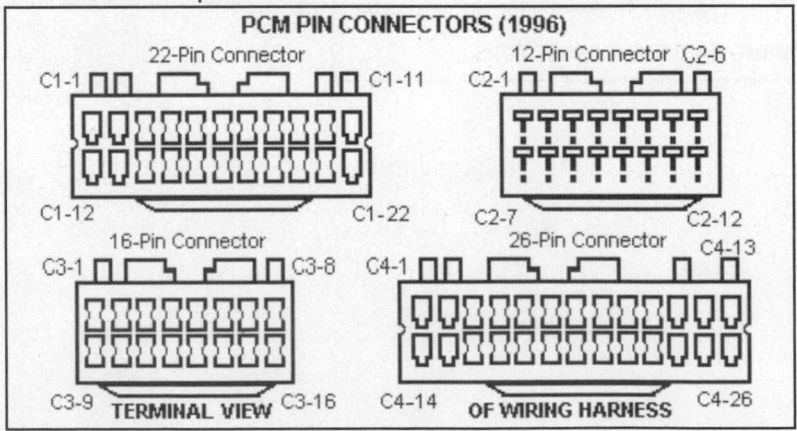

1996 Escort 1.8L I4 MFI VIN 8 (A/T) 16 Pin Connector

PCM Pin #	Wire Color	Circuit Description (16 Pin)	Value at Hot Idle
C3-1	---	Not Used	---
C3-2	RD/BL	HO2S-11 (B1 S1) Signal	0.1-1.1v
C3-3	WT/BK	TFT Sensor Signal	2.10-2.40v
C3-4	BL/WT	ECT Sensor Signal	0.5-0.6v
C3-5	LG/RD	Reference Voltage	4.9-5.1v
C3-6	RD/BK	IAT Sensor Signal	1.5-2.5v
C3-7	GN/WT	HO2S-11 (B1 S1) Heater	0v, Off: 12v
C3-8	BK/BL	Analog Return Signal	<0.050v
C3-9	RD	MAF Sensor Signal	0.6v
C3-10	RD/YL	HO2S-12 (B1 S2) Signal	0.1-1.1v
C3-11	LG/WT	TP Sensor Signal	0.4-1.0v
C3-12	---	Not Used	---
C3-13	---	Not Used	---
C3-14	RD/WT	Idle Switch Signal	Closed: 0v, Open: 12v
C3-15	---	Not Used	---
C3-16	BL/YL	PSP Switch Signal	Straight: 12v, Turned: 0v

1996 Escort 1.8L I4 MFI VIN 8 (A/T) 26 Pin Connector

PCM Pin #	Wire Color	Circuit Description (26 Pin)	Value at Hot Idle
C4-1	BK/LG	Power Ground	<0.1v
C4-2	BK/OR	Power Ground	<0.1v
C4-3	OR	CKP Sensor (-) Signal	435-475 Hz
C4-4	YL/BL	CMP 1 Sensor	5-7 Hz
C4-5	BL/RD	Keep Alive Power	12-14v
C4-6	BK/RD	High Speed Inlet Air Control	Over 5000 rpm: 12v
C4-7	---	Not Used	---
C4-8	---	Not Used	---
C4-9	BL/OR	Idle Air Control Solenoid	10.7v (30%)
C4-10	---	Not Used	---
C4-11	YL	Injector 1 Control	4.0-4.5 ms
C4-12	GN/RD	Injector 3 Control	4.0-4.5 ms
C4-13	---	Not Used	---
C4-14	WT/RD	Vehicle Power	12-14v
C4-15	BK/OR	Power Ground	<0.1v
C4-16	WT	CMP 2 Signal	5-7 Hz
C4-17	LG/WT	CKP Sensor (+) Signal	435-475 Hz
C4-18	GN/OR	Fuel Pressure Regulator	Startup: <3v, others: 12v
C4-19	---	Not Used	---
C4-20	BK/WT	Ignition SPOUT Signal	Varies: 0.1-0.3v
C4-21	---	Not Used	---
C4-22	---	Not Used	---
C4-23	WT/BL	EVAP Purge Solenoid	On: 0v, Off: 12v
C4-24	YL/BK	Injector 2 Control	4.0-4.5 ms
C4-25	GN/YL	Injector 4 Control	4.0-4.5 ms
C4-26	---	Not Used	---

Standard Colors and Abbreviations

Abbreviation	Color	Abbreviation	Color	Abbreviation	Color
BK	Black	GY	Gray	PK	Purple
BL	Blue	GN	Green	RD	Red
BR	Brown	LG	LT Green	TN	Tan
DB	Dark Blue	OR	Orange	WT	White
DG	DK Green	PK	Pink	YL	Yellow

1996 Escort 1.8L I4 MFI VIN 8 (M/T) 22 Pin Connector

PCM Pin #	Wire Color	Circuit Description (22 Pin)	Value at Hot Idle
C1-1	---	Not Used	---
C1-2	PK	Vehicle Start	KOEC: 9-11v
C1-3	YL/BK	MIL (lamp) Control	MIL On: 0v, Off: 12v
C1-4	BL/BK	A/C WOT Relay Control	On: 1v, Off: 12v
C1-5	LG/YL	Self-Test Input Signal	STI On: 0v, Off: 5v
C1-6	GN/BK	A/C High Pressure Switch	Switch On: 2.5v, Off: 12v
C1-7	GN	VSS (+) Signal	0 Hz, 55 mph: 125 Hz
C1-8	BL	VSS (-) Signal	0 Hz, 55 mph: 125 Hz
C1-9	GN	Brake Pedal Position Switch	Brake Off: 0v, On: 12v
C1-10	---	Not Used	---
C1-11	LG	Fuel Pump Control	On: 1v, Off: 12v
C1-12	---	Not Used	---
C1-13	BL/WT	ISO K-Line for OBD II	Digital Signals
C1-14	---	Not Used	---
C1-15	RD/BK	Headlamp Switch Input	Switch On: 12v, Off: 1v
C1-16	GY/RD	Rear Defroster Switch Signal	DEF On: 12v, Off: 1v
C1-17	BR/YL	A/T: Neutral Drive Switch	In 'N': 0v, Others: 5v
C1-17	BR/YL	M/T: Clutch Pedal Switch	Clutch In: 0v, Out: 5v
C1-18	BR/WT	Daytime Running Lamp Relay	Relay On: 0v, Off: 12v
C1-19	OR/BL	Blower Motor Control Switch	1st: 12v, others: 1v
C1-20	BK/GN	Cooling Fan Control	On: 1v, Off: 12v
C1-21	---	Not Used	---
C1-22	---	Not Used	---

1996 Escort 1.8L I4 MFI VIN 8 (M/T) 16 Pin Connector

PCM Pin #	Wire Color	Circuit Description (16 Pin)	Value at Hot Idle
C3-1	---	Not Used	---
C3-2	RD/BL	HO2S-11 (B1 S1) Signal	0.1-1.1v
C3-3	---	Not Used	---
C3-4	BL/WT	ECT Sensor Signal	0.5-0.6v
C3-5	LG/RD	Reference Voltage	4.9-5.1v
C3-6	RD/BK	IAT Sensor Signal	1.5-2.5v
C3-7	GN/WT	HO2S-11 (B1 S1) Heater	On: 1v, Off: 12v
C3-8	BK/BL	Analog Return Signal	<0.1v
C3-9	RD	MAF Sensor Signal	0.8, 55 mph: 1.9v
C3-10	RD/YL	HO2S-12 (B1 S2) Signal	0.1-1.1v
C3-11	LG/WT	TP Sensor Signal	0.1v, 55 mph: 2.1v
C3-12	---	Not Used	---
C3-13	---	Not Used	---
C3-14	RD/WT	Idle Switch Signal	Closed: 0v, Open: 12v
C3-15	---	Not Used	---
C3-16	BL/YL	PSP Switch Signal	Straight: 12v, Turned: 0v

Standard Colors and Abbreviations

Abbreviation	Color	Abbreviation	Color	Abbreviation	Color
BK	Black	GY	Gray	PK	Purple
BL	Blue	GN	Green	RD	Red
BR	Brown	LG	LT Green	TN	Tan
DB	Dark Blue	OR	Orange	WT	White
DG	DK Green	PK	Pink	YL	Yellow

1996 Escort 1.8L I4 MFI VIN 8 (M/T) 26 Pin Connector

PCM Pin #	Wire Color	Circuit Description (26 Pin)	Value at Hot Idle
C4-1	BK/LG	Power Ground	<0.1v
C4-2	BK/OR	Power Ground	<0.1v
C4-3	OR	CKP Sensor (-) Signal	435-475 Hz
C4-4	YL/BL	CMP 1 Signal	5-7 Hz
C4-5	BL/RD	Keep Alive Power	12-14v
C4-6	BK/RD	High Speed Inlet Air Solenoid	Below 5000 rpm: 1.5v
C4-7	---	Not Used	---
C4-8	---	Not Used	---
C4-9	BL/OR	Idle Air Control Solenoid	Varies: 8-10v
C4-10	---	Not Used	---
C4-11	YL	Injector 1 Control	4.0-4.5 ms
C4-12	GN/RD	Injector 3 Control	4.0-4.5 ms
C4-13	---	Not Used	---
C4-14	WT/RD	Vehicle Power	12-14v
C4-15	BK/OR	Power Ground	<0.1v
C4-16	WT	CMP 2 Signal	5-7 Hz
C4-17	LG/WT	CKP Sensor (+) Signal	435-475 Hz
C4-18	GN/OR	Fuel Pressure Regulator	Startup: <3v, others: 12v
C4-19	---	Not Used	---
C4-20	GN/WT	Spark Output Signal	50% duty cycle
C4-21	---	Not Used	---
C4-22	---	Not Used	---
C4-23	WT/BL	EVAP Purge Solenoid	12v, 55 mph: 1v
C4-24	YL/BK	Injector 2 Control	4.0-4.5 ms
C4-25	GN/YL	Injector 4 Control	4.0-4.5 ms
C4-26	---	Not Used	---

Pin Connector Graphic

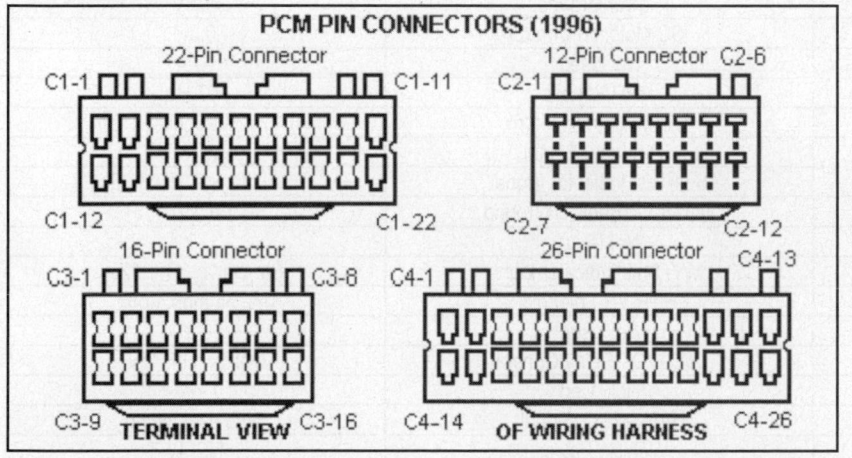

1990 Escort 1.9L I4 CFI VIN 9 (All) 60 Pin Connector

PCM Pin #	Wire Color	Circuit Description (60 Pin)	Value at Hot Idle
1	YL	Keep Alive Power	12-14v
2-3	---	Not Used	---
4	DG/YL	Ignition Diagnostic Monitor	20-31 Hz
5	RD/LG	Keep Alive Power	12-14v
6, 9	---	Not Used	---
7	LG/YL	ECT Sensor Signal	0.5-0.6v
8	PK/BK	Fuel Pump Monitor	On: 1v, Off: 12v
10	BK/YL	A/C Cycling Clutch Switch	A/C On: 12v, Off: 0v
11-15	---	Not Used	---
16	BK/OR	Ignition System Ground	<0.050v
17	TN/RD	STI Output, MIL Control	MIL On: 1v, Off: 12v
18-19	---	Not Used	---
20	BK	PCM Case Ground	<0.050v
21	BR/WT	ISC Motor (+) Control	0.3v
22	TN/LG	Fuel Pump Control	On: 1v, Off: 12v
23-24	---	Not Used	---
25	LG/PK	ACT Sensor Signal	1.5-2.5v
26	OR/WT	Reference Voltage	4.9-5.1v
27	BR/LG	PFE EGR Sensor	3.2v, 55 mph: 2.8v
28	LG/WT	Idle Tracking Switch	10.1v, 55 mph: 0v
29	DG/PK	HO2S-11 (B1 S1) Signal	0.1-1.1v
30	GY/OR	Neutral Drive Switch Signal	In 'N': 0v, Others: 5v
31-34	---	Not Used	---
35	GY/YL	EVAP Purge Solenoid	12v, 55 mph: 1v
36	YL/LG	Spark Output Signal	50% duty cycle
37	RD	Vehicle Power	12-14v
38-39	---	Not Used	---
40	BK/LG	Power Ground	<0.1v
41	WT/BL	ISC Motor (-) Control	0.3v
42-44	---	Not Used	---
45	LG/BK	MAP Sensor	107 Hz
46	BK/WT	Analog Signal Return	<0.050v
47	DG/LG	TP Sensor Signal	0.5-1.2v
48	WT/RD	Self-Test Indicator Signal	STI On: <0.1v, Off: 5v
49	OR	HO2S-11 (Bank 1) Ground	<0.050v
50	---	Not Used	---
51	TN/BL	M/T: Shift Indicator Light	SIL On: 1v, Off: 12v
52	YL	EGR VR Solenoid	0%, 55 mph: 45%
53	---	Not Used	---
54	BK/YL	A/C WOT Relay Control	On: 1v, Off: 12v
55, 59	---	Not Used	---
56	DB	PIP Sensor Signal	50% duty cycle
57	RD	Vehicle Power	12-14v
58	TN/RD	CFI Fuel Injector	1.5 ms
60	BK/LG	Power Ground	<0.1v

Pin Connector Graphic

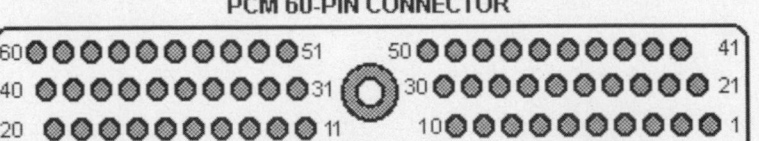

PCM 60-PIN CONNECTOR

Terminal View of 60-Pin PCM Harness Connector

1990 Escort 1.9L I4 MFI VIN J (All) 60 Pin Connector

PCM Pin #	Wire Color	Circuit Description (60 Pin)	Value at Hot Idle
1	YL	Keep Alive Power	12-14v
2-3	---	Not Used	---
4	DG/YL	Ignition Diagnostic Monitor	20-31 Hz
5-6	---	Not Used	---
7	LG/YL	ECT Sensor Signal	0.5-0.6v
8	PK/BK	Fuel Pump Monitor	On: 12v, Off: 0v
9, 11-15	---	Not Used	---
10	BK/YL	A/C Cycling Clutch Switch	A/C On: 12v, Off: 0v
16	BK/OR	Ignition System Ground	<0.050v
17	TN/BL	M/T: Shift Indicator Light	SIL On: 1v, Off: 12v
18-19	---	Not Used	---
20	BK	PCM Case Ground	<0.050v
21	OR/BK	IAC Motor Control	Idle: 9.5-10.3v
22	TN/LG	Fuel Pump Control	On: 1v, Off: 12v
23-24	---	Not Used	---
25	LG/PK	VAT Sensor	1.5-2.5v
26	OR/WT	Reference Voltage	4.9-5.1v
27-28	---	Not Used	---
29	DG/PK	HO2S-11 (B1 S1) Signal	0.1-1.1v
30	GY/OR	A/T: Neutral Drive Switch	In 'N': 0v, Others: 5v
30	GY/OR	M/T: Clutch Engage Switch	Clutch In: 0v, Out: 5v
31	---	Not Used	---
32	GY/YL	EVAP Purge Solenoid	12v, 55 mph: 1v
33-34	---	Not Used	---
35	YL	EGR Shutoff Solenoid Control	0%, 35 mph: 38%
36	YL/LG	Spark Output Signal	50% duty cycle
37	RD	Vehicle Power	12-14v
38-39	---	Not Used	---
40	BK/LG	Power Ground	<0.1v
41-42	---	Not Used	---
43	WT/BK	Vane Airflow Meter	0.7v
44	---	Not Used	---
45	LG/BK	BARO Sensor Signal	159 Hz (sea level)
46	BK/WT	Analog Signal Return	<0.050v
47	DG/LG	TP Sensor Signal	0.5-1.2v
48	WT/RD	Self-Test Indicator Signal	STI Open: 5v, Closed: 1v
49	OR	HO2S-11 (Bank 1) Ground	<0.050v
50-52, 55	---	Not Used	---
53 ('89-'90)	TN/RD	MIL (lamp) Control	MIL On: 1v, Off: 12v
54	OR/BL	A/C WOT Relay Control	On: 1v, Off: 12v
56	DB	PIP Sensor Signal	50% duty cycle
57	RD	Vehicle Power	12-14v
58	TN/RD	Injector Bank 1 (INJ 1 & 2)	4.8-5.0 ms
59	TN/OR	Injector Bank 2 (INJ 3 & 4)	4.8-5.0 ms
60	BK/LG	Power Ground	<0.1v

Pin Connector Graphic

PCM 60-PIN CONNECTOR

Terminal View of 60-Pin PCM Harness Connector

1991-92 Escort 1.9L I4 MFI VIN J (All) 60 Pin Connector

PCM Pin #	Wire Color	Circuit Description (60 Pin)	Value at Hot Idle
1	BL/RD	Keep Alive Power	12-14v
3	WT/BK	VSS (+) Signal	0 Hz, 55 mph: 125 Hz
4	RD	Ignition Diagnostic Monitor	20-31 Hz
2, 5	---	Not Used	
6 ('91)	BL	VSS (-) Signal	0 Hz, 55 mph: 125 Hz
6 ('92)	DB	VSS (-) Signal	0 Hz, 55 mph: 125 Hz
7	BL/WT	ECT Sensor Signal	0.5-0.6v
8	YL/GN	Data Bus (-) Signal	Digital Signals
9	GN/YL	Mass Airflow Return	<0.050v
10	GN/BK	A/C Cycling Clutch Switch	A/C On: 12v, Off: 0v
11	---	Not Used	---
12	YL/OR	Injector 3 Control	4.8-5.0 ms
13	GN/OR	Injector 4 Control	4.8-5.0 ms
14	---	Not Used	---
15 ('92)	OR/WT	EVAP Purge Solenoid	12v, 55 mph: 1v
16	RD/BL	Ignition System Ground	<0.050v
17	YL/BK	STI Output, MIL Control	MIL On: 1v, Off: 12v
18, 20	---	Not Used	---
19	BK/PK	Fuel Pump Monitor	On: 12v, Off: 0v
21	BL/OR	IAC Motor Control	9.7-12.0v
22	LG	Fuel Pump Control	On: 1v, Off: 12v
23-24	---	Not Used	---
25	WT/GN	ACT Sensor Signal	1.5-2.5v
26	LG/WT	Reference Voltage	4.9-5.1v
27	BL/YL	PFE EGR Sensor	3.2v, 55 mph: 2.8v
28	YL/BL	Data Bus (+) Signal	Digital Signals
29	GN/BL	HO2S-11 (B1 S1) Signal	0.1-1.1v
30	BR/YL	A/T: Neutral Drive Switch	In 'P': 0v, Others 5v
30	BR/YL	M/T: Clutch Engage Switch	Clutch In: 0v, Out: 12v
31	RD/BK	High Speed Fan Control	On: 1v, Off: 12v
32	---	Not Used	---
33	WT/BL	EGR VR Solenoid	0%, 55 mph: 45%
34	LG/BK	Octane Adjust Switch 2	Closed: 0v, Open: 9.1v
35	YL/WT	Low Speed Fan Control	On: 1v, Off: 12v
36	LG/WT	Spark Angle Word Signal	50% duty cycle
37	WT/RD	Vehicle Power	12-14v
38-39	---	Not Used	---
40	BK/GN	Power Ground	<0.1v

Pin Connector Graphic

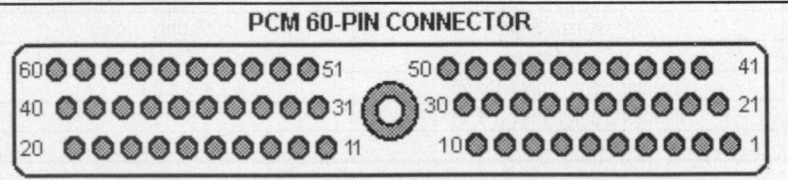

PCM 60-PIN CONNECTOR

Terminal View of 60-Pin PCM Harness Connector

Standard Colors and Abbreviations

Abbreviation	Color	Abbreviation	Color	Abbreviation	Color
BK	Black	GY	Gray	PK	Purple
BL	Blue	GN	Green	RD	Red
BR	Brown	LG	LT Green	TN	Tan
DB	Dark Blue	OR	Orange	WT	White
DG	DK Green	PK	Pink	YL	Yellow

1991-92 Escort 1.9L I4 MFI VIN J (All) 60 Pin Connector

PCM Pin #	Wire Color	Circuit Description (60 Pin)	Value at Hot Idle
41	OR	CID+ Sensor	400-425 Hz
42	BL/GN	CID- Sensor	400-425 Hz
43	GN/WT	Octane Adjust Switch 1	Closed: 0v, Open: 9.1v
44-45	---	Not Used	---
46	LG/BK	Analog Signal Return	<0.050v
47	RD/WT	TP Sensor Signal	Hot 0.5-1.2v
48	LG/YL	Self-Test Indicator Signal	STI Open: 5v, Closed: 1v
49	BK	HO2S-11 (Bank 1) Ground	<0.050v
49	OR/BK	HO2S-11 (Bank 1) Ground	<0.050v
50	BR/BK	MAF Sensor Signal	0.6-0.9v
51-52	---	Not Used	---
53	LG/RD	Shift Indicator Control	SIL On: 1v, Off 12v
54	BL/BK	A/C WOT Relay Control	On: 1v, Off: 12v
55	---	Not Used	---
56	GN/WT	PIP Sensor Signal	50% dwell
57	WT/RD	Vehicle Power	12-14v
58	YL	Injector Bank 1 (INJ 1 & 2)	4.8-5.0 ms
59	GN/RD	Injector Bank 2 (INJ 3 & 4)	4.8-5.0 ms
60	BK/GN	Power Ground	<0.1v

Pin Connector Graphic

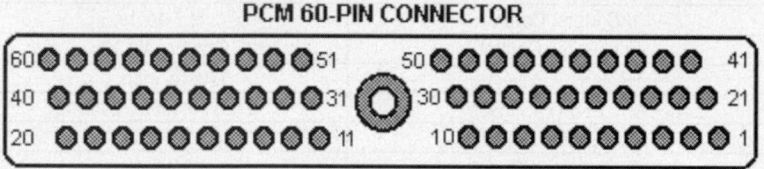

PCM 60-PIN CONNECTOR

Terminal View of 60-Pin PCM Harness Connector

Standard Colors and Abbreviations

Abbreviation	Color	Abbreviation	Color	Abbreviation	Color
BK	Black	GY	Gray	PK	Purple
BL	Blue	GN	Green	RD	Red
BR	Brown	LG	LT Green	TN	Tan
DB	Dark Blue	OR	Orange	WT	White
DG	DK Green	PK	Pink	YL	Yellow

1993-95 Escort 1.9L I4 MFI VIN J (All) 60 Pin Connector

PCM Pin #	Wire Color	Circuit Description (60 Pin)	Value at Hot Idle
1	BL/RD	Keep Alive Power	12-14v
2	WT	TOT Sensor Signal	2.10-2.40v
3	WT/BK	VSS (+) Signal	0 Hz, 55 mph: 125 Hz
4	RD	Ignition Diagnostic Monitor	20-31 Hz
5	GN	Brake Pedal Position Switch	Brake Off: 0v, On: 12v
6	BL	VSS (-) Signal	0 Hz, 55 mph: 125 Hz
6 ('94-'95)	DB	VSS (-) Signal	0 Hz, 55 mph: 125 Hz
7	BL/WT	ECT Sensor Signal	0.5-0.6v
8	YL/GN	Data Bus (-) Signal	Digital Signals
9	GN/YL	Mass Airflow Return	<0.050v
10	GN/BK	A/C Cycling Clutch Switch	A/C On: 12v, Off: 0v
11	LG/BL	Shift Solenoid 1 Control	12v, 55 mph: 1v
12	YL/OR	Injector 3 Control	4.8-5.1 ms
13	GN/OR	Injector 4 Control	4.8-5.1 ms
14	---	Not Used	---
15	OR/WT	EVAP Purge Solenoid	12v, 55 mph: 9v
16	RD/BL	Ignition System Ground	<0.1v
17	YL/BK	STI Output, MIL Control	MIL On: 1v, Off: 12v
18	BR/RD	TR Sensor Drive	12v, 55 mph: 12v
19	BK/PK	Fuel Pump Monitor	On: 12v, Off: 0v
20	BK	PCM Case Ground	<0.050v
21	OR/BK	IAC Motor Control	9.7-11.1v
22	LG	Fuel Pump Control	On: 1v, Off: 12v
23	RD/GN	TR Sensor Reverse	12v, 55 mph: 0v
24	WT/BL	A/T: TSS (+) Sensor	42-50 Hz (680-720 rpm)
25	WT/GN	IAT Sensor Signal	1.5-2.5v
26	LG/WT	Reference Voltage	4.9-5.1v
27	BL/YL	PFE EGR Sensor	3.2v, 55 mph: 3.4v
28	YL/BL	Data Bus (+) Signal	Digital Signals
29	GN/BL	HO2S-11 (B1 S1) Signal	0.1-1.1v
30	BK/BL	A/T: Neutral Drive Switch	In 'P': 0v, Others: 5v
30	BR/YL	A/T: Neutral Drive Switch	In 'P': 0v, Others: 5v
30	BK/BL	M/T: Clutch Engage Switch	Clutch In: 0v, Out 5v
30	BR/YL	M/T: Clutch Engage Switch	Clutch In: 0v, Out 5v
31	RD/BK	High Speed Fan Control	On: 1v, Off: 12v
32-34	---	Not Used	---
33	WT/BL	EGR VR Solenoid	0%, 55 mph: 45%
35	YL/WT	Low Speed Fan Control	On: 1v, Off: 12v
36	LG/WT	Spark Output Signal	50% duty cycle
37	WT/RD	Vehicle Power	12-14v
38	BR/BL	TR Sensor Overdrive	12v, 55 mph: 0v
39	---	Not Used	---
40	BK/GN	Power Ground	<0.1v

Pin Connector Graphic

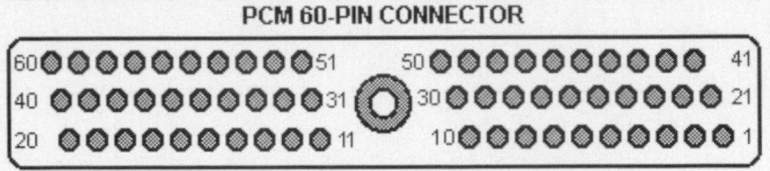

PCM 60-PIN CONNECTOR

Terminal View of 60-Pin PCM Harness Connector

1993-95 Escort 1.9L I4 MFI VIN J (All) 60 Pin Connector

PCM Pin #	Wire Color	Circuit Description (60 Pin)	Value at Hot Idle
41	OR	CID+ Sensor	5-7 Hz
42	BL/GN	CID- Sensor	<0.050v
43	GN/WT	Octane Adjust Switch 1	Closed: 0v, Open: 9.1v
44	YL/BL	A/T: TSS Sensor (-)	44-56 Hz (680-710 rpm)
45	BR/GN	TR Sensor Low	12v, 55 mph: 0v
46	LG/BK	Analog Signal Return	<0.050v
47	RD/WT	TP Sensor Signal	0.5-1.2v
48	LG/YL	Self-Test Indicator Signal	STI Open: 5v, Closed: 1v
49	BK	HO2S-11 (Bank 1) Ground	<0.050v
49	OR/BK	HO2S-11 (Bank 1) Ground	<0.050v
50	BR/BK	MAF Sensor Signal	0.7v
51	PK/BK	Shift Solenoid 2 Control	1v, 55 mph: 1v
52	OR/GN	Shift Solenoid 3 Control	12v, 55 mph: 12v
53	LG/RD	M/T: Shift Indicator Light	SIL On: 1v, Off: 12v
54	BL/BK	A/C WOT Relay Control	On: 1v, Off: 12v
55	PK/WT	TCC Solenoid Control	12v, 55 mph: 9v
56	GN/WT	PIP Sensor Signal	50% dwell
57	WT/RD	Vehicle Power	12-14v
58	YL	Injector 1 Control	4.8-5.1 ms
59	GN/RD	Injector 2 Control	4.8-5.1 ms
60	BK/GN	Power Ground	<0.1v
60	BK/GN	Power Ground	<0.1v

Pin Connector Graphic

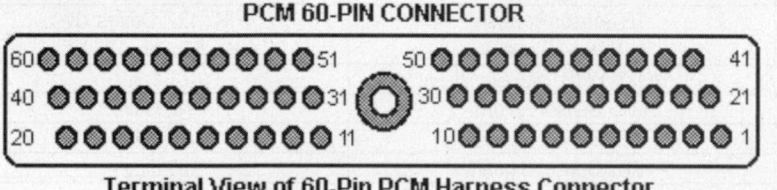

Standard Colors and Abbreviations

Abbreviation	Color	Abbreviation	Color	Abbreviation	Color
BK	Black	GY	Gray	PK	Purple
BL	Blue	GN	Green	RD	Red
BR	Brown	LG	LT Green	TN	Tan
DB	Dark Blue	OR	Orange	WT	White
DG	DK Green	PK	Pink	YL	Yellow

1996 Escort 1.9L I4 MFI VIN J (All) 104 Pin Connector

PCM Pin #	Wire Color	Circuit Description (104 Pin)	Value at Hot Idle
1	PK/BK	Shift Solenoid 2 Control	12v, 55 mph: 1v
2	YL/BK	MIL (lamp) Control	MIL On: 1v, Off: 12v
3-5	---	Not Used	---
6	BR/BK	TR Sensor Overdrive	In 'P': 0v, in O/D: 12v
7	BR/DG	TR Sensor Low	In 'P': 0v, in 1st: 12v
9	BR/YL	TR Sensor Drive	In 'P': 0v, in 2nd: 12v
8,10	---	Not Used	---
11	BL/PK	Purge Flow Sensor	1.55v, at 30 mph: 2.55v
13	PK/DG	Flash EEPROM Power	0.1v
12, 14	---	Not Used	---
15	BK/BL	Data Bus (-) Signal	Digital Signals
16	WT/BL	Data Bus (+) Signal	Digital Signals
17	RD/BK	High Speed Fan Control	On: 1v, Off: 12v
18-20	---	Not Used	---
21	BL/DG	CKP Sensor (+) Signal	435-475 Hz
22	OR	CKP Sensor (-) Signal	435-475 Hz
23	YL/BL	TSS Sensor (-)	340-380 Hz
24, 51	BK/YL	Power Ground	<0.1v
25	RD/BL	Power Ground	<0.1v
26	GY/YL	Coil Driver 1 Control	5° dwell
27	LG/BL	Shift Solenoid 1 Control	1v, 55 mph: 12v
28-29	---	Not Used	---
30	DG/WT	Octane Adjust Switch	Closed: 0v, Open: 9.3v
31, 34	---	Not Used	---
32	RD/GN	TR Sensor Reverse	In 'R': 12v, Others: 0v
33	BL	VSS (-) Signal	0 Hz, 55 mph: 125 Hz
35	GY	HO2S-12 (B1 S2) Signal	0.1-1.1v
36	OR/BL	MAF Sensor Return	<0.050v
37	WT	TFT Sensor Signal	2.10-2.40v
38	BL/WT	ECT Sensor Signal	0.5-0.6v
39	WT/GN	IAT Sensor Signal	1.5-2.5v
40	BK/PK	Fuel Pump Monitor	On: 12v, Off: 0v
41	GN/RD	A/C Cycling Clutch Switch	A/C On: 12v, Off: 0v
42-44, 46	---	Not Used	---
45	PK/WT	Low Speed Fan Control	On: 1v, Off: 12v
47	PK/RD	EGR VR Solenoid	0%, 55 mph: 45%
48	LG/RD	Clean Tachometer Output	25-38 Hz
49-50	---	Not Used	---
52	BR/RD	Coil Driver 2 Control	5° dwell
53	OR/GN	Shift Solenoid 3 Control	12v, 55 mph: 1v
54	GY/WT	TCC Solenoid Control	0%, 55 mph: 50%
55	BL/RD	Keep Alive Power	12-14v
56-57	---	Not Used	---
58	WT/BK	VSS (+) Signal	0 Hz, 55 mph: 125 Hz
59	---	Not Used	---
60	GY/DG	HO2S-11 (B1 S1) Signal	0.1-1.1v
61-63	---	Not Used	---
64	RD	A/T: Neutral Drive Switch	In 'N': 0v, in Gear: 12v
64	BR/YL	M/T: Clutch Pedal Position	Clutch In: 0v, Out: 5v

1996 Escort 1.9L I4 MFI VIN J (All) 104 Pin Connector

PCM Pin #	Wire Color	Circuit Description (104 Pin)	Value at Hot Idle
65	BL/YL	DPFE Sensor Signal	0.20-1.30v
66	---	Not Used	---
67	OR/WT	EVAP Purge Solenoid	0-100%
68	---	Not Used	---
69	BL/BK	A/C WOT Relay Control	On: 1v, Off: 12v
70	---	Not Used	---
71	WT/RD	Vehicle Power	12-14v
72-73	---	Not Used	---
74	BR/WT	Injector 3 Control	4.0-4.5 ms
75	YL	Injector 1 Control	4.0-4.5 ms
76	BL	CID- Sensor	5-7 Hz
77	BK/YL	Power Ground	<0.1v
78-79	---	Not Used	---
80	LG	Fuel Pump Control	On: 1v, Off: 12v
81-82	---	Not Used	---
83	BL/OR	IAC Motor Control	10v (30% on time)
84	WT/BL	TSS Sensor (+) Signal	340-380 Hz
85	DG	CID+ Sensor	5-7 Hz
86	DGN/BK	A/C Pressure Sensor Signal	A/C On: 1.5-1.9v
87	---	Not Used	---
88	BR/BK	MAF Sensor Signal	0.6-0.9v
89	RD/WT	TP Sensor Signal	0.53-1.27v
90	LG/WT	Reference Voltage	4.9-5.1v
91	LG/BK	Analog Signal Return	<0.050v
92	DG	Brake Pedal Position Switch	Brake Off: 0v, On: 12v
93	OR/BK	HO2S-11 (B1 S1) Heater	On: 1v, Off: 12v
94, 96	---	Not Used	---
95	PK	HO2S-12 (B1 S2) Heater	On: 1v, Off: 12v
97	WT/RD	Vehicle Power	12-14v
98-99	---	Not Used	---
100	DG/OR	Injector 4 Control	4.0-4.5 ms
101	PK/YL	Injector 2 Control	4.0-4.5 ms
102	---	Not Used	---
103	BK/YL	Power Ground	<0.1v
104	---	Not Used	---

Pin Connector Graphic

PCM 104-PIN CONNECTOR

Terminal View of 104-Pin PCM Wiring Harness Connector

Standard Colors and Abbreviations

Abbreviation	Color	Abbreviation	Color	Abbreviation	Color
BK	Black	GY	Gray	PK	Purple
BL	Blue	GN	Green	RD	Red
BR	Brown	LG	LT Green	TN	Tan
DB	Dark Blue	OR	Orange	WT	White
DG	DK Green	PK	Pink	YL	Yellow

1997 Escort 2.0L I4 MFI VIN P (All) 104 Pin Connector

PCM Pin #	Wire Color	Circuit Description (104 Pin)	Value at Hot Idle
1	PK/BK	Shift Solenoid 2 Control	12v, 55 mph: 1v
2	YL/BK	MIL (lamp) Control	MIL On: 1v, Off: 12v
3-5	---	Not Used	---
6	BR/BK	TR Sensor Overdrive	In 'P': 0v, in O/D: 11v
7	BR/GN	TR Sensor Low	In Manual 1st: 12v
8	PK/BL	IMRC Monitor Signal	5v
9	BR/YL	TR Sensor Drive	In Drive: 0v
10	---	Not Used	---
11	BL/PK	EVAP Purge Flow Sensor	0.8v, 55 mph: 3.0v
12, 14	---	Not Used	---
13	PK/GN	Flash EEPROM Power	0.1v
15	BK/BL	Data Bus (-) Signal	Digital Signals
16	WT/BL	Data Bus (+) Signal	Digital Signals
17	RD/BK	High Speed Fan Control	On: 1v, Off: 12v
18-20		Not Used	
21	BK	CKP Sensor (+) Signal	400-425 Hz
22	WT	CKP Sensor (-) Signal	400-425 Hz
23	RD	TSS Sensor (-)	340-380 Hz
24, 51	BK/YL	Power Ground	<0.1v
25	RD/BL	Shield Ground	<0.050v
26	GN/YL	Coil Driver 1 Control	5° dwell
27	LG/BL	Shift Solenoid 1 Control	1v, 55 mph: 12v
28-29	---	Not Used	---
30	GN/WT	Octane Adjust Switch	Closed: 0v, Open: 9.3v
31	YL/RD	PSP Switch Signal	Straight: 0.6v, turned: 2v
32	RD/GN	TR Sensor Reverse	In Reverse: 12v
33	BL	VSS (-) Signal	0 Hz, 55 mph: 125 Hz
34	---	Not Used	---
35	GY	HO2S-12 (B1 S2) Signal	0.1-1.1v
36	OR/BL	MAF Sensor Return	<0.050v
37	OR	TFT Sensor Signal	2.10-2.40v
38	BL/WT	ECT Sensor Signal	0.5-0.6v
39	WT/GN	IAT Sensor Signal	1.5-2.5v
40	BK/PK	Fuel Pump Monitor	On: 12v, Off: 0v
41	GN/RD	A/C Cycling Clutch Switch	A/C On: 12v, Off: 0v
42	RD/YL	IMRC Solenoid Control	12v (off)
43-44	---	Not Used	---
45	PK/WT	Low Speed Fan Control	On: 1v, Off: 12v
46	---	Not Used	---
47	PK/RD	EGR VR Solenoid	0%, 55 mph: 45%
48	LG/RD	Clean Tachometer Output	25-38 Hz
49-50	---	Not Used	---
52	BR/RD	Coil Driver 2 Control	5° dwell
53	GY/YL	Shift Solenoid 3 Control	6.7v, 55 mph: 6.7v
54	GY/WT	TCC Solenoid Control	0%, 55 mph: 95%
55	BL/RD	Keep Alive Power	12-14v
56-57	---	Not Used	---
58	WT/BK	VSS (+) Signal	0 Hz, 55 mph: 125 Hz
60	GY/DG	HO2S-11 (B1 S1) Signal	0.1-1.1v
61-63	---	Not Used	---
64	RD	TR Sensor Signal	In 'P': 0v, 55 mph: 1.7v
64	BR/YL	Clutch Pedal Position Switch	Clutch In: 0v, Out: 5v

1997 Escort 2.0L I4 MFI VIN P (All) 104 Pin Connector

PCM Pin #	Wire Color	Circuit Description (104 Pin)	Value at Hot Idle
65	BL/YL	DPFE Sensor Signal	0.20-1.30v
66	---	Not Used	---
67	OR/WT	EVAP Purge Solenoid	0%, 55 mph: 50%
68	---	Not Used	---
69	BL/BK	A/C WOT Relay Control	On: 1v, Off: 12v
70	---	Not Used	---
71	WT/RD	Vehicle Power	12-14v
72-73	---	Not Used	---
74	BR/WT	Injector 3 Control	3.3-3.7 ms
75	YL	Injector 1 Control	3.3-3.7 ms
76	BL	CID- Sensor	5-7 Hz
77	BK/YL	Power Ground	<0.1v
78-79	---	Not Used	---
80	LG	Fuel Pump Control	On: 1v, Off: 12v
81-82	---	Not Used	---
83	BL/OR	IAC Motor Control	10v (30% on time)
84	WT	TSS Sensor (+) Signal	340-380 Hz
85	GN	CID+ Sensor	5-9 Hz
86	GN/BK	A/C Pressure Sensor Signal	A/C On: 1.2-1.5v
87	---	Not Used	---
88	BR/BK	MAF Sensor Signal	0.6-0.9v
89	RD/WT	TP Sensor Signal	0.53-1.27v
90	LG/WT	Reference Voltage	4.9-5.1v
91	LG/BK	Analog Signal Return	<0.050v
92	DG	Brake Pedal Position Switch	Brake Off: 0v, On: 12v
93	OR/BK	HO2S-11 (B1 S1) Heater	On: 1v, Off: 12v
94	---	Not Used	---
95	PK	HO2S-12 (B1 S2) Heater	On: 1v, Off: 12v
96	---	Not Used	---
97	WT/RD	Vehicle Power	12-14v
98-99	---	Not Used	---
100	GN/OR	Injector 4 Control	3.3-3.7 ms
101	PK/YL	Injector 2 Control	3.3-3.7 ms
102	---	Not Used	---
103	BK/YL	Power Ground	<0.1v
104	---	Not Used	---

Pin Connector Graphic

```
            PCM 104-PIN CONNECTOR
 1 ●●●●●●●●●●●●●      ●●●●●●●●●●●●● 26
27 ●●●●●●●●●●●●●      ●●●●●●●●●●●●● 52
53 ●●●●●●●●●●●●●  ⬤  ●●●●●●●●●●●●● 78
79 ●●●●●●●●●●●●●      ●●●●●●●●●●●●● 104
```

Terminal View of 104-Pin PCM Wiring Harness Connector

Standard Colors and Abbreviations

Abbreviation	Color	Abbreviation	Color	Abbreviation	Color
BK	Black	GY	Gray	PK	Purple
BL	Blue	GN	Green	RD	Red
BR	Brown	LG	LT Green	TN	Tan
DB	Dark Blue	OR	Orange	WT	White
DG	DK Green	PK	Pink	YL	Yellow

1998-99 Escort 2.0L I4 MFI VIN P (All) 104 Pin Connector

PCM Pin #	Wire Color	Circuit Description (104 Pin)	Value at Hot Idle
1	PK/BK	Shift Solenoid 'B' Control	12v, 55 mph: 1v
2	YL/BK	MIL (lamp) Control	MIL On: 1v, Off: 12v
3-5	---	Not Used	---
6	BR/BL	TR Overdrive Switch	In O/D: 12v
7	BR/GN	TR Low Position Switch	In Manual 1st: 12v
8	PK/LB	IMRC Monitor Signal	5v
9	LG/BL	TR Drive Position Switch	In Drive: 0v
9-10	---	Not Used	---
11	BL/PK	EVAP Purge Flow Sensor	0.8v, 55 mph: 3.0v
12	WT	Fuel Level Indicator Signal	1.7v (1/2 full)
13	PK/GN	Flash EEPROM Power	0.1v
14	---	Not Used	---
15	BK/BL	Data Bus (-) Signal	Digital Signals
16	WT/BL	Data Bus (+) Signal	Digital Signals
17	RD/BK	High Speed Fan Control	On: 1v, Off: 12v
18-20	---	Not Used	---
21	BK	CKP Sensor (+) Signal	400-425 Hz
22	WT	CKP Sensor (-) Signal	400-425 Hz
23	RD	TSS Sensor (-)	340-380 Hz
24	BK/YL	Power Ground	<0.1v
25	RD/BL	Shield Ground	<0.050v
26	GN/YL	Coil 'A' Driver Control	5° dwell
27	LG/BL	Shift Solenoid 'A' Control	1v, 55 mph: 12v
28-29, 34	---	Not Used	---
30	GN/WT	Octane Adjust Switch	Closed: 0v, Open: 9.3v
31	YL/RD	PSP Switch Signal	Straight: 0.6v, turned: 2v
32	RD/GN	TR Reverse Position Switch	In Reverse: 12v
33	BL	VSS (-) Signal	0 Hz, 55 mph: 125 Hz
35	GY	HO2S-12 (B1 S2) Signal	0.1-1.1v
36	OR/BL	MAF Sensor Return	<0.050v
37	OR	TFT Sensor Signal	2.10-2.40v
38	BL/WT	ECT Sensor Signal	0.5-0.6v
39	WT/GN	IAT Sensor Signal	1.5-2.5v
40	BK/RD	Fuel Pump Monitor	On: 12v, Off: 0v
41	WT	A/C Cycling Clutch Switch	A/C On: 12v, Off: 0v
42	RD/YL	IMRC Solenoid Control	12v (off)
43	---	Not Used	---
45	PK/WT	Low Speed Fan Control	On: 1v, Off: 12v
46	---	Not Used	---
47	PK/RD	EGR VR Solenoid	0%, 55 mph: 45%
48	LG/RD	Clean Tachometer Output	25-38 Hz
49-50	---	Not Used	---
51	BK/YL	Power Ground	<0.1v
52	BR/RD	Coil 'B' Driver Control	5° dwell
53	OR/GN	Shift Solenoid 'C' Control	6.7v, 55 mph: 6.7v
54	GY/WT	TCC Solenoid Control	0%, 55 mph: 95%
55	BL/RD	Keep Alive Power	12-14v
56	OR/WT	EVAP Purge Solenoid	0-10 Hz (0-100%)
57	YL/RD	Knock Sensor 1 Signal	0v
58	WT/BK	VSS (+) Signal	0 Hz, 55 mph: 125 Hz
60	GY/GN	HO2S-11 (B1 S1) Signal	0.1-1.1v
61	---	Not Used	---
62	BK/WT	FTP Sensor Signal	2.6v (0" H2O - cap off)
63	WT/PK	Fuel Rail Pressure Sensor	2.8v (39 psi)

1998-99 Escort 2.0L I4 MFI VIN P (All) 104 Pin Connector

PCM Pin #	Wire Color	Circuit Description (104 Pin)	Value at Hot Idle
64	RD	TR Sensor Signal	In 'P': 0v, 55 mph: 1.7v
64	BR/YL	M/T: Clutch Pedal Switch	Clutch In: 0v, Out: 5v
65	BL/YL	DPFE Sensor Signal	0.20-1.30v
66	---	Not Used	---
67	BL/PK	EVAP CV Solenoid	0-10 Hz (0-100%)
68	---	Not Used	---
69	BL/BK	A/C WOT Relay Control	On: 1v, Off: 12v
71	WT/RD	Vehicle Power	12-14v
72	PK	Shift Indicator Control	SIL On: 1v, Off: 12v
73	---	Not Used	---
74	BR/WT	Injector 3 Control	3.3-3.7 ms
75	YL	Injector 1 Control	3.3-3.7 ms
76	BL	CMP Sensor Ground	5-7 Hz
77	BK/YL	Power Ground	<0.1v
78, 81, 87	---	Not Used	---
79	YL/GN	Fuel Cap Off Indicator	Digital Signal
80	WT/OR	Fuel Pump Control	On: 1v, Off: 12v
82	YL/PK	TOT Sensor Signal	2.10-2.40v
83	BL/OR	IAC Motor Control	10v (30% on time)
84	WT	TSS Sensor (+) Signal	340-380 Hz
85	GN	CMP Sensor Signal	5-9 Hz
86	GN/BK	A/C High Pressure Switch	A/C On: 1.2-1.5v
88	BR/BK	MAF Sensor Signal	0.6-0.9v
89	RD/WT	TP Sensor Signal	0.53-1.27v
90	LG/WT	Reference Voltage	4.9-5.1v
91	LG/BK	Analog Signal Return	<0.050v
92	DG	Brake Pedal Position Switch	Brake Off: 0v, On: 12v
93	OR/BK	HO2S-11 (B1 S1) Heater	On: 1v, Off: 12v
94, 96	---	Not Used	---
95	PK	HO2S-12 (B1 S2) Heater	On: 1v, Off: 12v
97	WT/RD	Vehicle Power	12-14v
98-99	---	Not Used	---
100	GN/OR	Injector 4 Control	3.3-3.7 ms
101	PK/YL	Injector 2 Control	3.3-3.7 ms
102	---	Not Used	---
103	BK/YL	Power Ground	<0.1v
104	---	Not Used	---

Pin Connector Graphic

PCM 104-PIN CONNECTOR

Terminal View of 104-Pin PCM Wiring Harness Connector

Standard Colors and Abbreviations

Abbreviation	Color	Abbreviation	Color	Abbreviation	Color
BK	Black	GY	Gray	PK	Purple
BL	Blue	GN	Green	RD	Red
BR	Brown	LG	LT Green	TN	Tan
DB	Dark Blue	OR	Orange	WT	White
DG	DK Green	PK	Pink	YL	Yellow

2000-02 Escort 2.0L I4 MFI VIN P (All) 104 Pin Connector

PCM Pin #	Wire Color	Circuit Description (104 Pin)	Value at Hot Idle
1	PK/BK	Shift Solenoid 'B' Control	12v, 55 mph: 1v
2	YL/BK	MIL (lamp) Control	MIL On: 1v, Off: 12v
3-5	---	Not Used	---
6	BR/BL	TR Overdrive Switch	In O/D: 12v
7	BR/GN	TR Low Position Switch	In Manual 1st: 12v
8	PK/LB	IMRC Monitor Signal	5v
9	BR/YL	TR Drive Position Switch	In Drive: 0v
10-11	---	Not Used	---
12	YL/BR	Fuel Level Indicator Signal	1.7v (1/2 full)
13	PK/GN	Flash EEPROM Power	0.1v
14	---	Not Used	---
15	BK/BL	Data Bus (-) Signal	Digital Signals
16	WT/BL	Data Bus (+) Signal	Digital Signals
17	RD/BK	High Speed Fan Control	On: 1v, Off: 12v
18-20	---	Not Used	---
21	BK	CKP Sensor (+) Signal	400-425 Hz
22	WT	CKP Sensor (-) Signal	400-425 Hz
23	RD	TSS Sensor (-)	340-380 Hz
24	BK/YL	Power Ground	<0.1v
25	RD/BL	Shield Ground	<0.050v
26	GN/YL	Coil 'A' Driver Control	5° dwell
27	LG/BL	Shift Solenoid 'A' Control	1v, 55 mph: 12v
28-30, 34	---	Not Used	---
31	YL/RD	PSP Switch Signal	Straight: 0.6v, Turned: 2v
32	RD/GN	TR Reverse Position Switch	In Reverse: 12v
33	BL	VSS (-) Signal	0 Hz, 55 mph: 125 Hz
35	GY	HO2S-12 (B1 S2) Signal	0.1-1.1v
36	OR/BL	MAF Sensor Return	<0.050v
37	OR	TFT Sensor Signal	2.10-2.40v
38	BL/WT	ECT Sensor Signal	0.5-0.6v
39	WT/GN	IAT Sensor Signal	1.5-2.5v
40	BK/RD	Fuel Pump Monitor	On: 12v, Off: 0v
41	WT	A/C Cycling Clutch Switch	A/C On: 12v, Off: 0v
42	RD/YL	IMRC Solenoid Control	12v (off)
43, 46	---	Not Used	---
45	PK/WT	Low Speed Fan Control	On: 1v, Off: 12v
47	PK/RD	EGR VR Solenoid	0%, 55 mph: 45%
48	LG/RD	Clean Tachometer Output	25-38 Hz
49-50	---	Not Used	---
51	BK/YL	Power Ground	<0.1v
52	BR/RD	Coil 'B' Driver Control	5° dwell
53	OR/GN	Shift Solenoid 'C' Control	6.7v, 55 mph: 6.7v
54	GY/WT	TCC Solenoid Control	0%, 55 mph: 95%
55	BL/RD	Keep Alive Power	12-14v
56	OR/WT	EVAP Purge Solenoid	0-10 Hz (0-100%)
57	RD	Knock Sensor 1 Signal	0v
58	WT/BK	VSS (+) Signal	0 Hz, 55 mph: 125 Hz
60	GY/GN	HO2S-11 (B1 S1) Signal	0.1-1.1v
61	---	Not Used	---
62	BK/WT	FTP Sensor Signal	2.6v (0" H2O - cap off)
63	WT/PK	Injection Pressure Switch	2.8v (39 psi)
64	RD	TR Sensor Signal	In 'P': 0v, 55 mph: 1.7v
64	BR/YL	M/T: Clutch Pedal Switch	Clutch In: 0v, Out: 5v

2000-02 Escort 2.0L I4 MFI VIN P (All) 104 Pin Connector

PCM Pin #	Wire Color	Circuit Description (104 Pin)	Value at Hot Idle
65 ('00)	BL/YL	DPFE Sensor Signal	0.20-1.30v
65 ('01-'02)	BL/YL	DPFE Sensor Signal	0.95-1.05v
67	BL/PK	EVAP CV Solenoid	0-10 Hz (0-100%)
68, 73	---	Not Used	---
69	BL/BK	A/C WOT Relay Control	On: 1v, Off: 12v
71	WT/RD	Vehicle Power	12-14v
72	PK	Shift Indicator Control	On: 1v, Off: 12v
74	BR/WT	Injector 3 Control	3.3-3.7 ms
75	YL	Injector 1 Control	3.3-3.7 ms
76	BL	CMP Sensor Ground	5-7 Hz
77	BK/YL	Power Ground	<0.1v
78, 81, 87	---	Not Used	---
79	YL/GN	Fuel Cap Off Indicator	On: 1v, Off: 12v
80	WT/OR	Fuel Pump Control	On: 1v, Off: 12v
82	YL/PK	TOT Sensor Signal	2.10-2.40v
83	BL/OR	IAC Motor Control	10v (30% on time)
84	WT	TSS Sensor (+) Signal	340-380 Hz
85	GN	CMP Sensor Signal	5-9 Hz
86	GN/BK	A/C High Pressure Switch	A/C On: 1.2-1.5v
88	BR/BK	MAF Sensor Signal	0.6-0.9v
89	RD/WT	TP Sensor Signal	0.53-1.27v
90	LG/WT	Reference Voltage	4.9-5.1v
91	LG/BK	Analog Signal Return	<0.050v
92	DG	Brake Pedal Position Switch	Brake Off: 0v, On: 12v
93	OR/BK	HO2S-11 (B1 S1) Heater	On: 1v, Off: 12v
94, 96	---	Not Used	---
95	PK	HO2S-12 (B1 S2) Heater	On: 1v, Off: 12v
97	WT/RD	Vehicle Power	12-14v
98-99	---	Not Used	---
100	GN/OR	Injector 4 Control	3.3-3.7 ms
101	PK/YL	Injector 2 Control	3.3-3.7 ms
103	BK/YL	Power Ground	<0.1v
102	---	Not Used	---
104	---	Not Used	---

Pin Connector Graphic

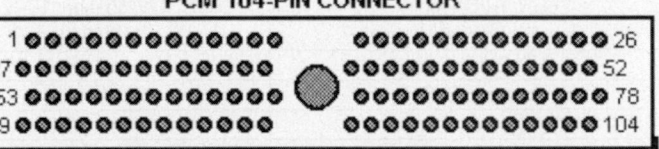

Terminal View of 104-Pin PCM Wiring Harness Connector

Standard Colors and Abbreviations

Abbreviation	Color	Abbreviation	Color	Abbreviation	Color
BK	Black	GY	Gray	PK	Purple
BL	Blue	GN	Green	RD	Red
BR	Brown	LG	LT Green	TN	Tan
DB	Dark Blue	OR	Orange	WT	White
DG	DK Green	PK	Pink	YL	Yellow

1998-99 Escort 2.0L I4 MFI VIN 3 (All) 104 Pin Connector

PCM Pin #	Wire Color	Circuit Description (104 Pin)	Value at Hot Idle
1	PK/BK	Shift Solenoid 2 or 'B' Control	1v, Others: 1v
2	YL/BL	MIL (lamp) Control	MIL On: 1v, Off: 12v
3-5, 8	---	Not Used	---
6	BR/BL	TR O/D Position Signal	In 'P': 1v, In O/D: 12v
7	BR/GN	TR Low Position Signal	In 'L': 12v, Others: 1v
9	BR/YL	TR Drive Position Signal	In 'D': 12v, Others: 1v
10-11, 14	---	Not Used	---
12	YL/BL	Fuel Level Indicator Signal	1.7v (1/2 full)
13	PK/GN	Flash EEPROM Power	0.1v
15	BK/BL	SCP Data Bus (-) Signal	Digital Signals
16	WT/BL	SCP Data Bus (+) Signal	Digital Signals
17	RD/GN	High Speed Fan Control	On: 1v, Off: 12v
18-20	---	Not Used	---
21	BK	CKP Sensor (+) Signal	450-480 Hz
22	WT	CKP Sensor (-) Signal	450-480 Hz
23	RD	TSS Sensor (-)	340-380 Hz (680-710 rpm)
24, 51	BK/YL	Power Ground	<0.1v
25	RD/BL	Shield Ground	<0.050v
26	GN/YL	Coil 'A' Driver Control	5° dwell
27	LG/BL	Shift Solenoid 'A' Control	1v, Others: 12v
28, 34	---	Not Used	---
29	PK/BK	TCS (switch) Signal	TCS & O/D On: 12v
30	WT/BK	Octane Adjust Switch	Closed: 0v, Open: 9.3v
31	YL/RD	PSP Switch Signal	Straight: 1v, turned: 12v
32	RD/GN	TR Reverse Position Signal	In 'R': 12v, Others: 1v
33	BL	VSS (-) Signal	0 Hz, 55 mph: 125 Hz
35	GY	HO2S-12 (B1 S2) Signal	0.1-1.1v
36	OR/BL	MAF Sensor Return	<0.1v
37	OR/GN	TFT Sensor Signal	2.10-2.40v
38	BL/WT	ECT Sensor Signal	0.5-0.6v
39	WT/GN	IAT Sensor Signal	1.5-2.5v
40	BK/RD	Fuel Pump Monitor	On: 12v, Off: 0v
41	WT	A/C Cycling Clutch Switch	Closed: 12v, Open: 0v
44	PK/GN	Variable Valve Timing Control	12v (Off)
45	PK/WT	Low Speed Fan Control	On: 1v, Off: 12v
46, 49, 51	---	Not Used	---
47	BK/GN	EGR VR Solenoid	12v, 55 mph: 10-12v
48	LG/RD	Clean Tachometer Output	20-31 Hz
52	BR/RD	Coil 'B' Driver Control	5° dwell
53	OR/GN	Shift Solenoid 'C' Control	12v, Others: 12v
54	GY/WT	TCC Solenoid Control	12v (0%)
55	BL/RD	Keep Alive Power	12-14v
56	OR/WT	EVAP Purge Solenoid	0-10 Hz (0-100%)
57	RD	Knock Sensor 1 Signal	0v
58	WT/BK	VSS (+) Signal	0 Hz, 55 mph: 125 Hz
59	---	Not Used	---
60	GY/GN	HO2S-11 (B1 S1) Signal	0.1-1.1v
61	---	Not Used	---
62	BK/WT	FTP Sensor Signal	2.6v (cap off)
63	WT/PK	FRP Sensor Signal	2.8v (39 psi)
64	RD	A/T: TR Sensor Signal	In 'P': 0v, 55 mph: 1.7v
64	BR/YL	M/T: Clutch Pedal Position	Clutch In: 1v, Out: 5v

1998-99 Escort 2.0L I4 MFI VIN 3 (All) 104 Pin Connector

PCM Pin #	Wire Color	Circuit Description (104 Pin)	Value at Hot Idle
65	WT/BL	DPFE Sensor Signal	0.20-1.30v
66	---	Not Used	---
67	BL/PK	EVAP CV Solenoid	0-10 Hz (0-100%)
69	BL/BK	A/C WOT Relay Control	On: 1v, Off: 12v
70	BK/YL	VCT Control (ZETECT)	VCT Off: 12v, On: 1v
71	WT/RD	Vehicle Power	12-14v
72	PK	Shift Indicator Control	On: 1v, Off: 12v
73, 78	---	Not Used	---
74	BK/WT	Injector 3 Control	2.3-2.9 ms
75	YL	Injector 1 Control	2.3-2.9 ms
76	LB	CMP Sensor Ground	<0.050v
77	BK/YL	Power Ground	<0.1v
79	WT/BL	TCIL (lamp) Control	On: 1v, Off: 12v
80	WT/OR	Fuel Pump Control	On: 1v, Off: 12v
81, 87	---	Not Used	---
82	YL/PK	EPC Solenoid Control	2.6v (72%)
83	WT/LB	IAC Motor Control	9v (33% duty cycle)
84	WT	TSS Sensor (+) Signal	340-380 Hz (680-710 rpm)
85	GN	CMP Sensor Signal	5-7 Hz
86	GN/BK	A/C Pressure Switch	A/C On: 12v (Open)
88	BR/BK	MAF Sensor Signal	0.6-0.9v
89	RD/WT	TP Sensor Signal	0.53-1.27v
90	LG/WT	Reference Voltage	4.9-5.1v
91	LG/BK	Analog Signal Return	<0.050v
92	GN	Brake Pedal Position Switch	Brake Off: 0v, On: 12v
93	OR/BK	HO2S-11 (B1 S1) Heater	On: 1v, Off: 12v
95	PK	HO2S-12 (B1 S2) Heater	On: 1v, Off: 12v
96	---	Not Used	---
97	WT/RD	Vehicle Power	12-14v
98-99	---	Not Used	---
100	GN/OR	Injector 4 Control	2.3-2.9 ms
101	PK/YL	Injector 2 Control	2.3-2.9 ms
102	BK/OR	Shift Solenoid 'D' Control	8v, 55 mph: 8-9v
103	BK/YL	Power Ground	<0.1v
104	---	Not Used	---

Pin Connector Graphic

PCM 104-PIN CONNECTOR

Terminal View of 104-Pin PCM Wiring Harness Connector

Standard Colors and Abbreviations

Abbreviation	Color	Abbreviation	Color	Abbreviation	Color
BK	Black	GY	Gray	PK	Purple
BL	Blue	GN	Green	RD	Red
BR	Brown	LG	LT Green	TN	Tan
DB	Dark Blue	OR	Orange	WT	White
DG	DK Green	PK	Pink	YL	Yellow

2000-02 Escort 2.0L I4 ZETEC-E VIN 3 (All) 104 Pin Connector

PCM Pin #	Wire Color	Circuit Description (104 Pin)	Value at Hot Idle
1	GN/BL	Shift Solenoid 'B' Control	1v, Others: 1v
2	---	Not Used	---
3	WT/VT	IMRC Solenoid Control	IMRC Off: 5v, On: 2.5v
4	GN/BL	TR Reverse Position Signal	In 'R': 12v, Others: 0v
5-6	---	Not Used	---
7	GN/WT	TR First Position Signal	In '1st': 12v, Others: 1v
8	GN/OR	TR Second Position Signal	In 2nd: 12v, Others: 1v
9-10	---	Not Used	---
11	GN/BK	TR OD Position Signal	In 'P': 1v, In O/D: 12v
12	BR/WT	Sensor Ground	<0.050v
13	OG/BK	Flash EEPROM Power	0.1v
14, 18	---	Not Used	---
15	BL/BK	SCP Bus (-) Signal	Digital Signals
16	GY/OR	SCP Bus (+) Signal	Digital Signals
17	BK/WT	High Speed Fan Control	On: 1v, Off: 12v
19	GY/OR	TX Signal	Digital Signals
21	BK	CKP Sensor (+) Signal	450-480 Hz
22	WT	CKP Sensor (-) Signal	450-480 Hz
23	RD	TSS Sensor (-)	340-380 Hz (680-710 rpm)
24	BK/YL	Power Ground	<0.1v
25	RD/BL	Power Ground	<0.1v
26	GN/YL	Coil 'A' Driver Control	5° dwell
27	LG/BL	Shift Solenoid 'A' Control	1v, Others: 12v
28-30	---	Not Used	---
31	YL/RD	PSP Switch Signal	Straight: 1v, Turned: 12v
32	RD/GN	TR Reverse Position Signal	In 'R': 12v, Others: 1v
33	BL	VSS (-) Signal	0 Hz, 55 mph: 125 Hz
34	---	Not Used	---
35	GY	HO2S-12 (B1 S2) Signal	0.1-1.1v
36	OR/BL	MAF Sensor Return	<0.1v
37	OR	TFT Sensor Signal	2.10-2.40v
38	BL/WT	ECT Sensor Signal	0.5-0.6v
39	WT/GN	IAT Sensor Signal	1.5-2.5v
40	BK/RD	Fuel Pump Monitor	On: 12v, Off: 0v
41	WT	A/C Cycling Clutch Switch	Closed: 12v, Open: 0v
42-43	---	Not Used	---
44	PK/GN	Variable Valve Timing Control	12v (Off)
45	PK/WT	Low Speed Fan Control	On: 1v, Off: 12v
46-47	---	Not Used	---
48	LG/RD	Clean Tachometer Output	20-31 Hz
49-50	---	Not Used	---
51	BK/YL	Power Ground	<0.1v
52	BR/RD	Coil 'B' Driver Control	5° dwell
53	OR/GN	Shift Solenoid 'C' Control	12v, Others: 12v
54	GY/WT	TCC Solenoid Control	12v (0%)
55	BK/RD	Keep Alive Power	12-14v
56	OR/WT	EVAP Purge Solenoid	0-10 Hz (0-100%)
57	RD	Knock Sensor 1 Signal	0v
58	WT/BK	VSS (+) Signal	0 Hz, 55 mph: 125 Hz
59	---	Not Used	---
60	GY/GN	HO2S-11 (B1 S1) Signal	0.1-1.1v
61	---	Not Used	---
62	BK/WT	FTP Sensor Signal	2.6v (cap off)
63	WT/VT	Injection Pressure Sensor	2.8v (39 psi)

2000-02 Escort 2.0L I4 MFI VIN 3 (All) 104 Pin Connector

PCM Pin #	Wire Color	Circuit Description (104 Pin)	Value at Hot Idle
64	RD	A/T: TR Sensor Signal	In 'P': 0v, 55 mph: 1.7v
64	BR/YL	M/T: Clutch Pedal Position	Clutch In: 1v, Out: 5v
65-66	---	Not Used	---
67	BL/PK	EVAP CV Solenoid	0-10 Hz (0-100%)
68	---	Not Used	---
69	BL/BK	A/C WOT Relay Control	On: 1v, Off: 12v
71	WT/RD	Vehicle Power	12-14v
70	---	Not Used	---
72	PK	Shift Indicator Control	On: 1v, Off: 12v
73	---	Not Used	---
74	BR/WT	Injector 3 Control	2.3-2.9 ms
75	YL	Injector 1 Control	2.3-2.9 ms
76	BL	CMP Sensor Ground	<0.050v
77	BK/YL	Power Ground	<0.1v
78	---	Not Used	---
79	YL/GN	Fuel Cap Off Indicator	Off: 12v, On: 1v
80	WT/OR	Fuel Pump Control	Off: 12v, On: 1v
81, 87	---	Not Used	---
82	YL/VT	EPC Solenoid Control	2.6v (72%)
83	BL/OR	IAC Motor Control	9v (33% duty cycle)
84	WT	TSS Sensor (+) Signal	340-380 Hz (680-710 rpm)
85	GN	CMP Sensor Signal	5-7 Hz
86	GN/BK	A/C Pressure Switch	A/C On: 12v (Open)
88	BR/BK	MAF Sensor Signal	0.6-0.9v
89	RD/WT	TP Sensor Signal	0.53-1.27v
90	LG/WT	Reference Voltage	4.9-5.1v
91	LG/BK	Analog Signal Return	<0.050v
92	GN	Brake Pedal Position Switch	Brake Off: 0v, On: 12v
93	OR/BK	HO2S-11 (B1 S1) Heater	On: 1v, Off: 12v
95	VT	HO2S-12 (B1 S2) Heater	On: 1v, Off: 12v
96	---	Not Used	---
97	WT/RD	Vehicle Power	12-14v
98-99	---	Not Used	---
100	GN/OR	Injector 4 Control	2.3-2.9 ms
101	VT/YL	Injector 2 Control	2.3-2.9 ms
102	---	Not Used	---
103	BK/YL	Power Ground	<0.1v
104	---	Not Used	---

Pin Connector Graphic

```
          PCM 104-PIN CONNECTOR
      1 ●●●●●●●●●●●●●    ●●●●●●●●●●●●● 26
     27 ●●●●●●●●●●●●●   ⬤ ●●●●●●●●●●●●● 52
     53 ●●●●●●●●●●●●●      ●●●●●●●●●●●●● 78
     79 ●●●●●●●●●●●●●    ●●●●●●●●●●●●● 104
```

Terminal View of 104-Pin PCM Wiring Harness Connector

Standard Colors and Abbreviations

Abbreviation	Color	Abbreviation	Color	Abbreviation	Color
BK	Black	GY	Gray	PK	Purple
BL	Blue	GN	Green	RD	Red
BR	Brown	LG	LT Green	TN	Tan
DB	Dark Blue	OR	Orange	WT	White
DG	DK Green	PK	Pink	YL	Yellow

FOCUS PIN TABLES

2000-2002 Focus 2.0L I4 DURATEC VIN P (All) 104 Pin Connector

PCM Pin #	Wire Color	Circuit Description (104 Pin)	Value at Hot Idle
1	GN/BL	Shift Solenoid 'B' Control	1v, at Cruise: 1v
2, 5-6	---	Not Used	---
3	WT/RD	IMRC Solenoid Control	IMRC Off: 5v, On: 2.5v
4	GN/BL	TR Reverse Position Signal	In 'R': 12v, Others: 0v
7	GN/WT	TR 1st Position Signal	Manual 1: 12v, Others: 1v
8	GN/OR	TR Drive Position Signal	In 'D': 12v, Others: 1v
9-10, 12	---	Not Used	---
11	GN/BK	TR Overdrive Position Signal	In O/D: 12v, Others: 1v
13	OR/BK	Flash EEPROM Power	0.1v
14, 18, 23	---	Not Used	---
15	BL/BK	Data Bus (-) Signal	Digital Signals
16	GY/OR	Data Bus (+) Signal	Digital Signals
17	BK/WT	High Speed Fan Control	On: 1v, Off: 12v
19	GY/OR	Passive Antitheft System	Digital Signals
20	BK/BL	Injector 3 Control	3.3-3.7 ms
21	WT/RD	CKP Sensor (+) Signal	400-425 Hz
22	BR/RD	CKP Sensor (-) Signal	400-425 Hz
24	BK/YL	Power Ground	<0.1v
25	BK	Shield Ground	<0.050v
26	BK/OR	Coil 'A' Driver Control	5° dwell
27	BK/RD	Vehicle Start Signal	KOEC: 9-11v
28	WT/VT	M/T: VSS (+) Signal	0 Hz, 55 mph: 125 Hz
29	GN/BK	TR Overdrive OFF Switch	O/D Off: 0v
30, 32	---	Not Used	---
31	BK/YL	PSP Switch Signal	Straight: 0.6v, turned: 2v
33	BR/BL	A/T: OSS (-) Sensor Signal	0 Hz, 55 mph: 120 Hz
34	WT/VT	A/T: TSS Sensor (+) Signal	340-380 Hz
35	WT/RD	HO2S-22 (B2 S2) Signal	0.1-1.1v
36	BR/BL	MAF Sensor Return	<0.050v
37	WT/GN	TFT Sensor Signal	2.10-2.40v
38	WT	ECT Sensor Signal	0.5-0.6v
39	WT/VT	IAT Sensor Signal	1.5-2.5v
40	GN/BK	Fuel Pump Driver Module	3.5v (100% on time)
41	GN/YL	A/C Cycling Clutch Switch	Closed: 12v, Open: 0v
42	BK/OR	Passive Antitheft System	Digital Signals
43, 45-46	---	Not Used	---
44	GN/OR	EPC Switch Signal	12v
47	BK/GN	EGR VR Solenoid	0%, 55 mph: 45%
48-50	---	Not Used	---
51	BK/YL	Power Ground	<0.1v
52	BK/GN	Coil 'B' Driver Control	5° dwell
53	WT/GN	Passive Antitheft System	Digital Signals
54	BK/WT	Fuel Pump Driver Module	1.5v (33% on time)
55	RD	Keep Alive Power	12-14v
56	BK/BL	EVAP Purge Solenoid	0-10 Hz (0-100%)
57	WT/BK	Knock Sensor 1 (-) Signal	0v
58	WT/BL	A/T: OSS (+) Sensor Signal	0 Hz, 55 mph: 120 Hz
59	GY	Generator Field Sense	130 Hz (30% on time)
60	WT	HO2S-21 (B2 S1) Signal	0.1-1.1v
61	---	Not Used	---
62	WT/VT	FTP Sensor Signal	2.6v (0" H2O - cap off)
63	WT/GN	FRP Sensor Signal	2.8v (39 psi)
64	GN/YL	A/T: Neutral Start Position	In 'N': 0v, Others: 5v
64	WT	M/T: Clutch Pedal Switch	Clutch In: 0v, Out: 5v

2000-2002 Focus 2.0L I4 DURATEC VIN P (All) 104 Pin Connector

PCM Pin #	Wire Color	Circuit Description (104 Pin)	Value at Hot Idle
65 ('00)	WT/BL	DPFE Sensor Signal	0.20-1.30v
65 ('01-'02)	WT/BL	DPFE Sensor Signal	0.95-1.05v
66	---	Not Used	---
67	BK/OR	EVAP CV Solenoid	0-10 Hz (0-100%)
68	BK/BL	Low Speed Fan Control	On: 1v, Off: 12v
69	BK/YL	A/C WOT Relay Control	On: 1v, Off: 12v
70	BK/WT	Injector 1 Control	3.3-3.7 ms
71, 97	GN/YL	Vehicle Power	12-14v
72	BL	Generator Field Control	0 Hz (0% on time)
73	GN/YL	Shift Solenoid 'A' Control	0v, 55 mph: 12v
74-75	---	Not Used	---
76	BR/WT	M/T: CMP Sensor Ground	5-7 Hz
76	BR/WT	A/T: TSS Sensor (-) Signal	340-380 Hz
77	BK/YL	Power Ground	<1v
78-79	---	Not Used	---
80	BK/WT	IMRC Solenoid Control	12v (off)
81	BK/RD	EPC Solenoid Control	2.5v (73% on time)
82	GN/BK	Shift Solenoid 'C' Control	0v, 55 mph: 4.10v
83	BK/YL	IAC Motor Control	10v (30% on time)
84, 98, 94	---	Not Used	---
85	WT/VT	CMP Sensor Signal	5-9 Hz
86	WT/RD	A/C Pressure Switch	A/C On: 12v (Open)
87	BR/YL	Knock Sensor 1 Signal	0v
88	WT/BL	MAF Sensor Signal	0.6-0.9v
89	WT	TP Sensor Signal	0.53-1.27v
90	YL	Reference Voltage	4.9-5.1v
91	BR	Analog Signal Return	<0.050v
92	GN/RD	Brake Pedal Position Switch	Brake Off: 0v, On: 12v
93	BK/YL	HO2S-21 (B2 S1) Heater	On: 1v, Off: 12v
95	BK/OR	Injector 4 Control	3.3-3.7 ms
96	BK/YL	Injector 2 Control	3.3-3.7 ms
99	GN/OR	Shift Solenoid 'D' Control	0v, 55 mph: 0v
100	BK/BL	HO2S-22 (B2 S2) Heater	On: 1v, Off: 12v
101	---	Not Used	---
102	GN/WT	Shift Solenoid 'E' Control	0v, 55 mph: 0v
103	BK/YL	Power Ground	<1v
104	---	Not Used	---

Pin Connector Graphic

```
PCM 104-PIN CONNECTOR

 1 ⊗⊗⊗⊗⊗⊗⊗⊗⊗⊗⊗⊗⊗   ⊗⊗⊗⊗⊗⊗⊗⊗⊗⊗⊗⊗⊗ 26
27 ⊗⊗⊗⊗⊗⊗⊗⊗⊗⊗⊗⊗⊗   ⊗⊗⊗⊗⊗⊗⊗⊗⊗⊗⊗⊗⊗ 52
53 ⊗⊗⊗⊗⊗⊗⊗⊗⊗⊗⊗⊗⊗ ⬤ ⊗⊗⊗⊗⊗⊗⊗⊗⊗⊗⊗⊗⊗ 78
79 ⊗⊗⊗⊗⊗⊗⊗⊗⊗⊗⊗⊗⊗   ⊗⊗⊗⊗⊗⊗⊗⊗⊗⊗⊗⊗⊗ 104
```

Terminal View of 104-Pin PCM Wiring Harness Connector

Standard Colors and Abbreviations

Abbreviation	Color	Abbreviation	Color	Abbreviation	Color
BK	Black	GY	Gray	PK	Purple
BL	Blue	GN	Green	RD	Red
BR	Brown	LG	LT Green	TN	Tan
DB	Dark Blue	OR	Orange	WT	White
DG	DK Green	PK	Pink	YL	Yellow

2003-05 Focus 2.0L I4 DURATEC VIN P (All) 104 Pin Connector

PCM Pin #	Wire Color	Circuit Description (104 Pin)	Value at Hot Idle
1	GN/BL	Shift Solenoid 'B' Control	In 2nd: 1v, Others: 1v
2	---	Not Used	---
3	WT/VT	IMRC Solenoid Monitor	5v
4	GN/BL	TR Reverse Position Signal	In 'R': 12v, Others: 0v
5-6	---	Not Used	---
7	GN/WT	TR First Position Signal	In 1st: 12v, Others: 1v
8	GN/OR	TR Second Position Signal	In 2nd: 12v, Others: 1v
9	WT/BK	Barometric Pressure Sensor	159 Hz (sea level)
10	---	Not Used	---
11	GN/BK	TR OD Position Signal	In 'P': 1v, In O/D: 12v
12	BR/WT	Sensor Ground	<0.050v
13	OG/BK	Flash EEPROM Power	0.1v
14, 18, 30	---	Not Used	---
15	BL/BK	SCP Bus (-) Signal	Digital Signals
16	GY/OR	SCP Bus (+) Signal	Digital Signals
17	BK/WT	High Speed Fan Control	On: 1v, Off: 12v
19	GY/OR	Passive Antitheft TX Signal	Digital Signals
20	BK/BL	Fuel Injector 3 Control	2.3-2.9 ms
21	WT/RD	CKP Sensor (+) Signal	450-480 Hz
22	BR/RD	CKP Sensor (-) Signal	450-480 Hz
23	BK/YL	Power Ground	<0.1v
24	BK/YL	Power Ground	<0.1v
25	BK	Power Ground	<0.1v
26	BK/OR	Coil 'A' Driver Control	5° dwell
27	BK/RD	Start Inhibit Relay Control	Relay Off: 12v, On: 1v
28	WT/VT	VSS (-) Signal	0 Hz, 55 mph: 125 Hz
29	GN/BK	Overdrive Cancel Switch	TCS & O/D On: 12v
31	WT/VT	PSP Switch Signal	Straight: 1v, Turned: 5v
32	BN/YL	Sensor Ground	<0.050v
33	BR/BL	Output Speed Sensor (-) Signal	Moving: AC pulse signals
34	WT/VT	Turbine Speed Sensor (+) Signal	340-380 Hz
35	WT/RD	HO2S-12 (B1 S2) Signal	0.1-1.1v
36	BR/BL	MAF Sensor Return	<0.1v
37	WT/GN	TFT Sensor Signal	2.10-2.40v
38	WT/VT	Cylinder Head Temperature Sensor	0.5-0.6v
39	WT/VT	MAF Sensor Signal	0.6-0.9v
40	GN/BK	Fuel Pump Monitor	On: 12v, Off: 0v
41	GN/YL	A/C Cycling Clutch Pressure Switch	Closed: 12v, Open: 0v
42	BK/OR	PATS Indicator Control	Indicator Off: 12v, On: 1v
43, 48	---	Not Used	---
44	GN/OR	Pressure Control Solenoid 'A'	Off: 12v, On: 1v
45	BK/RD	Camshaft Timing Adjuster Control	Off: 12v, On: 1v
46	BK/WT	Low Fuel Indicator	Digital Signals
47	BK/GN	Electric Vacuum Regulator Control	Off: 12v, On: 1v
49	BL/RD	Controller Area Network Bus 'L'	0-7-0v
50	GY/RD	Controller Area Network Bus 'H'	0-7-0v
51	BK/YL	Power Ground	<0.1v
52	BK/GN	Coil 'B' Driver Control	5° dwell
53	WT/GN	Passive Antitheft RX Signal	Digital Signals
54	BK/WT	Fuel Pump Control	Pump Off: 12v, On: 1v
55	RD	Keep Alive Power	12-14v
56	BK/BL	EVAP Purge Solenoid	0-10 Hz (0-100%)
57	WT/BK	Knock Sensor Signal	0v
58	WT/BL	VSS (+) Signal	Moving: AC pulse signals

2003-05 Focus 2.0L I4 DURATEC VIN P (All) 104 Pin Connector

PCM Pin #	Wire Color	Circuit Description (104 Pin)	Value at Hot Idle
59	GY	Generator Monitor Signal	130 Hz (45%)
60	WT	HO2S-11 (B1 S1) Signal	0.1-1.1v
61, 66	---	Not Used	---
62	WT/VT	FTP Sensor Signal	2.6v (cap off)
63	WT/GN	Fuel Rail Pressure Sensor	2.8v (39 psi)
64	GN/YL	A/T: TR Sensor Signal	In 'P': 0v, 55 mph: 1.7v
64	WT	M/T: Clutch Pedal Position	Clutch In: 1v, Out: 5v
65	WT/BL	EGR Pressure Transducer Signal	N/A
67	BK/OR	EVAP CV Solenoid	0-10 Hz (0-100%)
68	BK/YL	Ground Switched	0v
69	BK/WT	A/C WOT Relay Control	On: 1v, Off: 12v
70	BK/WT	Fuel Injector 1 Control	2.3-2.9 ms
71	GN/YL	Vehicle Power	12-14v
72	BL	Generator Communicator Control	1.5-10.5v (40-250 Hz)
73	GN/YL	Shift Solenoid A' Control	1v, Others: 1v
74	---	Not Used	---
75	BK/WT	Power Ground	<0.1v
76	BN	Turbine Speed Sensor (-) Signal	340-380 Hz
77	BK/YL	Power Ground	<0.1v
78, 84	---	Not Used	---
79	YL/GN	Fuel Cap Off Indicator	Off: 12v, On: 1v
80	BK/WT	IMRC Signal	IMRC Off: 5v, On: 2.5v
81	BK/RD	Pressure Control Solenoid 'A' (-)	<0.1v
82	GN/BK	Shift Solenoid 'C' Control	In Drive: 1v, Others: 12v
83	GN/YL	IAC Motor Control	9v (33% duty cycle)
85	WT/VT	CMP Sensor Signal	5-7 Hz
86	WT/RD	A/C Pressure Switch	A/C On: 12v (Open)
87	BR/YL	Knock Sensor Signal (-)	<0.050v
88	WT/BL	MAF Sensor Signal	0.6-0.9v
89	WT	TP Sensor Signal	0.53-1.27v
90	YL	Reference Voltage	4.9-5.1v
91	BR/GN	Analog Signal Return	<0.050v
92	GN/RD	Brake Pedal Position Switch	Brake Off: 0v, On: 12v
93	BK/YL	HO2S-11 (B1 S1) Heater	On: 1v, Off: 12v
94, 98	---	Not Used	---
95	BK/OR	Injector 4 Control	2.3-2.9 ms
96	BK/YL	Injector 3 Control	2.3-2.9 ms
97	GN/YL	Vehicle Power	12-14v
99	GN/OR	Shift Solenoid 'D' Control	In Drive: 1v, Others: 12v
100	BK/BL	HO2S-12 (B1 S2) Heater	On: 1v, Off: 12v
101	---	Not Used	---
102	GN/WT	Shift Solenoid 'E' Control	In O/D: 1v, Others: 12v
103	BK/YL	Power Ground	<0.1v
104	---	Not Used	---

Pin Connector Graphic

PCM 104-PIN CONNECTOR

Terminal View of 104-Pin PCM Wiring Harness Connector

2000-2002 Focus 2.0L I4 ZETEC-E VIN 3 (All) 104 Pin Connector

PCM Pin #	Wire Color	Circuit Description (104 Pin)	Value at Hot Idle
1	GN/BL	Shift Solenoid 'B' Control	1v, at Cruise: 1v
2, 5-6	---	Not Used	---
3	WT/RD	IMRC Solenoid Control	IMRC Off: 5v, On: 2.5v
4	GN/BL	TR Reverse Position Signal	In 'R': 12v, Others: 0v
7	GN/WT	TR 1st Position Signal	Manual 1: 12v, Others: 1v
8	GN/OR	TR Drive Position Signal	In 'D': 12v, Others: 1v
9-10, 12	---	Not Used	---
11	GN/BK	TR Overdrive Position Signal	In O/D: 12v, Others: 1v
13	OR/BK	Flash EEPROM Power	0.1v
14, 18, 23	---	Not Used	---
15	BL/BK	Data Bus (-) Signal	Digital Signals
16	GY/OR	Data Bus (+) Signal	Digital Signals
17	BK/WT	High Speed Fan Control	On: 1v, Off: 12v
19	GY/OR	Passive Antitheft System	Digital Signals
20	BK/BL	Injector 3 Control	3.3-3.7 ms
21	WT/RD	CKP Sensor (+) Signal	400-425 Hz
22	BR/RD	CKP Sensor (-) Signal	400-425 Hz
24	BK/YL	Power Ground	<0.1v
25	BK	Shield Ground	<0.050v
26	BK/OR	Coil 'A' Driver Control	5° dwell
27	BK/RD	Vehicle Start Signal	KOEC: 9-11v
28	WT/VT	M/T: VSS (+) Signal	0 Hz, 55 mph: 125 Hz
29	GN/BK	TR Overdrive OFF Switch	O/D Off: 0v
30, 32	---	Not Used	---
31	BK/YL	PSP Switch Signal	Straight: 0.6v, turned: 2v
33	BR/BL	A/T: OSS (-) Sensor Signal	0 Hz, 55 mph: 120 Hz
34	WT/VT	A/T: TSS Sensor (+) Signal	340-380 Hz
35	WT/RD	HO2S-22 (B2 S2) Signal	0.1-1.1v
36	BR/BL	MAF Sensor Return	<0.050v
37	WT/GN	TFT Sensor Signal	2.10-2.40v
38	WT	ECT Sensor Signal	0.5-0.6v
39	WT/VT	IAT Sensor Signal	1.5-2.5v
40	GN/BK	Fuel Pump Driver Module	3.5v (100% on time)
41	GN/YL	A/C Cycling Clutch Switch	Closed: 12v, Open: 0v
42	BK/OR	Passive Antitheft System	Digital Signals
43, 45-46	---	Not Used	---
44	GN/OR	EPC Switch Signal	12v
47	BK/GN	EGR VR Solenoid	0%, 55 mph: 45%
48-50	---	Not Used	---
51	BK/YL	Power Ground	<0.1v
52	BK/GN	Coil 'B' Driver Control	5° dwell
53	WT/GN	Passive Antitheft System	Digital Signals
54	BK/WT	Fuel Pump Driver Module	1.5v (33% on time)
55	RD	Keep Alive Power	12-14v
56	BK/BL	EVAP Purge Solenoid	0-10 Hz (0-100%)
57	WT/BK	Knock Sensor 1 (-) Signal	0v
58	WT/BL	A/T: OSS (+) Sensor Signal	0 Hz, 55 mph: 120 Hz
59	GY	Generator Field Sense	130 Hz (30% on time)
60	WT	HO2S-21 (B2 S1) Signal	0.1-1.1v
61	---	Not Used	---
62	WT/VT	FTP Sensor Signal	2.6v (0" H2O - cap off)
63	WT/GN	FRP Sensor Signal	2.8v (39 psi)
64	GN/YL	A/T: Neutral Start Position	In 'N': 0v, Others: 5v
64	WT	M/T: Clutch Pedal Switch	Clutch In: 0v, Out: 5v

2000-2002 Focus 2.0L I4 ZETEC-E VIN 3 (All) 104 Pin Connector

PCM Pin #	Wire Color	Circuit Description (104 Pin)	Value at Hot Idle
65 ('00)	WT/BL	DPFE Sensor Signal	0.20-1.30v
65 ('01-'02)	WT/BL	DPFE Sensor Signal	0.95-1.05v
66	---	Not Used	---
67	BK/OR	EVAP CV Solenoid	0-10 Hz (0-100%)
68	BK/BL	Low Speed Fan Control	On: 1v, Off: 12v
69	BK/YL	A/C WOT Relay Control	On: 1v, Off: 12v
70	BK/WT	Injector 1 Control	3.3-3.7 ms
71, 97	GN/YL	Vehicle Power	12-14v
72	BL	Generator Field Control	0 Hz (0% on time)
73	GN/YL	Shift Solenoid 'A' Control	0v, 55 mph: 12v
74-75	---	Not Used	---
76	BR/WT	M/T: CMP Sensor Ground	5-7 Hz
76	BR/WT	A/T: TSS Sensor (-) Signal	340-380 Hz
77	BK/YL	Power Ground	<1v
78-79	---	Not Used	---
80	BK/WT	IMRC Solenoid Control	12v (off)
81	BK/RD	EPC Solenoid Control	2.5v (73% on time)
82	GN/BK	Shift Solenoid 'C' Control	0v, 55 mph: 4.10v
83	BK/YL	IAC Motor Control	10v (30% on time)
84, 98, 94	---	Not Used	---
85	WT/VT	CMP Sensor Signal	5-9 Hz
86	WT/RD	A/C Pressure Switch	A/C On: 12v (Open)
87	BR/YL	Knock Sensor 1 Signal	0v
88	WT/BL	MAF Sensor Signal	0.6-0.9v
89	WT	TP Sensor Signal	0.53-1.27v
90	YL	Reference Voltage	4.9-5.1v
91	BR	Analog Signal Return	<0.050v
92	GN/RD	Brake Pedal Position Switch	Brake Off: 0v, On: 12v
93	BK/YL	HO2S-21 (B2 S1) Heater	On: 1v, Off: 12v
95	BK/OR	Injector 4 Control	3.3-3.7 ms
96	BK/YL	Injector 2 Control	3.3-3.7 ms
99	GN/OR	Shift Solenoid 'D' Control	0v, 55 mph: 0v
100	BK/BL	HO2S-22 (B2 S2) Heater	On: 1v, Off: 12v
101	---	Not Used	---
102	GN/WT	Shift Solenoid 'E' Control	0v, 55 mph: 0v
103	BK/YL	Power Ground	<1v
104	---	Not Used	---

Pin Connector Graphic

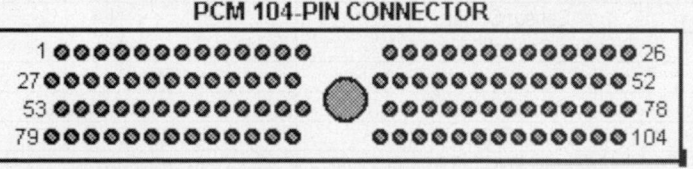

Terminal View of 104-Pin PCM Wiring Harness Connector

Standard Colors and Abbreviations

Abbreviation	Color	Abbreviation	Color	Abbreviation	Color
BK	Black	GY	Gray	PK	Purple
BL	Blue	GN	Green	RD	Red
BR	Brown	LG	LT Green	TN	Tan
DB	Dark Blue	OR	Orange	WT	White
DG	DK Green	PK	Pink	YL	Yellow

2003-05 Focus 2.0L I4 ZETEC-E VIN 3 (All) 104 Pin Connector

PCM Pin #	Wire Color	Circuit Description (104 Pin)	Value at Hot Idle
1	GN/BL	Shift Solenoid 'B' Control	In 2nd: 1v, Others: 1v
2	---	Not Used	---
3	WT/VT	IMRC Solenoid Monitor	5v
4	GN/BL	TR Reverse Position Signal	In 'R': 12v, Others: 0v
5-6	---	Not Used	---
7	GN/WT	TR First Position Signal	In 1st: 12v, Others: 1v
8	GN/OR	TR Second Position Signal	In 2nd: 12v, Others: 1v
9	WT/BK	Barometric Pressure Sensor	159 Hz (sea level)
10	---	Not Used	---
11	GN/BK	TR OD Position Signal	In 'P': 1v, In O/D: 12v
12	BR/WT	Sensor Ground	<0.050v
13	OG/BK	Flash EEPROM Power	0.1v
14, 18, 30	---	Not Used	---
15	BL/BK	SCP Bus (-) Signal	Digital Signals
16	GY/OR	SCP Bus (+) Signal	Digital Signals
17	BK/WT	High Speed Fan Control	On: 1v, Off: 12v
19	GY/OR	Passive Antitheft TX Signal	Digital Signals
20	BK/BL	Fuel Injector 3 Control	2.3-2.9 ms
21	WT/RD	CKP Sensor (+) Signal	450-480 Hz
22	BR/RD	CKP Sensor (-) Signal	450-480 Hz
23	BK/YL	Power Ground	<0.1v
24	BK/YL	Power Ground	<0.1v
25	BK	Power Ground	<0.1v
26	BK/OR	Coil 'A' Driver Control	5° dwell
27	BK/RD	Start Inhibit Relay Control	Relay Off: 12v, On: 1v
28	WT/VT	VSS (-) Signal	0 Hz, 55 mph: 125 Hz
29	GN/BK	Overdrive Cancel Switch	TCS & O/D On: 12v
31	WT/VT	PSP Switch Signal	Straight: 1v, Turned: 5v
32	BN/YL	Sensor Ground	<0.050v
33	BR/BL	Output Speed Sensor (-) Signal	0 Hz, 55 mph: 120 Hz
34	WT/VT	Turbine Speed Sensor (+) Signal	340-380 Hz
35	WT/RD	HO2S-12 (B1 S2) Signal	0.1-1.1v
36	BR/BL	MAF Sensor Return	<0.1v
37	WT/GN	TFT Sensor Signal	2.10-2.40v
38	WT/VT	Cylinder Head Temperature Sensor	0.5-0.6v
39	WT/VT	MAF Sensor Signal	0.6-0.9v
40	GN/BK	Fuel Pump Monitor	On: 12v, Off: 0v
41	GN/YL	A/C Cycling Clutch Pressure Switch	Closed: 12v, Open: 0v
42	BK/OR	PATS Indicator Control	Indicator Off: 12v, On: 1v
43	WT/BL	Sensor Signal	N/A
44	GN/OR	Pressure Control Solenoid 'A'	N/A
45	BK/RD	Camshaft Timing Adjuster Control	Off: 12v, On: 1v
46	BK/WT	Low Fuel Indicator	Digital Signals
47	BK/GN	Electric Vacuum Regulator Control	Off: 12v, On: 1v
48	---	Not Used	---
49	BL/RD	Controller Area Network Bus 'L'	0-7-0v
50	GY/RD	Controller Area Network Bus 'H'	0-7-0v
51	BK/YL	Power Ground	<0.1v
52	BK/GN	Coil 'B' Driver Control	5° dwell
53	WT/GN	Passive Antitheft RX Signal	Digital Signals
54	BK/WT	Fuel Pump Control	Pump Off: 12v, On: 1v
55	RD	Keep Alive Power	12-14v
56	BK/BL	EVAP Purge Solenoid	0-10 Hz (0-100%)
57	WT/BK	Knock Sensor Signal	0v

2003-05 Focus ZX2 2.0L I4 MFI VIN 3 (All) 104 Pin Connector

PCM Pin #	Wire Color	Circuit Description (104 Pin)	Value at Hot Idle
58	WT/BL	VSS (+) Signal	0 Hz, 55 mph: 125 Hz
59	GY	Generator Monitor Signal	130 Hz (45%)
60	WT	HO2S-11 (B1 S1) Signal	0.1-1.1v
61, 66	---	Not Used	---
62	WT/VT	FTP Sensor Signal	2.6v (cap off)
63	WT/GN	Fuel Rail Pressure Sensor	2.8v (39 psi)
64	GN/YL	A/T: TR Sensor Signal	In 'P': 0v, 55 mph: 1.7v
64	WT	M/T: Clutch Pedal Position	Clutch In: 1v, Out: 5v
65	WT/BL	EGR Pressure Transducer Signal	N/A
67	BK/OR	EVAP CV Solenoid	0-10 Hz (0-100%)
68	BK/YL	Ground Switched	0v
69	BK/WT	A/C WOT Relay Control	On: 1v, Off: 12v
70	BK/WT	Fuel Injector 1 Control	2.3-2.9 ms
71	GN/YL	Vehicle Power	12-14v
72	BL	Generator Communicator Control	1.5-10.5v (40-250 Hz)
73	GN/YL	Shift Solenoid A' Control	1v, Others: 1v
74	---	Not Used	---
75	BK/WT	Power Ground	<0.1v
76	BN	Turbine Speed Sensor (-) Signal	340-380 Hz
77	BK/YL	Power Ground	<0.1v
78, 84	---	Not Used	---
79	YL/GN	Fuel Cap Off Indicator	Off: 12v, On: 1v
80	BK/WT	IMRC Signal	IMRC Off: 5v, On: 2.5v
81	BK/RD	Pressure Control Solenoid 'A' (-)	DC pulse signals
82	GN/BK	Shift Solenoid 'C' Control	In Drive: 1v, Others: 12v
83	GN/YL	IAC Motor Control	9v (33% duty cycle)
85	WT/VT	CMP Sensor Signal	5-7 Hz
86	WT/RD	A/C Pressure Switch	A/C On: 12v (Open)
87	BR/YL	Knock Sensor Signal (-)	<0.050v
88	WT/BL	MAF Sensor Signal	0.6-0.9v
89	WT	TP Sensor Signal	0.53-1.27v
90	YL	Reference Voltage	4.9-5.1v
91	BR	Analog Signal Return	<0.050v
92	GN/RD	Brake Pedal Position Switch	Brake Off: 0v, On: 12v
93	BK/YL	HO2S-11 (B1 S1) Heater	On: 1v, Off: 12v
94, 98	---	Not Used	---
95	BK/OR	Injector 4 Control	2.3-2.9 ms
96	BK/YL	Injector 3 Control	2.3-2.9 ms
97	GN/YL	Vehicle Power	12-14v
99	GN/OR	Shift Solenoid 'D' Control	In Drive: 1v, Others: 12v
100	BK/BL	HO2S-12 (B1 S2) Heater	On: 1v, Off: 12v
101	---	Not Used	---
102	GN/WT	Shift Solenoid 'E' Control	In O/D: 1v, Others: 12v
103	BK/YL	Power Ground	<0.1v
104	---	Not Used	---

Pin Connector Graphic

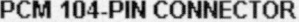

PCM 104-PIN CONNECTOR

Terminal View of 104-Pin PCM Wiring Harness Connector

2003-05 Focus 2.3L I4 DOHC MFI VIN Z (All) 104 Pin Connector

PCM Pin #	Wire Color	Circuit Description (104 Pin)	Value at Hot Idle
1	GN/BL	Shift Solenoid 'B' Control	In 2nd: 1v, Others: 1v
2	---	Not Used	---
3	WT/VT	IMRC Solenoid Monitor	5v
4	GN/BL	TR Reverse Position Signal	In 'R': 12v, Others: 0v
5-6	---	Not Used	---
7	GN/WT	TR First Position Signal	In 1st: 12v, Others: 1v
8	GN/OR	TR Second Position Signal	In 2nd: 12v, Others: 1v
9	WT/BK	Barometric Pressure Sensor	159 Hz (sea level)
10	---	Not Used	---
11	GN/BK	TR OD Position Signal	In 'P': 1v, In O/D: 12v
12	BR/WT	Sensor Ground	<0.050v
13	OG/BK	Flash EEPROM Power	0.1v
14, 18, 30	---	Not Used	---
15	BL/BK	SCP Bus (-) Signal	Digital Signals
16	GY/OR	SCP Bus (+) Signal	Digital Signals
17	BK/WT	High Speed Fan Control	On: 1v, Off: 12v
19	GY/OR	Passive Antitheft TX Signal	Digital Signals
20	BK/BL	Fuel Injector 3 Control	2.3-2.9 ms
21	WT/RD	CKP Sensor (+) Signal	450-480 Hz
22	BR/RD	CKP Sensor (-) Signal	450-480 Hz
23	BK/YL	Power Ground	<0.1v
24	BK/YL	Power Ground	<0.1v
25	BK	Power Ground	<0.1v
26	BK/OR	Coil 'A' Driver Control	5° dwell
27	BK/RD	Start Inhibit Relay Control	Relay Off: 12v, On: 1v
28	WT/VT	VSS (-) Signal	0 Hz, 55 mph: 125 Hz
29	GN/BK	Overdrive Cancel Switch	TCS & O/D On: 12v
31	WT/VT	PSP Switch Signal	Straight: 1v, Turned: 5v
32	BN/YL	Sensor Ground	<0.050v
33	BR/BL	Output Speed Sensor (-) Signal	0 Hz, 55 mph: 120 Hz
34	WT/VT	Turbine Speed Sensor (+) Signal	340-380 Hz
35	WT/RD	HO2S-12 (B1 S2) Signal	0.1-1.1v
36	BR/BL	MAF Sensor Return	<0.1v
37	WT/GN	TFT Sensor Signal	2.10-2.40v
38	WT	Engine Coolant Temperature Sensor	0.5-0.6v
39	WT/VT	MAF Sensor Signal	0.6-0.9v
40	GN/BK	Fuel Pump Monitor	On: 12v, Off: 0v
41	GN/YL	A/C Cycling Clutch Pressure Switch	Closed: 12v, Open: 0v
42	BK/OR	PATS Indicator Control	Indicator Off: 12v, On: 1v
43	WT/BL	Sensor Signal	N/A
44	GN/OR	Pressure Control Solenoid 'A'	N/A
45	BK/RD	Camshaft Timing Adjuster Control	Off: 12v, On: 1v
46	BK/WT	Low Fuel Indicator	Indicator Off: 12v, On: 1v
47	BK/GN	Electric Vacuum Regulator Control	Off: 12v, On: 1v
48	---	Not Used	---
49	BL/RD	Controller Area Network Bus 'L'	0-7-0v
50	GY/RD	Controller Area Network Bus 'H'	0-7-0v
51	BK/YL	Power Ground	<0.1v
52	BK/GN	Coil 'B' Driver Control	5° dwell
53	WT/GN	Passive Antitheft RX Signal	Digital Signals
54	BK/WT	Fuel Pump Control	Pump Off: 12v, On: 1v
55	RD	Keep Alive Power	12-14v
56	BK/BL	EVAP Purge Solenoid	0-10 Hz (0-100%)
57	WT/BK	Knock Sensor Signal	0v

2003-05 Focus 2.3L I4 DOHC MFI VIN Z (All) 104 Pin Connector

PCM Pin #	Wire Color	Circuit Description (104 Pin)	Value at Hot Idle
58	WT/BL	VSS (+) Signal	0 Hz, 55 mph: 125 Hz
59	GY	Generator Monitor Signal	130 Hz (45%)
60	WT	HO2S-11 (B1 S1) Signal	0.1-1.1v
61, 66	---	Not Used	---
62	WT/VT	FTP Sensor Signal	2.6v (cap off)
63	WT/GN	Fuel Rail Pressure Sensor	2.8v (39 psi)
64	GN/YL	A/T: TR Sensor Signal	In 'P': 0v, 55 mph: 1.7v
64	WT	M/T: Clutch Pedal Position	Clutch In: 1v, Out: 5v
65	WT/BL	EGR Pressure Transducer Signal	N/A
67	BK/OR	EVAP CV Solenoid	0-10 Hz (0-100%)
68	BK/YL	Ground Switched	0v
69	BK/WT	A/C WOT Relay Control	On: 1v, Off: 12v
70	BK/WT	Fuel Injector 1 Control	2.3-2.9 ms
71	GN/YL	Vehicle Power	12-14v
72	BL	Generator Communicator Control	1.5-10.5v (40-250 Hz)
73	GN/YL	Shift Solenoid A' Control	1v, Others: 1v
74	---	Not Used	---
75	BK/WT	Power Ground	<0.1v
76	BN	Turbine Speed Sensor (-) Signal	340-380 Hz
77	BK/YL	Power Ground	<0.1v
78, 84	---	Not Used	---
79	YL/GN	Fuel Cap Off Indicator	Off: 12v, On: 1v
80	BK/WT	IMRC Signal	IMRC Off: 5v, On: 2.5v
81	BK/RD	Pressure Control Solenoid 'A' (-)	DC pulse signals
82	GN/BK	Shift Solenoid 'C' Control	In Drive: 1v, Others: 12v
83	GN/YL	IAC Motor Control	9v (33% duty cycle)
85	WT/VT	CMP Sensor Signal	5-7 Hz
86	WT/RD	A/C Pressure Switch	A/C On: 12v (Open)
87	BR/YL	Knock Sensor Signal (-)	<0.050v
88	WT/BL	MAF Sensor Signal	0.6-0.9v
89	WT	TP Sensor Signal	0.53-1.27v
90	YL	Reference Voltage	4.9-5.1v
91	BR	Analog Signal Return	<0.050v
92	GN/RD	Brake Pedal Position Switch	Brake Off: 0v, On: 12v
93	BK/YL	HO2S-11 (B1 S1) Heater	On: 1v, Off: 12v
94, 98	---	Not Used	---
95	BK/OR	Injector 4 Control	2.3-2.9 ms
96	BK/YL	Injector 3 Control	2.3-2.9 ms
97	GN/YL	Vehicle Power	12-14v
99	GN/OR	Shift Solenoid 'D' Control	In Drive: 1v, Others: 12v
100	BK/BL	HO2S-12 (B1 S2) Heater	On: 1v, Off: 12v
101	---	Not Used	---
102	GN/WT	Shift Solenoid 'E' Control	In O/D: 1v, Others: 12v
103	BK/YL	Power Ground	<0.1v
104	---	Not Used	---

Pin Connector Graphic

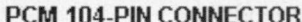

Terminal View of 104-Pin PCM Wiring Harness Connector

FIVE HUNDRED & FREESTYLE PIN TABLES

2005 Five Hundred, Freestyle 3.0L V6 MFI VIN 1 (A/T) C175A 58-Pin Connector

PCM Pin #	Wire Color	Circuit Description (58-Pin)	Value at Hot Idle
1	WT/RD	Accelerator Pedal Position Sensor	0.5-4.9v
2	---	Not Used	---
3	GY	SCP Data Bus (+)	Digital Signals
4	BL	SCP Data Bus (-)	<0.050v
5	BR	Signal Return	<0.050v
6-8	---	Not Used	---
9	BR/BL	A/C Clutch Relay Control	Relay Off: 12v, On: 1v
10-11	---	Not Used	---
12	BR/RD	EVAP Purge Solenoid Control	0-10 Hz (0-100%)
13	YL/GN	EEPROM Power	0.1v
14	WT/VT	Manual Transmission Switch (+) Signal	Switch Open: 12v, Closed: 1v
15	WT	Electronic Throttle Control Module	Digital Signals
16	WT/BL	Electronic Throttle Control Module	Digital Signals
17	BR	Electronic Throttle Control Module	Digital Signals
18-19	---	Not Used	---
20	YL	ETC Reference Voltage	4.9-5.1v
21	---	Not Used	---
22	WT/BK	Manual Transmission Switch (-) Signal	<0.050v
23	YL/RD	ETC Reference Voltage	4.9-5.1v
24	BK/RD	Power Ground	<0.1v
25	BK/RD	Power Ground	<0.1v
26	BK/RD	Power Ground	<0.1v
27	BK/RD	Power Ground	<0.1v
28	GN/OR	Brake Pressure Switch	Brake Off: 0v, On: 12v
29	---	Not Used	---
30	BR/BL	EVAP Vent Solenoid Control	0-10 Hz (0-100%)
31	WT/BL	MAF Sensor Signal	0.6v, 55 mph: 1.7v
32	GN/YL	Vehicle Power (Start-Run)	12-14v
33	GN/OR	Vehicle Power (Start-Run)	12-14v
34-37	---	Not Used	---
38	BR/BL	MAF Sensor Signal Return	<0.050v
39	---	Not Used	---
40	OG	Keep Alive Power	12-14v
41	WT/GN	Traction Control Disable Switch	Brake Off: 0v, On: 12v
42	WT/VT	A/C Pressure Switch Signal	0.9v (50 psi)
43	BR/YL	Power Ground	<0.1v
44	OG/YL	Keep Alive Power (CJB fuse)	12-14v
45-46	---	Not Used	---
47	WT/BK	Engine Cooling Fan Motor Control	Off: 12v, On: 1v
48	WT/BL	Restraint Control Module	Digital Signals
49	WT/VT	Rear Electronic Module	Digital Signals
50	---	Not Used	---
51	WT/VT	IAT Sensor Signal	1.5-2.5v
52	WT/VT	FTP Sensor Signal	2.6v (0" H2O - cap off)
53	GN/RD	Controller Area Network (+) Signal	0-7-0v
54	BL/RD	Controller Area Network (-) Signal	0-7-0v
55	YL	Reference Voltage	4.9-5.1v
56	BK/OR	Sensor Return	<0.050v
57	YL/BL	Speed Control Reference Voltage	4.9-5.1v
58	---	Not Used	---

2005 Five Hundred, Freestyle 3.0L V6 MFI VIN 1 (A/T) C175B 32-Pin Connector

PCM Pin #	Wire Color	Circuit Description (32-Pin)	Value at Hot Idle
1	BR	Shift Solenoid 'A' Control	12v, 55 mph: 1v
2	BR/RD	Shift Solenoid 'B' Control	12v, 55 mph: 1v
3-6	---	Not Used	---
7	BR/GN	Pressure Control Solenoid 'A' Control	8.8v
8	BR/WT	Shift Solenoid 'C' Control	12v, 55 mph: 12v
9	WT	Digital TR Sensor (TR3A)	0v, 55 mph: 1.7v
10	WT/RD	Digital TR Sensor (TR4)	0v, 55 mph: 11v
11	---	Not Used	---
12	BR/WT	Pressure Control 'C' Solenoid Control	Off: 12v, On: 1v
13	BR/YL	Pressure Control 'B' Solenoid Control	Off: 12v, On: 1v
14	BR/RD	Signal Return	<0.050v
15	BR/BL	HO2S-12 (B1 S2) Heater Control	Off: 12v, On: 1v
16	BK/GN	HO2S-22 (B2 S2) Heater Control	Off: 12v, On: 1v
17	BR/YL	Shift Solenoid 'D' Control	12v, 55 mph: 1v
18	WT/BL	Digital TR Sensor (TR2)	0v, 55 mph: 11v
19	---	Not Used	---
20	BR/WT	Torque Converter Clutch Control	12v, 55 mph: 0.1v
21	WT/BK	Intermediate Speed Shaft Signal	0 Hz, 55 mph: 1386 Hz
22	WT/GN	Digital TR Sensor (TR1)	0v, 55 mph: 11v
23	WT/RD	Transmission Fluid Temperature Sensor	2.10-2.40v
24-25	---	Not Used	---
26	WT/RD	Output Shaft Speed Sensor Signal	0 Hz, 975 Hz
27	WT/VT	Turbine Shaft Speed Sensor Signal	340 Hz, 1025 Hz
28	WT/BL	HO2S-12 (B1 S2) Signal	0.1-1.1v
29	WT/GN	HO2S-22 (B2 S2) Signal	0.1-1.1v
30-32	---	Not Used	---

Wire Harness Connectors Graphic

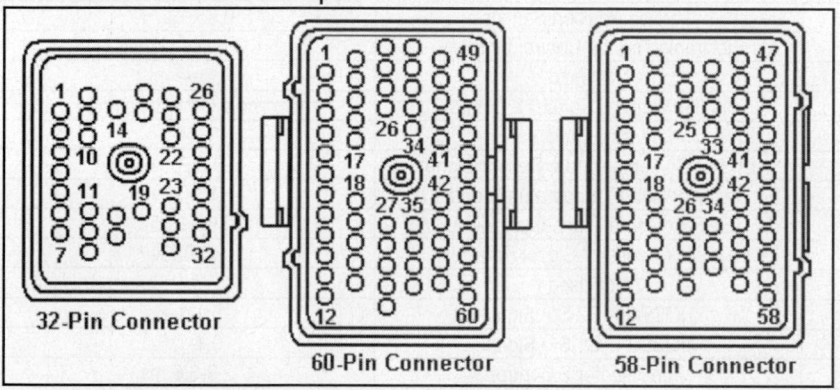

32-Pin Connector

60-Pin Connector

58-Pin Connector

Standard Colors and Abbreviations

Abbreviation	Color	Abbreviation	Color	Abbreviation	Color
BK	Black	GY	Gray	PK	Purple
BL	Blue	GN	Green	RD	Rcd
BR	Brown	LG	LT Green	TN	Tan
DB	Dark Blue	OR	Orange	WT	White
DG	DK Green	PK	Pink	YL	Yellow

2005 Five Hundred, Freestyle 3.0L V6 MFI VIN 1 (A/T) C175C 60 Pin Connector

PCM Pin #	Wire Color	Circuit Description (60 Pin)	Value at Hot Idle
1	WT/BK	Variable Valve Control Solenoid 1	DC pulse signals
2-3, 6	---	Not Used	---
4	WT/BL	Fuel Rail Temperature Sensor	1.7-3.5v (50-120ºF)
5	WT	Power Steering Pressure Switch	Straight: 0v, Turned: 5v
7	BR	HO2S-11 (B1 S1) Heater	On: 1v, Off: 12v
8	BR/RD	HO2S-21 (B2 S1) Signal	0.1-1.1v
9-10	---	Not Used	---
11	BR/BL	Fuel Injector 4 Control	2.9-3.6 ms
12	BR/GN	COP 4 Driver Control	5º, 55 mph: 8º
13	WT/VT	Variable Valve Timing Solenoid 2	DC pulse signals
14	YL/BK	Reference Voltage	4.9-5.1v
15	BR	Throttle Position Sensor Return	0.050v
16	BR/GN	EGR Pressure Transducer Return	<0.050v
17	BR/GN	Sensor Signal Return	<0.050v
18	---	Not Used	---
19	GY/RD	Generator Common Signal	0-130 Hz (varies)
20	BR/GN	Fuel Injector 5 Control	2.9-3.6 ms
21	BR/RD	Fuel Injector 3 Control	2.9-3.6 ms
22	BR/YL	COP 5 Driver Control	5º, 55 mph: 8º
23	WT/GN	EGR Pressure Transducer Signal	0-3.1v
24	YL	Battery Power (at all times)	12-14v
25-26	---	Not Used	---
27	YL/BL	Electronic Throttle Control Motor (-)	DC pulse signals
28	BR/YL	Fuel Injector 6 Control	2.9-3.6 ms
29	BR/WT	Intake Manifold Runner 1 Control	Solenoid Off: 12v, On: 1v
30	BR/WT	COP 6 Driver Control	5º, 55 mph: 8º
31, 33-34	---	Not Used	---
32	WT	Throttle Position Sensor	0.53-1.27v
35	WT/BL	Electronic Throttle Control Motor (+)	DC pulse signals
36	---	Not Used	---
37	BR/GN	Intake Manifold Runner 2 Control	Solenoid Off: 12v, On: 1v
38	---	Not Used	---
39	WT/GN	Engine Oil Temperature Sensor Signal	0.5-0.6v
40	WT/VT	Cylinder Head Temperature Sensor	0.5-0.6v
41	WT/BL	EGR Pressure Transducer Signal	0.95-1.05v
42	GY/BK	Knock Sensor 1 Signal Return	<0.050v
43	---	Not Used	---
44	WT/RD	HO2S-21 (B2 S1) Signal	0.1-1.1v
45	WT	HO2S-11 (B1 S1) Signal	0.1-1.1v
46	BR	Fuel Injector 2 Control	2.9-3.6 ms
47	BR/WT	Fuel Injector 1 Control	2.9-3.6 ms
48	BR/RD	COP 2 Driver Control	5º, 55 mph: 8º
49	WT/GN	Fuel Rail Pressure Sensor Signal	2.8v (39 psi)
50	WT/RD	Generator Monitor Signal	130 Hz (45%)
51	WT/BK	Knock Sensor 1 (+) Signal	0v
52	---	Not Used	---
53	WT/VT	CMP Sensor 1 Signal	6 Hz
54	WT/BK	CMP Sensor 2 Signal	6 Hz
55	GY/RD	CKP Sensor (-) Signal	390 Hz
56	WT/RD	CKP Sensor (+) Signal	390 Hz
57	WT/RD	Throttle Position Sensor	0.53-1.27v
58	BR	COP 1 Driver Control	5º, 55 mph: 8º
59	---	Not Used	---
60	BR/BL	COP 3 Driver Control	5º, 55 mph: 8º

MUSTANG PIN TABLES

1990 Mustang 2.3L I4 MFI VIN A (All) 60 Pin Connector

PCM Pin #	Wire Color	Circuit Description (60 Pin)	Value at Hot Idle
1	BK/OR	Keep Alive Power	12-14v
2	RD/LG	Brake Pedal Position Switch	Brake Off: 0v, On: 12v
3	YL/LG	PSP Switch Signal	Straight: 0v, Turned: 10v
4	DG/YL	Ignition Diagnostic Monitor	20-31 Hz
5-6, 8-9	---	Not Used	---
7	LG/YL	ECT Sensor Signal	0.5-0.6v
10	BK/YL	A/C Cycling Clutch Switch	A/C On: 12v, Off: 0v
11-15, 18-19	---	Not Used	---
16	BK/OR	Ignition System Ground	<0.050v
17	TN	STI Output, MIL Control	MIL On: 1v, Off: 12v
20	BK/LG	PCM Case Ground	<0.050v
21	WT/BL	IAC Motor Control	8-9v
22	TN/LG	Fuel Pump Control	On: 1v, Off: 12v
23	YL/RD	Knock Sensor Signal	0v
24	LG/PK	A/C Discharge Switch	A/C On: 12v, Off: 0v
25	LG/RD	ACT Sensor Signal	1.5-2.5v
25	LG/PK	ACT Sensor Signal	1.5-2.5v
26	OR/WT	Reference Voltage	4.9-5.1v
27	BR/LG	EGR EVP Sensor Signal	0.4v
28, 31-32	---	Not Used	---
29	DG/PKK	HO2S-11 (B1 S1) Signal	0.1-1.1v
29	DG/PK	HO2S-11 (B1 S1) Signal	0.1-1.1v
30	BK/WT	Neutral Drive Switch Signal	In 'N': 0v, Others: 5v
30	YL/RD	Neutral Drive Switch Signal	In 'N': 0v, Others: 5v
30	BK/WT	Clutch Engagement Switch	Clutch In: 0v, Out: 5v
30	YL/RD	Clutch Engagement Switch	Clutch In: 0v, Out: 5v
33	DG	EGR Vent Solenoid Control	12v, 55 mph: 1v
34-35	---	Not Used	---
36	YL/LG	Spark Output Signal	50% duty cycle
37, 57	RD	Vehicle Power	12-14v
38-39, 41-44	---	Not Used	---
40	BK/LG	Power Ground	<0.1v
45	LG/BK	MAP Sensor Signal	107 Hz
46	BK/WT	Analog Signal Return	<0.050v
47	DG/LG	TP Sensor Signal	0.5-1.2v
48	WT/RD	Self-Test Indicator Signal	STI Open: 5v, Closed: 1v
49	OR	HO2S-11 (Bank 1) Ground	<0.050v
50-51, 55	---	Not Used	---
52	YL	EGR Control Solenoid	Off-idle: 12v
53	OR/YL	Converter Clutch Override	Off: 12v, On: 1v
54	OR/BL	A/C WOT Relay Control	On: 1v, Off: 12v
56	DB	PIP Sensor Signal	50% duty cycle
58	TN/OR	Injector Bank 1 (INJ 1 & 4)	3.4-4.4 ms
59	TN/RD	Injector Bank 2 (INJ 2 & 3)	3.4-4.4 ms
60	BK	Power Ground	<0.1v

Pin Connector Graphic

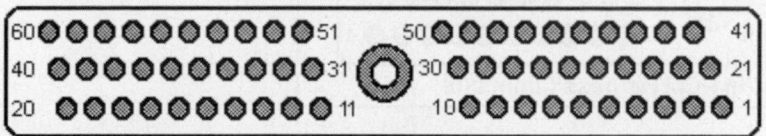

Terminal View of 60-Pin PCM Harness Connector

1991-93 Mustang 2.3L I4 MFI VIN M (All) 60 Pin Connector

PCM Pin #	Wire Color	Circuit Description (60 Pin)	Value at Hot Idle
1	YL	Keep Alive Power	12-14v
2	LG	Brake Pedal Position Switch	Brake Off: 0v, On: 12v
3	GY/BK	VSS (+) Signal	0 Hz, 55 mph: 125 Hz
4	TN/YL	Ignition Diagnostic Monitor	20-31 Hz
5	DB/OR	CID Sensor Signal	5-7 Hz
6	PK/OR	VSS (-) Signal	0 Hz, 55 mph: 125 Hz
7	LG/RD	ECT Sensor Signal	0.5-0.6v
8	DG/YL	Fuel Pump Monitor	On: 12v, Off: 0v
9, 28	PK, TN/OR	Data Bus (-), (+) Signals	Digital Signals
10	BK/YL	A/C Cycling Clutch Switch	A/C On: 12v, Off: 0v
11-13, 18-19, 23	---	Not Used	---
14	BL/RD	MAF Sensor Signal	0.6v
15	TN/BL	MAF Sensor Return	<0.050v
16	OR/RD	Ignition System Ground	<0.050v
17	PK/LG	STI Output, MIL Control	MIL On: 1v, Off: 12v
20	BK	PCM Case Ground	<0.050v
21	WT/BL	IAC Motor Control	8-9v
22	BL/OR	Fuel Pump Control	On: 1v, Off: 12v
24	YL/LG	PSP Switch Signal	Straight: 0v, Turned: 10v
25	GY	ACT Sensor Signal	1.5-2.5v
26	BR/WT	Reference Voltage	4.9-5.1v
27	BR/LG	EGR EVP Sensor	0.3v
29	GY/BL	HO2S-11 (B1 S1) Signal	0.1-1.1v
30	BL/YL	Neutral Drive Switch Signal	In 'N': 0v, Others: 5v
31	GY/YL	EVAP Purge Solenoid	0%, 35 mph: 40%
32	DB/YL	Dual Plug Inhibit Signal	DPI On: 12v, Off: 0v
33	BR/PK	EGR VR Solenoid	0%, 55 mph: 45%
34-35, 38-39	---	Not Used	---
36	PK	Spark Output Signal	50% dwell
37, 57	RD	Vehicle Power	12-14v
40, 60	BK	Power Ground	<0.1v
41-43, 50, 55	---	Not Used	---
44	DG	Octane Adjust Switch	Closed: 0v, Open: 9.1v
45	LG/BK	BARO Sensor Signal	159 Hz (sea level)
46	GY/RD	Analog Signal Return	<0.050v
47	GY/WT	TP Sensor Signal	0.5-1.2v
48	WT/PK	Self-Test Indicator Signal	STI Open: 5v, Closed: 1v
49	OR	HO2S-11 (Bank 1) Ground	<0.050v
51	LG/PK	Low Speed Fan Control	On: 1v, Off: 12v
52	OR/YL	Shift Solenoid 3-4 Control	On: 1v, Off: 12v
53	PK/YL	Converter Clutch Override	0%, 55 mph: 95%
54	PK/YL	A/C WOT Relay Control	On: 1v, Off: 12v
56	GY/OR	PIP Sensor Signal	50% duty cycle
58	TN	Injector Bank 1 (INJ 1 & 4)	3.0-3.8 ms
59	WT	Injector Bank 1 (INJ 2 & 3)	3.0-3.8 ms

Pin Connector Graphic

PCM 60-PIN CONNECTOR

Terminal View of 60-Pin PCM Harness Connector

1994-97 Mustang 3.8L V6 MFI VIN 4 (All) 104 Pin Connector

PCM Pin #	Wire Color	Circuit Description (104 Pin)	Value at Hot Idle
1	PK/OR	Shift Solenoid 2 Control	12v, 55 mph: 1v
2	PK/LG	MIL (lamp) Control	MIL On: 1v, Off: 12v
3-10, 12, 14	---	Not Used	---
11	PK/WT	Purge Flow Sensor	0.8v, 55 mph: 3v
13	P	Flash EEPROM Power	0.1v
15	PK/LB	Data Bus (-) Signal	Digital Signals
16	TN/OR	Data Bus (+) Signal	Digital Signals
17-20, 23	---	Not Used	---
21	DB	CKP Sensor (+) Signal	850-1120 Hz
22	GY	CKP Sensor (-) Signal	850-1120 Hz
24	BK/WT	Power Ground	<0.1v
25	BK	PCM Case Ground	<0.050v
26	DB/LG	Coil Driver 1 Control	5° dwell
27	OR/YL	Shift Solenoid 1 Control	1v, 55 mph: 12v
28, 31, 32	---	Not Used	---
29	TN/WT	TCS (switch) Signal	TCS & O/D On: 12v
30	DG	Octane Adjust Switch	Closed: 0v, Open: 9.3v
33	PK/OR	VSS (-) Signal	0 Hz, 55 mph: 125 Hz
34	---	Not Used	---
35	RD/LG	HO2S-12 (B1 S2) Signal	0.1-1.1v
36	TN/LB	MAF Sensor Return	<0.050v
37	OR/BK	TFT Sensor Signal	2.10-2.40v
38	LG/RD	ECT Sensor Signal	0.5-0.6v
39	GY	IAT Sensor Signal	1.5-2.5v
40	DG/YL	Fuel Pump Monitor	On: 12v, Off: 0v
41	DG/OR	A/C Cycling Clutch Switch	A/C On: 12v, Off: 0v
42-44, 46	---	Not Used	---
45	DB	Low Speed Fan Control	On: 1v, Off: 12v
47	BR/PK	EGR VR Solenoid	0%, 55 mph: 45%
48	OR/WT	Clean Tachometer Output	33-37 Hz
49-50, 53	---	Not Used	---
51	BK/WT	Power Ground	<0.1v
52	RD/LB	Coil Driver 2 Control	5° dwell
54	BR/OR	TCC Solenoid Control	TCC Off Idle: 0%
55	YL	Keep Alive Power	12-14v
56, 57, 59	---	Not Used	---
58	GY/BK	VSS (+) Signal	0 Hz, 55 mph: 125 Hz
60	GY/LB	HO2S-11 (B1 S1) Signal	0.1-1.1v
61	PK/LG	HO2S-22 (B2 S2) Signal	0.1-1.1v
62-63, 66	---	Not Used	---
64	LB/YL	TR Sensor Signal	In 'P': 0v, 55 mph: 1.7v
64	LB/YL	Neutral Position Switch	In 'N': 0v, Others: 5v
64	LB/YL	Clutch Pedal Position Switch	Clutch In: 0v, Out: 5v

1994-97 Mustang 3.8L V6 MFI VIN 4 (All) 104 Pin Connector

PCM Pin #	Wire Color	Circuit Description (104 Pin)	Value at Hot Idle
65	BR/LG	DPFE Sensor Signal	0.20-1.30v
67	GY/YL	EVAP Purge Solenoid	0-10 Hz (0-100%)
68	---	Not Used	---
69	PK/YL	A/C WOT Relay Control	On: 1v, Off: 12v
70, 72	---	Not Used	---
71	RD	Vehicle Power	12-14v
73	TN/BK	Injector 5 Control	4.9-5.1 ms
74	BR/YL	Injector 3 Control	4.9-5.1 ms
75	TN	Injector 1 Control	4.9-5.1 ms
76	BK/WT	Power Ground	<0.1v
77	BK/WT	Power Ground	<0.1v
78	PK/WT	Coil Driver 3 Control	5° dwell
79	WT/LG	TCIL (lamp) Control	On: 1v, Off: 12v
80	LB/OR	Fuel Pump Control	On: 1v, Off: 12v
81	WT/YL	EPC Solenoid Control	9.0v (15 psi)
82	PK/YL	TCC Solenoid	12v (0%)
83	WT/LB	IAC Motor Control	30% duty cycle
84	DG/WT	TSS Sensor Signal	43 Hz (650 rpm)
85	DB/OR	CID Sensor Signal	5-7 Hz
86	TN/LG	A/C Pressure Switch	A/C On: 12v (Open)
87	RD/BK	HO2S-21 (B2 S1) Signal	0.1-1.1v
88	LB/RD	MAF Sensor Signal	0.6v
89	GY/WT	TP Sensor Signal	0.53-1.27v
90	BR/WT	Reference Voltage	4.9-5.1v
91	GY/RD	Analog Signal Return	<0.050v
92	LG	Brake Pedal Position Switch	Brake Off: 0v, On: 12v
93	RD/WT	HO2S-11 (B1 S1) Heater	On: 1v, Off: 12v
94	YL/LB	HO2S-21 (B2 S1) Heater	On: 1v, Off: 12v
95	WT/BK	HO2S-12 (B1 S2) Heater	On: 1v, Off: 12v
96	TN/YL	HO2S-22 (B2 S2) Heater	On: 1v, Off: 12v
97	RD	Vehicle Power	12-14v
98	---	Not Used	---
99	LG/OR	Injector 6 Control	4.9-5.1 ms
100	BR/LB	Injector 4 Control	4.9-5.1 ms
101	WT	Injector 2 Control	4.9-5.1 ms
102	---	Not Used	---
103	BK/WT	Power Ground	<0.1v
104	---	Not Used	---

Pin Connector Graphic

PCM 104-PIN CONNECTOR

Terminal View of 104-Pin PCM Wiring Harness Connector

Standard Colors and Abbreviations

Abbreviation	Color	Abbreviation	Color	Abbreviation	Color
BK	Black	GY	Gray	PK	Purple
BL	Blue	GN	Green	RD	Red
BR	Brown	LG	LT Green	TN	Tan
DB	Dark Blue	OR	Orange	WT	White
DG	DK Green	PK	Pink	YL	Yellow

1998-99 Mustang 3.8L V6 MFI VIN 4 (All) 104 Pin Connector

PCM Pin #	Wire Color	Circuit Description (104 Pin)	Value at Hot Idle
1	PK/OR	Shift Solenoid 2 Control	12v, Others: 1v
2	PK/LG	MIL (lamp) Control	MIL On: 1v, Off: 12v
3	YL/BK	Digital TR1 Sensor	In 'P': 0v, 55 mph: 11v
4	---	Not Used	---
5	DG/YL	AIR Monitor Signal	Off: 0.1v
6-10	---	Not Used	---
11	PK/WT	Purge Flow Sensor	0.8v, 55 mph: 3v
12	---	Not Used	---
13	PK	Flash EEPROM Power	0.1v
14	---	Not Used	---
15	PK/LB	Data Bus (-) Signal	Digital Signals
16	TN/OR	Data Bus (+) Signal	Digital Signals
17-20	---	Not Used	---
21	DB	CKP Sensor (+) Signal	390-450 Hz
22	GY	CKP Sensor (-) Signal	390-450 Hz
23	---	Not Used	---
24	BK/WT	Power Ground	<0.1v
25	BK	PCM Case Ground	<0.050v
26	DB/LG	Coil Driver 1 Control	5° dwell
27	OR/YL	Shift Solenoid 1 Control	1v, Others: 1v
28	---	Not Used	---
29	TN/WT	TCS (switch) Signal	TCS & O/D On: 12v
30	DG	Octane Adjust Switch	Closed: 0v, Open: 9.3v
31-32	---	Not Used	---
33	PK/OR	VSS (-) Signal	0 Hz, 55 mph: 125 Hz
34	---	Not Used	---
35	RD/LG	HO2S-12 (B1 S2) Signal	0.1-1.1v
36	TN/LB	MAF Sensor Return	<0.050v
37	OR/BK	TFT Sensor Signal	2.10-2.40v
38	LG/RD	ECT Sensor Signal	0.5-0.6v
39	GY	IAT Sensor Signal	1.5-2.5v
40	DG/YL	Fuel Pump Monitor	On: 12v, Off: 0v
41	DG/OR	A/C High Pressure Switch	A/C On: 12v, Off: 1v
42-44	---	Not Used	---
45	DB	Low Speed Fan Control	On: 1v, Off: 12v
46	---	Not Used	---
47	BR/PK	EGR VR Solenoid	0%, 55 mph: 45%
48	OR/WT	Clean Tachometer Output	33-37 Hz
49	LB/BK	Digital TR2 Sensor	In 'P': 0v, 55 mph: 11v
50	WT/BK	Digital TR4 Sensor	In 'P': 0v, 55 mph: 11v
51	BK/WT	Power Ground	<0.1v
52	RD/LB	Coil Driver 2 Control	5° dwell
54	PK/YL	TCC Solenoid Control	0%, 55 mph: 95%
55	YL	Keep Alive Power	12-14v
56	LG/BK	EVAP CV Solenoid	0-10 Hz (0-100%)
57	---	Not Used	---
58	GY/BK	VSS (+) Signal	0 Hz, 55 mph: 125 Hz
59	---	Not Used	---
60	GY/LB	HO2S-11 (B1 S1) Signal	0.1-1.1v
61	PK/LG	HO2S-22 (B2 S2) Signal	0.1-1.1v
62-63	---	Not Used	---
64	LB/YL	TR Sensor	In 'P': 0v, in O/D: 1.7v
64	LB/YL	Neutral Position Switch	In 'N': 0v, Others: 5v
64	LB/YL	Clutch Pedal Position Switch	Clutch In: 0v, Out: 5v

1998-99 Mustang 3.8L V6 MFI VIN 4 (All) 104 Pin Connector

PCM Pin #	Wire Color	Circuit Description (104 Pin)	Value at Hot Idle
65	BR/LG	DPFE Sensor Signal	0.20-1.30v
66, 68	---	Not Used	---
67	GY/YL	EVAP Purge Solenoid	0-10 Hz (0-100%)
69	PK/YL	A/C WOT Relay Control	On: 1v, Off: 12v
70	WT/OR	AIR Bypass Solenoid	AIRB On: 1v, Off: 12v
71	RD	Vehicle Power	12-14v
72	---	Not Used	---
73	TN/BK	Injector 5 Control	4.9-5.1 ms
74	BR/YL	Injector 3 Control	4.9-5.1 ms
75	TN	Injector 1 Control	4.9-5.1 ms
76-77	BK/WT	Power Ground	<0.1v
78	PK/WT	Coil Driver 3 Control	6° dwell
79	WT/LG	TCIL (lamp) Control	On: 1v, Off: 12v
80	LB/OR	Fuel Pump Control	On: 1v, Off: 12v
81	WT/YL	EPC Solenoid Control	8.3v (8 psi)
82	PK/YL	TCC Solenoid Control	0%, 55 mph: 95%
83	WT/LB	IAC Motor Control	30% duty cycle
84	DG/WT	OSS Sensor Signal	0, 55 mph: 2500 rpm
85	DB/OR	CMP Sensor Signal	5-7 Hz
86	TN/LG	A/C Pressure Switch	A/C On: 12v (Open)
87	RD/BK	HO2S-21 (B2 S1) Signal	0.1-1.1v
88	LB/RD	MAF Sensor Signal	0.6v
89	GY/WT	TP Sensor Signal	0.53-1.27v
90	BR/WT	Reference Voltage	4.9-5.1v
91	GY/RD	Analog Signal Return	<0.050v
92	LG	Brake Pedal Position Switch	Brake Off: 0v, On: 12v
93	RD/WT	HO2S-11 (B1 S1) Heater	On: 1v, Off: 12v
94	YL/LB	HO2S-21 (B2 S1) Heater	On: 1v, Off: 12v
95	WT/BK	HO2S-12 (B1 S2) Heater	On: 1v, Off: 12v
96	TN/YL	HO2S-22 (B2 S2) Heater	On: 1v, Off: 12v
97	RD	Vehicle Power	12-14v
98	---	Not Used	---
99	LG/OR	Injector 6 Control	4.9-5.1 ms
100	BR/LB	Injector 4 Control	4.9-5.1 ms
101	W	Injector 2 Control	4.9-5.1 ms
102	---	Not Used	---
103	BK/WT	Power Ground	<0.1v
104	---	Not Used	---

Pin Connector Graphic

PCM 104-PIN CONNECTOR

```
 1 ooooooooooooo      ooooooooooooo 26
27 ooooooooooooo   ●  ooooooooooooo 52
53 ooooooooooooo      ooooooooooooo 78
79 ooooooooooooo      ooooooooooooo 104
```

Terminal View of 104-Pin PCM Wiring Harness Connector

Standard Colors and Abbreviations

Abbreviation	Color	Abbreviation	Color	Abbreviation	Color
BK	Black	GY	Gray	PK	Purple
BL	Blue	GN	Green	RD	Red
BR	Brown	LG	LT Green	TN	Tan
DB	Dark Blue	OR	Orange	WT	White
DG	DK Green	PK	Pink	YL	Yellow

2000 Mustang 3.8L V6 MFI VIN 4 (All) 104 Pin Connector

PCM Pin #	Wire Color	Circuit Description (104 Pin)	Value at Hot Idle
1	PK/OR	Shift Solenoid 'B' Control	12v, Others: 1v
3	YL/BK	Digital TR1 Sensor	In 'P': 0v, 55 mph: 11v
4	---	Not Used	---
5	WT	Air Injection Pump Monitor	1v, 55 mph: 1v
6-12	---	Not Used	---
13	PK	Flash EEPROM Power	0.1v
14	---	Not Used	---
15	PK/LB	Data Bus (-) Signal	Digital Signals
16	TN/OR	Data Bus (+) Signal	Digital Signals
17-20	---	Not Used	---
21	DB	CKP Sensor (+) Signal	390-450 Hz
22	GY	CKP Sensor (-) Signal	390-450 Hz
23-24	---	Not Used	---
25	BK	PCM Case Ground	<0.050v
26	DB/LG	Coil Driver 1 Control	5° dwell
27	OR/YL	Shift Solenoid 'A' Control	1v, Others: 1v
28	---	Not Used	---
29	TN/WT	TCS (switch) Signal	TCS & O/D On: 12v
30-34	---	Not Used	---
35	RD/LG	HO2S-12 (B1 S2) Signal	0.1-1.1v
36	TN/LB	MAF Sensor Return	<0.050v
37	OR/BK	TFT Sensor Signal	2.10-2.40v
38	LG/RD	ECT Sensor Signal	0.5-0.6v
39	GY	IAT Sensor Signal	1.5-2.5v
40	LB/OR	Fuel Pump Monitor	On: 12v, Off: 0v
41	DG/OR	A/C Cycling Switch	A/C On: 12v, Off: 1v
42	---	Not Used	---
45	DB	Low Speed Fan Control	On: 1v, Off: 12v
46	---	Not Used	---
47	BR/PK	EGR VR Solenoid	0%, 55 mph: 45%
48	---	Not Used	---
49	LB/BK	Digital TR2 Sensor	In 'P': 0v, 55 mph: 11v
50	WT/BK	Digital TR4 Sensor	In 'P': 0v, 55 mph: 11v
51	BK/WT	Power Ground	<0.1v
52	RD/LB	Coil Driver 2 Control	5° dwell
54	BR/OR	TCC Solenoid Control	0%, 55 mph: 95%
55	RD/WT	Keep Alive Power	12-14v
56	LG/BK	EVAP CV Solenoid	0-10 Hz (0-100%)
57-59	---	Not Used	---
60	GY/LB	HO2S-11 (B1 S1) Signal	0.1-1.1v
61	PK/LG	HO2S-22 (B2 S2) Signal	0.1-1.1v
62	RD/PK	FTP Sensor Signal	2.6v (0" H2O - cap off)
63	RD/PK	FRP Sensor Signal	2.8v (39 psi)
64	LB/YL	Digital TR3A Sensor	In 'P': 0v, in O/D: 1.7v

2000 Mustang 3.8L V6 MFI VIN 4 (All) 104 Pin Connector

PCM Pin #	Wire Color	Circuit Description (104 Pin)	Value at Hot Idle
65	BR/LG	DPFE Sensor Signal	0.20-1.30v
66	---	Not Used	---
67	GY/YL	EVAP Purge Solenoid	0-10 Hz (0-100%)
68	WT/OR	VSS (+) Signal	0 Hz, 55 mph: 125 Hz
69	PK/YL	A/C WOT Relay Control	On: 1v, Off: 12v
70	WT/OR	Electronic Air Management	On: 1v, Off: 12v
71	RD	Vehicle Power	12-14v
72	---	Not Used	---
73	TN/BK	Injector 5 Control	4.9-5.1 ms
74	BR/YL	Injector 3 Control	4.9-5.1 ms
75	T	Injector 1 Control	4.9-5.1 ms
76-77	BK/WT	Power Ground	<0.1v
78	PK/WT	Coil Driver 3 Control	6° dwell
79	---	Not Used	---
80	WT/RD	Fuel Pump Control	On: 1v, Off: 12v
81	WT/YL	EPC Solenoid Control	8.3v (8 psi)
82, 86	---	Not Used	---
83	WT/LB	IAC Motor Control	30% duty cycle
84	DG/WT	OSS Sensor Signal	0, 55 mph: 2500 rpm
85	DB/OR	CMP Sensor Signal	5-7 Hz
87	RD/BK	HO2S-21 (B2 S1) Signal	0.1-1.1v
88	LB/RD	MAF Sensor Signal	0.6v
89	GY/WT	TP Sensor Signal	0.53-1.27v
90	BR/WT	Reference Voltage	4.9-5.1v
91	GY/RD	Analog Signal Return	<0.050v
92	LG	Brake Pedal Position Switch	Brake Off: 0v, On: 12v
93	RD/WT	HO2S-11 (B1 S1) Heater	On: 1v, Off: 12v
94	YL/LB	HO2S-21 (B2 S1) Heater	On: 1v, Off: 12v
95	WT/BK	HO2S-12 (B1 S2) Heater	On: 1v, Off: 12v
96	TN/YL	HO2S-22 (B2 S2) Heater	On: 1v, Off: 12v
97	RD	Vehicle Power	12-14v
98	---	Not Used	---
99	LG/OR	Injector 6 Control	4.9-5.1 ms
100	BR/LB	Injector 4 Control	4.9-5.1 ms
101	W	Injector 2 Control	4.9-5.1 ms
102	---	Not Used	---
103	BK/WT	Power Ground	<0.1v
104	---	Not Used	---

Pin Connector Graphic

PCM 104-PIN CONNECTOR

```
 1 ●●●●●●●●●●●●●     ●●●●●●●●●●●●● 26
27 ●●●●●●●●●●●●●     ●●●●●●●●●●●●● 52
53 ●●●●●●●●●●●●●  ●  ●●●●●●●●●●●●● 78
79 ●●●●●●●●●●●●●     ●●●●●●●●●●●●● 104
```

Terminal View of 104-Pin PCM Wiring Harness Connector

Standard Colors and Abbreviations

Abbreviation	Color	Abbreviation	Color	Abbreviation	Color
BK	Black	GY	Gray	PK	Purple
BL	Blue	GN	Green	RD	Red
BR	Brown	LG	LT Green	TN	Tan
DB	Dark Blue	OR	Orange	WT	White
DG	DK Green	PK	Pink	YL	Yellow

2001 Mustang 3.8L V6 MFI VIN 4 (All) 104 Pin Connector

PCM Pin #	Wire Color	Circuit Description (104 Pin)	Value at Hot Idle
1	VT/OR	Shift Solenoid 2 Control	12v, Others: 1v
2	---	Not Used	---
3	YL/BK	Digital TR1 Sensor	In 'P': 0v, 55 mph: 11v
4-7	---	Not Used	---
8	DB/YL	IMRC Solenoid Control	5v (Off)
9-12	---	Not Used	---
13	PK	Flash EEPROM Power	0.1v
14	---	Not Used	---
15	PK/LB	Data Bus (-) Signal	Digital Signals
16	TN/OR	Data Bus (+) Signal	Digital Signals
17-20	---	Not Used	---
21	DB	CKP Sensor (+) Signal	390-450 Hz
22	GY	CKP Sensor (-) Signal	390-450 Hz
23-24	---	Not Used	---
25	BK	PCM Case Ground	<0.050v
26	DB/LG	Coil Driver 1 Control	5° dwell
27	OR/YL	Shift Solenoid 'A' Control	1v, Others: 1v
28	---	Not Used	---
29	TN/WT	TCS (switch) Signal	TCS & O/D On: 12v
30-34	---	Not Used	---
35	RD/LG	HO2S-12 (B1 S2) Signal	0.1-1.1v
36	TN/LB	MAF Sensor Return	<0.050v
37	OR/BK	TFT Sensor Signal	2.10-2.40v
38	LG/RD	ECT Sensor Signal	0.5-0.6v
39	GY	IAT Sensor Signal	1.5-2.5v
40	LB/OR	Fuel Pump Monitor	Pump On: 3.5v, Off: 0v
41	DG/OR	A/C Cycling Switch	A/C On: 12v, Off: 1v
42	BR	IMRC Solenoid Control	12v (Off)
43-44	---	Not Used	---
45	DB	Low Speed Fan Control	On: 1v, Off: 12v
46	---	Not Used	---
47	BR/PK	EGR VR Solenoid	0%, 55 mph: 45%
48	---	Not Used	---
49	LB/BK	Digital TR2 Sensor	In 'P': 0v, 55 mph: 11v
50	WT/BK	Digital TR4 Sensor	In 'P': 0v, 55 mph: 11v
51	BK/WT	Power Ground	<0.1v
52	RD/LB	Coil Driver 2 Control	5° dwell
54	BR/OR	TCC Solenoid Control	0%, 55 mph: 95%
55	RD/WT	Keep Alive Power	12-14v
56	LG/BK	EVAP CV Solenoid	0 Hz (0%)
57-59	---	Not Used	---
60	GY/LB	HO2S-11 (B1 S1) Signal	0.1-1.1v
61	PK/LG	HO2S-22 (B2 S2) Signal	0.1-1.1v
62	RD/PK	FTP Sensor Signal	2.6v (0" H2O - cap off)
63	RD/PK	FRP Sensor Signal	2.8v (39 psi)
64	LB/YL	Digital TR3A Sensor	In 'P': 0v, in O/D: 1.7v

2001 Mustang 3.8L V6 MFI VIN 4 (All) 104 Pin Connector

PCM Pin #	Wire Color	Circuit Description (104 Pin)	Value at Hot Idle
65	BR/LG	DPFE Sensor Signal	0.95-1.05v
66	YL/LG	CHT Sensor Signal	0.7v (194°F)
67	GY/YL	EVAP Purge Solenoid	0-10 Hz (0-100%)
68	WT/OR	VSS (+) Signal	0 Hz, 55 mph: 125 Hz
69	PK/YL	A/C WOT Relay Control	On: 1v, Off: 12v
71	RD	Vehicle Power	12-14v
72	---	Not Used	---
73	TN/BK	Injector 5 Control	3.8-4.9 ms
74	BR/YL	Injector 3 Control	3.8-4.9 ms
75	T	Injector 1 Control	3.8-4.9 ms
76-77	BK/WT	Power Ground	<0.1v
78	PK/WT	Coil Driver 3 Control	6° dwell
79	---	Not Used	---
80	WT/RD	Fuel Pump Control	On: 1v, Off: 12v
81	WT/YL	EPC Solenoid Control	8.3v (8 psi)
82	---	Not Used	---
83	WT/LB	IAC Motor Control	30% duty cycle
84	DG/WT	OSS Sensor Signal	0 Hz, 55 mph: 240 Hz
85	DB/OR	CMP Sensor Signal	5-7 Hz
86	---	Not Used	---
87	RD/BK	HO2S-21 (B2 S1) Signal	0.1-1.1v
88	LB/RD	MAF Sensor Signal	0.6-0.9v
89	GY/WT	TP Sensor Signal	0.53-1.27v
90	BR/WT	Reference Voltage	4.9-5.1v
91	GY/RD	Analog Signal Return	<0.050v
92	LG	Brake Pedal Position Switch	Brake Off: 0v, On: 12v
93	RD/WT	HO2S-11 (B1 S1) Heater	On: 1v, Off: 12v
94	YL/LB	HO2S-21 (B2 S1) Heater	On: 1v, Off: 12v
95	WT/BK	HO2S-12 (B1 S2) Heater	On: 1v, Off: 12v
96	TN/YL	HO2S-22 (B2 S2) Heater	On: 1v, Off: 12v
97	RD	Vehicle Power	12-14v
99	LG/OR	Injector 6 Control	3.8-4.9 ms
100	BR/LB	Injector 4 Control	3.8-4.9 ms
101	WT	Injector 2 Control	3.8-4.9 ms
102	---	Not Used	---
103	BK/WT	Power Ground	<0.1v
104	---	Not Used	---

Pin Connector Graphic

PCM 104-PIN CONNECTOR

Terminal View of 104-Pin PCM Wiring Harness Connector

Standard Colors and Abbreviations

Abbreviation	Color	Abbreviation	Color	Abbreviation	Color
BK	Black	GY	Gray	PK	Purple
BL	Blue	GN	Green	RD	Red
BR	Brown	LG	LT Green	TN	Tan
DB	Dark Blue	OR	Orange	WT	White
DG	DK Green	PK	Pink	YL	Yellow

2002-05 Mustang 3.8L V6 MFI VIN 4 (All) 104 Pin Connector

PCM Pin #	Wire Color	Circuit Description (104 Pin)	Value at Hot Idle
1	VT/OR	Shift Solenoid 'B' Control	12v, Others: 1v
2	---	Not Used	---
3	YL/BK	Digital TR1 Sensor	In 'P': 0v, 55 mph: 11v
4-6	---	Not Used	---
7	---	Not Used	---
8	DB/YL	IMRC Solenoid Control	5v (Off)
9	---	Not Used	---
10	DB/LG	Barometric Pressure Sensor Signal	159 Hz (sea level)
11-12	---	Not Used	---
13	VT	Flash EEPROM Power	0.1v
14	---	Not Used	---
15	PK/LB	SCP Bus (-) Signal	Digital Signals
16	TN/OR	SCP Bus (+) Signal	Digital Signals
17-20	---	Not Used	---
21	DB	CKP Sensor (+) Signal	390-450 Hz
22	GY	CKP Sensor (-) Signal	390-450 Hz
23-24	---	Not Used	---
25	BK	PCM Case Ground	<0.050v
26	DB/LG	Coil Driver 1 Control	5° dwell
27	OR/YL	Shift Solenoid 'A' Control	1v, Others: 1v
28	---	Not Used	---
29	TN/WT	TCS (switch) Signal	TCS & O/D On: 12v
30-34	---	Not Used	---
35	RD/LG	HO2S-12 (B1 S2) Signal	0.1-1.1v
36	TN/LB	MAF Sensor Return	<0.050v
37	OR/BK	TFT Sensor Signal	2.10-2.40v
38	---	Not Used	---
39	GY	IAT Sensor Signal	1.5-2.5v
40	LB/OR	Fuel Pump Monitor	Pump On: 3.5v, Off: 0v
41	DG/OR	A/C Cycling Switch	A/C On: 12v, Off: 1v
42	BR	IMRC Solenoid Control	12v (Off)
43-44	---	Not Used	---
45	DB	Low Speed Fan Control	On: 1v, Off: 12v
46	DB	Low Speed Fan Relay Control	Relay Off: 12v, On: 1v
47	BR/PK	EGR VR Solenoid	0%, 55 mph: 45%
48	---	Not Used	---
49	LB/BK	Digital TR2 Sensor	In 'P': 0v, 55 mph: 11v
50	WT/BK	Digital TR4 Sensor	In 'P': 0v, 55 mph: 11v
51	BK/WT	Power Ground	<0.1v
52	RD/LB	Coil Driver 2 Control	5° dwell
53	---	Not Used	---
54	BR/OR	TCC Solenoid Control	0%, 55 mph: 95%
55	RD/WT	Keep Alive Power	12-14v
56	LG/BK	EVAP CV Solenoid	0 Hz (0%)
57-59	---	Not Used	---
60	GY/LB	HO2S-11 (B1 S1) Signal	0.1-1.1v
61	VT/LG	HO2S-22 (B2 S2) Signal	0.1-1.1v
62	RD/PK	FTP Sensor Signal	2.6v (0" H2O - cap off)
63	RD/PK	FRP Sensor Signal	2.8v (39 psi)
64	LB/YL	Digital TR3A Sensor	In 'P': 0v, in O/D: 1.7v

2002-05 Mustang 3.8L V6 MFI VIN 4 (All) 104 Pin Connector

PCM Pin #	Wire Color	Circuit Description (104 Pin)	Value at Hot Idle
65	BR/LG	DPFE Sensor Signal	0.95-1.05v
66	YL/LG	CHT Sensor Signal	0.7v (194°F)
67	GY/YL	EVAP Purge Solenoid	0-10 Hz (0-100%)
68	WT/OR	VSS (+) Signal	0 Hz, 55 mph: 125 Hz
69	PK/YL	A/C WOT Relay Control	On: 1v, Off: 12v
70	---	Not Used	---
71	RD	Vehicle Power (Start-Run)	12-14v
72	---	Not Used	---
73	TN/BK	Injector 5 Control	3.8-4.9 ms
74	BR/YL	Injector 3 Control	3.8-4.9 ms
75	TN	Injector 1 Control	3.8-4.9 ms
76, 77	BK/WT	Power Ground	<0.1v
78	PK/WT	Coil Driver 3 Control	6° dwell
79	---	Not Used	---
80	WT/RD	Fuel Pump Control	Off: 12v, On: 1v
81	WT/YL	EPC Solenoid Control	8.3v (8 psi)
82	---	Not Used	---
83	WT/LB	IAC Motor Control	30% duty cycle
84	DG/WT	OSS Sensor Signal	0 Hz, 55 mph: 240 Hz
85	DB/OR	CMP Sensor Signal	5-7 Hz
86, 98	---	Not Used	---
87	RD/BK	HO2S-21 (B2 S1) Signal	0.1-1.1v
88	LB/RD	MAF Sensor Signal	0.6-0.9v
89	GY/WT	TP Sensor Signal	0.53-1.27v
90	BR/WT	Reference Voltage	4.9-5.1v
91	GY/RD	Sensor Return	<0.050v
92	LG	Brake Pedal Position Switch	Brake Off: 0v, On: 12v
93	RD/WT	HO2S-11 (B1 S1) Heater	On: 1v, Off: 12v
94	YL/LB	HO2S-21 (B2 S1) Heater	On: 1v, Off: 12v
95	WT/BK	HO2S-12 (B1 S2) Heater	On: 1v, Off: 12v
96	TN/YL	HO2S-22 (B2 S2) Heater	On: 1v, Off: 12v
97	RD	Vehicle Power	12-14v
99	LG/OR	Injector 6 Control	3.8-4.9 ms
100	BR/LB	Injector 4 Control	3.8-4.9 ms
101	WT	Injector 2 Control	3.8-4.9 ms
102	---	Not Used	---
103	BK/WT	Power Ground	<0.1v
104	---	Not Used	---

Pin Connector Graphic

PCM 104-PIN CONNECTOR

1 [row of pins] 26
27 [row of pins] 52
53 [row of pins] 78
79 [row of pins] 104

Terminal View of 104-Pin PCM Wiring Harness Connector

Standard Colors and Abbreviations

Abbreviation	Color	Abbreviation	Color	Abbreviation	Color
BK	Black	GY	Gray	PK	Purple
BL	Blue	GN	Green	RD	Red
BR	Brown	LG	LT Green	TN	Tan
DB	Dark Blue	OR	Orange	WT	White
DG	DK Green	PK	Pink	YL	Yellow

1996-97 Mustang 4.6L V8 MFI VIN V (All) 104 Pin Connector

PCM Pin #	Wire Color	Circuit Description (104 Pin)	Value at Hot Idle
1	---	Not Used	---
2	PK/LG	MIL (lamp) Control	MIL On: 1v, Off: 12v
3-4	---	Not Used	---
5	WT	EAM Monitor Signal	1v, 55 mph: 1v
6-7	---	Not Used	---
8	WT/OR	IMRC Monitor Signal	5v, 55 mph: 5v
9-10	---	Not Used	---
11	PK/WT	Purge Flow Sensor	0.8v, 55 mph: 3v
12	---	Not Used	---
13	PK	Flash EEPROM Power	0.1v
14	---	Not Used	---
15	PK/LB	Data Bus (-) Signal	Digital Signals
16	TN/OR	Data Bus (+) Signal	Digital Signals
17-20	---	Not Used	---
21	DB	CKP Sensor (+) Signal	850-1120 Hz
22	GY	CKP Sensor (-) Signal	850-1120 Hz
23	---	Not Used	---
24	BK/WT	Power Ground	<0.1v
25	BK	PCM Case Ground	<0.050v
26	DB/LG	Coil Driver 1 Control	5° dwell
27-29	---	Not Used	---
30	DG	Octane Adjust Switch	Closed: 0v, Open: 9.3v
31	---	Not Used	---
32	DG/PK	Knock Sensor 2 Signal	0v
33	PK/OR	VSS (-) Signal	0 Hz, 55 mph: 125 Hz
34	---	Not Used	---
35	RD/LG	HO2S-12 (B1 S2) Signal	0.1-1.1v
36	TN/LB	MAF Sensor Return	<0.050v
37	---	Not Used	---
38	LG/RD	ECT Sensor Signal	0.5-0.6v
39	GY	IAT Sensor Signal	1.5-2.5v
40	DG/YL	Fuel Pump Monitor	On: 12v, Off: 0v
41	DG/OR	A/C Cycling Clutch Switch	A/C On: 12v, Off: 0v
42	BR	IMRC Solenoid Control	12v, 55 mph: 12v
43-44	---	Not Used	---
45	DB	Low Speed Fan Control	On: 1v, Off: 12v
46	LG/PK	High Speed Fan Control	On: 1v, Off: 12v
47	BR/PK	EGR VR Solenoid	0%, 55 mph: 45%
48	OR/WT	Clean Tachometer Output	37-44 Hz
49-50	---	Not Used	---
51	BK/WT	Power Ground	<0.1v
52	RD/LB	Coil Driver 2 Control	5° dwell
53-54	---	Not Used	---
55	YL	Keep Alive Power	12-14v
56	---	Not Used	---
57	YL/RD	Knock Sensor 1 Signal	0v
58	GY/BK	VSS (+) Signal	0 Hz, 55 mph: 125 Hz
59	---	Not Used	---
60	GY/LB	HO2S-11 (B1 S1) Signal	0.1-1.1v
61	PK/LG	HO2S-22 (B2 S2) Signal	0.1-1.1v
62-64	---	Not Used	---

1996-97 Mustang 4.6L V8 MFI VIN V (All) 104 Pin Connector

PCM Pin #	Wire Color	Circuit Description (104 Pin)	Value at Hot Idle
65	BR/LG	DPFE Sensor Signal	0.20-1.30v
66	---	Not Used	---
67	GY/YL	EVAP Purge Solenoid	0-10 Hz (0-100%)
68	---	Not Used	---
69	PK/YL	A/C WOT Relay Control	On: 1v, Off: 12v
70	WT/OR	EAM System Control	12v, 55 mph: 12v
71	RD	Vehicle Power	12-14v
72	TN/RD	Injector 7 Control	2.4-2.8 ms
73	TN/BK	Injector 5 Control	2.4-2.8 ms
74	BR/YL	Injector 3 Control	2.4-2.8 ms
75	TN	Injector 1 Control	2.4-2.8 ms
76, 77	BK/WT	Power Ground	<0.1v
78	PK/WT	Coil Driver 3 Control	5° dwell
79	WT/RD	Low Fuel Pump Control	On: 1v, Off: 12v
80	LB/OR	High Fuel Pump Control	On: 1v, Off: 12v
81-82	---	Not Used	---
83	WT/LB	IAC Motor Control	10v (30% on time)
84	---	Not Used	---
85	DB/OR	CMP Sensor Signal	5-7 Hz
86	TN/LG	A/C Pressure Switch	A/C On: 12v (Open)
87	RD/BK	HO2S-21 (B2 S1) Signal	0.1-1.1v
88	LB/RD	MAF Sensor Signal	0.6v
89	GY/WT	TP Sensor Signal	0.53-1.27v
90	BR/WT	Reference Voltage	4.9-5.1v
91	GY/RD	Analog Signal Return	<0.050v
92	---	Not Used	---
93	RD/WT	HO2S-11 (B1 S1) Heater	On: 1v, Off: 12v
94	YL/LB	HO2S-21 (B2 S1) Heater	On: 1v, Off: 12v
95	WT/BK	HO2S-12 (B1 S2) Heater	On: 1v, Off: 12v
96	TN/YL	HO2S-22 (B2 S2) Heater	On: 1v, Off: 12v
97	RD	Vehicle Power	12-14v
98	LB	Injector 8 Control	2.4-2.8 ms
99	LG/OR	Injector 6 Control	2.4-2.8 ms
100	BR/LB	Injector 4 Control	2.4-2.8 ms
101	WT	Injector 2 Control	2.4-2.8 ms
102	---	Not Used	---
103	BK/WT	Power Ground	<0.1v
104	RD/YL	Coil Driver 4 Control	5° dwell

Pin Connector Graphic

PCM 104-PIN CONNECTOR

Terminal View of 104-Pin PCM Wiring Harness Connector

Standard Colors and Abbreviations

Abbreviation	Color	Abbreviation	Color	Abbreviation	Color
BK	Black	GY	Gray	PK	Purple
BL	Blue	GN	Green	RD	Red
BR	Brown	LG	LT Green	TN	Tan
DB	Dark Blue	OR	Orange	WT	White
DG	DK Green	PK	Pink	YL	Yellow

1998-99 Mustang 4.6L 4v V8 MFI VIN V (M/T) 104 Pin Connector

PCM Pin #	Wire Color	Circuit Description (104 Pin)	Value at Hot Idle
1	---	Not Used	---
2	PK/LG	MIL (lamp) Control	MIL On: 1v, Off: 12v
3-4	---	Not Used	---
5	WT	Electronic AIR Monitor	1v, 55 mph: 1v
6-7	---	Not Used	---
8	WT/OR	IMRC Monitor Signal	5v, 55 mph: 5v
9-10	---	Not Used	---
11 ('98)	PK/WT	PF Sensor Signal	1.1-1.6v
12	YL/WT	Fuel Level Indicator Signal	1.7v (1/2 full)
13	PK	Flash EEPROM Power	0.1v
14	---	Not Used	---
15	PK/LB	Data Bus (-) Signal	Digital Signals
16	TN/OR	Data Bus (+) Signal	Digital Signals
17-20	---	Not Used	---
21	DB	CKP Sensor (+) Signal	320-420 Hz
22	GY	CKP Sensor (-) Signal	320-420 Hz
23	---	Not Used	---
24	BK/WT	Power Ground	<0.1v
25	BK	PCM Case Ground	<0.050v
26	DB/LG	Coil Driver 1 Control	5° dwell
27-29	---	Not Used	---
30	DG	Octane Adjust Switch	Closed: 0v, Open: 9.3v
31	---	Not Used	---
32	DG/PK	Knock Sensor 2 Signal	0v
33	PK/OR	VSS (-) Signal	0 Hz, 55 mph: 125 Hz
34, 37	---	Not Used	---
35	RD/LG	HO2S-12 (B1 S2) Signal	0.1-1.1v
36	TN/LB	MAF Sensor Return	<0.050v
38	LG/RD	ECT Sensor Signal	0.5-0.6v
39	GY	IAT Sensor Signal	1.5-2.5v
40	DG/YL	Fuel Pump Monitor	On: 12v, Off: 0v
41	DG/OR	A/C High Pressure Cutout	A/C On: 12v, Off: 1v
42	BR	IMRC Solenoid Control	12v, 55 mph: 12v
43-44	---	Not Used	---
45	DB	Low Speed Fan Control	On: 1v, Off: 12v
46	LG/PK	High Speed Fan Control	On: 1v, Off: 12v
47	BR/PK	EGR VR Solenoid	0%, 55 mph: 45%
48	OR/WT	Clean Tachometer Output	37-44 Hz
49-50	---	Not Used	---
51	BK/WT	Power Ground	<0.1v
52	RD/LB	Coil Driver 2 Control	5° dwell
53-54	---	Not Used	---
55	YL	Keep Alive Power	12-14v
56	LG/BK	EVAP Purge Solenoid	0-10 Hz (0-100%)
57	YL/RD	Knock Sensor 1 Signal	0v
58	GY/BK	VSS (+) Signal	0 Hz, 55 mph: 125 Hz
59	---	Not Used	---
60	GY/LB	HO2S-11 (B1 S1) Signal	0.1-1.1v
61	PK/LG	HO2S-22 (B2 S2) Signal	0.1-1.1v
62-64	---	Not Used	---

1998-99 Mustang 4.6L 4v V8 MFI VIN V (M/T) 104 Pin Connector

PCM Pin #	Wire Color	Circuit Description (104 Pin)	Value at Hot Idle
65	BR/LG	DPFE Sensor Signal	0.20-1.30v
66, 68	---	Not Used	
67	GY/YL	EVAP Purge Solenoid	0-10 Hz (0-100%)
69	PK/YL	A/C WOT Relay Control	A/C On: 12v, Off: 1v
70	WT/OR	Electronic AIRB Solenoid	12v, Others: 12v
71	RD	Vehicle Power	12-14v
72	TN/RD	Injector 7 Control	2.4-2.8 ms
73	TN/BK	Injector 5 Control	2.4-2.8 ms
74	BR/YL	Injector 3 Control	2.4-2.8 ms
75	TN	Injector 1 Control	2.4-2.8 ms
76, 77	BK/WT	Power Ground	<0.1v
78	PK/WT	Coil Driver 3 Control	5° dwell
79	WT/RD	Low Fuel Pump Control	On: 1v, Off: 12v
80	LB/OR	Fuel Pump Control	On: 1v, Off: 12v
81-82	---	Not Used	---
82	RD/PK	FTP Sensor Signal	2.6v (0" H2O - cap off)
83	WT/LB	IAC Motor Control	9v (32% duty cycle)
84	---	Not Used	
85	DB/OR	CMP Sensor Signal	5-7 Hz
86	TN/LG	A/C High Pressure Switch	A/C On: 12v, Off: 1v
87	RD/BK	HO2S-21 (B2 S1) Signal	0.1-1.1v
88	LB/RD	MAF Sensor Signal	0.6v
89	GY/WT	TP Sensor Signal	0.53-1.27v
90	BR/WT	Reference Voltage	4.9-5.1v
91	GY/RD	Analog Signal Return	<0.050v
92	---	Not Used	---
93	RD/WT	HO2S-11 (B1 S1) Heater	On: 1v, Off: 12v
94	YL/LB	HO2S-21 (B2 S1) Heater	On: 1v, Off: 12v
95	WT/BK	HO2S-12 (B1 S2) Heater	On: 1v, Off: 12v
96	TN/YL	HO2S-22 (B2 S2) Heater	On: 1v, Off: 12v
97	RD	Vehicle Power	12-14v
98	LB	Injector 8 Control	2.4-2.8 ms
99	LG/OR	Injector 6 Control	2.4-2.8 ms
100	BR/LB	Injector 4 Control	2.4-2.8 ms
101	WT	Injector 2 Control	2.4-2.8 ms
102	---	Not Used	---
103	BK/WT	Power Ground	<0.1v
104	RD/YL	Coil Driver 4 Control	5° dwell

Pin Connector Graphic

Standard Colors and Abbreviations

Abbreviation	Color	Abbreviation	Color	Abbreviation	Color
BK	Black	GY	Gray	PK	Purple
BL	Blue	GN	Green	RD	Red
BR	Brown	LG	LT Green	TN	Tan
DB	Dark Blue	OR	Orange	WT	White
DG	DK Green	PK	Pink	YL	Yellow

2000-02 Mustang 4.6L V8 MFI VIN V (M/T) 104 Pin Connector

PCM Pin #	Wire Color	Circuit Description (104 Pin)	Value at Hot Idle
1	OR/YL	COP 6 (Integrated) Dwell	6° dwell
2	---	Not Used	---
3	BK/WT	Power Ground	<0.1v
4-12	---	Not Used	---
13	PK	Flash EEPROM Power	0.1v
14	---	Not Used	---
15	PK/LB	Data Bus (-) Signal	Digital Signals
16	TN/OR	Data Bus (+) Signal	Digital Signals
17-18	---	Not Used	---
19	DB	Low Speed Fan Control	On: 1v, Off: 12v
20	---	Not Used	---
21	DB	CKP Sensor (+) Signal	320-420 Hz
22	GY	CKP Sensor (-) Signal	320-420 Hz
23	DG/WT	Knock Sensor 2 (-) Signal	0v
24	---	Not Used	---
25	BK	PCM Case Ground	<0.050v
26	LG/WT	COP 1 (Integrated) Dwell	6° dwell
27	LG/YL	COP 5 (Integrated) Dwell	6° dwell
28-31	---	Not Used	---
32	YL	Knock Sensor 1 (-) Signal	0v
32-34	---	Not Used	---
35	RD/LG	HO2S-12 (B1 S2) Signal	0.1-1.1v
36	TN/LB	MAF Sensor Return	<0.050v
37	---	Not Used	---
38	LG/RD	ECT Sensor Signal	0.5-0.6v
39	GY	IAT Sensor Signal	1.5-2.5v
40	LB/OR	Fuel Pump Monitor	2.5-7.5v
41	DG/OR	A/C Cycling Switch Signal	A/C On: 12v, Off: 1v
42-45	---	Not Used	---
46	LG/PK	High Speed Fan Control	On: 1v, Off: 12v
47	BR/PK	EGR VR Solenoid	0%, 55 mph: 45%
48-50	---	Not Used	---
51	BK/WT	Power Ground	<0.1v
52	WT/PK	COP 3 (Integrated) Dwell	6° dwell
53	DG/PK	COP 4 (Integrated) Dwell	6° dwell
54	---	Not Used	---
55	RD/WT	Keep Alive Power	12-14v
56	LG/BK	EVAP Purge Solenoid	0-10 Hz (0-100%)
57	YL/RD	Knock Sensor 1 (+) Signal	0v
58-59	---	Not Used	---
60	GY/LB	HO2S-11 (B1 S1) Signal	0.1-1.1v
61	PK/LG	HO2S-22 (B2 S2) Signal	0.1-1.1v
62	RD/PK	FTP Sensor Signal	2.6v (0" H2O - cap off)
63	RD/PK	FRP Sensor Signal	2.8v (39 psi)
64	---	Not Used	---
65 ('00)	BR/LG	DPFE Sensor Signal	0.20-1.30v
65 ('01-'02)	BR/LG	DPFE Sensor Signal	0.95-1.05v
66	---	Not Used	---
67	GY/YL	EVAP CV Solenoid	0-10 Hz (0-100%)
68	WT/OR	VSS (+) Signal	0 Hz, 55 mph: 125 Hz
69	PK/YL	A/C WOT Relay Control	A/C On: 12v, Off: 1v
70	---	Not Used	---

2000-02 Mustang 4.6L V8 MFI VIN V (M/T) 104 Pin Connector

PCM Pin #	Wire Color	Circuit Description (104 Pin)	Value at Hot Idle
71	RD	Vehicle Power	12-14v
72	TN/RD	Injector 7 Control	2.4-2.8 ms
73	TN/BK	Injector 5 Control	2.4-2.8 ms
74	BR/YL	Injector 3 Control	2.4-2.8 ms
75	TN	Injector 1 Control	2.4-2.8 ms
76	---	Not Used	---
77	BK/WT	Power Ground	<0.1v
78	PK/LB	COP 7 (Integrated) Dwell	6° dwell
79	WT/RD	COP 8 (Integrated) Dwell	6° dwell
80	WT/RD	Fuel Pump Control	On: 1v, Off: 12v
81-82	---	Not Used	---
83	WT/LB	IAC Motor Control	9v (32% duty cycle)
84	DG/WT	OSS Sensor (+) Signal	0 Hz, 55 mph: 470 Hz
85	DB/OR	CMP Sensor Signal	5-7 Hz
86	TN/LG	A/C High Pressure Switch	A/C On: 12v, Off: 1v
87	RD/BK	HO2S-21 (B2 S1) Signal	0.1-1.1v
88	LB/RD	MAF Sensor Signal	0.6v
89	GY/WT	TP Sensor Signal	0.53-1.27v
90	BR/WT	Reference Voltage	4.9-5.1v
91	GY/RD	Analog Signal Return	<0.050v
92	LG	Brake Pedal Position Switch	Brake Off: 0v, On: 12v
93	RD/WT	HO2S-11 (B1 S1) Heater	On: 1v, Off: 12v
94	YL/LB	HO2S-21 (B2 S1) Heater	On: 1v, Off: 12v
95	WT/BK	HO2S-12 (B1 S2) Heater	On: 1v, Off: 12v
96	TN/YL	HO2S-22 (B2 S2) Heater	On: 1v, Off: 12v
97	RD	Vehicle Power	12-14v
98	LB	Injector 8 Control	2.4-2.8 ms
99	LG/OR	Injector 6 Control	2.4-2.8 ms
100	BR/LB	Injector 4 Control	2.4-2.8 ms
101	WT	Injector 2 Control	2.4-2.8 ms
102	DG/PK	Knock Sensor 2 (+) Signal	0v
103	BK/WT	Power Ground	<0.1v
104	PK/WT	COP 2 (Integrated) Dwell	6° dwell

Pin Connector Graphic

PCM 104-PIN CONNECTOR

1 ●●●●●●●●●●●● ●●●●●●●●●●●● 26
27 ●●●●●●●●●●●● ●●●●●●●●●●●● 52
53 ●●●●●●●●●●●● ●●●●●●●●●●●● 78
79 ●●●●●●●●●●●● ●●●●●●●●●●●● 104

Terminal View of 104-Pin PCM Wiring Harness Connector

Standard Colors and Abbreviations

Abbreviation	Color	Abbreviation	Color	Abbreviation	Color
BK	Black	GY	Gray	PK	Purple
BL	Blue	GN	Green	RD	Red
BR	Brown	LG	LT Green	TN	Tan
DB	Dark Blue	OR	Orange	WT	White
DG	DK Green	PK	Pink	YL	Yellow

2003-05 Mustang Mach 1 4.6L V8 DOHC VIN R (All) 104 Pin Connector

PCM Pin #	Wire Color	Circuit Description (104 Pin)	Value at Hot Idle
1	VT/YL	COP 6 (Integrated) Dwell	6° dwell
2, 4-5	---	Not Used	---
3	BK/WT	Power Ground	<0.1v
6	OR/YL	Shift Solenoid 'A' Control	1v, Others: 1v
7, 9	---	Not Used	---
8	WT/OR	IMRC Monitor Signal	5v, 55 mph: 5v
10	DB/LG	Barometric Pressure Sensor Signal	159 Hz
11	VT/OR	Shift Solenoid 'B' Control	12v, Others: 1v
12, 14	---	Not Used	---
13	VT	Flash EEPROM Power	0.1v
15	PK/LB	SCP Bus (-) Signal	Digital Signals
16	TN/OR	SCP Bus (+) Signal	Digital Signals
17-20	---	Not Used	---
21	DB	CKP Sensor (+) Signal	320-420 Hz
22	GY	CKP Sensor (-) Signal	320-420 Hz
23	DG/WT	Knock Sensor 1 (-) Signal	<0.050v
24	---	Not Used	---
25	BK	PCM Case Ground	<0.050v
26	LG/WT	COP 1 (Integrated) Dwell	6° dwell
27	LG/YL	COP 5 (Integrated) Dwell	6° dwell
28	DB	Low Speed Fan Relay Control	Relay Off: 12v, On: 1v
29	TN/WT	TCS (switch) Signal	TCS & O/D On: 12v
30-31, 33	---	Not Used	---
32	YL	Knock Sensor 2 (-) Signal	<0.050v
34	YL/BK	Digital TR1 Sensor	In 'P': 0v, 55 mph: 11v
35	RD/LG	HO2S-12 (B1 S2) Signal	0.1-1.1v
36	TN/LB	MAF Sensor Return	<0.050v
37	RD/YL	Intake Air Temperature Sensor 2 Signal	1.5-2.5v
38	LG/RD	Reference Voltage	4.9-5.1v
39	GY	Intake Air Temperature Sensor 1 Signal	1.5-2.5v
40	LB/OR	Fuel Pump Monitor	2.5-7.5v
41	DG/OR	A/C Cycling Switch Signal	A/C On: 12v, Off: 1v
42	BR	IMRC Solenoid Control	12v (Off)
43-44	---	Not Used	---
45	LG/VT	Supercharger Bypass Solenoid Control	Solenoid Off: 12v, On: 1v
46	LG/VT	High Speed Fan Relay Control	Relay Off: 12v, On: 1v
47	BR/PK	EGR VR Solenoid	0%, 55 mph: 45%
49	LB/BK	Digital TR2 Sensor	In 'P': 0v, 55 mph: 11v
50	WT/BK	Digital TR4 Sensor	In 'P': 0v, 55 mph: 11v
51	BK/WT	Power Ground	<0.1v
52	WT/PK	COP 3 (Integrated) Dwell	6° dwell
53	DG/PK	COP 4 (Integrated) Dwell	6° dwell
54	---	Not Used	---
55	RD/WT	Keep Alive Power	12-14v
56	LG/BK	EVAP Purge Solenoid	0-10 Hz (0-100%)
57	YE/RD	Knock Sensor 1 (+) Signal	0v
58	---	Not Used	---
60	GY/LB	HO2S-11 (B1 S1) Signal	0.1-1.1v
61	VT/LG	HO2S-22 (B2 S2) Signal	0.1-1.1v
62	RD/PK	FTP Sensor Signal	2.6v (0" H2O - cap off)
63	RD/PK	FRP Sensor Signal	2.8v (39 psi)
64	LB/YL	Digital TR3A Sensor	In 'P': 0v, in O/D: 1.7v
65	BR/LG	DPFE Sensor Signal	0.95-1.05v
66	YE/LG	Cylinder Head Temperature Sensor	0.5-0.6v

2003-05 Mustang Mach 1 4.6L V8 DOHC VIN R (All) 104 Pin Connector

PCM Pin #	Wire Color	Circuit Description (104 Pin)	Value at Hot Idle
67	GY/YL	EVAP CV Solenoid	0-10 Hz (0-100%)
68	WT/OR	VSS (+) Signal	Moving: AC pulse signals
69	PK/YL	A/C WOT Relay Control	A/C On: 12v, Off: 1v
70	---	Not Used	---
71	RD	Vehicle Power (Start-Run)	12-14v
72	TN/RD	Injector 7 Control	2.4-2.8 ms
73	TN/BK	Injector 5 Control	2.4-2.8 ms
74	BR/YL	Injector 3 Control	2.4-2.8 ms
75	TN	Injector 1 Control	2.4-2.8 ms
76, 77	BK/WT	Power Ground	<0.1v
78	PK/LB	COP 7 (Integrated) Dwell	6° dwell
79	WT/RD	COP 8 (Integrated) Dwell	6° dwell
80	WT/RD	Fuel Pump Control	Off: 12v, On: 1v
81	WT/YL	EPC Solenoid Control	8.3v (8 psi)
82	WT/OR	Charge Air Cooler Pump Relay Control	Relay Off: 12v, On: 1v
83	WT/LB	IAC Motor Control	9v (32% duty cycle)
84	DG/WT	OSS Sensor (+) Signal	0 Hz, 55 mph: 470 Hz
85	DB/OR	CMP Sensor Signal	5-7 Hz
86	TN/LG	A/C High Pressure Switch	A/C On: 12v, Off: 1v
87	RD/BK	HO2S-21 (B2 S1) Signal	0.1-1.1v
88	LB/RD	MAF Sensor Signal	0.6v
89	GY/WT	TP Sensor Signal	0.53-1.27v
90	BR/WT	Reference Voltage	4.9-5.1v
91	GY/RD	Sensor Ground	<0.050v
92	LG	Brake Pedal Position Switch	Brake Off: 0v, On: 12v
93	RD/WT	HO2S-11 (B1 S1) Heater	On: 1v, Off: 12v
94	YL/LB	HO2S-21 (B2 S1) Heater	On: 1v, Off: 12v
95	WT/BK	HO2S-12 (B1 S2) Heater	On: 1v, Off: 12v
96	TN/YL	HO2S-22 (B2 S2) Heater	On: 1v, Off: 12v
97	RD	Vehicle Power (Start-Run)	12-14v
98	LB	Injector 8 Control	2.4-2.8 ms
99	LG/OR	Injector 6 Control	2.4-2.8 ms
100	BR/LB	Injector 4 Control	2.4-2.8 ms
101	WT	Injector 2 Control	2.4-2.8 ms
102	DG/PK	Knock Sensor 2 (+) Signal	0v
103	BK/WT	Power Ground	<0.1v
104	PK/WT	COP 2 (Integrated) Dwell	6° dwell

Pin Connector Graphic

PCM 104-PIN CONNECTOR

Terminal View of 104-Pin PCM Wiring Harness Connector

Standard Colors and Abbreviations

Abbreviation	Color	Abbreviation	Color	Abbreviation	Color
BK	Black	GY	Gray	PK	Purple
BL	Blue	GN	Green	RD	Red
BR	Brown	LG	LT Green	TN	Tan
DB	Dark Blue	OR	Orange	WT	White
DG	DK Green	PK	Pink	YL	Yellow

1996-97 Mustang 4.6L 2v V8 MFI VIN W (All) 104 Pin Connector

PCM Pin #	Wire Color	Circuit Description (104 Pin)	Value at Hot Idle
1	PK/OR	Shift Solenoid 2 Control	12v, 55 mph: 1v
2	PK/LG	MIL (lamp) Control	MIL On: 1v, Off: 12v
3-10	---	Not Used	---
11	PK/WT	Purge Flow Sensor	0.8v, 55 mph: 3v
12	---	Not Used	---
13	PK	Flash EEPROM Power	0.1v
14	---	Not Used	---
15	PK/LB	Data Bus (-) Signal	Digital Signals
16	TN/OR	Data Bus (+) Signal	Digital Signals
17-20	---	Not Used	---
21	DB	CKP Sensor (+) Signal	850-1120 Hz
22	GY	CKP Sensor (-) Signal	850-1120 Hz
23	---	Not Used	---
24	BK/WT	Power Ground	<0.1v
25	BK	PCM Case Ground	<0.1v
26	DB/LG	Coil Driver 1 Control	5° dwell
27	OR/YL	Shift Solenoid 1 Control	1v, 55 mph: 12v
28	---	Not Used	---
29	TN/WT	TCS (switch) Signal	TCS & O/D On: 12v
30	DG	Octane Adjust Switch	Closed: 0v, Open: 9.3v
31-32	---	Not Used	---
33	PK/OR	VSS (-) Signal	0 Hz, 55 mph: 125 Hz
34	---	Not Used	---
35	RD/LG	HO2S-12 (B1 S2) Signal	0.1-1.1v
36	TN/LB	MAF Sensor Return	<0.1v
37	OR/BK	TFT Sensor Signal	2.10-2.40v
38	LG/RD	ECT Sensor Signal	0.5-0.6v
39	GY	IAT Sensor Signal	1.5-2.5v
40	DG/YL	Fuel Pump Monitor	On: 12v, Off: 0v
41	DG/OR	A/C Cycling Clutch Switch	A/C On: 12v, Off: 0v
42-44	---	Not Used	---
45	DB	Low Speed Fan Control	On: 1v, Off: 12v
46	LG/PK	High Speed Fan Control	On: 1v, Off: 12v
47	BR/PK	EGR VR Solenoid	0%, 55 mph: 45%
48	OR/WT	Clean Tachometer Output	33-37 Hz
49-50	---	Not Used	---
51	BK/WT	Power Ground	<0.1v
52	RD/LB	Coil Driver 2 Control	5° dwell
53	---	Not Used	---
54	BR/OR	TCC Solenoid Control	0%, 55 mph: 95%
55	YL	Keep Alive Power	12-14v
56-57	---	Not Used	---
58	GY/BK	VSS (+) Signal	0 Hz, 55 mph: 125 Hz
59	---	Not Used	---
60	GY/LB	HO2S-11 (B1 S1) Signal	0.1-1.1v
61	PK/LG	HO2S-22 (B2 S2) Signal	0.1-1.1v
62-63	---	Not Used	---
64	LB/YL	TR Sensor Signal	In 'P': 0v, 55 mph: 1.7v
65	BR/LG	DPFE Sensor Signal	0.20-1.30v
66	---	Not Used	---

1996-97 Mustang 4.6L 2v V8 MFI VIN W (All) 104 Pin Connector

PCM Pin #	Wire Color	Circuit Description (104 Pin)	Value at Hot Idle
67	GY/YL	EVAP Purge Solenoid	0-10 Hz (0-100%)
68	---	Not Used	---
69	PK/YL	A/C WOT Relay Control	On: 1v, Off: 12v
70	---	Not Used	---
71	RD	Vehicle Power	12-14v
72	TN/RD	Injector 7 Control	3.5-3-7 ms
73	TN/BK	Injector 5 Control	3.5-3-7 ms
74	BR/YL	Injector 3 Control	3.5-3-7 ms
75	TN	Injector 1 Control	3.5-3-7 ms
76-77	BK/WT	Power Ground	<0.1v
78	PK/WT	Coil Driver 3 Control	5° dwell
79	WT/LG	TCIL (lamp) Control	On: 1v, Off: 12v
80	LB/OR	Fuel Pump Control	On: 1v, Off: 12v
81	WT/YL	EPC Solenoid Control	9v (15 psi)
82	---	Not Used	---
83	WT/LB	IAC Motor Control	30% duty cycle
84	DG/WT	TSS Sensor Signal	43 Hz (650 rpm)
85	DB/OR	CID Sensor Signal	5-7 Hz
86	TN/LG	A/C Pressure Switch	A/C On: 12v (Open)
87	RD/BK	HO2S-21 (B2 S1) Signal	0.1-1.1v
88	LB/RD	MAF Sensor Signal	0.6v
89	GY/WT	TP Sensor Signal	0.53-1.27v
90	BR/WT	Reference Voltage	4.9-5.1v
91	GY/RD	Analog Signal Return	<0.050v
92	LG	Brake Pedal Position Switch	Brake Off: 0v, On: 12v
93	RD/WT	HO2S-11 (B1 S1) Heater	On: 1v, Off: 12v
94	YL/LB	HO2S-21 (B2 S1) Heater	On: 1v, Off: 12v
95	WT/BK	HO2S-12 (B1 S2) Heater	On: 1v, Off: 12v
96	TN/YL	HO2S-22 (B2 S2) Heater	On: 1v, Off: 12v
97	RD	Vehicle Power	12-14v
98	LB	Injector 8 Control	3.5-3-7 ms
99	LG/OR	Injector 6 Control	3.5-3-7 ms
100	BR/LB	Injector 4 Control	3.5-3-7 ms
101	WT	Injector 2 Control	3.5-3-7 ms
102	---	Not Used	---
103	BK/WT	Power Ground	<0.1v
104	RD/YL	Coil Driver 4 Control	5° dwell

Pin Connector Graphic

```
           PCM 104-PIN CONNECTOR
   1 ●●●●●●●●●●●●●    ●●●●●●●●●●●●● 26
  27 ●●●●●●●●●●●●●       ●●●●●●●●●●●●● 52
  53 ●●●●●●●●●●●●●    ●●●●●●●●●●●●● 78
  79 ●●●●●●●●●●●●●    ●●●●●●●●●●●●● 104
```

Terminal View of 104-Pin PCM Wiring Harness Connector

Standard Colors and Abbreviations

Abbreviation	Color	Abbreviation	Color	Abbreviation	Color
BK	Black	GY	Gray	PK	Purple
BL	Blue	GN	Green	RD	Red
BR	Brown	LG	LT Green	TN	Tan
DB	Dark Blue	OR	Orange	WT	White
DG	DK Green	PK	Pink	YL	Yellow

1998-99 Mustang 4.6L 2v V8 MFI VIN W (All) 104 Pin Connector

PCM Pin #	Wire Color	Circuit Description (104 Pin)	Value at Hot Idle
1	PK/OR	Shift Solenoid 2 Control	12v, 55 mph: 12v
2	PK/LG	MIL (lamp) Control	MIL On: 1v, Off: 12v
3	YL/BK	Digital TR1 Sensor	0v, 55 mph: 11v
4-10	---	Not Used	---
11 ('98)	PK/WT	PF Sensor Signal	1.1-1.6v
12	YL/WT	Fuel Level Indicator Signal	1.7v (1/2 full)
13	PK	Flash EEPROM Power	0.1v
14	---	Not Used	---
15	PK/LB	Data Bus (-) Signal	Digital Signals
16	TN/OR	Data Bus (+) Signal	Digital Signals
17-20	---	Not Used	---
21	DB	CKP Sensor (+) Signal	390-450 Hz
22	GY	CKP Sensor (-) Signal	390-450 Hz
23	---	Not Used	---
24	BK/WT	Power Ground	<0.1v
25	BK	PCM Case Ground	<0.050v
26	DB/LG	Coil Driver 1 Control	5° dwell
27	OR/YL	Shift Solenoid 1 Control	1v, 55 mph: 1v
28	---	Not Used	---
29	TN/WT	TCS (switch) Signal	TCS & O/D On: 12v
30	DG	Octane Adjust Switch	Closed: 0v, Open: 9.3v
31-32	---	Not Used	---
33	PK/OR	VSS (-) Signal	0 Hz, 55 mph: 125 Hz
34	---	Not Used	---
35	RD/LG	HO2S-12 (B1 S2) Signal	0.1-1.1v
36	TN/LB	MAF Sensor Return	<0.050v
37	OR/BK	TFT Sensor Signal	2.10-2.40v
38	LG/RD	ECT Sensor Signal	0.5-0.6v
39	GY	IAT Sensor Signal	1.5-2.5v
40	DG/YL	Fuel Pump Monitor	On: 12v, Off: 0v
41	DG/OR	A/C High Pressure Switch	AC On: 12v, Off: 0v
42, 44	---	Not Used	---
43	WT/RD	Fuel Pump Control	On: 1v, Off: 12v
45	DB	Low Speed Fan Control	On: 1v, Off: 12v
46	LG/PK	High Speed Fan Control	On: 1v, Off: 12v
47	BR/PK	EGR VR Solenoid	0%, 55 mph: 45%
48	OR/WT	Clean Tachometer Output	39-45 Hz
49	LB/BK	Digital TR2 Sensor	0v, 55 mph: 11v
50	WT/BK	Digital TR4 Sensor	0v, 55 mph: 11v
51	WT/YL	Power Ground	<0.1v
52	RD/LB	Coil Driver 2 Control	5° dwell
53	---	Not Used	---
54	PK/YL	TCC Solenoid Control	0%, 55 mph: 95%
55	Y	Keep Alive Power	12-14v
56	LG/BK	EVAP VMV Solenoid	0-10 Hz (0-100%)
57, 59	---	Not Used	---
58	GY/BK	VSS (+) Signal	0 Hz, 55 mph: 125 Hz
60	GY/LB	HO2S-11 (B1 S1) Signal	0.1-1.1v
61	PK/LG	HO2S-22 (B2 S2) Signal	0.1-1.1v
62	RD/PK	FTP Sensor Signal	2.6v (0" H2O - cap off)
63	---	Not Used	---
64	LB/YL	TR Sensor Signal	In 'P': 0v, in O/D: 1.7v
65	BR/LG	DPFE Sensor Signal	0.20-1.30v
66	---	Not Used	---

1998-99 Mustang 4.6L 2v V8 MFI VIN W (All) 104 Pin Connector

PCM Pin #	Wire Color	Circuit Description (104 Pin)	Value at Hot Idle
67	GY/YL	EVAP Purge Solenoid	0-10 Hz (0-100%)
68	---	Not Used	---
69	PK/YL	A/C WOT Relay Control	On: 1v, Off: 12v
70	---	Not Used	---
71	RD	Vehicle Power	12-14v
72	TN/RD	Injector 7 Control	3.5-3-7 ms
73	TN/BK	Injector 5 Control	3.5-3-7 ms
74	BR/YL	Injector 3 Control	3.5-3-7 ms
75	TN	Injector 1 Control	3.5-3-7 ms
76-77	BK/WT	Power Ground	<0.1v
78	PK/WT	Coil Driver 3 Control	5° dwell
79	WT/LG	TCIL (lamp) Control	On: 1v, Off: 12v
80	LB/OR	Fuel Pump Control	On: 1v, Off: 12v
81	WT/YL	EPC Solenoid Control	9v (15 psi)
82	---	Not Used	---
83	WT/LB	IAC Motor Control	20% duty cycle
84	DG/WT	TSS Sensor Signal	43 Hz (650 rpm)
85	DB/OR	CMP Sensor Signal	5-7 Hz
86	TN/LG	A/C Pressure Switch	On: 1v, Off: 12v
87	RD/BK	HO2S-21 (B2 S1) Signal	0.1-1.1v
88	LB/RD	MAF Sensor Signal	0.6v
89	GY/WT	TP Sensor Signal	0.53-1.27v
90	BR/WT	Reference Voltage	4.9-5.1v
91	GY/RD	Analog Signal Return	<0.050v
92	LG	Brake Pedal Position Switch	Brake Off: 0v, On: 12v
93	RD/WT	HO2S-11 (B1 S1) Heater	On: 1v, Off: 12v
94	YL/LB	HO2S-21 (B2 S1) Heater	On: 1v, Off: 12v
95	WT/BK	HO2S-12 (B1 S2) Heater	On: 1v, Off: 12v
96	TN/YL	HO2S-22 (B2 S2) Heater	On: 1v, Off: 12v
97	RD	Vehicle Power	12-14v
98	LB	Injector 8 Control	3.5-3-7 ms
99	LG/OR	Injector 6 Control	3.5-3-7 ms
100	LB	Injector 4 Control	3.5-3-7 ms
101	WT	Injector 2 Control	3.5-3-7 ms
102	---	Not Used	---
103	BK/WT	Power Ground	<0.1v
104	RD/YL	Coil Driver 4 Control	5° dwell

Pin Connector Graphic

PCM 104-PIN CONNECTOR

Terminal View of 104-Pin PCM Wiring Harness Connector

Standard Colors and Abbreviations

Abbreviation	Color	Abbreviation	Color	Abbreviation	Color
BK	Black	GY	Gray	PK	Purple
BL	Blue	GN	Green	RD	Red
BR	Brown	LG	LT Green	TN	Tan
DB	Dark Blue	OR	Orange	WT	White
DG	DK Green	PK	Pink	YL	Yellow

1996-97 Mustang GT 4.6L 2v V8 VIN X (All) 104 Pin Connector

PCM Pin #	Wire Color	Circuit Description (104 Pin)	Value at Hot Idle
1	PK/OR	Shift Solenoid 2 Control	12v, 55 mph: 1v
2	PK/LG	MIL (lamp) Control	MIL On: 1v, Off: 12v
3-10	---	Not Used	---
11	PK/WT	Purge Flow Sensor	0.8v, 55 mph: 3v
12	---	Not Used	---
13	PK	Flash EEPROM Power	0.1v
14	---	Not Used	---
15	PK/LB	Data Bus (-) Signal	Digital Signals
16	TN/OR	Data Bus (+) Signal	Digital Signals
17-20	---	Not Used	---
21	DB	CKP Sensor (+) Signal	850-1120 Hz
22	GY	CKP Sensor (-) Signal	850-1120 Hz
23	---	Not Used	---
24	BK/WT	Power Ground	<0.1v
25	BK	PCM Case Ground	<0.050v
26	DB/LG	Coil Driver 1 Control	5° dwell
27	OR/YL	Shift Solenoid 1 Control	1v, 55 mph: 12v
28	---	Not Used	---
29	TN/WT	TCS (switch) Signal	TCS & O/D On: 12v
30	DG	Octane Adjust Switch	Closed: 0v, Open: 9.3v
31-32	---	Not Used	---
33	PK/OR	VSS (-) Signal	0 Hz, 55 mph: 125 Hz
34	---	Not Used	---
35	RD/LG	HO2S-12 (B1 S2) Signal	0.1-1.1v
36	TN/LB	MAF Sensor Return	<0.050v
37	OR/BK	TFT Sensor Signal	2.10-2.40v
38	LG/RD	ECT Sensor Signal	0.5-0.6v
39	GY	IAT Sensor Signal	1.5-2.5v
40	DG/YL	Fuel Pump Monitor	On: 12v, Off: 0v
41	DG/OR	A/C High Pressure Switch	AC On: 12v, Off: 0v
42-44	---	Not Used	---
45	DB	Low Speed Fan Control	On: 1v, Off: 12v
46	LG/PK	High Speed Fan Control	On: 1v, Off: 12v
47	BR/PK	EGR VR Solenoid	0%, 55 mph: 45%
48	OR/WT	Clean Tachometer Output	33-37 Hz
49-50	---	Not Used	---
51	BK/WT	Power Ground	<0.1v
52	RD/LB	Coil Driver 2 Control	5° dwell
53	---	Not Used	---
54	BR/OR	TCC Solenoid Control	0%, 55 mph: 95%
55	YL	Keep Alive Power	12-14v
56-57	---	Not Used	---
58	GY/BK	VSS (+) Signal	0 Hz, 55 mph: 125 Hz
59	---	Not Used	---
60	GY/LB	HO2S-11 (B1 S1) Signal	0.1-1.1v
61	PK/LG	HO2S-22 (B2 S2) Signal	0.1-1.1v
62-63	---	Not Used	---
64	LB/YL	A/T: TR Sensor	In 'P': 0v, 55 mph: 1.7v
64	LB/YL	M/T: Clutch Pedal Position	Clutch In: 0v, Out: 5v
65	BR/LG	DPFE Sensor Signal	0.20-1.30v
66	---	Not Used	---

1996-97 Mustang GT 4.6L 2v V8 VIN X (All) 104 Pin Connector

PCM Pin #	Wire Color	Circuit Description (104 Pin)	Value at Hot Idle
67	GY/YL	EVAP Purge Solenoid	0-10 Hz (0-100%)
68	----	Not Used	---
69	PK/YL	A/C WOT Relay Control	On: 1v, Off: 12v
70	---	Not Used	---
71	RD	Vehicle Power	12-14v
72	TN/RD	Injector 7 Control	3.5-3-7 ms
73	TN/BK	Injector 5 Control	3.5-3-7 ms
74	TN/RD	Injector 3 Control	3.5-3-7 ms
75	TN	Injector 1 Control	3.5-3-7 ms
76-77	BK/WT	Power Ground	<0.1v
78	PK/WT	Coil Driver 3 Control	5° dwell
79	WT/LG	TCIL (lamp) Control	On: 1v, Off: 12v
80	LB/OR	Fuel Pump Control	On: 1v, Off: 12v
81	WT/YL	EPC Solenoid Control	9v (15 psi)
82	---	Not Used	---
83	WT/LB	IAC Motor Control	30% duty cycle
84	DG/WT	TSS Sensor Signal	43 Hz (650 rpm)
85	DB/OR	CID Sensor Signal	5-7 Hz
86	TN/LG	A/C Pressure Switch	A/C On: 12v (Open)
87	RD/BK	HO2S-21 (B2 S1) Signal	0.1-1.1v
88	LB/RD	MAF Sensor Signal	0.6v
89	GY/WT	TP Sensor Signal	0.53-1.27v
90	BR/WT	Reference Voltage	4.9-5.1v
91	GY/RD	Analog Signal Return	<0.050v
92	LG	Brake Pedal Position Switch	Brake Off: 0v, On: 12v
93	RD/WT	HO2S-11 (B1 S1) Heater	On: 1v, Off: 12v
94	YL/LB	HO2S-21 (B2 S1) Heater	On: 1v, Off: 12v
95	WT/BK	HO2S-12 (B1 S2) Heater	On: 1v, Off: 12v
96	TN/YL	HO2S-22 (B2 S2) Heater	On: 1v, Off: 12v
97	RD	Vehicle Power	12-14v
98	LB	Injector 8 Control	3.5-3-7 ms
99	LG/OR	Injector 6 Control	3.5-3-7 ms
100	LB	Injector 4 Control	3.5-3-7 ms
101	WT	Injector 2 Control	3.5-3-7 ms
102	---	Not Used	---
103	BK/WT	Power Ground	<0.1v
104	RD/YL	Coil Driver 4 Control	5° dwell

Pin Connector Graphic

PCM 104-PIN CONNECTOR

Terminal View of 104-Pin PCM Wiring Harness Connector

Standard Colors and Abbreviations

Abbreviation	Color	Abbreviation	Color	Abbreviation	Color
BK	Black	GY	Gray	PK	Purple
BL	Blue	GN	Green	RD	Red
BR	Brown	LG	LT Green	TN	Tan
DB	Dark Blue	OR	Orange	WT	White
DG	DK Green	PK	Pink	YL	Yellow

1998-99 Mustang GT 4.6L 2v V8 VIN X (All) 104 Pin Connector

PCM Pin #	Wire Color	Circuit Description (104 Pin)	Value at Hot Idle
1	PK/OR	Shift Solenoid 2 Control	12v, 55 mph: 0.1v
2	PK/LG	MIL (lamp) Control	MIL On: 1v, Off: 12v
3	YL/BK	Digital TR1 Sensor	0v, 55 mph: 11v
4-10	---	Not Used	---
11 ('88)	PK/WT	PF Sensor Signal	1.1-1.6v
12	YL/WT	Fuel Level Indicator Signal	1.7v (1/2 full)
13	PK	Flash EEPROM Power	0.1v
14	---	Not Used	---
15	PK/LB	Data Bus (-) Signal	Digital Signals
16	TN/OR	Data Bus (+) Signal	Digital Signals
17-20	---	Not Used	---
21	DB	CKP Sensor (+) Signal	390-450 Hz
22	GY	CKP Sensor (-) Signal	390-450 Hz
23	---	Not Used	---
24	BK/WT	Power Ground	<0.1v
25	BK	PCM Case Ground	<0.050v
26	DB/LG	Coil Driver 1 Control	5° dwell
27	OR/YL	Shift Solenoid 1 Control	1v, 55 mph: 1v
28	---	Not Used	---
29	TN/WT	TCS (switch) Signal	TCS & O/D On: 12v
30	DG	Octane Adjust Switch	Closed: 0v, Open: 9.3v
31-32	---	Not Used	---
33	PK/OR	VSS (-) Signal	0 Hz, 55 mph: 125 Hz
34	---	Not Used	---
35	RD/LG	HO2S-12 (B1 S2) Signal	0.1-1.1v
36	TN/LB	MAF Sensor Return	<0.050v
37	OR/BK	TFT Sensor Signal	2.10-2.40v
38	LG/RD	ECT Sensor Signal	0.5-0.6v
39	GY	IAT Sensor Signal	1.5-2.5v
40	DG/YL	Fuel Pump Monitor	On: 12v, Off: 0v
41	DG/OR	A/C High Pressure Switch	AC On: 12v, Off: 0v
42	---	Not Used	---
43	WT/RD	Fuel Pump Control	On: 1v, Off: 12v
44	---	Not Used	---
45	DB	Low Speed Fan Control	On: 1v, Off: 12v
46	LG/PK	High Speed Fan Control	On: 1v, Off: 12v
47	BR/PK	EGR VR Solenoid	0%, 55 mph: 45%
48	OR/WT	Clean Tachometer Output	39-45 Hz
49	LB/BK	Digital TR2 Sensor	0v, 55 mph: 11v
50	WT/BK	Digital TR4 Sensor	0v, 55 mph: 11v
51	WT/YL	Power Ground	<0.1v
52	RD/LB	Coil Driver 2 Control	5° dwell
53	---	Not Used	---
54	PK/YL	TCC Solenoid Control	0%, 55 mph: 95%
55	YL	Keep Alive Power	12-14v
56	LG/BK	EVAP VMV Solenoid	0-10 Hz (0-100%)
57, 59	---	Not Used	---
58	GY/BK	VSS (+) Signal	0 Hz, 55 mph: 125 Hz
60	GY/LB	HO2S-11 (B1 S1) Signal	0.1-1.1v
61	PK/LG	HO2S-22 (B2 S2) Signal	0.1-1.1v
62	RD/PK	FTP Sensor Signal	2.6v (0" H2O - cap off)
63	---	Not Used	---
64	DG/WT	TR Sensor Signal	In 'P': 0v, in O/D: 1.7v
65	BR/LG	DPFE Sensor Signal	0.20-1.30v

1998-99 Mustang GT 4.6L 2v V8 VIN X (All) 104 Pin Connector

PCM Pin #	Wire Color	Circuit Description (104 Pin)	Value at Hot Idle
66	---	Not Used	---
67	GY/YL	EVAP Purge Solenoid	0-10 Hz (0-100%)
68	---	Not Used	---
69	PK/YL	A/C WOT Relay Control	On: 1v, Off: 12v
70	---	Not Used	---
71	RD	Vehicle Power	12-14v
72	TN/RD	Injector 7 Signal (INJ 7)	3.5-3-7 ms
73	TN/BK	Injector 5 Control	3.5-3-7 ms
74	TN/RD	Injector 3 Control	3.5-3-7 ms
75	TN	Injector 1 Control	3.5-3-7 ms
76-77	BK/WT	Power Ground	<0.1v
78	PK/WT	Coil Driver 3 Control	5° dwell
79	WT/LG	TCIL (lamp) Control	On: 1v, Off: 12v
80	LB/OR	Fuel Pump Control	On: 1v, Off: 12v
81	WT/YL	EPC Solenoid Control	9v (15 psi)
82	---	Not Used	---
83	WT/LB	IAC Motor Control	20% duty cycle
84	DG/WT	TSS Sensor Signal	43 Hz (650 rpm)
85	DB/OR	CID Sensor Signal	5-7 Hz
86	TN/LG	A/C Pressure Switch	On: 1v, Off: 12v
87	RD/BK	HO2S-21 (B2 S1) Signal	0.1-1.1v
88	LB/RD	MAF Sensor Signal	0.6v
89	GY/WT	TP Sensor Signal	0.53-1.27v
90	BR/WT	Reference Voltage	4.9-5.1v
91	GY/RD	Analog Signal Return	<0.050v
92	LG	Brake Pedal Position Switch	Brake Off: 0v, On: 12v
93	RD/WT	HO2S-11 (B1 S1) Heater	On: 1v, Off: 12v
94	YL/LB	HO2S-21 (B2 S1) Heater	On: 1v, Off: 12v
95	WT/BK	HO2S-12 (B1 S2) Heater	On: 1v, Off: 12v
96	TN/YL	HO2S-22 (B2 S2) Heater	On: 1v, Off: 12v
97	RD	Vehicle Power	12-14v
98	LB	Injector 8 Signal (INJ 8)	3.5-3-7 ms
99	LG/OR	Injector 6 Control	3.5-3-7 ms
100	LB	Injector 4 Control	3.5-3-7 ms
101	WT	Injector 2 Control	3.5-3-7 ms
102	---	Not Used	---
103	BK/WT	Power Ground	<0.1v
104	RD/YL	Coil Driver 4 Control	5° dwell

Pin Connector Graphic

PCM 104-PIN CONNECTOR

Terminal View of 104-Pin PCM Wiring Harness Connector

Standard Colors and Abbreviations

Abbreviation	Color	Abbreviation	Color	Abbreviation	Color
BK	Black	GY	Gray	PK	Purple
BL	Blue	GN	Green	RD	Red
BR	Brown	LG	LT Green	TN	Tan
DB	Dark Blue	OR	Orange	WT	White
DG	DK Green	PK	Pink	YL	Yellow

2000-02 Mustang GT, GT Bullitt 4.6L 2v V8 VIN X (All) 104 Pin Connector

PCM Pin #	Wire Color	Circuit Description (104 Pin)	Value at Hot Idle
1	OR/YL	COP 6 (Integrated) Dwell	6° dwell
2	---	Not Used	---
3	BK/WT	Power Ground	<0.1v
4-5	---	Not Used	---
6	OR/YL	Shift Solenoid 'A' Control	1v, Others: 1v
7, 9	---	Not Used	---
8 ('02-'03)	WT/OR	IMRC Monitor Signal	5v, 55 mph: 5v
11	VT/OR	Shift Solenoid 'B' Control	12v, Others: 1v
12, 14	---	Not Used	---
13	VT	Flash EEPROM Power	0.1v
15	PK/LB	SCP Bus (-) Signal	Digital Signals
16	TN/OR	SCP Bus (+) Signal	Digital Signals
17-18, 20	---	Not Used	---
19	DB	Low Speed Fan Control	On: 1v, Off: 12v
21	DB	CKP Sensor (+) Signal	390-450 Hz
22	GY	CKP Sensor (-) Signal	390-450 Hz
23-24	---	Not Used	---
25	BK	PCM Case Ground	<0.050v
26	LG/WT	COP 1 (Integrated) Dwell	6° dwell
27	LG/YL	COP 5 (Integrated) Dwell	6° dwell
28	DB	Low Speed Fan Relay Control	Relay Off: 12v, On: 1v
29	TN/WT	TCS (switch) Signal	TCS & O/D On: 12v
30-33	---	Not Used	---
34	YL/BK	Digital TR1 Sensor	0v, 55 mph: 11v
35	RD/LG	HO2S-12 (B1 S2) Signal	0.1-1.1v
36	TN/LB	MAF Sensor Return	<0.050v
37	OR/BK	TFT Sensor Signal	2.10-2.40v
38	LG/RD	ECT Sensor Signal	0.5-0.6v
39	GY	IAT Sensor Signal	1.5-2.5v
40	LB/OR	Fuel Pump Monitor	3.5v (100%)
41	DG/OR	A/C Cycling Switch Signal	A/C On: 12v, Off: 1v
42	BR	IMRC Solenoid Control	12v (Off)
43-44	---	Not Used	---
45	LG/VT	Supercharger Bypass Solenoid Control	Solenoid Off: 12v, On: 1v
46	LG/PK	High Speed Fan Control	On: 1v, Off: 12v
47	BR/PK	EGR VR Solenoid	0%, 55 mph: 45%
49	LB/BK	Digital TR2 Sensor	0v, 55 mph: 11v
50	WT/BK	Digital TR4 Sensor	0v, 55 mph: 11v
51	BK/WT	Power Ground	<0.1v
52	WT/PK	COP 3 (Integrated) Dwell	6° dwell
53	DG/PK	COP 4 (Integrated) Dwell	6° dwell
54	BR/OR	TCC Solenoid Control	0%, 55 mph: 95%
55	RD/WT	Keep Alive Power	12-14v
56	LG/BK	EVAP Purge Solenoid	0-10 Hz (0-100%)
57-59	---	Not Used	---
60	GY/LB	HO2S-11 (B1 S1) Signal	0.1-1.1v
61	PK/LG	HO2S-22 (B2 S2) Signal	0.1-1.1v
62	RD/PK	FTP Sensor Signal	2.6v (0" H2O - cap off)
63	RD/PK	FRP Sensor Signal	3.0v (43 psi)
64	LB/YL	Digital TR3A Sensor	0v, 55 mph: 11v
65 ('00)	BR/LG	DPFE Sensor Signal	0.20-1.30v
65 ('01-'03)	BR/LG	DPFE Sensor Signal	0.95-1.05v
67	GY/YL	EVAP CV Solenoid	0-10 Hz (0-100%)
68	WT/OR	VSS (+) Signal	0 Hz, 55 mph: 125 Hz

2000-02 Mustang GT, GT Bullitt 4.6L 2v V8 VIN X (All) 104 Pin Connector

PCM Pin #	Wire Color	Circuit Description (104 Pin)	Value at Hot Idle
69	PK/YL	A/C WOT Relay Control	A/C On: 12v, Off: 1v
70	---	Not Used	---
71	RD	Vehicle Power	12-14v
72	TN/RD	Injector 7 Control	3.5-3.7 ms
73	TN/BK	Injector 5 Control	3.5-3.7 ms
74	BR/YL	Injector 3 Control	3.5-3.7 ms
75	TN	Injector 1 Control	3.5-3.7 ms
76, 77	BK/WT	Power Ground	<0.1v
78	PK/LB	COP 7 (Integrated) Dwell	6° dwell
79	WT/RD	COP 8 (Integrated) Dwell	6° dwell
80	WT/RD	Fuel Pump Control	1.3v (28%)
81	WT/YL	EPC Solenoid Control	9.5v (15-20 psi)
82	WT/OR	Charge Air Cooler Pump Relay Control	Relay Off: 12v, On: 1v
83	WT/LB	IAC Motor Control	10v (30% duty cycle)
84	DG/WT	OSS Sensor (+) Signal	0 Hz, 55 mph: 250 Hz
85	DB/OR	CMP Sensor Signal	5-7 Hz
86	TN/LG	A/C High Pressure Switch	A/C On: 12v, Off: 1v
87	RD/BK	HO2S-21 (B2 S1) Signal	0.1-1.1v
88	LB/RD	MAF Sensor Signal	0.6-0.9v
89	GY/WT	TP Sensor Signal	0.53-1.27v
90	BR/WT	Reference Voltage	4.9-5.1v
91	GY/RD	Analog Signal Return	<0.050v
92	LG	Brake Pedal Position Switch	Brake Off: 0v, On: 12v
93	RD/WT	HO2S-11 (B1 S1) Heater	On: 1v, Off: 12v
94	YL/LB	HO2S-21 (B2 S1) Heater	On: 1v, Off: 12v
95	WT/BK	HO2S-12 (B1 S2) Heater	On: 1v, Off: 12v
96	TN/YL	HO2S-22 (B2 S2) Heater	On: 1v, Off: 12v
97	RD	Vehicle Power	12-14v
98	LB	Injector 8 Control	3.5-3.7 ms
99	LG/OR	Injector 6 Control	3.5-3.7 ms
100	BR/LB	Injector 4 Control	3.5-3.7 ms
101	WT	Injector 2 Control	3.5-3.7 ms
102	---	Not Used	---
103	BK/WT	Power Ground	<0.1v
104	PK/WT	COP 2 (Integrated) Dwell	6° dwell

Pin Connector Graphic

PCM 104-PIN CONNECTOR

Terminal View of 104-Pin PCM Wiring Harness Connector

Standard Colors and Abbreviations

Abbreviation	Color	Abbreviation	Color	Abbreviation	Color
BK	Black	GY	Gray	PK	Purple
BL	Blue	GN	Green	RD	Red
BR	Brown	LG	LT Green	TN	Tan
DB	Dark Blue	OR	Orange	WT	White
DG	DK Green	PK	Pink	YL	Yellow

2003-05 Mustang Cobra 4.6L V8 DOHC VIN Y (M/T) 104 Pin Connector

PCM Pin #	Wire Color	Circuit Description (104 Pin)	Value at Hot Idle
1	VT/YL	COP 6 (Integrated) Dwell	6° dwell
2	---	Not Used	---
3	BK/WT	Power Ground	<0.1v
4-7	---	Not Used	---
8	WT/OR	IMRC Monitor Signal	5v, 55 mph: 5v
9	---	Not Used	---
10	DB/LG	Barometric Pressure Sensor Signal	159 Hz
11-12	---	Not Used	---
13	VT	Flash EEPROM Power	0.1v
14	---	Not Used	---
15	PK/LB	SCP Bus (-) Signal	Digital Signals
16	TN/OR	SCP Bus (+) Signal	Digital Signals
17-19	---	Not Used	---
20	TN/RD	Reverse Lockout Solenoid Control	Solenoid Off: 12v, On: 1v
21	DB	CKP Sensor (+) Signal	320-420 Hz
22	GY	CKP Sensor (-) Signal	320-420 Hz
23-24	---	Not Used	---
25	BK	PCM Case Ground	<0.050v
26	LG/WT	COP 1 (Integrated) Dwell	6° dwell
27	LG/YL	COP 5 (Integrated) Dwell	6° dwell
28-34	---	Not Used	---
35	RD/LG	HO2S-12 (B1 S2) Signal	0.1-1.1v
36	TN/LB	MAF Sensor Return	<0.050v
37	RD/YL	Intake Air Temperature Sensor 2 Signal	1.5-2.5v
38	---	Not Used	---
39	GY	Intake Air Temperature Sensor 1 Signal	1.5-2.5v
40	LB/OR	Fuel Pump Monitor	2.5-7.5v
41	DG/OR	A/C Cycling Switch Signal	A/C On: 12v, Off: 1v
42	BR	IMRC Solenoid Control	12v (Off)
43-44	---	Not Used	---
45	LG/VT	Supercharger Bypass Solenoid Control	Solenoid Off: 12v, On: 1v
46	---	Not Used	---
47	BR/PK	EGR VR Solenoid	0%, 55 mph: 45%
48-50	---	Not Used	---
51	BK/WT	Power Ground	<0.1v
52	WT/PK	COP 3 (Integrated) Dwell	6° dwell
53	DG/PK	COP 4 (Integrated) Dwell	6° dwell
54	---	Not Used	---
55	RD/WT	Keep Alive Power	12-14v
56	LG/BK	EVAP Purge Solenoid	0-10 Hz (0-100%)
57-59	---	Not Used	---
60	GY/LB	HO2S-11 (B1 S1) Signal	0.1-1.1v
61	VT/LG	HO2S-22 (B2 S2) Signal	0.1-1.1v
62	RD/PK	FTP Sensor Signal	2.6v (0" H2O - cap off)
63	RD/PK	FRP Sensor Signal	2.8v (39 psi)
64	---	Not Used	---
65	BR/LG	DPFE Sensor Signal	0.95-1.05v
66	YE/LG	Cylinder Head Temperature Sensor	0.5-0.6v
67	GY/YL	EVAP CV Solenoid	0-10 Hz (0-100%)
68	WT/OR	VSS (+) Signal	Moving: AC pulse signals
69	PK/YL	A/C WOT Relay Control	A/C On: 12v, Off: 1v
70	---	Not Used	---

2003-05 Mustang Cobra 4.6L V8 DOHC VIN Y (M/T) 104 Pin Connector

PCM Pin #	Wire Color	Circuit Description (104 Pin)	Value at Hot Idle
71	RD	Vehicle Power (Start-Run)	12-14v
72	TN/RD	Injector 7 Control	2.4-2.8 ms
73	TN/BK	Injector 5 Control	2.4-2.8 ms
74	BR/YL	Injector 3 Control	2.4-2.8 ms
75	TN	Injector 1 Control	2.4-2.8 ms
76	BK/WT	Power Ground	<0.1v
77	BK/WT	Power Ground	<0.1v
78	PK/LB	COP 7 (Integrated) Dwell	6° dwell
79	WT/RD	COP 8 (Integrated) Dwell	6° dwell
80	WT/RD	Fuel Pump Control	Off: 12v, On: 1v
81	---	Not Used	---
82	WT/DG	Charge Air Cooler Pump Relay Control	Relay Off: 12v, On: 1v
83	WT/LB	IAC Motor Control	9v (32% duty cycle)
84	DG/WT	OSS Sensor (+) Signal	0 Hz, 55 mph: 470 Hz
85	DB/OR	CMP Sensor Signal	5-7 Hz
86	TN/LG	A/C High Pressure Switch	A/C On: 12v, Off: 1v
87	RD/BK	HO2S-21 (B2 S1) Signal	0.1-1.1v
88	LB/RD	MAF Sensor Signal	0.6v
89	GY/WT	TP Sensor Signal	0.53-1.27v
90	BR/WT	Reference Voltage	4.9-5.1v
91	GY/RD	Sensor Ground	<0.050v
92	LG	Brake Pedal Position Switch	Brake Off: 0v, On: 12v
93	RD/WT	HO2S-11 (B1 S1) Heater	On: 1v, Off: 12v
94	YL/LB	HO2S-21 (B2 S1) Heater	On: 1v, Off: 12v
95	WT/BK	HO2S-12 (B1 S2) Heater	On: 1v, Off: 12v
96	TN/YL	HO2S-22 (B2 S2) Heater	On: 1v, Off: 12v
97	RD	Vehicle Power (Start-Run)	12-14v
98	LB	Injector 8 Control	2.4-2.8 ms
99	LG/OR	Injector 6 Control	2.4-2.8 ms
100	BR/LB	Injector 4 Control	2.4-2.8 ms
101	WT	Injector 2 Control	2.4-2.8 ms
102	DG/PK	Knock Sensor 2 (+) Signal	0v
103	BK/WT	Power Ground	<0.1v
104	PK/WT	COP 2 (Integrated) Dwell	6° dwell

Pin Connector Graphic

PCM 104-PIN CONNECTOR

Terminal View of 104-Pin PCM Wiring Harness Connector

Standard Colors and Abbreviations

Abbreviation	Color	Abbreviation	Color	Abbreviation	Color
BK	Black	GY	Gray	PK	Purple
BL	Blue	GN	Green	RD	Red
BR	Brown	LG	LT Green	TN	Tan
DB	Dark Blue	OR	Orange	WT	White
DG	DK Green	PK	Pink	YL	Yellow

1990-92 Mustang 5.0L V8 HO VIN E (All) 60 Pin Connector

PCM Pin #	Wire Color	Circuit Description (60 Pin)	Value at Hot Idle
1	YL	Keep Alive Power	12-14v
2, 5, 8, 11, 18, 28	---	Not Used	---
3	GY, PK	VSS (+), (-) Signal	0 Hz, 55 mph: 125 Hz
4	TN/YL	Ignition Diagnostic Monitor	20-31 Hz
7	LG/RD	ECT Sensor Signal	0.5-0.6v
9	TN/BL	MAF Sensor Return	<0.050v
10	BK/YL	A/C Cycling Clutch Switch	A/C On: 12v, Off: 0v
12	BR/YL	Injector 3 Control	4.6-4.8 ms
13	BR/BL	Injector 4 Control	4.6-4.8 ms
14	TN/BK	Injector 5 Control	4.6-4.8 ms
15	LG/OR	Injector 6 Control	4.6-4.8 ms
16	OR/RD	Ignition System Ground	<0.050v
17	PK/LG	STI Output, MIL Control	MIL On: 1v, Off: 12v
19	DG/YL	Fuel Pump Monitor	On: 12v, Off: 0v
20	BK	PCM Case Ground	<0.050v
21	WT/BL	IAC Motor Control	8-10v
22	BL/OR	Fuel Pump Control	On: 1v, Off: 12v
23-24, 34-35	---	Not Used	---
25	GY	ACT Sensor Signal	1.5-2.5v
26	BR/WT	Reference Voltage	4.9-5.1v
27	BR/LG	EGR EVP Sensor	Idle: 0.4 v
29	GY/BL	HO2S-21 (B2 S1) Signal	0.1-1.1v
30	BL/YL	Neutral Drive (A/T), Clutch (M/T) Switch	In 'P': 0v or with Clutch In: 0v
31	GY/YL	EVAP Purge Solenoid	0%, 55 mph: 40%
32	WT/LG	Thermactor Air Divert SOL	On: 1v, Off: 12v
33	BR/PK	EGR VR Solenoid	0%, 55 mph: 45%
36	PK	Spark Output Signal	50% duty cycle
37, 57	RD	Vehicle Power	12-14v
38	WT/YL	Thermactor Air Bypass SOL	TAB On: 1v, Off: 12v
39, 41, 51, 53, 55	---	Not Used	---
40, 60	BK/WT	Power Ground	<0.1v
42	TN/RD	Injector 7 Control	4.6-4.8 ms
43	RD/BK	HO2S-11 (B1 S1) Signal	0.1-1.1v
44	GY/RD	HO2S-11 (Bank 1) Ground	<0.050v
45	LG/BK	BARO Sensor Signal	159 Hz (sea level)
46	GY/RD	Analog Signal Return	<0.050v
47	GY/WT	TP Sensor Signal	0.5-1.2v
48	WT/PK	Self-Test Indicator Signal	STI Open: 5v, Closed: 1v
49	OR	HO2S-21 (Bank 2) Ground	<0.050v
50	LB/RD	MAF Sensor Signal	0.20-1.30v
52	LB	Injector 8 Control	4.6-4.8 ms
54	PK/YL	A/C WOT Cutoff Relay	On: 1v, Off: 12v
56	GY/OR	PIP Sensor Signal	50% duty cycle
58	TN	Injector 1 Control	4.6-4.8 ms
59	WT	Injector 2 Control	4.6-4.8 ms

Pin Connector Graphic

PCM 60-PIN CONNECTOR

Terminal View of 60-Pin PCM Harness Connector

1993 Mustang 5.0L V8 HO MFI VIN E (All) 60 Pin Connector

PCM Pin #	Wire Color	Circuit Description (60 Pin)	Value at Hot Idle
1	YL	Keep Alive Power	12-14v
2	---	Not Used	---
3	GY/BK	VSS (+) Signal	0 Hz, 55 mph: 125 Hz
4	TN/YL	Ignition Diagnostic Monitor	20-31 Hz
5	---	Not Used	---
6	PK/OR	VSS (-) Signal	0 Hz, 55 mph: 125 Hz
7	LG/RD	ECT Sensor Signal	0.5-0.6v
8	---	Not Used	---
9	TN/BL	MAF Sensor Return	<0.050v
10	BK/YL	A/C Cycling Clutch Switch	A/C On: 12v, Off: 0v
11	---	Not Used	---
12	BR/YL	Injector 3 Control	4.6-5.2 ms
13	BR/BL	Injector 4 Control	4.6-5.2 ms
14	TN/BK	Injector 5 Control	4.6-5.2 ms
15	LG/OR	Injector 6 Control	4.6-5.2 ms
16	OR/RD	Ignition System Ground	<0.050v
17	PK/LG	STI Output, MIL Control	MIL On: 1v, Off: 12v
18	---	Not Used	---
19	DG/YL	Fuel Pump Monitor	On: 12v, Off: 0v
20	BK	PCM Case Ground	<0.1v
21	WT/BL	IAC Motor Control	8-9v
22	BL/OR	Fuel Pump Control	On: 1v, Off: 12v
23-24	---	Not Used	---
25	GY	IAT Sensor Signal	1.5-2.5v
26	BR/WT	Reference Voltage	4.9-5.1v
27	BR/LG	EGR EVP Sensor	0.4v
28	---	Not Used	---
29	GY/BL	HO2S-21 (B2 S1) Signal	0.1-1.1v
30	BL/YL	A/T: Park/Neutral Position	In 'P': 0v, Others: 5v
30	BL/YL	M/T: Clutch Pedal Position	Clutch In: 0v, Out: 5v
31	GY/YL	EVAP Purge Solenoid	0%, 55 mph: 50%
32	WT/LG	Air Injection Diverter Solenoid	AIRD On: 1v, Off: 12v
33	BR/PK	EGR VR Solenoid	0%, 55 mph: 45%
34-35	---	Not Used	---
36	PK	Spark Output Signal	50% duty cycle
37	RD	Vehicle Power	12-14v
38	WT/YL	Air Injection Bypass Solenoid	AIRB On: 1v, Off: 12v
39	---	Not Used	---
40	BK/WT	Power Ground	<0.1v

1993 Mustang 5.0L V8 HO MFI VIN E (All) 60 Pin Connector

PCM Pin #	Wire Color	Circuit Description (60 Pin)	Value at Hot Idle
41	---	Not Used	---
42	TN/RD	Injector 7 Control	4.6-5.2 ms
43	RD/BK	HO2S-11 (B1 S1) Signal	0.1-1.1v
44	GY/RD	HO2S-11 (Bank 1) Ground	<0.050v
45	LG/BK	BARO Sensor Signal	159 Hz (sea level)
46	GY/RD	Analog Signal Return	<0.050v
47	GY/WT	TP Sensor Signal	0.5-1.2v
48	WT/PK	Self-Test Indicator Signal	STI Open: 5v, Closed: 1v
49	OR	HO2S-21 (Bank 2) Ground	<0.050v
50	LB/RD	MAF Sensor Signal	0.6-0.9v
51	---	Not Used	---
52	LB	Injector 8 Control	4.6-5.2 ms
53	---	Not Used	---
54	PK/YL	A/C WOT Cutoff Relay	On: 1v, Off: 12v
55	---	Not Used	---
56	GY/OR	PIP Sensor Signal	50% duty cycle
57	RD	Vehicle Power	12-14v
58	TN	Injector 1 Control	4.6-5.2 ms
59	WT	Injector 2 Control	4.6-5.2 ms
60	BK/WT	Power Ground	<0.1v

Pin Connector Graphic

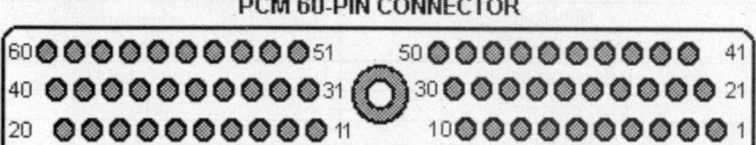

Terminal View of 60-Pin PCM Harness Connector

Standard Colors and Abbreviations

Abbreviation	Color	Abbreviation	Color	Abbreviation	Color
BK	Black	GY	Gray	PK	Purple
BL	Blue	GN	Green	RD	Red
BR	Brown	LG	LT Green	TN	Tan
DB	Dark Blue	OR	Orange	WT	White
DG	DK Green	PK	Pink	YL	Yellow

1994-95 Mustang 5.0L V8 MFI VIN D, VIN T (All) 60 Pin Connector

PCM Pin #	Wire Color	Circuit Description (60 Pin)	Value at Hot Idle
1	YL	Keep Alive Power	12-14v
2	LG	Brake Pedal Position Switch	Brake Off: 0v, On: 12v
3	GY/BK	VSS (+) Signal	0 Hz, 55 mph: 125 Hz
4	WT/PK	Ignition Diagnostic Monitor	20-31 Hz
5	DG/WT	TSS Sensor Signal	0 Hz, 35 mph: 115 Hz
6	PK/OR	VSS (-) Signal	0 Hz, 55 mph: 125 Hz
7	LG/RD	ECT Sensor Signal	0.5-0.6v
8	DG/YL	Fuel Pump Monitor	On: 12v, Off: 0v
9	TN/BL	MAF Sensor Return	<0.050v
10	DG/OR	A/C Cycling Clutch Signal	A/C On: 12v, Off: 0v
11	GY/YL	EVAP Purge Solenoid	0%, 55 mph: 60%
12	LG/OR	Injector 6 Control	5.0-5.2 ms
13	TN/RD	Injector 7 Control	5.0-5.2 ms
14	BL	Injector 8 Control	5.0-5.2 ms
15	TN/BK	Injector 5 Control	5.0-5.2 ms
16	OR/RD	Ignition System Ground	<0.050v
17	PK/LG	STI Output, MIL Control	MIL On: 1v, Off: 12v
18	TN/OR	Data Bus (-) Signal	Digital Signals
19	PK/BL	Data Bus (+) Signal	Digital Signals
20	BK	PCM Case Ground	<0.050v
21	WT/BL	IAC Motor Control	8-9v
22	BL/OR	Fuel Pump Control	On: 1v, Off: 12v
23-24	---	Not Used	---
25	GY	IAT Sensor Signal	1.5-2.5v
26	BR/WT	Reference Voltage	4.9-5.1v
27	BR/LG	EGR EVP Sensor	0.4v
28-29	---	Not Used	---
30	BL/YL	MLP Sensor Signal	In 'P': 0v, 55 mph: 1.7v
31	WT/OR	Air Injection Bypass Solenoid	AIRB On: 1v, Off: 12v
32	LG/PK	High Speed Fan Control	On: 1v, Off: 12v
33	BR/PK	EGR VR Solenoid	0%, 55 mph: 45%
34	BR	Air Injection Diverter Solenoid	AIRD On: 1v, Off: 12v
35	BR/BL	Injector 4 Control	5.0-5.2 ms
36	PK	Spark Output Signal	50% duty cycle
37	RD	Vehicle Power	12-14v
38	WT/YL	EPC Solenoid Control	Hot Idle: 9.1v (4 psi)
39	BR/YL	Injector 3 Control	5.0-5.2 ms
40	BK/WT	Power Ground	<0.1v

1994-95 Mustang 5.0L V8 MFI VIN D, VIN T (All) 60 Pin Connector

PCM Pin #	Wire Color	Circuit Description (60 Pin)	Value at Hot Idle
41	TN/WT	TCS (switch) Signal	TCS/OD On: 1v, Off: 12v
42	PK/BL	A/C Hi Pressure Cutoff Switch	On: 1v, Off: 12v
43	RD/BK	HO2S-11 (B1 S1) Signal	0.1-1.1v
44	GY/BL	HO2S-21 (B2 S1) Signal	0.1-1.1v
45	---	Not Used	---
46	GY/RD	Analog Signal Return	<0.050v
47	GY/WT	TP Sensor Signal	0.5-1.2v
48	WT/PK	Self-Test Indicator Signal	STI Open: 5v, Closed: 1v
49	OR/BK	TOT Sensor Signal	2.10-2.40v
50	LB/RD	MAF Sensor Signal	0.6-0.9v
51	OR/YL	Shift Solenoid 1 Control	1v, 55 mph: 12v
52	PK/OR	Shift Solenoid (SS2)	12v, 55 mph: 1v
53	PK/YL	Torque Converter Clutch	0%, 55 mph: 40%
53	BR/OR	Torque Converter Clutch	0%, 55 mph: 40%
54	PK/YL	A/C WOT Relay Control	On: 1v, Off: 12v
55	DB	Low Speed Fan Control	On: 1v, Off: 12v
56	GY/OR	PIP Sensor Signal	50% duty cycle
57	RD	Vehicle Power	12-14v
58	TN	Injector 1 Control	5.0-5.2 ms
59	WT	Injector 2 Control	5.0-5.2 ms
60	BK/WT	Power Ground	<0.1v

Pin Connector Graphic

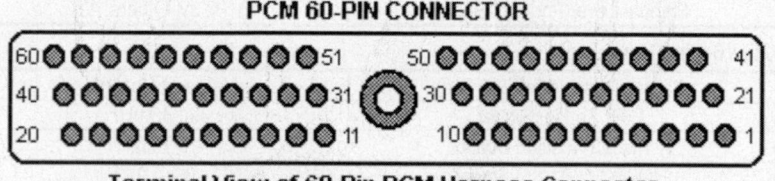

PCM 60-PIN CONNECTOR

Terminal View of 60-Pin PCM Harness Connector

Standard Colors and Abbreviations

Abbreviation	Color	Abbreviation	Color	Abbreviation	Color
BK	Black	GY	Gray	PK	Purple
BL	Blue	GN	Green	RD	Red
BR	Brown	LG	LT Green	TN	Tan
DB	Dark Blue	OR	Orange	WT	White
DG	DK Green	PK	Pink	YL	Yellow

PROBE PIN TABLES

1993-95 Probe 2.0L I4 MFI VIN A (A/T) 22 Pin Connector

PCM Pin #	Wire Color	Circuit Description (22 Pin)	Value at Hot Idle
22-A	BL/RD	Keep Alive Power	12- 14v
22-B	RD/BK	Vehicle Power (Power Relay)	12-14v
22-C	WT/RD	System Monitor Lamp	Lamp On: 5v, Off: 12v
22-D	BK/RD	Vehicle Start Signal	Cranking: 9-11v
22-E	WT/BL	MIL (lamp) Control	MIL On: 1v, MIL Off: 12v
22-F	WT/GN	Brake Pedal Position Switch	Brake Off: 0v, On: 12v
22-G	LG/RD	Self-Test Output Signal	STO On: 5v, Off: 12v
22-H	BR/BK	Overdrive Off Switch Signal	OD Switch On: 0.1v
22-I	BK/OR	Ignition Control Module Signal	0.2v, 55 mph: 0.3vv
22-J	WT	Headlamp Switch Signal	Lamps On: 12v, Off: 0.1v
22-K	GN	Daylight Running Lamp (DRL)	DRL On: 2.5v, Off: 12v
22-L	RD/WT	Self-Test Input	STI On: 0.1v, Off: 5v
22-M	GN/RD	VSS Signal to Cluster	Digital Signals
22-N	BR/YL	PSP Switch Signal	Straight: 12v, Turned: 0v
22-O	PK	Rear Defrost (DEF) Signal	Switch Off: 0.1v, On: 12v
22-P	OR/BK	Blower Motor Switch Signal	Off or 1st: 12v, 2nd on: 1v
22-Q	PK/GN	Overdrive Lamp (Cluster)	OD On: 1v, Off: 12v
22-R	PK/BL	A/C Cycling Clutch Switch	A/C On: 2v, Off: 12v
22-S	YL/WT	AC Relay Control	On: 1v, Off: 12v
22-T	BR	Idle Position Switch Signal	IDL closed: 0v, open: 12v
22-U	RD/BK	Vehicle Power	12-14v
22-V	GN	Manual Lever Position Signal	In 'P': 0v, Others: 12v

1993-95 Probe 2.0L I4 MFI VIN A (A/T) 16 Pin Connector

PCM Pin #	Wire Color	Circuit Description (16 Pin)	Value at Hot Idle
16-A	GN/OR	CKP Sensor Signal	390-450 Hz
16-B	LG/WT	Camshaft Position Sensor	5-7 Hz
16-C	BK/YL	HO2S-11 (B1 S1) Signal	0.1-1.1v
16-D	YL	TP Sensor Signal	0.5-1.3v
16-E	YL/BK	ECT Sensor Signal	0.5-0.6v
16-F	RD/WT	IAT Sensor Signal	1-5-2.5v
16-G	BL	Barometric Pressure Sensor	4.3v (28" Hg)
16-H	BL/YL	TOT Sensor Signal	2.10-2.40v
16-I	PK	Reference Voltage	4.9-5.1v
16-J	RD	EGR Temperature Sensor	2.25-2.55v (68-120°F)
16-K	BL/WT	Cooling Fan Temp. Sensor	0.8v
16-L	BK/RD	MAF Sensor Signal	1.9v
16-M	RD	MLP Sensor - Reverse	In 'R': 0v, Others: 0v
16-N	BL/OR	MLP Sensor - First	In 1st: 0v, Others: 0v
16-O	GN/WT	MLP Sensor - Second	In 2nd: 0v, Others: 0v
16-P	RD/BL	MLP Sensor - Drive	In 'P': 0v, Others: 12v

Standard Colors and Abbreviations

Abbreviation	Color	Abbreviation	Color	Abbreviation	Color
BK	Black	GY	Gray	PK	Purple
BL	Blue	GN	Green	RD	Red
BR	Brown	LG	LT Green	TN	Tan
DB	Dark Blue	OR	Orange	WT	White
DG	DK Green	PK	Pink	YL	Yellow

1993-95 Probe 2.0L I4 MFI VIN A (A/T) 26 Pin Connector

PCM Pin #	Wire Color	Circuit Description (26 Pin)	Value at Hot Idle
26-A	BK	Power Ground	<0.1v
26-B	BK	Power Ground	<0.1v
26-C	BK/RD	Power Ground	<0.1v
26-D	BK/BL	Analog Signal Return	<0.1v
26-E	---	Not Used	---
26-F	WT	TSS Sensor (+) Signal	680-720 rpm
26-G	BK/GN	Low Speed Fan Control	On: 1v, Off: 12v
26-H	RD	TSS Sensor (-)	680-720 rpm
26-I	BL/GN	High Speed Fan Control	On: 1v, Off: 12v
26-J	---	Not Used	---
26-K	WT/BL	EGR VR Solenoid	0%, 55 mph: 45%
26-L	LG	Fuel Pump Control	On: 1v, Off: 12v
26-M	GN	Fuel Pressure Regulator	Startup: <3v, other: 12v
26-N	WT/BK	EVAP Purge Solenoid	12v, 55 mph: 1v
26-O	LG/BK	Idle Air Control Valve	8-10v
26-P	BL	Shift Solenoid 1 Control	12v, 55 mph: 1v
26-Q	BL/BK	Shift Solenoid 2 Control	1v, 55 mph: 12v
26-R	GN/BK	Shift Solenoid 3 Control	12v, 55 mph: 1v
26-S	BL/WT	TCC Solenoid Control	12v, 55 mph: 1v
26-T	RD/WT	Downshift Timing Solenoid	DSS On: 1v, Off: 12v
26-U	RD/BK	TCC Solenoid Control	0v, 55 mph: 12v
26-V	RD/GN	Line Pressure Solenoid	9v, 55 mph: 8.4v
26-W	YL/BK	Injector 1 Control	3.5-3.7 ms
26-X	YL/WT	Injector 2 Control	3.5-3.7 ms
26-Y	YL/RD	Injector 3 Control	3.5-3.7 ms
26-Z	YL/GN	Injector 4 Control	3.5-3.7 ms

Pin Connector Graphic

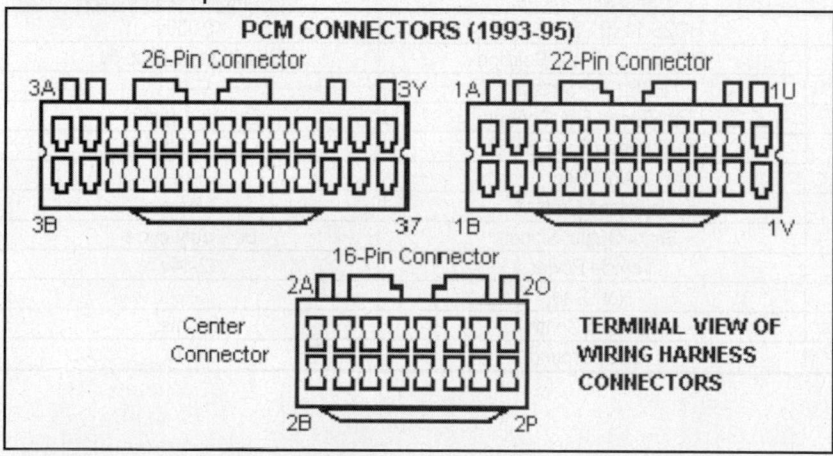

Standard Colors and Abbreviations

Abbreviation	Color	Abbreviation	Color	Abbreviation	Color
BK	Black	GY	Gray	PK	Purple
BL	Blue	GN	Green	RD	Red
BR	Brown	LG	LT Green	TN	Tan
DB	Dark Blue	OR	Orange	WT	White
DG	DK Green	PK	Pink	YL	Yellow

1993-95 Probe 2.0L I4 MFI VIN A (M/T) 60 Pin Connector

PCM Pin #	Wire Color	Circuit Description (60 Pin)	Value at Hot Idle
1	BL/RD	Keep Alive Power	12- 14v
2	---	Not Used	---
3	BL/WT	VSS (+) Signal	0 Hz, 55 mph: 125 Hz
4	BL/YL	Ignition Diagnostic Monitor	20-31 Hz
5	---	Not Used	---
6	BR/GN	VSS (-) Signal	0 Hz, 55 mph: 125 Hz
7	OR/GN	ECT Sensor Signal	0.5-0.6v
8	WT/YL	Fuel Pump Monitor	Pump On: 12v, Off; 0v
9	BR	MAF Sensor Return	<0.050v
10	---	Not Used	---
11	WT/BK	EVAP Purge Solenoid	12v, 55 mph: 1v
12	---	Not Used	---
13	OR/WT	Fuel Pressure Regulator	Startup: <3v, Others: 12v
14	OR/BK	Blower Motor Signal	Off or 1st: 12v, 2nd on: 1v
15	PK	Rear Defroster Signal	Switch Off: 0.1v, On: 12v
16	BK	Ignition Ground	<0.050v
17	BL	STI Output, MIL Control	MIL On: 1v, Off: 12v
18	---	Not Used	---
19	---	Not Used	---
20	---	Not Used	---
21	LG/BK	ISC Motor Control	8-10.1v
22	LG	Fuel Pump Control	On: 1v, Off: 12v
23	PK/BK	A/C Cycling Clutch Switch	On: 1v, Off: 12v
24	GN/WT	Camshaft Position Sensor	5-7 Hz
25	WT/LG	IAT Sensor Signal	1-5-2.5v
26	LG/PK	Reference Voltage	4.9-5.1v
27	---	Not Used	---
28	BR/YL	PSP Switch Signal	Straight: 12v, Turned: 0v
29	GN/BL	HO2S-11 (B1 S1) Ground	<0.050v
30	LG/BK	M/T: Clutch Pedal Position	Clutch In: 0v, Out: 5v
31	WT/BL	EGR EVP Sensor	0.4v
32	BL/OR	Low Speed Fan Control	On: 1v, Off: 12v
33	BL/GN	High Speed Fan Control	On: 1v, Off: 12v
34	---	Not Used	---
35	YL/GN	Injector 4 Control	3.8 ms
36	RD/BL	Spark Output Signal	50% duty cycle
37	WT/RD	Vehicle Power	12-14v
38	---	Not Used	---
39	YL/RD	Injector 3 Control	3.8 ms
40	BK/GN	Power Ground	<0.1v

1993-95 Probe 2.0L I4 MFI VIN A (M/T) 60 Pin Connector

PCM Pin #	Wire Color	Circuit Description (60 Pin)	Value at Hot Idle
41	---	Not Used	---
42	WT	Daytime Running Lamp	DRL On: 2v, Off: 12v
43	RD/GN	EGR Temperature Sensor	2.25-2.55v (68-120°F)
44	BK/YL	HO2S-11 (B1 S1) Signal	0.1-1.1v
45	W	Headlamp Switch Signal	Off or 1st: 12v, 2nd on: 1v
46	BK/BL	Analog Signal Return	<0.050v
47	LG/WT	TP Sensor Signal	0.5-1.1v
48	BL/PK	Self-Test Input	STI On: 0.1v, Off: 5v
49	---	Not Used	---
50	RD	MAF Sensor Signal	1.9v
51-53	---	Not Used	---
54	GN/BK	A/C Relay Control	On: 1v, Off: 12v
55	---	Not Used	---
56	RD/YL	PIP Sensor Signal	50% duty cycle
57	WT/RD	Vehicle Power	12-14v
58	YL/BK	Injector 1 Control	3.8 ms
59	YL/WT	Injector 2 Control	3.8 ms
60	BK/GN	Power Ground	<0.1v

Pin Connector Graphic

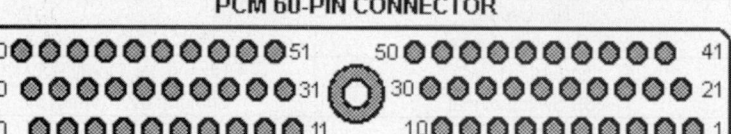

PCM 60-PIN CONNECTOR

60 ●●●●●●●●●● 51   50 ●●●●●●●●●● 41
40 ●●●●●●●●●● 31   30 ●●●●●●●●●● 21
20 ●●●●●●●●●● 11   10 ●●●●●●●●●● 1

Terminal View of 60-Pin PCM Harness Connector

Standard Colors and Abbreviations

Abbreviation	Color	Abbreviation	Color	Abbreviation	Color
BK	Black	GY	Gray	PK	Purple
BL	Blue	GN	Green	RD	Red
BR	Brown	LG	LT Green	TN	Tan
DB	Dark Blue	OR	Orange	WT	White
DG	DK Green	PK	Pink	YL	Yellow

1996-97 Probe 2.0L I4 MFI VIN A (All) 104 Pin Connector

PCM Pin #	Wire Color	Circuit Description (104 Pin)	Value at Hot Idle
1	BL/BK	Shift Solenoid 2 Control	1v, 55 mph: 12v
2	BL	MIL (lamp) Control	MIL On: 1v, Off: 12v
3-9	---	Not Used	---
10	YL/BK	Blower Motor Switch Signal	Motor On: 0.7v, Off: 12v
11-12	---	Not Used	---
13	GN/YL	Flash EEPROM Power	0.1v
14	GN	Daytime Running Lamps	Lamps Off: 12v, On: 1v
15	WT/RD	Data Bus (-) Signal	Digital Signals
16	OR/BK	Data Bus (+) Signal	Digital Signals
17	BL	High Speed Fan Control	On: 1v, Off: 12v
18	---	Not Used	---
19	GN/WT	Fuel Pressure Regulator	12v, 55 mph: 12v
20	---	Not Used	---
21	GN	CKP Sensor (+) Signal	390-450 Hz
22	BL	CKP Sensor (-) Signal	390-450 Hz
23	BK/GN	Power Ground	<0.1v
24	BK/GN	Power Ground	<0.1v
25	---	Not Used	---
26	GN	Coil Driver 1 Control	5° dwell
27	BL	Shift Solenoid 1 Control	12v, 55 mph: 1v
28	---	Not Used	---
29	BR/BK	TCS (switch) Signal	TCS & O/D On: 12v
30	OR/BL	Octane Adjust Switch	Closed: 0v, Open: 9.3v
31	BR/YL	PSP Switch Signal	Straight: 0v, Turned: 5v
32	---	Not Used	---
33	OR/BK	VSS (-) Signal	0 Hz, 55 mph: 125 Hz
34	LG	EGR BARO Pressure Sensor	Typical Range: 4.1-4.8v
35	BK/YL	HO2S-12 (B1 S2) Signal	0.1-1.1v
36	BR	MAF Sensor Return	<0.050v
37	PK/WT	TFT Sensor Signal	2.10-2.40v
38	OR/GN	ECT Sensor Signal	0.5-0.6v
39	WT/GN	IAT Sensor Signal	1.5-2.5v
40	WT/YL	Fuel Pump Monitor	On: 12v, Off: 0v
41	PK/BK	A/C Cycling Clutch Switch	A/C On: 12v, Off: 0v
42-44	---	Not Used	---
45	BL/OR	Low Speed Fan Control	On: 1v, Off: 12v
46-48	---	Not Used	---
49	WT	Headlight Switch Signal	Lights On: 12v, Off: 0.1v
46-48	---	Not Used	---
51	BK/GN	Power Ground	1v
52-53	---	Not Used	---
54	BL/LG	Modulated Converter Clutch	0%, 55 mph: 50%
55	BL/RD	Keep Alive Power	12-14v
56-57	---	Not Used	---
58	BL/WT	VSS (+) Signal	0 Hz, 55 mph: 125 Hz
59	---	Not Used	---
60	BL/WT	HO2S-11 (B1 S1) Signal	0.1-1.1v
61-63	---	Not Used	---
64	RD/BK	A/T CD4E: TR Sensor	In 'P': 0v, 55 mph: 1.7v
64	LG/BK	M/T: Clutch Pedal Position	Clutch In: 0v, Out: 5v

1996-97 Probe 2.0L I4 MFI VIN A (All) 104 Pin Connector

PCM Pin #	Wire Color	Circuit Description (104 Pin)	Value at Hot Idle
65	RD/GN	EGR Valve Position Sensor	0.20-1.30v
66	PK	Rear Defroster Switch Signal	Defrost On: 0.9v, Off: 12v
67	WT/PK	EVAP Purge Solenoid	12v, 55 mph: 1v
68	WT/BL	EGR Atmospheric Pressure	12v, 55 mph: 1v
69	GN/BK	A/C WOT Relay Control	On: 1v, Off: 12v
70	---	Not Used	---
71	WT/RD	Vehicle Power	12-14v
72	GN/BL	EGR VR Solenoid	12v, 55 mph: 1v
73	---	Not Used	---
74	YL/RD	Injector 3 Control	1.7-2.3 ms
75	YL/BK	Injector 1 Control	1.7-2.3 ms
76	BK/GN	Power Ground	<0.1v
77	BK/GN	Power Ground	<0.1v
78	---	Not Used	---
79	BR/YL	TCIL (lamp) Control	On: 1v, Off: 12v
80	LG	Fuel Pump Control	On: 1v, Off: 12v
81	BL/BR	EPC Solenoid Control	7-8v (8-9 psi)
82	---	Not Used	---
83	LG/BK	IAC Motor Control	9v (41% duty cycle)
84	BL/GN	TSS Sensor (+) Signal	42-50 Hz (680-720 rpm)
85	GY/WT	CID Sensor Signal	5-7 Hz
86	PK/YL	A/C Pressure Switch	A/C On: 12v (Open)
87	---	Not Used	---
88	RD	MAF Sensor Signal	0.6v
89	LG/WT	TP Sensor Signal	0.53-1.27v
90	LG/V	Reference Voltage	4.9-5.1v
91	BK/BL	Analog Signal Return	<0.050v
92	WT/BK	Brake Pedal Position Switch	Brake Off: 0v, On: 12v
93	YL	HO2S-11 (B1 S1) Heater	On: 1v, Off: 12v
94-96	---	Not Used	---
97	WT/RD	Vehicle Power	12-14v
98	OR/WT	EGR Check Solenoid	12v, 55 mph: 9-12v
99	---	Not Used	---
100	YL/GN	Injector 4 Control	1.7-2.3 ms
101	YL/WT	Injector 2 Control	1.7-2.3 ms
102	BL/V	Shift Solenoid 3 Control	12v, 55 mph: 1v
103	BK/GN	Power Ground	<0.1v
104	---	Not Used	---

Pin Connector Graphic

PCM 104-PIN CONNECTOR

```
 1 ○○○○○○○○○○○○○    ○○○○○○○○○○○○○ 26
27 ○○○○○○○○○○○○○    ○○○○○○○○○○○○○ 52
53 ○○○○○○○○○○○○○  ●  ○○○○○○○○○○○○○ 78
79 ○○○○○○○○○○○○○    ○○○○○○○○○○○○○ 104
```

Terminal View of 104-Pin PCM Wiring Harness Connector

Standard Colors and Abbreviations

Abbreviation	Color	Abbreviation	Color	Abbreviation	Color
BK	Black	GY	Gray	PK	Purple
BL	Blue	GN	Green	RD	Red
BR	Brown	LG	LT Green	TN	Tan
DB	Dark Blue	OR	Orange	WT	White
DG	DK Green	PK	Pink	YL	Yellow

1990-92 Probe 2.2L I4 MFI VIN C (A/T) 22 Pin Connector

PCM Pin #	Wire Color	Circuit Description (22 Pin)	Value at Hot Idle
22-A	BL/RD	Keep Alive Power	12- 14v
22-B	RD/BK	Vehicle Power (Power Relay)	12-14v
22-C	WT/YL	System Monitor Lamp	Lamp On: 5v, Off: 12v
22-D	BK/PK	Vehicle Start Signal	Cranking: 9-11v
22-E	WT/BL	MIL (lamp) Control	MIL On: 1v, MIL Off: 12v
22-F	WT/BK	Self-Test Output Signal	STO On: 5v, Off: 12v
22-G	---	Not Used	---
22-K	---	Not Used	---
22-H	WT/BK	Headlamp Switch Signal	Switch On: 12v, Off: 0.1v
22-I	RD/WT	Self-Test Input	STI On: 0.1v, Off: 5v
22-J	BK/BL	Rear Defroster Control	Switch On: 12v, Off: 0.1v
22-L	BL/BK	Condenser Fan Relay	On: 1v, Off: 12v
22-M	GN/RD	VSS Signal to Cluster	Digital Signals
22-N	BR/RD	PSP Switch Signal	Straight: 12v, Turned: 0v
22-O	BL/WT	A/C Cycling Clutch Switch	On: 1v, Off: 12v
22-P	BL/BK	Blower Motor Switch Signal	Off or 1st: 12v, 2nd on: 1v
22-Q	WT/GN	Brake Pedal Position Switch	Brake Off: 0v, On: 12v
22-R	BK/YL	A/T: PNP Position Switch	In 'P': 0v, Others: 12v
22-S	YL/WT	Speed Limiter Signal (Cluster)	Digital Signals
22-T	LG/WT	Idle Position Switch Signal	IDL closed: 0v, open: 12v
22-U	BK/YL	Vehicle Power	12-14v
22-V	YL/BK	Ignition Diagnostic Monitor	20-31 Hz

1990-92 Probe 2.2L I4 MFI VIN C (A/T) 16 Pin Connector

PCM Pin #	Wire Color	Circuit Description (16 Pin)	Value at Hot Idle
16-A	RD/WT	Vane Airflow Meter Reference	4.9-5.1v
16-B	RD/BK	Vane Airflow Meter Signal	3.3v
16-C	BK	O2S-11 (B1 S1) Signal	0.1-1.1v
16-D	BK/GN	Cooling Fan 1 Control	On: 1v, Off: 12v
16-E	YL/BK	ECT Sensor Signal	0.5-0.6v
16-F	LG/BK	TP Sensor Signal	0.5-1.3v
16-G	BL/RD	Radiator Temperature Switch	Switch closed: 0.1v
16-H	BL/WT	Manual Mode Switch	Closed: 0.1v, open: 5v
16-I	LG/RD	Reference Voltage	4.9-5.1v
16-J	YL/BL	EGR EVP Sensor	0.4v
16-K	RD	VAT Sensor Signal	2.23v (at 104ºF)
16-L	---	Not Used	---
16-M	YL/GN	VSS (+) Signal	0 Hz, 55 mph: 125 Hz
16-N	YL/BL	VSS (-) Signal	0 Hz, 55 mph: 125 Hz
16-O	WT/BK	EVAP Purge Solenoid	12v, 55 mph: 1v
16-P	BR/YL	Neutral Shift Signal to Cluster	Digital Signals

Standard Colors and Abbreviations

Abbreviation	Color	Abbreviation	Color	Abbreviation	Color
BK	Black	GY	Gray	PK	Purple
BL	Blue	GN	Green	RD	Red
BR	Brown	LG	LT Green	TN	Tan
DB	Dark Blue	OR	Orange	WT	White
DG	DK Green	PK	Pink	YL	Yellow

1990-92 Probe 2.2L I4 MFI VIN C (A/T) 26 Pin Connector

PCM Pin #	Wire Color	Circuit Description (26 Pin)	Value at Hot Idle
26-A	BK	Power Ground	<0.1v
26-B	BK	Power Ground	<0.1v
26-C	BK/LG	Power Ground	<0.1v
26-D	LG/YL	Analog Signal Return	<0.050v
26-E	YL	Manual Lever Position O/D	In O/D: 12v
26-F	---	Not Used	---
26-G	YL/RD	Manual Lever Position Low	In Low: 12v
26-H	YL/BK	Manual Lever Position Drive	In Drive: 12v
26-I	---	Not Used	---
26-J	---	Not Used	---
26-K	---	Not Used	---
26-L	---	Not Used	---
26-M	WT/RD	Fuel Pressure Regulator	Startup: <3v, Others: 12v
26-N	BL/LG	TOT Sensor Signal	2.10-2.40v
26-O	WT/BL	EGR VR Solenoid	0%, 55 mph: 45%
26-P	---	Not Used	---
26-Q	WT	Idle Air Control Valve	8-10v
26-R	---	Not Used	---
26-S	---	Not Used	---
26-T	---	Not Used	---
26-U	YL	Fuel Injector 1 Bank 1	3.6 ms
26-V	YL/BK	Fuel Injector 2 Bank 2	3.6 ms
26-W	BL	Shift Solenoid 1 Control	12v, 55 mph: 1v
26-X	BL/YL	Shift Solenoid 2 Control	1v, 55 mph: 12v
26-Y	BL/RD	Shift Solenoid 3 Control	12v, 55 mph: 1v
26-Z	BL/WT	TCC Solenoid Control	12v, 55 mph: 1v

Pin Connector Graphic

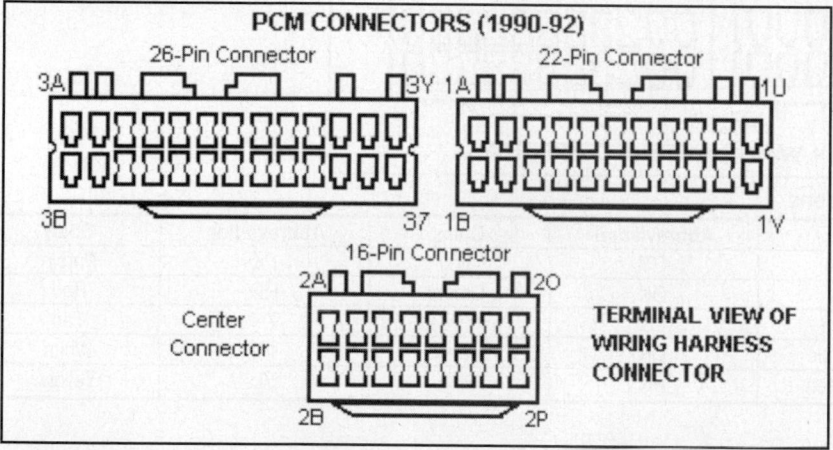

Standard Colors and Abbreviations

Abbreviation	Color	Abbreviation	Color	Abbreviation	Color
BK	Black	GY	Gray	PK	Purple
BL	Blue	GN	Green	RD	Red
BR	Brown	LG	LT Green	TN	Tan
DB	Dark Blue	OR	Orange	WT	White
DG	DK Green	PK	Pink	YL	Yellow

1990-92 Probe 2.2L I4 MFI VIN C (M/T) 22 Pin Connector

PCM Pin #	Wire Color	Circuit Description (22 Pin)	Value at Hot Idle
22-A	BL/RD	Keep Alive Power	12- 14v
22-B	RD/BK	Vehicle Power (Power Relay)	12-14v
22-C	WT/YL	System Monitor Lamp	Lamp On: 5v, Off: 12v
22-D	BK/PK	Vehicle Start Signal	Cranking: 9-11v
22-E	WT/BL	MIL (lamp) Control	MIL On: 1v, MIL Off: 12v
22-F	WT/BK	Self-Test Output Signal	STO On: 5v, Off: 12v
22-G	---	Not Used	---
22-H	---	Not Used	---
22-I	---	Not Used	---
22-J	BK/BL	Rear Defroster Control	Switch On: 12v, Off: 0.1v
22-K	---	Not Used	---
22-L	---	Not Used	---
22-M	---	Not Used	---
22-N	LG/WT	Idle Position Switch	IDL closed: 0v, Open: 12v
22-O	WT/GN	Brake Pedal Position Switch	Brake Off: 0v, On: 12v
22-P	BR/RD	PSP Switch Signal	Straight: 12v, Turned: 0v
22-Q	BL/WT	A/C Cycling Clutch Switch	On: 1v, Off: 12v
22-R	BK/GN	Cooling Fan 1 Control	On: 1v, Off: 12v
22-S	BL/BK	Blower Motor Switch	Off or 1st: 12v, 2nd on: 1v
22-T	BK/BL	Rear Defroster Switch Signal	AC On: 12v, Off: 0v
22-U	RD/BL	Clutch Engage Switch	Clutch In: 0v, Out: 11v
22-V	---	Not Used	---

Pin Connector Graphic

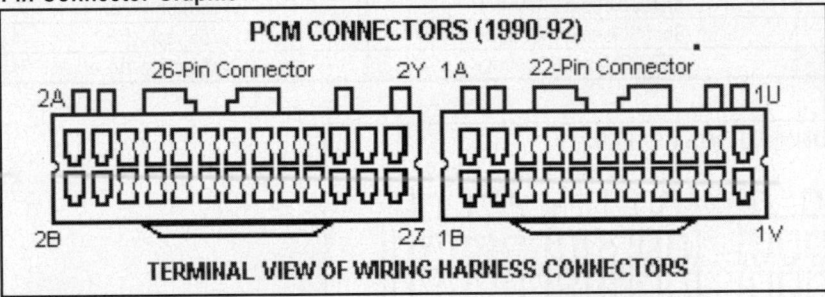

Standard Colors and Abbreviations

Abbreviation	Color	Abbreviation	Color	Abbreviation	Color
BK	Black	GY	Gray	PK	Purple
BL	Blue	GN	Green	RD	Red
BR	Brown	LG	LT Green	TN	Tan
DB	Dark Blue	OR	Orange	WT	White
DG	DK Green	PK	Pink	YL	Yellow

1990-92 Probe 2.2L I4 MFI VIN C (M/T) 26 Pin Connector

PCM Pin #	Wire Color	Circuit Description (26 Pin)	Value at Hot Idle
26-A	BK	Power Ground	<0.1v
26-B	BK	Power Ground	<0.1v
26-C	BK/LG	Power Ground	<0.1v
26-D	LG/YL	Analog Signal Return	<0.050v
26-E	---	Not Used	---
26-F	---	Not Used	---
26-G	YL/WT	Speed Limiter Signal	Digital Signals
26-H	---	Not Used	---
26-I	YL/BL	Ignition Diagnostic Monitor	20-31 Hz
26-J	RD/WT	Vane Airflow Meter Reference	4.9-5.1v
26-K	LG/RD	Vehicle Reference	4.9-5.1v
26-L	YL/BL	EGR EVP Sensor	0.4v
26-M	LG/BK	TP Sensor Signal	0.5-1.1v
26-N	BK	O2S-11 (B1 S1) Signal	0.1-1.1v
26-O	RD/BK	Vane Airflow Meter Signal	3.3v
26-P	RD	VAT Sensor Signal	2.23v (at 104°F)
26-Q	YL/BK	ECT Sensor Signal	0.5-0.6v
26-R	---	Not Used	---
26-S	---	Not Used	---
26-T	WT/RD	Fuel Pressure Regulator	Startup: <3v, Others: 12v
26-U	YL	Fuel Injector 1 Bank 1	3.6 ms
26-V	YL/BK	Fuel Injector 2 Bank 2	3.6 ms
26-W	WT	Idle Air Control Valve	8-10v
26-X	WT/BL	EVAP Purge Solenoid	12v, 55 mph: 1v
26-Y	WT/BL	EGR VR Solenoid	0%, 55 mph: 45%
26-Z	---	Not Used	---

Pin Connector Graphic

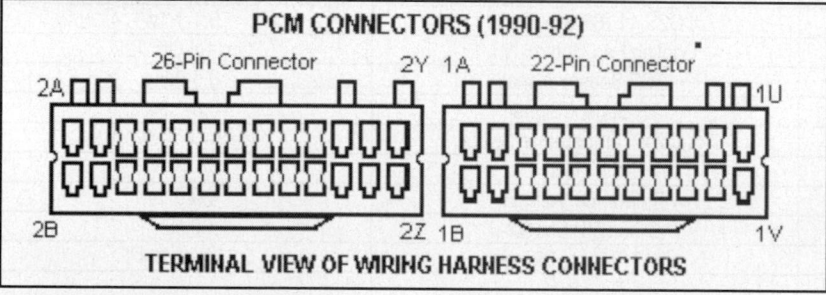

Standard Colors and Abbreviations

Abbreviation	Color	Abbreviation	Color	Abbreviation	Color
BK	Black	GY	Gray	PK	Purple
BL	Blue	GN	Green	RD	Red
BR	Brown	LG	LT Green	TN	Tan
DB	Dark Blue	OR	Orange	WT	White
DG	DK Green	PK	Pink	YL	Yellow

1990-92 Probe 2.2L I4 MFI TC VIN L (All) 22 Pin Connector

PCM Pin #	Wire Color	Circuit Description (22 Pin)	Value at Hot Idle
22-A	BL/RD	Keep Alive Power	12- 14v
22-B	RD/BK	Vehicle Power (Power Relay)	12-14v
22-C	WT/YL	System Monitor Lamp	Lamp On: 5v, Off: 12v
22-D	BK/PK	Vehicle Start Signal	Cranking: 9-11v
22-E	WT/BL	MIL (lamp) Control	MIL On: 1v, MIL Off: 12v
22-F	WT/BK	Self-Test Output Signal	STO On: 5v, Off: 12v
22-G	YL/BK	Spark Output Signal	50% duty cycle
22-H	WT/BK	Headlamp Switch Signal	Switch On: 12v, Off: 1v
22-I	RD/WT	Self-Test Input Signal	STI On: 1v, STI Off: 12v
22-J	BK/BL	Rear Defroster Control	Switch On: 12v, Off: 0.1v
22-K	---	Not Used	---
22-L	BL/BK	Condenser Fan Relay	On: 1v, Off: 12v
22-M	YL/WT	VSS Signal to Cluster	Digital Signals
22-N	BR/RD	PSP Switch Signal	Straight: 12v, Turned: 0v
22-O	BL/WT	A/C Cycling Clutch Switch	On: 1v, Off: 12v
22-P	BL/BK	Blower Motor Switch Signal	Off or 1st: 12v, 2nd on: 1v
22-Q	WT/GN	Brake Pedal Position Switch	Brake Off: 0v, On: 12v
22-R	RD/BK	A/T: Neutral Drive Switch	In 'P', 0v, Others: 12v
22-R	RD/BK	M/T: Clutch Engage Switch	Clutch In: 0v, Out: 12v
22-S	---	Not Used	---
22-T	LG/WT	Idle Switch Signal	IDL Closed: 0v, open: 12v
22-U	BK/YL	Vehicle Power	12-14v
22-V	BL/RD	Ignition Diagnostic Monitor	20-31 Hz

1990-92 Probe 2.2L I4 MFI TC VIN L (All) 16 Pin Connector

PCM Pin #	Wire Color	Circuit Description (16 Pin)	Value at Hot Idle
16-A	RD/WT	Vane Airflow Meter Reference	4.9-5.1v
16-B	RD/BK	Vane Airflow Meter Signal	3.3v
16-C	BK	O2S-11 (B1 S1) Signal	0.1-1.1v
16-D	BK/GN	Cooling Fan Control No. 1	On: 1v, Off: 12v
16-E	YL/BK	ECT Sensor Signal	0.5-0.6v
16-F	LG/BK	TP Sensor Signal	0.5-1.3v
16-G	BL/RD	Radiator Temperature Switch	Switch closed: 0.1v
16-H	---	Not Used	---
16-I	LG/RD	Reference Voltage	4.9-5.1v
16-J	YL/BL	EGR EVP Sensor	0.4v
16-K	RD	VAT Sensor Signal	2.23v (at 104°F)
16-L	---	Not Used	---
16-M	RD/YL	Knock Sensor Signal	0v
16-N	---	Not Used	---
16-O	---	Not Used	---
16-P	---	Not Used	---

Standard Colors and Abbreviations

Abbreviation	Color	Abbreviation	Color	Abbreviation	Color
BK	Black	GY	Gray	PK	Purple
BL	Blue	GN	Green	RD	Red
BR	Brown	LG	LT Green	TN	Tan
DB	Dark Blue	OR	Orange	WT	White
DG	DK Green	PK	Pink	YL	Yellow

1990-92 Probe 2.2L I4 MFI TC VIN L (All) 26 Pin Connector

PCM Pin #	Wire Color	Circuit Description (26 Pin)	Value at Hot Idle
26-A	BK	Power Ground	<0.1v
26-B	BK	Power Ground	<0.1v
26-C	BK/LG	Power Ground	<0.1v
26-D	LG/YL	Analog Signal Return	<0.050v
26-E	BL	CKP Sensor Signal	390-450 Hz
26-F	WT	CMP Sensor VREF	4.9-5.1v
26-G	GN	CMP Sensor 1 Signal	5-7 Hz
26-H	RD	CMP Sensor 2 Signal	22-28 Hz
26-I	---	Not Used	---
26-J	---	Not Used	---
26-K	---	Not Used	---
26-L	YL/RD	MIL (lamp) Control	MIL On: 1v, Off: 12v
26-M	WT/RD	Fuel Pressure Regulator	Startup: <3v, Others: 12v
26-N	---	Not Used	---
26-O	BL/WT	EGR Vent Solenoid Control	12v, at off-idle: 1v
26-P	EGRC	EGR Control Solenoid	12v, off-idle: 1v
26-Q	WT	Idle Air Control Valve	8-10v
26-R	BR/YL	Boost Control Solenoid	On: 1v, Off: 12v
26-S	---	Not Used	---
26-T	LG	Fuel Pump Control	On: 1v, Off: 12v
26-U	YL	Fuel Injector 1 Bank 1	3.6 ms
26-V	YL/BK	Fuel Injector 2 Bank 2	3.6 ms
26-W	---	Not Used	---
26-X	---	Not Used	---
26-Y	---	Not Used	---
26-Z	---	Not Used	---

Pin Connector Graphic

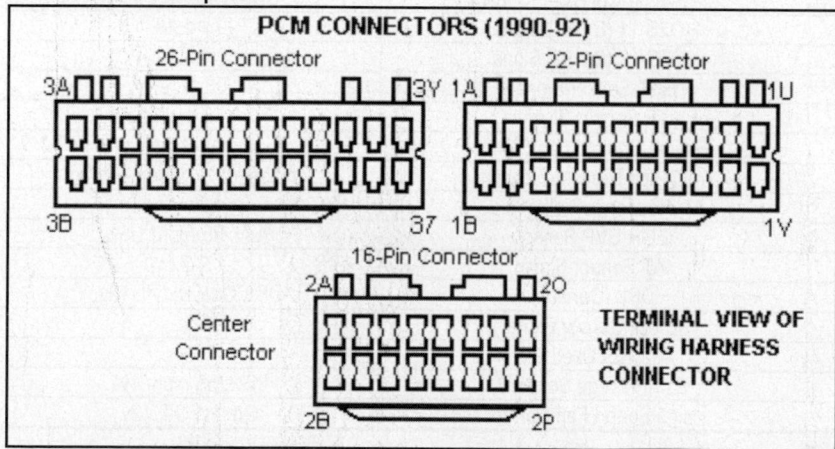

Standard Colors and Abbreviations

Abbreviation	Color	Abbreviation	Color	Abbreviation	Color
BK	Black	GY	Gray	PK	Purple
BL	Blue	GN	Green	RD	Red
BR	Brown	LG	LT Green	TN	Tan
DB	Dark Blue	OR	Orange	WT	White
DG	DK Green	PK	Pink	YL	Yellow

1993 Probe 2.5L V6 MFI VIN B (All) 22 Pin Connector

PCM Pin #	Wire Color	Circuit Description (22 Pin)	Value at Hot Idle
22-A	BL/RD	Keep Alive Power	12- 14v
22-B	RD/BK	Vehicle Power (Power Relay)	12-14v
22-C	BK/RD	Vehicle Start Signal	Cranking: 9-11v
22-D	WT/RD	System Monitor Lamp	Lamp On: 5v, Off: 12v
22-E	BL	MIL (lamp) Control	MIL On: 1v, MIL Off: 12v
22-F	LG/RD	Self-Test Output Signal	STO On: 5v, Off: 12v
22-G	BL/OR	Ignition Control Module Signal	0.1-0.3v
22-H	WT	Headlamp Switch Signal	Switch On: 12v, Off: 1v
22-I	RD/WT	Self-Test Input Signal	STI On: 1v, STI Off: 12v
22-J	PK	Rear Defroster Control	Switch On: 12v, Off: 0.1v
22-K	WT/BK	TR Switch Signal	In 'P': 0v, Others: 12v
22-L	GN/BK	A/C Heater Relay Control	On: 1v, Off: 12v
22-M	GN/RD	Vehicle Speed Sensor	0 Hz, 55 mph: 125 Hz
22-N	BK/YL	PSP Switch Signal	Straight: 12v, Turned: 0v
22-O	PK/BK	A/C Clutch Pressure Switch	Switch On: 0v, Off: 12v
22-P	OR/BK	Blower Motor Switch	Off: 12v, On: 1v
22-Q	WT/GN	Brake Pedal Position Switch	Brake Off: 0v, On: 12v
22-R	LG/BK	A/T: PNP Switch Signal	In 'P', 0v, Others: 12v
22-R	LG/BK	M/T: CPP Switch Signal	Clutch In: 0v, Out: 12v
22-S	GN	TCM RST1 Signal	Digital Signals
22-T	BR	Idle Switch Signal	Closed: 0V, Open: 12v
22-U	BK	Power Ground	<0.1v
22-V	LG/WT	TCM RST2 Signal	Digital Signals

1993 Probe 2.5L V6 MFI VIN B (All) 16 Pin Connector

PCM Pin #	Wire Color	Circuit Description (16 Pin)	Value at Hot Idle
16-A	GN/OR	Barometric Pressure Sensor	Hot Idle: 3.9v
16-B	RD	Measuring-Core VAF Sensor	3.06v, 55 mph: 1.7v
16-C	BK/YL	HO2S-11 (B1 S1) Signal	0.1-1.1v
16-D	BL/WT	HO2S-12 (B1 S2) Signal	0.1-1.1v
16-E	RD/GN	ECT Sensor Signal	0.5-0.6v
16-F	YL	TP Sensor Signal	0.7v, 55 mph: 2.1v
16-G	RD/WT	Engine Coolant Fan Signal	0.5-0.6v
16-H	PK/YL	A/C High Pressure Switch	Closed: 1v, Open: 12v
16-I	PK	Reference Voltage	4.9-5.1v
16-J	RD/BK	EGR EVP Sensor	0.4v, 55 mph: 1.8v
16-K	BK/RD	IAT Sensor Signal	1.5-2.5v
16-L	GN	DRL (Canada)	DRL On: 2v, Off: 12v
16-M	WT	Knock Sensor Signal	0v
16-N	---	Not Used	---
16-O	BL/BK	EVAP Purge Solenoid	12v, 55 mph: 1v
16-P	BL/GN	High Speed Fan Control	On: 1v, Off: 12v

Standard Colors and Abbreviations

Abbreviation	Color	Abbreviation	Color	Abbreviation	Color
BK	Black	GY	Gray	PK	Purple
BL	Blue	GN	Green	RD	Red
BR	Brown	LG	LT Green	TN	Tan
DB	Dark Blue	OR	Orange	WT	White
DG	DK Green	PK	Pink	YL	Yellow

1993 Probe 2.5L V6 MFI VIN B (All) 26 Pin Connector

PCM Pin #	Wire Color	Circuit Description (26 Pin)	Value at Hot Idle
26-A	BK	Power Ground	<0.1v
26-B	BK	Power Ground	<0.1v
26-C	BK/RD	Power Ground	<0.1v
26-D	BK/RD	Power Ground	<0.1v
26-E	LG/OR	CKP 1 Sensor	2.5v
26-F	BL	CKP Sensor Ground	<0.050v
26-G	BL/PK	CMP Sensor Signal	2.5v
26-H	GN	CKP 2 Sensor Signal	2.5v
26-I	WT/GN	Variable Resonance Induction	VRIS1 On: 1v, Off: 12v
26-J	BL/RD	Variable Resonance Induction	VRIS2 On: 1v, Off: 12v
26-K	---	Not Used	---
26-L	RD/WT	Low Speed Fan Control	On: 1v, Off: 12v
26-M	GN/BK	Fuel Pressure Regulator	Startup: <3v, other: 12v
26-N	BL/OR	Cooling Fan Control	On: 1v, Off: 12v
26-O	WT/BL	EGR Vent Solenoid	EGRV Off: 12v, On: 1v
26-P	GN/WT	EGR Control Solenoid	EGRC Off: 12v, On: 1v
26-Q	LG/BK	Idle Air Control Valve	8-10v
26-R	---	Not Used	---
26-S	---	Not Used	---
26-T	LG	Fuel Pump Control	On: 1v, Off: 12v
26-U	RD/LG	Injector 1 Control	3.6 ms
26-V	BL/WT	Injector 2 Control	3.6 ms
26-W	BR	Injector 3 Control	3.6 ms
26-X	RD/YL	Injector 4 Control	3.6 ms
26-Y	WT	Injector 5 Control	3.6 ms
26-Z	WT/BK	Injector 6 Control	3.6 ms

Pin Connector Graphic

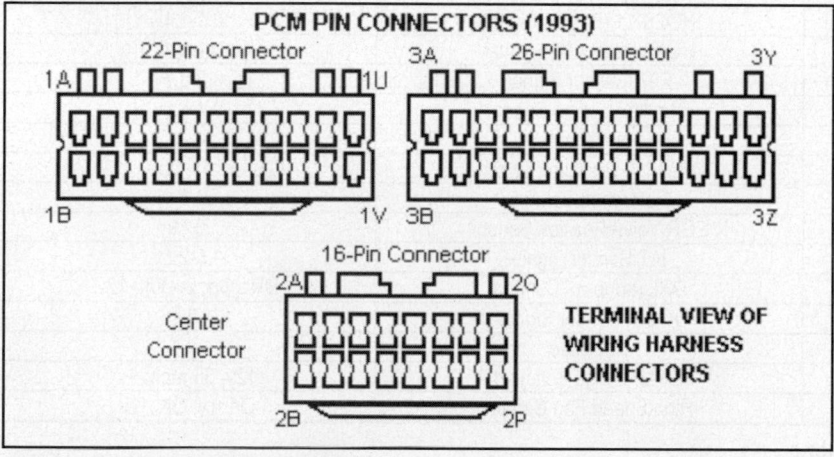

Standard Colors and Abbreviations

Abbreviation	Color	Abbreviation	Color	Abbreviation	Color
BK	Black	GY	Gray	PK	Purple
BL	Blue	GN	Green	RD	Red
BR	Brown	LG	LT Green	TN	Tan
DB	Dark Blue	OR	Orange	WT	White
DG	DK Green	PK	Pink	YL	Yellow

1994-95 Probe 2.5L V6 MFI VIN B (All) 22 Pin Connector

PCM Pin #	Wire Color	Circuit Description (22 Pin)	Value at Hot Idle
22-A	BL/RD	Keep Alive Power	12- 14v
22-B	RD/BK	Vehicle Power	12-14v
22-C	BK/RD	Vehicle Start Signal	KOEC: 9-11v
22-D	WT/RD	Switch Monitor Lamp Control	Lamp Off: 12v, On: 1v
22-E	BL	STO & MIL (lamp) Control	STO On: 5v, Off: 12v
22-F	-	Not Used	---
22-G	BL/OR	Ignition Control Module	Varies
22-H	WT	Headlamp Switch Signal	Switch On: 12v, Off: 1v
22-I	RD/WT	Self-Test Input	STI On: 1v, STI Off: 12v
22-J	PK	Rear Window Defroster	Switch On: 1v, Off: 12v
22-K	WT/BK	Torque Reduce & ECT Signal	Digital Signals
22-L	GN/BK	A/C Relay Control	On: 1v, Off: 12v
22-M	GN/RD	Vehicle Speed Sensor	0 Hz, 55 mph: 125 Hz
22-N	BK/YL	PSP Switch Signal	Straight: 12v, Turned: 0v
22-O	PK/BK	A/C Cycling Pressure Switch	Switch On: 12v, Off: 1v
22-P	OR/BK	Blower Motor Control Switch	Off or 1st: 12v, 2nd on: 1v
22-Q	WT/GN	Brake On/ Off Switch Signal	Brake Off: 0v, On: 12v
22-R	LG/BK	A/T: PNP Switch Signal	In 'P', 0v, Others: 12v
22-R	LG/BK	M/T: CPP Switch Signal	Clutch In: 0v, Out: 12v
22-S	GN	Torque Reduce Signal #1	Digital Signals
22-T	BR	Idle Switch Signal	Closed: 0v, Open: 12v
22-U	-	Not Used	---
22-V	LG/WT	Torque Reduce Signal #2	Digital Signals

1994-95 Probe 2.5L V6 MFI VIN B (All) 16 Pin Connector

PCM Pin #	Wire Color	Circuit Description (16 Pin)	Value at Hot Idle
16-A	GN/OR	Barometric Pressure Sensor	3.9v
16-B	RD	Measuring Core VAF Sensor	Idle: 3.06v
16-C	BK/YL	HO2S-11 (B1 S1) Signal	0.1-1.1v
16-D	BL/WT	HO2S-12 (B1 S2) Signal	0.1-1.1v
16-E	RD/GN	ECT Sensor Signal	0.5-0.6v
16-F	YL	Throttle Position Sensor	0.4-1.0v
16-G	-	Not Used	---
16-H	PK/YL	A/C High Pressure Switch	Closed: 1v, open: 12v
16-I	PK	Reference Voltage	4.9-5.1v
16-J	RD/BK	EGR Valve Position Sensor	0.4v
16-K	BK/RD	IAT Sensor Signal	1.5-2.5v
16-L	GN	DRL (lamps) - Canada	DRL On: 2v, Off: 12v
16-M	WT	Knock Sensor Signal	0v
16-N	-	Not Used	---
16-O	BL/BK	EVAP Purge Solenoid	12v, 55 mph: 1v
16-P	BL/GN	High Speed Fan Control	On: 1v, Off: 12v

Standard Colors and Abbreviations

Abbreviation	Color	Abbreviation	Color	Abbreviation	Color
BK	Black	GY	Gray	PK	Purple
BL	Blue	GN	Green	RD	Red
BR	Brown	LG	LT Green	TN	Tan
DB	Dark Blue	OR	Orange	WT	White
DG	DK Green	PK	Pink	YL	Yellow

1994-95 Probe 2.5L V6 MFI VIN B (All) 26 Pin Connector

PCM Pin #	Wire Color	Circuit Description (26 Pin)	Value at Hot Idle
26-A	BK	Power Ground	<0.1v
26-B	BK	Power Ground	<0.1v
26-C	BK/RD	Power Ground	<0.1v
26-D	BK/RD	Power Ground	<0.1v
26-E	LG/OR	CKP Sensor 1 Signal	2.5v
26-F	BL	CKP Sensor Ground	0.1v
26-G	BL/PK	CMP Sensor 1 Signal	2.5v
26-H	GN	CKP Sensor 2 Signal	2.5v
26-I	WT/GN	Variable Resonance Induction	VRIS1 On: 1v, Off: 12v
26-J	BL/RD	Variable Resonance Induction	VRIS2 On: 1v, Off: 12v
26-K	-	Not Used	---
26-L	RD/WT	Low Speed Fan Control	On: 1v, Off: 12v
26-M	GN/BK	Fuel Pressure Regulator	Startup: 3v, others: 12v
26-N	BL/OR	Condenser Fan Control	On: 1v, Off: 12v
26-O	WT/BL	EGR Vent Solenoid	12v, off-idle: 1v
26-P	GN/WT	EGR Control Solenoid	12v, off-idle: 1v
26-Q	LG/BK	Idle Air Control Valve	8-10v
26-R	-	Not Used	---
26-S	-	Not Used	---
26-T	LG	Fuel Pump Control	On: 1v, Off: 12v
26-U	RD/LG	Injector 1 Control	3.6 ms
26-V	BL/WT	Injector 2 Control	3.6 ms
26-W	BR	Injector 3 Control	3.6 ms
26-X	R/L	Injector 4 Control	3.6 ms
26-Y	W	Injector 5 Control	3.6 ms
26-Z	WT/BK	Injector 6 Control	3.6 ms

Pin Connector Graphic

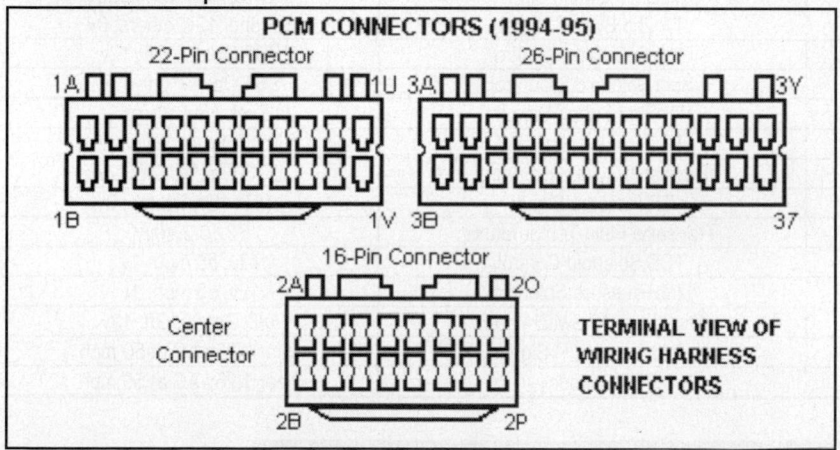

Standard Colors and Abbreviations

Abbreviation	Color	Abbreviation	Color	Abbreviation	Color
BK	Black	GY	Gray	PK	Purple
BL	Blue	GN	Green	RD	Red
BR	Brown	LG	LT Green	TN	Tan
DB	Dark Blue	OR	Orange	WT	White
DG	DK Green	PK	Pink	YL	Yellow

1996-97 Probe 2.5L V6 MFI VIN B (4EAT) 28 Pin Connector

PCM Pin #	Wire Color	Circuit Description (28 Pin)	Value at Hot Idle
C246-1	BL/GN	High Speed Fan Control	On: 1v, Off: 12v
C246-2	BR/LG	Condenser Fan Control	On: 1v, Off: 12v
C246-3	BL/OR	Low Speed Fan Control	On: 0v, Off: 12v
C246-4	GN/BK	Air Conditioning Relay	On: 1v, Off: 12v
C246-5	---	Not Used	---
C246-6	PK/YL	High Pressure Switch Signal	Closed: 1v, open: 12v
C246-7	RD/OR	ISO K-Line Signal for OBD II	Digital Signals
C246-8	---	Not Used	---
C246-9	GN/BK	FPRC Solenoid Control	ON: 1v, Off: 12v
C246-10	GN	Transmission Range Switch	AC On: 12v, Off: 0v
C246-13	BL	STI Output, MIL Control	MIL On: 1v, Off: 12v
C246-14	GN/RD	Vehicle Speed Sensor	Vehicle moving: 0-5v
C246-15	GN	DRL (lamps) - Canada	DRL On: 2v, Off: 12v
C246-16	YL/BK	Blower Motor Control Switch	Off - 1st: 12v, 2nd on: 1v
C246-17	WT/GN	Brake Pedal Position Switch	Brake Off: 0v, On: 12v
C246-20	GY	VRIS Solenoid 1 Control	VRIS1 On: 1v, Off: 12v
C246-21	BL/RD	VRIS Solenoid 2 Control	VRIS2 On: 1v, Off: 12v
C246-22	PK/BK	A/C Cycling Pressure Switch	On: 1v, Off: 12v
C246-23	BK/BR	Vehicle Start Signal	KOEC: 9-11v
C246-24	WT	Headlamp Switch Signal	Off or 1st: 12v, 2nd on: 1v
C246-25	PK	Rear Window Defrost Switch	DEF On: 1v, Off: 12v
C246-27	LG	Fuel Pump Control	On: 1v, Off: 12v

1996-97 Probe 2.5L V6 MFI VIN B (4EAT) 16 Pin Connector

PCM Pin #	Wire Color	Circuit Description (16 Pin)	Value at Hot Idle
C245-1	BL	Shift Solenoid 1 Control	SS1 On: 0v, Off: 12v
C245-2	BR/YL	Transaxle Control Indicator	On: 1v, Off: 12v
C245-3	RD/BL	TR Drive Switch Signal	In Drive: 12v, others: 0v
C245-4	GN/WT	TR 2nd Switch Signal	Second: 12v, others: 0v
C245-5	BL/BK	Shift Solenoid 2 Control	SS2 On: 1v, Off: 12v
C245-6	GN/BK	Shift Solenoid 3 Control	SS3 On: 1v, Off: 12v
C245-7	BL/OR	TR First Switch Signal	In First: 12v, others: 0v
C245-8	BL/WT	TCC Control Solenoid	TCC On: 1v, Off: 12v
C245-9	RD/WT	Downshift Solenoid Control	DSS On: 1v, Off: 12v
C245-10	RD/YL	TR Reverse Signal	In 'R': 12v, Others: 0v
C245-11	BL/YL	Transaxle Fluid Temperature	2.10-2.40v
C245-12	RD/BK	TCC Solenoid Control	12v, 55 mph: 1v
C245-13	RD/BK	Line Pressure Solenoid	12v, 55 mph: 1v
C245-14	BR/BK	Overdrive Off Switch Input	O/D On: 0v, Off: 12v
C245-15	WT	TSS Sensor (+) Signal	Near 1.25v AC at 50 mph
C245-16	RD	TSS Sensor (-)	Near 1.25v AC at 50 mph

Pin Connector Graphic

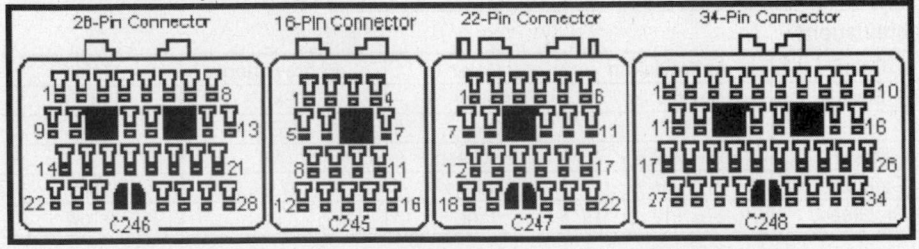

1996-97 Probe 2.5L V6 MFI VIN B (4EAT) 22 Pin Connector

PCM Pin #	Wire Color	Circuit Description (22 Pin)	Value at Hot Idle
C247-1	PK	Reference Voltage	4.9-5.1v
C247-2	RD	MC-VAF Sensor Signal	3.03v
C247-3	PK/WT	HO2S-21 (B2 S1) Signal	0.1-1.1v
C247-4	GN	HO2S-11 (B1 S1) Signal	0.1-1.1v
C247-5	RD/BL	ECT Sensor Signal	0.5-0.6v
C247-7	YL	TP Sensor Signal	0.5-1.2v
C247-8	GN	EGR Boost Sensor	4.3v (at 28" HG)
C247-9	RD/BK	EGR Valve Position Sensor	0.4v
C247-10	BK/RD	IAT Sensor Signal	1.5-2.5v
C247-11	RD/YL	HO2S-21 (B2 S1) Heater	On: 1v, Off: 12v
C247-12	RD/GN	HO2S-13 (B1 S3) Heater	On: 1v, Off: 12v
C247-16	WL	Knock Sensor Signal	0v
C247-17	BK/RD	Power Ground	<0.1v
C247-18	BK/YL	HO2S-12 (B1 S2) Signal	0.1-1.1v
C247-19	BL/WT	HO2S-13 (B1 S3) Signal	0.1-1.1v
C247-20	BL/YL	PSP Switch Signal Input	Straight: 12v, Turned: 0v
C247-21	BR	Idle Throttle Switch Signal	Closed: 0v, Open: 12v
C247-22	BK/BL	Analog Signal Return	<0.050v

1996-97 Probe 2.5L V6 MFI VIN B (4EAT) 34 Pin Connector

PCM Pin #	Wire Color	Circuit Description (34 Pin)	Value at Hot Idle
C248-1	RD/BK	Vehicle Power	12-14v
C248-2	PK/BL	HO2S-11 (B1 S1) Heater	On: 1v, Off: 12v
C248-3	PK	HO2S-21 (B2 S1) Heater	On: 1v, Off: 12v
C248-4	LG/BK	IAC Motor Control	8-10v
C248-5	RD/WT	Injector 1 Control	3.8-4.0 ms
C248-6	BL/WT	Injector 2 Control	3.8-4.0 ms
C248-7	BR	Injector 3 Control	3.8-4.0 ms
C248-8	RD/YL	Injector 4 Control	3.8-4.0 ms
C248-9	WT	Injector 5 Control	3.8-4.0 ms
C248-10	WT/BK	Injector 6 Control	3.8-4.0 ms
C248-11	BL/RD	Keep Alive Power	12-14v
C248-12	BL/PK	CMP Sensor Signal	5-7 Hz
C248-14	RD/WT	Self-Test Input Signal	STI On: 0v, Off: 12v
C248-15	GN	CKP Sensor Signal	410-445 Hz
C248-16	BL	CKP Sensor Ground	<0.050v
C248-17	RD/BK	Vehicle Power	12-14v
C248-18	PK/BK	EGR Vent Solenoid	EGRV On: 1v, Off: 12v
C248-19	GN/WT	EGR Control Solenoid	EGRC On: 1v, Off: 12v
C248-20	BL/BK	EVAP Purge Solenoid	CANP On: <1.v, Off: 12v
C248-21	BL/OR	Ignition Module Control	Varies: 0.1-0.3v
C248-27	BK/RD	Power Ground	<0.1v
C248-30	GY/GN	EGR Boost Check Solenoid	On: 1v, Off: 12v
C248-31, 32	BK	Power Ground	<0.1v
C248-33	BK	Power Ground	<0.1v
C248-34	BK	Power Ground	<0.1v

Standard Colors and Abbreviations

Abbreviation	Color	Abbreviation	Color	Abbreviation	Color
BK	Black	GY	Gray	PK	Purple
BL	Blue	GN	Green	RD	Red
BR	Brown	LG	LT Green	TN	Tan
DB	Dark Blue	OR	Orange	WT	White
DG	DK Green	PK	Pink	YL	Yellow

1996-97 Probe 2.5L V6 MFI VIN B (M/T) 28 & 22 Pin Connector

PCM Pin #	Wire Color	Circuit Description (28 Pin)	Value at Hot Idle
C246-1	BL/GN	High Speed Fan Control	On: 1v, Off: 12v
C246-2	BR/LG	Condenser Fan Control	On: 1v, Off: 12v
C246-3	BL/OR	Low Speed Fan Control	On: 1v, Off: 12v
C246-4	GN/BK	A/C Relay Control	On: 1v, Off: 12v
C246-5, 8	---	Not Used	---
C246-6	PK/YL	A/C High Pressure Switch	Closed: 1v, pen: 12v
C246-7	RD/OR	ISO K-Line for OBD II	Digital Signals
C246-9	GN/BK	FPRC Solenoid	FPRC On: 0v, Off: 12v
C246-10	GN	Neutral Position Switch	In 'N': 0v, Others: 5v
C246-10	GN	Clutch Pedal Position Switch	Clutch In: 0v, Out 5v
C246-11-12	---	Not Used	---
C246-13	BL	Self-Test Output & MIL	MIL On: 1v, Off: 12v
C246-14	GN/RD	Vehicle Speed Sensor	Vehicle moving: 0-5v
C246-15	GN	DRL (lamps) - Canada	DRL On: 2v, DRL Off: 12v
C246-16	YL/BK	Blower Motor Control Switch	Off or 1st: 12v, 2nd on: 1v
C246-17	WT/GN	Brake Pedal Position Switch	Brake Off: 0v, On: 12v
C246-18, 19	---	Not Used	---
C246-20	GY	VRIS Solenoid 1 Control	VRIS1 On: 1v, Off: 12v
C246-21	BL/RD	VRIS 2 Solenoid Control	VRIS2 On: 1v, Off: 12v
C246-22	PK/BK	A/C Cycling Pressure Switch	Closed: 1v, open: 12v
C246-23	BK/BR	Vehicle Start Signal	KOEC: 9-11v
C246-24	WT	Headlamp Switch Signal	Off or 1st: 12v, 2nd on: 1v
C246-25	PK	Rear Window Defrost Switch	Switch On: 1v, Off: 12v
C246-26, 28	---	Not Used	---
C246-27	LG	Fuel Pump Control	On: 1v, Off: 12v
PCM Pin #	**Wire Color**	**Circuit Description (22 Pin)**	**Value at Hot Idle**
C247-1	PK	Reference Voltage	4.9-5.1v
C247-2	RD	MCV Airflow Sensor Signal	3.03v
C247-3	PK/WT	HO2S-21 (B2 S1) Signal	0.1-1.1v
C247-4	GN	HO2S-11 (B1 S1) Signal	0.1-1.1v
C247-5	RD/BL	ECT Sensor Signal	0.5-0.6v
C247-6	---	Not Used	---
C247-7	YL	TP Sensor Signal	0.5-1.2v
C247-8	GN	EGR Boost Sensor	4.3v (at 28" HG)
C247-9	RD/BK	EGR EVP Sensor	0.4v
C247-10	BK/RD	IAT Sensor Signal	1.5-2.5v
C247-11	RD/YL	HO2S-12 (B1 S2) Heater	On: 1v, Off: 12v
C247-12	RD/GN	HO2S-13 (B1 S3) Heater	On: 1v, Off: 12v
C247-13-15	---	Not Used	---
C247-16	WT	Knock Sensor Signal	0v
C247-17	BK/RD	Power Ground	<0.1v
C247-18	BK/YL	HO2S-12 (B1 S2) Signal	0.1-1.1v
C247-19	BL/WT	HO2S-13 (B1 S3) Signal	0.1-1.1v
C247-20	BL/YL	PSP Switch Signal	Straight: 12v, Turned: 0v
C247-21	BR	Idle Throttle Switch Signal	IDL closed: 0v, Open: 12v
C247-22	BK/BL	Analog Signal Return	<0.050v

Standard Colors and Abbreviations

Abbreviation	Color	Abbreviation	Color	Abbreviation	Color
BK	Black	GY	Gray	PK	Purple
BL	Blue	GN	Green	RD	Red
BR	Brown	LG	LT Green	TN	Tan
DB	Dark Blue	OR	Orange	WT	White
DG	DK Green	PK	Pink	YL	Yellow

1996-97 Probe 2.5L V6 MFI VIN B (M/T) 'C248' 34 Pin Connector

PCM Pin #	Wire Color	Circuit Description (34 Pin)	Value at Hot Idle
C248-1	RD/BK	Vehicle Power	12-14v
C248-2	PK/BL	HO2S-11 (B1 S1) Heater	On: 1v, Off: 12v
C248-3	PK	HO2S-21 (B2 S1) Heater	On: 1v, Off: 12v
C248-4	LG/BK	IAC Motor Control	8-10v
C248-5	RD/WT	Injector 1 Control	3.8-4.0 ms
C248-6	BL/WT	Injector 2 Control	3.8-4.0 ms
C248-7	BR	Injector 3 Control	3.8-4.0 ms
C248-8	RD/YL	Injector 4 Control	3.8-4.0 ms
C248-9	WT	Injector 5 Control	3.8-4.0 ms
C248-10	WT/BK	Injector 6 Control	3.8-4.0 ms
C248-11	BL/RD	Keep Alive Power	12-14v
C248-12	BL/PK	Camshaft Position Sensor	5-7 Hz
C248-13	---	Not Used	---
C248-14	RD/WT	Self-Test Input Signal	STI On: 0v, Off: 12v
C248-15	GN	CKP Sensor Signal	410-440 Hz
C248-16	BL	CMP Sensor Ground	<0.050v
C248-17	---	Not Used	12-14v
C248-18	PK/BK	EGR Vent Solenoid Control	12v, 55 mph: 1v
C248-19	GN/WT	EGR Control Solenoid	12v, 55 mph: 1v
C248-20	BL/BK	EVAP Purge Solenoid	CANP On: <1.v, Off: 12v
C248-21	BL/OR	Ignition Control Module Signal	0.1-0.3v
C248-22	---	Not Used	---
C248-23	---	Not Used	---
C248-24	---	Not Used	---
C248-25	---	Not Used	---
C248-26	---	Not Used	---
C248-27	BK/RD	Power Ground	<0.1v
C248-28	---	Not Used	---
C248-29	---	Not Used	---
C248-30	GY/GN	EGR Boost Check Solenoid	On: 1v, Off: 12v
C248-31	BK	Power Ground	<0.1v
C248-32	BK	Power Ground	<0.1v
C248-33	BK	Power Ground	<0.1v
C248-34	BK	Power Ground	<0.1v

PCM Connector Graphic

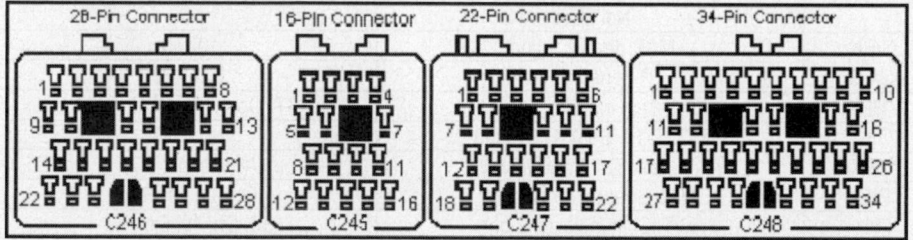

Standard Colors and Abbreviations

Abbreviation	Color	Abbreviation	Color	Abbreviation	Color
BK	Black	GY	Gray	PK	Purple
BL	Blue	GN	Green	RD	Red
BR	Brown	LG	LT Green	TN	Tan
DB	Dark Blue	OR	Orange	WT	White
DG	DK Green	PK	Pink	YL	Yellow

TAURUS PIN TABLES

1990 Taurus 2.5L I4 CFI VIN D (All) 60 Pin Connector

PCM Pin #	Wire Color	Circuit Description (60 Pin)	Value at Hot Idle
1	YL	Keep Alive Power	12-14v
2	RD/LG	Brake Pedal Position Switch	Brake Off: 0v, On: 12v
3	DG, OR	VSS (+), (-) Signal	0 Hz, 55 mph: 125 Hz
4	DG/YL	Ignition Diagnostic Monitor	20-31 Hz
5	RD/LG	Ignition Signal (Cranking)	KOEC: 9-11v
7	LG/YL	ECT Sensor Signal	0.5-0.6v
8	PK/BK	Fuel Pump Monitor	On: 12v, Off: 0v
9	BK, OR	Data Bus (-), (+) Signal	Digital Signals
10	PK	A/C Cycling Clutch Switch	A/C On: 12v, Off: 0v
11-12, 14-15	---	Not Used	---
13	OR/YL	Cruise Control Solenoid	Solenoid On: 1v
16	BK/OR	Ignition System Ground	<0.050v
17	TN/RD	STO & MIL Control	MIL On: 1v, Off: 12v
18-19, 32, 38	---	Not Used	---
20, 46	BK	PCM Case, Analog Signal Ground	<0.050v
21	YL, WT	ISC Motor (+), (-) Control	0.3v
22	TN/LG	Fuel Pump Control	On: 1v, Off: 12v
23	YL/LG	PSP Switch Signal	Straight: 0v, Turned: 5v
24	WT/RD	Idle Tracking Switch	11v, 55 mph: 0v
25	LG/PK	ACT Sensor Signal	1.5-2.5v
26	OR/WT	Reference Voltage	4.9-5.1v
27	GN or BR	EGR EVP Sensor	0.4v, 55 mph: 2.6v
29	GN/P	HO2S-11 (B1 S1) Signal	0.1-1.1v
29	DG/PK	HO2S-11 (B1 S1) Signal	0.1-1.1v
30	BL/YL	Neutral Drive (A/T), Clutch (M/T) Switch	In 'N' or Clutch "In": 0v
31	GY/YL	EVAP Purge Solenoid	12v, 55 mph: 1v
33	DG	EGR VR Solenoid	0%, 55 mph: 45%
34	BL/PK	Data Output Link	Digital Signals
35	WT/PK	Speed Control Vent Solenoid	Off: 0% duty cycle
36	YL/LG	Spark Output Signal	50% duty cycle
37, 57	RD	Vehicle Power	12-14v
39	OR	Speed Control Switch Ground	<0.050v
40, 60	BK/LG	Power Ground	<0.1v
42	GY/BK	S/C Vacuum Solenoid Control	Off: 0% duty cycle
43-44, 51, 59	---	Not Used	---
45	LG/BK	MAP Sensor Signal	107 Hz
47	DG/LG	TP Sensor Signal	0.5-1.2v
48	WT/BK	Self-Test Indicator Signal	STI Open: 5v, Closed: 1v
49	OR	HO2S-11 (Bank 1) Ground	<0.050v
50	BL/BK	Speed Control Switch	All speeds: 6.7v
52	PK	High Speed Fan Control	On: 1v, Off: 12v
53	PK	M/T: Shift Indicator Light	SIL On: 1v, Off: 12v
54	RD	A/C WOT Relay Control	On: 1v, Off: 12v
55	TN/OR	Low Speed Fan Control	On: 1v, Off: 12v
56	DB	PIP Sensor Signal	50% duty cycle
58	TN/RD	CFI Fuel Injector	1.6-1.7 ms

Pin Connector Graphic

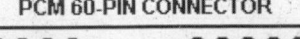
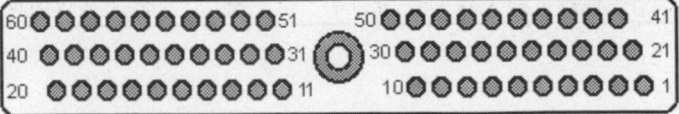

Terminal View of 60-Pin PCM Harness Connector

1991 Taurus 2.5L I4 MFI VIN D (A/T) 60 Pin Connector

PCM Pin #	Wire Color	Circuit Description (60 Pin)	Value at Hot Idle
1	YL	Keep Alive Power	12-14v
2	RD/LG	Brake Pedal Position Switch	Brake Off: 0v, On: 12v
3	DG/WT	VSS (+) Signal	0 Hz, 55 mph: 125 Hz
4	WT/PK	Ignition Diagnostic Monitor	20-31 Hz
5	GY/BK	TSS Sensor Signal	100 Hz (700 rpm)
6	OR/YL	VSS (+) Signal	0 Hz, 55 mph: 125 Hz
7	LG/RD	ECT Sensor Signal	0.5-0.6v
8	PK/BK	Fuel Pump Monitor	On: 12v, Off: 0v
9	TN/BL	MAF Sensor Return	<0.050v
10	PK/BL	A/C Cycling Clutch Switch	A/C On: 12v, Off: 0v
11	GN/YL	EVAP Purge Solenoid	12v, 55 mph: 1v
12	---	Not Used	---
13	TN/OR	Low Speed Fan Control	On: 1v, Off: 12v
14	LG/PK	High Speed Fan Control	On: 1v, Off: 12v
15	---	Not Used	---
16	OR/RD	Ignition System Ground	<0.050v
17	TN/RD	STI Output, MIL Control	MIL On: 1v, Off: 12v
18	OR/BK	Data Bus (+) Signal	Digital Signals
19	BK/OR	Data Bus (-) Signal	Digital Signals
20	BK	PCM Case Ground	<0.050v
21	OR/BK	IAC Motor Control	9.5-10.1v
22	TN/LG	Fuel Pump Control	On: 1v, Off: 12v
23	---	Not Used	---
24	DG	CID Sensor Signal	5-7 Hz
25	LG/PK	ACT Sensor Signal	1.5-2.5v
26	BR/WT	Reference Voltage	4.9-5.1v
27	BR/LG	PFE EGR Sensor	3.3v, 55 mph: 2.8v
28	YL/LG	PSP Sensor Signal	Straight: 0v, Turned: 11v
29	---	Not Used	---
30	BL/YL	MLP Sensor Signal	In 'P': 0v, 55 mph: 1.7v
31	---	Not Used	---
32	---	Not Used	---
33	BR/PK	EGR VR Solenoid	0%, 55 mph: 45%
34	---	Not Used	---
35	BR/BL	Injector 4 Control	4.8-5.0 ms
36	YL/LG	Spark Output Signal	50% duty cycle
37	RD	Vehicle Power	12-14v
38	WT/YL	EPC Solenoid Control	8-9v (9 psi)
39	BR/YL	Injector 3 Control	4.8-5.0 ms
40	BK/LG	Power Ground	<0.1v

Standard Colors and Abbreviations

Abbreviation	Color	Abbreviation	Color	Abbreviation	Color
BK	Black	GY	Gray	PK	Purple
BL	Blue	GN	Green	RD	Red
BR	Brown	LG	LT Green	TN	Tan
DB	Dark Blue	OR	Orange	WT	White
DG	DK Green	PK	Pink	YL	Yellow

1991 Taurus 2.5L I4 MFI VIN D (A/T) 60 Pin Connector

PCM Pin #	Wire Color	Circuit Description (60 Pin)	Value at Hot Idle
41-43	----	Not Used	---
44	RD/BK	HO2S-11 (B1 S1) Signal	0.1-1.1v
45	---	Not Used	---
46	GY/RD	Analog Signal Return	<0.050v
47	GN/WT	TP Sensor Signal	0.5-1.2v
48	BR	Self-Test Indicator Signal	STI Open: 5v, Closed: 1v
49	O/B	TOT Sensor Signal	2.10-2.40v
50	BL/RD	MAF Sensor Signal	0.6-0.9v
51	OR/YL	Shift Solenoid 1 Control	12v, 55 mph: 1v
52	PK/OR	Shift Solenoid 2 Control	1v, 55 mph: 12v
53	TN/WT	Lock-Up Solenoid	0%, 55 mph: 45%
54	PK/YL	A/C WOT Relay Control	On: 1v, Off: 12v
55	WT/RD	Shift Solenoid 3 Control	7-9v, 55 mph: 8-9v
56	DB	PIP Sensor Signal	50% duty cycle
57	RD	Vehicle Power	12-14v
58	TN	Injector 1 Control	4.8-5.0 ms
59	WT	Injector 2 Control	4.8-5.0 ms
60	BK/LG	Power Ground	<0.1v

Pin Connector Graphic

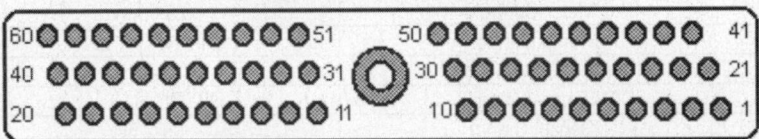

PCM 60-PIN CONNECTOR

Terminal View of 60-Pin PCM Harness Connector

Standard Colors and Abbreviations

Abbreviation	Color	Abbreviation	Color	Abbreviation	Color
BK	Black	GY	Gray	PK	Purple
BL	Blue	GN	Green	RD	Red
BR	Brown	LG	LT Green	TN	Tan
DB	Dark Blue	OR	Orange	WT	White
DG	DK Green	PK	Pink	YL	Yellow

1996-97 Taurus 3.0L V6 MFI VIN S (A/T) 104 Pin Connector

PCM Pin #	Wire Color	Circuit Description (104 Pin)	Value at Hot Idle
1	PK/OR	Shift Solenoid 2 Control	1v, 55 mph: 12v
2	PK/LG	MIL (lamp) Control	MIL On: 1v, Off: 12v
3-4, 6-7	---	Not Used	---
5	WT/OR	Electronic AIRB Monitor	5v
8	DB/YL	IMRC Monitor Signal	0.6v
9-12, 14	---	Not Used	---
13	PK	Flash EEPROM Power	0.1v
15	PK/LB	Data Bus (-) Signal	Digital Signals
16	TN/OR	Data Bus (+) Signal	Digital Signals
17-20, 23	---	Not Used	---
21	GY	CKP Sensor (+) Signal	850-1120 Hz
22	DB	CKP Sensor (-) Signal	850-1120 Hz
24, 51	BK/WT	Power Ground	<0.1v
25	BK	Power Ground	<0.1v
26	YL/BK	Coil Driver 1 Control	6° dwell
27	OR/YL	Shift Solenoid 1 Control	12v, 55 mph: 1v
28	TN/OR	Low Speed Fan Control	On: 1v, Off: 12v
29	TN/WT	Overdrive Cancel Switch	TCS & O/D On: 12v
30	DG	Octane Adjust Switch	Closed: 0v, Open: 9.3v
31	YL/LG	PSP Switch Signal	Straight: 0v, Turned: 12v
32, 34, 43	---	Not Used	---
33	PK/OR	VSS (-) Signal	0 Hz, 55 mph: 125 Hz
35	RD/BK	HO2S-12 (B1 S2) Signal	0.1-1.1v
36	TN/LB	MAF Sensor Return	<0.050v
37	OR/BK	TFT Sensor Signal	2.10-2.40v
38	LG/RD	ECT Sensor Signal	0.5-0.6v
39	GY	IAT Sensor Signal	1.5-2.5v
40	DG/YL	Fuel Pump Monitor	On: 12v, Off: 0v
41	PK/LB	A/C Cycling Clutch Switch	A/C On: 12v, Off: 0v
42	BR	IMRC Solenoid Control	12v (Off)
44-45, 49	---	Not Used	---
46	LG/PK	High Speed Fan Control	On: 1v, Off: 12v
47	BR/PK	EGR VR Solenoid	0%, 55 mph: 45%
48	TN/YL	Clean Tachometer Output	33-37 Hz
50, 54, 59	---	Not Used	---
52	YL/RD	Coil Driver 2 Control	6° dwell
53	PK/BK	Shift Solenoid 3 Control	12v, 55 mph: 1v
55	DG/OR	Keep Alive Power	12-14v
56	GY/YL	EVAP VMV Solenoid	0-10 Hz (0-100%)
57	YL/RD	Knock Sensor Signal	0v
58	GY/BK	VSS (+) Signal	0 Hz, 55 mph: 125 Hz
60	GY/LB	HO2S-11 (B1 S1) Signal	0.1-1.1v
61	PK/LG	HO2S-22 (B2 S2) Signal	0.1-1.1v
62	RD/PK	FTP Sensor Signal	2.6v (0" H2O - cap off)
63	---	Not Used	---
64	LB/YL	TR Sensor Signal	In 'P': 0v, 55 mph: 1.7v
65	BR/LG	DPFE Sensor Signal	0.20-1.30v
66	---	Not Used	---
67	PK/WT	EVAP Canister Vent Valve	0-10 Hz (0-100%)
68	---	Not Used	---
69	PK/YL	A/C WOT Relay Control	On: 1v, Off: 12v
70	BR	Electronic AIRB Solenoid	12v, 55 mph: 12v

1996-97 Taurus 3.0L V6 MFI VIN S (A/T) 104 Pin Connector

PCM Pin #	Wire Color	Circuit Description (104 Pin)	Value at Hot Idle
71	RD	Vehicle Power	12-14v
73	TN/BK	Injector 5 Control	2.3-5.5 ms
74	BR/YL	Injector 3 Control	2.3-5.5 ms
75	TN	Injector 1 Control	2.3-5.5 ms
76 & 77	BK/WT	Power Ground	<0.1v
78	YL/WT	Coil Driver 3 Control	6° dwell
79	WT/LG	TCIL (lamp) Control	On: 1v, Off: 12v
80	LB/OR	Fuel Pump Control	On: 1v, Off: 12v
81	WT/YL	EPC Solenoid Control	9.0v (15 psi)
82	PK/YL	TCC Solenoid Control	12v (0 psi)
83	WT/LB	IAC Motor Control	10v (55% duty cycle)
84	DG/WT	TSS Sensor Signal	50-65 Hz
85	DB/OR	CID Sensor Signal	5-7 Hz
86	BK/YL	A/C Pressure Switch	A/C On: 12v (Open)
87	RD/LG	HO2S-12 (B1 S2) Signal	0.1-1.1v
88	LB/RD	MAF Sensor Signal	0.6v
89	GY/WT	TP Sensor Signal	0.53-1.27v
90	BR/WT	Reference Voltage	4.9-5.1v
91	GY/RD	Analog Signal Return	<0.050v
92	RD/LG	Brake Pedal Position Switch	Brake Off: 0v, On: 12v
93	RD/WT	HO2S-11 (B1 S1) Heater	On: 1v, Off: 12v
94	WT/BK	HO2S-21 (B2 S1) Heater	On: 1v, Off: 12v
95	YL/LG	HO2S-12 (B1 S2) Heater	On: 1v, Off: 12v
96	TN/YL	HO2S-22 (B2 S2) Heater	On: 1v, Off: 12v
97	RD	Vehicle Power	12-14v
98	---	Not Used	---
99	LG/OR	Injector 6 Control	2.3-5.5 ms
100	BR/LB	Injector 4 Control	2.3-5.5 ms
101	WT	Injector 2 Control	2.3-5.5 ms
102	---	Not Used	---
103	BK/WT	Power Ground	<0.1v
104	---	Not Used	---

Pin Connector Graphic

PCM 104-PIN CONNECTOR

Terminal View of 104-Pin PCM Wiring Harness Connector

Standard Colors and Abbreviations

Abbreviation	Color	Abbreviation	Color	Abbreviation	Color
BK	Black	GY	Gray	PK	Purple
BL	Blue	GN	Green	RD	Red
BR	Brown	LG	LT Green	TN	Tan
DB	Dark Blue	OR	Orange	WT	White
DG	DK Green	PK	Pink	YL	Yellow

1998-99 Taurus 3.0L 4v V6 MFI VIN S (A/T) 104 Pin Connector

PCM Pin #	Wire Color	Circuit Description (104 Pin)	Value at Hot Idle
1	PK/OR	Shift Solenoid 2 Control	1v, 55 mph: 1v
2	PK/LG	MIL (lamp) Control	MIL On: 1v, Off: 12v
3-4, 6	---	Not Used	---
5	WT/OR	Electronic AIRB Monitor	5v
7, 9-11	---	Not Used	---
8	DB/YL	IMRC Monitor Signal	5v, 55 mph: 5v
12	YL/WT	Fuel Level Indicator Signal	1.7v (1/2 full)
13	PK	Flash EEPROM Power	0.1v
15	PK/LB	Data Bus (-) Signal	Digital Signals
14, 17-20	---	Not Used	---
16	TN/OR	Data Bus (+) Signal	Digital Signals
21	GY	CKP Sensor (+) Signal	390-520 Hz
22	DB	CKP Sensor (-) Signal	390-520 Hz
24	BK/WT	Power Ground	<0.1v
25	BK	Power Ground	<0.1v
26	YL/BK	Coil Driver 1 Control	5° dwell
27	OR/YL	Shift Solenoid 1 Control	12v, 55 mph: 0.1v
28	LB	Low Speed Fan Control	On: 1v, Off: 12v
29	TN/WT	Overdrive Cancel Switch	TCS & O/D On: 12v
30	DG	Octane Adjust Switch	Closed: 0v, Open: 9.3v
31	YL/LG	PSP Switch Signal	Straight: 0v, Turned: 12v
33	PK/OR	VSS (-) Signal	0 Hz, 55 mph: 125 Hz
35	RD/BK	HO2S-12 (B1 S2) Signal	0.1-1.1v
36	TN/LB	MAF Sensor Return	<0.050v
37	OR/BK	TFT Sensor Signal	2.10-2.40v
38	LG/RD	ECT Sensor Signal	0.5-0.6v
39	GY	IAT Sensor Signal	1.5-2.5v
40	DG/YL	Fuel Pump Monitor	On: 12v, Off: 0v
41	PK/LB	A/C Cycling Clutch Switch	On: 1v, Off: 12v
42	BR	IMRC Solenoid Control	12v (Off)
45	PK/WT	EVAP CV Solenoid	0-10 Hz (0-100%)
46	LG/PK	High Speed Fan Control	On: 1v, Off: 12v
47	BR/PK	EGR VR Solenoid	0%, 55 mph: 45%
48	TN/YL	Clean Tachometer Output	33-45 Hz
49-50	---	Not Used	---
51	BK/WT	Power Ground	<0.1v
52	YL/RD	Coil Driver 2 Control	5° dwell
53	PK/BK	Shift Solenoid 3 Control	12v, 55 mph: 0.1v
54	PK/YL	TCC Solenoid Control	0%, 55 mph: 95%
56	GY/YL	EVAP Purge Solenoid	0-10 Hz (0-100%)
55	DG/OR	Keep Alive Power	12-14v
57	YL/RD	Knock Sensor Signal	0v
58	GY/BK	VSS (+) Signal	0 Hz, 55 mph: 125 Hz
59	---	Not Used	---
60	GY/LB	HO2S-11 (B1 S1) Signal	0.1-1.1v
61	PK/LG	HO2S-22 (B2 S2) Signal	0.1-1.1v
62	RD/BK	FTP Sensor Signal	2.6v (0 in. H20)
63	---	Not Used	---
64	LB/YL	TR Sensor Signal	In 'P': 0v, 55 mph: 1.7v
65	BR/LG	DPFE Sensor Signal	0.20-1.30v
66	---	Not Used	---
67	PK/WT	EVAP Purge Solenoid	0-10 Hz (0-100%)
68	---	Not Used	---

1998-99 Taurus 3.0L 4v V6 MFI VIN S (A/T) 104 Pin Connector

PCM Pin #	Wire Color	Circuit Description (104 Pin)	Value at Hot Idle
69	PK/YL	A/C WOT Relay Control	On: 1v, Off: 12v
70	BR	Electronic AIRB Solenoid	12v, 55 mph: 12v
71	RD	Vehicle Power	12-14v
72	---	Not Used	---
73	TN/BK	Injector 5 Control	2.2-2.7 ms
74	BR/YL	Injector 3 Control	2.2-2.7 ms
75	TN	Injector 1 Control	2.2-2.7 ms
76, 77	BK/WT	Power Ground	<0.1v
78	YL/WT	Coil Driver 3 Control	5° dwell
79	WT/LG	TCIL (lamp) Control	On: 1v, Off: 12v
80	LB/OR	Fuel Pump Control	On: 1v, Off: 12v
81	WT/YL	EPC Solenoid Control	9.0v (15 psi)
82	PK/YL	TCC Solenoid Control	12v (0 psi)
83	WT/LB	IAC Motor Control	10v (55% duty cycle)
84	DG/WT	TSS Sensor Signal	43 Hz (700 rpm)
85	DB/OR	CID Sensor Signal	5-7 Hz
86	BK/YL	A/C High Pressure Switch	12v (Switch Open)
87	RD/LG	HO2S-21 (B2 S1) Signal	0.1-1.1v
88	LB/RD	MAF Sensor Signal	0.6v
89	GY/WT	TP Sensor Signal	0.53-1.27v
90	BR/WT	Reference Voltage	4.9-5.1v
91	GY/RD	Analog Signal Return	<0.050v
92	RD/LG	Brake Pedal Position Switch	Brake Off: 0v, On: 12v
93	RD/WT	HO2S-11 (B1 S1) Heater	On: 1v, Off: 12v
94	WT/BK	HO2S-21 (B2 S1) Heater	On: 1v, Off: 12v
95	YL/LG	HO2S-12 (B1 S2) Heater	On: 1v, Off: 12v
96	TN/YL	HO2S-22 (B2 S2) Heater	On: 1v, Off: 12v
71	RD	Vehicle Power	12-14v
98	---	Not Used	---
99	LG/OR	Injector 6 Control	2.2-2.7 ms
100	BR/LB	Injector 4 Control	2.2-2.7 ms
101	WT	Injector 2 Control	2.2-2.7 ms
102	---	Not Used	---
103	BK/WT	Power Ground	<0.1v
104	---	Not Used	---

Pin Voltage Connector

PCM 104-PIN CONNECTOR

Terminal View of 104-Pin PCM Wiring Harness Connector

Standard Colors and Abbreviations

Abbreviation	Color	Abbreviation	Color	Abbreviation	Color
BK	Black	GY	Gray	PK	Purple
BL	Blue	GN	Green	RD	Red
BR	Brown	LG	LT Green	TN	Tan
DB	Dark Blue	OR	Orange	WT	White
DG	DK Green	PK	Pink	YL	Yellow

2000 Taurus 3.0L 4v V6 MFI VIN S (A/T) 104 Pin Connector

PCM Pin #	Wire Color	Circuit Description (104 Pin)	Value at Hot Idle
1	DG/PK	COP 4 Driver Control	6° dwell
2	PK/LG	MIL (lamp) Control	MIL On: 1v, Off: 12v
3	BK/WT	Power Ground	<0.1v
5	WT/OR	Electronic AIRB Monitor	5v
6	OR/YL	Shift Solenoid 'A' Control	12v, 55 mph: 0.1v
12	YL/WT	Fuel Level Indicator Signal	1.7v (1/2 full)
13	PK	Flash EEPROM Power	0.1v
15	PK/LB	Data Bus (-) Signal	Digital Signals
16	TN/OR	Data Bus (+) Signal	Digital Signals
17	BR/YL	SCI Receive Signal	N/A
18	GY/RD	SCI Transmit Signal	N/A
20	PK/BK	Shift Solenoid 'A' Control	12v, 55 mph: 0.1v
21	GY	CKP Sensor (+) Signal	390-520 Hz
22	DB	CKP Sensor (-) Signal	390-520 Hz
24	BK/WT	Power Ground	<0.1v
25	BK	Power Ground	<0.1v
26	LG/WT	COP 1 Driver Control	6° dwell
27	OR/YL	COP 5 Driver Control	6° dwell
28	LB	Low Speed Fan Control	On: 1v, Off: 12v
30	DB/LG	Antitheft Indicator	N/A
31	YL/LG	PSP Switch Signal	Straight: 0v, Turned: 12v
32	YL	Knock Sensor (-) Signal	0v
34	YL/BK	Digital TR1 Sensor	In 'P': 0v, 55 mph: 11v
35	RD/BK	HO2S-12 (B1 S2) Signal	0.1-1.1v
36	TN/LB	MAF Sensor Return	<0.050v
37	OR/BK	TFT Sensor Signal	2.10-2.40v
38	LG/RD	ECT Sensor Signal	0.5-0.6v
39	GY	IAT Sensor Signal	1.5-2.5v
40	DG/YL	Fuel Pump Monitor	On: 12v, Off: 0v
41	DG/WT	A/C High Pressure Switch	On: 1v, Off: 12v
44	PK	Starter Relay Control	KOEC: 9-11v
46	LG/PK	High Speed Fan Control	On: 1v, Off: 12v
47	BR/PK	EGR VR Solenoid	0%, 55 mph: 45%
48	TN/YL	Clean Tachometer Output	33-45 Hz
49	LB/BK	Digital TR2 Sensor	In 'P': 0v, 55 mph: 11v
50	WT/BK	Digital TR4 Sensor	In 'P': 0v, 55 mph: 11v
51	BK/WT	Power Ground	<0.1v
52	PK/WT	COP 2 Driver Control	6° dwell
53	OR/YL	COP 6 Driver Control	6° dwell
55	PK/LB	Keep Alive Power	12-14v
54	PK/YL	TCC Solenoid Control	0%, 55 mph: 95%
56	GY/YL	EVAP Purge Solenoid	0-10 Hz (0-100%)
55	DG/OR	Keep Alive Power	12-14v
57	YL/RD	Knock Sensor (+) Signal	0v
59	DG/WT	TSS Sensor Signal	43 Hz (700 rpm)
60	GY/LB	HO2S-11 (B1 S1) Signal	0.1-1.1v
61	PK/LG	HO2S-22 (B2 S2) Signal	0.1-1.1v
62	RD/PK	FTP Sensor Signal	2.6v (0 in. H20)
63	WT/YL	EPC Solenoid Control	9.0v (15 psi)
64	RD/BK	Digital TR3A Sensor	In 'P': 0v, 55 mph: 1.7v

2000 Taurus 3.0L 4v V6 MFI VIN S (A/T) 104 Pin Connector

PCM Pin #	Wire Color	Circuit Description (104 Pin)	Value at Hot Idle
65	BR/LG	DPFE Sensor Signal	0.20-1.30v
67	PK/WT	EVAP CV Solenoid	0-10 Hz (0-100%)
68	GY/BK	VSS (+) Signal	0 Hz, 55 mph: 125 Hz
69	PK/YL	A/C WOT Relay Control	On: 1v, Off: 12v
70	BR	Electronic AIRB Solenoid	12v, 55 mph: 12v
71	RD	Vehicle Power	12-14v
73	TN/BK	Injector 5 Control	2.2-2.7 ms
74	BR/YL	Injector 3 Control	2.2-2.7 ms
75	TN	Injector 1 Control	2.2-2.7 ms
76	BK/WT	Power Ground	<0.1v
77	BK/WT	Power Ground	<0.1v
78	YL/PKK	COP 3 Driver Control	6° dwell
79	WT/LG	TCIL (lamp) Control	On: 1v, Off: 12v
80	LB/OR	Fuel Pump Control	On: 1v, Off: 12v
81	BR/OR	EPC Solenoid Control	9.0v (15 psi)
83	WT/LB	IAC Motor Control	10v (55% duty cycle)
84	DB/YL	OSS Sensor Signal	0 Hz, 55 mph: 131 Hz
85	DB/OR	CID Sensor Signal	5-7 Hz
86	BK/YL	A/C Clutch Relay Switch	12v (Switch Open)
87	RD/LG	HO2S-21 (B2 S1) Signal	0.1-1.1v
88	LB/RD	MAF Sensor Signal	0.6v
89	GY/WT	TP Sensor Signal	0.53-1.27v
90	BR/WT	Reference Voltage	4.9-5.1v
91	GY/RD	Analog Signal Return	<0.050v
92	RD/LG	Brake Pedal Position Switch	Brake Off: 0v, On: 12v
93	RD/WT	HO2S-11 (B1 S1) Heater	On: 1v, Off: 12v
94	WT/BK	HO2S-21 (B2 S1) Heater	On: 1v, Off: 12v
95	YL/LB	HO2S-12 (B1 S2) Heater	On: 1v, Off: 12v
96	TN/YL	HO2S-22 (B2 S2) Heater	On: 1v, Off: 12v
99	LG/OR	Injector 6 Control	2.2-2.7 ms
100	BR/LB	Injector 4 Control	2.2-2.7 ms
101	WT	Injector 2 Control	2.2-2.7 ms
102	---	Not Used	---
103	BK/WT	Power Ground	<0.1v
104	---	Not Used	---

Pin Connector Graphic

PCM 104-PIN CONNECTOR

Terminal View of 104-Pin PCM Wiring Harness Connector

Standard Colors and Abbreviations

Abbreviation	Color	Abbreviation	Color	Abbreviation	Color
BK	Black	GY	Gray	PK	Purple
BL	Blue	GN	Green	RD	Red
BR	Brown	LG	LT Green	TN	Tan
DB	Dark Blue	OR	Orange	WT	White
DG	DK Green	PK	Pink	YL	Yellow

2001 Taurus 3.0L 4v V6 MFI VIN S (A/T) 104 Pin Connector

PCM Pin #	Wire Color	Circuit Description (104 Pin)	Value at Hot Idle
1	PK/OR	Shift Solenoid 'B' Control	1v, 55 mph: 1v
2	PK/LG	MIL (lamp) Control	MIL On: 1v, Off: 12v
3	BK/WT	Power Ground	<0.1v
3	YL/BK	Digital TR1 Sensor	In 'P': 0v, 55 mph: 11v
4	BR/PK	Start System Voltage	Cranking: 9-11v
6	DG/WT	TSS Sensor Signal	50-65 Hz (790-820 rpm)
12	YL/WT	Fuel Level Indicator Signal	1.7v (1/2 full)
13	VT	Flash EEPROM Power	0.1v
15	PK/LB	SCP Data Bus (-) Signal	Digital Signals
16	TN/OR	SCP Data Bus (+) Signal	Digital Signals
17	BR/YL	SCI Receive (RX) Signal	Digital Signals
18	GY/RD	SCI Transmit (TX) Signal	Digital Signals
20	GY/YL	Generator Signal 2	130 Hz (45%)
21	GY	CKP Sensor (+) Signal	390-520 Hz
22	DB	CKP Sensor (-) Signal	390-520 Hz
24	BK/WT	Power Ground	<0.1v
25	BK	Chassis Ground	<0.1v
26	YL/BK	Coil 'A' Driver (Cyl 1 & 5)	6° dwell
27	OR/YL	Shift Solenoid 'A' Control	12v, 55 mph: 1v
28	LB	Low Speed Fan Control	On: 1v, Off: 12v
30	DB/LG	Antitheft Indicator	Refer to PAT System
31	YL/LG	PSP Switch Signal	Straight: 0v, Turned: 12v
32	YL	Knock Sensor (-) Signal	<0.050v
35	RD/BK	HO2S-12 (B1 S2) Signal	0.1-1.1v
36	TN/LB	MAF Sensor Return	<0.050v
37	OR/BK	TFT Sensor Signal	2.10-2.40v
38	LG/RD	ECT Sensor Signal	0.5-0.6v
39	GY	IAT Sensor Signal	1.5-2.5v
40	DG/YL	Fuel Pump Monitor	On: 12v, Off: 0v
41	DG/WT	A/C High Pressure Switch	On: 1v, Off: 12v
42	LG/YL	Generator Cutoff Relay	0.1v (relay on)
44	PK	Starter Relay Control	KOEC: 9-11v
45	LB/RD	Generator Sensor Signal	130 Hz (30%)
46	LG/VT	High Speed Fan Control	On: 1v, Off: 12v
47	BR/PK	EGR VR Solenoid	0%, 55 mph: 45%
48	TN/YL	Clean Tachometer Output	33-45 Hz
49	LB/BK	Digital TR2 Sensor	In 'P': 0v, 55 mph: 11v
50	WT/BK	Digital TR4 Sensor	In 'P': 0v, 55 mph: 11v
51	BK/WT	Power Ground	<0.1v
52	YL/RD	Coil 'B' Driver (Cyl 3 & 4)	6° dwell
53	PK/BK	Shift Solenoid 'C' Control	12v, 55 mph: 1v
54	VT/YL	TCC Solenoid Control	0%, 55 mph: 95%
55	PK/LB	Keep Alive Power	12-14v
56	GY/YL	EVAP Purge Solenoid	0-10 Hz (0-100%)
55	DG/OR	Keep Alive Power	12-14v
57	YL/RD	Knock Sensor (+) Signal	0v
60	GY/LB	HO2S-11 (B1 S1) Signal	0.1-1.1v
61	VT/LG	HO2S-22 (B2 S2) Signal	0.1-1.1v
62	RD/PK	FTP Sensor Signal	2.6v (0 in. H20)
63	WT/YL	Injector Pressure Sensor	2.74v (39 psi)
64	RD/BK	Digital TR3A Sensor	In 'P': 0v, 55 mph: 1.7v

2001 Taurus 3.0L 4v V6 MFI VIN S (A/T) 104 Pin Connector

PCM Pin #	Wire Color	Circuit Description (104 Pin)	Value at Hot Idle
65	BR/LG	DPFE Sensor Signal	0.95-1.05v
66	TN/OR	Evaporator Air Temp. Sensor	4.95-5.15v
67	VT/WT	EVAP CV Solenoid	0-10 Hz (0-100%)
68	GY/BK	VSS (+) Signal	0 Hz, 55 mph: 125 Hz
69	PK/YL	A/C WOT Relay Control	On: 1v, Off: 12v
71	RD	Vehicle Power	12-14v
73	TN/BK	Injector 5 Control	2.2-2.7 ms
74	BR/YL	Injector 3 Control	2.2-2.7 ms
75	TN	Injector 1 Control	2.2-2.7 ms
76	BK/WT	Power Ground	<0.1v
77	BK/WT	Power Ground	<0.1v
78	YL/WT	Coil 'C' Driver (Cyl 2 & 6)	6° dwell
79	WT/LG	TCIL (lamp) Control	On: 1v, Off: 12v
80	LB/OR	Fuel Pump Control	On: 1v, Off: 12v
81	BR/OR	EPC Solenoid Control	9.0v (15 psi)
82	BK/WT	Check Cap Indicator Light	On: 1v, Off: 12v
83	WT/LB	IAC Motor Control	10v (55% duty cycle)
84	DB/YL	OSS Sensor Signal	0 Hz, 55 mph: 131 Hz
85	DB/OR	CID Sensor Signal	5-7 Hz
86	BK/YL	A/C Clutch Relay Switch	12v (Switch Open)
87	RD/LG	HO2S-21 (B2 S1) Signal	0.1-1.1v
88	LB/RD	MAF Sensor Signal	0.6v
89	GY/WT	TP Sensor Signal	0.53-1.27v
90	BR/WT	Reference Voltage	4.9-5.1v
91	GY/RD	Analog Signal Return	<0.050v
92	RD/LG	Brake Pedal Position Switch	Brake Off: 0v, On: 12v
93	RD/WT	HO2S-11 (B1 S1) Heater	On: 1v, Off: 12v
94	WT/BK	HO2S-21 (B2 S1) Heater	On: 1v, Off: 12v
95	YL/LB	HO2S-12 (B1 S2) Heater	On: 1v, Off: 12v
96	TN/YL	HO2S-22 (B2 S2) Heater	On: 1v, Off: 12v
98	LG/RD	Generator Indicator Light	On: 1v, Off: 12v
99	LG/OR	Injector 6 Control	2.2-2.7 ms
100	BR/LB	Injector 4 Control	2.2-2.7 ms
101	W	Injector 2 Control	2.2-2.7 ms
103	BK/WT	Power Ground	<0.1v

Pin Voltage Connector

PCM 104-PIN CONNECTOR

1 26
27 52
53 78
79 104

Terminal View of 104-Pin PCM Wiring Harness Connector

Standard Colors and Abbreviations

Abbreviation	Color	Abbreviation	Color	Abbreviation	Color
BK	Black	GY	Gray	PK	Purple
BL	Blue	GN	Green	RD	Red
BR	Brown	LG	LT Green	TN	Tan
DB	Dark Blue	OR	Orange	WT	White
DG	DK Green	PK	Pink	YL	Yellow

2002-04 Taurus 3.0L 4v V6 MFI VIN S (A/T) 104 Pin Connector

PCM Pin #	Wire Color	Circuit Description (104 Pin)	Value at Hot Idle
1	VT/OR	Shift Solenoid 'B' Control	1v, 55 mph: 1v
2	PK/LG	MIL (lamp) Control	MIL On: 1v, Off: 12v
3	YL/BK	Digital TR1 Sensor	In 'P': 0v, 55 mph: 11v
4	BR/PK	System Voltage (Start)	Cranking: 9-11v
5, 7-11	---	Not Used	---
6	DG/WT	TSS Sensor Signal	50-65 Hz (790-820 rpm)
12	YL/WT	Fuel Level Indicator Signal	1.7v (1/2 full)
13	VT	Flash EEPROM Power	0.1v
14, 19	---	Not Used	---
15	PK/LB	SCP Bus (-) Signal	Digital Signals
16	TN/OR	SCP Bus (+) Signal	Digital Signals
17	BR/YL	Passive Antitheft RX Signal	Digital Signals
18	GY/RD	Passive Antitheft TX Signal	Digital Signals
20	GY/YL	Generator Monitor Signal	130 Hz (45%)
21	GY	CKP Sensor (+) Signal	390-520 Hz
22	DB	CKP Sensor (-) Signal	390-520 Hz
23, 29	---	Not Used	---
24	BK/WT	Power Ground	<0.1v
25	BK	Chassis Ground	<0.1v
26	YL/BK	Coil 1 Driver (Cyl 1 & 5)	6° dwell
27	OR/YL	Shift Solenoid 'A' Control	12v, 55 mph: 1v
28	LB	Engine Cooling Fan Brake Relay Control	Relay Off: 12v, On: 1v
30	DB/LG	Antitheft Indicator	Indicator Off: 12v, On: 1v
31	YL/LG	PSP Switch Signal	Straight: 0v, Turned: 12v
32	YL	Knock Sensor (-) Signal	<0.050v
33-34, 43	---	Not Used	---
35	RD/BK	HO2S-12 (B1 S2) Signal	0.1-1.1v
36	TN/LB	MAF Sensor Return	<0.050v
37	OR/BK	TFT Sensor Signal	2.10-2.40v
38	LG/RD	ECT Sensor Signal	0.5-0.6v
39	GY	IAT Sensor Signal	1.5-2.5v
40	DG/YL	Fuel Pump Monitor	On: 12v, Off: 0v
41	DG/WT	A/C High Pressure Switch	On: 1v, Off: 12v
42	LG/YL	Engine Cooling Fan Relay Control	Relay Off: 12v, On: 1v
44	PK	Starter Relay Control	Relay Off: 12v, On: 1v
45	LB/RD	Generator Sensor Signal	130 Hz (30%)
46	LG/VT	High Speed Fan Control	On: 1v, Off: 12v
47	BR/PK	EGR VR Solenoid Control	0%, 55 mph: 45%
48	TN/YL	Clean Tachometer Output	33-45 Hz
49	LB/BK	Digital TR2 Sensor	In 'P': 0v, 55 mph: 11v
50	WT/BK	Digital TR4 Sensor	In 'P': 0v, 55 mph: 11v
51	BK/WT	Power Ground	<0.1v
52	YL/RD	Coil 2 Driver (Cyl 3 & 4)	6° dwell
53	PK/BK	Shift Solenoid 'C' Control	12v, 55 mph: 1v
54	VT/YL	TCC Solenoid Control	0%, 55 mph: 95%
55	PK/LB	Keep Alive Power	12-14v
56	GY/YL	EVAP Purge Solenoid	0-10 Hz (0-100%)
57	YL/RD	Knock Sensor (+) Signal	0v
58-59	---	Not Used	---
60	GY/LB	HO2S-11 (B1 S1) Signal	0.1-1.1v
61	VT/LG	HO2S-22 (B2 S2) Signal	0.1-1.1v
62	RD/PK	FTP Sensor Signal	2.6v (0 in. H20)
63	WT/YL	Injector Pressure Sensor	2.74v (39 psi)
64	RD/BK	Digital TR3A Sensor	In 'P': 0v, 55 mph: 1.7v

2002-04 Taurus 3.0L 4v V6 MFI VIN S (A/T) 104 Pin Connector

PCM Pin #	Wire Color	Circuit Description (104 Pin)	Value at Hot Idle
65	BR/LG	EGR DPFE Sensor Signal	0.95-1.05v
66	TN/OR	Evaporator Air Temperature Sensor	4.95-5.15v
67	VT/WT	EVAP CV Solenoid	0-10 Hz (0-100%)
68	GY/BK	VSS (+) Signal	0 Hz, 55 mph: 125 Hz
69	PK/YL	A/C WOT Relay Control	On: 1v, Off: 12v
71	RD	Vehicle Power (Start-Run)	12-14v
72	---	Not Used	---
73	TN/BK	Injector 5 Control	2.2-2.7 ms
74	BR/YL	Injector 3 Control	2.2-2.7 ms
75	TN	Injector 1 Control	2.2-2.7 ms
76, 77	BK/WT	Power Ground	<0.1v
78	YL/WT	Coil 3 Driver (Cyl 2 & 6)	6° dwell
79	WT/LG	TCIL (lamp) Control	Off: 12v, On: 1v
80	LB/OR	Fuel Pump Control	Off: 12v, On: 1v
81	BR/OR	EPC Solenoid Control	9.0v (15 psi)
82	BK/WT	Check Cap Indicator Light	Off: 12v, On: 1v
83	WT/LB	IAC Motor Control	10v (55% duty cycle)
84	DB/YL	OSS Sensor Signal	0 Hz, 55 mph: 131 Hz
85	DB/OR	CMP Sensor Signal	5-7 Hz
86	BK/YL	A/C Clutch Relay Switch	12v (Switch Open)
87	RD/LG	HO2S-21 (B2 S1) Signal	0.1-1.1v
88	LB/RD	MAF Sensor Signal	0.6v
89	GY/WT	TP Sensor Signal	0.53-1.27v
90	BR/WT	Reference Voltage	4.9-5.1v
91	GY/RD	Sensor Ground	<0.050v
92	RD/LG	Brake Pedal Position Switch	Brake Off: 0v, On: 12v
93	RD/WT	HO2S-11 (B1 S1) Heater	On: 1v, Off: 12v
94	WT/BK	HO2S-21 (B2 S1) Heater	On: 1v, Off: 12v
95	YL/LB	HO2S-12 (B1 S2) Heater	On: 1v, Off: 12v
96	TN/YL	HO2S-22 (B2 S2) Heater	On: 1v, Off: 12v
97	---	Not Used	---
98	LG/RD	Generator Indicator Light	Off: 12v, On: 1v
99	LG/OR	Injector 6 Control	2.2-2.7 ms
100	BR/LB	Injector 4 Control	2.2-2.7 ms
101	WT	Injector 2 Control	2.2-2.7 ms
102, 104	---	Not Used	---
103	BK/WT	Power Ground	<0.1v

Pin Voltage Connector

PCM 104-PIN CONNECTOR

Terminal View of 104-Pin PCM Wiring Harness Connector

Standard Colors and Abbreviations

Abbreviation	Color	Abbreviation	Color	Abbreviation	Color
BK	Black	GY	Gray	PK	Purple
BL	Blue	GN	Green	RD	Red
BR	Brown	LG	LT Green	TN	Tan
DB	Dark Blue	OR	Orange	WT	White
DG	DK Green	PK	Pink	YL	Yellow

1990 Taurus 3.0L V6 MFI VIN U (A/T) 60 Pin Connector

PCM Pin #	Wire Color	Circuit Description (60 Pin)	Value at Hot Idle
1	YL	Keep Alive Power	12-14v
2	RD/LG	Brake Pedal Position Switch	Brake Off: 0v, On: 12v
3	DG, OR/YL	VSS (+), (-) Signal	0 Hz, 55 mph: 125 Hz
4	DG/YL	Ignition Diagnostic Monitor	20-31 Hz
5, 11-12, 14-15	---	Not Used	---
7	LG/YL	ECT Sensor Signal	0.5-0.6v
8	PK/BK	Fuel Pump Monitor	On: 12v, Off: 0v
9	PK, TN	Data Bus (-), (+) Signal	Digital Signals
10	PK/BL	A/C Cycling Clutch Switch	A/C On: 12v, Off: 0v
13	OR/YL	Speed Control Solenoid (-)	Solenoid On: 1v
16	BK/OR	Ignition System Ground	<0.050v
17	TN/RD	STI Output, MIL Control	MIL On: 1v, Off: 12v
18	DG/PK	Shift Solenoid 3-4 Control	12v, 55 mph: 1v
19	OR/YL	Shift Solenoid 2-3 Control	12v, 55 mph: 1v
20	BK	PCM Case Ground	<0.050v
21	OR/BK	IAC Motor Control	7-9v
22	TN/LG	Fuel Pump Control	On: 1v, Off: 12v
23	YL/RD	Knock Sensor Signal (49 States)	0v
24	YL/LG	PSP Switch Signal	Straight: 0v, Turned: 11v
25	LG/PK	ACT Sensor Signal	1.5-2.5v
26	OR/WT	Reference Voltage	4.9-5.1v
27	BR/LG	PFE EGR Sensor	3.2v, 55 mph: 2.9v
29	DG/PK	HO2S-11 (B1 S1) Signal	0.1-1.1v
30	PK/YL	Neutral Drive Switch Signal	In 'P': 0v, Others: 12v
31	GY/YL	EVAP Purge Solenoid	12v, 55 mph: 1v
32, 41, 43, 51	---	Not Used	---
33	DG	EGR VR Solenoid	0%, 55 mph: 45%
34	BL/PK	Data Output Link	Digital Signals
35	WT/PK	Speed Control Solenoid	0%, 55 mph: 97%
36	YL/LG	Spark Output Signal	6.93v (duty cycle)
37, 57	RD	Vehicle Power	12-14v
39, 50	OR, BL/BK	S/C Switch Ground, SC Signal	<0.050v, S/C On: 6.7v
40, 60	BK/LG	Power Ground	<0.1v
42	GY/BK	S/C Vacuum Solenoid Control	S/C Off: 0% duty cycle
44	GY/WT	AXOD Transmission Solenoid	On: 1v, Off: 12v
45	LG/BK	MAP Sensor Signal	107 Hz
46	BK/WT	Analog Signal Return	<0.050v
47	DG/LG	TP Sensor Signal	0.5-1.2v
48	WT/BK	Self-Test Indicator Signal	STI Open: 5v, Closed: 1v
49	OR	HO2S-21 (B2 S1) Signal	0.1-1.1v
52	PK	High Speed Fan Control	On: 1v, Off: 12v
53	TN/BL	Lock-Up Solenoid for AXOD	12v, 55 mph: 1v
54	RD	A/C WOT Relay Control	On: 1v, Off: 12v
55	TN/OR	Low Speed Fan Control	Fan On: 1v, Off: 12v
56	DB	PIP Sensor Signal	50% duty cycle
58	TN/RD	Injector Bank 1 (INJ 1, 2 & 4)	5.8-6.1 ms
59	TN/OR	Injector Bank 2 (INJ 3, 5 & 6)	5.8-6.1 ms

PCM 60-PIN CONNECTOR

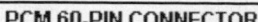

Terminal View of 60-Pin PCM Harness Connector

1991-95 Taurus 3.0L V6 MFI VIN U (A/T) 60 Pin Connector

PCM Pin #	Wire Color	Circuit Description (60 Pin)	Value at Hot Idle
1	YL	Keep Alive Power	12-14v
2	RD/LG	Brake Pedal Position Switch	Brake Off: 0v, On: 12v
3	DG/WT	VSS (+) Signal	0 Hz, 55 mph: 125 Hz
4	WT/PK	Ignition Diagnostic Monitor	20-31 Hz
4	DG/YL	Ignition Diagnostic Monitor	20-31 Hz
5	GY/BK	TSS Sensor Signal	100 Hz (700 rpm)
6	OR/YL	VSS (+) Signal	0 Hz, 55 mph: 125 Hz
7	LG/RD	ECT Sensor Signal	0.5-0.6v
8	PK/BK	Fuel Pump Monitor	On: 12v, Off: 0v
9	TN/BL	MAF Sensor Return	<0.050v
10	PK/BL	A/C Cycling Clutch Switch	A/C On: 12v, Off: 0v
11	GN/YL	EVAP Purge Solenoid	12v, 55 mph: 1v
12	LG/OR	Injector 6 Control	5.0-5.6 ms
13	TN/OR	Low Cooling Fan	On: 1v, Off: 12v
14	---	Not Used	---
15	TN/BK	Injector 5 Control	5.0-5.6 ms
16	OR/RD	Ignition System Ground	<0.050v
17	TN/RD	STI Output, MIL Control	MIL On: 1v, Off: 12v
18	TN/OR	Data Bus (+) Signal	Digital Signals
19	PK/BL	Data Bus (-) Signal	Digital Signals
20	BK	PCM Case Ground	<0.050v
21	BR/WT	IAC Motor Control	8-9v
22	TN/LG	Fuel Pump Control	On: 1v, Off: 12v
22	BL/OR	Fuel Pump Control	On: 1v, Off: 12v
23-24	---	Not Used	---
25	LG/PK	IAT Sensor Signal	1.5-2.5v
26	BR/WT	Reference Voltage	4.9-5.1v
27	BR/LG	PFE EGR Sensor	3.2v, 55 mph: 2.8v
28	YL/LG	PSP Switch Signal	Straight: 0v, Turned: 11v
29	---	Not Used	---
30	BL	MLP Sensor Signal	In 'P': 0v, 55 mph: 1.7v
30	BL/YL	MLP Sensor Signal	In 'P': 0v, 55 mph: 1.7v
31	LG/PK	High Cooling Fan	On: 1v, Off: 12v
32	---	Not Used	---
33	BR/PK	EGR VR Solenoid	0%, 55 mph: 45%
34	BL/BK	Data Output Link	Digital Signals
35	BR/BL	Injector 4 Control	5.0-5.6 ms
36	YL/LG	Spark Output Signal	50% duty cycle
37	RD	Vehicle Power	12-14v
38	WT/YL	EPC Solenoid Control	8-9v (9 psi)
39	BR/YL	Injector 3 Control	5.0-5.6 ms
40	BK/LG	Power Ground	<0.1v

1991-95 Taurus 3.0L V6 MFI VIN U (A/T) 60 Pin Connector

PCM Pin #	Wire Color	Circuit Description (60 Pin)	Value at Hot Idle
41	---	Not Used	---
42 ('94-'95)	TN/LG	A/C Pressure Sensor	A/C On: 12v (open)
43	RD/BK	HO2S-11 (B1 S1) Signal	0.1-1.1v
44	GY/BL	HO2S-21 (B2 S1) Signal	0.1-1.1v
45	---	Not Used	---
46	GY/RD	Analog Signal Return	<0.050v
47	GN/WT	TP Sensor Signal	0.5-1.2v
48	BR	Self-Test Indicator Signal	STI Open: 5v, Closed: 1v
49	OR/BK	TOT Sensor Signal	2.10-2.40v
50	BL/RD	MAF Sensor Signal	0.6-0.9v
51	OR/YL	Shift Solenoid 1 Control	12v, 55 mph: 1v
52	PK/OR	Shift Solenoid 2 Control	1v, 55 mph: 12v
53	TN/WT	TCC Solenoid Control	0%, 55 mph: 45%
54	PK/YL	A/C WOT Relay Control	On: 1v, Off: 12v
55	WT/RD	Shift Solenoid 3 Control	7-9v, 55 mph: 8-9v
56	DB	PIP Sensor Signal	50% duty cycle
56	GY/OR	PIP Sensor Signal	50% duty cycle
57	RD	Vehicle Power	12-14v
58	TN	Injector 1 Control	5.0-5.6 ms
59	WT	Injector 2 Control	5.0-5.6 ms
60	BK/LG	Power Ground	<0.1v

Pin Connector Graphic

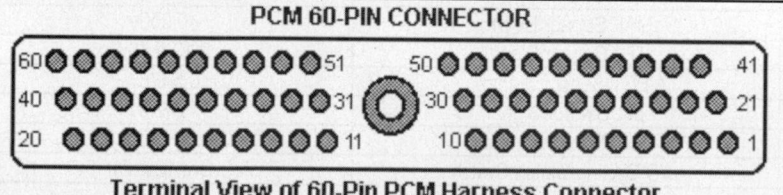

PCM 60-PIN CONNECTOR

Terminal View of 60-Pin PCM Harness Connector

Standard Colors and Abbreviations

Abbreviation	Color	Abbreviation	Color	Abbreviation	Color
BK	Black	GY	Gray	PK	Purple
BL	Blue	GN	Green	RD	Red
BR	Brown	LG	LT Green	TN	Tan
DB	Dark Blue	OR	Orange	WT	White
DG	DK Green	PK	Pink	YL	Yellow

1996-97 Taurus 3.0L V6 MFI VIN U (A/T) 104 Pin Connector

PCM Pin #	Wire Color	Circuit Description (104 Pin)	Value at Hot Idle
1	PK/OR	Shift Solenoid 2 Control	1v, 55 mph: 12v
2	PK/LG	MIL (lamp) Control	MIL On: 1v, Off: 12v
3-12	---	Not Used	---
13	PK	Flash EEPROM Power	0.1v
14	---	Not Used	---
15	PK/LB	Data Bus (-) Signal	Digital Signals
16	TN/OR	Data Bus (+) Signal	Digital Signals
17-20	---	Not Used	---
21	GY	CKP Sensor (+) Signal	510-540 Hz
22	DB	CKP Sensor (-) Signal	510-540 Hz
23	---	Not Used	---
24	BK/WT	Power Ground	<0.1v
25	BK	Power Ground	<0.1v
26	WT/BK	Coil Driver 1 Control	5° dwell
27	OR/YL	Shift Solenoid 1 Control	12v, 55 mph: 1v
28	TN/OR	Low Speed Fan Control	On: 1v, Off: 12v
29	---	Not Used	---
30	DG	Octane Adjust Switch	Closed: 0v, Open: 9.3v
31	YL/LG	PSP Switch Signal	Straight: 0v, Turned: 12v
32	---	Not Used	---
33	PK/OR	VSS (-) Signal	0 Hz, 55 mph: 125 Hz
34	---	Not Used	---
35	RD/BK	HO2S-12 (B1 S2) Signal	0.1-1.1v
36	TN/LB	MAF Sensor Return	<0.050v
37	OR/BK	TFT Sensor Signal	2.10-2.40v
38	LG/RD	ECT Sensor Signal	0.5-0.6v
39	GY	IAT Sensor Signal	1.5-2.5v
40	DG/YL	Fuel Pump Monitor	On: 12v, Off: 0v
41	PK/LB	A/C Cycling Clutch Switch	A/C On: 12v, Off: 0v
42-45	---	Not Used	---
46	LG/PK	High Speed Fan Control	On: 1v, Off: 12v
47	BR/PK	EGR VR Solenoid	0%, 55 mph: 45%
48	TN/YL	Clean Tachometer Output	42-50 Hz
49-50	---	Not Used	---
51	BK/WT	Power Ground	<0.1v
52	YL/RD	Coil Driver 2 Control	5° dwell
53	PK/BK	Shift Solenoid 3 Control	12v, 55 mph: 1v
54	---	Not Used	---
55	PK	Keep Alive Power	12-14v
56	GY/YL	EVAP VMV Solenoid	0-10 Hz (0-100%)
57	---	Not Used	---
58	GY/BK	VSS (+) Signal	0 Hz, 55 mph: 125 Hz
59	---	Not Used	---
60	GY/LB	HO2S-11 (B1 S1) Signal	0.1-1.1v
61	PK/LG	HO2S-22 (B2 S2) Signal	0.1-1.1v
62	RD/PK	FTP Sensor Signal	2.6v (0" H2O - cap off)
63	---	Not Used	---
64	LB/YL	TR Sensor Signal	In 'P': 0v, 55 mph: 1.7v

1996-97 Taurus 3.0L V6 MFI VIN U (A/T) 104 Pin Connector

PCM Pin #	Wire Color	Circuit Description (104 Pin)	Value at Hot Idle
65	BR/LG	DPFE Sensor Signal	0.20-1.30v
66, 68	---	Not Used	---
67	PK/WT	EVAP CV Solenoid	0-10 Hz (0-100%)
69	PK/YL	A/C WOT Relay Control	On: 1v, Off: 12v
70, 72	---	Not Used	---
71	RD	Vehicle Power	12-14v
73	TN/BK	Injector 5 Control	2.3-2.8 ms
74	BR/YL	Idle Air Control Valve	10v (32% duty cycle)
75	TN	Injector 1 Control	2.3-2.8 ms
76	BK/WT	Power Ground	<0.1v
77	BK/WT	Power Ground	<0.1v
78	YL/WT	Coil Driver 3 Control	5° dwell
79	---	Not Used	---
80	LB/OR	Fuel Pump Control	On: 1v, Off: 12v
81	WT/YL	EPC Solenoid Control	10.6v (40 psi)
82	PK/YL	TCC Solenoid Control	12v (0%)
83	WT/LB	Injector 3 Control	2.3-2.8 ms
84	DG/WT	TSS Sensor Signal	50-65 Hz (790-820 rpm)
85	DB/OR	CID Sensor Signal	6-8 Hz
86	BK/YL	A/C Pressure Switch	A/C On: 12v (Open)
87	RD/LG	HO2S-21 (B2 S1) Signal	0.1-1.1v
88	LB/RD	MAF Sensor Signal	0.9v
89	GY/WT	TP Sensor Signal	0.53-1.27v
90	BR/WT	Reference Voltage	4.9-5.1v
91	GY/RD	Analog Signal Return	<0.050v
92	RD/LG	Brake Pedal Position Switch	Brake Off: 0v, On: 12v
93	RD/WT	HO2S-11 (B1 S1) Heater	On: 1v, Off: 12v
94	WT/BK	HO2S-21 (B2 S1) Heater	On: 1v, Off: 12v
95	YL/LB	HO2S-12 (B1 S2) Heater	On: 1v, Off: 12v
96	TN/YL	HO2S-22 (B2 S2) Heater	On: 1v, Off: 12v
97	RD	Vehicle Power	12-14v
98	---	Not Used	---
99	LG/OR	Injector 6 Control	2.3-2.8 ms
100	BR/LB	Injector 4 Control	2.3-2.8 ms
101	WT	Injector 2 Control	2.3-2.8 ms
102	---	Not Used	---
103	BK/WT	Power Ground	<0.1v
104	---	Not Used	---

Pin Connector Graphic

Terminal View of 104-Pin PCM Wiring Harness Connector

Standard Colors and Abbreviations

Abbreviation	Color	Abbreviation	Color	Abbreviation	Color
BK	Black	GY	Gray	PK	Purple
BL	Blue	GN	Green	RD	Red
BR	Brown	LG	LT Green	TN	Tan
DB	Dark Blue	OR	Orange	WT	White
DG	DK Green	PK	Pink	YL	Yellow

1998-99 Taurus 3.0L V6 MFI VIN U (A/T) 104 Pin Connector

PCM Pin #	Wire Color	Circuit Description (104 Pin)	Value at Hot Idle
1	VT/OR	Shift Solenoid 'B' Control	1v, 55 mph: 12v
2	PK/LG	MIL (lamp) Control	MIL On: 1v, Off: 12v
3	YL/BK	Digital TR1 Sensor	In 'P': 0v, 55 mph: 11v
4-11	---	Not Used	---
12	YL/WT	Fuel Level Indicator Signal	1.7v (1/2 full)
13	PK	Flash EEPROM Power	0.1v
15	PK/LB	Data Bus (-) Signal	Digital Signals
16	TN/OR	Data Bus (+) Signal	Digital Signals
17-18	---	Not Used	---
21	GY	CKP Sensor (+) Signal	510-540 Hz
22	DB	CKP Sensor (-) Signal	510-540 Hz
24	BK/WT	Power Ground	<0.1v
25	BK	Chassis Ground	<0.1v
26	YL/BK	Coil Driver 1 Control	6° dwell
27	OR/YL	Shift Solenoid 'A' Control	12v, 55 mph: 1v
28	LB	Low Speed Fan Control	On: 1v, Off: 12v
30	---	Not Used	---
31	YL/LG	PSP Switch Signal	Straight: 0v, Turned: 12v
32	---	Not Used	---
33	PK/OR	VSS (-) Signal	0 Hz, 55 mph: 125 Hz
35	RD/BK	HO2S-12 (B1 S2) Signal	0.1-1.1v
36	TN/LB	MAF Sensor Return	<0.050v
37	OR/BK	TFT Sensor Signal	2.10-2.40v
38	LG/RD	ECT Sensor Signal	0.5-0.6v
39	GY	IAT Sensor Signal	1.5-2.5v
40	DG/YL	Fuel Pump Monitor	On: 12v, Off: 0v
41	DG/WT	A/C High Pressure Switch	AC On: 12v, Off: 0v
42-44	---	Not Used	---
45	PK/WT	EVAP CV Solenoid	0-10 Hz (0-100%)
46	LG/PK	High Speed Fan Control	On: 1v, Off: 12v
47	BR/PK	EGR VR Solenoid	0%, 55 mph: 45%
48	TN/YL	Clean Tachometer Output	42-50 Hz
49	LB/BK	Digital TR2 Sensor	In 'P': 0v, 55 mph: 11v
50	WT/BK	Digital TR4 Sensor	In 'P': 0v, 55 mph: 11v
51	BK/WT	Power Ground	<0.1v
52	YL/RD	Coil Driver 2 Control	6° dwell
53	PK/BK	Shift Solenoid 'C' Control	12v, 55 mph: 1v
54	DG/WT	TSS Sensor Signal	50-65 Hz (820-900 rpm)
55	DG/OR	Keep Alive Power	12-14v
56	GY/YL	EVAP Purge Solenoid	0-10 Hz (0-100%)
57, 59	---	Not Used	---
58	GY/BK	VSS (+) Signal	0 Hz, 55 mph: 125 Hz
60	GY/LB	HO2S-11 (B1 S1) Signal	0.1-1.1v
61	PK/LG	HO2S-22 (B2 S2) Signal	0.1-1.1v
62	RD/PK	FTP Sensor Signal	2.6v (0" H2O - cap off)
63	---	Not Used	---
64	LB/YL	Digital TR3A Sensor	In 'P': 0v, in O/D: 1.7v

1998-99 Taurus 3.0L V6 MFI VIN U (A/T) 104 Pin Connector

PCM Pin #	Wire Color	Circuit Description (104 Pin)	Value at Hot Idle
65	BR/LG	DPFE Sensor Signal	0.20-1.30v
66-68	---	Not Used	---
69	PK/YL	A/C WOT Relay Control	On: 1v, Off: 12v
70	---	Not Used	---
71	RD	Vehicle Power	12-14v
73	TN/BK	Injector 5 Control	2.3-2.8 ms
74	BR/YL	Injector 3 Control	2.3-2.8 ms
75	TN	Injector 1 Control	2.3-2.8 ms
76-77	BK/WT	Power Ground	<0.1v
78	YL/WT	Coil Driver 3 Control	6° dwell
80	LB/OR	Fuel Pump Control	On: 1v, Off: 12v
81	WT/YL	EPC Solenoid Control	10.6v (40 psi)
81-82	---	Not Used	---
83	WT/LB	IAC Motor Control	10v (32% duty cycle)
84	DG/WT	TSS Sensor Signal	50-65 Hz (790-820 rpm)
85	DB/OR	CID Sensor Signal	6-8 Hz
86	BK/YL	A/C Pressure Switch	A/C On: 12v (Open)
87	RD/LG	HO2S-21 (B2 S1) Signal	0.1-1.1v
88	LB/RD	MAF Sensor Signal	0.9v
89	GY/WT	TP Sensor Signal	0.53-1.27v
90	BR/WT	Reference Voltage	4.9-5.1v
91	GY/RD	Analog Signal Return	<0.050v
92	RD/LG	Brake Pedal Position Switch	Brake Off: 0v, On: 12v
93	RD/WT	HO2S-11 (B1 S1) Heater	On: 1v, Off: 12v
94	WT/BK	HO2S-21 (B2 S1) Heater	On: 1v, Off: 12v
95	YL/LB	HO2S-12 (B1 S2) Heater	On: 1v, Off: 12v
96	TN/YL	HO2S-22 (B2 S2) Heater	On: 1v, Off: 12v
97 ('98)	RD	Vehicle Power	12-14v
99	LG/OR	Injector 6 Control	2.3-2.8 ms
100	BR/LB	Injector 4 Control	2.3-2.8 ms
101	WT	Injector 2 Control	2.3-2.8 ms
102	---	Not Used	---
103	BK/WT	Power Ground	<0.1v
104	---	Not Used	---

Pin Connector Graphic

PCM 104-PIN CONNECTOR

Terminal View of 104-Pin PCM Wiring Harness Connector

Standard Colors and Abbreviations

Abbreviation	Color	Abbreviation	Color	Abbreviation	Color
BK	Black	GY	Gray	PK	Purple
BL	Blue	GN	Green	RD	Red
BR	Brown	LG	LT Green	TN	Tan
DB	Dark Blue	OR	Orange	WT	White
DG	DK Green	PK	Pink	YL	Yellow

2000 Taurus 3.0L V6 MFI VIN U (A/T) 104 Pin Connector

PCM Pin #	Wire Color	Circuit Description (104 Pin)	Value at Hot Idle
1	VT/OR	Shift Solenoid 'B' Control	1v, 55 mph: 12v
2	PK/LG	MIL (lamp) Control	MIL On: 1v, Off: 12v
3	YL/BK	Digital TR1 Sensor	In 'P': 0v, 55 mph: 11v
4	BR/PK	System Voltage (Start-Run)	Cranking: 9-11v
5	WT/OR	Secondary Air Injection Relay	On: 1v, Off: 12v
6	DG/WT	TSS Sensor Signal	50-65 Hz (790-820 rpm)
7-10	---	Not Used	---
12	YL/WT	Fuel Level Indicator Signal	1.7v (1/2 full)
13	VT	Flash EEPROM Power	0.1v
14	---	Not Used	---
15	PK/LB	SCP Data Bus (-) Signal	Digital Signals
16	TN/OR	SCP Data Bus (+) Signal	Digital Signals
17	BR/YL	SCI Receive (RX) Signal	Digital Signals
18	GY/RD	SCI Transmit (TX) Signal	Digital Signals
20	GY/YL	Generator Field Control	0 Hz
21	GY	CKP Sensor (+) Signal	510-540 Hz
22	DB	CKP Sensor (-) Signal	510-540 Hz
24	BK/WT	Power Ground	<0.1v
25	BK	Case Ground	<0.1v
26	YL/BK	Coil 1 Driver (Cyl 1 & 5)	6° dwell
27	OR/YL	Shift Solenoid 'A' Control	12v, 55 mph: 1v
28	LB	Low Speed Fan Control	On: 1v, Off: 12v
29	---	Not Used	---
30	DB/LG	Passive Antitheft Indicator	Lamp On: 1v, Off: 12v
31	YL/LG	PSP Switch Signal	Straight: 0v, Turned: 12v
32	YL	Knock Sensor 1 (-) Signal	0v
35	RD/BK	HO2S-12 (B1 S2) Signal	0.1-1.1v
36	TN/LB	MAF Sensor Return	<0.050v
37	OR/BK	TFT Sensor Signal	2.10-2.40v
38	LG/RD	ECT Sensor Signal	0.5-0.6v
39	GY	IAT Sensor Signal	1.5-2.5v
40	DG, PK	Fuel Pump Monitor	On: 12v, Off: 0v
41	DG/WT	A/C High Pressure Switch	AC On: 12v, Off: 0v
42	LG/YL	Generator Cutoff Relay	0.1v (with fan on)
44	PK	Starter Relay Control	KOEC: 9-11v
45	LB/RD	Generator Sensor Signal	TBD
46	LG/VT	High Speed Fan Control	On: 1v, Off: 12v
47	BR/PK	EGR VR Solenoid	0%, 55 mph: 45%
48	TN/YL	Clean Tachometer Output	42-50 Hz
49	LB/BK	Digital TR2 Sensor	In 'P': 0v, 55 mph: 11v
50	WT/BK	Digital TR4 Sensor	In 'P': 0v, 55 mph: 11v
51	BK/WT	Power Ground	<0.1v
52	YL/RD	Coil Driver 2 (Cyl 3 & 4)	6° dwell
53	PK/BK	Shift Solenoid 'C' Control	12v, 55 mph: 1v
54	VT/YL	TCC Solenoid Control	12v (0%)
55	PK/LB	Keep Alive Power	12-14v
56	GY/YL	EVAP Purge Solenoid	0-10 Hz (0-100%)
57	YL/RD	Knock Sensor 1 (+) Signal	0v
60	GY/LB	HO2S-11 (B1 S1) Signal	0.1-1.1v
61	VT/LG	HO2S-22 (B2 S2) Signal	0.1-1.1v
62	RD/PK	FTP Sensor Signal	2.6v (0" H2O - cap off)
63	WT/YL	Injector Pressure Sensor	TBD
64	LB/YL	Digital TR3A Sensor	In 'P': 0v, in O/D: 1.7v

2000 Taurus 3.0L V6 MFI VIN U (A/T) 104 Pin Connector

PCM Pin #	Wire Color	Circuit Description (104 Pin)	Value at Hot Idle
65	BR/LG	DPFE Sensor Signal	0.20-1.30v
66	TN/OR	Evaporator Air Temp. Sensor	4.95-5.15v
67	VT/WT	EVAP Purge Solenoid	0-10 Hz (0-100%)
68	GY/BK	VSS (+) Signal	0 Hz, 55 mph: 125 Hz
69	PK/YL	A/C WOT Relay Control	On: 1v, Off: 12v
70	BR	Secondary Air Injection Relay	On: 1v, Off: 12v
71	RD	Vehicle Power	12-14v
73	TN/BK	Injector 5 Control	2.3-2.8 ms
74	BR/YL	Injector 3 Control	2.3-2.8 ms
75	TN	Injector 1 Control	2.3-2.8 ms
76, 77	BK/WT	Power Ground	<0.1v
78	YL/WT	Coil 3 Driver (Cyl 2 & 6)	6° dwell
79	WT/LG	Check Transaxle Light	On: 1v, Off: 12v
80	LB/OR	Fuel Pump Control	On: 1v, Off: 12v
81	WT or BR	EPC Solenoid Control	10.6v (40 psi)
82	BK/WT	Check Cap Indicator Light	On: 1v, Off: 12v
83	WT/LB	IAC Motor Control	10v (32% duty cycle)
84	DB/YL	OSS Sensor Signal	0 Hz, 55 mph: 131 Hz
85	DB/OR	CID Sensor Signal	6-8 Hz
86	BK/YL	A/C Pressure Clutch Relay	A/C On: 12v, Off: 0v
87	RD/LG	HO2S-21 (B2 S1) Signal	0.1-1.1v
88	LB/RD	MAF Sensor Signal	0.9v
89	GY/WT	TP Sensor Signal	0.53-1.27v
90	BR/WT	Reference Voltage	4.9-5.1v
91	GY/RD	Analog Signal Return	<0.050v
92	RD/LG	Brake Pedal Position Switch	Brake Off: 0v, On: 12v
93	RD/WT	HO2S-11 (B1 S1) Heater	On: 1v, Off: 12v
94	WT/BK	HO2S-21 (B2 S1) Heater	On: 1v, Off: 12v
95	YL/LB	HO2S-12 (B1 S2) Heater	On: 1v, Off: 12v
96	TN/YL	HO2S-22 (B2 S2) Heater	On: 1v, Off: 12v
98	LG/RD	Generator Battery Indicator	Digital Signals
99	LG/OR	Injector 6 Control	2.3-2.8 ms
100	BR/LB	Injector 4 Control	2.3-2.8 ms
101	WT	Injector 2 Control	2.3-2.8 ms
102	---	Not Used	---
103	BK/WT	Power Ground	<0.1v
104	---	Not Used	---

Pin Connector Graphic

PCM 104-PIN CONNECTOR

Terminal View of 104-Pin PCM Wiring Harness Connector

Standard Colors and Abbreviations

Abbreviation	Color	Abbreviation	Color	Abbreviation	Color
BK	Black	GY	Gray	PK	Purple
BL	Blue	GN	Green	RD	Red
BR	Brown	LG	LT Green	TN	Tan
DB	Dark Blue	OR	Orange	WT	White
DG	DK Green	PK	Pink	YL	Yellow

2001-04 Taurus 3.0L V6 MFI VIN U (A/T) 104 Pin Connector

PCM Pin #	Wire Color	Circuit Description (104 Pin)	Value at Hot Idle
1	VT/OR	Shift Solenoid 'B' Control	1v, 55 mph: 12v
2	PK/LG	MIL (lamp) Control	MIL On: 1v, Off: 12v
3	YL/BK	Digital TR1 Sensor	In 'P': 0v, 55 mph: 11v
4	TN/RD	System Voltage (Start-Run)	Cranking: 9-11v
5	---	Not Used	---
6	DG/WT	TSS Sensor Signal	50-65 Hz (790-820 rpm)
7-10	---	Not Used	---
12	YL/WT	Fuel Level Indicator Signal	1.7v (1/2 full)
13	VT	Flash EEPROM Power	0.1v
14, 19	---	Not Used	---
15	PK/LB	SCP Bus (-) Signal	Digital Signals
16	TN/OR	SCP Bus (+) Signal	Digital Signals
17	BR/YL	Passive Antitheft RX Signal	Digital Signals
18	GY/RD	Passive Antitheft TX Signal	Digital Signals
20	GY/YL	Generator Field Control	130 Hz (45%)
21	GY	CKP Sensor (+) Signal	510-540 Hz
22	DB	CKP Sensor (-) Signal	510-540 Hz
23	---	Not Used	---
24	BK/WT	Power Ground	<0.1v
25	BK	Case Ground	<0.1v
26	YL/BK	Coil 1 Driver (Cyl 1 & 5)	6° dwell
27	OR/YL	Shift Solenoid 'A' Control	12v, 55 mph: 1v
28	LB	Engine Cooling Fan Brake Relay Control	Off: 12v, On: 1v
29, 33-34	---	Not Used	---
30	DB/LG	Passive Antitheft Indicator	Lamp On: 1v, Off: 12v
31	YL/LG	PSP Switch Signal	Straight: 0v, Turned: 12v
32	YL	Knock Sensor 1 (-) Signal	<0.050v
35	RD/BK	HO2S-12 (B1 S2) Signal	0.1-1.1v
36	TN/LB	MAF Sensor Return	<0.050v
37	OR/BK	TFT Sensor Signal	2.10-2.40v
38	LG/RD	ECT Sensor Signal	0.5-0.6v
39	GY	IAT Sensor Signal	1.5-2.5v
40	DG/YL	Fuel Pump Monitor	On: 12v, Off: 0v
41	DG/WT	A/C High Pressure Switch	AC On: 12v, Off: 0v
42	LG/YL	Generator Cutoff Relay	0.1v (with fan on)
44	PK	Starter Relay Control	KOEC: 9-11v
45	LB/RD	Generator Sensor Signal	130 Hz (30%)
46	LG/VT	High Cooling Speed Fan Control	Off: 12v, On: 1v
47	BR/PK	EGR VR Solenoid	0%, 55 mph: 45%
48	TN/YL	Clean Tachometer Output	42-50 Hz
49	LB/BK	Digital TR2 Sensor	In 'P': 0v, 55 mph: 11v
50	WT/BK	Digital TR4 Sensor	In 'P': 0v, 55 mph: 11v
51	BK/WT	Power Ground	<0.1v
52	YL/RD	Coil Driver 2 (Cyl 3 & 4)	6° dwell
53	PK/BK	Shift Solenoid 'C' Control	12v, 55 mph: 1v
54	VT/YL	TCC Solenoid Control	12v (0%)
55	PK/LB	Keep Alive Power	12-14v
56	GY/YL	EVAP Purge Solenoid	0-10 Hz (0-100%)
57	YL/RD	Knock Sensor (+) Signal	0v
60	GY/LB	HO2S-11 (B1 S1) Signal	0.1-1.1v
61	VT/LG	HO2S-22 (B2 S2) Signal	0.1-1.1v
62	RD/PK	FTP Sensor Signal	2.6v (0" H2O - cap off)
63	WT/YL	Injector Pressure Sensor	TBD
64	RD/BK	Digital TR3A Sensor	In 'P': 0v, in O/D: 1.7v

2001-04 Taurus 3.0L V6 MFI VIN U (A/T) 104 Pin Connector

PCM Pin #	Wire Color	Circuit Description (104 Pin)	Value at Hot Idle
65	BR/LG	DPFE Sensor Signal	0.95-1.05v
66	TN/OR	Evaporator Air Temperature Sensor	4.95-5.15v
67	VT/WT	EVAP Purge Solenoid	0-10 Hz (0-100%)
68	GY/BK	VSS (+) Signal	Moving: AC pulse signals
69	PK/YL	A/C WOT Relay Control	Off: 12v, On: 1v
70	---	Not Used	---
71	RD	Vehicle Power (Start-Run)	12-14v
73	TN/BK	Injector 5 Control	2.3-2.8 ms
74	BR/YL	Injector 3 Control	2.3-2.8 ms
75	TN	Injector 1 Control	2.3-2.8 ms
76, 77	BK/WT	Power Ground	<0.1v
78	YL/WT	Coil 3 Driver (Cyl 2 & 6)	6° dwell
79	WT/LG	Check Transaxle Light	Indicator Off: 12v, On: 1v
80	LB/OR	Fuel Pump Control	Off: 12v, On: 1v
81	BR/OR	EPC Solenoid Control	10.6v (40 psi)
82	BK/WT	Check Cap Indicator Light	Indicator Off: 12v, On: 1v
83	WT/LB	IAC Motor Control	10v (32% duty cycle)
84	DB/YL	OSS Sensor Signal	0 Hz, 55 mph: 131 Hz
85	DB/OR	CID Sensor Signal	6-8 Hz
86	BK/YL	A/C Pressure Clutch Relay	Relay Off: 12v, On: 1v
87	RD/LG	HO2S-21 (B2 S1) Signal	0.1-1.1v
88	LB/RD	MAF Sensor Signal	0.9v
89	GY/WT	TP Sensor Signal	0.53-1.27v
90	BR/WT	Reference Voltage	4.9-5.1v
91	GY/RD	Sensor Return	<0.050v
92	RD/LG	Brake Pedal Position Switch	Brake Off: 0v, On: 12v
93	RD/WT	HO2S-11 (B1 S1) Heater	On: 1v, Off: 12v
94	WT/BK	HO2S-21 (B2 S1) Heater	On: 1v, Off: 12v
95	YL/LB	HO2S-12 (B1 S2) Heater	On: 1v, Off: 12v
96	TN/YL	HO2S-22 (B2 S2) Heater	On: 1v, Off: 12v
97	RD	Vehicle Power (Start-Run)	12-14v
98	LG/RD	Generator Battery Indicator	Digital Signals
99	LG/OR	Injector 6 Control	2.3-2.8 ms
100	BR/LB	Injector 4 Control	2.3-2.8 ms
101	WT	Injector 2 Control	2.3-2.8 ms
102	---	Not Used	---
103	BK/WT	Power Ground	<0.1v
104	---	Not Used	---

Pin Connector Graphic

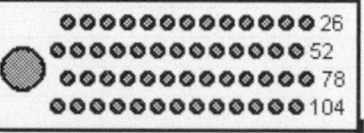

Standard Colors and Abbreviations

Abbreviation	Color	Abbreviation	Color	Abbreviation	Color
BK	Black	GY	Gray	PK	Purple
BL	Blue	GN	Green	RD	Red
BR	Brown	LG	LT Green	TN	Tan
DB	Dark Blue	OR	Orange	WT	White
DG	DK Green	PK	Pink	YL	Yellow

1990-93 Taurus 3.0L V6 MFI SHO VIN Y (M/T) 60 Pin Connector

PCM Pin #	Wire Color	Circuit Description (60 Pin)	Value at Hot Idle
1	YL	Keep Alive Power	12-14v
2	YL/LG	PSP Switch Signal	Straight: 0v, Turned: 11v
3	DG/WT	VSS (+) Signal	0 Hz, 55 mph: 125 Hz
4	GY/OR	Ignition Diagnostic Monitor	20-31 Hz
5	RD/LG	Brake Pedal Position Switch	Brake Off: 0v, On: 12v
6	OR/YL	VSS (-) Signal	0 Hz, 55 mph: 125 Hz
7	LG/YL	ECT Sensor Signal	0.5-0.6v
7	LG/RD	ECT Sensor Signal	0.5-0.6v
8	---	Not Used	---
9	TN/BL	MAF Sensor Return	<0.050v
10	PK/BL	A/C Cycling Clutch Switch	A/C On: 12v, Off: 0v
11	OR/YL	Speed Control Solenoid (+)	S/C On: 12-14v
12	BR/YL	Injector 3 Control	3.6-4.1 ms
13	BR/BL	Injector 4 Control	3.6-4.1 ms
14	TN/BL	Injector 5 Control	3.6-4.1 ms
14	TN/BK	Injector 5 Control	3.6-4.1 ms
15	LG	Injector 6 Control	3.6-4.1 ms
15	LG/OR	Injector 6 Control	3.6-4.1 ms
16	BK/OR	Ignition System Ground	<0.050v
16	OR/RD	Ignition System Ground	<0.050v
17	TN/RD	STI Output, MIL Control	MIL On: 1v, Off: 12v
18	PK	Octane Adjust Switch	Closed: 0v, Open: 9.1v
19	PK/BK	Fuel Pump Monitor	On: 12v, Off: 0v
20	BK	PCM Case Ground	<0.050v
21	OR/BK	IAC Motor Control	8-9v
22	TN/LG	Low Fuel Pump Control	On: 1v, Off: 12v
23	Y, YL/RD	Knock Sensor Signal	0v
24	DG	CID Sensor Signal	5-7 Hz
25	GY	ACT Sensor Signal	1.5-2.5v
25	LG/PK	ACT Sensor Signal	1.5-2.5v
26	BR/WT	Reference Voltage	4.9-5.1v
26	OR/WT	Reference Voltage	4.9-5.1v
27	BR/LG	PFE EGR Sensor	3.2v, 55 mph: 2.8v
28	BL/BK	S/C Command Switch Signal	S/C On: 6.7v
29	RD/BK	HO2S-21 (B2 S1) Signal	0.1-1.1v
29	DG/PK	HO2S-21 (B2 S1) Signal	0.1-1.1v
30	PK/YL	M/T: Clutch Engage Switch	Clutch In: 0v, Out: 5v
31	GY/YL	EVAP Purge Solenoid	12v, 55 mph: 1v
32	LG/PK	IAC Motor Control	8-10v
33	DG	EGR VR Solenoid	0%, 55 mph: 45%
33	BR/PK	EGR VR Solenoid	0%, 55 mph: 45%
34	---	Not Used	---
35	WT/PK	Speed Control Vent Solenoid	0%, 55 mph: 45%
36	YL/LG	Spark Output Signal	50% duty cycle
37	RD	Vehicle Power	12-14v
38	---	Not Used	---
39	OR	Speed Control Switch Ground	<0.050v
40	BK/LG	Power Ground	<0.1v

1990-93 Taurus 3.0L V6 MFI SHO VIN Y (M/T) 60 Pin Connector

PCM Pin #	Wire Color	Circuit Description (60 Pin)	Value at Hot Idle
41	BL/OR	High Fuel Pump Control	On: 1v, Off: 12v
42	---	Not Used	---
43	DB/LG	HO2S-11 (B1 S1) Signal	0.1-1.1v
44	---	Not Used	---
45	LG/BK	BARO Sensor Signal	159 Hz (sea level)
46	BK/WT	Analog Signal Return	<0.050v
46	GY/RD	Analog Signal Return	<0.050v
47	DG/LG	TP Sensor Signal	0.5-1.2v
47	GY/WT	TP Sensor Signal	0.5-1.2v
48	BR	STI Signal	STI Open: 5v, Closed: 1v
48	WT/BK	STI Signal	STI Open: 5v, Closed: 1v
49	OR	HO2S Ground	<0.050v
50 ('90-'91)	BL/RD	MAF Sensor Signal	0.8v, 55 mph: 1.3v
50 ('92-'93)	DB/OR	MAF Sensor Signal	0.8v, 55 mph: 1.3v
51	GY/BK	S/C Vacuum Solenoid Control	0%, 55 mph: 45%
52-53	---	Not Used	---
54	PK/YL	A/C WOT Relay Control	On: 1v, Off: 12v
54	RD	A/C WOT Relay Control	On: 1v, Off: 12v
55	TN/OR	Low Speed Fan Control	On: 1v, Off: 12v
56	DB	PIP Sensor Signal	50% duty cycle
57	RD	Vehicle Power	12-14v
58	TN	Injector 1 Control	3.6-4.1 ms
59	WT	Injector 2 Control	3.6-4.1 ms
60	BK/LG	Power Ground	<0.1v

Pin Connector Graphic

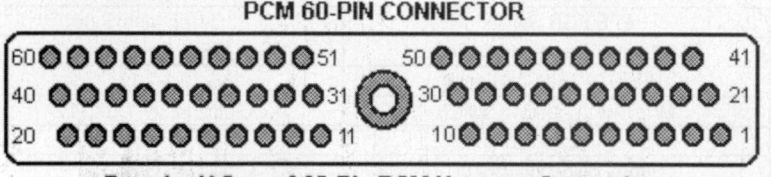

PCM 60-PIN CONNECTOR

Terminal View of 60-Pin PCM Harness Connector

Standard Colors and Abbreviations

Abbreviation	Color	Abbreviation	Color	Abbreviation	Color
BK	Black	GY	Gray	PK	Purple
BL	Blue	GN	Green	RD	Red
BR	Brown	LG	LT Green	TN	Tan
DB	Dark Blue	OR	Orange	WT	White
DG	DK Green	PK	Pink	YL	Yellow

1994-95 Taurus 3.0L V6 MFI SHO VIN Y (M/T) 60 Pin Connector

PCM Pin #	Wire Color	Circuit Description (60 Pin)	Value at Hot Idle
1	YL	Keep Alive Power	12-14v
2	YL/LG	PSP Switch Signal	Straight: 0v, Turned: 11v
3	DG/WT	VSS (+) Signal	0 Hz, 55 mph: 125 Hz
4	TN/YL	Ignition Diagnostic Monitor	20-31 Hz
5	RD/LG	Brake Pedal Position Switch	Brake Off: 0v, On: 12v
6	OR/YL	VSS (-) Signal	0 Hz, 55 mph: 125 Hz
7	LG/RD	ECT Sensor Signal	0.5-0.6v
8	---	Not Used	
9	TN/LB	MAF Sensor Return	<0.050v
10	PK/BL	A/C Cycling Clutch Switch	A/C On: 12v, Off: 0v
11	OR/YL	Speed Control Solenoid (+)	S/C On: 12-14v
12	BR/YL	Injector 3 Control	3.6-4.1 ms
13	BR/BL	Injector 4 Control	3.6-4.1 ms
14	TN/BK	Injector 5 Control	3.6-4.1 ms
15	LG/OR	Injector 6 Control	3.6-4.1 ms
16	OR/RD	Ignition System Ground	<0.050v
17	TN/RD	STI Output, MIL Control	MIL On: 1v, Off: 12v
18	PK	Octane Adjust Switch	Closed: 0v, Open: 9.1v
19	PK/BK	Fuel Pump Monitor	On: 12v, Off: 0v
20	BK	PCM Case Ground	<0.050v
21	OR/BK	IAC Motor Control	8-10v
22	TN/LG	Low Fuel Pump	On: 1v, Off: 12v
23	YL/RD	Knock Sensor Signal	0v
24	DG	CID Sensor Signal	5-7 Hz
25	GY	IAT Sensor Signal	1.5-2.5v
26	BR/WT	Reference Voltage	4.9-5.1v
27	BR/LG	PFE EGR Sensor	3.3v, 55 mph: 2.8v
28	BL/BK	S/Control Command Switch	S/C On: 6.7v
29	RD/BK	HO2S-21 (B2 S1) Signal	0.1-1.1v
30	PK/YL	M/T: Clutch Position Switch	Clutch In: 0v, Out: 11v
31	GY/YL	EVAP Purge Solenoid	12v, 55 mph: 1v
32	LG/PK	IMRC Solenoid Control	12v, 55 mph: 12v
33	BR/PK	EGR VR Solenoid	0%, 55 mph: 45%
34	---	Not Used	---
35	WT/PK	Speed Control Vent Solenoid	0%, 55 mph: 45%
36	YL/LG	Spark Output Signal	50% duty cycle
37	RD	Vehicle Power	12-14v
38	---	Not Used	---
39	OR	Speed Control Switch Ground	<0.050v
40	BK/LG	Power Ground	<0.1v

1994-95 Taurus 3.0L V6 MFI SHO VIN Y (M/T) 60 Pin Connector

PCM Pin #	Wire Color	Circuit Description (60 Pin)	Value at Hot Idle
41	BL/OR	High Fuel Pump Control	On: 1v, Off: 12v
41	TN/LG	High Fuel Pump Control	On: 1v, Off: 12v
42	---	Not Used	---
43	DB/LG	HO2S-11 (B1 S1) Signal	0.1-1.1v
44	---	Not Used	---
45	LG/BK	BARO Sensor Signal	159 Hz (sea level)
46	GY/RD	Analog Signal Return	<0.050v
47	GY/WT	TP Sensor Signal	0.5-2.1v
48	BR	Self-Test Indicator Signal	STI Open: 5v, Closed: 1v
49	OR	HO2S Ground	<0.050v
50	BL/RD	MAF Sensor Signal	0.6-0.9v
51	GY/BK	S/C Vacuum Solenoid Control	0%, 55 mph: 45%
52-53	---	Not Used	---
54	PK/YL	A/C WOT Relay Control	On: 1v, Off: 12v
55	TN/OR	Low Speed Fan Control	On: 1v, Off: 12v
56	DB	PIP Sensor Signal	50% duty cycle
57	RD	Vehicle Power	12-14v
58	TN	Injector 1 Control	3.6-4.1 ms
59	WT	Injector 2 Control	3.6-4.1 ms
60	BK/LG	Power Ground	<0.1v

Pin Connector Graphic

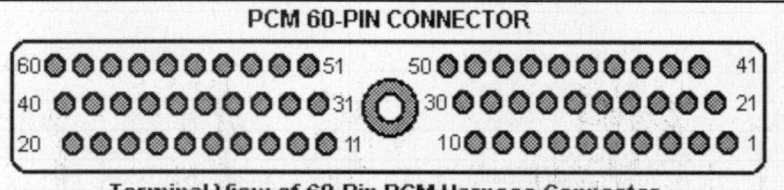

Terminal View of 60-Pin PCM Harness Connector

Standard Colors and Abbreviations

Abbreviation	Color	Abbreviation	Color	Abbreviation	Color
BK	Black	GY	Gray	PK	Purple
BL	Blue	GN	Green	RD	Red
BR	Brown	LG	LT Green	TN	Tan
DB	Dark Blue	OR	Orange	WT	White
DG	DK Green	PK	Pink	YL	Yellow

1993-95 Taurus 3.0L V6 FFV VIN 1 (A/T) 60 Pin Connector

PCM Pin #	Wire Color	Circuit Description (60 Pin)	Value at Hot Idle
1	YL	Keep Alive Power	12-14v
2	RD/LG	Brake Pedal Position Switch	Brake Off: 0v, On: 12v
3	DG/WT	VSS (+) Signal	0 Hz, 55 mph: 125 Hz
4	WT/PK	Ignition Diagnostic Monitor	20-31 Hz
5	GY/BK	TSS Sensor Signal	100 Hz (700 rpm)
6	OR/YL	VSS (-) Signal	0 Hz, 55 mph: 125 Hz
7	LG/RD	ECT Sensor Signal	0.5-0.6v
8	PK/BK	Fuel Pump Monitor	On: 12v, Off: 0v
9	TN/BL	MAF Sensor Return	<0.050v
10	PK/BL	A/C Cycling Clutch Switch	A/C On: 12v, Off: 0v
11	GN/YL	EVAP Purge Solenoid	12v, 55 mph: 1v
12	LG/OR	Injector 6 Control	3.8-6.1 ms
13	TN/OR	Low Cooling Fan	On: 1v, Off: 12v
14	BK/LG	Cold Start Injector (CSI)	KOEC only: 5 ms
15	TN/BK	Injector 5 Control	3.8-6.1 ms
16	OR/RD	Ignition System Ground	<0.050v
17	TN/RD	STI Output, MIL Control	MIL On: 1v, Off: 12v
18	TN/OR	Data Bus (+) Signal	Digital Signals
19	PK/BL	Data Bus (-) Signal	Digital Signals
20	BK	PCM Case Ground	<0.050v
21	BR/WT	IAC Motor Control	7-9v
22	BL/OR	Fuel Pump Control	On: 1v, Off: 12v
23	---	Not Used	---
24	DB/OR	CID Sensor Signal	5-7 Hz
25	GY	IAT Sensor Signal	1.5-2.5v
26	BR/WT	Reference Voltage	4.9-5.1v
27	BR/LG	PFE EGR Sensor	3.3v, 55 mph: 2.9v
28	YL/LG	PSP Switch Signal	Straight: 0v, Turned: 11v
29	---	Not Used	---
30	BL/YL	MLP Sensor Signal	In 'P': 0v, 55 mph: 1.7v
31	LG/PK	High Cooling Fan	On: 1v, Off: 12v
32	OR/LG	Low Fuel Pump Control	On: 1v, Off: 12v
33	BR/PK	EGR VR Solenoid	0%, 55 mph: 45%
34	BL/BK	Data Output Solenoid (DOL)	Digital Signals
34	BL/PK	Data Output Solenoid (DOL)	Digital Signals
35	BR/BL	Injector 4 Control	3.8-6.1 ms
36	PK	Spark Output Signal	6.93v (duty cycle)
37	RD	Vehicle Power	12-14v
38	WT/YL	EPC Solenoid Control	8-9v (9 psi)
39	BR/YL	Injector 3 Control	3.8-6.1 ms
40	BK/LG	Power Ground	<0.1v

1993-95 Taurus 3.0L V6 FFV VIN 1 (A/T) 60 Pin Connector

PCM Pin #	Wire Color	Circuit Description (60 Pin)	Value at Hot Idle
41-42	---	Not Used	---
43	RD/BK	HO2S-11 (B1 S1) Signal	0.1-1.1v
44	GY/BL	HO2S-21 (B2 S1) Signal	0.1-1.1v
45	DG/LG	Flexible Fuel Sensor Signal	30-70 Hz (100% gasoline)
46	GY/RD	Analog Signal Return	<0.050v
47	GN/WT	TP Sensor Signal	0.5-1.2v
48	BR	Self-Test Indicator Signal	STI Open: 5v, Closed: 1v
49	OR/BK	TOT Sensor Signal	2.10-2.40v
50	BL/RD	MAF Sensor Signal	0.6-0.9v
51	OR/YL	Shift Solenoid 1 Control	12v, 55 mph: 1v
52	PK/OR	Shift Solenoid 2 Control	1v, 55 mph: 12v
53	TN/WT	TCC Solenoid Control	0%, 55 mph: 45%
54	PK/YL	A/C WOT Relay Control	On: 1v, Off: 12v
55	WT/RD	Shift Solenoid 3 Control	7-9v, 55 mph: 8-9v
56	GY/OR	PIP Sensor Signal	50% duty cycle
57	RD	Vehicle Power	12-14v
58	TN	Injector 1 Control	3.8-6.1 ms
59	WT	Injector 2 Control	3.8-6.1 ms
60	BK/LG	Power Ground	<0.1v

Pin Connector Graphic

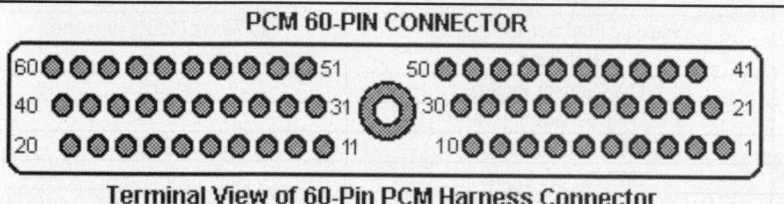

PCM 60-PIN CONNECTOR

Terminal View of 60-Pin PCM Harness Connector

Standard Colors and Abbreviations

Abbreviation	Color	Abbreviation	Color	Abbreviation	Color
BK	Black	GY	Gray	PK	Purple
BL	Blue	GN	Green	RD	Red
BR	Brown	LG	LT Green	TN	Tan
DB	Dark Blue	OR	Orange	WT	White
DG	DK Green	PK	Pink	YL	Yellow

1996-97 Taurus 3.0L V6 FFV VIN 1, VIN 2 (A/T) 104 Pin Connector

PCM Pin #	Wire Color	Circuit Description (104 Pin)	Value at Hot Idle
1	PK/OR	Shift Solenoid 2 Control	1v, 55 mph: 12v
2	PK/LG	MIL (lamp) Control	MIL On: 1v, Off: 12v
3-12	---	Not Used	---
13	WT/OR	Flash EEPROM Power	0.1v
14	---	Not Used	---
15	PK	Data Bus (-) Signal	Digital Signals
16	PK/LB	Data Bus (+) Signal	Digital Signals
17-20	---	Not Used	---
21	TN/OR	CKP Sensor (+) Signal	510-540 Hz
22	GY	CKP Sensor (-) Signal	510-540 Hz
23	---	Not Used	---
24	DB	Power Ground	<0.1v
25	BK/WT	Power Ground	<0.1v
26	BK	Coil Driver 1 Control	6° dwell
27	YL/BK	Shift Solenoid 1 Control	12v, 55 mph: 1v
28	OR/YL	Low Speed Fan Control	On: 1v, Off: 12v
29	---	Not Used	---
30	TN/OR	Octane Adjust Switch	Closed: 0v, Open: 9.3v
31	DG	PSP Switch Signal	Straight: 0v, Turned: 12v
32	---	Not Used	---
33	YL/LG	VSS (-) Signal	0 Hz, 55 mph: 125 Hz
34	PK/OR	Flexible Fuel Sensor Signal	30-70 Hz (100% gasoline)
35	---	Not Used	---
36	RD/BK	MAF Sensor Return	<0.050v
37	TN/LB	TFT Sensor Signal	2.10-2.40v
38	OR/BK	ECT Sensor Signal	0.5-0.6v
39	LG/RD	IAT Sensor Signal	1.5-2.5v
40	GY	Fuel Pump Monitor	On: 12v, Off: 0v
41	DG/YL	A/C High Pressure Switch	AC On: 12v, Off: 0v
42	---	Not Used	---
43	OR/LG	Data Output Link	Digital Signals
44-45	---	Not Used	---
46	LG/PK	High Speed Fan Control	On: 1v, Off: 12v
47	BR/PK	EGR VR Solenoid	0%, 55 mph: 45%
48	TN/YL	Clean Tachometer Output	42-50 Hz
49-50	---	Not Used	---
51	BK/WT	Power Ground	<0.1v
52	YL/RD	Coil Driver 2 Control	6° dwell
53	PK/BK	Shift Solenoid 3 Control	12v, 55 mph: 1v
54	---	Not Used	---
55	PK	Keep Alive Power	12-14v
56	GY/YL	EVAP VMV Solenoid	0-10 Hz (0-100%)
57	---	Not Used	---
58	GY/BK	VSS (+) Signal	0 Hz, 55 mph: 125 Hz
59	---	Not Used	---
60	GY/LB	HO2S-11 (B1 S1) Signal	0.1-1.1v
61	---	Not Used	---
62	RD/PK	FTP Sensor Signal	2.6v (0" H2O - cap off)
63	---	Not Used	---
64	LB/YL	Digital TR Sensor	In 'P': 0v, in O/D: 1.7v

1996-97 Taurus 3.0L V6 FFV VIN 1, VIN 2 (A/T) 104 Pin Connector

PCM Pin #	Wire Color	Circuit Description (104 Pin)	Value at Hot Idle
65	BR/LG	DPFE Sensor Signal	0.20-1.30v
66	---	Not Used	---
67	PK/WT	EVAP Purge Solenoid	0-10 Hz (0-100%)
68	---	Not Used	---
69	PK/YL	A/C WOT Relay Control	On: 1v, Off: 12v
70, 72	---	Not Used	---
71	RD	Vehicle Power	12-14v
73	TN/BK	Injector 5 Control	2.3-2.8 ms
74	BR/YL	Injector 3 Control	2.3-2.8 ms
75	TN	Injector 1 Control	2.3-2.8 ms
76	BK/WT	Power Ground	<0.1v
77	BK/WT	Power Ground	<0.1v
78	YL/WT	Coil Driver 3 Control	6° dwell
79	---	Not Used	---
80	LB/OR	Fuel Pump Control	On: 1v, Off: 12v
81	WT/YL	EPC Solenoid Control	10.6v (40 psi)
82	PK/YL	TCC Solenoid Control	0%, 55 mph: 95%
83	WT/LB	IAC Motor Control	10v (32% duty cycle)
84	DG/WT	TSS Sensor Signal	50-65 Hz (790-820 rpm)
85	DB/OR	CID Sensor Signal	6-8 Hz
86	BK/YL	A/C Pressure Switch	A/C On: 12v (Open)
87	RD/LG	HO2S-21 (B2 S1) Signal	0.1-1.1v
88	LB/RD	MAF Sensor Signal	0.9v
89	GY/WT	TP Sensor Signal	0.53-1.27v
90	BR/WT	Reference Voltage	4.9-5.1v
91	GY/RD	Analog Signal Return	<0.050v
92	RD/LG	Brake Pedal Position Switch	Brake Off: 0v, On: 12v
93	RD/WT	HO2S-11 (B1 S1) Heater	On: 1v, Off: 12v
94	WT/BK	HO2S-21 (B2 S1) Heater	On: 1v, Off: 12v
95, 96, 98	---	Not Used	---
97	RD	Vehicle Power	12-14v
99	LG/OR	Injector 6 Control	2.3-2.8 ms
100	BR/LB	Injector 4 Control	2.3-2.8 ms
101	WT	Injector 2 Control	2.3-2.8 ms
102	---	Not Used	---
103	BK/WT	Power Ground	<0.1v
104	---	Not Used	---

Pin Connector Graphic

PCM 104-PIN CONNECTOR

Terminal View of 104-Pin PCM Wiring Harness Connector

Standard Colors and Abbreviations

Abbreviation	Color	Abbreviation	Color	Abbreviation	Color
BK	Black	GY	Gray	PK	Purple
BL	Blue	GN	Green	RD	Red
BR	Brown	LG	LT Green	TN	Tan
DB	Dark Blue	OR	Orange	WT	White
DG	DK Green	PK	Pink	YL	Yellow

1998-99 Taurus 3.0L V6 FFV VIN 1, VIN 2 (A/T) 104 Pin Connector

PCM Pin #	Wire Color	Circuit Description (104 Pin)	Value at Hot Idle
1	PK/OR	Shift Solenoid 'B' Control	1v, 55 mph: 12v
2	PK/LG	MIL (lamp) Control	MIL On: 1v, Off: 12v
3	YL/BK	Digital TR1 Sensor	In 'P': 0v, 55 mph: 11v
5-12	---	Not Used	---
13	PK	Flash EEPROM Power	0.1v
15	PK/LB	Data Bus (-) Signal	Digital Signals
16	TN/OR	Data Bus (+) Signal	Digital Signals
17-18	---	Not Used	---
21	GY	CKP Sensor (+) Signal	510-540 Hz
22	DB	CKP Sensor (-) Signal	510-540 Hz
23	---	Not Used	---
24	BK/WT	Power Ground	<0.1v
25	BK	Power Ground	<0.1v
26	YL/BK	Coil Driver 1 Control	5° dwell
27	OR/YL	Shift Solenoid 'A' Control	12v, 55 mph: 0.1v
28	LB	Low Speed Fan Control	On: 1v, Off: 12v
29	---	Not Used	---
30	TN/OR	Octane Adjust Switch	Closed: 0v, Open: 9.3v
31	YL/LG	PSP Switch Signal	Straight: 0v, Turned: 12v
32	---	Not Used	---
33	PK/OR	VSS (-) Signal	0 Hz, 55 mph: 125 Hz
34	DG/LG	Flexible Fuel Sensor Signal	40-60 Hz (100% gasoline)
35	---	Not Used	---
36	TN/LB	MAF Sensor Return	<0.050v
37	OR/BK	TFT Sensor Signal	2.10-2.40v
38	LG/RD	ECT Sensor Signal	0.5-0.6v
39	GY	IAT Sensor Signal	1.5-2.5v
40	DG/YL	Fuel Pump Monitor	On: 12v, Off: 0v
41	DG/WT	A/C High Pressure Switch	AC On: 12v, Off: 0v
42	---	Not Used	---
43	OR/LG	Data Output Link	Digital Signals
44-45	---	Not Used	---
46	LG/PK	High Speed Fan Control	On: 1v, Off: 12v
47	BR/PK	EGR VR Solenoid	0%, 55 mph: 45%
48	TN/YL	Clean Tachometer Output	42-50 Hz
49	LB/BK	Digital TR2 Sensor	In 'P': 0v, 55 mph: 11v
50	WT/BK	Digital TR4 Sensor	In 'P': 0v, 55 mph: 11v
51	BK/WT	Power Ground	<0.1v
52	YL/RD	Coil Driver 2 Control	5° dwell
53	PK/BK	Shift Solenoid 'C' Control	12v, 55 mph: 0.1v
54	PK/YL	TCC Solenoid Control	0%, 55 mph: 95%
55	DG/OR	Keep Alive Power	12-14v
56	GY/YL	EVAP Purge Solenoid	0-10 Hz (0-100%)
57	---	Not Used	---
58	GY/BK	VSS (+) Signal	0 Hz, 55 mph: 125 Hz
59	---	Not Used	---
60	GY/LB	HO2S-11 (B1 S1) Signal	0.1-1.1v
61	---	Not Used	---
62	RD/PK	FTP Sensor Signal	2.6v (0" H2O - cap off)
63	---	Not Used	---
64	LB/YL	Digital TR Sensor	In 'P': 0v, in O/D: 1.7v

1998-99 Taurus 3.0L V6 FFV VIN 1, VIN 2 (A/T) 104 Pin Connector

PCM Pin #	Wire Color	Circuit Description (104 Pin)	Value at Hot Idle
65	BR/LG	DPFE Sensor Signal	0.20-1.30v
66	---	Not Used	---
67	PK/WT	EVAP Purge Solenoid	0-10 Hz (0-100%)
68	---	Not Used	---
69	PK/YL	A/C WOT Relay Control	On: 1v, Off: 12v
70	---	Not Used	---
71	RD	Vehicle Power	12-14v
73	TN/BK	Injector 5 Control	2.3-2.8 ms
74	BR/YL	Injector 3 Control	2.3-2.8 ms
75	TN	Injector 1 Control	2.3-2.8 ms
76	BK/WT	Power Ground	<0.1v
77	BK/WT	Power Ground	<0.1v
78	YL/WT	Coil Driver 3 Control	5° dwell
80	LB/OR	Fuel Pump Control	On: 1v, Off: 12v
81	WT/YL	EPC Solenoid Control	11v (38-42 psi)
82	PK/YL	TCC Solenoid Control	12v (0%)
83	WT/LB	IAC Motor Control	10v (32% duty cycle)
84	DG/WT	TSS Sensor Signal	50-65 Hz (790-850 rpm)
85	DB/OR	CID Sensor Signal	5-7 Hz
86	BK/YL	A/C Pressure Switch	A/C On: 12v (Open)
87	RD/LG	HO2S-21 (B2 S1) Signal	0.1-1.1v
88	LB/RD	MAF Sensor Signal	0.9v
89	GY/WT	TP Sensor Signal	0.53-1.27v
90	BR/WT	Reference Voltage	4.9-5.1v
91	GY/RD	Analog Signal Return	<0.050v
92	RD/LG	Brake Pedal Position Switch	Brake Off: 0v, On: 12v
93	RD/WT	HO2S-11 (B1 S1) Heater	On: 1v, Off: 12v
94	WT/BK	HO2S-21 (B2 S1) Heater	On: 1v, Off: 12v
95-96	YL/LB	HO2S-12 (B1 S2) Heater	On: 1v, Off: 12v
97	RD	Vehicle Power	12-14v
99	LG/OR	Injector 6 Control	2.3-2.8 ms
100	BR/LB	Injector 4 Control	2.3-2.8 ms
101	WT	Injector 2 Control	2.3-2.8 ms
102	---	Not Used	---
103	BK/WT	Power Ground	<0.1v
104	---	Not Used	---

Pin Connector Graphic

PCM 104-PIN CONNECTOR

Terminal View of 104-Pin PCM Wiring Harness Connector

Standard Colors and Abbreviations

Abbreviation	Color	Abbreviation	Color	Abbreviation	Color
BK	Black	GY	Gray	PK	Purple
BL	Blue	GN	Green	RD	Red
BR	Brown	LG	LT Green	TN	Tan
DB	Dark Blue	OR	Orange	WT	White
DG	DK Green	PK	Pink	YL	Yellow

2000 Taurus 3.0L V6 Flexible Fuel Vehicle VIN 2 (A/T) 104 Pin Connector

PCM Pin #	Wire Color	Circuit Description (104 Pin)	Value at Hot Idle
1	VT/OR	Shift Solenoid 'B' Control	1v, 55 mph: 12v
2	PK/LG	MIL (lamp) Control	MIL On: 1v, Off: 12v
3	YL/BK	Digital TR1 Sensor	In 'P': 0v, 55 mph: 11v
4	BR/PK	Start System Voltage	Cranking: 9-11v
5	---	Not Used	---
6	DG/WT	TSS Sensor Signal	50-65 Hz (790-820 rpm)
7-11	---	Not Used	---
12	YL/WT	Fuel Level Indicator Signal	1.7v (1/2 full)
13	VT	Flash EEPROM Power	0.1v
14	---	Not Used	---
15	PK/LB	SCP Bus (-) Signal	Digital Signals
16	TN/OR	SCP Bus (+) Signal	Digital Signals
17	BR/YL	Passive Antitheft RX Signal	Digital Signals
18	GY/RD	Passive Antitheft TX Signal	Digital Signals
19-20	---	Not Used	---
21	GY	CKP Sensor (+) Signal	510-540 Hz
22	DB	CKP Sensor (-) Signal	510-540 Hz
23	---	Not Used	---
24	BK/WT	Power Ground	<0.1v
25	BK	Chassis Ground	<0.1v
26	YL/BK	Coil 'A' Driver (Cyl 1 & 5)	5° dwell
27	OR/YL	Shift Solenoid 'A' Control	12v, 55 mph: 0.1v
28	LB	Low Speed Fan Control	On: 1v, Off: 12v
29, 33-34	---	Not Used	---
30	DB/LG	Antitheft Indicator Control	Indicator Off: 12v, On: 1v
31	YL/LG	PSP Switch Signal	Straight: 0v, Turned: 12v
32	YT	Knock Sensor 1 Ground	<0.050v
35	RD/BK	HO2S-12 (B1 S2) Signal	0.1-1.1v
36	TN/LB	MAF Sensor Return	<0.050v
37	OR/BK	TFT Sensor Signal	2.10-2.40v
38	LG/RD	ECT Sensor Signal	0.5-0.6v
39	GY	IAT Sensor Signal	1.5-2.5v
40	PK/BK	Fuel Pump Monitor	On: 12v, Off: 0v
41	DG/WT	A/C High Pressure Switch	AC On: 12v, Off: 0v
42-43, 45	---	Not Used	---
44	PK	Starter Relay Control	KOEC: 9-11v
46	LG/VT	High Speed Fan Control	On: 1v, Off: 12v
47	BR/PK	EGR VR Solenoid	0%, 55 mph: 45%
48	TN/YL	Clean Tachometer Output	42-50 Hz
49	LB/BK	Digital TR2 Sensor	In 'P': 0v, 55 mph: 11v
50	WT/BK	Digital TR4 Sensor	In 'P': 0v, 55 mph: 11v
51	BK/WT	Power Ground	<0.1v
52	YL/RD	Coil 'B' Driver (Cyl 3 & 4)	5° dwell
53	PK/BK	Shift Solenoid 'C' Control	12v, 55 mph: 0.1v
54	VT/YL	TCC Solenoid Control	0%, 55 mph: 95%
55	PK/LB	Keep Alive Power	12-14v
56	GY/YL	EVAP Purge Solenoid	0-10 Hz (0-100%)
57	YL/RD	Knock Sensor 1 (+) Signal	0v
58-59	---	Not Used	---
60	GY/LB	HO2S-11 (B1 S1) Signal	0.1-1.1v
61	VT/LG	HO2S-22 (B2 S2) Signal	0.1-1.1v
62	RD/PK	FTP Sensor Signal	2.6v (0" H2O - cap off)
63	WT/YL	Injector Pressure Sensor	2.7-3.7v (105-130 psi)
64	LB/YL	Digital TR3A Sensor	In 'P': 0v, in O/D: 1.7v

2000 Taurus 3.0L V6 Flexible Fuel Vehicle VIN 1, VIN 2 (A/T) 104 Pin Connector

PCM Pin #	Wire Color	Circuit Description (104 Pin)	Value at Hot Idle
65	BR/LG	DPFE Sensor Signal	0.20-1.30v
66	---	Not Used	---
67	PK/WT	EVAP CV Solenoid	0-10 Hz (0-100%)
68	GY/BK	VSS (+) Signal	0 Hz, 55 mph: 125 Hz
69	PK/YL	A/C WOT Relay Control	On: 1v, Off: 12v
71	RD	Vehicle Power	12-14v
72	---	Not Used	---
73	TN/BK	Injector 5 Control	2.3-2.8 ms
74	BR/YL	Injector 3 Control	2.3-2.8 ms
75	TN	Injector 1 Control	2.3-2.8 ms
76	BK/WT	Power Ground	<0.1v
77	BK/WT	Power Ground	<0.1v
78	YL/WT	Coil Driver 3 (Cyl 3 & 6)	5° dwell
80	LB/OR	Fuel Pump Control	On: 1v, Off: 12v
81	BR/OR	EPC Solenoid Control	11v (38-42 psi)
82	BK/WT	Check Cap Indicator Control	On: 1v, Off: 12v
83	WT/LB	IAC Motor Control	10v (32% duty cycle)
84	DB/YL	OSS Sensor Signal	0 Hz, 55 mph: 131 Hz
85	DB/OR	CID Sensor Signal	5-7 Hz
86	BK/YL	A/C Pressure Switch	A/C On: 12v (Open)
87	RD/LG	HO2S-21 (B2 S1) Signal	0.1-1.1v
88	LB/RD	MAF Sensor Signal	0.9v
89	GY/WT	TP Sensor Signal	0.53-1.27v
90	BR/WT	Reference Voltage	4.9-5.1v
91	GY/RD	Analog Signal Return	<0.050v
92	RD/LG	Brake Pedal Position Switch	Brake Off: 0v, On: 12v
93	RD/WT	HO2S-11 (B1 S1) Heater	On: 1v, Off: 12v
94	WT/BK	HO2S-21 (B2 S1) Heater	On: 1v, Off: 12v
95	YL/LB	HO2S-12 (B1 S2) Heater	On: 1v, Off: 12v
96	TN/YL	HO2S-22 (B2 S2) Heater	On: 1v, Off: 12v
97-98	---	Not Used	---
99	LG/OR	Injector 6 Control	2.3-2.8 ms
100	BR/LB	Injector 4 Control	2.3-2.8 ms
101	WT	Injector 2 Control	2.3-2.8 ms
102	---	Not Used	---
103	BK/WT	Power Ground	<0.1v
104	---	Not Used	---

Pin Connector Graphic

Terminal View of 104-Pin PCM Wiring Harness Connector

Standard Colors and Abbreviations

Abbreviation	Color	Abbreviation	Color	Abbreviation	Color
BK	Black	GY	Gray	PK	Purple
BL	Blue	GN	Green	RD	Red
BR	Brown	LG	LT Green	TN	Tan
DB	Dark Blue	OR	Orange	WT	White
DG	DK Green	PK	Pink	YL	Yellow

2001-04 Taurus 3.0L V6 Flexible Fuel Vehicle VIN 2 (A/T) 104 Pin Connector

PCM Pin #	Wire Color	Circuit Description (104 Pin)	Value at Hot Idle
1	VT/OR	Shift Solenoid 'B' Control	1v, 55 mph: 12v
2	PK/LG	MIL (lamp) Control	MIL On: 1v, Off: 12v
3	YL/BK	Digital TR1 Sensor	In 'P': 0v, 55 mph: 11v
4	TN/RD	System Voltage (Start-Run)	Cranking: 9-11v
5	---	Not Used	---
6	DG/WT	TSS Sensor Signal	50-65 Hz (790-820 rpm)
7-10	---	Not Used	---
12	YL/WT	Fuel Level Indicator Signal	1.7v (1/2 full)
13	VT	Flash EEPROM Power	0.1v
14, 19, 23	---	Not Used	---
15	PK/LB	SCP Bus (-) Signal	Digital Signals
16	TN/OR	SCP Bus (+) Signal	Digital Signals
17	BR/YL	Passive Antitheft RX Signal	Digital Signals
18	GY/RD	Passive Antitheft TX Signal	Digital Signals
20	GY/YL	Generator Field Control	130 Hz (45%)
21	GY	CKP Sensor (+) Signal	510-540 Hz
22	DB	CKP Sensor (-) Signal	510-540 Hz
24	BK/WT	Power Ground	<0.1v
25	BK	Case Ground	<0.1v
26	YL/BK	Coil 1 Driver (Cyl 1 & 5)	6° dwell
27	OR/YL	Shift Solenoid 'A' Control	12v, 55 mph: 1v
28	LB	Engine Cooling Fan Brake Relay Control	Off: 12v, On: 1v
29, 33-34	---	Not Used	---
30	DB/LG	Passive Antitheft Indicator	Lamp On: 1v, Off: 12v
31	YL/LG	PSP Switch Signal	Straight: 0v, Turned: 12v
32	YL	Knock Sensor 1 (-) Signal	<0.050v
35	RD/BK	HO2S-12 (B1 S2) Signal	0.1-1.1v
36	TN/LB	MAF Sensor Return	<0.050v
37	OR/BK	TFT Sensor Signal	2.10-2.40v
38	LG/RD	ECT Sensor Signal	0.5-0.6v
39	GY	IAT Sensor Signal	1.5-2.5v
40	DG/YL	Fuel Pump Monitor	On: 12v, Off: 0v
41	DG/WT	A/C High Pressure Switch	AC On: 12v, Off: 0v
42	LG/YL	Generator Cutoff Relay	0.1v (with fan on)
44	PK	Starter Relay Control	KOEC: 9-11v
45	LB/RD	Generator Sensor Signal	130 Hz (30%)
46	LG/VT	High Cooling Speed Fan Control	Off: 12v, On: 1v
47	BR/PK	EGR VR Solenoid Control	0%, 55 mph: 45%
48	TN/YL	Clean Tachometer Output	42-50 Hz
49	LB/BK	Digital TR2 Sensor	In 'P': 0v, 55 mph: 11v
50	WT/BK	Digital TR4 Sensor	In 'P': 0v, 55 mph: 11v
51	BK/WT	Power Ground	<0.1v
52	YL/RD	Coil Driver 2 (Cyl 3 & 4)	6° dwell
53	PK/BK	Shift Solenoid 'C' Control	12v, 55 mph: 1v
54	VT/YL	TCC Solenoid Control	12v (0%)
55	PK/LB	Keep Alive Power	12-14v
56	GY/YL	EVAP Purge Solenoid	0-10 Hz (0-100%)
57	YL/RD	Knock Sensor (+) Signal	0v
58-59	---	Not Used	---
60	GY/LB	HO2S-11 (B1 S1) Signal	0.1-1.1v
61	VT/LG	HO2S-22 (B2 S2) Signal	0.1-1.1v
62	RD/PK	FTP Sensor Signal	2.6v (0" H2O - cap off)
63	WT/YL	Injector Pressure Sensor	2.7-3.7v (105-130 psi)
64	RD/BK	Digital TR3A Sensor	In 'P': 0v, in O/D: 1.7v

2001-04 Taurus 3.0L V6 Flexible Fuel Vehicle VIN 2 (A/T) 104 Pin Connector

PCM Pin #	Wire Color	Circuit Description (104 Pin)	Value at Hot Idle
65	BR/LG	DPFE Sensor Signal	0.95-1.05v
66	TN/OR	Evaporator Air Temperature Sensor	4.95-5.15v
67	VT/WT	EVAP Purge Solenoid	0-10 Hz (0-100%)
68	GY/BK	VSS (+) Signal	Moving: AC pulse signals
69	PK/YL	A/C WOT Relay Control	Off: 12v, On: 1v
70	---	Not Used	---
71	RD	Vehicle Power (Start-Run)	12-14v
73	TN/BK	Injector 5 Control	2.3-2.8 ms
74	BR/YL	Injector 3 Control	2.3-2.8 ms
75	TN	Injector 1 Control	2.3-2.8 ms
76, 77	BK/WT	Power Ground	<0.1v
78	YL/WT	Coil 3 Driver (Cyl 2 & 6)	6° dwell
79	WT/LG	Check Transaxle Light	Indicator Off: 12v, On: 1v
80	LB/OR	Fuel Pump Control	Off: 12v, On: 1v
81	BR/OR	EPC Solenoid Control	10.6v (40 psi)
82	BK/WT	Check Cap Indicator Light	Indicator Off: 12v, On: 1v
83	WT/LB	IAC Motor Control	10v (32% duty cycle)
84	DB/YL	OSS Sensor Signal	0 Hz, 55 mph: 131 Hz
85	DB/OR	CID Sensor Signal	6-8 Hz
86	BK/YL	A/C Pressure Clutch Relay	Relay Off: 12v, On: 1v
87	RD/LG	HO2S-21 (B2 S1) Signal	0.1-1.1v
88	LB/RD	MAF Sensor Signal	0.9v
89	GY/WT	TP Sensor Signal	0.53-1.27v
90	BR/WT	Reference Voltage	4.9-5.1v
91	GY/RD	Sensor Return	<0.050v
92	RD/LG	Brake Pedal Position Switch	Brake Off: 0v, On: 12v
93	RD/WT	HO2S-11 (B1 S1) Heater	On: 1v, Off: 12v
94	WT/BK	HO2S-21 (B2 S1) Heater	On: 1v, Off: 12v
95	YL/LB	HO2S-12 (B1 S2) Heater	On: 1v, Off: 12v
96	TN/YL	HO2S-22 (B2 S2) Heater	On: 1v, Off: 12v
97	RD	Vehicle Power (Start-Run)	12-14v
98	LG/RD	Generator Battery Indicator	Digital Signals
99	LG/OR	Injector 6 Control	2.3-2.8 ms
100	BR/LB	Injector 4 Control	2.3-2.8 ms
101	WT	Injector 2 Control	2.3-2.8 ms
102	---	Not Used	---
103	BK/WT	Power Ground	<0.1v
104	---	Not Used	---

Pin Connector Graphic

PCM 104-PIN CONNECTOR

Terminal View of 104-Pin PCM Wiring Harness Connector

Standard Colors and Abbreviations

Abbreviation	Color	Abbreviation	Color	Abbreviation	Color
BK	Black	GY	Gray	PK	Purple
BL	Blue	GN	Green	RD	Red
BR	Brown	LG	LT Green	TN	Tan
DB	Dark Blue	OR	Orange	WT	White
DG	DK Green	PK	Pink	YL	Yellow

1993-95 Taurus 3.2L V6 MFI SHO VIN P (A/T) 60 Pin Connector

PCM Pin #	Wire Color	Circuit Description (60 Pin)	Value at Hot Idle
1	Y	Keep Alive Power	12-14v
2	RD/LG	Brake Pedal Position Switch	Brake Off: 0v, On: 12v
3	DG/WT	VSS (+) Signal	65 Hz at 55 mph
4	GY/OR	Ignition Diagnostic Monitor	20-31 Hz
5	GY/BK	TSS Sensor Signal	100 Hz (700 mph)
6	OR/YL	VSS (+) Signal	0 Hz, 55 mph: 125 Hz
7	LG/RD	ECT Sensor Signal	0.5-0.6v
8	PK/BK	Fuel Pump Monitor	On: 12v, Off: 0v
9	TN/BL	MAF Sensor Return	<0.050v
10	PK/BL	A/C Cycling Clutch Switch	A/C On: 12v, Off: 0v
11	GN/YL	EVAP Purge Solenoid	12v, 55 mph: 1v
12	LG/OR	Injector 6 Control	3.6-3.8 ms
13	TN/OR	Low Cooling Fan	On: 1v, Off: 12v
14	OR/YL	TCIL (lamp) Control	On: 1v, Off: 12v
14	WT/LG	TCIL (lamp) Control	On: 1v, Off: 12v
15	TN/BK	Injector 5 Control	3.6-3.8 ms
16	OR/RD	Ignition System Ground	<0.050v
17	TN/RD	STI Output, MIL Control	MIL On: 1v, Off: 12v
18	OR/BK	Data Bus (+) Signal	Digital Signals
19	BK/OR	Data Bus (-) Signal	Digital Signals
20	BK	PCM Case Ground	<0.050v
21	OR/BK	IAC Motor Control	8-9v
22	TN/LG	Fuel Pump Control	On: 1v, Off: 12v
23	YL/RD	Knock Sensor Signal	0v
24	DG	CID Sensor Signal	5-7 Hz
25	GY	IAT Sensor Signal	1.5-2.5v
26	BR/WT	Reference Voltage	4.9-5.1v
27	BR/LG	PFE EGR Sensor	0.6v
28	YL/LG	PSP Switch Signal	Straight: 0v, Turned: 11v
29	PK	Octane Adjust Switch	Closed: 0v, Open: 9.1v
30	BL/YL	MLP Sensor Signal	In 'P': 4.4v, in 'D': 2.0v
31	LG/PK	High Cooling Fan	On: 1v, Off: 12v
32	---	Not Used	---
33	BR/PK	EGR VR Solenoid	0%, 55 mph: 45%
34	LG/PK	IMRC Solenoid	12v, 55 mph: 12v
35	BR/BL	Injector 4 Control	3.6-3.8 ms
36	YL/LG	Spark Output Signal	50% duty cycle
37	RD	Vehicle Power	12-14v
38	WT/YL	EPC Solenoid Control	8.8v (9 psi)
39	BR/YL	Injector 3 Control	3.6-3.8 ms
40	BK/LG	Power Ground	<0.1v

1993-95 Taurus 3.2L V6 MFI SHO VIN P (A/T) 60 Pin Connector

PCM Pin #	Wire Color	Circuit Description (60 Pin)	Value at Hot Idle
41	WT/LG	TCS (switch) Signal	TCS & O/D On: 12v
41	TN/WT	TCS (switch) Signal	TCS & O/D On: 12v
42 ('94-'95)	DG/OR	Refrigerant Contain Switch	5v
43	DB/LG	HO2S-11 (B1 S1) Signal	0.1-1.1v
44	RD/BK	HO2S-21 (B2 S1) Signal	0.1-1.1v
45	---	Not Used	---
46	GY/RD	Analog Signal Return	<0.050v
47	GN/WT	TP Sensor Signal	0.5-2.1v
48	BR	Self-Test Indicator Signal	STI Open: 5v, Closed: 1v
49	OR/BK	TOT Sensor Signal	2.10-2.40v
50	BL/RD	MAF Sensor Signal	0.7v
51	OR/YL	Shift Solenoid 1 Control	12v, 55 mph: 1v
52	PK/OR	Shift Solenoid 2 Control	1v, 55 mph: 12v
53	TN/WT	TCC Solenoid Control	0%, 55 mph: 40%
53	PK/YL	TCC Solenoid Control	0%, 55 mph: 40%
54	PK/YL	A/C WOT Relay Control	On: 1v, Off: 12v
55	PK/BK	Shift Solenoid 3 Control	12v, 55 mph: 1v
55	WT/RD	Shift Solenoid 3 Control	12v, 55 mph: 1v
56	DB	PIP Sensor Signal	50% duty cycle
57	RD	Vehicle Power	12-14v
58	TN	Injector 1 Control	3.6-3.8 ms
59	WT	Injector 2 Control	3.6-3.8 ms
60	BK/LG	Power Ground	<0.1v

Pin Connector Graphic

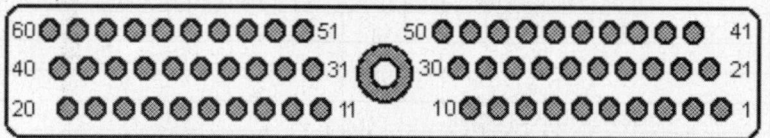

PCM 60-PIN CONNECTOR

Terminal View of 60-Pin PCM Harness Connector

Standard Colors and Abbreviations

Abbreviation	Color	Abbreviation	Color	Abbreviation	Color
BK	Black	GY	Gray	PK	Purple
BL	Blue	GN	Green	RD	Red
BR	Brown	LG	LT Green	TN	Tan
DB	Dark Blue	OR	Orange	WT	White
DG	DK Green	PK	Pink	YL	Yellow

1996-97 Taurus SHO 3.4L 4v V6 VIN N (A/T) 104 Pin Connector

PCM Pin #	Wire Color	Circuit Description (104 Pin)	Value at Hot Idle
1	OR/YL	COP 5 Driver Control	6° dwell
2	PK/LG	MIL (lamp) Control	MIL On: 1v, Off: 12v
3-4	---	Not Used	---
5	WT/OR	Electronic AIR Control	12v (Off)
6	OR/YL	Shift Solenoid 1 Control	12v, 55 mph: 1v
7	---	Not Used	---
8	DB/YL	IMRC Monitor	5v
9-10	---	Not Used	---
11	PK/OR	Shift Solenoid 2 Control	1v, 55 mph: 12v
12	WT/LG	TCIL (lamp) Control	On: 1v, Off: 12v
13	PK	Flash EEPROM Power	0.1v
14	---	Not Used	---
15	PK/LG	Data Bus (-) Signal	Digital Signals
16	TN/OR	Data Bus (+) Signal	Digital Signals
17-19	---	Not Used	---
20	PK/BK	Shift Solenoid 3 Control	1v, 55 mph: 12v
21	GY	CKP Sensor (+) Signal	380-420 Hz
22	DB	CKP Sensor (-) Signal	380-420 Hz
23, 24	BK/WT	Power Ground	<0.1v
25	BK	Power Ground	<0.1v
26	LG/WT	COP Driver 1	6° dwell
27	WT/PK	COP Driver 6	6° dwell
28	TN/OR	Low Speed Fan Control	On: 1v, Off: 12v
29	TN/WT	Overdrive Cancel Switch	OCS & OD On: 0.1v
30	DG	Octane Adjust Switch	Closed: 0v, Open: 9.3v
31	YL/LG	PSP Switch Signal	Straight: 0v, Turned: 12v
32	DG/PK	Knock Sensor 2 Signal	0v
33	PK/OR	VSS (-) Signal	0 Hz, 55 mph: 125 Hz
34, 43-45	---	Not Used	---
35	RD/BK	HO2S-12 (B1 S2) Signal	0.1-1.1v
36	TN/LB	MAF Sensor Return	<0.1v
37	OR/BK	TFT Sensor Signal	2.10-2.40v
38	LG/RD	ECT Sensor Signal	0.5-0.6v
39	GY	IAT Sensor Signal	1.5-2.5v
40	DG/YL	Fuel Pump Monitor	On: 12v, Off: 0v
41	PK/LB	A/C Cycling Clutch Switch	A/C On: 12v, Off: 0v
42	BR	IMRC Solenoid Control	12v (Off)
46	LG/PK	High Speed Fan Control	On: 1v, Off: 12v
47	BR/PK	EGR VR Solenoid	0%, 55 mph: 45%
48	TN/YL	Clean Tachometer Output	33-37 Hz
49-50, 54	---	Not Used	---
51	BK/WT	Power Ground	<0.1v
52	LG/YL	COP Driver 2	6° dwell
53	PK/LB	COP Driver 7	6° dwell
55	DG/OR	Keep Alive Power	12-14v
56	GY/YL	EVAP VMV Solenoid	0-10 Hz (0-100%)
57	YL/RD	Knock Sensor 1 Signal	0v
58	GY/BK	VSS (+) Signal	0 Hz, 55 mph: 125 Hz
60	GY/LB	HO2S-11 (B1 S1) Signal	0.1-1.1v
61	PK/LG	HO2S-22 (B2 S2) Signal	0.1-1.1v
62	RD/PK	FTP Sensor Signal	2.6v (0 in. H20)
63	---	Not Used	---
64	LB/YL	TR Sensor	In 'P': 0v, 55 mph: 1.7v

1996-97 Taurus SHO 3.4L 4v V6 VIN N (A/T) 104 Pin Connector

PCM Pin #	Wire Color	Circuit Description (104 Pin)	Value at Hot Idle
65	BR/LG	DPFE Sensor Signal	0.20-1.30v
66, 68	---	Not Used	---
67	PK/WT	EVAP Purge Solenoid	0-10 Hz (0-100%)
69	PK/YL	A/C WOT Relay Control	On: 1v, Off: 12v
70	BR	Electronic AIRB Solenoid	12v, 55 mph: 12v
71	RD	Vehicle Power	12-14v
72	TN/RD	Injector 7 Control	2.3-2-8 ms
73	TN/BK	Injector 5 Control	2.3-2-8 ms
74	BR/YL	Injector 3 Control	2.3-2-8 ms
75	TN	Injector 1 Control	2.3-2-8 ms
76, 77	BK/WT	Power Ground	<0.1v
78	DG/PK	COP Driver 3	6° dwell
79	WT/RD	COP Driver 8	6° dwell
80	LB/OR	Fuel Pump Control	On: 1v, Off: 12v
81	WT/YL	EPC Solenoid Control	9.0v (15 psi)
82	PK/YL	TCC Solenoid Control	12v (0 psi)
83	WT/LB	IAC Motor Control	10v (30% on time)
84	DG/WT	TSS Sensor Signal	43 Hz (650 rpm)
85	DB/OR	CID Sensor Signal	5-7 Hz
86	BK/YL	A/C Pressure Switch	A/C On: 12v (Open)
87	RD/LG	HO2S-21 (B2 S1) Signal	0.1-1.1v
88	LB/RD	MAF Sensor Signal	0.6v
89	GY/WT	TP Sensor Signal	0.53-1.27v
90	BR/WT	Reference Voltage	4.9-5.1v
91	GY/RD	Analog Signal Return	<0.050v
92	RD/LG	Brake Pedal Switch Signal	Brake Off: 0v, On: 12v
93	RD/WT	HO2S-11 (B1 S1) Heater	On: 1v, Off: 12v
94	WT/BK	HO2S-21 (B2 S1) Heater	On: 1v, Off: 12v
95	YL/LB	HO2S-12 (B1 S2) Heater	On: 1v, Off: 12v
96	TN/YL	HO2S-22 (B2 S2) Heater	On: 1v, Off: 12v
97	RD	Vehicle Power	12-14v
98	LB	Injector 8 Control	2.3-2-8 ms
99	LG/OR	Injector 6 Control	2.3-2-8 ms
100	BR/LB	Injector 4 Control	2.3-2-8 ms
101	WT	Injector 2 Control	2.3-2-8 ms
102	---	Not Used	---
103	BK/WT	Power Ground	<0.1v
104	PK/WT	Coil Driver 4 Control	6° dwell

Pin Connector Graphic

PCM 104-PIN CONNECTOR

Terminal View of 104-Pin PCM Wiring Harness Connector

Standard Colors and Abbreviations

Abbreviation	Color	Abbreviation	Color	Abbreviation	Color
BK	Black	GY	Gray	PK	Purple
BL	Blue	GN	Green	RD	Red
BR	Brown	LG	LT Green	TN	Tan
DB	Dark Blue	OR	Orange	WT	White
DG	DK Green	PK	Pink	YL	Yellow

1998-99 Taurus SHO 3.4L 4v V6 VIN N (A/T) 104 Pin Connector

PCM Pin #	Wire Color	Circuit Description (104 Pin)	Value at Hot Idle
1	OR/YL	COP 5 Driver Control	6° dwell
2	PK/LG	MIL (lamp) Control	MIL On: 1v, Off: 12v
3-4, 7	---	Not Used	---
5	WT/OR	Electronic AIR Monitor	1v, 55 mph: 1v
6	OR/YL	Shift Solenoid 1 Control	12v, 55 mph: 0.1v
8	DB/YL	IMRC Monitor	5v
9	YL/WT	Fuel Level Indicator Signal	1.7v (1/2 full)
10	---	Not Used	---
11	PK/OR	Shift Solenoid 2 Control	1v, 55 mph: 1v
12	WT/LG	TCIL (lamp) Control	On: 1v, Off: 12v
13	PK	Flash EEPROM Power	0.1v
14	---	Not Used	---
15	PK	Data Bus (-) Signal	Digital Signals
16	TN/OR	Data Bus (+) Signal	Digital Signals
17-19	---	Not Used	---
20	PK/BK	Shift Solenoid 3 Control	12v, 55 mph: 0.1v
21	GY	CKP Sensor (+) Signal	380-420 Hz
22	DB	CKP Sensor (-) Signal	380-420 Hz
23, 24, 51	BK/WT	Power Ground	<0.1v
25	BK	Power Ground	<0.1v
26	LG/WT	COP 1 Driver Control	6° dwell
27	WT/PK	COP 6 Driver Control	6° dwell
28	TN/OR	Low Speed Fan Control	On: 1v, Off: 12v
29	TN/WT	Overdrive Cancel Switch	OCS & O/D On: 0.1v
30	DG	Octane Adjust Switch	Closed: 0v, Open: 9.3v
31	YL/LG	PSP Switch Signal	Straight: 0v, Turned: 12v
32	DG/PK	Knock Sensor 2 Signal	0v
33	PK/OR	VSS (-) Signal	0 Hz, 55 mph: 125 Hz
35	RD/BK	HO2S-12 (B1 S2) Signal	0.1-1.1v
36	TN/LB	MAF Sensor Return	<0.050v
37	OR/BK	TFT Sensor Signal	2.10-2.40v
38	LG/RD	ECT Sensor Signal	0.5-0.6v
39	GY	IAT Sensor Signal	1.5-2.5v
40	DG/YL	Fuel Pump Monitor	On: 12v, Off: 0v
41	PK/LB	A/C Cycling Clutch Switch	AC On: 12v, Off: 0v
42	BR	IMRC Solenoid Control	12v (Off)
43, 49-50	---	Not Used	---
45	PK/WT	EVAP CV Solenoid	0-10 Hz (0-100 Hz)
46	LG/PK	High Speed Fan Control	On: 1v, Off: 12v
47	BR/PK	EGR VR Solenoid	0%, 55 mph: 45%
48	TN/YL	Clean Tachometer Output	33-37 Hz
52	LG/YL	COP 2 Driver Control	6° dwell
53	PK/LB	COP 7 Driver Control	6° dwell
54	PK/YL	TCC Solenoid Control	0%, 55 mph: 95%
55	DG/OR	Keep Alive Power	12-14v
56	GY/YL	EVAP Purge Solenoid	0-10 Hz (0-100%)
57	YL/RD	Knock Sensor Signal	0v
58	GY/BK	VSS (+) Signal	0 Hz, 55 mph: 125 Hz
59	DG/WT	TSS Sensor Signal	30-32 Hz (550-600 rpm)
60	GY/LB	HO2S-11 (B1 S1) Signal	0.1-1.1v
61	PK/LG	HO2S-22 (B2 S2) Signal	0.1-1.1v
62	RD/PK	FTP Sensor Signal	2.6v (0 in. H20)
63	---	Not Used	---
64	LB/YL	TR Sensor Signal	In 'P': 0v, 55 mph: 1.7v

1998-99 Taurus SHO 3.4L 4v V6 VIN N (A/T) 104 Pin Connector

PCM Pin #	Wire Color	Circuit Description (104 Pin)	Value at Hot Idle
65	BR/LG	DPFE Sensor Signal	0.20-1.30v
66, 68	---	Not Used	---
67	PK/WT	EVAP Purge Solenoid	0-10 Hz (0-100%)
69	PK/YL	A/C WOT Relay Control	On: 1v, Off: 12v
70	BR	Electronic AIRB Solenoid	12v, 55 mph: 12v
71	RD	Vehicle Power	12-14v
72	TN/RD	Injector 7 Control	2.3-3.2 ms
73	TN/BK	Injector 5 Control	2.3-3.2 ms
74	BR/YL	Injector 3 Control	2.3-3.2 ms
75	TN	Injector 1 Control	2.3-3.2 ms
77	BK/WT	Power Ground	<0.1v
78	DG/PK	COP 3 Driver Control	6° dwell
79	WT/RD	COP 8 Driver Control	6° dwell
80	LG/OR	Fuel Pump Control	On: 1v, Off: 12v
81	WT/YL	EPC Solenoid Control	9.3v (15 psi)
82	PK/YL	TCC Solenoid Control	12v (0 psi)
83	WT/LB	IAC Motor Control	10v (30% on time)
84	DG/WT	TSS Sensor Signal	43 Hz (650 rpm)
85	DB/OR	CID Sensor Signal	5-7 Hz
86	BK/YL	A/C High Pressure Switch	12v (Switch Open)
87	RD/LG	HO2S-21 (B2 S1) Signal	0.1-1.1v
88	LB/RD	MAF Sensor Signal	0.6v
89	GY/WT	TP Sensor Signal	0.53-1.27v
90	BR/WT	Reference Voltage	4.9-5.1v
91	GY/RD	Analog Signal Return	<0.050v
92	RD/LG	Brake Pedal Switch Signal	Brake Off: 0v, On: 12v
93	RD/WT	HO2S-11 (B1 S1) Heater	On: 1v, Off: 12v
94	WT/BK	HO2S-21 (B2 S1) Heater	On: 1v, Off: 12v
95	YL/LB	HO2S-12 (B1 S2) Heater	On: 1v, Off: 12v
96	TN/YL	HO2S-22 (B2 S2) Heater	On: 1v, Off: 12v
97	RD	Vehicle Power	12-14v
98	LB	Injector 8 Control	2.3-3.2 ms
99	LG/OR	Injector 6 Control	2.3-3.2 ms
100	BR/LB	Injector 4 Control	2.3-3.2 ms
101	WT	Injector 2 Control	2.3-3.2 ms
102	---	Not Used	---
103	BK/WT	Power Ground	<0.1v
104	PK/WT	COP 4 Driver Control	6° dwell

Pin Connector Graphic

```
        PCM 104-PIN CONNECTOR
  1 ooooooooooooooo    ooooooooooooooo 26
27 ooooooooooooooo     ooooooooooooooo 52
53 ooooooooooooooo  ●  ooooooooooooooo 78
79 ooooooooooooooo     ooooooooooooooo 104
```

Terminal View of 104-Pin PCM Wiring Harness Connector

Standard Colors and Abbreviations

Abbreviation	Color	Abbreviation	Color	Abbreviation	Color
BK	Black	GY	Gray	PK	Purple
BL	Blue	GN	Green	RD	Red
BR	Brown	LG	LT Green	TN	Tan
DB	Dark Blue	OR	Orange	WT	White
DG	DK Green	PK	Pink	YL	Yellow

1990 Taurus 3.8L V6 MFI VIN 4 (A/T) 60 Pin Connector

PCM Pin #	Wire Color	Circuit Description (60 Pin)	Value at Hot Idle
1	YL	Keep Alive Power	12-14v
2	RD/LG	PSP Switch Signal	Straight: 0v, Turned: 11v
3	DG/WT	VSS (+) Signal	0 Hz, 55 mph: 125 Hz
4	DG/YL	Ignition Diagnostic Monitor	20-31 Hz
5	---	Not Used	---
6	OR/YL	VSS (-) Signal	0 Hz, 55 mph: 125 Hz
7	LG/YL	ECT Sensor Signal	0.5-0.6v
8	BK/BL	Data Bus (-) Signal	Digital Signals
8	BK/OR	Data Bus (-) Signal	Digital Signals
9	GY/WT	AXOD Transmission Solenoid	On: 1v, Off: 12v
10	PK/BL	A/C Cycling Clutch Switch	A/C On: 12v, Off: 0v
10	PK	A/C Cycling Clutch Switch	A/C On: 12v, Off: 0v
11	OR/YL	Speed Control Solenoid (+)	S/C On: 12-14v
12	BR/YL	Injector 3 Control	6.0-6.8 ms
13	BR/BL	Injector 4 Control	6.0-6.8 ms
14	TN/BL	Injector 5 Control	6.0-6.8 ms
15	LG	Injector 6 Control	6.0-6.8 ms
16	GY	Ignition System Ground	<0.050v
17	TN/RD	STI Output, MIL Control	MIL On: 1v, Off: 12v
18	DG/PK	Transmission 3-4 Switch	3-4 Switch Closed: 0.1v
19	OR/YL	Transmission 2-3 Switch	2-3 Switch Closed: 0.1v
20	---	Not Used	---
21	OR/BK	IAC Motor Control	9-10v
22	TN/LG	Fuel Pump Control	On: 1v, Off: 12v
23-24	---	Not Used	---
25	LG/PK	ACT Sensor Signal	1.5-2.5v
26	OR/WT	Reference Voltage	4.9-5.1v
27	BR/LG	PFE EGR Sensor	3.2v, 55 mph: 2.8v
28	BL/BK	Speed Control Switch	Switch On: 6.7v
29	DB/LG	HO2S-21 (B2 S1) Signal	0.1-1.1v
30	PK/YL	Neutral Position Switch	In 'N': 0v, Others: 4.6v
31	GY/YL	EVAP Purge Solenoid	12v, 55 mph: 1v
32	---	Not Used	---
33	DG	EGR VR Solenoid	0%, 55 mph: 45%
34	BL/PK	Data Output Link	Digital Signals
35	WT/PK	Speed Control Vent Solenoid	98% duty cycle (55 mph)
36	YL/LG	Spark Output Signal	6.93v (duty cycle)
37	RD	Vehicle Power	12-14v
38	---	Not Used	---
39	OR	Speed Control Switch Ground	<0.050v
40	BK/LG	Power Ground	<0.1v

1990 Taurus 3.8L V6 MFI VIN 4 (A/T) 60 Pin Connector

PCM Pin #	Wire Color	Circuit Description (60 Pin)	Value at Hot Idle
41	PK	High Speed Fan Control	On: 1v, Off: 12v
42	---	Not Used	---
43	DG/PK	HO2S-11 (B1 S1) Signal	0.1-1.1v
44	TN/OR	Data Bus (+) Signal	Digital Signals
44	OR/BK	Data Bus (+) Signal	Digital Signals
45	LG/BK	MAP Sensor Signal	107 Hz
46	BK/WT	Analog Signal Return	<0.050v
47	DG/LG	TP Sensor Signal	0.5-1.2v
48	WT/BK	Self-Test Indicator Signal	STI Open: 5v, Closed: 1v
49	OR	HO2S-11 (Bank 1) Ground	<0.050v
50	PK/BK	Fuel Pump Monitor	On: 1v, Off: 12v
51	GY/BK	S/C Vacuum Solenoid	0%, 55 mph: 45%
52	---	Not Used	---
53	TN/BL	Lockup Solenoid for AXOD	12v, 55 mph: 1v
54	RD	A/C WOT Relay Control	On: 1v, Off: 12v
55	TN/OR	Low Speed Fan Control	On: 1v, Off: 12v
56	DB	PIP Sensor Signal	50% duty cycle
57	RD	Vehicle Power	12-14v
58	TN	Injector 1 Control	6.0-6.8 ms
59	WT	Injector 2 Control	6.0-6.8 ms
60	BK/LG	Power Ground	<0.1v

Pin Connector Graphic

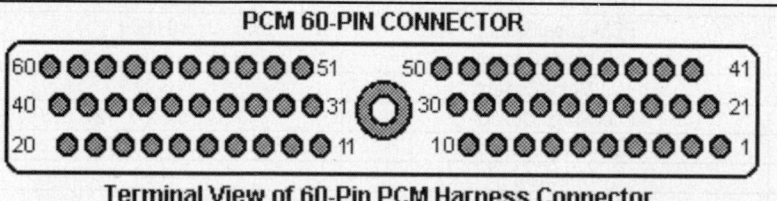

Terminal View of 60-Pin PCM Harness Connector

Standard Colors and Abbreviations

Abbreviation	Color	Abbreviation	Color	Abbreviation	Color
BK	Black	GY	Gray	PK	Purple
BL	Blue	GN	Green	RD	Red
BR	Brown	LG	LT Green	TN	Tan
DB	Dark Blue	OR	Orange	WT	White
DG	DK Green	PK	Pink	YL	Yellow

1991-95 Taurus 3.8L V6 MFI VIN 4 (A/T) 60 Pin Connector

PCM Pin #	Wire Color	Circuit Description (60 Pin)	Value at Hot Idle
1	YL	Keep Alive Power	12-14v
2	RD/LG	Brake Pedal Position Switch	Brake Off: 0v, On: 12v
3	DG/WT	VSS (-) Signal	0 Hz, 55 mph: 125 Hz
4	WT/PK	Ignition Diagnostic Monitor	20-31 Hz
4	DG/YL	Ignition Diagnostic Monitor	20-31 Hz
5	GY/BK	TSS Sensor Signal	100 Hz (700 rpm)
6	OR/YL	VSS (+) Signal	0 Hz, 55 mph: 125 Hz
7	LG/RD	ECT Sensor Signal	0.5-0.6v
8	PK/BK	Fuel Pump Monitor	On: 12v, Off: 0v
9	TN/BL	MAF Sensor Return	<0.050v
10	PK	A/C Cycling Clutch Signal	A/C On: 12v, Off: 0v
10	PK/BL	A/C Cycling Clutch Signal	A/C On: 12v, Off: 0v
11	GN/YL	EVAP Purge Solenoid	12v, 55 mph: 1v
12	LG/OR	Injector 6 Control	5.6-6.0 ms
13	TN/OR	Low Cooling Fan	On: 1v, Off: 12v
14	---	Not Used	---
15	TN/BK	Injector 5 Control	5.6-6.0 ms
16	GY	Ignition System Ground	<0.050v
17	TN/RD	STI Output, MIL Control	MIL On: 1v, Off: 12v
18	TN/OR	Data Bus (+) Signal	Digital Signals
18	OR/BK	Data Bus (+) Signal	Digital Signals
19	PK/BL	Data Bus (-) Signal	Digital Signals
19	BK/OR	Data Bus (-) Signal	Digital Signals
20	BK	PCM Case Ground	<0.050v
21	BL/OR	IAC Motor Control	8-9v
21	BR/WT	IAC Motor Control	8-9v
22	TN/LG	Fuel Pump Control	On: 1v, Off: 12v
22	BL/OR	Fuel Pump Control	On: 1v, Off: 12v
23-24	---	Not Used	---
25	GY	IAT Sensor Signal	1.5-2.5v
26	BR/WT	Reference Voltage	4.9-5.1v
27	BR/LG	PFE EGR Sensor	3.2v, 55 mph: 2.8v
28	YL/LG	PSP Sensor Signal	Straight: 0v, Turned: 11v
29	---	Not Used	---
30	BL	MLP Sensor Signal	In 'P': 0v, 55 mph: 1.7v
30	BL/YL	MLP Sensor Signal	In 'P': 0v, 55 mph: 1.7v
31	LG/PK	High Cooling Fan	On: 1v, Off: 12v
32	---	Not Used	---
33	BR/PK	EGR VR Solenoid	0%, 55 mph: 45%
34	BL/PK	Data Output Link	Digital Signals
35	BR/BL	Injector 4 Control	5.6-6.0 ms
36	YL/LG	Spark Output Signal	50% duty cycle
37	RD	Vehicle Power	12-14v
38	WT/YL	EPC Solenoid Control	7.8v (8-9 psi)
39	BR/YL	Injector 3 Control	5.6-6.0 ms
40	BK/LG	Power Ground	<0.1v

1991-95 Taurus 3.8L V6 MFI VIN 4 (A/T) 60 Pin Connector

PCM Pin #	Wire Color	Circuit Description (60 Pin)	Value at Hot Idle
41	---	Not Used	---
42 ('94-'95)	TN/LG	Refrigerant Contain Switch	5v
43	RD/BK	HO2S-11 (B1 S1) Signal	0.1-1.1v
44	DB/LG	HO2S-21 (B2 S1) Signal	0.1-1.1v
44	GY/BL	HO2S-21 (B2 S1) Signal	0.1-1.1v
45	---	Not Used	---
46	GY/RD	Analog Signal Return	<0.050v
47	GN/WT	TP Sensor Signal	0.5-1.2v
48	BR	Self-Test Indicator Signal	STI Open: 5v, Closed: 1v
49	OR/BK	TOT Sensor Signal	2.10-2.40v
50	BL/RD	MAF Sensor Signal	0.7v
51	OR/YL	Shift Solenoid 1 Control	12v, 55 mph: 1v
52	PK/OR	Shift Solenoid 2 Control	1v, 55 mph: 12v
53	TN/WT	TCC Solenoid Control	12v, 55 mph: 1v
54 ('91-'93)	PK/YL	A/C WOT Relay Control	On: 1v, Off: 12v
54 ('94-'95)	RD	A/C WOT Relay Control	On: 1v, Off: 12v
55	WT/RD	Shift Solenoid 3 Control	12v, 55 mph: 1v
56	DB	PIP Sensor Signal	50% duty cycle
56	GY/OR	PIP Sensor Signal	50% duty cycle
57	RD	Vehicle Power	12-14v
58	TN	Injector 1 Control	5.6-6.0 ms
59	WT	Injector 2 Control	5.6-6.0 ms
60	BK/LG	Power Ground	<0.1v

Pin Connector Graphic

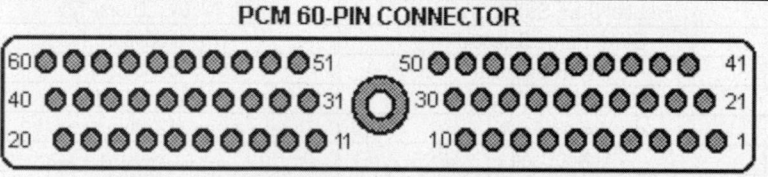

Terminal View of 60-Pin PCM Harness Connector

Standard Colors and Abbreviations

Abbreviation	Color	Abbreviation	Color	Abbreviation	Color
BK	Black	GY	Gray	PK	Purple
BL	Blue	GN	Green	RD	Red
BR	Brown	LG	LT Green	TN	Tan
DB	Dark Blue	OR	Orange	WT	White
DG	DK Green	PK	Pink	YL	Yellow

TEMPO PIN TABLES

1990 Tempo 2.3L I4 MFI VIN S, X (All) 60 Pin Connector

PCM Pin #	Wire Color	Circuit Description (60 Pin)	Value at Hot Idle
1	YL	Keep Alive Power	12-14v
2, 9	---	Not Used	---
3	DG/YL	VSS (+) Signal	0 Hz, 55 mph: 125 Hz
4	TN/YL	Ignition Diagnostic Monitor	20-31 Hz
5	RD/LG	Vehicle Power	12-14v
6	OR/YL	VSS (-) Signal	0 Hz, 55 mph: 125 Hz
7	LG/YL	ECT Sensor Signal	0.5-0.6v
8	PK/BK	Fuel Pump Monitor	On: 12v, Off: 0v
10	BK/YL	A/C Cycling Clutch Switch	A/C On: 12v, Off: 0v
11-12, 14-15	---	Not Used	---
13	TN/OR	Low Speed Fan Control	On: 1v, Off: 12v
16	BK	Ignition System Ground	<0.050v
17	TN/RD	STI Output, MIL Control	MIL On: 1v, Off: 12v
20	BK	PCM Case Ground	<0.050v
21	BR/WT	IAC Motor Control	8-10v
22	DG	Fuel Pump Control	On: 1v, Off: 12v
24	YL/LG	PSP Switch Signal	Straight 0v, Turned 11v
25	LG/PK	ACT Sensor Signal	1.5-2.5v
26	BR/WT	Reference Voltage	4.9-5.1v
26	OR/WT	Reference Voltage	4.9-5.1v
27	BR/LG	PFE EGR Sensor	3.2v, 55 mph: 2.8v
28, 32, 34-35, 41	---	Not Used	---
29	DG/PK	HO2S-11 (B1 S1) Signal	0.1-1.1v
30	WT/PK	Neutral Drive Switch Signal	In 'N': 0v, Others: 12v
31	GY/YL	EVAP Purge Solenoid	12v, at 55 mph: 1v
33	YL	EGR VR Solenoid	0%, 55 mph: 45%
36	YL/LG	Spark Output Signal	50% duty cycle
37, 57	RD	Vehicle Power	12-14v
38-39, 41-42	---	Not Used	---
40	BK/LG	Power Ground	<0.1v
43	PK	A/C Demand Signal	On: 1v, Off: 12v
43	LG/PK	A/C Demand Signal	On: 1v, Off: 12v
45	LG/BK	MAP Sensor Signal	107 Hz
46	BK/GY	Analog Signal Return	<0.050v
47	DG/LG	TP Sensor Signal	0.5-1.2v
48	WT/RD	Self-Test Indicator Signal	STI Open: 5v, Closed: 1v
49	OR	HO2S-11 (Bank 1) Ground	<0.050v
51	WT/RD	Air Management Solenoid	AM1 On: 1v, Off: 12v
52	YL	EGR VR Solenoid	0%, 55 mph: 45%
53	TN/WT	M/T: Shift Indicator Light	SIL On: 1v, Off: 12v
54	OR/BL	A/C WOT Relay Control	On: 1v, Off: 12v
56	DB	PIP Sensor Signal	50% duty cycle
58	TN/RD	Injector Bank 1 (INJ 1 & 4)	3.6-5.0 ms
59	TN/OR	Injector Bank 2 (INJ 2 & 3)	3.6-5.0 ms
60	BK/LG	Power Ground	<0.1v

Pin Connector Graphic

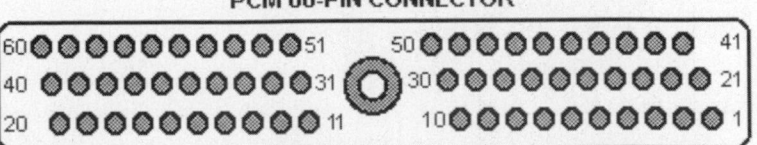

PCM 60-PIN CONNECTOR

Terminal View of 60-Pin PCM Harness Connector

1991 Tempo 2.3L I4 MFI VIN S, X (All) 60 Pin Connector

PCM Pin #	Wire Color	Circuit Description (60 Pin)	Value at Hot Idle
1	YL	Keep Alive Power	12-14v
2-5, 9	---	Not Used	---
3	DG/WT	VSS+ Signal	0 Hz, 55 mph: 125 Hz
4	TN/YL	Ignition Diagnostic Monitor	20-31 Hz
6	OR/YL	VSS- Signal	0 Hz, 55 mph: 125 Hz
7	LG/RD	ECT Sensor Signal	0.5-0.6v
8	PK/BK	Fuel Pump Monitor	On: 12v, Off: 0v
10	DG	A/C Cycling Clutch Switch	A/C On: 12v, Off: 0v
10	BK/YL	A/C Cycling Clutch Switch	A/C On: 12v, Off: 0v
11-15, 18-19	---	Not Used	---
16	OR/RD	Ignition System Ground	<0.1v
17	TN/RD	STI & MIL Control	MIL On: 1v, Off: 12v
20	BK	PCM Case Ground	<0.1v
21	BR/WT	IAC Motor Control	8-10v
22	DG/YL	Fuel Pump Control	On: 1v, Off: 12v
23, 28	---	Not Used	---
24	YL/LG	PSP Switch Signal	Straight 0v, Turned 11v
25	LG/PK	ACT Sensor Signal	1.5-2.5v
26	BR/WT	Reference Voltage	4.9-5.1v
27	BR/LG	PFE EGR Sensor Signal	3.2v, 55 mph: 2.8v
29	RD/BK	HO2S-11 (B1 S1) Signal	0.1-1.1v
30	WT/PK	A/T: Neutral Drive Switch	In 'N': 0v, Others: 12v
30	WT/PK	M/T: Clutch Pedal Position	Clutch In: 0v, Out: 12v
31	GY/YL	EVAP Purge Solenoid	12v, at 55 mph: 1v
32, 34-35	---	Not Used	---
33	YL	EGR VR Solenoid	0%, 55 mph: 45%
36	YL/LG	Spark Output Signal	50% duty cycle
37	RD	Vehicle Power	12-14v
38-39, 41-42	---	Not Used	---
40	BK/LG	Ground	<0.1v
43	PK	A/C Demand Signal	A/C Switch on: 12, Off: 0v
44, 50, 52, 55	---	Not Used	---
45	LG/BK	MAP Sensor Signal	107 Hz
46	GY/RD	Analog Signal Return	<0.1v
47	GN/WT	TP Sensor Signal	0.5-1.2v
48	WT/PK	Self-Test Indicator Signal	STI Open: 5v, Closed: 1v
49	OR	HO2S-11 (Bank 1) Ground	<0.1v
51	WT/RD	Air Management Solenoid	AM1 On: 1v, Off: 12v
53	TN/WT	M/T: Shift Indicator Light	SIL On: 1v, Off: 12v
54	OR/BL	A/C WOT Relay Control	On: 1v, Off: 12v
56	DB	PIP Signal	50% duty cycle
57	RD	Vehicle Power	12-14v
58	TN/RD	Injector Bank 1 (INJ 1 & 4)	3.6-5.0 ms
59	TN/OR	Injector Bank 2 (INJ 2 & 3)	3.6-5.0 ms
60	BK/LG	Power Ground	<0.1v

Pin Connector Graphic

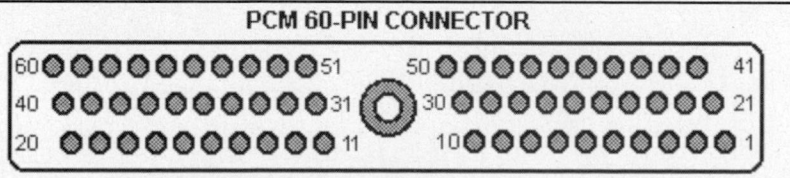

PCM 60-PIN CONNECTOR

Terminal View of 60-Pin PCM Harness Connector

1992-94 Tempo 2.3L I4 MFI VIN X (All) Pin Connector

PCM Pin #	Wire Color	Circuit Description (60 Pin)	Value at Hot Idle
1	YL	Keep Alive Power	12-14v
2	---	Not Used	---
3	DG/WT	VSS+ Signal	0 Hz, 55 mph: 125 Hz
4	WT/PK	Ignition Diagnostic Monitor	20-31 Hz
5	---	Not Used	---
6	OR/YL	VSS- Signal	0 Hz, 55 mph: 125 Hz
7	LG/RD	ECT Sensor Signal	0.5-0.6v
8	PK/BK	Fuel Pump Monitor	On: 12v, Off: 0v
9	TN/BL	MAF Sensor Return	<0.1v
10	PK/BL	A/C Cycling Clutch Switch	A/C On: 12v, Off: 0v
11	GN/YL	EVAP Purge Solenoid	12v, at 55 mph: 1v
12	---	Not Used	---
13	TN/OR	Low Speed Fan Control	On: 1v, Off: 12v
14	---	Not Used	---
15	---	Not Used	---
16	OR/RD	Ignition System Ground	<0.1v
17	TN/RD	STI & MIL Control	MIL On: 1v, Off: 12v
18	TN/OR	Data Bus (+) Signal	5v
19	PK/BL	Data Bus (-) Signal	Digital Signals
20	BK	PCM Case Ground	<0.1v
21	W	IAC Motor Control	8-10v
21	WT/BL	IAC Motor Control	8-10v
22	DG/YL	Fuel Pump Control	On: 1v, Off: 12v
23	---	Not Used	---
24	DG	CID Sensor Signal	5-7 Hz
25	GY	IAT Sensor Signal	1.5-2.5v
26	BR/WT	Reference Voltage	4.9-5.1v
27	BR/LG	EGR PFE Sensor Signal	3.2v, 55 mph: 2.8v
28	YL/LG	PSP Switch Signal	Straight: 0v, Turned: 11v
29	---	Not Used	---
30	PK/YL	A/T: Neutral Drive Switch	In 'N': 0v, Others: 5v
30	PK/YL	M/T: Clutch Engage Switch	Clutch In: 0v, Out: 5v
31	WT/RD	Air Management (1992)	12v, 55 mph: 1v
31	WT/RD	Air Bypass Solenoid (1993-94)	12v, 55 mph: 1v
32	---	Not Used	---
33	YL	EGR VR Solenoid	0%, 55 mph: 45%
34	---	Not Used	---
35	BR/BL	Injector 4 Control	5.1-5.3 ms
36	PK	Spark Output Signal	50% duty cycle
37	RD	Vehicle Power	12-14v
38	---	Not Used	---
39	BR/YL	Injector 3 Control	5.1-5.3 ms
40	BK/LG	Power Ground	<0.1v

1992-94 Tempo 2.3L I4 MFI VIN X (All) Pin Connector

PCM Pin #	Wire Color	Circuit Description (60 Pin)	Value at Hot Idle
41-43	---	Not Used	---
44	RD/BK	HO2S-11 (B1 S1) Signal	0.1-1.1v
45	---	Not Used	---
46	GY/RD	Analog Signal Return	<0.1v
47	GN/WT	TP Sensor (Volts)	0.5-1.2v
48	WT/PK	Self-Test Indicator Signal	STI Open: 5v, Closed: 1v
49	---	Not Used	---
50	LB/RD	MAF Sensor Signal	0.8v, 55 mph: 2.1v
50	BL/BK	MAF Sensor Signal	0.8v, 55 mph: 2.1v
51-52	---	Not Used	---
53	TN/WT	M/T: Shift Indicator Light	SIL On: 1v, Off: 12v
54	PK/YL	A/C WOT Relay Control	On: 1v, Off: 12v
55	---	Not Used	---
56	DB	PIP Signal	50% duty cycle
57	RD	Vehicle Power	12-14v
58	TN	Injector 1 Control	5.1-5.3 ms
59	WT	Injector 2 Signal	5.1-5.3 ms
60	BK/LG	Power Ground	<0.1v

Pin Connector Graphic

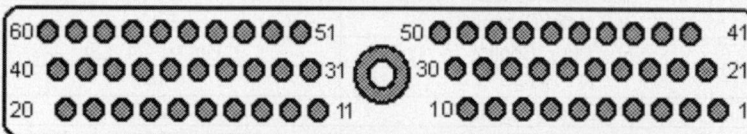

PCM 60-PIN CONNECTOR

Terminal View of 60-Pin PCM Harness Connector

Standard Colors and Abbreviations

Abbreviation	Color	Abbreviation	Color	Abbreviation	Color
BK	Black	GY	Gray	PK	Purple
BL	Blue	GN	Green	RD	Red
BR	Brown	LG	LT Green	TN	Tan
DB	Dark Blue	OR	Orange	WT	White
DG	DK Green	PK	Pink	YL	Yellow

1992-94 Tempo 3.0L V6 MFI VIN U (All) 60 Pin Connector

PCM Pin #	Wire Color	Circuit Description (60 Pin)	Value at Hot Idle
1	YL	Keep Alive Power	12-14v
2	---	Not Used	---
3	DG/WT	VSS (+) Signal	0 Hz, 55 mph: 125 Hz
4	WT/PK	Ignition Diagnostic Monitor	20-31 Hz
5	---	Not Used	---
6	OR/YL	VSS (-) Signal	0 Hz, 55 mph: 125 Hz
7	LG/RD	ECT Sensor Signal	0.5-0.6v
8	PK/BK	Fuel Pump Monitor	On: 12v, Off: 0v
9	TN/BL	MAF Sensor Return	<0.050v
10	PK/BL	A/C Cycling Clutch Switch	A/C On: 12v, Off: 0v
11	GN/YL	EVAP Purge Solenoid	12v, at 55 mph: 1v
12	LG/OR	Injector 6 Control	4.6-5.3 ms
13	TN/OR	Low Speed Fan Control	On: 1v, Off: 12v
14	---	Not Used	---
15	TN/BK	Injector 5 Control	4.6-5.3 ms
16	OR/RD	Ignition System Ground	<0.050v
17	TN/RD	STI Output, MIL Control	MIL On: 1v, Off: 12v
18	TN/OR	Data Bus (+) Signal	Digital Signals
19	PK/BL	Data Bus (-) Signal	Digital Signals
20	BK	PCM Case Ground	<0.050v
21	WT	IAC Motor Control	8-10v
21	WT/BL	IAC Motor Control	8-10v
22	DG/YL	Fuel Pump Control	On: 1v, Off: 12v
23-24	---	Not Used	---
25	GY	IAT Sensor Signal	1.5-2.5v
26	BR/WT	Reference Voltage	4.9-5.1v
27	BR/LG	PFE EGR Sensor	3.2v, 55 mph: 2.9v
28	YL/LG	PSP Switch Signal	Straight: 0v, Turned: 11v
29	---	Not Used	---
30	PK/YL	A/T: Neutral Drive Switch	In 'N': 0v, Others: 5v
30	PK/YL	M/T: Clutch Position Switch	Clutch In: 0v, Out: 5v
31	BL	High Speed Fan Control	On: 1v, Off: 12v
32	---	Not Used	---
33	YL	EGR VR Solenoid	0%, 55 mph: 45%
34	---	Not Used	---
35	BR/BL	Injector 4 Control	4.6-5.3 ms
36	PK	Spark Output Signal	50% duty cycle
37	RD	Vehicle Power	12-14v
38	---	Not Used	---
39	BR/YL	Injector 3 Control	4.6-5.3 ms
40	BK/LG	Power Ground	<0.1v

1992-94 Tempo 3.0L V6 MFI VIN U (All) 60 Pin Connector

PCM Pin #	Wire Color	Circuit Description (60 Pin)	Value at Hot Idle
41	---	Not Used	---
42 ('94)	TN/LG	A/C Cycling Clutch Switch	A/C On: 12v, Off: 0v
43	RD/BK	HO2S-11 (B1 S1) Signal	0.1-1.1v
44	GY/BL	HO2S-21 (B2 S1) Signal	0.1-1.1v
45	---	Not Used	---
46	GY/RD	Analog Signal Return	<0.050v
47	GN/WT	TP Sensor Signal	0.7v, 55 mph: 1.3v
48	WT/PK	Self-Test Indicator Signal	STI Open: 5v, Closed: 1v
49	---	Not Used	---
50	BL/RD	MAF Sensor Signal	0.7v, 55 mph: 2.1v
50	BL/BK	MAF Sensor Signal	0.7v, 55 mph: 2.1v
51-53	---	Not Used	---
54	PK/YL	A/C WOT Relay Control	On: 1v, Off: 12v
55	---	Not Used	---
56	DB	PIP Sensor Signal	50% duty cycle
57	RD	Vehicle Power	12-14v
58	TN	Injector 1 Control	4.6-5.3 ms
59	WT	Injector 2 Control	4.6-5.3 ms
60	BK/LG	Power Ground	<0.1v

Pin Connector Graphic

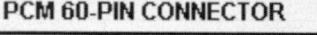
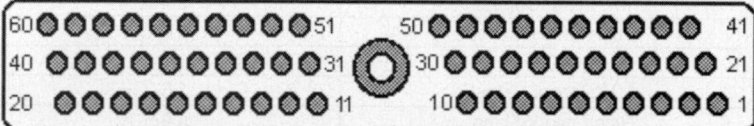

Terminal View of 60-Pin PCM Harness Connector

Standard Colors and Abbreviations

Abbreviation	Color	Abbreviation	Color	Abbreviation	Color
BK	Black	GY	Gray	PK	Purple
BL	Blue	GN	Green	RD	Red
BR	Brown	LG	LT Green	TN	Tan
DB	Dark Blue	OR	Orange	WT	White
DG	DK Green	PK	Pink	YL	Yellow

THUNDERBIRD TABLES

1990 Thunderbird 3.8L V6 MFI VIN 4 (A/T) 60 Pin Connector

PCM Pin #	Wire Color	Circuit Description (60 Pin)	Value at Hot Idle
1	YL	Keep Alive Power	12-14v
2	---	Not Used	---
3	DG/WT	VSS (+) Signal	0 Hz, 55 mph: 125 Hz
4	DG/YL	Ignition Diagnostic Monitor	20-31 Hz
5	LG	Brake Pedal Position Switch	Brake Off: 0v, On: 12v
6	BK/WT	VSS (-) Signal	0 Hz, 55 mph: 125 Hz
7	LG/YL	ECT Sensor Signal	0.5-0.6v
8	OR/BK	Data Bus (-) Signal	Digital Signals
8	BK/OR	Data Bus (-) Signal	Digital Signals
9	---	Not Used	---
10	PK/BL	A/C Cycling Clutch Switch	A/C On: 12v, Off: 0v
11	OR/YL	Speed Control Solenoid	Solenoid On: 12-14v
12	BR/YL	Injector 3 Control	6.0-6.2 ms
13	BR/BL	Injector 4 Control	6.0-6.2 ms
14	TN/BL	Injector 5 Control	6.0-6.2 ms
15	LG	Injector 6 Control	6.0-6.2 ms
16	DB, GY	Ignition System Ground	<0.050v
17	YL/BK	STI Output, MIL Control	MIL On: 1v, Off: 12v
18-19	--	Not Used	---
20	BK	PCM Case Ground	<0.050v
21	RD/LG	IAC Motor Control	20-40%
22	TN/LG	Fuel Pump Control	On: 1v, Off: 12v
23-24	---	Not Used	---
25	LG/PK	ACT Sensor Signal	1.5-2.5v
26	OR/WT	Reference Voltage	4.9-5.1v
27	BR/LG	PFE EGR Sensor	3.2v, 55 mph: 2.8v
28	BL/BK	S/C Command Switch	Switch On: 6.7v
29	TN/OR	HO2S-21 (B2 S1) Signal	0.1-1.1v
30	RD/BL	Neutral Drive Switch Signal	In 'N': 0v, Others: 4-5v
31	GY/YL	EVAP Purge Solenoid	12v, at 55 mph: 1v
32	---	Not Used	---
33	DG	EGR VR Solenoid	0%, 55 mph: 45%
34	BL/PK	Data Output Link	Digital Signals
35	WT/PK	Speed Control Vent Solenoid	98% (on at 55 mph)
36	YL/LG	Spark Output Signal	50% duty cycle
37	RD	Vehicle Power	12-14v
38	---	Not Used	---
39	WT/BL	Speed Control Switch Ground	<0.050v
40	BK/LG	Power Ground	<0.1v

1990 Thunderbird 3.8L V6 MFI VIN 4 (A/T) 60 Pin Connector

PCM Pin #	Wire Color	Circuit Description (60 Pin)	Value at Hot Idle
41-42	---	Not Used	---
43	TN/RD	HO2S-11 (B1 S1) Signal	0.1-1.1v
44	BK/OR	Data Bus (+) Signal	Digital Signals
44	OR/BK	Data Bus (+) Signal	Digital Signals
45	DB/LG	MAP Sensor Signal	107 Hz
46	BK/WT	Analog Signal Return	<0.050v
47	DG/LG	TP Sensor Signal	0.8v, 55 mph: 1.1v
48	WT/RD	Self-Test Indicator Signal	STI Open: 5v, Closed: 1v
49	OR	HO2S-11 (Bank 1) Ground	<0.050v
50	PK/BK	Fuel Pump Monitor	On: 12v, Off: 0v
51	GY/BK	S/C Vacuum Solenoid	50-90% (On at 55 mph)
52-53	---	Not Used	---
54	OR/BL	A/C WOT Relay Control	On: 1v, Off: 12v
55	---	Not Used	---
56	DB	PIP Sensor Signal	50% duty cycle
57	RD	Vehicle Power	12-14v
58	TN	Injector 1 Control	6.0-6.2 ms
59	WT	Injector 2 Control	6.0-6.2 ms
60	BK/LG	Power Ground	<0.1v

Pin Connector Graphic

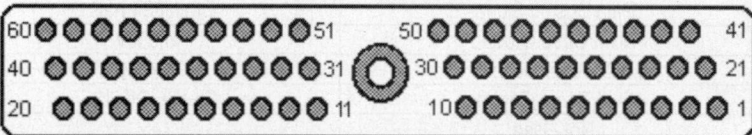

PCM 60-PIN CONNECTOR

Terminal View of 60-Pin PCM Harness Connector

Standard Colors and Abbreviations

Abbreviation	Color	Abbreviation	Color	Abbreviation	Color
BK	Black	GY	Gray	PK	Purple
BL	Blue	GN	Green	RD	Red
BR	Brown	LG	LT Green	TN	Tan
DB	Dark Blue	OR	Orange	WT	White
DG	DK Green	PK	Pink	YL	Yellow

1991-93 Thunderbird 3.8L V6 MFI VIN 4 (A/T) 60 Pin Connector

PCM Pin #	Wire Color	Circuit Description (60 Pin)	Value at Hot Idle
1	YL	Keep Alive Power	12-14v
2	---	Not Used	---
3	DG/WT	VSS (+) Signal	0 Hz, 55 mph: 125 Hz
4	RD/BL	Ignition Diagnostic Monitor	20-31 Hz
4	WT/PK	Ignition Diagnostic Monitor	20-31 Hz
5	---	Not Used	---
6	BK/WT	VSS (-) Signal	0 Hz, 55 mph: 125 Hz
6	GY/RD	VSS (-) Signal	0 Hz, 55 mph: 125 Hz
7	LG/YL	ECT Sensor Signal	0.5-0.6v
7	LG/RD	ECT Sensor Signal	0.5-0.6v
8	PK/BL	Fuel Pump Monitor	On: 12v, Off: 0v
9	TN/BL	MAF Sensor Return	<0.050v
10	PK/BK	A/C Cycling Clutch Switch	A/C On: 12v, Off: 0v
11	GY/YL	EVAP Purge Solenoid	12v, 55 mph: 1v
12	LG	Injector 6 Control	5.6-6.3 ms
12	LG/OR	Injector 6 Control	5.6-6.3 ms
13-14	---	Not Used	---
15	TN/BL	Injector 5 Control	5.6-6.3 ms
15	TN/BK	Injector 5 Control	5.6-6.3 ms
16	GY	Ignition System Ground	<0.050v
17	YL/BK	STI Output, MIL Control	MIL On: 1v, Off: 12v
18	TN/OR	Data Bus (+) Signal	Digital Signals
19	PK/BL	Data Bus (-) Signal	Digital Signals
20	BK	PCM Case Ground	<0.050v
21	RD/LG	IAC Motor Control	20-40%
22	TN/LG	Fuel Pump Control	On: 1v, Off: 12v
23-24	---	Not Used	---
25	LG/PK	IAT Sensor Signal	1.5-2.5v
25	GY	IAT Sensor Signal	1.5-2.5v
26	OR/WT	Reference Voltage	4.9-5.1v
26	BR/WT	Reference Voltage	4.9-5.1v
27	BR/LG	PFE EGR Sensor	3.2v, 55 mph: 2.9v
28-29	---	Not Used	---
30	RD/BL	Neutral Drive Switch Signal	In 'N': 0v, Others: 5v
31-32	---	Not Used	---
33	DG	EGR VR Solenoid	0%, 55 mph: 45%
33	BR/PK	EGR VR Solenoid	0%, 55 mph: 45%
34	BL/PK	Data Output Link	Digital Signals
35	BR/BL	Injector 4 Control	5.6-6.3 ms
36	YL/LG	Spark Output Signal	6.93v (50% duty cycle)
37	RD	Vehicle Power	12-14v
38	---	Not Used	---
39	BR/YL	Injector 3 Control	5.6-6.3 ms
40	BK/LG	Power Ground	<0.1v

1991-93 Thunderbird 3.8L V6 MFI VIN 4 (A/T) 60 Pin Connector

PCM Pin #	Wire Color	Circuit Description (60 Pin)	Value at Hot Idle
41-42	---	Not Used	---
43	TN/RD	HO2S-21 (B2 S1) Signal	0.1-1.1v
43	TN/WT	HO2S-21 (B2 S1) Signal	0.1-1.1v
44	TN/OR	HO2S-11 (B1 S1) Signal	0.1-1.1v
45	---	Not Used	---
46	BK/WT	Analog Signal Return	<0.050v
46	GY/RD	Analog Signal Return	<0.050v
47	DG/LG	TP Sensor Signal	0.5-1.2v
47	GY/WT	TP Sensor Signal	0.5-1.2v
48	WT/RD	Self-Test Indicator Signal	STI Open: 5v, Closed: 1v
48	WT/PK	Self-Test Indicator Signal	STI Open: 5v, Closed: 1v
49	---	Not Used	---
50	DB or BL	MAF Sensor Signal	0.7v, 55 mph: 1.3v
51-53	---	Not Used	---
54	OR/BL	A/C WOT Relay Control	On: 1v, Off: 12v
55	---	Not Used	---
56	DB	PIP Sensor Signal	50% duty cycle
57	RD	Vehicle Power	12-14v
58	TN	Injector 1 Control	5.6-6.3 ms
59	WT	Injector 2 Control	5.6-6.3 ms
60	BK/LG	Power Ground	<0.1v

Pin Connector Graphic

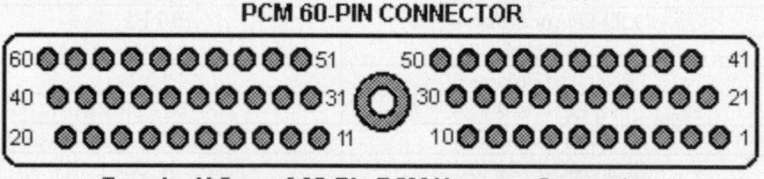

Standard Colors and Abbreviations

Abbreviation	Color	Abbreviation	Color	Abbreviation	Color
BK	Black	GY	Gray	PK	Purple
BL	Blue	GN	Green	RD	Red
BR	Brown	LG	LT Green	TN	Tan
DB	Dark Blue	OR	Orange	WT	White
DG	DK Green	PK	Pink	YL	Yellow

1994-95 Thunderbird 3.8L V6 MFI VIN 4 (A/T) 60 Pin Connector

PCM Pin #	Wire Color	Circuit Description (60 Pin)	Value at Hot Idle
1	YL	Keep Alive Power	12-14v
2	LG	Brake Pedal Position Switch	Brake Off: 0v, On: 12v
3	GY/BK	VSS (+) Signal	0 Hz, 55 mph: 125 Hz
4	OR/WT	Ignition Diagnostic Monitor	20-31 Hz
5	DG/WT	TSS Sensor Signal	At 55 mph: 205-220 Hz
6	PK/OR	VSS (-) Signal	0 Hz, 55 mph: 125 Hz
7	LG/RD	ECT Sensor Signal	0.5-0.6v
8	PK/BK	Fuel Pump Monitor	On: 12v, Off: 0v
9	TN/BL	MAF Sensor Return	<0.050v
10	PK/BL	A/C Cycling Clutch Switch	A/C On: 12v, Off: 0v
11	GY/YL	EVAP Purge Solenoid	12v, 55 mph: 1v
12	LG/OR	Injector 6 Control	5.6-6.3 ms
13	TN/OR	Low Speed Fan Control	On: 1v, Off: 12v
14	LG/PK	High Speed Fan Control	On: 1v, Off: 12v
15	TN/BK	Injector 5 Control	5.6-6.3 ms
16	OR/RD	Ignition System Ground	<0.050v
17	PK/LG	STO & MIL Control	MIL On: 1v, Off: 12v
18	TN/OR	Data Bus (+) Signal	Digital Signals
19	PK/BL	Data Bus (-) Signal	Digital Signals
20	BK	PCM Case Ground	<0.050v
21	WT/BL	IAC Motor Control	9.5-11v
22	BL/OR	Fuel Pump Control	On: 1v, Off: 12v
23	---	Not Used	---
24	DB/OR	CID Sensor Signal	5-7 Hz
25	GY	ATS Signal	1.5-2.5v
26	BR/WT	Reference Voltage	4.9-5.1v
27	BR/LG	PFE EGR Sensor	3.2v, 55 mph: 2.9v
28	---	Not Used	---
29	DG	Octane Adjust Switch	Closed: 0v, Open: 9.1v
30	BL/YL	MLP Sensor Signal	In P/N: 0v, in O/D: 2.1v
31	---	Not Used	---
32	---	Not Used	---
33	BR/PK	EGR VR Solenoid	0%, 55 mph: 45%
34	---	Not Used	---
35	BR/BL	Injector 4 Control	5.6-6.3 ms
36	PK	Spark Output Signal	50% duty cycle
37	RD	Vehicle Power	12-14v
38	YL/WT	EPC Solenoid Control	8.8v (20 psi)
39	BR/YL	Injector 3 Control	5.6-6.3 ms
40	BK/WT	Power Ground	<0.1v

1994-95 Thunderbird 3.8L V6 MFI VIN 4 (A/T) 60 Pin Connector

PCM Pin #	Wire Color	Circuit Description (60 Pin)	Value at Hot Idle
41	TN/WT	TCS (switch) Signal	TCS & O/D On: 12v
42	TN/LG	A/C High Pressure Switch	AC On: 12v, Off: 0v
43	RD/BK	HO2S-21 (B2 S1) Signal	0.1-1.1v
44	GY/BL	HO2S-11 (B1 S1) Signal	0.1-1.1v
45	---	Not Used	---
46	GY/RD	Analog Signal Return	<0.050v
47	GY/WT	TP Sensor Signal	0.5-1.2v
48	WT/PK	Self-Test Indicator Signal	STI Open: 5v, Closed: 1v
49	OR/BK	TOT Sensor Signal	2.10-2.40v
50	BL/RD	MAF Sensor Signal	0.6v, 55 mph: 1.3v
51	OR/YL	Shift Solenoid 1 Control	1v, 55 mph: 12v
52	PK/OR	Shift Solenoid 2 Control	12v, 55 mph: 1v
53	BK/OR	TCC Solenoid Control	0%, 55 mph: 45%
54	PK/YL	A/C WOT Relay Control	On: 1v, Off: 12v
55	WT/LG	TCIL (lamp) Control	On: 1v, Off: 12v
56	GY/OR	PIP Sensor Signal	50% duty cycle
56	PK	PIP Sensor Signal	50% duty cycle
57	RD	Vehicle Power	12-14v
58	TN	Injector 1 Control	5.6-6.3 ms
59	WT	Injector 2 Control	5.6-6.3 ms
60	BK/LG	Power Ground	<0.1v

Pin Connector Graphic

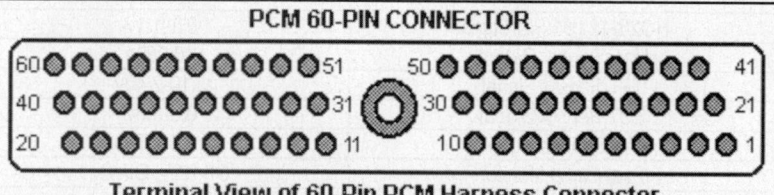

PCM 60-PIN CONNECTOR

Terminal View of 60-Pin PCM Harness Connector

Standard Colors and Abbreviations

Abbreviation	Color	Abbreviation	Color	Abbreviation	Color
BK	Black	GY	Gray	PK	Purple
BL	Blue	GN	Green	RD	Red
BR	Brown	LG	LT Green	TN	Tan
DB	Dark Blue	OR	Orange	WT	White
DG	DK Green	PK	Pink	YL	Yellow

1996-97 Thunderbird 3.8L V6 MFI VIN 4 (A/T) 104 Pin Connector

PCM Pin #	Wire Color	Circuit Description (104 Pin)	Value at Hot Idle
1	PK/OR	Shift Solenoid 2 Control	12v, 55 mph: 1v
2	PK/LG	MIL (lamp) Control	MIL On: 1v, Off: 12v
3-10	---	Not Used	---
11	PK/WT	Purge Flow Sensor	0.8v, 55 mph: 3v
12	---	Not Used	---
13	PK	Flash EEPROM Power	0.1v
14	---	Not Used	---
15	PK/LB	Data Bus (-) Signal	Digital Signals
16	TN/OR	Data Bus (+) Signal	Digital Signals
17-20	---	Not Used	---
21	DB	CKP Sensor (+) Signal	390-450 Hz
22	GY	CKP Sensor (-) Signal	390-450 Hz
23	---	Not Used	---
24	BK/WT	Power Ground	<0.1v
25	BK	PCM Case Ground	<0.050v
26	DB/LG	Coil Driver 1 Control	8°, 55 mph: 12° dwell
27	OR/YL	Shift Solenoid 1 Control	1v, 55 mph: 12v
28	---	Not Used	---
29	TN/WT	TCS (switch) Signal	TCS & O/D On: 12v
30	DG	Octane Adjust Switch	Closed: 0v, Open: 9.3v
31-32	---	Not Used	---
33	PK/OR	VSS (-) Signal	0 Hz, 55 mph: 125 Hz
34	---	Not Used	---
35	RD/LG	HO2S-12 (B1 S2) Signal	0.1-1.1v
36	TN/LB	MAF Sensor Return	<0.050v
37	OR/BK	TFT Sensor Signal	2.10-2.40v
38	LG/RD	ECT Sensor Signal	0.5-0.6v
39	GY	IAT Sensor Signal	1.5-2.5v
40	PK/BK	Fuel Pump Monitor	On: 12v, Off: 0v
41	PK/LB	A/C Cycling Clutch Switch	AC On: 12v, Off: 0v
42-44	---	Not Used	---
45	TN/OR	Low Speed Fan Control	On: 1v, Off: 12v
46	LG/PK	High Speed Fan Control	On: 1v, Off: 12v
47	BR/PK	EGR VR Solenoid	0%, 55 mph: 45%
48-50	---	Not Used	---
51	BK/WT	Power Ground	<0.1v
52	RD/LB	Coil Driver 2 Control	8°, 55 mph: 12° dwell
53	---	Not Used	---
54	BR/OR	TCC Solenoid Control	0%, 55 mph: 95%
55	YL	Keep Alive Power	12-14v
56-57	---	Not Used	---
58	GY/BK	VSS (+) Signal	0 Hz, 55 mph: 125 Hz
59	---	Not Used	---
60	GY/LB	HO2S-11 (B1 S1) Signal	0.1-1.1v
61	PK/LG	HO2S-22 (B2 S2) Signal	0.1-1.1v
62	---	Not Used	---
63	---	Not Used	---
64	LB/YL	TR Sensor Signal	In 'P': 0v, 55 mph: 1.7v

1996-97 Thunderbird 3.8L V6 MFI VIN 4 (A/T) 104 Pin Connector

PCM Pin #	Wire Color	Circuit Description (104 Pin)	Value at Hot Idle
65	BR/LG	DPFE Sensor Signal	0.20-1.30v
66, 68	---	Not Used	---
67	GY/YL	EVAP Purge Solenoid	0-10 Hz (0-100%)
69	PK/YL	A/C WOT Relay Control	On: 1v, Off: 12v
70, 72	---	Not Used	---
71	RD	Vehicle Power	12-14v
73	TN/BK	Injector 5 Control	4.5-4.8 ms
74	BR/YL	Injector 3 Control	4.5-4.8 ms
75	TN	Injector 1 Control	4.5-4.8 ms
76	BK/WT	Power Ground	<0.1v
77	BK/WT	Power Ground	<0.1v
78	PK/WT	Coil Driver 3 Control	8°, 55 mph: 12° dwell
79	WT/LG	TCIL (lamp) Control	On: 1v, Off: 12v
80	LB/OR	Fuel Pump Control	On: 1v, Off: 12v
81	WT/YL	EPC Solenoid Control	8.8v (20 psi)
82	---	Not Used	---
83	WT/LB	IAC Motor Control	10v (30% on time)
84	DG/WT	TSS Sensor Signal	43 Hz (650 rpm)
85	DG	CID Sensor Signal	5-7 Hz
86	TN/LG	A/C Pressure Switch	On: 1v, Off: 12v
87	RD/BK	HO2S-21 (B2 S1) Signal	0.1-1.1v
88	LB/RD	MAF Sensor Signal	0.6v, 55 mph: 1.9v
89	GY/WT	TP Sensor Signal	0.8v, 55 mph: 1.1v
90	BR/WT	Reference Voltage	4.9-5.1v
91	GY/RD	Analog Signal Return	<0.050v
92	LG	Brake Pedal Position Switch	Brake Off: 0v, On: 12v
93	RD/WT	HO2S-11 (B1 S1) Heater	On: 1v, Off: 12v
94	YL/LB	HO2S-21 (B2 S1) Heater	On: 1v, Off: 12v
95	WT/BK	HO2S-12 (B1 S2) Heater	On: 1v, Off: 12v
96	TN/YL	HO2S-22 (B2 S2) Heater	On: 1v, Off: 12v
97	RD	Vehicle Power	12-14v
98	---	Not Used	---
99	LG/OR	Injector 6 Control	4.5-4.8 ms
100	BR/LB	Injector 4 Control	4.5-4.8 ms
101	WT	Injector 2 Control	4.5-4.8 ms
102	---	Not Used	---
103	BK/WT	Power Ground	<0.1v
104	---	Not Used	---

Pin Connector Graphic

PCM 104-PIN CONNECTOR

Terminal View of 104-Pin PCM Wiring Harness Connector

Standard Colors and Abbreviations

Abbreviation	Color	Abbreviation	Color	Abbreviation	Color
BK	Black	GY	Gray	PK	Purple
BL	Blue	GN	Green	RD	Red
BR	Brown	LG	LT Green	TN	Tan
DB	Dark Blue	OR	Orange	WT	White
DG	DK Green	PK	Pink	YL	Yellow

1990-92 Thunderbird 3.8L V6 SC VIN C (A/T) 60 Pin Connector

PCM Pin #	Wire Color	Circuit Description (60 Pin)	Value at Hot Idle
1	YL	Keep Alive Power	12-14v
2	---	Not Used	---
3	DG/WT	VSS (+) Signal	0 Hz, 55 mph: 125 Hz
4	DG/YL	Ignition Diagnostic Monitor	20-31 Hz
5	LG	Brake Pedal Position Switch	Brake Off: 0v, On: 12v
6	BK/WT	VSS (-) Signal	0 Hz, 55 mph: 125 Hz
7	LG/YL	ECT Sensor Signal	0.5-0.6v
8	---	Not Used	---
9	TN/BL	MAF Sensor Return	<0.050v
10	PK/BL	A/C Cycling Clutch Switch	A/C On: 12v, Off: 0v
11	OR/YL	Speed Control Solenoid (+)	S/C On: 12-14v
12	LG	Injector 3 Control	4.0-4.2 ms
12	LG/OR	Injector 3 Control	4.0-4.2 ms
13	BR/BL	Injector 4 Control	4.0-4.2 ms
14	TN/BL	Injector 5 Control	4.0-4.2 ms
15	LG	Injector 6 Control	4.0-4.2 ms
16	BL	Ignition System Ground	<0.050v
17	YL/BK	STI Output, MIL Control	MIL On: 1v, Off: 12v
18	BK/WT	Octane Adjust Switch	Closed: 0v, Open: 9.1v
19	PK/BK	Fuel Pump Monitor	On: 12v, Off: 0v
20	BK	PCM Case Ground	<0.050v
21	RD/LG	IAC Motor Control	9.9v (40%)
22	TN/LG	Fuel Pump Control	On: 1v, Off: 12v
23	YL/RD	Knock Sensor Signal	0v
24	DG	CID Sensor Signal	5-7 Hz
25	LG/PK	ACT Sensor Signal	1.5-2.5v
26	OR/WT	Reference Voltage	4.9-5.1v
27	BR/LG	PFE EGR Sensor	3.2v, 55 mph: 2.8v
28	BL/BK	S/C Command Switch	Switch On: 6.7v
29	TN/OR	HO2S-21 (B2 S1) Signal	0.1-1.1v
30	RD/BL	Neutral Drive Switch Signal	In 'N': 0v, Others: 5v
31	GY/YL	EVAP Purge Solenoid	12v, 55 mph: 1v
32	OR/WT	Automatic Shock Control	4.3v
33	YL	EGR VR Solenoid	0%, 55 mph: 45%
34	---	Not Used	---
35	WT/PK	Speed Control Vent Solenoid	0%, 55 mph: 98%
36	YL/LG	Spark Output Signal	9.93v (50% duty cycle)
37	RD	Vehicle Power	12-14v
38 ('89)	LG/PK	Supercharger Bypass Solenoid	0%, 55 mph: 45%
39	WT/BL	Speed Control Switch Ground	<0.050v
40	BK/LG	Power Ground	<0.1v

1990-92 Thunderbird 3.8L V6 SC VIN C (A/T) 60 Pin Connector

PCM Pin #	Wire Color	Circuit Description (60 Pin)	Value at Hot Idle
41	PK	High Speed Fan Control	On: 1v, Off: 12v
42	---	Not Used	---
43	TN/RD	HO2S-11 (B1 S1) Signal	0.1-1.1v
44	---	Not Used	---
45	DB/LG	MAP Sensor Signal	107 Hz
46	BK/WT	Analog Signal Return	<0.050v
47	DG/LG	TP Sensor Signal	0.5-1.2v
48	WT/RD	Self-Test Indicator Signal	STI Open: 5v, Closed: 1v
49	OR	HO2S-11 (Bank 1) Ground	<0.050v
50	DB/OR	MAF Sensor Signal	0.7v
51	GY/BK	S/C Vacuum Solenoid	0%, 55 mph: 45%
52	---	Not Used	---
53	PK	Shift Indicator Control	SIL On: 1v, Off: 12v
54	RD	A/C WOT Relay Control	On: 1v, Off: 12v
55	TN/OR	Low Speed Fan Control	On: 1v, Off: 12v
56	DB	PIP Sensor Signal	50% duty cycle
57	RD	Vehicle Power	12-14v
58	TN	Injector 1 Control	4.0-4.2 ms
59	WT	Injector 2 Control	4.0-4.2 ms
60	BK/LG	Power Ground	<0.1v

Pin Connector Graphic

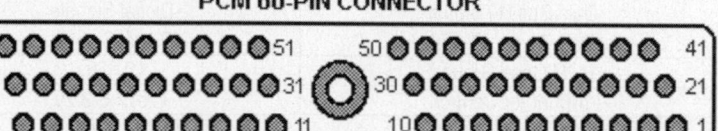

Terminal View of 60-Pin PCM Harness Connector

Standard Colors and Abbreviations

Abbreviation	Color	Abbreviation	Color	Abbreviation	Color
BK	Black	GY	Gray	PK	Purple
BL	Blue	GN	Green	RD	Red
BR	Brown	LG	LT Green	TN	Tan
DB	Dark Blue	OR	Orange	WT	White
DG	DK Green	PK	Pink	YL	Yellow

1990-93 Thunderbird 3.8L V6 SC VIN R (A/T) 60 Pin Connector

PCM Pin #	Wire Color	Circuit Description (60 Pin)	Value at Hot Idle
1	YL	Keep Alive Power	12-14v
2	---	Not Used	---
3	DG/WT	VSS (+) Signal	0 Hz, 55 mph: 125 Hz
4	DG/YL	Ignition Diagnostic Monitor	20-31 Hz
4	TN/YL	Ignition Diagnostic Monitor	20-31 Hz
5	---	Not Used	---
6	BK/WT	VSS (-) Signal	0 Hz, 55 mph: 125 Hz
6	GY/RD	VSS (-) Signal	0 Hz, 55 mph: 125 Hz
7	LG/YL	ECT Sensor Signal	0.5-0.6v
7	LG/RD	ECT Sensor Signal	0.5-0.6v
8 ('91-'93)	PK/BK	Fuel Pump Monitor	On: 12v, Off: 0v
9	TN/BL	MAF Sensor Return	<0.050v
10	PK/BL	A/C Cycling Clutch Switch	A/C On: 12v, Off: 0v
11	GY/YL	EVAP Purge Solenoid	12v, at 55 mph: 1v
12	TN/OR	Injector 6 Control	3.6-4.0 ms
12	LG/OR	Injector 6 Control	3.6-4.0 ms
13	TN/OR	Low Speed Fan Control	Fan On: 12v, Off: 0v
14	---	Not Used	---
15	TN/BL	Injector 5 Control	3.6-4.0 ms
15	TN/BK	Injector 5 Control	3.6-4.0 ms
16	BL, GY	Ignition System Ground	<0.050v
17	YL/BK	STI Output, MIL Control	MIL On: 1v, Off: 12v
18	TN/OR	Data Bus (+) Signal	Digital Signals
19	PK/BL	Data Bus (-) Signal	Digital Signals
20	BK	PCM Case Ground	<0.050v
21	RD/LG	IAC Motor Control	Idle: 7.6-9.2v
22	TN/LG	Fuel Pump Control	On: 1v, Off: 12v
23	YL/RD	Knock Sensor Signal	0v
24	DG/LG	CID Sensor Signal	5-7 Hz
25	LG/PK	IAT Sensor Signal	1.5-2.5v
25	GY	IAT Sensor Signal	1.5-2.5v
26	OR/WT	Reference Voltage	4.9-5.1v
26	BR/WT	Reference Voltage	4.9-5.1v
27-28	---	Not Used	---
29	BK/WT	Octane Adjust Switch	Closed: 0v, Open: 9.1v
29	GY/RD	Octane Adjust Switch	Closed: 0v, Open: 9.1v
30	RD/BL	Neutral Drive Switch Signal	In 'P': 0v, Others: 5v
31	PK	High Speed Fan Control	On: 1v, Off: 12v
31	LG/PK	High Speed Fan Control	On: 1v, Off: 12v
32	OR/WT	Automatic Ride Control	4.3v
33-34	---	Not Used	---
35	BR/BL	Injector 4 Control	3.6-4.0 ms
36	YL/LG	Spark Output Signal	50% duty cycle
37	RD	Vehicle Power	12-14v
38	---	Not Used	---
39	BR/YL	Injector 3 Control	3.6-4.0 ms
40	BK/LG	Power Ground	<0.1v

1990-93 Thunderbird 3.8L V6 SC VIN R (A/T) 60 Pin Connector

PCM Pin #	Wire Color	Circuit Description (60 Pin)	Value at Hot Idle
41-42	---	Not Used	---
43	TN/RD	HO2S-21 (B2 S1) Signal	0.1-1.1v
43	TN/WT	HO2S-21 (B2 S1) Signal	0.1-1.1v
44	TN/OR	HO2S-11 (B1 S1) Signal	0.1-1.1v
45	DB/LG	BARO Pressure Sensor	159 Hz (sea level)
46	BK/WT	Analog Signal Return	<0.050v
46	GY/RD	Analog Signal Return	<0.050v
47	DG/LG	TP Sensor Signal	Hot 0.5-1.2v
47	GY/WT	TP Sensor Signal	Hot 0.5-1.2v
48	WT/RD	Self-Test Indicator Signal	STI Open: 5v, Closed: 1v
48	WT/PK	Self-Test Indicator Signal	STI Open: 5v, Closed: 1v
49	---	Not Used	---
50	DB/OR	MAF Sensor Signal	Hot 0.7v
50	BL/RD	MAF Sensor Signal	Hot 0.7v
51-52	---	Not Used	---
52	---	Not Used	---
53	PK	Shift Indicator Lamp	SIL On: 1v, Off: 12v
54	RD or PK	A/C WOT Relay Control	On: 1v, Off: 12v
55	---	Not Used	---
56	DB	PIP Sensor Signal	50% duty cycle
57	RD	Vehicle Power	12-14v
58	TN	Injector 1 Control	3.6-4.0 ms
59	WT	Injector 2 Control	3.6-4.0 ms
60	BK/LG	Power Ground	<0.1v

Pin Connector Graphic

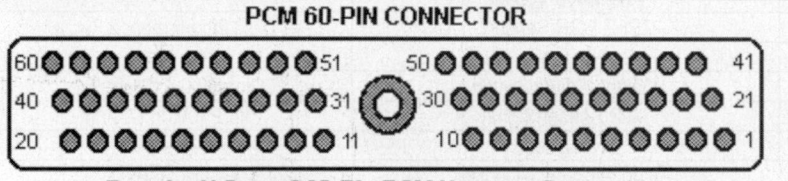

PCM 60-PIN CONNECTOR

Terminal View of 60-Pin PCM Harness Connector

Standard Colors and Abbreviations

Abbreviation	Color	Abbreviation	Color	Abbreviation	Color
BK	Black	GY	Gray	PK	Purple
BL	Blue	GN	Green	RD	Red
BR	Brown	LG	LT Green	TN	Tan
DB	Dark Blue	OR	Orange	WT	White
DG	DK Green	PK	Pink	YL	Yellow

1994-95 Thunderbird 3.8L V6 SC VIN R (A/T) 60 Pin Connector

PCM Pin #	Wire Color	Circuit Description (60 Pin)	Value at Hot Idle
1	YL	Keep Alive Power	12-14v
2	LG	Brake Pedal Position Switch	Brake Off: 0v, On: 12v
3	GY/BK	VSS (+) Signal	0 Hz, 55 mph: 125 Hz
4	OR/WT	Ignition Diagnostic Monitor	20-31 Hz
5	DG/WT	TSS Sensor Signal	215-230 Hz (55 mph)
6	PK/OR	VSS (-) Signal	0 Hz, 55 mph: 125 Hz
7	LG/RD	ECT Sensor Signal	0.5-0.6v
8	PK/BK	Fuel Pump Monitor	On: 12v, Off: 0v
9	TN/BL	MAF Sensor Return	<0.050v
10	PK/BL	A/C Cycling Clutch Switch	A/C On: 12v, Off: 0v
11	GY/YL	EVAP Purge Solenoid	12v, at 55 mph: 1v
12	LG/OR	Injector 6 Control	3.6-3.8 ms
13	TN/OR	Low Speed Fan Control	On: 1v, Off: 12v
14	LG/PK	High Speed Fan Control	On: 1v, Off: 12v
15	TN/BK	Injector 5 Control	3.6-3.8 ms
16	OR/RD	Ignition System Ground	<0.050v
17	PK/LG	STI Output, MIL Control	MIL On: 1v, Off: 12v
18	TN/OR	Data Bus (+) Signal	Digital Signals
19	PK/BL	Data Bus (-) Signal	Digital Signals
20	BK	PCM Case Ground	<0.050v
21	WT/BL	IAC Motor Control	9.2v (40%)
22	BL/OR	Fuel Pump Control	On: 1v, Off: 12v
23	YL/RD	Knock Sensor Signal	0v
24	DB/OR	CMP Sensor Signal	5-7 Hz
25	GY	IAT Sensor Signal	1.5-2.5v
26	BR/WT	Reference Voltage	4.9-5.1v
27	BR/LG	PFE EGR Sensor	3.2v, 55 mph: 2.8v
28	---	Not Used	---
29	DG	Octane Adjust Switch	Closed: 0v, Open: 9.1v
30	BL/YL	MLP Sensor Signal	In P: 0v, In D: 2.1v
31	---	Not Used	---
32	OR/WT	Automatic Shock Control	Hot 4.3v
33	BR/PK	EGR VR Solenoid	0%, 55 mph: 45%
34	OR	Cooling Fan System	Fan On: 1v, Off: 12v
35	BR/BL	Injector 4 Control	3.6-3.8 ms
36	PK	Spark Output Signal	50% duty cycle
37	RD	Vehicle Power	12-14v
38	WT/YL	EPC Solenoid Control	8.8v (20 psi)
39	BR/YL	Injector 3 Control	3.6-3.8 ms
40	BK/WT	Power Ground	<0.1v

1994-95 Thunderbird 3.8L V6 MFI SC VIN R (All) 60 Pin Connector

PCM Pin #	Wire Color	Circuit Description (60 Pin)	Value at Hot Idle
41	TN/WT	TCS (switch) Signal	TCS & O/D On: 12v
42	TN/LG	A/C Pressure Switch	A/C On: 12v, Off: 0v
43	RD/BK	HO2S-21 (B2 S1) Signal	0.1-1.1v
44	GY/BL	HO2S-11 (B1 S1) Signal	0.1-1.1v
45	DB/LG	BARO Sensor	159 Hz (sea level)
46	GY/RD	Analog Signal Return	<0.050v
47	GY/WT	TP Sensor Signal	0.5-1.2v
48	WT/PK	Self-Test Indicator Signal	STI Open: 5v, Closed: 1v
49	OR/BK	TOT Sensor Signal	2.10-2.40v
50	BL/RD	MAF Sensor Signal	0.7v, 55 mph: 2.1v
51	OR/YL	Shift Solenoid 1 Control	1v, 55 mph: 12v
52	PK/OR	Shift Solenoid 2 Control	12v, 55 mph: 1v
53	BK/OR	TCC Solenoid Control	0%, 55 mph: 95%
54	PK/YL	A/C WOT Relay Control	On: 1v, Off: 12v
55	WT/LG	TCIL (lamp) Control	On: 1v, Off: 12v
56	GY/OR	PIP Sensor Signal	50% duty cycle
57	RD	Vehicle Power	12-14v
58	TN	Injector 1 Control	3.6-3.8 ms
59	WT	Injector 2 Control	3.6-3.8 ms
60	BK/WT	Power Ground	<0.1v

Pin Connector Graphic

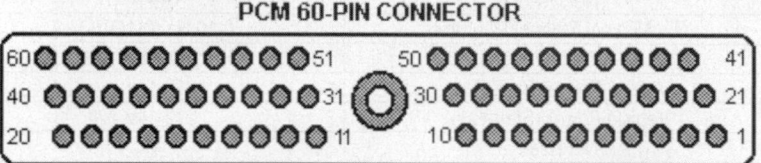

Terminal View of 60-Pin PCM Harness Connector

Standard Colors and Abbreviations

Abbreviation	Color	Abbreviation	Color	Abbreviation	Color
BK	Black	GY	Gray	PK	Purple
BL	Blue	GN	Green	RD	Red
BR	Brown	LG	LT Green	TN	Tan
DB	Dark Blue	OR	Orange	WT	White
DG	DK Green	PK	Pink	YL	Yellow

2002-05 Thunderbird 3.9L V8 VIN A (A/T) C175-B 58-Pin Connector

PCM Pin #	Wire Color	Circuit Description (58-Pin)	Value at Hot Idle
1	WT/RD	Electronic Throttle Control Module	Digital Signals
2	---	Not Used	---
3	GY	Data Bus (+) Signal	Digital Signals
4	BL	Data Bus (-) Signal	Digital Signals
5	BR	Signal Return	<0.050v
6-8	---	Not Used	---
9	BR/BL	A/C WOT Relay Control	On: 1v, Off: 12v
10-11	---	Not Used	---
12	BR/RD	EVAP Purge Solenoid Control	0-10 Hz (0-100%)
13	YL/GN	Flash EEPROM Power	0.1v
14	WT/VT	Manual Transmission Switch	N/A
15	WT	Electronic Throttle Control Module	Digital Signals
16	WT/VT	Electronic Throttle Control Module	Digital Signals
17	BR/WT	Electronic Throttle Control Module	Digital Signals
18-19	---	Not Used	---
20	YL/VT	ETC Reference Voltage	12-14v
21	---	Not Used	---
22	WT/BK	Manual Transmission Switch Signal	Switch Open: 12v, Closed: 1v
23	YL/RD	ETC Reference Voltage	12-14v
24-27	BK	Power Ground	<0.1v
28	OG/BL	Battery Power (Hot at all times)	12-14v
29	---	Not Used	---
30	BR/BL	EVAP Vent Solenoid Control	0-10 Hz (0-100%)
31	WT/BL	MAF Sensor Signal	0.6v, 55 mph: 3.3v
32	GN/YL	Vehicle Power (Start-Run)	12-14v
33	GN/OR	Vehicle Power (Start-Run)	12-14v
34-37	---	Not Used	---
38	BR/BL	MAF Sensor Return	<0.050v
39	---	Not Used	---
40	OR	Traction Control Disable Switch	Brake Off: 0v, On: 12v
41	WT/GN	Transmission Control Switch	TCS & O/D On: 12v
42	WT/VT	A/C Pressure Sensor Signal	0.9v (50 psi)
43	BN/YL	Power Ground	<0.1v
44	OR/YL	Keep Alive Power (CJB fuse)	12-14v
45-46	---	Not Used	---
47	WT/BK	Engine Cooling Fan Motor	Off: 12v, On: 1v
48	WT/BL	Restraint Control Module	Digital Signals
49	WT/VT	Rear Electronic Module	Digital Signals
50	---	Not Used	---
51	WT/GN	Intake Air Temperature Sensor	1.5-2.5v
52	WT/VT	Fuel Tank Pressure Sensor	2.6v (0" H2O - cap off)
53	GN/RD	Controller Area Network (+) Signal	0-7-0v
54	GN/RD	Controller Area Network (-) Signal	0-7-0v
55	YL	Reference Voltage	4.9-5.1v
56	BK/OR	Ground Switch Signal	<0.050v
57	YL/BL	Speed Control Reference Voltage	4.9-5.1v
58	---	Not Used	---

2002-05 Thunderbird 3.9L V8 VIN A (A/T) C175-T 32-Pin Connector

PCM Pin #	Wire Color	Circuit Description (32-Pin)	Value at Hot Idle
1	BR	Shift Solenoid 'A' Control	0.1, 55 mph: 12v
2	BR/RD	Shift Solenoid 'B' Control	12v, 55 mph: 12v
3-6	---	Not Used	---
7	BR/GN	Pressure Control 'A' Solenoid	8.8v
8	BR/WT	Shift Solenoid 'C' Control	12v, 55 mph: 12v
9	WT	DTR Sensor 3A Signal	In 'P': 0v, Others: 12v
10	WT/RD	DTR Sensor 4 Signal	In 'P': 0v, Others: 12v
11	---	Not Used	---
12	BR/WT	Pressure Control 'C' Solenoid	On: 1v, Off: 12v
13	BR/YL	Pressure Control 'B' Solenoid	On: 1v, Off: 12v
14	BR/RD	Sensor Signal Return	<0.050v
15	BR/BL	HO2S-12 (B1 S2) Heater	On: <1v, Off: 12v
16	BR/GN	HO2S-22 (B2 S2) Heater	On: <1v, Off: 12v
17	BR/YL	Shift Solenoid 'D' Control	0.1, 55 mph: 12v
18	WT/BL	DTR Sensor 2 Signal	In 'P': 0v, Others: 12v
19	---	Not Used	---
20	BR/WT	TCC Solenoid Control	12v, 55 mph: 0.1v
21	WT/BK	Intermediate Shaft Speed Sensor	0 Hz, 55 mph: 1320
22	WT/GN	DTR Sensor 1 Signal	In 'P': 0v, Others: 12v
23	WT/RD	TFT Sensor Signal	2.10-2.40v
24-25	---	Not Used	---
26	WT/RD	Output Shaft Speed Sensor	0 Hz, 55 mph: 1070 Hz
27	WT/VT	Turbine Shaft Speed Sensor	340 Hz, 55 mph: 1075
28	WT/BL	HO2S-12 (B1 S2) Signal	0.1-1.1v
29	WT/GN	HO2S-22 (B2 S2) Signal	0.1-1.1v
30-32	---	Not Used	---

PCM Pin Connectors Graphic

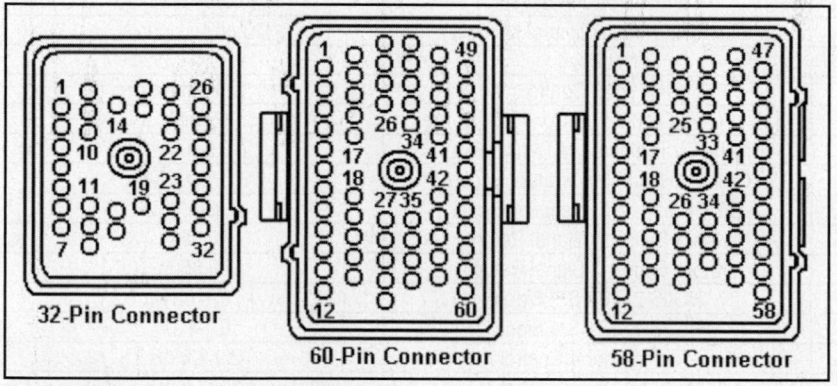

32-Pin Connector

60-Pin Connector

58-Pin Connector

Standard Colors and Abbreviations

Abbreviation	Color	Abbreviation	Color	Abbreviation	Color
BK	Black	GY	Gray	PK	Purple
BL	Blue	GN	Green	RD	Red
BR	Brown	LG	LT Green	TN	Tan
DB	Dark Blue	OR	Orange	WT	White
DG	DK Green	PK	Pink	YL	Yellow

2002-05 Thunderbird 3.9L V8 VIN A (A/T) C175-E 60 Pin Connector

PCM Pin #	Wire Color	Circuit Description (60 Pin)	Value at Hot Idle
1	WT/BK	Variable Valve Control Solenoid 1	DC pulse signals
2-3	---	Not Used	---
4	WT/BL	Fuel Rail Temperature Sensor	1.7-3.5v (50-120°F)
5	WT	Power Steering Pressure Switch	Straight: 0v, Turned: 5v
6	---	Not Used	---
7	BR	HO2S-11 (B1 S1) Heater	On: 1v, Off: 12v
8	BR/RD	HO2S-21 (B2 S1) Signal	0.1-1.1v
9-10, 18	---	Not Used	---
11	BR/GN	Fuel Injector 5 Control	2.9-3.6 ms
12	BR/YL	COP 5 Driver Control	5°, 55 mph: 8°
13	WT/VT	Variable Valve Timing Solenoid 2	DC pulse signals
14	YL/RD	Reference Voltage	4.9-5.1v
15	BR	Throttle Position Sensor Return	0.050v
16	BR/GN	EGR Pressure Transducer Return	<0.050v
17	BR/GN	Sensor Signal Return	<0.050v
19	GY/RD	Generator Common Signal	0-130 Hz (varies)
20	BR	Fuel Injector 2 Control	2.9-3.6 ms
21	BR/YL	Fuel Injector 6 Control	2.9-3.6 ms
22	BR/WT	COP 6 Driver Control	5°, 55 mph: 8°
23	WT/GN	EGR Pressure Transducer Signal	0-3.1v
24	YL	Battery Power (at all times)	12-14v
25-26	---	Not Used	---
27	YL/BL	Electronic Throttle Control Motor (-)	DC pulse signals
28	BR/RD	Fuel Injector 5 Control	2.9-3.6 ms
29	BR/WT	Fuel Injector 7 Control	2.9-3.6 ms
30	BR	COP 7 Driver Control	5°, 55 mph: 8°
31, 33-34	---	Not Used	---
32	WT	Throttle Position Sensor	0.53-1.27v
35	WT/BL	Electronic Throttle Control Motor (+)	DC pulse signals
36	---	Not Used	---
37	BR	Injector 8 Control	2.9-3.6 ms
38	BR/RD	COP 8 Driver Control	5°, 55 mph: 8°
39	WT/GN	Engine Oil Temperature Sensor Signal	0.5-0.6v
40	WT/VT	Cylinder Head Temperature Sensor	0.5-0.6v
41	WT/BL	EGR Pressure Transducer Signal	0.95-1.05v
42	GY/BK	Knock Sensor 1 Signal Return	<0.050v
43	GY/RD	Knock Sensor 2 Signal Return	<0.050v
44	WT/RD	HO2S-21 (B2 S1) Signal	0.1-1.1v
45	WT	HO2S-11 (B1 S1) Signal	0.1-1.1v
46	BR/BL	Fuel Injector 4 Control	2.9-3.6 ms
47	BR/WT	Fuel Injector 1 Control	2.9-3.6 ms
48	BR/GN	COP 4 Driver Control	5°, 55 mph: 8°
49	WT/GN	FRP Sensor Signal	2.8v (39 psi)
50	WT/GN	Generator Monitor Signal	130 Hz (45%)
51	WT/BK	Knock Sensor 1 (+) Signal	0v
52	WT/RD	Knock Sensor 2 (+) Signal	0v
53	WT/VT	CMP Sensor 1 Signal	6 Hz
54	WT/BK	CMP Sensor 2 Signal	6 Hz
55	GY/RD	CKP Sensor (-) Signal	390 Hz
56	WT/RD	CKP Sensor (+) Signal	390 Hz
57	WT/RD	Throttle Position Sensor	0.53-1.27v
58	BR	COP 1 Driver Control	5°, 55 mph: 8°
59	BR/RD	COP 2 Driver Control	5°, 55 mph: 8°
60	BR/BL	COP 3 Driver Control	5°, 55 mph: 8°

1994-97 Thunderbird 4.6L V8 MFI VIN W (A/T) 104 Pin Connector

PCM Pin #	Wire Color	Circuit Description (104 Pin)	Value at Hot Idle
1	PK/OR	Shift Solenoid 2 Control	12v, 55 mph: 1v
2	PK/LG	MIL (lamp) Control	MIL On: 1v, Off: 12v
3-10	---	Not Used	---
11	PK/WT	Purge Flow Sensor	0.8v, 55 mph: 3v
12	---	Not Used	---
13	PK	Flash EEPROM Power	0.1v
14	---	Not Used	---
15	PK/LB	Data Bus (-) Signal	Digital Signals
16	TN/OR	Data Bus (+) Signal	Digital Signals
17-20	---	Not Used	---
21	DB	CKP Sensor (+) Signal	450-480 Hz
22	GY	CKP Sensor (-) Signal	450-480 Hz
23	---	Not Used	---
24	BK/WT	Power Ground	<0.1v
25	BK	PCM Case Ground	<0.050v
26	DB/LG	Coil Driver 1 Control	6° dwell
27	OR/YL	Shift Solenoid 1 Control	1v, 55 mph: 12v
28	---	Not Used	---
29	TN/WT	TCS (switch) Signal	TCS & O/D On: 12v
30	DG	Octane Adjust Switch	Closed: 0v, Open: 9.3v
31-32	---	Not Used	---
33	PK/OR	VSS (-) Signal	0 Hz, 55 mph: 125 Hz
34	---	Not Used	---
35	RD/LG	HO2S-12 (B1 S2) Signal	0.1-1.1v
36	TN/LB	MAF Sensor Return	<0.050v
37	OR/BK	TFT Sensor Signal	2.10-2.40v
38	LG/RD	ECT Sensor Signal	0.5-0.6v
39	GY	IAT Sensor Signal	1.5-2.5v
40	PK/BK	Fuel Pump Monitor	On: 12v, Off: 0v
41	PK/LB	A/C Cycling Clutch Switch	AC On: 12v, Off: 0v
42-44	---	Not Used	---
45	TN/OR	Low Speed Fan Control	On: 1v, Off: 12v
46	LG/PK	High Speed Fan Control	On: 1v, Off: 12v
47	BR/PK	EGR VR Solenoid	0%, 55 mph: 45%
48-50	---	Not Used	---
51	BK/WT	Power Ground	<0.1v
52	RD/LB	Coil Driver 2 Control	6° dwell
53	---	Not Used	---
54	BR/OR	TCC Solenoid Control	0%, 55 mph: 95%
55	YL	Keep Alive Power	12-14v
56-57	---	Not Used	---
58	GY/BK	VSS (+) Signal	0 Hz, 55 mph: 125 Hz
59	---	Not Used	---
60	GY/LB	HO2S-11 (B1 S1) Signal	0.1-1.1v
61	PK/LG	HO2S-22 (B2 S2) Signal	0.1-1.1v
62-63	---	Not Used	---
64	LB/YL	TR Sensor Signal	In 'P': 0v, 55 mph: 1.7v

1994-97 Thunderbird 4.6L V8 MFI VIN W (A/T) 104 Pin Connector

PCM Pin #	Wire Color	Circuit Description (104 Pin)	Value at Hot Idle
65	BR/LG	DPFE Sensor Signal	0.20-1.30v
66, 68	---	Not Used	---
67	GY/YL	EVAP Purge Solenoid	0-10 Hz (0-100%)
69	PK/YL	A/C WOT Relay Control	On: 1v, Off: 12v
70	---	Not Used	---
71	RD	Vehicle Power	12-14v
72	TN/RD	Injector 7 Control	3.1-3.5 ms
73	TN/BK	Injector 5 Control	3.1-3.5 ms
74	BR/YL	Injector 3 Control	3.1-3.5 ms
75	TN	Injector 1 Control	3.1-3.5 ms
76-77	BK/WT	Power Ground	<0.1v
78	PK/WT	Coil Driver 3 Control	6° dwell
79	WT/LG	TCIL (lamp) Control	On: 1v, Off: 12v
80	LB/OR	Fuel Pump Control	On: 1v, Off: 12v
81	WT/YL	EPC Solenoid Control	9.0v (15 psi)
82	---	Not Used	---
83	WT/LB	IAC Motor Control	10v (30% on time)
84	DG/WT	TSS Sensor Signal	43 Hz (650 rpm)
85	DG	CID Sensor Signal	5-7 Hz
86	TN/LG	A/C Pressure Switch	A/C On: 12v (Open)
87	RD/BK	HO2S-21 (B2 S1) Signal	0.1-1.1v
88	LB/RD	MAF Sensor Signal	0.6v
89	GY/WT	TP Sensor Signal	0.53-1.27v
90	BR/WT	Reference Voltage	4.9-5.1v
91	GY/RD	Analog Signal Return	<0.050v
92	LG	Brake Pedal Position Switch	Brake Off: 0v, On: 12v
93	RD/WT	HO2S-11 (B1 S1) Heater	On: 1v, Off: 12v
94	YL/LB	HO2S-21 (B2 S1) Heater	On: 1v, Off: 12v
95	WT/BK	HO2S-12 (B1 S2) Heater	On: 1v, Off: 12v
96	TN/YL	HO2S-22 (B2 S2) Heater	On: 1v, Off: 12v
97	RD	Vehicle Power	12-14v
98	LB	Injector 8 Control	3.1-3.5 ms
99	LG/OR	Injector 6 Control	3.1-3.5 ms
100	BR/LB	Injector 4 Control	3.1-3.5 ms
101	WT	Injector 2 Control	3.1-3.5 ms
102	---	Not Used	---
103	BK/WT	Power Ground	<0.1v
104	RD/YL	Coil Driver 4 Control	6° dwell

Pin Connector Graphic

PCM 104-PIN CONNECTOR

Terminal View of 104-Pin PCM Wiring Harness Connector

Standard Colors and Abbreviations

Abbreviation	Color	Abbreviation	Color	Abbreviation	Color
BK	Black	GY	Gray	PK	Purple
BL	Blue	GN	Green	RD	Red
BR	Brown	LG	LT Green	TN	Tan
DB	Dark Blue	OR	Orange	WT	White
DG	DK Green	PK	Pink	YL	Yellow

1991-93 Thunderbird 5.0L V8 MFI VIN T (A/T) 60 Pin Connector

PCM Pin #	Wire Color	Circuit Description (60 Pin)	Value at Hot Idle
1	YL	Keep Alive Power	12-14v
2	---	Not Used	---
3	DG/WT	VSS (+) Signal	0 Hz, 55 mph: 125 Hz
4	RD/BL	Ignition Diagnostic Monitor	20-31 Hz
4	WT/PK	Ignition Diagnostic Monitor	20-31 Hz
5	---	Not Used	---
6	BK/WT	VSS (-) Signal	0 Hz, 55 mph: 125 Hz
6	GY/RD	VSS (-) Signal	0 Hz, 55 mph: 125 Hz
7	LG/YL	ECT Sensor Signal	0.5-0.6v
7	LG/RD	ECT Sensor Signal	0.5-0.6v
8	PK/BK	Fuel Pump Monitor	On: 12v, Off: 0v
9	TN/BL	MAF Sensor Return	<0.050v
10	PK/BL	A/C Cycling Clutch Switch	A/C On: 12v, Off: 0v
11	GY/YL	EVAP Purge Solenoid	12v, at 55 mph: 1v
12	LG	Injector 6 Control	4.9-5.2 ms
12	LG/OR	Injector 6 Control	4.9-5.2 ms
13	TN/OR	Injector 7 Control	4.9-5.2 ms
13	TN/RD	Injector 7 Control	4.9-5.2 ms
14	BL	Injector 8 Control	4.9-5.2 ms
15	TN/BL	Injector 5 Control	4.9-5.2 ms
15	TN/BK	Injector 5 Control	4.9-5.2 ms
16	GY	Ignition System Ground	<0.050v
17	YL/BK	STI Output, MIL Control	MIL On: 1v, Off: 12v
18	TN/OR	Data Bus (+) Signal	Digital Signals
19	PK/BL	Data Bus (-) Signal	Digital Signals
20	BK	PCM Case Ground	<0.050v
21	RD/LG	IAC Motor Control	9.3v (40%)
22	TN/LG	Fuel Pump Control	On: 1v, Off: 12v
23-24	---	Not Used	---
25	GY	IAT Sensor Signal	1.5-2.5v
25	LG/PK	IAT Sensor Signal	1.5-2.5v
26	OR/WT	Reference Voltage	4.9-5.1v
26	BR/WT	Reference Voltage	4.9-5.1v
27	BR/LG	EGR EVP Sensor Signal	0.4v
28-29	---	Not Used	---
30	RD/BL	Neutral Drive Switch Signal	In P/N: 0v, all others: 4-5v
31	WT/RD	Air Management 1 Solenoid	AM1 On: 1v, Off: 12v
32	OR/WT	Automatic Ride Control	Hot 4.3v
33	DG	EGR VR Solenoid	0%, 55 mph: 45%
33	BR/PK	EGR VR Solenoid	0%, 55 mph: 45%
34	BL/PK	Data Output Link	Digital Signals
35	BR/BL	Injector 4 Control	4.9-5.2 ms
36	YL/LG	Spark Output Signal	50% duty cycle
37	RD	Vehicle Power	12-14v
38	---	Not Used	---
39	BR/YL	Injector 3 Control	4.9-5.2 ms
40	BK/LG	Power Ground	<0.1v

1991-93 Thunderbird 5.0L V8 MFI VIN T (A/T) 60 Pin Connector

PCM Pin #	Wire Color	Circuit Description (60 Pin)	Value at Hot Idle
41-42	---	Not Used	---
43	TN/RD	HO2S-21 (B2 S1) Signal	0.1-1.1v
43	TN/WT	HO2S-21 (B2 S1) Signal	0.1-1.1v
44	TN/OR	HO2S-11 (B1 S1) Signal	0.1-1.1v
45	---	Not Used	---
46	BK/WT	Analog Signal Return	<0.050v
46	GY/RD	Analog Signal Return	<0.050v
47	DG/LG	TP Sensor Signal	1.0v, 55 mph: 1.2v
47	GY/WT	TP Sensor Signal	1.0v, 55 mph: 1.2v
48	WT/RD	Self-Test Indicator Signal	STI Open: 5v, Closed: 1v
48	WT/PK	Self-Test Indicator Signal	STI Open: 5v, Closed: 1v
49	---	Not Used	---
50	DB/OR	MAF Sensor Signal	0.7v, 55 mph: 2.1v
50	BL/RD	MAF Sensor Signal	0.7v, 55 mph: 2.1v
51-53	---	Not Used	---
54	OR/BL	A/C WOT Relay Control	On: 1v, Off: 12v
55	---	Not Used	---
56	DB	PIP Sensor Signal	50% duty cycle
57	RD	Vehicle Power	12-14v
58	TN	Injector 1 Control	4.9-5.2 ms
59	WT	Injector 2 Control	4.9-5.2 ms
60	BK/LG	Power Ground	<0.1v

Pin Connector Graphic

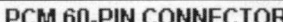

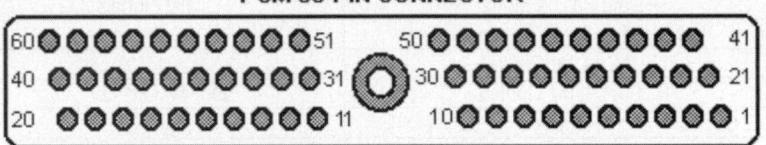

Terminal View of 60-Pin PCM Harness Connector

Standard Colors and Abbreviations

Abbreviation	Color	Abbreviation	Color	Abbreviation	Color
BK	Black	GY	Gray	PK	Purple
BL	Blue	GN	Green	RD	Red
BR	Brown	LG	LT Green	TN	Tan
DB	Dark Blue	OR	Orange	WT	White
DG	DK Green	PK	Pink	YL	Yellow

CONTINENTAL PIN TABLES

1990-94 Continental 3.8L V6 MFI VIN 4 (A/T) 60 Pin Connector

4	WT/PK	Ignition Diagnostic Monitor	20-31 Hz
5	DG/WT	TSS Sensor Signal	30 mph: 126-136 Hz
6	PK/OR	VSS (-) Signal	0 Hz, 55 mph: 125 Hz
7	LG/RD	ECT Sensor Signal	0.5-0.6v
8	PK/BK	Fuel Pump Monitor	12v, Off: 0v
8	DG/YL	Fuel Pump Monitor	12v, Off: 0v
9	TN/BL	MAF Sensor Return	<0.050v
10	PK	A/C Cycling Clutch Switch	Switch On: 12v, Off: 0v
10	LG/PK	A/C Cycling Clutch Switch	Switch On: 12v, Off: 0v
11	GY/YL	EVAP Purge Solenoid	12v, 55 mph: 1v
12	LG/OR	Injector 6 Control	6.0-6.8 ms
13	DB	Low Speed Fan Control	Fan On: 1v, Off: 12v
13	TN/OR	Low Speed Fan Control	Fan On: 1v, Off: 12v
14	---	Not Used	---
15	TN/BK	Injector 5 Control	6.0-6.8 ms
16	OR/RD	Ignition System Ground	<0.050v
17	PK/LG	Self-Test Output, MIL Control	MIL On: 1v, Off: 12v
18	TN/OR	Data Bus (+) Signal	Digital Signals
19	PK/BL	Data Bus (-) Signal	Digital Signals
20	BK	PCM Case Ground	<0.050v
21	WT	IAC Motor Control	Pulse Signals: 8-14v
21	WT/BL	IAC Motor Control	Pulse Signals: 8-14v
22	BL/OR	Fuel Pump Control	On: 1v, Off: 12v
22	TN/LG	Fuel Pump Control	On: 1v, Off: 12v
23	---	Not Used	---
24	---	Not Used	---
25	GY	IAT Sensor Signal	1.5-2.5v
26	BR/WT	Reference Voltage	4.9-5.1v
27	BR/LG	PFE EGR Sensor Signal	3.2v
28	YL/LG	PSP Switch Signal	Straight: 0v, Turned: 12v
29	---	Not Used	---
30	BL/YL	MLP Sensor Signal	In 'P': 4.4v, in 'D': 2.09v
31	LG/PK	High Speed Fan Control	Fan On: 0v, Off: 12v
32	DB/WT	Auto Suspension Control	5.5-9.0v
33	BR/PK	VR Solenoid Control	0%, 55 mph: 45%
34	---	Not Used	---
35	BR/BL	Injector 4 Control	6.0-6.8 ms
36	PK	Spark Output Signal	6.93v (50% dwell)
37	RD	Vehicle Power	12-14v
38	WT/YL	EPC Solenoid Control	8.8-9.9v
39	BR/YL	Injector 3 Control	6.0-6.8 ms
40	BK	Power Ground	<0.1v
40	BK/WT	Power Ground	<0.1v

1990-94 Continental 3.8L V6 MFI VIN 4 (A/T) 60 Pin Connector

PCM Pin #	Wire Color	Circuit Description (60 Pin)	Value at Hot Idle
41	---	Not Used	---
42 ('94)	TN/LG	AC High Pressure Switch	A/C Switch On: 0.8-1.9v
43	RD/BK	HO2S-21 (B2 S1) Signal	0.1-1.1v
44 ('91-'92)	GYL/BL	HO2S-11 (B1 S1) Signal	0.1-1.1v
45 ('93-'94)	GY/RD	HO2S-11 (B1 S1) Signal	0.1-1.1v
46	GY/RD	Analog Signal Return	<0.050v
47	GY/WT	TP Sensor Signal	0.8v, 55 mph: 1.1v
48	WT/PK	Self-Test Input Signal	STI On: 0v, Off: 5v
48	WT/PK	Self-Test Input Signal	STI On: 0v, Off: 5v
49	OR/BK	TOT Sensor Signal	2.10-2.40v
50	BL/RD	MAF Sensor Signal	0.7v, 55 mph: 2.1v
51	OR/YL	Shift Solenoid 1 Control	12v, 55 mph: 1v
52	PK/OR	Shift Solenoid 2 Control	1v, 55 mph: 12v
53	TN/WT	TCC Solenoid Control	30 mph: 11-12v (90%)
54	PK/YL	A/C WOT Relay Control	On: 1v, Off: 12v
55	PK/BK	Shift Solenoid 3 Control	12v, 55 mph: 1v
56	GY/OR	PIP Sensor Signal	6.93v (50% d/cycle)
57	RD	Vehicle Power	12-14v
58	TN	Injector 1 Control	6.0-6.8 ms
59	WT	Injector 2 Control	6.0-6.8 ms
60	BK/LG	Power Ground	<0.1v

Pin Connector Graphic

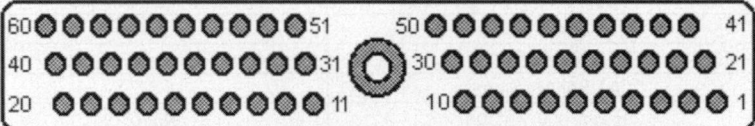

Standard Colors and Abbreviations

Abbreviation	Color	Abbreviation	Color	Abbreviation	Color
BK	Black	GY	Gray	PK	Purple
BL	Blue	GN	Green	RD	Red
BR	Brown	LG	LT Green	TN	Tan
DB	Dark Blue	OR	Orange	WT	White
DG	DK Green	PK	Pink	YL	Yellow

1995 Continental 4.6L V8 MFI VIN V (A/T) 60 Pin Connector

PCM Pin #	Wire Color	Circuit Description (60 Pin)	Value at Hot Idle
1	YL	Keep Alive Power	12-14v
2 ('93-'95)	LG	Brake Switch Signal	Brake Off: 0v, On: 12v
3	DG/WT	VSS (+) Signal	0 Hz, 55 mph: 125 Hz
4	TN/YL	Ignition Diagnostic Monitor	20-31 Hz
5	DG/WT	TSS+ Sensor	Idle: 790-820 Hz
6	PK/OR	VSS (-) Signal	0 Hz, 55 mph: 125 Hz
7	LG/RD	ECT Sensor Signal	0.5-0.6v
8	PK/BK	Fuel Pump Monitor	12v, Off: 0v
9	TN/LB	MAF Sensor Return	<0.050v
10	BK/YL	A/C Cycling Clutch Switch	Switch On: 12v, Off: 0v
11	GY/YL	EVAP Purge Solenoid	12v, 55 mph: 9v
12	LG/OR	Injector 6 Control	4.0-4.4 ms
13	TN/RD	Injector 7 Control	4.0-4.4 ms
14	LB	Injector 8 Control	4.0-4.4 ms
15	TN/BK	Injector 5 Control	4.0-4.4 ms
16	OR/RD	Ignition System Ground	<0.050v
17	PK/LG	Self-Test Output & MIL	MIL On: 1v, Off: 12v
18	TN/OR	Data Bus (+) Signal	Digital Signals
19	PK/LB	Data Bus (-) Signal	Digital Signals
20	BK	PCM Case Ground	<0.050v
21	WT/LB	IAC Motor Control	8.3-11.5v
22	LB/OR	Fuel Pump Control	On: 1v, Off: 12v
23	YL/RD	Knock Sensor Signal	0v
24	DB/OR	CID Sensor Signal	6-7 Hz
25	GY	ACT Sensor Signal	1.5-2.5v
26	BR/WT	Reference Voltage	4.9-5.1v
27	BR/LG	DPFE EGR Sensor Signal	0.4v, 55 mph: 2.9v
28	---	Not Used	---
29	DG	Octane Adjust Switch	Closed: 0v, Open: 9.1v
30	LB/YL	MLP Sensor Signal	In 'P': 0v, in O/D: 1.7v
31	---	Not Used	---
32	LG/BK	IMRC Solenoid	12v, 55 mph: 12v
33	BR/PK	VR Solenoid Control	0%, 55 mph: 45%
34	LB/PK	Data Output Link	Digital Signals
35	BR/LB	Injector 4 Control	4.0-4.4 ms
36	PK	Spark Angle Word Signal	6.93v (50% d/cycle)
37	RD	Vehicle Power	12-14v
38 ('93-'95)	WT/YL	EPC Solenoid Control	9.5v (20 psi)
39	BR/YL	Injector 3 Control	4.0-4.4 ms
40	BK/WT	Power Ground	<0.1v

1995 Continental 4.6L V8 MFI VIN V (A/T) 60 Pin Connector

PCM Pin #	Wire Color	Circuit Description (60 Pin)	Value at Hot Idle
41	TN/WT	TCS (switch) Signal	TCS & O/D On: 12v
42	---	Not Used	---
43	RD/BK	HO2S-21 (B2 S1) Signal	0.1-1.1v
44	GY/LB	HO2S-11 (B1 S1) Signal	0.1-1.1v
45	---	Not Used	---
46	GY/RD	Analog Signal Return	<0.050v
47	GY/WT	TP Sensor Signal	0.8v, 55 mph: 1.1v
48	WT/PK	Self-Test Input Signal	STI On: 0v, Off: 5v
49	OR/BK	TOT Sensor Signal	2.10-2.40v
50	LB/RD	MAF Sensor Signal	0.8v, 55 mph: 1.8v
51	OR/YL	Shift Solenoid 1 Control	1v, 55 mph: 12v
52	PK/OR	Shift Solenoid 2 Control	12v, 55 mph: 1v
53	PK/YL	TCC Solenoid Control	55 mph: 9-10v
54	PK/YL	A/C WOT Relay Control	On: 1v, Off: 12v
55	WT/LG	TCIL (lamp) Control	Lamp On: 1v, Off: 12v
56	GY/OR	PIP Sensor Signal	6.93v (50% dwell)
57	RD	Vehicle Power	12-14v
58	TN	Injector 1 Control	4.0-4.4 ms
59	WT	Injector 2 Control	4.0-4.4 ms
60	BK/WT	Power Ground	<0.1v

Pin Connector Graphic

PCM 60-PIN CONNECTOR

Terminal View of 60-Pin PCM Harness Connector

Standard Colors and Abbreviations

Abbreviation	Color	Abbreviation	Color	Abbreviation	Color
BK	Black	GY	Gray	PK	Purple
BL	Blue	GN	Green	RD	Red
BR	Brown	LG	LT Green	TN	Tan
DB	Dark Blue	OR	Orange	WT	White
DG	DK Green	PK	Pink	YL	Yellow

1996-97 Continental 4.6L V8 MFI VIN V (A/T) 104 Pin Connector

PCM Pin #	Wire Color	Circuit Description (104 Pin)	Value at Hot Idle
1	PK/OR	Shift Solenoid 2 Control	1v, 55 mph: 12v
2	PK/LG	MIL (lamp) Control	MIL On: 1v, Off: 12v
3-4	---	Not Used	---
5	BK/OR	Elect. Air Manage. System	1v, 55 mph: 1v
6-7	---	Not Used	---
8	LG/BK	IMRC Solenoid Monitor	All speeds: 5v
9-12, 14	---	Not Used	---
13	PK	Flash EEPROM Power	0.1v
15	PK/LB	Data Bus (-) Signal	Digital Signals
16	TN/OR	Data Bus (+) Signal	Digital Signals
17	---	Not Used	---
18	WT/PK	Traction Assist (PWM) Signal	Digital Signals
19-20	---	Not Used	---
21	BK/PK	CKP Sensor (+) Signal	310-330 Hz
22	GY/YL	CKP Sensor (-) Signal	310-330 Hz
23	---	Not Used	---
24	BK/WT	Power Ground	<0.1v
25	BK	PCM Case Ground	<0.050v
26	DB/LG	Coil Driver 1 Control	5°, 55 mph: 8° dwell
27	OR/YL	Shift Solenoid 1 Control	12v, 55 mph: 1v
28-29	---	Not Used	---
30	DG	Octane Adjust Switch	9.3v (Closed: 0v)
32	YL/LG	Knock Sensor 2 Signal	0v
33-34	---	Not Used	---
35	RD/LG	HO2S-12 (B1 S2) Signal	0.1-1.1v
36	TN/LB	MAF Sensor Return	<0.050v
37	OR/BK	TFT Sensor Signal	2.10-2.40v
38	LG/RD	ECT Sensor Signal	0.5-0.6v
39	GY	IAT Sensor Signal	1.5-2.5v
40	DG/YL	Fuel Pump Monitor	12v, Off: 0v
41	PK	A/C Cycling Clutch Switch	Switch On: 12v, Off: 0v
42-43	---	Not Used	---
44	BR	IMRC Solenoid Control	12v, 55 mph: 12v
46	LG/PK	High Speed Fan Control	Fan On: 1v, Off: 12v
47	BR/PK	VR Solenoid Control	0%, 55 mph: 45%
48-50	---	Not Used	---
51	BK/WT	Power Ground	<0.1v
52	RD/LB	Coil Driver 2 Control	5°, 55 mph: 8° dwell
53	PK/BK	Shift Solenoid 3 Control	12v, 55 mph: 1v
54	---	Not Used	---
55	YL	Keep Alive Power	12-14v
56	LG/BK	EVAP VMV (valve) Control	0-10 Hz (0-100%)
57	YL/RD	Knock Sensor 1 Signal	0v
58-59	---	Not Used	---
60	GY/LB	HO2S-11 (B1 S1) Signal	0.1-1.1v
61	PK/LG	HO2S-22 (B2 S2) Signal	0.1-1.1v
63	YL/WT	TP Sensor Signal 'B'	0.5-0.7v
64	LB/YL	TR Sensor Signal	In 'P': 0v, in O/D: 1.7v

1996-97 Continental 4.6L V8 MFI VIN V (A/T) 104 Pin Connector

PCM Pin #	Wire Color	Circuit Description (104 Pin)	Value at Hot Idle
65	BR/LG	DPFE EGR Sensor Signal	0.20-1.30v
68	DB	Low Speed Fan Control	Fan On: 1v, Off: 12v
69	PK/YL	A/C WOT Relay Control	On: 1v, Off: 12v
70	WT/OR	Air Mgmt. Solenoid Control	12v, 55 mph: 12v
71	RD	Vehicle Power	12-14v
72	TN/RD	Injector 7 Control	2.8-2-9 ms
73	TN/BK	Injector 5 Control	2.8-2-9 ms
74	BR/YL	Injector 3 Control	2.8-2-9 ms
75	TN	Injector 1 Control	2.8-2-9 ms
76, 77, 103	BK/WT	Power Ground	<0.1v
78	PK/WT	Coil Driver 3 Control	5°, 55 mph: 8° dwell
80	LB/OR	Fuel Pump Control	On: 1v, Off: 12v
81	WT/YL	EPC Solenoid Control	9.5v (17 psi)
82	DB/WT	TCC Solenoid Control	55 mph: 90-100%
83	WT/LB	IAC Motor Control	30%, 55 mph: 45%
84	DG/WT	TSS Sensor Signal	9.5v (17 psi)
85	DB/OR	CMP Sensor	4-6 Hz
86	TN/LG	A/C Pressure Switch Signal	Open: 12v, Closed: 0v
87	RD/BK	HO2S-21 (B2 S1) Signal	0.1-1.1v
88	LB/RD	MAF Sensor Signal	0.6v, 55 mph: 1.7v
89	GY/WT	TP Sensor Signal	0.53-1.27v
90	BR/WT	Reference Voltage	4.9-5.1v
91	GY/RD	Analog Signal Return	<0.050v
92	LG	Brake Switch Signal	Brake Off: 0v, On: 12v
93	RD/WT	HO2S-11 (B1 S1) Heater	1v, Off: 12v
94	YL/LB	HO2S-21 (B2 S1) Heater	1v, Off: 12v
95	WT/BK	HO2S-12 (B1 S2) Heater	1v, Off: 12v
96	TN/YL	HO2S-22 (B2 S2) Heater	1v, Off: 12v
97	RD	Vehicle Power	12-14v
98	LB	Injector 8 Control	2.8-2-9 ms
99	LG/OR	Injector 6 Control	2.8-2-9 ms
100	BR/LB	Injector 4 Control	2.8-2-9 ms
101	WT	Injector 2 Control	2.8-2-9 ms
102	---	Not Used	---
104	RD/YL	Coil Driver 4 Control	5°, 55 mph: 8° dwell

Pin Connector Graphic

PCM 104-PIN CONNECTOR

Terminal View of 104-Pin PCM Wiring Harness Connector

Standard Colors and Abbreviations

Abbreviation	Color	Abbreviation	Color	Abbreviation	Color
BK	Black	GY	Gray	PK	Purple
BL	Blue	GN	Green	RD	Red
BR	Brown	LG	LT Green	TN	Tan
DB	Dark Blue	OR	Orange	WT	White
DG	DK Green	PK	Pink	YL	Yellow

1998-99 Continental 4.6L V8 MFI VIN V (A/T) 104 Pin Connector

PCM Pin #	Wire Color	Circuit Description (104 Pin)	Value at Hot Idle
1	WT/PK	COP 5 Driver Control	5º, 55 mph: 8º dwell
2	PK/LG	MIL (lamp) Control	MIL On: 1v, Off: 12v
3	BK/WT	Power Ground	<0.1v
6	OR/YL	Shift Solenoid 1 Control	12v, 55 mph: 1v
8	LG/BK	IMRC Monitor	5v, 55 mph: 5v
9	YL/WT	Fuel Level Indicator Signal	1.7v (1/2 full)
11	PK/OR	Shift Solenoid 2 Control	1v, 55 mph: 1v
12	YL/WT	Fuel Level Indicator Signal	1.7v (1/2 full)
13	PK	Flash EEPROM Power	0.5v
14	LB/WT	EFTA Sensor	1.7-3.5v (50-120ºF)
15	PK	Data Bus (-) Signal	<0.050v
16	TN/OR	Data Bus (+) Signal	Digital Signals
18	RD/PK	Fuel Rail Pressure Sensor	2.8v (39 psi)
20	PK/BK	Shift Solenoid 3 Control	12v, 55 mph: 1v
21	BK	CKP (+) Sensor Signal	390-450 Hz
22	GY	CKP (-) Sensor Signal	390-450 Hz
25	BK	PCM Case Ground	<0.050v
26	LG/WT	COP 1 Driver Control	5º, 55 mph: 8º dwell
27	OR/YL	COP 6 Driver Control	5º, 55 mph: 8º dwell
28	DB	Low Cooling Fan Control	Fan On: 1v, Off: 12v
29	TN/WT	TCS (switch) Signal	TCS & O/D On: 12v
30	DG	Octane Adjust Switch	9.3v (Closed: 0v)
31	YL/LG	PSP Switch Signal	Straight: 5v, Turning: 0v
32	DG/PK	Knock Sensor 2 Signal	0v
33	PK/OR	VSS (-) Signal	0 Hz, 55 mph: 125 Hz
34	YL/BK	Digital TR1 Sensor Signal	0v, 55 mph: 11v
35	RD/LG	HO2S-12 (B1 S2) Signal	0.1-1.1v
36	TN/LB	MAF Sensor Return	<0.050v
37	OR/BK	TFT Sensor Signal	2.10-2.40v
38	LG/RD	ECT Sensor Signal	0.5-0.6v
39	GY	IAT Sensor Signal	1.5-2.5v
40	DG/YL	Fuel Pump Monitor	12v, Off: 0v
41	PK	A/C High Pressure Switch	Switch On: 12v, Off: 0v
42	BR	IMRC Solenoid Control	12v, 55 mph: 12v
43	WT/RD	Fuel Pump Control	On: 1v, Off: 12v
46	LG/PK	High Speed Fan Control	Fan On: 1v, Off: 12v
47	BR/PK	VR Solenoid Control	0%, 55 mph: 45%
48	OR/WT	Clean Tachometer Output	39-45 Hz
49	LB/BK	Digital TR2 Sensor Signal	0v, 55 mph: 11v
50	WT/BK	Digital TR4 Sensor Signal	0v, 55 mph: 11v
51	BK/WT	Power Ground	<0.1v
52	PK/LB	COP 2 Driver Control	5º, 55 mph: 8º dwell
53	PK/WT	COP 7 Driver Control	5º, 55 mph: 8º dwell
54	PK/YL	TCC Solenoid Control	12v, 55 mph: 1v
55	LG/RD	Keep Alive Power	12-14v
56	LG/BK	EVAP VMV (valve) Control	0-10 Hz (0-100%)
57	YL/RD	Knock Sensor 1 Signal	0v
58	GY/BK	VSS (+) Signal	0 Hz, 55 mph: 125 Hz
59	DG/WT	TSS Sensor Signal	40-45 Hz (645-650 rpm)
60	GY/LB	HO2S-11 (B1 S1) Signal	0.1-1.1v
61	PK/LG	HO2S-22 (B2 S2) Signal	0.1-1.1v
62	YL/WT	FTP Sensor Signal	2.6v (0" H2O - cap off)
64	LB/YL	Digital TR Sensor Signal	In 'P': 0v, in O/D: 1.7v

1998-99 Continental 4.6L V8 MFI VIN V (A/T) 104 Pin Connector

PCM Pin #	Wire Color	Circuit Description (104 Pin)	Value at Hot Idle
65	BR/LG	DPFE Sensor Signal	0.20-1.30v
67	PK/WT	EVAP Vent Valve	0-10 Hz (0-100%)
69	PK/YL	A/C WOT Relay Control	Relay On: 0v, Off: 12v
71, 97	RD	Vehicle Power	12-14v
72	TN/RD	Injector 7 Control	2.8-2-9 ms
73	TN/BK	Injector 5 Control	2.8-2-9 ms
74	BR/YL	Injector 3 Control	2.8-2-9 ms
75	TN	Injector 1 Control	2.8-2-9 ms
77, 103	BK/WT	Power Ground	<0.1v
78	DG/PK	COP 3 Driver Control	5º, 55 mph: 8º dwell
79	WT/RD	COP 8 Driver Control	5º, 55 mph: 8º dwell
80	LB/OR	Fuel Pump Control	On: 1v, Off: 12v
81	WT/YL	EPC Solenoid Control	9v (17 psi)
83	WT/LB	IAC Motor Control	35% duty cycle
84	DG/WT	TSS Sensor Signal	43 Hz (650 rpm)
85	DB/OR	CMP Sensor Signal	4-6 Hz
86	TN/LG	A/C Pressure Transducer	A/C On: 0.8-1.9v
87	RD/BK	HO2S-21 (B2 S1) Signal	0.1-1.1v
88	LB/RD	MAF Sensor Signal	0.6v, 55 mph: 1.7v
89	GY/WT	TP Sensor Signal	0.53-1.27v
90	BR/WT	Reference Voltage	4.9-5.1v
91	GY/RD	Analog Signal Return	<0.050v
92	LG	BPP Switch Signal	Brake On: 12v, Off: 0v
93	RD/WT	HO2S-11 (B1 S1) Heater	1v, Off: 12v
94	YL/LB	HO2S-21 (B2 S1) Heater	1v, Off: 12v
95	WT/BK	HO2S-12 (B1 S2) Heater	1v, Off: 12v
96	TN/YL	HO2S-22 (B2 S2) Heater	1v, Off: 12v
98	LB	Injector 8 Control	2.8-2-9 ms
99	LG/OR	Injector 6 Control	2.8-2-9 ms
100	BR/LB	Injector 4 Control	2.8-2-9 ms
101	WT	Injector 2 Control	2.8-2-9 ms
104	WT/PK	Coil Driver 4 Control	5º, 55 mph: 8º dwell

Pin Connector Graphic

Terminal View of 104-Pin PCM Wiring Harness Connector

Standard Colors and Abbreviations

Abbreviation	Color	Abbreviation	Color	Abbreviation	Color
BK	Black	GY	Gray	PK	Purple
BL	Blue	GN	Green	RD	Red
BR	Brown	LG	LT Green	TN	Tan
DB	Dark Blue	OR	Orange	WT	White
DG	DK Green	PK	Pink	YL	Yellow

2000-02 Continental 4.6L V8 MFI VIN V (A/T) 104 Pin Connector

PCM Pin #	Wire Color	Circuit Description (104 Pin)	Value at Hot Idle
1	LG/YL	COP 6 Driver Control	5°, 55 mph: 8° dwell
2, 4-5, 7-8	---	Not Used	---
3	BK/WT	Power Ground	<0.1v
6	OG/YL	Shift Solenoid 1 Control	12v, 55 mph: 1v
9	YL/WT	Fuel Level Input	1.7v (1/2 full)
10, 17-18	---	Not Used	---
11	VT/OR	Shift Solenoid 2 Control	1v, 55 mph: 1v
13	VT	Flash EEPROM Power	0.1v
15	PK/LB	Data Bus (-) Signal	Digital Signals
16	TN/OG	Data Bus (+) Signal	Digital Signals
20	PK/BK	Shift Solenoid 3 Control	12v, 55 mph: 1v
21	BK/PK	CKP Sensor (+) Signal	330-420 Hz
22	GY/YL	CKP Sensor (-) Signal	330-420 Hz
23	DG/WT	Knock Sensor (-) 1 Signal	0v
25	BK	Case Ground	<0.050v
26	LG/WT	COP 1 Driver Control	5°, 55 mph: 8° dwell
27	OG/YL	COP 5 Driver Control	5°, 55 mph: 8° dwell
28	DB	Low Cooling Fan Control	Fan On: 1v, Off: 12v
29-30, 33	---	Not Used	---
31	YL/LG	PSP Switch Signal	Straight: 5v, Turning: 0v
32	DG/VT	Knock Sensor 1 (+) Signal	0v
34	OG/BK	Digital TR1 Sensor Signal	0v, 55 mph: 11v
35	RD/LG	HO2S-12 (B1 S2) Signal	0.1-1.1v
36	TN/LB	MAF Sensor Return	<0.050v
37	OG/BK	TFT Sensor Signal	2.10-2.40v
38	LG/RD	ECT Sensor Signal	0.5-0.6v
39	GY	IAT Sensor Signal	1.5-2.5v
40	DG/YL	Fuel Pump Monitor	0-7v, Off: 0v
41	VT	A/C Clutch Switch Signal	A/C On: 12v, Off: 0v
42-45, 48	---	Not Used	---
46	LG/VT	High Speed Fan Control	Fan On: 1v, Off: 12v
47	BR/PK	VR Solenoid Control	0%, 55 mph: 45%
49	BK/WT	Digital TR2 Sensor Signal	0v, 55 mph: 11v
50	DG/OG	Digital TR4 Sensor Signal	0v, 55 mph: 11v
51	BK/WT	Power Ground	<0.1v
52	PK/LB	COP 3 Driver Control	5°, 55 mph: 8° dwell
53	PK/WT	COP 4 Driver Control	5°, 55 mph: 8° dwell
54	DB/WT	TCC Solenoid Control	12v, 55 mph: 1v
55	LG/RD	Keep Alive Power	12-14v
56	LG/BK	EVAP Purge Solenoid	0-10 Hz (0-100%)
57	YL/RD	Knock Sensor 1 (+) Signal	0v
58	---	Not Used	---
59	DG/WT	TSS Sensor Signal	40-45 Hz (645-650 rpm)
60	GY/LB	HO2S-11 (B1 S1) Signal	0.1-1.1v
61	VT/LG	HO2S-22 (B2 S2) Signal	0.1-1.1v
62	RD/PK	FTP Sensor Signal	2.6v (0" H2O - cap off)
63	RD/PK	Fuel Rail Pressure Sensor	2.8v (39 psi)
64	LB/YL	Digital TR3 Signal	In 'P': 0v, in O/D: 1.7v
65 ('00)	BR/LG	DPFE Sensor Signal	0.20-1.30v
65 ('01-'02)	BR/LG	DPFE Sensor Signal	0.95-1.05v
66	---	Not Used	---

2000-02 Continental 4.6L V8 MFI VIN V (A/T) 104 Pin Connector

PCM Pin #	Wire Color	Circuit Description (104 Pin)	Value at Hot Idle
67	VT/WT	EVAP CV Solenoid	0-10 Hz (0-100%)
69	PK/YL	A/C WOT Relay Control	12v (off)
70	---	Not Used	---
71	RD	Vehicle Power	12-14v
72	TN/RD	Injector 7 Control	2.8-2-9 ms
73	TN/BK	Injector 5 Control	2.8-2-9 ms
74	BR/YL	Injector 3 Control	2.8-2-9 ms
75	TN	Injector 1 Control	2.8-2-9 ms
77	BK/WT	Comm. Network Ground	<0.050v
78	DG/VT	COP 7 Driver Control	5º, 55 mph: 8º dwell
79	WT/RD	COP 8 Driver Control	5º, 55 mph: 8º dwell
80	LB/OR	Fuel Pump Control	1.5-1.7v (28-33%)
81	WT/YL	EPC Solenoid Control	8-9v (15-18 psi)
82	---	Not Used	---
83	WT/LB	IAC Motor Control	28-30%
84	---	Not Used	---
85	DB/OR	CMP Sensor Signal	4-6 Hz
86	TN/LG	A/C Pressure Sensor	A/C On: 0.6v (30 psi)
87	RD/BK	HO2S-21 (B2 S1) Signal	0.1-1.1v
88	LB/RD	MAF Sensor Signal	0.6v, 55 mph: 1.7v
89	GY/WT	TP Sensor Signal	0.53-1.27v
90	BR/WT	Reference Voltage	4.9-5.1v
91	GY/RD	Analog Signal Return	<0.050v
92	---	Not Used	---
93	RD/WT	HO2S-11 (B1 S1) Heater	1v, Off: 12v
94	YL/LB	HO2S-21 (B2 S1) Heater	1v, Off: 12v
95	WT/BK	HO2S-12 (B1 S2) Heater	1v, Off: 12v
96	TN/YL	HO2S-22 (B2 S2) Heater	1v, Off: 12v
97	RD	Vehicle Power	12-14v
98	LB	Injector 8 Control	2.8-2-9 ms
99	LG/OR	Injector 6 Control	2.8-2-9 ms
100	BR/LB	Injector 4 Control	2.8-2-9 ms
101	WT	Injector 2 Control	2.8-2-9 ms
102	YL	Knock Sensor 2 (-) Signal	0v
103	BK/WT	Communication Network GND	<0.050v
104	WT/PK	COP 2 Driver Control	5º, 55 mph: 8º dwell

Pin Connector Graphic

Standard Colors and Abbreviations

Abbreviation	Color	Abbreviation	Color	Abbreviation	Color
BK	Black	GY	Gray	PK	Purple
BL	Blue	GN	Green	RD	Red
BR	Brown	LG	LT Green	TN	Tan
DB	Dark Blue	OR	Orange	WT	White
DG	DK Green	PK	Pink	YL	Yellow

LS PIN TABLES

2000-01 LS 3.0L V6 VIN S (A/T) C175A 58-Pin Connector

PCM Pin #	Wire Color	Circuit Description (58-Pin)	Value at Hot Idle
1-2	---	Not Used	---
3	GY	SCP Data Bus (+)	Digital Signals
4	BL	SCP Data Bus (-)	<0.050v
6	BR/BL	EVAP CV Solenoid	0-10 Hz (0-100%)
7-8	---	Not Used	---
9	BR/BL	Ground, Switch	N/A
10-11	---	Not Used	---
12	BR/RD	EVAP Purge Solenoid	0-10 Hz (0-100%)
13	YL/GN	EEPROM Power	0.1v
14-16, 18	---	Not Used	---
17	BR	Analog Signal Return	<0.050v
19	BR/GN	Secondary AIR System Relay	On: 1v, Off: 12v
20	YL	Reference Voltage	4.9-5.1v
21-23	---	Not Used	---
24	BK/YL	Power Ground	<0.1v
25	BK/RD	Power Ground	<0.1v
26	BK/RD	Power Ground	<0.1v
27	BK/RD	Power Ground	<0.1v
28	OG/GN	Generator Field Signal	0.5-10.5v
29	GY/WT	Signal Return Normal	<0.050v
30, 35	---	Not Used	---
31	WT/BL	MAF Sensor Signal	0.6v, 55 mph: 1.7v
32	GN/YL	Vehicle Power (from relay)	12-14v
33	GN/OG	Vehicle Power (from relay)	12-14v
34	RD/BL	Secondary AIR Monitor	Pump On: 12v, Off: 0v
36	WT/GN	Normal	N/A
37	WT	PSP Switch Signal	Straight: 5v, Turning: 0v
38	BR/BL	MAF Sensor Return	<0.050v
39	---	Not Used	---
40	OG	BPP Signal from ABS	Brake Off: 0v, On: 12v
41	WT/GN	Sensor Signal	N/A
42	WT/VT	A/C Pressure Switch Signal	0.9v (50 psi)
43	BR/YL	Power Ground	<0.1v
44	OG/YL	Keep Alive Power (CJB fuse)	12-14v
45	WT/VT	Sensor Signal	N/A
46	GY/OR	Signal 2	N/A
47-48	---	Not Used	---
49	WT/VT	Sensor Signal	N/A
50	WT/BK	Generator Monitor Signal	130 Hz (45%)
51	WT/VT	IAT Sensor Signal	1.5-2.5v
52	WT/VT	FTP Sensor Signal	2.6v (0" H2O - cap off)
53-55	---	Not Used	---
56	BK/OR	Ground, Switched	N/A
57	YL/BL	Keep Alive Power	12-14v
58	WT/VT	Sensor Signal	N/A

2000-01 LS 3.0L V6 VIN S (A/T) C175B 32-Pin Connector

PCM Pin #	Wire Color	Circuit Description (32-Pin)	Value at Hot Idle
1	BR	Shift Solenoid 'A' Control	12v, 55 mph: 1v
2	BR/RD	Shift Solenoid 'B' Control	12v, 55 mph: 1v
3	BR/WT	Shift Solenoid 'C' Control	12v, 55 mph: 1v
4	BR/YL	Shift Solenoid 'D' Control	12v, 55 mph: 1v
5	BR/WT	TCC Solenoid Control	12v, 55 mph: 1v
6, 8	---	Not Used	---
7	BR/GN	Pressure Control A Solenoid	On: 1v, Off: 12v
9	WT	Digital TR Sensor (TR3A)	0v, 55 mph: 1.7v
10	WT/RD	Digital TR Sensor (TR4)	0v, 55 mph: 11v
11, 14	---	Not Used	---
12	BR/WT	Pressure Control C Solenoid	On: 1v, Off: 12v
13	BR/YL	Pressure Control B Solenoid	On: 1v, Off: 12v
15	BR/BL	HO2S-12 (B1 S2) Heater	1v, Off: 12v
16	BR/GN	HO2S-22 (B2 S2) Heater	1v, Off: 12v
17	BR	Sensor Signal Return	<0.050v
18	WT/BL	Digital TR Sensor (TR2)	0v, 55 mph: 11v
19	---	Not Used	---
21	WT/BK	Intermediate Speed Shaft	0 Hz, 55 mph: 1320 Hz
22	WT/GN	Digital TR Sensor (TR1)	0v, 55 mph: 11v
23	WT/RD	Trans. Fluid Temp. Sensor	2.10-2.40v
24-25	---	Not Used	---
26	WT/RD	Output Shaft Speed Sensor	0 Hz, 55 mph: 975 Hz
27	WT/VT	Turbine Shaft Speed Sensor	340 Hz, 55 mph: 1025 Hz
28	WT/BL	HO2S-12 (B1 S2) Signal	0.1-1.1v
29	WT/GN	HO2S-22 (B2 S2) Signal	0.1-1.1v
30	WT/BL	A/T Pressure Switch Signal	Open: 12v, Closed: 0v
31-32	---	Not Used	---

Pin Connectors Graphic

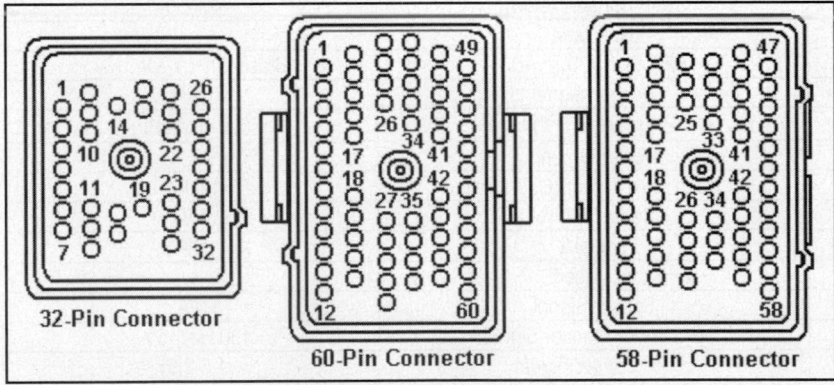

32-Pin Connector   60-Pin Connector   58-Pin Connector

Standard Colors and Abbreviations

Abbreviation	Color	Abbreviation	Color	Abbreviation	Color
BK	Black	GY	Gray	PK	Purple
BL	Blue	GN	Green	RD	Red
BR	Brown	LG	LT Green	TN	Tan
DB	Dark Blue	OR	Orange	WT	White
DG	DK Green	PK	Pink	YL	Yellow

2000-01 LS 3.0L V6 MFI VIN S (A/T) C175C 60 Pin Connector

PCM Pin #	Wire Color	Circuit Description (60 Pin)	Value at Hot Idle
1, 3-6	---	Not Used	---
2	BR/WT	Injector 1 Control	2.8-3.2 ms
7	BR	HO2S-11 (B1 S1) Heater	1v, Off: 12v
8	BR/RD	HO2S-21 (B2 S1) Heater	1v, Off: 12v
9	BR	IAC Solenoid Control	35% duty cycle
10, 15	---	Not Used	---
11	BR/GN	Injector 5 Control	2.8-3.2 ms
12	BR/GN	COP 4 Driver Control	5°, 55 mph: 8° dwell
13	BR/BL	COP 3 Driver Control	5°, 55 mph: 8° dwell
14	BR	Injector 2 Control	2.8-3.2 ms
16	BR/GN	VR Solenoid Control	0%, 55 mph: 45%
17	BR	Sensor Signal Return	<0.050v
18-19	---	Not Used	---
20	YL/BK	Sensor Voltage Reference	4.9-5.1v
21	BR/YL	Injector 6 Control	2.8-3.2 ms
22	BR/YL	COP 5 Driver Control	5°, 55 mph: 8° dwell
23	BR/RD	COP 2 Driver Control	5°, 55 mph: 8° dwell
24	BR/RD	Injector 3 Control	2.8-3.2 ms
25	---	Not Used	---
28	GY/RD	Generator Field Control	0-130 Hz (varies)
29	BR/WT	IMRC Monitor	5v, 55 mph: 5v
30	BR/WT	COP 6 Driver Control	5°, 55 mph: 8° dwell
31	BR	COP 1 Driver Control	5°, 55 mph: 8° dwell
32	BR/BL	Injector 4 Control	2.8-3.2 ms
33-35	---	Not Used	---
36	BR	Hydraulic Fan Solenoid	On: 1v, Off: 12v
37	BR	Power Steering Press. Switch	Straight: 5v, Turning: 0v
38	BR/RD	To Be Done	---
39	GY/BK	To Be Done	---
40	WT/VT	CHT Sensor Signal	0.7v (194°F)
41 ('00)	WT/BL	DPFE Sensor Signal	0.20-1.30v
41 ('01)	WT/BL	DPFE Sensor Signal	0.95-1.05v
42	GY/BK	Knock Sensor 1 Signal	0v
43	GY/RD	Knock Sensor 2 Signal	0v
44	WT/RD	HO2S-21 (B2 S1) Signal	0.1-1.1v
45	WT	HO2S-11 (B1 S1) Signal	0.1-1.1v
46-48	---	Not Used	---
49	WT/GN	Fuel Rail Pressure Sensor	2.8v (39 psi)
50	WT/RD	Generator Field Signal	130 Hz (40%)
51	WT/BK	Knock Sensor 1 Ground	0v
52	WT/RD	Knock Sensor 2 Ground	0v
53	WT/VT	CMP Sensor Signal	4-6 Hz
54	---	Not Used	---
55	GY/RD	CKP Sensor (+) Signal	390-450 Hz
56	WT/RD	CKP Sensor (-) Signal	390-450 Hz
57	WT	TP Sensor Signal	0.53-1.27v
58-60	---	Not Used	---

2002-05 LS 3.0L V6 VIN S (A/T) C175A 58-Pin Connector

PCM Pin #	Wire Color	Circuit Description (58-Pin)	Value at Hot Idle
1	WT/RD	Accelerator Pedal Position Sensor	0.5-4.9v
2	---	Not Used	---
3	GY	SCP Data Bus (+)	Digital Signals
4	BL	SCP Data Bus (-)	<0.050v
5	BR	Signal Return	<0.050v
6-8	---	Not Used	---
9	BR/BL	A/C Clutch Relay Control	Relay Off: 12v, On: 1v
10-11	---	Not Used	---
12	BR/RD	EVAP Purge Solenoid Control	0-10 Hz (0-100%)
13	YL/GN	EEPROM Power	0.1v
14	WT/VT	Manual Transmission Switch (+) Signal	Switch Open: 12v, Closed: 1v
15	WT	Electronic Throttle Control Module	Digital Signals
16	WT/BL	Electronic Throttle Control Module	Digital Signals
17	BR	Electronic Throttle Control Module	Digital Signals
18-19	---	Not Used	---
20	YL	ETC Reference Voltage	4.9-5.1v
21	---	Not Used	---
22	WT/BK	Manual Transmission Switch (-) Signal	<0.050v
23	YL/RD	ETC Reference Voltage	4.9-5.1v
24	BK/RD	Power Ground	<0.1v
25	BK/RD	Power Ground	<0.1v
26	BK/RD	Power Ground	<0.1v
27	BK/RD	Power Ground	<0.1v
28	GN/OR	Brake Pressure Switch	Brake Off: 0v, On: 12v
29	---	Not Used	---
30	BR/BL	EVAP Vent Solenoid Control	0-10 Hz (0-100%)
31	WT/BL	MAF Sensor Signal	0.6v, 55 mph: 1.7v
32	GN/YL	Vehicle Power (Start-Run)	12-14v
33	GN/OR	Vehicle Power (Start-Run)	12-14v
34-37	---	Not Used	---
38	BR/BL	MAF Sensor Signal Return	<0.050v
39	---	Not Used	---
40	OG	Keep Alive Power	12-14v
41	WT/GN	Traction Control Disable Switch	Brake Off: 0v, On: 12v
42	WT/VT	A/C Pressure Switch Signal	0.9v (50 psi)
43	BR/YL	Power Ground	<0.1v
44	OG/YL	Keep Alive Power (CJB fuse)	12-14v
45-46	---	Not Used	---
47	WT/BK	Engine Cooling Fan Motor Control	Off: 12v, On: 1v
48	WT/BL	Restraint Control Module	Digital Signals
49	WT/VT	Rear Electronic Module	Digital Signals
50	---	Not Used	---
51	WT/VT	IAT Sensor Signal	1.5-2.5v
52	WT/VT	FTP Sensor Signal	2.6v (0" H2O - cap off)
53	GN/RD	Controller Area Network (+) Signal	0-7-0v
54	BL/RD	Controller Area Network (-) Signal	0-7-0v
55	YL	Reference Voltage	4.9-5.1v
56	BK/OR	Sensor Return	<0.050v
57	YL/BL	Speed Control Reference Voltage	4.9-5.1v
58	---	Not Used	---

2002-05 LS 3.0L V6 VIN S (A/T) C175B 32-Pin Connector

PCM Pin #	Wire Color	Circuit Description (32-Pin)	Value at Hot Idle
1	BR	Shift Solenoid 'A' Control	12v, 55 mph: 1v
2	BR/RD	Shift Solenoid 'B' Control	12v, 55 mph: 1v
3-6	---	Not Used	---
7	BR/GN	Pressure Control Solenoid 'A' Control	8.8v
8	BR/WT	Shift Solenoid 'C' Control	12v, 55 mph: 12v
9	WT	Digital TR Sensor (TR3A)	0v, 55 mph: 1.7v
10	WT/RD	Digital TR Sensor (TR4)	0v, 55 mph: 11v
11	---	Not Used	---
12	BR/WT	Pressure Control 'C' Solenoid Control	Off: 12v, On: 1v
13	BR/YL	Pressure Control 'B' Solenoid Control	Off: 12v, On: 1v
14	BR/RD	Signal Return	<0.050v
15	BR/BL	HO2S-12 (B1 S2) Heater Control	Off: 12v, On: 1v
16	BK/GN	HO2S-22 (B2 S2) Heater Control	Off: 12v, On: 1v
17	BR/YL	Shift Solenoid 'D' Control	12v, 55 mph: 1v
18	WT/BL	Digital TR Sensor (TR2)	0v, 55 mph: 11v
19	---	Not Used	---
20	BR/WT	Torque Converter Clutch Control	12v, 55 mph: 0.1v
21	WT/BK	Intermediate Speed Shaft Signal	0 Hz, 55 mph: 1386 Hz
22	WT/GN	Digital TR Sensor (TR1)	0v, 55 mph: 11v
23	WT/RD	Transmission Fluid Temperature Sensor	2.10-2.40v
24-25	---	Not Used	---
26	WT/RD	Output Shaft Speed Sensor Signal	0 Hz, 975 Hz
27	WT/VT	Turbine Shaft Speed Sensor Signal	340 Hz, 1025 Hz
28	WT/BL	HO2S-12 (B1 S2) Signal	0.1-1.1v
29	WT/GN	HO2S-22 (B2 S2) Signal	0.1-1.1v
30-32	---	Not Used	---

Wire Harness Connectors Graphic

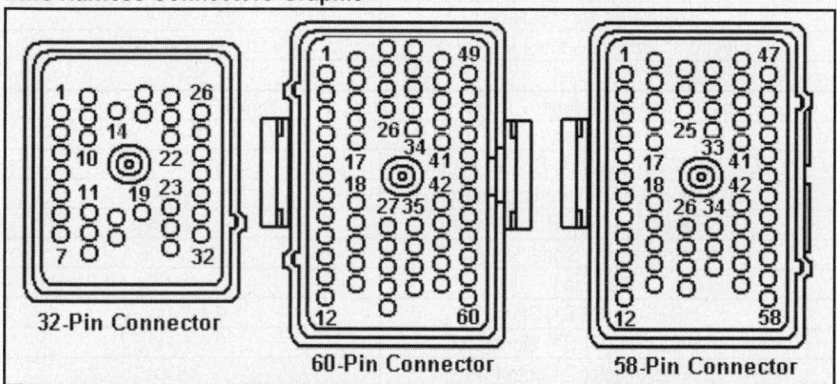

32-Pin Connector   60-Pin Connector   58-Pin Connector

Standard Colors and Abbreviations

Abbreviation	Color	Abbreviation	Color	Abbreviation	Color
BK	Black	GY	Gray	PK	Purple
BL	Blue	GN	Green	RD	Red
BR	Brown	LG	LT Green	TN	Tan
DB	Dark Blue	OR	Orange	WT	White
DG	DK Green	PK	Pink	YL	Yellow

2002-05 LS 3.0L V6 MFI VIN S (A/T) C175C 60 Pin Connector

PCM Pin #	Wire Color	Circuit Description (60 Pin)	Value at Hot Idle
1	WT/BK	Variable Valve Control Solenoid 1	DC pulse signals
2-3, 6	---	Not Used	---
4	WT/BL	Fuel Rail Temperature Sensor	1.7-3.5v (50-120ºF)
5	WT	Power Steering Pressure Switch	Straight: 0v, Turned: 5v
7	BR	HO2S-11 (B1 S1) Heater	On: 1v, Off: 12v
8	BR/RD	HO2S-21 (B2 S1) Signal	0.1-1.1v
9-10	---	Not Used	---
11	BR/BL	Fuel Injector 4 Control	2.9-3.6 ms
12	BR/GN	COP 4 Driver Control	5º, 55 mph: 8º
13	WT/VT	Variable Valve Timing Solenoid 2	DC pulse signals
14	YL/BK	Reference Voltage	4.9-5.1v
15	BR	Throttle Position Sensor Return	0.050v
16	BR/GN	EGR Pressure Transducer Return	<0.050v
17	BR/GN	Sensor Signal Return	<0.050v
18	---	Not Used	---
19	GY/RD	Generator Common Signal	0-130 Hz (varies)
20	BR/GN	Fuel Injector 5 Control	2.9-3.6 ms
21	BR/RD	Fuel Injector 3 Control	2.9-3.6 ms
22	BR/YL	COP 5 Driver Control	5º, 55 mph: 8º
23	WT/GN	EGR Pressure Transducer Signal	0-3.1v
24	YL	Battery Power (at all times)	12-14v
25-26	---	Not Used	---
27	YL/BL	Electronic Throttle Control Motor (-)	DC pulse signals
28	BR/YL	Fuel Injector 6 Control	2.9-3.6 ms
29	BR/WT	Intake Manifold Runner 1 Control	Solenoid Off: 12v, On: 1v
30	BR/WT	COP 6 Driver Control	5º, 55 mph: 8º
31, 33-34	---	Not Used	---
32	WT	Throttle Position Sensor	0.53-1.27v
35	WT/BL	Electronic Throttle Control Motor (+)	DC pulse signals
36	---	Not Used	---
37	BR/GN	Intake Manifold Runner 2 Control	Solenoid Off: 12v, On: 1v
38	---	Not Used	---
39	WT/GN	Engine Oil Temperature Sensor Signal	0.5-0.6v
40	WT/VT	Cylinder Head Temperature Sensor	0.5-0.6v
41	WT/BL	EGR Pressure Transducer Signal	0.95-1.05v
42	GY/BK	Knock Sensor 1 Signal Return	<0.050v
43	---	Not Used	---
44	WT/RD	HO2S-21 (B2 S1) Signal	0.1-1.1v
45	WT	HO2S-11 (B1 S1) Signal	0.1-1.1v
46	BR	Fuel Injector 2 Control	2.9-3.6 ms
47	BR/WT	Fuel Injector 1 Control	2.9-3.6 ms
48	BR/RD	COP 2 Driver Control	5º, 55 mph: 8º
49	WT/GN	Fuel Rail Pressure Sensor Signal	2.8v (39 psi)
50	WT/RD	Generator Monitor Signal	130 Hz (45%)
51	WT/BK	Knock Sensor 1 (+) Signal	0v
52	---	Not Used	---
53	WT/VT	CMP Sensor 1 Signal	6 Hz
54	WT/BK	CMP Sensor 2 Signal	6 Hz
55	GY/RD	CKP Sensor (-) Signal	390 Hz
56	WT/RD	CKP Sensor (+) Signal	390 Hz
57	WT/RD	Throttle Position Sensor	0.53-1.27v
58	BR	COP 1 Driver Control	5º, 55 mph: 8º
59	---	Not Used	---
60	BR/BL	COP 3 Driver Control	5º, 55 mph: 8º

2000-02 LS 3.9L V8 VIN A (A/T) C175A 58-Pin Connector

PCM Pin #	Wire Color	Circuit Description (58-Pin)	Value at Hot Idle
1-2	---	Not Used	---
3	GY	SCP Data Bus (+)	Digital Signals
4	BL	SCP Data Bus (-)	<0.050v
5	---	Not Used	---
6	BR/BL	EVAP CV Solenoid	0-10 Hz (0-100%)
7-8	---	Not Used	---
9	BR/BL	Ground, Switch	N/A
10-11	---	Not Used	---
12	BR/RD	EVAP Purge Solenoid	0-10 Hz (0-100%)
13	YL/GN	EEPROM Power	0.1v
14-16	---	Not Used	---
17	BR	Analog Signal Return	<0.050v
19	BR/GN	Secondary AIR System Relay	On: 1v, Off: 12v
20	YL	Reference Voltage	4.9-5.1v
21-23	---	Not Used	---
24	BK/YL	Power Ground	<0.1v
25	BK/RD	Power Ground	<0.1v
26	BK/RD	Power Ground	<0.1v
27	BK/RD	Power Ground	<0.1v
28	OG/GN	Generator Field Signal	0.5-10.5v
29	GY/WT	Signal Return Normal	<0.050v
30	---	Not Used	---
31	WT/BL	MAF Sensor Signal	0.6v, 55 mph: 1.7v
32	GN/YL	Vehicle Power (switched)	12-14v
33	GN/OG	Vehicle Power (switched)	12-14v
34	RD/BL	Secondary AIR Monitor	Pump On: 12v, Off: 0v
35	---	Not Used	---
36	WT/GN	Normal	N/A
37	WT	PSP Switch Signal	Straight: 5v, Turning: 0v
38	BR/BL	MAF Sensor Return	<0.050v
39	---	Not Used	---
40	OG	BPP Signal from ABS	Brake Off: 0v, On: 12v
41	WT/GN	Sensor Signal	N/A
42	WT/VT	A/C Pressure Switch Signal	0.9v (50 psi)
43	BR/YL	Power Ground	<0.1v
44	OG/YL	Keep Alive Power	12-14v
45	WT/VT	Sensor Signal	N/A
46	GY/OR	Signal 2	N/A
47-48	---	Not Used	---
49	WT/VT	Sensor Signal	N/A
50	WT/BK	Generator Monitor Signal	130 Hz (45%)
51	WT/VT	IAT Sensor Signal	1.5-2.5v
52	WT/VT	FTP Sensor Signal	2.6v (0" H2O - cap off)
53-55	---	Not Used	---
56	BK/OR	Ground, Switched	N/A
57	YL/BL	Keep Alive Power	12-14v
58	WT/VT	Sensor Signal	N/A

2000-02 LS 3.9L V8 VIN A (A/T) C175B 32-Pin Connector

PCM Pin #	Wire Color	Circuit Description (32-Pin)	Value at Hot Idle
1	BR	Shift Solenoid 'A' Control	12v, 55 mph: 1v
2	BR/RD	Shift Solenoid 'B' Control	12v, 55 mph: 1v
3	BR/WT	Shift Solenoid 'C' Control	12v, 55 mph: 1v
4	BR/YL	Shift Solenoid 'D' Control	12v, 55 mph: 1v
5	BR/WT	TCC Solenoid Control	12v, 55 mph: 1v
6	---	Not Used	---
7	BR/GN	Pressure Control A Solenoid	On: 1v, Off: 12v
9	WT	Digital TR Sensor (TR3A)	0v, 55 mph: 1.7v
10	WT/RD	Digital TR Sensor (TR4)	0v, 55 mph: 11v
11	---	Not Used	---
12	BR/WT	Pressure Control C Solenoid	On: 1v, Off: 12v
13	BR/YL	Pressure Control B Solenoid	On: 1v, Off: 12v
14, 19	---	Not Used	---
15	BR/BL	HO2S-12 (B1 S2) Heater	1v, Off: 12v
16	BR/GN	HO2S-22 (B2 S2) Heater	1v, Off: 12v
17	BR	Sensor Signal Return	<0.050v
18	WT/BL	Digital TR Sensor (TR2)	0v, 55 mph: 11v
21	WT/BK	Intermediate Speed Shaft	0 Hz, 55 mph: 1386 Hz
22	WT/GN	Digital TR Sensor (TR1)	0v, 55 mph: 11v
23	WT/RD	Trans. Fluid Temp. Sensor	2.10-2.40v
25	---	Not Used	---
26	WT/RD	Output Shaft Speed Sensor	0 Hz, 975 Hz
27	WT/VT	Turbine Shaft Speed Sensor	340 Hz, 1025 Hz
28	WT/BL	HO2S-12 (B1 S2) Signal	0.1-1.1v
29	WT/GN	HO2S-22 (B2 S2) Signal	0.1-1.1v
30	WT/BL	A/T Pressure Switch	Open: 12v, Closed: 0v
31-32	---	Not Used	---

Pin Connectors Graphic

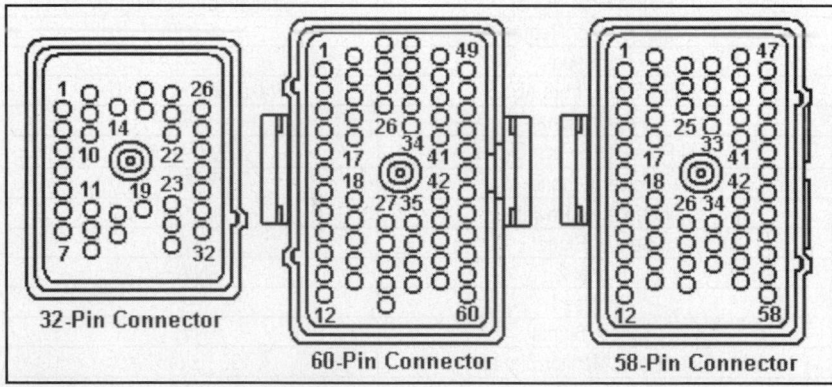

32-Pin Connector

60-Pin Connector

58-Pin Connector

Standard Colors and Abbreviations

Abbreviation	Color	Abbreviation	Color	Abbreviation	Color
BK	Black	GY	Gray	PK	Purple
BL	Blue	GN	Green	RD	Red
BR	Brown	LG	LT Green	TN	Tan
DB	Dark Blue	OR	Orange	WT	White
DG	DK Green	PK	Pink	YL	Yellow

2000-02 LS 3.9L V8 MFI VIN A (A/T) C175B 60 Pin Connector

PCM Pin #	Wire Color	Circuit Description (60 Pin)	Value at Hot Idle
1	BR/RD	COP 2 Driver Control	5°, 55 mph: 8° dwell
2	BR/WT	Injector 1 Control	2.8-3.6 ms
3-6	---	Not Used	---
7	BR	HO2S-11 (B1 S1) Heater	1v, Off: 12v
8	BR/RD	HO2S-21 (B2 S1) Heater	1v, Off: 12v
9	BR	IAC Solenoid Control	35% duty cycle
10, 15	---	Not Used	---
11	BR/GN	Injector 5 Control	2.8-3.6 ms
12	BR/WT	COP 6 Driver Control	5°, 55 mph: 8° dwell
13	BR/GN	COP 4 Driver Control	5°, 55 mph: 8° dwell
14	BR	Injector 2 Control	2.8-3.6 ms
16	BR/GN	VR Solenoid Control	0%, 55 mph: 45%
17	BR	Sensor Signal Return	<0.050v
18-19	---	Not Used	---
20	YL/BK	Sensor Voltage Reference	4.9-5.1v
21	BR/YL	Injector 6 Control	2.8-3.6 ms
22	BR/BL	COP 3 Driver Control	5°, 55 mph: 8° dwell
23	BR/YL	COP 5 Driver Control	5°, 55 mph: 8° dwell
24	BR/RD	Injector 3 Control	2.8-3.6 ms
25-27	---	Not Used	---
28	GY/RD	Generator Control Signal	0-130 Hz (varies)
29	BR/WT	Injector 7 Control	2.8-3.6 ms
30	BR	COP 7 Driver Control	5°, 55 mph: 8° dwell
31	BR	COP 1 Driver Control	5°, 55 mph: 8° dwell
32	BR/BL	Injector 4 Control	2.8-3.6 ms
33-35	---	Not Used	---
36	BR	Hydraulic Fan Solenoid	On: 1v, Off: 12v
37	BR	Injector 8 Control	2.8-3.6 ms
38	BR/RD	COP 8 Driver Control	5°, 55 mph: 8° dwell
39	---	Not Used	---
40	WT/VT	CHT Sensor Signal	0.7v (194°F)
41	WT/BL	DPFE Sensor Signal	0.95-1.05v
42	GY/BK	Knock Sensor 1 Signal	0v
43	GY/RD	Knock Sensor 2 Signal	0v
44	WT/RD	HO2S-21 (B2 S1) Signal	0.1-1.1v
45	WT	HO2S-11 (B1 S1) Signal	0.1-1.1v
46-48	---	Not Used	---
49	WT/GN	Fuel Rail Pressure Sensor	2.8v (39 psi)
50	WT/RD	Generator Field Signal	130 Hz (40%)
51	WT/BK	Knock Sensor 1 Ground	0v
52	WT/RD	Knock Sensor 2 Ground	0v
53	WT/VT	CMP Sensor Signal	6 Hz
54	---	Not Used	---
55	GY/RD	CKP Sensor (+) Signal	380 Hz
56	WT/RD	CKP Sensor (-) Signal	380 Hz
57	WT	TP Sensor Signal	0.53-1.27v
58-60	---	Not Used	---

2003-05 LS 3.9L V8 VIN A (A/T) C175A 58-Pin Connector

PCM Pin #	Wire Color	Circuit Description (58-Pin)	Value at Hot Idle
1	WT/RD	Accelerator Pedal Position Sensor	0.5-4.9v
2	---	Not Used	---
3	GY	SCP Data Bus (+)	Digital Signals
4	BL	SCP Data Bus (-)	<0.050v
5	BR	Signal Return	<0.050v
6-8	---	Not Used	---
9	BR/BL	A/C Clutch Relay Control	Relay Off: 12v, On: 1v
10-11	---	Not Used	---
12	BR/RD	EVAP Purge Solenoid Control	0-10 Hz (0-100%)
13	YL/GN	EEPROM Power	0.1v
14	WT/VT	Manual Transmission Switch (+) Signal	Switch Open: 12v, Closed: 1v
15	WT	Electronic Throttle Control Module	Digital Signals
16	WT/BL	Electronic Throttle Control Module	Digital Signals
17	BR	Electronic Throttle Control Module	Digital Signals
18-19	---	Not Used	---
20	YL	ETC Reference Voltage	4.9-5.1v
21	---	Not Used	---
22	WT/BK	Manual Transmission Switch (-) Signal	<0.050v
23	YL/RD	ETC Reference Voltage	4.9-5.1v
24	BK/RD	Power Ground	<0.1v
25	BK/RD	Power Ground	<0.1v
26	BK/RD	Power Ground	<0.1v
27	BK/RD	Power Ground	<0.1v
28	GN/OR	Brake Pressure Switch	Brake Off: 0v, On: 12v
29	---	Not Used	---
30	BR/BL	EVAP Vent Solenoid Control	0-10 Hz (0-100%)
31	WT/BL	MAF Sensor Signal	0.6v, 55 mph: 1.7v
32	GN/YL	Vehicle Power (Start-Run)	12-14v
33	GN/OR	Vehicle Power (Start-Run)	12-14v
34-37	---	Not Used	---
38	BR/BL	MAF Sensor Signal Return	<0.050v
39	---	Not Used	---
40	OG	Keep Alive Power	12-14v
41	WT/GN	Traction Control Disable Switch	Brake Off: 0v, On: 12v
42	WT/VT	A/C Pressure Switch Signal	0.9v (50 psi)
43	BR/YL	Power Ground	<0.1v
44	OG/YL	Keep Alive Power (CJB fuse)	12-14v
45-46	---	Not Used	---
47	WT/BK	Engine Cooling Fan Motor Control	Off: 12v, On: 1v
48	WT/BL	Restraint Control Module	Digital Signals
49	WT/VT	Rear Electronic Module	Digital Signals
50	---	Not Used	---
51	WT/VT	IAT Sensor Signal	1.5-2.5v
52	WT/VT	FTP Sensor Signal	2.6v (0" H2O - cap off)
53	GN/RD	Controller Area Network (+) Signal	0-7-0v
54	BL/RD	Controller Area Network (-) Signal	0-7-0v
55	YL	Reference Voltage	4.9-5.1v
56	BK/OR	Sensor Return	<0.050v
57	YL/BL	Speed Control Reference Voltage	4.9-5.1v
58	---	Not Used	---

2003-05 LS 3.9L V8 VIN A (A/T) C175B 32-Pin Connector

PCM Pin #	Wire Color	Circuit Description (32-Pin)	Value at Hot Idle
1	BR	Shift Solenoid 'A' Control	12v, 55 mph: 1v
2	BR/RD	Shift Solenoid 'B' Control	12v, 55 mph: 1v
3-6	---	Not Used	---
7	BR/GN	Pressure Control Solenoid 'A' Control	8.8v
8	BR/WT	Shift Solenoid 'C' Control	12v, 55 mph: 12v
9	WT	Digital TR Sensor (TR3A)	0v, 55 mph: 1.7v
10	WT/RD	Digital TR Sensor (TR4)	0v, 55 mph: 11v
11	---	Not Used	---
12	BR/WT	Pressure Control 'C' Solenoid Control	Off: 12v, On: 1v
13	BR/YL	Pressure Control 'B' Solenoid Control	Off: 12v, On: 1v
14	BR/RD	Signal Return	<0.050v
15	BR/BL	HO2S-12 (B1 S2) Heater Control	Off: 12v, On: 1v
16	BK/GN	HO2S-22 (B2 S2) Heater Control	Off: 12v, On: 1v
17	BR/YL	Shift Solenoid 'D' Control	12v, 55 mph: 1v
18	WT/BL	Digital TR Sensor (TR2)	0v, 55 mph: 11v
19	---	Not Used	---
20	BR/WT	Torque Converter Clutch Control	12v, 55 mph: 0.1v
21	WT/BK	Intermediate Speed Shaft Signal	0 Hz, 55 mph: 1386 Hz
22	WT/GN	Digital TR Sensor (TR1)	0v, 55 mph: 11v
23	WT/RD	Transmission Fluid Temperature Sensor	2.10-2.40v
24-25	---	Not Used	---
26	WT/RD	Output Shaft Speed Sensor Signal	0 Hz, 975 Hz
27	WT/VT	Turbine Shaft Speed Sensor Signal	340 Hz, 1025 Hz
28	WT/BL	HO2S-12 (B1 S2) Signal	0.1-1.1v
29	WT/GN	HO2S-22 (B2 S2) Signal	0.1-1.1v
30-32	---	Not Used	---

Wire Harness Connectors Graphic

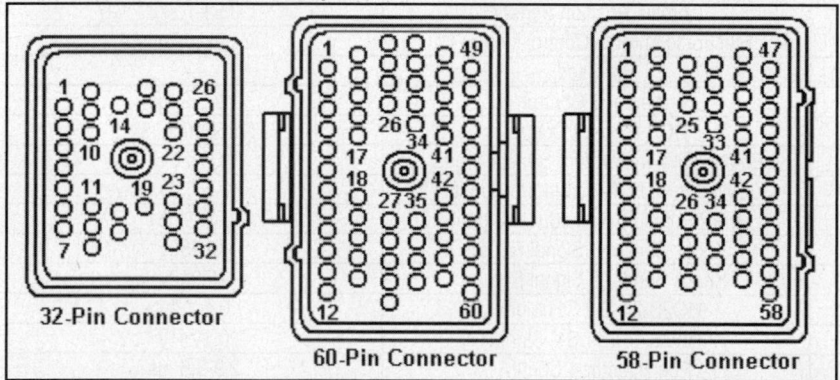

32-Pin Connector

60-Pin Connector

58-Pin Connector

Standard Colors and Abbreviations

Abbreviation	Color	Abbreviation	Color	Abbreviation	Color
BK	Black	GY	Gray	PK	Purple
BL	Blue	GN	Green	RD	Red
BR	Brown	LG	LT Green	TN	Tan
DB	Dark Blue	OR	Orange	WT	White
DG	DK Green	PK	Pink	YL	Yellow

2003-05 LS 3.9L V8 MFI VIN A (A/T) C175C 60 Pin Connector

PCM Pin #	Wire Color	Circuit Description (60 Pin)	Value at Hot Idle
1	WT/BK	Variable Valve Control Solenoid 1	DC pulse signals
2-3, 6	---	Not Used	---
4	WT/BL	Fuel Rail Temperature Sensor	1.7-3.5v (50-120ºF)
5	WT	Power Steering Pressure Switch	Straight: 0v, Turned: 5v
7	BR	HO2S-11 (B1 S1) Heater	On: 1v, Off: 12v
8	BR/RD	HO2S-21 (B2 S1) Signal	0.1-1.1v
9-10	---	Not Used	---
11	BR/GN	Fuel Injector 5 Control	2.9-3.6 ms
12	BR/YL	COP 5 Driver Control	5º, 55 mph: 8º
13	WT/VT	Variable Valve Timing Solenoid 2	DC pulse signals
14	YL/RD	Reference Voltage	4.9-5.1v
15	BR	Throttle Position Sensor Return	0.050v
16	BR/GN	EGR Pressure Transducer Return	<0.050v
17	BR/GN	Sensor Signal Return	<0.050v
18	---	Not Used	---
19	GY/RD	Generator Common Signal	0-130 Hz (varies)
20	BR	Fuel Injector 2 Control	2.9-3.6 ms
21	BR/YL	Fuel Injector 6 Control	2.9-3.6 ms
22	BR/WT	COP 6 Driver Control	5º, 55 mph: 8º
23	WT/GN	EGR Pressure Transducer Signal	0-3.1v
24	YL	Battery Power (at all times)	12-14v
25-26	---	Not Used	---
27	YL/BL	Electronic Throttle Control Motor (-)	DC pulse signals
28	BR/RD	Fuel Injector 5 Control	2.9-3.6 ms
29	BR/WT	Fuel Injector 7 Control	2.9-3.6 ms
30	BR	COP 7 Driver Control	5º, 55 mph: 8º
31, 33-34	---	Not Used	---
32	WT	Throttle Position Sensor	0.53-1.27v
35	WT/BL	Electronic Throttle Control Motor (+)	DC pulse signals
36	---	Not Used	---
37	BR	Injector 8 Control	2.9-3.6 ms
38	BR/RD	COP 8 Driver Control	5º, 55 mph: 8º
39	WT/GN	Engine Oil Temperature Sensor Signal	0.5-0.6v
40	WT/VT	Cylinder Head Temperature Sensor	0.5-0.6v
41	WT/BL	EGR Pressure Transducer Signal	0.95-1.05v
42	GY/BK	Knock Sensor 1 Signal Return	<0.050v
43	GY/RD	Knock Sensor 2 Signal Return	<0.050v
44	WT/RD	HO2S-21 (B2 S1) Signal	0.1-1.1v
45	WT	HO2S-11 (B1 S1) Signal	0.1-1.1v
46	BR/BL	Fuel Injector 4 Control	2.9-3.6 ms
47	BR/WT	Fuel Injector 1 Control	2.9-3.6 ms
48	BR/GN	COP 4 Driver Control	5º, 55 mph: 8º
49	WT/GN	Fuel Rail Pressure Sensor Signal	2.8v (39 psi)
50	WT/GN	Generator Monitor Signal	130 Hz (45%)
51	WT/BK	Knock Sensor 1 (+) Signal	0v
52	WT/RD	Knock Sensor 2 (+) Signal	0v
53	WT/VT	CMP Sensor 1 Signal	6 Hz
54	WT/BK	CMP Sensor 2 Signal	6 Hz
55	GY/RD	CKP Sensor (-) Signal	390 Hz
56	WT/RD	CKP Sensor (+) Signal	390 Hz
57	WT/RD	Throttle Position Sensor	0.53-1.27v
58	BR	COP 1 Driver Control	5º, 55 mph: 8º
59	BR/RD	COP 2 Driver Control	5º, 55 mph: 8º
60	BR/BL	COP 3 Driver Control	5º, 55 mph: 8º

MARK VII PIN TABLES

1990-92 Mark VII 5.0L V8 HO MFI VIN E (A/T) 60 Pin Connector

PCM Pin #	Wire Color	Circuit Description (60 Pin)	Value at Hot Idle
1	BK/OR	Keep Alive Power	12-14v
2	LG	Brake Switch Signal	Brake Off: 0v, On: 12v
3	DG/WT	VSS (+) Signal	0 Hz, 55 mph: 125 Hz
4	DG/YL	Ignition Diagnostic Monitor	20-31 Hz
4	BL/PK	Ignition Diagnostic Monitor	20-31 Hz
5	---	Not Used	---
6	PK/BL	VSS (-) Signal	0 Hz, 55 mph: 125 Hz
7	LG/YL	ECT Sensor Signal	0.5-0.6v
8	---	Not Used	---
9	TN/BL	MAF Sensor Return	<0.050v
10	LG/PK	A/C Cycling Clutch Switch	A/C On: 12v, Off: 0v
10	PK/BL	A/C Cycling Clutch Switch	A/C On: 12v, Off: 0v
11	LG/BK	Air Management 2 Solenoid	AM2 On: 1v, Off: 12v
12	TN/RD	Injector 3 Control	5.0-6.0 ms
13	TN/BK	Injector 4 Control	5.0-6.0 ms
14	TN/BK	Injector 5 Control	5.0-6.0 ms
15	TN/LG	Injector 6 Control	5.0-6.0 ms
16	BK/OR	Ignition System Ground	<0.050v
17	YL/BK	Self-Test Output, MIL Control	MIL On: 1v, Off: 12v
17	T/P	Self-Test Output, MIL Control	MIL On: 1v, Off: 12v
18-19	---	Not Used	---
20	BK	PCM Case Ground	<0.050v
21	WT/BL	IAC Motor Control	8-10v
22	BK/OR	Fuel Pump Control	On: 1v, Off: 12v
22	TN/LG	Fuel Pump Control	On: 1v, Off: 12v
23-24	---	Not Used	---
25	LG/PK	ACT Sensor Signal	1.5-2.5v
26	OR/WT	Reference Voltage	4.9-5.1v
27	BR/LG	EGR EVP Sensor Signal	0.4v
28	---	Not Used	---
29	DG/PK	HO2S-21 (B2 S1) Signal	0.1-1.1v
30	WT/PK	Neutral Drive Switch Signal	In 'N': 0v, Others: 5v
30	BL/YL	Neutral Drive Switch Signal	In 'N': 0v, Others: 5v
31	GY/YL	EVAP Purge Solenoid	12v, 55 mph: 1v
32	---	Not Used	---
33	DG	VR Solenoid Control	0%, 55 mph: 45%
34	BL/P	Data Output Link	Digital Signals
35	WT/PK	S/C Vent Solenoid Control	Vacuum Decreasing: 1v
36	YL/LG	Spark Output Signal	6.93v (50% d/cycle)
37	R, BK/YL	Vehicle Power	12-14v
38	GY/BK	S/C Vacuum Solenoid	Vacuum Increasing: 1v
39	BL/YL	Speed Control Switch Ground	<0.050v
40	BK/LG	Power Ground	<0.1v

1990-92 Mark VII 5.0L V8 HO MFI VIN E (A/T) 60 Pin Connector

PCM Pin #	Wire Color	Circuit Description (60 Pin)	Value at Hot Idle
41	OR/YL	Speed Control Solenoid (+)	S/C On at 55 mph: 12v
42	TN/OR	Injector 7 Control	5.0-6.0 ms
43	DB/LG	HO2S-11 (B1 S1) Signal	0.1-1.1v
43	BL/PK	HO2S-11 (B1 S1) Signal	0.1-1.1v
44	GY/RD	HO2S-11 (B1 S1) Ground	<0.050v
45	DB/LG	MAP Sensor Signal	107 Hz (sea level)
45	LG/BK	MAP Sensor Signal	107 Hz (sea level)
46	BK/WT	Analog Signal Return	<0.050v
47	GY/WT	TP Sensor Signal	0.8v, 55 mph: 1.1v
47	GY	TP Sensor Signal	0.8v, 55 mph: 1.1v
47	DG/BL	TP Sensor Signal	0.8v, 55 mph: 1.1v
48	WT/RD	Self-Test Input Signal	STI On: 0v, Off: 5v
48	WT/BK	Self-Test Input Signal	STI On: 0v, Off: 5v
48	TN/RD	Self-Test Input Signal	STI On: 0v, Off: 5v
49	OR	HO2S-21 (B2 S1) Ground	<0.050v
50	OR/RD	MAF Sensor Signal	0.8v, 55 mph: 1.8v
50	BL/BK	MAF Sensor Signal	0.8v, 55 mph: 1.8v
51	WT/RD	Air Management 1 Solenoid	AM1 On: 1v, Off: 12v
52	YL	Injector 8 Control	5.0-6.0 ms
53, 55	---	Not Used	---
54	OR/BL	A/C WOT Relay Control	On: 1v, Off: 12v
56	DB	PIP Sensor Signal	6.93v (50% d/cycle)
57	RD	Vehicle Power	12-14v
57	BK/YL	Vehicle Power	12-14v
58	TN	Injector 1 Control	5.0-6.0 ms
59	TN/RD	Injector 2 Control	5.0-6.0 ms
60	BK/LG	Power Ground	<0.1v

Pin Connector Graphic

PCM 60-PIN CONNECTOR

Terminal View of 60-Pin PCM Harness Connector

Standard Colors and Abbreviations

Abbreviation	Color	Abbreviation	Color	Abbreviation	Color
BK	Black	GY	Gray	PK	Purple
BL	Blue	GN	Green	RD	Red
BR	Brown	LG	LT Green	TN	Tan
DB	Dark Blue	OR	Orange	WT	White
DG	DK Green	PK	Pink	YL	Yellow

MARK VIII PIN TABLES

1993-95 Mark VIII 4.6L V8 MFI VIN V (A/T) 60 Pin Connector

PCM Pin #	Wire Color	Circuit Description (60 Pin)	Value at Hot Idle
1	YL	Keep Alive Power	12-14v
2	LG	Brake Switch Signal	Brake Off: 0v, On: 12v
3	DG/WT	VSS (+) Signal	0 Hz, 55 mph: 125 Hz
4	TN/YL	Ignition Diagnostic Monitor	20-31 Hz
5	DG/WT	TSS Sensor Signal	At 30 mph: 126-136 Hz
6	PK/OR	VSS (-) Signal	0 Hz, 55 mph: 125 Hz
7	LG/RD	ECT Sensor Signal	0.5-0.6v
8	PK/BK	Fuel Pump Monitor	12v, Off: 0v
9	TN/LB	MAF Sensor Return	<0.050v
10	BK/YL	A/C Cycling Clutch Switch	A/C On: 12v, Off: 0v
11	GY/YL	EVAP Purge Solenoid	12v, 55 mph: 1v
12	LG/OR	Injector 6 Control	4.0-4.2 ms
13	TN/RD	Injector 7 Control	4.0-4.4 ms
14	LB	Injector 8 Control	4.0-4.4 ms
15	TN/BK	Injector 5 Control	4.0-4.4 ms
16	OR/RD	Ignition System Ground	<0.050v
17	PK/LG	Self-Test Output & MIL	MIL On: 1v, Off: 12v
18	TN/OR	Data Bus (+) Signal	Digital Signals
19	PK/LB	Data Bus (-) Signal	Digital Signals
20	BK	PCM Case Ground	<0.050v
21	WT/LB	IAC Motor Control	8.3-11.5v
22	LB/OR	Fuel Pump Control	On: 1v, Off: 12v
23	YL/RD	Knock Sensor Signal	0v
24	DB/OR	CID Sensor Signal	6-7 Hz
25	GY	ACT Sensor Signal	1.5-2.5v
26	BR/WT	Reference Voltage	4.9-5.1v
27	BR/LG	DPFE EGR Sensor Signal	0.4v, 55 mph: 2.9v
28	---	Not Used	---
29	DG	Octane Adjust Switch	Closed: 0v, Open: 9.1v
30	LB/YL	MLP Sensor Signal	In 'P': 0v, in O/D: 1.7v
31	---	Not used	---
32	LG/BK	IMRC Solenoid Control	12v, 55 mph: 12v
33	BR/PK	VR Solenoid Control	0%, 55 mph: 45%
34	LB/PK	Data Output Link	Digital Signals
35	BR/LB	Injector 4 Control	4.0-4.4 ms
36	PK	Spark Angle Word Signal	6.93v (50% d/cycle)
37	RD	Vehicle Power	12-14v
38	WT/YL	EPC Solenoid Control	9.5v (20 psi)
39	BR/YL	Injector 3 Control	4.0-4.4 ms
40	BK/WT	Power Ground	<0.1v

1993-95 Mark VIII 4.6L V8 MFI VIN V (A/T) 60 Pin Connector

PCM Pin #	Wire Color	Circuit Description (60 Pin)	Value at Hot Idle
41	TN/WT	TCS (switch) Signal	TCS & O/D On: 12v
42	---	Not Used	---
43	RD/BK	HO2S-21 (B2 S1) Signal	0.1-1.1v
44	GY/LB	HO2S-11 (B1 S1) Signal	0.1-1.1v
45	---	Not Used	---
46	GY/RD	Analog Signal Return	<0.050v
47	GY/WT	TP Sensor Signal	0.8v, 55 mph: 1.1v
48	WT/PK	Self-Test Input Signal	STI On: 0v, Off: 5v
49	OR/BK	TOT Sensor Signal	2.10-2.40v
50	LB/RD	MAF Sensor Signal	0.8v, 55 mph: 1.8v
51	OR/YL	Shift Solenoid 1 Control	1v, 55 mph: 12v
52	PK/OR	Shift Solenoid 2 Control	12v, 55 mph: 1v
53	PK/YL	TCC Solenoid Control	12v, 55 mph: 9v
54	PK/YL	A/C WOT Relay Control	On: 1v, Off: 12v
55	WT/LG	TCIL (lamp) Control	Lamp On: 1v, Off: 12v
56	GY/OR	PIP Sensor Signal	6.93v (50% dwell)
57	RD	Vehicle Power	12-14v
58	TN	Injector 1 Control	4.0-4.4 ms
59	WT	Injector 2 Control	4.0-4.4 ms
60	BK/WT	Power Ground	<0.1v

Pin Connector Graphic

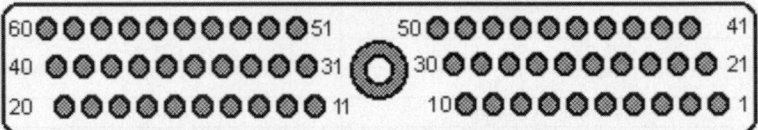

PCM 60-PIN CONNECTOR

Terminal View of 60-Pin PCM Harness Connector

Standard Colors and Abbreviations

Abbreviation	Color	Abbreviation	Color	Abbreviation	Color
BK	Black	GY	Gray	PK	Purple
BL	Blue	GN	Green	RD	Red
BR	Brown	LG	LT Green	TN	Tan
DB	Dark Blue	OR	Orange	WT	White
DG	DK Green	PK	Pink	YL	Yellow

1996-97 Mark VIII 4.6L V8 MFI VIN V (A/T) 104 Pin Connector

PCM Pin #	Wire Color	Circuit Description (104 Pin)	Value at Hot Idle
1	PK/OR	Shift Solenoid 2	1v, 55 mph: 12v
2	PK/LG	MIL (lamp) Control	MIL On: 1v, Off: 12v
3-4, 6-7	---	Not Used	---
5	W	EAM System Monitor	1v, 55 mph: 1v
8	DG/LG	IMRC Solenoid Monitor	5v, 55 mph: 5v
9-12	---	Not Used	---
13	YL/BK	Flash EEPROM Power	0.1v
15	PK/LB	Data Bus (-) Signal	Digital Signals
16	TN/OR	Data Bus (+) Signal	Digital Signals
14, 17-20	---	Not Used	---
21	DB	CKP Sensor (+) Signal	365-395 Hz
22	GY	CKP Sensor (-) Signal	365-395 Hz
23	---	Not Used	---
24	BK/WT	Power Ground	<0.1v
25	BK	PCM Case Ground	<0.050v
26	TN/WT	Coil Driver 1 Control	5°, 55 mph: 8° dwell
27	OR/YL	Shift Solenoid 1 Control	12v, 55 mph: 1v
28, 31, 34	---	Not Used	---
29	TN/WT	TCS (switch) Signal	TCS & O/D On: 12v
30	DG	Octane Adjust Switch	9.3v (Closed: 0v)
32	DG/PK	Knock Sensor 2 Signal	0v
33	PK/OR	VSS (-) Signal	0 Hz, 55 mph: 125 Hz
35	RD/LG	HO2S-12 (B1 S2) Signal	0.1-1.1v
36	TN/LB	MAF Sensor Return	<0.050v
37	OR/BK	TFT Sensor Signal	2.10-2.40v
38	LG/RD	ECT Sensor Signal	0.5-0.6v
39	GY	IAT Sensor Signal	1.5-2.5v
40	PK/BK	Fuel Pump Monitor	12v, Off: 0v
41	BK/YL	A/C Cycling Clutch Switch	A/C On: 12v
42	LG/BK	IMRC Solenoid Control	12v, 55 mph: 12v
43	LB/PK	Data Output Line	Digital Signals
44-46	---	Not Used	---
47	BR/PK	VR Solenoid Control	0%, 55 mph: 45%
48	TN/YL	Clean Tachometer Output	40-46 Hz
49-50	---	Not Used	---
51	BK/WT	Power Ground	<0.1v
52	TN/OR	Coil Driver 2 Control	5°, 55 mph: 8° dwell
53, 59	---	Not Used	---
54	PK/YL	TCC Solenoid Control	0%
55	Y	Keep Alive Power	12-14v
56	LG/BK	EVAP VMV (valve) Control	0-10 Hz (0-100%)
57	YL/RD	Knock Sensor 1 Signal	0v
58	DG/WT	VSS (+) Signal	0 Hz, 55 mph: 125 Hz
60	GY/LB	HO2S-12 (B1 S2) Signal	0.1-1.1v
61	PK/LG	HO2S-22 (B2 S2) Signal	0.1-1.1v
62-63	---	Not Used	---
64	LB/YL	TR Sensor Signal	In 'P': 0v, in O/D: 1.7v

1996-97 Mark VIII 4.6L V8 MFI VIN V (A/T) 104 Pin Connector

PCM Pin #	Wire Color	Circuit Description (104 Pin)	Value at Hot Idle
65	BR/LG	DPFE EGR Sensor Signal	0.20-1.30v
66-69	---	Not Used	---
70	OR/YL	EAM Solenoid Control	12v, 55 mph: 12v
71	RD	Vehicle Power	12-14v
72	TN/RD	Injector 7 Control	2.5-3.0 ms
73	TN/BK	Injector 5 Control	2.5-3.0 ms
74	BR/YL	Injector 3 Control	2.5-3.0 ms
75	TN	Injector 1 Control	2.5-3.0 ms
76, 77, 103	BK/WT	Power Ground	<0.1v
78	TN/LG	Coil Driver 3 Control	5°, 55 mph: 8° dwell
79	WT/LG	TCIL (lamp) Control	Lamp On: 1v, Off: 12v
80	LB/OR	Fuel Pump Control	On: 1v, Off: 12v
81	WT/YL	EPC Solenoid Control	8.3v (15 psi)
83	WT/LB	IAC Motor Control	30%, 55 mph: 45%
84	DG/WT	TSS Sensor Signal	43 Hz (647 rpm)
85	DB/OR	CMP Sensor	4-6 Hz
86	DG/WT	A/C Pressure Switch Signal	Open: 12v, Closed: 0v
87	RD/BK	HO2S-21 (B2 S1) Signal	0.1-1.1v
88	LB/RD	MAF Sensor Signal	0.6v, 55 mph: 1.7v
89	GY/WT	TP Sensor Signal	0.53-1.27v
90	BR/WT	Reference Voltage	4.9-5.1v
91	GY/RD	Analog Signal Return	<0.050v
92	LG	BPP Switch Signal	Brake Off: 0v, On: 12v
93	RD/WT	HO2S-11 (B1 S1) Heater	1v, Off: 12v
94	YL/LB	HO2S-21 (B2 S1) Heater	1v, Off: 12v
95	WT/BK	HO2S-12 (B1 S2) Heater	1v, Off: 12v
96	TN/YL	HO2S-22 (B2 S2) Heater	1v, Off: 12v
97	RD	Vehicle Power	12-14v
98	LB	Injector 8 Control	2.5-3.0 ms
99	LG/OR	Injector 6 Control	2.5-3.0 ms
100	BR/LB	Injector 4 Control	2.5-3.0 ms
101	WT	Injector 2 Control	2.5-3.0 ms
102	---	Not Used	---
104	TN/LB	Coil Driver 4 Control	5°, 55 mph: 8° dwell

Pin Connector Graphic

PCM 104-PIN CONNECTOR

```
1 ooooooooooooo     ooooooooooooo 26
27 ooooooooooooo    ooooooooooooo 52
53 ooooooooooooo  ⬤  ooooooooooooo 78
79 ooooooooooooo    ooooooooooooo 104
```

Terminal View of 104-Pin PCM Wiring Harness Connector

Standard Colors and Abbreviations

Abbreviation	Color	Abbreviation	Color	Abbreviation	Color
BK	Black	GY	Gray	PK	Purple
BL	Blue	GN	Green	RD	Red
BR	Brown	LG	LT Green	TN	Tan
DB	Dark Blue	OR	Orange	WT	White
DG	DK Green	PK	Pink	YL	Yellow

1998 Mark VIII 4.6L 4v V8 MFI VIN V (A/T) 104 Pin Connector

PCM Pin #	Wire Color	Circuit Description (104 Pin)	Value at Hot Idle
1	OR/YL	COP 5 Driver Control	5°, 55 mph: 8° dwell
2	PK/LG	MIL (lamp) Control	MIL On: 1v, Off: 12v
3, 24	BK/WT	Power Ground	<0.1v
5	BK/OR	Air Injection System Monitor	1v, 55 mph: 1v
6	OR/YL	Shift Solenoid 1 Control	1v, 55 mph: 1v
8	LG/BR	IMRC Monitor	5v, 55 mph: 5v
9	YL/WT	Fuel Level Indicator Signal	1.7v (1/2 full)
11	PK/OR	Shift Solenoid 2 Control	12v, 55 mph: 1v
13	PK	Flash EEPROM Power	0.5v
15	PK	Data Bus (-) Signal	Digital Signals
16	TN/OR	Data Bus (+) Signal	Digital Signals
20	PK/BK	Shift Solenoid 3 Control	12v, 55 mph: 1v
21	BK/PK	CKP Sensor (+) Signal	390-450 Hz
22	GY/YL	CKP Sensor (-) Signal	390-450 Hz
25	BK	PCM Case Ground	<0.050v
26	LG/WT	COP 1 Driver Control	5°, 55 mph: 8° dwell
27	LG/YL	COP 6 Driver Control	5°, 55 mph: 8° dwell
28	DB	Low Cooling Fan Control	Fan On: 1v, Off: 12v
29	TN/WT	TCS (switch) Signal	TCS & O/D On: 12v
30	DG	Octane Adjust Switch	9.3v (Closed: 0v)
31	YL/LG	PSP Switch Signal	Straight: 5v, Turning: 0v
32	DG/PK	Knock Sensor 2 Signal	0v
33	PK/OR	VSS (-) Signal	0 Hz, 55 mph: 125 Hz
34	YL/BK	Digital TR1 Sensor Signal	0v, 55 mph: 11v
35	RD/LG	HO2S-12 (B1 S2) Signal	0.1-1.1v
36	TN/LB	MAF Sensor Return	<0.050v
37	OR/BK	TFT Sensor Signal	2.10-2.40v
38	LG/RD	ECT Sensor Signal	0.5-0.6v
39	GY	IAT Sensor Signal	1.5-2.5v
40	DG/YL	Fuel Pump Monitor	12v, Off: 0v
41	PK	A/C High Pressure Switch	Open: 12v, Closed: 0v
42	BR	IMRC Solenoid Control	12v, 55 mph: 12v
43	WT/RD	Fuel Pump Control	On: 1v, Off: 12v
46	LG/PK	High Speed Fan Control	Fan On: 1v, Off: 12v
47	BR/PK	VR Solenoid Control	0%, 55 mph: 45%
48	OR/WT	Clean Tachometer Output	39-45 Hz
49	LB/BK	Digital TR2 Sensor Signal	0v, 55 mph: 11v
50	WT/BK	Digital TR4 Sensor Signal	0v, 55 mph: 11v
51, 77, 103	BK/WT	Power Ground	<0.1v
52	WT/PK	COP 2 Driver Control	5°, 55 mph: 8° dwell
53	DG/PK	COP 7 Driver Control	5°, 55 mph: 8° dwell
54	PK/YL	TCC Solenoid Control	12v, 55 mph: 1v
55	BR/PK	Keep Alive Power	12-14v
56	LG/BK	EVAP VMV (valve) Control	0-10 Hz (0-100%)
57	YL/RD	Knock Sensor 1 Signal	0v
58	GY/BK	VSS (+) Signal	0 Hz, 55 mph: 125 Hz
59	DG/WT	TSS Sensor Signal	40-45 Hz (645-650 rpm)
60	GY/LB	HO2S-11 (B1 S1) Signal	0.1-1.1v
61	PK/LG	HO2S-22 (B2 S2) Signal	0.1-1.1v
62	RD/PK	Fuel Tank Pressure Sensor	2.6v (0" H2O - cap off)
64	RD/LB	Digital TR Sensor Signal	In 'P': 0v, in O/D: 1.7v

1998 Mark VIII 4.6L 4v V8 MFI VIN V (A/T) 104 Pin Connector

PCM Pin #	Wire Color	Circuit Description (104 Pin)	Value at Hot Idle
65	BR/LG	DPFE Sensor Signal	0.5v, 55 mph: 3.1v
70	WT/OR	Air Mgmt. Solenoid Control	12v, 55 mph: 12v
67	PK/WT	EVAP Vent Valve	0-10 Hz (0-100%)
69	PK/YL	A/C WOT Relay Control	Relay On: 0v, Off: 12v
71, 97	RD	Vehicle Power	12-14v
72	TN/RD	Injector 7 Control	2.5-3-0 ms
73	TN/BK	Injector 5 Control	2.5-3-0 ms
74	BR/YL	Injector 3 Control	2.5-3-0 ms
75	TN	Injector 1 Control	2.5-3-0 ms
78	PK/LB	COP 3 Driver Control	5°, 55 mph: 8° dwell
79	WT/RD	COP 8 Driver Control	5°, 55 mph: 8° dwell
80	LB/OR	Fuel Pump Control	On: 1v, Off: 12v
81	WT/YL	EPC Solenoid Control	9v (17 psi)
83	WT/LB	IAC Motor Control	32% duty cycle
84	DG/YL	OSS Sensor Signal	132-136 (1355 rpm)
85	DB/OR	CMP Sensor Signal	4-6 Hz
86	TN/LG	A/C Pressure Transducer	0.8-1.4v
87	RD/BK	HO2S-21 (B2 S1) Signal	0.1-1.1v
88	LB/RD	MAF Sensor Signal	0.6v, 55 mph: 1.7v
89	GY/WT	TP Sensor Signal	0.53-1.27v
90	BR/WT	Reference Voltage	4.9-5.1v
91	GY/RD	Analog Signal Return	<0.050v
92	LG	Brake Switch Signal	Brake On: 12v, Off: 0v
93	RD/WT	HO2S-11 (B1 S1) Heater	1v, Off: 12v
94	YL/LB	HO2S-21 (B2 S1) Heater	1v, Off: 12v
95	WT/BK	HO2S-12 (B1 S2) Heater	1v, Off: 12v
96	TN/YL	HO2S-22 (B2 S2) Heater	1v, Off: 12v
98	LB	Injector 8 Control	2.5-3-0 ms
99	LG/OR	Injector 6 Control	2.5-3-0 ms
100	BR/LB	Injector 4 Control	2.5-3-0 ms
101	WT	Injector 2 Control	2.5-3-0 ms
104	PK/WT	Coil Driver 4 Control	5°, 55 mph: 8° dwell

Pin Connector Graphic

PCM 104-PIN CONNECTOR

Terminal View of 104-Pin PCM Wiring Harness Connector

Standard Colors and Abbreviations

Abbreviation	Color	Abbreviation	Color	Abbreviation	Color
BK	Black	GY	Gray	PK	Purple
BL	Blue	GN	Green	RD	Red
BR	Brown	LG	LT Green	TN	Tan
DB	Dark Blue	OR	Orange	WT	White
DG	DK Green	PK	Pink	YL	Yellow

TOWN CAR PIN TABLES

1991-95 Town Car 4.6L V8 MFI VIN W (A/T) 60 Pin Connector

PCM Pin #	Wire Color	Circuit Description (60 Pin)	Value at Hot Idle
1	YL	Keep Alive Power	12-14v
1	YL/BK	Keep Alive Power	12-14v
2 ('92-'95)	LG	Brake Switch Signal	Brake Off: 0v, On: 12v
3	YL/BL	VSS (+) Signal	0 Hz, 55 mph: 125 Hz
3	G/BK	VSS (+) Signal	0 Hz, 55 mph: 125 Hz
4	TN/YL	Ignition Diagnostic Monitor	20-31 Hz
4	WT/BL	Ignition Diagnostic Monitor	20-31 Hz
5	DG/WT	OSS Sensor (RPM)	30 mph: 260-1330 rpm
6	BK/BL	VSS (-) Signal	0 Hz, 55 mph: 125 Hz
6	PK/OR	VSS (-) Signal	0 Hz, 55 mph: 125 Hz
7	LG/RD	ECT Sensor Signal	0.5-0.6v
8	DG/YL	Fuel Pump Monitor	12v, Off: 0v
8	PK/BK	Fuel Pump Monitor	12v, Off: 0v
9	TN/BL	MAF Sensor Return	<0.050v
10	PK	A/C Cycling Clutch Switch	A/C On: 12v, Off: 0v
10	DG/OR	A/C Cycling Clutch Switch	A/C On: 12v, Off: 0v
10	PK/YL	A/C Cycling Clutch Switch	A/C On: 12v, Off: 0v
11	GY/YL	EVAP Purge Solenoid	On: 1v, Off: 12v
12	LG/OR	Injector 6 Control	4.0-4.4 ms
13	TN/RD	Injector 7 Control	4.0-4.4 ms
14	BL	Injector 8 Control	4.0-4.4 ms
15	TN/BK	Injector 5 Control	4.0-4.4 ms
15	TN	Injector 5 Control	4.0-4.4 ms
16	OR	Ignition System Ground	<01v
16	OR/RD	Ignition System Ground	<01v
17	PK/LG	Self-Test Output, MIL Control	MIL On: 1v, Off: 12v
17	TN/RD	Self-Test Output, MIL Control	MIL On: 1v, Off: 12v
18	TN/OR	Data Bus (+) Signal	Digital Signals
19	PK/BL	Data Bus (-) Signal	Digital Signals
20	BK	PCM Case Ground	<0.050v
21	WT	IAC Motor Control	9.8-11.8v
21	WT/BL	IAC Motor Control	9.8-11.8v
22	BL/OR	Fuel Pump Control	On: 1v, Off: 12v
23	---	Not Used	---
24	DG	CID Sensor Signal	5-9 Hz
24	DB/OR	CID Sensor Signal	5-9 Hz
25	LG/PK	ACT Sensor Signal	1.5-2.5v
25	GY	ACT Sensor Signal	1.5-2.5v
26	BR/WT	Reference Voltage	4.9-5.1v
27	BR/LG	DPFE EGR Sensor Signal	0.4v, 55 mph: 2.9v
28 ('91)	YL/LG	PSP Switch Signal	Straight: 0v, turned: 12v
29	WT/RD	Octane Adjust Switch	Closed: 0v, Open: 9.1v
29	DG	Octane Adjust Switch	Closed: 0v, Open: 9.1v
30 ('91)	WT/PK	NDS Signal	In 'N': 0v, Others: 5v
30 ('92-'95)	BL/YL	MLP Sensor Signal	In 'P': 0v, in O/D: 2.1v
31-32	---	Not Used	---
33	BR/PK	VR Solenoid Control	0%, 55 mph: 45%
34	DB/LG	Data Output Link	Digital Signals
35	BR/BL	Injector 4 Control	4.0-4.4 ms
36	PK	Spark Angle Word Signal	6.93v (50% d/cycle)
37, 57	RD	Vehicle Power	12-14v
38 ('92-'95)	WT/YL	EPC Solenoid Signal	9.5v (16-18 psi)

1991-95 Town Car 4.6L V8 MFI VIN W (A/T) 60 Pin Connector

PCM Pin #	Wire Color	Circuit Description (60 Pin)	Value at Hot Idle
39	BR/YL	Injector 3 Control	4.0-4.4 ms
40	BK/WT	Power Ground	<0.1v
41 ('93-'94)	TN/WT	TCS (switch) Signal	TCS & O/D On: 12v
42	---	Not Used	---
43	RD/BK	HO2S-11 (B1 S1) Signal	0.1-1.1v
44	GYL/BL	HO2S-21 (B2 S1) Signal	0.1-1.1v
45	---	Not Used	---
46	GY/RD	Analog Signal Return	<0.050v
47	GY/WT	TP Sensor Signal	0.8v, 55 mph: 1.1v
48	WT/PK	Self-Test Input Signal	STI On: 0v, Off: 5v
49	OR/BK	TOT Sensor Signal	2.10-2.40v
50	BL/RD	MAF Sensor Signal	0.7v, 55 mph: 2.1v
51 ('92-'95)	OR/YL	Shift Solenoid 1 Control	12v, 55 mph: 1v
52 ('92-'95)	PK/OR	Shift Solenoid 2 Control	1v, 55 mph: 12v
53 ('92-'95)	BR/OR	TCC Solenoid Control	12v, 30 mph: 11-12v
54	PK/YL	A/C WOT Relay Control	On: 1v, Off: 12v
54	OR/BL	A/C WOT Relay Control	On: 1v, Off: 12v
55 ('93-'95)	WT/LG	TCIL (lamp) Control	Lamp On: 1v, Off: 12v
56	GY/OR	PIP Sensor Signal	6.93v (50% dwell)
58	TN	Injector 1 Control	4.0-4.4 ms
59	WT	Injector 2 Control	4.0-4.4 ms
60	BK/WT	Power Ground	<0.1v

Pin Connector Graphic

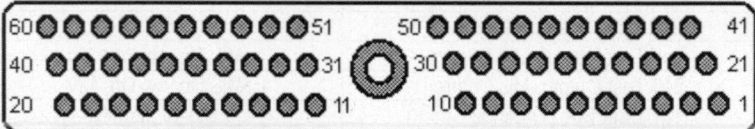

PCM 60-PIN CONNECTOR

Terminal View of 60-Pin PCM Harness Connector

Standard Colors and Abbreviations

Abbreviation	Color	Abbreviation	Color	Abbreviation	Color
BK	Black	GY	Gray	PK	Purple
BL	Blue	GN	Green	RD	Red
BR	Brown	LG	LT Green	TN	Tan
DB	Dark Blue	OR	Orange	WT	White
DG	DK Green	PK	Pink	YL	Yellow

1996-97 Town Car 4.6L V8 MFI VIN W (A/T) 104 Pin Connector

PCM Pin #	Wire Color	Circuit Description (104 Pin)	Value at Hot Idle
1	PK/OR	Shift Solenoid 2 Control	12v, 55 mph: 1v
2	PK/LG	MIL (lamp) Control	MIL On: 1v, Off: 12v
3-12, 14	---	Not Used	---
13	PK	Flash EEPROM Power	0.1v
15	PK/LB	Data Bus (-) Signal	Digital Signals
16	TN/OR	Data Bus (+) Signal	Digital Signals
17-20, 23	---	Not Used	---
21	BK/PK	CKP Sensor (+) Signal	440-490 Hz
22	GY/YL	CKP Sensor (-) Signal	440-490 Hz
24	BK/WT	Power Ground	<0.1v
25	BK	PCM Case Ground	<0.050v
26	TN/WT	Coil Driver 1 Control	5°, 55 mph: 8° dwell
27	OR/YL	Shift Solenoid 1 Control	1v, 55 mph: 12v
28, 31, 32	---	Not Used	---
29	TN/WT	TCS (switch) Signal	TCS & O/D On: 12v
30	DG	Octane Adjust Switch	9.3v (Closed: 0v)
33	PK/OR	VSS (-) Signal	0 Hz, 55 mph: 125 Hz
34	---	Not Used	---
35	RD/LG	HO2S-12 (B1 S2) Signal	0.1-1.1v
36	TN/LB	MAF Sensor Return	<0.050v
37	OR/BK	TFT Sensor Signal	2.10-2.40v
38	LG/RD	ECT Sensor Signal	0.5-0.6v
39	GY	IAT Sensor Signal	1.5-2.5v
40	DG/YL	Fuel Pump Monitor	12v, Off: 0v
41	BK/YL	A/C Cycling Clutch Switch	A/C On: 12v
42	---	Not Used	---
43	DB/LG	Data Output Link	Digital Signals
44	---	Not Used	---
45	DB	Low Speed Fan Control	Fan On: 1v, Off: 12v
46	---	Not Used	---
47	BR/PK	VR Solenoid Control	0%, 55 mph: 45%
48-50	---	Not Used	---
51	BK/WT	Power Ground	<0.1v
52	TN/OR	Coil Driver 2 Control	5°, 55 mph: 8° dwell
53	---	Not Used	---
54	PK/YL	TCC Solenoid Control	0%
55	YL/BK	Keep Alive Power	12-14v
56	LG/BK	EVAP VMV (valve) Control	0-10 Hz (0-100%)
57	---	Not Used	---
58	GY/BK	VSS (+) Signal	0 Hz, 55 mph: 125 Hz
59	---	Not Used	---
60	GY/LB	HO2S-11 (B1 S1) Signal	0.1-1.1v
61	PK/LG	HO2S-22 (B2 S2) Signal	0.1-1.1v
62	RD/PK	FTP Sensor Signal	2.6v (0" H2O - cap off)
63	---	Not Used	---
64	LB/YL	TR Sensor Signal	In 'P': 0v, in O/D: 1.7v

1996-97 Town Car 4.6L V8 MFI VIN W (A/T) 104 Pin Connector

PCM Pin #	Wire Color	Circuit Description (104 Pin)	Value at Hot Idle
65	BR/LG	DPFE EGR Sensor Signal	0.20-1.30v
66	---	Not Used	---
67	PK/WT	EVAP Vent Solenoid Control	0-10 Hz (0-100%)
68, 70	---	Not Used	---
69	PK/YL	A/C WOT Relay Control	On: 0.1v, Off: 12v
71	RD	Vehicle Power	12-14v
72	TN/RD	Injector 7 Control	3.3-3-5 ms
73	TN/BK	Injector 5 Control	3.3-3-5 ms
74	BR/YL	Injector 3 Control	3.3-3-5 ms
75	TN	Injector 1 Control	3.3-3-5 ms
76, 77	BK/WT	Power Ground	<0.1v
78	TN/LG	Coil Driver 3 Control	5°, 55 mph: 8° dwell
79	WT/LG	TCIL (lamp) Control	Lamp On: 1v, Off: 12v
80	LB/OR	Fuel Pump Control	On: 1v, Off: 12v
81	WT/YL	EPC Solenoid Control	9.5v (20 psi)
82, 86	---	Not Used	---
83	WT/LB	IAC Motor Control	30%, 55 mph: 45%
84	DG/WT	TSS Sensor Signal	43 Hz (650 rpm)
85	DG	CID Sensor Signal	5-7 Hz
87	RD/BK	HO2S-21 (B2 S1) Signal	0.1-1.1v
88	LB/RD	MAF Sensor Signal	0.6v, 55 mph: 1.7v
89	GY/WT	TP Sensor Signal	0.53-1.27v
90	BR/WT	Reference Voltage	4.9-5.1v
91	GY/RD	Analog Signal Return	<0.050v
92	LG	BPP Switch Signal	Brake Off: 0v, On: 12v
93	RD/WT	HO2S-11 (B1 S1) Heater	1v, Off: 12v
94	YL/LB	HO2S-21 (B2 S1) Heater	1v, Off: 12v
95	WT/BK	HO2S-12 (B1 S2) Heater	1v, Off: 12v
96	TN/YL	HO2S-22 (B2 S2) Heater	1v, Off: 12v
97	RD	Vehicle Power	12-14v
98	LB	Injector 8 Control	3.3-3-5 ms
99	LG/OR	Injector 6 Control	3.3-3-5 ms
100	BR/LB	Injector 4 Control	3.3-3-5 ms
101	WT	Injector 2 Control	3.3-3-5 ms
102	---	Not Used	---
103	BK/WT	Power Ground	<0.1v
104	TN/LB	Coil Driver 4 Control	5°, 55 mph: 8° dwell

Pin Connector Graphic

```
         PCM 104-PIN CONNECTOR
   1 ○○○○○○○○○○○○○    ○○○○○○○○○○○○○ 26
  27 ○○○○○○○○○○○○○    ○○○○○○○○○○○○○ 52
  53 ○○○○○○○○○○○○○  ⬤  ○○○○○○○○○○○○○ 78
  79 ○○○○○○○○○○○○○    ○○○○○○○○○○○○○ 104
```

Terminal View of 104-Pin PCM Wiring Harness Connector

Standard Colors and Abbreviations

Abbreviation	Color	Abbreviation	Color	Abbreviation	Color
BK	Black	GY	Gray	PK	Purple
BL	Blue	GN	Green	RD	Red
BR	Brown	LG	LT Green	TN	Tan
DB	Dark Blue	OR	Orange	WT	White
DG	DK Green	PK	Pink	YL	Yellow

1998-99 Town Car 4.6L 2v V8 MFI VIN W (A/T) 104 Pin Connector

PCM Pin #	Wire Color	Circuit Description (104 Pin)	Value at Hot Idle
1	OR/YL	COP 6 Driver Control	5°, 55 mph: 8° dwell
2	PK/LG	MIL (lamp) Control	MIL On: 1v, Off: 12v
3	BK/WT	Power Ground	<0.1v
6	OR/YL	Shift Solenoid 1 Control	1v, 55 mph: 1v
9	YL/WT	Fuel Level Indicator Signal	1.7v (1/2 full)
11	PK/OR	Shift Solenoid 2 Control	12v, 55 mph: 1v
12	WT/LG	TCIL (lamp) Control	Lamp On: 1v, Off: 12v
13	PK	Flash EEPROM Power	0.5v
15	PK/LB	Data Bus (-) Signal	Digital Signals
16	TN/OR	Data Bus (+) Signal	Digital Signals
21	BK/PK	CKP Sensor (+) Signal	440-490 Hz
22	GY/YL	CKP Sensor (-) Signal	440-490 Hz
24	BK/WT	Power Ground	<0.1v
25	BK	PCM Case Ground	<0.050v
26	LG/WT	COP 1 Driver Control	5°, 55 mph: 8° dwell
27	LG/YL	COP 5 Driver Control	1v, 30 mph: 12v
28	RD/OR	Low Speed Fan Control	Fan On: 1v, Off: 12v
29	TN/WT	TCS (switch) Signal	TCS & O/D On: 12v
30	WT/RD	Octane Adjust Switch	9.3v (Closed: 0v)
33	PK/OR	VSS (-) Signal	0 Hz, 55 mph: 125 Hz
34	YL/BK	Digital TR1 Sensor Signal	In 'P': 0v, 55 mph: 11v
35	RD/LG	HO2S-11 (B1 S1) Signal	0.1-1.1v
36	TN/LB	MAF Sensor Return	<0.050v
37	OR/BK	TFT Sensor Signal	2.10-2.40v
39	GY	IAT Sensor Signal	1.5-2.5v
40	DG/YL	Fuel Pump Monitor	12v, Off: 0v
41	BK/YL	A/C Cycling Clutch Switch	Switch On: 12v, Off: 0v
45	OR/RD	CHTIL (lamp) Control	Lamp On: 1v, Off: 12v
46	LG/RD	High Speed Fan Control	Fan On: 1v, Off: 12v
47	BR/PK	VR Solenoid Control	0%, 55 mph: 45%
49	LB/BK	Digital TR2 Sensor Signal	In 'P': 0v, 55 mph: 11v
50	WT/BK	Digital TR4 Sensor Signal	In 'P': 0v, 55 mph: 11v
51	BK/WT	Power Ground	<0.1v
52	WT/PK	COP 3 Driver Control	5°, 55 mph: 8° dwell
53	OR/PK	COP 4 Driver Control	5°, 55 mph: 8° dwell
54	PK/YL	TCC Solenoid Control	0%, 55 mph: 100%
55	RD/WT	Keep Alive Power	12-14v
56	LG/BK	EVAP Purge Valve Control	0-10 Hz (0-100%)
57	YL/RD	Knock Sensor 1 Signal	0v
58	GY/BK	VSS (+) Signal	0 Hz, 55 mph: 125 Hz
60	GY/LB	HO2S-11 (B1 S1) Signal	0.1-1.1v
61	PK/LG	HO2S-22 (B2 S2) Signal	0.1-1.1v
62	RD/PK	FTP Sensor Signal	2.6v (0" H2O - cap off)
64	DB/YL	Digital TR Sensor Signal	In 'P': 0v, in O/D: 1.7v

1998-99 Town Car 4.6L 2v V8 MFI VIN W (A/T) 104 Pin Connector

PCM Pin #	Wire Color	Circuit Description (104 Pin)	Value at Hot Idle
65	BR/LG	DPFE Sensor Signal	0.20-1.30v
66	YL/LG	CHT Sensor Signal	0.7v (194ºF)
67	PK/WT	EVAP Vent Solenoid Control	0-10 Hz (0-100%)
69	OR/LB	A/C WOT Cutoff Relay	On: 1v, Off: 12v
71	RD	Vehicle Power	12-14v
72	TN/RD	Injector 7 Control	3.4-3-7 ms
73	TN/BK	Injector 5 Control	3.4-3-7 ms
74	BR/YL	Injector 3 Control	3.4-3-7 ms
75	TN	Injector 1 Control	3.4-3-7 ms
78	PK/LB	COP 7 Driver Control	5º, 55 mph: 8º dwell
79	WT/RD	COP 8 Driver Control	5º, 55 mph: 8º dwell
80	LB/OR	Fuel Pump Control	On: 1v, Off: 12v
81	WT/YL	EPC Solenoid Control	9.5v (20 psi)
83	WT/LB	IAC Motor Control	34% duty cycle
84	DB/YL	OSS Sensor Signal	0 Hz, 55 mph: 216 Hz
85	DB/OR	CMP Sensor Signal	6-7 Hz
86	WT/BK	A/C High Pressure Switch	Open: 12v, Closed: 0v
87	RD/BK	HO2S-21 (B2 S1) Signal	0.1-1.1v
88	LB/RD	MAF Sensor Signal	0.6v, 55 mph: 1.8v
89	GY/WT	TP Sensor Signal	0.53-1.27v
90	BR/WT	Reference Voltage	4.9-5.1v
91	GY/RD	Analog Signal Return	<0.050v
92	LG	BPP Switch Signal	Brake On: 12v, Off: 0v
93	RD/WT	HO2S-11 (B1 S1) Heater	1v, Off: 12v
94	YL/LB	HO2S-21 (B2 S1) Heater	1v, Off: 12v
95	WT/BK	HO2S-12 (B1 S2) Heater	1v, Off: 12v
96	TN/YL	HO2S-22 (B2 S2) Heater	1v, Off: 12v
97	RD	Vehicle Power	12-14v
98	LB	Injector 8 Control	3.4-3-7 ms
99	LG/OR	Injector 6 Control	3.4-3-7 ms
100	BR/LB	Injector 4 Control	3.4-3-7 ms
101	WT	Injector 2 Control	3.4-3-7 ms
103	BK/WT	Power Ground	<0.1v
104	PK/WT	COP 2 Driver Control	5º, 55 mph: 8º dwell

Pin Connector Graphic

```
PCM 104-PIN CONNECTOR
 1 ◦◦◦◦◦◦◦◦◦◦◦◦◦      ◦◦◦◦◦◦◦◦◦◦◦◦◦ 26
27 ◦◦◦◦◦◦◦◦◦◦◦◦◦      ◦◦◦◦◦◦◦◦◦◦◦◦◦ 52
53 ◦◦◦◦◦◦◦◦◦◦◦◦◦   ●  ◦◦◦◦◦◦◦◦◦◦◦◦◦ 78
79 ◦◦◦◦◦◦◦◦◦◦◦◦◦      ◦◦◦◦◦◦◦◦◦◦◦◦◦ 104
```

Terminal View of 104-Pin PCM Wiring Harness Connector

Standard Colors and Abbreviations

Abbreviation	Color	Abbreviation	Color	Abbreviation	Color
BK	Black	GY	Gray	PK	Purple
BL	Blue	GN	Green	RD	Red
BR	Brown	LG	LT Green	TN	Tan
DB	Dark Blue	OR	Orange	WT	White
DG	DK Green	PK	Pink	YL	Yellow

2000-01 Town Car 4.6L 2v V8 VIN W (A/T) 104 Pin Connector

PCM Pin #	Wire Color	Circuit Description (104 Pin)	Value at Hot Idle
1	OG/YL	COP 6 Driver Control	5°, 55 mph: 8° dwell
2	---	Not Used	---
3	BK/WT	Power Ground	<0.1v
4-5	---	Not Used	---
6	OG/YL	Shift Solenoid 'A' Control	1v, 55 mph: 1v
7-10	---	Not Used	---
11	VT/OR	Shift Solenoid 'B' Control	12v, 55 mph: 1v
12, 14	---	Not Used	---
13	VT	Flash EEPROM Power	0.1v
15	PK/LB	Data Bus (-) Signal	Digital Signals
16	TN/OG	Data Bus (+) Signal	Digital Signals
17-20	---	Not Used	---
21	BK/PK	CKP Sensor (+) Signal	440-490 Hz
22	GY/YL	CKP Sensor (-) Signal	440-490 Hz
23-24	---	Not Used	---
25	BK	PCM Case Ground	<0.050v
26	LG/WT	COP 1 Driver Control	5°, 55 mph: 8° dwell
27	LG/YL	COP 5 Driver Control	1v, 30 mph: 12v
28	RD/OG	Low Speed Fan Control	Fan On: 1v, Off: 12v
29	TN/WT	TCS (switch) Signal	TCS & O/D On: 12v
30-33	---	Not Used	---
34	YL/BK	Digital TR1 Sensor Signal	In 'P': 0v, 55 mph: 11v
35	RD/LG	HO2S-12 (B1 S2) Signal	0.1-1.1v
36	TN/LB	MAF Sensor Return	<0.050v
37	OG/BK	TFT Sensor Signal	2.10-2.40v
38	---	Not Used	---
39	GY	IAT Sensor Signal	1.5-2.5v
40	DG/YL	Fuel Pump Monitor	12v, Off: 0v
41	BK/YL	A/C Cycling Clutch Switch	Switch On: 12v, Off: 0v
42-45	---	Not Used	---
46	LG/VT	High Speed Fan Control	Fan On: 1v, Off: 12v
47	BR/PK	VR Solenoid Control	0%, 55 mph: 45%
48	---	Not Used	---
49	LB/BK	Digital TR2 Sensor Signal	In 'P': 0v, 55 mph: 11v
50	WT/BK	Digital TR4 Sensor Signal	In 'P': 0v, 55 mph: 11v
51	BK/WT	Power Ground	<0.1v
52	WT/PK	COP 3 Driver Control	5°, 55 mph: 8° dwell
53	DG/VT	COP 4 Driver Control	5°, 55 mph: 8° dwell
54	VT/YL	TCC Solenoid Control	0%, 55 mph: 100%
55	RD/WT	Keep Alive Power	12-14v
56	LG/BK	EVAP Purge Valve Control	0-10 Hz (0-100%)
57-59	---	Not Used	---
60	GY/LB	HO2S-11 (B1 S1) Signal	0.1-1.1v
61	VT/LG	HO2S-22 (B2 S2) Signal	0.1-1.1v
62	RD/BK	FTP Sensor Signal	2.6v (0" H2O - cap off)
63	---	Not Used	---
64	LB/YL	Digital TR Sensor Signal	In 'P': 0v, in O/D: 1.7v
65 ('00)	BR/LG	DPFE Sensor Signal	0.20-1.30v
65 ('01)	BR/LG	DPFE Sensor Signal	0.95-1.05v
66	YL/LG	CHT Sensor Signal	0.7v (194°F)
67	VT/WT	EVAP CV Solenoid	0-10 Hz (0-100%)
68	---	Not Used	---

2000-01 Town Car 4.6L 2v V8 VIN W (A/T) 104 Pin Connector

PCM Pin #	Wire Color	Circuit Description (104 Pin)	Value at Hot Idle
69	OG/LB	A/C WOT Cutoff Relay	On: 1v, Off: 12v
70	---	Not Used	---
71	RD	Vehicle Power	12-14v
72	TN/RD	Injector 7 Control	3.4-3-7 ms
73	TN/BK	Injector 5 Control	3.4-3-7 ms
74	BR/YL	Injector 3 Control	3.4-3-7 ms
75	TN	Injector 1 Control	3.4-3-7 ms
76	---	Not Used	---
77	BK/WT	Power Ground	<0.1v
78	PK/LB	COP 7 Driver Control	5º, 55 mph: 8º dwell
79	WT/RD	COP 8 Driver Control	5º, 55 mph: 8º dwell
80	LB/OR	Fuel Pump Control	On: 1v, Off: 12v
81	WT/YL	EPC Solenoid Control	9.5v (20 psi)
82	---	Not Used	---
83	WT/LB	IAC Motor Control	34% duty cycle
84	DB/YL	OSS Sensor Signal	0 Hz, 55 mph: 216 Hz
85	DB/OR	CMP Sensor Signal	6-7 Hz
86	WT/BK	A/C High Pressure Switch	Open: 12v, Closed: 0v
87	RD/BK	HO2S-21 (B2 S1) Signal	0.1-1.1v
88	LB/RD	MAF Sensor Signal	0.6v, 55 mph: 1.8v
89	GY/WT	TP Sensor Signal	0.53-1.27v
90	BR/WT	Reference Voltage	4.9-5.1v
91	GY/RD	Analog Signal Return	<0.050v
92	---	Not Used	---
93	RD/WT	HO2S-11 (B1 S1) Heater	1v, Off: 12v
94	YL/LB	HO2S-21 (B2 S1) Heater	1v, Off: 12v
95	WT/BK	HO2S-12 (B1 S2) Heater	1v, Off: 12v
96	TN/YL	HO2S-22 (B2 S2) Heater	1v, Off: 12v
97	---	Not Used	---
98	LB	Injector 8 Control	3.4-3-7 ms
99	LG/OR	Injector 6 Control	3.4-3-7 ms
100	BR/LB	Injector 4 Control	3.4-3-7 ms
101	WT	Injector 2 Control	3.4-3-7 ms
102	---	Not Used	---
103	BK/WT	Power Ground	<0.1v
104	PK/WT	COP 2 Driver Control	5º, 55 mph: 8º dwell

Pin Connector Graphic

PCM 104-PIN CONNECTOR

Terminal View of 104-Pin PCM Wiring Harness Connector

Standard Colors and Abbreviations

Abbreviation	Color	Abbreviation	Color	Abbreviation	Color
BK	Black	GY	Gray	PK	Purple
BL	Blue	GN	Green	RD	Red
BR	Brown	LG	LT Green	TN	Tan
DB	Dark Blue	OR	Orange	WT	White
DG	DK Green	PK	Pink	YL	Yellow

2002-04 Town Car 4.6L 2v V8 VIN W (A/T) 104 Pin Connector

PCM Pin #	Wire Color	Circuit Description (104 Pin)	Value at Hot Idle
1	OG/YL	COP 6 Driver Control	5°, 55 mph: 8° dwell
2, 4-5	---	Not Used	---
3	BK/WT	Power Ground	<0.1v
6	OG/YL	Shift Solenoid 'A' Control	1v, 55 mph: 1v
7	YL/LG	Generator Regulator 'S' Terminal	0-130 Hz
8-9	---	Not Used	---
10	LG/RD	EGR System Module Signal	0.95-1.05v
11	VT/DG	Shift Solenoid 'B' Control	12v, 55 mph: 1v
12, 14	---	Not Used	---
13	VT	Flash EEPROM Power	0.1v
15	PK/LB	SCP Bus (-) Signal	Digital Signals
16	TN/OG	SCP Bus (+) Signal	Digital Signals
17	GY/OG	Passive Antitheft RX Signal	Digital Signals
18	WT/LG	Passive Antitheft TX Signal	Digital Signals
19-20	---	Not Used	---
21	BK/PK	CKP Sensor (+) Signal	440-490 Hz
22	GY/YL	CKP Sensor (-) Signal	440-490 Hz
23	---	Not Used	---
24	BK/WT	Power Ground	<0.1v
25	BK	PCM Case Ground	<0.050v
26	LG/WT	COP 1 Driver Control	5°, 55 mph: 8° dwell
27	LG/YL	COP 5 Driver Control	1v, 30 mph: 12v
28	RD/OG	Engine Cooling Fan Motor	Fan Off: 12v, On: 1v
29	TN/WT	TCS (switch) Signal	TCS & O/D On: 12v
30	OG/RD	Antitheft Indicator Control	Indicator Off: 12v, On: 1v
31	YL/LG	PSP Switch Signal	Straight: 0v, Turned: 12v
32	DG/VT	Knock Sensor (+) Signal	0v
33	---	Not Used	---
34	YL/BK	Digital TR1 Sensor Signal	In 'P': 0v, 55 mph: 11v
35	RD/LG	HO2S-12 (B1 S2) Signal	0.1-1.1v
36	TN/LB	MAF Sensor Return	<0.050v
37	OG/BK	TFT Sensor Signal	2.10-2.40v
38	BK/WT	Engine Air Temperature Sensor	1.5-2.5v
39	GY	Intake Air Temperature Sensor	1.5-2.5v
40	DG/YL	Fuel Pump Monitor	12v, Off: 0v
41	PK/LB	A/C Cycling Clutch Switch	Switch On: 12v, Off: 0v
42-43	---	Not Used	---
44	GY/RD	Starter Relay Control	Relay Off: 12v, On: 1v
45	LG/RD	Generator/Battery Indicator Control	Indicator Off: 12v, On: 1v
47	BR/PK	EGR System Module Signal	0.95-1.05v
46, 48	---	Not Used	---
49	LB/BK	Digital TR2 Sensor Signal	In 'P': 0v, 55 mph: 11v
50	WT/BK	Digital TR4 Sensor Signal	In 'P': 0v, 55 mph: 11v
51	BK/WT	Power Ground	<0.1v
52	WT/PK	COP 3 Driver Control	5°, 55 mph: 8° dwell
53	DG/VT	COP 4 Driver Control	5°, 55 mph: 8° dwell
54	VT/YL	TCC Solenoid Control	0%, 55 mph: 100%
55	RD/WT	Keep Alive Power	12-14v
56	LG/BK	EVAP Purge Valve Control	0-10 Hz (0-100%)
57	YL/RD	Knock Sensor (-) Signal	<0.050v
58-59	---	Not Used	---
60	GY/LB	HO2S-11 (B1 S1) Signal	0.1-1.1v

2002-04 Town Car 4.6L 2v V8 VIN W (A/T) 104 Pin Connector

PCM Pin #	Wire Color	Circuit Description (104 Pin)	Value at Hot Idle
61	VT/LG	HO2S-22 (B2 S2) Signal	0.1-1.1v
62	RD/BK	FTP Sensor Signal	2.6v (0" H2O - cap off)
63	OG/LG	Injector Pressure Sensor	2.7-3.7v (105-130 psi)
64	LB/YL	Digital TR Sensor Signal	In 'P': 0v, in O/D: 1.7v
65	BR/LG	EGR System Module Signal	0.95-1.05v
66	YL/LG	Cylinder Head Temperature Sensor Signal	0.7v (194°F)
67	VT/WT	EVAP CV Solenoid	0-10 Hz (0-100%)
68	GY/BK	Vehicle Speed Sensor (+) Signal	Moving: AC pulse signals
69	OG/LB	A/C WOT Cutoff Relay	Off: 12v, On: 1v
70	YL	Generator Battery Indicator	Indicator Off: 12v, On: 1v
71	RD	Vehicle Power (Start-Run)	12-14v
72	TN/RD	Injector 7 Control	3.4-3-7 ms
73	TN/BK	Injector 5 Control	3.4-3-7 ms
74	BR/YL	Injector 3 Control	3.4-3-7 ms
75	TN	Injector 1 Control	3.4-3-7 ms
76	BK/WT	Power Ground	<0.1v
77	BK/WT	Power Ground	<0.1v
78	PK/LB	COP 7 Driver Control	5°, 55 mph: 8° dwell
79	WT/RD	COP 8 Driver Control	5°, 55 mph: 8° dwell
80	LB/OR	Fuel Pump Control	On: 1v, Off: 12v
81	WT/YL	EPC Solenoid Control	9.5v (20 psi)
82	---	Not Used	---
83	WT/LB	IAC Motor Control	34% duty cycle
84	DB/YL	OSS Sensor Signal	0 Hz, 55 mph: 216 Hz
85	DB/OR	CMP Sensor Signal	6-7 Hz
86	WT/BK	A/C High Pressure Switch	Open: 12v, Closed: 0v
87	RD/BK	HO2S-21 (B2 S1) Signal	0.1-1.1v
88	LB/RD	MAF Sensor Signal	0.6v, 55 mph: 1.8v
89	GY/WT	TP Sensor Signal	0.53-1.27v
90	BR/WT	Reference Voltage	4.9-5.1v
91	GY/RD	Sensor Ground	<0.050v
92	---	Not Used	---
93	RD/WT	HO2S-11 (B1 S1) Heater	1v, Off: 12v
94	YL/LB	HO2S-21 (B2 S1) Heater	1v, Off: 12v
95	WT/BK	HO2S-12 (B1 S2) Heater	1v, Off: 12v
96	TN/YL	HO2S-22 (B2 S2) Heater	1v, Off: 12v
97	RD	System Power (Start-Run)	12-14v
98	LB	Injector 8 Control	3.4-3-7 ms
99	LG/OR	Injector 6 Control	3.4-3-7 ms
100	BR/LB	Injector 4 Control	3.4-3-7 ms
101	WT	Injector 2 Control	3.4-3-7 ms
102	---	Not Used	---
103	BK/WT	Power Ground	<0.1v
104	PK/WT	COP 2 Driver Control	5°, 55 mph: 8° dwell

Pin Connector Graphic

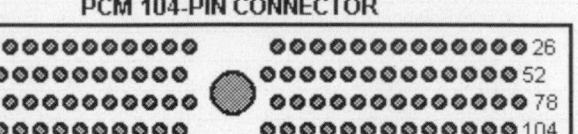

PCM 104-PIN CONNECTOR

Terminal View of 104-Pin PCM Wiring Harness Connector

1990 Town Car 5.0L V8 MFI VIN F (A/T) 60 Pin Connector

PCM Pin #	Wire Color	Circuit Description (60 Pin)	Value at Hot Idle
1	YL/BK	Keep Alive Power	12-14v
2	---	Not Used	---
3	YL/BL	VSS (+) Signal	0 Hz, 55 mph: 125 Hz
4	WT/BL	Ignition Diagnostic Monitor	20-31 Hz
5	DG/WT	Anti-Lock Brake Indicator	Digital Signals
6	BK/BL	VSS (-) Signal	0 Hz, 55 mph: 125 Hz
7	LG/YL	ECT Sensor Signal	0.5-0.6v
8	PK/BL	Data Bus (-) Signal	Digital Signals
9	TN/BL	MAF Sensor Return	<0.050v
10	PK	A/C Cycling Clutch Switch	A/C On: 12v, Off: 0v
11	OR/YL	Air Mgmt. 2 Solenoid Control	AM2 On: 1v, Off: 12v
12	BR/YL	Injector 3 Control	5.0-6.2 ms
13	BR/BL	Injector 4 Control	5.0-6.2 ms
14	TN/BL	Injector 5 Control	5.0-6.2 ms
15	LG	Injector 6 Control	5.0-6.2 ms
16	BK/OR	Ignition System Ground	<0.050v
17	TN/RD	Self-Test Output, MIL Control	MIL On: 1v, Off: 12v
18	---	Not Used	---
19	PK/BK	Fuel Pump Monitor	12v, Off: 0v
20	BK	PCM Case Ground	<0.050v
21	BL/OR	IAC Motor Control	5.0-11.5v
21	LG/WT	IAC Motor Control	5.0-11.5v
22	BL/OR	Fuel Pump Control	On: 1v, Off: 12v
22	TN/LG	Fuel Pump Control	On: 1v, Off: 12v
23	---	Not Used	---
24	---	Not Used	---
25	LG/PK	ACT Sensor Signal	1.5-2.5v
26	OR/WT	Reference Voltage	4.9-5.1v
27	BR/LG	EGR EVP Sensor Signal	0.4v, 55 mph: 2.9v
28	BL/BK	S/C Command Switch Signal	All speeds: 6.7v
29	DG/PK	HO2S-11 (B1 S1) Signal	0.1-1.1v
30	BL/YL	Neutral Drive Switch Signal	In 'N': 0v, Others: 5v
30	WT/PK	Neutral Drive Switch Signal	In 'N': 0v, Others: 5v
31	GY/YL	EVAP Purge Solenoid	12v, 55 mph: 1v
32	OR	Thermactor Air Divert Sol.	TAD On: 1v, Off: 12v
33	DG	VR Solenoid Control	0%, 55 mph: 45%
34	DB/LG	Data Output Link	Digital Signal
35	WT/BK	S/C Vent Solenoid Control	Vacuum Decreasing: 1v
35	WT/PK	S/C Vent Solenoid Control	Vacuum Decreasing: 1v
36	GY/OR	Spark Output Signal	6.93v (50% d/cycle)
37	RD	Vehicle Power	12-14v
38	YL	Thermactor Air Bypass SOL	TAB On: 1v, Off: 12v
39	GY/BK	Speed Control Switch Ground	<0.050v
40	BK/LG	Power Ground	<0.1v

1990 Town Car 5.0L V8 MFI VIN F (A/T) 60 Pin Connector

PCM Pin #	Wire Color	Circuit Description (60 Pin)	Value at Hot Idle
41	---	Not Used	---
42	TN/OR	Injector 7 Control	5.0-6.2 ms
43	DB/YL	HO2S-22 (B2 S2) Signal	0.1-1.1v
43	DG/YL	HO2S-22 (B2 S2) Signal	0.1-1.1v
44	TN/LG	Data Bus (+) Signal	Digital Signals
45	LG/BK	Barometric Pressure Sensor	± 4 159 Hz (Sea Level)
46	BK/WT	Analog Signal Return	<0.050v
47	DG/LG	TP Sensor Signal	0.8v, 55 mph: 1.1v
48	WT/RD	Self-Test Input Signal	STI On: 0v, Off: 5v
49	GYL/BL	HO2S-11 (B1 S1) Ground	<0.050v
50	DB/OR	MAF Sensor Signal	0.8v, 55 mph: 1.8v
51	GY/BK	Air Mgmt. 1 Solenoid Control	AM1 On: 1v, Off: 12v
52	BL	Injector 8 Control	5.0-6.2 ms
53	---	Not Used	---
54	OR/BL	A/C WOT Relay Control	On: 1v, Off: 12v
55	---	Not Used	---
56	DB	PIP Sensor Signal	6.93v (50% d/cycle)
57	RD	Vehicle Power	12-14v
58	TN	Injector 1 Control	5.0-6.2 ms
59	WT	Injector 2 Control	5.0-6.2 ms
60	BK/LG	Power Ground	<0.1v

Pin Connector Graphic

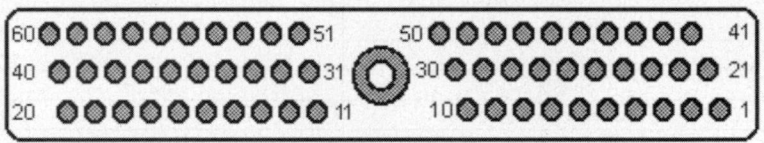

PCM 60-PIN CONNECTOR

Terminal View of 60-Pin PCM Harness Connector

Standard Colors and Abbreviations

Abbreviation	Color	Abbreviation	Color	Abbreviation	Color
BK	Black	GY	Gray	PK	Purple
BL	Blue	GN	Green	RD	Red
BR	Brown	LG	LT Green	TN	Tan
DB	Dark Blue	OR	Orange	WT	White
DG	DK Green	PK	Pink	YL	Yellow

COLONY PARK PIN TABLES

1990-91 Colony Park 5.0L V8 MFI VIN F (A/T) 60 Pin Connector

PCM Pin #	Wire Color	Circuit Description (60 Pin)	Value at Hot Idle
1	BK/OR	Keep Alive Power	12-14v
2	LG	Brake Switch Signal	Brake Off: 0v, On: 12v
3	DG/WT	VSS (+) Signal	0 Hz, 55 mph: 125 Hz
4	DG/YT	Ignition Diagnostic Monitor	20-31 Hz
5	---	Not Used	---
6	PK/LB	VSS (-) Signal	0 Hz, 55 mph: 125 Hz
7	LG/YL	ECT Sensor Signal	0.5-0.6v
8-9	---	Not Used	---
10	LG/PK	A/C Cycling Clutch Signal	A/C Off: 0v, On: 12v
11	LG/BK	Air Management 2 Solenoid	On: 1v, Off: 12v
12	BR/YL	Injector 3 Control	5.7-6.2 ms
13	BR/LB	Injector 4 Control	5.7-6.2 ms
14	TL/BL	Injector 5 Control	5.7-6.2 ms
15	LG	Injector 6 Control	5.7-6.2 ms
16	BK/OR	Ignition System Ground	<0.050v
17	YL/BK	Self-Test Output, MIL Control	MIL Off: 12v, On: 1v
18-19	---	Not Used	---
20	BK	Case Ground	<0.050v
21	WT/BL	IAC Motor Control	9.0-11.5v
22	TN/LG	Fuel Pump Control	Relay Off: 12v, On: 1v
23-24	---	Not Used	---
25	LG/PK	ACT Sensor Signal	1.5-2.5v
26	OR/WT	Reference Voltage	4.9-5.1v
27	BR/LG	EVP Sensor Signal	0.4v, 55 mph: 1.5v
28	---	Not Used	---
29	DG/PK	HO2S-21 (B1 S1) Signal	0.1-1.1v
30	WT/PK	Neutral Drive Switch	In 'N': 0v, Others: 5v
31	GY/YL	Canister Purge Solenoid	12v, 55 mph: 1v
32	---	Not Used	---
33	DG	VR Solenoid Control	0%, 55 mph: 0-45%
34	LB/PK	Data Output Link	Digital Signals
35	WT/PK	S/C Vent Solenoid Control	Vacuum Decreasing: 1v
36	YL/LG	Spark Output Signal	6.93v (50% d/cycle)
37	RD	Vehicle Power	12-14v
38	GY/BK	S/C Vacuum Solenoid Control	Vacuum Increasing: 1v
39	GY/BK	Speed Control Switch Ground	<0.050v
40	BK	Power Ground	<0.1v

1990-91 Colony Park 5.0L V8 MFI VIN F (A/T) 60 Pin Connector

PCM Pin #	Wire Color	Circuit Description (60 Pin)	Value at Hot Idle
41	OR/YL	Speed Control Servo (+) SIG	Solenoid On: 12v
42	TN/OR	Injector 7 Control	5.7-6.2 ms
42	TN/RD	Injector 7 Control	5.7-6.2 ms
43	DB/LG	HO2S-21 (B1 S1) Signal	0.1-1.1v
45	DB/LG	MAP Sensor Signal	107 Hz (sea level)
45	LG/BK	MAP Sensor Signal	107 Hz (sea level)
46	BK/WT	Analog Signal Return	<0.050v
47	DG/LG	TP Sensor Signal	0.8v, 55 mph: 1.1v
48	WT/RD	Self-Test Input Signal	STI On: 1v, Off: 5v
49	OR	HO2S-11 (Bank 1) Ground	<0.050v
50	OR/RD	Speed Command Control Sw.	6.7v
51	WT/RD	Air Management 1 Solenoid	AM1 On: 1v, Off: 12v
52	BL	Injector 8 Control	5.7-6.2 ms
52	TN/RD	Injector 8 Control	5.7-6.2 ms
53	---	Not Used	---
54	OR/BL	A/C WOT Relay Control	Relay Off: 12v, On: 1v
55	---	Not Used	---
56	DB	PIP Sensor Signal	6.93v (50% d/cycle)
57	RD	Vehicle Power	12-14v
58	TN	Injector 1 Control	5.7-6.2 ms
59	TN/WT	Injector 2 Control	5.7-6.2 ms
59	WT	Injector 2 Control	5.7-6.2 ms
60	BK/LG	Power Ground	<0.1v

Pin Connector Graphic

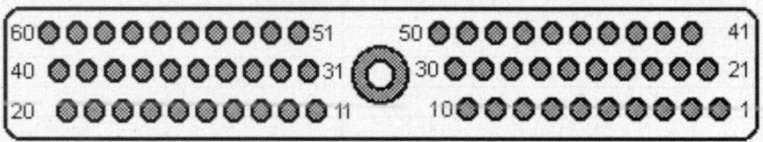

PCM 60-PIN CONNECTOR

Terminal View of 60-Pin PCM Harness Connector

Standard Colors and Abbreviations

Abbreviation	Color	Abbreviation	Color	Abbreviation	Color
BK	Black	GY	Gray	PK	Purple
BL	Blue	GN	Green	RD	Red
BR	Brown	LG	LT Green	TN	Tan
DB	Dark Blue	OR	Orange	WT	White
DG	DK Green	PK	Pink	YL	Yellow

COUGAR PIN TABLES

1999 Cougar 2.0L I4 4v MFI VIN 3 (All) 104 Pin Connector

PCM Pin #	Wire Color	Circuit Description (104 Pin)	Value at Hot Idle
1	BK/BL	Shift Solenoid 'B' Control	12v, 55 mph: 1v
2	BK/OR	MIL (lamp) Control	MIL On: 1v, Off: 12v
3-11	---	Not Used	---
12	WT	Fuel Level Indicator Signal	1.7v (1/2 full)
13	WT/BL	Flash EEPROM Power	0.5v
14	---	Not Used	---
15	BL	Data Bus (-) Signal	Digital Signals
16	GY	Data Bus (+) Signal	Digital Signals
17	GY/OR	Passive Anti-Theft System	Digital Signals
18	---	Not Used	---
19	BK/WT	High Speed Fan Control	Fan On: 1v, Off: 12v
20	BK/BL	Injector 3 Control	2.3-2.9 ms
21	WT/RD	CKP Sensor (+) Signal	450-480 Hz
22	BR/RD	CKP Sensor (-) Signal	450-480 Hz
24	BK/YL	Power Ground	<0.1v
25	BK/RD	Chassis Ground	<0.050v
26	BR/BL	Coil Driver 1 Control	5° dwell
27	BK/YL	Shift Solenoid 1 Control	1v, 55 mph: 12v
29	PK/BK	TCS (switch) Signal	TCS & O/D On: 12v
29	GN/BK	TCS (switch) Signal	TCS & O/D On: 12v
30	---	Not Used	---
31	WT	PSP Switch Signal	Straight: 1v, turned: 2-4v
32-33	---	Not Used	---
34	WT/GN	Passive Anti-Theft System	Digital Signals
35	WT/BL	HO2S-12 (B1 S2) Signal	0.1-1.1v
36	BR/BL	MAF Sensor Return	<0.050v
37	WT/RD	TFT Sensor Signal	2.10 to 2.40v
38	WT/GN	ECT Sensor Signal	0-5-0.6v
39	WT/PK	IAT Sensor Signal	1.5-2.5v
40	PK/BK	Fuel Pump Monitor	12v, Off: 0v
40	GN/BK	Fuel Pump Monitor	12v, Off: 0v
41	PK/BL	A/C High Pressure Cut-Off	Closed: 12v, Open: 0v
42	---	Not Used	---
43	WT	Fuel Flow Sensor	Digital Signal
44	BK/RD	VCT Actuator Control	VCT Off: 12v, On: 1v
45	BK/BL	Low Speed Fan Control	Fan On: 1v, Off: 12v
46-47	---	Not Used	---
48	WT/BK	Clean Tachometer Output	22-31 Hz
49-50	---	Not Used	---
51	BK/YL	Power Ground	<0.1v
52	BR/GN	Coil Driver 2 Control	5° dwell
53	---	Not Used	---
54	BK/BL	Fuel Pump Control	Relay Off: 12v, On: 1v
55	OR/YL	Keep Alive Power	12-14v
56	BK/OR	EVAP Purge Valve	0-10 Hz (0-100%)
57	WT/BK	Knock Sensor 1 Signal	0v
58	WT/BL	VSS (+) Signal	0 Hz, 55 mph: 125 Hz
59	---	Not Used	---
60	WT	HO2S-11 (B1 S1) Signal	0.1-1.1v
61	---	Not Used	---
62	WT/PK	FTP Sensor Signal	2.6v (0" H2O - cap off)
63	WT/GN	Fuel Pressure Transducer	2.8v (39 psi)
64	WT	Clutch Pedal Position Switch	Clutch In: 0v, Out: 5v

1999 Cougar 2.0L I4 4v MFI VIN 3 (All) 104 Pin Connector

PCM Pin #	Wire Color	Circuit Description (104 Pin)	Value at Hot Idle
65-66	---	Not Used	---
67	BK/OR	EVAP CV Solenoid	0-10 Hz (0-100%)
68	---	Not Used	---
69	BK/YL	A/C WOT Relay Control	Relay Off: 12v, On: 1v
70	BK/WT	Injector 1 Control	2.3-2.9 ms
71	GN/YL	Vehicle Power	12-14v
72, 75	---	Not Used	---
73	BK/YL	HO2S-11 (B1 S1) Signal	0.1-1.1v
74	BK/BL	Injector 3 Control	2.3-2.9 ms
76	BR/WT	CMP and TSS Ground	<0.050v
77	BK/YL	Power Ground	<0.1v
78	---	Not Used	---
79	WT/BK	TCIL (lamp) Control	Lamp On: 1v, Off: 12v
80	BK/WT	TCC Solenoid Control	12v, 55 mph: 1v
81	BK/RD	EPC Solenoid Control	8.9v
82	BK/BL	PATS Sensor Signal	Digital Signal
83	BK/YL	IAC Motor Control	9v (33% d/cycle)
84	WT/PK	TSS Sensor Signal	40-56 Hz
85	WT/PK	CMP Sensor Signal	5-7 Hz
86	BK/WT	A/C Pressure Switch	Switch On: 12v (Open)
87	BR/YL	Knock Sensor 1 (+) Signal	0v
88	WT/BL	MAF Sensor Signal	0.8v
89	WT	TP Sensor Signal	0.53-1.27v
90	YL	Reference Voltage	4.9-5.1v
91	BR	Analog Signal Return	<0.050v
92	PK	Brake Pedal Position Switch	Brake Off: 0v, On: 12v
93	BK/YL	HO2S-11 (B1 S1) Heater	1v, Off: 12v
94	---	Not Used	---
95	BK/OR	Injector 4 Control	2.3-2.9 ms
96	BK/YL	Injector 2 Control	2.3-2.9 ms
97	GN/YL	Vehicle Power	12-14v
98-99	---	Not Used	---
100	BK/OR	HO2S-12 (B1 S2) Heater	1v, Off: 12v
101	---	Not Used	---
102	BK/OR	Shift Solenoid 3 Control	8v, 55 mph: 8-9v
103	BK/YL	Power Ground	<0.1v
104	---	Not Used	---

Pin Connector Graphic

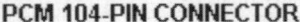

```
PCM 104-PIN CONNECTOR
 1 ●●●●●●●●●●●●    ●●●●●●●●●●●●● 26
27 ●●●●●●●●●●●●    ●●●●●●●●●●●●● 52
53 ●●●●●●●●●●●●  ●  ●●●●●●●●●●●● 78
79 ●●●●●●●●●●●●    ●●●●●●●●●●●● 104
```

Terminal View of 104-Pin PCM Wiring Harness Connector

Standard Colors and Abbreviations

Abbreviation	Color	Abbreviation	Color	Abbreviation	Color
BK	Black	GY	Gray	PK	Purple
BL	Blue	GN	Green	RD	Red
BR	Brown	LG	LT Green	TN	Tan
DB	Dark Blue	OR	Orange	WT	White
DG	DK Green	PK	Pink	YL	Yellow

2000-02 Cougar 2.0L I4 4v MFI VIN 3 (All) 104 Pin Connector

PCM Pin #	Wire Color	Circuit Description (104 Pin)	Value at Hot Idle
1	---	Not Used	---
2	BK/OR	MIL (lamp) Control	MIL On: 1v, Off: 12v
3-10	---	Not Used	---
6	BK/YL	Shift Solenoid 'B' Control	12v, 55 mph: 1v
8	BK/BL	IMRC Solenoid Control	0v
11	BK/BL	Shift Solenoid 'A' Control	1v, 55 mph: 12v
12	WT	Fuel Level Indicator Signal	1.7v (1/2 full)
13	WT/BL	Flash EEPROM Power	0.1v
14	---	Not Used	---
15	BL	Data Bus (-) Signal	Digital Signals
16	GY	Data Bus (+) Signal	Digital Signals
17	BK/WT	High Speed Cooling Fan	Relay Off: 12v, On: 1v
18	---	Not Used	---
19	GY/OR	PAT System	Digital Signals
20	BK/BL	Injector 3 Control	2.3-2.9 ms
21	WT/RD	CKP Sensor (+) Signal	450-480 Hz
22	BR/RD	CKP Sensor (-) Signal	450-480 Hz
24, 51	BK/YL	Power Ground	<0.1v
25	BK/RD	Chassis Ground	<0.050v
26	BR/BL	Coil Driver 1 Control	5° dwell
27	BK/YL	VTEC Solenoid	1v
28	WT/BK	VSS (-) Signal	0 Hz
29	GN/BK	TCS (switch) Signal	TCS & O/D On: 12v
30, 32-33	---	Not Used	---
31	WT	PSP Switch Signal	Straight: 1v, turned: 2-4v
34	WT/GN	Passive Anti-Theft System	Digital Signals
35	WT/BL	HO2S-12 (B1 S2) Signal	0.1-1.1v
36	BR/BL	MAF Sensor Return	<0.050v
37	WT/RD	TFT Sensor Signal	2.10 to 2.40v
38	WT/GN	ECT Sensor Signal	0-5-0.6v
39	WT/PK	IAT Sensor Signal	1.5-2.5v
40	PK/BK	Fuel Pump Monitor	12v, Off: 0v
40	GN/BK	Fuel Pump Monitor	12v, Off: 0v
41	PK/BL	A/C High Pressure Cut-Off	Closed: 12v, Open: 0v
42	---	Not Used	---
43	WT	Fuel Flow Sensor	Digital Signals
44	BK/RD	VCT Actuator Control	VCT Off: 12v, On: 1v
45	BK/BL	Low Speed Fan Control	Fan On: 1v, Off: 12v
46-47	---	Not Used	---
48	WT/BK	Clean Tachometer Output	22-31 Hz
49-50, 53	---	Not Used	---
52	BR/GN	Coil Driver 2 Control	5° dwell
54	BK/BL	Fuel Pump Control	Relay Off: 12v, On: 1v
55	OR/YL	Keep Alive Power	12-14v
56	BK/OR	EVAP Purge Valve	0-10 Hz (0-100%)
57	WT/BK	Knock Sensor 1 Signal	0v
58	WT/BL	VSS (+) Signal	0 Hz, 55 mph: 125 Hz
59, 61	---	Not Used	---
60	WT	HO2S-11 (B1 S1) Signal	0.1-1.1v
62	WT/PK	FTP Sensor Signal	2.6v (0" H2O - cap off)
63	WT/GN	Fuel Pressure Transducer	2.8v (39 psi)
64	WT	Clutch Pedal Position Switch	Clutch In: 0v, Out: 5v

2000-02 Cougar 2.0L I4 4v MFI VIN 3 (All) 104 Pin Connector

PCM Pin #	Wire Color	Circuit Description (104 Pin)	Value at Hot Idle
65 ('00)	WT/BL	DPFE Sensor Signal	0.20-1.30v
65 ('01-'02)	WT/BL	DPFE Sensor Signal	0.95-1.05v
65-66	---	Not Used	---
67	BK/OR	EVAP CV Solenoid	0-10 Hz (0-100%)
68	BK/BL	Cooling Fan Relay Control	Fan On: 1v, Off: 12v
69	BK/YL	A/C WOT Relay Control	Relay Off: 12v, On: 1v
70	BK/WT	Injector 1 Control	2.3-2.9 ms
71	GN/TN	Vehicle Power	12-14v
72, 74-75	---	Not Used	---
73	BK/YL	HO2S-11 (B1 S1) Signal	0.1-1.1v
76	BN/WT	CMP and TSS Ground	<0.050v
77	BK/YL	Power Ground	<0.1v
78	BN/YL	Ignition Transistor Ground	<0.050v
79	---	Not Used	---
80	BK/WT	TCC Solenoid Control	12v, 55 mph: 1v
81	BK/RD	EPC Solenoid Control	8.9v
82	BK/BL	PATS Sensor Signal	Digital Signals
83	BK/YL	IAC Motor Control	9v (33% d/cycle)
84	WT/VT	TSS Sensor Signal	40-56 Hz
85	WT/VT	CMP Sensor Signal	5-7 Hz
86	BK/BL	A/C Pressure Switch	Switch On: 12v (Open)
87	WT/RD	Knock Sensor 1 (+) Signal	0.1-1.1v
88	WT/BL	MAF Sensor Signal	0.8v
89	WT	TP Sensor Signal	0.53-1.27v
90	YL	Reference Voltage	4.9-5.1v
91	BN	Analog Signal Return	<0.050v
92	GN/RD	Brake Pedal Position Switch	Brake Off: 0v, On: 12v
93	BK/GN	HO2S-11 (B1 S1) Heater	1v, Off: 12v
94, 98-99	---	Not Used	---
95	BK/OR	Injector 4 Control	2.3-2.9 ms
96	BK/YL	Injector 2 Control	2.3-2.9 ms
97	GN/TN	Vehicle Power	12-14v
100	BK/OR	HO2S-12 (B1 S2) Heater	1v, Off: 12v
101	---	Not Used	---
102	BK/OR	Shift Solenoid 'C' Control	8v, 55 mph: 8-9v
103	BK/YL	Power Ground	<0.1v
104	---	Not Used	---

Pin Connector Graphic

PCM 104-PIN CONNECTOR

Terminal View of 104-Pin PCM Wiring Harness Connector

Standard Colors and Abbreviations

Abbreviation	Color	Abbreviation	Color	Abbreviation	Color
BK	Black	GY	Gray	PK	Purple
BL	Blue	GN	Green	RD	Red
BR	Brown	LG	LT Green	TN	Tan
DB	Dark Blue	OR	Orange	WT	White
DG	DK Green	PK	Pink	YL	Yellow

1999 Cougar 2.5L V6 4v MFI VIN L (All) 104 Pin Connector

PCM Pin #	Wire Color	Circuit Description (104 Pin)	Value at Hot Idle
1	BK/BL	Shift Solenoid 2 Control	1v, 55 mph: 12v
2	BK/OR	MIL (lamp) Control	MIL On: 1v, Off: 12v
4-7	---	Not Used	---
8	BK/BL	IMRC Solenoid Control	5v, 55 mph: 5v
9-11	---	Not Used	---
12	WT	Fuel Level Indicator Signal	1.7v (1/2 full)
13	WT/BL	Flash EEPROM Power	0.5v
14, 18	---	Not Used	---
15	BL	Data Bus (-) Signal	Digital Signals
16	GY	Data Bus (+) Signal	Digital Signals
17	GY/OR	Passive Anti-Theft System	Digital Signals
19	BK/WT	High Speed Fan Control	Fan On: 1v, Off: 12v
20	BK/BL	Injector 3 Control	2.1-2.5 ms
21	WT/RD	CKP Sensor (+) Signal	410-440 Hz
22	BR/RD	CKP Sensor (-) Signal	410-440 Hz
23, 28, 30	---	Not Used	---
24	BK/YL	Power Ground	<0.1v
25	BK/RD	Chassis Ground	<0.050v
26	BR/BL	Coil Driver 1 Control	5° dwell
27 ('99)	BK/YL	Shift Solenoid 1 Control	12v, 55 mph: 1v
29	PK/BK	TCS (switch) Signal	TCS & O/D On: 12v
31	WT	PSP Switch Signal	Straight: 0.1v, turned: 12v
32	BR/YL	Knock Sensor 1 Signal	0v
33	BR/BL	VSS (-) Signal	0 Hz, 55 mph: 125 Hz
34	WT/GN	Passive Anti-Theft System	Digital Signal
35	WT/BL	HO2S-12 (B1 S2) Signal	0.1-1.1v
36	BR/BL	MAF Sensor Return	<0.050v
37	WT/RD	TFT Sensor Signal	2.10 to 2.40v
38	WT/GN	ECT Sensor Signal	0-5-0.6v
39	WT/PK	IAT/ACT Sensor Signal	1.5-2.5v
40	PK/BK	Fuel Pump Monitor	12v, Off: 0v
40	GN/BK	Fuel Pump Monitor	12v, Off: 0v
41	PK/BL	A/C High Pressure Cut-Out	A/C Switch On: 12v
42	BK/RD	IMRC Solenoid Signal	12v, 55 mph: 12v
43	WT	Fuel Flow Sensor	Varies: 0-5.1v
45	BK/BL	Low Speed Fan Control	Fan On: 1v, Off: 12v
46, 50, 59	---	Not Used	---
47	BK/GN	VR Solenoid Control	0%, 55 mph: 40%
48	WT/BK	Clean Tachometer Output	35-42 Hz
51	BK/YL	Power Ground	<0.1v
52	BR/GN	Coil Driver 2 Control	5° dwell
54	BK/BL	Fuel Pump Control	Relay Off: 12v, On: 1v
55	OR/YL	Keep Alive Power	12-14v
56	BK/OR	EVAP Purge Solenoid	0-10 Hz (0-100%)
56	BK/BL	EVAP Purge Solenoid	0-10 Hz (0-100%)
57	WT/BK	Knock Sensor 1 Signal	0v
58	WT/PK	VSS (+) Signal	0 Hz, 55 mph: 125 Hz
60	WT	HO2S-11 (B1 S1) Signal	0.1-1.1v
61	WT/GN	HO2S-22 (B2 S2) Signal	0.1-1.1v
62	WT/PK	FTP Sensor Signal	2.6v (0" H2O - cap off)
63	WT/GN	Fuel Pressure Transducer	2.8v (39 psi)
64	WT/GN	A/T: TR Sensor Signal	In 'P': 0v, O/D: 1.7v
64	WT	M/T: Clutch Position Switch	Clutch In: 0v, Out: 5v

1999 Cougar 2.5L V6 4v MFI VIN L (All) 104 Pin Connector

PCM Pin #	Wire Color	Circuit Description (104 Pin)	Value at Hot Idle
65	WT/BL	DPFE Sensor Signal	0.20-1.30v
66, 68	---	Not Used	---
67	BK/OR	EVAP CV Solenoid	0-10 Hz (0-100%)
69	BK/YL	A/C WOT Relay Control	Relay Off: 12v, On: 1v
70	BK/WT	Injector 1 Control	2.1-2.5 ms
71	GN/YL	Vehicle Power	12-14v
72, 74-75	---	Not Used	---
73	BK/YL	HO2S-11 (B1 S1) Heater	1v, Off: 12v
76	BR/WT	CMP & TSS Ground	<0.050v
77	BK/YL	Power Ground	<0.1v
78	BR/YL	Coil Driver 3 Control	5° dwell
79	WT/BK	TCIL (lamp) Control	Lamp On: 1v, Off: 12v
80	BK/WT	TCC Solenoid Control	0%, 55 mph: 90-95%
81	BK/RD	EPC Solenoid Control	8.9v
82	BK/BL	PATS Sensor Signal	Digital Signal
83	BK/YL	IAC Motor Control	9v (33% d/cycle)
84	WT/PK	TSS Sensor Signal	46-50 Hz
85	WT/PK	CMP Sensor Signal	5-7 Hz
86	BK/WT	A/C Pressure Switch	Switch On: 12v (Open)
87	WT/RD	HO2S-21 (B1 S1) Signal	0.1-1.1v
88	WT/BL	MAF Sensor Signal	0.8v
89	WT	TP Sensor Signal	0.53-1.27v
90	YL	Reference Voltage	4.9-5.1v
91	BR	Analog Signal Return	<0.050v
92	OR	Brake Pedal Position Switch	Brake Off: 0v, On: 12v
92	GN/RD	Brake Pedal Position Switch	Brake Off: 0v, On: 12v
93	BK/GN	Injector 5 Control	2.1-2.5 ms
94	BK/RD	Injector 6 Control	2.1-2.5 ms
95	BK/OR	Injector 4 Control	2.1-2.5 ms
96	BK/YL	Injector 2 Control	2.1-2.5 ms
97	GN/YL	Vehicle Power	12-14v
98	---	Not Used	---
99	BK/BL	HO2S-21 (B2 S1) Heater	1v, Off: 12v
100	BK/OR	HO2S-12 (B1 S2) Heater	1v, Off: 12v
101	BK/GN	HO2S-22 (B2 S2) Heater	1v, Off: 12v
102	BK/OR	Shift Solenoid 3 Control	7-9v, 30 mph: 8-9v
103	BK/YL	Power Ground	<0.1v
104	---	Not Used	---

Pin Connector Graphic

Terminal View of 104-Pin PCM Wiring Harness Connector

Standard Colors and Abbreviations

Abbreviation	Color	Abbreviation	Color	Abbreviation	Color
BK	Black	GY	Gray	PK	Purple
BL	Blue	GN	Green	RD	Red
BR	Brown	LG	LT Green	TN	Tan
DB	Dark Blue	OR	Orange	WT	White
DG	DK Green	PK	Pink	YL	Yellow

2000-02 Cougar 2.5L V6 4v MFI VIN L (All) 104 Pin Connector

PCM Pin #	Wire Color	Circuit Description (104 Pin)	Value at Hot Idle
1	---	Not Used	---
2	BK/OR	MIL (lamp) Control	MIL On: 1v, Off: 12v
3-5	---	Not Used	---
6	BK/YL	Shift Solenoid 'A' Control	12v, 55 mph: 1v
8	BK/BL	IMRC Solenoid Control	5v, 55 mph: 5v
9-10	---	Not Used	---
11	BK/BL	Shift Solenoid 'B' Control	1v, 55 mph: 12v
12	WT	Fuel Level Indicator Signal	1.7v (1/2 full)
13	WT/BL	Flash EEPROM Power	0.5v
15	BL	Data Bus (-) Signal	Digital Signals
16	GY	Data Bus (+) Signal	Digital Signals
17	GY/OR	Passive Anti-Theft System	Digital Signals
19	BK/WT	High Speed Fan Control	Fan On: 1v, Off: 12v
20	BK/BL	Injector 3 Control	2.1-2.5 ms
21	WT/RD	CKP Sensor (+) Signal	410-440 Hz
22	BR/RD	CKP Sensor (-) Signal	410-440 Hz
24	BK/YL	Power Ground	<0.1v
25	BK/RD	Chassis Ground	<0.050v
26	BR/BL	Coil Driver 1 Control	5° dwell
29	PK/BK	TCS (switch) Signal	TCS & O/D On: 12v
31	WT	PSP Switch Signal	Straight: 0.1v, turned: 12v
32	BR/YL	Knock Sensor 1 Signal	0v
33	BR/BL	VSS (-) Signal	0 Hz, 55 mph: 125 Hz
34	WT/GN	Passive Anti-Theft System	Digital Signal
35	WT/BL	HO2S-12 (B1 S2) Signal	0.1-1.1v
36	BR/BL	MAF Sensor Return	<0.050v
37	WT/RD	TFT Sensor Signal	2.10 to 2.40v
38	WT/GN	ECT Sensor Signal	0-5-0.6v
39	WT/PK	IAT/ACT Sensor Signal	1.5-2.5v
40	PK/BK	Fuel Pump Monitor	12v, Off: 0v
40	GN/BK	Fuel Pump Monitor	12v, Off: 0v
41	PK/BL	A/C High Pressure Cut-Out	A/C Switch On: 12v
42	BK/RD	IMRC Solenoid Signal	12v, 55 mph: 12v
43	WT	Fuel Flow Sensor	Varies: 0-5.1v
45	BK/BL	Low Speed Fan Control	Fan On: 1v, Off: 12v
47	BK/GN	VR Solenoid Control	0%, 55 mph: 40%
48	WT/BK	Clean Tachometer Output	35-42 Hz
51	BK/YL	Power Ground	<0.1v
52	BR/GN	Coil Driver 2 Control	5° dwell
54	BK/BL	Fuel Pump Control	Relay Off: 12v, On: 1v
55	OR/YL	Keep Alive Power	12-14v
56	BK/OR	EVAP Purge Solenoid	0-10 Hz (0-100%)
56	BK/BL	EVAP Purge Solenoid	0-10 Hz (0-100%)
57	WT/BK	Knock Sensor 1 Signal	0v
58	WT/PK	VSS (+) Signal	0 Hz, 55 mph: 125 Hz
60	WT	HO2S-11 (B1 S1) Signal	0.1-1.1v
61	WT/GN	HO2S-22 (B2 S2) Signal	0.1-1.1v
62	WT/PK	FTP Sensor Signal	2.6v (0" H2O - cap off)
63	WT/GN	Fuel Pressure Transducer	2.8v (39 psi)
64	WT/GN	A/T: TR Sensor Signal	In 'P': 0v, O/D: 1.7v
64	WT	M/T: Clutch Position Switch	Clutch In: 0v, Out: 5v

2000-02 Cougar 2.5L V6 4v MFI VIN L (All) 104 Pin Connector

PCM Pin #	Wire Color	Circuit Description (104 Pin)	Value at Hot Idle
65	WT/BL	DPFE Sensor Signal	0.95-1.05v
66	---	Not Used	---
67	BK/OR	EVAP CV Solenoid	0-10 Hz (0-100%)
69	BK/YL	A/C WOT Relay Control	Relay Off: 12v, On: 1v
68	---	Not Used	---
70	BK/WT	Injector 1 Control	2.1-2.5 ms
71	GN/YL	Vehicle Power	12-14v
72, 74-75	---	Not Used	---
73	BK/YL	HO2S-11 (B1 S1) Heater	1v, Off: 12v
76	BR/WT	CMP & TSS Ground	<0.050v
77	BK/YL	Power Ground	<0.1v
78	BR/YL	Coil Driver 3 Control	5° dwell
79	WT/BK	TCIL (lamp) Control	Lamp On: 1v, Off: 12v
80	BK/WT	TCC Solenoid Control	0%, 55 mph: 90-95%
81	BK/RD	EPC Solenoid Control	8.9v
82	BK/BL	PATS Sensor Signal	Digital Signal
83	BK/YL	IAC Motor Control	9v (33% d/cycle)
84	WT/PK	TSS Sensor Signal	46-50 Hz
85	WT/PK	CMP Sensor Signal	5-7 Hz
86	BK/WT	A/C Pressure Switch	Switch On: 12v (Open)
87	WT/RD	HO2S-21 (B1 S1) Signal	0.1-1.1v
88	WT/BL	MAF Sensor Signal	0.8v
89	WT	TP Sensor Signal	0.53-1.27v
90	YL	Reference Voltage	4.9-5.1v
91	BR	Analog Signal Return	<0.050v
92	OR	Brake Pedal Position Switch	Brake Off: 0v, On: 12v
92	GN/RD	Brake Pedal Position Switch	Brake Off: 0v, On: 12v
93	BK/GN	Injector 5 Control	2.1-2.5 ms
94	BK/RD	Injector 6 Control	2.1-2.5 ms
95	BK/OR	Injector 4 Control	2.1-2.5 ms
96	BK/YL	Injector 2 Control	2.1-2.5 ms
97	GN/YL	Vehicle Power	12-14v
98	---	Not Used	---
99	BK/BL	HO2S-21 (B2 S1) Heater	1v, Off: 12v
100	BK/OR	HO2S-12 (B1 S2) Heater	1v, Off: 12v
101	BK/GN	HO2S-22 (B2 S2) Heater	1v, Off: 12v
102	BK/OR	Shift Solenoid 'C' Control	7-9v, 30 mph: 8-9v
103	BK/YL	Power Ground	<0.1v
104	---	Not Used	---

Pin Connector Graphic

PCM 104-PIN CONNECTOR

Terminal View of 104-Pin PCM Wiring Harness Connector

1990 Cougar 3.8L V6 MFI VIN 4 (A/T) 60 Pin Connector

PCM Pin #	Wire Color	Circuit Description (60 Pin)	Value at Hot Idle
1	YL	Keep Alive Power	12-14v
2	---	Not Used	---
3	DG/WT	VSS (+) Signal	0 Hz, 55 mph: 125 Hz
4	DG/YT	Ignition Diagnostic Monitor	20-31 Hz
5	LG	Brake Switch Signal	Brake Off: 0v, On: 12v
6	BK/WT	VSS (-) Signal	0 Hz, 55 mph: 125 Hz
7	LG/YL	ECT Sensor Signal	0.5-0.6v
8	OR/BK	Data Bus (-) Signal	Digital Signals
9	---	Not Used	---
10	PK/BL	A/C Cycling Clutch Signal	A/C Off: 0v, On: 12v
11	OR/YL	Speed Control Solenoid (+)	Solenoid On: 12-14V
12	BR/YL	Injector 3 Control	Hot Idle: 6.0-6.2 ms
13	BR/BL	Injector 4 Control	Hot Idle: 6.0-6.2 ms
14	TN/BL	Injector 5 Control	Hot Idle: 6.0-6.2 ms
15	LG	Injector 6 Control	Hot Idle: 6.0-6.2 ms
16	DB	Ignition System Ground	<0.050v
16	GY	Ignition System Ground	<0.050v
17	YL/BK	Self-Test Output, MIL Control	MIL Off: 12v, On: 1v
18	--	Not Used	---
19	---	Not Used	---
20	BK	Case Ground	<0.050v
21	RD/LG	IAC Motor Control	9-10v
22	TN/LG	Fuel Pump Control	Relay Off: 12v, On: 1v
23	---	Not Used	---
24	---	Not Used	---
25	LG/PK	ACT Sensor Signal	1.5-2.5v
26	OR/WT	Reference Voltage	4.9-5.1v
27	BR/LG	PFE Sensor Signal	3.2v, 55 mph: 2.8v
28	BL/BK	Speed Control Switch Signal	6.7v
29	TN/OR	HO2S-21 (B1 S1) Signal	0.1-1.1v
30	RD/BL	Neutral Drive Switch	In 'N': 0v, Others: 4-5v
31	GY/YL	Canister Purge Solenoid	12v, 55 mph: 1v
32	---	Not Used	---
33	DG	VR Solenoid Control	0%, 55 mph: 45%
34	BL/PK	Data Output Link	Digital Signals
35	WT/PK	S/C Vent Solenoid Control	Vacuum Decreasing: 1v
36	YL/LG	Spark Output Signal	6.93v (50% d/cycle)
37	RD	Vehicle Power	12-14v
38	---	Not Used	---
39	WT/BL	Speed Control Switch Ground	<0.050v
40	BK/LG	Power Ground	<0.1v

1990 Cougar 3.8L V6 MFI VIN 4 (A/T) 60 Pin Connector

PCM Pin #	Wire Color	Circuit Description (60 Pin)	Value at Hot Idle
41	---	Not Used	---
42	---	Not Used	---
43	TN/RD	HO2S-11 (B1 S1) Signal	0.1-1.1v
44	BK/OR	Data Bus (+) Signal	Digital Signals
45	DB/LG	MAP Sensor Signal	107 Hz (sea level)
46	BK/WT	Analog Signal Return	<0.050v
47	DG/LG	TP Sensor Signal	0.8v, 55 mph: 1.1v
48	WT/RD	Self-Test Input Signal	STI On: 1v, Off: 5v
49	OR	HO2S-11 (Bank 1) Ground	<0.050v
50	PK/BK	Fuel Pump Monitor	12v, Off: 0v
51	GY/BK	S/C Vacuum Solenoid Control	Vacuum Increasing: 1v
52	---	Not Used	---
53	---	Not Used	---
54	OR/BL	A/C WOT Relay Control	Relay Off: 12v, On: 1v
55	---	Not Used	---
56	GY	PIP Sensor Signal	6.93v (50% d/cycle)
56	DB	PIP Sensor Signal	6.93v (50% d/cycle)
57	RD	Vehicle Power	12-14v
58	TN	Injector 1 Control	6-6.2 ms
59	WT	Injector 2 Control	6-6.2 ms
60	BK/LG	Power Ground	<0.1v

Pin Connector Graphic

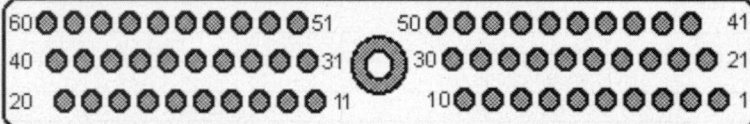

PCM 60-PIN CONNECTOR

Terminal View of 60-Pin PCM Harness Connector

Standard Colors and Abbreviations

Abbreviation	Color	Abbreviation	Color	Abbreviation	Color
BK	Black	GY	Gray	PK	Purple
BL	Blue	GN	Green	RD	Red
BR	Brown	LG	LT Green	TN	Tan
DB	Dark Blue	OR	Orange	WT	White
DG	DK Green	PK	Pink	YL	Yellow

1991-95 Cougar 3.8L V6 MFI VIN 4 (A/T) 60 Pin Connector

PCM Pin #	Wire Color	Circuit Description (60 Pin)	Value at Hot Idle
1	YL	Keep Alive Power	12-14v
2	---	Not Used	---
3	GY/BK	VSS (+) Signal	0 Hz, 55 mph: 125 Hz
4	RD/BL	Ignition Diagnostic Monitor	20-31 Hz
4	WT/PK	Ignition Diagnostic Monitor	20-31 Hz
5	---	Not Used	---
6	BK/WT	VSS (-) Signal	0 Hz, 55 mph: 125 Hz
6	GY/RD	VSS (-) Signal	0 Hz, 55 mph: 125 Hz
7	LG/YL	ECT Sensor Signal	0.5-0.6v
7	LG/RD	ECT Sensor Signal	0.5-0.6v
8	PK/BL	Fuel Pump Monitor	12v, Off: 0v
9	TN/BL	MAF Sensor Return	<0.050v
10	PK/BK	A/C Cycling Clutch Signal	A/C Off: 0v, On: 12v
11	GY/YL	Canister Purge SOL Control	12v, 55 mph: 9-11v
12	LG	Injector 6 Control	5.6-6.3 ms
12	LG/OR	Injector 6 Control	5.6-6.3 ms
13	---	Not Used	---
14	---	Not Used	---
15	TN/BL	Injector 5 Control	5.6-6.3 ms
15	TN/BK	Injector 5 Control	5.6-6.3 ms
16	GY	Ignition System Ground	<0.050v
17	YL/BK	Self-Test Output, MIL Control	MIL Off: 12v, On: 1v
18	TN/OR	Data Bus (+) Signal	Digital Signals
19	PK/BK	Data Bus (-) Signal	Digital Signals
20	BK	Case Ground	<0.050v
21	RD/LG	IAC Motor Control	9.5-11.0v
22	TN/LG	Fuel Pump Control	Relay Off: 12v, On: 1v
23	---	Not Used	---
24	---	Not Used	---
25	LG/PK	IAT Sensor Signal	1.5-2.5v
25	GY	IAT Sensor Signal	1.5-2.5v
26	OR/WT	Reference Voltage	4.9-5.1v
26	BR/WT	Reference Voltage	4.9-5.1v
27	BR/LG	PFE Sensor Signal	3.2v, 55 mph: 2.9v
28	---	Not Used	---
29	---	Not Used	---
30	RD/BL	Neutral Drive Switch	In 'N': 0v, Others: 5v
31	---	Not Used	---
32	---	Not Used	---
33	DG	VR Solenoid Control	0%, 55 mph: 0-45%
33	BR/PK	VR Solenoid Control	0%, 55 mph: 0-45%
34	BL/PK	Data Output Link	Digital Signals
35	BR/BL	Injector 4 Control	5.6-6.3 ms
36	YL/LG	Spark Output Signal	6.93v (50% d/cycle)
37	RD	Vehicle Power	12-14v
38	---	Not Used	---
39	BR/YL	Injector 3 Control	5.6-6.3 ms
40	BK/LG	Power Ground	<0.1v

1991-95 Cougar 3.8L V6 MFI VIN 4 (A/T) 60 Pin Connector

PCM Pin #	Wire Color	Circuit Description (60 Pin)	Value at Hot Idle
41-42	---	Not Used	---
43	TN/RD	HO2S-21 (B1 S1) Signal	0.1-1.1v
43	TN/WT	HO2S-21 (B1 S1) Signal	0.1-1.1v
44	TN/OR	HO2S-11 (B1 S1) Signal	0.1-1.1v
45	---	Not Used	---
46	BK/WT	Analog Signal Return	<0.050v
46	GY/RD	Analog Signal Return	<0.050v
47	DG/LG	TP Sensor Signal	0.8v, 55 mph: 1.1v
47	GY/WT	TP Sensor Signal	0.8v, 55 mph: 1.1v
48	WT/PK	Self-Test Input Signal	STI On: 1v, Off: 5v
48	WT/RD	Self-Test Input Signal	STI On: 1v, Off: 5v
49	---	Not Used	---
50	DB/OR	MAF Sensor Signal	0.7v, 55 mph: 2.0v
50	BL/RD	MAF Sensor Signal	0.7v, 55 mph: 2.0v
51-53	---	Not Used	---
54	OR/BL	A/C WOT Relay Control	Relay Off: 12v, On: 1v
55	---	Not Used	---
56	DB	PIP Sensor Signal	6.93v (50% d/cycle)
57	RD	Vehicle Power	12-14v
58	TN	Injector 1 Control	5.6-6.3 ms
59	WT	Injector 2 Control	5.6-6.3 ms
60	BK/LG	Power Ground	<0.1v

Pin Connector Graphic

PCM 60-PIN CONNECTOR

Terminal View of 60-Pin PCM Harness Connector

Standard Colors and Abbreviations

Abbreviation	Color	Abbreviation	Color	Abbreviation	Color
BK	Black	GY	Gray	PK	Purple
BL	Blue	GN	Green	RD	Red
BR	Brown	LG	LT Green	TN	Tan
DB	Dark Blue	OR	Orange	WT	White
DG	DK Green	PK	Pink	YL	Yellow

1996-97 Cougar 3.8L V6 MFI VIN 4 (A/T) 104 Pin Connector

PCM Pin #	Wire Color	Circuit Description (104 Pin)	Value at Hot Idle
1	PK/OR	Shift Solenoid 2 Control	12v, 55 mph: 1v
2	PK/LG	MIL (lamp) Control	MIL On: 1v, Off: 12v
3-10	---	Not Used	---
11	PK/WT	Purge Flow Sensor Signal	0.5-1.5v, 30 mph: 3v
12	---	Not Used	---
13	PK	Flash EEPROM Power	0.1v
14	---	Not Used	---
15	PK/LB	Data Bus (-) Signal	Digital Signals
16	TN/OR	Data Bus (+) Signal	Digital Signals
17-20	---	Not Used	---
21	DB	CKP Sensor (+) Signal	850-1120 Hz
22	GY	CKP Sensor (-) Signal	850-1120 Hz
23	---	Not Used	---
24	BK/WT	Power Ground	<0.1v
25	BK	Case Ground	<0.050v
26	DB/LG	Coil Driver 1 Control	5° dwell
27	OR/YL	Shift Solenoid 1 Control	1v, 55 mph: 12v
28	---	Not Used	---
29	TN/WT	TCS (switch) Signal	TCS & O/D On: 12v
30	DG	Octane Adjust Switch	9.3v (Closed: 0v)
31-32	---	Not Used	---
33	PK/OR	VSS (-) Signal	0 Hz, 55 mph: 125 Hz
34	---	Not Used	---
35	RD/LG	HO2S-12 (B1 S2) Signal	0.1-1.1v
36	TN/LB	MAF Sensor Return	<0.050v
37	OR/BK	TFT Sensor Signal	2.40-2.10v
38	LG/RD	ECT Sensor Signal	0.5-0.6v
39	GY	IAT Sensor Signal	1.5-2.5v
40	PK/BK	Fuel Pump Monitor	12v, Off: 0v
41	PK/LB	A/C Cycling Clutch Switch	AC On: 12v, Off: 0v
42-44	---	Not Used	---
45	TN/OR	Low Speed Fan Control	Fan On: 1v, Off: 12v
46	LG/PK	High Speed Fan Control	Fan On: 1v, Off: 12v
47	BR/PK	VR Solenoid Control	0%, 55 mph: 45%
48-50	---	Not Used	---
51	BK/WT	Power Ground	<0.1v
52	RD/LB	Coil Driver 2 Control	5° dwell
53	---	Not Used	---
54	BR/OR	TCC Solenoid Control	0%, 55 mph: 90%
55	YL	Keep Alive Power	12-14v
56-57	---	Not Used	---
58	GY/BK	VSS (+) Signal	0 Hz, 55 mph: 125 Hz
59	---	Not Used	---
60	GY/LB	HO2S-11 (B1 S1) Signal	0.1-1.1v
61	PK/LG	HO2S-22 (B2 S2) Signal	0.1-1.1v
62-63	---	Not Used	---
64	LBYL	TR Sensor Signal	In 'P': 0v, O/D: 1.7v

1996-97 Cougar 3.8L V6 MFI VIN 4 (A/T) 104 Pin Connector

PCM Pin #	Wire Color	Circuit Description (104 Pin)	Value at Hot Idle
65	BR/LG	DPFE Sensor Signal	0.20-1.30v
66, 68	---	Not Used	---
67	GY/YL	EVAP Purge Solenoid	0-10 Hz (0-100%)
69	PK/YL	A/C WOT Relay Control	Relay Off: 12v, On: 1v
70, 72	---	Not Used	---
71	RD	Vehicle Power	12-14v
73	TN/BK	Injector 5 Control	4.5-4-8 ms
74	BR/YL	Injector 3 Control	4.5-4-8 ms
75	TN	Injector 1 Control	4.5-4-8 ms
76	BK/WT	Power Ground	<0.1v
77	BK/WT	Power Ground	<0.1v
78	PK/WT	Coil Driver 3 Control	5° dwell
79	WT/LG	TCIL (lamp) Control	Lamp On: 1v, Off: 12v
80	LB/OR	Fuel Pump Control	Relay Off: 12v, On: 1v
81	WT/YL	EPC Solenoid Control	8.8v (20 psi)
82	---	Not Used	---
83	WT/LB	IAC Motor Control	10v (30% d/cycle)
84	DG/WT	TSS Sensor Signal	43 Hz (650 rpm)
85	DG	CMP Sensor Signal	5-7 Hz
86	TN/LG	A/C Pressure Switch	Switch On: 12v (Open)
87	RD/BK	HO2S-21 (B1 S1) Signal	0.1-1.1v
88	LB/RD	MAF Sensor Signal	0.6v
89	GY/WT	TP Sensor Signal	0.53-1.27v
90	BR/WT	Reference Voltage	4.9-5.1v
91	GY/RD	Analog Signal Return	<0.050v
92	LG	Brake Pedal Position Switch	Brake Off: 0v, On: 12v
93	RD/WT	HO2S-11 (B1 S1) Heater	1v, Off: 12v
94	YL/LB	HO2S-21 (B2 S1) Heater	1v, Off: 12v
95	WT/BK	HO2S-12 (B1 S2) Heater	1v, Off: 12v
96	TN/YL	HO2S-22 (B2 S2) Heater	1v, Off: 12v
97	RD	Vehicle Power	12-14v
98	---	Not Used	---
99	LG/OR	Injector 6 Control	4.5-4-8 ms
100	BR/LB	Injector 4 Control	4.5-4-8 ms
101	WT	Injector 2 Control	4.5-4-8 ms
102	---	Not Used	---
103	BK/WT	Power Ground	<0.1v
104	---	Not Used	---

Pin Connector Graphic

Standard Colors and Abbreviations

Abbreviation	Color	Abbreviation	Color	Abbreviation	Color
BK	Black	GY	Gray	PK	Purple
BL	Blue	GN	Green	RD	Red
BR	Brown	LG	LT Green	TN	Tan
DB	Dark Blue	OR	Orange	WT	White
DG	DK Green	PK	Pink	YL	Yellow

1990-92 Cougar 3.8L V6 Supercharged MFI VIN C (All) 60 Pin Connector

PCM Pin #	Wire Color	Circuit Description (60 Pin)	Value at Hot Idle
1	YL	Keep Alive Power	12-14v
2	---	Not Used	---
3	DG/WT	VSS (+) Signal	0 Hz, 55 mph: 125 Hz
4	DG/YT	Ignition Diagnostic Monitor	20-31 Hz
5	LG	Brake Switch Signal	Brake Off: 0v, On: 12v
6	BK/WT	VSS (-) Signal	0 Hz, 55 mph: 125 Hz
7	LG/YL	ECT Sensor Signal	0.5-0.6v
8	---	Not Used	---
9	TN/BL	MAF Sensor Return	<0.050v
10	PK/BL	A/C Cycling Clutch Signal	A/C Off: 0v, On: 12v
11	OR/YL	Speed Control Solenoid (+)	Solenoid On: 12-14v
12	BR/YL	Injector 3 Control	4.0-4.2 ms
13	BR/BL	Injector 4 Control	4.0-4.2 ms
14	TN/BL	Injector 5 Control	4.0-4.2 ms
15	LG	Injector 6 Control	4.0-4.2 ms
16	BL	Ignition System Ground	<0.050v
17	YL/BK	Self-Test Output, MIL Control	MIL Off: 12v, On: 1v
18	BK/WT	Octane Adjust Switch	Closed: 0v, Open: 9.1v
19	PK/BK	Fuel Pump Control	12v, Off: 0v
20	BK	Case Ground	<0.050v
21	RD/LG	IAC Motor Control	20-40%
22	TN/LG	Fuel Pump Control	Relay Off: 12v, On: 1v
23	YL/RD	Knock Sensor Signal	0v
24	DG	CID Sensor Signal	5-9 Hz
25	LG/PK	ACT Sensor Signal	1.5-2.5v
26	OR/WT	Reference Voltage	4.9-5.1v
27	BR/LG	PFE Sensor Signal	3.2v, 55 mph: 2.9v
28	BL/BK	Speed Control Switch Signal	6.7v
29	TN/OR	HO2S-21 (B1 S1) Signal	0.1-1.1v
30	RD/BL	Neutral Drive Switch	In 'N': 0v, Others: 5v
31	GY/YL	Canister Purge Solenoid	12v, 55 mph: 1v
32	OR/WT	Automatic Ride Control	4.3v
33	YL	VR Solenoid Control	0%, 55 mph: 45%
34	---	Not Used	---
35	WT/PK	S/C Vent Solenoid Control	0%, 55 mph: 98%
36	YL/LG	Spark Output Signal	6.93v (50% d/cycle)
37	RD	Vehicle Power	12-14v
38 ('89)	LG/PK	Supercharger Bypass Sol.	0% duty cycle
39	WT/BL	Speed Control Switch Ground	<0.050v
40	BK/LG	Power Ground	<0.1v

1990-92 Cougar 3.8L V6 SC MFI VIN C (All) 60 Pin Connector

PCM Pin #	Wire Color	Circuit Description (60 Pin)	Value at Hot Idle
41	PK	High Speed Fan Control	Fan On: 1v, Off: 12v
42	---	Not Used	---
43	TN/RD	HO2S-11 (B1 S1) Signal	0.1-1.1v
44	---	Not Used	---
45	DB/LG	MAP Sensor Signal	107 Hz (sea level)
46	BK/WT	Analog Signal Return	<0.050v
47	DG/LG	TP Sensor Signal	0.8v, 55 mph: 1.1v
48	WT/RD	Self-Test Input Signal	STI On: 1v, Off: 5v
49	OR	HO2S-11 (Bank 1) Ground	<0.050v
50	DB/OR	MAF Sensor Signal	0.7v, 55 mph: 2.0v
51	GY/BK	S/C Vacuum Solenoid Control	Vacuum Increasing: 1v
52	---	Not Used	---
53	PK	Shift Indicator Light Control	SIL On: 1v, Off: 12v
54	RD	A/C WOT Relay Control	Relay Off: 12v, On: 1v
55	TN/OR	Low Speed Fan Control	Fan On: 1v, Off: 12v
56	DB	PIP Sensor Signal	6.93v (50% d/cycle)
57	RD	Vehicle Power	12-14v
58	TN	Injector 1 Control	4.0-4.2 ms
59	WT	Injector 2 Control	4.0-4.2 ms
60	LG/BK	Power Ground	<0.1v

Pin Connector Graphic

PCM 60-PIN CONNECTOR

Terminal View of 60-Pin PCM Harness Connector

Standard Colors and Abbreviations

Abbreviation	Color	Abbreviation	Color	Abbreviation	Color
BK	Black	GY	Gray	PK	Purple
BL	Blue	GN	Green	RD	Red
BR	Brown	LG	LT Green	TN	Tan
DB	Dark Blue	OR	Orange	WT	White
DG	DK Green	PK	Pink	YL	Yellow

1990-94 Cougar 3.8L V6 SC MFI VIN R (All) 60 Pin Connector

PCM Pin #	Wire Color	Circuit Description (60 Pin)	Value at Hot Idle
1	YL	Keep Alive Power	12-14v
2	---	Not Used	---
3	DG/WT	VSS (+) Signal	0 Hz, 55 mph: 125 Hz
4	DG/YT	Ignition Diagnostic Monitor	20-31 Hz
5	LG	Brake Switch Signal	Brake Off: 0v, On: 12v
6	BK/WT	VSS (-) Signal	0 Hz, 55 mph: 125 Hz
7	LG/YL	ECT Sensor Signal	0.5-0.6v
8	---	Not Used	---
9	TN/BL	MAF Sensor Return	<0.050v
10	PK/BK	A/C Cycling Clutch Signal	A/C Off: 0v, On: 12v
11	OR/YL	Speed Control Solenoid (+)	Solenoid On: 12-14v
12	BR/YL	Injector 3 Control	4.0-4.2 ms
13	BR/BL	Injector 4 Control	4.0-4.2 ms
14	TN/BL	Injector 5 Control	4.0-4.2 ms
15	LG	Injector 6 Control	4.0-4.2 ms
16	BL	Ignition System Ground	<0.050v
17	YL/BK	Self-Test Output, MIL Control	MIL Off: 12v, On: 1v
18	BK/WT	Octane Adjust Switch	Closed: 0v, Open: 9.1v
19	PK/BK	Fuel Pump Control	12v, Off: 0v
20	BK	Case Ground	<0.050v
21	RD/LG	IAC Motor Control	Hot Idle: 9-9.4v
22	TN/LG	Fuel Pump Control	Relay Off: 12v, On: 1v
23	YL/RD	Knock Sensor Signal	0v
24	DG	CID Sensor Signal	5-9 Hz
25	LG/PK	ACT Sensor Signal	1.5-2.5v
26	OR/WT	Reference Voltage	4.9-5.1v
27	BR/LG	PFE Sensor Signal	3.2v, 55 mph: 2.9v
28	BL/BK	Speed Control Switch Signal	6.7v
29	TN/OR	HO2S-21 (B1 S1) Signal	0.1-1.1v
30	RD/BL	Neutral Drive Switch	In 'N': 0v, Others: 5v
31	GY/YL	Canister Purge Solenoid	12v, 55 mph: 1v
32	OR/WT	Automatic Ride Control	4.3v
33	YL	VR Solenoid Control	0%, 55 mph: 45%
34	---	Not Used	---
35	WT/PK	S/C Vent Solenoid Control	0%, 55 mph: 98%
36	YL/LG	Spark Output Signal	6.93v (50% d/cycle)
37	RD	Vehicle Power	12-14v
38	LG/PK	Supercharger Bypass Sol.	0% duty cycle
39	WT/BL	Speed Control Switch Ground	<0.050v
40	BK/LG	Power Ground	<0.1v

1990-94 Cougar 3.8L V6 SC MFI VIN R (All) 60 Pin Connector

PCM Pin #	Wire Color	Circuit Description (60 Pin)	Value at Hot Idle
41	PK	High Speed Fan Control	Fan On: 1v, Off: 12v
42	---	Not Used	---
43	TN/RD	HO2S-11 (B1 S1) Signal	0.1-1.1v
44	---	Not Used	---
45	DB/LG	MAP Sensor Signal	107 Hz (sea level)
46	BK/WT	Analog Signal Return	<0.050v
47	DG/LG	TP Sensor Signal	0.8v, 55 mph: 1.1v
48	WT/RD	Self-Test Input Signal	STI On: 1v, Off: 5v
49	OR	HO2S-11 (Bank 1) Ground	<0.050v
50	DB/OR	MAF Sensor Signal	0.7v, 55 mph: 2.0v
51	GY/BK	S/C Vacuum Solenoid Control	Vacuum Increasing: 1v
52	---	Not Used	---
53	PK	Shift Indicator Light Control	SIL On: 1v, Off: 12v
54	RD	A/C WOT Relay Control	Relay Off: 12v, On: 1v
55	TN/OR	Low Speed Fan Control	Fan On: 1v, Off: 12v
56	DB	PIP Sensor Signal	6.93v (50% d/cycle)
57	RD	Vehicle Power	12-14v
58	TN	Injector 1 Control	4.0-4.2 ms
59	WT	Injector 2 Control	4.0-4.2 ms
60	LG/BK	Power Ground	<0.1v

Pin Connector Graphic

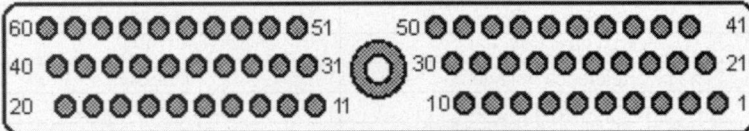

PCM 60-PIN CONNECTOR

Terminal View of 60-Pin PCM Harness Connector

Standard Colors and Abbreviations

Abbreviation	Color	Abbreviation	Color	Abbreviation	Color
BK	Black	GY	Gray	PK	Purple
BL	Blue	GN	Green	RD	Red
BR	Brown	LG	LT Green	TN	Tan
DB	Dark Blue	OR	Orange	WT	White
DG	DK Green	PK	Pink	YL	Yellow

1994-97 Cougar 4.6L V8 2v MFI VIN W (A/T) 104 Pin Connector

PCM Pin #	Wire Color	Circuit Description (104 Pin)	Value at Hot Idle
1	PK/OR	Shift Solenoid 2 Control	12v, 55 mph: 1v
2	PK/LG	MIL (lamp) Control	MIL On: 1v, Off: 12v
3-10	---	Not Used	---
11	PK/WT	Purge Flow Sensor Signal	0-10 Hz (0-100%)
12	---	Not Used	---
13	PK	Flash EEPROM Power	0.1v
14	---	Not Used	---
15	PK/LB	Data Bus (-) Signal	Digital Signals
16	TN/OR	Data Bus (+) Signal	Digital Signals
17-20	---	Not Used	---
21	DB	CKP Sensor (+) Signal	850-1120 Hz
22	GY	CKP Sensor (-) Signal	850-1120 Hz
23	---	Not Used	---
24	BK/WT	Power Ground	<0.1v
25	BK	Case Ground	<0.050v
26	DB/LG	Coil Driver 1 Control	5° dwell
27	OR/YL	Shift Solenoid 1 Control	1v, 55 mph: 12v
28	---	Not Used	---
29	TN/WT	TCS (switch) Signal	TCS & O/D On: 12v
30	DG	Octane Adjust Switch	9.3v (Closed: 0v)
31, 32	---	Not Used	---
33	PK/OR	VSS (-) Signal	0 Hz, 55 mph: 125 Hz
34	---	Not Used	---
35	RD/LG	HO2S-12 (B1 S2) Signal	0.1-1.1v
36	TN/LB	MAF Sensor Return	<0.050v
37	OR/BK	TFT Sensor Signal	2.40-2.10v
38	LG/RD	ECT Sensor Signal	0.5-0.6v
39	GY	IAT Sensor Signal	1.5-2.5v
40	PK/BK	Fuel Pump Monitor	12v, Off: 0v
41	PK/LB	A/C Cycling Clutch Switch	AC On: 12v, Off: 0v
42-44	---	Not Used	---
45	TN/OR	Low Speed Fan Control	Fan On: 1v, Off: 12v
46	LG/PK	High Speed Fan Control	Fan On: 1v, Off: 12v
47	BR/PK	VR Solenoid Control	0%, 55 mph: 45%
48-50	---	Not Used	---
51	BK/WT	Power Ground	<0.1v
52	RD/LB	Coil Driver 2 Control	5° dwell
53	---	Not Used	---
54	BR/OR	TCC Solenoid Control	0%, 55 mph: 95%
55	YL	Keep Alive Power	12-14v
56-57	---	Not Used	---
58	GY/BK	VSS (+) Signal	0 Hz, 55 mph: 125 Hz
59	---	Not Used	---
60	GY/LB	HO2S-11 (B1 S1) Signal	0.1-1.1v
61	PK/LG	HO2S-22 (B2 S2) Signal	0.1-1.1v
62-63	---	Not Used	---
64	LBYL	TR Sensor Signal	In 'P': 0v, O/D: 1.7v

1994-97 Cougar 4.6L V8 2v MFI VIN W (A/T) 104 Pin Connector

PCM Pin #	Wire Color	Circuit Description (104 Pin)	Value at Hot Idle
65	BR/LG	DPFE Sensor Signal	0.20-1.30v
66	---	Not Used	---
67	GY/YL	EVAP Purge Solenoid	0-10 Hz (0-100%)
68, 70, 82	---	Not Used	---
69	PK/YL	A/C WOT Relay Control	Relay Off: 12v, On: 1v
71	RD	Vehicle Power	12-14v
72	TN/RD	Injector 7 Control	3.1-3-5 ms
73	TN/BK	Injector 5 Control	3.1-3-5 ms
74	BR/YL	Injector 3 Control	3.1-3-5 ms
75	TN	Injector 1 Control	3.1-3-5 ms
76, 77	BK/WT	Power Ground	<0.1v
78	PK/WT	Coil Driver 3 Control	5° dwell
79	WT/LG	TCIL (lamp) Control	Lamp On: 1v, Off: 12v
80	LB/OR	Fuel Pump Control	Relay Off: 12v, On: 1v
81	WT/YL	EPC Solenoid Control	9.0v (15 psi)
83	WT/LB	IAC Motor Control	10v (30% d/cycle)
84	DG/WT	TSS Sensor Signal	43 Hz (650 rpm)
85	DG	CMP Sensor Signal	5-7 Hz
86	TN/LG	A/C Pressure Switch	Switch On: 12v (Open)
87	RD/BK	HO2S-21 (B1 S1) Signal	0.1-1.1v
88	LB/RD	MAF Sensor Signal	0.6v
89	GY/WT	TP Sensor Signal	0.53-1.27v
90	BR/WT	Reference Voltage	4.9-5.1v
91	GY/RD	Analog Signal Return	<0.050v
92	LG	Brake Pedal Position Switch	Brake Off: 0v, On: 12v
93	RD/WT	HO2S-11 (B1 S1) Heater	1v, Off: 12v
94	YL/LB	HO2S-21 (B2 S1) Heater	1v, Off: 12v
95	WT/BK	HO2S-12 (B1 S2) Heater	1v, Off: 12v
96	TN/YL	HO2S-22 (B2 S2) Heater	1v, Off: 12v
97	RD	Vehicle Power	12-14v
98	LB	Injector 8 Control	3.1-3-5 ms
99	LG/OR	Injector 6 Control	3.1-3-5 ms
100	BR/LB	Injector 4 Control	3.1-3-5 ms
101	WT	Injector 2 Control	3.1-3-5 ms
102	---	Not Used	---
103	BK/WT	Power Ground	<0.1v
104	RD/YL	Coil Driver 4 Control	5° dwell

Pin Connector Graphic

```
        PCM 104-PIN CONNECTOR
    1 ooooooooooooo    ooooooooooooo 26
  27 ooooooooooooo    ooooooooooooo 52
  53 ooooooooooooo  ⬤  ooooooooooooo 78
  79 ooooooooooooo    ooooooooooooo 104
```

Terminal View of 104-Pin PCM Wiring Harness Connector

Standard Colors and Abbreviations

Abbreviation	Color	Abbreviation	Color	Abbreviation	Color
BK	Black	GY	Gray	PK	Purple
BL	Blue	GN	Green	RD	Red
BR	Brown	LG	LT Green	TN	Tan
DB	Dark Blue	OR	Orange	WT	White
DG	DK Green	PK	Pink	YL	Yellow

1991-93 Cougar 5.0L V8 MFI VIN T (A/T) 60 Pin Connector

PCM Pin #	Wire Color	Circuit Description (60 Pin)	Value at Hot Idle
1	YL	Keep Alive Power	12-14v
2	---	Not Used	---
3	DG/WT	VSS (+) Signal	0 Hz, 55 mph: 125 Hz
4	RD/BL	Ignition Diagnostic Monitor	20-31 Hz
4	WT/PK	Ignition Diagnostic Monitor	20-31 Hz
5	---	Not Used	---
6	BK/WT	VSS (-) Signal	0 Hz, 55 mph: 125 Hz
6	GY/RD	VSS (-) Signal	0 Hz, 55 mph: 125 Hz
7	LG/YL	ECT Sensor Signal	0.5-0.6v
7	LG/RD	ECT Sensor Signal	0.5-0.6v
8	PK/BK	Fuel Pump Monitor	12v, Off: 0v
9	TN/BL	MAF Sensor Return	<0.050v
10	PK/BL	A/C Cycling Clutch Signal	A/C Off: 0v, On: 12v
11	GY/YL	Canister Purge Solenoid	12v, 55 mph: 1v
12	LG	Injector 6 Control	4.9-5.2 ms
12	LG/OR	Injector 6 Control	4.9-5.2 ms
13	TN/OR	Injector 7 Control	4.9-5.2 ms
13	TN/RD	Injector 7 Control	4.9-5.2 ms
14	BL	Injector 8 Control	4.9-5.2 ms
15	TN/BL	Injector 5 Control	4.9-5.2 ms
15	TN/BK	Injector 5 Control	4.9-5.2 ms
16	GY	Ignition System Ground	<0.050v
17	YL/BK	Self-Test Output, MIL Control	MIL Off: 12v, On: 1v
18	TN/OR	Data Bus (+) Signal	Digital Signals
19	PK/BL	Data Bus (-) Signal	Digital Signals
20	BK	Case Ground	<0.050v
21	RD/LG	IAC Motor Control	9.3-10.3v
22	TN/LG	Fuel Pump Control	Relay Off: 12v, On: 1v
23-24	---	Not Used	---
25	GY	IAT Sensor Signal	1.5-2.5v
25	LG/PK	IAT Sensor Signal	1.5-2.5v
26	OR/WT	Reference Voltage	4.9-5.1v
26	BR/WT	Reference Voltage	4.9-5.1v
27	BR/LG	EVP Sensor Signal	0.4v, 55 mph: 1.5v
28	---	Not Used	---
29	---	Not Used	---
30	RD/BL	Neutral Drive Switch	In 'P': 0v, Others: 4-5v
31	WT/RD	Air Management 1 Solenoid	AM1 On: 1v, Off: 12v
32	OR/WT	Automatic Shock Control	4.3v
33	DG	VR Solenoid Control	0%, 55 mph: 0-45%
33	BR/PK	VR Solenoid Control	0%, 55 mph: 0-45%
34	BL/PK	Data Output Link	Digital Signals
35	BR/BL	Injector 4 Control	4.9-5.2 ms
36	YL/LG	Spark Output Signal	6.93v (50% d/cycle)
37	RD	Vehicle Power	12-14v
38	---	Not Used	---
39	BR/YL	Injector 3 Control	4.9-5.2 ms
40	BK/LG	Power Ground	<0.1v

1991-93 Cougar 5.0L V8 MFI VIN T (A/T) 60 Pin Connector

PCM Pin #	Wire Color	Circuit Description (60 Pin)	Value at Hot Idle
41-42	---	Not Used	---
43	TN/RD	HO2S-21 (B1 S1) Signal	0.1-1.1v
43	TN/WT	HO2S-21 (B1 S1) Signal	0.1-1.1v
44	TN/OR	HO2S-11 (B1 S1) Signal	0.1-1.1v
45	---	Not Used	---
46	BK/WT	Analog Signal Return	<0.050v
46	GY/RD	Analog Signal Return	<0.050v
47	DG/LG	TP Sensor Signal	0.8v, 55 mph: 1.1v
47	GY/WT	TP Sensor Signal	0.8v, 55 mph: 1.1v
48	WT/RD	Self-Test Input Signal	STI On: 1v, Off: 5v
48	WT/PK	Self-Test Input Signal	STI On: 1v, Off: 5v
49	---	Not Used	---
50	DB/OR	MAF Sensor Signal	1.6v
50	BL/RD	MAF Sensor Signal	1.6v
51-53	---	Not Used	---
54	OR/BL	A/C WOT Relay Control	Relay Off: 12v, On: 1v
55	---	Not Used	---
56	DB	PIP Sensor Signal	6.93v (50% d/cycle)
57	RD	Vehicle Power	12-14v
58	TN	Injector 1 Control	4.9-5.2 ms
59	WT	Injector 2 Control	4.9-5.2 ms
60	BK/LG	Power Ground	<0.1v

Pin Connector Graphic

PCM 60-PIN CONNECTOR

Terminal View of 60-Pin PCM Harness Connector

Standard Colors and Abbreviations

Abbreviation	Color	Abbreviation	Color	Abbreviation	Color
BK	Black	GY	Gray	PK	Purple
BL	Blue	GN	Green	RD	Red
BR	Brown	LG	LT Green	TN	Tan
DB	Dark Blue	OR	Orange	WT	White
DG	DK Green	PK	Pink	YL	Yellow

GRAND MARQUIS PIN TABLES

1992-95 Grand Marquis 4.6L V8 VIN W (A/T) 60 Pin Connector

PCM Pin #	Wire Color	Circuit Description (60 Pin)	Value at Hot Idle
1	YL	Keep Alive Power	12-14v
2 ('93-'95)	LG	Brake Switch Signal	Brake Off: 0v, On: 12v
3	GY/BK	VSS (+) Signal	0 Hz, 55 mph: 125 Hz
4	TN/YL	Ignition Diagnostic Monitor	20-31 Hz
5	PK/BL	TSS Sensor (+) Signal	0 Hz, 30 mph: 130 Hz
6	PK/OR	VSS (-) Signal	0 Hz, 55 mph: 125 Hz
7	LG/RD	ECT Sensor Signal	0.5-0.6v
8	DG/YT	Fuel Pump Monitor	12v, Off: 0v
9	TN/BL	MAF Sensor Return	<0.050v
10	DG/OR	A/C Cycling Clutch Signal	A/C Off: 0v, On: 12v
11	GY/YL	Air Management 2 Solenoid	On: 1v, Off: 12v
12	LG/OR	Injector 6 Control	4.0-4.4 ms
13	TN/RD	Injector 7 Control	4.0-4.4 ms
14	BL	Injector 8 Control	4.0-4.4 ms
15	TN/BK	Injector 5 Control	4.0-4.4 ms
16	OR/RD	Ignition System Ground	<0.050v
16	OR	Ignition System Ground	<0.050v
17	TN/RD	Self-Test Output, MIL Control	MIL Off: 12v, On: 1v
18	TN/OR	Data Bus (+) Signal	Digital Signals
19	PK/BL	Data Bus (-) Signal	Digital Signals
20	BK	Case Ground	<0.050v
21	W	IAC Motor Control	8.3-11.5v
21	WT/BL	IAC Motor Control	8.3-11.5v
22	BL/OR	Fuel Pump Control	Relay Off: 12v, On: 1v
23	---	Not Used	---
24	DG	CID Sensor Signal	6-7 Hz
25	LG/PK	ACT Sensor Signal	1.5-2.5v
26	BR/WT	Reference Voltage	4.9-5.1v
27	BR/LG	DPFE Sensor Signal	0.20-1.30v
28	---	Not Used	---
29	WT/RD	Octane Adjust Switch	Closed: 0v, Open: 9.1v
30 ('93-'95)	WT/PK	Neutral Drive Switch	In 'P': 0v, Others: 5v
30 ('93-'95)	BL/YL	Neutral Drive Switch	In 'P': 0v, Others: 5v
30 ('93-'95)	WT/PK	MLP Sensor Signal	0v, O/D: 2.1v
30 ('93-'95)	BL/YL	MLP Sensor Signal	0v, O/D: 2.1v
31-32	---	Not Used	---
33	BR/PK	VR Solenoid Control	0%, 55 mph: 0-45%
34	BL/LG	Data Output Link	Digital Signals
35	BR/BL	Injector 4 Control	4.0-4.4 ms
36	PK	Spark Angle Word Signal	6.93v (50% d/cycle)
37	RD	Vehicle Power	12-14v
38 ('93-'95)	WT/YL	EPC Solenoid Control	9.5v (20 psi)
39	BR/YL	Injector 4 Control	4.0-4.4 ms
40	BK	Power Ground	<0.1v
40	BK/WT	Power Ground	<0.1v

1992-95 Grand Marquis 4.6L V8 VIN W (A/T) 60 Pin Connector

PCM Pin #	Wire Color	Circuit Description (60 Pin)	Value at Hot Idle
41 ('93-'95)	TN/WT	TCS Sensor Signal	TCS & O/D On: 12v
42	---	Not Used	---
43	RD/BK	HO2S-11 (B1 S1) Signal	0.1-1.1v
44	GY/BL	HO2S-21 (B1 S1) Signal	0.1-1.1v
45	---	Not Used	---
46	GY/RD	Analog Signal Return	<0.050v
47	GY/WT	TP Sensor Signal	0.8v, 55 mph: 1.1v
48	WT/PK	Self-Test Input Signal	STI On: 1v, Off: 5v
49 ('93-'95)	OR/BK	TOT Sensor Signal	2.40-2.10v
50	BL/RD	MAF Sensor Signal	0.8v
51	OR/YL	Shift Solenoid 1 Control	1v, 55 mph: 12v
52	PK/OR	Shift Solenoid 2 Control	12v, 55 mph: 1v
53	BR/OR	MCCC or TCC Solenoid	12v, 55 mph: 9-10v
54	OR/BL	A/C WOT Relay Control	Relay Off: 12v, On: 1v
55	DB/W	TCIL (lamp) Control	Lamp On: 1v, Off: 12v
55	WT/LG	TCIL (lamp) Control	Lamp On: 1v, Off: 12v
56	GY/OR	PIP Sensor Signal	6.93v (50% dwell)
57	RD	Vehicle Power	12-14v
58	TN	Injector 1 Control	4.0-4.4 ms
59	WT	Injector 2 Control	4.0-4.4 ms
60	BK	Power Ground	<0.1v

Pin Connector Graphic

PCM 60-PIN CONNECTOR

Terminal View of 60-Pin PCM Harness Connector

Standard Colors and Abbreviations

Abbreviation	Color	Abbreviation	Color	Abbreviation	Color
BK	Black	GY	Gray	PK	Purple
BL	Blue	GN	Green	RD	Red
BR	Brown	LG	LT Green	TN	Tan
DB	Dark Blue	OR	Orange	WT	White
DG	DK Green	PK	Pink	YL	Yellow

1996-97 Grand Marquis 4.6L V8 2v VIN W (A/T) 104 Pin Connector

PCM Pin #	Wire Color	Circuit Description (104 Pin)	Value at Hot Idle
1	PK/OR	Shift Solenoid 2 Control	12v, 55 mph: 1v
2	PK/LG	MIL (lamp) Control	MIL On: 1v, Off: 12v
3-12	---	Not Used	---
13	LG/YL	Flash EEPROM Power	0.1v
14	---	Not Used	---
15	PK/LB	Data Bus (-) Signal	Digital Signals
16	TN/OR	Data Bus (+) Signal	Digital Signals
17-20	---	Not Used	---
21	DB	CKP Sensor (+) Signal	850-1120 Hz
22	GY	CKP Sensor (-) Signal	850-1120 Hz
23	---	Not Used	---
24	BK/WT	Power Ground	<0.1v
25	BK	Case Ground	<0.050v
26	TN/WT	Coil Driver 1 Control	5° dwell
27	OR/YL	Shift Solenoid 1 Control	1v, 55 mph: 12v
28	---	Not Used	---
29	TN/WT	TCS (switch) Signal	TCS & O/D On: 12v
30	WT/RD	Octane Adjust Switch	9.3v (Closed: 0v)
31-32	---	Not Used	---
33	PK/OR	VSS (-) Signal	0 Hz, 55 mph: 125 Hz
34	---	Not Used	---
35	RD/LG	HO2S-12 (B1 S2) Signal	0.1-1.1v
36	TN/LB	MAF Sensor Return	<0.050v
37	OR/BK	TFT Sensor Signal	2.40-2.10v
38	LG/RD	ECT Sensor Signal	0.5-0.6v
39	GY	IAT Sensor Signal	1.5-2.5v
40	DG/YT	Fuel Pump Monitor	12v, Off: 0v
41	DG/OR	A/C Cycling Clutch Switch	A/C Switch On: 12v
42-44	---	Not Used	---
45	RD/OR	Low Speed Fan Control	Fan On: 1v, Off: 12v
46	---	Not Used	---
47	BR/PK	VR Solenoid Control	0%, 55 mph: 45%
48-50	---	Not Used	---
51	BK/WT	Power Ground	<0.1v
52	TN/OR	Coil Driver 2 Control	5° dwell
53	---	Not Used	---
54	RD/LB	TCC Solenoid Control	0%, 55 mph: 90%
55	YL/BK	Keep Alive Power	12-14v
56	LG/BK	EVAP VMV Solenoid	0-10 Hz (0-100%)
57	---	Not Used	---
58	GY/BK	VSS (+) Signal	0 Hz, 55 mph: 125 Hz
59	---	Not Used	---
60	GY/LB	HO2S-11 (B1 S1) Signal	0.1-1.1v
61	PK/LG	HO2S-22 (B2 S2) Signal	0.1-1.1v
62	RD/PK	FTP Sensor Signal	2.6v (0" H2O)
63	---	Not Used	---
64	LBYL	TR Sensor Signal	In 'P': 0v, O/D: 1.7v

1996-97 Grand Marquis 4.6L V8 2v VIN W (A/T) 104 Pin Connector

PCM Pin #	Wire Color	Circuit Description (104 Pin)	Value at Hot Idle
65	BR/LG	DPFE Sensor Signal	0.20-1.30v
66	---	Not Used	---
67	PK/WT	EVAP CV Solenoid	0-10 Hz (0-100%)
68	---	Not Used	---
69	OR/LB	A/C WOT Relay Control	Relay Off: 12v, On: 1v
70	---	Not Used	---
71	RD	Vehicle Power	12-14v
72	TN/RD	Injector 7 Control	3.4-3-7 ms
73	TN/BK	Injector 5 Control	3.4-3-7 ms
74	BR/YL	Injector 3 Control	3.4-3-7 ms
75	TN	Injector 1 Control	3.4-3-7 ms
76, 77	BK/WT	Power Ground	<0.1v
78	TN/LG	Coil Driver 3 Control	5° dwell
79	WT/LG	TCIL (lamp) Control	Lamp On: 1v, Off: 12v
80	LB/OR	Fuel Pump Control	Relay Off: 12v, On: 1v
81	WT/YL	EPC Solenoid Control	9.0v (15 psi)
82, 86	---	Not Used	---
83	WT/LB	IAC Motor Control	9.3v (34% duty cycle)
84	PK/LB	OSS Sensor Signal	0 Hz, 30 mph: 131 Hz
85	DG	CMP Sensor Signal	5-7 Hz
87	RD/BK	HO2S-21 (B1 S1) Signal	0.1-1.1v
88	LB/RD	MAF Sensor Signal	0.6v
89	GY/WT	TP Sensor Signal	0.53-1.27v
90	BR/WT	Reference Voltage	4.9-5.1v
91	GY/RD	Analog Signal Return	<0.050v
92	LG	Brake Pedal Position Switch	Brake Off: 0v, On: 12v
93	RD/WT	HO2S-11 (B1 S1) Heater	1v, Off: 12v
94	YL/LB	HO2S-21 (B2 S1) Heater	1v, Off: 12v
95	WT/BK	HO2S-12 (B1 S2) Heater	1v, Off: 12v
96	TN/YL	HO2S-22 (B2 S2) Heater	1v, Off: 12v
97	RD	Vehicle Power	12-14v
98	LB	Injector 8 Control	3.4-3-7 ms
99	LG/OR	Injector 6 Control	3.4-3-7 ms
100	BR/LB	Injector 4 Control	3.4-3-7 ms
101	WT	Injector 2 Control	3.4-3-7 ms
102	---	Not Used	---
103	BK/WT	Power Ground	<0.1v
104	RD/YL	Coil Driver 4 Control	5° dwell

Pin Connector Graphic

PCM 104-PIN CONNECTOR

Terminal View of 104-Pin PCM Wiring Harness Connector

Standard Colors and Abbreviations

Abbreviation	Color	Abbreviation	Color	Abbreviation	Color
BK	Black	GY	Gray	PK	Purple
BL	Blue	GN	Green	RD	Red
BR	Brown	LG	LT Green	TN	Tan
DB	Dark Blue	OR	Orange	WT	White
DG	DK Green	PK	Pink	YL	Yellow

1998-1999 Grand Marquis 4.6L V8 VIN W (A/T) 104 Pin Connector

PCM Pin #	Wire Color	Circuit Description (104 Pin)	Value at Hot Idle
1	OR/YL	COP 6 Driver Control	5° dwell
2	PK/LG	MIL (lamp) Control	MIL On: 1v, Off: 12v
3	BK/WT	Power Ground	<0.1v
4-5	---	Not Used	---
6	OR/YL	Shift Solenoid 'A' Control	1v, 55 mph: 12v
9	YL/WT	Fuel Level Indicator Signal	1.7v (1/2 full)
7-8, 10	---	Not Used	---
11	PK/OR	Shift Solenoid 'B' Control	12v, Cruise: 1v
12	WT/LG	TCIL (lamp) Control	Lamp On: 1v, Off: 5v
13	PK	Flash EEPROM Power	0.1v
15	PK/LB	Data Bus (-) Signal	Digital Signals
16	TN/OR	Data Bus (+) Signal	Digital Signals
17-20	---	Not Used	---
21	BK/PKK	CKP Sensor (+) Signal	440-490 Hz
22	GY/YL	CKP Sensor (-) Signal	440-490 Hz
23	---	Not Used	---
24	BK/WT	Power Ground	<0.1v
25	BK	PCM Case Ground	<0.050v
26	LG/WT	COP 1 Driver Control	5° dwell
27	LG/YL	COP 5 Driver Control	1v, 55 mph: 12v
28	RD/OR	Low Speed Fan Control	On: 1v, Off: 12v
29	TN/WT	TCS (switch) Signal	TCS & O/D On: 12v
30	WT/RD	Octane Adjust Switch	Closed: 0v, Open: 9.3v
31-32, 38	---	Not Used	---
33	PK/OR	VSS (-) Signal	0 Hz, 55 mph: 125 Hz
34	YL/BK	Digital TR1 Sensor	In 'P': 0v, 55 mph: 11v
35	RD/LG	HO2S-12 (B1 S2) Signal	0.1-1.1v
36	TN/LB	MAF Sensor Return	<0.050v
37	OR/BK	TFT Sensor Signal	2.10-2.40v
39	GY	IAT Sensor Signal	1.5-2.5v
40	DG/YT	Fuel Pump Monitor	On: 12v, Off: 0v
41	DG/OR	A/C Cycling Clutch Switch	AC On: 12v, Off: 0v
42	RD/WT	ECT Sensor Signal	0.5-0.6v
43	DB/LG	Fuel Flow Rate Signal	Digital Signals
45	OR/RD	CHTIL (lamp) Control	Lamp On: 1v, Off: 5v
46	LG/PK	High Speed Fan Control	On: 1v, Off: 12v
47	BR/PK	EGR VR Solenoid	0%, 55 mph: 45%
48	---	Not Used	---
49	RD/LB	Digital TR2 Sensor	In 'P': 0v, 55 mph: 11v
50	WT/BK	Digital TR4 Sensor	In 'P': 0v, 55 mph: 11v
51	BK/WT	Power Ground	<0.1v
52	WT/PK	COP 3 Driver Control	5° dwell
53	DG/PK	COP 4 Driver Control	5° dwell
54	PK/YL	TCC Solenoid Control	0%, 55 mph: 95%
55	YL/BK	Keep Alive Power	12-14v
56	LG/BK	EVAP Purge Solenoid	0-10 Hz (0-100%)
57, 59	---	Not Used	---
58	GY/BK	VSS + Sensor Signal	0 Hz, 55 mph: 125 Hz
60	GY/LB	HO2S-11 (B1 S1) Signal	0.1-1.1v
61	PK/LG	HO2S-22 (B2 S2) Signal	0.1-1.1v
62	RD/PK	FTP Sensor Signal	2.6v (0" H2O - cap off)
63	---	Not Used	---
64	LBYL	Digital TR3 Sensor	In 'P': 0v, in O/D: 1.7v

1998-1999 Grand Marquis 4.6L V8 VIN W (A/T) 104 Pin Connector

PCM Pin #	Wire Color	Circuit Description (104 Pin)	Value at Hot Idle
65	BR/LG	DPFE Sensor Signal	0.20-1.30v
66	YL/LG	CHT Sensor Signal	0.7v (194°F)
67	PK/WT	EVAP CV Solenoid	0-10 Hz (0-100%)
68, 70	---	Not Used	---
69	PK/YL	A/C WOT Relay Control	On: 1v, Off: 12v
82	---	Not Used	---
71	RD	Vehicle Power	12-14v
72	TN/RD	Injector 7 Control	3.4-3.7 ms
73	TN/BK	Injector 5 Control	3.4-3.7 ms
74	BR/YL	Injector 3 Control	3.4-3.7 ms
75	TN	Injector 1 Control	3.4-3.7 ms
77	BK/WT	Power Ground	<0.1v
78	PK/LB	COP 7 Driver Control	5° dwell
79	WT/RD	COP 8 Driver Control	5° dwell
80	LB/OR	Fuel Pump Control	On: 1v, Off: 12v
81	WT/YL	EPC Solenoid Control	9.5v (20 psi)
83	WT/LB	IAC Motor Control	34% duty cycle
84	DB/YL	OSS Sensor Signal	0 Hz
85	DB/OR	CMP Sensor Signal	6-7 Hz
86	WT/BK	A/C High Pressure Switch	Open: 12v, Closed: 0v
87	RD/BK	HO2S-21 (B2 S1) Signal	0.1-1.1v
88	LB/RD	MAF Sensor Signal	0.6v
89	GY/WT	TP Sensor Signal	0.53-1.27v
90	BR/WT	Reference Voltage	4.9-5.1v
91	GY/RD	Analog Signal Return	<0.050v
92	LG	Brake Pedal Position Switch	Brake Off: 0v, On: 12v
93	RD/WT	HO2S-11 (B1 S1) Heater	On: 1v, Off: 12v
94	YL/LB	HO2S-21 (B2 S1) Heater	On: 1v, Off: 12v
95	WT/BK	HO2S-12 (B1 S2) Heater	On: 1v, Off: 12v
96	TN/YL	HO2S-22 (B2 S2) Heater	On: 1v, Off: 12v
97 ('98 only)	RD	Vehicle Power	12-14v
98	LB	Injector 8 Control	3.4-3.7 ms
99	LG/OR	Injector 6 Control	3.4-3.7 ms
100	BR/LB	Injector 4 Control	3.4-3.7 ms
101	WT	Injector 2 Control	3.4-3.7 ms
102	---	Not Used	---
103	BK/WT	Power Ground	<0.1v
104	PK/WT	COP 2 Driver Control	5° dwell

Pin Connector Graphic

PCM 104-PIN CONNECTOR

```
1 ⊙⊙⊙⊙⊙⊙⊙⊙⊙⊙⊙⊙⊙    ⊙⊙⊙⊙⊙⊙⊙⊙⊙⊙⊙⊙⊙ 26
27 ⊙⊙⊙⊙⊙⊙⊙⊙⊙⊙⊙⊙⊙    ⊙⊙⊙⊙⊙⊙⊙⊙⊙⊙⊙⊙⊙ 52
53 ⊙⊙⊙⊙⊙⊙⊙⊙⊙⊙⊙⊙⊙  ●  ⊙⊙⊙⊙⊙⊙⊙⊙⊙⊙⊙⊙⊙ 78
79 ⊙⊙⊙⊙⊙⊙⊙⊙⊙⊙⊙⊙⊙    ⊙⊙⊙⊙⊙⊙⊙⊙⊙⊙⊙⊙⊙ 104
```

Terminal View of 104-Pin PCM Wiring Harness Connector

Standard Colors and Abbreviations

Abbreviation	Color	Abbreviation	Color	Abbreviation	Color
BK	Black	GY	Gray	PK	Purple
BL	Blue	GN	Green	RD	Red
BR	Brown	LG	LT Green	TN	Tan
DB	Dark Blue	OR	Orange	WT	White
DG	DK Green	PK	Pink	YL	Yellow

2000-02 Grand Marquis 4.6L V8 VIN W (A/T) 104 Pin Connector

PCM Pin #	Wire Color	Circuit Description (104 Pin)	Value at Hot Idle
1	OR/YL	COP 6 Driver Control	5° dwell
2	PK/LG	MIL (lamp) Control	MIL On: 1v, Off: 12v
3	BK/WT	Power Ground	<0.1v
4-5	---	Not Used	---
6	OR/YL	Shift Solenoid 'A' Control	1v, 55 mph: 12v
7-8	---	Not Used	---
9	YL/WT	Fuel Level Indicator Signal	1.7v (1/2 full)
10	---	Not Used	---
11	VT/OR	Shift Solenoid 'B' Control	12v, Cruise: 1v
12	WT/LG	TCIL (lamp) Control	Lamp On: 1v, Off: 5v
13	VT	Flash EEPROM Power	0.1v
15	PK/LB	Data Bus (-) Signal	Digital Signals
16	TN/OG	Data Bus (+) Signal	Digital Signals
17-20	---	Not Used	---
21	BK/PKK	CKP Sensor (+) Signal	440-490 Hz
22	GY/YL	CKP Sensor (-) Signal	440-490 Hz
23-24	---	Not Used	---
25	BK	PCM Case Ground	<0.050v
26	LG/WTT	COP 1 Driver Control	5° dwell
27	LG/YL	COP 5 Driver Control	1v, 55 mph: 12v
28	RD/OG	Low Speed Fan Control	On: 1v, Off: 12v
29	TN/WT	TCS (switch) Signal	TCS & O/D On: 12v
30-32	---	Not Used	---
33	PK/OR	VSS (-) Signal	0 Hz, 55 mph: 125 Hz
34	YL/BK	Digital TR1 Sensor	In 'P': 0v, 55 mph: 11v
35	RD/LG	HO2S-12 (B1 S2) Signal	0.1-1.1v
36	TN/LB	MAF Sensor Return	<0.050v
37	OG/BK	TFT Sensor Signal	2.10-2.40v
38	---	Not Used	---
39	GY	IAT Sensor Signal	1.5-2.5v
40	DG/YL	Fuel Pump Monitor	On: 12v, Off: 0v
41	BK/YL	A/C Pressure Cutout switch	AC On: 12v, Off: 0v
42	RD/WT	ECT Sensor Signal	0.5-0.6v
43	DB/LG	Fuel Flow Rate Signal	Digital Signals
45	OG/RD	CHTIL (lamp) Control	Lamp On: 1v, Off: 5v
46	LG/VT	High Speed Fan Control	On: 1v, Off: 12v
47	BR/VT	EGR VR Solenoid	0%, 55 mph: 45%
48	---	Not Used	---
49	LB/BK	Digital TR2 Sensor	In 'P': 0v, 55 mph: 11v
50	WT/BK	Digital TR4 Sensor	In 'P': 0v, 55 mph: 11v
51	BK/WT	Power Ground	<0.1v
52	WT/PK	COP 3 Driver Control	5° dwell
53	DG/VT	COP 4 Driver Control	5° dwell
54	VT/YL	TCC Solenoid Control	0%, 55 mph: 95%
55	YL/BK	Keep Alive Power	12-14v
56	LG/BK	EVAP Purge Solenoid	0-10 Hz (0-100%)
57	---	Not Used	---
58	GY/BK	VSS + Sensor Signal	0 Hz, 55 mph: 125 Hz
59	---	Not Used	---
60	GY/BK	HO2S-11 (B1 S1) Signal	0.1-1.1v
61	VT/LG	HO2S-22 (B2 S2) Signal	0.1-1.1v
62	RD/VT	FTP Sensor Signal	2.6v (0" H2O - cap off)
63	---	Not Used	---
64	LB/YL	Digital TR3 Sensor	In 'P': 0v, in O/D: 1.7v

2000-02 Grand Marquis 4.6L V8 VIN W (A/T) 104 Pin Connector

PCM Pin #	Wire Color	Circuit Description (104 Pin)	Value at Hot Idle
65 ('00)	BN/LG	DPFE Sensor Signal	0.20-1.30v
65 ('01-'02)	BN/LG	DPFE Sensor Signal	0.95-1.05v
66	YL/LB	CHT Sensor Signal	0.7v (194°F)
67	VT/WT	EVAP CV Solenoid	0-10 Hz (0-100%)
68	---	Not Used	---
69	PK/YL	A/C WOT Relay Control	On: 1v, Off: 12v
70, 82	---	Not Used	---
71	RD	Vehicle Power	12-14v
72	TN/RD	Injector 7 Control	3.4-3.7 ms
73	TN/BK	Injector 5 Control	3.4-3.7 ms
74	BN/YL	Injector 3 Control	3.4-3.7 ms
75	TN	Injector 1 Control	3.4-3.7 ms
77	BK/WT	Power Ground	<0.1v
78	PK/LB	COP 7 Driver Control	5° dwell
79	WT/RD	COP 8 Driver Control	5° dwell
80	LB/OR	Fuel Pump Control	On: 1v, Off: 12v
81	WT/YL	EPC Solenoid Control	9.5v (20 psi)
83	WT/LB	IAC Motor Control	34% duty cycle
84	DB/YL	OSS Sensor Signal	0 Hz
85	DB/OR	CMP Sensor Signal	6-7 Hz
86	WT/BK	A/C High Pressure Switch	Open: 12v, Closed: 0v
87	RD/BK	HO2S-21 (B2 S1) Signal	0.1-1.1v
88	LB/RD	MAF Sensor Signal	0.6v
89	GY/WT	TP Sensor Signal	0.53-1.27v
90	BN/WT	Reference Voltage	4.9-5.1v
91	GY/RD	Sensor Signal Return	<0.050v
92	LG	Brake Pedal Position Switch	Brake Off: 0v, On: 12v
93	RD/WT	HO2S-11 (B1 S1) Heater	On: 1v, Off: 12v
94	YL/LB	HO2S-21 (B2 S1) Heater	On: 1v, Off: 12v
95	WT/BK	HO2S-12 (B1 S2) Heater	On: 1v, Off: 12v
96	TN/YL	HO2S-22 (B2 S2) Heater	On: 1v, Off: 12v
98	LB	Injector 8 Control	3.4-3.7 ms
99	LG/OR	Injector 6 Control	3.4-3.7 ms
100	BN/LB	Injector 4 Control	3.4-3.7 ms
101	WT	Injector 2 Control	3.4-3.7 ms
102	---	Not Used	---
103	BK/WT	Power Ground	<0.1v
104	PK/WT	COP 2 Driver Control	5° dwell

Pin Connector Graphic

PCM 104-PIN CONNECTOR

Terminal View of 104-Pin PCM Wiring Harness Connector

Standard Colors and Abbreviations

Abbreviation	Color	Abbreviation	Color	Abbreviation	Color
BK	Black	GY	Gray	PK	Purple
BL	Blue	GN	Green	RD	Red
BR	Brown	LG	LT Green	TN	Tan
DB	Dark Blue	OR	Orange	WT	White
DG	DK Green	PK	Pink	YL	Yellow

2003-05 Grand Marquis 4.6L V8 VIN W (A/T) 104 Pin Connector

PCM Pin #	Wire Color	Circuit Description (104 Pin)	Value at Hot Idle
1	OR/YL	COP 6 Driver Control	5° dwell
2	PK/LG	MIL (lamp) Control	MIL On: 1v, Off: 12v
3	BK/WT	Power Ground	<0.1v
4-5, 14, 23	---	Not Used	---
6	OR/YL	Shift Solenoid 'A' Control	1v, 55 mph: 12v
7	YL/LG	Generator Regulator 'S' Terminal	0-130 Hz
8	RD/PK	Fuel Rail Temperature Sensor	1.7-3.5v (50-120°F)
9	YL/WT	Fuel Level Indicator Signal	1.7v (1/2 full)
10	LB/RD	MAP Sensor Signal	107 Hz
11	VT/OR	Shift Solenoid 'B' Control	12v, Cruise: 1v
12	WT/LG	TCIL (lamp) Control	Lamp On: 1v, Off: 5v
13	VT	Flash EEPROM Power	0.1v
15	PK/LB	SCP Bus (-) Signal	Digital Signals
16	TN/OR	SCP Bus (+) Signal	Digital Signals
17	GY/OR	RX Signal	Digital Signals
18	WT/LG	TX Signal	Digital Signals
19	OR/RD	Cylinder Head Temperature Lamp Control	Lamp Off: 12v, On: 1v
20	WT/LG	Fuel Door Release Solenoid Indicator	Solenoid Off: 12v, On: 1v
21	BK/PK	CKP Sensor (+) Signal	440-490 Hz
22	GY/YL	CKP Sensor (-) Signal	440-490 Hz
24, 51	BK/WT	Power Ground	<0.1v
25	BK	PCM Case Ground	<0.050v
26	LG/WT	COP 1 Driver Control	5° dwell
27	LG/YL	COP 5 Driver Control	1v, 55 mph: 12v
28	RD/OR	Low Speed Fan Control	On: 1v, Off: 12v
29	TN/WT	TCS (switch) Signal	TCS & O/D On: 12v
30	OG/RD	Light Emitting Diode Signal Ground	<0.050v
31	YL/LG	Power Steering Pressure Sensor	Straight: 0v, Turned: 12v
32	DG/VT	Knock Sensor Ground	<0.050v
33	---	Not Used	---
34	YL/BK	Digital TR1 Sensor	In 'P': 0v, 55 mph: 11v
35	RD/LG	HO2S-12 (B1 S2) Signal	0.1-1.1v
36	TN/LB	MAF Sensor Return	<0.050v
37	OR/BK	TFT Sensor Signal	2.10-2.40v
38	BK/WT	A/C Pressure Sensor Discharge Temp.	A/C On: 1.5-1.9v
39	GY	IAT Sensor Signal	1.5-2.5v
40	DG/YL	Fuel Pump Monitor	On: 12v, Off: 0v
41	PK/LB	A/C Pressure Cutout Switch	AC On: 12v, Off: 0v
42	RD/WT	ECT Sensor Signal	0.5-0.6v
43	DB/LG	Low Fuel Indicator	Digital Signals
44	GY/RD	Starter Relay Control	Relay Off: 12v, On: 9v
45	LG/RD	Generator Common	Digital Signals
46	---	Not Used	---
47	BR/PK	EGR VR Solenoid	0%, 55 mph: 45%
48	WT	Tachometer Output	DC pulse signals
49	LB/BK	Digital TR2 Sensor	In 'P': 0v, 55 mph: 11v
50	WT/BK	Digital TR4 Sensor	In 'P': 0v, 55 mph: 11v
52	WT/PK	COP 3 Driver Control	5° dwell
53	DG/VT	COP 4 Driver Control	5° dwell
54	VT/YL	TCC Solenoid Control	0%, 55 mph: 95%
55	YL/BK	Keep Alive Power	12-14v
56	LG/BK	EVAP Purge Solenoid Control	0-10 Hz (0-100%)
57	YL/RD	Knock Sensor Signal	0v
58-59	---	Not Used	---

2003-05 Grand Marquis 4.6L V8 VIN W (A/T) 104 Pin Connector

PCM Pin #	Wire Color	Circuit Description (104 Pin)	Value at Hot Idle
60	GY/BK	HO2S-11 (B1 S1) Signal	0.1-1.1v
61	VT/LG	HO2S-22 (B2 S2) Signal	0.1-1.1v
62	RD/PK	FTP Sensor Signal	2.6v (0" H2O - cap off)
63	OR/LG	Injection Pressure Sensor	2.8v (39 psi)
64	LB/YL	Digital TR3 Sensor	In 'P': 0v, in O/D: 1.7v
65	BR/LG	DPFE Sensor Signal	0.95-1.05v
66	YL/LG	Cylinder Head Temperature Sensor	0.7v (194ºF)
67	VT/WT	EVAP Canister Vent Solenoid	0-10 Hz (0-100%)
68	GY/BK	Vehicle Speed Sensor (+) Signal	0 Hz, 55 mph: 125 Hz
69	PK/YL	A/C WOT Relay Control	On: 1v, Off: 12v
70	YL	Generator Battery Indicator Control	Lamp Off: 12v, On: 1v
71	RD	Vehicle Power (Start-Run)	12-14v
72	TN/RD	Injector 7 Control	3.4-3.7 ms
73	TN/BK	Injector 5 Control	3.4-3.7 ms
74	BR/YL	Injector 3 Control	3.4-3.7 ms
75	TN	Injector 1 Control	3.4-3.7 ms
76	BK/WT	Power Ground	<0.1v
77	---	Not Used	---
78	PK/LB	COP 7 Driver Control	5º dwell
79	WT/RD	COP 8 Driver Control	5º dwell
80	LB/OR	Fuel Pump Control	On: 1v, Off: 12v
81	WT/YL	EPC Solenoid Control	9.5v (20 psi)
82	---	Not Used	---
83	WT/LB	IAC Motor Control	34% duty cycle
84	DB/YL	OSS Sensor Signal	0 Hz
85	DB/OR	CMP Sensor Signal	6-7 Hz
86	WT/BK	A/C High Pressure Switch	Open: 12v, Closed: 0v
87	RD/BK	HO2S-21 (B2 S1) Signal	0.1-1.1v
88	LB/RD	MAF Sensor Signal	0.6v
89	GY/WT	TP Sensor Signal	0.53-1.27v
90	BR/WT	Reference Voltage	4.9-5.1v
91	GY/RD	Analog Signal Return	<0.050v
92	DG	Brake Pedal Position Switch	Brake Off: 0v, On: 12v
93	RD/WT	HO2S-11 (B1 S1) Heater	On: 1v, Off: 12v
94	YL/LB	HO2S-21 (B2 S1) Heater	On: 1v, Off: 12v
95	WT/BK	HO2S-12 (B1 S2) Heater	On: 1v, Off: 12v
96	TN/YL	HO2S-22 (B2 S2) Heater	On: 1v, Off: 12v
97	RD	Vehicle Power (Start-Run)	12-14v
98	LB	Injector 8 Control	3.4-3.7 ms
99	LG/OR	Injector 6 Control	3.4-3.7 ms
100	BR/LB	Injector 4 Control	3.4-3.7 ms
101	WT	Injector 2 Control	3.4-3.7 ms
102	---	Not Used	---
103	BK/WT	Power Ground	<0.1v
104	PK/WT	COP 2 Driver Control	5º dwell

Pin Connector Graphic

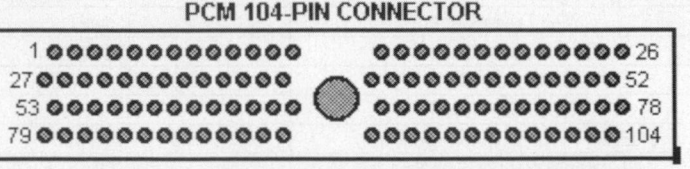

PCM 104-PIN CONNECTOR

Terminal View of 104-Pin PCM Wiring Harness Connector

1990-91 Grand Marquis 5.0L V8 MFI VIN F (A/T) 60 Pin Connector

PCM Pin #	Wire Color	Circuit Description (60 Pin)	Value at Hot Idle
1	BK/OR	Keep Alive Power	12-14v
1	YL	Keep Alive Power	12-14v
2 ('90-'91)	LG	Brake Switch Signal	Brake Off: 0v, On: 12v
3 ('90-'91)	DG/WT	VSS (+) Signal	0 Hz, 55 mph: 125 Hz
4	DG/YT	Ignition Diagnostic Monitor	20-31 Hz
4	BL/PK	Ignition Diagnostic Monitor	20-31 Hz
4	TN/YL	Ignition Diagnostic Monitor	20-31 Hz
5	---	Not Used	---
6 ('90-'91)	BK/WT	VSS (-) Signal	0 Hz, 55 mph: 125 Hz
6 ('90-'91)	GY/RD	VSS (-) Signal	0 Hz, 55 mph: 125 Hz
7	LG/YL	ECT Sensor Signal	0.5-0.6v
7	LG/RD	ECT Sensor Signal	0.5-0.6v
8-9	---	Not Used	---
10	LG/PK	A/C Cycling Clutch Signal	A/C Off: 0v, On: 12v
10	PK/BL	A/C Cycling Clutch Signal	A/C Off: 0v, On: 12v
11	LG/BK	Air Management 2 Solenoid	On: 1v, Off: 12v
12	TN/YL	Injector 3 Control	5.7-6.2 ms
12	BR/YL	Injector 3 Control	5.7-6.2 ms
13	TN/BK	Injector 4 Control	5.7-6.2 ms
13	BR/BL	Injector 4 Control	5.7-6.2 ms
14	TN/BL	Injector 5 Control	5.7-6.2 ms
14	TN	Injector 5 Control	5.7-6.2 ms
15	TN/LG	Injector 6 Control	5.7-6.2 ms
15	LG/OR	Injector 6 Control	5.7-6.2 ms
16	BK/OR	Ignition System Ground	<0.050v
16	OR	Ignition System Ground	<0.050v
17	YL	Self-Test Output, MIL Control	MIL Off: 12v, On: 1v
17	TN/RD	Self-Test Output, MIL Control	MIL Off: 12v, On: 1v
18-19	---	Not Used	---
20	BK	Case Ground	<0.050v
21	WT/BL	IAC Motor Control	9.0-11.5v
22	TN/LG	Fuel Pump Control	Relay Off: 12v, On: 1v
23-24	---	Not Used	---
25	LG/PK	ACT Sensor Signal	1.5-2.5v
26	OR/WT	Reference Voltage	4.9-5.1v
27	BR/LG	EVP Sensor Signal	0.4v, 55 mph: 1.5v
28, 32	---	Not Used	---
29	DG/PK	HO2S-11 (B1 S1) Signal	0.1-1.1v
29	RD/BK	HO2S-11 (B1 S1) Signal	0.1-1.1v
30	WT/PK	Neutral Drive Switch	In 'N': 0v, Others: 5v
30	BL/YL	Neutral Drive Switch	In 'N': 0v, Others: 5v
31	GY/YL	Canister Purge Solenoid	12v, 55 mph: 1v
33	DG	VR Solenoid Control	0%, 55 mph: 0-45%
33	BR/PK	VR Solenoid Control	0%, 55 mph: 0-45%
34	BL/P	Data Output Link	Digital Signals
34	BL/PK	Data Output Link	Digital Signals
35 ('90-'91)	WT/PK	S/C Vent Solenoid	Vacuum Decreasing: 1v
35 ('90-'91)	GY/BK	S/C Vent Solenoid	Vacuum Decreasing: 1v
36	YL/LG	Spark Output Signal	6.93v (50% d/cycle)
37, 57	RD	Vehicle Power	12-14v
38	GY/BK	S/C Vacuum Solenoid Control	Vacuum Increasing: 1v
38	LG	S/C Vacuum Solenoid Control	Vacuum Increasing: 1v
39 ('90-'91)	GY/BK	S/C Switch Ground	<0.050v
40	BK/LG	Power Ground	<0.1v

1990-91 Grand Marquis 5.0L V8 MFI VIN F (A/T) 60 Pin Connector

PCM Pin #	Wire Color	Circuit Description (60 Pin)	Value at Hot Idle
41 ('90-'91)	OR/YL	S/C Servo (+) Signal	Solenoid On: 12v
42	TN/OR	Injector 7 Control	5.7-6.2 ms
42	TN/RD	Injector 7 Control	5.7-6.2 ms
43	DB/LG	HO2S-21 (B1 S1) Signal	0.1-1.1v
44	---	Not Used	---
45	DB/LG	MAP Sensor Signal	107 Hz (sea level)
45	LG/BK	MAP Sensor Signal	107 Hz (sea level)
46	BK/WT	Analog Signal Return	<0.050v
46	GY/RD	Analog Signal Return	<0.050v
47	DG/LG	TP Sensor Signal	0.8v, 55 mph: 1.1v
47	GY/WT	TP Sensor Signal	0.8v, 55 mph: 1.1v
48	WT/RD	Self-Test Input Signal	STI On: 1v, Off: 5v
48	WT/BK	Self-Test Input Signal	STI On: 1v, Off: 5v
48	BR	Self-Test Input Signal	STI On: 1v, Off: 5v
49	OR	HO2S-11 (Bank 1) Ground	<0.050v
50 ('90-'91)	OR/RD	Speed Control Switch	6.7v
50 ('90-'91)	BL/BK	Speed Control Switch	6.7v
51	WT/RD	Air Management 1 Solenoid	AM1 On: 1v, Off: 12v
52	BL	Injector 8 Control	5.7-6.2 ms
52	TN/RD	Injector 8 Control	5.7-6.2 ms
53	---	Not Used	---
54	OR/BL	A/C WOT Relay Control	Relay Off: 12v, On: 1v
55	---	Not Used	---
56	DB	PIP Sensor Signal	6.93v (50% d/cycle)
57	BK/YL	Vehicle Power	12-14v
58	TN	Injector 1 Control	5.7-6.2 ms
59	TN/WT	Injector 2 Control	5.7-6.2 ms
59	WT	Injector 2 Control	5.7-6.2 ms
60	BK/LG	Power Ground	<0.1v

Pin Connector Graphic

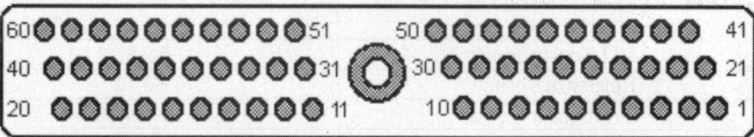

Terminal View of 60-Pin PCM Harness Connector

Standard Colors and Abbreviations

Abbreviation	Color	Abbreviation	Color	Abbreviation	Color
BK	Black	GY	Gray	PK	Purple
BL	Blue	GN	Green	RD	Red
BR	Brown	LG	LT Green	TN	Tan
DB	Dark Blue	OR	Orange	WT	White
DG	DK Green	PK	Pink	YL	Yellow

1990-91 Grand Marquis 5.0L V8 VIN F California 60 Pin Connector

PCM Pin #	Wire Color	Circuit Description (60 Pin)	Value at Hot Idle
1	BK/OR	Keep Alive Power	12-14v
1	YL	Keep Alive Power	12-14v
2	---	Not Used	---
3	DG/WT	VSS (+) Signal	0 Hz, 55 mph: 125 Hz
4	DG/YT	Ignition Diagnostic Monitor	20-31 Hz
5	LG	Brake On/Off Signal	Brake Off: 0v, On: 12v
6	BK/WT	VSS (-) Signal	0 Hz, 55 mph: 125 Hz
7	LG/YL	ECT Sensor Signal	0.5-0.6v
8	BK/OR	Data Bus (-) Signal	Digital Signals
9	---	Not Used	---
10	PK/BL	A/C Cycling Clutch Signal	A/C Off: 0v, On: 12v
11	OR/YL	Air Management 2 Solenoid	On: 1v, Off: 12v
12	BR/YL	Injector 3 Control	5.7-6.2 ms
13	BR/BL	Injector 4 Control	5.7-6.2 ms
14	TN/BL	Injector 5 Control	5.7-6.2 ms
15	LG	Injector 6 Control	5.7-6.2 ms
16	BK/OR	Ignition System Ground	<0.050v
17	TN/RD	Self-Test Output, MIL Control	MIL Off: 12v, On: 1v
18	---	Not Used	---
19	OR	Fuel Pump Monitor	12v, Off: 0v
20	BK	Case Ground	<0.050v
21	WT/BL	IAC Motor Control	9.0-11.5v
22	TN/LG	Fuel Pump Control	Relay Off: 12v, On: 1v
23-24	---	Not Used	---
25	LG/PK	ACT Sensor Signal	1.5-2.5v
26	OR/WT	Reference Voltage	4.9-5.1v
27	BR/LG	EVP Sensor Signal	0.4v, 55 mph: 1.5v
28	BL/BK	Speed Control Switch Signal	6.7v
29	DG/PK	HO2S-11 (B1 S1) Signal	0.1-1.1v
30	WT/PK	Neutral Drive Switch	In 'P': 0v, Others: 5v
31	GY/YL	Canister Purge Solenoid	12v, 55 mph: 1v
32	LG/BK	Thermactor Air Diverter SOL	On: 1v, Off: 12v
33	DG	VR Solenoid Control	0%, 55 mph: 0-45%
34	---	Not Used	---
35	WT/PK	S/C Vent Solenoid Control	Vacuum Decreasing: 1v
36	YL/LG	Spark Output Signal	6.93v (50% d/cycle)
37	RD	Vehicle Power	12-14v
38	WT/RD	Thermactor Air Bypass SOL	On: 1v, Off: 12v
39	GY/BK	Speed Control Switch Ground	<0.050v
40	BK/LG	Power Ground	<0.1v

1990-91 Grand Marquis 5.0L V8 VIN F California 60 Pin Connector

PCM Pin #	Wire Color	Circuit Description (60 Pin)	Value at Hot Idle
41	---	Not Used	---
42	TN/OR	Injector 7 Control	5.7-6.2 ms
43	DB/LG	HO2S-21 (B1 S1) Signal	0.1-1.1v
44	OR/BK	Data Bus (+) Signal	Digital Signals
45	LG/BK	MAP Sensor Signal	107 Hz (sea level)
46	BK/WT	Analog Signal Return	<0.050v
47	DG/LG	TP Sensor Signal	0.8v, 55 mph: 1.1v
48	WT/BK	Self-Test Input Signal	STI On: 1v, Off: 5v
49	OR	HO2S-11 (Bank 1) Ground	<0.050v
50	---	Not Used	---
51	GY/BK	Air Management 1 Solenoid	AM1 On: 1v, Off: 12v
52	BL	Injector 8 Control	5.7-6.2 ms
53	---	Not Used	---
54	OR/BL	A/C WOT Relay Control	Relay Off: 12v, On: 1v
55	---	Not Used	---
56	DB	PIP Sensor Signal	6.93v (50% dwell)
57	RD	Vehicle Power	12-14v
58	TN	Injector 1 Control	5.7-6.2 ms
59	WT	Injector 2 Control	5.7-6.2 ms
60	BK/LG	Power Ground	<0.1v

Pin Connector Graphic

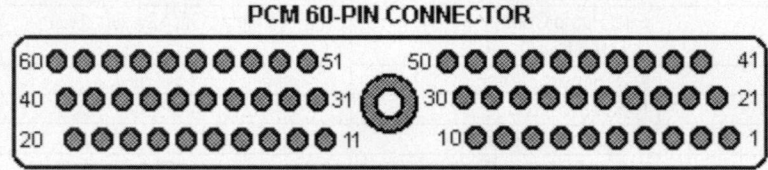

Standard Colors and Abbreviations

Abbreviation	Color	Abbreviation	Color	Abbreviation	Color
BK	Black	GY	Gray	PK	Purple
BL	Blue	GN	Green	RD	Red
BR	Brown	LG	LT Green	TN	Tan
DB	Dark Blue	OR	Orange	WT	White
DG	DK Green	PK	Pink	YL	Yellow

MARAUDER PIN TABLES

2003-04 Marauder 4.6L V8 MPI VIN V 104 Pin Connector

PCM Pin #	Wire Color	Circuit Description (104 Pin)	Value at Hot Idle
1	OR/YL	COP 6 Driver Control	5° dwell
2	PK/LG	MIL (lamp) Control	MIL On: 1v, Off: 12v
3	BK/WT	Power Ground	<0.1v
4-5, 14, 23	---	Not Used	---
6	OR/YL	Shift Solenoid 'A' Control	1v, 55 mph: 12v
7	YL/LG	Generator Regulator 'S' Terminal	0-130 Hz
8	RD/PK	Fuel Rail Temperature Sensor	1.7-3.5v (50-120°F)
9	YL/WT	Fuel Level Indicator Signal	1.7v (1/2 full)
10	LB/RD	MAP Sensor Signal	107 Hz
11	VT/OR	Shift Solenoid 'B' Control	12v, Cruise: 1v
12	WT/LG	TCIL (lamp) Control	Lamp On: 1v, Off: 5v
13	VT	Flash EEPROM Power	0.1v
15	PK/LB	SCP Bus (-) Signal	Digital Signals
16	TN/OR	SCP Bus (+) Signal	Digital Signals
17	GY/OR	RX Signal	Digital Signals
18	WT/LG	TX Signal	Digital Signals
19	OR/RD	Cylinder Head Temperature Lamp Control	Lamp Off: 12v, On: 1v
20	WT/LG	Fuel Door Release Solenoid Indicator	Solenoid Off: 12v, On: 1v
21	BK/PK	CKP Sensor (+) Signal	440-490 Hz
22	GY/YL	CKP Sensor (-) Signal	440-490 Hz
24	BK/WT	Power Ground	<0.1v
25	BK	PCM Case Ground	<0.050v
26	LG/WT	COP 1 Driver Control	5° dwell
27	LG/YL	COP 5 Driver Control	1v, 55 mph: 12v
28	RD/OR	Low Speed Fan Control	On: 1v, Off: 12v
29	TN/WT	TCS (switch) Signal	TCS & O/D On: 12v
30	OR/RD	Light Emitting Diode Signal Ground	<0.050v
31	YL/LG	Power Steering Pressure Sensor	Straight: 0v, Turned: 12v
32	DG/VT	Knock Sensor Ground	<0.050v
33	---	Not Used	---
34	YL/BK	Digital TR1 Sensor	In 'P': 0v, 55 mph: 11v
35	RD/LG	HO2S-12 (B1 S2) Signal	0.1-1.1v
36	TN/LB	MAF Sensor Return	<0.050v
37	OR/BK	TFT Sensor Signal	2.10-2.40v
38	BK/WT	A/C Pressure Sensor Discharge Temp.	A/C On: 1.5-1.9v
39	GY	IAT Sensor Signal	1.5-2.5v
40	DG/YL	Fuel Pump Monitor	On: 12v, Off: 0v
41	PK/LB	A/C Pressure Cutout Switch	AC On: 12v, Off: 0v
42	RD/WT	ECT Sensor Signal	0.5-0.6v
43	DB/LG	Low Fuel Indicator	Digital Signals
44	GY/RD	Starter Relay Control	Relay Off: 12v, On: 9v
45	LG/RD	Generator Common	Digital Signals
46	---	Not Used	---
47	BR/PK	EGR VR Solenoid	0%, 55 mph: 45%
48	LG/WT	Tachometer Output	DC pulse signals
49	LB/BK	Digital TR2 Sensor	In 'P': 0v, 55 mph: 11v
50	WT/BK	Digital TR4 Sensor	In 'P': 0v, 55 mph: 11v
51	BK/WT	Power Ground	<0.1v
52	WT/PK	COP 3 Driver Control	5° dwell
53	DG/VT	COP 4 Driver Control	5° dwell
54	VT/YL	TCC Solenoid Control	0%, 55 mph: 95%
55	YL/BK	Keep Alive Power	12-14v
56	LG/BK	EVAP Purge Solenoid Control	0-10 Hz (0-100%)
57	YL/RD	Knock Sensor Signal	0v

2003-04 Marauder 4.6L V8 MPI VIN V 104 Pin Connector

PCM Pin #	Wire Color	Circuit Description (104 Pin)	Value at Hot Idle
58-59	---	Not Used	---
60	GY/BK	HO2S-11 (B1 S1) Signal	0.1-1.1v
61	VT/LG	HO2S-22 (B2 S2) Signal	0.1-1.1v
62	RD/PK	FTP Sensor Signal	2.6v (0" H2O - cap off)
63	OR/LG	Injection Pressure Sensor	2.8v (39 psi)
64	LB/YL	Digital TR3 Sensor	In 'P': 0v, in O/D: 1.7v
65	BR/LG	DPFE Sensor Signal	0.95-1.05v
66	LG/RD	Engine Coolant Temperature Sensor	0.7v (194ºF)
67	VT/WT	EVAP Canister Vent Solenoid	0-10 Hz (0-100%)
68	GY/BK	Vehicle Speed Sensor (+) Signal	0 Hz, 55 mph: 125 Hz
69	PK/YL	A/C WOT Relay Control	On: 1v, Off: 12v
70	YL	Generator Battery Indicator Control	Lamp Off: 12v, On: 1v
71	RD	Vehicle Power (Start-Run)	12-14v
72	TN/RD	Injector 7 Control	3.4-3.7 ms
73	TN/BK	Injector 5 Control	3.4-3.7 ms
74	BR/YL	Injector 3 Control	3.4-3.7 ms
75	TN	Injector 1 Control	3.4-3.7 ms
76	BK/WT	Power Ground	<0.1v
77	---	Not Used	---
78	PK/LB	COP 7 Driver Control	5º dwell
79	WT/RD	COP 8 Driver Control	5º dwell
80	LB/OR	Fuel Pump Control	On: 1v, Off: 12v
81	WT/YL	EPC Solenoid Control	9.5v (20 psi)
82	---	Not Used	---
83	WT/LB	IAC Motor Control	34% duty cycle
84	DB/YL	OSS Sensor Signal	0 Hz
85	DB/OR	CMP Sensor Signal	6-7 Hz
86	WT/BK	A/C High Pressure Switch	Open: 12v, Closed: 0v
87	RD/BK	HO2S-21 (B2 S1) Signal	0.1-1.1v
88	LB/RD	MAF Sensor Signal	0.6v
89	GY/WT	TP Sensor Signal	0.53-1.27v
90	BR/WT	Reference Voltage	4.9-5.1v
91	GY/RD	Analog Signal Return	<0.050v
92	DG	Brake Pedal Position Switch	Brake Off: 0v, On: 12v
93	RD/WT	HO2S-11 (B1 S1) Heater	On: 1v, Off: 12v
94	YL/LB	HO2S-21 (B2 S1) Heater	On: 1v, Off: 12v
95	WT/BK	HO2S-12 (B1 S2) Heater	On: 1v, Off: 12v
96	TN/YL	HO2S-22 (B2 S2) Heater	On: 1v, Off: 12v
97	RD	Vehicle Power (Start-Run)	12-14v
98	LB	Injector 8 Control	3.4-3.7 ms
99	LG/OR	Injector 6 Control	3.4-3.7 ms
100	BR/LB	Injector 4 Control	3.4-3.7 ms
101	WT	Injector 2 Control	3.4-3.7 ms
102	---	Not Used	---
103	BK/WT	Power Ground	<0.1v
104	PK/WT	COP 2 Driver Control	5º dwell

Pin Connector Graphic

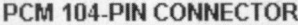

PCM 104-PIN CONNECTOR

Terminal View of 104-Pin PCM Wiring Harness Connector

MONTEGO PIN TABLES

2005 Montego 3.0L V6 MFI VIN 1 (A/T) C175A 58-Pin Connector

PCM Pin #	Wire Color	Circuit Description (58-Pin)	Value at Hot Idle
1	WT/RD	Accelerator Pedal Position Sensor	0.5-4.9v
2	---	Not Used	---
3	GY	SCP Data Bus (+)	Digital Signals
4	BL	SCP Data Bus (-)	<0.050v
5	BR	Signal Return	<0.050v
6-8	---	Not Used	---
9	BR/BL	A/C Clutch Relay Control	Relay Off: 12v, On: 1v
10-11	---	Not Used	---
12	BR/RD	EVAP Purge Solenoid Control	0-10 Hz (0-100%)
13	YL/GN	EEPROM Power	0.1v
14	WT/VT	Manual Transmission Switch (+) Signal	Switch Open: 12v, Closed: 1v
15	WT	Electronic Throttle Control Module	Digital Signals
16	WT/BL	Electronic Throttle Control Module	Digital Signals
17	BR	Electronic Throttle Control Module	Digital Signals
18-19	---	Not Used	---
20	YL	ETC Reference Voltage	4.9-5.1v
21	---	Not Used	---
22	WT/BK	Manual Transmission Switch (-) Signal	<0.050v
23	YL/RD	ETC Reference Voltage	4.9-5.1v
24	BK/RD	Power Ground	<0.1v
25	BK/RD	Power Ground	<0.1v
26	BK/RD	Power Ground	<0.1v
27	BK/RD	Power Ground	<0.1v
28	GN/OR	Brake Pressure Switch	Brake Off: 0v, On: 12v
29	---	Not Used	---
30	BR/BL	EVAP Vent Solenoid Control	0-10 Hz (0-100%)
31	WT/BL	MAF Sensor Signal	0.6v, 55 mph: 1.7v
32	GN/YL	Vehicle Power (Start-Run)	12-14v
33	GN/OR	Vehicle Power (Start-Run)	12-14v
34-37	---	Not Used	---
38	BR/BL	MAF Sensor Signal Return	<0.050v
39	---	Not Used	---
40	OG	Keep Alive Power	12-14v
41	WT/GN	Traction Control Disable Switch	Brake Off: 0v, On: 12v
42	WT/VT	A/C Pressure Switch Signal	0.9v (50 psi)
43	BR/YL	Power Ground	<0.1v
44	OG/YL	Keep Alive Power (CJB fuse)	12-14v
45-46	---	Not Used	---
47	WT/BK	Engine Cooling Fan Motor Control	Off: 12v, On: 1v
48	WT/BL	Restraint Control Module	Digital Signals
49	WT/VT	Rear Electronic Module	Digital Signals
50	---	Not Used	---
51	WT/VT	IAT Sensor Signal	1.5-2.5v
52	WT/VT	FTP Sensor Signal	2.6v (0" H2O - cap off)
53	GN/RD	Controller Area Network (+) Signal	0-7-0v
54	BL/RD	Controller Area Network (-) Signal	0-7-0v
55	YL	Reference Voltage	4.9-5.1v
56	BK/OR	Sensor Return	<0.050v
57	YL/BL	Speed Control Reference Voltage	4.9-5.1v
58	---	Not Used	---

2005 Montego 3.0L V6 MFI VIN 1 (A/T) C175B 32-Pin Connector

PCM Pin #	Wire Color	Circuit Description (32-Pin)	Value at Hot Idle
1	BR	Shift Solenoid 'A' Control	12v, 55 mph: 1v
2	BR/RD	Shift Solenoid 'B' Control	12v, 55 mph: 1v
3-6	---	Not Used	---
7	BR/GN	Pressure Control Solenoid 'A' Control	8.8v
8	BR/WT	Shift Solenoid 'C' Control	12v, 55 mph: 12v
9	WT	Digital TR Sensor (TR3A)	0v, 55 mph: 1.7v
10	WT/RD	Digital TR Sensor (TR4)	0v, 55 mph: 11v
11	---	Not Used	---
12	BR/WT	Pressure Control 'C' Solenoid Control	Off: 12v, On: 1v
13	BR/YL	Pressure Control 'B' Solenoid Control	Off: 12v, On: 1v
14	BR/RD	Signal Return	<0.050v
15	BR/BL	HO2S-12 (B1 S2) Heater Control	Off: 12v, On: 1v
16	BK/GN	HO2S-22 (B2 S2) Heater Control	Off: 12v, On: 1v
17	BR/YL	Shift Solenoid 'D' Control	12v, 55 mph: 1v
18	WT/BL	Digital TR Sensor (TR2)	0v, 55 mph: 11v
19	---	Not Used	---
20	BR/WT	Torque Converter Clutch Control	12v, 55 mph: 0.1v
21	WT/BK	Intermediate Speed Shaft Signal	0 Hz, 55 mph: 1386 Hz
22	WT/GN	Digital TR Sensor (TR1)	0v, 55 mph: 11v
23	WT/RD	Transmission Fluid Temperature Sensor	2.10-2.40v
24-25	---	Not Used	---
26	WT/RD	Output Shaft Speed Sensor Signal	0 Hz, 975 Hz
27	WT/VT	Turbine Shaft Speed Sensor Signal	340 Hz, 1025 Hz
28	WT/BL	HO2S-12 (B1 S2) Signal	0.1-1.1v
29	WT/GN	HO2S-22 (B2 S2) Signal	0.1-1.1v
30-32	---	Not Used	---

Wire Harness Connectors Graphic

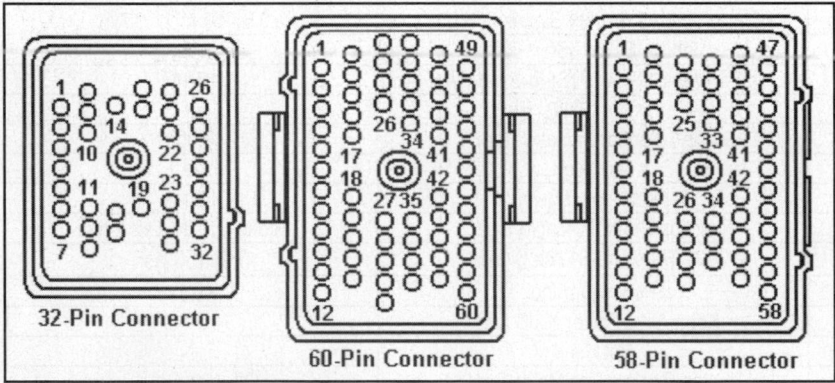

32-Pin Connector

60-Pin Connector

58-Pin Connector

Standard Colors and Abbreviations

Abbreviation	Color	Abbreviation	Color	Abbreviation	Color
BK	Black	GY	Gray	PK	Purple
BL	Blue	GN	Green	RD	Red
BR	Brown	LG	LT Green	TN	Tan
DB	Dark Blue	OR	Orange	WT	White
DG	DK Green	PK	Pink	YL	Yellow

2005 Montego 3.0L V6 MFI VIN 1 (A/T) C175C 60 Pin Connector

PCM Pin #	Wire Color	Circuit Description (60 Pin)	Value at Hot Idle
1	WT/BK	Variable Valve Control Solenoid 1	DC pulse signals
2-3, 6	---	Not Used	---
4	WT/BL	Fuel Rail Temperature Sensor	1.7-3.5v (50-120ºF)
5	WT	Power Steering Pressure Switch	Straight: 0v, Turned: 5v
7	BR	HO2S-11 (B1 S1) Heater	On: 1v, Off: 12v
8	BR/RD	HO2S-21 (B2 S1) Signal	0.1-1.1v
9-10	---	Not Used	---
11	BR/BL	Fuel Injector 4 Control	2.9-3.6 ms
12	BR/GN	COP 4 Driver Control	5º, 55 mph: 8º
13	WT/VT	Variable Valve Timing Solenoid 2	DC pulse signals
14	YL/BK	Reference Voltage	4.9-5.1v
15	BR	Throttle Position Sensor Return	0.050v
16	BR/GN	EGR Pressure Transducer Return	<0.050v
17	BR/GN	Sensor Signal Return	<0.050v
18	---	Not Used	---
19	GY/RD	Generator Common Signal	0-130 Hz (varies)
20	BR/GN	Fuel Injector 5 Control	2.9-3.6 ms
21	BR/RD	Fuel Injector 3 Control	2.9-3.6 ms
22	BR/YL	COP 5 Driver Control	5º, 55 mph: 8º
23	WT/GN	EGR Pressure Transducer Signal	0-3.1v
24	YL	Battery Power (at all times)	12-14v
25-26	---	Not Used	---
27	YL/BL	Electronic Throttle Control Motor (-)	DC pulse signals
28	BR/YL	Fuel Injector 6 Control	2.9-3.6 ms
29	BR/WT	Intake Manifold Runner 1 Control	Solenoid Off: 12v, On: 1v
30	BR/WT	COP 6 Driver Control	5º, 55 mph: 8º
31, 33-34	---	Not Used	---
32	WT	Throttle Position Sensor	0.53-1.27v
35	WT/BL	Electronic Throttle Control Motor (+)	DC pulse signals
36	---	Not Used	---
37	BR/GN	Intake Manifold Runner 2 Control	Solenoid Off: 12v, On: 1v
38	---	Not Used	---
39	WT/GN	Engine Oil Temperature Sensor Signal	0.5-0.6v
40	WT/VT	Cylinder Head Temperature Sensor	0.5-0.6v
41	WT/BL	EGR Pressure Transducer Signal	0.95-1.05v
42	GY/BK	Knock Sensor 1 Signal Return	<0.050v
43	---	Not Used	---
44	WT/RD	HO2S-21 (B2 S1) Signal	0.1-1.1v
45	WT	HO2S-11 (B1 S1) Signal	0.1-1.1v
46	BR	Fuel Injector 2 Control	2.9-3.6 ms
47	BR/WT	Fuel Injector 1 Control	2.9-3.6 ms
48	BR/RD	COP 2 Driver Control	5º, 55 mph: 8º
49	WT/GN	Fuel Rail Pressure Sensor Signal	2.8v (39 psi)
50	WT/RD	Generator Monitor Signal	130 Hz (45%)
51	WT/BK	Knock Sensor 1 (+) Signal	0v
52	---	Not Used	---
53	WT/VT	CMP Sensor 1 Signal	6 Hz
54	WT/BK	CMP Sensor 2 Signal	6 Hz
55	GY/RD	CKP Sensor (-) Signal	390 Hz
56	WT/RD	CKP Sensor (+) Signal	390 Hz
57	WT/RD	Throttle Position Sensor	0.53-1.27v
58	BR	COP 1 Driver Control	5º, 55 mph: 8º
59	---	Not Used	---
60	BR/BL	COP 3 Driver Control	5º, 55 mph: 8º

MYSTIQUE PIN TABLES

1995 Mystique 2.0L I4 MFI VIN 3 (All) 60 Pin Connector

PCM Pin #	Wire Color	Circuit Description (60 Pin)	Value at Hot Idle
1	OR/YL	Keep Alive Power	12- 14v
2	OR	Brake On/Off Signal	Brake Off: 0v, On: 12v
3	WT/PK	VSS (+) Signal	0 Hz, 55 mph: 125 Hz
4	WT/GN	Ignition Diagnostic Monitor	20-31 Hz
5	WT/PK	A/T: TSS Sensor (+) Signal	42-50 Hz (680-720 rpm)
6	---	Not Used	---
7	WT/GN	ECT Sensor Signal	0.5-0.6v
8	PK/BK	Fuel Pump Monitor	Pump On: 12v, Off; 0v
9	BR/BL	MAF Sensor Return	<0.050v
10	PK/BL	A/C Cycling Clutch Switch	A/C Switch On: 12v
11	BK/OR	EVAP Canister Purge Control	12v, 55 mph: 1v
12	---	Not Used	---
13	BK/BL	Low Speed Fan Control	Fan On: 1v, Off: 12v
14-15	---	Not Used	---
16	BK/BL	Ignition Ground	<0.050v
17	BK/OR	STI Output, MIL Control	MIL On: 1v, Off: 12v
18	WT/BK	Data Bus (+) Signal	Digital Signals
19	BN/BK	Data Bus (-) Signal	Digital Signals
20	BK/RD	Case Ground	<0.050v
21	BK/OR	IAC Solenoid Control	Varies: 8-11v
22	BK/BL	Fuel Pump Control	Relay Off: 12v, On: 1v
23	WT/BK	Knock Sensor Signal	No Knock: 2.5v
24	WT/PK	Camshaft Position Sensor	5-7 Hz
25	WT/PK	IAT Sensor Signal	1.5-2.5v
26	YL	Reference Voltage	4.9-5.1v
27	WT/BL	DPFE Sensor Signal	0.20-1.30v
28	WT/PK	PSP Switch Signal	Straight: 12v, Turned: 0v
29	WT/BK	Octane Adjust Switch Signal	9.3v (Closed: 0v)
30	WT	A/T: PNP Sensor Signal	In P/N: 0v, 55 mph: 1.7v
30	WT	M/T: Clutch Pedal Position	Clutch In: 0v, Out: 5v
31	BK/WT	High Speed Fan Control	Fan On: 1v, Off: 12v
32	---	Not Used	---
33	GN/WT	VR Solenoid Control	12v, 30 mph: 10-12v
34	---	Not Used	---
35	BK/OR	Injector 4 Control	2.3-2.9 ms
36	WT/PK	Spark Output Signal	50% duty cycle
37	GN/YL	Vehicle Power	12-14v
38	BK/RD	A/T: EPC Solenoid Control	7-8v (8-9 psi)
39	BK/BL	Injector 3 Control	2.3-2.9 ms
40	BK/YL	Power Ground	<0.1v

1995 Mystique 2.0L I4 MFI VIN 3 (All) 60 Pin Connector

PCM Pin #	Wire Color	Circuit Description (60 Pin)	Value at Hot Idle
41	WT/BK	A/T: TCS (switch) Signal	TCS & O/D On: 12v
42-43	---	Not Used	---
44	WT	HO2S-11 (B1 S1) Signal	0.1-1.1v
45	---	Not Used	---
46	BN	Analog Signal Return	<0.050v
47	WT	TP Sensor Signal	0.5-1.1v
48	WT/BK	Self-Test Input	STI On: 0.1v, Off: 5v
49	WT/RD	A/T: TFT Sensor Signal	2.40-2.10v
50	WT/BL	MAF Sensor Signal	1.9v
51	BK/YL	A/T: Shift Solenoid 1 Control	12v, 55 mph: 1v
52	BK/BL	A/T: Shift Solenoid 2 Control	1v, 55 mph: 12v
53	BK/WT	A/T: TCC Solenoid Control	TCC On: 1v, Off: 12v
54	BK/YL	A/C WOT Relay Control	Relay Off: 12v, On: 1v
55	BK/OR	A/T: Shift Solenoid 3 Control	12v, 55 mph: 1v
56	WT/BK	PIP Sensor Signal	50% duty cycle
57	GN/YL	Vehicle Power	12-14v
58	BK/WT	Injector 1 Control	2.3-2.9 ms
59	BK/YL	Injector 2 Control	2.3-2.9 ms
60	BK/YL	Power Ground	<0.1v

Pin Connector Graphic

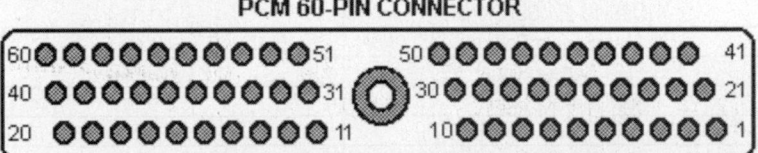

Terminal View of 60-Pin PCM Harness Connector

Standard Colors and Abbreviations

Abbreviation	Color	Abbreviation	Color	Abbreviation	Color
BK	Black	GY	Gray	PK	Purple
BL	Blue	GN	Green	RD	Red
BR	Brown	LG	LT Green	TN	Tan
DB	Dark Blue	OR	Orange	WT	White
DG	DK Green	PK	Pink	YL	Yellow

1996-97 Mystique 2.0L I4 MFI VIN 3 (All) 104 Pin Connector

PCM Pin #	Wire Color	Circuit Description (104 Pin)	Value at Hot Idle
1	BK/BL	Shift Solenoid 2 Control	12v, 55 mph: 1v
2	BK/OR	MIL (lamp) Control	MIL On: 1v, Off: 12v
3-12	---	Not Used	---
13	WT/BL	Flash EEPROM Power	0.5v
14	---	Not Used	---
15	BR/BL	Data Bus (-) Signal	Digital Signals
16	WT/BL	Data Bus (+) Signal	Digital Signals
17-20	---	Not Used	---
21	WT/RD	CKP Sensor (+) Signal	450-480 Hz
22	BR/WT	CKP Sensor (-) Signal	450-480 Hz
23	---	Not Used	---
24	BK/YL	Power Ground	<0.1v
25	BK/RD	Chassis Ground	<0.050v
26	BR/BL	Coil Driver 1 Control	5° dwell
27	BK/YL	Shift Solenoid 1 Control	1v, 55 mph: 12v
28	---	Not Used	---
29	PK/BK	TCS (switch) Signal	TCS & O/D On: 12v
30	WT/BK	Octane Adjust Switch Signal	9.3v (Closed: 0v)
31	WT/PK	PSP Sensor Signal	Straight: 1v, Turning: 12v
31	WT	PSP Sensor Signal	Straight: 1v, Turning: 12v
32-34	---	Not Used	---
35	WT/BL	HO2S-12 (B1 S2) Signal	0.1-1.1v
36	BR/BL	MAF Sensor Return	<0.050v
37	WT/RD	TFT Sensor Signal	2.10 to 0.40v
38	WT/GN	ECT Sensor Signal	0.5-0.6v
39	WT/PK	IAT Sensor Signal	1.5-2.5v
40	PK/BK	Fuel Pump Monitor	12v, Off: 0v
41	PK/BL	A/C Cycling Clutch Switch	Closed: 12v, Open: 0v
42-44	---	Not Used	---
45	BK/BL	Low Speed Fan Control	Fan On: 1v, Off: 12v
46	BK/WT	High Fan Control	Fan On: 1v, Off: 12v
47	BK/GN	VR Solenoid Control	0 Hz, 30 mph: 40 Hz
48	WT/BK	Clean Tachometer Output	20-31 Hz
49-50	---	Not Used	---
51	BK/YL	Power Ground	<0.1v
52	BR/GN	Coil Driver 2 Control	5° dwell
53	---	Not Used	---
54	BK/WT	TCC Solenoid Control	0%, 30 mph: 80%
55	OR/YL	Keep Alive Power	12-14v
56	BK/OR	EVAP VMV Solenoid	0-10 Hz (0-100%)
57	WT/BK	Knock Sensor 1 Signal	0v
58	WT/PK	VSS (+) Signal	0 Hz, 55 mph: 125 Hz
59	---	Not Used	---
60	WT	HO2S-11 (B1 S1) Signal	0.1-1.1v
61-63	---	Not Used	---
64	WT/GN	A/T: TR Sensor Signal	In 'P': 0v, O/D: 1.7v
64	WT	M/T: Clutch Pedal Position	Clutch In: 0v, Out: 5v

1996-97 Mystique 2.0L I4 MFI VIN 3 (All) 104 Pin Connector

PCM Pin #	Wire Color	Circuit Description (104 Pin)	Value at Hot Idle
65	WT/BL	DPFE Sensor Signal	0.20-1.30v
66-68	---	Not Used	---
69	BK/YL	A/C WOT Relay Control	Relay Off: 12v, On: 1v
70	BK/YL	ZETECT VCT Control	VCT Off: 12v, On: 1v
71	GN/YL	Vehicle Power	12-14v
72-73	---	Not Used	---
74	BK/BL	Injector 3 Control	2.3-2.9 ms
75	BK/WT	Injector 1 Control	2.3-2.9 ms
76	BK/YL	CMP and TSS Sensor Ground	<0.050v
77	BK/YL	Power Ground	<0.1v
78, 82	---	Not Used	---
79	WT/BL	TCIL (lamp) Control	Lamp On: 1v, Off: 12v
80	BK/BL	Fuel Pump Control	Pump On: 1v, Off: 12v
81	BK/RD	EPC Solenoid Control	8.9v
83	BK/YL	IAC Motor Control	9v (33% d/cycle)
84	WT/PK	TSS Sensor Signal	40-56 Hz
85	WT/PK	CID Sensor Signal	5-7 Hz
86	BK/WT	A/C Pressure Switch	A/C On: 12v (Open)
87	---	Not Used	---
88	WT/BL	MAF Sensor Signal	0.8v
89	WT	TP Sensor Signal	0.53-1.27v
90	YL	Reference Voltage	4.9-5.1v
91	BR	Analog Signal Return	<0.050v
92	O	Brake Pedal Position Switch	Brake Off: 0v, On: 12v
93	BK/YL	HO2S-11 (B1 S1) Heater	1v, Off: 12v
94	---	Not Used	---
95	BK/OR	HO2S-12 (B1 S2) Heater	1v, Off: 12v
96	---	Not Used	---
97	GN/YL	Vehicle Power	12-14v
98-99	---	Not Used	---
100	BK/OR	Injector 4 Control	2.3-2.9 ms
101	BK/YL	Injector 2 Control	2.3-2.9 ms
102	BK/OR	Shift Solenoid 3 Control	8v, 55 mph: 8-9v
103	BK/YL	Power Ground	<0.1v
104	---	Not Used	---

Pin Connector Graphic

```
        PCM 104-PIN CONNECTOR
   1 0000000000000   00000000000000 26
  27 0000000000000   00000000000000 52
  53 0000000000000 ⬤ 00000000000000 78
  79 0000000000000   00000000000000 104
```

Terminal View of 104-Pin PCM Wiring Harness Connector

Standard Colors and Abbreviations

Abbreviation	Color	Abbreviation	Color	Abbreviation	Color
BK	Black	GY	Gray	PK	Purple
BL	Blue	GN	Green	RD	Red
BR	Brown	LG	LT Green	TN	Tan
DB	Dark Blue	OR	Orange	WT	White
DG	DK Green	PK	Pink	YL	Yellow

1998-2000 Mystique 2.0L I4 MFI VIN 3 (All) 104 Pin Connector

PCM Pin #	Wire Color	Circuit Description (104 Pin)	Value at Hot Idle
1	BK/BL	Shift Solenoid 2	12v, 55 mph: 1v
2	BK/OR	MIL (lamp) Control	MIL On: 1v, Off: 12v
3-11, 14	---	Not Used	---
12	WT	Fuel Level Indicator Signal	1.7v (1/2 full)
13	WT/BL	Flash EEPROM Power	0.5v
15	BL	Data Bus (-) Signal	Digital Signals
16	GY	Data Bus (+) Signal	Digital Signals
17-19, 23	---	Not Used	---
20	WT/BK	Transmission Indicator Signal	In 'P': 0v, O/D: 1.7v
21	WT/RD	CKP Sensor (+) Signal	450-480 Hz
22	BR/WT	CKP Sensor (-) Signal	450-480 Hz
22	BR/RD	CKP Sensor (-) Signal	450-480 Hz
24	BK/YL	Power Ground	<0.1v
25	BK/RD	Chassis Ground	<0.050v
26	BR/BL	Coil Driver 1 Control	5° dwell
27	BK/YL	Shift Solenoid 1 Control	1v, 55 mph: 12v
28	WT/BL	Speedometer Indicator Signal	Digital Signals
29	PK/BK	TCS (switch) Signal	TCS & O/D On: 12v
29	GN/BK	TCS (switch) Signal	TCS & O/D On: 12v
30	WT/BK	Octane Adjust Switch	9.3v (Closed: 0v)
31	WT/PK	PSP Sensor Signal	Straight: 1v, Turning: 12v
31	WT	PSP Sensor Signal	Straight: 1v, Turning: 12v
33	WT/PK	VSS (-) Signal	0 Hz, 55 mph: 125 Hz
34	WT/PK	TSS Sensor Signal (2000)	40-56 Hz
35	WT/BL	HO2S-12 (B1 S2) Signal	0.1-1.1v
36	BR/BL	MAF Sensor Return	<0.050v
37	WT/RD	TFT Sensor Signal	2.10 to 0.40v
38	WT/GN	ECT Sensor Signal	0.5-0.6v
39	WT/PK	IAT Sensor Signal	1.5-2.5v
40	PK/BK	Fuel Pump Monitor	12v, Off: 0v
40	GN/BK	Fuel Pump Monitor	12v, Off: 0v
41	PK/BL	A/C Cycling Clutch Switch	Closed: 12v, Open: 0v
41	GN/BL	A/C Cycling Clutch Switch	Closed: 12v, Open: 0v
44	BK/RD	VCT Actuator	VCT Off: 12v, On: 1v
45	BK/BL	Low Speed Fan Control	Fan On: 1v, Off: 12v
46	BK/WT	High Fan Control	Fan On: 1v, Off: 12v
47	BK/GN	VR Solenoid Control	12v, 30 mph: 10-12v
48	WT/BK	Clean Tachometer Output	20-31 Hz
50	---	Not Used	---
51	BK/YL	Power Ground	<0.1v
52	BR/GN	Coil Driver 2 Control	5° dwell
54	BK/WT	TCC Solenoid Control	0%, 30 mph: 80%
55	OR/YL	Keep Alive Power	12-14v
55	RD	Keep Alive Power	12-14v
56	BK/OR	EVAP Purge Valve	0-10 Hz (0-100%)
57	WT/BK	Knock Sensor (+) Signal	0v
58	WT/BK	VSS (+) Signal	0 Hz, 55 mph: 125 Hz
58	WT/PK	VSS (+) Signal	0 Hz, 55 mph: 125 Hz
59 ('20)	WT/GN	Generator Load Indicator	1.5-10.5v (40-250 Hz)
60	WT	HO2S-11 (B1 S1) Signal	0.1-1.1v
62	WT/PK	FTP Sensor Signal	2.6v (0" H2O - cap off)
63	WT/PK	FRP Sensor Signal	2.8v (39 psi)
63	WT/GN	FRP Sensor Signal	2.8v (39 psi)

1998-2000 Mystique 2.0L I4 MFI VIN 3 (All) 104 Pin Connector

PCM Pin #	Wire Color	Circuit Description (104 Pin)	Value at Hot Idle
64	WT/GN	A/T: TR Sensor Signal	In 'P': 0v, O/D: 1.7v
64	WT	M/T: Clutch or Neutral Switch	Clutch In: 0v, Out: 5v
65	WT/BL	DPFE Sensor Signal	0.20-1.30v
66, 68	---	Not Used	---
67	BK/OR	EVAP CV Solenoid	0-10 Hz (0-100%)
69	BK/YL	A/C WOT Relay Control	Relay Off: 12v, On: 1v
70	BK/YL	VCT Control (ZETECT)	VCT Off: 12v, On: 1v
71, 97	GN/YL	Vehicle Power	12-14v
73 ('20)	BK/YL	Shift Solenoid 1 Control	1v, 55 mph: 12v
74	BK/BL	Injector 3 Control	2.3-2.9 ms
75	BK/WT	Injector 1 Control	2.3-2.9 ms
76	BR	CMP and TSS Sensor Ground	<0.050v
76	BR/WT	CMP and TSS Sensor Ground	<0.050v
77	BK/YL	Power Ground	<0.1v
79	WT/BL	TCIL (lamp) Control	Lamp On: 1v, Off: 12v
80	BK/BL	Fuel Pump Control	Pump On: 1v, Off: 12v
80	BK/RD	Fuel Pump Control	Pump On: 1v, Off: 12v
81	BK/RD	EPC Solenoid Control	8.9v
83	BK/YL	IAC Motor Control	9v (33% d/cycle)
84	WT/PK	TSS Sensor Signal	40-56 Hz
85	WT/PK	CMP Sensor Signal	5-7 Hz
86	BK/WT	A/C Pressure Switch	A/C On: 12v (Open)
87	BR/YL	Knock Sensor (-) Signal	0v
88	WT/BL	MAF Sensor Signal	0.8v
89	WT	TP Sensor Signal	0.53-1.27v
90	YL	Reference Voltage	4.9-5.1v
91	BR	Analog Signal Return	<0.050v
92	OR	Brake Pedal Position Switch	Brake Off: 0v, On: 12v
92	GN/RD	Brake Pedal Position Switch	Brake Off: 0v, On: 12v
93	BK/YL	HO2S-11 (B1 S1) Heater	1v, Off: 12v
95	BK/OR	HO2S-12 (B1 S2) Heater	1v, Off: 12v
99	BK/WT	Modulated Lockup Solenoid	On: 1v, Off: 12v
100	BK/OR	Injector 4 Control	2.3-2.9 ms
101	BK/YL	Injector 2 Control	2.3-2.9 ms
102	BK/OR	Shift Solenoid 3 Control	8v, 55 mph: 8-9v
103	BK/YL	Power Ground	<0.1v
104	---	Not Used	---

Pin Connector Graphic

PCM 104-PIN CONNECTOR

Terminal View of 104-Pin PCM Wiring Harness Connector

Standard Colors and Abbreviations

Abbreviation	Color	Abbreviation	Color	Abbreviation	Color
BK	Black	GY	Gray	PK	Purple
BL	Blue	GN	Green	RD	Red
BR	Brown	LG	LT Green	TN	Tan
DB	Dark Blue	OR	Orange	WT	White
DG	DK Green	PK	Pink	YL	Yellow

1995 Mystique 2.5L V6 MFI VIN L (All) 22 Pin Connector

PCM Pin #	Wire Color	Circuit Description (22 Pin)	Value at Hot Idle
22-A	BL/RD	Keep Alive Power	12- 14v
22-B	RD/BK	Vehicle Power	12-14v
22-C	BK/RD	Vehicle Start Signal	KOEC: 9-11v
22-D	WT/RD	Switch Monitor Lamp Control	Lamp Off: 12v, On: 1v
22-E	BL	STO & MIL (lamp) Control	STO On: 5v, Off: 12v
22-F	---	Not Used	---
22-G	BL/OR	Ignition Control Module	Varies
22-H	WT	Headlamp Switch Signal	Switch On: 12v, Off: 1v
22-I	RD/WT	Self-Test Input	STI On: 1v, STI Off: 12v
22-J	PK	Rear Window Defroster	Switch On: 1v, Off: 12v
22-K	WT/BK	Torque Reduce & ECT Signal	Digital Signal
22-L	GN/BK	A/C Relay Control	Relay Off: 12v, On: 1v
22-M	GN/RD	Vehicle Speed Sensor	0 Hz, 55 mph: 125 Hz
22-N	BL/YL	PSP Switch Signal	Straight: 12v, Turned: 0v
22-O	PK/BK	A/C Cycling Pressure Switch	Switch On: 12v, Off: 1v
22-P	OR/BK	Blower Motor Control Switch	Off or 1st: 12v, 2nd on: 1v
22-Q	WT/GN	Brake On/ Off Switch Signal	Brake Off: 0v, On: 12v
22-R	LG/BK	A/T: PNP Switch Signal	In 'P', 0v, Others: 12v
22-R	LG/BK	M/T: CPP Switch Signal	Clutch In: 0v, Out: 12v
22-S	GN	Torque Reduce Signal #1	Digital Signal
22-T	BR	Idle Switch Signal	Closed: 0v, Open: 12v
22-U	---	Not Used	---
22-V	LG/WT	Torque Reduce Signal #2	Digital Signal

1995 Mystique 2.5L V6 MFI VIN L (All) 16 Pin Connector

PCM Pin #	Wire Color	Circuit Description (16 Pin)	Value at Hot Idle
16-A	GN/OR	Barometric Pressure Sensor	3.9v
16-B	RD	Measuring Core VAF Sensor	Idle: 3.06v
16-C	BK/YL	HO2S-11 (B1 S1) Signal	0.1-1.1v
16-D	BL/WT	HO2S-12 (B1 S2) Signal	0.1-1.1v
16-E	R/GN	ECT Sensor Signal	0.5-0.6v
16-F	YL	Throttle Position Sensor	0.4-1.0v
16-G	-	Not Used	---
16-N	-	Not Used	---
16-H	PK/YL	A/C High Pressure Switch	Closed: 1v, open: 12v
16-I	PK	Reference Voltage	4.9-5.1v
16-J	RD/BK	EGR Valve Position Sensor	Idle: 0.4v
16-K	BK/RD	IAT Sensor Signal	1.5-3.5v
16-L	GN	DRL Signal - Canada	DRL On: 2v, Off: 12v
16-M	WT	Knock Sensor Signal	0v
16-O	BL/BK	EVAP Canister Purge Control	12v, 55 mph: 1v
16-P	BL/GN	High Speed Fan Control	Fan On: 1v, Off: 12v

1995 Mystique 2.5L V6 MFI VIN L (All) 26 Pin Connector

PCM Pin #	Wire Color	Circuit Description (26 Pin)	Value at Hot Idle
26-A	BK	Power Ground	<0.1v
26-B	BK	Power Ground	<0.1v
26-C	BK/RD	Power Ground	<0.1v
26-D	BK/RD	Power Ground	<0.1v
26-E	LG/OR	CKP Sensor 1 Signal	2.5v
26-F	BL	CKP Sensor Ground	0.1v
26-G	BL/PK	CMP Sensor 1 Signal	2.5v
26-H	GN	CKP Sensor 2 Signal	2.5v
26-I	WT/GN	Variable Resonance Induction	VRIS1 On: 1v, Off: 12v
26-J	BL/RD	Variable Resonance Induction	VRIS2 On: 1v, Off: 12v
26-K	---	Not Used	---
26-L	RD/WT	Low Speed Fan Control	Relay Off: 12v, On: 1v
26-M	GN/BK	Fuel Press. Regulator Control	Startup: 3v, others: 12v
26-N	BL/OR	Condenser Fan Control	Fan On: 1v, Off: 12v
26-O	WT/BL	EGR Vent Solenoid	12v, Cruise: 1v
26-P	GN/WT	EGR Control Solenoid	12v, Cruise: 1v
26-Q	LG/BK	Idle Air Control Valve	8-10v
26-R	---	Not Used	---
26-S	---	Not Used	---
26-T	LG	Fuel Pump Control	Relay Off: 12v, On: 1v
26-U	RD/LG	Injector 1 Control	3.6 ms
26-V	BL/WT	Injector 2 Control	3.6 ms
26-W	BR	Injector 3 Control	3.6 ms
26-X	RD/YL	Injector 4 Control	3.6 ms
26-Y	WT	Injector 5 Control	3.6 ms
26-Z	WT/BK	Injector 6 Control	3.6 ms

Pin Connector Graphic

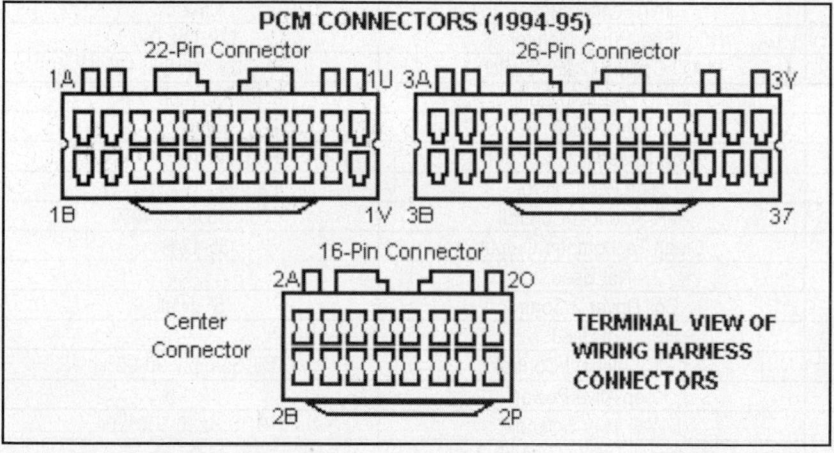

Standard Colors and Abbreviations

Abbreviation	Color	Abbreviation	Color	Abbreviation	Color
BK	Black	GY	Gray	PK	Purple
BL	Blue	GN	Green	RD	Red
BR	Brown	LG	LT Green	TN	Tan
DB	Dark Blue	OR	Orange	WT	White
DG	DK Green	PK	Pink	YL	Yellow

1996-97 Mystique 2.5L V6 MFI VIN L (All) 104 Pin Connector

PCM Pin #	Wire Color	Circuit Description (104 Pin)	Value at Hot Idle
1	BK/BL	Shift Solenoid 2	1v, 55 mph: 12v
2	BK/OR	MIL (lamp) Control	MIL On: 1v, Off: 12v
3-7	---	Not Used	---
8	BK/BL	IMRC Solenoid Control	5v, 55 mph: 5v
9-12, 14	---	Not Used	---
13	WT/BL	Flash EEPROM Power	0.5v
15	BR/BL	Data Bus (-) Signal	Digital Signals
16	WT/BL	Data Bus (+) Signal	Digital Signals
17	GY/OR	Passive Anti-Theft System	Digital Signal
18-20	---	Not Used	---
21	WT/RD	CKP Sensor (+) Signal	410-440 Hz
22	BR/RD	CKP Sensor (-) Signal	410-440 Hz
23	---	Not Used	---
24, 51	BK/YL	Power Ground	<0.1v
25	BK/RD	Chassis Ground	<0.050v
26	BR/BL	Coil Driver 1 Control	5° dwell
27	BK/YL	CD4E A/T: Shift Solenoid 1	12v, 55 mph: 1v
28	---	Not Used	---
29	PK/BK	CD4E A/T: TCS (switch)	TCS & O/D On: 12v
30	WT/BK	Octane Adjust Switch	9.3v (Closed: 0v)
31	WT	PSP Sensor Signal	Straight: 1v, Turning: 12v
32-34	---	Not Used	---
34	WT/GN	Passive Anti-Theft System	Digital Signal
35	WT/BL	HO2S-12 (B1 S2) Signal	0.1-1.1v
36	BR/BL	MAF Sensor Return	<0.050v
37	WT/RD	TFT Sensor Signal	2.10 to 0.40v
38	WT/GN	ECT Sensor Signal	0.5-0.6v
39	WT/PK	IAT Sensor Signal	1.5-2.5v
40	PK/BK	Fuel Pump Monitor	12v, Off: 0v
41	PK/BL	A/C Cycling Clutch Switch	A/C Switch On: 12v
42	BK/RD	IMRC Solenoid Control	12v, 55 mph: 12v
43-44	---	Not Used	---
45	BK/BL	Low Speed Fan Control	Fan On: 1v, Off: 12v
46	BK/WT	High Fan Control	Fan On: 1v, Off: 12v
47	BK/GN	VR Solenoid Control	0%, 55 mph: 40%
48	WT/BK	Clean Tachometer Output	35-42 Hz
49-50	---	Not Used	---
52	BR/GN	Coil Driver 2 Control	5° dwell
53	---	Not Used	---
54	BK/WT	TCC Solenoid Control	0%, 55 mph: 90-95%
55	OR/YL	Keep Alive Power	12-14v
56	BK/OR	EVAP VMV Solenoid	0-10 Hz (0-100%)
57	WT/BK	Knock Sensor Signal	0v
58	WT/PK	VSS (+) Signal	0 Hz, 55 mph: 125 Hz
59	---	Not Used	---
60	WT	HO2S-11 (B1 S1) Signal	0.1-1.1v
61	WT/GN	HO2S-22 (B2 S2) Signal	0.1-1.1v
62-63	---	Not Used	---
64	WT/GN	CD4E A/T: TR Sensor Signal	In 'P': 0v, O/D: 1.7v
64	WT	M/T: Clutch Position Switch	Clutch In: 0v, Out: 5v

1996-97 Mystique 2.5L V6 MFI VIN L (All) 104 Pin Connector

PCM Pin #	Wire Color	Circuit Description (104 Pin)	Value at Hot Idle
65	WT/BL	DPFE Sensor Signal	0.20-1.30v
66, 68	---	Not Used	---
67	BK/OR	EVAP CV Solenoid	0-10 Hz (0-100%)
69	BK/YL	A/C WOT Relay Control	Relay Off: 12v, On: 1v
70	---	Not Used	---
71	GN/YL	Vehicle Power	12-14v
72, 82	---	Not Used	---
73	BK/GN	Injector 5 Control	2.1-2.5 ms
74	BK/BL	Injector 3 Control	2.1-2.5 ms
75	BK/WT	Injector 1 Control	2.1-2.5 ms
76	BK/YL	CMP & TSS Ground	<0.050v
77	BK/YL	Power Ground	<0.1v
78	BR/YL	Coil Driver 3 Control	5° dwell
79	WT/BL	TCIL (lamp) Control	Lamp On: 1v, Off: 12v
80	BK/BL	Fuel Pump Control	Pump On: 1v, Off: 12v
81	BK/RD	EPC Solenoid Control	8.9v
83	BK/YL	IAC Motor Control	9v (33% d/cycle)
84	WT/PK	TSS Sensor Signal	46-50 Hz
85	WT/PK	CMP Sensor Signal	5-7 Hz
86	BK/WT	A/C Pressure Switch	A/C On: 12v (Open)
87	WT/RD	HO2S-21 (B1 S1) Signal	0.1-1.1v
88	WT/BL	MAF Sensor Signal	0.8v
89	WT	TP Sensor Signal	0.53-1.27v
90	YL	Reference Voltage	4.9-5.1v
91	BR	Analog Signal Return	<0.050v
92	OR	Brake Pedal Position Switch	Brake Off: 0v, On: 12v
93	BK/YL	HO2S-11 (B1 S1) Heater	1v, Off: 12v
94	BK/BL	HO2S-21 (B2 S1) Heater	1v, Off: 12v
95	BK/OR	HO2S-12 (B1 S2) Heater	1v, Off: 12v
96	BK/GN	HO2S-22 (B2 S2) Heater	1v, Off: 12v
97	GN/YL	Vehicle Power	12-14v
98	---	Not Used	---
99	BK/RD	Injector 6 Control	2.1-2.5 ms
100	BK/OR	Injector 4 Control	2.1-2.5 ms
101	BK/YL	Injector 2 Control	2.1-2.5 ms
102	BK/OR	Shift Solenoid 3	7-9v, 30 mph: 8-9v
103	BK/YL	Power Ground	<0.1v
98, 104	---	Not Used	---

Pin Connector Graphic

```
           PCM 104-PIN CONNECTOR
  1 ○○○○○○○○○○○○○    ○○○○○○○○○○○○○ 26
 27 ○○○○○○○○○○○○○         ○○○○○○○○○○○○○ 52
 53 ○○○○○○○○○○○○○    ●    ○○○○○○○○○○○○○ 78
 79 ○○○○○○○○○○○○○    ○○○○○○○○○○○○○ 104
```

Terminal View of 104-Pin PCM Wiring Harness Connector

Standard Colors and Abbreviations

Abbreviation	Color	Abbreviation	Color	Abbreviation	Color
BK	Black	GY	Gray	PK	Purple
BL	Blue	GN	Green	RD	Red
BR	Brown	LG	LT Green	TN	Tan
DB	Dark Blue	OR	Orange	WT	White
DG	DK Green	PK	Pink	YL	Yellow

1998-99 Mystique 2.5L V6 MFI VIN L (All) 104 Pin Connector

PCM Pin #	Wire Color	Circuit Description (104 Pin)	Value at Hot Idle
1	BK/BL	Shift Solenoid 2 Control	1v, 55 mph: 12v
2	BK/OR	MIL (lamp) Control	MIL On: 1v, Off: 12v
3-7	---	Not Used	---
8	BK/BL	IMRC Solenoid Control	5v, 55 mph: 5v
9-11, 14	---	Not Used	---
12	WT	Fuel Level Indicator Signal	1.7v (1/2 full)
12	WT/BK	Fuel Level Indicator Signal	1.7v (1/2 full)
13	WT/BL	Flash EEPROM Power	0.5v
15	BL	Data Bus (+) Signal	Digital Signals
16	GY	Data Bus (-) Signal	Digital Signals
17	GY/OR	Passive Anti-Theft System	Digital Signals
18-19	---	Not Used	---
20	BK/BL	Injector 3 Control	2.1-2.5 ms
21	WT/RD	CKP Sensor (+) Signal	410-440 Hz
22	BR/RD	CKP Sensor (-) Signal	410-440 Hz
23	---	Not Used	---
24	BK/YL	Power Ground	<0.1v
25	BK/RD	Chassis Ground	<0.050v
26	BR/BL	Coil Driver 1 Control	5° dwell
27	BK/YL	Shift Solenoid 1 Control	12v, 55 mph: 1v
28	---	Not Used	---
29	PK/BK	TCS (switch) Signal	TCS & O/D On: 12v
30	WT/BK	Octane Adjust Switch	9.3v (Closed: 0v)
31	WT	PSP Switch Signal	Straight: 1v, Turning: 12v
32-33	---	Not Used	---
34	WT/GN	Passive Anti-Theft System	Digital Signal
35	WT/BL	HO2S-12 (B1 S2) Signal	0.1-1.1v
36	BR/BL	MAF Sensor Return	<0.050v
37	WT/RD	TFT Sensor Signal	2.10 to 0.40v
38	WT/GN	ECT Sensor Signal	0.5-0.6v
39	WT/PK	IAT Sensor Signal	1.5-2.5v
40	PK/BK	Fuel Pump Monitor	12v, Off: 0v
41	PK/BL	A/C Cycling Clutch Switch	A/C Switch On: 12v
42	BK/RD	IMRC Solenoid Signal	12v, 55 mph: 12v
43-44	---	Not Used	---
45	BK/BL	Low Speed Fan Control	Fan On: 1v, Off: 12v
46	BK/WT	High Fan Control	Fan On: 1v, Off: 12v
47	BK/GN	VR Solenoid Control	0%, 55 mph: 40%
48	WT/BK	Clean Tachometer Output	35-42 Hz
49-50	---	Not Used	---
51	BK/YL	Power Ground	<0.1v
52	BR/GN	Coil Driver 2 Control	5° dwell
53, 59	---	Not Used	---
54	BK/BL	Fuel Pump Control	Pump On: 1v, Off: 12v
55	OR/YL	Keep Alive Power	12-14v
56	BK/OR	EVAP Purge Valve	0-10 Hz (0-100%)
57	WT/BK	Knock Sensor 1 Signal	0v
58	WT/PK	VSS (+) Signal	0 Hz, 55 mph: 125 Hz
60	WT	HO2S-11 (B1 S1) Signal	0.1-1.1v
61	WT/GN	HO2S-22 (B2 S2) Signal	0.1-1.1v
62	WT/PK	FTP Sensor Signal	2.6v (cap off)
63 ('99)	WT/PK	FRP Sensor Signal	2.8v (39 psi)
64	WT/GN	A/T: TR Sensor Signal	In 'P': 0v, O/D: 1.7v
64	WT	M/T: Clutch Position Switch	Clutch In: 0v, Out: 5v

1998-99 Mystique 2.5L V6 MFI VIN L (All) 104 Pin Connector

PCM Pin #	Wire Color	Circuit Description (104 Pin)	Value at Hot Idle
65	WT/BL	DPFE Sensor Signal	0.20-1.30v
66, 68, 72	---	Not Used	---
67	BK/OR	EVAP CV Solenoid	0-10 Hz (0-100%)
69	BK/YL	A/C WOT Relay Control	Relay Off: 12v, On: 1v
70	BK/YL	ZETECT VCT Control	VCT Off: 12v, On: 1v
71	GN/YL	Vehicle Power	12-14v
73	BK/YL	HO2S-11 (B1 S1) Heater	1v, Off: 12v
74	BK/BL	Injector 3 Control	2.1-2.5 ms
75	BK/WT	Injector 1 Control	2.1-2.5 ms
76	BK/YL	CMP & TSS Sensor Ground	<0.050v
77	BK/YL	Power Ground	<0.1v
78	BR/YL	Coil Driver 3 Control	5° dwell
79	WT/BK	TCIL (lamp) Control	Lamp On: 1v, Off: 12v
80	BK/WT	TCC Solenoid Control	0%, 55 mph: 90-95%
81	BK/RD	EPC Solenoid Control	8.9v
82	---	Not Used	---
83	BK/YL	IAC Motor Control	9v (33% d/cycle)
84	WT/PK	TSS Sensor Signal	46-50 Hz
85	WT/PK	CMP Sensor Signal	5-7 Hz
86	BK/WT	A/C Pressure Switch	A/C On: 12v (Open)
87	WT/RD	HO2S-21 (B1 S1) Signal	0.1-1.1v
88	WT/BL	MAF Sensor Signal	0.8v
89	WT	TP Sensor Signal	0.53-1.27v
90	YL	Reference Voltage	4.9-5.1v
91	BR	Analog Signal Return	<0.050v
92	OR	Brake Pedal Position Switch	Brake Off: 0v, On: 12v
93	BK/YL	Injector 5 Control	2.1-2.5 ms
94	BK/BL	Injector 6 Control	2.1-2.5 ms
95	BK/OR	Injector 4 Control	2.1-2.5 ms
96	BK/GN	Injector 2 Control	2.1-2.5 ms
97	GN/YL	Vehicle Power	12-14v
98	---	Not Used	---
99	BK/BL	HO2S-21 (B2 S1) Heater	1v, Off: 12v
100	BK/OR	HO2S-12 (B1 S2) Heater	1v, Off: 12v
101	BK/GN	HO2S-22 (B2 S2) Heater	1v, Off: 12v
102	BK/OR	Shift Solenoid 3 Control	7-9v, 30 mph: 8-9v
103	BK/YL	Power Ground	<0.1v
104	---	Not Used	---

Pin Connector Graphic

Terminal View of 104-Pin PCM Wiring Harness Connector

Standard Colors and Abbreviations

Abbreviation	Color	Abbreviation	Color	Abbreviation	Color
BK	Black	GY	Gray	PK	Purple
BL	Blue	GN	Green	RD	Red
BR	Brown	LG	LT Green	TN	Tan
DB	Dark Blue	OR	Orange	WT	White
DG	DK Green	PK	Pink	YL	Yellow

2000 Mystique 2.5L V6 MFI VIN L (All) 104 Pin Connector

PCM Pin #	Wire Color	Circuit Description (104 Pin)	Value at Hot Idle
1	---	Not Used	---
2	BK/OR	MIL (lamp) Control	MIL On: 1v, Off: 12v
3-5, 7	---	Not Used	---
6	BK/YL	Shift Solenoid 1 Control	12v, 55 mph: 1v
8	BK/BL	IMRC Solenoid Control	5v, 55 mph: 5v
9	WT	Fuel Level Indicator Signal	1.7v (1/2 full)
7, 10	---	Not Used	---
11	BK/BL	Shift Solenoid 2 Control	12v, 55 mph: 1v
12	WT/BK	Gear Indicator Signal	Digital Signal
13	WT/BL	Flash EEPROM Power	0.5v
14, 17-19	---	Not Used	---
15	BL	Data Bus (-) Signal	Digital Signals
16	GY	Data Bus (+) Signal	Digital Signals
20	BK/BL	Injector 3 Control	2.1-2.5 ms
21	WT/RD	CKP Sensor (+) Signal	410-440 Hz
22	BR/RD	CKP Sensor (-) Signal	410-440 Hz
23, 30, 34	---	Not Used	---
24	BK/YL	Power Ground	<0.1v
25	BK/RD	Chassis Ground	<0.050v
26	BR/BL	Coil Driver 1 Control	5° dwell
27	BK/YL	Shift Solenoid 1 Control	12v, 55 mph: 1v
28	WT/BL	Speedometer Indicator Signal	Digital Signals
29	GN/BK	TCS (switch) Signal	TCS & O/D On: 12v
31	WT	PSP Switch Signal	Straight: 1v, Turning: 12v
32	BR/YL	Knock Sensor (-) Signal	0v
33	BR/BL	VSS (-) Signal	0 Hz, 55 mph: 125 Hz
35	WT/BL	HO2S-12 (B1 S2) Signal	0.1-1.1v
36	BR/BL	MAF Sensor Return	<0.050v
37	WT/RD	TFT Sensor Signal	2.10 to 0.40v
38	WT/GN	ECT Sensor Signal	0.5-0.6v
39	WT/PK	ACT Sensor Signal	1.5-2.5v
40	GN/BK	Fuel Pump Monitor	12v, Off: 0v
41	GN/BL	A/C Cycling Clutch Switch	A/C Switch On: 12v
42	BK/RD	IMRC Motor Control Signal	12v, 55 mph: 12v
43-44, 49-50	---	Not Used	---
45	BK/BL	Low Speed Fan Control	Fan On: 1v, Off: 12v
46	BK/WT	High Fan Control	Fan On: 1v, Off: 12v
47	BK/GN	VR Solenoid Control	0%, 55 mph: 40%
48	WT/BK	Clean Tachometer Output	35-42 Hz
51	BK/YL	Power Ground	<0.1v
52	BR/GN	Coil Driver 2 Control	5° dwell
53, 59	---	Not Used	---
54	BK/RD	Fuel Pump Control	Pump On: 1v, Off: 12v
55	RD	Keep Alive Power	12-14v
56	BK/OR	EVAP Purge Valve	0-10 Hz (0-100%)
57	WT/BK	Knock Sensor + Signal	0v
58	WT/BL	VSS (+) Signal	0 Hz, 55 mph: 125 Hz
60	WT	HO2S-11 (B1 S1) Signal	0.1-1.1v
61	WT/GN	HO2S-22 (B2 S2) Signal	0.1-1.1v
62	WT/PK	FTP Sensor Signal	2.6v (cap off)
63	WT/GN	FRP Sensor Signal	2.8v (39 psi)
64	WT/GN	A/T: TR Sensor Signal	In 'P': 0v, O/D: 1.7v
64	WT	M/T: Clutch Position Switch	Clutch In: 0v, Out: 5v

2000 Mystique 2.5L V6 MFI VIN L (All) 104 Pin Connector

PCM Pin #	Wire Color	Circuit Description (104 Pin)	Value at Hot Idle
65	WT/BL	DPFE Sensor Signal	0.20-1.30v
66, 68	---	Not Used	---
67	BK/OR	EVAP CV Solenoid	0-10 Hz (0-100%)
69	BK/YL	A/C WOT Relay Control	Relay Off: 12v, On: 1v
70	BK/WT	Injector 1 Control	2.1-2.5 ms
71	GN/YL	Vehicle Power	12-14v
72, 75	---	Not Used	---
73	BK/YL	HO2S-11 (B1 S1) Heater	1v, Off: 12v
74	BK/BL	Injector 3 Control	2.1-2.5 ms
76	BR/WT	CMP & TSS Ground	<0.050v
77	BK/YL	Power Ground	<0.1v
78	BR/YL	Coil Driver 3 Control	5° dwell
79	WT/BK	TCIL (lamp) Control	Lamp On: 1v, Off: 12v
80	BK/WT	TCC Solenoid Control	0%, 55 mph: 90-95%
81	BK/RD	EPC Solenoid Control	8.9v
82	---	Not Used	---
83	BK/YL	IAC Motor Control	9v (33% d/cycle)
84	WT/PK	TSS Sensor Signal	46-50 Hz
85	WT/PK	CMP Sensor Signal	5-7 Hz
86	BK/WT	A/C Pressure Switch	A/C On: 12v (Open)
87	WT/RD	HO2S-21 (B1 S1) Signal	0.1-1.1v
88	WT/BL	MAF Sensor Signal	0.8v
89	WT	TP Sensor Signal	0.53-1.27v
90	YL	Reference Voltage	4.9-5.1v
91	BR	Analog Signal Return	<0.050v
92	GN/RD	Brake Pedal Position Switch	Brake Off: 0v, On: 12v
93	BK/GN	Injector 5 Control	2.1-2.5 ms
94	BK/RD	Injector 6 Control	2.1-2.5 ms
95	BK/OR	Injector 4 Control	2.1-2.5 ms
96	BK/YL	Injector 2 Control	2.1-2.5 ms
97	GN/YL	Vehicle Power	12-14v
98	---	Not Used	---
99	BK/BL	HO2S-21 (B2 S1) Heater	1v, Off: 12v
100	BK/OR	HO2S-12 (B1 S2) Heater	1v, Off: 12v
101	BK/GN	HO2S-22 (B2 S2) Heater	1v, Off: 12v
102	BK/OR	Shift Solenoid 3 Control	7-9v, 30 mph: 8-9v
103	BK/YL	Power Ground	<0.1v
104	---	Not Used	---

Pin Connector Graphic

PCM 104-PIN CONNECTOR

Terminal View of 104-Pin PCM Wiring Harness Connector

Standard Colors and Abbreviations

Abbreviation	Color	Abbreviation	Color	Abbreviation	Color
BK	Black	GY	Gray	PK	Purple
BL	Blue	GN	Green	RD	Red
BR	Brown	LG	LT Green	TN	Tan
DB	Dark Blue	OR	Orange	WT	White
DG	DK Green	PK	Pink	YL	Yellow

SABLE PIN TABLES

1996-97 Sable 3.0L V6 4v MFI VIN S (A/T) 104 Pin Connector

PCM Pin #	Wire Color	Circuit Description (104 Pin)	Value at Hot Idle
1	PK/OR	Shift Solenoid 2 Control	1v, 55 mph: 12v
2	PK/LG	MIL (lamp) Control	MIL On: 1v, Off: 12v
3-4	---	Not Used	---
5	WT/OR	Electronic AIR System	0.1v (Off)
6-7	---	Not Used	---
8	DB/Y	IMRC Monitor Signal	5v, 55 mph: 5v
9-12, 14	---	Not Used	---
13	PK	Flash EEPROM Power	0.1v
15	PK/LB	Data Bus (-) Signal	Digital Signals
16	TN/OR	Data Bus (+) Signal	Digital Signals
17-20, 23	---	Not Used	---
21	GY	CKP Sensor (+) Signal	850-1120 Hz
22	DB	CKP Sensor (-) Signal	850-1120 Hz
24	BK/WT	Power Ground	<0.1v
25	BK	Chassis Ground	<0.050v
26	YL/BK	Coil Driver 1 Control	5° dwell
27	OR/YL	Shift Solenoid 1 Control	12v, 55 mph: 1v
28	TN/OR	Low Speed Fan Control	Fan On: 1v, Off: 12v
29	TN/WT	Overdrive Cancel Switch	TCS & O/D On: 12v
30	DG	Octane Adjust Switch	9.3v (Closed: 0v)
31	YL/LG	PSP Switch Signal	Straight: 0v, turned: 12v
32, 34	---	Not Used	---
33	PK/OR	VSS (-) Signal	0 Hz, 55 mph: 125 Hz
35	RD/BK	HO2S-12 (B1 S2) Signal	0.1-1.1v
36	TN/LB	MAF Sensor Return	<0.050v
37	OR/BK	TFT Sensor Signal	2.40-2.10v
38	LG/RD	ECT Sensor Signal	0.5-0.6v
39	GY	IAT Sensor Signal	1.5-2.5v
40	DG/YT	Fuel Pump Monitor	12v, Off: 0v
41	PK/LB	A/C Cycling Clutch Switch	A/C Switch On: 12v
42	BR	IMRC Solenoid Control	12v, 55 mph: 12v
43-45	---	Not Used	---
46	LG/PK	High Speed Fan Control	Fan On: 1v, Off: 12v
47	BR/PK	VR Solenoid Control	0%, 55 mph: 45%
48	TN/YL	Clean Tachometer Output	33-37 Hz
49-50	---	Not Used	---
51	BK/WT	Power Ground	<0.1v
52	YL/RD	Coil Driver 2 Control	5° dwell
53	PK/BK	Shift Solenoid 3 Control	12v, 55 mph: 1v
54	---	Not Used	---
55	DG/OR	Keep Alive Power	12-14v
56	GY/YL	EVAP VMV Solenoid	0-10 Hz (0-100%)
57	YL/RD	Knock Sensor Signal	0v
58	GY/BK	VSS (+) Signal	0 Hz, 55 mph: 125 Hz
59	---	Not Used	---
60	GY/LB	HO2S-11 (B1 S1) Signal	0.1-1.1v
61	PK/LG	HO2S-22 (B2 S2) Signal	0.1-1.1v
62	RD/PK	FTP Sensor Signal	2.6v (0" H20)
63	---	Not Used	---
64	LBYL	TR Sensor Signal	In 'P': 0v, O/D: 1.7v

1996-97 Sable 3.0L V6 4v MFI VIN S (A/T) 104 Pin Connector

PCM Pin #	Wire Color	Circuit Description (104 Pin)	Value at Hot Idle
65	BR/LG	DPFE Sensor Signal	0.20-1.30v
66, 68, 72	---	Not Used	---
64	LBYL	TR Sensor Signal	In 'P': 0v, O/D: 1.7v
67	PK/WT	EVAP CV Solenoid	0-10 Hz (0-100%)
69	PK/YL	A/C WOT Relay Control	Relay Off: 12v, On: 1v
70	BR	Elect/Secondary Air Injection	12v, 55 mph: 12v
71	RD	Vehicle Power	12-14v
73	TN/BK	Injector 5 Control	2.3-5.5 ms
74	BR/YL	Injector 3 Control	2.3-5.5 ms
75	TN	Injector 1 Control	2.3-5.5 ms
76, 77	BK/WT	Power Ground	<0.1v
78	YL/WT	Coil Driver 3 Control	5° dwell
79	WT/LG	TCIL (lamp) Control	Lamp On: 1v, Off: 12v
80	LB/OR	Fuel Pump Control	Relay Off: 12v, On: 1v
81	WT/YL	EPC Solenoid Control	9.0v (15 psi)
82	PK/YL	TCC Solenoid Control	12v (0 psi)
83	WT/LB	IAC Motor Control	10v (30% d/cycle)
84	DG/WT	TSS Sensor Signal	43 Hz (650 rpm)
85	DB/OR	CID Sensor Signal	5-7 Hz
86	BK/YL	A/C Pressure Switch	Switch On: 12v (Open)
87	RD/LG	HO2S-21 (B1 S1) Signal	0.1-1.1v
88	LB/RD	MAF Sensor Signal	0.6v
89	GY/WT	TP Sensor Signal	0.53-1.27v
90	BR/WT	Reference Voltage	4.9-5.1v
91	GY/RD	Analog Signal Return	<0.050v
92	RD/LG	Brake Switch Signal	Brake Off: 0v, On: 12v
93	RD/WT	HO2S-11 (B1 S1) Heater	1v, Off: 12v
94	WT/BK	HO2S-21 (B2 S1) Heater	1v, Off: 12v
95	YL/LG	HO2S-12 (B1 S2) Heater	1v, Off: 12v
96	TN/YL	HO2S-22 (B2 S2) Heater	1v, Off: 12v
97	RD	Vehicle Power	12-14v
98	---	Not Used	---
99	LG/OR	Injector 6 Control	2.3-5.5 ms
100	BR/LB	Injector 4 Control	2.3-5.5 ms
101	WT	Injector 2 Control	2.3-5.5 ms
102	---	Not Used	---
103	BK/WT	Power Ground	<0.1v
104	---	Not Used	---

Pin Connector Graphic

PCM 104-PIN CONNECTOR

Terminal View of 104-Pin PCM Wiring Harness Connector

Standard Colors and Abbreviations

Abbreviation	Color	Abbreviation	Color	Abbreviation	Color
BK	Black	GY	Gray	PK	Purple
BL	Blue	GN	Green	RD	Red
BR	Brown	LG	LT Green	TN	Tan
DB	Dark Blue	OR	Orange	WT	White
DG	DK Green	PK	Pink	YL	Yellow

1998-99 Sable 3.0L 4v V6 MFI VIN S (A/T) 104 Pin Connector

PCM Pin #	Wire Color	Circuit Description (104 Pin)	Value at Hot Idle
1	PK/OR	Shift Solenoid 2 Control	1v, 55 mph: 1v
2	PK/LG	MIL (lamp) Control	MIL On: 1v, Off: 12v
3-4	---	Not Used	---
5	WT/OR	Electronic AIRB Monitor	5v
8	DB/Y	IMRC Monitor Signal	5v, 55 mph: 5v
9-11	---	Not Used	---
12	YL/WT	Fuel Level Indicator Signal	1.7v (1/2 full)
13	PK	Flash EEPROM Power	0.1v
14	---	Not Used	---
15	PK/LB	Data Bus (-) Signal	Digital Signals
16	TN/OR	Data Bus (+) Signal	Digital Signals
17-20	---	Not Used	---
21	GY	CKP Sensor (+) Signal	390-520 Hz
22	DB	CKP Sensor (-) Signal	390-520 Hz
24	BK/WT	Power Ground	<0.1v
25	BK	Power Ground	<0.1v
26	YL/BK	Coil Driver 1 Control	5° dwell
27	OR/YL	Shift Solenoid 1 Control	12v, 55 mph: 0.1v
28	LB	Low Speed Fan Control	On: 1v, Off: 12v
29	TN/WT	Overdrive Cancel Switch	TCS & O/D On: 12v
30	DG	Octane Adjust Switch	Closed: 0v, Open: 9.3v
31	YL/LG	PSP Switch Signal	Straight: 0v, Turned: 12v
32	---	Not Used	---
33	PK/OR	VSS (-) Signal	0 Hz, 55 mph: 125 Hz
35	RD/BK	HO2S-12 (B1 S2) Signal	0.1-1.1v
36	TN/LB	MAF Sensor Return	<0.050v
37	OR/BK	TFT Sensor Signal	2.10-2.40v
38	LG/RD	ECT Sensor Signal	0.5-0.6v
39	GY	IAT Sensor Signal	1.5-2.5v
40	DG/YT	Fuel Pump Monitor	On: 12v, Off: 0v
41	PK/LB	A/C Cycling Clutch Switch	On: 1v, Off: 12v
42	BR	IMRC Solenoid Control	12v (Off)
43	---	Not Used	---
45	PK/WT	EVAP CV Solenoid	0-10 Hz (0-100%)
46	LG/PK	High Speed Fan Control	On: 1v, Off: 12v
47	BR/PK	EGR VR Solenoid	0%, 55 mph: 45%
48	TN/YL	Clean Tachometer Output	33-45 Hz
49-50	---	Not Used	---
51	BK/WT	Power Ground	<0.1v
52	YL/RD	Coil Driver 2 Control	5° dwell
53	PK/BK	Shift Solenoid 3 Control	12v, 55 mph: 0.1v
54	PK/YL	TCC Solenoid Control	0%, 55 mph: 95%
56	GY/YL	EVAP Purge Solenoid	0-10 Hz (0-100%)
55	DG/OR	Keep Alive Power	12-14v
57	YL/RD	Knock Sensor Signal	0v
58	GY/BK	VSS (+) Signal	0 Hz, 55 mph: 125 Hz
59	---	Not Used	---
60	GY/LB	HO2S-11 (B1 S1) Signal	0.1-1.1v
61	PK/LG	HO2S-22 (B2 S2) Signal	0.1-1.1v
62	RD/BK	FTP Sensor Signal	2.6v (0 in. H20)
63	---	Not Used	---
64	LBYL	TR Sensor Signal	In 'P': 0v, 55 mph: 1.7v

1998-99 Sable 3.0L 4v V6 MFI VIN S (A/T) 104 Pin Connector

PCM Pin #	Wire Color	Circuit Description (104 Pin)	Value at Hot Idle
65	BR/LG	DPFE Sensor Signal	0.20-1.30v
66, 68, 72	---	Not Used	---
64	LBYL	TR Sensor Signal	In 'P': 0v, 55 mph: 1.7v
67	PK/WT	EVAP Purge Solenoid	0-10 Hz (0-100%)
69	PK/YL	A/C WOT Relay Control	On: 1v, Off: 12v
70	BR	Electronic AIRB Solenoid	12v, 55 mph: 12v
71	RD	Vehicle Power	12-14v
73	TN/BK	Injector 5 Control	2.2-2.7 ms
74	BR/YL	Injector 3 Control	2.2-2.7 ms
75	TN	Injector 1 Control	2.2-2.7 ms
76, 77	BK/WT	Power Ground	<0.1v
78	YL/WT	Coil Driver 3 Control	5° dwell
79	WT/LG	TCIL (lamp) Control	On: 1v, Off: 12v
80	LB/OR	Fuel Pump Control	On: 1v, Off: 12v
81	WT/YL	EPC Solenoid Control	9.0v (15 psi)
82	VT/YL	TCC Solenoid Control	12v (0 psi)
83	WT/LB	IAC Motor Control	10v (55% duty cycle)
84	DG/WT	TSS Sensor Signal	43 Hz (700 rpm)
85	DB/OR	CID Sensor Signal	5-7 Hz
86	BK/YL	A/C High Pressure Switch	12v (Switch Open)
87	RD/LG	HO2S-21 (B2 S1) Signal	0.1-1.1v
88	LB/RD	MAF Sensor Signal	0.6v
89	GY/WT	TP Sensor Signal	0.53-1.27v
90	BR/WT	Reference Voltage	4.9-5.1v
91	GY/RD	Analog Signal Return	<0.050v
92	RD/LG	Brake Pedal Position Switch	Brake Off: 0v, On: 12v
93	RD/WT	HO2S-11 (B1 S1) Heater	On: 1v, Off: 12v
94	WT/BK	HO2S-21 (B2 S1) Heater	On: 1v, Off: 12v
95	YL/LG	HO2S-12 (B1 S2) Heater	On: 1v, Off: 12v
96	TN/YL	HO2S-22 (B2 S2) Heater	On: 1v, Off: 12v
97	RD	Vehicle Power	12-14v
98	---	Not Used	---
99	LG/OR	Injector 6 Control	2.2-2.7 ms
100	BR/LB	Injector 4 Control	2.2-2.7 ms
101	WT	Injector 2 Control	2.2-2.7 ms
98, 102	---	Not Used	---
103	BK/WT	Power Ground	<0.1v
104	---	Not Used	---

Pin Connector Graphic

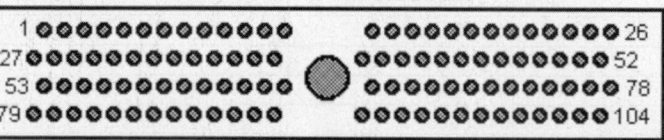

PCM 104-PIN CONNECTOR

Terminal View of 104-Pin PCM Wiring Harness Connector

Standard Colors and Abbreviations

Abbreviation	Color	Abbreviation	Color	Abbreviation	Color
BK	Black	GY	Gray	PK	Purple
BL	Blue	GN	Green	RD	Red
BR	Brown	LG	LT Green	TN	Tan
DB	Dark Blue	OR	Orange	WT	White
DG	DK Green	PK	Pink	YL	Yellow

2000 Sable 3.0L 4v V6 MFI VIN S (A/T) 104 Pin Connector

PCM Pin #	Wire Color	Circuit Description (104 Pin)	Value at Hot Idle
1	DG/PK	COP 4 Driver Control	6° dwell
2	PK/LG	MIL (lamp) Control	MIL On: 1v, Off: 12v
3	BK/WT	Power Ground	<0.1v
5	WT/OR	Electronic AIRB Monitor	5v
6	OR/YL	Shift Solenoid 'A' Control	12v, 55 mph: 0.1v
4, 7-11	---	Not Used	---
12	YL/WT	Fuel Level Indicator Signal	1.7v (1/2 full)
13	PK	Flash EEPROM Power	0.1v
14, 19	---	Not Used	---
15	PK/LB	Data Bus (-) Signal	Digital Signals
16	TN/OR	Data Bus (+) Signal	Digital Signals
17	BR/YL	SCI Receive Signal	N/A
18	GY/RD	SCI Transmit Signal	N/A
20	PK/BK	Shift Solenoid 'A' Control	12v, 55 mph: 0.1v
21	GY	CKP Sensor (+) Signal	390-520 Hz
22	DB	CKP Sensor (-) Signal	390-520 Hz
24	BK/WT	Power Ground	<0.1v
25	BK	Power Ground	<0.1v
26	LG/WT	COP 1 Driver Control	6° dwell
27	OR/YL	COP 5 Driver Control	6° dwell
28	LB	Low Speed Fan Control	On: 1v, Off: 12v
29	---	Not Used	---
30	DB/LG	Anti-theft Indicator	N/A
31	YL/LG	PSP Switch Signal	Straight: 0v, Turned: 12v
32	YL	Knock Sensor (-) Signal	0v
34	YL/BK	Digital TR1 Sensor	In 'P': 0v, 55 mph: 11v
35	RD/BK	HO2S-12 (B1 S2) Signal	0.1-1.1v
36	TN/LB	MAF Sensor Return	<0.050v
37	OR/BK	TFT Sensor Signal	2.10-2.40v
38	LG/RD	ECT Sensor Signal	0.5-0.6v
39	GY	IAT Sensor Signal	1.5-2.5v
40	DG/YT	Fuel Pump Monitor	On: 12v, Off: 0v
41	DG/WT	A/C High Pressure Switch	On: 1v, Off: 12v
42-43	---	Not Used	---
44	PK	Starter Relay Control	KOEC: 9-11v
46	LG/PK	High Speed Fan Control	On: 1v, Off: 12v
47	BR/PK	EGR VR Solenoid	0%, 55 mph: 45%
48	TN/YL	Clean Tachometer Output	33-45 Hz
49	LB/BK	Digital TR2 Sensor	In 'P': 0v, 55 mph: 11v
50	WT/BK	Digital TR4 Sensor	In 'P': 0v, 55 mph: 11v
51	BK/WT	Power Ground	<0.1v
52	PK/WT	COP 2 Driver Control	6° dwell
53	OR/YL	COP 6 Driver Control	6° dwell
55	PK/LB	Keep Alive Power	12-14v
54	PK/YL	TCC Solenoid Control	0%, 55 mph: 95%
56	GY/YL	EVAP Purge Solenoid	0-10 Hz (0-100%)
55	DG/OR	Keep Alive Power	12-14v
57	YL/RD	Knock Sensor (+) Signal	0v
59	DG/WT	TSS Sensor Signal	43 Hz (700 rpm)
60	GY/LB	HO2S-11 (B1 S1) Signal	0.1-1.1v
61	PK/LG	HO2S-22 (B2 S2) Signal	0.1-1.1v
62	RD/PK	FTP Sensor Signal	2.6v (0 in. H20)
63	WT/YL	EPC Solenoid Control	9.0v (15 psi)
64	RD/BK	Digital TR3A Sensor	In 'P': 0v, 55 mph: 1.7v

2000 Sable 3.0L 4v V6 MFI VIN S (A/T) 104 Pin Connector

PCM Pin #	Wire Color	Circuit Description (104 Pin)	Value at Hot Idle
65	BR/LG	DPFE Sensor Signal	0.20-1.30v
66, 72	---	Not Used	---
67	PK/WT	EVAP CV Solenoid	0-10 Hz (0-100%)
68	GY/BK	VSS (+) Signal	0 Hz, 55 mph: 125 Hz
69	PK/YL	A/C WOT Relay Control	On: 1v, Off: 12v
70	BR	Electronic AIRB Solenoid	12v, 55 mph: 12v
71	RD	Vehicle Power	12-14v
73	TN/BK	Injector 5 Control	2.2-2.7 ms
74	BR/YL	Injector 3 Control	2.2-2.7 ms
75	TN	Injector 1 Control	2.2-2.7 ms
76, 77	BK/WT	Power Ground	<0.1v
78	YLPK	COP 3 Driver Control	6° dwell
79	WT/LG	TCIL (lamp) Control	On: 1v, Off: 12v
80	LB/OR	Fuel Pump Control	On: 1v, Off: 12v
81	BR/OR	EPC Solenoid Control	9.0v (15 psi)
82	VT/YL	TCC Solenoid Control	12v (0 psi)
83	WT/LB	IAC Motor Control	10v (55% duty cycle)
84	DB/Y	OSS Sensor Signal	0 Hz, 55 mph: 131 Hz
85	DB/OR	CID Sensor Signal	5-7 Hz
86	BK/YL	A/C Clutch Relay Switch	12v (Switch Open)
87	RD/LG	HO2S-21 (B2 S1) Signal	0.1-1.1v
88	LB/RD	MAF Sensor Signal	0.6v
89	GY/WT	TP Sensor Signal	0.53-1.27v
90	BR/WT	Reference Voltage	4.9-5.1v
91	GY/RD	Analog Signal Return	<0.050v
92	RD/LG	Brake Pedal Position Switch	Brake Off: 0v, On: 12v
93	RD/WT	HO2S-11 (B1 S1) Heater	On: 1v, Off: 12v
94	WT/BK	HO2S-21 (B2 S1) Heater	On: 1v, Off: 12v
95	YL/LB	HO2S-12 (B1 S2) Heater	On: 1v, Off: 12v
96	TN/YL	HO2S-22 (B2 S2) Heater	On: 1v, Off: 12v
97-98	---	Not Used	---
99	LG/OR	Injector 6 Control	2.2-2.7 ms
100	BR/LB	Injector 4 Control	2.2-2.7 ms
101	WT	Injector 2 Control	2.2-2.7 ms
102	---	Not Used	---
103	BK/WT	Power Ground	<0.1v
104	---	Not Used	---

Pin Connector Graphic

Standard Colors and Abbreviations

Abbreviation	Color	Abbreviation	Color	Abbreviation	Color
BK	Black	GY	Gray	PK	Purple
BL	Blue	GN	Green	RD	Red
BR	Brown	LG	LT Green	TN	Tan
DB	Dark Blue	OR	Orange	WT	White
DG	DK Green	PK	Pink	YL	Yellow

2001 Sable 3.0L 4v V6 MFI VIN S (A/T) 104 Pin Connector

PCM Pin #	Wire Color	Circuit Description (104 Pin)	Value at Hot Idle
1	PK/OR	Shift Solenoid 'B' Control	1v, 55 mph: 1v
2	PK/LG	MIL (lamp) Control	MIL On: 1v, Off: 12v
3	YL/BK	Digital TR1 Sensor	In 'P': 0v, 55 mph: 11v
4	BR/PK	Start System Voltage	Cranking: 9-11v
5	---	Not Used	---
6	DG/WT	TSS Sensor Signal	50-65 Hz (790-820 rpm)
7-11	---	Not Used	---
12	YL/WT	Fuel Level Indicator Signal	1.7v (1/2 full)
13	VT	Flash EEPROM Power	0.1v
15	PK/LB	SCP Data Bus (-) Signal	Digital Signals
16	TN/OR	SCP Data Bus (+) Signal	Digital Signals
17	BR/YL	SCI Receive (RX) Signal	Digital Signals
18	GY/RD	SCI Transmit (TX) Signal	Digital Signals
20	GY/YL	Generator Signal 2	TBD
21	GY	CKP Sensor (+) Signal	390-520 Hz
22	DB	CKP Sensor (-) Signal	390-520 Hz
24	BK/WT	Power Ground	<0.1v
25	BK	Chassis Ground	<0.1v
26	YL/BK	Coil 'A' Driver (Cyl 1 & 5)	6° dwell
27	OR/YLEL	Shift Solenoid 'A' Control	12v, 55 mph: 1v
28	LB	Low Speed Fan Control	On: 1v, Off: 12v
30	DB/LG	Anti-theft Indicator	Refer to PAT System
31	YL/LG	PSP Switch Signal	Straight: 0v, Turned: 12v
32	YT	Knock Sensor (-) Signal	<0.050v
35	RD/BK	HO2S-12 (B1 S2) Signal	0.1-1.1v
36	TN/LB	MAF Sensor Return	<0.050v
37	OR/BK	TFT Sensor Signal	2.10-2.40v
38	LG/RD	ECT Sensor Signal	0.5-0.6v
39	GY	IAT Sensor Signal	1.5-2.5v
40	DG/YT	Fuel Pump Monitor	On: 12v, Off: 0v
41	DG/WT	A/C High Pressure Switch	On: 1v, Off: 12v
42	LG/YL	Generator Cutoff Relay	0.1v (relay on)
44	PK	Starter Relay Control	KOEC: 9-11v
45	LB/RD	Generator Sensor Signal	130 Hz (30%)
46	LG/VT	High Speed Fan Control	On: 1v, Off: 12v
47	BR/PK	EGR VR Solenoid	0%, 55 mph: 45%
48	TN/YL	Clean Tachometer Output	33-45 Hz
49	LB/BK	Digital TR2 Sensor	In 'P': 0v, 55 mph: 11v
50	WT/BK	Digital TR4 Sensor	In 'P': 0v, 55 mph: 11v
51	BK/WT	Power Ground	<0.1v
52	YL/RD	Coil 'B' Driver (Cyl 3 & 4)	6° dwell
53	PK/BK	Shift Solenoid 'C' Control	12v, 55 mph: 1v
54	VT/YL	TCC Solenoid Control	0%, 55 mph: 95%
55	PK/LB	Keep Alive Power	12-14v
56	GY/YL	EVAP Purge Solenoid	0-10 Hz (0-100%)
55	DG/OR	Keep Alive Power	12-14v
57	YL/RD	Knock Sensor (+) Signal	0v
60	GY/LB	HO2S-11 (B1 S1) Signal	0.1-1.1v
61	VT/LG	HO2S-22 (B2 S2) Signal	0.1-1.1v
62	RD/PK	FTP Sensor Signal	2.6v (0 in. H20)
63	WT/YL	Injector Pressure Sensor	2.74v (39 psi)
64	RD/BK	Digital TR3A Sensor	In 'P': 0v, 55 mph: 1.7v

2001 Sable 3.0L 4v V6 MFI VIN S (A/T) 104 Pin Connector

PCM Pin #	Wire Color	Circuit Description (104 Pin)	Value at Hot Idle
65	BR/LG	DPFE Sensor Signal	0.95-1.05v
66	TN/OR	Evaporator Air Temp. Sensor	4.95-5.15v
67	VT/WT	EVAP CV Solenoid	0-10 Hz (0-100%)
68	GY/BK	VSS (+) Signal	0 Hz, 55 mph: 125 Hz
69	PK/YL	A/C WOT Relay Control	On: 1v, Off: 12v
70, 72	---	Not Used	---
71	RD	Vehicle Power	12-14v
73	TN/BK	Injector 5 Control	2.2-2.7 ms
74	BR/YL	Injector 3 Control	2.2-2.7 ms
75	TN	Injector 1 Control	2.2-2.7 ms
76, 77	BK/WT	Power Ground	<0.1v
78	YL/WT	Coil 'C' Driver (Cyl 2 & 6)	6° dwell
79	WT/LG	TCIL (lamp) Control	On: 1v, Off: 12v
80	LB/OR	Fuel Pump Control	On: 1v, Off: 12v
81	BR/OR	EPC Solenoid Control	9.0v (15 psi)
82	BK/WT	Check Cap Indicator Light	On: 1v, Off: 12v
83	WT/LB	IAC Motor Control	10v (55% duty cycle)
84	DB/YL	OSS Sensor Signal	0 Hz, 55 mph: 131 Hz
85	DB/OR	CID Sensor Signal	5-7 Hz
86	BK/YL	A/C Clutch Relay Switch	12v (Switch Open)
87	RD/LG	HO2S-21 (B2 S1) Signal	0.1-1.1v
88	LB/RD	MAF Sensor Signal	0.6v
89	GY/WT	TP Sensor Signal	0.53-1.27v
90	BR/WT	Reference Voltage	4.9-5.1v
91	GY/RD	Analog Signal Return	<0.050v
92	RD/LG	Brake Pedal Position Switch	Brake Off: 0v, On: 12v
93	RD/WT	HO2S-11 (B1 S1) Heater	On: 1v, Off: 12v
94	WT/BK	HO2S-21 (B2 S1) Heater	On: 1v, Off: 12v
95	YL/LB	HO2S-12 (B1 S2) Heater	On: 1v, Off: 12v
96	TN/YL	HO2S-22 (B2 S2) Heater	On: 1v, Off: 12v
97	---	Not Used	---
98	LG/RD	Generator Indicator Light	On: 1v, Off: 12v
99	LG/OR	Injector 6 Control	2.2-2.7 ms
100	BR/LB	Injector 4 Control	2.2-2.7 ms
101	WT	Injector 2 Control	2.2-2.7 ms
102	---	Not Used	---
103	BK/WT	Power Ground	<0.1v
104	---	Not Used	---

Pin Connector Graphic

```
PCM 104-PIN CONNECTOR
1 ○○○○○○○○○○○○○    ○○○○○○○○○○○○○ 26
27 ○○○○○○○○○○○○○   ○○○○○○○○○○○○○ 52
53 ○○○○○○○○○○○○○ ⬤ ○○○○○○○○○○○○○ 78
79 ○○○○○○○○○○○○○   ○○○○○○○○○○○○○ 104
```
Terminal View of 104-Pin PCM Wiring Harness Connector

Standard Colors and Abbreviations

Abbreviation	Color	Abbreviation	Color	Abbreviation	Color
BK	Black	GY	Gray	PK	Purple
BL	Blue	GN	Green	RD	Red
BR	Brown	LG	LT Green	TN	Tan
DB	Dark Blue	OR	Orange	WT	White
DG	DK Green	PK	Pink	YL	Yellow

2002-04 Sable 3.0L 4v V6 MFI VIN S (A/T) 104 Pin Connector

PCM Pin #	Wire Color	Circuit Description (104 Pin)	Value at Hot Idle
1	VT/OR	Shift Solenoid 'B' Control	1v, 55 mph: 1v
2	PK/LG	MIL (lamp) Control	MIL On: 1v, Off: 12v
3	YL/BK	Digital TR1 Sensor	In 'P': 0v, 55 mph: 11v
4	TN/RD	System Voltage (Start)	Cranking: 9-11v
5, 7-11	---	Not Used	---
6	DG/WT	TSS Sensor Signal	50-65 Hz (790-820 rpm)
12	YL/WT	Fuel Level Indicator Signal	1.7v (1/2 full)
13	VT	Flash EEPROM Power	0.1v
14, 19	---	Not Used	---
15	PK/LB	SCP Bus (-) Signal	Digital Signals
16	TN/OR	SCP Bus (+) Signal	Digital Signals
17	BR/YL	Passive Antitheft RX Signal	Digital Signals
18	GY/RD	Passive Antitheft TX Signal	Digital Signals
20	GY/YL	Generator Monitor Signal	130 Hz (45%)
21	GY	CKP Sensor (+) Signal	390-520 Hz
22	DB	CKP Sensor (-) Signal	390-520 Hz
23, 29	---	Not Used	---
24	BK/WT	Power Ground	<0.1v
25	BK	Chassis Ground	<0.1v
26	YL/BK	Coil 1 Driver (Cyl 1 & 5)	6° dwell
27	OR/YL	Shift Solenoid 'A' Control	12v, 55 mph: 1v
28	LB	Engine Cooling Fan Brake Relay Control	Relay Off: 12v, On: 1v
30	DB/LG	Antitheft Indicator	Indicator Off: 12v, On: 1v
31	YL/LG	PSP Switch Signal	Straight: 0v, Turned: 12v
32	YL	Knock Sensor (-) Signal	<0.050v
33-34, 43	---	Not Used	---
35	RD/BK	HO2S-12 (B1 S2) Signal	0.1-1.1v
36	TN/LB	MAF Sensor Return	<0.050v
37	OR/BK	TFT Sensor Signal	2.10-2.40v
38	LG/RD	ECT Sensor Signal	0.5-0.6v
39	GY	IAT Sensor Signal	1.5-2.5v
40	DG/YL	Fuel Pump Monitor	On: 12v, Off: 0v
41	DG/WT	A/C High Pressure Switch	On: 1v, Off: 12v
42	LG/YL	Engine Cooling Fan Relay Control	Relay Off: 12v, On: 1v
44	PK	Starter Relay Control	Relay Off: 12v, On: 1v
45	LB/RD	Generator Sensor Signal	130 Hz (30%)
46	LG/VT	High Speed Fan Control	On: 1v, Off: 12v
47	BR/PK	EGR VR Solenoid Control	0%, 55 mph: 45%
48	TN/YL	Clean Tachometer Output	33-45 Hz
49	LB/BK	Digital TR2 Sensor	In 'P': 0v, 55 mph: 11v
50	WT/BK	Digital TR4 Sensor	In 'P': 0v, 55 mph: 11v
51	BK/WT	Power Ground	<0.1v
52	YL/RD	Coil 2 Driver (Cyl 3 & 4)	6° dwell
53	PK/BK	Shift Solenoid 'C' Control	12v, 55 mph: 1v
54	VT/YL	TCC Solenoid Control	0%, 55 mph: 95%
55	PK/LB	Keep Alive Power	12-14v
56	GY/YL	EVAP Purge Solenoid	0-10 Hz (0-100%)
57	YL/RD	Knock Sensor (+) Signal	0v
58-59	---	Not Used	---
60	GY/LB	HO2S-11 (B1 S1) Signal	0.1-1.1v
61	VT/LG	HO2S-22 (B2 S2) Signal	0.1-1.1v
62	RD/PK	FTP Sensor Signal	2.6v (0 in. H20)
63	WT/YL	Injector Pressure Sensor	2.74v (39 psi)
64	RD/BK	Digital TR3A Sensor	In 'P': 0v, 55 mph: 1.7v

2002-04 Sable 3.0L 4v V6 MFI VIN S (A/T) 104 Pin Connector

PCM Pin #	Wire Color	Circuit Description (104 Pin)	Value at Hot Idle
65	BR/LG	EGR DPFE Sensor Signal	0.95-1.05v
66	TN/OR	Evaporator Air Temperature Sensor	4.95-5.15v
67	VT/WT	EVAP CV Solenoid	0-10 Hz (0-100%)
68	GY/BK	VSS (+) Signal	0 Hz, 55 mph: 125 Hz
69	PK/YL	A/C WOT Relay Control	On: 1v, Off: 12v
71	RD	Vehicle Power (Start-Run)	12-14v
72	---	Not Used	---
73	TN/BK	Injector 5 Control	2.2-2.7 ms
74	BR/YL	Injector 3 Control	2.2-2.7 ms
75	TN	Injector 1 Control	2.2-2.7 ms
76, 77	BK/WT	Power Ground	<0.1v
78	YL/WT	Coil 3 Driver (Cyl 2 & 6)	6° dwell
79	WT/LG	TCIL (lamp) Control	Off: 12v, On: 1v
80	LB/OR	Fuel Pump Control	Off: 12v, On: 1v
81	BR/OR	EPC Solenoid Control	9.0v (15 psi)
82	BK/WT	Check Cap Indicator Light	Off: 12v, On: 1v
83	WT/LB	IAC Motor Control	10v (55% duty cycle)
84	DB/YL	OSS Sensor Signal	0 Hz, 55 mph: 131 Hz
85	DB/OR	CMP Sensor Signal	5-7 Hz
86	BK/YL	A/C Clutch Relay Switch	12v (Switch Open)
87	RD/LG	HO2S-21 (B2 S1) Signal	0.1-1.1v
88	LB/RD	MAF Sensor Signal	0.6v
89	GY/WT	TP Sensor Signal	0.53-1.27v
90	BR/WT	Reference Voltage	4.9-5.1v
91	GY/RD	Sensor Ground	<0.050v
92	RD/LG	Brake Pedal Position Switch	Brake Off: 0v, On: 12v
93	RD/WT	HO2S-11 (B1 S1) Heater	On: 1v, Off: 12v
94	WT/BK	HO2S-21 (B2 S1) Heater	On: 1v, Off: 12v
95	YL/LB	HO2S-12 (B1 S2) Heater	On: 1v, Off: 12v
96	TN/YL	HO2S-22 (B2 S2) Heater	On: 1v, Off: 12v
97	---	Not Used	---
98	LG/RD	Generator Indicator Light	Off: 12v, On: 1v
99	LG/OR	Injector 6 Control	2.2-2.7 ms
100	BR/LB	Injector 4 Control	2.2-2.7 ms
101	WT	Injector 2 Control	2.2-2.7 ms
102, 104	---	Not Used	---
103	BK/WT	Power Ground	<0.1v

Pin Voltage Connector

PCM 104-PIN CONNECTOR

Terminal View of 104-Pin PCM Wiring Harness Connector

Standard Colors and Abbreviations

Abbreviation	Color	Abbreviation	Color	Abbreviation	Color
BK	Black	GY	Gray	PK	Purple
BL	Blue	GN	Green	RD	Red
BR	Brown	LG	LT Green	TN	Tan
DB	Dark Blue	OR	Orange	WT	White
DG	DK Green	PK	Pink	YL	Yellow

1991-95 Sable 3.0L V6 2v MFI VIN U (A/T) 60 Pin Connector

PCM Pin #	Wire Color	Circuit Description (60 Pin)	Value at Hot Idle
1	YL	Keep Alive Power	12-14v
2	RD/LG	Brake Switch Signal	Brake Off: 0v, On: 12v
3	DG/WT	VSS (+) Signal	0 Hz, 55 mph: 125 Hz
4	WT/PK	Ignition Diagnostic Monitor	20-31 Hz
4	DG/YT	Ignition Diagnostic Monitor	20-31 Hz
5	GY/BK	TSS Sensor Signal	100 Hz (700 rpm)
6	OR/YL	VSS (+) Signal	0 Hz, 55 mph: 125 Hz
7	LG/RD	ECT Sensor Signal	0.5-0.6v
8	PK/BK	Fuel Pump Monitor	12v, Off: 0v
9	TN/BL	MAF Sensor Return	<0.050v
10	PK/BL	A/C Cycling Clutch Signal	A/C Off: 0v, On: 12v
10	PK	A/C Cycling Clutch Signal	A/C Off: 0v, On: 12v
11	GN/YL	Canister Purge Solenoid	12v, Cruise: 1v
12	LG/OR	Injector 6 Control	5.0-5.6 ms
13	TN/OR	Low Speed Cooling Fan	Fan On: 1v, Off: 12v
14	---	Not Used	---
15	TN/BK	Injector 5 Control	5.0-5.6 ms
16	OR/RD	Ignition System Ground	<0.050v
17	TN/RD	Self-Test Output, MIL Control	MIL Off: 12v, On: 1v
18	TN/OR	Data Bus (+) Signal	Digital Signals
19	PK/BL	Data Bus (-) Signal	Digital Signals
20	BK	Case Ground	<0.050v
21	BR/WT	IAC Motor Control	8-9v
22	BL/OR	Fuel Pump Control	Relay Off: 12v, On: 1v
22	TN/LG	Fuel Pump Control	Relay Off: 12v, On: 1v
23-24	---	Not Used	---
25	LG/PK	IAT Sensor Signal	1.5-2.5v
26	BR/WT	Reference Voltage	4.9-5.1v
27	BR/LG	PFE Sensor Signal	3.2v, 55 mph: 2.9v
28	YL/LG	PSP Switch Signal	Straight: 0v, Turned: 11v
29	---	Not Used	---
30	BL	A/T: MLP Sensor Signal	In 'P': 0v, O/D: 1.7v
30	BL/YL	A/T: MLP Sensor Signal	In 'P': 0v, O/D: 1.7v
31	LG/PK	High Speed Cooling Fan	Fan On: 1v, Off: 12v
32	---	Not Used	---
33	BR/PK	VR Solenoid Control	0%, 55 mph: 45%
34	BL/BK	Data Output Link	Digital Signal
34	BL/PK	Data Output Link	Digital Signal
35	BR/BL	Injector 4 Control	5.0-5.6 ms
36	YL/LG	Spark Output Signal	6.93v (50% d/cycle)
37	RD	Vehicle Power	12-14v
38	WT/YL	EPC Solenoid Control	8-9v (9 psi)
39	BR/YL	Injector 3 Control	5.0-5.6 ms
40	BK/LG	Power Ground	<0.1v

1991-95 Sable 3.0L V6 2v MFI VIN U (A/T) 60 Pin Connector

PCM Pin #	Wire Color	Circuit Description (60 Pin)	Value at Hot Idle
41	---	Not Used	---
42 ('94-'95)	TN/LG	A/C Pressure Sensor	A/C On: 12v, Off: 0v
43	RD/BK	HO2S-11 (B1 S1) Signal	0.1-1.1v
44	GY/BL	HO2S-21 (B1 S1) Signal	0.1-1.1v
45	---	Not Used	---
46	GY/RD	Analog Signal Return	<0.050v
47	GN/WT	TP Sensor Signal	0.8v, 55 mph: 1.1v
48	BR	Self-Test Input Signal	STI On: 1v, Off: 5v
49	OR/BK	TOT Sensor Signal	2.40-2.10v
50	BL/RD	MAF Sensor Signal	0.8v
51	OR/YL	Shift Solenoid 1 Control	12v, 55 mph: 1v
52	PK/OR	Shift Solenoid 2 Control	1v, 55 mph: 12v
53	TN/WT	TCC Solenoid Control	Idle: 0%, 30 mph: 0-45%
54	PK/YL	A/C WOT Relay Control	Relay Off: 12v, On: 1v
55	WT/RD	Shift Solenoid 3 Control	7-9v, 30 mph: 8-9v
56	DB	PIP Sensor Signal	6.93v (50% d/cycle)
56	GY/OR	PIP Sensor Signal	6.93v (50% d/cycle)
57	RD	Vehicle Power	12-14v
58	TN	Injector 1 Control	5.0-5.6 ms
59	WT	Injector 2 Control	5.0-5.6 ms
60	BK/LG	Power Ground	<0.1v

Pin Connector Graphic

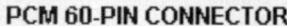

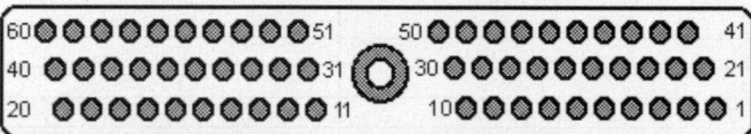

Terminal View of 60-Pin PCM Harness Connector

Standard Colors and Abbreviations

Abbreviation	Color	Abbreviation	Color	Abbreviation	Color
BK	Black	GY	Gray	PK	Purple
BL	Blue	GN	Green	RD	Red
BR	Brown	LG	LT Green	TN	Tan
DB	Dark Blue	OR	Orange	WT	White
DG	DK Green	PK	Pink	YL	Yellow

1996-97 Sable 3.0L V6 2v MFI VIN U (A/T) 104 Pin Connector

PCM Pin #	Wire Color	Circuit Description (104 Pin)	Value at Hot Idle
1	PK/OR	Shift Solenoid 2 Control	1v, 55 mph: 12v
2	PK/LG	MIL (lamp) Control	MIL On: 1v, Off: 12v
3-12	---	Not Used	---
13	PK	Flash EEPROM Power	0.1v
14	---	Not Used	---
15	PK/LB	Data Bus (-) Signal	Digital Signals
16	TN/OR	Data Bus (+) Signal	Digital Signals
17-20	---	Not Used	---
21	GY	CKP Sensor (+) Signal	510-540 Hz
22	DB	CKP Sensor (-) Signal	510-540 Hz
23	---	Not Used	---
24	BK/WT	Power Ground	<0.1v
25	BK	Chassis Ground	<0.050v
26	WT/BK	Coil Driver 1 Control	5° dwell
27	OR/YL	Shift Solenoid 1 Control	12v, 55 mph: 1v
28	TN/OR	Low Speed Fan Control	Fan On: 1v, Off: 12v
29	---	Not Used	---
30	DG	Octane Adjust Switch	9.3v (Closed: 0v)
31	YL/LG	PSP Switch Signal	Straight: 0v, Turned: 12v
32	---	Not Used	---
33	PK/OR	VSS (-) Signal	0 Hz, 55 mph: 125 Hz
34	---	Not Used	---
35	RD/BK	HO2S-12 (B1 S2) Signal	0.1-1.1v
36	TN/LB	MAF Sensor Return	<0.050v
37	OR/BK	TFT Sensor Signal	2.40-2.10v
38	LG/RD	ECT Sensor Signal	0.5-0.6v
39	GY	IAT Sensor Signal	1.5-2.5v
40	DG/YT	Fuel Pump Monitor	12v, Off: 0v
41	PK/LB	A/C Cycling Clutch Switch	A/C Switch On: 12v
42-45	---	Not Used	---
46	LG/PK	High Speed Fan Control	Fan On: 1v, Off: 12v
47	BR/PK	VR Solenoid Control	0%, 55 mph: 45%
48	TN/YL	Clean Tachometer Output	42-50 Hz
49-50	---	Not Used	---
51	BK/WT	Power Ground	1v
52	YL/RD	Coil Driver 2 Control	5° dwell
53	PK/BK	Shift Solenoid 3 Control	12v, 55 mph: 1v
54	---	Not Used	---
55	PK	Keep Alive Power	12-14v
56	GY/YL	EVAP VMV Solenoid	0-10 Hz (0-100%)
57	---	Not Used	---
58	GY/BK	VSS (+) Signal	0 Hz, 55 mph: 125 Hz
59	---	Not Used	---
60	GY/LB	HO2S-11 (B1 S1) Signal	0.1-1.1v
61	PK/LG	HO2S-22 (B2 S2) Signal	0.1-1.1v
62	RD/PK	FTP Sensor Signal	2.6v (0" H2O - cap off)
63	---	Not Used	---
64	LBYL	TR Sensor Signal	In 'P': 0v, O/D: 1.7v

1996-97 Sable 3.0L V6 2v MFI VIN U (A/T) 104 Pin Connector

PCM Pin #	Wire Color	Circuit Description (104 Pin)	Value at Hot Idle
65	BR/LG	DPFE Sensor Signal	0.20-1.30v
66, 68	---	Not Used	---
67	PK/WT	EVAP CV Solenoid	0-10 Hz (0-100%)
69	PK/YL	A/C WOT Relay Control	Relay Off: 12v, On: 1v
70	---	Not Used	---
71	RD	Vehicle Power	12-14v
72	---	Not Used	---
73	TN/BK	Injector 5 Control	3.8-4.7 ms
74	BR/YL	Injector 3 Control	3.8-4.7 ms
75	TN	Injector 1 Control	3.8-4.7 ms
76-77	BK/WT	Power Ground	<0.1v
78	YL/WT	Coil Driver 3 Control	5° dwell
79	---	Not Used	---
80	LB/OR	Fuel Pump Control	Relay Off: 12v, On: 1v
81	WT/YL	EPC Solenoid Control	10.6v (40 psi)
82	PK/YL	TCC Solenoid Control	12v (0%)
83	WT/LB	IAC Motor Control	10v (45% d/cycle)
84	DG/WT	TSS Sensor Signal	50-65 Hz (790-820 rpm)
85	DB/OR	CID Sensor Signal	6-8 Hz
86	BK/YL	A/C Pressure Switch	Switch On: 12v (Open)
87	RD/LG	HO2S-21 (B1 S1) Signal	0.1-1.1v
88	LB/RD	MAF Sensor Signal	0.9v
89	GY/WT	TP Sensor Signal	0.53-1.27v
90	BR/WT	Reference Voltage	4.9-5.1v
91	GY/RD	Analog Signal Return	<0.050v
92	RD/LG	Brake Pedal Position Switch	Brake Off: 0v, On: 12v
93	RD/WT	HO2S-11 (B1 S1) Heater	1v, Off: 12v
94	WT/BK	HO2S-21 (B2 S1) Heater	1v, Off: 12v
95	YL/LB	HO2S-12 (B1 S2) Heater	1v, Off: 12v
96	TN/YL	HO2S-22 (B2 S2) Heater	1v, Off: 12v
97	RD	Vehicle Power	12-14v
98	---	Not Used	---
99	LG/OR	Injector 6 Control	3.8-4.7 ms
100	BR/LB	Injector 4 Control	3.8-4.7 ms
101	WT	Injector 2 Control	3.8-4.7 ms
102	---	Not Used	---
103	BK/WT	Power Ground	<0.1v
104	---	Not Used	---

Pin Connector Graphic

```
           PCM 104-PIN CONNECTOR
  1 ●●●●●●●●●●●●●   ●●●●●●●●●●●●● 26
27 ●●●●●●●●●●●●●   ●●●●●●●●●●●●● 52
53 ●●●●●●●●●●●●●      ●   ●●●●●●●●●●●●● 78
79 ●●●●●●●●●●●●●   ●●●●●●●●●●●●● 104
   Terminal View of 104-Pin PCM Wiring Harness Connector
```

Standard Colors and Abbreviations

Abbreviation	Color	Abbreviation	Color	Abbreviation	Color
BK	Black	GY	Gray	PK	Purple
BL	Blue	GN	Green	RD	Red
BR	Brown	LG	LT Green	TN	Tan
DB	Dark Blue	OR	Orange	WT	White
DG	DK Green	PK	Pink	YL	Yellow

1998-99 Sable 3.0L V6 MFI VIN U (A/T) 104 Pin Connector

PCM Pin #	Wire Color	Circuit Description (104 Pin)	Value at Hot Idle
1	VTN/OR	Shift Solenoid 'B' Control	1v, 55 mph: 12v
2	PK/LG	MIL (lamp) Control	MIL On: 1v, Off: 12v
3	YL/BK	Digital TR1 Sensor	In 'P': 0v, 55 mph: 11v
4-11	---	Not Used	---
12	YL/WT	Fuel Level Indicator Signal	1.7v (1/2 full)
13	PK	Flash EEPROM Power	0.1v
15	PK/LB	Data Bus (-) Signal	Digital Signals
16	TN/OR	Data Bus (+) Signal	Digital Signals
17-18	---	Not Used	---
21	GY	CKP Sensor (+) Signal	510-540 Hz
22	DB	CKP Sensor (-) Signal	510-540 Hz
24	BK/WT	Power Ground	<0.1v
25	BK	Chassis Ground	<0.1v
26	YL/BK	Coil Driver 1 Control	6° dwell
27	OR/YL	Shift Solenoid 'A' Control	12v, 55 mph: 1v
28	LB	Low Speed Fan Control	On: 1v, Off: 12v
30	---	Not Used	---
31	YL/LG	PSP Switch Signal	Straight: 0v, Turned: 12v
32	---	Not Used	---
33	PK/OR	VSS (-) Signal	0 Hz, 55 mph: 125 Hz
35	RD/BK	HO2S-12 (B1 S2) Signal	0.1-1.1v
36	TN/LB	MAF Sensor Return	<0.050v
37	OR/BK	TFT Sensor Signal	2.10-2.40v
38	LG/RD	ECT Sensor Signal	0.5-0.6v
39	GY	IAT Sensor Signal	1.5-2.5v
40	DG/YT	Fuel Pump Monitor	On: 12v, Off: 0v
41	DG/WT	A/C High Pressure Switch	AC On: 12v, Off: 0v
42-44	---	Not Used	---
45	PK/WT	EVAP CV Solenoid	0-10 Hz (0-100%)
46	LG/PK	High Speed Fan Control	On: 1v, Off: 12v
47	BR/PK	EGR VR Solenoid	0%, 55 mph: 45%
48	TN/YL	Clean Tachometer Output	42-50 Hz
49	LB/BK	Digital TR2 Sensor	In 'P': 0v, 55 mph: 11v
50	WT/BK	Digital TR4 Sensor	In 'P': 0v, 55 mph: 11v
51	BK/WT	Power Ground	<0.1v
52	YL/RD	Coil Driver 2 Control	6° dwell
53	PK/BK	Shift Solenoid 'C' Control	12v, 55 mph: 1v
54	DG/WT	TSS Sensor Signal	50-65 Hz (820-900 rpm)
55	DG/OR	Keep Alive Power	12-14v
56	GY/YL	EVAP Purge Solenoid	0-10 Hz (0-100%)
57	---	Not Used	---
58	GY/BK	VSS (+) Signal	0 Hz, 55 mph: 125 Hz
59	---	Not Used	---
60	GY/LB	HO2S-11 (B1 S1) Signal	0.1-1.1v
61	PK/LG	HO2S-22 (B2 S2) Signal	0.1-1.1v
62	RD/PK	FTP Sensor Signal	2.6v (0" H2O - cap off)
63	---	Not Used	---
64	LBYL	Digital TR3A Sensor	In 'P': 0v, in O/D: 1.7v

1998-99 Sable 3.0L V6 MFI VIN U (A/T) 104 Pin Connector

PCM Pin #	Wire Color	Circuit Description (104 Pin)	Value at Hot Idle
65	BR/LG	DPFE Sensor Signal	0.20-1.03v
67-68	---	Not Used	---
69	PK/YL	A/C WOT Relay Control	On: 1v, Off: 12v
70	---	Not Used	---
71	RD	Vehicle Power	12-14v
72	---	Not Used	---
73	TN/BK	Injector 5 Control	2.3-2.8 ms
74	BR/YL	Injector 3 Control	2.3-2.8 ms
75	TN	Injector 1 Control	2.3-2.8 ms
76, 77	BK/WT	Power Ground	<0.1v
78	YL/WT	Coil Driver 3 Control	6° dwell
80	LB/OR	Fuel Pump Control	On: 1v, Off: 12v
81	WT/YL	EPC Solenoid Control	10.6v (40 psi)
81-82	---	Not Used	---
83	WT/LB	IAC Motor Control	10v (32% duty cycle)
84	DG/WT	TSS Sensor Signal	50-65 Hz (790-820 rpm)
85	DB/OR	CID Sensor Signal	6-8 Hz
86	BK/YL	A/C Pressure Switch	A/C On: 12v (Open)
87	RD/LG	HO2S-21 (B2 S1) Signal	0.1-1.1v
88	LB/RD	MAF Sensor Signal	0.9v
89	GY/WT	TP Sensor Signal	0.53-1.27v
90	BR/WT	Reference Voltage	4.9-5.1v
91	GY/RD	Analog Signal Return	<0.050v
92	RD/LG	Brake Pedal Position Switch	Brake Off: 0v, On: 12v
93	RD/WT	HO2S-11 (B1 S1) Heater	On: 1v, Off: 12v
94	WT/BK	HO2S-21 (B2 S1) Heater	On: 1v, Off: 12v
95	YL/LB	HO2S-12 (B1 S2) Heater	On: 1v, Off: 12v
96	TN/YL	HO2S-22 (B2 S2) Heater	On: 1v, Off: 12v
97 ('98)	RD	Vehicle Power	12-14v
99	LG/OR	Injector 6 Control	2.3-2.8 ms
100	BR/LB	Injector 4 Control	2.3-2.8 ms
101	WT	Injector 2 Control	2.3-2.8 ms
102	---	Not Used	---
103	BK/WT	Power Ground	<0.1v
104	---	Not Used	---

Pin Connector Graphic

PCM 104-PIN CONNECTOR

Terminal View of 104-Pin PCM Wiring Harness Connector

Standard Colors and Abbreviations

Abbreviation	Color	Abbreviation	Color	Abbreviation	Color
BK	Black	GY	Gray	PK	Purple
BL	Blue	GN	Green	RD	Red
BR	Brown	LG	LT Green	TN	Tan
DB	Dark Blue	OR	Orange	WT	White
DG	DK Green	PK	Pink	YL	Yellow

2000 Sable 3.0L V6 MFI VIN U (A/T) 104 Pin Connector

PCM Pin #	Wire Color	Circuit Description (104 Pin)	Value at Hot Idle
1	VTN/OR	Shift Solenoid 'B' Control	1v, 55 mph: 12v
2	PK/LG	MIL (lamp) Control	MIL On: 1v, Off: 12v
3	YL/BK	Digital TR1 Sensor	In 'P': 0v, 55 mph: 11v
4	BN/PK	Start System Voltage	Cranking: 9-11v
5	WT/OR	Secondary Air Injection Relay	On: 1v, Off: 12v
6	DG/WT	TSS Sensor Signal	50-65 Hz (790-820 rpm)
7-11, 14	---	Not Used	---
12	YL/WT	Fuel Level Indicator Signal	1.7v (1/2 full)
13	VT	Flash EEPROM Power	0.1v
15	PK/LB	SCP Data Bus (-) Signal	Digital Signals
16	TN/OR	SCP Data Bus (+) Signal	Digital Signals
17	BN/Y	SCI Receive (RX) Signal	Digital Signals
18	GY/RD	SCI Transmit (TX) Signal	Digital Signals
19	---	Not Used	---
20	GY/YL	Generator Field Control	0 Hz
21	GY	CKP Sensor (+) Signal	510-540 Hz
22	DB	CKP Sensor (-) Signal	510-540 Hz
24	BK/WT	Power Ground	<0.1v
25	BK	Case Ground	<0.1v
26	YL/BK	Coil 1 Driver (Cyl 1 & 5)	6° dwell
27	OR/YL	Shift Solenoid 'A' Control	12v, 55 mph: 1v
28	LB	Low Speed Fan Control	On: 1v, Off: 12v
29	---	Not Used	---
30	DB/LG	Passive Anti-theft Indicator	Lamp On: 1v, Off: 12v
31	YL/LG	PSP Switch Signal	Straight: 0v, Turned: 12v
32	YL	Knock Sensor 1 (-) Signal	0v
33-34, 43	---	Not Used	---
35	RD/BK	HO2S-12 (B1 S2) Signal	0.1-1.1v
36	TN/LB	MAF Sensor Return	<0.050v
37	OR/BK	TFT Sensor Signal	2.10-2.40v
38	LG/RD	ECT Sensor Signal	0.5-0.6v
39	GY	IAT Sensor Signal	1.5-2.5v
40	DG, PK	Fuel Pump Monitor	On: 12v, Off: 0v
41	DG/WT	A/C High Pressure Switch	AC On: 12v, Off: 0v
42	LG/YL	Generator Cutoff Relay	0.1v (with fan on)
44	PK	Starter Relay Control	KOEC: 9-11v
45	LB/RD	Generator Sensor Signal	TBD
46	LG/VT	High Speed Fan Control	On: 1v, Off: 12v
47	BR/PK	EGR VR Solenoid	0%, 55 mph: 45%
48	TN/YL	Clean Tachometer Output	42-50 Hz
49	LB/BK	Digital TR2 Sensor	In 'P': 0v, 55 mph: 11v
50	WT/BK	Digital TR4 Sensor	In 'P': 0v, 55 mph: 11v
51	BK/WT	Power Ground	<0.1v
52	YL/RD	Coil Driver 2 (Cyl 3 & 4)	6° dwell
53	PK/BK	Shift Solenoid 'C' Control	12v, 55 mph: 1v
54	VT/YL	TCC Solenoid Control	12v (0%)
55	PK/LB	Keep Alive Power	12-14v
56	GY/YL	EVAP Purge Solenoid	0-10 Hz (0-100%)
57	YL/RD	Knock Sensor 1 (+) Signal	0v
60	GY/LB	HO2S-11 (B1 S1) Signal	0.1-1.1v
61	VT/LG	HO2S-22 (B2 S2) Signal	0.1-1.1v
62	RD/PK	FTP Sensor Signal	2.6v (0" H2O - cap off)
63	WT/YL	Injector Pressure Sensor	2.74v (39 psi)
64	LBYL	Digital TR3A Sensor	In 'P': 0v, in O/D: 1.7v

2000 Sable 3.0L V6 MFI VIN U (A/T) 104 Pin Connector

PCM Pin #	Wire Color	Circuit Description (104 Pin)	Value at Hot Idle
65	BN/LG	DPFE Sensor Signal	0.20-1.30v
66	TN/OR	Evaporator Air Temp. Sensor	4.95-5.15v
67	VT/WT	EVAP Purge Solenoid	0-10 Hz (0-100%)
68	GY/BK	VSS (+) Signal	0 Hz, 55 mph: 125 Hz
69	PK/YL	A/C WOT Relay Control	On: 1v, Off: 12v
70	BN	Secondary Air Injection Relay	On: 1v, Off: 12v
71	RD	Vehicle Power	12-14v
73	TN/BK	Injector 5 Control	2.3-2.8 ms
74	BR/YL	Injector 3 Control	2.3-2.8 ms
75	TN	Injector 1 Control	2.3-2.8 ms
76, 77	BK/WT	Power Ground	<0.1v
78	YL/WT	Coil 3 Driver (Cyl 2 & 6)	6° dwell
79	WT/LG	Check Transaxle Light	On: 1v, Off: 12v
80	LB/OR	Fuel Pump Control	On: 1v, Off: 12v
81	BR/OR	EPC Solenoid Control	10.6v (40 psi)
82	BK/WT	Check Cap Indicator Light	On: 1v, Off: 12v
83	WT/LB	IAC Motor Control	10v (32% duty cycle)
84	DB/YL	OSS Sensor Signal	0 Hz, 55 mph: 131 Hz
85	DB/OR	CID Sensor Signal	6-8 Hz
86	BK/YL	A/C Pressure Clutch Relay	A/C On: 12v, Off: 0v
87	RD/LG	HO2S-21 (B2 S1) Signal	0.1-1.1v
88	LB/RD	MAF Sensor Signal	0.9v
89	GY/WT	TP Sensor Signal	0.53-1.27v
90	BR/WT	Reference Voltage	4.9-5.1v
91	GY/RD	Analog Signal Return	<0.050v
92	RD/LG	Brake Pedal Position Switch	Brake Off: 0v, On: 12v
93	RD/WT	HO2S-11 (B1 S1) Heater	On: 1v, Off: 12v
94	WT/BK	HO2S-21 (B2 S1) Heater	On: 1v, Off: 12v
95	YL/LB	HO2S-12 (B1 S2) Heater	On: 1v, Off: 12v
96	TN/YL	HO2S-22 (B2 S2) Heater	On: 1v, Off: 12v
98	LG/RD	Generator Battery Indicator	Digital Signals
99	LG/OR	Injector 6 Control	2.3-2.8 ms
100	BR/LB	Injector 4 Control	2.3-2.8 ms
101	WT	Injector 2 Control	2.3-2.8 ms
102	---	Not Used	---
103	BK/WT	Power Ground	<0.1v
104	---	Not Used	---

Pin Connector Graphic

PCM 104-PIN CONNECTOR

Terminal View of 104-Pin PCM Wiring Harness Connector

Standard Colors and Abbreviations

Abbreviation	Color	Abbreviation	Color	Abbreviation	Color
BK	Black	GY	Gray	PK	Purple
BL	Blue	GN	Green	RD	Red
BR	Brown	LG	LT Green	TN	Tan
DB	Dark Blue	OR	Orange	WT	White
DG	DK Green	PK	Pink	YL	Yellow

2001-04 Sable 3.0L V6 MFI VIN U (A/T) 104 Pin Connector

PCM Pin #	Wire Color	Circuit Description (104 Pin)	Value at Hot Idle
1	VT/OR	Shift Solenoid 'B' Control	1v, 55 mph: 12v
2	PK/LG	MIL (lamp) Control	MIL On: 1v, Off: 12v
3	YL/BK	Digital TR1 Sensor	In 'P': 0v, 55 mph: 11v
4	TN/RD	System Voltage (Start-Run)	Cranking: 9-11v
5	---	Not Used	---
6	DG/WT	TSS Sensor Signal	50-65 Hz (790-820 rpm)
7-10	---	Not Used	---
12	YL/WT	Fuel Level Indicator Signal	1.7v (1/2 full)
13	VT	Flash EEPROM Power	0.1v
14, 19	---	Not Used	---
15	PK/LB	SCP Bus (-) Signal	Digital Signals
16	TN/OR	SCP Bus (+) Signal	Digital Signals
17	BR/YL	Passive Antitheft RX Signal	Digital Signals
18	GY/RD	Passive Antitheft TX Signal	Digital Signals
20	GY/YL	Generator Field Control	130 Hz (45%)
21	GY	CKP Sensor (+) Signal	510-540 Hz
22	DB	CKP Sensor (-) Signal	510-540 Hz
23	---	Not Used	---
24	BK/WT	Power Ground	<0.1v
25	BK	Case Ground	<0.1v
26	YL/BK	Coil 1 Driver (Cyl 1 & 5)	6° dwell
27	OR/YL	Shift Solenoid 'A' Control	12v, 55 mph: 1v
28	LB	Engine Cooling Fan Brake Relay Control	Off: 12v, On: 1v
29, 33-34	---	Not Used	---
30	DB/LG	Passive Antitheft Indicator	Lamp On: 1v, Off: 12v
31	YL/LG	PSP Switch Signal	Straight: 0v, Turned: 12v
32	YL	Knock Sensor 1 (-) Signal	<0.050v
35	RD/BK	HO2S-12 (B1 S2) Signal	0.1-1.1v
36	TN/LB	MAF Sensor Return	<0.050v
37	OR/BK	TFT Sensor Signal	2.10-2.40v
38	LG/RD	ECT Sensor Signal	0.5-0.6v
39	GY	IAT Sensor Signal	1.5-2.5v
40	DG/YL	Fuel Pump Monitor	On: 12v, Off: 0v
41	DG/WT	A/C High Pressure Switch	AC On: 12v, Off: 0v
42	LG/YL	Generator Cutoff Relay	0.1v (with fan on)
44	PK	Starter Relay Control	KOEC: 9-11v
45	LB/RD	Generator Sensor Signal	130 Hz (30%)
46	LG/VT	High Cooling Speed Fan Control	Off: 12v, On: 1v
47	BR/PK	EGR VR Solenoid	0%, 55 mph: 45%
48	TN/YL	Clean Tachometer Output	42-50 Hz
49	LB/BK	Digital TR2 Sensor	In 'P': 0v, 55 mph: 11v
50	WT/BK	Digital TR4 Sensor	In 'P': 0v, 55 mph: 11v
51	BK/WT	Power Ground	<0.1v
52	YL/RD	Coil Driver 2 (Cyl 3 & 4)	6° dwell
53	PK/BK	Shift Solenoid 'C' Control	12v, 55 mph: 1v
54	VT/YL	TCC Solenoid Control	12v (0%)
55	PK/LB	Keep Alive Power	12-14v
56	GY/YL	EVAP Purge Solenoid	0-10 Hz (0-100%)
57	YL/RD	Knock Sensor (+) Signal	0v
60	GY/LB	HO2S-11 (B1 S1) Signal	0.1-1.1v
61	VT/LG	HO2S-22 (B2 S2) Signal	0.1-1.1v
62	RD/PK	FTP Sensor Signal	2.6v (0" H2O - cap off)
63	WT/YL	Injector Pressure Sensor	TBD
64	RD/BK	Digital TR3A Sensor	In 'P': 0v, in O/D: 1.7v

2001-04 Sable 3.0L V6 MFI VIN U (A/T) 104 Pin Connector

PCM Pin #	Wire Color	Circuit Description (104 Pin)	Value at Hot Idle
65	BR/LG	DPFE Sensor Signal	0.95-1.05v
66	TN/OR	Evaporator Air Temperature Sensor	4.95-5.15v
67	VT/WT	EVAP Purge Solenoid	0-10 Hz (0-100%)
68	GY/BK	VSS (+) Signal	Moving: AC pulse signals
69	PK/YL	A/C WOT Relay Control	Off: 12v, On: 1v
70	---	Not Used	---
71	RD	Vehicle Power (Start-Run)	12-14v
73	TN/BK	Injector 5 Control	2.3-2.8 ms
74	BR/YL	Injector 3 Control	2.3-2.8 ms
75	TN	Injector 1 Control	2.3-2.8 ms
76, 77	BK/WT	Power Ground	<0.1v
78	YL/WT	Coil 3 Driver (Cyl 2 & 6)	6° dwell
79	WT/LG	Check Transaxle Light	Indicator Off: 12v, On: 1v
80	LB/OR	Fuel Pump Control	Off: 12v, On: 1v
81	BR/OR	EPC Solenoid Control	10.6v (40 psi)
82	BK/WT	Check Cap Indicator Light	Indicator Off: 12v, On: 1v
83	WT/LB	IAC Motor Control	10v (32% duty cycle)
84	DB/YL	OSS Sensor Signal	0 Hz, 55 mph: 131 Hz
85	DB/OR	CID Sensor Signal	6-8 Hz
86	BK/YL	A/C Pressure Clutch Relay	Relay Off: 12v, On: 1v
87	RD/LG	HO2S-21 (B2 S1) Signal	0.1-1.1v
88	LB/RD	MAF Sensor Signal	0.9v
89	GY/WT	TP Sensor Signal	0.53-1.27v
90	BR/WT	Reference Voltage	4.9-5.1v
91	GY/RD	Sensor Return	<0.050v
92	RD/LG	Brake Pedal Position Switch	Brake Off: 0v, On: 12v
93	RD/WT	HO2S-11 (B1 S1) Heater	On: 1v, Off: 12v
94	WT/BK	HO2S-21 (B2 S1) Heater	On: 1v, Off: 12v
95	YL/LB	HO2S-12 (B1 S2) Heater	On: 1v, Off: 12v
96	TN/YL	HO2S-22 (B2 S2) Heater	On: 1v, Off: 12v
97	RD	Vehicle Power (Start-Run)	12-14v
98	LG/RD	Generator Battery Indicator	Digital Signals
99	LG/OR	Injector 6 Control	2.3-2.8 ms
100	BR/LB	Injector 4 Control	2.3-2.8 ms
101	WT	Injector 2 Control	2.3-2.8 ms
102	---	Not Used	---
103	BK/WT	Power Ground	<0.1v
104	---	Not Used	---

Pin Connector Graphic

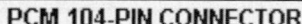

```
        PCM 104-PIN CONNECTOR
   1 ●●●●●●●●●●●●   ●●●●●●●●●●●●● 26
  27 ●●●●●●●●●●●●●      ●●●●●●●●●●●●● 52
  53 ●●●●●●●●●●●●●  ⬤  ●●●●●●●●●●●●● 78
  79 ●●●●●●●●●●●●●      ●●●●●●●●●●●●●● 104
```

Terminal View of 104-Pin PCM Wiring Harness Connector

Standard Colors and Abbreviations

Abbreviation	Color	Abbreviation	Color	Abbreviation	Color
BK	Black	GY	Gray	PK	Purple
BL	Blue	GN	Green	RD	Red
BR	Brown	LG	LT Green	TN	Tan
DB	Dark Blue	OR	Orange	WT	White
DG	DK Green	PK	Pink	YL	Yellow

1990 Sable 3.8L V6 2v MFI VIN 4 (A/T) 60 Pin Connector

PCM Pin #	Wire Color	Circuit Description (60 Pin)	Value at Hot Idle
1	YL	Keep Alive Power	12-14v
2	RD/LG	PSP Switch Signal	Straight: 0v, Turned: 11v
3	DG/WT	VSS (+) Signal	0 Hz, 55 mph: 125 Hz
4	DG/YT	Ignition Diagnostic Monitor	20-31 Hz
5	---	Not Used	---
6	OR/YL	VSS (-) Signal	0 Hz, 55 mph: 125 Hz
7	LG/YL	ECT Sensor Signal	0.5-0.6v
8	BK/OR	Data Bus (-) Signal	Digital Signals
9	GY/WT	AXOD Solenoid	On: 1v, Off: 12v
10	PK/BL	A/C Cycling Clutch Signal	A/C Off: 0v, On: 12v
10	PK	A/C Cycling Clutch Signal	A/C Off: 0v, On: 12v
11	OR/YL	Speed Control Solenoid (+)	S/C On: 12-14v
12	BR/YL	Injector 3 Control	6.0-6.8 ms
13	BR/BL	Injector 4 Control	6.0-6.8 ms
14	TN/BL	Injector 5 Control	6.0-6.8 ms
15	LG	Injector 6 Control	6.0-6.8 ms
16	GY	Ignition System Ground	<0.050v
17	TN/RD	Self-Test Output, MIL Control	MIL Off: 12v, On: 1v
18	DG/PK	Transmission 3-4 Switch	3-4 Switch Closed: 0.1v
19	OR/YL	Transmission 2-3 Switch	2-3 Switch Closed: 0.1v
20 ('89)	BK	Case Ground	<0.050v
21	OR/BK	IAC Motor Control	9-10v
22	TN/LG	Fuel Pump Control	Relay Off: 12v, On: 1v
23-24	---	Not Used	---
25	LG/PK	ACT Sensor Signal	1.5-2.5v
26	OR/WT	Reference Voltage	4.9-5.1v
27	BR/LG	PFE Sensor Signal	3.2v, 55 mph: 2.9v
28	BL/BK	Speed Control Switch Signal	S/C On: 6.7v
29	DB/LG	HO2S-21 (B1 S1) Signal	0.1-1.1v
30	PK/YL	Neutral Drive Switch	In 'N': 0v, Others: 4.6v
31	GY/YL	Canister Purge Solenoid	12v, Cruise: 1v
32	---	Not Used	---
33	DG	VR Solenoid Control	0%, 55 mph: 45%
34	BL/PK	Data Output Link	Digital Signals
35	WT/PK	S/C Vent Solenoid Control	Vacuum Decreasing: 1v
36	YL/LG	Spark Output Signal	6.93v (duty cycle)
37	RD	Vehicle Power	12-14v
38	---	Not Used	---
39	OR	Speed Control Switch Ground	<0.050v
40	BK/LG	Power Ground	<0.1v

1990 Sable 3.8L V6 2v MFI VIN 4 (A/T) 60 Pin Connector

PCM Pin #	Wire Color	Circuit Description (60 Pin)	Value at Hot Idle
41	PK	High Speed Fan Control	Fan On: 1v, Off: 12v
42	---	Not Used	---
43	DG/PK	HO2S-11 (B1 S1) Signal	0.1-1.1v
44	OR/BK	Data Bus (+) Signal	Digital Signals
45	LG/BK	MAP Sensor Signal	107 Hz (sea level)
46	BK/WT	Analog Signal Return	<0.050v
47	DG/LG	TP Sensor Signal	0.8v, 55 mph: 1.1v
48	WT/BK	Self-Test Input Signal	STI On: 1v, Off: 5v
49	OR	HO2S-11 (Bank 1) Ground	<0.050v
50	PK/BK	Fuel Pump Monitor	Relay Off: 12v, On: 1v
51	GY/BK	S/C Vacuum Solenoid Control	Vacuum Increasing: 1v
52	---	Not Used	---
53	TN/BL	AXOD Lockup Solenoid	12v, 55 mph: 1v
54	RD	A/C WOT Relay Control	Relay Off: 12v, On: 1v
55	TN/OR	Low Speed Fan Control	Fan On: 1v, Off: 12v
56	DB	PIP Sensor Signal	6.93v (50% d/cycle)
57	RD	Vehicle Power	12-14v
58	TN	Injector 1 Control	6.0-6.8 ms
59	WT	Injector 2 Control	6.0-6.8 ms
60	BK/LG	Power Ground	<0.1v

Pin Connector Graphic

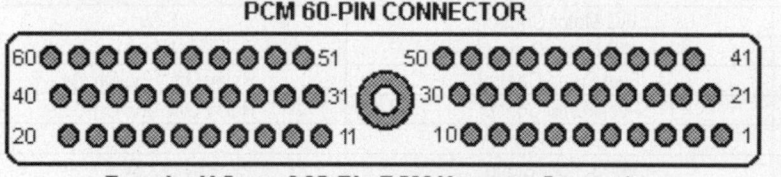

Standard Colors and Abbreviations

Abbreviation	Color	Abbreviation	Color	Abbreviation	Color
BK	Black	GY	Gray	PK	Purple
BL	Blue	GN	Green	RD	Red
BR	Brown	LG	LT Green	TN	Tan
DB	Dark Blue	OR	Orange	WT	White
DG	DK Green	PK	Pink	YL	Yellow

1991-95 Sable 3.8L V6 2v MFI VIN 4 (A/T) 60 Pin Connector

PCM Pin #	Wire Color	Circuit Description (60 Pin)	Value at Hot Idle
1	YL	Keep Alive Power	12-14v
2	RD/LG	Brake Switch Signal	Brake Off: 0v, On: 12v
3	DG/WT	VSS (-) Signal	0 Hz, 55 mph: 125 Hz
4	WT/PK	Ignition Diagnostic Monitor	20-31 Hz
4	DG/YT	Ignition Diagnostic Monitor	20-31 Hz
5	GY/BK	TSS Sensor Signal	100 Hz (700 rpm)
6	OR/YL	VSS (+) Signal	0 Hz, 55 mph: 125 Hz
7	LG/RD	ECT Sensor Signal	0.5-0.6v
8	PK/BK	Fuel Pump Monitor	12v, Off: 0v
9	TN/BL	MAF Sensor Return	<0.050v
10	PK	A/C Cycling Clutch Signal	A/C Off: 0v, On: 12v
10	PK/BL	A/C Cycling Clutch Signal	A/C Off: 0v, On: 12v
11	GN/YL	Canister Purge Solenoid	12v, Cruise: 1v
12	LG/OR	Injector 6 Control	5.6-6.0 ms
13	TN/OR	Low Speed Fan Control	Fan On: 1v, Off: 12v
14	---	Not Used	---
15	TN/BK	Injector 5 Control	5.6-6.0 ms
16	GY	Ignition System Ground	<0.050v
17	TN/RD	Self-Test Output, MIL Control	MIL Off: 12v, On: 1v
18	OR/BK	Data Bus (+) Signal	Digital Signals
19	PK/BL	Data Bus (-) Signal	Digital Signals
20	BK	Case Ground	<0.050v
21	BL/OR	IAC Motor Control	8-9v
21	BR/WT	IAC Motor Control	8-9v
22	TN/LG	Fuel Pump Control	Relay Off: 12v, On: 1v
22	BL/OR	Fuel Pump Control	Relay Off: 12v, On: 1v
23-24	---	Not Used	---
25	GY	IAT Sensor Signal	1.5-2.5v
26	BR/WT	Reference Voltage	4.9-5.1v
27	BR/LG	PFE Sensor Signal	3.2v, 55 mph: 2.9v
28	YL/LG	PSP Switch Signal	Straight: 0v, Turned: 11v
29	---	Not Used	---
30	BL	MLP Sensor Signal	In 'P': 0v, O/D: 1.7v
30	BL/YL	MLP Sensor Signal	In 'P': 0v, O/D: 1.7v
31	LG/PK	High Speed Fan Control	Fan On: 1v, Off: 12v
32	---	Not Used	---
33	BR/PK	VR Solenoid Control	0%, 55 mph: 45%
34	BL/PK	Data Output Link	Digital Signals
35	BR/BL	Injector 4 Control	5.6-6.0 ms
36	YL/LG	Spark Output Signal	6.93v (50% d/cycle)
37	RD	Vehicle Power	12-14v
38	WT/YL	EPC Solenoid Control	7.8v (8-9 psi)
39	BR/YL	Injector 3 Control	5.6-6.0 ms
40	BK/LG	Power Ground	<0.1v

1991-95 Sable 3.8L V6 2v MFI VIN 4 (A/T) 60 Pin Connector

PCM Pin #	Wire Color	Circuit Description (60 Pin)	Value at Hot Idle
41	---	Not Used	---
42 ('94-'95)	TN/LG	Refrigerant Containment Sw.	5v
43	RD/BK	HO2S-11 (B1 S1) Signal	0.1-1.1v
44	DB/LG	HO2S-21 (B1 S1) Signal	0.1-1.1v
44	GY/BL	HO2S-21 (B1 S1) Signal	0.1-1.1v
45	---	Not Used	---
46	GY/RD	Analog Signal Return	<0.050v
47	GN/WT	TP Sensor Signal	0.8v, 55 mph: 1.1v
48	BR	Self-Test Input Signal	STI On: 1v, Off: 5v
49	OR/BK	TOT Sensor Signal	2.40-2.10v
50	BL/RD	MAF Sensor Signal	0.7v, 55 mph: 2.0v
51	OR/YL	Shift Solenoid 1 Control	12v, 55 mph: 1v
52	PK/OR	Shift Solenoid 2 Control	1v, 55 mph: 12v
53	TN/WT	TCC Solenoid Control	12v, 55 mph: 1v
54	PK/, R	A/C WOT Relay Control	Relay Off: 12v, On: 1v
55	WT/RD	Shift Solenoid 3 Control	12v, 55 mph: 1v
56	DB	PIP Sensor Signal	6.93v (50% d/cycle)
56	GY/OR	PIP Sensor Signal	6.93v (50% d/cycle)
57	RD	Vehicle Power	12-14v
58	TN	Injector 1 Control	5.6-6.0 ms
59	WT	Injector 2 Control	5.6-6.0 ms
60	BK/LG	Power Ground	<0.1v

Pin Connector Graphic

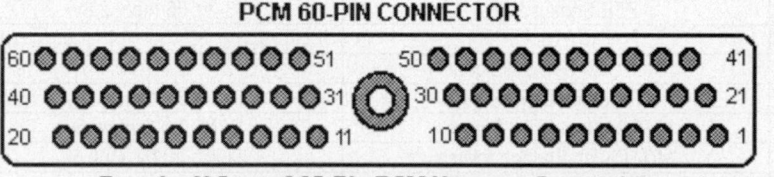

Standard Colors and Abbreviations

Abbreviation	Color	Abbreviation	Color	Abbreviation	Color
BK	Black	GY	Gray	PK	Purple
BL	Blue	GN	Green	RD	Red
BR	Brown	LG	LT Green	TN	Tan
DB	Dark Blue	OR	Orange	WT	White
DG	DK Green	PK	Pink	YL	Yellow

TOPAZ PIN TABLES

1990 Topaz 2.3L I4 MFI VIN S, VIN X (All) 60 Pin Connector

PCM Pin #	Wire Color	Circuit Description (60 Pin)	Value at Hot Idle
1	YL	Keep Alive Power	12-14v
2	---	Not Used	---
3	DG/WT	VSS (+) Signal	0 Hz, 55 mph: 125 Hz
4	DG/YT	Ignition Diagnostic Monitor	20-31 Hz
5	RD/LG	Key Power	12-14v
6	OR/YL	VSS (-) Signal	0 Hz, 55 mph: 125 Hz
7	LG/YL	ECT Sensor Signal	0.5-0.6v
8	PK/BK	Fuel Pump Monitor	12v, Off: 0v
9	---	Not Used	---
10	TN/YL	A/C Cycling Clutch Signal	A/C Off: 0v, On: 12v
11-15	---	Not Used	---
16	BK/OR	Ignition System Ground	<0.050v
16	BK/WT	Ignition System Ground	<0.050v
17	TN/RD	Self-Test Output, MIL Control	MIL Off: 12v, On: 1v
18-19	---	Not Used	---
20	BK	Case Ground	<0.050v
21	BR/WT	IAC Motor Control	8-10v
22	OR/BL	Fuel Pump Control	Relay Off: 12v, On: 1v
23	---	Not Used	---
24	YL/LG	PSP Switch Signal	Straight: 0v, Turned: 5v
25	LG/PK	ACT Sensor Signal	1.5-2.5v
26	OR/WT	Reference Voltage	4.9-5.1v
27	BR/LG	PFE Sensor Signal	3.2v, 55 mph: 2.8v
28	---	Not Used	---
29	DG/PK	HO2S-11 (B1 S1) Signal	0.1-1.1v
30	GY/OR	A/T: Neutral Drive Switch	In 'N': 0v, Others: 5v
30	WT/PK	A/T: Neutral Drive Switch	In 'N': 0v, Others: 5v
30	GY/OR	M/T: Clutch Engage Switch	Clutch In: 0v, Out: 5v
30	WT/PK	M/T: Clutch Engage Switch	Clutch In: 0v, Out: 5v
31	GY/YL	Canister Purge Solenoid	12v, 55 mph: 1v
32	---	Not Used	---
33	YL	VR Solenoid Control	0%, 55 mph: 0-45%
34-35	---	Not Used	---
36	YL/LG	Spark Output Signal	6.93v (50% d/cycle)
37	RD	Vehicle Power	12-14v
38-39	---	Not Used	---
40	BK/LG	Power Ground	<0.1v

1990 Topaz 2.3L I4 MFI VIN S, VIN X (All) 60 Pin Connector

PCM Pin #	Wire Color	Circuit Description (60 Pin)	Value at Hot Idle
41-42	---	Not Used	---
43	LG/PK	A/C Demand Switch Signal	A/C On: 12v, Off: 0v
44	---	Not Used	---
45	LG/BK	MAP Sensor Signal	107 Hz (sea level)
46	BK/WT	Analog Signal Return	<0.050v
47	DG/LG	TP Sensor Signal	0.8v, 55 mph: 1.1v
48	WT/RD	Self-Test Input Signal	STI On: 1v, Off: 5v
49	OR	HO2S-11 (Bank 1) Ground	<0.050v
50	---	Not Used	---
51	WT/RD	Air Management 1 Solenoid	AM1 On: 1v, Off: 12v,
52	---	Not Used	---
53	TN/BL	M/T: Shift Indicator Light	SIL On: 1v, Off: 12v
54	OR/LB	A/C WOT Relay Control	Relay Off: 12v, On: 1v
55	---	Not Used	---
56	DB	PIP Sensor Signal	6.93v (50% d/cycle)
57	RD	Vehicle Power	12-14v
58	TN/RD	Injector Bank 1	3.8-3.9 ms
59	TN/OR	Injector Bank 2	3.8-3.9 ms
60	BK/LG	Power Ground	<0.1v

Pin Connector Graphic

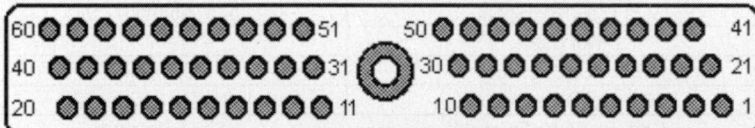

PCM 60-PIN CONNECTOR

60 — 51   50 — 41
40 — 31   30 — 21
20 — 11   10 — 1

Terminal View of 60-Pin PCM Harness Connector

Standard Colors and Abbreviations

Abbreviation	Color	Abbreviation	Color	Abbreviation	Color
BK	Black	GY	Gray	PK	Purple
BL	Blue	GN	Green	RD	Red
BR	Brown	LG	LT Green	TN	Tan
DB	Dark Blue	OR	Orange	WT	White
DG	DK Green	PK	Pink	YL	Yellow

1991 Topaz 2.3L I4 MFI VIN S, VIN X (All) 60 Pin Connector

PCM Pin #	Wire Color	Circuit Description (60 Pin)	Value at Hot Idle
1	YL	Keep Alive Power	12-14v
2	---	Not Used	---
3	DG/WT	VSS (+) Signal	0 Hz, 55 mph: 125 Hz
4	TN/YL	Ignition Diagnostic Monitor	20-31 Hz
5	---	Not Used	---
6	OR/YL	VSS (-) Signal	0 Hz, 55 mph: 125 Hz
7	LG/RD	ECT Sensor Signal	0.5-0.6v
8	PK/BK	Fuel Pump Monitor	12v, Off: 0v
9	---	Not Used	---
10	DG	A/C Cycling Clutch Signal	A/C Off: 0v, On: 12v
10	BK/YL	A/C Cycling Clutch Signal	A/C Off: 0v, On: 12v
11-15	---	Not Used	---
16	OR/RD	Ignition System Ground	<0.050v
17	TN/RD	Self-Test Output, MIL Control	MIL Off: 12v, On: 1v
18-19	---	Not Used	---
20	BK	Case Ground	<0.050v
21	BR/WT	IAC Motor Control	8-10v
22	DG/YT	Fuel Pump Control	Relay Off: 12v, On: 1v
23	---	Not Used	---
24	YL/LG	PSP Switch Signal	Straight 0v, Turned 11v
25	LG/PK	ACT Sensor Signal	1.5-2.5v
26	BR/WT	Reference Voltage	4.9-5.1v
27	BR/LG	PFE Sensor Signal	3.2v, 55 mph: 2.9v
28	---	Not Used	---
29	RD/BK	HO2S-11 (B1 S1) Signal	0.1-1.1v
30	WT/PK	Neutral Drive Switch	In 'N': 0v, Others: 12v
31	GY/YL	Canister Purge Solenoid	12v, 55 mph: 1v
32	---	Not Used	---
33	YL	VR Solenoid Control	0%, 55 mph: 0-45%
34-35	---	Not Used	---
36	YL/LG	Spark Output Signal	6.93v (50% d/cycle)
37	RD	Vehicle Power	12-14v
38-39	---	Not Used	---
40	BK/LG	Power Ground	<0.1v

1991 Topaz 2.3L I4 MFI VIN S, VIN X (All) 60 Pin Connector

PCM Pin #	Wire Color	Circuit Description (60 Pin)	Value at Hot Idle
41-42	---	Not Used	---
43	PK	A/C Demand Switch Signal	A/C On: 12v, AC Off: 0v
44	---	Not Used	---
45	LG/BK	MAP Sensor Signal	107 Hz (sea level)
46	GY/RD	Analog Signal Return	<0.050v
47	GN/WT	TP Sensor Signal	0.8v, 55 mph: 1.1v
48	WT/PK	Self-Test Input Signal	STI On: 1v, Off: 5v
49	OR	HO2S-11 (Bank 1) Ground	<0.050v
50	---	Not Used	---
51	WT/RD	Air Management 1 Solenoid	AM1 On: 1v, Off: 12v
52	---	Not Used	---
53	TN/WT	M/T: Shift Indicator Light	SIL On: 1v, Off: 12v
54	OR/BL	A/C WOT Relay Control	Relay Off: 12v, On: 1v
55	---	Not Used	---
56	DB	PIP Sensor Signal	6.93v (50% d/cycle)
57	RD	Vehicle Power	12-14v
58	TN/RD	Injector Bank 1	3.6-5.0 ms
59	TN/OR	Injector Bank 2	3.6-5.0 ms
60	BK/LG	Power Ground	<0.1v

Pin Connector Graphic

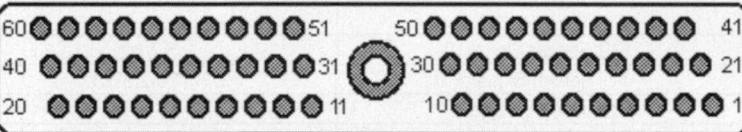

Terminal View of 60-Pin PCM Harness Connector

Standard Colors and Abbreviations

Abbreviation	Color	Abbreviation	Color	Abbreviation	Color
BK	Black	GY	Gray	PK	Purple
BL	Blue	GN	Green	RD	Red
BR	Brown	LG	LT Green	TN	Tan
DB	Dark Blue	OR	Orange	WT	White
DG	DK Green	PK	Pink	YL	Yellow

1992-94 Topaz 2.3L I4 MFI VIN X (All) 60 Pin Connector

PCM Pin #	Wire Color	Circuit Description (60 Pin)	Value at Hot Idle
1	YL	Keep Alive Power	12-14v
2	---	Not Used	---
3	DG/WT	VSS (+) Signal	0 Hz, 55 mph: 125 Hz
4	WT/PK	Ignition Diagnostic Monitor	20-31 Hz
5	---	Not Used	---
6	OR/YL	VSS (-) Signal	0 Hz, 55 mph: 125 Hz
7	LG/RD	ECT Sensor Signal	0.5-0.6v
8	PK/BK	Fuel Pump Monitor	12v, Off: 0v
9	TN/BL	MAF Sensor Return	<0.050v
10	PK/BL	A/C Cycling Clutch Signal	A/C Off: 0v, On: 12v
11	GN/YL	Canister Purge Solenoid	12v, 55 mph: 1v
12	---	Not Used	---
13	TN/OR	Low Speed Fan Control	Fan On: 1v, Off: 12v
14-15	---	Not Used	---
16	OR/RD	Ignition System Ground	<0.050v
17	TN/RD	Self-Test Output, MIL Control	MIL Off: 12v, On: 1v
18	TN/OR	Data Bus (+) Signal	Digital Signals
19	PK/BL	Data Bus (-) Signal	Digital Signals
20	BK	Case Ground	<0.050v
21	WT	IAC Motor Control	8-10v
21	WT/BL	IAC Motor Control	8-10v
22	DG/YT	Fuel Pump Control	Relay Off: 12v, On: 1v
23	---	Not Used	---
24	DG	CID Sensor Signal	5-7 Hz
25	GY	IAT Sensor Signal	1.5-2.5v
26	BR/WT	Reference Voltage	4.9-5.1v
27	BR/LG	PFE Sensor Signal	3.2v, 55 mph: 2.9v
28	YL/LG	PSP Switch Signal	Straight: 0v, Turned: 11v
29	---	Not Used	---
30	PK/YL	A/T: Neutral Drive Switch	In 'N': 0v, Others: 5v
30	PK/YL	M/T: Clutch Engage Switch	Clutch In: 0v, Out: 5v
31	WT/RD	Air Mgmt. Bypass Solenoid	0v, at 55 mph: 12v
32	---	Not Used	---
33	YL	VR Solenoid Control	0%, 55 mph: 45%
34	---	Not Used	---
35	BR/BL	Injector 4 Control	5.1-5.3 ms
36	PK	Spark Output Signal	6.93v (50% d/cycle)
37	RD	Vehicle Power	12-14v
38	---	Not Used	---
39	BR/YL	Injector 3 Control	5.1-5.3 ms
40	BK/LG	Power Ground	<0.1v

1992-94 Topaz 2.3L I4 MFI VIN X (All) 60 Pin Connector

PCM Pin #	Wire Color	Circuit Description (60 Pin)	Value at Hot Idle
41-43	---	Not Used	---
44	RD/BK	HO2S-11 (B1 S1) Signal	0.1-1.1v
45	---	Not Used	---
46	GY/RD	Analog Signal Return	<0.050v
47	GN/WT	TP Sensor Signal	0.8v, 55 mph: 1.1v
48	WT/PK	Self-Test Input Signal	STI On: 1v, Off: 5v
49	---	Not Used	---
50	LB/RD	MAF Sensor Signal	0.8v
50	BL/BK	MAF Sensor Signal	0.8v
51-52	---	Not Used	---
53	TN/WT	M/T: Shift Indicator Light	SIL On: 1v, Off: 12v
54	PK/YL	A/C WOT Relay Control	Relay Off: 12v, On: 1v
55	---	Not Used	---
56	DB	PIP Sensor Signal	6.93v (50% d/cycle)
57	RD	Vehicle Power	12-14v
58	TN	Injector 1 Control	5.1-5.3 ms
59	WT	Injector 2 Control	5.1-5.3 ms
60	BK/LG	Power Ground	<0.1v

Pin Connector Graphic

PCM 60-PIN CONNECTOR

Terminal View of 60-Pin PCM Harness Connector

Standard Colors and Abbreviations

Abbreviation	Color	Abbreviation	Color	Abbreviation	Color
BK	Black	GY	Gray	PK	Purple
BL	Blue	GN	Green	RD	Red
BR	Brown	LG	LT Green	TN	Tan
DB	Dark Blue	OR	Orange	WT	White
DG	DK Green	PK	Pink	YL	Yellow

1992-94 Topaz 3.0L V6 MFI VIN U (All) 60 Pin Connector

PCM Pin #	Wire Color	Circuit Description (60 Pin)	Value at Hot Idle
1	YL	Keep Alive Power	12-14v
2	---	Not Used	---
3	DG/WT	VSS (+) Signal	0 Hz, 55 mph: 125 Hz
4	WT/PK	Ignition Diagnostic Monitor	20-31 Hz
5	---	Not Used	---
6	OR/YL	VSS (-) Signal	0 Hz, 55 mph: 125 Hz
7	LG/RD	ECT Sensor Signal	0.5-0.6v
8	PK/BK	Fuel Pump Monitor	12v, Off: 0v
9	TN/BL	MAF Sensor Return	<0.050v
10	PK/BL	A/C Cycling Clutch Signal	A/C Off: 0v, On: 12v
11	GN/YL	Canister Purge Solenoid	12v, 55 mph: 1v
12	LG/OR	Injector 6 Control	4.6-5.3 ms
13	TN/OR	Low Speed Fan Control	Fan On: 1v, Off: 12v
14	---	Not Used	---
15	TN/BK	Injector 5 Control	4.6-5.3 ms
16	OR/RD	Ignition System Ground	<0.050v
17	TN/RD	Self-Test Output, MIL Control	MIL Off: 12v, On: 1v
18	TN/OR	Data Bus (+) Signal	Digital Signals
19	PK/BL	Data Bus (-) Signal	Digital Signals
20	BK	Case Ground	<0.050v
21	WT	IAC Motor Control	8-10v
21	WT/BK	IAC Motor Control	8-10v
22	DG/YT	Fuel Pump Control	Relay Off: 12v, On: 1v
23-24	---	Not Used	---
25	GY	IAT Sensor Signal	1.5-2.5v
26	BR/WT	Reference Voltage	4.9-5.1v
27	BR/LG	PFE Sensor Signal	3.2v, 55 mph: 2.9v
28	YL/LG	PSP Switch Signal	Straight: 0v, Turned: 11v
29	---	Not Used	---
30	PK/YL	A/T: Neutral Drive Switch	In 'N': 0v, Others: 5v
30	PK/YL	M/T: Clutch Position Switch	Clutch In: 0v, Out: 5v
31	BL	High Speed Fan Control	Fan On: 1v, Off: 12v
32	---	Not Used	---
33	YL	VR Solenoid Control	0%, 55 mph: 45%
34	---	Not Used	---
35	BR/BL	Injector 4 Control	4.6-5.3 ms
36	PK	Spark Output Signal	6.93v (50% d/cycle)
37	RD	Vehicle Power	12-14v
38	---	Not Used	---
39	BR/YL	Injector 3 Control	4.6-5.3 ms
40	BK/LG	Power Ground	<0.1v

1992-94 Topaz 3.0L V6 MFI VIN U (All) 60 Pin Connector

PCM Pin #	Wire Color	Circuit Description (60 Pin)	Value at Hot Idle
41	---	Not Used	---
42 ('94)	TN/LG	A/C Cycling Clutch Signal	A/C Off: 0v, On: 12v
43	RD/BK	HO2S-11 (B1 S1) Signal	0.1-1.1v
44	GY/BL	HO2S-21 (B1 S1) Signal	0.1-1.1v
45	---	Not Used	---
46	GY/RD	Analog Signal Return	<0.050v
47	GN/WT	TP Sensor Signal	0.5v-1.2v
48	WT/PK	Self-Test Input Signal	STI On: 1v, Off: 5v
49	---	Not Used	---
50	BL/RD	MAF Sensor Signal	0.8v
50	BL/BK	MAF Sensor Signal	0.8v
51-53	---	Not Used	---
54	PK/YL	A/C WOT Relay Control	Relay Off: 12v, On: 1v
55	---	Not Used	---
56	DB	PIP Sensor Signal	6.93v (50% d/cycle)
57	RD	Vehicle Power	12-14v
58	TN	Injector 1 Control	4.6-5.3 ms
59	WT	Injector 2 Control	4.6-5.3 ms
60	BK/LG	Power Ground	<0.1v

Pin Connector Graphic

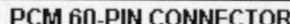

Standard Colors and Abbreviations

Abbreviation	Color	Abbreviation	Color	Abbreviation	Color
BK	Black	GY	Gray	PK	Purple
BL	Blue	GN	Green	RD	Red
BR	Brown	LG	LT Green	TN	Tan
DB	Dark Blue	OR	Orange	WT	White
DG	DK Green	PK	Pink	YL	Yellow

TRACER PIN TABLES

1991-92 Tracer 1.8L I4 MFI VIN 8 (All) 22 Pin Connector

PCM Pin #	Wire Color	Circuit Description (22 Pin)	Value at Hot Idle
22-A	BL/RD	Keep Alive Power	12-14v
22-B	WT/RD	Vehicle Power	12-14v
22-C	PK	Start Signal	KOEC: 9-11v
22-D	WT/YL	Switch Monitor Lamp Signal	SML On: 5v, Off: 12v
22-E	YL/BK	MIL (lamp) Control	MIL On: 2.5v, Off: 12v
22-F	WT/BK	Self-Test Output, MIL Control	STO On: 5v, Off: 12v
22-G	GN/WT	Spark Output Signal	6.93v (50% d/cycle)
22-H	---	Not Used	---
22-I	---	Not Used	---
22-J	BL/BK	A/C WOT Relay Control	A/C On: 12v, Off: 2.5v
22-K	LG/YL	Self-Test Input Signal	STI On: 0v, Off: 5v
22-L	BR/WT	DRL (Canada)	DRL Off: 12.5v, On: 2.5v
22-M	---	Not Used	---
22-N	RD/WT	IDL Switch Signal	Closed: 0v, Open: 12v
22-O	GN	M/T: Brake Switch Signal	Brake Off: 0v, On: 12v
22-O	GN/RD	4EAT A/T: Down Shift Signal	DSS Off: 1v, On: 12v
22-P	BL/YL	PSP Switch Signal	Straight: 0v, Turned: 12v
22-Q	GN/BK	A/C Switch Signal	A/C On: 2.5v, Off: 12v
22-R	BK/GN	Cooling Fan Control	Fan On: 1v, Off: 12v
22-S	OR/BL	Blower Motor Switch Signal	Off or 1st: 12v, 2nd>: 1v
22-T	BK/BL	Rear Defroster Switch Signal	Switch Off: 1v, On: 12v
22-U	RD/BK	Headlamp Switch Signal	HDL On: 12v, Off: 1v
22-V	BR/YL	M/T: Neutral Drive Switch	In 'N': 0v, Others: 11v
22-V	BR/YL	M/T: Clutch Engage Switch	Clutch In: 0v, Out: 12v
22-V	BK/BL	A/T: MLP Switch Signal	In 'N': 0v, Others: 12v

Pin Connector Graphic

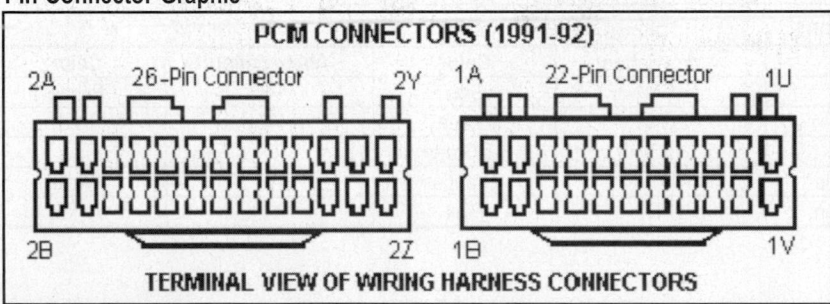

Standard Colors and Abbreviations

Abbreviation	Color	Abbreviation	Color	Abbreviation	Color
BK	Black	GY	Gray	PK	Purple
BL	Blue	GN	Green	RD	Red
BR	Brown	LG	LT Green	TN	Tan
DB	Dark Blue	OR	Orange	WT	White
DG	DK Green	PK	Pink	YL	Yellow

1991-92 Tracer 1.8L I4 MFI VIN 8 (All) 26 Pin Connector

PCM Pin #	Wire Color	Circuit Description (26 Pin)	Value at Hot Idle
26-A	BK/OR	Power Ground	<0.1v
26-B	BK/OR	Power Ground	<0.1v
26-C	BK/LG	Power Ground	<0.1v
26-D	BK/WT	Analog Signal Return	<0.050v
26-E	WT	CKP Sensor Signal	400-425 Hz
26-F	---	Not Used	---
26-G	YL/BL	CID Sensor Signal	5-7 Hz
26-H	BK	Power Ground (California)	<0.1v
26-H	BK/YL	Vehicle Power (Canada)	12-14v
26-I	---	Not Used	---
26-J	WT/RD	A/T: Vehicle Power Relay	12-14v
26-K	LG/RD	Reference Voltage	4.9-5.1v
26-L	LG/WT	M/T: WOT Switch Signal	12v, 55 mph: 12v
26-M	LG/WT	4EAT A/T: TP Sensor Signal	0.5-2.1v
26-N	RD/BL	O2S-11 (Bank 1 Sensor 1)	0.1-1.1v
26-O	RD	VAF Sensor Signal	3.3v
26-P	RD/BK	VAT Sensor Signal	1.5-2.5v
26-Q	BL/WT	ECT Sensor Signal	0.5-0.6v
26-R	---	Not Used	---
26-S	BK/RD	High Speed Inlet Air Solenoid	Below 5000 rpm: 1.5v
26-T	GN/OR	Fuel Press. Regulator Control	149 sec after startup: <3v
26-U	YL	Injector Bank 1	3.6 ms
26-V	YL/BK	Injector Bank 2	3.6 ms
26-W	BL/OR	IAC Motor Control	8-10v
26-X	WT/BL	Canister Purge Solenoid	12v, 55 mph: 1v
26-Y	---	Not Used	---
26-Z	BL/GN	ECT Signal to 4EAT Module	ECT at 162°F: <2.5v

Pin Connector Graphic

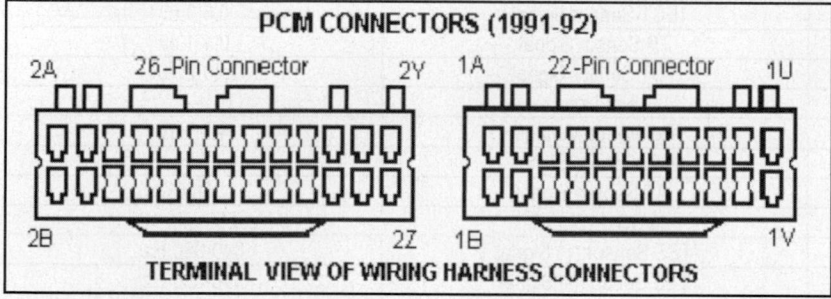

Standard Colors and Abbreviations

Abbreviation	Color	Abbreviation	Color	Abbreviation	Color
BK	Black	GY	Gray	PK	Purple
BL	Blue	GN	Green	RD	Red
BR	Brown	LG	LT Green	TN	Tan
DB	Dark Blue	OR	Orange	WT	White
DG	DK Green	PK	Pink	YL	Yellow

1993 Tracer 1.8L I4 4v MFI VIN 8 (A/T) 22 Pin Connector

PCM Pin #	Wire Color	Circuit Description (22 Pin)	Value at Hot Idle
22-A	DB/RD	Keep Alive Power	12- 14v
22-B	WT/RD	Vehicle Power	12-14v
22-C	PK	Vehicle Start Signal	Cranking: 9-11v
22-D	WT/YL	System Monitor Lamp	Lamp On: 5v, Off: 12v
22-E	YL/BK	MIL (lamp) Control	MIL On: 2.5v, Off: 12v
22-F	WT/BK	Self-Test Output, MIL Control	STO On: 5v, Off: 12v
22-G	DG/WT	Ignition Control Module Signal	0.1-0.3v
22-H	RD/BK	Headlamp Switch Signal	Switch On: 12v, Off: 1v
22-I	LG/YL	Self-Test Input Signal	STI On: 0v, Off: 5v
22-J	BK/DB	Rear Defroster Control	Switch Off: 1v, On: 12v
22-K	BK	Power Ground	<0.1v
22-L	DB/BK	A/C & Heater Relay Control	A/C On: 2.5v, AC Off: 12v
22-M	DG	VSS (+) Signal	Not moving: 0v
22-N	DB/YL	PSP Switch Signal	Straight: 12v, Turned: 0v
22-O	DG/DB	A/C Switch Signal	A/C On: 2.5v, AC Off: 12v
22-P	OR/DB	Blower Motor Switch	Off or 1st: 12v, 2nd on: 1v
22-Q	DG	Brake Switch Signal	Brake Off: 0v, On: 12v
22-R	BK/DB	MLP Switch Signal	In 'P': 0v, Others: 12v
22-S	---	Not Used	---
22-T	RD/WT	Idle Switch Signal	Closed: 0V, Open: 12v
22-U	---	Not Used	---
22-V	---	Not Used	---

1993 Tracer 1.8L I4 4v MFI VIN 8 (A/T) 16 Pin Connector

PCM Pin #	Wire Color	Circuit Description (16 Pin)	Value at Hot Idle
16-A	WT	CKP Sensor Signal	2.5v
16-B	RD	VAF Sensor Signal	0.2v
16-C	RD/DB	O2S-11 (Bank 1 Sensor 1)	0.1-1.1v
16-D	BK/DG	Cooling Fan Control	Fan On: 1v, Off: 12v
16-E	DB/W	ECT Sensor Signal	0.5-0.6v
16-F	LG/WT	TP Sensor Signal	0.4-1.0v
16-G	WT/BK	TOT Sensor Signal	2.40-2.10v
16-H	---	Not Used	---
16-I	LG/RD	Reference Voltage	4.9-5.1v
16-J	YL/DB	CID Sensor Signal	2.5v
16-K	RD/BK	IAT Sensor Signal	1.5-2.5v
16-L	DB	VSS (-) Signal	0 Hz, 55 mph: 125 Hz
16-M	WT/DB	TSS Sensor (+) Signal	340-380 Hz
16-N	YL/DB	TSS Sensor (-) Signal	340-380 Hz
16-O	WT/DB	Canister Purge Solenoid	12v, Cruise: 1v
16-P	---	Not Used	---

1993 Tracer 1.8L I4 4v MFI VIN 8 (A/T) 26 Pin Connector

PCM Pin #	Wire Color	Circuit Description (26 Pin)	Value at Hot Idle
26-A	BK/OR	Power Ground	<0.1v
26-B	BK/OR	Power Ground	<0.1v
26-C	BK/LG	Power Ground	<0.1v
26-D	BK/WT	Analog Signal Return	<0.050v
26-E	YL	MLP Overdrive Switch Signal	Switch On: 1v, Off: 12v
26-F	BR/WT	DRL (Canada)	DRL On: 2v, Off: 12v
26-G	YL/WT	MLP Sensor Low Signal	In 'L': 12v, Others: 0v
26-H	YL/RD	MLP Sensor Drive Signal	Drive: 12v, Others: 0v
26-I	BK/RD	High Speed Inlet Air Solenoid	Less than 4900 rpm: 1v
26-J	---	Not Used	---
26-K	---	Not Used	---
26-L	---	Not Used	---
26-M	DG/OR	Fuel Press. Regulator Control.	149 sec after startup: <3v
26-N	---	Not Used	---
26-O	---	Not Used	---
26-P	---	Not Used	---
26-Q	DB/OR	IAC Valve Control	8-10v
26-R	---	Not Used	---
26-S	---	Not Used	---
26-T	---	Not Used	---
26-U	YL	Injector Bank 1 (INJ 1 & 3)	3.6 ms
26-V	YL/BK	Injector Bank 2 (INJ 2 & 4)	3.6 ms
26-W	DB/OR	Shift Solenoid 1 Control	Shifting: 12v, other: 1v
26-X	DB/YL	Shift Solenoid 2 Control	Shifting: 12v, other: 1v
26-Y	OR	Shift Solenoid 3 Control	Shifting: 12v, other: 1v
26-Z	DB	TCC Solenoid Control	0v, Cruise: 12v

Pin Connector Graphic

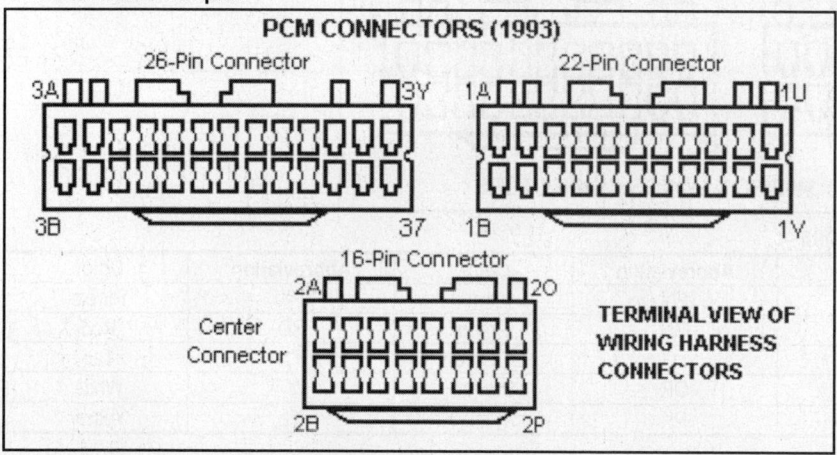

PCM CONNECTORS (1993)

26-Pin Connector

22-Pin Connector

16-Pin Connector

Center Connector

TERMINAL VIEW OF WIRING HARNESS CONNECTORS

Standard Colors and Abbreviations

Abbreviation	Color	Abbreviation	Color	Abbreviation	Color
BK	Black	GY	Gray	PK	Purple
BL	Blue	GN	Green	RD	Red
BR	Brown	LG	LT Green	TN	Tan
DB	Dark Blue	OR	Orange	WT	White
DG	DK Green	PK	Pink	YL	Yellow

1993 Tracer 1.8L I4 4v MFI VIN 8 (M/T) 22 Pin Connector

PCM Pin #	Wire Color	Circuit Description (22 Pin)	Value at Hot Idle
22-1	DB/RD	Keep Alive Power	12- 14v
22-2	WT/RD	Vehicle Power (Main Relay)	12-14v
22-3	PK	Vehicle Start Signal	Cranking: 9-11v
22-4	WT/YL	System Monitor Lamp	Lamp On: 5v, Off: 12v
22-5	YL/BK	MIL (lamp) Control	MIL On: 1v, Off: 12v
22-6	WT/BK	Self-Test Output, MIL Control	Buzzer On: 5v, Off: 12v
22-7	DG/WT	Spark Output Signal	6.93v (50% d/cycle)
22-8	---	Not Used	---
22-9	---	Not Used	---
22-10	DB/BK	AC WOT Relay Control	Relay Off: 12v, On: 1v
22-11	LG/YL	Self-Test Input Signal	STI On: 0v, STI Off: 5v
22-12	BR/WT	DRL (Canada)	DRL Off: 12v, On: 2.5v
22-13	---	Not Used	---
22-14	RD/WT	Idle Switch Input	Closed: 0V, Open: 12v
22-15	DG	Brake Switch Signal	Brake Off: 0v, On: 12v
22-16	DB/YL	PSP Switch Signal	Straight: 12v, Turned: 0v
22-17	DB/DB	A/C Switch Signal	A/C On: 2.5v, Off: 12v
22-18	BK/DG	Fan Motor Control	Fan On: 1v, Off: 12v
22-19	OR/DB	Blower Motor Switch Signal	Off or 1st: 12v, 2nd on: 1v
22-20	BK/DB	Rear Defroster Signal	Switch On: 12v, Off: 0.1v
22-21	RD/BK	Headlamp Switch Signal	Switch On: 12v, Off: 1v
22-22	BR/YL	M/T: Neutral Drive Switch	In 'N': 0v, Others: 12v
22-22	BR/YL	M/T: Clutch Pedal Switch	Clutch In: 0v, Out: 12v

Pin Connector Graphic

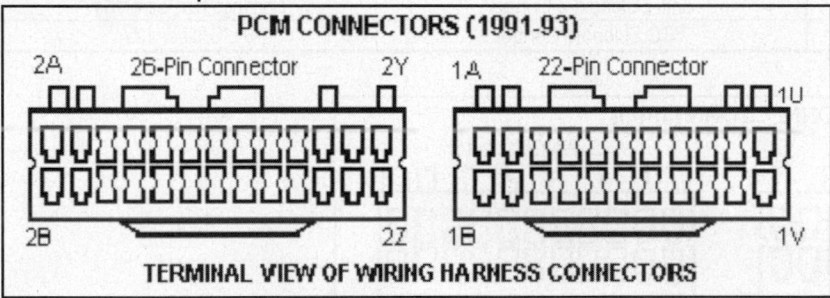

PCM CONNECTORS (1991-93)

2A 26-Pin Connector 2Y 1A 22-Pin Connector 1U

2B 2Z 1B 1V

TERMINAL VIEW OF WIRING HARNESS CONNECTORS

Standard Colors and Abbreviations

Abbreviation	Color	Abbreviation	Color	Abbreviation	Color
BK	Black	GY	Gray	PK	Purple
BL	Blue	GN	Green	RD	Red
BR	Brown	LG	LT Green	TN	Tan
DB	Dark Blue	OR	Orange	WT	White
DG	DK Green	PK	Pink	YL	Yellow

1993 Tracer 1.8L I4 4v MFI VIN 8 (M/T) 26 Pin Connector

PCM Pin #	Wire Color	Circuit Description (26 Pin)	Value at Hot Idle
26-1	BK/OR	Power Ground	<0.1v
26-2	BK/OR	Power Ground	<0.1v
26-3	BK/LG	Power Ground	<0.1v
26-4	BK/WT	Analog Signal Return	<0.050v
26-5	WT	CKP Sensor Signal	2.5v
26-6	---	Not Used	---
26-7	YL/DB	CID Sensor Signal	2.5v
26-8	BK	Power Ground	<0.1v
26-9	---	Not Used	---
26-10	---	Not Used	---
26-11	LG/RD	Reference Voltage	4.9-5.1v
26-12	LG/WT	A/C WOT Relay Control	Relay On: 0.1v, 12v
26-13	---	Not Used	---
26-14	RD/DB	O2S-11 (Bank 1 Sensor 1)	0.1-1.1v
26-15	RD	VAF Meter Signal	0.3v
26-16	RD/BK	IAT Sensor Signal	1.5-2.5v
26-17	DB/W	ECT Sensor Signal	0.5-0.6v
26-18	---	Not Used	---
26-19	BK/RD	High Speed Inlet Air Solenoid	More than 5000 rpm: 12v
26-20	DG/OR	Fuel Pressure Regulator	149 sec after startup: <3v
26-21	YL	Injector Bank 1 (INJ 1 & 3)	3.6 ms
26-22	YL/BK	Injector Bank 2 (INJ 2 & 4)	3.6 ms
26-23	DB/OR	IAC Valve Control	8-10v
26-24	WT/DB	Canister Purge Solenoid	12v, Cruise: 1v
26-25	---	Not Used	---
26-26	---	Not Used	---

Pin Connector Graphic

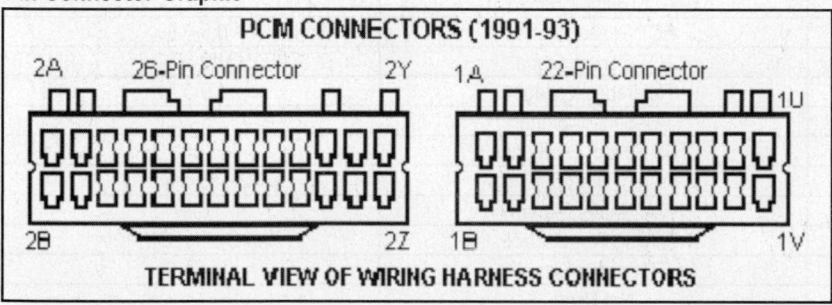

Standard Colors and Abbreviations

Abbreviation	Color	Abbreviation	Color	Abbreviation	Color
BK	Black	GY	Gray	PK	Purple
BL	Blue	GN	Green	RD	Red
BR	Brown	LG	LT Green	TN	Tan
DB	Dark Blue	OR	Orange	WT	White
DG	DK Green	PK	Pink	YL	Yellow

1994-95 Tracer 1.8L I4 4v MFI VIN 8 (A/T) 22 Pin Connector

PCM Pin #	Wire Color	Circuit Description (22 Pin)	Value at Hot Idle
22-1	DB/RD	Keep Alive Power	12- 14v
22-2	PK	Vehicle Start Signal	Cranking: 9-11v
22-3	YL/BK	MIL (lamp) Control	MIL On: 1v, Off: 12v
22-4	DG/WT	CKP Sensor Signal	2.5v
22-5	LG/YL	Data Link Connector	Digital Signals
22-6	LG/YL	Data Link Connector	Digital Signals
22-6	BK/YL	Vehicle Power (Canada)	12-14v
22-7	DG	VSS (+) Signal	0 Hz, 55 mph: 125 Hz
22-8	DG/BK	A/C Switch Signal	A/C On: 2.5v, Off: 12v
22-9	DG	Brake Switch Signal	Brake Off: 0v, On: 12v
22-10	---	Not Used	---
22-11	---	Not Used	---
22-12	WT/RD	Vehicle Power (Main Relay)	12-14v
22-13	WT/YL	Data Link Connector	Digital Signals
22-14	WT/BK	Data Link Connector	Digital Signals
22-15	RD/BK	Park Lamp Signal	Switch On: 12v, Off: 1v
22-16	BK/DB	Rear Defroster Switch Signal	Switch On: 12v, Off: 1v
22-17	DB/BK	A/C WOT Relay Control	Relay Off: 12v, On: 1v
22-18	DB/YL	PSP Switch Signal	Straight: 12v, Turned: 0v
22-19	OR/DB	Cooling Fan Control	Fan On: 1v, Off: 12v
22-20	BK/DB	Ignition Start Signal	KOEC: 9-11v
22-21	RD/WT	Throttle Position Idle Switch	Closed: 0V, Open: 12v
22-22	---	Not Used	---

1994-95 Tracer 1.8L I4 4v MFI VIN 8 (A/T) 16 Pin Connector

PCM Pin #	Wire Color	Circuit Description (16 Pin)	Value at Hot Idle
16-1	WT	CKP Sensor Signal	2.5v
16-2	RD/DB	O2S-11 (Bank 1 Sensor 1)	0.1-1.1v
16-3	DB/W	ECT Sensor Signal	0.5-0.6v
16-4	WT/BK	TOT Sensor Signal	2.40-2.10v
16-5	LG/RD	Reference Voltage	4.9-5.1v
16-6	RD/BK	IAT Sensor Signal	1.5-2.5v
16-7	WT/DB	TSS Sensor (+) Signal	340-380 Hz
16-8	WT/DB	Canister Purge Solenoid	12v, 55 mph: 1v
16-9	RD	VAF Meter Signal	0.3v
16-10	BK/DG	Cooling Fan Control	Fan On: 1v, Off: 12v
16-11	LG/WT	TP Sensor Signal	0.4-1.0v
16-12	---	Not Used	---
16-13	YL/DB	CID Sensor Signal	2.5v
16-14	DB	VSS (-) Signal	0 Hz, 55 mph: 125 Hz
16-15	YL/DB	TSS Sensor (-) Signal	340-380 Hz
16-16	---	Not Used	---

1994-95 Tracer 1.8L I4 4v MFI VIN 8 (A/T) 26 Pin Connector

PCM Pin #	Wire Color	Circuit Description (26 Pin)	Value at Hot Idle
26-1	BK/OR	Power Ground	<0.1v
26-2	BK/LG	Power Ground	<0.1v
26-3	YL	TRS Overdrive Signal	Switch On: 1v, Off: 12v
26-4	YL/WT	TRS Low Signal	In 'L': 12v, Others: 1v
26-5	BK/RD	High Speed Inlet Air Solenoid	Less than 4900 rpm: 1v
26-6	---	Not Used	---
26-7	DG/OR	Fuel Pressure Regulator	149 sec after startup: <3v
26-8	---	Not Used	---
26-9	DB/OR	IAC Valve Control	8-10v
26-10	---	Not Used	---
26-11	YL	Injector Bank 1 (INJ 1 & 3)	3.6 ms
26-12	DB/OR	Shift Solenoid 1 Control	SS1 On: 1v, Off: 12v
26-13	OR	Shift Solenoid 3 Control	SS3 On: 1v, Off: 12v
26-14	BK/OR	Power Ground	<0.1v
26-15	BK/WT	Analog Signal Return	<0.050v
26-16	BR/WT	DRL (Canada)	DRL On: 1v, Off: 12v
26-17	YL/RD	TR Sensor Drive Signal	In 'D': Others: 1v
26-18	---	Not Used	---
26-19	---	Not Used	---
26-20	---	Not Used	---
26-21	---	Not Used	---
26-22	---	Not Used	---
26-23	---	Not Used	---
26-24	YL/BK	Injector Bank 2 (INJ 2 & 4)	3.6 ms
26-25	DB/YL	Shift Solenoid 2 Control	SS2 On: 1v, Off: 12v
26-26	DB	TCC Solenoid Control	During Shifting: 12v

Pin Connector Graphic

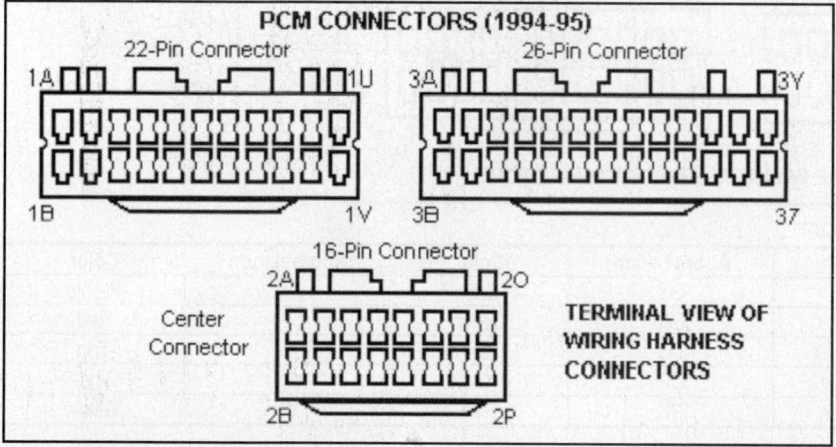

PCM CONNECTORS (1994-95)
22-Pin Connector
1A ... 1U
1B ... 1V
26-Pin Connector
3A ... 3Y
3B ... 37
16-Pin Connector
2A ... 2O
Center Connector
2B ... 2P
TERMINAL VIEW OF WIRING HARNESS CONNECTORS

Standard Colors and Abbreviations

Abbreviation	Color	Abbreviation	Color	Abbreviation	Color
BK	Black	GY	Gray	PK	Purple
BL	Blue	GN	Green	RD	Red
BR	Brown	LG	LT Green	TN	Tan
DB	Dark Blue	OR	Orange	WT	White
DG	DK Green	PK	Pink	YL	Yellow

1994-95 Tracer 1.8L I4 4v MFI VIN 8 (M/T) 22 Pin Connector

PCM Pin #	Wire Color	Circuit Description (22 Pin)	Value at Hot Idle
22-A	DB/RD	Keep Alive Power	12- 14v
22-B	PK	Vehicle Start Signal	Cranking: 9-11v
22-C	YL/BK	MIL (lamp) Control	MIL On: 1v, Off: 12v
22-D	DG/WT	Spark Output Signal	6.93v (50% d/cycle)
22-E	LG/YL	Self-Test Input Signal	STI On: 0v, STI Off: 5v
22-F	---	Not Used	---
22-G	---	Not Used	---
22-H	DG	Brake Switch Signal	Brake Off: 0v, On: 12v
22-I	DG/BK	A/C Switch Signal	A/C On: 2.5v, Off: 12v
22-J	OR/DB	Blower Motor Switch Signal	Off or 1st: 12v, 2nd on: 1v
22-K	RD/BK	Headlamp Switch Signal	Switch On: 12v, Off: 1v
22-L	WT/RD	Vehicle Power (Main Relay)	12-14v
22-M	WT/YL	Data Link Connector	Digital Signals
22-N	WT/BK	Data Link Connector	Digital Signals
22-O	---	Not Used	---
22-P	DB/BK	A/C WOT Relay Control	Relay Off: 12v, On: 1v
22-Q	BR/WT	DRL (Canada)	DRL Off: 12v, On: 2.5v
22-R	RD/WT	Idle Position Switch Signal	Closed: 0V, Open: 12v
22-S	DB/YL	PSP Switch Signal	Straight: 12v, Turned: 0v
22-T	BK/DG	Cooling Fan Control	Fan On: 1v, Off: 12v
22-U	BK/DB	Rear Defroster Switch Signal	AC On: 12v, Off: 0v
22-V	BR/YL	M/T: Neutral Drive Switch	In 'N': 0v, Others: 12v
22-V	BR/YL	M/T: Clutch Engage Switch	Clutch In: 0v, Out: 12v

Pin Connector Graphic

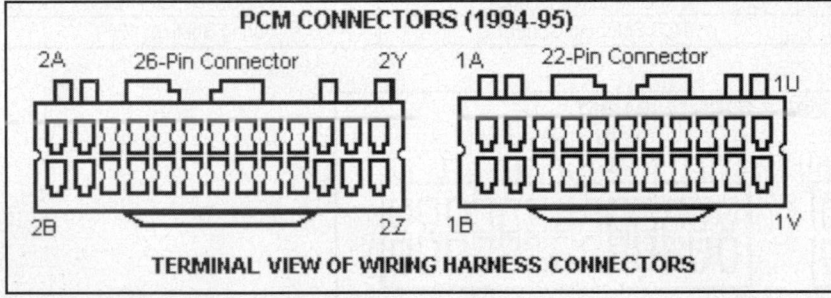

Standard Colors and Abbreviations

Abbreviation	Color	Abbreviation	Color	Abbreviation	Color
BK	Black	GY	Gray	PK	Purple
BL	Blue	GN	Green	RD	Red
BR	Brown	LG	LT Green	TN	Tan
DB	Dark Blue	OR	Orange	WT	White
DG	DK Green	PK	Pink	YL	Yellow

1994-95 Tracer 1.8L I4 4v MFI VIN 8 (M/T) 26 Pin Connector

PCM Pin #	Wire Color	Circuit Description (26 Pin)	Value at Hot Idle
26-A	BK/OR	Power Ground	<0.1v
26-B	BK/LG	Power Ground	<0.1v
26-C	W	CKP Sensor Signal	2.5v
26-D	YL/DB	CID Sensor Signal	2.5v
26-E	---	Not Used	---
26-F	LG/RD	Reference Voltage	4.9-5.1v
26-G	---	Not Used	---
26-H	RD	VAF Meter Signal	0.3v
26-I	DB/W	ECT Sensor Signal	0.5-0.6v
26-J	BK/RD	High Speed Inlet Air Solenoid	More than 5000 rpm: 12v
26-K	YL	Injector Bank 1 (INJ 1 & 3)	3.6 ms
26-L	DB/OR	IAC Valve Control	8-10v
26-M	---	Not Used	---
26-N	BK/OR	Power Ground	<0.1v
26-O	BK/WT	Analog Signal Return	<0.050v
26-P	---	Not Used	---
26-Q ('94)	BK	Power Ground	<0.1v
26-Q ('95)	BK/YL	Ignition Power	12-14v
26-R	---	Not Used	---
26-S	LG/WT	TP Sensor Signal	0.4-1.0v
26-T	RD/DB	O2S-11 (Bank 1 Sensor 1)	0.1-1.1v
26-U	RD/BK	IAT Sensor Signal	1.5-2.5v
26-V	---	Not Used	---
26-W	DG/OR	Fuel Pressure Regulator	149 sec after startup: <3v
26-X	YL/BK	Injector Bank 2 (INJ 2 & 4)	3.6 ms
26-Y	WT/DB	Canister Purge Solenoid	12v, 55 mph: 1v
26-Z	---	Not Used	---

Pin Connector Graphic

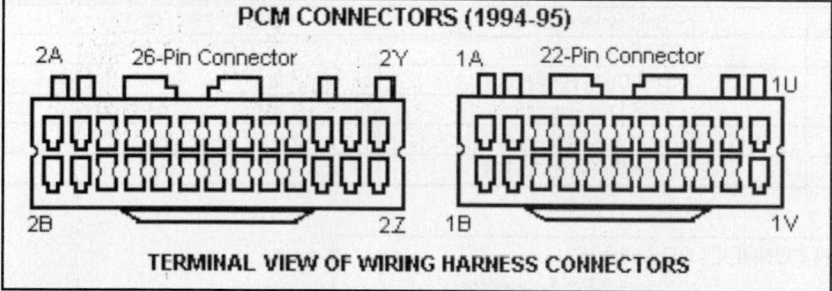

Standard Colors and Abbreviations

Abbreviation	Color	Abbreviation	Color	Abbreviation	Color
BK	Black	GY	Gray	PK	Purple
BL	Blue	GN	Green	RD	Red
BR	Brown	LG	LT Green	TN	Tan
DB	Dark Blue	OR	Orange	WT	White
DG	DK Green	PK	Pink	YL	Yellow

1996 Tracer 1.8L I4 4v MFI VIN 8 (A/T) 22 Pin Connector

PCM Pin #	Wire Color	Circuit Description (22 Pin)	Value at Hot Idle
C1-2	PK	Vehicle Start	KOEC: 9-11v
C1-3	YL/BK	MIL (lamp) Control	MIL On: 0v, Off: 12v
C1-4	BL/BK	A/C WOT Relay Control	A/C On: 2.5v, Off: 12v
C1-5	LG/YL	Self-Test Input Signal	STI On: 0v, Off: 5v
C1-6	GN/BK	A/C High Pressure Switch	A/C On: 2.5v, Off: 12v
C1-7	GN	VSS (+) Signal	0 Hz, 55 mph: 125 Hz
C1-8	BL	VSS (-) Signal	0 Hz, 55 mph: 125 Hz
C1-9	GN	Brake Switch Signal	Brake On: 12v. Off: 0v
C1-10	---	Not Used	---
C1-11	LG	Fuel Pump Control	Relay On, 0v, Off: 12v
C1-12	---	Not Used	---
C1-13	BL/WT	ISO K-Line (OBD II)	Digital Signals: 0-7-0-7v
C1-14	---	Not Used	---
C1-15	RD/BK	Headlamp Switch Signal	Switch On: 12v, Off: 1v
C1-16	GY/RD	Rear Defroster Switch Signal	DEF On: 12v, Off: 1v
C1-17	---	Not Used	---
C1-18	BR/WT	DRL Switch (Canada)	Switch On: 0v, Off: 12v
C1-19	OR/BL	Blower Motor Control Switch	Off/1st: 12v, 2nd on: 1v
C1-20	BK/GN	Cooling Fan Relay	Relay On: 0v, Off: 12v
C1-21-22	---	Not Used	---

1996 Tracer 1.8L I4 4v MFI VIN 8 (A/T) 12 Pin Connector

PCM Pin #	Wire Color	Circuit Description (12 Pin)	Value at Hot Idle
C2-1	BL/OR	Shift Solenoid 1-2	SS1 On: 1v, Off: 12 v
C2-2	OR	Shift Solenoid 3-4	SS3 On: 1v, Off: 12 v
C2-3	---	Not Used	---
C2-4	YL/RD	TRS 2nd Position	In 2nd: 12v, Others: 0v
C2-5	---	Not Used	---
C2-6	WT/BL	TSS Sensor (+) Signal	340-380 Hz
C2-7	BL/YL	Shift Solenoid 3-4	SS3 On: 1v, Off: 12 v
C2-8	BL	TCC Solenoid Control	TCC On: 0v, Off: 12v
C2-9	YL	TRS Drive Position	In 'D': 12v, Others: 0v
C2-10	BR/WT	TRS Low Position	In 'L': 12v, Others: 0v
C2-11	---	Not Used	---
C2-12	YL/BL	ECT Sensor Signal to TCM	0.5-0.6v

Pin Connector Graphic

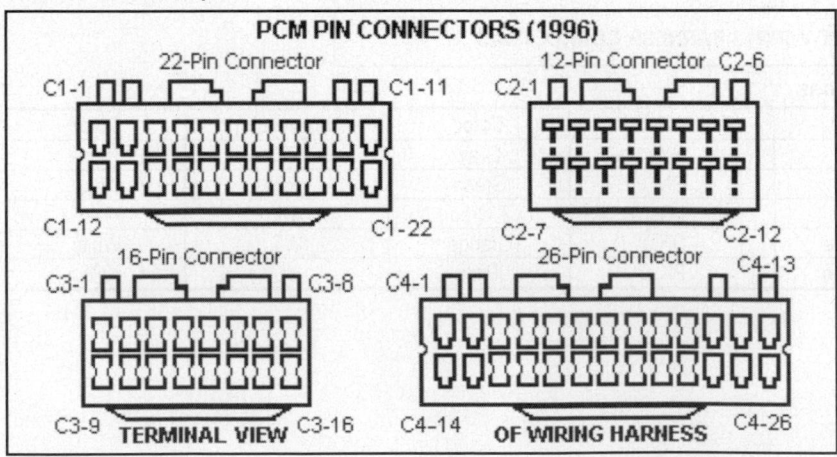

1996 Tracer 1.8L I4 4v MFI VIN 8 (A/T) 16 Pin Connector

PCM Pin #	Wire Color	Circuit Description (16 Pin)	Value at Hot Idle
C3-1	---	Not Used	---
C3-2	RD/BL	HO2S-11 (B1 S1) Signal	0.1-1.1v
C3-3	WT/BK	TFT Sensor Signal	2.40-2.10v
C3-4	BL/WT	ECT Sensor Signal	0.6v
C3-5	LG/RD	Reference Voltage	4.9-5.1v
C3-6	RD/BK	IAT Sensor Signal	1.5-2.5v
C3-7	GN/WT	HO2S-11 (B1 S1) Heater	1v, Off: 12v
C3-8	BK/BL	Analog Return Signal	<0.050v
C3-9	RD	MAF Sensor Signal	0.6v
C3-10	RD/YL	HO2S-12 (B1 S2) Signal	0.1-1.1v
C3-11	LG/WT	TP Sensor Signal	0.4-1.0v
C3-12	---	Not Used	---
C3-13	---	Not Used	---
C3-14	RD/WT	Idle Position Switch Signal	Closed: 0V, Open: 12v
C3-15	---	Not Used	---
C3-16	BL/YL	PSP Switch Signal	Straight: 12v, Turned: 0v

1996 Tracer 1.8L I4 4v MFI VIN 8 (A/T) 26 Pin Connector

PCM Pin #	Wire Color	Circuit Description (26 Pin)	Value at Hot Idle
C4-1	BK/LG	Power Ground	<0.1v
C4-2	BK/OR	Power Ground	<0.1v
C4-3	OR	CKP Sensor (-) Signal	435-475 Hz
C4-4	YL/BL	CMP 1 Sensor Signal	5-7 Hz
C4-5	BL/RD	Keep Alive Power	12-14v
C4-6	BK/RD	High Speed Inlet Air Solenoid	More than 5000 rpm: 12v
C4-7	---	Not Used	---
C4-8	---	Not Used	---
C4-9	BL/OR	IAC Motor Control	10.7v (30%)
C4-10	---	Not Used	---
C4-11	YL	Injector 1 Control	4-4.5 ms
C4-12	GN/RD	Injector 3 Control	4-4.5 ms
C4-13	---	Not Used	---
C4-14	WT/RD	Vehicle Power	12-14v
C4-15	BK/OR	Power Ground	<0.1v
C4-16	W	CMP 2 Sensor Signal	5-7 Hz
C4-17	LG/WT	CKP Sensor (+) Signal	435-475 Hz
C4-18	GN/OR	Fuel Pressure Regulator	149 sec after start: <3v
C4-19	---	Not Used	---
C4-20	BK/WT	Ignition Spark Output Signal	Varies: 0.1-0.3v
C4-21	---	Not Used	---
C4-22	---	Not Used	---
C4-23	WT/BL	Canister Purge Solenoid	Purge On: 0v, Off: 12v
C4-24	YL/BK	Injector 2 Control	4-4.5 ms
C4-25	GN/YL	Injector 4 Control	4-4.5 ms
C4-26	---	Not Used	---

1996 Tracer 1.8L I4 4v MFI VIN 8 (M/T) 22 Pin Connector

PCM Pin #	Wire Color	Circuit Description (22 Pin)	Value at Hot Idle
C1-1	---	Not Used	---
C1-2	PK	Vehicle Start	KOEC: 9-11v
C1-3	YL/BK	MIL (lamp) Control	MIL On: 0v, Off: 12v
C1-4	BL/BK	A/C WOT Relay Control	A/C On: 2.5v, Off: 12v
C1-5	LG/YL	Self-Test Input Signal	STI On: 0v, Off: 5v
C1-6	GN/BK	A/C High Pressure Switch	A/C On: 2.5v, Off: 12v
C1-7	GN	VSS (+) Signal	0 Hz, 55 mph: 125 Hz
C1-8	BL	VSS (-) Signal	0 Hz, 55 mph: 125 Hz
C1-9	GN	Brake Switch Signal	Brake On: 12v. Off: 0v
C1-10	---	Not Used	---
C1-11	LG	Fuel Pump Control	Relay On, 0v, Off: 12v
C1-12	---	Not Used	---
C1-13	BL/WT	ISO K-Line (OBD II)	Digital Signals: 0-7-0-7v
C1-14	---	Not Used	---
C1-15	RD/BK	Headlamp Switch Signal	Switch On: 12v, Off: 1v
C1-16	GY/RD	Rear Defroster Switch Signal	Switch On: 12v, Off: 1v
C1-17	BR/YL	M/T: Neutral Drive Switch	In 'N': 0v, Others: 5v
C1-17	BR/YL	M/T: Clutch Pedal Switch	Clutch In: 0v, Out: 5v
C1-18	BR/WT	DRL (Canada)	Relay On: 0v, Off: 12v
C1-19	OR/BL	Blower Motor Control Switch	Off or 1st: 12v, 2nd on: 1v
C1-20	BK/GN	Cooling Fan Control	Relay On: 0v, Off: 12v
C1-21	---	Not Used	---
C1-22	---	Not Used	---

1996 Tracer 1.8L I4 4v MFI VIN 8 (M/T) 16 Pin Connector

PCM Pin #	Wire Color	Circuit Description (16 Pin)	Value at Hot Idle
C3-1	---	Not Used	---
C3-2	RD/BL	HO2S-11 (B1 S1) Signal	0.1-1.1v
C3-3	---	Not Used	---
C3-4	BL/WT	ECT Sensor Signal	0.5-0.6v
C3-5	LG/RD	Reference Voltage	4.9-5.1v
C3-6	RD/BK	IAT Sensor Signal	1.5-2.5v
C3-7	GN/WT	HO2S-11 (B1 S1) Heater	1v, Off: 12v
C3-8	BK/BL	Analog Return Signal	<0.050v
C3-9	RD	MAF Sensor Signal	0.6v
C3-10	RD/YL	HO2S-12 (B1 S2) Signal	0.1-1.1v
C3-11	LG/WT	TP Sensor Signal	0.5v
C3-12, 13	---	Not Used	---
C3-14	RD/WT	Idle Position Switch Signal	Closed: 0v, Open: 12v
C3-15	---	Not Used	---
C3-16	BL/YL	PSP Switch Signal	Straight: 12v, Turned: 0v

1996 Tracer 1.8L I4 4v MFI VIN 8 (M/T) 26 Pin Connector

PCM Pin #	Wire Color	Circuit Description (26 Pin)	Value at Hot Idle
C4-1	BK/LG	Power Ground	<0.1v
C4-2	BK/OR	Power Ground	<0.1v
C4-3	OR	CKP Sensor (-) Signal	435-475 Hz
C4-4	YL/BL	CMP 1 Sensor Signal	5-7 Hz
C4-5	BL/RD	Keep Alive Power	12-14v
C4-6	BK/RD	High Speed Inlet Air Solenoid	More than 5000 rpm: 12v
C4-7	---	Not Used	---
C4-8	---	Not Used	---
C4-9	BL/OR	IAC Motor Control	Varies: 8-10v
C4-10	---	Not Used	---
C4-11	YL	Injector 1 Control	4.0-4.5 ms
C4-12	GN/RD	Injector 3 Control	4.0-4.5 ms
C4-13	---	Not Used	---
C4-14	WT/RD	Vehicle Power	12-14v
C4-15	BK/OR	Power Ground	<0.1v
C4-16	WT	CMP 2 Sensor Signal	5-7 Hz
C4-17	LG/WT	CKP Sensor (+) Signal	435-475 Hz
C4-18	GN/OR	Fuel Pressure Regulator	149 sec after startup: <3v
C4-19	---	Not Used	---
C4-20	GN/WT	Spark Output Signal	6.93v (50% d/cycle)
C4-21	---	Not Used	---
C4-22	---	Not Used	---
C4-23	WT/BL	Canister Purge Solenoid	12v, 55 mph: 1v
C4-24	YL/BK	Injector 2 Control	4.0-4.5 ms
C4-25	GN/YL	Injector 4 Control	4.0-4.5 ms
C4-26	---	Not Used	---

Pin Connector Graphic

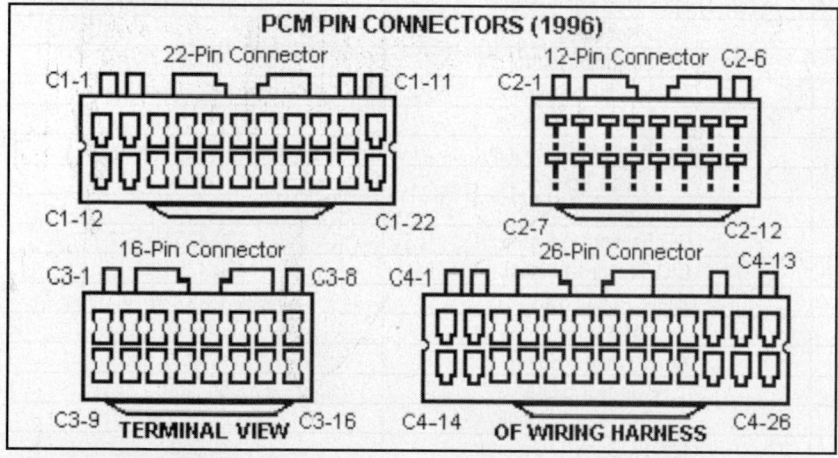

Standard Colors and Abbreviations

Abbreviation	Color	Abbreviation	Color	Abbreviation	Color
BK	Black	GY	Gray	PK	Purple
BL	Blue	GN	Green	RD	Red
BR	Brown	LG	LT Green	TN	Tan
DB	Dark Blue	OR	Orange	WT	White
DG	DK Green	PK	Pink	YL	Yellow

1991-92 Tracer 1.9L I4 MFI VIN J (All) 60 Pin Connector

PCM Pin #	Wire Color	Circuit Description (60 Pin)	Value at Hot Idle
1	BL/RD	Keep Alive Power	12-14v
2, 5, 11, 14	---	Not Used	---
3	WT/BK	VSS (+) Signal	0 Hz, 55 mph: 125 Hz
4	RD	Ignition Diagnostic Monitor	20-31 Hz
6	BL	VSS (-) Signal	0 Hz, 55 mph: 125 Hz
6	DB	VSS (-) Signal	0 Hz, 55 mph: 125 Hz
7	BL/WT	ECT Sensor Signal	0.5-0.6v
8	Y/GN	Data Bus (-) Signal	Digital Signals
9	GN/YL	MAF Sensor Return	<0.050v
10	GN/BK	A/C Cycling Clutch Signal	A/C Off: 0v, On: 12v
12	YL/OR	Injector 3 Control	4.8-5.0 ms
13	GN/OR	Injector 4 Control	4.8-5.0 ms
15 ('92)	OR/WT	Canister Purge Solenoid	12v, 55 mph: 9-11v
16	RD/BL	Ignition System Ground	<0.050v
17	YL/BK	Self-Test Output, MIL Control	MIL On: 1v, Off: 12v
18, 23-24	---	Not Used	---
19	BK/PKK	Fuel Pump Monitor	12v, Off: 0v
20	---	Not Used	---
21	BL/OR	IAC Motor Control	9.7-12.0v
22	LG	Fuel Pump Control	Relay Off: 12v, On: 1v
25	WT/GN	ACT Sensor Signal	1.5-2.5v
26	LG/WT	Reference Voltage	4.9-5.1v
27	BL/YL	PFE Sensor Signal	3.2v, 55 mph: 2.8v
28	YL/BL	Data Bus (+) Signal	Digital Signals
29	GN/BL	HO2S-11 (B1 S1) Signal	0.1-1.1v
30	BR/YL	A/T: Neutral Drive Switch	In 'P': 0v, Others 5v
30	BR/YL	M/T: Clutch Engage Switch	Clutch In: 0v, Out: 12v
31	RD/BK	High Speed Fan Control	Fan On: 1v, Off: 12v
32, 38-39	---	Not Used	---
33	WT/BL	VR Solenoid Control	0%, 55 mph: 45%
34	LG/BK	Octane Adjust Switch 2	Closed: 0v, Open: 9.1v
35	YL/WT	Low Speed Fan Control	Fan On: 1v, Off: 12v
36	LG/WT	Spark Angle Word Signal	6.93v (50% d/cycle)
37	WT/RD	Vehicle Power	12-14v
40	BK/GN	Power Ground	<0.1v
41	OR	CID Sensor (+) Signal	400-425 Hz
42	BL/GN	CID Sensor (-) Signal	400-425 Hz
43	GN/WT	Octane Adjust Switch 1	Closed: 0v, Open: 9.1v
44-45	---	Not Used	---
46	LG/BK	Analog Signal Return	<0.050v
47	RD/WT	TP Sensor Signal	0.8v, 55 mph: 1.1v
48	LG/YL	Self-Test Input Signal	STI On: 1v, Off: 5v
49	BK	HO2S-11 (Bank 1) Ground	<0.050v
49	OR/BK	HO2S-11 (Bank 1) Ground	<0.050v
50	BR/BK	MAF Sensor Signal	0.8v
51-52, 55	---	Not Used	---
53	LG/RD	Shift Indicator Light Control	SIL On: 1v, Off 12v
54	BL/BK	A/C WOT Relay Control	Relay Off: 12v, On: 1v
56	GN/WT	PIP Sensor Signal	6.93v (50% dwell)
57	WT/RD	Vehicle Power	12-14v
58	YL	Injector Bank 1 (INJ 1 & 2)	4.8-5.0 ms
59	GN/RD	Injector Bank 2 (INJ 3 & 4)	4.8-5.0 ms
60	BK/GN	Power Ground	<0.1v

1993-95 Tracer 1.9L I4 MFI VIN J (All) 60 Pin Connector

PCM Pin #	Wire Color	Circuit Description (60 Pin)	Value at Hot Idle
1	BL/RD	Keep Alive Power	12-14v
2	WT	TOT Sensor Signal	2.40-2.10v
3	WT/BK	VSS (+) Signal	0 Hz, 55 mph: 125 Hz
4	RD	Ignition Diagnostic Monitor	22-31 Hz
5	GN	Brake Switch Signal	Brake Off: 0v, On: 12v
6	BL	VSS (-) Signal	0 Hz, 55 mph: 125 Hz
6	DB	VSS (-) Signal	0 Hz, 55 mph: 125 Hz
7	BL/WT	ECT Sensor Signal	0.5-0.6v
8	Y/GN	Data Bus (-) Signal	Digital Signals
9	GN/YL	MAF Sensor Return	<0.050v
10	GN/BK	A/C Cycling Clutch Signal	A/C Off: 0v, On: 12v
11	LG/BL	Shift Solenoid 1 Control	12v, 55 mph: 1v
12	YL/OR	Injector 3 Control	4.8-5.1 ms
13	GN/OR	Injector 4 Control	4.8-5.1 ms
14	---	Not Used	---
15	OR/WT	Canister Purge Solenoid	12v, 55 mph: 9v
16	RD/BL	Ignition System Ground	<0.050v
17	YL/BK	Self-Test Output, MIL Control	MIL On: 1v, Off: 12v
18	BR/RD	A/T: TR Sensor Drive	12v, 55 mph: 12v
19	BK/PKK	Fuel Pump Monitor	12v, Off: 0v
20	BK	Case Ground	<0.050v
21	OR/BK	IAC Motor Control	9.7-11.1v
22	LG	Fuel Pump Control	Relay Off: 12v, On: 1v
23	R/GN	TR Sensor Reverse	12v, 55 mph: 10v
24	WT/BL	A/T: TSS Sensor Signal	42-50 Hz (680-720 rpm)
25	WT/GN	IAT Sensor Signal	1.5-2.5v
26	LG/WT	Reference Voltage	4.9-5.1v
27	BL/YL	PFE Sensor Signal	3.2v, 55 mph: 3.4v
28	YL/BL	Data Bus (+) Signal	Digital Signals
29	GN/BL	HO2S-11 (B1 S1) Signal	0.1-1.1v
30	BK/BL	A/T: Neutral Drive Switch	In P: 0v, all others: 5v
30	BR/YL	A/T: Neutral Drive Switch	In P: 0v, all others: 5v
30	BK/BL	M/T: Clutch Engage Switch	Clutch In: 0v, Out 5v
30	BR/YL	M/T: Clutch Engage Switch	Clutch In: 0v, Out 5v
31	RD/BK	High Speed Fan Control	Fan On: 1v, Off: 12v
32	---	Not Used	---
33	WT/BL	VR Solenoid Control	0%, 55 mph: 45%
34	---	Not Used	---
35	YL/WT	Low Speed Fan Control	Fan On: 1v, Off: 12v
36	LG/WT	Spark Output Signal	6.93v (50% d/cycle)
37	WT/RD	Vehicle Power	12-14v
38	BR/BL	TRS Overdrive Position	O/D Off: 12v, O/D On: 1v
39	---	Not Used	---
40	BK/GN	Power Ground	<0.1v

1993-95 Tracer 1.9L I4 MFI VIN J (All) 60 Pin Connector

PCM Pin #	Wire Color	Circuit Description (60 Pin)	Value at Hot Idle
41	OR	CID Sensor (+) Signal	5-7 Hz
42	BL/GN	CID Sensor (-) Signal	5-7 Hz
43	GN/WT	Octane Adjust Switch 1	Closed: 0v, Open: 9.1v
44	YL/BL	A/T: TSS Sensor Signal	44-56 Hz (680-710 rpm)
45	BR/GN	A/T: Transmission Low Signal	12v, 55 mph: 0v
46	LG/BK	Analog Signal Return	<0.050v
47	RD/WT	TP Sensor Signal	0.8v, 55 mph: 1.1v
48	LG/YL	Self-Test Input Signal	STI On: 1v, Off: 5v
49	BK	HO2S-11 (Bank 1) Ground	<0.050v
49	OR/BK	HO2S-11 (Bank 1) Ground	<0.050v
50	BR/BK	MAF Sensor Signal	0.7v, 55 mph: 2.0v
51	PK/BK	A/T: Shift Solenoid 2 Control	1v, 35 mph: 1v
52	OR/GN	A/T: Shift Solenoid 3 Control	12v, 35 mph: 12v
53	LG/RD	M/T: Shift Indicator Light	SIL On: 1v, Off: 12v
54	BL/BK	A/C WOT Relay Control	Relay Off: 12v, On: 1v
55	PK/WT	TCC Solenoid Control	12v, 55 mph: 9v
56	GN/WT	PIP Sensor Signal	6.93v (50% dwell)
57	WT/RD	Vehicle Power	12-14v
58	YL	Injector 1 Control	4.8-5.1 ms
59	GN/RD	Injector 2 Control	4.8-5.1 ms
60	BK/GN	Power Ground	<0.1v

Pin Connector Graphic

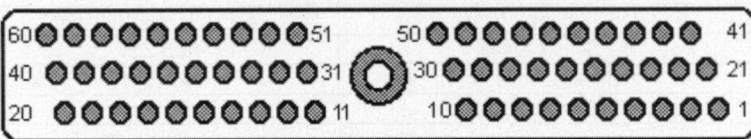

PCM 60-PIN CONNECTOR

Terminal View of 60-Pin PCM Harness Connector

Standard Colors and Abbreviations

Abbreviation	Color	Abbreviation	Color	Abbreviation	Color
BK	Black	GY	Gray	PK	Purple
BL	Blue	GN	Green	RD	Red
BR	Brown	LG	LT Green	TN	Tan
DB	Dark Blue	OR	Orange	WT	White
DG	DK Green	PK	Pink	YL	Yellow

1996 Tracer 1.9L I4 MFI VIN J (All) 104 Pin Connector

PCM Pin #	Wire Color	Circuit Description (104 Pin)	Value at Hot Idle
1	PK/BK	Shift Solenoid 2 Control	12v, 55 mph: 1v
2	YL/BK	MIL (lamp) Control	MIL On: 1v, Off: 12v
3-5	---	Not Used	---
6	BR/BK	TR Sensor Overdrive	In 'P': 0v, in OD: 12v
7	BR/DG	TR Sensor Low	In 'P': 0v, in 1st: 12v
8	---	Not Used	
9	BR/YL	TR Sensor Drive	In 'P': 0v, in 2nd: 12v
10, 12	---	Not Used	
11	BL/PK	Purge Flow Sensor	0.5-1.5v, 30 mph: 3v
13	PK/DG	Flash EEPROM Power	0.1v
14	---	Not Used	---
15	BK/BL	Data Bus (-) Signal	Digital Signals
16	WT/BL	Data Bus (+) Signal	Digital Signals
17	RD/BK	High Speed Fan Control	Fan On: 1v, Off: 12v
18, 20	---	Not Used	---
21	BL/DG	CKP Sensor (+) Signal	435-475 Hz
22	OR	CKP Sensor (-) Signal	435-475 Hz
23	YL/BL	TSS Sensor (-) Signal	340-380 Hz
24	BK/YL	Power Ground	<0.1v
25	RD/BL	Chassis Ground	<0.050v
26	GY/YL	Coil Driver 1 Control	5° dwell
27	LG/BL	Shift Solenoid 1 Control	1v, 55 mph: 12v
28-29	---	Not Used	---
30	DG/WT	Octane Adjust Switch	9.3v (Closed: 0v)
31, 34	---	Not Used	---
32	R/GN	TR Sensor Reverse	Reverse: 12v, Others: 0v
33	BL	VSS (-) Signal	0 Hz, 55 mph: 125 Hz
35	GY	HO2S-12 (B1 S2) Signal	0.1-1.1v
36	OR/BL	MAF Sensor Return	<0.050v
37	WT	TFT Sensor Signal	2.40-2.10v
38	BL/WT	ECT Sensor Signal	0.5-0.6v
39	WT/GN	IAT Sensor Signal	1.5-2.5v
40	BK/PKK	Fuel Pump Monitor	12v, Off: 0v
41	GN/RD	A/C Cycling Clutch Switch	A/C Switch On: 12v
42-44	---	Not Used	---
45	PK/WT	Low Speed Fan Control	Fan On: 1v, Off: 12v
46	---	Not Used	---
47	PK/RD	VR Solenoid Control	0%, 55 mph: 45%
48	LG/RD	Clean Tachometer Output	25-38 Hz
49-50	---	Not Used	---
51	BK/YL	Power Ground	<0.1v
52	BR/RD	Coil Driver 2 Control	5° dwell
53	OR/GN	Shift Solenoid 3 Control	12v, 55 mph: 1v
54	GY/WT	TCC Solenoid Control	25-38 Hz
55	BL/RD	Keep Alive Power	12-14v
56-57	---	Not Used	---
58	WT/BK	VSS (+) Signal	0 Hz, 55 mph: 125 Hz
60	GY/DG	HO2S-11 (B1 S1) Signal	0.1-1.1v
61-63	---	Not Used	---
64	RD	A/T: Neutral Drive Switch	In 'N': 0v, in Gear: 12v
64	BR/YL	M/T: Clutch Position Switch	Clutch In: 0v, Out: 5v

1996 Tracer 1.9L I4 MFI VIN J (All) 104 Pin Connector

PCM Pin #	Wire Color	Circuit Description (104 Pin)	Value at Hot Idle
65	BL/YL	DPFE Sensor Signal	0.20-1.30v
66	---	Not Used	---
67	OR/WT	Canister Purge Solenoid	0-10 Hz (0-100%)
68	---	Not Used	---
69	BL/BK	A/C WOT Relay Control	Relay Off: 12v, On: 1v
70	---	Not Used	---
71	WT/RD	Vehicle Power	12-14v
72-73	---	Not Used	---
74	BR/WT	Injector 3 Control	4.0-4.5 ms
75	Y	Injector 1 Control	4.0-4.5 ms
76	BL	CMP Sensor (-) Signal	5-7 Hz
77	BK/YL	Power Ground	<0.1v
78-79	---	Not Used	---
80	LG	Fuel Pump Control	Relay Off: 12v, On: 1v
81-82	---	Not Used	---
83	BL/OR	IAC Motor Control	10v (30% d/cycle)
84	WT/BL	TSS Sensor (+) Signal	340-380 Hz
85	DG	CMP Sensor (+) Signal	5-7 Hz
86	DG/BK	A/C Pressure Sensor	A/C On: 1.5-1.9v
87	---	Not Used	---
88	BR/BK	MAF Sensor Signal	0.8v
89	RD/WT	TP Sensor Signal	0.53-1.27v
90	LG/WT	Reference Voltage	4.9-5.1v
91	LG/BK	Analog Signal Return	<0.050v
92	DG	Brake Switch Signal	Brake Off: 0v, On: 12v
93	OR/BK	HO2S-11 (B1 S1) Heater	1v, Off: 12v
94	---	Not Used	---
95	PK	HO2S-12 (B1 S2) Heater	1v, Off: 12v
96	---	Not Used	---
97	WT/RD	Vehicle Power	12-14v
98-99	---	Not Used	---
100	DG/OR	Injector 4 Control	4.0-4.5 ms
101	PK/YL	Injector 2 Control	4.0-4.5 ms
102	---	Not Used	---
103	BK/YL	Power Ground	<0.1v
104	---	Not Used	---

Pin Connector Graphic

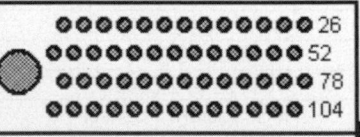

PCM 104-PIN CONNECTOR

Terminal View of 104-Pin PCM Wiring Harness Connector

Standard Colors and Abbreviations

Abbreviation	Color	Abbreviation	Color	Abbreviation	Color
BK	Black	GY	Gray	PK	Purple
BL	Blue	GN	Green	RD	Red
BR	Brown	LG	LT Green	TN	Tan
DB	Dark Blue	OR	Orange	WT	White
DG	DK Green	PK	Pink	YL	Yellow

1997 Tracer 2.0L I4 2v MFI VIN P (All) 104 Pin Connector

PCM Pin #	Wire Color	Circuit Description (104 Pin)	Value at Hot Idle
1	PK/BK	Shift Solenoid 2 Control	12v, 55 mph: 1v
2	YL/BK	MIL (lamp) Control	MIL On: 1v, Off: 12v
3-5	---	Not Used	---
6	BR/BK	TR Sensor Overdrive	In O/D: 12v, Others: 0v
7	BR/GN	TR Sensor Low	In 1st: 12v, Others: 0v
8	PK/BL	IMRC Solenoid Monitor	All speeds: 5v
9	BR/YL	TR Sensor Drive	In 'D': 0v, Others: 12v
10	---	Not Used	---
11	BL/PK	EVAP Purge Flow Sensor	0.8v, 30 mph: 3.0v
12	---	Not Used	---
14	---	Not Used	---
13	PK/GN	Flash EEPROM Power	0.1v
15	BK/BL	Data Bus (-) Signal	Digital Signals
16	WT/BL	Data Bus (+) Signal	Digital Signals
17	RD/BK	High Speed Fan Control	Fan On: 1v, Off: 12v
18-20	---	Not Used	---
21	BK	CKP Sensor (+) Signal	400-425 Hz
22	WT	CKP Sensor (-) Signal	400-425 Hz
23	RD	TSS Sensor (-) Signal	340-380 Hz
24	BK/YL	Power Ground	<0.1v
25	RD/BL	Case Ground	<0.050v
26	GN/YL	Coil Driver 1 Control	5° dwell
27	LG/BL	Shift Solenoid 1 Control	1v, 55 mph: 12v
28-29, 34	---	Not Used	---
30	GN/WT	Octane Adjust Switch	9.3v (Closed: 0v)
31	YL/RD	PSP Switch Signal	Straight: 1v, Turning: 12v
32	R/GN	TR Sensor Reverse	Reverse: 12v, others: 0v
33	BL	VSS (-) Signal	0 Hz, 55 mph: 125 Hz
35	GY	HO2S-12 (B1 S2) Signal	0.1-1.1v
36	OR/BL	MAF Sensor Return	<0.050v
37	OR	TFT Sensor Signal	2.40-2.10v
38	BL/WT	ECT Sensor Signal	0.5-0.6v
39	WT/GN	IAT Sensor Signal	1.5-2.5v
40	BK/PKK	Fuel Pump Monitor	12v, Off: 0v
41	GN/RD	A/C Cycling Clutch Switch	A/C Switch On: 12v
42	RD/YL	IMRC Solenoid Control	12v, 55 mph: 12v
43-44, 46	---	Not Used	---
45	PK/WT	Low Speed Fan Control	Fan On: 1v, Off: 12v
47	PK/RD	VR Solenoid Control	0%, 55 mph: 45%
48	LG/RD	Clean Tachometer Output	25-38 Hz
49-50	---	Not Used	---
51	BK/YL	Power Ground	<0.1v
52	BR/RD	Coil Driver 2 Control	5° dwell
53	GY/YL	Shift Solenoid 3 Control	6.7v, 55 mph: 6.7v
54	GY/WT	TCC Solenoid Control	0%, 55 mph: 95%
55	BL/RD	Keep Alive Power	12-14v
56-57	---	Not Used	---
58	WT/BK	VSS (+) Signal	0 Hz, 55 mph: 125 Hz
60	GY/DG	HO2S-12 (B1 S2) Signal	0.1-1.1v
61-63	---	Not Used	---
64	RD	A/T: TR Position Sensor	In 'P': 0v, O/D: 1.7v
64	BR/YL	M/T: Clutch Pedal Position	Clutch In: 0v, Out: 5v

1997 Tracer 2.0L I4 2v MFI VIN P (All) 104 Pin Connector

PCM Pin #	Wire Color	Circuit Description (104 Pin)	Value at Hot Idle
65	BL/YL	DPFE Sensor Signal	0.20-1.30v
66	---	Not Used	---
67	OR/WT	EVAP Purge Solenoid	0-10 Hz (0-100%)
68	---	Not Used	---
69	BL/BK	A/C WOT Relay Control	Relay Off: 12v, On: 1v
70	---	Not Used	---
71	WT/RD	Vehicle Power	12-14v
72-73	---	Not Used	---
74	BR/WT	Injector 3 Control	3.3-3.7 ms
75	YL	Injector 1 Control	3.3-3.7 ms
76	BL	CMP Sensor (-) Signal	5-7 Hz
77	BK/YL	Power Ground	<0.1v
78-79	---	Not Used	---
80	LG	Fuel Pump Control	Relay Off: 12v, On: 1v
81-82	---	Not Used	---
83	BL/OR	IAC Motor Control	10v (30% d/cycle)
84	WT	TSS Sensor (+) Signal	340-380 Hz
85	GN	CMP Sensor (+) Signal	5-9 Hz
86	GN/BK	A/C Pressure Sensor	A/C On: 1.2-1.5v
87	---	Not Used	---
88	BR/BK	MAF Sensor Signal	0.8v
89	RD/WT	TP Sensor Signal	0.53-1.27v
90	LG/WT	Reference Voltage	4.9-5.1v
91	LG/BK	Analog Signal Return	<0.050 v
92	DG	Brake Switch Signal	Brake Off: 0v, On: 12v
93	OR/BK	HO2S-11 (B1 S1) Heater	1v, Off: 12v
94	---	Not Used	---
95	PK	HO2S-12 (B1 S2) Heater	1v, Off: 12v
96	---	Not Used	---
97	WT/RD	Vehicle Power	12-14v
98-99	---	Not Used	---
100	GN/OR	Injector 4 Control	3.3-3.7 ms
101	PK/YL	Injector 2 Control	3.3-3.7 ms
102	---	Not Used	---
103	BK/YL	Power Ground	<0.1v
104	---	Not Used	---

Pin Connector Graphic

PCM 104-PIN CONNECTOR

Terminal View of 104-Pin PCM Wiring Harness Connector

Standard Colors and Abbreviations

Abbreviation	Color	Abbreviation	Color	Abbreviation	Color
BK	Black	GY	Gray	PK	Purple
BL	Blue	GN	Green	RD	Red
BR	Brown	LG	LT Green	TN	Tan
DB	Dark Blue	OR	Orange	WT	White
DG	DK Green	PK	Pink	YL	Yellow

1998-99 Tracer 2.0L I4 MFI VIN P (All) 104 Pin Connector

PCM Pin #	Wire Color	Circuit Description (104 Pin)	Value at Hot Idle
1	PK/BK	Shift Solenoid 2 Control	0.1v, 55 mph: 0.1v
2	YL/BK	MIL (lamp) Control	MIL On: 1v, Off: 12v
3-5	---	Not Used	---
6	BR/BL	TR Sensor Overdrive	In O/D: 12v, others: 0v
7	BR/GN	TR Sensor Low	In 1st: 12v, others: 0v
8	PK/LB	IMRC Monitor Signal	All speeds: 5v
9	LG/BL	TR Drive Position	In 'D': 0v, 55 mph: 11.5v
10	---	Not Used	---
11	BL/PK	EVAP Purge Flow Sensor	0.8v, 30 mph: 3.0v
12	Y/BR	Fuel Level Indicator Signal	11.7v (1/2 full)
13	PK/GN	Flash EEPROM Power	0.1v
14	---	Not Used	---
15	BK/BL	Data Bus (-) Signal	Digital Signals
16	WT/BL	Data Bus (+) Signal	Digital Signals
17	RD/BK	High Speed Fan Control	Fan On: 1v, Off: 12v
18-20	---	Not Used	---
21	BK	CKP Sensor (+) Signal	400-425 Hz
22	WT	CKP Sensor (-) Signal	400-425 Hz
23	RD	TSS Sensor (-) Signal	340-380 Hz
24, 51	BK/YL	Power Ground	<0.1v
25	RD/BL	Shield Ground	<0.050v
26	GN/YL	Coil Driver 1 Control	5° dwell
27	LG/BL	Shift Solenoid 1 Control	0.1v, 55 mph: 12v
28-29, 34	---	Not Used	---
30	GN/WT	Octane Adjust Switch	9.3v (Closed: 0v)
31	YL/RD	PSP Sensor Signal	Straight: 0.5v, turned: 2v
32	R/GN	TR Sensor Reverse	Reverse: 12v, others: 0v
33	BL	VSS (-) Signal	0 Hz, 55 mph: 125 Hz
35	GY	HO2S-12 (B1 S2) Signal	0.1-1.1v
36	OR/BL	MAF Sensor Return	<0.050v
37	OR	TFT Sensor Signal	2.10 to 2.40v
38	BL/WT	ECT Sensor Signal	0.5-0.6v
39	WT/GN	IAT Sensor Signal	1.5-2.5v
40	BK/RD	Fuel Pump Monitor	12v, Off: 0v
41	W	A/C Cycling Clutch Switch	A/C Switch On: 12v
42	RD/YL	IMRC Solenoid Control	12v, 55 mph: 12v
43-44, 46	---	Not Used	---
45	PK/WT	Low Speed Cooling Fan	Fan On: 1v, Off: 12v
47	PK/RD	VR Solenoid Control	0%, 55 mph: 45%
48	LG/RD	Clean Tachometer Output	25-38 Hz
49-50	---	Not Used	---
52	BR/RD	Coil Driver 2 Control	5° dwell
53	OR/GN	Shift Solenoid 3 Control	12v, 55 mph: 12v
54	GY/WT	TCC Solenoid Control	0%, 55 mph: 95%
55	BL/RD	Keep Alive Power	12-14v
56	OR/WT	EVAP Purge Solenoid	0-10 Hz (0-100%)
57	RD	Knock Sensor 1 Signal	0v
58	WT/BK	VSS (+) Signal	0 Hz, 55 mph: 125 Hz
59, 61	---	Not Used	---
60	GY/GN	HO2S-11 (B1 S1) Signal	0.1-1.1v
62	BK/WT	FTP Sensor Signal	2.6v (0" H2O - cap off)
63	WT/PK	Fuel Rail Pressure Sensor	2.8v (39 psi)
64	RD	A/T: TR Sensor Signal	In 'P': 0.1v, O/D: 5v
64	BR/YL	M/T: CPP Switch Signal	Clutch In: 0v, Out: 5v

1998-99 Tracer 2.0L I4 MFI VIN P (All) 104 Pin Connector

PCM Pin #	Wire Color	Circuit Description (104 Pin)	Value at Hot Idle
65	BL/YL	DPFE Sensor Signal	0.20-1.30v
66, 68	---	Not Used	---
67	BL/PK	EVAP CV Solenoid	0-10 Hz (0-100%)
69	BL/BK	A/C WOT Relay Control	Relay Off: 12v, On: 1v
70	---	Not Used	---
71	WT/RD	Vehicle Power	12-14v
72	PK	M/T: Shift Indicator Lamp	SIL On: 1v, Off: 12v
73	---	Not Used	---
74	BR/WT	Injector 3 Control	3.3-3.7 ms
75	YL	Injector 1 Control	3.3-3.7 ms
76	BL	CMP Sensor (-) Signal	5-7 Hz
77	BK/YL	Power Ground	<0.1v
78-79	---	Not Used	---
80	WT/OR	Fuel Pump Control	Relay Off: 12v, On: 1v
81	---	Not Used	---
82	YL/PK	EPC Solenoid Control	2.5v (73%)
83	BL/OR	IAC Motor Control	10v (30% d/cycle)
84	WT	TSS Sensor (+) Signal	340-380 Hz
85	GN	CMP Sensor (+) Signal	5-7 Hz
86	GN/BK	A/C High Pressure Switch	AC On: 1.2-1.5v
87	---	Not Used	---
88	BR/BK	MAF Sensor Signal	0.8v
89	RD/WT	TP Sensor Signal	0.53-1.27v
90	LG/WT	Reference Voltage	4.9-5.1v
91	LG/BK	Analog Signal Return	<0.050v
92	GN	Brake Pedal Position Switch	Brake Off: 0v, On: 12v
93	OR/BK	HO2S-11 (B1 S1) Heater	1v, Off: 12v
94	---	Not Used	---
95	PK	HO2S-12 (B1 S2) Heater	1v, Off: 12v
96	---	Not Used	---
97	WT/RD	Vehicle Power	12-14v
98-99	---	Not Used	---
100	GN/OR	Injector 4 Control	3.3-3.7 ms
101	PK/YL	Injector 2 Control	3.3-3.7 ms
102	---	Not Used	---
103	BK/YL	Power Ground	<0.1v
104	---	Not Used	---

Pin Connector Graphic

PCM 104-PIN CONNECTOR

Terminal View of 104-Pin PCM Wiring Harness Connector

Standard Colors and Abbreviations

Abbreviation	Color	Abbreviation	Color	Abbreviation	Color
BK	Black	GY	Gray	PK	Purple
BL	Blue	GN	Green	RD	Red
BR	Brown	LG	LT Green	TN	Tan
DB	Dark Blue	OR	Orange	WT	White
DG	DK Green	PK	Pink	YL	Yellow

AVIATOR PIN TABLES

2003-04 Aviator 4.6L V8 DOHC 4v VIN H C175B 58 Pin Connector

PCM Pin #	Wire Color	Circuit Description (58 Pin)	Value at Hot Idle
1, 5	---	Not Used	---
2	DG/YL	Fuel Pump Power	Pump On: 12v, Off: 0v
3	TN/OR	SCP Data Bus (+) Signal	Digital Signals
4	PK/LB	SCP Data Bus (-) Signal	<0.050v
6	VT/WT	EVAP CV Solenoid	0-10 Hz (0-100%)
7	GY/BK	Vehicle Speed Sensor (+)	0 Hz, 55 mph: 125 Hz
8	LG/RD	Generator Monitor Signal	130 Hz (45%)
9	PK/YL	A/C Clutch Relay	12v (relay on: 1v)
10	YL/WT	Throttle Position Sensor 2 Signal	0.5-4.1v
11	WT/LG	Passive Antitheft TX Signal	Digital Signals
12	LG/BK	EVAP Canister Purge Solenoid Control	0-10 Hz (0-100%)
13	WT/VT	Module Programming Signal	0.1v
14	GY/OR	Passive Antitheft RX Signal	Digital Signals
15-16, 18	---	Not Used	---
17	GY/RD	Sensor Return	<0.050v
19	WT/BK	Auxiliary Condenser Cooling Fan Relay	Relay Off: 12v, On: 1v
20	BR/WT	Reference Voltage	4.9-5.1v
21	---	Not Used	---
22	DB/LG	Passive Antitheft Indicator Control	Indicator Off: 12v, On: 1v
23-24	LB/BK	4WD Indicator Low Signal	12v (switch on: 1v)
25-26	BK/WT	Communication Network Ground	<0.050v
27	BK/WT	Communication Network Ground	<0.050v
28	BK/YL	Brake Pressure Switch	12v (Brake On: 0v)
29	OR/LB	Speed Control Motor 'A' Control	DC pulse signals
30	BK/YL	A/C Cyclic Pressure Switch	A/C Off: 0v, On: 12v
31	LB/RD	MAF Sensor Signal	0.7v, 55 mph: 1.8v
32	VT	Vehicle Power (Start-Run)	12-14v
33	VT	Vehicle Power (Start-Run)	12-14v
34-35	---	Not Used	---
36	LG/WT	Speed Control Motor 'B' Control	DC pulse signals
37	YL/LG	Power Steering Pressure Switch	Straight: 0v, Turned: 12v
38	TN/LB	MAF Sensor Return	<0.050v
39	OR	Starter Motor Relay Circuit	Relay Off: 0v, On: 12v
40	LG	Brake Position Switch	Brake Off: 12v, On: 1v
41	TN/WT	Transmission Overdrive Cancel Switch	Switch Off: 0v, On: 12v
42, 48-50	---	Not Used	---
43	BK	Power Ground	<0.1v
44	RD/WT	Battery Power	12-14v
45	BK	Speed Control Switch Ground	<0.050v
46	BR/WT	Speed Control Motor 'B' Control	DC pulse signals
47	RD/YL	A/C Head Pressure Switch	Switch Open: 12v, Closed: 0v
51	GY	Air Charge Temperature Sensor	1.5-2.5v
52	RD/PK	Fuel Tank Pressure Sensor	2.6v (0" H20 - cap off)
53-55	---	Not Used	---
56	DG/OR	Speed Control Switch	Off: 0v, On: 6.7v
57	LB/BK	Speed Control On/Off to Amplifier	0v (switch on: 6.7v)
58	LB/OR	Fuel Pump Relay Control	On: 1v, Off: 12v

2003-04 Aviator 4.6L V8 DOHC 4v VIN H C175T 32 Pin Connector

PCM Pin #	Wire Color	Circuit Description (32 Pin)	Value at Hot Idle
1	OR/YL	Shift Solenoid 'A' Control	12v, 55 mph: 1v
2	VT/OR	Shift Solenoid 'B' Control	12v, 55 mph: 1v
3	PK/BK	Shift Solenoid 'C' Control	12v, 55 mph: 1v
4	BR/OR	Shift Solenoid 'D' Control	12v, 55 mph: 1v
5	VT/YL	TCC Solenoid Control	12v, 55 mph: 1v
6	---	Not Used	---
7	WT/YL	Electronic Pressure Solenoid Control	On: 1v, Off: 12v
8	---	Not Used	---
9	LB/YL	Digital TR Sensor (TR3A)	0v, 55 mph: 1.7v
10	WT/BK	Digital TR Sensor (TR4)	0v, 55 mph: 11v
11	---	Not Used	---
12	WT	Pressure Control Solenoid 'C'	On: 1v, Off: 12v
13	LB/PK	Pressure Control Solenoid 'B'	On: 1v, Off: 12v
14	---	Not Used	---
15	WT/BK	HO2S-12 (B1 S2) Heater Control	Heater Off: 12v, On: 1v
16	TN/YL	HO2S-22 (B2 S2) Heater Control	Heater Off: 12v, On: 1v
17	GY/RD	Sensor Return	<0.050v
18	LB/BK	Digital TR Sensor (TR2)	0v, 55 mph: 11v
19-20	---	Not Used	---
21	YL/LG	Overdrive Drum Speed Sensor	200 Hz, 55 mph: 1185 Hz
22	YL/BK	Digital TR Sensor (TR1)	0v, 55 mph: 11v
23	OR/BK	Transmission Fluid Temperature Sensor	2.10-2.40v
24-25	---	Not Used	---
26	DB/YL	Output Shaft Speed Sensor	0 Hz, 985 Hz
27	DG/WT	Turbine Shaft Speed Sensor	360 Hz, 890 Hz
28	RD/LG	HO2S-12 (B1 S2) Signal	0.1-1.1v
29	VT/LG	HO2S-22 (B2 S2) Signal	0.1-1.1v
30-32	---	Not Used	---

Pin Connector Graphic

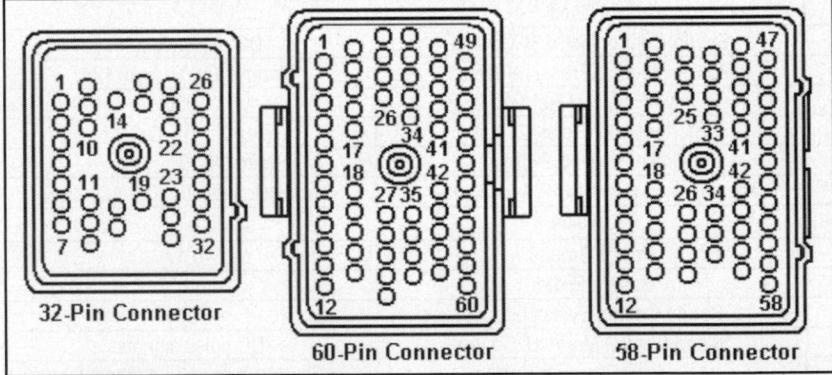

32-Pin Connector

60-Pin Connector

58-Pin Connector

Standard Colors and Abbreviations

Abbreviation	Color	Abbreviation	Color	Abbreviation	Color
BK	Black	GY	Gray	PK	Purple
BL	Blue	GN	Green	RD	Red
BR	Brown	LG	LT Green	TN	Tan
DB	Dark Blue	OR	Orange	WT	White
DG	DK Green	PK	Pink	YL	Yellow

2003-04 Aviator 4.6L V8 DOHC 4v VIN H C175E 60 Pin Connector

PCM Pin #	Wire Color	Circuit Description (60 Pin)	Value at Hot Idle
1	PK/WT	Coil On Plug (COP) 2 Driver (dwell)	5°, 55 mph: 7°
2	TN	Fuel Injector 1 Control	3.3-3.8 ms
3	DB/LG	Intake Manifold Communication Control	0%, over 3500 rpm: 100%
4-6	---	Not Used	---
7	RD/WT	HO2S-11 (B1 S1) Heater Control	Heater Off: 12v, On: 1v
8	YL/LB	HO2S-21 (B2 S1) Heater Control	Heater Off: 12v, On: 1v
9	WT/LB	IAC Solenoid Control	10.7v (33%)
10	---	Not Used	---
11	TN/BK	Fuel Injector 5 Control	3.3-3.8 ms
12	OR/YL	Coil On Plug (COP) 6 Driver (dwell)	5°, 55 mph: 7°
13	PK/LB	Coil On Plug (COP) 7 Driver (dwell)	5°, 55 mph: 7°
13	PK/WT	Coil 1 Driver (dwell)	5°, 55 mph: 7°
14	WT	Fuel Injector 2 Control	3.3-3.8 ms
15	---	Not Used	---
16	BR/PK	Electric VR Solenoid Control	0%, 55 mph: 45%
17	GY/RD	Sensor Return	<0.050v
18-19	---	Not Used	---
20	BR/WT	Sensor Voltage Reference	4.9-5.1v
21	LG/OR	Fuel Injector 6 Control	3.3-3.8 ms
22	LG/YL	Coil On Plug (COP) 5 Driver (dwell)	5°, 55 mph: 7°
23	WT/PK	Coil On Plug (COP) 3 Driver (dwell)	5°, 55 mph: 7°
24	BR/YL	Fuel Injector 3 Control	3.3-3.8 ms
25-28	---	Not Used	---
29	TN/RD	Fuel Injector 3 Control	3.3-3.8 ms
30	DG/VT	Coil On Plug (COP) 4 Driver (dwell)	5°, 55 mph: 7°
31	LG/WT	Coil On Plug (COP) 1 Driver (dwell)	5°, 55 mph: 7°
32	BR/LB	Fuel Injector 4 Control	3.3-3.8 ms
33-37	---	Not Used	---
38	WT/RD	Fuel Injector 8 Control	3.3-3.8 ms
39	---	Not Used	---
40	YL/LG	Cylinder Head Temperature Sensor	0.5-0.6v
41	BR/LG	EGR DPFE Sensor Signal	0.95-1.05v
42	YL	Knock Sensor 1 Ground	<0.050v
43	DG/WT	Knock Sensor 2 Ground	0v
44	RD/BK	HO2S-21 (B2 S1) Signal	0.1-1.1v
45	GY/LB	HO2S-11 (B1 S1) Signal	0.1-1.1v
46	LG/RD	Engine Coolant Temperature Sensor	0.5-0.6v
47-48	---	Not Used	---
51	YL/RD	Knock Sensor 1 Signal	0v
52	DG/VT	Knock Sensor 2 Signal	0v
53	DB/OR	Camshaft Position Sensor Signal	6 Hz
54	---	Not Used	---
55	DB	CKP Sensor (+) Signal	400 Hz
56	GY	CKP Sensor (-) Signal	400 Hz
57	GY/WT	TP Sensor Signal	0.53-1.27v
58	LG/WT	Intake Manifold Communication Sensor	0.5-4.5v
59	LG/BK	Manifold Absolute Pressure Sensor	4v, Over 3500 rpm: 1v
60	---	Not Used	---

Pin Connector Graphic

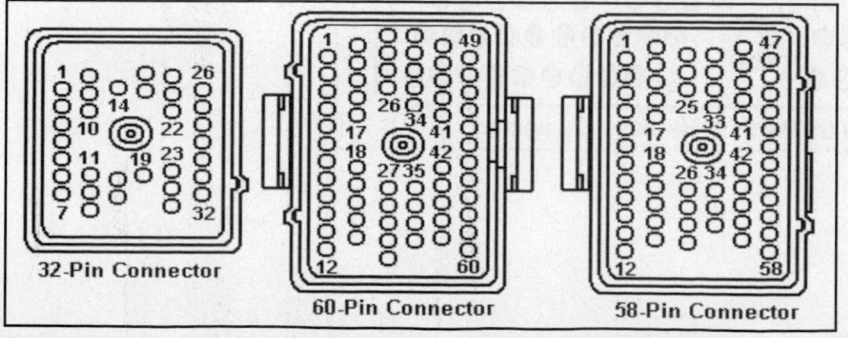

32-Pin Connector   60-Pin Connector   58-Pin Connector

BRONCO PIN TABLES

1990 Bronco 4.9L I6 MFI VIN Y 60 Pin Connector

PCM Pin #	Wire Color	Circuit Description (60 Pin)	Value at Hot Idle
1	YL	Keep Alive Power	12-14v
3	GY/BK	VSS (+) Signal	0 Hz, 55 mph: 125 Hz
4	DG/YL	Ignition Diagnostic Monitor	20-31 Hz
6	BK	VSS (-) Signal	0 Hz, 55 mph: 125 Hz
7	LG/YL	ECT Sensor Signal	0.5-0.6v
8	BR	Fuel Pump Monitor	On: 12v, Off: 0v
9	BK/OR	Data Bus (-) Signal	Digital Signals
10	BK/YL	A/C Switch Signal	A/C On: 12v, Off: 0v
11	WT/BK	Air Management 2 Solenoid	AM2 On: 1v, Off: 12v
16	BK/OR	Ignition System Ground	<0.050v
17	PK/LG	Self-Test Output & MIL	MIL On: 1v, Off: 12v
18	TN/LG	Inferred Mileage Sensor	Digital Signals
20	BK	PCM Case Ground	<0.050v
21	GY/WT	IAC Solenoid Control	8.8-10.0v
22	TN/LG	Fuel Pump Control	On: 1v, Off: 12v
23	LG/BK	Knock Sensor Signal	0v
24	YL/LG	PSP Switch Signal	Straight: 0v, Turned: 12v
25	YL/RD	ACT Sensor Signal	1.5-2.5v
26	OR/WT	Reference Voltage	4.9-5.1v
27	BR/LG	EVP Sensor Signal	0.4v
28	OR/BK	Data Bus (+) Signal	Digital Signals
29	GY/BL	HO2S-11 (B1 S1) Signal	0.1-1.1v
30	GY/YL	A/T: Neutral Drive Switch	In 'N': 0v, Others: 5v
30	GY/YL	M/T: CPP Switch Signal	Clutch In: 0v, Out: 5v
31	GY/YL	EVAP Purge Solenoid	12v, 55 mph: 1v
33	DG	VR Solenoid Control	0%, 55 mph: 45%
36	YL/LG	Spark Output Signal	6.93v (50%)
37	RD	Vehicle Power	12-14v
40	BK/LG	Power Ground	<0.1v
41-42	---	Not Used	---
43	LG/PK	A/C Demand Switch	A/C On: 12v, Off: 0v
45	LG/BK	MAP Sensor Signal	107 Hz (sea level)
46	BK/WT	Analog Signal Return	<0.050v
47	DG/LG	TP Sensor Signal	0.5-1.2v
48	WT/RD	Self-Test Indicator Signal	STI On: 1v, Off: 5v
49	OR	HO2S-11 (B1 S1) Ground	<0.050v
50	---	Not Used	---
51	WT/RD	Air Management 1 Solenoid	1v, 55 mph: 12v
52-55	---	Not Used	---
56	DB	PIP Sensor Signal	6.93v (50%)
57	RD	Vehicle Power	12-14v
58	TN/OR	Injector Bank 1 (INJ 1, 3 & 5)	6.4-6.8 ms
59	TN/RD	Injector Bank 2 (INJ 2, 4 & 6)	6.4-6.8 ms
60	BK/LG	Power Ground	<0.1v

Pin Connector Graphic

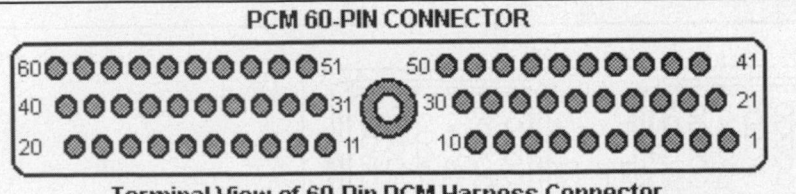

PCM 60-PIN CONNECTOR

Terminal View of 60-Pin PCM Harness Connector

1990 Bronco 4.9L I6 MFI VIN Y (E4OD) 60 Pin Connector

PCM Pin #	Wire Color	Circuit Description (60 Pin)	Value at Hot Idle
1	YL	Keep Alive Power	12-14v
2	LG	Brake Pedal Switch	Brake Off: 12v, On: 1v
3	GY/BK	VSS (+) Signal	0 Hz, 55 mph: 125 Hz
4	DG/YL	Ignition Diagnostic Monitor	20-31 Hz
6	BK	VSS (-) Signal	0 Hz, 55 mph: 125 Hz
7	LG/YL	ECT Sensor Signal	0.5-0.6v
8	BR	Fuel Pump Monitor	On: 12v, Off: 0v
9	BK/OR	Data Bus (-) Signal	Digital Signals
10	BK/YL	A/C Switch Signal	A/C On: 12v, Off: 0v
11	WT/BK	Air Management 2 Solenoid	AM2 On: 1v, Off: 12v
12	BL/BK	4x4 Switch Signal	12v (switch closed: 0v)
16	BK/OR	Ignition System Ground	<0.050v
17	PK/LG	Self-Test Output & MIL	MIL On: 1v, Off: 12v
18	TN/LG	Inferred Mileage Sensor	Digital Signals
19	PK/OR	Shift Solenoid 2 Control	1v, 55 mph: 12v
20	BK/LG	Case Ground	<0.1v
21	GY/WT	IAC Solenoid Control	9.8-10.6v
22	TN/LG	Fuel Pump Control	On: 1v, Off: 12v
23	LG/BK	Knock Sensor Signal	0v
24	YL/LG	PSP Switch Signal	Straight: 0v, Turned: 12v
25	YL/RD	ACT Sensor Signal	1.5-2.5v
26	OR/WT	Reference Voltage	4.9-5.1v
27	BR/LG	EVP Sensor Signal	0.4v
28	OR/BK	Data Bus (+) Signal	Digital Signals
29	GY/BL	HO2S-11 (B1 S1) Signal	0.1-1.1v
30	BL/YL	MLP Sensor Signal	In 'P': 0v, in O/D: 5v
31	GY/YL	EVAP Purge Solenoid	12v, 55 mph: 1v
32	LG/WT	Overdrive Cancel Indicator	OCIL On: 1v, Off: 12v
33	DG	VR Solenoid Control	0%, 55 mph: 45%
36	YL/LG	Spark Output Signal	6.93v (50%)
37	RD	Vehicle Power	12-14v
38	WT/YL	EPC Solenoid Control	9.5v (5 psi)
40	BK/LG	Power Ground	<0.1v
41	WT	Overdrive Cancel Switch	OCS & O/D On: 0.1v
42	OR/BK	TOT Sensor Signal	2.10-2.40v
43	LG/PK	A/C Demand Switch	A/C On: 12v, Off: 0v
45	LG/BK	MAP Sensor Signal	107 Hz (sea level)
46	BK/WT	Analog Signal Return	<0.050v
47	DG/LG	TP Sensor Signal	0.5-1.2v
48	WT/RD	Self-Test Indicator Signal	STI On: 1v, Off: 5v
49	OR	HO2S-11 (B1 S1) Ground	<0.050v
51	WT/RD	Air Management 1 Solenoid	1v, 55 mph: 12v
52	OR/YL	Shift Solenoid 1 Control	12v, 55 mph: 1v
53	PK/YL	TCC Solenoid Control	12v, 55 mph: 1v
55	BR	Coast Clutch Switch Signal	12v, 55 mph: 12v
56	DB	PIP Sensor Signal	6.93v (50%)
57	RD	Vehicle Power	12-14v
58	TN/OR	Injector Bank 1 (INJ 1, 3 & 5)	6.4-6.8 ms
59	TN/RD	Injector Bank 2 (INJ 2, 4 & 6)	6.4-6.8 ms
60	BK/LG	Power Ground	<0.1v

Pin Connector Graphic

PCM 60-PIN CONNECTOR

Terminal View of 60-Pin PCM Harness Connector

1991-94 Bronco 4.9L I6 MFI VIN Y 60 Pin Connector

PCM Pin #	Wire Color	Circuit Description (60 Pin)	Value at Hot Idle
1	YL	Keep Alive Power	12-14v
3	GY/BK	VSS (+) Signal	0 Hz, 55 mph: 125 Hz
4	TN/YL	Ignition Diagnostic Monitor	20-31 Hz
4	YL/BK	Ignition Diagnostic Monitor	20-31 Hz
6	PK/OR	VSS (-) Signal	0 Hz, 55 mph: 125 Hz
7	LG/RD	ECT Sensor Signal	0.5-0.6v
8	DG/YL	Fuel Pump Monitor	On: 12v, Off: 0v
9	PK/BL	Data Bus (-) Signal	Digital Signals
10	BK/YL	A/C Switch Signal	A/C On: 12v, Off: 0v
10	PK/BL	A/C Switch Signal	A/C On: 12v, Off: 0v
11	BR	Air Management 2 Solenoid	AM2 On: 1v, Off: 12v
16	OR/RD	Ignition System Ground	<0.050v
17	PK/LG	Self-Test Output & MIL	MIL On: 1v, Off: 12v
20	BK	PCM Case Ground	<0.050v
21	WT/BL	IAC Solenoid Control	10.7v (33%)
21	BL/WT	IAC Solenoid Control	10.7v (33%)
22	BL/OR	Fuel Pump Control	On: 1v, Off: 12v
23	YL/RD	Knock Sensor Signal	0v
25	GN/YL	IAT Sensor Signal	1.5-2.5v
25	GY	IAT Sensor Signal	1.5-2.5v
26	BR/WT	Reference Voltage	4.9-5.1v
27	BR/LG	EVP Sensor Signal	0.4v
28	TN/OR	Data Bus (+) Signal	Digital Signals
29	GY/BL	HO2S-11 (B1 S1) Signal	0.1-1.1v
30	BL/YL	A/T: Neutral Drive Switch	In 'N': 0v, Others: 5v
30	BL/YL	M/T: CPP Switch Signal	Clutch In: 0v, Out: 5v
31	GY/YL	EVAP Purge Solenoid	12v, 55 mph: 1v
32	WT/LG	Overdrive Cancel Indicator	OCIL On: 1v, Off: 12v
33	BR/PK	VR Solenoid Control	0%, 55 mph: 45%
36	PK	Spark Output Signal	6.93v (50%)
37	RD	Vehicle Power	12-14v
40	BK/WT	Power Ground	<0.1v
43	PK	A/C Demand Switch	A/C On: 12v, Off: 0v
45	LG/BK	MAP Sensor Signal	107 Hz (sea level)
46	GY/RD	Analog Signal Return	<0.050v
47	GY/WT	TP Sensor Signal	1.5-1.2v
48	WT/PK	Self-Test Indicator Signal	STI On: 1v, Off: 5v
49	OR	HO2S-11 (B1 S1) Ground	<0.050v
51	WT/OR	Air Management 1 Solenoid	1v, 55 mph: 12v
56	GY/OR	PIP Sensor Signal	6.93v (50%)
57	RD	Vehicle Power	12-14v
58	TN	Injector Bank 1 (INJ 1, 3 & 5)	5.6-6.4 ms
59	WT	Injector Bank 2 (INJ 2, 4 & 6)	5.6-6.4 ms
60	BK/WT	Power Ground	<0.1v

Pin Connector Graphic

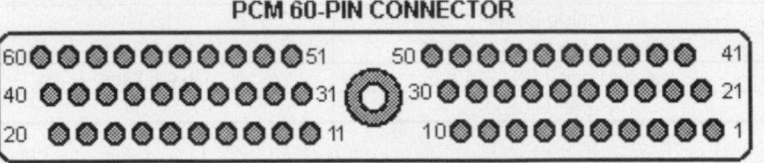

1991-94 Bronco 4.9L I6 MFI VIN Y (E4OD) 60 Pin Connector

PCM Pin #	Wire Color	Circuit Description (60 Pin)	Value at Hot Idle
1	YL	Keep Alive Power	12-14v
2	LG	Brake Pedal Switch	Brake Off: 12v, On: 1v
3	GY/BK	VSS (+) Signal	0 Hz, 55 mph: 125 Hz
4	TN or YL/BK	Ignition Diagnostic Monitor	20-31 Hz
5, 13-15	---	Not Used	---
6	PK/OR	VSS (-) Signal	0 Hz, 55 mph: 125 Hz
7	LG/RD	ECT Sensor Signal	0.5-0.6v
8	DG/YL	Fuel Pump Monitor	On: 12v, Off: 0v
9	PK/BL	Data Bus (-) Signal	Digital Signals
10	BK or PK/BL	A/C Switch Signal	A/C On: 12v, Off: 0v
11	BR	Air Management 2 Solenoid	AM2 On: 1v, Off: 12v
12	BL or PK	4x4 Switch Signal	12v (switch closed: 0v)
16	OR/RD	Ignition System Ground	<0.050v
17	PK/LG	Self-Test Output & MIL	MIL On: 1v, Off: 12v
18, 24	---	Not Used	---
19	PK/OR	Shift Solenoid 2 Control	1v, Off: 12v
20	BK	PCM Case Ground	<0.050v
21	WT or BK	IAC Solenoid Control	10.7v (33%)
22	BL/OR	Fuel Pump Control	On: 1v, Off: 12v
23	YL/RD	Knock Sensor Signal	0v
25	GN/YL or GY	ACT Sensor Signal	1.5-2.5v
26	BR/WT	Reference Voltage	4.9-5.1v
27	BR/LG	EVP Sensor Signal	0.4v
28	TN/OR	Data Bus (+) Signal	Digital Signals
29	GY/BL	HO2S-11 (B1 S1) Signal	0.1-1.1v
30	BL/YL	MLP Sensor Signal	In 'P': 0v, in O/D: 5v
31	GY/YL	EVAP Purge Solenoid	12v, 55 mph: 1v
32	WT/LG	Overdrive Cancel Indicator	OCIL On: 1v, Off: 12v
33	BR/PK	VR Solenoid Control	0%, 55 mph: 45%
34-35, 39, 44	---	Not Used	---
36	PK	Spark Output Signal	6.93v (50%)
37	RD	Vehicle Power	12-14v
38	WT/YL	EPC Solenoid Control	9.5v (5 psi)
40	BK/WT	Power Ground	<0.1v
41	TN/WT	Overdrive Cancel Switch	OCS Closed: 0.1v
42	OR/BK	TOT Sensor Signal	2.10-2.40v
43	PK	A/C Demand Switch	A/C On: 12v, Off: 0v
45	LG/BK	MAP Sensor Signal	107 Hz (sea level)
46	GY/RD	Analog Signal Return	<0.050v
47	GY/WT	TP Sensor Signal	0.5-1.2v
48	WT/PK	Self-Test Indicator Signal	STI On: 1v, Off: 5v
49	OR	HO2S-11 (B1 S1) Ground	<0.050v
50, 54	---	Not Used	---
51	WT/OR	Air Management 1 Solenoid	1v, 55 mph: 12v
52	OR/YL	Shift Solenoid 1 Control	12v, 55 mph: 1v
53	PK/YL	TCC Solenoid Control	12v, 55 mph: 9v
55	BR/OR	Coast Clutch Switch Signal	12v, 55 mph: 12v
56	GY/OR	PIP Sensor Signal	6.93v (50%)
57	RD	Vehicle Power	12-14v
58	TN	Injector Bank 1 (INJ 1, 3 & 5)	5.6-6.0 ms
59	W	Injector Bank 2 (INJ 2, 4 & 6)	5.6-6.0 ms
60	BK/WT	Power Ground	<0.1v

PCM 60-PIN CONNECTOR

Terminal View of 60-Pin PCM Harness Connector

1995 Bronco 4.9L I6 MFI VIN Y California 60 Pin Connector

PCM Pin #	Wire Color	Circuit Description (60 Pin)	Value at Hot Idle
2	LG	Brake Pedal Switch	Brake Off: 12v, On: 1v
3	GY/BK	VSS (+) Signal	0 Hz, 55 mph: 125 Hz
4	YL/BK	Ignition Diagnostic Monitor	20-31 Hz
6	PK/OR	VSS (-) Signal	0 Hz, 55 mph: 125 Hz
7	LG/RD	ECT Sensor Signal	0.5-0.6v
8	DG/YL	Fuel Pump Monitor	On: 12v, Off: 0v
9	TN/BL	MAF Sensor Return	<0.050v
10	DG/OR	A/C Switch Signal	A/C On: 12v, Off: 0v
11	GY/YL	EVAP Purge Solenoid	12v, 55 mph: 1v
12	LG/OR	Injector 6 Control	5.5-7.1 ms
15	TN/BK	Injector 5 Control	5.5-7.1 ms
16	TN/YL	Ignition System Ground	<0.050v
17	PK/LG	Self-Test Output & MIL	MIL On: 1v, Off: 12v
18	TN/OR	Data Bus (-) Signal	Digital Signals
19	PK/BK	Data Bus (+) Signal	Digital Signals
20	BK	PCM Case Ground	<0.050v
21	WT/BL	IAC Solenoid Control	10.7v (33%)
22	BL/OR	Fuel Pump Control	On: 1v, Off: 12v
23	YL/RD	Knock Sensor Signal	0v
25	GY	IAT Sensor Signal	1.5-2.5v
26	BR/WT	Reference Voltage	4.9-5.1v
27	BR/LG	EVP Sensor Signal	0.4v
30	BL/YL	A/T: Neutral Drive Switch	In 'N': 0v, Others: 5v
30	GY/YL	M/T: CPP Switch Signal	Clutch In: 0v, Out: 5v
31	WT/OR	Thermactor Air Bypass	TAB On: 1v, Off: 12v
32	WT/LG	TCIL (lamp) Control	TCIL On: 1v, Off: 12v
33	BR/PK	VR Solenoid Control	0%, 55 mph: 45%
34	BR	Thermactor Air Diverter	TAD On: 1v, Off: 12v
35	BR/BL	Injector 4 Control	5.5-7.1 ms
36	PK	Spark Output Signal	6.93v (50%)
37, 57	RD	Vehicle Power	12-14v
39	BR/YL	Injector 3 Control	5.5-7.1 ms
40	BK/WT	Power Ground	<0.1v
41	TN/WT	TCS (switch) Signal	TCS & O/D On: 12v
42	BL/YL	4x4 Switch Signal	12v (switch closed: 0v)
43	RD/BK	HO2S-21 (B2 S1) Signal	0.1-1.1v
44	GY/BL	HO2S-11 (B1 S1) Signal	0.1-1.1v
46	GY/RD	Analog Signal Return	<0.050v
47	GY/WT	TP Sensor Signal	0.5-1.2v
48	WT/PK	Self-Test Indicator Signal	STI On: 1v, Off: 5v
50	BL/RD	MAF Sensor Signal	0.8v
54	PK/YL	A/C WOT Relay Control	On: 1v, Off: 12v
56	GY/OR	PIP Sensor Signal	6.93v (50%)
58	TN	Injector 1 Control	5.5-7.1 ms
59	WT	Injector 2 Control	5.5-7.1 ms
60	BK/WT	Power Ground	<0.1v

Pin Connector Graphic

Terminal View of 60-Pin PCM Harness Connector

1995 Bronco 4.9L I6 VIN Y (All) 60 Pin Connector

PCM Pin #	Wire Color	Circuit Description (60 Pin)	Value at Hot Idle
1	YL	Keep Alive Power	12-14v
2	LG	Brake Switch Signal	Brake Off: 12v, On: 1v
3	GY/BK	VSS (+) Signal	0 Hz, 55 mph: 125 Hz
4	YL/BK	Ignition Diagnostic Monitor	20-31 Hz
6	PK/OR	VSS (-) Signal	0 Hz, 55 mph: 125 Hz
7	LG/RD	ECT Sensor Signal	0.5-0.6v
8	DG/YL	Fuel Pump Monitor	On: 12v, Off: 0v
9	PK/BL	Data Bus (-) Signal	Digital Signals
10	PK/BL	A/C Switch Signal	A/C On: 12v, Off: 0v
11	BR	Air Management 2 Solenoid	AM2 On: 1v, Off: 12v
12	BL/BK	4x4 Indicator Lamp	12v (switch closed: 0v)
16	OR/RD	Ignition System Ground	<0.050v
17	PK/LG	Self-Test Output & MIL	MIL On: 1v, Off: 12v
20	BK	PCM Case Ground	<0.050v
21	WT/BL	IAC Solenoid Control	10.7v (33%)
22	BL/OR	Fuel Pump Control	On: 1v, Off: 12v
23	YL/RD	Knock Sensor Signal	0v
25	GY	IAT Sensor Signal	1.5-2.5v
26	BR/WT	Reference Voltage	4.9-5.1v
27	BR/LG	EVP Sensor Signal	0.4v
28	TN/OR	Data Bus (+) Signal	Digital Signals
29	GY/BL	HO2S-11 (B1 S1) Signal	0.1-1.1v
30	GY/YL	A/T: Neutral Drive Switch	In 'N': 0v, Others: 5v
30	BL/YL	M/T: CPP Switch Signal	Clutch In: 0v, Out: 5v
31	GY/YL	EVAP Purge Solenoid	12v, 55 mph: 1v
32	WT/LG	TCIL (lamp) Control	TCIL On: 1v, Off: 12v
33	BR/PK	VR Solenoid Control	0%, 55 mph: 45%
36	PK	Spark Output Signal	6.93v (50%)
37	RD	Vehicle Power	12-14v
40	BK/WT	Power Ground	<0.1v
41	TN/WT	TCS (switch) Signal	0.1-0.2v
43	PK	A/C Demand Switch	A/C On: 12v, Off: 0v
45	LG/BK	MAP Sensor Signal	107 Hz (sea level)
46	GY/RD	Analog Signal Return	<0.050v
47	GY/WT	TP Sensor Signal	0.53-1.27v
48	WT/PK	Self-Test Indicator Signal	STI On: 1v, Off: 5v
49	OR	HO2S-11 (B1 S1) Ground	<0.050v
51	WT/OR	Air Management 1 Solenoid	AM1 On: 1v, 55 mph: 12v
56	GY/OR	PIP Sensor Signal	6.93v (50%)
57	RD	Vehicle Power	12-14v
58	TN	Injector Bank 1 (INJ 1, 3 & 5)	6.2-7.4 ms
59	WT	Injector Bank 2 (INJ 2, 4 & 6)	6.2-7.4 ms
60	BK/WT	Power Ground	<0.1v

Pin Connector Graphic

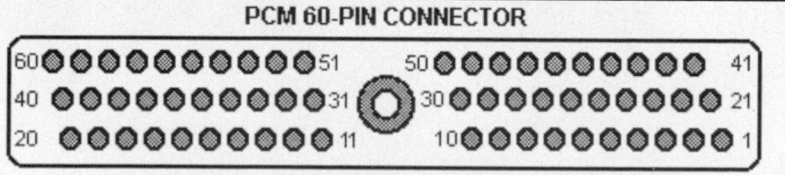

PCM 60-PIN CONNECTOR

Terminal View of 60-Pin PCM Harness Connector

1995 Bronco 4.9L I6 VIN Y (E4OD) California 60 Pin Connector

PCM Pin #	Wire Color	Circuit Description (60 Pin)	Value at Hot Idle
1	YL	Keep Alive Power	12-14v
2	LG	Brake Pedal Switch	Brake Off: 12v, On: 1v
3	GY/BK	VSS (+) Signal	0 Hz, 55 mph: 125 Hz
4	YL/BK	Ignition Diagnostic Monitor	20-31 Hz
5	---	Not Used	---
6	PK/OR	VSS (-) Signal	0 Hz, 55 mph: 125 Hz
7	LG/RD	ECT Sensor Signal	0.5-0.6v
8	DG/YL	Fuel Pump Monitor	On: 12v, Off: 0v
9	TN/BL	MAF Sensor Return	<0.050v
10	DG/OR	A/C Switch Signal	A/C On: 12v, Off: 0v
11	GY/YL	EVAP Purge Solenoid	12v, 55 mph: 1v
12	LG/OR	Injector 6 Control	5.5-7.1 ms
13-14	---	Not Used	---
15	TN/BK	Injector 5 Control	5.5-7.1 ms
16	TN/YL	Ignition System Ground	<0.050v
17	PK/LG	Self-Test Output & MIL	MIL On: 1v, Off: 12v
18	TN/OR	Data Bus (-) Signal	Digital Signals
19	PK/BK	Data Bus (+) Signal	Digital Signals
20	BK	PCM Case Ground	<0.050v
21	WT/BL	IAC Solenoid Control	10.7v (33%)
22	BL/OR	Fuel Pump Control	On: 1v, Off: 12v
23	YL/RD	Knock Sensor Signal	0v
24	---	Not Used	---
25	GY	IAT Sensor Signal	1.5-2.5v
26	BR/WT	Reference Voltage	4.9-5.1v
27	BR/LG	EVP Sensor Signal	0.4v
28-29	---	Not Used	---
30	BL/YL	TR Sensor Signal	In 'P': 0v, in O/D: 5v
30	GY/YL	TR Sensor Signal	In 'P': 0v, in O/D: 5v
31	WT/OR	Thermactor Air Bypass	TAB On: 1v, Off: 12v
32	WT/LG	TCIL (lamp) Control	TCIL On: 1v, Off: 12v
33	BR/PK	VR Solenoid Control	0%, 55 mph: 45%
34	BR	Thermactor Air Diverter	TAD On: 1v, Off: 12v
35	BR/BL	Injector 4 Control	5.5-7.1 ms
36	PK	Spark Output Signal	6.93v (50%)
37	RD	Vehicle Power	12-14v
38	WT/YL	EPC Solenoid Control	9.5v (5 psi)
39	BR/YL	Injector 3 Control	5.5-7.1 ms
40	BK/WT	Power Ground	<0.1v

1995 Bronco 4.9L I6 VIN Y (E4OD) California 60 Pin Connector

PCM Pin #	Wire Color	Circuit Description (60 Pin)	Value at Hot Idle
41	TN/WT	TCS (switch) Signal	TCS & O/D On: 12v
42	BL/YL	4x4 Switch Signal	12v (switch closed: 0v)
43	RD/BK	HO2S-21 (B2 S1) Signal	0.1-1.1v
44	GY/BL	HO2S-11 (B1 S1) Signal	0.1-1.1v
45	---	Not Used	---
46	GY/RD	Analog Signal Return	<0.050v
47	GY/WT	TP Sensor Signal	0.5-1.2v
48	WT/PK	Self-Test Indicator Signal	STI On: 1v, Off: 5v
49	OR/BK	TOT Sensor Signal	2.10-2.40v
50	BL/RD	MAF Sensor Signal	0.8v
51	OR/YL	Shift Solenoid 1 Control	12v, 55 mph: 1v
52	PK/OR	Shift Solenoid 2 Control	1v, 30 mph: 12v
53	PK/YL	TCC Solenoid Control	12v, 55 mph: 9v
54	PK/YL	A/C WOT Relay Control	On: 1v, Off: 12v
55	BR/OR	Coast Clutch Switch Signal	12v, 55 mph: 12v
56	GY/OR	PIP Sensor Signal	6.93v (50%)
57	RD	Vehicle Power	12-14v
58	TN	Injector 1 Control	5.5-7.1 ms
59	WT	Injector 2 Control	5.5-7.1 ms
60	BK/WT	Power Ground	<0.1v

Pin Connector Graphic

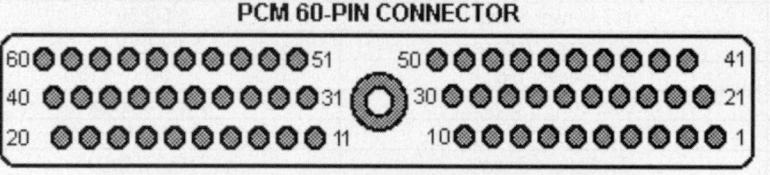

PCM 60-PIN CONNECTOR

Terminal View of 60-Pin PCM Harness Connector

Standard Colors and Abbreviations

Abbreviation	Color	Abbreviation	Color	Abbreviation	Color
BK	Black	GY	Gray	PK	Purple
BL	Blue	GN	Green	RD	Red
BR	Brown	LG	LT Green	TN	Tan
DB	Dark Blue	OR	Orange	WT	White
DG	DK Green	PK	Pink	YL	Yellow

1995 Bronco 4.9L I6 VIN Y (E4OD) 60 Pin Connector

PCM Pin #	Wire Color	Circuit Description (60 Pin)	Value at Hot Idle
1	YL	Keep Alive Power	12-14v
2	LG	Brake Switch Signal	Brake Off: 12v, On: 1v
3	GY/BK	VSS (+) Signal	0 Hz, 55 mph: 125 Hz
4	YL/BK	Ignition Diagnostic Monitor	20-31 Hz
5	---	Not Used	---
6	PK/OR	VSS (-) Signal	0 Hz, 55 mph: 125 Hz
7	LG/RD	ECT Sensor Signal	0.5-0.6v
8	DG/YL	Fuel Pump Monitor	On: 12v, Off: 0v
9	PK/BL	Data Bus (-) Signal	Digital Signals
10	PK/BL	A/C Switch Signal	A/C On: 12v, Off: 0v
11	BR	Air Management 2 Solenoid	AM2 On: 1v, Off: 12v
12	BL/BK	4x4 Indicator Lamp	12v (switch closed: 0v)
13-15	---	Not Used	---
16	OR/RD	Ignition System Ground	<0.050v
17	PK/LG	Self-Test Output & MIL	MIL On: 1v, Off: 12v
18	---	Not Used	---
19	PK/OR	Shift Solenoid 2 Control	12v, 55 mph: 1v
20	BK	PCM Case Ground	<0.050v
21	WT/BL	IAC Solenoid Control	10.7v (33%)
22	BL/OR	Fuel Pump Control	On: 1v, Off: 12v
23	YL/RD	Knock Sensor Signal	0v
24	---	Not Used	---
25	GY	IAT Sensor Signal	1.5-2.5v
26	BR/WT	Reference Voltage	4.9-5.1v
27	BR/LG	EVP Sensor Signal	0.4v
28	TN/OR	Data Bus (+) Signal	Digital Signals
29	GY/BL	HO2S-11 (B1 S1) Signal	0.1-1.1v
30	GY/YL	TR Sensor Signal	In 'P': 0v, in O/D: 5v
30	BL/YL	TR Sensor Signal	In 'P': 0v, in O/D: 5v
31	GY/YL	EVAP Purge Solenoid	12v, 55 mph: 1v
32	WT/LG	TCIL (lamp) Control	TCIL On: 1v, Off: 12v
33	BR/PK	VR Solenoid Control	0%, 55 mph: 45%
34-35	---	Not Used	---
36	PK	Spark Output Signal	6.93v (50%)
37	RD	Vehicle Power	12-14v
38	WT/YL	EPC Solenoid Control	9.5v (5 psi)
39	---	Not Used	---
40	BK/WT	Power Ground	<0.1v

1995 Bronco 4.9L I6 VIN Y (E4OD) 60 Pin Connector

PCM Pin #	Wire Color	Circuit Description (60 Pin)	Value at Hot Idle
41	TN/WT	TCS (switch) Signal	TCS & O/D On: 12v
42	OR/BK	TOT Sensor Signal	2.10-2.40v
43	PK	A/C Demand Switch	A/C On: 12v, Off: 0v
44	---	Not Used	---
45	LG/BK	MAP Sensor Signal	107 Hz (sea level)
46	GY/RD	Analog Signal Return	<0.050v
47	GY/WT	TP Sensor Signal	0.5-1.2v
48	WT/PK	Self-Test Indicator Signal	STI On: 1v, Off: 5v
49	OR	HO2S-11 (B1 S1) Ground	<0.050v
50	---	Not Used	---
51	WT/OR	Air Management 1 Solenoid	1v, 55 mph: 12v
52	OR/YL	Shift Solenoid 1 Control	1v, 30 mph: 12v
53	PK/YL	TCC Solenoid Control	12v, 55 mph: 9v
54	---	Not Used	---
55	BR/OR	Coast Clutch Solenoid	12v, 55 mph: 12v
56	GY/OR	PIP Sensor Signal	6.93v (50%)
57	RD	Vehicle Power	12-14v
58	TN	Injector Bank 1 (INJ 1, 3 & 5)	6.2-7.4 ms
59	WT	Injector Bank 2 (INJ 2, 4 & 6)	6.2-7.4 ms
60	BK/WT	Power Ground	<0.1v

Pin Connector Graphic

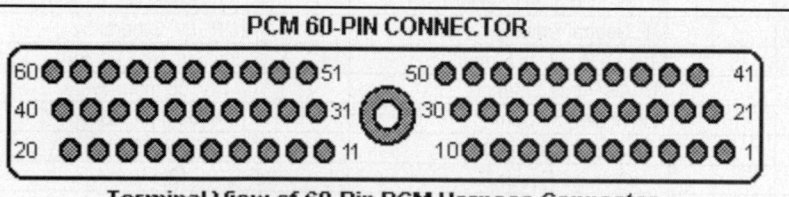

PCM 60-PIN CONNECTOR

Terminal View of 60-Pin PCM Harness Connector

Standard Colors and Abbreviations

Abbreviation	Color	Abbreviation	Color	Abbreviation	Color
BK	Black	GY	Gray	PK	Purple
BL	Blue	GN	Green	RD	Red
BR	Brown	LG	LT Green	TN	Tan
DB	Dark Blue	OR	Orange	WT	White
DG	DK Green	PK	Pink	YL	Yellow

1990 Bronco 5.0L V8 MFI VIN N (All) 60 Pin Connector

PCM Pin #	Wire Color	Circuit Description (60 Pin)	Value at Hot Idle
1	YL	Keep Alive Power	12-14v
3	GY/BK	VSS (+) Signal	0 Hz, 55 mph: 125 Hz
3	DG/WT	VSS (+) Signal	0 Hz, 55 mph: 125 Hz
4	DG/YL	Ignition Diagnostic Monitor	20-31 Hz
6	BK	VSS (-) Signal	0 Hz, 55 mph: 125 Hz
7	LG/YL	ECT Sensor Signal	0.5-0.6v
8	BR	Fuel Pump Monitor	On: 12v, Off: 0v
10	BK/YL	A/C Switch Signal	A/C On: 12v, Off: 0v
11	WT/BK	Air Management 2 Solenoid	AM2 On: 1v, Off: 12v
16	BK/OR	Ignition System Ground	<0.050v
17	PK/LG	Self-Test Output & MIL	MIL On: 1v, Off: 12v
18	TN/LG	Inferred Mileage Sensor	Digital Signals
20	BK	PCM Case Ground	<0.050v
21	GY/WT	IAC Solenoid Control	9.0v
22	TN/LG	Fuel Pump Control	On: 1v, Off: 12v
23	LG/BK	Knock Sensor Signal	0v
24	YL/LG	PSP Switch Signal	Straight: 0v, Turned: 12v
25	YL/RD	ACT Sensor Signal	1.5-2.5v
26	OR/WT	Reference Voltage	4.9-5.1v
27	BR/LG	EVP Sensor Signal	0.4v
29	GY/BK	HO2S-11 (B1 S1) Signal	0.1-1.1v
29	DG/PK	HO2S-11 (B1 S1) Signal	0.1-1.1v
30	GY/YL	A/T: Neutral Drive Switch	In 'P': 0v, Others: 5v
30	BL/YL	M/T: CPP Switch Signal	Clutch In: 0v, Out: 5v
31	GY/YL	EVAP Purge Solenoid	12v, 55 mph: 1v
33	DG	VR Solenoid Control	0%, 55 mph: 45%
36	YL/LG	Spark Output Signal	6.93v (50%)
37	RD	Vehicle Power	12-14v
40	BK/LG	Power Ground	<0.1v
45	DG/LG	MAP Sensor Signal	107 Hz (sea level)
45	DB/LG	MAP Sensor Signal	107 Hz (sea level)
46	BK/WT	Analog Signal Return	<0.050v
47	DG/LG	TP Sensor Signal	0.5-1.2v
48	WT/RD	Self-Test Indicator Signal	STI On: 1v, Off: 5v
49	OR	HO2S-11 (B1 S1) Ground	<0.050v
51	WT/RD	Air Management 1 Solenoid	1v, 55 mph: 12v
56	DB	PIP Sensor Signal	6.93v (50%)
57	RD	Vehicle Power	12-14v
58	TN/OR	Injector Bank 1 (INJ 1, 4, 5, 8)	5.0-5.7 ms
59	TN/RD	Injector Bank 2 (INJ 2, 3, 6, 7)	5.0-5.7 ms
60	BK/LG	Power Ground	<0.1v

Pin Connector Graphic

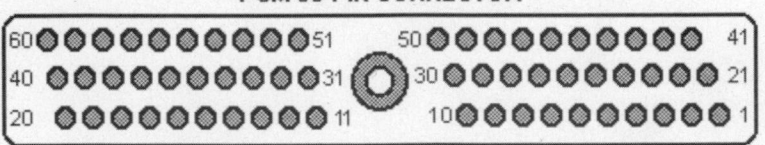

Terminal View of 60-Pin PCM Harness Connector

Standard Colors and Abbreviations

Abbreviation	Color	Abbreviation	Color	Abbreviation	Color
BK	Black	GY	Gray	PK	Purple
BL	Blue	GN	Green	RD	Red
BR	Brown	LG	LT Green	TN	Tan
DB	Dark Blue	OR	Orange	WT	White
DG	DK Green	PK	Pink	YL	Yellow

1991-93 Bronco 5.0L V8 MFI VIN N (A/T) 60 Pin Connector

PCM Pin #	Wire Color	Circuit Description (60 Pin)	Value at Hot Idle
1	YL	Keep Alive Power	12-14v
2, 5	---	Not Used	---
3	GY/BK	VSS (+) Signal	0 Hz, 55 mph: 125 Hz
4	TN/YL	Ignition Diagnostic Monitor	20-31 Hz
6	PK/OR	VSS (-) Signal	0 Hz, 55 mph: 125 Hz
7	LG/RD	ECT Sensor Signal	0.5-0.6v
8	DG/YL	Fuel Pump Monitor	On: 12v, Off: 0v
9	BK/OR	Data Bus (-) Signal	Digital Signals
9	PK/BL	Data Bus (-) Signal	Digital Signals
10	BK/YL	A/C Switch Signal	A/C On: 12v, Off: 0v
10	PK/BL	A/C Switch Signal	A/C On: 12v, Off: 0v
11	BR	Air Management 2 Solenoid	AM2 On: 1v, Off: 12v
12-15, 18-19	---	Not Used	---
16	OR/RD	Ignition System Ground	<0.050v
17	PK/LG	Self-Test Output & MIL	MIL On: 1v, Off: 12v
20	BK	PCM Case Ground	<0.050v
21	WT/BL	IAC Solenoid Control	10.7v (33%)
22	BL/OR	Fuel Pump Control	On: 1v, Off: 12v
23	YL/RD	Knock Sensor Signal	0v
24	YL/LG	PSP Switch Signal	Straight: 0v, Turned: 12v
25	GY	ACT Sensor Signal	1.5-2.5v
26	BR/WT	Reference Voltage	4.9-5.1v
27	BR/LG	EVP Sensor Signal	0.4v
28	TN/OR	Data Bus (+) Signal	Digital Signals
29	GY/BL	HO2S-11 (B1 S1) Signal	0.1-1.1v
30	GY/YL	A/T: Neutral Drive Switch	In 'P': 0v, Others: 5v
30	BL/YL	M/T: CPP Switch Signal	Clutch In: 0v, Out: 5v
31 ('92-'93)	GY/YL	EVAP Purge Solenoid	12v, 55 mph: 1v
32, 41-44	---	Not Used	---
33	BR/PK	VR Solenoid Control	0%, 55 mph: 45%
34-35, 38-39	---	Not Used	---
36	PK	Spark Output Signal	6.93v (50%)
37	RD	Vehicle Power	12-14v
40	BK/WT	Power Ground	<0.1v
45	LG/BK	MAP Sensor Signal	107 Hz (sea level)
46	GY/RD	Analog Signal Return	<0.050v
47	GY/WT	TP Sensor Signal	0.5-1.2v
48	WT/PK	Self-Test Indicator Signal	STI On: 1v, Off: 5v
49	OR	HO2S-11 (B1 S1) Ground	<0.050v
50	---	Not Used	---
51	WT/OR	Air Management 1 Solenoid	1v, 55 mph: 12v
52-55	---	Not Used	---
56	GY/OR	PIP Sensor Signal	6.93v (50%)
57	RD	Vehicle Power	12-14v
58	TN	Injector Bank 1 (INJ 1, 4, 5, 8)	4.4-5.6 ms
59	WT	Injector Bank 2 (INJ 2, 3, 6, 7)	4.4-5.6 ms
60	BK/WT	Power Ground	<0.1v

Pin Connector Graphic

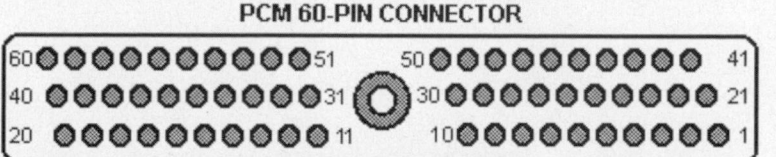

PCM 60-PIN CONNECTOR

Terminal View of 60-Pin PCM Harness Connector

1994-95 Bronco 5.0L V8 VIN N [4R70W] 60 Pin Connector

PCM Pin #	Wire Color	Circuit Description (60 Pin)	Value at Hot Idle
1	YL	Keep Alive Power	12-14v
2	LG	Brake Pedal Switch	Brake Off: 12v, On: 1v
3	GY/BK	VSS (+) Signal	0 Hz, 55 mph: 125 Hz
4	YL/BK	Ignition Diagnostic Monitor	20-31 Hz
5	DG/WT	TSS (+) Sensor Signal	30 mph: 126-136 Hz
6	PK/OR	VSS (-) Signal	0 Hz, 55 mph: 125 Hz
7	LG/RD	ECT Sensor Signal	0.5-0.6v
8	DG/YL	Fuel Pump Monitor	On: 12v, Off: 0v
9	TN/BL	MAF Sensor Return	<0.050v
10	DG/OR	A/C Switch Signal	A/C On: 12v, Off: 0v
11	GY/YL	EVAP Purge Solenoid	12v, 55 mph: 1v
12	LG/OR	Injector 6 Control	5.0-5.5 ms
13	TN/RD	Injector 7 Control	5.0-5.5 ms
14	BL	Injector 8 Control	5.0-5.5 ms
15	TN/BK	Injector 5 Control	5.0-5.5 ms
16	OR/RD	Ignition System Ground	<0.050v
17	PK/LG	Self-Test Output & MIL	MIL On: 1v, Off: 12v
18	TN/OR	Data Bus (+) Signal	Digital Signals
19	PK/BL	Data Bus (-) Signal	Digital Signals
20	BK	PCM Case Ground	<0.050v
21	WT/BL	IAC Solenoid Control	10.7v (33%)
22	BL/OR	Fuel Pump Control	On: 1v, Off: 12v
23	YL/RD	Knock Sensor Signal	0v
24	---	Not Used	---
25	GY	IAT Sensor Signal	1.5-2.5v
26	BR/WT	Reference Voltage	4.9-5.1v
27	BR/LG	EVP Sensor Signal	0.4v
28	---	Not Used	---
29	BK /LG	TSS (-) Sensor Signal	30 mph: 126-136 Hz
30	BL/YL	MLP Sensor Signal	In 'P': 0v, in O/D: 5v
31	WT/OR	Air Bypass Solenoid Control	AIRB On 1v, Off: 12v
32	WT/LG	TCIL (lamp) Control	TCIL On: 1v, Off: 12v
33	BR/PK	VR Solenoid Control	0%, 55 mph: 45%
34	BR	Air Diverter Solenoid Control	AIRD On 1v, Off: 12v
35	BR/BL	Injector 4 Control	5.0-5.5 ms
36	PK	Spark Output Signal	6.93v (50%)
37	RD	Vehicle Power	12-14v
38	WT/YL	EPC Solenoid Control	9.5v (5 psi)
39	BR/YL	Injector 3 Control	5.0-5.5 ms
40	BK/WT	Power Ground	<0.1v

1994-95 Bronco 5.0L V8 VIN N [4R70W] 60 Pin Connector

PCM Pin #	Wire Color	Circuit Description (60 Pin)	Value at Hot Idle
41	TN/WT	TCS (switch) Signal	TCS Closed: 0.1v
42	BL/BK	4x4 Indicator Light	12v (switch closed: 0v)
43	---	Not Used	---
44	GY/BL	HO2S-11 (B1 S1) Signal	0.1-1.1v
45	---	Not Used	---
46	GY/RD	Analog Signal Return	<0.050v
47	GY/WT	TP Sensor Signal	0.5-1.2v
48	WT/PK	Self-Test Indicator Signal	STI On: 1v, Off: 5v
49	OR/BK	TOT Sensor Signal	2.10-2.40v
50	BL/RD	MAF Sensor Signal	0.7v
51	OR/YL	Shift Solenoid 1 Control	12v, 55 mph: 1v
52	PK/OR	Shift Solenoid 2 Control	1v, 30 mph: 12v
53	PK/YL	TCC Solenoid Control	12v, 55 mph: 1v
54-55	---	Not Used	---
56	GY/OR	PIP Sensor Signal	6.93v (50%)
57	RD	Vehicle Power	12-14v
58	TN	Injector 1 Control	5.0-5.5 ms
59	WT	Injector 2 Control	5.0-5.5 ms
60	BK/WT	Power Ground	<0.1v

Pin Connector Graphic

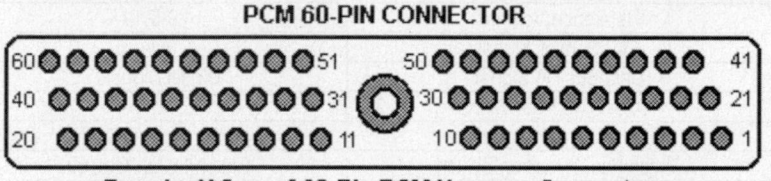

Terminal View of 60-Pin PCM Harness Connector

Standard Colors and Abbreviations

Abbreviation	Color	Abbreviation	Color	Abbreviation	Color
BK	Black	GY	Gray	PK	Purple
BL	Blue	GN	Green	RD	Red
BR	Brown	LG	LT Green	TN	Tan
DB	Dark Blue	OR	Orange	WT	White
DG	DK Green	PK	Pink	YL	Yellow

1994-95 Bronco 5.0L MFI VIN N (All) 60 Pin Connector

PCM Pin #	Wire Color	Circuit Description (60 Pin)	Value at Hot Idle
1	YL	Keep Alive Power	12-14v
2	---	Not Used	---
3	GY/BK	VSS (+) Signal	0 Hz, 55 mph: 125 Hz
4	WT/PK	Ignition Diagnostic Monitor	20-31 Hz
5	---	Not Used	---
6	PK/OR	VSS (-) Signal	0 Hz, 55 mph: 125 Hz
7	LG/RD	ECT Sensor Signal	0.5-0.6v
8	DG/YL	Fuel Pump Monitor	On: 12v, Off: 0v
9	PK/BL	Data Bus (-) Signal	Digital Signals
10	PK/BL	A/C Switch Signal	A/C On: 12v, Off: 0v
11	BR	Air Management 2 Solenoid	AM2 On: 1v, Off: 12v
12-15	---	Not Used	---
16	OR/RD	Ignition System Ground	<0.050v
17	PK/LG	Self-Test Output & MIL	MIL On: 1v, Off: 12v
18-19	---	Not Used	---
20	BK	PCM Case Ground	<0.050v
21	WT/BL	IAC Solenoid Control	10.7v (33%)
22	BL/OR	Fuel Pump Control	On: 1v, Off: 12v
23	YL/RD	Knock Sensor Signal	0v
24	YL/LG	PSP Switch Signal	Straight: 0v, Turned: 12v
25	GY	IAT Sensor Signal	1.5-2.5v
26	BR/WT	Reference Voltage	4.9-5.1v
27	BR/LG	EVP Sensor Signal	0.4v
28	TN/OR	Data Bus (+) Signal	Digital Signals
29	GY/YL	HO2S-11 (B1 S1) Signal	0.1-1.1v
30	BL/YL	MLP Sensor Signal	In 'P': 0v, in O/D: 5v
30	BL/YL	Clutch Pedal Position Switch	Clutch In: 0v, Out: 5v
31	GY/YL	EVAP Purge Solenoid	12v, 55 mph: 1v
32	---	Not Used	---
33	BR/PK	VR Solenoid Control	0%, 55 mph: 45%
36	PK	Spark Output Signal	6.93v (50%)
37	RD	Vehicle Power	12-14v
40	BK/WT	Power Ground	<0.1v
41	TN/WT	TCS (switch) Signal	TCS Closed: 0.1v
42	OR/BK	TOT Sensor Signal	2.10-2.40v
43	PK	A/C Demand Switch	A/C On: 12v, Off: 0v
44	---	Not Used	---
45	LG/BK	MAP Sensor Signal	107 Hz (sea level)
46	GY/RD	Analog Signal Return	<0.050v
47	GY/WT	TP Sensor Signal	0.5-1.2v
48	WT/PK	Self-Test Indicator Signal	STI On: 1v, Off: 5v
49	OR	HO2S-11 (B1 S1) Ground	<0.050v
50	---	Not Used	---
51	WT/OR	Air Management 1 Solenoid	1v, 55 mph: 12v
52-55	---	Not Used	---
56	GY/OR	PIP Sensor Signal	6.93v (50%)
57	RD	Vehicle Power	12-14v
58	TN	Injector Bank 1 (INJ 1, 4, 5, 8)	4.4-5.4 ms
59	WT	Injector Bank 2 (INJ 2, 3, 6, 7)	4.4-5.4 ms
60	BK/WT	Power Ground	<0.1v

PCM 60-PIN CONNECTOR

Terminal View of 60-Pin PCM Harness Connector

1994-95 Bronco 5.0L V8 VIN N (E4OD) 60 Pin Connector

PCM Pin #	Wire Color	Circuit Description (60 Pin)	Value at Hot Idle
1	YL	Keep Alive Power	12-14v
2	LG	Brake On/Off Switch	Brake Off: 12v, On: 1v
3	GY/BK	VSS (+) Signal	0 Hz, 55 mph: 125 Hz
4	WT/PK	Ignition Diagnostic Monitor	20-31 Hz
5	---	Not Used	---
6	PK/OR	VSS (-) Signal	0 Hz, 55 mph: 125 Hz
7	LG/RD	ECT Sensor Signal	0.5-0.6v
8	DG/YL	Fuel Pump Monitor	On: 12v, Off: 0v
9	TN/BL	MAF Sensor Return	<0.050v
10	BK/YL	A/C Switch Signal	A/C On: 12v, Off: 0v
10	DG/OR	A/C Switch Signal	A/C On: 12v, Off: 0v
11	GY/YL	EVAP Purge Solenoid	12v, 55 mph: 1v
12	LG/OR	Injector 6 Control	4.7-5.1 ms
13	TN/RD	Injector 7 Control	4.7-5.1 ms
14	BL	Injector 8 Control	4.7-5.1 ms
15	TN/BK	Injector 5 Control	4.7-5.1 ms
16	OR/RD	Ignition System Ground	<0.050v
17	PK/LG	Self-Test Output & MIL	MIL On: 1v, Off: 12v
18	TN/OR	Data Bus (+) Signal	Digital Signals
19	PK/BL	Data Bus (-) Signal	Digital Signals
20	BK	PCM Case Ground	<0.050v
21	WT/BL	IAC Solenoid Control	10.7v (33%)
22	BL/OR	Fuel Pump Control	On: 1v, Off: 12v
23	YL/RD	Knock Sensor Signal	0v
24	---	Not Used	---
25	GY	IAT Sensor Signal	1.5-2.5v
26	BR/WT	Reference Voltage	4.9-5.1v
27	BR/LG	EVP Sensor Signal	0.4v
28-29	---	Not Used	---
30	BL/YL	MLP Sensor Signal	In 'P': 0v, in O/D: 5v
31	WT/OR	Air Bypass Solenoid Control	AIRB On: 1v, Off: 12v
32	WT/LG	TCIL (lamp) Control	TCIL On: 1v, Off: 12v
33	BR/PK	VR Solenoid Control	0%, 55 mph: 45%
34	BR	Air Diverter Solenoid	AIRD On: 1v, Off: 12v
35	BR/BL	Injector 4 Control	4.7-5.1 ms
36	PK	Spark Output Signal	6.93v (50%)
37	RD	Vehicle Power	12-14v
38	WT/YL	EPC Solenoid Control	9.5v (5 psi)
39	BR/YL	Injector 3 Control	4.7-5.1 ms
40	BK/WT	Power Ground	<0.1v

1994-95 Bronco 5.0L V8 VIN N (E4OD) 60 Pin Connector

PCM Pin #	Wire Color	Circuit Description (60 Pin)	Value at Hot Idle
41	TN/WT	TCS (switch) Signal	TCS Closed: 0.1v
42	BL/BK	4x4 Indicator Light	4x4 Switch On: 0.1v
43	---	Not Used	---
44	GY/BL	HO2S-11 (B1 S1) Signal	0.1-1.1v
45	---	Not Used	---
46	GY/RD	Analog Signal Return	<0.050v
47	GY/WT	TP Sensor Signal	0.5-1.2v
48	WT/PK	Self-Test Indicator Signal	STI On: 1v, Off: 5v
49	OR/BK	HO2S-11 (B1 S1) Ground	<0.050v
50	BL/RD	MAF Sensor Signal	0.8v
51	OR/YL	Shift Solenoid 1 Control	12v, 55 mph: 1v
52	PK/OR	Shift Solenoid 2 Control	1v, 30 mph: 12v
53	PK/YL	TCC Solenoid Control	12v, 55 mph: 9v
54	---	Not Used	---
55	BR/OR	Coast Clutch Solenoid	12v, 55 mph: 12v
56	GY/OR	PIP Sensor Signal	6.93v (50%)
57	RD	Vehicle Power	12-14v
58	TN	Injector 1 Control	4.7-5.1 ms
59	WT	Injector 2 Control	4.7-5.1 ms
60	BK/WT	Power Ground	<0.1v

Pin Connector Graphic

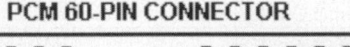

PCM 60-PIN CONNECTOR

Terminal View of 60-Pin PCM Harness Connector

Standard Colors and Abbreviations

Abbreviation	Color	Abbreviation	Color	Abbreviation	Color
BK	Black	GY	Gray	PK	Purple
BL	Blue	GN	Green	RD	Red
BR	Brown	LG	LT Green	TN	Tan
DB	Dark Blue	OR	Orange	WT	White
DG	DK Green	PK	Pink	YL	Yellow

1996 Bronco 5.0L MFI VIN N (All) 104 Pin Connector

PCM Pin #	Wire Color	Circuit Description (104 Pin)	Value at Hot Idle
1	---	Not Used	---
2	PK/LG	MIL (lamp) Control	MIL On: 1v, Off: 12v
4	LG/RD	Power Take-Off Signal	0.1v (off)
5-12	---	Not Used	---
13	PK	Flash EPROM Power	0.1v
14	LG/BK	4x4 Low Switch Signal	12v (Closed: 0v)
15	PK/LB	Data Bus (-) Signal	Digital Signals
16	TN/OR	Data Bus (+) Signal	Digital Signals
18-22	---	Not Used	---
23	OR/RD	Ignition Ground	<0.050v
24	BK/WT	Power Ground	<0.1v
25	BK/LB	Case Ground	<0.050v
26-28	---	Not Used	---
29	TN/WT	TCS (switch) Signal	TCS & O/D On: 12v
30-32	---	Not Used	---
33	PK/OR	PSOM (-) Signal	<0.050v
34	---	Not Used	---
35	RD/LG	HO2S-12 (B1 S2) Signal	0.1-1.1v
36	TN/LB	MAF Sensor Return	<0.050v
37	OR/BK	TFT Sensor Signal	2.10-2.40v
38	LG/RD	ECT Sensor Signal	0.5-0.6v
39	GY	IAT Sensor Signal	1.5-2.5v
40	DG/YL	Fuel Pump Monitor	On: 12v, Off: 0v
41	DG/OR	A/C Switch Signal	A/C On: 12v, Off: 0v
42-43	---	Not Used	---
44	BR	Secondary AIR Diverter	AIRD On: 1v, Off: 12v
45-46	---	Not Used	---
47	BK/PK	VR Solenoid Control	0%, 55 mph: 45%
48	YL/BK	Clean Tachometer Output	39-45 Hz
50	---	Not Used	---
51	BK/WT	Power Ground	<0.1v
52-54	---	Not Used	---
55	YL	Keep Alive Power	12-14v
56	LG/BK	EVAP Purge Solenoid	0-10 Hz (0-100%)
57	---	Not Used	---
58	GY/BK	PSOM (+) Signal	0 Hz, 55 mph: 125 Hz
59	DG/LG	Misfire Detection Sensor	45-55 Hz
60	GY/LB	HO2S-11 (B1 S1) Signal	0.1-1.1v
61-63	---	Not Used	---
64	LB/YL	A/T: Neutral Position Switch	In 'P': 0v, Others: 5v
64	LB/YL	M/T: Clutch Engage Switch	Clutch In: 0v, Out: 5v
65	BR/LG	DPFE Sensor Signal	0.20-1.30v
66-69	---	Not Used	---

1996 Bronco 5.0L MFI VIN N (All) 104 Pin Connector

PCM Pin #	Wire Color	Circuit Description (104 Pin)	Value at Hot Idle
70	WT/OR	Secondary AIR Bypass	AIRB On: 1v, Off: 12v
71	RD	Vehicle Power	12-14v
72	TN/RD	Injector 7 Control	3.2-4.5 ms
73	TN/BK	Injector 5 Control	3.2-4.5 ms
74	BR/YL	Injector 3 Control	3.2-4.5 ms
75	TN	Injector 1 Control	3.2-4.5 ms
76	BK/WT	Power Ground	<0.1v
77	BK/WT	Power Ground	<0.1v
78	---	Not Used	---
79	WT/LG	TCIL (lamp) Control	TCIL On: 1v, Off: 12v
80	LB/OR	Fuel Pump Control	On: 1v, Off: 12v
81-82	---	Not Used	---
83	WT/LB	IAC Solenoid Control	10.8v (33%)
84	DB/YL	OSS (+) Sensor Signal	0 Hz, 30 mph: 130 Hz
85	---	Not Used	---
87	RD/BK	HO2S-21 (B2 S1) Signal	0.1-1.1v
88	LB/RD	MAF Sensor Signal	0.8v, 55 mph: 2.1v
89	GY/WT	TP Sensor Signal	0.53-1.27v
90	BR/WT	Reference Voltage	4.9-5.1v
91	GY/RD	Analog Signal Return	<0.050v
92	LG	Brake Position Switch	Brake Off: 12v, On: 1v
93	RD/WT	HO2S-11 (B1 S1) Heater	On: 1v, Off: 12v
94	YL/LB	HO2S-21 (B2 S1) Heater	On: 1v, Off: 12v
95	WT/BK	HO2S-12 (B1 S2) Heater	On: 1v, Off: 12v
96	---	Not Used	---
97	RD	Vehicle Power	12-14v
98	LG	Injector 8 Control	3.2-4.5 ms
99	LG/OR	Injector 6 Control	3.2-4.5 ms
100	BR/LB	Injector 4 Control	3.2-4.5 ms
101	W	Injector 2 Control	3.2-4.5 ms
102	---	Not Used	---
103	BK/WT	Power Ground	<0.1v
104	---	Not Used	---

Pin Connector Graphic

PCM 104-PIN CONNECTOR

Terminal View of 104-Pin PCM Wiring Harness Connector

Standard Colors and Abbreviations

Abbreviation	Color	Abbreviation	Color	Abbreviation	Color
BK	Black	GY	Gray	PK	Purple
BL	Blue	GN	Green	RD	Red
BR	Brown	LG	LT Green	TN	Tan
DB	Dark Blue	OR	Orange	WT	White
DG	DK Green	PK	Pink	YL	Yellow

1996 Bronco 5.0L VIN N (E4OD) 104 Pin Connector

PCM Pin #	Wire Color	Circuit Description (104 Pin)	Value at Hot Idle
1	PK/OR	Shift Solenoid 2 Control	12v, 55 mph: 12v
2	PK/LG	MIL (lamp) Control	MIL On: 1v, Off: 12v
3, 5, 7, 12	---	Not Used	---
4	LG/RD	Power Take-Off Signal	0.1v (off)
6	BK/YL	OSS (-) Sensor Signal	At 30 mph: 125-131 Hz
13	PK	Flash EPROM Power	0.1v
14	LG/BK	4x4 Low Switch Signal	12v (Closed: 0v)
15	PK/LB	Data Bus (-) Signal	Digital Signals
16	TN/OR	Data Bus (+) Signal	Digital Signals
17-22, 26	---	Not Used	---
23	OR/RD	Ignition Ground	<0.050v
24	BK/WT	Power Ground	<0.1v
25	BK/LB	Case Ground	<0.050v
27	OR/YL	Shift Solenoid 1 Control	1v, 55 mph: 12v
28	---	Not Used	---
29	TN/WT	TCS (switch) Signal	TCS & O/D On: 12v
30-32	---	Not Used	---
33	PK/OR	PSOM (-) Signal	<0.050v
34	---	Not Used	---
35	RD/LG	HO2S-12 (B1 S2) Signal	0.1-1.1v
36	TN/LB	MAF Sensor Return	<0.050v
37	OR/BK	TFT Sensor Signal	2.10-2.40v
38	LG/RD	ECT Sensor Signal	0.5-0.6v
39	GY	IAT Sensor Signal	1.5-2.5v
40	DG/YL	Fuel Pump Monitor	On: 12v, Off: 0v
41	DG/OR	A/C Switch Signal	A/C On: 12v, Off: 0v
42-43	---	Not Used	---
44	BR	Secondary AIR Diverter	AIRD On: 1v, Off: 12v
45-46	---	Not Used	---
47	BK/PK	VR Solenoid Control	0%, 55 mph: 45%
48	YL/BK	Clean Tachometer Output	39-45 Hz
49	GY/OR	PIP Sensor Signal	6.93v (50%)
50	PK	Spark Output Signal	6.93v (50%)
51	BK/WT	Power Ground	<0.1v
52, 57	---	Not Used	---
53	BR/OR	Coast Clutch Solenoid	12v, 55 mph: 12v
54	PK/YL	TCC Solenoid Control	0%, 55 mph: 95%
55	YL	Keep Alive Power	12-14v
56	LG/BK	EVAP Purge Solenoid	0-10 Hz (0-100%)
58	GY/BK	PSOM (+) Signal	0 Hz, 55 mph: 125 Hz
59	DG/LG	Misfire Detection Sensor	45-55 Hz
60	GY/LB	HO2S-11 (B1 S1) Signal	0.1-1.1v
61-63	---	Not Used	---
64	LB/YL	TR Sensor Signal	In 'P': 0v, in O/D: 5v
65	BR/LG	DPFE Sensor Signal	0.20-1.30v
66-69	---	Not Used	---

1996 Bronco 5.0L VIN N (E4OD) 104 Pin Connector

PCM Pin #	Wire Color	Circuit Description (104 Pin)	Value at Hot Idle
70	WT/OR	Secondary AIR Bypass	AIRB On: 1v, Off: 12v
71	RD	Vehicle Power	12-14v
72	TN/RD	Injector 7 Control	3.2-4.5 ms
73	TN/BK	Injector 5 Control	3.2-4.5 ms
74	BR/YL	Injector 3 Control	3.2-4.5 ms
75	TN	Injector 1 Control	3.2-4.5 ms
76	BK/WT	Power Ground	<0.1v
77	BK/WT	Power Ground	<0.1v
78	---	Not Used	---
79	WT/LG	TCIL (lamp) Control	TCIL On: 1v, Off: 12v
80	LB/OR	Fuel Pump Control	On: 1v, Off: 12v
81	WT/YL	EPC Solenoid Control	10v (26 psi)
82	---	Not Used	---
83	WT/LB	IAC Solenoid Control	10.8v (33%)
84-86	---	Not Used	---
87	RD/BK	HO2S-21 (B2 S1) Signal	0.1-1.1v
88	LB/RD	MAF Sensor Signal	0.8v, 55 mph: 2.1v
89	GY/WT	TP Sensor Signal	0.53-1.27v
90	BR/WT	Reference Voltage	4.9-5.1v
91	GY/RD	Analog Signal Return	<0.050v
92	LG	Brake Position Switch	Brake Off: 12v, On: 1v
93	RD/WT	HO2S-11 (B1 S1) Heater	On: 1v, Off: 12v
94	YL/LB	HO2S-21 (B2 S1) Heater	On: 1v, Off: 12v
95	WT/BK	HO2S-12 (B1 S2) Heater	On: 1v, Off: 12v
96	---	Not Used	---
97	RD	Vehicle Power	12-14v
98	LG	Injector 8 Control	3.2-4.5 ms
99	LG/OR	Injector 6 Control	3.2-4.5 ms
100	BR/LB	Injector 4 Control	3.2-4.5 ms
101	WT	Injector 2 Control	3.2-4.5 ms
102	---	Not Used	---
103	BK/WT	Power Ground	<0.1v
104	---	Not Used	---

Pin Connector Graphic

PCM 104-PIN CONNECTOR

Terminal View of 104-Pin PCM Wiring Harness Connector

Standard Colors and Abbreviations

Abbreviation	Color	Abbreviation	Color	Abbreviation	Color
BK	Black	GY	Gray	PK	Purple
BL	Blue	GN	Green	RD	Red
BR	Brown	LG	LT Green	TN	Tan
DB	Dark Blue	OR	Orange	WT	White
DG	DK Green	PK	Pink	YL	Yellow

1990 Bronco 5.8L MFI VIN H (All) 60 Pin Connector

PCM Pin #	Wire Color	Circuit Description (60 Pin)	Value at Hot Idle
1	YL	Keep Alive Power	12-14v
2, 5, 9	---	Not Used	---
3	DG/WT	VSS (+) Signal	0 Hz, 55 mph: 125 Hz
4	DG/YL	Ignition Diagnostic Monitor	20-31 Hz
6	BK	VSS (-) Signal	0 Hz, 55 mph: 125 Hz
7	LG/YL	ECT Sensor Signal	0.5-0.6v
8	BR	Fuel Pump Monitor	On: 12v, Off: 0v
10	BK/YL	A/C Switch Signal	A/C On: 12v, Off: 0v
11	WT/BK	Air Management 2 Solenoid	AM2 On: 1v, Off: 12v
12-15, 19	---	Not Used	---
16	BK/OR	Ignition System Ground	<0.050v
17	PK/LG	Self-Test Output & MIL	MIL On: 1v, Off: 12v
18	TN/LG	Inferred Mileage Sensor	Digital Signals
20	BK	PCM Case Ground	0.050v
21	GY/WT	IAC Solenoid Control	8.0-9.1v
22	TN/LG	Fuel Pump Control	On: 1v, Off: 12v
23	LG/BK	Knock Sensor Signal	0v
24	YL/LG	PSP Switch Signal	Straight: 0v, Turned: 12v
25	YL/RD	ACT Sensor Signal	1.5-2.5v
26	OR/WT	Reference Voltage	4.9-5.1v
27	BR/LG	EVP Sensor Signal	0.4v
28, 32	---	Not Used	---
29	GY/BL	HO2S-11 (B1 S1) Signal	0.1-1.1v
29	DG/PK	HO2S-11 (B1 S1) Signal	0.1-1.1v
30	GY/YL	A/T: Neutral Drive Switch	In 'P': 0v, Others: 5v
30	GY/YL	M/T: Clutch Engage Switch	Clutch In: 0v, Out: 5v
31	GY/YL	EVAP Purge Solenoid	12v, 55 mph: 1v
33	DG	VR Solenoid Control	0%, 55 mph: 45%
34-35, 38-39	---	Not Used	---
36	YL/LG	Spark Output Signal	6.93v (50%)
37	R	Vehicle Power	12-14v
40	BK/LG	Power Ground	<0.1v
42-45, 50	BK/LG	Power Ground	<0.1v
45	DG/LG	MAP Sensor Signal	107 Hz (sea level)
45	DB/LG	MAP Sensor Signal	107 Hz (sea level)
46	BK/WT	Analog Signal Return	<0.050v
47	DG/LG	TP Sensor Signal	0.5-1.2v
48	WT/RD	Self-Test Indicator Signal	STI On: 1v, Off: 5v
49	OR	HO2S-11 (B1 S1) Ground	<0.050v
51	WT/RD	Air Management 1 Solenoid	1v, 55 mph: 12v
52-56	---	Not Used	---
57	RD	Vehicle Power	12-14v
56	DB	PIP Sensor Signal	6.93v (50%)
58	TN/OR	Injector Bank 1 (INJ 1, 4, 5, 8)	5.8-6.4 ms
59	TN/RD	Injector Bank 2 (INJ 2, 3, 6, 7)	5.8-6.4 ms
60	BK/LG	Power Ground	<0.1v

Pin Connector Graphic

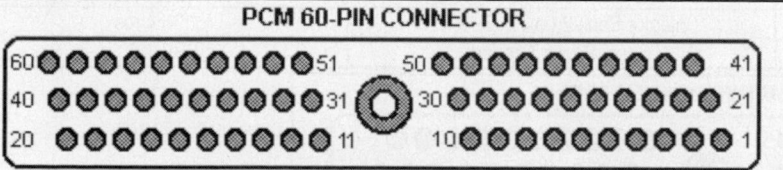

PCM 60-PIN CONNECTOR

Terminal View of 60-Pin PCM Harness Connector

1990 Bronco 5.8L V8 VIN H (E4OD) 60 Pin Connector

PCM Pin #	Wire Color	Circuit Description (60 Pin)	Value at Hot Idle
1	YL	Keep Alive Power	12-14v
2	LG	Brake Pedal Switch	Brake Off: 12v, On: 1v
3	DG/WT	VSS (+) Signal	0 Hz, 55 mph: 125 Hz
4	DG/YL	Ignition Diagnostic Monitor	20-31 Hz
5	---	Not Used	---
6	BK	VSS (-) Signal	0 Hz, 55 mph: 125 Hz
7	LG/YL	ECT Sensor Signal	0.5-0.6v
8	BR	Fuel Pump Monitor	On: 12v, Off: 0v
9	---	Not Used	---
10	BK/YL	A/C Switch Signal	A/C On: 12v, Off: 0v
11	WT/BK	Air Management 2 Solenoid	AM2 On: 1v, Off: 12v
12	BL/BK	4x4 Indicator Light	12v (Closed: 0v)
16	BK/OR	Ignition System Ground	<0.050v
17	PK/LG	Self-Test Output & MIL	MIL On: 1v, Off: 12v
18	TN/LG	Inferred Mileage Sensor	Digital Signals
19	DG/PK	Shift Solenoid 2 Control	1v, 55 mph: 1v
20	BK/LG	Case, Power Ground	<0.1v
21	GY/WT	IAC Solenoid Control	8.0-9.1v
22	TN/LG	Fuel Pump Control	On: 1v, Off: 12v
23	LG/BK	Knock Sensor Signal	0v
24	YL/LG	PSP Switch Signal	Straight: 0v, Turned: 12v
25	YL/RD	ACT Sensor Signal	1.5-2.5v
26	OR/WT	Reference Voltage	4.9-5.1v
27	BR/LG	EVP Sensor Signal	0.4v
28	---	Not Used	---
29	DG/PK	HO2S-11 (B1 S1) Signal	0.1-1.1v
30	BL/WT	MLP Sensor Signal	In 'P': 0v, in O/D: 5v
31	GY/YL	EVAP Purge Solenoid	12v, 55 mph: 1v
32	LG/WT	Overdrive Cancel Indicator	OCIL On: 1v, Off: 12v
33	DG	VR Solenoid Control	0%, 55 mph: 45%
36	YL/LG	Spark Output Signal	6.93v (50%)
37	RD	Vehicle Power	12-14v
38	BL/YL	EPC Solenoid Control	9.2v (5 psi)
40	BK/LG	Case, Power Ground	<0.1v
41	TN/WT	Overdrive Cancel Switch	OCS Closed: 0.1v
42	OR/BK	TOT Sensor Signal	2.10-2.40v
45	DB/LG	MAP Sensor Signal	107 Hz (sea level)
46	BK/WT	Analog Signal Return	<0.050v
47	DG/LG	TP Sensor Signal	0.5-1.2v
48	WT/RD	Self-Test Indicator Signal	STI On: 1v, Off: 5v
49	OR	HO2S-11 (B1 S1) Ground	<0.050v
51	WT/RD	Air Management 1 Solenoid	1v, 55 mph: 12v
52	OR/YL	Shift Solenoid 1 Control	1v, 55 mph: 12v
53	PK/YL	TCC Solenoid Control	12v, 55 mph: 1v
55	BR	Coast Clutch Solenoid	12v, 55 mph: 12v
56	DB	PIP Sensor Signal	6.93v (50%)
57	RD	Vehicle Power	12-14v
58	TN/OR	Injector Bank 1 (INJ 1, 4, 5, 8)	5.8-6.4 ms
59	TN/RD	Injector Bank 2 (INJ 2, 3, 6, 7)	5.8-6.4 ms
60	BK/LG	Case, Power Ground	<0.1v

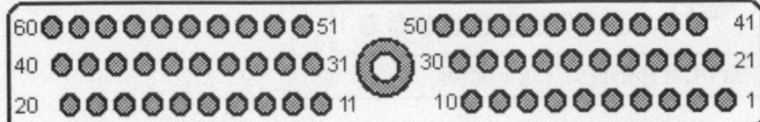

PCM 60-PIN CONNECTOR

Terminal View of 60-Pin PCM Harness Connector

1991-93 Bronco 5.8L MFI VIN H (All) 60 Pin Connector

PCM Pin #	Wire Color	Circuit Description (60 Pin)	Value at Hot Idle
1	YL	Keep Alive Power	12-14v
2	---	Not Used	---
3	GY/BK	VSS (+) Signal	0 Hz, 55 mph: 125 Hz
4	YL/BK	Ignition Diagnostic Monitor	20-31 Hz
4	TN/YL	Ignition Diagnostic Monitor	20-31 Hz
5	---	Not Used	---
6	PK/OR	VSS (-) Signal	0 Hz, 55 mph: 125 Hz
7	LG/RD	ECT Sensor Signal	0.5-0.6v
8	DG/YL	Fuel Pump Monitor	On: 12v, Off: 0v
9	PK/BL	Data Bus (-) Signal	Digital Signals
9	BK/OR	Data Bus (-) Signal	Digital Signals
10	BK/YL	A/C Switch Signal	A/C On: 12v, Off: 0v
10	PK/BL	A/C Switch Signal	A/C On: 12v, Off: 0v
11	BR	Air Management 2 Solenoid	AM2 On: 1v, Off: 12v
12-15	---	Not Used	---
16	OR/RD	Ignition System Ground	<0.050v
17	PK/LG	Self-Test Output & MIL	MIL On: 1v, Off: 12v
18-19	---	Not Used	---
20	BK	PCM Case Ground	<0.050v
21	WT/BL	IAC Solenoid Control	7.2-9.2v
22	BL/OR	Fuel Pump Control	On: 1v, Off: 12v
23 ('92-'93)	YL/RD	Knock Sensor Signal	0v
24 ('92-'93)	YL/LG	PSP Switch Signal	Straight: 0v, Turned: 12v
25	GY	ACT Sensor Signal	1.5-2.5v
26	BR/WT	Reference Voltage	4.9-5.1v
27	BR/LG	EVP Sensor Signal	0.4v
28	TN/OR	Data Bus (+) Signal	Digital Signals
29	GY/BL	HO2S-11 (B1 S1) Signal	0.1-1.1v
29	GY/YL	HO2S-11 (B1 S1) Signal	0.1-1.1v
30	BL/YL	A/T: Neutral Drive Switch	In 'P': 0v, Others: 5v
30	BL/YL	M/T: Clutch Engage Switch	Clutch In: 0v, Out: 5v
31-32	---	Not Used	---
33	BR/PK	VR Solenoid Control	0%, 55 mph: 45%
34-35	---	Not Used	---
36	PK	Spark Output Signal	6.93v (50%)
37	RD	Vehicle Power	12-14v
38-39	---	Not Used	---
40	BK/WT	Power Ground	<0.1v
41-44	---	Not Used	---
45	LG/BK	MAP Sensor Signal	107 Hz (sea level)
46	GY/RD	Analog Signal Return	<0.050v
47	GY/WT	TP Sensor Signal	0.5-1.2v
48	WT/PK	Self-Test Indicator Signal	STI On: 1v, Off: 5v
49	Or	HO2S-11 (B1 S1) Ground	<0.050v
50	---	---	---
51	WT/OR	Air Management 1 Solenoid	AM1 On: 1v, 55 mph: 12v
52-55	---	Not Used	---
56	GY/OR	PIP Sensor Signal	6.93v (50%)
57	RD	Vehicle Power	12-14v
58	TN	Injector Bank 1 (INJ 1, 4, 5, 8)	3.5-5.0 ms
59	WT	Injector Bank 2 (INJ 2, 3, 6, 7)	3.5-5.0 ms
60	BK/WT	Power Ground	<0.1v

Pin Connector Graphic

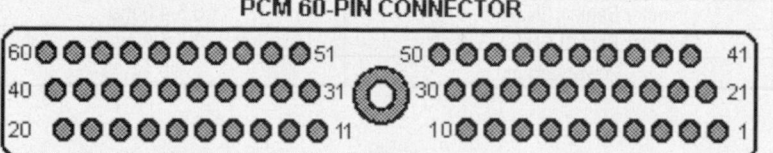

PCM 60-PIN CONNECTOR

Terminal View of 60-Pin PCM Harness Connector

1991-93 Bronco 5.8L V8 VIN H (E4OD) 60 Pin Connector

PCM Pin #	Wire Color	Circuit Description (60 Pin)	Value at Hot Idle
1	YL	Keep Alive Power	12-14v
2	LG	Brake Pedal Switch	Brake Off: 12v, On: 1v
3	GY/BK	VSS (+) Signal	0 Hz, 55 mph: 125 Hz
4	YL/BK	Ignition Diagnostic Monitor	20-31 Hz
4	TN/YL	Ignition Diagnostic Monitor	20-31 Hz
5, 13-15, 18	---	Not Used	---
6	PK/OR	VSS (-) Signal	0 Hz, 55 mph: 125 Hz
7	LG/RD	ECT Sensor Signal	0.5-0.6v
8	DG/YL	Fuel Pump Monitor	On: 12v, Off: 0v
9	PK/BL	Data Bus (+), (+) Signals	Digital Signals, <0.050v
10	BK/YL	A/C Switch Signal	A/C On: 12v, Off: 0v
10	PK/BL	A/C Switch Signal	A/C On: 12v, Off: 0v
11	BR	Air Management 2 Solenoid	AM2 On: 1v, Off: 12v
12	PK/BL	4x4 Indicator Light	12v (switch closed: 0v)
16	OR/RD	Ignition System Ground	<0.050v
17	PK/LG	Self-Test Output & MIL	MIL On: 1v, Off: 12v
19	PK/OR	Shift Solenoid 2 Control	12v, 55 mph: 1v
20	BK	PCM Case Ground	<0.050v
21	WT/BL	IAC Solenoid Control	7.0-10.1v
22	BL/OR	Fuel Pump Control	On: 1v, Off: 12v
23 ('92-'93)	YL/RD	Knock Sensor Signal	0v
24 ('92-'93)	YL/LG	PSP Switch Signal	Straight: 0v, Turned: 12v
25	GY	ACT Sensor Signal	1.5-2.5v
26	BR/WT	Reference Voltage	4.9-5.1v
27	BR/LG	EVP Sensor Signal	0.4v
28	TN/OR	Data Bus (+) Signal	Digital Signals
29	GY/BL	HO2S-11 (B1 S1) Signal	0.1-1.1v
29	GY/YL	HO2S-11 (B1 S1) Signal	0.1-1.1v
30	BL/YL	MLP Sensor Signal	In 'P': 0v, in O/D: 5v
31	GY/YL	EVAP Purge Solenoid	12v, 55 mph: 1v
32	WT/LG	Overdrive Cancel Indicator	OCIL On: 1v, Off: 12v
33	BR/PK	VR Solenoid Control	0%, 55 mph: 45%
34-35, 39, 44	---	Not Used	---
36	PK	Spark Output Signal	6.93v (50%)
37, 57	RD	Vehicle Power	12-14v
38	WT/YL	EPC Solenoid Control	9.5v (5 psi)
40	BK/WT	Power Ground	<0.1v
41	TN/WT	Overdrive Cancel Switch	OCS Closed: 0.1v
42	OR/BK	TOT Sensor Signal	2.10-2.40v
43 ('91)	PK	A/C Demand Switch	A/C On: 12v, Off: 0v
45	LG/BK	MAP Sensor Signal	107 Hz (sea level)
46	GY/RD	Analog Signal Return	<0.050v
47	GY/WT	TP Sensor Signal	0.5-1.2v
48	WT/PK	Self-Test Indicator Signal	STI On: 1v, Off: 5v
49	OR	HO2S-11 (B1 S1) Ground	<0.050v
50, 54	---	Not Used	---
51	WT/OR	Air Management 1 Solenoid	1v, 55 mph: 12v
52	OR/YL	Shift Solenoid 1 Control	1v, 55 mph: 12v
53	PK/YL	TCC Solenoid Control	1v, 55 mph: 12v
55	BR/OR	Coast Clutch Switch Signal	12v, 55 mph: 12v
56	GY/OR	PIP Sensor Signal	6.93v (50%)
58	TN	Injector Bank 1 (INJ 1, 4, 5, 8)	3.5-4.0 ms
59	WT	Injector Bank 2 (INJ 2, 3, 6, 7)	3.5-4.0 ms
60	BK/WT	Power Ground	<0.1v

1994 Bronco 5.8L V8 MFI VIN H (E4OD) 60 Pin Connector

PCM Pin #	Wire Color	Circuit Description (60 Pin)	Value at Hot Idle
1	YL	Keep Alive Power	12-14v
2	LG	Brake Pedal Switch	Brake Off: 12v, On: 1v
3	GY/BK	VSS (+) Signal	0 Hz, 55 mph: 125 Hz
4	YL/BK	Ignition Diagnostic Monitor	20-31 Hz
5, 13-15	---	Not Used	---
6	PK/OR	VSS (-) Signal	0 Hz, 55 mph: 125 Hz
7	LG/RD	ECT Sensor Signal	0.5-0.6v
8	DG/YL	Fuel Pump Monitor	On: 12v, Off: 0v
9	PK/BL	Data Bus (-) Signal	Digital Signals
9	BK/OR	Data Bus (-) Signal	Digital Signals
10	BK/YL	A/C Switch Signal	A/C On: 12v, Off: 0v
10	PK/BL	A/C Switch Signal	A/C On: 12v, Off: 0v
11	BR	Air Management 2 Solenoid	AM2 On: 1v, Off: 12v
12	PK/BL	4x4 Indicator Light	12v (switch closed: 0v)
16	OR/RD	Ignition System Ground	<0.050v
17	PK/LG	Self-Test Output & MIL	MIL On: 1v, Off: 12v
18, 23-24	---	Not Used	---
19	PK/OR	Shift Solenoid 2 Control	12v, 55 mph: 1v
20	BK	PCM Case Ground	<0.050v
21	WT/BL	IAC Solenoid Control	7.0-10.1v
22	BL/OR	Fuel Pump Control	On: 1v, Off: 12v
25	GY	ACT Sensor Signal	1.5-2.5v
26	BR/WT	Reference Voltage	4.9-5.1v
27	BR/LG	EVP Sensor Signal	0.4v
28	TN/OR	Data Bus (+) Signal	Digital Signals
29	GY/BL	HO2S-11 (B1 S1) Signal	0.1-1.1v
29	GY/YL	HO2S-11 (B1 S1) Signal	0.1-1.1v
30	BL/YL	MLP Sensor Signal	In 'P': 0v, in O/D: 5v
31	GY/YL	EVAP Purge Solenoid	12v, 55 mph: 1v
32	WT/LG	Overdrive Cancel Indicator	OCIL On: 1v, Off: 12v
33	BR/PK	VR Solenoid Control	0%, 55 mph: 45%
34-35, 39, 44	---	Not Used	---
36	PK	Spark Output Signal	6.93v (50%)
37	RD	Vehicle Power	12-14v
38	WT/YL	EPC Solenoid Control	9.5v (5 psi)
40	BK/WT	Power Ground	<0.1v
41	TN/WT	TCS (switch) Signal	TCS Closed: 0.1v
42	OR/BK	TOT Sensor Signal	2.10-2.40v
43	PK	A/C Demand Switch	A/C On: 12v, Off: 0v
45	LG/BK	MAP Sensor Signal	107 Hz (sea level)
46	GY/RD	Analog Signal Return	<0.050v
47	GY/WT	TP Sensor Signal	0.5-1.2v
48	WT/PK	Self-Test Indicator Signal	STI On: 1v, Off: 5v
49	OR	HO2S-11 (B1 S1) Ground	<0.050v
50, 54	---	Not Used	---
51	WT/OR	Air Management 1 Solenoid	1v, 55 mph: 12v
52	OR/YL	Shift Solenoid 1 Control	1v, 30 mph: 12v
53	PK/YL	TCC Solenoid Control	12v, 55 mph: 9v
55	BR/OR	Coast Clutch Switch Signal	12v, 55 mph: 12v
56	GY/OR	PIP Sensor Signal	6.93v (50%)
57	RD	Vehicle Power	12-14v
58	TN	Injector Bank 1 (INJ 1, 4, 5, 8)	3.5-5.0 ms
59	WT	Injector Bank 2 (INJ 2, 3, 6, 7)	3.5-5.0 ms
60	BK/WT	Power Ground	<0.1v

1995 Bronco 5.8L V8 VIN H (E4OD) California 60 Pin Connector

PCM Pin #	Wire Color	Circuit Description (60 Pin)	Value at Hot Idle
1	YL	Keep Alive Power	12-14v
2	LG	Brake Pedal Switch	Brake Off: 12v, On: 1v
3	GY/BK	VSS (+) Signal	0 Hz, 55 mph: 125 Hz
4	YL/BK	Ignition Diagnostic Monitor	20-31 Hz
5	---	Not Used	---
6	PK/OR	VSS (-) Signal	0 Hz, 55 mph: 125 Hz
7	LG/RD	ECT Sensor Signal	0.5-0.6v
8	DG/YL	Fuel Pump Monitor	On: 12v, Off: 0v
9	TN/BL	MAF Sensor Return	<0.050v
10	DG/OR	A/C Switch Signal	A/C On: 12v, Off: 0v
11	GY/YL	EVAP Purge Solenoid	12v, 55 mph: 1v
12	LG/OR	Injector 6 Control	5.5-6.1 ms
13	TN/RD	Injector 7 Control	5.5-6.1 ms
14	BL	Injector 8 Control	5.5-6.1 ms
15	TN/BK	Injector 5 Control	5.5-6.1 ms
16	OR/RD	Ignition System Ground	<0.050v
17	PK/LG	Self-Test Output & MIL	MIL On: 1v, Off: 12v
18	TN/OR	Data Bus (+) Signal	Digital Signals
19	PK/BL	Data Bus (-) Signal	Digital Signals
20	BK	PCM Case Ground	<0.050v
21	WT/BL	IAC Solenoid Control	10.7v (33%)
22	BL/OR	Fuel Pump Control	On: 1v, Off: 12v
23-24	---	Not Used	---
25	GY	IAT Sensor Signal	1.5-2.5v
26	BR/WT	Reference Voltage	4.9-5.1v
27	BR/LG	EVP Sensor Signal	0.4v
28-29	---	Not Used	---
30	BL/YL	MLP Sensor Signal	In 'P': 0v, in O/D: 5v
31	---	Not Used	---
32	WT/LG	TCIL (lamp) Control	TCIL On: 1v, Off: 12v
33	BR/PK	VR Solenoid Control	0%, 55 mph: 45%
34	BR	Air Diverter Solenoid Control	12v, 55 mph: 12v
35	BR/BL	Injector 4 Control	5.5-6.1 ms
36	PK	Spark Output Signal	6.93v (50%)
37	RD	Vehicle Power	12-14v
38	WT/YL	EPC Solenoid Control	9.5v (5 psi)
39	BR/YL	Injector 3 Control	5.5-6.1 ms
40	BK/WT	Power Ground	<0.1v

Pin Connector Graphic

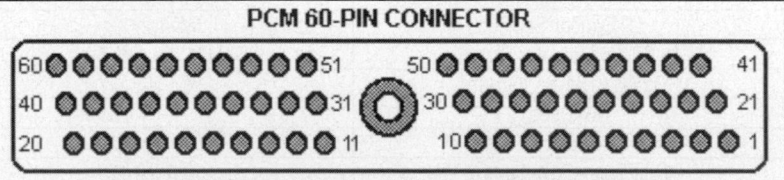

Terminal View of 60-Pin PCM Harness Connector

1995 Bronco 5.8L V8 VIN H (E4OD) California 60 Pin Connector

PCM Pin #	Wire Color	Circuit Description (60 Pin)	Value at Hot Idle
41	TN/WT	TCS (switch) Signal	TCS Closed: 0.1v
42	BL/PK	4x4 Indicator Light	12v (switch closed: 0v)
43	RD/BK	HO2S-21 (B2 S1) Signal	0.1-1.1v
44	GY/BL	HO2S-11 (B1 S1) Signal	0.1-1.1v
45	---	Not Used	---
46	GY/RD	Analog Signal Return	<0.050v
47	GY/WT	TP Sensor Signal	0.5-1.2v
48	WT/PK	Self-Test Indicator Signal	STI On: 1v, Off: 5v
49	OR/BK	TOT Sensor Signal	2.10-2.40v
50	BL/RD	MAF Sensor Signal	0.8v
51	OR/YL	Shift Solenoid 1 Control	1v, 30 mph: 12v
52	PK/OR	Shift Solenoid 2 Control	12v, 55 mph: 1v
53	PK/YL	TCC Solenoid Control	12v, 55 mph: 9v
54	---	Not Used	---
55	BR/OR	Coast Clutch Solenoid	12v, 55 mph: 12v
56	GY/OR	PIP Sensor Signal	6.93v (50%)
57	RD	Vehicle Power	12-14v
58	TN	Injector 1 Control	5.5-6.1 ms
59	WT	Injector 2 Control	5.5-6.1 ms
60	BK/WT	Power Ground	<0.1v

Pin Connector Graphic

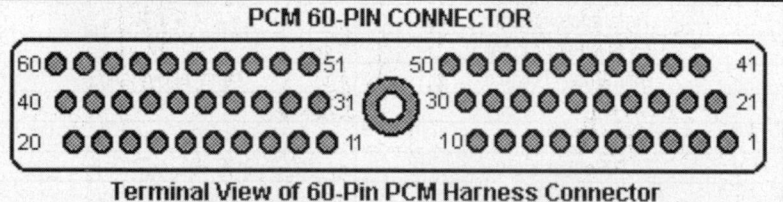

PCM 60-PIN CONNECTOR

Terminal View of 60-Pin PCM Harness Connector

Standard Colors and Abbreviations

Abbreviation	Color	Abbreviation	Color	Abbreviation	Color
BK	Black	GY	Gray	PK	Purple
BL	Blue	GN	Green	RD	Red
BR	Brown	LG	LT Green	TN	Tan
DB	Dark Blue	OR	Orange	WT	White
DG	DK Green	PK	Pink	YL	Yellow

1995 Bronco 5.8L V8 VIN H (E4OD) 60 Pin Connector

PCM Pin #	Wire Color	Circuit Description (60 Pin)	Value at Hot Idle
1	YL	Keep Alive Power	12-14v
2	LG	Brake Pedal Switch	Brake Off: 12v, On: 1v
3	GY/BK	VSS (+) Signal	0 Hz, 55 mph: 125 Hz
4	YL/BK	Ignition Diagnostic Monitor	20-31 Hz
5, 13-15, 18	---	Not Used	---
6	PK/OR	VSS (-) Signal	0 Hz, 55 mph: 125 Hz
7	LG/RD	ECT Sensor Signal	0.5-0.6v
8	DG/YL	Fuel Pump Monitor	On: 12v, Off: 0v
9	PK/BL	Data Bus (-) Signal	Digital Signals
9	BK/OR	Data Bus (-) Signal	Digital Signals
10	BK/YL	A/C Switch Signal	A/C On: 12v, Off: 0v
10	PK/BL	A/C Switch Signal	A/C On: 12v, Off: 0v
11	BR	Air Management 2 Solenoid	AM2 On: 1v, Off: 12v
12	PK/BL	4x4 Indicator Light	12v (switch closed: 0v)
16	OR/RD	Ignition System Ground	<0.050v
17	PK/LG	Self-Test Output & MIL	MIL On: 1v, Off: 12v
19	PK/OR	Shift Solenoid 2 Control	12v, 55 mph: 1v
20	BK	PCM Case Ground	<0.050v
21	WT/BL	IAC Solenoid Control	7.0-10.1v
22	BL/OR	Fuel Pump Control	On: 1v, Off: 12v
23-24, 34-35	---	Not Used	---
25	GY	ACT Sensor Signal	1.5-2.5v
26	BR/WT	Reference Voltage	4.9-5.1v
27	BR/LG	EVP Sensor Signal	0.4v
28	TN/OR	Data Bus (+) Signal	Digital Signals
29	GY/BL	HO2S-11 (B1 S1) Signal	0.1-1.1v
29	GY/YL	HO2S-11 (B1 S1) Signal	0.1-1.1v
30	BL/YL	MLP Sensor Signal	In 'P': 0v, in O/D: 5v
31	GY/YL	EVAP Purge Solenoid	12v, 55 mph: 1v
32	WT/LG	Overdrive Cancel Indicator	OCIL On: 1v, Off: 12v
33	BR/PK	VR Solenoid Control	0%, 55 mph: 45%
36	PK	Spark Output Signal	6.93v (50%)
37	RD	Vehicle Power	12-14v
38	WT/YL	EPC Solenoid Control	9.5v (5 psi)
39, 44, 50, 54	---	Not Used	---
40	BK/WT	Power Ground	<0.1v
41	TN/WT	TCS (switch) Signal	TCS Closed: 0.1v
42	OR/BK	TOT Sensor Signal	2.10-2.40v
43	PK	A/C Demand Switch	A/C On: 12v, Off: 0v
45	LG/BK	MAP Sensor Signal	107 Hz (sea level)
46	GY/RD	Analog Signal Return	<0.050v
47	GY/WT	TP Sensor Signal	0.5-1.2v
48	WT/PK	Self-Test Indicator Signal	STI On: 1v, Off: 5v
49	OR	HO2S-11 (B1 S1) Ground	<0.050v
51	WT/OR	Air Management 1 Solenoid	1v, 55 mph: 12v
52	OR/YL	Shift Solenoid 1 Control	1v, 30 mph: 12v
53	PK/YL	TCC Solenoid Control	12v, 55 mph: 9v
55	BR/OR	Coast Clutch Switch Signal	12v, 55 mph: 12v
56	GY/OR	PIP Sensor Signal	6.93v (50%)
57	RD	Vehicle Power	12-14v
58	TN	Injector Bank 1 (INJ 1, 4, 5, 8)	3.5-5.0 ms
59	WT	Injector Bank 2 (INJ 2, 3, 6, 7)	3.5-5.0 ms
60	BK/WT	Power Ground	<0.1v

1996 Bronco 5.8L V8 MFI VIN H (E4OD) 104 Pin Connector

PCM Pin #	Wire Color	Circuit Description (104 Pin)	Value at Hot Idle
1	PK/OR	Shift Solenoid 2 Control	12v, 55 mph: 12v
2	PK/LG	MIL (lamp) Control	MIL On: 1v, Off: 12v
4	LG/RD	Power Take-Off Signal	0.1v (off)
5-12, 17, 22	---	Not Used	---
13	PK	Flash EPROM Power	0.1v
14	LG/BK	4x4 Low Switch Signal	12v (Closed: 0v)
15	PK/LB	Data Bus (-) Signal	Digital Signals
16	TN/OR	Data Bus (+) Signal	Digital Signals
23	OR/RD	Ignition Ground	<0.050v
24	BK/WT	Power Ground	<0.1v
25	BK/LB	Case Ground	<0.050v
26, 28	---	Not Used	---
27	OR/YL	Shift Solenoid 1 Control	1v, 55 mph: 12v
29	TN/WT	TCS (switch) Signal	TCS & O/D On: 12v
30-32, 34	---	Not Used	---
33	PK/OR	PSOM (+) Signal	<0.050v
35	RD/LG	HO2S-12 (B1 S2) Signal	0.1-1.1v
36	TN/LB	MAF Sensor Return	<0.050v
37	OR/BK	TFT Sensor Signal	2.10-2.40v
38	LG/RD	ECT Sensor Signal	0.5-0.6v
39	GY	IAT Sensor Signal	1.5-2.5v
40	DG/YL	Fuel Pump Monitor	On: 12v, Off: 0v
41	BK/YL	A/C Switch Signal	A/C On: 12v, Off: 0v
42-43, 46	---	Not Used	---
44	BR	Secondary AIR Diverter	AIRD On: 1v, Off: 12v
47	BK/PK	VR Solenoid Control	0%, 55 mph: 45%
48	YL/BK	Clean Tachometer Output	39-45 Hz
49	GY/OR	PIP Sensor Signal	6.93v (50%)
50	PK	Spark Output Signal	6.93v (50%)
51	BK/WT	Power Ground	<0.1v
52, 57	---	Not Used	---
53	BR/OR	Coast Clutch Solenoid	12v, 55 mph: 12v
54	PK/YL	TCC Solenoid Control	0%, 55 mph: 95%
55	Y	Keep Alive Power	12-14v
56	LG/BK	EVAP Purge Solenoid	0-10 Hz (0-100%)
58	GY/BK	PSOM (-) Signal	0 Hz, 55 mph: 125 Hz
59	DG/LG	Misfire Detection Sensor	45-55 Hz
60	GY/LB	HO2S-11 (B1 S1) Signal	0.1-1.1v
61-63	---	Not Used	---
64	LB/YL	TR Sensor Signal	In 'P': 0v, in O/D: 5v

Pin Connector Graphic

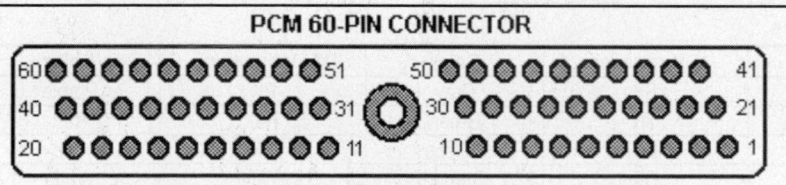

PCM 60-PIN CONNECTOR

Terminal View of 60-Pin PCM Harness Connector

1996 Bronco 5.8L V8 MFI VIN H (E4OD) 104 Pin Connector

PCM Pin #	Wire Color	Circuit Description (104 Pin)	Value at Hot Idle
65	BR/LG	DPFE Sensor Signal	0.20-1.30v
66-69	---	Not Used	---
70	WT/OR	Secondary AIR Bypass	AIRB On: 1v, Off: 12v
71	RD	Vehicle Power	12-14v
72	TN/RD	Injector 7 Control	4.2-4.6 ms
73	TN/BK	Injector 5 Control	4.2-4.6 ms
74	BR/YL	Injector 3 Control	4.2-4.6 ms
75	TN	Injector 1 Control	4.2-4.6 ms
76	BK/WT	Power Ground	<0.1v
77	BK/WT	Power Ground	<0.1v
78, 82	---	Not Used	---
79	WT/LG	TCIL (lamp) Control	TCIL On: 1v, Off: 12v
80	LB/OR	Fuel Pump Control	On: 1v, Off: 12v
81	WT/YL	EPC Solenoid Control	9.2v (5 psi)
83	WT/LB	IAC Solenoid Control	10.7v (33%)
84-86	---	Not Used	---
87	RD/BK	HO2S-21 (B2 S1) Signal	0.1-1.1v
88	LB/RD	MAF Sensor Signal	0.8v, 55 mph: 2.1v
89	GY/WT	TP Sensor Signal	0.53-1.27v
90	BR/WT	Reference Voltage	4.9-5.1v
91	GY/RD	Analog Signal Return	<0.050v
92	LG	Brake Position Switch	Brake Off: 12v, On: 1v
93	RD/WT	HO2S-11 (B1 S1) Heater	On: 1v, Off: 12v
94	YL/LB	HO2S-21 (B2 S1) Heater	On: 1v, Off: 12v
95	WT/BK	HO2S-12 (B1 S2) Heater	On: 1v, Off: 12v
96	---	Not Used	---
97	RD	Vehicle Power	12-14v
98	LG	Injector 8 Control	4.2-4.6 ms
99	LG/OR	Injector 6 Control	4.2-4.6 ms
100	BR/LB	Injector 4 Control	4.2-4.6 ms
101	WT	Injector 2 Control	4.2-4.6 ms
102	---	Not Used	---
103	BK/WT	Power Ground	<0.1v
104	---	Not Used	---

Pin Connector Graphic

PCM 104-PIN CONNECTOR

Terminal View of 104-Pin PCM Wiring Harness Connector

Standard Colors and Abbreviations

Abbreviation	Color	Abbreviation	Color	Abbreviation	Color
BK	Black	GY	Gray	PK	Purple
BL	Blue	GN	Green	RD	Red
BR	Brown	LG	LT Green	TN	Tan
DB	Dark Blue	OR	Orange	WT	White
DG	DK Green	PK	Pink	YL	Yellow

ESCAPE PIN TABLES

2001-03 Escape 2.0L I4 4v ZETEC VIN B (All) 104 Pin Connector

PCM Pin #	Wire Color	Circuit Description (104 Pin)	Value at Hot Idle
1	VT/OR	Shift Solenoid 'B' Control	12, 55 mph: 1v
2-12	---	Not Used	---
13	VT	Flash EEPROM Power	0.1v
14	---	Not Used	---
15	PK/LB	Data Bus (-) Signal	<0.050v
16	TN/OR	Data Bus (+) Signal	Digital Signals
17	DB	High Speed Fan Control	Fan Off: 12v, On: 1v
18	---	Not Used	---
19	BR/OR	Passive Antitheft TX Signal	Digital Signals
20	BR/YL	Fuel Injector 3 Control	2.5-3.0 ms
21	BK/PK	CKP Sensor (+) Signal	400-500 Hz
22	GY/YL	CKP Sensor (-) Signal	400-500 Hz
23	BR/LG	Power Ground	<0.1v
24	BK	Power Ground	<0.1v
25	BR/WT	Power Ground	<0.1v
26	DG/VT	Coil 1 Driver (dwell)	5°, 55 mph: 7°
27	DB/OR	Starter Relay Circuit	Relay Off: 12v, On: 1v
28	GY/BK	VSS (+) Signal	0 Hz, 55 mph: 125 Hz
29	OR/YL	Overdrive Cancel Switch	OCS Off: 0v, On: 12v
30	---	Not Used	---
31	YL/LG	Power Steering Pressure Switch	Straight: 0v, Turning: 12v
32	---	Not Used	---
33	BR/LG	M/T: Power Ground	<0.1v
34	WT/LB	Turbine Shaft Speed Sensor	340 Hz, 55 mph: 1090 Hz
35	RD/LG	HO2S-12 (B1 S2) Signal	0.1-1.1v
36	TN/BK	MAF Sensor Return	<0.050v
37	OR/BK	Transmission Fluid Temperature Sensor	2.10-2.40v
38	LG/RD	Cylinder Head Temperature Sensor	0.6 (194°F)
39	PK/BK	Sensor Return	<0.050v
40	DG/YL	Fuel Pump Relay Control	Off: 12v, On: 1v
41	DG/OR	A/C Dual Pressure Switch	A/C Off: 0v, On: 12v
43	BK	Medium Speed Fan Control Relay	Relay Off: 12v, On: 1v
44-50	---	Not Used	---
51	BK	Power Ground	<0.1v
52	LG/WT	Coil 2 Driver (dwell)	5°, 55 mph: 7°
53	RD/LG	Passive Antitheft RX Signal	Digital Signals
54	LB/OR	Fuel Pump Relay Control	Relay Off: 12v, On: 1v
55	RD/LG	Battery Power	12-14v
56	LG/BK	EVAP Vapor Management Valve	0-10 Hz (0-100%)
57	YL/RD	Knock Sensor 1 (+) Signal	0v
58	DB/YL	A/T: Output Shaft Speed Sensor	0 Hz, 55 mph: 720 Hz
58	DB/YL	M/T: VSS (-) Signal	0 Hz, 55 mph: 125 Hz
59	GY/OR	Generator Load Indicator	1.5-10.v (40-250 Hz)
60	GY/LG	HO2S-11 (B1 S1) Signal	0.1-1.1v
61	---	Not Used	---
62	RD/PK	Fuel Tank Pressure Sensor	2.6v (0" HG - cap off)
63	---	Not Used	---
64	LB/YL	Digital TR Sensor	0v, in O/D: 1.7v

2001-03 Escape 2.0L I4 4v ZETEC VIN B (All) 104 Pin Connector

PCM Pin #	Wire Color	Circuit Description (104 Pin)	Value at Hot Idle
65	BR/LG	EGR DPFE Sensor Signal	0.95-1.05v
66	---	Not Used	---
67	VT/WT	EVAP Canister Vent Solenoid Control	0-10 Hz (0-100%)
68	LB	Low Speed Fan Relay Control	Fan Off: 12v, On: 1v
69	PK/YL	A/C WOT Relay Control	Relay Off: 12v, On: 1v
70	TN	Fuel Injector 1 Control	2.5-3.0 ms
71	WT/RD	Vehicle Power (Start-Run)	12-14v
72	BR/WT	Generator Control Signal	0-130 Hz
73	OR/YL	Shift Solenoid 'A' Control	1v, 55 mph: 1v
74-75	---	Not Used	---
76	BK	Power Ground	<0.1v
77	BK	Power Ground	<0.1v
78	---	Not Used	---
79	VT	Electric VR Solenoid Control	0%, 55 mph: 45%
80	---	Not Used	---
81	WT/YL	EPC Solenoid Control	9.5v
82	---	Not Used	---
83	WT/LB	IAC Solenoid Control	10.7v (33%)
84	---	Output Shaft Speed (+) Signal	0 Hz, 55 mph: 720 Hz
85	DB/OR	CMP Sensor (+) Signal	5-7 Hz
86	BK/YL	A/C Dual Pressure Switch	A/C On: 12v (Open)
87	DG/VT	Knock Sensor 1 (-) Signal	<0.050v
88	LB/RD	MAF Sensor Signal	0.6-0.9v
89	GY/WT	TP Sensor Signal	0.53-1.27v
90	BR/WT	Reference Voltage	4.9-5.1v
91	OR	Signal Return	<0.050v
92	LG	Brake Position Switch	Brake Off: 12v, On: 1v
93	RD/WT	HO2S-11 (B1 S1) Heater Control	Off: 12v, On: 1v
94	---	Not Used	---
95	BR/LB	Injector 4 Control	2.5-3.0 ms
96	WT	Injector 2 Control	2.5-3.0 ms
97	WT/RD	Vehicle Power	12-14v
98	---	Not Used	---
99	VT/YL	Torque Converter Clutch	0%, 55 mph: 95%
100	WT/BK	HO2S-12 (B1 S2) Heater	On: 1v, Off: 12v
101	---	Not Used	---
102	PK/BK	Shift Solenoid 3 (3-2T/TCCS)	12v, 55 mph: 8.8v
103	BK	Power Ground	<0.1v
104	---	Not Used	---

Pin Connector Graphic

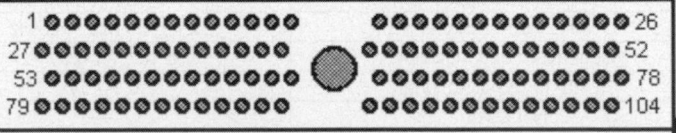

PCM 104-PIN CONNECTOR

Terminal View of 104-Pin PCM Wiring Harness Connector

Standard Colors and Abbreviations

Abbreviation	Color	Abbreviation	Color	Abbreviation	Color
BK	Black	GY	Gray	PK	Purple
BL	Blue	GN	Green	RD	Red
BR	Brown	LG	LT Green	TN	Tan
DB	Dark Blue	OR	Orange	WT	White
DG	DK Green	PK	Pink	YL	Yellow

2001-03 Escape 3.0L V6 4v DURATEC VIN 1 (A/T) 104 Pin Connector

PCM Pin #	Wire Color	Circuit Description (104 Pin)	Value at Hot Idle
1	WT/PK	Coil On Plug (COP) 4 Driver (dwell)	6°, 55 mph: 8°
2	---	Not Used	---
3	BR/LG	Power Ground	<0.1v
4-5	---	Not Used	---
6	OR/YL	Shift Solenoid 'A' Control	12v, 55 mph: 12v
7-10	---	Not Used	---
11	VT/OR	Shift Solenoid 'B' Control	1v, 55 mph: 12v
13	VT	Flash EEPROM Power	0.1v
14		Not Used	---
15	PK/LB	Data Bus (-) Signal	<0.050v
16	TN/OR	Data Bus (+) Signal	Digital Signals
17	RD/BK	Passive Antitheft RX Signal	Digital Signals
18	BR/OR	Passive Antitheft Output	0.8v
19	---	Not Used	---
20	PK/BK	Shift Solenoid 3 (3-2T/TCCS)	12v, 55 mph: 8.8v
21	BK/PK	CKP Sensor (+) Signal	400-450 Hz
22	GY/YL	CKP Sensor (-) Signal	400-450 Hz
23	---	Not Used	---
24	BK	Power Ground	<0.1v
25	BK/WT	Power Ground	<0.1v
26	LG/WT	Coil On Plug (COP) 1 Driver (dwell)	6°, 55 mph: 8°
27	PK/WT	Coil On Plug (COP) 5 Driver (dwell)	6°, 55 mph: 8°
28	DB	Low Speed Fan Relay Control	Relay Off: 12v, On: 1v
29	OR/YL	Overdrive Cancel Switch	OCS Off: 0v, On: 12v
30	---	Not Used	---
31	YL/LG	Power Steering Pressure Switch	Straight: 0v, Turning: 12v
32	DG/VT	Knock Sensor (-) Signal	<0.050v
33-34	---	Not Used	---
35	RD/LG	HO2S-12 (B1 S2) Signal	0.1-1.1v
36	TN/BK	MAF Sensor Return	<0.050v
37	OR/BK	Transmission Fluid Temperature Sensor	2.10-2.40v
38	LG/RD	Engine Coolant Temperature Sensor	0.5-0.6v
39	PK/BK	Sensor Ground	<0.050v
40	DG/YL	Fuel Pump Relay	Relay Off: 12v, On: 1v
41	DG/OR	A/C Dual Pressure Switch Signal	A/C Off: 0v, On: 12v
42	BK	Medium Speed Fan Relay Control	Relay Off: 12v, On: 1v
43	---	Not Used	---
44	DB/OR	Starter Relay Control Circuit	Relay Off: 12v, On: 1v
45	---	Not Used	---
46	LB	High Speed Fan Relay Control	Relay Off: 12v, On: 1v
47	VT	Electric VR Solenoid Control	0%, 55 mph: 45%
48-50	---	Not Used	---
51	BK	Power Ground	<0.1v
52	DG/VT	Coil On Plug (COP) 2 Driver (dwell)	6°, 55 mph: 8°
53	LG/YL	Coil On Plug (COP) 6 Driver (dwell)	6°, 55 mph: 8°
54	VT/YL	Torque Converter Clutch Control	0%, 55 mph: 95%
55	RD/LG	Battery Power	12-14v
56	LG/BK	EVAP Vapor Management Valve	0-10 Hz (0-100%)
57	YL/RD	Knock Sensor 1 (+) Signal	0v
58	DB/YL	A/T: Output Shaft Speed Sensor (+)	0 Hz, 55 mph: 720 Hz
58	DB/YL	M/T: VSS (+) Signal	0 Hz, 55 mph: 125 Hz
59	WT/LB	Turbine Shaft Speed Sensor	50 Hz, 55 mph: 120 Hz
60	GY/LG	HO2S-11 (B1 S1) Signal	0.1-1.1v
61	VT/LG	HO2S-22 (B2 S2) Signal	0.1-1.1v
62	RD/PK	Fuel Tank Pressure Sensor	2.6v (0" HG - cap off)
63	---	Not Used	---
64	LB/YL	Digital TR Sensor	0v, in O/D: 1.7v

2001-02 Escape 3.0L V6 4v DURATEC VIN 1 (A/T) 104 Pin Connector

PCM Pin #	Wire Color	Circuit Description (104 Pin)	Value at Hot Idle
65	BR/LG	EGR DPFE Sensor Signal	0.95-1.05v
66	---	Not Used	---
67	VT/WT	EVAP Canister Vent Solenoid Control	0-10 Hz (0-100%)
68	GY/BK	Vehicle Speed Sensor (+) Signal	0 Hz, 55 mph: 125 Hz
69	PK/YL	A/C WOT Relay Control	Off: 12v, On: 1v
70	---	Not Used	---
71	WT/RD	Vehicle Power (Start-Run)	12-14v
72	---	Not Used	---
73	TN/BK	Fuel Injector 5 Control	2.6-3.2 ms
74	BR/YL	Fuel Injector 3 Control	2.5-3.2 ms
75	TN	Fuel Injector 1 Control	2.5-3.2 ms
76	BK	Power Ground	<0.1v
77	---	Not Used	---
78	OR/YL	Coil On Plug (COP) 3 Driver (dwell)	6°, 55 mph: 8°
79	---	Not Used	---
80	LB/OR	Fuel Pump Relay	Relay Off: 12v, On: 1v
81	WT/YL	EPC Solenoid Control	9.5v
82	---	Not Used	---
83	WT/LB	IAC Solenoid Control	10.7v (33%)
84	DB/YL	Output Shaft Speed (-) Signal	0 Hz, 55 mph: 720 Hz
84	DB/YL	M/T: VSS (-) Signal	0 Hz, 55 mph: 125 Hz
85	DB/OR	Camshaft Position Sensor	5-7 Hz
86	BK/YL	A/C Dual Pressure Switch	A/C On: 12v (Open)
87	RD/BK	HO2S-21 (B2 S1) Signal	0.1-1.1v
88	LB/RD	MAF Sensor Signal	0.6-0.9v
89	GY/WT	TP Sensor Signal	0.53-1.27v
90	BR/WT	Reference Voltage	4.9-5.1v
91	OR	Signal Return	<0.050v
92	LG	Brake Position Switch	Brake Off: 12v, On: 1v
93	RD/WT	HO2S-11 (B1 S1) Heater	On: 1v, Off: 12v
94	YL/LB	HO2S-21 (B2 S1) Heater	On: 1v, Off: 12v
95	WT/BK	HO2S-12 (B1 S2) Heater	On: 1v, Off: 12v
96	TN/YL	HO2S-22 (B2 S2) Heater	On: 1v, Off: 12v
97	WT/RD	Vehicle Power	12-14v
98	---	Not Used	---
99	LG/OR	Injector 6 Control	2.5-3.2 ms
100	BR/LB	Injector 4 Control	2.5-3.2 ms
101	WT	Injector 2 Control	2.5-3.2 ms
102, 104	---	Not Used	---
103	BK	Power Ground	<0.1v

Pin Connector Graphic

PCM 104-PIN CONNECTOR

Terminal View of 104-Pin PCM Wiring Harness Connector

Standard Colors and Abbreviations

Abbreviation	Color	Abbreviation	Color	Abbreviation	Color
BK	Black	GY	Gray	PK	Purple
BL	Blue	GN	Green	RD	Red
BR	Brown	LG	LT Green	TN	Tan
DB	Dark Blue	OR	Orange	WT	White
DG	DK Green	PK	Pink	YL	Yellow

EXCURSION PIN TABLES
2000-02 Excursion 5.4L V8 VIN L (A/T) 104 Pin Connector

PCM Pin #	Wire Color	Circuit Description (104 Pin)	Value at Hot Idle
1	OR/YL	COP 6 Driver (dwell)	5°, 55 mph: 8°
2	PK/LG	MIL (lamp) Control	MIL On: 1v, Off: 12v
3	BK/WT	Power Ground	<0.1v
4-5	---	Not Used	---
6	OR/YL	Shift Solenoid 1 Control	1v, 55 mph: 12v
7-8	---	Not Used	---
9	YL/WT	Fuel Level Indicator Signal	1.7v (1/2 full)
10	---	Not Used	---
11	PK/ORG	Shift Solenoid 2 Control	12v, 55 mph: 12v
12	WT/LG	TCIL (lamp) Control	TCIL On: 1v, Off: 12v
13	VT	Flash EEPROM Power	0.1v
14	LB/BK	4x4 Low Indicator Switch	Switch On: 0v, Off: 12v
15	PK/LB	Data Bus (-) Signal	Digital Signals
16	TN/OR	Data Bus (+) Signal	Digital Signals
17-19	---	Not Used	---
20	BR/O	Coast Clutch Solenoid	12v, 55 mph: 12v
21	DB	CKP (+) Sensor Signal	380-410 Hz
22	GY	CKP (-) Sensor Signal	380-410 Hz
23-24	BK/WT	Power Ground	<0.1v
25	LB/YL	Case Ground	<0.050v
26	LG/WT	COP 1 Driver (dwell)	5°, 55 mph: 8°
27	LG/YL	COP 5 Driver (dwell)	5°, 55 mph: 8°
28	---	Not Used	---
29	TN/WT	TCS (switch) Signal	TCS & O/D On: 12v
30-31	---	Not Used	---
32	DG/PK	Knock Sensor (-) Signal	<0.050v
33	PK/OR	VSS (-) Signal	0 Hz, 55 mph: 125 Hz
34	WT/YL	Digital TR1 Sensor	0v, 55 mph: 11.5v
35	RD/LG	HO2S-12 (B1 S2) Signal	0.1-1.1v
36	TN/LB	MAF Sensor Return	<0.050v
37	OR/BK	TFT Sensor Signal	2.10-2.40v
38	---	Not Used	---
39	GY	IAT Sensor Signal	1.5-2.5v
40	DG/YL	Fuel Pump Monitor	On: 12v, Off: 0v
41	TN/LG	A/C Head Pressure Switch	A/C On: 12v, Off: 0v
42	---	Not Used	---
43	OR/LG	Data Output Link	Digital Signals
44	---	Not Used	---
45	RD/WT	ECT Signal to Dash	Digital Signals
46	---	Not Used	---
47	BR/PK	VR Solenoid Control	0%, 55 mph: 45%
48	WT/PK	Clean Tachometer Output	65 Hz, 55 mph: 175 Hz
49	DB/WT	Digital TR2 Sensor	0v, 55 mph: 11.5v
50	DG/YL	Digital TR4 Sensor	0v, 55 mph: 11.5v
51	BK/WT	Power Ground	<0.1v
52	WT/PK	COP 3 Driver (dwell)	5°, 55 mph: 8°
53	DG/PK	COP 4 Driver (dwell)	5°, 55 mph: 8°
54	PK/YL	TCC Solenoid Control	0%, 55 mph: 95%
55	RD/WT	Keep Alive Power	12-14v
56	LG/BK	EVAP Purge Solenoid	0-10 Hz (0-100%)
57	YL/RD	Knock Sensor (+) Signal	0v
58	GY/BK	VSS (+) Signal	0 Hz, 55 mph: 125 Hz
59	DG/WT	TSS Sensor Signal	0 Hz, 700 Hz
60	GY/LB	HO2S-11 (B1 S1) Signal	0.1-1.1v
61	---	Not Used	---
62	RD/PK	FTP Sensor Signal	2.6v at 0" H20 (cap off)
63	---	Not Used	---
64	LB/YL	Digital TR3A Sensor	In 'P': 0v, in O/D: 1.7v

2000-02 Excursion 5.4L V8 VIN L (A/T) 104 Pin Connector

PCM Pin #	Wire Color	Circuit Description (104 Pin)	Value at Hot Idle
68	---	Not Used	---
65 ('00)	BR/LG	DPFE Sensor Signal	0.20-1.30v
65 ('01-'02)	BR/LG	DPFE Sensor Signal	0.95-1.05v
66	YL/LG	CHT Sensor Signal	0.6 (194ºF)
67	PK/WT	EVAP CV Solenoid	0-10 Hz (0-100%)
69	PK/YL	AC Switch Signal	A/C On: 12v, Off: 0v
70	---	Not Used	---
71	RD	Vehicle Power	12-14v
72	TN/RD	Injector 7 Control	3.8-4.6 ms
73	TN/BK	Injector 5 Control	3.8-4.6 ms
74	BR/YL	Injector 3 Control	3.8-4.6 ms
75	TN	Injector 1 Control	3.8-4.6 ms
76, 77	BK/WT	Power Ground	<0.1v
78	PK/LB	COP 7 Driver (dwell)	5º, 55 mph: 8º
79	WT/RD	COP 8 Driver (dwell)	5º, 55 mph: 8º
80	LB/OR	Fuel Pump Relay Control	On: 1v, Off: 12v
81	WT/YL	EPC Solenoid Control	9.2v (5 psi)
82	---	Not Used	---
83	WT/LB	IAC Solenoid Control	10.7v (33%)
84	DB/YL	OSS Sensor (+) Signal	0 Hz, 55 mph: 2050 Hz
85	DG	CMP Sensor Signal	9 Hz, 55 mph: 16 Hz
86	---	Not Used	---
87	RD/BK	HO2S-21 (B2 S1) Signal	0.1-1.1v
88	LB/RD	MAF Sensor Signal	0.8v, 55 mph: 1.6v
89	GY/WT	TP Sensor Signal	0.9v, 55 mph: 1.3v
90	BR/WT	Reference Voltage	4.9-5.1v
91	GY/RD	Sensor Ground	<0.050v
92	RD/LG	Brake Position Switch	Brake Off: 12v, On: 1v
93	RD/WT	HO2S-11 (B1 S1) Heater	On: 1v, Off: 12v
94	YL/LB	HO2S-21 (B2 S1) Heater	On: 1v, Off: 12v
95	WT/BK	HO2S-12 (B1 S2) Heater	On: 1v, Off: 12v
97	RD	Vehicle Power	12-14v
98	LB	Injector 8 Control	3.8-4.6 ms
99	LG/OR	Injector 6 Control	3.8-4.6 ms
100	WT/LB	Injector 4 Control	3.8-4.6 ms
101	WT	Injector 2 Control	3.8-4.6 ms
102	---	Not Used	---
103	BK/WT	Power Ground	<0.1v
104	PK/WT	COP 2 Driver (dwell)	5º, 55 mph: 8º

Pin Connector Graphic

PCM 104-PIN CONNECTOR

Terminal View of 104-Pin PCM Wiring Harness Connector

Standard Colors and Abbreviations

Abbreviation	Color	Abbreviation	Color	Abbreviation	Color
BK	Black	GY	Gray	PK	Purple
BL	Blue	GN	Green	RD	Red
BR	Brown	LG	LT Green	TN	Tan
DB	Dark Blue	OR	Orange	WT	White
DG	DK Green	PK	Pink	YL	Yellow

2003-05 Excursion 5.4L V8 VIN L (A/T) 104 Pin Connector

PCM Pin #	Wire Color	Circuit Description (104 Pin)	Value at Hot Idle
1	OR/YL	COP 6 Driver (dwell)	5°, 55 mph: 8°
2	---	Not Used	---
3	BK/WT	Power Ground	<0.1v
4	LB/YL	Customer Access Signal	Digital Signals
5	---	Not Used	---
6	OR/YL	Shift Solenoid 'A' Control	1v, 55 mph: 12v
7-10	---	Not Used	---
11	VT/OR	Shift Solenoid 'B' Control	12v, 55 mph: 1v
12	WT/LG	TCIL (lamp) Control	7.7v (Switch On: 0v)
13	VT	Flash EEPROM Power	0.1v
14	---	Not Used	---
15	PK/LB	SCP Data Bus (-) Signal	<0.050v
16	TN/OR	SCP Data Bus (+) Signal	Digital Signals
17-19	---	Not Used	---
20	BR/OR	Coast Clutch Solenoid	12v, 55 mph: 12v
21	DB	CKP (-) Sensor Signal	411 Hz
22	GY	CKP (+) Sensor Signal	411 Hz
23	BK/WT	Power Ground	<0.1v
24	BK/WT	Power Ground	<0.1v
25	LB/YL	Chassis Ground	<0.050v
26	LG/WT	COP 1 Driver (dwell)	5°, 55 mph: 8°
27	LG/YL	COP 5 Driver (dwell)	5°, 55 mph: 8°
28	---	Not Used	---
29	TN/WT	TCS (switch) Signal	TCS & O/D On: 12v
30-31	---	Not Used	---
32	DG/VT	Knock Sensor (-) Signal	<0.050v
33	PK/OR	Power Ground	<0.1v
34	YL/BK	Digital TR1 Sensor	0v, 55 mph: 11.5v
35	RD/LG	HO2S-12 (B1 S2) Signal	0.1-1.1v
36	TN/LB	MAF Sensor Return	<0.050v
37	OR/BK	Transmission Fluid Temperature Sensor	2.10-2.40v
38	---	Not Used	---
39	GY	Intake Air Temperature Sensor	1.5-2.5v
40	DG/YL	Fuel Pump Monitor	Relay Off: 12v, On: 1v
41	TN/LG	A/C Pressure Switch Signal	Switch Closed: 12v
42	---	Not Used	---
43	OR/LG	Data Output Link	5v
44-46	---	Not Used	---
47	BR/PK	Electric VR Solenoid Control	0%, 55 mph: 45%
48	LG/WT	Customer Access (Tachometer)	DC signals
49	LB/BK	Digital TR2 Sensor	0v, 55 mph: 11.5v
50	WT/BK	Digital TR4 Sensor	0v, 55 mph: 11.5v
51	BK/WT	Power Ground	<0.1v
52	WT/PK	COP 3 Driver (dwell)	5°, 55 mph: 8°
53	DG/VT	COP 4 Driver (dwell)	5°, 55 mph: 8°
54	VT/YL	TCC Solenoid Control	0%, 55 mph: 95%
55	RD/WT	Keep Alive Power	12-14v
56	LG/BK	EVAP Vapor Management Valve	0-10 Hz (0-100%)
57	YL/RD	Knock Sensor 1 Signal	0v
58	GY/BK	Vehicle Speed Sensor Signal	Moving: DC signals
59	DG/WT	TSS Sensor Signal	300 Hz, 55 mph: 980
60	GY/LB	HO2S-11 (B1 S1) Signal	0.1-1.1v
61	---	Not Used	---
62	RD/PK	Fuel Tank Pressure Sensor	2.6v (0" H2O - cap off)
63	---	Not Used	---
64	LB/YL	Digital TR3A Sensor Signal	In 'P': 0v, in O/D: 1.7v

2003-05 Excursion 5.4L V8 VIN L (A/T) 104 Pin Connector

PCM Pin #	Wire Color	Circuit Description (104 Pin)	Value at Hot Idle
65	BR/LG	EGR DPFE Sensor Signal	0.95-1.05v
66	YL/LG	Cylinder Head Temperature Sensor	0.6 (194°F)
67	VT/WT	EVAP Canister Vent Solenoid Control	0-10 Hz (0-100%)
68	---	Not Used	---
69	PK/YL	AC Switch Signal	A/C On: 12v, Off: 0v
70	---	Not Used	---
71	RD	Vehicle Power (Start-Run)	12-14v
72	TN/RD	Fuel Injector 7 Control	3.8-4.6 ms
73	TN/BK	Fuel Injector 5 Control	3.8-4.6 ms
74	BR/YL	Fuel Injector 3 Control	3.8-4.6 ms
75	TN	Fuel Injector 1 Control	3.8-4.6 ms
76	BK/WT	Power Ground	<0.1v
77	BK/WT	Power Ground	<0.1v
78	PK/LB	COP 7 Driver (dwell)	5°, 55 mph: 8°
79	WT/RD	COP 8 Driver (dwell)	5°, 55 mph: 8°
80	LB/OR	Fuel Pump Relay Control	Relay Off: 12v, On: 1v
81	WT/YL	EPC Solenoid Control	9.2v (5 psi)
82	---	Not Used	---
83	WT/LB	IAC Solenoid Control	10.7v (33%)
84	DB/YL	OSS Sensor (+) Signal	0 Hz, 55 mph: 2050 Hz
85	DG	CMP Sensor Signal	9 Hz, 55 mph: 16 Hz
86	---	Not Used	---
87	RD/BK	HO2S-21 (B2 S1) Signal	0.1-1.1v
88	LB/RD	MAF Sensor Signal	0.8v, 55 mph: 1.6v
89	GY/WT	TP Sensor Signal	0.9v, 55 mph: 1.3v
90	BR/WT	Reference Voltage	4.9-5.1v
91	GY/RD	Sensor Return	<0.050v
92	RD/LG	Brake Position Switch	Brake Off: 12v, On: 1v
93	RD/WT	HO2S-11 (B1 S1) Heater Control	On: 1v, Off: 12v
94	YL/LB	HO2S-21 (B2 S1) Heater Control	On: 1v, Off: 12v
95	WT/BK	HO2S-12 (B1 S2) Heater Control	On: 1v, Off: 12v
97	RD	Vehicle Power (Start-Run)	12-14v
98	LB	Fuel Injector 8 Control	3.8-4.6 ms
99	LG/OR	Fuel Injector 6 Control	3.8-4.6 ms
100	BR/LB	Fuel Injector 4 Control	3.8-4.6 ms
101	WT	Fuel Injector 2 Control	3.8-4.6 ms
102	---	Not Used	---
103	BK/WT	Power Ground	<0.1v
104	PK/WT	COP 2 Driver (dwell)	5°, 55 mph: 8°

Pin Connector Graphic

PCM 104-PIN CONNECTOR

Terminal View of 104-Pin PCM Wiring Harness Connector

Standard Colors and Abbreviations

Abbreviation	Color	Abbreviation	Color	Abbreviation	Color
BK	Black	GY	Gray	PK	Purple
BL	Blue	GN	Green	RD	Red
BR	Brown	LG	LT Green	TN	Tan
DB	Dark Blue	OR	Orange	WT	White
DG	DK Green	PK	Pink	YL	Yellow

2000-01 Excursion 6.8L V10 VIN S (A/T) 104 Pin Connector

PCM Pin #	Wire Color	Circuit Description (104 Pin)	Value at Hot Idle
1	OR/YL	COP 6 Driver (dwell)	5°, 55 mph: 8°
2	PK/LG	MIL (lamp) Control	MIL On: 1v, Off: 12v
3	BK/WT	Power Ground	<0.1v
4	LB/YL	Power Takeoff Signal	PTO Off: 0v, On: 12v
5	---	Not Used	---
6	OR/YL	Shift Solenoid 1 Control	1v, 55 mph: 12v
7-10	---	Not Used	---
9	YL/WT	Fuel Level Indicator Signal	1.7v (1/2 full)
11	PK/OR	Shift Solenoid 2 Control	12v, 55 mph: 12v
12	WT/LG	TCIL (lamp) Control	TCIL On: 1v, Off: 12v
13	PK	Flash EEPROM Power	0.1v
14	LB/BK	4x4 Low Switch Signal	Switch On: 0v, Off: 12v
15	PK/LB	Data Bus (-) Signal	Digital Signals
16	TN/OR	Data Bus (+) Signal	Digital Signals
17-19	---	Not Used	---
20	BR/OR	Coast Clutch Solenoid	12v, 55 mph: 12v
21	DB	CKP (-) Sensor Signal	500-525 Hz
22	GY	CKP (+) Sensor Signal	500-525 Hz
23-24	BK/WT	Power Ground	<0.1v
25	LB/YL	Case Ground	<0.050v
26	LG/WT	COP 1 Driver (dwell)	5°, 55 mph: 8°
27	LG/YL	COP 10 Driver (dwell)	5°, 55 mph: 8°
28	---	Not Used	---
29	TN/WT	TCS (switch) Signal	TCS & O/D On: 12v
30-31	---	Not Used	---
32	DG/PK	Knock Sensor 1 Return	<0.050v
33	PK/OR	VSS (-) Signal	0 Hz, 55 mph: 125 Hz
34	WT/YL	Digital TR1 Sensor	0v, 55 mph: 11.5v
35	RD/LG	HO2S-12 (B1 S2) Signal	0.1-1.1v
36	TN/LB	MAF Sensor Return	<0.050v
37	OR/BK	TFT Sensor Signal	2.10-2.40v
38	---	Not Used	---
39	GY	IAT Sensor Signal	1.5-2.5v
40	DG/YL	Fuel Pump Monitor	On: 12v, Off: 0v
41	TN/LG	A/C Head Pressure Switch	A/C On: 12v, Off: 0v
42	GY/RD	Injector 10 Control	3.8-4.6 ms
43	OR/LG	Data Output Link	Digital Signals
45	RD/WT	ECT Sensor Signal to Dash	0.6v
46	---	Not Used	---
47	BR/PK	VR Solenoid Control	0%, 55 mph: 45%
48	WT/PK	Clean Tachometer Output	65 Hz, 55 mph: 175 Hz
49	DB/WT	Digital TR2 Sensor	0v, 55 mph: 11.5v
50	DG/YL	Digital TR4 Sensor	0v, 55 mph: 11.5v
51	BK/WT	Power Ground	<0.1v
52	WT/PK	COP 5 Driver (dwell)	5°, 55 mph: 8°
53	DG/PK	COP 7 Driver (dwell)	5°, 55 mph: 8°
54	PK/YL	TCC Solenoid Control	0%, 55 mph: 95%
55	RD/WT	Keep Alive Power	12-14v
56	LG/BK	EVAP Purge Solenoid	0-10 Hz (0-100%)
57	YL/RD	Knock Sensor (+) Signal	0v
58	GY/BK	VSS (+) Signal	0 Hz, 55 mph: 125 Hz
59	DG/WT	TSS Sensor Signal	100 Hz, 55 mph: 270 Hz
60	GY/LB	HO2S-11 (B1 S1) Signal	0.1-1.1v
61	---	Not Used	---
62	RD/PK	Fuel Tank Pressure Sensor	2.6v (0" H2O - cap off)
63	---	Not Used	---
64	LB/YL	Digital TR3A Sensor	In 'P': 0v, in O/D: 1.7v

2000-01 Excursion 6.8L V10 VIN S (A/T) 104 Pin Connector

PCM Pin #	Wire Color	Circuit Description (104 Pin)	Value at Hot Idle
65	BR/LG	DPFE Sensor Signal	0.9v, 55 mph: 3.1v
66	YL/LG	CHT Sensor Signal	0.6 (194°F)
68	GY/BK	Injector 9 Control	3.8-4.6 ms
69	PK/YL	AC Switch Signal	A/C On: 12v, Off: 0v
70	---	Not Used	---
71	RD	Vehicle Power (Start-Run)	12-14v
72	TN/RD	Injector 7 Control	3.8-4.6 ms
73	TN/BK	Injector 5 Control	3.8-4.6 ms
74	BR/YL	Injector 3 Control	3.8-4.6 ms
75	TN	Injector 1 Control	3.8-4.6 ms
76	BK/WT	Power Ground	<0.1v
77	BK/WT	Power Ground	<0.1v
78	PK/LB	COP 2 Driver (dwell)	5°, 55 mph: 8°
79	WT/RD	COP 8 Driver (dwell)	5°, 55 mph: 8°
80	LB/OR	Fuel Pump Control	On: 1v, Off: 12v
81	WT/YL	EPC Solenoid Control	9.2v (5 psi)
82	WT/RD	COP 9 Driver (dwell)	5°, 55 mph: 8°
83	WT/LB	IAC Solenoid Control	10.7v (33%)
84	DB/YL	OSS Sensor Signal	0 Hz, 55 mph: 2050 Hz
85	DG	CMP Sensor Signal	9 Hz, 55 mph: 16 Hz
86	---	Not Used	---
87	RD/BK	HO2S-21 (B2 S1) Signal	0.1-1.1v
88	LB/RD	MAF Sensor Signal	0.8v, 55 mph: 1.6v
89	GY/WT	TP Sensor Signal	0.9v, 55 mph: 1.3v
90	BR/WT	Reference Voltage	4.9-5.1v
91	GY/RD	Sensor Return	<0.050v
92	RD/LG	Brake Position Switch	Brake Off: 12v, On: 1v
93	RD/WT	HO2S-11 (B1 S1) Heater	On: 1v, Off: 12v
94	YL/LB	HO2S-21 (B2 S1) Heater	On: 1v, Off: 12v
97	RD	Vehicle Power	12-14v
98	LB	Injector 8 Control	3.8-4.6 ms
99	LG/OR	Injector 6 Control	3.8-4.6 ms
100	BR/LB	Injector 4 Control	3.8-4.6 ms
101	WT	Injector 2 Control	3.8-4.6 ms
102	YL/BK	COP 4 Driver (dwell)	5°, 55 mph: 8°
103	BK/WT	Power Ground	<0.1v
104	PK/WT	COP 3 Driver (dwell)	5°, 55 mph: 8°

Pin Connector Graphic

```
PCM 104-PIN CONNECTOR
 1 ●●●●●●●●●●●●   ●●●●●●●●●●●●● 26
27 ●●●●●●●●●●●●   ●●●●●●●●●●●●● 52
53 ●●●●●●●●●●●●  ●  ●●●●●●●●●●●●● 78
79 ●●●●●●●●●●●●   ●●●●●●●●●●●●● 104
```

Terminal View of 104-Pin PCM Wiring Harness Connector

Standard Colors and Abbreviations

Abbreviation	Color	Abbreviation	Color	Abbreviation	Color
BK	Black	GY	Gray	PK	Purple
BL	Blue	GN	Green	RD	Red
BR	Brown	LG	LT Green	TN	Tan
DB	Dark Blue	OR	Orange	WT	White
DG	DK Green	PK	Pink	YL	Yellow

2002-05 Excursion 6.8L V10 VIN S (A/T) 104 Pin Connector

PCM Pin #	Wire Color	Circuit Description (104 Pin)	Value at Hot Idle
1	OR/YL	COP 6 Driver (dwell)	5°, 55 mph: 8°
2	---	Not Used	---
3	BK/WT	Power Ground	<0.1v
4	LB/YL	Power Takeoff Signal	PTO Off: 0v, On: 12v
5	---	Not Used	---
6	OR/YL	Shift Solenoid 1 Control	1v, 55 mph: 12v
7-8	---	Not Used	---
9	YL/WT	Fuel Level Indicator Signal	1.7v (1/2 full)
10	---	Not Used	---
11	PK/OR	Shift Solenoid 2 Control	12v, 55 mph: 12v
12	WT/LG	TCIL (lamp) Control	TCIL On: 1v, Off: 12v
13	PK	Flash EEPROM Power	0.1v
14	LB/BK	4x4 Low Switch Signal	Switch On: 0v, Off: 12v
15	PK/LB	Data Bus (-) Signal	Digital Signals
16	TN/OR	Data Bus (+) Signal	Digital Signals
17-19	---	Not Used	---
20	BR/OR	Coast Clutch Solenoid	12v, 55 mph: 12v
21	DB	CKP (-) Sensor Signal	500-525 Hz
22	GY	CKP (+) Sensor Signal	500-525 Hz
23	BK/WT	Power Ground	<0.1v
24	BK/WT	Power Ground	<0.1v
25	LB/YL	Case Ground	<0.050v
26	LG/WT	COP 1 Driver (dwell)	5°, 55 mph: 8°
27	LG/YL	COP 10 Driver (dwell)	5°, 55 mph: 8°
28	---	Not Used	---
29	TN/WT	TCS (switch) Signal	TCS & O/D On: 12v
30-31	---	Not Used	---
32	DG/VT	Knock Sensor 1 Return	<0.050v
33	PK/OR	VSS (-) Signal	0 Hz, 55 mph: 125 Hz
34	WT/YL	Digital TR1 Sensor	0v, 55 mph: 11.5v
35	RD/LG	HO2S-12 (B1 S2) Signal	0.1-1.1v
36	TN/LB	MAF Sensor Return	<0.050v
37	OR/BK	TFT Sensor Signal	2.10-2.40v
38	---	Not Used	---
39	GY	IAT Sensor Signal	1.5-2.5v
40	DG/YL	Fuel Pump Monitor	On: 12v, Off: 0v
41	TN/LG	A/C Head Pressure Switch	A/C On: 12v, Off: 0v
42	GY/RD	Injector 10 Control	3.8-4.6 ms
43	OR/LG	Data Output Link	Digital Signals
44-46	---	Not Used	---
47	BR/PK	VR Solenoid Control	0%, 55 mph: 45%
48	LG/WT	Clean Tachometer Output	65 Hz, 55 mph: 175 Hz
49	LB/BK	Digital TR2 Sensor	0v, 55 mph: 11.5v
50	WT/BK	Digital TR4 Sensor	0v, 55 mph: 11.5v
51	BK/WT	Power Ground	<0.1v
52	WT/PK	COP 5 Driver (dwell)	5°, 55 mph: 8°
53	DG/VT	COP 7 Driver (dwell)	5°, 55 mph: 8°
54	VT/YL	TCC Solenoid Control	0%, 55 mph: 95%
55	RD/WT	Keep Alive Power	12-14v
56	LG/BK	EVAP Vapor Management Valve	0-10 Hz (0-100%)
57	YL/RD	Knock Sensor (+) Signal	0v
58	YL/BK	VSS (+) Signal	0 Hz, 55 mph: 125 Hz
59	DG/WT	TSS Sensor Signal	100 Hz, 55 mph: 270 Hz
60	GY/LB	HO2S-11 (B1 S1) Signal	0.1-1.1v
61	---	Not Used	---
62	RD/PK	Fuel Tank Pressure Sensor	2.6v (0" H2O - cap off)
63	---	Not Used	---
64	LB/YL	Digital TR3A Sensor	In 'P': 0v, in O/D: 1.7v

2002-05 Excursion 6.8L V10 VIN S (A/T) 104 Pin Connector

PCM Pin #	Wire Color	Circuit Description (104 Pin)	Value at Hot Idle
65	BR/LG	EGR DPFE Sensor Signal	0.9v, 55 mph: 3.1v
66	YL/LG	Cylinder Head Temperature Sensor	0.6 (194°F)
68	GY/BK	Injector 9 Control	3.8-4.6 ms
69	PK/YL	AC Switch Signal	A/C On: 12v, Off: 0v
70	---	Not Used	---
71	RD	Vehicle Power (Start-Run)	12-14v
72	TN/RD	Injector 7 Control	3.8-4.6 ms
73	TN/BK	Injector 5 Control	3.8-4.6 ms
74	BR/YL	Injector 3 Control	3.8-4.6 ms
75	TN	Injector 1 Control	3.8-4.6 ms
76	BK/WT	Power Ground	<0.1v
77	BK/WT	Power Ground	<0.1v
78	PK/LB	COP 2 Driver (dwell)	5°, 55 mph: 8°
79	WT/RD	COP 8 Driver (dwell)	5°, 55 mph: 8°
80	LB/OR	Fuel Pump Control	On: 1v, Off: 12v
81	WT/YL	EPC Solenoid Control	9.2v (5 psi)
82	WT/RD	COP 9 Driver (dwell)	5°, 55 mph: 8°
83	WT/LB	IAC Solenoid Control	10.7v (33%)
84	DB/YL	OSS Sensor Signal	0 Hz, 55 mph: 2050 Hz
85	DG	CMP Sensor Signal	9 Hz, 55 mph: 16 Hz
86	---	Not Used	---
87	RD/BK	HO2S-21 (B2 S1) Signal	0.1-1.1v
88	LB/RD	MAF Sensor Signal	0.8v, 55 mph: 1.6v
89	GY/WT	TP Sensor Signal	0.9v, 55 mph: 1.3v
90	BR/WT	Reference Voltage	4.9-5.1v
91	GY/RD	Sensor Return	<0.050v
92	RD/LG	Brake Position Switch	Brake Off: 12v, On: 1v
93	RD/WT	HO2S-11 (B1 S1) Heater	Heater Off: 12v, On: 1v
94	YL/LB	HO2S-21 (B2 S1) Heater	Heater Off: 12v, On: 1v
95	WT/BK	HO2S-12 (B1 S2) Heater	Heater Off: 12v, On: 1v
96	---	Not Used	---
97	RD	Vehicle Power	12-14v
98	LB	Injector 8 Control	3.8-4.6 ms
99	LG/OR	Injector 6 Control	3.8-4.6 ms
100	BR/LB	Injector 4 Control	3.8-4.6 ms
101	WT	Injector 2 Control	3.8-4.6 ms
102	YL/BK	COP 4 Driver (dwell)	5°, 55 mph: 8°
103	BK/WT	Power Ground	<0.1v
104	PK/WT	COP 3 Driver (dwell)	5°, 55 mph: 8°

Pin Connector Graphic

PCM 104-PIN CONNECTOR

Terminal View of 104-Pin PCM Wiring Harness Connector

Standard Colors and Abbreviations

Abbreviation	Color	Abbreviation	Color	Abbreviation	Color
BK	Black	GY	Gray	PK	Purple
BL	Blue	GN	Green	RD	Red
BR	Brown	LG	LT Green	TN	Tan
DB	Dark Blue	OR	Orange	WT	White
DG	DK Green	PK	Pink	YL	Yellow

2000-01 Excursion 6.8L V10 Bi-Fuel VIN S (A/T) 104 Pin Connector

PCM Pin #	Wire Color	Circuit Description (104 Pin)	Value at Hot Idle
1	OR/YL	COP 6 Driver (dwell)	5°, 55 mph: 8°
2	PK/LG	MIL (lamp) Control	MIL On: 1v, Off: 12v
3	BK/WT	Power Ground	<0.1v
4	LB/YL	Power Takeoff Signal	PTO Off: 0v, On: 12v
5	---	Not Used	---
6	OR/YL	Shift Solenoid 1 Control	1v, 55 mph: 12v
7-8	---	Not Used	---
9	YL/WT	Fuel Level Indicator Signal	1.7v (1/2 full)
10	---	Not Used	---
11	PK/OR	Shift Solenoid 2 Control	12v, 55 mph: 12v
12	WT/LG	TCIL (lamp) Control	TCIL On: 1v, Off: 12v
13	PK	Flash EEPROM Power	0.1v
14	LB/BK	4x4 Low Switch Signal	Switch On: 0v, Off: 12v
15	PK/LB	Data Bus (-) Signal	Digital Signals
16	TN/OR	Data Bus (+) Signal	Digital Signals
17-18	---	Not Used	---
19	OR/LG	Fuel Pump Speed Relay	Relay On: 12v, Off: 0v
20	BR/OR	Coast Clutch Solenoid	12v, 55 mph: 12v
21	DB	CKP (-) Sensor Signal	500-525 Hz
22	GY	CKP (+) Sensor Signal	500-525 Hz
23-24	BK/WT	Power Ground	<0.1v
25	LB/YL	Case Ground	<0.050v
26	LG/WT	COP 1 Driver (dwell)	5°, 55 mph: 8°
27	LG/YL	COP 10 Driver (dwell)	5°, 55 mph: 8°
28	---	Not Used	---
29	TN/WT	TCS (switch) Signal	TCS & O/D On: 12v
30-31	---	Not Used	---
32	DG/PK	Knock Sensor 1 Return	<0.050v
33	PK/OR	VSS (-) Signal	0 Hz, 55 mph: 125 Hz
34	WT/YL	Digital TR1 Sensor	0v, 55 mph: 11.5v
35	RD/LG	HO2S-12 (B1 S2) Signal	0.1-1.1v
36	TN/LB	MAF Sensor Return	<0.050v
37	OR/BK	TFT Sensor Signal	2.10-2.40v
38	---	Not Used	---
39	GY	IAT Sensor Signal	1.5-2.5v
40	DG/YL	Fuel Pump Monitor	On: 12v, Off: 0v
41	TN/LG	A/C Head Pressure Switch	A/C On: 12v, Off: 0v
42	GY/RD	Injector 10 Control	3.8-4.6 ms
43	OR/LG	Data Output Link	Digital Signals
45	RD/WT	ECT Sensor Signal to Dash	0.6v
46	GY/BK	Vehicle Speed Sensor	0 Hz, 55 mph: 125 Hz
47	BR/PK	VR Solenoid Control	0%, 55 mph: 45%
48	WT/PK	Clean Tachometer Output	65 Hz, 55 mph: 175 Hz
49	DB/WT	Digital TR2 Sensor	0v, 55 mph: 11.5v
50	DG/YL	Digital TR4 Sensor	0v, 55 mph: 11.5v
51	BK/WT	Power Ground	<0.1v
52	WT/PK	COP 5 Driver (dwell)	5°, 55 mph: 8°
53	DG/PK	COP 7 Driver (dwell)	5°, 55 mph: 8°
54	PK/YL	TCC Solenoid Control	0%, 55 mph: 95%
55	RD/WT	Keep Alive Power	12-14v
56	LG/BK	EVAP Purge Solenoid	0-10 Hz (0-100%)
57	YL/RD	Knock Sensor (+) Signal	0v
58	GY/BK	VSS (+) Signal	0 Hz, 55 mph: 125 Hz
59	DG/WT	TSS Sensor Signal	100 Hz, 55 mph: 270 Hz
60	GY/LB	HO2S-11 (B1 S1) Signal	0.1-1.1v
61	---	Not Used	---
62	RD/PK	Fuel Tank Pressure Sensor	2.6v (0" H2O - cap off)
63	---	Not Used	---
64	LB/YL	Digital TR3A Sensor	In 'P': 0v, in O/D: 1.7v

2000-01 Excursion 6.8L V10 Bi-Fuel VIN S (A/T) 104 Pin Connector

PCM Pin #	Wire Color	Circuit Description (104 Pin)	Value at Hot Idle
65	BR/LG	EGR DPFE Sensor Signal	0.9v, 55 mph: 3.1v
66	YL/LG	Cylinder Head Temperature Sensor	0.6 (194°F)
67	VT/WT	EVAP Canister Vent Solenoid Control	0-10 Hz (0-100%)
68	GY/BK	Injector 9 Control	3.8-4.6 ms
69	PK/YL	AC Switch Signal	A/C On: 12v, Off: 0v
70	---	Not Used	---
71	RD	Vehicle Power	12-14v
72	TN/RD	Injector 7 Control	3.8-4.6 ms
73	TN/BK	Injector 5 Control	3.8-4.6 ms
74	BR/YL	Injector 3 Control	3.8-4.6 ms
75	TN	Injector 1 Control	3.8-4.6 ms
76	BK/WT	Power Ground	<0.1v
77	BK/WT	Power Ground	<0.1v
78	PK/LB	COP 2 Driver (dwell)	5°, 55 mph: 8°
79	WT/RD	COP 8 Driver (dwell)	5°, 55 mph: 8°
80	LB/OR	Fuel Pump Control	On: 1v, Off: 12v
81	WT/YL	EPC Solenoid Control	9.2v (5 psi)
82	WT/RD	COP 9 Driver (dwell)	5°, 55 mph: 8°
83	WT/LB	IAC Solenoid Control	10.7v (33%)
84	DB/YL	OSS Sensor Signal	0 Hz, 55 mph: 2050 Hz
85	DG	CMP Sensor Signal	9 Hz, 55 mph: 16 Hz
86	---	Not Used	---
87	RD/BK	HO2S-21 (B2 S1) Signal	0.1-1.1v
88	LB/RD	MAF Sensor Signal	0.8v, 55 mph: 1.6v
89	GY/WT	TP Sensor Signal	0.9v, 55 mph: 1.3v
90	BR/WT	Reference Voltage	4.9-5.1v
91	GY/RD	Sensor Ground	<0.050v
92	RD/LG	Brake Position Switch	Brake Off: 12v, On: 1v
93	RD/WT	HO2S-11 (B1 S1) Heater	Heater Off: 12v, On: 1v
94	YL/LB	HO2S-21 (B2 S1) Heater	Heater Off: 12v, On: 1v
95	WT/BK	HO2S-12 (B1 S2) Heater	Heater Off: 12v, On: 1v
96	---	Not Used	---
97	RD	Vehicle Power (Start-Run)	12-14v
98	LB	Injector 8 Control	3.8-4.6 ms
99	LG/OR	Injector 6 Control	3.8-4.6 ms
100	BR/LB	Injector 4 Control	3.8-4.6 ms
101	WT	Injector 2 Control	3.8-4.6 ms
102	YL/BK	COP 4 Driver (dwell)	5°, 55 mph: 8°
103	BK/WT	Power Ground	<0.1v
104	PK/WT	COP 3 Driver (dwell)	5°, 55 mph: 8°

Pin Connector Graphic

Standard Colors and Abbreviations

Abbreviation	Color	Abbreviation	Color	Abbreviation	Color
BK	Black	GY	Gray	PK	Purple
BL	Blue	GN	Green	RD	Red
BR	Brown	LG	LT Green	TN	Tan
DB	Dark Blue	OR	Orange	WT	White
DG	DK Green	PK	Pink	YL	Yellow

2002-05 Excursion 6.8L V10 Bi-Fuel VIN S (A/T) 104 Pin Connector

PCM Pin #	Wire Color	Circuit Description (104 Pin)	Value at Hot Idle
1	OR/YL	COP 6 Driver (dwell)	5°, 55 mph: 8°
2	---	Not Used	---
3	BK/WT	Power Ground	<0.1v
4	LB/YL	Power Takeoff Signal	PTO Off: 0v, On: 12v
5	---	Not Used	---
6	OR/YL	Shift Solenoid 1 Control	1v, 55 mph: 12v
7-8	---	Not Used	---
9	YL/WT	Fuel Level Indicator Signal	1.7v (1/2 full)
10	---	Not Used	---
11	PK/OR	Shift Solenoid 2 Control	12v, 55 mph: 12v
12	WT/LG	TCIL (lamp) Control	TCIL On: 1v, Off: 12v
13	PK	Flash EEPROM Power	0.1v
14	LB/BK	4x4 Low Switch Signal	Switch On: 0v, Off: 12v
15	PK/LB	Data Bus (-) Signal	Digital Signals
16	TN/OR	Data Bus (+) Signal	Digital Signals
17-18	---	Not Used	---
19	OR/LG	Fuel Pump Speed Relay	Relay On: 12v, Off: 0v
20	BR/OR	Coast Clutch Solenoid	12v, 55 mph: 12v
21	DB	CKP (-) Sensor Signal	500-525 Hz
22	GY	CKP (+) Sensor Signal	500-525 Hz
23-24	BK/WT	Power Ground	<0.1v
25	LB/YL	Case Ground	<0.050v
26	LG/WT	COP 1 Driver (dwell)	5°, 55 mph: 8°
27	LG/YL	COP 10 Driver (dwell)	5°, 55 mph: 8°
28	---	Not Used	---
29	TN/WT	TCS (switch) Signal	TCS & O/D On: 12v
30-31	---	Not Used	---
32	DG/PK	Knock Sensor 1 Return	<0.050v
33	PK/OR	VSS (-) Signal	0 Hz, 55 mph: 125 Hz
34	WT/YL	Digital TR1 Sensor	0v, 55 mph: 11.5v
35	RD/LG	HO2S-12 (B1 S2) Signal	0.1-1.1v
36	TN/LB	MAF Sensor Return	<0.050v
37	OR/BK	TFT Sensor Signal	2.10-2.40v
39	GY	IAT Sensor Signal	1.5-2.5v
40	DG/YL	Fuel Pump Monitor	On: 12v, Off: 0v
41	TN/LG	A/C Head Pressure Switch	A/C On: 12v, Off: 0v
42	GY/RD	Injector 10 Control	3.8-4.6 ms
43	OR/LG	Data Output Link	Digital Signals
45	RD/WT	ECT Sensor Signal to Dash	0.6v
46	GY/BK	Vehicle Speed Sensor	0 Hz, 55 mph: 125 Hz
47	BR/PK	VR Solenoid Control	0%, 55 mph: 45%
48	WT/PK	Clean Tachometer Output	65 Hz, 55 mph: 175 Hz
49	DB/WT	Digital TR2 Sensor	0v, 55 mph: 11.5v
50	DG/YL	Digital TR4 Sensor	0v, 55 mph: 11.5v
51	BK/WT	Power Ground	<0.1v
52	WT/PK	COP 5 Driver (dwell)	5°, 55 mph: 8°
53	DG/PK	COP 7 Driver (dwell)	5°, 55 mph: 8°
54	PK/YL	TCC Solenoid Control	0%, 55 mph: 95%
55	RD/WT	Keep Alive Power	12-14v
56	LG/BK	EVAP Purge Solenoid	0-10 Hz (0-100%)
57	YL/RD	Knock Sensor (+) Signal	0v
58	GY/BK	VSS (+) Signal	0 Hz, 55 mph: 125 Hz
59	DG/WT	TSS Sensor Signal	100 Hz, 55 mph: 270 Hz
60	GY/LB	HO2S-11 (B1 S1) Signal	0.1-1.1v
61	---	Not Used	---
62	RD/PK	Fuel Tank Pressure Sensor	2.6v (0" H2O - cap off)
63	---	Not Used	---
64	LB/YL	Digital TR3A Sensor	In 'P': 0v, in O/D: 1.7v

2002-05 Excursion 6.8L V10 Bi-Fuel VIN S (A/T) 104 Pin Connector

PCM Pin #	Wire Color	Circuit Description (104 Pin)	Value at Hot Idle
65	BR/LG	DPFE Sensor Signal	0.9v, 55 mph: 3.1v
66	YL/LG	CHT Sensor Signal	0.6 (194ºF)
67	VT/WT	EVAP Canister Vent Solenoid Control	0-10 Hz (0-100%)
68	GY/BK	Injector 9 Control	3.8-4.6 ms
69	PK/YL	AC Switch Signal	A/C On: 12v, Off: 0v
70	---	Not Used	---
71	RD	Vehicle Power (Start-Run)	12-14v
72	TN/RD	Injector 7 Control	3.8-4.6 ms
73	TN/BK	Injector 5 Control	3.8-4.6 ms
74	BR/YL	Injector 3 Control	3.8-4.6 ms
75	TN	Injector 1 Control	3.8-4.6 ms
76	BK/WT	Power Ground	<0.1v
77	BK/WT	Power Ground	<0.1v
78	PK/LB	COP 2 Driver (dwell)	5º, 55 mph: 8º
79	WT/RD	COP 8 Driver (dwell)	5º, 55 mph: 8º
80	LB/OR	Fuel Pump Control	On: 1v, Off: 12v
81	WT/YL	EPC Solenoid Control	9.2v (5 psi)
82	WT/RD	COP 9 Driver (dwell)	5º, 55 mph: 8º
83	WT/LB	IAC Solenoid Control	10.7v (33%)
84	DB/YL	OSS Sensor Signal	0 Hz, 55 mph: 2050 Hz
85	DG	CMP Sensor Signal	9 Hz, 55 mph: 16 Hz
86	DB/YL	OSS Sensor Signal	0 Hz, 55 mph: 250 Hz
87	RD/BK	HO2S-21 (B2 S1) Signal	0.1-1.1v
88	LB/RD	MAF Sensor Signal	0.8v, 55 mph: 1.6v
89	GY/WT	TP Sensor Signal	0.9v, 55 mph: 1.3v
90	BR/WT	Reference Voltage	4.9-5.1v
91	GY/RD	Sensor Ground	<0.050v
92	RD/LG	Brake Position Switch	Brake Off: 12v, On: 1v
93	RD/WT	HO2S-11 (B1 S1) Heater	Heater Off: 12v, On: 1v
94	YL/LB	HO2S-21 (B2 S1) Heater	Heater Off: 12v, On: 1v
95	WT/BK	HO2S-12 (B1 S2) Heater	Heater Off: 12v, On: 1v
96	---	Not Used	---
97	RD	Vehicle Power (Start-Run)	12-14v
98	LB	Injector 8 Control	3.8-4.6 ms
99	LG/OR	Injector 6 Control	3.8-4.6 ms
100	BR/LB	Injector 4 Control	3.8-4.6 ms
101	WT	Injector 2 Control	3.8-4.6 ms
102	YL/BK	COP 4 Driver (dwell)	5º, 55 mph: 8º
103	BK/WT	Power Ground	<0.1v
104	PK/WT	COP 3 Driver (dwell)	5º, 55 mph: 8º

Pin Connector Graphic

PCM 104-PIN CONNECTOR

Terminal View of 104-Pin PCM Wiring Harness Connector

Standard Colors and Abbreviations

Abbreviation	Color	Abbreviation	Color	Abbreviation	Color
BK	Black	GY	Gray	PK	Purple
BL	Blue	GN	Green	RD	Red
BR	Brown	LG	LT Green	TN	Tan
DB	Dark Blue	OR	Orange	WT	White
DG	DK Green	PK	Pink	YL	Yellow

2000-01 Excursion 7.3L V8 Diesel VIN F 104 Pin Connector

PCM Pin #	Wire Color	Circuit Description (104 Pin)	Value at Hot Idle
1	---	Not Used	---
2	PK/LG	MIL (lamp) Control	MIL On: 1v, Off: 12v
3	RD/YL	Low Current Sensor Return	<0.050v
4	LB/YL	Neutral Gear Switch	Switch On: 0v, Off: 12v
5	LG/RD	Parking Brake Applied Switch	Park Brake Applied: 0v
6	OR/YL	Shift Solenoid 1 Control	SS1 On: 1v, Off: 12v
7	---	Not Used	---
8	WT/LG	Glow Plug Monitor Left Bank	Digital Signal: 0-12-0v
9	WT/PK	Glow Plug Monitor Right Bank	Digital Signal: 0-12-0v
10	RD/OR	Idle Validation Switch	12v, off-idle: 0v
11	RD/GN	Shift Solenoid 'B' Control	SS2 On: 1v, Off: 12v
12	WT/LG	TCIL (lamp) Control	Lamp On: 1v, Off: 12v
13	PK	Generic Scan Tool Input	0.1v
14	OR/LB	4x4 Low Switch Input	Switch On: 0v, Off: 12v
15	PK/LB	Data Bus (-)	<0.050v
16	TN/OR	Data Bus (+)	Digital Signals
17	WT/YL	Digital TR1 Sensor	In 'P': 0v, in 'D': 10.7v
18	---	Not Used	---
19	WT/PK	Tachometer Reflected Signal	6.5v / 130 Hz
20	BR/OR	Coast Clutch Solenoid	CCS On: 0v, Off: 12v
21	DG	CMP Sensor Signal	Digital Signal: 0-12-0v
22-23	---	Not Used	---
24	YL/WT	APP Sensor Ground	<0.050v
25	LB/YL	Speedometer Ground	<0.050v
26-27	---	Not Used	---
28	RD	Water In Fuel Indicator Lamp	Lamp on: 1v, Off: 12v
29	TN/WT	TCS (switch) Signal	TCS & O/D On: 12v
30	PK/LB	Exhaust Back Pressure	0.9v, off-idle: 2.5v
31	RD/LG	Brake Pressure Switch	Brake On: 0v, Off: 12v
32-34	---	Not Used	---
35	WT/YL	Generator Power Switch	12-14v
36	GY/RD	Fuel Heater/Water in Fuel	Heater On: 12v
37	OR/BK	TFT Sensor Signal	0.3-4.5v
38	LG/RD	Engine Oil Temperature	0.3-4.7v
39	GY	IAT Sensor Signal	0.2-4.5v
40	PK/BK	Fuel Pump Power	12-14v
41	TN/LG	A/C Head Pressure Switch	0v or 12v
42	GY/RD	Exhaust Backpressure Signal	Digital Signal: 0-12-0v
43-46	---	Not Used	---
47	WT/RD	Wastegate Control Solenoid	Solenoid Off: 12v, On: 1v
48	GY/WT	Electronic Feedback Line	Digital Signals
49	DB/WT	Digital TR2 Sensor	In 'P': 0v, in 'D': 10.7v
50	DG/YL	Digital TR4 Sensor	In 'P': 0v, in 'D': 10.7v
51	BK/WT	Power Ground	<0.1v
52-53	---	Not Used	---
54	PK/YL	TCC Solenoid Control	TCC On: 1v, Off: 12v
55	RD/WT	Keep Alive Power	12-14v
56-57	---	Not Used	---
58	GY/BK	VSS (+) Signal	0 Hz, 55 mph: 125 Hz
59	DB/YL	OSS Sensor Signal	Varying Signal
60	---	Not Used	---
61	LB/BK	Speed Control On/Off Switch	Switch On: 12v, Off: 0v
62	RD/YL	ACT Sensor Signal	1.5-2.5v
63	---	Not Used	---
64	LB/YL	Digital TR3A Sensor3	In Park: 4.5v, in Drive: 2.2v

2000-01 Excursion 7.3L V8 Diesel VIN F 104 Pin Connector

PCM Pin #	Wire Color	Circuit Description (104 Pin)	Value at Hot Idle
65	LB	CMP Sensor Ground	<0.050v
66-69	---	Not Used	---
70	BK/PK	Wait To Start Indicator Control	Indicator Off: 12v, On: 1v
71	RD	Vehicle Power	12-14v
72-76	---	Not Used	---
77	BK/WT	Power Ground	<0.1v
78	---	Not Used	---
79	LG/BK	MAP Sensor Signal	1-3v
80	WT/BK	IDM Enable Relay Control	On: 1v, Off: 12v
81	WT/YL	EPC Solenoid Control	7.5v (8 psi)
82	---	Not Used	---
83	YL/RD	Injector Pressure Regulator Control	Duty Cycle: 0-12-0v
84	DG/WT	TSS Sensor Signal	At 30 mph: 126-136 Hz
85-86	---	Not Used	---
87	DB/LG	Injection Control Pressure	0.75v (Min: 0.83v startup)
88	---	Not Used	---
89	GY/WT	APP Sensor Signal	0.5-1.6v
90	BR/WT	Reference Voltage	4.9-5.1v
91	GY/RD	Sensor Return Signal	<0.050v
92	RD/LG	Brake Position Switch	Brake Off: 12v, On: 1v
93	---	Not Used	---
94	LB/OR	Fuel Pump Relay Output	12-14v
95	BR/OR	Fuel Delivery Control Signal	49 Hz
96	YL/LB	CMP Sensor Signal	6v (5 Hz)
97	RD	Vehicle Power (Start-Run)	12-14v
98	PK	Manifold Intake Air Heater Relay Control	Relay Off: 12v, On: 1v
99	---	Not Used	---
100	YL/WT	Fuel Level Indicator Signal	1.7v (1/2 full)
101	PK/OR	Glow Plug Relay Control	Relay Off: 12v, On: 1v
102	---	Not Used	---
103	BK/WT	Power Ground	<0.1v
104	---	Not Used	---

Pin Connector Graphic

PCM 104-PIN CONNECTOR

Terminal View of 104-Pin PCM Wiring Harness Connector

Standard Colors and Abbreviations

Abbreviation	Color	Abbreviation	Color	Abbreviation	Color
BK	Black	GY	Gray	PK	Purple
BL	Blue	GN	Green	RD	Red
BR	Brown	LG	LT Green	TN	Tan
DB	Dark Blue	OR	Orange	WT	White
DG	DK Green	PK	Pink	YL	Yellow

2002-05 Excursion 7.3L V8 Diesel VIN F (A/T) 104 Pin Connector

PCM Pin #	Wire Color	Circuit Description (104 Pin)	Value at Hot Idle
1-2	---	Not Used	---
3	RD/YL	Manifold Intake Air Heater Monitor	Off: 12v, On: 1v
4	---	Not Used	---
5	LG/RD	Parking Brake Applied Switch	Brake Off: 12v, Applied: 0v
6	OR/YL	Shift Solenoid 1 Control	SS1 Off: 12v, On: 1v
7	---	Not Used	---
8	WT/LG	Glow Plug Control Module Communication	Digital Signal: 0-12-0v
9	---	Not Used	---
10	RD/LG	Idle Validation Switch	12v, off-idle: 0v
11	VT/OG	Shift Solenoid 2 Control	SS2 Off: 12v, On: 1v
12	WT/LG	TCIL (lamp) Control	7.7v (Switch On: 0v)
13	VT	Flash EEPROM Power Supply	0.1v
14	---	Not Used	---
15	PK/LB	SCP Data Bus (-) Signal	<0.050v
16	TN/OR	SCP Data Bus (+) Signal	Digital Signals
17	YL/BK	Digital TR1 Sensor	In 'P': 0v, in Drive: 10.7v
18	PK/YL	A/C Relay Control	Relay Off: 12v, On: 1v
19	LG/WT	Tachometer Reflected Signal	6.5v / 130 Hz
20	BR/OR	Coast Clutch Solenoid Control	CCS On: 0v, Off: 12v
21	DG	CMP Sensor Signal	Digital Signal: 0-12-0v
22-23	---	Not Used	---
24	YL/WT	APP Sensor Ground	<0.050v
25	LB/YL	Case Ground	<0.050v
26-28	---	Not Used	---
29	TN/WT	TCS (switch) Signal	TCS & O/D On: 12v
30	VT/LB	Exhaust Back Pressure Sensor	0.9v, off-idle: 2.5v
31	BK/YL	Brake Pressure Switch	Brake Off: 12v, On: 1v
32-34	---	Not Used	---
35	WT/YL	Alternator No. 1 (Top) Monitor	6-10v
36	GY/RD	Fuel Heater / Water in Fuel	Heater On: 12v, Off: 0v
37	OR/BK	Transmission Fluid Temperature Sensor	2.10-2.40v
38	LG/RD	Engine Oil Temperature Sensor	0.3-4.7v
39	GY	Intake Air Temperature Sensor	1.5-2.5v
40	PK/BK	Fuel Pump Monitor	Pump Off: 12v, On: 1v
41	TN/LG	A/C Head Pressure Switch	0v or 12v
42	GY/RD	Exhaust Backpressure Regulator Control	Duty Cycle (0-12-0v)
43	OR/LG	Overhead Console Fuel Consumption	0-5-0v
44	---	Not Used	---
45	OG/LB	Cruise Control Set Indicator	Indicator Off: 12v, On: 1v
47	WT/RD	Wastegate Control Solenoid Control	Solenoid Off: 12v, On: 1v
48	GY/WT	Injector Driver Module	Digital Signal: 0.9-3.0v
49	LB/PK	Digital TR2 Sensor	In 'P': 0v, in Drive: 10.7v
50	WT/BK	Digital TR4 Sensor	In 'P': 0v, in Drive: 10.7v
51	BK/WT	Power Ground	<0.1v
52-53	---	Not Used	---
54	VT/YL	TCC Solenoid Control	TCC Off: 12v, On: 1v
55	RD/WT	Direct Battery	12-14v
56-57	---	Not Used	---
58	GY/BK	VSS (+) Signal	0 Hz, 55 mph: 125 Hz
59	DB/YL	OSS Sensor Signal	AC pulse signals
60	YL	Alternator No. 2 (Bottom) Control	6-10v
61	LB/BK	Speed Control On/Off Switch	Switch On: 12v, Off: 0v
62	RD/YL	MAT Sensor Signal	1.5-2.5v
63	---	Not Used	---
64	LB/YL	Digital TR3 Sensor	In Park: 4.5v, in Drive: 2.2v

2002-05 Excursion 7.3L V8 Diesel VIN F (A/T) 104 Pin Connector

PCM Pin #	Wire Color	Circuit Description (104 Pin)	Value at Hot Idle
65	LB	CMP Sensor Ground	<0.050v
66	LB/YL	Power Take-Off Enable	0v (Off)
67	LG/RD	Generator/Battery Indicator Control	Indicator Off: 12v, On: 1v
68-70	---	Not Used	---
71	RD	Vehicle Power (Start-Run)	12-14v
72-76	---	Not Used	---
77	BK/WT	Power Ground	<0.1v
78	---	Not Used	---
79	LG/BK	Manifold Absolute Pressure Sensor	1-3v
80	WT/BK	Injector Drive Module Relay Control	Relay Off: 12v, On: 1v
81	WT/YL	EPC Solenoid Control	7.5v (8 psi)
82	---	Not Used	---
83	YL/RD	Injector Pressure Regulator Control	Duty Cycle: 0-12-0v
84	DG/WT	Turbine Shaft Speed Sensor	30 mph: 130 Hz
85-86	---	Not Used	---
87	DB/LG	Injection Control Pressure Sensor	0.1-3.0v
88	---	Not Used	---
89	GY/WT	APP Sensor Signal	0.5-1.6v
90	BR/WT	Reference Voltage	4.9-5.1v
91	GY/RD	Sensor Ground	<0.050v
92	RD/LG	Brake Pedal Switch	Brake Off: 0v, On: 12v
93	---	Not Used	---
94	LB/OR	Fuel Pump Relay Control	Relay Off: 12v, On: 1v
95	BR/OR	Fuel Delivery Control	40-240 Hz
96	YL/LB	CID Sensor Signal	5-30 Hz
97	RD	Vehicle Power (Start-Run)	12-14v
98	VT	Manifold Intake Air Heater Relay Control	Relay Off: 12v, On: 1v
99-100	---	Not Used	---
101	VT/OR	Glow Plug Relay Control	Relay Off: 12v, On: 1v
102	---	Not Used	---
103	BK/WT	Power Ground	<0.1v
104	---	Not Used	---

Pin Connector Graphic

PCM 104-PIN CONNECTOR

```
 1 ●●●●●●●●●●●●●      ●●●●●●●●●●●●● 26
27 ●●●●●●●●●●●●●      ●●●●●●●●●●●●● 52
53 ●●●●●●●●●●●●●  ⬤  ●●●●●●●●●●●●● 78
79 ●●●●●●●●●●●●●      ●●●●●●●●●●●●● 104
```

Terminal View of 104-Pin PCM Wiring Harness Connector

Standard Colors and Abbreviations

Abbreviation	Color	Abbreviation	Color	Abbreviation	Color
BK	Black	GY	Gray	PK	Purple
BL	Blue	GN	Green	RD	Red
BR	Brown	LG	LT Green	TN	Tan
DB	Dark Blue	OR	Orange	WT	White
DG	DK Green	PK	Pink	YL	Yellow

EXPEDITION PIN TABLES

1997 Expedition 4.6L V8 VIN W, VIN 6 (A/T) 104 Pin Connector

PCM Pin #	Wire Color	Circuit Description (104 Pin)	Value at Hot Idle
1	PK/OR	Shift Solenoid 2 Control	12v, 55 mph: 1v
2	PK/LG	MIL (lamp) Control	MIL On: <1v, Off: 12v
3	YL/BK	Digital TR1 Sensor	0v, 55 mph: 11.5v
12	YL/WT	Fuel Level Indicator Signal	1.7v (1/2 full)
13	PK	Flash EEPROM Power	0.1v
14	LB/BK	4x4 Indicator Signal	4x4 On: 1v, Off: 12v
15	PK	Data Bus (-) Signal	Digital Signals
16	TN/OR	Data Bus (+) Signal	Digital Signals
19	DB/WT	Automatic Ride Control	Digital Signals
21	DB	CKP (-) Sensor Signal	430-475 Hz
22	GY	CKP (+) Sensor Signal	430-475 Hz
24	BK/WT	Power Ground	<0.1v
25	LG/YL	Case Ground	<0.050v
26	DB/LG	Coil 1 Driver (dwell)	5°, 55 mph: 8°
27	OR/YL	Shift Solenoid 1 Control	1v, 55 mph: 1v
29	TN/WT	TCS (switch) Signal	TCS & O/D On: 12v
30	DG	Octane Adjustment	9.3v (switch shorted: 0v)
33	PK/OR	VSS (-) Signal	0 Hz, 55 mph: 125 Hz
35	RD/LG	HO2S-12 (B1 S2) Signal	0.1-1.1v
36	TN/LB	MAF Sensor Return	<0.050v
37	OR/BK	TFT Sensor Signal	2.10-2.40v
38	LG/RD	ECT Sensor Signal	0.5-0.6v
39	GY	IAT Sensor Signal	1.5-2.5v
40	DG/YL	Fuel Pump Monitor	On: 12v, Off: 0v
41	BK/YL	A/C Switch Signal	A/C On: 12v, Off: 0v
43	OR/LG	Overhead Trip Computer	Digital Signals
45	RD/WT	CHTIL (lamp) Control	Lamp On: 1v, Off: 12v
46	BR	Intake Manifold Tuning Valve	12v, 55 mph: 12v
47	BR/PK	VR Solenoid Control	0%, 55 mph: 45%
48	WT/PK	Clean Tachometer Output Signal	39-49 Hz
49	LB/BK	Digital TR2 Sensor	0v, 55 mph: 11.5v
50	WT/BK	Digital TR4 Sensor	0v, 55 mph: 11.5v
51	BK/WT	Power Ground	<0.1v
52	RD/LB	Coil 2 Driver (dwell)	5°, 55 mph: 8°
54	PK/YL	TCC Solenoid Control	0%, 55 mph: 95%
55	BK/LG	Keep Alive Power	12-14v
56	LG/BK	EVAP Purge Solenoid	0-10 Hz (0-100%)
57	YL/RD	Knock Sensor 1 Signal	0v
58	GY/BK	VSS (+) Signal	0 Hz, 55 mph: 125 Hz
60	GY/LB	HO2S-11 (B1 S1) Signal	0.1-1.1v
61	PK/LG	HO2S-22 (B2 S2) Signal	0.1-1.1v
62	RD/PK	FTP Sensor Signal	2.6v (0" H2O - cap off)
64	LB/YL	TR Sensor Signal	In 'P': 0v, in O/D: 1.7v

1997 Expedition 4.6L V8 VIN W, VIN 6 (A/T) 104 Pin Connector

PCM Pin #	Wire Color	Circuit Description (104 Pin)	Value at Hot Idle
65	BR/LG	DPFE Sensor Signal	0.20-1.30v
66	YL/LG	CHT Sensor Signal	0.6 to 1.7v (194°F)
67	PK/WT	EVAP CV Solenoid	0-10 Hz (0-100%)
69	PK/YL	A/C WOT Relay Control	Relay On: 0.1v, Off: 12v
71	RD	Vehicle Power	12-14v
72	TN/RD	Injector 7 Control	2.7-4.1 ms
73	TN/BK	Injector 5 Control	2.7-4.1 ms
74	BR/YL	Injector 3 Control	2.7-4.1 ms
75	TN	Injector 1 Control	2.7-4.1 ms
76, 77	BK/WT	Power Ground	<0.1v
78	PK/WT	Coil 3 Driver (dwell)	5°, 55 mph: 8°
79	WT/LG	TCIL (lamp) Control	TCIL On: 1v, Off: 12v
80	LB/OR	Fuel Pump Control	On: 1v, Off: 12v
81	WT/YL	EPC Solenoid Control	9.1v (6 psi)
83	WT/LB	IAC Solenoid Control	10.7v (33%)
84	DB/YL	OSS Sensor Signal	0 Hz, 55 mph: 250 Hz
85	DG	CID Sensor Signal	6 Hz, 55 mph: 15 Hz
87	RD/BK	HO2S-21 (B2 S1) Signal	0.1-1.1v
88	LB/RD	MAF Sensor Signal	0.6v, 55 mph: 2.1v
89	GY/WT	TP Sensor Signal	0.53-1.27v
90	BR/WT	Reference Voltage	4.9-5.1v
91	GY/RD	Analog Signal Return	<0.050v
92	LG	Brake Switch Signal	Brake Off: 12v, On: 1v
93	RD/WT	HO2S-11 (B1 S1) Heater	On: 1v, Off: 12v
94	YL/LB	HO2S-21 (B2 S1) Heater	On: 1v, Off: 12v
95	WT/BK	HO2S-12 (B1 S2) Heater	On: 1v, Off: 12v
96	TN/YL	HO2S-22 (B2 S2) Heater	On: 1v, Off: 12v
97	RD	Vehicle Power	12-14v
98	LB	Injector 8 Control	2.7-4.1 ms
99	LG/OR	Injector 6 Control	2.7-4.1 ms
100	BR/LB	Injector 4 Control	2.7-4.1 ms
101	WT	Injector 2 Control	2.7-4.1 ms
103	BK/WT	Power Ground	<0.1v
104	RD/YL	Coil 4 Driver (dwell)	5°, 55 mph: 8°

Pin Connector Graphic

```
         PCM 104-PIN CONNECTOR
   1 ●●●●●●●●●●●●●    ●●●●●●●●●●●●● 26
  27 ●●●●●●●●●●●●●    ●●●●●●●●●●●●● 52
  53 ●●●●●●●●●●●●●  ●  ●●●●●●●●●●●●● 78
  79 ●●●●●●●●●●●●●    ●●●●●●●●●●●●● 104
```

Terminal View of 104-Pin PCM Wiring Harness Connector

Standard Colors and Abbreviations

Abbreviation	Color	Abbreviation	Color	Abbreviation	Color
BK	Black	GY	Gray	PK	Purple
BL	Blue	GN	Green	RD	Red
BR	Brown	LG	LT Green	TN	Tan
DB	Dark Blue	OR	Orange	WT	White
DG	DK Green	PK	Pink	YL	Yellow

1998-99 Expedition 4.6L V8 VIN 6, VIN W (A/T) 104 Pin Connector

PCM Pin #	Wire Color	Circuit Description (104 Pin)	Value at Hot Idle
1	PK/OR	Shift Solenoid 2 Control	12v, 55 mph: 1v
2	PK/LG	MIL (lamp) Control	MIL On: <1v, Off: 12v
3	YL/BK	Digital TR1 Sensor	0v, 55 mph: 11.5v
4-11	---	Not Used	---
12	YL/WT	Fuel Level Indicator Signal	1.7v (1/2 full)
13	PK	Flash EEPROM Power	0.1v
14	LB/BK	4x4 Indicator Signal	4x4 On: 1v, Off: 12v
15	PK	Data Bus (-) Signal	Digital Signals
16	TN/OR	Data Bus (+) Signal	Digital Signals
17	---	Not Used	---
19	DB/WT	Automatic Ride Control	Digital Signals
20	---	Not Used	---
21	DB	CKP (-) Sensor Signal	430-475 Hz
22	GY	CKP (+) Sensor Signal	430-475 Hz
24	BK/WT	Power Ground	<0.1v
25	LG/YL	Case Ground	<0.050v
26	DB/LG	Coil 1 Driver (dwell)	5°, 55 mph: 8°
27	OR/YL	Shift Solenoid 1 Control	1v, 55 mph: 1v
28	---	Not Used	---
29	TN/WT	TCS (switch) Signal	TCS & O/D On: 12v
30	DG	Octane Adjustment	9.3v (switch shorted: 0v)
31-32	---	Not Used	---
33	PK/OR	VSS (-) Signal	0 Hz, 55 mph: 125 Hz
35	RD/LG	HO2S-12 (B1 S2) Signal	0.1-1.1v
36	TN/LB	MAF Sensor Return	<0.050v
37	OR/BK	TFT Sensor Signal	2.10-2.40v
38	LG/RD	ECT Sensor Signal	0.5-0.6v
39	GY	IAT Sensor Signal	1.5-2.5v
40	DG/YL	Fuel Pump Monitor	On: 12v, Off: 0v
41	BK/YL	A/C Switch Signal	A/C On: 12v, Off: 0v
42	---	Not Used	---
43	OR/LG	Overhead Trip Computer	Digital Signals
44	---	Not Used	---
45	RD/WT	CHTIL (lamp) Control	Lamp On: 1v, Off: 12v
46	BR	Intake Manifold Tuning Valve	12v, 55 mph: 12v
47	BR/PK	VR Solenoid Control	0%, 55 mph: 45%
48	WT/PK	Clean Tachometer Output Signal	39-49 Hz
49	LB/BK	Digital TR2 Sensor	0v, 55 mph: 11.5v
50	WT/BK	Digital TR4 Sensor	0v, 55 mph: 11.5v
51	BK/WT	Power Ground	<0.1v
52	RD/LB	Coil 2 Driver (dwell)	5°, 55 mph: 8°
54	PK/YL	TCC Solenoid Control	0%, 55 mph: 95%
55	BK/LG	Keep Alive Power	12-14v
56	LG/BK	EVAP Purge Solenoid	0-10 Hz (0-100%)
57	YL/RD	Knock Sensor 1 Signal	0v
58	GY/BK	VSS (+) Signal	0 Hz, 55 mph: 125 Hz
59	---	Not Used	---
60	GY/LB	HO2S-11 (B1 S1) Signal	0.1-1.1v
61	PK/LG	HO2S-22 (B2 S2) Signal	0.1-1.1v
62	RD/PK	FTP Sensor Signal	2.6v (0" H2O - cap off)
63	---	Not Used	---
64	LB/YL	TR Sensor Signal	In 'P': 0v, in O/D: 1.7v

1998-99 Expedition 4.6L V8 VIN 6, VIN W (A/T) 104 Pin Connector

PCM Pin #	Wire Color	Circuit Description (104 Pin)	Value at Hot Idle
65	BR/LG	DPFE Sensor Signal	0.20-1.30v
66	YL/LG	CHT Sensor Signal	0.6 to 1.7v (194°F)
67	PK/WT	EVAP CV Solenoid	0-10 Hz (0-100%)
69	PK/YL	A/C WOT Relay Control	Relay On: 0.1v, Off: 12v
71	RD	Vehicle Power	12-14v
72	TN/RD	Injector 7 Control	2.7-3.5 ms
73	TN/BK	Injector 5 Control	2.7-3.5 ms
74	BR/YL	Injector 3 Control	2.7-3.5 ms
75	TN	Injector 1 Control	2.7-3.5 ms
76, 77	BK/WT	Power Ground	<0.1v
78	PK/WT	Coil 3 Driver (dwell)	5°, 55 mph: 8°
79	WT/LG	TCIL (lamp) Control	TCIL On: 1v, Off: 12v
80	LB/OR	Fuel Pump Control	On: 1v, Off: 12v
81	WT/YL	EPC Solenoid Control	9.1v (6 psi)
83	WT/LB	IAC Solenoid Control	10.7v (33%)
84	DB/YL	OSS Sensor Signal	0 Hz, 55 mph: 250 Hz
85	DG	CID Sensor Signal	6 Hz, 55 mph: 15 Hz
87	RD/BK	HO2S-21 (B2 S1) Signal	0.1-1.1v
88	LB/RD	MAF Sensor Signal	0.6v, 55 mph: 2.1v
89	GY/WT	TP Sensor Signal	0.53-1.27v
90	BR/WT	Reference Voltage	4.9-5.1v
91	GY/RD	Analog Signal Return	<0.050v
92	LG	Brake Switch Signal	Brake Off: 12v, On: 1v
93	RD/WT	HO2S-11 (B1 S1) Heater	On: 1v, Off: 12v
94	YL/LB	HO2S-21 (B2 S1) Heater	On: 1v, Off: 12v
95	WT/BK	HO2S-12 (B1 S2) Heater	On: 1v, Off: 12v
96	TN/YL	HO2S-22 (B2 S2) Heater	On: 1v, Off: 12v
97	RD	Vehicle Power	12-14v
98	LB	Injector 8 Control	2.7-3.5 ms
99	LG/OR	Injector 6 Control	2.7-3.5 ms
100	BR/LB	Injector 4 Control	2.7-3.5 ms
101	WT	Injector 2 Control	2.7-3.5 ms
103	BK/WT	Power Ground	<0.1v
104	RD/YL	Coil 4 Driver (dwell)	5°, 55 mph: 8°

Pin Connector Graphic

PCM 104-PIN CONNECTOR

Terminal View of 104-Pin PCM Wiring Harness Connector

Standard Colors and Abbreviations

Abbreviation	Color	Abbreviation	Color	Abbreviation	Color
BK	Black	GY	Gray	PK	Purple
BL	Blue	GN	Green	RD	Red
BR	Brown	LG	LT Green	TN	Tan
DB	Dark Blue	OR	Orange	WT	White
DG	DK Green	PK	Pink	YL	Yellow

2000-02 Expedition 4.6L V8 VIN 6, VIN W (A/T) 104 Pin Connector

PCM Pin #	Wire Color	Circuit Description (104 Pin)	Value at Hot Idle
1	OR/YL	COP 6 Driver (dwell)	5°, 55 mph: 8°
2	---	Not Used	---
3	BK/WT	Power Ground	<0.1v
4-5	---	Not Used	---
6	OR/YL	Shift Solenoid 'A' Control	1v, 55 mph: 1v
7-10	---	Not Used	---
11	PK/OR	Shift Solenoid 'B' Control	12v, 55 mph: 1v
12	WT/LG	TCIL (lamp) Control	Lamp On: 1v, Off: 12v
13	PK	Flash EEPROM Power	0.1v
14	LB/BK	4x4 Low Switch	Switch On: 0v, Off: 12v
15	PK/LB	Data Bus (-) Signal	Digital Signals
16	TN/OR	Data Bus (+) Signal	Digital Signals
17-18	---	Not Used	---
19	DB/WT	S/C Module to Acceleration	Digital Signals
20	---	Not Used	---
21	DB	CKP (-) Sensor Signal	430-475 Hz
22	GY	CKP (+) Sensor Signal	430-475 Hz
23	DG/WT	Dual Knock Sensor (-) Signal	<0.050v
24	---	Not Used	---
25	LB/YL	Case Ground	<0.050v
26	LG/WT	COP 1 Driver (dwell)	5°, 55 mph: 8°
27	LG/YL	COP 5 Driver (dwell)	5°, 55 mph: 8°
28	---	Not Used	---
29	TN/WT	TCS (switch) Signal	TCS & O/D On: 12v
30-31	---	Not Used	---
32	DG/PK	Dual Knock Sensor (-) Signal	<0.050v
33	---	Not Used	---
34	YL/BK	TR Sensor Signal	In 'P': 0v, in O/D: 1.7v
35	RD/LG	HO2S-12 (B1 S2) Signal	0.1-1.1v
36	TN/LB	MAF Sensor Return	<0.050v
37	OR/BK	TFT Sensor Signal	2.10-2.40v
38	---	Not Used	---
39	GY	IAT Sensor Signal	1.5-2.5v
40	DG/YL	Fuel Pump Monitor	On: 12v, Off: 0v
41	BK/YL	A/C High Pressure Switch	A/C On: 12v, Off: 0v
42	PK/OR	TP Sensor Signal	0.9v, 55 mph: 1.3v
43	OR/LG	Data Output Link	Digital Signals
44-45	---	Not Used	---
46	BR	IMRC (valve) Control	12v
47	BR/PK	VR Solenoid Control	0%, 55 mph: 45%
48	---	Not Used	---
49	LB/BK	Digital TR2 Sensor	0v, 55 mph: 11.5v
50	WT/BK	Digital TR4 Sensor	0v, 55 mph: 11.5v
51	BK/WT	Power Ground	<0.1v
52	WT/PK	COP 3 Driver (dwell)	5°, 55 mph: 8°
53	DG/PK	COP 4 Driver (dwell)	5°, 55 mph: 8°
54	PK/YL	TCC Solenoid Control	0%, 55 mph: 95%
55	RD/WT	Keep Alive Power	12-14v
56	LG/BK	EVAP Purge Solenoid	0-10 Hz (0-100%)
57	YL/RD	Dual Knock Sensor (+) Signal	0v
58-59	---	Not Used	---
60	GY/LB	HO2S-11 (B1 S1) Signal	0.1-1.1v
61	PK/LG	HO2S-22 (B2 S2) Signal	0.1-1.1v
62	RD/PK	FTP Sensor Signal	2.6v (0" H2O - cap off)
63	---	Not Used	---
64	LB/YL	Digital TR3A Sensor	0v, 55 mph: 11.5v

2000-02 Expedition 4.6L V8 VIN 6, VIN W (A/T) 104 Pin Connector

PCM Pin #	Wire Color	Circuit Description (104 Pin)	Value at Hot Idle
65 ('00)	BR/LG	DPFE Sensor Signal	0.20-1.30v
65 ('01-'02)	BR/LG	DPFE Sensor Signal	0.95-1.05v
66	YL/LG	CHT Sensor Signal	0.6 to 1.7v (194°F)
67	PK/WT	EVAP CV Solenoid	0-10 Hz (0-100%)
68	GY/BK	VSS Out Signal	Digital Signals
69	PK/YL	A/C WOT Relay Control	On: 1v, Off: 12v
70	---	Not Used	---
71	RD	Vehicle Power	12-14v
72	TN/RD	Injector 7 Control	2.7-4.1 ms
73	TN/BK	Injector 5 Control	2.7-4.1 ms
74	BR/YL	Injector 3 Control	2.7-4.1 ms
75	TN	Injector 1 Control	2.7-4.1 ms
76	---	Not Used	---
77	BK/WT	Power Ground	<0.1v
78	PK/LB	COP 7 Driver (dwell)	5°, 55 mph: 8°
79	WT/RD	COP 8 Driver (dwell)	5°, 55 mph: 8°
80	LB/OR	Fuel Pump Relay Control	On: 1v, Off: 12v
81	WT/YL	EPC Solenoid Control	9.1v (6 psi)
83	WT/LB	IAC Solenoid Control	10.7v (33%)
84	DB/YL	OSS Sensor Signal	0 Hz, 55 mph: 250 Hz
85	DG	CMP Sensor Signal	6 Hz, 55 mph: 15 Hz
87	RD/BK	HO2S-21 (B2 S1) Signal	0.1-1.1v
88	LB/RD	MAF Sensor Signal	0.6v, 55 mph: 2.1v
89	GY/WT	TP Sensor Signal	0.53-1.27v
90	BR/WT	Reference Voltage	4.9-5.1v
91	GY/RD	Sensor Ground	<0.050v
92	RD/LG	Brake Position Switch	Brake Off: 12v, On: 1v
93	RD/WT	HO2S-11 (B1 S1) Heater	On: 1v, Off: 12v
94	YL/LB	HO2S-21 (B2 S1) Heater	On: 1v, Off: 12v
95	WT/BK	HO2S-12 (B1 S2) Heater	On: 1v, Off: 12v
96	TN/YL	HO2S-22 (B2 S2) Heater	On: 1v, Off: 12v
97	RD	Vehicle Power	12-14v
98	LB	Injector 8 Control	2.7-4.1 ms
99	LG/OR	Injector 6 Control	2.7-4.1 ms
100	BR/LB	Injector 4 Control	2.7-4.1 ms
101	WT	Injector 2 Control	2.7-4.1 ms
102	DG/WT	Dual Knock Sensor (-) Signal	0v
103	BK/WT	Power Ground	<0.1v
104	PK/WT	COP 2 Driver (dwell)	5°, 55 mph: 8°

Pin Connector Graphic

PCM 104-PIN CONNECTOR

Terminal View of 104-Pin PCM Wiring Harness Connector

Standard Colors and Abbreviations

Abbreviation	Color	Abbreviation	Color	Abbreviation	Color
BK	Black	GY	Gray	PK	Purple
BL	Blue	GN	Green	RD	Red
BR	Brown	LG	LT Green	TN	Tan
DB	Dark Blue	OR	Orange	WT	White
DG	DK Green	PK	Pink	YL	Yellow

2003-05 Expedition 4.6L V8 MFI VIN W (A/T) C138b Pin Connector

PCM Pin #	Wire Color	Circuit Description (46 Pin)	Value at Hot Idle
1	BK/WT	Power Ground	<0.1v
2	PK/YL	A/C WOT Relay Control	Relay Off: 12v, On: 1v
3	PK/OR	TP Sensor Signal	0.9v, 55 mph: 1.3v
4	DB/OG	Starter Relay Control Circuit	Relay Off: 12v, On: 1v
5	---	Not Used	---
6	OG/LB	Cooling Fan Speed Signal	0.5-4.9v
7	TN/LG	Generator Communicator Command	1.5-10.5v (40-250 Hz)
8	RD/PK	Fuel Tank Pressure Sensor	2.6v (0" H2O - cap off)
9	OR/RD	Antitheft Indicator Control	Indicator Off: 12v, On: 1v
10	LB/YL	Power Ground	<0.1v
11	BK/WT	Power Ground	<0.1v
12	VT/YL	Speed Control Servo Signal 'C'	DC pulse signals
13	GY/BK	VSS (+) Signal	0 Hz, 55 mph: 125 Hz
14	BR/PK	Generator Monitor Signal	130 Hz (45%)
15	BK/YL	Brake Pedal Position Switch	Brake Off: 0v, On: 12v
16	---	Not Used	---
17	DG/OR	Steering Wheel Control Ground	<0.050v
18	YL/LG	Power Steering Pressure Switch	Straight: 0v, Turned: 12v
19	GY	IAT Sensor Signal	1.5-2.5v
20	DG/YL	Fuel Pump Relay Output	Off: 0v, On: 12v
21	---	Not Used	---
22	TN/WT	TCS (switch) Signal	TCS & O/D On: 12v
23	BK/WT	Power Ground	<0.1v
24	PK/BK	Speed Control Servo Signal 'B'	DC pulse signals
25	LG/VT	Cooling Fan Control	Digital Signal: 0-12-0v
26	TN	Speed Control Servo Common	<0.050v
27	LB/OR	Fuel Pump Relay Control	Relay Off: 12v, On: 1v
28-29	---	Not Used	---
30	TN/LB	MAF Sensor Return	<0.050v
31	LB/RD	MAF Sensor Signal	0.6v, 55 mph: 2.1v
32	TN/OR	SCP Data Bus (+) Signal	Digital Signals
33	GY/RD	Sensor Ground	<0.050v
34	RD	Vehicle Power (Start-Run)	12-14v
35	DG/WT	Speed Control Servo Signal 'A'	DC pulse signals
36	VT/WT	EVAP Canister Vent Solenoid Control	0-10 Hz (0-100%)
37	LB/BK	Speed Control Switch	0v (switch on: 6.7v)
38	LG/BK	EVAP Purge Solenoid Control	0-10 Hz (0-100%)
39	VT	Module Programming Signal	0.1v
40	RD/WT	Direct Battery	12-14v
41	OR/LB	A/C Pressure Switch Signal	0v (switch on: 12v)
42	GY/OR	Passive Antitheft RX Signal	Digital Signals
43	WT/LG	Passive Antitheft TX Signal	Digital Signals
44	PK/LB	Data Bus (-) Signal	<0.050v
45	BR/WT	Reference Voltage	4.9-5.1v
46	RD	Vehicle Power (Start-Run)	12-14v

2003-05 Expedition 4.6L V8 MFI VIN W (A/T) C138t Pin Connector

PCM Pin #	Wire Color	Circuit Description (30 Pin)	Value at Hot Idle
1	---	Not Used	---
2	VT/LG	HO2S-22 (B2 S2) Signal	0.1-1.1v
3	RD/LG	HO2S-12 (B1 S2) Signal	0.1-1.1v
4-10	---	Not Used	---
11	VT/YL	TCC Solenoid Control	0%, 55 mph: 95%
12	OR/YL	Shift Solenoid 'A' Control	1v, 55 mph: 1v
13	VT/OR	Shift Solenoid 'B' Control	12v, 55 mph: 1v
14	VT/OR	Coast Clutch Solenoid Control	12v, 55 mph: 12v
15-16	---	Not Used	---
17	LB/YL	Digital TR3A Sensor	0v, 55 mph: 11.5v
18	LB/BK	Digital TR2 Sensor	0v, 55 mph: 11.5v
19	WT/BK	Digital TR4 Sensor	0v, 55 mph: 11.5v
20	YL/BK	Digital TR1 Sensor	In 'P': 0v, in O/D: 1.7v
21	WT/BK	HO2S-12 (B1 S2) Heater	On: 1v, Off: 12v
22	---	Not Used	---
23	WT/YL	EPC Solenoid Control	9.1v (6 psi)
24	---	Not Used	---
25	DB/YL	OSS Sensor Signal	0 Hz, 55 mph: 250 Hz
26	DG/WT	Turbine Shaft Speed Sensor	360 Hz, 890 Hz
27	GY/RD	Sensor Ground	<0.050v
28	OR/BK	Transmission Fluid Temperature Sensor	2.10-2.40v
29	TN/YL	HO2S-22 (B2 S2) Heater	On: 1v, Off: 12v
30	---	Not Used	---

Pin Connector Graphic

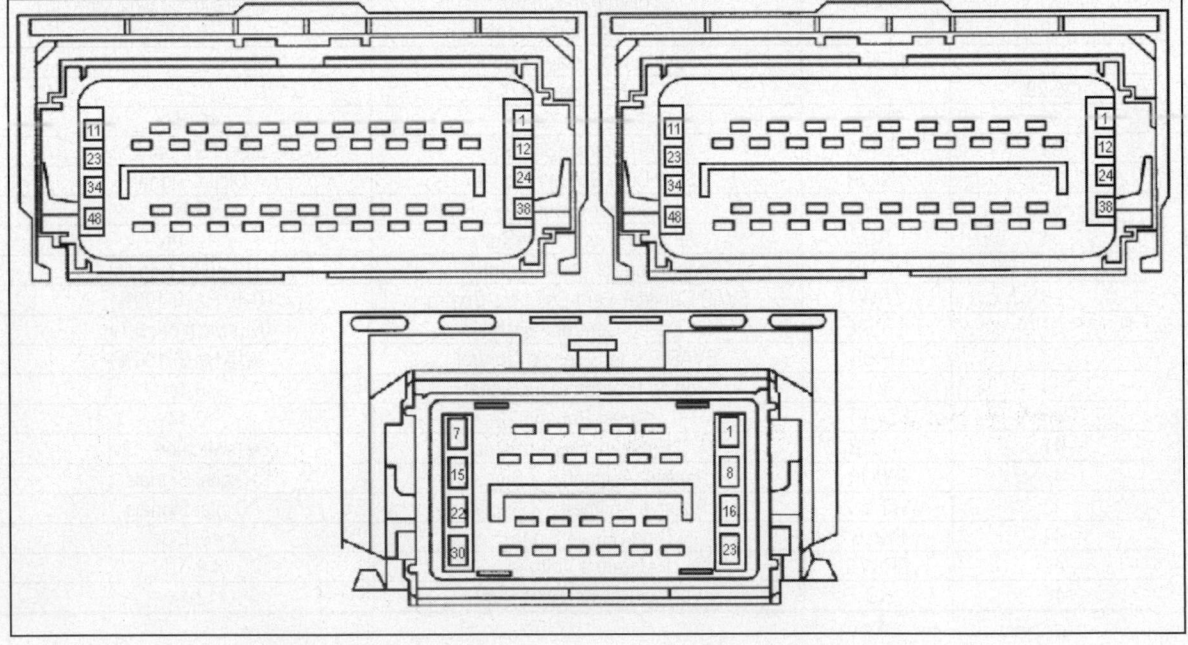

2003-05 Expedition 4.6L V8 MFI VIN W (A/T) C1381e Pin Connector

PCM Pin #	Wire Color	Circuit Description (46 Pin)	Value at Hot Idle
1	PK/WT	COP 2 Driver (dwell)	5°, 55 mph: 8°
2	WT/LB	IAC Solenoid Control	10.7v (33%)
3	TN/BK	Injector 5 Control	2.7-4.1 ms
4	LG/OR	Injector 6 Control	2.7-4.1 ms
5	TN/RD	Injector 7 Control	2.7-4.1 ms
6	LB	Injector 8 Control	2.7-4.1 ms
7	BR/YL	Injector 3 Control	2.7-4.1 ms
8	BR/LB	Injector 4 Control	2.7-4.1 ms
9	GY/YL	Vehicle Power (Start-Run)	12-14v
10	---	Not Used	---
11	WT/RD	COP 8 Driver (dwell)	5°, 55 mph: 8°
12	PK/LB	COP 7 Driver (dwell)	5°, 55 mph: 8°
13	BR	Intake Manifold Runner Control Module	12v
14	TN	Injector 1 Control	2.7-4.1 ms
15	WT	Injector 2 Control	2.7-4.1 ms
16-19	---	Not Used	---
20	YL/LB	HO2S-21 (B2 S1) Heater	On: 1v, Off: 12v
21	RD/WT	HO2S-11 (B1 S1) Heater	On: 1v, Off: 12v
22	BR/PK	EGR Vacuum Regulator Control	0%, 55 mph: 45%
23	DG/VT	COP 4 Driver (dwell)	5°, 55 mph: 8°
24	WT/PK	COP 3 Driver (dwell)	5°, 55 mph: 8°
25	GY/RD	Sensor Ground	<0.050v
26	GY/LB	HO2S-11 (B1 S1) Signal	0.1-1.1v
27	RD/BK	HO2S-21 (B2 S1) Signal	0.1-1.1v
28	DG/VT	Knock Sensor 2 (+) Signal	0v
29	YL/RD	Knock Sensor 1 (+) Signal	0v
30	DB	CKP (+) Sensor Signal	400-420 Hz
31	DG	Camshaft Position Sensor	6 Hz, 55 mph: 15 Hz
32	DG/PK	Dual Knock Sensor (-) Signal	<0.050v
33	BR/LG	ECT Sensor Signal	0.5-0.6v
34	LG/YL	COP 5 Driver (dwell)	5°, 55 mph: 8°
35	LG/WT	COP 1 Driver (dwell)	5°, 55 mph: 8°
36	BR/WT	Reference Voltage	4.9-5.1v
37-38	---	Not Used	---
39	DG/WT	Knock Sensor 2 (-) Signal	<0.050v
40	YL	Knock Sensor 1 (-) Signal	<0.050v
41	GY	CKP (-) Sensor Signal	400-420 Hz
42-43	---	Not Used	---
44	GY/WT	TP Sensor Signal	0.53-1.27v
45	YL/LG	CHT Sensor Signal	0.6 to 1.7v (194°F)
46	OR/YL	COP 6 Driver (dwell)	5°, 55 mph: 8°

1997 Expedition 5.4L V8 VIN L (E4OD) 104 Pin Connector

PCM Pin #	Wire Color	Circuit Description (104 Pin)	Value at Hot Idle
1	LG/YL	COP 6 Driver (dwell)	5°, 55 mph: 8°
2	PK/LG	MIL (lamp) Control	MIL On: 1v, Off: 12v
3	BK/WT	Power Ground	<0.1v
6	OR/YL	Shift Solenoid 1 Control	1v, 30 mph: 12v
9	YL/WT	Fuel Level Indicator Signal	1.7v at 50%
11	PK/OR	Shift Solenoid 2 Control	12v, 55 mph: 1v
12	WT/LG	TCIL (lamp) Control	TCIL On: 1v, Off: 12v
13	PK	Flash EEPROM Power	0.1v
14	LB/BK	4x4 Indicator Signal	4x4 On: 1v, Off: 12v
15	PK	Data Bus (-) Signal	Digital Signals
16	TN/OR	Data Bus (+) Signal	Digital Signals
19	DB/WT	Suspension Control Module	Digital Signals
20	BR/OR	Coast Clutch Solenoid	12v, 55 mph: 12v
21	DB	CKP (+) Sensor Signal	400-420 Hz
22	GY	CKP (-) Sensor Signal	400-420 Hz
24	BK/WT	Power Ground	<0.1v
25	BK	Case Ground	<0.050v
26	LG/WT	COP 1 Driver (dwell)	5°, 55 mph: 8°
27	OR/YL	COP 5 Driver (dwell)	5°, 55 mph: 8°
29	TN/WT	TCS (switch) Signal	TCS & O/D On: 12v
30	DG	Octane Adjustment Signal	9.3v (Closed: 0v)
33	PK/OR	VSS (-) Signal	0 Hz, 55 mph: 125 Hz
34	YL/BK	Digital TR1 Sensor	0v, 55 mph: 11.5v
35	RD/LG	HO2S-12 (B1 S2) Signal	0.1-1.1v
36	TN/LB	MAF Sensor Return	<0.050v
37	OR/BK	TFT Sensor Signal	2.10-2.40v
39	GY	IAT Sensor Signal	1.5-2.5v
40	DG/YL	Fuel Pump Monitor	On: 12v, Off: 0v
41	BK/YL	A/C High Pressure Switch	A/C On: 12v, Off: 0v
43	OR/LG	Overhead Trip Computer	Digital Signals
45	RD/WT	CHIL (lamp) Control	CHT On: 1v, Off: 12v
46	BR	IMTV (tuning valve) Control	1v, 55 mph: 1v
47	BR/PK	VR Solenoid Control	0%, 55 mph: 45%
48	WT/PK	Clean Tachometer Output	46 Hz
49	LB/BK	Digital TR2 Sensor	0v, 55 mph: 11.5v
50	WT/BK	Digital TR4 Sensor	0v, 55 mph: 11.5v
51	BK/WT	Power Ground	<0.1v
52	PK/WT	COP 3 Driver (dwell)	5°, 55 mph: 8°
53	PK/B	COP 4 Driver (dwell)	5°, 55 mph: 8°
54	PK/YL	TCC Solenoid Control	0%, 55 mph: 95%
55	BK/LG	Keep Alive Power	12-14v
56	LG/BK	EVAP Purge Solenoid	0-10 Hz (0-100%)
57	YL/RD	Knock Sensor Signal	0v
58	GY/BK	VSS (+) Signal	0 Hz, 55 mph: 125 Hz
60	GY/LB	HO2S-11 (B1 S1) Signal	0.1-1.1v
61	PK/LG	HO2S-22 (B2 S2) Signal	0.1-1.1v
62	RD/PK	FTP Sensor Signal	2.6v (0" H2O - cap off)
64	LB/YL	Digital TR3A Sensor	In 'P': 0v, in O/D: 1.7v

1997 Expedition 5.4L V8 VIN L (E4OD) 104 Pin Connector

PCM Pin #	Wire Color	Circuit Description (104 Pin)	Value at Hot Idle
65	BR/LG	DPFE Sensor Signal	0.20-1.30v
66	LG	CHT Sensor Signal	4.52v (68°F)
67	PK/WT	EVAP CV Solenoid	0-10 Hz (0-100%)
71	RD	Vehicle Power	12-14v
72	TN/RD	Injector 7 Control	3.2-4.0 ms
73	TN/BK	Injector 5 Control	3.2-4.0 ms
74	BR/YL	Injector 3 Control	3.2-4.0 ms
75	TN	Injector 1 Control	3.2-4.0 ms
77	BK/WT	Power Ground	<0.1v
78	WT/PK	COP 7 Driver (dwell)	5°, 55 mph: 8°
79	WT/RD	COP 8 Driver (dwell)	5°, 55 mph: 8°
80	LB/OR	Fuel Pump Control	On: 1v, Off: 12v
81	WT/YL	EPC Solenoid Control	9.1v (4 psi)
83	WT/LB	IAC Solenoid Control	10.7v (33%)
85	DG	CMP Sensor Signal	5-7 Hz
87	RD/BK	HO2S-21 (B2 S1) Signal	0.1-1.1v
88	LB/RD	MAF Sensor Signal	0.8v
89	GY/WT	TP Sensor Signal	0.53-1.27v
90	BR/WT	Reference Voltage	4.9-5.1v
91	GY/RD	Analog Signal Return	<0.050v
92	LG	Brake Pedal Switch Signal	Brake Off: 12v, On: 1v
93	RD/WT	HO2S-11 (B1 S1) Heater	On: 1v, Off: 12v
94	YL/LB	HO2S-21 (B2 S1) Heater	On: 1v, Off: 12v
95	WT/BK	HO2S-12 (B1 S2) Heater	On: 1v, Off: 12v
96	TN/YL	HO2S-22 (B2 S2) Heater	On: 1v, Off: 12v
97	RD	Vehicle Power	12-14v
98	LB	Injector 8 Control	3.2-4.0 ms
99	LG/OR	Injector 6 Control	3.2-4.0 ms
100	BR/LB	Injector 4 Control	3.2-4.0 ms
101	WT	Injector 2 Control	3.2-4.0 ms
103	BK/WT	Power Ground	<0.1v
104	DG/PK	COP 2 Driver (dwell)	5°, 55 mph: 8°

Pin Connector Graphic

PCM 104-PIN CONNECTOR

Terminal View of 104-Pin PCM Wiring Harness Connector

Standard Colors and Abbreviations

Abbreviation	Color	Abbreviation	Color	Abbreviation	Color
BK	Black	GY	Gray	PK	Purple
BL	Blue	GN	Green	RD	Red
BR	Brown	LG	LT Green	TN	Tan
DB	Dark Blue	OR	Orange	WT	White
DG	DK Green	PK	Pink	YL	Yellow

1998-99 Expedition 5.4L VIN L (E4OD) 104 Pin Connector

PCM Pin #	Wire Color	Circuit Description (104 Pin)	Value at Hot Idle
1	LG/YL	COP 6 Driver (dwell)	5°, 55 mph: 8°
2	PK/LG	MIL (lamp) Control	MIL On: 1v, Off: 12v
3, 24	BK/WT	Power Ground	<0.1v
6	OR/YL	Shift Solenoid 1 Control	1v, 30 mph: 12v
9	YL/WT	Fuel Level Indicator Signal	1.7v (1/2 full)
11	PK/OR	Shift Solenoid 2 Control	12v, 55 mph: 1v
12	WT/LG	TCIL (lamp) Control	TCIL On: 1v, Off: 12v
13	PK	Flash EEPROM Power	0.1v
14	LB/BK	4x4 Indicator Signal	4x4 Switch Off: 0.1v
15	PK	Data Bus (-) Signal	Digital Signals
16	TN/OR	Data Bus (+) Signal	Digital Signals
19	DB/WT	Suspension Control Module	Digital Signals
20	BR/OR	Coast Clutch Solenoid	12v, 55 mph: 12v
21	DB	CKP (+) Sensor Signal	430-475 Hz
22	GY	CKP (-) Sensor Signal	430-475 Hz
25	BK	Case Ground	<0.050v
26	LG/WT	COP 1 Driver (dwell)	5°, 55 mph: 8°
27	OR/YL	COP 5 Driver (dwell)	5°, 55 mph: 8°
29	TN/WT	TCS (switch) Signal	TCS & O/D On: 12v
30	DG	Octane Adjustment Signal	9.3v (Closed: 0v)
33	PK/OR	VSS (-) Signal	0 Hz, 55 mph: 125 Hz
34	YL/BK	Digital TR1 Sensor	0v, 55 mph: 11.5v
35	RD/LG	HO2S-12 (B1 S2) Signal	0.1-1.1v
36	TN/LB	MAF Sensor Return	<0.050v
37	OR/BK	TFT Sensor Signal	2.10-2.40v
39	GY	IAT Sensor Signal	1.5-2.5v
40	DG/YL	Fuel Pump Monitor	On: 12v, Off: 0v
41	BK/YL	A/C Request Signal	A/C On: 12v, Off: 0v
43	OR/LG	Overhead Trip Computer	Digital Signals
45	RD/WT	CHIL (lamp) Control	CHT On: 1v, Off: 12v
46	BR	IMTV (tuning valve) Control	12v, 55 mph: 12v
47	BR/PK	VR Solenoid Control	0%, 55 mph: 45%
48	WT/PK	Clean Tachometer Output	46 Hz
49	LB/BK	Digital TR2 Sensor	0v, 55 mph: 11.5v
50	WT/BK	Digital TR4 Sensor	0v, 55 mph: 11.5v
51	BK/WT	Power Ground	<0.1v
52	PK/WT	COP 3 Driver (dwell)	5°, 55 mph: 8°
53	PK/LB	COP 4 Driver (dwell)	5°, 55 mph: 8°
54	PK/YL	TCC Solenoid Control	0%, 55 mph: 95%
55	BK/LG	Keep Alive Power	12-14v
56	LG/BK	EVAP Purge Solenoid	0-10 Hz (0-100%)
57	YL/RD	Knock Sensor 1 Signal	0v
58	GY/BK	VSS (+) Signal	0 Hz, 55 mph: 125 Hz
60	GY/LB	HO2S-11 (B1 S1) Signal	0.1-1.1v
61	PK/LG	HO2S-22 (B2 S2) Signal	0.1-1.1v
62	RD/PK	FTP Sensor Signal	2.6v (0" H2O - cap off)
64	LB/YL	Digital TR3A Sensor	In 'P': 0v, in O/D: 1.7v

1998-99 Expedition 5.4L VIN L (E4OD) 104 Pin Connector

PCM Pin #	Wire Color	Circuit Description (104 Pin)	Value at Hot Idle
65	BR/LG	DPFE Sensor Signal	0.2-1.30v
66	YL/LG	CHT Sensor Signal	0.6v (194ºF)
67	PK/WT	EVAP CV Solenoid	0-10 Hz (0-100%)
71	RD	Vehicle Power	12-14v
72	TN/RD	Injector 7 Control	2.7-3.5 ms
73	TN/BK	Injector 5 Control	2.7-3.5 ms
74	BR/YL	Injector 3 Control	2.7-3.5 ms
75	TN	Injector 1 Control	2.7-3.5 ms
77	BK/WT	Power Ground	<0.1v
78	WT/PK	COP 7 Driver (dwell)	5º, 55 mph: 8º
79	WT/RD	COP 8 Driver (dwell)	5º, 55 mph: 8º
80	LB/OR	Fuel Pump Control	On: 1v, Off: 12v
81	WT/YL	EPC Solenoid Control	9.1v (4 psi)
83	WT/LB	IAC Solenoid Control	10.7v (33%)
85	DG	CMP Sensor Signal	6 Hz, 55 mph: 13-17 Hz
87	RD/BK	HO2S-21 (B2 S1) Signal	0.1-1.1v
88	LB/RD	MAF Sensor Signal	0.8v, 55 mph: 1.9v
89	GY/WT	TP Sensor Signal	0.53-1.27v
90	BR/WT	Reference Voltage	4.9-5.1v
91	GY/RD	Analog Signal Return	<0.050v
92	LG	Brake Pedal Switch Signal	Brake Off: 12v, On: 1v
93	RD/WT	HO2S-11 (B1 S1) Heater	On: 1v, Off: 12v
94	YL/LB	HO2S-21 (B2 S1) Heater	On: 1v, Off: 12v
95	WT/BK	HO2S-12 (B1 S2) Heater	On: 1v, Off: 12v
96	TN/YL	HO2S-22 (B2 S2) Heater	On: 1v, Off: 12v
97	RD	Vehicle Power	12-14v
98	LB	Injector 8 Control	2.7-3.5 ms
99	LG/OR	Injector 6 Control	2.7-3.5 ms
100	BR/LB	Injector 4 Control	2.7-3.5 ms
101	WT	Injector 2 Control	2.7-3.5 ms
103	BK/WT	Power Ground	<0.1v
104	DG/PK	COP 2 Driver (dwell)	5º, 55 mph: 8º

Pin Connector Graphic

```
          PCM 104-PIN CONNECTOR
  1 ●●●●●●●●●●●●●        ●●●●●●●●●●●●● 26
 27 ●●●●●●●●●●●●●        ●●●●●●●●●●●●● 52
 53 ●●●●●●●●●●●●●   ⬤   ●●●●●●●●●●●●● 78
 79 ●●●●●●●●●●●●●        ●●●●●●●●●●●●● 104
```

Terminal View of 104-Pin PCM Wiring Harness Connector

Standard Colors and Abbreviations

Abbreviation	Color	Abbreviation	Color	Abbreviation	Color
BK	Black	GY	Gray	PK	Purple
BL	Blue	GN	Green	RD	Red
BR	Brown	LG	LT Green	TN	Tan
DB	Dark Blue	OR	Orange	WT	White
DG	DK Green	PK	Pink	YL	Yellow

2000-02 Expedition 5.4L V8 VIN L (E4OD) 104 Pin Connector

PCM Pin #	Wire Color	Circuit Description (104 Pin)	Value at Hot Idle
1	OR/YL	COP 6 Driver (dwell)	5°, 55 mph: 8°
3	BK/WT	Power Ground	<0.1v
6	OR/YL	Shift Solenoid 1 Control	1v, 30 mph: 12v
11	PK/OR	Shift Solenoid 2 Control	12v, 55 mph: 1v
12	WT/LG	TCIL (lamp) Control	TCIL On: 1v, Off: 12v
13	PK	Flash EEPROM Power	0.1v
14	LB/BK	4x4 Low Indicator Switch	4x4 Switch Off: 0.1v
15	PK/LB	Data Bus (-) Signal	Digital Signals
16	TN/OR	Data Bus (+) Signal	Digital Signals
19	DB/WT	Suspension Control Module	Digital Signals
20	BR/OR	Coast Clutch Solenoid	12v, 55 mph: 12v
21	DB	CKP (+) Sensor Signal	430-475 Hz
22	GY	CKP (-) Sensor Signal	430-475 Hz
25	LB/YL	Case Ground	<0.050v
26	LG/WT	COP 1 Driver (dwell)	5°, 55 mph: 8°
27	LG/YL	COP 5 Driver (dwell)	5°, 55 mph: 8°
29	TN/WT	TCS (switch) Signal	TCS & O/D On: 12v
32	DG/PK	Knock Sensor 1 Return	<0.050v
34	YL/BK	Digital TR1 Sensor	0v, 55 mph: 11.5v
35	RD/LG	HO2S-12 (B1 S2) Signal	0.1-1.1v
36	TN/LB	MAF Sensor Return	<0.050v
37	OR/BK	TFT Sensor Signal	2.10-2.40v
38	LG/RD	ECT Sensor Signal	0.5-0.6v
39	GY	IAT Sensor Signal	1.5-2.5v
40	DG/YL	Fuel Pump Monitor	On: 12v, Off: 0v
41	BK/YL	A/C Request Signal	A/C On: 12v, Off: 0v
42	PK/OR	TP Sensor Signal	0.9v, 55 mph: 1.3v
43	OR/LG	Data Output Link	5v
46	BR	IMRC (valve) Control	12v
47	BR/PK	VR Solenoid Control	0%, 55 mph: 45%
49	LB/BK	Digital TR2 Sensor	0v, 55 mph: 11.5v
50	WT/BK	Digital TR4 Sensor	0v, 55 mph: 11.5v
51	BK/WT	Power Ground	<0.1v
52	WT/PK	COP 3 Driver (dwell)	5°, 55 mph: 8°
53	DG/PK	COP 4 Driver (dwell)	5°, 55 mph: 8°
54	PK/YL	TCC Solenoid Control	0%, 55 mph: 95%
55	RD/WT	Keep Alive Power	12-14v
56	LG/BK	EVAP Purge Solenoid	0-10 Hz (0-100%)
57	YL/RD	Knock Sensor 1 Signal	0v
59	DG/WT	TSS Sensor Signal	100 Hz, 30 mph: 700 Hz
60	GY/LB	HO2S-11 (B1 S1) Signal	0.1-1.1v
61	PK/LG	HO2S-22 (B2 S2) Signal	0.1-1.1v
62	RD/PK	FTP Sensor Signal	2.6v (0" H2O - cap off)
64	LB/YL	Digital TR3A Sensor	In 'P': 0v, in O/D: 1.7v

2000-02 Expedition 5.4L V8 VIN L (E4OD) 104 Pin Connector

PCM Pin #	Wire Color	Circuit Description (104 Pin)	Value at Hot Idle
65 ('00)	BR/LG	DPFE Sensor Signal	0.2-1.30v
65 ('01-'02)	BR/LG	DPFE Sensor Signal	0.95-1.05v
66	YL/LG	CHT Sensor Signal	0.6v (194°F)
67	PK/WT	EVAP CV Solenoid	0-10 Hz (0-100%)
68	GY/BK	Vehicle Speed Sensor	0 Hz, 55 mph: 125 Hz
71	RD	Vehicle Power	12-14v
72	TN/RD	Injector 7 Control	2.7-3.5 ms
73	TN/BK	Injector 5 Control	2.7-3.5 ms
74	BR/YL	Injector 3 Control	2.7-3.5 ms
75	TN	Injector 1 Control	2.7-3.5 ms
77	BK/WT	Power Ground	<0.1v
78	PK/LB	COP 7 Driver (dwell)	5°, 55 mph: 8°
79	WT/RD	COP 8 Driver (dwell)	5°, 55 mph: 8°
80	LB/OR	Fuel Pump Control	On: 1v, Off: 12v
81	WT/YL	EPC Solenoid Control	9.1v (4 psi)
83	WT/LB	IAC Solenoid Control	10.7v (33%)
84	DB/YL	OSS Sensor Signal	0 Hz, 55 mph: 350 Hz
85	DG	CMP Sensor Signal	6 Hz, 55 mph: 13-17 Hz
87	RD/BK	HO2S-21 (B2 S1) Signal	0.1-1.1v
88	LB/RD	MAF Sensor Signal	0.8v, 55 mph: 1.9v
89	GY/WT	TP Sensor Signal	0.53-1.27v
90	BR/WT	Reference Voltage	4.9-5.1v
91	GY/RD	Analog Signal Return	<0.050v
92	RD/LG	Brake Pedal Switch Signal	Brake Off: 12v, On: 1v
93	RD/WT	HO2S-11 (B1 S1) Heater	On: 1v, Off: 12v
94	YL/LB	HO2S-21 (B2 S1) Heater	On: 1v, Off: 12v
95	WT/BK	HO2S-12 (B1 S2) Heater	On: 1v, Off: 12v
96	TN/YL	HO2S-22 (B2 S2) Heater	On: 1v, Off: 12v
97	RD	Vehicle Power	12-14v
98	LB	Injector 8 Control	2.7-3.5 ms
99	LG/OR	Injector 6 Control	2.7-3.5 ms
100	BR/LB	Injector 4 Control	2.7-3.5 ms
101	W	Injector 2 Control	2.7-3.5 ms
103	BK/WT	Power Ground	<0.1v
104	PK/WT	COP 2 Driver (dwell)	5°, 55 mph: 8°

Pin Connector Graphic

PCM 104-PIN CONNECTOR

Terminal View of 104-Pin PCM Wiring Harness Connector

Standard Colors and Abbreviations

Abbreviation	Color	Abbreviation	Color	Abbreviation	Color
BK	Black	GY	Gray	PK	Purple
BL	Blue	GN	Green	RD	Red
BR	Brown	LG	LT Green	TN	Tan
DB	Dark Blue	OR	Orange	WT	White
DG	DK Green	PK	Pink	YL	Yellow

2003-05 Expedition 5.4L V8 MFI VIN L (A/T) C175a Pin Connector

PCM Pin #	Wire Color	Circuit Description (46 Pin)	Value at Hot Idle
1	BK/WT	Power Ground	<0.1v
2	PK/YL	A/C WOT Relay Control	Relay Off: 12v, On: 1v
3	PK/OR	TP Sensor Signal	0.9v, 55 mph: 1.3v
4	DB/OG	Starter Relay Control Circuit	Relay Off: 12v, On: 1v
5	---	Not Used	---
6	OG/LB	Cooling Fan Speed Signal	0.5-4.9v
7	TN/LG	Generator Communicator Command	1.5-10.5v (40-250 Hz)
8	RD/PK	Fuel Tank Pressure Sensor	2.6v (0" H2O - cap off)
9	OR/RD	Antitheft Indicator Control	Indicator Off: 12v, On: 1v
10	LB/YL	Power Ground	<0.1v
11	BK/WT	Power Ground	<0.1v
12	VT/YL	Speed Control Servo Signal 'C'	DC pulse signals
13	GY/BK	VSS (+) Signal	0 Hz, 55 mph: 125 Hz
14	BR/PK	Generator Monitor Signal	130 Hz (45%)
15	BK/YL	Brake Pedal Position Switch	Brake Off: 0v, On: 12v
16	---	Not Used	---
17	DG/OR	Steering Wheel Control Ground	<0.050v
18	YL/LG	Power Steering Pressure Switch	Straight: 0v, Turned: 12v
19	GY	IAT Sensor Signal	1.5-2.5v
20	DG/YL	Fuel Pump Relay Output	Off: 0v, On: 12v
21	---	Not Used	---
22	TN/WT	TCS (switch) Signal	TCS & O/D On: 12v
23	BK/WT	Power Ground	<0.1v
24	PK/BK	Speed Control Servo Signal 'B'	DC pulse signals
25	LG/VT	Cooling Fan Control	Digital Signal: 0-12-0v
26	TN	Speed Control Servo Common	<0.050v
27	LB/OR	Fuel Pump Relay Control	Relay Off: 12v, On: 1v
28-29	---	Not Used	---
30	TN/LB	MAF Sensor Return	<0.050v
31	LB/RD	MAF Sensor Signal	0.6v, 55 mph: 2.1v
32	TN/OR	SCP Data Bus (+) Signal	Digital Signals
33	GY/RD	Sensor Ground	<0.050v
34	RD	Vehicle Power (Start-Run)	12-14v
35	DG/WT	Speed Control Servo Signal 'A'	DC pulse signals
36	VT/WT	EVAP Canister Vent Solenoid Control	0-10 Hz (0-100%)
37	LB/BK	Speed Control Switch	N/A
38	LG/BK	EVAP Purge Solenoid Control	0-10 Hz (0-100%)
39	VT	Module Programming Signal	0.1v
40	RD/WT	Direct Battery	12-14v
41	OR/LB	A/C Pressure Switch Signal	0v (switch on: 12v)
42	GY/OR	Passive Antitheft RX Signal	Digital Signals
43	WT/LG	Passive Antitheft TX Signal	Digital Signals
44	PK/LB	Data Bus (-) Signal	<0.050v
45	BR/WT	Reference Voltage	4.9-5.1v
46	RD	Vehicle Power (Start-Run)	12-14v

2003-05 Expedition 5.4L V8 MFI VIN L (A/T) C138t Pin Connector

PCM Pin #	Wire Color	Circuit Description (30 Pin)	Value at Hot Idle
1	---	Not Used	---
2	VT/LG	HO2S-22 (B2 S2) Signal	0.1-1.1v
3	RD/LG	HO2S-12 (B1 S2) Signal	0.1-1.1v
4-10	---	Not Used	---
11	VT/YL	TCC Solenoid Control	0%, 55 mph: 95%
12	OR/YL	Shift Solenoid 'A' Control	1v, 55 mph: 1v
13	VT/OR	Shift Solenoid 'B' Control	12v, 55 mph: 1v
14	BR/OR	Coast Clutch Solenoid Control	12v, 55 mph: 12v
15-16	---	Not Used	---
17	LB/YL	Digital TR3A Sensor	0v, 55 mph: 11.5v
18	LB/BK	Digital TR2 Sensor	0v, 55 mph: 11.5v
19	WT/BK	Digital TR4 Sensor	0v, 55 mph: 11.5v
20	YL/BK	Digital TR1 Sensor	In 'P': 0v, in O/D: 1.7v
21	WT/BK	HO2S-12 (B1 S2) Heater	On: 1v, Off: 12v
22	---	Not Used	---
23	WT/YL	EPC Solenoid Control	9.1v (6 psi)
24	---	Not Used	---
25	DB/YL	OSS Sensor Signal	0 Hz, 55 mph: 250 Hz
26	DG/WT	Turbine Shaft Speed Sensor	360 Hz, 890 Hz
27	GY/RD	Sensor Ground	<0.050v
28	OR/BK	Transmission Fluid Temperature Sensor	2.10-2.40v
29	TN/YL	HO2S-22 (B2 S2) Heater	On: 1v, Off: 12v
30	---	Not Used	---

Pin Connector Graphic

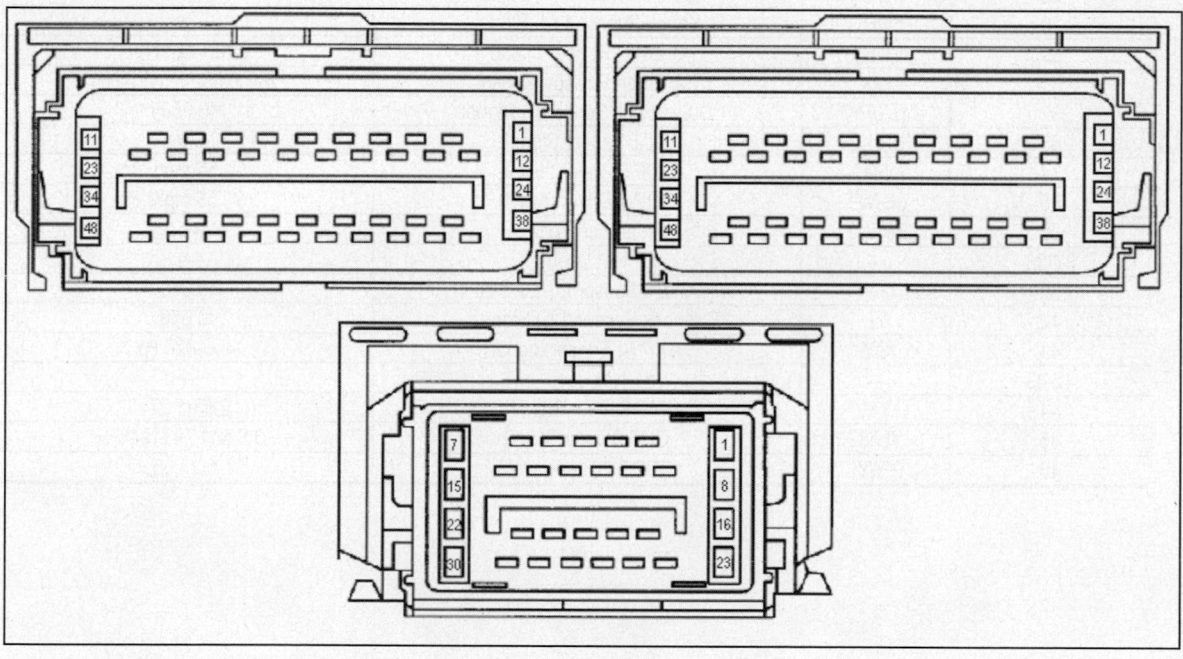

2003-05 Expedition 5.4L V8 MFI VIN L (A/T) C1381e Pin Connector

PCM Pin #	Wire Color	Circuit Description (46 Pin)	Value at Hot Idle
1	PK/WT	COP 2 Driver (dwell)	5°, 55 mph: 8°
2	WT/LB	IAC Solenoid Control	10.7v (33%)
3	TN/BK	Injector 5 Control	2.7-4.1 ms
4	LG/OR	Injector 6 Control	2.7-4.1 ms
5	TN/RD	Injector 7 Control	2.7-4.1 ms
6	LB	Injector 8 Control	2.7-4.1 ms
7	BR/YL	Injector 3 Control	2.7-4.1 ms
8	BR/LB	Injector 4 Control	2.7-4.1 ms
9	GY/YL	Vehicle Power (Start-Run)	12-14v
10	---	Not Used	---
11	WT/RD	COP 8 Driver (dwell)	5°, 55 mph: 8°
12	PK/LB	COP 7 Driver (dwell)	5°, 55 mph: 8°
13	BR	Intake Manifold Runner Control Module	12v
14	TN	Injector 1 Control	2.7-4.1 ms
15	WT	Injector 2 Control	2.7-4.1 ms
16-19	---	Not Used	---
20	YL/LB	HO2S-21 (B2 S1) Heater	On: 1v, Off: 12v
21	RD/WT	HO2S-11 (B1 S1) Heater	On: 1v, Off: 12v
22	BR/PK	EGR Vacuum Regulator Control	0%, 55 mph: 45%
23	DG/VT	COP 4 Driver (dwell)	5°, 55 mph: 8°
24	WT/PK	COP 3 Driver (dwell)	5°, 55 mph: 8°
25	GY/RD	Sensor Ground	<0.050v
26	GY/LB	HO2S-11 (B1 S1) Signal	0.1-1.1v
27	RD/BK	HO2S-21 (B2 S1) Signal	0.1-1.1v
28	DG/VT	Knock Sensor 2 (+) Signal	0v
29	YL/RD	Knock Sensor 1 (+) Signal	0v
30	DB	CKP (+) Sensor Signal	400-420 Hz
31	DG	Camshaft Position Sensor	6 Hz, 55 mph: 15 Hz
32	DG/PK	Dual Knock Sensor (-) Signal	<0.050v
33	BR/LG	ECT Sensor Signal	0.5-0.6v
34	LG/YL	COP 5 Driver (dwell)	5°, 55 mph: 8°
35	LG/WT	COP 1 Driver (dwell)	5°, 55 mph: 8°
36	BR/WT	Reference Voltage	4.9-5.1v
37-38	---	Not Used	---
39	DG/WT	Knock Sensor 2 (-) Signal	<0.050v
40	YL	Knock Sensor 1 (-) Signal	<0.050v
41	GY	CKP (-) Sensor Signal	400-420 Hz
42-43	---	Not Used	---
44	GY/WT	TP Sensor Signal	0.53-1.27v
45	YL/LG	CHT Sensor Signal	0.6 to 1.7v (194°F)
46	OR/YL	COP 6 Driver (dwell)	5°, 55 mph: 8°

EXPLORER PIN TABLES

1997 Explorer 4.0L V6 MFI VIN E (A/T) 104 Pin Connector

PCM Pin #	Wire Color	Circuit Description (104 Pin)	Value at Hot Idle
1	PK/OR	Shift Solenoid 2 Control	12v, 55 mph: 12v
2	PK/LG	MIL (lamp) Control	MIL On: 1v, Off: 12v
3	YL/BK	Digital TR1 Sensor	In 'P': 0v, 55 mph: 11.5v
6	DB/YL	OSS Sensor Signal	0 Hz, 55 mph: 350 Hz
12	YL/WT	Fuel Level Indicator Signal	1.7v (1/2 full)
13	PK	Flash EPROM Power	0.1v
14	LB/BK	4x4 Low Switch Signal	12v (Switch On: 0v)
15	PK	Data Bus (-) Signal	Digital Signals
16	TN/OR	Data Bus (+) Signal	Digital Signals
19	OR/BK	Accelerator Ride Control	0.1v (Off)
21	DB	CKP (+) Sensor Signal	430-460 Hz
22	GY	CKP (-) Sensor Signal	430-460 Hz
24	BK/WT	Power Ground	<0.1v
25	BK	Case Ground	<0.050v
26	YL/BK	Coil 1 Driver (dwell)	5°, 55 mph: 8°
27	OR/YL	Shift Solenoid 1 Control	1v, 55 mph: 12v
28	BR/OR	Coast Clutch Solenoid	12v, 55 mph: 12v
29	TN/WT	TCS (switch) Signal	TCS & O/D On: 12v
30	DG	Octane Adjustment	9.3v (Closed: 0v)
33	PK/OR	VSS (-) Signal	0 Hz, 55 mph: 125 Hz
35	RD/LG	HO2S-12 (B1 S2) Signal	0.1-1.1v
36	TN/LB	MAF Sensor Return	<0.050v
37	OR/BK	TFT Sensor Signal	2.10-2.40v
38	LG/RD	ECT Sensor Signal	0.5-0.6v
39	DG	IAT Sensor Signal	1.5-2.5v
40	DG/YL	Fuel Pump Monitor	On: 12v, Off: 0v
41	DG/OR	A/C Switch Signal	A/C On: 12v, Off: 0v
43	LB/PK	Neutral Tow Connector Signal	N/A
46	BR	Inlet Air Control Valve	0.9v (15%)
47	BK/PK	VR Solenoid Control	0%, 55 mph: 45%
48	TN/YL	Clean Tachometer Output	39-45 Hz
49	LB/BK	Digital TR2 Sensor	0v, 55 mph: 11.5v
50	YL/LG	Digital TR4 Sensor	0v, 55 mph: 11.5v
51	BK/WT	Power Ground	<0.1v
52	YL/RD	Coil 2 Driver (dwell)	5°, 55 mph: 8°
53	PK/BK	Shift Solenoid 3 Control	12v, 55 mph: 1v
54	PK/YL	TCC Solenoid Control	0%, 55 mph: 95%
55	Y	Keep Alive Power	12-14v
56	LG/BK	EVAP Purge Solenoid	0-10 Hz (0-100%)
57	YL/RD	Knock Sensor 1 Signal	0v
58	GY/BK	VSS (+) Signal	0 Hz, 55 mph: 125 Hz
60	GY/LB	HO2S-11 (B1 S1) Signal	0.1-1.1v
61	PK/LG	HO2S-22 (B2 S2) Signal	0.1-1.1v
62	RD/PK	FTP Sensor Signal	2.6v (0" H2O - cap off)
64	LB/YL	Digital TR3A Sensor	In 'P': 0v, 30 mph: 1.7v

1997 Explorer 4.0L V6 MFI VIN E (A/T) 104 Pin Connector

PCM Pin #	Wire Color	Circuit Description (104 Pin)	Value at Hot Idle
65	BR/LG	DPFE Sensor Signal	0.20-1.30v
67	PK/WT	EVAP CV Solenoid	0-10 Hz (0-100%)
69	PK/YL	A/C WOT Relay Control	On: 1v, Off: 12v
71	RD	Vehicle Power	12-14v
73	TN/BK	Injector 5 Control	3.6-7.5 ms
74	BR/YL	Injector 3 Control	3.6-7.5 ms
75	TN	Injector 1 Control	3.6-7.5 ms
76, 77	BK/WT	Power Ground	<0.1v
78	YL/WT	Coil 3 Driver (dwell)	5°, 55 mph: 8°
79	WT/LG	TCIL (lamp) Control	TCIL On: 1v, Off: 12v
80	LB/OR	Fuel Pump Control	On: 1v, Off: 12v
81	WT/YL	EPC Solenoid Control	10v (26 psi)
83	WT/LB	IAC Solenoid Control	10.8v (33%)
84	DG/WT	TSS (+) Sensor Signal	100-125 Hz
85	DB/OR	CMP Sensor Signal	7 Hz, 55 mph: 18 Hz
87	RD/BK	HO2S-21 (B2 S1) Signal	0.1-1.1v
88	LB/RD	MAF Sensor Signal	0.8v, 55 mph: 2.1v
89	GY/WT	TP Sensor Signal	0.53-1.27v
90	BR/WT	Reference Voltage	4.9-5.1v
91	GY/RD	Analog Signal Return	<0.050v
92	LG	Brake Position Switch	Brake Off: 12v, On: 1v
93	RD/WT	HO2S-11 (B1 S1) Heater	On: 1v, Off: 12v
94	YL/LB	HO2S-21 (B2 S1) Heater	On: 1v, Off: 12v
95	WT/BK	HO2S-12 (B1 S2) Heater	On: 1v, Off: 12v
96	TN/YL	HO2S-22 (B2 S2) Heater	On: 1v, Off: 12v
97	RD	Vehicle Power	12-14v
99	LG/OR	Injector 6 Control	3.6-7.5 ms
100	BR/LB	Injector 4 Control	3.6-7.5 ms
101	WT	Injector 2 Control	3.6-7.5 ms
103	BK/WT	Power Ground	<0.1v

Pin Connector Graphic

PCM 104-PIN CONNECTOR

Terminal View of 104-Pin PCM Wiring Harness Connector

Standard Colors and Abbreviations

Abbreviation	Color	Abbreviation	Color	Abbreviation	Color
BK	Black	GY	Gray	PK	Purple
BL	Blue	GN	Green	RD	Red
BR	Brown	LG	LT Green	TN	Tan
DB	Dark Blue	OR	Orange	WT	White
DG	DK Green	PK	Pink	YL	Yellow

1998-99 Explorer 4.0L V6 VIN E 104 Pin Connector

PCM Pin #	Wire Color	Circuit Description (104 Pin)	Value at Hot Idle
1	PK/OR	Shift Solenoid 'B' Control	12v, 55 mph: 12v
2	PK/LG	MIL (lamp) Control	MIL On: 1v, Off: 12v
3	YL/BK	Digital TR1 Sensor	In 'P': 0v, 55 mph: 11.5v
4-5, 7-11	---	Not Used	---
6	DB/YL	OSS Sensor Signal	0 Hz, 55 mph: 350 Hz
12	YL/WT	Fuel Level Indicator Signal	1.7v (1/2 full)
13	VT	Flash EPROM Power	0.1v
14	LB/BK	4x4 Indicator Signal	4x4 On: 1v, Off: 12v
15	PK/LB	Data Bus (-) Signal	Digital Signals
16	TN/OR	Data Bus (+) Signal	Digital Signals
17-20	---	Not Used	---
19	OR/BK	Accelerator Ride Control	0.1v (Off)
21	DB	CKP (+) Sensor Signal	430-460 Hz
22	GY	CKP (-) Sensor Signal	<0.050v
24, 51	BK/WT	Power Ground	<0.1v
25	BK	Case Ground	<0.050v
26	TN/WT	Coil 1 Driver (dwell)	5°, 55 mph: 8°
27	OR/YL	Shift Solenoid 'A' Control	1v, 55 mph: 12v
28	BR/OR	Coast Clutch Solenoid	12v, 55 mph: 12v
28 (2000-01)	BR/OR	Shift Solenoid 'D' Control	12v, 55 mph: 12v
29	TN/WT	TCS (switch) Signal	TCS & O/D On: 12v
30 ('98-'99)	DG	Octane Adjustment	9.3v (Closed: 0v)
31-32, 34	---	Not Used	---
32 ('98-'99)	DG/VT	Knock Sensor 1 Ground	<0.050v
33 ('98-'99)	PK/OR	VSS (-) Signal	0 Hz, 55 mph: 125 Hz
35	RD/LG	HO2S-12 (B1 S2) Signal	0.1-1.1v
36	TN/LB	MAF Sensor Return	<0.050v
37	OR/BK	TFT Sensor Signal	2.10-2.40v
38	LG/RD	ECT Sensor Signal	0.5-0.6v
39	GY	IAT Sensor Signal	1.5-2.5v
40	DG/YL	Fuel Pump Monitor	On: 12v, Off: 0v
41	PK	A/C Switch Signal	A/C On: 12v, Off: 0v
42	---	Not Used	---
43 ('98-'99)	LB/PK	Neutral Tow Connector Signal	N/A
44	---	Not Used	---
45	YL/WT	PCM Signal to the GEM	Digital Signals
46 ('98-'99)	BR	Inlet Air Control Valve	0.9v (15%)
47	BRD/PK	VR Solenoid Control	0%, 55 mph: 45%
48	TN/YL	Clean Tachometer Output	39-45 Hz
49	LB/BK	Digital TR2 Sensor	0v, 55 mph: 11.5v
50	WT/BK	Digital TR4 Sensor	0v, 55 mph: 11.5v
52	TN/OR	Coil 2 Driver (Control)	5°, 55 mph: 8°
53	P/BK	Shift Solenoid 'C' Control	12v, 55 mph: 1v
54	PK/YL	TCC Solenoid Control	0%, 55 mph: 95%
55	YL	Keep Alive Power	12-14v
56	LG/BK	EVAP Purge Solenoid	0-10 Hz (0-100%)
57	---	Not Used	---
58	GY/BK	VSS (+) Signal	0 Hz, 55 mph: 125 Hz
59	---	Not Used	---
60	GY/LB	HO2S-11 (B1 S1) Signal	0.1-1.1v
61	---	Not Used	---
62	RD/PK	FTP Sensor Signal	2.6v (0" H2O - cap off)
64	LB/YL	Digital TR3 Sensor	In 'P': 0v, 30 mph: 1.7v

1998-99 Explorer 4.0L V6 VIN E 104 Pin Connector

PCM Pin #	Wire Color	Circuit Description (104 Pin)	Value at Hot Idle
65	BR/LG	DPFE Sensor Signal	0.20-1.30v
66	---	Not Used	---
67	PK/WT	EVAP CV Solenoid	0-10 Hz (0-100%)
68	---	Not Used	---
69	PK/YL	A/C WOT Relay Control	On: 1v, Off: 12v
71	RD	Vehicle Power	12-14v
70, 72	---	Not Used	---
73	TN/BK	Injector 5 Control	3.6-7.5 ms
74	BR/YL	Injector 3 Control	3.6-7.5 ms
75	TN	Injector 1 Control	3.6-7.5 ms
76, 77	BK/WT	Power Ground	<0.1v
78	TN/LG	Coil 3 Driver (dwell)	5°, 55 mph: 8°
79	WT/LG	TCIL (lamp) Control	TCIL On: 1v, Off: 12v
80	LB/OR	Fuel Pump Control	On: 1v, Off: 12v
81	WT/YL	EPC Solenoid Control	10v (26 psi)
82	---	Not Used	---
83	WT/LB	IAC Solenoid Control	10.8v (33%)
84	DG/WT	TSS (+) Sensor Signal	100-125 Hz
85	DB/OR	CMP Sensor Signal	7 Hz, 55 mph: 18 Hz
86	BK/YL	A/C Pressure Switch Signal	Switch Open: 12v
87	RD/BK	HO2S-21 (B2 S1) Signal	0.1-1.1v
88	LB/RD	MAF Sensor Signal	0.8v, 55 mph: 2.1v
89	GY/WT	TP Sensor Signal	0.53-1.27v
90	BR/WT	Reference Voltage	4.9-5.1v
91	GY/RD	Sensor Ground	<0.050v
92	RD/LG	Brake Position Switch	Brake Off: 12v, On: 1v
93	RD/WT	HO2S-11 (B1 S1) Heater	On: 1v, Off: 12v
94	YL/LB	HO2S-21 (B2 S1) Heater	On: 1v, Off: 12v
95	WT/BK	HO2S-12 (B1 S2) Heater	On: 1v, Off: 12v
96, 98	---	Not Used	---
97	RD	Vehicle Power	12-14v
99	LG/OR	Injector 6 Control	3.6-7.5 ms
100	BR/LB	Injector 4 Control	3.6-7.5 ms
101	WT	Injector 2 Control	3.6-7.5 ms
102	---	Not Used	---
103	BK/WT	Power Ground	<0.1v
104	---	Not Used	---

Pin Connector Graphic

PCM 104-PIN CONNECTOR

Terminal View of 104-Pin PCM Wiring Harness Connector

Standard Colors and Abbreviations

Abbreviation	Color	Abbreviation	Color	Abbreviation	Color
BK	Black	GY	Gray	PK	Purple
BL	Blue	GN	Green	RD	Red
BR	Brown	LG	LT Green	TN	Tan
DB	Dark Blue	OR	Orange	WT	White
DG	DK Green	PK	Pink	YL	Yellow

2000-02 Explorer 4.0L V6 SOHC MFI VIN E 104 Pin Connector

PCM Pin #	Wire Color	Circuit Description (104 Pin)	Value at Hot Idle
1	PK/OR	Shift Solenoid 'B' Control	12v, 55 mph: 12v
2	PK/LG	MIL (lamp) Control	MIL On: 1v, Off: 12v
3	YL/BK	Digital TR1 Sensor	In 'P': 0v, 55 mph: 11.5v
4-5	---	Not Used	---
6	DB/YL	OSS Sensor Signal	0 Hz, 55 mph: 350 Hz
7-11	---	Not Used	---
12	YL/WT	Fuel Level Indicator Signal	1.7v (1/2 full)
13	PK	Flash EPROM Power	0.1v
14	LB/BK	4x4 Low Indicator Switch	Switch On: 1v, Off: 12v
15	PK/LB	Data Bus (-) Signal	Digital Signals
16	TN/OR	Data Bus (+) Signal	Digital Signals
17-18	---	Not Used	---
19	OR/BK	Load Leveling Acceleration	Digital Signals
20	---	Not Used	---
21	DB	CKP (+) Sensor Signal	430-460 Hz
22	GY	CKP (-) Sensor Signal	<0.050v
23	---	Not Used	---
24	BK/WT	Power Ground	<0.1v
25	BK	Chassis Ground	<0.050v
26	TN/WT	Coil 1 Driver (dwell)	5°, 55 mph: 8°
27	OR/YL	Shift Solenoid 'A' Control	1v, 55 mph: 12v
28	BR/OR	Shift Solenoid 'D' Control	12v, 55 mph: 12v
29	TN/WT	TCS (switch) Signal	TCS & O/D On: 12v
30	---	Not Used	---
32	DG/VT	Knock Sensor (-) Return	<0.050v
33-34	---	Not Used	---
35	RD/LG	HO2S-12 (B1 S2) Signal	0.1-1.1v
36	TN/LB	MAF Sensor Return	<0.050v
37	OR/BK	TFT Sensor Signal	2.10-2.40v
38	LG/RD	ECT Sensor Signal	0.5-0.6v
39	GY	IAT Sensor Signal	1.5-2.5v
40	DG/YL	Fuel Pump Monitor	On: 12v, Off: 0v
41	PK	A/C High Pressure Switch	Switch On: 12v (open)
42	---	Not Used	---
43	LB/PK	Message Center Fuel Flow	Digital Signals
44	---	Not Used	---
45	YL/WT	PCM Signal to GEM	Digital Signals
46	BK/WT	Fuel Cap Indicator (Cluster)	Digital Signals
47	BR/PK	EGR VR Regulator Solenoid	0%, 55 mph: 45%
48	TN/YL	Clean Tachometer Output	39-45 Hz
49	LB/BK	Digital TR2 Sensor	0v, 55 mph: 11.5v
50	WT/BK	Digital TR4 Sensor	0v, 55 mph: 11.5v
51	BK/WT	Power Ground	<0.1v
52	TN/OR	Coil 2 Driver (Control)	5°, 55 mph: 8°
53	PK/BK	Shift Solenoid 'C' Control	12v, 55 mph: 1v
54	PK/YL	TCC Solenoid Control	0%, 55 mph: 95%
55	YL	Keep Alive Power	12-14v
56	LG/BK	EVAP Purge Solenoid	0-10 Hz (0-100%)
57	YL/RD	Knock Sensor (+) Signal	0v
58	GY/BK	VSS (+) Signal	0 Hz, 55 mph: 125 Hz
59	---	Not Used	---
60	GY/LB	HO2S-11 (B1 S1) Signal	0.1-1.1v
61	PK/LG	HO2S-22 (B2 S2) Signal	0.1-1.1v
63	---	Not Used	---
62	RD/PK	FTP Sensor Signal	2.6v (0" H2O - cap off)
64	LB/YL	Digital TR3 Sensor	In 'P': 0v, 30 mph: 1.7v

2000-02 Explorer 4.0L V6 SOHC MFI VIN E 104 Pin Connector

PCM Pin #	Wire Color	Circuit Description (104 Pin)	Value at Hot Idle
65 ('00)	BR/LG	DPFE Sensor Signal	0.20-1.30v
65	BR/LG	DPFE Sensor Signal	0.95-1.05v
66	---	Not Used	---
67	VT/WT	EVAP CV Solenoid	0-10 Hz (0-100%)
69	PK/YL	A/C WOT Relay Control	On: 1v, Off: 12v
70	---	Not Used	---
71	RD	Vehicle Power	12-14v
73	TN/BK	Injector 5 Control	3.6-7.5 ms
74	BR/YL	Injector 3 Control	3.6-7.5 ms
75	TN	Injector 1 Control	3.6-7.5 ms
76, 77	BK/WT	Power Ground	<0.1v
78	TN/LG	Coil 3 Driver (dwell)	5°, 55 mph: 8°
79	WT/LG	TCIL (lamp) Control	TCIL On: 1v, Off: 12v
80	LB/OR	Fuel Pump Relay Control	On: 1v, Off: 12v
81	WT/YL	EPC Solenoid Control	10v (26 psi)
82	---	Not Used	---
83	WT/LB	IAC Solenoid Control	10.8v (33%)
84	DG/WT	TSS Sensor Signal	100-125 Hz
85	DB/OR	CMP Sensor Signal	7 Hz, 55 mph: 18 Hz
86	BK/YL	A/C Pressure Sensor	Switch open: 12v
87	RD/BK	HO2S-21 (B2 S1) Signal	0.1-1.1v
88	LB/RD	MAF Sensor Signal	0.8v, 55 mph: 2.1v
89	GY/WT	TP Sensor Signal	0.53-1.27v
90	BR/WT	Reference Voltage	4.9-5.1v
91	GY/RD	Sensor Ground	<0.050v
92	RD/LG	Brake Position Switch	Brake Off: 12v, On: 1v
93	RD/WT	HO2S-11 (B1 S1) Heater	On: 1v, Off: 12v
94	YL/LB	HO2S-21 (B2 S1) Heater	On: 1v, Off: 12v
95	WT/BK	HO2S-12 (B1 S2) Heater	On: 1v, Off: 12v
96	TN/YL	HO2S-22 (B2 S2) Heater	On: 1v, Off: 12v
97	RD	Vehicle Power	12-14v
98	---	Not Used	---
99	LG/OR	Injector 6 Control	3.6-7.5 ms
100	BR/LB	Injector 4 Control	3.6-7.5 ms
101	WT	Injector 2 Control	3.6-7.5 ms
102	---	Not Used	---
103	BK/WT	Power Ground	<0.1v
104	---	Not Used	---

Pin Connector Graphic

PCM 104-PIN CONNECTOR

1 ●●●●●●●●●●●●● ●●●●●●●●●●●●● 26
27 ●●●●●●●●●●●●● ●●●●●●●●●●●●● 52
53 ●●●●●●●●●●●●● ●●●●●●●●●●●●● 78
79 ●●●●●●●●●●●●● ●●●●●●●●●●●●● 104

Terminal View of 104-Pin PCM Wiring Harness Connector

Standard Colors and Abbreviations

Abbreviation	Color	Abbreviation	Color	Abbreviation	Color
BK	Black	GY	Gray	PK	Purple
BL	Blue	GN	Green	RD	Red
BR	Brown	LG	LT Green	TN	Tan
DB	Dark Blue	OR	Orange	WT	White
DG	DK Green	PK	Pink	YL	Yellow

2002-05 Explorer 4.0L V6 MFI VIN K 58 Pin Connector

PCM Pin #	Wire Color	Circuit Description (58 Pin)	Value at Hot Idle
1, 5, 15-16	---	Not Used	---
2	PK/BK	Fuel Pump Power	Pump On: 12v, Off: 0v
3	TN/OR	SCP Data Bus (+)	Digital Signals
4	PK/LB	SCP Data Bus (-)	Digital Signals
6	VT/WT	EVAP Canister Vent Valve Control	0-10 Hz (0-100%)
7	GY/BK	Vehicle Speed Sensor (+) Signal	0 Hz, 55 mph: 125 Hz
8	LG/RD	Generator Monitor Signal	130 Hz (45%)
9	PK/YL	A/C Clutch Relay Control	12v (relay on: 1v)
10	YL/WT	4WD Indicator Control	Indicator Off: 12v, On: 1v
11	WT/LG	Passive Antitheft TX Signal	Digital Signals
12	LG/BK	EVAP Purge Solenoid	0-10 Hz (0-100%)
13	VT	Module Programming Signal	0.1v
14	GY/OR	Passive Antitheft RX Signal	Digital Signals
18-19, 21	---	Not Used	---
17	GY/RD	Sensor Ground	<0.050v
20	BR/WT	Reference Voltage	4.9-5.1v
22	DB/LG	Passive Antitheft Indicator Control	Indicator Off: 12v, On: 1v
23	LB/BK	4WD Indicator Low Signal	12v (switch on: 1v)
24	BK/WT	Power Ground	<0.1v
25	BK/WT	Power Ground	<0.1v
26	BK/WT	Power Ground	<0.1v
27	BK/WT	Power Ground	<0.1v
28	BK/YL	Brake Pressure Switch	12v (Brake On: 0v)
29	OR/LB	Cruise Set Indicator Control	Indicator Off: 12v, On: 1v
30	BK/YL	A/C Clutch Relay (switched)	12v (relay on: 1v)
31	LB/RD	MAF Sensor Signal	0.7v, 55 mph: 1.8v
32	VT	Vehicle Power (Start-Run)	12-14v
33	VT	Vehicle Power (Start-Run)	12-14v
34-35	---	Not Used	---
36	OR/LB	Cruise Set Indicator Control	12v (switch set: 0v)
37	---	Not Used	---
38	TN/LB	MAF Sensor Return	<0.050v
39	OR	Starter Motor Relay Circuit	Relay Off: 12v, On: 1v
40	RD/LG	Brake Position Switch	Brake Off: 12v, On: 1v
41	TN/WT	Overdrive Cancel Switch	0v (switch on: 12v)
42, 48-49	---	Not Used	---
43	BK	Power Ground	<0.1v
44	RD/WT	Keep Alive Power (CJB fuse)	12-14v
45	BK	Speed Control Switch Ground	<0.050v
46	OR/LB	Cruise Switch Indicator Control	Indicator Off: 12v, On: 1v
47	RD/YL	A/C Pressure Switch Signal	0v (switch on: 12v)
50	VT	A/C Demand Switch	0v (A/C On: 12v)
51	GY	Intake Air Temperature Sensor	1.5-2.5v
52	RD/PK	Fuel Tank Pressure Sensor	2.6v (0" H20 - cap off)
53-55	---	Not Used	---
56	DG/OR	Speed Control Switch Ground	<0.050v
57	LB/BK	Speed Control Switch Input	0v (switch on: 6.7v)
58	LB/OR	Fuel Pump Relay Control	Relay Off: 12v, On: 1v

2002-05 Explorer 4.0L V6 MFI VIN K 32 Pin Connector

PCM Pin #	Wire Color	Circuit Description (32 Pin)	Value at Hot Idle
1	OR/YL	Shift Solenoid 'A' Control	12v, 55 mph: 1v
2	VT/OR	Shift Solenoid 'B' Control	12v, 55 mph: 1v
3	PK/BK	Shift Solenoid 'C' Control	12v, 55 mph: 1v
4	BR/OR	Shift Solenoid 'D' Control	12v, 55 mph: 1v
5	VT/YL	TCC Solenoid Control	12v, 55 mph: 1v
6	---	Not Used	---
7	WT/YL	Electronic Pressure Control	On: 1v, Off: 12v
8	---	Not Used	---
9	LB/YL	Digital TR Sensor (TR3A)	0v, 55 mph: 1.7v
10	WT/BK	Digital TR Sensor (TR4)	0v, 55 mph: 11v
11	---	Not Used	---
12	WT	Motor Position #4	On: 1v, Off: 12v
13	LB/PK	Pressure Control 'B' Solenoid	On: 1v, Off: 12v
14	---	Not Used	---
15	WT/BK	HO2S-12 (B1 S2) Heater	1v, Off: 12v
16	TN/YL	HO2S-22 (B2 S2) Heater	1v, Off: 12v
17	GY/RD	Sensor Reference Ground	<0.050v
18	LB/BK	Digital TR Sensor (TR2)	0v, 55 mph: 11v
19-20	---	Not Used	---
21	YL/LG	Overdrive Drum Speed Input	200 Hz, 55 mph: 1185 Hz
22	YL/BK	Digital TR Sensor (TR1)	0v, 55 mph: 11v
23	OR/BK	Transmission Fluid Temperature Sensor	2.10-2.40v
24-25	---	Not Used	---
26	DB/YL	Output Shaft Speed Sensor	0 Hz, 985 Hz
27	DG/WT	Turbine Shaft Speed Sensor	360 Hz, 890 Hz
28	RD/LG	HO2S-12 (B1 S2) Signal	0.1-1.1v
29	VT/LG	HO2S-22 (B2 S2) Signal	0.1-1.1v
30-32	---	Not Used	---

Pin Connector Graphic

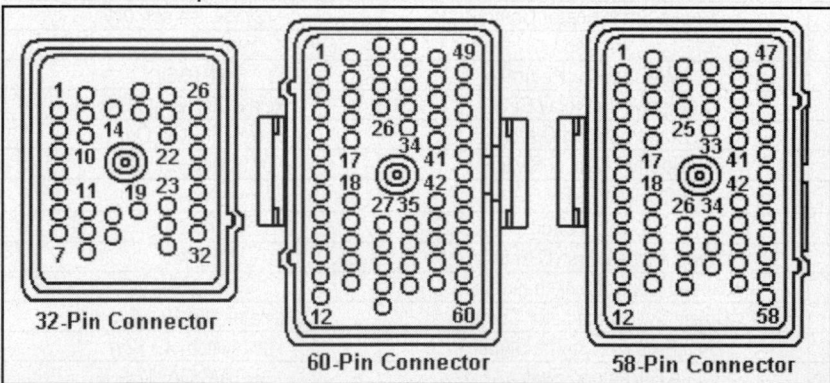

32-Pin Connector   60-Pin Connector   58-Pin Connector

Standard Colors and Abbreviations

Abbreviation	Color	Abbreviation	Color	Abbreviation	Color
BK	Black	GY	Gray	PK	Purple
BL	Blue	GN	Green	RD	Red
BR	Brown	LG	LT Green	TN	Tan
DB	Dark Blue	OR	Orange	WT	White
DG	DK Green	PK	Pink	YL	Yellow

2002-05 Explorer 4.0L V6 MFI VIN K 60 Pin Connector

PCM Pin #	Wire Color	Circuit Description (60 Pin)	Value at Hot Idle
1	---	Not Used	---
2	TN	Injector 1 Control	3.3-3.8 ms
3-6	---	Not Used	---
7	RD/WT	HO2S-11 (B1 S1) Heater	1v, Off: 12v
8	YL/LB	HO2S-21 (B2 S1) Heater	1v, Off: 12v
9	WT/LB	IAC Solenoid Control	10.7v (33%)
10	---	Not Used	---
11	TN/BK	Injector 5 Control	3.3-3.8 ms
12	---	Not Used	---
13	PK/WT	Coil 1 Driver (dwell)	5°, 55 mph: 7°
14	WT	Injector 2 Control	3.3-3.8 ms
15	---	Not Used	---
16	BR/PK	VR Solenoid Control	0%, 55 mph: 45%
17	GY/RD	Sensor Reference Ground	<0.050v
18-19	---	Not Used	---
20	BR/WT	Sensor Voltage Reference	4.9-5.1v
21	LG/OR	Injector 6 Control	3.3-3.8 ms
22	---	Not Used	---
23	RD/LB	Coil 2 Driver (dwell)	5°, 55 mph: 7°
24	BR/YL	Injector 3 Control	3.3-3.8 ms
25-30	---	Not Used	---
31	DB/LG	Coil 3 Driver (dwell)	5°, 55 mph: 7°
32	BR/LB	Injector 4 Control	3.3-3.8 ms
33-40	---	Not Used	---
41	BR/LG	DPFE Sensor Signal	0.95-1.05v
42	YL	Knock Sensor 1 Signal	0v
43	DG/WT	Knock Sensor 2 Signal	0v
44	RD/BK	HO2S-21 (B2 S1) Signal	0.1-1.1v
45	GY/LB	HO2S-11 (B1 S1) Signal	0.1-1.1v
46	LG/RD	ECT Sensor Signal	0.5-0.6v
47-50	---	Not Used	---
51	YL/RD	Knock Sensor 1 Ground	0v
52	---	Not Used	---
53	DG	CMP Sensor Signal	6 Hz
54	---	Not Used	---
55	DB	CKP Sensor (-) Signal	400 Hz
56	GY	CKP Sensor (+) Signal	400 Hz
57	GY/WT	TP Sensor Signal	0.53-1.27v
58-60	---	Not Used	---

1991-94 Explorer 4.0L V6 MFI VIN X 60 Pin Connector

PCM Pin #	Wire Color	Circuit Description (60 Pin)	Value at Hot Idle
1	YL	Keep Alive Power	12-14v
2	LG	Brake Pedal Switch	Brake Off: 12v, On: 1v
3	GY/BK	VSS (+) Signal	0 Hz, 55 mph: 125 Hz
4	TN/YL	Ignition Diagnostic Monitor	20-31 Hz
6	PK/OR	VSS (-) Signal	0 Hz, 55 mph: 125 Hz
7	LG/RD	ECT Sensor Signal	0.5-0.6v
8	DG/YL	Fuel Pump Monitor	On: 12v, Off: 0v
9 ('92-'94)	PK/BL	Data Bus (-) Signal	Digital Signals
10	DG/OR	A/C Switch Signal	A/C On: 12v, Off: 0v
14	BL/RD	MAF Sensor Signal	0.7v
15	TN/BL	MAF Sensor Return	<0.050v
16	OR/RD	Ignition System Ground	<0.050v
17	PK/LG	Self-Test Output & MIL	MIL On: 1v, Off: 12v
20	BK	PCM Case Ground	<0.050v
21	WT/BL	IAC Solenoid Control	10.7v (33%)
22	BL/OR	Fuel Pump Control	On: 1v, Off: 12v
25	GY	IAT Sensor Signal	1.5-2.5v
26	BR/WT	Reference Voltage	4.9-5.1v
28 ('92-'94)	TN/OR	Data Bus (+) Signal	Digital Signals
29	GY/BL	HO2S-11 (B1 S1) Signal	0.1-1.1v
30	BL/YL	A/T: Neutral Drive Switch	In 'P': 0v, Others: 5v
30	BL/YL	M/T: Clutch Engage Switch	Clutch In: 0v, Out: 5v
31	GY/DB	EVAP Purge Solenoid	12v, 55 mph: 1v
31	GY/YL	EVAP Purge Solenoid	12v, 55 mph: 1v
36	PK	Spark Output Signal	6.93v (50%)
37	RD	Vehicle Power	12-14v
39 ('93-'94)	RD/BK	HO2S-12 (B1 S2) Signal	0.1-1.1v
40	BK/WT	Power Ground	<0.1v
44	DG	Octane Adjustment	9.3v (Closed: 0v)
45 ('91)	LG/BK	BARO Sensor Signal	159 Hz (sea level)
46	GY/RD	Analog Signal Return	<0.050v
47	GY/WT	TP Sensor Signal	0.5-1.2v
48	WT/PK	Self-Test Input Signal	STI On: 1v, Off: 5v
49	OR	HO2S-11 (B1 S1) Ground	<0.050v
49	GY/RD	HO2S-11 (B1 S1) Ground	<0.050v
52	OR/YL	Shift Solenoid 3/4	1v, 55 mph: 12v
53	PK/YL	TCC Solenoid Control	12v, 55 mph: 1v
54	PK/YL	A/C WOT Relay Control	On: 1v, Off: 12v
55	---	Not Used	---
56	GY/OR	PIP Sensor Signal	6.93v (50%)
57	RD	Vehicle Power	12-14v
58	TN	Injector Bank 1 (INJ 1, 2, 4)	3.3-5.7 ms
59	WT	Injector Bank 2 (INJ 3, 5, 6)	3.3-5.7 ms
60	BK/LG	Power Ground	<0.1v
60	BK/WT	Power Ground	<0.1v

1993-94 Explorer 4.0L V6 MFI VIN X California 60 Pin Connector

PCM Pin #	Wire Color	Circuit Description (60 Pin)	Value at Hot Idle
1	YL	Keep Alive Power	12-14v
2	LG	Brake Pedal Switch	Brake Off: 12v, On: 1v
3	GY/BK	VSS (+) Signal	0 Hz, 55 mph: 125 Hz
4	TN/YL	Ignition Diagnostic Monitor	20-31 Hz
5	---	Not Used	---
6	PK/OR	VSS (-) Signal	0 Hz, 55 mph: 125 Hz
7	LG/RD	ECT Sensor Signal	0.5-0.6v
8	DG/YL	Fuel Pump Monitor	On: 12v, Off: 0v
9	TN/BL	MAF Sensor Return	<0.050v
10	DG/OR	A/C Switch Signal	A/C On: 12v, Off: 0v
11	GY/YL	EVAP Purge Solenoid	12v, 55 mph: 1v
12	LG/OR	Injector 6 Control	3.3-5.7 ms
13-14	---	Not Used	---
15	TN/BK	Injector 5 Control	3.3-5.7 ms
16	OR/RD	Ignition System Ground	<0.050v
17	PK/LG	Self-Test Output & MIL	MIL On: 1v, Off: 12v
18	TN/OR	Data Bus (+) Signal	Digital Signals
19	PK/BL	Data Bus (-) Signal	Digital Signals
20	BK/LG	PCM Case Ground	<0.050v
21	WT/BL	IAC Solenoid Control	10.7v (33%)
22	BL/OR	Fuel Pump Control	On: 1v, Off: 12v
23	---	Not Used	---
24	DB/OR	CMP Sensor Signal	5-7 Hz
25	GY	IAT Sensor Signal	1.5-2.5v
26	BR/WT	Reference Voltage	4.9-5.1v
27	BR/LG	DPFE Sensor Signal	0.20-1.20v
28	---	Not Used	---
29	DG	Octane Adjustment	9.3v (Closed: 0v)
30	BL/YL	A/T: Neutral Drive Switch	In 'P': 0v, Others: 5v
30	BL/YL	M/T: Clutch Engage Switch	Clutch In: 0v, Out: 5v
31-32	---	Not Used	---
33	BR/PK	VR Solenoid Control	0%, 55 mph: 45%
34	---	Not Used	---
35	BR/BL	Injector 4 Control	3.3-5.7 ms
36	PK	Spark Output Signal	6.93v (50%)
37	RD	Vehicle Power	12-14v
38	---	Not Used	---
39	BR/YL	Injector 3 Control	3.3-5.7 ms
40	BK/WT	Power Ground	<0.1v

1993-94 Explorer 4.0L V6 MFI VIN X California 60 Pin Connector

PCM Pin #	Wire Color	Circuit Description (60 Pin)	Value at Hot Idle
41-42	---	Not Used	---
43	RD/BK	HO2S-21 (B2 S1) Signal	0.1-1.1v
44	GY/BL	HO2S-11 (B1 S1) Signal	0.1-1.1v
45	---	Not Used	---
46	GY/RD	Analog Signal Return	<0.050v
47	GY/WT	TP Sensor Signal	0.5-1.2v
48	WT/PK	Self-Test Input Signal	STI On: 1v, Off: 5v
49	---	Not Used	---
50	BL/RD	MAF Sensor Signal	0.7v
51	OR/YL	Shift Solenoid 3-4 Control	1v, 55 mph: 12v
52	---	Not Used	---
53	PK/YL	TCC Solenoid Control	12v, 55 mph: 1v
54	PK/YL	A/C WOT Relay Control	On: 1v, Off: 12v
55	---	Not Used	---
56	GY/OR	PIP Sensor Signal	6.93v (50%)
57	RD	Vehicle Power	12-14v
58	TN	Injector 1 Control	3.3-5.7 ms
59	WT	Injector 2 Control	3.3-5.7 ms
60	BK/WT	Power Ground	<0.1v

Pin Connector Graphic

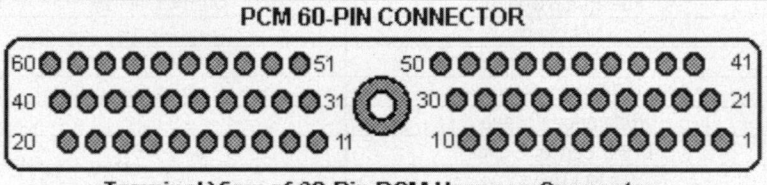

PCM 60-PIN CONNECTOR

Terminal View of 60-Pin PCM Harness Connector

Standard Colors and Abbreviations

Abbreviation	Color	Abbreviation	Color	Abbreviation	Color
BK	Black	GY	Gray	PK	Purple
BL	Blue	GN	Green	RD	Red
BR	Brown	LG	LT Green	TN	Tan
DB	Dark Blue	OR	Orange	WT	White
DG	DK Green	PK	Pink	YL	Yellow

1995 Explorer 4.0L V6 MFI VIN X 60 Pin Connector

PCM Pin #	Wire Color	Circuit Description (60 Pin)	Value at Hot Idle
1	YL	Keep Alive Power	12-14v
2	LG	Brake Pedal Switch	Brake Off: 12v, On: 1v
3	GY/BK	VSS (+) Signal	0 Hz, 55 mph: 125 Hz
4	TN/YL	Ignition Diagnostic Monitor	20-31 Hz
5	DG/WT	TSS (+) Sensor Signal	100 Hz, 55 mph: 270 Hz
6	PK/OR	VSS (-) Signal	0 Hz, 55 mph: 125 Hz
7	LG/RD	ECT Sensor Signal	0.5-0.6v
8	DG/YL	Fuel Pump Monitor	On: 12v, Off: 0v
9	TN/BL	MAF Sensor Return	<0.050v
10	DG/OR	A/C Switch Signal	A/C On: 12v, Off: 0v
11	GY/YL	EVAP Purge Solenoid	12v, 55 mph: 1v
12	LG/OR	Injector 6 Control	3.3-5.7 ms
13	WT/LG	TCIL (lamp) Control	TCIL On: 1v, Off: 12v
14	---	Not Used	---
15	TN/BK	Injector 5 Control	3.3-5.7 ms
16	OR/RD	Ignition System Ground	<0.050v
17	PK/LG	Self-Test Output & MIL	MIL On: 1v, Off: 12v
18	TN/OR	Data Bus (+) Signal	Digital Signals
19	PK/BL	Data Bus (-) Signal	Digital Signals
20	BK	PCM Case Ground	<0.050v
21	WT/BL	IAC Solenoid Control	10.7v (33%)
22	BL/OR	Fuel Pump Control	On: 1v, Off: 12v
23	---	Not Used	---
24	DB/OR	CMP Sensor Signal	5-7 Hz
25	GY	IAT Sensor Signal	1.5-2.5v
26	BR/WT	Reference Voltage	4.9-5.1v
27	BR/LG	DPFE Sensor Signal	0.4v
28	---	Not Used	---
29	DG	Octane Adjustment	9.3v (Closed: 0v)
30	BL/YL	A/T: Neutral Drive Switch	In 'P': 0v, Others: 5v
30	BL/YL	M/T: Clutch Engage Switch	Clutch In: 0v, Out: 5v
31	BL/PK	Fuel Flow Rate Signal	Digital Signal
32	BR/OR	Coasting Clutch Solenoid	12v, 55 mph: 12v
33	BR/PK	VR Solenoid Control	0%, 55 mph: 45%
34	OR/BK	Automatic Ride Control	4.3v
35	BR/BL	Injector 4 Control	3.3-5.7 ms
36	PK	Spark Output Signal	6.93v (50%)
37	RD	Vehicle Power	12-14v
38	WT/YL	EPC Solenoid Control	10.9v (26 psi)
39	BR/YL	Injector 3 Control	3.3-5.7 ms
40	BK/WT	Power Ground	<0.1v

1995 Explorer 4.0L V6 MFI VIN X 60 Pin Connector

PCM Pin #	Wire Color	Circuit Description (60 Pin)	Value at Hot Idle
41	TN/WT	TCS (switch) Signal	TCS & O/D On: 12v
42	BL/BK	Low Range Indicator Signal	N/A
43	RD/BK	HO2S-22 (B2 S2) Signal	0.1-1.1v
44	GY/BL	HO2S-11 (B1 S1) Signal	0.1-1.1v
45	---	Not Used	---
46	GY/RD	Analog Signal Return	<0.050v
47	GY/WT	TP Sensor Signal	0.5-1.2v
48	WT/PK	Self-Test Input Signal	STI On: 1v, Off: 5v
49	OR/BK	TFT Sensor Signal	2.10-2.40v
50	BL/RD	MAF Sensor Signal	0.7v
51	OR/YL	Shift Solenoid 1 Control	1v, 30 mph: 12v
52	PK/OR	Shift Solenoid 2 Control	12v, 55 mph: 1v
53	PK/BK	TCC Solenoid Control	12v, 55 mph: 9v
54	PK/YL	A/C WOT Relay Control	On: 1v, Off: 12v
55	PK/OR	Shift Solenoid 3 Control	12v, 55 mph: 1v
56	GY/OR	PIP Sensor Signal	6.93v (50%)
57	RD	Vehicle Power	12-14v
58	TN	Injector 1 Control	3.3-5.7 ms
59	WT	Injector 2 Control	3.3-5.7 ms
60	BK/WT	Power Ground	<0.1v

Pin Connector Graphic

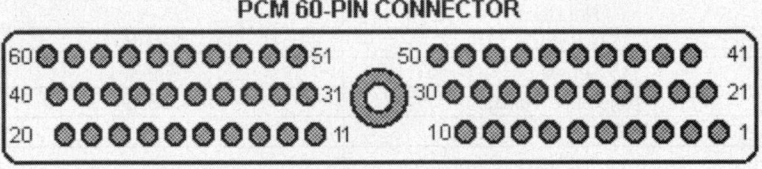

PCM 60-PIN CONNECTOR

Terminal View of 60-Pin PCM Harness Connector

Standard Colors and Abbreviations

Abbreviation	Color	Abbreviation	Color	Abbreviation	Color
BK	Black	GY	Gray	PK	Purple
BL	Blue	GN	Green	RD	Red
BR	Brown	LG	LT Green	TN	Tan
DB	Dark Blue	OR	Orange	WT	White
DG	DK Green	PK	Pink	YL	Yellow

1996-97 Explorer 4.0L V6 MFI VIN X 104 Pin Connector

PCM Pin #	Wire Color	Circuit Description (104 Pin)	Value at Hot Idle
1	PK/OR	Shift Solenoid 2 Control	12v, 55 mph: 12v
2	PK/LG	MIL (lamp) Control	MIL On: 1v, Off: 12v
13	VT	Flash EPROM Power	0.1v
14	LB/BK	4x4 Low Switch Signal	12v (switch closed: 0v)
15	PK/LB	Data Bus (-) Signal	Digital Signals
16	TN/OR	Data Bus (+) Signal	Digital Signals
19	OR/BK	Accelerator Ride Control	0.1v (Off)
21	DB	CKP (+) Sensor Signal	430-460 Hz
22	GY	CKP (-) Sensor Signal	430-460 Hz
24	BK/WT	Power Ground	<0.1v
25	BK	Case Ground	<0.050v
26	YL/BK	Coil 1 Driver (dwell)	5º, 55 mph: 8º
27	OR/YL	Shift Solenoid 1 Control	1v, 55 mph: 12v
28	BR/OR	Coast Clutch Solenoid	12v, 55 mph: 12v
29	TN/WT	TCS (switch) Signal	TCS & O/D On: 12v
30	DG	Octane Adjustment	9.3v (Closed: 0v)
33	PK/OR	VSS (-) Signal	0 Hz, 55 mph: 125 Hz
35	RD/LG	HO2S-12 (B1 S2) Signal	0.1-1.1v
36	TN/LB	MAF Sensor Return	<0.050v
37	OR/BK	TFT Sensor Signal	2.10-2.40v
38	LG/RD	ECT Sensor Signal	0.5-0.6v
39	GY	IAT Sensor Signal	1.5-2.5v
40	DG/YL	Fuel Pump Monitor	On: 12v, Off: 0v
41	DG/OR	A/C Switch Signal	A/C On: 12v, Off: 0v
43	LB/PK	Data Output Link	5v
47	BK/PK	VR Solenoid Control	0%, 55 mph: 45%
48	TN/YL	Clean Tachometer Output	39-45 Hz
51	BK/WT	Power Ground	<0.1v
52	WT/RD	Coil 2 Driver (dwell)	5º, 55 mph: 8º
53	PK/BK	Shift Solenoid 3 Control	12v, 55 mph: 1v
54	PK/YL	TCC Solenoid Control	0%, 55 mph: 95%
55	YL	Keep Alive Power	12-14v
56	LG/BK	EVAP Purge Solenoid	0-10 Hz (0-100%)
58	GY/BK	VSS (+) Signal	0 Hz, 55 mph: 125 Hz
60	GY/LB	HO2S-11 (B1 S1) Signal	0.1-1.1v
62	RD/PK	FTP Sensor Signal	2.6v (0" H2O - cap off)
64	LB/YL	A/T: TR Sensor Signal	In 'P': 0v, 55 mph: 1.7v
64	LB/YL	M/T: Clutch Pedal Switch	Clutch In: 0v, Out: 12v

1996-97 Explorer 4.0L V6 MFI VIN X 104 Pin Connector

PCM Pin #	Wire Color	Circuit Description (104 Pin)	Value at Hot Idle
65	BR/LG	DPFE Sensor Signal	0.20-1.30v
67	PK/WT	EVAP CV Solenoid	0-10 Hz (0-100%)
69	PK/YL	A/C WOT Relay Control	On: 1v, Off: 12v
71	RD	Vehicle Power	12-14v
73	TN/BK	Injector 5 Control	3.4-3.8 ms
74	BR/YL	Injector 3 Control	3.4-3.8 ms
75	TN	Injector 1 Control	3.4-3.8 ms
76, 77	BK/WT	Power Ground	<0.1v
78	YL/WT	Coil 3 Driver (dwell)	5°, 55 mph: 8°
79	WT/LG	TCIL (lamp) Control	TCIL On: 1v, Off: 12v
80	LB/OR	Fuel Pump Control	On: 1v, Off: 12v
81	WT/YL	EPC Solenoid Control	10v (26 psi)
83	WT/LB	IAC Solenoid Control	10.8v (33%)
84	DG/WT	TSS (+) Sensor Signal	100-125 Hz
85	DB/OR	CMP Sensor Signal	7 Hz, 55 mph: 18 Hz
87	RD/BK	HO2S-21 (B2 S1) Signal	0.1-1.1v
88	LB/RD	MAF Sensor Signal	0.8v, 55 mph: 2.1v
89	GY/WT	TP Sensor Signal	0.53-1.27v
90	BR/WT	Reference Voltage	4.9-5.1v
91	GY/RD	Analog Signal Return	<0.050v
92	LG	Brake Position Switch	Brake Off: 12v, On: 1v
93	RD/WT	HO2S-11 (B1 S1) Heater	On: 1v, Off: 12v
94	YL/LB	HO2S-21 (B2 S1) Heater	On: 1v, Off: 12v
95	WT/BK	HO2S-12 (B1 S2) Heater	On: 1v, Off: 12v
97	RD	Vehicle Power	12-14v
99	LG/OR	Injector 6 Control	3.4-3.8 ms
100	BR/LB	Injector 4 Control	3.4-3.8 ms
101	WT	Injector 2 Control	3.4-3.8 ms
103	BK/WT	Power Ground	<0.1v

Pin Connector Graphic

PCM 104-PIN CONNECTOR

Terminal View of 104-Pin PCM Wiring Harness Connector

Standard Colors and Abbreviations

Abbreviation	Color	Abbreviation	Color	Abbreviation	Color
BK	Black	GY	Gray	PK	Purple
BL	Blue	GN	Green	RD	Red
BR	Brown	LG	LT Green	TN	Tan
DB	Dark Blue	OR	Orange	WT	White
DG	DK Green	PK	Pink	YL	Yellow

1998-2002 Explorer 4.0L V6 MFI VIN X 104 Pin Connector

PCM Pin #	Wire Color	Circuit Description (104 Pin)	Value at Hot Idle
1	PK/OR	Shift Solenoid 'B' Control	12v, 55 mph: 12v
2	PK/LG	MIL (lamp) Control	MIL On: 1v, Off: 12v
3	YL/BK	Digital TR1 Sensor	In 'P': 0v, 55 mph: 11.5v
4-5	---	Not Used	---
6	DB/YL	OSS Sensor Signal	0 Hz, 55 mph: 350 Hz
7-11	---	Not Used	---
12	YL/WT	Fuel Level Indicator Signal	1.7v (1/2 full)
13	VT	Flash EPROM Power	0.1v
14	LB/BK	4x4 Low Indicator Control	4x4 On: 1v, Off: 12v
15	PK/LB	Data Bus (-) Signal	Digital Signals
16	TN/OR	Data Bus (+) Signal	Digital Signals
17-18	---	Not Used	---
19	OR/BK	Load Leveling Acceleration	Digital Signals
20, 23	---	Not Used	---
21	DB	CKP (+) Sensor Signal	430-460 Hz
22	GY	CKP (-) Sensor Signal	430-460 Hz
24, 51	BK/WT	Power Ground	<0.1v
25	BK	Case Ground	<0.050v
26	TN/WT	Coil 1 Driver (dwell)	5°, 55 mph: 8°
27	OR/YL	Shift Solenoid 'A' Control	1v, 55 mph: 12v
28 ('98-'99)	BR/OR	Coast Clutch Solenoid	12v, 55 mph: 12v
28 ('00-'02)	BR/OR	Shift Solenoid 'D' Control	12v, 55 mph: 12v
29	TN/WT	TCS (switch) Signal	TCS & O/D On: 12v
30 ('98-'99)	DG	Octane Adjustment	9.3v (Closed: 0v)
31, 34	---	Not Used	---
32 ('00-'02)	DG/PK	Knock Sensor (-) Signal	<0.050v
33 ('98-'99)	PK/OR	VSS (-) Signal	0 Hz, 55 mph: 125 Hz
35	RD/LG	HO2S-12 (B1 S2) Signal	0.1-1.1v
36	TN/LB	MAF Sensor Return	<0.050v
37	OR/BK	TFT Sensor Signal	2.10-2.40v
38	LG/RD	ECT Sensor Signal	0.5-0.6v
39	GY	IAT Sensor Signal	1.5-2.5v
40	DG/YL	Fuel Pump Monitor	On: 12v, Off: 0v
41	PK	A/C Switch Signal	A/C On: 12v, Off: 0v
42, 44	---	Not Used	---
43	LB/PK	Message Center (Fuel Flow)	Digital Signals
45	YL/WT	PCM Signals to the GEM	Digital Signals
46	BK/WT	Fuel Cap Indicator (Cluster)	Digital Signals
47	BR/PK	VR Solenoid Control	0%, 55 mph: 45%
48	TN/YL	Clean Tachometer Output	39-45 Hz
49	LB/BK	Digital TR2 Sensor	In 'P': 0v, 55 mph: 11.5v
50	WT/BK	Digital TR4 Sensor	In 'P': 0v, 55 mph: 11.5v
52	TN/OR	Coil 2 Driver (Control)	5°, 55 mph: 8°
53	PK/BK	Shift Solenoid 'C' Control	12v, 55 mph: 1v
54	PK/YL	TCC Solenoid Control	0%, 55 mph: 95%
55	YL	Keep Alive Power	12-14v
56	LG/BK	EVAP Purge Solenoid	0-10 Hz (0-100%)
57 ('00-'02)	YL/RD	Knock Sensor (+) Signal	0v
58	GY/BK	VSS (+) Signal	0 Hz, 55 mph: 125 Hz
59	---	Not Used	---
60	GY/LB	HO2S-11 (B1 S1) Signal	0.1-1.1v
61	PK/LG	HO2S-22 (B2 S2) Signal	0.1-1.1v
62	RD/PK	FTP Sensor Signal	2.6v (0" H2O - cap off)
63	---	Not Used	---
64	LB/YL	A/T: Digital TR3 Sensor	In 'P': 0v, 55 mph: 1.7v

1998-2002 Explorer 4.0L V6 MFI VIN X 104 Pin Connector

PCM Pin #	Wire Color	Circuit Description (104 Pin)	Value at Hot Idle
65	BR/LG	DPFE Sensor Signal	0.20-1.30v
66	---	Not Used	---
67	PK/WT	EVAP CV Solenoid	0-10 Hz (0-100%)
68	---	Not Used	---
69	PK/YL	A/C WOT Relay Control	On: 1v, Off: 12v
70	---	Not Used	---
71	RD	Vehicle Power	12-14v
72	---	Not Used	---
73	TN/BK	Injector 5 Control	3.4-3.8 ms
74	BR/YL	Injector 3 Control	3.4-3.8 ms
75	TN	Injector 1 Control	3.4-3.8 ms
76	BK/WT	Power Ground	<0.1v
77	BK/WT	Power Ground	<0.1v
78	TN/LG	Coil 3 Driver (dwell)	5°, 55 mph: 8°
79	WT/LG	TCIL (lamp) Control	TCIL On: 1v, Off: 12v
80	LB/OR	Fuel Pump Control	On: 1v, Off: 12v
81	WT/YL	EPC Solenoid Control	10v (26 psi)
82	---	Not Used	---
83	WT/LB	IAC Solenoid Control	10.7v (33%)
84	DG/WT	TSS Sensor Signal	100-125 Hz
85	DB/OR	CMP Sensor Signal	7 Hz, 55 mph: 18 Hz
86	BK/YL	A/C Pressure Sensor	Switch open: 12v
87	RD/BK	HO2S-21 (B2 S1) Signal	0.1-1.1v
88	LB/RD	MAF Sensor Signal	0.8v, 55 mph: 2.1v
89	GY/WT	TP Sensor Signal	0.53-1.27v
90	BR/WT	Reference Voltage	4.9-5.1v
91	GY/RD	Sensor Ground	<0.050v
92	RD/LG	Brake Position Switch	Brake Off: 12v, On: 1v
93	RD/WT	HO2S-11 (B1 S1) Heater	On: 1v, Off: 12v
94	YL/LB	HO2S-21 (B2 S1) Heater	On: 1v, Off: 12v
95	WT/BK	HO2S-12 (B1 S2) Heater	On: 1v, Off: 12v
97	RD	Vehicle Power	12-14v
98	---	Not Used	---
99	LG/OR	Injector 6 Control	3.4-3.8 ms
100	BR/LB	Injector 4 Control	3.4-3.8 ms
101	WT	Injector 2 Control	3.4-3.8 ms
102	---	Not Used	---
103	BK/WT	Power Ground	<0.1v
104	---	Not Used	---

Pin Connector Graphic

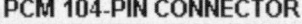

PCM 104-PIN CONNECTOR

Terminal View of 104-Pin PCM Wiring Harness Connector

Standard Colors and Abbreviations

Abbreviation	Color	Abbreviation	Color	Abbreviation	Color
BK	Black	GY	Gray	PK	Purple
BL	Blue	GN	Green	RD	Red
BR	Brown	LG	LT Green	TN	Tan
DB	Dark Blue	OR	Orange	WT	White
DG	DK Green	PK	Pink	YL	Yellow

2002-05 Explorer 4.6L V8 VIN W 58 Pin Connector

PCM Pin #	Wire Color	Circuit Description (58 Pin)	Value at Hot Idle
1, 5	---	Not Used	---
2	PK/BK	Fuel Pump Power	Pump On: 12v, Off: 0v
3	TN/OR	SCP Data Bus (+)	Digital Signals
4	PK/LB	SCP Data Bus (-)	Digital Signals
6	VT/WT	EVAP Canister Vent Solenoid Control	0-10 Hz (0-100%)
7	GY/BK	Vehicle Speed Sensor (+) Signal	0 Hz, 55 mph: 125 Hz
8	LG/RD	Generator Monitor Signal	130 Hz (45%)
9	PK/YL	A/C Clutch Relay	12v (relay on: 1v)
11	WT/LG	TX Signal	Digital Signals
12	LG/BK	EVAP Purge Solenoid	0-10 Hz (0-100%)
13	VT	Module Programming Signal	0.1v
14	GY/OR	RX Signal	Digital Signals
15-16, 21	---	Not Used	---
17	GY/RD	Sensor Ground	<0.050v
18-19	---	Not Used	---
20	BR/WT	Reference Voltage	4.9-5.1v
22	DB/LG	Passive Antitheft Indicator Control	Indicator Off: 12v, On: 1v
23	LB/BK	4WD Indicator Low Signal	12v (switch on: 1v)
24-27	BK/WT	Power Ground	<0.1v
28	BK/YL	Brake Pressure Switch	12v (Brake On: 0v)
29	OR/LB	Cruise Set Indicator Control	N/A
30	BK/YL	A/C Clutch Relay (switched)	12v (relay on: 1v)
31	LB/RD	MAF Sensor Signal	0.7v, 55 mph: 1.8v
32	VT	Vehicle Power (Start-Run)	12-14v
33	VT	Vehicle Power (Start-Run)	12-14v
34-35	---	Not Used	---
36	OR/LB	Cruise Set Indicator Control	12v (switch set: 0v)
37	---	Not Used	---
38	TN/LB	MAF Sensor Return	<0.050v
39	OR	Starter Motor Relay Signal	0v
40	RD/LG	Brake Position Switch	Brake Off: 12v, On: 1v
41	TN/WT	Overdrive Cancel Switch	0v (switch on: 12v)
42	---	Not Used	---
43	BK	Power Ground	<0.1v
44	RD/WT	Keep Alive Power (CJB fuse)	12-14v
45	BK	Speed Control Switch Ground	<0.050v
46	OR/LB	Cruise Switch Indicator	12v
47	RD/YL	A/C Pressure Switch Signal	0v (switch on: 12v)
48-49	---	Not Used	---
50	VT	A/C Demand Switch	0v (A/C On: 12v)
51	GY	IAT Sensor Signal	1.5-2.5v
52	RD/PK	FTP Sensor Signal	2.6v (0" H20 - cap off)
53-55	---	Not Used	---
56	DG/OR	Speed Control Switch Ground	<0.050v
57	LB/BK	Speed Control Switch Input	0v (switch on: 6.7v)
58	LB/OR	Fuel Pump Relay Control	On: 1v, Off: 12v

2002-05 Explorer 4.6L V8 VIN W 32 Pin Connector

PCM Pin #	Wire Color	Circuit Description (32 Pin)	Value at Hot Idle
1	OR/YL	Shift Solenoid 'A' Control	12v, 55 mph: 1v
2	VT/OR	Shift Solenoid 'B' Control	12v, 55 mph: 1v
3	PK/BK	Shift Solenoid 'C' Control	12v, 55 mph: 1v
4	BR/OR	Shift Solenoid 'D' Control	12v, 55 mph: 1v
5	VT/YL	TCC Solenoid Control	12v, 55 mph: 1v
6	---	Not Used	---
7	WT/YL	Electronic Pressure Control	On: 1v, Off: 12v
8	---	Not Used	---
9	LB/YL	Digital TR Sensor (TR3A)	0v, 55 mph: 1.7v
10	WT/BK	Digital TR Sensor (TR4)	0v, 55 mph: 11v
11	---	Not Used	---
12	WT	Motor Position #4	On: 1v, Off: 12v
13	LB/PK	Pressure Control 'B' Solenoid	On: 1v, Off: 12v
14	---	Not Used	---
15	WT/BK	HO2S-12 (B1 S2) Heater	1v, Off: 12v
16	TN/YL	HO2S-22 (B2 S2) Heater	1v, Off: 12v
17	GY/RD	Sensor Reference Ground	<0.050v
18	LB/BK	Digital TR Sensor (TR2)	0v, 55 mph: 11v
19-20	---	Not Used	---
21	YL/LG	Overdrive Drum Speed Input	200 Hz, 55 mph: 1185 Hz
22	YL/BK	Digital TR Sensor (TR1)	0v, 55 mph: 11v
23	OR/BK	Transmission Fluid Temperature Sensor	2.10-2.40v
24-25	---	Not Used	---
26	DB/YL	Output Shaft Speed Sensor	0 Hz, 985 Hz
27	DG/WT	Turbine Shaft Speed Sensor	360 Hz, 890 Hz
28	RD/LG	HO2S-12 (B1 S2) Signal	0.1-1.1v
29	VT/LG	HO2S-22 (B2 S2) Signal	0.1-1.1v
30-32	---	Not Used	---

Pin Connector Graphic

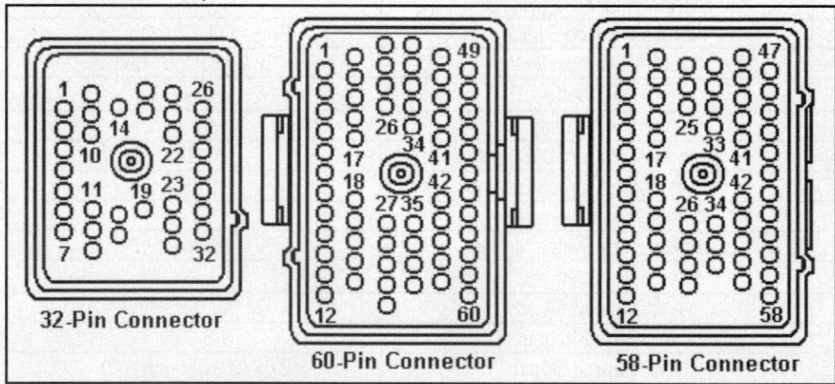

32-Pin Connector

60-Pin Connector

58-Pin Connector

Standard Colors and Abbreviations

Abbreviation	Color	Abbreviation	Color	Abbreviation	Color
BK	Black	GY	Gray	PK	Purple
BL	Blue	GN	Green	RD	Red
BR	Brown	LG	LT Green	TN	Tan
DB	Dark Blue	OR	Orange	WT	White
DG	DK Green	PK	Pink	YL	Yellow

2002-05 Explorer 4.6L V8 VIN W 60 Pin Connector

PCM Pin #	Wire Color	Circuit Description (60 Pin)	Value at Hot Idle
1	PK/WT	COP 2 Driver (dwell)	5°, 55 mph: 7°
2	TN	Injector 1 Control	3.3-3.8 ms
3-6	---	Not Used	---
7	RD/WT	HO2S-11 (B1 S1) Heater	1v, Off: 12v
8	YL/LB	HO2S-21 (B2 S1) Heater	1v, Off: 12v
9	WT/LB	IAC Solenoid Control	10.7v (33%)
10, 15	---	Not Used	---
11	TN/BK	Injector 5 Control	3.3-3.8 ms
12	OR/YL	COP 6 Driver Control	5°, 55 mph: 7°
13	PK/LB	COP 7 Driver Control	5°, 55 mph: 7°
14	WT	Injector 2 Control	3.3-3.8 ms
16	BR/PK	VR Solenoid Control	0%, 55 mph: 45%
17	GY/RD	Sensor Reference Ground	<0.050v
18-19	---	Not Used	---
20	BR/WT	Sensor Voltage Reference	4.9-5.1v
21	LG/OR	Injector 6 Control	3.3-3.8 ms
22	LG/YL	COP 5 Driver Control	5°, 55 mph: 7°
23	BR/YL	COP 5 Driver Control	5°, 55 mph: 7°
24	BR/YL	Injector 3 Control	3.3-3.8 ms
25-28	---	Not Used	---
29	TN/RD	Injector 7 Control	3.3-3.8 ms
30	DG/VT	COP 4 Driver Control	5°, 55 mph: 7°
31	LG/WT	COP 1 Driver (dwell)	5°, 55 mph: 7°
32	BR/LB	Injector 4 Control	3.3-3.8 ms
33-36	---	Not Used	---
37	LB	Injector 8 Control	3.3-3.8 ms
38	WT/RD	COP 8 Driver Control	5°, 55 mph: 7°
39	---	Not Used	---
40	YL/LB	CHT Sensor Signal	0.7v (194°F)
41	BR/LG	DPFE Sensor Signal	0.95-1.05v
42	YL	Knock Sensor 1 Signal	0v
43	DG/WT	Knock Sensor 2 Signal	0v
44	RD/BK	HO2S-21 (B2 S1) Signal	0.1-1.1v
45	GY/LB	HO2S-11 (B1 S1) Signal	0.1-1.1v
46-50	---	Not Used	---
51	YL/RD	Knock Sensor 1 Ground	0v
52	DG/VT	Knock Sensor 2 Ground	0v
53	DG	CMP Sensor Signal	6 Hz
54	---	Not Used	---
55	DB	CKP Sensor (+) Signal	400 Hz
56	GY	CKP Sensor (-) Signal	400 Hz
57	GY/WT	TP Sensor Signal	0.53-1.27v
58-60	---	Not Used	---

Pin Connector Graphic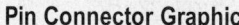

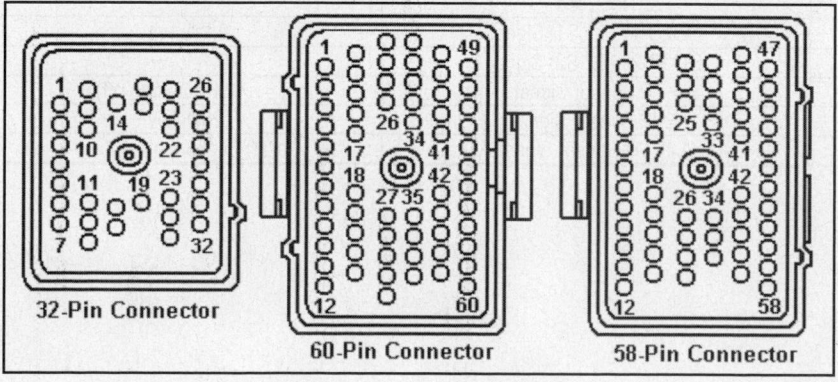

32-Pin Connector   60-Pin Connector   58-Pin Connector

1996-97 Explorer 5.0L V8 VIN P 104 Pin Connector

PCM Pin #	Wire Color	Circuit Description (104 Pin)	Value at Hot Idle
1	PK/OR	Shift Solenoid 2 Control	12v, 55 mph: 12v
2	PK/LG	MIL (lamp) Control	MIL On: 1v, Off: 12v
3-12	---	Not Used	---
13	VT	Flash EPROM Power	0.1v
14	---	Not Used	---
15	PK/LB	Data Bus (-) Signal	Digital Signals
16	TN/OR	Data Bus (+) Signal	Digital Signals
17-18	---	Not Used	---
19	OR/BK	Accelerator Ride Control	0.1v (Off)
20	---	Not Used	---
21	BK/PK	CKP (+) Sensor Signal	430-460 Hz
22	GY/YL	CKP (-) Sensor Signal	430-460 Hz
23	---	Not Used	---
24	BK/WT	Power Ground	<0.1v
25	BK	Case Ground	<0.050v
26	DB/LG	Coil 1 Driver (dwell)	5°, 55 mph: 8°
27	OR/YL	Shift Solenoid 1 Control	1v, 55 mph: 12v
28	---	Not Used	---
29	TN/WT	TCS (switch) Signal	TCS & O/D On: 12v
30	DG	Octane Adjustment	9.3v (Closed: 0v)
31-32	---	Not Used	---
33	PK/OR	VSS (-) Signal	0 Hz, 55 mph: 125 Hz
34	---	Not Used	---
35	RD/LG	HO2S-12 (B1 S2) Signal	0.1-1.1v
36	TN/LB	MAF Sensor Return	<0.050v
37	OR/BK	TFT Sensor Signal	2.10-2.40v
38	LG/RD	ECT Sensor Signal	0.5-0.6v
39	GY	IAT Sensor Signal	1.5-2.5v
40	DG/YL	Fuel Pump Monitor	On: 12v, Off: 0v
41	DG/OR	A/C Switch Signal	A/C On: 12v, Off: 0v
42	---	Not Used	---
43	LB/PK	Data Output Link	5v
44-46	---	Not Used	---
47	BK/PK	VR Solenoid Control	0%, 55 mph: 45%
48	TN/YL	Clean Tachometer Output	39-45 Hz
49-50	---	Not Used	---
51	BK/WT	Power Ground	<0.1v
52	RD/LB	Coil 2 Driver (dwell)	5°, 55 mph: 8°
53	---	Not Used	---
54	DB/WT	TCC Solenoid Control	0%, 55 mph: 95%
55	Y	Keep Alive Power	12-14v
56	LG/BK	EVAP Purge Solenoid	0-10 Hz (0-100%)
57	---	Not Used	---
58	GY/BK	VSS (+) Signal	0 Hz, 55 mph: 125 Hz
59	---	Not Used	---
60	GY/LB	HO2S-11 (B1 S1) Signal	0.1-1.1v
61	PK/LG	HO2S-22 (B2 S2) Signal	0.1-1.1v
62	RD/PK	FTP Sensor Signal	2.6v (0" H2O - cap off)
63	OR/YL	EVP Sensor Signal	0.20-1.30v
64	LB/YL	A/T: TR Sensor Signal	In 'P': 0v, 55 mph: 1.7v

1996-97 Explorer 5.0L V8 VIN P 104 Pin Connector

PCM Pin #	Wire Color	Circuit Description (104 Pin)	Value at Hot Idle
65-66	---	Not Used	---
67	PK/WT	EVAP CV Solenoid	0-10 Hz (0-100%)
68	---	Not Used	---
69	PK/YL	A/C WOT Relay Control	On: 1v, Off: 12v
70	---	Not Used	---
71	RD	Vehicle Power	12-14v
72	TN/RD	Injector 7 Control	3.2-4.5 ms
73	TN/BK	Injector 5 Control	3.2-4.5 ms
74	BR/YL	Injector 3 Control	3.2-4.5 ms
75	TN	Injector 1 Control	3.2-4.5 ms
76	BK/WT	Power Ground	<0.1v
77	BK/WT	Power Ground	<0.1v
78	PK/WT	Coil 4 Driver (dwell)	5°, 55 mph: 8°
79	WT/LG	TCIL (lamp) Control	TCIL On: 1v, Off: 12v
80	LB/OR	Fuel Pump Control	On: 1v, Off: 12v
81	WT/YL	EPC Solenoid Control	10v (26 psi)
83	WT/LB	IAC Solenoid Control	10.8v (33%)
84	DB/YL	OSS Sensor Signal	0 Hz, 55 mph: 230 Hz
85	DB/OR	CMP Sensor Signal	5-10 Hz
86	---	Not Used	---
87	RD/BK	HO2S-21 (B2 S1) Signal	0.1-1.1v
88	LB/RD	MAF Sensor Signal	0.8v, 55 mph: 2.1v
89	GY/WT	TP Sensor Signal	0.53-1.27v
90	BR/WT	Reference Voltage	4.9-5.1v
91	GY/RD	Analog Signal Return	<0.050v
92	LG	Brake Position Switch	Brake Off: 12v, On: 1v
93	RD/WT	HO2S-11 (B1 S1) Heater	On: 1v, Off: 12v
94	YL/LB	HO2S-21 (B2 S1) Heater	On: 1v, Off: 12v
95	WT/BK	HO2S-12 (B1 S2) Heater	On: 1v, Off: 12v
96	TN/YL	HO2S-22 (B2 S2) Heater	On: 1v, Off: 12v
97	RD	Vehicle Power	12-14v
98	LB	Injector 8 Control	3.2-4.5 ms
99	LG/OR	Injector 6 Control	3.2-4.5 ms
100	BR/LB	Injector 4 Control	3.2-4.5 ms
101	WT	Injector 2 Control	3.2-4.5 ms
102	---	Not Used	---
103	BK/WT	Power Ground	<0.1v
104	RD/YL	Coil 4 Driver (dwell)	5°, 55 mph: 8°

Pin Connector Graphic

PCM 104-PIN CONNECTOR

Terminal View of 104-Pin PCM Wiring Harness Connector

Standard Colors and Abbreviations

Abbreviation	Color	Abbreviation	Color	Abbreviation	Color
BK	Black	GY	Gray	PK	Purple
BL	Blue	GN	Green	RD	Red
BR	Brown	LG	LT Green	TN	Tan
DB	Dark Blue	OR	Orange	WT	White
DG	DK Green	PK	Pink	YL	Yellow

1998-2001 Explorer 5.0L V8 VIN P 104 Pin Connector

PCM Pin #	Wire Color	Circuit Description (104 Pin)	Value at Hot Idle
1	PK/OR	Shift Solenoid 'B' Control	12v, 55 mph: 1v
2	PK/LG	MIL (lamp) Control	MIL On: 1v, Off: 12v
3	YL/BK	Digital TR1 Sensor	In 'P': 0v, Others: 11.5v
4-11	---	Not Used	---
12	YL/WT	Fuel Level Indicator Signal	1.7v (1/2 full)
13	VT	Flash EPROM Power	0.1v
14	---	Not Used	---
15	PK/LB	Data Bus (-) Signal	Digital Signals
16	TN/OR	Data Bus (+) Signal	Digital Signals
17-18	---	Not Used	---
19	OR/BK	Air Suspension Control Signal	Digital Signals
20	---	Not Used	---
21	BK/PK	CKP (+) Sensor Signal	460-500 Hz
22	GY/YL	CKP (-) Sensor Signal	460-500 Hz
24	BK/WT	Power Ground	<0.1v
25	BK	Case Ground	<0.050v
26	DB/LG	Coil 1 Driver (dwell)	5°, 55 mph: 8°
27	OR/YL	Shift Solenoid 'A' Control	1v, 55 mph: 1v
28	---	Not Used	---
29	TN/WT	TCS or Manual Switch	TCS or M/D pressed: 12v
30 ('98-'99)	DG	Octane Adjustment	9.3v (Closed: 0v)
31-32	---	Not Used	---
33 ('98-'99)	PK/OR	VSS (-) Signal	0 Hz, 55 mph: 125 Hz
35	RD/LG	HO2S-12 (B1 S2) Signal	0.1-1.1v
36	TN/LB	MAF Sensor Return	<0.050v
37	OR/BK	TFT Sensor Signal	2.10-2.40v
38	LG/RD	ECT Sensor Signal	0.5-0.6v
39	GY	IAT Sensor Signal	1.5-2.5v
40	DG/YL	Fuel Pump Monitor	On: 12v, Off: 0v
41	VT	A/C Switch Signal	Switch On: 12v (open)
42, 45-46	---	Not Used	---
43 ('98-'99)	LB/PK	Neutral Tow Connector Signal	N/A
47	BR/PK	VR Solenoid Control	0%, 55 mph: 45%
48	TN/YL	Clean Tachometer Output	40-55 Hz
49	LB/BK	Digital TR2 Sensor	In 'P': 0v, Others: 11.5v
50	WT/BK	Digital TR4 Sensor	In 'P': 0v, Others: 11.5v
51	BK/WT	Power Ground	<0.1v
52	RD/LB	Coil 2 Driver (Control)	5°, 55 mph: 8°
53	---	Not Used	---
54	DB/WT	TCC Solenoid Control	0%, 55 mph: 95%
55	YL	Keep Alive Power	12-14v
56	LG/BK	EVAP Purge Solenoid	0-10 Hz (0-100%)
57	---	Not Used	---
58	GY/BK	VSS (+) Signal	0 Hz, 55 mph: 125 Hz
59	---	Not Used	---
60	GY/LB	HO2S-11 (B1 S1) Signal	0.1-1.1v
61	PK/LG	HO2S-22 (B2 S2) Signal	0.1-1.1v
62	RD/PK	FTP Sensor Signal	2.6v (0" H2O - cap off)
63	---	Not Used	---
64	LB/YL	Digital TR3 Sensor	In 'P': 0v, 55 mph: 1.7v

1998-2001 Explorer 5.0L V8 VIN P 104 Pin Connector

PCM Pin #	Wire Color	Circuit Description (104 Pin)	Value at Hot Idle
65 ('98-'00)	BR/LG	DPFE Sensor Signal	0.20-1.30v
65 ('01-'00)	BR/LG	DPFE Sensor Signal	0.95-1.05v
66	---	Not Used	---
67	PK/WT	EVAP CV Solenoid	0-10 Hz (0-100%)
68	---	Not Used	---
69	PK/YL	A/C WOT Relay Control	On: 1v, Off: 12v
70	---	Not Used	---
71	RD	Vehicle Power	12-14v
72	TN/RD	Injector 7 Control	3.2-4.5 ms
73	TN/BK	Injector 5 Control	3.2-4.5 ms
74	BR/YL	Injector 3 Control	3.2-4.5 ms
75	TN	Injector 1 Control	3.2-4.5 ms
76	BK/WT	Power Ground	<0.1v
77	BK/WT	Power Ground	<0.1v
78	PK/WT	Coil 3 Driver (dwell)	5°, 55 mph: 8°
79	WT/LG	TCIL (lamp) Control	TCIL On: 1v, Off: 12v
80	LB/OR	Fuel Pump Control	On: 1v, Off: 12v
81	WT/YL	EPC Solenoid Control	9.5v (10 psi)
82	---	Not Used	---
83	WT/LB	IAC Solenoid Control	10.8v (12-30%)
84	DB/YL	OSS Sensor Signal	At 55 mph: 230-280 Hz
85	DB/OR	CMP Sensor Signal	5-10 Hz
86	BK/YL	A/C Pressure Switch Signal	Open: 12v (<24 psi)
87	RD/BK	HO2S-21 (B2 S1) Signal	0.1-1.1v
88	LB/RD	MAF Sensor Signal	0.8v, 55 mph: 2.1v
89	GY/WT	TP Sensor Signal	0.53-1.27v
90	BR/WT	Reference Voltage	4.9-5.1v
91	GY/RD	Analog Signal Return	<0.050v
92	RD/LG	Brake Position Switch	Brake Off: 12v, On: 1v
93	RD/WT	HO2S-11 (B1 S1) Heater	On: 1v, Off: 12v
94	YL/LB	HO2S-21 (B2 S1) Heater	On: 1v, Off: 12v
95	WT/BK	HO2S-12 (B1 S2) Heater	On: 1v, Off: 12v
96	TN/YL	HO2S-22 (B2 S2) Heater	On: 1v, Off: 12v
97	RD	Vehicle Power	12-14v
98	LB	Injector 8 Control	3.2-4.5 ms
99	LG/OR	Injector 6 Control	3.2-4.5 ms
100	BR/LB	Injector 4 Control	3.2-4.5 ms
101	WT	Injector 2 Control	3.2-4.5 ms
102	---	Not Used	---
103	BK/WT	Power Ground	<0.1v
104	RD/YL	Coil 4 Driver (dwell)	5°, 55 mph: 8°

Pin Connector Graphic

Standard Colors and Abbreviations

Abbreviation	Color	Abbreviation	Color	Abbreviation	Color
BK	Black	GY	Gray	PK	Purple
BL	Blue	GN	Green	RD	Red
BR	Brown	LG	LT Green	TN	Tan
DB	Dark Blue	OR	Orange	WT	White
DG	DK Green	PK	Pink	YL	Yellow

MOUNTAINEER PIN TABLES

1998-99 Mountaineer 4.0L V6 VIN E 104 Pin Connector

PCM Pin #	Wire Color	Circuit Description (104 Pin)	Value at Hot Idle
1	PK/OR	Shift Solenoid 'B' Control	12v, 55 mph: 12v
2	PK/LG	MIL (lamp) Control	MIL On: 1v, Off: 12v
3	YL/BK	Digital TR1 Sensor	In 'P': 0v, 55 mph: 11.5v
4-5	---	Not Used	---
6	DB/YL	OSS Sensor Signal	0 Hz, 55 mph: 350 Hz
7-11	---	Not Used	---
12	YL/WT	Fuel Level Indicator Signal	1.7v (1/2 full)
13	VT	Flash EPROM Power	0.1v
14	LB/BK	4x4 Indicator Signal	4x4 On: 1v, Off: 12v
15	PK/LB	Data Bus (-) Signal	Digital Signals
16	TN/OR	Data Bus (+) Signal	Digital Signals
17-20	---	Not Used	---
19	OR/BK	Accelerator Ride Control	0.1v (Off)
21	DB	CKP (+) Sensor Signal	430-460 Hz
22	GY	CKP (-) Sensor Signal	<0.050v
24	BK/WT	Power Ground	<0.1v
25	BK	Case Ground	<0.050v
26	TN/WT	Coil 1 Driver (dwell)	5°, 55 mph: 8°
27	OR/YL	Shift Solenoid 'A' Control	1v, 55 mph: 12v
28	BR/OR	Coast Clutch Solenoid	12v, 55 mph: 12v
29	TN/WT	TCS (switch) Signal	TCS & O/D On: 12v
30	DG	Octane Adjustment	9.3v (Closed: 0v)
31-32	---	Not Used	---
32	DG/VT	Knock Sensor 1 Ground	<0.050v
33	PK/OR	VSS (-) Signal	0 Hz, 55 mph: 125 Hz
34	---	Not Used	---
35	RD/LG	HO2S-12 (B1 S2) Signal	0.1-1.1v
36	TN/LB	MAF Sensor Return	<0.050v
37	OR/BK	TFT Sensor Signal	2.10-2.40v
38	LG/RD	ECT Sensor Signal	0.5-0.6v
39	GY	IAT Sensor Signal	1.5-2.5v
40	DG/YL	Fuel Pump Monitor	On: 12v, Off: 0v
41	PK	A/C Switch Signal	A/C On: 12v, Off: 0v
42	---	Not Used	---
43	LB/PK	Neutral Tow Connector Signal	N/A
44	---	Not Used	---
45	YL/WT	PCM Signal to the GEM	Digital Signals
46	BR	Inlet Air Control Valve	0.9v (15%)
47	BRD/PK	VR Solenoid Control	0%, 55 mph: 45%
48	TN/YL	Clean Tachometer Output	39-45 Hz
49	LB/BK	Digital TR2 Sensor	0v, 55 mph: 11.5v
50	WT/BK	Digital TR4 Sensor	0v, 55 mph: 11.5v
51	BK/WT	Power Ground	<0.1v
52	TN/OR	Coil 2 Driver (Control)	5°, 55 mph: 8°
53	P/BK	Shift Solenoid 'C' Control	12v, 55 mph: 1v
54	PK/YL	TCC Solenoid Control	0%, 55 mph: 95%
55	YL	Keep Alive Power	12-14v
56	LG/BK	EVAP Purge Solenoid	0-10 Hz (0-100%)
57	---	Not Used	---
58	GY/BK	VSS (+) Signal	0 Hz, 55 mph: 125 Hz
59	---	Not Used	---
60	GY/LB	HO2S-11 (B1 S1) Signal	0.1-1.1v
61	---	Not Used	---
62	RD/PK	FTP Sensor Signal	2.6v (0" H2O - cap off)
63	---	Not Used	---
64	LB/YL	Digital TR3 Sensor	In 'P': 0v, 30 mph: 1.7v

1998-99 Mountaineer 4.0L V6 VIN E 104 Pin Connector

PCM Pin #	Wire Color	Circuit Description (104 Pin)	Value at Hot Idle
65	BR/LG	DPFE Sensor Signal	0.20-1.30v
66	---	Not Used	---
67	PK/WT	EVAP CV Solenoid	0-10 Hz (0-100%)
68	---	Not Used	---
69	PK/YL	A/C WOT Relay Control	On: 1v, Off: 12v
71	RD	Vehicle Power	12-14v
70	---	Not Used	---
72	---	Not Used	---
73	TN/BK	Injector 5 Control	3.6-7.5 ms
74	BR/YL	Injector 3 Control	3.6-7.5 ms
75	TN	Injector 1 Control	3.6-7.5 ms
76	BK/WT	Power Ground	<0.1v
77	BK/WT	Power Ground	<0.1v
78	TN/LG	Coil 3 Driver (dwell)	5º, 55 mph: 8º
79	WT/LG	TCIL (lamp) Control	TCIL On: 1v, Off: 12v
80	LB/OR	Fuel Pump Control	On: 1v, Off: 12v
81	WT/YL	EPC Solenoid Control	10v (26 psi)
82	---	Not Used	---
83	WT/LB	IAC Solenoid Control	10.8v (33%)
84	DG/WT	TSS (+) Sensor Signal	100-125 Hz
85	DB/OR	CMP Sensor Signal	7 Hz, 55 mph: 18 Hz
86	BK/YL	A/C Pressure Switch Signal	Switch Open: 12v
87	RD/BK	HO2S-21 (B2 S1) Signal	0.1-1.1v
88	LB/RD	MAF Sensor Signal	0.8v, 55 mph: 2.1v
89	GY/WT	TP Sensor Signal	0.53-1.27v
90	BR/WT	Reference Voltage	4.9-5.1v
91	GY/RD	Sensor Ground	<0.050v
92	RD/LG	Brake Position Switch	Brake Off: 12v, On: 1v
93	RD/WT	HO2S-11 (B1 S1) Heater	On: 1v, Off: 12v
94	YL/LB	HO2S-21 (B2 S1) Heater	On: 1v, Off: 12v
95	WT/BK	HO2S-12 (B1 S2) Heater	On: 1v, Off: 12v
96	---	Not Used	---
97	RD	Vehicle Power	12-14v
98	---	Not Used	---
99	LG/OR	Injector 6 Control	3.6-7.5 ms
100	BR/LB	Injector 4 Control	3.6-7.5 ms
101	WT	Injector 2 Control	3.6-7.5 ms
102	---	Not Used	---
103	BK/WT	Power Ground	<0.1v
104	---	Not Used	---

Pin Connector Graphic

PCM 104-PIN CONNECTOR

Terminal View of 104-Pin PCM Wiring Harness Connector

Standard Colors and Abbreviations

Abbreviation	Color	Abbreviation	Color	Abbreviation	Color
BK	Black	GY	Gray	PK	Purple
BL	Blue	GN	Green	RD	Red
BR	Brown	LG	LT Green	TN	Tan
DB	Dark Blue	OR	Orange	WT	White
DG	DK Green	PK	Pink	YL	Yellow

2000-02 Mountaineer 4.0L V6 SOHC MFI VIN E 104 Pin Connector

PCM Pin #	Wire Color	Circuit Description (104 Pin)	Value at Hot Idle
1	PK/OR	Shift Solenoid 'B' Control	12v, 55 mph: 12v
2	PK/LG	MIL (lamp) Control	MIL On: 1v, Off: 12v
3	YL/BK	Digital TR1 Sensor	In 'P': 0v, 55 mph: 11.5v
4-5	---	Not Used	---
6	DB/YL	OSS Sensor Signal	0 Hz, 55 mph: 350 Hz
7-11	---	Not Used	---
12	YL/WT	Fuel Level Indicator Signal	1.7v (1/2 full)
13	PK	Flash EPROM Power	0.1v
14	LB/BK	4x4 Low Indicator Switch	Switch On: 1v, Off: 12v
15	PK/LB	Data Bus (-) Signal	Digital Signals
16	TN/OR	Data Bus (+) Signal	Digital Signals
17-18	---	Not Used	---
19	OR/BK	Load Leveling Acceleration	Digital Signals
20	---	Not Used	---
21	DB	CKP (+) Sensor Signal	430-460 Hz
22	GY	CKP (-) Sensor Signal	<0.050v
23	---	Not Used	---
24	BK/WT	Power Ground	<0.1v
25	BK	Chassis Ground	<0.050v
26	TN/WT	Coil 1 Driver (dwell)	5°, 55 mph: 8°
27	OR/YL	Shift Solenoid 'A' Control	1v, 55 mph: 12v
28	BR/OR	Shift Solenoid 'D' Control	12v, 55 mph: 12v
29	TN/WT	TCS (switch) Signal	TCS & O/D On: 12v
30	---	Not Used	---
32	DG/VT	Knock Sensor (-) Return	<0.050v
33-34	---	Not Used	---
35	RD/LG	HO2S-12 (B1 S2) Signal	0.1-1.1v
36	TN/LB	MAF Sensor Return	<0.050v
37	OR/BK	TFT Sensor Signal	2.10-2.40v
38	LG/RD	ECT Sensor Signal	0.5-0.6v
39	GY	IAT Sensor Signal	1.5-2.5v
40	DG/YL	Fuel Pump Monitor	On: 12v, Off: 0v
41	PK	A/C High Pressure Switch	Switch On: 12v (open)
42	---	Not Used	---
43	LB/PK	Message Center Fuel Flow	Digital Signals
44	---	Not Used	---
45	YL/WT	PCM Signal to GEM	Digital Signals
46	BK/WT	Fuel Cap Indicator (Cluster)	Digital Signals
47	BR/PK	EGR VR Regulator Solenoid	0%, 55 mph: 45%
48	TN/YL	Clean Tachometer Output	39-45 Hz
49	LB/BK	Digital TR2 Sensor	0v, 55 mph: 11.5v
50	WT/BK	Digital TR4 Sensor	0v, 55 mph: 11.5v
51	BK/WT	Power Ground	<0.1v
52	TN/OR	Coil 2 Driver (Control)	5°, 55 mph: 8°
53	PK/BK	Shift Solenoid 'C' Control	12v, 55 mph: 1v
54	PK/YL	TCC Solenoid Control	0%, 55 mph: 95%
55	YL	Keep Alive Power	12-14v
56	LG/BK	EVAP Purge Solenoid	0-10 Hz (0-100%)
57	YL/RD	Knock Sensor (+) Signal	0v
58	GY/BK	VSS (+) Signal	0 Hz, 55 mph: 125 Hz
59	---	Not Used	---
60	GY/LB	HO2S-11 (B1 S1) Signal	0.1-1.1v
61	PK/LG	HO2S-22 (B2 S2) Signal	0.1-1.1v
63	---	Not Used	---
62	RD/PK	FTP Sensor Signal	2.6v (0" H2O - cap off)
64	LB/YL	Digital TR3 Sensor	In 'P': 0v, 30 mph: 1.7v

2000-02 Mountaineer 4.0L V6 SOHC MFI VIN E 104 Pin Connector

PCM Pin #	Wire Color	Circuit Description (104 Pin)	Value at Hot Idle
65 ('00)	BR/LG	DPFE Sensor Signal	0.20-1.30v
65	BR/LG	DPFE Sensor Signal	0.95-1.05v
66	---	Not Used	---
67	VT/WT	EVAP CV Solenoid	0-10 Hz (0-100%)
69	PK/YL	A/C WOT Relay Control	On: 1v, Off: 12v
70	---	Not Used	---
71	RD	Vehicle Power	12-14v
73	TN/BK	Injector 5 Control	3.6-7.5 ms
74	BR/YL	Injector 3 Control	3.6-7.5 ms
75	TN	Injector 1 Control	3.6-7.5 ms
76	BK/WT	Power Ground	<0.1v
77	BK/WT	Power Ground	<0.1v
78	TN/LG	Coil 3 Driver (dwell)	5°, 55 mph: 8°
79	WT/LG	TCIL (lamp) Control	TCIL On: 1v, Off: 12v
80	LB/OR	Fuel Pump Relay Control	On: 1v, Off: 12v
81	WT/YL	EPC Solenoid Control	10v (26 psi)
82	---	Not Used	---
83	WT/LB	IAC Solenoid Control	10.8v (33%)
84	DG/WT	TSS Sensor Signal	100-125 Hz
85	DB/OR	CMP Sensor Signal	7 Hz, 55 mph: 18 Hz
86	BK/YL	A/C Pressure Sensor	Switch open: 12v
87	RD/BK	HO2S-21 (B2 S1) Signal	0.1-1.1v
88	LB/RD	MAF Sensor Signal	0.8v, 55 mph: 2.1v
89	GY/WT	TP Sensor Signal	0.53-1.27v
90	BR/WT	Reference Voltage	4.9-5.1v
91	GY/RD	Sensor Ground	<0.050v
92	RD/LG	Brake Position Switch	Brake Off: 12v, On: 1v
93	RD/WT	HO2S-11 (B1 S1) Heater	On: 1v, Off: 12v
94	YL/LB	HO2S-21 (B2 S1) Heater	On: 1v, Off: 12v
95	WT/BK	HO2S-12 (B1 S2) Heater	On: 1v, Off: 12v
96	TN/YL	HO2S-22 (B2 S2) Heater	On: 1v, Off: 12v
97	RD	Vehicle Power	12-14v
98	---	Not Used	---
99	LG/OR	Injector 6 Control	3.6-7.5 ms
100	BR/LB	Injector 4 Control	3.6-7.5 ms
101	WT	Injector 2 Control	3.6-7.5 ms
102	---	Not Used	---
103	BK/WT	Power Ground	<0.1v
104	---	Not Used	---

Pin Connector Graphic

PCM 104-PIN CONNECTOR

Terminal View of 104-Pin PCM Wiring Harness Connector

Standard Colors and Abbreviations

Abbreviation	Color	Abbreviation	Color	Abbreviation	Color
BK	Black	GY	Gray	PK	Purple
BL	Blue	GN	Green	RD	Red
BR	Brown	LG	LT Green	TN	Tan
DB	Dark Blue	OR	Orange	WT	White
DG	DK Green	PK	Pink	YL	Yellow

2002-05 Mountaineer 4.0L V6 MFI VIN K 58 Pin Connector

PCM Pin #	Wire Color	Circuit Description (58 Pin)	Value at Hot Idle
1, 5, 15-16	---	Not Used	---
2	PK/BK	Fuel Pump Power	Pump On: 12v, Off: 0v
3	TN/OR	SCP Data Bus (+)	Digital Signals
4	PK/LB	SCP Data Bus (-)	Digital Signals
6	VT/WT	EVAP Canister Vent Valve Control	0-10 Hz (0-100%)
7	GY/BK	Vehicle Speed Sensor (+) Signal	0 Hz, 55 mph: 125 Hz
8	LG/RD	Generator Monitor Signal	130 Hz (45%)
9	PK/YL	A/C Clutch Relay Control	12v (relay on: 1v)
10	YL/WT	4WD Indicator Control	Indicator Off: 12v, On: 1v
11	WT/LG	Passive Antitheft TX Signal	Digital Signals
12	LG/BK	EVAP Purge Solenoid	0-10 Hz (0-100%)
13	VT	Module Programming Signal	0.1v
14	GY/OR	Passive Antitheft RX Signal	Digital Signals
18-19, 21	---	Not Used	---
17	GY/RD	Sensor Ground	<0.050v
20	BR/WT	Reference Voltage	4.9-5.1v
22	DB/LG	Passive Antitheft Indicator Control	Indicator Off: 12v, On: 1v
23	LB/BK	4WD Indicator Low Signal	12v (switch on: 1v)
24	BK/WT	Power Ground	<0.1v
25	BK/WT	Power Ground	<0.1v
26	BK/WT	Power Ground	<0.1v
27	BK/WT	Power Ground	<0.1v
28	BK/YL	Brake Pressure Switch	12v (Brake On: 0v)
29	OR/LB	Cruise Set Indicator Control	Indicator Off: 12v, On: 1v
30	BK/YL	A/C Clutch Relay (switched)	12v (relay on: 1v)
31	LB/RD	MAF Sensor Signal	0.7v, 55 mph: 1.8v
32	VT	Vehicle Power (Start-Run)	12-14v
33	VT	Vehicle Power (Start-Run)	12-14v
34-35	---	Not Used	---
36	OR/LB	Cruise Set Indicator Control	12v (switch set: 0v)
37	---	Not Used	---
38	TN/LB	MAF Sensor Return	<0.050v
39	OR	Starter Motor Relay Circuit	Relay Off: 12v, On: 1v
40	RD/LG	Brake Position Switch	Brake Off: 12v, On: 1v
41	TN/WT	Overdrive Cancel Switch	0v (switch on: 12v)
42, 48-49	---	Not Used	---
43	BK	Power Ground	<0.1v
44	RD/WT	Keep Alive Power (CJB fuse)	12-14v
45	BK	Speed Control Switch Ground	<0.050v
46	OR/LB	Cruise Switch Indicator Control	Indicator Off: 12v, On: 1v
47	RD/YL	A/C Pressure Switch Signal	0v (switch on: 12v)
50	VT	A/C Demand Switch	0v (A/C On: 12v)
51	GY	Intake Air Temperature Sensor	1.5-2.5v
52	RD/PK	Fuel Tank Pressure Sensor	2.6v (0" H20 - cap off)
53-55	---	Not Used	---
56	DG/OR	Speed Control Switch Ground	<0.050v
57	LB/BK	Speed Control Switch Input	0v (switch on: 6.7v)
58	LB/OR	Fuel Pump Relay Control	Relay Off: 12v, On: 1v

2002-05 Mountaineer 4.0L V6 MFI VIN K 32 Pin Connector

PCM Pin #	Wire Color	Circuit Description (32 Pin)	Value at Hot Idle
1	OR/YL	Shift Solenoid 'A' Control	12v, 55 mph: 1v
2	VT/OR	Shift Solenoid 'B' Control	12v, 55 mph: 1v
3	PK/BK	Shift Solenoid 'C' Control	12v, 55 mph: 1v
4	BR/OR	Shift Solenoid 'D' Control	12v, 55 mph: 1v
5	VT/YL	TCC Solenoid Control	12v, 55 mph: 1v
6	---	Not Used	---
7	WT/YL	Electronic Pressure Control	On: 1v, Off: 12v
8	---	Not Used	---
9	LB/YL	Digital TR Sensor (TR3A)	0v, 55 mph: 1.7v
10	WT/BK	Digital TR Sensor (TR4)	0v, 55 mph: 11v
11	---	Not Used	---
12	WT	Motor Position #4	On: 1v, Off: 12v
13	LB/PK	Pressure Control 'B' Solenoid	On: 1v, Off: 12v
14	---	Not Used	---
15	WT/BK	HO2S-12 (B1 S2) Heater	1v, Off: 12v
16	TN/YL	HO2S-22 (B2 S2) Heater	1v, Off: 12v
17	GY/RD	Sensor Reference Ground	<0.050v
18	LB/BK	Digital TR Sensor (TR2)	0v, 55 mph: 11v
19-20	---	Not Used	---
21	YL/LG	Overdrive Drum Speed Input	200 Hz, 55 mph: 1185 Hz
22	YL/BK	Digital TR Sensor (TR1)	0v, 55 mph: 11v
23	OR/BK	Transmission Fluid Temperature Sensor	2.10-2.40v
24-25	---	Not Used	---
26	DB/YL	Output Shaft Speed Sensor	0 Hz, 985 Hz
27	DG/WT	Turbine Shaft Speed Sensor	360 Hz, 890 Hz
28	RD/LG	HO2S-12 (B1 S2) Signal	0.1-1.1v
29	VT/LG	HO2S-22 (B2 S2) Signal	0.1-1.1v
30-32	---	Not Used	---

Pin Connector Graphic

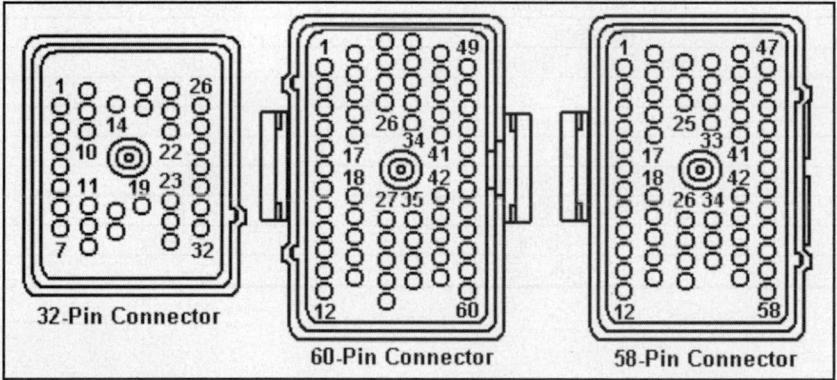

32-Pin Connector · 60-Pin Connector · 58-Pin Connector

Standard Colors and Abbreviations

Abbreviation	Color	Abbreviation	Color	Abbreviation	Color
BK	Black	GY	Gray	PK	Purple
BL	Blue	GN	Green	RD	Red
BR	Brown	LG	LT Green	TN	Tan
DB	Dark Blue	OR	Orange	WT	White
DG	DK Green	PK	Pink	YL	Yellow

2002-05 Mountaineer 4.0L V6 MFI VIN K 60 Pin Connector

PCM Pin #	Wire Color	Circuit Description (60 Pin)	Value at Hot Idle
1	---	Not Used	---
2	TN	Injector 1 Control	3.3-3.8 ms
3-6	---	Not Used	---
7	RD/WT	HO2S-11 (B1 S1) Heater	1v, Off: 12v
8	YL/LB	HO2S-21 (B2 S1) Heater	1v, Off: 12v
9	WT/LB	IAC Solenoid Control	10.7v (33%)
10	---	Not Used	---
11	TN/BK	Injector 5 Control	3.3-3.8 ms
12	---	Not Used	---
13	PK/WT	Coil 1 Driver (dwell)	5°, 55 mph: 7°
14	WT	Injector 2 Control	3.3-3.8 ms
15	---	Not Used	---
16	BR/PK	VR Solenoid Control	0%, 55 mph: 45%
17	GY/RD	Sensor Reference Ground	<0.050v
18-19	---	Not Used	---
20	BR/WT	Sensor Voltage Reference	4.9-5.1v
21	LG/OR	Injector 6 Control	3.3-3.8 ms
22	---	Not Used	---
23	RD/LB	Coil 2 Driver (dwell)	5°, 55 mph: 7°
24	BR/YL	Injector 3 Control	3.3-3.8 ms
25-30	---	Not Used	---
31	DB/LG	Coil 3 Driver (dwell)	5°, 55 mph: 7°
32	BR/LB	Injector 4 Control	3.3-3.8 ms
33-40	---	Not Used	---
41	BR/LG	DPFE Sensor Signal	0.95-1.05v
42	YL	Knock Sensor 1 Signal	0v
43	DG/WT	Knock Sensor 2 Signal	0v
44	RD/BK	HO2S-21 (B2 S1) Signal	0.1-1.1v
45	GY/LB	HO2S-11 (B1 S1) Signal	0.1-1.1v
46	LG/RD	ECT Sensor Signal	0.5-0.6v
47-50	---	Not Used	---
51	YL/RD	Knock Sensor 1 Ground	0v
52	---	Not Used	---
53	DG	CMP Sensor Signal	6 Hz
54	---	Not Used	---
55	DB	CKP Sensor (-) Signal	400 Hz
56	GY	CKP Sensor (+) Signal	400 Hz
57	GY/WT	TP Sensor Signal	0.53-1.27v
58-60	---	Not Used	---

2002-05 Mountaineer 4.6L V8 VIN W 58 Pin Connector

PCM Pin #	Wire Color	Circuit Description (58 Pin)	Value at Hot Idle
1, 5	---	Not Used	---
2	PK/BK	Fuel Pump Power	Pump On: 12v, Off: 0v
3	TN/OR	SCP Data Bus (+)	Digital Signals
4	PK/LB	SCP Data Bus (-)	Digital Signals
6	VT/WT	EVAP Canister Vent Solenoid Control	0-10 Hz (0-100%)
7	GY/BK	Vehicle Speed Sensor (+) Signal	0 Hz, 55 mph: 125 Hz
8	LG/RD	Generator Monitor Signal	130 Hz (45%)
9	PK/YL	A/C Clutch Relay	12v (relay on: 1v)
11	WT/LG	TX Signal	Digital Signals
12	LG/BK	EVAP Purge Solenoid	0-10 Hz (0-100%)
13	VT	Module Programming Signal	0.1v
14	GY/OR	RX Signal	Digital Signals
15-16, 21	---	Not Used	---
17	GY/RD	Sensor Ground	<0.050v
18-19	---	Not Used	---
20	BR/WT	Reference Voltage	4.9-5.1v
22	DB/LG	Passive Antitheft Indicator Control	Indicator Off: 12v, On: 1v
23	LB/BK	4WD Indicator Low Signal	12v (switch on: 1v)
24-27	BK/WT	Power Ground	<0.1v
28	BK/YL	Brake Pressure Switch	12v (Brake On: 0v)
29	OR/LB	Cruise Set Indicator Control	N/A
30	BK/YL	A/C Clutch Relay (switched)	12v (relay on: 1v)
31	LB/RD	MAF Sensor Signal	0.7v, 55 mph: 1.8v
32	VT	Vehicle Power (Start-Run)	12-14v
33	VT	Vehicle Power (Start-Run)	12-14v
34-35	---	Not Used	---
36	OR/LB	Cruise Set Indicator Control	12v (switch set: 0v)
37	---	Not Used	---
38	TN/LB	MAF Sensor Return	<0.050v
39	OR	Starter Motor Relay Signal	0v
40	RD/LG	Brake Position Switch	Brake Off: 12v, On: 1v
41	TN/WT	Overdrive Cancel Switch	0v (switch on: 12v)
42	---	Not Used	---
43	BK	Power Ground	<0.1v
44	RD/WT	Keep Alive Power (CJB fuse)	12-14v
45	BK	Speed Control Switch Ground	<0.050v
46	OR/LB	Cruise Switch Indicator	12v
47	RD/YL	A/C Pressure Switch Signal	0v (switch on: 12v)
48-49	---	Not Used	---
50	VT	A/C Demand Switch	0v (A/C On: 12v)
51	GY	IAT Sensor Signal	1.5-2.5v
52	RD/PK	FTP Sensor Signal	2.6v (0" H20 - cap off)
53-55	---	Not Used	---
56	DG/OR	Speed Control Switch Ground	<0.050v
57	LB/BK	Speed Control Switch Input	0v (switch on: 6.7v)
58	LB/OR	Fuel Pump Relay Control	On: 1v, Off: 12v

2002-05 Mountaineer 4.6L V8 VIN W 32 Pin Connector

PCM Pin #	Wire Color	Circuit Description (32 Pin)	Value at Hot Idle
1	OR/YL	Shift Solenoid 'A' Control	12v, 55 mph: 1v
2	VT/OR	Shift Solenoid 'B' Control	12v, 55 mph: 1v
3	PK/BK	Shift Solenoid 'C' Control	12v, 55 mph: 1v
4	BR/OR	Shift Solenoid 'D' Control	12v, 55 mph: 1v
5	VT/YL	TCC Solenoid Control	12v, 55 mph: 1v
6	---	Not Used	---
7	WT/YL	Electronic Pressure Control	On: 1v, Off: 12v
8	---	Not Used	---
9	LB/YL	Digital TR Sensor (TR3A)	0v, 55 mph: 1.7v
10	WT/BK	Digital TR Sensor (TR4)	0v, 55 mph: 11v
11	---	Not Used	---
12	WT	Motor Position #4	On: 1v, Off: 12v
13	LB/PK	Pressure Control 'B' Solenoid	On: 1v, Off: 12v
14	---	Not Used	---
15	WT/BK	HO2S-12 (B1 S2) Heater	1v, Off: 12v
16	TN/YL	HO2S-22 (B2 S2) Heater	1v, Off: 12v
17	GY/RD	Sensor Reference Ground	<0.050v
18	LB/BK	Digital TR Sensor (TR2)	0v, 55 mph: 11v
19-20	---	Not Used	---
21	YL/LG	Overdrive Drum Speed Input	200 Hz, 55 mph: 1185 Hz
22	YL/BK	Digital TR Sensor (TR1)	0v, 55 mph: 11v
23	OR/BK	Transmission Fluid Temperature Sensor	2.10-2.40v
24-25	---	Not Used	---
26	DB/YL	Output Shaft Speed Sensor	0 Hz, 985 Hz
27	DG/WT	Turbine Shaft Speed Sensor	360 Hz, 890 Hz
28	RD/LG	HO2S-12 (B1 S2) Signal	0.1-1.1v
29	VT/LG	HO2S-22 (B2 S2) Signal	0.1-1.1v
30-32	---	Not Used	---

Pin Connector Graphic

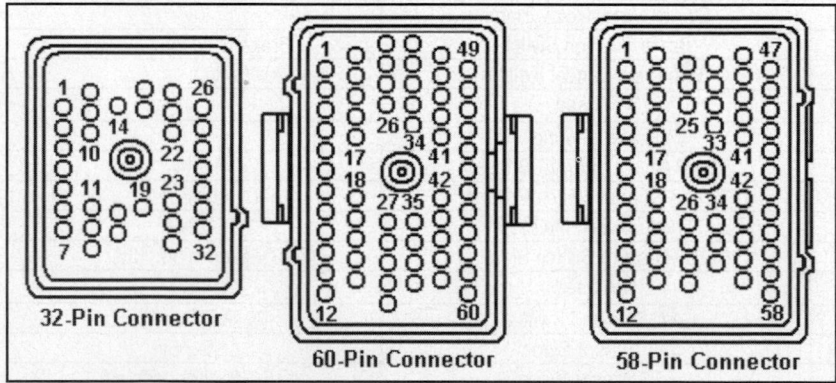

32-Pin Connector   60-Pin Connector   58-Pin Connector

Standard Colors and Abbreviations

Abbreviation	Color	Abbreviation	Color	Abbreviation	Color
BK	Black	GY	Gray	PK	Purple
BL	Blue	GN	Green	RD	Red
BR	Brown	LG	LT Green	TN	Tan
DB	Dark Blue	OR	Orange	WT	White
DG	DK Green	PK	Pink	YL	Yellow

2002-05 Mountaineer 4.6L V8 VIN W 60 Pin Connector

PCM Pin #	Wire Color	Circuit Description (60 Pin)	Value at Hot Idle
1	PK/WT	COP 2 Driver (dwell)	5°, 55 mph: 7°
2	TN	Injector 1 Control	3.3-3.8 ms
3-6	---	Not Used	---
7	RD/WT	HO2S-11 (B1 S1) Heater	1v, Off: 12v
8	YL/LB	HO2S-21 (B2 S1) Heater	1v, Off: 12v
9	WT/LB	IAC Solenoid Control	10.7v (33%)
10, 15	---	Not Used	---
11	TN/BK	Injector 5 Control	3.3-3.8 ms
12	OR/YL	COP 6 Driver Control	5°, 55 mph: 7°
13	PK/LB	COP 7 Driver Control	5°, 55 mph: 7°
14	WT	Injector 2 Control	3.3-3.8 ms
16	BR/PK	VR Solenoid Control	0%, 55 mph: 45%
17	GY/RD	Sensor Reference Ground	<0.050v
18-19	---	Not Used	---
20	BR/WT	Sensor Voltage Reference	4.9-5.1v
21	LG/OR	Injector 6 Control	3.3-3.8 ms
22	LG/YL	COP 5 Driver Control	5°, 55 mph: 7°
23	BR/YL	COP 5 Driver Control	5°, 55 mph: 7°
24	BR/YL	Injector 3 Control	3.3-3.8 ms
25-28	---	Not Used	---
29	TN/RD	Injector 7 Control	3.3-3.8 ms
30	DG/VT	COP 4 Driver Control	5°, 55 mph: 7°
31	LG/WT	COP 1 Driver (dwell)	5°, 55 mph: 7°
32	BR/LB	Injector 4 Control	3.3-3.8 ms
33-36	---	Not Used	---
37	LB	Injector 8 Control	3.3-3.8 ms
38	WT/RD	COP 8 Driver Control	5°, 55 mph: 7°
39	---	Not Used	---
40	YL/LB	CHT Sensor Signal	0.7v (194°F)
41	BR/LG	DPFE Sensor Signal	0.95-1.05v
42	YL	Knock Sensor 1 Signal	0v
43	DG/WT	Knock Sensor 2 Signal	0v
44	RD/BK	HO2S-21 (B2 S1) Signal	0.1-1.1v
45	GY/LB	HO2S-11 (B1 S1) Signal	0.1-1.1v
46-50	---	Not Used	---
51	YL/RD	Knock Sensor 1 Ground	0v
52	DG/VT	Knock Sensor 2 Ground	0v
53	DG	CMP Sensor Signal	6 Hz
54	---	Not Used	---
55	DB	CKP Sensor (+) Signal	400 Hz
56	GY	CKP Sensor (-) Signal	400 Hz
57	GY/WT	TP Sensor Signal	0.53-1.27v
58-60	---	Not Used	---

Pin Connector Graphic

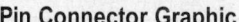
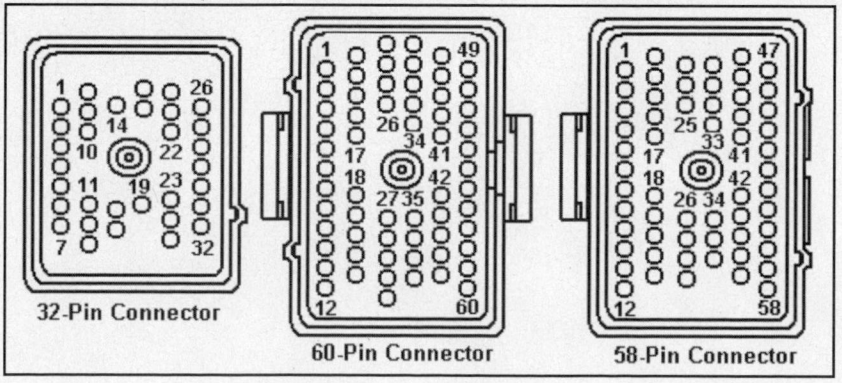

32-Pin Connector

60-Pin Connector

58-Pin Connector

1996-97 Mountaineer 5.0L V8 VIN P 104 Pin Connector

PCM Pin #	Wire Color	Circuit Description (104 Pin)	Value at Hot Idle
1	PK/OR	Shift Solenoid 2 Control	12v, 55 mph: 12v
2	PK/LG	MIL (lamp) Control	MIL On: 1v, Off: 12v
3-12, 14, 17	---	Not Used	---
13	VT	Flash EPROM Power	0.1v
15	PK/LB	Data Bus (-) Signal	Digital Signals
16	TN/OR	Data Bus (+) Signal	Digital Signals
18, 20, 23	---	Not Used	---
19	OR/BK	Accelerator Ride Control	0.1v (Off)
21	BK/PK	CKP (+) Sensor Signal	430-460 Hz
22	GY/YL	CKP (-) Sensor Signal	430-460 Hz
24	BK/WT	Power Ground	<0.1v
25	BK	Case Ground	<0.050v
26	DB/LG	Coil 1 Driver (dwell)	5º, 55 mph: 8º
27	OR/YL	Shift Solenoid 1 Control	1v, 55 mph: 12v
28	---	Not Used	---
29	TN/WT	TCS (switch) Signal	TCS & O/D On: 12v
30	DG	Octane Adjustment	9.3v (Closed: 0v)
31-32, 34	---	Not Used	---
33	PK/OR	VSS (-) Signal	0 Hz, 55 mph: 125 Hz
35	RD/LG	HO2S-12 (B1 S2) Signal	0.1-1.1v
36	TN/LB	MAF Sensor Return	<0.050v
37	OR/BK	TFT Sensor Signal	2.10-2.40v
38	LG/RD	ECT Sensor Signal	0.5-0.6v
39	GY	IAT Sensor Signal	1.5-2.5v
40	DG/YL	Fuel Pump Monitor	On: 12v, Off: 0v
41	DG/OR	A/C Switch Signal	A/C On: 12v, Off: 0v
43	LB/PK	Data Output Link	5v
42, 44-46	---	Not Used	---
47	BK/PK	VR Solenoid Control	0%, 55 mph: 45%
48	TN/YL	Clean Tachometer Output	39-45 Hz
49-50, 53	---	Not Used	---
51	BK/WT	Power Ground	<0.1v
52	RD/LB	Coil 2 Driver (dwell)	5º, 55 mph: 8º
54	DB/WT	TCC Solenoid Control	0%, 55 mph: 95%
55	Y	Keep Alive Power	12-14v
56	LG/BK	EVAP Purge Solenoid	0-10 Hz (0-100%)
57	---	Not Used	---
58	GY/BK	VSS (+) Signal	0 Hz, 55 mph: 125 Hz
59	---	Not Used	---
60	GY/LB	HO2S-11 (B1 S1) Signal	0.1-1.1v
61	PK/LG	HO2S-22 (B2 S2) Signal	0.1-1.1v
62	RD/PK	FTP Sensor Signal	2.6v (0" H2O - cap off)
63	OR/YL	EVP Sensor Signal	0.20-1.30v
64	LB/YL	A/T: TR Sensor Signal	In 'P': 0v, 55 mph: 1.7v

1996-97 Mountaineer 5.0L V8 VIN P 104 Pin Connector

PCM Pin #	Wire Color	Circuit Description (104 Pin)	Value at Hot Idle
65-66	---	Not Used	---
67	PK/WT	EVAP CV Solenoid	0-10 Hz (0-100%)
68	---	Not Used	---
69	PK/YL	A/C WOT Relay Control	On: 1v, Off: 12v
70	---	Not Used	---
71	RD	Vehicle Power	12-14v
72	TN/RD	Injector 7 Control	3.2-4.5 ms
73	TN/BK	Injector 5 Control	3.2-4.5 ms
74	BR/YL	Injector 3 Control	3.2-4.5 ms
75	TN	Injector 1 Control	3.2-4.5 ms
76	BK/WT	Power Ground	<0.1v
77	BK/WT	Power Ground	<0.1v
78	PK/WT	Coil 4 Driver (dwell)	5°, 55 mph: 8°
79	WT/LG	TCIL (lamp) Control	TCIL On: 1v, Off: 12v
80	LB/OR	Fuel Pump Control	On: 1v, Off: 12v
81	WT/YL	EPC Solenoid Control	10v (26 psi)
83	WT/LB	IAC Solenoid Control	10.8v (33%)
84	DB/YL	OSS Sensor Signal	0 Hz, 55 mph: 230 Hz
85	DB/OR	CMP Sensor Signal	5-10 Hz
86	---	Not Used	---
87	RD/BK	HO2S-21 (B2 S1) Signal	0.1-1.1v
88	LB/RD	MAF Sensor Signal	0.8v, 55 mph: 2.1v
89	GY/WT	TP Sensor Signal	0.53-1.27v
90	BR/WT	Reference Voltage	4.9-5.1v
91	GY/RD	Analog Signal Return	<0.050v
92	LG	Brake Position Switch	Brake Off: 12v, On: 1v
93	RD/WT	HO2S-11 (B1 S1) Heater	On: 1v, Off: 12v
94	YL/LB	HO2S-21 (B2 S1) Heater	On: 1v, Off: 12v
95	WT/BK	HO2S-12 (B1 S2) Heater	On: 1v, Off: 12v
96	TN/YL	HO2S-22 (B2 S2) Heater	On: 1v, Off: 12v
97	RD	Vehicle Power	12-14v
98	LB	Injector 8 Control	3.2-4.5 ms
99	LG/OR	Injector 6 Control	3.2-4.5 ms
100	BR/LB	Injector 4 Control	3.2-4.5 ms
101	WT	Injector 2 Control	3.2-4.5 ms
102	---	Not Used	---
103	BK/WT	Power Ground	<0.1v
104	RD/YL	Coil 4 Driver (dwell)	5°, 55 mph: 8°

Pin Connector Graphic

```
        PCM 104-PIN CONNECTOR
  1 ◎◎◎◎◎◎◎◎◎◎◎◎◎   ◎◎◎◎◎◎◎◎◎◎◎◎◎ 26
 27 ◎◎◎◎◎◎◎◎◎◎◎◎◎    ●    ◎◎◎◎◎◎◎◎◎◎◎◎◎ 52
 53 ◎◎◎◎◎◎◎◎◎◎◎◎◎         ◎◎◎◎◎◎◎◎◎◎◎◎◎ 78
 79 ◎◎◎◎◎◎◎◎◎◎◎◎◎   ◎◎◎◎◎◎◎◎◎◎◎◎◎ 104
```

Terminal View of 104-Pin PCM Wiring Harness Connector

Standard Colors and Abbreviations

Abbreviation	Color	Abbreviation	Color	Abbreviation	Color
BK	Black	GY	Gray	PK	Purple
BL	Blue	GN	Green	RD	Red
BR	Brown	LG	LT Green	TN	Tan
DB	Dark Blue	OR	Orange	WT	White
DG	DK Green	PK	Pink	YL	Yellow

1998-2001 Mountaineer 5.0L V8 VIN P 104 Pin Connector

PCM Pin #	Wire Color	Circuit Description (104 Pin)	Value at Hot Idle
1	PK/OR	Shift Solenoid 'B' Control	12v, 55 mph: 1v
2	PK/LG	MIL (lamp) Control	MIL On: 1v, Off: 12v
3	YL/BK	Digital TR1 Sensor	In 'P': 0v, Others: 11.5v
4-11	---	Not Used	---
12	YL/WT	Fuel Level Indicator Signal	1.7v (1/2 full)
13	VT	Flash EPROM Power	0.1v
14	---	Not Used	---
15	PK/LB	Data Bus (-) Signal	Digital Signals
16	TN/OR	Data Bus (+) Signal	Digital Signals
17-18	---	Not Used	---
19	OR/BK	Air Suspension Control Signal	Digital Signals
20	---	Not Used	---
21	BK/PK	CKP (+) Sensor Signal	460-500 Hz
22	GY/YL	CKP (-) Sensor Signal	460-500 Hz
24	BK/WT	Power Ground	<0.1v
25	BK	Case Ground	<0.050v
26	DB/LG	Coil 1 Driver (dwell)	5°, 55 mph: 8°
27	OR/YL	Shift Solenoid 'A' Control	1v, 55 mph: 1v
28	---	Not Used	---
29	TN/WT	TCS or Manual Switch	TCS or M/D pressed: 12v
30 ('98-'99)	DG	Octane Adjustment	9.3v (Closed: 0v)
31-32	---	Not Used	---
33 ('98-'99)	PK/OR	VSS (-) Signal	0 Hz, 55 mph: 125 Hz
35	RD/LG	HO2S-12 (B1 S2) Signal	0.1-1.1v
36	TN/LB	MAF Sensor Return	<0.050v
37	OR/BK	TFT Sensor Signal	2.10-2.40v
38	LG/RD	ECT Sensor Signal	0.5-0.6v
39	GY	IAT Sensor Signal	1.5-2.5v
40	DG/YL	Fuel Pump Monitor	On: 12v, Off: 0v
41	VT	A/C Switch Signal	Switch On: 12v (open)
42, 45-46	---	Not Used	---
43 ('98-'99)	LB/PK	Neutral Tow Connector Signal	N/A
47	BR/PK	VR Solenoid Control	0%, 55 mph: 45%
48	TN/YL	Clean Tachometer Output	40-55 Hz
49	LB/BK	Digital TR2 Sensor	In 'P': 0v, Others: 11.5v
50	WT/BK	Digital TR4 Sensor	In 'P': 0v, Others: 11.5v
51	BK/WT	Power Ground	<0.1v
52	RD/LB	Coil 2 Driver (Control)	5°, 55 mph: 8°
53, 57	---	Not Used	---
54	DB/WT	TCC Solenoid Control	0%, 55 mph: 95%
55	YL	Keep Alive Power	12-14v
56	LG/BK	EVAP Purge Solenoid	0-10 Hz (0-100%)
58	GY/BK	VSS (+) Signal	0 Hz, 55 mph: 125 Hz
59	---	Not Used	---
60	GY/LB	HO2S-11 (B1 S1) Signal	0.1-1.1v
61	PK/LG	HO2S-22 (B2 S2) Signal	0.1-1.1v
62	RD/PK	FTP Sensor Signal	2.6v (0" H2O - cap off)
63	---	Not Used	---
64	LB/YL	Digital TR3 Sensor	In 'P': 0v, 55 mph: 1.7v

1998-2001 Mountaineer 5.0L V8 VIN P 104 Pin Connector

PCM Pin #	Wire Color	Circuit Description (104 Pin)	Value at Hot Idle
65 ('98-'00)	BR/LG	DPFE Sensor Signal	0.20-1.30v
65 ('01-'00)	BR/LG	DPFE Sensor Signal	0.95-1.05v
66, 68, 70	---	Not Used	---
67	PK/WT	EVAP CV Solenoid	0-10 Hz (0-100%)
69	PK/YL	A/C WOT Relay Control	On: 1v, Off: 12v
71	RD	Vehicle Power	12-14v
72	TN/RD	Injector 7 Control	3.2-4.5 ms
73	TN/BK	Injector 5 Control	3.2-4.5 ms
74	BR/YL	Injector 3 Control	3.2-4.5 ms
75	TN	Injector 1 Control	3.2-4.5 ms
76, 77	BK/WT	Power Ground	<0.1v
78	PK/WT	Coil 3 Driver (dwell)	5°, 55 mph: 8°
79	WT/LG	TCIL (lamp) Control	TCIL On: 1v, Off: 12v
80	LB/OR	Fuel Pump Control	On: 1v, Off: 12v
81	WT/YL	EPC Solenoid Control	9.5v (10 psi)
82	---	Not Used	---
83	WT/LB	IAC Solenoid Control	10.8v (12-30%)
84	DB/YL	OSS Sensor Signal	At 55 mph: 230-280 Hz
85	DB/OR	CMP Sensor Signal	5-10 Hz
86	BK/YL	A/C Pressure Switch Signal	Open: 12v (<24 psi)
87	RD/BK	HO2S-21 (B2 S1) Signal	0.1-1.1v
88	LB/RD	MAF Sensor Signal	0.8v, 55 mph: 2.1v
89	GY/WT	TP Sensor Signal	0.53-1.27v
90	BR/WT	Reference Voltage	4.9-5.1v
91	GY/RD	Analog Signal Return	<0.050v
92	RD/LG	Brake Position Switch	Brake Off: 12v, On: 1v
93	RD/WT	HO2S-11 (B1 S1) Heater	On: 1v, Off: 12v
94	YL/LB	HO2S-21 (B2 S1) Heater	On: 1v, Off: 12v
95	WT/BK	HO2S-12 (B1 S2) Heater	On: 1v, Off: 12v
96	TN/YL	HO2S-22 (B2 S2) Heater	On: 1v, Off: 12v
97	RD	Vehicle Power	12-14v
98	LB	Injector 8 Control	3.2-4.5 ms
99	LG/OR	Injector 6 Control	3.2-4.5 ms
100	BR/LB	Injector 4 Control	3.2-4.5 ms
101	WT	Injector 2 Control	3.2-4.5 ms
102	---	Not Used	---
103	BK/WT	Power Ground	<0.1v
104	RD/YL	Coil 4 Driver (dwell)	5°, 55 mph: 8°

Pin Connector Graphic

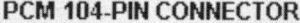
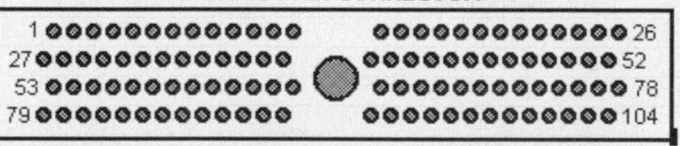

PCM 104-PIN CONNECTOR

Terminal View of 104-Pin PCM Wiring Harness Connector

Standard Colors and Abbreviations

Abbreviation	Color	Abbreviation	Color	Abbreviation	Color
BK	Black	GY	Gray	PK	Purple
BL	Blue	GN	Green	RD	Red
BR	Brown	LG	LT Green	TN	Tan
DB	Dark Blue	OR	Orange	WT	White
DG	DK Green	PK	Pink	YL	Yellow

NAVIGATOR PIN TABLES

1998-99 Navigator 5.4L V8 MFI VIN L (4R100) 104 Pin Connector

PCM Pin #	Wire Color	Circuit Description (104 Pin)	Value at Hot Idle
1	LG/YL	COP 5 Driver (dwell)	5°, 55 mph: 8°
2	PK/LG	MIL (lamp) Control	MIL On: 1v, Off: 12v
3	BK/WT	Power Ground	<0.1v
6	OR/YL	Shift Solenoid 1 Control	1v, 55 mph: 12v
9	YL/WT	Fuel Level Indicator Signal	1.7v (1/2 full)
11	PK/OR	Shift Solenoid 2 Control	12v, 55 mph: 12v
12	WT/LG	TCIL (lamp) Control	TCIL On: 1v, Off: 12v
13	VT	Flash EEPROM Power	0.1v
14	LB/BK	4x4 Indicator Control	4x4 lamp On: 1v, Off: 0v
15	PK	Data Bus (-) Signal	Digital Signals
16	TN/OR	Data Bus (+) Signal	Digital Signals
19	DB/WT	Air Suspension Control	Digital Signals
20	BR/OR	Coast Clutch Solenoid	12v, 55 mph: 12v
21	DB	CKP (+) Sensor Signal	400-420 Hz
22	GY	CKP (-) Sensor Signal	400-420 Hz
24	BK/WT	Power Ground	<0.1v
25	LG/YL	Case Ground	<0.050v
26	LG/WT	COP 1 Driver (dwell)	5°, 55 mph: 8°
27	OR/YL	COP 6 Driver (dwell)	5°, 55 mph: 8°
29	TN/WT	TCS (switch) Signal	TCS & O/D On: 12v
30	DG	Octane Adjust Shorting Bar	9.3v (Closed: 0v)
33	PK/OR	VSS (-) Signal	0 Hz, 55 mph: 125 Hz
34	YL/BK	Digital TR1 Sensor	0v, 55 mph: 11.5v
35	RD/LG	HO2S-12 (B1 S2) Signal	0.1-1.1v
36	TN/LB	MAF Sensor Return	<0.050v
37	OR/BK	TFT Sensor Signal	2.10-2.40v
39	GY	IAT Sensor Signal	1.5-2.5v
40	DG/YL	Fuel Pump Monitor	On: 12v, Off: 0v
41	BK/YL	A/C High Pressure Switch	Switch On: 12v (open)
43	OR/LG	Overhead Console Module	Digital Signals
45	RD/WT	CHIL (lamp) Control	CHT On: 1v, Off: 12v
46	BR	IMTV (tuning valve) Control	12v
47	BR/PK	VR Solenoid Control	0%, 55 mph: 45%
48	WT/PK	Clean Tachometer Output	46 Hz, 55 mph: 115 Hz
49	LB/BK	Digital TR2 Sensor	0v, 55 mph: 11.5v
50	WT/BK	Digital TR4 Sensor	0v, 55 mph: 11.5v
51	BK/WT	Power Ground	<0.1v
52	PK/WT	COP Driver 2 Driver	5°, 55 mph: 8°
53	PK/B	COP 7 Driver (dwell)	5°, 55 mph: 8°
54	PK/YL	TCC Solenoid Control	0%, 55 mph: 95%
55	BK/LG	Keep Alive Power	12-14v
56	LG/BK	EVAP Purge Solenoid	0-10 Hz (0-100%)
57	YL/RD	Knock Sensor 1 Signal	0v
58	GY/BK	VSS (+) Signal	0 Hz, 55 mph: 125 Hz
60	GY/LB	HO2S-11 (B1 S1) Signal	0.1-1.1v
61	PK/LG	HO2S-22 (B2 S2) Signal	0.1-1.1v
62	RD/PK	FTP Sensor Signal	2.6v (0" H2O - cap off)
64	LB/YL	Digital TR3A Sensor	In 'P': 0v, in O/D: 1.7v

1998-99 Navigator 5.4L V8 MFI VIN L (4R100) 104 Pin Connector

PCM Pin #	Wire Color	Circuit Description (104 Pin)	Value at Hot Idle
65	BR/LG	DPFE Sensor Signal	0.20-1.30v
66	YL/LG	CHT Sensor Signal	0.6 to 1.7v (194ºF)
67	PK/WT	EVAP CV Solenoid	0-10 Hz (0-100%)
71	RD	Vehicle Power	12-14v
72	TN/RD	Injector 7 Control	3.2-3.8 ms
73	TN/BK	Injector 5 Control	3.2-3.8 ms
74	BR/YL	Injector 3 Control	3.2-3.8 ms
75	TN	Injector 1 Control	3.2-3.8 ms
77	BK/WT	Power Ground	<0.1v
78	WT/PK	COP 3 Driver (dwell)	5º, 55 mph: 8º
79	WT/RD	COP 8 Driver (dwell)	5º, 55 mph: 8º
80	LB/OR	Fuel Pump Control	On: 1v, Off: 12v
81	WT/YL	EPC Solenoid Control	9.1v (5 psi)
83	WT/LB	IAC Solenoid Control	10.7v (33%)
85	DG	CMP Sensor Signal	6 Hz, 55 mph: 15-17 Hz
87	RD/BK	HO2S-21 (B2 S1) Signal	0.1-1.1v
88	LB/RD	MAF Sensor Signal	0.8v, 55 mph: 2.3v
89	GY/WT	TP Sensor Signal	0.53-1.27v
90	BR/WT	Reference Voltage	4.9-5.1v
91	GY/RD	Analog Signal Return	<0.050v
92	LG	Brake Pedal Switch Signal	Brake Off: 12v, On: 1v
93	RD/WT	HO2S-11 (B1 S1) Heater	On: 1v, Off: 12v
94	YL/LB	HO2S-21 (B2 S1) Heater	On: 1v, Off: 12v
95	WT/BK	HO2S-12 (B1 S2) Heater	On: 1v, Off: 12v
96	TN/YL	HO2S-22 (B2 S2) Heater	On: 1v, Off: 12v
97	RD	Vehicle Power	12-14v
98	LB	Injector 8 Control	3.2-3.8 ms
99	LG/OR	Injector 6 Control	3.2-3.8 ms
100	BR/LB	Injector 4 Control	3.2-3.8 ms
101	WT	Injector 2 Control	3.2-3.8 ms
103	BK/WT	Power Ground	<0.1v
104	DG/PK	COP 4 Driver (dwell)	5º, 55 mph: 8º

Pin Connector Graphic

PCM 104-PIN CONNECTOR

Terminal View of 104-Pin PCM Wiring Harness Connector

Standard Colors and Abbreviations

Abbreviation	Color	Abbreviation	Color	Abbreviation	Color
BK	Black	GY	Gray	PK	Purple
BL	Blue	GN	Green	RD	Red
BR	Brown	LG	LT Green	TN	Tan
DB	Dark Blue	OR	Orange	WT	White
DG	DK Green	PK	Pink	YL	Yellow

1999-2001 Navigator 5.4L V8 VIN A (A/T) 104 Pin Connector

PCM Pin #	Wire Color	Circuit Description (104 Pin)	Value at Hot Idle
1	OR/YL	COP 6 Driver (dwell)	5°, 55 mph: 8°
3	BK/WT	Power Ground	<0.1v
6	OR/YL	Shift Solenoid 1 Control	1v, 55 mph: 12v
11	VT/O	Shift Solenoid 2 Control	12v, 55 mph: 12v
12	WT/LG	TCIL (lamp) Control	TCIL On: 1v, Off: 12v
13	VT	Flash EEPROM Power	0.1v
14	LB/BK	4x4 Low Indicator Switch	Switch On: 0v, Off: 12v
15	PK/LB	Data Bus (-) Signal	Digital Signals
16	TN/OR	Data Bus (+) Signal	Digital Signals
19	DB/WT	Air Suspension Control	Digital Signals
20	BR/OR	Coast Clutch Solenoid	12v, 55 mph: 12v
21	DB	CKP (+) Sensor Signal	400-420 Hz
22	GY	CKP (-) Sensor Signal	400-420 Hz
24	BK/WT	Power Ground	<0.1v
25	LB/YL	Case Ground	<0.050v
26	LG/WT	COP 1 Driver (dwell)	5°, 55 mph: 8°
27	OR/YL	COP 5 Driver (dwell)	5°, 55 mph: 8°
29	TN/WT	TCS (switch) Signal	TCS & O/D On: 12v
32	DG/VT	Knock Sensor 1 Signal	0v
34	YL/BK	Digital TR1 Sensor	0v, 55 mph: 11.5v
35	RD/LG	HO2S-12 (B1 S2) Signal	0.1-1.1v
36	TN/LB	MAF Sensor Return	<0.050v
37	OR/BK	TFT Sensor Signal	2.10-2.40v
38	LG/RD	ECT Sensor Signal	0.5-0.6v
39	GY	IAT Sensor Signal	1.5-2.5v
40	DG/YL	Fuel Pump Monitor	On: 12v, Off: 0v
41	BK/YL	A/C High Pressure Switch	Switch On: 12v (open)
42	PK/OR	TP Sensor Signal	0.53-1.27v
43	OR/LG	Overhead Console Module	Digital Signals
43 ('00-'02)	OR/LG	Data Output Link	Digital Signals
46	BR	IMRC (valve) Control	12v
47	BR/PK	VR Solenoid Control	0%, 55 mph: 45%
49	LB/BK	Digital TR2 Sensor	0v, 55 mph: 11.5v
50	WT/BK	Digital TR4 Sensor	0v, 55 mph: 11.5v
51	BK/WT	Power Ground	<0.1v
52	WT/PK	COP 3 Driver (dwell)	5°, 55 mph: 8°
53	DG/VT	COP 4 Driver (dwell)	5°, 55 mph: 8°
54	VTR/YL	TCC Solenoid Control	0%, 55 mph: 95%
55	RD/WTH	Keep Alive Power	12-14v
56	LG/BK	EVAP Purge Solenoid	0-10 Hz (0-100%)
57	YL/RD	Knock Sensor 1 Signal	0v
59	DG/WT	TSS Sensor Signal	370 Hz (700 rpm)
60	GY/LB	HO2S-11 (B1 S1) Signal	0.1-1.1v
61	VTR/LG	HO2S-22 (B2 S2) Signal	0.1-1.1v
62	RD/PK	FTP Sensor Signal	2.6v (0" H2O - cap off)
64	LB/YL	Digital TR3A Sensor	In 'P': 0v, in O/D: 1.7v

1999-2001 Navigator 5.4L V8 VIN A (A/T) 104 Pin Connector

PCM Pin #	Wire Color	Circuit Description (104 Pin)	Value at Hot Idle
65 ('99-'00)	BR/LG	DPFE Sensor Signal	0.20-1.30v
65 ('01-'02)	BR/LG	DPFE Sensor Signal	0.95-1.05v
66	YL/LG	CHT Sensor Signal	0.6 to 1.7v (194°F)
67	VT/WT	EVAP CV Solenoid	0-10 Hz (0-100%)
68	GY/BK	Vehicle Speed Out Signal	0 Hz, 55 mph: 125 Hz
69	PK/YL	A/C WOT Relay Control	On: 1v, Off: 12v
71	RD	Vehicle Power	12-14v
72	TN/RD	Injector 7 Control	3.2-3.8 ms
73	TN/BK	Injector 5 Control	3.2-3.8 ms
74	BR/YL	Injector 3 Control	3.2-3.8 ms
75	TN	Injector 1 Control	3.2-3.8 ms
77	BK/WT	Power Ground	<0.1v
78	PK/LB	COP 7 Driver (dwell)	5°, 55 mph: 8°
79	WT/RD	COP 8 Driver (dwell)	5°, 55 mph: 8°
80	LB/OR	Fuel Pump Relay Control	On: 1v, Off: 12v
81	WT/YL	EPC Solenoid Control	9.1v (5 psi)
83	WT/LB	IAC Solenoid Control	10.7v (33%)
84	DB/YL	OSS Sensor Signal	0 Hz, 55 mph: 125 Hz
85	DG	CMP Sensor Signal	6 Hz, 55 mph: 14-17 Hz
87	RD/BK	HO2S-21 (B2 S1) Signal	0.1-1.1v
88	LB/RD	MAF Sensor Signal	0.8v, 55 mph: 2.3v
89	GY/WT	TP Sensor Signal	0.53-1.27v
90	BR/WT	Reference Voltage	4.9-5.1v
91	GY/RD	Analog Signal Return	<0.050v
92	RD/LG	Brake Pedal Switch Signal	Brake Off: 12v, On: 1v
93	RD/WT	HO2S-11 (B1 S1) Heater	On: 1v, Off: 12v
94	YL/LB	HO2S-21 (B2 S1) Heater	On: 1v, Off: 12v
95	WT/BK	HO2S-12 (B1 S2) Heater	On: 1v, Off: 12v
96	TN/YL	HO2S-22 (B2 S2) Heater	On: 1v, Off: 12v
97	RD	Vehicle Power	12-14v
98	LB	Injector 8 Control	3.2-3.8 ms
99	LG/OR	Injector 6 Control	3.2-3.8 ms
100	BR/LB	Injector 4 Control	3.2-3.8 ms
101	W	Injector 2 Control	3.2-3.8 ms
103	BK/WT	Power Ground	<0.1v
104	PK/WT	COP 2 Driver (dwell)	5°, 55 mph: 8°

Pin Connector Graphic

```
      PCM 104-PIN CONNECTOR
 1 ●●●●●●●●●●●●●    ●●●●●●●●●●●●● 26
27 ●●●●●●●●●●●●●    ●●●●●●●●●●●●● 52
53 ●●●●●●●●●●●●●  ●  ●●●●●●●●●●●●● 78
79 ●●●●●●●●●●●●●    ●●●●●●●●●●●●● 104
```

Terminal View of 104-Pin PCM Wiring Harness Connector

Standard Colors and Abbreviations

Abbreviation	Color	Abbreviation	Color	Abbreviation	Color
BK	Black	GY	Gray	PK	Purple
BL	Blue	GN	Green	RD	Red
BR	Brown	LG	LT Green	TN	Tan
DB	Dark Blue	OR	Orange	WT	White
DG	DK Green	PK	Pink	YL	Yellow

2001-02 Expedition 5.4L V8 VIN R (E4OD) 104 Pin Connector

PCM Pin #	Wire Color	Circuit Description (104 Pin)	Value at Hot Idle
1	OR/YL	COP 6 Driver (dwell)	5º, 55 mph: 8º
2	---	Not Used	---
3	BK/WT	Power Ground	<0.1v
6	OR/YL	Shift Solenoid 1 Control	1v, 30 mph: 12v
7-10	---	Not Used	---
11	VT/OR	Shift Solenoid 2 Control	12v, 55 mph: 1v
12	WT/LG	TCIL (lamp) Control	TCIL On: 1v, Off: 12v
13	VT	Flash EEPROM Power	0.1v
14	LB/BK	4x4 Low Indicator Switch	4x4 Switch Off: 0.1v
15	PK/LB	Data Bus (-) Signal	Digital Signals
16	TN/OR	Data Bus (+) Signal	Digital Signals
17-18	---	Not Used	---
19	DB/WT	Suspension Control Module	Digital Signals
20	BR/OR	Coast Clutch Solenoid	12v, 55 mph: 12v
21	DB	CKP (+) Sensor Signal	430-475 Hz
22	GY	CKP (-) Sensor Signal	430-475 Hz
25	LB/YL	Case Ground	<0.050v
26	LG/WT	COP 1 Driver (dwell)	5º, 55 mph: 8º
27	LG/YL	COP 5 Driver (dwell)	5º, 55 mph: 8º
28	---	Not Used	---
29	TN/WT	TCS (switch) Signal	TCS & O/D On: 12v
30-31	---	Not Used	---
32	DG/PK	Knock Sensor 1 Return	<0.050v
33	---	Not Used	---
34	YL/BK	Digital TR1 Sensor	0v, 55 mph: 11.5v
35	RD/LG	HO2S-12 (B1 S2) Signal	0.1-1.1v
36	TN/LB	MAF Sensor Return	<0.050v
37	OR/BK	TFT Sensor Signal	2.10-2.40v
38	---	Not Used	---
39	GY	IAT Sensor Signal	1.5-2.5v
40	DG/YL	Fuel Pump Monitor	On: 12v, Off: 0v
41	BK/YL	A/C Request Signal	A/C On: 12v, Off: 0v
42	PK/OR	TP Sensor Signal	0.9v, 55 mph: 1.3v
43	OR/LG	Data Output Link	5v
44-45	---	Not Used	---
46	BR	IMRC (valve) Control	12v
47	BR/PK	VR Solenoid Control	0%, 55 mph: 45%
48	---	Not Used	---
49	LB/BK	Digital TR2 Sensor	0v, 55 mph: 11.5v
50	WT/BK	Digital TR4 Sensor	0v, 55 mph: 11.5v
51	BK/WT	Power Ground	<0.1v
52	WT/PK	COP 3 Driver (dwell)	5º, 55 mph: 8º
53	DG/VT	COP 4 Driver (dwell)	5º, 55 mph: 8º
54	VT/YL	TCC Solenoid Control	0%, 55 mph: 95%
55	RD/WT	Keep Alive Power	12-14v
56	LG/BK	EVAP Purge Solenoid	0-10 Hz (0-100%)
57	YL/RD	Knock Sensor 1 Signal	0v
58	---	Not Used	---
59	DG/WT	TSS Sensor Signal	100 Hz, 30 mph: 700 Hz
60	GY/LB	HO2S-11 (B1 S1) Signal	0.1-1.1v
61	VT/LG	HO2S-22 (B2 S2) Signal	0.1-1.1v
62	RD/PK	FTP Sensor Signal	2.6v (0" H2O - cap off)
63	---	Not Used	---
64	LB/YL	Digital TR3A Sensor	In 'P': 0v, in O/D: 1.7v

2001-02 Navigator 5.4L V8 VIN R (E4OD) 104 Pin Connector

PCM Pin #	Wire Color	Circuit Description (104 Pin)	Value at Hot Idle
65	BR/LG	DPFE Sensor Signal	0.95-1.05v
66	YL/LG	CHT Sensor Signal	0.6v (194°F)
67	VT/WT	EVAP CV Solenoid	0-10 Hz (0-100%)
68	GY/BK	Vehicle Speed Sensor	0 Hz, 55 mph: 125 Hz
69	PK/YL	A/C WOT Relay Control	Relay Off: 12v, On: 1v
70	---	Not Used	---
71	RD	Vehicle Power	12-14v
72	TN/RD	Injector 7 Control	2.7-3.5 ms
73	TN/BK	Injector 5 Control	2.7-3.5 ms
74	BR/YL	Injector 3 Control	2.7-3.5 ms
75	TN	Injector 1 Control	2.7-3.5 ms
76	BK/WT	Power Ground	<0.1v
77	---	Not Used	---
78	PK/LB	COP 7 Driver (dwell)	5°, 55 mph: 8°
79	WT/RD	COP 8 Driver (dwell)	5°, 55 mph: 8°
80	LB/OR	Fuel Pump Control	On: 1v, Off: 12v
81	WT/YL	EPC Solenoid Control	9.1v (4 psi)
82	---	Not Used	---
83	WT/LB	IAC Solenoid Control	10.7v (33%)
84	DB/YL	OSS Sensor Signal	0 Hz, 55 mph: 350 Hz
85	DG	CMP Sensor Signal	6 Hz, 55 mph: 13-17 Hz
86	---	Not Used	---
87	RD/BK	HO2S-21 (B2 S1) Signal	0.1-1.1v
88	LB/RD	MAF Sensor Signal	0.8v, 55 mph: 1.9v
89	GY/WT	TP Sensor Signal	0.53-1.27v
90	BR/WT	Reference Voltage	4.9-5.1v
91	GY/RD	Analog Signal Return	<0.050v
92	RD/LG	Brake Pedal Switch Signal	Brake Off: 12v, On: 1v
93	RD/WT	HO2S-11 (B1 S1) Heater	On: 1v, Off: 12v
94	YL/LB	HO2S-21 (B2 S1) Heater	On: 1v, Off: 12v
95	WT/BK	HO2S-12 (B1 S2) Heater	On: 1v, Off: 12v
96	TN/YL	HO2S-22 (B2 S2) Heater	On: 1v, Off: 12v
97	RD	Vehicle Power	12-14v
98	LB	Injector 8 Control	2.7-3.5 ms
99	LG/OR	Injector 6 Control	2.7-3.5 ms
100	BR/LB	Injector 4 Control	2.7-3.5 ms
101	WT	Injector 2 Control	2.7-3.5 ms
102	---	Not Used	---
103	BK/WT	Power Ground	<0.1v
104	PK/WT	COP 2 Driver (dwell)	5°, 55 mph: 8°

Pin Connector Graphic

```
        PCM 104-PIN CONNECTOR
    1 ●●●●●●●●●●●●●      ●●●●●●●●●●●●● 26
  27 ●●●●●●●●●●●●●      ●●●●●●●●●●●●● 52
  53 ●●●●●●●●●●●●●   ⬡  ●●●●●●●●●●●●● 78
  79 ●●●●●●●●●●●●●      ●●●●●●●●●●●●● 104
    Terminal View of 104-Pin PCM Wiring Harness Connector
```

Standard Colors and Abbreviations

Abbreviation	Color	Abbreviation	Color	Abbreviation	Color
BK	Black	GY	Gray	PK	Purple
BL	Blue	GN	Green	RD	Red
BR	Brown	LG	LT Green	TN	Tan
DB	Dark Blue	OR	Orange	WT	White
DG	DK Green	PK	Pink	YL	Yellow

2003-05 Navigator 5.4L V8 MFI VIN R (E4OD) C175a Pin Connector

PCM Pin #	Wire Color	Circuit Description (46 Pin)	Value at Hot Idle
1	BK/WT	Power Ground	<0.1v
2	PK/YL	A/C WOT Relay Control	Relay Off: 12v, On: 1v
3	PK/OR	TP Sensor Signal	0.9v, 55 mph: 1.3v
4	DB/OG	Starter Relay Control Circuit	Relay Off: 12v, On: 1v
5	---	Not Used	---
6	OG/LB	Cooling Fan Speed Signal	0.5-4.9v
7	TN/LG	Generator Communicator Command	1.5-10.5v (40-250 Hz)
8	RD/PK	Fuel Tank Pressure Sensor	2.6v (0" H2O - cap off)
9	OR/RD	Antitheft Indicator Control	Indicator Off: 12v, On: 1v
10	LB/YL	Power Ground	<0.1v
11	BK/WT	Power Ground	<0.1v
12	VT/YL	Speed Control Servo Signal 'C'	DC pulse signals
13	GY/BK	VSS (+) Signal	0 Hz, 55 mph: 125 Hz
14	BR/PK	Generator Monitor Signal	130 Hz (45%)
15	BK/YL	Brake Pedal Position Switch	Brake Off: 0v, On: 12v
16	---	Not Used	---
17	DG/OR	Steering Wheel Control Ground	<0.050v
18	YL/LG	Power Steering Pressure Switch	Straight: 0v, Turned: 12v
19	GY	IAT Sensor Signal	1.5-2.5v
20	DG/YL	Fuel Pump Relay Output	Off: 0v, On: 12v
21	---	Not Used	---
22	TN/WT	TCS (switch) Signal	TCS & O/D On: 12v
23	BK/WT	Power Ground	<0.1v
24	PK/BK	Speed Control Servo Signal 'B'	DC pulse signals
25	LG/VT	Cooling Fan Control	Digital Signal: 0-12-0v
26	TN	Speed Control Servo Common	<0.050v
27	LB/OR	Fuel Pump Relay Control	Relay Off: 12v, On: 1v
28-29	---	Not Used	---
30	TN/LB	MAF Sensor Return	<0.050v
31	LB/RD	MAF Sensor Signal	0.6v, 55 mph: 2.1v
32	TN/OR	SCP Data Bus (+) Signal	Digital Signals
33	GY/RD	Sensor Ground	<0.050v
34	RD	Vehicle Power (Start-Run)	12-14v
35	DG/WT	Speed Control Servo Signal 'A'	DC pulse signals
36	VT/WT	EVAP Canister Vent Solenoid Control	0-10 Hz (0-100%)
37	LB/BK	Speed Control Switch	N/A
38	LG/BK	EVAP Purge Solenoid Control	0-10 Hz (0-100%)
39	VT	Module Programming Signal	0.1v
40	RD/WT	Direct Battery	12-14v
41	OR/LB	A/C Pressure Switch Signal	0v (switch on: 12v)
42	GY/OR	Passive Antitheft RX Signal	Digital Signals
43	WT/LG	Passive Antitheft TX Signal	Digital Signals
44	PK/LB	Data Bus (-) Signal	<0.050v
45	BR/WT	Reference Voltage	4.9-5.1v
46	RD	Vehicle Power (Start-Run)	12-14v

2003-05 Navigator 5.4L V8 MFI VIN R (E4OD) C138t Pin Connector

PCM Pin #	Wire Color	Circuit Description (30 Pin)	Value at Hot Idle
1	---	Not Used	---
2	VT/LG	HO2S-22 (B2 S2) Signal	0.1-1.1v
3	RD/LG	HO2S-12 (B1 S2) Signal	0.1-1.1v
4-10	---	Not Used	---
11	VT/YL	TCC Solenoid Control	0%, 55 mph: 95%
12	OR/YL	Shift Solenoid 'A' Control	1v, 55 mph: 1v
13	VT/OR	Shift Solenoid 'B' Control	12v, 55 mph: 1v
14	BR/OR	Coast Clutch Solenoid Control	12v, 55 mph: 12v
15-16	---	Not Used	---
17	LB/YL	Digital TR3A Sensor	0v, 55 mph: 11.5v
18	LB/BK	Digital TR2 Sensor	0v, 55 mph: 11.5v
19	WT/BK	Digital TR4 Sensor	0v, 55 mph: 11.5v
20	YL/BK	Digital TR1 Sensor	In 'P': 0v, in O/D: 1.7v
21	WT/BK	HO2S-12 (B1 S2) Heater	On: 1v, Off: 12v
22	---	Not Used	---
23	WT/YL	EPC Solenoid Control	9.1v (6 psi)
24	---	Not Used	---
25	DB/YL	OSS Sensor Signal	0 Hz, 55 mph: 250 Hz
26	DG/WT	Turbine Shaft Speed Sensor	360 Hz, 890 Hz
27	GY/RD	Sensor Ground	<0.050v
28	OR/BK	Transmission Fluid Temperature Sensor	2.10-2.40v
29	TN/YL	HO2S-22 (B2 S2) Heater	On: 1v, Off: 12v
30	---	Not Used	---

Pin Connector Graphic

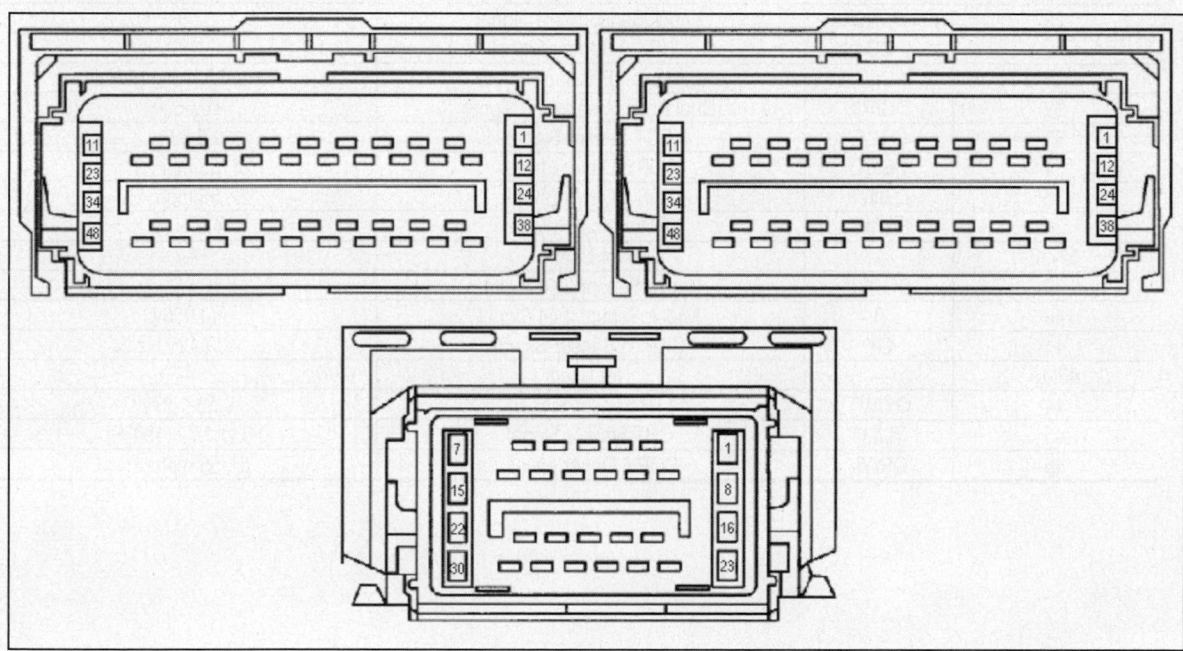

2003-05 Navigator 5.4L V8 MFI VIN R (E4OD) C1381e Pin Connector

PCM Pin #	Wire Color	Circuit Description (46 Pin)	Value at Hot Idle
1	PK/WT	COP 2 Driver (dwell)	5°, 55 mph: 8°
2	WT/LB	IAC Solenoid Control	10.7v (33%)
3	TN/BK	Injector 5 Control	2.7-4.1 ms
4	LG/OR	Injector 6 Control	2.7-4.1 ms
5	TN/RD	Injector 7 Control	2.7-4.1 ms
6	LB	Injector 8 Control	2.7-4.1 ms
7	BR/YL	Injector 3 Control	2.7-4.1 ms
8	BR/LB	Injector 4 Control	2.7-4.1 ms
9	GY/YL	Vehicle Power (Start-Run)	12-14v
10	---	Not Used	---
11	WT/RD	COP 8 Driver (dwell)	5°, 55 mph: 8°
12	PK/LB	COP 7 Driver (dwell)	5°, 55 mph: 8°
13	BR	Intake Manifold Runner Control Module	12v
14	TN	Injector 1 Control	2.7-4.1 ms
15	WT	Injector 2 Control	2.7-4.1 ms
16-19	---	Not Used	---
20	YL/LB	HO2S-21 (B2 S1) Heater	On: 1v, Off: 12v
21	RD/WT	HO2S-11 (B1 S1) Heater	On: 1v, Off: 12v
22	BR/PK	EGR Vacuum Regulator Control	0%, 55 mph: 45%
23	DG/VT	COP 4 Driver (dwell)	5°, 55 mph: 8°
24	WT/PK	COP 3 Driver (dwell)	5°, 55 mph: 8°
25	GY/RD	Sensor Ground	<0.050v
26	GY/LB	HO2S-11 (B1 S1) Signal	0.1-1.1v
27	RD/BK	HO2S-21 (B2 S1) Signal	0.1-1.1v
28	DG/VT	Knock Sensor 2 (+) Signal	0v
29	YL/RD	Knock Sensor 1 (+) Signal	0v
30	DB	CKP (+) Sensor Signal	400-420 Hz
31	DG	Camshaft Position Sensor	6 Hz, 55 mph: 15 Hz
32	DG/PK	Dual Knock Sensor (-) Signal	<0.050v
33	BR/LG	ECT Sensor Signal	0.5-0.6v
34	LG/YL	COP 5 Driver (dwell)	5°, 55 mph: 8°
35	LG/WT	COP 1 Driver (dwell)	5°, 55 mph: 8°
36	BR/WT	Reference Voltage	4.9-5.1v
37-38	---	Not Used	---
39	DG/WT	Knock Sensor 2 (-) Signal	<0.050v
40	YL	Knock Sensor 1 (-) Signal	<0.050v
41	GY	CKP (-) Sensor Signal	400-420 Hz
42-43	---	Not Used	---
44	GY/WT	TP Sensor Signal	0.53-1.27v
45	YL/LG	CHT Sensor Signal	0.6 to 1.7v (194°F)
46	OR/YL	COP 6 Driver (dwell)	5°, 55 mph: 8°

AEROSTAR PIN TABLES

1990 Aerostar 3.0L V6 MFI VIN U (A/T) 60 Pin Connector

PCM Pin #	Wire Color	Circuit Description (60 Pin)	Value at Hot Idle
1	YL	Keep Alive Power	12-14v
2	RD/LG	Brake Switch Signal	Brake Off: 12v, On: 1v
3	DG/WT	VSS (+) Signal	0 Hz, at 55 mph: 125 Hz
4	DG/YL	Ignition Diagnostic Monitor	20-31 Hz
6	BK/YL	VSS (-) Signal	0 Hz, at 55 mph: 125 Hz
6	BK/WT	VSS (-) Signal	0 Hz, at 55 mph: 125 Hz
7	LG/YL	Engine Coolant Temperature Sensor	0.5-0.6v
8	PK/BK	Fuel Pump Monitor	On: 12v, Off: 0v
8	OR/BL	Fuel Pump Monitor	On: 12v, Off: 0v
10	BK/YL	A/C Pressure Switch Signal	A/C On: 12v, off: 0v
16	BK/OR	Ignition System Ground	<0.050v
17	TN/RD	Self-Test Output & MIL	Lamp On: 1v, Off: 12v
20	BK	PCM Case Ground	<0.050v
21	OR/BK	IAC Solenoid Control	10.6-11.8v
22	TN/LG	Fuel Pump Control	On: 1v, Off: 12v
23	YL/RD	Knock Sensor Signal	0v
24	YL/LG	Power Steering Pressure Switch	Straight: 0v,Turned: 12v
25	LG/PK	ACT Sensor Signal	1.5-2.5v
26	OR/WT	Reference Voltage	4.9-5.1v
29	DG/PK	HO2S-11 (B1 S1) Signal	0.1-1.1v
30	BL/YL	A/T: Neutral Drive Switch	In 'N': 0v, Others: 5v
30	WT/BK	A/T: Neutral Drive Switch	In 'N': 0v, Others: 5v
30	WT/BK	M/T: Clutch Engage Switch	Clutch Out: 5v, In: 0v
30	BL/BK	M/T: Clutch Engage Switch	Clutch Out: 5v, In: 0v
31	GY/YL	Canister Purge Solenoid	12v, 55 mph: 1v
34	BL/PK	Data Output Link	5v
36	YL/LG	Spark Output Signal	6.93v (50%)
37	RD	Vehicle Power	12-14v
40	BK/LG	Power Ground	<0.1v
45	DB/LG	MAP Sensor Signal	107 Hz
46	BK/WT	Analog Signal Return	<0.050v
47	DG/LG	TP Sensor Signal	0.7v, 55 mph: 2.1v
48	WT/RD	Self-Test Input Signal	STI On: 1v, Off: 5v
49	OR	HO2S-11 (B1 S1) Ground	<0.050v
52	TN/BL	Shift Solenoid 3-4 Control	1v, 55 mph: 12v
53	OR/YL	TCC Solenoid Control	12v, 55 mph: 1v
53	WT	TCC Solenoid Control	12v, 55 mph: 1v
54	GY/WT	A/C WOT Relay Control	On: 1v, Off: 12v
54	RD	A/C WOT Relay Control	On: 1v, Off: 12v
56	DB	PIP Sensor Signal	6.93v (50%)
57	RD	Vehicle Power	12-14v
58	TN/OR	Injector Bank 1 (INJ 1, 2 & 4)	5.6-6.0 ms
59	TN/RD	Injector Bank 2 (INJ 3, 5 & 6)	5.6-6.0 ms
60	BK/LG	Power Ground	<0.1v

Pin Connector Graphic

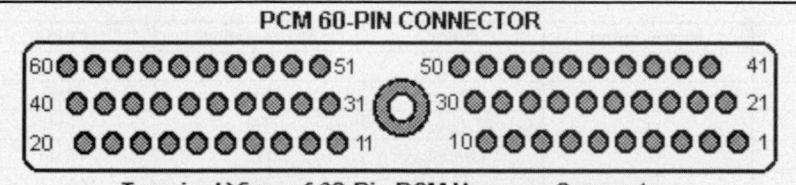

Terminal View of 60-Pin PCM Harness Connector

1991 Aerostar 3.0L V6 MFI VIN U (A/T) 60 Pin Connector

PCM Pin #	Wire Color	Circuit Description (60 Pin)	Value at Hot Idle
1	YL	Keep Alive Power	12-14v
2	LG	Brake Switch Signal	Brake Off: 12v, On: 1v
3	GY/BK	VSS (+) Signal	0 Hz, at 55 mph: 125 Hz
4	TN/YL	Ignition Diagnostic Monitor	20-31 Hz
6	BK/YL	VSS (-) Signal	0 Hz, at 55 mph: 125 Hz
7	LG/YL	Engine Coolant Temperature Sensor	0.5-0.6v
8	PK/BK	Fuel Pump Monitor	On: 12v, Off: 0v
10	BK/YL	A/C Pressure Switch Signal	A/C On: 12v, off: 0v
16	OR/RD	Ignition System Ground	<0.050v
17	TN/RD	Self-Test Output & MIL	MIL On: 1v, Off: 12v
20	BK	PCM Case Ground	<0.050v
21	WT/BL	IAC Solenoid Control	10.5-11.8v
22	BL/OR	Fuel Pump Control	On: 1v, Off: 12v
24	YL/LG	Power Steering Pressure Switch	Straight: 0v, Turned: 12v
25	LG/PK	ACT Sensor Signal	1.5-2.5v
26	BR/WT	Reference Voltage	4.9-5.1v
29	RD/BK	HO2S-11 (B1 S1) Signal	0.1-1.1v
30	BL/YL	A/T: Neutral Drive Switch	In 'N': 0v, Others: 5v
30	BL/YL	M/T: Clutch Engage Switch	Clutch Out: 5v, In: 0v
31	GY/YL	Canister Purge Solenoid	12v, 55 mph: 1v
34	BL/PK	Data Output Link	5v
36	PK	Spark Output Signal	6.93v (50%)
37	RD	Vehicle Power	12-14v
40	BK/LG	Power Ground	<0.1v
45	DB/LG	MAP Sensor Signal	107 Hz
46	GY/RD	Analog Signal Return	<0.050v
47	GY/WT	TP Sensor Signal	0.7v, 55 mph: 2.1v
48	WT/PK	Self-Test Input Signal	STI On: 1v, Off: 5v
49	OR	HO2S-11 (B1 S1) Ground	<0.050v
52	OR/YL	Shift Solenoid 3-4 Control	1v, 55 mph: 12v
53	WT	TCC Solenoid Control	12v, 55 mph: 1v
54	GY/WT	A/C WOT Relay Control	On: 1v, Off: 12v
56	GY/OR	PIP Sensor Signal	6.93v (50%)
57	RD	Vehicle Power	12-14v
58	TN	Injector Bank 1 (INJ 1, 2 & 4)	5.6-6.0 ms
59	WT	Injector Bank 2 (INJ 3, 5 & 6)	5.6-6.0 ms
60	BK/LG	Power Ground	<0.1v

Pin Connector Graphic

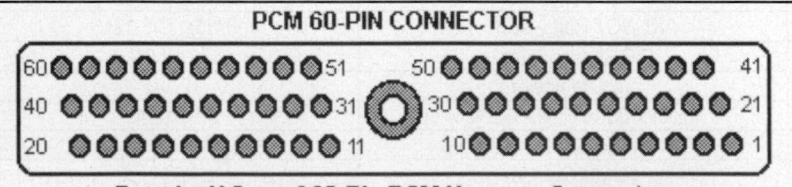

PCM 60-PIN CONNECTOR

Terminal View of 60-Pin PCM Harness Connector

Standard Colors and Abbreviations

Abbreviation	Color	Abbreviation	Color	Abbreviation	Color
BK	Black	GY	Gray	PK	Purple
BL	Blue	GN	Green	RD	Red
BR	Brown	LG	LT Green	TN	Tan
DB	Dark Blue	OR	Orange	WT	White
DG	DK Green	PK	Pink	YL	Yellow

1992-95 Aerostar 3.0L V6 MFI VIN U (A/T) 60 Pin Connector

PCM Pin #	Wire Color	Circuit Description (60 Pin)	Value at Hot Idle
1	YL	Keep Alive Power	12-14v
2	RD/LG	Brake Switch Signal	Brake Off: 12v, On: 1v
3	GY/BK	VSS (+) Signal	0 Hz, at 55 mph: 125 Hz
4	WT/PK	Ignition Diagnostic Monitor	20-31 Hz
5	---	Not Used	---
6	PK/OR	VSS (-) Signal	0 Hz, at 55 mph: 125 Hz
7	LG/RD	Engine Coolant Temperature Sensor	0.5-0.6v
8	DG/YL	Fuel Pump Monitor	On: 12v, Off: 0v
9	TN/BL	MAF Sensor Return	<0.050v
10	DG/OR	A/C Pressure Switch Signal	A/C On: 12v, off: 0v
11	GY/YL	Canister Purge Solenoid	12v, 55 mph: 1v
12	LB/OR	Injector 6 Control	5.0-5.5 ms
13	---	Not Used	---
14 ('92)	PK/BK	TCC Solenoid Control	0%, at 55 mph: 95%
15	TN/BK	Injector 5 Control	5.0-5.5 ms
16	BK/OR	Ignition System Ground	<0.050v
16	OR/RD	Ignition System Ground	<0.050v
17	PK/LG	Self-Test Output & MIL	MIL On: 1v, Off: 12v
18	TN/OR	Data Bus (+)	Digital Signals
19	PK/BL	Data Bus (-)	Digital Signals
20	BK	PCM Case Ground	<0.050v
21	WT/BL	IAC Solenoid Control	10.8-12.6v
22	BL/OR	Fuel Pump Control	On: 1v, Off: 12v
23-24	---	Not Used	---
25	GY	ACT Sensor Signal	1.5-2.5v
26	BR/WT	Reference Voltage	4.9-5.1v
27-29	---	Not Used	---
30	BL/YL	A/T: Neutral Drive Switch	In 'N': 0v, Others: 5v
30	BL/YL	M/T: Clutch Engage Switch	Clutch Out: 5v, In: 0v
31-33	---	Not Used	---
34	BL/PK	Data Output Link	5v
35	BR/LB	Injector 4 Control	5.0-5.5 ms
36	PK	Spark Output Signal	6.93v (50%)
37	RD	Vehicle Power	12-14v
38	---	Not Used	---
39	BR/YL	Injector 3 Control	5.0-5.5 ms
40	BK/LG	Power Ground	<0.1v

1992-95 Aerostar 3.0L V6 MFI VIN U (A/T) 60 Pin Connector

PCM Pin #	Wire Color	Circuit Description (60 Pin)	Value at Hot Idle
41-43	---	Not Used	---
44	GY/BL	HO2S-11 (B1 S1) Signal	0.1-1.1v
45	---	Not Used	---
46	GY/RD	Analog Signal Return	<0.050v
47	GY/WT	TP Sensor Signal	0.7v, 55 mph: 2.1v
48	WT/PK	Self-Test Input Signal	STI On: 1v, Off: 5v
49	---	Not Used	---
50	BL/RD	MAF Sensor Signal	0.9v, 55 mph: 2.2v
51	OR/YL	Shift Solenoid 3-4 Control	1v, 55 mph: 12v
52	---	Not Used	---
53 ('93-'94)	PK/BK	TCC Solenoid Control	12v, 55 mph: 1v
53 ('93-'94)	PK/BK	TCC Solenoid Control	12v, 55 mph: 1v
54	PK/YL	A/C WOT Relay Control	On: 1v, Off: 12v
55	---	Not Used	---
56	GY/OR	PIP Sensor Signal	6.93v (50%)
57	RD	Vehicle Power	12-14v
58	TN	Injector 1 Control	5.0-5.5 ms
59	WT	Injector 2 Control	5.0-5.5 ms
60	BK/LG	Power Ground	<0.1v

Pin Connector Graphic

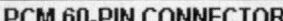
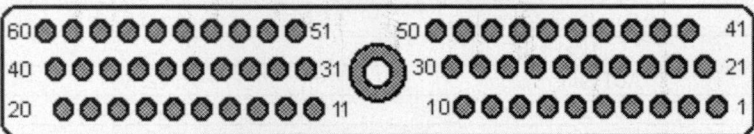

PCM 60-PIN CONNECTOR

Terminal View of 60-Pin PCM Harness Connector

Standard Colors and Abbreviations

Abbreviation	Color	Abbreviation	Color	Abbreviation	Color
BK	Black	GY	Gray	PK	Purple
BL	Blue	GN	Green	RD	Red
BR	Brown	LG	LT Green	TN	Tan
DB	Dark Blue	OR	Orange	WT	White
DG	DK Green	PK	Pink	YL	Yellow

1996-97 Aerostar 3.0L V6 VIN U (A/T) 104 Pin Connector

PCM Pin #	Wire Color	Circuit Description (104 Pin)	Value at Hot Idle
1	BK/WT	Shift Solenoid 2 Control	1v, 55 mph: 12v
2	PK/LG	MIL (lamp) Control	MIL On: 1v, Off: 12v
3-10, 12, 14	---	Not Used	---
11	BL/LG	Purge Flow Sensor	0.8v, 55 mph: 3.0v
13	PK	Flash EEPROM	0.1v
15	PK/LB	Data Bus (-)	Digital Signals
16	TN/OR	Data Bus (+)	Digital Signals
17-20, 23	---	Not Used	---
21	DB	CKP (+) Sensor	518-540 Hz
22	GY	CKP (-) Sensor	518-540 Hz
24	BK/WT	Power Ground	<0.050v
25	BK	Case Ground	<0.050v
26	YL/BK	Coil 1 Driver (Dwell)	5°, at 55 mph: 8°
27	OR/YL	Shift Solenoid 1 Control	12v, 55 mph: 1v
28	BR/OR	Coast Clutch Solenoid	12v, 55 mph: 12v
29	TN/WT	TCS (switch) Signal	TCS & O/D On: 12v
30	DG	Octane Adjustment	9.3v (switch shorted: 0v)
31-32	---	Not Used	---
33	PK/OR	VSS (-) Signal	0 Hz, at 55 mph: 125 Hz
35	RD/LG	HO2S-12 (B1 S2) Signal	0.1-1.1v
36	TN/LB	MAF Sensor Return	<0.050v
37	OR/BK	Transmission Fluid Temperature Sensor	2.10-2.40v
38	LG/RD	Engine Coolant Temperature Sensor	0.5-0.6v
39	GY	Intake Air Temperature Sensor	1.5-2.5v
40	DG/YL	Fuel Pump Monitor	On: 12v, Off: 0v
41	TN/YL	A/C Pressure Switch Signal	A/C On: 12v, off: 0v
42, 44-46	---	Not Used	---
43	LB/PK	Data Output Link Signal	5v
47	BR/PK	VR Solenoid Control	0%, at 55 mph: 45%
48	TN/YL	Clean Tachometer Output	42-50 Hz
49-50	---	Not Used	---
51	BK/WT	Power Ground	<0.1v
52	YL/RD	Coil 2 Driver (Dwell)	5°, at 55 mph: 8°
53	PK/BK	Shift Solenoid 3 Control	12v, 55 mph: 1v
54	PK/YL	TCC Solenoid Control	0%, at 55 mph: 95%
55	YL	Keep Alive Power	12-14v
56-57, 59	---	Not Used	---
58	GY/BK	VSS (+) Signal	0 Hz, at 55 mph: 125 Hz
60	GY/LB	HO2S-11 (B1 S1) Signal	0.1-1.1v
61-63	---	Not Used	---
64	LB/YL	Transmission Range Sensor	In 'P': 0v, in O/D: 1.7v
65	BR/LG	EGR DPFE Sensor Signal	0.20-1.30v
66, 68	---	Not Used	---
67	GY/YL	EVAP Purge Solenoid	0-10 Hz (0-100%)
69	PK/YL	A/C WOT Relay Control	On: 1v, Off: 12v
70	---	Not Used	---

1996-97 Aerostar 3.0L V6 VIN U (A/T) 104 Pin Connector

PCM Pin #	Wire Color	Circuit Description (104 Pin)	Value at Hot Idle
71	RD	Vehicle Power	12-14v
72	---	Not Used	---
73	TN/BK	Injector 5 Control	4.5-4.8 ms
74	BR/YL	Injector 3 Control	4.5-4.8 ms
75	TN	Injector 1 Control	4.5-4.8 ms
76, 77	BK/WT	Power Ground	<0.1v
78	YL/WT	Coil 3 Driver (Dwell)	5°, at 55 mph: 8°
79	WT/LG	TCIL (lamp) Control	Lamp On: 1v, Off: 12v
80	LB/OR	Fuel Pump Control	On: 1v, Off: 12v
81	WT/YL	EPC Solenoid Control	10.7v (27 psi)
82	---	Not Used	---
83	WT/LB	IAC Solenoid Control	10v (45% duty cycle)
84	DG/WT	Turbine Shaft Speed Sensor	120 Hz, 30 mph: 300 Hz
85	DB/OR	Camshaft Position Sensor	7 Hz, 55 mph: 13-17 Hz
86-87	---	Not Used	---
88	LB/RD	MAF Sensor Signal	0.9v, 55 mph: 2.2v
89	GY/WT	TP Sensor Signal	0.53-1.27v
90	BR/WT	Reference Voltage	4.9-5.1v
91	GY/RD	Analog Signal Return	<0.050v
92	LG	Brake Position Switch	Brake Off: 12v, On: 1v
93	RD/WT	HO2S-11 (B1 S1) Heater	1v, Off: 12v
94	---	Not Used	---
95	WT/BK	HO2S-12 (B1 S2) Heater	1v, Off: 12v
96	---	Not Used	---
97	RD	Vehicle Power	12-14v
98	---	Not Used	---
99	LG/OR	Injector 6 Control	4.5-4.8 ms
100	BR/LB	Injector 4 Control	4.5-4.8 ms
101	WT	Injector 2 Control	4.5-4.8 ms
102	---	Not Used	---
103	BK/WT	Power Ground	<0.1v
104	---	Not Used	---

Pin Connector Graphic

PCM 104-PIN CONNECTOR

Terminal View of 104-Pin PCM Wiring Harness Connector

Standard Colors and Abbreviations

Abbreviation	Color	Abbreviation	Color	Abbreviation	Color
BK	Black	GY	Gray	PK	Purple
BL	Blue	GN	Green	RD	Red
BR	Brown	LG	LT Green	TN	Tan
DB	Dark Blue	OR	Orange	WT	White
DG	DK Green	PK	Pink	YL	Yellow

1990 Aerostar 4.0L V6 OHV VIN X (A/T) 60 Pin Connector

PCM Pin #	Wire Color	Circuit Description (60 Pin)	Value at Hot Idle
1	YL	Keep Alive Power	12-14v
2	RD/LG	Brake Switch Signal	Brake Off: 12v, On: 1v
3	DG/WT	VSS (+) Signal	0 Hz, at 55 mph: 125 Hz
4	DG/YL	Ignition Diagnostic Monitor	20-31 Hz
6	BK/YL	VSS (-) Signal	0 Hz, at 55 mph: 125 Hz
7	LG/YL	Engine Coolant Temperature Sensor	0.5-0.6v
8	PK/BK	Fuel Pump Monitor	On: 12v, Off: 0v
9	BK/OR	Data Bus (+)	Digital Signals
10	BK/YL	A/C Pressure Switch Signal	A/C On: 12v, off: 0v
14	DB/OR	MAF Sensor Signal	0.7v, 55 mph: 2.4v
15	TN/BL	MAF Sensor Return	<0.050v
16	BK/OR	Ignition System Ground	<0.050v
17	TN/RD	Self-Test Output & MIL	MIL On: 1v, Off: 12v
20	BK	PCM Case Ground	<0.050v
21	OR/BK	IAC Solenoid Control	10.2v (39%)
22	TN/LG	Fuel Pump Control	On: 1v, Off: 12v
24	YL/LG	Power Steering Pressure Switch	Straight: 0v, Turned: 12v
25	LG/PK	ACT Sensor Signal	1.5-2.5v
26	OR/WT	Reference Voltage	4.9-5.1v
28	OR/BK	Data Bus (-)	Digital Signals
29	DG/PK	HO2S-11 (B1 S1) Signal	0.1-1.1v
30	BL/YL	A/T: Neutral Drive Switch	In 'N': 0v, Others: 5v
31	GY/YL	Canister Purge Solenoid	12v, 55 mph: 1v
34	BL/PK	Data Output Link Signal	Digital Signals
36	YL/LG	Spark Output Signal	6.93v (50%)
37	RD	Vehicle Power	12-14v
40	BK/LG	Power Ground	<0.1v
44	OR	Octane Adjustment	9.3v (switch shorted: 0v)
45	DB/LG	BARO Sensor Signal	159 Hz (Sea Level)
46	BK/WT	Analog Signal Return	<0.050v
47	DG/LG	TP Sensor Signal	0.7v, 55 mph: 2.1v
48	WT/RD	Self-Test Input Signal	STI On: 1v, Off: 5v
49	OR	HO2S-11 (B1 S1) Ground	<0.050v
52	TN/BL	Shift Solenoid 3-4 Control	1v, 55 mph: 12v
53	WT	TCC Solenoid Control	12v, 55 mph: 1v
54	GY/WT	A/C WOT Relay Control	On: 1v, Off: 12v
56	DB	PIP Sensor Signal	6.93v (50%)
57	RD	Vehicle Power	12-14v
58	TN/OR	Injector Bank 1 (INJ 1, 2 & 4)	3.0-3.2 ms
59	TN	Injector Bank 2 (INJ 3, 5 & 6)	3.0-3.2 ms
60	BK/LG	Power Ground	<0.1v

Pin Connector Graphic

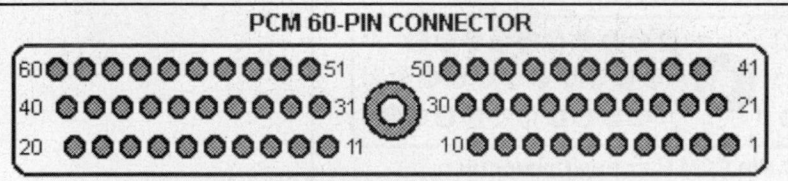

PCM 60-PIN CONNECTOR

Terminal View of 60-Pin PCM Harness Connector

1991 Aerostar 4.0L V6 OHV VIN X (A/T) 60 Pin Connector

PCM Pin #	Wire Color	Circuit Description (60 Pin)	Value at Hot Idle
1	YL	Keep Alive Power	12-14v
2	RD/LG	Brake Switch Signal	Brake Off: 12v, On: 1v
3	DG/WT	VSS (+) Signal	0 Hz, at 55 mph: 125 Hz
4	TN/YL	Ignition Diagnostic Monitor	20-31 Hz
6	BK/YL	VSS (-) Signal	0 Hz, at 55 mph: 125 Hz
7	LG/RD	Engine Coolant Temperature Sensor	0.5-0.6v
8	PK/BK	Fuel Pump Monitor	On: 12v, Off: 0v
9	BK/OR	Data Bus (-)	Digital Signals
10	BK/YL	A/C Pressure Switch Signal	A/C On: 12v, off: 0v
14	BL/RD	MAF Sensor Signal	0.7v, 55 mph: 2.4v
15	TN/BL	MAF Sensor Return	<0.050v
16	OR/RD	Ignition System Ground	<0.050v
17	TN/RD	Self-Test Output & MIL	MIL On: 1v, Off: 12v
20	BK	PCM Case Ground	<0.050v
21	OR/BK	IAC Solenoid Control	10.2v (39%)
22	TN/LG	Fuel Pump Control	On: 1v, Off: 12v
25	LG/PK	ACT Sensor Signal	1.5-2.5v
26	BR/WT	Reference Voltage	4.9-5.1v
28	OR/BK	Data Bus (+)	Digital Signals
29	RD/BK	HO2S-11 (B1 S1) Signal	0.1-1.1v
30	BL/YL	A/T: Neutral Drive Switch	In 'P': 0v, Others: 5v
31	GY/YL	Canister Purge Solenoid	12v, 55 mph: 1v
34	BL/PK	Data Output Link Signal	Digital Signals
36	YL/LG	Spark Output Signal	6.93v (50%)
37	RD	Vehicle Power	12-14v
40	BK/LG	Power Ground	<0.1v
44	OR	Octane Adjustment	9.3v (switch shorted: 0v)
45	DB/LG	BARO Sensor Signal	159 Hz (Sea Level)
46	GY/RD	Analog Signal Return	<0.050v
47	GY/WT	TP Sensor Signal	0.7v, 55 mph: 2.1v
48	WT/RD	Self-Test Input Signal	STI On: 1v, Off: 5v
49	OR	HO2S-11 (B1 S1) Ground	<0.050v
52	TN/WT	Shift Solenoid 3-4 Control	1v, 55 mph: 12v
53	WT	TCC Solenoid Control	12v, 55 mph: 1v
54	GY/WT	A/C WOT Relay Control	On: 1v, Off: 12v
56	DB	PIP Sensor Signal	6.93v (50%)
57	RD	Vehicle Power	12-14v
58	TN/OR	Injector Bank 1 (INJ 1, 2 & 4)	3.3-3.5 ms
59	TN/RD	Injector Bank 2 (INJ 3, 5 & 6)	3.3-3.5 ms
60	BK/LG	Power Ground	<0.1v

Pin Connector Graphic

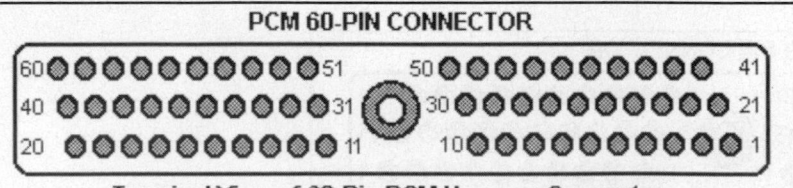

PCM 60-PIN CONNECTOR

Terminal View of 60-Pin PCM Harness Connector

1992-95 Aerostar 4.0L V6 VIN X (A/T) 60 Pin Connector

PCM Pin #	Wire Color	Circuit Description (60 Pin)	Value at Hot Idle
1	YL	Keep Alive Power	12-14v
2	RD/LG	Brake Switch Signal	Brake Off: 12v, On: 1v
3	GY/BK	VSS (+) Signal	0 Hz, at 55 mph: 125 Hz
3	DG/WT	VSS (+) Signal	0 Hz, at 55 mph: 125 Hz
4	TN/YL	Ignition Diagnostic Monitor	20-31 Hz
5	---	Not Used	----
6	PK/OR	VSS (-) Signal	0 Hz, at 55 mph: 125 Hz
7	LG/RD	Engine Coolant Temperature Sensor	0.5-0.6v
8	DG/YL	Fuel Pump Monitor	On: 12v, Off: 0v
9	PK/BL	Data Bus (-)	Digital Signals
10	DG/OR	A/C Pressure Switch Signal	A/C On: 12v, off: 0v
11-13	---	Not Used	---
14	BL/RD	MAF Sensor Signal	0.7v, 55 mph: 2.4v
15	TN/BL	MAF Sensor Return	<0.050v
16	OR/RD	Ignition System Ground	<0.050v
17	PK/LG	Self-Test Output & MIL	MIL On: 1v, Off: 12v
18-19	---	Not Used	---
20	BK	PCM Case Ground	<0.050v
21	WT/BL	IAC Solenoid Control	10.2v (39%)
22	BL/OR	Fuel Pump Control	On: 1v, Off: 12v
23-24	---	Not Used	---
25	GY	Intake Air Temperature Sensor	1.5-2.5v
26	BR/WT	Reference Voltage	4.9-5.1v
27	---	Not Used	---
28	TN/OR	Data Bus (+)	Digital Signals
29	GY/BL	HO2S-11 (B1 S1) Signal	0.1-1.1v
29	WT/BL	HO2S-11 (B1 S1) Signal	0.1-1.1v
30	BL/YL	A/T: Neutral Drive Switch	In 'N': 0v, Others: 5v
31	GY/YL	Canister Purge Solenoid	12v, 55 mph: 1v
32-33	---	Not Used	---
34	BL/PK	Data Output Link Signal	Digital Signals
35	---	Not Used	---
36 ('92-'93)	PK	Spark Angle Word	6.93v (50%)
36 ('93-'94)	PK	Spark Output Signal	6.93v (50%)
37	RD	Vehicle Power	12-14v
38-39	---	Not Used	---
40	BK/LG	Power Ground	<0.1v

1992-95 Aerostar 4.0L V6 VIN X (A/T) 60 Pin Connector

PCM Pin #	Wire Color	Circuit Description (60 Pin)	Value at Hot Idle
41-43	---	Not Used	---
44	DG	Octane Adjustment	9.3v (switch shorted: 0v)
45	---	Not Used	---
46	GY/RD	Analog Signal Return	<0.050v
47	GY/WT	TP Sensor Signal	0.7v, 55 mph: 2.1v
48	WT/PK	Self-Test Input Signal	STI On: 1v, Off: 5v
49	OR	HO2S-11 (B1 S1) Ground	<0.050v
50-51	---	Not Used	---
52	OR/YL	Shift Solenoid 3-4 Control	1v, 55 mph: 12v
53	PK/BK	TCC Solenoid Control	12v, 55 mph: 1v
54	PK/YL	A/C WOT Relay Control	On: 1v, Off: 12v
55	---	Not Used	---
56	GY/OR	PIP Sensor Signal	6.93v (50%)
57	RD	Vehicle Power	12-14v
58	TN	Injector Bank 1 (INJ 1, 2 & 4)	3.3-3.6 ms
59	WT	Injector Bank 2 (INJ 3, 5 & 6)	3.3-3.6 ms
60	BK/LG	Power Ground	<0.1v

Pin Connector Graphic

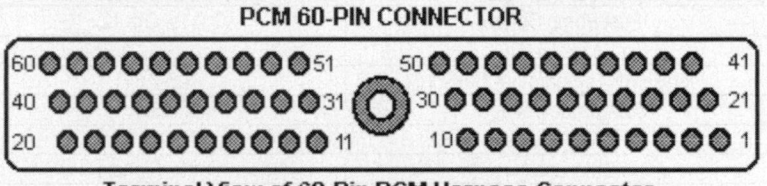

Standard Colors and Abbreviations

Abbreviation	Color	Abbreviation	Color	Abbreviation	Color
BK	Black	GY	Gray	PK	Purple
BL	Blue	GN	Green	RD	Red
BR	Brown	LG	LT Green	TN	Tan
DB	Dark Blue	OR	Orange	WT	White
DG	DK Green	PK	Pink	YL	Yellow

1996-97 Aerostar 4.0L V6 VIN X (A/T) 104 Pin Connector

PCM Pin #	Wire Color	Circuit Description (104 Pin)	Value at Hot Idle
1	BK/WT	Shift Solenoid 2 Control	12v, 55 mph: 12v
2	PK/LG	MIL (lamp) Control	MIL On: 1v, Off: 12v
3-10	---	Not Used	---
11	BL/LG	Purge Flow Sensor	0.8v, 30 mph: 3v
12	---	Not Used	---
13	VT	Flash EPROM Power	0.1v
14	---	Not Used	---
15	PK/LB	Data Bus (-)	Digital Signals
16	TN/OR	Data Bus (+)	Digital Signals
17-20	---	Not Used	---
21	DB	CKP (+) Sensor	450-480 Hz
22	GY	CKP (-) Sensor	450-480 Hz
23	---	Not Used	---
24	BK/WT	Power Ground	<0.1v
25	BK	Case Ground	<0.050v
26	YL/BK	Coil 1 Driver (Dwell)	5°, at 55 mph: 8°
27	OR/YL	Shift Solenoid 1 Control	1v, 55 mph: 12v
28	BR/OR	Coast Clutch Solenoid	12v, 55 mph: 12v
29	TN/WT	TCS (switch) Signal	TCS & O/D On: 12v
30	DG	Octane Adjustment	9.3v (switch shorted: 0v)
31-32	---	Not Used	---
33	PK/OR	PSOM (-) Signals	<0.050v
34	---	Not Used	---
35	RD/LG	HO2S-12 (B1 S2) Signal	0.1-1.1v
36	TN/LB	MAF Sensor Return	<0.050v
37	OR/BK	Transmission Fluid Temperature Sensor	2.10-2.40v
38	LG/RD	Engine Coolant Temperature Sensor	0.5-0.6v
39	G/Y	Intake Air Temperature Sensor	1.5-2.5v
40	DG/YL	Fuel Pump Monitor	On: 12v, Off: 0v
41	TN/YL	A/C Pressure Switch Signal	A/C On: 12v, off: 0v
42	---	Not Used	---
43	LB/PK	Overhead Trip Computer	Digital Signals
44-46	---	Not Used	---
47	BR/PK	VR Solenoid Control	0%, at 55 mph: 45%
48	TN/YL	Clean Tachometer Output	38-42 Hz
49-50	---	Not Used	---
51	BK/WT	Power Ground	<0.1v
52	YL/RD	Coil 2 Driver (Dwell)	5°, at 55 mph: 8°
53	PK/BK	Shift Solenoid 3 Control	12v, 55 mph: 1v
54	PK/YL	TCC Solenoid Control	0%, at 55 mph: 95%
55	YL	Keep Alive Power	12-14v
56-57	---	Not Used	---
58	GY/BK	PSOM (+) Signals	Digital signals
59	---	Not Used	---
60	GY/LB	HO2S-11 (B1 S1) Signal	0.1-1.1v
61-63	---	Not Used	---
64	LB/YL	Digital Transmission Range Sensor	In 'P': 0v, in O/D: 1.7v

1996-97 Aerostar 4.0L V6 VIN X (A/T) 104 Pin Connector

PCM Pin #	Wire Color	Circuit Description (104 Pin)	Value at Hot Idle
65	BR/LG	EGR DPFE Sensor Signal	0.20-1.30v
66	---	Not Used	---
67	GY/YL	EVAP Purge Solenoid	0-10 Hz (0-100%)
68	---	Not Used	---
69	PK/YL	A/C WOT Relay Control	On: 1v, Off: 12v
70	---	Not Used	---
71	RD	Vehicle Power	12-14v
72	---	Not Used	---
73	TN/BK	Injector 5 Control	3.9-4.1 ms
74	BR/YL	Injector 3 Control	3.9-4.1 ms
75	TN	Injector 1 Control	3.9-4.1 ms
76	BK/WT	Power Ground	<0.1v
77	BK/WT	Power Ground	<0.1v
78	YL/WT	Coil 3 Driver (Dwell)	5°, at 55 mph: 8°
79	WT/LG	TCIL (lamp) Control	Lamp On: 1v, Off: 12v
80	LB/OR	Fuel Pump Control	On: 1v, Off: 12v
81	WT/YL	EPC Solenoid Control	10.2v (24 psi)
82	---	Not Used	---
83	WT/LB	IAC Solenoid Control	10.7v (33%)
84	DG/WT	Turbine Shaft Speed Sensor	120 Hz, 30 mph: 300 Hz
85	DB/OR	Camshaft Position Sensor	7 Hz, 55 mph: 18 Hz
86-87	---	Not Used	---
88	LB/RD	MAF Sensor Signal	0.8v, 55 mph: 1.6v
89	GY/WT	TP Sensor Signal	0.53-1.27v
90	BR/WT	Reference Voltage	4.9-5.1v
91	GY/RD	Analog Signal Return	<0.050v
92	LG	Brake Position Switch	Brake Off: 12v, On: 1v
93	RD/WT	HO2S-11 (B1 S1) Heater	0.1-1.1v
94	---	Not Used	---
95	WT/BK	HO2S-22 (B2 S2) Heater	0.1-1.1v
96	---	Not Used	---
97	RD	Vehicle Power	12-14v
98	---	Not Used	---
99	LG/OR	Injector 6 Control	4.5-4.8 ms
100	BR/LB	Injector 4 Control	4.5-4.8 ms
101	W	Injector 2 Control	4.5-4.8 ms
102	---	Not Used	---
103	BK/WT	Power Ground	<0.1v
104	---	Not Used	---

Pin Connector Graphic

PCM 104-PIN CONNECTOR

```
 1 ●●●●●●●●●●●●●   ●●●●●●●●●●●●●● 26
27 ●●●●●●●●●●●●●       ●●●●●●●●●●●●● 52
53 ●●●●●●●●●●●●●   ⬤   ●●●●●●●●●●●●● 78
79 ●●●●●●●●●●●●●   ●●●●●●●●●●●●●● 104
```

Terminal View of 104-Pin PCM Wiring Harness Connector

Standard Colors and Abbreviations

Abbreviation	Color	Abbreviation	Color	Abbreviation	Color
BK	Black	GY	Gray	PK	Purple
BL	Blue	GN	Green	RD	Red
BR	Brown	LG	LT Green	TN	Tan
DB	Dark Blue	OR	Orange	WT	White
DG	DK Green	PK	Pink	YL	Yellow

E-SERIES VAN PIN TABLES

1997-99 E-Series Van 4.2L V6 VIN 2 (A/T) 104-P Connector

PCM Pin #	Wire Color	Circuit Description (104 Pin)	Value at Hot Idle
1	PK/OR	Shift Solenoid 2 Control	12v, 55 mph: 1v
2	PK/LG	MIL (lamp) Control	MIL On: <1v, Off: 12v
3	YL/BK	Digital TR1 Sensor Signal	0v, at 55 mph: 11.5v
8	DB/YL	Electric IMRC Monitor	5v, 55 mph: 5v
12	YL/WT	Fuel Level Indicator Signal	1.7v (1/2 full)
13	VT	Flash EEPROM	0.1v
14	LB/BK	4x4 Indicator Signal	Off: 7.7v, On: 0v
15	PK/LB	Data Bus (-) Signal	Digital Signals
16	TN/OR	Data Bus (+) Signal	Digital Signals
19	DB/WT	Automatic Ride Control	Digital Signals
21	DB	CKP (+) Sensor	430-500 Hz
22	GY	CKP (-) Sensor	430-500 Hz
24	BK/WT	Power Ground	<0.1v
25	LG/YL	Case Ground	<0.050v
25	BK/LB	Case Ground	<0.050v
26	DB/LG	Coil 1 Driver (Dwell)	5°, at 55 mph: 8°
27	OR/YL	Shift Solenoid 1 Control	1v, 55 mph: 1v
29	TN/WT	TCS (switch) Signal	TCS & O/D On: 12v
30 ('97-'99)	DG	Octane Adjustment	9.3v (switch shorted: 0v)
32	DG/PK	Knock Sensor 1 Ground	<0.050v
33	PK/OR	VSS (-) Signal	0 Hz, at 55 mph: 125 Hz
35	RD/LG	HO2S-12 (B1 S2) Signal	0.1-1.1v
36	TN/LB	MAF Sensor Return	<0.050v
37	OR/BK	Transmission Fluid Temperature Sensor	2.10-2.40v
38	LG/RD	Engine Coolant Temperature Sensor	0.5-0.6v
39	GY	Intake Air Temperature Sensor	1.5-2.5v
40	DG/YL	Fuel Pump Monitor	On: 12v, Off: 0v
41	TN/LG	A/C Cycling Clutch Switch	A/C On: 12v, off: 0v
42	DB/LG	Electric IMRC Control	12v, 55 mph: 12v
43	OR/LG	Overhead Trip Computer	Digital Signals
45	RD/WT	CHTIL (lamp) Control	Lamp On: 1v, Off: 12v
46	BR	Intake Manifold Tuning Valve	12v, 55 mph: 12v
47	BR/PK	VR Solenoid Control	0%, at 55 mph: 45%
48	WT/PK	Clean Tachometer Output Signal	35-49 Hz
49	LB/BK	Digital TR2 Sensor	0v, at 55 mph: 11.5v
50	LG/RD	Digital TR4 Sensor	0v, at 55 mph: 11.5v
51	BK/WT	Power Ground	<0.1v
52	RD/LB	Coil 2 Driver (Dwell)	5°, at 55 mph: 8°
54	PKWT	TCC Solenoid Control	0%, at 55 mph: 95%
54	PK/YL	TCC Solenoid Control	0%, at 55 mph: 95%
55	YL	Keep Alive Power	12-14v
56	LG/BK	EVAP Purge Solenoid	0-10 Hz (0-100%)
57	YL/RD	Knock Sensor 1 Signal	0v
58	GY/BK	VSS (+) Signal	0 Hz, at 55 mph: 125 Hz
60	GY/LB	HO2S-11 (B1 S1) Signal	0.1-1.1v
61	PK/LG	HO2S-22 (B2 S2) Signal	0.1-1.1v
62	RD/PK	FTP Sensor Signal	2.6v (0" H2O - cap off)
64	LB/YL	Transmission Range Sensor	In 'P': 0v, in O/D: 1.7v

1997-99 E-Series Van 4.2L V6 VIN 2 (A/T) 104-P Connector

PCM Pin #	Wire Color	Circuit Description (104 Pin)	Value at Hot Idle
65	BR/LG	EGR DPFE Sensor Signal	0.20-1.30v
66	YL/LG	Cylinder Head Temperature Sensor	0.6 to 1.7v (194ºF)
67	PKWT	EVAP CV Solenoid	0-10 Hz (0-100%)
69	BK/LB	A/C WOT Relay Control	On: 1v, Off: 12v
70-71	RD	Vehicle Power	12-14v
73	TN/BK	Injector 5 Control	2.7-4.1 ms
74	BR/YL	Injector 3 Control	2.7-4.1 ms
75	TN	Injector 1 Control	2.7-4.1 ms
76-77	BK/WT	Power Ground	<0.1v
78	PKWT	Coil 3 Driver (Dwell)	5º, at 55 mph: 8º
79	WT/LG	TCIL (lamp) Control	Lamp On: 1v, Off: 12v
80	LB/OR	Fuel Pump Control	On: 1v, Off: 12v
81	WT/OR	EPC Solenoid Control	10v (15-20 psi)
83	WT/LB	IAC Solenoid Control	10v (31% duty cycle)
84	DB/YL	Output Shaft Speed Sensor	0 Hz, 55 mph: 250 Hz
85	DG	CID Sensor Signal	6 Hz, 55 mph: 16 Hz
87	RD/BK	HO2S-21 (B2 S1) Signal	0.1-1.1v
88	LB/RD	MAF Sensor Signal	0.6v, 55 mph: 2.2v
89	GY/WT	TP Sensor Signal	0.9v, 55 mph: 1.1v
90	BR/WT	Reference Voltage	4.9-5.1v
91	GY/RD	Analog Signal Return	<0.050v
92	LG	Brake Switch Signal	Brake Off: 12v, On: 1v
93	RD/WT	HO2S-11 (B1 S1) Heater	1v, Off: 12v
94	YL/LB	HO2S-21 (B2 S1) Heater	1v, Off: 12v
95	WT/BK	HO2S-12 (B1 S2) Heater	1v, Off: 12v
96	TN/YL	HO2S-22 (B2 S2) Heater	1v, Off: 12v
97	RD	Vehicle Power	12-14v
99	LG/OR	Injector 6 Control	2.7-4.1 ms
100	BR/LB	Injector 4 Control	2.7-4.1 ms
101	WT	Injector 2 Control	2.7-4.1 ms
102	---	Not Used	---
103	BK/WT	Power Ground	<0.1v
104	---	Not Used	---

Pin Connector Graphic

PCM 104-PIN CONNECTOR

Terminal View of 104-Pin PCM Wiring Harness Connector

Standard Colors and Abbreviations

Abbreviation	Color	Abbreviation	Color	Abbreviation	Color
BK	Black	GY	Gray	PK	Purple
BL	Blue	GN	Green	RD	Red
BR	Brown	LG	LT Green	TN	Tan
DB	Dark Blue	OR	Orange	WT	White
DG	DK Green	PK	Pink	YL	Yellow

2000-05 E-Series Van 4.2L V6 VIN 2 (A/T) 104-P Connector

PCM Pin #	Wire Color	Circuit Description (104 Pin)	Value at Hot Idle
1	VT/OR	Shift Solenoid 2 Control	12v, 55 mph: 1v
2	---	Not Used	---
3	ORBK	Digital TR1 Sensor Signal	0v, at 55 mph: 11.5v
4-5	---	Not Used	---
6	DG/WT	Turbine Shaft Speed Sensor	120 Hz, 30 mph: 300 Hz
7	---	Not Used	---
8	DB/YL	Intake Manifold Runner Control	5v, 55 mph: 5v
9-12	---	Not Used	---
13	VT	Flash EEPROM	0.1v
14	---	Not Used	---
15	PK/LB	SCP Bus (-) Signal	<0.050v
16	TN/OR	SCP Bus (+) Signal	Digital Signals
17	GY/OR	Passive Antilock RX Signal	Digital Signals
18	WT/LG	Passive Antilock TX Signal	Digital Signals
19	---	Not Used	---
20	VT	Generator 'S' Terminal	Frequency Signal
21	BK/PK	CKP (+) Sensor	430-500 Hz
22	GY/YL	CKP (-) Sensor	430-500 Hz
23	---	Not Used	---
24	BK/WT	Power Ground	<0.1v
25	---	Not Used	---
26	DB/LG	Coil 1 Driver (Dwell)	5°, at 55 mph: 8°
27	OR/YL	Shift Solenoid 1 Control	1v, 55 mph: 1v
28	DB/LG	Low Speed Fan Relay Control	Relay Off: 12v, On: 1v
29	---	Not Used	---
30	DB/LG	Passive Antitheft Indicator Control	Indicator Off: 12v, On: 1v
31	YL/LG	Power Steering Pressure Switch	Straight: 0v, Turned: 12v
32-34	---	Not Used	---
35	RD/LG	HO2S-12 (B1 S2) Signal	0.1-1.1v
36	TN/LB	MAF Sensor Return	<0.050v
37	OR/BK	Transmission Fluid Temperature Sensor	2.10-2.40v
38	LG/RD	Engine Coolant Temperature Sensor	0.5-0.6v
39	GY	Intake Air Temperature Sensor	1.5-2.5v
40	DG/YL	Fuel Pump Monitor	Off: 0v, On: 12v
41	BK/YL	A/C Pressure Switch Signal	A/C On: 12v, Off: 0v
42	BR	Intake Manifold Runner Control Module	12v, 55 mph: 12v
43	---	Not Used	---
44	LG/YL	Starter Relay Circuit	Off: 12v, On: 1v
45	RD/PK	Generator 'L' Terminal	0v or 12v
46	LG/VT	High Speed Cooling Fan Relay Control	Relay Off: 12v, On: 1v
47	BR/PK	EGR VR Solenoid Control	0%, at 55 mph: 45%
48	---	Not Used	---
49	LB/BK	Digital TR2 Sensor	0v, at 55 mph: 11.5v
50	WT/BK	Digital TR4 Sensor	0v, at 55 mph: 11.5v
51	BK/WT	Power Ground	<0.1v
52	RD/LB	Coil 2 Driver (Dwell)	5°, at 55 mph: 8°
53	PK/BK	Shift Solenoid 'C' Control	12v, 55 mph: 1v
54	RD/LB	TCC Solenoid Control	0%, at 55 mph: 95%
55	RD	Direct Battery	12-14v
56	LG/BK	EVAP Canister Purge Solenoid	0-10 Hz (0-100%)
57-59	---	Not Used	---
60	GY/LB	HO2S-11 (B1 S1) Signal	0.1-1.1v
61	VT/LG	HO2S-22 (B2 S2) Signal	0.1-1.1v
62	RD/PK	Fuel Tank Pressure Sensor	2.6v (0" H2O - cap off)
63	---	Not Used	---
64	LB/YL	Digital Transmission Range Sensor	In 'P': 0v, in O/D: 1.7v

2000-02 E-Series Van 4.2L V6 VIN 2 (A/T) 104-P Connector

PCM Pin #	Wire Color	Circuit Description (104 Pin)	Value at Hot Idle
65 ('00)	BR/LG	EGR DPFE Sensor Signal	0.20-1.30v
65 ('01-'02)	BR/LG	EGR DPFE Sensor Signal	0.95-1.05v
66	---	Not Used	---
67	VT/WT	EVAP Canister Vent Solenoid Control	0-10 Hz (0-100%)
68	---	Not Used	---
69	PK/YL	A/C Relay Control	On: 1v, Off: 12v
70	---	Not Used	---
71	RD	Vehicle Power (Start-Run)	12-14v
73	TN/BK	Injector 5 Control	2.7-4.1 ms
74	BR/YL	Injector 3 Control	2.7-4.1 ms
75	TN	Injector 1 Control	2.7-4.1 ms
76	BK/WT	Power Ground	<0.1v
77	---	Not Used	---
78	PK/WT	Coil 3 Driver (Dwell)	5°, at 55 mph: 8°
79	---	Not Used	---
80	LB/OR	Fuel Pump Relay Control	Off: 12v, On: 1v
81	WT/YL	EPC Solenoid Control	10v (15-20 psi)
82	---	Not Used	---
83	WT/LB	IAC Solenoid Control	10v (31% duty cycle)
84	DB/YL	Output Shaft Speed Sensor	0 Hz, 55 mph: 250 Hz
85	DG/OR	Camshaft Position Sensor	6 Hz, 55 mph: 16 Hz
86	TN/LG	A/C Pressure Switch	Switch On: 12v (Open)
87	RD/BK	HO2S-21 (B2 S1) Signal	0.1-1.1v
88	LB/RD	MAF Sensor Signal	0.6v, 55 mph: 2.2v
89	GY/WT	TP Sensor Signal	0.9v, 55 mph: 1.1v
90	BR/WT	Reference Voltage	4.9-5.1v
91	GY/RD	Sensor Ground	<0.050v
92	---	Not Used	---
93	RD/WT	HO2S-11 (B1 S1) Heater Control	1v, Off: 12v
94	YL/LB	HO2S-21 (B2 S1) Heater Control	1v, Off: 12v
95	WT/BK	HO2S-12 (B1 S2) Heater Control	1v, Off: 12v
96	TN/YL	HO2S-22 (B2 S2) Heater Control	1v, Off: 12v
97-98	---	Not Used	---
99	LG/OR	Injector 6 Control	2.7-4.1 ms
100	BR/LB	Injector 4 Control	2.7-4.1 ms
101	WT	Injector 2 Control	2.7-4.1 ms
102	---	Not Used	---
103	BK/WT	Power Ground	<0.1v
104	---	Not Used	---

Pin Connector Graphic

```
                  PCM 104-PIN CONNECTOR
    1 ●●●●●●●●●●●●●     ●●●●●●●●●●●●● 26
   27 ●●●●●●●●●●●●●          ●●●●●●●●●●●●● 52
   53 ●●●●●●●●●●●●●    ●    ●●●●●●●●●●●●● 78
   79 ●●●●●●●●●●●●●     ●●●●●●●●●●●●● 104
```

Terminal View of 104-Pin PCM Wiring Harness Connector

Standard Colors and Abbreviations

Abbreviation	Color	Abbreviation	Color	Abbreviation	Color
BK	Black	GY	Gray	PK	Purple
BL	Blue	GN	Green	RD	Red
BR	Brown	LG	LT Green	TN	Tan
DB	Dark Blue	OR	Orange	WT	White
DG	DK Green	PK	Pink	YL	Yellow

1997-99 E-Series Van 4.6L V8 VIN 6 (A/T) 104-P Connector

PCM Pin #	Wire Color	Circuit Description (104 Pin)	Value at Hot Idle
1	VTN/ORG	Shift Solenoid 2 Control	12v, 55 mph: 1v
2	PK/LG	MIL (lamp) Control	MIL On: <1v, Off: 12v
3	YL/BK	Digital TR1 Sensor Signal	0v, at 55 mph: 11.5v
12	YL/WT	Fuel Level Indicator Signal	1.7v (1/2 full)
13	VT	Flash EEPROM	0.1v
14	LB/BK	4x4 Indicator Signal	Off: 12v, On: 1v
15	PK/LB	Data Bus (-) Signal	Digital Signals
16	TN/OR	Data Bus (+) Signal	Digital Signals
19	DB/WT	Automatic Ride Control	Digital Signals
21	DB	CKP (+) Sensor	430-475 Hz
22	GY	CKP (-) Sensor	430-475 Hz
24	BK/WT	Power Ground	<0.1v
25	BK/LB	Case Ground	<0.050v
26	DB/LG	Coil 1 Driver (Dwell)	5°, at 55 mph: 8°
27	OR/YL	Shift Solenoid 1 Control	1v, 55 mph: 1v
29	TN/WT	TCS (switch) Signal	TCS & O/D On: 12v
30	DG	Octane Adjustment	9.3v (switch shorted: 0v)
33	PK/OR	VSS (-) Signal	0 Hz, at 55 mph: 125 Hz
35	RD/LG	HO2S-12 (B1 S2) Signal	0.1-1.1v
36	TN/LB	MAF Sensor Return	<0.050v
37	OR/BK	Transmission Fluid Temperature Sensor	2.10-2.40v
38	LG/RD	Engine Coolant Temperature Sensor	0.5-0.6v
39	GY	Intake Air Temperature Sensor	1.5-2.5v
40	DG/YL	Fuel Pump Monitor	On: 12v, Off: 0v
41	BK/YL	A/C Cycling Clutch Switch	A/C On: 12v, off: 0v
42	BR	Inlet Air Control Solenoid	12v, 55 mph: 12v
43	OR/LG	Overhead Trip Computer	Digital Signals
45	RD/WT	CHTIL (lamp) Control	Lamp On: 1v, Off: 12v
46	BR	Intake Manifold Tuning Valve	12v, 55 mph: 12v
47	BR/PK	VR Solenoid Control	0%, at 55 mph: 45%
48	WT/PK	Clean Tachometer Output Signal	39-49 Hz
49	LB/BK	Digital TR2 Sensor	0v, at 55 mph: 11.5v
50	LG/RD	Digital TR4 Sensor	0v, at 55 mph: 11.5v
51	BK/WT	Power Ground	<0.1v
52	RD/LB	Coil 2 Driver (Dwell)	5°, at 55 mph: 8°
54	PK/YL	TCC Solenoid Control	0%, at 55 mph: 95%
55	Y	Keep Alive Power	12-14v
56	LG/BK	EVAP Purge Solenoid	0-10 Hz (0-100%)
57	YL/RD	Knock Sensor 1 Signal	0v
58	GY/BK	VSS (+) Signal	0 Hz, at 55 mph: 125 Hz
60	GY/LB	HO2S-11 (B1 S1) Signal	0.1-1.1v
61	PK/LG	HO2S-22 (B2 S2) Signal	0.1-1.1v
62	RD/PK	Fuel Tank Pressure Sensor	2.6v (0" H2O - cap off)

1997-99 E-Series Van 4.6L V8 VIN 6 (A/T) 104-P Connector

PCM Pin #	Wire Color	Circuit Description (104 Pin)	Value at Hot Idle
64	LB/YL	Clutch Pedal Position Switch	5v (Clutch In: 0v)
64	LB/YL	Transmission Range Sensor	In 'P': 0v, in O/D: 1.7v
65	BR/LG	EGR DPFE Sensor Signal	0.20-1.30v
66	YL/LG	Cylinder Head Temperature Sensor	0.6 to 1.7v (194°F)
67	PKWT	EVAP CV Solenoid	0-10 Hz (0-100%)
69	PK/YL	A/C WOT Relay Control	On: 1v, Off: 12v
71, 97	RD	Vehicle Power	12-14v
72	TN/RD	Injector 7 Control	2.7-4.1 ms
73	TN/BK	Injector 5 Control	2.7-4.1 ms
74	BR/YL	Injector 3 Control	2.7-4.1 ms
75	TN	Injector 1 Control	2.7-4.1 ms
76, 77, 103	BK/WT	Power Ground	<0.1v
78	PK/WT	Coil 3 Driver (Dwell)	5°, at 55 mph: 8°
79	WT/LG	TCIL (lamp) Control	Lamp On: 1v, Off: 12v
80	LB/OR	Fuel Pump Control	On: 1v, Off: 12v
81	WT/OR	EPC Solenoid Control	9.1v (6 psi)
83	WT/LB	IAC Solenoid Control	10v (31% duty cycle)
84	DG/WT	Output Shaft Speed Sensor	0 Hz, 55 mph: 250 Hz
85	DG	CID Sensor Signal	6 Hz, 55 mph: 15 Hz
87	RD/BK	HO2S-21 (B2 S1) Signal	0.1-1.1v
88	LB/RD	MAF Sensor Signal	0.6v, 55 mph: 2.1v
89	GY/WT	TP Sensor Signal	0.53-1.27v
90	BR/WT	Reference Voltage	4.9-5.1v
91	GY/RD	Analog Signal Return	<0.050v
92	LG	Brake Switch Signal	Brake Off: 12v, On: 1v
93	RD/WT	HO2S-11 (B1 S1) Heater	1v, Off: 12v
94	YL/LB	HO2S-21 (B2 S1) Heater	1v, Off: 12v
95	WT/BK	HO2S-12 (B1 S2) Heater	1v, Off: 12v
96	TN/YL	HO2S-22 (B2 S2) Heater	1v, Off: 12v
98	LB	Injector 8 Control	2.7-4.1 ms
99	LG/OR	Injector 6 Control	2.7-4.1 ms
100	BR/LB	Injector 4 Control	2.7-4.1 ms
101	WT	Injector 2 Control	2.7-4.1 ms
102	---	Not Used	---
104	RD/YL	Coil 4 Driver (Dwell)	5°, at 55 mph: 8°

Pin Connector Graphic

PCM 104-PIN CONNECTOR

Terminal View of 104-Pin PCM Wiring Harness Connector

Standard Colors and Abbreviations

Abbreviation	Color	Abbreviation	Color	Abbreviation	Color
BK	Black	GY	Gray	PK	Purple
BL	Blue	GN	Green	RD	Red
BR	Brown	LG	LT Green	TN	Tan
DB	Dark Blue	OR	Orange	WT	White
DG	DK Green	PK	Pink	YL	Yellow

2000-05 E-Series 4.6L V8 VIN W (All) 104 Pin Connector

PCM Pin #	Wire Color	Circuit Description (104 Pin)	Value at Hot Idle
1	OR/YL	COP 6 Driver (Dwell)	5°, at 55 mph: 8°
2	PK/LG	MIL (lamp) Control	MIL On: <1v, Off: 12v
3	BK/WT	Power Ground	<0.1v
4-5	---	Not Used	---
6	OR/YL	Shift Solenoid 1 Control	1v, 55 mph: 1v
9	YL/WT	Fuel Pump Monitor	On: 12v, Off: 0v
10	---	Not Used	---
11	VT/OR	Shift Solenoid 2 Control	12v, 55 mph: 1v
12	WT/LG	TCIL (lamp) Control	Indicator Off: 12v, On: 1v
13	VT	Flash EEPROM	0.1v
14	OG/LB	4WD Low Signal	0v or 12v
15	PK/LB	SCP Bus (-) Signal	<0.050v
16	TN/OR	SCP Bus (+) Signal	Digital Signals
17-19	---	Not Used	---
20	BR/OR	Coast Clutch Solenoid Control	12v, 55 mph: 12v
21	DB	CKP (+) Sensor	390-430 Hz
22	GY	CKP (-) Sensor	390-430 Hz
23	---	Not Used	---
24	BK/WT	Power Ground	<0.1v
25	BK/LB	Case Ground	<0.050v
26	LG/WT	COP 1 Driver (Dwell)	5°, at 55 mph: 8°
27	LG/YL	COP 5 Driver (Dwell)	5°, at 55 mph: 8°
28	---	Not Used	---
29	TN/WT	PTO (switch) Signal	PTO On: 12v
30-31	---	Not Used	---
32	DG/VT	Knock Sensor 1 Ground	<0.050v
33	---	Not Used	---
34	YL/BK	Transmission Range Sensor	In 'P': 0v, in O/D: 1.7v
35	RD/LG	HO2S-12 (B1 S2) Signal	0.1-1.1v
36	TN/LB	MAF Sensor Return	<0.050v
37	OR/BK	Transmission Fluid Temperature Sensor	2.10-2.40v
38	LG/RD	Engine Oil Temperature Sensor	-40°F: 4.7v, 230°F: 0.36v
39	GY	Intake Air Temperature Sensor	1.5-2.5v
40	DG/YL	Fuel Pump Monitor	On: 12v, Off: 0v
41	TN/LG	A/C Pressure Switch Signal	A/C On: 12v, Off: 0v
42-44	---	Not Used	---
45	YL/LB	Engine Coolant Temperature Sensor	0.5-0.6v
46	BR	Intake Manifold Tuning Valve	12v, 55 mph: 12v
47	BR/PK	EGR VR Solenoid Control	0%, at 55 mph: 45%
48	WT/PK	Clean Tachometer Output Signal	DC pulse signals
49	LB/BK	Digital TR2 Sensor	0v, at 55 mph: 11.5v
50	LG/RD	Digital TR4 Sensor	0v, at 55 mph: 11.5v
51	BK/WT	Power Ground	<0.1v
52	WT/PK	COP 3 Driver (Dwell)	5°, at 55 mph: 8°
53	DG/VT	COP 4 Driver (Dwell)	5°, at 55 mph: 8°
54	VT/YL	TCC Solenoid Control	0%, at 55 mph: 95%
55	YL	Keep Alive Power	12-14v
56	LG/BK	EVAP Purge Solenoid	0-10 Hz (0-100%)
57	YL/RD	Knock Sensor (-) Signal	<0.050v
58	---	Not Used	---
59	DG/WT	Turbine Shaft Speed Sensor	120 Hz, 30 mph: 300 Hz
60	GY/LB	HO2S-11 (B1 S1) Signal	0.1-1.1v
61	VT/LG	HO2S-22 (B2 S2) Signal	0.1-1.1v
62	RD/PK	Fuel Tank Pressure Sensor	2.6v (0" H2O - cap off)
63	---	Not Used	---
64	LB/YL	Digital TR3A Sensor	0v, at 55 mph: 11.5v

2000-05 E-Series Van 4.6L V8 VIN W (All) 104 Pin Connector

PCM Pin #	Wire Color	Circuit Description (104 Pin)	Value at Hot Idle
65 '00)	BR/LG	EGR DPFE Sensor Signal	0.20-1.30v
65 ('01-'02)	BR/LG	EGR DPFE Sensor Signal	0.95-1.05v
66	YL/LG	Cylinder Head Temperature Sensor	0.6 to 1.7v (194°F)
67	VT/WT	EVAP CV Solenoid	0-10 Hz (0-100%)
68	GY/BK	Vehicle Speed Output	0 Hz, at 55 mph: 125 Hz
69	BK/LB	A/C Relay Control	On: 1v, Off: 12v
71	RD	Vehicle Power (Start Run)	12-14v
72	TN/RD	Injector 7 Control	2.7-4.1 ms
73	TN/BK	Injector 5 Control	2.7-4.1 ms
74	BR/YL	Injector 3 Control	2.7-4.1 ms
75	TN	Injector 1 Control	2.7-4.1 ms
76	BK/WT	Power Ground	<0.1v
77	BK/WT	Power Ground	<0.1v
78	PK/LB	COP 7 Driver (Dwell)	5°, at 55 mph: 8°
79	WT/RD	COP 8 Driver (Dwell)	5°, at 55 mph: 8°
80	LB/OR	Fuel Pump Control	On: 1v, Off: 12v
81	WT/YL	EPC Solenoid Control	9.1v (6 psi)
83	WT/LB	IAC Solenoid Control	10v (31% duty cycle)
84	DG/YL	Output Shaft Speed Sensor	0 Hz, 55 mph: 250 Hz
85	DG	Camshaft Position Sensor	6 Hz, 55 mph: 15 Hz
87	RD/BK	HO2S-21 (B2 S1) Signal	0.1-1.1v
88	LB/RD	MAF Sensor Signal	0.6v, 55 mph: 2.1v
89	GY/WT	TP Sensor Signal	0.53-1.27v
90	BR/WT	Reference Voltage	4.9-5.1v
91	GY/RD	Analog Signal Return	<0.050v
92	LG	Brake Position Switch	Brake Off: 12v, On: 1v
93	RD/WT	HO2S-11 (B1 S1) Heater	1v, Off: 12v
94	YL/LB	HO2S-21 (B2 S1) Heater	1v, Off: 12v
95	WT/BK	HO2S-12 (B1 S2) Heater	1v, Off: 12v
96	TN/YL	HO2S-22 (B2 S2) Heater	1v, Off: 12v
97	RD	Vehicle Power (Start Run)	12-14v
98	LB	Injector 8 Control	2.7-4.1 ms
99	LG/OR	Injector 6 Control	2.7-4.1 ms
100	BR/LB	Injector 4 Control	2.7-4.1 ms
101	WT	Injector 2 Control	2.7-4.1 ms
103	BK/WT	Power Ground	<0.1v
104	PK/WT	COP 2 Driver (Dwell)	5°, at 55 mph: 8°

Pin Connector Graphic

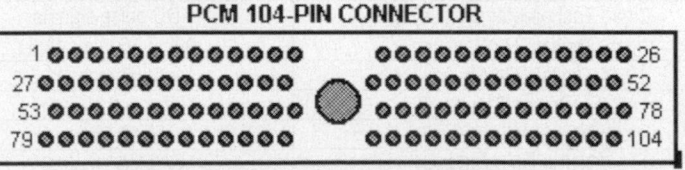

Standard Colors and Abbreviations

Abbreviation	Color	Abbreviation	Color	Abbreviation	Color
BK	Black	GY	Gray	PK	Purple
BL	Blue	GN	Green	RD	Red
BR	Brown	LG	LT Green	TN	Tan
DB	Dark Blue	OR	Orange	WT	White
DG	DK Green	PK	Pink	YL	Yellow

1990 E-Series 4.9L I6 OHV MFI VIN Y (All) 60 Pin Connector

PCM Pin #	Wire Color	Circuit Description (60 Pin)	Value at Hot Idle
1	BK/OR	Keep Alive Power	12-14v
3	DG/WT	VSS (+) Signal	0 Hz, at 55 mph: 125 Hz
4	DG/YL	Ignition Diagnostic Monitor	20-31 Hz
6	OR/YL	VSS (-) Signal	0 Hz, at 55 mph: 125 Hz
7	LG/YL	Engine Coolant Temperature Sensor	0.5-0.6v
8	OR/BL	Fuel Pump Monitor	On: 12v, Off: 0v
9	BK/OR	Data Bus (-)	Digital Signals
10	BK/YL	A/C Pressure Switch Signal	A/C On: 12v, off: 0v
11	WT/BK	Air Management 2 Solenoid	On: 1v, Off: 12v
16	BK/OR	Ignition System Ground	<0.050v
17	TN/RD	Self-Test Output & MIL	MIL On: 1v, Off: 12v
20	BK	PCM Case Ground	<0.050v
21	GY/WT	IAC Solenoid Control	9.3-10.1v
22	TN/LG	Fuel Pump Control	On: 1v, Off: 12v
23	LG/BK	Knock Sensor 1 Signal	0v
24	YL/LG	Power Steering Pressure Switch	Straight: 0v,Turned: 12v
25	YL/RD	ACT Sensor Signal	1.5-2.5v
26	OR/WT	Reference Voltage	4.9-5.1v
27	BR/LG	EVP Sensor Signal	0.4v, 55 mph: 3.1v
28	OR/BK	Data Bus (+)	Digital Signals
29	DG/PK	HO2S-11 (B1 S1) Signal	0.1-1.1v
30	BL/YL	A/T: Neutral Drive Switch	In 'P': 0v, Others: 5v
30	BL/YL	M/T: Clutch Engage Switch	Clutch Out: 5v, In: 0v
31	GY/YL	Canister Purge Solenoid	12v, 55 mph: 1v
33	DG	VR Solenoid Control	0%, at 55 mph: 45%
36	YL/LG	Spark Output Signal	6.93v (50%)
37	RD	Vehicle Power	12-14v
40	BK/LG	Power Ground	<0.1v
43	LG/PK	A/C Demand Switch	A/C On: 12v, Off: 0v
45	DB/LG	MAP Sensor Signal	107 Hz
46	BK/WT	Analog Signal Return	<0.050v
47	DG/LG	TP Sensor Signal	1v, 55 mph: 2.1v
48	WT/RD	Self-Test Indicator Signal	STI On: 1v, Off: 5v
49	OR	HO2S-11 (B1 S1) Ground	<0.050v
51	WT/RD	Air Management 1 Solenoid	On: 1v, Off: 12v
56	DB	PIP Sensor Signal	6.93v (50%)
57	RD	Vehicle Power	12-14v
58	TN/OR	Injector Bank 1 (INJ 1, 3 & 5)	6.4-6.8 ms
59	TN/RD	Injector Bank 2 (INJ 2, 4 & 6)	6.4-6.8 ms
60	BK/LG	Power Ground	<0.1v

Pin Connector Graphic

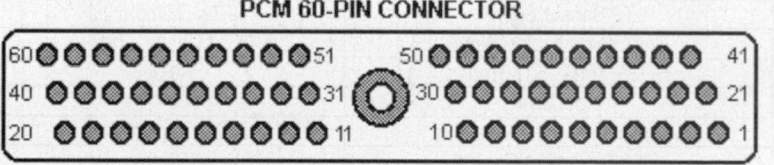

Standard Colors and Abbreviations

Abbreviation	Color	Abbreviation	Color	Abbreviation	Color
BK	Black	GY	Gray	PK	Purple
BL	Blue	GN	Green	RD	Red
BR	Brown	LG	LT Green	TN	Tan
DB	Dark Blue	OR	Orange	WT	White
DG	DK Green	PK	Pink	YL	Yellow

1990 E-Series 4.9L I6 OHV VIN Y (E4OD) 60 Pin Connector

PCM Pin #	Wire Color	Circuit Description (60 Pin)	Value at Hot Idle
1	BK/OR	Keep Alive Power	12-14v
2	LG	Brake Switch Signal	Brake Off: 12v, On: 1v
3	DG/WT	VSS (+) Signal	0 Hz, at 55 mph: 125 Hz
4	DG/YL	Ignition Diagnostic Monitor	20-31 Hz
5	---	Not Used	---
6	OR/YL	VSS (-) Signal	0 Hz, at 55 mph: 125 Hz
7	LG/YL	Engine Coolant Temperature Sensor	0.5-0.6v
8	OR/BL	Fuel Pump Monitor	On: 12v, Off: 0v
9	BK/OR	Data Bus (-)	Digital Signals
10	BK/YL	A/C Pressure Switch Signal	A/C On: 12v, off: 0v
11	WT/BK	Air Management 2 Solenoid	On: 1v, Off: 12v
12-15	---	Not Used	---
16	BK/OR	Ignition System Ground	<0.050v
17	TN/RD	Self-Test Output & MIL	MIL On: 1v, Off: 12v
18	---	Not Used	---
19	DG/PK	Shift Solenoid 2 Control	12v, 55 mph: 1v
20	BK	PCM Case Ground	<0.050v
21	GY/WT	IAC Solenoid Control	9.3-10.1v
22	TN/LG	Fuel Pump Control	On: 1v, Off: 12v
23	LG/BK	Knock Sensor 1 Signal	0v
24	YL/LG	Power Steering Pressure Switch	Straight: 0v, Turned: 12v
25	YL/RD	Intake Air Temperature Sensor	1.5-2.5v
26	OR/WT	Reference Voltage	4.9-5.1v
27	BR/LG	EVP Sensor Signal	0.4v, 55 mph: 3.1v
28	OR/BK	Data Bus (+)	Digital Signals
29	DG/PK	HO2S-11 (B1 S1) Signal	0.1-1.1v
30	BL/YL	MLP Sensor Signal	In 'P': 0v, in O/D: 5v
31	GY/YL	Canister Purge Solenoid	12v, 55 mph: 1v
32	LG/WT	OCIL (lamp) Control	Lamp On: 1v, Off: 12v
33	DG	VR Solenoid Control	0%, at 55 mph: 45%
34-35	---	Not Used	---
36	YL/LG	Spark Output Signal	6.93v (50%)
37	RD	Vehicle Power	12-14v
38	BL/YL	EPC Solenoid Control	9.1v (4 psi)
39	---	Not Used	---
40	BK/LG	Power Ground	<0.1v
41	TN/BL	Overdrive Cancel Switch	OCS & O/D On: 0.1v
42	OR/BK	TOT Sensor Signal	2.10-2.40v
43	LG/PK	A/C Demand Switch	A/C On: 12v, Off: 0v
44	---	Not Used	---
45	DB/LG	MAP Sensor Signal	107 Hz
46	BK/WT	Analog Signal Return	<0.050v
47	DG/LG	TP Sensor Signal	0.7v, 55 mph: 2.1v
48	WT/RD	Self-Test Indicator Signal	STI On: 1v, Off: 5v
49	OR	HO2S-11 (B1 S1) Ground	<0.050v
50	---	Not Used	---
51	WT/RD	Air Management 1 Solenoid	On: 1v, Off: 12v
52	OR/YL	Shift Solenoid 1 Control	1v, 55 mph: 1v
53	PK/YL	TCC Solenoid Control	12v, 55 mph: 1v
54	---	Not Used	---
55	BR	Coast Clutch Solenoid	12v, 55 mph: 12v
56	DB	PIP Sensor Signal	6.93v (50%)
57	R	Vehicle Power	12-14v
58	TN/OR	Injector Bank 1 (INJ 1, 3 & 5)	6.4-6.8 ms
59	TN/RD	Injector Bank 2 (INJ 2, 4 & 6)	6.4-6.8 ms
60	BK/LG	Power Ground	<0.1v

1991 E-Series 4.9L I6 OHV VIN Y (All) 60 Pin Connector

PCM Pin #	Wire Color	Circuit Description (60 Pin)	Value at Hot Idle
1	BK/OR	Keep Alive Power	12-14v
2, 5	---	Not Used	---
3	DG/WT	VSS (+) Signal	0 Hz, at 55 mph: 125 Hz
4	TN/YL	Ignition Diagnostic Monitor	20-31 Hz
6	OR/YL	VSS (-) Signal	0 Hz, at 55 mph: 125 Hz
7	LG/PK	Engine Coolant Temperature Sensor	0.5-0.6v
8	DG/YL	Fuel Pump Monitor	On: 12v, Off: 0v
9	BK/OR	Data Bus (-) California	Digital Signals
10	BK/YL	A/C Pressure Switch Signal	A/C On: 12v, off: 0v
11	BR	Air Management 2 Solenoid	On: 1v, Off: 12v
12-15	---	Not Used	---
16	OR/RD	Ignition System Ground	<0.050v
17	TN/RD	Self-Test Output & MIL	MIL On: 1v, Off: 12v
18-19	---	Not Used	---
20	BK	PCM Case Ground	<0.050v
21	GY/WT	IAC Solenoid Control	8.5-10.2v
22	TN/LG	Fuel Pump Control	On: 1v, Off: 12v
23	LG/BK	Knock Sensor 1 Signal	0v
24	---	Not Used	---
25	YL/RD	ACT Sensor Signal	1.5-2.5v
26	BR/WT	Reference Voltage	4.9-5.1v
27	BR/LG	EVP Sensor Signal	0.4v, 55 mph: 3.1v
28	OR/BK	Data Bus (+)	Digital Signals
29	RD/BK	HO2S-11 (B1 S1) Signal	0.1-1.1v
30	WT/RD	A/T: Neutral Drive Switch	In 'P': 0v, Others: 5v
30	WT/RD	M/T: Clutch Engage Switch	Clutch Out: 5v, In: 0v
31	GY/YL	Canister Purge Solenoid	12v, 55 mph: 1v
32	---	Not Used	---
33	BR/PK	VR Solenoid Control	0%, at 55 mph: 45%
34-35	---	Not Used	---
36	YL/LG	Spark Output Signal	6.93v (50%)
37	Rd	Vehicle Power	12-14v
38-39	---	Not Used	---
40	BK/LG	Power Ground	<0.1v
41-42	---	Not Used	---
43	PK	A/C Demand Switch	A/C On: 12v, Off: 0v
44	---	Not Used	---
45	DB/LG	MAP Sensor Signal	107 Hz
46	GY/RD	Analog Signal Return	<0.050v
47	GY/WT	TP Sensor Signal	0.7v, 55 mph: 2.1v
48	WT/RD	Self-Test Indicator Signal	STI On: 1v, Off: 5v
49	OR	HO2S-11 (B1 S1) Ground	<0.050v
50	---	Not Used	---
51	OR	Air Management 1 Solenoid	On: 1v, Off: 12v
52-55	---	Not Used	---
56	DB	PIP Sensor Signal	6.93v (50%)
57	RD	Vehicle Power	12-14v
58	TN/OR	Injector Bank 1 (INJ 1, 3 & 5)	5.0-6.4 ms
59	TN/RD	Injector Bank 2 (INJ 2, 4 & 6)	5.0-6.4 ms
60	BK/LG	Power Ground	<0.1v

Pin Connector Graphic

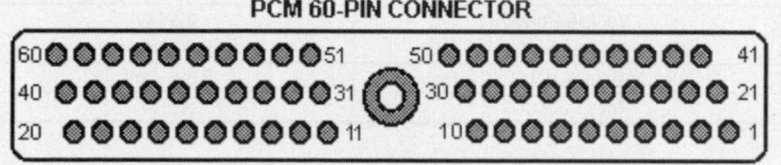

Terminal View of 60-Pin PCM Harness Connector

1991 E-Series 4.9L I6 OHV VIN Y (E4OD) 60 Pin Connector

Y	Wire Color	Circuit Description (60 Pin)	Value at Hot Idle
1	BK/OR	Keep Alive Power	12-14v
2	LG	Brake Switch Signal	Brake Off: 12v, On: 1v
3	DG/WT	VSS (+) Signal	0 Hz, at 55 mph: 125 Hz
4	TN/YL	Ignition Diagnostic Monitor	20-31 Hz
5	---	Not Used	---
6	OR/YL	VSS (-) Signal	0 Hz, at 55 mph: 125 Hz
7	LG/PK	Engine Coolant Temperature Sensor	0.5-0.6v
8	DG/YL	Fuel Pump Monitor	On: 12v, Off: 0v
9	BK/OR	Data Bus (-)	Digital Signals
10	BK/YL	A/C Pressure Switch Signal	A/C On: 12v, off: 0v
11	BR	Air Management 2 Solenoid	On: 1v, Off: 12v
12-15	---	Not Used	---
16	OR/RD	Ignition System Ground	<0.050v
17	TN/RD	Self-Test Output & MIL	MIL On: 1v, Off: 12v
18	---	Not Used	---
19	PK	Shift Solenoid 2 Control	12v, 55 mph: 1v
20	BK	PCM Case Ground	<0.050v
21	GY/WT	IAC Solenoid Control	8.5-10.3v
22	TN/LG	Fuel Pump Control	On: 1v, Off: 12v
23	LG/BK	Knock Sensor 1 Signal	0v
24	---	Not Used	---
25	YL/RD	ACT Sensor Signal	1.5-2.5v
26	BR/WT	Reference Voltage	4.9-5.1v
27	BR/LG	EVP Sensor Signal	0.4v, 55 mph: 3.1v
28	OR/BK	Data Bus (+)	Digital Signals
29	RD/BK	HO2S-11 (B1 S1) Signal	0.1-1.1v
30	WT/RD	MLP Sensor Signal	In 'P': 0v, in O/D: 5v
31	GY/YL	Canister Purge Solenoid	12v, 55 mph: 1v
32	WT/LG	OCIL (lamp) Control	Lamp On: 1v, Off: 12v
33	BR/PK	VR Solenoid Control	0%, at 55 mph: 45%
34-35	---	Not Used	---
36	YL/LG	Spark Output Signal	6.93v (50%)
37	RD	Vehicle Power	12-14v
38	BL/YL	EPC Solenoid Control	7.7v (5 psi)
39	---	Not Used	---
40	BK/LG	Power Ground	<0.1v
41	TN/WT	Overdrive Cancel Switch	OCS & O/D On: 0.1v
42	OR/BK	TOT Sensor Signal	2.10-2.40v
43	PK	A/C Demand Switch	A/C On: 12v, Off: 0v
44	---	Not Used	---
45	DB/LG	MAP Sensor Signal	107 Hz
46	GY/RD	Analog Signal Return	<0.050v
47	GY/WT	TP Sensor Signal	0.7v, 55 mph: 2.1v
48	WT/RD	Self-Test Indicator Signal	STI On: 1v, Off: 5v
49	OR	HO2S-11 (B1 S1) Ground	<0.050v
50	---	Not Used	---
51	OR	Air Management 1 Solenoid	On: 1v, Off: 12v
52	OR	Shift Solenoid 1 Control	1v, 55 mph: 12v
53	PK/YL	TCC Solenoid Control	0%, at 55 mph: 95%
54	---	Not Used	---
55	BR/OR	Coast Clutch Switch Signal	12v, 55 mph: 12v
56	DB	PIP Sensor Signal	6.93v (50%)
57	RD	Vehicle Power	12-14v
58	TN/OR	Injector Bank 1 (INJ 1, 3 & 5)	5.6-6.4 ms
59	TN/RD	Injector Bank 2 (INJ 2, 4 & 6)	5.6-6.4 ms
60	BK/LG	Power Ground	<0.1v

1992-93 E-Series 4.9L I6 OHV VIN Y (All) 60 Pin Connector

PCM Pin #	Wire Color	Circuit Description (60 Pin)	Value at Hot Idle
1	YL	Keep Alive Power	12-14v
2, 5	---	Not Used	---
3	GY/BK	VSS (+) Signal	0 Hz, at 55 mph: 125 Hz
4	TN/YL	Ignition Diagnostic Monitor	20-31 Hz
6	PK/OR	VSS (-) Signal	0 Hz, at 55 mph: 125 Hz
7	LG/RD	Engine Coolant Temperature Sensor	0.5-0.6v
8	DG/YL	Fuel Pump Monitor	On: 12v, Off: 0v
9	PK/BL	Data Bus (-)	Digital Signals
10	BK/YL	A/C Pressure Switch Signal	A/C On: 12v, off: 0v
11	BR	Air Management 2 Solenoid	On: 1v, Off: 12v
12-15	---	Not Used	---
16	OR/RD	Ignition System Ground	<0.050v
17	PK/LG	Self-Test Output & MIL	MIL On: 1v, Off: 12v
18-19, 24	---	Not Used	---
20	BK	PCM Case Ground	<0.050v
21	WT/BL	IAC Solenoid Control	8.2-10.1v
22	BL/OR	Fuel Pump Control	On: 1v, Off: 12v
23	YL/RD	Knock Sensor 1 Signal	0v
25	GY	ACT Sensor Signal	1.5-2.5v
26	BR/WT	Reference Voltage	4.9-5.1v
27	BR/LG	EVP Sensor Signal	0.4v, 55 mph: 3.1v
28	TN/OR	Data Bus (+)	Digital Signals
29	GY/BL	HO2S-11 (B1 S1) Signal	0.1-1.1v
30	BL/YL	A/T: Neutral Drive Switch	In 'P': 0v, Others: 5v
30	BL/YL	M/T: Clutch Engage Switch	Clutch Out: 5v, In: 0v
31	GY/YL	Canister Purge Solenoid	12v, 55 mph: 1v
32	---	Not Used	---
33	BR/PK	VR Solenoid Control	0%, at 55 mph: 45%
34-35	---	Not Used	---
36	PK	Spark Output Signal	6.93v (50%)
37	RD	Vehicle Power	12-14v
38-39	---	Not Used	---
40	BK/WT	Power Ground	<0.1v
41-42	---	Not Used	---
43	PK	A/C Demand Switch	A/C On: 12v, Off: 0v
44	---	Not Used	---
45	LG/BK	MAP Sensor Signal	107 Hz
46	GY/RD	Analog Signal Return	<0.050v
47	GY/WT	TP Sensor Signal	0.7v, 55 mph: 2.1v
48	WT/PK	Self-Test Indicator Signal	STI On: 1v, Off: 5v
49	OR	HO2S-11 (B1 S1) Ground	<0.050v
50	---	Not Used	---
51	WT/OR	Air Management 1 Solenoid	On: 1v, Off: 12v
52-54	---	Not Used	---
55	---	Not Used	---
56	GY/OR	PIP Sensor Signal	6.93v (50%)
57	RD	Vehicle Power	12-14v
58	TN	Injector Bank 1 (INJ 1, 3 & 5)	6.8-7.0 ms
59	WT	Injector Bank 2 (INJ 2, 4 & 6)	6.8-7.0 ms
60	BK/WT	Power Ground	<0.1v

1992-93 E-Series 4.9L I6 VIN Y (E4OD) 60 Pin Connector

PCM Pin #	Wire Color	Circuit Description (60 Pin)	Value at Hot Idle
1	YL	Keep Alive Power	12-14v
2	LG	Brake Switch Signal	Brake Off: 12v, On: 1v
3	GY/BK	VSS (+) Signal	0 Hz, at 55 mph: 125 Hz
4	TN/YL	Ignition Diagnostic Monitor	20-31 Hz
5	---	Not Used	---
6	PK/OR	VSS (-) Signal	0 Hz, at 55 mph: 125 Hz
7	LG/RD	Engine Coolant Temperature Sensor	0.5-0.6v
8	DG/YL	Fuel Pump Monitor	On: 12v, Off: 0v
9	PK/BL	Data Bus (-)	Digital Signals
10	BK/YL	A/C Pressure Switch Signal	A/C On: 12v, off: 0v
11	BR	Air Management 2 Solenoid	On: 1v, Off: 12v
12-15	---	Not Used	---
16	OR/RD	Ignition System Ground	<0.050v
17	PK/LG	Self-Test Output & MIL	MIL On: 1v, Off: 12v
18	---	Not Used	---
19	PK/OR	Shift Solenoid 2 Control	12v, 55 mph: 1v
20	BK	PCM Case Ground	<0.050v
21	WT/BL	IAC Solenoid Control	8.2-10.1v
22	BL/OR	Fuel Pump Control	On: 1v, Off: 12v
23	YL/RD	Knock Sensor 1 Signal	0v
24	---	Not Used	---
25	GY	ACT Sensor Signal	1.5-2.5v
26	BR/WT	Reference Voltage	4.9-5.1v
27	BR/LG	EVP Sensor Signal	0.4v, 55 mph: 3.1v
28	TN/OR	Data Bus (+)	Digital Signals
29	GY/BL	HO2S-11 (B1 S1) Signal	0.1-1.1v
30	BL/YL	MLP Sensor Signal	In 'P': 0v, in O/D: 5v
31	GY/YL	Canister Purge Solenoid	12v, 55 mph: 1v
32	WT/LG	OCIL (lamp) Control	Lamp On: 1v, Off: 12v
33	BR/PK	VR Solenoid Control	0%, at 55 mph: 45%
34-35	---	Not Used	---
36	PK	Spark Output Signal	6.93v (50%)
37	RD	Vehicle Power	12-14v
38	WT/YL	EPC Solenoid Control	7.7-8.7v
39	---	Not Used	---
40	BK/WT	Power Ground	<0.1v
41	TN/WT	Overdrive Cancel Switch	OCS & O/D On: 0.1v
42	OR/BK	TOT Sensor Signal	2.10-2.40v
43	PK	A/C Demand Switch	A/C On: 12v, Off: 0v
44	---	Not Used	---
45	LG/BK	MAP Sensor Signal	107 Hz
46	GY/RD	Analog Signal Return	<0.050v
47	GY/WT	TP Sensor Signal	0.7v, 55 mph: 2.1v
48	WT/PK	Self-Test Indicator Signal	STI On: 1v, Off: 5v
49	OR	HO2S-11 (B1 S1) Ground	<0.050v
50	---	Not Used	---
51	WT/OR	Air Management 1 Solenoid	On: 1v, Off: 12v
52	OR/YL	Shift Solenoid 1 Control	1v, 55 mph: 12v
53	PK/YL	TCC Solenoid Control	0%, at 55 mph: 95%
54	---	Not Used	---
55	BR/OR	Coast Clutch Solenoid	12v, 55 mph: 12v
56	GY/OR	PIP Sensor Signal	6.93v (50%)
57	RD	Vehicle Power	12-14v
58	TN	Injector Bank 1 (INJ 1, 3 & 5)	6.2-7.4 ms
59	WT	Injector Bank 2 (INJ 2, 4 & 6)	6.2-7.4 ms
60	BK/WT	Power Ground	<0.1v

1994-95-Series 4.9L I6 OHV VIN Y (All) 60 Pin Connector

PCM Pin #	Wire Color	Circuit Description (60 Pin)	Value at Hot Idle
1	YL	Keep Alive Power	12-14v
2	LG	Brake Switch Signal	Brake Off: 12v, On: 1v
3	GY/BK	VSS (+) Signal	0 Hz, at 55 mph: 125 Hz
4	WT/PK	Ignition Diagnostic Monitor	20-31 Hz
5	---	Not Used	---
6	PK/OR	VSS (-) Signal	0 Hz, at 55 mph: 125 Hz
7	LG/RD	Engine Coolant Temperature Sensor	0.5-0.6v
8	DG/YL	Fuel Pump Monitor	On: 12v, Off: 0v
9	TN/BL	MAF Sensor Return	<0.050v
10	DG/OR	A/C Pressure Switch Signal	A/C On: 12v, off: 0v
11	GY/YL	Canister Purge Solenoid	12v, 55 mph: 1v
12	LG/OR	Injector 6 Control	5.5-7.1 ms
13-14	---	Not Used	---
15	TN/BK	Injector 5 Control	5.5-7.1 ms
16	OR/RD	Ignition System Ground	<0.1v
17	PK/LG	Self-Test Output & MIL	MIL On: 1v, Off: 12v
18	TN/OR	Data Bus (-)	Digital Signals
19	PK/OR	Data Bus (+)	Digital Signals
20	BK	PCM Case Ground	<0.050v
21	WT/BL	IAC Solenoid Control	8.6-10.7v
22	BL/OR	Fuel Pump Control	On: 1v, Off: 12v
23	YL/RD	Knock Sensor 1 Signal	0v
24	---	Not Used	---
25	GY	Intake Air Temperature Sensor	1.5-2.5v
26	BR/WT	Reference Voltage	4.9-5.1v
27	BR/LG	EVP Sensor Signal	0.4v, 55 mph: 3.1v
28-29	---	Not Used	---
30	BL/YL	A/T: PNP Sensor Signal	In 'P': 0v, Others: 5v
30	BL/YL	M/T: Clutch Pedal Position	Clutch Out: 5v, In: 0v
31	WT/OR	Thermactor Air Bypass	TAB On: 1v, Off: 12v
32	WT/LG	TCIL (lamp) Control	Lamp On: 1v, Off: 12v
33	BR/PK	VR Solenoid Control	0%, al 55 mph: 45%
34	BR	Thermactor Air Diverter	TAD On: 1v, Off: 12v
35	BR/BL	Injector 4 Control	5.5-7.1 ms
36	PK	Spark Output Signal	6.93v (50%)
37	RD	Vehicle Power	12-14v
38	---	Not Used	---
39	BR/YL	Injector 3 Control	5.5-7.1 ms
40	BK/WT	Power Ground	<0.1v
41	TN/WT	TCS (switch) Signal	TCS & O/D On: 12v
42	---	Not Used	---
43	RD/BR	HO2S-21 (B2 S1) Signal	0.1-1.1v
44	GY/BL	HO2S-11 (B1 S1) Signal	0.1-1.1v
45	---	Not Used	---
46	GY/RD	Analog Signal Return	<0.050v
47	GY/WT	TP Sensor Signal	0.7v, 55 mph: 2.1v
48	WT/PK	Self-Test Indicator Signal	STI On: 1v, Off: 5v
49	---	Not Used	---
50	BL/RD	MAF Sensor Signal	0.8v, 55 mph: 1.6v
51-53	---	Not Used	---
54	PK/YL	A/C WOT Relay Control	On: 1v, Off: 12v
55	---	Not Used	---
56	GY/OR	PIP Sensor Signal	6.93v (50%)
57	RD	Vehicle Power	12-14v
58	TN	Injector 1 Control	5.5-7.1 ms
59	WT	Injector 2 Control	5.5-7.1 ms
60	BK/WT	Power Ground	<0.1v

1994-95 E-Series 4.9L I6 VIN Y (E4OD) 60 Pin Connector

PCM Pin #	Wire Color	Circuit Description (60 Pin)	Value at Hot Idle
1	YL	Keep Alive Power	12-14v
2	LG	Brake Switch Signal	Brake Off: 12v, On: 1v
3	GY/BK	VSS (+) Signal	0 Hz, at 55 mph: 125 Hz
4	WT/PK	Ignition Diagnostic Monitor	20-31 Hz
5	---	Not Used	---
6	PK/OR	VSS (-) Signal	0 Hz, at 55 mph: 125 Hz
7	LG/RD	Engine Coolant Temperature Sensor	0.5-0.6v
8	DG/YL	Fuel Pump Monitor	On: 12v, Off: 0v
9	PK/BL	Data Bus (-)	Digital Signals
10	BK/YL	A/C Pressure Switch Signal	A/C On: 12v, off: 0v
11	BR	Air Management 2 Solenoid	On: 1v, Off: 12v
12-15	---	Not Used	---
16	OR/RD	Ignition System Ground	<0.050v
17	PK/LG	Self-Test Output & MIL	MIL On: 1v, Off: 12v
18	---	Not Used	---
19	PK/OR	Shift Solenoid 2 Control	12v, 55 mph: 1v
20	BK	PCM Case Ground	<0.050v
21	WT/BL	IAC Solenoid Control	10.7v (32%)
22	BL/OR	Fuel Pump Control	On: 1v, Off: 12v
23	YL/RD	Knock Sensor 1 Signal	0v
24	---	Not Used	---
25	GY	Intake Air Temperature Sensor	1.5-2.5v
26	BR/WT	Reference Voltage	4.9-5.1v
27	BR/LG	EVP Sensor Signal	0.4v, 55 mph: 3.1v
28	TN/OR	Data Bus (+)	Digital Signals
29	GY/BL	HO2S-11 (B1 S1) Signal	0.1-1.1v
30	BL/YL	MLP Sensor Signal	In 'P': 0v, in Drive: 5v
31	GY/YL	Canister Purge Solenoid	12v, 55 mph: 1v
32	WT/LG	TCIL (lamp) Control	Lamp On: 1v, Off: 12v
33	BR/PK	VR Solenoid Control	0%, at 55 mph: 45%
34-35	---	Not Used	---
36	PK	Spark Output Signal	6.93v (50%)
37	RD	Vehicle Power	12-14v
38	WT/YL	EPC Solenoid Control	8.1v (4 psi)
39	---	Not Used	---
40	BK/WT	Power Ground	<0.1v
41	TN/WT	TCS (switch) Signal	TCS & O/D On: 12v
42	OR/BK	TOT Sensor Signal	2.10-2.40v
43	PK	A/C Demand Switch	A/C On: 12v, Off: 0v
44	---	Not Used	---
45	LG/BK	MAP Sensor Signal	107 Hz (sea level)
46	GY/RD	Analog Signal Return	<0.050v
47	GY/WT	TP Sensor Signal	0.7v, 55 mph: 2.1v
48	WT/PK	Self-Test Indicator Signal	STI On: 1v, Off: 5v
49	OR	HO2S-11 (B1 S1) Ground	<0.050v
50	---	Not Used	---
51	WT/OR	Air Management 1 Solenoid	On: 1v, Off: 12v
52	OR/YL	Shift Solenoid 1 Control	1v, 55 mph: 1v
53	PK/YL	TCC Solenoid Control	0%, at 55 mph: 95%
54	---	Not Used	---
55	BR/OR	Coast Clutch Solenoid	12v, 55 mph: 12v
56	GY/OR	PIP Sensor Signal	6.93v (50%)
57	RD	Vehicle Power	12-14v
58	TN	Injector Bank 1 (INJ 1, 3 & 5)	6.2-7.4 ms
59	WT	Injector Bank 2 (INJ 2, 4 & 6)	6.2-7.4 ms
60	BK/WT	Power Ground	<0.1v

1995 E-Series 4.9L I6 MFI VIN Y California 60 Pin Connector

PCM Pin #	Wire Color	Circuit Description (60 Pin)	Value at Hot Idle
1	YL	Keep Alive Power	12-14v
2	LG	Brake Switch Signal	Brake Off: 12v, On: 1v
3	GY/BK	VSS (+) Signal	0 Hz, at 55 mph: 125 Hz
4	WT/PK	Ignition Diagnostic Monitor	20-31 Hz
5	---	Not Used	---
6	PK/OR	VSS (-) Signal	0 Hz, at 55 mph: 125 Hz
7	LG/RD	Engine Coolant Temperature Sensor	0.5-0.6v
8	DG/YL	Fuel Pump Monitor	On: 12v, Off: 0v
9	PK/BL	Data Bus (-)	Digital Signals
10	BK/YL	A/C Pressure Switch Signal	A/C On: 12v, off: 0v
11	BR	Air Management 2 Solenoid	On: 1v, Off: 12v
12-15	---	Not Used	---
16	OR/RD	Ignition System Ground	<0.050v
17	PK/LG	Self-Test Output & MIL	MIL On: 1v, Off: 12v
18	---	Not Used	---
19	PK/BL	Shift Solenoid 2 Control	12v, 55 mph: 1v
20	BK	PCM Case Ground	<0.050v
21	WT/BL	IAC Solenoid Control	10.7v (33%)
22	BL/OR	Fuel Pump Control	On: 1v, Off: 12v
23	YL/RD	Knock Sensor 1 Signal	0v
24	---	Not Used	---
25	GY	Intake Air Temperature Sensor	1.5-2.5v
26	BR/WT	Reference Voltage	4.9-5.1v
27	BR/LG	EVP Sensor Signal	0.4v, 55 mph: 3.1v
28	TN/OR	Data Bus (+)	Digital Signals
29	GY/BL	HO2S-11 (B1 S1) Signal	0.1-1.1v
30	BL/YL	MLP Sensor Signal	In 'P': 0v, in Drive: 5v
31	GY/YL	Canister Purge Solenoid	12v, 55 mph: 1v
32	WT/LG	TCIL (lamp) Control	Lamp On: 1v, Off: 12v
33	BR/PK	VR Solenoid Control	0%, at 55 mph: 45%
34-35	---	Not Used	---
36	PK	Spark Output Signal	6.93v (50%)
37	RD	Vehicle Power	12-14v
38	WT/YL	EPC Solenoid Control	8.1v (4 psi)
39	---	Not Used	---
40	BK/WT	Power Ground	<0.1v

1995 E-Series 4.9L I6 MFI VIN Y (Cal) 60 Pin Connector

PCM Pin #	Wire Color	Circuit Description (60 Pin)	Value at Hot Idle
41	TN/WT	TCS (switch) Signal	TCS & O/D On: 12v
42	OR/BK	TOT Sensor Signal	2.10-2.40v
43	PK	A/C Demand Switch	A/C On: 12v, Off: 0v
44	---	Not Used	---
45	LG/BK	MAP Sensor Signal	107 Hz (sea level)
46	GY/RD	Analog Signal Return	<0.050v
47	GY/WT	TP Sensor Signal	0.7v, 55 mph: 2.1v
48	WT/PK	Self-Test Indicator Signal	STI On: 1v, Off: 5v
49	OR	HO2S-11 (B1 S1) Ground	<0.050v
50	---	Not Used	---
51	WT/OR	Air Management 1 Solenoid	On: 1v, Off: 12v
52	OR/YL	Shift Solenoid 1 Control	1v, 55 mph: 12v
53	PK/YL	TCC Solenoid Control	12v, 55 mph: 1v
54	---	Not Used	---
55	BR/OR	Coast Clutch Solenoid	12v, 55 mph: 12v
56	GY/OR	PIP Sensor Signal	6.93v (50%)
57	RD	Vehicle Power	12-14v
58	TN	Injector Bank 1 (INJ 1, 3 & 5)	6.2-7.4 ms
59	WT	Injector Bank 2 (INJ 2, 4 & 6)	6.2-7.4 ms
60	BK/WT	Power Ground	<0.1v

Pin Connector Graphic

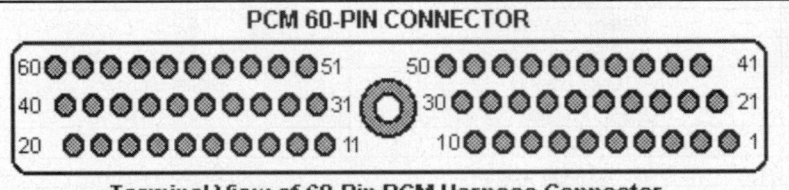

PCM 60-PIN CONNECTOR

Terminal View of 60-Pin PCM Harness Connector

Standard Colors and Abbreviations

Abbreviation	Color	Abbreviation	Color	Abbreviation	Color
BK	Black	GY	Gray	PK	Purple
BL	Blue	GN	Green	RD	Red
BR	Brown	LG	LT Green	TN	Tan
DB	Dark Blue	OR	Orange	WT	White
DG	DK Green	PK	Pink	YL	Yellow

1996 E-Series 4.9L I6 OHV VIN Y (A/T) 104 Pin Connector

PCM Pin #	Wire Color	Circuit Description (104 Pin)	Value at Hot Idle
1	---	Not Used	---
2	PK/LG	MIL (lamp) Control	MIL On: 1v, Off: 12v
3-12	---	Not Used	---
13	PK	Flash EEPROM	0.1v
14	---	Not Used	---
15	PK/LB	Data Bus (-)	Digital Signals
16	TN/OR	Data Bus (+)	Digital Signals
17-22	---	Not Used	---
23	OR/RD	Ignition System Ground	<0.050v
24	BK/WT	Power Ground	<0.1v
25	BK	Chassis Ground	<0.050v
26-28	---	Not Used	---
29	TN/WT	Transmission Cancel Switch	12v (Closed: 1v)
30-32	---	Not Used	---
33	OR/YL	PSOM (-) Signal	<0.050v
34	---	Not Used	---
35	RD/LG	HO2S-12 (B1 S2) Signal	0.1-1.1v
36	TN/LB	MAF Sensor Return	<0.050v
37	---	Not Used	---
38	LG/RD	Engine Coolant Temperature Sensor	0.5-0.6v
39	GY	Intake Air Temperature Sensor	1.5-2.5v
40	DG/YL	Fuel Pump Monitor	On: 12v, Off: 0v
41	BK/YL	A/C Pressure Switch Signal	A/C On: 12v, off: 0v
42-43	---	Not Used	---
44	BR	Secondary AIR Diverter	On: 1v, Off: 12v
45-46	---	Not Used	---
47	BR/PK	VR Solenoid Control	0%, at 55 mph: 45%
48	WT/PK	Ignition Diagnostic Monitor	20-31 Hz
49	GY/OR	PIP Sensor Signal	6.93v (50%)
50	PK	Spark Output Signal	6.93v (50%)
51	BK/WT	Power Ground	<0.1v
52-54	---	Not Used	---
55	YL	Keep Alive Power	12-14v
56	LG/BK	EVAP VMV Solenoid	0-10 Hz (0-100%)
57	YL/RD	Knock Sensor 1 Signal	0v
58	GY/BK	PSOM (+) Signal	0 Hz, at 55 mph: 125 Hz
59	DG/LG	Misfire Detection Sensor	35 Hz (150 mv AC)
60	GY/LB	HO2S-11 (B1 S1) Signal	0.1-1.1v
61-63	---	Not Used	---
64	LB/YL	PNP Switch Signal	In 'P': 0v, Others: 5v

1996 E-Series 4.9L I6 OHV VIN Y (A/T) 104 Pin Connector

PCM Pin #	Wire Color	Circuit Description (104 Pin)	Value at Hot Idle
65	BR/LG	EGR DPFE Sensor Signal	0.20-1.30v
66-68	---	Not Used	---
69	PK/YL	A/C WOT Relay Control	On: 1v, Off: 12v
70	WT/OR	Secondary Air Bypass Solenoid	AIRB On: 1v, Off: 12v
71	RD	Vehicle Power	12-14v
72	---	Not Used	---
73	TN/BK	Injector 5 Control	3.8-4.7 ms
74	BR/YL	Injector 3 Control	3.8-4.7 ms
75	TN	Injector 1 Control	3.8-4.7 ms
76	BK/WT	Power Ground	<0.1v
77	BK/WT	Power Ground	<0.1v
78	---	Not Used	---
79	WT/LG	TCIL (lamp) Control	Lamp On: 1v, Off: 12v
80	LB/OR	Fuel Pump Control	On: 1v, Off: 12v
81-82	---	Not Used	---
83	WT/LB	IAC Solenoid Control	10.7v (33%)
84-86	---	Not Used	---
87	RD/BK	HO2S-21 (B2 S1) Signal	0.1-1.1v
88	LB/RD	MAF Sensor Signal	0.8v, 55 mph: 1.6v
89	GY/WT	TP Sensor Signal	0.53-1.27v
90	BR/WT	Reference Voltage	4.9-5.1v
91	GY/RD	Analog Signal Return	<0.050v
92	LG	Brake Switch Signal	Brake Off: 12v, On: 1v
93	RD/WT	HO2S-11 (B1 S1) Heater	1v, Off: 12v
94	YL/LB	HO2S-21 (B2 S1) Heater	1v, Off: 12v
95	WT/BK	HO2S-12 (B1 S2) Heater	1v, Off: 12v
96, 98	---	Not Used	---
97	R	Vehicle Power	12-14v
99	LG/OR	Injector 6 Control	3.8-4.7 ms
100	BR/LB	Injector 4 Control	3.8-4.7 ms
101	W	Injector 2 Control	3.8-4.7 ms
102	---	Not Used	---
103	BK/WT	Power Ground	<0.1v
104	---	Not Used	---

Pin Connector Graphic

```
        PCM 104-PIN CONNECTOR
  1 ●●●●●●●●●●●●    ●●●●●●●●●●●● 26
 27 ●●●●●●●●●●●●    ●●●●●●●●●●●● 52
 53 ●●●●●●●●●●●●  ⬤ ●●●●●●●●●●●● 78
 79 ●●●●●●●●●●●●    ●●●●●●●●●●●● 104
```

Terminal View of 104-Pin PCM Wiring Harness Connector

Standard Colors and Abbreviations

Abbreviation	Color	Abbreviation	Color	Abbreviation	Color
BK	Black	GY	Gray	PK	Purple
BL	Blue	GN	Green	RD	Red
BR	Brown	LG	LT Green	TN	Tan
DB	Dark Blue	OR	Orange	WT	White
DG	DK Green	PK	Pink	YL	Yellow

1996 E-Series 4.9L I6 OHV VIN Y (E4OD) 104 Pin Connector

PCM Pin #	Wire Color	Circuit Description (104 Pin)	Value at Hot Idle
1	PK/OR	Shift Solenoid 2 Control	12v, 55 mph: 1v
2	PK/LG	MIL (lamp) Control	MIL On: 1v, Off: 12v
3-12	---	Not Used	---
13	PK	Flash EEPROM	0.1v
14	---	Not Used	---
15	PK/LB	Data Bus (-)	Digital Signals
16	TN/OR	Data Bus (+)	Digital Signals
17-22	---	Not Used	---
23	OR/RD	Ignition System Ground	<0.050v
24	BK/WT	Power Ground	<0.1v
25	BK	Chassis Ground	<0.050v
26	---	Not Used	---
27	OR/YL	Shift Solenoid 1 Control	1v, 55 mph: 12v
28	---	Not Used	---
29	TN/WT	Transmission Cancel Switch	12v (Closed: 1v)
30-32	---	Not Used	---
33	OR/YL	PSOM (-) Signal	<0.050v
34	---	Not Used	---
35	RD/LG	HO2S-12 (B1 S2) Signal	0.1-1.1v
36	TN/LB	MAF Sensor Return	<0.050v
37	OR/BK	Transmission Fluid Temperature Sensor	2.10-2.40v
38	LG/RD	Engine Coolant Temperature Sensor	0.5-0.6v
39	GY	Intake Air Temperature Sensor	1.5-2.5v
40	DG/YL	Fuel Pump Monitor	On: 12v, Off: 0v
41	BK/YL	A/C Pressure Switch Signal	A/C On: 12v, off: 0v
42-43	---	Not Used	---
44	BR	Secondary AIR Diverter	On: 1v, Off: 12v
45-46	---	Not Used	---
47	BR/PK	VR Solenoid Control	0%, at 55 mph: 45%
48	WT/PK	Ignition Diagnostic Monitor	20-31 Hz
49	GY/OR	PIP Sensor Signal	6.93v (50%)
50	PK	Spark Output Signal	6.93v (50%)
51	BK/WT	Power Ground	<0.1v
52	---	Not Used	---
53	BR/OR	Coast Clutch Solenoid	12v, 55 mph: 12v
54	DB/WT	Torque Converter Clutch	0%, at 55 mph: 95%
55	Y	Keep Alive Power	12-14v
56	LG/BK	EVAP VMV Solenoid	0-10 Hz (0-100%)
57	YL/RD	Knock Sensor 1 Signal	0v
58	GY/BK	PSOM (+) Signal	0 Hz, at 55 mph: 125 Hz
59	DG/LG	Misfire Detection Sensor	35 Hz (150 mv AC)
60	GY/LB	HO2S-11 (B1 S1) Signal	0.1-1.1v
61-63	---	Not Used	---
64	LB/YL	Transmission Range Sensor	In 'P': 0v, in O/D: 1.7v

1996 E-Series 4.9L I6 OHV VIN Y (E4OD) 104 Pin Connector

PCM Pin #	Wire Color	Circuit Description (104 Pin)	Value at Hot Idle
65	BR/LG	EGR DPFE Sensor Signal	0.20-1.30v
66-68	---	Not Used	---
69	PK/YL	A/C WOT Relay Control	On: 1v, Off: 12v
70	WT/OR	Secondary Air Bypass Solenoid	AIRB On: 1v, Off: 12v
71	RD	Vehicle Power	12-14v
72	---	Not Used	---
73	TN/BK	Injector 5 Control	3.8-4.7 ms
74	BR/YL	Injector 3 Control	3.8-4.7 ms
75	TN	Injector 1 Control	3.8-4.7 ms
76	BK/WT	Power Ground	<0.1v
77	BK/WT	Power Ground	<0.1v
78	---	Not Used	---
79	WT/LG	TCIL (lamp) Control	Lamp On: 1v, Off: 12v
80	LB/OR	Fuel Pump Control	On: 1v, Off: 12v
81	WT/YL	EPC Solenoid Control	10.6v (40 psi)
82	---	Not Used	---
83	WT/LB	IAC Solenoid Control	10.7v (33%)
84-86	---	Not Used	---
87	RD/BK	HO2S-21 (B2 S1) Signal	0.1-1.1v
88	LB/RD	MAF Sensor Signal	0.8v, 55 mph: 1.6v
89	GY/WT	TP Sensor Signal	0.53-1.27v
90	BR/WT	Reference Voltage	4.9-5.1v
91	GY/RD	Analog Signal Return	<0.050v
92	LG	Brake Switch Signal	Brake Off: 12v, On: 1v
93	RD/WT	HO2S-11 (B1 S1) Heater	1v, Off: 12v
94	YL/LB	HO2S-21 (B2 S1) Heater	1v, Off: 12v
95	WT/BK	HO2S-12 (B1 S2) Heater	1v, Off: 12v
96	---	Not Used	---
97	RD	Vehicle Power	12-14v
98	---	Not Used	---
99	LG/OR	Injector 6 Control	3.8-4.7 ms
100	BR/LB	Injector 4 Control	3.8-4.7 ms
101	WT	Injector 2 Control	3.8-4.7 ms
102	---	Not Used	---
103	BK/WT	Power Ground	<0.1v
104	---	Not Used	---

Pin Connector Graphic

PCM 104-PIN CONNECTOR

Terminal View of 104-Pin PCM Wiring Harness Connector

Standard Colors and Abbreviations

Abbreviation	Color	Abbreviation	Color	Abbreviation	Color
BK	Black	GY	Gray	PK	Purple
BL	Blue	GN	Green	RD	Red
BR	Brown	LG	LT Green	TN	Tan
DB	Dark Blue	OR	Orange	WT	White
DG	DK Green	PK	Pink	YL	Yellow

1990 E-Series 5.0L V8 VIN N (A/T) 60 Pin Connector

PCM Pin #	Wire Color	Circuit Description (60 Pin)	Value at Hot Idle
1	YL	Keep Alive Power	12-14v
2, 5	---	Not Used	---
3	GY/BK	VSS (+) Signal	0 Hz, at 55 mph: 125 Hz
3	DG/WT	VSS (+) Signal	0 Hz, at 55 mph: 125 Hz
4	DG/YL	Ignition Diagnostic Monitor	20-31 Hz
6	BK	VSS (-) Signal	0 Hz, at 55 mph: 125 Hz
7	LG/YL	Engine Coolant Temperature Sensor	0.5-0.6v
8	BR	Fuel Pump Monitor	On: 12v, Off: 0v
9	---	Not Used	---
10	BK/YL	A/C Pressure Switch Signal	A/C On: 12v, off: 0v
11	WT/BK	Air Management 2 Solenoid	On: 1v, Off: 12v
12-15	---	Not Used	---
16	BK/OR	Ignition System Ground	<0.050v
17	PK/LG	Self-Test Output & MIL	MIL On: 1v, Off: 12v
18-19	---	Not Used	---
20	BK	PCM Case Ground	<0.050v
21	GY/WT	IAC Solenoid Control	9v (31%)
22	TN/LG	Fuel Pump Control	On: 1v, Off: 12v
23	LG/BK	Knock Sensor 1 Signal	0v
24	YL/LG	Power Steering Pressure Switch	Straight: 0v, Turned: 12v
25	YL/RD	ACT Sensor Signal	1.5-2.5v
26	OR/WT	Reference Voltage	4.9-5.1v
27	BR/LG	EVP Sensor Signal	0.4v, 55 mph: 3.1v
28	---	Not Used	---
29	GY/BK	HO2S-11 (B1 S1) Signal	0.1-1.1v
29	DG/PK	HO2S-11 (B1 S1) Signal	0.1-1.1v
30	GY/YL	A/T: Neutral Drive Switch	In 'P': 0v, Others: 5v
30	BL/YL	A/T: Neutral Drive Switch	In 'P': 0v, Others: 5v
31	GY/YL	Canister Purge Solenoid	12v, 55 mph: 1v
32, 34-35	---	Not Used	---
33	DG	VR Solenoid Control	0%, at 55 mph: 45%
36	YL/LG	Spark Output Signal	6.93v (50%)
37	RD	Vehicle Power	12-14v
38-39	---	Not Used	---
40	BK/LG	Power Ground	<0.1v
41-44	---	Not Used	---
45	DG/LG	MAP Sensor Signal	107 Hz (sea level)
45	DB/LG	MAP Sensor Signal	107 Hz (sea level)
46	BK/WT	Analog Signal Return	<0.50v
47	DG/LG	TP Sensor Signal	0.7v, 55 mph: 2.1v
48	WT/RD	Self-Test Indicator Signal	STI On: 1v, Off: 5v
49	OR	HO2S-11 (B1 S1) Ground	<0.050v
50, 52-55	---	Not Used	---
51	WT/RD	Air Management 1 Solenoid	On: 1v, Off: 12v
56	DB	PIP Sensor Signal	6.93v (50%)
57	RD	Vehicle Power	12-14v
58	TN/OR	Injector Bank 1 (INJ 1, 4, 5, 8)	5.0-5.7 ms
59	TN/RD	Injector Bank 2 (INJ 2, 3, 6, 7)	5.0-5.7 ms
60	BK/LG	Power Ground	<0.1v

1991 E-Series 5.0L V8 OHV VIN N (A/T) 60 Pin Connector

PCM Pin #	Wire Color	Circuit Description (60 Pin)	Value at Hot Idle
1	BK/OR	Keep Alive Power	12-14v
2	---	Not Used	---
3	DG/WT	VSS (+) Signal	0 Hz, at 55 mph: 125 Hz
4	TN/YL	Ignition Diagnostic Monitor	20-31 Hz
5	---	Not Used	---
6	OR/YL	VSS (-) Signal	0 Hz, at 55 mph: 125 Hz
7	LG/RD	Engine Coolant Temperature Sensor	0.5-0.6v
8	DG/YL	Fuel Pump Monitor	On: 12v, Off: 0v
9	BK/OR	Data Bus (-)	Digital Signals
10	BK/YL	A/C Pressure Switch Signal	A/C On: 12v, off: 0v
11	BR	Air Management 2 Solenoid	On: 1v, Off: 12v
12-15	---	Not Used	---
16	OR/RD	Ignition System Ground	<0.050v
17	TN/RD	Self-Test Output & MIL	MIL On: 1v, Off: 12v
18-19	---	Not Used	---
20	BK	PCM Case Ground	<0.050v
21	GY/WT	IAC Solenoid Control	10.7v (33%)
22	TN/LG	Fuel Pump Control	On: 1v, Off: 12v
23	LG/BK	Knock Sensor 1 Signal	0v
24	YL/LG	Power Steering Pressure Switch	Straight: 0v, Turned: 12v
25	YL/RD	ACT Sensor Signal	1.5-2.5v
26	BR/WT	Reference Voltage	4.9-5.1v
27	BR/LG	EVP Sensor Signal	0.4v, 55 mph: 3.1v
28	OR/BK	Data Bus (+)	Digital Signals
29	RD/BK	HO2S-11 (B1 S1) Signal	0.1-1.1v
30	---	Not Used	---
31	GY/YL	Canister Purge Solenoid	12v, 55 mph: 1v
32	---	Not Used	---
33	BR/PK	VR Solenoid Control	0%, at 55 mph: 45%
34-35	---	Not Used	---
36	YL/LG	Spark Output Signal	6.93v (50%)
37	RD	Vehicle Power	12-14v
38-39	---	Not Used	---
40	BK/LG	Power Ground	<0.1v
41-44	---	Not Used	---
45	DB/LG	MAP Sensor Signal	107 Hz (sea level)
46	GY/RD	Analog Signal Return	<0.050v
47	GY/WT	TP Sensor Signal	0.7v, 55 mph: 2.1v
48	WT/RD	Self-Test Indicator Signal	STI On: 1v, Off: 5v
49	OR	HO2S-11 (B1 S1) Ground	<0.050v
50	---	Not Used	---
51	WT/OR	Air Management 1 Solenoid	On: 1v, Off: 12v
52-55	---	Not Used	---
56	DB	PIP Sensor Signal	6.93v (50%)
57	RD	Vehicle Power	12-14v
58	TN/OR	Injector Bank 1 (INJ 1, 4, 5, 8)	4.4-5.6 ms
59	TN/RD	Injector Bank 2 (INJ 2, 3, 6, 7)	4.4-5.6 ms
60	BK/LG	Power Ground	<0.1v

Pin Connector Graphic

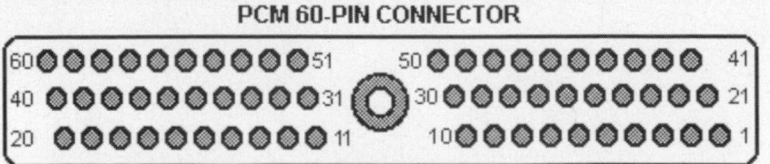

Terminal View of 60-Pin PCM Harness Connector

1992-93 E-Series 5.0L V8 VIN N (A/T) 60 Pin Connector

PCM Pin #	Wire Color	Circuit Description (60 Pin)	Value at Hot Idle
1	YL	Keep Alive Power	12-14v
2	---	Not Used	---
3	GY/BK	PSOM (-) Signal	<0.050v
4	TN/YL	Ignition Diagnostic Monitor	20-31 Hz
5	---	Not Used	---
6	PK/OR	PSOM (+) Signal	0 Hz, at 55 mph: 125 Hz
7	LG/RD	Engine Coolant Temperature Sensor	0.5-0.6v
8	DG/YL	Fuel Pump Monitor	On: 12v, Off: 0v
9	PK/LB	Data Bus (-)	Digital Signals
10	BK/YL	A/C Pressure Switch Signal	A/C On: 12v, off: 0v
11	BR	Air Management 2 Solenoid	On: 1v, Off: 12v
12-15	---	Not Used	---
16	OR/RD	Ignition System Ground	<0.050v
17	PK/LG	Self-Test Output & MIL	MIL On: 1v, Off: 12v
18-19	---	Not Used	---
20	BK	Case Ground	<0.050v
21	WT/LB	IAC Solenoid Control	10.7v (33%)
22	LB/OR	Fuel Pump Control	On: 1v, Off: 12v
23	YL/RD	Knock Sensor 1 Signal	0v
24	---	Not Used	---
25	GY	ACT Sensor Signal	1.5-2.5v
26	BR/WT	Reference Voltage	4.9-5.1v
27	BR/LG	EVP Sensor Signal	0.4v, 55 mph: 3.1v
28	TN/OR	Data Bus (+)	Digital Signals
29	GY/LB	HO2S-11 (B1 S1) Signal	0.1-1.1v
30	LB/YL	Neutral Drive Switch	In 'P': 0v, Others: 5v
31	GY/YL	Canister Purge Solenoid	12v, 55 mph: 1v
32	---	Not Used	---
33	BR/PK	VR Solenoid Control	0%, at 55 mph: 45%
34-35	---	Not Used	---
36	PK	Spark Output Signal	6.93v (50%)
37	RD	Vehicle Power	12-14v
38-39	---	Not Used	---
40	BK/WT	Power Ground	<0.1v
41-44	---	Not Used	---
45	LG/BK	MAP Sensor Signal	107 Hz (sea level)
46	GY/RD	Analog Signal Return	<0.050v
47	GY/WT	TP Sensor Signal	0.7v, 55 mph: 2.1v
48	WT/PK	Self-Test Indicator Signal	STI On: 1v, Off: 5v
49	OR	HO2S-11 (B1 S1) Ground	<0.050v
50	---	Not Used	---
51	WT/OR	Air Management 1 Solenoid	On: 1v, Off: 12v
52-55	---	Not Used	---
56	GY/OR	PIP Sensor Signal	6.93v (50%)
57	RD	Vehicle Power	12-14v
58	TN	Injector Bank 1 (INJ 1, 4, 5, 8)	4.4-5.6 ms
59	WT	Injector Bank 2 (INJ 2, 3, 6, 7)	4.4-5.6 ms
60	BK/WT	Power Ground	<0.1v

Pin Connector Graphic

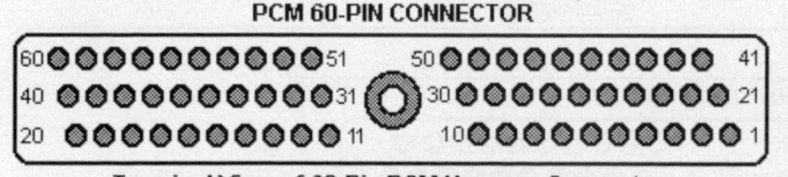

Terminal View of 60-Pin PCM Harness Connector

1992-93 E-Series 5.0L V8 VIN N (E4OD) 60 Pin Connector

PCM Pin #	Wire Color	Circuit Description (60 Pin)	Value at Hot Idle
1	YL	Keep Alive Power	12-14v
2	LG	Brake Switch Signal	Brake Off: 12v, On: 1v
3	GY/BK	PSOM (-) Signal	<0.050v
4	TN/YL	Ignition Diagnostic Monitor	20-31 Hz
5, 12-15	---	Not Used	---
6	PK/OR	PSOM (+) Signal	0 Hz, at 55 mph: 125 Hz
7	LG/RD	Engine Coolant Temperature Sensor	0.5-0.6v
8	DG/YL	Fuel Pump Monitor	On: 12v, Off: 0v
9	PK/LB	Data Bus (-)	Digital Signals
10	BK/YL	A/C Pressure Switch Signal	A/C On: 12v, off: 0v
11	BR	Air Management 2 Solenoid	On: 1v, Off: 12v
16	OR/RD	Ignition System Ground	<0.050v
17	PK/LG	Self-Test Output & MIL	MIL On: 1v, Off: 12v
18, 24	---	Not Used	---
19	PK/OR	Shift Solenoid 1 Control	1v, 55 mph: 12v
20	BK	Case Ground	<0.050v
21	WT/LB	IAC Solenoid Control	10.7v (33%)
22	LB/OR	Fuel Pump Control	On: 1v, Off: 12v
23	YL/RD	Knock Sensor 1 Signal	0v
25	GY	ACT Sensor Signal	1.5-2.5v
26	BR/WT	Reference Voltage	4.9-5.1v
27	BR/LG	EVP Sensor Signal	0.4v, 55 mph: 3.1v
28	TN/OR	Data Bus (+)	Digital Signals
29	GY/LB	HO2S-11 (B1 S1) Signal	0.1-1.1v
30	LB/YL	Neutral Drive Switch	In 'P': 0v, Others: 5v
31	GY/YL	Canister Purge Solenoid	12v, 55 mph: 1v
32	WT/LG	TCIL (lamp) Control	Lamp On: 1v, Off: 12v
33	BR/PK	VR Solenoid Control	0%, at 55 mph: 45%
34-35, 39	---	Not Used	---
36	PK	Spark Output Signal	6.93v (50%)
37	RD	Vehicle Power	12-14v
38	WT/YL	EPC Solenoid Control	9.5v (5 psi)
40	BK/WT	Power Ground	<0.1v
41	TN/WT	Overdrive Cancel Switch	OCS Switch On: 0.1v
42	OR/BK	TOT Sensor Signal	2.10-2.40v
43-44	---	Not Used	---
45	LG/BK	MAP Sensor Signal	107 Hz (sea level)
46	GY/RD	Analog Signal Return	<0.050v
47	GY/WT	TP Sensor Signal	0.7v, 55 mph: 2.1v
48	WT/PK	Self-Test Indicator Signal	STI On: 1v, Off: 5v
49	OR	HO2S-11 (B1 S1) Ground	<0.050v
50, 54	---	Not Used	---
51	WT/OR	Air Management 1 Solenoid	On: 1v, Off: 12v
52	OR/YL	Shift Solenoid 2 Control	12v, 55 mph: 1v
53	PK/YL	TCC Solenoid Control	0%, at 55 mph: 95%
55	BR/OR	Coast Clutch Solenoid	12v, 55 mph: 12v
56	GY/OR	PIP Sensor Signal	6.93v (50%)
57	RD	Vehicle Power	12-14v
58	TN	Injector Bank 1 (INJ 1, 4, 5, 8)	4.4-5.6 ms
59	WT	Injector Bank 2 (INJ 2, 3, 6, 7)	4.4-5.6 ms
60	BK/WT	Power Ground	<0.1v

Pin Connector Graphic

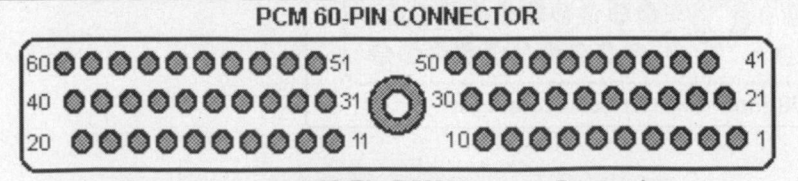

Terminal View of 60-Pin PCM Harness Connector

1994-95 E-Series 5.0L V8 VIN N (A/T) 60 Pin Connector

PCM Pin #	Wire Color	Circuit Description (60 Pin)	Value at Hot Idle
1	YL	Keep Alive Power	12-14v
2	LG	Brake Switch Signal	Brake Off: 12v, On: 1v
3	GY/BK	VSS (+) Signal	0 Hz, at 55 mph: 125 Hz
4	WT/PK	Ignition Diagnostic Monitor	20-31 Hz
5	DG/WT	Turbine Shaft Speed Sensor	120 Hz, 30 mph: 300 Hz
6	PK/OR	VSS (-) Signal	0 Hz, at 55 mph: 125 Hz
7	LG/RD	Engine Coolant Temperature Sensor	0.5-0.6v
8	DG/YL	Fuel Pump Monitor	On: 12v, Off: 0v
9	TN/BL	MAF Sensor Return	<0.050v
10	BK/YL	A/C Pressure Switch Signal	A/C On: 12v, off: 0v
11	GY/YL	Canister Purge Solenoid	12v, 55 mph: 1v
12	LG/OR	Injector 6 Control	4.7-5.5 ms
13	TN/RD	Injector 7 Control	4.7-5.5 ms
14	BL	Injector 8 Control	4.7-5.5 ms
15	TN/BK	Injector 5 Control	4.7-5.5 ms
16	OR/RD	Ignition System Ground	<0.050v
17	PK/LG	Self-Test Output & MIL	MIL On: 1v, Off: 12v
18	TN/OR	Data Bus (+)	Digital Signals
19	PK/BL	Data Bus (-)	Digital Signals
20	BK	PCM Case Ground	<0.050v
21	WT/BL	IAC Solenoid Control	10.7v (33%)
22	BL/OR	Fuel Pump Control	On: 1v, Off: 12v
23	YL/RD	Knock Sensor 1 Signal	0v
24	---	Not Used	---
25	GY	Intake Air Temperature Sensor	1.5-2.5v
26	BR/WT	Reference Voltage	4.9-5.1v
27	BR/LG	EVP Sensor Signal	0.4v, 55 mph: 3.1v
28	---	Not Used	---
29	WT/BL	Turbine Shaft Speed Sensor	120 Hz, 30 mph: 300 Hz
30	BL/YL	MLP Sensor Signal	In 'P': 0v, in O/D: 5v
31	WT/OR	Air Bypass Solenoid Control	AIRB On: 1v, Off: 12v
32	WT/LG	TCIL (lamp) Control	Lamp On: 1v, Off: 12v
33	BR/PK	VR Solenoid Control	0%, at 55 mph: 45%
34	BR	Air Diverter Solenoid Control	On: 1v, Off: 12v
35	BR/BL	Injector 4 Control	4.7-5.5 ms
36	PK	Spark Output Signal	6.93v (50%)
37	RD	Vehicle Power	12-14v
38	WT/YL	EPC Solenoid Control	9.5v (5 psi)
39	BR/YL	Injector 3 Control	4.7-5.5 ms
40	BK/WT	Power Ground	<0.1v

1994-95 E-Series 5.0L V8 VIN N (A/T) 60 Pin Connector

Y	Wire Color	Circuit Description (60 Pin)	Value at Hot Idle
41	TN/WT	TCS (switch) Signal	TCS & O/D On: 12v
42-43	---	Not Used	---
44	GY/BL	HO2S-11 (B1 S1) Signal	0.1-1.1v
45	---	Not Used	---
46	GY/RD	Analog Signal Return	<0.050v
47	GY/WT	TP Sensor Signal	0.7v, 55 mph: 2.1v
48	WT/PK	Self-Test Indicator Signal	STI On: 1v, Off: 5v
49	OR/BK	TOT Sensor Signal	2.10-2.40v
50	BL/RD	MAF Sensor Signal	0.8v, 55 mph: 1.6v
51	OR/YL	Shift Solenoid 1 Control	1v, 55 mph: 12v
52	PK/OR	Shift Solenoid 2 Control	12v, 55 mph: 1v
53	PK/YL	TCC Solenoid Control	0%, at 55 mph: 95%
54-55	---	Not Used	---
56	GY/OR	PIP Sensor Signal	6.93v (50%)
57	RD	Vehicle Power	12-14v
58	TN	Injector 1 Control	4.7-5.5 ms
59	WT	Injector 2 Control	4.7-5.5 ms
60	BK/WT	Power Ground	<0.1v

Pin Connector Graphic

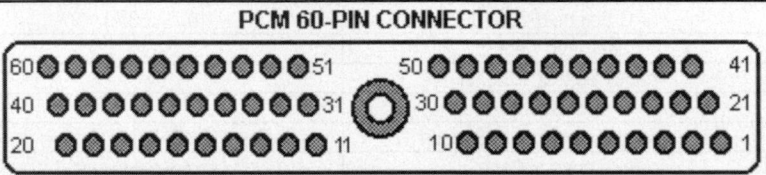

Standard Colors and Abbreviations

Abbreviation	Color	Abbreviation	Color	Abbreviation	Color
BK	Black	GY	Gray	PK	Purple
BL	Blue	GN	Green	RD	Red
BR	Brown	LG	LT Green	TN	Tan
DB	Dark Blue	OR	Orange	WT	White
DG	DK Green	PK	Pink	YL	Yellow

1996 E-Series 5.0L V8 OHV VIN N (A/T) 104 Pin Connector

PCM Pin #	Wire Color	Circuit Description (104 Pin)	Value at Hot Idle
1	---	Not Used	---
2	PK/LG	MIL (lamp) Control	MIL On: 1v, Off: 12v
3, 5	---	Not Used	---
4	LG/RD	Power Take-Off Signal	0.1v (Off)
6	GY	Output Shaft Speed Sensor	0 Hz, 55 mph: 130 Hz
7-12	---	Not Used	---
13	PK	Flash EPROM Power	0.1v
14	---	Not Used	---
15	PK/LB	Data Bus (-)	Digital Signals
16	TN/OR	Data Bus (+)	Digital Signals
17-22	---	Not Used	---
23	OR/RD	Ignition System Ground	<0.050v
24	BK/WT	Power Ground	<0.1v
25	BK	Case Ground	<0.050v
26	---	Not Used	---
27	OR/YL	Shift Solenoid 1 Control	1v, 55 mph: 12v
28	---	Not Used	---
29	TN/WT	TCS (switch) Signal	TCS & O/D On: 12v
30-32	---	Not Used	---
33	PK/OR	PSOM (-) Signal	<0.050v
34	---	Not Used	---
35	RD/LG	HO2S-12 (B1 S2) Signal	0.1-1.1v
36	TN/LB	MAF Sensor Return	<0.050v
37	---	Not Used	---
38	LG/RD	Engine Coolant Temperature Sensor	0.5-0.6v
39	GY	Intake Air Temperature Sensor	1.5-2.5v
40	DG/YL	Fuel Pump Monitor	On: 12v, Off: 0v
41	BK/YL	A/C Pressure Switch Signal	A/C On: 12v, off: 0v
42-43	---	Not Used	---
44	BR	Secondary AIR Diverter	On: 1v, Off: 12v
45-46	---	Not Used	---
47	BK/PK	VR Solenoid Control	0%, at 55 mph: 45%
48	WT/PK	Clean Tachometer Output	38-42 Hz
49	GY/OR	PIP Sensor Signal	6.93v (50%)
50	PK	Spark Output Signal	6.93v (50%)
51	BK/WT	Power Ground	<0.1v
52-54	---	Not Used	---
55	YL	Keep Alive Power	12-14v
56	LG/BK	EVAP VMV Solenoid	0-10 Hz (0-100%)
57	---	Not Used	---
58	GY/BK	PSOM (+) Signal	0 Hz, at 55 mph: 125 Hz
59	DB	Misfire Detection Sensor	45-55 Hz
60	GY/LB	HO2S-11 (B1 S1) Signal	0.1-1.1v
61-63	---	Not Used	---
64	LB/YL	PNP Switch Signal	In 'P': 0v, Others: 5v

1996 E-Series 5.0L V8 OHV VIN N (A/T) 104 Pin Connector

PCM Pin #	Wire Color	Circuit Description (104 Pin)	Value at Hot Idle
65	BR/LG	EGR DPFE Sensor Signal	0.20-1.30v
66-69	---	Not Used	---
70	WT/OR	Secondary Air Bypass Solenoid	AIRB On: 1v, Off: 12v
71	RD	Vehicle Power	12-14v
72	TN/RD	Injector 7 Control	3.2-4.5 ms
73	TN/BK	Injector 5 Control	3.2-4.5 ms
74	BR/YL	Injector 3 Control	3.2-4.5 ms
75	TN	Injector 1 Control	3.2-4.5 ms
76	BK/WT	Power Ground	<0.1v
77	BK/WT	Power Ground	<0.1v
78	---	Not Used	---
79	WT/LG	TCIL (lamp) Control	Lamp On: 1v, Off: 12v
80	LB/OR	Fuel Pump Control	On: 1v, Off: 12v
81-82	---	Not Used	---
83	WT/LB	IAC Solenoid Control	10.7v (33%)
84	DG/YL	Output Shaft Speed Sensor	0 Hz, 55 mph: 130 Hz
85-86	---	Not Used	---
87	RD/BK	HO2S-21 (B2 S1) Signal	0.1-1.1v
88	LB/RD	MAF Sensor Signal	0.8v, 55 mph: 2.3v
89	GY/WT	TP Sensor Signal	0.9v, 55 mph: 1.2v
90	BR/WT	Reference Voltage	4.9-5.1v
91	GY/RD	Analog Signal Return	<0.050v
92	LG	Brake Position Switch	Brake Off: 12v, On: 1v
93	RD/WT	HO2S-11 (B1 S1) Heater	1v, Off: 12v
94	YL/LB	HO2S-21 (B2 S1) Heater	1v, Off: 12v
95	WT/BK	HO2S-12 (B1 S2) Heater	1v, Off: 12v
96	---	Not Used	---
97	RD	Vehicle Power	12-14v
98	LG	Injector 8 Control	3.2-4.5 ms
99	LG/OR	Injector 6 Control	3.2-4.5 ms
100	BR/LB	Injector 4 Control	3.2-4.5 ms
101	WT	Injector 2 Control	3.2-4.5 ms
102	---	Not Used	---
103	BK/WT	Power Ground	<0.1v
104	---	Not Used	---

Pin Connector Graphic

PCM 104-PIN CONNECTOR

Terminal View of 104-Pin PCM Wiring Harness Connector

Standard Colors and Abbreviations

Abbreviation	Color	Abbreviation	Color	Abbreviation	Color
BK	Black	GY	Gray	PK	Purple
BL	Blue	GN	Green	RD	Red
BR	Brown	LG	LT Green	TN	Tan
DB	Dark Blue	OR	Orange	WT	White
DG	DK Green	PK	Pink	YL	Yellow

1996 E-Series 5.0L V8 VIN N (E4OD) 104 Pin Connector

PCM Pin #	Wire Color	Circuit Description (104 Pin)	Value at Hot Idle
1	PK/OR	Shift Solenoid 2 Control	12v, 55 mph: 12v
2	PK/LG	MIL (lamp) Control	MIL On: 1v, Off: 12v
3, 5	---	Not Used	---
4	LG/RD	Power Take-Off Signal	0.1v (Off)
6	GY	Output Shaft Speed Sensor	0 Hz, 30 mph: 130 Hz
7-12	---	Not Used	---
13	VT	Flash EPROM Power	0.1v
14	---	Not Used	---
15	PK/LB	Data Bus (-)	Digital Signals
16	TN/OR	Data Bus (+)	Digital Signals
17-22	---	Not Used	---
23	OR/RD	Ignition System Ground	<0.050v
24	BK/WT	Power Ground	<0.1v
25	BK	Case Ground	<0.050v
26	---	Not Used	---
27	OR/YL	Shift Solenoid 1 Control	1v, 55 mph: 12v
28	---	Not Used	---
29	TN/WT	TCS (switch) Signal	TCS & O/D On: 12v
30-32	---	Not Used	---
33	PK/OR	PSOM (-) Signal	<0.050v
34	---	Not Used	---
35	RD/LG	HO2S-12 (B1 S2) Signal	0.1-1.1v
36	TN/LB	MAF Sensor Return	<0.050v
37	OR/BK	Transmission Fluid Temperature Sensor	2.10-2.40v
38	LG/RD	Engine Coolant Temperature Sensor	0.5-0.6v
39	GY	Intake Air Temperature Sensor	1.5-2.5v
40	DG/YL	Fuel Pump Monitor	On: 12v, Off: 0v
41	BK/YL	A/C Pressure Switch Signal	A/C On: 12v, off: 0v
42-43	---	Not Used	---
44	BR	Secondary AIR Diverter	On: 1v, Off: 12v
45-46	---	Not Used	---
47	BK/PK	VR Solenoid Control	0%, at 55 mph: 45%
48	WT/PK	Clean Tachometer Output	38-42 Hz
49	GY/OR	PIP Sensor Signal	6.93v (50%)
50	PK	Spark Output Signal	6.93v (50%)
51	BK/WT	Power Ground	<0.1v
52	---	Not Used	---
53	BR/OR	Coast Clutch Solenoid	12v, 55 mph: 12v
54	DB/WT	TCC Solenoid Control	0%, at 55 mph: 95%
55	YL	Keep Alive Power	12-14v
56	LG/BK	EVAP VMV Solenoid	0-10 Hz (0-100%)
57	---	Not Used	---
58	GY/BK	PSOM (+) Signal	0 Hz, at 55 mph: 125 Hz
59	DB	Misfire Detection Sensor	45-55 Hz
60	GY/LB	HO2S-11 (B1 S1) Signal	0.1-1.1v
61-63	---	Not Used	---
64	LB/YL	Digital Transmission Range Sensor	In 'P': 0v, in O/D: 1.7v

1996 E-Series 5.0L V8 VIN N (E4OD) 104 Pin Connector

PCM Pin #	Wire Color	Circuit Description (104 Pin)	Value at Hot Idle
65	BR/LG	EGR DPFE Sensor Signal	0.20-1.30v
66-69	---	Not Used	---
70	WT/OR	Secondary Air Bypass Solenoid	AIRB On: 1v, Off: 12v
71	RD	Vehicle Power	12-14v
72	TN/RD	Injector 7 Control	3.2-4.5 ms
73	TN/BK	Injector 5 Control	3.2-4.5 ms
74	BR/YL	Injector 3 Control	3.2-4.5 ms
75	TN	Injector 1 Control	3.2-4.5 ms
76	BK/WT	Power Ground	<0.1v
77	BK/WT	Power Ground	<0.1v
78	---	Not Used	---
79	WT/LG	TCIL (lamp) Control	Lamp On: 1v, Off: 12v
80	LB/OR	Fuel Pump Control	On: 1v, Off: 12v
81	WT/YL	EPC Solenoid Control	10v (26 psi)
82	---	Not Used	---
83	WT/LB	IAC Solenoid Control	10.7v (33%)
84	DG/YL	Output Shaft Speed Sensor	0 Hz, 30 mph: 130 Hz
85-86	---	Not Used	---
87	RD/BK	HO2S-21 (B2 S1) Signal	0.1-1.1v
88	LB/RD	MAF Sensor Signal	0.8v, 55 mph: 2.3v
89	GY/WT	TP Sensor Signal	0.9v, 55 mph: 1.2v
90	BR/WT	Reference Voltage	4.9-5.1v
91	GY/RD	Analog Signal Return	<0.050v
92	LG	Brake Position Switch	Brake Off: 12v, On: 1v
93	RD/WT	HO2S-11 (B1 S1) Heater	1v, Off: 12v
94	YL/LB	HO2S-21 (B2 S1) Heater	1v, Off: 12v
95	WT/BK	HO2S-12 (B1 S2) Heater	1v, Off: 12v
96	---	Not Used	---
97	RD	Vehicle Power	12-14v
98	LG	Injector 8 Control	3.2-4.5 ms
99	LG/OR	Injector 6 Control	3.2-4.5 ms
100	BR/LB	Injector 4 Control	3.2-4.5 ms
101	WT	Injector 2 Control	3.2-4.5 ms
102	---	Not Used	---
103	BK/WT	Power Ground	<0.1v
104	---	Not Used	---

Pin Connector Graphic

```
               PCM 104-PIN CONNECTOR
   1 ●●●●●●●●●●●●●    ●●●●●●●●●●●●● 26
  27 ●●●●●●●●●●●●●    ●●●●●●●●●●●●● 52
  53 ●●●●●●●●●●●●●   ▨   ●●●●●●●●●●●●● 78
  79 ●●●●●●●●●●●●●    ●●●●●●●●●●●●● 104
```

Terminal View of 104-Pin PCM Wiring Harness Connector

Standard Colors and Abbreviations

Abbreviation	Color	Abbreviation	Color	Abbreviation	Color
BK	Black	GY	Gray	PK	Purple
BL	Blue	GN	Green	RD	Red
BR	Brown	LG	LT Green	TN	Tan
DB	Dark Blue	OR	Orange	WT	White
DG	DK Green	PK	Pink	YL	Yellow

1997 E-Series 5.4L V8 VIN L (E4OD) 104 Pin Connector

PCM Pin #	Wire Color	Circuit Description (104 Pin)	Value at Hot Idle
1	PK/LB	COP 6 Driver (Dwell)	5°, at 55 mph: 8°
2	PK/LG	MIL (lamp) Control	MIL On: 1v, Off: 12v
3	BK/WT	Power Ground	<0.1v
4-5	---	Not Used	---
6	OR/YL	Shift Solenoid 1 Control	1v, 55 mph: 12v
7-8	---	Not Used	---
9	YL/WT	Fuel Level Indicator Signal	1.7v (1/2 full)
10	---	Not Used	---
11	PK/OR	Shift Solenoid 2 Control	12v, 55 mph: 1v
12	WT/LG	TCIL (lamp) Control	Lamp On: 1v, Off: 12v
13	PK	Flash EEPROM	0.1v
14	---	Not Used	---
15	PK/LB	Data Bus (-)	Digital Signals
16	TN/OR	Data Bus (+)	Digital Signals
17-19	---	Not Used	---
20	BR/OR	Coast Clutch Solenoid	12v, 55 mph: 12v
21	DB	CKP (+) Sensor	400-420 Hz
22	GY	CKP (-) Sensor	400-420 Hz
23	---	Not Used	---
24	BK/WT	Power Ground	<0.1v
25	BK/LB	Case Ground	<0.050v
26	DB/YL	COP 1 Driver (Dwell)	5°, at 55 mph: 8°
27	PKWT	COP 5 Driver (Dwell)	5°, at 55 mph: 8°
28	---	Not Used	---
29	TN/WT	TCS (switch) Signal	TCS & O/D On: 12v
30	DG	Octane Adjustment	9.3v (switch shorted: 0v)
31-32	---	Not Used	---
33	PK/OR	VSS (-) Signal	0 Hz, at 55 mph: 125 Hz
34	OR/BK	Transmission Range Sensor 1	0v, at 55 mph: 11.5v
35	RD/LG	HO2S-12 (B1 S2) Signal	0.1-1.1v
36	TN/LB	MAF Sensor Return	<0.050v
37	OR/BK	Transmission Fluid Temperature Sensor	2.10-2.40v
38	---	Not Used	---
39	GY	Intake Air Temperature Sensor	1.5-2.5v
40	DG/YL	Fuel Pump Monitor	On: 12v, Off: 0v
41	BK/YL	A/C High Pressure Switch	Switch Closed: 12v
42-44	---	Not Used	---
45	YL/LB	CHIL (lamp) Control	Lamp On: 1v, Off: 12v
46	---	Not Used	---
47	BR/PK	VR Solenoid Control	0%, at 55 mph: 45%
48	---	Not Used	---
49	LB/BK	Digital TR2 Sensor Signal	0v, at 55 mph: 11.5v
50	GY/BK	Digital TR4 Sensor Signal	0v, at 55 mph: 11.5v
51	BK/WT	Power Ground	<0.1v
52	LG/WT	COP 3 Driver (Dwell)	5°, at 55 mph: 8°
53	OR/YL	COP 4 Driver (Dwell)	5°, at 55 mph: 8°
54	PK/YL	TCC Solenoid Control	0%, at 55 mph: 95%
55	YL	Keep Alive Power	12-14v
56	LG/BK	EVAP VMV Solenoid	0-10 Hz (0-100%)
57	YL/RD	Knock Sensor 1 Signal	0v
58	GY/BK	VSS (+) Signal	0 Hz, at 55 mph: 125 Hz
59	---	Not Used	---
60	GY/LB	HO2S-11 (B1 S1) Signal	0.1-1.1v
61, 63	---	Not Used	---
62	RD/PK	Fuel Tank Pressure Sensor	2.6v (0" H2O - cap off)
64	LB/YL	Transmission Range Sensor 3	In 'P': 0v, in O/D: 1.7v

1997 E-Series 5.4L V8 VIN L (E4OD) 104 Pin Connector

PCM Pin #	Wire Color	Circuit Description (104 Pin)	Value at Hot Idle
65	BR/LG	EGR DPFE Sensor Signal	0.20-1.30v
66	YL/LG	Cylinder Head Temperature Sensor	0.6v (194°F)
67	PKWT	EVAP CV Solenoid	0-10 Hz (0-100%)
68-70	---	Not Used	---
71	RD	Vehicle Power	12-14v
72	TN/RD	Injector 7 Control	3.2-3.8 ms
73	TN/BK	Injector 5 Control	3.2-3.8 ms
74	BR/YL	Injector 3 Control	3.2-3.8 ms
75	TN	Injector 1 Control	3.2-3.8 ms
76	---	Not Used	---
77	BK/WT	Power Ground	<0.1v
78	WT/PK	COP 7 Driver (Dwell)	5°, at 55 mph: 8°
79	WT/RD	COP 8 Driver (Dwell)	5°, at 55 mph: 8°
80	LB/OR	Fuel Pump Control	On: 1v, Off: 12v
82	---	Not Used	---
81	WT/YL	EPC Solenoid Control	9.1v (4 psi)
83	WT/LB	IAC Solenoid Control	10.7v (33%)
84	---	Not Used	---
85	DB/OR	Camshaft Position Sensor	6 Hz, 55 mph: 14-18 Hz
86	---	Not Used	---
87	RD/BK	HO2S-21 (B2 S1) Signal	0.1-1.1v
88	LB/RD	MAF Sensor Signal	0.8v, 55 mph: 1.6v
89	GY/WT	TP Sensor Signal	0.53-1.27v
90	BR/WT	Reference Voltage	4.9-5.1v
91	GY/RD	Analog Signal Return	<0.050v
92	LG	Brake Position Switch	Brake Off: 12v, On: 1v
93	RD/WT	HO2S-11 (B1 S1) Heater	1v, Off: 12v
94	YL/LB	HO2S-21 (B2 S1) Heater	1v, Off: 12v
95	WT/BK	HO2S-12 (B1 S2) Heater	1v, Off: 12v
96	---	Not Used	---
97	RD	Vehicle Power	12-14v
98	LB	Injector 8 Control	3.2-3.8 ms
99	LG/OR	Injector 6 Control	3.2-3.8 ms
100	BR/LG	Injector 4 Control	3.2-3.8 ms
101	WT	Injector 2 Control	3.2-3.8 ms
102	---	Not Used	---
103	BK/WT	Power Ground	<0.1v
104	DB/LG	COP 2 Driver (Dwell)	5°, at 55 mph: 8°

Pin Connector Graphic

Terminal View of 104-Pin PCM Wiring Harness Connector

Standard Colors and Abbreviations

Abbreviation	Color	Abbreviation	Color	Abbreviation	Color
BK	Black	GY	Gray	PK	Purple
BL	Blue	GN	Green	RD	Red
BR	Brown	LG	LT Green	TN	Tan
DB	Dark Blue	OR	Orange	WT	White
DG	DK Green	PK	Pink	YL	Yellow

1998-99 E-Series 5.4L V8 VIN L (E4OD) 104 Pin Connector

PCM Pin #	Wire Color	Circuit Description (104 Pin)	Value at Hot Idle
1	PK/LB	COP 6 Driver (Dwell)	5°, at 55 mph: 8°
2	PK/LG	MIL (lamp) Control	MIL On: 1v, Off: 12v
3	BK/WT	Power Ground	<0.1v
4-5	---	Not Used	---
6	OR/YL	Shift Solenoid 1 Control	1v, 55 mph: 12v
7-8	---	Not Used	---
9	YL/WT	Fuel Level Indicator Signal	1.7v (1/2 full)
10	---	Not Used	---
11	PK/OR	Shift Solenoid 2 Control	12v, 55 mph: 1v
12	WT/LG	TCIL (lamp) Control	Lamp On: 1v, Off: 12v
13	PK	Flash EEPROM	0.1v
14	---	Not Used	---
15	PK/LB	Data Bus (-)	Digital Signals
16	TN/OR	Data Bus (+)	Digital Signals
17-19	---	Not Used	---
20	BR/OR	Coast Clutch Solenoid	12v, 55 mph: 12v
21	DB	CKP (+) Sensor	410 Hz, 55 mph: 1 KHz
22	GY	CKP (-) Sensor	410 Hz, 55 mph: 1 KHz
23	---	Not Used	---
24	BK/WT	Case Ground	<0.050v
25	BK/LB	Case Ground	<0.050v
26	DB/YL	COP 1 Driver (Dwell)	5°, at 55 mph: 8°
27	PKWT	COP 5 Driver (Dwell)	5°, at 55 mph: 8°
28	---	Not Used	---
29	TN/WT	TCS (switch) Signal	TCS & O/D On: 12v
30	DG	Octane Adjustment	9.3v (switch shorted: 0v)
31-32	---	Not Used	---
33	PK/OR	VSS (-) Signal	0 Hz, at 55 mph: 125 Hz
34	YL/BK	Digital TR1 Sensor	0v, at 55 mph: 11.5v
35	RD/LG	HO2S-12 (B1 S2) Signal	0.1-1.1v
36	TN/LB	MAF Sensor Return	<0.050v
37	OR/BK	Transmission Fluid Temperature Sensor	2.10-2.40v
38	---	Not Used	---
39	GY	Intake Air Temperature Sensor	1.5-2.5v
40	DG/YL	Fuel Pump Monitor	On: 12v, Off: 0v
41	BK/YL	A/C High Pressure Switch	Switch Closed: 12v
42-44	---	Not Used	---
45	YL/LB	CHIL (lamp) Control	Lamp On: 1v, Off: 12v
46	BR	Intake Manifold Tuning Valve	0.1v
47	BR/PK	VR Solenoid Control	0%, at 55 mph: 45%
48	WT/PK	Clean Tachometer Output Signal	46 Hz, 55 mph: 115 Hz
49	LB/BK	Digital TR2 Sensor	0v, at 55 mph: 11.5v
50	LG/RD	Digital TR4 Sensor	0v, at 55 mph: 11.5v
51	BK/WT	Power Ground	<0.1v
52	LG/WT	COP 3 Driver (Dwell)	5°, at 55 mph: 8°
53	OR/YL	COP 4 Driver (Dwell)	5°, at 55 mph: 8°
54	PK/YL	TCC Solenoid Control	0%, at 55 mph: 95%
55	Y	Keep Alive Power	12-14v
56	LG/BK	EVAP Purge Solenoid	0-10 Hz (0-100%)
57	YL/RD	Knock Sensor 1 Signal	0v
58	GY/BK	VSS (+) Signal	0 Hz, at 55 mph: 125 Hz
60	GY/LB	HO2S-11 (B1 S1) Signal	0.1-1.1v
61	---	Not Used	---
62	RD/PK	Fuel Tank Pressure Sensor	2.6v (0" H2O - cap off)
63	---	Not Used	---
64	LB/YL	Digital TR Sensor	In 'P': 0v, in O/D: 1.7v

1998-99 E-Series 5.4L V8 VIN L (E4OD) 104 Pin Connector

PCM Pin #	Wire Color	Circuit Description (104 Pin)	Value at Hot Idle
65	BR/LG	EGR DPFE Sensor Signal	0.20-1.30v
66	YL/LG	Cylinder Head Temperature Sensor	0.6v (194ºF)
67	PKWT	EVAP CV Solenoid	0-10 Hz (0-100%)
68-70	---	Not Used	---
71	RD	Vehicle Power	12-14v
72	TN/RD	Injector 7 Control	3.2-3.8 ms
73	TN/BK	Injector 5 Control	3.2-3.8 ms
74	BR/YL	Injector 3 Control	3.2-3.8 ms
75	TN	Injector 1 Control	3.2-3.8 ms
76	---	Not Used	---
77	BK/WT	Power Ground	<0.1v
78	WT/PK	COP 7 Driver (Dwell)	5º, at 55 mph: 8º
79	WT/RD	COP 8 Driver (Dwell)	5º, at 55 mph: 8º
80	LB/OR	Fuel Pump Control	On: 1v, Off: 12v
81	WT/YL	EPC Solenoid Control	9.1v (5 psi)
82	---	Not Used	---
83	WT/LB	IAC Solenoid Control	10.7v (33%)
84	DB/YL	Output Shaft Speed Sensor	0 Hz, 55 mph: 228 Hz
85	DG	Camshaft Position Sensor	6 Hz, 55 mph: 14-18 Hz
86	---	Not Used	---
87	RD/BK	HO2S-21 (B2 S1) Signal	0.1-1.1v
88	LB/RD	MAF Sensor Signal	0.8v, 55 mph: 1.6v
89	GY/WT	TP Sensor Signal	0.53-1.27v
90	BR/WT	Reference Voltage	4.9-5.1v
91	GY/RD	Analog Signal Return	<0.050v
92	LG	Brake Position Switch	Brake Off: 12v, On: 1v
93	RD/WT	HO2S-11 (B1 S1) Heater	1v, Off: 12v
94	YL/LB	HO2S-21 (B2 S1) Heater	1v, Off: 12v
95	WT/BK	HO2S-12 (B1 S2) Heater	1v, Off: 12v
96	---	Not Used	---
97	RD	Vehicle Power	12-14v
98	LB	Injector 8 Control	3.2-3.8 ms
99	LG/OR	Injector 6 Control	3.2-3.8 ms
100	BR/LG	Injector 4 Control	3.2-3.8 ms
101	WT	Injector 2 Control	3.2-3.8 ms
102	---	Not Used	---
103	BK/WT	Power Ground	<0.1v
104	DB/LG	COP 2 Driver (Dwell)	5º, at 55 mph: 8º

Pin Connector Graphic

```
PCM 104-PIN CONNECTOR
 1 ●●●●●●●●●●●●   ●●●●●●●●●●●● 26
27 ●●●●●●●●●●●●   ●●●●●●●●●●●● 52
53 ●●●●●●●●●●●●  ◉  ●●●●●●●●●●●● 78
79 ●●●●●●●●●●●●   ●●●●●●●●●●●● 104
```

Terminal View of 104-Pin PCM Wiring Harness Connector

Standard Colors and Abbreviations

Abbreviation	Color	Abbreviation	Color	Abbreviation	Color
BK	Black	GY	Gray	PK	Purple
BL	Blue	GN	Green	RD	Red
BR	Brown	LG	LT Green	TN	Tan
DB	Dark Blue	OR	Orange	WT	White
DG	DK Green	PK	Pink	YL	Yellow

2000-02 E-Series 5.4L V8 VIN L (A/T) 104 Pin Connector

PCM Pin #	Wire Color	Circuit Description (104 Pin)	Value at Hot Idle
1	OR/YL	COP 6 Driver (Dwell)	5°, at 55 mph: 8°
2	PK/LG	MIL (lamp) Control	MIL On: 1v, Off: 12v
3	BK/WT	Power Ground	<0.1v
4	PK/LB	Power Takeoff Signal	PTO Off: 0v, On: 12v
5	---	Not Used	---
6	OR/YL	Shift Solenoid 1 Control	1v, 55 mph: 1v
7-8	---	Not Used	---
9	YL/WT	Fuel Pump Monitor	On: 1v, Off: 12v
10	---	Not Used	---
11	PK/OR	Shift Solenoid 2 Control	12v, 55 mph: 1v
12	WT/LG	TCIL (lamp) Control	Lamp On: 1v, Off: 12v
13	PK	Flash EEPROM	0.1v
14	OR/LB	4x4 Low Switch	Switch On: 0v, Off: 12v
15	PK/LB	Data Bus (-)	Digital Signals
16	TN/OR	Data Bus (+)	Digital Signals
17-19	---	Not Used	---
20	BR/OR	Coast Clutch Solenoid	12v, 55 mph: 12v
21	DB	CKP (+) Sensor	410 Hz, 55 mph: 1 KHz
22	GY	CKP (-) Sensor	410 Hz, 55 mph: 1 KHz
24	BK/WT	Case Ground	<0.050v
25	BK/LB	Case Ground	<0.050v
26	DB/YL	COP 1 Driver (Dwell)	5°, at 55 mph: 8°
27	PKWT	COP 5 Driver (Dwell)	5°, at 55 mph: 8°
28	---	Not Used	---
29	TN/WT	TCS (switch) Signal	TCS & O/D On: 12v
30-31	---	Not Used	---
32	DG/PK	Knock Sensor 1 Return	<0.050v
33	---	Not Used	---
34	YL/BK	Digital TR1 Sensor	0v, at 55 mph: 11.5v
35	RD/LG	HO2S-12 (B1 S2) Signal	0.1-1.1v
36	TN/LB	MAF Sensor Return	<0.050v
37	OR/BK	Transmission Fluid Temperature Sensor	2.10-2.40v
38	---	Not Used	---
39	GY	Intake Air Temperature Sensor	1.5-2.5v
40	DG/YL	Fuel Pump Monitor	On: 12v, Off: 0v
41	BK/YL	A/C Pressure Switch Signal	A/C On: 12v, Off: 0v
42-44	---	Not Used	---
45	YL/LB	Engine Coolant Temperature Sensor	0.5-0.6v
46	---	Not Used	---
47	BR/PK	VR Solenoid Control	0%, at 55 mph: 45%
48	---	Not Used	---
49	LB/BK	Digital TR2 Sensor	0v, at 55 mph: 11.5v
50	LG/RD	Digital TR4 Sensor	0v, at 55 mph: 11.5v
51	BK/WT	Power Ground	<0.1v
52	WT/PK	COP 3 Driver (Dwell)	5°, at 55 mph: 8°
53	DG/PK	COP 4 Driver (Dwell)	5°, at 55 mph: 8°
54	PK/YL	TCC Solenoid Control	0%, at 55 mph: 95%
55	YL	Keep Alive Power	12-14v
56	LG/BK	EVAP Purge Solenoid	0-10 Hz (0-100%)
57	YL/RD	Knock Sensor 1 Signal	0v
58	---	Not Used	---
59	DG/WT	Turbine Shaft Speed Sensor	120 Hz, 30 mph: 300 Hz
60	GY/LB	HO2S-11 (B1 S1) Signal	0.1-1.1v
61	---	Not Used	---
62	RD/PK	Fuel Tank Pressure Sensor	2.6v (0" H2O - cap off)
63	---	Not Used	---
64	LB/YL	Digital TR3A Sensor	In 'P': 0v, in O/D: 1.7v

2000-02 E-Series 5.4L V8 VIN L (A/T) 104 Pin Connector

PCM Pin #	Wire Color	Circuit Description (104 Pin)	Value at Hot Idle
65 ('00)	BR/LG	EGR DPFE Sensor Signal	0.20-1.30v
65 ('01-'02)	BR/LG	EGR DPFE Sensor Signal	0.95-1.05v
66	YL/LG	Cylinder Head Temperature Sensor	0.6v (194°F)
67	PKWT	EVAP CV Solenoid	0-10 Hz (0-100%)
68	GY/BK	Vehicle Speed Signal	0 Hz, 55 mph: 124 Hz
69-70	---	Not Used	---
71	RD	Vehicle Power	12-14v
72	TN/RD	Injector 7 Control	3.2-3.8 ms
73	TN/BK	Injector 5 Control	3.2-3.8 ms
74	BR/YL	Injector 3 Control	3.2-3.8 ms
75	TN	Injector 1 Control	3.2-3.8 ms
76	BK/WT	Power Ground	<0.1v
77	BK/WT	Power Ground	<0.1v
78	PK/LB	COP 7 Driver (Dwell)	5°, at 55 mph: 8°
79	WT/RD	COP 8 Driver (Dwell)	5°, at 55 mph: 8°
80	LB/OR	Fuel Pump Control	On: 1v, Off: 12v
81	WT/YL	EPC Solenoid Control	8.0v (3.5 psi)
82	---	Not Used	---
83	WT/LB	IAC Solenoid Control	8.7v (43%)
84	DB/YL	Output Shaft Speed Sensor	0 Hz, 55 mph: 228 Hz
85	DG	Camshaft Position Sensor	7 Hz, 55 mph: 13-17 Hz
86	---	Not Used	---
87	RD/BK	HO2S-21 (B2 S1) Signal	0.1-1.1v
88	LB/RD	MAF Sensor Signal	0.8v, 55 mph: 1.9v
89	GY/WT	TP Sensor Signal	0.53-1.27v
90	BR/WT	Reference Voltage	4.9-5.1v
91	GY/RD	Analog Signal Return	<0.050v
92	LG	Brake Position Switch	Brake Off: 12v, On: 1v
93	RD/WT	HO2S-11 (B1 S1) Heater	1v, Off: 12v
94	YL/LB	HO2S-21 (B2 S1) Heater	1v, Off: 12v
95	WT/BK	HO2S-12 (B1 S2) Heater	1v, Off: 12v
96	---	Not Used	---
97	RD	Vehicle Power	12-14v
98	LB	Injector 8 Control	3.2-3.8 ms
99	LG/OR	Injector 6 Control	3.2-3.8 ms
100	BR/LG	Injector 4 Control	3.2-3.8 ms
101	WT	Injector 2 Control	3.2-3.8 ms
102	---	Not Used	---
103	BK/WT	Power Ground	<0.1v
104	PKWT	COP 2 Driver (Dwell)	5°, at 55 mph: 8°

Pin Connector Graphic

PCM 104-PIN CONNECTOR

Terminal View of 104-Pin PCM Wiring Harness Connector

Standard Colors and Abbreviations

Abbreviation	Color	Abbreviation	Color	Abbreviation	Color
BK	Black	GY	Gray	PK	Purple
BL	Blue	GN	Green	RD	Red
BR	Brown	LG	LT Green	TN	Tan
DB	Dark Blue	OR	Orange	WT	White
DG	DK Green	PK	Pink	YL	Yellow

2003-05 E-Series 5.4L V8 VIN L (A/T) 104 Pin Connector

PCM Pin #	Wire Color	Circuit Description (104 Pin)	Value at Hot Idle
1	OR/YL	COP 6 Driver (Dwell)	5º, at 55 mph: 8º
2	PK/LG	MIL (lamp) Control	MIL On: 1v, Off: 12v
3	BK/WT	Power Ground	<0.1v
4-5	---	Not Used	---
6	OR/YL	Shift Solenoid 'A' Control	1v, 55 mph: 1v
7-8	---	Not Used	---
9	YL/WT	Low Fuel Indicator	Indicator Off: 12v, On: 1v
10	---	Not Used	---
11	VT/OR	Shift Solenoid 'B' Control	12v, 55 mph: 1v
12	WT/LG	TCIL (lamp) Control	Lamp Off: 12v, On: 1v
13	VT	Flash EEPROM	0.1v
14	OR/LB	4x4 Low Switch	Switch Off: 12v, On: 1v
15	PK/LB	SCP Data Bus (-)	<0.050v
16	TN/OR	SCP Data Bus (+)	Digital Signals
17-19	---	Not Used	---
20	BR/OR	Coast Clutch Solenoid Control	12v, 55 mph: 12v
21	DB	CKP (+) Sensor	410 Hz, 55 mph: 1 KHz
22	GY	CKP (-) Sensor	410 Hz, 55 mph: 1 KHz
23	---	Not Used	---
24	BK/WT	Case Ground	<0.050v
25	BK/LB	Power Ground	<0.1v
26	LG/WT	COP 1 Driver (Dwell)	5º, at 55 mph: 8º
27	LG/YL	COP 5 Driver (Dwell)	5º, at 55 mph: 8º
28	---	Not Used	---
29	TN/WT	TCS (switch) Signal	TCS & O/D On: 12v
30-31	---	Not Used	---
32	DG/VT	Knock Sensor 1 Ground	<0.050v
33	---	Not Used	---
34	YL/BK	Digital TR1 Sensor	0v, at 55 mph: 11.5v
35	RD/LG	HO2S-12 (B1 S2) Signal	0.1-1.1v
36	TN/LB	MAF Sensor Return	<0.050v
37	OR/BK	Transmission Fluid Temperature Sensor	2.10-2.40v
38	---	Not Used	---
39	GY	Intake Air Temperature Sensor	1.5-2.5v
40	DG/YL	Fuel Pump Monitor	On: 12v, Off: 0v
41	BK/YL	A/C Pressure Switch Signal	A/C On: 12v, Off: 0v
42-44	---	Not Used	---
45	YL/LB	Engine Coolant Temperature Sensor	0.5-0.6v
46	---	Not Used	---
47	BR/PK	VR Solenoid Control	0%, at 55 mph: 45%
48	---	Not Used	---
49	LB/BK	Digital TR2 Sensor	0v, at 55 mph: 11.5v
50	LG/RD	Digital TR4 Sensor	0v, at 55 mph: 11.5v
51	BK/WT	Power Ground	<0.1v
52	WT/PK	COP 3 Driver (Dwell)	5º, at 55 mph: 8º
53	DG/VT	COP 4 Driver (Dwell)	5º, at 55 mph: 8º
54	VT/YL	TCC Solenoid Control	0%, at 55 mph: 95%
55	YL	Keep Alive Power	12-14v
56	LG/BK	EVAP Canister Purge Valve	0-10 Hz (0-100%)
57	YL/RD	Knock Sensor 1 Ground	<0.050v
58	---	Not Used	---
59	DG/WT	Turbine Shaft Speed Sensor	120 Hz, 30 mph: 300 Hz
60	GY/LB	HO2S-11 (B1 S1) Signal	0.1-1.1v
61	---	Not Used	---
62	RD/PK	Fuel Tank Pressure Sensor	2.6v (0" H2O - cap off)
63	---	Not Used	---
64	LB/YL	Digital TR3A Sensor	In 'P': 0v, in O/D: 1.7v

2003-05 E-Series 5.4L V8 VIN L (A/T) 104 Pin Connector

PCM Pin #	Wire Color	Circuit Description (104 Pin)	Value at Hot Idle
65	BR/LG	EGR DPFE Sensor Signal	0.95-1.05v
66	YL/LG	Cylinder Head Temperature Sensor	0.6v (194°F)
67	VT/WT	EVAP Canister Vent Valve	0-10 Hz (0-100%)
68	GY/BK	Vehicle Speed Sensor	0 Hz, 55 mph: 124 Hz
69-70	---	Not Used	---
71	RD	Vehicle Power (Start-Run)	12-14v
72	TN/RD	Injector 7 Control	3.2-3.8 ms
73	TN/BK	Injector 5 Control	3.2-3.8 ms
74	BR/YL	Injector 3 Control	3.2-3.8 ms
75	TN	Injector 1 Control	3.2-3.8 ms
76	BK/WT	Power Ground	<0.1v
77	BK/WT	Power Ground	<0.1v
78	PK/LB	COP 7 Driver (Dwell)	5°, at 55 mph: 8°
79	WT/RD	COP 8 Driver (Dwell)	5°, at 55 mph: 8°
80	LB/OR	Fuel Pump Relay Control	Off: 12v, On: 1v
81	WT/YL	EPC Solenoid Control	8.0v (3.5 psi)
82	---	Not Used	---
83	WT/LB	IAC Solenoid Control	8.7v (43%)
84	DB/YL	Output Shaft Speed Sensor	0 Hz, 55 mph: 228 Hz
85	DG	Camshaft Position Sensor	7 Hz, 55 mph: 13-17 Hz
86	---	Not Used	---
87	RD/BK	HO2S-21 (B2 S1) Signal	0.1-1.1v
88	LB/RD	Mass Airflow Sensor	0.8v, 55 mph: 1.9v
89	GY/WT	Throttle Position Sensor	0.53-1.27v
90	BR/WT	Reference Voltage	4.9-5.1v
91	GY/RD	Sensor Ground	<0.050v
92	LG	Brake Position Switch	Brake Off: 12v, On: 1v
93	RD/WT	HO2S-11 (B1 S1) Heater Control	1v, Off: 12v
94	YL/LB	HO2S-21 (B2 S1) Heater Control	1v, Off: 12v
95	WT/BK	HO2S-12 (B1 S2) Heater Control	1v, Off: 12v
96	---	Not Used	---
97	RD	Vehicle Power (Start-Run)	12-14v
98	LB	Injector 8 Control	3.2-3.8 ms
99	LG/OR	Injector 6 Control	3.2-3.8 ms
100	BR/LG	Injector 4 Control	3.2-3.8 ms
101	WT	Injector 2 Control	3.2-3.8 ms
102	---	Not Used	---
103	BK/WT	Power Ground	<0.1v
104	PKWT	COP 2 Driver (Dwell)	5°, at 55 mph: 8°

Pin Connector Graphic

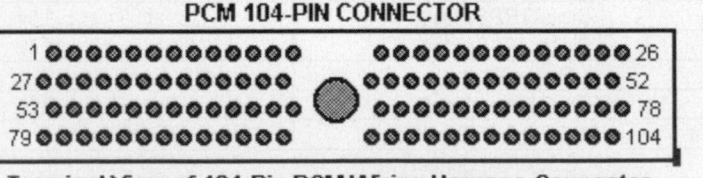

PCM 104-PIN CONNECTOR

Terminal View of 104-Pin PCM Wiring Harness Connector

Standard Colors and Abbreviations

Abbreviation	Color	Abbreviation	Color	Abbreviation	Color
BK	Black	GY	Gray	PK	Purple
BL	Blue	GN	Green	RD	Red
BR	Brown	LG	LT Green	TN	Tan
DB	Dark Blue	OR	Orange	WT	White
DG	DK Green	PK	Pink	YL	Yellow

1998-89 E-Series 5.4L V8 CNG VIN M (A/T) 104 Pin Connector

PCM Pin #	Wire Color	Circuit Description (104 Pin)	Value at Hot Idle
1	OR/YL	COP 6 Driver (Dwell)	5°, at 55 mph: 8°
2	PK/LG	MIL (lamp) Control	MIL On: 1v, Off: 12v
3	BK/WT	Power Ground	<0.1v
4	WT/PK	Power Takeoff Signal	PTO Off: 0v, On: 12v
5	---	Not Used	---
6	OG/YL	Shift Solenoid 'A' Control	1v, 55 mph: 12v
7-10	---	Not Used	---
11	VTN/ORG	Shift Solenoid 'B' Control	12v, 55 mph: 1v
12	WT/LG	TCIL (lamp) Control	Lamp On: 1v, Off: 12v
13	VT	Flash EEPROM	0.1v
14	---	Not Used	---
15	PK/LB	Data Bus (-)	Digital Signals
16	TN/OR	Data Bus (+)	Digital Signals
17-20	---	Not Used	---
20	BR/OG	Coast Clutch Solenoid	12v, 55 mph: 12v
21	DB	CKP (+) Sensor	440-490 Hz
22	GY	CKP (-) Sensor	440-490 Hz
24	BK/WT	Power Ground	<0.1v
25	BK/LB	Case Ground	<0.050v
26	LG/WTH	COP 1 Driver (Dwell)	5°, at 55 mph: 8°
27	LG/OR	COP 5 Driver (Dwell)	5°, at 55 mph: 8°
29	TN/WT	TCS (switch) Signal	TCS & O/D On: 12v
30-32	---	Not Used	---
33	PK/ORG	VSS (-) Signal Return	<0.050v
34	YL/BK	Transmission Range Sensor 1	0v, at 55 mph: 11.5v
36	TN/LB	MAF Sensor Return	<0.050v
37	OG/BK	Transmission Fluid Temperature Sensor	2.10-2.40v
38	---	Not Used	---
39	GY	Intake Air Temperature Sensor	1.5-2.5v
40	DG/YL	Fuel Pump Monitor	On: 12v, Off: 0v
41	BK/YL	A/C Pressure Switch Signal	A/C On: 12v, Off: 0v
42-44	---	Not Used	---
45	YL/LB	Engine Coolant Temperature Sensor	0.5-0.6v
48	WT/PK	Clean Tachometer Output	55 Hz, 55 mph: 120 Hz
49	LB/BK	Digital TR2 Sensor Signal	0v, at 55 mph: 11.5v
50	LG/RD	Digital TR4 Sensor Signal	0v, at 55 mph: 11.5v
51	BK/WT	Power Ground	<0.1v
52	WT/PK	COP 3 Driver (Dwell)	5°, at 55 mph: 8°
53	DG/PK	COP 4 Driver (Dwell)	5°, at 55 mph: 8°
54	VT/YL	TCC Solenoid Control	0%, at 55 mph: 95%
55	YL	Keep Alive Power	12-14v
56-57	---	Not Used	---
58	GY/BK	VSS (+) Signal	0 Hz, at 55 mph: 125 Hz
59	DG/WT	Turbine Shaft Speed Sensor	120 Hz, 30 mph: 300 Hz
60	GY/LB	HO2S-11 (B1 S1) Signal	0.1-1.1v
62	LB	Fuel Rail Temperature Sensor	1.7-3.5v (50-120°F)
63	RD/PK	Injection Pressure Sensor	2-3.7v (90-100 psi)
64	LB/YL	Digital TR3A Sensor Signal	In 'P': 0v, in O/D: 1.7v

1998-99 E-Series 5.4L V8 CNG VIN M 104 Pin Connector

PCM Pin #	Wire Color	Circuit Description (104 Pin)	Value at Hot Idle
65	---	Not Used	---
66	YL/LG	Cylinder Head Temperature Sensor	0.6v (194°F)
68-70	---	Not Used	---
71	RD	Vehicle Power	12-14v
72	TN/RD	Injector 7 (to NGV Module)	3.9-4.6 ms
73	TN/BK	Injector 5 (to NGV Module)	3.9-4.6 ms
74	BR/YL	Injector 3 (to NGV Module)	3.9-4.6 ms
75	TN	Injector 1 (to NGV Module)	3.9-4.6 ms
77	BK/WT	Power Ground	<0.1v
78	PK/LB	COP 7 Driver (Dwell)	5°, at 55 mph: 8°
79	WT/RD	COP 8 Driver (Dwell)	5°, at 55 mph: 8°
80	LB/OR	Fuel Pump Control	On: 1v, Off: 12v
81	WT/YL	EPC Solenoid Control	9.1v (4 psi)
82	---	Not Used	---
83	WT/LB	IAC Solenoid Control	10.7v (33%)
84	---	Not Used	---
85	DG	Camshaft Position Sensor	6 Hz, 55 mph: 14-18 Hz
86	---	Not Used	---
87	RD/BK	HO2S-21 (B2 S1) Signal	0.1-1.1v
88	LB/RD	MAF Sensor Signal	0.8v, 55 mph: 1.6v
89	GY/WT	TP Sensor Signal	0.53-1.27v
90	BR/WT	Reference Voltage	4.9-5.1v
91	GY/RD	Sensor Signal Return	<0.050v
92	LG	Brake Position Switch	Brake Off: 12v, On: 1v
93	RD/WT	HO2S-11 (B1 S1) Heater	1v, Off: 12v
94	YL/LB	HO2S-21 (B2 S1) Heater	1v, Off: 12v
95-96	---	Not Used	---
97	RD	Vehicle Power	12-14v
98	LB	Injector 8 (to NGV Module)	3.9-4.6 ms
99	LG/OR	Injector 6 (to NGV Module)	3.9-4.6 ms
100	BR/LB	Injector 4 (to NGV Module)	3.9-4.6 ms
101	WT	Injector 2 (to NGV Module)	3.9-4.6 ms
102	---	Not Used	---
103	BK/WT	Power Ground	<0.1v
104	PK/WT	COP 2 Driver (Dwell)	5°, at 55 mph: 8°

Pin Connector Graphic

PCM 104-PIN CONNECTOR

Terminal View of 104-Pin PCM Wiring Harness Connector

Standard Colors and Abbreviations

Abbreviation	Color	Abbreviation	Color	Abbreviation	Color
BK	Black	GY	Gray	PK	Purple
BL	Blue	GN	Green	RD	Red
BR	Brown	LG	LT Green	TN	Tan
DB	Dark Blue	OR	Orange	WT	White
DG	DK Green	PK	Pink	YL	Yellow

2000-02 E-Series 5.4L V8 CNG VIN M 104 Pin Connector

PCM Pin #	Wire Color	Circuit Description (104 Pin)	Value at Hot Idle
1	OR/YL	COP 6 Driver (Dwell)	5°, at 55 mph: 8°
2	PK/LG	MIL (lamp) Control	MIL On: 1v, Off: 12v
3	BK/WT	Power Ground	<0.1v
4	VTN/LB	Auxiliary Power Feed	PTO Off: 0v, On: 12v
5	---	Not Used	---
6	OG/YL	Shift Solenoid 'A' Control	1v, 55 mph: 12v
7-10	---	Not Used	---
11	VTN/ORG	Shift Solenoid 'B' Control	12v, 55 mph: 1v
12	WT/LG	TCIL (lamp) Control	Lamp On: 1v, Off: 12v
13	VT	Flash EEPROM	0.1v
14	---	Not Used	---
15	PK/LB	Data Bus (-)	Digital Signals
16	TN/OR	Data Bus (+)	Digital Signals
17-19	---	Not Used	---
20	BR/OG	Coast Clutch Solenoid	12v, 55 mph: 12v
21	DB	CKP (+) Sensor	440-490 Hz
22	GY	CKP (-) Sensor	440-490 Hz
23	---	Not Used	---
24	BK/WT	Power Ground	<0.1v
25	BK/LB	Case Ground	<0.050v
26	LG/WTH	COP 1 Driver (Dwell)	5°, at 55 mph: 8°
27	LG/YL	COP 5 Driver (Dwell)	5°, at 55 mph: 8°
28	LG/OR	COP 5 Driver (Dwell)	5°, at 55 mph: 8°
29	TN/WT	TCS (switch) Signal	TCS & O/D On: 12v
30-33	LG/OR	COP 5 Driver (Dwell)	5°, at 55 mph: 8°
34	YL/BK	Transmission Range Sensor 1	0v, at 55 mph: 11.5v
36	TN/LB	MAF Sensor Return	<0.050v
37	OG/BK	Transmission Fluid Temperature Sensor	2.10-2.40v
38	---	Not Used	---
39	GY	Intake Air Temperature Sensor	1.5-2.5v
40	DG/YL	Fuel Pump Monitor	On: 12v, Off: 0v
41	BK/YL	A/C Pressure Switch Signal	A/C On: 12v, Off: 0v
42-44	---	Not Used	---
45	YL/LB	Engine Coolant Temperature Sensor	0.5-0.6v
46-47	---	Not Used	---
48	WT/PK	Clean Tachometer Output	55 Hz, 55 mph: 120 Hz
49	LB/BK	Digital TR2 Sensor Signal	0v, at 55 mph: 11.5v
50	LG/RD	Digital TR4 Sensor Signal	0v, at 55 mph: 11.5v
51	BK/WT	Power Ground	<0.1v
52	WT/PK	COP 3 Driver (Dwell)	5°, at 55 mph: 8°
53	DG/PK	COP 4 Driver (Dwell)	5°, at 55 mph: 8°
54	VT/YL	TCC Solenoid Control	0%, at 55 mph: 95%
55	YL	Keep Alive Power	12-14v
56-57	---	Not Used	---
59	DG/WT	Turbine Shaft Speed Sensor	120 Hz, 30 mph: 300 Hz
60	GY/LB	HO2S-11 (B1 S1) Signal	0.1-1.1v
61	---	Not Used	---
62	LB	Fuel Rail Temperature Sensor	1.7-3.5v (50-120°F)
63	RD/PK	Injection Pressure Sensor	2-3.7v (90-100 psi)
64	LB/YL	Digital TR3A Sensor Signal	In 'P': 0v, in O/D: 1.7v
65	---	Not Used	---

2000-02 E-Series 5.4L V8 CNG VIN M 104 Pin Connector

PCM Pin #	Wire Color	Circuit Description (104 Pin)	Value at Hot Idle
66	YL/LG	Cylinder Head Temperature Sensor	0.6v (194ºF)
67	---	Not Used	---
68	GY/BK	Vehicle Speed Signal	0 Hz, at 55 mph: 125 Hz
69-70	---	Not Used	---
71	RD	Vehicle Power	12-14v
72	TN/RD	Injector 7 (to NGV Module)	3.9-4.6 ms
73	TN/BK	Injector 5 (to NGV Module)	3.9-4.6 ms
74	BR/YL	Injector 3 (to NGV Module)	3.9-4.6 ms
75	TN	Injector 1 (to NGV Module)	3.9-4.6 ms
76	---	Not Used	---
77	BK/WT	Power Ground	<0.1v
78	PK/LB	COP 7 Driver (Dwell)	5º, at 55 mph: 8º
79	WT/RD	COP 8 Driver (Dwell)	5º, at 55 mph: 8º
80	LB/OR	Fuel Pump Control	On: 1v, Off: 12v
81	WT/YL	EPC Solenoid Control	9.1v (4 psi)
82	---	Not Used	---
83	WT/LB	IAC Solenoid Control	10.7v (33%)
84	DB/YL	Output Shaft Speed Sensor	0 Hz, 55 mph: 228 Hz
85	DG	Camshaft Position Sensor	6 Hz, 55 mph: 14-18 Hz
86	---	Not Used	---
87	RD/BK	HO2S-21 (B2 S1) Signal	0.1-1.1v
88	LB/RD	MAF Sensor Signal	0.8v, 55 mph: 1.6v
89	GY/WT	TP Sensor Signal	0.53-1.27v
90	BR/WT	Reference Voltage	4.9-5.1v
91	GY/RD	Sensor Signal Return	<0.050v
92	LG	Brake Position Switch	Brake Off: 12v, On: 1v
93	RD/WT	HO2S-11 (B1 S1) Heater	1v, Off: 12v
94	YL/LB	HO2S-21 (B2 S1) Heater	1v, Off: 12v
95-96	---	Not Used	---
97	RD	Vehicle Power	12-14v
98	LB	Injector 8 (to NGV Module)	3.9-4.6 ms
99	LG/OR	Injector 6 (to NGV Module)	3.9-4.6 ms
100	BR/LB	Injector 4 (to NGV Module)	3.9-4.6 ms
101	WT	Injector 2 (to NGV Module)	3.9-4.6 ms
102	---	Not Used	---
103	BK/WT	Power Ground	<0.1v
104	PK/WT	COP 2 Driver (Dwell)	5º, at 55 mph: 8º

Pin Connector Graphic

PCM 104-PIN CONNECTOR

Terminal View of 104-Pin PCM Wiring Harness Connector

Standard Colors and Abbreviations

Abbreviation	Color	Abbreviation	Color	Abbreviation	Color
BK	Black	GY	Gray	PK	Purple
BL	Blue	GN	Green	RD	Red
BR	Brown	LG	LT Green	TN	Tan
DB	Dark Blue	OR	Orange	WT	White
DG	DK Green	PK	Pink	YL	Yellow

2003-05 E-Series 5.4L V8 CNG VIN M (A/T) 104 Pin Connector

PCM Pin #	Wire Color	Circuit Description (104 Pin)	Value at Hot Idle
1	OR/YL	COP 6 Driver (Dwell)	5º, at 55 mph: 8º
2	PK/LG	MIL (lamp) Control	MIL On: 1v, Off: 12v
3	BK/WT	Power Ground	<0.1v
4-5	---	Not Used	---
6	OR/YL	Shift Solenoid 'A' Control	1v, 55 mph: 1v
7-8	---	Not Used	---
9	YL/WT	Low Fuel Indicator	Indicator Off: 12v, On: 1v
10	---	Not Used	---
11	VT/OR	Shift Solenoid 'B' Control	12v, 55 mph: 1v
12	WT/LG	TCIL (lamp) Control	Lamp Off: 12v, On: 1v
13	VT	Flash EEPROM	0.1v
14	OR/LB	4x4 Low Switch	Switch Off: 12v, On: 1v
15	PK/LB	SCP Data Bus (-)	<0.050v
16	TN/OR	SCP Data Bus (+)	Digital Signals
17-19	---	Not Used	---
20	BR/OR	Coast Clutch Solenoid Control	12v, 55 mph: 12v
21	DB	CKP (+) Sensor	410 Hz, 55 mph: 1 KHz
22	GY	CKP (-) Sensor	410 Hz, 55 mph: 1 KHz
23	---	Not Used	---
24	BK/WT	Case Ground	<0.050v
25	BK/LB	Power Ground	<0.1v
26	LG/WT	COP 1 Driver (Dwell)	5º, at 55 mph: 8º
27	LG/YL	COP 5 Driver (Dwell)	5º, at 55 mph: 8º
28	---	Not Used	---
29	TN/WT	TCS (switch) Signal	TCS & O/D On: 12v
30-31	---	Not Used	---
32	DG/VT	Knock Sensor 1 Ground	<0.050v
33	---	Not Used	---
34	YL/BK	Digital TR1 Sensor	0v, at 55 mph: 11.5v
35	RD/LG	HO2S-12 (B1 S2) Signal	0.1-1.1v
36	TN/LB	MAF Sensor Return	<0.050v
37	OR/BK	Transmission Fluid Temperature Sensor	2.10-2.40v
38	---	Not Used	---
39	GY	Intake Air Temperature Sensor	1.5-2.5v
40	DG/YL	Fuel Pump Monitor	On: 12v, Off: 0v
41	BK/YL	A/C Pressure Switch Signal	A/C On: 12v, Off: 0v
42-44	---	Not Used	---
45	YL/LB	Engine Coolant Temperature Sensor	0.5-0.6v
46	---	Not Used	---
47	BR/PK	VR Solenoid Control	0%, at 55 mph: 45%
48	---	Not Used	---
49	LB/BK	Digital TR2 Sensor	0v, at 55 mph: 11.5v
50	LG/RD	Digital TR4 Sensor	0v, at 55 mph: 11.5v
51	BK/WT	Power Ground	<0.1v
52	WT/PK	COP 3 Driver (Dwell)	5º, at 55 mph: 8º
53	DG/VT	COP 4 Driver (Dwell)	5º, at 55 mph: 8º
54	VT/YL	TCC Solenoid Control	0%, at 55 mph: 95%
55	YL	Keep Alive Power	12-14v
56	LG/BK	EVAP Canister Purge Valve	0-10 Hz (0-100%)
57	YL/RD	Knock Sensor 1 Ground	<0.050v
58	---	Not Used	---
59	DG/WT	Turbine Shaft Speed Sensor	120 Hz, 30 mph: 300 Hz
60	GY/LB	HO2S-11 (B1 S1) Signal	0.1-1.1v
61	---	Not Used	---
62	LB	Fuel Rail Temperature Sensor	1.7-3.5v (50-120ºF)
63	RD/PK	Injection Pressure Sensor	2-3.7v (90-100 psi)
64	LB/YL	Digital TR3A Sensor	In 'P': 0v, in O/D: 1.7v

2003-05 E-Series 5.4L V8 CNG VIN M (A/T) 104 Pin Connector

PCM Pin #	Wire Color	Circuit Description (104 Pin)	Value at Hot Idle
65	BR/LG	EGR DPFE Sensor Signal	0.95-1.05v
66	YL/LG	Cylinder Head Temperature Sensor	0.6v (194ºF)
67	VT/WT	EVAP Canister Vent Valve	0-10 Hz (0-100%)
68	GY/BK	Vehicle Speed Sensor	0 Hz, 55 mph: 124 Hz
69-70	---	Not Used	---
71	RD	Vehicle Power (Start-Run)	12-14v
72	TN/RD	Injector 7 Control	3.2-3.8 ms
73	TN/BK	Injector 5 Control	3.2-3.8 ms
74	BR/YL	Injector 3 Control	3.2-3.8 ms
75	TN	Injector 1 Control	3.2-3.8 ms
76	BK/WT	Power Ground	<0.1v
77	BK/WT	Power Ground	<0.1v
78	PK/LB	COP 7 Driver (Dwell)	5º, at 55 mph: 8º
79	WT/RD	COP 8 Driver (Dwell)	5º, at 55 mph: 8º
80	LB/OR	Fuel Pump Relay Control	Off: 12v, On: 1v
81	WT/YL	EPC Solenoid Control	8.0v (3.5 psi)
82	---	Not Used	---
83	WT/LB	IAC Solenoid Control	8.7v (43%)
84	DB/YL	Output Shaft Speed Sensor	0 Hz, 55 mph: 228 Hz
85	DG	Camshaft Position Sensor	7 Hz, 55 mph: 13-17 Hz
86	---	Not Used	---
87	RD/BK	HO2S-21 (B2 S1) Signal	0.1-1.1v
88	LB/RD	Mass Airflow Sensor	0.8v, 55 mph: 1.9v
89	GY/WT	Throttle Position Sensor	0.53-1.27v
90	BR/WT	Reference Voltage	4.9-5.1v
91	GY/RD	Sensor Ground	<0.050v
92	LG	Brake Position Switch	Brake Off: 12v, On: 1v
93	RD/WT	HO2S-11 (B1 S1) Heater Control	1v, Off: 12v
94	YL/LB	HO2S-21 (B2 S1) Heater Control	1v, Off: 12v
95	WT/BK	HO2S-12 (B1 S2) Heater Control	1v, Off: 12v
96	---	Not Used	---
97	RD	Vehicle Power (Start-Run)	12-14v
98	LB	Injector 8 Control	3.2-3.8 ms
99	LG/OR	Injector 6 Control	3.2-3.8 ms
100	BR/LG	Injector 4 Control	3.2-3.8 ms
101	WT	Injector 2 Control	3.2-3.8 ms
102	---	Not Used	---
103	BK/WT	Power Ground	<0.1v
104	PKWT	COP 2 Driver (Dwell)	5º, at 55 mph: 8º

Pin Connector Graphic

Terminal View of 104-Pin PCM Wiring Harness Connector

Standard Colors and Abbreviations

Abbreviation	Color	Abbreviation	Color	Abbreviation	Color
BK	Black	GY	Gray	PK	Purple
BL	Blue	GN	Green	RD	Red
BR	Brown	LG	LT Green	TN	Tan
DB	Dark Blue	OR	Orange	WT	White
DG	DK Green	PK	Pink	YL	Yellow

1998-02 E-Series 5.4L V8 NGV Module 60 Pin Connector

PCM Pin #	Wire Color	Circuit Description (60 Pin)	Value at Hot Idle
1	YL	Keep Alive Power	12-14v
2	---	Not Used	---
3	TN	Injector 1 Signal from PCM	3.9-4.6 ms
4	WT	Injector 2 Signal from PCM	3.9-4.6 ms
5	BR/YL	Injector 3 Signal from PCM	3.9-4.6 ms
6	---	Not Used	---
7	RD/PK	Fuel Tank Pressure Signal	2.6v (0" HG - cap off)
8-15	---	Not Used	---
16	BK/LB	NGV Timer	N/A
17	---	Not Used	---
18	PK/LB	SCP Data Bus (-)	<0.050V
19	TN/OG	SCP Data Bus (+)	Digital Signals
20	BK/LB	Power Ground	<0.1v
21-22	---	Not Used	--
23	BR/LB	Injector 4 Signal from PCM	3.9-6.5 ms
24	TN/BK	Injector 5 Signal from PCM	3.9-4.6 ms
25	LG/OR	Injector 6 Signal from PCM	3.9-4.6 ms
26	BR/WT	Reference Voltage	4.9-5.1v
27-32	---	Not Used	---
33	TN/BK	Injector 5 Control	3.9-6.5 ms
34	---	Not Used	---
35	BR/LB	Injector 4 Control	3.9-6.5 ms
36	---	Not Used	---
37	RD	Vehicle Power Input	12-14v
38	YL/WT	Fuel Gauge Signal	Varies
39	BR/YL	Injector 3 Control	2.9-6.5 ms
40	BK	Power Ground	<0.1v
41	---	Not Used	---
42	LG/OR	Injector 6 Control	3.9-6.5 ms
43	TN/RD	Injector 7 Signal from PCM	3.9-6.5 ms
44	LB	Injector 8 Signal from PCM	3.9-6.5 ms
45	---	Not Used	---
46	GY/RD	Sensor Signal Return	<0.050v
47	OG/LG	Fuel Tank Temperature	Varies: 0.1-4.9v
48-52	---	Not Used	---
53	TN/RD	Injector 7 Control	3.9-6.5 ms
54	LB	Injector 8 Control	3.9-6.5 ms
55-56	---	Not Used	---
57	RD	Vehicle Power	12-14v
58	TN	Injector 1 Control	3.9-6.5 ms
59	WT	Injector 2 Control	3.9-6.5 ms
60	BK	Power Ground	<01v

Pin Connector Graphic

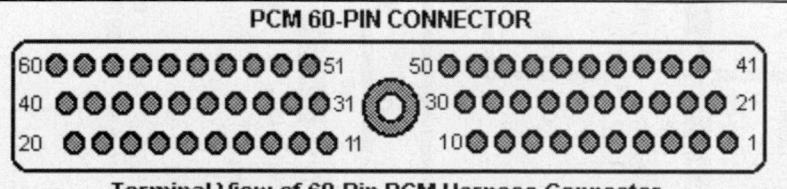

PCM 60-PIN CONNECTOR

Terminal View of 60-Pin PCM Harness Connector

Standard Colors and Abbreviations

Abbreviation	Color	Abbreviation	Color	Abbreviation	Color
BK	Black	GY	Gray	PK	Purple
BL	Blue	GN	Green	RD	Red
BR	Brown	LG	LT Green	TN	Tan
DB	Dark Blue	OR	Orange	WT	White
DG	DK Green	PK	Pink	YL	Yellow

1998-02 E-Series 5.4L V8 NGV (A/T) Module Wiring

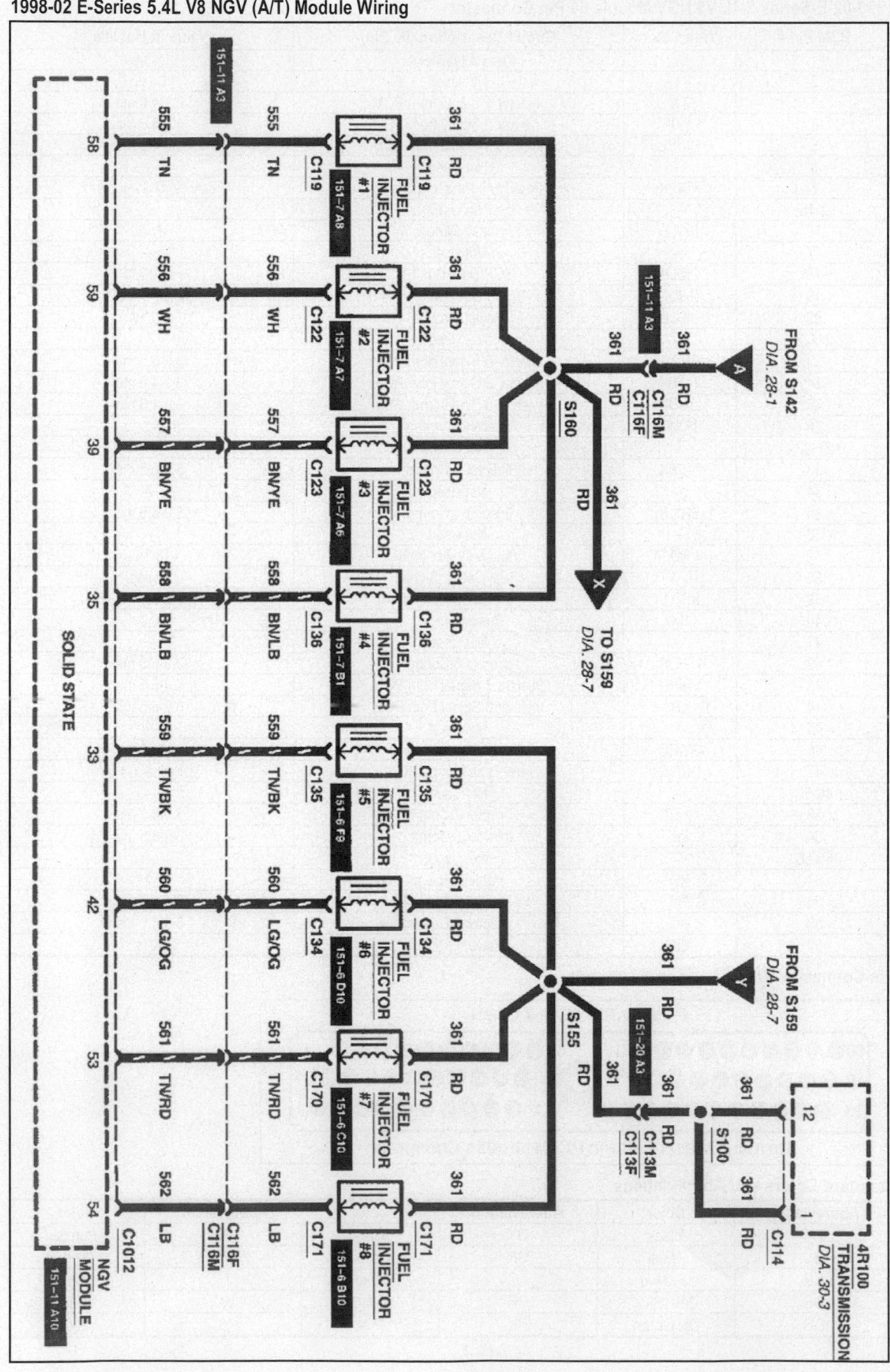

1990 E-Series 5.8L V8 VIN H (A/T) 60 Pin Connector

PCM Pin #	Wire Color	Circuit Description (60 Pin)	Value at Hot Idle
1	BK/OR	Keep Alive Power	12-14v
2, 5	---	Not Used	---
3	DG/WT	VSS (+) Signal	0 Hz, at 55 mph: 125 Hz
4	DG/YL	Ignition Diagnostic Monitor	20-31 Hz
6	OR/YL	VSS (-) Signal	0 Hz, at 55 mph: 125 Hz
6	GY/BK	VSS (-) Signal	0 Hz, at 55 mph: 125 Hz
7	LG/YL	Engine Coolant Temperature Sensor	0.5-0.6v
7	GN/YL	Engine Coolant Temperature Sensor	0.5-0.6v
8	OR/BL	Fuel Pump Monitor	On: 12v, Off: 0v
9, 12-15	---	Not Used	---
10	BK/YL	A/C Pressure Switch Signal	A/C On: 12v, off: 0v
11	WT/BK	Air Management 2 Solenoid	On: 1v, Off: 12v
16	BK/OR	Ignition System Ground	<0.050v
17	TN/RD	Self-Test Output & MIL	MIL On: 1v, Off: 12v
18	TN/LG	Inferred Mileage Sensor	Digital Signals
19, 23-24	---	Not Used	---
20	BK	PCM Case Ground	<0.050v
21	GY/WT	IAC Solenoid Control	8.3v
22	TN/LG	Fuel Pump Control	On: 1v, Off: 12v
25	YL/RD	ACT Sensor Signal	1.5-2.5v
26	OR/WT	Reference Voltage	4.9-5.1v
27	BR/LG	EVP Sensor Signal	0.4v, 55 mph: 3.1v
27	DG/LG	EVP Sensor Signal	0.4v, 55 mph: 3.1v
28	---	Not Used	---
29	DG/PK	HO2S-11 (B1 S1) Signal	0.1-1.1v
30	BL/WT	A/T: Neutral Drive Switch	In 'P': 0v, Others: 5v
31	GY/YL	Canister Purge Solenoid	12v, 55 mph: 1v
32, 34-35	---	Not Used	---
33	DG	VR Solenoid Control	0%, at 55 mph: 45%
36	YL/GN	Spark Output Signal	6.93v (50%)
37	RD	Vehicle Power	12-14v
38-39	---	Not Used	---
40	BK/LG	Power Ground	<0.1v
41-44	---	Not Used	---
45	DB/LG	MAP Sensor Signal	107 Hz (sea level)
46	BK/WT	Analog Signal Return	<0.050v
47	DG/LG	TP Sensor Signal	0.7v, 55 mph: 2.1v
48	WT/RD	Self-Test Indicator Signal	STI On: 1v, Off: 5v
49	OR	HO2S-11 (B1 S1) Ground	<0.050v
50	---	Not Used	---
51	WT/RD	Air Management 1 Solenoid	On: 1v, Off: 12v
52-55	---	Not Used	---
56	DB	PIP Sensor Signal	6.93v (50%)
57	R	Vehicle Power	12-14v
58	TN/OR	Injector Bank 1 (INJ 1, 4, 5, 8)	5.6-6.4 ms
59	TN/RD	Injector Bank 2 (INJ 2, 3, 6, 7)	5.6-6.4 ms
60	BK/LG	Power Ground	<0.1v

Pin Connector Graphic

Terminal View of 60-Pin PCM Harness Connector

1990 E-Series 5.8L V8 VIN H (E4OD) 60 Pin Connector

PCM Pin #	Wire Color	Circuit Description (60 Pin)	Value at Hot Idle
1	BK/OR	Keep Alive Power	12-14v
2	LG	Brake Switch Signal	Brake Off: 12v, On: 1v
3	DG/WT	VSS (+) Signal	0 Hz, at 55 mph: 125 Hz
4	DG/YL	Ignition Diagnostic Monitor	20-31 Hz
5, 9, 12-15	---	Not Used	---
6	OR/YL	VSS (-) Signal	0 Hz, at 55 mph: 125 Hz
7	LG/YL	Engine Coolant Temperature Sensor	0.5-0.6v
8	OR/BL	Fuel Pump Monitor	On: 12v, Off: 0v
10	BK/YL	A/C Pressure Switch Signal	A/C On: 12v, off: 0v
11	WT/BK	Air Management 2 Solenoid	On: 1v, Off: 12v
16	BK/OR	Ignition System Ground	<0.050v
17	TN/RD	Self-Test Output & MIL	MIL On: 1v, Off: 12v
18	TN/LG	Inferred Mileage Sensor	Digital Signals
19	DG/PK	Shift Solenoid 2 Control	12v, 30 mph: 1v
20	BK	PCM Case Ground	<0.050v
21	GY/WT	IAC Solenoid Control	8.3v
22	TN/LG	Fuel Pump Control	On: 1v, Off: 12v
23, 28	---	Not Used	---
24	YL/LG	Power Steering Pressure Switch	Straight: 0v, Turned: 12v
25	YL/RD	ACT Sensor Signal	1.5-2.5v
26	OR/WT	Reference Voltage	4.9-5.1v
27	BR/LG	EVP Sensor Signal	0.4v, 55 mph: 3.1v
29	DG/PK	HO2S-11 (B1 S1) Signal	0.1-1.1v
30	BL/WT	MLP Sensor Signal	In 'P': 0v, in O/D: 5v
31	GY/YL	Canister Purge Solenoid	12v, 55 mph: 1v
32	LG/WT	OCIL (lamp) Control	Lamp On: 1v, Off: 12v
33	DG	VR Solenoid Control	0%, at 55 mph: 45%
34-35	---	Not Used	---
36	YL/LG	Spark Output Signal	6.93v (50%)
37	RD	Vehicle Power	12-14v
38	BL/YL	EPC Solenoid Control	9.2v (5 psi)
39, 43-44	---	Not Used	---
40	BK/LG	Power Ground	<0.1v
41	TN/BL	Overdrive Cancel Switch	OCS Closed: 0.1v
42	OR/BK	TOT Sensor Signal	2.10-2.40v
45	DB/LG	MAP Sensor Signal	107 Hz
46	BK/WT	Analog Signal Return	<0.050v
47	DG/LG	TP Sensor Signal	0.7v, 55 mph: 2.1v
48	WT/RD	Self-Test Indicator Signal	STI On: 1v, Off: 5v
49	OR	HO2S-11 (B1 S1) Ground	<0.050v
50, 54	---	Not Used	---
51	WT/RD	Air Management 1 Solenoid	On: 1v, Off: 12v
52	OR/YL	Shift Solenoid 1 Control	In 'P': 1v, 30 mph 12v
53	PK/YL	TCC Solenoid Control	12v, 55 mph: 1v
55	BR	Coast Clutch Solenoid	12v, 55 mph: 12v
56	DB	PIP Sensor Signal	6.93v (50%)
57	RD	Vehicle Power	12-14v
58	TN/OR	Injector Bank 1 (INJ 1, 4, 5, 8)	5.8-6.4 ms
59	TN/RD	Injector Bank 2 (INJ 2, 3, 6, 7)	5.8-6.4 ms
60	BK/LG	Power Ground	<0.1v

Pin Connector Graphic

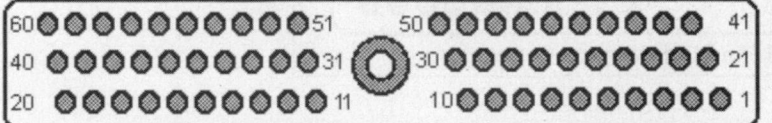

PCM 60-PIN CONNECTOR

Terminal View of 60-Pin PCM Harness Connector

1991 E-Series 5.8L V8 OHV VIN H (All) 60 Pin Connector

PCM Pin #	Wire Color	Circuit Description (60 Pin)	Value at Hot Idle
1	BK/OR	Keep Alive Power	12-14v
2	LG	Brake Switch Signal	Brake Off: 12v, On: 1v
3	DG/WT	VSS (+) Signal	0 Hz, at 55 mph: 125 Hz
4	TN/YL	Ignition Diagnostic Monitor	20-31 Hz
5, 12-15	---	Not Used	---
6	OR/YL	VSS (-) Signal	0 Hz, at 55 mph: 125 Hz
7	LG/RD	Engine Coolant Temperature Sensor	0.5-0.6v
8	DG/YL	Fuel Pump Monitor	On: 12v, Off: 0v
9	BK/OR	Data Bus (-)	Digital Signals
10	BK/YL	A/C Pressure Switch Signal	A/C On: 12v, off: 0v
11	BR	Air Management 2 Solenoid	On: 1v, Off: 12v
16	OR/RD	Ignition System Ground	<0.050v
17	TN/RD	Self-Test Output & MIL	MIL On: 1v, Off: 12v
18, 24	---	Not Used	---
19	PK	Shift Solenoid 2 Control	12v, 55 mph: 1v
20	BK	PCM Case Ground	<0.050v
21	GY/WT	IAC Solenoid Control	7.0-8.5v
22	TN/LG	Fuel Pump Control	On: 1v, Off: 12v
23	LG/BK	Knock Sensor 1 Signal	0v
25	YL/RD	ACT Sensor Signal	1.5-2.5v
26	BR/WT	Reference Voltage	4.9-5.1v
27	BR/LG	EVP Sensor Signal	0.3v, 55 mph: 2.3v
28	OR/BK	Data Bus (+)	Digital Signals
29	RD/BK	HO2S-11 (B1 S1) Signal	0.1-1.1v
30	WT/RD	MLP Sensor Signal	In 'P': 0v, in O/D: 5v
31	GY/YL	Canister Purge Solenoid	12v, 55 mph: 1v
32	WT/LG	OCIL (lamp) Control	Lamp On: 1v, Off: 12v
33	BR/PK	VR Solenoid Control	0%, at 55 mph: 45%
34-35, 39	---	Not Used	---
36	YL/LG	Spark Output Signal	6.93v (50%)
37	RD	Vehicle Power	12-14v
38	BL/YL	EPC Solenoid Control	9.5v (5 psi)
40	BK/LG	Power Ground	<0.1v
41	TN/WT	Overdrive Cancel Switch	Switch Closed: 0.1v
42	OR/BK	TOT Sensor Signal	2.10-2.40v
43-44, 50, 54	---	Not Used	---
44	---	Not Used	---
45	DB/LG	MAP Sensor Signal	107 Hz
46	GY/RD	Analog Signal Return	<0.050v
47	GY/WT	TP Sensor Signal	0.7v, 55 mph: 2.1v
48	WT/RD	Self-Test Indicator Signal	STI On: 1v, Off: 5v
49	OR	HO2S-11 (B1 S1) Ground	<0.050v
51	WT/RD	Air Management 1 Solenoid	On: 1v, Off: 12v
52	OR	Shift Solenoid 1 Control	1v, 55 mph: 12v
53	PK/YL	TCC Solenoid Control	0%, at 55 mph: 95%
55	BR/OR	Coast Clutch Solenoid	12v, 55 mph: 12v
56	DB	PIP Sensor Signal	6.93v (50%)
57	RD	Vehicle Power	12-14v
58	TN/OR	Injector Bank 1 (INJ 1, 4, 5, 8)	5.0-5.8 ms
59	TN/RD	Injector Bank 2 (INJ 2, 3, 6, 7)	5.0-5.8 ms
60	BK/LG	Power Ground	<0.1v

Pin Connector Graphic

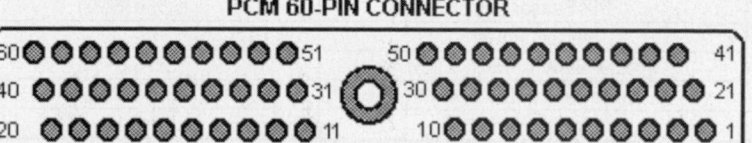

Terminal View of 60-Pin PCM Harness Connector

1992-94 E-Series 5.8L V8 VIN H (A/T) 60 Pin Connector

PCM Pin #	Wire Color	Circuit Description (60 Pin)	Value at Hot Idle
1	YL	Keep Alive Power	12-14v
2	LG	Brake Switch Signal	Brake Off: 12v, On: 1v
3	GY/BK	VSS (+) Signal	0 Hz, at 55 mph: 125 Hz
4	TN/YL	Ignition Diagnostic Monitor	20-31 Hz
4	WT/PK	Ignition Diagnostic Monitor	20-31 Hz
5	---	Not Used	---
6	PK/OR	VSS (-) Signal	0 Hz, at 55 mph: 125 Hz
7	LG/RD	Engine Coolant Temperature Sensor	0.5-0.6v
8	DG/YL	Fuel Pump Monitor	On: 12v, Off: 0v
9	PK/BL	Data Bus (-)	Digital Signals
10	BK/YL	A/C Pressure Switch Signal	A/C On: 12v, off: 0v
11	BR	Air Management 2 Solenoid	On: 1v, Off: 12v
12-15	---	Not Used	---
16	OR/RD	Ignition System Ground	<0.050v
17	PK/LG	Self-Test Output & MIL	MIL On: 1v, Off: 12v
18	---	Not Used	---
19	PK/OR	Shift Solenoid 2 Control	12v, 55 mph: 1v
20	BK	PCM Case Ground	<0.050v
21	WT/BL	IAC Solenoid Control	10.7v (33%)
22	BL/OR	Fuel Pump Control	On: 1v, Off: 12v
23	YL/RD	Knock Sensor 1 Signal	0v
24	---	Not Used	---
25	GY	ACT Sensor Signal	1.5-2.5v
26	BR/WT	Reference Voltage	4.9-5.1v
27	BR/LG	EVP Sensor Signal	0.4v, 55 mph: 3.1v
28	TN/OR	Data Bus (+)	Digital Signals
29	GY/BL	HO2S-11 (B1 S1) Signal	0.1-1.1v
30	BL/YL	MLP Sensor Signal	In 'P': 0v, in O/D: 5v
31	GY/YL	Canister Purge Solenoid	12v, 55 mph: 1v
32	WT/LG	OCIL (lamp) Control	Lamp On: 1v, Off: 12v
33	BR/PK	VR Solenoid Control	0%, at 55 mph: 45%
34-35	---	Not Used	---
36	PK	Spark Output Signal	6.93v (50%)
37	RD	Vehicle Power	12-14v
38	WT/YL	EPC Solenoid Control	9.5v (5 psi)
39	---	Not Used	---
40	BK/WT	Power Ground	<0.1v
41 ('92-'93)	TN/WT	Overdrive Cancel Switch	OCS On: 12v, Off: 0v
41 ('94)	TN/WT	TCS (switch) Signal	TCS & O/D On: 12v
42	OR/BK	TOT Sensor Signal	2.10-2.40v
43 ('94)	PK	A/C Demand Switch	A/C On: 12v, Off: 0v
44	---	Not Used	---
45	LG/BK	MAP Sensor Signal	107 Hz (sea level)
46	GY/RD	Analog Signal Return	<0.050v
47	GY/WT	TP Sensor Signal	0.7v, 55 mph: 2.1v
48	WT/PK	Self-Test Indicator Signal	STI On: 1v, Off: 5v
49	OR	HO2S-11 (B1 S1) Ground	<0.050v
50	---	Not Used	---
51	WT/OR	Air Management 1 Solenoid	AM1 On: 1v, Off: 12v
52	OR/YL	Shift Solenoid 1 Control	1v, 55 mph: 12v
53	PK/YL	TCC Solenoid Control	0%, at 55 mph: 95%
54	---	Not Used	---
55	BR/OR	Coast Clutch Solenoid	12v, 55 mph: 12v
56	GY/OR	PIP Sensor Signal	6.93v (50%)
57	RD	Vehicle Power	12-14v
58	TN	Injector Bank 1 (INJ 1, 3 & 5)	3.5-4.0 ms
59	WT	Injector Bank 2 (INJ 2, 4 & 6)	3.5-4.0 ms
60	BK/WT	Power Ground	<0.1v

1995 E-Series 5.8L V8 VIN H (A/T) California 60 Pin Connector

PCM Pin #	Wire Color	Circuit Description (60 Pin)	Value at Hot Idle
1	Y	Keep Alive Power	12-14v
2	LG	Brake Switch Signal	Brake Off: 12v, On: 1v
3	GY/BK	VSS (+) Signal	0 Hz, at 55 mph: 125 Hz
4	WT/PK	Ignition Diagnostic Monitor	20-31 Hz
5	---	Not Used	---
6	PK/OR	VSS (-) Signal	0 Hz, at 55 mph: 125 Hz
7	LG/RD	Engine Coolant Temperature Sensor	0.5-0.6v
8	DG/YL	Fuel Pump Monitor	On: 12v, Off: 0v
9	TN/BL	MAF Sensor Return	<0.050v
10	BK/YL	A/C Pressure Switch Signal	A/C On: 12v, off: 0v
11	GY/YL	Canister Purge Solenoid	12v, 55 mph: 1v
12	LG/OR	Injector 6 Control	5.5-6.1 ms
13	TN/RD	Injector 7 Control	5.5-6.1 ms
14	BL	Injector 8 Control	5.5-6.1 ms
15	TN/BK	Injector 5 Control	5.5-6.1 ms
16	OR/RD	Ignition System Ground	<0.050v
17	PK/LG	Self-Test Output & MIL	MIL On: 1v, Off: 12v
18	TN/OR	Data Bus (+)	Digital Signals
19	PK/BL	Data Bus (-)	Digital Signals
20	BK	PCM Case Ground	<0.050v
21	WT/BL	IAC Solenoid Control	10.7v (33%)
22	BL/OR	Fuel Pump Control	On: 1v, Off: 12v
23	---	Not Used	---
24	---	Not Used	---
25	GY	Intake Air Temperature Sensor	1.5-2.5v
26	BR/WT	Reference Voltage	4.9-5.1v
27	BR/LG	EVP Sensor Signal	0.4v, 55 mph: 3.1v
28	---	Not Used	---
29	---	Not Used	---
30	BL/YL	MLP Sensor Signal	In 'P': 0v, in O/D: 1.7v
31	---	Not Used	---
32	WT/LG	TCIL (lamp) Control	Lamp On: 1v, Off: 12v
33	BR/PK	VR Solenoid Control	0%, at 55 mph: 45%
34	BR	Air Diverter Solenoid Control	On: 1v, Off: 12v
35	BR/BL	Injector 4 Control	5.5-6.1 ms
36	PK	Spark Output Signal	6.93v (50%)
37	R	Vehicle Power	12-14v
38	WT/YL	EPC Solenoid Control	9.5v (5 psi)
39	BR/YL	Injector 3 Control	5.5-6.1 ms
40	BK/WT	Power Ground	<0.1v

1995 E-Series 5.8L V8 VIN H (A/T) California 60 Pin Connector

PCM Pin #	Wire Color	Circuit Description (60 Pin)	Value at Hot Idle
41	TN/WT	TCS (switch) Signal	TCS & O/D On: 12v
42	---	Not Used	---
43	RD/BK	HO2S-21 (B2 S1) Signal	0.1-1.1v
44	GY/BL	HO2S-11 (B1 S1) Signal	0.1-1.1v
45	---	Not Used	---
46	GY/RD	Analog Signal Return	<0.050v
47	GY/WT	TP Sensor Signal	0.7v, 55 mph: 2.1v
48	WT/PK	Self-Test Indicator Signal	STI On: 1v, Off: 5v
49	OR/BK	TOT Sensor Signal	2.10-2.40v
50	BL/RD	MAF Sensor Signal	0.8v, 55 mph: 1.6v
51	OR/YL	Shift Solenoid 1 Control	12v, 55 mph: 1v
52	PK/OR	Shift Solenoid 2 Control	1v, 55 mph: 12v
53	PK/YL	TCC Solenoid Control	0%, at 55 mph: 95%
54	---	Not Used	---
55	BR/OR	Coasting Clutch Solenoid	12v, 55 mph: 12v
56	GY/OR	PIP Sensor Signal	6.93v (50%)
57	RD	Vehicle Power	12-14v
58	TN	Injector 1 Control	5.5-6.1 ms
59	WT	Injector 2 Control	5.5-6.1 ms
60	BK/WT	Power Ground	<0.1v

Pin Connector Graphic

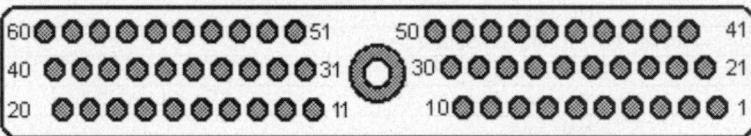

PCM 60-PIN CONNECTOR

Terminal View of 60-Pin PCM Harness Connector

Standard Colors and Abbreviations

Abbreviation	Color	Abbreviation	Color	Abbreviation	Color
BK	Black	GY	Gray	PK	Purple
BL	Blue	GN	Green	RD	Red
BR	Brown	LG	LT Green	TN	Tan
DB	Dark Blue	OR	Orange	WT	White
DG	DK Green	PK	Pink	YL	Yellow

1995 E-Series 5.8L V8 VIN H (E4OD) 60 Pin Connector

PCM Pin #	Wire Color	Circuit Description (60 Pin)	Value at Hot Idle
1	YL	Keep Alive Power	12-14v
2	LG	Brake Switch Signal	Brake On: 12v: Off: 0v
3	GY/BK	VSS (+) Signal	0 Hz, at 55 mph: 125 Hz
4	WT/PK	Ignition Diagnostic Monitor	20-31 Hz
5	---	Not Used	---
6	PK/OR	VSS (-) Signal	0 Hz, at 55 mph: 125 Hz
7	LG/RD	Engine Coolant Temperature Sensor	0.5-0.6v
8	DG/YL	Fuel Pump Monitor	On: 12v, Off: 0v
9	PK/BL	Data Bus (-)	Digital Signals
10	BK/YL	A/C Pressure Switch Signal	A/C On: 12v, off: 0v
11	BR	Air Management 2 Solenoid	On: 1v, Off: 12v
12-15	---	Not Used	---
16	OR/RD	Ignition System Ground	<0.050v
17	PK/LG	Self-Test Output & MIL	MIL On: 1v, Off: 12v
18	---	Not Used	---
19	PK/OR	Shift Solenoid 2 Control	12v, 55 mph: 1v
20	BK	PCM Case Ground	<0.050v
21	WT/BL	IAC Solenoid Control	10.7v (33%)
22	BL/OR	Fuel Pump Control	On: 1v, Off: 12v
23	YL/RD	Knock Sensor 1 Signal	0v
24	---	Not Used	---
25	GY	ACT Sensor Signal	1.5-2.5v
26	BR/WT	Reference Voltage	4.9-5.1v
27	BR/LG	EVP Sensor Signal	0.4v, 55 mph: 3.1v
28	TN/OR	Data Bus (+)	Digital Signals
29	GY/BL	HO2S-11 (B1 S1) Signal	0.1-1.1v
30	BL/YL	MLP Sensor Signal	In 'P': 0v, in O/D: 5v
31	GY/YL	Canister Purge Solenoid	12v, 55 mph: 1v
32	WT/LG	OCIL (lamp) Control	Lamp On: 1v, Off: 12v
33	BR/PK	VR Solenoid Control	0%, at 55 mph: 45%
34-35	---	Not Used	---
36	PK	Spark Output Signal	6.93v (50%)
37	RD	Vehicle Power	12-14v
38	WT/YL	EPC Solenoid Control	9.5v (5 psi)
39	---	Not Used	---
40	BK/WT	Power Ground	<0.1v

1995 E-Series 5.8L V8 VIN H (E4OD) 60 Pin Connector

PCM Pin #	Wire Color	Circuit Description (60 Pin)	Value at Hot Idle
41	TN/WT	TCS (switch) Signal	TCS & O/D On: 12v
42	OR/BK	TOT Sensor Signal	2.10-2.40v
43	PK	A/C Demand Switch	A/C On: 12v, Off: 0v
44	---	Not Used	---
45	LG/BK	MAP Sensor Signal	107 Hz (sea level)
46	GY/RD	Analog Signal Return	<0.050v
47	GY/WT	TP Sensor Signal	0.7v, 55 mph: 2.1v
48	WT/PK	Self-Test Indicator Signal	STI On: 1v, Off: 5v
49	OR	HO2S-11 (B1 S1) Ground	<0.050v
50	---	Not Used	---
51	WT/OR	Air Management 1 Solenoid	On: 1v, Off: 12v
52	OR/YL	Shift Solenoid 1 Control	1v, 55 mph: 12v
53	PK/YL	TCC Solenoid Control	0%, at 55 mph: 95%
54	---	Not Used	---
55	BR/OR	Coast Clutch Solenoid	12v, 55 mph: 12v
56	GY/OR	PIP Sensor Signal	6.93v (50%)
57	RD	Vehicle Power	12-14v
58	TN	Injector Bank 1 (INJ 1, 4, 5, 8)	3.5-5.0 ms
59	WT	Injector Bank 2 (INJ 2, 4, 6, 7)	3.5-5.0 ms
60	BK/WT	Power Ground	<0.1v

Pin Connector Graphic

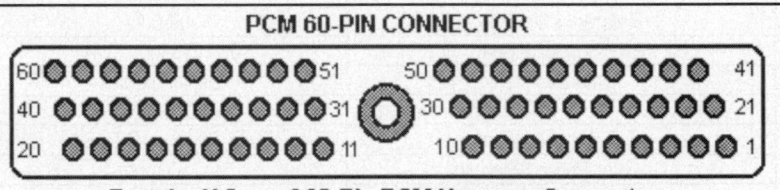

Standard Colors and Abbreviations

Abbreviation	Color	Abbreviation	Color	Abbreviation	Color
BK	Black	GY	Gray	PK	Purple
BL	Blue	GN	Green	RD	Red
BR	Brown	LG	LT Green	TN	Tan
DB	Dark Blue	OR	Orange	WT	White
DG	DK Green	PK	Pink	YL	Yellow

1996 E-Series 5.8L V8 VIN H (E4OD) 104 Pin Connector

PCM Pin #	Wire Color	Circuit Description (104 Pin)	Value at Hot Idle
1	PK/OR	Shift Solenoid 2 Control	12v, 55 mph: 12v
2	PK/LG	MIL (lamp) Control	MIL On: 1v, Off: 12v
3	---	Not Used	---
4	LG/RD	Power Take-Off Signal	PTO On: 12v, Off: 0v
5-12	---	Not Used	---
13	PK	Flash EPROM Power	0.1v
14	---	Not Used	---
15	PK/LB	Data Bus (-)	Digital Signals
16	TN/OR	Data Bus (+)	Digital Signals
17-22	---	Not Used	---
23	OR/RD	Ignition System Ground	<0.050v
24	BK/WT	Power Ground	<0.1v
25	BK	Case Ground	<0.050v
26	---	Not Used	---
27	OR/YL	Shift Solenoid 1 Control	1v, 55 mph: 12v
28	---	Not Used	
29	TN/WT	TCS (switch) Signal	TCS & O/D On: 12v
30-32	---	Not Used	---
33	PK/OR	PSOM (-) Signal	<0.050v
34	---	Not Used	---
35	RD/LG	HO2S-12 (B1 S2) Signal	0.1-1.1v
36	TN/LB	MAF Sensor Return	<0.050v
37	OR/BK	Transmission Fluid Temperature Sensor	2.10-2.40v
38	LG/RD	Engine Coolant Temperature Sensor	0.5-0.6v
39	GY	Intake Air Temperature Sensor	1.5-2.5v
40	DG/YL	Fuel Pump Monitor	On: 12v, Off: 0v
41	BK/YL	A/C Pressure Switch Signal	A/C On: 12v, off: 0v
42-46	---	Not Used	---
47	BK/PK	VR Solenoid Control	0%, at 55 mph: 45%
48	WT/PK	Clean Tachometer Output	38-42 Hz
49	GY/OR	PIP Sensor Signal	6.93v (50%)
50	PK	Spark Output Signal	6.93v (50%)
51	BK/WT	Power Ground	<0.1v
52	---	Not Used	---
53	BR/OR	Coast Clutch Solenoid	12v, 55 mph: 12v
54	DB/WT	TCC Solenoid Control	0%, at 55 mph: 95%
55	YL	Keep Alive Power	12-14v
56	LG/BK	EVAP VMV Solenoid	0-10 Hz (0-100%)
57	---	Not Used	---
58	GY/BK	PSOM (+) Signal	0 Hz, at 55 mph: 125 Hz
59	DB	Misfire Detection Sensor	45-55 Hz
60	GY/LB	HO2S-11 (B1 S1) Signal	0.1-1.1v
61-63	---	Not Used	---
64	LB/YL	Digital Transmission Range Sensor	In 'P': 0v, in O/D: 1.7v

1996 E-Series 5.8L V8 VIN H (E4OD) 104 Pin Connector

PCM Pin #	Wire Color	Circuit Description (104 Pin)	Value at Hot Idle
65	BR/LG	EGR DPFE Sensor Signal	1v, 55 mph: 1v
66-70	---	Not Used	---
71	RD	Vehicle Power	12-14v
72	TN/RD	Injector 7 Control	4.2-4.6 ms
73	TN/BK	Injector 5 Control	4.2-4.6 ms
74	BR/YL	Injector 3 Control	4.2-4.6 ms
75	TN	Injector 1 Control	4.2-4.6 ms
76	BK/WT	Power Ground	<0.1v
77	BK/WT	Power Ground	<0.1v
78	---	Not Used	---
79	WT/LG	TCIL (lamp) Control	Lamp On: 1v, Off: 12v
80	LB/OR	Fuel Pump Control	On: 1v, Off: 12v
81	WT/YL	EPC Solenoid Control	10v (26 psi)
82	---	Not Used	---
83	WT/LB	IAC Solenoid Control	10.7v (33%)
84-86	---	Not Used	---
87	RD/BK	HO2S-21 (B2 S1) Signal	0.1-1.1v
88	LB/RD	MAF Sensor Signal	0.8v, 55 mph: 2.3v
89	GY/WT	TP Sensor Signal	0.9v, 55 mph: 1.2v
90	BR/WT	Reference Voltage	4.9-5.1v
91	GY/RD	Analog Signal Return	<0.050v
92	LG	Brake Position Switch	Brake Off: 12v, On: 1v
93	RD/WT	HO2S-11 (B1 S1) Heater	1v, Off: 12v
94	YL/LB	HO2S-21 (B2 S1) Heater	1v, Off: 12v
95	WT/BK	HO2S-12 (B1 S2) Heater	1v, Off: 12v
96	---	Not Used	---
97	R	Vehicle Power	12-14v
98	LG	Injector 8 Control	4.2-4.6 ms
99	LG/OR	Injector 6 Control	4.2-4.6 ms
100	BR/LB	Injector 4 Control	4.2-4.6 ms
101	W	Injector 2 Control	4.2-4.6 ms
102	---	Not Used	---
103	BK/WT	Power Ground	<0.1v
104	---	Not Used	---

Pin Connector Graphic

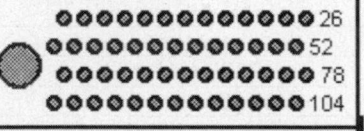

PCM 104-PIN CONNECTOR

Terminal View of 104-Pin PCM Wiring Harness Connector

Standard Colors and Abbreviations

Abbreviation	Color	Abbreviation	Color	Abbreviation	Color
BK	Black	GY	Gray	PK	Purple
BL	Blue	GN	Green	RD	Red
BR	Brown	LG	LT Green	TN	Tan
DB	Dark Blue	OR	Orange	WT	White
DG	DK Green	PK	Pink	YL	Yellow

1997 E-Series 6.8L V10 VIN S (E4OD) 104 Pin Connector

PCM Pin #	Wire Color	Circuit Description (104 Pin)	Value at Hot Idle
1	PK/LB	COP 6 Driver (Dwell)	5°, at 55 mph: 8°
2	PK/LG	MIL (lamp) Control	MIL On: 1v, Off: 12v
3	BK/WT	Power Ground	<0.1v
4-5	---	Not Used	---
6	OR/YL	Shift Solenoid 1 Control	1v, 55 mph: 12v
7-8	---	Not Used	---
9	YL/WT	Fuel Level Indicator Signal	1.7v (1/2 full)
10	---	Not Used	---
11	PK/OR	Shift Solenoid 2 Control	12v, 55 mph: 12v
12	---	Not Used	---
13	VT	Flash EEPROM	0.1v
14	---	Not Used	---
15	PK/LB	Data Bus (-)	Digital Signals
16	TN/OR	Data Bus (+)	Digital Signals
17-19	---	Not Used	---
20	BR/OR	Coast Clutch Solenoid	12v, 55 mph: 12v
21	DB	CKP (+) Sensor	400-420 Hz
22	GY	CKP (-) Sensor	400-420 Hz
23-24	BK/WT	Power Ground	<0.1v
25	BK/LB	Case Ground	<0.050v
26	DB/YL	COP 1 Driver (Dwell)	5°, at 55 mph: 8°
27	PKWT	COP 10 Driver (Dwell)	5°, at 55 mph: 8°
28	WT/LG	TCIL (lamp) Control	Lamp On: 1v, Off: 12v
29	TN/WT	TCS (switch) Signal	TCS & O/D On: 12v
30	DG	Octane Adjustment	9.3v (switch shorted: 0v)
31-32	---	Not Used	---
33	PK/OR	VSS (-) Signal	0 Hz, at 55 mph: 125 Hz
34	YL/BK	Digital TR1 Sensor	0v, at 55 mph: 11.5v
35	RD/LG	HO2S-12 (B1 S2) Signal	0.1-1.1v
36	TN/LB	MAF Sensor Return	<0.1v
37	OR/BK	Transmission Fluid Temperature Sensor	2.10-2.40v
38	---	Not Used	---
39	GY	Intake Air Temperature Sensor	1.5-2.5v
40	DG/YL	Fuel Pump Monitor	On: 12v, Off: 0v
41	BK/YL	A/C High Pressure Switch	A/C On: 12v, off: 0v
42	GY/RD	Injector 10 Control	4.2-4.6 ms
43-44	---	Not Used	---
45	YL/LB	CHIL (lamp) Control	CHT Off: 12v, On: 1v
46	---	Not Used	---
47	BR/PK	VR Solenoid Control	0%, at 55 mph: 45%
48	PK/BL	Clean Tachometer Output	60-70 Hz
49	LB/BK	Digital TR2 Sensor	0v, at 55 mph: 11.5v
50	LG/RD	Digital TR4 Sensor	0v, at 55 mph: 11.5v
51	BK/WT	Power Ground	<0.1v
52	PKWT	COP 5 Driver (Dwell)	5°, at 55 mph: 8°
53	WT/PK	COP 7 Driver (Dwell)	5°, at 55 mph: 8°
54	PK/YL	TCC Solenoid Control	0%, at 55 mph: 95%
55	Y	Keep Alive Power	12-14v
56	LG/BK	EVAP VMV Solenoid	0-10 Hz (0-100%)
57	YL/RD	Knock Sensor 1 Signal	0v
58	GY/BK	VSS (+) Signal	0 Hz, at 55 mph: 125 Hz
59	---	Not Used	---
60	GY/LB	HO2S-11 (B1 S1) Signal	0.1-1.1v
61	---	Not Used	---
62	RD/PK	Fuel Tank Pressure Sensor	2.6v (0" H2O - cap off)
63	---	Not Used	---
64	LB/YL	Digital Transmission Range Sensor	In 'P': 0v, in O/D: 1.7v

1997 E-Series 6.8L V10 VIN S (E4OD) 104 Pin Connector

PCM Pin #	Wire Color	Circuit Description (104 Pin)	Value at Hot Idle
65	BR/LG	EGR DPFE Sensor Signal	0.20-1.30v
66	YL/LG	Cylinder Head Temperature Sensor	0.6v (194°F)
67	PKWT	EVAP Purge Solenoid	0-10 Hz (0-100%)
68	GY/BK	Injector 9 Control	4.2-4.6 ms
69-70	---	Not Used	---
71	RD	Vehicle Power (Start-Run)	12-14v
72	TN/RD	Injector 7 Control	4.2-4.6 ms
73	TN/BK	Injector 5 Control	4.2-4.6 ms
74	BR/YL	Injector 3 Control	4.2-4.6 ms
75	TN	Injector 1 Control	4.2-4.6 ms
76	BK/WT	Power Ground	<0.1v
77	BK/WT	Power Ground	<0.1v
78	DB/LG	COP 2 Driver (Dwell)	5°, at 55 mph: 8°
79	WT/RD	COP 8 Driver (Dwell)	5°, at 55 mph: 8°
80	LB/OR	Fuel Pump Control	On: 1v, Off: 12v
81	WT/YL	EPC Solenoid Control	9.2v (5 psi)
82	DG/PK	COP 9 Driver (Dwell)	5°, at 55 mph: 8°
83	WT/LB	IAC Solenoid Control	10.7v (33%)
84	---	Not Used	---
85	DG	Camshaft Position Sensor	9 Hz, 55 mph: 16 Hz
86	---	Not Used	---
87	RD/BK	HO2S-21 (B2 S1) Signal	0.1-1.1v
88	LB/RD	MAF Sensor Signal	0.8v, 55 mph: 1.6v
89	GY/WT	TP Sensor Signal	0.9v, 55 mph: 1.3v
90	BR/WT	Reference Voltage	4.9-5.1v
91	GY/RD	Analog Signal Return	<0.050v
92	LG	Brake Position Switch	Brake Off: 12v, On: 1v
93	RD/WT	HO2S-11 (B1 S1) Heater	1v, Off: 12v
94	YL/LB	HO2S-21 (B2 S1) Heater	1v, Off: 12v
95	WT/BK	HO2S-12 (B1 S2) Heater	1v, Off: 12v
96	---	Not Used	---
97	RD	Vehicle Power (Start-Run)	12-14v
98	LB	Injector 8 Control	4.2-4.6 ms
99	LG/OR	Injector 6 Control	4.2-4.6 ms
100	BR/LB	Injector 4 Control	4.2-4.6 ms
101	W	Injector 2 Control	4.2-4.6 ms
102	OR/YL	COP 4 Driver (Dwell)	5°, at 55 mph: 8°
103	BK/WT	Power Ground	<0.1v
104	LG/WT	COP 3 Driver (Dwell)	5°, at 55 mph: 8°

Pin Connector Graphic

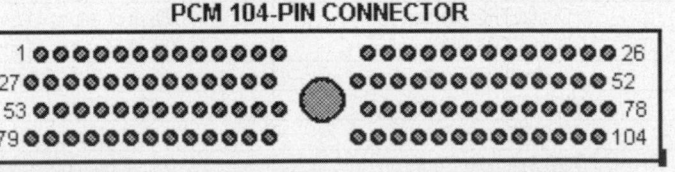

PCM 104-PIN CONNECTOR

Terminal View of 104-Pin PCM Wiring Harness Connector

Standard Colors and Abbreviations

Abbreviation	Color	Abbreviation	Color	Abbreviation	Color
BK	Black	GY	Gray	PK	Purple
BL	Blue	GN	Green	RD	Red
BR	Brown	LG	LT Green	TN	Tan
DB	Dark Blue	OR	Orange	WT	White
DG	DK Green	PK	Pink	YL	Yellow

1998-99 E-Series 6.8L V10 VIN S (E4OD) 104 Pin Connector

PCM Pin #	Wire Color	Circuit Description (104 Pin)	Value at Hot Idle
1	PK/LB	COP 6 Driver (Dwell)	5°, at 55 mph: 8°
2	PK/LG	MIL (lamp) Control	MIL On: 1v, Off: 12v
3	BK/WT	Power Ground	<0.1v
4	PK/LB	Power Takeoff Signal	PTO Off: 0v, On: 12v
5	---	Not Used	---
6	OR/YL	Shift Solenoid 1 Control	1v, 55 mph: 12v
7-8	---	Not Used	---
9	YL/WT	Fuel Level Indicator Signal	1.7v (1/2 full)
10	---	Not Used	---
11	PK/OR	Shift Solenoid 2 Control	12v, 55 mph: 12v
12	---	Not Used	---
13	VT	Flash EEPROM	0.1v
14	---	Not Used	---
15	PK/LB	Data Bus (-)	Digital Signals
16	TN/OR	Data Bus (+)	Digital Signals
17-19	---	Not Used	---
20	BR/OR	Coast Clutch Solenoid	12v, 55 mph: 12v
21	DB	CKP (+) Sensor	500-525 Hz
22	GY	CKP (-) Sensor	500-525 Hz
23-24	BK/WT	Power Ground	<0.1v
25	BK/LB	Case Ground	<0.050v
26	DB/YL	COP 1 Driver (Dwell)	5°, at 55 mph: 8°
27	PKWT	COP 10 Driver (Dwell)	5°, at 55 mph: 8°
28	WT/LG	TCIL (lamp) Control	Lamp On: 1v, Off: 12v
29	TN/WT	TCS (switch) Signal	TCS & O/D On: 12v
30	DG	Octane Adjustment	9.3v (switch shorted: 0v)
31-32	---	Not Used	---
33	PK/OR	VSS (-) Signal	0 Hz, at 55 mph: 125 Hz
34	YL/BK	Digital TR1 Sensor	0v, at 55 mph: 11.5v
35	RD/LG	HO2S-12 (B1 S2) Signal	0.1-1.1v
36	TN/LB	MAF Sensor Return	<0.050v
37	OR/BK	Transmission Fluid Temperature Sensor	2.10-2.40v
38	---	Not Used	---
39	GY	Intake Air Temperature Sensor	1.5-2.5v
40	DG/YL	Fuel Pump Monitor	On: 12v, Off: 0v
41	BK/YL	A/C Pressure Switch Signal	A/C On: 12v, off: 0v
42	GY/RD	Injector 10 Control	3.8-4.6 ms
43-44	---	Not Used	---
45	YL/LB	CHIL (lamp) Control	CHT Off: 12v, On: 1v
46	---	Not Used	---
47	BR/PK	VR Solenoid Control	0%, at 55 mph: 45%
48	PK/BL	Clean Tachometer Output	65 Hz, 55 mph: 175 Hz
49	LB/BK	Digital TR2 Sensor	0v, at 55 mph: 11.5v
50	LG/RD	Digital TR4 Sensor	0v, at 55 mph: 11.5v
51	BK/WT	Power Ground	<0.1v
52	PKWT	COP 5 Driver (Dwell)	5°, at 55 mph: 8°
53	WT/PK	COP 7 Driver (Dwell)	5°, at 55 mph: 8°
54	PK/YL	TCC Solenoid Control	0%, at 55 mph: 95%
55	YL	Keep Alive Power	12-14v
56	LG/BK	EVAP Purge Solenoid	0-10 Hz (0-100%)
57	YL/RD	Knock Sensor 1 Signal	0v
58	GY/BK	VSS (+) Signal	0 Hz, at 55 mph: 125 Hz
59	---	Not Used	---
60	GY/LB	HO2S-11 (B1 S1) Signal	0.1-1.1v
61-63	---	Not Used	---
62	RD/PK	Fuel Tank Pressure Sensor	2.6v (0" H2O - cap off)
64	LB/YL	Digital Transmission Range Sensor	In 'P': 0v, in O/D: 1.7v

1998-99 E-Series 6.8L V10 VIN S (E4OD) 104 Pin Connector

PCM Pin #	Wire Color	Circuit Description (104 Pin)	Value at Hot Idle
65	BR/LG	EGR DPFE Sensor Signal	0.20-1.30v
66	YL/LG	Cylinder Head Temperature Sensor	0.6v (194ºF)
67	PKWT	EVAP Purge Solenoid	0-10 Hz (0-100%)
68	GY/BK	Injector 9 Control	3.8-4.6 ms
69-70	---	Not Used	---
71	RD	Vehicle Power	12-14v
72	TN/RD	Injector 7 Control	3.8-4.6 ms
73	TN/BK	Injector 5 Control	3.8-4.6 ms
74	BR/YL	Injector 3 Control	3.8-4.6 ms
75	TN	Injector 1 Control	3.8-4.6 ms
76	BK/WT	Power Ground	<0.1v
77	BK/WT	Power Ground	<0.1v
78	DB/LG	COP 2 Driver (Dwell)	5º, at 55 mph: 8º
79	WT/RD	COP 8 Driver (Dwell)	5º, at 55 mph: 8º
80	LB/OR	Fuel Pump Control	On: 1v, Off: 12v
81	WT/YL	EPC Solenoid Control	9.2v (5 psi)
82	DG/PK	COP 9 Driver (Dwell)	5º, at 55 mph: 8º
83	WT/LB	IAC Solenoid Control	10.7v (33%)
84	---	Not Used	---
85	DG	Camshaft Position Sensor	9 Hz, 55 mph: 16 Hz
86	---	Not Used	---
87	RD/BK	HO2S-21 (B2 S1) Signal	0.1-1.1v
88	LB/RD	MAF Sensor Signal	0.8v, 55 mph: 1.6v
89	GY/WT	TP Sensor Signal	0.9v, 55 mph: 1.3v
90	BR/WT	Reference Voltage	4.9-5.1v
91	GY/RD	Analog Signal Return	<0.050v
92	LG	Brake Position Switch	Brake Off: 12v, On: 1v
93	RD/WT	HO2S-11 (B1 S1) Heater	1v, Off: 12v
94	YL/LB	HO2S-21 (B2 S1) Heater	1v, Off: 12v
95	WT/BK	HO2S-12 (B1 S2) Heater	1v, Off: 12v
96	---	Not Used	---
97	RD	Vehicle Power	12-14v
98	LB	Injector 8 Control	3.8-4.6 ms
99	LG/OR	Injector 6 Control	3.8-4.6 ms
100	BR/LB	Injector 4 Control	3.8-4.6 ms
101	WT	Injector 2 Control	3.8-4.6 ms
102	OR/YL	COP 4 Driver (Dwell)	5º, at 55 mph: 8º
103	BK/WT	Power Ground	<0.1v
104	LG/WT	COP 3 Driver (Dwell)	5º, at 55 mph: 8º

Pin Connector Graphic

PCM 104-PIN CONNECTOR

Terminal View of 104-Pin PCM Wiring Harness Connector

Standard Colors and Abbreviations

Abbreviation	Color	Abbreviation	Color	Abbreviation	Color
BK	Black	GY	Gray	PK	Purple
BL	Blue	GN	Green	RD	Red
BR	Brown	LG	LT Green	TN	Tan
DB	Dark Blue	OR	Orange	WT	White
DG	DK Green	PK	Pink	YL	Yellow

2000-2002 E-Series 6.8L V10 VIN S (A/T) 104 Pin Connector

PCM Pin #	Wire Color	Circuit Description (104 Pin)	Value at Hot Idle
1	OR/YL	COP 6 Driver (Dwell)	5°, at 55 mph: 8°
2	PK/LG	MIL (lamp) Control	MIL On: 1v, Off: 12v
3	BK/WT	Power Ground	<0.1v
4	PK/LB	Power Takeoff Signal	PTO Off: 0v, On: 12v
5	---	Not Used	---
6	OR/YL	Shift Solenoid 'A' Control	1v, 55 mph: 12v
7-10	---	Not Used	---
9	YL/WT	Fuel Pump Sender	TBD
11	PK/OR	Shift Solenoid 'B' Control	12v, 55 mph: 12v
12	WT/LG	Overdrive Cancel Switch	TBD
13	VT	Flash EEPROM	0.1v
14	OR/LB	4x4 Low Switch	Switch On: 0v, Off: 12v
15	PK/LB	Data Bus (-)	Digital Signals
16	TN/OR	Data Bus (+)	Digital Signals
17-19	---	Not Used	---
20	BR/OR	Coast Clutch Solenoid	12v, 55 mph: 12v
21	DB	CKP (+) Sensor	500-525 Hz
22	GY	CKP (-) Sensor	500-525 Hz
23-24	BK/WT	Power Ground	<0.1v
25	BK/LB	Case Ground	<0.050v
26	LG/WT	COP 1 Driver (Dwell)	5°, at 55 mph: 8°
27	WT/RD	COP 10 Driver (Dwell)	5°, at 55 mph: 8°
28	---	Not Used	---
29	TN/WT	TCS (switch) Signal	TCS & O/D On: 12v
30-31	---	Not Used	---
32	DG/PK	Knock Sensor 1 Return	<0.050v
33	---	Not Used	---
34	YL/BK	Digital TR1 Sensor	0v, at 55 mph: 11.5v
35	RD/LG	HO2S-12 (B1 S2) Signal	0.1-1.1v
36	TN/LB	MAF Sensor Return	<0.050v
37	OR/BK	Transmission Fluid Temperature Sensor	2.10-2.40v
38	---	Not Used	---
39	GY	Intake Air Temperature Sensor	1.5-2.5v
40	DG/YL	Fuel Pump Monitor	On: 12v, Off: 0v
41	BK/YL	A/C Pressure Switch Signal	A/C On: 12v, Off: 0v
42	GY/RD	Injector 10 Control	3.8-4.6 ms
43-44	---	Not Used	---
45	YL/LB	Engine Coolant Temperature Sensor	0.5-0.6v
46	GY/BK	Vehicle Speed Signal	0 Hz, at 55 mph: 125 Hz
47	BR/PK	VR Solenoid Control	0%, at 55 mph: 45%
48	WT/PK	Clean Tachometer Output	65 Hz, 55 mph: 175 Hz
49	LB/BK	Digital TR2 Sensor	0v, at 55 mph: 11.5v
50	LG/RD	Digital TR4 Sensor	0v, at 55 mph: 11.5v
51	BK/WT	Power Ground	<0.1v
52	LG/YL	COP 5 Driver (Dwell)	5°, at 55 mph: 8°
53	PK/LB	COP 7 Driver (Dwell)	5°, at 55 mph: 8°
54	PK/YL	TCC Solenoid Control	0%, at 55 mph: 95%
55	Y	Keep Alive Power	12-14v
56	LG/BK	EVAP Purge Solenoid	0-10 Hz (0-100%)
57	YL/RD	Knock Sensor 1 Signal	0v
58-59	---	Not Used	---
60	GY/LB	HO2S-11 (B1 S1) Signal	0.1-1.1v
61	---	Not Used	---
62	RD/PK	Fuel Tank Pressure Sensor	2.6v (0" H2O - cap off)
63	---	Not Used	---
64	LB/YL	Digital TR3A Sensor Signal	In 'P': 0v, in O/D: 1.7v

2000-2002 E-Series 6.8L V10 VIN S (A/T) 104 Pin Connector

PCM Pin #	Wire Color	Circuit Description (104 Pin)	Value at Hot Idle
65 ('00)	BR/LG	EGR DPFE Sensor Signal	0.20-1.30v
65 ('01-'02)	BR/LG	EGR DPFE Sensor Signal	0.95-1.05v
66	YL/LG	Cylinder Head Temperature Sensor	0.6v (194°F)
67	PKWT	EVAP Purge Solenoid	0-10 Hz (0-100%)
68	GY/BK	Injector 9 Control	3.8-4.6 ms
71	RD	Vehicle Power (Start-Run)	12-14v
72	TN/RD	Injector 7 Control	3.8-4.6 ms
73	TN/BK	Injector 5 Control	3.8-4.6 ms
74	BR/YL	Injector 3 Control	3.8-4.6 ms
75	TN	Injector 1 Control	3.8-4.6 ms
76	BK/WT	Power Ground	<0.1v
77	BK/WT	Power Ground	<0.1v
78	PKWT	COP 2 Driver (Dwell)	5°, at 55 mph: 8°
79	WT/RD	COP 8 Driver (Dwell)	5°, at 55 mph: 8°
80	LB/OR	Fuel Pump Control	On: 1v, Off: 12v
81	WT/YL	EPC Solenoid Control	9.2v (5 psi)
82	YL/BK	COP 9 Driver (Dwell)	5°, at 55 mph: 8°
83	WT/LB	IAC Solenoid Control	10.7v (33%)
84	DB/YL	Output Shaft Speed Sensor	0 Hz, 55 mph: 2 KHz
85	DG	Camshaft Position Sensor	9 Hz, 55 mph: 16 Hz
87	RD/BK	HO2S-21 (B2 S1) Signal	0.1-1.1v
88	LB/RD	MAF Sensor Signal	0.8v, 55 mph: 1.6v
89	GY/WT	TP Sensor Signal	0.9v, 55 mph: 1.3v
90	BR/WT	Reference Voltage	4.9-5.1v
91	GY/RD	Analog Signal Return	<0.050v
92	LG	Brake Position Switch	Brake Off: 12v, On: 1v
93	RD/WT	HO2S-11 (B1 S1) Heater	1v, Off: 12v
94	YL/LB	HO2S-21 (B2 S1) Heater	1v, Off: 12v
95	WT/BK	HO2S-12 (B1 S2) Heater	1v, Off: 12v
96	---	Not Used	---
97	RD	Vehicle Power (Start-Run)	12-14v
98	LB	Injector 8 Control	3.8-4.6 ms
99	LG/OR	Injector 6 Control	3.8-4.6 ms
100	BR/LB	Injector 4 Control	3.8-4.6 ms
101	WT	Injector 2 Control	3.8-4.6 ms
102	DG/PK	COP 4 Driver (Dwell)	5°, at 55 mph: 8°
103	BK/WT	Power Ground	<0.1v
104	WT/PK	COP 3 Driver (Dwell)	5°, at 55 mph: 8°

Pin Connector Graphic

Standard Colors and Abbreviations

Abbreviation	Color	Abbreviation	Color	Abbreviation	Color
BK	Black	GY	Gray	PK	Purple
BL	Blue	GN	Green	RD	Red
BR	Brown	LG	LT Green	TN	Tan
DB	Dark Blue	OR	Orange	WT	White
DG	DK Green	PK	Pink	YL	Yellow

2003-05 E-Series 6.8L V10 VIN S (All) 104 Pin Connector

PCM Pin #	Wire Color	Circuit Description (104 Pin)	Value at Hot Idle
1	OR/YL	COP 6 Driver (dwell)	5°, 55 mph: 9°
2	---	Not Used	---
3	BK/WT	Power Ground	<0.1v
4	LB/YL	Power Take-Off (if equipped)	0v (Off)
5	---	Not Used	---
6	OR/YL	Shift Solenoid 1 Control	1v, 55 mph: 12v
7-10	---	Not Used	---
11	PK/OR	Shift Solenoid 2 Control	12v, 55 mph: 12v
12	WT/LG	TCIL (lamp) Control	TCIL On: 1v, Off: 2.2v
13	VT	Flash EEPROM Power	0.1v
14	LB/BK	4x4 Low Indicator Switch	12v (switch closed: 0v)
15	PK/LB	SCP Data Bus (-) Signal	<0.050v
16	TN/OR	SCP Data Bus (+) Signal	Digital Signals
17-19	---	Not Used	---
20	BR/OR	Coast Clutch Solenoid Control	12v, 55 mph: 12v
21	DB	CKP (-) Sensor Signal	400-420 Hz
22	GY	CKP (+) Sensor Signal	400-420 Hz
23-24	BK/WT	Power Ground	<0.1v
25	LB/YL	Case Ground	<0.050v
26	LG/WT	COP 1 Driver (dwell)	5°, 55 mph: 9°
27	LG/YL	COP 10 Driver (dwell)	5°, 55 mph: 9°
28	---	Not Used	---
29	TN/WT	Overdrive Cancel Switch) Signal	Switch Off: 12v, On: 1v
30-31	---	Not Used	---
32	DG/VT	Knock Sensor 1 (-) Signal	<0.050v
33	PK/OR	Ground	<0.050v
34	YL/BK	Digital TR1 Sensor	0v, 55 mph: 11.5v
35	RD/LG	HO2S-12 (B1 S2) Signal	0.1-1.1v
36	TN/LB	MAF Sensor Return	<0.050v
37	OR/BK	TFT Sensor Signal	2.10-2.40v
38	---	Not Used	---
39	GY	IAT Sensor Signal	1.5-2.5v
40	DG/YL	Fuel Pump Monitor	Off: 12v, On: 1v
41	TN/LG	A/C Switch Signal	A/C On: 12v, Off: 0v
42	GY/RD	Injector 10 Control	4.2-4.6 ms
43	OR/LG	Data Output Link	Digital Signals
44-46	---	Not Used	---
47	BR/PK	VR Solenoid Control	0%, 55 mph: 45%
48	LG/WT	Clean Tachometer Output	60-70 Hz
49	LB/BK	Digital TR2 Sensor	0v, 55 mph: 11.5v
50	WT/BK	Digital TR4 Sensor	0v, 55 mph: 11.5v
51	BK/WT	Power Ground	<0.1v
52	WT/PK	COP 5 Driver (dwell)	5°, 55 mph: 9°
53	DGVT	COP 7 Driver (dwell)	5°, 55 mph: 9°
54	VT/YL	TCC Solenoid Control	0%, 55 mph: 95%
55	RD/WT	Keep Alive Power	12-14v
56	LG/BK	EVAP Purge Solenoid	0-10 Hz (0-100%)
57	YL/RD	Knock Sensor 1 (+) Signal	0v
58	GY/BK	VSS (+) Signal	0 Hz, 55 mph: 125 Hz
59	DG/WT	TSS Sensor Signal	300 Hz, 55 mph: 980
60	GY/LB	HO2S-11 (B1 S1) Signal	0.1-1.1v
61	---	Not Used	---
62	RD/PK	FTP Sensor Signal	2.6v at 0" H20 (cap off)
63	---	Not Used	---
64	LB/YL	A/T: Digital TR3 Sensor	In 'P': 0v, in O/D: 1.7v
64	LB/YL	M/T: CPP Switch Signal	5v (clutch "in": 0v)

2003-05 E-Series 6.8L V10 VIN S (All) 104 Pin Connector

PCM Pin #	Wire Color	Circuit Description (104 Pin)	Value at Hot Idle
65	BR/LG	DPFE Sensor Signal (California)	0.95-1.05v
66	YL/LG	Cylinder Head Temperature Sensor	0.6 (194°F)
67	VT/WT	EVAP Canister Vent Solenoid Control	0-10 Hz (0-100%)
68	GY/BK	Injector 9 Control	4.2-4.6 ms
69	PK/YL	A/C Clutch Relay Control	A/C On: 12v, Off: 0v
70	---	Not Used	---
71	RD	Vehicle Power	12-14v
72	TN/RD	Injector 7 Control	4.2-4.6 ms
73	TN/BK	Injector 5 Control	4.2-4.6 ms
74	BR/YL	Injector 3 Control	4.2-4.6 ms
75	TN	Injector 1 Control	4.2-4.6 ms
76	BK/WT	Power Ground	<0.1v
77	BK/WT	Power Ground	<0.1v
78	PK/LB	COP 2 Driver (dwell)	5°, 55 mph: 9°
79	WT/RD	COP 8 Driver (dwell)	5°, 55 mph: 9°
80	LB/OR	Fuel Pump Relay Control	Off: 12v, On: 1v
81	WT/YL	EPC Solenoid Control	9.2v (5 psi)
82	WT/RD	COP 9 Driver (dwell)	5°, 55 mph: 9°
83	WT/LB	IAC Solenoid Control	10.7v (33%)
84	DB/YL	Output Shaft Speed Sensor	0 Hz, 55 mph: 470 Hz
85	DG	CMP Sensor Signal	10 Hz, 55 mph: 16 Hz
86	---	Not Used	---
87	RD/BK	HO2S-21 (B2 S1) Signal	0.1-1.1v
88	LB/RD	MAF Sensor Signal	0.8v, 55 mph: 1.6v
89	GY/WT	TP Sensor Signal	0.53-1.27v
90	BR/WT	Reference Voltage	4.9-5.1v
91	GY/RD	Sensor Return	<0.050v
92	RD/LG	Brake Pedal Switch	0v (Brake On: 12v)
93	RD/WT	HO2S-11 (B1 S1) Heater	1v (Heater Off: 12v)
94	YL/LB	HO2S-21 (B2 S1) Heater	1v (Heater Off: 12v)
95	WT/BK	HO2S-12 (B1 S2) Heater	1v (Heater Off: 12v)
96	---	Not Used	---
97	RD	Vehicle Power	12-14v
98	LB	Injector 8 Control	4.2-4.6 ms
99	LG/OR	Injector 6 Control	4.2-4.6 ms
100	BR/LB	Injector 4 Control	4.2-4.6 ms
101	W	Injector 2 Control	4.2-4.6 ms
102	YL/BK	COP 4 Driver (dwell)	5°, 55 mph: 9°
103	BK/WT	Power Ground	<0.1v
104	PK/WT	COP 3 Driver (dwell)	5°, 55 mph: 9°

Pin Connector Graphic

PCM 104-PIN CONNECTOR

```
 1 ●●●●●●●●●●●●      ●●●●●●●●●●●● 26
27 ●●●●●●●●●●●●   ●  ●●●●●●●●●●●● 52
53 ●●●●●●●●●●●●      ●●●●●●●●●●●● 78
79 ●●●●●●●●●●●●      ●●●●●●●●●●●● 104
```

Terminal View of 104-Pin PCM Wiring Harness Connector

Standard Colors and Abbreviations

Abbreviation	Color	Abbreviation	Color	Abbreviation	Color
BK	Black	GY	Gray	PK	Purple
BL	Blue	GN	Green	RD	Red
BR	Brown	LG	LT Green	TN	Tan
DB	Dark Blue	OR	Orange	WT	White
DG	DK Green	PK	Pink	YL	Yellow

1996-99 E-Series 7.3L V8 Diesel VIN F (A/T) 104 Pin Connector

PCM Pin #	Wire Color	Circuit Description (104 Pin)	Value at Hot Idle
1	PK/OR	Shift Solenoid 2 Control	12v, 55 mph: 12v
2	PK/LG	MIL (lamp) Control	MIL On: 1v, Off: 12v
3	RD/YL	Manifold Intake Air Heater Relay	Heater On: 12v, Off: 0v
4	OR/GR	Parking Brake Applied Switch	Switch Up: 12, Down: 0v
5	RD/OR	Idle Validation Switch	Switch Up: 0v, Down: 12v
5 (Cal)	LG/RD	Parking Brake Applied Switch	Park Brake Applied: 0v
6 (Cal)	OR/YL	Shift Solenoid 1 Control	SS1 On: 1v, Off: 12v
7	---	Not Used	---
8 (Cal)	GY	Glow Plug Monitor High Side	Plugs Off: 0v, On: 12v
9 (Cal)	OR	Glow Plug Monitor (Bank 1)	Plugs Off: 0v, On: 12v
10 (Cal)	RD/OR	Idle Validation Switch	0v, off-12v
11 (Cal)	PK/OR	Shift Solenoid 2 Control	SS2 On: 1v, Off: 12v
12 (Cal)	WT/LG	TCIL (lamp) Control	Lamp On: 1v, Off: 12v
13	VT	Flash EPROM Power	0.1v
14	LG/BK	4x4 Low Switch	Switch On: 0v, Off: 12v
15	PK/LB	Data Bus (-)	Digital Signals
16	TN/OR	Data Bus (+)	Digital Signals
17 (Cal)	OR/BK	Digital TR1 Sensor Signal	Manual 1 or 2: 10.7v
19 (Cal)	WT/PK	Tachometer Reflected Signal	6.5v (130 Hz)
20	---	Not Used	---
21 (Cal)	DG	Camshaft Position Sensor	7 Hz
22-23	---	Not Used	---
24	YL/BK	ACP Sensor Ground	<0.050v
25	BK	Case Ground	<0.050v
26	---	Not Used	---
27	OR/YL	Shift Solenoid 1 Control	1v, 55 mph: 12v
28	PK/YL	TCC Solenoid Control	12v, 55 mph: 1v
29	TN/LB	Clutch Pedal Position Switch	Clutch In: 0v, Out: 12v
29	TN/WT	TCS (switch) Signal	TCS & O/D On: 12v
30	PK/LB	Exhaust Back Pressure	0.9v, Cruise: 2.5v
31	RD/LG	Brake Pedal Applied Switch	Pedal Up: 12v, Down: 0v
32	---	Not Used	---
33	PK/OR	VSS (-) Signal	0 Hz, at 55 mph: 125 Hz
34	LG/BK	MAP Sensor Signal	107 Hz
35-36	---	Not Used	---
37	OR/BK	Transmission Fluid Temperature Sensor	2.10-2.40v
38	LG/RD	Engine Oil Temperature	-40°F: 4.7v, 230°F: 0.36v
39	GY	Intake Air Temperature Sensor	1.5-2.5v
40	DG/OR	Speed Control Ground	<0.050v
41	BK/YL	AC Clutch (ACC) Signal	AC & Clutch On: 12v
42	GY/RD	Exhaust Backpressure Signal	0-12v (duty cycle)
43-47	---	Not Used	---
48	GY/WT	Electronic Feedback Line	Digital Signals
49	DG	Camshaft Position Sensor	6.5v (130 Hz0
50	WT/PK	Tachometer Signal (CMP)	6.5v (130 Hz)
51	BK/WT	Power Ground	<0.1v
52	---	Not Used	---
53	BR/OR	Coast Clutch Solenoid	12v, 55 mph: 12v
54	---	Not Used	---
55	YL	Keep Alive Power	12-14v
56-57	---	Not Used	---
58	GY/BK	VSS (+) Signal	0 Hz, at 55 mph: 125 Hz
61	LB/BK	Speed Control Cruise Signal	0 or 12v
62	---	Not Used	---
63	DB/LG	BARO Sensor	4.6v (sea level)
64	LB/YL	Digital Transmission Range Sensor	In 'P': 0v, in O/D: 1.7v

1996-99 E-Series 7.3L V8 Diesel VIN F (A/T) 104 Pin Connector

PCM Pin #	Wire Color	Circuit Description (104 Pin)	Value at Hot Idle
65	LB	Camshaft Position Sensor Ground	<0.050v
66-69	---	Not Used	---
70	WT/BK	Glow Plug Lamp Control	Lamp On: 0v, Off: 12v
71	RD	Vehicle Power (Start-Run)	12-14v
72-75	---	Not Used	---
76	BK/WT	Power Ground	<0.1v
77	BK/WT	Power Ground	<0.1v
78	---	Not Used	---
79	WT/LG	TCIL (lamp) Control	Lamp On: 1v, Off: 12v
80	BK/PK	IDM Relay Control	On: 1v, Off: 12v
81	WT/YL	EPC Solenoid Control	7.5v (8 psi)
82	---	Not Used	---
83	YL/RD	Injector Pressure Regulator	12v, 55 mph: 12v
84	DB/LG	BARO Sensor Signal	At Sea Level: 4.64v
85-86	---	Not Used	---
87	DB/LG	Injection Control Pressure	0.75v (Min: 0.83v startup)
88	---	Not Used	---
89	GY/WT	ACP Sensor Signal	0.5-0.9v
90	BR/WT	Reference Voltage	4.9-5.1v
91	GY/RD	Analog Signal Return	<0.050v
92	LG	Brake Position Switch	Brake Off: 12v, On: 1v
93-94	---	Not Used	---
95	BR/OR	Fuel Delivery Control Signal	1v / 49 Hz
96	YL/LB	Camshaft Position Sensor	5 Hz (6v)
97	RD	Vehicle Power (Start-Run)	12-14v
98-99	---	Not Used	---
100	YL/WT	Fuel Level Indicator Signal	1.7v (1/2 full)
101	PK/OR	Glow Plug Relay Control	On: 1v, Off: 12v
103	BK/WT	Power Ground	<0.1v
104	---	Not Used	---

Pin Connector Graphic

PCM 104-PIN CONNECTOR

Terminal View of 104-Pin PCM Wiring Harness Connector

Standard Colors and Abbreviations

Abbreviation	Color	Abbreviation	Color	Abbreviation	Color
BK	Black	GY	Gray	PK	Purple
BL	Blue	GN	Green	RD	Red
BR	Brown	LG	LT Green	TN	Tan
DB	Dark Blue	OR	Orange	WT	White
DG	DK Green	PK	Pink	YL	Yellow

2000-02 E-Series 7.3L V8 Diesel VIN F (A/T) 104 Pin Connector

PCM Pin #	Wire Color	Circuit Description (104 Pin)	Value at Hot Idle
1	---	Not Used	---
2	PK/LG	MIL (lamp) Control	MIL On: 1v, Off: 12v
3	RD/YL	Manifold Intake Air Heater Monitor Signal	Heater On: 12v, Off: 0v
4	OR/GR	Driveline Disconnect Switch	Switch On: 0v, Off: 12v
5	LG/RD	Brake Warning Switch Input	Applied: 0v, Off: 12v
6	OR/YL	Shift Solenoid 1/A Control	SS1 On: 1v, Off: 12v
7	---	Not Used	---
8	WT/PK	Glow Plug Monitor High Side	Plugs Off: 0v, On: 12v
8 (Cal)	WT/LG	Glow Plug Control Module	Digital Signals
9	WT	Glow Plug Monitor Right Bank	Plugs Off: 0v, On: 12v
10	RD/OR	Idle Validation Switch	12, Cruise: 0v
11	PK/OR	Shift Solenoid 'B' Control	SS2 On: 1v, Off: 12v
12	WT/LG	TCIL (lamp) Control	Lamp On: 1v, Off: 12v
13	PK	Generic Scan Tool Input	0.1v
14	OR/LB	4x4 Low Switch	Switch On: 0v, Off: 12v
15	PK/LB	Data Bus (-)	Digital Signals
16	TN/OR	Data Bus (+)	Digital Signals
17	OR/BK	Digital TR1 Sensor	In 'P': 0v, Drive: 10.7v
18	---	Not Used	---
19	WT/PK	Clean Tachometer Output	6.5v (130 Hz)
20	BR/OR	Coast Clutch Solenoid	CCS On: 0v, Off: 12v
21	DG	Camshaft Position Sensor	7 Hz
22-23	---	Not Used	---
24	YL/BK	APP Sensor Ground	<0.050v
25	BK/LB	Case Ground	<0.050v
26-27	---	Not Used	---
28	RD	Water In Fuel Indicator Lamp	Lamp on: 1v, Off: 12v
29	TN/WT	A/T: TCS (switch) Signal	TCS & O/D On: 12v
30	PK/LB	Exhaust Back Pressure	0.9v, off-idle: 2.5v
31	RD/LG	Brake Pressure Switch	Brake On: 0v, Off: 12v
32	---	Not Used	---
33	PK/OR	VSS (-) Signal	<0.050v
34	WT/LG	Glow Plug Monitor Left Bank	Plugs Off: 0v, On: 12v
35	WT/YL	Generator Power Switch	12-14v
36	GY/RD	Alternator #1 (top) Monitor	6-10v
36	GY/RD	Fuel Line Heater	Heater On: 12v, Off: 0v
37	OR/BK	Transmission Fluid Temperature Sensor	0.3-4.5v
38	LG/RD	Engine Oil Temperature	0.3-4.7v
39	GY	Intake Air Temperature Sensor	0.2-4.5v
40	PK/BK	Speed Control Ground	<0.050v
41	BK/YL	A/C Pressure Switch Signal	A/C On: 12v, Off: 0v
42	GY/RD	Exhaust Backpressure Signal	Digital Signals
43-47	---	Not Used	---
48	GY/WT	Electronic Feedback Line	Digital Signals
49	WT/PK	Digital TR2 Sensor	In 'P': 0v, Drive: 10.7v
50	GY/BK	Digital TR4 Sensor	In 'P': 0v, Drive: 10.7v
51	BK/WT	Power Ground	<0.1v
52-53	---	Not Used	---
54	PK/YL	TCC Solenoid Control	TCC On: 1v, Off: 12v
55	YL	Keep Alive Power	12-14v
56-57	---	Not Used	---
58	GY/BK	VSS (+) Signal	0 Hz, at 55 mph: 125 Hz
59	---	Not Used	---
60	YL	Alternator #2 (bottom) Monitor	6-10v
61	LB/BK	Speed Control Cruise Signal	S/C Switch On: 12v
62-63	---	Not Used	---
64	LB/YL	Digital TR Sensor3	In 'P': 4.5v, Drive: 2.2v

2000-02 E-Series 7.3L V8 Diesel VIN F (A/T) 104 Pin Connector

PCM Pin #	Wire Color	Circuit Description (104 Pin)	Value at Hot Idle
65	LB	Camshaft Position Sensor Ground	<0.050v
66	TN/LG	A/C Pressure Switch	0-4.9v
67	GY/LB	Battery Direct	12-14v
68-69	---	Not Used	---
70	BK/PK	Glow Plug Lamp Control	Lamp On: 1v, Off: 12v
71	RD	Vehicle Power (Start-Run)	12-14v
72-75	---	Not Used	---
76	BK/WT	Power Ground	<0.1v
77	BK/WT	Power Ground	<0.1v
78	---	Not Used	---
79	LG/BK	MAP Sensor Signal	1-3v
80	WT/BK	IDM Relay Control	On: 1v, Off: 12v
81	WT/YL	EPC Solenoid Control	7.5v (8 psi)
82	---	Not Used	---
83	YL/RD	Injector Pressure Regulator	12v
84	DG/WT	Turbine Shaft Speed Sensor	30 mph: 126-136 Hz
85-86	---	Not Used	---
87	DB/LG	Injection Control Pressure	0.75v (Min: 0.83v startup)
88	---	Not Used	---
89	GY/WT	APP Sensor Signal	0.5-1.6v
90	BR/WT	Reference Voltage	4.9-5.1v
91	GY/RD	Analog Signal Return	<0.050v
92	LG	Brake Position Switch	Brake Off: 12v, On: 1v
93	---	Not Used	---
94	LB/OR	Fuel Pump Control	On: 1v, Off: 12v
95	BR/OR	Fuel Delivery Control Signal	49 Hz
96	YL/LB	Camshaft Position Sensor	6v (5 Hz)
97	RD	Vehicle Power (Start-Run)	12-14v
98	PK	Manifold Intake Air Heater Control	12-14v
99	---	Not Used	---
100	YL/WT	Fuel Level Indicator Signal	1.7v (1/2 full)
101	PK/OR	Glow Plug Relay Control	On: 1v, Off: 12v
102	---	Not Used	---
103	BK/WT	Power Ground	<0.1v
104	---	Not Used	---

Pin Connector Graphic

PCM 104-PIN CONNECTOR

1 ●●●●●●●●●●●● ●●●●●●●●●●●●● 26
27 ●●●●●●●●●●●● ●●●●●●●●●●●●● 52
53 ●●●●●●●●●●●● ●●●●●●●●●●●●● 78
79 ●●●●●●●●●●●● ●●●●●●●●●●●●● 104

Terminal View of 104-Pin PCM Wiring Harness Connector

Standard Colors and Abbreviations

Abbreviation	Color	Abbreviation	Color	Abbreviation	Color
BK	Black	GY	Gray	PK	Purple
BL	Blue	GN	Green	RD	Red
BR	Brown	LG	LT Green	TN	Tan
DB	Dark Blue	OR	Orange	WT	White
DG	DK Green	PK	Pink	YL	Yellow

2003 E-Series 7.3L V8 Diesel VIN F (All) 104 Pin Connector

PCM Pin #	Wire Color	Circuit Description (104 Pin)	Value at Hot Idle
1	---	Not Used	---
2	PK/LG	MIL (lamp) Control	MIL Off: 12v, On: 1v
3	RD/YL	Manifold Intake Air Heater Monitor	Off: 12v, On: 1v
4	---	Not Used	---
5	LG/RD	Parking Brake Applied Switch	Brake Off: 12v, Applied: 0v
6	OR/YL	Shift Solenoid 1 Control	SS1 Off: 12v, On: 1v
7	---	Not Used	---
8	WT/LG	Glow Plug Monitor Left Bank	Plugs Off: 0v, On: 12v
9	---	Not Used	---
10	RD/LG	Idle Validation Switch	12v, off-idle: 0v
11	VT/OG	Shift Solenoid 2 Control	SS2 Off: 12v, On: 1v
12	WT/LG	TCIL (lamp) Control	7.7v (Switch On: 0v)
13	VT	Flash EEPROM Power Supply	0.1v
14	---	Not Used	---
15	PK/LB	SCP Data Bus (-) Signal	<0.050v
16	TN/OR	SCP Data Bus (+) Signal	Digital Signals
17	OR/BK	Digital TR1 Sensor	In 'P': 0v, in Drive: 10.7v
18	PK/YL	A/C Relay Control	Relay Off: 12v, On: 1v
19	WT/PK	Tachometer Reflected Signal	6.5v / 130 Hz
20	BR/OR	Coast Clutch Solenoid Control	CCS On: 0v, Off: 12v
21	DG	CMP Sensor Signal	Digital Signal: 0-12-0v
22-23	---	Not Used	---
24	YL/WT	APP Sensor Ground	<0.050v
25	BK/LB	Case Ground	<0.050v
26-27	---	Not Used	---
28	RD	Water In Fuel Indicator Control	Indicator Off: 12v, On: 1v
29	TN/WT	A/T: TCS (switch) Signal	TCS & O/D On: 12v
29	TN/LB	M/T: CCP Switch Signal	Clutch In: 0v, Out: 12v
30	VT/LB	Exhaust Back Pressure Sensor	0.9v, off-idle: 2.5v
31	BK/YL	Brake Pressure Switch	Brake Off: 12v, On: 1v
32	---	Not Used	---
33	PK/OG	VSS (-) Signal	<0.050v
34	---	Not Used	---
35	WT/YL	Alternator No. 1 (Top) Monitor	6-10v
36	GY/RD	Fuel Heater / Water in Fuel	Heater On: 12v
37	OR/BK	TFT Sensor Signal	2.10-2.40v
38	LG/RD	Engine Oil Temperature Sensor	0.3-4.7v
39	GY	Intake Air Temperature Sensor	1.5-2.5v
40	PK/BK	Fuel Pump Monitor	Pump Off: 12v, On: 1v
41	TN/LG	A/C Head Pressure Switch	0v or 12v
42	GY/RD	Exhaust Backpressure Signal	Digital: 0-12-0-12v
43	OR/LG	Overhead Console Fuel Consumption	0-5-0v
44-46	---	Not Used	---
47	WT/RD	Wastegate Control Solenoid Control	Solenoid Off: 12v, On: 1v
48	GY/WT	Electronic Feedback Line	0.9-3.0v
49	WT/PK	Digital TR2 Sensor	In 'P': 0v, in Drive: 10.7v
50	GY/BK	Digital TR4 Sensor	In 'P': 0v, in Drive: 10.7v
51	BK/WT	Power Ground	<0.1v
52-53	---	Not Used	---
54	VT/YL	TCC Solenoid Control	TCC Off: 12v, On: 1v
55	RD/WT	Keep Alive Power	12-14v
56-57	---	Not Used	---
58	GY/BK	VSS (+) Signal	0 Hz, 55 mph: 125 Hz
59	DB/YL	OSS Sensor Signal	AC pulse signals
60	YL	Alternator No. 2 (Bottom) Control	6-10v
61	LB/BK	Speed Control On/Off Switch	Switch On: 12v, Off: 0v
62	RD/YL	MAT Sensor Signal	1.5-2.5v
63	---	Not Used	---

2003 E-Series 7.3L V8 Diesel VIN F 104 Pin Connector

PCM Pin #	Wire Color	Circuit Description (104 Pin)	Value at Hot Idle
64	LB/YL	Digital TR3 Sensor	In P: 4.5v, in Drive: 2.2v
65	LB	CMP Sensor Ground	<0.050v
66	LB/YL	Power Take-Off Signal	0v (Off)
67	LG/RD	Battery Direct	12-14v
68-69	---	Not Used	---
70	BK/PK	Glow Plug Lamp Control	7.7v (Switch On: 0v)
71	RD	Vehicle Power	12-14v
72-75	---	Not Used	---
76	BK/WT	Power Ground	<0.1v
77	BK/WT	Power Ground	<0.1v
78	---	Not Used	---
79	LG/BK	MAP Sensor Signal	1-3v
80	WT/BK	IDM Enable Relay Control	Off: 12v, On: 1v
81	WT/YL	EPC Solenoid Control	7.5v (8 psi)
82	---	Not Used	---
83	YL/RD	Injector Pressure Regulator	12v
84	DG/WT	TSS Sensor Signal	30 mph: 130 Hz
85-86	---	Not Used	---
87	DB/LG	Injection Control Pressure	0.1-3.0v
88	---	Not Used	---
89	GY/WT	APP Sensor Signal	0.5-1.6v
90	BR/WT	Reference Voltage	4.9-5.1v
91	GY/RD	Sensor Ground	<0.050v
92	RD/LG	Brake Pedal Switch	0v (Brake On: 12v)
93	---	Not Used	---
94	LB/OR	Fuel Pump Relay Control	Relay Off: 12v, On: 1v
95	BR/OR	Fuel Delivery Control	40-240 Hz
96	YL/LB	CID Sensor Signal	5-30 Hz
97	RD	Vehicle Power	12-14v
98	VT	PCM Signal to Relay	12v
99	---	Not Used	---
100	YL/WT	Fuel Level Indicator Signal	1.7v (1/2 full)
101	VT/OR	Glow Plug Relay Control	Off: 12v, On: 1v
102	---	Not Used	---
103	BK/WT	Power Ground	<0.1v
104	---	Not Used	---

Pin Connector Graphic

PCM 104-PIN CONNECTOR

Terminal View of 104-Pin PCM Wiring Harness Connector

Standard Colors and Abbreviations

Abbreviation	Color	Abbreviation	Color	Abbreviation	Color
BK	Black	GY	Gray	PK	Purple
BL	Blue	GN	Green	RD	Red
BR	Brown	LG	LT Green	TN	Tan
DB	Dark Blue	OR	Orange	WT	White
DG	DK Green	PK	Pink	YL	Yellow

1990 E-Series 7.5L V8 MFI VIN G (A/T) 60 Pin Connector

PCM Pin #	Wire Color	Circuit Description (60 Pin)	Value at Hot Idle
1	BK/OR	Keep Alive Power	12-14v
2, 5, 9	---	Not Used	---
3	DG/WT	VSS (+) Signal	0 Hz, at 55 mph: 125 Hz
4	DG/YL	Ignition Diagnostic Monitor	20-31 Hz
6	OR/YL	VSS (-) Signal	0 Hz, at 55 mph: 125 Hz
6	GY/BK	VSS (-) Signal	0 Hz, at 55 mph: 125 Hz
7	LG/YL	Engine Coolant Temperature Sensor	0.5-0.6v
7	GN/YL	Engine Coolant Temperature Sensor	0.5-0.6v
8	OR/BL	Fuel Pump Monitor	On: 12v, Off: 0v
8	RD	Fuel Pump Monitor	On: 12v, Off: 0v
10	BK/YL	A/C Pressure Switch Signal	A/C On: 12v, off: 0v
11	WT/BK	Air Management 2 Solenoid	On: 1v, Off: 12v
12-15, 19	---	Not Used	---
16	BK/OR	Ignition System Ground	<0.050v
17	PK/LG	Self-Test Output & MIL	MIL On: 1v, Off: 12v
17	TN/RD	Self-Test Output & MIL	MIL On: 1v, Off: 12v
18	TN/LG	Inferred Mileage Sensor	Digital Signals
20	BK	PCM Case Ground	<0.050v
21	GY/WT	IAC Solenoid Control	8.1-10.1v
22	TN/LG	Fuel Pump Control	On: 1v, Off: 12v
23	LG/BK	Knock Sensor 1 Signal	0v
24	YL/LG	Power Steering Pressure Switch	Straight: 0v, Turned: 12v
25	YL/RD	ACT Sensor Signal	1.5-2.5v
26	OR/WT	Reference Voltage	4.9-5.1v
27	BR/LG	EVP Sensor Signal	0.4v, 55 mph: 3.1v
27	DG/LG	EVP Sensor Signal	0.4v, 55 mph: 3.1v
28, 32, 34-35	---	Not Used	---
29	DG/PK	HO2S-11 (B1 S1) Signal	0.1-1.1v
30	BL/WT	A/T: Neutral Drive Switch	In 'P': 0v, Others: 5v
30	BR/WT	A/T: Neutral Drive Switch	In 'P': 0v, Others: 5v
31	GY/YL	Canister Purge Solenoid	12v, 55 mph: 1v
33	DG	VR Solenoid Control	0%, at 55 mph: 45%
36	YL/GN	Spark Output Signal	6.93v (50%)
37	RD	Vehicle Power	12-14v
38-39	---	Not Used	---
40	BK/LG	Power Ground	<0.1v
41-44	---	Not Used	---
45	DB/LG	MAP Sensor Signal	107 Hz
46	BK/WT	Analog Signal Return	<0.050v
47	DG/LG	TP Sensor Signal	1.0v
48	WT/RD	Self-Test Indicator Signal	STI On: 1v, Off: 5v
49	OR	HO2S-11 (B1 S1) Ground	<0.050v
50	---	Not Used	---
51	WT/RD	Air Management 1 Solenoid	On: 1v, Off: 12v
52-55	---	Not Used	---
56	DB	PIP Sensor Signal	6.93v (50%)
57	RD	Vehicle Power	12-14v
58	TN/OR	Injector Bank 1 (INJ 1, 4, 5, 8)	5.7-7.0 ms
59	TN/RD	Injector Bank 2 (INJ 2, 3, 6, 7)	5.7-7.0 ms
60	BK/LG	Power Ground	<0.1v

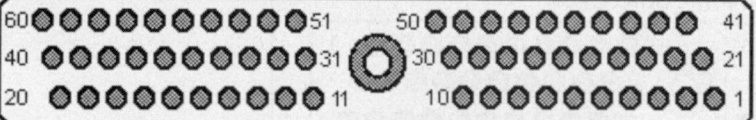

PCM 60-PIN CONNECTOR

Terminal View of 60-Pin PCM Harness Connector

1990 E-Series 7.5L V8 VIN G (E4OD) 60 Pin Connector

PCM Pin #	Wire Color	Circuit Description (60 Pin)	Value at Hot Idle
1	BK/OR	Keep Alive Power	12-14v
2	LG	Brake Switch Signal	Brake Off: 12v, On: 1v
3	DG/WT	VSS (+) Signal	0 Hz, at 55 mph: 125 Hz
4	DG/YL	Ignition Diagnostic Monitor	20-31 Hz
5, 9, 12-15	---	Not Used	---
6	OR/YL	VSS (-) Signal	0 Hz, at 55 mph: 125 Hz
7	LG/YL	Engine Coolant Temperature Sensor	0.5-0.6v
8	OR/BL	Fuel Pump Monitor	On: 12v, Off: 0v
10	BK/YL	A/C Pressure Switch Signal	A/C On: 12v, off: 0v
11	WT/BK	Air Management 2 Solenoid	On: 1v, Off: 12v
16	BK/OR	Ignition System Ground	<0.050v
17	PK/LG	Self-Test Output & MIL	MIL On: 1v, Off: 12v
17	TN/RD	Self-Test Output & MIL	MIL On: 1v, Off: 12v
18	TN/LG	Inferred Mileage Sensor	Digital Signals
19	DG/PK	Shift Solenoid 2 Control	12v, 55 mph: 1v
20	BK	PCM Case Ground	<0.050v
21	GY/WT	IAC Solenoid Control	8.1-10.1v
22	TN/LG	Fuel Pump Control	On: 1v, Off: 12v
24	YL/LG	Power Steering Pressure Switch	Straight: 0v, Turned: 12v
25	YL/RD	ACT Sensor Signal	1.5-2.5v
26	OR/WT	Reference Voltage	4.9-5.1v
27	BR/LG	EVP Sensor Signal	0.4v, 55 mph: 3.1v
28, 34-35, 39	---	Not Used	---
29	DG/PK	HO2S-11 (B1 S1) Signal	0.1-1.1v
30	BL/WT	MLP Sensor Signal	In 'P': 0v, in O/D: 5v
31	GY/YL	Canister Purge Solenoid	12v, 55 mph: 1v
32	LG/WT	OCIL (lamp) Control	Lamp On: 1v, Off: 12v
33	DG	VR Solenoid Control	0%, at 55 mph: 45%
36	YL/LG	Spark Output Signal	6.93v (50%)
37	RD	Vehicle Power	12-14v
38	BL/YL	EPC Solenoid Control	9.5v (5 psi)
40	BK/LG	Power Ground	<0.1v
41	TN/BL	Overdrive Cancel Switch	OCS Closed: 1v
42	OR/BK	TOT Sensor Signal	2.10-2.40v
43-44	---	Not Used	---
45	DB/LG	MAP Sensor Signal	107 Hz (sea level)
46	BK/WT	Analog Signal Return	<0.050v
47	DG/LG	TP Sensor Signal	0.7v, 55 mph: 2.1v
48	WT/RD	Self-Test Indicator Signal	STI On: 1v, Off: 5v
49	OR	HO2S-11 (B1 S1) Ground	<0.050v
50, 54	---	Not Used	---
51	WT/RD	Air Management 1 Solenoid	AM1 On: 1v, Off: 12v
52	OR/YL	Shift Solenoid 1 Control	1v, 55 mph: 12v
53	PK/YL	TCC Solenoid Control	0%, at 55 mph: 95%
55	BR	Coast Clutch Solenoid	12v, 55 mph: 12v
56	DB	PIP Sensor Signal	6.93v (50%)
57	RD	Vehicle Power	12-14v
58	TN/OR	Injector Bank 1 (INJ 1, 4, 5, 8)	5.7-7.0 ms
59	TN/RD	Injector Bank 2 (INJ 2, 3, 6, 7)	5.7-7.0 ms
60	BK/LG	Power Ground	<0.1v

PCM 60-PIN CONNECTOR

Terminal View of 60-Pin PCM Harness Connector

1991 E-Series 7.5L V8 VIN G (E4OD) 60 Pin Connector

PCM Pin #	Wire Color	Circuit Description (60 Pin)	Value at Hot Idle
1	BK/OR	Keep Alive Power	12-14v
2	LG	Brake Switch Signal	Brake Off: 12v, On: 1v
3	DG/WT	VSS (+) Signal	0 Hz, at 55 mph: 125 Hz
4	TN/YL	Ignition Diagnostic Monitor	20-31 Hz
5, 9, 12-15	---	Not Used	---
6	OR/YL	VSS (-) Signal	0 Hz, at 55 mph: 125 Hz
7	LG/RD	Engine Coolant Temperature Sensor	0.5-0.6v
8	DG/YL	Fuel Pump Monitor	On: 12v, Off: 0v
9	---	Not Used	---
10	BK/YL	A/C Pressure Switch Signal	A/C On: 12v, off: 0v
16	OR/RD	Ignition System Ground	<0.050v
17	TN/RD	Self-Test Output & MIL	MIL On: 1v, Off: 12v
18, 23-24	---	Not Used	---
19	PK	Shift Solenoid 2 Control	1v, 55 mph: 12v
20	BK	PCM Case Ground	<0.050v
21	GY/WT	IAC Solenoid Control	8.1-10.1v
22	TN/LG	Fuel Pump Control	On: 1v, Off: 12v
25	YL/RD	ACT Sensor Signal	1.5-2.5v
26	BR/WT	Reference Voltage	4.9-5.1v
27	BR/LG	EVP Sensor Signal	0.4v, 55 mph: 3.1v
28, 34-35	---	Not Used	---
29	RD/BK	HO2S-11 (B1 S1) Signal	0.1-1.1v
30	WT/RD	MLP Sensor Signal	In 'P': 0v, in O/D: 5v
31	GY/YL	Canister Purge Solenoid	12v, 55 mph: 1v
32	WT/LG	OCIL (lamp) Control	Lamp On: 1v, Off: 12v
33	BR/PK	VR Solenoid Control	0%, at 55 mph: 45%
36	YL/LG	Spark Output Signal	6.93v (50%)
37	RD	Vehicle Power	12-14v
38	BL/YL	EPC Solenoid Control	9.5v (5 psi)
39, 43-44	---	Not Used	---
40	BK/LG	Power Ground	<0.1v
41	TN/WT	Overdrive Cancel Switch	OCS pressed: 12v
42	OR/BK	TOT Sensor Signal	2.10-2.40v
45	DB/LG	MAP Sensor Signal	107 Hz (sea level)
46	GY/RD	Analog Signal Return	<0.050v
47	GY/WT	TP Sensor Signal	0.7v, 55 mph: 2.1v
48	WT/OR	Self-Test Indicator Signal	STI On: 1v, Off: 5v
49	OR	HO2S-11 (B1 S1) Ground	<0.050v
50, 54	---	Not Used	---
51	WT/RD	Air Management 1 Solenoid	On: 1v, Off: 12v
52	OR	Shift Solenoid 1 Control	1v, 55 mph: 12v
53	PK/YL	TCC Solenoid Control	0%, at 55 mph: 95%
55	BR/OR	Coast Clutch Solenoid	12v, 55 mph: 12v
56	DB	PIP Sensor Signal	6.93v (50%)
57	RD	Vehicle Power	12-14v
58	TN/OR	Injector Bank 1 (INJ 1, 4, 5, 8)	6.0-6.6 ms
59	TN/RD	Injector Bank 2 (INJ 2, 3, 6, 7)	6.0-6.6 ms
60	BK/LG	Power Ground	<0.1v

Pin Connector Graphic

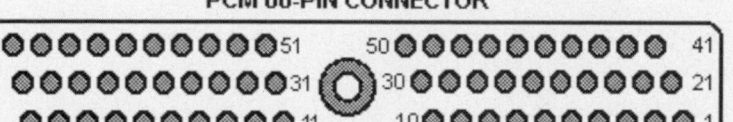

Terminal View of 60-Pin PCM Harness Connector

1992-95 E-Series 7.5L V8 VIN G (E4OD) 60 Pin Connector

PCM Pin #	Wire Color	Circuit Description (60 Pin)	Value at Hot Idle
1	YL	Keep Alive Power	12-14v
2	LG	Brake Switch Signal	Brake Off: 12v, On: 1v
3	GY/BK	VSS (+) Signal	0 Hz, at 55 mph: 125 Hz
4	TN/LG	Ignition Diagnostic Monitor	20-31 Hz
4	WT/PK	Ignition Diagnostic Monitor	20-31 Hz
5	---	Not Used	---
6	PK/OR	VSS (-) Signal	0 Hz, at 55 mph: 125 Hz
7	LG/RD	Engine Coolant Temperature Sensor	0.5-0.6v
8	DG/YL	Fuel Pump Monitor	On: 12v, Off: 0v
9	PK/BL	Data Bus (-)	Digital Signals
10	BK/YL	A/C Pressure Switch Signal	A/C On: 12v, off: 0v
11	BR	Air Management 2 Solenoid	On: 1v, Off: 12v
12-15	---	Not Used	---
16	OR/RD	Ignition System Ground	<0.050v
17	PK/LG	Self-Test Output & MIL	MIL On: 1v, Off: 12v
18	---	Not Used	---
19	PK/OR	Shift Solenoid 2 Control	1v, 55 mph: 12v
20	BK	PCM Case Ground	<0.050v
21	WT/BL	IAC Solenoid Control	8.1-10.1v
22	BL/OR	Fuel Pump Control	On: 1v, Off: 12v
23-24	---	Not Used	---
25	GY	ACT Sensor Signal	1.5-2.5v
26	BR/WT	Reference Voltage	4.9-5.1v
27	BR/LG	EVP Sensor Signal	0.4v, 55 mph: 3.1v
28	TN/OR	Data Bus (+)	Digital Signals
29	GY/BL	HO2S-11 (B1 S1) Signal	0.1-1.1v
30	BL/YL	MLP Sensor Signal	In 'P': 0v, in O/D: 5v
31	GY/YL	Canister Purge Solenoid	12v, 55 mph: 1v
32	WT/LG	OCIL (lamp) Control	Lamp On: 1v, Off: 12v
33	BR/PK	VR Solenoid Control	0%, at 55 mph: 45%
34-35	---	Not Used	---
36	PK	Spark Output Signal	6.93v (50%)
37	RD	Vehicle Power	12-14v
38	WT/YL	EPC Solenoid Control	9.5v (5 psi)
39	---	Not Used	---
40	BK/WT	Power Ground	<0.1v

1992-95 E-Series 7.5L V8 VIN G (E4OD) 60 Pin Connector

PCM Pin #	Wire Color	Circuit Description (60 Pin)	Value at Hot Idle
41	TN/WT	Overdrive Cancel Switch	OCS Closed: 0.1v
42	OR/BK	TOT Sensor Signal	2.10-2.40v
43-44	---	Not Used	---
45	LG/BK	MAP Sensor Signal	107 Hz (sea level)
46	GY/RD	Analog Signal Return	<0.050v
47	GY/WT	TP Sensor Signal	0.7v, 55 mph: 2.1v
48	WT/PK	Self-Test Indicator Signal	STI On: 1v, Off: 5v
49	OR	HO2S-11 (B1 S1) Ground	<0.050v
50	---	Not Used	---
51	WT/OR	Air Management 1 Solenoid	On: 1v, Off: 12v
52	OR/YL	Shift Solenoid 1 Control	1v, 55 mph: 12v
53	PK/YL	TCC Solenoid Control	0%, at 55 mph: 95%
54	---	Not Used	---
55	BR/OR	Coast Clutch Solenoid	12v, 55 mph: 12v
56	GY/OR	PIP Sensor Signal	6.93v (50%)
57	RD	Vehicle Power	12-14v
58	TN	Injector Bank 1 (INJ 1, 4, 5, 8)	6.0-6.6 ms
59	WT	Injector Bank 2 (INJ 2, 3, 6, 7)	6.0-6.6 ms
60	BK/WT	Power Ground	<0.1v

Pin Connector Graphic

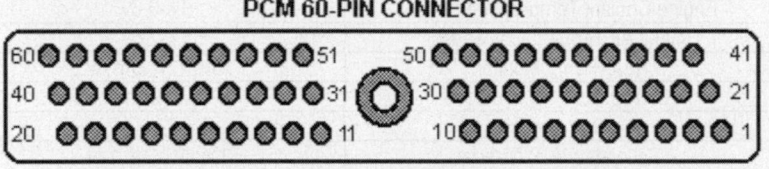

PCM 60-PIN CONNECTOR

Terminal View of 60-Pin PCM Harness Connector

Standard Colors and Abbreviations

Abbreviation	Color	Abbreviation	Color	Abbreviation	Color
BK	Black	GY	Gray	PK	Purple
BL	Blue	GN	Green	RD	Red
BR	Brown	LG	LT Green	TN	Tan
DB	Dark Blue	OR	Orange	WT	White
DG	DK Green	PK	Pink	YL	Yellow

1996 E-Series 7.5L V8 VIN G (E4OD) 104 Pin Connector

PCM Pin #	Wire Color	Circuit Description (104 Pin)	Value at Hot Idle
1	PK/OR	Shift Solenoid 2 Control	12v, 55 mph: 12v
2	PK/LG	MIL (lamp) Control	MIL On: 1v, Off: 12v
3-12	---	Not Used	---
13	VT	Flash EPROM Power	0.1v
14	---	Not Used	---
15	PK/LB	Data Bus (-)	Digital Signals
16	TN/OR	Data Bus (+)	Digital Signals
17-22	---	Not Used	---
23	OR/RD	Ignition System Ground	<0.050v
24	BK/WT	Power Ground	<0.1v
25	BK	Case Ground	<0.050v
26	---	Not Used	---
27	OR/YL	Shift Solenoid 1 Control	1v, 55 mph: 12v
28	---	Not Used	---
29	TN/WT	TCS (switch) Signal	TCS & O/D On: 12v
30-32	---	Not Used	---
33	PK/OR	PSOM (-) Signal	<0.050v
34	---	Not Used	---
35	RD/LG	HO2S-12 (B1 S2) Signal	0.1-1.1v
36	TN/LB	MAF Sensor Return	<0.050v
37	OR/BK	Transmission Fluid Temperature Sensor	2.10-2.40v
38	LG/RD	Engine Coolant Temperature Sensor	0.5-0.6v
39	GY	Intake Air Temperature Sensor	1.5-2.5v
40	DG/YL	Fuel Pump Monitor	On: 12v, Off: 0v
41	DG/OR	A/C Pressure Switch Signal	AC On: 12v, Off: 0v
42-43	---	Not Used	---
44	BR	Secondary AIR Diverter	On: 1v, Off: 12v
45-46	---	Not Used	---
47	BK/PK	VR Solenoid Control	0%, at 55 mph: 45%
48	WT/PK	Clean Tachometer Output	38-42 Hz
49	GY/OR	PIP Sensor Signal	6.93v (50%)
50	PK	Spark Output Signal	6.93v (50%)
51	BK/WT	Power Ground	<0.1v
52	---	Not Used	---
53	YL/BK	Coast Clutch Solenoid	12v, 55 mph: 12v
54	PK/YL	TCC Solenoid Control	0%, at 55 mph: 95%
55	Y	Keep Alive Power	12-14v
56	LG/BK	EVAP VMV Solenoid	0-10 Hz (0-100%)
57	---	Not Used	---
58	GY/BK	PSOM (+) Signal	0 Hz, at 55 mph: 125 Hz
59	DG/LG	Misfire Detection Sensor	45-55 Hz
60	GY/LB	HO2S-11 (B1 S1) Signal	0.1-1.1v
61-63	---	Not Used	---
64	LB/YL	Digital Transmission Range Sensor	In 'P': 0v, in O/D: 1.7v

1996 E-Series 7.5L V8 VIN G (E4OD) 104 Pin Connector

PCM Pin #	Wire Color	Circuit Description (104 Pin)	Value at Hot Idle
65	BR/LG	EGR DPFE Sensor Signal	1v, 55 mph: 1v
66-69	---	Not Used	---
70	WT/OR	Secondary Air Bypass Solenoid	AIRB On: 1v, Off: 12v
71	RD	Vehicle Power (Start-Run)	12-14v
72	TN/RD	Injector 7 Control	4.3-4.6 ms
73	TN/BK	Injector 5 Control	4.3-4.6 ms
74	BR/YL	Injector 3 Control	4.3-4.6 ms
75	TN	Injector 1 Control	4.3-4.6 ms
76	BK/WT	Power Ground	<0.1v
77	BK/WT	Power Ground	<0.1v
78	---	Not Used	---
79	WT/LG	TCIL (lamp) Control	Lamp On: 1v, Off: 12v
80	LB/OR	Fuel Pump Control	On: 1v, Off: 12v
81	WT/YL	EPC Solenoid Control	10v (26 psi)
82	---	Not Used	---
83	WT/LB	IAC Solenoid Control	10.7v (33%)
84-86	---	Not Used	---
87	RD/BK	HO2S-21 (B2 S1) Signal	0.1-1.1v
88	LB/RD	MAF Sensor Signal	0.8v, 55 mph: 2.3v
89	GY/WT	TP Sensor Signal	0.9v, 55 mph: 1.2v
90	BR/WT	Reference Voltage	4.9-5.1v
91	GY/RD	Analog Signal Return	<0.050v
92	LG	Brake Position Switch	Brake Off: 12v, On: 1v
93	RD/WT	HO2S-11 (B1 S1) Heater	1v, Off: 12v
94	YL/LB	HO2S-21 (B2 S1) Heater	1v, Off: 12v
95	WT/BK	HO2S-12 (B1 S2) Heater	1v, Off: 12v
96	---	Not Used	---
97	RD	Vehicle Power	12-14v
98	LG	Injector 8 Control	4.3-4.6 ms
99	LG/OR	Injector 6 Control	4.3-4.6 ms
100	BR/LB	Injector 4 Control	4.3-4.6 ms
101	W	Injector 2 Control	4.3-4.6 ms
102	---	Not Used	---
103	BK/WT	Power Ground	<0.1v
104	---	Not Used	---

Pin Connector Graphic

```
PCM 104-PIN CONNECTOR

 1 ●●●●●●●●●●●●●    ●●●●●●●●●●●●● 26
27 ●●●●●●●●●●●●●    ●●●●●●●●●●●●● 52
53 ●●●●●●●●●●●●●  ⬤ ●●●●●●●●●●●●● 78
79 ●●●●●●●●●●●●●    ●●●●●●●●●●●●● 104
```

Terminal View of 104-Pin PCM Wiring Harness Connector

Standard Colors and Abbreviations

Abbreviation	Color	Abbreviation	Color	Abbreviation	Color
BK	Black	GY	Gray	PK	Purple
BL	Blue	GN	Green	RD	Red
BR	Brown	LG	LT Green	TN	Tan
DB	Dark Blue	OR	Orange	WT	White
DG	DK Green	PK	Pink	YL	Yellow

VILLAGER PIN TABLES

1993-95 Villager 3.0L V6 VIN W (A/T) 116-Pin Connector

PCM Pin #	Wire Color	Circuit Description (116-Pin)	Value at Hot Idle
1	BL	Ignition Control Signal	2000 rpm: 1.2-1.3v
2	GN/WT	Tachometer Output Signal	2000 rpm: 1.2-1.3v
3	RD	Ignition Check Signal	9.0-12.0v
4	WT/GN	PCM Relay Control	Key off: 12-14v for 2 sec.
5	OR/BK	Fuel Flow Signal to IPC	Digital Signals
6	BK	Power Ground	<0.1v
7	YL/RD	Data Link Connector	5v
8	WT/PK	EGR Temperature Sensor	0.4-5v, 55 mph: 0.2-1v
9	BL/OR	Low Speed Fan Control	Fan On: 1v, Off: 12v
10	BR/WT	High Speed Fan Control	Fan On: 1v, Off: 12v
11	GY/RD	A/C Relay Control	On: 1v, Off: 12v
12	PK	Power Steering Pressure Switch	Straight: 5v, Turned: 0v
13	BK	Power Ground	<0.1v
14	YL/BL	Data Link Connector Signal	5v
15	YL/BK	Data Link Connector Signal	5v
16	WT/BL	MAF Sensor Signal	1.7v, 55 mph: 2.8v
17	OR/BL	MAF Sensor Ground	<0.050v
18	LG/RD	Engine Coolant Temperature Sensor	0.5-0.6v
19	LG/BK	HO2S-11 (B1 S1) Signal	0.1-1.1v
20	RD	TP Sensor Signal	0.4-0.1v
21	WT/BK	Analog Signal Return	<0.050v
22	RD/WT	Camshaft Position Sensor	0.2-0.4v
23	BL/WT	Diagnostic Test Connector	5v
24	GY/RD	MIL (lamp) Control	Lamp On: 1v, Off: 12v
25-26	---	Not Used	---
27	WT	Knock Sensor 1 Signal	0v
28	RD/GN	TP Sensor Signal to TCM	0.4-0.1v
29	WT/BK	Analog Signal Return	<0.050v
30	RD/WT	CKP Sensor Reference	0.2-0.4v
31	RD/BL	CKP Sensor Signal	2.5-2.7v
32	GN/YL	Speedometer Signal to PCM	Moving: Digital Signals
33	BR/YL	IDL Switch (-) Signal	Throttle Closed: 8-10v
34	BL/BK	Vehicle Start Signal	KOEC: 9-11v
35	GN/BK	PNP Signal to TCM	In 'P': 0v, Others: 5v
36	BL/YL	Ignition Power	12-14v
37	BR	Reference Voltage	4.9-5.1v
38	RD/BK	Vehicle Power	12-14v
39	BK/RD	Power Ground	<0.1v
40	RD/BL	CKP Sensor Signal	2.5-2.7v

1993-95 Villager 3.0L V6 VIN W (A/T) 116-Pin Connector

PCM Pin #	Wire Color	Circuit Description (116-Pin)	Value at Hot Idle
41	WT/RD	A/C High Pressure Switch	A/C On: 2.5v, Off: 12-14v
42-43	---	Not Used	---
44	RD/YL	IDL Switch (+) Signal	Closed: 8-10v, Open: 0v
45	---	Not Used	---
46	YL	Keep Alive Power	12-14v
47	RD/BK	Vehicle Power	12-14v
48	BK/RD	Power Ground	<0.1v
101	GN/OR	Injector 1 Control	4.2-4.6 ms
102	GY	EGR Control Solenoid	0.7v, 2000 rpm: 12v
103	GN/RD	Injector 3 Control	4.2-4.6 ms
104	BL/RD	Fuel Pump Control	On: 1v, Off: 12v
105	YL/GN	Injector 5 Control	4.2-4.6 ms
106 ('94-'95)	BK	HO2S-11 (B1 S1) Signal	0.1-1.1v
107	BK	Power Ground	<0.1v
108	BK	Power Ground	<0.1v
109	RD/BK	Vehicle Power	12-14v
110	GN	Injector 2 Control	4.2-4.6 ms
111	---	Not Used	---
112	YL/PK	Injector 4 Control	4.2-4.6 ms
113	LB	Auxiliary Air Valve Control	8-11v, with load: 4-7v
114	GY/BL	Injector 6 Control	4.2-4.6 ms
115	---	Not Used	---
116	BK	Power Ground	<0.1v

Pin Connector Graphic

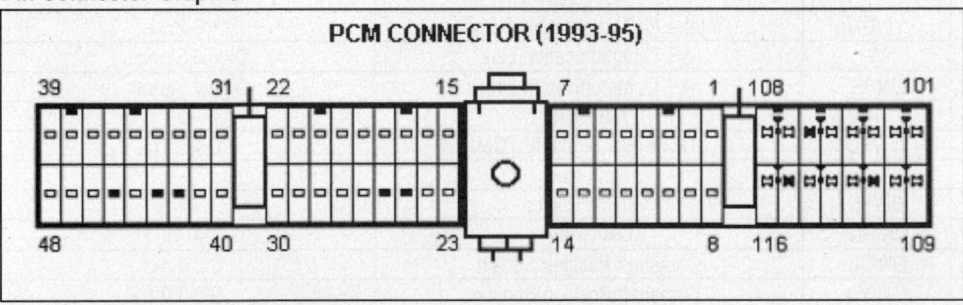

PCM CONNECTOR (1993-95)

Standard Colors and Abbreviations

Abbreviation	Color	Abbreviation	Color	Abbreviation	Color
BK	Black	GY	Gray	PK	Purple
BL	Blue	GN	Green	RD	Red
BR	Brown	LG	LT Green	TN	Tan
DB	Dark Blue	OR	Orange	WT	White
DG	DK Green	PK	Pink	YL	Yellow

1996-98 Villager 3.0L V6 VIN W, 1 (A/T) 88-Pin Connector

PCM Pin #	Wire Color	Circuit Description (88-Pin)	Value at Hot Idle
1	BL	Ignition Control Signal	0.4-0.6v
2	RD	Ignition Check Signal	9.0-9.5v
3	GN/WT	Tachometer Output Signal	1.0v
4	WT/GN	Main Power Relay Control	On: 1v, Off: 12v
5-6	---	Not Used	---
7	BR/BK	TCM Communication Line	Digital Signals
8	BL/RD	Fuel Pump Control	On: 1v, Off: 12v
9	BL/BK	A/C High Pressure Switch	5v
10	BK	Power Ground	<0.1v
11	---	Not Used	
12	BL/WT	Self-Test Input Signal	STI On: 1v, Off: 5v
13	BR/WT	High Speed Fan 1 & 2 Control	Fan On: 1v, Off: 12v
14	BL/OR	Low Speed Fan Control	Fan On: 1v, Off: 12v
15	GY/RD	A/C Relay Control	A/C & BLMT On: 1v
16-17	---	Not Used	---
18	PK	MIL (lamp) Control	MIL On: 1v, Off: 12v
19	BK	Power Ground	<0.1v
20	BL/BK	Vehicle Start Signal	KOEC: 9-11v
21	WT/RD	A/C Dual Pressure Switch	A/C & BLMT On: 2-2.5v
22	GN/BK	PNP Switch Signal	In 'P': 0v, Others: 5v
23	RD	TP Sensor Signal	0.4-1.1v
24	GN/WT	A/T Communication Line	Digital Signals
25	PK	Power Steering Pressure Switch	Straight: 5v, Turned: 0v
26	GN/YL	Speedometer Signal to PCM	Moving: Digital Signals
27	---	Not Used	---
28	YL/GN	Intake Air Temperature Sensor	1.5-2.5v
29	W	A/T Communication Line	Digital Signals
30	GN/YL	A/T Communication Line	Digital Signals
31-32	---	Not Used	---
33	WT/RD	TP Sensor Signal to TCM	0.4-1.1v
34-37	---	Not Used	---
38	BL/YL	Vehicle Power	12-14v
39	BK/RD	Main Ground	<0.1v
40	GN/BK	Camshaft Position Sensor	0.2-0.4v
41	GN/YL	Camshaft Position Sensor	2.0-3.0v
42	---	Not Used	---
43	BK/RD	Power Ground	<0.1v
44	GN/BK	Camshaft Position Sensor	0.2-0.4v
45	GN/YL	Camshaft Position Sensor	2.0-3.0v
46	LG	HO2S-11 (B1 S1) Signal	0.1-1.1v
47	WT/BL	MAF Sensor Signal	1.6v, 55 mph: 2.8v
48	OR/BL	MAF Sensor Reference	12-14v
49	BR	Reference Voltage	4.9-5.1v
50	BK/YL	Analog Signal Return	<0.050v
51	LG/RD	Engine Coolant Temperature Sensor	0.5-0.6v
52	WT	HO2S-12 (B1 S2) Signal	0.1-1.1v
53	LG	CKP (+) Sensor	0.4v
54	WT	Knock Sensor 1 Signal	0v
55	LB	IAC or AAC Valve Control	With Load: 4-7v
56	BK/WT	Vehicle Power	12-14v
57	---	Not Used	---
58	YL/GN	OBD II DLC ISO K-Line	5v
59-60	---	Not Used	---

1996-98 Villager 3.0L V6 VIN W, 1 (A/T) 88-Pin Connector

PCM Pin #	Wire Color	Circuit Description (88-Pin)	Value at Hot Idle
61	BK/WT	Vehicle Power	12-14v
62	WT/PK	EGR Temperature Sensor	0.4-5v, 55 mph: 0.2-1v
63	---	Not Used	---
64	YL/RD	DLC Signal to Scan Tool	0v
65	YL/BK	DLC Signal to Scan Tool	0v
66-67	---	Not Used	---
68	YL/BL	DLC Signal to Scan Tool	0v
69	---	Not Used	---
70	YL	Keep Alive Power	12-14v
71	---	Not Used	---
72	GN/OR	Injector 1 Control	4.2-4.6 ms
73	GY	EGR/EVAP Solenoid Control	12v, >3200 rpm: 1v
74	GN/RD	Injector 3 Control	4.2-4.6 ms
75	---	Not Used	---
76	BK	Power Ground	<0.1v
77	GN	Injector 2 Control	4.2-4.6 ms
78	---	Not Used	---
79	YL/PK	Injector 4 Control	4.2-4.6 ms
80	---	Not Used	---
81	YL/GN	Injector 5 Control	4.2-4.6 ms
82	BK	Power Ground	<0.1v
83	BK/WT	Vehicle Power	12-14v
84	GY/BL	Injector 6 Control	4.2-4.6 ms
85	BK	HO2S-11 (B1 S1) Heater	Heater Off: 12v, On: 1v
86	YL	HO2S-12 (B1 S2) Heater	Heater Off: 12v, On: 1v
87	---	Not Used	---
88	BK	Power Ground	<0.1v

Pin Connector Graphic

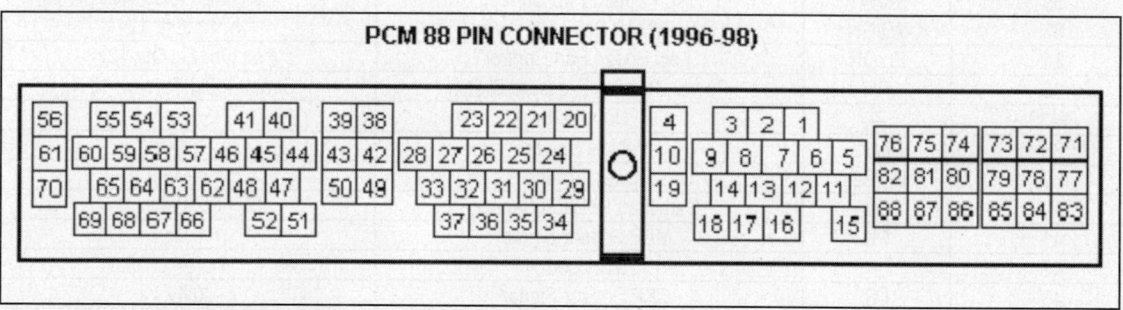

Standard Colors and Abbreviations

Abbreviation	Color	Abbreviation	Color	Abbreviation	Color
BK	Black	GY	Gray	PK	Purple
BL	Blue	GN	Green	RD	Red
BR	Brown	LG	LT Green	TN	Tan
DB	Dark Blue	OR	Orange	WT	White
DG	DK Green	PK	Pink	YL	Yellow

1999-2000 Villager 3.3L V6 SOHC VIN T 104 Pin Connector

PCM Pin #	Wire Color	Circuit Description (104 Pin)	Value at Hot Idle
1	BL	Ignition Control Signal	0.4-0.6v
2	W	IDM Sensor Signal	5-7 Hz
3	GN/WT	Tachometer Output Signal	30-35 Hz
4	WT/GN	Main Power Relay Control	On: 1v, Off: 12v
5	GN/BK	EVAP Purge Solenoid	0-10 Hz (0-100%)
6	---	Not Used	---
7	GN/PK	TCM Communication Line	Digital Signals
8	OR/BK	Instrument Cluster Signal	N/A
9	BL/BK	A/C High Pressure Switch	Any speed: 5v
10	BK	Power Ground	<0.1v
11	BL/RD	Main Relay Control	On: 1v, Off: 12v
12	LG	Self-Test Input Signal	STI On: 1v, Off: 5v
13-16	---	Not Used	---
17	BL/WT	DLC (connector) Signals	Digital Signals
18	PK	MIL (lamp) Control	MIL On: 1v, Off: 12v
19	BK	Power Ground	<0.1v
20	BL/BK	Vehicle Start Signal	KOEC: 9-11v
21	WT/RD	A/C Dual Pressure Switch	A/C & BLMT On: 2-2.5v
22	GY/BK	PNP Switch Signal	In 'P': 0v, Others: 5v
23	RD	TP Sensor Signal	0.4-1.1v
24	GN/WT	A/T Communication Line	Digital Signals
25	WT	TCM Signal (TCM Pin 6)	Digital Signals
26	BL/YL	Wide Open Throttle Signal	12v, WOT: 0v
27	GN/YL	Vehicle Speed Sensor	Digital Signals
28	---	Not Used	---
29	BL/RD	A/T Communication Line	Digital Signals
30	BL/WT	A/T Communication Line	Digital Signals
31	PK/BK	TCM Signal (TCM Pin 7)	Digital Signals
32	BR/WT	High Speed Fan Control	Fan On: 1v, Off: 12v
33	RD/GN	BARO/MAP Solenoid Control	On: 1v, Off: 12v
34	BL/OR	Low Speed Fan Control	Fan On: 1v, Off: 12v
35	PK	Power Steering Pressure Switch	Straight: 0v, Turned: 12v
36-37	---	Not Used	---
38	BL/YL	Vehicle Power	12-14v
39	BK/RD	Power Ground	<0.1v
40	WT	Camshaft Position Sensor	0.2-0.4v
41	PK	MAP Sensor Signal	0.9v
42	---	Not Used	---
43	LG	CKP Sensor Signal	0.4v
44	---	Not Used	---
45	BK/RD	Power Ground	<0.1v
46	WT	Camshaft Position Sensor	0.2-0.4v
47	OR	Camshaft Position Sensor	2.0-3.0v
48	LG	HO2S-11 (B1 S1) Signal	0.1-1.1v
49-51	BR	Reference Voltage	4.9-5.1v
52	WT/BL	MAF Sensor Signal	1.6v, 55 mph: 2.8v
53	BL/OR	MAF Sensor Return	<0.050v
54	WT	HO2S-12 (B1 S2) Heater	1v, Off: 12v

1999-2000 Villager 3.3L V6 SOHC VIN T 104 Pin Connector

PCM Pin #	Wire Color	Circuit Description (104 Pin)	Value at Hot Idle
55-56	---	Not Used	---
57	BR	Vehicle Reference	4.9-5.1v
58	BK/YL	Analog Sensor Ground	<0.050v
59	LG/RD	Engine Coolant Temperature Sensor	0.5-0.6v
60	RD/YL	Fuel Pump Control	On: 1v, Off: 12v
61	YL/GN	Intake Air Temperature Sensor	1.5-2.5v
62	RD	EVAP Pressure Sensor	2.6v at 0" Hg (cap off)
63	WT	Knock Sensor Signal	0v
64-65	---	Not Used	---
66	BK/WT	Main Relay Power to PCM	12-14v
67	---	Not Used	---
68	YL/GN	DLC Signal to Scan Tool	No Scan Tool: 0v
69-70	---	Not Used	---
71	BK/WT	Vehicle Power (Main Relay)	12-14v
72-73	---	Not Used	---
74	YL/RD	DLC (connector) Signal	Digital Signals
75	YL/RD	DLC (connector) Signal	Digital Signals
76	WT/PK	EGR Temperature Sensor	0.4-5v, 55 mph: 0.2-1v
77-79	---	Not Used	---
80	YL	Direct Battery	12-14v
81	LB	ISC Solenoid Control	7-9v
82	GN/OR	Injector 1 Control	4.0-4.6 ms
83	GY	VR Solenoid Control	0%, at 55 mph: 45%
84	GN/RD	Injector 3 Control	4.0-4.6 ms
85, 87	---	Not Used	---
86	YL/GN	Injector 5 Control	4.0-4.6 ms
88	LG/BK	EVAP CV Solenoid	0-10 Hz (0-100%)
89	GN	Injector 2 Control	4.0-4.6 ms
90	---	Not Used	---
91	YL/PK	Injector 4 Control	4.0-4.6 ms
92	---	Not Used	---
93	GY/BL	Injector 6 Control	4.0-4.6 ms
94-95	---	Not Used	---
96	BK	Power Ground	<0.1v
97	BK/WT	Vehicle Power (Main Relay)	12-14v
98	PK	BARO/MAP Solenoid Control	On: 1v, Off: 12v
99	BK	HO2S-11 (B1 S1) Heater	1v, Off: 12v
100	BL/GN	EVAP Vacuum Cut Valve	0-10 Hz (0-100%)
101	---	Not Used	---
102	YL	HO2S-12 (B1 S2) Signal	0.1-1.1v
103	---	Not Used	---
104	BK	Power Ground	<0.1v

Pin Connector Graphic

97	98	99	100		101	102	103	104
89	90	91	92		93	94	95	96
81	82	83	84		85	86	87	88

15	16	17	18	19
11	12	13	14	10
5	6	7	8	9
1	2	3		4

34	35		36	37	57	58
29	30	31	32	33	44	45
24	25	26	27	28		
20	21	22		23	38	39

59	60	61	62	76		77	78	79	80
52	53	54	55	56	72	73	74	75	71
46	47	48	49	50	51	67	68	69	70
40	41	42	43		63	64	65		88

Standard Colors and Abbreviations

Abbreviation	Color	Abbreviation	Color	Abbreviation	Color
BK	Black	GY	Gray	PK	Purple
BL	Blue	GN	Green	RD	Red
BR	Brown	LG	LT Green	TN	Tan
DB	Dark Blue	OR	Orange	WT	White
DG	DK Green	PK	Pink	YL	Yellow

2001-02 Villager 3.3L V6 SOHC VIN T 104 Pin Connector

PCM Pin #	Wire Color	Circuit Description (104 Pin)	Value at Hot Idle
1	BL	Ignition Control Signal	0.4-0.6v
2	WT	IDM Sensor Signal	5-7 Hz
3	GN/WT	Tachometer Output Signal	30-35 Hz
4	WT/GN	Main Power Relay Control	On: 1v, Off: 12v
5	GN/BK	EVAP Purge Solenoid	0-10 Hz (0-100%)
6	---	Not Used	---
7	GN/PK	TCM Communication Line	Digital Signals
8	OR/BK	Instrument Cluster Signal	N/A
9	BL/BK	A/C High Pressure Switch	Any speed: 5v
10	BK	Power Ground	<0.1v
11	BL/RD	Main Relay Control	On: 1v, Off: 12v
12	LG	Self-Test Input Signal	STI On: 1v, Off: 5v
13-16	---	Not Used	---
17	BL/WT	DLC (connector) Signals	Digital Signals
18	PK	MIL (lamp) Control	MIL On: 1v, Off: 12v
19	BK	Power Ground	<0.1v
20	BL/BK	Vehicle Start Signal	KOEC: 9-11v
21	WT/RD	A/C Dual Pressure Switch	A/C & BLMT On: 2-2.5v
22	GY/BK	PNP Switch Signal	In 'P': 0v, Others: 5v
23	RD	TP Sensor Signal	0.4-1.1v
24	GN/WT	A/T Communication Line	Digital Signals
25	WT	TCM Signal (TCM Pin 6)	Digital Signals
26	BL/YL	Wide Open Throttle Signal	12v, WOT: 0v
27	GN/YL	Vehicle Speed Sensor	Digital Signals
28	---	Not Used	---
29	BL/RD	A/T Communication Line	Digital Signals
30	BL/WT	A/T Communication Line	Digital Signals
31	PK/BK	TCM Signal (TCM Pin 7)	Digital Signals
32	BR/WT	High Speed Fan Control	Fan On: 1v, Off: 12v
33	RD/GN	BARO/MAP Solenoid Control	On: 1v, Off: 12v
34	BL/OR	Low Speed Fan Control	Fan On: 1v, Off: 12v
35	PK	Power Steering Pressure Switch	Straight: 0v, Turned: 12v
36-37	---	Not Used	---
38	BL/YL	Vehicle Power	12-14v
39	BK/RD	Power Ground	<0.1v
40	WT	Camshaft Position Sensor	0.2-0.4v
41	PK	MAP Sensor Signal	0.9v
42	---	Not Used	---
43	LG	CKP Sensor Signal	0.4v
44	---	Not Used	---
45	BK/RD	Power Ground	<0.1v
46	WT	Camshaft Position Sensor	0.2-0.4v
47	OR	Camshaft Position Sensor	2.0-3.0v
48	LG	HO2S-11 (B1 S1) Signal	0.1-1.1v
49-51	BR	Reference Voltage	4.9-5.1v
52	WT/BL	MAF Sensor Signal	1.6v, 55 mph: 2.8v
53	BL/OR	MAF Sensor Return	<0.050v
54	WT	HO2S-12 (B1 S2) Heater	1v, Off: 12v

2001-02 Villager 3.3L V6 SOHC VIN T 104 Pin Connector

PCM Pin #	Wire Color	Circuit Description (104 Pin)	Value at Hot Idle
55-56	---	Not Used	---
57	BR	Vehicle Reference	4.9-5.1v
58	BK/YL	Analog Sensor Ground	<0.050v
59	LG/RD	Engine Coolant Temperature Sensor	0.5-0.6v
60	RD/YL	Fuel Pump Control	On: 1v, Off: 12v
61	YL/GN	Intake Air Temperature Sensor	1.5-2.5v
62	RD	EVAP Pressure Sensor	2.6v at 0" Hg (cap off)
63	WT	Knock Sensor Signal	0v
64-65	---	Not Used	---
66	BK/WT	Main Relay Power to PCM	12-14v
67	---	Not Used	---
68	YL/GN	DLC Signal to Scan Tool	No Scan Tool: 0v
69-70	---	Not Used	---
71	BK/WT	Vehicle Power (Main Relay)	12-14v
72-73	---	Not Used	---
74	YL/RD	DLC (connector) Signal	Digital Signals
75	YL/RD	DLC (connector) Signal	Digital Signals
76	WT/PK	EGR Temperature Sensor	0.4-5v, 55 mph: 0.2-1v
77-79	---	Not Used	---
80	YL	Direct Battery	12-14v
81	LB	ISC Solenoid Control	7-9v
82	GN/OR	Injector 1 Control	4.0-4.6 ms
83	GY	VR Solenoid Control	0%, at 55 mph: 45%
84	GN/RD	Injector 3 Control	4.0-4.6 ms
85	---	Not Used	---
86	YL/GN	Injector 5 Control	4.0-4.6 ms
87	---	Not Used	---
88	LG/BK	EVAP CV Solenoid	0-10 Hz (0-100%)
89	GN	Injector 2 Control	4.0-4.6 ms
90	---	Not Used	---
91	YL/PK	Injector 4 Control	4.0-4.6 ms
92	---	Not Used	---
93	GY/BL	Injector 6 Control	4.0-4.6 ms
94-95	---	Not Used	---
96	BK	Power Ground	<0.1v
97	BK/WT	Vehicle Power (Main Relay)	12-14v
98	PK	BARO/MAP Solenoid Control	On: 1v, Off: 12v
99	BK	HO2S-11 (B1 S1) Heater	1v, Off: 12v
100	BL/GN	EVAP Vacuum Cut Valve	0-10 Hz (0-100%)
101	---	Not Used	---
102	YL	HO2S-12 (B1 S2) Signal	0.1-1.1v
103	---	Not Used	---
104	BK	Power Ground	<0.1v

Pin Connector Graphic

97	98	99	100		101	102	103	104		15	16	17	18		19			34	35		36	37	57	58		59	60	61	62	76		77	78	79		80
89	90	91	92		93	94	95	96		11	12	13	14		10			29	30	31	32	33	44	45		52	53	54	55	56	72	73	74	75		71
81	82	83	84		85	86	87	88		5	6	7	8	9				24	25	26	27	28				46	47	48	49	50	51	67	68	69	70	
									1	2	3			4			20	21	22		23	38	39		40	41	42	43			63	64	65		88

Standard Colors and Abbreviations

Abbreviation	Color	Abbreviation	Color	Abbreviation	Color
BK	Black	GY	Gray	PK	Purple
BL	Blue	GN	Green	RD	Red
BR	Brown	LG	LT Green	TN	Tan
DB	Dark Blue	OR	Orange	WT	White
DG	DK Green	PK	Pink	YL	Yellow

<u>WINDSTAR PIN TABLES</u>

1995-97 Windstar 3.0L V6 VIN U (A/T) 104 Pin Connector

PCM Pin #	Wire Color	Circuit Description (104 Pin)	Value at Hot Idle
1	PK/OR	Shift Solenoid 2 Control	1v, 55 mph: 12v
2	PK/LG	MIL (lamp) Control	MIL On: 1v, Off: 12v
3	---	Not Used	---
4 ('95)	RD/OR	Cooling Fan Monitor	Fan On: 12v, Off: 0v
5-9	---	Not Used	---
10 ('96-'97)	YL/BK	Coil 1 Driver (Dwell) (Cyl 1, 5)	5°, at 55 mph: 8°
11 ('96-'97)	YL/WT	Coil 2 Driver (Dwell) (Cyl 3, 4)	5°, at 55 mph: 8°
12 ('96-'97)	YL/RD	Coil 3 Driver (Dwell) (Cyl 2, 6)	5°, at 55 mph: 8°
13	YL/BK	Flash EEPROM	0.1v
14	---	Not Used	---
15	PK/LB	Data Bus (-)	Digital Signals
16	TN/OR	Data Bus (+)	Digital Signals
17	LG/PK	High Speed Fan Control	Fan On: 1v, Off: 12v
18-22	---	Not Used	---
23 ('95)	OR/RD	Ignition System Ground	<0.050v
24	BK/WT	Power Ground	<0.1v
25	BK	Case Ground	<0.050v
26	---	Not Used	---
27	OR/YL	Shift Solenoid 1 Control	12v, 55 mph: 1v
28	---	Not Used	---
29	TN/WT	TCS (switch) Signal	12v (Closed: 0v)
30	DG	Octane Adjustment	9.3v (switch shorted: 0v)
31-32	---	Not Used	---
33	OR/YL	VSS (-) Signal	0 Hz, at 55 mph: 125 Hz
34	---	Not Used	---
35	RD/LG	HO2S-12 (B1 S2) Signal	0.1-1.1v
36	TN/LB	MAF Sensor Return	<0.050v
37	OR/BK	Transmission Fluid Temperature Sensor	2.10-2.40v
38	LG/RD	Engine Coolant Temperature Sensor	0.5-0.6v
39	GY	Intake Air Temperature Sensor	1.5-2.5v
40	DG/YL	Fuel Pump Monitor	On: 12v, Off: 0v
41	BK/YL	A/C Pressure Switch Signal	A/C On: 12v, off: 0v
42	---	Not Used	---
43	OR/LG	Fuel Flow Rate Signal	Digital Signals
44-46	---	Not Used	---
47	BK/PK	VR Solenoid Control	0%, at 55 mph: 45%
48	WT/PK	Clean Tachometer Output	33-37 Hz
49 ('95)	GY/OR	Profile Ignition Pickup	6.93v (50%)
50 ('95)	PK	Spark Output Signal	6.93v (50%)
51	BK/WT	Power Ground	<0.1v
52	---	Not Used	---
53	PK/BK	Shift Solenoid 3 Control	12v, 55 mph: 1v
54	---	Not Used	---
55	YL	Keep Alive Power	12-14v
56	LG/BK	EVAP VMV Solenoid	0-10 Hz (0-100%)
57	---	Not Used	---
58	DG/WT	VSS (+) Signal	0 Hz, at 55 mph: 125 Hz
59	---	Not Used	---
60	GY/LB	HO2S-11 (B1 S1) Signal	0.1-1.1v
61	PK/LG	HO2S-22 (B2 S2) Signal	0.1-1.1v
62-63	---	Not Used	---
64	LB/YL	Transmission Range Sensor	In 'P': 0v, in O/D: 1.7v

1995-97 Windstar 3.0L V6 VIN U (A/T) 104 Pin Connector

PCM Pin #	Wire Color	Circuit Description (104 Pin)	Value at Hot Idle
65	BR/LG	EGR DPFE Sensor Signal	0.20-1.30v
66-68	---	Not Used	---
69	PK/YL	A/C WOT Relay Control	On: 1v, Off: 12v
70	---	Not Used	---
71	RD	Vehicle Power	12-14v
72	---	Not Used	---
73	TN/BK	Injector 5 Control	5.0-5.2 ms
74	BR/YL	Injector 3 Control	5.0-5.2 ms
75	TN	Injector 1 Control	5.0-5.2 ms
76	BK/WT	Power Ground	<0.1v
77	BK/WT	Power Ground	<0.1v
78	---	Not Used	---
79	WT/LG	TCIL (lamp) Control	Lamp On: 1v, Off: 12v
80	LB/OR	Fuel Pump Control	On: 1v, Off: 12v
81	WT/YL	EPC Solenoid Control	9.9v (15 psi)
82	RD/LB	TCC Solenoid Control	0%, at 55 mph: 95%
83	WT/LB	IAC Solenoid Control	10.7v (33%)
84	GY/BK	Turbine Shaft Speed Sensor	120 Hz, 30 mph: 300 Hz
85	DB/OR	Camshaft Position Sensor	6 Hz, 55 mph: 14-18 Hz
86	TN/LG	A/C Pressure Switch Signal	Switch On: 12v (Open)
87	RD/BK	HO2S-21 (B2 S1) Signal	0.1-1.1v
88	LB/RD	MAF Sensor Signal	0.8v, 55 mph: 1.6v
89	GY/WT	TP Sensor Signal	0.53-1.27v
90	BR/WT	Reference Voltage	4.9-5.1v
91	GY/RD	Analog Signal Return	<0.050v
92	RD/LG	Brake Position Switch	Brake Off: 12v, On: 1v
93	RD/WT	HO2S-11 (B1 S1) Heater	1v, Off: 12v
94	YL/LB	HO2S-21 (B2 S1) Heater	1v, Off: 12v
95	WT/BK	HO2S-12 (B1 S2) Heater	1v, Off: 12v
96	TN/YL	HO2S-22 (B2 S2) Heater	1v, Off: 12v
97	RD	Vehicle Power	12-14v
98	DB	Low Speed Fan Control	Fan On: 1v, Off: 12v
99	LG/OR	Injector 6 Control	5.0-5.2 ms
100	BR/LB	Injector 4 Control	5.0-5.2 ms
101	WT	Injector 2 Control	5.0-5.2 ms
102	---	Not Used	---
103	BK/WT	Power Ground	<0.1v
104	---	Not Used	---

Pin Connector Graphic

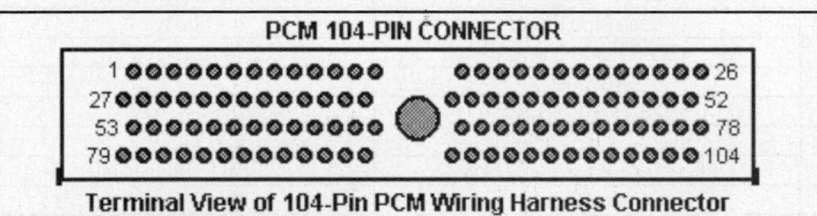

Standard Colors and Abbreviations

Abbreviation	Color	Abbreviation	Color	Abbreviation	Color
BK	Black	GY	Gray	PK	Purple
BL	Blue	GN	Green	RD	Red
BR	Brown	LG	LT Green	TN	Tan
DB	Dark Blue	OR	Orange	WT	White
DG	DK Green	PK	Pink	YL	Yellow

1998-2000 Windstar 3.0L V6 VIN U (A/T) 104 Pin Connector

PCM Pin #	Wire Color	Circuit Description (104 Pin)	Value at Hot Idle
1	PK/OR	Shift Solenoid 2/B Control	1v, 55 mph: 12v
2 ('98-'99)	PK/LG	MIL (lamp) Control	MIL On: 1v, Off: 12v
3	OR/BK	Digital TR1 Sensor	In 'P': 0v, 55 mph: 11v
4-5	---	Not Used	---
6	DB/YL	Output Shaft Speed Sensor	0 Hz, 55 mph: 2 KHz
7-12	---	Not Used	---
13	VT	Flash EEPROM	0.1v
14	---	Not Used	---
15	PK/LB	Data Bus (-)	Digital Signals
16	TN/OR	Data Bus (+)	Digital Signals
17 ('98-'99)	LG/PK	High Speed Fan Control	Fan On: 1v, Off: 12v
18-20	---	Not Used	---
21	BK/PK	CKP (+) Sensor	518-540 Hz
22	GY/YL	CKP (-) Sensor	518-540 Hz
23	---	Not Used	---
24	BK/WT	Power Ground	<0.1v
25	BK	Case Ground	<0.050v
26	DB/LG	Coil 1 Driver (Dwell)	5°, at 55 mph: 8°
27	OR/YL	Shift Solenoid 1/A Control	12v, 55 mph: 1v
28	DB	Low Speed Fan Control	Fan On: 1v, Off: 12v
29-34	---	Not Used	---
35	RD/LG	HO2S-12 (B1 S2) Signal	0.1-1.1v
36	TN/LB	MAF Sensor Return	<0.050v
37	OR/BK	Transmission Fluid Temperature Sensor	2.10-2.40v
38	LG/RD	Engine Coolant Temperature Sensor	0.5-0.6v
39	GY	Intake Air Temperature Sensor	1.5-2.5v
40	DG/YL	Fuel Pump Monitor	On: 12v, Off: 0v
41	BK/YL	A/C High Pressure Switch	A/C On: 12v, off: 0v
42	---	Not Used	---
43	OR/LG	Fuel Flow Output Signal	Digital Signals
44-45	---	Not Used	---
46	LG/PK	High Speed Fan Control	Fan On: 1v, Off: 12v
47	BR/PK	VR Solenoid Control	0%, at 55 mph: 45%
48	---	Not Used	---
49	LB/BK	Digital TR2 Sensor	In 'P': 0v, 55 mph: 11v
50	WT/BK	Digital TR4 Sensor	In 'P': 0v, 55 mph: 11v
51	BK/WT	Power Ground	<0.1v
52	RD/LB	Coil 2 Driver (Dwell)	5°, at 55 mph: 8°
53	PK/BK	Shift Solenoid 3/C Control	12v, 55 mph: 1v
54	RD/LB	TCC Solenoid Control	0%, at 55 mph: 95%
55	Y	Keep Alive Power	12-14v
56	LG/BK	EVAP Purge Solenoid	0-10 Hz (0-100%)
57-59	---	Not Used	---
60	GY/LB	HO2S-11 (B1 S1) Signal	0.1-1.1v
61	PK/LG	HO2S-22 (B2 S2) Signal	0.1-1.1v
62	RD/PK	Fuel Tank Pressure Sensor	2.6v at 0" Hg (cap off)
63	---	Not Used	---
64	LB/YL	Transmission Range Sensor	In 'P': 0v, in O/D: 1.7v

1998-2000 Windstar 3.0L V6 VIN U (A/T) 104 Pin Connector

PCM Pin #	Wire Color	Circuit Description (104 Pin)	Value at Hot Idle
65	BR/LG	EGR DPFE Sensor Signal	0.20-1.30v
66	---	Not Used	---
67	PKWT	EVAP CV Solenoid	0-10 Hz (0-100%)
68	---	Not Used	---
69	PK/YL	A/C WOT Relay Control	On: 1v, Off: 12v
70	---	Not Used	---
71	RD	Vehicle Power	12-14v
72	---	Not Used	---
73	TN/BK	Injector 5 Control	5.0-5.2 ms
74	BR/YL	Injector 3 Control	5.0-5.2 ms
75	TN	Injector 1 Control	5.0-5.2 ms
76	BK/WT	Power Ground	<0.1v
77	BK/WT	Power Ground	<0.1v
78	PKWT	Coil 3 Driver (Dwell)	5°, at 55 mph: 8°
79	---	Not Used	---
80	LB/OR	Fuel Pump Control	On: 1v, Off: 12v
81	WT/YL	EPC Solenoid Control	9.9v (15 psi)
82	---	Not Used	---
83	WT/LB	IAC Solenoid Control	10.7v (33%)
84	GY/BK	Turbine Shaft Speed Sensor	120 Hz, 30 mph: 300 Hz
85	DB/OR	Camshaft Position Sensor	6 Hz, 55 mph: 15-20 Hz
86	TN/LG	A/C Pressure Switch Signal	Switch On: 12v (Open)
87	RD/BK	HO2S-21 (B2 S1) Signal	0.1-1.1v
88	LB/RD	MAF Sensor Signal	0.8v, 55 mph: 1.9v
89	GY/WT	TP Sensor Signal	0.9v, 55 mph: 1.2v
90	BR/WT	Reference Voltage	4.9-5.1v
91	GY/RD	Analog Signal Return	<0.050v
92	---	Not Used	---
93	RD/WT	HO2S-11 (B1 S1) Heater	1v, Off: 12v
94	YL/LB	HO2S-21 (B2 S1) Heater	1v, Off: 12v
95	WT/BK	HO2S-12 (B1 S2) Heater	1v, Off: 12v
96	TN/YL	HO2S-22 (B2 S2) Heater	1v, Off: 12v
97	RD	Vehicle Power	12-14v
98	---	Not Used	---
99	LG/OR	Injector 6 Control	5.0-5.2 ms
100	BR/LB	Injector 4 Control	5.0-5.2 ms
101	WT	Injector 2 Control	5.0-5.2 ms
102	---	Not Used	---
103	BK/WT	Power Ground	<0.1v
104	---	Not Used	---

Pin Connector Graphic

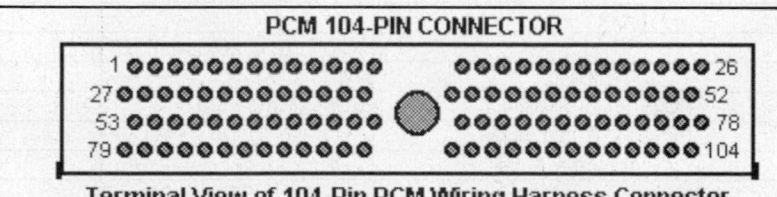

PCM 104-PIN CONNECTOR

Terminal View of 104-Pin PCM Wiring Harness Connector

Standard Colors and Abbreviations

Abbreviation	Color	Abbreviation	Color	Abbreviation	Color
BK	Black	GY	Gray	PK	Purple
BL	Blue	GN	Green	RD	Red
BR	Brown	LG	LT Green	TN	Tan
DB	Dark Blue	OR	Orange	WT	White
DG	DK Green	PK	Pink	YL	Yellow

1995 Windstar 3.8L V6 VIN 4 (A/T) 104 Pin Connector

PCM Pin #	Wire Color	Circuit Description (104 Pin)	Value at Hot Idle
1	PK/OR	Shift Solenoid 2 Control	1v, 55 mph: 12v
2	PK/LG	MIL (lamp) Control	MIL On: 1v, Off: 12v
3	---	Not Used	---
4	RD/OR	Low Speed Fan Control	Fan On: 12v, Off: 0v
5-12	---	Not Used	---
13	YL/BK	Flash EEPROM	0.1v
14	---	Not Used	---
15	PK/LB	Data Bus (-)	Digital Signals
16	TN/OR	Data Bus (+)	Digital Signals
17	LG/PK	High Speed Fan Control	Fan On: 1v, Off: 12v
18-22	---	Not Used	---
23	OR/RD	Ignition System Ground	<0.050v
24	BK/WT	Power Ground	<0.1v
25	BK	Case Ground	<0.050v
26	---	Not Used	---
27	OR/YL	Shift Solenoid 1 Control	12v, 55 mph: 1v
28	---	Not Used	---
29	TN/WT	TCS (switch) Signal	0v (Closed: 12v)
30	DG	Octane Adjustment	9.3v (switch shorted: 0v)
31-32	---	Not Used	---
33	OR/YL	VSS (-) Signal	0 Hz, at 55 mph: 125 Hz
34	---	Not Used	---
35	RD/LG	HO2S-12 (B1 S2) Signal	0.1-1.1v
36	TN/LB	MAF Sensor Return	<0.050v
37	OR/BK	Transmission Fluid Temperature Sensor	2.10-2.40v
38	LG/RD	Engine Coolant Temperature Sensor	0.5-0.6v
39	GY	Intake Air Temperature Sensor	1.5-2.5v
40	DG/YL	Fuel Pump Monitor	On: 12v, Off: 0v
41	BK/YL	A/C High Pressure Switch	A/C On: 12v, off: 0v
42	---	Not Used	---
43	OR/LG	Fuel Flow Rate Signal	Digital Signals
44-46	---	Not Used	---
47	BR/PK	VR Solenoid Control	0%, at 55 mph: 45%
48	WT/PK	Clean Tachometer Output	33-37 Hz
49	GY/OR	Profile Ignition Pickup	6.93v (50%)
50	PK	Spark Output Signal	6.93v (50%)
51	BK/WT	Power Ground	<0.1v
52	---	Not Used	---
53	PK/BK	Shift Solenoid 3 Control	12v, 55 mph: 1v
54	---	Not Used	---
55	YL	Keep Alive Power	12-14v
56	LG/BK	EVAP Purge Solenoid	0-10 Hz (0-100%)
57	---	Not Used	---
58	DG/WT	VSS (+) Signal	0 Hz, at 55 mph: 125 Hz
59	---	Not Used	---
60	GY/LB	HO2S-11 (B1 S1) Signal	0.1-1.1v
61	PK/LG	HO2S-12 (B1 S2) Signal	0.1-1.1v
62-63	---	Not Used	---
64	LB/YL	Digital Transmission Range Sensor	In 'P': 0v, in O/D: 1.7v

1995 Windstar 3.8L V6 VIN 4 (A/T) 104 Pin Connector

PCM Pin #	Wire Color	Circuit Description (104 Pin)	Value at Hot Idle
65	BR/LG	EGR DPFE Sensor Signal	0.20-1.30v
66-68	---	Not Used	---
69	PK/YL	A/C WOT Relay Control	On: 1v, Off: 12v
70	---	Not Used	---
71	RD	Vehicle Power	12-14v
72	---	Not Used	---
73	TN/BK	Injector 5 Control	3.5-3.8 ms
74	BR/YL	Injector 3 Control	3.5-3.8 ms
75	TN	Injector 1 Control	3.5-3.8 ms
76	BK/WT	Power Ground	<0.1v
77	BK/WT	Power Ground	<0.1v
78	---	Not Used	---
79	WT/LG	TCIL (lamp) Control	Lamp On: 1v, Off: 12v
80	LB/OR	Fuel Pump Control	On: 1v, Off: 12v
81	WT/YL	EPC Solenoid Control	9v (15 psi)
82	RD/LB	TCC Solenoid Control	0%, at 55 mph: 95%
83	WT/LB	IAC Solenoid Control	10.7v (33%)
84	GY/BK	Turbine Shaft Speed Sensor	120 Hz, 30 mph: 300 Hz
85	DB/OR	Camshaft Position Sensor	6 Hz, 55 mph: 14-18 Hz
86	TN/LG	A/C Pressure Switch	Switch On: 12v (Open)
87	RD/BK	HO2S-21 (B2 S1) Signal	0.1-1.1v
88	LB/RD	MAF Sensor Signal	0.8v, 55 mph: 1.6v
89	GY/WT	TP Sensor Signal	0.53-1.27v
90	BR/WT	Reference Voltage	4.9-5.1v
91	GY/RD	Analog Signal Return	<0.050v
92	RD/LG	Brake Position Switch	Brake Off: 12v, On: 1v
93	RD/WT	HO2S-11 (B1 S1) Heater	1v, Off: 12v
94	YL/LB	HO2S-21 (B2 S1) Heater	1v, Off: 12v
95	WT/BK	HO2S-12 (B1 S2) Heater	1v, Off: 12v
96	TN/YL	HO2S-22 (B2 S2) Heater	1v, Off: 12v
97	RD	Vehicle Power	12-14v
98	DB	Low Speed Fan Control	Fan On: 1v, Off: 12v
99	LG/OR	Injector 6 Control	3.5-3.8 ms
100	BK/LB	Injector 4 Control	3.5-3.8 ms
101	WT	Injector 2 Control	3.5-3.8 ms
102	---	Not Used	---
103	BK/WT	Power Ground	<0.1v
104	---	Not Used	---

Pin Connector Graphic

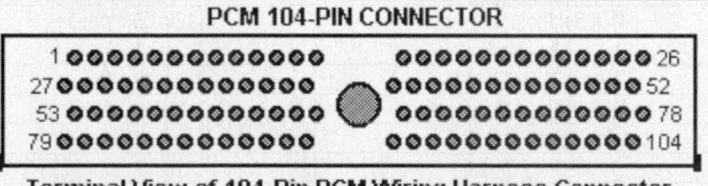

Standard Colors and Abbreviations

Abbreviation	Color	Abbreviation	Color	Abbreviation	Color
BK	Black	GY	Gray	PK	Purple
BL	Blue	GN	Green	RD	Red
BR	Brown	LG	LT Green	TN	Tan
DB	Dark Blue	OR	Orange	WT	White
DG	DK Green	PK	Pink	YL	Yellow

1996-97 Windstar 3.8L V6 VIN 4 (A/T) 104 Pin Connector

PCM Pin #	Wire Color	Circuit Description (104 Pin)	Value at Hot Idle
1	PK/OR	Shift Solenoid 2 Control	1v, 55 mph: 12v
2	PK/LG	MIL (lamp) Control	MIL On: 1v, Off: 12v
3-7	---	Not Used	---
8	OR/WT	IMRC Monitor 2	6.9v, 55 mph: 6.9v
9	TN	IMRC Monitor 1	6.9v, 55 mph: 6.9v
10-12	---	Not Used	---
13	YL/BK	Flash EEPROM	0.1v
14	---	Not Used	---
15	PK/LB	Data Bus (-)	Digital Signals
16	TN/OR	Data Bus (+)	Digital Signals
17-20	---	Not Used	---
21	DB	CKP (+) Sensor	390-450 Hz
22	GY	CKP (-) Sensor	390-450 Hz
23	---	Not Used	---
24	BK/WT	Power Ground	<0.1v
25	BK	Case Ground	<0.050v
26	DB/LG	Coil 1 Driver (Dwell)	5°, at 55 mph: 8°
27	OR/YL	Shift Solenoid 1 Control	12v, 55 mph: 1v
28	DB	Low Speed Fan Control	Fan On: 1v, Off: 12v
29	TN/WT	TCS (switch) Signal	0v (Closed: 12v)
30	DG	Octane Adjustment	9.3v (switch shorted: 0v)
31-32	---	Not Used	---
33	OR/YL	VSS (-) Signal	0 Hz, at 55 mph: 125 Hz
34	---	Not Used	---
35	RD/LG	HO2S-12 (B1 S2) Signal	0.1-1.1v
36	TN/LB	MAF Sensor Return	<0.050v
37	OR/BK	Transmission Fluid Temperature Sensor	2.10-2.40v
38	LG/RD	Engine Coolant Temperature Sensor	0.5-0.6v
39	GY	Intake Air Temperature Sensor	1.5-2.5v
40	DG/YL	Fuel Pump Monitor	On: 12v, Off: 0v
41	BK/YL	A/C High Pressure Switch	A/C On: 12v, off: 0v
42	BR	IMRC Solenoid Control	12v, 55 mph: 12v
43	OR/LG	Fuel Flow Rate Signal	Digital Signals
44-45	---	Not Used	---
46	LG/PK	High Speed Fan Control	Fan On: 1v, Off: 12v
47	BR/PK	VR Solenoid Control	0%, at 55 mph: 45%
48	WT/PK	Clean Tachometer Output	33-37 Hz
49-50	---	Not Used	---
51	BK/WT	Power Ground	<0.1v
52	RD/LB	Coil 2 Driver (Dwell)	5°, at 55 mph: 8°
53	PK/BK	Shift Solenoid 3 Control	12v, 55 mph: 1v
54	---	Not Used	---
55	YL	Keep Alive Power	12-14v
56	LG/BK	EVAP Purge Solenoid	0-10 Hz (0-100%)
57	---	Not Used	---
58	DG/WT	VSS (+) Signal	0 Hz, at 55 mph: 125 Hz
59	---	Not Used	---
60	GY/LB	HO2S-11 (B1 S1) Signal	0.1-1.1v
61	PK/LG	HO2S-12 (B1 S2) Signal	0.1-1.1v
62-63	---	Not Used	---
64	LB/YL	Digital Transmission Range Sensor	In 'P': 0v, in O/D: 1.7v

1995-97 Windstar 3.8L V6 VIN 4 (A/T) 104 Pin Connector

PCM Pin #	Wire Color	Circuit Description (104 Pin)	Value at Hot Idle
65	BR/LG	EGR DPFE Sensor Signal	0.20-1.30v
66-68	---	Not Used	---
69	PK/YL	A/C WOT Relay Control	On: 1v, Off: 12v
70	---	Not Used	---
71	RD	Vehicle Power (Start-Run)	12-14v
72	---	Not Used	---
73	TN/BK	Injector 5 Control	3.5-3.8 ms
74	BR/YL	Injector 3 Control	3.5-3.8 ms
75	TN	Injector 1 Control	3.5-3.8 ms
76	BK/WT	Power Ground	<0.1v
77	BK/WT	Power Ground	<0.1v
78	RD/LB	Coil 3 Driver (Dwell)	5°, at 55 mph: 8°
79	WT/LG	TCIL (lamp) Control	Lamp On: 1v, Off: 12v
80	LB/OR	Fuel Pump Control	On: 1v, Off: 12v
81	WT/YL	EPC Solenoid Control	9v (15 psi)
82	RD/LB	TCC Solenoid Control	0%, at 55 mph: 95%
83	WT/LB	IAC Solenoid Control	10.7v (33%)
84	GY/BK	Turbine Shaft Speed Sensor	120 Hz, 30 mph: 300 Hz
85	DB/OR	Camshaft Position Sensor	6 Hz, 55 mph: 14-18 Hz
86	TN/LG	A/C Pressure Switch	Switch On: 12v (Open)
87	RD/BK	HO2S-21 (B2 S1) Signal	0.1-1.1v
88	LB/RD	MAF Sensor Signal	0.8v, 55 mph: 1.6v
89	GY/WT	TP Sensor Signal	0.53-1.27v
90	BR/WT	Reference Voltage	4.9-5.1v
91	GY/RD	Analog Signal Return	<0.050v
92	RD/LG	Brake Position Switch	Brake Off: 12v, On: 1v
93	RD/WT	HO2S-11 (B1 S1) Heater	1v, Off: 12v
94	YL/LB	HO2S-21 (B2 S1) Heater	1v, Off: 12v
95	WT/BK	HO2S-12 (B1 S2) Heater	1v, Off: 12v
96	TN/YL	HO2S-22 (B2 S2) Heater	1v, Off: 12v
97	RD	Vehicle Power (Start-Run)	12-14v
98	---	Not Used	---
99	LG/OR	Injector 6 Control	3.5-3.8 ms
100	BK/LB	Injector 4 Control	3.5-3.8 ms
101	WT	Injector 2 Control	3.5-3.8 ms
103	BK/WT	Power Ground	<0.1v

Pin Connector Graphic

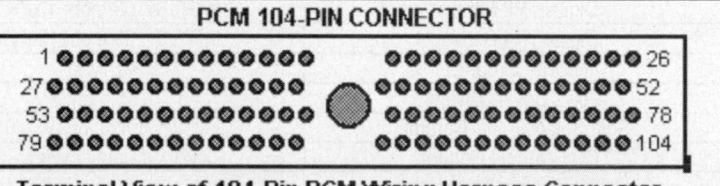

PCM 104-PIN CONNECTOR

Terminal View of 104-Pin PCM Wiring Harness Connector

Standard Colors and Abbreviations

Abbreviation	Color	Abbreviation	Color	Abbreviation	Color
BK	Black	GY	Gray	PK	Purple
BL	Blue	GN	Green	RD	Red
BR	Brown	LG	LT Green	TN	Tan
DB	Dark Blue	OR	Orange	WT	White
DG	DK Green	PK	Pink	YL	Yellow

1998-99 Windstar 3.8L V6 VIN 4 (A/T) 104 Pin Connector

PCM Pin #	Wire Color	Circuit Description (104 Pin)	Value at Hot Idle
1	PK/OR	Shift Solenoid 2 Control	1v, 55 mph: 12v
2	PK/LG	MIL (lamp) Control	MIL On: 1v, Off: 12v
3	OR/BK	Digital TR1 Sensor	In 'P': 0v, 55 mph: 10.7v
4-5	---	Not Used	---
6	DG/WT	Turbine Shaft Speed Sensor	43 Hz, 55 mph: 108 Hz
7	---	Not Used	---
8	OR/WT	IMRC Monitor Signal	5v, 55 mph: 5v
9-12	---	Not Used	---
13	VT	Flash EEPROM	0.1v
14	---	Not Used	---
15	PK/LB	Data Bus (-)	Digital Signals
16	TN/OR	Data Bus (+)	Digital Signals
19	---	Not Used	---
20	PK	Generator Field Signal	130 Hz (37%)
21	BK/PK	CKP (-) Sensor	390-450 Hz
22	GY/YL	CKP (+) Sensor	390-450 Hz
23	---	Not Used	---
24	BK/WT	Power Ground	<0.1v
25	BK	Case Ground	<0.050v
26	DB/LG	Coil 1 Driver (Dwell)	5°, at 55 mph: 8°
27	OR/YL	Shift Solenoid 1 Control	12v, 55 mph: 1v
28	DB	Low Speed Fan Control	Fan On: 1v, Off: 12v
29	---	Not Used	---
30	DB/LG	PATSIL Control	Indicator On: 1v
31	YL/LG	Power Steering Pressure Switch	Straight: 0v, Turned: 12v
32-34	---	Not Used	---
35	RD/LG	HO2S-12 (B1 S2) Signal	0.1-1.1v
36	TN/LB	MAF Sensor Return	<0.050v
37	OR/BK	Transmission Fluid Temperature Sensor	2.10-2.40v
38	LG/RD	Engine Coolant Temperature Sensor	0.5-0.6v
39	GY	Intake Air Temperature Sensor	1.5-2.5v
40	DG/YL	Fuel Pump Monitor	On: 12v, Off: 0v
41	BK/YL	A/C Pressure Switch Signal	A/C On: 12v, off: 0v
42	BR	IMRC Solenoid Control	12v, 55 mph: 12v
43-45	---	Not Used	---
46	LG/PK	High Speed Fan Control	Fan On: 1v, Off: 12v
47	BR/PK	VR Solenoid Control	0%, at 55 mph: 45%
48	---	Not Used	---
49	LG/BK	Digital TR2 Sensor	In 'P': 0v, 55 mph: 10.7v
50	WT/BK	Digital TR4 Sensor	In 'P': 0v, 55 mph: 10.7v
51	BK/WT	Power Ground	<0.1v
52	RD/LB	Coil 2 Driver (Dwell)	5°, at 55 mph: 8°
53	PK/BK	Shift Solenoid 3/C Control	12v, 55 mph: 1v
54	RD/LB	TCC Solenoid Control	12v (0% duty cycle)
55	RD	Keep Alive Power	12-14v
56	LG/BK	EVAP Purge Solenoid	0-10 Hz (0-100%)
57-59	---	Not Used	---
60	GY/LB	HO2S-11 (B1 S1) Signal	0.1-1.1v
61	PK/LG	HO2S-22 (B2 S2) Signal	0.1-1.1v
62	RD/PK	Fuel Tank Pressure Sensor	2.6v at 0" Hg (cap off)
63	---	Not Used	---
64	LB/YL	Digital TR3 Sensor Signal	In 'P': 0v, in O/D: 1.7v

1998-99 Windstar 3.8L V6 VIN 4 (A/T) 104 Pin Connector

PCM Pin #	Wire Color	Circuit Description (104 Pin)	Value at Hot Idle
65	BR/LG	EGR DPFE Sensor Signal	0.20-1.30v
66	---	Not Used	---
67	PKWT	EVAP CV Solenoid	0-10 Hz (0-100%)
68	---	Not Used	---
69	PK/YL	A/C WOT Relay Control	On: 1v, Off: 12v
70	---	Not Used	---
71	RD	Vehicle Power	12-14v
72	---	Not Used	---
73	TN/BK	Injector 5 Control	3.0-4.0 ms
74	BR/YL	Injector 3 Control	3.0-4.0 ms
75	TN	Injector 1 Control	3.0-4.0 ms
76-77	BK/WT	Power Ground	<0.1v
78	PKWT	Coil 3 Driver (Dwell)	5°, at 55 mph: 8°
79	---	Not Used	---
80	LB/OR	Fuel Pump Control	On: 1v, Off: 12v
81	WT/YL	EPC Solenoid Control	9.2v (15 psi)
82	---	Not Used	---
83	WT/LB	IAC Solenoid Control	10.7v (33%)
84	GY/BK	Output Shaft Speed Sensor	0 Hz, 55 mph: 475 Hz
85	DB/OR	Camshaft Position Sensor	6 Hz, 55 mph: 13-15 Hz
86	TN/LG	A/C Pressure Switch Signal	1v (54 psi)
87	RD/BK	HO2S-21 (B2 S1) Signal	0.1-1.1v
88	LB/RD	MAF Sensor Signal	0.8v, 55 mph: 1.8v
89	GY/WT	TP Sensor Signal	0.53-1.27v
90	BR/WT	Reference Voltage	4.9-5.1v
91	GY/RD	Analog Signal Return	<0.050v
92	---	Not Used	---
93	RD/WT	HO2S-11 (B1 S1) Heater	1v, Off: 12v
94	YL/LB	HO2S-21 (B2 S1) Heater	1v, Off: 12v
95	WT/BK	HO2S-12 (B1 S2) Heater	1v, Off: 12v
96	TN/YL	HO2S-22 (B2 S2) Heater	1v, Off: 12v
97	RD	Vehicle Power	12-14v
98	---	Not Used	---
99	LG/OR	Injector 6 Control	3.0-4.0 ms
100	BR/LB	Injector 4 Control	3.0-4.0 ms
101	WT	Injector 2 Control	3.0-4.0 ms
102	---	Not Used	---
103	BK/WT	Power Ground	<0.1v
104	---	Not Used	---

Pin Connector Graphic

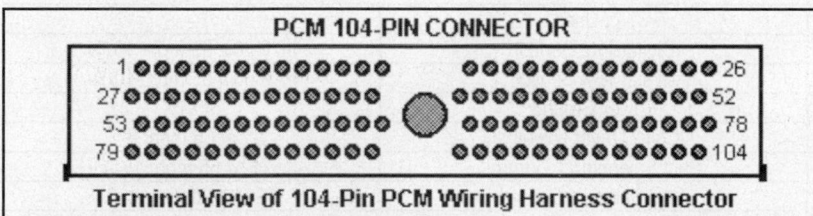

PCM 104-PIN CONNECTOR

Terminal View of 104-Pin PCM Wiring Harness Connector

Standard Colors and Abbreviations

Abbreviation	Color	Abbreviation	Color	Abbreviation	Color
BK	Black	GY	Gray	PK	Purple
BL	Blue	GN	Green	RD	Red
BR	Brown	LG	LT Green	TN	Tan
DB	Dark Blue	OR	Orange	WT	White
DG	DK Green	PK	Pink	YL	Yellow

2000 Windstar 3.8L V6 VIN 4 (A/T) 104 Pin Connector

PCM Pin #	Wire Color	Circuit Description (104 Pin)	Value at Hot Idle
1	VT/OR	Shift Solenoid 2 Control	1v, 55 mph: 12v
2	---	Not Used	---
3	OR/BK	Digital TR1 Sensor	In 'P': 0v, 55 mph: 10.7v
4-5	---	Not Used	---
6	DG/WT	Turbine Shaft Speed Sensor	43 Hz, 55 mph: 108 Hz
7	---	Not Used	---
8	OR/WT	IMRC Monitor Signal	5v, 55 mph: 5v
9-12	---	Not Used	---
13	VT	Flash EEPROM	0.1v
14	---	Not Used	---
15	PK/LB	SCP Data Bus (-)	<0.050v
16	TN/OR	SCP Data Bus (+)	Digital Signals
17	GN/OR	PATSIN Signal	12v
18	WT/LG	PATSOUT Signal	12v
19	---	Not Used	---
20	VT	Generator Field Signal	130 Hz (37%)
21	BK/PK	CKP (-) Sensor	390-450 Hz
22	GY/YL	CKP (+) Sensor	390-450 Hz
23	---	Not Used	---
24	BK/WT	Power Ground	<0.1v
25	---	Not Used	---
26	DB/LG	Coil 1 Driver (Dwell)	5°, at 55 mph: 8°
27	OR/YL	Shift Solenoid 'A' Control	12v, 55 mph: 1v
28	DB	Low Speed Fan Control	Fan On: 1v, Off: 12v
29	---	Not Used	---
30	DB/LG	Passive Antitheft Indicator Control	Indicator Off: 12v, On: 1v
31	YL/LG	Power Steering Pressure Switch	Straight: 0v, Turned: 12v
32-34	---	Not Used	---
35	RD/LG	HO2S-12 (B1 S2) Signal	0.1-1.1v
36	TN/LB	MAF Sensor Return	<0.050v
37	OR/BK	Transmission Fluid Temperature Sensor	2.10-2.40v
38	LG/RD	Engine Coolant Temperature Sensor	0.5-0.6v
39	GY	Intake Air Temperature Sensor	1.5-2.5v
40	DG/YL	Fuel Pump Monitor	Off: 0v, On: 12v
41	BK/YL	A/C Pressure Switch Signal	A/C On: 12v, off: 0v
42	BR	IMRC Solenoid Control	12v, 55 mph: 12v
43	---	Not Used	---
44	LG/YL	PATSTRT	0v
45	RD/PK	Generator Communication	0-130 Hz
46	LG/VT	High Speed Fan Relay Control	Relay Off: 12v, On: 1v
47	BR/PK	EGR VR Solenoid Control	0%, at 55 mph: 45%
48	---	Not Used	---
49	LG/BK	Digital TR2 Sensor	In 'P': 0v, 55 mph: 10.7v
50	WT/BK	Digital TR4 Sensor	In 'P': 0v, 55 mph: 10.7v
51	BK/WT	Power Ground	<0.1v
52	RD/LB	Coil 2 Driver (Dwell)	5°, at 55 mph: 8°
53	PK/BK	Shift Solenoid 'C' Control	12v, 55 mph: 1v
54	RD/LB	TCC Solenoid Control	12v (0% duty cycle)
55	RD	Direct Battery	12-14v
56	LG/BK	EVAP Canister Purge Solenoid	0-10 Hz (0-100%)
57-59	---	Not Used	---
60	GY/LB	HO2S-11 (B1 S1) Signal	0.1-1.1v
61	PK/LG	HO2S-22 (B2 S2) Signal	0.1-1.1v
62	RD/PK	Fuel Tank Pressure Sensor	2.6v at 0" Hg (cap off)
63	---	Not Used	---
64	LB/YL	Digital TR3 Sensor Signal	In 'P': 0v, in O/D: 1.7v

2000 Windstar 3.8L V6 VIN 4 (A/T) 104 Pin Connector

PCM Pin #	Wire Color	Circuit Description (104 Pin)	Value at Hot Idle
65	BR/LG	EGR DPFE Sensor Signal	0.20-1.30v
66	---	Not Used	---
67	VT/WT	EVAP CV Solenoid	0-10 Hz (0-100%)
68	---	Not Used	---
69	PK/YL	A/C WOT Relay Control	On: 1v, Off: 12v
70	---	Not Used	---
71	RD	Vehicle Power (Start-Run)	12-14v
72	---	Not Used	---
73	TN/BK	Injector 5 Control	3.0-4.0 ms
74	BR/YL	Injector 3 Control	3.0-4.0 ms
75	TN	Injector 1 Control	3.0-4.0 ms
76	BK/WT	Power Ground	<0.1v
77	---	Not Used	---
78	PKWT	Coil 3 Driver (Dwell)	5°, at 55 mph: 8°
79	---	Not Used	---
80	LB/OR	Fuel Pump Control	On: 1v, Off: 12v
81	WT/YL	EPC Solenoid Control	9.2v (15 psi)
82	---	Not Used	---
83	WT/LB	IAC Solenoid Control	10.7v (33%)
84	DB/YL	Output Shaft Speed Sensor	0 Hz, 55 mph: 475 Hz
85	DB/OR	Camshaft Position Sensor	6 Hz, 55 mph: 13-15 Hz
86	TN/LG	A/C Pressure Switch Signal	1v (54 psi)
87	RD/BK	HO2S-21 (B2 S1) Signal	0.1-1.1v
88	LB/RD	MAF Sensor Signal	0.8v, 55 mph: 1.8v
89	GY/WT	TP Sensor Signal	0.53-1.27v
90	BR/WT	Reference Voltage	4.9-5.1v
91	GY/RD	Sensor Ground	<0.050v
92	---	Not Used	---
93	RD/WT	HO2S-11 (B1 S1) Heater	1v, Off: 12v
94	YL/LB	HO2S-21 (B2 S1) Heater	1v, Off: 12v
95	WT/BK	HO2S-12 (B1 S2) Heater	1v, Off: 12v
96	TN/YL	HO2S-22 (B2 S2) Heater	1v, Off: 12v
97-98	---	Not Used	---
99	LG/OR	Injector 6 Control	3.0-4.0 ms
100	BR/LB	Injector 4 Control	3.0-4.0 ms
101	WT	Injector 2 Control	3.0-4.0 ms
102	---	Not Used	---
103	BK/WT	Power Ground	<0.1v
104	---	Not Used	---

Pin Connector Graphic

PCM 104-PIN CONNECTOR

Terminal View of 104-Pin PCM Wiring Harness Connector

Standard Colors and Abbreviations

Abbreviation	Color	Abbreviation	Color	Abbreviation	Color
BK	Black	GY	Gray	PK	Purple
BL	Blue	GN	Green	RD	Red
BR	Brown	LG	LT Green	TN	Tan
DB	Dark Blue	OR	Orange	WT	White
DG	DK Green	PK	Pink	YL	Yellow

2001-03 Windstar 3.8L V6 VIN 4 (A/T) 104 Pin Connector

PCM Pin #	Wire Color	Circuit Description (104 Pin)	Value at Hot Idle
1	VT/OR	Shift Solenoid 2 Control	1v, 55 mph: 12v
2	---	Not Used	---
3	OR/BK	Digital TR1 Sensor	In 'P': 0v, 55 mph: 10.7v
4-5	---	Not Used	---
6	DG/WT	Turbine Shaft Speed Sensor	43 Hz, 55 mph: 108 Hz
7	---	Not Used	---
8	OR/WT	IMRC Monitor Signal	5v, 55 mph: 5v
9-12	---	Not Used	---
13	VT	Flash EEPROM	0.1v
14	---	Not Used	---
15	PK/LB	SCP Data Bus (-)	<0.050v
16	TN/OR	SCP Data Bus (+)	Digital Signals
17	GY/OR	PATSIN Signal	12v
18	WT/LG	PATSOUT Signal	12v
19	---	Not Used	---
20	VT	Generator Field Signal	130 Hz (37%)
21	BK/PK	CKP (-) Sensor	390-450 Hz
22	GY/YL	CKP (+) Sensor	390-450 Hz
23	---	Not Used	---
24	BK/WT	Power Ground	<0.1v
25	---	Not Used	---
26	DB/LG	Coil 1 Driver (Dwell)	5°, at 55 mph: 8°
27	OR/YL	Shift Solenoid 'A' Control	12v, 55 mph: 1v
28	DB	Low Speed Fan Control	Fan On: 1v, Off: 12v
29	---	Not Used	---
30	DB/LG	Passive Antitheft Indicator Control	Indicator Off: 12v, On: 1v
31	YL/LG	Power Steering Pressure Switch	Straight: 0v, Turned: 12v
32-34	---	Not Used	---
35	RD/LG	HO2S-12 (B1 S2) Signal	0.1-1.1v
36	TN/LB	MAF Sensor Return	<0.050v
37	OR/BK	Transmission Fluid Temperature Sensor	2.10-2.40v
38	LG/RD	Engine Coolant Temperature Sensor	0.5-0.6v
39	GY	Intake Air Temperature Sensor	1.5-2.5v
40	DG/YL	Fuel Pump Monitor	Off: 0v, On: 12v
41	BK/YL	A/C Pressure Switch Signal	A/C On: 12v, off: 0v
42	BR	IMRC Solenoid Control	12v, 55 mph: 12v
43	---	Not Used	---
44	LG/YL	PATSTRT	0v
45	RD/PK	Generator Communication	0-130 Hz
46	LG/VT	High Speed Fan Relay Control	Relay Off: 12v, On: 1v
47	BR/PK	EGR VR Solenoid Control	0%, at 55 mph: 45%
48	---	Not Used	---
49	LG/BK	Digital TR2 Sensor	In 'P': 0v, 55 mph: 10.7v
50	WT/BK	Digital TR4 Sensor	In 'P': 0v, 55 mph: 10.7v
51	BK/WT	Power Ground	<0.1v
52	RD/LB	Coil 2 Driver (Dwell)	5°, at 55 mph: 8°
53	PK/BK	Shift Solenoid 'C' Control	12v, 55 mph: 1v
54	RD/LB	TCC Solenoid Control	12v (0% duty cycle)
55	RD	Direct Battery	12-14v
56	LG/BK	EVAP Canister Purge Solenoid	0-10 Hz (0-100%)
57-59	---	Not Used	---
60	GY/LB	HO2S-11 (B1 S1) Signal	0.1-1.1v
61	VT/LG	HO2S-22 (B2 S2) Signal	0.1-1.1v
62	RD/PK	Fuel Tank Pressure Sensor	2.6v at 0" Hg (cap off)
63	---	Not Used	---
64	LB/YL	Digital TR3 Sensor Signal	In 'P': 0v, in O/D: 1.7v

2001-03 Windstar 3.8L V6 VIN 4 (A/T) 104 Pin Connector

PCM Pin #	Wire Color	Circuit Description (104 Pin)	Value at Hot Idle
65	BR/LG	EGR DPFE Sensor Signal	0.95-1.05v
66	---	Not Used	---
67	VT/WT	EVAP CV Solenoid	0-10 Hz (0-100%)
68	---	Not Used	---
69	PK/YL	A/C WOT Relay Control	On: 1v, Off: 12v
70	---	Not Used	---
71	RD	Vehicle Power (Start-Run)	12-14v
72	---	Not Used	---
73	TN/BK	Injector 5 Control	3.0-4.0 ms
74	BR/YL	Injector 3 Control	3.0-4.0 ms
75	TN	Injector 1 Control	3.0-4.0 ms
76	BK/WT	Power Ground	<0.1v
77	---	Not Used	---
78	PKWT	Coil 3 Driver (Dwell)	5°, at 55 mph: 8°
79	---	Not Used	---
80	LB/OR	Fuel Pump Relay Control	Off: 12v, On: 1v
81	WT/YL	EPC Solenoid Control	9.2v (15 psi)
82	---	Not Used	---
83	WT/LB	IAC Solenoid Control	10.7v (33%)
84	DB/YL	Output Shaft Speed Sensor	0 Hz, 55 mph: 475 Hz
85	DB/OR	Camshaft Position Sensor	6 Hz, 55 mph: 13-15 Hz
86	TN/LG	A/C Pressure Switch Signal	1v (54 psi)
87	RD/BK	HO2S-21 (B2 S1) Signal	0.1-1.1v
88	LB/RD	MAF Sensor Signal	0.8v, 55 mph: 1.8v
89	GY/WT	TP Sensor Signal	0.53-1.27v
90	BR/WT	Reference Voltage	4.9-5.1v
91	GY/RD	Sensor Ground	<0.050v
92	---	Not Used	---
93	RD/WT	HO2S-11 (B1 S1) Heater	1v, Off: 12v
94	YL/LB	HO2S-21 (B2 S1) Heater	1v, Off: 12v
95	WT/BK	HO2S-12 (B1 S2) Heater	1v, Off: 12v
96	TN/YL	HO2S-22 (B2 S2) Heater	1v, Off: 12v
97-98	---	Not Used	---
99	LG/OR	Injector 6 Control	3.0-4.0 ms
100	BR/LB	Injector 4 Control	3.0-4.0 ms
101	WT	Injector 2 Control	3.0-4.0 ms
102	---	Not Used	---
103	BK/WT	Power Ground	<0.1v
104	---	Not Used	---

Pin Connector Graphic

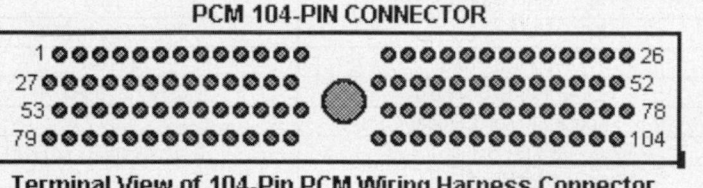

Standard Colors and Abbreviations

Abbreviation	Color	Abbreviation	Color	Abbreviation	Color
BK	Black	GY	Gray	PK	Purple
BL	Blue	GN	Green	RD	Red
BR	Brown	LG	LT Green	TN	Tan
DB	Dark Blue	OR	Orange	WT	White
DG	DK Green	PK	Pink	YL	Yellow

F-SERIES TRUCK PIN TABLES

1997-99 F-Series 4.2L V6 MFI VIN 2 (All) 104 Pin Connector

PCM Pin #	Wire Color	Circuit Description (104 Pin)	Value at Hot Idle
1	PK/OR	Shift Solenoid 2 Control	12v, 55 mph: 1v
2	PK/LG	MIL (lamp) Control	MIL Off: 12v, On: 1v
3	YL/BK	Digital TR1 Sensor	0v, 55 mph: 11.5v
4-7	---	Not Used	---
8	WT/OR	Intake Manifold Runner No. 1	55 mph: 5v or 2.5v
9	BR	Intake Manifold Runner No. 2	55 mph: 5v or 2.5v
11	---	Not Used	---
12	YL/WT	Fuel Level Indicator Signal	1.7v (1/2 full)
13	PK	Flash EEPROM Power	0.1v
14	LB/BK	4x4 Low Indicator Lamp	7.7v (Switch On: 0v)
15	PK/LB	SCP Data Bus (-) Signal	<0.050v
16	TN/OR	SCP Data Bus (+) Signal	Digital Signals
17-20	---	Not Used	---
21	DB	CKP (-) Sensor Signal	430-475 Hz
22	GY	CKP (+) Sensor Signal	430-475 Hz
23	---	Not Used	---
24	BK/WT	Power Ground	<0.1v
25	LG/YL	Case Ground	<0.050v
26	DB/LG	Coil 1 Driver (dwell)	6°, 55 mph: 9°
27	OR/YL	Shift Solenoid 1 Control	1v, 55 mph: 1v
29	TN/WT	TCS (switch) Signal	TCS & O/D On: 12v
30	DG	Octane Adjustment	Closed: 1v, Open: 9.3v
31-32	---	Not Used	---
33	PK/OR	VSS (-) Signal	0 Hz, 55 mph: 125 Hz
35	RD/LG	HO2S-12 (B1 S2) Signal	0.1-1.1v
36	TN/LB	MAF Sensor Return	<0.050v
37	OR/BK	TFT Sensor Signal	2.10-2.40v
38	LG/RD	ECT Sensor Signal	0.5-0.6v
39	GY	IAT Sensor Signal	1.5-2.5v
40	DG/YL	Fuel Pump Monitor	On: 12v, Off: 0v
41	BK/YL	A/C Switch Signal	A/C On: 12v, Off: 0v
42	BR	Intake Manifold Runner	12v, 55 mph: 12v
43-46	---	Not Used	---
47	BR/PK	VR Solenoid Control	0%, 55 mph: 45%
48	WT/PK	Clean Tachometer Output	65 Hz, 55 mph: 175 Hz
49	LB/BK	Digital TR2 Sensor	0v, 55 mph: 11.5v
50	WT/BK	Digital TR4 Sensor	0v, 55 mph: 11.5v
51	BK/WT	Power Ground	<0.1v
52	RD/LB	Coil 3 Driver (dwell)	6°, 55 mph: 9°
53	---	Not Used	---
54	PK/YL	TCC Solenoid Control	0%, 55 mph: 95%
55	BK/LG	Keep Alive Power	12-14v
56	LG/BK	EVAP Purge Solenoid	0-10 Hz (0-100%)
57	YL/RD	Knock Sensor 1 Signal	0v
58	GY/BK	VSS (+) Signal	0 Hz, 55 mph: 125 Hz
59	---	Not Used	---
60	GY/LB	HO2S-11 (B1 S1) Signal	0.1-1.1v
61	PK/LG	HO2S-22 (B2 S2) Signal	0.1-1.1v
62	RD/PK	FTP Sensor Signal	2.6v (0" H2O - cap off)
73	---	Not Used	---
64	LB/YL	A/T: Digital TR3 Sensor	0v, 55 mph: 11.5v
64	LB/YL	M/T: CPP Switch Signal	5v (clutch "in": 0v)

1997-99 F-Series 4.2L V6 MFI VIN 2 (All) 104 Pin Connector

PCM Pin #	Wire Color	Circuit Description (104 Pin)	Value at Hot Idle
65	BR/LG	DPFE Sensor Signal	0.20-1.30v
66	YL/LG	CHT Sensor Signal	0.6v (194°F)
67	PK/WT	EVAP CV Solenoid	0-10 Hz (0-100%)
69	BK/LB	A/C WOT Relay Control	Off: 12v, On: 1v
70, 72	---	Not Used	---
71	RD	Vehicle Power	12-14v
73	TN/BK	Injector 5 Control	2.7-4.1 ms
74	BR/YL	Injector 3 Control	2.7-4.1 ms
75	TN	Injector 1 Control	2.7-4.1 ms
76, 77	BK/WT	Power Ground	<0.1v
78	PK/WT	Coil 2 Driver (dwell)	6°, 55 mph: 9°
79	WT/LG	TCIL (lamp) Control	7.7v (Switch On: 0v)
80	LB/OR	Fuel Pump Control	Off: 12v, On: 1v
81	WT/YL	EPC Solenoid Control	10v (15-20 psi)
83	WT/LB	IAC Solenoid Control	10.7v (33%)
84	DB/YL	OSS Sensor Signal	0 Hz, 55 mph: 250 Hz
85	DG	CMP Sensor Signal	6 Hz, 55 mph: 15 Hz
86	---	Not Used	---
87	RD/BK	HO2S-21 (B2 S1) Signal	0.1-1.1v
88	LB/RD	MAF Sensor Signal	0.6v, 55 mph: 2.2v
89	GY/WT	TP Sensor Signal	0.53-1.27v
90	BR/WT	Reference Voltage	4.9-5.1v
91	GY/RD	Analog Signal Return	<0.050v
92	LG	Brake Pedal Switch	0v (Brake On: 12v)
93	RD/WT	HO2S-11 (B1 S1) Heater	1v (Heater Off: 12v)
94	YL/LB	HO2S-21 (B2 S1) Heater	1v (Heater Off: 12v)
95	WT/BK	HO2S-12 (B1 S2) Heater	1v (Heater Off: 12v)
96	TN/YL	HO2S-22 (B2 S2) Heater	1v (Heater Off: 12v)
97	RD	Vehicle Power	12-14v
98	---	Not Used	---
99	LG/OR	Injector 6 Control	2.7-4.1 ms
100	BR/LB	Injector 4 Control	2.7-4.1 ms
101	WT	Injector 2 Control	2.7-4.1 ms
102	---	Not Used	---
103	BK/WT	Power Ground	<0.1v
104	RD/YL	Coil 4 Driver (dwell)	6°, 55 mph: 9°

Pin Connector Graphic

PCM 104-PIN CONNECTOR

Terminal View of 104-Pin PCM Wiring Harness Connector

Standard Colors and Abbreviations

Abbreviation	Color	Abbreviation	Color	Abbreviation	Color
BK	Black	GY	Gray	PK	Purple
BL	Blue	GN	Green	RD	Red
BR	Brown	LG	LT Green	TN	Tan
DB	Dark Blue	OR	Orange	WT	White
DG	DK Green	PK	Pink	YL	Yellow

2000-02 F-Series 4.2L V6 MFI VIN 2 (All) 104 Pin Connector

PCM Pin #	Wire Color	Circuit Description (104 Pin)	Value at Hot Idle
1	VT/OR	Shift Solenoid 'B' Control	12v, 55 mph: 1v
2	---	Not Used	---
3	YL/BK	Digital TR1 Sensor	0v, 55 mph: 11.5v
4-7	---	Not Used	---
8	WT/OR	Intake Manifold Runner No. 1	55 mph: 5v or 2.5v
13	PK	Flash EEPROM Power	0.1v
14	LB/BK	4x4 Low Indicator Switch	12v (Switch On: 0v)
15	PK/LB	SCP Data Bus (-) Signal	<0.050v
16	TN/OR	SCP Data Bus (+) Signal	Digital Signals
17-18	---	Not Used	---
21	DB	CKP (-) Sensor Signal	430-475 Hz
22	GY	CKP (+) Sensor Signal	430-475 Hz
23	---	Not Used	---
24	BK/WT	Power Ground	<0.1v
25	LB/YL	Case Ground	<0.050v
26	DB/LG	Coil 1 Driver (dwell)	6°, 55 mph: 9°
27	OR/YL	Shift Solenoid 'A' Control	1v, 55 mph: 1v
29	TN/WT	TCS (switch) Signal	TCS & O/D On: 12v
30-34	---	Not Used	---
35	RD/LG	HO2S-12 (B1 S2) Signal	0.1-1.1v
36	TN/LB	MAF Sensor Return	<0.050v
37	OR/BK	TFT Sensor Signal	2.10-2.40v
38	---	Not Used	---
39	GY	IAT Sensor Signal	1.5-2.5v
40	DG/YL	Fuel Pump Monitor	On: 12v, Off: 0v
41	BK/YL	A/C Switch Signal	Switch on: 12v, off: 1v
42	BR	Intake Manifold Runner	12v, 55 mph: 12v
43-46	---	Not Used	---
47	BR/PK	VR Solenoid Control	0%, 55 mph: 45%
48	---	Not Used	---
49	LB/BK	Digital TR2 Sensor	0v, 55 mph: 11.5v
50	WT/BK	Digital TR4 Sensor	0v, 55 mph: 11.5v
51	BK/WT	Power Ground	<0.1v
52	RD/LB	Coil 3 Driver (dwell)	6°, 55 mph: 9°
53	---	Not Used	---
54	VT/YL	TCC Solenoid Control	0%, 55 mph: 95%
55	RD/WT	Keep Alive Power	12-14v
56	LG/BK	EVAP Purge Solenoid	0-10 Hz (0-100%)
57	YL/RD	Knock Sensor 1 Signal	0v
58	PK	Transfer Case Speed Sensor	0 Hz, 55 mph: 471 Hz
60	GY/LB	HO2S-11 (B1 S1) Signal	0.1-1.1v
61	VT/LG	HO2S-22 (B2 S2) Signal	0.1-1.1v
62	RD/PK	FTP Sensor Signal	2.6v (0" H2O - cap off)
64	LB/YL	A/T: Digital TR3 Sensor	0v, 55 mph: 11.5v
64	LB/YL	M/T: CPP Switch Signal	5v (clutch "in": 0v)

2000-02 F-SERIES 4.2L V6 MFI VIN 2 (ALL) 104 Pin Connector

PCM Pin #	Wire Color	Circuit Description (104 Pin)	Value at Hot Idle
65	BR/LG	DPFE Sensor Signal	0.20-1.30v
66	YL/LG	CHT Sensor Signal	0.6v (194°F)
67	VT/WT	EVAP CV Solenoid	0-10 Hz (0-100%)
68	GY/BK	VSS (+) Signal	0 Hz, 55 mph: 125 Hz
69	PK/YL	A/C WOT Relay Control	Off: 12v, On: 1v
71	RD	Vehicle Power	12-14v
72	---	Not Used	---
73	TN/BK	Injector 5 Control	2.7-4.1 ms
74	BR/YL	Injector 3 Control	2.7-4.1 ms
75	TN	Injector 1 Control	2.7-4.1 ms
76	---	Not Used	---
77	BK/WT	Power Ground	<0.1v
78	PK/WT	Coil 2 Driver (dwell)	6°, 55 mph: 9°
79	WT/LG	TCIL (lamp) Control	7.7v (Switch On: 0v)
80	LB/OR	Fuel Pump Control	Off: 12v, On: 1v
81	WT/YL	EPC Solenoid Control	10v (15-20 psi)
82	---	Not Used	---
83	WT/LB	IAC Solenoid Control	10.7v (33%)
84	DB/YL	OSS Sensor Signal	0 Hz, 55 mph: 250 Hz
85	DG	CMP Sensor Signal	6 Hz, 55 mph: 15 Hz
86	---	Not Used	---
87	RD/BK	HO2S-21 (B2 S1) Signal	0.1-1.1v
88	LB/RD	MAF Sensor Signal	0.6v, 55 mph: 2.2v
89	GY/WT	TP Sensor Signal	0.53-1.27v
90	BR/WT	Reference Voltage	4.9-5.1v
91	GY/RD	Analog Signal Return	<0.050v
92	RD/LG	Brake Pedal Switch	0v (Brake On: 12v)
93	RD/WT	HO2S-11 (B1 S1) Heater	1v (Heater Off: 12v)
94	YL/LB	HO2S-21 (B2 S1) Heater	1v (Heater Off: 12v)
95	WT/BK	HO2S-12 (B1 S2) Heater	1v (Heater Off: 12v)
96	TN/YL	HO2S-22 (B2 S2) Heater	1v (Heater Off: 12v)
97	RD	Vehicle Power	12-14v
99	LG/OR	Injector 6 Control	2.7-4.1 ms
100	BR/LB	Injector 4 Control	2.7-4.1 ms
101	WT	Injector 2 Control	2.7-4.1 ms
102	---	Not Used	---
103	BK/WT	Power Ground	<0.1v
104	---	Not Used	---

Pin Connector Graphic

```
PCM 104-PIN CONNECTOR
1  ⊙⊙⊙⊙⊙⊙⊙⊙⊙⊙⊙⊙⊙    ⊙⊙⊙⊙⊙⊙⊙⊙⊙⊙⊙⊙⊙ 26
27 ⊙⊙⊙⊙⊙⊙⊙⊙⊙⊙⊙⊙⊙    ⊙⊙⊙⊙⊙⊙⊙⊙⊙⊙⊙⊙⊙ 52
53 ⊙⊙⊙⊙⊙⊙⊙⊙⊙⊙⊙⊙⊙  ●  ⊙⊙⊙⊙⊙⊙⊙⊙⊙⊙⊙⊙⊙ 78
79 ⊙⊙⊙⊙⊙⊙⊙⊙⊙⊙⊙⊙⊙    ⊙⊙⊙⊙⊙⊙⊙⊙⊙⊙⊙⊙⊙ 104
```

Terminal View of 104-Pin PCM Wiring Harness Connector

Standard Colors and Abbreviations

Abbreviation	Color	Abbreviation	Color	Abbreviation	Color
BK	Black	GY	Gray	PK	Purple
BL	Blue	GN	Green	RD	Red
BR	Brown	LG	LT Green	TN	Tan
DB	Dark Blue	OR	Orange	WT	White
DG	DK Green	PK	Pink	YL	Yellow

2003-05 F-Series 4.2L V6 MFI VIN 2 (All) 104 Pin Connector

PCM Pin #	Wire Color	Circuit Description (104 Pin)	Value at Hot Idle
1	VT/OR	Shift Solenoid 'B' Control	12v, 55 mph: 1v
2	---	Not Used	---
3	YL/BK	Digital TR1 Sensor	0v, 55 mph: 11.5v
4-7	---	Not Used	---
8	WT/OR	Intake Manifold Runner No. 1	55 mph: 5v or 2.5v
13	VT	Flash EEPROM Power	0.1v
14	LB/BK	4x4 Low Indicator Switch	12v (Switch On: 0v)
15	PK/LB	Data Bus (-) Signal	Digital Signals
16	TN/OR	Data Bus (+) Signal	Digital Signals
17-18	---	Not Used	---
19	OG/LG	Fuel Pump High/Low Relay Control	Relay Off: 12v, On: 1v
20	BR/OR	Coast Clutch Solenoid Control (MSOF)	12v, 55 mph: 12v
21	DB	CKP (-) Sensor Signal	430-475 Hz
22	GY	CKP (+) Sensor Signal	430-475 Hz
23	---	Not Used	---
24	BK/WT	Power Ground	<0.1v
25	LB/YL	Case Ground	<0.050v
26	DB/LG	Coil 1 Driver (dwell)	6°, 55 mph: 9°
27	OR/YL	Shift Solenoid 'A' Control	1v, 55 mph: 1v
28	---	Not Used	---
29	TN/WT	TCS (switch) Signal	TCS & O/D On: 12v
30-34	---	Not Used	---
35	RD/LG	HO2S-12 (B1 S2) Signal	0.1-1.1v
36	TN/LB	MAF Sensor Return	<0.050v
37	OR/BK	TFT Sensor Signal	2.10-2.40v
38	---	Not Used	---
39	GY	IAT Sensor 2 Signal	1.5-2.5v
40	DG/YL	Fuel Pump Monitor	Off: 12v, On: 1v
41	BK/YL	A/C Pressure Switch Signal	Switch on: 12v, off: 1v
42	BR	Intake Manifold Runner Monitor	12v, 55 mph: 12v
43-46	---	Not Used	---
47	BR/PK	VR Solenoid Control	0%, 55 mph: 45%
48	---	Not Used	---
49	LB/BK	Digital TR2 Sensor	0v, 55 mph: 11.5v
50	WT/BK	Digital TR4 Sensor	0v, 55 mph: 11.5v
51	BK/WT	Power Ground	<0.1v
52	RD/LB	Coil 3 Driver (dwell)	6°, 55 mph: 9°
53	---	Not Used	---
54	VT/YL	TCC Solenoid Control	0%, 55 mph: 95%
55	RD/WT	Keep Alive Power	12-14v
56	LG/BK	EVAP Purge Solenoid	0-10 Hz (0-100%)
57	YL/RD	Knock Sensor 1 Signal	0v
58	---	Not Used	---
59	PK	Transfer Case Speed Sensor (MSOF)	0 Hz, 55 mph: 471 Hz
60	GY/LB	HO2S-11 (B1 S1) Signal	0.1-1.1v
61	VT/LG	HO2S-22 (B2 S2) Signal	0.1-1.1v
62	RD/PK	FTP Sensor Signal	2.6v (0" H2O - cap off)
63	---	Not Used	---
64	LB/YL	A/T: Digital TR3 Sensor	0v, 55 mph: 11.5v
64	LB/YL	M/T: CPP Switch Signal	5v (clutch "in": 0v)

2003-05 F-Series 4.2L V6 MFI VIN 2 (All) 104 Pin Connector

PCM Pin #	Wire Color	Circuit Description (104 Pin)	Value at Hot Idle
65	BR/LG	DPFE Sensor Signal	0.20-1.30v
66	YL/LG	Cylinder Head Temperature Sensor	0.6v (194°F)
67	VT/WT	EVAP CV Solenoid	0-10 Hz (0-100%)
68	GY/BK	VSS (+) Signal	0 Hz, 55 mph: 125 Hz
69	PK/YL	A/C WOT Relay Control	Off: 12v, On: 1v
70	BK/WT	Check Fuel Cap Indicator Control	Indicator Off: 12v, On: 1v
71	RD	Vehicle Power	12-14v
72	---	Not Used	---
73	TN/BK	Injector 5 Control	2.7-4.1 ms
74	BR/YL	Injector 3 Control	2.7-4.1 ms
75	TN	Injector 1 Control	2.7-4.1 ms
76	---	Not Used	---
77	BK/WT	Power Ground	<0.1v
78	PK/WT	Coil 2 Driver (dwell)	6°, 55 mph: 9°
79	WT/LG	TCIL (lamp) Control	7.7v (Switch On: 0v)
80	LB/OR	Fuel Pump Control	Off: 12v, On: 1v
81	WT/YL	EPC Solenoid Control	10v (15-20 psi)
82	---	Not Used	---
83	WT/LB	IAC Solenoid Control	10.7v (33%)
84	DB/YL	OSS Sensor Signal	0 Hz, 55 mph: 250 Hz
85	DG	CMP Sensor Signal	6 Hz, 55 mph: 15 Hz
86	---	Not Used	---
87	RD/BK	HO2S-21 (B2 S1) Signal	0.1-1.1v
88	LB/RD	MAF Sensor Signal	0.6v, 55 mph: 2.2v
89	GY/WT	TP Sensor Signal	0.53-1.27v
90	BR/WT	Reference Voltage	4.9-5.1v
91	GY/RD	Sensor Return	<0.050v
92	RD/LG	Brake Pedal Switch	0v (Brake On: 12v)
93	RD/WT	HO2S-11 (B1 S1) Heater	1v (Heater Off: 12v)
94	YL/LB	HO2S-21 (B2 S1) Heater	1v (Heater Off: 12v)
95	WT/BK	HO2S-12 (B1 S2) Heater	1v (Heater Off: 12v)
96	TN/YL	HO2S-22 (B2 S2) Heater	1v (Heater Off: 12v)
97	RD	Vehicle Power	12-14v
98	---	Not Used	---
99	LG/OR	Injector 6 Control	2.7-4.1 ms
100	BR/LB	Injector 4 Control	2.7-4.1 ms
101	WT	Injector 2 Control	2.7-4.1 ms
102	---	Not Used	---
103	BK/WT	Power Ground	<0.1v
104	---	Not Used	---

Pin Connector Graphic

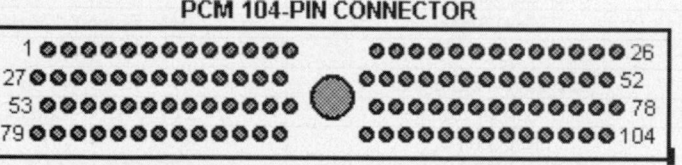

PCM 104-PIN CONNECTOR

Terminal View of 104-Pin PCM Wiring Harness Connector

Standard Colors and Abbreviations

Abbreviation	Color	Abbreviation	Color	Abbreviation	Color
BK	Black	GY	Gray	PK	Purple
BL	Blue	GN	Green	RD	Red
BR	Brown	LG	LT Green	TN	Tan
DB	Dark Blue	OR	Orange	WT	White
DG	DK Green	PK	Pink	YL	Yellow

1997-99 F-Series 4.6L V8 VIN W, VIN 6 (All) 104 Pin Connector

PCM Pin #	Wire Color	Circuit Description (104 Pin)	Value at Hot Idle
1	PK/OR	Shift Solenoid 2 Control	12v, 55 mph: 1v
2 ('97-'98)	PK/LG	MIL (lamp) Control	MIL Off: 12v, On: 1v
3	YL/BK	Digital TR1 Sensor	0v, 55 mph: 11.5v
4-11, 17-20	---	Not Used	---
12 ('97-'98)	YL/WT	Fuel Level Indicator Signal	1.7v (1/2 full)
13	VT	Flash EEPROM Power	0.1v
14	LB/BK	4x4 Indicator Switch	7.7v (Switch On: 0v)
15	PK/LB	Data Bus (-) Signal	Digital Signals
16	TN/OG	Data Bus (+) Signal	Digital Signals
21	DB	CKP (+) Sensor Signal	430-475 Hz
22	GY	CKP (-) Sensor Signal	430-475 Hz
23, 28	---	Not Used	---
24	BK/WT	Power Ground	<0.1v
25	BK/LB	Case Ground	<0.050v
26	DB/LG	Coil 1 Driver (dwell)	6°, 55 mph: 9°
27	OG/YL	Shift Solenoid 1 Control	1v, 55 mph: 12v
29	TN/WT	TCS (switch) Signal	TCS & O/D On: 12v
30 ('97-'98)	DG	Octane Adjustment	Closed: 1v, Open: 9.3v
31-32, 34	---	Not Used	---
33 ('97-'98)	PK/OR	VSS (-) Signal	0 Hz, 55 mph: 125 Hz
35	RD/LG	HO2S-12 (B1 S2) Signal	0.1-1.1v
36	TN/LB	MAF Sensor Return	<0.050v
37	OG/BK	TFT Sensor Signal	2.10-2.40v
38 ('97-'98)	LG/RD	ECT Sensor Signal	0.5-0.6v
39	GY	IAT Sensor Signal	1.5-2.5v
40	DG/YL	Fuel Pump Monitor	On: 12v, Off: 0v
41	BK/YL	A/C Switch Signal	A/C On: 12v, Off: 0v
42	BR	Inlet Air Control Valve	12v, 55 mph: 12v
43-44	---	Not Used	---
45 ('97-'98)	RD/WT	CHTIL (lamp) Control	7.7v (Switch On: 0v)
46	BR	Intake Manifold Tuning Valve	12v, 55 mph: 12v
47	BR/PK	VR Solenoid Control	0%, 55 mph: 45%
48 ('97-'98)	WT/PK	Clean Tachometer Output	39-49 Hz
49	LB/BK	Digital TR2 Sensor	0v, 55 mph: 11.5v
50	WT/BK	Digital TR4 Sensor	0v, 55 mph: 11.5v
51	BK/WT	Power Ground	<0.1v
52	RD/LB	Coil 2 Driver (dwell)	6°, 55 mph: 9°
53, 59	---	Not Used	---
54	VT/YL	TCC Solenoid Control	0%, 55 mph: 95%
55	RD/WT	Keep Alive Power	12-14v
56	LG/BK	EVAP Purge Solenoid	0-10 Hz (0-100%)
57	YL/RD	Knock Sensor 1 Signal	0v
58	PK	VSS (+) Signal	0 Hz, 55 mph: 125 Hz
60	GY/LB	HO2S-11 (B1 S1) Signal	0.1-1.1v
61	VT/LG	HO2S-22 (B2 S2) Signal	0.1-1.1v
62	RD/PK	FTP Sensor Signal	2.6v (0" H2O - cap off)
63	---	Not Used	---
64	LB/YL	A/T: Digital TR Sensor 3	0v, 55 mph: 11.5v
64	LB/YL	M/T: CPP Switch Signal	5v (clutch "in": 0v)

1997-99 F-Series 4.6L V8 VIN W, VIN 6 (All) 104 Pin Connector

PCM Pin #	Wire Color	Circuit Description (104 Pin)	Value at Hot Idle
65	BR/LG	DPFE Sensor Signal	0.20-1.30v
66	YL/LG	CHT Sensor Signal	0.6v (194°F)
67	VT/WT	EVAP CV Solenoid	0-10 Hz (0-100%)
68	GY/BK	Vehicle Speed Sensor Output	0 Hz, 55 mph: 125 Hz
69	PK/YL	A/C WOT Relay Control	12v (Relay On: 1v)
70	---	Not Used	---
71, 97	RD	Vehicle Power	12-14v
72	TN/RD	Injector 7 Control	2.7-4.1 ms
73	TN/BK	Injector 5 Control	2.7-4.1 ms
74	BR/YL	Injector 3 Control	2.7-4.1 ms
75	TN	Injector 1 Control	2.7-4.1 ms
76, 77, 103	BK/WT	Power Ground	<0.1v
78	PK/WT	Coil 3 Driver (dwell)	6°, 55 mph: 9°
79	WT/LG	TCIL (lamp) Control	12v (Lamp On: 1v)
80	LB/OR	Fuel Pump Control	Off: 12v, On: 1v
81	WT/YL	EPC Solenoid Control	9.1v (6 psi)
82, 86, 102	---	Not Used	---
83	WT/LB	IAC Solenoid Control	10.7v (33%)
84	DB/YL	OSS Sensor Signal	0 Hz, 55 mph: 250 Hz
85	DG	CMP Sensor Signal	6 Hz, 55 mph: 15 Hz
87	RD/BK	HO2S-21 (B2 S1) Signal	0.1-1.1v
88	LB/RD	MAF Sensor Signal	0.6v, 55 mph: 2.1v
89	GY/WT	TP Sensor Signal	0.53-1.27v
90	BR/WT	Reference Voltage	4.9-5.1v
91	GY/RD	Analog Signal Return	<0.050v
92	RD/LG	Brake Pedal Switch	0v (Brake On: 12v)
93	RD/WT	HO2S-11 (B1 S1) Heater	1v (Heater Off: 12v)
94	YL/LB	HO2S-21 (B2 S1) Heater	1v (Heater Off: 12v)
95	WT/BK	HO2S-12 (B1 S2) Heater	1v (Heater Off: 12v)
96	TN/YL	HO2S-22 (B2 S2) Heater	1v (Heater Off: 12v)
98	LB	Injector 8 Control	2.7-4.1 ms
99	LG/OR	Injector 6 Control	2.7-4.1 ms
100	BR/LB	Injector 4 Control	2.7-4.1 ms
101	WT	Injector 2 Control	2.7-4.1 ms
103	BK/WT	Power Ground	<0.1v
104	RD/YL	Coil 4 Driver (dwell)	6°, 55 mph: 9°

Pin Connector Graphic

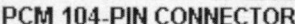

PCM 104-PIN CONNECTOR

Terminal View of 104-Pin PCM Wiring Harness Connector

Standard Colors and Abbreviations

Abbreviation	Color	Abbreviation	Color	Abbreviation	Color
BK	Black	GY	Gray	PK	Purple
BL	Blue	GN	Green	RD	Red
BR	Brown	LG	LT Green	TN	Tan
DB	Dark Blue	OR	Orange	WT	White
DG	DK Green	PK	Pink	YL	Yellow

2000-02 F-Series 4.6L V8 VIN 6, VIN W (All) 104 Pin Connector

PCM Pin #	Wire Color	Circuit Description (104 Pin)	Value at Hot Idle
1	OG/YL	COP 6 Driver (dwell)	5°, 55 mph: 8°
2	---	Not Used	---
3	BK/WT	Power Ground	<0.1v
4	PK	Transfer Case Speed Sensor	0 Hz, 55 mph:
6	OG/YL	Shift Solenoid 'A' Control	1v, 55 mph: 1v
7	PK	Transfer Case VSS Signal	0 Hz, 55 mph: 471 Hz
11	VT/OR	Shift Solenoid 'B' Control	12v, 55 mph: 1v
12	WT/LG	TCIL (lamp) Control	12v (Lamp On: 1v)
13	VT	Flash EEPROM Power	0.1v
14	LB/BK	4x4 Indicator Switch	7.7v (Switch On: 0v)
15	PK/LB	SCP Data Bus (-) Signal	<0.050v
16	TN/OG	SCP Data Bus (+) Signal	Digital Signals
21	DB	CKP (-) Sensor Signal	430-475 Hz
22	GY	CKP (+) Sensor Signal	430-475 Hz
25	LB/YL	Case Ground	<0.050v
26	LG/WT	COP 1 Driver (dwell)	5°, 55 mph: 8°
27	LG/YL	COP 5 Driver (dwell)	5°, 55 mph: 8°
29	TN/WT	TCS (switch) Signal	TCS & O/D On: 12v
32	DG/VT	Knock Sensor 1 Return	<0.050v
34	YL/BK	TR1 Sensor Signal	In 'P': 0v, in O/D: 1.7v
35	RD/LG	HO2S-12 (B1 S2) Signal	0.1-1.1v
36	TN/LB	MAF Sensor Return	<0.050v
37	OG/BK	TFT Sensor Signal	2.10-2.40v
39	GY	IAT Sensor Signal	1.5-2.5v
40	DB/YL	Fuel Pump Monitor	On: 12v, Off: 0v
41	BK/YL	A/C Switch Signal	Switch on: 12v, off: 1v
46	BR	Intake Manifold Tuning Valve	12v, 55 mph: 12v
47	BR/PK	VR Solenoid Control	0%, 55 mph: 45%
49	LB/BK	Digital TR2 Sensor	0v, 55 mph: 11.5v
50	WT/BK	Digital TR4 Sensor	0v, 55 mph: 11.5v
51	BK/WT	Power Ground	<0.1v
52	WT/PK	COP 3 Driver (dwell)	5°, 55 mph: 8°
53	DG/VT	COP 4 Driver (dwell)	5°, 55 mph: 8°
54	VT/YL	TCC Solenoid Control	0%, 55 mph: 95%
55	RD/WT	Keep Alive Power	12-14v
56	LG/BK	EVAP Purge Solenoid	0-10 Hz (0-100%)
57	YL/RD	Knock Sensor 1 Signal	0v
60	GY/LB	HO2S-11 (B1 S1) Signal	0.1-1.1v
61	VT/LG	HO2S-22 (B2 S2) Signal	0.1-1.1v
62	RD/PK	FTP Sensor Signal	2.6v (0" H2O - cap off)
64	LB/YL	A/T: Digital TR3 Sensor	0v, 55 mph: 11.5v
64	LB/YL	M/T: CPP Switch Signal	5v (clutch "in": 0v)

2000-02 F-Series 4.6L V8 VIN 6, VIN W (All) 104 Pin Connector

PCM Pin #	Wire Color	Circuit Description (104 Pin)	Value at Hot Idle
65	BR/LG	EGR DPFE Sensor Signal	0.20-1.30v
66	YL/LG	CHT Sensor Signal	0.6v (194°F)
67	VT/WT	EVAP CV Solenoid	0-10 Hz (0-100%)
68	GY/BK	Vehicle Speed Control	0v
69	PK/YL	A/C WOT Relay Control	Off: 12v, On: 1v
71, 97	RD	Vehicle Power	12-14v
72	TN/RD	Injector 7 Control	2.7-4.1 ms
73	TN/BK	Injector 5 Control	2.7-4.1 ms
74	BR/YL	Injector 3 Control	2.7-4.1 ms
75	TN	Injector 1 Control	2.7-4.1 ms
77	BK/WT	Power Ground	<0.1v
78	PK/LB	COP 7 Driver (dwell)	5°, 55 mph: 8°
79	WT/RD	COP 8 Driver (dwell)	5°, 55 mph: 8°
80	LB/OR	Fuel Pump Relay Control	Off: 12v, On: 1v
81	WT/YL	EPC Solenoid Control	9.1v (6 psi)
83	WT/LB	IAC Solenoid Control	10.7v (33%)
84	DB/YL	OSS Sensor Signal	0 Hz, 55 mph: 250 Hz
85	DG	CMP Sensor Signal	6 Hz, 55 mph: 15 Hz
87	RD/BK	HO2S-21 (B2 S1) Signal	0.1-1.1v
88	LB/RD	MAF Sensor Signal	0.6v, 55 mph: 2.1v
89	GY/WT	TP Sensor Signal	0.53-1.27v
90	BR/WT	Reference Voltage	4.9-5.1v
91	GY/RD	Analog Signal Return	<0.050v
92	RD/LG	Brake Pedal Switch	0v (Brake On: 12v)
93	RD/WT	HO2S-11 (B1 S1) Heater	1v (Heater Off: 12v)
94	YL/LB	HO2S-21 (B2 S1) Heater	1v (Heater Off: 12v)
95	WT/BK	HO2S-12 (B1 S2) Heater	1v (Heater Off: 12v)
96	TN/YL	HO2S-22 (B2 S2) Heater	1v (Heater Off: 12v)
98	LB	Injector 8 Control	2.7-4.1 ms
99	LG/OR	Injector 6 Control	2.7-4.1 ms
100	BR/LB	Injector 4 Control	2.7-4.1 ms
101	WT	Injector 2 Control	2.7-4.1 ms
103	BK/WT	Power Ground	<0.1v
104	PK/WT	COP 2 Driver (dwell)	5°, 55 mph: 8°

Pin Connector Graphic

PCM 104-PIN CONNECTOR

Terminal View of 104-Pin PCM Wiring Harness Connector

Standard Colors and Abbreviations

Abbreviation	Color	Abbreviation	Color	Abbreviation	Color
BK	Black	GY	Gray	PK	Purple
BL	Blue	GN	Green	RD	Red
BR	Brown	LG	LT Green	TN	Tan
DB	Dark Blue	OR	Orange	WT	White
DG	DK Green	PK	Pink	YL	Yellow

2003-05 F-Series 4.6L V8 VIN 6, VIN W (All) 104 Pin Connector

PCM Pin #	Wire Color	Circuit Description (104 Pin)	Value at Hot Idle
1	OG/YL	COP 6 Driver (dwell)	5°, 55 mph: 8°
2	---	Not Used	---
3	BK/WT	Power Ground	<0.1v
4	PK	Transfer Case Speed Sensor	0 Hz, 55 mph:
5	---	Not Used	---
6	OG/YL	Shift Solenoid 'A' Control	1v, 55 mph: 1v
7-10	---	Not Used	---
11	VT/OR	Shift Solenoid 'B' Control	12v, 55 mph: 1v
12	WT/LG	TCIL (lamp) Control	12v (Lamp On: 1v)
13	VT	Flash EEPROM Power	0.1v
14	LB/BK	4x4 Indicator Switch	7.7v (Switch On: 0v)
15	PK/LB	SCP Data Bus (-) Signal	<0.050v
16	TN/OG	SCP Data Bus (+) Signal	Digital Signals
17-18	---	Not Used	---
19	OG/LG	Fuel Pump High/Low Relay Control	Relay Off: 12v, On: 1v
20	BR/OR	Coast Clutch Solenoid Control (MSOF)	12v, 55 mph: 12v
21	DB	CKP (-) Sensor Signal	430-475 Hz
22	GY	CKP (+) Sensor Signal	430-475 Hz
23-24	---	Not Used	---
25	LB/YL	Case Ground	<0.050v
26	LG/WT	COP 1 Driver (dwell)	5°, 55 mph: 8°
27	LG/YL	COP 5 Driver (dwell)	5°, 55 mph: 8°
28	---	Not Used	---
29	TN/WT	TCS (switch) Signal	TCS & O/D On: 12v
30-31	---	Not Used	---
32	DG/VT	Knock Sensor 1 Return	<0.050v
33	---	Not Used	---
34	YL/BK	TR1 Sensor Signal	In 'P': 0v, in O/D: 1.7v
35	RD/LG	HO2S-12 (B1 S2) Signal	0.1-1.1v
36	TN/LB	MAF Sensor Return	<0.050v
37	OG/BK	TFT Sensor Signal	2.10-2.40v
38	---	Not Used	---
39	GY	IAT Sensor 2 Signal	1.5-2.5v
40	DB/YL	Fuel Pump Monitor	On: 12v, Off: 0v
41	BK/YL	A/C Pressure Switch Signal	Switch on: 12v, off: 1v
42-45	---	Not Used	---
46	BR	Intake Manifold Tuning Valve	12v, 55 mph: 12v
47	BR/PK	VR Solenoid Control	0%, 55 mph: 45%
48	---	Not Used	---
49	LB/BK	Digital TR2 Sensor	0v, 55 mph: 11.5v
50	WT/BK	Digital TR4 Sensor	0v, 55 mph: 11.5v
51	BK/WT	Power Ground	<0.1v
52	WT/PK	COP 3 Driver (dwell)	5°, 55 mph: 8°
53	DG/VT	COP 4 Driver (dwell)	5°, 55 mph: 8°
54	VT/YL	TCC Solenoid Control	0%, 55 mph: 95%
55	RD/WT	Keep Alive Power	12-14v
56	LG/BK	EVAP Purge Solenoid	0-10 Hz (0-100%)
57	YL/RD	Knock Sensor 1 Signal	0v
58-59	---	Not Used	---
60	GY/LB	HO2S-11 (B1 S1) Signal	0.1-1.1v
61	VT/LG	HO2S-22 (B2 S2) Signal	0.1-1.1v
62	RD/PK	FTP Sensor Signal	2.6v (0" H2O - cap off)
63	---	Not Used	---
64	LB/YL	A/T: Digital TR3 Sensor	0v, 55 mph: 11.5v
64	LB/YL	M/T: CPP Switch Signal	5v (clutch "in": 0v)

2003-05 F-Series 4.6L V8 VIN 6, VIN W (All) 104 Pin Connector

PCM Pin #	Wire Color	Circuit Description (104 Pin)	Value at Hot Idle
65	BR/LG	EGR DPFE Sensor Signal	0.20-1.30v
66	YL/LG	CHT Sensor Signal	0.6v (194ºF)
67	VT/WT	EVAP CV Solenoid	0-10 Hz (0-100%)
68	GY/BK	Vehicle Speed Control	0v
69	PK/YL	A/C WOT Relay Control	Off: 12v, On: 1v
70	BK/WT	Check Fuel Cap Indicator Control	Indicator Off: 12v, On: 1v
71	RD	Vehicle Power	12-14v
72	TN/RD	Injector 7 Control	2.7-4.1 ms
73	TN/BK	Injector 5 Control	2.7-4.1 ms
74	BR/YL	Injector 3 Control	2.7-4.1 ms
75	TN	Injector 1 Control	2.7-4.1 ms
76	---	Not Used	---
77	BK/WT	Power Ground	<0.1v
78	PK/LB	COP 7 Driver (dwell)	5º, 55 mph: 8º
79	WT/RD	COP 8 Driver (dwell)	5º, 55 mph: 8º
80	LB/OR	Fuel Pump Relay Control	Off: 12v, On: 1v
81	WT/YL	EPC Solenoid Control	9.1v (6 psi)
82, 86	---	Not Used	---
83	WT/LB	IAC Solenoid Control	10.7v (33%)
84	DB/YL	OSS Sensor Signal	0 Hz, 55 mph: 250 Hz
85	DG	CMP Sensor Signal	6 Hz, 55 mph: 15 Hz
87	RD/BK	HO2S-21 (B2 S1) Signal	0.1-1.1v
88	LB/RD	MAF Sensor Signal	0.6v, 55 mph: 2.1v
89	GY/WT	TP Sensor Signal	0.53-1.27v
90	BR/WT	Reference Voltage	4.9-5.1v
91	GY/RD	Sensor Return	<0.050v
92	RD/LG	Brake Pedal Switch	0v (Brake On: 12v)
93	RD/WT	HO2S-11 (B1 S1) Heater	1v (Heater Off: 12v)
94	YL/LB	HO2S-21 (B2 S1) Heater	1v (Heater Off: 12v)
95	WT/BK	HO2S-12 (B1 S2) Heater	1v (Heater Off: 12v)
96	TN/YL	HO2S-22 (B2 S2) Heater	1v (Heater Off: 12v)
97	RD	Vehicle Power	12-14v
98	LB	Injector 8 Control	2.7-4.1 ms
99	LG/OR	Injector 6 Control	2.7-4.1 ms
100	BR/LB	Injector 4 Control	2.7-4.1 ms
101	WT	Injector 2 Control	2.7-4.1 ms
102	---	Not Used	---
103	BK/WT	Power Ground	<0.1v
104	PK/WT	COP 2 Driver (dwell)	5º, 55 mph: 8º

Pin Connector Graphic

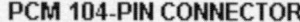

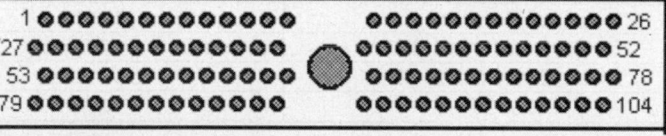

Terminal View of 104-Pin PCM Wiring Harness Connector

Standard Colors and Abbreviations

Abbreviation	Color	Abbreviation	Color	Abbreviation	Color
BK	Black	GY	Gray	PK	Purple
BL	Blue	GN	Green	RD	Red
BR	Brown	LG	LT Green	TN	Tan
DB	Dark Blue	OR	Orange	WT	White
DG	DK Green	PK	Pink	YL	Yellow

1990 F-Series 4.9L I6 MFI VIN Y (All) 60 Pin Connector

PCM Pin #	Wire Color	Circuit Description (60 Pin)	Value at Hot Idle
1	YL	Keep Alive Power	12-14v
2, 5	---	Not Used	---
3	DG/WT	VSS (+) Signal	0 Hz, 55 mph: 125 Hz
3	GY/BK	VSS (+) Signal	0 Hz, 55 mph: 125 Hz
4	DG/YL	IDM Sensor Signal	20-31 Hz
6	BK	VSS (-) Signal	0 Hz, 55 mph: 125 Hz
7	LG/YL	ECT Sensor Signal	0.5-0.6v
8	BR	Fuel Pump Monitor	On: 12v, Off: 0v
9	BK/OR	Data Bus (-) Signal	Digital Signals
10	BK/YL	A/C Switch Signal	A/C On: 12v, Off: 0v
11	WT/BK	Air Management 2 Solenoid	AM2 Off: 12v, On: 1v
16	BK/OR	Ignition System Ground	<0.050v
17	TN/RD	Self Test Output & MIL	MIL Off: 12v, On: 1v
17	PK/LG	Self Test Output & MIL	MIL Off: 12v, On: 1v
18	TN/LG	Inferred Mileage Sensor	Digital Signals
20	BK	PCM Case Ground	<0.050v
21	GY/WT	IAC Solenoid Control	8.8-10.0v
22	TN/LG	Fuel Pump Control	Off: 12v, On: 1v
23	LG/BK	Knock Sensor Signal	0v
24	YL/LG	PSP Switch Signal	0v (turning: 12v)
25	YL/RD	ACT Sensor Signal	1.5-2.5v
26	OR/WT	Reference Voltage	4.9-5.1v
27	BR/LG	EVP Sensor Signal	0.4v, 55 mph: 2.6v
28	OR/BK	Data Bus (+) Signal	Digital Signals
29	DG/PK	HO2S-11 (B1 S1) Signal	0.1-1.1v
29	GY/BL	HO2S-11 (B1 S1) Signal	0.1-1.1v
30	GY/YL	A/T: Neutral Drive Switch	In 'N': Others: 5v
30	GY/YL	M/T: CPP Switch Signal	5v (clutch "in": 0v)
31	GY/YL	Canister Purge Solenoid	12v, 55 mph: 1v
33	DG	VR Solenoid Control	0%, 55 mph: 45%
36	YL/LG	Spark Output Signal	6.93v (50%)
37	RD	Vehicle Power	12-14v
40	BK/LG	Power Ground	<0.1v
43	LG/PK	A/C Demand Switch	A/C On: 12v, Off: 0v
45	DB/LG	MAP Sensor Signal	107 Hz (sea level)
45	LG/BK	MAP Sensor Signal	107 Hz (sea level)
46	BK/WT	Analog Signal Return	<0.050v
47	DG/LG	TP Sensor Signal	1v, 55 mph: 1.4v
48	WT/RD	Self Test Indicator Signal	STI On: 1v, Off: 5v
49	OR	HO2S-11 (B1 S1) Ground	<0.050v
51	WT/RD	Air Management 1 Solenoid	AM1 Off: 12v, On: 1v
56	DB	PIP Sensor Signal	6.93v (50%)
57	RD	Vehicle Power	12-14v
58	TN/OR	Injector Bank 1 (INJ 1, 3 & 5)	6.4-6.8 ms
59	TN/RD	Injector Bank 2 (INJ 2, 4 & 6)	6.4-6.8 ms
60	BK/LG	Power Ground	<0.1v

Pin Connector Graphic

Terminal View of 60-Pin PCM Harness Connector

1990 F-Series 4.9L I6 MFI VIN Y (E4OD) 60 Pin Connector

PCM Pin #	Wire Color	Circuit Description (60 Pin)	Value at Hot Idle
1	YL	Keep Alive Power	12-14v
2	LG	Brake On/Off Switch	0v (Brake On: 12v)
3	GY/BK	VSS (+), (-) Signals	0 Hz, 55 mph: 125 Hz
4	DG/YL	IDM Sensor Signal	20-31 Hz
5, 13-15	---	Not Used	---
7	LG/YL	ECT Sensor Signal	0.5-0.6v
8	BR	Fuel Pump Monitor	On: 12v, Off: 0v
9, 28	BK, OR/BK	Data Bus (-), (+) Signals	Digital Signal
10	BK/YL	A/C Switch Signal	A/C On: 12v, Off: 0v
11	WT/BK	Air Management 2 Solenoid	AM2 Off: 12v, On: 1v
12	BL/BK	4x4 Switch Signal	12v (Switch On: 0v)
16	BK/OR	Ignition System Ground	<0.050v
17	PK/LG	Self Test Output & MIL	MIL Off: 12v, On: 1v
18	TN/LG	Inferred Mileage Sensor	Digital Signals
19	PK/OR	Shift Solenoid 2 Control	1v, 55 mph: 12v
20	BK	PCM Case Ground	<0.050v
21	GY/WT	IAC Solenoid Control	9.8-10.6v
22	TN/LG	Fuel Pump Control	Off: 12v, On: 1v
23	LG/BK	Knock Sensor Signal	0v
24	YL/LG	PSP Switch Signal	0v (turning: 12v)
25	YL/RD	ACT Sensor Signal	1.5-2.5v
26	OR/WT	Reference Voltage	4.9-5.1v
27	BR/LG	EVP Sensor Signal	0.4v, 55 mph: 2.6v
29	GY/BL	HO2S-11 (B1 S1) Signal	0.1-1.1v
30	BL/YL	MLP Sensor Signal	In 'P': 0v, in O/D: 5v
31	GY/YL	Canister Purge Solenoid	12v, 55 mph: 1v
32	LG/WT	OCIL (lamp) Control	7.7v (Switch On: 0v)
33	DG	VR Solenoid Control	0%, 55 mph: 45%
34-35, 39	---	Not Used	---
36	YL/LG	Spark Output Signal	6.93v (50%)
37, 57	RD	Vehicle Power	12-14v
38	WT/YL	EPC Solenoid Control	9.5v (5 psi)
40, 60	BK/LG	Power Ground	<0.1v
41	TN/WT	OCS (switch) Signal	OCS & O/D On: 12
42	OR/BK	TOT Sensor Signal	2.10-2.40v
43	LG/PK	A/C Demand Switch	A/C On: 12v, Off: 0v
44, 50, 54	---	Not Used	---
45	LG/BK	MAP Sensor Signal	107 Hz (sea level)
46	BK/WT	Analog Signal Return	<0.050v
47	DG/LG	TP Sensor Signal	1v, 55 mph: 1.4v
48	WT/RD	Self Test Indicator Signal	STI On: 1v, Off: 5v
49	OR	HO2S-11 (B1 S1) Ground	<0.050v
51	WT/RD	Air Management 1 Solenoid	AM1 Off: 12v, On: 1v
52	OR/YL	Shift Solenoid 1 Control	12v, 55 mph: 1v
53	PK/YL	TCC Solenoid Control	0v, On 55 mph: 12v
55	BR	Coast Clutch Switch	12v, 55 mph: 12v
56	DB	PIP Sensor Signal	6.93v (50%)
58	TN/OR	Injector Bank 1 (INJ 1, 3 & 5)	6.4-6.8 ms
59	TN/RD	Injector Bank 2 (INJ 2, 4 & 6)	6.4-6.8 ms

Pin Connector Graphic

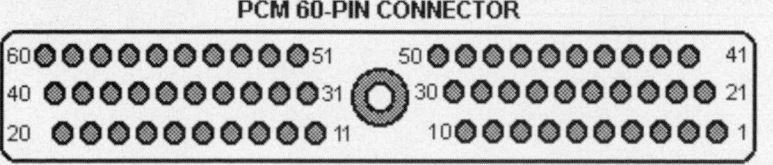

PCM 60-PIN CONNECTOR

Terminal View of 60-Pin PCM Harness Connector

1991-94 F-Series 4.9L I6 VIN Y (All) 60 Pin Connector

PCM Pin #	Wire Color	Circuit Description (60 Pin)	Value at Hot Idle
1	YL	Keep Alive Power	12-14v
3	GY/BK	VSS (+) Signal	0 Hz, 55 mph: 125 Hz
4	TN/YL	IDM Sensor Signal	20-31 Hz
4	YL/BK	IDM Sensor Signal	20-31 Hz
6	PK/OR	VSS (-) Signal	0 Hz, 55 mph: 125 Hz
7	LG/RD	ECT Sensor Signal	0.5-0.6v
8	DG/YL	Fuel Pump Monitor	On: 12v, Off: 0v
9	PK/BL	Data Bus (-) Signal	Digital Signals
10	BK/YL	A/C Switch Signal	A/C On: 12v, Off: 0v
10	PK/BL	A/C Switch Signal	A/C On: 12v, Off: 0v
11	BR	Air Management 2 Solenoid	AM2 Off: 12v, On: 1v
16	OR/RD	Ignition System Ground	<0.050v
17	PK/LG	Self Test Output & MIL	MIL Off: 12v, On: 1v
20	BK	PCM Case Ground	<0.050v
21	WT/BL	IAC Solenoid Control	10.7v (33%)
22	BL/OR	Fuel Pump Control	Off: 12v, On: 1v
23	YL/RD	Knock Sensor Signal	0v
25	GY	IAT Sensor Signal	1.5-2.5v
26	BR/WT	Reference Voltage	4.9-5.1v
27	BR/LG	EVP Sensor Signal	0.4v, 55 mph: 2.6v
28	TN/OR	Data Bus (+) Signal	Digital Signals
29	GY/BL	HO2S-11 (B1 S1) Signal	0.1-1.1v
30	BL/YL	A/T: Neutral Drive Switch	In 'N': Others: 5v
30	BL/YL	M/T: CPP Switch Signal	5v (clutch "in": 0v)
31	GY/YL	Canister Purge Solenoid	12v, 55 mph: 1v
32	WT/LG	OCIL (lamp) Control	7.7v (Switch On: 0v)
33	BR/PK	VR Solenoid Control	0%, 55 mph: 45%
36	PK	Spark Output Signal	6.93v (50%)
37	RD	Vehicle Power	12-14v
38	WT/YL	EPC Solenoid Control	9.5v (5 psi)
40	BK/WT	Power Ground	<0.1v
43	PK	A/C Demand Switch	A/C On: 12v, Off: 0v
45	LG/BK	MAP Sensor Signal	107 Hz (sea level)
46	GY/RD	Analog Signal Return	<0.050v
47	GY/WT	TP Sensor Signal	1.5-1.2v
48	WT/PK	Self Test Indicator Signal	STI On: 1v, Off: 5v
49	OR	HO2S-11 (B1 S1) Ground	<0.050v
51	WT/OR	Air Management 1 Solenoid	AM1 Off: 12v, On: 1v
56	GY/OR	PIP Sensor Signal	6.93v (50%)
57	RD	Vehicle Power	12-14v
58	TN	Injector Bank 1 (INJ 1, 3 & 5)	5.6-6.4 ms
59	WT	Injector Bank 2 (INJ 2, 4 & 6)	5.6-6.4 ms
60	BK/WT	Power Ground	<0.1v

Pin Connector Graphic

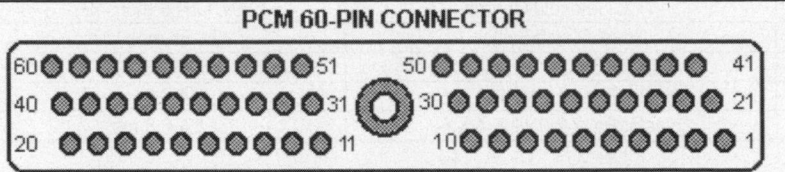

1991-94 F-Series 4.9L I6 VIN Y (E4OD) 60 Pin Connector

PCM Pin #	Wire Color	Circuit Description (60 Pin)	Value at Hot Idle
1	YL	Keep Alive Power	12-14v
2	LG	Brake Position Switch	0v (Brake On: 12v)
3	GY/BK	VSS (+) Signal	0 Hz, 55 mph: 125 Hz
4	TN/YL	IDM Sensor Signal	20-31 Hz
4	YL/BK	IDM Sensor Signal	20-31 Hz
5	---	Not Used	---
6	PK/OR	VSS (-) Signal	0 Hz, 55 mph: 125 Hz
7	LG/RD	ECT Sensor Signal	0.5-0.6v
8	DG/YL	Fuel Pump Monitor	On: 12v, Off: 0v
9	PK/BL	Data Bus (-) Signal	Digital Signals
10	BK/YL	A/C Switch Signal	A/C On: 12v, Off: 0v
10	PK/BL	A/C Switch Signal	A/C On: 12v, Off: 0v
11	BR	Air Management 2 Solenoid	AM2 Off: 12v, On: 1v
12	BL/BK	4x4 Switch Signal	12v (switch closed: 0v)
12	PK/BL	4x4 Switch Signal	12v (switch closed: 0v)
13-15	---	Not Used	---
16	OR/RD	Ignition System Ground	<0.050v
17	PK/LG	Self Test Output & MIL	MIL Off: 12v, On: 1v
18	---	Not Used	---
19	PK/OR	Shift Solenoid 2 Control	1v, 55 mph: 12v
20	BK	PCM Case Ground	<0.050v
21	WT/BL	IAC Solenoid Control	10.7v (33%)
22	BL/OR	Fuel Pump Control	Off: 12v, On: 1v
23	YL/RD	Knock Sensor Signal	0v
24	---	Not Used	---
25	GN/YL	ACT Sensor Signal	1.5-2.5v
26	BR/WT	Reference Voltage	4.9-5.1v
27	BR/LG	EVP Sensor Signal	0.4v, 55 mph: 2.6v
28	TN/OR	Data Bus (+) Signal	Digital Signals
29	GY/BL	HO2S-11 (B1 S1) Signal	0.1-1.1v
30	BL/YL	MLP Sensor Signal	In 'P': 0v, in O/D: 5v
31	GY/YL	Canister Purge Solenoid	12v, 55 mph: 1v
32	WT/LG	OCIL (lamp) Control	7.7v (Switch On: 0v)
33	BR/PK	VR Solenoid Control	0%, 55 mph: 45%
34	---	Not Used	---
35	---	Not Used	---
36	PK	Spark Output Signal	6.93v (50%)
37	RD	Vehicle Power	12-14v
38	WT/YL	EPC Solenoid Control	9.5v (5 psi)
39	---	Not Used	---
40	BK/WT	Power Ground	<0.1v

1991-94 F-Series 4.9L I6 VIN Y (E4OD) 60 Pin Connector

PCM Pin #	Wire Color	Circuit Description (60 Pin)	Value at Hot Idle
41	TN/WT	OCS (switch) Signal	12v (switch closed: 0v)
42	OR/BK	TOT Sensor Signal	2.10-2.40v
43	PK	A/C Demand Switch	A/C On: 12v, Off: 0v
44	---	Not Used	---
45	LG/BK	MAP Sensor Signal	107 Hz (sea level)
46	GY/RD	Analog Signal Return	<0.050v
47	GY/WT	TP Sensor Signal	1v, 55 mph: 1.4v
48	WT/PK	Self Test Indicator Signal	STI On: 1v, Off: 5v
49	OR	HO2S-11 (B1 S1) Ground	<0.050v
50	---	Not Used	---
51	WT/OR	Air Management 1 Solenoid	AM1 Off: 12v, On: 1v
52	OR/YL	Shift Solenoid 1 Control	12v, 55 mph: 1v
53	PK/YL	TCC Solenoid Control	0v, On 55 mph: 12v
54	---	Not Used	---
55	BR/OR	Coast Clutch Switch	12v, 55 mph: 12v
56	GY/OR	PIP Sensor Signal	6.93v (50%)
57	RD	Vehicle Power	12-14v
58	TN	Injector Bank 1 (INJ 1, 3 & 5)	5.6-6.0 ms
59	WT	Injector Bank 2 (INJ 2, 4 & 6)	5.6-6.0 ms
60	BK/WT	Power Ground	<0.1v

Pin Connector Graphic

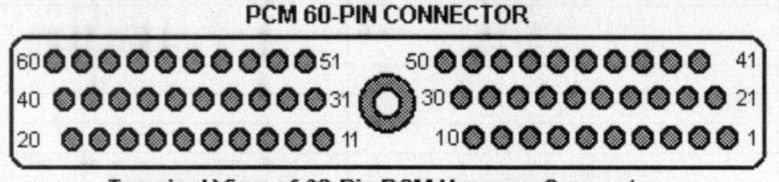

PCM 60-PIN CONNECTOR

Terminal View of 60-Pin PCM Harness Connector

Standard Colors and Abbreviations

Abbreviation	Color	Abbreviation	Color	Abbreviation	Color
BK	Black	GY	Gray	PK	Purple
BL	Blue	GN	Green	RD	Red
BR	Brown	LG	LT Green	TN	Tan
DB	Dark Blue	OR	Orange	WT	White
DG	DK Green	PK	Pink	YL	Yellow

1995 F-Series 4.9L VIN Y (All) California 60 Pin Connector

PCM Pin #	Wire Color	Circuit Description (60 Pin)	Value at Hot Idle
1	---	Not Used	---
2	LG	Brake Pedal Switch	0v (Brake On: 12v)
3	GY/BK	VSS (+) Signal	0 Hz, 55 mph: 125 Hz
4	YL/BK	IDM Sensor Signal	20-31 Hz
5	---	Not Used	---
6	PK/OR	VSS (-) Signal	0 Hz, 55 mph: 125 Hz
7	LG/RD	ECT Sensor Signal	0.5-0.6v
8	DG/YL	Fuel Pump Monitor	On: 12v, Off: 0v
9	TN/BL	MAF Sensor Return	<0.050v
10	DG/OR	A/C Switch Signal	A/C On: 12v, Off: 0v
11	GY/YL	Canister Purge Solenoid	12v, 55 mph: 1v
12	LG/OR	Injector 6 Control	5.5-7.1 ms
13	---	Not Used	---
14	---	Not Used	---
15	TN/BK	Injector 5 Control	5.5-7.1 ms
16	TN/YL	Ignition System Ground	<0.050v
17	PK/LG	Self Test Output & MIL	MIL Off: 12v, On: 1v
18	TN/OR	Data Bus (-) Signal	Digital Signals
19	PK/BK	Data Bus (+) Signal	Digital Signals
20	BK	PCM Case Ground	<0.050v
21	WT/BL	IAC Solenoid Control	10.7v (33%)
22	BL/OR	Fuel Pump Control	Off: 12v, On: 1v
23	YL/RD	Knock Sensor Signal	0v
24	---	Not Used	---
25	GY	IAT Sensor Signal	1.5-2.5v
26	BR/WT	Reference Voltage	4.9-5.1v
27	BR/LG	EVP Sensor Signal	0.4v, 55 mph: 2.6v
28-29	---	Not Used	---
30	BL/YL	A/T: Neutral Drive Switch	In 'P': 0v, Others: 5v
30	GY/YL	A/T: Neutral Drive Switch	In 'P': 0v, Others: 5v
30	BL/YL	M/T: CPP Switch Signal	5v (clutch "in": 0v)
30	GY/YL	M/T: CPP Switch Signal	5v (clutch "in": 0v)
31	WT/OR	Thermactor Air Bypass	TAB Off: 12v, On: 1v
32	WT/LG	TCIL (lamp) Control	12v (TCIL On: 0v)
33	BR/PK	VR Solenoid Control	0%, 55 mph: 45%
34	BR	Thermactor Air Diverter	TAD Off: 12v, On: 1v
35	BR/BL	Injector 4 Control	5.5-7.1 ms
36	PK	Spark Output Signal	6.93v (50%)
37	RD	Vehicle Power	12-14v
38	---	Not Used	---
39	BR/YL	Injector 3 Control	5.5-7.1 ms
40	BK/WT	Power Ground	<0.1v

1995 F-Series 4.9L VIN Y (All) California 60 Pin Connector

PCM Pin #	Wire Color	Circuit Description (60 Pin)	Value at Hot Idle
41	TN/WT	TCS (switch) Signal	TCS & O/D On: 12v
42	BL/YL	4x4 Switch Signal	12v (switch closed: 0v)
43	RD/BK	HO2S-21 (B2 S1) Signal	0.1-1.1v
44	GY/BL	HO2S-11 (B1 S1) Signal	0.1-1.1v
45	---	Not Used	---
46	GY/RD	Analog Signal Return	<0.1v
47	GY/WT	TP Sensor Signal	1v, 55 mph: 1.4v
48	WT/PK	Self Test Indicator Signal	STI On: 1v, Off: 5v
49	---	Not Used	---
50	BL/RD	MAF Sensor Signal	0.8v, 55 mph: 1.6v
51	---	Not Used	---
52	---	Not Used	---
53	---	Not Used	---
54	PK/YL	A/C WOT Cutout Control	Off: 12v, On: 1v
55	---	Not Used	---
56	GY/OR	PIP Sensor Signal	6.93v (50%)
57	RD	Vehicle Power	12-14v
58	TN	Injector 1 Control	5.5-7.1 ms
59	WT	Injector 2 Control	5.5-7.1 ms
60	BK/WT	Power Ground	<0.1v

Pin Connector Graphic

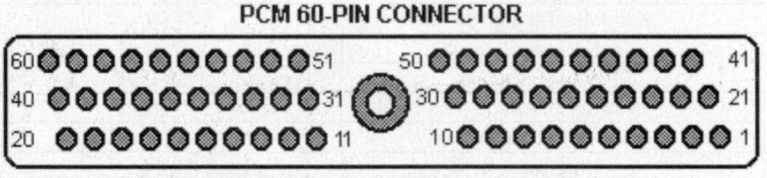

PCM 60-PIN CONNECTOR

Terminal View of 60-Pin PCM Harness Connector

Standard Colors and Abbreviations

Abbreviation	Color	Abbreviation	Color	Abbreviation	Color
BK	Black	GY	Gray	PK	Purple
BL	Blue	GN	Green	RD	Red
BR	Brown	LG	LT Green	TN	Tan
DB	Dark Blue	OR	Orange	WT	White
DG	DK Green	PK	Pink	YL	Yellow

1995 F-Series 4.9L I6 VIN Y (All) 60 Pin Connector

PCM Pin #	Wire Color	Circuit Description (60 Pin)	Value at Hot Idle
1	YL	Keep Alive Power	12-14v
2	LG	Brake Pedal Switch	0v (Brake On: 12v)
3	GY/BK	VSS (+) Signal	0 Hz, 55 mph: 125 Hz
4	YL/BK	IDM Sensor Signal	20-31 Hz
6	PK/OR	VSS (-) Signal	0 Hz, 55 mph: 125 Hz
7	LG/RD	ECT Sensor Signal	0.5-0.6v
8	DG/YL	Fuel Pump Monitor	On: 12v, Off: 0v
9	PK/BL	Data Bus (-) Signal	Digital Signals
10	PK/BL	A/C Switch Signal	A/C On: 12v, Off: 0v
11	BR	Air Management 2 Solenoid	AM2 Off: 12v, On: 1v
12	BL/BK	4x4 Indicator Lamp	12v (switch closed: 0v)
16	OR/RD	Ignition System Ground	<0.050v
17	PK/LG	Self Test Output & MIL	MIL Off: 12v, On: 1v
20	BK	PCM Case Ground	<0.050v
21	WT/BL	IAC Solenoid Control	10.7v (33%)
22	BL/OR	Fuel Pump Control	Off: 12v, On: 1v
23	YL/RD	Knock Sensor Signal	0v
25	GY	IAT Sensor Signal	1.5-2.5v
26	BR/WT	Reference Voltage	4.9-5.1v
27	BR/LG	EVP Sensor Signal	0.4v, 55 mph: 2.6v
28	TN/OR	Data Bus (+) Signal	Digital Signals
29	GY/BL	HO2S-11 (B1 S1) Signal	0.1-1.1v
30	GY/YL	M/T: CPP Switch Signal	5v (clutch "in": 0v)
30	BL/YL	M/T: CPP Switch Signal	5v (clutch "in": 0v)
31	GY/YL	Canister Purge Solenoid	12v, 55 mph: 1v
32	WT/LG	TCIL (lamp) Control	7.7v (Switch On: 0v)
33	BR/PK	VR Solenoid Control	0%, 55 mph: 45%
36	PK	Spark Output Signal	6.93v (50%)
37	RD	Vehicle Power	12-14v
40	BK/WT	Power Ground	<0.1v
41	TN/WT	TCS (switch) Signal	0.1-0.2v
43	PK	A/C Demand Switch	A/C On: 12v, Off: 0v
45	LG/BK	MAP Sensor Signal	107 Hz (sea level)
46	GY/RD	Analog Signal Return	<0.050v
47	GY/WT	TP Sensor Signal	0.9-1.0v
48	WT/PK	Self Test Indicator Signal	STI on: 1v, Off: 5v
49	OR	HO2S-11 (B1 S1) Ground	<0.050v
51	WT/OR	Air Management 1 Solenoid	AM1 Off: 12v, On: 1v
56	GY/OR	Profile Ignition Truck Signal	6.93v (50%)
57	RD	Vehicle Power	12-14v
58	TN	Injector Bank 1 (INJ 1, 3 & 5)	6.2-7.4 ms
59	WT	Injector Bank 2 (INJ 2, 4 & 6)	6.2-7.4 ms
60	BK/WT	Power Ground	<0.1v

Pin Connector Graphic

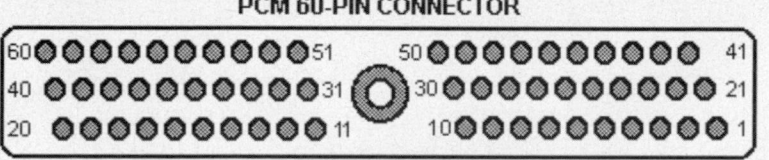

Terminal View of 60-Pin PCM Harness Connector

1995 F-Series 4.9L I6 VIN Y (E4OD) Calif. 60 Pin Connector

PCM Pin #	Wire Color	Circuit Description (60 Pin)	Value at Hot Idle
1	YL	Keep Alive Power	12-14v
2	LG	Brake Pedal Switch	0v (Brake On: 12v)
3	GY/BK	VSS (+) Signal	0 Hz, 55 mph: 125 Hz
4	YL/BK	IDM Sensor Signal	20-31 Hz
5	---	Not Used	---
6	PK/OR	VSS (-) Signal	0 Hz, 55 mph: 125 Hz
7	LG/RD	ECT Sensor Signal	0.5-0.6v
8	DG/YL	Fuel Pump Monitor	On: 12v, Off: 0v
9	TN/BL	MAF Sensor Return	<0.050v
10	DG/OR	A/C Switch Signal	A/C On: 12v, Off: 0v
11	GY/YL	Canister Purge Solenoid	12v, 55 mph: 1v
12	LG/OR	Injector 6 Control	5.5-7.1 ms
13-14	---	Not Used	---
15	TN/BK	Injector 5 Control	5.5-7.1 ms
16	OR/RD	Ignition System Ground	<0.050v
17	PK/LG	Self Test Output & MIL	MIL Off: 12v, On: 1v
18	TN/OR	Data Bus (-) Signal	Digital Signals
19	PK/BL	Data Bus (+) Signal	Digital Signals
20	BK	PCM Case Ground	<0.050v
21	WT/BL	IAC Solenoid Control	10.7v (33%)
22	BL/OR	Fuel Pump Control	Off: 12v, On: 1v
23	YL/RD	Knock Sensor Signal	0v
24	---	Not Used	---
25	GY	Idle Air Temperature Sensor	0.5-0.6v
26	BR/WT	Reference Voltage	4.9-5.1v
27	BR/LG	EVP Sensor Signal	0.4v, 55 mph: 2.6v
28-29	---	Not Used	---
30	BL/YL	TR Sensor Signal	In 'P': 0v, in O/D: 5v
30	GY/YL	TR Sensor Signal	In 'P': 0v, in O/D: 5v
31	WT/OR	Thermactor Air Bypass	TAB Off: 12v, On: 1v
32	WT/LG	TCIL (lamp) Control	7.7v (Switch On: 0v)
33	BR/PK	VR Solenoid Control	0%, 55 mph: 45%
34	BR	Thermactor Air Diverter	TAD Off: 12v, On: 1v
35	BR/BL	Injector 4 Control	5.5-7.1 ms
36	PK	Spark Output Signal	6.93v (50%)
37	RD	Vehicle Power	12-14v
38	WT/YL	EPC Solenoid Control	9.5v (5 psi)
39	BR/YL	Injector 3 Control	5.5-7.1 ms
40	BK/WT	Power Ground	<0.1v

1995 F-Series 4.9L I6 VIN Y (E4OD) Calif. 60 Pin Connector

PCM Pin #	Wire Color	Circuit Description (60 Pin)	Value at Hot Idle
41	TN/WT	TCS (switch) Signal	TCS & O/D On: 12v
42	BL/YL	4x4 Switch Signal	12v (switch closed: 0v)
43	RD/BK	HO2S-21 (B2 S1) Signal	0.1-1.1v
44	GY/BL	HO2S-11 (B1 S1) Signal	0.1-1.1v
45	---	Not Used	---
46	GY/RD	Analog Signal Return	<0.050v
47	GY/WT	TP Sensor Signal	1v, 55 mph: 1.4v
48	WT/PK	Self Test Indicator Signal	STI On: 1v, Off: 5v
49	OR/BK	TOT Sensor Signal	2.10-2.40v
50	BL/RD	MAF Sensor Signal	0.8v, 55 mph: 1.6v
51	OR/YL	Shift Solenoid 1 Control	12v, 55 mph: 1v
52	PK/OR	Shift Solenoid 2 Control	1v, 55 mph: 12v
53	PK/YL	TCC Solenoid Control	0v, On 55 mph: 12v
54	PK/YL	A/C WOT Relay Control	Off: 12v, On: 1v
55	BR/OR	Coast Clutch Solenoid	12v, 55 mph: 12v
56	GY/OR	PIP Sensor Signal	6.93v (50%)
57	RD	Vehicle Power	12-14v
58	TN	Injector 1 Control	5.5-7.1 ms
59	WT	Injector 2 Control	5.5-7.1 ms
60	BK/WT	Power Ground	<0.1v

Pin Connector Graphic

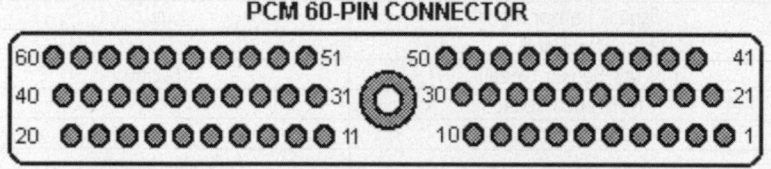

Terminal View of 60-Pin PCM Harness Connector

Standard Colors and Abbreviations

Abbreviation	Color	Abbreviation	Color	Abbreviation	Color
BK	Black	GY	Gray	PK	Purple
BL	Blue	GN	Green	RD	Red
BR	Brown	LG	LT Green	TN	Tan
DB	Dark Blue	OR	Orange	WT	White
DG	DK Green	PK	Pink	YL	Yellow

1995 F-Series 4.9L I6 VIN Y (E4OD) Federal 60 Pin Connector

PCM Pin #	Wire Color	Circuit Description (60 Pin)	Value at Hot Idle
1	YL	Keep Alive Power	12-14v
2	LG	Brake Pedal Switch	0v (Brake On: 12v)
3	GY/BK	VSS (+) Signal	0 Hz, 55 mph: 125 Hz
4	YL/BK	IDM Sensor Signal	20-31 Hz
5	---	Not Used	---
6	PK/OR	VSS (-) Signal	0 Hz, 55 mph: 125 Hz
7	LG/RD	ECT Sensor Signal	0.5-0.6v
8	DG/YL	Fuel Pump Monitor	On: 12v, Off: 0v
9	PK/BL	Data Bus (-) Signal	Digital Signals
10	PK/BL	A/C Switch Signal	A/C On: 12v, Off: 0v
11	BR	Air Management 2 Solenoid	AM2 Off: 12v, On: 1v
12	BL/BK	4x4 Indicator Lamp	12v (switch closed: 0v)
13	---	Not Used	---
14	---	Not Used	---
15	---	Not Used	---
16	OR/RD	Ignition System Ground	<0.050v
17	PK/LG	Self Test Output & MIL	MIL Off: 12v, On: 1v
18	---	Not Used	---
19	PK/OR	Shift Solenoid 2 Control	12v, 55 mph: 1v
20	BK	PCM Case Ground	<0.050v
21	WT/BL	IAC Solenoid Control	10.7v (33%)
22	BL/OR	Fuel Pump Control	Off: 12v, On: 1v
23	YL/RD	Knock Sensor Signal	0v
24	---	Not Used	---
25	GY	IAT Sensor Signal	1.5-2.5v
26	BR/WT	Reference Voltage	4.9-5.1v
27	BR/LG	EVP Sensor Signal	0.4v, 55 mph: 2.6v
28	TN/OR	Data Bus (+) Signal	Digital Signals
29	GY/BL	HO2S-11 (B1 S1) Signal	0.1-1.1v
30	GY/YL	TR Sensor Signal	In 'P': 0v, in O/D: 5v
30	BL/YL	TR Sensor Signal	In 'P': 0v, in O/D: 5v
31	GY/YL	Canister Purge Solenoid	12v, 55 mph: 1v
32	WT/LG	TCIL (lamp) Control	7.7v (Switch On: 0v)
33	BR/PK	VR Solenoid Control	0%, 55 mph: 45%
34	---	Not Used	---
35	---	Not Used	---
36	PK	Spark Output Signal	6.93v (50%)
37	RD	Vehicle Power	12-14v
38	WT/YL	EPC Solenoid Control	9.5v (5 psi)
39	---	Not Used	---
40	BK/WT	Power Ground	<0.1v

1995 F-Series 4.9L I6 VIN Y (E4OD) Federal 60-P Connector

PCM Pin #	Wire Color	Circuit Description (60 Pin)	Value at Hot Idle
41	TN/WT	TCS (switch) Signal	TCS & O/D On: 12v
42	OR/BK	TOT Sensor Signal	2.10-2.40v
43	PK	A/C Demand Switch	A/C On: 12v, Off: 0v
44	---	Not Used	---
45	LG/BK	MAP Sensor Signal	107 Hz (sea level)
46	GY/RD	Analog Signal Return	<0.050v
47	GY/WT	TP Sensor Signal	1v, 55 mph: 1.4v
48	WT/PK	Self Test Indicator Signal	STI On: 1v, Off: 5v
49	OR	HO2S-11 (B1 S1) Ground	<0.050v
50	---	Not Used	---
51	WT/OR	Air Management 1 Solenoid	AM1 Off: 12v, On: 1v
52	OR/YL	Shift Solenoid 1 Control	1v, 55 mph: 12v
53	PK/YL	TCC Solenoid Control	0v, On 55 mph: 12v
54	---	Not Used	---
55	BR/OR	Coast Clutch Solenoid	12v, 55 mph: 12v
56	GY/OR	PIP Sensor Signal	6.93v (50%)
57	RD	Vehicle Power	12-14v
58	TN	Injector Bank 1 (INJ 1, 3 & 5)	6.2-7.4 ms
59	WT	Injector Bank 2 (INJ 2, 4 & 6)	6.2-7.4 ms
60	BK/WT	Power Ground	<0.1v

Pin Connector Graphic

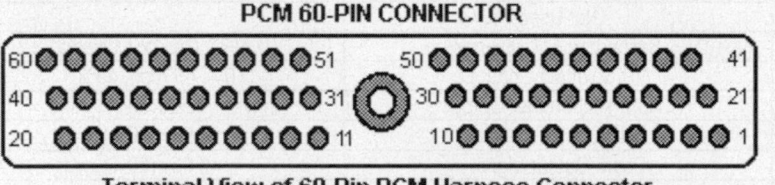

Terminal View of 60-Pin PCM Harness Connector

Standard Colors and Abbreviations

Abbreviation	Color	Abbreviation	Color	Abbreviation	Color
BK	Black	GY	Gray	PK	Purple
BL	Blue	GN	Green	RD	Red
BR	Brown	LG	LT Green	TN	Tan
DB	Dark Blue	OR	Orange	WT	White
DG	DK Green	PK	Pink	YL	Yellow

1996 Truck 4.9L I6 MFI VIN Y (All) 104 Pin Connector

PCM Pin #	Wire Color	Circuit Description (104 Pin)	Value at Hot Idle
1	---	Not Used	---
2	PK/LG	MIL (lamp) Control	MIL Off: 12v, On: 1v
3-12	---	Not Used	---
13	PK	Flash EPROM Power	0.1v
14	LB/BK	4x4 Low Switch Signal	12v (switch closed: 0v)
15	PK/LB	Data Bus (-) Signal	Digital Signals
16	TN/OR	Data Bus (+) Signal	Digital Signals
17-22	---	Not Used	---
23	OR/RD	Ignition Ground	<0.050v
24	BK/WT	Power Ground	<0.1v
25	BK	Case Ground	<0.050v
26-28	---	Not Used	---
29	TN/WT	TCS (switch) Signal	TCS & O/D On: 12v
30-32, 34	---	Not Used	---
33	PK/OR	PSOM (-) Signal	<0.050v
35	RD/LG	HO2S-12 (B1 S2) Signal	0.1-1.1v
36	TN/LB	MAF Sensor Return	<0.050v
37	OR/BK	TFT Sensor Signal	2.10-2.40v
38	LG/RD	ECT Sensor Signal	0.5-0.6v
39	GY	IAT Sensor Signal	1.5-2.5v
40	DG/YL	Fuel Pump Monitor	On: 12v, Off: 0v
41	DG/OR	A/C Switch Signal	A/C On: 12v, Off: 0v
42-43, 46	---	Not Used	---
44	BR	Secondary AIR Diverter	AIRD Off: 12v, On: 1v
47	BK/PK	VR Solenoid Control	0%, 55 mph: 45%
48	YL/BK	Clean Tachometer Output	39-45 Hz
49	GY/OR	PIP Sensor Signal	6.93v (50%)
50	PK	Spark Output Signal	6.93v (50%)
51	BK/WT	Power Ground	<0.1v
52-54	---	Not Used	---
55	YL	Keep Alive Power	12-14v
56	LG/BK	EVAP Purge Solenoid	0-10 Hz (0-100%)
57	YL/RD	Knock Sensor Signal	0v
58	GY/BK	PSOM (+) Signal	0 Hz, 55 mph: 125 Hz
59	DG/LG	Misfire Detection Sensor	45-55 Hz
60	GY/LB	HO2S-11 (B1 S1) Signal	0.1-1.1v
61-63	---	Not Used	---
64	LB/YL	A/T: PNP Switch Signal	In 'P': 0v, Others: 5v
64	LB/YL	M/T: CPP Switch Signal	5v (clutch "in": 0v)

1996 Truck 4.9L I6 MFI VIN Y (All) 104 Pin Connector

PCM Pin #	Wire Color	Circuit Description (104 Pin)	Value at Hot Idle
65	BR/LG	DPFE Sensor Signal	0.95-1.05v
66-68	---	Not Used	---
69	PK/YL	A/C WOT Relay Control	At idle with A/C on: 12v
70	WT/OR	Secondary AIR Bypass	AIRB Off: 12v, On: 1v
71	RD	Vehicle Power	12-14v
72	---	Not Used	---
73	TN/BK	Injector 5 Control	3.2-4.5 ms
74	BR/YL	Injector 3 Control	3.2-4.5 ms
75	TN	Injector 1 Control	3.2-4.5 ms
76	BK/WT	Power Ground	<0.1v
77	BK/WT	Power Ground	<0.1v
78	---	Not Used	---
79	WT/LG	TCIL (lamp) Control	7.7v (Switch On: 0v)
80	LB/OR	Fuel Pump Control	Off: 12v, On: 1v
81-82	---	Not Used	---
83	WT/LB	IAC Solenoid Control	10.7v (33%)
84-86	---	Not Used	---
87	RD/BK	HO2S-21 (B2 S1) Signal	0.1-1.1v
88	LB/RD	MAF Sensor Signal	0.8v, 55 mph: 1.8v
89	GY/WT	TP Sensor Signal	0.53-1.27v
90	BR/WT	Reference Voltage	4.9-5.1v
91	GY/RD	Analog Signal Return	<0.050v
92	LG	Brake Pedal Switch	0v (Brake On: 12v)
93	RD/WT	HO2S-11 (B1 S1) Heater	1v (Heater Off: 12v)
94	YL/LB	HO2S-21 (B2 S1) Heater	1v (Heater Off: 12v)
95	WT/BK	HO2S-12 (B1 S2) Heater	1v (Heater Off: 12v)
96, 98	---	Not Used	---
97	RD	Vehicle Power	12-14v
99	LG/OR	Injector 6 Control	3.2-4.5 ms
100	BR/LB	Injector 4 Control	3.2-4.5 ms
101	W	Injector 2 Control	3.2-4.5 ms
102	---	Not Used	---
103	BK/WT	Power Ground	<0.1v
104	---	Not Used	---

Pin Connector Graphic

PCM 104-PIN CONNECTOR

Terminal View of 104-Pin PCM Wiring Harness Connector

Standard Colors and Abbreviations

Abbreviation	Color	Abbreviation	Color	Abbreviation	Color
BK	Black	GY	Gray	PK	Purple
BL	Blue	GN	Green	RD	Red
BR	Brown	LG	LT Green	TN	Tan
DB	Dark Blue	OR	Orange	WT	White
DG	DK Green	PK	Pink	YL	Yellow

1996 Truck 4.9L I6 MFI VIN Y (E4OD) 104 Pin Connector

PCM Pin #	Wire Color	Circuit Description (104 Pin)	Value at Hot Idle
1	PK/OR	Shift Solenoid 2 Control	1v, 55 mph: 12v
2	PK/LG	MIL (lamp) Control	MIL Off: 12v, On: 1v
3-12	---	Not Used	---
13	PK	Flash EPROM Power	0.1v
14	LB/BK	4x4 Low Switch	12v (switch closed: 0v)
15	PK/LB	Data Bus (-) Signal	Digital Signals
16	TN/OR	Data Bus (+) Signal	Digital Signals
17-22	---	Not Used	---
23	OR/RD	Ignition Ground	<0.050v
24, 51	BK/WT	Power Ground	<0.1v
25	BK	Case Ground	<0.050v
26, 28	---	Not Used	---
27	OR/YL	Shift Solenoid 1 Control	1v, 55 mph: 12v
29	TN/WT	TCS (switch) Signal	TCS & O/D On: 12v
30-32, 34	---	Not Used	---
33	PK/OR	PSOM (-) Signal	<0.050v
35	RD/LG	HO2S-12 (B1 S2) Signal	0.1-1.1v
36	TN/LB	MAF Sensor Return	050v
37	OR/BK	TFT Sensor Signal	2.10-2.40v
38	LG/RD	ECT Sensor Signal	0.5-0.6v
39	GY	IAT Sensor Signal	1.5-2.5v
40	DG/YL	Fuel Pump Monitor	On: 12v, Off: 0v
41	DG/OR	A/C Switch Signal	A/C On: 12v, Off: 0v
42-43	---	Not Used	---
44	BR	Secondary AIR Diverter	AIRD Off: 12v, On: 1v
45-46	---	Not Used	---
47	BK/PK	VR Solenoid Control	0%, 55 mph: 45%
48	YL/BK	Clean Tachometer Output	39-45 Hz
49	GY/OR	PIP Sensor Signal	6.93v (50%)
50	PK	Spark Output Signal	6.93v (50%)
52	---	Not Used	---
53	BR/OR	Coast Clutch Solenoid	12v, 55 mph: 12v
54	PK/YL	TCC Solenoid Control	0%, 55 mph: 95%
55	YL	Keep Alive Power	12-14v
56	LG/BK	EVAP Purge Solenoid	0-10 Hz (0-100%)
57	YL/RD	Knock Sensor Signal	0v
58	GY/BK	PSOM (+) Signal	0 Hz, 55 mph: 125 Hz
59	DG/LG	Misfire Detection Sensor	45-55 Hz
60	GY/LB	HO2S-11 (B1 S1) Signal	0.1-1.1v
61-63	---	Not Used	---
64	LB/YL	TR Sensor Signal	In 'P': 0v, in O/D: 5v

1996 Truck 4.9L I6 MFI VIN Y (E4OD) 104 Pin Connector

PCM Pin #	Wire Color	Circuit Description (104 Pin)	Value at Hot Idle
65	BR/LG	DPFE Sensor Signal	0.95-1.05v
66-68	---	Not Used	---
69	PK/YL	A/C WOT Relay Control	Off: 12v, On: 1v
70	WT/OR	Secondary AIR Bypass	AIRB Off: 12v, On: 1v
71	RD	Vehicle Power	12-14v
72	---	Not Used	---
73	TN/BK	Injector 5 Control	3.2-4.5 ms
74	BR/YL	Injector 3 Control	3.2-4.5 ms
75	TN	Injector 1 Control	3.2-4.5 ms
76	BK/WT	Power Ground	<0.1v
77	BK/WT	Power Ground	<0.1v
78, 82	---	Not Used	---
79	WT/LG	TCIL (lamp) Control	7.7v (Switch On: 0v)
80	LB/OR	Fuel Pump Control	Off: 12v, On: 1v
81	WT/YL	EPC Solenoid Control	10v (26 psi)
83	WT/LB	IAC Solenoid Control	10.7v (33%)
84-86	---	Not Used	---
87	RD/BK	HO2S-21 (B2 S1) Signal	0.1-1.1v
88	LB/RD	MAF Sensor Signal	0.8v, 55 mph: 1.8v
89	GY/WT	TP Sensor Signal	0.53-1.27v
90	BR/WT	Reference Voltage	4.9-5.1v
91	GY/RD	Analog Signal Return	<0.050v
92	LG	Brake Pedal Switch	0v (Brake On: 12v)
93	RD/WT	HO2S-11 (B1 S1) Heater	1v (Heater Off: 12v)
94	YL/LB	HO2S-21 (B2 S1) Heater	1v (Heater Off: 12v)
95	WT/BK	HO2S-12 (B1 S2) Heater	1v (Heater Off: 12v)
96, 98	---	Not Used	---
97	RD	Vehicle Power	12-14v
99	LG/OR	Injector 6 Control	3.2-4.5 ms
100	BR/LB	Injector 4 Control	3.2-4.5 ms
101	WT	Injector 2 Control	3.2-4.5 ms
102	---	Not Used	---
103	BK/WT	Power Ground	<0.1v
104	---	Not Used	---

Pin Connector Graphic

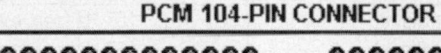

PCM 104-PIN CONNECTOR

Terminal View of 104-Pin PCM Wiring Harness Connector

Standard Colors and Abbreviations

Abbreviation	Color	Abbreviation	Color	Abbreviation	Color
BK	Black	GY	Gray	PK	Purple
BL	Blue	GN	Green	RD	Red
BR	Brown	LG	LT Green	TN	Tan
DB	Dark Blue	OR	Orange	WT	White
DG	DK Green	PK	Pink	YL	Yellow

1990 F-Series 5.0L V8 VIN N (All) 60 Pin Connector

PCM Pin #	Wire Color	Circuit Description (60 Pin)	Value at Hot Idle
1	Y, BK/R	Keep Alive Power	12-14v
3	GY	VSS (+) Signal	0 Hz, 55 mph: 125 Hz
3	DG	VSS (+) Signal	0 Hz, 55 mph: 125 Hz
4	DG, BK	IDM Sensor Signal	20-31 Hz
5	RD	Inferred Mileage Sensor	Digital Signals
6	BK	VSS (-) Signal	0 Hz, 55 mph: 125 Hz
7	LG/YL	ECT Sensor Signal	0.5-0.6v
8 ('89-'90)	BR	Fuel Pump Monitor	On: 12v, Off: 0v
9, 12, 15	---	Not Used	---
10	BK, LG	A/C Switch Signal	A/C On: 12v, Off: 0v
11	WT, BL	Air Management 2 Solenoid	AM2 Off: 12v, On: 1v
16	BK	Ignition System Ground	<0.050v
16	BK/OR	Ignition System Ground	<0.050v
17	PK, TN	Self Test Output & MIL	MIL Off: 12v, On: 1v
17	BL/YL	Self Test Output & MIL	MIL Off: 12v, On: 1v
18 ('88-'89)	TN/LG	Inferred Mileage Sensor	Digital Signals
20	BK	PCM Case Ground	<0.050v
20	BK/WT	PCM Case Ground	<0.050v
21	GY, GN	IAC Solenoid Control	9.2-10.3v
22	TN/LG, OR	Fuel Pump Control	Off: 12v, On: 1v
23	LG/BK	Knock Sensor Signal	0v
23	BK	Knock Sensor Signal	0v
24	YL, TN	PSP Switch Signal	0v (turning: 12v)
24	OR/BR	PSP Switch Signal	0v (turning: 12v)
25	YL, RD	ACT Sensor Signal	1.5-2.5v
26	OR/WT	Reference Voltage	4.9-5.1v
27	BR/LG	EVP Sensor Signal	0.4v, 55 mph: 2.6v
29	GY, DG	HO2S-11 (B1 S1) Signal	0.1-1.1v
30	DG, GY	A/T: Neutral Drive Switch	In 'P': 0v, Others: 5v
30	DG/WT	M/T: Clutch Engage Switch	5v (clutch "in": 0v)
30	GY/YL	M/T: Clutch Engage Switch	5v (clutch "in": 0v)
31 ('88-'90)	GY/YL	Canister Purge Solenoid	12v, 55 mph: 1v
33	DG, WT	VR Solenoid Control	0%, 55 mph: 45%
36	YL/LG	Spark Output Signal	6.93v (50%)
37	RD	Vehicle Power	12-14v
40	BK/WT	Power Ground	<0.1v
45	DG/LG	MAP Sensor Signal	107 Hz (sea level)
45	BL/LG	MAP Sensor Signal	107 Hz (sea level)
46	BK/WT	Analog Signal Return	<0.050v
47	DG, GN	TP Sensor Signal	1v, 55 mph: 1.4v
48	WT, BK	Self Test Indicator Signal	STI On: 1v, Off: 5v
49	OR	HO2S-11 (B1 S1) Ground	<0.050v
51	WT/RD, YL	Air Management 1 Solenoid	AM1 Off: 12v, On: 1v
56	DB, BR	PIP Sensor Signal	6.93v (50%)
57	RD	Vehicle Power	12-14v
58	TN/OR, WT	Injector Bank 1 (INJ 1, 4, 5, 8)	5.0-5.7 ms
59	TN/RD, WT	Injector Bank 2 (INJ 2, 3, 6, 7)	5.0-5.7 ms
60	BK/WT	Power Ground	<0.1v

Pin Connector Graphic

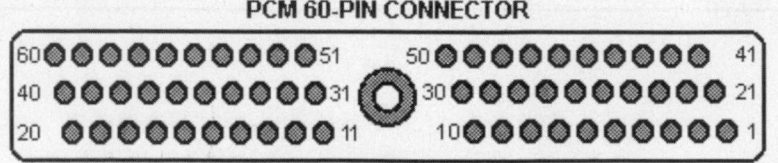

Terminal View of 60-Pin PCM Harness Connector

1991-93 F-Series 5.0L V8 VIN N (All) 60 Pin Connector

PCM Pin #	Wire Color	Circuit Description (60 Pin)	Value at Hot Idle
1	YL	Keep Alive Power	12-14v
3	GY/BK	VSS (+) Signal	0 Hz, 55 mph: 125 Hz
4	TN/YL	IDM Sensor Signal	20-31 Hz
6	PK/OR	VSS (-) Signal	0 Hz, 55 mph: 125 Hz
7	LG/RD	ECT Sensor Signal	0.5-0.6v
8	DG/YL	Fuel Pump Monitor	On: 12v, Off: 0v
9	BK, PK	Data Bus (-) Signal	Digital Signals
10	BK/YL	A/C Switch Signal	A/C On: 12v, Off: 0v
10	PK/BL	A/C Switch Signal	A/C On: 12v, Off: 0v
11	BR	Air Management 2 Solenoid	AM2 Off: 12v, On: 1v
16	OR/RD	Ignition System Ground	<0.050v
17	PK/LG	Self Test Output & MIL	MIL Off: 12v, On: 1v
20	BK	PCM Case Ground	<0.050v
21	WT/BL	IAC Solenoid Control	10.7v (33%)
22	BL/OR	Fuel Pump Control	Off: 12v, On: 1v
23	YL/RD	Knock Sensor Signal	0v
24	YL/LG	PSP Switch Signal	0v (turning: 12v)
25	GY	ACT Sensor Signal	1.5-2.5v
26	BR/WT	Reference Voltage	4.9-5.1v
27	BR/LG	EVP Sensor Signal	0.4v, 55 mph: 2.6v
28	TN/OR	Data Bus (+) Signal	Digital Signals
29	GY/BL	HO2S-11 (B1 S1) Signal	0.1-1.1v
31 ('92-'93)	GY/YL	Canister Purge Solenoid	12v, 55 mph: 1v
33	BR/PK	VR Solenoid Control	0%, 55 mph: 45%
36	PK	Spark Output Signal	6.93v (50%)
37	RD	Vehicle Power	12-14v
40	BK/WT	Power Ground	<0.1v
45	LG/BK	MAP Sensor Signal	107 Hz (sea level)
46	GY/RD	Analog Signal Return	<0.050v
47	GY/WT	TP Sensor Signal	1v, 55 mph: 1.4v
48	WT/PK	Self Test Indicator Signal	STI On: 1v, Off: 5v
49	OR	HO2S-11 (B1 S1) Ground	<0.050v
51	WT/OR	Air Management 1 Solenoid	AM1 Off: 12v, On: 1v
56	GY/OR	PIP Sensor Signal	6.93v (50%)
57	RD	Vehicle Power	12-14v
58	TN	Injector Bank 1 (INJ 1, 4, 5, 8)	4.4-5.6 ms
59	WT	Injector Bank 2 (INJ 2, 3, 6, 7)	4.4-5.6 ms
60	BK/WT	Power Ground	<0.1v

Pin Connector Graphic

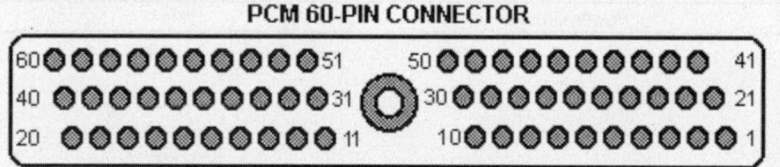

Terminal View of 60-Pin PCM Harness Connector

Standard Colors and Abbreviations

Abbreviation	Color	Abbreviation	Color	Abbreviation	Color
BK	Black	GY	Gray	PK	Purple
BL	Blue	GN	Green	RD	Red
BR	Brown	LG	LT Green	TN	Tan
DB	Dark Blue	OR	Orange	WT	White
DG	DK Green	PK	Pink	YL	Yellow

1994-95 F-Series 5.0L V8 VIN N (All) 60 Pin Connector

PCM Pin #	Wire Color	Circuit Description (60 Pin)	Value at Hot Idle
1	YL	Keep Alive Power	12-14v
2	LG	Brake Position Switch	0v (Brake On: 12v)
3	GY/BK	VSS (+) Signal	0 Hz, 55 mph: 125 Hz
4	YL/BK	IDM Sensor Signal	20-31 Hz
5	DG/WT	TSS (+) Sensor Signal	At 55 mph: 126-136 Hz
6	PK/OR	VSS (-) Signal	0 Hz, 55 mph: 125 Hz
7	LG/RD	ECT Sensor Signal	0.5-0.6v
8	DG/YL	Fuel Pump Monitor	On: 12v, Off: 0v
9	TN/BL	MAF Sensor Return	<0.050v
10	DG/OR	A/C Switch Signal	A/C On: 12v, Off: 0v
11	GY/YL	Canister Purge Solenoid	12v, 55 mph: 1v
12	LG/OR	Injector 6 Control	5.0-5.5 ms
13	TN/RD	Injector 7 Control	5.0-5.5 ms
14	BL	Injector 8 Control	5.0-5.5 ms
15	TN/BK	Injector 5 Control	5.0-5.5 ms
16	OR/RD	Ignition System Ground	<0.050v
17	PK/LG	Self Test Output & MIL	MIL Off: 12v, On: 1v
18	TN/OR	Data Bus (+) Signal	Digital Signals
19	PK/BL	Data Bus (-) Signal	Digital Signals
20	BK	PCM Case Ground	Digital Signals
21	WT/BL	IAC Solenoid Control	10.7v (33%)
22	BL/OR	Fuel Pump Control	Off: 12v, On: 1v
23	YL/RD	Knock Sensor Signal	0v
24	---	Not Used	---
25	GY	IAT Sensor Signal	1.5-2.5v
26	BR/WT	Reference Voltage	4.9-5.1v
27	BR/LG	EVP Sensor Signal	0.4v, 55 mph: 2.6v
28	---	Not Used	---
29	BK /LG	TSS (-) Sensor Signal	<0.050v
30	BL/YL	A/T: MLP Sensor Signal	In 'P': 0v, in O/D: 5v
30	GY/YL	M/T: CPP Switch Signal	5v (clutch "in": 0v)
31	WT/OR	Air Bypass Solenoid Control	AIRB On 1v, Off: 12v
32	WT/LG	TCIL (lamp) Control	7.7v (Switch On: 0v)
33	BR/PK	VR Solenoid Control	0%, 55 mph: 45%
34	BR	Air Diverter Solenoid Control	AIRD On 1v, Off: 12v
35	BR/BL	Injector 4 Control	5.0-5.5 ms
36	PK	Spark Output Signal	6.93v (50%)
37	RD	Vehicle Power	12-14v
38	WT/YL	EPC Solenoid Control	9.5v (5 psi)
39	BR/YL	Injector 3 Control	5.0-5.5 ms
40	BK/WT	Power Ground	<0.1v

1994-95 F-Series 5.0L V8 VIN N (All) 60 Pin Connector

PCM Pin #	Wire Color	Circuit Description (60 Pin)	Value at Hot Idle
41	TN/WT	TCS (switch) Signal	TCS closed: 12, open: 0v
42	BL/BK	4x4 Indicator Light	12v (switch closed: 0v)
43	---	Not Used	---
44	GY/BL	HO2S-11 (B1 S1) Signal	0.1-1.1v
45	---	Not Used	---
46	GY/RD	Analog Signal Return	<0.050v
47	GY/WT	TP Sensor Signal	1v, 55 mph: 1.4v
48	WT/PK	Self Test Indicator Signal	STI On: 1v, Off: 5v
49	OR/BK	TOT Sensor Signal	2.10-2.40v
50	BL/RD	MAF Sensor Signal	0.7v, 55 mph: 1.9v
51	OR/YL	Shift Solenoid 1 Control	12v, 55 mph: 1v
52	PK/OR	Shift Solenoid 2 Control	1v, 55 mph: 12v
53	PK/YL	TCC Solenoid Control	0v, On 55 mph: 12v
54	---	Not Used	---
55	---	Not Used	---
56	GY/OR	PIP Sensor Signal	6.93v (50%)
57	RD	Vehicle Power	12-14v
58	TN	Injector 1 Control	5.0-5.5 ms
59	WT	Injector 2 Control	5.0-5.5 ms
60	BK/WT	Power Ground	<0.1v

Pin Connector Graphic

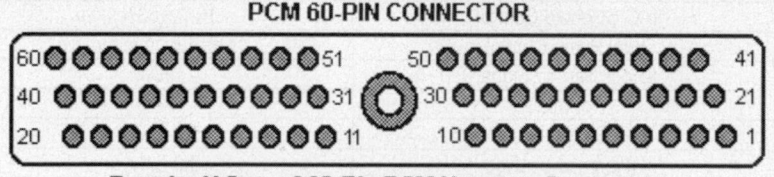

Standard Colors and Abbreviations

Abbreviation	Color	Abbreviation	Color	Abbreviation	Color
BK	Black	GY	Gray	PK	Purple
BL	Blue	GN	Green	RD	Red
BR	Brown	LG	LT Green	TN	Tan
DB	Dark Blue	OR	Orange	WT	White
DG	DK Green	PK	Pink	YL	Yellow

1994-95 F-Series 5.0L V8 VIN N (E4OD) 60 Pin Connector

PCM Pin #	Wire Color	Circuit Description (60 Pin)	Value at Hot Idle
1	YL	Keep Alive Power	12-14v
2	LG	Brake Position Switch	0v (Brake On: 12v)
3	GY/BK	VSS (+) Signal	0 Hz, 55 mph: 125 Hz
4	WT/PK	IDM Sensor Signal	20-31 Hz
4	YL/BK	IDM Sensor Signal	20-31 Hz
5	---	Not Used	---
6	PK/OR	VSS (-) Signal	0 Hz, 55 mph: 125 Hz
7	LG/RD	ECT Sensor Signal	0.5-0.6v
8	DG/YL	Fuel Pump Monitor	On: 12v, Off: 0v
9	TN/BL	MAF Sensor Return	0.050v
10	BK/YL	A/C Switch Signal	A/C On: 12v, Off: 0v
10	DG/OR	A/C Switch Signal	A/C On: 12v, Off: 0v
11	GY/YL	Canister Purge Solenoid	12v, 55 mph: 1v
12	LG/OR	Injector 6 Control	4.7-5.1 ms
13	TN/RD	Injector 7 Control	4.7-5.1 ms
14	BL	Injector 8 Control	4.7-5.1 ms
15	TN/BK	Injector 5 Control	4.7-5.1 ms
16	OR/RD	Ignition System Ground	<0.050v
17	PK/LG	Self Test Output & MIL	MIL Off: 12v, On: 1v
18	TN/OR	Data Bus (+) Signal	Digital Signals
19	PK/BL	Data Bus (-) Signal	Digital Signals
20	BK	PCM Case Ground	<0.050v
21	WT/BL	IAC Solenoid Control	10.7v (33%)
22	BL/OR	Fuel Pump Control	Off: 12v, On: 1v
23	YL/RD	Knock Sensor Signal	0v
24	---	Not Used	---
25	GY	IAT Sensor Signal	1.5-2.5v
26	BR/WT	Reference Voltage	4.9-5.1v
27	BR/LG	EVP Sensor Signal	0.4v, 55 mph: 2.6v
28	---	Not Used	---
29	---	Not Used	---
30	BL/YL	MLP Sensor Signal	In 'P': 0v, in O/D: 5v
31	WT/OR	Air Bypass Solenoid Control	AIRB Off: 12v, On: 1v
32	WT/LG	TCIL (lamp) Control	7.7v (Switch On: 0v)
33	BR/PK	VR Solenoid Control	0%, 55 mph: 45%
34	BR	Air Diverter Solenoid Control	AIRD Off: 12v, On: 1v
35	BR/BL	Injector 4 Control	4.7-5.1 ms
36	PK	Spark Output Signal	6.93v (50%)
37	RD	Vehicle Power	12-14v
38	WT/YL	EPC Solenoid Control	9.5v (5 psi)
39	BR/YL	Injector 3 Control	4.7-5.1 ms
40	BK/WT	Power Ground	<0.1v

1994-95 F-Series 5.0L V8 VIN N (E4OD) 60 Pin Connector

PCM Pin #	Wire Color	Circuit Description (60 Pin)	Value at Hot Idle
41	TN/WT	TCS (switch) Signal	TCS closed: 12, open: 0v
42	BL/BK	4x4 Indicator Light	4x4 Switch On: 0.1v
43	---	Not Used	---
44	GY/BL	HO2S-11 (B1 S1) Signal	0.1-1.1v
45	---	Not Used	---
46	GY/RD	Analog Signal Return	<0.050v
47	GY/WT	TP Sensor Signal	1v, 55 mph: 1.4v
48	WT/PK	Self Test Indicator Signal	STI On: 1v, Off: 5v
49	OR/BK	HO2S-11 (B1 S1) Ground	<0.050v
50	BL/RD	MAF Sensor Signal	0.8v, 55 mph: 1.6v
51	OR/YL	Shift Solenoid 1 Control	12v, 55 mph: 1v
52	PK/OR	Shift Solenoid 2 Control	1v, 55 mph: 12v
53	PK/YL	TCC Solenoid Control	0v, On 55 mph: 12v
54	---	Not Used	---
55	BR/OR	Coast Clutch Solenoid	12v, 55 mph: 12v
56	GY/OR	PIP Sensor Signal	6.93v (50%)
57	RD	Vehicle Power	12-14v
58	TN	Injector 1 Control	4.7-5.1 ms
59	WT	Injector 2 Control	4.7-5.1 ms
60	BK/WT	Power Ground	<0.1v

Pin Connector Graphic

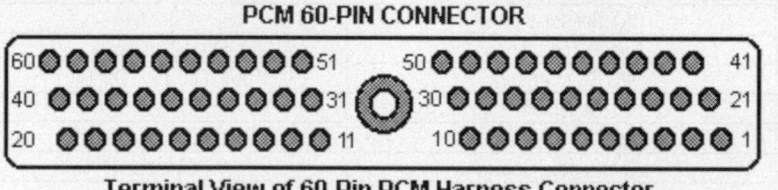

Terminal View of 60-Pin PCM Harness Connector

Standard Colors and Abbreviations

Abbreviation	Color	Abbreviation	Color	Abbreviation	Color
BK	Black	GY	Gray	PK	Purple
BL	Blue	GN	Green	RD	Red
BR	Brown	LG	LT Green	TN	Tan
DB	Dark Blue	OR	Orange	WT	White
DG	DK Green	PK	Pink	YL	Yellow

1996 F-Series 5.0L V8 VIN N (All) 104 Pin Connector

PCM Pin #	Wire Color	Circuit Description (104 Pin)	Value at Hot Idle
1 (E40D)	PK/OR	Shift Solenoid 2 Control	1v, 55 mph: 12v
2	PK/LG	MIL (lamp) Control	MIL Off: 12v, On: 1v
3, 5	---	Not Used	---
4	LG/RD	Power Take-Off (if equipped)	0v (Off)
6 (E40D)	BK/YL	OSS (-) Sensor Signal	<0.050 v
7-12, 17-22	---	Not Used	---
13	PK	Flash EPROM Power	0.1v
14	LG/BK	4x4 Low Switch	12v (switch closed: 0v)
15	PK/LB	Data Bus (-) Signal	Digital Signals
16	TN/OR	Data Bus (+) Signal	Digital Signals
23	OR/RD	Ignition Ground	<0.050v
24, 51	BK/WT	Power Ground	<0.1v
25	BK/LB	Case Ground	<0.050v
26, 28	---	Not Used	---
27 (E40D)	OR/YL	Shift Solenoid 1 Control	1v, 55 mph: 12v
29	TN/WT	TCS (switch) Signal	TCS & O/D On: 12v
30-32, 34	---	Not Used	---
33	PK/OR	PSOM (-) Signal	<0.050v
35	RD/LG	HO2S-12 (B1 S2) Signal	0.1-1.1v
36	TN/LB	MAF Sensor Return	<0.050v
37	OR/BK	TFT Sensor Signal	2.10-2.40v
38	LG/RD	ECT Sensor Signal	0.5-0.6v
39	GY	IAT Sensor Signal	1.5-2.5v
40	DG/YL	Fuel Pump Monitor	On: 12v, Off: 0v
41	DG/OR	A/C Switch Signal	A/C On: 12v, Off: 0v
42-43, 45-46	---	Not Used	---
44	BR	Secondary AIR Diverter	AIRD Off: 12v, On: 1v
47	BK/PK	VR Solenoid Control	0%, 55 mph: 45%
48	YL/BK	Clean Tachometer Output	39-45 Hz
49	GY/OR	PIP Sensor Signal	6.93v (50%)
50	PK	Spark Output Signal	6.93v (50%)
52, 57	---	Not Used	---
53 (E40D)	BR/OR	CCS Solenoid Control	12v, 55 mph: 12v
54 (E40D)	PK/YL	TCC Solenoid Control	0%, 55 mph: 95%
55	YL	Keep Alive Power	12-14v
56	LG/BK	EVAP Purge Solenoid	0-10 Hz (0-100%)
58	GY/BK	PSOM (+) Signal	0 Hz, 55 mph: 125 Hz
59	DG/LG	Misfire Detection Sensor	45-55 Hz
60	GY/LB	HO2S-11 (B1 S1) Signal	0.1-1.1v
61-63	---	Not Used	---
64 (E40D)	LB/YL	TR Sensor Signal	In 'P': 0v, in O/D: 5v
64 (4R70W)	LB/YL	PNP Switch Signal	In 'P': 0v, Others: 5v
64	LB/YL	M/T: CPP Switch Signal	5v (clutch "in": 0v)

1996 F-Series 5.0L V8 VIN N (All) 104 Pin Connector

PCM Pin #	Wire Color	Circuit Description (104 Pin)	Value at Hot Idle
65	BR/LG	DPFE Sensor Signal	0.95-1.05v
66-69	---	Not Used	---
70	WT/OR	Secondary AIR Bypass	AIRB Off: 12v, On: 1v
71	RD	Vehicle Power	12-14v
72	TN/RD	Injector 7 Control	3.2-4.5 ms
73	TN/BK	Injector 5 Control	3.2-4.5 ms
74	BR/YL	Injector 3 Control	3.2-4.5 ms
75	TN	Injector 1 Control	3.2-4.5 ms
76	BK/WT	Power Ground	<0.1v
77	BK/WT	Power Ground	<0.1v
78, 82	---	Not Used	---
79	WT/LG	TCIL (lamp) Control	7.7v (Switch On: 0v)
80	LB/OR	Fuel Pump Control	Off: 12v, On: 1v
81 (E40D)	WT/YL	EPC Solenoid	10v (26 psi)
83	WT/LB	IAC Solenoid Control	10.7v (33%)
84 (E40D)	DB/YL	OSS (+) Sensor Signal	0 Hz, 55 mph: 128 Hz
85-86	---	Not Used	---
87	RD/BK	HO2S-21 (B2 S1) Signal	0.1-1.1v
88	LB/RD	MAF Sensor Signal	0.8v, 55 mph: 1.8v
89	GY/WT	TP Sensor Signal	0.53-1.27v
90	BR/WT	Reference Voltage	4.9-5.1v
91	GY/RD	Analog Signal Return	<0.050 v
92	LG	Brake Pedal Switch	0v (Brake On: 12v)
93	RD/WT	HO2S-11 (B1 S1) Heater	1v (Heater Off: 12v)
94	YL/LB	HO2S-21 (B2 S1) Heater	1v (Heater Off: 12v)
95	WT/BK	HO2S-12 (B1 S2) Heater	1v (Heater Off: 12v)
96	---	Not Used	---
97	RD	Vehicle Power	12-14v
98	LG	Injector 8 Control	3.2-4.5 ms
99	LG/OR	Injector 6 Control	3.2-4.5 ms
100	BR/LB	Injector 4 Control	3.2-4.5 ms
101	WT	Injector 2 Control	3.2-4.5 ms
102	---	Not Used	---
103	BK/WT	Power Ground	<0.1v
104	---	Not Used	---

Pin Connector Graphic

PCM 104-PIN CONNECTOR

Terminal View of 104-Pin PCM Wiring Harness Connector

Standard Colors and Abbreviations

Abbreviation	Color	Abbreviation	Color	Abbreviation	Color
BK	Black	GY	Gray	PK	Purple
BL	Blue	GN	Green	RD	Red
BR	Brown	LG	LT Green	TN	Tan
DB	Dark Blue	OR	Orange	WT	White
DG	DK Green	PK	Pink	YL	Yellow

1997 F-Series 5.4L V8 VIN L (E4OD) 104 Pin Connector

PCM Pin #	Wire Color	Circuit Description (104 Pin)	Value at Hot Idle
1	PK/LB	COP 6 Driver (dwell)	5°, 55 mph: 8°
2	PK/LG	MIL (lamp) Control	MIL Off: 12v, On: 1v
3	BK/WT	Power Ground	<0.1v
4-5, 7-8, 10	---	Not Used	---
6	OR/YL	Shift Solenoid 1 Control	1v, 55 mph: 12v
9	YL/WT	Fuel Level Indictor Signal	1.7v (1/2 full)
11	PK/OR	Shift Solenoid 2 Control	12v, 55 mph: 1v
12	WT/LG	TCIL (lamp) Control	7.7v (Switch On: 0v)
13	PK	Flash EEPROM Power	0.1v
14, 17-19	---	Not Used	---
15	PK/LB	Data Bus (-) Signal	Digital Signals
16	TN/OR	Data Bus (+) Signal	Digital Signals
20	BR/OR	Coast Clutch Solenoid	12v, 55 mph: 12v
21	DB	CKP (+) Sensor Signal	400-420 Hz
22	GY	CKP (-) Sensor Signal	400-420 Hz
23, 28	---	Not Used	---
24, 51	BK/WT	Power Ground	<0.1v
25	BK/LB	Case Ground	<0.050v
26	DB/YL	COP 1 Driver (dwell)	5°, 55 mph: 8°
27	PK/WT	COP 5 Driver (dwell)	5°, 55 mph: 8°
31-32, 38	---	Not Used	---
29	TN/WT	TCS (switch) Signal	TCS & O/D On: 12v
30	DG	Octane Adjustment	9.3v (shorted: 0v)
33	PK/OR	VSS (-) Signal	At 55 mph; 125 Hz
34	OR/BK	Digital TR1 Sensor	0v, 55 mph: 11.5v
35	RD/LG	HO2S-12 (B1 S2) Signal	0.1-1.1v
36	TN/LB	MAF Sensor Return	<0.050v
37	OR/BK	TFT Sensor Signal	2.10-2.40v
39	GY	IAT Sensor Signal	1.5-2.5v
40	DG/YL	Fuel Pump Monitor	On: 12v, Off: 0v
41	BK/YL	A/C Switch Signal	Switch On: 12v, Off: 0v
42-44, 48	---	Not Used	---
45	YL/LB	CHIL (lamp) Control	CHT Off: 12v, On: 1v
46	BR	Intake Manifold Tuning Valve	12v, 55 mph: 12v
47	BR/PK	VR Solenoid Control	0%, 55 mph: 45%
49	LB/BK	Digital TR2 Sensor	0v, 55 mph: 11.5v
50	GY/BK	Digital TR4 Sensor	0v, 55 mph: 11.5v
52	LG/WT	COP 3 Driver (dwell)	5°, 55 mph: 8°
53	OR/YL	COP 4 Driver (dwell)	5°, 55 mph: 8°
54	PK/YL	TCC Solenoid Control	0%, 55 mph: 95%
55	Y	Keep Alive Power	12-14v
56	LG/BK	EVAP Purge Solenoid	0-10 Hz (0-100%)
57	YL/RD	Knock Sensor Signal	0v
58	GY/BK	VSS (+) Signal	0 Hz, 55 mph: 125 Hz
59	---	Not Used	---
60	GY/LB	HO2S-11 (B1 S1) Signal	0.1-1.1v
61	---	Not Used	---
62	RD/PK	FTP Sensor Signal	2.6v (0" H2O - cap off)
63	---	Not Used	---
64	LB/YL	Digital TR3 Sensor	In 'P': 0v, in O/D: 1.7v

1997 F-Series 5.4L V8 VIN L (E4OD) 104 Pin Connector

PCM Pin #	Wire Color	Circuit Description (104 Pin)	Value at Hot Idle
65	BR/LG	DPFE Sensor Signal	0.95-1.05v
66	YL/LG	CHT Sensor Signal	0.7v (194ºF)
67	PK/WT	EVAP CV Solenoid	0-10 Hz (0-100%)
68-70, 76	---	Not Used	---
71, 97	RD	Vehicle Power	12-14v
72	TN/RD	Injector 7 Control	2.7-3.5 ms
73	TN/BK	Injector 5 Control	2.7-3.5 ms
74	BR/YL	Injector 3 Control	2.7-3.5 ms
75	TN	Injector 1 Control	2.7-3.5 ms
82, 84, 86	---	Not Used	---
77	BK/WT	Power Ground	<0.1v
78	WT/PK	COP 7 Driver (dwell)	5º, 55 mph: 8º
79	WT/RD	COP 8 Driver (dwell)	5º, 55 mph: 8º
80	LB/OR	Fuel Pump Control	Off: 12v, On: 1v
81	WT/YL	EPC Solenoid Control	9.1v (4 psi)
83	WT/LB	IAC Solenoid Control	10.7v (33%)
85	DB/OR	CMP Sensor Signal	7 Hz, 55 mph: 15 Hz
87	RD/BK	HO2S-21 (B2 S1) Signal	0.1-1.1v
88	LB/RD	MAF Sensor Signal	0.8v, 55 mph: 1.6v
89	GY/WT	TP Sensor Signal	0.53-1.27v
90	BR/WT	Reference Voltage	4.9-5.1v
91	GY/RD	Analog Signal Return	<0.050v
92	LG	Brake Pedal Switch	0v (Brake On: 12v)
93	RD/WT	HO2S-11 (B1 S1) Heater	1v (Heater Off: 12v)
94	YL/LB	HO2S-21 (B2 S1) Heater	1v (Heater Off: 12v)
95	WT/BK	HO2S-12 (B1 S2) Heater	1v (Heater Off: 12v)
96	---	Not Used	---
98	LB	Injector 8 Control	2.7-3.5 ms
99	LG/OR	Injector 6 Control	2.7-3.5 ms
100	BR/LG	Injector 4 Control	2.7-3.5 ms
101	WT	Injector 2 Control	2.7-3.5 ms
102	---	Not Used	---
103	BK/WT	Power Ground	<0.1v
104	DB/LG	COP 2 Driver (dwell)	5º, 55 mph: 8º

Pin Connector Graphic

PCM 104-PIN CONNECTOR

Terminal View of 104-Pin PCM Wiring Harness Connector

Standard Colors and Abbreviations

Abbreviation	Color	Abbreviation	Color	Abbreviation	Color
BK	Black	GY	Gray	PK	Purple
BL	Blue	GN	Green	RD	Red
BR	Brown	LG	LT Green	TN	Tan
DB	Dark Blue	OR	Orange	WT	White
DG	DK Green	PK	Pink	YL	Yellow

1998-99 F-Series 5.4L V8 VIN L (A/T) 104 Pin Connector

PCM Pin #	Wire Color	Circuit Description (104 Pin)	Value at Hot Idle
1	LG/YL	COP 6 Driver (dwell)	5°, 55 mph: 8°
2 ('98)	PK/LG	MIL (lamp) Control	MIL Off: 12v, On: 1v
3, 24, 51	BK/WT	Power Ground	<0.1v
6	OR/YL	Shift Solenoid 1 Control	1v, 55 mph: 12v
7 ('99)	PK	Transfer Case Speed Sensor	0 Hz, 55 mph: 471 Hz
9 ('98)	YL/WT	Fuel Level Indicator Signal	1.7v (1/2 full)
11	PK/OR	Shift Solenoid 2 Control	12v, 55 mph: 1v
12	WT/LG	TCIL (lamp) Control	7.7v (Switch On: 0v)
13	PK	Flash EEPROM Power	0.1v
14	LB/BK	4x4 Low Indicator Switch	4x4 Switch On: 0.1v
15	PK	Data Bus (-) Signal	Digital Signals
16	TN/OR	Data Bus (+) Signal	Digital Signals
20	BR/OR	Coast Clutch Solenoid	12v, 55 mph: 12v
21	DB	CKP (+) Sensor Signal	411 Hz
22	GY	CKP (-) Sensor Signal	411 Hz
25	BK/LB	Case Ground	<0.050v
26	LG/WT	COP 1 Driver (dwell)	5°, 55 mph: 8°
27	LG/YL	COP 5 Driver (dwell)	5°, 55 mph: 8°
29	TN/WT	TCS (switch) Signal	TCS & O/D On: 12v
30 ('98)	DG	Octane Adjustment	9.3v (shorted: 0v)
32	DG/PK	Knock Sensor 1 Signal	0v
33 ('98)	PK/OR	VSS (-) Signal	0 Hz, 55 mph: 125 Hz
34	YL/BK	Digital TR1 Sensor	0v, 55 mph: 11.5v
35	RD/LG	HO2S-12 (B1 S2) Signal	0.1-1.1v
36	TN/LB	MAF Sensor Return	<0.050v
37	OR/BK	TFT Sensor Signal	2.10-2.40v
39	GY	IAT Sensor Signal	1.5-2.5v
40	DG/YL	Fuel Pump Monitor	On: 12v, Off: 0v
41	BK/YL	A/C Switch Signal	Switch Closed: 12v
45 ('98)	RD/WT	CHIL (lamp) Control	7.7v (Switch On: 0v)
46	BR	Intake Manifold Tuning Valve	12v, 55 mph: 12v
47	BR/PK	VR Solenoid Control	0%, 55 mph: 45%
48 ('98)	WT/PK	Clean Tachometer Output	55 Hz, 55 mph: 120 Hz
49	LB/BK	Digital TR2 Sensor	0v, 55 mph: 11.5v
50	WT/BK	Digital TR4 Sensor	0v, 55 mph: 11.5v
52	LG/WT	COP 3 Driver (dwell)	5°, 55 mph: 8°
53	PK/WT	COP 4 Driver (dwell)	5°, 55 mph: 8°
54	PK/YL	TCC Solenoid Control	0%, 55 mph: 95%
55	RD/WT	Keep Alive Power	12-14v
56	LG/BK	EVAP Purge Solenoid	0-10 Hz (0-100%)
57	YL/RD	Knock Sensor 1 Signal	0v
58 ('99)	GY/BK	VSS (+) Signal	0 Hz, 55 mph: 125 Hz
60	GY/LB	HO2S-11 (B1 S1) Signal	0.1-1.1v
61	PK/LG	HO2S-22 (B2 S2) Signal	0.1-1.1v
62	RD/PK	FTP Sensor Signal	2.6v (0" H2O - cap off)
64	LB/YL	Digital TR3 Sensor	In 'P': 0v, in O/D: 1.7v

1998-99 F-Series 5.4L V8 VIN L (A/T) 104 Pin Connector

PCM Pin #	Wire Color	Circuit Description (104 Pin)	Value at Hot Idle
65	BR/LG	DPFE Sensor Signal	0.95-1.05v
66	YL/LG	CHT Sensor Signal	0.7v (194°F)
67	PK/WT	EVAP CV Solenoid	0-10 Hz (0-100%)
68 ('99)	GY/BK	Transfer Case Speed Sensor	0 Hz, 55 mph: 471 Hz
69 ('99)	PK/YL	A/C WOT Relay Control	Off: 12v, On: 1v
71, 97	RD	Vehicle Power	12-14v
72	TN/RD	Injector 7 Control	2.7-3.5 ms
73	TN/BK	Injector 5 Control	2.7-3.5 ms
74	BR/YL	Injector 3 Control	2.7-3.5 ms
75	TN	Injector 1 Control	2.7-3.5 ms
77, 103	BK/WT	Power Ground	<0.1v
78	WT/PK	COP 7 Driver (dwell)	5°, 55 mph: 8°
79	WT/RD	COP 8 Driver (dwell)	5°, 55 mph: 8°
80	LB/OR	Fuel Pump Control	Off: 12v, On: 1v
81	WT/YL	EPC Solenoid Control	9.1v (4 psi)
83	WT/LB	IAC Solenoid Control	10.7v (33%)
84	DB/YL	OSS Sensor Signal	0 Hz, 55 mph: 700 Hz
85	DG	CMP Sensor Signal	7 Hz, 55 mph: 15 Hz
87	RD/BK	HO2S-21 (B2 S1) Signal	0.1-1.1v
88	LB/RD	MAF Sensor Signal	0.8v, 55 mph: 1.6v
89	GY/WT	TP Sensor Signal	0.53-1.27v
90	BR/WT	Reference Voltage	4.9-5.1v
91	GY/RD	Analog Signal Return	<0.050v
92	RD/LG	Brake Position Switch	0v (Brake On: 12v)
93	RD/WT	HO2S-11 (B1 S1) Heater	1v (Heater Off: 12v)
94	YL/LB	HO2S-21 (B2 S1) Heater	1v (Heater Off: 12v)
95	WT/BK	HO2S-12 (B1 S2) Heater	1v (Heater Off: 12v)
96	TN/YL	HO2S-22 (B2 S2) Heater	1v (Heater Off: 12v)
98	LB	Injector 8 Control	2.7-3.5 ms
99	LG/OR	Injector 6 Control	2.7-3.5 ms
100	BR/LB	Injector 4 Control	2.7-3.5 ms
101	WT	Injector 2 Control	2.7-3.5 ms
104	PK/WT	COP 2 Driver (dwell)	5°, 55 mph: 8°

Pin Connector Graphic

PCM 104-PIN CONNECTOR

1 ◦◦◦◦◦◦◦◦◦◦◦◦◦ ◦◦◦◦◦◦◦◦◦◦◦◦◦ 26
27 ◦◦◦◦◦◦◦◦◦◦◦◦◦ ◦◦◦◦◦◦◦◦◦◦◦◦◦ 52
53 ◦◦◦◦◦◦◦◦◦◦◦◦◦ ◦◦◦◦◦◦◦◦◦◦◦◦◦ 78
79 ◦◦◦◦◦◦◦◦◦◦◦◦◦ ◦◦◦◦◦◦◦◦◦◦◦◦◦ 104

Terminal View of 104-Pin PCM Wiring Harness Connector

Standard Colors and Abbreviations

Abbreviation	Color	Abbreviation	Color	Abbreviation	Color
BK	Black	GY	Gray	PK	Purple
BL	Blue	GN	Green	RD	Red
BR	Brown	LG	LT Green	TN	Tan
DB	Dark Blue	OR	Orange	WT	White
DG	DK Green	PK	Pink	YL	Yellow

2000-02 F-Series 5.4L V8 VIN L (All) 104 Pin Connector

PCM Pin #	Wire Color	Circuit Description (104 Pin)	Value at Hot Idle
1	OR/YL	COP 6 Driver (dwell)	5°, 55 mph: 8°
3	BK/WT	Power Ground	<0.1v
6	OR/YL	Shift Solenoid 'A' Control	1v, 55 mph: 12v
7	PK	Transfer Case Speed Sensor	0 Hz, 55 mph: 471 Hz
11	PK/OR	Shift Solenoid 'B' Control	12v, 55 mph: 1v
12	WT/LG	TCIL (lamp) Control	7.7v (Switch On: 0v)
13	PK	Flash EEPROM Power	0.1v
14	LB/BK	4x4 Low Indicator Switch	12v (Switch On: 0v)
15	PK/LB	Data Bus (-) Signal	Digital Signals
16	TN/OR	Data Bus (+) Signal	Digital Signals
20	BR/OR	Coast Clutch Solenoid	12v, 55 mph: 12v
21	DB	CKP (-) Sensor Signal	411 Hz
22	GY	CKP (+) Sensor Signal	411 Hz
25	LB/YL	Chassis Ground	<0.050v
26	LG/WT	COP 1 Driver (dwell)	5°, 55 mph: 8°
27	LG/YL	COP 5 Driver (dwell)	5°, 55 mph: 8°
29	TN/WT	TCS (switch) Signal	TCS & O/D On: 12v
32	DG/PK	Knock Sensor (-) Signal	0v
34	YL/BK	Digital TR1 Sensor	0v, 55 mph: 11.5v
35	RD/LG	HO2S-12 (B1 S2) Signal	0.1-1.1v
36	TN/LB	MAF Sensor Return	<0.050v
37	OR/BK	TFT Sensor Signal	2.10-2.40v
39	GY	IAT Sensor Signal	1.5-2.5v
40	DG/YL	Fuel Pump Monitor	On: 12v, Off: 0v
41	BK/YL	A/C Switch Signal	Switch Closed: 12v
46	BR	Intake Manifold Tuning Valve	12v, 55 mph: 12v
47	BR/PK	VR Solenoid Control	0%, 55 mph: 45%
49	LB/BK	Digital TR2 Sensor	0v, 55 mph: 11.5v
50	WT/BK	Digital TR4 Sensor	0v, 55 mph: 11.5v
51	BK/WT	Power Ground	<0.1v
52	WT/PK	COP 3 Driver (dwell)	5°, 55 mph: 8°
53	DG/PK	COP 4 Driver (dwell)	5°, 55 mph: 8°
54	PK/YL	TCC Solenoid Control	0%, 55 mph: 95%
55	RD/WT	Keep Alive Power	12-14v
56	LG/BK	EVAP Purge Solenoid	0-10 Hz (0-100%)
57	YL/RD	Knock Sensor 1 Signal	0v
59	DG/WT	TSS Sensor Signal	300 Hz, 55 mph: 980
60	GY/LB	HO2S-11 (B1 S1) Signal	0.1-1.1v
61	PK/LG	HO2S-22 (B2 S2) Signal	0.1-1.1v
62	RD/PK	FTP Sensor Signal	2.6v (0" H2O - cap off)
64	LB/YL	Digital TR3 Sensor	In 'P': 0v, in O/D: 1.7v

2000-02 F-Series 5.4L V8 VIN L (All) 104 Pin Connector

PCM Pin #	Wire Color	Circuit Description (104 Pin)	Value at Hot Idle
65 ('00)	BR/LG	DPFE Sensor Signal	0.20-1.30v
65 ('01-'02)	BR/LG	DPFE Sensor Signal	0.95-1.05v
66	YL/LG	CHT Sensor Signal	0.7v (194ºF)
67	PK/WT	EVAP CV Solenoid	0-10 Hz (0-100%)
68	GY/BK	VSS (+) Signal	0 Hz, 55 mph: 125 Hz
69	PK/YL	A/C WOT Relay Control	Off: 12v, On: 1v
71	RD	Vehicle Power	12-14v
72	TN/RD	Injector 7 Control	2.7-3.5 ms
73	TN/BK	Injector 5 Control	2.7-3.5 ms
74	BR/YL	Injector 3 Control	2.7-3.5 ms
75	TN	Injector 1 Control	2.7-3.5 ms
76	---	Not Used	---
77	BK/WT	Power Ground	<0.1v
78	PK/LB	COP 7 Driver (dwell)	5º, 55 mph: 8º
79	WT/RD	COP 8 Driver (dwell)	5º, 55 mph: 8º
80	LB/OR	Fuel Pump Relay Control	Off: 12v, On: 1v
81	WT/YL	EPC Solenoid Control	9.1v (4 psi)
82,86	---	Not Used	---
83	WT/LB	IAC Valve Control	10.7v (33%)
84	DB/YL	OSS Sensor Signal	0 Hz, 55 mph: 228 Hz
85	DG	CMP Sensor Signal	7 Hz, 55 mph: 15 Hz
87	RD/BK	HO2S-21 (B2 S1) Signal	0.1-1.1v
88	LB/RD	MAF Sensor Signal	0.8v, 55 mph: 1.6v
89	GY/WT	TP Sensor Signal	0.53-1.27v
90	BR/WT	Reference Voltage	4.9-5.1v
91	GY/RD	Sensor Return	<0.050v
92	RD/LG	Brake Pedal Switch	0v (Brake On: 12v)
93	RD/WT	HO2S-11 (B1 S1) Heater	1v (Heater Off: 12v)
94	YL/LB	HO2S-21 (B2 S1) Heater	1v (Heater Off: 12v)
95	WT/BK	HO2S-12 (B1 S2) Heater	1v (Heater Off: 12v)
96	TN/YL	HO2S-22 (B2 S2) Heater	1v (Heater Off: 12v)
97	RD	Vehicle Power	12-14v
98	YL	Injector 8 Control	2.7-3.5 ms
99	YL/LB	Injector 6 Control	2.7-3.5 ms
100	YL/BK	Injector 4 Control	2.7-3.5 ms
101	YL/RD	Injector 2 Control	2.7-3.5 ms
102	---	Not Used	---
103	BK/WT	Power Ground	<0.1v
104	PK/WT	COP 2 Driver (dwell)	5º, 55 mph: 8º

Pin Connector Graphic

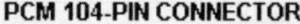

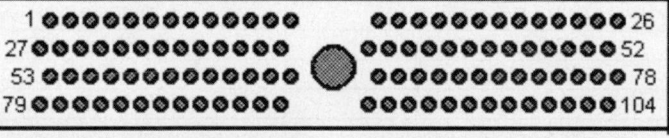

PCM 104-PIN CONNECTOR

Terminal View of 104-Pin PCM Wiring Harness Connector

Standard Colors and Abbreviations

Abbreviation	Color	Abbreviation	Color	Abbreviation	Color
BK	Black	GY	Gray	PK	Purple
BL	Blue	GN	Green	RD	Red
BR	Brown	LG	LT Green	TN	Tan
DB	Dark Blue	OR	Orange	WT	White
DG	DK Green	PK	Pink	YL	Yellow

2003-05 F-Series 5.4L V8 VIN L (All) 104 Pin Connector

PCM Pin #	Wire Color	Circuit Description (104 Pin)	Value at Hot Idle
1	OR/YL	COP 6 Driver (dwell)	5°, 55 mph: 8°
2	---	Not Used	---
3	BK/WT	Power Ground	<0.1v
4	PK	Transfer Case Speed Sensor (MSOF)	12v, 55 mph: 12v
5	---	Not Used	---
6	OR/YL	Shift Solenoid 'A' Control	1v, 55 mph: 12v
7-10	---	Not Used	---
11	VT/OR	Shift Solenoid 'B' Control	12v, 55 mph: 1v
12	WT/LG	TCIL (lamp) Control	7.7v (Switch On: 0v)
13	VT	Flash EEPROM Power	0.1v
14	LB/BK	4x4 Low Indicator Switch	12v (Switch On: 0v)
15	PK/LB	SCP Data Bus (-) Signal	<0.050v
16	TN/OR	SCP Data Bus (+) Signal	Digital Signals
17-19	---	Not Used	---
20	BR/OR	Coast Clutch Solenoid	12v, 55 mph: 12v
21	DB	CKP (-) Sensor Signal	411 Hz
22	GY	CKP (+) Sensor Signal	411 Hz
23-24	---	Not Used	---
25	LB/YL	Chassis Ground	<0.050v
26	LG/WT	COP 1 Driver (dwell)	5°, 55 mph: 8°
27	LG/YL	COP 5 Driver (dwell)	5°, 55 mph: 8°
28	---	Not Used	---
29	TN/WT	TCS (switch) Signal	TCS & O/D On: 12v
30-31	---	Not Used	---
32	DG/VT	Knock Sensor (-) Signal	<0.050v
33	---	Not Used	---
34	YL/BK	Digital TR1 Sensor	0v, 55 mph: 11.5v
35	RD/LG	HO2S-12 (B1 S2) Signal	0.1-1.1v
36	TN/LB	MAF Sensor Return	<0.050v
37	OR/BK	TFT Sensor Signal	2.10-2.40v
38	---	Not Used	---
39	GY	IAT Sensor Signal	1.5-2.5v
40	DG/YL	Fuel Pump Monitor	On: 12v, Off: 0v
41	BK/YL	A/C Pressure Switch Signal	Switch Closed: 12v
42-45	---	Not Used	---
46	BR	Intake Manifold Tuning Valve Control	12v, 55 mph: 12v
47	BR/PK	VR Solenoid Control	0%, 55 mph: 45%
48	---	Not Used	---
49	LB/BK	Digital TR2 Sensor	0v, 55 mph: 11.5v
50	WT/BK	Digital TR4 Sensor	0v, 55 mph: 11.5v
51	BK/WT	Power Ground	<0.1v
52	WT/PK	COP 3 Driver (dwell)	5°, 55 mph: 8°
53	DG/VT	COP 4 Driver (dwell)	5°, 55 mph: 8°
54	VT/YL	TCC Solenoid Control	0%, 55 mph: 95%
55	RD/WT	Keep Alive Power	12-14v
56	LG/BK	EVAP Purge Solenoid	0-10 Hz (0-100%)
57	YL/RD	Knock Sensor 1 Signal	0v
58	---	Not Used	---
59	DG/WT	TSS Sensor Signal	300 Hz, 55 mph: 980
60	GY/LB	HO2S-11 (B1 S1) Signal	0.1-1.1v
61	VT/LG	HO2S-22 (B2 S2) Signal	0.1-1.1v
62	RD/PK	FTP Sensor Signal	2.6v (0" H2O - cap off)
63	---	Not Used	---
64	LB/YL	Digital TR3A Sensor Signal	In 'P': 0v, in O/D: 1.7v

2003-05 F-Series 5.4L V8 VIN L (A/T) 104 Pin Connector

PCM Pin #	Wire Color	Circuit Description (104 Pin)	Value at Hot Idle
65	BR/LG	EGR DPFE Sensor Signal	0.95-1.05v
66	YL/LG	Cylinder Head Temperature Sensor	0.7v (194°F)
67	VT/WT	EVAP Canister Vent Solenoid	0-10 Hz (0-100%)
68	GY/BK	VSS (+) Signal	0 Hz, 55 mph: 125 Hz
69	PK/YL	A/C WOT Relay Control	Off: 12v, On: 1v
70	BK/WT	Check Fuel Cap Indicator Control	Indicator Off: 12v, On: 1v
71	RD	Vehicle Power	12-14v
72	TN/RD	Injector 7 Control	2.7-3.5 ms
73	TN/BK	Injector 5 Control	2.7-3.5 ms
74	BR/YL	Injector 3 Control	2.7-3.5 ms
75	TN	Injector 1 Control	2.7-3.5 ms
76	---	Not Used	---
77	BK/WT	Power Ground	<0.1v
78	PK/LB	COP 7 Driver (dwell)	5°, 55 mph: 8°
79	WT/RD	COP 8 Driver (dwell)	5°, 55 mph: 8°
80	LB/OR	Fuel Pump Relay Control	Off: 12v, On: 1v
81	WT/YL	EPC Solenoid Control	9.1v (4 psi)
82	---	Not Used	---
83	WT/LB	IAC Valve Control	10.7v (33%)
84	DB/YL	OSS Sensor Signal	0 Hz, 55 mph: 228 Hz
85	DG	CMP Sensor Signal	7 Hz, 55 mph: 15 Hz
86	---	Not Used	---
87	RD/BK	HO2S-21 (B2 S1) Signal	0.1-1.1v
88	LB/RD	MAF Sensor Signal	0.8v, 55 mph: 1.6v
89	GY/WT	TP Sensor Signal	0.53-1.27v
90	BR/WT	Reference Voltage	4.9-5.1v
91	GY/RD	Sensor Return	<0.050v
92	RD/LG	Brake Pedal Switch	0v (Brake On: 12v)
93	RD/WT	HO2S-11 (B1 S1) Heater	1v (Heater Off: 12v)
94	YL/LB	HO2S-21 (B2 S1) Heater	1v (Heater Off: 12v)
95	WT/BK	HO2S-12 (B1 S2) Heater	1v (Heater Off: 12v)
96	TN/YL	HO2S-22 (B2 S2) Heater	1v (Heater Off: 12v)
97	RD	Vehicle Power	12-14v
98	YL	Injector 8 Control	2.7-3.5 ms
99	YL/LB	Injector 6 Control	2.7-3.5 ms
100	BR/LB	Injector 4 Control	2.7-3.5 ms
101	YL/RD	Injector 2 Control	2.7-3.5 ms
102	---	Not Used	---
103	BK/WT	Power Ground	<0.1v
104	PK/WT	COP 2 Driver (dwell)	5°, 55 mph: 8°

Pin Connector Graphic

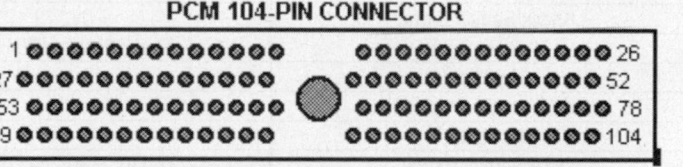

PCM 104-PIN CONNECTOR

Terminal View of 104-Pin PCM Wiring Harness Connector

Standard Colors and Abbreviations

Abbreviation	Color	Abbreviation	Color	Abbreviation	Color
BK	Black	GY	Gray	PK	Purple
BL	Blue	GN	Green	RD	Red
BR	Brown	LG	LT Green	TN	Tan
DB	Dark Blue	OR	Orange	WT	White
DG	DK Green	PK	Pink	YL	Yellow

1997 F-Series 5.4L CNG VIN M (E4OD) 104 Pin Connector

PCM Pin #	Wire Color	Circuit Description (104 Pin)	Value at Hot Idle
1	PK/LB	COP 6 Driver (dwell)	5°, 55 mph: 8°
2	PK/LG	MIL (lamp) Control	MIL Off: 12v, On: 1v
3	BK/WT	Power Ground	<0.1v
4-5, 7-8, 10	---	Not Used	---
6	OR/YL	Shift Solenoid 1 Control	1v, 55 mph: 12v
9	YL/WT	Fuel Level Indicator Signal	1.7v (1/2 full)
11	PK/OR	Shift Solenoid 2 Control	12v, 55 mph: 1v
12	WT/LG	TCIL (lamp) Control	7.7v (Switch On: 0v)
13	PK	Flash EEPROM Power	0.1v
14, 17-19	---	Not Used	---
15	PK/LB	Data Bus (-) Signal	Digital Signals
16	TN/OR	Data Bus (+) Signal	Digital Signals
20	BR/OR	Coast Clutch Solenoid	12v, 55 mph: 12v
21	DB	CKP (+) Sensor Signal	460-500 Hz
22	GY	CKP (-) Sensor Signal	460-500 Hz
23	---	Not Used	---
24, 51	BK/WT	Power Ground	<0.1v
25	BK/LB	Case Ground	<0.050v
26	DB/YL	COP 1 Driver (dwell)	5°, 55 mph: 8°
27	PK/WT	COP 5 Driver (dwell)	5°, 55 mph: 8°
28, 31-32	---	Not Used	---
29	TN/WT	TCS (switch) Signal	TCS & O/D On: 12v
30	DG	Octane Adjustment	9.3v (shorted: 0v)
33	PK/OR	VSS (-) Signal	At 55 mph; 125 Hz
34	OR/BK	Digital TR1 Sensor	0v, 55 mph: 11.5v
36	TN/LB	MAF Sensor Return	<0.050v
37	OR/BK	TFT Sensor Signal	2.10-2.40v
38	---	Not Used	---
39	GY	IAT Sensor Signal	1.5-2.5v
40	DG/YL	Fuel Pump Monitor	On: 12v, Off: 0v
41	BK/YL	A/C Hi Pressure Cutoff Switch	AC & Switch On: 12v
42-44, 48	---	Not Used	---
45	YL/LB	CHIL (lamp) Control	CHT Off: 12v, On: 1v
46	BR	Intake Manifold Tuning Valve	12v, 55 mph: 12v
47	BR/PK	VR Solenoid Control	0%, 55 mph: 45%
49	LB/BK	Digital TR2 Sensor	0v, 55 mph: 11.5v
50	GY/BK	Digital TR4 Sensor	0v, 55 mph: 11.5v
52	LG/WT	COP 3 Driver (dwell)	5°, 55 mph: 8°
53	OR/YL	COP 4 Driver (dwell)	5°, 55 mph: 8°
54	PK/YL	TCC Solenoid Control	0%, 55 mph: 95%
55	YL	Keep Alive Power	12-14v
56	LG/BK	EVAP Purge Solenoid	0-10 Hz (0-100%)
57	YL/RD	Knock Sensor Signal	0v
58	GY/BK	VSS (+) Signal	0 Hz, 55 mph: 125 Hz
59	---	Not Used	---
60	GY/LB	HO2S-11 (B1 S1) Signal	0.1-1.1v
61	---	Not Used	---
62	RD/PK	FTP Sensor Signal	2.6v (0" H2O - cap off)
63	---	FRP Sensor Signal	2-3.7v (90-100 psi)
64	LB/YL	Digital TR3 Sensor	In 'P': 0v, in O/D: 1.7v

1997 F-Series 5.4L CNG VIN M (E4OD) 104 Pin Connector

PCM Pin #	Wire Color	Circuit Description (104 Pin)	Value at Hot Idle
65	BR/LG	DPFE Sensor Signal	0.95-1.05v
66	YL/LG	CHT Sensor Signal	0.6v (194°F)
67	PK/WT	EVAP CV Solenoid	0-10 Hz (0-100%)
68, 70	---	Not Used	---
71, 97	RD	Vehicle Power	12-14v
72	TN/RD	Injector 7 Control	2.7-3.5 ms
73	TN/BK	Injector 5 Control	2.7-3.5 ms
74	BR/YL	Injector 3 Control	2.7-3.5 ms
75	TN	Injector 1 Control	2.7-3.5 ms
76, 82, 84	---	Not Used	---
77	BK/WT	Power Ground	<0.1v
78	WT/PK	COP 7 Driver (dwell)	5°, 55 mph: 8°
79	WT/RD	COP 8 Driver (dwell)	5°, 55 mph: 8°
80	LB/OR	Fuel Pump Control	Off: 12v, On: 1v
81	WT/YL	EPC Solenoid Control	9.1v (4 psi)
83	WT/LB	IAC Solenoid Control	10.7v (33%)
85	DB/OR	CMP Sensor Signal	7 Hz, 55 mph: 15 Hz
87	RD/BK	HO2S-21 (B2 S1) Signal	0.1-1.1v
88	LB/RD	MAF Sensor Signal	0.8v, 55 mph: 1.6v
89	GY/WT	TP Sensor Signal	0.53-1.27v
90	BR/WT	Reference Voltage	4.9-5.1v
91	GY/RD	Analog Signal Return	<0.050v
92	LG	Brake Pedal Switch	0v (Brake On: 12v)
93	RD/WT	HO2S-11 (B1 S1) Heater	1v (Heater Off: 12v)
94	YL/LB	HO2S-21 (B2 S1) Heater	1v (Heater Off: 12v)
86, 96	---	Not Used	---
98	LB	Injector 8 Control	2.7-3.5 ms
99	LG/OR	Injector 6 Control	2.7-3.5 ms
100	BR/LG	Injector 4 Control	2.7-3.5 ms
101	WT	Injector 2 Control	2.7-3.5 ms
102	---	Not Used	---
103	BK/WT	Power Ground	<0.1v
104	DB/LG	COP 2 Driver (dwell)	5°, 55 mph: 8°

Pin Connector Graphic

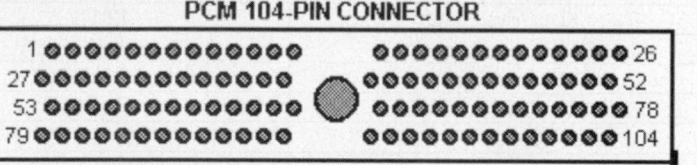

Standard Colors and Abbreviations

Abbreviation	Color	Abbreviation	Color	Abbreviation	Color
BK	Black	GY	Gray	PK	Purple
BL	Blue	GN	Green	RD	Red
BR	Brown	LG	LT Green	TN	Tan
DB	Dark Blue	OR	Orange	WT	White
DG	DK Green	PK	Pink	YL	Yellow

1998-99 F-Series 5.4L CNG VIN M, VIN Z (E4OD) 104-P Connector

PCM Pin #	Wire Color	Circuit Description (104 Pin)	Value at Hot Idle
1	LG/YL	COP 6 Driver (dwell)	5°, 55 mph: 8°
2 ('98)	PK/LG	MIL (lamp) Control	MIL Off: 12v, On: 1v
3, 24	BK/WT	Power Ground	<0.1v
4	WT/PK	Power Takeoff Signal	PTO Off: 0v, On: 12v
6	OR/YL	Shift Solenoid 1 Control	1v, 55 mph: 12v
9	YL/WT	Fuel Level Indicator Signal	1.7v (1/2 full)
11	PK/OR	Shift Solenoid 2 Control	12v, 55 mph: 1v
12	WT/LG	TCIL (lamp) Control	7.7v (Switch On: 0v)
13	PK	Flash EEPROM Power	0.1v
15	PK/LB	Data Bus (-)	<0.050v
16	TN/OR	Data Bus (+)	Digital Signals
20	BR/OR	Coast Clutch Solenoid	12v, 55 mph: 12v
21	DB	CKP (-) Sensor Signal	440-490 Hz
22	GY	CKP (+) Sensor Signal	440-490 Hz
25	BK/LB	Case Ground	<0.050v
26	LG/WT	COP 1 Driver (dwell)	5°, 55 mph: 8°
27	LG/OR	COP 5 Driver (dwell)	5°, 55 mph: 8°
29	TN/WT	TCS (switch) Signal	TCS & O/D On: 12v
30	DG	Octane Adjust Shorting Bar	9.3v (shorted: 0v)
33 ('98)	PK/OR	VSS (-) Signal	0 Hz, 55 mph: 125 Hz
34	YL/BK	Digital TR1 Sensor	0v, 55 mph: 11.5v
36	TN/LB	MAF Sensor Return	<0.050v
37	OR/BK	TFT Sensor Signal	2.10-2.40v
39	GY	IAT Sensor Signal	1.5-2.5v
40	DG/YL	Fuel Pump Monitor	On: 12v, Off: 0v
41	BK/YL	A/C Switch Signal	Switch Closed: 12v
45 ('98)	YL/LB	CHIL (lamp) Control	7.7v (Switch On: 0v)
46	BR	Intake Manifold Tuning Valve	12v, 55 mph: 12v
48 ('98)	WT/PK	Clean Tachometer Output	55 Hz, 55 mph: 120 Hz
49	LB/BK	Digital TR2 Sensor	0v, 55 mph: 11.5v
50	LG/RD	Digital TR4 Sensor	0v, 55 mph: 11.5v
51	BK/WT	Power Ground	<0.1v
52	WT/PK	COP 3 Driver (dwell)	5°, 55 mph: 8°
53	OR/YL	COP 4 Driver (dwell)	5°, 55 mph: 8°
54	PK/YL	TCC Solenoid Control	0%, 55 mph: 95%
55	RD/WT	Keep Alive Power	12-14v
56	LG/BK	EVAP Purge Solenoid	0-10 Hz (0-100%)
58 ('99)	GY/BK	VSS (+) Signal	0 Hz, 55 mph: 125 Hz
60	GY/LB	HO2S-11 (B1 S1) Signal	0.1-1.1v
62	LB	EFT Sensor Signal	1.7-3.5v (50-120°F)
63	RD/PK	Fuel Rail Pressure Sensor	2-3.7v (90-100 psi)
64	LB/YL	Digital TR 3A Sensor	In 'P': 0v, in O/D: 1.7v

1998-99 F-Series 5.4L CNG VIN M, VIN Z (E4OD) 104-P Connector

PCM Pin #	Wire Color	Circuit Description (104 Pin)	Value at Hot Idle
66	YL/LG	CHT Sensor Signal	0.6 (194°F)
68	GY/BK	VSS (-) Signal	0 Hz, 55 mph: 125 Hz
69	PK/YL	A/C WOT Relay Control	Off: 12v, On: 1v
71, 97	RD	Vehicle Power	12-14v
72	WT	Injector 7 Signal NGV Module	3.9-4.6 ms
73	WT/LB	Injector 5 Signal NGV Module	3.9-4.6 ms
74	WT/BK	Injector 3 Signal NGV Module	3.9-4.6 ms
75	WT/RD	Injector 1 Signal NGV Module	3.9-4.6 ms
77, 103	BK/WT	Power Ground	<0.1v
78	PK/LB	COP 7 Driver (dwell)	5°, 55 mph: 8°
79	WT/RD	COP 8 Driver (dwell)	5°, 55 mph: 8°
80	LB/OR	Fuel Shutoff Valve Control	Off: 12v, On: 1v
81	WT/YL	EPC Solenoid Control	9.1v (4 psi)
83	WT/LB	IAC Solenoid Control	10.7v (33%)
84	DB/YL	OSS Sensor Signal	0 Hz, 55 mph: 720 Hz
85	DG	CMP Sensor Signal	6 Hz, 55 mph: 16 Hz
87	RD/BK	HO2S-21 (B2 S1) Signal	0.1-1.1v
88	LB/RD	MAF Sensor Signal	0.8v, 55 mph: 1.6v
89	GY/WT	TP Sensor Signal	0.53-1.27v
90	BR/WT	Reference Voltage	4.9-5.1v
91	GY/RD	Analog Signal Return	<0.050v
92	RD/LG	Brake Pedal Switch	0v (Brake On: 12v)
93	RD/WT	HO2S-11 (B1 S1) Heater	1v (Heater Off: 12v)
94	YL/LB	HO2S-21 (B2 S1) Heater	1v (Heater Off: 12v)
98	YL	Injector 8 Signal NGV Module	3.9-4.6 ms
99	YL/LB	Injector 6 Signal NGV Module	3.9-4.6 ms
100	YL/BK	Injector 4 Signal NGV Module	3.9-4.6 ms
101	YL/RD	Injector 2 Signal NGV Module	3.9-4.6 ms
104	PK/WT	COP 2 Driver (dwell)	5°, 55 mph: 8°

Pin Connector Graphic

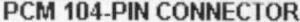

PCM 104-PIN CONNECTOR

Terminal View of 104-Pin PCM Wiring Harness Connector

Standard Colors and Abbreviations

Abbreviation	Color	Abbreviation	Color	Abbreviation	Color
BK	Black	GY	Gray	PK	Purple
BL	Blue	GN	Green	RD	Red
BR	Brown	LG	LT Green	TN	Tan
DB	Dark Blue	OR	Orange	WT	White
DG	DK Green	PK	Pink	YL	Yellow

2000-02 F-Series 5.4L V8 CNG VIN M, VIN Z (E4OD) 104-P Connector

PCM Pin #	Wire Color	Circuit Description (104 Pin)	Value at Hot Idle
1	OR/YL	COP 6 Driver (dwell)	5°, 55 mph: 8°
3	BK/WT	Power Ground	<0.1v
6	OR/YL	Shift Solenoid 'A' Control	1v, 55 mph: 12v
11	PK/OR	Shift Solenoid 'B' Control	12v, 55 mph: 1v
12	WT/LG	TCIL (lamp) Control	7.7v (Switch On: 0v)
13	VT	Flash EEPROM Power	0.1v
15	PK/LB	Data Bus (-)	<0.050v
16	TN/OR	Data Bus (+)	Digital Signals
20	BR/OR	Coast Clutch Solenoid	12v, 55 mph: 12v
21	DB	CKP (-) Sensor Signal	440-490 Hz
22	GY	CKP (+) Sensor Signal	440-490 Hz
25	LB/YL	Chassis Ground	<0.050v
26	LG/WT	COP 1 Driver (dwell)	5°, 55 mph: 8°
27	LG/YL	COP 5 Driver (dwell)	5°, 55 mph: 8°
29	TN/WT	TCS (switch) Signal	TCS & O/D On: 12v
34	YL/BK	Digital TR1 Sensor	0v, 55 mph: 11.5v
36	TN/LB	MAF Sensor Return	<0.050v
37	OR/BK	TFT Sensor Signal	2.10-2.40v
39	GY	IAT Sensor Signal	1.5-2.5v
40	DG/YL	Fuel Pump Monitor	On: 12v, Off: 0v
41	BK/YL	A/C Switch Signal	Switch Closed: 12v
49	LB/BK	Digital TR2 Sensor	0v, 55 mph: 11.5v
50	WT/BK	Digital TR4 Sensor	0v, 55 mph: 11.5v
51	BK/WT	Power Ground	<0.1v
52	WT/PK	COP 3 Driver (dwell)	5°, 55 mph: 8°
53	DG/PK	COP 4 Driver (dwell)	5°, 55 mph: 8°
54	PK/YL	TCC Solenoid Control	0%, 55 mph: 95%
55	RD/WT	Keep Alive Power	12-14v
59	DG/WT	TSS Sensor Signal	300 Hz, 55 mph: 980
60	GY/LB	HO2S-11 (B1 S1) Signal	0.1-1.1v
62	LB	EFT Sensor Signal	1.7-3.5v (50-120°F)
63	RD/PK	Fuel Rail Pressure Sensor	2-3.7v (90-100 psi)
64	LB/YL	Digital TR 3 Sensor	In 'P': 0v, in O/D: 1.7v

2000-02 F-Series 5.4L V8 CNG VIN M, VIN Z (E4OD) 104-P Connector

PCM Pin #	Wire Color	Circuit Description (104 Pin)	Value at Hot Idle
66	YL/LG	CHT Sensor Signal	0.6 (194°F)
67	PK/WT	EVAP CV Solenoid	0-10 Hz (0-100%)
68	GY/BK	VSS (-) Signal	0 Hz, 55 mph: 125 Hz
69	PK/YL	A/C WOT Relay Control	Off: 12v, On: 1v
71, 97	RD	Vehicle Power	12-14v
72	WT	Injector 7 Signal NGV Module	3.9-4.6 ms
73	WT/LB	Injector 5 Signal NGV Module	3.9-4.6 ms
74	WT/BK	Injector 3 Signal NGV Module	3.9-4.6 ms
75	WT/RD	Injector 1 Signal NGV Module	3.9-4.6 ms
77, 103	BK/WT	Power Ground	<0.1v
78	PK/LB	COP 7 Driver (dwell)	5°, 55 mph: 8°
79	WT/RD	COP 8 Driver (dwell)	5°, 55 mph: 8°
80	LB/OR	Fuel Pump Control	Off: 12v, On: 1v
81	WT/YL	EPC Solenoid Control	9.1v (4 psi)
83	WT/LB	IAC Solenoid Control	10.7v (33%)
84	DB/YL	OSS Sensor Signal	0 Hz, 55 mph: 720 Hz
85	DG	CMP Sensor Signal	6 Hz, 55 mph: 16 Hz
87	RD/BK	HO2S-21 (B2 S1) Signal	0.1-1.1v
88	LB/RD	MAF Sensor Signal	0.8v, 55 mph: 1.6v
89	GY/WT	TP Sensor Signal	0.53-1.27v
90	BR/WT	Reference Voltage	4.9-5.1v
91	GY/RD	Analog Signal Return	<0.050v
92	RD/LG	Brake Pedal Switch	0v (Brake On: 12v)
93	RD/WT	HO2S-11 (B1 S1) Heater	1v (Heater Off: 12v)
94	YL/LB	HO2S-21 (B2 S1) Heater	1v (Heater Off: 12v)
98	YL	Injector 8 Signal NGV Module	3.9-4.6 ms
99	YL/LB	Injector 6 Signal NGV Module	3.9-4.6 ms
100	YL/BK	Injector 4 Signal NGV Module	3.9-4.6 ms
101	YL/RD	Injector 2 Signal NGV Module	3.9-4.6 ms
104	PK/WT	COP 2 Driver (dwell)	5°, 55 mph: 8°

Pin Connector Graphic

Standard Colors and Abbreviations

Abbreviation	Color	Abbreviation	Color	Abbreviation	Color
BK	Black	GY	Gray	PK	Purple
BL	Blue	GN	Green	RD	Red
BR	Brown	LG	LT Green	TN	Tan
DB	Dark Blue	OR	Orange	WT	White
DG	DK Green	PK	Pink	YL	Yellow

2003-05 F-Series 5.4L V8 CNG VIN M, VIN Z (E4OD) 104-P Connector

PCM Pin #	Wire Color	Circuit Description (104 Pin)	Value at Hot Idle
1	OR/YL	COP 6 Driver (dwell)	5°, 55 mph: 8°
2	---	Not Used	---
3	BK/WT	Power Ground	<0.1v
4	PK	Transfer Case Speed Sensor (MSOF)	0 Hz, 55 mph: 471 Hz
4-5	---	Not Used	---
6	OR/YL	Shift Solenoid 'A' Control	1v, 55 mph: 12v
7-10	---	Not Used	---
11	PK/OR	Shift Solenoid 'B' Control	12v, 55 mph: 1v
12	WT/LG	TCIL (lamp) Control	7.7v (Switch On: 0v)
13	VT	Flash EEPROM Power	0.1v
14	---	Not Used	---
15	PK/LB	SCP Data Bus (-)	<0.050v
16	TN/OR	SCP Data Bus (+)	Digital Signals
17-19	---	Not Used	---
20	BR/OR	Coast Clutch Solenoid Control (MSOF)	12v, 55 mph: 12v
21	DB	CKP (-) Sensor Signal	440-490 Hz
22	GY	CKP (+) Sensor Signal	440-490 Hz
23-24	---	Not Used	---
25	LB/YL	Chassis Ground	<0.050v
26	LG/WT	COP 1 Driver (dwell)	5°, 55 mph: 8°
27	LG/YL	COP 5 Driver (dwell)	5°, 55 mph: 8°
28	---	Not Used	---
29	TN/WT	TCS (switch) Signal	TCS & O/D On: 12v
34	YL/BK	Digital TR1 Sensor	0v, 55 mph: 11.5v
31-35	---	Not Used	---
36	TN/LB	MAF Sensor Return	<0.050v
37	OR/BK	TFT Sensor Signal	2.10-2.40v
38	---	Not Used	---
39	GY	IAT Sensor Signal	1.5-2.5v
40	DG/YL	Fuel Pump Monitor	On: 12v, Off: 0v
41	BK/YL	A/C Switch Signal	Switch Closed: 12v
42-48	---	Not Used	---
49	LB/BK	Digital TR2 Sensor	0v, 55 mph: 11.5v
50	WT/BK	Digital TR4 Sensor	0v, 55 mph: 11.5v
51	BK/WT	Power Ground	<0.1v
52	WT/PK	COP 3 Driver (dwell)	5°, 55 mph: 8°
53	DG/PK	COP 4 Driver (dwell)	5°, 55 mph: 8°
54	PK/YL	TCC Solenoid Control	0%, 55 mph: 95%
55	RD/WT	Keep Alive Power	12-14v
59	DG/WT	TSS Sensor Signal	300 Hz, 55 mph: 980
60	GY/LB	HO2S-11 (B1 S1) Signal	0.1-1.1v
61	---	Not Used	---
62	LB	Fuel Rail Temperature Sensor	1.7-3.5v (50-120°F)
63	RD/PK	Fuel Rail Pressure Sensor	2-3.7v (90-100 psi)
64	LB/YL	Digital TR 3 Sensor	In 'P': 0v, in O/D: 1.7v

2003-05 F-Series 5.4L V8 CNG VIN M, VIN Z (E4OD) 104-P Connector

PCM Pin #	Wire Color	Circuit Description (104 Pin)	Value at Hot Idle
65	---	Not Used	---
66	YL/LG	CHT Sensor Signal	0.6 (194°F)
67	PK/WT	EVAP CV Solenoid	0-10 Hz (0-100%)
68	GY/BK	VSS (-) Signal	0 Hz, 55 mph: 125 Hz
69	PK/YL	A/C WOT Relay Control	Off: 12v, On: 1v
70	BK/WT	Check Fuel Cap Indicator Control	Indicator Off: 12v, On: 1v
71	RD	Vehicle Power	12-14v
72	WT	Injector 7 Signal (NGV Module)	3.9-4.6 ms
73	WT/LB	Injector 5 Signal (NGV Module)	3.9-4.6 ms
74	WT/BK	Injector 3 Signal (NGV Module)	3.9-4.6 ms
75	WT/RD	Injector 1 Signal (NGV Module)	3.9-4.6 ms
76	---	Not Used	---
77	BK/WT	Power Ground	<0.1v
78	PK/LB	COP 7 Driver (dwell)	5°, 55 mph: 8°
79	WT/RD	COP 8 Driver (dwell)	5°, 55 mph: 8°
80	LB/OR	Fuel Pump Control	Off: 12v, On: 1v
81	WT/YL	EPC Solenoid Control	9.1v (4 psi)
82, 86	---	Not Used	---
83	WT/LB	IAC Solenoid Control	10.7v (33%)
84	DB/YL	OSS Sensor Signal	0 Hz, 55 mph: 720 Hz
85	DG	CMP Sensor Signal	6 Hz, 55 mph: 16 Hz
87	RD/BK	HO2S-21 (B2 S1) Signal	0.1-1.1v
88	LB/RD	MAF Sensor Signal	0.8v, 55 mph: 1.6v
89	GY/WT	TP Sensor Signal	0.53-1.27v
90	BR/WT	Reference Voltage	4.9-5.1v
91	GY/RD	Sensor Return	<0.050v
92	RD/LG	Brake Pedal Switch	0v (Brake On: 12v)
93	RD/WT	HO2S-11 (B1 S1) Heater	1v (Heater Off: 12v)
94	YL/LB	HO2S-21 (B2 S1) Heater	1v (Heater Off: 12v)
95-96	---	Not Used	---
97	RD	Vehicle Power	12-14v
98	YL	Injector 8 Signal (NGV Module)	3.9-4.6 ms
99	YL/LB	Injector 6 Signal (NGV Module)	3.9-4.6 ms
100	YL/BK	Injector 4 Signal (NGV Module)	3.9-4.6 ms
101	YL/RD	Injector 2 Signal (NGV Module)	3.9-4.6 ms
102	---	Not Used	---
103	BK/WT	Power Ground	<0.1v
104	PK/WT	COP 2 Driver (dwell)	5°, 55 mph: 8°

Pin Connector Graphic

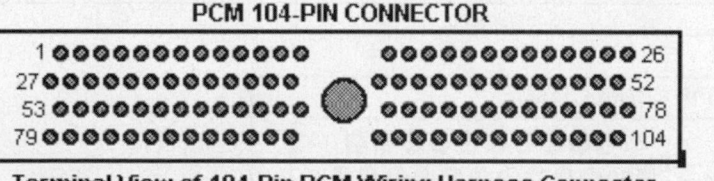

PCM 104-PIN CONNECTOR

Terminal View of 104-Pin PCM Wiring Harness Connector

Standard Colors and Abbreviations

Abbreviation	Color	Abbreviation	Color	Abbreviation	Color
BK	Black	GY	Gray	PK	Purple
BL	Blue	GN	Green	RD	Red
BR	Brown	LG	LT Green	TN	Tan
DB	Dark Blue	OR	Orange	WT	White
DG	DK Green	PK	Pink	YL	Yellow

1990 F-Series 5.8L V8 VIN H (All) 60 Pin Connector

PCM Pin #	Wire Color	Circuit Description (60 Pin)	Value at Hot Idle
1	YL	Keep Alive Power	12-14v
3	GY/BK	VSS (+) Signal	0 Hz, 55 mph: 125 Hz
3	DG/WT	VSS (+) Signal	0 Hz, 55 mph: 125 Hz
4	DG/YL	IDM Sensor Signal	20-31 Hz
6	BK	VSS (-) Signal	0 Hz, 55 mph: 125 Hz
7	LG/YL	ECT Sensor Signal	0.5-0.6v
8 ('89-'90)	BR	Fuel Pump Monitor	On: 12v, Off: 0v
10	BK/YL	A/C Switch Signal	A/C On: 12v, Off: 0v
11	WT/BK	Air Management 2 Solenoid	AM2 Off: 12v, On: 1v
16	BK/OR	Ignition System Ground	<0.050v
17	PK/LG	Self Test Output & MIL	MIL Off: 12v, On: 1v
18	TN/LG	Inferred Mileage Sensor	Digital Signals
20	BK	PCM Case Ground	0.050v
21	GY/WT	IAC Solenoid Control	8.3-9.1v
22	TN/LG	Fuel Pump Control	Off: 12v, On: 1v
23	LG/BK	Knock Sensor Signal	0v
24	YL/LG	PSP Switch Signal	0v (turning: 12v)
25	YL/RD	ACT Sensor Signal	1.5-2.5v
26	OR/WT	Reference Voltage	4.9-5.1v
27	BR/LG	EVP Sensor Signal	0.4v, 55 mph: 2.6v
29	GY/BL	HO2S-11 (B1 S1) Signal	0.1-1.1v
29	DG/PK	HO2S-11 (B1 S1) Signal	0.1-1.1v
30	GY/YL	A/T: Neutral Drive Switch	In 'P': 0v, Others: 5v
30	GY/YL	M/T: Clutch Engage Switch	5v (clutch "in": 0v)
31	GY/YL	Canister Purge Solenoid	12v, 55 mph: 1v
33	DG	VR Solenoid Control	0%, 55 mph: 45%
36	YL/LG	Spark Output Signal	6.93v (50%)
37	RD	Vehicle Power	12-14v
40	BK/LG	Power Ground	<0.1v
45	DG/LG	MAP Sensor Signal	107 Hz (sea level)
45	DB/LG	MAP Sensor Signal	107 Hz (sea level)
46	BK/WT	Analog Signal Return	<0.050v
47	DG/LG	TP Sensor Signal	1v, 55 mph: 1.4v
48	WT/RD	Self Test Indicator Signal	STI On: 1v, Off: 5v
49	OR	HO2S-11 (B1 S1) Ground	<0.050v
51	WT/RD	Air Management 1 Solenoid	AM1 Off: 12v, On: 1v
56	DB	PIP Sensor Signal	6.93v (50%)
57	RD	Vehicle Power	12-14v
58	TN/OR	Injector Bank 1 (INJ 1, 4, 5, 8)	5.8-6.4 ms
59	TN/RD	Injector Bank 2 (INJ 2, 3, 6, 7)	5.8-6.4 ms
60	BK/LG	Power Ground	<0.1v

Pin Connector Graphic

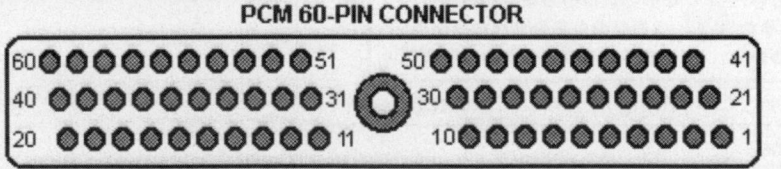

PCM 60-PIN CONNECTOR

Terminal View of 60-Pin PCM Harness Connector

1990 F-Series 5.8L V8 VIN H (E4OD) 60 Pin Connector

PCM Pin #	Wire Color	Circuit Description (60 Pin)	Value at Hot Idle
1	YL	Keep Alive Power	12-14v
2	LG	Brake Position Switch	0v (Brake On: 12v)
3	DG, GY	VSS (+) Signal	0 Hz, 55 mph: 125 Hz
4	DG/YL	IDM Sensor Signal	20-31 Hz
6	BK	VSS (-) Signal	0 Hz, 55 mph: 125 Hz
7	LG/YL	ECT Sensor Signal	0.5-0.6v
8	BR	Fuel Pump Monitor	On: 12v, Off: 0v
10	BK/YL	A/C Switch Signal	A/C On: 12v, Off: 0v
11	WT/BK	Air Management 2 Solenoid	AM2 Off: 12v, On: 1v
12	BL/BK	4x4 Indicator Light	Light Off: 12v, On: 1v
16	BK/OR	Ignition System Ground	<0.050v
17	PK/LG	Self Test Output & MIL	MIL Off: 12v, On: 1v
18	TN/LG	Inferred Mileage Sensor	Digital Signals
19	DG/PK	Shift Solenoid 2 Control	1v, 55 mph: 1v
20	BK	PCM Case Ground	<0.050v
21	GY/WT	IAC Solenoid Control	8.3-9.1v
22	TN/LG	Fuel Pump Control	Off: 12v, On: 1v
23	LG/BK	Knock Sensor Signal	0v
24	YL/LG	PSP Switch Signal	0v (turning: 12v)
25	YL/RD	ACT Sensor Signal	1.5-2.5v
26	OR/WT	Reference Voltage	4.9-5.1v
27	BR/LG	EVP Sensor Signal	0.4v, 55 mph: 2.6v
29	DG, GY	HO2S-11 (B1 S1) Signal	0.1-1.1v
30	BL/WT	MLP Sensor Signal	In 'P': 0v, in O/D: 5v
31	GY/YL	Canister Purge Solenoid	12v, 55 mph: 1v
32	LG/WT	OCIL (lamp) Control	7.7v (Switch On: 0v)
33	DG	VR Solenoid Control	0%, 55 mph: 45%
36	YL/LG	Spark Output Signal	6.93v (50%)
37, 57	RD	Vehicle Power	12-14v
38	BL/YL	EPC Solenoid Control	9.2v (5 psi)
40	BK/LG	Power Ground	<0.1v
41	TN/WT	OCS (switch) Signal	12v (switch closed: 0v)
42	OR/BK	TOT Sensor Signal	2.10-2.40v
43-44	---	Not Used	---
45	DB/LG	MAP Sensor Signal	107 Hz (sea level)
46	BK/WT	Analog Signal Return	<0.050v
47	DG/LG	TP Sensor Signal	1v, 55 mph: 1.4v
48	WT/RD	Self Test Indicator Signal	STI On: 1v, Off: 5v
49	OR	HO2S-11 (B1 S1) Ground	<0.050v
51	WT/RD	Air Management 1 Solenoid	AM1 Off: 12v, On: 1v
52	OR/YL	Shift Solenoid 1 Control	1v, 55 mph: 12v
53	PK/YL	TCC Solenoid Control	0v, On 55 mph: 12v
55	BR	Coast Clutch Solenoid	12v, 55 mph: 12v
56	DB	PIP Sensor Signal	6.93v (50%)
58	TN/OR	Injector Bank 1 (INJ 1, 4, 5, 8)	5.8-6.4 ms
59	TN/RD	Injector Bank 2 (INJ 2, 3, 6, 7)	5.8-6.4 ms
60	BK/LG	Power Ground	<0.1v

Pin Connector Graphic

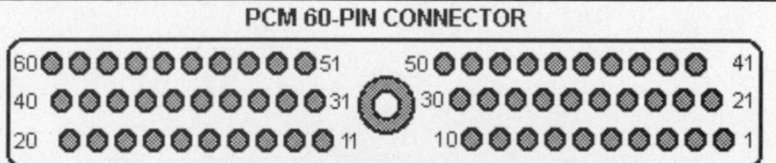

PCM 60-PIN CONNECTOR

60 ... 51   50 ... 41
40 ... 31   30 ... 21
20 ... 11   10 ... 1

Terminal View of 60-Pin PCM Harness Connector

1991-93 F-Series 5.8L V8 VIN H (All) 60 Pin Connector

PCM Pin #	Wire Color	Circuit Description (60 Pin)	Value at Hot Idle
1	YL	Keep Alive Power	12-14v
3	GY/BK	VSS (+) Signal	0 Hz, 55 mph: 125 Hz
4	YL/BK	IDM Sensor Signal	20-31 Hz
4	TN/YL	IDM Sensor Signal	20-31 Hz
6	PK/OR	VSS (-) Signal	0 Hz, 55 mph: 125 Hz
7	LG/RD	ECT Sensor Signal	0.5-0.6v
8	DG/YL	Fuel Pump Monitor	On: 12v, Off: 0v
9	PK, BK	Data Bus (-) Signal	Digital Signals
10	BK, PK	A/C Switch Signal	A/C On: 12v, Off: 0v
11	BR	Air Management 2 Solenoid	AM2 Off: 12v, On: 1v
16	OR/RD	Ignition System Ground	<0.050v
17	PK/LG	Self Test Output & MIL	MIL Off: 12v, On: 1v
20	BK	PCM Case Ground	<0.050v
21	WT/BL	IAC Solenoid Control	7.2-9.2v
22	BL/OR	Fuel Pump Control	Off: 12v, On: 1v
23 ('92-'93)	YL/RD	Knock Sensor Signal	0v
24 ('92-'93)	YL/LG	PSP Switch Signal	0v (turning: 12v)
25	GY	ACT Sensor Signal	1.5-2.5v
26	BR/WT	Reference Voltage	4.9-5.1v
27	BR/LG	EVP Sensor Signal	0.4v, 55 mph: 2.6v
28	TN/OR	Data Bus (+) Signal	Digital Signals
29	GY/BL	HO2S-11 (B1 S1) Signal	0.1-1.1v
30	BL/YL	A/T: Neutral Drive Switch	In 'P': 0v, Others: 5v
30	BL/YL	M/T: Clutch Engage Switch	5v (clutch "in": 0v)
33	BR/PK	VR Solenoid Control	0%, 55 mph: 45%
36	PK	Spark Output Signal	6.93v (50%)
37	RD	Vehicle Power	12-14v
40	BK/WT	Power Ground	<0.1v
45	LG/BK	MAP Sensor Signal	107 Hz (sea level)
46	GY/RD	Analog Signal Return	<0.050v
47	GY/WT	TP Sensor Signal	1v, 55 mph: 1.4v
48	WT/PK	Self Test Indicator Signal	STI On: 1v, Off: 5v
49	OR	HO2S-11 (B1 S1) Ground	<0.050v
51	WT/OR	Air Management 1 Solenoid	AM1 Off: 12v, On: 1v
56	GY/OR	PIP Sensor Signal	6.93v (50%)
57	RD	Vehicle Power	12-14v
58	TN	Injector Bank 1 (INJ 1, 4, 5, 8)	3.5-5.0 ms
59	WT	Injector Bank 2 (INJ 2, 3, 6, 7)	3.5-5.0 ms
60	BK/WT	Power Ground	<0.1v

Pin Connector Graphic

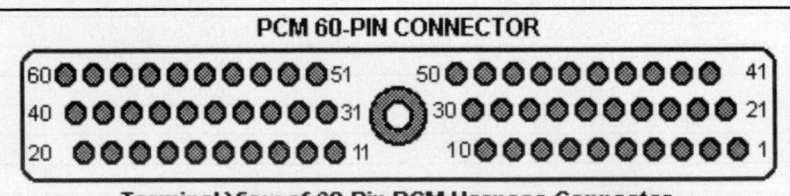

PCM 60-PIN CONNECTOR

Terminal View of 60-Pin PCM Harness Connector

1991-93 F-Series 5.8L V8 VIN H (E4OD) 60 Pin Connector

PCM Pin #	Wire Color	Circuit Description (60 Pin)	Value at Hot Idle
1	YL	Keep Alive Power	12-14v
2	LG	Brake Position Switch	0v (Brake On: 12v)
3	GY/BK	VSS (+) Signal	0 Hz, 55 mph: 125 Hz
4	YL/BK	IDM Sensor Signal	20-31 Hz
4	TN/YL	IDM Sensor Signal	20-31 Hz
5	---	Not Used	---
6	PK/OR	VSS (-) Signal	0 Hz, 55 mph: 125 Hz
7	LG/RD	ECT Sensor Signal	0.5-0.6v
8	DG/YL	Fuel Pump Monitor	On: 12v, Off: 0v
9	PK/BL	Data Bus (-) Signal	<0.050v
9	BK/OR	Data Bus (+) Signal	Digital Signals
10	BK/YL	A/C Switch Signal	A/C On: 12v, Off: 0v
10	PK/BL	A/C Switch Signal	A/C On: 12v, Off: 0v
11	BR	Air Management 2 Solenoid	AM2 Off: 12v, On: 1v
12	BL/BK	4x4 Indicator Light	Light On: 0.1v, Off: 12v
13-15	---	Not Used	---
16	OR/RD	Ignition System Ground	<0.050v
17	PK/LG	Self Test Output & MIL	MIL Off: 12v, On: 1v
18	---	Not Used	---
19	PK/OR	Shift Solenoid 2 Control	12v, 55 mph: 1v
20	BK	PCM Case Ground	<0.050v
21	WT/BL	IAC Solenoid Control	10.7v (33%)
22	BL/OR	Fuel Pump Control	Off: 12v, On: 1v
23 ('92-'93)	YL/RD	Knock Sensor Signal	0v
24 ('92-'93)	YL/LG	PSP Switch Signal	0v (turning: 12v)
25	GY	ACT Sensor Signal	1.5-2.5v
26	BR/WT	Reference Voltage	4.9-5.1v
27	BR/LG	EVP Sensor Signal	0.4v, 55 mph: 2.6v
28	TN/OR	Data Bus (+) Signal	Digital Signals
29	GY/BL	HO2S-11 (B1 S1) Signal	0.1-1.1v
29	GY/YL	HO2S-11 (B1 S1) Signal	0.1-1.1v
30	BL/YL	MLP Sensor Signal	In 'P': 0v, in O/D: 5v
31	GY/YL	Canister Purge Solenoid	12v, 55 mph: 1v
32	WT/LG	OCIL (lamp) Control	7.7v (Switch On: 0v)
33	BR/PK	VR Solenoid Control	0%, 55 mph: 45%
34	---	Not Used	---
35	---	Not Used	---
36	PK	Spark Output Signal	6.93v (50%)
37	RD	Vehicle Power	12-14v
38	WT/YL	EPC Solenoid Control	9.5v (5 psi)
39	---	Not Used	---
40	BK/WT	Power Ground	<0.1v

1991-93 F-Series 5.8L V8 VIN H (E4OD) 60 Pin Connector

PCM Pin #	Wire Color	Circuit Description (60 Pin)	Value at Hot Idle
41	TN/WT	OCS (switch) Signal	12v (switch closed: 0v)
42	OR/BK	TOT Sensor Signal	2.10-2.40v
43 ('91)	PK	A/C Demand Switch	A/C On: 12v, Off: 0v
44	---	Not Used	---
45	LG/BK	MAP Sensor Signal	107 Hz (sea level)
46	GY/RD	Analog Signal Return	<0.050v
47	GY/WT	TP Sensor Signal	1v, 55 mph: 1.4v
48	WT/PK	Self Test Indicator Signal	STI On: 1v, Off: 5v
49	OR	HO2S-11 (B1 S1) Ground	<0.050v
50	---	Not Used	---
51	WT/OR	Air Management 1 Solenoid	Solenoid Off: 12v, On: 1v
52	OR/YL	Shift Solenoid 1 Control	0.3-0.4v
53	PK/YL	TCC Solenoid Control	0v, On 55 mph: 12v
54	---	Not Used	---
55	BR/OR	Coast Clutch Solenoid	12v, 55 mph: 12v
56	GY/OR	PIP Sensor Signal	6.93v (50%)
57	RD	Vehicle Power	12-14v
58	TN	Injector Bank 1 (INJ 1, 4, 5, 8)	3.5-4.0 ms
59	WT	Injector Bank 2 (INJ 2, 3, 6, 7)	3.5-4.0 ms
60	BK/WT	Power Ground	<0.1v

Pin Connector Graphic

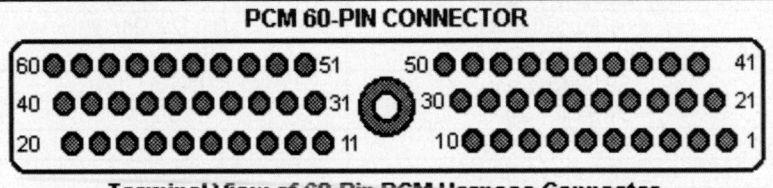

PCM 60-PIN CONNECTOR

Terminal View of 60-Pin PCM Harness Connector

Standard Colors and Abbreviations

Abbreviation	Color	Abbreviation	Color	Abbreviation	Color
BK	Black	GY	Gray	PK	Purple
BL	Blue	GN	Green	RD	Red
BR	Brown	LG	LT Green	TN	Tan
DB	Dark Blue	OR	Orange	WT	White
DG	DK Green	PK	Pink	YL	Yellow

1994 F-Series 5.8L V8 VIN H (All) 60 Pin Connector

PCM Pin #	Wire Color	Circuit Description (60 Pin)	Value at Hot Idle
1	YL	Keep Alive Power	12-14v
2	LG	Brake Position Switch	0v (Brake On: 12v)
3	GY/BK	VSS (+) Signal	0 Hz, 55 mph: 125 Hz
4	YL/BK	IDM Sensor Signal	20-31 Hz
5	---	Not Used	---
6	PK/OR	VSS (-) Signal	0 Hz, 55 mph: 125 Hz
7	LG/RD	ECT Sensor Signal	0.5-0.6v
8	DG/YL	Fuel Pump Monitor	On: 12v, Off: 0v
9	PK/BL	Data Bus (-) Signal	<0.050v
9	BK/OR	Data Bus (+) Signal	Digital Signals
10	BK/YL	A/C Switch Signal	A/C On: 12v, Off: 0v
10	PK/BL	A/C Switch Signal	A/C On: 12v, Off: 0v
11	BR	Air Management 2 Solenoid	AM2 Off: 12v, On: 1v
12	PK/BL	4x4 Indicator Light	12v (switch closed: 0v)
13-15	---	Not Used	---
16	OR/RD	Ignition System Ground	<0.050v
17	PK/LG	Self Test Output & MIL	MIL Off: 12v, On: 1v
18-19	---	Not Used	---
20	BK	PCM Case Ground	<0.050v
21	WT/BL	IAC Solenoid Control	10.7v (33%)
22	BL/OR	Fuel Pump Control	Off: 12v, On: 1v
23	---	Not Used	---
24	---	Not Used	---
25	GY	ACT Sensor Signal	1.5-2.5v
26	BR/WT	Reference Voltage	4.9-5.1v
27	BR/LG	EVP Sensor Signal	0.4v, 55 mph: 2.6v
28	TN/OR	Data Bus (+) Signal	Digital Signals
29	GY/BL	HO2S-11 (B1 S1) Signal	0.1-1.1v
29	GY/YL	HO2S-11 (B1 S1) Signal	0.1-1.1v
30	BL/YL	A/T: Neutral Drive Switch	In 'N': Others: 5v
30	BL/YL	M/T: Clutch Engage Switch	5v (clutch "in": 0v)
31	GY/YL	Canister Purge Solenoid	12v, 55 mph: 1v
32	WT/LG	OCIL (lamp) Control	7.7v (Switch On: 0v)
33	BR/PK	VR Solenoid Control	0%, 55 mph: 45%
34	---	Not Used	---
35	---	Not Used	---
36	PK	Spark Output Signal	6.93v (50%)
37	RD	Vehicle Power	12-14v
38	---	Not Used	---
39	---	Not Used	---
40	BK/WT	Power Ground	<0.1v

1994 F-Series 5.8L V8 VIN H (All) 60 Pin Connector

PCM Pin #	Wire Color	Circuit Description (60 Pin)	Value at Hot Idle
41-42	---	Not Used	---
43	PK	A/C Demand Switch	A/C On: 12v, Off: 0v
44	---	Not Used	---
45	LG/BK	MAP Sensor Signal	107 Hz (sea level)
46	GY/RD	Analog Signal Return	<0.050v
47	GY/WT	TP Sensor Signal	1v, 55 mph: 1.4v
48	WT/PK	Self Test Indicator Signal	STI On: 1v, Off: 5v
49	OR	HO2S-11 (B1 S1) Ground	<0.050v
50	---	Not Used	---
51	WT/OR	Air Management 1 Solenoid	Solenoid Off: 12v, On: 1v
52-55	---	Not Used	---
56	GY/OR	PIP Sensor Signal	6.93v (50%)
57	RD	Vehicle Power	12-14v
58	TN	Injector Bank 1 (INJ 1, 4, 5, 8)	3.5-5.0 ms
59	WT	Injector Bank 2 (INJ 2, 3, 6, 7)	3.5-5.0 ms
60	BK/WT	Power Ground	<0.1v

Pin Connector Graphic

PCM 60-PIN CONNECTOR

Terminal View of 60-Pin PCM Harness Connector

Standard Colors and Abbreviations

Abbreviation	Color	Abbreviation	Color	Abbreviation	Color
BK	Black	GY	Gray	PK	Purple
BL	Blue	GN	Green	RD	Red
BR	Brown	LG	LT Green	TN	Tan
DB	Dark Blue	OR	Orange	WT	White
DG	DK Green	PK	Pink	YL	Yellow

1994 F-Series 5.8L V8 MFI VIN H (E4OD) 60 Pin Connector

PCM Pin #	Wire Color	Circuit Description (60 Pin)	Value at Hot Idle
1	YL	Keep Alive Power	12-14v
2	LG	Brake Position Switch	0v (Brake On: 12v)
3	GY/BK	VSS (+) Signal	0 Hz, 55 mph: 125 Hz
4	YL/BK	IDM Sensor Signal	20-31 Hz
5	---	Not Used	---
6	PK/OR	VSS (-) Signal	0 Hz, 55 mph: 125 Hz
7	LG/RD	ECT Sensor Signal	0.5-0.6v
8	DG/YL	Fuel Pump Monitor	On: 12v, Off: 0v
9	PK/BL	Data Bus (-) Signal	<0.050v
9	BK/OR	Data Bus (+) Signal	Digital Signals
10	BK/YL	A/C Switch Signal	A/C On: 12v, Off: 0v
10	PK/BL	A/C Switch Signal	A/C On: 12v, Off: 0v
11	BR	Air Management 2 Solenoid	AM2 Off: 12v, On: 1v
12	PK/BL	4x4 Indicator Light	12v (switch closed: 0v)
13	---	Not Used	---
14	---	Not Used	---
15	---	Not Used	---
16	OR/RD	Ignition System Ground	<0.050v
17	PK/LG	Self Test Output & MIL	MIL Off: 12v, On: 1v
18	---	Not Used	---
19	PK/OR	Shift Solenoid 2 Control	12v, 55 mph: 1v
20	BK	PCM Case Ground	<0.050v
21	WT/BL	IAC Solenoid Control	10.7v (33%)
22	BL/OR	Fuel Pump Control	Off: 12v, On: 1v
23	---	Not Used	---
24	---	Not Used	---
25	GY	ACT Sensor Signal	1.5-2.5v
26	BR/WT	Reference Voltage	4.9-5.1v
27	BR/LG	EVP Sensor Signal	0.4v, 55 mph: 2.6v
28	TN/OR	Data Bus (+) Signal	Digital Signals
29	GY/BL	HO2S-11 (B1 S1) Signal	0.1-1.1v
29	GY/YL	HO2S-11 (B1 S1) Signal	0.1-1.1v
30	BL/YL	MLP Sensor Signal	In 'P': 0v, in O/D: 5v
31	GY/YL	Canister Purge Solenoid	12v, 55 mph: 1v
32	WT/LG	OCIL (lamp) Control	7.7v (Switch On: 0v)
33	BR/PK	VR Solenoid Control	0%, 55 mph: 45%
34	---	Not Used	---
35	---	Not Used	---
36	PK	Spark Output Signal	6.93v (50%)
37	RD	Vehicle Power	12-14v
38	WT/YL	EPC Solenoid Control	9.5v (5 psi)
39	---	Not Used	---
40	BK/WT	Power Ground	<0.1v

1994 F-Series 5.8L V8 MFI VIN H (E4OD) 60 Pin Connector

PCM Pin #	Wire Color	Circuit Description (60 Pin)	Value at Hot Idle
41	TN/WT	TCS (switch) Signal	TCS closed: 12, open: 0v
42	OR/BK	TOT Sensor Signal	2.10-2.40v
43	PK	A/C Demand Switch	A/C On: 12v, Off: 0v
44	---	Not Used	---
45	LG/BK	MAP Sensor Signal	107 Hz (sea level)
46	GY/RD	Analog Signal Return	<0.050v
47	GY/WT	TP Sensor Signal	1v, 55 mph: 1.4v
48	WT/PK	Self Test Indicator Signal	STI On: 1v, Off: 5v
49	OR	HO2S-11 (B1 S1) Ground	<0.050v
50	---	Not Used	---
51	WT/OR	Air Management 1 Solenoid	Solenoid Off: 12v, On: 1v
52	OR/YL	Shift Solenoid 1 Control	1v, 55 mph: 12v
53	PK/YL	TCC Solenoid Control	0v, On 55 mph: 12v
54	---	Not Used	---
55	BR/OR	Coast Clutch Switch	12v, 55 mph: 12v
56	GY/OR	PIP Sensor Signal	6.93v (50%)
57	RD	Vehicle Power	12-14v
58	TN	Injector Bank 1 (INJ 1, 4, 5, 8)	3.5-5.0 ms
59	WT	Injector Bank 2 (INJ 2, 3, 6, 7)	3.5-5.0 ms
60	BK/WT	Power Ground	<0.1v

Pin Connector Graphic

PCM 60-PIN CONNECTOR

Terminal View of 60-Pin PCM Harness Connector

Standard Colors and Abbreviations

Abbreviation	Color	Abbreviation	Color	Abbreviation	Color
BK	Black	GY	Gray	PK	Purple
BL	Blue	GN	Green	RD	Red
BR	Brown	LG	LT Green	TN	Tan
DB	Dark Blue	OR	Orange	WT	White
DG	DK Green	PK	Pink	YL	Yellow

1995 F-Series 5.8L V8 VIN H (E4OD) California 60-P Connector

PCM Pin #	Wire Color	Circuit Description (60 Pin)	Value at Hot Idle
1	YL	Keep Alive Power	12-14v
2	LG	Brake Pedal Switch	0v (Brake On: 12v)
3	GY/BK	VSS (+) Signal	At 55 mph; 125 Hz
4	YL/BK	IDM Sensor Signal	20-31 Hz
5	---	Not Used	---
6	PK/OR	VSS (-) Signal	0 Hz, 55 mph: 125 Hz
7	LG/RD	ECT Sensor Signal	0.5-0.6v
8	DG/YL	Fuel Pump Monitor	On: 12v, Off: 0v
9	TN/BL	MAF Sensor Return	<0.050v
10	DG/OR	A/C Switch Signal	A/C On: 12v, Off: 0v
11	LB/BK	EVAP Purge Solenoid	12v, 55 mph: 1v
12	LG/OR	Injector 6 Control	5.5-6.1 ms
13	TN/RD	Injector 7 Control	5.5-6.1 ms
14	LB	Injector 8 Control	5.5-6.1 ms
15	TN/BK	Injector 5 Control	5.5-6.1 ms
16	OR/RD	Ignition System Ground	<0.050v
17	PK/LG	Self Test Output & MIL	MIL Off: 12v, On: 1v
18	TN/OR	Data Bus (+) Signal	Digital Signals
19	PK/BL	Data Bus (-) Signal	Digital Signals
20	BK	PCM Case Ground	Digital Signals
21	WT/BL	IAC Solenoid Control	10.6-10.7v
22	BL/OR	Fuel Pump Control	Off: 12v, On: 1v
23	---	Not Used	---
24	---	Not Used	---
25	GY	IAT Sensor Signal	1.5-2.5v
26	BR/WT	Reference Voltage	4.9-5.1v
27	BR/LG	EVP Sensor Signal	0.4v, 55 mph: 2.6v
28	---	Not Used	---
29	BK/LG	Not Used	---
30	LB/YL	MLP Sensor Signal	In 'P': 0v, in O/D: 5v
31	---	Not Used	---
32	WT/LG	TCIL (lamp) Control	7.7v (Switch On: 0v)
33	BR/PK	VR Solenoid Control	0%, 55 mph: 45%
34	BR	Air Diverter Solenoid Control	12v
35	BR/BL	Injector 4 Control	5.5-6.1 ms
36	PK	Spark Output Signal	6.93v (50%)
37	RD	Vehicle Power	12-14v
38	WT/YL	EPC Solenoid Control	9.5v (5 psi)
39	BR/YL	Injector 3 Control	5.5-6.1 ms
40	BK/WT	Power Ground	<0.1v

1995 F-Series 5.8L V8 VIN H (E4OD) California 60-P Connector

PCM Pin #	Wire Color	Circuit Description (60 Pin)	Value at Hot Idle
41	TN/WT	TCS (switch) Signal	TCS closed: 12, open: 0v
42	BL/PK	4x4 Indicator Light	12v (switch closed: 0v)
43	RD/BK	HO2S-21 (B2 S1) Signal	0.1-1.1v
44	GY/BL	HO2S-11 (B1 S1) Signal	0.1-1.1v
45	---	Not Used	---
46	GY/RD	Analog Signal Return	<0.050v
47	GY/WT	TP Sensor Signal	1v, 55 mph: 1.4v
48	WT/PK	Self Test Indicator Signal	STI On: 1v, Off: 5v
49	OR/BK	TOT Sensor Signal	2.10-2.40v
50	LB/RD	MAF Sensor Signal	0.8v, 55 mph: 1.6v
51	OR/YL	Shift Solenoid 1 Control	1v, 55 mph: 12v
52	PK/OR	Shift Solenoid 2 Control	12v, 55 mph: 1v
53	PK/YL	TCC Solenoid Control	0v, On 55 mph: 12v
54	---	Not Used	---
55	BR/OR	Coast Clutch Solenoid	1v, 55 mph: 12v
56	GY/OR	PIP Sensor Signal	6.93v (50%)
57	RD	Vehicle Power	12-14v
58	TN	Injector 1 Control	5.5-6.1 ms
59	WT	Injector 2 Control	5.5-6.1 ms
60	BK/WT	Power Ground	<0.1v

Pin Connector Graphic

PCM 60-PIN CONNECTOR

60 ●●●●●●●●●● 51   50 ●●●●●●●●●● 41
40 ●●●●●●●●●● 31   ○ 30 ●●●●●●●●●● 21
20 ●●●●●●●●●● 11   10 ●●●●●●●●●● 1

Terminal View of 60-Pin PCM Harness Connector

Standard Colors and Abbreviations

Abbreviation	Color	Abbreviation	Color	Abbreviation	Color
BK	Black	GY	Gray	PK	Purple
BL	Blue	GN	Green	RD	Red
BR	Brown	LG	LT Green	TN	Tan
DB	Dark Blue	OR	Orange	WT	White
DG	DK Green	PK	Pink	YL	Yellow

1995 F-Series 5.8L V8 VIN H (All) 60 Pin Connector

PCM Pin #	Wire Color	Circuit Description (60 Pin)	Value at Hot Idle
1	YL	Keep Alive Power	12-14v
2	LG	Brake Pedal Switch	0v (Brake On: 12v)
3	GY/BK	VSS (+) Signal	0 Hz, 55 mph: 125 Hz
4	YL/BK	IDM Sensor Signal	20-31 Hz
5	---	Not Used	---
6	PK/OR	VSS (-) Signal	0 Hz, 55 mph: 125 Hz
7	LG/RD	ECT Sensor Signal	0.5-0.6v
8	DG/YL	Fuel Pump Monitor	On: 12v, Off: 0v
9	PK/BL	Data Bus (-) Signal	Digital Signals
10	PK/BL	A/C Switch Signal	A/C On: 12v, Off: 0v
11	BR	Air Management 2 Solenoid	AM2 Off: 12v, On: 1v
12	PK/BL	4x4 Indicator Light	12v (switch closed: 0v)
13-15	---	Not Used	---
16	OR/RD	Ignition System Ground	<0.050v
17	PK/LG	Self Test Output & MIL	MIL Off: 12v, On: 1v
18	---	Not Used	---
19	PK/OR	Shift Solenoid 2 Control	12v, 55 mph: 1v
20	BK	PCM Case Ground	<0.050v
21	WT/BL	IAC Solenoid Control	10.7v (33%)
22	BL/OR	Fuel Pump Control	Off: 12v, On: 1v
23-24	---	Not Used	---
25	GY	ACT Sensor Signal	1.5-2.5v
26	BR/WT	Reference Voltage	4.9-5.1v
27	BR/LG	EVP Sensor Signal	0.4v, 55 mph: 2.6v
28	TN/OR	Data Bus (+) Signal	Digital Signals
29	GY/BL	HO2S-11 (B1 S1) Signal	0.1-1.1v
30	BL/YL	MLP Sensor Signal	In 'P': 0v, in O/D: 5v
31	GY/YL	Canister Purge Solenoid	12v, 55 mph: 1v
32	WT/LG	OCIL (lamp) Control	7.7v (Switch On: 0v)
33	BR/PK	VR Solenoid Control	0%, 55 mph: 45%
34-35	---	Not Used	---
36	PK	Spark Output Signal	6.93v (50%)
37	R	Vehicle Power	12-14v
38	WT/YL	EPC Solenoid Control	9.5v (5 psi)
39	---	Not Used	---
40	BK/WT	Power Ground	<0.1v

1995 F-Series 5.8L V8 VIN H (All) 60 Pin Connector

PCM Pin #	Wire Color	Circuit Description (60 Pin)	Value at Hot Idle
41	TN/WT	TCS (switch) Signal	TCS closed: 12, open: 0v
42	OR/BK	TOT Sensor Signal	2.10-2.40v
43	PK	A/C Demand Switch	A/C On: 12v, Off: 0v
44	---	Not Used	---
45	LG/BK	MAP Sensor Signal	107 Hz (sea level)
46	GY/RD	Analog Signal Return	<0.050v
47	GY/WT	TP Sensor Signal	1v, 55 mph: 1.4v
48	WT/PK	Self Test Indicator Signal	STI On: 1v, Off: 5v
49	OR	HO2S-11 (B1 S1) Ground	<0.050v
50	---	Not Used	---
51	WT/OR	Air Management 1 Solenoid	AM1 Off: 12v, On: 1v
52	OR/YL	Shift Solenoid 1 Control	1v, 55 mph: 12v
53	PK/YL	TCC Solenoid Control	0v, On 55 mph: 12v
54	---	Not Used	---
55	BR/OR	Coast Clutch Solenoid	12v, 55 mph: 12v
56	GY/OR	PIP Sensor Signal	6.93v (50%)
57	RD	Vehicle Power	12-14v
58	TN	Injector Bank 1 (INJ 1, 4, 5, 8)	3.5-5.0 ms
59	WT	Injector Bank 2 (INJ 2, 3, 6, 7)	3.5-5.0 ms
60	BK/WT	Power Ground	<0.1v

Pin Connector Graphic

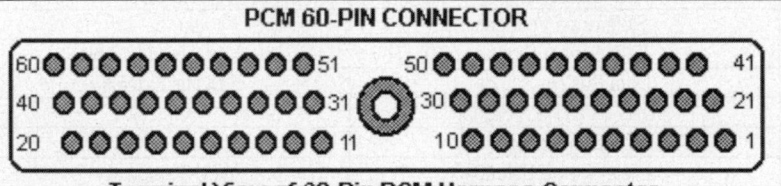

PCM 60-PIN CONNECTOR

Terminal View of 60-Pin PCM Harness Connector

Standard Colors and Abbreviations

Abbreviation	Color	Abbreviation	Color	Abbreviation	Color
BK	Black	GY	Gray	PK	Purple
BL	Blue	GN	Green	RD	Red
BR	Brown	LG	LT Green	TN	Tan
DB	Dark Blue	OR	Orange	WT	White
DG	DK Green	PK	Pink	YL	Yellow

1996-97 F-Series 5.8L V8 VIN H (E4OD) 104 Pin Connector

PCM Pin #	Wire Color	Circuit Description (104 Pin)	Value at Hot Idle
1	PK/OR	Shift Solenoid 2 Control	1v, 55 mph: 12v
2	PK/LG	MIL (lamp) Control	MIL Off: 12v, On: 1v
4	LG/RD	Power Take-Off (if equipped)	0v (Off)
5-12	---	Not Used	---
13	PK	Flash EPROM Power	0.1v
14	LG/BK	4x4 Low Indicator Switch	12v (switch closed: 0v)
15	PK/LB	Data Bus (-) Signal	Digital Signals
16	TN/OR	Data Bus (+) Signal	Digital Signals
17-22	---	Not Used	---
23	OR/RD	Ignition Ground	<0.050v
24, 51	BK/WT	Power Ground	<0.1v
25	BK/LB	Case Ground	<0.050v
26, 28	---	Not Used	---
27	OR/YL	Shift Solenoid 1 Control	1v, 55 mph: 12v
29	TN/WT	TCS (switch) Signal	TCS & O/D On: 12v
30-32, 34	---	Not Used	---
33	PK/OR	PSOM (-) Signal	<0.050v
35	RD/LG	HO2S-12 (B1 S2) Signal	0.1-1.1v
36	TN/LB	MAF Sensor Return	<0.050v
37	OR/BK	TFT Sensor Signal	2.10-2.40v
38	LG/RD	ECT Sensor Signal	0.5-0.6v
39	GY	IAT Sensor Signal	1.5-2.5v
40	DG/YL	Fuel Pump Monitor	On: 12v, Off: 0v
41	BK/YL	A/C Switch Signal	A/C On: 12v, Off: 0v
42-43, 46	---	Not Used	---
44	BR	Secondary AIR Diverter	AIRD Off: 12v, On: 1v
47	BK/PK	VR Solenoid Control	0%, 55 mph: 45%
48	YL/BK	Clean Tachometer Output	39-45 Hz
49	GY/OR	PIP Sensor Signal	6.93v (50%)
50	PK	Spark Output Signal	6.93v (50%)
52, 57	---	Not Used	---
53	BR/OR	Coast Clutch Solenoid	12v, 55 mph: 12v
54	PK/YL	TCC Solenoid Control	0%, 55 mph: 95%
55	YL	Keep Alive Power	12-14v
56	LG/BK	EVAP Purge Solenoid	0-10 Hz (0-100%)
58	GY/BK	PSOM (+) Signal	0 Hz, 55 mph: 125 Hz
59	DG/LG	Misfire Detection Sensor	45-55 Hz
60	GY/LB	HO2S-11 (B1 S1) Signal	0.1-1.1v
61-63	---	Not Used	---
64	LB/YL	TR Sensor Signal	In 'P': 0v, in O/D: 5v

1996-97 Truck 5.8L V8 VIN H (E4OD) 104 Pin Connector

PCM Pin #	Wire Color	Circuit Description (104 Pin)	Value at Hot Idle
65	BR/LG	DPFE Sensor Signal	0.95-1.05v
66-69	---	Not Used	---
70	WT/OR	Secondary AIR Bypass	AIRB Off: 12v, On: 1v
71	RD	Vehicle Power	12-14v
72	TN/RD	Injector 7 Control	4.2-4.6 ms
73	TN/BK	Injector 5 Control	4.2-4.6 ms
74	BR/YL	Injector 3 Control	4.2-4.6 ms
75	TN	Injector 1 Control	4.2-4.6 ms
76	BK/WT	Power Ground	<0.1v
77	BK/WT	Power Ground	<0.1v
78, 82	---	Not Used	---
79	WT/LG	TCIL (lamp) Control	7.7v (Switch On: 0v)
80	LB/OR	Fuel Pump Control	Off: 12v, On: 1v
81	WT/YL	EPC Solenoid Control	9.2v (5 psi)
83	WT/LB	IAC Solenoid Control	10.7v (33%)
84-86	---	Not Used	---
87	RD/BK	HO2S-21 (B2 S1) Signal	0.1-1.1v
88	LB/RD	MAF Sensor Signal	0.8v, 55 mph: 1.8v
89	GY/WT	TP Sensor Signal	0.53-1.27v
90	BR/WT	Reference Voltage	4.9-5.1v
91	GY/RD	Analog Signal Return	<0.050v
92	LG	Brake Pedal Switch	0v (Brake On: 12v)
93	RD/WT	HO2S-11 (B1 S1) Heater	1v (Heater Off: 12v)
94	YL/LB	HO2S-21 (B2 S1) Heater	1v (Heater Off: 12v)
95	WT/BK	HO2S-12 (B1 S2) Heater	1v (Heater Off: 12v)
96	---	Not Used	---
97	RD	Vehicle Power	12-14v
98	LG	Injector 8 Control	4.2-4.6 ms
99	LG/OR	Injector 6 Control	4.2-4.6 ms
100	BR/LB	Injector 4 Control	4.2-4.6 ms
101	WT	Injector 2 Control	4.2-4.6 ms
102	---	Not Used	---
103	BK/WT	Power Ground	<0.1v
104	---	Not Used	---

Pin Connector Graphic

PCM 104-PIN CONNECTOR

Terminal View of 104-Pin PCM Wiring Harness Connector

Standard Colors and Abbreviations

Abbreviation	Color	Abbreviation	Color	Abbreviation	Color
BK	Black	GY	Gray	PK	Purple
BL	Blue	GN	Green	RD	Red
BR	Brown	LG	LT Green	TN	Tan
DB	Dark Blue	OR	Orange	WT	White
DG	DK Green	PK	Pink	YL	Yellow

2003-05 F-Series 6.0L V8 Diesel VIN P (All) C138a Pin Connector

PCM Pin #	Wire Color	Circuit Description (46 Pin)	Value at Hot Idle
1	---	Not Used	---
2	PK/YL	A/C Relay Control	Relay Off: 12v, On: 1v
3	OG/LG	Data Output Link	Digital Signals
4	BR/PK	Starter Relay Circuit Control	Relay Off: 12v, On: 1v
5	LB/OR	Fuel Pump Relay Control	Relay Off: 12v, On: 1v
6-7	---	Not Used	---
8	TN/LG	A/C Pressure Switch Signal	Switch on: 12v, off: 1v
9	OG/LB	Speed Control Indicator Control	Indicator Off: 12v, On: 1v
10	LB/YL	Power Ground	<0.1v
11	BK/WT	Power Ground	<0.1v
12	---	Not Used	---
13	PK/YL	CAN Bus 1H Signal	0-7-0v
14	RD/WT	CAN Bus 1L Signal	0-7-0v
15	RD	Water In Fuel Indicator Control	Indicator Off: 12v, On: 1v
16	TN/LG	A/C Head Pressure Switch	0v or 12v
17	LG/RD	Parking Brake Applied Switch	Brake Off: 12v, Applied: 0v
18	RD/LG	Brake Pedal Switch	0v (Brake On: 12v)
19	PK/BK	Fuel Pump Monitor	Pump Off: 12v, On: 1v
20	LB/RD	APP Sensor 2 Ground	<0.050v
21	TN/LB	MAF Sensor Ground	<0.050v
22	---	Not Used	---
23	BK/WT	Power Ground	<0.1v
24	BK	Speed Control Ground	<0.050v
25	YL/WT	APP Sensor 3 Signal	0.8-3.5v
26	LB/BK	APP Sensor 1 Signal	0.7-4.2v
27	---	Not Used	---
28	BK/YL	Brake Pedal Switch	0v (Brake On: 12v)
29	WT/LB	APP Sensor 2 Reference Voltage	4.9-5.1v
30	VT	Generator/Battery Indicator Control	Indicator Off: 12v, On: 1v
31	LB/BK	Speed Control On/Off Switch	Switch On: 12v, Off: 0v
32	TN/OR	SCP Data Bus (+) Signal	Digital Signals
33	GY/RD	Ground	<0.050v
34	RD	Vehicle Power (Start-Run)	12-14v
35	GY/BK	VSS (+) Signal	0 Hz, 55 mph: 125 Hz
36	TN/WT	Tow Haul Switch	Off: 0v, On: 12v
37	YL	APP Sensor 2 Signal	1.4-4.1v
38	DB/LG	Barometric Absolute Pressure Sensor	159 Hz
39	VT	Flash EEPROM Power	0.1v
40	RD/WT	Battery Direct	12-14v
41	LG/BK	Manifold Absolute Pressure Sensor	0.5-4.5v
42	LB/RD	MAF Sensor Signal	0.1-4.7v
43	GY	Intake Air Temperature Sensor	1.5-2.5v
44	PK/LB	SCP Data Bus (-) Signal	<0.050v
45	BR/WT	Reference Voltage	4.9-5.1v
46	RD	Vehicle Power (Start-Run)	12-14v

2003-05 F-Series 6.0L V8 Diesel VIN P (All) C138b Pin Connector

PCM Pin #	Wire Color	Circuit Description (30 Pin)	Value at Hot Idle
1	PK/WT	Reference Voltage	4.9-5.1v
2	VT/YL	Pressure Control Solenoid 'A'	Solenoid Off: 12v, On: 1v
3	YL/LG	Reverse Lamps Relay Control	Relay Off: 12v, On: 1v
4	RD/WT	Transfer Case Neutral Signal	N/A
5	WT/LG	TCIL (lamp) Control	7.7v (Switch On: 0v)
6	---	Not Used	---
7	YL/WT	EPC Solenoid Control	7.5v (8 psi)
8	---	Not Used	---
9	OG/YL	Shift Solenoid Pressure Control 'A'	Solenoid Off: 12v, On: 1v
10	VT/OG	Shift Solenoid Pressure Control 'B'	Solenoid Off: 12v, On: 1v
11	PK/BK	Shift Solenoid Pressure Control 'C'	Solenoid Off: 12v, On: 1v
12	BK/LG	Shift Solenoid Pressure Control 'D'	Solenoid Off: 12v, On: 1v
13	DB/WT	Shift Solenoid Pressure Control 'E'	Solenoid Off: 12v, On: 1v
14	BR/OG	TCC Solenoid Control	TCC Off: 12v, On: 1v
15-16	---	Not Used	---
17	OG/YL	Shift Solenoid Pressure Control 'A'	Solenoid Off: 12v, On: 1v
18	LB/PK	Shift Solenoid Pressure Control 'B'	Solenoid Off: 12v, On: 1v
19	LB/RD	Shift Solenoid Pressure Control 'C'	Solenoid Off: 12v, On: 1v
20	WT/RD	Shift Solenoid Pressure Control 'D'	Solenoid Off: 12v, On: 1v
21	PK/LB	Shift Solenoid Pressure Control 'E'	Solenoid Off: 12v, On: 1v
22	BK/WT	TR Sensor TR-P Ground	<0.050v
23-24	---	Not Used	---
25	LB/YL	TR Sensor TR-P Signal	0-12-0v
26	OG/BK	TFT Sensor Signal	2.10-2.40v
27	GY/OG	ISS Sensor Signal	AC pulse signals
28	DB/YL	OSS Sensor Signal	AC pulse signals
29	DG/WT	TSS Sensor Signal	AC pulse signals
30	OG/WT	Sensor Ground	<0.050v

Pin Connector Graphic

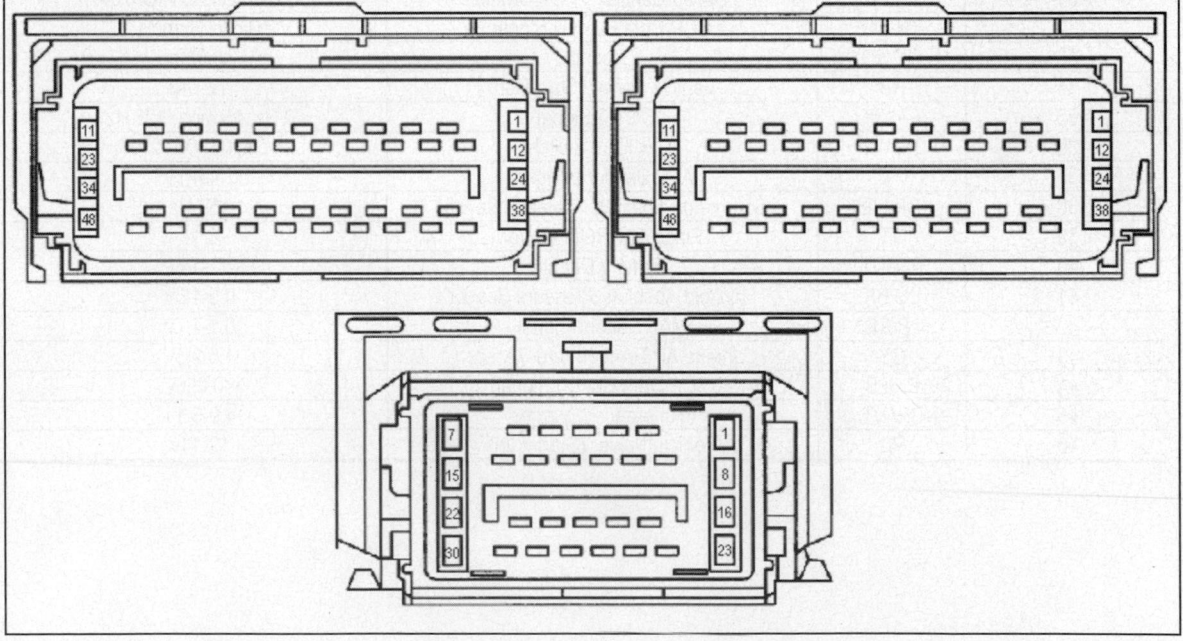

2003-05 F-Series 6.0L V8 Diesel VIN P C1381c Pin Connector

PCM Pin #	Wire Color	Circuit Description (46 Pin)	Value at Hot Idle
1	WT/YL	Charge Indicator Monitor	12v (Fault Detected: 0v)
2	YL/RD	Injector Pressure Regulator	12v
3	VT/OR	Glow Plug Enable	Relay Off: 12v, On: 1v
4	YL	Generator 2 Monitor	12v (Fault Detected: 0v)
5	---	Not Used	---
6	RD/OR	Electric Fan Speed Signal	0.1-4.9v
7-9	---	Not Used	---
10	BK/LB	Variable Geometry Actuator Signal	0-12-0v
11	DB	Variable Geometry Actuator Power	12-14v
12	GY/WT	Electronic Throttle Control (-) Signal	<0.050v
13	---	Not Used	---
14	DB	Electric Fan Speed Control	0-12-0v
15-16	---	Not Used	---
17	WT/LG	Glow Plug Module Signal	Off: 12v, On: 1v
18	---	Not Used	---
19	LG/YL	Injector Driver Module Command	0-6000 Hz
20	DB/OG	Injector Driver Module CID Signal	0.5-50 Hz
21	---	Not Used	---
22	BR/PK	Engine Cooling Fan Ground	<0.1v
29	TN/LB	M/T: CCP Switch Signal	Clutch In: 0v, Out: 12v
30	VT/LB	Exhaust Back Pressure Sensor	0.9v, off-idle: 2.5v
31	BK/YL	Brake Pressure Switch	Brake Off: 12v, On: 1v
32	YL/WT	Coolant Temperature Sensor	0.6 (194ºF)
33	LB/OR	EGR Valve Actuator Position Sense	0.6-3.5v
34-35	---	Not Used	---
36	BR/WT	Reference Voltage	4.9-5.1v
37	RD/LG	CAN Bus 2H Signal	0-7-0v
38	OG/LB	EGR Throttle Position Sensor	0.1-5.1v
39-40	---	Not Used	---
41	GY	Crankshaft Position Sensor (-) Signal	0-6000 Hz
42	BK	Ground (Drain Wire)	<0.050v
43	OR	Camshaft Position Sensor (-) Signal	0.5-50 Hz
44	LG/RD	Engine Oil Temperature Sensor	40ºF: 4.7v, 230ºF: 0.358v
45	RD/WT	Manifold Air Temperature Sensor	0.2-4.7v
46	BR/LG	Electric Fan Clutch Reference Voltage	12-14v

1999-2002 F-Series 6.8L V10 VIN S (All) 104 Pin Connector

PCM Pin #	Wire Color	Circuit Description (104 Pin)	Value at Hot Idle
1	OR/YL	COP 6 Driver (dwell)	5°, 55 mph: 9°
2	PK/LG	MIL (lamp) Control	MIL Off: 12v, On: 1v
3	BK/WT	Power Ground	<0.1v
4	LB/YL	Power Take-Off (if equipped)	0v (Off)
6	OR/YL	Shift Solenoid 1 Control	1v, 55 mph: 12v
9	YL/WT	Fuel Level Indicator Signal	1.7v (1/2 full)
11	PK/OR	Shift Solenoid 2 Control	12v, 55 mph: 12v
12	WT/LG	TCIL (lamp) Control	TCIL On: 1v, Off: 2.2v
13	VT	Flash EEPROM Power	0.1v
14	LB/BK	4x4 Low Indicator Switch	12v (switch closed: 0v)
15	PK/LB	Data Bus (-)	Digital Signals
16	TN/OR	Data Bus (+)	Digital Signals
19	OR/LG	Fuel Pump High Speed Relay	Off: 12v, On: 1v
20	BR/OR	Coast Clutch Solenoid	12v, 55 mph: 12v
21	DB	CKP (-) Sensor Signal	400-420 Hz
22	GY	CKP (+) Sensor Signal	400-420 Hz
23-24	BK/WT	Power Ground	<0.1v
25	LB/YL	Case Ground	<0.050v
26	LG/WT	COP 1 Driver (dwell)	5°, 55 mph: 9°
27	LG/YL	COP 10 Driver (dwell)	5°, 55 mph: 9°
29	TN/WT	TCS (switch) Signal	TCS & O/D On: 12v
32 ('00-'02)	DG/PK	Knock Sensor 1 (-) Signal	<0.050v
33	PK/OR	VSS (-) Signal	0 Hz, 55 mph: 125 Hz
34	YL/BK	Digital TR1 Sensor	0v, 55 mph: 11.5v
35	RD/LG	HO2S-12 (B1 S2) Signal	0.1-1.1v
36	TN/LB	MAF Sensor Return	<0.050v
37	OR/BK	TFT Sensor Signal	2.10-2.40v
39	GY	IAT Sensor Signal	1.5-2.5v
40	DG/YL	Fuel Pump Monitor	On: 12v, Off: 0v
41	TN/LG	A/C Switch Signal	A/C On: 12v, Off: 0v
42	GY/RD	Injector 10 Control	4.2-4.6 ms
43	OR/LG	Data Output Link	Digital Signals
45	RD/WT	CHIL (lamp) Control	CHT Off: 12v, On: 1v
46	GY/BK	VSS (+) Signal	0 Hz, 55 mph: 125 Hz
47	BR/PK	VR Solenoid Control	0%, 55 mph: 45%
48	WT/PK	Clean Tachometer Output	60-70 Hz
49	LB/BK	Digital TR2 Sensor	0v, 55 mph: 11.5v
50	WT/BK	Digital TR4 Sensor	0v, 55 mph: 11.5v
51	BK/WT	Power Ground	<0.1v
52	WT/PK	COP 5 Driver (dwell)	5°, 55 mph: 9°
53	DGVT	COP 7 Driver (dwell)	5°, 55 mph: 9°
54	PK/YL	TCC Solenoid Control	0%, 55 mph: 95%
55	RD/WT	Keep Alive Power	12-14v
56	LG/BK	EVAP Purge Solenoid	0-10 Hz (0-100%)
57	YL/RD	Knock Sensor 1 (+) Signal	0v
58	GY/BK	VSS (+) Signal	0 Hz, 55 mph: 125 Hz
59	DG/WT	TSS Sensor Signal	300 Hz, 55 mph: 980
60	GY/LB	HO2S-11 (B1 S1) Signal	0.1-1.1v
61	---	Not Used	---
62	RD/PK	FTP Sensor Signal	2.6v at 0" H20 (cap off)
63	---	Not Used	---
64	LB/YL	A/T: Digital TR3 Sensor	In 'P': 0v, in O/D: 1.7v
64	LB/YL	M/T: CPP Switch Signal	5v (clutch "in": 0v)

1999-2002 F-Series 6.8L V10 VIN S (All) 104 Pin Connector

PCM Pin #	Wire Color	Circuit Description (104 Pin)	Value at Hot Idle
65 ('99-'00)	BR/LG	DPFE Sensor Signal	0.20-1.30v
65 ('01-'02)	BR/LG	DPFE Sensor Signal	0.95-1.05v
66	YL/LG	CHT Sensor Signal	0.6 (194°F)
67	VT/WT	EVAP Canister Vent Solenoid Control	0-10 Hz (0-100%)
68	GY/BK	Injector 9 Control	4.2-4.6 ms
69 ('00-'02)	PK/YL	A/C Clutch Relay Control	A/C On: 12v, Off: 0v
71, 97	RD	Vehicle Power	12-14v
72	TN/RD	Injector 7 Control	4.2-4.6 ms
73	TN/BK	Injector 5 Control	4.2-4.6 ms
74	BR/YL	Injector 3 Control	4.2-4.6 ms
75	TN	Injector 1 Control	4.2-4.6 ms
76, 77, 103	BK/WT	Power Ground	<0.1v
78	PK/LB	COP 2 Driver (dwell)	5°, 55 mph: 9°
79	WT/RD	COP 8 Driver (dwell)	5°, 55 mph: 9°
80	LB/OR	Fuel Pump Relay Control	Off: 12v, On: 1v
81	WT/YL	EPC Solenoid Control	9.2v (5 psi)
82	DG/PK	COP 9 Driver (dwell)	5°, 55 mph: 9°
82 ('00-'02)	WT/RD	COP 9 Driver (dwell)	5°, 55 mph: 9°
83	WT/LB	IAC Solenoid Control	10.7v (33%)
84	DB/YL	Output Shaft Speed Sensor	0 Hz, 55 mph: 470 Hz
85	DG	CMP Sensor Signal	10 Hz, 55 mph: 16 Hz
86	---	Not Used	---
87	RD/BK	HO2S-21 (B2 S1) Signal	0.1-1.1v
88	LB/RD	MAF Sensor Signal	0.8v, 55 mph: 1.6v
89	GY/WT	TP Sensor Signal	0.53-1.27v
90	BR/WT	Reference Voltage	4.9-5.1v
91	GY/RD	Analog Signal Return	<0.050v
92	RD/LG	Brake Pedal Switch	0v (Brake On: 12v)
93	RD/WT	HO2S-11 (B1 S1) Heater	1v (Heater Off: 12v)
94	YL/LB	HO2S-21 (B2 S1) Heater	1v (Heater Off: 12v)
95	WT/BK	HO2S-12 (B1 S2) Heater	1v (Heater Off: 12v)
98	LB	Injector 8 Control	4.2-4.6 ms
99	LG/OR	Injector 6 Control	4.2-4.6 ms
100	BR/LB	Injector 4 Control	4.2-4.6 ms
101	WT	Injector 2 Control	4.2-4.6 ms
102	YL/BK	COP 4 Driver (dwell)	5°, 55 mph: 9°
104	PK/WT	COP 3 Driver (dwell)	5°, 55 mph: 9°

Pin Connector Graphic

PCM 104-PIN CONNECTOR

Terminal View of 104-Pin PCM Wiring Harness Connector

Standard Colors and Abbreviations

Abbreviation	Color	Abbreviation	Color	Abbreviation	Color
BK	Black	GY	Gray	PK	Purple
BL	Blue	GN	Green	RD	Red
BR	Brown	LG	LT Green	TN	Tan
DB	Dark Blue	OR	Orange	WT	White
DG	DK Green	PK	Pink	YL	Yellow

2003-05 F-Series 6.8L V10 VIN S (All) 104 Pin Connector

PCM Pin #	Wire Color	Circuit Description (104 Pin)	Value at Hot Idle
1	OR/YL	COP 6 Driver (dwell)	5°, 55 mph: 9°
2	---	Not Used	---
3	BK/WT	Power Ground	<0.1v
4	LB/YL	Power Take-Off (if equipped)	0v (Off)
5	---	Not Used	---
6	OR/YL	Shift Solenoid 1 Control	1v, 55 mph: 12v
7-10	---	Not Used	---
11	PK/OR	Shift Solenoid 2 Control	12v, 55 mph: 12v
12	WT/LG	TCIL (lamp) Control	TCIL On: 1v, Off: 2.2v
13	VT	Flash EEPROM Power	0.1v
14	LB/BK	4x4 Low Indicator Switch	12v (switch closed: 0v)
15	PK/LB	SCP Data Bus (-) Signal	<0.050v
16	TN/OR	SCP Data Bus (+) Signal	Digital Signals
17-19	---	Not Used	---
20	BR/OR	Coast Clutch Solenoid Control	12v, 55 mph: 12v
21	DB	CKP (-) Sensor Signal	400-420 Hz
22	GY	CKP (+) Sensor Signal	400-420 Hz
23-24	BK/WT	Power Ground	<0.1v
25	LB/YL	Case Ground	<0.050v
26	LG/WT	COP 1 Driver (dwell)	5°, 55 mph: 9°
27	LG/YL	COP 10 Driver (dwell)	5°, 55 mph: 9°
28	---	Not Used	---
29	TN/WT	Overdrive Cancel Switch) Signal	Switch Off: 12v, On: 1v
30-31	---	Not Used	---
32	DG/VT	Knock Sensor 1 (-) Signal	<0.050v
33	PK/OR	Ground	<0.050v
34	YL/BK	Digital TR1 Sensor	0v, 55 mph: 11.5v
35	RD/LG	HO2S-12 (B1 S2) Signal	0.1-1.1v
36	TN/LB	MAF Sensor Return	<0.050v
37	OR/BK	TFT Sensor Signal	2.10-2.40v
38	---	Not Used	---
39	GY	IAT Sensor Signal	1.5-2.5v
40	DG/YL	Fuel Pump Monitor	Off: 12v, On: 1v

2003-05 F-Series 6.8L V10 VIN S (All) 104 Pin Connector

PCM Pin #	Wire Color	Circuit Description (104 Pin)	Value at Hot Idle
41	TN/LG	A/C Switch Signal	A/C On: 12v, Off: 0v
42	GY/RD	Injector 10 Control	4.2-4.6 ms
43	OR/LG	Data Output Link	Digital Signals
44-46	---	Not Used	---
47	BR/PK	VR Solenoid Control	0%, 55 mph: 45%
48	LG/WT	Clean Tachometer Output	60-70 Hz
49	LB/BK	Digital TR2 Sensor	0v, 55 mph: 11.5v
50	WT/BK	Digital TR4 Sensor	0v, 55 mph: 11.5v
51	BK/WT	Power Ground	<0.1v
52	WT/PK	COP 5 Driver (dwell)	5°, 55 mph: 9°
53	DGVT	COP 7 Driver (dwell)	5°, 55 mph: 9°
54	VT/YL	TCC Solenoid Control	0%, 55 mph: 95%
55	RD/WT	Keep Alive Power	12-14v
56	LG/BK	EVAP Purge Solenoid	0-10 Hz (0-100%)
57	YL/RD	Knock Sensor 1 (+) Signal	0v
58	GY/BK	VSS (+) Signal	0 Hz, 55 mph: 125 Hz
59	DG/WT	TSS Sensor Signal	300 Hz, 55 mph: 980
60	GY/LB	HO2S-11 (B1 S1) Signal	0.1-1.1v
61	---	Not Used	---
62	RD/PK	FTP Sensor Signal	2.6v at 0" H20 (cap off)
63	---	Not Used	---
64	LB/YL	A/T: Digital TR3 Sensor	In 'P': 0v, in O/D: 1.7v
64	LB/YL	M/T: CPP Switch Signal	5v (clutch "in": 0v)

2003-05 F-Series 6.8L V10 VIN S (All) 104 Pin Connector

PCM Pin #	Wire Color	Circuit Description (104 Pin)	Value at Hot Idle
65	BR/LG	DPFE Sensor Signal (California)	0.95-1.05v
66	YL/LG	Cylinder Head Temperature Sensor	0.6 (194ºF)
67	VT/WT	EVAP Canister Vent Solenoid Control	0-10 Hz (0-100%)
68	GY/BK	Injector 9 Control	4.2-4.6 ms
69	PK/YL	A/C Clutch Relay Control	A/C On: 12v, Off: 0v
70	---	Not Used	
71	RD	Vehicle Power	12-14v
72	TN/RD	Injector 7 Control	4.2-4.6 ms
73	TN/BK	Injector 5 Control	4.2-4.6 ms
74	BR/YL	Injector 3 Control	4.2-4.6 ms
75	TN	Injector 1 Control	4.2-4.6 ms
76, 77	BK/WT	Power Ground	<0.1v
78	PK/LB	COP 2 Driver (dwell)	5º, 55 mph: 9º
79	WT/RD	COP 8 Driver (dwell)	5º, 55 mph: 9º
80	LB/OR	Fuel Pump Relay Control	Off: 12v, On: 1v
81	WT/YL	EPC Solenoid Control	9.2v (5 psi)
82	WT/RD	COP 9 Driver (dwell)	5º, 55 mph: 9º
83	WT/LB	IAC Solenoid Control	10.7v (33%)
84	DB/YL	Output Shaft Speed Sensor	0 Hz, 55 mph: 470 Hz
85	DG	CMP Sensor Signal	10 Hz, 55 mph: 16 Hz
86	---	Not Used	---
87	RD/BK	HO2S-21 (B2 S1) Signal	0.1-1.1v
88	LB/RD	MAF Sensor Signal	0.8v, 55 mph: 1.6v
89	GY/WT	TP Sensor Signal	0.53-1.27v
90	BR/WT	Reference Voltage	4.9-5.1v
91	GY/RD	Sensor Return	<0.050v
92	RD/LG	Brake Pedal Switch	0v (Brake On: 12v)
93	RD/WT	HO2S-11 (B1 S1) Heater	1v (Heater Off: 12v)
94	YL/LB	HO2S-21 (B2 S1) Heater	1v (Heater Off: 12v)
95	WT/BK	HO2S-12 (B1 S2) Heater	1v (Heater Off: 12v)
96	---	Not Used	---
97	RD	Vehicle Power	12-14v
98	LB	Injector 8 Control	4.2-4.6 ms
99	LG/OR	Injector 6 Control	4.2-4.6 ms
100	BR/LB	Injector 4 Control	4.2-4.6 ms
101	W	Injector 2 Control	4.2-4.6 ms
102	YL/BK	COP 4 Driver (dwell)	5º, 55 mph: 9º
103	BK/WT	Power Ground	<0.1v
104	PK/WT	COP 3 Driver (dwell)	5º, 55 mph: 9º

Pin Connector Graphic

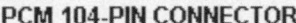

PCM 104-PIN CONNECTOR

Terminal View of 104-Pin PCM Wiring Harness Connector

Standard Colors and Abbreviations

Abbreviation	Color	Abbreviation	Color	Abbreviation	Color
BK	Black	GY	Gray	PK	Purple
BL	Blue	GN	Green	RD	Red
BR	Brown	LG	LT Green	TN	Tan
DB	Dark Blue	OR	Orange	WT	White
DG	DK Green	PK	Pink	YL	Yellow

1996-97 F-Series 7.3L V8 Diesel VIN F 104 Pin Connector

PCM Pin #	Wire Color	Circuit Description (104 Pin)	Value at Hot Idle
1	PK/OR	Shift Solenoid 2 Control	1v, 55 mph: 12v
2	PK/LG	MIL (lamp) Control	MIL Off: 12v, On: 1v
4	PK/WT	Parking Brake Applied Switch	Parking Brake Applied: 0v
5	RD/OR	Idle Validation Switch Signal	Switch up: 0v, down: 12v
5	LG/RD	Parking Brake Switch (Cal)	Parking Brake Applied: 0v
6	OR/YL	Shift Solenoid 1 Control (Cal)	SS1 Off: 12v, On: 1v
8	GY	Glow Plug Monitor High (Cal)	Plugs Off: 0v, On: 12v
9	OR	Glow Plug Monitor R/S (Cal)	Plugs Off: 0v, On: 12v
10	RD/OR	Idle Validation Switch (Cal)	0v, off-12v
11	PK/OR	Shift Solenoid 2 Control (Cal)	SS2 Off: 12v, On: 1v
12	WT/LG	TCIL (lamp) Control (Cal)	7.7v (Switch On: 0v)
13	PK	Flash EPROM Power	0.1v
14	LG/BK	4x4 Low Indicator Switch	Switch On: 0v, Off: 12v
15	PK/LB	Data Bus (-) Signal	Digital Signals
16	TN/OR	Data Bus (+) Signal	Digital Signals
17	OR/BK	TR Sensor Signal 1 (Cal)	Manual 1 or 2: 10.7v
19	WT/PK	Tachometer Signal Cal)	6.5v / 130 Hz
21	DG	CMP Sensor Signal (Cal)	6 Hz, 55 mph: 15 Hz
24	YL/BK	Accelerator Pedal (-) Sensor	<0.050v
25	BK	Case Ground	<0.050v
27	OR/YL	Shift Solenoid 1 Control	0.1v, 35 mph: 12v
28	PK/YL	TCC Solenoid Control	0v, On 55 mph: 12v
29	TN/LB	Clutch Pedal Position Switch	Clutch In: 0v, Out: 12v
29	TN/WT	TCS (switch) Signal	TCS & O/D On: 12v
30	PK/LB	Exhaust Back Press. Sensor	0.80-0.95v
31	BK/YL	Brake Pedal Applied Switch	Brake Off: 12, On: 1v
33	PK/OR	VSS (-) Signal	0 Hz, 55 mph: 125 Hz
34	LG/BK	MAP Sensor Signal	159 Hz (sea level)
37	OR/BK	TFT Sensor Signal	2.10-2.40v
38	LG/RD	EOT Sensor Signal	0.36v at 230ºF
39	GY	IAT Sensor Signal	1.5-2.5v
40	DG/OR	Speed Control Ground	<0.050v
41	DG/OR	A/C Clutch Signal	A/C On: 12v, Off: 0v
42	GY/RD	Exhaust Back Pressure	6-8v (when enabled)
48	GY/WT	Electronic Feedback Line	Digital Signals
49	DG	CMP (+) Sensor Signal	0.7v
50	WT/PK	Tachometer Signal from CMP	6.5v / 130 Hz
51	BK/WT	Power Ground	<0.1v
53	BR/OR	Coast Clutch Solenoid	CCS Off: 12v, On: 1v
55	YL	Keep Alive Power	12-14v
58	GY/BK	VSS (+) Signal	0 Hz, 55 mph: 125 Hz
61	LB/BK	Speed Control Cruise Signal	Varies 0-12v
64	LB/YL	A/T: TR Sensor Signal	In 'P': 0v, in O/D: 5v

1996-97 F-Series 7.3L V8 Diesel VIN F 104 Pin Connector

PCM Pin #	Wire Color	Circuit Description (104 Pin)	Value at Hot Idle
65	LB	CMP (-) Sensor Signal	<0.050v
70	WT/BK	IDM Relay Signal	IDM Off: 12v, On: 0v
71	RD	Vehicle Power	12-14v
76	BK/WT	Power Ground	<0.1v
77	BK/WT	Power Ground	<0.1v
79	WT/LG	TCIL (lamp) Control	7.7v (Switch On: 0v)
80	BK/PK	Glow Plug Lamp Control	7.7v (Switch On: 0v)
81	WT/YL	EPC Solenoid Control	7.5v (8 psi)
83	YL/RD	Injector Pressure Regulator	12v duty cycle signal
84	DB/LG	BARO Sensor Signal	At Sea Level: 4.64v
87	DB/LG	Injection Control Pressure	0.75v (Min: 0.83v startup)
89	GY/WT	APP Sensor Signal	0.5-0.9v
90	BR/WT	Reference Voltage	4.9-5.1v
91	GY/RD	Analog Signal Return	<0.050v
92	LG	Brake Pedal Switch	0v (Brake On: 12v)
95	BR/OR	Fuel Demand Command	49 Hz
96	YL/LB	CID Sensor Signal	5 Hz
97	RD	Vehicle Power	12-14v
101	PK/OR	Glow Plug Relay Control	Off: 12v, On: 1v
103	BK/WT	Power Ground	<0.1v

Pin Connector Graphic

PCM 104-PIN CONNECTOR

Terminal View of 104-Pin PCM Wiring Harness Connector

Standard Colors and Abbreviations

Abbreviation	Color	Abbreviation	Color	Abbreviation	Color
BK	Black	GY	Gray	PK	Purple
BL	Blue	GN	Green	RD	Red
BR	Brown	LG	LT Green	TN	Tan
DB	Dark Blue	OR	Orange	WT	White
DG	DK Green	PK	Pink	YL	Yellow

1998-99 F-Series 7.3L V8 Diesel VIN F 104 Pin Connector

PCM Pin #	Wire Color	Circuit Description (104 Pin)	Value at Hot Idle
1, 3, 7, 9	---	Not Used	---
2	PK/LG	MIL (lamp) Control	MIL Off: 12v, On: 1v
4	LB/YL	Neutral Gear Switch	Switch On: 0v, Off: 12v
5	LG/RD	Brake Warning Indicator	Parking Brake Applied: 0v
6	OR/YL	Shift Solenoid 1 Control	SS1 Off: 12v, On: 1v
8	WT/LG	Glow Plug Monitor	Plugs Off: 0v, On: 12v
10	RD/LG	Idle Validation Switch	0v, off-12v
11	PK/OR	Shift Solenoid 2 Control	SS2 Off: 12v, On: 1v
12	WT/LG	TCIL (lamp) Control	7.7v (Switch On: 0v)
13	PK	Flash EPROM Power	0.1v
14	LG/BK	4x4 Low Indicator Switch	Switch On: 0v, Off: 12v
15	PK/LB	Data Bus (-) Signal	Digital Signals
16	TN/OR	Data Bus (+) Signal	Digital Signals
17	WT/YL	Digital TR1 Sensor	0v, 55 mph: 11.5v
18, 22-23	---	Not Used	---
19	WT/PK	Tachometer Reflected Signal	6.5v / 130 Hz
20	BR/OR	Coast Clutch Solenoid	12v, 55 mph: 12v
21	DG	CMP Sensor Signal	5-7 Hz
24	YL/WT	TP Sensor Signal	0.53-1.27v
25	LB/YL	Tachometer VSS Ground	<0.050v
26-27	---	Not Used	---
28	RD	Water In Fuel Indicator Lamp	7.7v (Switch On: 0v)
29	TN/WT	A/T: TCS (switch) Signal	TCS & O/D On: 12v
29	TN/LB	M/T: CPP (switch) Signal	Clutch In: 0v, Out: 12v
30	PK/LB	Exhaust Back Pressure	0.9v, off-idle: 2.5v
31	BK/YL	Brake Pedal Applied Switch	Brake Off: 12, On: 1v
32-34	---	Not Used	---
35	WT/YL	Generator Power Switch	12-14v
36	GY/RD	Fuel Heater Signal	Heater On: 12v
37	OR/BK	TFT Sensor Signal	2.10-2.40v
38	LG/RD	Engine Oil Temperature	-40ºF: 4.7v, 230ºF: 0.358v
39	GY	IAT Sensor Signal	1.5-2.5v
40	OR/YL	Fuel Pump Relay	Off: 12v, On: 1v
41	TN/LG	A/C Switch Signal	Switch On: 0v, Off: 12v
42	GY/RD	Exhaust Pressure Regulator	Digital: 0-12-0-12v
43	OR/LG	DLC Signal	Digital Signals
44-46, 52-53	---	Not Used	---
47	WT/RD	Wastegate Control Solenoid	Solenoid Off: 12v, On: 1v
48	GY/WT	Electronic Feedback Line	0.4-2.2v -12v digital signal
49	DB/WT	Digital TR2 Sensor	0v, 55 mph: 11.5v
50	DG/YL	Digital TR4 Sensor	0v, 55 mph: 11.5v
51, 77, 103	BK/WT	Power Ground	<0.1v
54	PK/YL	TCC Solenoid Control	Clutch Off: 12v, On: 1v
55	RD/WT	Keep Alive Power	12-14v
56-57, 68-69	---	Not Used	---
58	GY/BK	VSS (+) Signal	0 Hz, 55 mph: 125 Hz
59	DB/YL	OSS Sensor Signal	0 Hz, 55 mph: 470 Hz
60	YL	Alternator No. 2 (bottom) Motor	6-10v
61	LB/BK	Speed Control Cruise Signal	Signal varies: 0-12v
62	RD/YL	ACT Sensor Signal	1.5-2.5v
63	DB/LG	BARO Sensor Signal	4.71v (sea level)
64	LB/YL	Digital TR3 Sensor	0v, 55 mph: 11.5v

1998-99 F-Series 7.3L V8 Diesel VIN F 104 Pin Connector

PCM Pin #	Wire Color	Circuit Description (104 Pin)	Value at Hot Idle
65	LB	CMP (-) Sensor Signal	<0.050v
66	LB/YL	Take Off Power Input	N/A
67	LG/RD	Generator Indicator	Indicator Off: 12v, On: 1v
70	WT/BK	Glow Plug Lamp Control	7.7v (Switch On: 0v)
71, 97	RD	Vehicle Power	12-14v
72-76, 78	---	Not Used	---
79	LG/BK	MAP Sensor Signal	0.9v
80	WT/BK	IDM Enable	Off: 12v, On: 1v
81	WT/YL	EPC Solenoid Control	7.5v (8 psi)
82, 85-86	---	Not Used	---
83	YL/RD	Injector Pressure Regulator	12v (Duty Cycle)
84	DG/WT	TSS Sensor Signal	300 Hz, 55 mph: 980
87	DB/LG	Injection Control Pressure	0.75v (Min: 0.83v startup)
88, 93	---	Not Used	---
89	GY/WT	APP Sensor Signal	0.5-0.9v
90	BR/WT	Reference Voltage	4.9-5.1v
91	GY/RD	Analog Signal Return	<0.050v
92	RD/LG	Brake Pedal Switch	0v (Brake On: 12v)
94	LB/OR	Fuel Pump Relay	
95	BR/OR	Fuel Demand Command Signal	49 Hz
96	YL/LB	CMP (+) Sensor Signal	5 Hz
98-99	---	Not Used	---
100	YL/WT	Fuel Level Indicator Signal	1.7v (1/2 full)
102, 104	---	Not Used	---
101	PK/OR	Glow Plug Relay Control	Off: 12v, On: 1v

Pin Connector Graphic

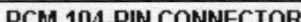

Standard Colors and Abbreviations

Abbreviation	Color	Abbreviation	Color	Abbreviation	Color
BK	Black	GY	Gray	PK	Purple
BL	Blue	GN	Green	RD	Red
BR	Brown	LG	LT Green	TN	Tan
DB	Dark Blue	OR	Orange	WT	White
DG	DK Green	PK	Pink	YL	Yellow

2000-02 F-Series 7.3L V8 Diesel VIN F 104 Pin Connector

PCM Pin #	Wire Color	Circuit Description (104 Pin)	Value at Hot Idle
1	---	Not Used	---
2	PK/LG	MIL (lamp) Control	MIL Off: 12v, On: 1v
3	RD/YL	Low Current Sensor Return	<0.050v
4	LB/YL	Neutral Gear Switch	Switch On: 0v, Off: 12v
5	LG/RD	Parking Brake Applied Switch	Parking Brake Applied: 0v
6	OR/YL	Shift Solenoid 1 Control	SS1 Off: 12v, On: 1v
7, 9	---	Not Used	---
8	WT/LG	Glow Plug Monitor Left Bank	Plugs Off: 0v, On: 12v
10	RD/LG	Idle Validation Switch	12v, off-idle: 0v
11	PK/OR	Shift Solenoid 2 Control	SS2 Off: 12v, On: 1v
12	WT/LG	TCIL (lamp) Control	7.7v (Switch On: 0v)
13	PK	Generic Scan Tool Input	0.1v
14	LB/BK	4x4 Low Switch Input	Switch On: 0v, Off: 12v
15	PK/LB	Data Bus (-)	<0.050v
16	TN/OR	Data Bus (+)	Digital Signals
17	WT/YL	Digital TR1 Sensor	In 'P': 0v, in Drive: 10.7v
18	---	Not Used	---
19	WT/PK	Tachometer Reflected Signal	6.5v / 130 Hz
20	BR/OR	Coast Clutch Solenoid	CCS On: 0v, Off: 12v
21	DG	CMP Sensor Signal	Digital Signal: 0-12-0v
22-23	---	Not Used	---
24	YL/WT	APP Sensor Ground	<0.050v
25	LB/YL	Speedometer Ground	<0.050v
26-27	---	Not Used	---
28	RD	Water In Fuel Indicator Lamp	7.7v (Switch On: 0v)
29	TN/WT	A/T: TCS (switch) Signal	TCS & O/D On: 12v
29	TN/LB	M/T: CCP Switch Signal	Clutch In: 0v, Out: 12v
30	PK/LB	Exhaust Back Pressure	0.9v, off-idle: 2.5v
31	BK/YL	Brake Pressure Switch	Brake Off: 12v, On: 1v
32-34	---	Not Used	---
35	WT/YL	Generator Power Switch	12-14v
36	GY/RD	Fuel Heater / Water in Fuel	Heater On: 12v
37	OR/BK	TFT Sensor Signal	2.10-2.40v
38	LG/RD	Engine Oil Temperature	0.3-4.7v
39	GY	IAT Sensor Signal	1.5-2.5v
40	PK/BK	Fuel Pump Power	12-14v

2000-02 F-Series 7.3L V8 Diesel VIN F 104 Pin Connector

41	TN/LG	A/C Head Pressure Switch	0v or 12v
42	GY/RD	Exhaust Backpressure Signal	Digital: 0-12-0-12v
43	OR/LG	Data Output Link	Digital Signals
44	---	Not Used	---
45	OR/LB	Speed Control Indicator	0v
46	---	Not Used	---
47	WT/RD	Wastegate Control Solenoid	Solenoid Off: 12v, On: 1v
48	GY/WT	Electronic Feedback Line	Digital Signals
49	DB/WT	Digital TR2 Sensor	In 'P': 0v, in Drive: 10.7v
50	DG/YL	Digital TR4 Sensor	In 'P': 0v, in Drive: 10.7v
51	BK/WT	Power Ground	<0.1v
52-53	---	Not Used	---
54	PK/YL	TCC Solenoid Control	TCC Off: 12v, On: 1v
55	RD/WT	Keep Alive Power	12-14v
56-57	---	Not Used	---
58	GY/BK	VSS (+) Signal	0 Hz, 55 mph: 125 Hz
59	DB/YL	OSS Sensor Signal	Varying Signal
60	YL	Generator Power	12-14v
61	LB/BK	Speed Control On/Off Switch	Switch On: 12v, Off: 0v
62	RD/YL	MAT Sensor Signal	1.5-2.5v

2000-02 F-Series 7.3L V8 Diesel VIN F 104 Pin Connector

PCM Pin #	Wire Color	Circuit Description (104 Pin)	Value at Hot Idle
63	---	Not Used	---
64	LB/YL	Digital TR3 Sensor	In P: 4.5v, in Drive: 2.2v
65	LB	CMP Sensor Ground	<0.050v
66	LB/YL	Power Take-Off Signal	0v (Off)
67	LG/RD	Battery Direct	12-14v
68-69	---	Not Used	---
70	BK/PK	Glow Plug Lamp Control	7.7v (Switch On: 0v)
71	RD	Vehicle Power	12-14v
72-76	---	Not Used	---
77	BK/WT	Power Ground	<0.1v
78	---	Not Used	---
79	LG/BK	MAP Sensor Signal	1-3v
80	WT/BK	IDM Enable Relay Control	Off: 12v, On: 1v
81	WT/YL	EPC Solenoid Control	7.5v (8 psi)
82	---	Not Used	---
83	YL/RD	Injector Pressure Regulator	12v
84	DG/WT	TSS Sensor Signal	30 mph: 130 Hz
85-86	---	Not Used	---
87	DB/LG	Injection Control Pressure	0.75v (Min: 0.83v startup)
88	---	Not Used	---
89	GY/WT	APP Sensor Signal	0.5-1.6v
90	BR/WT	Reference Voltage	4.9-5.1v
91	GY/RD	Sensor Return Signal	<0.050v
92	RD/LG	Brake Pedal Switch	0v (Brake On: 12v)
93	---	Not Used	---
94	LB/OR	Fuel Pump Relay Output	12-14v
95	BR/OR	Fuel Delivery Control Signal	49 Hz
96	YL/LB	CID Sensor Signal	5 Hz
97	RD	Vehicle Power	12-14v
98	PK	PCM Signal to Relay	Digital Signals
99	---	Not Used	---
100	YL/WT	Fuel Level Indicator Signal	1.7v (1/2 full)
101	PK/OR	Glow Plug Relay Control	Off: 12v, On: 1v
102	---	Not Used	---
103	BK/WT	Power Ground	<0.1v
104	---	Not Used	---

Pin Connector Graphic

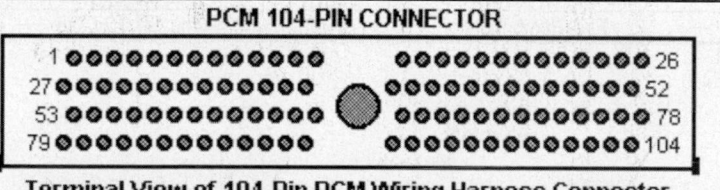

Terminal View of 104-Pin PCM Wiring Harness Connector

Standard Colors and Abbreviations

Abbreviation	Color	Abbreviation	Color	Abbreviation	Color
BK	Black	GY	Gray	PK	Purple
BL	Blue	GN	Green	RD	Red
BR	Brown	LG	LT Green	TN	Tan
DB	Dark Blue	OR	Orange	WT	White
DG	DK Green	PK	Pink	YL	Yellow

2003 F-Series 7.3L V8 Diesel VIN F (All) 104 Pin Connector

PCM Pin #	Wire Color	Circuit Description (104 Pin)	Value at Hot Idle
1	---	Not Used	---
2	PK/LG	MIL (lamp) Control	MIL Off: 12v, On: 1v
3	RD/YL	Manifold Intake Air Heater Monitor	Off: 12v, On: 1v
4	---	Not Used	---
5	LG/RD	Parking Brake Applied Switch	Brake Off: 12v, Applied: 0v
6	OR/YL	Shift Solenoid 1 Control	SS1 Off: 12v, On: 1v
7	---	Not Used	---
8	WT/LG	Glow Plug Monitor Left Bank	Plugs Off: 0v, On: 12v
9	---	Not Used	---
10	RD/LG	Idle Validation Switch	12v, off-idle: 0v
11	VT/OG	Shift Solenoid 2 Control	SS2 Off: 12v, On: 1v
12	WT/LG	TCIL (lamp) Control	7.7v (Switch On: 0v)
13	VT	Flash EEPROM Power Supply	0.1v
14	---	Not Used	---
15	PK/LB	SCP Data Bus (-) Signal	<0.050v
16	TN/OR	SCP Data Bus (+) Signal	Digital Signals
17	OR/BK	Digital TR1 Sensor	In 'P': 0v, in Drive: 10.7v
18	PK/YL	A/C Relay Control	Relay Off: 12v, On: 1v
19	WT/PK	Tachometer Reflected Signal	6.5v / 130 Hz
20	BR/OR	Coast Clutch Solenoid Control	CCS On: 0v, Off: 12v
21	DG	CMP Sensor Signal	Digital Signal: 0-12-0v
22-23	---	Not Used	---
24	YL/WT	APP Sensor Ground	<0.050v
25	BK/LB	Case Ground	<0.050v
26-27	---	Not Used	---
28	RD	Water In Fuel Indicator Control	Indicator Off: 12v, On: 1v
29	TN/WT	A/T: TCS (switch) Signal	TCS & O/D On: 12v
29	TN/LB	M/T: CCP Switch Signal	Clutch In: 0v, Out: 12v
30	VT/LB	Exhaust Back Pressure Sensor	0.9v, off-idle: 2.5v
31	BK/YL	Brake Pressure Switch	Brake Off: 12v, On: 1v
32	---	Not Used	---
33	PK/OG	VSS (-) Signal	<0.050v
34	---	Not Used	---
35	WT/YL	Alternator No. 1 (Top) Monitor	6-10v
36	GY/RD	Fuel Heater / Water in Fuel	Heater On: 12v
37	OR/BK	TFT Sensor Signal	2.10-2.40v
38	LG/RD	Engine Oil Temperature Sensor	0.3-4.7v
39	GY	Intake Air Temperature Sensor	1.5-2.5v
40	PK/BK	Fuel Pump Monitor	Pump Off: 12v, On: 1v

2003 F-Series 7.3L V8 Diesel VIN F (All) 104 Pin Connector

41	TN/LG	A/C Head Pressure Switch	0v or 12v
42	GY/RD	Exhaust Backpressure Signal	Digital: 0-12-0-12v
43	OR/LG	Overhead Console Fuel Consumption	0-5-0v
44-46	---	Not Used	---
47	WT/RD	Wastegate Control Solenoid Control	Solenoid Off: 12v, On: 1v
48	GY/WT	Electronic Feedback Line	0.9-3.0v
49	WT/PK	Digital TR2 Sensor	In 'P': 0v, in Drive: 10.7v
50	GY/BK	Digital TR4 Sensor	In 'P': 0v, in Drive: 10.7v
51	BK/WT	Power Ground	<0.1v
52-53	---	Not Used	---
54	VT/YL	TCC Solenoid Control	TCC Off: 12v, On: 1v
55	RD/WT	Keep Alive Power	12-14v
56-57, 63	---	Not Used	---
58	GY/BK	VSS (+) Signal	0 Hz, 55 mph: 125 Hz
59	DB/YL	OSS Sensor Signal	AC pulse signals
60	YL	Alternator No. 2 (Bottom) Control	6-10v
61	LB/BK	Speed Control On/Off Switch	Switch On: 12v, Off: 0v
62	RD/YL	MAT Sensor Signal	1.5-2.5v

2003 F-Series 7.3L V8 Diesel VIN F 104 Pin Connector

PCM Pin #	Wire Color	Circuit Description (104 Pin)	Value at Hot Idle
63	---	Not Used	---
64	LB/YL	Digital TR3 Sensor	In P: 4.5v, in Drive: 2.2v
65	LB	CMP Sensor Ground	<0.050v
66	LB/YL	Power Take-Off Signal	0v (Off)
67	LG/RD	Battery Direct	12-14v
68-69	---	Not Used	---
70	BK/PK	Glow Plug Lamp Control	7.7v (Switch On: 0v)
71	RD	Vehicle Power	12-14v
72-75	---	Not Used	---
76	BK/WT	Power Ground	<0.1v
77	BK/WT	Power Ground	<0.1v
78	---	Not Used	---
79	LG/BK	MAP Sensor Signal	1-3v
80	WT/BK	IDM Enable Relay Control	Off: 12v, On: 1v
81	WT/YL	EPC Solenoid Control	7.5v (8 psi)
82	---	Not Used	---
83	YL/RD	Injector Pressure Regulator	12v
84	DG/WT	TSS Sensor Signal	30 mph: 130 Hz
85-86	---	Not Used	---
87	DB/LG	Injection Control Pressure	0.1-3.0v
88	---	Not Used	---
89	GY/WT	APP Sensor Signal	0.5-1.6v
90	BR/WT	Reference Voltage	4.9-5.1v
91	GY/RD	Sensor Ground	<0.050v
92	RD/LG	Brake Pedal Switch	0v (Brake On: 12v)
93	---	Not Used	---
94	LB/OR	Fuel Pump Relay Control	Relay Off: 12v, On: 1v
95	BR/OR	Fuel Delivery Control	40-240 Hz
96	YL/LB	CID Sensor Signal	5-30 Hz
97	RD	Vehicle Power	12-14v
98	VT	PCM Signal to Relay	12v
99	---	Not Used	---
100	YL/WT	Fuel Level Indicator Signal	1.7v (1/2 full)
101	VT/OR	Glow Plug Relay Control	Off: 12v, On: 1v
102	---	Not Used	---
103	BK/WT	Power Ground	<0.1v
104	---	Not Used	---

Pin Connector Graphic

PCM 104-PIN CONNECTOR

Terminal View of 104-Pin PCM Wiring Harness Connector

Standard Colors and Abbreviations

Abbreviation	Color	Abbreviation	Color	Abbreviation	Color
BK	Black	GY	Gray	PK	Purple
BL	Blue	GN	Green	RD	Red
BR	Brown	LG	LT Green	TN	Tan
DB	Dark Blue	OR	Orange	WT	White
DG	DK Green	PK	Pink	YL	Yellow

1990 F-Series 7.5L V8 VIN G (All) 60 Pin Connector

PCM Pin #	Wire Color	Circuit Description (60 Pin)	Value at Hot Idle
1	YL	Keep Alive Power	12-14v
3	GY/BK	VSS (+) Signal	0 Hz, 55 mph: 125 Hz
3	DG/WT	VSS (+) Signal	0 Hz, 55 mph: 125 Hz
4	DG/YL	IDM Sensor Signal	20-31 Hz
6	BK	VSS (-) Signal	0 Hz, 55 mph: 125 Hz
7	LG/YL	ECT Sensor Signal	0.5-0.6v
8	BR	Fuel Pump Monitor	On: 12v, Off: 0v
10	BK/YL	A/C Switch Signal	A/C On: 12v, Off: 0v
11	WT/BK	Air Management 2 Solenoid	AM2 Off: 12v, On: 1v
16	BK/OR	Ignition System Ground	<0.050v
17	PK/LG	Self Test Output & MIL	MIL Off: 12v, On: 1v
18	TN/LG	Inferred Mileage Sensor	Digital Signals
20	BK	PCM Case Ground	<0.050v
21	GY/WT	IAC Solenoid Control	8.0-10.0v
22	TN/LG	Fuel Pump Control	Off: 12v, On: 1v
23 ('89-'90)	LG/BK	Knock Sensor Signal	0v
24 ('89-'90)	YL/LG	PSP Switch Signal	0v (turning: 12v)
25	YL/RD	ACT Sensor Signal	1.5-2.5v
26	OR/WT	Reference Voltage	4.9-5.1v
27	BR/LG	EVP Sensor Signal	0.4v, 55 mph: 2.6v
29	DG/PK	HO2S-11 (B1 S1) Signal	0.1-1.1v
29	GY/BL	HO2S-11 (B1 S1) Signal	0.1-1.1v
30	GY/YL	A/T: Neutral Drive Switch	In 'P': 0v, Others: 5v
30	GY/YL	M/T: Clutch Engage Switch	5v (clutch "in": 0v)
31	GY/YL	Canister Purge Solenoid	12v, 55 mph: 1v
33	DG	VR Solenoid Control	0%, 55 mph: 45%
36	YL/LG	Spark Output Signal	6.93v (50%)
37	RD	Vehicle Power	12-14v
40	BK/LG	Power Ground	<0.1v
45	DG/LG	MAP Sensor Signal	107 Hz (sea level)
45	DB/LG	MAP Sensor Signal	107 Hz (sea level)
46	BK/WT	Analog Signal Return	<0.050v
47	DG/LG	TP Sensor Signal	1v, 55 mph: 1.4v
48	WT/RD	Self Test Indicator Signal	STI On: 1v, Off: 5v
49	OR	HO2S-11 (B1 S1) Ground	<0.050v
51	WT/RD	Air Management 1 Solenoid	AM1 Off: 12v, On: 1v
56	DB	PIP Sensor Signal	6.93v (50%)
57	RD	Vehicle Power	12-14v
58	TN/OR	Injector Bank 1 (INJ 1, 4, 5, 8)	5.7-7.0 ms
59	TN/RD	Injector Bank 2 (INJ 2, 3, 6, 7)	5.7-7.0 ms
60	BK/LG	Power Ground	<0.1v

Pin Connector Graphic

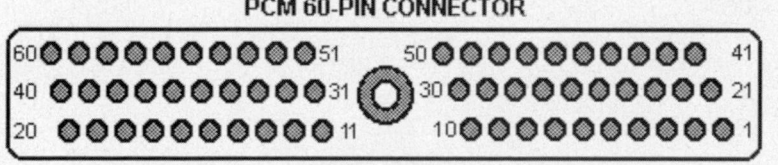

PCM 60-PIN CONNECTOR

Terminal View of 60-Pin PCM Harness Connector

1990 F-Series 7.5L V8 VIN G (E4OD) 60 Pin Connector

PCM Pin #	Wire Color	Circuit Description (60 Pin)	Value at Hot Idle
1	YL	Keep Alive Power	12-14v
2	LG	Brake Position Switch	0v (Brake On: 12v)
3	GY, DG	VSS (+) Signal	0 Hz, 55 mph: 125 Hz
4	DG/YL	IDM Sensor Signal	20-31 Hz
6	BK	VSS (-) Signal	0 Hz, 55 mph: 125 Hz
7	LG/YL	ECT Sensor Signal	0.5-0.6v
8	BR	Fuel Pump Monitor	On: 12v, Off: 0v
10	BK/YL	A/C Switch Signal	A/C On: 12v, Off: 0v
11 ('89)	WT/BK	Air Management 2 Solenoid	AM2 Off: 12v, On: 1v
12	BL/BK	4x4 Indicator Light	4x4 Switch Closed: 1v
16	BK/OR	Ignition System Ground	<0.050v
17	PK/LG	Self Test Output & MIL	MIL Off: 12v, On: 1v
18 ('89)	TN/LG	Inferred Mileage Sensor	Digital Signals
19	DG/PK	Shift Solenoid Control	1v, 55 mph: 1v
20	BK	PCM Case Ground	<0.050v
21	GY/WT	IAC Solenoid Control	8.0-10.0v
22	TN/LG	Fuel Pump Control	Off: 12v, On: 1v
23 ('89)	LG/BK	Knock Sensor Signal	0v
24 ('89)	YL/LG	PSP Switch Signal	0v (turning: 12v)
25	YL/RD	ACT Sensor Signal	1.5-2.5v
26	OR/WT	Reference Voltage	4.9-5.1v
27	BR/LG	EVP Sensor Signal	0.4v, 55 mph: 2.6v
29	DG/PK	HO2S-11 (B1 S1) Signal	0.1-1.1v
29	GY/BL	HO2S-11 (B1 S1) Signal	0.1-1.1v
30	BL/WT	MLP Sensor Signal	In 'P': 0v, in O/D: 5v
30	BL/YL	MLP Sensor Signal	In 'P': 0v, in O/D: 5v
31	GY/YL	Canister Purge Solenoid	12v, 55 mph: 1v
32	LG/WT	OCIL (lamp) Control	7.7v (Switch On: 0v)
33	DG	VR Solenoid Control	0%, 55 mph: 45%
36	YL/LG	Spark Output Signal	6.93v (50%)
37, 57	RD	Vehicle Power	12-14v
38	BL/YL	EPC Solenoid Control	8.8v
40	BK/LG	Power Ground	<0.1v

1990 F-Series 7.5L V8 VIN G (E4OD) 60 Pin Connector

PCM Pin #	Wire Color	Circuit Description (60 Pin)	Value at Hot Idle
41	TN/WT	OCS (switch) Signal	12v (switch closed: 0v)
42	OR/BK	TOT Sensor Signal	2.10-2.40v
45	DB/LG	MAP Sensor Signal	107 Hz (sea level)
46	BK/WT	Analog Signal Return	<0.050v
47	DG/LG	TP Sensor Signal	1v, 55 mph: 1.4v
48	WT/RD	Self Test Indicator Signal	STI On: 1v, Off: 5v
49	OR	HO2S-11 (B1 S1) Ground	<0.050v
51	WT/RD	Air Management 1 Solenoid	Solenoid Off: 12v, On: 1v
52	OR/YL	Shift Solenoid 1 Control	1v, 55 mph: 12v
53	PK/YL	TCC Solenoid Control	0v, On 55 mph: 12v
55	BR	Coast Clutch Solenoid	12v, 55 mph: 12v
56	DB	PIP Sensor Signal	6.93v (50%)
58	TN/OR	Injector Bank 1 (INJ 1, 4, 5, 8)	5.7-7.0 ms
59	TN/RD	Injector Bank 2 (INJ 2, 3, 6, 7)	5.7-7.0 ms
40, 60	BK/LG	Power Ground	<0.1v

Pin Connector Graphic

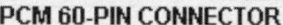

PCM 60-PIN CONNECTOR

60 ●●●●●●●●●●● 51   50 ●●●●●●●●●● 41
40 ●●●●●●●●●●● 31   ◎ 30 ●●●●●●●●●● 21
20 ●●●●●●●●●●● 11   10 ●●●●●●●●●● 1

Terminal View of 60-Pin PCM Harness Connector

Standard Colors and Abbreviations

Abbreviation	Color	Abbreviation	Color	Abbreviation	Color
BK	Black	GY	Gray	PK	Purple
BL	Blue	GN	Green	RD	Red
BR	Brown	LG	LT Green	TN	Tan
DB	Dark Blue	OR	Orange	WT	White
DG	DK Green	PK	Pink	YL	Yellow

1991 F-Series 7.5L V8 VIN G (All) 60 Pin Connector

PCM Pin #	Wire Color	Circuit Description (60 Pin)	Value at Hot Idle
1	YL	Keep Alive Power	12-14v
3	GY/BK	VSS (+) Signal	0 Hz, 55 mph: 125 Hz
4	TN/YL	IDM Sensor Signal	20-31 Hz
6	PK/OR	VSS (-) Signal	0 Hz, 55 mph: 125 Hz
7	LG/RD	ECT Sensor Signal	0.5-0.6v
8	DG/YL	Fuel Pump Monitor	On: 12v, Off: 0v
10	BK/YL	A/C Switch Signal	A/C On: 12v, Off: 0v
16	BK/OR	Ignition System Ground	<0.050v
17	PK/LG	Self Test Output & MIL	MIL Off: 12v, On: 1v
20	BK	PCM Case Ground	<0.050v
21	WT/BL	IAC Solenoid Control	8.0-10.5v
22	BL/OR	Fuel Pump Control	Off: 12v, On: 1v
25	GY	ACT Sensor Signal	1.5-2.5v
26	BR/WT	Reference Voltage	4.9-5.1v
27	BR/LG	EVP Sensor Signal	0.3v
29	GY/BL	HO2S-11 (B1 S1) Signal	0.1-1.1v
30	BL/YL	A/T: Neutral Drive Switch	In 'P': 0v, Others: 5v
30	BL/YL	M/T: Clutch Position Switch	Clutch In 0v, Out: 5v
31	GY/YL	Canister Purge Solenoid	12v, 55 mph: 1v
33	BR/PK	VR Solenoid Control	0%, 55 mph: 45%
36	PK	Spark Output Signal	6.93v (50%)
37	RD	Vehicle Power	12-14v
40	BK/WT	Power Ground	<0.1v
45	LG/BK	MAP Sensor Signal	107 Hz (sea level)
46	GY/RD	Analog Signal Return	<0.050v
47	GY/WT	TP Sensor Signal	1v, 55 mph: 1.4v
48	WT/PK	Self Test Indicator Signal	STI On: 1v, Off: 5v
49	OR	HO2S-11 (B1 S1) Ground	<0.050v
51	WT/RD	Air Management 1 Solenoid	AM1 Off: 12v, On: 1v
56	GY/OR	PIP Sensor Signal	6.93v (50%)
57	RD	Vehicle Power	12-14v
58	TN	Injector Bank 1 (INJ 1, 4, 5, 8)	5.6-6.2 ms
59	WT	Injector Bank 2 (INJ 2, 3, 6, 7)	5.6-6.2 ms
60	BK/LG	Power Ground	<0.1v

Pin Connector Graphic

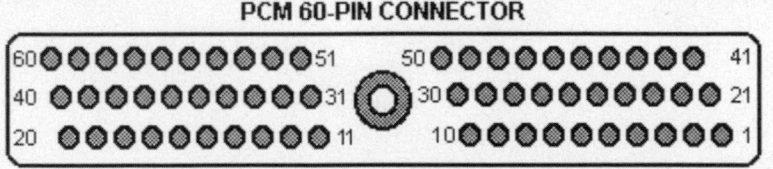

PCM 60-PIN CONNECTOR

Terminal View of 60-Pin PCM Harness Connector

1991 F-Series 7.5L V8 MFI VIN G (E4OD) 60 Pin Connector

PCM Pin #	Wire Color	Circuit Description (60 Pin)	Value at Hot Idle
1	YL	Keep Alive Power	12-14v
2	LG	Brake Position Switch	0v (Brake On: 12v)
3	GY/BK	VSS (+) Signal	0 Hz, 55 mph: 125 Hz
4	TN/YL	IDM Sensor Signal	20-31 Hz
6	PK/OR	VSS (-) Signal	0 Hz, 55 mph: 125 Hz
7	LG/RD	ECT Sensor Signal	0.5-0.6v
8	DG/YL	Fuel Pump Monitor	On: 12v, Off: 0v
10	BK/YL	A/C Switch Signal	A/C On: 12v, Off: 0v
12	BL/BK	4x4 Indicator Light Signal	4x4 Switch Closed: 1v
16	BK/OR	Ignition System Ground	<0.050v
17	PK/LG	Self Test Output & MIL	MIL Off: 12v, On: 1v
19	PK/OR	Shift Solenoid 2 Control	1v, 55 mph: 1v
20	BK	PCM Case Ground	<0.050v
21	WT/BL	IAC Solenoid Control	8.0-10.5v
22	BL/OR	Fuel Pump Control	Off: 12v, On: 1v
25	GY	ACT Sensor Signal	1.5-2.5v
26	BR/WT	Reference Voltage	4.9-5.1v
27	BR/LG	EVP Sensor Signal	0.3v
29	GY/BL	HO2S-11 (B1 S1) Signal	0.1-1.1v
30	BL/YL	MLP Sensor Signal	In 'P': 4.4v, in OD: 2.1v
31	GY/YL	Canister Purge Solenoid	12v, 55 mph: 1v
32	WT/LG	OCIL (lamp) Control	7.7v (Switch On: 0v)
33	BR/PK	VR Solenoid Control	0%, 55 mph: 45%
36	PK	Spark Output Signal	6.93v (50%)
37	RD	Vehicle Power	12-14v
38	WT/YL	EPC Solenoid Control	9.5v (5 psi)
40	BK/WT	Power Ground	<0.1v
41	TN/WT	OCS (switch) Signal	12v (switch closed: 0v)
42	OR/BK	TOT Sensor Signal	2.10-2.40v
45	LG/BK	MAP Sensor Signal	107 Hz (sea level)
46	GY/RD	Analog Signal Return	<0.050v
47	GY/WT	TP Sensor Signal	1v, 55 mph: 1.4v
48	WT/PK	Self Test Indicator Signal	STI On: 1v, Off: 5v
49	OR	HO2S-11 (B1 S1) Ground	<0.050v
51	WT/RD	Air Management 1 Solenoid	AM1 Off: 12v, On: 1v
52	OR/YL	Shift Solenoid 1 Control	1v, 55 mph: 12v
53	PK/YL	TCC Solenoid Control	0v, On 55 mph: 12v
54	---	Not Used	---
55	BR/OR	Coast Clutch Solenoid	12v, 55 mph: 12v
56	GY/OR	PIP Sensor Signal	6.93v (50%)
57	RD	Vehicle Power	12-14v
58	TN	Injector Bank 1 (INJ 1, 4, 5, 8)	6.0-6.6 ms
59	WT	Injector Bank 2 (INJ 2, 3, 6, 7)	6.0-6.6 ms
60	BK/LG	Power Ground	<0.1v

Pin Connector Graphic

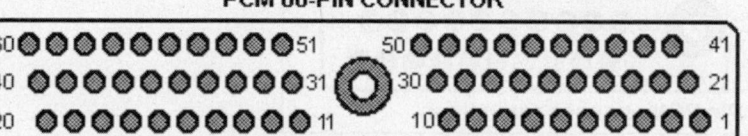

PCM 60-PIN CONNECTOR

Terminal View of 60-Pin PCM Harness Connector

1992-95 F-Series 7.5L V8 VIN G (All) 60 Pin Connector

PCM Pin #	Wire Color	Circuit Description (60 Pin)	Value at Hot Idle
1	YL	Keep Alive Power	12-14v
2	LG	Brake Position Switch	0v (Brake On: 12v)
3	GY/BK	VSS (+) Signal	0 Hz, 55 mph: 125 Hz
4	YL/BK	IDM Sensor Signal	20-31 Hz
4	WT/PK	IDM Sensor Signal	20-31 Hz
4	TN/YL	IDM Sensor Signal	20-31 Hz
6	PK/OR	VSS (-) Signal	0 Hz, 55 mph: 125 Hz
7	LG/RD	ECT Sensor Signal	0.5-0.6v
8	DG/YL	Fuel Pump Monitor	On: 12v, Off: 0v
9	PK/BL	Data Bus (-) Signal	Digital Signals
10	PK/BL	A/C Switch Signal	A/C On: 12v, Off: 0v
11 ('93-'95)	BR	Air Management 2 Solenoid	AM2 Off: 12v, On: 1v
16	OR/RD	Ignition System Ground	<0.050v
17	PK/LG	Self Test Output & MIL	MIL Off: 12v, On: 1v
20	BK	PCM Case Ground	<0.050v
21	WT/BL	IAC Solenoid Control	8.0-10.0v
22 ('92-'93)	BL/OR	Fuel Pump Control	Off: 12v, On: 1v
23 ('92-'93)	YL/RD	Knock Sensor Signal	0v
25	GY	IAT Sensor Signal	1.5-2.5v
26	BR/WT	Reference Voltage	4.9-5.1v
27	BK/YL	EVP Sensor Signal	0.4v, 55 mph: 2.6v
27	BR/LG	EVP Sensor Signal	0.4v, 55 mph: 2.6v
28	TN/OR	Data Bus (+) Signal	Digital Signals
29	GY/BL	HO2S-11 (B1 S1) Signal	0.1-1.1v
30	GY/YL	A/T: Neutral Drive Switch	In 'P': 0v, Others: 5v
30	GY/YL	M/T: Clutch Engage Switch	5v (clutch "in": 0v)
31	GY/YL	Canister Purge Solenoid	1v, 55 mph: 12v
33	BR/PK	VR Solenoid Control	0%, 55 mph: 45%
36	PK	Spark Output Signal	6.93v (50%)
37	RD	Vehicle Power	12-14v
40	BK/WT	Power Ground	<0.1v
45	LG/BK	MAP Sensor Signal	107 Hz (sea level)
46	GY/RD	Analog Signal Return	<0.050v
47	GY/WT	TP Sensor Signal	1v, 55 mph: 1.4v
48	WT/PK	Self Test Indicator Signal	STI On: 1v, Off: 5v
49	OR	HO2S-11 (B1 S1) Ground	<0.050v
51	WT/OR	Air Management 1 Solenoid	AM1 Off: 12v, On: 1v
56	GY/OR	PIP Sensor Signal	6.93v (50%)
57	RD	Vehicle Power	12-14v
58	TN	Injector Bank 1 (INJ 1, 4, 5, 8)	5.0-5.8 ms
59	WT	Injector Bank 2 (INJ 2, 3, 6, 7)	5.0-5.8 ms
60	BK/WT	Power Ground	<0.1v

Pin Connector Graphic

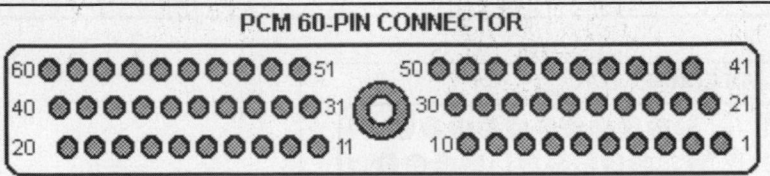

1992-95 F-Series 7.5L V8 VIN G (E4OD) 60 Pin Connector

PCM Pin #	Wire Color	Circuit Description (60 Pin)	Value at Hot Idle
1	YL	Keep Alive Power	12-14v
2	LG	Brake Position Switch	0v (Brake On: 12v)
3	GY/BK	VSS (+) Signal	0 Hz, 55 mph: 125 Hz
4	YL, WT	IDM Sensor Signal	20-31 Hz
6	PK/OR	VSS (-) Signal	0 Hz, 55 mph: 125 Hz
7	LG/RD	ECT Sensor Signal	0.5-0.6v
8	DG/YL	Fuel Pump Monitor	On: 12v, Off: 0v
9	PK/BL	Data Bus (-) Signal	Digital Signals
10	PK/BL	A/C Switch Signal	A/C On: 12v, Off: 0v
11 ('93-'95)	BR	Air Management 2 Solenoid	AM2 Off: 12v, On: 1v
12	BL/BK	4x4 Indicator Light Signal	12v (switch closed: 0v)
16	OR/RD	Ignition System Ground	<0.050v
17	PK/LG	Self Test Output & MIL	MIL Off: 12v, On: 1v
19	PK/OR	Shift Solenoid 2 Control	1v, 55 mph: 12v
20	BK	PCM Case Ground	<0.050v
21	WT/BL	IAC Solenoid Control	8.0-10.0v
22	BL/OR	Fuel Pump Control	Off: 12v, On: 1v
25	GY	IAT Sensor Signal	1.5-2.5v
26	BR/WT	Reference Voltage	4.9-5.1v
27	BK/YL	EVP Sensor Signal	0.4v, 55 mph: 2.6v
27	BR/LG	EVP Sensor Signal	0.4v, 55 mph: 2.6v
28	TN/OR	Data Bus (+) Signal	Digital Signals
29	GY/BL	HO2S-11 (B1 S1) Signal	0.1-1.1v
30	BL/YL	MLP Sensor Signal	In 'P': 0v, in O/D: 5v
31	GY/YL	Canister Purge Solenoid	12v, 55 mph: 1v
32	WT/LG	OCIL or TCIL (lamp) Control	7.7v (Switch On: 0v)
33	BR/PK	VR Solenoid Control	0%, 55 mph: 45%
36	PK	Spark Output Signal	6.93v (50%)
37, 57	RD	Vehicle Power	12-14v
38	WT/YL	EPC Solenoid Control	9.5v (5 psi)
40	BK/WT	Power Ground	<0.1v
41	TN/WT	OCS or TCS (switch) Signal	12v (switch closed: 0v)
42	OR/BK	TOT Sensor Signal	2.10-2.40v
45	LG/BK	MAP Sensor Signal	107 Hz (sea level)
46	GY/RD	Analog Signal Return	<0.050v
47	GY/WT	TP Sensor Signal	1v, 55 mph: 1.4v
48	WT/PK	Self Test Indicator Signal	STI On: 1v, Off: 5v
49	OR	HO2S-11 (B1 S1) Ground	<0.050v
51	WT/OR	Air Management 1 Solenoid	AM1 Off: 12v, On: 1v
52	OR/YL	Shift Solenoid 1 Control	12v, 55 mph: 1v
53	PK/YL	TCC Solenoid Control	0v, On 55 mph: 12v
55	BR/OR	Coast Clutch Solenoid	12v, 55 mph: 12v
56	GY/OR	PIP Sensor Signal	6.93v (50%)
58	TN	Injector Bank 1 (INJ 1, 4, 5, 8)	6.0-6.8 ms
59	WT	Injector Bank 2 (INJ 2, 3, 6, 7)	6.0-6.8 ms
60	BK/WT	Power Ground	<0.1v

Pin Connector Graphic

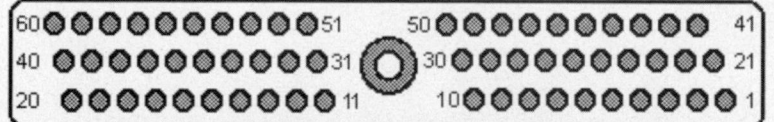

Terminal View of 60-Pin PCM Harness Connector

1996-97 F-Series 7.5L V8 MFI VIN G (All) 104 Pin

PCM Pin #	Wire Color	Circuit Description (104 Pin)	Value at Hot Idle
1	---	Not Used	---
2	PK/LG	MIL (lamp) Control	MIL Off: 12v, On: 1v
4	LG/RD	Power Take-Off (if equipped)	0v (Off)
5-12	---	Not Used	---
13	PK	Flash EPROM Power	0.1v
14	LG/BK	4x4 Low Switch	12v (switch closed: 0v)
15	PK/LB	Data Bus (-) Signal	Digital Signals
16	TN/OR	Data Bus (+) Signal	Digital Signals
17-22	---	Not Used	---
23	OR/RD	Ignition Ground	<0.050v
24	BK/WT	Power Ground	<0.1v
25	BK/LB	Case Ground	<0.050v
26-28	---	Not Used	---
29	TN/WT	TCS (switch) Signal	TCS & O/D On: 12v
30-32	---	Not Used	---
33	PK/OR	PSOM (-) Signal	<0.050v
34, 37	---	Not Used	---
35	RD/LG	HO2S-12 (B1 S2) Signal	0.1-1.1v
36	TN/LB	MAF Sensor Return	<0.050v
38	LG/RD	ECT Sensor Signal	0.5-0.6v
39	GY	IAT Sensor Signal	1.5-2.5v
40	DG/YL	Fuel Pump Monitor	On: 12v, Off: 0v
41	DG/OR	A/C Switch Signal	A/C On: 12v, Off: 0v
42-43, 46	---	Not Used	---
44	BR	Secondary AIR Diverter	AIRD Off: 12v, On: 1v
47	BK/PK	VR Solenoid Control	0%, 55 mph: 45%
48	YL/BK	Clean Tachometer Output	39-45 Hz
49	GY/OR	PIP Sensor Signal	6.93v (50%)
50	PK	Spark Output Signal	6.93v (50%)
51	BK/WT	Power Ground	<0.1v
52-54, 57	---	Not Used	---
55	YL	Keep Alive Power	12-14v
56	LG/BK	EVAP Purge Solenoid	0-10 Hz (0-100%)
58	GY/BK	PSOM (+) Signal	0 Hz, 55 mph: 125 Hz
59	DG/LG	Misfire Detection Sensor	45-55 Hz
60	GY/LB	HO2S-11 (B1 S1) Signal	0.1-1.1v
61-63	---	Not Used	---
64	LB/YL	A/T: PNP Switch Signal	In 'P': 0v, Others: 5v
64	LB/YL	M/T: CPP Switch Signal	5v (clutch "in": 0v)

1996-97 F-Series 7.5L V8 MFI VIN G (All) 104 Pin

PCM Pin #	Wire Color	Circuit Description (104 Pin)	Value at Hot Idle
65	BR/LG	DPFE Sensor Signal	0.95-1.05v
66-69	---	Not Used	---
70	WT/OR	Secondary AIR Bypass	AIRB Off: 12v, On: 1v
71	RD	Vehicle Power	12-14v
72	TN/RD	Injector 7 Control	4.3-4.6 ms
73	TN/BK	Injector 5 Control	4.3-4.6 ms
74	BR/YL	Injector 3 Control	4.3-4.6 ms
75	TN	Injector 1 Control	4.3-4.6 ms
76	BK/WT	Power Ground	<0.1v
77	BK/WT	Power Ground	<0.1v
78, 81-82	---	Not Used	---
79	WT/LG	TCIL (lamp) Control	7.7v (Switch On: 0v)
80	LB/OR	Fuel Pump Control	Off: 12v, On: 1v
83	WT/LB	IAC Solenoid Control	10.7v (33%)
84-86	---	Not Used	---
87	RD/BK	HO2S-21 (B2 S1) Signal	0.1-1.1v
88	LB/RD	MAF Sensor Signal	0.8v, 55 mph: 1.8v
89	GY/WT	TP Sensor Signal	1v, 55 mph: 1.6v
90	BR/WT	Reference Voltage	4.9-5.1v
91	GY/RD	Analog Signal Return	<0.050v
92	LG	Brake Pedal Switch	0v (Brake On: 12v)
93	RD/WT	HO2S-11 (B1 S1) Heater	1v (Heater Off: 12v)
94	YL/LB	HO2S-21 (B2 S1) Heater	1v (Heater Off: 12v)
95	WT/BK	HO2S-12 (B1 S2) Heater	1v (Heater Off: 12v)
96	---	Not Used	---
97	RD	Vehicle Power	12-14v
98	LG	Injector 8 Control	4.3-4.6 ms
99	LG/OR	Injector 6 Control	4.3-4.6 ms
100	BR/LB	Injector 4 Control	4.3-4.6 ms
101	W	Injector 2 Control	4.3-4.6 ms
102	---	Not Used	---
103	BK/WT	Power Ground	<0.1v
104	---	Not Used	---

Pin Connector Graphic

PCM 104-PIN CONNECTOR

Terminal View of 104-Pin PCM Wiring Harness Connector

Standard Colors and Abbreviations

Abbreviation	Color	Abbreviation	Color	Abbreviation	Color
BK	Black	GY	Gray	PK	Purple
BL	Blue	GN	Green	RD	Red
BR	Brown	LG	LT Green	TN	Tan
DB	Dark Blue	OR	Orange	WT	White
DG	DK Green	PK	Pink	YL	Yellow

1996-97 F-Series 7.5L V8 VIN G (E4OD) 104 Pin Connector

PCM Pin #	Wire Color	Circuit Description (104 Pin)	Value at Hot Idle
1	PK/OR	Shift Solenoid 2 Control	1v, 55 mph: 12v
2	PK/LG	MIL (lamp) Control	MIL Off: 12v, On: 1v
4	LG/RD	Power Take-Off (if equipped)	0v (Off)
5-12	---	Not Used	---
13	PK	Flash EPROM Power	0.1v
14	LG/BK	4x4 Low Switch Signal	12v (switch closed: 0v)
15	PK/LB	Data Bus (-) Signal	Digital Signals
16	TN/OR	Data Bus (+) Signal	Digital Signals
17-22	---	Not Used	---
23	OR/RD	Ignition Ground	<0.050v
24, 51	BK/WT	Power Ground	<0.1v
25	BK/LB	Case Ground	<0.050v
26, 28	---	Not Used	---
27	OR/YL	Shift Solenoid 1 Control	1v, 55 mph: 12v
29	TN/WT	TCS (switch) Signal	TCS & O/D On: 12v
30-32, 34	---	Not Used	---
33	PK/OR	PSOM (-) Signal	<0.050v
35	RD/LG	HO2S-12 (B1 S2) Signal	0.1-1.1v
36	TN/LB	MAF Sensor Return	<0.050v
37	OR/BK	TFT Sensor Signal	2.10-2.40v
38	LG/RD	ECT Sensor Signal	0.5-0.6v
39	GY	IAT Sensor Signal	1.5-2.5v
40	DG/YL	Fuel Pump Monitor	On: 12v, Off: 0v
41	DG/OR	A/C Switch Signal	A/C On: 12v, Off: 0v
42-43, 46	---	Not Used	---
44	BR	Secondary AIR Diverter	AIRD Off: 12v, On: 1v
47	BK/PK	VR Solenoid Control	0%, 55 mph: 45%
48	YL/BK	Clean Tachometer Output	39-45 Hz
49	GY/OR	PIP Sensor Signal	6.93v (50%)
50	PK	Spark Output Signal	6.93v (50%)
52, 57	---	Not Used	---
53	YL/BK	Coast Clutch Solenoid	12v, 55 mph: 12v
54	PK/YL	TCC Solenoid Control	0%, 55 mph: 95%
55	Y	Keep Alive Power	12-14v
56	LG/BK	EVAP Purge Solenoid	0-10 Hz (0-100%)
58	GY/BK	PSOM (+) Signal	0 Hz, 55 mph: 125 Hz
59	DG/LG	Misfire Detection Sensor	45-55 Hz
60	GY/LB	HO2S-11 (B1 S1) Signal	0.1-1.1v
61-63	---	Not Used	---
64	LB/YL	TR Sensor Signal	In 'P': 0v, in O/D: 5v

1996-97 F-Series 7.5L V8 VIN G (E4OD) 104 Pin Connector

PCM Pin #	Wire Color	Circuit Description (104 Pin)	Value at Hot Idle
65	BR/LG	DPFE Sensor Signal	0.95-1.05v
66-69	---	Not Used	---
70	WT/OR	Secondary AIR Bypass	AIRB Off: 12v, On: 1v
71	RD	Vehicle Power	12-14v
72	TN/RD	Injector 7 Control	4.3-4.6 ms
73	TN/BK	Injector 5 Control	4.3-4.6 ms
74	BR/YL	Injector 3 Control	4.3-4.6 ms
75	TN	Injector 1 Control	4.3-4.6 ms
76	BK/WT	Power Ground	<0.1v
77	BK/WT	Power Ground	<0.1v
78, 82	---	Not Used	---
79	WT/LG	TCIL (lamp) Control	7.7v (Switch On: 0v)
80	LB/OR	Fuel Pump Control	Off: 12v, On: 1v
81	WT/YL	EPC Solenoid Control	9.2v (5 psi)
83	WT/LB	IAC Solenoid Control	10.7v (33%)
84-86	---	Not Used	---
87	RD/BK	HO2S-21 (B2 S1) Signal	0.1-1.1v
88	LB/RD	MAF Sensor Signal	0.8v, 55 mph: 1.8v
89	GY/WT	TP Sensor Signal	0.53-1.27v
90	BR/WT	Reference Voltage	4.9-5.1v
91	GY/RD	Analog Signal Return	<0.050v
92	LG	Brake Pedal Switch	0v (Brake On: 12v)
93	RD/WT	HO2S-11 (B1 S1) Heater	1v (Heater Off: 12v)
94	YL/LB	HO2S-21 (B2 S1) Heater	1v (Heater Off: 12v)
95	WT/BK	HO2S-12 (B1 S2) Heater	1v (Heater Off: 12v)
96	---	Not Used	---
97	RD	Vehicle Power	12-14v
98	LG	Injector 8 Control	4.3-4.6 ms
99	LG/OR	Injector 6 Control	4.3-4.6 ms
100	BR/LB	Injector 4 Control	4.3-4.6 ms
101	WT	Injector 2 Control	4.3-4.6 ms
102	---	Not Used	---
103	BK/WT	Power Ground	<0.1v
104	---	Not Used	---

Pin Connector Graphic

PCM 104-PIN CONNECTOR

Terminal View of 104-Pin PCM Wiring Harness Connector

Standard Colors and Abbreviations

Abbreviation	Color	Abbreviation	Color	Abbreviation	Color
BK	Black	GY	Gray	PK	Purple
BL	Blue	GN	Green	RD	Red
BR	Brown	LG	LT Green	TN	Tan
DB	Dark Blue	OR	Orange	WT	White
DG	DK Green	PK	Pink	YL	Yellow

LIGHTNING PIN TABLES

1999-2003-05 Pickup 5.4L V8 VIN 3 (E4OD) 104 Pin Connector

PCM Pin #	Wire Color	Circuit Description (104 Pin)	Value at Hot Idle
1	OR/YL	COP 6 Driver (dwell)	5°, 55 mph: 8°
2	---	Not Used	---
3	BK/WT	Power Ground	<0.1v
4	PK	Transfer Case Speed Sensor (MSOF)	12v, 55 mph: 12v
5	---	Not Used	---
6	OR/YL	Shift Solenoid 1 Control	1v, 55 mph: 12v
7-10	---	Not Used	---
11	VT/OR	Shift Solenoid 2 Control	12v, 55 mph: 1v
12	WT/LG	TCIL (lamp) Control	7.7v (Switch On: 0v)
13	VT	Flash EEPROM Power	Digital Signals
14	LB/BK	4x4 Low Indicator Switch	12v (Switch On: 0v)
15	PK/LB	SCP Data Bus (-) Signal	<0.050v
16	TN/OR	SCP Data Bus (+) Signal	Digital Signals
17-18	---	Not Used	---
19	OR/LG	Fuel Pump High/Low Control	Off: 12v, On: 1v
20	BR/OR	Coast Clutch Solenoid	12v, 55 mph: 12v
21	DB	CKP (+) Sensor Signal	411 Hz
22	GY	CKP (-) Sensor Signal	411 Hz
23-24	---	Not Used	---
25	LB/YL	Case Ground	<0.050v
26	LG/WT	COP 1 Driver (dwell)	5°, 55 mph: 8°
27	LG/YL	COP 5 Driver (dwell)	5°, 55 mph: 8°
28	---	Not Used	---
29	TN/WT	TCS (switch) Signal	TCS & O/D On: 12v
30-31	---	Not Used	---
32	DG/PK	Knock Sensor (-) Signal	0v
33	---	Not Used	---
34	YL/BK	Digital TR1 Sensor	0v, 55 mph: 11.5v
35	RD/LG	HO2S-12 (B1 S2) Signal	0.1-1.1v
36	TN/LB	MAF Sensor Return	<0.050v
37	OR/BK	TFT Sensor Signal	2.10-2.40v
38	RD/YL	Intake Air Temperature 2 Sensor	1.5-2.5v
39	GY	Intake Air Temperature 1 Sensor	1.5-2.5v
40	RD	Fuel Pump Monitor	On: 12v, Off: 0v
41	BK/YL	A/C Switch Signal	Switch Closed: 12v
42	LG/VT	Supercharger Bypass Solenoid	Solenoid Off: 12v, On: 1v
43-45	---	Not Used	---
47	BR/PK	VR Solenoid Control	0%, 55 mph: 45%
48	---	Not Used	---
49	LB/BK	Digital TR2 Sensor	0v, 55 mph: 11.5v
50	WT/BK	Digital TR4 Sensor	0v, 55 mph: 11.5v
51	BK/WT	Power Ground	<0.1v
52	WT/PK	COP 3 Driver (dwell)	5°, 55 mph: 8°
53	DG/PK	COP 4 Driver Dwell	5°, 55 mph: 8°
54	PK/YL	TCC Solenoid Control	0%, 55 mph: 95%
55	RD/WT	Keep Alive Power	12-14v
56	LG/BK	EVAP Purge Solenoid	0-10 Hz (0-100%)
57	YL/RD	Knock Sensor (+) Signal	0v
58	---	Not Used	---
59	DG/WT	TSS (+) Sensor Signal	300 Hz, 55 mph: 980
60	GY/LB	HO2S-11 (B1 S1) Signal	0.1-1.1v
61	PK/LG	HO2S-22 (B2 S2) Signal	0.1-1.1v
62	RD/PK	FTP Sensor Signal	2.6v (0" H2O - cap off)
63	---	Not Used	---
64	LB/YL	Digital TR3 Sensor	In 'P': 0v, in O/D: 1.7v

1999-2003-05 Pickup 5.4L V8 VIN 3 (E4OD) 104 Pin Connector

PCM Pin #	Wire Color	Circuit Description (104 Pin)	Value at Hot Idle
65	BR/LG	EGR DPFE Sensor Signal	0.95-1.05v
66	YL/LG	Cylinder Head Temperature Sensor	0.6v (194°F)
67	PK/WT	EVAP CV Solenoid	0-10 Hz (0-100%)
68	GY/BK	VSS (+) Signal	0 Hz, 55 mph: 125 Hz
69	PK/YL	A/C WOT Relay Control	Off: 12v, On: 1v
70	BK/WT	Check Fuel Cap Indicator Control	Indicator Off: 12v, On: 1v
71	RD	Vehicle Power	12-14v
72	WT	Injector 7 Control	2.7-3.5 ms
73	WT/LB	Injector 5 Control	2.7-3.5 ms
74	WT/BK	Injector 3 Control	2.7-3.5 ms
75	WT/RD	Injector 1 Control	2.7-3.5 ms
70	---	Not Used	---
77	BK/WT	Power Ground	<0.1v
78	PK/LB	COP 7 Driver Dwell	5°, 55 mph: 8°
79	WT/RD	COP 8 Driver Dwell	5°, 55 mph: 8°
80	LB/OR	Fuel Pump Relay Control	Off: 12v, On: 1v
81	WT/YL	EPC Solenoid Control	9.1v (4 psi)
82, 86	---	Not Used	---
83	WT/LB	IAC Solenoid Control	10.7v (33%)
84	DB/YL	OSS Sensor Signal	0 Hz, 55 mph: 700 Hz
85	DG	CMP Sensor Signal	7 Hz, 55 mph: 15 Hz
87	RD/BK	HO2S-21 (B2 S1) Signal	0.1-1.1v
88	LB/RD	MAF Sensor Signal	0.8v, 55 mph: 1.6v
89	GY/WT	TP Sensor Signal	0.53-1.27v
90	BR/WT	Reference Voltage	4.9-5.1v
91	GY/RD	Analog Signal Return	<0.050v
92	RD/LG	Brake Pedal Switch	0v (Brake On: 12v)
93	RD/WT	HO2S-11 (B1 S1) Heater	1v (Heater Off: 12v)
94	YL/LB	HO2S-21 (B2 S1) Heater	1v (Heater Off: 12v)
95	WT/BK	HO2S-12 (B1 S2) Heater	1v (Heater Off: 12v)
96	TN/YL	HO2S-22 (B2 S2) Heater	1v (Heater Off: 12v)
97	RD	Vehicle Power	12-14v
98	YL	Injector 8 Control	2.7-3.5 ms
99	YL/LB	Injector 6 Control	2.7-3.5 ms
100	YL/BK	Injector 4 Control	2.7-3.5 ms
101	YL/RD	Injector 2 Control	2.7-3.5 ms
102	---	Not Used	---
103	BK/WT	Power Ground	<0.1v
104	PK/WT	COP 2 Driver (dwell)	5°, 55 mph: 8°

Pin Connector Graphic

Terminal View of 104-Pin PCM Wiring Harness Connector

Standard Colors and Abbreviations

Abbreviation	Color	Abbreviation	Color	Abbreviation	Color
BK	Black	GY	Gray	PK	Purple
BL	Blue	GN	Green	RD	Red
BR	Brown	LG	LT Green	TN	Tan
DB	Dark Blue	OR	Orange	WT	White
DG	DK Green	PK	Pink	YL	Yellow

RANGER PIN TABLES

1990 Pickup 2.3L I4 VIN A (All) 60 Pin Connector

PCM Pin #	Wire Color	Circuit Description (60 Pin)	Value at Hot Idle
1	YL/BK	Keep Alive Power	12-14v
2	LG	Brake Position Switch	0v (Brake On: 12v)
3	DG/WT	VSS (+) Signal	0 Hz, 55 mph: 125 Hz
4	BK/YL	IDM Sensor Signal	20-31 Hz
6	OR/YL	VSS (-) Signal	0 Hz, 55 mph: 125 Hz
7	LG/YL	ECT Sensor Signal	0.5-0.6v
8	OR/BL	Fuel Pump Monitor	On: 12v, Off: 0v
9	TN/OR	Data Bus (-) Signal (California)	Digital Signals
10	TN/YL	A/C Switch Signal	A/C On: 12v, Off: 0v
14	BK/LG	Mass Airflow Sensor (California)	0.6v
15	TN/BL	Mass Airflow Return (California)	<0.050v
16	BK/OR	Ignition System Ground	<0.050v
17	TN/RD	Self Test Output & MIL	MIL Off: 12v, On: 1v
18	WT/RD	Octane Adjust Sensor Signal	9.3v (shorted: 0v)
20	BK	PCM Case Ground	<0.050v
21	GY/WT	IAC Solenoid Control	10.6-11.0v
22	TN/LG	Fuel Pump Control	Off: 12v, On: 1v
24	YL/LG	PSP Switch Signal	0v (turning: 12v)
25	YL/RD	ACT Sensor Signal	1.5-2.5v
26	OR/WT	Reference Voltage	4.9-5.1v
27	BR/LG	EVP Sensor Signal	0.4v
28 (Cal)	TN/RD	Data Bus (+) Signal	Digital Signals
29	DG/PK	HO2S-11 (B1 S1) Signal	0.1-1.1v
30	WT/BK	A/T: Neutral Drive Switch	In 'P': 0v, Others: 5v
30	WT/BK	M/T: Clutch Engage Switch	5v (clutch "in": 0v)
32	GY/OR	Dual Plug Inhibit	0.1v
33	DG	VR Solenoid Control	0%, 55 mph: 45%
36	YL/LG	Spark Output Signal	6.93v (50% dwell)
37	RD	Vehicle Power	12-14v
40	BK/LG	Power Ground	<0.1v
43	LG/PK	A/C Demand Signal	A/C On: 12v
45	DG/BL	MAP Sensor Signal	107 Hz (sea level)
46	BK/WT	Analog Signal Return	<0.050v
47	DG/LG	TP Sensor Signal	1v, 55 mph: 1.4v
48	WT/RD	Self Test Input Signal	STI On: 1v, Off: 5v
49	OR	HO2S-11 (B1 S1) Ground	<0.050v
52	TN/BL	Shift Solenoid 3-4	Solenoid Off: 12v, On: 1v
53	WT	TCC Solenoid Control	0v, On 55 mph: 12v
54	PK	A/C WOT Cutout Control	Off: 12v, On: 1v
56	DB	PIP Sensor Signal	6.93v (50%)
57	RD	Vehicle Power	12-14v
58	TN	Injector Bank 1 (INJ 1 & 4)	3.7-4.4 ms
59	WT	Injector Bank 2 (INJ 2 & 3)	3.7-4.4 ms
60	BK/LG	Power Ground	<0.1v

Pin Connector Graphic

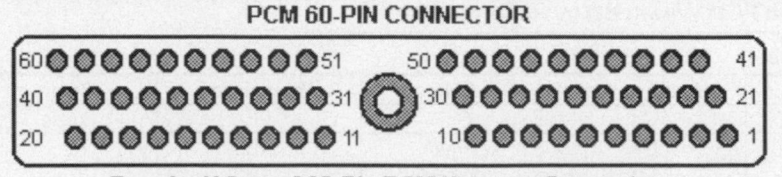

1991-94 Pickup 2.3L I4 VIN A (All) 60 Pin Connector

PCM Pin #	Wire Color	Circuit Description (60 Pin)	Value at Hot Idle
1	YL	Keep Alive Power	12-14v
2	LG	Brake Position Switch	0v (Brake On: 12v)
3	GY/BK	VSS (+) Signal	0 Hz, 55 mph: 125 Hz
4	TN/YL	IDM Sensor Signal	20-31 Hz
5 ('93-'94)	GY	CMP Sensor Signal	7 Hz, 55 mph: 15 Hz
6	PK/OR	VSS (-) Signal	0 Hz, 55 mph: 125 Hz
7	LG/RD	ECT Sensor Signal	0.5-0.6v
8	DG/YL	Fuel Pump Monitor	On: 12v, Off: 0v
9 ('92-'94)	BK/BL	Data Bus (-) Signal	Digital Signals
10	DG/OR	A/C Switch Signal	A/C On: 12v, Off: 0v
14	BL/RD	MAF Sensor Signal	0.7v, 55 mph: 1.9v
15	TN/BL	MAF Sensor Return	<0.050v
16	OR/RD	Ignition System Ground	<0.050v
17	PK/LG	Self Test Output & MIL	MIL Off: 12v, On: 1v
20	BK	PCM Case Ground	<0.050v
20	BK/LG	PCM Case Ground	<0.050v
21	WT/BL	IAC Solenoid Control	10.7v (33%)
22	BL/OR	Fuel Pump Control	Off: 12v, On: 1v
24	YL/LG	PSP Switch Signal	0v (turning: 12v)
25	GY	ACT Sensor Signal	1.5-2.5v
26	BR/WT	Reference Voltage	4.9-5.1v
27	BR/LG	EVP Sensor Signal	0.5v, 55 mph: 0.8v
28 ('92-'94)	TN/OR	Data Bus (+) Signal	Digital Signals
29	GY/BL	HO2S-11 (B1 S1) Signal	0.1-1.1v
30	BL/YL	A/T: Neutral Drive Switch	In 'P': 0v, Others: 5v
30	BL/YL	M/T: Clutch Engage Switch	5v (clutch "in": 0v)
32	DB/YL	Dual Plug Inhibit	0.1v
33	BR/PK	VR Solenoid Control	0%, 55 mph: 45%
36	PK	Spark Output Signal	6.93v (50% dwell)
37	RD	Vehicle Power	12-14v
40	BK/WT	Power Ground	<0.1v
43	PK	A/C Demand Signal	A/C On: 12v
44	DG	Octane Adjustment	9.3v (shorted: 0v)
46	GY/RD	Analog Signal Return	<0.050v
47	GY/WT	TP Sensor Signal	0.9v, 55 mph: 1.5v
48	WT/PK	Self Test Input Signal	STI On: 1v, Off: 5v
49	O, GY/RD	HO2S-11 (B1 S1) Ground	<0.050v
52	OR/YL	Shift Solenoid 3-4	Solenoid Off: 12v, On: 1v
53	PK/YL	TCC Solenoid Control	0v, On 55 mph: 12v
54	PK/YL	A/C WOT Cutout Control	Off: 12v, On: 1v
56	GY/OR	PIP Sensor Signal	6.93v (50%)
57	RD	Vehicle Power	12-14v
58	TN	Injector Bank 1 (INJ 1 & 4)	3.3-3.7 ms
59	WT	Injector Bank 2 (INJ 2 & 3)	3.3-3.7 ms
60	BK/WT	Power Ground	<0.1v

Pin Connector Graphic

Terminal View of 60-Pin PCM Harness Connector

1994 Pickup 2.3L VIN A California (All) 60 Pin Connector

PCM Pin #	Wire Color	Circuit Description (60 Pin)	Value at Hot Idle
1	YL	Keep Alive Power	12-14v
2	LG	Brake Position Switch	0v (Brake On: 12v)
3	GY/BK	VSS (+) Signal	0 Hz, 55 mph: 125 Hz
4	TN/YL	IDM Sensor Signal	20-31 Hz
6	PK/OR	VSS (-) Signal	0 Hz, 55 mph: 125 Hz
7	LG/RD	ECT Sensor Signal	0.5-0.6v
8	DG/YL	Fuel Pump Monitor	On: 12v, Off: 0v
9	TN/BL	MAF Sensor Return	<0.050v
10	DG/OR	A/C Switch Signal	A/C On: 12v, Off: 0v
16	OR/RD	Ignition System Ground	<0.050v
17	PK/LG	Self-Test Output & MIL	MIL Off: 12v, On: 1v
18	TN/OR	Data Bus (+) Signal	Digital Signals
19	PK/BL	Data Bus (-) Signal	Digital Signals
20	BK/LG	PCM Case Ground	<0.050v
21	WT/BL	IAC Solenoid Control	9.5-11.5v
22	BL/OR	Fuel Pump Control	Off: 12v, On: 1v
24	DB/OR	CMP Sensor Signal	7 Hz, 55 mph: 15 Hz
25	GY	IAT Sensor Signal	1.5-2.5v
26	BR/WT	Reference Voltage	4.9-5.1v
27	BR/LG	EVP Sensor Signal	0.4v, 55 mph: 1.3v
28	YL/LG	PSP Switch Signal	0v (turning: 12v)
29	DG	HO2S-11 (B1 S1) Signal	0.1-1.1v
30	BL/YL	A/T: Park Neutral Switch	In 'P': 0v, Others: 5v
30	BL/YL	M/T: Clutch Engage Switch	5v (clutch "in": 0v)
31	DB/YL	Dual Plug Inhibit Switch	0.1v
33	BR/PK	VR Solenoid Control	0%, 55 mph: 45%
35	BR/BK	Injector 4 Control	3.3-3.5 ms
36	PK	Spark Output Signal	6.93v (50% dwell)
37	RD	Vehicle Power	12-14v
39	BR/YL	Injector 3 Control	3.3-3.5 ms
40	BK/WT	Power Ground	<0.1v
44	GY/BK	Octane Adjustment	9.3v (shorted: 0v)
46	GY/RD	Analog Signal Return	<0.050v
47	GY/WT	TP Sensor Signal	1v, 55 mph: 1.4v
48	WT/PK	Self Test Input Signal	STI On: 1v, Off: 5v
50	BL/RD	MAF Sensor Signal	0.7v, 55 mph: 1.9v
52	OR/YL	Shift Solenoid 3-4	Solenoid Off: 12v, On: 1v
53	PK/YL	TCC Solenoid Control	0v, On 55 mph: 12v
54	PK/YL	A/C WOT Cutout Control	Off: 12v, On: 1v
56	GY/OR	PIP Sensor Signal	6.93v (50%)
57	RD	Vehicle Power	12-14v
58	TN	Injector 1 Control	3.3-3.5 ms
59	WT	Injector 2 Control	3.3-3.5 ms
60	BK/WT	Power Ground	<0.1v

Pin Connector Graphic

PCM 60-PIN CONNECTOR

Terminal View of 60-Pin PCM Harness Connector

1995-97 Pickup 2.3L I4 VIN A (All) 104 Pin Connector

PCM Pin #	Wire Color	Circuit Description (104 Pin)	Value at Hot Idle
1	BK/WT	Shift Solenoid 2 Control	12v, 55 mph: 1v
2	PK/LG	MIL (lamp) Control	MIL Off: 12v, On: 1v
3-10, 12	---	Not Used	---
11	BL/LG	EVAP Purge Flow Sensor	0.8v, at 55 mph: 3v
13	PK	Flash EEPROM Power	0.1v
14	GY/BK	4x4 Low Switch Signal	12v (switch closed: 0v)
15	PK/LB	Data Bus (-) Signal	Digital Signals
16	TN/OR	Data Bus (+) Signal	Digital Signals
17-20, 23	---	Not Used	---
21	DB	CKP (+) Sensor Signal	390-450 Hz
22	GY	CKP (-) Sensor Signal	390-450 Hz
24, 51	BK/WT	Power Ground	<0.1v
25	BK	Case Ground	<0.050v
26	TN/WT	Coil 1 Driver (dwell)	6°, 55 mph: 9°
27	OR/YL	Shift Solenoid 1 Control	1v, 55 mph: 12v
28	BR/OR	Coast Clutch Solenoid	12v, 55 mph: 12v
29	TN/WT	TCS (switch) Signal	TCS & O/D On: 12v
30	DG	Octane Adjustment	9.3v (shorted: 0v)
31	YL/LG	PSP Switch Signal	0v (turning: 12v)
32, 34	---	Not Used	---
33	PK/OR	VSS (-) Signal	0 Hz, 55 mph: 125 Hz
35	RD/LG	HO2S-12 (B1 S2) Signal	0.1-1.1v
36	TN/LB	MAF Sensor Return	<0.050v
37	OR/BK	TFT Sensor Signal	2.10-2.40v
38	LG/RD	ECT Sensor Signal	0.5-0.6v
39	GY	IAT Sensor Signal	1.5-2.5v
40	DG/YL	Fuel Pump Monitor	On: 12v, Off: 0v
41	TN/YL	A/C Switch Signal	A/C On: 12v, Off: 0v
42-46	---	Not Used	---
47	BR/PK	VR Solenoid Control	0%, 55 mph: 45%
48	TN/YL	Clean Tachometer Output	25-38 Hz
49-50	---	Not Used	---
52	TN/OR	Coil 2 Driver (dwell)	5°, 55 mph: 8°
53	PK/BK	Shift Solenoid 3 Control	1v, 55 mph: 12v
54	PK/YL	TCC Solenoid Control	0%, 55 mph: 95%
55	YL	Keep Alive Power	12-14v
56-57	---	Not Used	---
58	GY/BK	VSS (+) Signal	0 Hz, 55 mph: 125 Hz
59	---	Not Used	---
60	GY/LB	HO2S-11 (B1 S1) Signal	0.1-1.1v
61-63	---	Not Used	---
64	LB/YL	A/T: TR Sensor Signal	In 'P': 0v, in O/D: 5v
64	LB/YL	M/T: CPP Switch Signal	5v (clutch "in": 0v)
65	BR/LG	DPFE Sensor Signal	0.20-1.30v
66	---	Not Used	---
67	GY/YL	EVAP Purge Solenoid	0-10 Hz (0-100%)

1995-97 Pickup 2.3L I4 VIN A (All) 104 Pin Connector

PCM Pin #	Wire Color	Circuit Description (104 Pin)	Value at Hot Idle
68	---	Not Used	---
69	PK/YL	A/C WOT Cutout Control	Off: 12v, On: 1v
70	---	Not Used	---
71	RD	Vehicle Power	12-14v
72-73	---	Not Used	---
74	BR/YL	Injector 3 Control	4.0-4.5 ms
75	TN	Injector 1 Control	4.0-4.5 ms
76	BK/WT	Power Ground	<0.1v
77	BK/WT	Power Ground	<0.1v
78	TN/LB	Coil 3 Driver (dwell)	5°, 55 mph: 8°
79	WT/LG	TCIL (lamp) Control	7.7v (Switch On: 0v)
80	LB/OR	Fuel Pump Control	Off: 12v, On: 1v
81	WT/YL	EPC Solenoid Control	11v (24 psi)
82	---	Not Used	---
83	WT/LB	IAC Solenoid Control	10.7v (33%)
84	DG/WT	TSS Sensor Signal	105 Hz (775 rpm)
85	DB/OR	CMP Sensor Signal	7 Hz, 55 mph: 15 Hz
86-87	---	Not Used	---
88	LB/RD	MAF Sensor Signal	0.8v, 55 mph: 1.6v
89	GY/WT	TP Sensor Signal	1v, 55 mph: 1.7v
90	BR/WT	Reference Voltage	4.9-5.1v
91	GY/RD	Analog Signal Return	<0.050v
92	LG	Brake Pedal Switch	0v (Brake On: 12v)
93	RD/WT	HO2S-11 (B1 S1) Heater	1v (Heater Off: 12v)
94	---	---	---
95	WT/BK	HO2S-12 (B1 S2) Heater	1v (Heater Off: 12v)
96	---	---	---
97	RD	Vehicle Power	12-14v
98-99	---	Not Used	---
100	BR/LB	Injector 4 Control	4.0-4.5 ms
101	WT	Injector 2 Control	4.0-4.5 ms
102	---	Not Used	---
103	BK/WT	Power Ground	<0.1v
104	TN/LG	Coil 3 Driver (dwell)	5°, 55 mph: 8°

Pin Connector Graphic

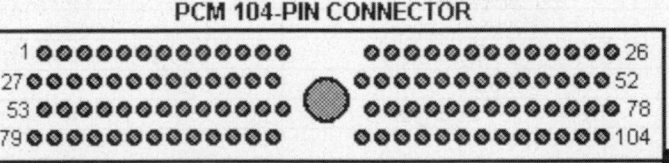

PCM 104-PIN CONNECTOR

Terminal View of 104-Pin PCM Wiring Harness Connector

Standard Colors and Abbreviations

Abbreviation	Color	Abbreviation	Color	Abbreviation	Color
BK	Black	GY	Gray	PK	Purple
BL	Blue	GN	Green	RD	Red
BR	Brown	LG	LT Green	TN	Tan
DB	Dark Blue	OR	Orange	WT	White
DG	DK Green	PK	Pink	YL	Yellow

2001-05 Pickup 2.3L VIN D (All) 104 Pin Connector

PCM Pin #	Wire Color	Circuit Description (104 Pin)	Value at Hot Idle
1	VT/OR	Shift Solenoid 'B' Control	1v, 55 mph: 12v
2	PK/LG	MIL (lamp) Control	MIL Off: 12v, On: 1v
3	YL/BK	Digital TR1 Sensor	0v, 55 mph: 11.5v
4-5, 7	---	Not Used	---
6	DG/WT	TSS Sensor Signal	0 Hz, 55 mph: 385 Hz
8	WT	Swirl Control Motor Signal	Digital Signals
9-11	---	Not Used	---
12	YL/WT	Fuel Level Indicator Signal	1.7v (1/2 full)
13	VT	Flash EEPROM Power	0.1v
14	LB/BK	4WD Indicator Low Signal	12v (switch closed: 0v)
15	PK/LB	SCP Data Bus (-) Signal	<0.050v
16	TN/OR	SCP Data Bus (+) Signal	Digital Signals
17	GY/OR	Passive Antitheft RX Signal	Digital Signals
18	WT/LG	Passive Antitheft TX Signal	Digital Signals
19	BK/PK	ECT Sensor Signal	0.5-0.6v
20, 23	---	Not Used	---
21	DB	CKP (+) Sensor Signal	440-490 Hz
22	GY	CKP (-) Sensor Signal	440-490 Hz
24	BK/WT	Power Ground	<0.1v
25	BK	Chassis Ground	<0.050v
26	TN/WT	Coil 1 Driver (dwell)	5°, 55 mph: 8°
27	OR/YL	Shift Solenoid 'A' Control	1v, 55 mph: 12v
28	BR/OR	Coast Clutch Solenoid Control	1v, 55mph: 12v
29	TN/WT	TCS (switch) Signal	TCS & O/D On: 12v
30	DB/LG	Anti-Theft Indicator Signal	Digital Signals
31	YL/LG	Power Steering Pressure Switch	0v (turning: 12v)
32	YL	Knock Sensor Signal	0v
33-34	---	Not Used	---
35	RD/LG	HO2S-12 (B1 S2) Signal	0.1-1.1v
36	TN/LB	MAF Sensor Return	<0.050v
37	OR/BK	TFT Sensor Signal	2.10-2.40v
38	---	Not Used	---
39	GY	IAT Sensor Signal	1.5-2.5v
40	DG/YL	Fuel Pump Monitor	On: 12v, Off: 0v
41	VT	A/C Demand Signal	A/C On: 12v, Off: 0v
42	WT/OR	Swirl Control Motor Signal	Digital Signals
43	---	Not Used	---
44	DB/OR	Starter Relay Control Circuit	Relay Off: 12v, On: 1v
45	OR/WT	Engine Cooling Fan Control	Fan Off: 12v, On: 1v
46	DB	Electric Thermostat to PCM	0.1-4.9v
47	---	Not Used	---
48	TN/YL	Clean Tachometer Output	25-38 Hz
49	LB/BK	Digital TR2 Sensor	0v, 55 mph: 11.5v
50	WT/BK	Digital TR4 Sensor	0v, 55 mph: 11.5v
51	BK/WT	Power Ground	<0.1v
52	TN/OR	Coil 2 Driver (dwell)	5°, 55 mph: 8°
53	PK/BK	Shift Solenoid 'C' Control	1v, 55 mph: 12v
54	VT/YL	TCC Solenoid Control	0%, 55 mph: 95%
55	YL	Battery Power	12-14v
56	LG/BK	EVAP Purge Control Valve	0-10 Hz (0-100%)
57	YL/RD	Knock Sensor Signal	0v
58	---	Not Used	---
59	GY/OR	ISS Sensor Signal	0 Hz, 55 mph: 1150 Hz
60	GY/LB	HO2S-11 (B1 S1) Signal	0.1-1.1v
61	---	Not Used	---
62	RD/PK	FTP Sensor Signal	2.6v (1/2 full)
63	LG/BK	MAP Sensor Signal	1-2v

2001-02 Pickup 2.3L VIN D (All) 104 Pin Connector

PCM Pin #	Wire Color	Circuit Description (104 Pin)	Value at Hot Idle
64	LB/YL	A/T: TR Sensor Signal	In 'P': 0v, in O/D: 5v
64	LB/YL	M/T: CPP Switch Signal	5v (clutch "in": 0v)
65	---	Not Used	
66	YL/LG	CHT Sensor Signal	0.6v (194°F)
67	VT/WT	EVAP Canister Vent Solenoid	0-10 Hz (0-100%)
68	GY/BK	VSS (+) Signal	0 Hz, 55 mph: 125 Hz
69	PK/YL	A/C Clutch Relay Control	Off: 12v, On: 1v
70	---	Not Used	---
71	RD	Vehicle Power	12-14v
72	GY/RD	EGR Stepper Motor (B1) Control	Digital Signals
73	VT/OR	EGR Stepper Motor (A1) Control	Digital Signals
74	BR/YL	Injector 3 Control	2.7-3.5 ms
75	TN	Injector 1 Control	2.7-3.5 ms
76, 77	BK/WT	Power Ground	<0.1v
78	---	Not Used	---
79	WT/LG	TCIL (lamp) Control	7.7v (Switch On: 0v)
80	LB/OR	Fuel Pump Relay Control	Off: 12v, On: 1v
81	WT/YL	EPC Solenoid Control	9.1v (4 psi)
82	BK/WT	Check Gas Cap Indicator Control	Indicator Off: 12v, On: 1v
83	WT/LB	IAC Solenoid Control	10.7v (33%)
84	DB/YL	OSS Sensor Signal	0 Hz, 55 mph: 385 Hz
85	DB/OR	CMP Sensor Signal	7 Hz, 55 mph: 15 Hz
86	BK/YL	A/C Switch Signal	A/C On: 12v, Off: 0v
87	---	Not Used	---
88	LB/RD	MAF Sensor Signal	0.8v, 55 mph: 1.6v
89	GY/WT	TP Sensor Signal	0.53-1.27v
90	BR/WT	Reference Voltage	4.9-5.1v
91	GY/RD	Sensor Return	<0.050v
92	RD/LG	Brake Pedal Switch	0v (Brake On: 12v)
93	RD/WT	HO2S-11 (B1 S1) Heater	1v (Heater Off: 12v)
94	---	Not Used	---
95	WT/BK	HO2S-12 (B1 S2) Heater	1v (Heater Off: 12v)
96	---	Not Used	---
97	RD	Vehicle Power	12-14v
98	TN/RD	EGR Stepper Motor (B2) Control	Digital Signals
99	DG	EGR Stepper Motor (A2) Control	Digital Signals
100	BR/LB	Injector 4 Control	2.7-3.5 ms
101	WT	Injector 2 Control	2.7-3.5 ms
102	---	Not Used	---
103	BK/WT	Power Ground	<0.1v
104	---	Not Used	---

Pin Connector Graphic

PCM 104-PIN CONNECTOR

Terminal View of 104-Pin PCM Wiring Harness Connector

Standard Colors and Abbreviations

Abbreviation	Color	Abbreviation	Color	Abbreviation	Color
BK	Black	GY	Gray	PK	Purple
BL	Blue	GN	Green	RD	Red
BR	Brown	LG	LT Green	TN	Tan
DB	Dark Blue	OR	Orange	WT	White
DG	DK Green	PK	Pink	YL	Yellow

1998-2001 Pickup 2.5L VIN C (All) 104 Pin Connector

PCM Pin #	Wire Color	Circuit Description (104 Pin)	Value at Hot Idle
1	PK/OR	Shift Solenoid 2/B Control	1v, 55 mph: 12v
2	PK/LG	MIL (lamp) Control	MIL Off: 12v, On: 1v
3	YL/BK	Digital TR1 Sensor	0v, 55 mph: 11.5v
12	YL/WT	Fuel Level Indicator Signal	1.7v (1/2 full)
13	PK	Flash EEPROM Power	0.1v
15	PK/LB	Data Bus (-) Signal	Digital Signals
16	TN/OR	Data Bus (+) Signal	Digital Signals
21	DB	CKP (+) Sensor Signal	440-490 Hz
22	GY	CKP (-) Sensor Signal	440-490 Hz
24	BK/WT	Power Ground	<0.1v
25	BK	Case Ground	<0.050v
26	TN/WT	Coil 1 Driver (dwell)	5°, 55 mph: 8°
27	OR/YL	Shift Solenoid 1/A Control	1v, 55 mph: 12v
28	BR/OR	Coast Clutch Solenoid	1v, 55mph: 12v
29	TN/WT	TCS (switch) Signal	TCS & O/D On: 12v
31	YL/LG	PSP Switch Signal	0v (turning: 12v)
35	RD/LG	HO2S-12 (B1 S2) Signal	0.1-1.1v
36	TN/LB	MAF Sensor Return	<0.050v
37	OR/BK	TFT Sensor Signal	2.10-2.40v
38	LG/RD	ECT Sensor Signal	0.5-0.6v
39	GY	IAT Sensor Signal	1.5-2.5v
40	DG/YL	Fuel Pump Monitor	On: 12v, Off: 0v
41	PK	A/C Demand Signal	A/C On: 12v, Off: 0v
47	BR/PK	VR Solenoid Control	0%, 55 mph: 45%
48	TN/YL	Clean Tachometer Output	25-38 Hz
49	LB/BK	Digital TR2 Sensor	0v, 55 mph: 11.5v
50	WT/BK	Digital TR4 Sensor	0v, 55 mph: 11.5v
51	BK/WT	Power Ground	1v
52	TN/OR	Coil 2 Driver (dwell)	5°, 55 mph: 8°
53	PK/BK	Shift Solenoid 'C' Control	1v, 55 mph: 12v
54	PK/YL	TCC Solenoid Control	0%, 55 mph: 95%
55	YL	Keep Alive Power	12-14v
56	LG/BK	EVAP Purge Solenoid	0-10 Hz (0-100%)
58	GY/BK	VSS (+) Signal	0 Hz, 55 mph: 125 Hz
60	GY/LB	HO2S-11 (B1 S1) Signal	0.1-1.1v
62	RD/PK	FTP Sensor Signal	2.6v (1/2 full)
64	LB/YL	A/T: TR Sensor Signal	In 'P': 0v, In O/D: 5v
64	LB/YL	M/T: CPP Switch Signal	5v (clutch "in": 0v)
65	BR/LG	DPFE Sensor Signal	0.95-1.05v
67	PK/WT	EVAP CV Solenoid	0-10 Hz (0-100%)

1998-2001 Pickup 2.5L VIN C (All) 104 Pin Connector

PCM Pin #	Wire Color	Circuit Description (104 Pin)	Value at Hot Idle
69	PK/YL	A/C WOT Relay Control	Off: 12v, On: 1v
71	RD	Vehicle Power	12-14v
74	BR/YL	Injector 3 Control	2.7-3.5 ms
75	TN	Injector 1 Control	2.7-3.5 ms
76	BK/WT	Power Ground	<0.1v
77	BK/WT	Power Ground	<0.1v
78	TN/LG	Coil 3 Driver (dwell)	5°, 55 mph: 8°
79	WT/LG	TCIL (lamp) Control	7.7v (Switch On: 0v)
80	LB/OR	Fuel Pump Control	Off: 12v, On: 1v
81	WT/YL	EPC Solenoid Control	9.1v (4 psi)
83	WT/LB	IAC Solenoid Control	10.7v (33%)
84	DG/WT	TSS Sensor Signal	0 Hz, 55 mph: 385 Hz
85	DB/OR	CMP Sensor Signal	7 Hz, 55 mph: 15 Hz
86	BK/YL	A/C Switch Signal	A/C On: 12v, Off: 0v
88	LB/RD	MAF Sensor Signal	0.8v, 55 mph: 1.6v
89	GY/WT	TP Sensor Signal	0.53-1.27v
90	BR/WT	Reference Voltage	4.9-5.1v
91	GY/RD	Analog Signal Return	<0.050v
92	RD/LG	Brake Pedal Switch	0v (Brake On: 12v)
93	RD/WT	HO2S-11 (B1 S1) Heater	1v (Heater Off: 12v)
95	WT/BK	HO2S-12 (B1 S2) Heater	1v (Heater Off: 12v)
97	RD	Vehicle Power	12-14v
100	BR/LB	Injector 4 Control	2.7-3.5 ms
101	WT	Injector 2 Control	2.7-3.5 ms
103	BK/WT	Power Ground	<0.1v
104	TN/LB	Coil 4 Driver (dwell)	5°, 55 mph: 8°

Pin Connector Graphic

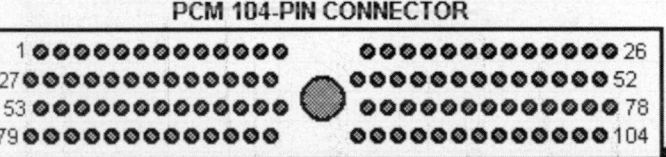

PCM 104-PIN CONNECTOR

Terminal View of 104-Pin PCM Wiring Harness Connector

Standard Colors and Abbreviations

Abbreviation	Color	Abbreviation	Color	Abbreviation	Color
BK	Black	GY	Gray	PK	Purple
BL	Blue	GN	Green	RD	Red
BR	Brown	LG	LT Green	TN	Tan
DB	Dark Blue	OR	Orange	WT	White
DG	DK Green	PK	Pink	YL	Yellow

1990 Pickup 2.9L V6 VIN T (All) 60 Pin Connector

PCM Pin #	Wire Color	Circuit Description (60 Pin)	Value at Hot Idle
1	YL/BK	Keep Alive Power	12-14v
2	LG	Brake Position Switch	0v (Brake On: 12v)
3	DG/WT	VSS (+) Signal	0 Hz, 55 mph: 125 Hz
4	DG/YL	IDM Sensor Signal	20-31 Hz
5, 9	---	Not Used	---
6	OR/YL	VSS (-) Signal	0 Hz, 55 mph: 125 Hz
7	LG/YL	ECT Sensor Signal	0.5-0.6v
8	OR/BL	Fuel Pump Monitor	On: 12v, Off: 0v
10	TN/YL	A/C Switch Signal	A/C On: 12v, Off: 0v
11-13	---	Not Used	---
14	DB/OR	MAF Sensor Signal (California)	0.8v, 55 mph: 1.6v
15	TN/BL	MAF Sensor Return	<0.050v
16	BK/OR	Ignition System Ground	<0.050v
17	TN/RD	Self Test Output & MIL	MIL Off: 12v, On: 1v
18-19	---	Not Used	---
20	BK	PCM Case Ground	<0.050v
21	GY/WT	IAC Solenoid Control	9.3-11.0v
22	TN/LG	Fuel Pump Control	Off: 12v, On: 1v
25	LG/PK	ACT Sensor Signal	1.5-2.5v
26	OR/WT	Reference Voltage	4.9-5.1v
29	DG/PK	HO2S-11 (B1 S1) Signal	0.1-1.1v
30	WT/BK	A/T: Neutral Drive Switch	In 'P': 0v, Others: 5v
30	WT/BK	M/T: Clutch Engage Switch	5v (clutch "in": 0v)
36	YL/LG	Spark Output Signal	6.93v (50% dwell)
37	RD	Vehicle Power	12-14v
40	BK/LG	Power Ground	<0.1v
45	DB/LG	MAP Sensor Signal	107 Hz (sea level)
46	BK/WT	Analog Signal Return	<0.050v
47	DG/LG	TP Sensor Signal	1v, 55 mph: 1.4v
48	WT/RD	Self-Test Input Signal	STI On: 1v, Off: 5v
49	OR	HO2S-11 (B1 S1) Ground	<0.050v
52	TN/BL	Shift Solenoid 3-4 Control	Solenoid Off: 12v, On: 1v
53	WT	TCC Solenoid Control	0v, On 55 mph: 12v
54	PK	A/C WOT Relay Control	Off: 12v, On: 1v
56	DB	PIP Sensor Signal	6.93v (50%)
57	RD	Vehicle Power	12-14v
58	LG/WT	Injector Bank 1 (INJ 1, 2 & 4)	3.3-4.5 ms
58	TN	Injector Bank 1 (INJ 1, 2 & 4)	3.3-4.5 ms
59	TN/RD, WT	Injector Bank 2 (INJ 3, 5 & 6)	3.3-4.5 ms
60	BK/LG	Power Ground	<0.1v

Pin Connector Graphic

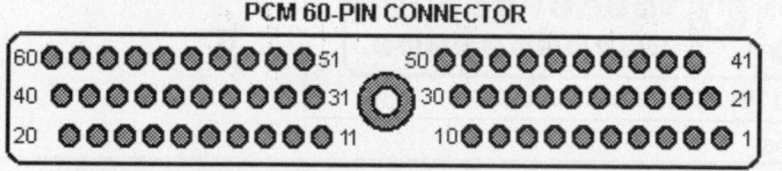

Terminal View of 60-Pin PCM Harness Connector

Standard Colors and Abbreviations

Abbreviation	Color	Abbreviation	Color	Abbreviation	Color
BK	Black	GY	Gray	PK	Purple
BL	Blue	GN	Green	RD	Red
BR	Brown	LG	LT Green	TN	Tan
DB	Dark Blue	OR	Orange	WT	White
DG	DK Green	PK	Pink	YL	Yellow

1991-92 Pickup 2.9L V6 VIN T (All) 60 Pin Connector

PCM Pin #	Wire Color	Circuit Description (60 Pin)	Value at Hot Idle
1	YL	Keep Alive Power	12-14v
2	LG	Brake Position Switch	0v (Brake On: 12v)
3	GY/BK	VSS (+) Signal	0 Hz, 55 mph: 125 Hz
4	TN/YL	IDM Sensor Signal	20-31 Hz
6	PK/OR	VSS (-) Signal	0 Hz, 55 mph: 125 Hz
7	LG/RD	ECT Sensor Signal	0.5-0.6v
8	DG/YL	Fuel Pump Monitor	On: 12v, Off: 0v
9 ('92)	PK/BL	Data Bus (-) Signal	Digital Signals
10	DG/OR	A/C Switch Signal	A/C On: 12v, Off: 0v
11-15, 18-19	---	Not Used	---
16	OR/RD	Ignition System Ground	<0.050v
17	PK/LG	Self Test Output & MIL	MIL Off: 12v, On: 1v
20	BK	PCM Case Ground	<0.050v
21	WT/BL	IAC Solenoid Control	10.7v (33%)
22	BL/OR	Fuel Pump Control	Off: 12v, On: 1v
25	GY	ACT Sensor Signal	1.5-2.5v
26	BR/WT	Reference Voltage	4.9-5.1v
28 ('92)	TN/OR	Data Bus (+) Signal	Digital Signals
29	GY/BL	HO2S-11 (B1 S1) Signal	0.1-1.1v
30	BL/YL	A/T: Neutral Drive Switch	In 'P': 0v, Others: 5v
30	BL/YL	M/T: Clutch Engage Switch	5v (clutch "in": 0v)
36	PK	Spark Output Signal	6.93v (50% dwell)
37	RD	Vehicle Power	12-14v
40	BK/WT	Power Ground	<0.1v
45	LG/BK	MAP Sensor Signal	107 Hz (sea level)
46	GY/RD	Analog Signal Return	<0.050v
47	GY/WT	TP Sensor Signal	1v, 55 mph: 1.4v
48	WT/PK	Self-Test Input Signal	STI On: 1v, Off: 5v
49	OR	HO2S-11 (B1 S1) Ground	<0.050v
52	OR/YL	Shift Solenoid 3-4 Control	Solenoid Off: 12v, On: 1v
53	PK/YL	TCC Solenoid Control	0v, On 55 mph: 12v
54	PK/YL	A/C WOT Relay Control	Off: 12v, On: 1v
56	GY/OR	PIP Sensor Signal	6.93v (50%)
57	RD	Vehicle Power	12-14v
58	TN	Injector Bank 1 (INJ 1, 2 & 4)	3.3-3.5 ms
59	WT	Injector Bank 2 (INJ 3, 5 & 6)	3.3-3.5 ms
60	BK/WT	Power Ground	<0.1v

Pin Connector Graphic

PCM 60-PIN CONNECTOR

Terminal View of 60-Pin PCM Harness Connector

Standard Colors and Abbreviations

Abbreviation	Color	Abbreviation	Color	Abbreviation	Color
BK	Black	GY	Gray	PK	Purple
BL	Blue	GN	Green	RD	Red
BR	Brown	LG	LT Green	TN	Tan
DB	Dark Blue	OR	Orange	WT	White
DG	DK Green	PK	Pink	YL	Yellow

1991 Pickup 3.0L V6 VIN U (All) 60 Pin Connector

PCM Pin #	Wire Color	Circuit Description (60 Pin)	Value at Hot Idle
1	YL	Keep Alive Power	12-14v
2	LG	Brake Position Switch	0v (Brake On: 12v)
3	GY/BK	VSS (+) Signal	0 Hz, 55 mph: 125 Hz
4	WT/PK	IDM Sensor Signal	20-31 Hz
6	PK/OR	VSS (-) Signal	0 Hz, 55 mph: 125 Hz
7	LG/RD	ECT Sensor Signal	0.5-0.6v
8	DG/YL	Fuel Pump Monitor	On: 12v, Off: 0v
10	DG/OR	A/C Switch Signal	A/C On: 12v, Off: 0v
14	BL/RD	MAF Sensor Signal	0.8v, 55 mph: 1.6v
15	TN/BL	MAF Sensor Return	<0.050v
16	OR/RD	Ignition System Ground	<0.050v
17	PK/LG	Self Test Output & MIL	MIL Off: 12v, On: 1v
20	BK	PCM Case Ground	<0.050v
21	WT/BL	IAC Solenoid Control	9.3-11.0v
22	BL/OR	Fuel Pump Control	Off: 12v, On: 1v
25	GY	ACT Sensor Signal	1.5-2.5v
26	BR/WT	Reference Voltage	4.9-5.1v
29	GY/BL	HO2S-11 (B1 S1) Signal	0.1-1.1v
30	BL/YL	A/T: Neutral Drive Switch	In 'P': 0v, Others: 5v
30	BL/YL	M/T: Clutch Engage Switch	5v (clutch "in": 0v)
31	GY/YL	Canister Purge Solenoid	CANP On at 55 mph: 1v
36	PK	Spark Output Signal	6.93v (50% dwell)
37	RD	Vehicle Power	12-14v
40	BK/WT	Power Ground	<0.1v
45	LG/BK	BARO Sensor Signal	159 Hz (sea level)
46	GY/RD	Analog Signal Return	<0.050v
47	GY/WT	TP Sensor Signal	0.7v, 55 mph: 1.1v
48	WT/PK	Self-Test Input Signal	STI On: 1v, Off: 5v
49	OR	HO2S-11 (B1 S1) Ground	<0.050v
53	PK/YL	Converter Clutch Override	0v, On 55 mph: 12v
53	PK/YL	TCC Solenoid Control	0v, On 55 mph: 12v
54	PK/YL	A/C WOT Relay Control	Off: 12v, On: 1v
56	GY/OR	PIP Sensor Signal	6.93v (50%)
57	RD	Vehicle Power	12-14v
58	TN	Injector Bank 1 (INJ 1, 2 & 4)	3.7-3.9 ms
59	WT	Injector Bank 2 (INJ 3, 5 & 6)	3.7-3.9 ms
60	BK/LG	Power Ground	<0.1v

Pin Connector Graphic

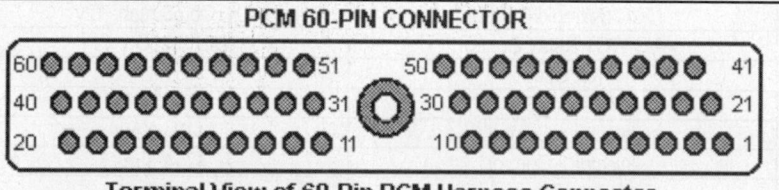

PCM 60-PIN CONNECTOR

Terminal View of 60-Pin PCM Harness Connector

Standard Colors and Abbreviations

Abbreviation	Color	Abbreviation	Color	Abbreviation	Color
BK	Black	GY	Gray	PK	Purple
BL	Blue	GN	Green	RD	Red
BR	Brown	LG	LT Green	TN	Tan
DB	Dark Blue	OR	Orange	WT	White
DG	DK Green	PK	Pink	YL	Yellow

1992-94 Pickup 3.0L V6 VIN U (All) 60 Pin Connector

PCM Pin #	Wire Color	Circuit Description (60 Pin)	Value at Hot Idle
1	YL	Keep Alive Power	12-14v
2	LG	Brake Position Switch	0v (Brake On: 12v)
3	GY/BK	VSS (+) Signal	0 Hz, 55 mph: 125 Hz
4	WT/PK	IDM Sensor Signal	20-31 Hz
6	PK/OR	VSS (-) Signal	0 Hz, 55 mph: 125 Hz
7	LG/RD	ECT Sensor Signal	0.5-0.6v
8	DG/YL	Fuel Pump Monitor	On: 12v, Off: 0v
9	TN/BL	MAF Sensor Return	<0.050v
10	DG/OR	A/C Switch Signal	A/C On: 12v, Off: 0v
11	GY/YL	Canister Purge Solenoid	12v, 55 mph: 1v
12	LG/OR	Injector 6 Control	4.0-4.3 ms
14 ('92-'93)	PK/YL	TCC Solenoid Control	0v, On 55 mph: 12v
15	TN/BK	Injector 6 Control	4.0-4.3 ms
16	OR/RD	Ignition System Ground	<0.050v
17	PK/LG	Self Test Output & MIL	MIL Off: 12v, On: 1v
18	TN/OR	Data Bus (+) Signal	Digital Signals
19	PK/BL	Data Bus (-) Signal	Digital Signals
20	BK	PCM Case Ground	<0.050v
20	BK/LG	PCM Case Ground	<0.050v
21	WT/BL	IAC Solenoid Control	10.8-12.6v
22	BL/OR	Fuel Pump Control	Off: 12v, On: 1v
25	GY	ACT Sensor Signal	1.5-2.5v
26	BR/WT	Reference Voltage	4.9-5.1v
27 ('93-'94)	BR/LG	DPFE Sensor Signal	0.5v, 55 mph: 0.8v
30	BL/YL	A/T: Park Neutral Switch	In 'P': 0v, Others: 5v
30	BL/YL	M/T: Clutch Engage Switch	5v (clutch "in": 0v)
33 ('93-'94)	BR/PK	VR Solenoid Control	0%, 55 mph: 45%
35	BR/BL	Injector 4 Control	4.0-4.3 ms
36	PK	Spark Output Signal	6.93v (50%)
37, 57	RD	Vehicle Power	12-14v
39	BR/YL	Injector 3 Control	4.0-4.3 ms
40	BK/WT	Power Ground	<0.1v
43	RD/BK	HO2S-21 (B2 S1) Signal	0.1-1.1v
44	GY/BL	HO2S-11 (B1 S1) Signal	0.1-1.1v
46	GY/RD	Analog Signal Return	<0.050v
47	GY/WT	TP Sensor Signal	1v, 55 mph: 1.4v
48	WT/PK	Self-Test Input Signal	STI On: 1v, Off: 5v
50	BL/RD	MAF Sensor Signal	0.8v, 55 mph: 1.8v
51	OR/YL	Shift Solenoid 3-4 Control	Solenoid Off: 12v, On: 1v
53 ('94)	PK/YL	TCC Solenoid Control	0v, On 55 mph: 12v
54	PK/YL	A/C WOT Relay Control	Off: 12v, On: 1v
55	---	Not Used	---
56	GY/OR	PIP Sensor Signal	6.93v (50%)
58	TN	Injector 1 Control	4.0-4.3 ms
59	WT	Injector 2 Control	4.0-4.3 ms
60	BK/WT	Power Ground	<0.1v

Pin Connector Graphic

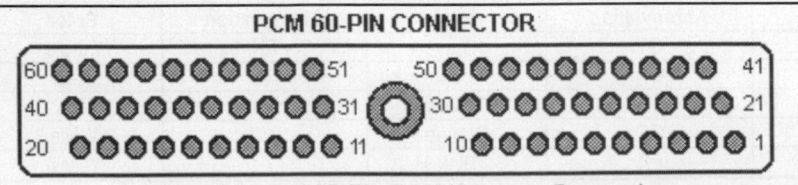

Terminal View of 60-Pin PCM Harness Connector

1995-97 Pickup 3.0L V6 VIN U (All) 104 Pin Connector

PCM Pin #	Wire Color	Circuit Description (104 Pin)	Value at Hot Idle
1	BK/WT	Shift Solenoid 2 Control	1v, 55 mph: 12v
2	PK/LG	MIL (lamp) Control	MIL Off: 12v, On: 1v
3-10, 12	---	Not Used	---
11	BL/LG	EVAP Purge Flow Sensor	0.8v, 55 mph: 3.0v
13	PK	Flash EEPROM Power	0.1v
14	GY/BK	4x4 Low Switch Signal	12v (switch closed: 0v)
15	PK/LB	Data Bus (-) Signal	Digital Signals
16	TN/OR	Data Bus (+) Signal	Digital Signals
17-20, 23	---	Not Used	---
21	DB	CKP (+) Sensor Signal	518-540 Hz
22	GY	CKP (-) Sensor Signal	518-540 Hz
24	BK/WT	Power Ground	<0.1v
25	BK	Chassis Ground	<0.050v
26	TN/WT	Coil 1 Driver (dwell)	6°, 55 mph: 9°
27	OR/YL	Shift Solenoid 1 Control	12v, 55 mph: 1v
28	BR/OR	Coast Clutch Solenoid	1v, 55 mph: 12v
29	TN/WT	TCS (switch) Signal	TCS & O/D On: 12v
30	DG	Octane Adjustment	9.3v (shorted: 0v)
31-32	---	Not Used	---
33	PK/OR	VSS (-) Signal	0 Hz, 55 mph: 125 Hz
35	RD/LG	HO2S-12 (B1 S2) Signal	0.1-1.1v
36	TN/LB	MAF Sensor Return	<0.050v
37	OR/BK	TFT Sensor Signal	2.10-2.40v
38	LG/RD	ECT Sensor Signal	0.5-0.6v
39	GY	IAT Sensor Signal	1.5-2.5v
40	DG/YL	Fuel Pump Monitor	On: 12v, Off: 0v
41	TN/YL	A/C Switch Signal	A/C On: 12v, Off: 0v
42-46	---	Not Used	---
47	BR/PK	VR Solenoid Control	0%, 55 mph: 45%
48	TN/YL	Clean Tachometer Output	42-50 Hz
49-50	---	Not Used	---
51	BK/WT	Power Ground	1v
52	TN/OR	Coil 2 Driver (dwell)	6°, 55 mph: 9°
53	PK/BK	Shift Solenoid 3 Control	12v, 55 mph: 1v
54	PK/YL	TCC Solenoid Control	0% (TCC Off)
55	YL	Keep Alive Power	12-14v
56-57, 59	---	Not Used	---
58	GY/BK	VSS (+) Signal	0 Hz, 55 mph: 125 Hz
60	GY/LB	HO2S-11 (B1 S1) Signal	0.1-1.1v
61-63	---	Not Used	---
64	LB/YL	A/T: TR Sensor Signal	In 'P': 0v, in O/D: 5v
64	LB/YL	M/T: Clutch Pedal Switch	5v (clutch "in": 0v)

1995-97 Pickup 3.0L V6 VIN U (All) 104 Pin Connector

PCM Pin #	Wire Color	Circuit Description (104 Pin)	Value at Hot Idle
65	BR/LG	DPFE Sensor Signal	0.20-1.30v
66	---	Not Used	---
67	GY/YL	Canister Purge Solenoid	12v, 55 mph: 1v
68	---	Not Used	---
69	PK/YL	A/C WOT Cutoff Relay	Off: 12v, On: 1v
70	---	Not Used	---
71	RD	Vehicle Power	12-14v
72	---	Not Used	---
73	TN/BK	Injector 5 Control	4.5-4.8 ms
74	BR/YL	Injector 3 Control	4.5-4.8 ms
75	TN	Injector 1 Control	4.5-4.8 ms
76-77	BK/WT	Power Ground	<0.1v
78	TN/LB	Coil 3 Driver (dwell)	6°, 55 mph: 9°
79	WT/LG	TCIL (lamp) Control	7.7v (Switch On: 0v)
80	LB/OR	Fuel Pump Control	Off: 12v, On: 1v
81	WT/YL	EPC Solenoid Control	10.6v (40 psi)
82	---	Not Used	---
83	WT/LB	IAC Solenoid Control	10.7v (33%)
84	DG/WT	TSS Sensor Signal	50-65 Hz
85	DB/OR	CMP Sensor Signal	7 Hz, 55 mph: 15 Hz
86	---	Not Used	---
87	RD/BK	HO2S-11 (B1 S1) Signal	0.1-1.1v
88	LB/RD	MAF Sensor Signal	0.9v
89	GY/WT	TP Sensor Signal	0.53-1.27v
90	BR/WT	Reference Voltage	4.9-5.1v
91	GY/RD	Analog Signal Return	<0.050v
92	LG	Brake Pedal Switch	0v (Brake On: 12v)
93	RD/WT	HO2S-11 (B1 S1) Heater	1v (Heater Off: 12v)
94	YL/LB	HO2S-21 (B2 S1) Heater	1v (Heater Off: 12v)
95	WT/BK	HO2S-12 (B1 S2) Heater	1v (Heater Off: 12v)
96	---	Not Used	---
97	RD	Vehicle Power	12-14v
98	---	Not Used	---
99	LG/OR	Injector 6 Control	4.5-4.8 ms
100	BR/LB	Injector 4 Control	4.5-4.8 ms
101	WT	Injector 2 Control	4.5-4.8 ms
102	---	Not Used	---
103	BK/WT	Power Ground	<0.1v
104	---	Not Used	---

Pin Connector Graphic

Terminal View of 104-Pin PCM Wiring Harness Connector

Standard Colors and Abbreviations

Abbreviation	Color	Abbreviation	Color	Abbreviation	Color
BK	Black	GY	Gray	PK	Purple
BL	Blue	GN	Green	RD	Red
BR	Brown	LG	LT Green	TN	Tan
DB	Dark Blue	OR	Orange	WT	White
DG	DK Green	PK	Pink	YL	Yellow

1998-2002 Pickup 3.0L V6 MFI VIN U (All) 104 Pin Connector

PCM Pin #	Wire Color	Circuit Description (104 Pin)	Value at Hot Idle
1	PK/OR	Shift Solenoid 2/B Control	1v, 55 mph: 12v
2	PK/LG	MIL (lamp) Control	MIL Off: 12v, On: 1v
3	YL/BK	Digital TR1 Sensor	0v, 55 mph: 11.5v
4-11	---	Not Used	---
12	YL/WT	Fuel Level Indicator Signal	1.7v (1/2 full)
13	PK	Flash EEPROM Power	0.1v
14	LB/BK	4x4 Low Indicator Switch	12v (switch closed: 0v)
15	PK/LB	Data Bus (-) Signal	Digital Signals
16	TN/OR	Data Bus (+) Signal	Digital Signals
17 ('02)	GY/OR	Passive Antitheft RX Signal	Digital Signals
18 ('02)	WT/LG	Passive Antitheft TX Signal	Digital Signals
19-20	---	Not Used	---
21	DB	CKP (+) Sensor Signal	440-490 Hz
22	GY	CKP (-) Sensor Signal	440-490 Hz
23	---	Not Used	---
24	BK/WT	Power Ground	<0.1v
25	BK	Case Ground	<0.050v
26	TN/WT	Coil 1 Driver (dwell)	6°, 55 mph: 9°
27	OR/YL	Shift Solenoid 'A' Control	1v, 55 mph: 12v
28	BR/OR	Shift Solenoid 'D' Control	1v, 55mph: 12v
29	TN/WT	TCS (switch) Signal	TCS & O/D On: 12v
30 ('02)	DB/LG	Antitheft Indicator Control	Indicator Off: 12v, On: 1v
31-34	---	Not Used	---
35	RD/LG	HO2S-12 (B1 S2) Signal	0.1-1.1v
36	TN/LB	MAF Sensor Return	<0.050v
37	OR/BK	TFT Sensor Signal	2.10-2.40v
38	LG/RD	ECT Sensor Signal	0.5-0.6v
39	GY	IAT Sensor Signal	1.5-2.5v
40	DG/YL	Fuel Pump Monitor	On: 12v, Off: 0v
41	PK	A/C Demand Signal	A/C On: 12v, Off: 0v
42-47	---	Not Used	---
48	TN/YL	Clean Tachometer Output	25-38 Hz
49	LB/BK	Digital TR2 Sensor	0v, 55 mph: 11.5v
50	WT/BK	Digital TR4 Sensor	0v, 55 mph: 11.5v
51	BK/WT	Power Ground	1v
52	TN/OR	Coil 3 Driver (dwell)	6°, 55 mph: 9°
53	PK/BK	Shift Solenoid 3/C Control	1v, 55 mph: 12v
54	PK/YL	TCC Solenoid Control	0%, 55 mph: 95%
55	YL	Keep Alive Power	12-14v
56	LG/BK	EVAP Purge Solenoid	0-10 Hz (0-100%)
57	---	Not Used	---
58	GY/BK	VSS (+) Signal	0 Hz, 55 mph: 125 Hz
59	---	Not Used	---
60	GY/LB	HO2S-11 (B1 S1) Signal	0.1-1.1v
61, 63, 66	---	Not Used	---
62	RD/PK	FTP Sensor Signal	2.6v (1/2 full)
64	LB/YL	A/T: TR Sensor Signal	In 'P': 0v, in O/D: 5v
64	LB/YL	M/T: CPP Switch Signal	5v (clutch "in": 0v)

1998-2002 Pickup 3.0L V6 MFI VIN U (All) 104 Pin Connector

PCM Pin #	Wire Color	Circuit Description (104 Pin)	Value at Hot Idle
65 ('98-'00)	BR/LG	DPFE Sensor Signal	0.20-1.30v
65 ('01-'02)	BR/LG	DPFE Sensor Signal	0.95-1.05v
67	PK/WT	EVAP CV Solenoid	0-10 Hz (0-100%)
68	---	Not Used	---
69	PK/YL	A/C WOT Relay Control	Off: 12v, On: 1v
70	---	Not Used	---
71	RD	Vehicle Power	12-14v
72	---	Not Used	---
73	TN/BK	Injector 5 Control	2.7-3.5 ms
74	BR/YL	Injector 3 Control	2.7-3.5 ms
75	TN	Injector 1 Control	2.7-3.5 ms
76-77	BK/WT	Power Ground	<0.1v
78	TN/LG	Coil 2 Driver (dwell)	6°, 55 mph: 9°
79	WT/LG	TCIL (lamp) Control	7.7v (Switch On: 0v)
80	LB/OR	Fuel Pump Control	Off: 12v, On: 1v
81	WT/YL	EPC Solenoid Control	9.1v (4 psi)
82	---	Not Used	---
83	WT/LB	IAC Solenoid Control	10.7v (33%)
84	DG/WT	TSS Sensor Signal	120 Hz, 55 mph: 260 Hz
85	DB/OR	CMP Sensor Signal	7 Hz, 55 mph: 15 Hz
86	BK/YL	A/C Switch Signal	A/C On: 12v, Off: 0v
87	RD/BK	HO2S-12 (B1 S2) Signal	0.1-1.1v
88	LB/RD	MAF Sensor Signal	0.8v, 55 mph: 1.6v
89	GY/WT	TP Sensor Signal	0.53-1.27v
90	BR/WT	Reference Voltage	4.9-5.1v
91	GY/RD	Analog Signal Return	<0.050v
92	RD/LG	Brake Pedal Switch	0v (Brake On: 12v)
93	RD/WT	HO2S-11 (B1 S1) Heater	1v (Heater Off: 12v)
94	YL/LB	HO2S-12 (B1 S2) Heater	1v (Heater Off: 12v)
95	WT/BK	HO2S-13 (B1 S3) Heater	1v (Heater Off: 12v)
96	---	Not Used	---
97	RD	Vehicle Power	12-14v
98	---	Not Used	---
99	LG/OR	Injector 6 Control	2.7-3.5 ms
100	BR/LB	Injector 4 Control	2.7-3.5 ms
101	WT	Injector 2 Control	2.7-3.5 ms
102	---	Not Used	---
103	BK/WT	Power Ground	<0.1v
104	---	Not Used	---

Pin Connector Graphic

PCM 104-PIN CONNECTOR

Terminal View of 104-Pin PCM Wiring Harness Connector

Standard Colors and Abbreviations

Abbreviation	Color	Abbreviation	Color	Abbreviation	Color
BK	Black	GY	Gray	PK	Purple
BL	Blue	GN	Green	RD	Red
BR	Brown	LG	LT Green	TN	Tan
DB	Dark Blue	OR	Orange	WT	White
DG	DK Green	PK	Pink	YL	Yellow

2003-05 Pickup 3.0L V6 MFI VIN U (All) 104 Pin Connector

PCM Pin #	Wire Color	Circuit Description (104 Pin)	Value at Hot Idle
1	VT/OR	Shift Solenoid 'B' Control	1v, 55 mph: 12v
2	PK/LG	MIL (lamp) Control	MIL Off: 12v, On: 1v
3	YL/BK	Digital TR1 Sensor	0v, 55 mph: 11.5v
4-5	---	Not Used	---
6	DG/WT	TSS Sensor Signal	120 Hz, 55 mph: 260 Hz
7-11	---	Not Used	---
12	YL/WT	Fuel Level Indicator Signal	1.7v (1/2 full)
13	VT	Flash EEPROM Power	0.1v
14	LB/BK	4x4 Low Indicator Switch	12v (switch closed: 0v)
15	PK/LB	SCP Data Bus (-) Signal	<0.050v
16	TN/OR	SCP Data Bus (+) Signal	Digital Signals
17	GY/OR	Passive Antitheft RX Signal	Digital Signals
18	WT/LG	Passive Antitheft TX Signal	Digital Signals
19-20	---	Not Used	---
21	DB	CKP (+) Sensor Signal	440-490 Hz
22	GY	CKP (-) Sensor Signal	440-490 Hz
23	---	Not Used	---
24	BK/WT	Power Ground	<0.1v
25	BK	Case Ground	<0.050v
26	TN/WT	Coil 1 Driver (dwell)	6°, 55 mph: 9°
27	OR/YL	Shift Solenoid 'A' Control	1v, 55 mph: 12v
28	BR/OR	Coast Clutch Solenoid	1v, 55mph: 12v
29	TN/WT	TCS (switch) Signal	TCS & O/D On: 12v
30	DB/LG	Antitheft Indicator Control	Indicator Off: 12v, On: 1v
31-34	---	Not Used	---
35	RD/LG	HO2S-12 (B1 S2) Signal	0.1-1.1v
36	TN/LB	MAF Sensor Return	<0.050v
37	OR/BK	TFT Sensor Signal	2.10-2.40v
38	LG/RD	ECT Sensor Signal	0.5-0.6v
39	GY	IAT Sensor Signal	1.5-2.5v
40	DG/YL	Fuel Pump Monitor	Off: 12v, On: 1v
41	VT	A/C Demand Signal	A/C On: 12v, Off: 0v
42-43	---	Not Used	---
44	DB/OR	Starter Relay Control Circuit	Relay Off: 12v, On: 1v
45-47	---	Not Used	---
48	TN/YL	Clean Tachometer Output	25-38 Hz
49	LB/BK	Digital TR2 Sensor	0v, 55 mph: 11.5v
50	WT/BK	Digital TR4 Sensor	0v, 55 mph: 11.5v
51	BK/WT	Power Ground	1v
52	TN/OR	Coil 3 Driver (dwell)	6°, 55 mph: 9°
53	PK/BK	Shift Solenoid 'C' Control	1v, 55 mph: 12v
54	VT/YL	TCC Solenoid Control	0%, 55 mph: 95%
55	YL	Keep Alive Power	12-14v
56	LG/BK	EVAP Purge Solenoid Control	0-10 Hz (0-100%)
57	YL/RD	Knock Sensor Signal	0v
58	---	Not Used	---
59	GY/OR	Intermediate Speed Shaft Sensor	0 Hz, 55 mph: 1150 Hz
60	GY/LB	HO2S-11 (B1 S1) Signal	0.1-1.1v
61	---	Not Used	---
62	RD/PK	FTP Sensor Signal	2.6v (1/2 full)
63	---	Not Used	---
64	LB/YL	A/T: TR Sensor Signal	In 'P': 0v, in O/D: 5v
64	LB/YL	M/T: CPP Switch Signal	5v (clutch "in": 0v)

2003-05 Pickup 3.0L V6 MFI VIN U (All) 104 Pin Connector

PCM Pin #	Wire Color	Circuit Description (104 Pin)	Value at Hot Idle
65-66	---	Not Used	---
67	VT/WT	EVAP Canister Purge Solenoid	0-10 Hz (0-100%)
68	GY/BK	Vehicle Speed Sensor Signal	0 Hz, 55 mph: 125 Hz
69	PK/YL	A/C WOT Relay Control	Off: 12v, On: 1v
70	---	Not Used	---
71	RD	Vehicle Power (Start-Run)	12-14v
72	---	Not Used	---
73	TN/BK	Injector 5 Control	2.7-3.5 ms
74	BR/YL	Injector 3 Control	2.7-3.5 ms
75	TN	Injector 1 Control	2.7-3.5 ms
76, 77	BK/WT	Power Ground	<0.1v
78	TN/LG	Coil 2 Driver (dwell)	6°, 55 mph: 9°
79	WT/LG	TCIL (lamp) Control	7.7v (Switch On: 0v)
80	LB/OR	Fuel Pump Control	Off: 12v, On: 1v
81	WT/YL	EPC Solenoid Control	9.1v (4 psi)
82	BK/WT	Check Fuel Cap Indicator Control	Indicator Off: 12v, On: 1v
83	WT/LB	IAC Solenoid Control	10.7v (33%)
84	DG/YL	OSS Sensor Signal	120 Hz, 55 mph: 260 Hz
85	DB/OR	CMP Sensor Signal	7 Hz, 55 mph: 15 Hz
86	BK/YL	A/C Switch Signal	A/C On: 12v, Off: 0v
87	RD/BK	HO2S-12 (B1 S2) Signal	0.1-1.1v
88	LB/RD	MAF Sensor Signal	0.8v, 55 mph: 1.6v
89	GY/WT	TP Sensor Signal	0.53-1.27v
90	BR/WT	Reference Voltage	4.9-5.1v
91	GY/RD	Analog Signal Return	<0.050v
92	RD/LG	Brake Pedal Switch	0v (Brake On: 12v)
93	RD/WT	HO2S-11 (B1 S1) Heater	1v (Heater Off: 12v)
94	YL/LB	HO2S-12 (B1 S2) Heater	1v (Heater Off: 12v)
95	WT/BK	HO2S-13 (B1 S3) Heater	1v (Heater Off: 12v)
96	---	Not Used	---
97	RD	Vehicle Power (Start-Run)	12-14v
98	---	Not Used	---
99	LG/OR	Injector 6 Control	2.7-3.5 ms
100	BR/LB	Injector 4 Control	2.7-3.5 ms
101	WT	Injector 2 Control	2.7-3.5 ms
102	---	Not Used	---
103	BK/WT	Power Ground	<0.1v
104	---	Not Used	---

Pin Connector Graphic

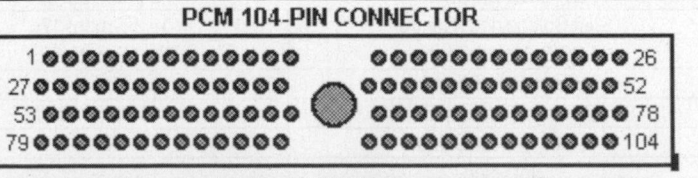

PCM 104-PIN CONNECTOR

Terminal View of 104-Pin PCM Wiring Harness Connector

Standard Colors and Abbreviations

Abbreviation	Color	Abbreviation	Color	Abbreviation	Color
BK	Black	GY	Gray	PK	Purple
BL	Blue	GN	Green	RD	Red
BR	Brown	LG	LT Green	TN	Tan
DB	Dark Blue	OR	Orange	WT	White
DG	DK Green	PK	Pink	YL	Yellow

1999-2002 Pickup 3.0L V6 Flexible Fuel Vehicle VIN V (All) 104 Pin Connector

PCM Pin #	Wire Color	Circuit Description (104 Pin)	Value at Hot Idle
1	VT/OR	Shift Solenoid 'B' Control	1v, 55 mph: 12v
2	PK/LG	MIL (lamp) Control	MIL Off: 12v, On: 1v
3	YL/BK	Digital TR1 Sensor	0v, 55 mph: 11.5v
4-11	---	Not Used	---
12	YL/WT	Fuel Level Indicator Signal	1.7v (1/2 full)
13	PK	Flash EEPROM Power	0.1v
14 ('99)	LB/BK	4x4 Low Indicator Switch	12v (switch closed: 0v)
15	PK/LB	Data Bus (-) Signal	<0.050v
16	TN/OR	Data Bus (+) Signal	Digital Signals
17-20	---	Not Used	---
21	DB	CKP (+) Sensor Signal	440-490 Hz
22	GY	CKP (-) Sensor Signal	440-490 Hz
23	---	Not Used	---
24	BK/WT	Power Ground	<0.1v
25	BK	Case Ground	<0.050v
26	TN/WT	Coil 1 Driver (dwell)	6°, 55 mph: 9°
27	OR/YL	Shift Solenoid 1/A Control	1v, 55 mph: 12v
28	BR/OR	Coast Clutch Solenoid	1v, 55mph: 12v
29	TN/WT	TCS (switch) Signal	TCS & O/D On: 12v
30	---	Not Used	---
31 ('99)	YL/LG	PSP Switch Signal	0v (turning: 12v)
32-33	---	Not Used	---
34	DG/LG	Fuel Composition Sensor	40-60 Hz
35	RD/LG	HO2S-13 (B1 S3) Signal	0.1-1.1v
36	TN/LB	MAF Sensor Return	<0.050v
37	OR/BK	TFT Sensor Signal	2.10-2.40v
38	LG/RD	ECT Sensor Signal	0.5-0.6v
39	GY	IAT Sensor Signal	1.5-2.5v
40	DG/YL	Fuel Pump Monitor	On: 12v, Off: 0v
41	PK	A/C Demand Signal	A/C On: 12v, Off: 0v
42-46	---	Not Used	---
47	BR/PK	VR Solenoid Control	0%, 55 mph: 45%
48	TN/YL	Clean Tachometer Output	25-38 Hz
49	LB/BK	Digital TR2 Sensor	0v, 55 mph: 11.5v
50	WT/BK	Digital TR4 Sensor	0v, 55 mph: 11.5v
51	BK/WT	Power Ground	<0.1v
52	TN/OR	Coil 3 Driver (dwell)	6°, 55 mph: 9°
53	PK/BK	Shift Solenoid 3 Control	1v, 55 mph: 12v
54	PK/YL	TCC Solenoid Control	0%, 55 mph: 95%
55	Y	Keep Alive Power	12-14v
56	LG/BK	EVAP Purge Solenoid	0-10 Hz (0-100%)
58	GY/BK	VSS (+) Signal	0 Hz, 55 mph: 125 Hz
60	GY/LB	HO2S-11 (B1 S1) Signal	0.1-1.1v
62	RD/PK	FTP Sensor Signal	2.6v (1/2 full)
62	RD/PK	FTP Sensor Signal	2.6v (1/2 full)
64	LB/YL	A/T: TR3A Sensor Signal	In 'P': 0v, in O/D: 5v
64	LB/YL	M/T: CPP Switch Signal	5v (clutch "in": 0v)

1999-2002 Pickup 3.0L V6 Flexible Fuel Vehicle VIN V (All) 104 Pin Connector

PCM Pin #	Wire Color	Circuit Description (104 Pin)	Value at Hot Idle
65 ('99-'00)	BR/LG	DPFE Sensor Signal	0.20-1.30v
65 ('01-'02)	BR/LG	DPFE Sensor Signal	0.95-1.05v
67	PK/WT	EVAP CV Solenoid	0-10 Hz (0-100%)
69	PK/YL	A/C WOT Relay Control	Off: 12v, On: 1v
69 ('00-'02)	PK/YL	A/C Clutch Relay Control	Off: 12v, On: 1v
71	RD	Vehicle Power	12-14v
73	TN/BK	Injector 5 Control	2.7-3.5 ms
74	BR/YL	Injector 3 Control	2.7-3.5 ms
75	TN	Injector 1 Control	2.7-3.5 ms
76-77	BK/WT	Power Ground	<0.1v
78	TN/LG	Coil 2 Driver (dwell)	6°, 55 mph: 9°
79	WT/LG	TCIL (lamp) Control	7.7v (Switch On: 0v)
80	LB/OR	Fuel Pump Control	Off: 12v, On: 1v
81	WT/YL	EPC Solenoid Control	9.1v (4 psi)
83	WT/LB	IAC Solenoid Control	10.7v (33%)
84	DG/WT	OSS Sensor Signal	0 Hz, 530 Hz
85	DB/OR	CMP Sensor Signal	7 Hz, 55 mph: 15 Hz
86	BK/YL	A/C Switch Signal	A/C On: 12v, Off: 0v
87	RD/BK	HO2S-12 (B1 S2) Signal	0.1-1.1v
88	LB/RD	MAF Sensor Signal	0.8v, 55 mph: 1.6v
89	GY/WT	TP Sensor Signal	0.53-1.27v
90	BR/WT	Reference Voltage	4.9-5.1v
91	GY/RD	Analog Signal Return	<0.050v
92	RD/LG	Brake Pedal Switch	0v (Brake On: 12v)
93	RD/WT	HO2S-11 (B1 S1) Heater	1v (Heater Off: 12v)
94	YL/LB	HO2S-12 (B1 S2) Heater	1v (Heater Off: 12v)
95	WT/BK	HO2S-13 (B1 S3) Heater	1v (Heater Off: 12v)
97	RD	Vehicle Power	12-14v
99	LG/OR	Injector 6 Control	2.7-3.5 ms
100	BR/LB	Injector 4 Control	2.7-3.5 ms
101	WT	Injector 2 Control	2.7-3.5 ms
103	BK/WT	Power Ground	<0.1v

Pin Connector Graphic

PCM 104-PIN CONNECTOR

Terminal View of 104-Pin PCM Wiring Harness Connector

Standard Colors and Abbreviations

Abbreviation	Color	Abbreviation	Color	Abbreviation	Color
BK	Black	GY	Gray	PK	Purple
BL	Blue	GN	Green	RD	Red
BR	Brown	LG	LT Green	TN	Tan
DB	Dark Blue	OR	Orange	WT	White
DG	DK Green	PK	Pink	YL	Yellow

2003-05 Pickup 3.0L V6 Flexible Fuel Vehicle VIN V (All) 104 Pin Connector

PCM Pin #	Wire Color	Circuit Description (104 Pin)	Value at Hot Idle
1	VT/OR	Shift Solenoid 'B' Control	1v, 55 mph: 12v
2	PK/LG	MIL (lamp) Control	MIL Off: 12v, On: 1v
3	YL/BK	Digital TR1 Sensor	0v, 55 mph: 11.5v
4-5	---	Not Used	---
6	DG/WT	TSS Sensor Signal	120 Hz, 55 mph: 260 Hz
7-11	---	Not Used	---
12	YL/WT	Fuel Level Indicator Signal	1.7v (1/2 full)
13	VT	Flash EEPROM Power	0.1v
14	LB/BK	4x4 Low Indicator Switch	12v (switch closed: 0v)
15	PK/LB	SCP Data Bus (-) Signal	<0.050v
16	TN/OR	SCP Data Bus (+) Signal	Digital Signals
17	GY/OR	Passive Antitheft RX Signal	Digital Signals
18	WT/LG	Passive Antitheft TX Signal	Digital Signals
19	---	Not Used	---
20	YL/WT	Fuel Composition Sensor	40-60 Hz
21	DB	CKP (+) Sensor Signal	440-490 Hz
22	GY	CKP (-) Sensor Signal	440-490 Hz
23	---	Not Used	---
24	BK/WT	Power Ground	<0.1v
25	BK	Case Ground	<0.050v
26	TN/WT	Coil 1 Driver (dwell)	6°, 55 mph: 9°
27	OR/YL	Shift Solenoid 'A' Control	1v, 55 mph: 12v
28	BR/OR	Coast Clutch Solenoid	1v, 55mph: 12v
29	TN/WT	TCS (switch) Signal	TCS & O/D On: 12v
30	DB/LG	Antitheft Indicator Control	Indicator Off: 12v, On: 1v
31-34	---	Not Used	---
35	RD/LG	HO2S-12 (B1 S2) Signal	0.1-1.1v
36	TN/LB	MAF Sensor Return	<0.050v
37	OR/BK	TFT Sensor Signal	2.10-2.40v
38	LG/RD	ECT Sensor Signal	0.5-0.6v
39	GY	IAT Sensor Signal	1.5-2.5v
40	DG/YL	Fuel Pump Monitor	Off: 12v, On: 1v
41	VT	A/C Demand Signal	A/C On: 12v, Off: 0v
42-43	---	Not Used	---
44	DB/OR	Starter Relay Control Circuit	Relay Off: 12v, On: 1v
45-47	---	Not Used	---
48	TN/YL	Clean Tachometer Output	25-38 Hz
49	LB/BK	Digital TR2 Sensor	0v, 55 mph: 11.5v
50	WT/BK	Digital TR4 Sensor	0v, 55 mph: 11.5v
51	BK/WT	Power Ground	1v
52	TN/OR	Coil 3 Driver (dwell)	6°, 55 mph: 9°
53	PK/BK	Shift Solenoid 'C' Control	1v, 55 mph: 12v
54	VT/YL	TCC Solenoid Control	0%, 55 mph: 95%
55	YL	Keep Alive Power	12-14v
56	LG/BK	EVAP Purge Solenoid Control	0-10 Hz (0-100%)
57	YL/RD	Knock Sensor Signal	0v
58	---	Not Used	---
59	GY/OR	Intermediate Speed Shaft Sensor	0 Hz, 55 mph: 1150 Hz
60	GY/LB	HO2S-11 (B1 S1) Signal	0.1-1.1v
61	---	Not Used	---
62	RD/PK	FTP Sensor Signal	2.6v (1/2 full)
63	---	Not Used	---
64	LB/YL	A/T: TR Sensor Signal	In 'P': 0v, in O/D: 5v
64	LB/YL	M/T: CPP Switch Signal	5v (clutch "in": 0v)

2003-05 Pickup 3.0L V6 Flexible Fuel Vehicle VIN V (All) 104 Pin Connector

PCM Pin #	Wire Color	Circuit Description (104 Pin)	Value at Hot Idle
65-66	---	Not Used	---
67	VT/WT	EVAP Canister Purge Solenoid	0-10 Hz (0-100%)
68	GY/BK	Vehicle Speed Sensor Signal	0 Hz, 55 mph: 125 Hz
69	PK/YL	A/C WOT Relay Control	Off: 12v, On: 1v
70	---	Not Used	---
71	RD	Vehicle Power (Start-Run)	12-14v
72	---	Not Used	---
73	TN/BK	Injector 5 Control	2.7-3.5 ms
74	BR/YL	Injector 3 Control	2.7-3.5 ms
75	TN	Injector 1 Control	2.7-3.5 ms
76, 77	BK/WT	Power Ground	<0.1v
78	TN/LG	Coil 2 Driver (dwell)	6°, 55 mph: 9°
79	WT/LG	TCIL (lamp) Control	7.7v (Switch On: 0v)
80	LB/OR	Fuel Pump Control	Off: 12v, On: 1v
81	WT/YL	EPC Solenoid Control	9.1v (4 psi)
82	BK/WT	Check Fuel Cap Indicator Control	Indicator Off: 12v, On: 1v
83	WT/LB	IAC Solenoid Control	10.7v (33%)
84	DG/YL	OSS Sensor Signal	120 Hz, 55 mph: 260 Hz
85	DB/OR	CMP Sensor Signal	7 Hz, 55 mph: 15 Hz
86	BK/YL	A/C Switch Signal	A/C On: 12v, Off: 0v
87	RD/BK	HO2S-12 (B1 S2) Signal	0.1-1.1v
88	LB/RD	MAF Sensor Signal	0.8v, 55 mph: 1.6v
89	GY/WT	TP Sensor Signal	0.53-1.27v
90	BR/WT	Reference Voltage	4.9-5.1v
91	GY/RD	Analog Signal Return	<0.050v
92	RD/LG	Brake Pedal Switch	0v (Brake On: 12v)
93	RD/WT	HO2S-11 (B1 S1) Heater	1v (Heater Off: 12v)
94	YL/LB	HO2S-21 (B2 S1) Heater	1v (Heater Off: 12v)
95	WT/BK	HO2S-12 (B1 S2) Heater	1v (Heater Off: 12v)
96	---	Not Used	---
97	RD	Vehicle Power (Start-Run)	12-14v
98	---	Not Used	---
99	LG/OR	Injector 6 Control	2.7-3.5 ms
100	BR/LB	Injector 4 Control	2.7-3.5 ms
101	WT	Injector 2 Control	2.7-3.5 ms
102	---	Not Used	---
103	BK/WT	Power Ground	<0.1v
104	---	Not Used	---

Pin Connector Graphic

PCM 104-PIN CONNECTOR

Terminal View of 104-Pin PCM Wiring Harness Connector

Standard Colors and Abbreviations

Abbreviation	Color	Abbreviation	Color	Abbreviation	Color
BK	Black	GY	Gray	PK	Purple
BL	Blue	GN	Green	RD	Red
BR	Brown	LG	LT Green	TN	Tan
DB	Dark Blue	OR	Orange	WT	White
DG	DK Green	PK	Pink	YL	Yellow

1990 Pickup 4.0L V6 MFI VIN X (All) 60 Pin Connector

PCM Pin #	Wire Color	Circuit Description (60 Pin)	Value at Hot Idle
1	YL/BK	Keep Alive Power	12-14v
2	LG	Brake Position Switch	0v (Brake On: 12v)
3	DG/WT	VSS (+) Signal	0 Hz, 55 mph: 125 Hz
4	DG/YL	IDM Sensor Signal	20-31 Hz
6	OR/YL	VSS (-) Signal	0 Hz, 55 mph: 125 Hz
7	LG/YL	ECT Sensor Signal	0.5-0.6v
8	OR/BL	Fuel Pump Monitor	On: 12v, Off: 0v
10	TN/YL	A/C Switch Signal	A/C On: 12v, Off: 0v
14	DB/OR	MAF Sensor Signal	0.8v
15	TN/BL	MAF Sensor Return	<0.050v
16	BK/OR	Ignition System Ground	<0.050v
17	TN/RD	Self Test Output & MIL	MIL Off: 12v, On: 1v
20	BK	PCM Case Ground	<0.050v
21	GY/WT	IAC Solenoid Control	8.0-11.5v
22	TN/LG	Fuel Pump Control	Off: 12v, On: 1v
25	LG/PK	ACT Sensor Signal	1.5-2.5v
26	OR/WT	Reference Voltage	4.9-5.1v
29	DG/PK	HO2S-11 (B1 S1) Signal	0.1-1.1v
30	WT/BK	A/T: Neutral Drive Switch	In 'P': 0v, Others: 5v
30	WT/BK	M/T: Clutch Engage Switch	5v (clutch "in": 0v)
36	YL/LG	Spark Output Signal	6.93v (50%)
37	RD	Vehicle Power	12-14v
40	BK/LG	Power Ground	<0.1v
45	DB/LG	BARO Sensor Signal	159 Hz (sea level)
46	BK/WT	Analog Signal Return	<0.050v
47	DG/LG	TP Sensor Signal	1v, 55 mph: 1.4v
48	WT/RD	Self-Test Input Signal	STI On: 1v, Off: 5v
49	OR	HO2S-11 (B1 S1) Ground	<0.050v
52	TN/BL	Shift Solenoid 3-4 Control	Solenoid Off: 12v, On: 1v
53	WT	TCC Solenoid Control	0v, On 55 mph: 12v
54	PK	A/C WOT Relay Control	Off: 12v, On: 1v
56	BK/BL	PIP Sensor Signal	6.93v (50%)
57	RD	Vehicle Power	12-14v
58	TN	Injector Bank 1 (INJ 1, 2 & 4)	3.0-3.2 ms
59	WT	Injector Bank 2 (INJ 3, 5 & 6)	3.0-3.2 ms
60	BK/LG	Power Ground	<0.1v

Pin Connector Graphic

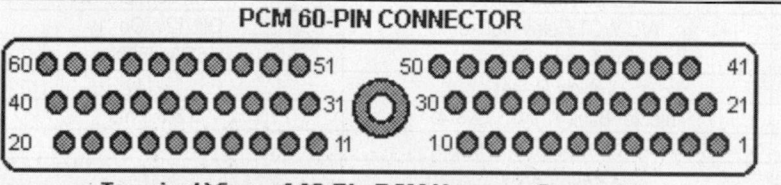

PCM 60-PIN CONNECTOR

Terminal View of 60-Pin PCM Harness Connector

Standard Colors and Abbreviations

Abbreviation	Color	Abbreviation	Color	Abbreviation	Color
BK	Black	GY	Gray	PK	Purple
BL	Blue	GN	Green	RD	Red
BR	Brown	LG	LT Green	TN	Tan
DB	Dark Blue	OR	Orange	WT	White
DG	DK Green	PK	Pink	YL	Yellow

1991-94 Pickup 4.0L V6 VIN X (All) 60 Pin Connector

PCM Pin #	Wire Color	Circuit Description (60 Pin)	Value at Hot Idle
1	YL	Keep Alive Power	12-14v
2	LG	Brake Position Switch	0v (Brake On: 12v)
3	GY/BK	VSS (+) Signal	0 Hz, 55 mph: 125 Hz
4	TN/YL	IDM Sensor Signal	20-31 Hz
6	PK/OR	VSS (-) Signal	0 Hz, 55 mph: 125 Hz
7	LG/RD	ECT Sensor Signal	0.5-0.6v
8	DG/YL	Fuel Pump Monitor	On: 12v, Off: 0v
9 ('92-'94)	PK/BL	Data Bus (-) Signal	Digital Signals
10	DG/OR	A/C Switch Signal	A/C On: 12v, Off: 0v
14	BL/RD	MAF Sensor Signal	0.8v
15	TN/BL	MAF Sensor Return	<0.050v
16	OR/RD	Ignition System Ground	<0.050v
17	PK/LG	Self Test Output & MIL	MIL Off: 12v, On: 1v
20	BK	PCM Case Ground	<0.050v
20	BK/LG	PCM Case Ground	<0.050v
21	WT/BL	IAC Solenoid Control	8.0-11.5v
22	BL/OR	Fuel Pump Control	Off: 12v, On: 1v
25	GY	IAT Sensor Signal	1.5-2.5v
26	BR/WT	Reference Voltage	4.9-5.1v
28 ('92-'94)	TN/OR	Data Bus (+) Signal	Digital Signals
29	GY/BL	HO2S-11 (B1 S1) Signal	0.1-1.1v
30	BL/YL	A/T: Neutral Drive Switch	In 'P': 0v, Others: 5v
30	BL/YL	M/T: Clutch Engage Switch	5v (clutch "in": 0v)
31	GY/DB	Canister Purge Solenoid	CANP On at 55 mph: 1v
31	GY/YL	Canister Purge Solenoid	CANP On at 55 mph: 1v
36	PK	Spark Output Signal	6.93v (50%)
37	RD	Vehicle Power	12-14v
39 ('93-'94)	RD/BK	HO2S-21 (B2 S1) Signal	0.1-1.1v
40	BK/WT	Power Ground	<0.1v
44	DG	Octane Adjustment	9.3v (shorted: 0v)
45 ('91)	LG/BK	BARO Sensor Signal	159 Hz (sea level)
46	GY/RD	Analog Signal Return	<0.050v
47	GY/WT	TP Sensor Signal	1v, 55 mph: 1.4v
48	WT/PK	Self-Test Input Signal	STI On: 1v, Off: 5v
49	OR	HO2S-11 (B1 S1) Ground	<0.050v
49	GY/RD	HO2S-11 (B1 S1) Ground	<0.050v
52	OR/YL	Shift Solenoid 3-4 Control	Solenoid Off: 12v, On: 1v
53	PK/YL	TCC Solenoid Control	0v, On 55 mph: 12v
54	PK/YL	A/C WOT Relay Control	Off: 12v, On: 1v
56	GY/OR	PIP Sensor Signal	6.93v (50%)
57	RD	Vehicle Power	12-14v
58	TN	Injector Bank 1 (INJ 1, 2 & 4)	3.3-3.5 ms
59	WT	Injector Bank 2 (INJ 3, 5 & 6)	3.3-3.5 ms
60	BK/LG	Power Ground	<0.1v
60	BK/WT	Power Ground	<0.1v

Pin Connector Graphic

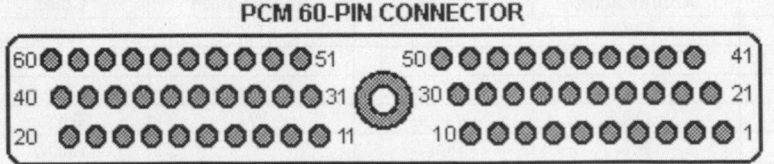

PCM 60-PIN CONNECTOR

Terminal View of 60-Pin PCM Harness Connector

1993-94 Pickup 4.0L V6 VIN X MFI (All) California 60 Pin

PCM Pin #	Wire Color	Circuit Description (60 Pin)	Value at Hot Idle
1	YL	Keep Alive Power	12-14v
2	LG	Brake Position Switch	0v (Brake On: 12v)
3	GY/BK	VSS (+) Signal	0 Hz, 55 mph: 125 Hz
4	TN/YL	IDM Sensor Signal	20-31 Hz
5, 13-14	---	Not Used	---
6	PK/OR	VSS (-) Signal	0 Hz, 55 mph: 125 Hz
7	LG/RD	ECT Sensor Signal	0.5-0.6v
8	DG/YL	Fuel Pump Monitor	On: 12v, Off: 0v
9	TN/BL	MAF Sensor Return	<0.050v
10	DG/OR	A/C Switch Signal	A/C On: 12v, Off: 0v
11	GY/YL	Canister Purge Solenoid	CANP On at 55 mph: 1v
12	BR/BL	Injector 6 Control	3.3-5.7 ms
12	LG/OR	Injector 6 Control	3.3-5.7 ms
15	TN/BK	Injector 5 Control	3.3-5.7 ms
16	OR/RD	Ignition System Ground	<0.050v
17	PK/LG	Self-Test Output & MIL	MIL Off: 12v, On: 1v
18	TN, PK	Data Bus (+), (-) Signals	Digital Signals, <0.050v
20	BK/LG	PCM Case Ground	<0.050v
21	WT/BL	IAC Solenoid Control	10.7v (33%)
22	BL/OR	Fuel Pump Control	Off: 12v, On: 1v
23, 28, 31-32	---	Not Used	---
24	DB/OR	CMP Sensor Signal	7 Hz, 55 mph: 15 Hz
25	GY	IAT Sensor Signal	1.5-2.5v
26	BR/WT	Reference Voltage	4.9-5.1v
27	BR/LG	DPFE Sensor Signal	0.20-1.30v
29	DG	Octane Adjustment	9.3v (shorted: 0v)
30	BL/YL	A/T: Neutral Drive Switch	In 'P': 0v, Others: 5v
30	BL/YL	M/T: Clutch Engage Switch	5v (clutch "in": 0v)
33	BR/PK	VR Solenoid Control	0%, 55 mph: 45%
34, 38, 41-42	---	Not Used	---
35	BR/BL	Injector 4 Control	3.3-5.7 ms
36	PK	Spark Output Signal	6.93v (50%)
37, 57	RD	Vehicle Power (Start-Run)	12-14v
39	BR/YL	Injector 3 Control	3.3-5.7 ms
40	BK/WT	Power Ground	<0.1v
43	RD/BK	HO2S-21 (B2 S1) Signal	0.1-1.1v
44	GY/BL	HO2S-11 (B1 S1) Signal	0.1-1.1v
45, 49, 52, 55	---	Not Used	---
46	GY/RD	Analog Signal Return	<0.050v
47	GY/WT	TP Sensor Signal	1v, 55 mph: 1.4v
48	WT/PK	Self-Test Input Signal	STI On: 1v, Off: 5v
50	BL/RD	MAF Sensor Signal	0.8v
51	OR/YL	Shift Solenoid 3-4 Control	Solenoid Off: 12v, On: 1v
53	PK/YL	TCC Solenoid Control	0v, On 55 mph: 12v
54	PK/YL	A/C WOT Relay Control	Off: 12v, On: 1v
56	GY/OR	PIP Sensor Signal	6.93v (50%)
58	TN	Injector 1 Control	3.3-5.7 ms
59	WT	Injector 2 Control	3.3-5.7 ms
60	BK/WT	Power Ground	<0.1v

Pin Connector Graphic

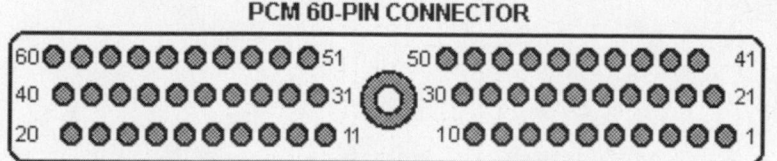

Terminal View of 60-Pin PCM Harness Connector

1995-97 Pickup 4.0L V6 VIN X (All) 104 Pin Connector

PCM Pin #	Wire Color	Circuit Description (104 Pin)	Value at Hot Idle
1	BK/WT	Shift Solenoid 2 Control	1v, 55 mph: 12v
2	PK/LG	MIL (lamp) Control	MIL Off: 12v, On: 1v
3-10, 12	---	Not Used	---
11	BL/LG	Purge Flow Sensor	0.8v, at 55 mph: 3v
13	PK	Flash EPROM Power	0.1v
14	GY/BK	4x4 Low Switch Signal	12v (Switch On: 0v)
15	PK/LB	Data Bus (-) Signal	Digital Signals
16	TN/OR	Data Bus (+) Signal	Digital Signals
17-20, 23	---	Not Used	---
21	DB	CKP (+) Sensor Signal	460-500 Hz
22	GY	CKP (-) Sensor Signal	460-500 Hz
24	BK/WT	Power Ground	<0.1v
25	BK	Case Ground	<0.050v
26	TN/WT	Coil 1 Driver (dwell)	5°, 55 mph: 8°
27	OR/YL	Shift Solenoid 1 Control	1v, 55 mph: 12v
28	BR/OR	Coast Clutch Solenoid	1v, 55 mph: 12v
29	TN/WT	TCS (switch) Signal	TCS & O/D On: 12v
30	DG	Octane Adjustment	9.3v (shorted: 0v)
31-32, 34	---	Not Used	---
33	PK/OR	VSS (-) Signal	0 Hz, 55 mph: 125 Hz
35	RD/LG	HO2S-12 (B1 S2) Signal	0.1-1.1v
36	TN/LB	MAF Sensor Return	<0.050v
37	OR/BK	TFT Sensor Signal	2.10-2.40v
38	LG/RD	ECT Sensor Signal	0.5-0.6v
39	G/Y	IAT Sensor Signal	1.5-2.5v
40	DG/YL	Fuel Pump Monitor	On: 12v, Off: 0v
41	TN/YL	A/C Switch Signal	A/C On: 12v, Off: 0v
42-46	---	Not Used	---
47	BR/PK	VR Solenoid Control	0%, 55 mph: 45%
48	TN/YL	Clean Tachometer Output	39-45 Hz
49-50	---	Not Used	---
51	BK/WT	Power Ground	<0.1v
52	TN/OR	Coil 2 Driver (dwell)	5°, 55 mph: 8°
53	PK/BK	Shift Solenoid 3 Control	12v, 55 mph: 1v
54	PK/YL	TCC Solenoid Control	0%, 55 mph: 95%
55	YL	Keep Alive Power	12-14v
56-57	---	Not Used	---
58	GY/BK	VSS (+) Signal	0 Hz, 55 mph: 125 Hz
59	---	Not Used	---
60	GY/LB	HO2S-11 (B1 S1) Signal	0.1-1.1v
61-63	---	Not Used	---
64	LB/YL	A/T: MLP Sensor Signal	In 'P': 0v, in O/D: 5v
64	LB/YL	M/T: Clutch Pedal Switch	Clutch In: 0v, Out: 12v
65	BR/LG	DPFE Sensor Signal	0.20-1.30v
66	---	Not Used	---
67	GY/YL	Canister Purge Solenoid	12v, 55 mph: 1v
68	---	Not Used	---

1995-97 Pickup 4.0L V6 VIN X (All) 104 Pin Connector

PCM Pin #	Wire Color	Circuit Description (104 Pin)	Value at Hot Idle
69	PK/YL	A/C WOT Relay Control	Off: 12v, On: 1v
70	---	Not Used	---
71	RD	Vehicle Power	12-14v
72	---	Not Used	---
73	TN/BK	Injector 5 Control	4.5-4.8 ms
74	BR/YL	Injector 3 Control	4.5-4.8 ms
75	TN	Injector 1 Control	4.5-4.8 ms
76	BK/WT	Power Ground	<0.1v
77	BK/WT	Power Ground	<0.1v
78	TN/LB	Coil 3 Driver (dwell)	5°, 55 mph: 8°
79	WT/LG	TCIL (lamp) Control	7.7v (Switch On: 0v)
80	LB/OR	Fuel Pump Control	Off: 12v, On: 1v
81	WT/YL	EPC Solenoid Control	10.2v (24 psi)
82	---	Not Used	---
83	WT/LB	IAC Solenoid Control	10.7v (33%)
84	DG/WT	TSS Sensor Signal	115 Hz, 55 mph: 260 Hz
85	DB/OR	CMP Sensor Signal	7 Hz, 55 mph: 15 Hz
86	---	Not Used	---
87	RD/BK	HO2S-21 (B2 S1) Signal	0.1-1.1v
88	LB/RD	MAF Sensor Signal	0.8v
89	GY/WT	TP Sensor Signal	0.7v, 55 mph: 1.1v
90	BR/WT	Reference Voltage	4.9-5.1v
91	GY/RD	Analog Signal Return	<0.050v
92	LG	Brake Pedal Switch	0v (Brake On: 12v)
93	RD/WT	HO2S-11 (B1 S1) Heater	1v (Heater Off: 12v)
94	YL/LB	HO2S-21 (B2 S1) Heater	1v (Heater Off: 12v)
95	WT/BK	HO2S-22 (B2 S2) Heater	1v (Heater Off: 12v)
96	---	Not Used	---
97	RD	Vehicle Power	12-14v
98	---	Not Used	---
99	LG/OR	Injector 6 Control	4.5-4.8 ms
100	BR/LB	Injector 4 Control	4.5-4.8 ms
101	W	Injector 2 Control	4.5-4.8 ms
102	---	Not Used	---
103	BK/WT	Power Ground	<0.1v
104	---	Not Used	---

Pin Connector Graphic

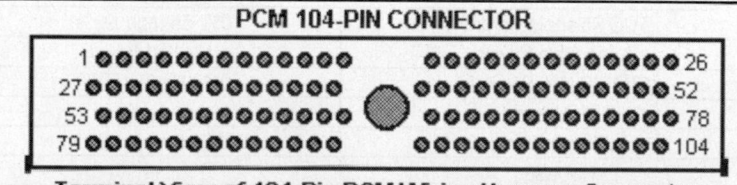

Terminal View of 104-Pin PCM Wiring Harness Connector

Standard Colors and Abbreviations

Abbreviation	Color	Abbreviation	Color	Abbreviation	Color
BK	Black	GY	Gray	PK	Purple
BL	Blue	GN	Green	RD	Red
BR	Brown	LG	LT Green	TN	Tan
DB	Dark Blue	OR	Orange	WT	White
DG	DK Green	PK	Pink	YL	Yellow

1998 Pickup 4.0L V6 VIN X (All) 104 Pin Connector

PCM Pin #	Wire Color	Circuit Description (104 Pin)	Value at Hot Idle
1	PK/OR	Shift Solenoid 2 Control	1v, 55 mph: 12v
2	PK/LG	MIL (lamp) Control	MIL Off: 12v, On: 1v
3	YL/BK	Digital TR1 Sensor	0v, 55 mph: 11.5v
4-5	---	Not Used	---
6	DB/YL	TSS Sensor Signal	110 Hz, 55 mph: 260 Hz
7-11	---	Not Used	---
12	YL/WT	Fuel Level Indicator Signal	1.7v (1/2 full)
13	PK	Flash EEPROM Power	0.1v
14	LB/BK	4x4 Low Indicator Switch	12v (switch closed: 0v)
15	PK/LB	Data Bus (-) Signal	Digital Signals
16	TN/OR	Data Bus (+) Signal	Digital Signals
17-20	---	Not Used	---
21	DB	CKP (+) Sensor Signal	440-490 Hz
22	GY	CKP (-) Sensor Signal	440-490 Hz
23	---	Not Used	---
24	BK/WT	Power Ground	<0.1v
25	BK	Case Ground	<0.050v
26	TN/WT	Coil 1 Driver (dwell)	6°, 55 mph: 9°
27	OR/YL	Shift Solenoid 1 Control	1v, 55 mph: 12v
28	BR/OR	Coast Clutch Solenoid	1v, 55mph: 12v
29	TN/WT	TCS (switch) Signal	TCS & O/D On: 12v
30	---	Not Used	---
31	YL/LG	PSP Switch Signal	0v (turning: 12v)
35	RD/LG	HO2S-13 (B1 S3) Signal	0.1-1.1v
36	TN/LB	MAF Sensor Return	<0.050v
37	OR/BK	TFT Sensor Signal	2.10-2.40v
38	LG/RD	ECT Sensor Signal	0.5-0.6v
39	GY	IAT Sensor Signal	1.5-2.5v
40	DG/YL	Fuel Pump Monitor	On: 12v, Off: 0v
41	PK	A/C Demand Signal	A/C On: 12v, Off: 0v
42-46	---	Not Used	---
47	BR/PK	VR Solenoid Control	0%, 55 mph: 45%
48	TN/YL	Clean Tachometer Output	25-38 Hz
49	LB/BK	Digital TR2 Sensor	0v, 55 mph: 11.5v
50	WT/BK	Digital TR4 Sensor	0v, 55 mph: 11.5v
51	BK/WT	Power Ground	<0.1v
52	TN/OR	Coil 2 Driver (dwell)	6°, 55 mph: 9°
53	PK/BK	Shift Solenoid 3 Control	1v, 55 mph: 12v
54	PK/YL	TCC Solenoid Control	0%, 55 mph: 95%
55	YL	Keep Alive Power	12-14v
56	LG/BK	EVAP Purge Solenoid	0-10 Hz (0-100%)
57	---	Not Used	---
58	GY/BK	VSS (+) Signal	0 Hz, 55 mph: 125 Hz
59, 61	---	Not Used	---
60	GY/LB	HO2S-11 (B1 S1) Signal	0.1-1.1v
62	RD/PK	FTP Sensor Signal	2.6v (1/2 full)
63	---	Not Used	---
64	LB/YL	A/T: TR Sensor Signal	In 'P': 0v, in O/D: 5v
64	LB/YL	M/T: CPP Switch Signal	5v (clutch "in": 0v)

1998 Pickup 4.0L V6 VIN X (All) 104 Pin Connector

PCM Pin #	Wire Color	Circuit Description (104 Pin)	Value at Hot Idle
65	BR/LG	DPFE Sensor Signal	0.20-1.30v
67	PK/WT	EVAP CV Solenoid	0-10 Hz (0-100%)
68, 70, 72	---	Not Used	---
69	PK/YL	A/C WOT Relay Control	Off: 12v, On: 1v
71	RD	Vehicle Power	12-14v
73	TN/BK	Injector 5 Control	2.7-3.5 ms
74	BR/YL	Injector 3 Control	2.7-3.5 ms
75	TN	Injector 1 Control	2.7-3.5 ms
76-77	BK/WT	Power Ground	<0.1v
78	TN/LG	Coil 3 Driver (dwell)	6°, 55 mph: 9°
79	WT/LG	TCIL (lamp) Control	7.7v (Switch On: 0v)
80	LB/OR	Fuel Pump Control	Off: 12v, On: 1v
81	WT/YL	EPC Solenoid Control	9.1v (4 psi)
82	---	Not Used	---
83	WT/LB	IAC Solenoid Control	10.7v (33%)
84	DG/WT	TSS Sensor Signal	115 Hz, 55 mph: 260 Hz
85	DB/OR	CMP Sensor Signal	7 Hz, 55 mph: 15 Hz
86	BK/YL	A/C Switch Signal	A/C On: 12v, Off: 0v
87	RD/BK	HO2S-12 (B1 S2) Signal	0.1-1.1v
88	LB/RD	MAF Sensor Signal	0.8v, 55 mph: 1.6v
89	GY/WT	TP Sensor Signal	0.53-1.27v
90	BR/WT	Reference Voltage	4.9-5.1v
91	GY/RD	Analog Signal Return	<0.050v
92	RD/LG	Brake Pedal Switch	0v (Brake On: 12v)
93	RD/WT	HO2S-11 (B1 S1) Heater	1v (Heater Off: 12v)
94	YL/LB	HO2S-12 (B1 S2) Heater	1v (Heater Off: 12v)
95	WT/BK	HO2S-13 (B1 S3) Heater	1v (Heater Off: 12v)
96, 98	---	Not Used	---
97	RD	Vehicle Power	12-14v
99	LG/OR	Injector 6 Control	2.7-3.5 ms
100	BR/LB	Injector 4 Control	2.7-3.5 ms
101	W	Injector 2 Control	2.7-3.5 ms
102	---	Not Used	---
103	BK/WT	Power Ground	<0.1v
104	---	Not Used	---

Pin Connector Graphic

PCM 104-PIN CONNECTOR

Terminal View of 104-Pin PCM Wiring Harness Connector

Standard Colors and Abbreviations

Abbreviation	Color	Abbreviation	Color	Abbreviation	Color
BK	Black	GY	Gray	PK	Purple
BL	Blue	GN	Green	RD	Red
BR	Brown	LG	LT Green	TN	Tan
DB	Dark Blue	OR	Orange	WT	White
DG	DK Green	PK	Pink	YL	Yellow

1999-2000 Pickup 4.0L V6 MFI VIN X (All) 104 Pin Connector

PCM Pin #	Wire Color	Circuit Description (104 Pin)	Value at Hot Idle
1	PK/OR	Shift Solenoid 'B' Control	1v, 55 mph: 12v
2	PK/LG	MIL (lamp) Control	MIL Off: 12v, On: 1v
3	YL/BK	Digital TR1 Sensor	0v, 55 mph: 11.5v
4-5	---	Not Used	---
6	DB/YL	OSS Sensor Signal	115 Hz, 55 mph: 260 Hz
7-11	---	Not Used	---
12	YL/WT	Fuel Level Indicator Signal	1.7v (1/2 full)
13	PK	Flash EEPROM Power	0.1v
14	LB/BK	4x4 Low Indicator Switch	12v (switch closed: 0v)
15	PK/LB	Data Bus (-) Signal	Digital Signals
16	TN/OR	Data Bus (+) Signal	Digital Signals
17-20	---	Not Used	---
21	DB	CKP (+) Sensor Signal	440-490 Hz
22	GY	CKP (-) Sensor Signal	440-490 Hz
23	---	Not Used	---
24	BK/WT	Power Ground	<0.1v
25	BK	Case Ground	<0.050v
26	TN/WT	Coil 1 Driver (dwell)	6°, 55 mph: 9°
27	OR/YL	Shift Solenoid 'A' Control	1v, 55 mph: 12v
28	BR/OR	Shift Solenoid D Control	1v, 55 mph: 12v
29	TN/WT	TCS (switch) Signal	TCS & O/D On: 12v
30-34	---	Not Used	---
35	RD/LG	HO2S-13 (B1 S3) Signal	0.1-1.1v
36	TN/LB	MAF Sensor Return	<0.050v
37	OR/BK	TFT Sensor Signal	2.10-2.40v
38	LG/RD	ECT Sensor Signal	0.5-0.6v
39	GY	IAT Sensor Signal	1.5-2.5v
40	DG/YL	Fuel Pump Monitor	On: 12v, Off: 0v
41	PK	A/C Demand Signal	A/C On: 12v, Off: 0v
42-47	---	Not Used	---
48	TN/YL	Clean Tachometer Output	25-38 Hz
49	LB/BK	Digital TR2 Sensor	0v, 55 mph: 11.5v
50	WT/BK	Digital TR4 Sensor	0v, 55 mph: 11.5v
51	BK/WT	Power Ground	<0.1v
52	TN/OR	Coil 2 Driver (dwell)	6°, 55 mph: 9°
53	PK/BK	Shift Solenoid 'C' Control	1v, 55 mph: 12v
54	PK/YL	TCC Solenoid Control	0%, 55 mph: 95%
55	Y	Keep Alive Power	12-14v
56	LG/BK	EVAP Purge Solenoid	0-10 Hz (0-100%)
57	---	Not Used	---
58	GY/BK	VSS (+) Signal	0 Hz, 55 mph: 125 Hz
59	---	Not Used	---
60	GY/LB	HO2S-11 (B1 S1) Signal	0.1-1.1v
61	---	Not Used	---
62	RD/PK	FTP Sensor Signal	2.6v (1/2 full)
63	---	Not Used	---
64	LB/YL	A/T: TR3A Sensor Signal	In 'P': 0v, in O/D: 5v
64	LB/YL	M/T: CPP Switch Signal	5v (clutch "in": 0v)

1999-2000 Pickup 4.0L V6 MFI VIN X (All) 104 Pin Connector

PCM Pin #	Wire Color	Circuit Description (104 Pin)	Value at Hot Idle
65	BR/LG	DPFE Sensor Signal	0.20-1.30v
66	---	Not Used	---
67	PK/WT	EVAP CV Solenoid	0-10 Hz (0-100%)
68	---	Not Used	---
69	PK/YL	A/C Clutch Relay Control	Off: 12v, On: 1v
70	---	Not Used	---
71	RD	Vehicle Power	12-14v
72	---	Not Used	---
73	TN/BK	Injector 5 Control	2.7-3.5 ms
74	BR/YL	Injector 3 Control	2.7-3.5 ms
75	TN	Injector 1 Control	2.7-3.5 ms
76-77	BK/WT	Power Ground	<0.1v
78	TN/LG	Coil 3 Driver (dwell)	6º, 55 mph: 9º
79	WT/LG	TCIL (lamp) Control	7.7v (Switch On: 0v)
80	LB/OR	Fuel Pump Relay Control	Off: 12v, On: 1v
81	WT/YL	EPC Solenoid Control	9.1v (4 psi)
82	---	Not Used	---
83	WT/LB	IAC Solenoid Control	10.7v (33%)
84	DG/WT	TSS Sensor Signal	115 Hz, 55 mph: 260 Hz
85	DB/OR	CMP Sensor Signal	7 Hz, 55 mph: 15 Hz
86	BK/YL	A/C Switch Signal	A/C On: 12v, Off: 0v
87	RD/BK	HO2S-12 (B1 S2) Signal	0.1-1.1v
88	LB/RD	MAF Sensor Signal	0.8v, 55 mph: 1.6v
89	GY/WT	TP Sensor Signal	0.53-1.27v
90	BR/WT	Reference Voltage	4.9-5.1v
91	GY/RD	Analog Signal Return	<0.050v
92	RD/LG	Brake Pedal Switch	0v (Brake On: 12v)
93	RD/WT	HO2S-11 (B1 S1) Heater	1v (Heater Off: 12v)
94	YL/LB	HO2S-12 (B1 S2) Heater	1v (Heater Off: 12v)
95	WT/BK	HO2S-13 (B1 S3) Heater	1v (Heater Off: 12v)
96	---	Not Used	---
97	RD	Vehicle Power	12-14v
98	---	Not Used	---
99	LG/OR	Injector 6 Control	2.7-3.5 ms
100	BR/LB	Injector 4 Control	2.7-3.5 ms
101	WT	Injector 2 Control	2.7-3.5 ms
102	---	Not Used	---
103	BK/WT	Power Ground	<0.1v
104	---	Not Used	---

Pin Connector Graphic

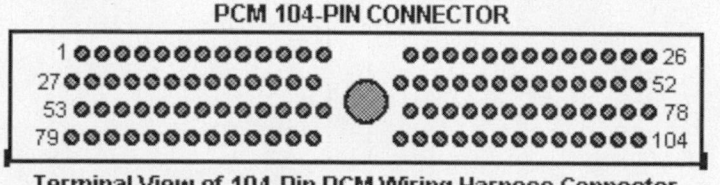

Terminal View of 104-Pin PCM Wiring Harness Connector

Standard Colors and Abbreviations

Abbreviation	Color	Abbreviation	Color	Abbreviation	Color
BK	Black	GY	Gray	PK	Purple
BL	Blue	GN	Green	RD	Red
BR	Brown	LG	LT Green	TN	Tan
DB	Dark Blue	OR	Orange	WT	White
DG	DK Green	PK	Pink	YL	Yellow

2002-05 Pickup 4.0L V6 MFI VIN E (All) 104 Pin Connector

PCM Pin #	Wire Color	Circuit Description (104 Pin)	Value at Hot Idle
1	VT/OR	Shift Solenoid 'B' Control	1v, 55 mph: 12v
2	PK/LG	MIL (lamp) Control	MIL Off: 12v, On: 1v
3	YL/BK	Digital TR1 Sensor	0v, 55 mph: 11.5v
4-5	---	Not Used	---
6	DG/WT	TSS Sensor Signal	115 Hz, 55 mph: 260 Hz
7-11	---	Not Used	---
12	YL/WT	Fuel Level Indicator Signal	1.7v (1/2 full)
13	VT	Flash EEPROM Power	0.1v
14	LB/BK	4x4 Low Indicator Switch	12v (switch closed: 0v)
15	PK/LB	SCP Data Bus (-) Signal	<0.050v
16	TN/OR	SCP Data Bus (+) Signal	Digital Signals
17	GY/OR	Passive Antitheft RX Signal	Digital Signals
18	WT/LG	Passive Antitheft TX Signal	Digital Signals
19-20	---	Not Used	---
21	DB	CKP (+) Sensor Signal	440-490 Hz
22	GY	CKP (-) Sensor Signal	440-490 Hz
23	---	Not Used	---
24	BK/WT	Power Ground	<0.1v
25	BK	Case Ground	<0.050v
26	TN/WT	Coil 1 Driver (dwell)	6°, 55 mph: 9°
27	OR/YL	Shift Solenoid 'A' Control	1v, 55 mph: 12v
28	BR/OR	Shift Solenoid 'D' Control	1v, 55 mph: 12v
29	TN/WT	TCS (switch) Signal	TCS & O/D On: 12v
30	DB/LG	Antitheft Indicator Control	Indicator Off: 12v, On: 1v
31	---	Not Used	---
32	DG/VT	Knock Sensor Signal	0v
33-34	---	Not Used	---
35	RD/LG	HO2S-13 (B1 S3) Signal	0.1-1.1v
36	TN/LB	MAF Sensor Return	<0.050v
37	OR/BK	TFT Sensor Signal	2.10-2.40v
38	LG/RD	ECT Sensor Signal	0.5-0.6v
39	GY	IAT Sensor Signal	1.5-2.5v
40	DG/YL	Fuel Pump Monitor	On: 12v, Off: 0v

2002-05 Pickup 4.0L V6 MFI VIN E (All) 104 Pin Connector

41	VT	A/C Demand Signal	A/C On: 12v, Off: 0v
42-43	---	Not Used	---
44	DB/OR	Starter Relay Control Circuit	Relay Off: 12v, On: 1v
45-46	---	Not Used	---
47	BN/PK	EGR Vacuum Regulator Control	0%, 55 mph: 45%
48	TN/YL	Clean Tachometer Output	25-38 Hz
49	LB/BK	Digital TR2 Sensor	0v, 55 mph: 11.5v
50	WT/BK	Digital TR4 Sensor	0v, 55 mph: 11.5v
51	BK/WT	Power Ground	<0.1v
52	TN/OR	Coil 2 Driver (dwell)	6º, 55 mph: 9º
53	PK/BK	Shift Solenoid 'C' Control	1v, 55 mph: 12v
54	VT/YL	TCC Solenoid Control	0%, 55 mph: 95%
55	BK/LG	Keep Alive Power	12-14v
56	LG/BK	EVAP Purge Solenoid Control	0-10 Hz (0-100%)
57	YL/RD	Knock Sensor Signal	0v
58	---	Not Used	---
59	GY/OR	Intermediate Speed Sensor Signal	0 Hz, 55 mph: 125 Hz
60	GY/LB	HO2S-11 (B1 S1) Signal	0.1-1.1v
61	---	Not Used	---
62	RD/PK	FTP Sensor Signal	2.6v (1/2 full)
63	---	Not Used	---
64	LB/YL	A/T: TR3A Sensor Signal	In 'P': 0v, in O/D: 5v
64	LB/YL	M/T: CPP Switch Signal	5v (clutch "in": 0v)

2002-05 Pickup 4.0L V6 MFI VIN X (All) 104 Pin Connector

PCM Pin #	Wire Color	Circuit Description (104 Pin)	Value at Hot Idle
65	BR/LG	EGR DPFE Sensor Signal	0.95-1.05v
66	---	Not Used	---
67	VT/WT	EVAP Canister Vent Solenoid Control	0-10 Hz (0-100%)
68	GY/BK	Vehicle Speed Sensor Signal	0 Hz, 55 mph: 125 Hz
69	PK/YL	A/C Clutch Relay Control	Off: 12v, On: 1v
70	---	Not Used	---
71	RD	Vehicle Power (Start-Run)	12-14v
72	---	Not Used	---
73	TN/BK	Injector 5 Control	2.7-3.5 ms
74	BR/YL	Injector 3 Control	2.7-3.5 ms
75	TN	Injector 1 Control	2.7-3.5 ms
76-77	BK/WT	Power Ground	<0.1v
78	TN/LG	Coil 3 Driver (dwell)	6°, 55 mph: 9°
79	WT/LG	TCIL (lamp) Control	7.7v (Switch On: 0v)
80	LB/OR	Fuel Pump Relay Control	Off: 12v, On: 1v
81	WT/YL	EPC Solenoid Control	9.1v (4 psi)
82	BK/WT	Check Fuel Cap Indicator Control	Indicator Off: 12v, On: 1v
83	WT/LB	IAC Solenoid Control	10.7v (33%)
84	DB/YL	OSS Sensor Signal	115 Hz, 55 mph: 260 Hz
85	DB/OR	CMP Sensor Signal	7 Hz, 55 mph: 15 Hz
86	BK/YL	A/C Switch Signal	A/C On: 12v, Off: 0v
87	RD/BK	HO2S-12 (B1 S2) Signal	0.1-1.1v
88	LB/RD	MAF Sensor Signal	0.8v, 55 mph: 1.6v
89	GY/WT	TP Sensor Signal	0.53-1.27v
90	BR/WT	Reference Voltage	4.9-5.1v
91	GY/RD	Analog Signal Return	<0.050v
92	RD/LG	Brake Pedal Switch	0v (Brake On: 12v)
93	RD/WT	HO2S-11 (B1 S1) Heater	1v (Heater Off: 12v)
94	YL/LB	HO2S-21 (B2 S1) Heater	1v (Heater Off: 12v)
95	WT/BK	HO2S-12 (B1 S2) Heater	1v (Heater Off: 12v)
96	---	Not Used	---
97	RD	Vehicle Power	12-14v
98	---	Not Used	---
99	LG/OR	Injector 6 Control	2.7-3.5 ms
100	BR/LB	Injector 4 Control	2.7-3.5 ms
101	WT	Injector 2 Control	2.7-3.5 ms
102	---	Not Used	---
103	BK/WT	Power Ground	<0.1v
104	---	Not Used	---

Pin Connector Graphic

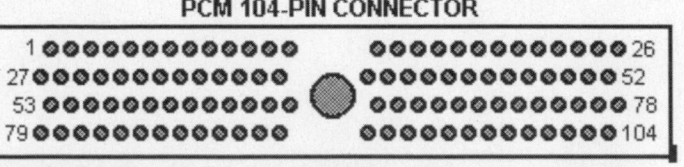

PCM 104-PIN CONNECTOR

Terminal View of 104-Pin PCM Wiring Harness Connector

Standard Colors and Abbreviations

Abbreviation	Color	Abbreviation	Color	Abbreviation	Color
BK	Black	GY	Gray	PK	Purple
BL	Blue	GN	Green	RD	Red
BR	Brown	LG	LT Green	TN	Tan
DB	Dark Blue	OR	Orange	WT	White
DG	DK Green	PK	Pink	YL	Yellow

BLACKWOOD PIN TABLES
2002-05 Pickup 5.4L V8 4v VIN A (A/T) 104 Pin Connector

PCM Pin #	Wire Color	Circuit Description (104 Pin)	Value at Hot Idle
1	OR/YL	COP 6 Driver (dwell)	5°, 55 mph: 8°
2	---	Not Used	---
3	BK/WT	Power Ground	<0.1v
4-5	---	Not Used	---
7-10	OR/YL	Shift Solenoid 1 Control	1v, 55 mph: 12v
4-5	---	Not Used	---
11	VT/OR	Shift Solenoid 2 Control	12v, 55 mph: 12v
12	WT/LG	TCIL (lamp) Control	TCIL Off: 12v, On: 1v
13	VT	Flash EEPROM Power	0.1v
14	LB/BK	4x4 Low Indicator Switch	Switch On: 0v, Off: 12v
15	PK/LB	Data Bus (-) Signal	Digital Signals
16	TN/OR	Data Bus (+) Signal	Digital Signals
17-18	---	Not Used	---
19	DB/WT	Air Suspension Control	Digital Signals
20	BR/OR	Coast Clutch Solenoid	12v, 55 mph: 12v
21	DB	CKP (+) Sensor Signal	400-420 Hz
22	GY	CKP (-) Sensor Signal	400-420 Hz
23	---	Not Used	---
24	BK/WT	Power Ground	<0.1v
25	LB/YL	Case Ground	<0.050v
26	LG/WT	COP 1 Driver (dwell)	5°, 55 mph: 8°
27	OR/YL	COP 5 Driver (dwell)	5°, 55 mph: 8°
28	---	Not Used	---
29	TN/WT	TCS (switch) Signal	TCS & O/D On: 12v
32	DG/VT	Knock Sensor 1 Signal	0v
33	---	Not Used	---
34	YL/BK	Digital TR1 Sensor	0v, 55 mph: 11.5v
35	RD/LG	HO2S-12 (B1 S2) Signal	0.1-1.1v
36	TN/LB	MAF Sensor Return	<0.050v
37	OR/BK	TFT Sensor Signal	2.10-2.40v
38	LG/RD	ECT Sensor Signal	0.5-0.6v
39	GY	IAT Sensor Signal	1.5-2.5v
40	DG/YL	Fuel Pump Monitor	On: 12v, Off: 0v

2002-05 Pickup 5.4L V8 4v VIN A (A/T) 104 Pin Connector

41	BK/YL	A/C High Pressure Switch	Switch On: 12v (open)
42	PK/OR	TP Sensor Signal	0.53-1.27v
43	OR/LG	Overhead Console Module	Digital Signals
43	OR/LG	Data Output Link	Digital Signals
46	BR	IMRC (valve) Control	12v
47	BR/PK	VR Solenoid Control	0%, 55 mph: 45%
48	---	Not Used	---
49	LB/BK	Digital TR2 Sensor	0v, 55 mph: 11.5v
50	WT/BK	Digital TR4 Sensor	0v, 55 mph: 11.5v
51	BK/WT	Power Ground	<0.1v
52	WT/PK	COP 3 Driver (dwell)	5°, 55 mph: 8°
53	DG/VT	COP 4 Driver (dwell)	5°, 55 mph: 8°
54	VT/YL	TCC Solenoid Control	0%, 55 mph: 95%
55	RD/WT	Keep Alive Power	12-14v
56	LG/BK	EVAP Purge Solenoid	0-10 Hz (0-100%)
57	YL/RD	Knock Sensor 1 Signal	0v
58	---	Not Used	---
59	DG/WT	TSS Sensor Signal	370 Hz (700 rpm)
60	GY/LB	HO2S-11 (B1 S1) Signal	0.1-1.1v
61	VT/LG	HO2S-22 (B2 S2) Signal	0.1-1.1v
62	RD/PK	FTP Sensor Signal	2.6v (0" H2O - cap off)
63	---	Not Used	---
64	LB/YL	Digital TR3A Sensor	In 'P': 0v, in O/D: 1.7v

2002-05 Pickup 5.4L V8 4v VIN A (A/T) 104 Pin Connector

PCM Pin #	Wire Color	Circuit Description (104 Pin)	Value at Hot Idle
65	BR/LG	DPFE Sensor Signal	0.95-1.05v
66	YL/LG	CHT Sensor Signal	0.6 to 1.7v (194°F)
67	VT/WT	EVAP CV Solenoid	0-10 Hz (0-100%)
68	GY/BK	Vehicle Speed Out Signal	0 Hz, 55 mph: 125 Hz
69	PK/YL	A/C WOT Relay Control	Off: 12v, On: 1v
70	BK/WT	Check Fuel Cap Indicator Control	Indicator Off: 12v, On: 1v
71	RD	Vehicle Power (Start-Run)	12-14v
72	TN/RD	Injector 7 Control	3.2-3.8 ms
73	TN/BK	Injector 5 Control	3.2-3.8 ms
74	BR/YL	Injector 3 Control	3.2-3.8 ms
75	TN	Injector 1 Control	3.2-3.8 ms
77	BK/WT	Power Ground	<0.1v
78	PK/LB	COP 7 Driver (dwell)	5°, 55 mph: 8°
79	WT/RD	COP 8 Driver (dwell)	5°, 55 mph: 8°
80	LB/OR	Fuel Pump Relay Control	Off: 12v, On: 1v
81	WT/YL	EPC Solenoid Control	9.1v (5 psi)
82, 86	---	Not Used	---
83	WT/LB	IAC Solenoid Control	10.7v (33%)
84	DB/YL	OSS Sensor Signal	0 Hz, 55 mph: 125 Hz
85	DG	CMP Sensor Signal	6 Hz, 55 mph: 14-17 Hz
87	RD/BK	HO2S-21 (B2 S1) Signal	0.1-1.1v
88	LB/RD	MAF Sensor Signal	0.8v, 55 mph: 2.3v
89	GY/WT	TP Sensor Signal	0.53-1.27v
90	BR/WT	Reference Voltage	4.9-5.1v
91	GY/RD	Analog Signal Return	<0.050v
92	RD/LG	Brake Pedal Switch Signal	0v (brake on: 12v)
93	RD/WT	HO2S-11 (B1 S1) Heater	Off: 12v, On: 1v
94	YL/LB	HO2S-21 (B2 S1) Heater	Off: 12v, On: 1v
95	WT/BK	HO2S-12 (B1 S2) Heater	Off: 12v, On: 1v
96	TN/YL	HO2S-22 (B2 S2) Heater	Off: 12v, On: 1v
97	RD	Vehicle Power (Start-Run)	12-14v
98	LB	Injector 8 Control	3.2-3.8 ms
99	LG/OR	Injector 6 Control	3.2-3.8 ms
100	BR/LB	Injector 4 Control	3.2-3.8 ms
101	WT	Injector 2 Control	3.2-3.8 ms
102	---	Not Used	---
103	BK/WT	Power Ground	<0.1v
104	PK/WT	COP 2 Driver (dwell)	5°, 55 mph: 8°

Pin Connector Graphic

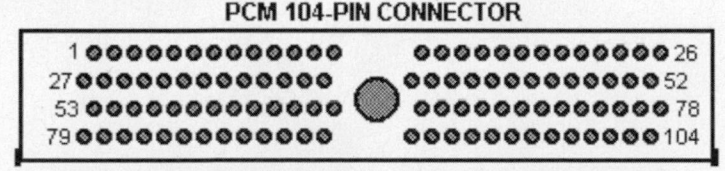

PCM 104-PIN CONNECTOR

Terminal View of 104-Pin PCM Wiring Harness Connector

Standard Colors and Abbreviations

Abbreviation	Color	Abbreviation	Color	Abbreviation	Color
BK	Black	GY	Gray	PK	Purple
BL	Blue	GN	Green	RD	Red
BR	Brown	LG	LT Green	TN	Tan
DB	Dark Blue	OR	Orange	WT	White
DG	DK Green	PK	Pink	YL	Yellow

Domestic Manual ISBN 1-4180-0606-8/Part No. 130606
Import Manual ISBN 1-4180-1537-7/Part No. 131537

Chilton has added so much to its labor guide manual that we've had to put it in two volumes! We've added hundreds of new labor operations—including maintenance services and electronic system diagnosis—to the **Chilton® 2006 Import** and **Domestic Labor Guide Manuals**. All labor times for 1981 through 2006 vehicles consider the real world environment in which technicians work: worn, rusted, or dirty components, being serviced with tools commonly used in the aftermarket. Chilton labor times are accepted by most insurance and extended warranty companies. Vehicle makes and models conform to current Automotive Aftermarket Industry Association standards.

Labor Guide Manual Benefits:

- parts terminology is more standardized across different OEMs to simplify reference
- a total of more than 2,500 pages of updated Chilton labor times appear in these two volumes
- make sure your students have this latest edition because our experts have updated hundreds of labor times for earlier models

Hardcover Manuals are 8 1/2" x 11", ©2006

Labor Guide CD-ROM Benefits:

- easy-to-use software to create and print professional-quality estimates and invoices
- three user-defined levels of labor rates correspond to different types of job scenarios, for "real-world" application
- functions as a database of aftermarket labor times for monitoring warranty and insurance claims
- software keeps track of customers and prior estimates for time-saving recall
- customizable application allows service writers to add labor operations and times, and parts companies to add labor times to existing parts ordering systems

CD ISBN 1-4180-0605-X/Part No. 130605
©2006

Previous Year Editions:
Chilton 2005 Labor Guide Manual, ISBN 1-4018-7412-6/Part No. 27412
Chilton 2005 Labor Guide CD-ROM, ISBN 1-4018-7818-0/Part No. 27818

The **Chilton® 2006 Mechanical Service Manuals** provide updated coverage through 2005 models and even many 2006 models, as made available from original equipment manufacturers (OEMs). Chilton is still your reliable source for fast, accurate repairs and reassembly and it still provides the lowest-priced professional repair manuals on the market! These manuals are organized by make, model, and system so information gathering is easier. Now with even more illustrations and a streamlined index, it's no wonder more automotive professionals turn to Chilton Professional Manuals for their mechanical service and repair information.

Mechanical Service Manual Benefits:

- access up-to-date service and repair information covering model years 2002-2006, all logically arranged by manufacturer
- follow clear, step-by-step procedures—from drive train to chassis—to yield fast, accurate results
- service more mechanical systems, including brakes, engines, suspensions, steering, and related components
- know what special tools are required for specific jobs, as Chilton editors describe and illustrate them to make repair work go more smoothly

2006 Editions

Chilton 2006 DaimlerChrysler Mechanical Service Manual—ISBN 1-4180-0600-9/Part No. 130600
Chilton 2006 Ford Mechanical Service Manual—ISBN 1-4180-0601-7/Part No. 130601
Chilton 2006 General Motors Mechanical Service Manual—ISBN 1-4180-0602-5/Part No. 130602
Chilton 2006 Asian Mechanical Service Manual—Volume I—ISBN 1-4180-0947-4/Part No. 130947
Chilton 2006 Asian Mechanical Service Manual—Volume II—ISBN 1-4180-0948-2/Part No. 130948
Chilton 2006 Asian Mechanical Service Manual—Volume III—ISBN 1-4180-0949-0/Part No. 130949
Chilton 2006 Asian Mechanical Service Manual—3 Volume Set—ISBN 1-4180-0603-3/Part No. 130603
Chilton 2006 European Mechanical Service Manual—ISBN 1-4180-0604-1/Part No. 130604

Asian Manuals Expected Release Date—Summer 2006
European Manuals Expected Release Date—Fall 2006
Manuals are 8 1/2" x 11", ©2006

2005 Editions

Chilton 2005 General Motors Mechanical Service Manual—ISBN 1-4018-7146-1/Part No. 27146
Chilton 2005 Chrysler Mechanical Service Manual—ISBN 1-4018-6718-9/Part No. 26718
Chilton 2005 Ford Mechanical Service Manual—ISBN 1-4018-6719-7/Part No. 26719
Chilton 2005 European Mechanical Service Manual—ISBN 1-4018-6720-0/Part No. 126720
Chilton 2005 Asian Mechanical Service Manual – Volume I—(Acura-Mazda) ISBN 1-4018-6716-2/Part No. 26716
Chilton 2005 Asian Mechanical Service Manual – Volume II—(Mitsubishi-Toyota)
    ISBN 1-4018-6717-0/Part No. 26717
Chilton 2005 Asian Mechanical Service Manual – 2 Volume Set—ISBN 1-4018-7180-1/Part No. 27180

Manuals are 8 1/2" x 11", ©2005

 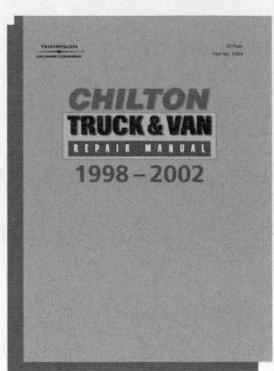  

The *Chilton® Perennial Editions* contain repair and maintenance information for popular mechanical systems that may not be available elsewhere. They offer a wide range of repair information on cars, trucks, vans, and SUVs dating back to the early 1960s, and as current as 2002. Information for 1993 and later model years includes scheduled maintenance interval charts.

Benefits:

- covers the most common vehicle models found in the repair aftermarket today
- gain quick understanding of systems using exploded-view illustrations, diagrams, and charts
- simplify tough jobs with easy-to-follow removal and installation instructions for heater core and other components
- obtain complete coverage of repair procedures from drive train to chassis and associated components

Auto Repair Manual, 1998-2002, 1,426 pages
　　ISBN 0-8019-9362-8/Part No. 9362
Auto Repair Manual, 1993-1997, 2,064 pages
　　ISBN 0-8019-7919-6/Part No. 7919
Auto Repair Manual, 1988-1992, 1,284 pages
　　ISBN 0-8019-7906-4/Part No. 7906
Auto Repair Manual, 1980-1987, 1,344 pages
　　ISBN 0-8019-7670-7/Part No. 7670

Import Car Repair Manual, 1998-2002, 1,792 pps
　　ISBN 0-8019-9363-6/Part No. 9363
Import Car Repair Manual, 1993-1997, 2,080 pps
　　ISBN 0-8019-7920-X/Part No. 7920
Import Car Repair Manual, 1988-1992, 1,632 pages
　　ISBN 0-8019-7907-2/Part No. 7907
Import Car Repair Manual, 1980-1987, 1,488 pages
　　ISBN 0-8019-7672-3/Part No. 7672

Truck & Van Repair Manual, 1998-2002, 1,408 pages
　　ISBN 0-8019-9364-4/Part No. 9364
Truck & Van Repair Manual, 1993-1997, 2,096 pages
　　ISBN 0-8019-7921-8/Part No. 7921
Truck & Van Repair Manual, 1991-1995, 1,664 pages
　　ISBN 0-8019-7911-0/Part No. 7911
Truck & Van Repair Manual, 1986-1990, 1,536 pages
　　ISBN 0-8019-7902-1/Part No. 7902
Truck & Van Repair Manual, 1979-1986, 1,440 pages
　　ISBN 0-8019-7655-3/Part No. 7655

SUV Repair Manual, 1998-2002, 1,292 pages
　　ISBN 0-8019-9365-2/Part No. 9365

Hardcover manuals are 8 1/2" x 11".

Chilton Collector's Editions—*Reference Manuals for Vintage Vehicles*
Auto Repair Manual, 1964-1971, ISBN 0-8019-5974-8/Part No. 5974,
Truck & Van Repair Manual, 1961-1971, ISBN 0-8019-6198-X/Part No. 6198
Truck & Van Repair Manual, 1971-1978, ISBN 0-8019-7012-1/Part No. 7012